Development Through the Lifespan

THIRD EDITION

Laura E. Berk
Illinois State University

ALLYN AND BACON

Boston New York San Francisco
Mexico City Montreal Toronto London Madrid Munich
Paris Hong Kong Singapore Tokyo Cape Town Sydney

Dedication

In loving memory of my parents,
Sofie Lentschner Eisenberg and Philip Vernon Eisenberg

Managing Editor: Tom Pauken
Director of Sales Specialists: Joyce Nilsen
Marketing Manager: Wendy Gordon
Composition Buyer: Linda Cox
Manufacturing Manager: Megan Cochran
Senior Production Editor: Elizabeth Gale Napolitano
Text Design and Composition: Schneck-DePippo Graphics
Cover Coordinator: Linda Knowles
Photo Researchers: Sarah Evertson—ImageQuest; Laurie Frankenthaler
Copyeditor: Connie Day
Proofreader: Bill Heckman

For related titles and support materials, visit our online catalog at www.ablongman.com

Library of Congress Cataloging-in-Publication Data

Berk, Laura E.
Development through the lifespan / Laura E. Berk. — 3rd ed.
p. cm.
Includes bibliographical references and index.
ISBN 0-205-39157-5
1. Developmental psychology. I. Title
BF713.B465 2003
155—dc21 2002043913

Printed in the United States of America
10 9 8 7 6 5 4 3 VH 07 06 05 04

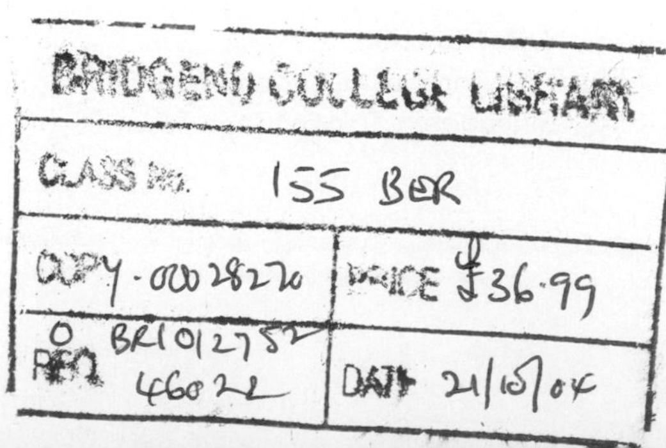

Milestones of Development Tables Credits
(credit sequence keyed to top-to-bottom placement of photos): **Page 200:** column 1: Laura Dwight; column 3: Laura Dwight; Laura Dwight; column 4: Myrleen Ferguson/PhotoEdit; **page 201:** column 1: Laura Dwight; David J. Sams/Stock Boston; column 3: Laura Dwight; Laura Dwight; **page 272:** column 1: David Young-Wolff/PhotoEdit; column 3: Laura Dwight; Nancy Sheehan/PhotoEdit; column 4: Laura Dwight; **page 273:** column 1: Laura Dwight; column 2: Myrleen Ferguson Cate/PhotoEdit; column 4: B & C Alexander; **page 340:** column 1: Patricia Selfe; column 2: Bob Daemmrich/The Image Works; column 3: Fuji Photos/The Image Works; column 4: Jeff Greenberg/PhotoEdit; **page 341:** column 1: Patricia Selfe; column 2: Mary Kate Denny/PhotoEdit; column 3: Charles Gupton/Corbis; MM Flash!Light/Stock Boston; column 4: Charles Gupton/Stock Boston; **page 408:** column 1: Patricia Selfe; David Young-Wolff/PhotoEdit; column 2: Tony Freeman/PhotoEdit; Jose L. Pelaez/Corbis; column 3: Robert Harbison; Michael Newman/PhotoEdit; **page 409:** column 1: Patricia Selfe; column 2: Nancy Richmond/The Image Works; column 3: Syracuse Newspapers/John Berry/The Image Works; **page 478:** column 1: Michelle Bridwell/PhotoEdit; Jon Feingersh/Corbis; column 2: Rob Lewine/Corbis; column 3: Yang Liu/Corbis; Jose Luis Pelaez, Inc./Corbis; **page 479:** column 1: Ed Bock/Corbis; column 2: Tony Freeman/PhotoEdit; Lawrence Migdale/Stock Boston, LLC.; column 3: Tony Freeman/PhotoEdit; **page 542:** column 1: Yann Layma/Getty Images/Stone; Robin Sachs/PhotoEdit; column 2: Adam Smith/Getty Images/Taxi; column 3: AP/Wide World Photos; **page 543:** column 1: Tony Savino/The Image Works; column 2: Spencer Grant/PhotoEdit; column 3: Norbert Schaefer/Corbis; **page 616:** column 1: Robert Brennan/PhotoEdit; Mike Yamashita/Woodfin Camp & Associates; column 2: LWA-Dan Tardif/Corbis; Jakob Halaska/Index Stock Imagery; column 3: Paula Bronstein/Getty Images; David Turnley/Corbis; **page 617:** column 1: David Young-Wolff/PhotoEdit; AP/Wide World Photos; column 2: AP/Wide World Photos; Chris Steele-Perkins/Magnum Photos; column 3: David Young Wolff/Photo Edit; Michael Newman/PhotoEdit

Development Through the Lifespan

About the Author

Laura E. Berk is a distinguished professor of psychology at Illinois State University, where she teaches human development to both undergraduate and graduate students. She received her bachelor's degree in psychology from the University of California, Berkeley, and her master's and doctoral degrees in educational psychology from the University of Chicago. She has been a visiting scholar at Cornell University, UCLA, Stanford University, and the University of South Australia. Berk has published widely on the effects of school environments on children's development, the development of private speech, and most recently the role of make-believe play in the development of self-regulation. Her research has been funded by the U.S. Office of Education and the National Institute of Child Health and Human Development. It has appeared in many prominent journals, including *Child Development, Developmental Psychology, Merrill-Palmer Quarterly, Journal of Abnormal Child Psychology, Development and Psychopathology,* and *Early Childhood Research Quarterly.* Her empirical studies have attracted the attention of the general public, leading to contributions to *Psychology Today* and *Scientific American.* Berk has served as research editor for *Young Children* and consulting editor for *Early Childhood Research Quarterly.* She is author of the chapter on the extracurriculum for the *Handbook of Research on Curriculum* (American Educational Research Association), the chapter on development for *The Many Faces of Psychological Research in the Twenty-First Century* (Society for the Teaching of Psychology), and the article on Vygotsky for the *Encyclopedia of Cognitive Science.* Her books include *Private Speech: From Social Interaction to Self-Regulation, Scaffolding Children's Learning: Vygotsky and Early Childhood Education,* and *Landscapes of Development: An Anthology of Readings.* In addition to *Development Through the Lifespan,* she is author of the best-selling texts *Child Development* and *Infants, Children, and Adolescents,* published by Allyn and Bacon. Her recently published book for parents and teachers is *Awakening Children's Minds: How Parents and Teachers Can Make a Difference.*

Brief Contents

Part I
THEORY AND RESEARCH IN HUMAN DEVELOPMENT 2

1 History, Theory, and Research Strategies 2

Part II
FOUNDATIONS OF DEVELOPMENT 42

2 Biological and Environmental Foundations 42
3 Prenatal Development, Birth, and the Newborn Baby 74

Part III
INFANCY AND TODDLERHOOD: THE FIRST TWO YEARS 112

4 Physical Development in Infancy and Toddlerhood 112
5 Cognitive Development in Infancy and Toddlerhood 142
6 Emotional and Social Development in Infancy and Toddlerhood 172

Part IV
EARLY CHILDHOOD: TWO TO SIX YEARS 202

7 Physical and Cognitive Development in Early Childhood 202
8 Emotional and Social Development in Early Childhood 242

Part V
MIDDLE CHILDHOOD: SIX TO ELEVEN YEARS 274

9 Physical and Cognitive Development in Middle Childhood 274
10 Emotional and Social Development in Middle Childhood 312

Part VI
ADOLESCENCE: THE TRANSITION TO ADULTHOOD 342

11 Physical and Cognitive Development in Adolescence 342
12 Emotional and Social Development in Adolescence 380

Part VII
EARLY ADULTHOOD 410

13 Physical and Cognitive Development in Early Adulthood 410
14 Emotional and Social Development in Early Adulthood 444

Part VIII
MIDDLE ADULTHOOD 480

15 Physical and Cognitive Development in Middle Adulthood 480
16 Emotional and Social Development in Middle Adulthood 510

Part IX
LATE ADULTHOOD 544

17 Physical and Cognitive Development in Late Adulthood 544
18 Emotional and Social Development in Late Adulthood 582

Part X
THE END OF LIFE 618

19 Death, Dying, and Bereavement 618

Glossary G-1
References R-1
Name Index NI-1
Subject Index SI-1

List of Features

A LIFESPAN VISTA

- Impact of Historical Times on the Life Course: The Great Depression and World War II *36*
- The Prenatal Environment and Health in Later Life *84*
- Brain Plasticity: Insights from Research on Brain-Damaged Children and Adults *120*
- Infantile Amnesia *156*
- Maternal Depression and Children's Development *176*
- David: A Boy Who Was Reared as a Girl *261*
- Children of War *335*
- Extracurricular Activities: Contexts for Positive Youth Development *376*
- Two Routes to Adolescent Delinquency *404*
- The Obesity Epidemic: How Americans Became the Heaviest People in the World *422*
- Childhood Attachment Patterns and Adult Romantic Relationships *452*
- Generative Adults Tell Their Life Stories *514*
- What Can We Learn About Aging from Centenarians? *550*
- World War II Refugee and Evacuee Children Look Back from the Vantage Point of Old Age *588*

SOCIAL ISSUES

- The Access Program: A Community–Researcher Partnership *28*
- The Pros and Cons of Reproductive Technologies *54*
- A Cross-National Perspective on Health Care and Other Policies for Parents and Newborn Babies *98*
- Does Child Care in Infancy Threaten Attachment Security and Later Adjustment? *190*
- Chronic Middle Ear Infection in Early Childhood: Consequences for Development *211*
- School Readiness and Grade Retention *303*
- Children's Eyewitness Testimony *337*
- Homosexuality: Coming Out to Oneself and Others *357*
- Development of Civic Responsibility *393*
- Sex Differences in Attitudes Toward Sexuality *427*
- Partner Abuse *427*
- Grandparents Rearing Grandchildren: The Skipped-Generation Family *528*
- Interventions for Caregivers of Elders with Dementia *568*
- Elder Suicide *592*
- Voluntary Active Euthanasia: Lessons from Australia and the Netherlands *637*

CULTURAL INFLUENCES

- Immigrant Youths: Amazing Adaptation *30*
- The African-American Extended Family *63*
- Cultural Variation in Infant Sleeping Arrangements *122*
- Caregiver–Toddler Interaction and Early Make-Believe Play *158*
- Young Children's Daily Life in a Yucatec Mayan Village *226*
- Cultural Variations in Personal Storytelling: Implications for Early Self-Concept *246*
- Education in Japan, Taiwan, and the United States *308*
- Identity Development Among Ethnic Minority Adolescents *386*
- Work-Study Apprenticeships in Germany *440*
- A Global Perspective on Family Planning *464*
- Menopause as a Biocultural Event *488*
- Cultural Variations in the Experience of Aging *558*
- Cultural Variations in Mourning Behavior *642*

BIOLOGY & ENVIRONMENT

- Resiliency *10*
- Uncoupling Genetic–Environmental Correlations for Mental Illness and Antisocial Behavior *70*
- The Mysterious Tragedy of Sudden Infant Death Syndrome *103*
- Development of Infants with Severe Visual Impairments *136*
- Do Infants Have Built-In Numerical Knowledge? *152*
- Biological Basis of Shyness and Sociability *182*
- "Mindblindness" and Autism *228*
- Temperament and Conscience Development in Young Children *254*
- Children with Attention-Deficit Hyperactivity Disorder *288*
- Bullies and Their Victims *325*
- Sex Differences in Spatial Abilities *370*
- Anti-Aging Effects of Dietary Calorie Restriction: Relevant to Humans? *485*
- What Factors Promote Psychological Well-Being in Midlife? *520*
- Taking Time Seriously: Time Perception and Social Goals *597*

Contents

A Personal Note to Students xviii
Preface for Instructors xix

Part I

THEORY AND RESEARCH IN HUMAN DEVELOPMENT 2

1 History, Theory, and Research Strategies 2

Human Development as a Scientific, Interdisciplinary, and Applied Field 5
Basic Issues 5
- Continuous or Discontinuous Development? 6
- One Course of Development or Many? 6
- Nature or Nurture as More Important? 7

The Lifespan Perspective: A Balanced Point of View 7
- Development as Lifelong 8
- Development as Multidimensional and Multidirectional 9
- Development as Plastic 9
- ■ BIOLOGY & ENVIRONMENT: Resiliency 10
- Development as Embedded in Multiple Contexts 10

Historical Foundations 13
- Philosophies of Childhood 13
- Philosophies of Adulthood and Aging 13
- Scientific Beginnings 14

Mid-Twentieth-Century Theories 15
- The Psychoanalytic Perspective 15
- Behaviorism and Social Learning Theory 18
- Piaget's Cognitive-Developmental Theory 19

Recent Theoretical Perspectives 20
- Information Processing 21
- Ethology and Evolutionary Psychology 22
- Vygotsky's Sociocultural Theory 23
- Ecological Systems Theory 24

Comparing and Evaluating Theories 26

Studying Development 27
- Common Research Methods 27
- General Research Designs 31
- Designs for Studying Development 34
- ■ SOCIAL ISSUES: The Access Program: A Community–Researcher Partnership 28
- ■ CULTURAL INFLUENCES: Immigrant Youths: Amazing Adaptation 30
- ■ A LIFESPAN VISTA: Impact of Historical Times on the Life Course: The Great Depression and World War II 36

Ethics in Lifespan Research 37

Summary 39
Important Terms and Concepts 41

Part II

FOUNDATIONS OF DEVELOPMENT 42

2 Biological and Environmental Foundations 42

Genetic Foundations 44
- The Genetic Code 44
- The Sex Cells 45
- Male or Female? 46
- Multiple Births 46
- Patterns of Genetic Inheritance 46
- Chromosomal Abnormalities 50

Reproductive Choices 51
- Genetic Counseling 52
- Prenatal Diagnosis and Fetal Medicine 52
- ■ SOCIAL ISSUES: The Pros and Cons of Reproductive Technologies 54
- Genetic Testing 56
- Adoption 58

Environmental Contexts for Development 58
- The Family 58
- Socioeconomic Status and Family Functioning 59
- The Impact of Poverty 60
- Beyond the Family: Neighborhoods, Towns, and Cities 61
- The Cultural Context 62
- ■ CULTURAL INFLUENCES: The African-American Extended Family 63

Understanding the Relationship Between Heredity and Environment 66
- The Question, "How Much?" 67
- The Question, "How?" 68

■ BIOLOGY & ENVIRONMENT: Uncoupling Genetic–Environmental Correlations for Mental Illness and Antisocial Behavior 70
Summary 71
Important Terms and Concepts 73
FYI: For Further Information and Help 73

3 Prenatal Development, Birth, and the Newborn Baby 74

Prenatal Development 76
Conception 76
Period of the Zygote 78
Period of the Embryo 79
Period of the Fetus 79
Prenatal Environmental Influences 81
Teratogens 81
■ A LIFESPAN VISTA: The Prenatal Environment and Health in Later Life 84
Other Maternal Factors 88
The Importance of Prenatal Health Care 89
Childbirth 91
The Stages of Childbirth 91
The Baby's Adaptation to Labor and Delivery 91
The Newborn Baby's Appearance 92
Assessing the Newborn's Physical Condition: The Apgar Scale 93
Approaches to Childbirth 93
Natural, or Prepared, Childbirth 94
Home Delivery 94
Medical Interventions 95
Fetal Monitoring 95
Labor and Delivery Medication 95
Cesarean Delivery 96
Preterm and Low-Birth-Weight Infants 96
Preterm versus Small for Date 97
Consequences for Caregiving 97
Interventions for Preterm Infants 97
■ SOCIAL ISSUES: A Cross-National Perspective on Health Care and Other Policies for Parents and Newborn Babies 98
Understanding Birth Complications 99
The Newborn Baby's Capacities 100
Newborn Reflexes 100
Newborn States 102
■ BIOLOGY & ENVIRONMENT: The Mysterious Tragedy of Sudden Infant Death Syndrome 103
Sensory Capacities 105
Neonatal Behavioral Assessment 107
Adjusting to the New Family Unit 108
Summary 109
Important Terms and Concepts 111
FYI: For Further Information and Help 111

Part III

INFANCY AND TODDLERHOOD: THE FIRST TWO YEARS 112

4 Physical Development in Infancy and Toddlerhood 112

Body Growth 114
Changes in Body Size and Muscle-Fat Makeup 114
Individual and Group Differences 114
Changes in Body Proportions 115
Brain Development 116
Development of Neurons 116
Development of the Cerebral Cortex 117
Sensitive Periods of Development 119
■ A LIFESPAN VISTA: Brain Plasticity: Insights from Research on Brain-Damaged Children and Adults 120
Changing States of Arousal 120
■ CULTURAL INFLUENCES: Cultural Variation in Infant Sleeping Arrangements 122
Influences on Early Physical Growth 123
Heredity 123
Nutrition 123
Malnutrition 125
Emotional Well-Being 126
Learning Capacities 126
Classical Conditioning 126
Operant Conditioning 127
Habituation 128
Imitation 128
Motor Development 129
The Sequence of Motor Development 130
Motor Skills as Dynamic Systems 130
Cultural Variations in Motor Development 131
Fine Motor Development: Reaching and Grasping 132
Perceptual Development 133
Hearing 133
Vision 133

■ BIOLOGY & ENVIRONMENT: Development of Infants with Severe Visual Impairments 136
Intermodal Perception 138
Understanding Perceptual Development 139

Summary 140
Important Terms and Concepts 141
FYI: For Further Information and Help 141

5 Cognitive Development in Infancy and Toddlerhood 142

Piaget's Cognitive-Developmental Theory 144
Piaget's Ideas About Cognitive Change 144
The Sensorimotor Stage 145
Follow-Up Research on Sensorimotor Development 147
Evaluation of the Sensorimotor Stage 150
■ BIOLOGY & ENVIRONMENT: Do Infants Have Built-In Numerical Knowledge? 152

Information Processing 153
Structure of the Information-Processing System 153
Attention 153
Memory 154
Categorization 155
■ A LIFESPAN VISTA: Infantile Amnesia 156
Evaluation of Information-Processing Findings 157

The Social Context of Early Cognitive Development 157
■ CULTURAL INFLUENCES: Caregiver–Toddler Interaction and Early Make-Believe Play 158

Individual Differences in Early Mental Development 160
Infant Intelligence Tests 160
Early Environment and Mental Development 161
Early Intervention for At-Risk Infants and Toddlers 162

Language Development 164
Three Theories of Language Development 164
Getting Ready to Talk 165
First Words 167
The Two-Word Utterance Phase 167
Individual and Cultural Differences 168
Supporting Early Language Development 168

Summary 170
Important Terms and Concepts 171
FYI: For Further Information and Help 171

6 Emotional and Social Development in Infancy and Toddlerhood 172

Erikson's Theory of Infant and Toddler Personality 174
Basic Trust versus Mistrust 174
Autonomy versus Shame and Doubt 174

Emotional Development 174
Development of Some Basic Emotions 175
■ A LIFESPAN VISTA: Maternal Depression and Children's Development 176
Understanding and Responding to the Emotions of Others 178
Emergence of Self-Conscious Emotions 179
Beginnings of Emotional Self-Regulation 179

Temperament and Development 180
The Structure of Temperament 180
Measuring Temperament 181
■ BIOLOGY & ENVIRONMENT: Biological Basis of Shyness and Sociability 182
Stability of Temperament 182
Genetic Influences 183
Environmental Influences 184
Temperament and Child Rearing: The Goodness-of-Fit Model 184

Development of Attachment 185
Ethological Theory of Attachment 186
Measuring the Security of Attachment 187
Stability of Attachment and Cultural Variations 188
Factors That Affect Attachment Security 188
■ SOCIAL ISSUES: Does Child Care in Infancy Threaten Attachment Security and Later Adjustment? 190
Multiple Attachments 192
Attachment and Later Development 193

Self-Development During the First Two Years 194
Self-Awareness 195
Categorizing the Self 196
Emergence of Self-Control 196

Summary 198
Important Terms and Concepts 199
FYI: For Further Information and Help 199
MILESTONES: Development in Infancy and Toddlerhood 200

Part IV

EARLY CHILDHOOD: TWO TO SIX YEARS 202

7 Physical and Cognitive Development in Early Childhood 202

PHYSICAL DEVELOPMENT 204
Body Growth 204
Skeletal Growth 204
Asynchronies in Physical Growth 205
Brain Development 206
Handedness 206
Other Advances in Brain Development 207
Influences on Physical Growth and Health 208
Heredity and Hormones 208
Emotional Well-Being 208
Nutrition 208
Infectious Disease 209
■ SOCIAL ISSUES: Chronic Middle Ear Infection in Early Childhood: Consequences for Development 211
Childhood Injuries 210
Motor Development 213
Gross Motor Development 213
Fine Motor Development 213
Individual Differences in Motor Skills 215

COGNITIVE DEVELOPMENT 216
Piaget's Theory: The Preoperational Stage 216
Advances in Mental Representation 216
Make-Believe Play 216
Limitations of Preoperational Thought 217
Follow-Up Research on Preoperational Thought 218
Evaluation of the Preoperational Stage 221
Piaget and Education 222
Vygotsky's Sociocultural Theory 223
Private Speech 223
Social Origins of Early Childhood Cognition 223
Vygotsky and Education 234
Evaluation of Vygotsky's Theory 225
■ CULTURAL INFLUENCES: Young Children's Daily Life in a Yucatec Mayan Village 226
Information Processing 225
Attention 225
Memory 226
The Young Child's Theory of Mind 228
■ BIOLOGY & ENVIRONMENT: "Mindblindness" and Autism 229
Early Childhood Literacy 230
Young Children's Mathematical Reasoning 231
Individual Differences in Mental Development 232
Home Environment and Mental Development 232
Preschool, Kindergarten, and Child Care 233
Educational Television 234
Language Development 236
Vocabulary 236
Grammar 236
Conversation 237
Supporting Language Learning in Early Childhood 237

Summary 238
Important Terms and Concepts 240
FYI: For Further Information and Help 241

8 Emotional and Social Development in Early Childhood 242

Erikson's Theory: Initiative versus Guilt 244
Self-Understanding 245
Foundations of Self-Concept 245
■ CULTURAL INFLUENCES: Cultural Variations in Personal Storytelling: Implications for Early Self-Concept 246
Emergence of Self-Esteem 245
Emotional Development 247
Understanding Emotion 247
Emotional Self-Regulation 248
Self-Conscious Emotions 248
Empathy 249
Peer Relations 249
Advances in Peer Sociability 249
First Friendships 251
Parental Influences on Early Peer Relations 251
Foundations of Morality 252
The Psychoanalytic Perspective 253
■ BIOLOGY & ENVIRONMENT: Temperament and Conscience Development in Young Children 254
Social Learning Theory 254
The Cognitive-Developmental Perspective 256
The Other Side of Morality: Development of Aggression 257
Gender Typing 260
Gender-Stereotyped Beliefs and Behavior 260
Genetic Influences on Gender Typing 260
■ A LIFESPAN VISTA: David: A Boy Who Was Reared as a Girl 261

Environmental Influences on Gender Typing 260
Gender Identity 263
Reducing Gender Stereotyping in Young Children 264

Child Rearing and Emotional and Social Development 264
Child-Rearing Styles 265
Cultural Variations 266
Child Maltreatment 267

Summary 269
Important Terms and Concepts 271
FYI: For Further Information and Help 271
MILESTONES: Development in Early Childhood 272

Part V

MIDDLE CHILDHOOD: SIX TO ELEVEN YEARS 274

9 Physical and Cognitive Development in Middle Childhood 274

PHYSICAL DEVELOPMENT 276

Body Growth 276

Common Health Problems 277
Vision and Hearing 277
Malnutrition 278
Obesity 278
Illnesses 280
Unintentional Injuries 280

Motor Development and Play 281
Gross Motor Development 281
Fine Motor Development 281
Sex Differences 282
Games with Rules 282
Physical Education 283

COGNITIVE DEVELOPMENT 284

Piaget's Theory: The Concrete Operational Stage 284
Achievements of the Concrete Operational Stage 285
Limitations of Concrete Operational Thought 286
Recent Research on Concrete Operational Thought 286
Evaluation of the Concrete Operational Stage 287

Information Processing 287
Attention 288
■ BIOLOGY & ENVIRONMENT: Children with Attention-Deficit Hyperactivity Disorder 288
Memory Strategies 289
The Knowledge Base and Memory Performance 290
Culture, Schooling, and Memory Strategies 290
The School-Age Child's Theory of Mind 291
Cognitive Self-Regulation 291
Applications of Information Processing to Academic Learning 291

Individual Differences in Mental Development 293
Defining and Measuring Intelligence 293
Recent Efforts to Define Intelligence 294
Explaining Individual and Group Differences in IQ 296

Language Development 299
Vocabulary 299
Grammar 299
Pragmatics 299
Learning Two Languages at a Time 300

Learning in School 301
Class Size 301
Educational Philosophies 302
■ SOCIAL ISSUES: School Readiness and Grade Retention 303
Teacher–Student Interaction 304
Grouping Practices 304
Teaching Children with Special Needs 305
How Well Educated Are North American Children? 307
■ CULTURAL INFLUENCES: Education in Japan, Taiwan, and the United States 308

Summary 309
Important Terms and Concepts 311
FYI: For Further Information and Help 311

10 Emotional and Social Development in Middle Childhood 312

Erikson's Theory: Industry versus Inferiority 320

Self-Understanding 314
Changes in Self-Concept 315
Development of Self-Esteem 315
Influences on Self-Esteem 316

Emotional Development 319
Self-Conscious Emotions 319
Emotional Understanding 319
Emotional Self-Regulation 320

Understanding Others: Perspective Taking 320

Moral Development 321

Learning About Justice Through Sharing 321
Changes in Moral and Social-Conventional Understanding 322

Peer Relations 322
Peer Groups 322
Friendships 323
Peer Acceptance 324
■ BIOLOGY & ENVIRONMENT: Bullies and Their Victims 325

Gender Typing 326
Gender-Stereotyped Beliefs 326
Gender Identity and Behavior 326
Cultural Influences on Gender Typing 327

Family Influences 327
Parent–Child Relationships 328
Siblings 328
Only Children 329
Divorce 329
Blended Families 331
Maternal Employment and Dual-Earner Families 333

Some Common Problems of Development 334
Fears and Anxieties 334
■ A LIFESPAN VISTA: Children of War 335
Child Sexual Abuse 334
■ SOCIAL ISSUES: Children's Eyewitness Testimony 337
Fostering Resiliency in Middle Childhood 336

Summary 338
Important Terms and Concepts 339
FYI: For Further Information and Help 339
MILESTONES: Development in Middle Childhood 340

Part VI

ADOLESCENCE: THE TRANSITION TO ADULTHOOD 342

11 Physical and Cognitive Development in Adolescence 342

PHYSICAL DEVELOPMENT 344

Conceptions of Adolescence 344
The Biological Perspective 344
The Social Perspective 344
A Balanced Point of View 345

Puberty: The Physical Transition to Adulthood 345
Hormonal Changes 345
Body Growth 346
Motor Development and Physical Activity 347
Sexual Maturation 347
Individual and Group Differences 348

The Psychological Impact of Pubertal Events 349
Reactions to Pubertal Changes 349
Pubertal Change, Emotion, and Social Behavior 350
Early versus Late Maturation 351

Health Issues 352
Nutritional Needs 352
Eating Disorders 352
Sexual Activity 354
■ SOCIAL ISSUES: Homosexuality: Coming Out to Oneself and Others 357
Sexually Transmitted Disease 356
Adolescent Pregnancy and Parenthood 358
Substance Use and Abuse 361

COGNITIVE DEVELOPMENT 363

Piaget's Theory: The Formal Operational Stage 363
Hypothetico-Deductive Reasoning 363
Propositional Thought 364
Recent Research on Formal Operational Thought 364

An Information-Processing View of Adolescent Cognitive Development 365
Scientific Reasoning: Coordinating Theory with Evidence 365
How Scientific Reasoning Develops 366

Consequences of Abstract Thought 367
Argumentativeness 367
Self-Consciousness and Self-Focusing 367
Idealism and Criticism 368
Planning and Decision Making 368

Sex Differences in Mental Abilities 369
■ BIOLOGY & ENVIRONMENT: Sex Differences in Spatial Abilities 370

Learning in School 371
School Transitions 371
Academic Achievement 372
Dropping Out 374
■ A LIFESPAN VISTA: Extracurricular Activities: Contexts for Positive Youth Development 376

Summary 377
Important Terms and Concepts 379
FYI: For Further Information and Help 379

12 Emotional and Social Development in Adolescence 380

Erikson's Theory: Identity versus Identity Confusion 382

Self-Understanding 383
Changes in Self-Concept 383
Changes in Self-Esteem 383

Paths to Identity 384
Identity Status and Psychological Well-Being 385
Factors That Affect Identity Development 385
CULTURAL INFLUENCES: Identity Development Among Ethnic Minority Adolescents 386

Moral Development 387
Piaget's Theory of Moral Development 387
Kohlberg's Extension of Piaget's Theory 388
Are There Sex Differences in Moral Reasoning? 390
Environmental Influences on Moral Reasoning 391
Moral Reasoning and Behavior 392
SOCIAL ISSUES: Development of Civic Responsibility 393

Gender Typing 394

The Family 394
Parent–Child Relationships 395
Family Circumstances 396
Siblings 396

Peer Relations 396
Friendships 396
Cliques and Crowds 398
Dating 399
Peer Conformity 399

Problems of Development 400
Depression 400
Suicide 401
Delinquency 403
A LIFESPAN VISTA: Two Routes to Adolescent Delinquency 404

Summary 405
Important Terms and Concepts 407
FYI: For Further Information and Help 407
MILESTONES: Development in Adolescence 408

Part VII

EARLY ADULTHOOD *410*

13 *Physical and Cognitive Development in Early Adulthood* 410

PHYSICAL DEVELOPMENT 412
Biological Aging Begins in Early Adulthood 413
Aging at the Level of DNA and Body Cells 413
Aging at the Level of Organs and Tissues 414

Physical Changes 414
Cardiovascular and Respiratory Systems 414
Motor Performance 416
Immune System 417
Reproductive Capacity 418

Health and Fitness 419
Nutrition 419
A LIFESPAN VISTA: The Obesity Epidemic: How Americans Became the Heaviest People in the World 422
Exercise 423
Substance Abuse 424
Sexuality 426
SOCIAL ISSUES: Sex Differences in Attitudes Toward Sexuality 427
Psychological Stress 430

COGNITIVE DEVELOPMENT 432
Changes in the Structure of Thought 432
Perry's Theory 432
Schaie's Theory 432
Labouvie-Vief's Theory 433

Information Processing: Expertise and Creativity 433

Changes in Mental Abilities 434

The College Experience 435
Psychological Impact of Attending College 436
Dropping Out 436

Vocational Choice 436
Selecting a Vocation 436
Factors Influencing Vocational Choice 437
Vocational Preparation of Non-College-Bound Young Adults 439
CULTURAL INFLUENCES: Work–Study Apprenticeships in Germany 440

Summary 441
Important Terms and Concepts 443
FYI: For Further Information and Help 443

14 *Emotional and Social Development in Early Adulthood* 444

Erikson's Theory: Intimacy versus Isolation 446

Other Theories of Adult Psychosocial Development 447
Levinson's Seasons of Life 447
Vaillant's Adaptation to Life 449
Limitations of Levinson's and Vaillant's Theories 449
The Social Clock 449

Close Relationships 450
Romantic Love 450
A LIFESPAN VISTA: Childhood Attachment Patterns and Adult Romantic Relationships 452
Friendships 452
Loneliness 455

The Family Life Cycle 456
Leaving Home 456
Joining of Families in Marriage 457
■ SOCIAL ISSUES: Partner Abuse 460
Parenthood 460
■ CULTURAL INFLUENCES: A Global Perspective on Family Planning 464

The Diversity of Adult Lifestyles 466
Singlehood 466
Cohabitation 466
Childlessness 466
Divorce and Remarriage 468
Variant Styles of Parenthood 469

Career Development 471
Establishing a Career 471
Women and Ethnic Minorities 472
Combining Work and Family 473

Summary 475
Important Terms and Concepts 477
FYI: For Further Information and Help 477
MILESTONES: Development in Early Adulthood 478

Part VIII

MIDDLE ADULTHOOD 480

15 Physical and Cognitive Development in Middle Adulthood 480

PHYSICAL DEVELOPMENT 483
Physical Changes 483
Vision 483
Hearing 484
Skin 484
Muscle–Fat Makeup 484
■ BIOLOGY & ENVIRONMENT: Anti-Aging Effects of Dietary Calorie Restriction: Relevant to Humans? 485
Skeleton 486
Reproductive System 486
■ CULTURAL INFLUENCES: Menopause as a Biocultural Event 488

Health and Fitness 489
Sexuality 489
Illness and Disability 489
Hostility and Anger 493

Adapting to the Physical Challenges of Midlife 494
Stress Management 494
Exercise 496
An Optimistic Outlook 496
Gender and Aging: A Double Standard 497

COGNITIVE DEVELOPMENT 498
Changes in Mental Abilities 498
Crystallized and Fluid Intelligence 498
Individual and Group Differences 500

Information Processing 501
Speed of Processing 501
Attention 502
Memory 502
Practical Problem Solving and Expertise 503
Creativity 504
Information Processing in Context 504

Vocational Life and Cognitive Development 505

Adult Learners: Becoming a College Student in Midlife 506
Characteristics of Returning Students 506
Supporting Returning Students 506

Summary 508
Important Terms and Concepts 509
FYI: For Further Information and Help 509

16 Emotional and Social Development in Middle Adulthood 510

Erikson's Theory: Generativity versus Stagnation 512
■ A LIFESPAN VISTA: Generative Adults Tell Their Life Stories 514

Other Theories of Psychosocial Development in Midlife 515
Levinson's Seasons of Life 515
Vaillant's Adaptation to Life 516
Is There a Midlife Crisis? 517
Stage or Life Events Approach 517

Stability and Change in Self-Concept and Personality 518
Possible Selves 518
Self-Acceptance, Autonomy, and Environmental Mastery 519
Coping Strategies 519
■ BIOLOGY & ENVIRONMENT: What Factors Promote Psychological Well-Being in Midlife? 520
Gender Identity 520
Individual Differences in Personality Traits 522

Relationships at Midlife 524
Marriage and Divorce 524
Changing Parent–Child Relationships 525
Grandparenthood 526

■ SOCIAL ISSUES: Grandparents Rearing Grandchildren: The Skipped-Generation Family 528
Middle-Aged Children and Their Aging Parents 529
Siblings 531
Friendships 532
Relationships Across Generations 533

Vocational Life 533
Job Satisfaction 534
Career Development 535
Career Change at Midlife 536
Unemployment 537
Planning for Retirement 537

Summary 539
Important Terms and Concepts 540
FYI: For Further Information and Help 541
MILESTONES: Development in Middle Adulthood 542

Part IX

LATE ADULTHOOD 544

17 Physical and Cognitive Development in Late Adulthood 544

PHYSICAL DEVELOPMENT 546

Life Expectancy 547
Variations in Life Expectancy 547
Life Expectancy in Late Adulthood 547
■ A LIFESPAN VISTA: What Can We Learn About Aging from Centenarians? 550
Maximum Lifespan 548

Physical Changes 549
Nervous System 549
Sensory Systems 550
Cardiovascular and Respiratory Systems 553
Immune System 554
Sleep 554
Physical Appearance and Mobility 555
Adapting to Physical Changes of Late Adulthood 556
■ CULTURAL INFLUENCES: Cultural Variations in the Experience of Aging 558

Health, Fitness, and Disability 558
Nutrition and Exercise 560
Sexuality 560
Physical Disabilities 561
Mental Disabilities 564
■ SOCIAL ISSUES: Interventions for Caregivers of Elders with Dementia 568
Health Care 567

COGNITIVE DEVELOPMENT 570

Memory 571
Deliberate versus Automatic Memory 571
Associative Memory 571
Remote Memory 572
Prospective Memory 573

Language Processing 573

Problem Solving 574

Wisdom 575

Factors Related to Cognitive Change 576

Cognitive Interventions 577

Lifelong Learning 577
Types of Programs 577
Benefits of Continuing Education 578

Summary 579
Important Terms and Concepts 581
FYI: For Further Information and Help 581

18 Emotional and Social Development in Late Adulthood 582

Erikson's Theory: Ego Integrity versus Despair 584

Other Theories of Psychosocial Development in Late Adulthood 585
Peck's Theory: Three Tasks of Ego Integrity 585
Labouvie-Vief's Theory: Emotional Expertise 585
Reminiscence and Life Review 586
■ A LIFESPAN VISTA: World War II Refugee and Evacuee Children Look Back from the Vantage Point of Old Age 588

Stability and Change in Self-Concept and Personality 587
Secure and Multifaceted Self-Concept 587
Agreeableness, Sociability, and Acceptance of Change 587
Spirituality and Religiosity 587

Individual Differences in Psychological Well-Being 591
Control versus Dependency 591
Health 592
■ SOCIAL ISSUES: Elder Suicide 592
Negative Life Changes 593
Social Support and Social Interaction 594

A Changing Social World 595
Social Theories of Aging 595
■ BIOLOGY & ENVIRONMENT: Taking Time Seriously: Time Perception and Social Goals 597

Social Contexts of Aging: Communities, Neighborhoods, and Housing 596

Relationships in Late Adulthood 600
Marriage 601
Gay and Lesbian Partnerships 601
Divorce and Remarriage 602
Widowhood 602
Never-Married, Childless Older Adults 603
Siblings 604
Friendships 605
Relationships with Adult Children 606
Relationships with Adult Grandchildren and Great-Grandchildren 606
Elder Maltreatment 607

Retirement and Leisure 609
The Decision to Retire 609
Adjustment to Retirement 610
Leisure Activities 611

Successful Aging 612

Summary 613
Important Terms and Concepts 615
FYI: For Further Information and Help 615
MILESTONES: Development in Late Adulthood 616

Part X

THE END OF LIFE 618

19 Death, Dying, and Bereavement 618

How We Die 620
Physical Changes 620
Defining Death 621
Death with Dignity 622

Understanding of and Attitudes Toward Death 622
Childhood 623
Adolescence 624
Adulthood 625
Death Anxiety 626

Thinking and Emotions of Dying People 627
Do Stages of Dying Exist? 627
Contextual Influences on Adaptations to Dying 628

A Place to Die 631
Home 631
Hospital 631
The Hospice Approach 632

The Right to Die 633
Passive Euthanasia 634
Voluntary Active Euthanasia 636
■ SOCIAL ISSUES: Voluntary Active Euthanasia: Lessons from Australia and the Netherlands 637
Assisted Suicide 636

Bereavement: Coping with the Death of a Loved One 639
Grief Process 639
Personal and Situational Variations 640
■ CULTURAL INFLUENCES: Cultural Variations in Mourning Behavior 642
Bereavement Interventions 644

Death Education 645

Summary 646
Important Terms and Concepts 647
FYI: For Further Information and Help 647

Glossary G-1
References R-1
Name Index NI-1
Subject Index SI-1

A Personal Note to Students

My 33 years of teaching child development have brought me in contact with thousands of students like you—students with diverse college majors, future goals, interests, and needs. Some are affiliated with my own department, psychology, but many come from other related fields—education, sociology, anthropology, family studies, nursing, and biology, to name just a few. Each semester, my students' aspirations have proved to be as varied as their fields of study. Many look toward careers in applied work—caregiving, nursing, counseling, social work, school psychology, and program administration. Some plan to teach, and a few want to do research. Most hope someday to become parents, whereas others are already parents who come with a desire to better understand and rear their children. And almost all arrive with a deep curiosity about how they themselves developed from tiny infants into the complex human beings they are today.

My goal in preparing this third edition of *Development Through the Lifespan* is to provide a textbook that meets the instructional goals of your course as well as your personal interests and needs. To achieve these objectives, I have grounded this book in a carefully selected body of classic and current theory and research. In addition, the text highlights the lifespan perspective on development and the interacting contributions of biology and environment to the developing person. It also illustrates commonalities and differences between ethnic groups and cultures, and discusses the broader social contexts in which we develop. I have provided a unique pedagogical program that will assist you in mastering information, integrating various aspects of development, critically examining controversial issues, and applying what you have learned.

I hope that learning about human development will be as rewarding for you as I have found it over the years. I would like to know what you think about both the field of human development and this book. I welcome your comments; please feel free to send them to me at Department of Psychology, Box 4620, Illinois State University, Normal, IL 61790, or care of the publisher, who will forward them to me.

—Laura E. Berk

Preface for Instructors

My decision to write *Development Through the Lifespan* was inspired by a wealth of professional and personal experiences. First and foremost were the interests and concerns of hundreds of students of human development with whom I have worked in three decades of college teaching. Each semester, their insights and questions have revealed how an understanding of any single period of development is enriched by an appreciation of the entire lifespan. Second, as I moved through adult development myself, I began to think more intensely about factors that have shaped and reshaped my own life course—family, friends, mentors, coworkers, community, and larger society. My career well established, my marriage having stood the test of time, and my children launched into their adult lives, I felt that a deeper grasp of these multiple, interacting influences would help me better appreciate where I had been and where I would be going in the years ahead. I was also convinced that such knowledge could contribute to my becoming a better teacher, scholar, family member, and citizen. And because teaching has been so central and gratifying a part of my work life, I wanted to bring to others a personally meaningful understanding of lifespan development.

The years since *Development Through the Lifespan* first appeared have been a period of considerable expansion and change in theory and research. This third edition represents these rapidly transforming aspects of the field, with a wealth of new content and teaching tools:

- *Diverse pathways of change are highlighted.* Investigators have reached broad consensus that variations in biological makeup and everyday tasks lead to wide individual differences in paths of change and resulting competencies. This edition pays more attention to variability in development and to recent theories—including ecological, sociocultural, and dynamic systems—that attempt to explain it. Multicultural and cross-cultural findings, including international comparisons, are enhanced throughout the text. Biology and Environment and Cultural Influences boxes also accentuate the theme of diversity in development.
- *The lifespan perspective is emphasized.* As in previous editions, the lifespan perspective—development as lifelong, multidimensional, multidirectional, plastic, and embedded in multiple contexts—continues to serve as a unifying approach to understanding human change and is woven thoroughly into the text. In addition, special Lifespan Vista boxes discuss lifespan-perspective assumptions and consider development across a wide age span.
- *The complex, bidirectional relationship between biology and environment is given greater attention.* Accumulating evidence on development of the brain, motor skills, cognitive competencies, temperament and personality, and developmental problems underscores the way biological factors emerge in, are modified by, and share power with experience. The interconnection between biology and environment is integral to the lifespan perspective and is revisited throughout the text narrative.
- *Inclusion of interdisciplinary research is expanded.* The move toward viewing thoughts, feelings, and behavior as an integrated whole, affected by a wide array of influences in biology, social context, and culture, has motivated developmental researchers to strengthen their links with other fields of psychology and with other disciplines. Topics and findings included in this edition increasingly reflect the contributions of educational psychology, social psychology, health psychology, clinical psychology, neuropsychology, biology, pediatrics, geriatrics, sociology, anthropology, social welfare, and other fields.
- *The links between theory, research, and applications are strengthened.* As researchers intensify their efforts to generate findings that can be applied to real-life situations, I have placed greater weight on social policy issues and sound theory- and research-based applications.
- The role of active student learning is made more explicit. Ask Yourself questions at the end of each major section have been expanded to promote three approaches to engaging actively with the subject matter—*Review, Apply,* and *Connect.* This feature assists students in reflecting on what they have read from multiple vantage points. In addition, definitions of important terms have been highlighted within the text, allowing students to see these terms in a meaningful context.

Text Philosophy

The basic approach of this book has been shaped by my own professional and personal history as a teacher, researcher, and parent. It consists of seven philosophical ingredients that I regard as essential for students to emerge from a course with a thorough understanding of lifespan development. Each theme is woven into every chapter:

1. An understanding of the diverse array of theories in the field and the strengths and shortcomings of each. The first chapter begins by emphasizing that only knowledge of multiple theories can do justice to the richness of human development.

As I take up each age period and domain of development, I present a variety of theoretical perspectives, indicate how each highlights previously overlooked aspects of development, and discuss research that evaluates it. Consideration of contrasting theories also serves as the context for an evenhanded analysis of many controversial issues.

2. **A grasp of the lifespan perspective as an integrative approach to development.** I introduce the lifespan perspective as an organizing framework in the first chapter and continually refer to and illustrate its assumptions throughout the text, in an effort to help students construct an overall vision of development from conception to death.

3. **Knowledge of both the sequence of human development and the processes that underlie it.** Students are provided with a discussion of the organized sequence of development along with processes of change. An understanding of process—how complex combinations of biological and environmental events produce development—has been the focus of most recent research. Accordingly, the text reflects this emphasis. But new information about the timetable of change has also emerged. In many ways, the very young and the old have proved to be far more competent than they were believed to be in the past. In addition, many milestones of adult development, such as finishing formal education, entering a career, getting married, having children, and retiring, have become less predictable. Current evidence on the sequence and timing of development, along with its implications for process, is presented for all periods of the lifespan.

4. **An appreciation of the impact of context and culture on human development.** A wealth of research indicates that people live in rich physical and social contexts that affect all domains of development. Throughout the book, students travel to distant parts of the world as I review a growing body of cross-cultural evidence. The text narrative also discusses many findings on socioeconomically and ethnically diverse people within the United States and Canada. Furthermore, the impact of historical time period and cohort membership receives continuous attention. In this vein, gender issues—the distinctive but continually evolving experiences, roles, and life paths of males and females—are granted substantial emphasis. Besides highlighting the effects of immediate settings, such as family, neighborhood, and school, I make a concerted effort to underscore the influence of larger social structures—societal values, laws, and government programs—on lifelong well-being.

5. **An understanding of the joint contributions of biology and environment to development.** The field recognizes more powerfully than ever before the joint roles of hereditary/constitutional and environmental factors—that these contributions to development combine in complex ways and cannot be separated in a simple manner. Numerous examples of how biological dispositions can be maintained as well as transformed by social contexts are presented throughout the book.

6. **A sense of the interdependency of all domains of development—physical, cognitive, emotional, and social.** Every chapter emphasizes an integrated approach to human development. I show how physical, cognitive, emotional, and social development are interwoven. Within the text narrative, and in a special series of Ask Yourself questions at the end of major sections, students are referred to other sections of the book to deepen their grasp of relationships among various aspects of change.

7. **An appreciation of the interrelatedness of theory, research, and applications.** Throughout this book, I emphasize that theories of human development and the research stimulated by them provide the foundation for sound, effective practices with children, adolescents, and adults. The link between theory, research, and applications is reinforced by an organizational format in which theory and research are presented first, followed by practical implications. In addition, a current focus in the field—harnessing knowledge of human development to shape social policies that support human needs throughout the lifespan—is reflected in every chapter. The text addresses the current condition of children, adolescents, and adults in the United States, Canada, and around the world and shows how theory and research have combined with public interest to spark successful interventions. Many important applied topics are considered, such as family planning, infant mortality, maternal employment and child care, teenage pregnancy and parenthood, domestic violence, exercise and adult health, lifelong learning, grandparents rearing grandchildren, adjustment to retirement, and adapting to widowhood.

Text Organization

I have chosen a chronological organization for *Development Through the Lifespan.* The book begins with an introductory chapter that describes the history of the field, contemporary theories, and research strategies. It is followed by two chapters on the foundations of development. Chapter 2 combines an overview of biological and environmental contexts into a single, integrated discussion of these multifaceted influences on development. Chapter 3 is devoted to prenatal development, birth, and the newborn baby. With this foundation, students are ready to look closely at seven major age periods: infancy and toddlerhood (Chapters 4, 5, and 6), early childhood (Chapters 7 and 8), middle childhood (Chapters 9 and 10), adolescence (Chapters 11 and 12), early adulthood (Chapters 13 and 14), middle adulthood (Chapters 15 and 16), and late adulthood (Chapters 17 and 18). Topical chapters within each chronological division cover physical development, cognitive development, and emotional and social development. The book concludes with a chapter on death, dying, and bereavement (Chapter 19).

The chronological approach assists students in thoroughly understanding each age period. It also eases the task of integrating the various domains of development because each is discussed in close proximity. At the same time, a chronologically organized book requires that theories covering several age periods be presented piecemeal. This creates a challenge for students, who must link the various parts together. To assist with this task, I frequently remind students of important earlier achievements before discussing new developments, referring back to related sections with page references. Also, chapters or sections devoted to the same topic (for example, cognitive development) are similarly organized, making it easier for students to draw connections across age periods and construct an overall view of developmental change.

New Coverage in the Third Edition

Lifespan development is a fascinating and ever-changing field of study, with constantly emerging new discoveries and refinements in existing knowledge. The third edition represents this burgeoning contemporary literature, with over 1,700 new citations. Cutting-edge topics throughout the text underscore the book's major themes. Here is a sampling:

- **Chapter 1:** Updated Biology and Environment box on resiliency • New illustration of information-processing research—a study of children's problem solving • New section on evolutionary developmental psychology • Updated consideration of Vygotsky's view of development • New, clearer illustration of the structure of the environment in ecological systems theory • New Social Issues box on the Access Program, a community–researcher partnership • Enhanced discussion of cohort effects, with examples.
- **Chapter 2:** Expanded and updated discussion of basic genetics, including resemblance of the human genome to the genomes of other species and gene–environment exchanges within the cell • Updated Social Issues box on the pros and cons of reproductive technologies, including new techniques and related ethical concerns • Updated consideration of the implications of the Human Genome Project for lifespan development, with emphasis on the status of genetic treatments • Revised and updated section on the development of adopted children • Updated section on environmental contexts for development, with special attention to family and neighborhood influences and the overall well-being of children, families, and the aged in the United States and Canada • New section on environmental influences on gene expression, including discussion of epigenesis • New Biology and Environment box on uncoupling genetic–environmental correlations for mental illness and antisocial behavior.
- **Chapter 3:** Updated discussion of teratogens and other maternal influences on birth outcomes and later development • Enhanced Lifespan Vista box on prenatal development and health in later life • Updated statistics and findings on medical interventions during childbirth • Revised and updated Social Issues box on a cross-national perspective on health care and other policies for parents and newborn babies • Updated Biology and Environment box on sudden infant death syndrome • Enhanced section on infant crying, including cultural variations in infant soothing and new evidence on abnormal crying • New findings on infant pain perception • Updated consideration of the role of newborn smell in locating food and identifying the caregiver.
- **Chapter 4:** Enhanced discussion of brain development, including production and migration of neurons, formation of synapses, and synaptic pruning • Updated research on the role of early experience in inducing brain lateralization • New Lifespan Vista box on brain plasticity throughout the lifespan, with insights from research on children and adults with brain damage • Updated Cultural Influences box on cultural variation in infant sleeping arrangements, with information on the protective value of cosleeping for at-risk infants • Enhanced coverage of sensitive periods in brain development, including discussion of appropriate stimulation in infancy • Updated information on the benefits of breastfeeding • New research on the negative consequences of malnutrition for health and psychological development • Updated findings on the development of speech perception • New findings on the relationship of motor development to depth perception • Expanded and updated consideration of early face perception.
- **Chapter 5:** Reorganized and condensed presentation of Piaget's sensorimotor stage • Clarified definition of mental representation and its functions in cognition • Description and critique of the violation-of-expectation method • Updated research on development of object permanence • Inclusion of the core knowledge perspective on cognitive development • New Biology and Environment box addressing the question, Do infants have built-in numerical knowledge? • Revised presentation of the information-processing system, including the central executive • Updated research on development of infant attention • Updated findings on development of representation, memory, and categorization in infancy • Updated Lifespan Vista box on infantile amnesia • Expanded treatment of the social context of early cognitive development, with attention to cultural variations in mental strategies • New evidence on long-term effects of the Carolina Abecedarian Project • Updated findings on the role of babbling in early language development.
- **Chapter 6:** Updated Lifespan Vista box on maternal depression and children's development • Cross-cultural research on development of stranger anxiety • Enhanced discussion of emotional self-regulation, including cultural variations • Expanded discussion of Rothbart's model of temperament • New research on the stability of temperament • Enhanced treatment of cultural influences on temperament • Updated discussion of the stability of attachment • New findings on the relationship of disorganized/disoriented attachment to later development • Enhanced discussion of the ecological context of attachment • Updated consideration of early sibling relationships •

Updated Social Issues box on whether child care in infancy threatens attachment security and later adjustment, including new findings from the NICHD Study of Early Child Care • Clarified description of the I-self and the me-self • Enhanced discussion of individual differences in early development of self-control, with emphasis on the importance of sensitive, supportive parenting.

- **Chapter 7:** New evidence on lateralization and handedness • Expanded discussion of environmental influences on preschoolers' eating behaviors • Updated statistics on child health indicators, including nutrition, immunization, and infectious disease • New Social Issues box on consequences of chronic middle ear infection for development • International comparisons of rates of childhood unintentional injuries • Updated consideration of development of drawing • Inclusion of research on scaffolding and cognitive development • Revised section evaluating Vygotsky's theory • New Cultural Influences box on young children's daily life in a Yucatec Mayan village, illustrating diversity in preschool cognitive development • New section on advances in autobiographical memory, including research on adult communication styles for prompting children's autobiographical narratives • Updated discussion of the young child's theory of mind, including an illustration of a typical false-belief task • Enhanced consideration of early literacy experiences, with emphasis on the consequences of storybook reading for language, literacy, and academic skills • Updated discussion of children's early math problem-solving strategies • New evidence on outcomes associated with child-centered versus academic preschools and kindergartens • Expanded and updated discussion of the long-term benefits of preschool intervention for poverty-stricken children • New findings on the impact of educational television on cognitive and social skills • Updated research on vocabulary development, emphasizing the diverse strategies preschoolers use to figure out word meanings.
- **Chapter 8:** New Cultural Influences box on implications of cultural variations in personal storytelling for preschoolers' self-concepts • Updated research on preschoolers' self-esteem • New evidence on understanding of emotions, emotional self-regulation, and self-conscious emotions • Revised and updated section on parenting and children's peer relations, including direct and indirect influences • New evidence on North American parents' use of corporal punishment • New research on moral understanding, with special attention to preschoolers' capacity to distinguish among moral imperatives, social conventions, and matters of personal choice • Updated research on development of aggression, including new findings on the effects of TV violence • Enhanced discussion of genetic influences on gender typing, including a new Lifespan Vista box describing a case study of a boy reared as a girl • Updated treatment of child-rearing styles and their impact on development • New findings on ethnic variations in child-rearing styles • Updated section on child maltreatment.
- **Chapter 9:** Current findings on development of myopia • Updated and expanded discussion of childhood obesity • Enhanced consideration of childhood illnesses, including children with chronic diseases • Current information on extent of physical education • Expanded discussion of spatial reasoning, including children's understanding of maps • Enhanced treatment of an information-processing view of concrete operational thought, with greater attention to Case's neo-Piagetian theory • Attention to parents' and teachers' influences on children's planning and cognitive self-regulation • Updated findings on reading and mathematics development • Revised section on Sternberg's triarchic theory of intelligence • New research on culture, communication styles, and children's mental test performance • Discussion of dynamic testing in reducing cultural bias in intelligence testing • Revised and updated sections on bilingual development and bilingual education • New Caregiving Concerns table on signs of high-quality education in elementary school • New Social Issues box on school readiness and early grade retention • Consideration of reciprocal teaching, a Vygotsky-inspired educational innovation • Updated research on gifted children, with attention to the distinction between talent and creativity and family contributions to development of talent • New evidence on academic achievement of American and Canadian students in cross-national perspective.
- **Chapter 10:** Enhanced discussion of self-concept and self-esteem, including cultural variations and parenting influences • Expanded treatment of self-conscious emotions, emotional understanding, and emotional self-regulation • Updated consideration of children's understanding of links between moral rules and social conventions • Expanded discussion of peer groups • New evidence on peer acceptance, including two subtypes of popular children • Updated Biology and Environment box on bullies and their victims • Updated section on divorce, with special attention to long-term consequences • New evidence on child care for school-age children • Updated Lifespan Vista box on children of war.
- **Chapter 11:** Updated statistics on physical activity rates of American and Canadian teenagers • Expanded treatment of ethnic variations in pubertal growth • New findings on the contribution of family experiences to pubertal timing • Enhanced discussion of changes in parent–child relationships at puberty • Updated research on anorexia nervosa and bulimia nervosa • Expanded discussion of factors related to adolescent parenthood • Updated findings on substance use and abuse, including international comparisons • Expanded treatment of information-processing research on adolescent cognition • Updated coverage of sex differences in mathematical and spatial abilities • Current evidence on the impact of school transitions on adolescent adjustment • Enhanced discussion of parenting and adolescent achievement, including a new section on parent–school partnerships • Updated research on factors leading up to dropping out

of high school and long-term consequences • New Lifespan Vista box on extracurricular activities and positive youth development.

- **Chapter 12:** Updated research on self-concept and self-esteem • Enhanced discussion of identity development, including the role of close friends • Updated Cultural influences box on ethnic identity • Expanded treatment of sex differences in moral reasoning, including cross-cultural research • Enhanced discussion of influences on moral reasoning, including a revised section on the impact of culture • New Social Issues box on development of civic responsibility • Updated discussion of parent–child relationships in adolescence • Expanded treatment of parent–child relationships and peer-group membership • Updated consideration of adolescent dating, including special challenges for homosexual adolescents • New evidence on adolescent depression, including sex differences • Special emphasis on the joint influence of personal an contextual factors on adolescent problem behavior, as illustrated by depression, suicide, and delinquency.
- **Chapter 13:** Updated discussion of theories of biological aging • Expanded consideration of SES differences in health, including, cross-national variations • Updated discussion of factors contributing to the decline in female reproductive capacity after age 35 • Current evidence on implications of overweight and obesity for health • New Lifespan Vista box on why the United States exceeds all other nations in prevalence of obesity • Updated findings on exercise intensity and health, including a comparison of American and Canadian government recommendations • Current evidence on the prevalence of smoking in young adults • New research on attitudes toward homosexuality • Updated discussion of factors linked to sexual coercion • Expanded consideration of psychological stress, including changes from early to middle adulthood • Enhanced discussion of psychological changes during the college years • New findings on gender differences in vocational choice • Updated Cultural Influences box on work–study apprenticeships in Germany.
- **Chapter 14:** Updated research on gender differences and similarities in mate selection • Updated Lifespan Vista box on associations between childhood attachment patterns and adult romantic relationships • New research on factors related to partner abuse, including international comparisons illustrating the importance of traditional gender roles and poverty • Enhanced discussion of changes in loneliness during adulthood • Updated findings on factors linked to timing of leaving home in early adulthood • New research on marital roles, with international comparisons of men's and women's participation in housework • New findings on factors linked to marital satisfaction, including the role of religiosity • Revised section on the transition to parenthood, including factors that predict continued marital happiness versus marital strain • Updated research on family size and parents' and children's well-being • New Cultural Influences box providing a global perspective on family planning • Expanded discussion of the relationship of cohabitation to marital success, including the role of cultural values and the cohabitation experience • Enhanced consideration of factors linked to divorce • New evidence on adjustment of stepparents and never-married single parents • Expanded and updated consideration of career development of women and ethnic minorities • Updated discussion of factors linked to work–family role overload and its consequences for parents' and children's well-being.
- **Chapter 15:** Updated discussion of changes in vision, including risk of glaucoma • Revised and updated discussion of hormone therapy, including recent evidence on benefits and risks • Updated statistics on illness and disability in midlife • Expanded discussion of stress management in adulthood • Enhanced consideration of physical and mental health benefits of exercise in midlife, and ways to encourage middle-aged adults to exercise • New research on gender and the double standard of aging • Expanded discussion of changes in mental abilities during adulthood, including Kaufman's research on verbal and performance IQ, with implications for development of crystallized and verbal intelligence • Enhanced consideration of changes in the quality of creativity in middle adulthood • Revised section on supporting returning students.
- **Chapter 16:** Expanded discussion of generativity, including ethnic variations • New Lifespan Vista box on life stories of generative adults • New findings on midlife crisis • Updated consideration of midlife gains in coping strategies • New evidence on stability of personality traits during adulthood • Enhanced discussion of changing parent–child relationships in midlife, including cultural variations in "launching children and moving on" • Updated section on caring for aging parents, including cultural variations in sense of obligation and in caregiver stress • Revised and updated Social Issues box on grandparents rearing grandchildren • New evidence on development of midlife sibling relationships • Updated research on midlife job satisfaction • Enhanced consideration of planning for retirement.
- **Chapter 17:** Revised and expanded discussion of average life expectancy and active lifespan, including international comparisons • Updated Lifespan Vista box on centenarians • Updated research on aging of the nervous system and declines in vision and hearing • Revised and expanded consideration adapting to physical changes of late adulthood, including new sections on assistive technology and overcoming stereotypes of aging • Enhanced treatment of cultural variations in the experience of aging • New findings on the impact of diet, vitamin-mineral supplements, and exercise on physical aging and health in late adulthood • Updated discussion of barriers to regular exercise in late life, with implications for effective intervention • Updated statistics on leading causes of death in late adulthood • Updated statistics on SES and ethnic variations in late-life health • New findings

on risk of dementia in African Americans • Updated research on genetic and environmental risk factors for Alzheimer's disease • New Social Issues box on interventions for caregivers of elders with dementia • Enhanced discussion of long-term care, including nursing home placement and ethnic and international comparisons • Expanded discussion of changes in memory, including a new section on associative memory • New research on competence in everyday problem solving in late adulthood • Expanded and updated discussion of the cognitive and emotional ingredients of wisdom, and its development.

- **Chapter 18:** Updated research on the functions of reminiscence • Enhanced discussion of religion and spirituality, with new findings on consequences for elders' well-being • Updated Lifespan Vista box on the influence of a shortened time perspective on elders' social goals • New section on elderly gay and lesbian partners • Updated consideration of adjustment to widowhood • New findings on changes in sibling relationships in late adulthood • Updated research on grandparent–adult grandchild relationships • New evidence on adaptation to retirement • Increased emphasis on elders' active efforts to sustain a sense of personal continuity in coping with diverse life challenges • Expanded treatment of gender differences in late-life adjustment • New research on successful aging.
- **Chapter 19:** Updated research on death anxiety • Revised and updated discussion of thinking and emotions of dying people, emphasizing individual and cultural differences • Expanded explanation of appropriate death • New findings on cultural variations in candidness with dying people • Updated consideration of hospital death, pointing out the need for comprehensive treatment programs aimed at easing physical, emotional, and spiritual pain • Inclusion of research on the benefits of hospice for dying patients and their families • Updated discussion of euthanasia and assisted suicide, including recent opinion polls and legal status in Western nations • Enhanced consideration of legalization of doctor-assisted suicide in Oregon • Updated Social Issues box on legalization of voluntary active euthanasia, with special emphasis on the Netherlands • Inclusion of the dual-process model of grieving • Expanded discussion of bereavement overload, including risks posed by random school murders and terrorist attacks • Updated coverage of bereavement interventions • Updated Cultural Influences box on cultural variations in mourning behavior, including discussion of website "cemeteries."

Pedagogical Features

Maintaining a highly accessible writing style—one that is lucid and engaging without being simplistic—continues to be one of the text's goals. I frequently converse with students, encouraging them to relate what they read to their own lives. In doing so, I hope to make the study of human development involving and pleasurable.

- **Stories and Vignettes About Real People.** To help students construct a clear image of development and to enliven the text narrative, each chronological age division is unified by case examples woven throughout that set of chapters. For example, the middle childhood section highlights the experiences and concerns of 10-year-old Joey; 8-year-old Lizzie; their divorced parents, Rena and Drake; and their classmates Mona, Terry, and Jermaine. In the chapters on late adulthood, students get to know Walt and Ruth, a vibrant retired couple, along with Walt's older brother Dick and his wife Goldie and Ruth's sister Ida, a victim of Alzheimer's disease. Besides a set of main characters, who bring unity to each age period, many additional vignettes offer vivid examples of development and diversity among children, adolescents, and adults.

- **Chapter Introductions and End-of-Chapter Summaries.** To provide a helpful preview of chapter content, I include an outline and overview in each chapter introduction. Especially comprehensive end-of-chapter summaries, organized according to the major divisions of each chapter and highlighting important terms, remind students of key points in the text discussion. Review questions are included in the summary to encourage active study.

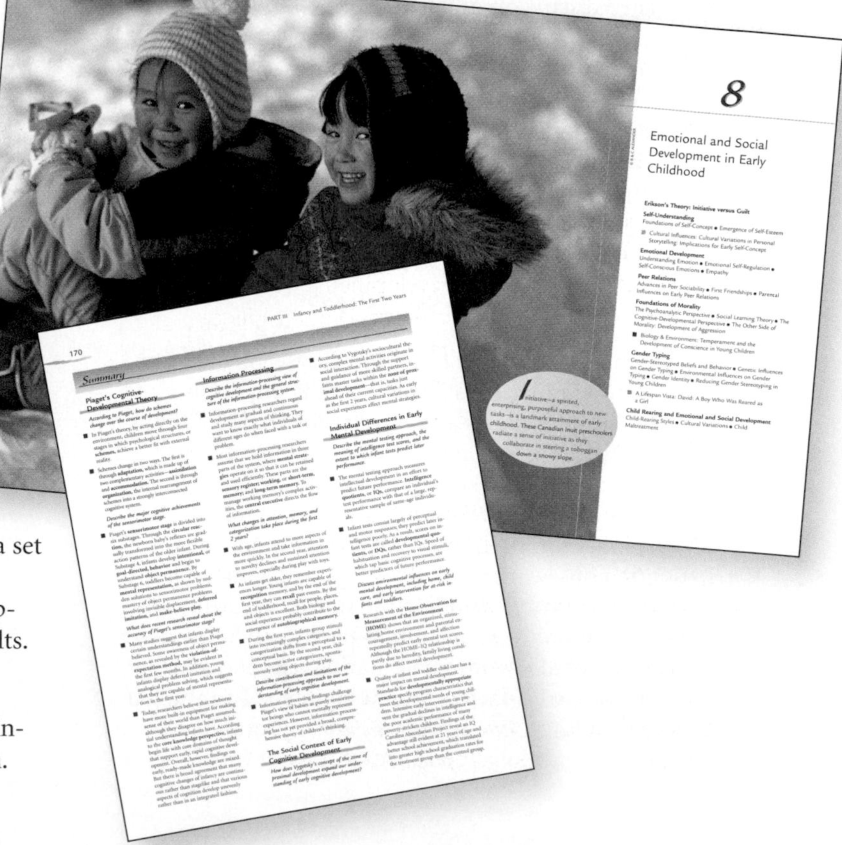

● **Ask Yourself Questions.** Active engagement with the subject matter is also supported by Ask Yourself questions at the end of each major section. Three types of questions prompt students to think about human development in diverse ways: ***Review*** questions help students recall and comprehend information they have just read; ***Apply*** questions encourage the application of knowledge to controversial issues and problems faced by children, adolescents, and adults; ***Connect*** questions help students build an image of the whole person by integrating what they have learned across age periods and domains of development. An icon (*www.*) in the text indicates that each question is answered on the text's companion website. Students may compare their reasoning to a model response.

Four types of thematic boxes accentuate the philosophical themes of this book:

● **A Lifespan Vista boxes** are devoted to topics that have long-term implications for development or involve intergenerational issues. Examples include: *Brain Plasticity: Insights from Research on Brain-Damaged Children and Adults; David: A Boy Who Was Reared as a Girl; Extracurricular Activities: Contexts for Positive Youth Development; The Obesity Epidemic: How Americans Became the Heaviest People in the World;* and *Generative Adults Tell Their Life Stories.*

● **Social Issues boxes** discuss the impact of social conditions on children, adolescents, and adults and emphasize the need for sensitive social policies to ensure their well-being—for example, *The Access Program: A Community–Researcher Partnership; School Readiness and Early Grade Retention; Chronic Middle Ear Infection in Early Childhood and Consequences for Development; Development of Civic Responsibility;* and *Interventions for Caregivers of Elders with Dementia.*

● **Cultural Influences boxes** have been expanded and updated to deepen attention to culture threaded throughout the text. They highlight both cross-cultural and multicultural variations in human development—for example, *Young Children's Daily Life in a Yucatec Mayan Village; Implications of Cultural Variations in Personal Storytelling for Preschoolers' Self-Concepts; A Global Perspective on Family Planning;* and *Cultural Variations in the Experience of Aging.*

● **Biology & Environment boxes.** New to this edition, this special feature highlights the growing attention to the complex, bidirectional relationship between biology and environment. Examples include *Uncoupling Genetic–Environmental Correlations for Mental Illness and Antisocial Behavior; Do Infants Have Built-In Numerical Knowledge?; Children with Attention-Deficit Hyperactivity Disorder; Sex Differences in Spatial Abilities;* and *Anti-Aging Effects of Dietary Calorie Restriction: Relevant to Humans?*

● **Caregiving Concerns Tables.** The relationship of theory and research to practice is woven throughout the text narrative. To accentuate this linkage, Caregiving Concerns tables provide easily accessible practical advice on the importance of caring for oneself and others throughout the lifespan. They include: *Dos and Don'ts for a Healthy Pregnancy; Building a Foundation for Good Eating Habits; Signs of High-Quality Education in Elementary School; Ways to Foster a Mastery-Oriented Approach to Learning and Prevent Learned Helplessness; Helping Children Adjust to Their Parents' Divorce; Keeping Love Alive in a Romantic Partnership; Ways Middle-Aged Parents Can Promote Positive Ties with Their Adult Children; Fostering Adaptation to Widowhood in Late Adulthood;* and *Resolving Grief After a Loved One Dies.*

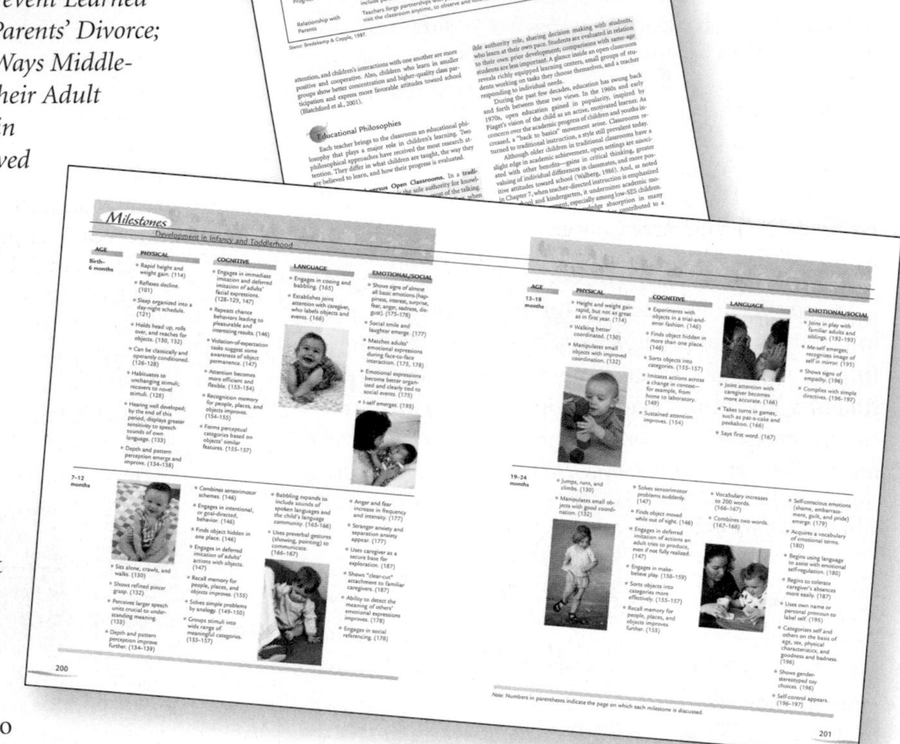

● **Milestones Tables.** A Milestones table appears at the end of each age division of the text. These tables summarize major physical, cognitive, language, emotional, and social attainments, providing a convenient aid for reviewing the chronology of lifespan development.

● **Additional Tables, Illustrations, and Photographs.** Tables are liberally included to help students grasp essential points in the text narrative and extend information on a topic. The many full-color figures and illustrations depict important theories, methods, and research findings. Photos have been carefully selected to portray human development and to represent the diversity of people in the United States, Canada, and around the world.

● **In-Text Key Terms with Definitions, End-of-Chapter Term List and End-of-Book Glossary.** Mastery of terms that make up the central vocabulary of the field is promoted through highlighted key-term and concept definitions, which appear in the text narrative, an end-of-chapter term list, and an end-of-book glossary.

Acknowledgments

The dedicated contributions of many individuals helped make this book a reality and contributed to refinements and improvements in this third edition. An impressive cast of reviewers provided many helpful suggestions, constructive criticisms, and enthusiasm for the organization and content of the text. I am grateful to each one of them.

Reviewers for the First and Second Editions

Paul C. Amrhein, University of New Mexico
Doreen Arcus, University of Massachusetts, Lowell
René L. Babcock, Central Michigan University
W. Keith Berg, University of Florida
James A. Bird, Weber State University
Joyce Bishop, Golden West College
Ed Brady, Belleville Area College
Michele Y. Breault, Truman State University
Joan B. Cannon, University of Massachusetts, Lowell
Michael Caruso, University of Toledo
Gary Creasey, Illinois State University
Rhoda Cummings, University of Nevada, Reno
Rita M. Curl, Minot State University
Carol Lynn Davis, University of Maine
Maria P. Fracasso, Towson University
Elizabeth F. Garner, University of North Florida
Clifford Gray, Pueblo Community College
Laura Hanish, Arizona State University
Traci Haynes, Columbus State Community College
Vernon Haynes, Youngstown State University
Paula Hillman, University of Wisconsin, Whitewater
Lera Joyce Johnson, Centenary College of Louisiana
Janet Kalinowski, Ithaca College
Kevin Keating, Broward Community College
Wendy Kliewer, Virginia Commonwealth University
Robert B. McLaren, California State University, Fullerton
Randy Mergler, California State University
Karla K. Miley, Black Hawk College
Teri Miller, Milwaukee Area Technical College
Gary T. Montgomery, University of Texas, Pan American
Feleccia Moore-Davis, Houston Community College
Karen Nelson, Austin College
Bob Newby, Tarleton State University
Jill Norvilitis, Buffalo State College
Patricia O'Brien, University of Illinois at Chicago
Peter Oliver, University of Hartford
Ellen Pastorino, Gainesville College
Leslee K. Polina, Southeast Missouri State University
Leon Rappaport, Kansas State University
Stephanie J. Rowley, University of North Carolina
Randall Russac, University of North Florida
Marie Saracino, Stephen F. Austin State University
Bonnie Seegmiller, City University of New York, Hunter College
Richard Selby, Southeast Missouri State University
Paul S. Silverman, University of Montana
Glenda Smith, North Harris College
Jeanne Spaulding, Houston Community College
Thomas Spencer, San Francisco State University
Bruce Stam, Chemeketa Community College
Vince Sullivan, Pensacola Junior College
Mojisola F. Tiamiyu, University of Toledo
Joe Tinnin, Richland College
L. Monique Ward, University of Michigan
Rob Weisskirch, California State University, Fullerton
Ursula M. White, El Paso Community College
Lois J. Willoughby, Miami-Dade Community College
Deborah R. Winters, New Mexico State University

Reviewers for the Third Edition

Doreen Arcus, University of Massachusetts, Lowell
Renée L. Babcock, Central Michigan University
Dilek Buchholz, Weber State University
Lanthan D. Camblin Jr., University of Cincinnati
Judith W. Cameron, Ohio State University
Susan L. Churchill, University of Nebraska-Lincoln
Laurie Gottlieb, McGill University
Karl Hennig, St. Francis Xavier University
Rebecca A. López, California State University-Long Beach
Dale A. Lund, University of Utah
Pamela Manners, Troy State University
Steve Mitchell, Somerset Community College
Nancy Ogden, Mount Royal College
Verna C. Pangman, University of Manitoba
Robert Pasnak, George Mason University
Warren H. Phillips, Iowa State University
Dana Plude, University of Maryland
Dolores Pushkar, Concordia University
Pamela Roberts, California State University, Long Beach
Elmer Ruhnke, Manatee Community College
Edythe H. Schwartz, California State University-Sacramento
Catya von Károlyi, University of Wisconsin-Eau Claire
Nancy White, Youngstown State University
Carol L. Wilkinson, Whatcom Community College

In addition, I thank the following individuals for responding to a survey that provided vital feedback for the new edition:

Glen Adams, Harding University
Kenneth Anderson, Calhoun Community College
Drusilla D. Glascoe, Salt Lake Community College
Barbara Hunter, Southwestern College
Mona Ibrahim, Concordia College
Margaret Kasimatis, Carroll College
Richard L. McWhorter, Texas A&M University
Michelle Moriarty, Johnson County Community College
Gayle Pitman, Sacramento City College
Mellis Schmidt, Northern New Mexico Community College
Donna Seagle, Chattanooga State Technical Community College
Romona Smith, Holmes Community College

Colleagues and students at Illinois State University aided my research and contributed significantly to the text's supplements. Richard Payne, Department of Politics and Government, is a kind and devoted friend with whom I have shared many profitable discussions about the writing process, the condition of children and the elderly, and other topics that significantly influenced my perspective on lifespan development and social policy. JoDe Paladino and Sara Harris joined me in preparing a thoroughly revised Instructor's Resource

Manual and Study Guide. In addition, JoDe's outstanding, dedicated work in helping conduct literature searches in preparation for the revision has been invaluable. Tara Kindelberger, Denise Shafer, and Lisa Sowa spent countless hours gathering and organizing library materials, bringing to the task unmatched organizational skills and dependability.

The supplements package also benefited from the talents and diligence of several other individuals. Leslie Barnes-Young of Francis Marion University prepared the excellent Lecture Enhancements included in the Instructor's Resource Manual. Gabrielle Principe of Ursinus College and Naomi Tyler of Vanderbilt University authored a superb Test Bank. Joyce Munsch of California State University, Northridge assisted in creating outstanding materials for the student web site.

I have been fortunate to work with an exceptionally capable editorial team at Allyn and Bacon. It has been a privilege to author this book under the sponsorship of Tom Pauken, Managing Editor. His careful editing of the manuscript, organizational skills, responsive day-to-day communication, insights and suggestions, astute problem solving, and thoughtfulness have greatly enhanced the quality of this edition and made its preparation enjoyable and rewarding. Tom has been a pleasure to work with in all respects, and I look forward to working with him on future projects. Thanks, also, to Sean Wakely, editor of the first edition, for his keen sense of vision and commitment to forging a true editor–author partnership, which made the creation of this text possible. The rich tree images that grace the covers of all three editions, radiant metaphors of lifespan development, were Sean's inspiration.

Liz Napolitano, Senior Production Editor, coordinated the complex production tasks that resulted in an exquisitely beautiful third edition. I am grateful for her exceptional aesthetic sense, attention to detail, flexibility, efficiency, and thoughtfulness. I thank Sarah Evertson and Laurie Frankenthaler for obtaining the exceptional photographs that so aptly illustrate the text narrative. Connie Day and Bill Heckman provided outstanding copyediting and proofreading.

I would like to express a heartfelt thank you to Joyce Nilsen, Director of Sales Specialists, and Brad Parkins and Wendy Gordon, Marketing Managers, for the outstanding work they have done in marketing my texts. Joyce, Brad, and Wendy have made sure that accurate and clear information about my books and their ancillaries reached Allyn and Bacon's sales force and that the needs of prospective and current adopters were met. Marcie Mealia, Field Marketing Specialist, has also devoted much time and energy to marketing activities, and I greatly appreciate the lovely social occasions she has planned and the kind greetings that she sends from time to time.

A final word of gratitude goes to my family, whose love, patience, and understanding have enabled me to be wife, mother, teacher, researcher, and text author at the same time. My sons, David and Peter, have taken a special interest in this project. Their reflections on events and progress in their own lives, conveyed over telephone and e-mail and during holiday visits, helped mold the early adulthood chapters. My husband, Ken, willingly made room for yet another time-consuming endeavor in our life together and communicated his belief in its importance in a great many unspoken, caring ways.

—Laura E. Berk

Supplementary Materials

● **Instructor's Supplements.** A variety of teaching tools are available to assist instructors in organizing lectures, planning demonstrations and examinations, and ensuring student comprehension:

- **Instructor's Resource Manual (IRM).** Prepared by Laura Berk and Sara Harris of Illinois State University and Leslie Barnes-Young of Francis Marion University, this thoroughly revised IRM contains additional material to enrich your class presentations. For each chapter, the IRM provides a Chapter-at-a-Glance grid, Brief Chapter Summary, Learning Objectives, detailed Lecture Outline, Lecture Enhancements, Learning Activities, Ask Yourself questions with answers, Suggested Student Readings, Transparency listing, and Media Materials.
- **Test Bank.** Prepared by Gabrielle Principe of Ursinus College and Naomi Tyler of Vanderbilt University, the Test Bank contains over 2,000 multiple-choice questions, each of which is cross-referenced to a Learning Objective, page-referenced to chapter content, and classified by type (factual, applied, or conceptual). Each chapter also includes a selection of essay questions and sample answers.
- **Computerized Test Bank.** This computerized version of the Test Bank, in easy-to-use software, lets you prepare tests for printing as well as for network and online testing. It has full editing capability for Windows and Macintosh.
- **Transparencies.** Two hundred full-color transparencies taken from the text and other sources are referenced in the IRM for the most appropriate use in your classroom presentations.
- **"Development Through the Lifespan in Action" Observation Program.** This revised and expanded real-life videotape is over two hours in length and contains hundreds of observation segments that illustrate the many theories, concepts, and milestones of human development. New additions include Childbirth, Adolescent Friendship, and Confronting a Life-Threatening Illness in Early Adulthood. An Observation Guide helps students use the video in conjunction with the textbook, deepening their understanding and applying what they have learned to everyday life. The videotape and Observation Guide are free to instructors who adopt the text and are available to students at a discount when packaged with the text.
- **"A Window on Lifespan Development" Running Observational Footage Video.** This second video complements the Observation Program above, through two hours of unscripted footage on many aspects of development through the lifespan. An accompanying Video Guide is also available.
- **PowerPoint CD-ROM.** This CD offers electronic slides of lecture outlines and illustrations from the textbook and allows you to customize content.
- **Allyn & Bacon Digital Media Archive for Berk.** This collection of media products—including charts, graphs, tables, figures, audio, and video clips—assists you in meeting your classroom goals.
- **Online Course Management.** This tool permits the integration of text-specific content, testing materials, and your own material. Available in CourseCompass, Blackboard, and WebCT formats.

● **Student Supplements** Beyond the study aids found in the textbook, Allyn and Bacon offers a number of supplements for students:

- **Grade Aid with Practice Tests.** Prepared by JoDe Paladino, Laura Berk, and Sara Harris of Illinois State University, this helpful guide offers Chapter Summaries, Learning Objectives, Study Questions organized according to major headings in the text, Ask Yourself questions that also appear in the text, Crossword Puzzles for mastering important terms, and two multiple-choice Self-Tests per chapter.
- **Website.** The companion website, *http://www.ablongman.com/berk* offers support for students through chapter-specific learning objectives, annotated web resources, flashcard vocabulary building, activities, practice tests, model answers to the book's Ask Yourself questions, and milestones tables. The site also features an *eThemes of the Times* section with full text articles from the *New York Times.*
- **Video Workshop for *Development Through the Lifespan.*** This complete teaching and learning system includes quality video footage on an easy-to-use CD-ROM plus a Student Learning Guide and an Instructor's Teaching Guide—both with text-specific Correlation Grids. VideoWorkshop is available FREE when packaged with the text. Contact your Allyn & Bacon sales representative for additional details and ordering information or visit *www.ablongman.com/videoworkshop*.
- **ResearchNavigator™.** Through three exclusive databases, this intuitive search interface allows students to efficiently make the most of their research time and provides extensive help in the research process. EBSCO's *ContentSelect* Academic Journal Database permits a discipline-specific search through professional and popular journals. Also featured are the *New York Times* Search-by-Subject Archive, and *Best of the Web* Link Library. (A required access code is contained in *Research Navigator Guide*).

- **Research Navigator Guide: Human Development, with Research Navigator Access Code.** Designed to help students select and evaluate research from the Web, this booklet contains a practical discussion of search engines, detailed information on evaluating online sources, citation guidelines for web resources, additional web activities and links for psychology, and an access code and guide to *ResearchNavigator* (available only when packaged with the text).

- **Tutor Center.** (Access Code Required) The Tutor Center provides students free, one-on-one, interactive tutoring from qualified instructors on all material in the text. Tutors offer help with understanding major developmental principles and suggest effective study techniques. Tutoring assistance is available by phone, FAX, Internet, and e-mail during Tutor Center hours. For more details and ordering information, please contact your Allyn & Bacon publisher's representative or visit *www.aw.com/tutorcenter.*

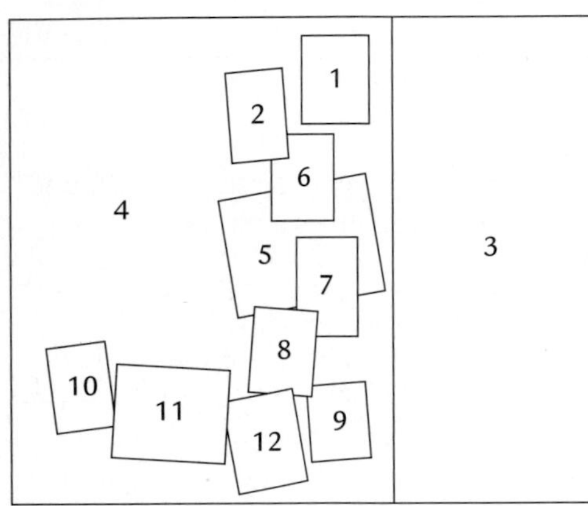

Legend for Photos Accompanying Sofie's Story Sofie's story is told in Chapters 1 and 19, from her birth to her death. The photos that appear at the beginning of Chapter 1 follow her through her lifespan. They are:

1. Sofie as a baby, with her mother in 1908.
2. Sofie, age 6, with her brother, age 8, in 1914.
3. Sofie, age 10, before a birthday party in 1919.
4. Sofie, age 18, high school graduation in 1926.
5. Sofie's German passport.
6. Sofie and Phil in their mid-thirties, during World War II, when they became engaged.
7. Sofie, age 60, and daughter Laura on Laura's wedding day in 1968.
8. Sofie and Phil in 1968, less than 2 years before Sofie died.
9. Sofie, age 61, and her first grandchild, Ellen, October 1969, less than 3 months before Sofie died.
10. Sofie's grandsons, David and Peter, ages 5 and 2, children of Laura and Ken.
11. Laura, Ken, and sons Peter and David, ages 10 and 13, on the occasion of David's Bar Mitzvah in 1985.
12. Laura and sons Peter and David, as young adults, in 2001.

Development Through the Lifespan

Ehefrau
Gestalt
Gesicht oval
Farbe der Augen grau
Unterschrift des Paßinhabers
Geschlecht
KAUAI

PHOTOS COURTESY OF LAURA E. BERK

History, Theory, and Research Strategies

Human Development as a Scientific, Applied, and Interdisciplinary Field

Basic Issues
Continuous or Discontinuous Development? • One Course of Development or Many? • Nature or Nurture as More Important?

The Lifespan Perspective: A Balanced Point of View
Development as Lifelong • Development as Multidimensional and Multidirectional • Development as Plastic • Development as Embedded in Multiple Contexts

■ Biology & Environment: Resiliency

Historical Foundations
Philosophies of Childhood • Philosophies of Adulthood and Aging • Scientific Beginnings

Mid-Twentieth-Century Theories
The Psychoanalytic Perspective • Behaviorism and Social Learning Theory • Piaget's Cognitive-Developmental Theory

Recent Theoretical Perspectives
Information Processing • Ethology and Evolutionary Developmental Psychology • Vygotsky's Sociocultural Theory • Ecological Systems Theory

Comparing and Evaluating Theories

Studying Development
Common Research Methods • General Research Designs • Designs for Studying Development

■ Social Issues: The Access Program: A Community–Researcher Partnership

■ Cultural Influences: Immigrant Youths: Amazing Adaptation

■ A Lifespan Vista: Impact of Historical Times on the Life Course: The Great Depression and World War II

Ethics in Lifespan Research

This photo essay chronicles the life course and family legacy of Sofie Lentschner. It begins in 1908 with Sofie's infancy and concludes in 2001, 30 years after Sofie's death, with her young adult grandsons, Peter and David. For a description of each photo, see the legend on page xxx.

Sofie Lentschner was born in 1908, the second child of Jewish parents who made their home in Leipzig, Germany, a city of thriving commerce and cultural vitality. Her father was a successful businessman and community leader. Her mother was a socialite well known for her charm, beauty, and hospitality. As a baby, Sofie displayed the determination and persistence that would be sustained throughout her life. She sat for long periods inspecting small objects with her eyes and hands. The single event that consistently broke her gaze was the sound of the piano in the parlor. As soon as Sofie could crawl, she steadfastly pulled herself up to finger its keys and marveled at the tinkling sound.

By the time Sofie entered elementary school, she was an introspective child, often ill at ease at the festive parties that girls of her family's social standing were expected to attend. She immersed herself in her schoolwork, especially in mastering the foreign languages that were a regular part of German elementary and secondary education. Twice a week, she took piano lessons from the finest teacher in Leipzig. By the time Sofie graduated from high school, she spoke English and French fluently and had become an accomplished pianist. Whereas most German girls of her time married by age 20, Sofie postponed serious courtship in favor of entering the university. Her parents began to wonder whether their intense, studious daughter would ever settle into family life.

Sofie wanted marriage as well as education, but her plans were thwarted by the political turbulence of her times. When Hitler rose to power in the early 1930s, Sofie's father feared for the safety of his wife and children and moved the family to Belgium. Conditions for Jews in Europe quickly worsened. The Nazis plundered Sofie's family home and confiscated her father's business. By the end of the 1930s, Sofie had lost contact with all but a handful of her aunts, uncles, cousins, and childhood friends, many of whom (she later learned) were herded into cattle cars and transported to the slave labor and death camps at Auschwitz-Birkenau. In 1939, as anti-Jewish laws and atrocities intensified, Sofie's family fled to the United States.

As Sofie turned 30, her parents concluded she would never marry and would need a career for financial security. They agreed to support her return to school, and Sofie earned two master's degrees, one in music and the other in librarianship. Then, on a blind date, she met Philip, a U.S. army officer. Philip's calm, gentle nature complemented Sofie's intensity and worldliness. Within 6 months they married. During the next 4 years, two daughters and a son were born. Soon Sofie's father became ill. The strain of uprooting his family and losing his home and business had shattered his health. After months of being bedridden, he died of heart failure.

When World War II ended, Philip left the army and opened a small men's clothing store. Sofie divided her time between caring for the children and helping Philip in the store. Now in her forties, she was a devoted mother, but few women her age were still rearing young children. As Philip struggled with the business, he spent longer hours at work, and Sofie often felt lonely. She rarely touched the piano, which brought back painful memories of youthful life plans shattered by war. Sofie's sense of isolation and lack of fulfillment frequently left her short-tempered. Late at night, she and Philip could be heard arguing.

As Sofie's children grew older and parenting took less time, she returned to school once more, this time to earn a teaching credential. Finally, at age 50, she launched a career. For the next decade, Sofie taught German and French to high school students and English to newly arrived immigrants. Besides easing her family's financial difficulties, she felt a gratifying sense of accomplishment and creativity. These years were among the most energetic and satisfying of Sofie's life. She had an unending enthusiasm for teaching—for transmitting her facility with language, her firsthand knowledge of the consequences of hatred and oppression, and her practical understanding of how to adapt to life in a new land. She watched her children, whose young lives were free of the trauma of war, adopt many of her values and commitments and begin their marital and vocational lives at the expected time.

Sofie approached age 60 with an optimistic outlook. As she and Philip were released from the financial burden of paying for their children's college education, they looked forward to greater leisure. Their affection and respect for one another deepened. Once again, Sofie began to play the piano. But this period of contentment was short-lived.

One morning, Sofie awoke and felt a hard lump under her arm. Several days later, her doctor diagnosed cancer. Sofie's spirited disposition and capacity to adapt to radical life changes helped her meet the illness head on. She defined it as an enemy—to be fought and overcome. As a result, she lived 5 more years. Despite the exhaustion of chemotherapy, Sofie maintained a full schedule of teaching duties and continued to visit and run errands for her elderly mother. But as she weakened physically, she no longer had the stamina to meet her classes. Gradually, she gave in to the ravaging illness. Bedridden for the last few weeks, she slipped quietly into death with Philip at her side. The funeral chapel overflowed with hundreds of Sofie's students. She had granted each a memorable image of a woman of courage and caring.

One of Sofie's three children, Laura, is the author of this book. Married a year before Sofie died, Laura and her husband, Ken, often think of Sofie's message, spoken privately to them on the eve of their wedding day: "I learned from my own life and marriage that you must build a life together but also a life apart. You must grant each other the time, space, and support to forge your own identities, your own ways of expressing yourselves and giving to others. The most important ingredient of your relationship must be respect."

Laura and Ken settled in a small midwestern city, near Illinois State University, where they continue to teach today—Laura in the Department of Psychology, Ken in the Department of Mathematics. They have two sons, David and Peter, to whom Laura has related many stories about Sofie's life and who carry her legacy forward. David shares his grandmother's penchant for teaching; he is a first-grade teacher in California. Peter practices law in Chicago; he shares his grandmother's love of music, playing violin, viola, and mandolin in his spare time.

Sofie's story raises a wealth of fascinating issues about human life histories:

- What determines the features that Sofie shares with others and those that make her unique—in physical characteristics, mental capacities, interests, and behaviors?
- What led Sofie to retain the same persistent, determined disposition throughout her life but to change in other essential ways?
- How do historical and cultural conditions—for Sofie, the persecution that destroyed her childhood home, caused the death of family members and friends, and engendered her flight to the United States—affect well-being throughout life?
- How does the timing of events—for example, Sofie's early exposure to foreign languages and her delayed entry into marriage, parenthood, and career—affect development?
- What factors—both personal and environmental—led Sofie to die sooner than expected?

These are central questions addressed by **human development,** a field of study devoted to understanding constancy and change throughout the lifespan. Great diversity characterizes the interests and concerns of investigators who study human development. But all share a single goal: to describe and identify those factors that influence consistencies and transformations in people from conception to death.

Human Development as a Scientific, Applied, and Interdisciplinary Field

Look again at the questions just listed, and you will see that they are not just of scientific interest. Each is of *applied,* or practical, importance as well. In fact, scientific curiosity is just one factor that led human development to become the exciting field of study it is today. Research about development has also been stimulated by social pressures to better people's lives. For example, the beginning of public education in the early part of this century led to a demand for knowledge about what and how to teach children of different ages. The interest of the medical profession in improving people's health required an understanding of physical development, nutrition, and disease. The social service profession's desire to treat anxieties and behavior problems and to help people adjust to major life events, such as divorce, job loss, or the death of a loved one, required information about personality and social development. And parents have continually asked for expert advice about child-rearing practices and experiences that would foster happy and successful lives for their children.

Our large storehouse of information about human development is *interdisciplinary.* It grew through the combined efforts of people from many fields of study. Because of the need for solutions to everyday problems at all ages, researchers from psychology, sociology, anthropology, and biology joined forces in research with professionals from education, family studies, medicine, public health, and social service, to name just a few. Today, the field of human development is a melting pot of contributions. Its body of knowledge is not just scientifically important but also relevant and useful.

Basic Issues

Research on human development is a relatively recent endeavor. Studies of children did not begin until the early part of the twentieth century. Investigations into adult development, aging, and change over the life course emerged only in the 1960s and 1970s (Elder, 1998). Nevertheless, ideas about how people grow and change have existed for centuries. As these speculations combined with research, they inspired the construction of *theories* of development. A **theory** is an orderly, integrated set of statements that describes, explains, and predicts behavior. For example, a good theory of infant–caregiver attachment would (1) *describe* the behaviors of babies of 6 to 8 months of age as they seek the affection and comfort of a familiar adult, (2) *explain* how and why infants develop this strong desire to bond with a caregiver, and (3) *predict* the consequences of this emotional bond for relationships throughout life.

Theories are vital tools for two reasons. First, they provide organizing frameworks for our observations of people. In other words, they *guide and give meaning to what we see.* Second, theories that are verified by research provide a sound basis for practical action. Once a theory helps us *understand* development, we are in a much better position to know *what to do* in our efforts to improve the welfare and treatment of children and adults.

As we will see later, theories are influenced by the cultural values and belief systems of their times. But theories differ in one important way from mere opinion and belief: A theory's continued existence depends on *scientific verification.* This means that the theory must be tested with a fair set of research procedures agreed on by the scientific community and that its findings must endure, or be replicated, over time.

In the field of human development, there are many theories with very different ideas about what people are like and how they change. The study of development provides no ultimate truth because investigators do not always agree on the meaning of what they see. In addition, humans are complex beings; they change physically, mentally, emotionally, and socially. As yet, no single theory has explained all these aspects. However, the existence of many theories helps advance knowledge because researchers are continually trying to support, contradict, and integrate these different points of view.

This chapter introduces you to major theories of human development and research strategies used to test them. We will return to each theory in greater detail, as well as introduce other important but less grand theories, in later chapters. Although there are many theories, we can easily organize them, since almost all take a stance on three basic issues: (1) Is the course of development continuous or discontinuous? (2) Does one course of development characterize all people, or are there many possible courses? (3) Are genetic or environmental factors more important in determining development? Let's look closely at each of these issues.

Continuous or Discontinuous Development?

How can we best describe the differences in capacities between small infants, young children, adolescents, and adults? As Figure 1.1 illustrates, major theories recognize two possibilities.

One view holds that infants and preschoolers respond to the world in much the same way as adults do. The difference between the immature and mature being is simply one of *amount or complexity*. For example, when Sofie was a baby, her perception of a piano melody, memory for past events, and ability to sort objects into categories may have been much like our own. Perhaps her only limitation was that she could not perform these skills with as much information and precision as we can. If this is so, then change in her thinking must be **continuous**—a process of gradually augmenting the same types of skills that were there to begin with.

A second view regards infants and children as having *unique ways of thinking, feeling, and behaving,* ones quite different from adults'. In other words, development is **discontinuous**—a process in which new and different ways of interpreting and responding to the world emerge at particular time periods. From this perspective, infant Sofie was not yet able to perceive and organize events and objects as a mature person could. Instead, she moved through a series of developmental steps, each with unique features, until she reached the highest level of functioning.

Theories that accept the discontinuous perspective regard development as taking place in **stages**—*qualitative changes* in thinking, feeling, and behaving that characterize specific periods of development. In stage theories, development is much like climbing a staircase, with each step corresponding to a more mature, reorganized way of functioning. The stage concept also assumes that people undergo periods of rapid transformation as they step up from one stage to the next. In other words, change is fairly sudden rather than gradual and ongoing.

Does development actually take place in a neat, orderly sequence of stages? For now, let's note that this is a very ambitious assumption that has not gone unchallenged. We will review some very influential stage theories later in this chapter.

One Course of Development or Many?

Stage theorists assume that people everywhere follow the same sequence of development. Yet the field of human development is becoming increasingly aware that children and adults live in distinct **contexts,** or unique combinations of personal and environmental circumstances that can result in different paths of change. For example, a shy individual who

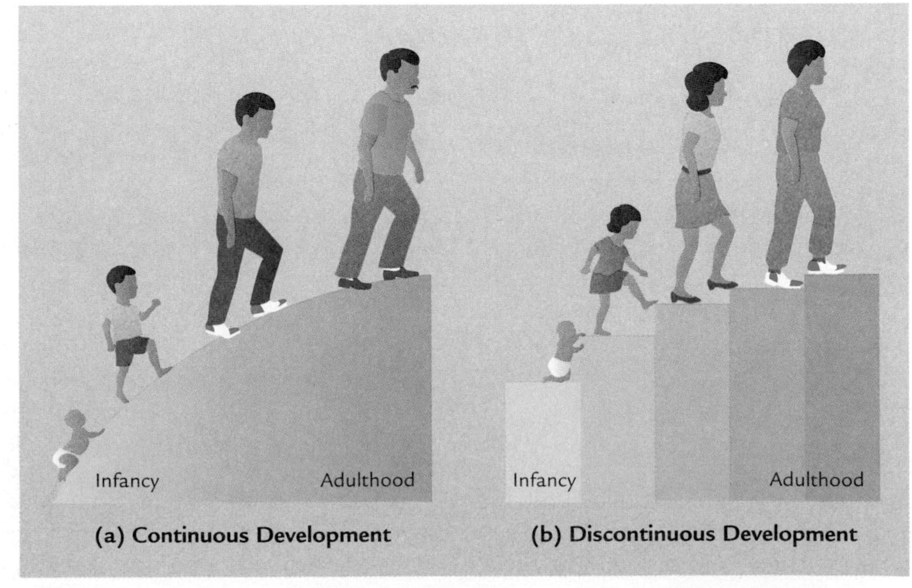

FIGURE 1.1 Is development continuous or discontinuous? (a) Some theorists believe that development is a smooth, continuous process. Individuals gradually add more of the same types of skills. (b) Other theorists think that development takes place in discontinuous stages. People change rapidly as they step up to a new level and then change very little for a while. With each new step, the person interprets and responds to the world in a qualitatively different way.

fears social encounters develops in very different contexts from those of a social agemate who readily seeks out other people (Rubin & Coplan, 1998). Children and adults in non-Western village societies encounter experiences in their families and communities that differ sharply from those of people in large Western cities. These different circumstances foster different intellectual capacities, social skills, and feelings about the self and others (Shweder et al., 1998).

As we will see, contemporary theorists regard the contexts that shape development as many-layered and complex. On the personal side, they include heredity and biological makeup. On the environmental side, they include immediate settings, such as home, school, and neighborhood, as well as circumstances more remote from people's everyday lives—community resources, societal values, and historical time period. Finally, a special interest in culture has made researchers more conscious than ever before of diversity in development.

Nature or Nurture as More Important?

In addition to describing the course of human development, each theory takes a stance on a major question about its underlying causes: Are genetic or environmental factors more important in determining development? This is the age-old **nature–nurture controversy.** By *nature,* we mean inborn biological givens—the hereditary information we receive from our parents at the moment of conception. By *nurture,* we mean the complex forces of the physical and social world that influence our biological makeup and psychological experiences before and after birth.

Although all theories grant at least some role to both nature and nurture, they vary in emphasis. For example, consider the following questions: Is the developing person's ability to think in more complex ways largely the result of an inborn timetable of growth? Or is it primarily influenced by stimulation from parents and teachers? Do children acquire language rapidly because they are genetically predisposed to do so or because parents tutor them from an early age? And what accounts for the vast individual differences among people—in height, weight, physical coordination, intelligence, personality, and social skills? Is nature or nurture more responsible?

The stances theories take on nature versus nurture affect their explanations of individual differences. Some theorists emphasize *stability*—that individuals who are high or low in a characteristic (such as verbal ability, anxiety, or sociability) will remain so at later ages. These theorists typically stress the importance of *heredity.* If they regard environment as important, they usually point to *early experiences* as establishing a lifelong pattern of behavior. Powerful negative events in the first few years, they argue, cannot be fully overcome by later, more positive ones (Bowlby, 1980; Sroufe, Egeland, & Kreutzer, 1990). Other theorists take a more optimistic view (Chess & Thomas, 1984; Nelson, 2002; Werner & Smith, 2001). They emphasize *plasticity*—that change is possible and likely if new experiences support it.

Throughout this chapter and the remainder of this book, we will see that investigators disagree, at times sharply, on the question of *stability or change.* Their answers often vary across *domains,* or aspects, of development. Think back to Sofie's story, and you will see that her linguistic ability and persistent approach to challenges were stable over the lifespan. In contrast, her psychological well-being and life satisfaction fluctuated considerably.

The Lifespan Perspective: A Balanced Point of View

So far, we have discussed basic issues of human development in terms of extremes—solutions on one side or the other. As we trace the unfolding of the field in the rest of this chapter, you will see that the positions of many theories have softened. Modern ones, especially, recognize the merits of both sides. Some theorists believe that both continuous and discontinuous changes occur and alternate with one another. And some acknowledge that development can have both universal features and features unique to the individual and his or her contexts. Furthermore, an increasing number of investigators regard heredity and environment as inseparably interwoven, each affecting the potential of the other to modify the child's traits and capacities (de Waal, 1999; Wachs, 2000).

These balanced visions owe much to the expansion of research from a nearly exclusive focus on the first 2 decades to include adulthood. In the first half of the twentieth century, it was widely assumed that development stopped at adolescence. Infancy and childhood were viewed as periods of rapid transformation, adulthood as a plateau, and aging as a period of decline. The changing character of the North American population awakened researchers to the idea that development is lifelong. Due to improvements in nutrition, sanitation, and medical knowledge, the *average life expectancy* (the number of years an individual born in a particular year can expect to live) gained more in the twentieth century than in the preceding five thousand years. In 1900, it was just under age 50; today, it is around age 77 in the United States and 79 in Canada. As a result, there are more older adults, a trend that has occurred in most of the world but is especially striking in industrialized nations. People age 65 and older accounted for about 4 percent of the North American population in 1900, 7 percent in 1950, and 12 percent in 2000. Growth in sheer numbers of elderly American and Canadian adults during the twentieth century has been even more dramatic, increasing more than eleven-fold, from 3 million to 34 million in the United States and from 290,000 to 4 million in Canada (U.S. Bureau of the Census, 2002a; Statistics Canada, 2002k).

■ Today, older adults are healthier and more active than in previous generations. These women, who are in their seventies, easily hike for miles while enjoying a day of bird-watching.

Older adults are not just more numerous; they are also healthier and more active. They challenge the earlier stereotype of the withering person and have sparked a profound shift in our view of human development. Compared with other approaches, the **lifespan perspective** offers a more complex vision of change and the factors that underlie it. Four assumptions make up this broader view: (1) development as lifelong, (2) development as multidimensional and multidirectional, (3) development as highly plastic, and (4) development as embedded in multiple contexts (Baltes, Lindenberger, & Staudinger, 1998; Smith & Baltes, 1999).

Development as Lifelong

According to the lifespan perspective, no age period is supreme in its impact on the life course. Instead, events occurring during each major period, summarized in Table 1.1, can have equally powerful effects on the future path of change. Within each period, change occurs in three broad domains: *physical, cognitive,* and *social,* which we separate for convenience of discussion (refer to Figure 1.2 for a description of each). Yet, as you are already aware from reading the first part of this chapter, the domains are not really distinct; they overlap and interact a great deal.

Every age period has its own agenda, bringing with it unique demands and opportunities that yield some similarities in development across many individuals. Nevertheless, throughout life, the challenges people face and the adjustments they make are highly diverse in timing and pattern, as the remaining assumptions make clear.

Table 1.1 Major Periods of Human Development

Period	Approximate Age Range	Brief Description
Prenatal	Conception to birth	The one-celled organism transforms into a human baby with remarkable capacities to adjust to life outside the womb.
Infancy and toddlerhood	Birth–2 years	Dramatic changes in the body and brain support emergence of a wide array of motor, perceptual, and intellectual capacities and first intimate ties to others.
Early childhood	2–6 years	The play years, in which motor skills are refined, thought and language expand at an astounding pace, a sense of morality is evident, and children begin to establish ties to peers.
Middle childhood	6–11 years	The school years, marked by advances in athletic abilities; logical thought processes; basic literacy skills; understanding of self, morality, and friendship; and peer-group membership.
Adolescence	11–20 years	Puberty leads to an adult-sized body and sexual maturity. Thought becomes abstract and idealistic and school achievement more serious. Adolescents focus on defining personal values and goals and establishing autonomy from the family.
Early adulthood	20–40 years	Most young people leave home, complete their education, and begin full-time work. Major concerns are developing a career; forming an intimate partnership; and marrying, rearing children, or establishing other lifestyles.
Middle adulthood	40–60 years	Many people are at the height of their careers and attain leadership positions. They must also help their children begin independent lives and their parents adapt to aging. They become more aware of their own mortality.
Late adulthood	60 years–death	People adjust to retirement, to decreased physical strength and health, and often to the death of a spouse. They reflect on the meaning of their lives.

■ **FIGURE 1.2 Major domains of development.** The three domains are not really distinct. Rather, they overlap and interact.

Development as Multidimensional and Multidirectional

Think back to Sofie's life and how she was continually faced with new demands and opportunities. The lifespan perspective regards the challenges and adjustments of development as *multidimensional*—affected by an intricate blend of biological, psychological, and social forces.

Lifespan development is also *multidirectional*—in at least two ways. First, development is not limited to improved performance. Instead, at all periods, it is a joint expression of growth and decline. When Sofie directed her energies toward mastering languages and music as a school-age child, she gave up refining other skills to their full potential. When she chose to become a teacher in adulthood, she let go of other career options. Although gains are especially evident early in life, and losses during the final years, people of all ages can improve current skills and develop new skills, including ones that compensate for reduced functioning (Freund & Baltes, 2000). One elderly psychologist who noticed his difficulty remembering people's names devised graceful ways of explaining his memory failure. Often he flattered his listener by remarking that he tended to forget only the names of important people! Under these conditions, he reflected, "forgetting may even be a pleasure" (Skinner, 1983, p. 240).

Second, besides being multidirectional over time, change is multidirectional within the same domain of development. Although some qualities of Sofie's cognitive functioning (such as memory) probably declined in her mature years, her knowledge of English and French undoubtedly grew throughout her life. And she also developed new forms of thinking. For example, Sofie's wealth of experience and ability to cope with diverse problems led her to become expert in practical matters—a quality of reasoning called wisdom. Recall the wise advice that Sofie gave Laura and Ken on the eve of their wedding day. We will consider the development of wisdom in Chapter 17. Notice, in these examples, how the lifespan perspective includes both continuous and discontinuous change.

Development as Plastic

Lifespan researchers emphasize that development is *plastic* at all ages. For example, consider Sofie's social reserve in

Biology & Environment

Resiliency

John and his best friend Gary grew up in a run-down, crime-ridden, inner-city neighborhood. By age 10, each had experienced years of family conflict followed by parental divorce. Reared for the rest of childhood and adolescence in mother-headed households, John and Gary rarely saw their fathers. Both achieved poorly, dropped out of high school, and were in and out of trouble with the police.

Then John and Gary's paths diverged. By age 30, John had fathered two children with women he never married, had spent time in prison, was unemployed, and drank alcohol heavily. In contrast, Gary had returned to finish high school, had studied auto mechanics at a community college, and had become manager of a gas station and repair shop. Married with two children, he had saved his earnings and bought a home. He was happy, healthy, and well adapted to life.

A wealth of evidence shows that environmental risks—poverty, negative family interactions and parental divorce, job loss, mental illness, and drug abuse—predispose children to future problems (Masten & Coatsworth, 1998). Why did Gary "beat the odds" and come through unscathed?

New evidence on **resiliency**—the ability to adapt effectively in the face of threats to development—is receiving increased attention because investigators want to find ways to protect young people from the damaging effects of stressful life conditions (Masten, 2001). This interest has been inspired by several long-term studies on the relationship of life stressors in childhood to competence and adjustment in adolescence and adulthood (Garmezy, 1993; Rutter, 1987; Werner & Smith, 2001). In each study, some individuals were shielded from negative outcomes, whereas others had lasting problems. Four broad factors seemed to offer protection from the damaging effects of stressful life events. Let's briefly consider each.

Personal Characteristics. A child's biologically endowed characteristics can reduce exposure to risk or lead to experiences that compensate for early stressful events. Intellectual ability, for example, is a protective factor. It increases the chances that a child will have rewarding experiences in school that offset the impact of a stressful home life and that further enhance mental ability (Masten et al., 1999). Temperament is particularly powerful. Children with easygoing, sociable dispositions have a special capacity to adapt to change and elicit positive responses from others. Children who are emotionally reactive and irritable often strain the patience of people around them (Milgram & Palti, 1993; Smith & Prior, 1995). For example, both John and Gary moved several times during their childhoods. Each time, John became anxious and angry. Gary, however, looked forward to making new friends and exploring new parts of the neighborhood.

A Warm Parental Relationship. A close relationship with at least one parent who provides affection and assistance and introduces order and organization into the child's life fosters resiliency. But note that this factor (as well as the next one) is not independent of children's personal characteristics. Children who are relaxed, socially responsive, and able to deal with change are easier to rear and more likely to enjoy positive relationships with parents and other people. At the

childhood and her decision to study rather than marry as a young adult. As new opportunities arose, Sofie moved easily into marriage and childbearing in her thirties. And although parenthood and financial difficulties posed challenges to Sofie and Philip's happiness, their relationship gradually became richer and more fulfilling. In Chapter 17, we will see that intellectual performance also remains flexible with advancing age. Elderly people respond to special training with substantial (but not unlimited) gains in a wide variety of mental abilities (Schaie, 1996).

Evidence on plasticity makes it clear that aging is not an eventual "shipwreck," as has often been assumed. Instead, the metaphor of a "butterfly"—of metamorphosis and continued potential—provides a far more accurate picture of lifespan change (Lemme, 2002). Still, development gradually becomes less plastic, as both capacity and opportunity for change are reduced. And plasticity varies greatly across individuals. Some children and adults experience more diverse life circumstances. And as the Biology and Environment box above reveals, some adapt more easily to changing conditions than do others.

Development as Embedded in Multiple Contexts

According to the lifespan perspective, pathways of change are highly diverse because development is *embedded in multiple contexts.* Although these wide-ranging influences can be organized into three categories, they work together, combining in unique ways to fashion each life course.

● **Age-Graded Influences.** Events that are strongly related to age and therefore fairly predictable in when they occur and how long they last are called **age-graded influences.** For example, most individuals walk shortly after their

same time, some children may develop more attractive dispositions as a result of parental warmth and attention (Smith & Prior, 1995; Wyman et al., 1999).

Social Support Outside the Immediate Family. A person outside the immediate family—perhaps a grandparent, teacher, or close friend—who forms a special relationship with the child can promote resiliency. Gary received support in adolescence from his grandfather, who listened to Gary's concerns and helped him solve problems constructively. In addition, Gary's grandfather had a stable marriage and work life and handled stressors skillfully. Consequently, he served as a model of effective coping (Zimmerman & Arunkumar, 1994).

A Strong Community. Opportunities to participate in community life increase the likelihood that older children and adolescents will overcome adversity. Extracurricular activities at school, religious youth groups, scouting, and other organizations teach important social skills, such as cooperation, leadership, and contributing to others' welfare. As participants acquire these competencies, they gain in self-reliance, self-esteem, and community commitment. As a college student, Gary volunteered with Habitat for Humanity, joining a team building affordable housing in low-income neighborhoods. Community involvement gave Gary additional opportunities to form meaningful relationships, which further strengthened his resiliency (Seccombe, 2002).

Research on resiliency highlights the complex connections between heredity and environment. Armed with positive characteristics, which stem from innate endowment, favorable rearing experiences, or both, children and adolescents take action to reduce stressful situations. Nevertheless, when many risks pile up, they are increasingly difficult to overcome (Quyen et al., 1998). Therefore, interventions must reduce risks and enhance relationships at home, in school, and in the community that protect young people against the negative effects of risk. This means attending to both the person and the environment—strengthening the individual's capacities as well as fixing problems.

© ALAN HICKS/GETTY IMAGES/STONE

■ This child's special relationship with her grandfather provides the social support she needs to cope with stress and solve problems constructively. A warm tie with a person outside the immediate family can promote resiliency.

first birthday, acquire their native language during the preschool years, reach puberty around ages 12 to 14, and (for women) experience menopause in their late forties or early fifties. These milestones are influenced by biology, but social customs can create age-graded influences as well. Starting school around age 6, getting a driver's license at age 16, and entering college around age 18 are events of this kind. Age-graded influences are especially prevalent during childhood and adolescence, when biological changes are rapid and cultures impose many age-related experiences to ensure that young people acquire the skills they need to participate in their society.

● **History-Graded Influences.** Development is also profoundly affected by forces unique to a particular historical era. Examples include epidemics, wars, and periods of economic prosperity or depression; technological advances, such as the introduction of television, computers, and the Internet; and changing cultural values, such as revised attitudes toward women and ethnic minorities. These **history-graded influences** explain why people born around the same time—called a *cohort*—tend to be alike in ways that set them apart from people born at other times.

● **Nonnormative Influences.** *Normative* means typical, or average. Age-graded and history-graded influences are normative because each affects large numbers of people in a similar way. **Nonnormative influences** are events that are irregular, in that they happen to just one or a few people and do not follow a predictable timetable. Consequently, they enhance the multidirectionality of development. Piano lessons in childhood with an inspiring teacher; a blind date with Philip; delayed marriage, parenthood, and career entry; and a battle with cancer are nonnormative influences that had a major impact on the direction of Sofie's life. Because they occur haphazardly, nonnormative events are among the most difficult for researchers to capture

■ Starting kindergarten marks a major transition in the lives of these 5- and 6-year-olds. Because the first day of school occurs at about the same age for most children in industrialized nations, it is an *age-graded influence.*

and study. Yet, as each of us can attest from our own experiences, they can affect us in powerful ways.

Lifespan investigators point out that nonnormative influences have become more powerful and age-graded influences less so in contemporary adult development. Compared with Sofie's era, the ages at which people finish their education, enter careers, get married, have children, and retire are much more diverse (Schroots & Birren, 1990). Indeed, Sofie's "off-time" accomplishments would have been less unusual had she been born a generation or two later! Age remains a powerful organizer of everyday experiences, and age-related expectations have certainly not disappeared. But age markers have blurred, and they vary across ethnic groups and cultures. The increasing role of nonnormative events in the life course adds to the fluid nature of lifespan development.

Notice that instead of a single line of development, the lifespan perspective emphasizes many potential pathways and outcomes—an image more like tree branches extending in diverse directions, which may undergo continuous and stagewise transformations (see Figure 1.3). Now let's turn to the historical foundations of the field as a prelude to discussing major theories that address various aspects of change.

Ask Yourself

REVIEW

Distinguish among age-graded, history-graded, and nonnormative influences on lifespan development. Cite an example of each in Sofie's story at the beginning of this chapter.

CONNECT

What stance does the lifespan perspective take on the issue of *one course of development or many*? How about *nature or nurture* and *stability or change*? Explain.

www.

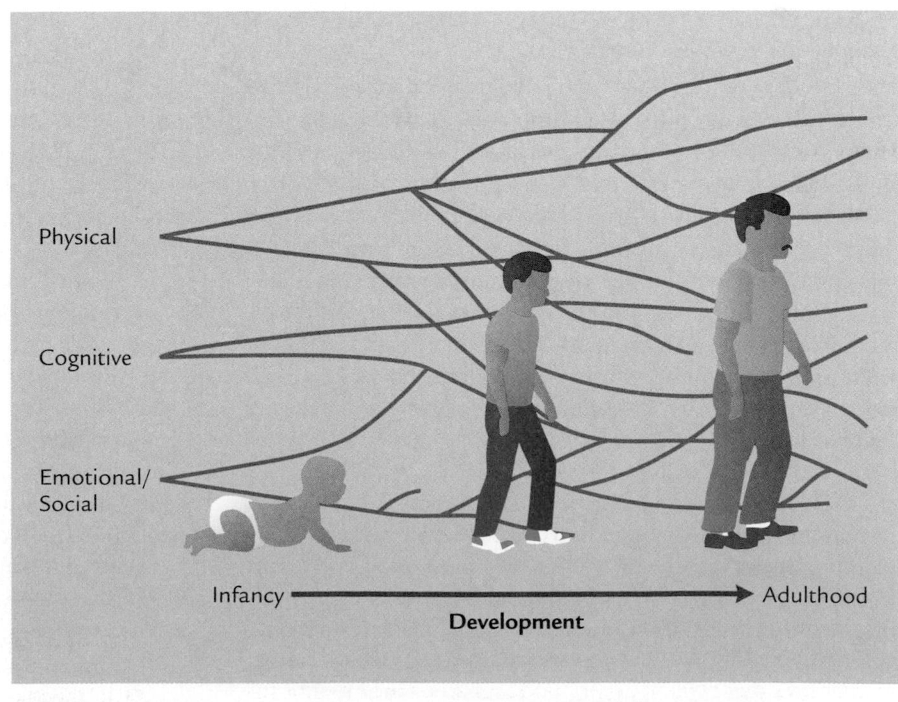

■ **FIGURE 1.3 The lifespan view of development.** Rather than envisioning a single line of stagewise or continuous change (see Figure 1.1 on page 6), lifespan theorists conceive of development as more like tree branches extending in diverse directions. Many potential pathways are possible, depending on the contexts that influence the individual's life course. Each branch in this treelike image represents a possible skill within one of the major domains of development. The crossing of the branches signifies that the domains—physical, cognitive, emotional, and social—are interrelated.

Historical Foundations

Contemporary theories of human development are the result of centuries of change in Western cultural values, philosophical thinking, and scientific progress. To understand the field as it exists today, we must return to its early beginnings—to influences that long preceded scientific study. We will see that many early ideas linger on as important forces in current theory and research.

Philosophies of Childhood

In medieval Europe (the sixth through the fifteenth centuries), little importance was placed on childhood as a separate phase of life. Once children emerged from infancy, they were regarded as miniature, already-formed adults, a view called **preformationism** (Ariès, 1962). Certain laws did recognize that children needed protection from people who might mistreat them, and medical works provided special instructions for their care. However, despite practical awareness of the vulnerability of children, there were no philosophies of the uniqueness of childhood or separate developmental periods (Borstelmann, 1983).

In the sixteenth century, a revised image of children sprang from the Puritan belief in original sin. Harsh, restrictive parenting practices were recommended as the most efficient means of taming the depraved child. Although punitiveness was the prevailing child-rearing philosophy, affection for their children prevented most Puritan parents from using extremely repressive measures. Instead, they tried to promote reason in their sons and daughters so they could tell right from wrong and resist temptation (Clarke-Stewart, 1998).

● **John Locke.** The seventeenth-century Enlightenment brought philosophies that emphasized ideals of human dignity and respect. The writings of John Locke (1632–1704), a leading British philosopher, served as the forerunner of a twentieth-century perspective that we will discuss shortly: behaviorism. Locke viewed the child as a **tabula rasa,** or, translated from Latin, "blank slate." According to this idea, children are, to begin with, nothing at all, and all kinds of experiences can shape their characters. Locke (1690/1892) described parents as rational tutors who can mold the child in any way they wish through careful instruction, effective example, and rewards for good behavior. His philosophy led to a change from harshness toward children to kindness and compassion.

Look carefully at Locke's ideas, and you will see that he took a firm stand on basic issues discussed earlier in this chapter. Lock regarded development as *continuous;* adultlike behaviors are gradually built up through the warm, consistent teachings of parents. Furthermore, Locke's view of the child as a tabula rasa led him to champion *nurture*—the power of the environment to shape the child. And his faith in nurture suggests the possibility of *many courses of development* and *change at later ages* due to new experiences. Finally, Locke's philosophy characterizes children as passive—as doing little to shape their own destiny, which is written on blank slates by others. This vision has been discarded. All contemporary theories view the developing person as an active, purposeful being who contributes substantially to his or her own development.

● **Jean Jacques Rousseau.** In the eighteenth century, a new theory of childhood was introduced by French philosopher Jean Jacques Rousseau (1712–1778). Children, Rousseau (1762/1955) thought, were not blank slates to be filled by adult instruction. Instead, they were **noble savages,** naturally endowed with a sense of right and wrong and with an innate plan for orderly, healthy growth. Unlike Locke, Rousseau thought children's built-in moral sense and unique ways of thinking and feeling would only be harmed by adult training. His was a child-centered philosophy in which the adult should be receptive to the child's needs at each of four stages of development: infancy, childhood, late childhood, and adolescence.

Rousseau's philosophy includes two influential concepts. The first is the concept of *stage,* which we discussed earlier in this chapter. The second is the concept of **maturation,** which refers to a genetically determined, naturally unfolding course of growth. Unlike Locke, Rousseau saw children as determining their own destinies. And he took a different stand on basic developmental issues. He saw development as a discontinuous, *stagewise* process that follows a *single, unified course* mapped out by *nature.*

Philosophies of Adulthood and Aging

Shortly after Rousseau devised his conception of childhood, the first lifespan views appeared. In the eighteenth and early nineteenth centuries, two German philosophers—John Nicolaus Tetens (1736–1807) and Friedrich August Carus

© ERIC LESSING/ART RESOURCE, NY

■ In this medieval painting, young children are depicted as miniature adults. Their dress and expressions resemble those of their elders. Through the fifteenth century, little emphasis was placed on childhood as a unique phase of the life cycle.

(1770–1808)—urged that attention to development be extended through adulthood. Each asked important questions about aging.

Tetens (1777) addressed the origins and extent of individual differences, the degree to which behavior can be changed in adulthood, and the impact of historical eras on the life course. He was ahead of his time in recognizing that older people can compensate for intellectual declines that, at times, may reflect hidden gains. For example, Tetens suggested that some memory difficulties are due to searching for a word or name among a lifetime of accumulated information—a possibility acknowledged by current research (Maylor & Valentine, 1992).

Carus (1808) moved beyond Rousseau's stages by identifying four periods that span the life course: childhood, youth, adulthood, and senescence. Like Tetens, Carus viewed aging not only as decline but also as progression. His writings reflect a remarkable awareness of multidirectionality and plasticity, which are at the heart of the lifespan perspective.

Scientific Beginnings

The study of development evolved quickly during the late nineteenth and early twentieth centuries. Early observations of human change were soon followed by improved methods and theories. Each advance contributed to the firm foundation on which the field rests today.

Darwin: Forefather of Scientific Child Study. Charles Darwin (1809–1882), a British naturalist, is often considered the forefather of scientific child study. Darwin (1859/1936) observed the infinite variation among plant and animal species. He also saw that within a species, no two individuals are exactly alike. From these observations, he constructed his famous theory of evolution.

The theory emphasized two related principles: *natural selection* and *survival of the fittest.* Darwin explained that certain species survived in particular environments because they have characteristics that fit with, or are adapted to, their surroundings. Other species die off because they are not well suited to their environments. Individuals within a species who best meet the survival requirements of the environment live long enough to reproduce and pass their more favorable characteristics to future generations. Darwin's emphasis on the adaptive value of physical characteristics and behavior found its way into important developmental theories.

During his explorations, Darwin discovered that the early prenatal growth of many species is strikingly similar. Other scientists concluded from Darwin's observation that the development of the human child followed the same general plan as the evolution of the human species. Although this belief eventually proved inaccurate, efforts to chart parallels between child growth and human evolution prompted researchers to make careful observations of all aspects of children's behavior. Out of these first attempts to document an idea about development, scientific child study was born.

The Normative Period. G. Stanley Hall (1846–1924), one of the most influential American psychologists of the early twentieth century, is generally regarded as the founder of the child study movement (Dixon & Lerner, 1999). He also foreshadowed lifespan research by writing one of the few books of his time on aging. Inspired by Darwin's work, Hall and his well-known student Arnold Gesell (1880–1961) devised theories of childhood and adolescence based on evolutionary ideas. These early leaders regarded development as a genetically determined process that unfolds automatically, much like a flower (Gesell, 1933; Hall, 1904).

Hall and Gesell are remembered less for their one-sided theories than for their intensive efforts to describe all aspects of development. They launched the **normative approach,** in which measures of behavior are taken on large numbers of individuals, and age-related averages are computed to represent

Darwin's theory of evolution emphasizes the adaptive value of physical characteristics and behavior. Affection and care in families is adaptive throughout the lifespan, promoting survival and psychological well-being. Here a daughter helps her elderly mother with medication.

typical development. Using this method, Hall constructed elaborate questionnaires asking children of different ages almost everything they could tell about themselves—interests, fears, imaginary playmates, dreams, friendships, everyday knowledge, and more (White, 1992). In the same fashion, Gesell collected detailed normative information on the motor achievements, social behaviors, and personality characteristics of infants and children (Gesell & Ilg, 1946/1949a, 1943/1949b).

Gesell was also among the first to make knowledge about child development meaningful to parents. If, as he believed, the timetable of development is the product of millions of years of evolution, then children are naturally knowledgeable about their needs. His child-rearing advice, in the tradition of Rousseau, recommended sensitivity to children's cues (Thelen & Adolph, 1992). Along with Benjamin Spock's *Baby and Child Care,* Gesell's books became a central part of a rapidly expanding popular literature for parents.

● **The Mental Testing Movement.** While Hall and Gesell were developing their theories and methods in the United States, French psychologist Alfred Binet (1857–1911) was also taking a normative approach to child development, but for a different reason. In the early 1900s, Binet and his colleague Theodore Simon were asked to find a way to identify children with learning problems who needed to be placed in special classes. The first successful intelligence test, which they constructed for this purpose, grew out of practical educational concerns.

In 1916, at Stanford University, Binet's test was adapted for use with English-speaking children. Since then the English version has been known as the Stanford-Binet Intelligence Scale. Besides providing a score that could successfully predict school achievement, the Binet test sparked tremendous interest in individual differences in development. Comparisons of the scores of people who vary in gender, ethnicity, birth order, family background, and other characteristics became a major focus of research. Intelligence tests also rose quickly to the forefront of the controversy over nature versus nurture that has continued to this day.

Ask Yourself

REVIEW

Explain how central assumptions of the lifespan perspective are reflected in Tetens's and Carus's philosophies of adulthood and aging.

APPLY

Suppose we could arrange a debate between John Locke and Jean Jacques Rousseau on the nature–nurture controversy. Summarize the argument that each historical figure would be likely to present.

www.

Mid-Twentieth-Century Theories

In the mid-twentieth century, human development expanded into a legitimate discipline. As it attracted increasing interest, a variety of mid-twentieth-century theories emerged, each of which continues to have followers today.

The Psychoanalytic Perspective

In the 1930s and 1940s, as more people sought help from professionals in dealing with emotional difficulties, a new question had to be addressed: How and why did people become the way they are? To treat psychological problems, psychiatrists and social workers turned to an emerging approach to personality development because of its emphasis on understanding the unique life history of each person.

According to the **psychoanalytic perspective,** people move through a series of stages in which they confront conflicts between biological drives and social expectations. The way these conflicts are resolved determines the individual's ability to learn, to get along with others, and to cope with anxiety. Although many individuals contributed to the psychoanalytic perspective, two were especially influential: Sigmund Freud, founder of the psychoanalytic movement, and Erik Erikson.

● **Freud's Theory.** Freud (1856–1939), a Viennese physician, saw patients in his practice with a variety of nervous symptoms, such as hallucinations, fears, and paralyses, that appeared to have no physical basis. Seeking a cure for these troubled adults, Freud found that their symptoms could be relieved by having patients talk freely about painful events of their childhoods. On the basis of adult remembrances, he examined the unconscious motivations of his patients and constructed his **psychosexual theory,** which emphasized that how parents manage their child's sexual and aggressive drives in the first few years is crucial for healthy personality development.

Three Parts of the Personality. In Freud's theory, three parts of the personality—id, ego, and superego—become integrated during five stages, summarized in Table 1.2 on page 16. The *id,* the largest portion of the mind, is the source of basic biological needs and desires. The *ego*—the conscious, rational part of personality—emerges in early infancy to redirect the id's impulses so they are discharged on appropriate objects at acceptable times and places. For example, aided by the ego, the hungry baby of a few months of age stops crying when he sees his mother unfasten her clothing for breast-feeding. And the more competent preschooler goes into the kitchen and gets a snack on her own.

Between 3 and 6 years of age, the *superego,* or conscience, develops from interactions with parents, who insist that children conform to the values of society. Now the ego faces the

Table 1.2 Freud's Psychosexual Stages

Psychosexual Stage	Period of Development	Description
Oral	Birth–1 year	The new ego directs the baby's sucking activities toward breast or bottle. If oral needs are not met appropriately, the individual may develop such habits as thumb sucking, fingernail biting, and pencil chewing in childhood and overeating and smoking in later life.
Anal	1–3 years	Toddlers and preschoolers enjoy holding and releasing urine and feces. Toilet training becomes a major issue between parent and child. If parents insist that children be trained before they are ready or make too few demands, conflicts about anal control may appear in the form of extreme orderliness and cleanliness or messiness and disorder.
Phallic	3–6 years	Id impulses transfer to the genitals, and the child finds pleasure in genital stimulation. Freud's *Oedipus conflict* for boys and *Electra conflict* for girls arise, and young children feel a sexual desire for the other-sex parent. To avoid punishment, they give up this desire and, instead, adopt the same-sex parent's characteristics and values. As a result, the superego is formed, and children feel guilty each time they violate its standards. The relations among id, ego, and superego established at this time determine the individual's basic personality.
Latency	6–11 years	Sexual instincts die down, and the superego develops further. The child acquires new social values from adults outside the family and from play with same-sex peers.
Genital	Adolescence	Puberty causes the sexual impulses of the phallic stage to reappear. If development has been successful during earlier stages, it leads to marriage, mature sexuality, and the birth and rearing of children.

■ In psychoanalytic theory, the ego redirects the id's impulses so the child's needs are satisfied in socially acceptable ways. Here a 3-year-old who wants to play with a puzzle refrains from grabbing and, instead, asks to help her classmate.

increasingly complex task of reconciling the demands of the id, the external world, and conscience (Freud, 1923/1974). For example, when the ego is tempted to gratify an id impulse by hitting a playmate to get an attractive toy, the superego may warn that such behavior is wrong. The ego must decide which of the two forces (id or superego) will win this inner struggle, or it must work out a compromise, such as asking for a turn with the toy. According to Freud, the relations established among the id, ego, and superego during the preschool years determine the individual's basic personality.

Psychosexual Development. Freud (1938/1973) believed that during childhood, sexual impulses shift their focus from the oral to the anal to the genital regions of the body. In each stage of development, parents walk a fine line between permitting too much or too little gratification of their child's basic needs. If parents strike an appropriate balance, then children grow into well-adjusted adults with the capacity for mature sexual behavior, investment in family life, and rearing of the next generation.

Freud's psychosexual theory highlighted the importance of family relationships for children's development. It was the first theory to stress the role of early experience. But Freud's perspective was eventually criticized. First, the theory overemphasized the influence of sexual feelings in development. Second, because it was based on the problems of sexually repressed, well-to-do adults, it did not apply in cultures differing from nineteenth-century Victorian society. Finally, Freud had not studied children directly.

● **Erikson's Theory.** Several of Freud's followers took what was useful from his theory and improved on his vision.

The most important of these neo-Freudians is Erik Erikson (1902–1994).

Although Erikson (1950) accepted Freud's basic psychosexual framework, he expanded the picture of development at each stage. In his **psychosocial theory,** Erikson emphasized that the ego does not just mediate between id impulses and superego demands. At each stage, it acquires attitudes and skills that make the individual an active, contributing member of society. A basic psychological conflict, which is resolved along a continuum from positive to negative, determines healthy or maladaptive outcomes at each stage. As Table 1.3 shows, Erikson's first five stages parallel Freud's stages, but Erikson added three adult stages.

Finally, unlike Freud, Erikson pointed out that normal development must be understood in relation to each culture's life situation. For example, among the Yurok Indians on the Northwest coast of the United States, babies are deprived of breast-feeding for the first 10 days after birth and instead are fed a thin soup from a small shell. At age 6 months, infants are abruptly weaned—if necessary, by having the mother leave for a few days. These experiences, from our cultural vantage point, might seem cruel. But Erikson explained that the Yurok live in a world in which salmon fill the river just once a year, a circumstance that requires considerable self-restraint for survival. In this way, he showed that child rearing can be understood only by making reference to the competencies valued and needed by the individual's society.

● **Contributions and Limitations of Psychoanalytic Theory.** A special strength of the psychoanalytic perspective is its emphasis on the individual's unique life history as worthy of study and understanding (Emde, 1992). Consistent with this view, psychoanalytic theorists accept the *clinical method,* which synthesizes information from a variety of sources into a detailed picture of the personality functioning of a single person. (We will discuss the clinical method further at the end of this chapter.) Psychoanalytic theory has also inspired a wealth of research on many aspects of emotional and social development, including infant–caregiver attachment, aggression, sibling relationships, child-rearing practices, morality, gender roles, and adolescent identity.

Despite its extensive contributions, the psychoanalytic perspective is no longer in the mainstream of human development research (Cairns, 1998). Psychoanalytic theorists may have become isolated from the rest of the field because they were so strongly committed to the clinical approach that they failed to consider other methods. In addition, many psychoanalytic ideas, such as psychosexual stages and ego functioning, are so vague that they are difficult or impossible to test empirically (Thomas, 2000; Westen & Gabbard, 1999).

Table 1.3 Erikson's Psychosocial Stages, with Corresponding Psychosexual Stages Indicated

Psychosocial Stage	Period of Development	Description
Basic trust versus mistrust (Oral)	Birth–1 year	From warm, responsive care, infants gain a sense of trust, or confidence, that the world is good. Mistrust occurs when infants have to wait too long for comfort and are handled harshly.
Autonomy versus shame and doubt (Anal)	1–3 years	Using new mental and motor skills, children want to choose and decide for themselves. Autonomy is fostered when parents permit reasonable free choice and do not force or shame the child.
Initiative versus guilt (Phallic)	3–6 years	Through make-believe play, children experiment with the kind of person they can become. Initiative—a sense of ambition and responsibility—develops when parents support their child's new sense of purpose. The danger is that parents will demand too much self-control, which leads to overcontrol, meaning too much guilt.
Industry versus diffusion (Latency)	6–11 years	At school, children develop the capacity to work and cooperate with others. Inferiority develops when negative experiences at home, at school, or with peers lead to feelings of incompetence.
Identity versus identity confusion (Genital)	Adolescence	The adolescent tries to answer the question, Who am I, and what is my place in society? Self-chosen values and vocational goals lead to a lasting personal identity. The negative outcome is confusion about future adult roles.
Intimacy versus isolation	Young adulthood	Young people work on establishing intimate ties to others. Because of earlier disappointments, some individuals cannot form close relationships and remain isolated.
Generativity versus stagnation	Middle adulthood	Generativity means giving to the next generation through child rearing, caring for other people, or productive work. The person who fails in these ways feels an absence of meaningful accomplishment.
Ego integrity versus despair	Old age	In this final stage, individuals reflect on the kind of person they have been. Integrity results from feeling that life was worth living as it happened. Older people who are dissatisfied with their lives fear death.

Nevertheless, Erikson's broad outline of lifespan change captures the essence of personality development during each major period of the life course, so we will return to it in later chapters. We will also encounter additional perspectives that clarify the attainments of early, middle, and late adulthood and that are within the tradition of stage models of psychosocial development (Levinson, 1978, 1996; Vaillant, 1977).

Behaviorism and Social Learning Theory

As psychoanalytic theory gained in prominence, human development was also influenced by a very different perspective. According to **behaviorism,** directly observable events—stimuli and responses—are the appropriate focus of study. American behaviorism began with the work of psychologist John Watson (1878–1958) in the early twentieth century. Watson wanted to create an objective science of psychology and rejected the psychoanalytic concern with the unseen workings of the mind (Horowitz, 1992).

● **Traditional Behaviorism.** Watson was inspired by studies of animal learning carried out by famous Russian physiologist Ivan Pavlov. Pavlov knew that dogs release saliva as an innate reflex when they are given food. But he noticed that his dogs were salivating before they tasted any food—when they saw the trainer who usually fed them. The dogs, Pavlov reasoned, must have learned to associate a neutral stimulus (the trainer) with another stimulus (food) that produces a reflexive response (salivation). As a result of this association, the neutral stimulus could bring about a response resembling the reflex. Eager to test this idea, Pavlov successfully taught dogs to salivate at the sound of a bell by pairing it with the presentation of food. He had discovered *classical conditioning.*

Watson wanted to find out if classical conditioning could be applied to children's behavior. In a historic experiment, he taught Albert, an 11-month-old infant, to fear a neutral stimulus—a soft white rat—by presenting it several times with a sharp, loud sound, which naturally scared the baby. Little Albert, who at first had reached out eagerly to touch the furry rat, began to cry and turn his head away when he caught sight of it (Watson & Raynor, 1920). In fact, Albert's fear was so intense that researchers eventually challenged the ethics of studies like this one. Consistent with Locke's tabula rasa, Watson concluded that environment is the supreme force in development. Adults can mold children's behavior, he thought, by carefully controlling stimulus–response associations. And development is a continuous process, consisting of a gradual increase in the number and strength of these associations.

Another form of behaviorism was B. F. Skinner's (1904–1990) *operant conditioning theory.* According to Skinner, behavior can be increased by following it with a wide variety of *reinforcers,* such as food, praise, or a friendly smile. It can also be decreased through *punishment,* such as disapproval or withdrawal of privileges. As a result of Skinner's work, operant conditioning became a broadly applied learning principle. We will consider these conditioning techniques further in Chapter 4.

■ Social learning theory recognizes that children acquire many skills through modeling. By observing and imitating her mother's behavior, this Vietnamese preschooler is becoming a skilled user of chopsticks.

● **Social Learning Theory.** Psychologists quickly became interested in whether behaviorism might offer a more direct and effective explanation of the development of social behavior than the less precise concepts of psychoanalytic theory. This concern sparked the emergence of several approaches that built on the principles of conditioning that came before them, offering expanded views of how children and adults acquire new responses.

Several kinds of **social learning theory** emerged. The most influential, devised by Albert Bandura, emphasized *modeling,* otherwise known as *imitation* or *observational learning,* as a powerful source of development. Bandura (1977) recognized that children acquire many favorable and unfavorable responses simply by watching and listening to others around them. The baby who claps her hands after her mother does so, the child who angrily hits a playmate in the same way that he has been punished at home, and the teenager who wears the same clothes and hairstyle as her friends at school are all displaying observational learning.

Bandura's work continues to influence much research on social development. However, like the field of human development as a whole, today his theory stresses the importance of *cognition,* or thinking. In fact, the most recent revision of Bandura's (1989, 1992) theory places such strong emphasis on how we think about ourselves and other people that he calls it a *social-cognitive* rather than a social learning approach.

According to this view, children gradually become more selective in what they imitate. From watching others engage in self-praise and self-blame and through feedback about the worth of their own actions, children develop *personal standards* for behavior and a *sense of self-efficacy*—the belief that their own abilities and characteristics will help them succeed. These cognitions guide responses in particular situations (Bandura, 1999). For example, imagine a parent who often remarks, "I'm glad I kept working on that task, even though it was hard," who explains the value of persistence, and who encourages her child by saying, "I know you can do a good job on that homework!" Soon the child starts to view himself as hardworking and high achieving and selects people with these characteristics as models. In this way, as individuals acquire attitudes, values, and convictions about themselves, they control their own learning and behavior.

● **Contributions and Limitations of Behaviorism and Social Learning Theory.** Like psychoanalytic theory, behaviorism and social learning theory have been helpful in treating emotional and behavior problems. Yet the techniques are decidedly different. **Behavior modification** consists of procedures that combine conditioning and modeling to eliminate undesirable behaviors and increase desirable responses. It has been used to relieve a wide range of difficulties in children and adults, such as persistent aggression, language delays, and extreme fears (Pierce & Epling, 1995; Wolpe & Plaud, 1997).

Nevertheless, modeling and reinforcement do not provide a complete account of development. Many theorists believe that behaviorism and social learning theory offer too narrow a view of important environmental influences. These extend beyond immediate reinforcements and modeled behaviors to the richness of the physical and social worlds. Finally, behaviorism and social learning theory have been criticized for underestimating people's contributions to their own development. In emphasizing cognition, Bandura is unique among theorists whose work grew out of the behaviorist tradition in granting children an active role in their own learning.

Piaget's Cognitive-Developmental Theory

If one individual has influenced research on child development more than any other, it is Swiss cognitive theorist Jean Piaget (1896–1980). North American investigators had been aware of Piaget's work since 1930. However, they did not grant it much attention until 1960, mainly because his ideas were very much at odds with behaviorism, which dominated psychology during the middle of the twentieth century (Zigler & Gilman, 1998). Piaget did not believe that knowledge could be imposed on a reinforced child. According to his **cognitive-developmental theory,** children actively construct knowledge as they manipulate and explore their world.

● **Piaget's Stages.** Piaget's view of development was greatly influenced by his early training in biology. Central to his theory is the biological concept of *adaptation* (Piaget, 1971). Just as the structures of the body are adapted to fit with the environment, so the structures of the mind develop to better fit with, or represent, the external world. In infancy and early childhood, children's understanding is different from adults'. For example, Piaget believed that young babies do not realize that an object hidden from view—a favorite toy or even the mother—continues to exist. He also concluded that preschoolers' thinking is full of faulty logic. For example, children younger than age 7 commonly say that the amount of milk or lemonade changes when it is poured into a different-shaped container. According to Piaget, children eventually revise these incorrect ideas in their ongoing efforts to achieve an equilibrium, or balance, between internal structures and information they encounter in their everyday worlds.

In Piaget's theory, as the brain develops and children's experiences expand, they move through four broad stages, each characterized by qualitatively distinct ways of thinking. Table 1.4 provides a brief description of Piaget's stages. In the *sensorimotor stage,* cognitive development begins with the baby's use of the senses and movements to explore the world. These action patterns evolve into the symbolic but illogical thinking of the preschooler in the *preoperational stage.* Then cognition is transformed into the more organized reasoning of the school-age child in the *concrete operational stage.* Finally, in the *formal*

Table 1.4 Piaget's Stages of Cognitive Development

Stage	Period of Development	Description
Sensorimotor	Birth–2 years	Infants "think" by acting on the world with their eyes, ears, hands, and mouth. As a result, they invent ways of solving sensorimotor problems, such as pulling a lever to hear the sound of a music box, finding hidden toys, and putting objects in and taking them out of containers.
Preoperational	2–7 years	Preschool children use symbols to represent their earlier sensorimotor discoveries. Development of language and make-believe play takes place. However, thinking lacks the logical qualities of the two remaining stages.
Concrete operational	7–11 years	Children's reasoning becomes logical. School-age children understand that a certain amount of lemonade or play dough remains the same even after its appearance changes. They also organize objects into hierarchies of classes and subclasses. However, thinking falls short of adult intelligence. It is not yet abstract.
Formal operational	11 years on	The capacity for abstraction permits adolescents to reason with symbols that do not refer to objects in the real world, as in advanced mathematics. They can also think of all possible outcomes in a scientific problem, not just the most obvious ones.

operational stage, thought becomes the complex, abstract reasoning system of the adolescent and adult.

● **Piaget's Methods of Study.** Piaget devised special methods for investigating how children think. In the early part of his career, he carefully observed his three infant children and also presented them with everyday problems, such as an attractive object that could be grasped, mouthed, kicked, or searched for. From their reactions, Piaget derived his ideas about cognitive changes during the first 2 years. In studying childhood and adolescent thought, Piaget took advantage of children's ability to describe their thinking. He adapted the clinical method of psychoanalysis, conducting open-ended *clinical interviews* in which a child's initial response to a task served as the basis for the next question Piaget would ask. We will look more closely at this technique when we discuss research methods later in this chapter.

● **Contributions and Limitations of Piaget's Theory.** Piaget's cognitive-developmental perspective convinced the field that children are active learners whose minds consist of rich structures of knowledge. Besides investigating children's understanding of the physical world, Piaget explored their reasoning about the social world. His stages have sparked a wealth of research on children's conceptions of themselves, other people, and human relationships. Practically speaking, Piaget's theory encouraged the development of educational philosophies and programs that emphasize discovery learning and direct contact with the environment.

Despite Piaget's overwhelming contributions, his theory has been challenged. Research indicates that Piaget underestimated the competencies of infants and preschoolers. When young children are given tasks scaled down in difficulty, their understanding appears closer to that of the older child and adult than Piaget assumed. Furthermore, many studies show that children's performance on Piagetian problems can be improved with training. This finding raises questions about his assumption that discovery learning rather than adult teaching is the best way to foster development. Critics also point out that Piaget's stagewise account pays insufficient attention to social and cultural influences on development. Finally, some lifespan theorists take issue with Piaget's conclusion that no major cognitive changes occur after adolescence. Several have proposed accounts of postformal thought that stress important transformations in adulthood (Arlin, 1989; Labouvie-Vief, 1985).

Today, the field of human development is divided over its loyalty to Piaget's ideas. Those who continue to find merit in Piaget's stage approach accept a modified view—one in which changes in thinking are not sudden and abrupt but take place gradually (Case, 1992, 1998; Fischer & Bidell, 1998). Others have turned to an approach that emphasizes continuous gains in children's cognition: information processing. And still others have been drawn to theories that focus on the role of children's social and cultural contexts. We take up these approaches in the next section.

■ Through careful observations of and clinical interviews with children, Jean Piaget developed his comprehensive theory of cognitive development. His work has inspired more research on children than any other theory.

Ask Yourself

REVIEW
Cite similarities and differences between Freud's and Erikson's views of development.

REVIEW
What aspect of behaviorism made it attractive to critics of psychoanalytic theory? How did Piaget's theory respond to a major limitation of behaviorism?

APPLY
A 4-year-old becomes frightened of the dark and refuses to go to sleep at night. How would a psychoanalyst and a behaviorist differ in their views of how this problem developed?

www.

Recent Theoretical Perspectives

New ways of understanding the developing person are constantly emerging—questioning, building on, and enhancing the discoveries of earlier theories. Today, a burst of fresh approaches and research emphases is broadening our understanding of the lifespan.

Information Processing

During the 1970s, researchers turned to the field of cognitive psychology for ways to understand the development of thinking. The design of digital computers that use mathematically specified steps to solve problems suggested to psychologists that the human mind might also be viewed as a symbol-manipulating system through which information flows—a perspective called **information processing** (Klahr & MacWhinney, 1998). From presentation to the senses at input to behavioral responses at output, information is actively coded, transformed, and organized.

Information-processing researchers often use flowcharts to map the precise series of steps individuals use to solve problems and complete tasks, much like the plans devised by programmers to get computers to perform a series of "mental operations." Let's look at an example to clarify the usefulness of this approach. In a study of problem solving, a researcher provided a pile of blocks varying in size, shape, and weight and asked school-age children to build a bridge across a "river" (painted on a floor mat) that was too wide for any single block to span (Thornton, 1999). Figure 1.4 shows one solution to the problem: two plank-like blocks span the water, each held in place by the counterweight of heavy blocks on the bridge's towers. Whereas older children easily built successful bridges, only one 5-year-old did. Careful tracking of her efforts revealed that she repeatedly tried unsuccessful strategies, such as pushing two planks together and pressing down on their ends to hold them in place. But eventually, her experimentation triggered the idea of using the blocks as counterweights. Her mistaken procedures helped her understand why the counterweight approach worked.

A wide variety of information-processing models exist. Some, like the one just considered, track children's mastery of one or a few tasks. Others describe the human cognitive system as a whole (Atkinson & Shiffrin, 1968; Lockhart & Craik, 1990). These general models are used as guides for asking questions about broad changes in thinking. For example, does a child's ability to solve problems become more organized and "planful" with age? Why is information processing slower among older than younger adults? Are declines in memory during old age evident on only some or on all types of tasks?

Like Piaget's cognitive-developmental theory, information processing regards people as active, sense-making beings. But unlike Piaget's theory, there are no stages of development. Rather, the thought processes studied—perception, attention, memory, planning strategies, categorization of information, and comprehension of written and spoken prose—are regarded as similar at all ages but present to a lesser or greater extent. Therefore, the view of development is one of continuous change.

A great strength of the information-processing approach is its commitment to careful, rigorous research methods. Because it has provided precise accounts of how children and adults tackle many cognitive tasks, its findings have important implications for education (Geary, 1994; Siegler, 1998). But information processing has fallen short in some respects. Although good at analyzing thinking into its components, information processing has had difficulty putting them back together into a comprehensive theory. In addition, aspects of cognition that are not linear and logical, such as imagination and creativity, are all but ignored by this approach (Lutz & Sternberg, 1999). Finally, much information-processing research has been conducted in

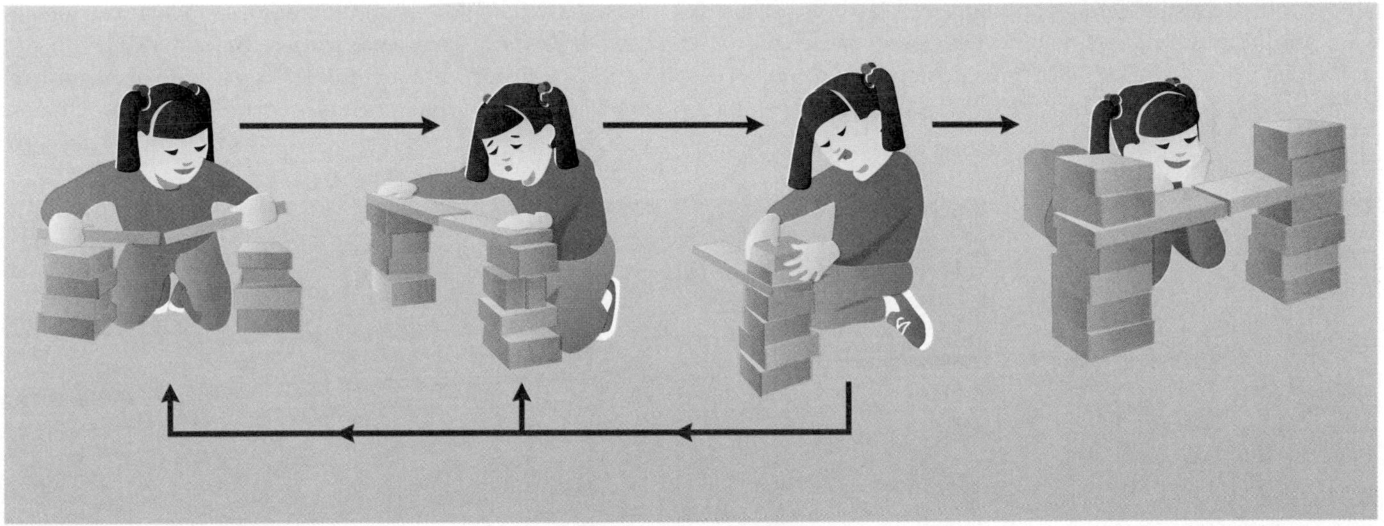

FIGURE 1.4 Information-processing flowchart showing the steps that a 5-year-old used to solve a bridge-building problem. Her task was to use blocks varying in size, shape, and weight, some of which were planklike, to construct a bridge across a "river" (painted on a floor mat) too wide for any single block to span. The child discovered how to counterweight and balance the bridge. The arrows reveal that even after building a successful counterweight, she returned to earlier, unsuccessful strategies, which seemed to help her understand why the counterweight approach worked. (Adapted from Thornton, 1999.)

laboratories rather than in real-life situations. Recently, investigators have addressed this concern by studying conversations, stories, memory for everyday events, and academic problem solving.

An advantage of having many theories is that they encourage one another to attend to previously neglected dimensions of people's lives. A unique feature of the final three perspectives we will discuss is their focus on *contexts for development.* The first of these views emphasizes that development of many capacities is influenced by our long evolutionary history.

Ethology and Evolutionary Developmental Psychology

Ethology is concerned with the adaptive, or survival, value of behavior and its evolutionary history (Dewsbury, 1992; Hinde, 1989). Its roots can be traced to the work of Darwin. Two European zoologists, Konrad Lorenz and Niko Tinbergen, laid its modern foundations. Watching diverse animal species in their natural habitats, Lorenz and Tinbergen observed behavior patterns that promote survival. The best known of these is *imprinting,* the early following behavior of certain baby birds, such as geese, that ensures that the young will stay close to the mother and be fed and protected from danger. Imprinting takes place during an early, restricted time period of development. If the mother goose is not present during this time but an object resembling her in important features is, young goslings may imprint on it instead (Lorenz, 1952).

Observations of imprinting led to a major concept in human development: the *critical period.* It refers to a limited time span during which the individual is biologically prepared to acquire certain adaptive behaviors but needs the support of an appropriately stimulating environment. Many researchers have conducted studies to find out whether complex cognitive and social behaviors must be learned during certain time periods. For example, if children are deprived of adequate food or physical and social stimulation during their early years, will their intelligence be impaired? If language is not mastered during early childhood, is the capacity to acquire it reduced?

In later chapters, we will discover that the term *sensitive period* applies better to human development than does the strict notion of a critical period (Bornstein, 1989). A **sensitive period** is a time that is optimal for certain capacities to emerge and in which the individual is especially responsive to environmental influences. However, its boundaries are less well defined than are those of a critical period. Development may occur later, but it is harder to induce.

Inspired by observations of imprinting, British psychoanalyst John Bowlby (1969) applied ethological theory to the understanding of the human infant–caregiver relationship. He argued that infant smiling, babbling, grasping, and crying are built-in social signals that encourage the parent to approach, care for, and interact with the baby. By keeping the mother near, these behaviors help ensure that the infant will be fed, protected from danger, and provided with the stimulation and affection necessary for healthy growth. The development of attachment in humans is a lengthy process involving changes in psychological structures that lead the baby to form a deep affectional tie with the caregiver. Bowlby (1979) believed that this bond has lifelong consequences, affecting relationships "from cradle to grave" (p. 129). In later chapters, we will consider research that evaluates this assumption.

Observations by ethologists have shown that many aspects of social behavior, including emotional expressions, aggression, cooperation, and social play, resemble those of our primate relatives. Recently, researchers have extended this effort in a new area of research called **evolutionary developmental psychology.** It seeks to understand the adaptive value of species-wide cognitive, emotional, and social competencies as those competencies change with age. Evolutionary developmental psychologists ask such questions as, What role does the newborn's visual preference for facelike stimuli play in survival? Does it support older infants' capacity to distinguish familiar caregivers from unfamiliar people? Why do children

■ Konrad Lorenz, one of the founders of ethology and a keen observer of animal behavior, developed the concept of imprinting. Here, young geese who were separated from their mother and placed in the company of Lorenz during an early, sensitive period show that they have imprinted on him. They follow as he swims through the water, a response that promotes survival.

play in sex-segregated groups? What do they learn from such play that might lead to adult gender-typed behaviors, such as male dominance and female investment in caregiving?

As these examples suggest, evolutionary psychologists are not just concerned with the biological basis of development. They are also interested in how individuals learn because learning lends flexibility and greater adaptiveness to behavior (Bjorklund & Pellegrini, 2000; Geary, 1999). The evolutionary selection benefits of behavior are believed to be strongest in the first half of life—to ensure survival, reproduction, and effective parenting. As people age, social and cultural factors become increasingly important in generating and maintaining high levels of functioning (Smith & Baltes, 1999). The next contextual perspective we will discuss, Vygotsky's sociocultural theory, serves as an excellent complement to ethology because it highlights social and cultural contexts for development.

Vygotsky's Sociocultural Theory

The field of human development has recently seen a dramatic increase in studies addressing the cultural context of people's lives. Investigations that make comparisons across cultures, and among ethnic groups within cultures, provide insight into whether developmental pathways apply to all people or are limited to particular environmental conditions. As a result, cross-cultural and multicultural research helps us untangle the contributions of biological and environmental factors to the timing, order of appearance, and diversity of children's and adults' behaviors (Greenfield, 1994).

In the past, cross-cultural studies focused on broad cultural differences in development—for example, whether children in one culture are more advanced in motor development or do better on intellectual tasks than children in another. However, this approach can lead us to conclude incorrectly that one culture is superior in enhancing development, whereas another is deficient. In addition, it does not help us understand the precise experiences that contribute to cultural differences in behavior.

Today, more research is examining the relationship of *culturally specific practices* to development. The contributions of Russian psychologist Lev Vygotsky (1896–1934) have played a major role in this trend. Vygotsky's (1934/1987) perspective is called **sociocultural theory.** It focuses on how *culture*—the values, beliefs, customs, and skills of a social group—is transmitted to the next generation. According to Vygotsky, *social interaction*—in particular, cooperative dialogues with more knowledgeable members of society—is necessary for children to acquire the ways of thinking and behaving that make up a community's culture (Wertsch & Tulviste, 1992). Vygotsky believed that as adults and more expert peers help children master culturally meaningful activities, the communication between them becomes part of children's thinking. As children internalize the essential features of these dialogues, they can use the language within them to guide their own thought and actions and to acquire new skills (Berk, 2001).

■ This girl of Bali, Indonesia, is learning traditional dance steps through the guidance of an adult expert. According to Vygotsky's sociocultural theory, social interaction between children and more knowledgeable members of their culture leads to ways of thinking and behaving essential for success in that culture.

Vygotsky's theory has been especially influential in the study of cognitive development. Vygotsky agreed with Piaget that children are active, constructive beings. But unlike Piaget, who emphasized children's independent efforts to make sense of their world, Vygotsky viewed cognitive development as a *socially mediated process*—as dependent on the support that adults and more mature peers provide as children try new tasks.

In Vygotsky's theory, children undergo certain stagewise changes. For example, when they acquire language, their ability to participate in dialogues with others is greatly enhanced, and mastery of culturally valued competencies surges forward. When children enter school, they spend much time discussing language, literacy, and other academic concepts—experiences that encourage them to reflect on their own thinking. As a result, they show dramatic gains in reasoning and problem solving.

Although most research inspired by Vygotsky's theory focuses on children, his ideas apply to people of any age. A central theme is that cultures select tasks for their members, and social interaction surrounding those tasks leads to competencies essential for success in a particular culture. For example, in industrialized nations, teachers can be seen helping people learn to read, drive a car, or use a computer (Schwebel, Maher, & Fagley, 1990). Among the Zinacanteco Indians of southern Mexico, adult experts guide young girls as they master complicated weaving techniques (Childs & Greenfield, 1982). In Brazil, child candy sellers with little or no schooling develop sophisticated mathematical abilities as the result of buying candy from wholesalers, pricing it in collaboration with adults and experienced peers, and bargaining with customers on city streets (Saxe, 1988).

Vygotsky's theory, and the research stimulated by it, reveal that children in every culture develop unique strengths. At the same time, Vygotsky's emphasis on culture and social experience led him to neglect the biological side of development. Although he recognized the importance of heredity and brain growth, he said little about their role in cognitive change. Furthermore, Vygotsky's emphasis on social transmission of knowledge meant that he placed less emphasis than other theorists on children's capacity to shape their own development. Contemporary followers of Vygotsky grant the individual and society more balanced roles (Gauvain, 1999; Rogoff, 1998).

Ecological Systems Theory

Urie Bronfenbrenner, an American psychologist, is responsible for an approach to human development that has moved to the forefront of the field over the past 2 decades because it offers the most differentiated and thorough account of contextual influences on development. **Ecological systems theory** views the person as developing within a complex *system* of relationships affected by multiple levels of the surrounding environment. Since the child's biological dispositions join with environmental forces to mold development, Bronfenbrenner recently characterized his perspective as a *bioecological model* (Bronfenbrenner & Morris, 1998).

As Figure 1.5 shows, Bronfenbrenner envisions the environment as a series of nested structures that includes but extends beyond the home, school, neighborhood, and workplace settings in which people spend their everyday lives. Each layer of the environment is viewed as having a powerful impact on development.

● The Microsystem.

The innermost level of the environment is the **microsystem,** which consists of activities and interaction patterns in the person's immediate surroundings. Bronfenbrenner emphasizes that to understand development at this level, we must keep in mind that all relationships are *bidirectional.* For example, adults affect children's behavior, but children's biologically and socially influenced characteristics—their physical attributes, personalities, and capacities—also affect adults' behavior. For example, a friendly, attentive child is likely to evoke positive and patient reactions from parents, whereas an active, distractible youngster is more likely to receive restriction and punishment. When these bidirectional interactions occur often over time, they have an enduring impact on development (Bronfenbrenner, 1995; Collins et al., 2000).

At the same time, other individuals in the microsystem affect the quality of any two-person relationship. If they are supportive, then interaction is enhanced. For example, when parents encourage one another in their child-rearing roles, each

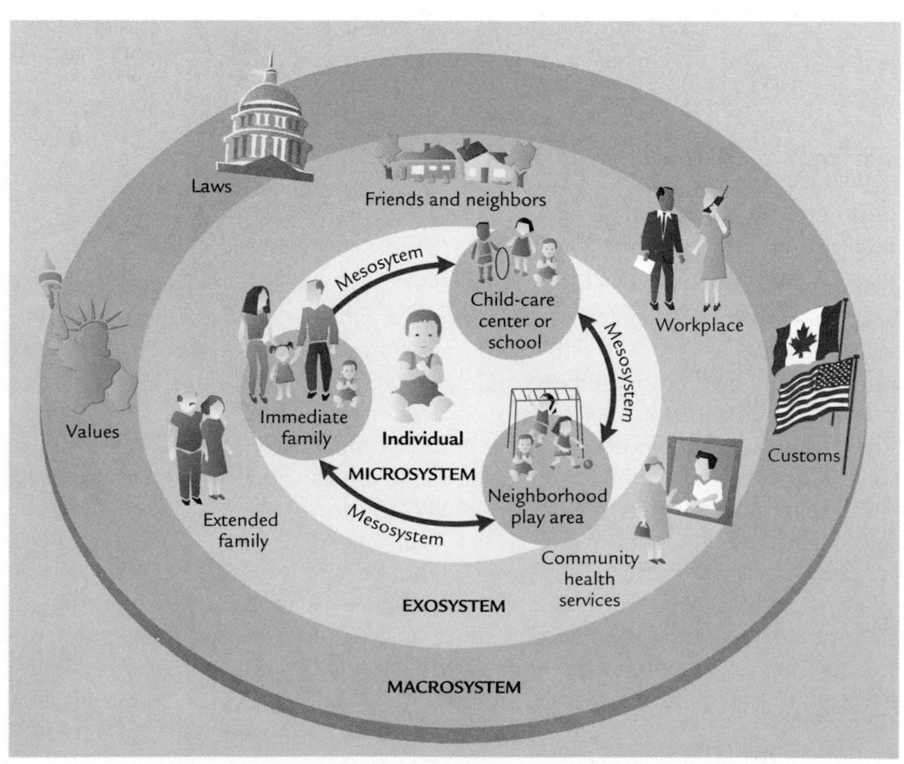

■ **FIGURE 1.5 Structure of the environment in ecological systems theory.** The *microsystem* concerns relations between the developing person and the immediate environment; the *mesosystem,* connections among immediate settings; the *exosystem,* social settings that affect but do not contain the developing person; and the *macrosystem,* the values, laws, customs, and resources of the culture that affect activities and interactions at all inner layers. The *chronosystem* (not pictured) is not a specific context. Instead, it refers to the dynamic, ever-changing nature of the person's environment.

engages in more effective parenting (Cowan, Powell, & Cowan, 1998). In contrast, marital conflict is associated with inconsistent discipline and hostile reactions toward children. In response, children typically become hostile, and both parent and child adjustment suffers (Hetherington & Stanley-Hagen, 2002).

● **The Mesosystem.** The second level of Bronfenbrenner's model, the **mesosystem,** encompasses connections between microsystems. For example, a child's academic progress depends not just on activities that take place in classrooms. It is also promoted by parent involvement in school life and by the extent to which academic learning is carried over into the home (Connors & Epstein, 1996). Among adults, how well a person functions as spouse and parent at home is affected by relationships in the workplace, and vice versa (Gottfried, Gottfried, & Bathurst, 2002).

● **The Exosystem.** The **exosystem** refers to social settings that do not contain the developing person but nevertheless affect experiences in immediate settings. These can be formal organizations, such as the board of directors in the individual's workplace or health and welfare services in the community. For example, flexible work schedules, paid maternity and paternity leave, and sick leave for parents whose children are ill are ways that work settings can help parents rear children and, indirectly, enhance the development of both adult and child. Exosystem supports can also be informal, such as social networks—friends and extended-family members who provide advice, companionship, and even financial assistance. Research confirms the negative impact of a breakdown in exosystem activities. Families who are socially isolated because they have few personal or community-based ties or who are affected by unemployment show increased rates of marital conflict and child abuse (Emery & Laumann-Billings, 1998).

● **The Macrosystem.** The outermost level of Bronfenbrenner's model, the **macrosystem,** is not a specific context. Instead, it consists of cultural values, laws, customs, and resources. The priority that the macrosystem gives to the needs of children and adults affects the support they receive at inner levels of the environment. For example, in countries that require high-quality standards for child care and workplace benefits for employed parents, children are more likely to have favorable experiences in their immediate settings. And when the government provides a generous pension plan for retirees, it supports the well-being of the elderly.

● **A Dynamic, Ever-Changing System.** According to Bronfenbrenner, the environment is not a static force that affects people in a uniform way. Instead, it is dynamic and ever-changing. Whenever individuals add or let go of roles or settings in their lives, the breadth of their microsystems changes. These shifts in contexts, or ecological transitions, as Bronfenbrenner calls them, take place throughout life and are often important turning points in development. Starting school, entering the workforce, marrying, becoming a parent, getting divorced, moving, and retiring are examples.

■ In ecological systems theory, development occurs within a complex system of relationships affected by multiple levels of the environment. This father greets his daughter at the end of the school day. The girl's experiences at school (microsystem) and the father's experiences at work (exosystem) affect father–daughter interaction.

Bronfenbrenner refers to the temporal dimension of his model as the **chronosystem** (the prefix *chrono* means "time"). Changes in life events can be imposed externally. Alternatively, they can arise from within the person, since individuals select, modify, and create many of their own settings and experiences. How they do so depends on their age; their physical, intellectual, and personality characteristics; and their environmental opportunities. Therefore, in ecological systems theory, development is neither controlled by environmental circumstances nor driven by inner dispositions. Instead, people are products and producers of their environments, so both people and their environments form a network of interdependent effects. We will see many more examples of these principles in later chapters.

Ask Yourself

REVIEW

Explain how each recent theoretical perspective regards children and adults as active contributors to their own development.

REVIEW

What features of Vygotsky's sociocultural theory distinguish it from Piaget's theory and from information processing?

CONNECT

Is Bronfenbrenner's ecological systems theory compatible with assumptions of the lifespan perspective: development as lifelong, multidirectional, highly plastic, and embedded in multiple contexts? Explain.

www.

Comparing and Evaluating Theories

In the preceding sections, we reviewed theoretical perspectives that are major forces in human development research. They differ in many respects. First, they focus on different domains of development. Some, such as the psychoanalytic perspective and ethology, emphasize emotional and social development. Others, such as Piaget's cognitive-developmental theory, information processing, and Vygotsky's sociocultural theory, stress changes in thinking. The remaining approaches—behaviorism, social learning theory, ecological systems theory, and the lifespan perspective—discuss many aspects of human functioning.

Second, every theory contains a point of view about development. As we conclude our review of theoretical perspectives, identify the stand each theory takes on the controversial issues presented at the beginning of this chapter. Then check your analysis against Table 1.5.

Table 1.5 Stances of Major Theories on Basic Issues in Human Development

Theory	Continuous or Discontinuous Development?	One Course of Development or Many?	Nature or Nurture as More Important?
Psychoanalytic perspective	*Discontinuous:* Stages of psychosexual and psychosocial development are emphasized.	*One course:* Stages are assumed to be universal.	*Both nature and nurture:* Innate impulses are channeled and controlled through child-rearing experiences. *The individual as stable:* Early experiences set the course of later development.
Behaviorism and social learning theory	*Continuous:* Development involves an increase in learned behaviors.	*Many possible courses:* Behaviors reinforced and modeled may vary from person to person.	*Emphasis on nurture:* Development is the result of conditioning and modeling. *Both early and later experiences* are important.
Piaget's cognitive-developmental theory	*Discontinuous:* Stages of cognitive development are emphasized.	*One course:* Stages are assumed to be universal.	*Both nature and nurture:* Development occurs as the brain matures and children exercise their innate drive to discover reality in a generally stimulating environment. *Both early and later experiences* are important.
Information processing	*Continuous:* Children and adults gradually improve in perception, attention, memory, and problem-solving skills.	*One course:* Changes studied characterize most or all children and adults.	*Both nature and nurture:* Children and adults are active, sense-making beings who modify their thinking as the brain matures and they confront new environmental demands. *Both early and later experiences* are important.
Ethology and evolutionary developmental psychology	*Both continuous and discontinuous:* Children and adults gradually develop a wider range of adaptive behaviors. Sensitive periods occur, in which qualitatively distinct capacities emerge fairly suddenly.	*One course:* Adaptive behaviors and sensitive periods apply to all members of a species.	*Both nature and nurture:* Evolution and heredity influence behavior, and learning lends greater flexibility and adaptiveness to it. In sensitive periods, *early experiences* set the course of later development.
Vygotsky's sociocultural theory	*Both continuous and discontinuous:* Language development and schooling lead to stagewise changes. Dialogues with more expert members of society also lead to continuous changes that vary from culture to culture.	*Many possible courses:* Socially mediated changes in thought and behavior vary from culture to culture.	*Both nature and nurture:* Heredity, brain growth, and dialogues with more expert members of society jointly contribute to development. *Both early and later experiences* are important.
Ecological systems theory	*Not specified.*	*Many possible courses:* Biological dispositions join with environmental forces at multiple levels to mold development in unique ways.	*Both nature and nurture:* The individual's characteristics and the reactions of others affect each other in a bidirectional fashion. *Both early and later experiences* are important.
Lifespan perspective	*Both continuous and discontinuous:* Continuous gains and declines, as well as discontinuous, stagewise emergence of new skills, occur during all age periods.	*Many possible courses:* Development is embedded in multiple contexts that vary from person to person, leading to diverse pathways of change.	*Both nature and nurture:* Development is multidimensional, affected by an intricate blend of biological and social forces. Emphasizes plasticity at all ages. *Both early and later experiences* are important.

Finally, we have seen that theories have strengths and limitations. Perhaps you found that you were attracted to some theories, but you had doubts about others. As you read more about development in later chapters, you may find it useful to keep a notebook in which you test your theoretical likes and dislikes against the evidence. Don't be surprised if you revise your ideas many times, just as theorists have done throughout this century.

Studying Development

In every science, theories, like those we have just reviewed, guide the collection of information, its interpretation, and its application to real-life situations. In fact, research usually begins with a prediction about behavior drawn from a theory, or what we call a *hypothesis.* But theories and hypotheses are only the beginning of the many activities that result in sound evidence on human development. Conducting research according to scientifically accepted procedures involves many steps and choices. Investigators must decide which participants, and how many, to include. Then they must figure out what the participants will be asked to do and when, where, and how many times each will have to be seen. Finally, they must examine and draw conclusions from their data.

In the following sections, we look at research strategies commonly used to study human development. We begin with *research methods*—the specific activities of participants, such as taking tests, answering questionnaires, responding to interviews, or being observed. Then we turn to *research designs*—overall plans for research studies that permit the best possible test of the investigator's hypothesis. Finally, we discuss ethical issues involved in doing research with human participants.

At this point, you may be wondering, Why learn about research strategies? Why not leave these matters to research specialists and concentrate on what is already known about the developing person and how this knowledge can be applied? There are two reasons. First, each of us must be a wise and critical consumer of knowledge. Knowing the strengths and limitations of various research strategies becomes important in separating dependable information from misleading results. Second, individuals who work directly with children or adults may be in a unique position to build bridges between research and practice by carrying out research, either on their own or in partnership with experienced investigators. Currently, communities and researchers are collaborating in designing, implementing, and evaluating interventions that enhance lifespan development (Lerner, Fisher, & Weinberg, 2000). For an inspiring example of this new wave of "outreach" research, consult the Social Issues box on page 28. To broaden these efforts, a basic understanding of the research process is essential.

Common Research Methods

How does a researcher choose a basic approach to gathering information? Common methods include systematic observation, self-reports (such as questionnaires and interviews), clinical or case studies of a single individual, and ethnographies of the life circumstances of a specific group of people.

● **Systematic Observation.** To find out how people actually behave, a researcher may choose systematic observation. Observations can be made in different ways. One approach is to go into the field, or natural environment, and observe the behavior of interest—a method called **naturalistic observation.**

A study of preschoolers' responses to their peers' distress provides a good example of this technique (Farver & Branstetter, 1994). Observing 3- and 4-year-olds in child-care centers, the researchers recorded each instance of crying and the reactions of nearby children—whether they ignored, watched curiously, commented on the child's unhappiness, scolded or teased, or shared, helped, or expressed sympathy. Caregiver behaviors, such as explaining why a child was crying, mediating conflict, or offering comfort, were noted to see if adult sensitivity was related to children's caring responses. A strong relationship emerged. The great strength of naturalistic observation is that investigators can see directly the everyday behaviors they hope to explain.

Naturalistic observation also has a major limitation: Not all individuals have the same opportunity to display a particular behavior in everyday life. In the study just described, some children might have witnessed a child crying more often than others had. For this reason, they might have displayed more compassion.

Researchers commonly deal with this difficulty by making **structured observations,** in which the investigator sets up a laboratory situation that evokes the behavior of interest so that every participant has an equal opportunity to display the response. In one study, children's comforting behavior was observed by playing a tape recording of a baby crying in the next room. Using an intercom, children could either talk to the baby or push a button so they did not have to listen (Eisenberg et al., 1993). Notice how structured observation permits more control over the research situation. But its great disadvantage is that people do not necessarily behave in the laboratory as they do in everyday life.

The procedures used to collect systematic observations vary, depending on the nature of the research problem. Some investigators must describe the entire stream of behavior—everything said and done over a certain time period. In one of my own studies, I wanted to find out how sensitive, responsive, and verbally stimulating caregivers were when they interacted with children in child-care centers (Berk, 1985). In this case, everything each caregiver said and did—even the amount of time she spent away from the children, taking coffee breaks and talking on the phone—was important. In other studies, only one or a few kinds of behavior are needed, and it is not necessary to preserve the entire behavior stream. In these instances, researchers use more efficient observation procedures in which they record only certain events or mark off behaviors on checklists.

Systematic observation provides invaluable information on how children and adults actually behave, but it tells us little

Social Issues

The Access Program: A Community–Researcher Partnernership

In Lorain County, Ohio, local businesses launched the Access Program, an extraordinary intervention that aims to increase the number of students who enroll in college. The program targets young people who otherwise would be unlikely to seek higher education. Alex, an eighth grader, and his parents (who work on a factory assembly line and do not have college degrees) participate. Access Program administrators believe that the earlier students and parents plan for postsecondary education and career development, the more successful young people will be in getting into college, graduating, and reaching their career goals.

The Access Program views getting ready for college as a collaboration between schools and families. Program representatives visit parents' workplaces to give presentations on planning for education beyond high school, including financial aid opportunities. At school, Alex meets with a volunteer advisor to discuss his interests and to complete a time line, which maps out what he ought to do from now through high school to prepare for college. Because of his interest in the medical field, Alex has been matched with a mentor who is a surgical nurse at a local hospital. She explained to Alex how she prepared for her job and introduced him to other hospital-related careers. After Alex enters high school, he will have a specially trained advisor who will assist with the details of college planning.

When program managers wanted to find out how well the Access Program was achieving its goals, they contacted a nearby child development research foundation, which worked closely with the program to carry out an evaluation (Oden, 2000). Researchers conducted interviews with students and parents about their program experiences and looked at educational outcomes. Findings revealed that of participating high school seniors, 94 percent had been accepted by one or more colleges, and 88 percent had enrolled—far more than would have been expected, given statistics for students of noncollege-educated parents. For students, a warm, supportive mentoring relationship had the strongest impact on educational aspirations. Parents especially appreciated the program's visits to worksites and attention to both general and specific information, including training directed at financial planning and filling out application forms. Many reported that their viewpoints had changed from not expecting their child to go to college to making definite college plans. Program leaders used the study's findings to acquire additional funding and hire a full-time mentor recruiter to expand the program's mentoring services.

© DAVID YOUNG-WOLFF/PHOTOEDIT

■ This high school student, whose parents did not go to college, explores higher education options with a specially trained advisor. In Lorain County, Ohio, the Access Program has greatly increased the number of high school students who enroll in college.

The Access Program illustrates how researchers can help communities make programs more effective. And when findings highlight program characteristics that work, those features can be replicated in other communities. In this way, researcher–community partnerships can lead to widespread benefits, along with increased understanding of development.

about the reasoning behind their responses. For this kind of information, researchers must turn to self-report techniques.

● **Self-Reports.** Self-reports are instruments that ask participants to provide information on their perceptions, thoughts, abilities, feelings, attitudes, beliefs, and past experiences. They range from relatively unstructured interviews to highly structured interviews, questionnaires, and tests.

In a **clinical interview,** a flexible, conversational style is used to probe for the participant's point of view. Consider the following example, in which Piaget questioned a 5-year-old child about his understanding of dreams:

> *Where does the dream come from?*—I think you sleep so well that you dream.—*Does it come from us or from outside?*—From outside.—*What do we dream with?*—I don't know.—*With the hands?... With nothing?*—Yes, with nothing.—*When you are in bed and you dream, where is the dream?*—In my bed, under the blanket. I don't really know. If it was in my stomach, the bones would be in the way and I shouldn't see it.—*Is the dream there when you sleep?*—Yes, it is in the bed beside me. (Piaget, 1926/1930, pp. 97–98)

Notice how Piaget encouraged the child to expand his ideas. Although a researcher conducting clinical interviews with more than one participant would typically ask the same first question to ensure a common task, individualized prompts are given to evoke a fuller picture of each person's reasoning (Ginsburg, 1997).

The clinical interview has two major strengths. First, it permits people to display their thoughts in terms that are as close as possible to the way they think in everyday life. Second, the clinical interview can provide a large amount of information in a fairly brief period. For example, in an hour-long session, we can obtain a wide range of information on child rearing from a parent or on life circumstances from an elder—much more than we could capture by observing for the same amount of time.

A major limitation of the clinical interview has to do with the accuracy with which people report their thoughts, feelings, and experiences. Some participants, desiring to please the interviewer, may make up answers that do not represent their actual thinking. When asked about past events, they may have trouble recalling exactly what happened. And because the clinical interview depends on verbal ability and expressiveness, it may underestimate the capacities of individuals who have difficulty putting their thoughts into words.

The clinical interview has also been criticized because of its flexibility. When questions are phrased differently for each participant, responses may be due to the manner of interviewing rather than to real differences in the way people think about a certain topic. **Structured interviews,** in which each participant is asked the same set of questions in the same way, can eliminate this problem. In addition, these instruments are much more efficient. Answers are briefer, and researchers can obtain written responses from an entire group of children or adults at the same time. Also, when structured interviews use multiple-choice, yes/no, and true/false formats, as is done on many tests and questionnaires, a computer can tabulate the answers. However, these approaches do not yield the same depth of information as a clinical interview. And they can still be affected by the problem of inaccurate reporting.

● The Clinical, or Case Study, Method. An outgrowth of psychoanalytic theory, which stresses the importance of understanding a single life history, the **clinical, or case study, method** brings together a wide range of information on one person, including interviews, observations, and sometimes test scores. The aim is to obtain as complete a picture as possible of that individual's psychological functioning and the experiences that led up to it.

The clinical method is well suited to studying the development of individuals who are few in number and who vary widely in characteristics. For example, the method has been used to find out what contributes to the accomplishments of *prodigies*—extremely gifted children who attain adult competence in a field before age 10 (Gardner, 1998b). Consider Adam, a boy who read, wrote, and composed musical pieces before he was out of diapers. By age 4, Adam was deeply involved in mastering human symbol systems—BASIC for the computer, French, German, Russian, Sanskrit, Greek, ancient hieroglyphs, music, and mathematics. Adam's parents provided a home rich in stimulation and reared him with affection, firmness, and humor. They searched for schools in which he could both develop his abilities and form rewarding social relationships. He graduated from college at age 18 and continues to pursue musical composition. Would Adam have realized his abilities without the chance combination of his special gift with nurturing, committed parents? Probably not, researchers concluded (Goldsmith, 2000).

■ Using the clinical interview, this researcher asks a mother to describe her child's development. The method permits a large amount of information to be gathered in a relatively short period. However, a drawback is that participants do not always report information accurately.

The clinical method yields richly detailed case narratives that offer valuable insights into the multiplicity of factors that affect development. Nevertheless, like all other methods, it has drawbacks. Information is often collected unsystematically and subjectively, permitting too much leeway for researchers' theoretical preferences to bias their observations and interpretations. In addition, investigators cannot assume that their conclusions apply to anyone other than the person studied. Even when patterns emerge across several cases, it is wise to try to confirm them with other research strategies.

● Methods for Studying Culture. A growing interest in the impact of culture has led researchers to adjust the methods just considered as well as tap procedures specially devised for cross-cultural and multicultural research. Which approach investigators choose depends on their research goals (Triandis, 1995, 1998).

Sometimes researchers are interested in characteristics that are believed to be universal but that vary in degree from one society to the next. These investigators might ask, Do parents make greater maturity demands of children in some cultures than in others? How strong are gender stereotypes in different nations? In each instance, several cultural groups will be compared, and all participants must be questioned or observed in the same way. Therefore, researchers draw on the self-report and observational procedures we have already considered, adapting them through translation so they can be understood in each cultural context. For example, to study cultural variation in parenting attitudes, the same questionnaire, asking for ratings on such items as "If my child gets into trouble, I expect him or her to handle the problem mostly by himself or herself," is given to all participants (Chen et al., 1998).

Cultural Influences

Immigrant Youths: Amazing Adaptation

During the past quarter century, a rising tide of immigrants has come to North America, fleeing war and persecution in their homelands or otherwise seeking better life chances. Today, nearly 20 percent of the U.S. youth population has foreign-born parents; almost 30 percent of these youths are foreign-born themselves. Similarly, immigrant youths are the fastest-growing segment of the Canadian population (Fuligni, 1998a; Statistics Canada, 2000). They are ethnically diverse. In the United States, most come from Asia and Latin America; in Canada, from Asia, Africa, and the Middle East.

Academic Achievement and Adjustment. Although educators and lay people often assume that the transition to a new country has a negative impact on psychological well-being, recent evidence reveals that children of immigrant parents adapt amazingly well. Students who are first-generation (foreign-born) or second-generation (American-born with immigrant parents) achieve in school as well as or better than students of native-born parents. Their success is evident in many academic subjects, including English, even though they are likely to come from non-English-speaking homes (Fuligni, 1997; Rumbaut, 1997).

Findings on psychological adjustment resemble those on achievement. Compared with their agemates, adolescents from immigrant families are less likely to commit delinquent and violent acts, to use drugs and alcohol, and to have early sex. They are also in better health—less likely to be obese and to have missed school because of illness. And in terms of self-esteem, they feel as positively about themselves as young people with native-born parents and report less emotional distress. These successes do not depend on having extensive time to adjust to a new way of life. The school performance and psychological well-being of recently arrived high school students is as high as—and sometimes higher than—that of students who come at younger ages (Fuligni, 1997, 1998; Rumbaut, 1997).

The outcomes just described are strongest for Asian youths, less dramatic for other ethnicities (Fuligni, 1997; Kao & Tienda, 1995). Variations in parental education and income account for these differences. Still, even first- and second-generation youths from ethnic groups that face considerable economic hardship (such as Mexican and Southeast Asian) are remarkably successful (Harris, 2000; Kao, 2000). Factors other than income are responsible.

Family and Community Influences. Ethnographies of immigrant populations reveal that uniformly, parents express the belief that education is the surest way to improve life chances. Consequently, they place a high value on their children's academic achievement (Suarez-Orozco & Suarez-Orozco, 1995; Zhou & Bankston, 1998). Aware of the challenges their children face, immigrant parents underscore the importance of trying hard. They remind their children that educational opportunities were not available in their native countries and that as a

At other times, researchers want to uncover the *cultural meanings* of children's and adults' behaviors by becoming as familiar as possible with their way of life (Shweder et al., 1998). To achieve this goal, researchers rely on a method borrowed from the field of anthropology—**ethnography.** Like the clinical method, ethnographic research is largely a descriptive, qualitative technique. But instead of aiming to understand a single individual, it is directed toward understanding a culture or a distinct social group, achieving its goals through *participant observation.* Typically, the researcher lives with the cultural community for a period of months or years, participating in all aspects of its daily life. Extensive field notes are gathered, consisting of a mix of observations, self-reports from members of the culture, and careful interpretations by the investigator (Jessor, 1996; Shweder, 1996). Later, these notes are put together into a description of the community that tries to capture its unique values and social processes.

The ethnographic approach assumes that by entering into close contact with a social group, researchers can understand the beliefs and behaviors of its members in a way not possible with an observational visit, interview, or questionnaire. In some ethnographies, investigators focus on many aspects of experience, as one team of researchers did in describing what it is like to grow up in a small American town (Peshkin, 1978). In other instances, the research is limited to one or a few settings, such as home, school, or neighborhood life (LeVine et al., 1994; Peshkin, 1997; Valdés, 1998). Researchers interested in cultural comparisons may supplement traditional self-report and observational methods with ethnography if they suspect that unique meanings underlie cultural differences, as the Cultural Influences box above reveals.

Ethnographers strive to minimize their influence on the culture they are studying by becoming part of it. Nevertheless, at times their presence does alter the situation. And as with clinical research, investigators' cultural values and theoretical commitments sometimes lead them to observe selectively or misinterpret what they see. In addition, the findings of ethnographic studies cannot be assumed to apply, or generalize, beyond the people and settings in which the research was conducted.

Before moving on to research designs, you may find it helpful to refer to Table 1.6 on page 32, which summarizes the strengths and limitations of the research methods just considered.

result, they themselves are often limited to menial jobs.

Adolescents from immigrant families internalize their parents' valuing of education, endorsing it more strongly than agemates with native-born parents (Asakawa, 2001; Fuligni, 1997). Because minority ethnicities usually stress allegiance to family and community over individual goals, first- and second-generation young people feel a strong sense of obligation to their parents (Fuligni et al., 1999). They view school success as one of the most important ways they can repay their parents for the hardships they endured in coming to a new land. Both family relationships and school achievement protect these youths from risky behaviors, such as delinquency, early pregnancy, and drug use (refer to the Biology and Environment box on pages 10–11).

Immigrant parents typically develop close ties to an ethnic community. It exerts additional control through a high consensus on values and constant monitoring of young people's activities. Consider Versailles Village, a low-income Vietnamese neighborhood in New Orleans, where the overwhelming majority of high school students say that obedience to parents and working hard are very important (Zhou & Bankston, 1998). A local education association promotes achievement by offering after-school homework tutoring sessions and English- and Vietnamese-language classes. Almost 70 percent of Vietnamese adolescents enroll, and attendance is positively related to school performance.

The comments of Vietnamese teenagers capture the power of these family and community forces:

> *Thuy Trang, age 14, middle-school Student of the Year:* "When my parents first immigrated from Vietnam, they spent every waking hour working hard to support a family. They have sacrificed for me, and I am willing to do anything for them."

> *Elizabeth, age 16, straight-A student, like her two older sisters:* "My parents know pretty much all the kids in the neighborhood. . . . Everybody here knows everybody else. It's hard to get away with much." (Zhou & Bankston, 1998, pp. 93, 130)

■ This family recently immigrated from Ecuador to the United States. Ethnographic research reveals that parents in immigrant families typically place a high value on academic achievement and emphasize family and community over individual goals. Consequently, their adolescent children are likely to feel a strong sense of obligation to meeting their parents' expectations.

Ask Yourself

REVIEW

Why might a researcher choose structured observation over naturalistic observation? How about the reverse? What might lead the researcher to opt for clinical interviewing over systematic observation?

APPLY

A researcher is interested in how elders experience daily life in different cultures. Which method should she use? Explain.

CONNECT

The lifespan perspective points out that nonnormative events can affect development in powerful ways. Reread the description of nonnormative influences on pages 11–12. Which method would be most likely to tap them?

www.

General Research Designs

In deciding on a research design, investigators choose a way of setting up a study that permits them to test their hypotheses with the greatest certainty possible. Two main types of designs are used in all research on human behavior: correlational and experimental.

● **Correlational Design.** In a **correlational design,** researchers gather information on already-existing groups of individuals, generally in natural life circumstances, and make no effort to alter their experiences. Then they look at relationships between participants' characteristics and their behavior or development. Suppose we want to answer such questions as, Do parents' styles of interacting with children have any bearing on children's intelligence? Does the arrival of a baby influence a couple's marital satisfaction? Does the death of a spouse in old age affect the surviving partner's physical health and psychological well-being? In these and many other instances, the conditions of interest are difficult or impossible to arrange and control and must be studied as they currently exist.

Table 1.6 Strengths and Limitations of Common Research Methods

Method	Description	Strengths	Limitations
Systematic Observation			
Naturalistic observation	Observation of behavior in natural contexts	Reflects participants' everyday lives.	Cannot control conditions under which participants are observed.
Structured observation	Observation of behavior in a laboratory	Grants each participant an equal opportunity to display the behavior of interest.	May not yield observations typical of participants' behavior in everyday life.
Self-Reports			
Clinical interview	Flexible interviewing procedure in which the investigator obtains a complete account of the participant's thoughts	Comes as close as possible to the way participants think in everyday life. Great breath and depth of information can be obtained in a short time.	May not result in accurate reporting of information. Flexible procedure makes comparing individuals' responses difficult.
Structured interview, questionnaires, and tests	Self-report instruments in which each participant is asked the same questions in the same way	Permits comparisons of participants' responses and efficient data collection and scoring.	Does not yield the same depth of information as a clinical interview. Responses are still subject to inaccurate reporting.
Clinical Method (Case Study)	A full picture of a single individual's psychological functioning, obtained by combining interviews, observations, and test scores	Provides rich, descriptive insights into factors that affect development.	May be biased by researchers' theoretical preferences. Findings cannot be applied to individuals other than the participant.
Ethnography	Participant observation of a culture or distinct social group. By making extensive field notes, the researcher tries to capture the culture's unique values and social processes	Provides a more thorough and accurate description than can be derived from a single observational visit, interview, or questionnaire.	May be biased by researchers' values and theoretical preferences. Findings cannot be applied to individuals and settings other than the ones studied.

Correlational studies have one major limitation: We cannot infer cause and effect. For example, if we find that parental interaction is related to children's intelligence, we still do not know whether parents' behavior actually *causes* intellectual differences among children. In fact, the opposite is certainly possible. The behaviors of highly intelligent children may be so attractive that they cause parents to interact more favorably. Or a third variable that we did not even think about studying, such as amount of noise and distraction in the home, may be causing both maternal interaction and children's intelligence to change.

In correlational studies, and in other types of research designs, investigators often examine relationships by using a **correlation coefficient,** a number that describes how two measures, or variables, are associated with one another. We will encounter the correlation coefficient in discussing research findings throughout this book, so let's look at what it is and how it is interpreted. A correlation coefficient can range in value from +1.00 to –1.00. The *magnitude, or size, of the number* shows the *strength of the relationship.* A zero correlation indicates no relationship, but the closer the value is to +1.00 or –1.00, the stronger the relationship. For instance, a correlation of –.78 is high, –.52 is moderate, and –.18 is low. Note, however, that correlations of +.52 and –.52 are equally strong. The *sign of the number* (+ or –) refers to the *direction of the relationship.* A positive sign (+) means that as one variable increases, the other also increases. A negative sign (–) indicates that as one variable increases, the other decreases.

Let's take some examples to illustrate how a correlation coefficient works. In one study, a researcher found that a measure of maternal language stimulation at 13 months was positively correlated with the size of children's vocabularies at 20 months, at +.50 (Tamis-LeMonda & Bornstein, 1994). This is a moderate correlation, which indicates that the more mothers spoke to their infants, the more advanced their children were in language development. In another study, a researcher reported that the extent to which mothers ignored their 10-month-olds' bids for attention was negatively correlated with children's willingness to comply with parental demands 1 year later—at –.46 for boys and –.36 for girls (Martin, 1981). These moderate correlations reveal that the more mothers ignored their babies, the less cooperative their children were.

Both of these investigations found a relationship between maternal behavior and children's early development. Although the researchers suspected that maternal behavior affected the children's responses, in neither study could they really be sure about cause and effect. However, finding a relationship in a correlational study suggests that tracking down its cause—with a more powerful experimental strategy, if possible—would be worthwhile.

● **Experimental Design.** An **experimental design** permits inferences about cause and effect because researchers use an evenhanded procedure to assign people to two or more treatment conditions. In an experiment, the events and behaviors of interest are divided into two types: independent and dependent variables. The **independent variable** is the one the investigator expects to cause changes in another variable. The **dependent variable** is the one the investigator expects to be influenced by the independent variable. Cause-and-effect relationships can be detected because the researcher directly *controls or manipulates* changes in the independent variable by exposing participants to the treatment conditions. Then the researcher compares their performance on measures of the dependent variable.

In one *laboratory experiment,* investigators explored the impact of adults' angry interactions on children's adjustment (El-Sheikh, Cummings, & Reiter, 1996). They hypothesized that the way angry encounters end (independent variable) affects children's emotional reactions (dependent variable). Four- and 5-year-olds were brought one at a time to a laboratory, accompanied by their mothers. One group was exposed to an unresolved-anger treatment, in which two adult actors entered the room and argued but did not work out their disagreements. The other group witnessed a resolved-anger treatment, in which the adults ended their disputes by apologizing and compromising. During a follow-up adult conflict, children in the resolved-anger treatment showed less distress, as measured by anxious facial expressions, freezing in place, and seeking closeness to their mothers. The experiment revealed that anger resolution can reduce the stressful impact of adult conflict on children.

In experimental studies, investigators must take special precautions to control for participants' characteristics that could reduce the accuracy of their findings. For example, in the study just described, if a greater number of children from homes high in parental conflict ended up in the unresolved-anger treatment, we could not tell whether the independent variable or the children's backgrounds produced the results. To protect against this problem, researchers engage in **random assignment** of participants to treatment conditions. By using an unbiased procedure, such as drawing numbers out of a hat or flipping a coin, investigators increase the likelihood that participants' characteristics will be equally distributed across treatment groups.

● **Modified Experimental Designs: Field and Natural Experiments.** Most experiments are conducted in laboratories where researchers can achieve the maximum possible control over treatment conditions. But as we have already indicated, findings obtained in laboratories may not always apply to everyday situations. The ideal solution to this problem is to do experiments in the field as a complement to laboratory investigations. In *field experiments,* investigators capitalize on rare opportunities to assign people randomly to treatment conditions in natural settings. In the laboratory experiment just described, we can conclude that the emotional climate established by adults affects children's behavior in the laboratory. But does it also do so in daily life?

© FRANK MURRAY

■ Does the death of a spouse in old age affect the surviving partner's physical health and psychological well-being? A correlational design can be used to answer this question, but it does not permit researchers to determine the precise cause of their findings.

Another study helps answer this question (Yarrow, Scott, & Waxler, 1973). This time, the research was carried out in a child-care center. A caregiver deliberately interacted differently with two groups of preschoolers. In one condition (the *nurturant treatment*), she modeled many instances of warmth and helpfulness. In the second condition (the *control,* since it involved no treatment), she behaved as usual, with no special emphasis on concern for others. Two weeks later, the researchers created several situations that called for helpfulness. For example, a visiting mother asked each child to watch her baby for a few moments, but the baby's toys had fallen out of the playpen. The investigators found that children exposed to the nurturant treatment were much more likely to return toys to the baby than were those in the control condition.

Often researchers cannot randomly assign participants and manipulate conditions in the real world. Sometimes they can compromise by conducting *natural experiments.* Treatments that already exist, such as different child-care centers, schools, workplaces, or retirement villages, are compared. These studies differ from correlational research only in that groups of participants are carefully chosen to ensure that their characteristics are as much alike as possible. In this way, investigators

Table 1.7 Strengths and Limitations of Research Designs

Method	Description	Strengths	Limitations
General			
Correlational	The investigator obtains information on already-existing groups without altering participants' experiences.	Permits study of relationships between variables.	Does not permit inferences about cause-and-effect relationships.
Experimental	Through random assignment of participants to treatment conditions, the investigator manipulates an independent variable and looks at its effect on a dependent variable. Can be conducted in the laboratory or the natural environment.	Permits inferences about cause-and-effect relationships.	When conducted in the laboratory, findings may not apply to the real world. When conducted in the field, control over the treatment is usually weaker than in the laboratory.
Developmental			
Longitudinal	The investigator studies the same group of participants repeatedly at different ages.	Permits study of common patterns and individual differences in development and relationships between early and later events and behaviors.	Age-related changes may be distorted because of participant dropout, practice effects, and cohort effects.
Cross-sectional	The investigator studies groups of participants differing in age at the same point in time.	More efficient than the longitudinal design. Not plagued by such problems as participant dropout and practice effects.	Does not permit study of individual developmental trends. Age differences may be distorted because of cohort effects.
Longitudinal-sequential	The investigator studies two or more groups of participants born in different years, following each group longitudinally.	Permits both longitudinal and cross-sectional comparisons. Reveals cohort effects.	May have the same problems as longitudinal and cross-sectional strategies, but the design itself helps identify difficulties.

rule out as best they can alternative explanations for their treatment effects. But despite these efforts, natural experiments are unable to achieve the precision and rigor of true experimental research.

To help you compare correlational and experimental designs, Table 1.7 summarizes their strengths and limitations. It also includes an overview of designs for studying development, to which we now turn.

Designs for Studying Development

Scientists interested in human development require information about the way research participants change over time. To answer questions about development, they must extend correlational and experimental approaches to include measurements at different ages. Longitudinal and cross-sectional designs are special *developmental research strategies.* In each, age comparisons form the basis of the research plan.

● **The Longitudinal Design.** In a **longitudinal design,** a group of participants is studied repeatedly at different ages, and changes are noted as the participants mature. The time spanned may be relatively short (a few months to several years) or very long (a decade or even a lifetime). The longitudinal approach has two major strengths. First, because it tracks the performance of each person over time, researchers can identify common patterns of development as well as individual differences. Second, longitudinal studies permit investigators to examine relationships between early and later events and behaviors. Let's take an example to illustrate these ideas.

A group of researchers wondered whether children who display extreme personality styles—either angry and explosive or shy and withdrawn—retain the same dispositions when they become adults. In addition, the researchers wanted to know what kinds of experiences promote stability or change in personality and what consequences explosiveness and shyness have for long-term adjustment. To answer these questions, the researchers delved into the archives of the Guidance Study, a well-known longitudinal investigation initiated in 1928 at the University of California, Berkeley, and continued for several decades (Caspi, Elder, & Bem, 1987, 1988).

Results revealed that the two personality styles were only moderately stable. Between ages 8 and 30, a good number of individuals remained the same, whereas others changed substantially. When stability did occur, it appeared to be due to a "snowballing effect," in which children evoked responses from adults and peers that acted to maintain their dispositions (Caspi, 1998). In other words, explosive youngsters were likely to be treated with anger and hostility (to which they reacted with even greater unruliness), whereas shy children were apt to be ignored.

Persistence of extreme personality styles affected many areas of adult adjustment. For men, the results of early explosiveness were most apparent in their work lives, in the form of conflicts with supervisors, frequent job changes, and unemployment. Since few women in this sample of an earlier generation worked after marriage, their family lives were most affected. Explosive girls grew up to be hotheaded wives and parents who were especially prone to divorce. Sex differences in the long-term consequences of shyness were even greater. Men who had been withdrawn in childhood were delayed in marrying, becoming fathers, and developing stable careers. Because a withdrawn, unassertive style was socially acceptable for females, women who had shy personalities showed no special adjustment problems.

• Problems in Conducting Longitudinal Research. Despite their strengths, longitudinal investigations pose a number of problems. For example, participants may move away or drop out of the research for other reasons. This changes the original sample so it no longer represents the population to whom researchers would like to generalize their findings. Also, from repeated study, people may become "test-wise." Their performance may improve as a result of *practice effects*—better test-taking skills and increased familiarity with the test—not because of factors commonly associated with development.

But the most widely discussed threat to longitudinal findings is **cohort effects** (see page 11): Individuals born in the same time period are influenced by a particular set of historical and cultural conditions. Results based on one cohort may not apply to people developing in other times. For example, unlike the findings on female shyness described in the preceding section, which were gathered in the 1950s, today's shy young women tend to be poorly adjusted—a difference that may be due to changes in gender roles in Western societies. Shy adults, whether male or female, frequently feel depressed and have few sources of social support (Caspi et al., 2000). Similarly, a longitudinal study of the lifespan would probably result in quite different findings if it were carried out in the first decade of the twenty-first century, around the time of World War II, or during the Great Depression of the 1930s. (See the Lifespan Vista box on page 36.)

• The Cross-Sectional Design. The length of time it takes for many behaviors to change, even in limited longitudinal studies, has led researchers to turn toward a more convenient strategy for studying development. In the **cross-sectional design,** groups of people differing in age are studied at the same point in time.

A study in which students in grades 3, 6, 9, and 12 filled out a questionnaire about their sibling relationships provides a good illustration (Buhrmester & Furman, 1990). Findings revealed that sibling interaction was characterized by greater equality and less power assertion with age. Also, feelings of sibling companionship declined during adolescence. The researchers thought that several factors contributed to these age differences. As later-born children become more competent and independent, they no longer need, and are probably less willing to accept, direction from older siblings. In addition, as adolescents move from psychological dependence on the family to greater involvement with peers, they may have less time and emotional need to invest in siblings. These intriguing ideas about the development of sibling relationships, as we will see in Chapter 12, have been confirmed in subsequent research.

• Problems in Conducting Cross-Sectional Research. The cross-sectional design is an efficient strategy for describing age-related trends. Because participants are measured only once, researchers need not be concerned about such difficulties as participant dropout or practice effects. But evidence about change at the level at which it actually occurs—the individual—is not available (Kraemer et al., 2000). For example, in the cross-sectional study of sibling relationships just discussed, comparisons are limited to age-group averages. We cannot tell if important individual differences exist. Indeed, longitudinal findings reveal that adolescents vary considerably in the changing quality of their sibling relationships, many becoming more distant but some becoming more supportive and intimate (Dunn, Slomkowski, & Beardsall, 1994).

Cross-sectional studies—especially those that cover a wide age span—have another problem. Like longitudinal research, they can be threatened by cohort effects. For example, comparisons of 10-year-old cohorts, 20-year-old cohorts, and 30-year-old cohorts—groups born and reared in different years—may not really represent age-related changes. Instead, they may reflect unique experiences associated with the historical period in which each age group grew up.

• Improving Developmental Designs. To overcome some of the limitations of longitudinal and cross-sectional research, investigators sometimes combine the two approaches. One way of doing so is the **longitudinal-sequential design,** in which a sequence of samples (two or more age groups) are followed for a number of years.

The design has two advantages. First, it permits us to find out whether cohort effects are operating by comparing people of the same age who were born in different years. Using the example shown in Figure 1.6 on page 37, we can compare the three samples at ages 20, 30, and 40. If they do not differ, we can rule out cohort effects. Second, we can make longitudinal and cross-sectional comparisons. If outcomes are similar in both, then we can be especially confident about our findings.

In a study that used the design in Figure 1.6, researchers wanted to find out whether adult personality development progresses as Erikson's psychosocial theory predicts (Whitbourne et al., 1992). Questionnaires measuring Erikson's stages were given to three cohorts of 20-year-olds, each born a decade apart. The cohorts were reassessed at 10-year intervals. Consistent with Erikson's theory, longitudinal and cross-sectional gains in identity and intimacy occurred between ages 20 and 30—a trend unaffected by historical time period. But a powerful cohort effect emerged for consolidation of the sense of industry: At age 20, Cohort 1 scored substantially below Cohorts 2 and 3. Look at Figure 1.6 again, and notice that

A Lifespan Vista

Impact of Historical Times on the Life Course: The Great Depression and World War II

Economic disaster, wars, and periods of rapid social change profoundly affect people's lives. Yet their impact depends on when they strike during the life course. Glen Elder (1999) capitalized on the economic hardship families experienced during the Great Depression of the 1930s to study its influence on lifespan development. He delved into the vast archives of two major longitudinal studies: (1) the Oakland Growth Study, an investigation of individuals born in the early 1920s who were adolescents when the Depression took its toll, and (2) the Guidance Study, whose participants were born in the late 1920s and were young children when their families faced severe economic losses.

In both cohorts, relationships changed when economic deprivation struck. As unemployed fathers lost status, mothers took greater control over family affairs. This reversal of traditional gender roles often sparked conflict. Fathers sometimes became explosive and punitive toward their children. At other times, they withdrew into passivity and depression. Mothers often became frantic with worry over the well-being of their husbands and children, and many entered the labor force to make ends meet (Elder, Liker, & Cross, 1984).

Outcomes for Adolescents. Although unusual burdens were placed on them as family lives changed, the Oakland Growth Study cohort—especially the boys—weathered economic hardship quite well. As adolescents, they were too old to be wholly dependent on their highly stressed parents. Boys spent less time at home as they searched for part-time jobs, and many turned toward adults and peers outside the family for emotional support. Girls took over household chores and cared for younger siblings. Their greater involvement in family affairs exposed them to more parental conflict and unhappiness. Consequently, adolescent girls' adjustment in economically deprived homes was somewhat less favorable than adolescent boys' (Elder, Van Nguyen, & Caspi, 1985).

These changes had major consequences for adolescents' future aspirations and adult lives. As girls focused on home and family, they were less likely to think about college and careers and more likely to marry early. Boys learned that economic resources could not be taken for granted, and they tended to make a very early commitment to an occupational choice. And the chance to become a parent was especially important to men whose lives had been disrupted by the Depression. Perhaps because they believed that a rewarding career could not be guaranteed, they viewed children as the most enduring benefit of their adult lives.

Outcomes for Children. Unlike the Oakland Growth Study cohort, the Guidance Study participants were within the years of intense family dependency when the Depression struck. For young boys (who, as we will see in later chapters, are especially prone to adjustment problems in the face of family stress), the impact of economic strain was severe. They showed emotional difficulties and poor attitudes toward school and work that persisted through the teenage years (Elder & Caspi, 1988).

But as the Guidance Study sample became adolescents, another major historical event occurred: In 1941, the United States entered World War II. As a result, thousands of men left their communities for military bases, leading to dramatic life changes. Some combat veterans came away with symptoms of emotional trauma that persisted for decades. Yet for most young soldiers, war mobilization broadened their range of knowledge and experience. It also granted time out from civilian responsibilities, giving many soldiers a chance to consider where their lives were going. And the GI Bill of Rights enabled them to expand their education and acquire new skills after the war. By middle adulthood, the Guidance Study war veterans had reversed the early negative impact of the Great Depression. They were more successful educationally and occupationally than their counterparts who had not entered the service (Elder & Hareven, 1993).

Clearly, cultural-historical change does not have a uniform impact on development. Outcomes can vary considerably, depending on the pattern of historical events and the age at which people experience them.

■ Historical time period has profound implications for development. The Great Depression of the 1930s left this farm family without a steady income. Children were more negatively affected than adolescents, who were no longer entirely dependent on their highly stressed parents.

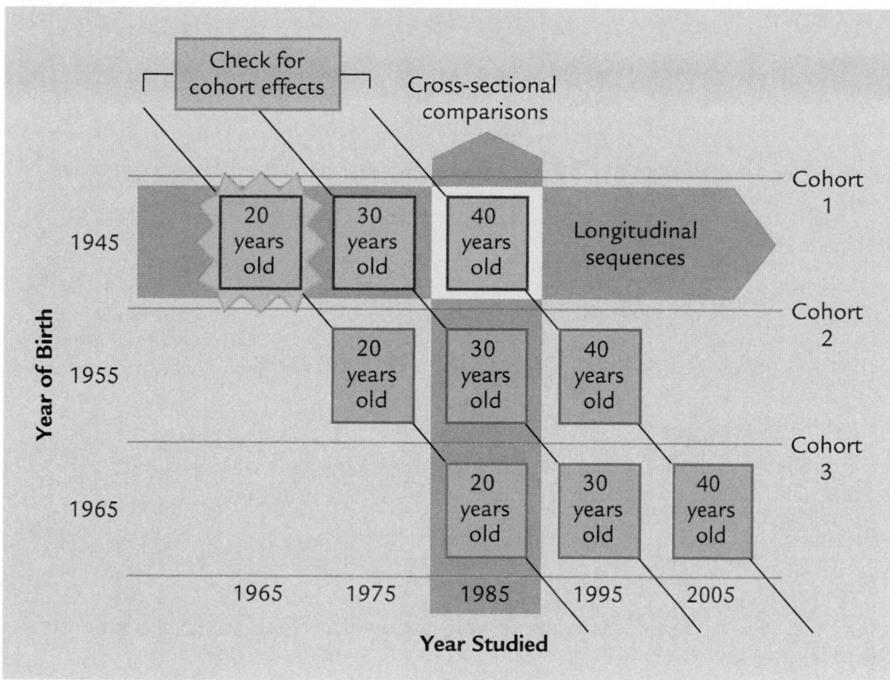

FIGURE 1.6 Example of a longitudinal-sequential design. Three cohorts, born in 1945 (blue), 1955 (pink), and 1965 (green), are followed longitudinally from 20 to 40 years of age. The design permits the researcher to check for cohort effects by comparing people of the same age who were born in different years. In a study that used this design, the Cohort 1 20-year-olds differed substantially from the 20-year-olds in Cohorts 2 and 3, indicating powerful history-graded influences. The longitudinal-sequential design also permits longitudinal and cross-sectional comparisons. Similar outcomes lend additional confidence in the results.

members of Cohort 1 reached age 20 in the mid-1960s. As college students, they were part of a political protest movement that reflected disenchantment with the work ethic. Once out of college, they caught up with the other cohorts, perhaps as a result of experiencing the pressures of the work world.

To date, only a handful of longitudinal-sequential studies have been conducted. Yet the design permits researchers to profit from the strengths of both longitudinal and cross-sectional strategies. And in uncovering cohort effects, it also helps explain diversity in development (Magnusson & Stattin, 1998).

Ask Yourself

REVIEW

Explain how cohort effects can distort the findings of both longitudinal and cross-sectional studies. How does the longitudinal-sequential design reveal cohort effects?

APPLY

A researcher compares older adults with chronic heart disease to those who are free of major health problems and finds that the first group scores lower on mental tests. What type of research design was used? Should the researcher conclude that heart disease causes a decline in intellectual functioning in late adulthood? Explain.

APPLY

A researcher wants to find out if preschoolers enrolled in child-care centers achieve as well in elementary school as preschoolers not in child care. Which developmental design is appropriate for answering this question? Why?

www.

Ethics in Lifespan Research

Research into human behavior creates ethical issues because, unfortunately, the quest for scientific knowledge can sometimes exploit people. For this reason, special guidelines for research have been developed by the federal government, funding agencies, and research-oriented associations, such as the American Psychological Association (1994). Table 1.8 on page 38 presents a summary of basic research rights drawn from these guidelines. Once you have examined them, read the following research situations, each of which poses a serious ethical dilemma. What precautions do you think should be taken in each instance?

- In a study of moral development, an investigator wants to assess children's ability to resist temptation by videotaping their behavior without their knowledge. Seven-year-olds are promised a prize for solving difficult puzzles. They are also told not to look at a classmate's correct solutions, which are deliberately placed at the back of the room. Telling children ahead of time that cheating is being studied or that their behavior is being closely monitored will destroy the purpose of the study.
- A researcher wants to study the impact of mild daily exercise on the physical and mental health of elderly patients in nursing homes. He consults each resident's doctor to make sure that the exercise routine will not be harmful. But when he seeks the residents' consent, he finds that many do not comprehend the purpose of the

Table 1.8 Rights of Research Participants

Research Right	Description
Protection from harm	Participants have the right to be protected from physical or psychological harm in research. If in doubt about the harmful effects of research, investigators should seek the opinion of others. When harm seems possible, investigators should find other means for obtaining the desired information or abandon the research.
Informed consent	Participants, including children and the elderly, have the right to have explained to them, in language appropriate to their level of understanding, all aspects of the research that may affect their willingness to participate. When children are participants, informed consent of parents as well as of others who act on the child's behalf (such as school officials) should be obtained, preferably in writing. Older adults who are cognitively impaired should be asked to appoint a surrogate decision maker. If they cannot do so, then someone should be named by an ethics committee that includes relatives and professionals who know the person well. All participants have the right to discontinue participation in the research at any time.
Privacy	Participants have the right to concealment of their identity on all information collected in the course of research. They also have this right with respect to written reports and any informal discussions about the research.
Knowledge of results	Participants have the right to be informed of the results of research in language appropriate to their level of understanding.
Beneficial treatments	If experimental treatments believed to be beneficial are under investigation, participants in control groups have the right to alternative beneficial treatments if they are available.

Sources: American Psychological Association, 1994; Cassel, 1988; Society for Research in Child Development, 1993.

research. And some appear to agree simply to relieve feelings of isolation and loneliness.

As these examples indicate, when children or the aged take part in research, the ethical concerns are especially complex. Immaturity makes it difficult or impossible for children to evaluate for themselves what participation in research will mean. And because mental impairment rises with very advanced age, some older adults cannot make voluntary and informed choices. The life circumstances of others make them unusually vulnerable to pressure for participation (Kimmel & Moody, 1990; Society for Research in Child Development, 1993).

Virtually every committee that has worked on developing ethical principles for research has concluded that conflicts arising in research situations often cannot be resolved with simple right or wrong answers (Stanley & Seiber, 1992). The ultimate responsibility for the ethical integrity of research lies with the investigator. However, researchers are advised or, in the case of federally funded research, required to seek advice from others. Special committees exist in colleges, universities, and other institutions for this purpose. These committees weigh the costs of the research to participants in terms of inconvenience and possible psychological or physical injury against the study's value for advancing knowledge and improving conditions of life. If there are any risks to participants' safety and welfare that the worth of the research does not justify, then preference is always given to the interests of the participants.

The ethical principle of *informed consent* requires special interpretation when participants cannot fully appreciate the research goals and activities. Parental consent is meant to protect the safety of children whose ability to decide is not yet mature. For children 7 years and older, their own informed consent should be obtained in addition to parental consent. Around age 7, changes in children's thinking permit them to better understand simple scientific principles and the needs of others. Researchers should respect and enhance these new capacities by providing school-age children with a full explanation of research activities in language they can understand (Fisher, 1993; Thompson, 1992). Extra care must be taken when telling children that the information they provide will be kept confidential and that they can end their participation at any time. Children may not understand, and sometimes do not believe, these promises (Abramovitch et al., 1995; Ondrusek et al., 1998).

Most older adults require no more than the usual informed-consent procedures. Yet many investigators set upper age limits in studies relevant to the elderly, thereby excluding the oldest adults (Bayer & Tadd, 2000). Researchers should not stereotype the elderly as incompetent to decide about participation or to engage in the research activities. Nevertheless, extra measures must be taken to protect those who are cognitively impaired or in settings for care of the chronically ill. Sometimes these individuals may agree to participate simply to obtain rewarding social interaction. Yet participation should not be automatically withheld, since it can result in personal as well as scientific benefits (High & Doole, 1995). In these instances, potential participants should be asked to appoint a surrogate decision maker. If they cannot do so, then someone should be named by an ethics committee that includes relatives and professionals who know the person well. As an added precaution, if the elderly person is incapable of consenting and the risks of the research are more than minimal, then the study should not be done unless it is likely to directly benefit the participant (Cassel, 1988).

Finally, all ethical guidelines advise that special precautions be taken in the use of deception and concealment, as occurs when researchers observe people from behind one-way mirrors, give them false feedback about their performance, or do not tell them the truth regarding what the research is about. When these kinds of procedures are used, *debriefing,* in which the investigator provides a full account and justification of the activities, occurs after the research session is over. Debriefing should also take place with children, but it does not always work well. Despite explanations, children may come away from the situation with their belief in the honesty of adults undermined. Ethical standards permit deception if investigators satisfy institutional committees that such practices are necessary. Nevertheless, because deception may have serious emotional consequences for some youngsters, investigators should try to come up with other research strategies when children are involved.

Summary

Human Development as a Scientific, Applied, and Interdisciplinary Field

What is human development, and what factors stimulated expansion of the field?

- **Human development** is an interdisciplinary field devoted to understanding human constancy and change throughout the lifespan. Research on human development has been stimulated by both scientific curiosity and social pressures to better people's lives.

Basic Issues

Identify three basic issues on which theories of human development take a stand.

- **Theories** of human development take a stance on three basic issues: (1) Is development a **continuous** process, or does it follow a series of **discontinuous stages**? (2) Does one general course of development characterize all individuals, or do many possible courses exist, depending on the **contexts** in which children and adults live? (3) Is development determined primarily by **nature** or **nurture,** and is it stable or open to change?

The Lifespan Perspective: A Balanced Point of View

Describe the lifespan perspective on development.

- The **lifespan perspective** is a balanced view that recognizes great complexity in human change and the factors that underlie it. According to this view, development is lifelong, multidimensional (affected by a blend of biological, psychological, and social forces), multidirectional (a joint expression of growth and decline), and plastic (open to change with the support of new experiences).
- Furthermore, the lifespan perspective regards the life course as embedded in multiple contexts. Although these contexts operate in an interconnected fashion, they can be organized into three categories: (1) **age-graded influences** that are predictable in timing and duration; (2) **history-graded influences,** forces unique to a particular historical era; and (3) **nonnormative influences,** events unique to one or a few individuals.

Historical Foundations

Describe major historical influences on theories of development.

- Contemporary theories of human development have roots extending far into the past. In medieval times, children were regarded as miniature adults, a view called **preformationism.** In the sixteenth century, childhood came to be viewed as a distinct phase of life. However, the Puritan belief in original sin led to a harsh philosophy of child rearing. The Enlightenment brought new ideas favoring more humane child treatment. Locke's **tabula rasa** furnished the basis for twentieth-century behaviorism. Rousseau's **noble savage** foreshadowed the concepts of stage and **maturation.**
- In the eighteenth and early nineteenth centuries, two German philosophers extended conceptions of development through adulthood. Tetens and Carus anticipated many aspects of the contemporary lifespan perspective.
- Darwin's theory of evolution influenced important twentieth-century theories and inspired scientific child study. In the early twentieth century, Hall and Gesell introduced the **normative approach,** which produced a large body of descriptive facts about development. Binet and Simon constructed the first successful intelligence test, initiating the mental testing movement.

Mid-Twentieth-Century Theories

What theories influenced human development research in the mid-twentieth century?

- In the 1930s and 1940s, psychiatrists and social workers turned to the **psychoanalytic perspective** for help in treating people's psychological problems. In Freud's **psychosexual theory,** the individual moves through five stages, during which three portions of the personality—id, ego, and superego—become integrated. Erikson's **psychosocial theory** builds on Freud's theory by emphasizing the development of culturally relevant attitudes and skills and the lifespan nature of development.
- As psychoanalytic theory gained in prominence, **behaviorism** and **social learning theory** emerged, emphasizing principles of conditioning and modeling and practical procedures of **behavior modification** to eliminate undesirable behaviors and increase desirable responses.
- In contrast to behaviorism, Piaget's **cognitive-developmental theory** emphasizes an active individual whose mind consists of rich structures of knowledge. According to Piaget, children move through four stages, beginning with the baby's sensorimotor action patterns and ending with the elaborate, abstract reasoning system of the adolescent. Piaget's work has stimulated a wealth of research on children's thinking and has encouraged educational philosophies and programs that emphasize discovery learning.

Recent Theoretical Perspectives

Describe recent theoretical perspectives on human development.

- **Information processing** views the mind as a complex, symbol-manipulating system much like a computer. This approach helps investigators achieve a detailed understanding of what individuals of different ages do when faced with tasks and problems. Information processing regards development as a matter of continuous change. Its findings have important implications for education.

- Three contemporary perspectives place special emphasis on contexts for development. **Ethology** stresses the evolutionary origins and adaptive value of behavior and inspired the **sensitive period** concept. In **evolutionary developmental psychology,** researchers have extended this emphasis, seeking to understand the adaptiveness of species-wide competencies as they change over time.

- Vygotsky's **sociocultural theory** has enhanced our understanding of cultural influences, especially in the area of cognitive development. Through cooperative dialogues with more expert members of society, children come to use language to guide their own thought and actions and acquire culturally relevant knowledge and skills.

- In **ecological systems theory,** nested layers of the environment—**microsystem, mesosystem, exosystem,** and **macrosystem**—are seen as major influences on the developing person. The **chronosystem** represents the dynamic, ever-changing nature of individuals and their experiences.

Comparing and Evaluating Theories

Identify the stand taken by each major theory on the three basic issues of human development.

- Theories that are major forces in human development research vary in their focus on different domains of development, in their view of development, and in their strengths and weaknesses. (For a full summary, see Table 1.5 on page 26.)

Studying Development

Describe methods commonly used in research on human development.

- **Naturalistic observations,** gathered in everyday environments, permit researchers to see directly the everyday behaviors they hope to explain. In contrast, **structured observations** take place in laboratories, where every participant has an equal opportunity to display the behaviors of interest.

- Self-report methods can be flexible and open-ended like the **clinical interview.** Alternatively, **structured interviews,** tests, and questionnaires, which permit efficient administration and scoring, can be given. Investigators use the **clinical, or case study, method** when they desire an in-depth understanding of a single individual.

- A growing interest in the impact of culture has prompted researchers to adapt observational and self-report methods to permit direct comparisons of cultures. To uncover the cultural meanings of children's and adults' behaviors, researchers rely on a method borrowed from the field of anthropology—**ethnography.** It uses participant observation to understand the unique values and social processes of a culture or distinct social group.

Distinguish correlational and experimental research designs, noting the strengths and limitations of each.

- The **correlational design** examines relationships between variables as they happen to occur, without altering people's experiences. The **correlation coefficient** is often used to measure the association between variables. Correlational studies do not permit inferences about cause and effect. However, their use is justified when it is difficult or impossible to control the variables of interest.

- An **experimental design** permits inferences about cause and effect. Researchers manipulate an **independent variable** by exposing participants to two or more treatment conditions. Then they determine what effect this variable has on a **dependent variable. Random assignment** reduces the chances that characteristics of participants will affect the accuracy of experimental findings.

- To achieve high degrees of control, most experiments are conducted in laboratories, but their findings may not apply to everyday life. Field and natural experiments compare treatments in natural environments. These approaches, however, are less rigorous than laboratory experiments.

Describe designs for studying development, noting the strengths and limitations of each.

- The **longitudinal design** permits study of common patterns as well as individual differences in development and of the relationship between early and later events and behaviors. Among problems researchers face in conducting longitudinal research are biased samples, participant dropout, practice effects, and **cohort effects**—difficulty generalizing to people developing in other historical times.

- The **cross-sectional design** offers an efficient approach to studying development. However, it is limited to comparisons of age-group averages. Findings of cross-sectional studies also can be distorted by cohort effects, especially when they cover a wide age span.

- To overcome some of the limitations of these designs, investigators sometimes combine the two approaches. The **longitudinal-sequential** design permits researchers to test for cohort effects and to compare longitudinal and cross-sectional findings.

Ethics in Lifespan Research

What special ethical concerns arise in research on human development?

- Research creates ethical issues, since the quest for scientific knowledge can sometimes exploit people. The ethical principle of informed consent requires special safeguards for children and for elderly people who are cognitively impaired or in settings for the care of the chronically ill. The use of deception in research with children is especially risky because it may undermine their basic faith in the trustworthiness of adults.

Important Terms and Concepts

age-graded influences (p. 10)
behavior modification (p. 19)
behaviorism (p. 18)
chronosystem (p. 25)
clinical interview (p. 28)
clinical, or case study, method (p. 29)
cognitive-developmental theory (p. 19)
cohort effects (p. 35)
contexts (p. 6)
continuous development (p. 6)
correlation coefficient (p. 32)
correlational design (p. 31)
cross-sectional design (p. 35)
dependent variable (p. 33)
discontinuous development (p. 6)
ecological systems theory (p. 24)
ethnography (p. 30)
ethology (p. 22)
evolutionary developmental psychology (p. 22)
exosystem (p. 25)
experimental design (p. 33)
history-graded influences (p. 11)
human development (p. 5)
independent variable (p. 33)
information processing (p. 21)
lifespan perspective (p. 8)
longitudinal design (p. 34)
longitudinal-sequential design (p. 35)
macrosystem (p. 25)
maturation (p. 13)
mesosystem (p. 25)
microsystem (p. 24)
naturalistic observation (p. 27)
nature–nurture controversy (p. 7)
noble savage (p. 13)
nonnormative influences (p. 11)
normative approach (p. 14)
preformationism (p. 13)
psychoanalytic perspective (p. 15)
psychosexual theory (p. 15)
psychosocial theory (p. 17)
random assignment (p. 33)
resiliency (p. 10)
sensitive period (p. 22)
social learning theory (p. 18)
sociocultural theory (p. 23)
stage (p. 6)
structured interview (p. 29)
structured observation (p. 27)
tabula rasa (p. 13)
theory (p. 5)

FYI For Further Information and Help

Consult the Companion Website for *Development Through the Lifespan, Third Edition,* (www.ablongman.com/berk) where you will find the following resources for this chapter:

- **Chapter Objectives**
- **Flashcards** for studying important terms and concepts
- **Annotated Weblinks** to guide you in further research
- **Ask Yourself** questions, which you can answer and then check against a sample response
- **Suggested Readings**
- **Practice Test** with immediate scoring and feedback

2

Biological and Environmental Foundations

Genetic Foundations
The Genetic Code ● The Sex Cells ● Male or Female? ● Multiple Births ● Patterns of Genetic Inheritance ● Chromosomal Abnormalities

Reproductive Choices
Genetic Counseling ● Prenatal Diagnosis and Fetal Medicine ● Genetic Testing ● Adoption

■ Social Issues: The Pros and Cons of Reproductive Technologies

Environmental Contexts for Development
The Family ● Socioeconomic Status and Family Functioning ● The Impact of Poverty ● Beyond the Family: Neighborhoods, Towns, and Cities ● The Cultural Context

■ Cultural Influences: The African-American Extended Family

Understanding the Relationship Between Heredity and Environment
The Question, "How Much?" ● The Question, "How?"

■ Biology & Environment: Uncoupling Genetic–Environmental Correlations for Mental Illness and Antisocial Behavior

A complex blend of genetic and environmental influences leads members of this three-generation Inupiat family of western Alaska to be both alike and different in physical characteristics and behavior.

"It's a girl," announces the doctor, who holds up the squalling little creature while her new parents gaze with amazement at their miraculous creation.

"A girl! We've named her Sarah!" exclaims the proud father to eager relatives waiting by the telephone for word about their new family member.

As we join these parents in thinking about how this wondrous being came into existence and imagining her future, we are struck by many questions. How could this baby, equipped with everything necessary for life outside the womb, have developed from the union of two tiny cells? What ensures that Sarah will, in due time, roll over, reach for objects, walk, talk, make friends, learn, imagine, and create—just like every other normal child born before her? Why is she a girl and not a boy, dark-haired rather than blond, calm and cuddly instead of wiry and energetic? What difference will it make that Sarah is given a name and place in one family, community, nation, and culture rather than another?

To answer these questions, this chapter takes a close look at the foundations of development: heredity and environment. Because nature has prepared us for survival, all humans have features in common. Yet each human being is also unique. Take a moment to jot down the most obvious similarities and differences in physical characteristics and behavior for several of your friends and their parents. Did you find that one person shows combined features of both parents, another resembles just one parent, whereas a third is not like either parent? These directly observable characteristics are called **phenotypes.** They depend in part on the individual's **genotype**—the complex blend of genetic information that determines our species and influences all our unique characteristics. Yet throughout life, phenotypes are also affected by the person's history of experiences in the environment.

We begin our discussion of development at the moment of conception, an event that establishes the hereditary makeup of the new individual. In the first section of this chapter, we review basic genetic principles that help explain similarities and differences between us in appearance and behavior. Next, we turn to aspects of the environment that play powerful roles throughout the lifespan. In the final section of this chapter, we take up the question of how nature and nurture *work together* to shape the course of development.

Genetic Foundations

Each of us is made up of trillions of separate units called *cells.* Inside every cell is a control center, or *nucleus,* that contains rodlike structures called **chromosomes,** which store and transmit genetic information. Human chromosomes come in 23 matching pairs (an exception is the XY pair in males, which we will discuss shortly). Each pair member corresponds to the other in size, shape, and genetic functions. One is inherited from the mother and one from the father (see Figure 2.1).

The Genetic Code

Chromosomes are made up of a chemical substance called **deoxyribonucleic acid,** or **DNA.** As Figure 2.2 shows, DNA is a long, double-stranded molecule that looks like a twisted ladder. Each rung of the ladder consists of a specific pair of chemical substances called *bases,* joined together between the two sides. It is this sequence of bases that provides genetic instructions. A **gene** is a segment of DNA along the length of the chromosome. Genes can be of different lengths—perhaps 100 to several thousand ladder rungs long. An estimated 30,000 genes lie along the human chromosomes.

We share some of our genetic makeup with even the simplest organisms, such as bacteria and molds, and most of it with other mammals, especially primates. Between 98 and 99 percent of chimpanzee and human DNA is identical. This means that only a small portion of our heredity is responsible

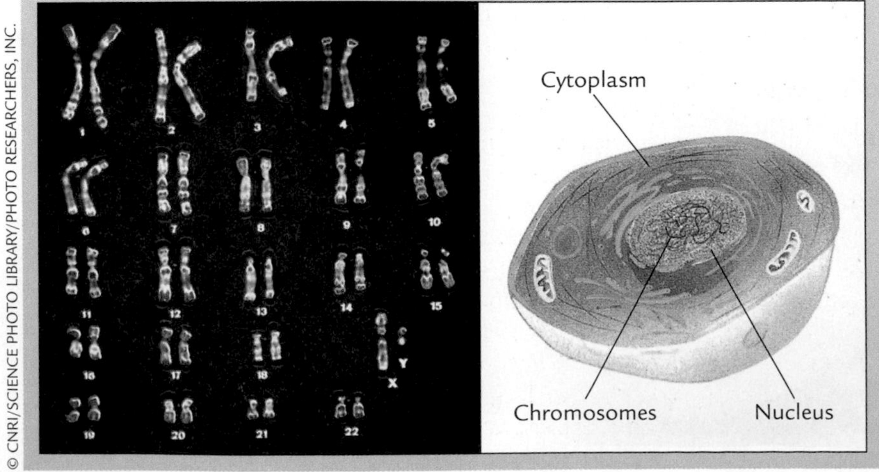

FIGURE 2.1 A karyotype, or photograph, of human chromosomes. The 46 chromosomes on the left were isolated from a body cell (shown on the right), stained, greatly magnified, and arranged in pairs according to decreasing size of the upper "arm" of each chromosome. Note the twenty-third pair, XY. The cell donor is a male. In a female, the twenty-third pair would be XX.

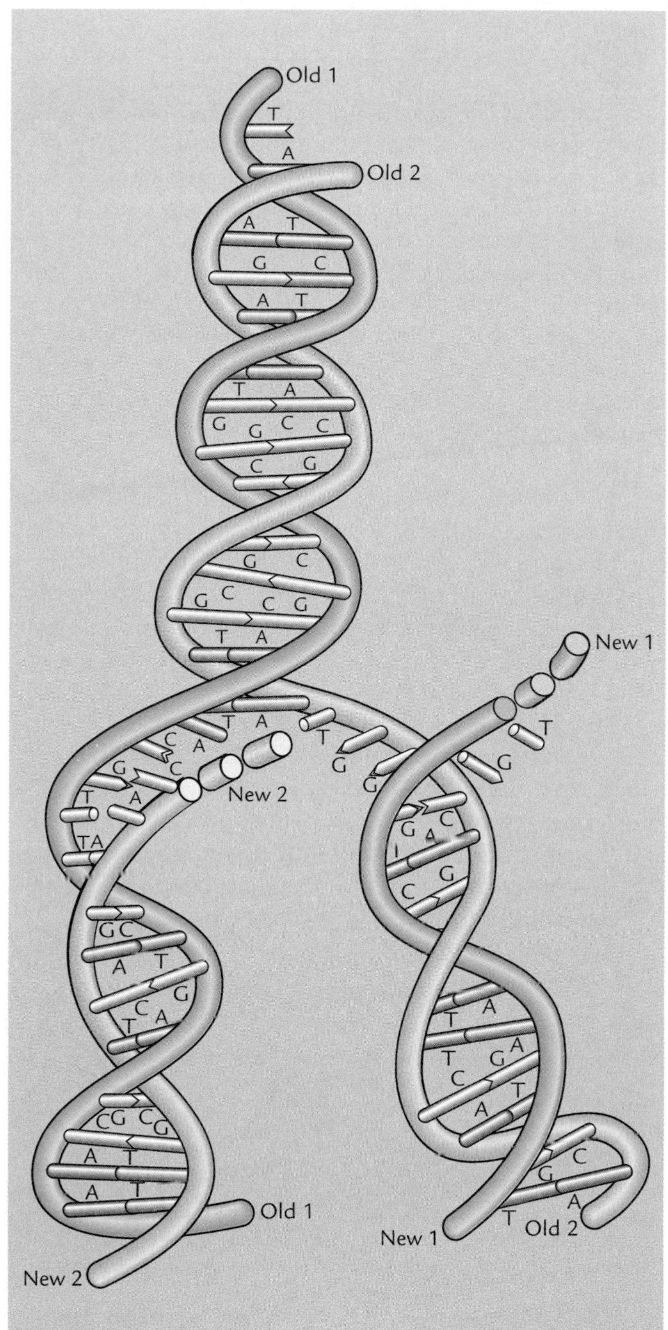

■ **FIGURE 2.2 DNA's ladderlike structure.** This figure shows that the pairings of bases across the rungs of the ladder are very specific: adenine (A) always appears with thymine (T), and cytosine (C) always appears with guanine (G). Here, the DNA ladder duplicates by splitting down the middle of its ladder rungs. Each free base picks up a new complementary partner from the area surrounding the cell nucleus.

for the traits that make us human, from our upright gait to our extraordinary language and cognitive capacities. And the genetic variation from one human to the next is even less! Individuals around the world are about 99.1 percent genetically identical (Gibbons, 1998). Only a tiny quantity of DNA contributes to human variation in traits and capacities.

A unique feature of DNA is that it can duplicate itself through a process called **mitosis.** This special ability permits the one-celled fertilized ovum to develop into a complex human being composed of a great many cells. Refer again to Figure 2.2, and you will see that during mitosis, the chromosomes copy themselves. As a result, each new body cell contains the same number of chromosomes and identical genetic information.

Genes accomplish their task by sending instructions for making a rich assortment of proteins to the *cytoplasm,* the area surrounding the cell nucleus. Proteins, which trigger chemical reactions throughout the body, are the biological foundation on which our characteristics are built. How do humans, with far fewer genes than scientists once thought (only twice as many as the worm or fly), manage to develop into such complex beings? The answer lies in the proteins our genes make, which break up and reassemble in staggering variety—about 10 to 20 million altogether. In simpler species, the number of proteins is far more limited. Furthermore, the communication system between the cell nucleus and cytoplasm, which fine-tunes gene activity, is more intricate in humans than in simpler organisms. Within the cell, a wide range of environmental factors modify gene expression (Davies, Howell, & Gardner, 2001). So even at this microscopic level, biological events are the result of *both* genetic and nongenetic forces.

The Sex Cells

New individuals are created when two special cells called **gametes,** or sex cells—the sperm and ovum—combine. A gamete contains only 23 chromosomes, half as many as a regular body cell. Gametes are formed through a cell division process called **meiosis,** which ensures that a constant quantity of genetic material is transmitted from one generation to the next. When sperm and ovum unite at conception, the cell that results, called a **zygote,** will again have 46 chromosomes.

In *meiosis,* the chromosomes pair up and exchange segments, so that genes from one are replaced by genes from another. Then chance determines which member of each pair will gather with others and end up in the same gamete. These events make the likelihood of nontwin offspring of the same two parents being genetically the same extremely slim—about 1 in 700 trillion (Gould & Keeton, 1997). Therefore, meiosis helps us understand why siblings differ from each other, even though they have features in common because their genotypes come from the same pool of parental genes.

In the male, four sperm are produced when meiosis is complete. Also, the cells from which sperm arise are produced continuously throughout life. For this reason, a healthy man can father a child at any age after sexual maturity. In the female, gamete production results in just one ovum. In addition, the female is born with all her ova already present in her ovaries, and she can bear children for only 3 to 4 decades. Still, there are plenty of female sex cells. About 1 to 2 million are present at birth, 40,000 remain at adolescence, and approximately 350 to 450 will mature during a woman's childbearing years (Moore & Persaud, 1998).

Male or Female?

Return to Figure 2.1, and note that 22 of the 23 pairs of chromosomes are matching pairs, called **autosomes.** The twenty-third pair consists of **sex chromosomes.** In females, this pair is called XX; in males, it is called XY. The X is a relatively long chromosome, whereas the Y is short and carries little genetic material. When gametes are formed in males, the X and Y chromosomes separate into different sperm cells. In females, all gametes carry an X chromosome. Therefore, the sex of the new organism is determined by whether an X-bearing or a Y-bearing sperm fertilizes the ovum.

Multiple Births

Ruth and Peter, a couple I know well, tried for several years to have a child, without success. When Ruth reached age 33, her doctor prescribed a fertility drug, and twins—Jeannie and Jason—were born. Jeannie and Jason are **fraternal,** or **dizygotic, twins,** the most common type of multiple birth, resulting from the release and fertilization of two ova. Therefore, Jeannie and Jason are genetically no more alike than ordinary siblings. Older maternal age (up to 35 to 39 years) and fertility drugs are major causes of multiple births—factors responsible for the dramatic rise in twinning since the 1970s (Bortolus et al., 1999). As Table 2.1 shows, other genetic and environmental factors are also involved.

Twins can be created in another way. Sometimes a zygote that has started to duplicate separates into two clusters of cells that develop into two individuals. These are called **identical,** or **monozygotic, twins** because they have the same genetic makeup. The frequency of identical twins is unrelated to the factors listed in Table 2.1. It is about the same around the world—about 3 of every 1,000 births (Tong, Caddy, & Short, 1997). Animal research has uncovered a variety of environmental influences that prompt this type of twinning, including temperature changes, variation in oxygen levels, and late fertilization of the ovum.

During their early years, children of single births are often healthier and develop more rapidly than twins (Mogford-Bevan, 1999). Jeannie and Jason were born early (as are most twins)—3 weeks before Ruth's due date. As we will see in Chapter 3, like other premature infants, they required special care after birth. When the twins came home from the hospital, Ruth and Peter had to divide time between them. Perhaps because neither baby got quite as much attention as the average single infant, Jeannie and Jason walked and talked several months later than other children their age, although both caught up in development by middle childhood (Lytton & Gallagher, 2002).

Patterns of Genetic Inheritance

Jeannie has her parents' dark, straight hair, whereas Jason is curly-haired and blond. Patterns of genetic inheritance—the way genes from each parent interact—explain these outcomes. Recall that except for the XY pair in males, all chromosomes come in corresponding pairs. Two forms of each gene occur at the same place on the autosomes, one inherited from the mother and one from the father. If the genes from both parents are alike, the child is **homozygous** and will display the inherited trait. If the genes are different, then the child is **heterozygous,** and relationships between the genes determine the trait that will appear.

● **Dominant–Recessive Inheritance.** In many heterozygous pairings, **dominant–recessive inheritance** occurs: Only one gene affects the child's characteristics. It is called *dominant;* the second gene, which has no effect, is called *recessive.* Hair color is an example. The gene for dark hair is dominant (we can represent it with a capital *D),* whereas the one for blond hair is recessive (symbolized by a lowercase *b).* A child who inherits a homozygous pair of dominant genes *(DD)* and a child who inherits a heterozygous pair *(Db)* will be dark-haired, even though their genotypes differ. Blond hair (like Jason's) can result only from having two recessive genes *(bb).* Still, heterozygous individuals with just one recessive gene *(Db)* can pass that trait to their children. Therefore, they are called **carriers** of the trait.

Table 2.1 Maternal Factors Linked to Fraternal Twinning

Factor	Description
Ethnicity	Occurs in 4 per 1,000 births among Asians, 8 per 1,000 births among whites, 12 to 16 per 1,000 births among blacks[a]
Family history of twinning	Occurs more often among women whose mothers and sisters gave birth to fraternal twins
Age	Rises with maternal age, peaking between 35 and 39 years, and then rapidly falls
Nutrition	Occurs less often among women with poor diets; occurs more often among women who are tall and overweight or of normal weight as opposed to slight body build
Number of births	Is more likely with each additional birth
Fertility drugs and in vitro fertilization	Is more likely with fertility hormones and in vitro fertilization (see page 54), which also increase the chances of triplets to quintuplets

[a]Worldwide rates, not including multiple births resulting from use of fertility drugs.
Source: Bortolus et al., 1999; Mange & Mange, 1998.

■ These identical, or monozygotic, twins were created when a duplicating zygote separated into two clusters of cells, and two individuals with the same genetic makeup developed. Identical twins look alike, and as we will see later in this chapter, tend to resemble each other in a variety of psychological characteristics.

Some human characteristics that follow the rules of dominant–recessive inheritance are given in Table 2.2 on the right and Table 2.3 on page 48. As you can see, many disabilities and diseases are the product of recessive genes. One of the most frequently occurring recessive disorders is *phenylketonuria,* or *PKU.* It affects the way the body breaks down proteins contained in many foods. Infants born with two recessive genes lack an enzyme that converts one of the basic amino acids that make up proteins (phenylalanine) into a by-product essential for body functioning (tyrosine). Without this enzyme, phenylalanine quickly builds to toxic levels that damage the central nervous system. By 1 year, infants with PKU are permanently retarded.

Despite its potentially damaging effects, PKU provides an excellent illustration of the fact that inheriting unfavorable genes does not always lead to an untreatable condition. All U.S. states and Canadian provinces require that each newborn be given a blood test for PKU. If the disease is found, doctors place the baby on a diet low in phenylalanine. Children who receive this treatment show delayed development of higher-order cognitive skills, such as planning and problem solving, in infancy and childhood because even small amounts of phenylalanine interfere with brain functioning. But as long as dietary treatment begins early and continues, children with PKU usually attain an average level of intelligence and have a normal lifespan (Pietz et al., 1998; Smith, Klim, & Hanley, 2000).

In dominant–recessive inheritance, if we know the genetic makeup of the parents, we can predict the percentage of children in a family who are likely to display or carry a trait. Figure 2.3 (see page 49) illustrates this for PKU. Notice that for a child to inherit the condition, each parent must carry a recessive gene.

Only rarely are serious diseases due to dominant genes. Think about why this is so. Children who inherit the dominant gene always develop the disorder. They seldom live long enough to reproduce, and the harmful dominant gene is eliminated from the family's heredity in a single generation. Some dominant disorders, however, do persist. One of them is *Huntington disease,* a condition in which the central nervous system degenerates. Why has this disease endured? Its symptoms usually do not appear until age 35 or later, after the person has passed the dominant gene to his or her children.

● **Codominance.** In some heterozygous circumstances, the dominant–recessive relationship does not hold completely. Instead, we see **codominance,** a pattern of inheritance in which both genes influence the person's characteristics.

The *sickle cell trait,* a heterozygous condition present in many black Africans, provides an example. *Sickle cell anemia* (see Table 2.3) occurs in full form when a child inherits two recessive genes. They cause the usually round red blood cells to become sickle (or crescent-moon) shaped, especially under low-oxygen conditions. The sickled cells clog the blood vessels and block the flow of blood. Individuals who have the disorder suffer severe attacks involving intense pain, swelling, and tissue damage. They generally die in the first 20 years of life; few live past age 40. Heterozygous individuals are protected from the disease under most circumstances. However, when

Table 2.2 Examples of Dominant and Recessive Characteristics

Dominant	Recessive
Dark hair	Blond hair
Normal hair	Pattern baldness
Curly hair	Straight hair
Nonred hair	Red hair
Facial dimples	No dimples
Normal hearing	Some forms of deafness
Normal vision	Nearsightedness
Farsightedness	Normal vision
Normal vision	Congenital eye cataracts
Normally pigmented skin	Albinism
Double-jointedness	Normal joints
Type A blood	Type O blood
Type B blood	Type O blood
Rh-positive blood	Rh-negative blood

Note: Many normal characteristics that were previously thought to be due to dominant–recessive inheritance, such as eye color, are now regarded as due to multiple genes. For the characteristics listed here, there still seems to be general agreement that the simple dominant–recessive relationship holds.

Source: McKusick, 1998.

Table 2.3 Examples of Dominant and Recessive Diseases

Disease	Description	Mode of Inheritance	Incidence	Treatment
Autosomal Diseases				
Cooley's anemia	Pale appearance, retarded physical growth, and lethargic behavior begin in infancy.	Recessive	1 in 500 births to parents of Mediterranean descent	Frequent blood transfusion; death from complications usually occurs by adolescence.
Cystic fibrosis	Lungs, liver, and pancreas secrete large amounts of thick mucus, leading to breathing and digestive difficulties.	Recessive	1 in 2,000 to 2,500 Caucasian births; 1 in 16,000 births to North Americans of African descent	Bronchial drainage, prompt treatment of respiratory infection, dietary management. Advances in medical care allow survival with good life quality into adulthood.
Phenylketonuria (PKU)	Inability to metabolize the amino acid phenylalanine, contained in many proteins, causes severe central nervous system damage in the first year of life.	Recessive	1 in 8,000 births	Placing the child on a special diet results in average intelligence and normal lifespan. Subtle difficulties with planning and problem-solving are often present.
Sickle cell anemia	Abnormal sickling of red blood cells causes oxygen deprivation, pain, swelling, and tissue damage. Anemia and susceptibility to infections, especially pneumonia, occur.	Recessive	1 in 500 births to North Americans of African descent	Blood transformations, painkillers, prompt treat-ment of infection. No known cure; 50 percent die by age 20.
Tay-Sachs disease	Central nervous system degeneration, with onset at about 6 months, leads to poor muscle tone, blindness, deafness, and convulsions.	Recessive	1 in 3,600 births to Jews of European descent and to French Canadians	None. Death by 3 to 4 years of age.
Huntington disease	Central nervous system degeneration leads to muscular coordination difficulties, mental deterioration, and personality changes. Symptoms usually do not appear until age 35 or later.	Dominant	1 in 18,000 to 25,000 births	None. Death occurs 10 to 20 years after symptom onset.
Marfan syndrome	Tall, slender build; thin, elongated arms and legs. Heart defects and eye abnormalities, especially of the lens. Excessive lengthening of the body results in a variety of skeletal defects.	Dominant	1 in 20,000 births	Correction of heart and eye defects sometimes possible. Death from heart failure in young adulthood is common.
X-Linked Diseases				
Duchenne muscular dystrophy	Degenerative muscle disease. Abnormal gait, loss of ability to walk between 7 and 13 years of age.	Recessive	1 in 3,000 to 5,000 male births	None. Death from respiratory infection or weakening of the heart muscle usually occurs in adolescence.
Hemophilia	Blood fails to clot normally. Can lead to severe internal bleeding and tissue damage.	Recessive	1 in 4,000 to 7,000 male births	Blood transfusions. Safety precautions to prevent injury.
Diabetes insipidus	Insufficient production of the hormone vasopressin results in excessive thirst and urination. Dehydration can cause central nervous system damage.	Recessive	1 in 2,500 male births	Hormone replacement.

Note: For recessive disorders listed, carrier status can be detected in prospective parents through a blood test or genetic analyses. For all disorders listed, prenatal diagnosis is available (see page 53).

Sources: Behrman, Kliegman, & Arvin, 1996; Chodirker et al., 2001; Gott, 1998; Grody, 1999; Knoers et al., 1993; McKusick, 1998; Schulman & Black, 1997.

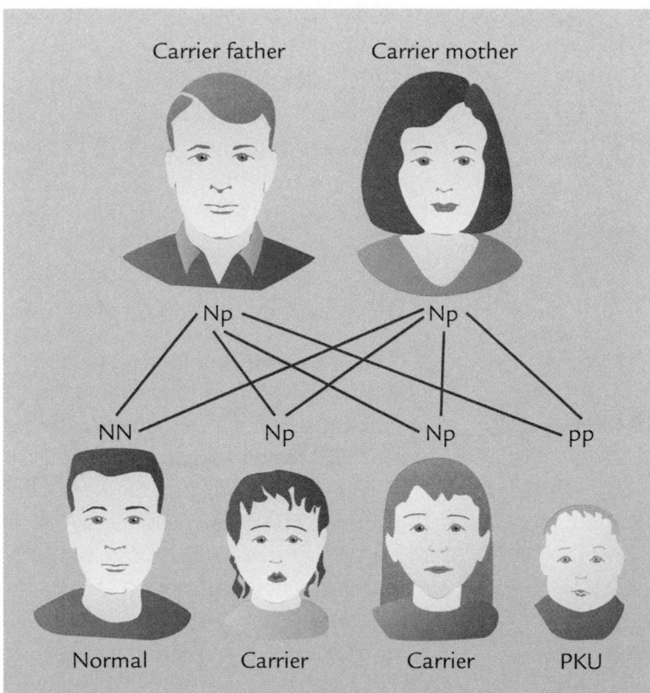

■ **FIGURE 2.3 Dominant–recessive mode of inheritance, as illustrated by PKU.** When both parents are heterozygous carriers of the recessive gene (*p*), we can predict that 25 percent of their offspring are likely to be normal (*NN*), 50 percent are likely to be carriers (*Np*), and 25 percent are likely to inherit the disorder (*pp*). Notice that the PKU-affected child, in contrast to his siblings, has light hair. The recessive gene for PKU affects more than one trait. It also leads to fair coloring.

they experience oxygen deprivation—for example, at high altitudes or after intense physical exercise—the single recessive gene asserts itself, and a temporary, mild form of the illness occurs.

The sickle cell gene is common among black Africans for a special reason. Carriers of it are more resistant to malaria than are individuals with two genes for normal red blood cells. In Africa, where malaria is common, these carriers survived and reproduced more frequently than others, leading the gene to be maintained in the black population.

● **X-Linked Inheritance.** Males and females have an equal chance of inheriting recessive disorders carried on the autosomes, such as PKU and sickle cell anemia. But when a harmful recessive gene is carried on the X chromosome, **X-linked inheritance** applies. Males are more likely to be affected because their sex chromosomes do not match. In females, any recessive gene on one X chromosome has a good chance of being suppressed by a dominant gene on the other X. But the Y chromosome is only about one-third as long and therefore lacks many corresponding genes to override those on the X. A well-known example is *hemophilia,* a disease in which the blood fails to clot normally. Figure 2.4 on page 50 shows its greater likelihood of inheritance by male children whose mothers carry the abnormal gene.

Besides X-linked disorders, many sex differences reveal the male to be at a disadvantage. Rates of miscarriage, infant and childhood deaths, birth defects, learning disabilities, behavior disorders, and mental retardation are greater for boys (Halpern, 1997). It is possible that these sex differences can be traced to the genetic code. The female, with two X chromosomes, benefits from a greater variety of genes. Nature, however, seems to have adjusted for the male's disadvantage. Worldwide, about 106 boys are born for every 100 girls, and judging from miscarriage and abortion statistics, an even greater number of males are conceived (Pyeritz, 1998).

Nevertheless, in recent decades the proportion of male births has declined in many industrialized countries, including Canada, Denmark, Germany, Finland, the Netherlands, Norway, and the United States (Jongbloet et al., 2001). Some researchers blame increased occupational and community exposure to pesticides for a reduction in sperm counts overall, especially Y-bearing sperm. But the precise cause is unknown.

● **Genetic Imprinting.** More than 1,000 human characteristics follow the rules of dominant–recessive and codominant inheritance (McKusick, 1998). In these cases, whichever parent contributes a gene to the new individual, the gene responds in the same way. Geneticists, however, have identified some exceptions. In **genetic imprinting,** genes are *imprinted,* or chemically marked, in such a way that one member of the pair (either the mother's or the father's) is activated, regardless of its makeup. The imprint is often temporary: It may be erased in the next generation, and it may not occur in all individuals (Everman & Cassidy, 2000).

Imprinting helps us understand certain puzzling genetic patterns. For example, children are more likely to develop diabetes if their father, rather than their mother, suffers from it. And people with asthma or hay fever tend to have mothers, not fathers, with the illness. Imprinting may also explain why Huntington disease, when inherited from the father, tends to emerge at an earlier age and to progress more rapidly (Navarrete, Martinez, & Salamanca, 1994).

Genetic imprinting can also operate on the sex chromosomes, as *fragile X syndrome* reveals. In this disorder, an abnormal repetition of a sequence of DNA bases occurs on the X chromosome, damaging a particular gene. Fragile X syndrome is the most common inherited cause of mild to moderate mental retardation. It has also been linked to 2 to 3 percent of cases of autism, a serious emotional disorder of early childhood involving bizarre, self-stimulating behavior and delayed or absent language and communication. Research reveals that the defective gene at the fragile site is expressed only when it is passed from mother to child (Ashley-Koch et al., 1998).

● **Mutation.** How are harmful genes created in the first place? The answer is **mutation,** a sudden but permanent change in a segment of DNA. A mutation may affect only one or two genes, or it may involve many genes, as in the chromosomal disorders we will discuss shortly. Some mutations occur

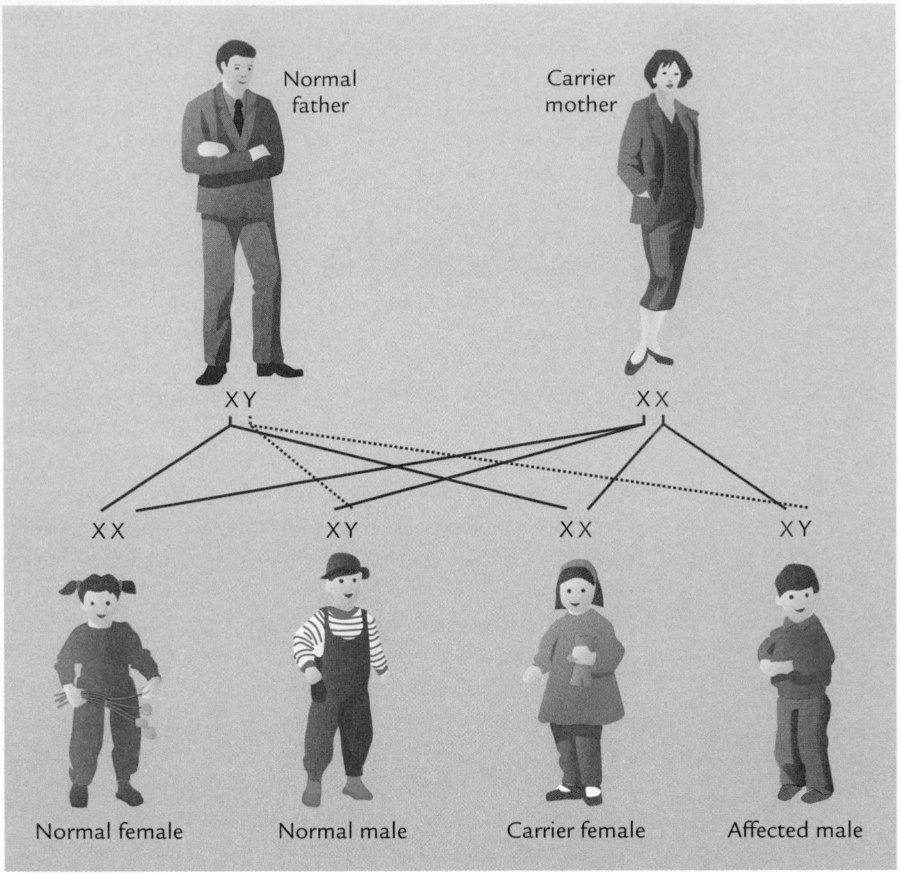

FIGURE 2.4 X-linked inheritance. In the example shown here, the gene on the father's X chromosome is normal. The mother has one normal and one abnormal recessive gene on her X chromosomes. By looking at the possible combinations of the parents' genes, we can predict that 50 percent of male children are likely to have the disorder and 50 percent of female children are likely to be carriers of it.

spontaneously, simply by chance. Others are caused by hazardous environmental agents in our food supply or the air we breathe.

Although nonionizing forms of radiation—electromagnetic waves and microwaves—have no demonstrated impact on DNA, ionizing (high-energy) radiation is an established cause of mutation. Women who receive repeated doses before conception are more likely to miscarry or give birth to children with hereditary defects. Genetic abnormalities, such as physical malformations and childhood cancer, are also higher when fathers are exposed to radiation in their occupations (Brent, 1999). However, infrequent and mild exposure to radiation does not cause genetic damage. Instead, high doses over a long period impair DNA.

● **Polygenic Inheritance.** So far, we have discussed patterns of inheritance in which people either display a particular trait or do not. These cut-and-dried individual differences are much easier to trace to their genetic origins than are characteristics that vary continuously among people, such as height, weight, intelligence, and personality. These traits are due to **polygenic inheritance,** in which many genes determine the characteristic in question. Polygenic inheritance is complex, and much about it is still unknown. In the final section of this chapter, we will discuss how researchers infer the influence of heredity on human attributes when they do not know the precise patterns of inheritance.

Chromosomal Abnormalities

Besides harmful recessive genes, abnormalities of the chromosomes are a major cause of serious developmental problems. Most chromosomal defects result from mistakes during meiosis, when the ovum and sperm are formed. A chromosome pair does not separate properly, or part of a chromosome breaks off. Because these errors involve more DNA than problems due to single genes, they usually produce many physical and mental symptoms.

● **Down Syndrome.** The most common chromosomal disorder, occurring in 1 out of every 800 live births, is *Down syndrome.* In 95 percent of cases, it results from a failure of the twenty-first pair of chromosomes to separate during meiosis, so the new individual inherits three of these chromosomes rather than the normal two. In other less frequent forms, an extra broken piece of a twenty-first chromosome is present. Or an error occurs during the early stages of mitosis, causing some but not all body cells to have the defective chromosomal makeup (called a *mosaic* pattern). In these instances, because less genetic material is involved, symptoms of the disorder are less extreme (Hodapp, 1996).

The consequences of Down syndrome include mental retardation, memory and speech problems, limited vocabulary, and slow motor development (Chapman & Hesketh, 2000). Affected individuals also have distinct physical features—a short, stocky build, a flattened face, a protruding tongue, almond-shaped eyes, and an unusual crease running across the palm of the hand. In addition, infants with Down syndrome are often born with eye cataracts and heart and intestinal defects. Three decades ago, most died by early adulthood. Today, because of medical advances, many survive into their sixties and beyond (Selikowitz, 1997).

Infants with Down syndrome smile less readily, show poor eye-to-eye contact, and explore objects less persistently. But when parents encourage them to engage with their surroundings, Down syndrome children develop more favorably (Sigman, 1999). They also benefit from infant and preschool

■ The facial features of the 9-year-old boy on the left are typical of Down syndrome. Although his intellectual development is impaired, this child is doing well because he is growing up in a stimulating home where his special needs are met and he is loved and accepted. Here he collaborates with his normally developing 4-year-old brother in making won ton dumplings.

intervention programs, although emotional, social, and motor skills improve more than intellectual performance (Hines & Bennett, 1996). Thus, environmental factors affect how well children with Down syndrome fare.

The risk of Down syndrome rises dramatically with maternal age, from 1 in 1,900 births at age 20, to 1 in 300 at age 35, to 1 in 30 at age 45 (Halliday et al., 1995; Meyers et al., 1997). Why is this so? Geneticists believe that the ova, present in the woman's body since her own prenatal period, weaken over time. As a result, chromosomes do not separate properly as they complete the process of meiosis at conception. In about 5 to 10 percent of cases, the extra genetic material originates with the father. However, Down syndrome and other chromosomal abnormalities are not related to advanced paternal age (Muller et al., 2000; Savage et al., 1998). In these instances, the mutation occurs for other unknown reasons.

● **Abnormalities of the Sex Chromosomes.** Disorders of the autosomes other than Down syndrome usually disrupt development so severely that miscarriage occurs. When such babies are born, they rarely survive beyond early childhood. In contrast, abnormalities of the sex chromosomes usually lead to fewer problems. In fact, sex chromosome disorders often are not recognized until adolescence when, in some of the deviations, puberty is delayed. The most common problems involve the presence of an extra chromosome (either X or Y) or the absence of one X in females.

A variety of myths exist about individuals with sex chromosome disorders. For example, males with *XYY syndrome* are not necessarily more aggressive and antisocial than XY males. And most children with sex chromosome disorders do not suffer from mental retardation. Instead, their intellectual problems are usually very specific. Verbal difficulties—for example, with reading and vocabulary—are common among girls with *triple X syndrome (XXX)* and boys with *Klinefelter syndrome (XXY),* both of whom inherit an extra X chromosome. In contrast, girls with *Turner syndrome (XO),* who are missing an X, have trouble with spatial relationships—for example, drawing pictures, telling right from left, following travel directions, and noticing changes in facial expressions (Geschwind et al., 2000; Money, 1993; Ross, Zinn, & McCauley, 2000). These findings tell us that adding to or subtracting from the usual number of X chromosomes results in particular intellectual deficits. At present, geneticists do not know why.

Ask Yourself

REVIEW

Explain the genetic origins of PKU and Down syndrome. Cite evidence that both heredity and environment contribute to the development of individuals with these disorders.

APPLY

Gilbert's genetic makeup is homozygous for dark hair. Jan's is homozygous for blond hair. What color is Gilbert's hair? How about Jan's? What proportion of their children are likely to be dark-haired? Explain.

CONNECT

Referring to ecological systems theory (Chapter 1, pages 24–25), explain why parents of children with genetic disorders often experience increased stress. What factors, within and beyond the family, can help these parents support their children's development?

www.

Reproductive Choices

Two years after they were married, Ted and Marianne gave birth to their first child. Kendra appeared to be a healthy infant, but by 4 months her growth slowed. Diagnosed as having Tay-Sachs disease (see Table 2.3), Kendra died at 2 years of age. Ted and Marianne were devastated by

Kendra's death. Although they did not want to bring another infant into the world who would endure such suffering, they badly wanted to have a child. They began to avoid family get-togethers, where little nieces and nephews were constant reminders of the void in their lives.

In the past, many couples with genetic disorders in their families chose not to bear a child at all rather than risk the birth of an abnormal baby. Today, genetic counseling and prenatal diagnosis help people make informed decisions about conceiving, carrying a pregnancy to term, or adopting a child.

Genetic Counseling

Genetic counseling is a communication process designed to help couples assess their chances of giving birth to a baby with a hereditary disorder and choose the best course of action in view of risks and family goals (Shiloh, 1996). Individuals likely to seek counseling are those who have had difficulties bearing children, such as repeated miscarriages, or who know that genetic problems exist in their families. In addition, women who delay childbearing past age 35 are candidates for genetic counseling. After this time, the overall rate of chromosomal abnormalities rises sharply, from 1 in every 190 to as many as 1 in every 10 pregnancies at age 48 (Meyers et al., 1997).

If a family history of mental retardation, physical defects, or inherited diseases exists, the genetic counselor interviews the couple and prepares a *pedigree,* a picture of the family tree in which affected relatives are identified. The pedigree is used to estimate the likelihood that parents will have an abnormal child, using the same genetic principles discussed earlier in this chapter. In the case of many disorders, blood tests or genetic analyses can reveal whether the parent is a carrier of the harmful gene. Carrier detection is possible for all of the recessive diseases listed in Table 2.3, as well as others, and for fragile X syndrome.

When all the relevant information is in, the genetic counselor helps people consider appropriate options. These include "taking a chance" and conceiving, choosing from among a variety of reproductive technologies (see the Social Issues box on pages 54–55), or adopting a child.

Prenatal Diagnosis and Fetal Medicine

If couples who might bear an abnormal child decide to conceive, several **prenatal diagnostic methods**—medical procedures that permit detection of problems before birth—are available (see Table 2.4). Women of advanced maternal age are prime candidates for *amniocentesis* or *chorionic villus sampling* (see Figure 2.5). Except for *ultrasound* and *maternal blood analysis,* prenatal diagnostic methods should not be used routinely, as they have some chance of injuring the developing organism.

Prenatal diagnosis has led to advances in fetal medicine. For example, by inserting a needle into the uterus, doctors can

Table 2.4 Prenatal Diagnostic Methods

Method	Description
Amniocentesis	The most widely used technique. A hollow needle is inserted through the abdominal wall to obtain a sample of fluid in the uterus. Cells are examined for genetic defects. Can be performed by 11 to 14 weeks after conception but safest after 15 weeks; 1 to 2 more weeks are required for test results. Small risk of miscarriage.
Chorionic villus sampling	A procedure that can be used if results are desired or needed very early in pregnancy. A thin tube is inserted into the uterus through the vagina, or a hollow needle is inserted through the abdominal wall. A small plug of tissue is removed from the end of one or more chorionic villi, the hairlike projections on the membrane surrounding the developing organism. Cells are examined for genetic defects. Can be performed at 6 to 8 weeks after conception, and results are available within 24 hours. Entails a slightly greater risk of miscarriage than does amniocentesis. Also associated with a small risk of limb deformities, which increases the earlier the procedure is performed.
Fetoscopy	A small tube with a light source at one end is inserted into the uterus to inspect the fetus for defects of the limbs and face. Also allows a sample of fetal blood to be obtained, permitting diagnosis of such disorders as hemophilia and sickle cell anemia as well as neural defects (see below). Usually performed between 15 and 18 weeks after conception, but can be done as early as 5 weeks. Entails some risk of miscarriage.
Ultrasound	High-frequency sound waves are beamed at the uterus; their reflection is translated into a picture on a videoscreen that reveals the size, shape, and placement of the fetus. By itself, permits assessment of fetal age, detection of multiple pregnancies, and identification of gross physical defects. Also used to guide amniocentesis, chorionic villus sampling, and fetoscopy. When used five or more times, may increase the chances of low birth weight.
Maternal blood analysis	By the second month of pregnancy, some of the developing organism's cells enter the maternal bloodstream. An elevated level of alpha-fetoprotein may indicate kidney disease, abnormal closure of the esophagus, or neural tube defects, such as anencephaly (absence of most of the brain) and spina bifida (bulging of the spinal cord from the spinal column). Isolated cells can be examined for genetic defects, such as Down syndrome.
Preimplantation genetic diagnosis	After in vitro fertilization and duplication of the zygote into a cluster of about eight to ten cells, one or two cells are removed and examined for hereditary defects. Only if that sample is free of detectable genetic disorders is the fertilized ovum implanted in the woman's uterus.

Sources: Eiben et al., 1997; Lissens & Sermon, 1997; Moore & Persaud, 1998; Newnham et al., 1993; Quintero, Puder, & Cotton, 1993; Sutcliffe, 2002; Wapner, 1997; Willner, 1998.

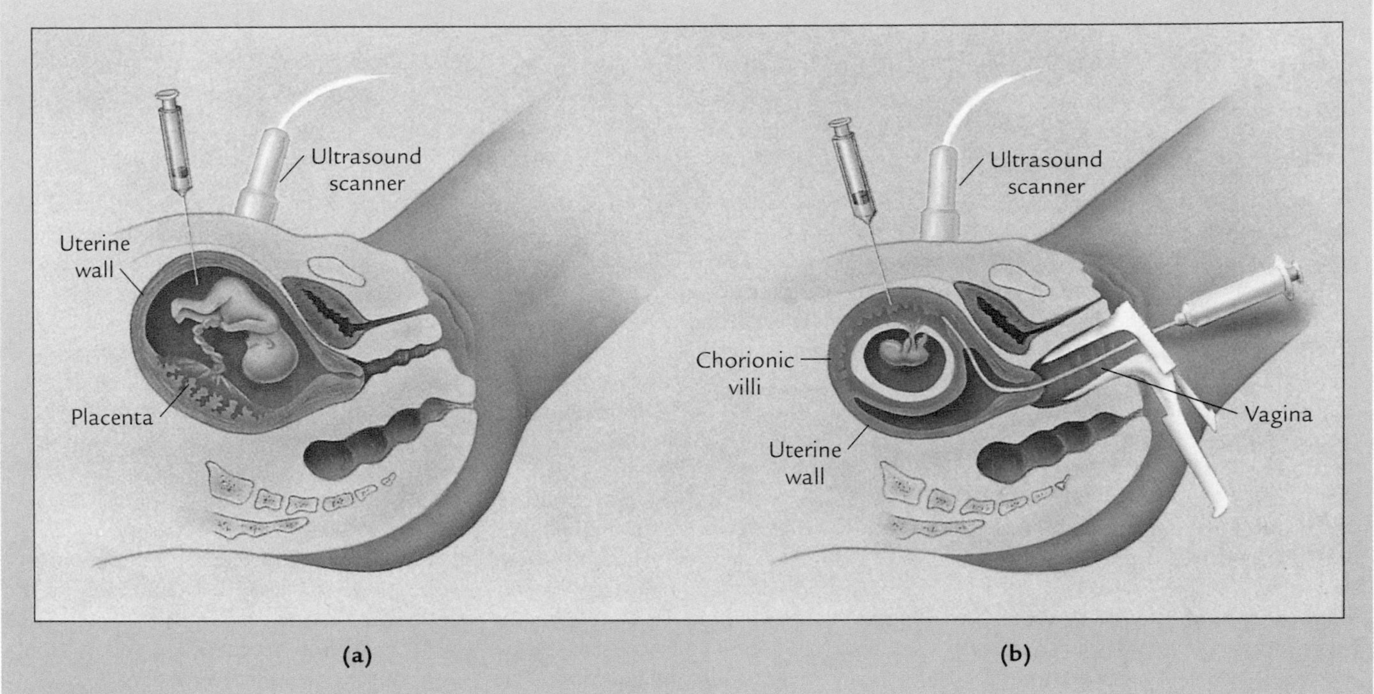

FIGURE 2.5 Amniocentesis and chorionic villus sampling. Today, more than 250 defects and diseases can be detected before birth using these two procedures. (a) In amniocentesis, a hollow needle is inserted through the abdominal wall into the uterus. Fluid is withdrawn and fetal cells are cultured, a process that takes about 3 weeks. (b) Chorionic villus sampling can be performed much earlier in pregnancy, at 6 to 8 weeks after conception, and results are available within 24 hours. Two approaches to obtaining a sample of chorionic villus are shown: inserting a thin tube through the vagina into the uterus and inserting a needle through the abdominal wall. In both amniocentesis and chorionic villus sampling, an ultrasound scanner is used for guidance. (From K. L. Moore & T. V. N. Persaud, 1998, *Before We Are Born,* 5th ed., Philadelphia: Saunders, p. 115. Adapted by permission of the publisher and author.)

administer drugs to the fetus. Surgery has been performed to repair such problems as heart and lung malformations, urinary tract obstructions, and neural defects. A fetus with a hereditary immune deficiency received a bone marrow transplant from his father that succeeded in creating a normally functioning immune system (Flake et al., 1996).

These techniques frequently result in complications, the most common being premature labor and miscarriage (James, 1998). Yet parents may be willing to try almost any option, even if there is only a slim chance of success. Currently, the medical profession is struggling with how to help parents make informed decisions about fetal surgery. One suggestion is that the advice of an independent counselor be provided—a doctor or nurse who understands the risks but is not involved in doing research on or performing the procedure.

Advances in *genetic engineering* also offer hope for correcting hereditary defects. As part of the Human Genome Project—an ambitious, international research program aimed at deciphering the chemical makeup of human genetic material (genome)—researchers have mapped the sequence of all human DNA base pairs. Using that information, they are "annotating" the genome—identifying all its genes and their functions, including their protein products and what they do. A major goal is to understand the estimated 4,000 human disorders, those due to single genes and those resulting from a complex interplay of multiple genes and environmental factors.

Already, thousands of genes have been identified, including those involved in hundreds of diseases, such as cystic fibrosis, Huntington disease, Duchenne muscular dystrophy, Marfan syndrome, and some forms of cancer (Jimeniz-Sanchez, Childs, & Valle, 2001). As a result, new treatments are being explored, such as *gene therapy*—delivering DNA carrying a functional

The Human Genome Project is leading to new gene-based treatments for hereditary disorders such as Alzheimer's disease. When the daughter and granddaughter of this Alzheimer's victim reach late adulthood, they may be spared this devastating illness.

Social Issues

The Pros and Cons of Reproductive Technologies

Some couples decide not to risk pregnancy because of a history of genetic disease. Many others—in fact, one-sixth of all couples who try to conceive—discover that they are sterile. And some never-married adults and gay and lesbian partners want to bear children. Today, increasing numbers of individuals are turning to alternative methods of conception—technologies that, although they fulfill the wish of parenthood, have become the subject of heated debate.

Donor Insemination and In Vitro Fertilization. For several decades, *donor insemination*—injection of sperm from an anonymous man into a woman—has been used to overcome male reproductive difficulties. In recent years, it has also permitted women without a heterosexual partner to become pregnant. Donor insemination is 70 to 80 percent successful, resulting in 30,000 to 50,000 births in North America each year (Cooper & Glazer, 1999).

In vitro fertilization is another reproductive technology that has become increasingly common. Since the first "test tube" baby was born in England in 1978, 1 percent of all children in developed countries—about 39,000 babies in the United States and 3,500 babies in Canada—have been conceived through this technique annually (Sutcliffe, 2002). With in vitro fertilization, hormones are given to a woman, stimulating ripening of several ova. These are removed surgically and placed in a dish of nutrients, to which sperm are added. Usually, in vitro fertilization is used to treat women whose fallopian tubes are permanently damaged. But a recently developed technique permits a single sperm to be injected directly into an ovum, thereby overcoming most male fertility problems. And a new "sex sorter" method helps ensure that couples who carry X-linked diseases (which usually affect males) have a daughter. Once an ovum is fertilized and begins to duplicate into several cells, it is injected into the mother's uterus.

The overall success rate of in vitro fertilization is about 25 to 30 percent. However, success declines steadily with age, from 37 percent in women younger than age 35 to 3 percent in women age 43 and older (U.S. Department of Health and Human Services, 1999a). By mixing and matching gametes, pregnancies can be brought about when either or both partners have a reproductive problem. Fertilized ova and sperm can even be frozen and stored in embryo banks for use at some future time, thereby guaranteeing healthy zygotes should age or illness lead to fertility problems.

Children conceived through these methods may be genetically unrelated to one or both of their parents. In addition, most parents who have used in vitro fertilization do not tell their children about their origins, although health professionals now encourage them to do so. Does lack of genetic ties or secrecy surrounding these techniques interfere with parent–child relationships? Perhaps because of a strong desire for parenthood, caregiving is actually somewhat warmer for young children conceived through in vitro fertilization or donor insemination. And in vitro infants are as securely attached to their parents, and children and adolescents as well adjusted, as their counterparts who were naturally conceived (Chan, Raboy, & Patterson, 1998; Gibson et al., 2000; Golombok, MacCallum, & Goodman, 2001).

Although donor insemination and in vitro fertilization have many benefits, serious questions have arisen about their use. Most U.S. states and Canadian provinces have no legal guidelines for these procedures. As a result, donors are not always screened for genetic or sexually transmitted diseases. Furthermore, in many countries (including the United States and Canada), doctors are not required to keep records of donor characteristics. Yet the resulting children may someday want to know their genetic background or need to know it for medical reasons. Also, critics worry that the in vitro "sex sorter" method will lead to parental sex selection, thereby eroding the moral value that children of both sexes are equally precious. And finally, in some instances, frozen embryos have been stolen—that is, taken and sold without permission of the couple owning them (Blum, 2000).

Surrogate Motherhood. An even more controversial form of medically assisted conception is *surrogate motherhood.* Typically in this procedure, sperm from a man whose wife is infertile are used to inseminate a woman, called a surrogate, who is paid a fee for her childbearing services. In return, the surrogate agrees to turn the baby over to the man (who is the natural father). The child is then adopted by his wife.

Although most of these arrangements proceed smoothly, those that end up in court highlight serious risks for all concerned. In one case, both parties rejected the infant with severe disabilities that resulted from the pregnancy. In others, the surrogate mother wanted to keep the baby or the couple changed their mind during the pregnancy. These children came into the world in the midst of conflict that threatened to last for years.

Because surrogacy favors the wealthy as contractors for infants and the less economically advantaged as surrogates, it may promote exploitation of financially needy women (Sureau, 1997). In addition, most surrogates already have children of their own who may be deeply affected by the pregnancy. Knowledge that their mother would give away a baby for profit may cause these youngsters to worry about the security of their own family circumstances.

New Reproductive Frontiers. Reproductive technologies are evolving faster than societies can weigh the ethics of these procedures. Doctors have used donor ova from younger women in combination with in vitro fertilization to help postmenopausal women become pregnant. Most recipients are in their 40s, but a 62-year-old has given birth in Italy and a 63-year-old in the United States. Even though candidates for postmenopausal-assisted childbirth are selected on the basis of good health, serious questions arise about bringing children into the world whose parents may not live to see them reach adulthood. Based on U.S. life expectancy data, 1 in 3 mothers and 1 in 2 fathers having a baby at age 55 will die before their child enters college (U.S. Bureau of the Census, 2002c).

Currently, experts are debating other reproductive options. In one instance, a woman with a busy stage career who could have become pregnant naturally chose to combine in vitro fertilization (using her own ova and her husband's sperm) and surrogate motherhood. This permitted the woman to continue her career while the surrogate carried her biological child (Wood, 2001). At donor banks, customers can select ova or sperm on the basis of physical characteristics and even IQ. Some worry that this practice is a dangerous step toward selective breeding through "designer babies"—controlling offspring characteristics by manipulating the genetic makeup of fertilized ova.

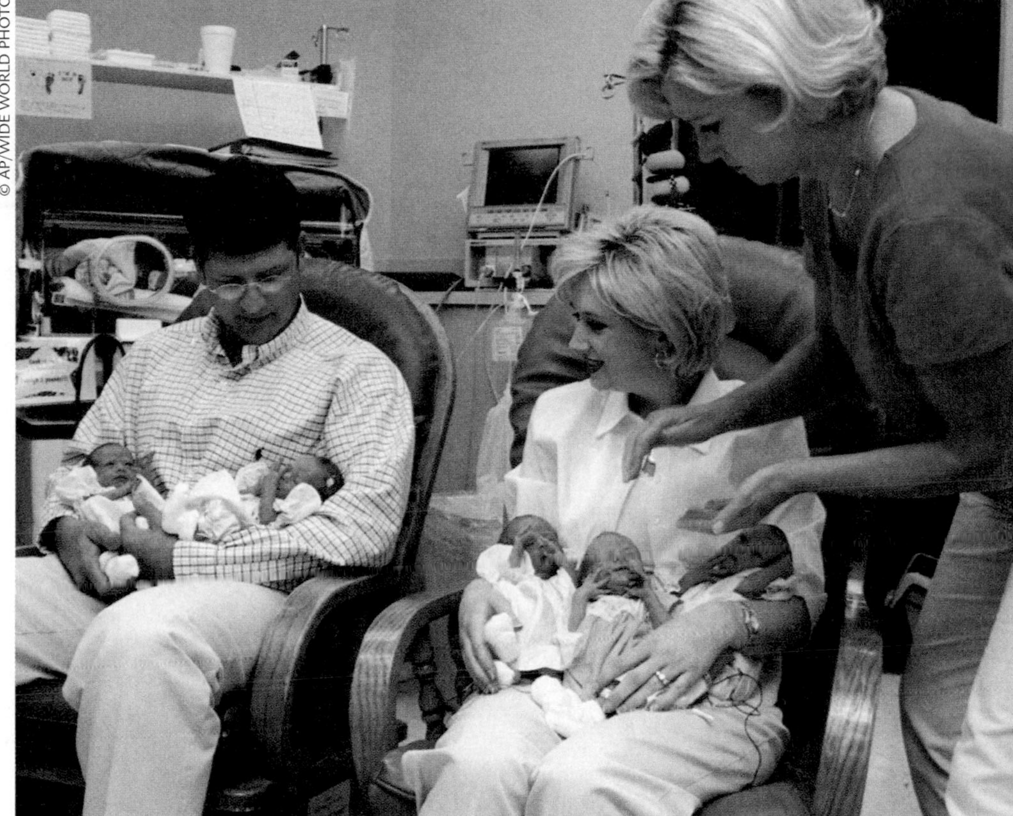

■ Although reproductive technologies permit many barren couples to have healthy newborns, they can pose grave ethical dilemmas. Fertility drugs and in vitro fertilization often lead to multiple fetuses. When three or more fill the uterus, pregnancy complications are often so severe that doctors recommend aborting one or more to save the others. These 3-week-old babies, being held by their parents in the intensive care nursery, are the only documented quintuplets to have all been born alive in Mississippi.

Finally, scientists have successfully cloned (made multiple copies of) fertilized ova in sheep, cattle, and monkeys, and they are working on effective ways to do so in humans. By providing extra ova for injection, cloning might improve the success rate of in vitro fertilization. But it also opens the possibility of mass-producing genetically identical people. Therefore, it is widely condemned (Fasouliotis & Schenker, 2000).

Although new reproductive technologies permit many barren couples to rear healthy newborn babies, laws are needed to regulate such practices. In Australia, New Zealand, and Sweden, individuals conceived with donated gametes have a right to information about their genetic origins (Hunter, Salter-Ling, & Glover, 2000). Pressure from those working in the field of assisted reproduction may soon lead to similar policies in the United States and Canada.

In the case of paying surrogate mothers to bear another's child, Australia, Canada, many European nations, and some U.S. states have banned the practice, arguing that the status of a baby should not be a matter of commercial arrangement and that a part of the body should not be rented or sold (McGee, 1997). England, France, and Italy have prohibited in vitro fertilization for women past menopause (Andrews & Elster, 2000). At present, nothing is known about the long-term health and psychological consequences of being a product of these procedures. Research on how such children grow up, including later-appearing medical conditions and knowledge and feelings about their origins, is important for weighing the pros and cons of these techniques.

Steps Before Conception to Increase the Chances of Having a Healthy Baby

SUGGESTION	RATIONALE
Arrange for a physical exam.	A physical exam before conception permits detection of diseases and other medical problems that might reduce fertility, be difficult to treat after the onset of pregnancy, or affect the developing organism.
Reduce or eliminate toxins under your control.	Because the developing organism is highly sensitive to damaging environmental agents during the early weeks of pregnancy (see Chapter 3), drugs, alcohol, cigarette smoke, radiation, pollution, chemical substances in the home and workplace, and infectious diseases should be avoided while trying to conceive. Furthermore, ionizing radiation and some industrial chemicals are known to cause mutations.
Consider your genetic makeup.	Find out if anyone in your family has had a child with a genetic disease or disability. If so, seek genetic counseling before conception.
Consult your doctor after 12 months of unsuccessful efforts at conception.	Long periods of infertility may be due to undiagnosed spontaneous abortions, which can be caused by genetic defects in either partner. If a physical exam reveals a healthy reproductive system, seek genetic counseling.

gene to the cells, thereby correcting a genetic abnormality. In recent experiments, gene therapy relieved symptoms in hemophilia patients and patients with severe immune system dysfunction (Cavazzana-Calvo et al., 2000; Kay et al., 2000). Another approach is *proteomics,* modifying gene-specified proteins involved in biological aging and disease (Blumenthal, 2001).

Genetic treatments, however, seem some distance away for most single-gene defects—and far off for diseases involving multiple genes that combine in complex ways with each other and the environment (Collins & McKusick, 2001). The Caregiving Concerns table above summarizes steps that prospective parents can take before conception to protect the genetic health of their child.

Genetic Testing

Fortunately, over 90 percent of pregnancies in the United States result in the birth of healthy infants with a good chance of a life free of genetic disease. For those who harbor genes that might lead to later-emerging disorders, a few predictive tests exist, such as those for breast and colon cancer. Scientists predict that by the year 2010 many more such tests will be available, permitting people to find out about their genetic risks and, hopefully, take steps to reduce them—through medical monitoring, lifestyle changes, or drug therapy (Druker & Lydon, 2000).

Although its potential benefits are great, at present genetic testing raises serious social, ethical, and legal concerns. A major controversy involves testing children and adults who are at risk but who do not yet show disease symptoms. Delay between the availability of predictive tests and effective interventions means that people must live with the knowledge that they might become seriously ill. A related concern is the need for greater knowledge of genetics, by both health professionals and the general public. Without this understanding, doctors and patients may misinterpret genetic risk. For example, genes associated with breast cancer were first identified in families with a high incidence of the disease. In those families, 85 percent of individuals harboring the genes developed cancer. But in the general population, the risk is reduced to 35 to 50 percent (Burke, Atkins, & Gwinn, 2002). Furthermore, some people whose tests revealed abnormal genes have encountered discrimination in the workplace, where they have lost their health insurance and even their jobs.

Experts in medical ethics recommend that genetic testing be offered only to high-risk individuals who undergo extensive education and counseling and agree to close medical and psychological follow-up if they test positive (Haddad et al., 1999). And procedures have been devised in the United States and Canada to ensure the privacy of genetic information and to prohibit mandatory genetic testing by employers (Privacy Commissioner of Canada, 2002; Yang, Flake, & Adzick, 1999). As the Human Genome Project continues to make invaluable contributions to human health, scientists are debating the best ways to protect the public interest while reaping the project's monumental rewards.

Adoption

Adults who cannot bear children, who are likely to pass along a genetic disorder, or who are older and single but want a family are turning to adoption in increasing numbers. Adoption agencies usually try to find parents of the same ethnic and religious background as the child. Where possible, they also try to choose parents who are the same age as most natural parents. Because the availability of healthy babies has declined (fewer young unwed mothers give up their babies today than in the past), more people are adopting from other countries or taking children who are older or who have developmental problems.

Adopted children and adolescents—whether born in another country or in the country of their adoptive parents—have more learning and emotional difficulties than other children, a difference that increases with the child's age at time of adoption (Levy-Shiff, 2001; Miller et al., 2000; Sharma, McCue, & Benson, 1998). There are many possible reasons for adoptees' more problematic childhoods. The biological mother may have been unable to care for the child because of emotional problems believed to be partly genetic, such as alcoholism or severe depression. She may have passed this tendency to her offspring. Or perhaps she experienced stress, poor diet, or inadequate medical care during pregnancy—factors that can affect the child (as we will see in Chapter 3). Furthermore, children adopted after infancy are more likely than their nonadopted peers to have a history of conflict-ridden family relationships, lack of parental affection, and neglect and abuse. Finally, adoptive parents and children, who are genetically unrelated, are less alike in intelligence and personality than biological relatives—differences that may threaten family harmony.

But despite these risks, most adopted children fare surprisingly well. In a Swedish study, researchers followed over 600 infant candidates for adoption into adolescence. Some were adopted shortly after birth; some were reared in foster homes; and some were reared by their biological mothers, who changed their minds about giving them up. As Figure 2.6 shows, adoptees developed much more favorably than children reared in foster families or returned to their birth mothers (Bohman & Sigvardsson, 1990). And in a study of internationally adopted children in the Netherlands, sensitive maternal care and secure attachment in infancy predicted cognitive and social competence at age 7 (Stams, Juffer, & van IJzendoorn, 2002). So even when children are not genetically related to their parents, an early warm, trusting parent–child relationship fosters development. Older-placed adoptive children with troubled family histories also develop feelings of trust and affection for their adoptive parents as they come to feel loved and supported in their new families (Sherrill & Pinderhughes, 1999).

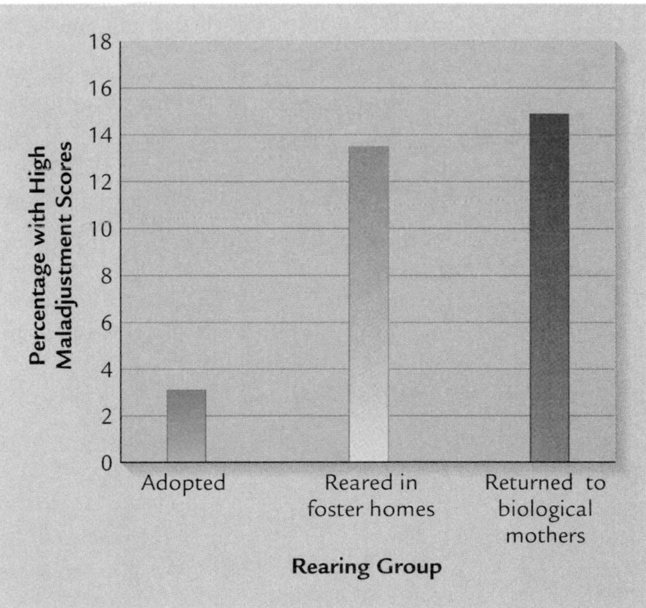

FIGURE 2.6 Relationship of type of rearing to maladjustment among a sample of Swedish adolescents who had been candidates for adoption at birth. Compared with the other two groups, adopted young people were rated by teachers as having far fewer problems, including anxiety, withdrawal, aggression, inability to concentrate, peer difficulties, and poor school motivation. (Adapted from Bohman & Sigvardsson, 1990.)

By adolescence, adoptees' lives are often complicated by unresolved curiosity about their roots. Some have difficulty accepting the possibility that they may never know their birth parents. Others worry about what they would do if their birth parents suddenly reappeared (Grotevant & Kohler, 1999). Nevertheless, the decision to search for birth parents is usually postponed until early adulthood, when marriage and childbirth may trigger it (Schaffer & Kral, 1988). Despite concerns about their origins, most adoptees appear well adjusted as adults. And as long as their parents took steps to help them learn about their heritage in childhood, transracially or transculturally adopted young people generally develop identities that are healthy blends of their birth and rearing backgrounds (Brooks & Barth, 1999).

As we conclude our discussion of reproductive choices, perhaps you are wondering how things turned out for Ted and Marianne. Through genetic counseling, Marianne discovered a history of Tay-Sachs disease on her mother's side of the family. Ted had a distant cousin who died of the disorder. The genetic counselor explained that the chances of giving birth to another affected baby were 1 in 4. Ted and Marianne took the risk. Their son Douglas is now 12 years old. Although Douglas is a carrier of the recessive allele, he is a normal, healthy boy. In a few years, Ted and Marianne will tell Douglas about his genetic history and explain the importance of genetic counseling and testing before he has children of his own.

Ask Yourself

REVIEW

Why is genetic counseling called a *communication process*? Who should seek it?

REVIEW

Describe the ethical pros and cons of fetal surgery, surrogate motherhood, and postmenopausal-assisted childbearing.

CONNECT

How does research on adoption reveal resiliency? Which of the factors related to resiliency (see Chapter 1, pages 10–11) is central in positive outcomes for adoptees?

www.

Environmental Contexts for Development

Just as complex as the genetic inheritance that sets the stage for development is the surrounding environment—a many-layered set of influences that combine to help or hinder physical and psychological well-being. Take a moment to reflect on your own childhood, and jot down a brief description of events and people that you regard as having had a significant impact on your development. Next, do the same for your adult life. When I ask my students to do this, most items they list involve their families. This emphasis is not surprising, since the family is the first and longest-lasting context for development. But other settings turn out to be important as well. Friends, neighbors, school, workplace, community organizations, and church, synagogue, or mosque generally make the top ten.

Think back to Bronfenbrenner's ecological systems theory, discussed in Chapter 1. It emphasizes that the environments extending beyond the *microsystem,* or the immediate settings just mentioned, powerfully affect development. Indeed, my students rarely mention one very important context. Its impact is so pervasive that we seldom stop to think about it in our daily lives. This is the *macrosystem,* or broad social climate of society—its values and programs that support and protect human development. All people need help with the demands of each phase of the lifespan—through well-designed housing, safe neighborhoods, good schools, well-equipped recreational facilities, affordable health services, and high-quality child care and other services that permit them to meet both work and family responsibilities. And some people, because of poverty or special tragedies, need considerably more help than others.

In the following sections, we take up these contexts for development. Because they affect every age and aspect of change, we will return to them in later chapters. For now, our discussion emphasizes that besides heredity, environments can enhance or create risks for development. And when a vulnerable child or adult—an individual with physical or psychological problems—is exposed to unfavorable contexts, development is seriously threatened.

The Family

In power and breadth of influence, no context equals the family. The family creates bonds between people that are unique. Attachments to parents and siblings usually last a lifetime and serve as models for relationships in the wider world of neighborhood, school, and community. Within the family, children learn the language, skills, and social and moral values of their culture. And at all ages, people turn to family members for information, assistance, and interesting and pleasurable interaction. Warm, gratifying family ties predict physical and psychological health throughout development. In contrast, isolation or alienation from the family is often associated with developmental problems (Parke & Buriel, 1998).

Contemporary researchers view the family as a network of interdependent relationships. Recall from ecological systems theory that *bidirectional influences* exist in which the behaviors of each family member affect those of others (Bronfenbrenner, 1989, 1995). Indeed, the very term *system* implies that the responses of all family members are related. These system influences operate both directly and indirectly.

Direct Influences. The next time you have a chance to observe family members interacting, watch carefully. You are likely to see that kind, patient communication evokes cooperative, harmonious responses, whereas harshness and impatience engender angry, resistive behavior. Each of these reactions, in turn, forges a new link in the interactive chain. In the first instance, a positive message tends to follow; in the second, a negative or avoidant one tends to occur.

These observations fit with a wealth of research on the family system. For example, many studies show that when parents' requests are accompanied by warmth and affection, children tend to cooperate. And when children willingly comply, their parents are likely to be warm and gentle in the future. In contrast, parents who discipline with harshness and impatience have children who refuse and rebel. And because children's misbehavior is stressful for parents, they may increase their use of punishment, leading to more unruliness by the child (Dodge, Pettit, & Bates, 1994; Stormshak et al., 2000). This principle applies to other two-person family relationships, such as brother and sister, husband and wife, and

■ The family is a complex social system in which each person's behavior influences the behavior of others, in both direct and indirect ways. The positive mealtime atmosphere in this Saudi Arabian family is a product of many sources, including parents who respond to children with warmth and patience, aunts and uncles who support parents in their child-rearing roles, and children who have developed cooperative dispositions.

parent and adult child. In each case, the behavior of one family member helps sustain a form of interaction in the other that either promotes or undermines psychological well-being.

● **Indirect Influences.** The impact of family relationships on development becomes even more complicated when we consider that interaction between any two members is affected by others present in the setting. Bronfenbrenner calls these indirect influences the effect of *third parties.*

Third parties can serve as supports for development. For example, when parents' marital relationship is warm and considerate, mothers and fathers praise and stimulate their children more and nag and scold them less. In contrast, when a marriage is tense and hostile, parents tend to be less responsive to their children's needs and to criticize, express anger, and punish (Cox, Paley, & Harter, 2001; Erel & Burman, 1995). Similarly, children can affect their parents' relationship in powerful ways. For example, as we will see in Chapter 10, some children show lasting emotional problems when their parents divorce. But longitudinal research reveals that long before the marital breakup, some children of divorcing couples were impulsive and defiant. These behaviors may have contributed to as well as been caused by their parents' marital problems (Amato & Booth, 1996; Hetherington, 1999).

Yet even when family relationships are strained by third parties, other members may help restore effective interaction. Grandparents are a case in point. They can promote children's development in many ways—both directly, by responding warmly to the child, and indirectly, by providing parents with child-rearing advice, models of child-rearing skill, and even financial assistance. Of course, like any indirect influence, grandparents can sometimes be harmful. When quarrelsome relations exist between grandparents and parents, parent–child communication may suffer.

● **Adapting to Change.** Think back to the *chronosystem* in Bronfenbrenner's theory (see page 25). The interplay of forces within the family is dynamic and ever-changing. Important events, such as the birth of a baby, a change of jobs, or an elderly parent joining the household due to declining health, create challenges that modify existing relationships. The way such events affect family interaction depends on the support provided by other family members as well as on the developmental status of each participant. For example, the arrival of a new baby prompts very different reactions in a toddler than in a school-age child. And caring for an ill elderly parent is more stressful for a middle-aged adult still rearing young children than for an adult of the same age without child-rearing responsibilities.

Historical time period also contributes to a dynamic family system. In recent decades, a declining birth rate, a high divorce rate, and expansion of women's roles have led to a smaller family size. This, combined with a longer lifespan, means that more generations are alive with fewer members in the youngest ones, leading to a "top-heavy" family structure. Consequently, young people today are more likely to have older relatives than at any time in history—a circumstance that can be enriching as well as a source of tension. In sum, as this complex intergenerational system moves through time, relationships are constantly revised as members adjust to their own and others' development as well as to external pressures.

Despite these variations, some general patterns in family functioning do exist. In the United States, Canada, and other Western nations, one important source of these consistencies is socioeconomic status.

Socioeconomic Status and Family Functioning

People in industrialized nations are stratified on the basis of what they do at work and how much they earn for doing it—factors that determine their social position and economic well-being. Researchers assess a family's standing on this continuum through an index called **socioeconomic status (SES).** It combines three interrelated, but not completely overlapping, variables: (1) years of education and (2) the prestige of and skill required by one's job, both of which measure social status, and (3) income, which measures economic status. As socioeconomic status rises and falls, people face changing circumstances that profoundly affect family functioning.

SES affects the timing and duration of phases of the family life cycle. People who work in skilled and semiskilled manual occupations (for example, machinists, truck drivers, and custodians) tend to marry and have children earlier as well as give birth to more children than people in white-collar and professional occupations. The two groups also differ in values and expectations. For example, when asked about personal qualities they desire for their children, lower-SES parents tend to emphasize external characteristics, such as obedience, politeness, neatness, and cleanliness. In contrast, higher-SES parents emphasize psychological traits, such as curiosity, happiness, self-direction, and cognitive and social maturity (Hoff, Laursen, & Tardiff, 2002; Tudge et al., 2000). In addition, fathers in higher-SES families tend to be more involved in child rearing and household responsibilities. Lower-SES fathers, partly because of their gender-stereotyped beliefs and partly because of economic necessity, focus more on their provider role (Rank, 2000).

These differences are reflected in family interaction. Parents higher in SES talk to and stimulate their infants and preschoolers more and grant them greater freedom to explore. When their children are older, higher-SES parents use more warmth, explanations, and verbal praise. Commands, such as "You do that because I told you to," as well as criticism and physical punishment occur more often in low-SES households (Hoff, Laursen, & Tardiff, 2002).

The life conditions of families help explain these findings. Lower-SES adults often feel a sense of powerlessness and lack of influence in their relationships beyond the home. For example, at work, they must obey the rules of others in positions of power and authority. When they get home, their parent–child interaction seems to duplicate these experiences, with them in the authority role. Higher levels of stress combined with a stronger belief in the value of physical punishment contribute to low-SES parents' greater use of coercive discipline (Pinderhughes et al., 2000). Higher-SES parents, in contrast, have more control over their own lives. At work, they are used to making independent decisions and convincing others of their point of view. At home, they teach these skills to their children (Greenberger, O'Neil, & Nagel, 1994).

Education also contributes to SES differences in family interaction. Higher-SES parents' interest in providing verbal stimulation and nurturing inner traits is supported by years of schooling, during which they learned to think about abstract, subjective ideas (Uribe, LeVine, & LeVine, 1994). Furthermore, the greater economic security of higher-SES families permits them to devote more time, energy, and material resources to nurturing their children's psychological characteristics.

As early as the second year of life, SES is positively correlated with cognitive and language development. Throughout childhood and adolescence, higher-SES youngsters do better in school (Brody, 1997b; Walker et al., 1994). And they attain higher levels of education, which greatly enhances their opportunities for a prosperous adult life. Researchers believe that differences in family functioning have much to do with these outcomes.

The Impact of Poverty

When families slip into poverty, development is seriously threatened. Consider the case of Zinnia Mae, who grew up in Trackton, a close-knit black community located in a small southeastern American city (Heath, 1990). As unemployment struck Trackton in the 1980s and citizens moved away, 16-year-old Zinnia Mae caught a ride to Atlanta. Two years later, Zinnia Mae was the mother of three children—a daughter and twin boys. She had moved into high-rise public housing.

Each of Zinnia Mae's days was much the same. She watched TV and talked with girlfriends on the phone. The children had only one set meal (breakfast) and otherwise ate whenever they were hungry or bored. Their play space was limited to the living room sofa and a mattress on the floor. Toys consisted of scraps of a blanket, spoons and food cartons, a small rubber ball, a few plastic cars, and a roller skate abandoned in the building. Zinnia Mae's most frequent words were "I'm so tired." She worried about where to find baby-sitters so she could go to the laundry or grocery, and about what she would do if she located the twins' father, who had stopped sending money.

Over the past 30 years, economic changes in the United States and Canada have caused the poverty rate to climb substantially; in recent years, it has dropped and then risen again. Today, nearly 12 percent of the population in the United States and Canada are affected. Those hit hardest are parents under age 25 with young children and elderly people who live alone. Poverty is also magnified among ethnic minorities and women. For example, 17 percent of American and Canadian children are poor, a rate that climbs to 32 percent for Native American children, 34 percent for African-American and Hispanic children, and 60 percent for Canadian Aboriginal children. (Aboriginal peoples in Canada include First Nations, Inuit, and Métis.) For single mothers with preschool children and elderly women on their own, the poverty rate in both countries is nearly 50 percent (Canadian National Council of Welfare, 2002; U.S. Bureau of the Census, 2002c).

Joblessness, a high divorce rate, a lower remarriage rate among women than men, widowhood, and (as we will see later) inadequate government programs to meet family needs are responsible for these disheartening statistics. The child poverty rate is higher than that of any age group. And of all Western nations, the United States has a higher percentage of extremely poor children. These circumstances are particularly worrisome because the earlier poverty begins, the deeper it is, and the longer it lasts, the more devastating its effects on physical and mental health and school achievement (Children's Defense Fund, 2003; Zigler & Hall, 2000).

The constant stresses that accompany poverty gradually weaken the family system. Poor families have many daily hassles—bills to pay, the car breaking down, loss of welfare and unemployment payments, something stolen from the house, to name just a few. When daily crises arise, family members become depressed, irritable, and distracted, and hostile interactions increase (McLoyd, 1998). These outcomes are especially severe in families that must live in poor housing and dangerous neighborhoods—conditions that make everyday existence even more difficult, while reducing social supports that help people cope with economic hardship (Brooks-Gunn & Duncan, 1997).

Besides poverty, another problem—one uncommon 25 years ago—has reduced the life chances of many children and adults. On any given night, approximately 35,000 people in Canada and 350,000 people in the United States have no place

■ Homelessness in the United States and Canada has risen over the past 2 decades. Families like this one travel from place to place in search of employment and a safe and secure place to live. Because of constant stress and few social supports, homeless children are usually behind in development, have health problems, and show poor psychological adjustment.

to live (Pohl, 2001; Wright, 1999). The majority are adults on their own, many of whom suffer from serious mental illness. But nearly 29 percent of the homeless in Canada and 40 percent in the United States are families with children. The rise in homelessness is due to a number of factors, the most important of which is a decline in the availability of government-supported low-cost housing and the release of large numbers of mentally ill people from hospitals and institutions, without an increase in community treatment programs aimed at helping them adjust to ordinary life and get better.

Most homeless families consist of women with children under age 5. Besides health problems (which affect most homeless people), homeless children suffer from developmental delays and serious emotional stress (Bratt, 2002). An estimated 25 to 30 percent of those who are old enough do not go to school. Those who do enroll achieve less well than other poverty-stricken children due to poor attendance and health and emotional difficulties (Vostanis, Grattan, & Cumella, 1997).

© ROBERT BRENNER/PHOTOEDIT

■ The resources neighborhoods offer are important for development and well-being at all ages. Here, adults and children participate in a barn raising for a needy family living in their neighborhood.

Beyond the Family: Neighborhoods, Towns, and Cities

In ecological systems theory, the mesosystem and exosystem underscore that ties between family and community are vital for psychological well-being throughout the lifespan. From our discussion of poverty, perhaps you can see why. In poverty-stricken urban areas, community life is usually disrupted. Families move often, parks and playgrounds are in disarray, and community centers providing organized leisure time activities do not exist. Family violence and child abuse and neglect are greatest in neighborhoods where residents are dissatisfied with their community, describing it as a socially isolated place to live. In contrast, when family ties to the community are strong—as indicated by regular church attendance and frequent contact with friends and relatives—family stress and adjustment problems are reduced (Garbarino & Kostelny, 1993; Magnuson & Duncan, 2002).

● **Neighborhoods.** Let's look closely at the functions of communities in the lives of children and adults by beginning with the neighborhood. What were your childhood experiences like in the yards, streets, and parks surrounding your home? How did you spend your time, whom did you get to know, and how important were these moments to you?

The resources and social ties offered by neighborhoods play an important part in children's development. In several studies, low-SES families were randomly assigned vouchers to move out of public housing into neighborhoods varying widely in affluence. Compared with their peers who remained in poverty-stricken areas, children and youths who moved into low-poverty neighborhoods showed substantially better physical and mental health and school achievement (Goering, 2003; Rubinowitz & Rosenbaum, 2000).

Neighborhood resources have a greater impact on economically disadvantaged than well-to-do young people (McLeod & Shanahan, 1996). Affluent families are not as dependent on their immediate surroundings for social support, education, and leisure pursuits. They can afford to reach beyond the streets near their homes, transporting their children to lessons and entertainment and, if necessary, to better-quality schools in distant parts of the community (Elliott et al., 1996). In low-income neighborhoods, after-school programs that substitute for lack of resources by providing enrichment activities are associated with improved school performance and psychological adjustment in middle childhood (Posner & Vandell, 1994; Vandell & Posner, 1999). Neighborhood organizations and informal social activities predict favorable development in adolescence, including increased self-confidence, school achievement, and educational aspirations (Gonzales et al., 1996).

Neighborhoods also affect adults' well-being. For example, an employed parent who can rely on a neighbor to assist her school-age child in her absence and who lives in an area safe for walking to and from school gains the peace of mind essential for productive work. During late adulthood, neighborhoods become increasingly important because elders spend more time in their homes. Despite the availability of planned housing for older people, about 90 percent remain in regular housing, usually in the same neighborhood where they lived during their working lives (Parmelee & Lawton, 1990; U.S. Bureau of the Census, 2002c). Proximity to relatives and friends is a significant factor in the decision to move or stay put late in life. In the absence of nearby family members, the elderly mention neighbors and nearby friends as resources they rely on most for physical and social support (Hooyman & Kiyak, 2002).

● **Towns and Cities.** Neighborhoods are embedded in towns and cities, which also mold children's and adults' daily lives. A well-known study examined the kinds of community settings children entered and the roles they played in a Midwestern town with a population of 700 (Barker, 1955). Many settings existed, and children were granted important responsibilities—stocking shelves at Kane's Grocery Store, playing in the town band, and operating the snow plow when help was short. They did so alongside adults, who taught them skills needed to become responsible members of the community. Compared with large urban areas, small towns offer stronger connections between settings that influence children's lives. For example, since most citizens know each other and schools serve as centers of community life, contact between teachers and parents occurs often—an important factor in promoting children's academic achievement (Eccles & Harold, 1996).

Like children, adults in small towns penetrate more settings and are more likely to occupy positions of leadership because a greater proportion of residents are needed to meet community needs—for example, by serving on the town council, on the school board, or in other civic roles. In late adulthood, people residing in small towns and suburbs have neighbors who are more willing to provide assistance. As a result, they develop a greater number of warm relationships with nonrelatives (Lawton, 1980). As one 99-year-old resident of a small Midwestern community, living alone and leading an active life, commented, "I don't think I could get along if I didn't have good neighbors." The family next door helps him with grocery shopping, checks each night to make sure his basement light is off (the signal that he is out of the shower and into bed), and looks out in the morning to see that his garage door is raised (the signal that he is up and okay) (Fergus, 1995).

Of course, children and adults in small towns cannot visit museums, go to professional baseball games, or attend orchestra concerts on a regular basis. The variety of settings is reduced compared to large cities. In small towns, however, active involvement in the community is likely to be greater throughout the lifespan. Also, public places in small towns are relatively safe and secure. Responsible adults are present in almost all settings to keep an eye on children. And the elderly feel safer—a strong contributor to how satisfied they are with their place of residence (Parmelee & Lawton, 1990). These conditions are hard to match in today's urban environments.

Think back to the case of Zinnia Mae and her three young children described on page 60. It reveals that community life is especially undermined in high-rise urban housing projects, which are heavily populated by young single mothers who are separated from family and friends by the cost and inconvenience of cross-town transportation. They report intense feelings of loneliness in the small, cramped apartments. Zinnia Mae agreed to tape-record her family interactions over a 2-year period (Heath, 1990). In 500 hours of tape, Zinnia Mae started a conversation with her daughter and twin sons only 18 times. Cut off from community ties and overwhelmed by financial worries and feelings of helplessness, she found herself unable to join in activities with her children. The result was a barren, understimulating environment—one very different from the home and community in which Zinnia Mae herself had grown up.

The Cultural Context

Our discussion in Chapter 1 emphasized that human development can be fully understood only when viewed in its larger cultural context. In the following sections, we expand on this important theme by taking up the macrosystem's role in development. First, we discuss ways that cultural values and practices affect environmental contexts for development. Second, we consider how healthy development depends on laws and government programs that shield people from harm and foster their well-being.

● **Cultural Values and Practices.** Cultures shape family interaction and community settings beyond the home—in short, all aspects of daily life. Many of us remain blind to aspects of our own cultural heritage until we see them in relation to the practices of others.

Each semester, I ask my students to think about the question: Who should be responsible for rearing young children? Here are some typical answers: "If parents decide to have a baby, then they should be ready to care for it." "Most people are not happy about others intruding into family life." These statements reflect a widely held opinion in the United States—that the care and rearing of children, and paying for that care, are the duty of parents, and only parents (Rickel & Becker, 1997; Scarr, 1996). This autonomous view has a long history—one in which independence, self-reliance, and the privacy of family life emerged as central North American values. It is one reason, among others, that the public has been slow to accept the idea of government-supported benefits for all families, such as high-quality child care. This strong emphasis on individualism has also contributed to the large number of American and Canadian families that remain poor, despite the fact that their members are gainfully employed (Pohl, 2002; Zigler & Hall, 2000).

Although many people value independence and privacy, not all citizens share the same values. Some are part of **subcultures**—groups of people with beliefs and customs that differ from those of the larger culture. Many ethnic minority groups in the United States and Canada have cooperative family structures, which help protect their members from the harmful effects of poverty. As the Cultural Influences box on the following page indicates, the African-American tradition of **extended family households,** in which three or more generations live together, is a vital feature of black family life that has enabled its members to survive, despite a long history of prejudice and economic deprivation. Within the extended family, grandparents play meaningful roles in guiding younger generations; adults with employment, marital, or child-rearing difficulties receive assistance and emotional support; and caregiving is enhanced for children and the elderly. Active and involved extended families also characterize other minorities, such as Asian, Native American, Hispanic, and Canadian Aboriginal subcultures (Harrison et al., 1994).

Consider our discussion so far, and you will see that it reflects a broad dimension on which cultures and subcultures

Cultural Influences

The African-American Extended Family

The African-American extended family can be traced to the African heritage of most black Americans. In many African societies, newly married couples do not start their own households. Instead, they live with a large extended family, which assists its members with all aspects of daily life. This tradition of maintaining a broad network of kin ties traveled to the United States during the period of slavery. Since then, it has served as a protective shield against the destructive impact of poverty and racial prejudice on African-American family life. Today, more black than white adults have relatives other than their own children living in the same household. African-American parents also live closer to kin, often establish familylike relationships with friends and neighbors, see more relatives during the week, and perceive them as more important figures in their lives (Wilson et al., 1995).

© MICHAEL SCHWARTZ/THE IMAGE WORKS

■ Strong bonds with extended-family members have helped protect many African-American children growing up under conditions of poverty and single parenthood. This extended family gathers to celebrate the eighty-fifth birthday of their oldest member.

By providing emotional support and sharing income and essential resources, the African-American extended family helps reduce the stress of poverty and single parenthood. In addition, extended-family members often help with child rearing. Furthermore, black adolescent mothers living in extended families are more likely to complete high school and get a job and less likely to be on welfare than are mothers living on their own—factors that in turn benefit children's well-being (Trent & Harlan, 1994).

For single mothers who were very young at the time of their child's birth, extended-family living continues to be associated with more positive adult–child interaction during the preschool years. Otherwise, establishing an independent household with the help of nearby relatives is related to improved child rearing. Perhaps this arrangement permits the more mature mother who has developed effective parenting skills to implement them (Chase-Lansdale, Brooks-Gunn, & Zamsky, 1994). In families rearing adolescents, kinship support increases the likelihood of effective parenting, which is related to adolescents' self-reliance, emotional well-being, and reduced delinquency (Taylor & Roberts, 1995).

Finally, the extended family plays an important role in transmitting African-American culture. Compared with nuclear-family households (which include only parents and their children), extended-family arrangements place more emphasis on cooperation and moral and religious values. And older black adults, such as grandparents and great-grandparents, regard educating children about their African heritage as especially important (Taylor, 2000). These influences strengthen family bonds, protect children's development, and increase the chances that the extended-family lifestyle will carry over to the next generation.

differ: the extent to which *collectivism versus individualism* is emphasized. In **collectivist societies,** people define themselves as part of a group and stress group goals over individual goals. In **individualistic societies,** people think of themselves as separate entities and are largely concerned with their own personal needs (Triandis, 1995). Although individualism tends to increase as cultures become more complex, cross-national differences remain. The United States is strongly individualistic, and Canada falls between the United States and most Western European countries. As we will see in the next section, collectivist versus individualistic values have a powerful impact on a nation's approach to protecting human development and well-being.

● **Public Policies and Lifespan Development.** When widespread social problems arise, such as poverty, homelessness, hunger, and disease, nations attempt to solve them by developing **public policies**—laws and government programs designed to improve current conditions. For example, when poverty increases and families become homeless, a country might decide to build more low-cost housing, raise the minimum wage, and increase welfare benefits. When reports indicate that many children are not achieving well in school, federal and state governments might grant more tax money to school districts and make sure that help reaches pupils who need it most. And when senior citizens have difficulty making ends meet because of inflation, a nation might increase its social security benefits.

Nevertheless, American and Canadian public policies safeguarding children and youths have lagged behind policies for the elderly. And both sets of policies have been especially slow to emerge in the United States.

Policies for Children, Youths, and Families. We have already seen in previous sections that although many North American children fare well, a large number grow up in environments that threaten their development. As Table 2.5 reveals, the United States does not rank well on any key measure of children's health and well-being. Canada fares better, devoting considerably more of its resources to education and health. For example, it grants all its citizens government-funded health care.

The problems of children and youths extend beyond the indicators in the table. For example, approximately 14 percent of American children have no health insurance, making them the largest segment of the U.S. uninsured population (Children's Defense Fund, 2002). Furthermore, the United States and Canada have been slow to move toward national standards and funding for child care. In both countries, much child care is substandard in quality (Goelman et al., 2000; NICHD Early Child Care Research Network, 2000a). In families affected by divorce, weak enforcement of child support payments heightens poverty in mother-headed families. By the time they finish high school, many North-American non-college-bound young people do not have the vocational preparation they need to contribute fully to society. And about 11 percent of U.S. and Canadian adolescents leave high school without a diploma. Those who do not finish their education are at risk for lifelong poverty (Children's Defense Fund, 2002; Statistics Canada, 2002o).

Why have attempts to help children and youths been especially difficult to realize in the United States and, to a lesser extent, in Canada? A complex set of political and economic forces is involved. North American values of self-reliance and privacy have made government hesitant to become involved in family matters. Furthermore, good social programs are expensive, and they must compete for a fair share of a country's economic resources. Children can easily remain unrecognized in this process because they cannot vote or speak out to protect their own interests, as adult citizens do (Zigler & Finn-Stevenson, 1999). Instead, they must rely on the goodwill of others to become an important government priority.

Policies for the Elderly. Until well into the twentieth century, the United States had few policies in place to protect its aging population. For example, social security benefits, which address the income needs of retired citizens who contributed to society through prior employment, were not awarded until the late 1930s. Yet most Western nations had social security systems in place a decade or more earlier; Canada's began in 1927 (DiNitto, 2002). In the 1960s, U.S. federal spending on programs for the elderly rapidly expanded. Medicare, a national health insurance program for older people that pays partial health care costs, was initiated. Canadian elders, like other residents of Canada, benefit from government-supported full health coverage, a system initiated in the 1950s (also called Medicare).

Table 2.5 How Do the United States and Canada Compare to Other Nations on Indicators of Children's Health and Well-Being?

Indicator	U.S. Rank[a]	Canadian Rank[a]	Some Countries the United States and Canada Trail
Childhood poverty[b] (among 23 industrialized nations considered)	19th	19th	Australia, Czech Republic, Germany, Norway, Sweden, Spain
Infant deaths in the first year of life (worldwide)	24th	15th	Hong Kong, Ireland, Singapore, Spain
Teenage pregnancy rate (among 45 industrialized nations considered)	45th	25th	Albania, Australia, Czech Republic, Denmark, Poland, Netherlands
Expenditures on education as percentage of gross domestic product (among 22 industrialized nations considered)	10th	6th	*For Canada:* Israel, Sweden *For the U.S.:* Australia, France, New Zealand, Sweden
Expenditures on health as a percentage of gross domestic product[c] (among 22 industrialized nations considered)	16th	4th	*For Canada:* Iceland, Switzerland, France *For the U.S.:* Austria, Australia, Hungary, New Zealand

[a]1 = highest rank

[b]The U.S. and Canadian childhood poverty rates of 17 percent greatly exceed those of any of these nations. For example, the rate is 12 percent in Australia, 6 percent in the Czech Republic, 4 percent in Norway, and 2.5 percent in Sweden.

[c]Gross domestic product is the value of all goods and services produced by a nation during a specified time period. It provides an overall measure of a nation's wealth.

Sources: Perie et al., 2000; Singh & Darroch, 2000; United Nations Children's Fund, 2000; United Nations Development Programme, 2002; U.S. Bureau of the Census, 2002b.

■ Overall, senior citizens in the United States are better off economically than are children. Nevertheless, many older adults—especially women and ethnic minorities—are poverty stricken. American elders are less well off than Canadian elders, who benefit from Canada's more generous income supplements and full health-care coverage.

Social security and Medicare consume 96 percent of the U.S. federal budget for the elderly; only 4 percent is devoted to other programs. Consequently, U.S. programs for the elderly have been criticized for neglecting social services (Hooyman & Kiyak, 2002). To meet this need, a national network for planning, coordinating, and delivering assistance to the aged has been established. Approximately 700 Area Agencies on Aging operate at regional and local levels, assessing community needs and offering communal and home-delivered meals, self-care education, elder abuse prevention, and a wide range of other social services. However, limited funding means that the Area Agencies help far too few people in need.

As noted earlier, many senior citizens—especially women, ethnic minorities, and those living alone—remain in dire economic straits. Those who had interrupted employment histories, held jobs without benefits, or suffered lifelong poverty are not eligible for social security. Although all Americans age 65 and older are guaranteed a minimum income, it is less than the poverty line—the amount judged necessary for bare subsistence by the federal government. Furthermore, social security benefits are rarely enough to serve as a sole source of retirement income; they must be supplemented through other pensions and family savings. But a substantial percentage of U.S. aging citizens do not have access to these resources. Therefore, they are more likely to be among the "near poor" than are other age groups (Koff & Park, 1999). Because of Canada's more generous income supplements as part of its Old Age Security Program, far fewer Canadian than U.S. elders are poverty stricken.

Nevertheless, the U.S. aging population is financially much better off now than in the past. Today, the elderly are a large, powerful, well-organized constituency, far more likely than children or low-income families to attract the support of politicians. As a result, the number of aging poor has declined from 1 out of 3 people in 1960 to 1 out of 10 in the early twenty-first century (U.S. Bureau of the Census, 2002c). And senior citizens are healthier and more independent than ever before. Still, as Figure 2.7 shows, the elderly in the United States are not as well off as those in other Western nations.

● **Looking Toward the Future.** Despite the worrisome state of many children, families, and aging citizens, progress is being made in improving their condition. Throughout this book, we will discuss many successful programs that could be expanded. Also, growing awareness of the gap between what we know and what we do to better people's lives has led experts in human development to join with concerned citizens as advocates for more effective policies. As a result, several influential interest groups with the well-being of children or the elderly as their central purpose have emerged. Among the most vigorous in the United States are the Children's Defense Fund and the American Association of Retired Persons.

A private, nonprofit organization founded by Marion Wright Edelman in 1973, the *Children's Defense Fund* engages in research, public education, legal action, drafting of legislation, congressional testimony, and community organizing. Each year, it publishes *The State of America's Children,* which provides a comprehensive analysis of the current condition of children, government-sponsored programs serving them, and proposals for improving child and family programs.

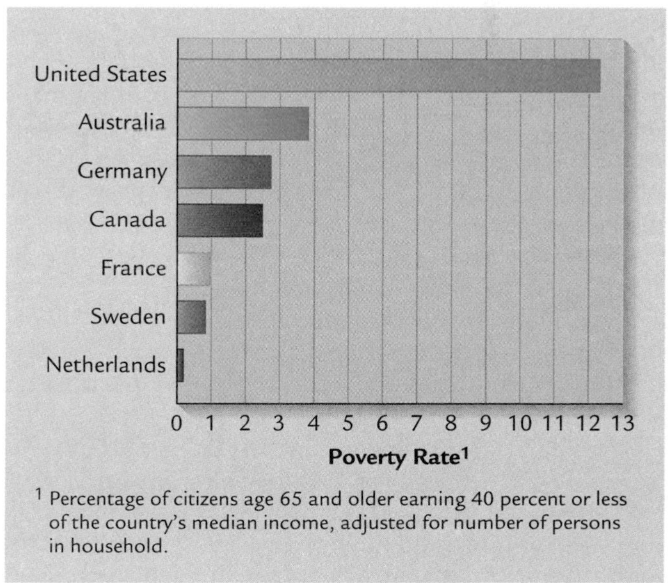

■ **FIGURE 2.7 Percentage of citizens age 65 and older living in poverty in seven industrialized nations.** Among countries listed, public expenditures on social security and other income guarantees for senior citizens are highest in the Netherlands, lowest in the United States (United Nations Development Programme, 2002). Consequently, poverty among the elderly is virtually nonexistent in the Netherlands but greatly exceeds that of other nations in the United States.

■ In 1973, Marion Wright Edelman founded the Children's Defense Fund, a private, nonprofit organization that provides a strong, effective voice for American children, who cannot vote, lobby, or speak for themselves. Edelman continues to serve as president of the Children's Defense Fund today.

Nearly half of Americans over age 50 (both retired and employed)—more than 35 million people—are members of the *American Association of Retired Persons (AARP).* Founded by Ethel Percy Andrus in 1958, AARP has a large and energetic lobbying staff that works for increased government benefits of all kinds for the aged. Each year, it releases the *AARP Public Policy Agenda,* which forms the basis for advocacy activities in diverse areas, including income, health care, social services, housing, and personal and legal rights. Among AARP's programs is an effort to mobilize elderly voters, an initiative that keeps lawmakers highly sensitive to policy proposals affecting older Americans.

Besides strong advocacy, the design and implementation of public policies that enhance human development depend on policy-relevant research that documents needs and evaluates programs to spark improvements. Today, more researchers are collaborating with community and government agencies to enhance the social relevance of their investigations. They are also doing a better job of disseminating their findings to the public, through television documentaries, newspaper stories, magazine articles, and direct reports to government officials (Denner et al., 1999). As a result, they are helping to create a sense of immediacy about the condition of children, families, and the aged that is necessary to spur a society into action.

Ask Yourself

REVIEW

Links between family and community foster development throughout the lifespan. Cite examples and research findings from our discussion that support this idea.

APPLY

On one of your trips to the local shopping center, you see a father getting very angry at his young son. Using ecological systems theory, list as many factors as you can that might account for the father's behavior.

APPLY

Check your local newspaper and one or two national news magazines to see how often articles appear on the condition of children, families, and the aged. Why is it important for researchers to communicate with the general public about the well-being of these sectors of the population?

CONNECT

How does poverty affect the functioning of the family system, thereby placing all domains of development at risk?

www.

Understanding the Relationship Between Heredity and Environment

So far in this chapter, we have discussed a wide variety of hereditary and environmental influences, each of which has the power to alter the course of development. Yet people who are born into the same family (and who therefore share genes and environments) often are quite different in characteristics. We also know that some individuals are affected more than others by their homes, neighborhoods, and communities. Cases exist in which a person provided with all the advantages in life does poorly, whereas another exposed to the worst of rearing conditions does well. How do scientists explain the impact of heredity and environment when they seem to work in so many different ways?

All contemporary researchers agree that both heredity and environment are involved in every aspect of development. But for polygenic traits (due to many genes) such as intelligence and personality, scientists are a long way from knowing the precise hereditary influences involved. They must study the impact of genes on these complex characteristics indirectly.

Some believe that it is useful and possible to answer the question of *how much* each factor contributes to differences among people. A growing consensus, however, regards that question as unanswerable. These investigators view genetic and environmental influences as inseparable. The important question, they maintain, is how nature and nurture work together. Let's consider each position in turn.

The Question, "How Much?"

Two methods—heritability estimates and concordance rates—are used to infer the role of heredity in complex human characteristics. Let's look closely at the information these procedures yield, along with their limitations.

● **Heritability.** **Heritability estimates** measure the extent to which individual differences in complex traits, such as intelligence and personality, in a specific population are due to genetic factors. Researchers have obtained heritabilities for intelligence and a variety of personality characteristics. We will take a brief look at their findings here and return to them in later chapters, when we consider these topics in greater detail. Heritability estimates are obtained from **kinship studies,** which compare the characteristics of family members. The most common type of kinship study compares identical twins, who share all their genes, with fraternal twins, who share only some. If people who are genetically more alike are also more similar in intelligence and personality, then the researcher assumes that heredity plays an important role.

Kinship studies of intelligence provide some of the most controversial findings in the field of human development. Some experts claim a strong genetic influence, whereas others believe that heredity is barely involved. Currently, most kinship findings support a moderate role for heredity. When many twin studies are examined, correlations between the scores of identical twins are consistently higher than those of fraternal twins. In a summary of more than 13,000 twin pairs, the correlation for intelligence was .86 for identical twins and .55 for fraternal twins (Scarr, 1997).

Researchers use a complex statistical procedure to compare these correlations, arriving at a heritability estimate ranging from 0 to 1.00. The value for intelligence is about .50 for child and adolescent twin samples in Western industrialized nations. This indicates that differences in genetic makeup can explain half the variation in intelligence (Plomin, 1994b). However, heritability increases in adulthood, with some estimates as high as .80. As we will see later, one explanation is that adults exert greater control over their intellectual experiences than do children (McClearn et al., 1997; McGue & Christensen, 2002). The intelligence of adopted children is more strongly related to the scores of their biological parents than to those of their adoptive parents, offering further support for the role of heredity (Horn, 1983; Scarr & Weinberg, 1983).

Heritability research also reveals that genetic factors are important in temperament and personality. For frequently studied traits, such as sociability, emotional expressiveness, and activity level, heritability estimates obtained on child, adolescent, and young adult twins are moderate, at .40 to .50 (Rothbart & Bates, 1998). Unlike intelligence, however, heritability of personality does not increase over the lifespan (Brody, 1997).

© T. K. WANSTALL/THE IMAGE WORKS

■ Identical twins Bob and Bob were separated by adoption shortly after birth and not reunited until adulthood. The two Bobs discovered they were alike in many ways. Both hold bachelor's degrees in engineering, are married to teachers named Brenda, wear glasses, have mustaches, smoke pipes, and are volunteer firemen. The study of identical twins reared apart reveals that heredity contributes to many psychological characteristics. Nevertheless, not all separated twins match up as well as this pair, and generalizing from twin evidence to the population is controversial.

● **Concordance.** A second measure used to infer the contribution of heredity to complex characteristics is the **concordance rate.** It refers to the percentage of instances in which both twins show a trait when it is present in one twin. Researchers typically use concordance to study the contribution of heredity to emotional and behavior disorders, which can be judged as either present or absent.

A concordance rate ranges from 0 to 100 percent. A score of 0 indicates that if one twin has the trait, the other one never has it. A score of 100 means that if one twin has the trait, the other one always has it. When a concordance rate is much higher for identical twins than for fraternal twins, heredity is believed to play a major role. Twin studies of schizophrenia (a disorder involving delusions and hallucinations, difficulty distinguishing fantasy from reality, and irrational and inappropriate behaviors) and severe depression show this pattern of findings. In the case of schizophrenia, the concordance rate for identical twins is 50 percent, that for fraternal twins only 18 percent. For severe depression, the figures are 69 percent and 25 percent (Gottesman, 1991; McGuffin & Sargeant, 1991). Once again, adoption studies support these results. Biological relatives of schizophrenic and depressed adoptees are more likely than adoptive relatives to share the same disorder (Bock & Goode, 1996; Loehlin, Willerman, & Horn, 1988).

Taken together, concordance and adoption research suggests that the tendency for schizophrenia and depression to run in families is partly due to genetic factors. However, we also know that environment is involved, since the concordance rate for identical twins would have to be 100 percent for heredity to be the only influence.

● **Limitations of Heritability and Concordance.** Serious questions have been raised about the accuracy of heritability estimates and concordance rates. First, each value refers only to the particular population studied and its unique range of genetic and environmental influences. For example, imagine a country in which people's home, school, and community experiences are very similar. Under these conditions, individual differences in intelligence and personality would be largely genetic, and heritability estimates would be close to 1.00. Conversely, the more environments vary, the greater their opportunity to account for individual differences, and the lower heritability estimates are likely to be (Plomin, 1994a).

Second, the accuracy of heritability estimates and concordance rates depends on the extent to which the twin pairs used reflect genetic and environmental variation in the population. Yet most twins studied are reared together under highly similar conditions. Even when separated twins are available, social service agencies often place them in advantaged homes that are alike in many ways (Eisenberg, 1998). Because the environments of most twin pairs are less diverse than those of the general population, heritability estimates are likely to exaggerate the role of heredity.

Heritability estimates are controversial because they can easily be misapplied. For example, high heritabilities have been used to suggest that ethnic differences in intelligence, such as the poorer performance of black children compared to white children, have a genetic basis (Jensen, 1969, 1985, 1998). Yet this line of reasoning is widely regarded as incorrect. Heritabilities computed on mostly white twin samples do not tell us what is responsible for test score differences between ethnic groups. We have already seen that large economic and cultural differences are involved. In Chapter 9, we will discuss research indicating that when black children are adopted into economically advantaged homes at an early age, their scores are well above average and substantially higher than those of children growing up in impoverished families.

Perhaps the most serious criticism of heritability estimates and concordance rates has to do with their usefulness. Although they are interesting statistics, they give us no precise information about how intelligence and personality develop or how individuals might respond to environments designed to help them develop as far as possible (Bronfenbrenner & Ceci, 1994; Wachs, 1999). Indeed, the heritability of intelligence is higher in advantaged homes and communities, which permit children to actualize their genetic endowment. In disadvantaged environments, children are prevented from realizing their potential. Consequently, enhancing their experiences through interventions—such as parent education and high-quality preschool or child care—has a greater impact on development (Bronfenbrenner & Morris, 1998).

According to one group of experts, heritability estimates have too many problems to yield any firm conclusions about the relative strength of nature and nurture (Collins et al., 2000). Although these statistics confirm that heredity contributes to complex traits, they do not tell us how environment can modify genetic influences.

The Question, "How?"

Today, most researchers view development as the result of a dynamic interplay between heredity and environment. How do nature and nurture work together? Several concepts shed light on this question.

● **Reaction Range.** The first of these ideas is **range of reaction,** or each person's unique, genetically determined response to the environment (Gottesman, 1963). Let's explore this idea in Figure 2.8. Reaction range can apply to any characteristic; here it is illustrated for intelligence. Notice that when environments vary from extremely unstimulating to highly enriched, Ben's intelligence increases steadily, Linda's rises sharply and then falls off, and Ron's begins to increase only after the environment becomes modestly stimulating.

Reaction range highlights two important points. First, it shows that because each of us has a unique genetic makeup, we respond differently to the same environment. Note in Figure 2.8 how a poor environment results in similarly low scores for all three individuals. But Linda is by far the best-performing child when environments provide an intermediate level of stimulation. And when environments are highly enriched, Ben does best, followed by Ron, both of whom now outperform

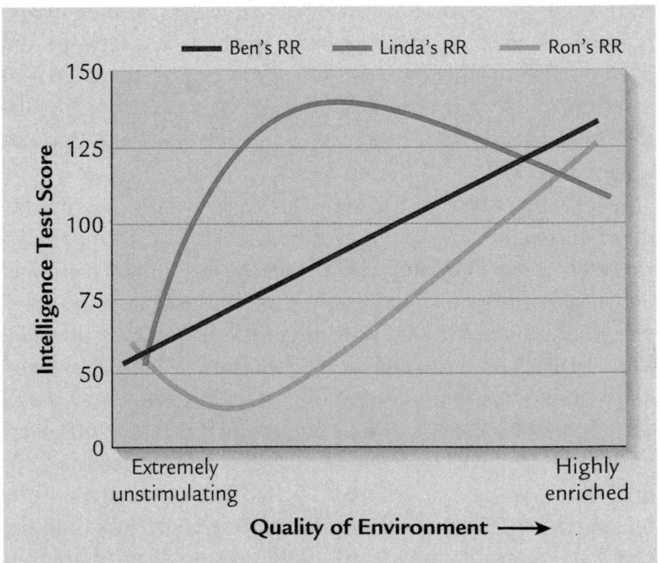

■ **FIGURE 2.8 Intellectual ranges of reaction (RR) for three children in environments that vary from extremely unstimulating to highly enriched.** Each child, due to his or her genetic makeup, responds differently as quality of the environment changes. Ben's intelligence increases steadily, Linda's rises sharply and then falls off, and Ron's begins to increase only after the environment becomes modestly stimulating. (Adapted from Wahlsten, 1994.)

Linda. Second, sometimes different genetic–environmental combinations can make two people look the same! For example, if Linda is reared in a minimally stimulating environment, her score will be about 100—average for people in general. Ben and Ron can also obtain this score, but to do so, they must grow up in a fairly enriched home. In sum, range of reaction reveals that unique blends of heredity and environment lead to both similarities and differences in behavior (Wahlsten, 1994).

● **Canalization.** The concept of canalization provides another way of understanding how heredity and environment combine. **Canalization** is the tendency of heredity to restrict the development of some characteristics to just one or a few outcomes. A behavior that is strongly canalized follows a genetically set growth plan, and only powerful environmental forces can change it (Waddington, 1957). For example, infant perceptual and motor development seems to be strongly canalized because all normal human babies eventually roll over, reach for objects, sit up, crawl, and walk. It takes extreme conditions to modify these behaviors. In contrast, intelligence and personality are less strongly canalized, since they vary much more with changes in the environment.

When we look at the behaviors constrained by heredity, we can see that canalization is highly adaptive. Through it, nature ensures that children will develop certain species-typical skills under a wide range of rearing conditions, thereby promoting survival.

● **Genetic–Environmental Correlation.** Nature and nurture work together in still another way. Several investigators point out that a major problem in trying to separate heredity and environment is that they are often correlated (Plomin, 1994b; Scarr & McCartney, 1983). According to the concept of **genetic–environmental correlation,** our genes influence the environments to which we are exposed. The way this happens changes with age.

Passive and Evocative Correlation. At younger ages, two types of genetic–environmental correlation are common. The first is called *passive* correlation because the child has no control over it. Early on, parents provide environments influenced by their own heredity. For example, parents who are good athletes emphasize outdoor activities and enroll their children in swimming and gymnastics. Besides getting exposed to an "athletic environment," the children may have inherited their parents' athletic ability. As a result, they are likely to become good athletes for both genetic and environmental reasons.

The second type of genetic–environmental correlation is *evocative.* Children evoke responses that are influenced by the child's heredity, and these responses strengthen the child's original style. For example, an active, friendly baby is likely to receive more social stimulation than a passive, quiet infant. And a cooperative, attentive child probably receives more patient and sensitive interactions from parents than an inattentive, distractible child.

©MITCH WOJNAROWICZ/THE IMAGE WORKS

■ This mother is an accomplished skier who exposes her children to skiing. In addition, the children may have inherited their mother's athletic talent. When heredity and environment are correlated, they jointly foster the same capacities, and the influence of one cannot be separated from the influence of the other.

Active Correlation. At older ages, *active* genetic–environmental correlation becomes common. As children extend their experiences beyond the immediate family and are given the freedom to make more choices, they actively seek environments that fit with their genetic tendencies. The well-coordinated, muscular child spends more time at after-school sports, the musically talented youngster joins the school orchestra and practices his violin, and the intellectually curious child is a familiar patron at her local library.

This tendency to actively choose environments that complement our heredity is called **niche-picking** (Scarr & McCartney, 1983). Infants and young children cannot do much niche-picking because adults select environments for them. In contrast, older children, adolescents, and adults are much more in charge of their environments. The niche-picking idea explains why pairs of identical twins reared apart during childhood and later reunited may find, to their great surprise, that they have similar hobbies, food preferences, and vocations—a trend that is especially marked when twins' environmental opportunities are similar (Bouchard et al., 1990; Plomin, 1994b). Niche-picking also helps us understand longitudinal findings that identical twins become somewhat more alike, and fraternal twins and adopted siblings less alike, in intelligence with age (Loehlin, Horn, & Willerman, 1997; McGue & Bouchard, 1998). The influence of heredity and environment is not constant but changes over time. With age, genetic factors may become more important in determining the environments we experience and choose for ourselves.

Biology & Environment

Uncoupling Genetic–Environmental Correlations for Mental Illness and Antisocial Behavior

Diagnosed with schizophrenia, Lars's and Sven's biological mothers had such difficulty functioning in everyday life that each gave up her infant son for adoption. Lars had the misfortune of being placed with adoptive parents who, like his biological mother, were mentally ill. His home life was chaotic, and his parents were punitive and neglectful. Sven's adoptive parents, in contrast, were psychologically healthy and reared him with love, patience, and consistency.

Lars displays a commonly observed genetic–environmental correlation: a predisposition for schizophrenia coupled with maladaptive parenting. Will he be more likely than Sven, whose adoption *uncoupled* this adverse genotype–environment link, to develop mental illness? In a large Finnish adoption study, nearly 200 adopted children of schizophrenic mothers were followed into adulthood (Tienari et al., 1994). Those (like Sven) who were reared by healthy adoptive parents showed little mental illness—no more than a control group with healthy biological and adoptive parents. In contrast, psychological impairments piled up in adoptees (like Lars) with both disturbed biological and adoptive parents. These children were considerably more likely to develop mental illness than were controls whose biological parents were healthy but who were being reared by severely disturbed adoptive parents.

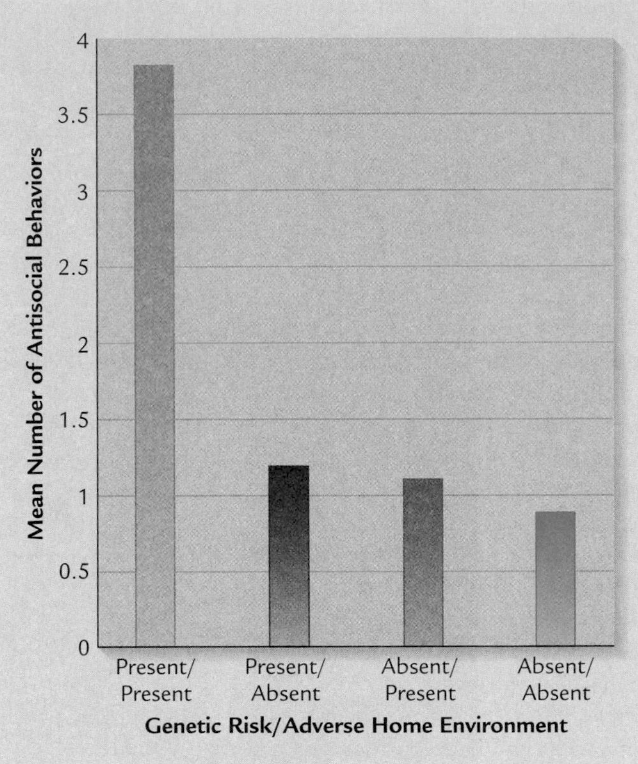

■ **FIGURE 2.9**
Antisocial behavior of adoptees varying in genetic and home-environment risk for criminality. Adolescent adoptees at genetic risk for criminality displayed a high rate of antisocial behavior only when reared in unfavorable homes. When reared in favorable homes, they did not differ from adoptees at no genetic risk. (Adapted from Cadoret, Cain, & Crowe, 1983.)

Similar findings emerged in several American and Swedish adoption studies addressing genetic and environmental contributions to antisocial behavior (Bohman, 1996; Yates, Cadoret, & Troughton, 1999). As Figure 2.9 shows, adopted infants whose biological mothers were imprisoned criminal offenders displayed a high rate of antisocial behavior in adolescence only when reared in unfavorable homes, as indicated by adoptive parents or siblings with severe adjustment problems. In families free of psychological disturbance, adoptees with a predisposition to criminality did not differ from those without this genetic background.

In sum, the chances that genes for psychological disorder will be expressed are far greater when child rearing is maladaptive. Healthy-functioning families seem to promote healthy development in children, despite a genetic risk associated with mental illness or criminality in a biological parent.

● **Environmental Influences on Gene Expression.** Notice how, in the concepts just considered, heredity is granted priority. In range of reaction, it *limits* responsiveness to varying environments. In canalization, it *restricts* the development of certain behaviors. Similarly, some theorists regard genetic–environmental correlation as entirely driven by genetics (Harris, 1998; Rowe, 1994). They believe that children's genetic makeup causes them to receive, evoke, or seek experiences that realize their inborn tendencies. Others argue that heredity does not dictate children's experiences or development in a rigid way. For example, as the Biology and Environment box above illustrates, parents and other caring adults can *uncouple* unfavorable genetic–environmental correlations. They often provide children with experiences that modify the expression of heredity, yielding favorable outcomes.

Accumulating evidence reveals that the relationship between heredity and environment is not a one-way street, from genes to environment to behavior. Instead, like other system

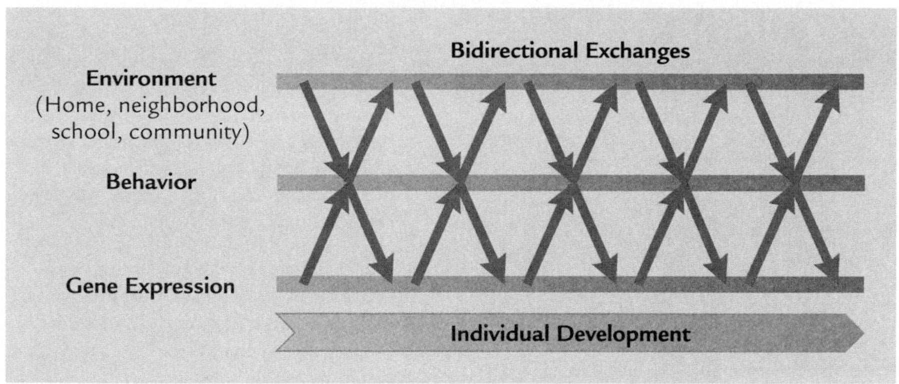

FIGURE 2.10 The epigenetic framework. Development takes place through ongoing, bidirectional exchanges between heredity and all levels of the environment. Genes affect behavior and experiences. Experiences and behavior also affect gene expression. (Adapted from Gottlieb, 2000.)

influences considered in this and the previous chapter, it is bidirectional: Genes affect people's behavior and experiences, but their experiences and behavior also affect gene expression (Gottlieb, 2000). Stimulation—both internal to the person (activity within the cytoplasm of the cell) and external to the person (home, neighborhood, school, and community)—triggers gene activity.

Researchers call this view of the relationship between heredity and environment the *epigenetic framework* (Gottlieb, 1998, 2002). It is depicted in Figure 2.10. **Epigenesis** means development resulting from ongoing, bidirectional exchanges between heredity and all levels of the environment. To illustrate, granting a baby a healthy diet increases brain growth, leading to new connections between nerve cells, which transform gene expression. This opens the door to new gene–environment exchanges—for example, advanced exploration of objects and interaction with caregivers, which further enhance brain growth and gene expression. These ongoing, bidirectional influences foster cognitive and social development. In contrast, harmful environments can dampen gene expression, at times so profoundly that later experiences can do little to change characteristics (such as intelligence) that were flexible to begin with.

A major reason that researchers are interested in the nature–nurture issue is that they want to improve environments so people can develop as far as possible. The concept of epigenesis reminds us that development is best understood as a series of complex exchanges between nature and nurture. Although people cannot be changed in any way we might desire, environments can modify genetic influences. The success of any attempt to improve development depends on the characteristics we want to change, the genetic makeup of the individual, and the type and timing of our intervention.

Ask Yourself

REVIEW

What is epigenesis, and how does it differ from range of reaction and genetic–environmental correlation? Provide an example of epigenesis.

APPLY

Bianca's parents are accomplished musicians. At age 4, Bianca began taking piano lessons. By age 10, she was accompanying the school choir. At age 14, she asked if she could attend a special music high school. Explain how genetic–environmental correlation promoted Bianca's talent.

CONNECT

One possible reason for the increase in heritability of intelligence from childhood to adulthood is that compared with children, adults exert more control over their own intellectual experiences (see page 68). What type of genetic–environmental correlation is involved? What light does this explanation shed on whether heritability estimates yield "pure" measures of genetic influences on human traits?

www.

Summary

Genetic Foundations

What are genes, and how are they transmitted from one generation to the next?

- Each individual's **phenotype,** or directly observable characteristics, is a product of both **genotype** and environment.

Chromosomes, rodlike structures within the cell nucleus, contain our hereditary endowment. Along their length are **genes,** segments of **DNA** that send instructions for making a rich assortment of proteins to the cytoplasm of the cell—a process that makes us distinctly human and influences our development and characteristics. We share most of our genetic makeup with other mammals, especially primates.

- **Gametes,** or sex cells, are produced by the process of cell division known as **meiosis.** Because each individual receives a unique set of genes from each parent, meiosis ensures that children will be genetically different. Once sperm and ovum unite, the resulting **zygote** starts to develop into a complex human being through cell duplication, or **mitosis.**

- If the fertilizing sperm carries an X chromosome, the child will be a girl; if it contains a Y chromosome, a boy will be born. **Fraternal,** or **dizygotic, twins** result when two ova are released from the mother's ovaries and each is fertilized. In contrast, **identical,** or **monozygotic,** twins develop when a zygote divides in two during the early stages of cell duplication.

Describe various patterns of genetic inheritance.

- **Dominant–recessive** and **codominant** relationships are patterns of inheritance that apply to traits controlled by single genes. In dominant–recessive inheritance, **heterozygous** individuals with one recessive gene are **carriers** of the recessive trait.

- When recessive disorders are **X-linked** (carried on the X chromosome), males are more likely to be affected. **Genetic imprinting** is a newly discovered pattern of inheritance in which one parent's gene is activated, regardless of its makeup.

- Unfavorable genes arise from **mutations,** which can occur spontaneously or be induced by hazardous environmental agents.

- Human traits that vary continuously, such as intelligence and personality, are **polygenic,** or influenced by many genes. Because the genetic principles involved are unknown, scientists must study the influence of heredity on these characteristics indirectly.

Describe major chromosomal abnormalities, and explain how they occur.

- Most chromosomal abnormalities are due to errors in meiosis. The most common chromosomal disorder is Down syndrome, which results in physical defects and mental retardation. Disorders of the **sex chromosomes** are generally milder than defects of the **autosomes.** Contrary to popular belief, males with XYY syndrome are not prone to aggression. Studies of children with triple X, Klinefelter, and Turner syndromes reveal that adding to or subtracting from the usual number of X chromosomes leads to specific intellectual problems.

Reproductive Choices

What procedures can assist prospective parents in having healthy children?

- **Genetic counseling** helps couples at risk for giving birth to children with genetic abnormalities decide whether or not to conceive. **Prenatal diagnostic methods** make early detection of genetic problems possible. Although reproductive technologies, such as donor insemination, in vitro fertilization, surrogate motherhood, and postmenopausal-assisted childbirth, permit many individuals to become parents who otherwise would not, they raise serious legal and ethical concerns.

- Many parents who cannot conceive or who have a high likelihood of transmitting a genetic disorder decide to adopt. Adopted children have more learning and emotional problems than children in general. However, warm, sensitive parenting predicts favorable development, and in the long run most adopted children fare well.

Environmental Contexts For Development

Describe the social systems perspective on family functioning, along with aspects of the environment that support family well-being and development.

- Just as complex as heredity are the environments in which human development takes place. The family is the first and foremost context for development. Ecological systems theory emphasizes that the behaviors of each family member affect those of others. The family system is also dynamic, constantly adjusting to new events, to developmental changes in its members, and to societal change.

- Despite these variations, one source of consistency in family functioning is **socioeconomic status (SES).** Higher-SES families tend to be smaller, to place greater emphasis on nurturing psychological traits, and to promote warm, verbally stimulating interaction with children. Lower-SES families often stress external characteristics and engage in more restrictive child rearing. Development is seriously undermined by poverty and homelessness.

- Supportive ties to the surrounding environment foster well-being throughout the lifespan. Communities that encourage constructive leisure activities, warm interactions among residents, connections between settings, and children and adults' active participation enhance development.

- The values and practices of cultures and **subcultures** affect all aspects of daily life. **Extended family households,** in which three or more generations live together, are common among ethnic minorities. They protect development under conditions of high life stress.

- In the complex world in which we live, favorable development depends on **public policies.** Effective social programs are influenced by many factors, including cultural values that stress **collectivism** over **individualism,** a nation's economic resources, and organizations and individuals that work for an improved quality of life. American and Canadian policies safeguarding children and their families are not as well developed as those safeguarding the elderly.

Understanding the Relationship Between Heredity and Environment

Explain the various ways heredity and environment may combine to influence complex traits.

- Some researchers believe it is useful and possible to determine "how much" heredity and environment contribute to individual differences. These investigators compute **heritability estimates** and **concordance rates** from **kinship studies.** Although these measures show that genetic factors influence such traits as intelligence and personality, their accuracy and usefulness have been challenged.

- Most researchers view development as the result of a dynamic interplay between nature and nurture and ask "how" heredity and environment work together. The concepts of **range of reaction, canalization, genetic–environmental correlation, niche-picking,** and **epigenesis** remind us that development is best understood as a series of complex exchanges between nature and nurture that change over the lifespan.

Important Terms and Concepts

autosomes (p. 46)
canalization (p. 69)
carrier (p. 46)
chromosomes (p. 44)
codominance (p. 47)
collectivist societies (p. 63)
concordance rate (p. 67)
deoxyribonucleic acid (DNA) (p. 44)
dominant–recessive inheritance (p. 46)
epigenesis (p. 71)
extended family household (p. 62)
fraternal, or dizygotic, twins (p. 46)
gametes (p. 45)
gene (p. 44)
genetic counseling (p. 52)
genetic–environmental correlation (p. 69)
genetic imprinting (p. 49)
genotype (p. 44)
heritability estimate (p. 67)
heterozygous (p. 46)
homozygous (p. 46)
identical, or monozygotic, twins (p. 46)
individualistic societies (p. 63)
kinship studies (p. 67)
meiosis (p. 45)
mitosis (p. 45)
mutation (p. 49)
niche-picking (p. 69)
phenotype (p. 44)
polygenic inheritance (p. 50)
prenatal diagnostic methods (p. 52)
public policies (p. 63)
range of reaction (p. 68)
sex chromosomes (p. 46)
socioeconomic status (SES) (p. 59)
subculture (p. 62)
X-linked inheritance (p. 49)
zygote (p. 45)

FYI For Further Information and Help

Consult the Companion Website for *Development Through the Lifespan, Third Edition,* (www.ablongman.com/berk) where you will find the following resources for this chapter:

- **Chapter Objectives**
- **Flashcards** for studying important terms and concepts
- **Annotated Weblinks** to guide you in further research
- **Ask Yourself** questions, which you can answer and then check against a sample response
- **Suggested Readings**
- **Practice Test** with immediate scoring and feedback

3

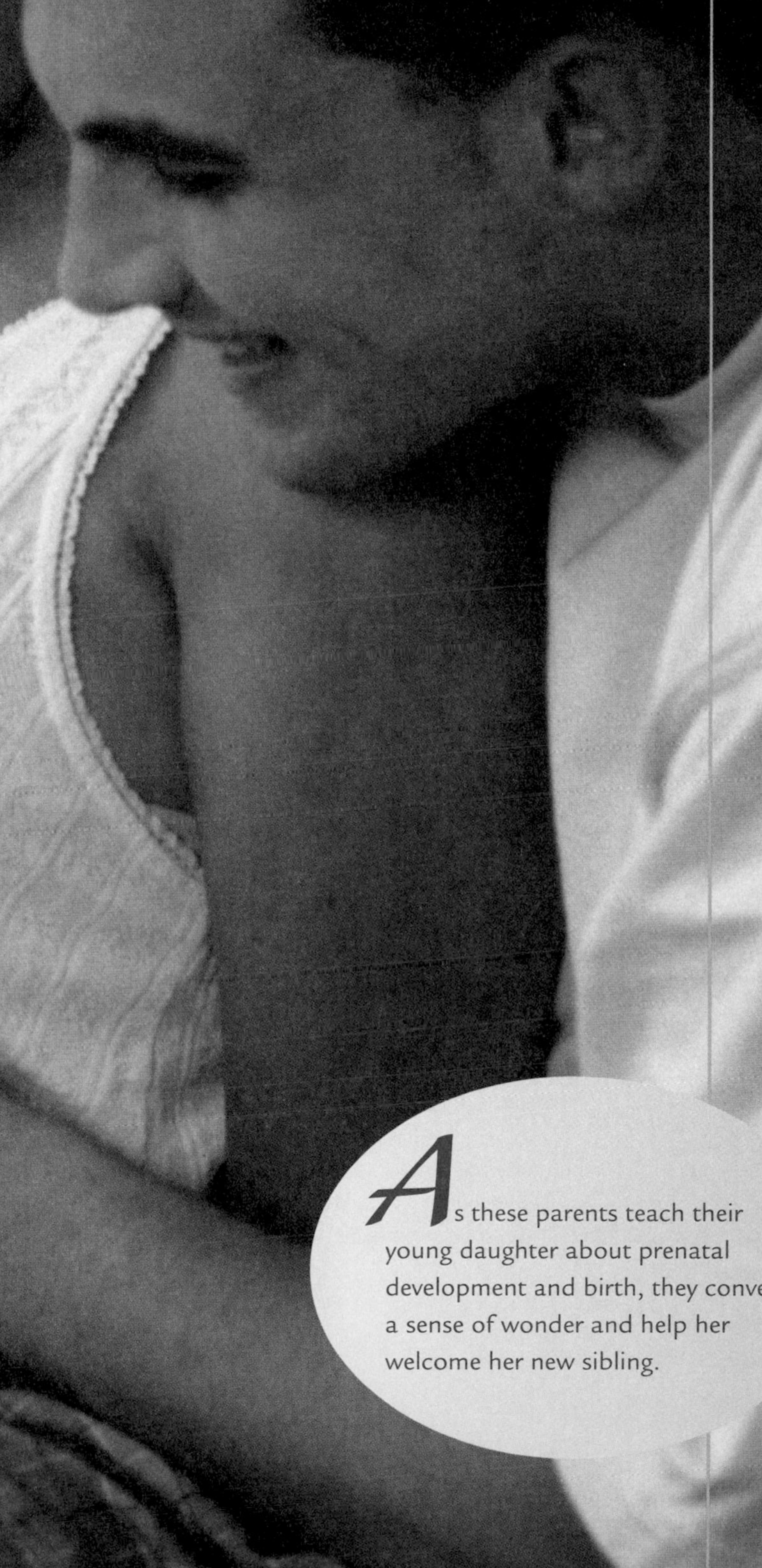

©JON FEINGERSH/CORBIS

As these parents teach their young daughter about prenatal development and birth, they convey a sense of wonder and help her welcome her new sibling.

Prenatal Development, Birth, and the Newborn Baby

Prenatal Development
Conception ● Period of the Zygote ● Period of the Embryo ● Period of the Fetus

Prenatal Environmental Influences
Teratogens ● Other Maternal Factors ● The Importance of Prenatal Health Care

■ A Lifespan Vista: The Prenatal Environment and Health in Later Life

Childbirth
The Stages of Childbirth ● The Baby's Adaptation to Labor and Delivery ● The Newborn Baby's Appearance ● Assessing the Newborn's Physical Condition: The Apgar Scale

Approaches to Childbirth
Natural, or Prepared, Childbirth ● Home Delivery

Medical Interventions
Fetal Monitoring ● Labor and Delivery Medication ● Cesarean Delivery

Preterm and Low-Birth-Weight Infants
Preterm versus Small for Date ● Consequences for Caregiving ● Interventions for Preterm Infants

■ Social Issues: A Cross-National Perspective on Health Care and Other Policies for Parents and Newborn Babies

Understanding Birth Complications

The Newborn Baby's Capacities
Newborn Reflexes ● Newborn States ● Sensory Capacities ● Neonatal Behavioral Assessment

■ Biology & Environment: The Mysterious Tragedy of Sudden Infant Death Syndrome

Adjusting to the New Family Unit

After months of wondering if the time in their own lives was right, Yolanda and Jay decided to have a baby. I met them one fall in my child development class, when Yolanda was just 2 months pregnant. Both were full of questions: "How does the baby grow before birth? When are different organs formed? Has its heart begun to beat? Can it hear, feel, or sense our presence?"

Most of all, Yolanda and Jay wanted to do everything possible to make sure their baby would be born healthy. At first, they believed that the uterus completely shielded the developing organism from any dangers in the environment. All babies born with problems, they thought, had unfavorable genes. After browsing through several pregnancy books, Yolanda and Jay realized they were wrong. Yolanda started to wonder about her diet. And she asked me whether an aspirin for a headache, a glass of wine at dinner, or a few cups of coffee during study hours might be harmful.

In this chapter, we answer Yolanda's and Jay's questions, along with a great many more that scientists have asked about the events before birth. First, we trace prenatal development, paying special attention to environmental supports for healthy growth, as well as damaging influences that threaten the child's health and survival. Next, we turn to the events of childbirth. Today, women in industrialized nations have many more choices than ever before about where and how they give birth, and modern hospitals often go to great lengths to make the arrival of a new baby a rewarding, family-centered event.

Yolanda and Jay's son Joshua reaped the benefits of their careful attention to his needs during pregnancy. He was strong, alert, and healthy at birth. Nevertheless, the birth process does not always go smoothly. We will consider the pros and cons of medical interventions, such as pain-relieving drugs and surgical deliveries, designed to ease a difficult birth and protect the health of mother and baby. Our discussion also addresses the development of infants born underweight or too early, before the prenatal period is complete. We conclude with a close look at the remarkable capacities of newborns.

Prenatal Development

The sperm and ovum that unite to form the new individual are uniquely suited for the task of reproduction. The ovum is a tiny sphere, measuring $1/175$ inch in diameter, that is barely visible to the naked eye as a dot the size of the period at the end of this sentence. But in its microscopic world, it is a giant—the largest cell in the human body. The ovum's size makes it a perfect target for the much smaller sperm, which measure only $1/500$ inch.

Conception

About once every 28 days, in the middle of a woman's menstrual cycle, an ovum bursts from one of her *ovaries,* two walnut-sized organs located deep inside her abdomen, and is drawn into one of two *fallopian tubes*—long, thin structures that lead to the hollow, soft-lined uterus (see Figure 3.1). While the ovum trav-

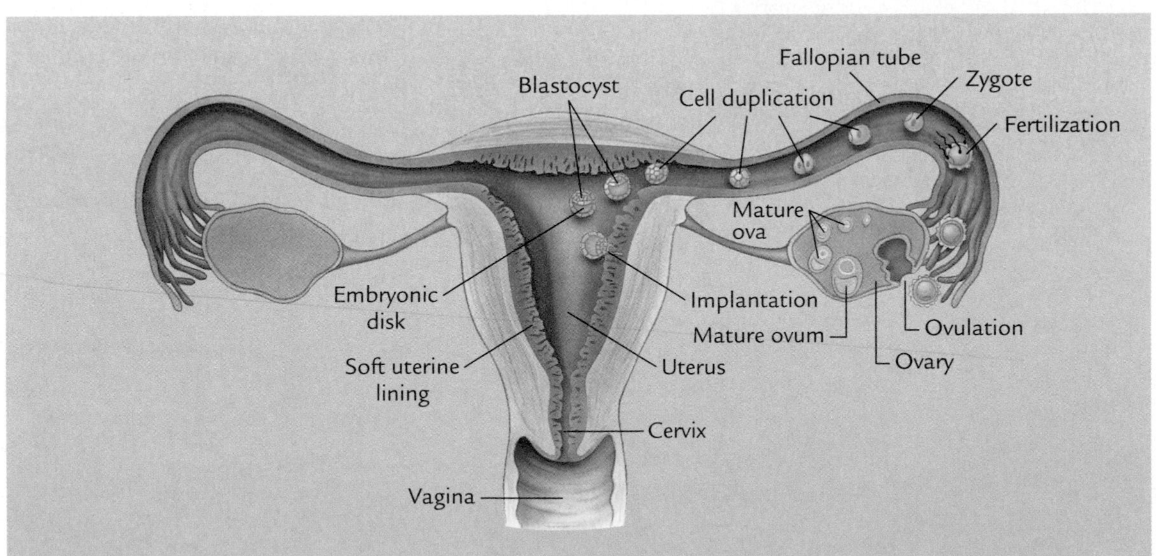

■ **FIGURE 3.1 Female reproductive organs, showing fertilization, early cell duplication, and implantation.** As the zygote moves down the fallopian tube, it begins to duplicate, at first slowly and then more rapidly. By the fourth day, it forms a hollow, fluid-filled ball called a blastocyst. The inner cells will become the new organism. The outer cells, or trophoblast, will provide protective covering. At the end of the first week, the blastocyst begins to implant in the uterine lining. (From K. L. Moore and T. V. N. Persaud, 1998, *Before We Are Born,* 5th ed., Philadelphia: Saunders, p. 44. Reprinted by permission of the publisher and the authors.)

els, the spot on the ovary from which it was released, now called the *corpus luteum,* secretes hormones that prepare the lining of the uterus to receive a fertilized ovum. If pregnancy does not occur, the corpus luteum shrinks, and the lining of the uterus is discarded 2 weeks later with menstruation.

The male produces sperm in vast numbers—an average of 300 million a day—in the *testes,* two glands located in the *scrotum,* a sac that lies just behind the penis. In the final process of maturation, each sperm develops a tail that permits it to swim long distances, upstream in the female reproductive tract, through the *cervix* (opening of the uterus) and into the *fallopian tube,* where fertilization usually takes place. The journey is difficult, and many sperm die. Only 300 to 500 reach the ovum, if one happens to be present. Sperm live for up to 6 days and can lie in wait for the ovum, which survives for only 1 day after being released into the fallopian tube. However, most conceptions result from intercourse occurring during a 3-day period—on the day of ovulation or during the 2 days preceding it (Wilcox, Weinberg, & Baird, 1995).

With conception, the story of prenatal development begins to unfold. The vast changes that take place during the 38 weeks of pregnancy are usually divided into three phases: (1) the period of the zygote, (2) the period of the embryo, and (3) the period of the fetus. As we look at what happens in each, you may find it useful to refer to Table 3.1, which summarizes the milestones of prenatal development.

Table 3.1 Major Milestones of Prenatal Development

Trimester	Period	Weeks	Length and Weight	Major Events
First	Zygote	1		The one-celled zygote multiplies and forms a blastocyst.
		2		The blastocyst burrows into the uterine lining. Structures that feed and protect the developing organism begin to form—amnion, chorion, yolk sac, placenta, and umbilical cord.
	Embryo	3–4	¼ inch (6 mm.)	A primitive brain and spinal cord appear. Heart, muscles, ribs, backbone, and digestive tract begin to develop.
		5–8	1 inch (2.5 cm.); 1/7 ounce (4 g.)	Many external body structures (face, arms, legs, toes, fingers) and internal organs form. The sense of touch begins to develop, and the embryo can move.
	Fetus	9–12	3 inches (7.6 cm.); less than 1 ounce (28 g.)	Rapid increase in size begins. Nervous system, organs, and muscles become organized and connected, and new behavioral capacities (kicking, thumb sucking, mouth opening, and rehearsal of breathing) appear. External genitals are well formed, and the fetus's sex is evident.
Second		13–24	12 inches (30 cm.); 1.8 pounds (820 g.)	The fetus continues to enlarge rapidly. In the middle of this period, fetal movements can be felt by the mother. Vernix and lanugo keep the fetus's skin from chapping in the amniotic fluid. Most of the brain's neurons are present by 24 weeks. Eyes are sensitive to light, and the fetus reacts to sound.
Third		25–38	20 inches (50 cm.); 7.5 pounds (3,400 g.)	The fetus has a chance of survival if born during this time. Size increases. Lungs mature. Rapid brain development causes sensory and behavioral capacities to expand. In the middle of this period, a layer of fat is added under the skin. Antibodies are transmitted from mother to fetus to protect against disease. Most fetuses rotate into an upside-down position in preparation for birth.

Sources: Moore & Persaud, 1998; Nilsson & Hamberger, 1990.

Period of the Zygote

The period of the zygote lasts about 2 weeks, from fertilization until the tiny mass of cells drifts down and out of the fallopian tube and attaches itself to the wall of the uterus. The zygote's first cell duplication is long and drawn out; it is not complete until about 30 hours after conception. Gradually new cells are added at a faster rate. By the fourth day, 60 to 70 cells form a hollow, fluid-filled ball called a *blastocyst* (refer again to Figure 3.1). The cells on the inside, called the *embryonic disk,* will become the new organism; the outer ring of cells, termed the *trophoblast,* will provide protective covering.

● **Implantation.** Between the seventh and ninth days after fertilization, **implantation** occurs: the blastocyst burrows deep into the uterine lining. Surrounded by the woman's nourishing blood, it starts to grow in earnest. At first, the trophoblast (protective outer layer) multiplies fastest. It forms a membrane, called the **amnion,** that encloses the developing organism in *amniotic fluid,* which helps keep the temperature of the prenatal world constant and provides a cushion against any jolts caused by the woman's movements. A *yolk sac* emerges that produces blood cells until the liver, spleen, and bone marrow are mature enough to take over this function (Moore & Persaud, 1998).

The events of these first 2 weeks are delicate and uncertain. As many as 30 percent of zygotes do not make it through this phase. In some, the sperm and ovum do not join properly. In others, cell duplication never begins. By preventing implantation in these cases, nature quickly eliminates most prenatal abnormalities (Sadler, 2000).

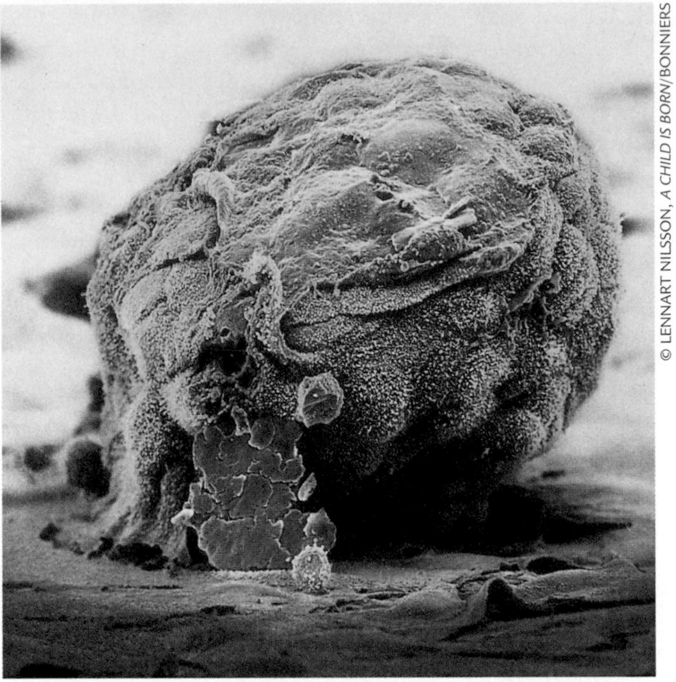

© LENNART NILSSON, *A CHILD IS BORN*/BONNIERS

■ **Period of the zygote: seventh to ninth day.** The fertilized ovum duplicates at an increasingly rapid rate, forming a hollow ball of cells, or blastocyst, by the fourth day after fertilization. Here the blastocyst, magnified thousands of times, burrows into the uterine lining between the seventh and ninth day.

■ In this photo taken with the aid of a powerful microscope, sperm have completed their journey up the female reproductive tract and are beginning to penetrate the surface of the enormous-looking ovum, the largest cell in the human body. When one of the sperm is successful at fertilizing the ovum, the resulting zygote will begin to duplicate.

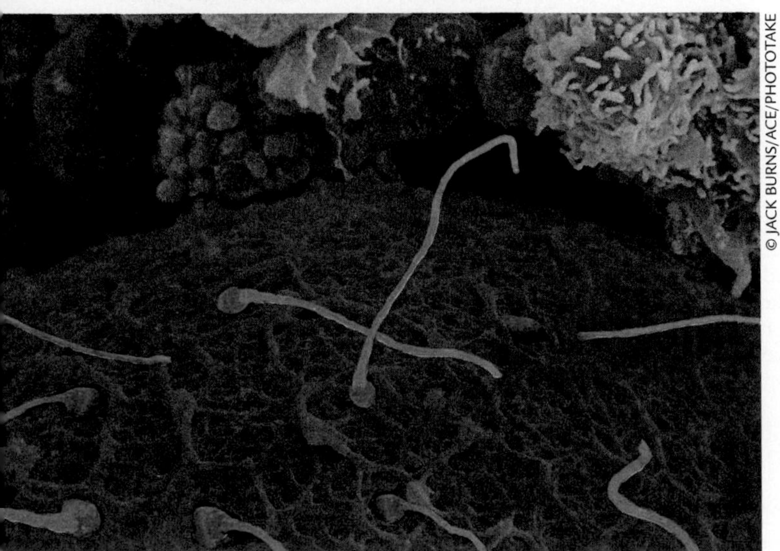

© JACK BURNS/ACE/PHOTOTAKE

● **The Placenta and Umbilical Cord.** By the end of the second week, cells of the trophoblast form another protective membrane—the **chorion,** which surrounds the amnion. From the chorion, tiny fingerlike *villi,* or blood vessels, begin to emerge.[1] As these villi burrow into the uterine wall, the placenta starts to develop. By bringing the mother's and embryo's blood close together, the **placenta** permits food and oxygen to reach the developing organism and waste products to be carried away. A membrane forms that allows these substances to be exchanged but prevents the mother's and embryo's blood from mixing directly.

The placenta is connected to the developing organism by the **umbilical cord,** which first appears as a tiny stalk and eventually grows to a length of 1 to 3 feet. The umbilical cord contains one large vein that delivers blood loaded with nutrients and two arteries that remove waste products. The force of blood flowing through the cord keeps it firm, so it seldom tangles while the embryo, like a space-walking astronaut, floats freely in its fluid-filled chamber (Moore & Persaud, 1998).

By the end of the period of the zygote, the developing organism has found food and shelter. These dramatic beginnings take place before most mothers know they are pregnant.

[1]Recall from Chapter 2 that *chorionic villus sampling* is the prenatal diagnostic method that can be performed earliest, by 6 to 8 weeks after conception.

Period of the Embryo

The period of the **embryo** lasts from the second through the eighth week of pregnancy. During these brief 6 weeks, the most rapid prenatal changes take place as the groundwork is laid for all body structures and internal organs.

● **Last Half of the First Month.** In the first week of this period, the embryonic disk forms three layers of cells: (1) the *ectoderm,* which will become the nervous system and skin; (2) the *mesoderm,* from which will develop the muscles, skeleton, circulatory system, and other internal organs; and (3) the *endoderm,* which will become the digestive system, lungs, urinary tract, and glands. These three layers give rise to all parts of the body.

At first, the nervous system develops fastest. The ectoderm folds over to form a **neural tube,** which will become the spinal cord and brain. While the nervous system is developing, the heart begins to pump blood, and muscles, backbone, ribs, and digestive tract start to appear. At the end of the first month, the curled embryo—only 1/4 inch long—consists of millions of organized groups of cells with specific functions.

● **The Second Month.** In the second month, growth continues rapidly. The eyes, ears, nose, jaw, and neck form. Tiny buds become arms, legs, fingers, and toes. Internal organs are more distinct: the intestines grow, the heart develops separate chambers, and the liver and spleen take over production of blood cells so that the yolk sac is no longer needed. Changing body proportions cause the embryo's posture to become more upright. Now 1 inch long and 1/7 of an ounce in weight, the embryo can sense its world. It responds to touch, particularly in the mouth area and on the soles of the feet. And it can move, although its tiny flutters are still too light to be felt by the mother (Nilsson & Hamberger, 1990).

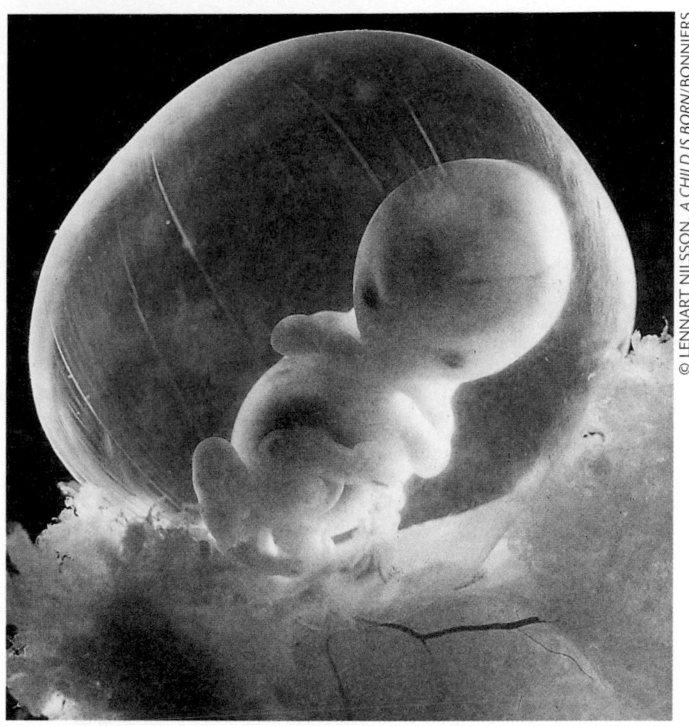

■ **Period of the embryo: seventh week.** The embryo's posture is more upright. Body structures—eyes, nose, arms, legs, and internal organs—are more distinct. An embryo of this age responds to touch. It can also move, although at less than one inch long and one ounce in weight, it is still too tiny to be felt by the mother.

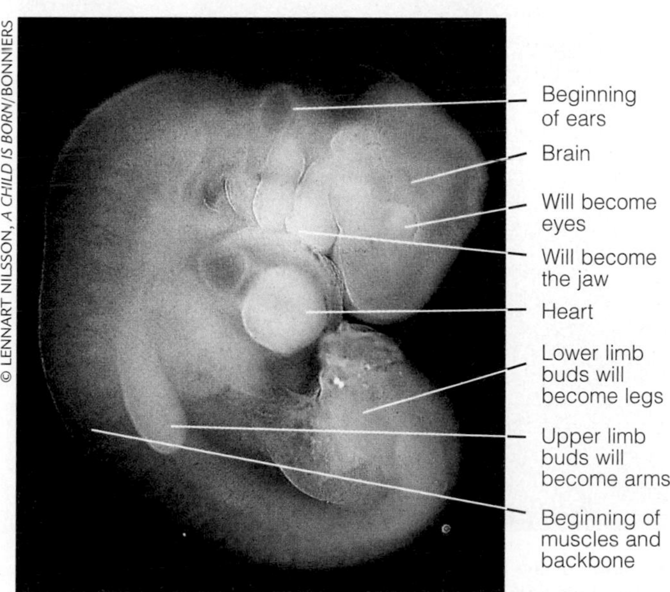

■ **Period of the embryo: fourth week.** In actual size, this 4-week-old embryo is only 1/4-inch long, but many body structures have begun to form. The primitive tail will disappear by the end of the embryonic period.

Period of the Fetus

Lasting from the ninth week to the end of pregnancy, the period of the **fetus** is the "growth and finishing" phase. During this longest prenatal period, the organism increases rapidly in size.

● **The Third Month.** In the third month, the organs, muscles, and nervous system start to become organized and connected. The fetus kicks, bends its arms, forms a fist, curls its toes, opens its mouth, and even sucks its thumb. The tiny lungs begin to expand and contract in an early rehearsal of breathing movements. By the twelfth week, the external genitals are well formed, and the sex of the fetus can be detected with ultrasound. Other finishing touches appear, such as fingernails, toenails, tooth buds, and eyelids. The heartbeat can now be heard through a stethoscope.

Prenatal development is sometimes divided into **trimesters,** or three equal time periods. At the end of the third month, the *first trimester* is complete. Two more must pass before the fetus is fully prepared to survive outside the womb.

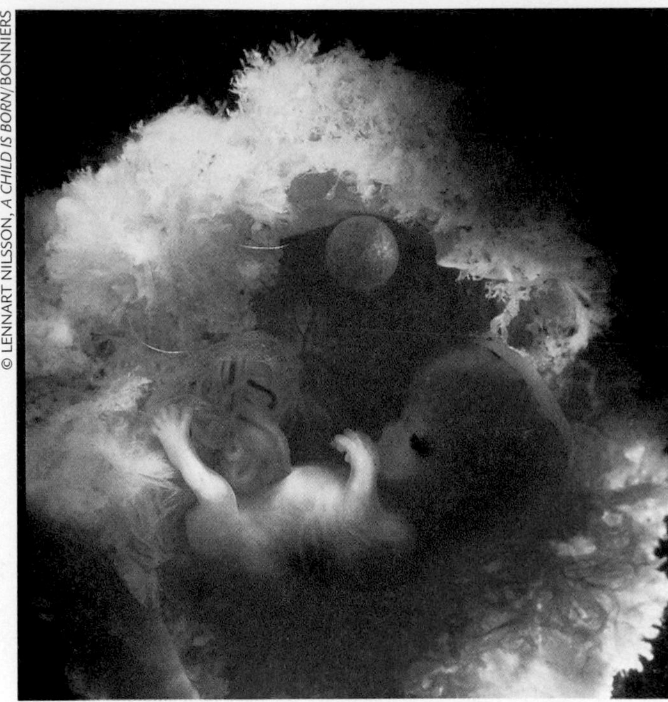

■ **Period of the fetus: eleventh week.** The organism is increasing rapidly in size. At 11 weeks, the brain and muscles are better connected. The fetus can kick, bend its arms, open and close its hands and mouth, and suck its thumb. Notice the yolk sac, which shrinks as pregnancy advances. The internal organs have taken over its function of producing blood cells.

● **The Second Trimester.** By the middle of the second trimester, between 17 and 20 weeks, the new being has grown large enough that the mother can feel its movements. A white, cheeselike substance called **vernix** protects its skin from chapping during the long months spent bathing in the amniotic fluid. White, downy hair called **lanugo** also appears over the entire body, helping the vernix stick to the skin.

At the end of the second trimester, many organs are well developed. And most of the brain's *neurons* (nerve cells that store and transmit information) are in place; few will be produced after this time. However, *glial cells,* which support and feed the neurons, continue to increase at a rapid rate throughout pregnancy, as well as after birth.

Brain growth means new behavioral capacities. The 20-week-old fetus can be stimulated as well as irritated by sounds. And if a doctor looks inside the uterus with fetoscopy (see Chapter 2, page 52), fetuses try to shield their eyes from the light with the hands, indicating that sight has begun to emerge (Nilsson & Hamberger, 1990). Still, a fetus born at this time cannot survive. Its lungs are immature, and the brain cannot yet control breathing and body temperature.

● **The Third Trimester.** During the final trimester, a fetus born early has a chance for survival. The point at which the fetus can first survive, called the **age of viability,** occurs sometime between 22 and 26 weeks (Moore & Persaud, 1998). If born between the seventh and eighth month, however, the baby usually needs help breathing. Although the brain's respiratory center is mature, tiny air sacs in the lungs are not yet ready to inflate and exchange carbon dioxide for oxygen.

The brain continues to make great strides. The *cerebral cortex,* the seat of human intelligence, enlarges. As neurological organization improves, the fetus spends more time awake. At 20 weeks, heart rate variability reveals no periods of alertness. But by 28 weeks, fetuses are awake about 11 percent of the time, a figure that rises to 16 percent just before birth (DiPietro et al., 1996a).

The fetus also takes on the beginnings of a personality. In one study, pattern of fetal activity just before birth predicted infant temperament at 3 and 6 months of age. Fetuses who cycled between quiet and active periods tended to become calm babies with predictable sleep–wake schedules. In contrast, fetuses who were highly active for long stretches were more likely to become difficult babies—fussy, upset by new experiences, irregular in eating and sleeping, and highly active (DiPietro et al., 1996b). Although these relationships are only modest, they suggest that parents with highly active fetuses should prepare for extra challenges! As we will see in Chapter 10, sensitive care can modify a difficult baby's temperament.

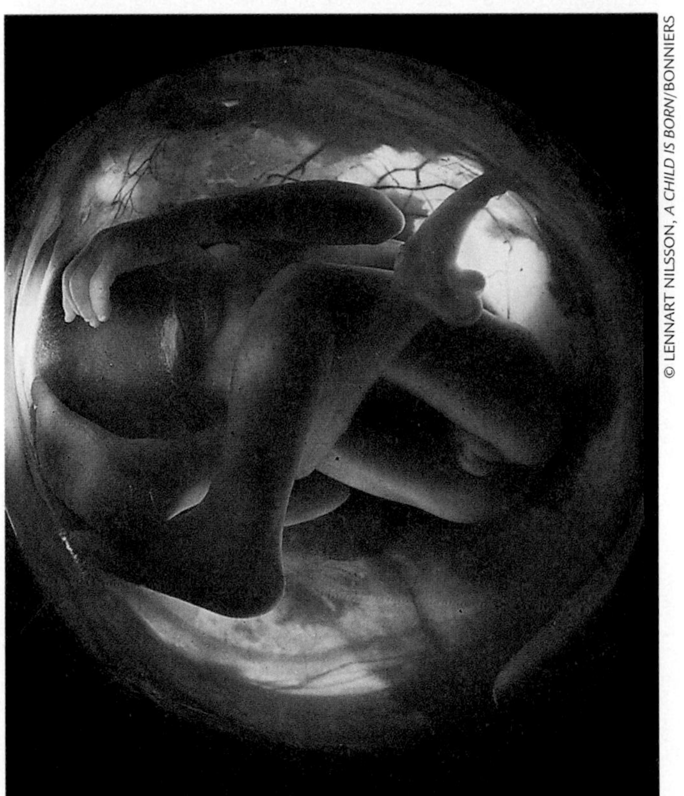

■ **Period of the fetus: twenty-second week.** This fetus is almost a foot long and weights slightly more than 1 pound. Its movements can be felt easily by the mother and by other family members who place a hand on her abdomen. The fetus has reached the age of viability. If born, it has a slim chance of surviving.

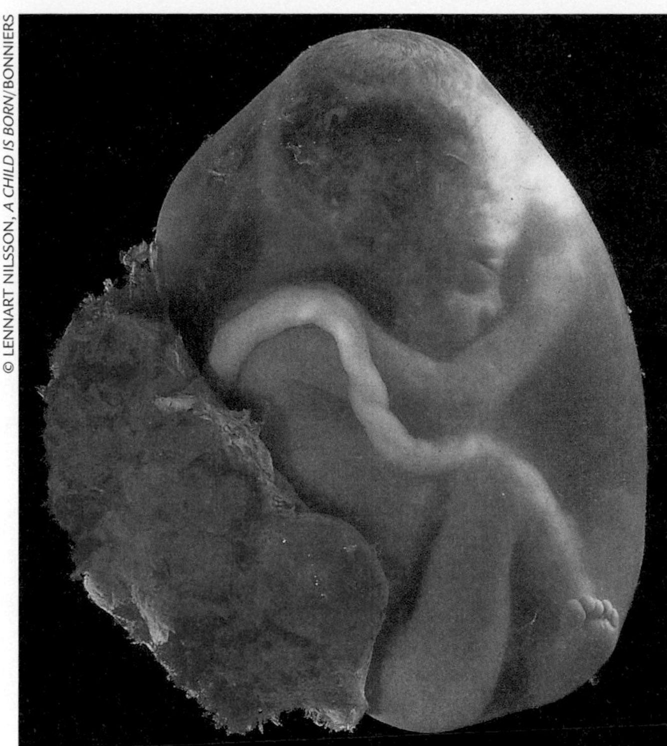

■ **Period of the fetus: thirty-sixth week.** This fetus fills the uterus. To support its need for nourishment, the umbilical cord and placenta have grown large. Notice the vernix (cheeselike substance) on the skin, which protects it from chapping. The fetus has accumulated a layer of fat to assist with temperature regulation after birth. In 2 more weeks, it will be full term.

The third trimester brings greater responsiveness to external stimulation. Around 24 weeks, fetuses can first feel pain, so after this time painkillers should be used in any surgical procedures (Royal College of Obstetricians and Gynecologists, 1997). By 25 weeks, fetuses react to nearby sounds with body movements (Kisilevsky & Low, 1998). And in the last weeks of pregnancy, they learn to prefer the tone and rhythm of their mother's voice. In one clever study, mothers read aloud Dr. Seuss's lively book *The Cat in the Hat* for the last 6 weeks of pregnancy. After birth, their infants learned to turn on recordings of the mother's voice by sucking on nipples. They showed a preference for the familiar poem by sucking harder to hear *The Cat in the Hat* than other rhyming stories (DeCasper & Spence, 1986).

During the final 3 months, the fetus gains more than 5 pounds and grows 7 inches. In the eighth month, a layer of fat is added to assist with temperature regulation. The fetus also receives antibodies from the mother's blood that protect against illnesses, since the newborn's immune system will not work well until several months after birth. In the last weeks, most fetuses assume an upside-down position, partly because of the shape of the uterus and because the head is heavier than the feet. Growth slows, and birth is about to take place.

Ask Yourself

REVIEW

Why is the period of the embryo regarded as the most dramatic prenatal phase? Why is the period of the fetus called the "growth and finishing" phase?

APPLY

Amy, 2 months pregnant, wonders how the developing organism is being fed. "I don't look pregnant yet, so does that mean not much development has occurred?" How would you respond to Amy?

CONNECT

How does brain development relate to fetal behavior?

www.

Prenatal Environmental Influences

Although the prenatal environment is far more constant than the world outside the womb, many factors can affect the developing embryo and fetus. Yolanda and Jay learned that there was much they could do to create a safe environment for development before birth.

Teratogens

The term **teratogen** refers to any environmental agent that causes damage during the prenatal period. It comes from the Greek word *teras,* meaning "malformation" or "monstrosity." This label was selected because scientists first learned about harmful prenatal influences from infants who had been profoundly damaged. Yet the harm done by teratogens is not always simple and straightforward. It depends on the following factors:

- *Dose.* We will see as we discuss particular teratogens that larger doses over longer time periods usually have more negative effects.
- *Heredity.* The genetic makeup of the mother and the developing organism plays an important role. Some individuals are better able to withstand harmful environments.
- *Other negative influences.* The presence of several negative factors at once, such as poor nutrition, lack of medical care, and additional teratogens, can worsen the impact of a single harmful agent.
- *Age.* The effects of teratogens vary with the age of the organism at time of exposure.

We can best understand this last idea if we think of the *sensitive period* concept introduced in Chapter 1. A sensitive period is a limited time span during which a part of the body or a behavior is biologically prepared to develop rapidly. During that time, it is especially sensitive to its surroundings. If the environment is harmful, then damage occurs, and recovery is difficult and sometimes impossible.

Figure 3.2 summarizes prenatal sensitive periods. In the period of the zygote, before implantation, teratogens rarely have any impact. If they do, the tiny mass of cells is usually so badly damaged that it dies. The embryonic period is the time when serious defects are most likely to occur because the foundations for all body parts are being laid down. During the fetal period, teratogenic damage is usually minor. However, organs such as the brain, eyes, and genitals can still be strongly affected.

The effects of teratogens are not limited to immediate physical damage. Some health effects are subtle and delayed (see the Lifespan Vista box on pages 84–85). Furthermore, psychological consequences may occur indirectly, as a result of physical damage. For example, a defect resulting from drugs the mother took during pregnancy can change the reactions of others to the child as well as the child's ability to explore the environment. Over time, parent–child interaction, peer relations, and cognitive and social development may suffer.

Notice how an important idea about development that we discussed in earlier chapters is at work here: *bidirectional influences* between child and environment. Now let's take a look at what scientists have discovered about a variety of teratogens.

● **Prescription and Nonprescription Drugs.** In the early 1960s, the world learned a tragic lesson about drugs and prenatal development. At that time, a sedative called *thalidomide* was widely available in Canada, Europe, and South America. When taken by mothers 4 to 6 weeks after conception, thalidomide produced gross deformities of the embryo's arms and legs and, less frequently, damage to the ears, heart, kidneys, and genitals. About 7,000 infants worldwide were affected (Moore & Persaud, 1998). As children exposed to thalidomide

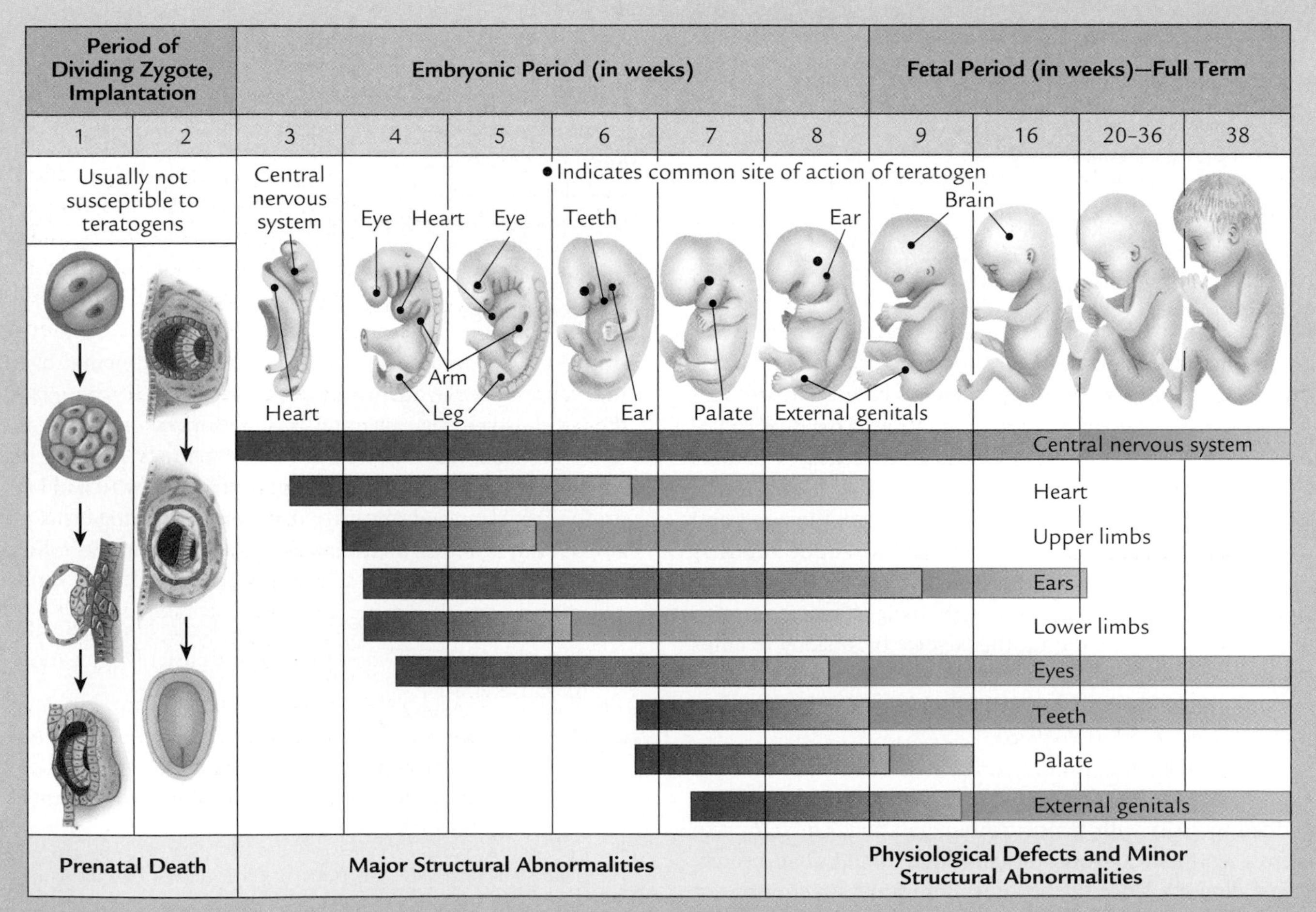

■ **FIGURE 3.2 Sensitive periods in prenatal development.** Each organ or structure has a sensitive period, during which its development may be disturbed. Blue horizontal bars indicate highly sensitive periods. Green horizontal bars indicate periods that are somewhat less sensitive to teratogens, although damage can occur. (From K. L. Moore & T. V. N. Persaud, 1998, *Before We Are Born,* 5th ed., Philadelphia: Saunders, p. 166. Adapted by permission of the publisher and authors.)

grew older, many scored below average in intelligence. Perhaps the drug damaged the central nervous system directly. Or perhaps the child-rearing conditions of these severely deformed youngsters impaired their intellectual development.

Another medication, a synthetic hormone called *diethylstilbestrol (DES),* was widely prescribed between 1945 and 1970 to prevent miscarriages. As daughters of these mothers reached adolescence and young adulthood, they showed unusually high rates of cancer of the vagina and malformations of the uterus. When they tried to have children, their pregnancies more often resulted in prematurity, low birth weight, and miscarriage than those of non-DES-exposed women. Young men showed an increased risk of genital abnormalities and cancer of the testes (Giusti, Iwamoto, & Hatch, 1995; Palmlund, 1996).

Any drug taken by the mother that has a molecule small enough to penetrate the placental barrier can enter the embryonic or fetal bloodstream. Yet many pregnant women continue to take over-the-counter drugs without consulting their doctors. Aspirin is one of the most common. Several studies suggest that regular aspirin use is linked to low birth weight, infant death around the time of birth, poorer motor development, and lower intelligence scores in early childhood, although other research fails to confirm these findings (Barr et al., 1990; Hauth et al., 1995; Streissguth et al., 1987). Coffee, tea, cola, and cocoa contain another frequently consumed drug, caffeine. Heavy caffeine intake (more than 3 cups of coffee per day) is associated with low birth weight, miscarriage, and newborn withdrawal symptoms, such as irritability and vomiting (Fernandes et al., 1998; Gilbert-Barness, 2000).

Because children's lives are involved, we must take findings like these seriously. At the same time, we cannot be sure that these drugs actually cause the problems mentioned. Often mothers take more than one drug. If the prenatal organism is injured, it is hard to tell which drug might be responsible or whether other factors correlated with drug taking are really at fault. Until we have more information, the safest course of action is the one Yolanda took: cut down or avoid these drugs entirely.

● **Illegal Drugs.** Use of highly addictive mood-altering drugs, such as cocaine and heroin, has become more widespread, especially in poverty-stricken inner cities where drugs provide a temporary escape from a daily life of hopelessness. The number of "cocaine babies" born in the United States has reached crisis levels in recent years, amounting to hundreds of thousands annually (Cornelius et al., 1999; Landry & Whitney, 1996).

Babies born to users of cocaine, heroin, or methadone (a less addictive drug used to wean people away from heroin) are at risk for a wide variety of problems, including prematurity, low birth weight, physical defects, breathing difficulties, and death around the time of birth (Datta-Bhutada, Johnson, & Rosen, 1998; Walker, Rosenberg, & Balaban-Gil, 1999). In addition, these infants arrive drug-addicted. They are often feverish and irritable at birth and have trouble sleeping, and their cries are abnormally shrill and piercing—a common symptom among stressed newborns (Friedman, 1996; Ostrea, Ostrea, & Simpson, 1997). When mothers with many problems of their own must take care of these babies, who are difficult to calm down, cuddle, and feed, behavior problems are likely to persist.

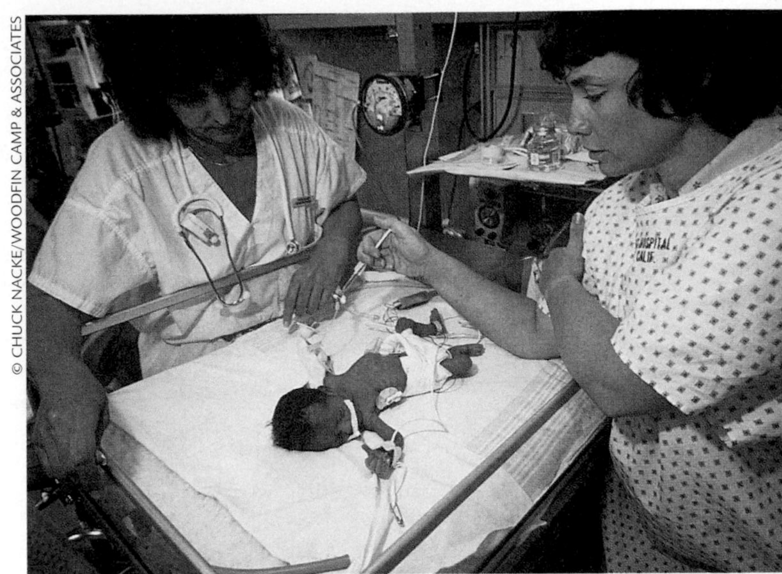

■ This baby, whose mother took crack during pregnancy, was born many weeks before his due date. He breathes with the aid of a respirator. His central nervous system may be damaged. Researchers are not yet sure if these outcomes are caused by crack or by the many other high-risk behaviors of drug users.

Throughout the first year of life, heroin- and methadone-exposed infants are less attentive to the environment, and their motor development is slow. After infancy, some children get better, whereas others remain jittery and inattentive. The kind of parenting these youngsters receive may explain why problems last for some but not for others (Cosden, Peerson, & Elliott, 1997).

Growing evidence on cocaine suggests that many prenatally exposed babies have lasting difficulties. Cocaine constricts the blood vessels, causing oxygen delivered to the developing organism to fall dramatically for 15 minutes following a high dose. It also alters the production and functioning of neurons and the chemical balance in the fetus's brain. These effects may contribute to a specific set of cocaine-linked physical defects, including eye, bone, genital, urinary tract, kidney, and heart deformities, as well as brain hemorrhages and seizures (Espy, Kaufmann, & Glisky, 1999; Mayes, 1999; Plessinger & Woods, 1998). Motor, visual, attention, memory, and language problems appear in infancy and persist into the preschool years (Mayes et al., 1996). Babies born to mothers who smoke crack (a cheap form of cocaine that delivers high doses quickly through the lungs) seem worst off in terms of low birth weight and damage to the central nervous system (Bender et al., 1995; Richardson et al., 1996).

Still, it is difficult to isolate the precise damage caused by cocaine, since many users take several drugs, display other high-risk behaviors, and engage in insensitive caregiving (Lester, 2000). The joint impact of these factors worsens outcomes for children (Alessandri, Bendersky, & Lewis, 1998;

A Lifespan Vista

The Prenatal Environment and Health in Later Life

When Michael entered the world 55 years ago, 6 weeks premature and weighing only 4 pounds, the doctor delivering him wasn't sure he would make it. Michael not only survived but enjoyed good health until his mid-forties when, during a routine medical checkup, he was diagnosed with high blood pressure and adult-onset diabetes. Michael wasn't overweight, didn't smoke, and didn't eat high-fat foods—risk factors for these conditions. Nor did the illnesses run in his family. Could the roots of Michael's health problems date back to his prenatal development? Increasing evidence suggests that prenatal environmental factors—ones not toxic (like tobacco or alcohol) but rather fairly subtle, such as the flow of nutrients and hormones across the placenta—can affect health decades later (Wheeler, Barker, & O'Brien, 1999).

Low Birth Weight and Heart Disease, Stroke, and Diabetes. Carefully controlled animal experiments reveal that a poorly nourished, underweight fetus experiences changes in body structure and function that result in cardiovascular disease in adulthood (Franco et al., 2002). To explore this relationship in humans, researchers tapped public records, gathering information on the birth weights of 15,000 British men and women and the occurrence of disease in middle adulthood. Those weighing less than 5 pounds at birth had a 50 percent greater chance of dying of heart disease and stroke, after SES and a variety of other health risks were controlled. The connection between birth weight and cardiovascular disease was strongest for people whose weight-to-length ratio at birth was very low—a sign of prenatal growth stunting (Godfrey & Barker, 2000; Martyn, Barker, & Osmond, 1996).

In other large-scale studies, a consistent link between low birth weight and heart disease, stroke, and diabetes in middle adulthood has emerged—for both sexes and in diverse countries, including Finland, India, Jamaica, and the United States (Eriksson et al., 2001; Fall et al., 1998; Forsén et al., 2000; Godfrey & Barker, 2001). Smallness itself does not cause later health problems; rather, researchers believe, complex factors associated with it are involved.

Some speculate that a poorly nourished fetus diverts large amounts of blood to the brain, causing organs in the abdomen such as the liver and kidney (involved in controlling cholesterol and blood pressure) to be undersized (Barker, 2002). The result is heightened later risk for heart disease and stroke. In the case of diabetes, inadequate prenatal nutrition may permanently impair functioning of the pancreas, leading glucose intolerance to rise as the person ages (Rich-Edwards et al., 1999). Yet another hypothesis, supported by both animal and human research, is that the malfunctioning placentas of some expectant mothers permit high levels of stress hormones to reach the fetus, which retards fetal growth, increases fetal blood pressure, and promotes hyperglycemia (excess blood sugar). These outcomes predispose the developing person to later disease (Osmond & Barker, 2000). And finally, slow fetal growth leads many children to compensate by gaining excessive weight that, when sustained in adulthood, promotes heart disease and diabetes (Barker, 1999).

High Birth Weight and Breast Cancer. The other prenatal growth extreme—high birth weight—is related to breast cancer, the most common malignancy in adult women (Andersson et al.,

Carta et al., 2001). And because cocaine-exposed babies have not been followed beyond early childhood, long-term consequences are unknown.

Another illegal drug, marijuana, is used more widely than heroin and cocaine. Studies examining its relationship to low birth weight and prematurity reveal mixed findings (Fried, 1993). Several researchers have linked prenatal marijuana exposure to smaller head size (a measure of brain growth), newborn startle reactions, disturbed sleep, and inattention in infancy and childhood (Dahl et al., 1995; Fried, Watkinson, & Gray, 1999; Lester & Dreher, 1989). As with cocaine, however, long-term effects have not been established.

● **Tobacco.** Although smoking has declined in Western nations, an estimated 12 percent of American women and 19 percent of Canadian women smoke during their pregnancies (Health Canada, 2001b; Matthews, 2001). The most well-known effect of smoking during the prenatal period is low birth weight. But the likelihood of other serious consequences, such as miscarriage, prematurity, impaired breathing during sleep, infant death, and cancer later in childhood is also increased (Franco et al., 2000, Walker, Rosenberg, & Balaban-Gil, 1999). The more cigarettes a mother smokes, the greater the chances that her baby will be affected. If a pregnant woman decides to stop smoking at any time, even during the last trimester, she reduces the likelihood that her infant will be born underweight and suffer from future problems (Klesges et al., 2001).

Even when a baby of a smoking mother appears to be born in good physical condition, slight behavioral abnormalities may threaten the child's development. Newborns of smoking mothers are less attentive to sounds and display more muscle tension (Fried & Makin, 1987). An unresponsive, restless baby may not

2001; Vatten et al., 2002). In one investigation, the mothers of 589 nurses with invasive breast cancer and 1,569 nurses who did not have breast cancer were asked to provide their daughters' birth weights, early life exposures (for example, smoking during pregnancy), and a family health history (such as relatives diagnosed with breast cancer). The nurses themselves provided information on adult health. After other risk factors were controlled, high birth weight—especially more than 8.7 pounds—emerged as a clear predictor of breast cancer (see Figure 3.3) (Michels et al., 1996). Researchers think that the culprit is excessive maternal estrogen during pregnancy, which promotes large fetal size and alters beginning breast tissue so that it may respond to estrogen in adulthood by becoming malignant.

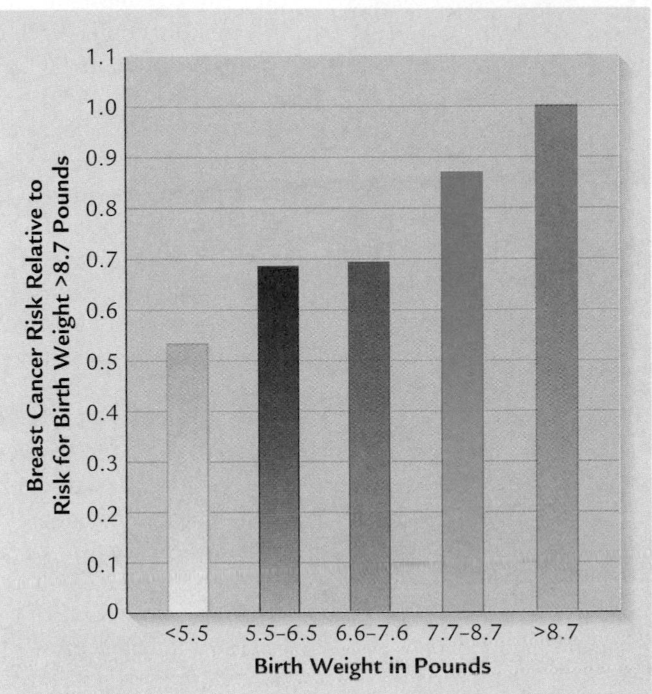

■ **FIGURE 3.3 Relationship of birth weight to breast-cancer risk in adulthood.** In a study of 589 nurses with invasive breast cancer and 1,569 nurses who did not have breast cancer, birth weight predicted breast cancer incidence after many other prenatal and postnatal health risks were controlled. The breast cancer risk was especially high for women whose birth weights were greater than 8.7 pounds. (Adapted from Michels et al., 1996.)

Prevention. The prenatal development-later-life illness relationships emerging in research do not mean that the illnesses are inevitable. Rather, prenatal environmental conditions *influence* adult health, and the steps we take to protect our health can prevent prenatal risks from becoming reality. Researchers advise individuals who were low-weight at birth to get regular medical checkups and to be attentive to diet, weight, fitness, and stress—controllable factors that contribute to heart disease and adult-onset diabetes. And high-birth-weight women should be conscientious about breast self-exams and mammograms, which permit breast cancer to be detected early and, in many instances, cured.

evoke the kind of interaction from adults that promotes healthy psychological development. Some studies report that prenatally exposed children have shorter attention spans, poorer memories, lower mental test scores, and more behavior problems in childhood and adolescence, even after many other factors have been controlled (Cornelius et al., 2001; Trasti et al., 1999; Wasserman et al., 2001). But other researchers have not confirmed these findings (Barr et al., 1990; Streissguth et al., 1989).

Exactly how can smoking harm the fetus? Nicotine, the addictive substance in tobacco, constricts blood vessels, lessens blood flow to the uterus, and causes the placenta to grow abnormally. This reduces transfer of nutrients, so the fetus gains weight poorly. Also, nicotine raises the concentration of carbon monoxide in the bloodstreams of both mother and fetus. Carbon monoxide displaces oxygen from red blood cells. It damages the central nervous system and reduces birth weight in the fetuses of laboratory animals. Similar effects may occur in humans (Friedman, 1996).

Finally, from one-third to one-half of nonsmoking pregnant women are "passive smokers" because their husbands, relatives, and co-workers use cigarettes. Passive smoking is also related to low birth weight, infant death, and possible long-term impairments in attention and learning (Dejin-Karlsson et al., 1998; Makin, Fried, & Watkinson, 1991). Clearly, expectant mothers should avoid smoke-filled environments.

● **Alcohol.** In a moving story, Michael Dorris (1989), a Dartmouth University anthropology professor, described what it was like to raise his adopted son, Adam, whose biological mother drank heavily throughout pregnancy and died of alcohol poisoning shortly after his birth. A Sioux Indian,

■ The mother of this severely retarded boy drank heavily during pregnancy. His widely spaced eyes, thin upper lip, and short eyelid openings are typical of fetal alcohol syndrome.

Adam was born with **fetal alcohol syndrome (FAS).** Mental retardation; impaired motor coordination, attention, memory, and language; and overactivity are typical of children with the disorder (Connor et al., 2001; Schonfeld et al., 2001). Distinct physical symptoms also accompany it, including slow physical growth and a particular pattern of facial abnormalities: widely spaced eyes, short eyelid openings, a small upturned nose, a thin upper lip, and a small head, indicating that the brain has not developed fully. In a related condition, known as **fetal alcohol effects (FAE),** individuals display only some of these abnormalities. Usually, their mothers drank alcohol in smaller quantities (Goodlet & Johnson, 1999; Mattson et al., 1998).

Even when provided with enriched diets, FAS babies fail to catch up in physical size during infancy or childhood. Mental impairment is also permanent: In his teens and twenties, Adam's intelligence remained below average, and he had trouble concentrating and keeping a routine job. He also suffered from poor judgment. He might buy something and not wait for change or wander off in the middle of a task. The more alcohol consumed by a woman during pregnancy, the poorer the child's motor coordination, information processing, reasoning, and scores on intelligence and achievement tests during the preschool and school years (Aronson, Hagberg, & Gillberg, 1997; Hunt et al., 1995; Streissguth et al., 1994). In adolescence, prenatal alcohol exposure is associated with poor school performance, trouble with the law, inappropriate sexual behavior, alcohol and drug abuse, depression, and other lasting mental health problems (Kelly, Day, & Streissguth, 2000).

How does alcohol produce its devastating effects? First, it interferes with brain development during the early months of pregnancy, resulting in structural damage and abnormalities in brain functioning, including transfer of messages from one part of the brain to another (Bookstein et al., 2002; Roebuck, Mattson, & Riley, 1999). Second, the body uses large quantities of oxygen to metabolize alcohol. A pregnant woman's heavy drinking draws away oxygen that the developing organism needs for cell growth.

About 25 percent of American and Canadian mothers reported drinking at some time during their pregnancies. As with heroin and cocaine, alcohol abuse is highest among poverty-stricken women (Health Canada, 2002a; U.S. Department of Health and Human Services, 2002). On the reservation where Adam was born, many children show symptoms of prenatal alcohol exposure. Unfortunately, when girls with FAS or FAE later become pregnant, the poor judgment caused by the syndrome often prevents them from understanding why they should avoid alcohol themselves. Thus, the tragic cycle is likely to be repeated in the next generation.

How much alcohol is safe during pregnancy? Even mild drinking, below one drink per day, is associated with reduced head size and body growth among children followed into adolescence (Day et al., 2002). And as little as 2 ounces of alcohol a day, taken very early in pregnancy, is linked to FAS-like facial features (Astley et al., 1992). Recall that other factors—both genetic and environmental—can make some fetuses more vulnerable to teratogens. Therefore, no amount of alcohol is safe, and pregnant women should avoid it entirely.

● **Radiation.** Defects due to radiation were tragically apparent in the children born to pregnant women who survived the bombing of Hiroshima and Nagasaki during World War II. Similar abnormalities surfaced in the 9 months following the 1986 Chernobyl, Ukraine, nuclear power plant accident. After each disaster, the incidence of miscarriage and babies born with underdeveloped brains, physical deformities, and slow physical growth rose dramatically (Schull & Otake, 1999; Terestchenko, Lyaginskaya, & Burtzeva, 1991).

Even when a radiation-exposed baby appears normal, problems may appear later. For example, even low-level radiation, as the result of industrial leakage or medical X-rays, can increase the risk of childhood cancer (Fattibene et al., 1999). In middle childhood, prenatally exposed Chernobyl children had lower intelligence test scores and had rates of language and emotional disorders two to three times greater than those of nonexposed Russian children. Furthermore, Chernobyl children's parents were highly anxious, due to forced evacuation from their homes and worries about living in irradiated areas. The more tension parents reported, the poorer their children's emotional functioning (Kolominsky, Igumnov, & Drozdovitch, 1999). Stressful rearing conditions seemed to combine

© CHARLTON-BOURDILLON/GETTY IMAGES

■ This child's mother was just a few weeks pregnant during the Chernobyl nuclear power plant disaster. Radiation exposure is probably responsible for his limb deformities. He is also at risk for low intelligence and language and emotional disorders.

with the damaging effects of prenatal radiation to impair children's development.

● **Environmental Pollution.** In industrialized nations, an astounding number of potentially dangerous chemicals are released into the environment. Over 100,000 are in common use in the United States, and many new pollutants are introduced each year.

Mercury is an established teratogen. In the 1950s, an industrial plant released waste containing high levels of mercury into a bay providing food and water for the town of Minamata, Japan. Many children born at the time displayed physical deformities, mental retardation, abnormal speech, difficulty in chewing and swallowing, and uncoordinated movements. Autopsies of those who died revealed widespread brain damage (Dietrich, 1999).

Another teratogen, *lead,* is present in paint flaking off the walls of old buildings and in certain materials used in industrial occupations. High levels of prenatal lead exposure are consistently related to prematurity, low birth weight, brain damage, and a wide variety of physical defects (Dye-White, 1986). Even low levels seem to be dangerous. Affected babies show slightly poorer mental and motor development (Dietrich, Berger, & Succop, 1993; Wasserman et al., 1994).

For many years, *polychlorinated biphenyls (PCBs)* were used to insulate electrical equipment until research showed that, like mercury, they found their way into waterways and entered the food supply. In Taiwan, prenatal exposure to very high levels of PCBs in rice oil resulted in low birth weight, discolored skin, deformities of the gums and nails, brain-wave abnormalities, and delayed cognitive development (Chen & Hsu, 1994; Chen et al., 1994). Steady, low-level PCB exposure is also harmful. Women who frequently ate PCB-contaminated Great Lakes fish, compared with those who ate little or no fish, had infants with lower birth weights, smaller heads, and less interest in their surroundings (Jacobson et al., 1984; Stewart et al., 2000). Follow-ups later in the first year and in early childhood revealed persisting memory difficulties and lower verbal intelligence (Jacobson, 1998; Jacobson et al., 1992).

● **Infectious Disease.** On her first prenatal visit, Yolanda's doctor asked if she and Jay had already had certain infectious diseases, such as measles, mumps, and chicken pox. Although most illnesses, such as the common cold, have little or no impact on the embryo or fetus, a few are major causes of miscarriage and birth defects.

In the mid-1960s, a worldwide epidemic of *rubella* (German, or 3-day, measles) led to the birth of over 20,000 American babies with serious defects. Consistent with the sensitive period concept, the greatest damage occurs when rubella strikes during the embryonic period. Over 50 percent of infants whose mothers become ill during that time show heart defects; eye cataracts; deafness; genital, urinary, and intestinal abnormalities; and mental retardation. Infection during the fetal period is less harmful, but low birth weight, hearing loss, and bone defects may still occur (Eberhart-Phillips, Frederick, & Baron, 1993). Even though vaccination in infancy and childhood is routine, about 10 to 20 percent of women in North America and Western Europe lack the rubella antibody, so new disease outbreaks are possible (Lee et al., 1992; Pebody et al., 2000).

The *human immunodeficiency virus (HIV),* which leads to *acquired immune deficiency syndrome (AIDS),* a disease that destroys the immune system, has infected increasing numbers of women over the past decade. When they become pregnant, they pass the deadly virus to the developing organism about 20 to 30 percent of the time (Nourse & Butler, 1998). AIDS progresses rapidly in infants. By 6 months, weight loss, diarrhea, and repeated respiratory illnesses are common. The virus also causes brain damage. Most prenatal AIDS babies survive for only 5 to 8 months after the appearance of these symptoms (Parks, 1996). The antiviral drug zidovudine (ZDV) reduces prenatal AIDS transmission as much as 95 percent, with no harmful consequences of drug treatment for children (Culnane et al., 1999). It has led to a dramatic decline in prenatally acquired AIDS in Western nations.

The developing organism is especially sensitive to the family of herpes viruses, for which there is no vaccine or treatment. Among these, *cytomegalovirus* (the most frequent prenatal infection, transmitted through respiratory or sexual contact) and *herpes simplex 2* (which is sexually transmitted) are especially dangerous. In both, the virus invades the mother's genital tract, infecting babies either at birth or during pregnancy and resulting in miscarriage, low birth weight, physical malformations, and mental retardation (Behrman, Kliegman, & Jenson, 2000).

Several bacterial and parasitic diseases are also teratogens. Among the most common is *toxoplasmosis,* caused by a parasite found in many animals. Pregnant women may become infected from eating raw or undercooked meat or from contact with the feces of infected cats. About 40 percent of women who have the disease transmit it to the developing organism. If it strikes during the first trimester, it is likely to cause eye and brain damage. Later infection is linked to mild visual and cognitive impairments (Jones et al., 2001). Expectant mothers can avoid toxoplasmosis by making sure the meat they eat is well cooked, having pet cats checked for the disease, and turning over the care of litter boxes to other family members.

Other Maternal Factors

Besides avoiding teratogens, expectant parents can support the development of the embryo or fetus in other ways. In healthy, physically fit women, regular moderate exercise, such as walking, swimming, hiking, and aerobics, is related to increased birth weight (Hatch et al., 1993). However, very frequent, vigorous exercise predicts the opposite outcome—lower birth weight than in healthy, nonexercising controls (Pivarnik, 1998). (Note, also, that pregnant women with health problems, such as circulatory difficulties or previous miscarriages, should consult their doctors about a physical fitness routine.) We examine other maternal factors—nutrition, emotional stress, blood type, and age and previous births—in the following sections.

● **Nutrition.** Children grow more rapidly during the prenatal period than at any other phase of development. During this time, they depend totally on the mother for nutrients. A healthy diet that results in a weight gain of 25 to 30 pounds helps ensure the health of mother and baby.

Prenatal malnutrition can cause serious damage to the central nervous system. The poorer the mother's diet, the greater the loss in brain weight, especially if malnutrition occurred during the last trimester. During that time, the brain is increasing rapidly in size, and a maternal diet high in all the basic nutrients is necessary for it to reach its full potential (Morgane et al., 1993). An inadequate diet during pregnancy can also distort the structure of other organs, including the pancreas, liver, and blood vessels, resulting in lifelong health problems (Barker, 1994).

Because poor nutrition suppresses development of the immune system, prenatally malnourished babies frequently catch respiratory illnesses (Chandra, 1991). In addition, they often are irritable and unresponsive to stimulation. The behavioral effects of poor nutrition quickly combine with an impoverished, stressful home life. With age, low intelligence and serious learning problems become more apparent (Pollitt, 1996).

■ This government-sponsored nutrition class in a village in India prevents prenatal malnutrition by promoting a proper diet for pregnant women. Mothers also learn how breast-feeding can protect their newborn baby's healthy growth (see Chapter 4, page 124).

Many studies show that providing pregnant women with adequate food has a substantial impact on the health of their newborn babies. Yet the growth demands of the prenatal period require more than just increasing the quantity of a typical diet. Vitamin and mineral enrichment is also crucial.

For example, folic acid supplementation around the time of conception greatly reduces abnormalities of the neural tube, such as anencephaly and spina bifida (see Table 2.4 on page 53). In addition, adequate folate intake during the last 10 weeks of pregnancy cuts in half the risk of premature delivery and low birth weight (MCR Vitamin Study Research Group, 1991; Scholl, Heidiger, & Belsky, 1996). Because of these findings, U.S. and Canadian government guidelines recommend that all women of childbearing age consume at least 0.4 but not more than 1 milligram of folic acid per day (excessive intake can be harmful). Currently, bread, flour, rice, pasta, and other grain products are being fortified with folic acid.

When poor nutrition persists throughout the prenatal period, infants usually require more than dietary improvement. Successful interventions must also break the cycle of apathetic mother–baby interactions. Some do so by teaching parents how to interact effectively with their infants, whereas others focus on stimulating infants to promote active engagement with their physical and social surroundings (Grantham-McGregor et al., 1994; Grantham-McGregor, Schofield, & Powell, 1987).

Although prenatal malnutrition is highest in poverty-stricken regions of the world, it is not limited to developing countries. The U.S. Special Supplemental Food Program for Women, Infants, and Children provides food packages to low-income pregnant women, but funding is limited, and only 70 percent of those eligible are served (Children's Defense Fund, 2002). Besides food, the Canada Prenatal Nutrition Program provides counseling, social support, access to health care, and shelter to all pregnant women at risk for poor birth outcomes (Health Canada, 2002c).

● **Emotional Stress.** When women experience severe emotional stress during pregnancy, their babies are at risk for a wide variety of difficulties. Intense anxiety is associated with a higher rate of miscarriage, prematurity, low birth weight, and newborn irritability, respiratory illness, and digestive disturbances. It is also related to certain physical defects, such as cleft lip and palate and pyloric stenosis—tightening of the infant's stomach outlet, which must be treated surgically (Carmichael & Shaw, 2000; Hoffman & Hatch, 1996).

When we experience fear and anxiety, stimulant hormones released into our bloodstream cause us to be "poised for action." Large amounts of blood are sent to parts of the body involved in the defensive response—the brain, the heart, and muscles in the arms, legs, and trunk. Blood flow to other organs, including the uterus, is reduced. As a result, the fetus receives less oxygen and nutrients. Stress hormones also cross the placenta, causing the fetus's heart rate and activity level to rise dramatically. Finally, stress weakens the immune system, making pregnant women more susceptible to infectious disease (Cohen & Williamson, 1991; Monk et al., 2000).

But stress-related prenatal complications are greatly reduced when mothers have husbands, other family members, and friends who offer social support (McLean et al., 1993; Nuckolls, Cassel, & Kaplan, 1972). The link between social support and positive pregnancy outcomes is particularly strong for low-income women, who often lead highly stressful daily lives (Hoffman & Hatch, 1996).

● **Rh Factor Incompatibility.** When the inherited blood types of mother and fetus differ, serious problems sometimes result. The most common cause of these difficulties is **Rh factor incompatibility.** When the mother is Rh-negative (lacks the Rh blood protein) and the father is Rh-positive (has the protein), the baby may inherit the father's Rh-positive blood type. If even a little of a fetus's Rh-positive blood crosses the placenta into the Rh-negative mother's bloodstream, she begins to form antibodies to the foreign Rh protein. If these enter the fetus's system, they destroy red blood cells, reducing the oxygen supply to organs and tissues. Mental retardation, miscarriage, heart damage, and infant death can occur.

Since it takes time for the mother to produce Rh antibodies, first born children are rarely affected. The danger increases with each additional pregnancy. Fortunately, the harmful effects of Rh incompatibility can be prevented in most cases. After the birth of each Rh-positive baby, Rh-negative mothers are routinely given a vaccine to prevent the buildup of antibodies. In emergency cases, blood transfusions can be performed immediately after delivery or, if necessary, even before birth.

● **Maternal Age and Previous Births.** In Chapter 2, we noted that women who delay childbearing until their thirties or forties face increased risk of infertility, miscarriage, and babies born with chromosomal defects. Are other pregnancy complications more common for older mothers? For many years, scientists thought so. But healthy women in their forties have no more prenatal difficulties than women in their twenties (Bianco et al., 1996; Dildy et al., 1996; Prysak, Lorenz, & Kisly, 1995).

In the case of teenage mothers, does physical immaturity cause prenatal complications? Again, research shows that it does not. As we will see in Chapter 11, nature tries to ensure that once a girl can conceive, she is physically ready to carry and give birth to a baby. Infants of teenagers are born with a higher rate of problems for quite different reasons. Many pregnant adolescents do not have access to medical care or are afraid to seek it. In addition, most pregnant teenagers come from low-income backgrounds, where stress, poor nutrition, and health problems are common (Coley & Chase-Lansdale, 1998).

The Importance of Prenatal Health Care

Yolanda had her first prenatal appointment 3 weeks after missing her menstrual period. After that, she visited the doctor's office once a month until she was 7 months pregnant, then twice during the eighth month. As birth grew near, Yolanda's appointments increased to once a week. The doctor kept track of her general health, her weight gain, and the capacity of her uterus and cervix to support the fetus. The fetus's growth was also carefully monitored.

Yolanda's pregnancy, like most others, was free of complications. But unexpected difficulties can arise, especially when mothers have health problems to begin with. For example, women with diabetes need careful monitoring. Extra sugar in the mother's bloodstream causes the fetus to grow larger than average, making pregnancy and birth problems more common.

■ During a routine prenatal visit, this doctor uses ultrasound to show an expectant mother an image of her fetus and to evaluate its development. All pregnant women should receive early and regular prenatal care to protect their own health and the health of their babies.

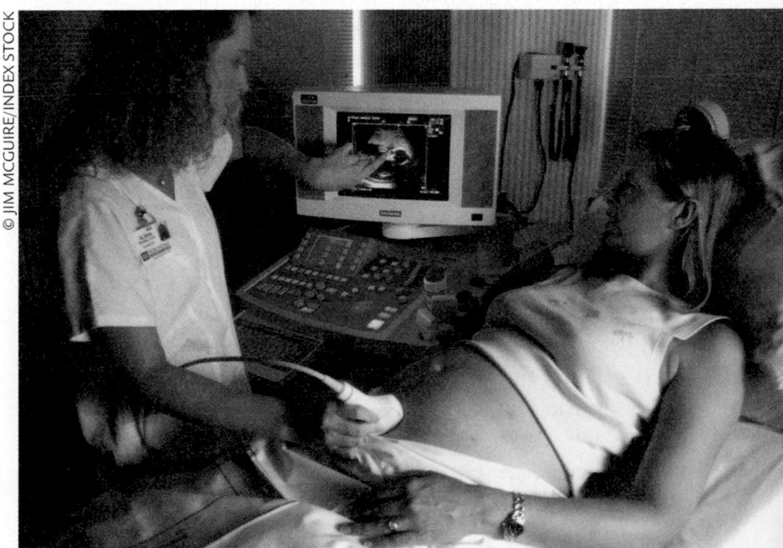

Dos and Don'ts for a Healthy Pregnancy

DO	DON'T
Do make sure that you have been vaccinated against infectious diseases dangerous to the embryo and fetus, such as rubella, before you get pregnant. Most vaccinations are not safe during pregnancy.	Don't take any drugs without consulting your doctor.
Do see a doctor as soon as you suspect that you are pregnant—within a few weeks after a missed menstrual period.	Don't smoke. If you have already smoked during part of your pregnancy, cut down or (better yet) quit. If other members of your family are smokers, ask them to quit or to smoke outside.
Do continue to get regular medical checkups throughout pregnancy.	Don't drink alcohol from the time you decide to get pregnant. If you find it difficult to give up alcohol, ask for help from your doctor, local family service agency, or nearest chapter of Alcoholics Anonymous.
Do obtain literature from your doctor, local library, and bookstore about prenatal development and care. Ask questions about anything you do not understand.	Don't engage in activities that might expose your baby to environmental hazards, such as chemical pollutants or radiation. If you work in an occupation that involves these agents, ask for a safer assignment or a leave of absence.
Do eat a well-balanced diet and take vitamin-mineral supplements as prescribed by your doctor. On average, a woman should increase her intake by 100 calories a day in the first trimester, 265 in the second, and 430 in the third. Gain 25 to 30 pounds gradually.	Don't engage in activities that might expose your baby to harmful infectious diseases, such as childhood illnesses and toxoplasmosis.
Do keep physically fit through mild exercise. If possible, join a special exercise class for expectant mothers.	Don't choose pregnancy as a time to go on a diet.
Do avoid emotional stress. If you are a single parent, find a relative or friend on whom you can count for emotional support.	Don't overeat and gain too much weight during pregnancy. A very large weight gain is associated with complications.
Do get plenty of rest. An overtired mother is at risk for pregnancy complications.	
Do enroll in a prenatal and childbirth education class along with your partner. When parents know what to expect, the 9 months before birth can be one of the most joyful times of life.	

Another complication, *toxemia* (sometimes called *eclampsia*), in which blood pressure increases sharply and the face, hands, and feet swell in the last half of pregnancy, is experienced by 5 to 10 percent of pregnant women. If untreated, toxemia can cause convulsions in the mother and fetal death. Usually, hospitalization, bed rest, and drugs can lower blood pressure to a safe level. If not, the baby must be delivered at once (Carlson, Eisenstat, & Ziporyn, 1996).

Unfortunately, 18 percent of pregnant women in the United States wait until after the first trimester to seek prenatal care, and 4 percent delay until the end of pregnancy or never get any at all. Most of these mothers are adolescents, unmarried, or poverty stricken. Their infants are far more likely to be born underweight and to die before birth or during the first year of life than are babies of mothers who receive early medical attention (Children's Defense Fund, 2002).

Why do these mothers delay going to the doctor? One reason is a lack of health insurance. Although the very poorest of these mothers are eligible for government-sponsored health services, many low-income women do not qualify. As we will see when we take up birth complications later in this chapter, in nations where affordable medical care is universally available, late-care pregnancies and maternal and infant health problems are greatly reduced.

Besides financial hardship, some mothers have other reasons for not seeking prenatal care. When researchers asked women who first went to the doctor late in pregnancy why they waited so long, they mentioned a wide variety of obstacles. These included situational barriers, such as difficulty finding a doctor and getting an appointment and lack of transportation. The women also mentioned many personal barriers—psychological stress, the demands of taking care of other young children, ambivalence about the pregnancy, and family crises (Maloni et al., 1996; Rogers & Shiff, 1996). Many were also engaging in high-risk behaviors, such as smoking and drug abuse. These women, who had no medical attention for most of their pregnancies, were among those who need it most!

Clearly, public education about the importance of early and sustained prenatal care for all pregnant women is badly needed. The Caregiving Concerns table above summarizes "dos and don'ts" for a healthy pregnancy, based on our discussion of the prenatal environment.

Ask Yourself

REVIEW

Why is it difficult to determine the effects of some environmental agents, such as over-the-counter drugs and pollution, on the embryo and fetus?

APPLY

Nora, pregnant for the first time, has heard about the teratogenic impact of alcohol and tobacco. Nevertheless, she believes that a few cigarettes and a glass of wine a day won't be harmful. Provide Nora with research-based reasons for not smoking or drinking.

CONNECT

List teratogens and other maternal factors that affect brain development during the prenatal period. Why is the central nervous system often affected when the prenatal environment is compromised?

CONNECT

What is a sensitive period? How is this concept relevant to understanding the impact of teratogens?

www.

Childbirth

Although Yolanda and Jay completed my course 3 months before their baby was born, both agreed to return the following spring to share their experiences with my next class. Two-week-old Joshua came along as well. Yolanda and Jay's story revealed that the birth of a baby is one of the most dramatic and emotional events in human experience. Jay was present throughout Yolanda's labor and delivery. Yolanda explained,

> By morning, we knew I was in labor. It was Thursday, so we went in for my usual weekly appointment. The doctor said, yes, the baby was on the way, but it would be a while. He told us to go home and relax or take a leisurely walk and come to the hospital in 3 or 4 hours. We checked in at 3 in the afternoon; Joshua arrived at 2 o'clock the next morning. When, finally, I was ready to deliver, it went quickly; a half-hour or so and some good hard pushes, and there he was! His body had stuff all over it, his face was red and puffy, and his head was misshapen, but I thought, "Oh! he's beautiful. I can't believe he's really here!"

Jay was also elated by Joshua's birth. "I wanted to support Yolanda and to experience as much as I could. It was awesome, indescribable," he said, holding Joshua over his shoulder and patting and kissing him gently. In the following sections, we explore the experience of childbirth, from both the parents' and the baby's point of view.

The Stages of Childbirth

It is not surprising that childbirth is often referred to as *labor.* It is the hardest physical work a woman may ever do. A complex series of hormonal changes initiates the process, which naturally divides into three stages (see Figure 3.4 on page 92).

1. *Dilation and effacement of the cervix.* This is the longest stage of labor, lasting, on the average, 12 to 14 hours in a first birth and 4 to 6 hours in later births. Contractions of the uterus gradually become more frequent and powerful, causing the cervix, or uterine opening, to open (dilate) and thin (efface). When the cervix opens completely, contractions reach a peak called *transition.* As a result, a clear channel from the uterus into the vagina, or birth canal, is formed.
2. *Birth of the baby.* This stage is much shorter, lasting about 50 minutes in a first birth and 20 minutes in later births. Strong contractions of the uterus continue, but the mother also feels a natural urge to squeeze and push with her abdominal muscles. As she does so with each contraction, she forces the baby down and out.
3. *Delivery of the placenta.* Labor comes to an end with a few final contractions and pushes. These cause the placenta to separate from the wall of the uterus and be delivered in about 5 to 10 minutes.

The Baby's Adaptation to Labor and Delivery

At first glance, labor and delivery seem like a dangerous ordeal for the baby. The strong contractions of Yolanda's uterus exposed Joshua's head to a great deal of pressure, and they squeezed the placenta and the umbilical cord repeatedly. Each time, Joshua's supply of oxygen was temporarily reduced.

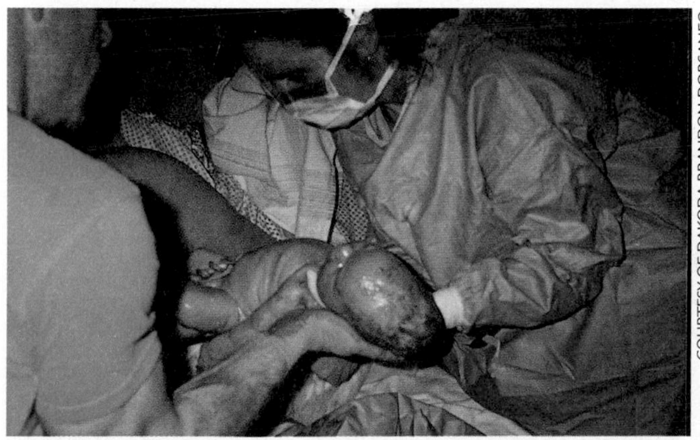

COURTESY OF DAKODA BRANDON DORSANEO

■ This newborn baby is held by his mother's birthing coach (on the left) and midwife (on the right) just after delivery. The umbilical cord has not yet been cut. Notice how the infant's head is molded from being squeezed through the birth canal for many hours. It is also very large in relation to his body. As the infant takes his first few breaths, his body turns from blue to pink. He is wide awake and ready to get to know his new surroundings.

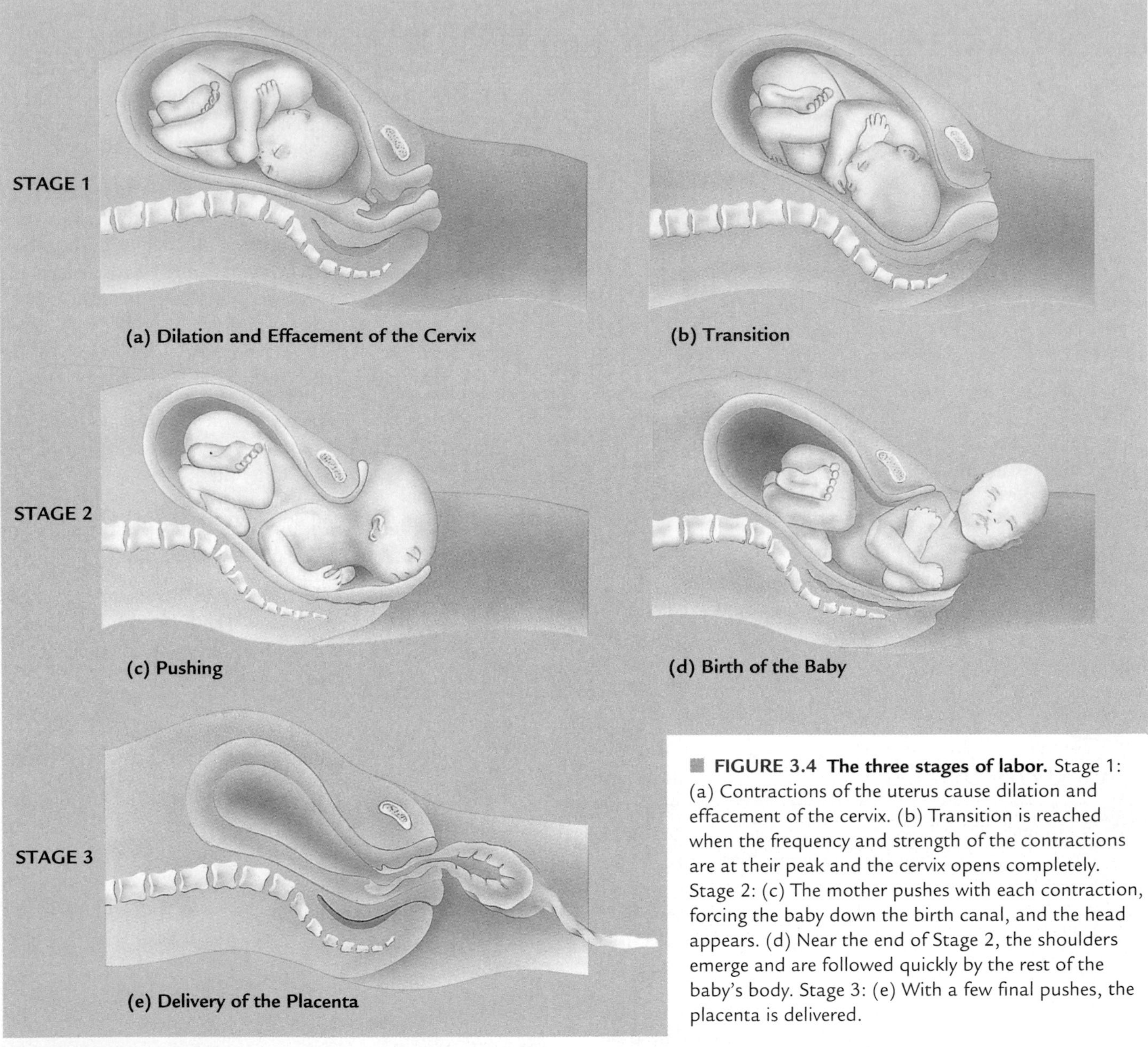

FIGURE 3.4 The three stages of labor. Stage 1: (a) Contractions of the uterus cause dilation and effacement of the cervix. (b) Transition is reached when the frequency and strength of the contractions are at their peak and the cervix opens completely. Stage 2: (c) The mother pushes with each contraction, forcing the baby down the birth canal, and the head appears. (d) Near the end of Stage 2, the shoulders emerge and are followed quickly by the rest of the baby's body. Stage 3: (e) With a few final pushes, the placenta is delivered.

Fortunately, healthy babies are well equipped to withstand the trauma of childbirth. The force of the contractions causes the infant to produce high levels of stress hormones. Recall that during pregnancy, the effects of maternal stress can endanger the baby. In contrast, during childbirth the infant's production of stress hormones is adaptive. It helps the baby withstand oxygen deprivation by sending a rich supply of blood to the brain and heart. In addition, it prepares the baby to breathe effectively by causing the lungs to absorb any remaining fluid and by expanding the bronchial tubes (passages leading to the lungs). Finally, stress hormones arouse the infant into alertness. Joshua was born wide awake, ready to interact with the surrounding world (Lagercrantz & Slotkin, 1986).

The Newborn Baby's Appearance

What do babies look like after birth? Jay smiled when my students asked this question. "Yolanda and I are probably the only people in the world who thought Joshua was beautiful!" The average newborn is 20 inches long and 7½ pounds in weight; boys tend to be slightly longer and heavier than girls. The head is very large in comparison to the trunk and legs, which are short and bowed. As we will see in later chapters, the combination of a big head (with its well-developed brain) and a small body means that human infants learn quickly in the first few months of life. But unlike most other mammals, they cannot get around on their own until much later.

Table 3.2 The Apgar Scale

Sign[a]	Score 0	1	2
Heart rate	No heartbeat	Under 100 beats per minute	100 to 140 beats per minute
Respiratory effort	No breathing for 60 seconds	Irregular, shallow breathing	Strong breathing and crying
Reflex irritability (sneezing, coughing, and grimacing)	No response	Weak reflexive response	Strong reflexive response
Muscle tone	Completely limp	Weak movements of arms and legs	Strong movements of arms and legs
Color[b]	Blue body, arms, and legs	Body pink with blue arms and legs	Body, arms, and legs completely pink

[a] To remember these signs, you may find it helpful to use a technique in which the original labels are reordered and renamed as follows: color = **A**ppearance, heart rate = **P**ulse, reflex irritability = **G**rimace, muscle tone = **A**ctivity, and respiratory effort = **R**espiration. Together, the first letters of the new labels spell **Apgar**.

[b] Color is the least reliable of the Apgar signs. The skin tone of nonwhite babies makes it difficult to apply the "pink" criterion. However, newborns of all races can be rated for pinkish glow resulting from the flow of oxygen through body tissues.

Source: Apgar, 1953.

Even though newborn babies may look strange, some features do make them attractive. Their round faces, chubby cheeks, large foreheads, and big eyes make adults feel like picking them up and cuddling them (Berman, 1980; Lorenz, 1943).

Assessing the Newborn's Physical Condition: The Apgar Scale

Infants who have difficulty making the transition to life outside the uterus must be given special help at once. To assess the infant's physical condition quickly, doctors and nurses use the **Apgar Scale.** As Table 3.2 shows, a rating of 0, 1, or 2 on each of five characteristics is made at 1 and 5 minutes after birth. A combined Apgar score of 7 or better indicates that the infant is in good physical condition. If the score is between 4 and 6, the baby requires assistance in establishing breathing and other vital signs. If the score is 3 or below, the infant is in serious danger and requires emergency medical attention. Two Apgar ratings are given, since some babies have trouble adjusting at first but do quite well after a few minutes (Apgar, 1953).

Approaches to Childbirth

Childbirth practices, like other aspects of family life, are molded by the society of which mother and baby are a part. In many village and tribal cultures, expectant mothers are well acquainted with the childbirth process. For example, the Jarara of South America and the Pukapukans of the Pacific Islands treat birth as a vital part of daily life. The Jarara mother gives birth in full view of the entire community, including small children. The Pukapukan girl is so familiar with the events of labor and delivery that she can frequently be seen playing at it. Using a coconut to represent the baby, she stuffs it inside her dress, imitates the mother's pushing, and lets the nut fall at the proper moment. In most nonindustrialized cultures, women are assisted during the birth process. Among the Mayans of the Yucatán, the mother leans against the body of a woman called the "head helper," who supports her weight and breathes with her during each contraction (Jordan, 1993; Mead & Newton, 1967).

In large Western nations, childbirth has changed dramatically over the centuries. Before the late 1800s, birth usually took place at home and was a family-centered event. The industrial revolution brought greater crowding to cities, along with new health problems. As a result, childbirth moved from home to the hospital, where the health of mothers and babies could be protected. Once doctors assumed responsibility for childbirth, women's knowledge of it declined, and relatives and friends were no longer welcome to participate (Borst, 1995).

By the 1950s and 1960s, women started to question the medical procedures that came to be used routinely during labor and delivery. Many felt that frequent use of strong drugs and delivery instruments had robbed them of a precious experience and were often not necessary or safe for the baby. Gradually, a natural childbirth movement arose in Europe and spread to North America. Its purpose was to make hospital birth as comfortable and rewarding for mothers as possible. Today, most hospitals carry this theme further by offering birth centers that are family centered and homelike. *Freestanding birth centers,* which encourage early contact between parents and baby but offer less backup medical care, also exist. And a small but growing number of North American women are rejecting institutional birth entirely and choosing to have their babies at home.

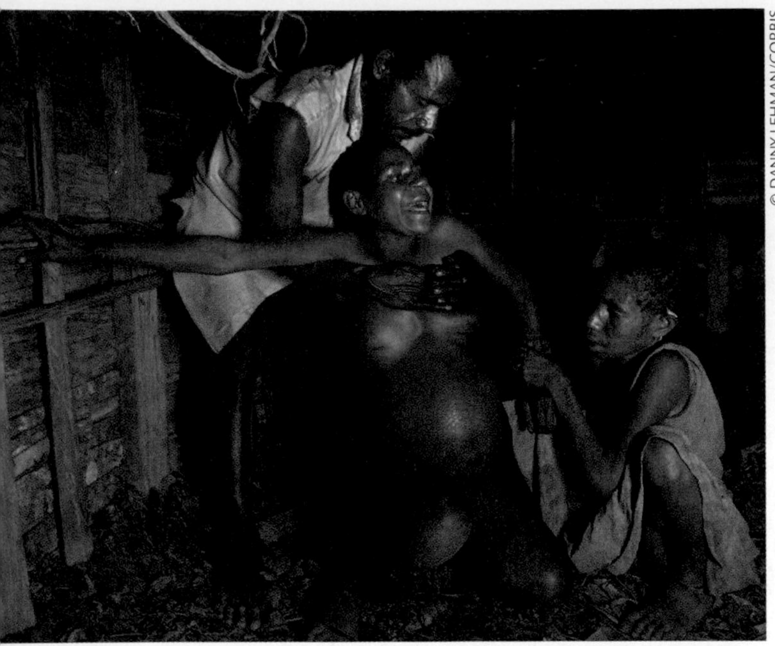

■ In this village society in Papua New Guinea, a woman gives birth in an upright squatting position. Her husband supports her body while an elderly woman helper soothes and encourages.

Natural, or Prepared, Childbirth

Yolanda and Jay chose **natural,** or **prepared, childbirth**—a group of techniques aimed at reducing pain and medical intervention and making childbirth as rewarding an experience as possible. Although many natural childbirth programs exist, most draw on methods developed by Grantly Dick-Read (1959) in England and Ferdinand Lamaze (1958) in France. These physicians recognized that cultural attitudes had taught women to fear the birth experience. An anxious, frightened woman in labor tenses her muscles, turning the mild pain that sometimes accompanies strong contractions into a great deal of pain.

In a typical natural childbirth program, the expectant mother and a companion (the father, a relative, or a friend) participate in three activities:

- *Classes.* Yolanda and Jay attended a series of classes in which they learned about the anatomy and physiology of labor and delivery. Knowledge about the birth process reduces a mother's fear.
- *Relaxation and breathing techniques.* During each class, Yolanda was taught relaxation and breathing exercises aimed at counteracting the pain of uterine contractions.
- *Labor coach.* Jay learned how to help Yolanda during childbirth by reminding her to relax and breathe, massaging her back, supporting her body, and offering words of encouragement and affection.

Studies comparing mothers who experience natural childbirth with those who do not reveal more positive attitudes toward the childbirth experience, less pain, and less use of medication—usually very little or none at all (Hetherington, 1990; Mackey, 1995). Social support is an important part of the success of natural childbirth techniques. In Guatemalan and American hospitals that routinely isolated patients during childbirth, some mothers were randomly assigned a companion who stayed with them throughout labor and delivery, talking to them, holding their hands, and rubbing their backs to promote relaxation. These mothers had fewer birth complications and shorter labors than women with no companionship. Guatemalan mothers receiving support also interacted more positively with their babies during the first hour after delivery, talking, smiling, and gently stroking (Kennell et al., 1991; Sosa et al., 1980). Furthermore, social support makes Western hospital-birth customs more acceptable to women from parts of the world where assistance from family and community members is the norm (Granot et al., 1996).

Home Delivery

Home birth has always been popular in certain industrialized nations, such as England, the Netherlands, and Sweden. The number of North American women choosing to have their babies at home has grown in recent years, although it remains small, at about 1 percent (Curtin & Park, 1999). Some home births are attended by doctors, but most are handled by certified *nurse–midwives* who have degrees in nursing and additional training in childbirth management.

The joys and perils of home delivery are well illustrated by the story that Don, who painted my house as I worked on this

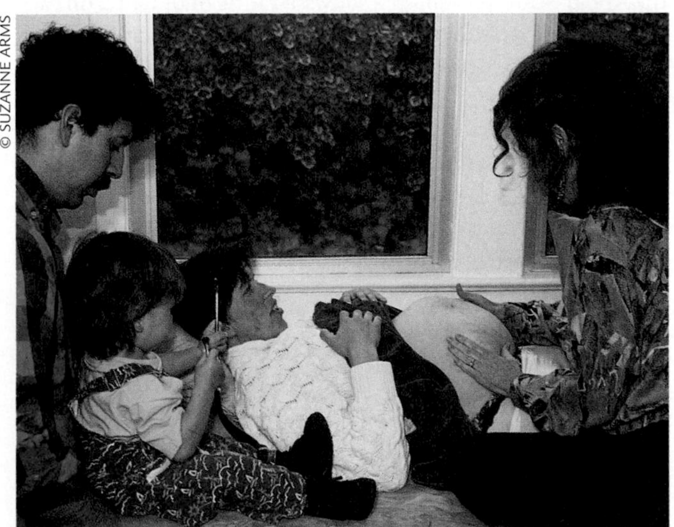

■ About to give birth at home, this mother discusses the progress of her labor with the midwife while her husband and their older child look on. Mothers who choose home birth want to make the experience an important part of family life, avoid unnecessary medical procedures, and exercise greater control over their own care and that of their babies.

book, related to me. "Our first child was delivered in the hospital," he said. "Even though I was present, Kathy and I found the atmosphere to be rigid and insensitive. We wanted a warmer, more personal birth environment." With the coaching of a nurse–midwife, Don delivered their second child, Cindy, at their farmhouse, 3 miles out of town. Three years later, when Kathy went into labor with Marnie, a heavy snowstorm prevented the midwife from reaching the house on time. Don delivered the baby alone, but the birth was difficult. Marnie failed to breathe for several minutes; with great effort, Don revived her. The frightening memory of Marnie's limp, blue body convinced Don and Kathy to return to the hospital to have their last child. By then, hospital practices had changed, and the event was a rewarding one for both parents.

Don and Kathy's experience raises the question: Is it just as safe to give birth at home as in a hospital? For healthy women who are assisted by a well-trained doctor or midwife, it seems so, since complications rarely occur (Olsen, 1997). However, if attendants are not carefully trained and prepared to handle emergencies, the rate of infant death is high (Mehlmadrona & Madrona, 1997). When mothers are at risk for any kind of complication, the appropriate place for labor and delivery is the hospital, where lifesaving treatment is available.

Medical Interventions

Two-year-old Melinda walks with a halting, lumbering gait and has difficulty keeping her balance. She has *cerebral palsy,* a general term for a variety of impairments in muscle coordination that result from brain damage before, during, or just after birth.

Like 10 percent of youngsters with cerebral palsy, Melinda's brain damage was caused by **anoxia,** or inadequate oxygen supply, during labor and delivery (Anslow, 1998). Her mother got pregnant accidentally, was frightened and alone, and arrived at the hospital at the last minute. Melinda was in **breech position,** turned so that the buttocks or feet would be delivered first, and the umbilical cord was wrapped around her neck. Had her mother come to the hospital earlier, doctors could have monitored Melinda's condition and delivered her surgically as soon as squeezing of the umbilical cord led to distress, reducing the damage or preventing it entirely.

In cases like Melinda's, medical interventions during childbirth are clearly justified. But in others, they can interfere with delivery and even pose new risks. In the following sections, we examine some commonly used medical techniques.

Fetal Monitoring

Fetal monitors are electronic instruments that track the baby's heart rate during labor. An abnormal heartbeat may indicate that the baby is in distress due to anoxia and needs to be delivered immediately. Most American hospitals require continuous fetal monitoring; it is used in over 80 percent of American births. In Canada, continuous monitoring is usually reserved for babies at risk for birth complications (Banta & Thacker, 2001; Liston et al., 2002). The most popular type of monitor is strapped across the mother's abdomen throughout labor. A second, more accurate method involves threading a recording device through the cervix and placing it directly under the baby's scalp.

Fetal monitoring is a safe medical procedure that has saved the lives of many babies in high-risk situations. Nevertheless, the practice is controversial. In healthy pregnancies, it does not reduce the rate of infant brain damage or death. Critics also worry that fetal monitors identify many babies as in danger who, in fact, are not (Berkus et al., 1999; Thacker, Stroup, & Chang, 2001). Monitoring is linked to an increased rate of cesarean (surgical) deliveries, which we will discuss shortly. In addition, some women complain that the devices are uncomfortable, prevent them from moving easily, and interfere with the normal course of labor.

Still, it is likely that fetal monitors will continue to be used routinely in the United States, even though they are not necessary in most cases. Today, doctors can be sued for malpractice if an infant dies or is born with problems and they cannot show that they did everything possible to protect the baby.

Labor and Delivery Medication

Some form of medication is used in 80 to 95 percent of North American births (Glosten, 1998). *Analgesics,* drugs used to relieve pain, may be given in mild doses during labor to help a mother relax. *Anesthetics* are a stronger type of painkiller that blocks sensation. A regional anesthetic may be injected into the spinal column to numb the lower half of the body.

Although pain-relieving drugs enable doctors to perform essential life-saving medical interventions, they can cause problems when used routinely. Anesthesia weakens uterine contractions during the first stage of labor and interferes with the mother's ability to feel contractions and push during the second stage. As a result, labor is prolonged (Alexander et al., 1998). In addition, since labor and delivery medication rapidly crosses the placenta, the newborn baby may be sleepy and withdrawn, suck poorly during feedings, and be irritable when awake (Emory, Schlackman, & Fiano, 1996).

Does use of medication during childbirth have a lasting impact on physical and mental development? Some researchers claim so (Brackbill, McManus, & Woodward, 1985). However, their findings have been challenged (Golub, 1996; Riordan et al., 2000). Anesthesia may be related to other risk factors that could account for long-term consequences in some studies, but more research is needed to sort out these effects.

Cesarean Delivery

A **cesarean delivery** is a surgical birth; the doctor makes an incision in the mother's abdomen and lifts the baby out of the uterus. Thirty years ago, cesarean delivery was rare. Since then, the cesarean rate has climbed. Today, it is about 20 percent in the United States and Canada (Health Canada, 2000b; U.S. Bureau of the Census, 2002c). Some countries, such as Japan and the Netherlands, have cesarean rates of less than 7 percent (Samuels & Samuels, 1996). Yet these nations, as we will see shortly, have very low infant death rates.

Cesareans have always been warranted by medical emergencies, such as Rh incompatibility, premature separation of the placenta from the uterus, or serious maternal illness or infection (for example, the herpes simplex 2 virus, which can infect the baby during a vaginal delivery). However, surgical delivery is not always needed in other instances. For example, although the most common reason for a cesarean is a previous cesarean, the technique used today—a small horizontal cut in the lower part of the uterus—makes vaginal birth safe in later pregnancies. Cesareans are often justified in breech births, in which the baby risks head injury or anoxia (as in Melinda's case). But the infant's exact position makes a difference. Certain breech babies fare just as well with a normal delivery as with a cesarean (Ismail et al., 1999). Sometimes the doctor can gently turn the baby into a head-down position during the early part of labor (Flamm & Quilligan, 1995).

When a cesarean delivery does occur, both mother and baby need extra support. Although the operation is quite safe, it requires more time for recovery. Because anesthetic may have crossed the placenta, newborns are more likely to be sleepy and unresponsive and to have breathing difficulties (Cox & Schwartz, 1990). These factors can negatively affect the early mother–infant relationship.

Ask Yourself

REVIEW

Describe the elements and benefits of natural childbirth. What aspect contributes greatly to favorable outcomes, and why?

REVIEW

Use of one medical intervention during childbirth increases the likelihood that others will also be used. Provide as many examples as you can to illustrate this idea.

CONNECT

How have history-graded influences (see Chapter 1, page 11) affected approaches to childbirth in Western nations? What effects have these changes had on the health and adjustment of mothers and newborn babies?

www.

Preterm and Low-Birth-Weight Infants

Babies born 3 weeks or more before the end of a full 38-week pregnancy or who weigh less than $5\frac{1}{2}$ pounds have, for many years, been referred to as "premature." A wealth of research indicates that premature babies are at risk for many problems. Birth weight is the best available predictor of infant survival and healthy development. Many newborns who weigh less than $3\frac{1}{3}$ pounds experience difficulties that are not overcome, an effect that becomes stronger as birth weight decreases (see Figure 3.5) (Minde, 2000; Palta et al., 2000). Frequent illness, inattention, overactivity, language delays, low intelligence test scores, and deficits in motor coordination and school learning are some of the difficulties that extend into childhood (Hack et al., 1994, 1995; Mayes & Bornstein, 1997).

About 1 in 14 American infants and 1 in 18 Canadian infants is born underweight. Although the problem can strike unexpectedly, it is highest among poverty-stricken women. Consequently, many ethnic minority and teenage mothers have low-birth-weight newborns (Children's Defense Fund, 2002; Statistics Canada, 2001a). These mothers, as noted earlier, are more likely to be undernourished and to be exposed to other harmful

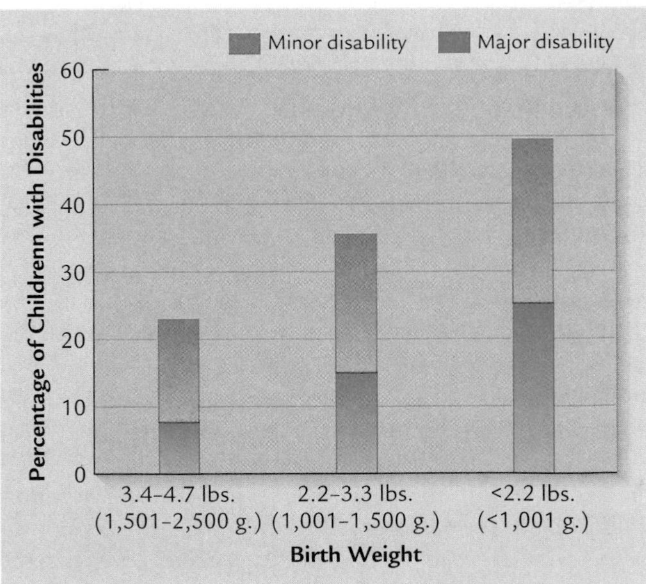

FIGURE 3.5 Incidence of major and minor disabilities by birth weight, obtained from studies of low-birth-weight children at school age. *Major disabilities* include cerebral palsy, mental retardation, and vision and hearing impairments. *Minor disabilities* include slightly below-average intelligence, learning disabilities (usually in reading, spelling, and math), mild difficulties in motor control, and behavior problems (including poor attention and impulse control, aggressiveness, noncompliance, depression, passivity, anxiety, and difficulty separating from parents). (Adapted from D'Agostino & Clifford, 1998.)

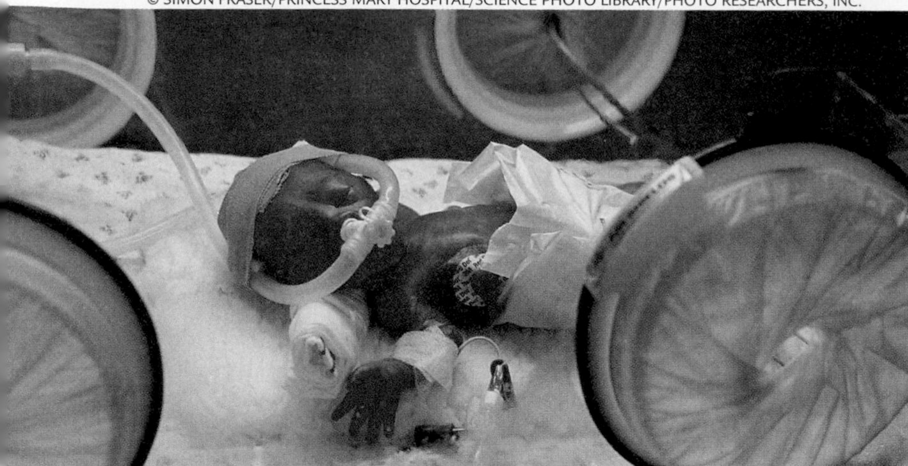

■ This baby was born 13 weeks before her due date and weighs little more than 2 pounds. Because her lungs are too immature to function independently, she breathes with the aid of a respirator. Her survival and development are seriously at risk.

environmental influences. In addition, they often do not receive the prenatal care necessary to protect their vulnerable babies.

Recall from Chapter 2 that prematurity is also common among twins. Because space inside the uterus is restricted, twins gain less weight after the twentieth week of pregnancy than do singletons.

Preterm versus Small for Date

Although low-birth-weight infants face many obstacles to healthy development, most go on to lead normal lives; half of those who weighed only a few pounds at birth have no disability (refer again to Figure 3.5). To better understand why some babies do better than others, researchers divide them into two groups. **Preterm** infants are born several weeks or more before their due date. Although they are small, their weight may still be appropriate, based on time spent in the uterus. **Small-for-date** babies are below their expected weight when length of the pregnancy is taken into account. Some small-for-date infants are actually full term. Others are preterm infants who are especially underweight.

Of the two types of babies, small-for-date infants usually have more serious problems. During the first year, they are more likely to die, catch infections, and show evidence of brain damage. By middle childhood, they have lower intelligence test scores, are less attentive, achieve less well in school, and are socially immature (Hediger et al., 2002; Schothorst & van Engeland, 1996). Small-for-date infants probably experienced inadequate nutrition before birth. Perhaps their mothers did not eat properly, the placenta did not function normally, or the babies themselves had defects that prevented them from growing as they should.

Consequences for Caregiving

Imagine a scrawny, thin-skinned infant whose body is only a little larger than the size of your hand. You try to play with the baby by stroking and talking softly, but he is sleepy and unresponsive. When you feed him, he sucks poorly. He is usually irritable during the short, unpredictable periods in which he is awake.

The appearance and behavior of preterm babies can lead parents to be less sensitive and responsive in caring for them. Compared with full-term infants, preterm babies—especially those who are very ill at birth—are less often held close, touched, and talked to gently. At times, mothers of these infants resort to interfering pokes and verbal commands in an effort to obtain a higher level of response from the baby (Barratt, Roach, & Leavitt, 1996). This may explain why preterm babies as a group are at risk for child abuse. When they are born to isolated, poverty-stricken mothers who cannot provide good nutrition, health care, and parenting, the likelihood of unfavorable outcomes is increased. In contrast, parents with stable life circumstances and social supports usually can overcome the stresses of caring for a preterm infant. In these cases, even sick preterm babies have a good chance of catching up in development by middle childhood (Liaw & Brooks-Gunn, 1993).

These findings suggest that how well preterm babies develop has a great deal to do with the parent–child relationship. Consequently, interventions directed at supporting both sides of this tie are more likely to help these infants recover.

Interventions for Preterm Infants

A preterm baby is cared for in a special Plexiglas-enclosed bed called an *isolette.* Temperature is carefully controlled because these babies cannot yet regulate their own body temperature effectively. To help protect the baby from infection, air is filtered before it enters the isolette. Infants born more than 6 weeks early commonly have a disorder called *respiratory distress syndrome* (otherwise known as *hyaline membrane disease*). Their tiny lungs are so poorly developed that the air sacs collapse, causing serious breathing difficulties. When a preterm infant breathes with the aid of a respirator, is fed through a stomach tube, and receives medication through an intravenous needle, the isolette can be very isolating indeed! Physical needs that otherwise would lead to close contact and other human stimulation are met mechanically.

● **Special Infant Stimulation.** At one time, doctors believed that stimulating such a fragile baby could be harmful. Now we know that in proper doses, certain kinds of stimulation can help preterm infants develop. In some intensive care nurseries, preterm babies can be seen rocking in suspended hammocks or lying on waterbeds designed to replace the gentle motion they would have received while still in the mother's uterus. Other forms of stimulation have also been used—an attractive mobile or a tape recording of a heartbeat, soft music, or the mother's voice. These experiences promote faster weight gain, more predictable sleep patterns, and greater alertness (Marshall-Baker, Lickliter, & Cooper, 1998; Standley, 1998).

Touch is an especially important form of stimulation. In baby animals, touching the skin releases certain brain chemicals

Social Issues

A Cross-National Perspective on Health Care and Other Policies for Parents and Newborn Babies

Infant mortality is an index used around the world to assess the overall health of a nation's children. It refers to the number of deaths in the first year of life per 1,000 live births. Although the United States has the most up-to-date health care technology in the world, it has made less progress than many other countries in reducing infant deaths. Over the past three decades, it has slipped in the international rankings, from 7th in the 1950s to 24th in the year 2001. Members of America's poor ethnic minorities, African-American babies especially, are at greatest risk. Black infants are more than twice as likely as white infants to die in the first year of life (U.S. Bureau of the Census, 2002c). Canada, in contrast, has achieved one of the lowest infant mortality rates in the world. It ranks 15th and differs only slightly from top-ranked countries. Still, infant mortality rates among Canada's lowest-income groups are two to four times higher than the nation's as a whole (Health Canada, 2000b).

Neonatal mortality, the rate of death within the first month of life, accounts for 67 percent of the infant death rate in the United States and for 80 percent in Canada. Two factors are largely responsible for neonatal mortality. The first is serious physical defects, most of which cannot be prevented. The percentage of babies born with physical defects is about the same in all ethnic and income groups. The second leading cause of neonatal mortality is low birth weight, which is largely preventable. African-American and Canadian Aboriginal babies are four times more likely to be born early and underweight than are white infants (Children's Defense Fund, 2002; Health Canada, 2002g).

Widespread poverty and, in the United States, weak health care programs for mothers and young children are largely responsible for these trends. Each country in Figure 3.6 that outranks the United States in infant survival provides all its citizens with government-sponsored health care benefits. And each takes extra steps to make sure that pregnant mothers and babies have access to good nutrition, high-quality medical care, and social and economic supports that promote effective parenting.

For example, all Western European nations guarantee women a certain number of prenatal visits at very low or no cost. After a baby is born, a health professional routinely visits the home to provide counseling about infant care and to arrange continuing medical services. Home assistance is especially extensive in the Netherlands. For a token fee, each mother is granted a specially trained maternity helper, who assists with infant care, shopping, housekeeping, meal preparation, and the care of other children during the days after delivery (Buekens et al., 1993; Kamerman, 1993).

Paid, job-protected employment leave is another vital societal intervention for new parents. Canadian mothers or fathers are eligible for up to 1 year of parental leave. Paid leave is widely available in other industrialized nations as well. Sweden has the most generous parental leave program in the world. Parents have the right to paid birth leave of 2 weeks for fathers plus 15 months of paid leave to share between them (Seward, Yeats, & Zottarelli, 2002). Even less-developed nations provide parental leave benefits. For example, in the People's Republic of China, a new mother is granted 3 months' leave at regular pay. Furthermore, many countries supplement basic paid leave. In Germany, for example, after a fully paid 3-month leave, a parent may

---

that support physical growth—effects believed to occur in humans as well (Field, 1998). When preterm infants were massaged several times each day in the hospital, they gained weight faster and, at the end of the first year, were advanced in mental and motor development over preterm babies not given this stimulation (Field, 2001; Field et al., 1986). In developing countries where hospitalization is not always possible, skin-to-skin "kangaroo baby care," in which the preterm infant is tucked between the mother's breasts, is being encouraged. The technique is used often in Western nations as a supplement to hospital intensive care. It fosters oxygenation of the baby's body, temperature regulation, improved sleep and feeding, and infant survival. In addition, mothers who practice kangaroo care feel more confident about handling and meeting the needs of their infants (Gale & VandenBerg, 1998).

● **Training Parents in Infant Caregiving Skills.** When preterm infants develop more quickly, parents are likely to feel encouraged and to interact with the baby more effectively. Interventions that support the parenting side of this relationship generally teach parents about the infant's characteristics and promote caregiving skills. For parents with the economic and personal resources to care for a low-birth-weight infant, just a few sessions of coaching in recognizing and responding to the baby's needs are linked to steady gains in mental test performance that, after several years, equal those of full-term children (Achenbach et al., 1990). Warm parenting that helps preterm infants sustain attention (for example, gently commenting on and showing the baby features of a toy) is especially helpful in promoting favorable early cognitive and language development (Smith et al., 1996).

When preterm infants live in stressed, low-income households, long-term, intensive intervention is necessary. In the Infant Health and Development Project, preterm babies born into poverty received a comprehensive intervention that combined medical follow-up, weekly parent training sessions, and enrollment in cognitively stimulating child care. More than four times as many intervention children as controls (39 versus

take 2 more years at a modest flat rate and a third year at no pay (Kamerman, 2000). Yet in the United States, the federal government mandates *only 12 weeks of unpaid leave* for employees in businesses with at least 50 workers. In 2002, California became the first state to guarantee a mother or father paid leave—up to 6 weeks at half salary.

But research indicates that 6 weeks of childbirth leave (the norm in the United States) is too short. When a family is stressed by a baby's arrival, leaves of 6 weeks or less are linked to maternal anxiety and depression and to negative interactions with the baby. Longer leaves of 12 weeks or more predict favorable maternal mental health and sensitive, responsive caregiving (Clark et al., 1997; Hyde et al., 1995). Single women and their babies are most hurt by the absence of a generous national paid leave policy. These mothers are usually the sole source of support for their families and can least afford to take time from their jobs.

In countries with low infant mortality rates, expectant parents need not wonder how they will get health care and other resources to support their baby's development. The powerful impact of universal, high-quality medical and social services on maternal and infant well-being provides strong justification for implementing similar programs in the United States.

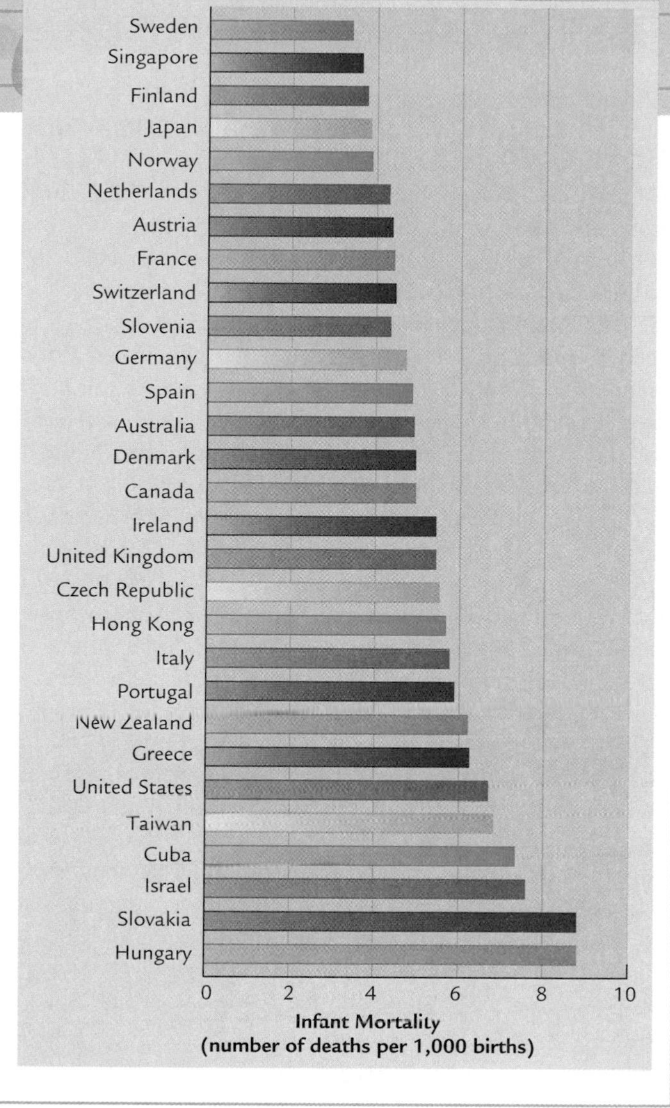

FIGURE 3.6 Infant mortality in 29 nations. Despite its advanced health care technology, the United States ranks poorly. It is 24th in the world, with a death rate of 6.7 infants per 1,000 births. Canada grants all its citizens government-funded health care and ranks 15th. Its infant death rate is 4.9 per 1,000 births. (Adapted from U.S. Bureau of the Census, 2002a.)

9 percent) were within normal range at age 3 in intelligence, psychological adjustment, and physical growth (Bradley et al., 1994). In addition, mothers in the intervention group were more affectionate and more often encouraged play and cognitive mastery in their children—one reason their 3-year-olds may have been developing so favorably (McCarton, 1998).

Yet by age 5, the intervention children had lost ground. And by age 8, the development of intervention and control children no longer differed (Brooks-Gunn et al., 1994; McCarton et al., 1997). These very vulnerable children need high-quality intervention well beyond age 3—even into the school years. And special strategies, such as extra adult–child interaction, may be necessary to achieve lasting changes in children with the lowest birth weights (Berlin et al., 1998).

Finally, the high rate of underweight babies in the United States—one of the worst in the industrialized world—could be greatly reduced by improving the health and social conditions described in the Social Issues box above. Fortunately, today we can save many preterm infants. But an even better course of action would be to prevent this serious threat to infant survival and development before it happens.

Understanding Birth Complications

In the preceding sections, we discussed a variety of birth complications. Now let's try to put the evidence together. Can any general principles help us understand how infants who survive a traumatic birth are likely to develop? A landmark study carried out in Hawaii provides answers to this question.

In 1955, Emmy Werner began to follow the development of nearly 700 infants on the island of Kauai who experienced either mild, moderate, or severe birth complications. Each was matched, on the basis of SES and ethnicity, with a healthy newborn (Werner & Smith, 1982). Findings revealed that the likelihood of long-term difficulties increased if birth trauma had been severe. But among mildly to moderately stressed children, the best predictor of how well they did in later years was the quality of their home environments. Children growing up in stable families did almost as well on measures of intelligence and psychological adjustment as those with no birth problems. Those exposed to poverty, family disorganization, and mentally ill parents often developed serious learning difficulties, behavior problems, and emotional disturbance.

The Kauai study tells us that as long as birth injuries are not overwhelming, a supportive home environment can restore children's growth. But the most intriguing cases in this study were the handful of exceptions. A few children with fairly serious birth complications and troubled family environments grew into competent adults who fared as well as controls in career attainment and psychological adjustment. Werner found that these children relied on factors outside the family and within themselves to overcome stress. Some had especially attractive personalities that caused them to receive positive responses from relatives, neighbors, and peers. In other instances, a grandparent, aunt, uncle, or baby-sitter established a warm relationship with the child and provided the needed emotional support (Werner, 1989, 1993, 2001; Werner & Smith, 1992).

Do these outcomes remind you of the characteristics of resilient children, discussed in Chapter 1? The Kauai study—and other similar investigations—reveal that the impact of early biological risks often wanes as children's personal characteristics and social experiences contribute increasingly to their functioning (Laucht, Esser, & Schmidt, 1997; Resnick et al., 1999). In sum, when the overall balance of life events tips toward the favorable side, children with serious birth problems can develop successfully.

Ask Yourself

REVIEW

Sensitive care can help preterm infants recover, but unfortunately they are less likely to receive this kind of care than are full-term newborns. Explain why.

APPLY

Cecilia and Adena each gave birth to a 3-pound baby 7 weeks preterm. Cecilia is single and on welfare. Adena and her husband are happily married and earn a good income. Plan an intervention for helping each baby develop.

CONNECT

List all the factors discussed in this chapter that increase the chances that an infant will be born underweight. How many of these factors could be prevented by better health care for mothers and babies?

www.

The Newborn Baby's Capacities

Newborn infants have a remarkable set of capacities that are crucial for survival and for evoking attention and care from parents. In relating to the physical world and building their first social relationships, babies are active from the very start.

Newborn Reflexes

A **reflex** is an inborn, automatic response to a particular form of stimulation. Reflexes are the newborn baby's most obvious organized patterns of behavior. As Jay placed Joshua on a table in my classroom, we saw several. When Jay bumped the side of the table, Joshua reacted by flinging his arms wide and bringing them back toward his body. As Yolanda stroked Joshua's cheek, he turned his head in her direction. When she put her finger in Joshua's palm, he grabbed on tightly. Look at Table 3.3 and see if you can name the newborn reflexes that Joshua displayed.

Some reflexes have survival value. For example, the rooting reflex helps the infant locate the mother's nipple. And if sucking were not automatic, our species would be unlikely to survive for a single generation!

A few reflexes form the basis for complex motor skills that will develop later. The stepping reflex looks like a primitive walking response. In infants who gain weight quickly in the weeks after birth, the stepping reflex drops out because thigh

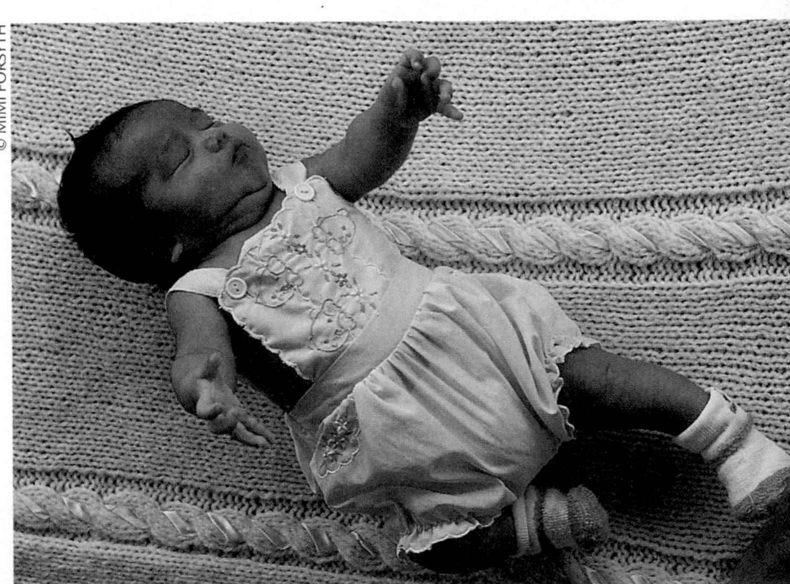

In the Moro reflex, loss of support or a sudden loud sound causes this baby to arch her back, extend her arms outward, and then bring them in toward her body.

Table 3.3 Some Newborn Reflexes

Reflex	Stimulation	Response	Age of Disappearance	Function
Eye blink	Shine bright light at eyes or clap hand near head	Infant quickly closes eyelids	Permanent	Protects infant from strong stimulation
Withdrawal	Prick sole of foot with pin	Foot withdraws, with flexion of knee and hip	Weakens after 10 days	Protects infant from unpleasant tactile stimulation
Rooting	Stroke cheek near corner of mouth	Head turns toward source of stimulation	3 weeks (becomes voluntary head turning at this time)	Helps infant find the nipple
Sucking	Place finger in infant's mouth	Infant sucks finger rhythmically	Replaced by voluntary sucking after 4 months	Permits feeding
Swimming	Place infant face down in pool of water	Baby paddles and kicks in swimming motion	4–6 months	Helps infant survive if dropped into water
Moro	Hold infant horizontally on back and let head drop slightly, or produce a sudden loud sound against surface supporting infant	Infant makes an "embracing" motion by arching back, extending legs, throwing arms outward, and then bringing arms in toward the body	6 months	In human evolutionary past, may have helped infant cling to mother
Palmar grasp	Place finger in infant's hand and press against palm	Spontaneous grasp of finger	3–4 months	Prepares infant for voluntary grasping
Tonic neck	Turn baby's head to one side while lying awake on back	Infant lies in a "fencing position." One arm is extended in front of eyes on side to which head is turned, other arm is flexed	4 months	May prepare infant for voluntary reaching
Stepping	Hold infant under arms and permit bare feet to touch a flat surface	Infant lifts one foot after another in stepping response	2 months in infants who gain weight quickly; sustained in lighter infants	Prepares infant for voluntary walking
Babinski	Stroke sole of foot from toe toward heel	Toes fan out and curl as foot twists in	8-12 months	Unknown

Sources: Knobloch & Pasamanick, 1974; Prechtl & Beintema, 1965; Thelen, Fisher, & Ridley-Johnson, 1984.

and calf muscles are not strong enough to lift the baby's chubby legs. But if the infant's lower body is dipped in water, the reflex reappears because the buoyancy of the water lightens the load on the baby's muscles (Thelen, Fisher, & Ridley-Johnson, 1984). When the stepping reflex is exercised regularly, babies make more reflexive stepping movements and are likely to walk several weeks earlier than if stepping is not practiced (Zelazo, 1983; Zelazo et al., 1993). However, there is no special need for infants to practice the stepping reflex because all normal babies walk in due time.

Some reflexes help parents and infants establish gratifying interaction. A baby who searches for and successfully finds the nipple, sucks easily during feedings, and grasps when the hand is touched encourages parents to respond lovingly and feel competent as caregivers. Reflexes can also help caregivers comfort the baby because they permit infants to control stress and amount of stimulation. For example, on short trips with Joshua to the grocery store, Yolanda brought along a pacifier. If he became fussy, sucking helped quiet him until she could feed, change, or hold and rock him.

Look at Table 3.3 again, and you will see that most newborn reflexes disappear during the first 6 months. Researchers believe that this is due to a gradual increase in voluntary control over behavior as the cerebral cortex matures. Pediatricians test reflexes carefully, especially if a newborn has experienced birth trauma, because reflexes can reveal the health of the baby's nervous system. Weak or absent reflexes, overly rigid or exaggerated reflexes, and reflexes that persist beyond the point in development when they should normally disappear can signal brain damage (Zafeiriou, 2000).

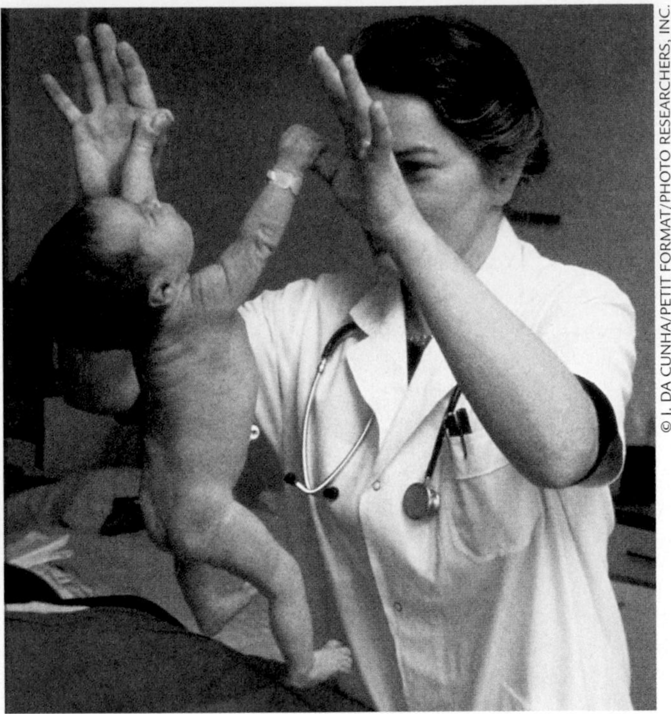
© J. DA CUNHA/PETIT FORMAT/PHOTO RESEARCHERS, INC.

■ The palmar grasp reflex is so strong during the first week after birth that many infants can use it to support their entire weight.

Newborn States

Throughout the day and night, newborn infants move in and out of five different **states of arousal,** or degrees of sleep and wakefulness, described in Table 3.4. During the first month, these states alternate frequently. Quiet alertness is the most fleeting. It moves relatively quickly toward fussing and crying. Much to the relief of their fatigued parents, newborns spend the greatest amount of time asleep—about 16 to 18 hours a day. Although newborns sleep more at night than during the day, their sleep–wake cycles are affected more by fullness–hunger than by darkness–light (Goodlin-Jones, Burnham, & Anders, 2000).

However, striking individual differences in daily rhythms exist that affect parents' attitudes toward and interaction with the baby. A few newborns sleep for long periods, increasing the energy their well-rested parents have for sensitive, responsive care. Other babies cry a great deal, and their parents must exert great effort to soothe them. If they do not succeed, parents may feel less competent and less positive toward their infant. Of the five states listed in Table 3.4, the two extremes of sleep and crying have been of greatest interest to researchers. Each tells us something about normal and abnormal early development.

● **Sleep.** One day, Yolanda and Jay watched Joshua while he slept and wondered why his eyelids and body twitched and his rate of breathing varied. Sleep is made up of at least two states. During irregular, or **rapid-eye-movement (REM), sleep,** electrical brain-wave activity is remarkably similar to that of the waking state. The eyes dart beneath the lids; heart rate, blood pressure, and breathing are uneven; and slight muscle twitches occur. In contrast, during **non-rapid-eye-movement (NREM) sleep,** the body is almost motionless, and heart rate, breathing, and brain-wave activity are slow and regular.

Like children and adults, newborns alternate between REM and NREM sleep. However, they spend far more time in the REM state than they ever will again. REM sleep accounts for 50 percent of a newborn baby's sleep time. By 3 to 5 years, it has declined to an adultlike level of 20 percent (Roffwarg, Muzio, & Dement, 1966).

Why do young infants spend so much time in REM sleep? In older children and adults, the REM state is associated with dreaming. Babies probably do not dream, at least not in the same way we do. Young infants are believed to have a special need for the stimulation of REM sleep because they spend little time in an alert state, when they can get input from the environment. REM sleep seems to be a way in which the brain stimulates itself. Sleep researchers believe that this stimulation is vital for growth of the central nervous system. In support of this idea, the percentage of

Table 3.4 Infant States of Arousal

State	Description	Daily Duration in Newborn
Regular sleep	The infant is at full rest and shows little or no body activity. The eyelids are closed, no eye movements occur, the face is relaxed, and breathing is slow and regular.	8–9 hours
Irregular sleep	Gentle limb movements, occasional stirring, and facial grimacing occur. Although the eyelids are closed, occasional rapid eye movements can be seen beneath them. Breathing is irregular.	8–9 hours
Drowsiness	The infant is either falling asleep or waking up. Body is less active than in irregular sleep but more active than in regular sleep. The eyes open and close; when open, they have a glazed look. Breathing is even but somewhat faster than in regular sleep.	Varies
Quiet alertness	The infant's body is relatively inactive, with eyes open and attentive. Breathing is even.	2–3 hours
Waking activity and crying	The infant shows frequent bursts of uncoordinated body activity. Breathing is very irregular. Face may be relaxed or tense and wrinkled. Crying may occur.	1–4 hours

Source: Wolff, 1966.

Biology & Environment

The Mysterious Tragedy of Sudden Infant Death Syndrome

Millie awoke with a start one morning and looked at the clock. It was 7:30, and Sasha had missed both her night waking and her early morning feeding. Wondering if she was all right, Millie and her husband Stuart tiptoed into the room. Sasha lay still, curled up under her blanket. She had died silently during her sleep.

Sasha was a victim of **sudden infant death syndrome (SIDS),** the unexpected death, usually during the night, of an infant under 1 year of age that remains unexplained after thorough investigation. In industrialized nations, SIDS is the leading cause of infant mortality between 1 and 12 months of age (Health Canada, 2001c; MacDorman & Atkinson, 1999).

Although the precise cause of SIDS is not known, its victims usually show physical problems from the very beginning. Early medical records of SIDS babies reveal higher rates of prematurity and low birth weight, poor Apgar scores, and limp muscle tone. Abnormal heart rate and respiration and disturbances in sleep–wake activity are also involved (Leach et al., 1999; Malloy & Hoffman, 1995). At the time of death, over half of SIDS babies have a mild respiratory infection (Kohlendorfer, Kiechl, & Sperl, 1998). This seems to increase the chances of respiratory failure in an already vulnerable baby.

One hypothesis about the cause of SIDS is that impaired brain functioning prevents these infants from learning how to respond when their survival is threatened—for example, when respiration is suddenly interrupted (Horne et al., 2000). Between 2 and 4 months, when SIDS is most likely, reflexes decline and are replaced by voluntary, learned responses. Respiratory and muscular weaknesses may stop SIDS babies from acquiring behaviors that replace defensive reflexes. As a result, when breathing difficulties occur during sleep, infants do not wake up, shift their position, or cry out for help. Instead, they simply give in to oxygen deprivation and death.

In an effort to reduce the occurrence of SIDS, researchers are studying environmental factors related to it. Maternal cigarette smoking, both during and after pregnancy, as well as smoking by other caregivers, strongly predicts the disorder. Babies exposed to cigarette smoke have more respiratory infections and are two to three times more likely than nonexposed infants to die of SIDS (Dybing & Sanner, 1999; Sundell, 2001). Prenatal abuse of drugs that depress central nervous system functioning (opiates and barbiturates) increases the risk of SIDS tenfold (Kandall & Gaines, 1991). SIDS babies are also more likely to sleep on their stomachs than on their backs and often are wrapped very warmly in clothing and blankets (Gallard, Taylor, & Bolton, 2002).

Some researchers suspect that nicotine, depressant drugs, excessive body warmth, and respiratory infection lead to physiological stress, which disrupts the normal sleep pattern. When sleep-deprived infants experience a sleep "rebound," they sleep more deeply, which results in loss of vital muscle tone in the airway passages. In at-risk babies, the airway may collapse, and the infant may fail to arouse sufficiently to reestablish breathing (Simpson, 2001). In other cases, healthy babies sleeping face down in soft bedding may die from continually breathing their own exhaled breath.

Quitting smoking, changing an infant's sleeping position, and removing a few bedclothes can reduce the incidence of SIDS. For example, if women refrained from smoking while pregnant, an estimated 30 percent of SIDS cases would be prevented. Public education campaigns that discourage parents from putting babies down on their stomachs have led to dramatic reductions in SIDS in many industrialized countries (American Academy of Pediatrics, 2000; Schlaud et al., 1999).

When SIDS does occur, surviving family members require a great deal of help to overcome a sudden and unexpected death. As Millie commented 6 months after Sasha's death, "It's the worst crisis we've ever been through. What's helped us most are the comforting words of others who've experienced the same tragedy."

REM sleep is especially great in the fetus and in preterm babies, who are even less able than full-term newborns to take advantage of external stimulation (DiPietro et al., 1996a; Sahni et al., 1995).

Whereas the brain-wave activity of REM sleep safeguards the central nervous system, rapid eye movements protect the health of the eye. Eye movements cause the vitreous (gelatinlike substance within the eye) to circulate, thereby delivering oxygen to parts of the eye that do not have their own blood supply. During sleep, when the eyes and the vitreous are still, visual structures are at risk for anoxia. As the brain cycles through periods of REM sleep, rapid eye movements stir up the vitreous, ensuring that the eye is fully oxygenated (Blumberg & Lucas, 1996).

Because the normal sleep behavior of a newborn baby is organized and patterned, observations of sleep states can help identify central nervous system abnormalities. In infants who are brain damaged or have experienced serious birth trauma, disturbed REM–NREM sleep cycles are often present. Babies with poor sleep organization are likely to be behaviorally disorganized and, therefore, to have difficulty learning and evoking caregiver interactions that enhance their development (Groome et al., 1997; Halpern, MacLean, & Baumeister, 1995). And the brain-functioning problems that underlie newborn sleep irregularities may culminate in sudden infant death syndrome, a major cause of infant mortality (see the Biology and Environment box above).

● **Crying.** Crying is the first way that babies communicate, letting parents know they need food, comfort, and stimulation.

During the weeks after birth, all infants have some fussy periods when they are difficult to console. But most of the time, the nature of the cry, combined with the experiences that led up to it, helps guide parents toward its cause. The baby's cry is a complex stimulus that varies in intensity, from a whimper to a message of all-out distress (Gustafson, Wood, & Green, 2000). As early as the first few weeks of life, infants can be identified by the unique vocal "signature" of their cry, which helps parents locate their baby from a distance (Gustafson, Green, & Cleland, 1994).

Young infants usually cry because of physical needs. Hunger is the most common cause, but babies may also cry in response to temperature change when undressed, a sudden noise, or a painful stimulus. Newborns (as well as older babies) often cry at the sound of another crying baby (Dondi, Simion, & Caltran, 1999). Some researchers believe this response reflects an inborn capacity to react to the suffering of others. Furthermore, crying typically increases during the early weeks, peaks at about 6 weeks, and then declines. Because this trend appears in many cultures with vastly different infant care practices, researchers believe that normal readjustments of the central nervous system underlie it (Barr, 2001).

The next time you hear an infant cry, note your own reaction. The sound stimulates strong feelings of arousal and discomfort in just about anyone (Boukydis & Burgess, 1982; Murray, 1985). This powerful response is probably innately programmed in humans to make sure that babies receive the care and protection they need to survive.

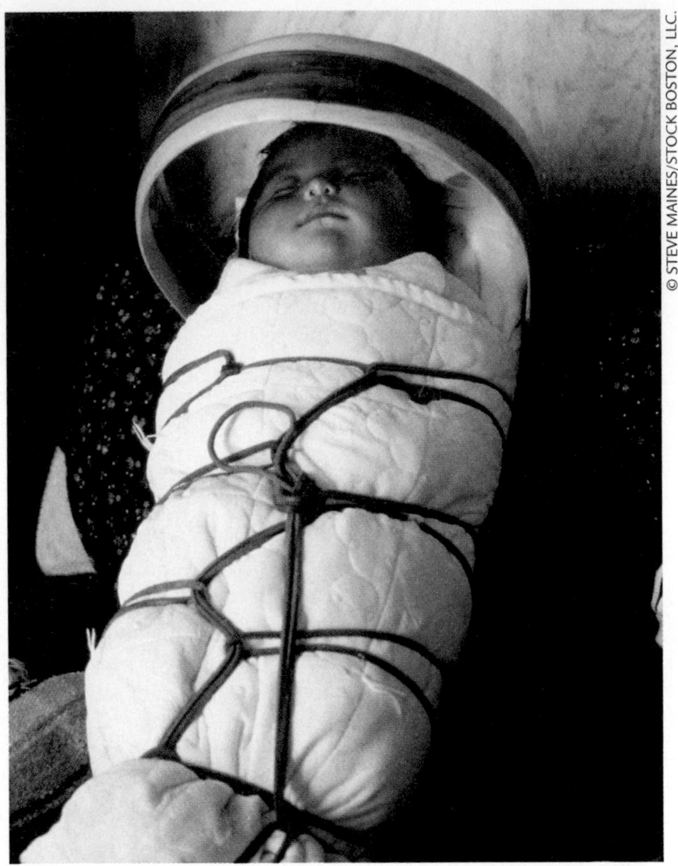

■ Some cultures routinely swaddle young infants, restricting movement and increasing warmth by wrapping blankets tightly around the body. This Navajo baby rests on a traditional cradle board that can be strapped to the mother's back. Swaddling reduces crying and promotes sleep.

Soothing Crying Infants. Although parents do not always interpret their baby's cry correctly, experience improves their accuracy (Thompson & Leger, 1999). Fortunately, as the Caregiving Concerns table on the following page indicates, there are many ways to soothe a crying baby when feeding and diaper changing do not work. The technique that Western parents usually try first, lifting the baby to the shoulder, is the most effective.

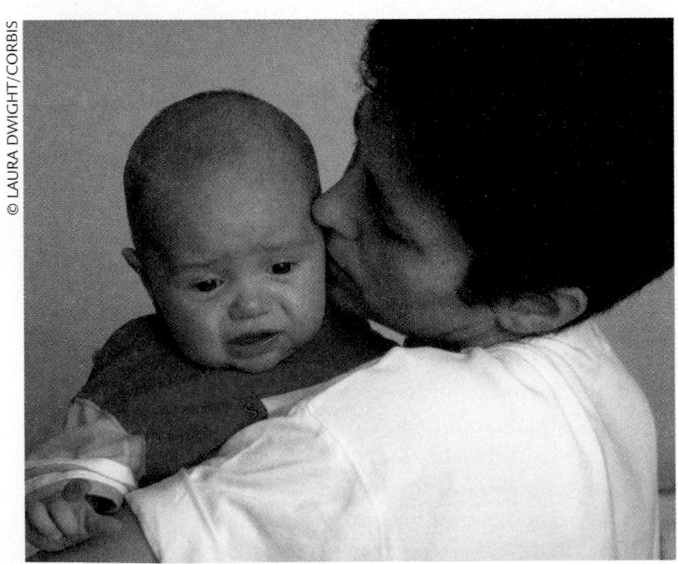

■ To soothe her crying infant, this mother holds her baby upright against her gently moving body. Besides encouraging infants to stop crying, this technique causes them to become quietly alert and attentive to the environment.

Another common soothing method is swaddling—wrapping the baby snugly in a blanket. Among the Quechua, who live in the cold, high-altitude desert regions of Peru, young babies are dressed in layers of clothing and blankets that cover the head and body. The result—a nearly sealed, warm pouch placed on the mother's back that moves rhythmically as she walks—reduces crying and promotes sleep. As a result, the baby conserves energy for early growth in the harsh Peruvian highlands (Tronick, Thomas, & Daltabuit, 1994).

In many tribal and village societies and non-Western developed nations, infants spend most of the day and night in close physical contact with their caregivers. Among the !Kung of the desert regions of Bostwana, Africa, mothers carry their young babies in grass-lined, animal-skin slings hung on their hips, so the infants can see their surroundings and can nurse at will. The Gusii, an agricultural and herding society of Kenya, care for infants by placing them on the backs of their mothers, who respond to crying with ready feeding. Japanese mothers also spend much time in close body contact with their babies (Small, 1998). Infants in these cultures show shorter bouts of crying than do North American babies (Barr, 2001).

Ways to Soothe a Crying Newborn

METHOD	EXPLANATION
Lift the baby to the shoulder and rock or walk.	This provides a combination of physical contact, upright posture, and motion. It is the most effective soothing technique.
Swaddle the baby.	Restricting movement and increasing warmth often soothe a young infant.
Offer a pacifier.	Sucking helps babies control their own level of arousal. Sucking a sweetened pacifier relieves pain and quiets a crying infant.
Talk softly or play rhythmic sounds.	Continuous, monotonous, rhythmic sounds, such as a clock ticking, a fan whirring, or peaceful music, are more effective than intermittent sounds.
Take the baby for a short car ride or walk in a baby carriage; swing the baby in a cradle.	Gentle, rhythmic motion of any kind helps lull the baby to sleep.
Massage the baby's body.	Stroke the baby's torso and limbs with continuous, gentle motions. This technique is used in some non-Western cultures to relax the baby's muscles.
Combine several of the methods listed above.	Stimulating several of the baby's senses at once is often more effective than stimulating only one.
If these methods do not work, permit the baby to cry for a short period of time.	Occasionally, a baby responds well just to being put down and will, after a few minutes, fall asleep.

Sources: Blass, 1999; Campos, 1989; Heinl, 1983; Lester, 1985; Reisman, 1987.

Abnormal Crying. Like reflexes and sleep patterns, the infant's cry offers a clue to central nervous system distress. The cries of brain-damaged babies and babies who have experienced prenatal and birth complications are often shrill, piercing, and shorter in duration than the cries of healthy infants (Boukydis & Lester, 1998; Green, Irwin, & Gustafson, 2000). Even newborns with a fairly common problem—*colic,* or persistent crying—tend to have high-pitched, harsh-sounding cries (Zeskind & Barr, 1997). Although the cause of colic is unknown, some researchers believe it is due to disturbed brain regulation of sleep–wake cycles (Papousek & Papousek, 1996). Others think it results from a temporary difficulty in calming down after becoming upset, which subsides between 3 and 6 months of age (Barr & Gunnar, 2000).

Most parents try to respond to a crying baby with extra care and sensitivity, but sometimes the cry is so unpleasant and the infant so difficult to soothe that parents become frustrated, resentful, and angry. Preterm and ill babies are more likely to be abused by highly stressed parents, who frequently mention a high-pitched, grating cry as one factor that caused them to lose control and harm the baby (Frodi, 1985). We will discuss a host of additional influences on child abuse in Chapter 8.

Sensory Capacities

On his visit to my class, Joshua looked wide-eyed at my bright pink blouse and turned to the sound of his mother's voice. During feedings, he lets Yolanda know by the way he sucks that he prefers the taste of breast milk to a bottle of plain water. Clearly, Joshua has some well-developed sensory capacities. In the following sections, we explore the newborn baby's responsiveness to touch, taste, smell, sound, and visual stimulation.

● Touch.

In our discussion of preterm infants, we saw that touch helps stimulate early physical growth. As we will see in Chapter 6, it is vital for emotional development as well. Therefore, it is not surprising that sensitivity to touch is well developed at birth. The reflexes listed in Table 3.3 reveal that the newborn baby responds to touch, especially around the mouth, on the palms, and on the soles of the feet. During the prenatal period, these areas, along with the genitals, are the first to become sensitive to touch (Humphrey, 1978).

At birth, infants are quite sensitive to pain. If male newborns are circumcised, anesthetic is sometimes not used because of the risk of giving drugs to a very young infant. Babies often respond with a high-pitched, stressful cry and a dramatic rise in heart rate, blood pressure, palm sweating, pupil dilation, and muscle tension (Jorgensen, 1999). Recent research establishing the safety of certain local anesthetics for newborns promises to ease the pain of these procedures. Offering a nipple that delivers a sugar solution is also helpful; it quickly reduces crying and discomfort in young babies. And combining the sweet liquid with gentle holding by the parent lessens pain even more (Gormally et al., 2000; Overgaard & Knudsen, 1999). Allowing a newborn to endure severe pain can affect later behavior. In one study, newborns who were not given a local anesthetic during circumcision reacted more intensely to a routine vaccination at 4 to 6 months of age than did their anesthetized counterparts (Taddio et al., 1997).

● **Taste and Smell.** All infants come into the world with the ability to communicate their taste preferences to caregivers. Facial expressions reveal that newborns can distinguish several basic tastes. Like adults, they relax their facial muscles in response to sweetness, purse their lips when the taste is sour, and show a distinct archlike mouth opening when it is bitter (Steiner, 1979). These reactions are important for survival, since (as we will see in Chapter 4) the food that is ideally suited to support the infant's early growth is the sweet-tasting milk of the mother's breast. Not until 4 months of age will babies prefer the salty taste to plain water, a change that may prepare them to accept solid foods (Mennella & Beauchamp, 1998).

Nevertheless, newborns can readily learn to like a taste that at first evoked either a neutral or a negative response. For example, babies allergic to cow's-milk formula who are given a soy or other vegetable-based substitute (typically very strong and bitter-tasting) soon prefer it to regular formula. A taste previously disliked can come to be preferred when it is paired with relief of hunger (Harris, 1997).

Like taste, certain odor preferences are innate. For example, the smell of bananas or chocolate causes a relaxed, pleasant facial expression, whereas the odor of rotten eggs makes the infant frown (Steiner, 1979). Newborns can also identify the location of an odor and, if it is unpleasant, defend themselves by turning their heads in the other direction (Reiser, Yonas, & Wikner, 1976).

In many mammals, the sense of smell plays an important role in feeding and in protecting the young from predators by helping mothers and babies identify each other. Although smell is less well developed in humans, traces of its survival value remain. Newborns given a choice between the smell of their own mother's amniotic fluid and that of another mother spend more time oriented toward the familiar fluid (Marlier, Schaal, & Soussignan, 1998). The smell of the mother's amniotic fluid is comforting; babies exposed to it cry less than babies who are not (Varendi et al., 1998).

Immediately after birth, babies placed face down between their mother's breasts spontaneously latch on to a nipple and begin sucking within an hour. If one breast is washed to remove its natural scent, most newborns grasp the unwashed breast, indicating that they are guided by smell (Porter & Winberg, 1999). At 4 days of age, breast-fed babies prefer the smell of their own mother's breast to that of an unfamiliar lactating mother (Cernoch & Porter, 1985). Bottle-fed babies orient to the smell of any lactating woman over the smell of formula or of a nonlactating woman (Marlier & Schaal, 1997; Porter et al., 1992). Newborns' dual attraction to the odors of their mother and the lactating breast helps them locate an appropriate food source and, in the process, distinguish their caregiver from other people.

● **Hearing.** Newborn infants can hear a wide variety of sounds, but they prefer complex sounds, such as noises and voices, to pure tones (Bench et al., 1976). In the first few days, infants can already tell the difference between a few sound patterns—a series of tones arranged in ascending versus descending order, utterances with two as opposed to three syllables, the stress patterns of words, such as *ma*-ma versus ma-*ma*, and happy-sounding speech as opposed to speech with angry, sad, or neutral emotional qualities (Bijeljac-Babic, Bertoncini, & Mehler, 1993; Mastropieri & Turkewitz, 1999; Sansavini, Bertoncini, & Giovanelli, 1997).

Newborns are especially sensitive to the sounds of human speech, and they are biologically prepared to respond to the sounds of any human language. Young infants can make fine-grained distinctions among a wide variety of speech sounds. For example, when given a nipple that turns on a recording of the "ba" sound, babies suck vigorously and then slow down as the novelty wears off. When the sound switches to "ga," sucking picks up, indicating that infants detect this subtle difference. Using this method, researchers have found only a few speech sounds that newborn infants cannot discriminate (Jusczyk, 1995). These capacities reveal that the baby is marvelously prepared for the awesome task of acquiring language. In Chapter 4, we will see that babies' speech perception changes markedly by the middle of the first year, after much exposure to their native tongue.

Listen carefully to yourself the next time you talk to a young baby. You will probably speak in a high-pitched, expressive voice and use a rising tone at the ends of phrases and sentences. Adults probably communicate this way because they notice that infants are more attentive when they do so. Indeed, newborns prefer speech with these characteristics (Aslin, Jusczyk, & Pisoni, 1998). They will also suck more on a nipple to hear a recording of their mother's voice than that of an unfamiliar woman, and to hear their native language as opposed to a foreign language (Moon, Cooper, & Fifer, 1993; Spence & DeCasper, 1987). These preferences may have developed from hearing the muffled sounds of the mother's voice before birth.

● **Vision.** Vision is the least developed of the senses at birth. Visual structures in both the eye and the brain are not yet fully formed. For example, cells in the retina, a membrane lining the inside of the eye that captures light and transforms it into messages that are sent to the brain, are not as mature or densely packed as they will be in several months. Also, muscles of the lens, which permits us to adjust our visual focus to varying distances, are weak (Banks & Bennett, 1988).

As a result, newborn babies cannot focus their eyes as well as an adult can, and **visual acuity,** or fineness of discrimination, is limited. At birth, infants perceive objects at a distance of 20 feet about as clearly as adults do at 400 feet (Gwiazda & Birch, 2001). In addition, unlike adults (who see nearby objects most clearly), newborn babies see equally unclearly across a wide range of distances (Banks, 1980). As a result, images such as the parent's face, even from close up, look quite blurred.

Although newborn babies cannot see well, they actively explore their environment by scanning it for interesting sights and tracking moving objects. However, their eye movements are slow and inaccurate (Aslin, 1993). Joshua's captivation with my pink blouse reveals that he is attracted to bright objects. Although newborns prefer to look at colored rather than gray stimuli, they are not yet good at discriminating colors. It will take a month or two for color vision to improve (Adams, Courage, & Mercer, 1994; Teller, 1998).

Neonatal Behavioral Assessment

A variety of instruments permit doctors, nurses, and researchers to assess the behavior of newborn babies. The most widely used of these tests, T. Berry Brazelton's **Neonatal Behavioral Assessment Scale (NBAS),** looks at the baby's reflexes, state changes, responsiveness to physical and social stimuli, and other reactions (Brazelton & Nugent, 1995).

The NBAS has been given to many infants around the world. As a result, researchers have learned about individual and cultural differences in newborn behavior and how child-rearing practices can maintain or change a baby's reactions. For example, NBAS scores of Asian and Native-American babies reveal that they are less irritable than Caucasian infants. Mothers in these cultures often encourage their babies' calm dispositions through holding and nursing at the first signs of discomfort (Chisholm, 1989; Muret-Wagstaff & Moore, 1989). In contrast, the poor NBAS scores of undernourished infants in Zambia, Africa, are quickly changed by the way their mothers care for them. The Zambian mother carries her baby on her hip all day, providing a rich variety of sensory stimulation. By 1 week of age, a once unresponsive newborn has been transformed into an alert, contented baby (Brazelton, Koslowski, & Tronick, 1976).

Can you tell from these examples why a single NBAS score is not a good predictor of later development? Because newborn behavior and parenting styles combine to shape development, *changes in NBAS scores* over the first week or two of life (rather than a single score) provide the best estimate of the baby's ability to recover from the stress of birth. NBAS "recovery curves" predict intelligence with moderate success well into the preschool years (Brazelton, Nugent, & Lester, 1987).

The NBAS has also been used to help parents get to know their infants. In some hospitals, health professionals discuss with or demonstrate to parents the newborn capacities assessed by the NBAS. Parents of both preterm and full-term newborns who participate in these programs interact more confidently and effectively with their babies (Eiden & Reifman, 1996). In one study, Brazilian mothers who experienced a 50-minute NBAS-based discussion a few days after delivery were more likely than controls who received only health care information to establish eye contact, smile, vocalize, and soothe in response to infant signals a month later (Wendland-Carro, Piccinini, & Millar, 1999). Although lasting effects on development have not been demonstrated, NBAS-based interventions are clearly useful in helping the parent–infant relationship get off to a good start.

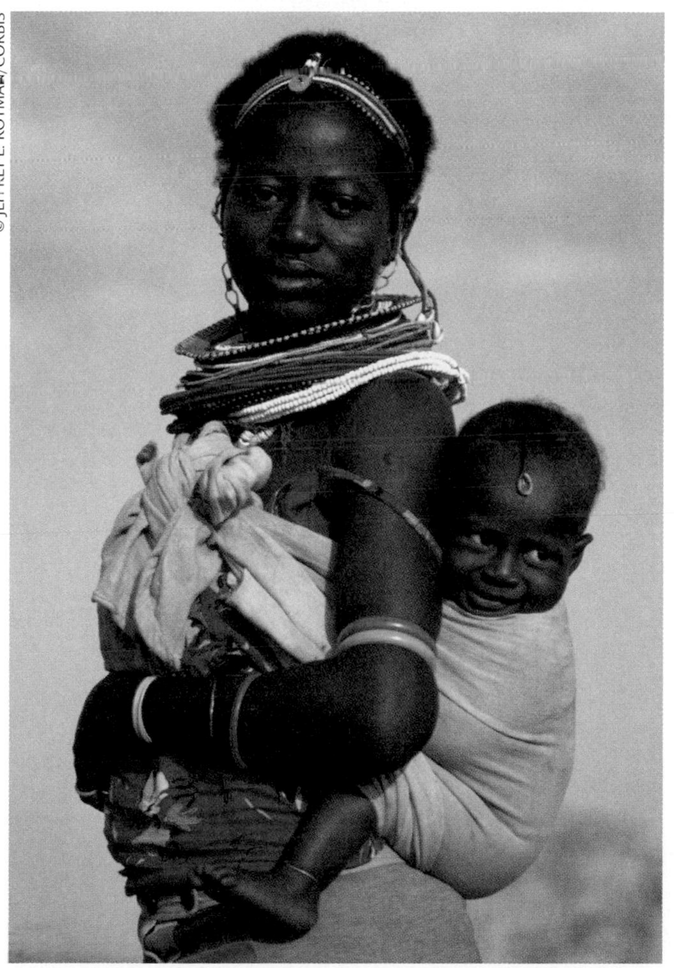

© JEFFREY L. ROTMAN/CORBIS

■ Similar to women in the Zambian culture, this mother of the El Molo people of northern Kenya carries her baby about all day, providing close physical contact, a rich variety of stimulation, and ready feeding.

Ask Yourself

REVIEW

What functions does REM sleep serve in young infants? Can sleep tell us anything about the health of the newborn's central nervous system? Explain.

APPLY

Jackie, who had a difficult delivery, observes her 2-day-old daughter Kelly being given the NBAS. Kelly scores poorly on many items. Jackie wonders if this means Kelly will not develop normally. How would you respond to Jackie's concern?

CONNECT

How do the diverse capacities of newborn babies contribute to their first social relationships? Provide as many examples as you can.

www.

Adjusting to the New Family Unit

The early weeks after a new baby enters the family are full of profound changes. The mother needs to recover from childbirth and adjust to massive hormone shifts in her body. If she is breastfeeding, energies must be devoted to working out this intimate relationship. The father needs to become a part of this new threesome while supporting the mother in her recovery. At times, he may feel ambivalent about the baby, who constantly demands and gets the mother's attention. And as we will see in Chapter 6, siblings—especially those who are young and first born—understandably feel displaced. They sometimes react with jealousy and anger.

While all this is going on, the tiny infant is very assertive about his urgent physical needs, demanding to be fed, changed, and comforted at odd times of the day and night. A family schedule that was once routine and predictable is now irregular and uncertain. Yolanda spoke candidly about the changes that she and Jay experienced:

> When we brought Joshua home, we had to deal with the realities of our new responsibility. Joshua seemed so small and helpless, and we worried about whether we would be able to take proper care of him. It took us 20 minutes to change the first diaper. I rarely feel rested because I'm up two to four times every night, and I spend a good part of my waking hours trying to anticipate Joshua's rhythms and needs. If Jay weren't so willing to help by holding and walking Joshua, I think I'd find it much harder.

How long does this time of adjustment to parenthood last? In Chapter 14, we will see that when parents try to support each other's needs, the stress caused by the birth of a baby remains manageable, and family relationships and responsibilities are worked out after a few months. Nevertheless, as one pair of counselors who have worked with many new parents point out, "As long as children are dependent on their parents, those parents find themselves preoccupied with thoughts of their children. This does not keep them from enjoying other aspects of their lives, but it does mean that they never return to being quite the same people they were before they became parents" (Colman & Colman, 1991, p. 198).

Summary

Prenatal Development

List the three phases of prenatal development, and describe the major milestones of each.

- The first prenatal phase, the period of the zygote, lasts about 2 weeks, from fertilization until the blastocyst becomes deeply **implanted** in the uterine lining. During this time, structures that will support prenatal growth begin to form. The embryonic disk is surrounded by the **amnion,** which fills with amniotic fluid to regulate temperature and cushion against the mother's movements. From the **chorion,** villi emerge that burrow into the uterine wall, and the **placenta** starts to develop. The developing organism is connected to the placenta by the **umbilical cord.**

- The period of the **embryo** lasts from 2 to 8 weeks, during which the foundations for all body structures are laid down. In the first week of this period, the **neural tube** forms, and the nervous system starts to develop. Other organs follow and grow rapidly. At the end of this phase, the embryo responds to touch and can move.

- The period of the **fetus,** lasting until the end of pregnancy, involves a dramatic increase in body size and the completion of physical structures. By the middle of the second **trimester,** the mother can feel movement. The fetus becomes covered with **vernix,** which protects the skin from chapping. White, downy hair called **lanugo** helps the vernix stick to the skin. At the end of the second trimester, production of neurons in the brain is complete.

- The **age of viability** occurs at the beginning of the third trimester, sometime between 22 and 26 weeks. The brain continues to develop rapidly, and new sensory and behavioral capacities emerge. The lungs gradually mature, the fetus fills the uterus, and birth is near.

Prenatal Environmental Influences

What are teratogens, and what factors influence their impact?

- **Teratogens** are environmental agents that cause damage during the prenatal period. Their effects conform to the sensitive period concept. The impact of teratogens varies with the amount and length of exposure, the genetic makeup of mother and fetus, the presence or absence of other harmful agents, and the age of the organism at time of exposure. The developing organism is especially vulnerable during the embryonic period because all essential body structures are emerging. Although immediate physical damage is easy to notice, serious health and psychological consequences may not show up until later in development.

List agents known to be or suspected of being teratogens, and discuss evidence supporting the harmful impact of each.

- The prenatal impact of many commonly used medications, such as aspirin and caffeine, is hard to separate from other factors correlated with drug taking. Babies whose mothers took heroin, methadone, or cocaine during pregnancy have withdrawal symptoms after birth and are jittery and inattentive. Using cocaine is especially risky because it is associated with physical defects and central nervous system damage.

- Infants of parents who use tobacco are often born underweight and may have attention, learning, and behavior problems in childhood. When mothers consume alcohol in large quantities, **fetal alcohol syndrome (FAS),** a disorder involving mental retardation, poor attention, overactivity, slow physical growth, and facial abnormalities, often results. Smaller amounts of alcohol may lead to some of these problems—a condition known as **fetal alcohol effects (FAE).**

- High levels of radiation, mercury, lead, and PCBs lead to physical malformations and severe brain damage. Low-level exposure has also been linked to diverse impairments, including lower intelligence test scores and, in the case of radiation, language and emotional disorders.

- Among infectious diseases, rubella causes a wide variety of abnormalities, which vary with the time the disease strikes during pregnancy. The human immunodeficiency virus (HIV), responsible for AIDS, results in brain damage, delayed mental and motor development, and early death. Toxoplasmosis in the first trimester may lead to eye and brain damage.

Describe the impact of other maternal factors on prenatal development.

- In healthy, physically fit women, regular moderate exercise is related to increased birth weight. However, very frequent, vigorous exercise results in lower birth weight. When the mother's diet is inadequate, low birth weight and damage to the brain and other organs are major concerns.

- Severe emotional stress is linked to many pregnancy complications, although its impact can be reduced by providing the mother with emotional support. **Rh factor incompatibility**—an Rh-negative mother and Rh-positive fetus—can lead to oxygen deprivation, brain and heart damage, and infant death.

- Aside from the risk of chromosomal abnormalities in older women, maternal age and number of previous births are not major causes of prenatal problems. Instead, poor health and environmental risks associated with poverty are the strongest predictors of pregnancy complications.

Why is early and regular health care vital during the prenatal period?

- Unexpected difficulties, such as toxemia, can arise, especially when mothers have health problems to begin with. Prenatal health care is especially critical for women unlikely to seek it—in particular, those who are young, single, and poor.

Childbirth

Describe the three stages of childbirth, the baby's adaptation to labor and delivery, and the newborn baby's appearance.

- Childbirth takes place in three stages, beginning with contractions that open the cervix so the mother can push the baby through the birth canal and ending with delivery of the placenta. During labor, infants produce high levels of stress hormones, which help them withstand oxygen deprivation, clear the lungs for breathing, and arouse them into alertness at birth.

- Newborn infants have large heads, small bodies, and facial features that make adults feel like cuddling them. The **Apgar Scale** assesses the baby's physical condition at birth.

Approaches to Childbirth

Describe natural childbirth and home delivery, noting any benefits and concerns associated with each.

- **Natural,** or **prepared, childbirth** involves classes in which prospective parents learn about labor and delivery, relaxation and breathing techniques to counteract pain, and coaching during childbirth. The method helps reduce the stress, pain, and use of medication. Social support, a vital part of natural childbirth, is linked to fewer birth complications and shorter labors. As long as mothers are healthy and are assisted by a well-trained doctor or midwife, giving birth at home is just as safe as giving birth in a hospital.

Medical Interventions

List common medical interventions during childbirth, circumstances that justify their use, and any dangers associated with each.

- When pregnancy and birth complications make **anoxia** likely, **fetal monitors** help save the lives of many babies. However, when used routinely, they may identify infants as in danger who, in fact, are not.

- Medication to relieve pain is necessary in complicated deliveries. When given in large doses, it may prolong labor and produce a depressed state in the newborn that affects the early mother–infant relationship.

- **Cesarean deliveries** are justified in cases of medical emergency and serious maternal illness and sometimes when babies are in **breech position.** Many unnecessary cesareans are performed in the United States.

Preterm and Low-Birth-Weight Infants

What are the risks of preterm birth and low birth weight, and what factors can help infants who survive a traumatic birth?

- **Preterm** and low-birth-weight babies are especially likely to be born to poverty-stricken mothers. Compared with preterm babies whose weight is appropriate for time spent in the uterus, **small-for-date** infants are more likely to develop poorly. The fragile appearance and unresponsive, irritable behavior of preterm infants can lead parents to be less sensitive and responsive in caring for them.

- Some interventions provide special stimulation in the intensive care nursery. Others teach parents how to care for and interact with their babies. When preterm infants live in stressed, low-income households, long-term, intensive intervention is required. A major cause of **infant mortality** is low birth weight.

Understanding Birth Complications

- When infants experience birth trauma, a supportive home environment can help restore their growth. Even children with fairly serious birth complications can recover with the help of favorable life events.

The Newborn Baby's Capacities

Describe the newborn baby's reflexes and states of arousal, including sleep characteristics and ways to soothe a crying baby.

- Infants begin life with remarkable skills for relating to their physical and social worlds. **Reflexes** are the newborn baby's most obvious organized patterns of behavior. Some have survival value, others provide the foundation for voluntary motor skills, and still others contribute to early social relationships.

- Although newborns move in and out of five different **states of arousal,** they spend most of their time asleep. Sleep consists of at least two states: **rapid-eye-movement (REM)** sleep and **non-rapid-eye-movement (NREM) sleep.** REM sleep provides young infants with stimulation essential for central nervous system development. Rapid eye movements ensure that structures of the eye remain oxygenated during sleep. Disturbed REM–NREM cycles are a sign of central nervous system abnormalities, which may contribute to **sudden infant death syndrome (SIDS).**

- A crying baby stimulates strong feelings of discomfort in nearby adults. The intensity of the cry and the experiences that led up to it help parents identify what is wrong. Once feeding and diaper changing have been tried, lifting the baby to the shoulder is the most effective soothing technique. In societies where babies spend most of the day and night in close physical contact with their caregivers, crying is greatly reduced.

Describe the newborn baby's sensory capacities.

- The senses of touch, taste, smell, and sound are well developed at birth. Newborns are sensitive to pain, prefer sweet tastes and smells, and orient toward the odor of their own mother's amniotic fluid and the lactating breast. Already they can distinguish a few sound patterns as well as almost all speech sounds. They are especially responsive to high-pitched, expressive voices, their own mother's voice, and speech in their native tongue.

- Vision is the least mature of the newborn's senses. At birth, focusing ability and **visual acuity** are limited. In exploring the visual field, newborn babies are attracted to bright objects but have difficulty discriminating colors.

Why is neonatal behavioral assessment useful?

- The most widely used instrument for assessing the behavior of the newborn infant is Brazelton's **Neonatal Behavioral Assessment Scale (NBAS).** The NBAS has helped researchers understand individual and cultural differences in newborn behavior. Sometimes it is used to teach parents about their baby's capacities.

Adjusting to the New Family Unit

Describe typical changes in the family after the birth of a new baby.

- The new baby's arrival is exciting but stressful. When parents are sensitive to each other's needs, adjustment problems are usually temporary, and the transition to parenthood goes well.

Important Terms and Concepts

age of viability (p. 80)
amnion (p. 78)
anoxia (p. 95)
Apgar Scale (p. 93)
breech position (p. 95)
cesarean delivery (p. 96)
chorion (p. 78)
embryo (p. 79)
fetal alcohol effects (FAE) (p. 86)
fetal alcohol syndrome (FAS) (p. 86)
fetal monitors (p. 95)
fetus (p. 79)
implantation (p. 78)
infant mortality (p. 98)
lanugo (p. 80)
natural, or prepared, childbirth (p. 94)
Neonatal Behavioral Assessment Scale (NBAS) (p. 107)
neural tube (p. 79)
non-rapid-eye-movement (NREM) sleep (p. 102)
placenta (p. 78)
preterm (p. 97)
rapid-eye-movement (REM) sleep (p. 102)
reflex (p. 100)
Rh factor incompatibility (p. 89)
small for date (p. 97)
states of arousal (p. 102)
sudden infant death syndrome (SIDS) (p. 103)
teratogen (p. 81)
trimesters (p. 79)
umbilical cord (p. 78)
vernix (p. 80)
visual acuity (p. 106)

FYI For Further Information and Help

Consult the Companion Website for *Development Through the Lifespan, Third Edition,* (www.ablongman.com/berk) where you will find the following resources for this chapter:

- **Chapter Objectives**
- **Flashcards** for studying important terms and concepts
- **Annotated Weblinks** to guide you in further research
- **Ask Yourself** questions, which you can answer and then check against a sample response
- **Suggested Readings**
- **Practice Test** with immediate scoring and feedback

Physical Development in Infancy and Toddlerhood

Body Growth
Changes in Body Size and Muscle-Fat Makeup ● Individual and Group Differences ● Changes in Body Proportions

Brain Development
Development of Neurons ● Development of the Cerebral Cortex ● Sensitive Periods of Brain Development ● Changing States of Arousal

- A Lifespan Vista: Brain Plasticity: Insights from Research on Brain-Damaged Children and Adults
- Cultural Influences: Cultural Variation in Infant Sleeping Arrangements

Influences on Early Physical Growth
Heredity ● Nutrition ● Malnutrition ● Emotional Well-Being

Learning Capacities
Classical Conditioning ● Operant Conditioning ● Habituation ● Imitation

Motor Development
The Sequence of Motor Development ● Motor Skills as Dynamic Systems ● Cultural Variations in Motor Development ● Fine Motor Development: Reaching and Grasping

Perceptual Development
Hearing ● Vision ● Intermodal Perception ● Understanding Perceptual Development

- Biology & Environment: Development of Infants with Severe Visual Impairments

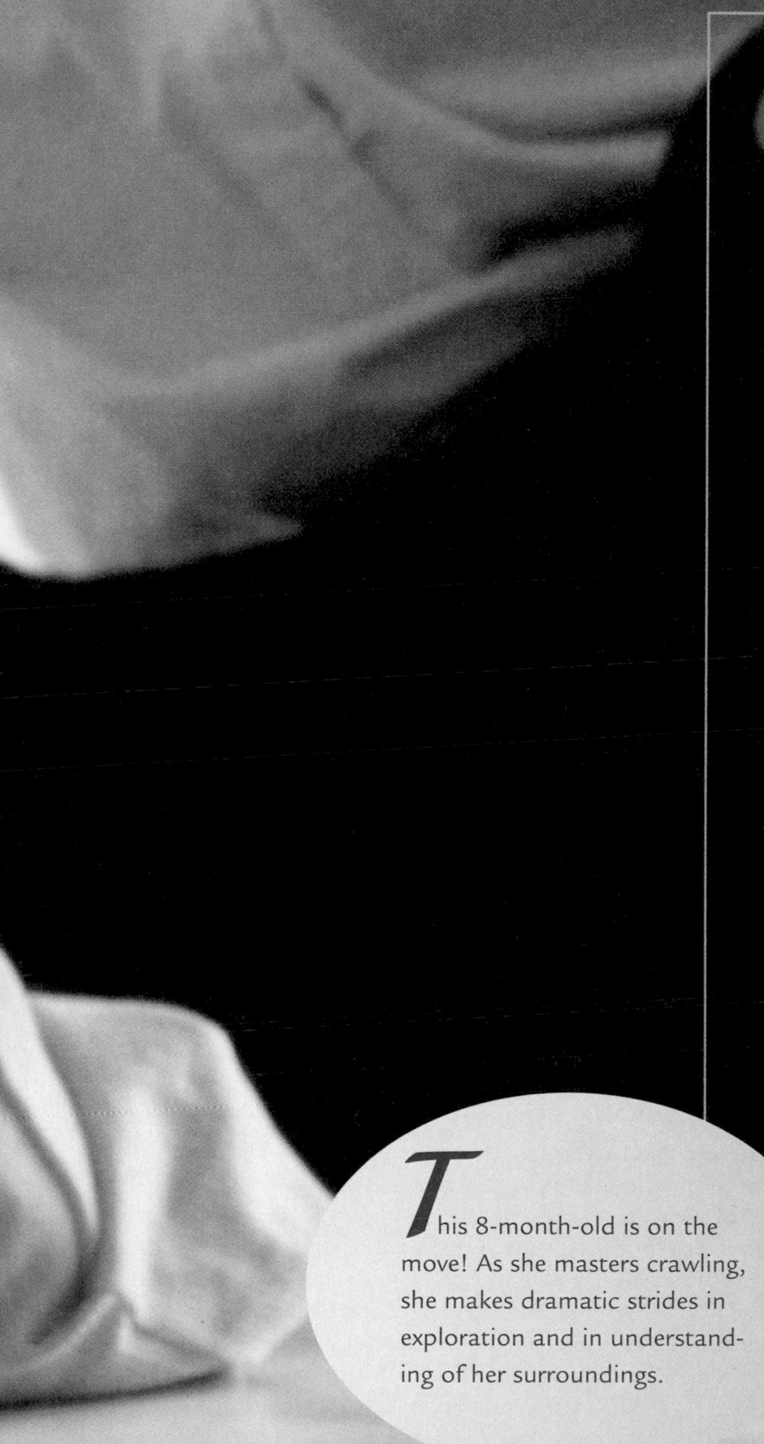

This 8-month-old is on the move! As she masters crawling, she makes dramatic strides in exploration and in understanding of her surroundings.

On a brilliant June morning, 16-month-old Caitlin emerged from her front door, ready for the short drive to the child-care home where she spent her weekends while her mother, Carolyn, and her father, David, worked. Clutching a teddy bear in one hand and her mother's arm with the other, Caitlin descended the front steps. "One! Two! Threeee!" Carolyn counted as she helped Caitlin down. "How much she's changed," Carolyn thought to herself, looking at the child who, not long ago, had been a newborn cradled in her arms. With her first steps, Caitlin had passed from infancy to toddlerhood—a period spanning the second year of life. At first, Caitlin did, indeed, "toddle" with an awkward gait, rocking from side to side and tipping over frequently. But her face reflected the thrill of conquering a new skill.

As they walked toward the car, Carolyn and Caitlin caught sight of 3-year-old Eli and his father, Kevin, in the neighboring yard. Eli dashed toward them, waving a bright yellow envelope. Carolyn bent down to open the envelope and took out a card. It read, "Announcing the arrival of Grace Ann. Born: Cambodia. Age: 16 months." Carolyn turned toward Kevin and Eli. "That is wonderful news! When can we see her?"

Let's wait a few days," Kevin suggested. "Monica's taken Grace to the doctor this morning. She's underweight and malnourished." Kevin described Monica's first night with Grace in a hotel room in Phnom Penh before they flew to the United States. Grace lay on the bed, withdrawn and fearful. Eventually she fell asleep, clutching crackers in both hands.

Carolyn felt a tug at her sleeve. Caitlin was impatient. Off they drove to child care, where Vanessa had just dropped off her 18-month-old son, Timmy. Within moments, Caitlin and Timmy were in the sandbox, shoveling sand into plastic cups and buckets with the help of their caregiver Ginette. A few weeks later, Grace joined Caitlin and Timmy at Ginette's child-care home. Although still tiny and unable to crawl or walk, she had grown taller and heavier, and her sad, vacant gaze had given way to a ready smile and an enthusiastic desire to imitate and explore. When Caitlin headed for the sandbox, Grace stretched out her arms, asking Ginette to carry her there, too. Soon Grace was pulling herself up at every opportunity. Finally, at age 18 months, she walked!

This chapter traces physical growth during the first 2 years—one of the most remarkable and busiest times of development. We will see how rapid changes in the infant's body and brain support learning, motor skills, and perceptual capacities. Caitlin, Grace, and Timmy will join us along the way to illustrate individual differences and environmental influences on physical development.

Body Growth

The next time you have a chance, briefly observe several infants and toddlers while walking in your neighborhood or at a shopping center. You will see that their capabilities differ vastly. One reason for the change in what children can do over the first 2 years is that their bodies change enormously—faster than at any other time after birth.

Changes in Body Size and Muscle–Fat Makeup

As Figure 4.1 illustrates, by the end of the first year a typical infant's height is 50 percent greater than it was at birth, and by 2 years of age it is 75 percent greater. Weight shows similar dramatic gains. By 5 months of age, birth weight has doubled, at 1 year it has tripled, and at 2 years it has quadrupled.

Rather than making steady gains, infants and toddlers grow in little spurts. In one study, children who were followed over the first 21 months of life went for periods of 7 to 63 days with no growth. Then they added as much as half an inch in a 24-hour period! Almost always, parents described their babies as irritable and very hungry on the day before the spurt (Lampl, 1993; Lampl, Veldhuis, & Johnson, 1992).

One of the most obvious changes in infants' appearance is their transformation into round, plump babies by the middle of the first year. This early rise in "baby fat," which peaks at about 9 months, helps the small infant maintain a constant body temperature (Tanner, 1990). During the second year, most toddlers slim down, a trend that continues into middle childhood. In contrast, muscle tissue increases very slowly during infancy and will not reach a peak until adolescence. Babies are not very muscular creatures, and their strength and physical coordination are limited.

Individual and Group Differences

As in all aspects of development, differences among children in body size and muscle–fat makeup exist. In infancy, girls are slightly shorter and lighter and have a higher ratio of fat to muscle than boys. These small sex differences remain throughout early and middle childhood and will be greatly magnified at adolescence. Ethnic differences in body size are apparent as well. Grace was below the growth norms (height and weight averages for children her age). Although early malnutrition contributed to Grace's small size, even after substantial catch-up she remained below North American norms, a trend typical for Asian children. In contrast, Timmy is slightly above average, as African-American children tend to be (Tanner, 1990).

Children of the same age also differ in *rate* of physical growth. In other words, some make faster progress than others toward a mature body size. We cannot tell how quickly a child's physical growth is moving along just by looking at current body size, since children grow to different heights and

■ **FIGURE 4.1 Body growth during the first 2 years.** Andy and Amy are brother and sister, born 2 years apart. These photos, taken by their parents, depict the dramatic changes in body size and proportions during infancy and toddlerhood. In the first year, the head is quite large in proportion to the rest of the body, and height and weight gain are especially rapid. During the second year, the lower portion of the body catches up. Notice, also, how Andy and Amy added "baby fat" in the early months of life and then slimmed down, a trend that continues into middle childhood. From birth on, Andy was slightly taller and heavier than Amy—a typical sex difference. We will revisit Andy's and Amy's growth in Chapters 7, 9, and 11.

weights in adulthood. For example, Timmy is larger and heavier than Caitlin and Grace, but he is not physically more mature. In a moment, you will see why.

The best way of estimating a child's physical maturity is to use *skeletal age,* a measure of bone development. It is determined by X-raying the long bones of the body to see the extent to which soft, pliable cartilage has hardened into bone, a gradual process that is completed in adolescence. When the skeletal ages of infants and children are examined, African-American children tend to be slightly ahead of Caucasian children at all ages. And girls are considerably ahead of boys. At birth, the sexes differ by about 4 to 6 weeks, a gap that widens over infancy and childhood and is responsible for the fact that girls reach their full body size several years before boys (Tanner, Healy, & Cameron, 2001). Girls' greater physical maturity may contribute to their greater resistance to harmful environmental influences. As noted in Chapter 2, girls experience fewer developmental problems than boys, and infant and childhood mortality for girls is also lower.

Changes in Body Proportions

As the child's overall size increases, different parts of the body grow at different rates. Two growth patterns describe these changes in body proportions. The first, depicted in Figure 4.2 on page 116, is called the **cephalocaudal trend.** Translated from Latin, it means "head to tail." As you can see, during the prenatal period the head develops more rapidly than the lower part of the body. At birth, the head takes up one-fourth of total body length, the legs only one-third. Notice how the lower portion of the body catches up. By age 2, the head accounts for only one-fifth and the legs for nearly one-half of body length.

The second pattern is the **proximodistal trend,** meaning growth proceeds, literally, from "near to far," or from the center of the body outward. In the prenatal period, the head, chest, and trunk grow first, followed by the arms and legs, and finally by the hands and feet. During infancy and childhood, the arms and legs continue to grow somewhat ahead of the hands and feet. As we will see later, motor development follows these same developmental trends.

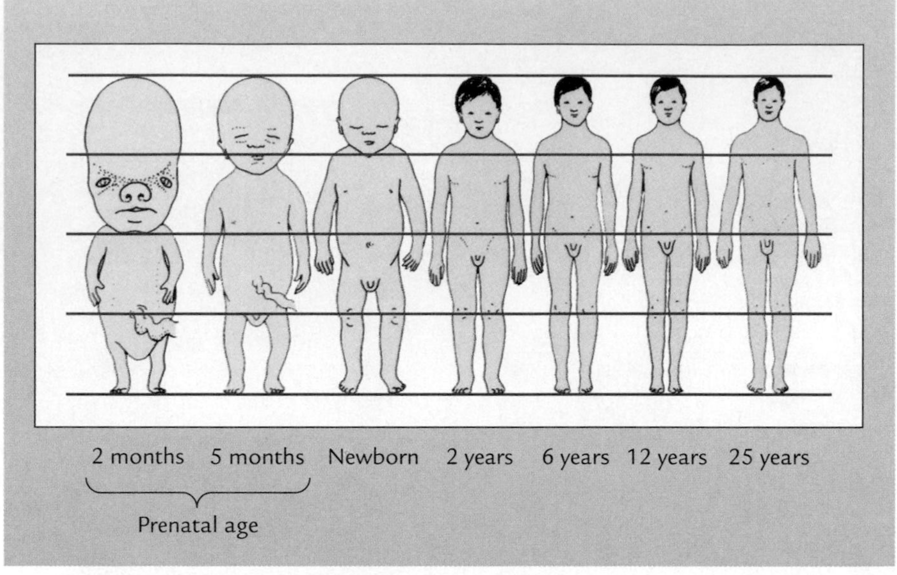

FIGURE 4.2 Changes in body proportions from the early prenatal period to adulthood. This figure illustrates the cephalocaudal trend of physical growth. The head gradually becomes smaller, and the legs longer, in proportion to the rest of the body.

Brain Development

At birth, the brain is nearer than any other physical structure to its adult size, and it continues to develop at an astounding pace throughout infancy and toddlerhood. To best understand brain growth, we need to look at it from two vantage points: (1) the microscopic level of individual brain cells and (2) the larger structural level of the cerebral cortex, which is responsible for the highly developed intelligence of our species.

Development of Neurons

The human brain has 100 to 200 billion **neurons,** or nerve cells that store and transmit information, many of which have thousands of direct connections with other neurons. Neurons differ from other body cells in that they are not tightly packed together. They have tiny gaps, or **synapses,** between them, where fibers from different neurons come close together but do not touch. Neurons release chemicals that cross the synapse, sending messages to one another.

The basic story of brain growth concerns how neurons develop and form this elaborate communication system. Major milestones of brain development are summarized in Figure 4.3. During the prenatal period, neurons are produced in the primitive neural tube of the embryo. From there, they migrate to form the major parts of the brain, traveling along threads produced by a network of guiding cells. By the end of the second trimester of pregnancy, production and migration of neurons are largely complete. Once neurons are in place, they differentiate, establishing their unique functions by extending their fibers to form synaptic connections with neighboring cells. As Figure 4.4 reveals, during the first 2 years, growth of neural fibers and synapses increases at an astounding pace (Huttenlocher, 1994; Moore & Persaud, 1998). Because developing neurons require space for these connective structures, a surprising aspect of brain growth is that when synapses are formed, many surrounding neurons die—20 to 80 percent depending on the brain region (Diamond & Hopson, 1999; Stiles, 2001a). Fortunately, during the prenatal period, the neural tube produces far more neurons than the brain will ever need.

As neurons form connections, *stimulation* becomes necessary for their survival. Neurons that are stimulated by input from the surrounding environment continue to establish new synapses, forming increasingly elaborate systems of communication that lead to more complex abilities. At first, stimulation results in an overabundance of synapses, many of which serve identical functions, thereby helping to ensure that the child will acquire certain skills. Neurons seldom stimulated soon lose their synapses, a process called **synaptic pruning.** It returns neurons not needed at the moment to an uncommitted state so they can support future development (Johnson, 1998). Notice how, for this process to go forward, appropriate stimulation of the child's brain is vital during periods in which the formation of synapses is at its peak (Eisenberg, 1999; Greenough et al., 1993).

Perhaps you are wondering, if few neurons are produced after the prenatal period, what causes the dramatic increase in brain size during the first 2 years? About half the brain's volume is made up of **glial cells,** which do not carry messages. Instead, they are responsible for **myelination,** the coating of neural fibers with an insulating fatty sheath (called *myelin*) that improves the efficiency of message transfer. Glial cells multiply dramatically from the end of pregnancy through the second year of life, a process that continues at a slower pace through middle childhood. Dramatic increases in neural fibers and

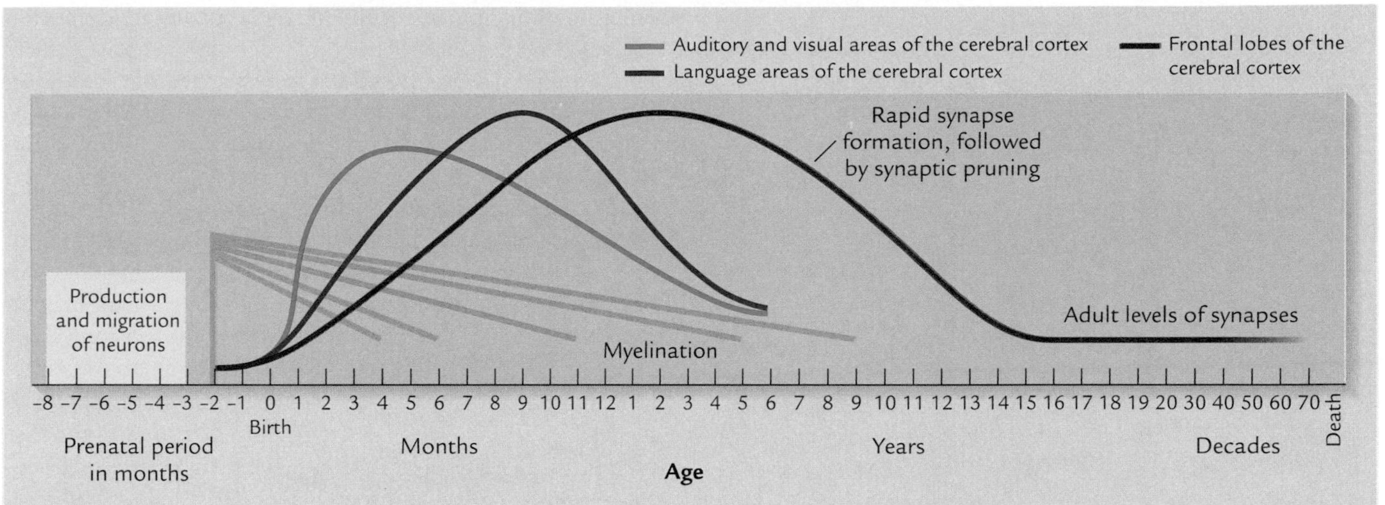

FIGURE 4.3 Major milestones of brain development. Formation of synapses is rapid during the first 2 years of life, especially in the auditory, visual, and language areas of the cerebral cortex. The frontal lobes, responsible for thought, undergo more extended synaptic growth. In each area, overproduction of synapses is followed by synaptic pruning as stimulation strengthens needed connections and returns neurons not needed at the moment to an uncommitted state so that they can support future skills. The frontal lobes are among the last regions to attain adult levels of synaptic connections—in mid- to late adolescence. (Adapted from Thompson & Nelson, 2001.)

myelination are responsible for the rapid gain in overall size of the brain. At birth, the brain is nearly 30 percent of its adult weight; by age 2, it reaches 70 percent (Thatcher et al., 1996).

Development of the Cerebral Cortex

The **cerebral cortex** surrounds the brain, looking much like a half-shelled walnut. It is the largest, most complex brain structure—accounting for 85 percent of the brain's weight, containing the greatest number of neurons and synapses, and responsible for the unique intelligence of our species. The cerebral cortex is the last brain structure to stop growing. For this reason, it is believed to be sensitive to environmental influences for a much longer period than any other part of the brain.

● **Regions of the Cortex.** As Figure 4.5 on page 118 shows, different regions of the cerebral cortex have specific functions, such as receiving information from the senses, instructing the body to move, and thinking. The order in which cortical regions develop corresponds to the order in which various capacities emerge in the infant and growing child. A burst of synaptic growth in the auditory and visual areas occurs during the first year—a period of dramatic gains in auditory and visual perception. Among areas responsible for body movement, neurons that control the head, arms, and chest form connections before those that control the trunk and legs. (Can you name this growth trend?) And areas that support language show dramatic synaptic growth during infancy and toddlerhood, when language development begins to flourish.

One of the last regions of the cortex to develop is the *frontal lobes,* which are responsible for thought—in particular, consciousness, inhibition of impulses, and regulation of behavior through planning. From age 2 months on, this area functions more effectively. Formation and pruning of synapses in the frontal lobes continue for many years, yielding an adult level

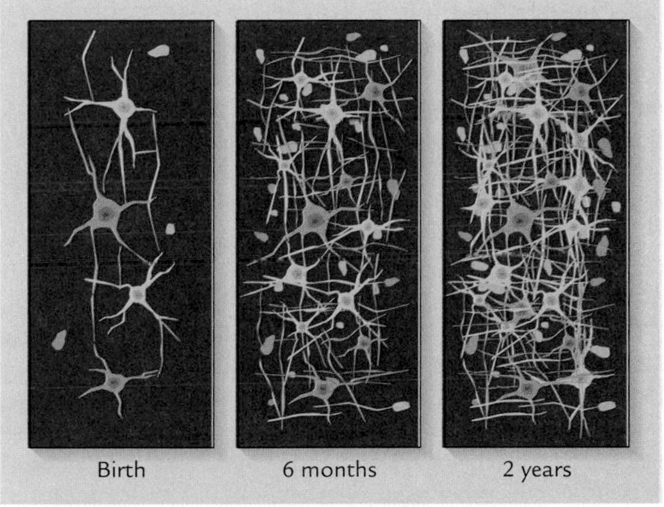

FIGURE 4.4 Development of synaptic connections in the brain. Growth of neural fibers takes place rapidly from birth to 2 years. During this time, new synapses form at an astounding pace, supporting the emergence of many new capacities. Stimulation is vitally important for maintaining and increasing this complex communication network. (Reprinted by permission of the publisher from *The Postnatal Development of the Human Cerebral Cortex,* vols. I-III, by Jesse LeRoy Conel, Cambridge, MA: Harvard University Press. Copyright ©1939, 1975 by the President and Fellows of Harvard College.)

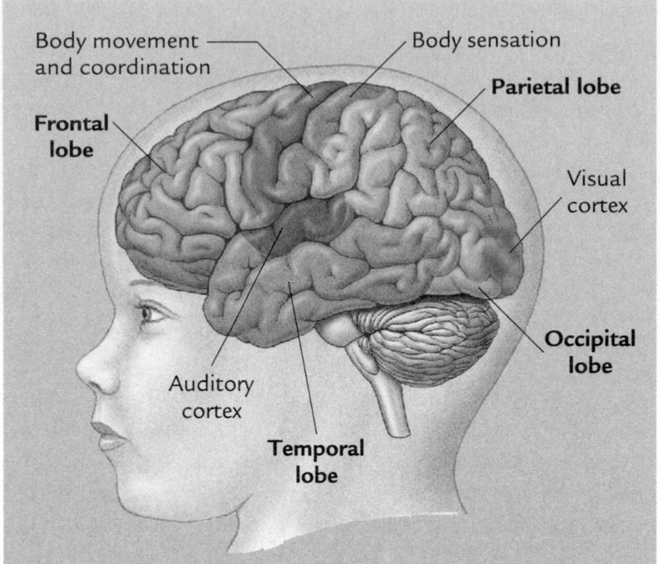

■ **FIGURE 4.5 The left side of the human brain, showing the cerebral cortex.** The cortex is divided into different lobes, each of which contains a variety of regions with specific functions. Some major regions are labeled here.

of synaptic connections around mid- to late adolescence (Nelson, 2002; Thompson et al., 2000).

● **Lateralization and Plasticity of the Cortex.** The cerebral cortex has two *hemispheres,* or sides—left and right—that differ in their functions. Some tasks are done mostly by one hemisphere and some by the other. For example, each hemisphere receives sensory information from and controls only one side of the body—the one opposite to it.[1] For most of us, the left hemisphere is largely responsible for verbal abilities (such as spoken and written language) and positive emotion (for example, joy). The right hemisphere handles spatial abilities (judging distances, reading maps, and recognizing and producing geometric shapes) and negative emotion (such as distress) (Banish & Heller, 1998; Nelson & Bosquet, 2000). This pattern may be reversed in left-handed people, but more often, the cortex of left-handers is less clearly specialized than that of right-handers.

Specialization of the two hemispheres is called **lateralization.** Why are behaviors and abilities lateralized? According to one view, the left hemisphere is better at processing information in a sequential, analytic (piece-by-piece) way, which is a good approach for dealing with communicative information—both verbal (language) and emotional (a joyful smile). In contrast, the right hemisphere is specialized for processing information in a holistic, integrative manner, ideal for making sense of spatial information and regulating negative emotion (Banish, 1998). A lateralized brain is certainly adaptive. It permits a wider array of functions to be carried out effectively than if both sides processed information exactly the same way.

Researchers are interested in when brain lateralization occurs because they want to know more about **brain plasticity.** In a highly *plastic* cortex, many areas are not yet committed to specific functions. If a part of the brain is damaged, other parts can take over tasks that it would have handled. But once the hemispheres lateralize, damage to a particular region means that the abilities controlled by it will be lost forever.

At birth, the hemispheres have already begun to specialize. Most newborns show greater electrical brain-wave activity in the left hemisphere while listening to speech sounds and displaying positive emotions. In contrast, the right hemisphere reacts more strongly to nonspeech sounds as well as to stimuli (such as a sour-tasting fluid) that cause infants to display negative emotion (Davidson, 1994; Fox & Davidson, 1986).

Nevertheless, dramatic evidence for substantial plasticity in the young brain comes from research on brain-damaged children and adults, summarized in the Lifespan Vista box on pages 120–121. Furthermore, a growing body of research reveals that early experience molds brain organization, inducing specialization of certain regions of the cerebral cortex. For example, electrical brain-wave recordings reveal that deaf adults who, as infants and children, learned sign language (a spatial skill) depend more on the right hemisphere for language processing than do hearing individuals (Neville & Bruer, 2001).

■ This 12-month-old, growing up in the Santa Clara Pueblo community of New Mexico, beats the ceremonial drum she has seen her father play during festival dances. Her more advanced actions on objects suggest that she might be in the midst of a brain growth spurt. How brain development can best be supported through stimulation during such periods is a challenging research question.

[1]The eyes are an exception. Messages from the right half of each retina go to the right hemisphere; messages from the left half of each retina go to the left hemisphere. Thus, visual information from *both* eyes is received by *both* hemispheres.

Also, toddlers advanced in language development show greater left-hemispheric specialization for language than their more slowly developing agemates. Apparently, the very process of acquiring language promotes lateralization (Bates, 1999; Mills, Coffey-Corina, & Neville, 1997).

During the first few years, the brain is more plastic than at any later time of life. Its flexibility protects young children's ability to learn, which is fundamental to survival (Nelson, 2000). And although the cortex is programmed from the start for hemispheric specialization, experience greatly influences the rate and success of this genetic program.

Sensitive Periods of Brain Development

Recall that stimulation of the brain is vital during periods in which it is growing most rapidly. The existence of sensitive periods in development of the cerebral cortex has been amply demonstrated in studies of animals exposed to extreme forms of sensory deprivation. For example, there seems to be a time when rich and varied visual experiences must occur for the visual centers of the brain to develop normally. If a month-old kitten is deprived of light for as brief a time as 3 or 4 days, these areas of the brain degenerate. If the kitten is kept in the dark during the fourth week of life and longer, the damage is severe and permanent (Crair, Gillespie, & Stryker, 1998). Enriched versus deprived early environments also affect overall brain growth. When animals reared in physically and socially stimulating surroundings are compared with animals reared in isolation, the brains of the stimulated animals show much denser synaptic connections (Greenough & Black, 1992).

● **Brain Growth Spurts.** Because we cannot ethically expose children to such experiments, researchers interested in identifying sensitive periods for human brain development must rely on less direct evidence. They have identified intermittent brain growth spurts from infancy to early adulthood, based on gains in brain weight and skull size as well as changes in electrical brain-wave activity of the cerebral cortex. For example, several surges in frontal-lobe activity, which gradually spread to other cortical regions, occur during the first 2 years of life: at 3 to 4 months, when infants typically reach for objects; around 8 months, when they begin to crawl and search for hidden objects; around 12 months, when they walk and display more advanced object-search behaviors; and between 1½ and 2 years, when language flourishes (Bell & Fox, 1994, 1998; Fischer & Bidell, 1998). Later electrical activity spurts, at ages 9, 12, 15, and 18 to 20, may reflect the emergence and refinement of abstract thought (Fischer & Rose, 1995).

Massive production of synapses may underlie brain growth spurts in the first 2 years. Development of more complex and efficient neural networks, due to synaptic pruning and longer-distance connections between the frontal lobes and other cortical regions, may account for the later ones. Researchers are convinced that what "wires" a child's brain during each of these periods is experience. But they still have many questions to answer about just how brain and behavioral development might best be supported during each growth spurt.

● **Appropriate Stimulation.** The evidence we do have confirms that the brain is particularly spongelike during the first few years of life; children learn new skills rapidly. As we will see later in this chapter and in chapters to come, understimulating infants and young children by depriving them of rich and varied experiences available in caring family environments impairs their development. Studies of infants from Eastern European orphanages show that the earlier they are removed from these barren settings, the greater their catch-up in development. When children spend their first 2 years or more in deprived institutional care with little social contact and sensory stimulation, all domains of development usually remain delayed (Ames, 1997; Johnson, 2000). Deficits in concentration, attention, and control of anger and other impulses are especially severe (Gunnar, 2001).

Unlike the orphanage children just described, Grace, whom Monica and Kevin had adopted in Cambodia, showed favorable progress. Two years earlier, they had adopted Grace's

■ This boy has spent his first 2 years in a Romanian orphanage, with little adult contact and stimulation. The longer he remains in a barren environment, the more he will withdraw and wither and display permanent impairments in all domains of development.

A Lifespan Vista

Brain Plasticity: Insights from Research on Brain-Damaged Children and Adults

In the first few years of life, the brain is highly plastic. It can reorganize areas committed to specific functions in a way that the mature brain cannot. Consequently, adults who suffered brain injuries in infancy and early childhood have fewer cognitive impairments than adults with later-occurring injuries. Nevertheless, the young brain is not totally plastic. When it is injured, its functioning is compromised. The extent of plasticity depends on several factors, including age at time of injury, site of damage, and skill area. Furthermore, plasticity is not restricted to childhood. Some reorganization after injury also occurs in the mature brain.

Brain Plasticity in Infancy and Early Childhood. In a large study of children with injuries to the cerebral cortex that occurred before birth or in the first 6 months of life, language and spatial skills were assessed repeatedly into the school years (Stiles, 2001a; Stiles et al., 1998). All the children had experienced early brain seizures or hemorrhages. Brain-imaging techniques revealed the precise site of damage.

Regardless of whether injury occurred in the left or right cerebral hemisphere, the children showed delays in language development that persisted until about 3½ years of age. The fact that damage to either hemisphere affected early language competence indicates that at first, language functioning is broadly distributed in the brain. But by age 5, the children caught up in both vocabulary and grammatical skills. Undamaged areas—either in the left or the right hemisphere—had taken over these language functions. Still, mild language deficits persist in children with early brain injuries. When asked to tell a story, they use less complex language than their normal agemates (Reilly, Bates, & Marchman, 1998).

Compared with language, spatial skills are more impaired after early brain injury. When 5- and 6-year-olds who had suffered brain injury before 6 months of age were asked to copy designs, children with right-hemispheric damage had trouble with holistic processing—accurately representing the overall shape. In contrast, children with left-hemispheric damage captured the basic shape but omitted fine-grained details (see Figure 4.6) (Akshoomoff & Stiles, 1995). The brain-injured children showed some improvement in their drawings over the school years—gains that do not occur in brain-injured adults. Still, the children had considerable difficulty drawing accurately (Akshoomoff et al., 2001).

Clearly, recovery after early brain injury is greater for language than spatial skills. Why is this so? Researchers speculate that spatial processing is the older of the two capacities in our evolutionary history and, therefore, more lateralized at birth (Stiles, 2001b). But early brain injury has far less impact on *both* language and spatial skills than does later injury. In sum, the young brain is remarkably plastic.

Brain Plasticity in Adulthood. Although more limited, brain plasticity is also evident in adulthood. For example, in a study of an adult born with fused fingers, brain-imaging techniques revealed that the motor area of the cerebral cortex lacked defined areas for finger control. Several weeks after surgery to separate the fingers, distinct cortical areas for each finger had emerged (Mogilner et al., 1993). Furthermore, adult stroke victims often display considerable recovery, especially in response to stimulation of language and motor skills. Brain-imaging techniques reveal that structures adjacent to the permanently damaged area or in the opposite cerebral hemi-

older brother, Eli. When Eli was 2 years old, Monica and Kevin sent a letter and a photo of Eli to his biological mother, describing a bright, happy child. The next day, she tearfully asked an adoption agency to send her baby daughter to join Eli and his American family. Although Grace's early environment was very deprived, her biological mother's loving care—gentle holding, soft speaking, and breastfeeding—may have prevented irreversible damage to her brain.

Besides impoverished environments, ones that overwhelm children with expectations beyond their current capacities interfere with the brain's potential. In recent years, expensive early-learning centers have sprung up, in which infants are trained with letter and number flash cards and slightly older toddlers are given a full curriculum of reading, math, science, art, music, gym, and more. There is no evidence that these programs yield smarter, better "superbabies." Instead, trying to prime infants with stimulation for which they are not ready can cause them to withdraw, thereby threatening their interest in learning and creating conditions much like stimulus deprivation! In addition, when such programs do not produce young geniuses, they can lead to disappointed parents who view their children as failures at a very tender age. Thus, they rob infants of a psychologically healthy start, and they deprive parents of relaxed, pleasurable participation in their children's early growth.

Changing States of Arousal

Rapid brain growth means that the organization of sleep and wakefulness changes substantially between birth and 2 years, and fussiness and crying also decline. The newborn

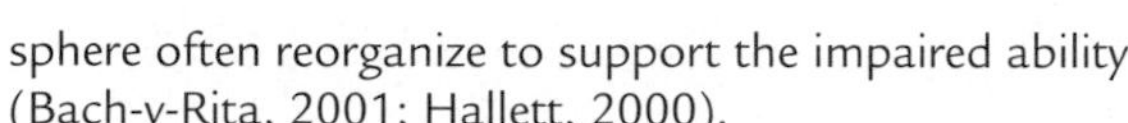

sphere often reorganize to support the impaired ability (Bach-y-Rita, 2001; Hallett, 2000).

In infancy and childhood, the goal of brain growth is to form neural connections that support essential skills. Animal research indicates that plasticity is greatest while the brain is forming many new synapses; plasticity declines during synaptic pruning (Kolb & Gibb, 2001). At older ages, specialized brain structures are in place, but after injury they can still reorganize to some degree. Plasticity seems to be a basic property of the nervous system throughout the lifespan. Researchers hope to discover how experience and brain plasticity work together to promote learning at each age period so they can help children and adults—with and without brain injuries—develop at their best.

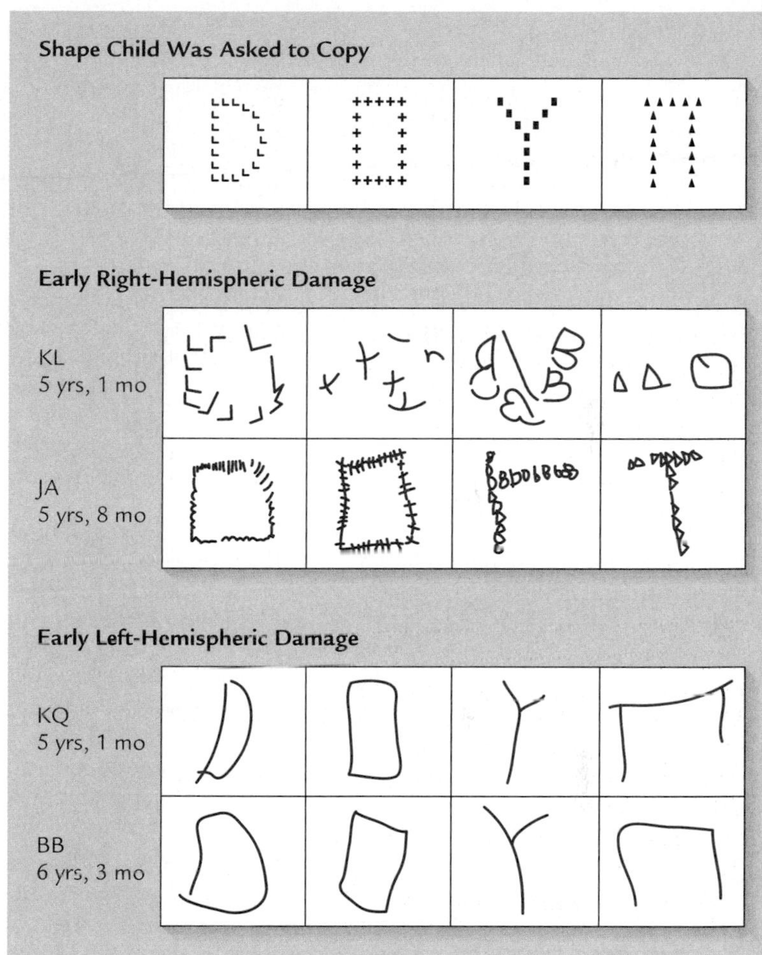

■ **FIGURE 4.6 Impairments in spatial skills in 5- and 6-year-olds who had experienced brain injury before birth or in the first 6 months of life.** Compared with language skills, spatial skills are more impaired after early brain injury. When researchers had children copy designs, those with right-hemispheric damage had difficulty representing the overall shape. Those with left-hemispheric damage captured the basic shape but omitted fine-grained details. Although drawings improved over the school years, difficulties with spatial processing remained. (Adapted from J. Stiles, 2001a, "Neural Plasticity and Cognitive Development," *Developmental Neuropsychology, 18,* p. 261. Reprinted by permission.)

baby takes round-the-clock naps that total about 16 to 18 hours. The decline in total sleep time over the first 2 years is not great; the average 2-year-old still needs 12 to 13 hours. The greatest changes are that periods of sleep and wakefulness become fewer but longer, and the sleep–wake pattern increasingly conforms to a night–day schedule (Whitney & Thoman, 1994). By the second year, children generally need only one or two naps during the day (Blum & Carey, 1996).

Although these changing arousal patterns are due to brain development, they are affected by the social environment. In most Western nations, parents usually succeed in getting their babies to sleep through the night around 4 months of age by offering an evening feeding before putting them down in a separate, quiet room. In this way, they push young infants to the limits of their neurological capacities. Not until the middle of the first year is the secretion of *melatonin,* a hormone within the brain that promotes drowsiness, much greater at night than during the day (Sadeh, 1997).

As the Cultural Influences box on page 122 reveals, the practice of isolating infants to promote sleep is rare elsewhere in the world. When babies sleep with their parents, their average sleep period remains constant at 3 hours from 1 to 8 months of age. Only at the end of the first year, as REM sleep (the state that usually prompts waking) declines, do infants move in the direction of an adultlike sleep–waking schedule (Ficca et al., 1999).

Even after infants sleep through the night, they continue to wake at night occasionally. In surveys carried out in Australia, Great Britain, Israel, and the United States, parent reports indicated that night wakings peaked between 1½ and 2 years and then declined (Armstrong, Quinn, & Dadds, 1994; Scher et al., 1995). As Chapter 6 will reveal, the challenges of this period—the ability to range farther from the caregiver

Cultural Influences

Cultural Variation in Infant Sleeping Arrangements

While awaiting the birth of a new baby, North American parents typically furnish a room as the infant's sleeping quarters. For decades, child-rearing advice from experts strongly encouraged the nighttime separation of baby from parent. For example, the most recent edition of Dr. Spock's *Baby and Child Care* recommends that infants be moved out of their parents' rooms early in the first year, explaining, "Otherwise, there is a chance that they may become dependent on this arrangement" (Spock & Parker, 1998, p. 102).

Yet parent–infant "cosleeping" is common around the globe. Japanese children usually lie next to their mothers throughout infancy and early childhood and continue to sleep with a parent or other family member until adolescence (Takahashi, 1990). Among the Maya of rural Guatemala, mother–infant cosleeping is interrupted only by the birth of a new baby, at which time the older child is moved beside the father or to another bed in the same room (Morelli et al., 1992). Cosleeping is also frequent in some North American subcultures. African-American infants and children frequently fall asleep with their parents and remain with them for part or all of the night (Brenner et al., 2003; Lozoff et al., 1995). Appalachian children of eastern Kentucky typically sleep with their parents for the first 2 years of life (Abbott, 1992).

Cultural values—specifically, collectivism versus individualism (see Chapter 2, page 63)—strongly influence infant sleeping arrangements. In one study, researchers interviewed American middle-SES mothers and Guatemalan Mayan mothers about their sleeping practices. American mothers conveyed an individualistic perspective, mentioning the importance of early independence, preventing bad habits, and protecting their own privacy. In contrast, Mayan mothers stressed a collectivist perspective, explaining that cosleeping helps build a close parent– child bond, which is necessary for children to learn the ways of people around them (Morelli et al., 1992).

Perhaps because more mothers are breastfeeding, during the past decade, cosleeping among American mothers and their infants increased from 6 to 13 percent (McKenna, 2002; Willinger et al., 2003). Research suggests that bedsharing evolved to protect infants' survival and health. Cosleeping babies breastfeed three times longer at night than do infants who sleep alone. Because infants arouse to nurse more often when sleeping next to their mothers, some researchers believe that cosleeping may help safeguard babies at risk for sudden infant death syndrome (see page 103) (Mosko, Richard, & McKenna, 1997a). And contrary to popular belief, mothers' total sleep time is not decreased by cosleeping, although they experience a great number of brief awakenings, which permits them to check on their baby (Mosko, Richard, & McKenna, 1997b).

Infant sleeping practices affect other aspects of family life. Sleep problems are not an issue for Mayan parents. Babies doze off in the midst of ongoing family activities and are carried to bed by their mothers. In the United States, getting young children ready for bed often requires an elaborate ritual that takes a good part of the evening. Perhaps bedtime struggles, so common in Western homes but rare elsewhere in the world, are related to the stress young children feel when they are required to fall asleep without assistance (Latz, Wolf, & Lozoff, 1999).

Critics of cosleeping warn that infants might become trapped under the parent's body or in soft covers and suffocate. Use of quilts and comforters while cosleeping is hazardous, and unknowingly, too many North American families use them (Willinger et al., 2003). But with appropriate precautions, parents and infants can cosleep safely (McKenna, 2001). In cultures where cosleeping is widespread, parents and infants usually sleep with light covering on hard surfaces, such as firm mattresses, floor mats, and wooden planks (Nelson, Schiefenhoevel, & Haimerl, 2000).

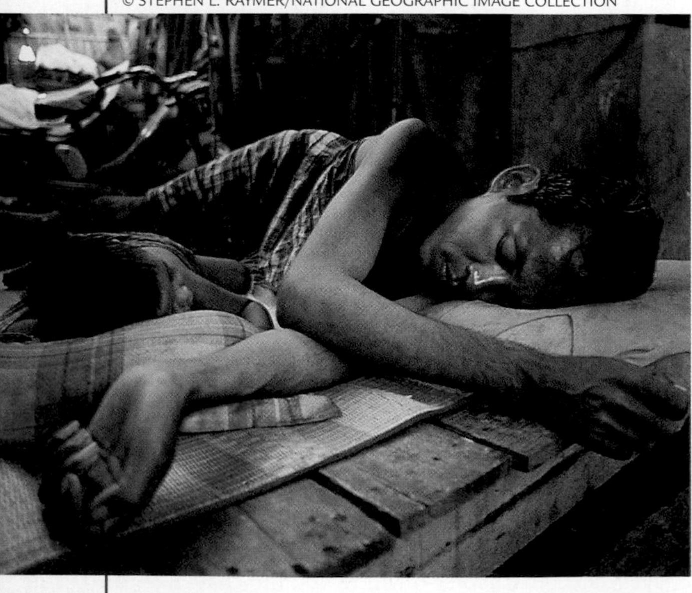

■ This Cambodian father and child sleep together—a practice common around the globe. Many parents who practice cosleeping believe that it helps build a close parent–child bond. Notice that the family sleeps on hard wooden surfaces, which protect cosleeping children from entrapment in soft bedding.

and awareness of the self as separate from others—often prompt anxiety, evident in disturbed sleep and clinginess. When parents offer comfort, these behaviors subside.

Ask Yourself

REVIEW

How does stimulation affect early brain development? Cite evidence at the level of neurons and at the level of the cerebral cortex.

REVIEW

Explain why overproduction of synapses and synaptic pruning are adaptive processes that foster brain development.

APPLY

Which infant enrichment program would you choose: one that emphasizes gentle talking and touching, exposure to sights and sounds, and simple social games, or one that includes word and number drills and classical music lessons? Explain.

CONNECT

Is research on brain plasticity consistent with the assumption of the lifespan perspective that development is plastic at all ages? (See Chapter 1, page 9.) Explain, providing evidence on how brain plasticity changes with development.

www.

Influences on Early Physical Growth

Physical growth, like other aspects of development, results from a continuous and complex interplay between genetic and environmental factors. Heredity, nutrition, and emotional well-being all affect early physical growth.

Heredity

Since identical twins are much more alike in body size than fraternal twins, we know that heredity is important in physical growth. When diet and health are adequate, height and rate of physical growth are largely determined by heredity (Tanner, 1990). In fact, as long as negative environmental influences such as poor nutrition and illness are not severe, children and adolescents typically show *catch-up growth*—a return to a genetically determined growth path.

Genetic makeup also affects body weight, since the weights of adopted children correlate more strongly with those of their biological than adoptive parents (Stunkard et al., 1986). However, as far as weight is concerned, environment—in particular, nutrition—plays an especially important role.

Nutrition

Nutrition is important at any time of development, but it is especially crucial during the first 2 years because the baby's brain and body are growing so rapidly. Pound for pound, an infant's energy needs are twice those of an adult. Twenty-five percent of the infant's total caloric intake is devoted to growth, and extra calories are needed to keep rapidly developing organs functioning properly (Pipes, 1996).

● **Breast- versus Bottle-Feeding.** Babies not only need enough food, they need the right kind of food. In early infancy, breast milk is especially suited to their needs, and bottled formulas try to imitate it. The Caregiving Concerns table on page 124 summarizes the major nutritional and health advantages of breastfeeding.

Because of these benefits, breastfed babies in poverty-stricken regions of the world are much less likely to be malnourished and 6 to 14 times more likely to survive the first year of life. Breastfeeding exclusively for the first 6 months, followed by combining breastfeeding with solid foods until the end of the first year, would save the lives of more than one million infants annually. Even breastfeeding for just a few weeks would offer some protection against respiratory and intestinal infections that are devastating to young children in developing countries. Furthermore, because a mother is less likely to get pregnant while she is nursing, breastfeeding helps increase spacing among siblings, a major factor in reducing infant and childhood deaths in developing countries (Darton-Hill & Coyne, 1998). (Note, however, that breastfeeding is not a reliable method of birth control.)

Yet many mothers in the developing world do not know about the benefits of breastfeeding. Consequently, they give their babies commercial formula or low-grade nutrients, such as rice water or highly dilute cow or goat milk. These foods often lead to illness because they are contaminated due to poor sanitation. The United Nations has encouraged all hospitals and maternity units in developing countries to promote breastfeeding as long as mothers do not have viral or bacterial infections (such as HIV and tuberculosis) that can be transmitted to the baby. Today, most developing countries have banned the practice of giving free or subsidized formula to any new mother who desires it.

Partly as a result of the natural childbirth movement, breastfeeding has become more common in industrialized nations, especially among well-educated women. Today, 65 percent of American mothers and 73 percent of Canadian mothers breastfeed. However, about two-thirds of breastfeeding American mothers and nearly half of Canadian mothers stop

Nutritional and Health Advantages of Breastfeeding

ADVANTAGE	DESCRIPTION
Correct balance of fat and protein	Compared with the milk of other mammals, human milk is higher in fat and lower in protein. This balance, as well as the unique proteins and fats contained in human milk, is ideal for a rapidly myelinating nervous system.
Nutritional completeness	A mother who breastfeeds need not add other foods to her infant's diet until the baby is 6 months old. The milks of all mammals are low in iron, but the iron contained in breast milk is much more easily absorbed by the baby's system. Consequently, bottle-fed infants need iron-fortified formula.
Protection against disease	Breastfeeding transfers antibodies and other infection-fighting agents from mother to child and enhances functioning of the immune system. As a result, breastfed babies have far fewer respiratory and intestinal illnesses and allergic reactions than bottle-fed infants. Components of human milk that protect against disease can be added to formulas, but breastfeeding provides superior immunity.
Healthy physical growth	Breastfed infants gain less weight and are leaner at 1 year of age than bottle-fed infants, a growth pattern that may help prevent later overweight and obesity.
Protection against faulty jaw development and tooth decay	Sucking the mother's nipple instead of an artificial nipple helps avoid malocclusion, a condition in which the upper and lower jaws do not meet properly. It also protects against tooth decay due to sweet liquid remaining in the mouths of infants who fall asleep while sucking on a bottle.
Digestibility	The composition of breast milk makes it more digestible than cow's milk. Because breastfed babies have a different kind of bacteria growing in their intestines than do bottle-fed infants, they rarely become constipated or have diarrhea.
Smoother transition to solid foods	Breastfed infants accept new solid foods more easily than bottle-fed infants, perhaps because of their greater experience with a variety of flavors, which pass from the maternal diet into the mother's milk.

Sources: Dewey, 2001; Pickering et al., 1998; Raisler, 1999; U.S. Department of Health and Human Services, 2000a.

■ Breastfeeding is especially important in developing countries, where infants are at risk for malnutrition and early death due to widespread poverty. This baby of Rajasthan, India, is likely to grow normally during the first year because her mother decided to breastfeed.

after a few months (Health Canada, 2002b; U.S. Department of Health and Human Services, 2002a). Not surprisingly, mothers who return to work sooner wean their babies from the breast earlier (Arora et al., 2000). However, a mother who cannot be with her baby all the time can pump her milk into a bottle or combine breast- and bottle-feeding.

Women who cannot or do not want to breastfeed sometimes worry that they are depriving their baby of an experience essential for healthy psychological development. Yet breast- and bottle-fed youngsters in industrialized nations do not differ in emotional adjustment (Fergusson & Woodward, 1999). Some studies report a slight advantage in mental test performance for children and adolescents who are breastfed, after many factors are controlled. Others, however, find no cognitive benefits (Jain, Concat, & Leventhal, 2002; Mortensen et al., 2002). Notice in the Caregiving Concerns table that breast milk provides nutrients ideally suited for early brain development.

● Are Chubby Babies at Risk for Later Overweight and Obesity? Timmy was an enthusiastic eater from early infancy. He nursed vigorously and gained weight quickly. By 5 months, he began reaching for food on his parents' plates. Vanessa wondered: Was she overfeeding Byron and increasing his chances of being permanently overweight?

A slight correlation exists between rapid weight gain in infancy and overweight at older ages (Stettler et al., 2002). But most chubby babies thin out during toddlerhood and early childhood, as weight gain slows and they become more active. Infants and toddlers can eat nutritious foods freely, without risk of becoming too fat.

How can concerned parents prevent their infants from becoming overweight children and adults? One way is to encourage good eating habits. Candy, soft drinks, sweetened juices, French fries, and other high-calorie foods loaded with sugar, salt, and saturated fats should be avoided. When given such foods regularly, young children start to prefer them (Birch & Fisher, 1995). Physical exercise is another safeguard against excessive weight gain. Once toddlers learn to walk, climb, and run, parents should encourage their natural delight at being able to control their bodies by providing opportunities for energetic play.

Malnutrition

Osita is an Ethiopian 2-year-old whose mother has never had to worry about his gaining too much weight. When she weaned him at 1 year, there was little for him to eat besides starchy rice flour cakes. Soon his belly enlarged, his feet swelled, his hair began to fall out, and a rash appeared on his skin. His bright-eyed, curious behavior vanished, and he became irritable and listless.

In developing countries and war-torn areas where food resources are limited, malnutrition is widespread. Recent evidence indicates that 40 to 60 percent of the world's children do not get enough to eat (Bellamy, 1998). Among the 4 to 7 percent who are severely affected, malnutrition leads to two dietary diseases: marasmus and kwashiorkor.

Marasmus is a wasted condition of the body caused by a diet low in all essential nutrients. It usually appears in the first year of life when a baby's mother is too malnourished to produce enough breast milk and bottle-feeding is also inadequate. Her starving baby becomes painfully thin and is in danger of dying.

Osita has **kwashiorkor,** caused by an unbalanced diet very low in protein. Kwashiorkor usually strikes after weaning, between 1 and 3 years of age. It is common in areas of the world where children get just enough calories from starchy foods, but protein resources are scarce. The child's body responds by breaking down its own protein reserves, which causes the swelling and other symptoms that Osita experienced.

■ The swollen abdomen and listless behavior of this Honduran child are classic symptoms of kwashiorkor, a nutritional illness that results from a diet very low in protein.

Children who survive these extreme forms of malnutrition grow to be smaller in all body dimensions (Galler, Ramsey, & Solimano, 1985a). And when their diets improve, they are at risk for excessive weight gain. Nationwide surveys in Russia, China, and South Africa reveal that growth-stunted children are far more likely to be overweight than their nonstunted agemates (Popkin, Richards, & Montiero, 1996). To protect itself, a malnourished body establishes a low basal metabolism rate, which may endure after nutrition improves. Also, malnutrition may disrupt appetite control centers in the brain, causing the child to overeat when food becomes plentiful.

Learning and behavior are also seriously affected. One long-term study of marasmic children revealed that an improved diet led to some catch-up growth in height but to little improvement in head size (Stoch et al., 1982). The malnutrition probably interfered with growth of neural fibers and myelination, causing a permanent loss in brain weight. These children score low on intelligence tests, show poor fine-motor coordination, and have difficulty paying attention (Galler et al., 1990; Galler, Ramsey, & Solimano, 1985b). They also display a more intense stress response to fear-arousing situations, perhaps caused by the constant, gnawing pain of hunger (Fernald & Grantham-McGregor, 1998).

Malnutrition is not confined to developing countries. Some poverty-stricken North American children go to bed hungry (Children's Defense Fund, 2002; Wachs, 1995). Although few have marasmus or kwashiorkor, their physical growth and ability to learn are still affected.

Ask Yourself

REVIEW

Explain why breastfeeding can have lifelong consequences for the development of babies born in poverty-stricken regions of the world.

APPLY

Ten-month-old Shaun is below average in height and painfully thin. He has one of two serious growth disorders. Name them, and indicate what clues you would look for to determine which one Shaun has.

www.

Emotional Well-Being

We are not used to thinking of affection and stimulation as necessary for healthy physical growth, but they are just as vital as food. **Nonorganic failure to thrive** is a growth disorder resulting from lack of parental love that is usually present by 18 months of age. Infants who have it show all the signs of marasmus: Their bodies look wasted, and they are withdrawn and apathetic. But no organic (or biological) cause for the baby's wasted appearance can be found. Enough food is offered, and the infant does not have a serious illness.

Lana, an observant nurse at a public health clinic, became concerned about 8-month-old Melanie, who was 3 pounds lighter than she had been at her last checkup. Her mother claimed to feed her often. Lana noted Melanie's behavior. She showed little interest in toys but, instead, kept her eyes on nearby adults, anxiously watching their every move. She rarely smiled when her mother came near or cuddled when picked up (Steward, 2001).

Family circumstances surrounding failure to thrive help explain these typical reactions. During feeding, diaper changing, and play, Melanie's mother sometimes seemed cold and distant, at other times impatient and hostile (Hagekull, Bohlin, & Rydell, 1997). Melanie tried to protect herself by tracking her mother's whereabouts and, when she approached, avoiding her gaze. Often an unhappy marriage and parental psychological disturbance contribute to these serious caregiving problems (Drotar, Pallotta, & Eckerle, 1994; Duniz et al., 1996). Sometimes the baby is irritable and displays abnormal feeding behaviors, such as poor sucking or vomiting—circumstances that stress the parent–child relationship further (Wooster, 1999).

In Melanie's case, her alcoholic father was out of work, and her parents argued constantly. Melanie's mother had little energy to meet Melanie's psychological needs. When treated early, by helping parents or placing the baby in a caring foster home, failure-to-thrive infants show quick catch-up growth. But if the disorder is not corrected in infancy, most children remain small and show lasting cognitive and emotional difficulties (Wooster, 2000).

Learning Capacities

Learning refers to changes in behavior as the result of experience. Babies come into the world with built-in learning capacities that permit them to profit from experience immediately. Infants are capable of two basic forms of learning, which were introduced in Chapter 1: classical and operant conditioning. They also learn through their natural preference for novel stimulation. Finally, shortly after birth, babies learn by observing others; they can soon imitate the facial expressions and gestures of adults.

Classical Conditioning

Newborn reflexes, discussed in Chapter 3, make **classical conditioning** possible in the young infant. In this form of learning, a new stimulus is paired with a stimulus that leads to a reflexive response. Once the baby's nervous system makes the connection between the two stimuli, the new stimulus produces the behavior by itself.

Classical conditioning is of great value to infants because it helps them recognize which events usually occur together in the everyday world. As a result, they can anticipate what is about to happen next, and the environment becomes more orderly and predictable. Let's take a closer look at the steps of classical conditioning.

As Carolyn settled down in the rocking chair to nurse Caitlin, she often stroked Caitlin's forehead. Soon Carolyn noticed that each time Caitlin's forehead was stroked, she made sucking movements. Caitlin had been classically conditioned. Here is how it happened (see Figure 4.7):

1. Before learning takes place, an **unconditioned stimulus (UCS)** must consistently produce a reflexive, or **unconditioned, response (UCR).** In Caitlin's case, sweet breast milk (UCS) resulted in sucking (UCR).

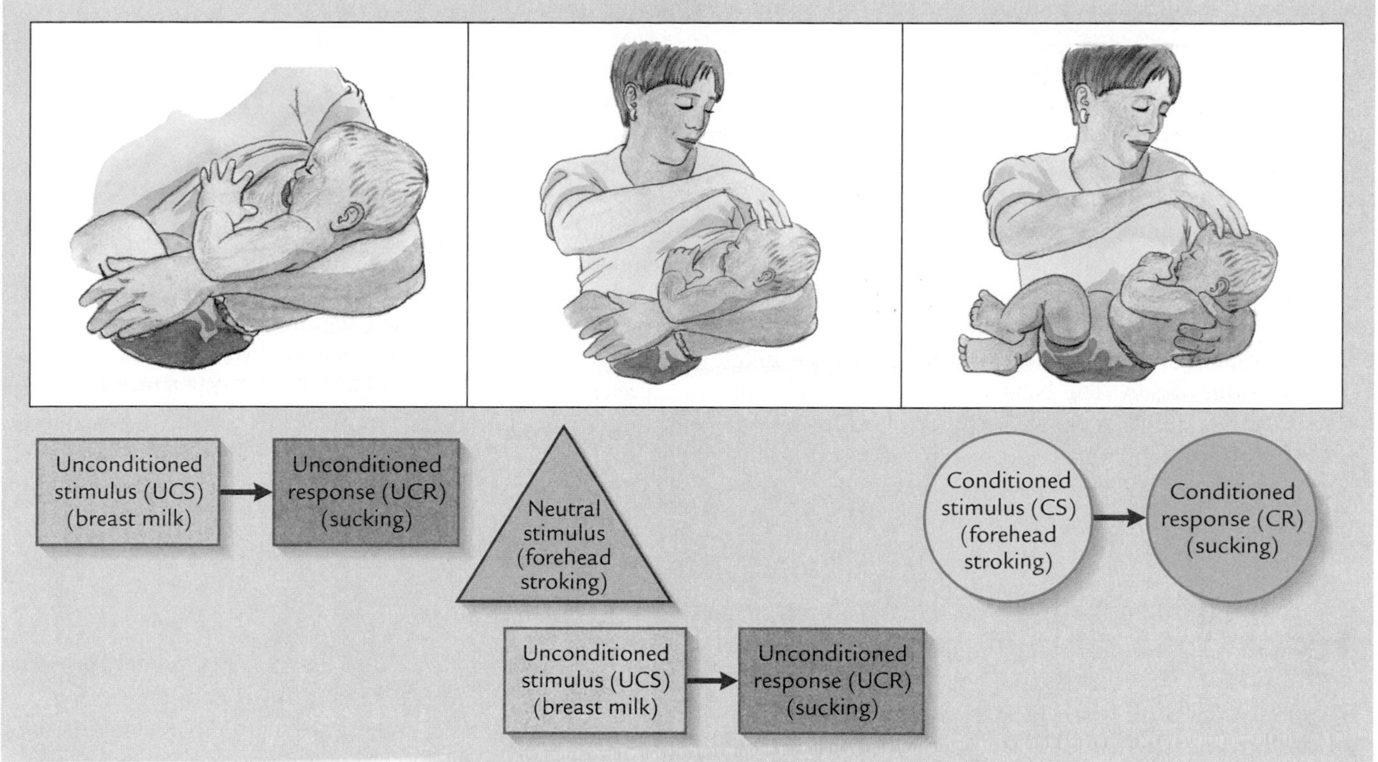

■ **FIGURE 4.7 The steps of classical conditioning.** The example here shows how Caitlin's mother classically conditioned her to make sucking movements by stroking her forehead at the beginning of feedings.

2. To produce learning, a *neutral stimulus* that does not lead to the reflex is presented just before, or at about the same time as, the UCS. Carolyn stroked Caitlin's forehead as each nursing period began. The stroking (neutral stimulus) was paired with the taste of milk (UCS).
3. If learning has occurred, the neutral stimulus by itself produces a response similar to the reflexive response. The neutral stimulus is then called a **conditioned stimulus (CS),** and the response it elicits is called a **conditioned response (CR).** We know that Caitlin has been classically conditioned because stroking her forehead outside the feeding situation (CS) results in sucking (CR).

If the CS is presented alone enough times, without being paired with the UCS, the CR will no longer occur. In other words, if Carolyn strokes Caitlin's forehead again and again without feeding her, Caitlin will gradually stop sucking in response to stroking. This is referred to as *extinction.*

Young infants can be classically conditioned most easily when the association between two stimuli has survival value. Caitlin learned quickly in the feeding situation, since learning the stimuli that accompany feeding improves the infant's ability to get food and survive (Blass, Ganchrow, & Steiner, 1984). In contrast, some responses are very difficult to classically condition in young babies. Fear is one of them. Until infants have the motor skills to escape from unpleasant events, they do not have a biological need to form these associations. But after 6 months of age, fear is easy to condition. In Chapter 6, we will discuss the development of fear, as well as other emotional reactions.

Operant Conditioning

In classical conditioning, babies build expectations about stimulus events in the environment, but they do not influence the stimuli that occur. In **operant conditioning,** infants act (or operate) on the environment, and stimuli that follow their behavior change the probability that the behavior will occur again. A stimulus that increases the occurrence of a response is called a **reinforcer.** For example, sweet liquid *reinforces* the sucking response in newborn babies. Removing a desirable stimulus or presenting an unpleasant one to decrease the occurrence of a response is called **punishment.** A sour-tasting fluid *punishes* newborn babies' sucking response. It causes them to purse their lips and stop sucking entirely.

Because the young infant can control only a few behaviors, successful operant conditioning in the early weeks of life is limited to sucking and head-turning responses. However, many stimuli besides food can serve as reinforcers. For example, researchers have created special laboratory conditions in which the baby's rate of sucking on a nipple produces a variety of interesting sights and sounds. Newborns will suck faster to see visual designs or hear music and human voices (Floccia, Christophe, & Bertoncini, 1997). As these findings suggest, operant conditioning has become a powerful tool for finding out what stimuli babies can perceive and which ones they prefer.

As infants get older, operant conditioning includes a wider range of responses and stimuli. For example, special mobiles have been hung over the cribs of 2- to 6-month-olds. When the baby's foot is attached to the mobile with a long cord, the infant can, by kicking, make the mobile turn. Under these conditions, it takes only a few minutes for infants to start kicking vigorously (Rovee-Collier, 1999; Shields & Rovee-Collier, 1992).

Operant conditioning soon modifies parents' and babies' reactions to each other. As the infant gazes into the adult's eyes, the adult looks and smiles back, and then the infant looks and smiles again. The behavior of each partner reinforces the other, and as a result, both parent and baby continue their pleasurable interaction. In Chapter 6, we will see that this contingent responsiveness plays an important role in the development of infant–caregiver attachment.

Look carefully at the findings just described, and you will see that young babies are active learners; they use any means they can to explore and control their surroundings in an effort to meet their needs for nutrition, stimulation, and social contact (Rovee-Collier, 1996). In fact, when infants' environments are so disorganized that their behavior does not lead to predictable outcomes, serious difficulties ranging from intellectual retardation to apathy and depression can result (Cicchetti & Aber, 1986; Seligman, 1975).

Habituation

At birth, the human brain is set up to be attracted to novelty. Infants tend to respond more strongly to a new element that has entered their environment. **Habituation** refers to a gradual reduction in the strength of a response due to repetitive stimulation. Looking, heart rate, and respiration rate may all decline, indicating a loss of interest. Once this has occurred, a new stimulus—some kind of change in the environment—causes responsiveness to return to a high level, an increase called **recovery.** For example, when you walk through a familiar space, you notice things that are new and different, such as a recently purchased picture on the wall or a piece of furniture that has been moved. Habituation and recovery enable us to focus our attention on those aspects of the environment we know the least about. As a result, learning is more efficient.

By studying infants' habituation and recovery, researchers can explore their understanding of the world. For example, a baby who first *habituates* to a visual pattern (a photo of a baby) and then *recovers* to a new one (a photo of a bald man) appears to remember the first stimulus and perceive the second one as new and different from it. This method of studying infant perception and cognition, illustrated in Figure 4.8, can be used with newborn babies, including those who are preterm. It has even been used to study the fetus's sensitivity to external stimuli—for example, by measuring changes in fetal heart rate when various repeated sounds are presented. The capacity to habituate and recover is evident in the third trimester of pregnancy (Sandman et al., 1997).

Imitation

Newborn babies come into the world with a primitive ability to learn through **imitation**—by copying the behavior of another person. For example, Figure 4.9 shows infants from 2 days to several weeks old imitating several adult facial expressions (Field et al., 1982; Meltzoff & Moore, 1977). The newborn's capacity to imitate extends to certain gestures, such as head movements, and has been demonstrated in many ethnic groups and cultures (Meltzoff & Kuhl, 1994).

But a few studies have failed to reproduce these findings (see, for example, Anisfeld et al., 2001). Therefore, some researchers regard the capacity as little more than an automatic response, much like a reflex. Others claim that newborns imitate many facial expressions even after short delays—when the adult is no longer demonstrating the behavior. These observations suggest that the capacity is flexible and voluntary (Hayne, 2002; Meltzoff & Moore, 1999).

As we will see in Chapter 5, infants' capacity to imitate improves greatly over the first 2 years. But however limited it is at birth, imitation is a powerful means of learning. Using imita-

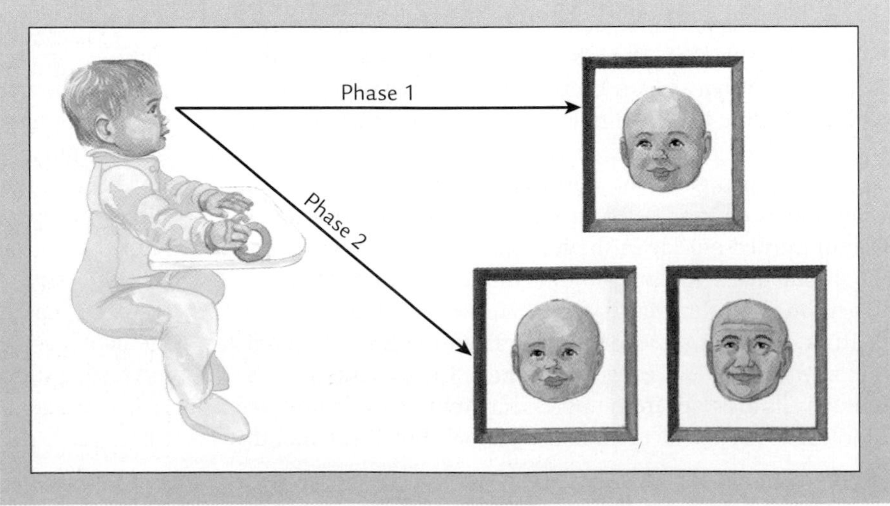

■ **FIGURE 4.8 Example of how the habituation/recovery sequence can be used to study infant perception and cognition.** In Phase 1, infants are shown (habituated to) a photo of a baby. In Phase 2, infants are again shown the baby photo, but this time it appears alongside a photo of a bald-headed man. Infants recovered to (spent more time looking at) the photo of the man, indicating that they remembered the baby and perceived the man's face as different from it. (Adapted from Fagan & Singer, 1979.)

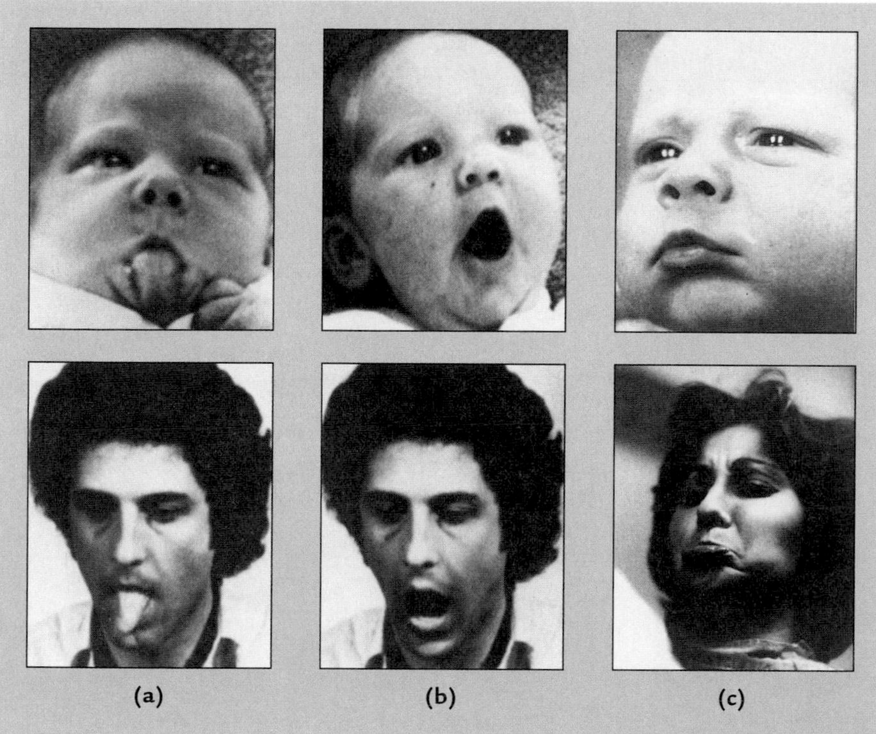

■ **FIGURE 4.9 Photographs from two of the first studies of newborn imitation.** Those on the left show 2- to 3-week-old infants imitating tongue protrusion (a), and mouth opening (b). The one on the right shows a 2-day-old infant imitating a sad (c) adult facial expression. (From A. N. Meltzoff & M. K. Moore, 1977, "Imitation of Facial and Manual Gestures by Human Neonates," *Science, 198,* p. 75; and T. M. Field et al., 1982, "Discrimination and Imitation of Facial Expressions by Neonates," *Science, 218,* p. 180. Copyright 1977 and 1982, respectively, by the AAAS. Reprinted by permission.)

tion, young infants explore their social world, getting to know people by matching their behavioral states. In the process, babies notice similarities between their own actions and those of others and start to find out about themselves. Furthermore, by tapping into infants' ability to imitate, adults can get infants to express desirable behaviors, and once they do, adults can encourage these further. Finally, caregivers take great pleasure in a baby who imitates their facial gestures and actions. Imitation seems to be one of those capacities that help get the infant's relationship with parents off to a good start.

Ask Yourself

REVIEW

Provide an example of classical conditioning, operant conditioning, and habituation/recovery in young infants. Why is each type of learning useful?

APPLY

Nine-month-old Byron has a toy with large, colored push buttons on it. Each time he pushes a button, he hears a nursery tune. Which learning capacity is the manufacturer of this toy taking advantage of? What can Byron's play with the toy reveal about his perception—specifically, his ability to distinguish sound patterns?

CONNECT

Infants with nonorganic failure to thrive rarely smile at friendly adults. Also, they anxiously keep track of nearby adults. Explain these reactions using the learning capacities discussed in the preceding sections.

www.

Motor Development

Carolyn, Monica, and Vanessa each kept baby books, filling them with proud notations about when their children held up their heads, reached for objects, sat by themselves, and walked alone. Parents' enthusiasm for these achievements makes perfect sense. With each new motor skill, babies master their bodies and the environment in a new way. For example, sitting upright gives infants an entirely different perspective on the world. Reaching permits babies to find out about objects by acting on them. And when infants can move on their own, their opportunities for exploration multiply.

Babies' motor achievements have a powerful effect on their social relationships. When at 7½ months Caitlin crawled, Carolyn and David began to restrict her movements by saying no, expressing mild impatience, and picking her up and moving her—strategies that were unnecessary before. When Caitlin walked 3 days after her first birthday, first "testing of wills" occurred (Biringen et al., 1995). Despite her mother's warnings, she sometimes pulled items from shelves that were "off limits."

At the same time, expressions of affection and playful activities expanded as Caitlin sought out her parents for greetings, hugs, and a gleeful game of hide-and-seek (Campos, Kermoian, & Zumbahlen, 1992). Soon after, Caitlin turned the pages of a book while she and her parents named the pictures. Caitlin's delight as she worked on new motor competencies triggered pleasurable reactions in others, which encouraged her efforts further (Mayes & Zigler, 1992). Motor skills, social competencies, cognition, and language developed together and supported one another.

Table 4.1 Gross and Fine Motor Development in the First Two Years

Motor Skill	Average Age Achieved	Age Range in Which 90 Percent of Infants Achieve the Skill
When held upright, holds head erect and steady	6 weeks	3 weeks–4 months
When prone, lifts self by arms	2 months	3 weeks–4 months
Rolls from side to back	2 months	3 weeks–5 months
Grasps cube	3 months, 3 weeks	2–7 months
Rolls from back to side	4½ months	2–7 months
Sits alone	7 months	5–9 months
Crawls	7 months	5–11 months
Pulls to stand	8 months	5–12 months
Plays pat-a-cake	9 months, 3 weeks	7–15 months
Stands alone	11 months	9–16 months
Walks alone	11 months, 3 weeks	9–17 months
Builds tower of two cubes	11 months, 3 weeks	10–19 months
Scribbles vigorously	14 months	10–21 months
Walks up stairs with help	16 months	12–23 months
Jumps in place	23 months, 2 weeks	17–30 months
Walks on tiptoe	25 months	16–30 months

Sources: Bayley, 1969, 1993.
Photos: (top) © Laura Dwight; (lower left) © Barbara Peacock/Getty Images/Taxi; (lower right) © Elizabeth Crews/The Image Works.

The Sequence of Motor Development

Gross motor development refers to control over actions that help infants get around in the environment, such as crawling, standing, and walking. In contrast, *fine motor development* has to do with smaller movements, such as reaching and grasping. Table 4.1 shows the average age at which infants and toddlers achieve a variety of gross and fine motor skills. Most (but not all) children follow this sequence.

Notice that the table also presents the age ranges during which the majority of babies accomplish each skill. These indicate that although the *sequence* of motor development is fairly uniform across children, large individual differences exist in *rate* of motor progress. We would be concerned about a child's development only if many motor skills were seriously delayed.

Look at Table 4.1 once more, and you will see that there is organization and direction to the infant's motor achievements. The *cephalocaudal trend* is evident. Motor control of the head comes before control of the arms and trunk, which comes before control of the legs. You can also see the *proximodistal trend:* Head, trunk, and arm control precedes coordination of the hands and fingers. These similarities between physical and motor development suggest a genetic contribution to motor progress.

But we must be careful not to think of motor skills as following a fixed, maturational timetable. Instead, each skill is a product of earlier motor attainments and a contributor to new ones. Furthermore, children acquire motor skills in highly individual ways. For example, Grace, who spent most of her days lying in a hammock until her adoption, did not try to crawl because she rarely spent time on her tummy or on firm surfaces that enabled her to move on her own. As a result, she pulled to a stand and walked before she crawled! Many influences—both internal and external to the child—join together to support the vast transformations in motor competencies of the first 2 years.

Motor Skills as Dynamic Systems

According to **dynamic systems theory of motor development,** mastery of motor skills involves acquiring increasingly complex *systems of action.* When motor skills work as a *system,* separate abilities blend together, each cooperating with others to produce more effective ways of exploring and controlling the environment. For example, control of the head and upper chest are combined into sitting with support. Kicking, rocking on all fours, and reaching are gradually put together into crawling. Then crawling, standing, and stepping are united into walking alone (Thelen, 1989).

Each new skill is a joint product of the following factors: (1) central nervous system development, (2) movement capacities of the body, (3) goals the child has in mind, and (4) environmental supports for the skill. Change in any element makes the system less stable, and the child starts to explore and select new, more effective motor patterns. The broader physical environment also has a profound impact on motor skills. For example, if children were reared in the moon's reduced gravity, they would prefer jumping to walking or running!

When a skill is first acquired, it is tentative and unstable. Infants must refine it so that it becomes smooth and accurate. In one study, a baby just starting to crawl often collapsed on her tummy and moved backward instead of forward. Gradually, she figured out how to propel herself along by alternately pulling with her arms and pushing with her feet. As she experimented with muscle patterns and observed the consequences of her movements, she perfected the crawling motion (Adolph, Vereijken, & Denny, 1998).

Look carefully at dynamic systems theory, and you will see why motor development cannot be genetically determined. Because exploration and the desire to master new tasks motivate it, heredity can map it out only at a general level (Hopkins & Butterworth, 1997; Thelen & Smith, 1998). Each skill is acquired by combining early accomplishments into a more complex system that permits the child to reach a goal. Consequently, different paths to the same motor skill exist.

Cultural Variations in Motor Development

Cross-cultural research shows how early movement opportunities and a stimulating environment contribute to motor development. Several decades ago, Wayne Dennis (1960) observed infants in Iranian orphanages who were deprived of the tantalizing surroundings that induce infants to acquire motor skills. The Iranian babies spent their days lying on their backs in cribs, without toys to play with. As a result, most did not move on their own until after 2 years of age. When they finally did move, the constant experience of lying on their backs led them to scoot in a sitting position rather than crawl on their hands and knees. Because babies who scoot come up against furniture with their feet, not their hands, they are far less likely to pull themselves to a standing position in preparation for walking. Therefore, walking was delayed.

Cultural variations in infant-rearing practices also affect motor development. Take a quick survey of parents you know, asking this question: Should sitting, crawling, and walking be deliberately encouraged? Answers vary widely from culture to culture. Japanese mothers believe such efforts are unnecessary. Among the Zinacanteco Indians of Southern Mexico, rapid motor progress is actively discouraged. Babies who walk before they know enough to keep away from cooking fires and weaving looms are viewed as dangerous to themselves and disruptive to others (Greenfield, 1992).

In contrast, among the Kipsigis of Kenya and the West Indians of Jamaica, babies hold their heads up, sit alone, and walk considerably earlier than North American infants. Kipsigi parents deliberately teach these motor skills. In the first few months, babies are seated in holes dug in the ground, and rolled blankets are used to keep them upright. Walking is promoted by frequently bouncing babies on their feet (Super, 1981). As Figure 4.10 shows, the West Indian mothers use a

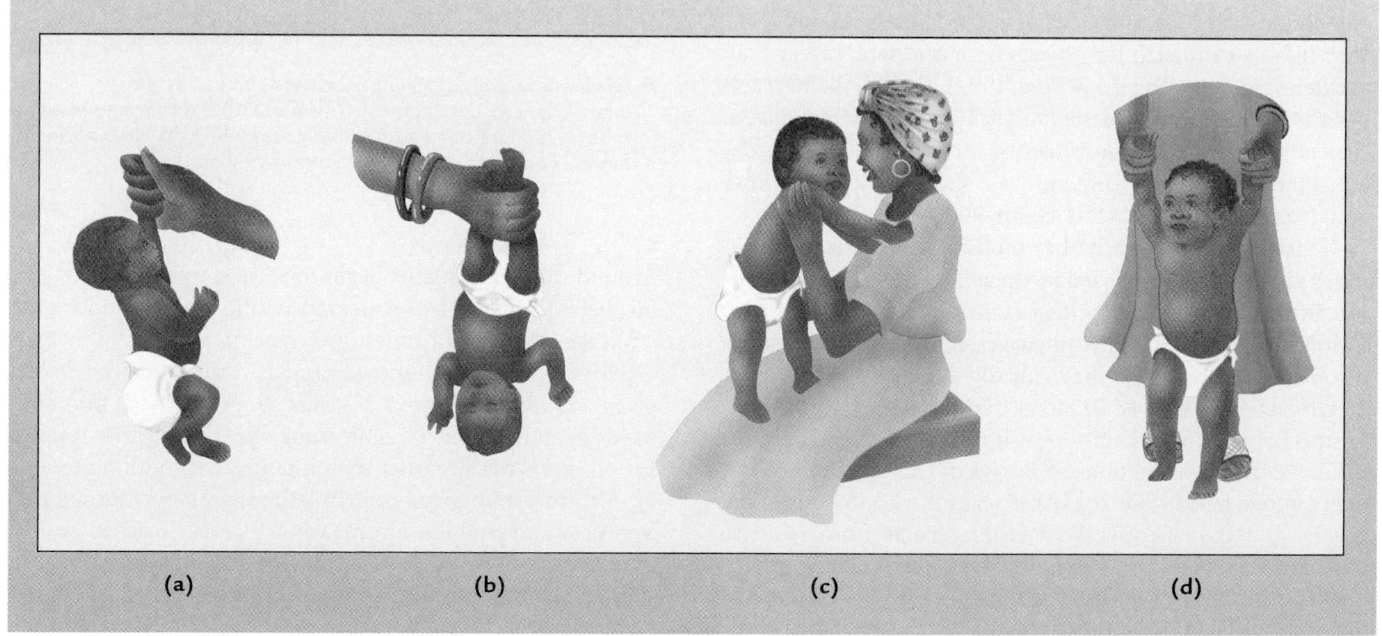

■ **FIGURE 4.10 West Indians of Jamaica use a formal handling routine with their babies.** Exercises practiced in the first few months include (a) stretching each arm while suspending the baby and (b) holding the infant upside-down by the ankles. Later in the first year, the baby is (c) "walked" up the mother's body and (d) encouraged to take steps on the floor while supported. (Adapted from B. Hopkins & T. Westra, 1988, "Maternal Handling and Motor Development: An Intracultural Study," *Genetic, Social and General Psychology Monographs, 14,* pp. 385, 388, 389. Reprinted by permission of the Helen Dwight Reid Educational Foundation. Published by Heldref Publications, 1319 Eighteenth St., N.W., Washington, DC 20036-1802.)

highly stimulating formal handling routine with their infants, explaining that exercise helps babies grow up strong, healthy, and physically attractive (Hopkins & Westra, 1988).

Fine Motor Development: Reaching and Grasping

Of all motor skills, reaching may play the greatest role in infant cognitive development because it opens up a whole new way of exploring the environment. By grasping things, turning them over, and seeing what happens when they are released, infants learn a great deal about the sights, sounds, and feel of objects.

Reaching and grasping, like many other motor skills, start out as gross, diffuse activity and move toward mastery of fine movements. Newborns make poorly coordinated swipes or swings, called *prereaching,* toward an object in front of them. Because they cannot control their arms and hands, they rarely contact the object. Like newborn reflexes, prereaching eventually drops out, around 7 weeks of age. Yet these early behaviors suggest that babies are biologically prepared to coordinate hand with eye in the act of reaching (Thelen, 2001).

At about 3 months, as infants develop the necessary head and shoulder control, voluntary reaching appears and gradually improves in accuracy (Spencer et al., 2000). By 5 to 6 months, infants can reach for and grasp an object in a room that has been darkened during the reach by switching off the lights—a skill that improves over the first year (Clifton et al., 1994; McCarty & Ashmead, 1999). Early on, vision is freed from the basic act of reaching so it can focus on more complex adjustments. By 7 months, the arms become more independent; infants can reach for objects with one arm, rather than by extending both (Fagard & Pezé, 1997). During the next few months, infants become better at reaching for moving objects. They adjust their hand movements, accurately obtaining objects that spin, change direction, or move closer or farther away (Wentworth, Benson, & Haith, 2000).

Once infants can reach, they modify their grasp. The newborn's grasp reflex is replaced by the *ulnar grasp,* a clumsy motion in which the fingers close against the palm. Still, even 3-month-olds readily adjust their grasp to the size and shape of objects—a capacity that improves over the first year (Newman, Atkinson, & Braddick, 2001). At around 4 to 5 months, when infants begin to sit up, both hands become coordinated in exploring objects. Babies of this age can hold an object in one hand while the other scans it with the tips of the fingers, and they frequently transfer objects from hand to hand (Rochat & Goubet, 1995). By the end of the first year, infants use the thumb and index finger in a well-coordinated *pincer grasp.* Then the ability to manipulate objects greatly expands. The 1-year-old can pick up raisins and blades of grass, turn knobs, and open and close small boxes.

Between 8 and 11 months, reaching and grasping are well practiced. As a result, attention is released from the motor skill itself to events that occur before and after attaining the object.

■ Of all motor skills, reaching is believed to play the greatest role in infant cognitive development. This 6-month-old enjoys squeezing, rubbing, and batting soft toys hung above his crib. He is beginning to coordinate both hands in exploring objects.

Around this time, infants begin to solve simple problems that involve reaching, such as searching for and finding a hidden toy.

Like other motor milestones, reaching is affected by early experience. In a well-known study, institutionalized babies given a moderate amount of visual stimulation—at first, simple designs and later, a mobile hung over their cribs—reached for objects 6 weeks earlier than infants given nothing to look at. A third group given massive stimulation—patterned crib bumpers and mobiles at an early age—also reached sooner than unstimulated babies. But this heavy enrichment took its toll. These infants looked away and cried a great deal, and they were not as advanced in reaching as the moderately stimulated group (White & Held, 1966). Recall from our discussion of brain development that more stimulation is not necessarily better. Trying to push infants beyond their readiness to handle stimulation can undermine the development of important motor skills.

Ask Yourself

REVIEW

Cite evidence that motor development is not genetically determined but rather is a joint product of biological, psychological, and environmental factors.

APPLY

Roseanne hung mobiles and pictures above her newborn baby's crib, hoping this would stimulate her infant's motor development. Is Roseanne doing the right thing? Why or why not?

CONNECT

Provide several examples of how motor development influences infants' social experiences. How do social experiences, in turn, influence motor development?

www.

Perceptual Development

In Chapter 3, you learned that touch, taste, smell, and hearing—but not vision—are remarkably well developed at birth. Now let's turn to a related question: How does perception change over the first year of life?

Our discussion will focus on hearing and vision because almost all research addresses these two aspects of perceptual development. Unfortunately, little research exists on how touch, taste, and smell develop after birth. Also, in Chapter 3, we used the word *sensation* to talk about these capacities. Now we use the word *perception*. The reason is that sensation suggests a fairly passive process—what the baby's receptors detect when they are exposed to stimulation. In contrast, perception is much more active. When we perceive, we organize and interpret what we see.

As we review the perceptual achievements of infancy, you will probably find it hard to tell where perception leaves off and thinking begins. Thus, the research we are about to discuss provides an excellent bridge to the topic of Chapter 5, cognitive development during the first 2 years.

Hearing

On Timmy's first birthday, Vanessa bought several tapes of nursery songs, and she turned one on each afternoon at naptime. Soon Timmy let her know his favorite tune. If she put on "Twinkle, Twinkle," he stood up in his crib and whimpered until she replaced it with "Jack and Jill." Timmy's behavior illustrates the greatest change in hearing over the first year of life: Babies start to organize sounds into complex patterns. If two slightly different melodies are played, 1-year-olds can tell that they are not the same (Morrongiello, 1986).

Responsiveness to sound provides support for the young baby's exploration of the environment through sight and touch. Infants as young as 3 days old turn their eyes and head in the general direction of a sound, an ability that improves greatly over the first 6 months (Litovsky & Ashmead, 1997). By this time, infants make judgments about how far away a sound is. They are less likely to try to retrieve a sounding object in the dark if it is beyond their reach (Clifton, Perris, & Bullinger, 1991).

As we will see in the next chapter, throughout the first year babies are preparing to acquire language. Recall from Chapter 3 that newborns can distinguish nearly all sounds in human languages and that they prefer listening to their native tongue. As infants continue to listen to the talk of people around them, they learn to focus on meaningful sound variations in their own language. By 6 months of age, they "screen out" sounds not used in their own language (Kuhl et al., 1992; Polka & Werker, 1994).

In the second half of the first year, infants focus on larger speech units that are critical to figuring out the meaning of what they hear. They recognize familiar words in spoken passages and can detect clauses and phrases in sentences (Jusczyk & Aslin, 1995; Jusczyk & Hohne, 1997). Around 7 to 9 months, infants extend this rhythmic sensitivity to individual words. They listen much longer to speech with stress patterns common in their own language, and they perceive it in wordlike segments (Jusczyk, Houston, & Newsome, 1999; Jusczyk, 2001).

How do infants begin to detect the structure of sentences and words—information vital for linking speech units with their meanings? Research reveals that they are remarkable analyzers of sound patterns. In detecting words, for example, they distinguish syllables that frequently occur together (indicating they belong to the same word) from those that seldom occur together (indicating a word boundary) (Saffran, Aslin, & Newport, 1996). Also, they notice patterns in word sequences. In a study using nonsense words, 7-month-olds discriminated the ABA structure of "ga ti ga" and "li na li" from the ABB structure of "wo fe fe" and "ta la la" (Marcus et al., 1999). The infants seemed to detect a simple word-order pattern—a capacity that may help them figure out the basic grammar of their language.

Vision

More than any other sense, humans depend on vision for exploring their environment. Although at first a baby's visual world is fragmented, it undergoes extraordinary changes during the first 7 to 8 months of life.

Visual development is supported by rapid maturation of the eye and visual centers in the cerebral cortex. Recall from Chapter 3 that the newborn baby focuses and perceives color poorly. Around 2 months, infants can focus on objects and discriminate colors about as well as adults can (Teller, 1998). Visual acuity (fineness of discrimination) improves steadily throughout the first year, reaching a near-adult level of about 20/20 by 6 months (Gwiazda & Birch, 2001). Scanning the environment

FIGURE 4.11 The visual cliff. Plexiglas covers the deep and shallow sides. By refusing to cross the deep side and showing a preference for the shallow side, this infant demonstrates the ability to perceive depth.

and tracking moving objects improve over the first half-year as eye movements come under voluntary control (von Hofsten & Rosander, 1998; Johnson, 1995).

As babies see more clearly and explore their visual field more adeptly, they figure out the characteristics of objects and how they are arranged in space. We can best understand how they do so by examining the development of two aspects of vision: depth and pattern perception.

● **Depth Perception.** *Depth perception* is the ability to judge the distance of objects from one another and from ourselves. It is important for understanding the layout of the environment and for guiding motor activity. To reach for objects, babies must have some sense of depth. Later, when infants crawl, depth perception helps them avoid bumping into furniture and falling down stairs.

Figure 4.11 shows the well-known *visual cliff,* designed by Eleanor Gibson and Richard Walk (1960) and used in the earliest studies of depth perception. It consists of a Plexiglas-covered table with a platform at the center, a "shallow" side with a checkerboard pattern just under the glass, and a "deep" side with a checkerboard several feet below the glass. The researchers found that crawling babies readily crossed the shallow side, but most reacted with fear to the deep side. They concluded that around the time infants crawl, most distinguish deep and shallow surfaces and avoid drop-offs that look dangerous.

Gibson and Walk's research shows that crawling and avoidance of drop-offs are linked, but it does not tell us how they are related or when depth perception first appears. To better understand the development of depth perception, researchers have turned to babies' ability to detect specific depth cues, using methods that do not require that they crawl.

Motion provides a great deal of information about depth, and it is the first depth cue to which infants are sensitive. Babies 3 to 4 weeks of age blink their eyes defensively when an object is moved toward their face as though it is going to hit (Nánez & Yonas, 1994). *Binocular* depth cues arise because our two eyes have slightly different views of the visual field. The brain blends these two images but also registers the difference between them. Research in which infants wear special goggles, like those for 3-D movies, reveals that sensitivity to binocular cues emerges between 2 and 3 months and improves rapidly over the first half-year (Birch, 1993). Finally, around 6 to 7 months, babies develop sensitivity to *pictorial* depth cues—the same ones that artists use to make a painting look three-dimensional. Examples include lines that create the illusion of perspective, changes in texture (nearby textures are more detailed than faraway ones), and overlapping objects (an object partially hidden by another object is perceived to be more distant) (Sen, Yonas, & Knill, 2001; Yonas et al., 1986).

Why does perception of depth cues emerge in the order just described? Researchers speculate that motor development is involved. For example, control of the head during the early weeks of life may help babies notice motion and binocular cues. And around 5 to 6 months, the ability to turn, poke, and feel the surface of objects may promote perception of pictorial cues (Bushnell & Boudreau, 1993). Indeed, as we will see next, one aspect of motor progress—independent movement—plays a vital role in refinement of depth perception.

● **Crawling and Depth Perception.** At 6 months, Timmy started crawling. "He's like a fearless daredevil," exclaimed Vanessa. "If I put him down in the middle of our bed, he crawls right over the edge. The same thing's happened by the stairs." Will Timmy become more wary of the side of the bed and the staircase as he becomes a more experienced crawler? Research suggests that he will. Infants with more crawling experience (regardless of when they start to crawl) are far more likely to refuse to cross the deep side of the visual cliff (Bertenthal, Campos, & Barrett, 1984).

From extensive everyday experience, babies gradually figure out how to use depth cues to detect the danger of falling. But because the loss of body control that leads to falling differs greatly for each body position, babies must undergo this

Crawling promotes three-dimensional understanding, such as wariness of drop-offs and memory for object locations. As this baby moves about, he takes note of how to get from place to place, where objects are in relation to himself and to other objects, and what they look like from different points of view.

learning separately for each posture. In one study, 9 month-olds who were experienced sitters but novice crawlers were placed on the edge of a shallow drop-off that could be widened (Adolph, 2000). While in the familiar sitting position, infants avoided leaning out for an attractive toy at distances likely to result in falling. But in the unfamiliar crawling position, they headed over the edge, even when the distance was extremely wide! As infants discover how to avoid falling in different postures and situations, their understanding of depth expands.

Crawling experience promotes other aspects of three-dimensional understanding. For example, seasoned crawlers are better than their inexperienced agemates at remembering object locations and finding hidden objects (Bai & Bertenthal, 1992; Campos et al., 2000). Why does crawling make such a difference? Compare your own experience of the environment when you are driven from one place to another as opposed to when you walk or drive yourself. When you move on your own, you are much more aware of landmarks and routes of travel, and you take more careful note of what things look like from different points of view. The same is true for infants. In fact, crawling promotes a new level of brain organization, as indicated by more organized brain-wave activity in the cerebral cortex. Perhaps crawling strengthens certain neural connections, especially those involved in vision and understanding of space (Bell & Fox, 1996). As the Biology and Environment box on page 136 reveals, the link between independent movement and spatial knowledge is also evident in a population with a very different perceptual experience: infants with severe visual impairments.

● **Pattern and Face Perception.** Even newborns prefer to look at patterned rather than plain stimuli—for example, a drawing of the human face or one with scrambled facial features rather than a black-and-white oval (Fantz, 1961). As infants get older, they prefer more complex patterns. For example, 3-week-old infants look longest at black-and-white checkerboards with a few large squares, whereas 8- and 14-week-olds prefer those with many squares (Brennan, Ames, & Moore, 1966).

A general principle, called **contrast sensitivity,** explains these early pattern preferences (Banks & Ginsburg, 1985). *Contrast* refers to the difference in the amount of light between adjacent regions in a pattern. If babies *are sensitive to* (can detect) the contrast in two or more patterns, they prefer the one with more contrast. To understand this idea, look at the checkerboards in the top row of Figure 4.12. To us, the one with many small squares has more contrasting elements. Now look at the bottom row, which shows how these checkerboards appear to infants in the first few weeks of life. Because of their poor vision, very young babies cannot resolve the features in more complex patterns, so they prefer to look at the large, bold checkerboard. Around 2 months of age, when detection of fine-grained detail has improved, infants become sensitive to the contrast in complex patterns and spend more time looking at them (Gwiazda & Birch, 2001; Teller, 1997).

FIGURE 4.12 The way two checkerboards differing in complexity look to infants in the first few weeks of life. Because of their poor vision, very young infants cannot resolve the fine detail in the complex checkerboard. It appears blurred, like a gray field. The large, bold checkerboard appears to have more contrast, so babies prefer to look at it. (Adapted from M. S. Banks & P. Salapatek, 1983, "Infant Visual Perception," in M. M. Haith & J. J. Campos [Eds.], *Handbook of Child Psychology: Vol. 2. Infancy and Developmental Psychobiology* [4th ed.], New York: Wiley, p. 504. Copyright © 1983 by John Wiley & Sons. Reprinted by permission.)

Biology & Environment

Development of Infants with Severe Visual Impairments

Research on infants who can see little or nothing at all dramatically illustrates the interdependence of vision, motor exploration, social interaction, and understanding of the world. In a longitudinal study, infants with a visual acuity of 20/800 or worse (they had only dim light perception or were blind) were followed through the preschool years. Compared to agemates with less severe visual impairments, they showed serious delays in all aspects of development. Motor and cognitive functioning suffered the most; with age, performance in both domains became increasingly distant from that of other children (Hatton et al., 1997).

What explains these profound developmental delays? Minimal or absent vision seems to alter the child's experiences in at least two crucial, interrelated ways.

Impact on Motor Exploration and Spatial Understanding. Infants with severe visual impairments attain gross and fine motor milestones many months later than their sighted counterparts (Levtzion-Korach et al., 2000). For example, on average, blind infants do not reach for and manipulate objects until 12 months, crawl until 13 months, and walk until 19 months (compare these averages to the norms in Table 4.1 on page 130). Why is this so?

Infants with severe visual impairments must rely on sound to identify the whereabouts of objects. But sound does not function as a precise clue to object location until much later than vision—around the middle of the first year. And because infants who cannot see have difficulty engaging their caregivers, adults may not provide them with rich, early exposure to sounding objects. As a result, the baby comes to understand relatively late that there is a world of interesting objects to explore.

Until "reaching on sound" is achieved, infants with severe visual impairments are not motivated to move independently. Because of their own uncertainty and parents' protection and restraint to prevent injury, blind infants are typically tentative in their movements. These factors delay motor development further.

Motor and cognitive development are closely linked, especially for infants with little or no vision. These babies build an understanding of the location and arrangement of objects in space only after reaching and crawling (Bigelow, 1992). Inability to imitate the motor actions of others presents additional challenges as these children get older, contributing to declines in motor and cognitive progress relative to peers with better vision (Hatton et al., 1997).

Impact on the Caregiver–Infant Relationship. Infants who see poorly have great difficulty evoking stimulating caregiver interaction. They cannot make eye contact, imitate, or pick up nonverbal social cues. Their emotional expressions are muted; for example, their smile is fleeting and unpredictable. Consequently, these infants may receive little adult attention, play, and other stimulation vital for all aspects of development (Tröster & Brambring, 1992).

When a visually impaired child does not learn how to participate in social interaction during infancy, communication is compromised in early childhood. In an observational study of blind children enrolled in preschools with sighted agemates, the blind children seldom initiated contact with peers and teachers. When they did interact, they had trouble interpreting the meaning of others' reactions and responding appropriately (Preisler, 1991, 1993).

Interventions. Parents, teachers, and professional caregivers can help infants with minimal vision overcome early developmental delays through stimulating, responsive interaction. Until a close emotional bond with an adult is forged, visually impaired babies cannot establish vital links with their environments.

Techniques that help infants become aware of their physical and social surroundings include heightened sensory input through combining sound and touch (holding, touching, or bringing the baby's hands to the adult's face while talking or singing), engaging in many repetitions, and consistently reinforcing the infant's efforts to make contact. Manipulative play with objects that make sounds is also vital. Finally, rich language stimulation can compensate for visual loss (Conti-Ramsden & Pérez-Pereira, 1999). It grants young children a ready means of finding out about objects, events, and behaviors they cannot see. Once language emerges, many children with limited or no vision show impressive rebounds. Some acquire a unique capacity for abstract thinking, and most master social and practical skills that permit them to lead productive, independent lives (Warren, 1994).

■ This 20-month-old, who has no vision, reacts with glee as her father guides her exploration of a novel object through touch and sound. Adults who encourage and reinforce children's efforts to make contact with their physical and social surroundings prevent the developmental delays typically associated with severely impaired vision.

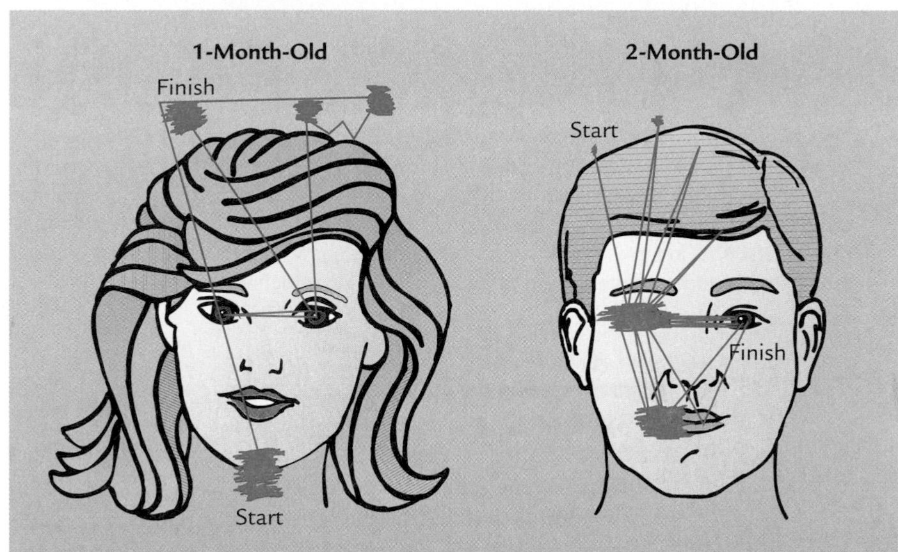

FIGURE 4.13 Visual scanning of the pattern of the human face by 1- and 2-month-old infants. One-month-olds limit their scanning to single features on the border of the stimulus, whereas 2-month-olds explore internal features. (From P. Salapatek, 1975, "Pattern Perception in Early Infancy," in L. B. Cohen & P. Salapatek [Eds.], *Infant Perception: From Sensation to Cognition,* New York: Academic Press, p. 201. Reprinted by permission.)

In the early weeks of life, infants respond to the separate parts of a pattern. For example, when shown drawings of human faces, 1-month-olds limit their visual exploration to the border of the stimulus and stare at single, high-contrast features, such as the hairline or chin (see Figure 4.13). At about 2 months, when scanning ability and contrast sensitivity improve, infants thoroughly explore a pattern's internal features, pausing briefly to look at each part (Bronson, 1991).

Once babies can take in all aspects of a pattern, they integrate them into a unified whole. By 4 months, they are so good at detecting pattern organization that they even perceive subjective boundaries that are not really present. For example, they perceive a square in the center of Figure 4.14a, just as you do (Ghim, 1990). Older infants carry this responsiveness to subjective form even further. For example, 9-month-olds show a special preference for an organized series of moving lights that resembles a human being walking, in that they look much longer at this display than at upside-down or scrambled versions (Bertenthal, 1993). By 12 months, infants can detect objects represented by incomplete drawings, even when as much as two-thirds of the drawing is missing (see Figure 4.14b) (Rose, Jankowski, & Senior, 1997).

The baby's tendency to search for structure in a patterned stimulus applies to face perception. Newborns prefer to look at simple, facelike stimuli with features arranged naturally (upright) rather than unnaturally (upside-down or sideways) (see Figure 4.15a on page 138) (Mondloch et al., 1999). They also track a facial pattern moving across their visual field farther than they track other stimuli (Johnson, 1999). Some researchers claim that these behaviors reflect a built-in capacity to orient toward members of one's own species, just as many newborn animals do (Johnson, 2001). Others argue that newborns are exposed to faces more often than other stimuli—early experiences that could quickly "wire" the brain to detect faces.

Although newborns respond to a simple facelike structure, they cannot discriminate a complex facial pattern from other,

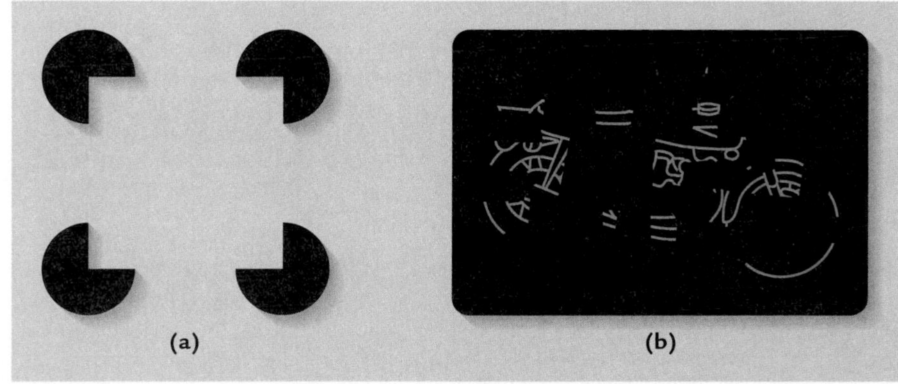

FIGURE 4.14 Subjective boundaries in visual patterns. (a) Do you perceive a square in the middle of the figure on the left? By 4 months of age, infants do, too. (b) What does the image on the right, missing two-thirds of its outline, look like to you? By 12 months, infants detect the image of a motorcycle. After habituating to the incomplete motorcycle image, they were shown an intact motorcycle figure paired with a novel form. Twelve-month-olds recovered to (looked longer at) the novel figure, indicating that they recognized the motorcycle pattern on the basis of very little visual information. (Adapted from Ghim, 1990; Rose, Jankowski, & Senior, 1997.)

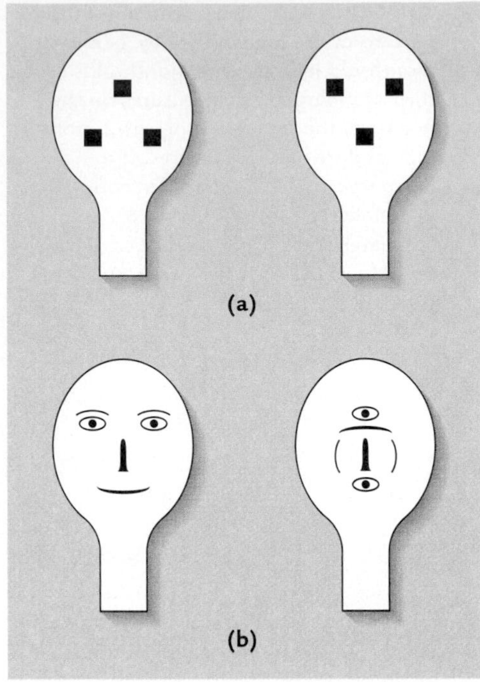

■ **FIGURE 4.15 Early face perception.** (a) Newborns prefer to look at the simple pattern resembling a face on the left over the upside-down version on the right. This preference for a facelike stimulus disappears by age 6 weeks. Some researchers believe it is innate, orients newborns toward people, and is replaced by more complex perceptual learning as the cerebral cortex develops and visual capacities improve. (b) When the complex drawing of a face on the left and the equally complex, scrambled version on the right are moved across newborns' visual field, they follow the face longer—another finding that suggests a built-in capacity to orient toward people. But if the two stimuli are presented side by side, infants show no preference for the face until 2 to 3 months of age. (From Johnson, 1999; Mondloch et al., 1999.)

equally complex patterns (see Figure 4.15b). At 2 to 3 months, when infants can combine stimulus elements into an organized whole, they do prefer a drawing of the human face to other, equally complex configurations (Dannemiller & Stephens, 1988). The baby's tendency to search for pattern structure is quickly applied to face perception. By 2 months, infants recognize aspects of their mother's facial features; they look longer at her face than at an unfamiliar woman's face (Bartrip, Morton, & de Schonen, 2001).

Around 3 months, infants make fine distinctions among the features of different faces. For example, they can tell the difference between the photos of two strangers, even when the faces are moderately similar. And between 7 and 10 months, infants start to perceive emotional expressions as meaningful wholes. They treat positive faces (happy and surprised) as different from negative ones (sad and fearful) (Ludemann, 1991). As infants recognize and respond to the expressive behavior of others, face perception supports their earliest social relationships.

The development of depth and pattern perception is summarized in Table 4.2. Up to this point, we have considered the infant's sensory systems one by one. Now let's examine their coordination.

Intermodal Perception

When we take in information from the environment, we often use **intermodal perception.** That is, we combine stimulation from more than one *modality,* or sensory system. Recent evidence indicates that from the start, babies perceive the world in an intermodal fashion (Meltzoff, 1990; Spelke, 1987). Recall that newborns turn in the general direction of a sound and reach for objects in a primitive way. These behaviors suggest that infants expect sight, sound, and touch to go together.

Table 4.2 Development of Visual Perception

	Birth–1 Month	2–4 Months	5–12 Months
Depth perception	Sensitivity to motion cues	Sensitivity to binocular cues	Sensitivity to pictorial cues; wariness of heights
Pattern perception	Preference for patterns with large elements Visual exploration limited to the border of a stimulus and single features	Visual exploration of entire stimulus, including internal features Combining pattern elements into an organized whole	Detection of increasingly complex, meaningful patterns
Face perception	Preference for a simple, facelike pattern	Preference for a complex facial pattern over other, equally complex patterns Preference for the mother's face over an unfamiliar woman's face Ability to distinguish photos of strangers' faces	Ability to perceive emotional expressions as meaningful wholes

Experiencing the integration of sensory modalities in these ways prepares young babies for detecting the wealth of intermodal associations in their everyday worlds (Slater et al., 1999).

Within a few months, infants make impressive intermodal matches. Three- and 4-month-olds can relate a child's or adult's moving lips to the corresponding sounds in speech. And 7-month-olds can link a happy or angry voice with the appropriate face of a speaking person (Bahrick, Netto, & Hernandez-Reif, 1998; Soken & Pick, 1992). Of course, many intermodal associations, such as the way a train sounds or a teddy bear feels, must be based on direct exposure. Yet even newborns acquire these relationships remarkably quickly, often after just one contact with a new situation (Morrongiello, Fenwick, & Chance, 1998). In addition, when researchers try to teach intermodal matches by pairing sights and sounds that do not naturally go together, babies will not learn them (Bahrick, 1992).

How does intermodal perception begin so early and develop so quickly? Young infants seem biologically primed to focus on intermodal information. They better detect changes in stimulation that occur simultaneously in two modalities (sight and sound) than those that occur in only one (Lewkowicz, 1996). Furthermore, detection of *amodal relations*—for example, the common tempo and rhythm in the sight and sound of clapping hands—develops first and may provide a basis for detecting other intermodal matches (Bahrick, 2001). Finally, early parent–infant interaction presents the baby with a rich context—consisting of many concurrent sights, sounds, touches, and smells—for expanding intermodal knowledge (Lickliter & Bahrich, 2000). Intermodal perception is yet another capacity that illustrates infants' active efforts to build an orderly, predictable world.

Understanding Perceptual Development

Now that we have reviewed the development of infant perceptual capacities, how can we put together this diverse array of amazing achievements? Eleanor and James Gibson provide widely accepted answers. According to the Gibsons' **differentiation theory,** infants actively search for **invariant features** of the environment—those that remain stable—in a constantly changing perceptual world. For example, in pattern perception, at first babies are confronted with a confusing mass of stimulation. But very quickly, they search for features that stand out along the border of a stimulus and orient toward images that crudely represent a face. Soon they explore internal features, noticing *stable relationships* between those features. As a result, they detect patterns, such as squares and complex faces. The development of intermodal perception also reflects this principle. Babies seek out invariant relationships—at first a common tempo and rhythm in concurrent sights and sounds, later more detailed associations—that unite information across modalities.

The Gibsons use the word *differentiation* (meaning analyze or break down) to describe their theory because over time, the baby detects finer and finer invariant features among stimuli. In addition to pattern perception, differentiation applies to depth perception; recall how sensitivity to motion and binocular cues precedes detection of fine-grained pictorial features. So one way of understanding perceptual development is to think of it as a built-in tendency to search for order and consistency, a capacity that becomes more fine-tuned with age (Gibson, 1970; Gibson, 1979).

Acting on the environment is vital in perceptual differentiation. Think back to the links between motor milestones and perceptual development discussed in this chapter. Infants constantly look for ways in which the environment affords opportunities for action (Gibson, 2000). By moving about and exploring the environment, they figure out which objects can be grasped, squeezed, bounced, or stroked and when a surface is safe to cross or presents the possibility of falling. As a result, they differentiate the world in new ways and act more competently (Adolph, 1997).

As we conclude this chapter, it is only fair to note that some researchers believe that babies do more than make sense of experience by searching for invariant features. Instead, they *impose meaning* on what they perceive, constructing categories of objects and events in the surrounding environment. We have seen the glimmerings of this cognitive point of view in this chapter. For example, older babies interpret a familiar face as a source of pleasure and affection and a pattern of blinking lights as a moving human being. This cognitive perspective also has merit in understanding the achievements of infancy. In fact, many researchers combine these two positions, regarding infant development as proceeding from a perceptual to a cognitive emphasis over the first year of life (Haith & Benson, 1998; Mandler, 1998).

Ask Yourself

REVIEW

Using research on crawling, show how motor and perceptual development support one another.

REVIEW

How does face perception change over the first year, and what gains in pattern perception support it?

APPLY

After several weeks of crawling, Benji learned to avoid going headfirst down the stairs. Now he has started to walk. Can his parents trust him not to try walking down a steep surface? Explain.

CONNECT

According to differentiation theory, perceptual development reflects infants' active search for invariant features. Provide examples from research on hearing, pattern perception, and intermodal perception.

www.

Summary

Body Growth

Describe major changes in body growth over the first 2 years.

- Changes in height and weight are rapid during the first 2 years. Body fat is laid down quickly in the first 9 months, whereas muscle development is slow and gradual. Skeletal age is the best way to estimate the child's physical maturity; girls are advanced over boys. Growth of parts of the body follows **cephalocaudal** and **proximodistal trends,** resulting in changing body proportions.

Brain Development

What changes in brain development occur during infancy and toddlerhood, at the level of individual brain cells and the level of the cerebral cortex?

- Early in development, the brain grows faster than any other organ of the body. Once **neurons** are in place, they form **synapses,** or connections, at a rapid rate. During the peak period of synaptic growth in any brain area, many surrounding neurons die to make room for synaptic connections. Stimulation determines which neurons will survive and continue to establish new synapses and which will lose their connective fibers through **synaptic pruning. Glial cells,** which are responsible for **myelination,** multiply rapidly into the second year and contribute to large gains in brain weight.

- Different regions of the **cerebral cortex** develop in the same order in which various capacities emerge in the growing child, with the frontal lobes among the last to develop. **Lateralization** refers to specialization of the hemispheres of the cerebral cortex. Although some brain specialization exists at birth, in the first few years of life there is high **brain plasticity.** Both heredity and early experience contribute to brain organization.

- Changes in the electrical brain-wave activity of the cortex, along with gains in brain weight and skull size, indicate that brain growth spurts occur intermittently from infancy through adolescence. These coincide with major cognitive changes and may be sensitive periods in which appropriate stimulation is necessary for full development.

How does the organization of sleep and wakefulness change over the first 2 years?

- The infant's changing arousal patterns are primarily affected by brain growth, but the social environment also plays a role. Short periods of sleep and wakefulness are put together and better coincide with a night–day schedule. Infants in Western nations sleep through the night much earlier than babies throughout most of the world, who sleep with their parents.

Influences on Early Physical Growth

Cite evidence that heredity, nutrition, and affection and stimulation contribute to early physical growth.

- Twin and adoption studies reveal that heredity contributes to body size and rate of physical growth.

- Breast milk is ideally suited to infants' growth needs and offers protection against disease. Breastfeeding prevents malnutrition and infant death in poverty-stricken areas of the world. Although breast- and bottle-fed babies do not differ in emotional adjustment, some studies report a slight advantage in mental test performance for children and adolescents who were breastfed.

- Most chubby babies slim down during toddlerhood and early childhood. Infants and toddlers can eat nutritious foods freely, without risk of becoming overweight.

- **Marasmus** and **kwashiorkor** are dietary diseases caused by malnutrition that affect many children in developing countries. If these conditions continue, body growth and brain development can be permanently stunted.

- **Nonorganic failure to thrive** illustrates the importance of stimulation and affection for normal physical growth.

Learning Capacities

Describe four infant learning capacities, the conditions under which they occur, and the unique value of each.

- Classical conditioning permits infants to recognize which events usually occur together in the everyday world. Infants can be **classically conditioned** when the pairing of an **unconditioned stimulus** (**UCS**) and a **conditioned stimulus** (**CS**) has survival value. Young babies are easily conditioned in the feeding situation. Classical conditioning of fear is difficult before 6 months of age.

- **Operant conditioning** helps infants to explore and control their surroundings. In addition to food, interesting sights and sounds serve as effective **reinforcers,** increasing the occurrence of a preceding behavior. **Punishment** involves removing a desirable stimulus or presenting an unpleasant one to decrease the occurrence of a response.

- **Habituation** and **recovery** reveal that at birth, babies are attracted to novelty. They tend to respond more strongly to a new element that has entered their environment. Newborn infants also have a primitive ability to imitate the facial expressions and gestures of adults. **Imitation** is a powerful means of learning and contributes to the parent–infant bond.

Motor Development

Describe the general course of motor development during the first 2 years, along with factors that influence it.

- Like physical development, motor development follows the cephalocaudal and proximodistal trends. According to **dynamic systems theory of motor development,** new motor skills are achieved by combining existing skills into increasingly complex systems of action. Each new skill is a joint product of central nervous system development, movement possibilities of the body, goals the child has in mind, and environmental supports for the skill.

- Movement opportunities and a stimulating environment profoundly affect motor development, as shown by research on infants raised in institutions in which they were deprived of stimulation. Cultural values and child-rearing customs contribute to the emergence and refinement of early motor skills.

- During the first year, infants perfect their reaching and grasping. The poorly coordinated prereaching of the newborn period drops out. Voluntary reaching gradually becomes more accurate and flexible, and the clumsy ulnar grasp is transformed into a refined pincer grasp.

Perceptual Development

What changes in hearing, depth and pattern perception, and intermodal perception take place during infancy?

- Over the first year, infants organize sounds into more complex patterns and can identify their precise location. Infants also become more sensitive to the sounds of their own language. During the second half of the first year, they use their remarkable ability to analyze sound patterns to detect meaningful speech units.

- Rapid development of the eye and visual centers in the brain supports the development of focusing, color discrimination, and visual acuity during the first half-year.

The ability to scan the environment and track moving objects also improves.

- Research on depth perception reveals that responsiveness to motion cues develops first, followed by sensitivity to binocular and then to pictorial cues. Experience in crawling enhances depth perception and other aspects of three-dimensional understanding. However, babies must learn to avoid drop-offs for each body position.
- **Contrast sensitivity** accounts for infants' early pattern preferences. At first, babies look at the border of a stimulus and at single features. Around 2 months, they explore the internal features of a pattern and start to detect pattern organization. Over time, they discriminate increasingly complex, meaningful patterns.
- Newborns prefer to look at and track simple, facelike stimuli, which suggests an adaptive capacity to orient toward members of the human species. At 2 to 3 months, they can discriminate a complex facial pattern and make fine distinctions between the features of different faces. In the second half-year, they perceive emotional expressions as meaningful wholes.
- From the start, infants are capable of **intermodal perception.** During the first year, they quickly combine information across sensory modalities, often after just one exposure to a new situation. Detection of amodal relations (such as common tempo and rhythm in sights and sounds) precedes, and may provide a basis for detecting, other intermodal matches.

Explain differentiation theory of perceptual development.

- According to **differentiation theory,** perceptual development is a matter of detecting **invariant features** in a constantly changing perceptual world. Acting on the world plays a major role in perceptual differentiation. Other researchers take a more cognitive view, suggesting that at an early age, infants impose meaning on what they perceive. Many researchers combine these two ideas.

Important Terms and Concepts

brain plasticity (p. 118)
cephalocaudal trend (p. 115)
cerebral cortex (p. 117)
classical conditioning (p. 126)
conditioned response (CR) (p. 127)
conditioned stimulus (CS) (p. 127)
contrast sensitivity (p. 135)
differentiation theory (p. 139)
dynamic systems theory of motor development (p. 130)
glial cells (p. 116)
habituation (p. 128)
imitation (p. 128)
intermodal perception (p. 138)
invariant features (p. 139)
kwashiorkor (p. 125)
lateralization (p. 118)
marasmus (p. 125)
myelination (p. 116)
neurons (p. 116)
nonorganic failure to thrive (p. 126)
operant conditioning (p. 127)
proximodistal trend (p. 115)
punishment (p. 127)
recovery (p. 128)
reinforcer (p. 127)
synapses (p. 116)
synaptic pruning (p. 116)
unconditioned response (UCR) (p. 126)
unconditioned stimulus (UCS) (p. 126)

FYI For Further Information and Help

Consult the Companion Website for *Development Through the Lifespan, Third Edition,* (www.ablongman.com/berk) where you will find the following resources for this chapter:

- **Chapter Objectives**
- **Flashcards** for studying important terms and concepts
- **Annotated Weblinks** to guide you in further research
- **Ask Yourself** questions, which you can answer and then check against a sample response
- **Suggested Readings**
- **Practice Test** with immediate scoring and feedback

5

Cognitive Development in Infancy and Toddlerhood

© JOSE LUIS PELAEZ, INC./CORBIS

Piaget's Cognitive-Developmental Theory
Piaget's Ideas About Cognitive Change • The Sensorimotor Stage • Follow-up Research on Infant Cognitive Development • Evaluation of the Sensorimotor Stage

■ Biology & Environment: Do Infants Have Built-In Numerical Knowledge?

Information Processing
Structure of the Information-Processing System • Attention • Memory • Categorization • Evaluation of Information-Processing Findings

■ A Lifespan Vista: Infantile Amnesia

The Social Context of Early Cognitive Development

■ Cultural Influences: Caregiver–Toddler Interaction and Early Make-Believe Play

Individual Differences in Early Mental Development
Infant Intelligence Tests • Early Environment and Mental Development • Early Intervention for At-Risk Infants and Toddlers

Language Development
Three Theories of Language Development • Getting Ready to Talk • First Words • The Two-Word Utterance Phase • Individual and Cultural Differences • Supporting Early Language Development

Infants' and toddlers' cognition develops rapidly through seeing, hearing, and touching and the sensitive support of caring adults. This 7-month-old radiates a sense of wonder as his mother helps him explore a summer wildflower.

When Caitlin, Grace, and Timmy gathered at Ginette's child-care home, the playroom was alive with activity. The three spirited explorers, each nearly 18 months old, were bent on discovery. Grace dropped shapes through holes in a plastic box that Ginette held and adjusted so the harder ones would fall smoothly into place. Once a few shapes were inside, Grace grabbed and shook the box, squealing with delight as the lid fell open and the shapes scattered around her. The clatter attracted Timmy, who picked up a shape, carried it to the railing at the top of the basement steps, and dropped it overboard, then followed with a teddy bear, a ball, his shoe, and a spoon. In the meantime, April pulled open a drawer, unloaded a set of wooden bowls, stacked them in a pile, knocked it over, and then banged two bowls together.

As the toddlers experimented, I could see the beginnings of language—a whole new way of influencing the world. Caitlin was the most vocal. "All gone baw!" she exclaimed as Timmy tossed the bright red ball down the basement steps. "Bye-bye," Grace chimed in, waving as the ball disappeared from sight. Later in the day, Grace revealed that she could use words and gestures to pretend. "Night-night," she said as she put her head down and closed her eyes, ever so pleased that in make-believe, she could decide for herself when and where to go to bed.

Over the first 2 years, the small, reflexive newborn baby becomes a self-assertive, purposeful being who solves simple problems and has started to master the most amazing human ability: language. Parents often wonder, "How does all this happen so quickly?" This question has also captivated researchers, yielding a wealth of findings along with vigorous debate over how to explain the astonishing pace of infant and toddler cognition.

In this chapter, we take up three perspectives on early cognitive development: *Piaget's cognitive-developmental theory, information processing,* and *Vygotsky's sociocultural theory.* We will also consider the usefulness of tests that measure infants' and toddlers' intellectual progress. Our discussion concludes with the beginnings of language. We will see how toddlers' first words build on early cognitive achievements and how, very soon, new words and expressions greatly increase the speed and flexibility of their thinking. Throughout development, cognition and language mutually support one another.

Piaget's Cognitive-Developmental Theory

Swiss theorist Jean Piaget inspired a vision of children as busy, motivated explorers whose thinking develops as they act directly on the environment. Influenced by his background in biology, he believed that the child's mind forms and modifies psychological structures so they achieve a better fit with external reality. Recall from Chapter 1 that in Piaget's theory, children move through four stages between infancy and adolescence. During those stages, all aspects of cognition develop in an integrated fashion, changing in a similar way at about the same time. The first stage is the **sensorimotor stage,** which spans the first 2 years of life.

As the name of this stage implies, Piaget believed that infants and toddlers "think" with their eyes, ears, hands, and other sensorimotor equipment. They cannot yet carry out many activities inside their heads. But by the end of toddlerhood, children can solve practical, everyday problems and represent their experiences in speech, gesture, and play. To appreciate Piaget's view of how these vast changes take place, let's consider some important concepts.

Piaget's Ideas About Cognitive Change

According to Piaget, specific psychological structures—organized ways of making sense of experience called **schemes**—change with age. At first, schemes are sensorimotor action patterns. For example, at 6 months, Timmy dropped objects in a fairly rigid way, simply by letting go of a rattle or teething ring and watching with interest. By age 18 months, his "dropping scheme" had become much more deliberate and creative. He tossed all sorts of objects down the basement stairs, throwing some up in the air, bouncing others off walls, releasing some gently and others forcefully. Soon his schemes will move from an *action-based level* to a *mental level.* Instead of just acting on objects, he will show evidence of thinking before he acts. This change, as we will see later, marks the transition from sensorimotor to preoperational thought.

In Piaget's theory, two processes account for changes in schemes: *adaptation* and *organization.*

● **Adaptation.** The next time you have a chance, notice how infants and toddlers tirelessly repeat actions that lead to interesting effects. **Adaptation** involves building schemes through direct interaction with the environment. It consists of two complementary activities: *assimilation* and *accommodation.* During **assimilation,** we use our current schemes to interpret the external world. For example, when Timmy dropped objects, he was assimilating them all into his sensorimotor "dropping scheme." In **accommodation,** we create new schemes or adjust old ones after noticing that our current ways of thinking do not fit the environment completely. When Timmy dropped objects in different ways, he modified his dropping scheme to take account of the varied properties of objects.

According to Piaget, the balance between assimilation and accommodation varies over time. When children are not changing much, they assimilate more than they accommodate. Piaget called this a state of cognitive *equilibrium,* implying a steady, comfortable condition. During rapid cognitive change, however, children are in a state of *disequilibrium,* or cognitive discomfort. They realize that new information does not match their current schemes, so they shift away from assimilation toward accommodation. Once they modify their schemes, they move back toward

■ According to Piaget's theory, at first schemes are motor action patterns. As this 8-month-old takes apart, turns, and bangs these pots and pans, he discovers that his movements have predictable effects on objects and that objects influence one another in predictable ways.

assimilation, exercising their newly changed structures until they are ready to be modified again.

Each time this back-and-forth movement between equilibrium and disequilibrium occurs, more effective schemes are produced. Because the times of greatest accommodation are the earliest ones, the sensorimotor stage is Piaget's most complex period of development.

● **Organization.** Schemes also change through **organization,** a process that takes place internally, apart from direct contact with the environment. Once children form new schemes, they rearrange them, linking them with other schemes to create a strongly interconnected cognitive system. For example, eventually Timmy will relate "dropping" to "throwing" and to his developing understanding of "nearness" and "farness." According to Piaget, schemes reach a true state of equilibrium when they become part of a broad network of structures that can be jointly applied to the surrounding world (Piaget, 1936/1952).

In the following sections, we will first describe infant development as Piaget saw it, noting research that supports his observations. Then we will consider evidence demonstrating that in some ways, the cognitive competence of babies is more advanced than Piaget believed it to be.

The Sensorimotor Stage

The difference between the newborn baby and the 2-year-old child is so vast that the sensorimotor stage is divided into six substages (see Table 5.1 for a summary). Piaget's observations of his own three children served as the basis for this sequence of development. Although this is a very small sample, Piaget watched carefully and also presented his son and two daughters with everyday problems (such as hidden objects) that helped reveal their understanding of the world.

According to Piaget, at birth infants know so little about the world that they cannot purposefully explore their surroundings. The **circular reaction** provides a special means of adapting first schemes. It involves stumbling onto a new experience caused by the baby's own motor activity. The reaction is "circular" because the infant tries to repeat the event again and again. As a result, a sensorimotor response that first occurred by chance becomes strengthened into a new scheme. Consider Caitlin, who at age 2 months accidentally made a smacking sound after a feeding. The sound was new and intriguing, so

Table 5.1 Summary of Piaget's Sensorimotor Substages

Sensorimotor Substage	Adaptive Behaviors
1. Reflexive schemes (birth to 1 month)	Newborn reflexes (see Chapter 3, page 101)
2. Primary circular reactions (1–4 months)	Simple motor habits centered around the infant's own body; limited anticipation of events
3. Secondary circular reactions (4–8 months)	Actions aimed at repeating interesting effects in the surrounding world; imitation of familiar behaviors
4. Coordination of secondary circular reactions (8–12 months)	Intentional, or goal-directed, behavior; ability to find a hidden object in the first location in which it is hidden (object permanence); improved anticipation of events; imitation of behaviors slightly different from those the infant usually performs
5. Tertiary circular reactions (12–18 months)	Exploration of the properties of objects by acting on them in novel ways; imitation of unfamiliar behaviors; ability to search in several locations for a hidden object (accurate A-B search)
6. Mental representation (18 months–2 years)	Internal depictions of objects and events, as indicated by sudden solutions to problems, ability to find an object that has been moved while out of sight (invisible displacement), deferred imitation, and make-believe play

Caitlin tried to repeat it until, after a few days, she became quite expert at smacking her lips.

During the first 2 years, the circular reaction changes in several ways. At first it centers around the infant's own body. Later, it turns outward, toward manipulation of objects. Finally, it becomes experimental and creative, aimed at producing novel effects in the environment. Young children's difficulty inhibiting new and interesting behaviors may underlie the circular reaction. But this immaturity in inhibition seems to be adaptive! It helps ensure that new skills will not be interrupted before they consolidate (Carey & Markman, 1999). Piaget considered revisions in the circular reaction so important that he named the sensorimotor substages after them (refer again to Table 5.1).

● **Repeating Chance Behaviors.** For Piaget, newborn reflexes are the building blocks of sensorimotor intelligence. At first, in Substage 1, babies suck, grasp, and look in much the same way, no matter what experiences they encounter. Carolyn reported an amusing example of her daughter Caitlin's indiscriminate sucking at age 2 weeks. She lay next to her father while he took a nap. Suddenly, he awoke with a start. Caitlin had latched on and begun to suck on his back!

Around 1 month, as babies enter Substage 2, they start to gain voluntary control over their actions through the *primary circular reaction,* by repeating chance behaviors largely motivated by basic needs. This leads to some simple motor habits, such as sucking their fists or thumbs. Babies of this substage also begin to vary their behavior in response to environmental demands. For example, they open their mouths differently for a nipple than for a spoon. Young infants also start to anticipate events. For example, at 3 months, when Timmy awoke from his nap, he cried out with hunger. But as soon as Vanessa entered the room, his crying stopped. He knew that feeding time was near.

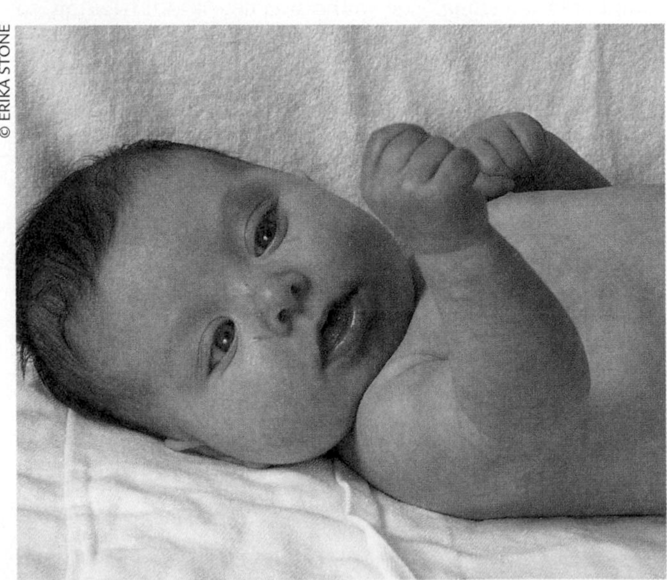

■ During Piaget's Substage 2, infants' adaptations are oriented toward their own bodies. This young baby carefully watches the movements of her hands, a primary circular reaction that helps her gain voluntary control over her behavior.

During Substage 3, which lastis from 4 to 8 months, infants sit up and reach for and manipulate objects (see Chapter 4). These motor achievements play a major role in turning their attention outward toward the environment. Using the *secondary circular reaction,* they try to repeat interesting events caused by their own actions. For example, 4-month-old Caitlin accidentally knocked a toy hung in front of her, producing a fascinating swinging motion. Over the next 3 days, Caitlin tried to repeat this effect, at first by grasping and then by waving her arms. Finally she succeeded in hitting the toy and gleefully repeated the motion. She had built the sensorimotor scheme of "hitting." Improved control over their own behavior permits infants to imitate the behavior of others more effectively. However, 4- to 8-month-olds cannot adapt flexibly and quickly enough to imitate novel behaviors (Kaye & Marcus, 1981). Therefore, although they enjoy watching an adult demonstrate a game of pat-a-cake, they are not yet able to participate.

● **Intentional Behavior.** In Substage 4, 8- to 12-month-olds combine schemes into new, more complex action sequences. As a result, actions that lead to new schemes no longer have a random, hit-or-miss quality—*accidentally* bringing the thumb to the mouth or *happening* to hit the toy. Instead, 8- to 12-month-olds can engage in **intentional,** or **goal-directed, behavior,** coordinating schemes deliberately to solve simple problems. The clearest example is provided by Piaget's famous object-hiding task, in which he shows the baby an attractive toy and then hides it behind his hand or under a cover. Infants of this substage can find the object. In doing so, they coordinate two schemes—"pushing" aside the obstacle and "grasping" the toy. Piaget regarded these action sequences as the foundation for all problem solving.

Retrieving hidden objects reveals that infants have begun to master **object permanence,** the understanding that objects continue to exist when out of sight. But awareness of object permanence is not yet complete. If the baby reaches several times for an object at a first hiding place (A) and sees it moved to a second (B), she will still search for it in the first hiding place (A). Because babies make this *A-not-B search error,* Piaget concluded that they do not have a clear image of the object as persisting when hidden from view.

Substage 4 brings additional advances. First, infants can better anticipate events, so they sometimes use their capacity for intentional behavior to try to change those events. At 10 months, Timmy crawled after Vanessa when she put on her coat, whimpering to keep her from leaving. Second, babies can imitate behaviors slightly different from those they usually perform. After watching someone else, they try to stir with a spoon, push a toy car, or drop raisins in a cup. Once again, they draw on intentional behavior, purposefully modifying schemes to fit an observed action (Piaget, 1945/1951).

In Substage 5, which lasts from 12 to 18 months, the *tertiary circular reaction* emerges. Toddlers repeat behaviors with variation, provoking new results. Recall how Timmy dropped objects over the basement steps, trying this, then that, and then another action. Because they approach the world in this delib-

■ The capacity to search for and find hidden objects between 8 and 12 months of age marks a major advance in cognitive development. This infant displays intentional, or goal-directed, behavior and coordinates schemes in obtaining a toy—capacities that are the foundation for all problem solving.

erately exploratory way, 12- to 18-month-olds are far better sensorimotor problem solvers than they were before. For example, Grace figured out how to fit a shape through a hole in a container by turning and twisting it until it fell through, and she discovered how to use a stick to get toys that were out of reach. According to Piaget, this capacity to experiment leads to a more advanced understanding of object permanence. Toddlers look in several locations to find a hidden toy, displaying an accurate A–B search. Their more flexible action patterns also permit them to imitate many more behaviors, such as stacking blocks, scribbling on paper, and making funny faces.

● **Mental Representation.** Substage 6 culminates with the ability to create **mental representations**—internal depictions of information that the mind can manipulate. Our most powerful mental representations are of two kinds: (1) *images,* or mental pictures of objects, people, and spaces, and (2) *concepts,* or categories in which similar objects or events are grouped together. Using a mental image, we can retrace our steps when we've misplaced something. Or we can imitate another's behavior long after we've observed it. And by thinking in concepts and labeling them (for example, *ball* for all rounded, movable objects used in play), we become more efficient thinkers, organizing our diverse experiences into meaningful, manageable, and memorable units.

Piaget noted that in arriving at solutions suddenly rather than through trial-and-error behavior, 18- to 24-month-olds seem to experiment with actions inside their heads—evidence that they can mentally represent their experiences. For example, at 19 months Grace received a new push toy. As she played with it for the first time, she rolled it over the carpet and ran into the sofa. She paused for a moment, as if to "think," and then immediately turned the toy in a new direction. Representation results in several other capacities. First, it enables older toddlers to solve advanced object permanence problems involving *invisible displacement*—finding a toy moved while out of sight, such as into a small box while under a cover. Second, it permits **deferred imitation**—the ability to remember and copy the behavior of models who are not present. Finally, it makes possible **make-believe play,** in which children act out everyday and imaginary activities. Like Grace's pretending to go to sleep at the beginning of this chapter, the toddler's make-believe is very simple. Make-believe expands greatly in early childhood and is so important for psychological development that we will return to it again. In sum, as the sensorimotor stage draws to a close, mental symbols have become major instruments of thinking.

Follow-Up Research on Infant Cognitive Development

Many studies suggest that infants display a wide array of understandings earlier than Piaget believed. For example, recall the operant conditioning research reviewed in Chapter 4, in which newborns sucked vigorously on a nipple to gain access to interesting sights and sounds. This behavior, which closely resembles Piaget's secondary circular reaction, indicates that babies try to explore and control the external world long before 4 to 8 months. In fact, they do so as soon as they are born.

A major method used to find out what infants know about hidden objects and other aspects of physical reality capitalizes on habituation/recovery, which we discussed in Chapter 4. In the **violation-of-expectation method,** researchers habituate babies to a physical event. Then they determine whether infants recover (look longer at) a *possible event* (a variation of the first event that follows physical laws) or an *impossible event* (a variation that violates physical laws). Recovery to the impossible event suggests that the infant is surprised at a deviation from physical reality and, therefore, is aware of that aspect of the physical world.

But as we will see, the violation-of-expectation method is controversial. Some critics believe that it indicates only limited awareness of physical events, not the full-blown understandings that Piaget detected when he observed infants acting on their surroundings, such as searching for hidden objects (Bremner, 1998). Other critics are convinced that the method is flawed—that it reveals only babies' perceptual preference for novelty, not their understanding of experience (Haith, 1999). Let's examine this debate in light of recent evidence.

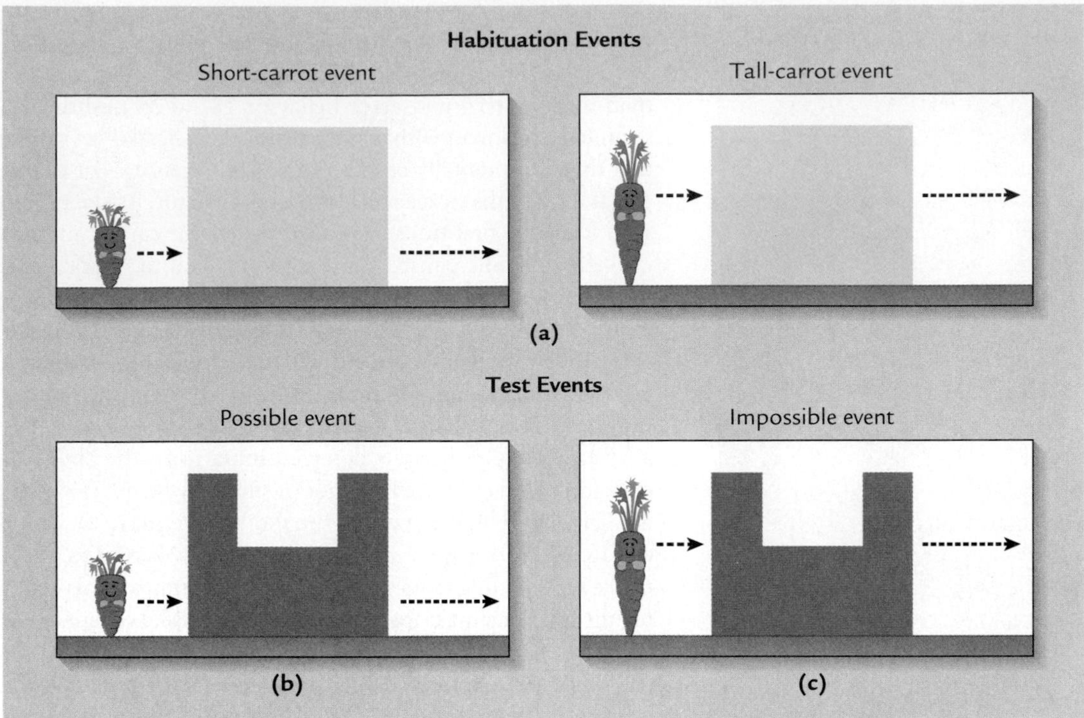

■ **FIGURE 5.1 Testing infants for understanding of object permanence using the violation-of-expectation method.** (a) First, infants were habituated to two events: a short carrot and a tall carrot moving behind a yellow screen, on alternative trials. Next the researchers presented two test events. The color of the screen was changed to help infants notice its window. (b) In the *possible event,* the carrot shorter than the window's lower edge moved behind the blue screen and reappeared on the other side. (c) In the *impossible event,* the carrot taller than the window's lower edge moved behind the screen, did not appear in the window, but then emerged intact on the other side. Infants as young as 3½ months recovered to (looked longer at) the impossible event, suggesting that they had some understanding of object permanence. (Adapted from R. Baillargeon & J. DeVos, 1991, "Object Permanence in Young Infants: Further Evidence," *Child Development, 62,* p. 1230. © The Society for Research in Child Development. Reprinted with permission of the Society for Research in Child Development.)

● **Object Permanence.** In a series of studies using the violation-of-expectation method, Renée Baillargeon and her collaborators claimed to have found evidence for object permanence in the first few months of life. One of Baillargeon's studies is illustrated in Figure 5.1 (Aguiar & Baillargeon, 1999, 2002; Baillargeon & DeVos, 1991). After habituating to a short and a tall carrot moving behind a screen, infants were given two test events: (1) a *possible event,* in which the short carrot moved behind a screen, could not been seen in its window, and reappeared on the other side, and (2) an *impossible event,* in which the tall carrot moved behind a screen, could not be seen in its window (although it was taller than the window's lower edge), and reappeared. Three-month-olds looked longer at the *impossible event,* suggesting that they expected an object moved behind a screen to continue to exist.

But studies using procedures similar to Baillargeon's failed to confirm some of her findings (Bogartz, Shinskey, & Shilling, 2000; Cashon & Cohen, 2000; Rivera, Wakeley, & Langer, 1999). Baillargeon and others answer that these opposing investigations did not include crucial controls. And they emphasize that infants look longer at a wide variety of impossible events that make it appear as though an object covered by a screen no longer exists (Aslin, 2000; Baillargeon, 2000). Still, critics question what babies' looking preferences actually tell us about their understanding.

If 3-month-olds do have some notion of object permanence, then what explains Piaget's finding that much older infants (who are quite capable of voluntary reaching) do not try to search for hidden objects? Consistent with Piaget's theory, research suggests that searching for hidden objects represents a true advance in understanding of object permanence because infants solve some object-hiding tasks before others. Ten-month-olds search for an object placed on a table and covered by a cloth before they search for an object that a hand deposits under a cloth (Moore & Meltzoff, 1999). In the second, more difficult task, infants seem to expect the object to reappear in the hand because that is where the object initially disappeared. When the hand emerges without the object, they conclude that there is no other place the object could be. Not until 14 months can most infants infer that the hand deposited the object under the cloth.

Once 8- to 12-month-olds actively search for hidden objects, they make the A-not-B search error. Some research suggests that they search at A (where they found the object on previous reaches) instead of B (its most recent location) because they have trouble inhibiting a previously rewarded response (Diamond, Cruttenden, & Neiderman, 1994). Another possibility is that after finding the object several times at A, they do not attend closely when it is hidden at B (Ruffman & Langman, 2002). A more comprehensive explanation is that a complex, dynamic system of factors—having built a habit of reaching toward A, continuing to look at A, having the hiding place at B look similar to the one at A, and maintaining a constant body posture—increases the chances that the baby will make the A-not-B search error. Research shows that disrupting any one of these factors increases 10-month-olds' accurate searching at B (Smith et al., 1999).

In sum, before 12 months, infants have difficulty translating what they know about object location into a successful search strategy. The ability to engage in an accurate A–B search coincides with rapid development of the frontal lobes of the cerebral cortex at the end of the first year (Bell, 1998). Also crucial are a wide variety of experiences perceiving, acting on, and remembering objects.

■ Deferred imitation greatly enriches young children's adaptations to their surrounding world. This toddler probably watched an adult watering flowers. Later, he imitates the behavior, having learned through observation the function of sprinkling cans.

● **Mental Representation.** In Piaget's theory, infants lead purely sensorimotor lives; they cannot represent experience until about 18 months of age. Yet 8-month-olds' ability to recall the location of a hidden object, even after delays of more than a minute, indicates that they mentally represent objects (McDonough, 1999). And new studies of deferred imitation and problem solving reveal that representational thought is evident even earlier.

Deferred Imitation. Piaget studied imitation by noting when his own three children demonstrated it in their everyday behavior. Under these conditions, a great deal must be known about the infant's daily life to be sure that deferred imitation—which requires infants to represent another's past behavior—has occurred.

Laboratory research reveals that deferred imitation is present at 6 weeks of age! Infants who watched an unfamiliar adult's facial expression imitated it when exposed to the same adult the next day (Meltzoff & Moore, 1994). Perhaps young babies use this imitation to identify and communicate with people they have seen before. As motor capacities improve, infants start to copy actions with objects. In one study, 6- and 9-month-olds were shown an "activity" board with 12 novel objects fastened to it—for example, a frog whose legs jump when a cord is pulled. An adult modeled the actions of 6 objects. When tested a day later, babies of both ages were far more likely to produce the actions they had seen than actions associated with objects that had not been demonstrated (Collie & Hayne, 1999).

Between 12 and 18 months, toddlers use deferred imitation skillfully to enrich their range of sensorimotor schemes. They retain modeled behaviors for several months and imitate across a change in context. For example, they enact in the laboratory behaviors learned at home and generalize actions to similar objects varying in size and color (Hayne, Boniface, & Barr, 2000; Klein & Meltzoff, 1999). At the end of the second year, toddlers can imitate actions an adult *tries* to produce, even if these are not fully realized (Meltzoff, 1995). On one occasion, a mother attempted to pour some raisins into a small bag but missed, spilling them. A moment later, her 18-month-old son began dropping the raisins into the bag, indicating that he had inferred his mother's intention. By age 2, children mimic entire social roles—such as mommy, daddy, or baby—during make-believe play.

Problem Solving. As Piaget indicated, infants develop intentional action sequences around 7 to 8 months, using them to solve simple problems, such as pulling on a cloth to obtain a toy resting on its far end (Willatts, 1999). Soon after, infants' representational skills permit more effective problem solving than Piaget's theory suggests.

By 10 to 12 months, infants can *solve problems by analogy*—take a strategy from one problem and apply it to other relevant problems. In one study, babies were given three similar problems, each requiring them to overcome a barrier, grasp a string, and pull it to get an attractive toy. The problems differed in all aspects of their specific features (see Figure 5.2 on page 150). On the first problem, the parent demonstrated the solution and encouraged the infant to imitate. Babies obtained the toy more readily on each additional problem, suggesting that they had

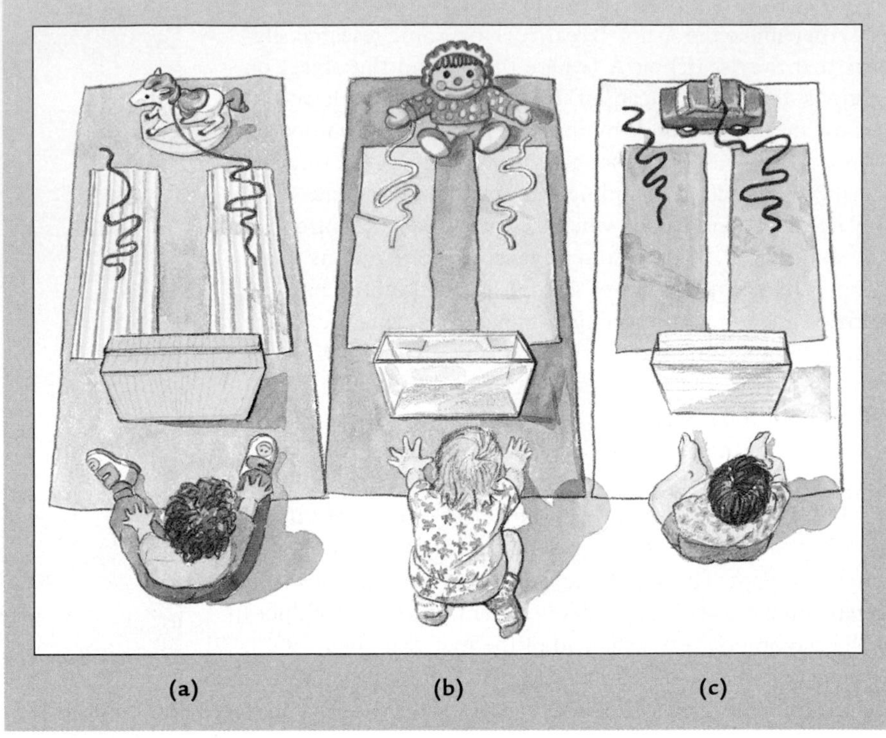

■ **FIGURE 5.2 Analogical problem solving by 10- to 12-month-olds.** After the parent demonstrated the solution to problem (a), infants solved (b) and (c) with increasing efficiency, even though those problems differed in all aspects of their superficial features.(From Z. Chen, R. P. Sanchez, & T. Campbell, 1997, "From Beyond to Within Their Grasp: The Rudiments of Analogical Problem Solving in 10- to 13-Month-Olds," *Developmental Psychology, 33,* p. 792. Copyright © 1997 by the American Psychological Association. Reprinted by permission of the publisher and author.)

formed a flexible mental representation of actions that access an out-of-reach object (Chen, Sanchez, & Campbell, 1997).

With age, children become better at reasoning by analogy, applying relevant strategies across increasingly dissimilar situations (Goswami, 1996). But even in the first year, infants have some ability to move beyond trial-and-error experimentation, represent a solution mentally, and use it in new contexts.

Evaluation of the Sensorimotor Stage

Table 5.2 summarizes the remarkable cognitive attainments we have just considered. Compare this table with the description of Piaget's sensorimotor substages on page 145. You will see that infants anticipate events, actively search for hidden objects, master A–B object search, flexibly vary their sensorimotor schemes, and engage in make-believe play within Piaget's time frame. Yet many other capacities—including secondary circular reactions, understanding of object properties, first signs of object permanence, deferred imitation, and problem solving by analogy—emerge earlier than Piaget expected. These findings show that the cognitive attainments of infancy do not develop together in the neat, stepwise fashion that Piaget assumed.

Disagreements between Piaget's observations and those of recent research raise controversial questions about how infant development takes place. Consistent with Piaget's ideas, sensorimotor action helps infants construct some forms of knowledge. For example, in Chapter 4, we saw that experience in crawling enhances depth perception and ability to find hidden objects. Yet we have also seen that infants comprehend a great deal before they are capable of the motor behaviors that Piaget assumed led to those understandings. How can we account for babies' amazing cognitive accomplishments?

Most researchers believe that young babies have more built-in cognitive equipment for making sense of experience than granted by Piaget, who thought they constructed all mental representations out of sensorimotor activity. But intense disagreement exists over how much initial understanding infants have. As we have seen, much evidence on young infants' cognition rests on the violation-of-expectation method. Researchers who lack confidence in this method argue that babies' cognitive starting point is limited. For example, some believe that newborns begin life with a set of biases, or learning procedures—such as powerful techniques for analyzing complex, perceptual information—that grant infants a means for constructing schemes (Elman et al., 1996; Haith & Benson, 1998; Karmiloff-Smith, 1992).

Others, convinced by violation-of-expectation findings, believe that infants start out with impressive understandings. According to this **core knowledge perspective,** babies are born with a set of innate knowledge systems, or *core domains of thought.* Each of these prewired understandings permits a ready grasp of new, related information and therefore supports early, rapid development (Carey & Markman, 1999; Spelke & Newport, 1998). Core knowledge theorists argue that infants could not make sense of the complex stimulation around them without having been "set up" in the course of evolution to comprehend crucial aspects of it. Researchers have conducted many studies of 2- to 6-month-olds' *physical knowledge,* including object permanence, object solidity (that

Table 5.2 Some Cognitive Attainments of Infancy and Toddlerhood

Age	Cognitive Attainments
Birth–1 month	Secondary circular reactions using limited motor skills, such as sucking a nipple to gain access to interesting sights and sounds
1–4 months	Violation-of-expectation findings suggest awareness of many object properties, including object permanence, object solidity, and gravity; deferred imitation of an adult's facial expression over a short delay (1 day)
4–8 months	Violation-of-expectation findings suggest basic numerical knowledge and improved physical knowledge; deferred imitation of an adult's novel actions on objects over a short delay (1 day)
8–12 months	Ability to search for a hidden object in diverse situations—when covered by a cloth, when a hand deposits it under a cloth, and when it is moved from one location to another (accurate A–B search); ability to solve sensorimotor problems by analogy to a previous, similar problem
12–18 months	Deferred imitation of an adult's novel actions on an object over a long delay (several months) and across a change in context (from home to laboratory and using similar objects varying in size and color)
18 months–2 years	Deferred imitation of actions an adult tries to produce, even if these are not fully realized, indicating a beginning capacity to infer others' intentions; imitation of social roles, such as mommy, daddy, or baby, in make-believe play

Note: Which of the capacities listed in the table indicate that mental representation emerges earlier than predicted by Piaget's sensorimotor substages?

■ Did this toddler figure out that each block placed on the tower will fall without support through extensive sensorimotor activity, as Piaget assumed? Or did she begin life with prewired physical knowledge—a core domain of thought that promotes early, rapid understanding?

one object cannot move through another object), and gravity (that an object will fall without support) (Baillargeon, 1994, Hespos & Baillargeon, 2001). They have also investigated infants' *numerical knowledge,* or ability to distinguish small quantities (see the Biology and Environment box on page 152). Furthermore, core knowledge theorists assume that *linguistic knowledge* is etched into the structure of the human brain—a possibility we will consider when we take up language development. And infants' early orientation toward people initiates swift development of *psychological knowledge*—in particular, understanding of mental states, such as intentions, emotions, desires, and beliefs, which we will begin to address in Chapter 6.

But as the Biology and Environment box reveals, studies of young infants' knowledge yield mixed results. Nevertheless, broad agreement exists on two issues. First, many cognitive changes of infancy are gradual and continuous rather than abrupt and stagelike (Flavell, Miller, & Miller, 2002). Second, rather than developing together, various aspects of infant cognition change unevenly because of the challenges posed by different types of tasks and infants' varying experience with them. These ideas serve as the basis for another major approach to cognitive development—*information processing*—which we take up next.

Before we turn to this alternative point of view, let's conclude our discussion of the sensorimotor stage by recognizing Piaget's enormous contributions. His work inspired a wealth of research on infant cognition, including studies that challenged his theory. In addition, his observations have been of great practical value. Teachers and caregivers continue to look to the sensorimotor stage for guidelines on how to create developmentally appropriate environments for infants and toddlers, an issue we address later in this chapter.

Biology & Environment

Do Infants Have Built-In Numerical Knowledge?

How does cognition develop so rapidly during the early years of life? According to the *core knowledge perspective,* infants have an inherited foundation of knowledge that quickly becomes more elaborate as they explore, play, and interact with others (Geary & Bjorklund, 2000). Do infants display numerical understandings so early that some knowledge must be innate? The violation-of-expectation method has been used to answer this question.

In the best known of these investigations, 5-month-olds saw a screen raised to hide a single toy animal (see Figure 5.3). Then they watched a hand place a second toy behind the screen. Finally, the screen was removed to reveal either one toy or two toys. If infants kept track of the two objects, then they should look longer at the one-toy display (*impossible outcome*)—which is what they did. In additional experiments, 5-month-olds given this task looked longer at three objects than at two. These results, and those of other similar studies, suggest that in the first half-year, babies can discriminate quantities up to three and use that knowledge to perform simple arithmetic—not just addition, but also subtraction, in which two objects are covered by a screen and one is removed (Wynn, 1992; Wynn, Bloom, & Chiang, 2002).

These findings, like other violation-of-expectation results, are controversial. Critics question what looking preferences actually tell us about infants' numerical knowledge. And some researchers report that 5-month-olds cannot add and subtract small quantities. In experiments similar to those just described, looking preferences were inconsistent (Feigenson, Carey, & Spelke, 2002; Wakeley, Rivera, & Langer, 2000). Furthermore, infants' knowledge of number is surprising, given that toddlers usually do not show these understandings. Before 14 to 16 months, toddlers have difficulty with less-than and greater-than relationships between small sets. And not until the preschool years do children add and subtract small sets correctly.

Overall, in some studies, infants display amazing knowledge; in others, they do not. And if such knowledge is innate, older children should reason in the same way as infants, yet they do not always do so. Core knowledge theorists respond that infants' looking behaviors may be more reliable indicators of understanding than older children's counting behaviors, which may not tap their true competencies (Wynn, 2002). Still, skeptics claim that human evolution may not have equipped infants with ready-made knowledge, which could limit their ability to adapt to changes in the environment (Haith, 1999; Meltzoff & Moore, 1998). At present, just what babies start out with—innate understandings or general learning strategies that permit them to discover knowledge quickly—continues to be hotly debated.

Original Event

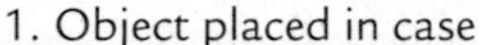

(a)

Test Events

Possible outcome

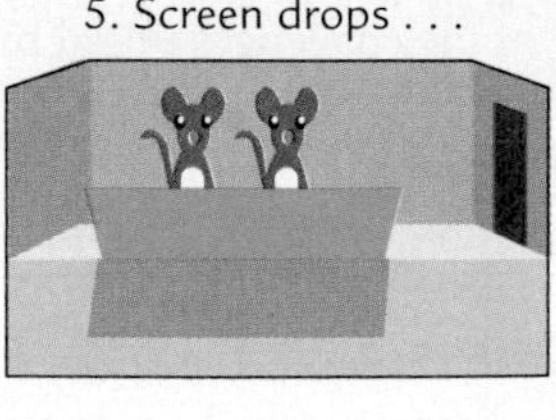

(b)

Impossible outcome

(c)

■ **FIGURE 5.3 Testing infants for basic number concepts.** (a) First, infants saw a screen raised in front of a toy animal. Then an identical toy was added behind the screen. Next, the researchers presented two outcomes. (b) In the *possible outcome,* the screen dropped to reveal two toy animals. (c) In the *impossible outcome,* the screen dropped to reveal one toy animal. Five-month-olds shown the impossible outcome looked longer than did 5-month-olds shown the possible outcome. The researchers concluded that infants can discriminate the quantities "one" and "two" and use that knowledge to perform simple addition: 1 + 1 = 2. A variation of this procedure suggested that 5-month-olds could also do simple subtraction: 2 – 1 = 1. (From K. Wynn, 1992, "Addition and Subtraction by Human Infants," *Nature, 358,* p. 749. Reprinted by permission.)

Ask Yourself

REVIEW
Explain how cognition changes in Piaget's theory, giving examples of assimilation, accommodation, and organization.

REVIEW
Using the text discussion on pages 145–150, construct your own table providing an overview of infant and toddler cognitive milestones. Which entries in the table are consistent with Piaget's sensorimotor stage? Which ones develop earlier than Piaget anticipated?

APPLY
Ten-month-old Mimi's father holds up her favorite teething biscuit, deposits it under a napkin, and shows Mimi his empty hand. Mimi looks puzzled and fails to search for the biscuit. Explain why Mimi finds this object-hiding task difficult.

www.

Information Processing

Information-processing theorists agree with Piaget that children are active, inquiring beings, but they do not provide a single, unified theory of cognitive development. Instead, they focus on many aspects of thinking, from attention, memory, and categorization skills to complex problem solving.

Recall from Chapter 1 that the information-processing approach frequently relies on computerlike flowcharts to describe the human cognitive system. The computer model of human thinking is attractive because it is explicit and precise. Information-processing theorists are not satisfied with general concepts, such as assimilation and accommodation, to describe how children think. Instead, they want to know exactly what individuals of different ages do when faced with a task or problem (Klahr & MacWhinney, 1998; Siegler, 1998).

Structure of the Information-Processing System

Most information-processing researchers assume that we hold information in three parts of the mental system for processing: *the sensory register*; *working,* or *short-term, memory*; and *long-term memory* (see Figure 5.4 on page 154). As information flows through each, we can operate on and transform it using **mental strategies,** increasing the chances that we will retain information and use it efficiently. To understand this more clearly, let's look at each aspect of the mental system.

First, information enters the **sensory register.** Here, sights and sounds are represented directly and stored briefly. Look around you, and then close your eyes. An image of what you saw persists for a few seconds, but then it decays, or disappears, unless you use mental strategies to preserve it. For example, you can *attend to* some information more carefully than to other information, increasing the chances that it will transfer to the next step of the information-processing system.

The second part of the mind is **working, or short-term, memory,** where we actively "work" on a limited amount of information, applying mental strategies. For example, if you are studying this book effectively, you are taking notes, repeating information to yourself, or grouping pieces of information together. Think, for a moment, about why you apply these strategies. The sensory register, although limited, can take in a wide panorama of information. The capacity of working memory is more restricted. But by meaningfully connecting pieces of information into a single representation, we reduce the number of pieces we must attend to, thereby making room in working memory for more. Also, the more thoroughly we learn information, the more *automatically* we use it. Automatic cognitive processing expands working memory by permitting us to focus on other information simultaneously.

To manage its complex activities, a special part of working memory called the **central executive** directs the flow of information. It decides what to attend to, coordinates incoming information with information already in the system, and selects, applies, and monitors strategies (Baddeley, 1993, 2000). The central executive is the conscious, reflective part of our mental system.

The longer we hold information in working memory, the greater the likelihood that it will transfer to the third, and largest, storage area—**long-term memory,** our permanent knowledge base, which is limitless. In fact, we store so much in long-term memory that we sometimes have problems with *retrieval,* or getting information back from the system. To aid retrieval, we apply strategies, just as we do in working memory. Information in long-term memory is *categorized* according to a master plan based on contents, much like a library shelving system. As a result, we can retrieve it easily by following the same network of associations used to store it in the first place.

Information-processing researchers believe that the basic structure of the mental system is similar throughout life. However, the *capacity* of the system—the amount of information that can be retained and processed at once and the speed with which it can be processed—increases, making possible more complex forms of thinking with age (Case, 1998; Miller & Vernon, 1997). Gains in information-processing capacity are partly due to brain development and partly due to improvements in strategies, such as attending to information and categorizing it effectively. The development of these strategies is already under way in the first 2 years of life.

Attention

How does attention develop in early infancy? Recall from our discussion of perceptual development in Chapter 4 that between 1 and 2 months of age, infants shift from attending to a single high-contrast feature of their visual world to exploring objects and patterns more thoroughly. Besides attending

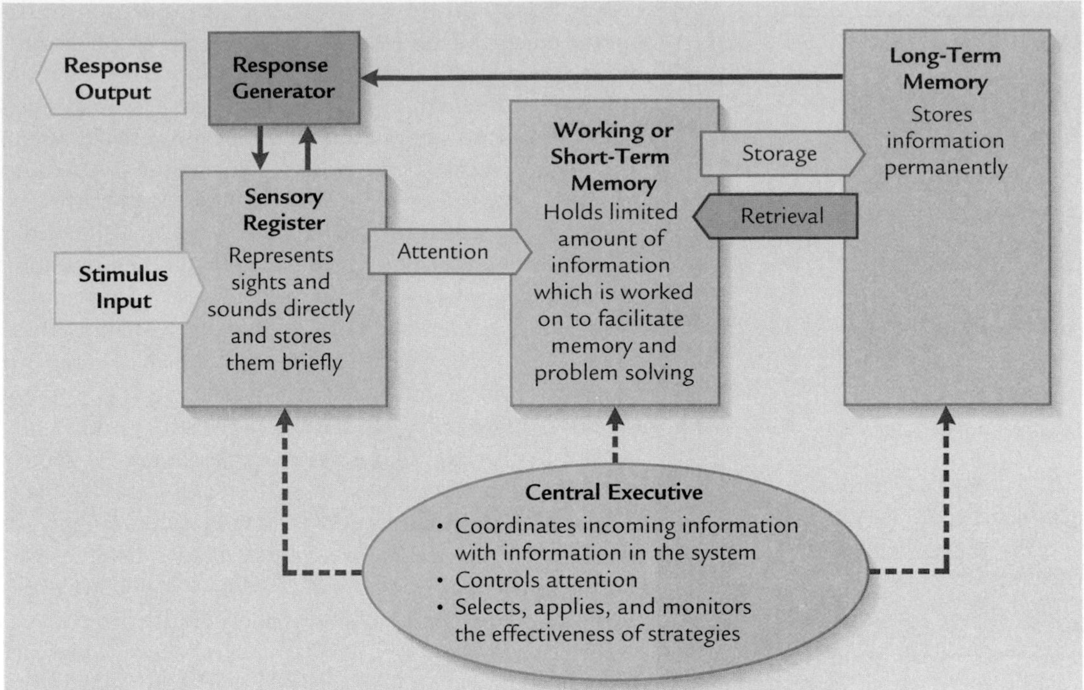

FIGURE 5.4 Store model of the human information-processing system. Information flows through three parts of the mental system: the *sensory register; working,* or *short-term, memory;* and *long-term memory.* In each, mental strategies can be used to manipulate information, increasing the efficiency of thinking and the chances that information will be retained. Strategies also permit us to think flexibly, adapting information to changing circumstances. The *central executive* is the conscious, reflective part of working memory. It coordinates incoming information already in the system, decides what to attend to, and oversees the use of strategies.

to more aspects of the environment, infants gradually become more efficient at managing their attention, taking in information more quickly with age. Habituation research reveals that preterm and newborn babies require a long time to habituate and recover to novel visual stimuli—about 3 or 4 minutes. But by 4 or 5 months, infants require as little as 5 to 10 seconds to take in a complex visual stimulus and recognize that it differs from a previous one (Slater et al., 1996).

One reason that very young babies' habituation times are so long is that they have difficulty disengaging their attention from interesting stimuli (Frick, Colombo, & Saxon, 1999). Once, Carolyn held a doll dressed in red-and-white checked overalls in front of 2-month-old Caitlin, who stared intently until, unable to break her gaze, she burst into tears. Just as important as attending to a stimulus is the ability to shift attention from one stimulus to another. By 4 to 6 months, infants' attention becomes more flexible (Hood, Atkinson, & Braddick, 1998).

During the first year, infants attend to novel and eye-catching events (Richards & Holley, 1999). With the transition to toddlerhood, children become increasingly capable of intentional behavior (refer back to Piaget's Substage 4). Consequently, attraction to novelty declines (but does not disappear) and *sustained attention* improves, especially when children play with toys. When a toddler engages in goal-directed behavior even in a limited way, such as stacking blocks or putting them in a container, attention must be maintained to reach the goal. As plans and activities become more complex, so does the duration of attention (Ruff & Lawson, 1990; Ruff & Rothbart, 1996).

Memory

Habituation research provides a window into infant memory. Studies show that infants gradually make finer distinctions among visual stimuli and remember them longer—at 3 months, for about 24 hours; by the end of the first year, for several days and, in the case of some stimuli (such as a photo of the human face), even weeks (Fagan, 1973; Pascalis, de Haan, & Nelson, 1998). Yet recall that what babies know about the stimuli to which they habituate and recover is not always clear. Some researchers argue that infants' understanding is best revealed through their active efforts to master their environment (Rovee-Collier, 2001). Consistent with this view, habituation research greatly underestimates infants' memory when compared with methods that rely on their active exploration of objects.

Using operant conditioning, Carolyn Rovee-Collier studied infant memory by teaching babies to move a mobile by kicking a foot tied to it with a long cord. She found that 2- to 3-month-olds remembered how to activate the mobile 1 week after training, and with a prompt (the experimenter briefly rotates the mobile for the baby), as long as 4 weeks. By 6 months of age, retention increases to 2 weeks and, with prompting, to 6 weeks (Rovee-Collier & Bhatt, 1993; Rovee-Collier, 1999). Around the middle of the first year, tasks in which babies control stimulation by manipulating buttons or switches work well for studying memory. When infants and toddlers pressed a lever to make a toy train move around a track, duration of memory continued to increase with age; 13 weeks after training, 18-month-olds remembered how to press the lever (Hartshorn et al., 1998).

So far, we have discussed only **recognition**—noticing when a stimulus is identical or similar to one previously experienced. This is the simplest form of memory because all babies have to do is indicate (by looking or kicking) that a new stimulus is identical or similar to a previous one. **Recall** is more challenging because it involves remembering something in the absence of perceptual support. To recall, you must generate a mental image of the past experience. Can infants engage in recall? By the end of the first year, they can. We know because they find hidden objects and imitate the actions of others hours or days after they observed the behavior.

Between 1 and 2 years, children's recall for people, places, and objects is excellent. Think back to findings on deferred imitation, which show that 14-month-olds recall and reproduce highly unusual behaviors several months after having observed them. Yet a puzzling finding is that as adults, we no longer recall our earliest experiences. The Lifespan Vista box on page 156 helps explain why.

Categorization

As infants gradually remember more information, they store it in a remarkably orderly fashion. Some creative variations of the operant conditioning research described earlier have been used to find out about infant categorization. One such study is described and illustrated in Figure 5.5. In fact, young babies categorize stimuli on the basis of shape, size, and other physical properties at such an early age that categorization is among the strongest evidence that babies' brains are set up from the start to represent and organize experience in adultlike ways (Mandler, 1998).

Habituation/recovery has also been used to study infant categorization. Researchers show babies a series of stimuli belonging to one category and then see whether they recover to (look longer at) a picture that is not a member of the category. Findings reveal that 7- to 12-month-olds structure objects into an impressive array of meaningful categories—food items, furniture, birds, animals, vehicles, kitchen utensils, plants, and more (Mandler & McDonough, 1993, 1996, 1998; Oakes, Coppage, & Dingel, 1997). Besides organizing the physical world, infants of this age categorize their emotional and social worlds. They sort people and their voices by gender and age (Bahrick, Netto, & Hernandez-Reif, 1998; Poulin-DuBois et al., 1994), have begun to distinguish emotional expressions, and can separate the natural movements of people from other motions (see Chapter 4, pages 136–138).

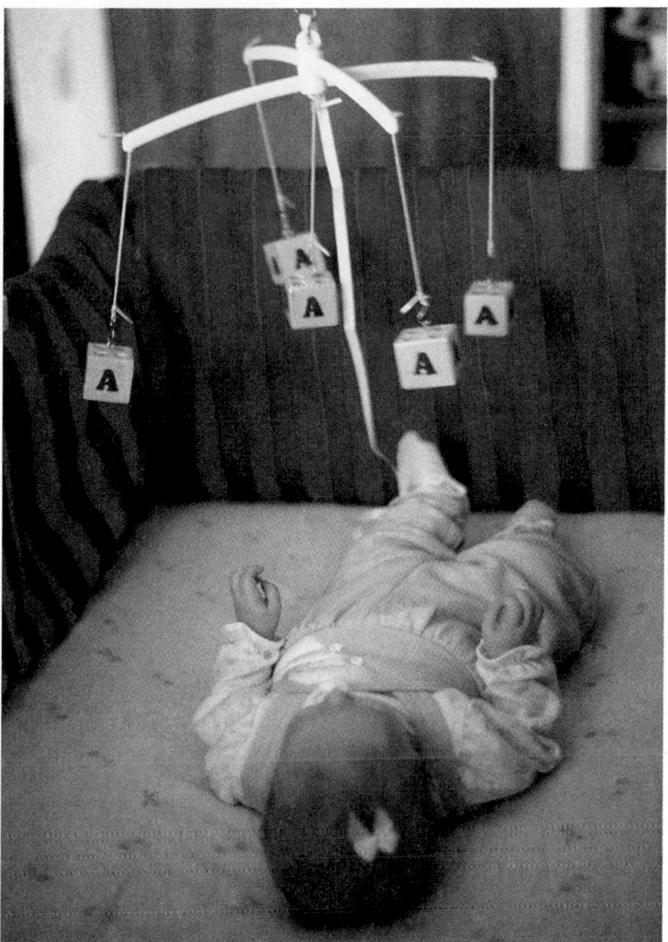

■ **FIGURE 5.5 Investigating infant categorization using operant conditioning.** Three-month-olds were taught to kick to move a mobile that was made of small blocks, all with the letter A on them. After a delay, kicking returned to a high level only if the babies were shown a mobile whose elements were labeled with the same form (the letter A). If the form was changed (from As to 2s), infants no longer kicked vigorously. While making the mobile move, the babies had grouped together its features. They associated the kicking response with the category A and, at later testing, distinguished it from the category 2 (Bhatt, Rovee-Collier, & Weiner, 1994; Hayne, Rovee-Collier, & Perris, 1987).

The earliest categories are *perceptual*—based on similar overall appearance or prominent object part, such as legs for animals and wheels for vehicles (Rakison & Butterworth, 1998). But by the end of the first year, more categories are *conceptual*—based on common function and behavior. For example, 1-year-olds group together kitchen utensils because each is used to prepare food for eating.

In the second year, children become active categorizers. Around 12 months, they touch objects that go together, without grouping them. Sixteen-month-olds can group objects into a single category. For example, when given four balls and four boxes, they put all the balls together but not the boxes. Around 18 months, they sort objects into two classes (Gopnik & Meltzoff, 1987). Compared with habituation/recovery,

A Lifespan Vista

Infantile Amnesia

If toddlers remember many aspects of their everyday lives, then what explains *infantile amnesia*—the fact that practically none of us can retrieve events that happened to us before age 3? This forgetting cannot be due merely to the passage of time because we can recall many events that happened long ago (Eacott, 1999). At present, several explanations of infantile amnesia exist.

One theory credits brain development. Growth of the frontal lobes of the cerebral cortex, along with other structures, may be necessary before experiences can be stored in ways that permit them to be retrieved many years later (Boyer & Diamond, 1992).

Yet the idea of vastly different approaches to remembering in younger and older individuals has been questioned because toddlers can describe memories verbally and retain them for extensive periods. A growing number of researchers believe that rather than a radical change in the way experience is represented, the decline of infantile amnesia requires the emergence of a special form of recall—**autobiographical memory,** or representations of special, one-time events that are long-lasting because they are imbued with personal meaning. For example, perhaps you recall the day a sibling was born, the first time you took an airplane, or a move to a new house.

For memories to become autobiographical, at least two developments are necessary. First, the child must have a well-developed image of the self. Yet in the first few years, the sense of self is not yet mature enough to serve as an anchor for one-time events (Howe, 2003). Second, autobiographical memory requires that children integrate personal experiences into a meaningful life story. Recent evidence reveals that preschoolers learn to structure memories in narrative form by talking about them with adults, who expand on their recollections by explaining what happened when, where, and with whom (Nelson, 1993).

Between 2 and 5 years of age, children's interest in memory-related conversations increases greatly—a change that may support the rise in autobiographical memories during this period. Interestingly, parents talk about the past in more detail with daughters (Bruce, Dolan, & Phillips-Grant, 2000; Reese, Haden, & Fivush, 1996). And collectivist cultural values lead Asian parents to discourage children from talking about themselves (Han, Leichtman, & Wang, 1998). Perhaps because women's early experiences were integrated into more coherent narratives, they report an earlier age of first memory and more vivid early memories than do men. Similarly, first memories of Caucasian-American adults are, on average, 6 months earlier than those of Asians (Mullen, 1994).

The decline of infantile amnesia probably represents a change to which both biology and social experience contribute. One speculation is that vital changes in the frontal lobes of the cerebral cortex during toddlerhood pave the way for an *explicit memory* system—one in which children remember consciously rather than *implicitly,* without conscious awareness (Rovee-Collier & Barr, 2001). In Chapters 7 and 9, we will see that deliberate recall of information and events improves greatly during childhood. It undoubtedly supports the success of conversations about the past in structuring children's autobiographical memories.

■ This toddler probably will not recall the exciting celebration of the Hindu Holi, or "bright," festival, at which she and family members splashed one another with colored powders and joined in dancing and folk singing. For autobiographical memory to develop, young children must have a well-developed self-image to which to attach special experiences and, with assistance from adults, integrate those experiences into a meaningful life story.

touching, sorting, and other play behaviors better reveal the meanings that toddlers attach to categories because they are applying those meanings in their everyday activities. For example, after having watched an adult give a toy dog a drink from a cup, 14-month-olds shown a rabbit and a motorcycle usually offer the drink only to the rabbit (Mandler & McDonough, 1998). Their behavior reveals a clear understanding that certain actions are appropriate for some categories of items (animals) and not others (vehicles).

How does this perceptual-to-conceptual change take place? Although researchers disagree on whether this shift requires a new approach to analyzing experience, most acknowledge that exploration of objects and expanding knowledge of the world contribute to older infants' capacity to group objects by their functions and behaviors (Mandler, 2000; Quinn et al., 2000). Finally, as adults label objects for children ("This one's a car, and that one's a bicycle"), they help toddlers refine their earliest categories (Waxman, 1995).

Evaluation of Information-Processing Findings

Information-processing research underscores the continuity of human thinking from infancy into adult life. In attending to the environment, remembering everyday events, and categorizing objects, Caitlin, Grace, and Timmy think in ways that are remarkably similar to our own, even though they are far from being the proficient mental processors we are. Findings on infant memory and categorization join with other research in challenging Piaget's view of early cognitive development. If 3-month-olds can remember events for as long as 4 weeks and categorize stimuli, then they must have some ability to represent their experiences mentally.

Information processing has contributed greatly to our view of young babies as sophisticated cognitive beings. Still, its greatest drawback stems from its central strength: After analyzing cognition into its components (such as perception, attention, and memory), information processing has had difficulty putting them back together into a broad, comprehensive theory. One approach to overcoming this weakness has been to combine Piaget's theory with the information-processing approach, an effort we will take up in Chapter 9. A more recent trend has been the application of a *dynamic systems view* to early cognition (see Chapter 4, page 130). Researchers analyze each cognitive attainment to see how it results from a complex system of prior accomplishments and the child's current goals (Thelen & Smith, 1998; Courage & Howe, 2002). These ideas have yet to be fully tested, but they may move the field closer to a more powerful view of how the mind of the infant and child develops.

The Social Context of Early Cognitive Development

Take a moment to review the short episode at the beginning of this chapter in which Grace dropped shapes into a container. Notice that Grace learns about the toy with Ginette's help. With adult support, Grace will gradually become better at matching shapes to openings and dropping them into the container. Then she will be able to perform the activity (and others like it) on her own.

Vygotsky's sociocultural theory has helped researchers realize that children live in rich social contexts that affect the way their cognitive world is structured (Rogoff, 1998; Wertsch & Tulviste, 1992). Vygotsky believed that complex mental activities, such as voluntary attention, deliberate memory, and problem solving, have their origins in social interaction. Through

■ This father assists his young son in putting together a puzzle through gentle physical support and simple words. By bringing the task within the child's zone of proximal development (see page 158) and adjusting his communication to suit the child's needs, the father transfers mental strategies to the child and promotes his cognitive development.

Cultural Influences

Caregiver–Toddler Interaction and Early Make-Believe Play

One of my husband Ken's activities with our two sons when they were young was to bake pineapple upside-down cake, a favorite treat. One Sunday afternoon when a cake was in the making, 21-month-old Peter stood on a chair at the kitchen sink, busily pouring water from one cup to another.

"He's in the way, Dad!" complained 4-year-old David, trying to pull Peter away from the sink.

"Maybe if we let him help, he'll give us room," Ken suggested. As David stirred the batter, Ken poured some into a small bowl for Peter, moved his chair to the side of the sink, and handed him a spoon.

"Here's how you do it, Peter," instructed David, with an air of superiority. Peter watched as David stirred, and then he tried to copy his motion. When it was time to pour the batter, Ken helped Peter hold and tip the small bowl.

"Time to bake it," said Ken.

"Bake it, bake it," repeated Peter, as he watched Ken slip the pan into the oven.

Several hours later, we observed one of Peter's earliest instances of make-believe play. He got his pail from the sandbox and, after filling it with a handful of sand, carried it into the kitchen and put it down on the floor in front of the oven. "Bake it, bake it," Peter called to Ken. Together, father and son lifted the pretend cake inside the oven.

Until recently, most researchers studied make-believe play apart from the social environment in which it occurs, while children played alone. Probably for this reason, Piaget and his followers concluded that toddlers discover make-believe independently, as soon as they are capable of representational schemes. Vygotsky's theory has challenged this view. He believed that society provides children with opportunities to represent culturally meaningful activities in play. Make-believe, like other complex mental activities, is first learned under the guidance of experts. In the example just described, Peter's capacity to represent daily events was extended when Ken drew him into the baking task and helped him act it out in play.

Current evidence supports the idea that early make-believe is the combined result of children's readiness to engage in it and social experiences that promote it. In one observational study of middle-SES American toddlers, 75 to 80 percent of make-believe involved mother–child interaction (Haight & Miller, 1993). At 12 months, make-believe was fairly one-sided; almost all play episodes were initiated by mothers. By the end of the second year, mothers and children displayed mutual interest in getting make-believe started; half of pretend episodes were initiated by each.

When adults participate, toddlers' make-believe is more elaborate (O'Reilly & Bornstein, 1993). For example, play themes are more varied. And toddlers are more likely to com-

joint activities with more mature members of their society, children come to master activities and think in ways that have meaning in their culture.

A special Vygotskian concept explains how this happens. The **zone of proximal** (or potential) **development** refers to a range of tasks that the child cannot yet handle alone but can do with the help of more skilled partners. To understand this idea, think of a sensitive adult (such as Ginette) who introduces a child to a new activity. The adult picks a task that the child can master but that is challenging enough that the child cannot do it by herself. Or the adult capitalizes on an activity that the child has chosen. Such tasks are especially suited for fostering development. Then, as the adult guides and supports, the child joins in the interaction and picks up mental strategies, and her competence increases. When this happens, the adult steps back, permitting the child to take more responsibility for the task.

As we will see in Chapters 7 and 9, Vygotsky's ideas have been applied mostly at older ages, when children are more skilled in language and social communication. But recently, Vygotsky's theory has been extended to infancy and toddlerhood. Recall that babies are equipped with capacities that ensure that caregivers will interact with them. Then adults adjust the environment and their communication in ways that promote learning adapted to their cultural circumstances.

A study by Barbara Rogoff and her collaborators (1984) illustrates this process. The researchers watched how several adults played with Rogoff's son and daughter over the first 2 years, while a jack-in-the-box toy was nearby. In the early months, adults tried to focus the baby's attention by working the toy, and as the bunny popped out, saying something like "My, what happened?" By the end of the first year (when the baby's cognitive and motor skills had improved), interaction centered on how to use the jack-in-the-box. When the infant reached for the toy, adults guided the baby's hand in turning the crank and putting the bunny back in the box. During the second year, adults helped from a distance. They used gestures and verbal prompts, such as rotating a hand in a turning motion near the crank. Research indicates that this fine-tuned support is related to advanced play, language, and problem solving during the second year (Bornstein et al., 1992; Tamis-LeMonda & Bornstein, 1989).

As early as the first 2 years, cultural variations in social experiences affect mental strategies. Note how, in the example

bine schemes into complex sequences, as Peter did when he put sand in the bucket ("making the batter"), carried it into the kitchen, and (with Ken's help) put it in the oven ("baking the cake"). The more parents pretend with their toddlers, the more time their children devote to make-believe. And in certain collectivist societies, such as Argentina and Japan, mother–toddler other-directed pretending, as in feeding or putting a doll to sleep, is particularly frequent and rich in maternal expressions of affection and praise (Bornstein et al., 1999).

In some cultures, older siblings are toddlers' first play partners. For example, in Indonesia and Mexico, where extended family households and sibling caregiving are common, make-believe is more frequent and more complex with older siblings than with mothers. As early as 3 to 4 years of age, children provide rich, challenging stimulation to their younger brothers and sisters. The fantasy play of these toddlers is just as well developed as that of their middle-SES American counterparts (Farver, 1993; Farver & Wimbarti, 1995).

As we will see in Chapter 7, make-believe is a major means through which children extend their cognitive skills and learn about important activities in their culture. Vygotsky's theory, and the findings that support it, tell us that providing a stimulating physical environment is not enough to promote early cognitive development. In addition, toddlers must be invited and encouraged by more skilled members of their culture to participate in the social world around them. Parents and teachers can enhance early make-believe by playing often with toddlers, responding, guiding, and elaborating on their make-believe themes.

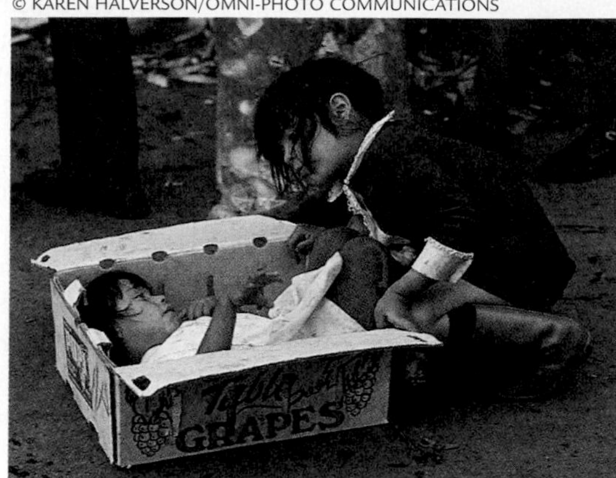

© KAREN HALVERSON/OMNI-PHOTO COMMUNICATIONS

■ In Mexico, where sibling caregiving is common, make-believe play is more frequent and complex with older siblings than with mothers. In creating a pretend scene, this 5-year-old provides her younger sister with enjoyable and challenging stimulation.

just described, adults and children focused their attention on a single activity. This strategy, common in Western middle-SES homes, is well suited to lessons in which children learn skills apart from the everyday situations in which those skills will later be used. In contrast, Guatemalan Mayan adults and toddlers often attend to several events at once. For example, one 12-month-old skillfully put objects in a jar while watching a passing truck and whistling on a toy whistle his mother had slipped in his mouth (Chavajay & Rogoff, 1999). Processing several competing events simultaneously may be vital in cultures where children largely learn not through lessons but through keen observation of others.

Earlier we saw how infants and toddlers create new schemes by acting on the physical world (Piaget) and how certain skills become better developed as children represent their experiences more efficiently and meaningfully (information processing). Vygotsky adds a third dimension to our understanding by emphasizing that many aspects of cognitive development are socially mediated. The Cultural Influences box above presents additional evidence for this idea. And we will see even more evidence in the next section, as we look at individual differences in mental development during the first 2 years.

Ask Yourself

REVIEW

Cite evidence that categorization becomes less perceptual and more conceptual with age. What factors support this shift? How can adults foster toddlers' categorization?

APPLY

Caitlin played with toys in a more intentional, goal-directed way as a toddler than as an infant. What impact is Caitlin's more advanced toy play likely to have on the development of attention? How is her cultural background likely to affect her attention?

APPLY

At age 18 months, Timmy's mother stood behind him, helping him throw a large ball into a box. When Timmy showed that he could throw the ball, his mother stepped back and let him try on his own. Using Vygotsky's ideas, explain how Timmy's mother is supporting his cognitive development.

www.

Individual Differences in Early Mental Development

Because of Grace's deprived early environment, Kevin and Monica had a psychologist give her one of many tests available for assessing mental development in infants and toddlers. Worried about Timmy's progress, Vanessa also arranged for him to be tested. At age 22 months, he had only a handful of words in his vocabulary, played in a less mature way than Caitlin and Grace, and seemed restless and overactive.

The testing approach is very different from the cognitive theories we have just discussed, which try to explain the *process* of development—how children's thinking changes over time. In contrast, designers of mental tests focus on cognitive *products.* They seek to measure behaviors that reflect mental development and to arrive at scores that predict future performance, such as later intelligence, school achievement, and adult vocational success. This concern with prediction arose nearly a century ago, when French psychologist Alfred Binet designed the first successful intelligence test, which predicted school achievement (see Chapter 1). It inspired the design of many new tests, including ones that measure intelligence at very early ages.

Infant Intelligence Tests

Accurately measuring the intelligence of infants is a challenge because they cannot answer questions or follow directions. All we can do is present them with stimuli, coax them to respond, and observe their behavior. As a result, most infant tests emphasize perceptual and motor responses, along with a few tasks that tap early language and cognition. For example, the *Bayley Scales of Infant Development,* a commonly used test for children between 1 month and 3½ years, consists of two parts: (1) the Mental Scale, which includes such items as turning to a sound, looking for a fallen object, building a tower of cubes, and naming pictures; and (2) the Motor Scale, which assesses gross and fine motor skills, such as grasping, sitting, drinking from a cup, and jumping (Bayley, 1993).

● **Computing Intelligence Test Scores.** Intelligence tests for infants, children, and adults are scored in much the same way. When a test is constructed, it is given to a large, representative sample of individuals. Performances of people at each age level form a *normal* or *bell-shaped curve* in which most scores fall near the center (the mean or average) and progressively fewer fall toward the extremes. On the basis of this distribution, the test designer computes *norms,* or standards against which future test takers can be compared. For example, if Timmy does better than 50 percent of his agemates, his score will be 100, an average test score. If he exceeds most children his age, his score will be much higher. If he does better than only a small percentage of 2-year-olds, his score will be much lower.

A score computed this way, permitting an individual's test performance to be compared with those of other same-age individuals, is called an **intelligence quotient,** or **IQ**—a term you have undoubtedly heard before. Table 5.3 describes the meaning of a range of IQ scores. Notice how the IQ offers a way of finding out whether an individual is ahead, behind, or on time (average) in mental development in relation to others of the same age. The great majority of individuals (96 percent) have IQs between 70 and 130. Only a very few achieve higher or lower scores.

● **Predicting Later Performance from Infant Tests.** Many people assume, incorrectly, that IQ is a measure of inborn ability that does not change with age. Despite careful construction, most infant tests predict later intelligence poorly. Longitudinal research reveals that the majority of children show substantial fluctuations in IQ between toddlerhood and adolescence—10 to 20 points in most cases and sometimes much more (McCall, 1993).

Because infants and toddlers are especially likely to become distracted, fatigued, or bored during testing, their scores often do not reflect their true abilities. In addition, the perceptual and motor items on infant tests differ from the tasks given to older children, which emphasize verbal, conceptual, and problem-solving skills. Because of concerns that infant test scores do not tap the same dimensions of intelligence measured at older ages, they are conservatively labeled **developmental quotients,** or **DQs,** rather than IQs.

■ A trained examiner tests this baby with the Bayley Scales of Infant Development while his father looks on. The perceptual and motor items on most infant tests differ from the tasks given to older children, which emphasize verbal, conceptual, and problem-solving skills. For normally developing children, infant tests predict later intelligence poorly.

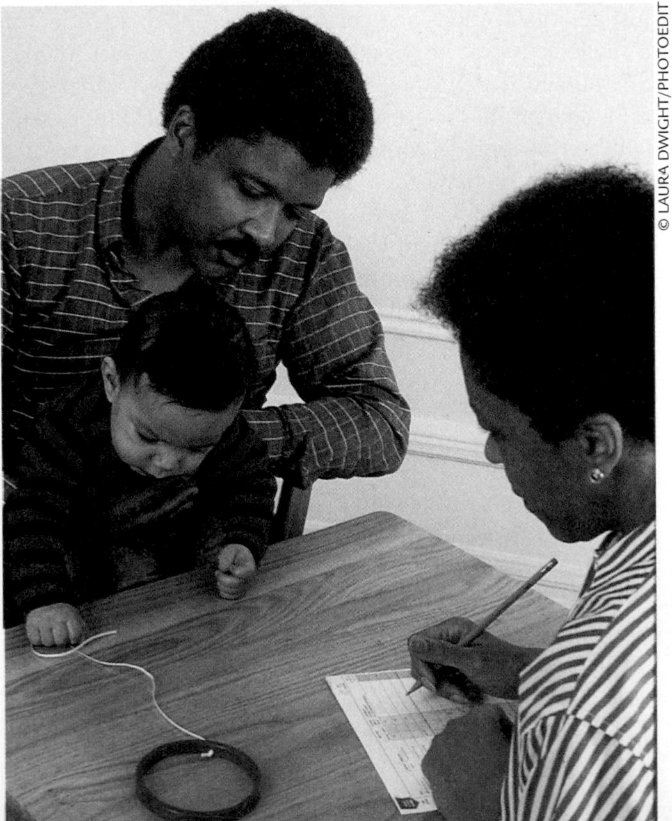

Table 5.3 Meaning of Different IQ Scores

Score	Percentile Rank (Child Does Better Than . . . Percent of Same-Age Children)
70	2
85	16
100 (average IQ)	50
115	84
130	98

Infant tests are somewhat better at making long-term predictions for extremely low-scoring babies. Today, they are largely used for *screening*—helping to identify for further observation and intervention babies whose very low scores mean that they are likely to have developmental problems in the future (Kopp, 1994).

Because infant tests do not predict later IQ for most children, researchers have turned to the information-processing approach to assess early mental progress. Their findings show that speed of habituation and recovery to visual stimuli are among the best available infant predictors of intelligence from early childhood into adolescence (McCall & Carriger, 1993; Sigman, Cohen, & Beckwith, 1997). Habituation and recovery seem to predict later IQ more effectively than traditional infant tests because they assesses quickness of thinking, a characteristic of bright individuals. They also tap basic cognitive processes—attention, memory, and response to novelty—that underlie intelligent behavior at all ages (Colombo, 1995; Rose & Feldman, 1997). The consistency of these findings prompted designers of the most recent edition of the Bayley test to include several items that tap cognitive skills, such as habituation/recovery, object permanence, and categorization.

Early Environment and Mental Development

In Chapter 2, we indicated that intelligence is a complex blend of hereditary and environmental influences. Many studies have examined the relationship of environmental factors to infant and toddler mental test scores. As we consider this evidence, you will encounter findings that highlight the role of heredity as well.

● **Home Environment.** The **Home Observation for Measurement of the Environment (HOME)** is a checklist for gathering information about the quality of children's home lives through observation and interviews with parents (Caldwell & Bradley, 1994). The Caregiving Concerns table below lists factors measured by HOME during the first 3 years. Each is positively related to toddlers' mental test performance. Regardless of SES and ethnicity, an organized, stimulating physical setting and parental encouragement, involvement, and affection repeatedly predict infant and early childhood IQ (Espy, Molfese, & DiLalla, 2001; Klebanov et al., 1998; Roberts, Burchinal, & Durham, 1999). The extent to which parents talk to infants and toddlers is particularly important. As the final section of this chapter will reveal, it contributes strongly to early language progress. Language progress, in turn, predicts intelligence and academic achievement in elementary school (Hart & Risley, 1995).

Yet we must interpret these correlational findings with caution. In all the studies, children were reared by their biological parents, with whom they share not just a common environment but also a common heredity. Parents who are genetically

Home Observation for Measurement of the Environment (HOME): Infancy and Toddler Subscales

SUBSCALE	SAMPLE ITEM
Emotional and verbal responsiveness of the parent	Parent caresses or kisses child at least once during observer's visit.
Acceptance of the child	Parent does not interfere with child's actions or restrict child's movements more than three times during observer's visit.
Organization of the physical environment	Child's play environment appears safe and free of hazards.
Provision of appropriate play materials	Parent provides toys or interesting activities for child during observer's visit.
Parental involvement with the child	Parent tends to keep child within visual range and to look at child often during observer's visit.
Variety in daily stimulation	Child eats at least one meal per day with mother or father, according to parental report.

Source: Elardo & Bradley, 1981.

more intelligent might provide better experiences as well as give birth to genetically brighter children, who also evoke more stimulation from their parents. Note that this hypothesis refers to *genetic–environmental correlation* (see Chapter 2, page 69), and research supports it (Cherney, 1994). But heredity does not account for all of the correlation between home environment and mental test scores. Family living conditions continue to predict children's IQ beyond the contribution of parental IQ and education (Klebanov et al., 1998; Chase-Lansdale et al., 1997). In one study, infants and children growing up in less crowded homes had parents who were far more verbally responsive to them—a major contributor to language, intellectual, and academic progress (Evans, Maxwell, & Hart, 1999).

Can the research summarized so far help us understand Vanessa's concern about Timmy's development? Indeed, it can. Ben, the psychologist who tested Timmy, found that he scored only slightly below average. Ben talked with Vanessa about her child-rearing practices and watched her play with Timmy. A single parent, Vanessa worked long hours and had little energy for Timmy at the end of the day. Ben also noticed that Vanessa, anxious about how well Timmy was doing, tended to pressure him. She constantly tried to dampen his active behavior and bombarded him with directions, such as "That's enough ball play. Stack these blocks." Ben explained that when parents are intrusive in these ways, infants and toddlers are likely to be distractible, play immaturely, and do poorly on mental tests (Bradley et al., 1989; Fiese, 1990). He coached Vanessa in how to interact sensitively with Timmy. At the same time, he assured her that Timmy's current performance need not forecast his future development. Warm, responsive parenting that builds on toddlers' current capacities is a much better indicator of how they will do later than is an early mental test score.

● **Infant and Toddler Child Care.** Home environments are not the only influential settings in which young children spend their days. Today, more than 60 percent of North American mothers with a child under age 2 are employed (Statistics Canada, 2002g; U.S. Bureau of the Census, 2002c). Child care for infants and toddlers has become common, and its quality has a major impact on mental development. Research consistently shows that infants and young children exposed to poor-quality child care, regardless of whether they come from middle- or low-SES homes, score lower on measures of cognitive and social skills (Hausfather et al., 1997; Kohen et al., 2000; NICHD Early Child Care Research Network, 2000b).

In contrast, good child care can reduce the negative impact of a stressed, poverty-stricken home life, and it sustains the benefits of growing up in an economically advantaged family (Lamb, 1998). In Swedish longitudinal research, entering high-quality child care in infancy and toddlerhood was associated with cognitive, emotional, and social competence in middle childhood and adolescence (Andersson, 1989, 1992; Broberg et al., 1997).

Visit some child-care settings, and take notes on what you see. In contrast to most European countries and to Australia and New Zealand, where child care is nationally regulated and funded to ensure its quality, reports on American and Canadian child care are cause for concern. Standards are set by the states and provinces and vary greatly across each nation. In some places, caregivers need no special training in child development, and one adult is permitted to care for 6 to 12 babies at once (Children's Defense Fund, 2002). In nationwide studies of child-care quality in the United States and Canada, only 20 to 25 percent of child-care centers and homes provided infants and toddlers with sufficiently positive, stimulating experiences to promote healthy psychological development; most settings offered substandard care (Doherty et al., 2000; Goelman et al., 2000; NICHD Early Childhood Research Network, 2000a).

The Caregiving Concerns table on the following page lists signs of high-quality care that can be used in choosing a child-care setting for an infant or toddler, based on standards for **developmentally appropriate practice** devised by the U.S. National Association for the Education of Young Children. These standards specify program characteristics that meet the developmental and individual needs of young children as established by current research and consensus among experts. Caitlin, Grace, and Timmy are fortunate to be in a child-care home that meets these standards. Children from low-income and poverty-stricken families are especially likely to have inadequate child care (Pungello & Kurtz-Costes, 1999).

Child care in the United States and Canada is affected by a macrosystem of individualistic values and weak government regulation and funding. Furthermore, many parents think that their children's child-care experiences are better than they actually are (Helburn, 1995). Inability to identify good care means that many parents do not demand it. Yet nations that invest in child care have selected a highly cost-effective means of protecting children's well-being. Much like the programs we are about to consider, excellent child care can also serve as effective early intervention for children whose development is at risk.

Early Intervention for At-Risk Infants and Toddlers

Many studies indicate that children living in poverty are likely to show gradual declines in intelligence test scores and to achieve poorly when they reach school age (Brody, 1997b). These problems are largely due to stressful home environments that undermine children's ability to learn and increase the chances that they will remain poor throughout their lives. A variety of intervention programs have been developed to break this tragic cycle of poverty. Although most begin during the preschool years (we will discuss these in Chapter 7), a few start during infancy and continue through early childhood.

Some interventions are center-based; children attend an organized child-care or preschool program where they receive educational, nutritional, and health services, and child-rearing and other social-service supports are provided to parents as well. Other interventions are home-based. A skilled adult visits the home and works with parents, teaching them how to stimulate a very young child's development. In most programs, participating children score higher on mental tests by age 2

Signs of Developmentally Appropriate Infant and Toddler Child Care

PROGRAM CHARACTERISTIC	SIGNS OF QUALITY
Physical setting	Indoor environment is clean, in good repair, well lighted, and well ventilated. Fenced outdoor play space is available. Setting does not appear overcrowded when children are present.
Toys and equipment	Play materials are appropriate for infants and toddlers and stored on low shelves within easy reach. Cribs, high chairs, infant seats, and child-sized tables and chairs are available. Outdoor equipment includes small riding toys, swings, slide, and sandbox.
Caregiver–child ratio	In child-care centers, caregiver–child ratio is no greater than 1 to 3 for infants and 1 to 6 for toddlers. Group size (number of children in one room) is no greater than 6 infants with 2 caregivers and 12 toddlers with 2 caregivers. In child-care homes, caregiver is responsible for no more than 6 children; within this group, no more than 2 are infants and toddlers. Staffing is consistent, so infants and toddlers can form relationships with particular caregivers.
Daily activities	Daily schedule includes times for active play, quiet play, naps, snacks, and meals. It is flexible rather than rigid, to meet the needs of individual children. Atmosphere is warm and supportive, and children are never left unsupervised.
Interactions among adults and children	Caregivers respond promptly to infants' and toddlers' distress; hold, talk to, sing, and read to them; and interact with them in a contingent manner that respects the individual child's interests and tolerance for stimulation.
Caregiver qualifications	Caregiver has some training in child development, first aid, and safety.
Relationships with parents	Parents are welcome anytime. Caregivers talk frequently with parents about children's behavior and development.
Licensing and accreditation	Child-care setting, whether a center or a home, is licensed by the state or province. In the United States, accreditation by the National Academy of Early Childhood Programs or the National Association for Family Child Care is evidence of an especially high-quality program.

Source: Bredekamp & Copple, 1997; National Association for the Education of Young Children, 1998.

than do untreated controls. These gains persist as long as the program lasts and occasionally longer. The more intense the intervention (for example, full-day, year-round high-quality child care plus support services for parents), the greater children's cognitive and academic performance throughout childhood and adolescence (Ramey, Campbell, & Ramey, 1999).

The Carolina Abecedarian Project illustrates these positive outcomes. In the 1970s, more than one hundred 3-week- to 3-month-old infants from poverty-stricken families were randomly assigned to either a treatment group or a control group. Treatment infants were enrolled in full-time, year-round child care through the preschool years. There they received stimulation aimed at promoting motor, cognitive, language, and social skills and, after age 3, prereading and math concepts. At all ages, special emphasis was placed on rich, responsive adult–child verbal communication. All children received nutrition and health services; the primary difference between treatment and controls was the child-care experience.

As Figure 5.6 on page 164 shows, by 12 months of age, the IQs of the two groups diverged. Treatment children maintained their IQ advantage when last tested—at 21 years of age. In addition, throughout their years of schooling, treatment youths achieved considerably better in reading and math. These gains translated into greater high school graduation rates for the treatment group than for the control group (Campbell et al., 2001; Ramey & Ramey, 1999). While the children were in elementary school, the researchers conducted a second experiment to compare the impact of early and later intervention. From kindergarten through second grade, half the treatment group and half the control group were provided a resource teacher, who introduced educational activities addressing the child's specific learning needs into the home. School-age intervention had no impact on IQ. And although it enhanced children's academic achievement, the effects were much weaker than the impact of very early intervention (Campbell & Ramey, 1995).

Without some form of early intervention, many children born into economically disadvantaged families will not reach their potential. Recognition of this reality led the U.S. Congress to provide limited funding for intervention services directed at infants and toddlers who already have serious developmental problems or who are at risk for problems because of poverty. At present, available programs are not nearly enough to meet the need (Children's Defense Fund, 2002). Nevertheless, those that exist are a promising beginning.

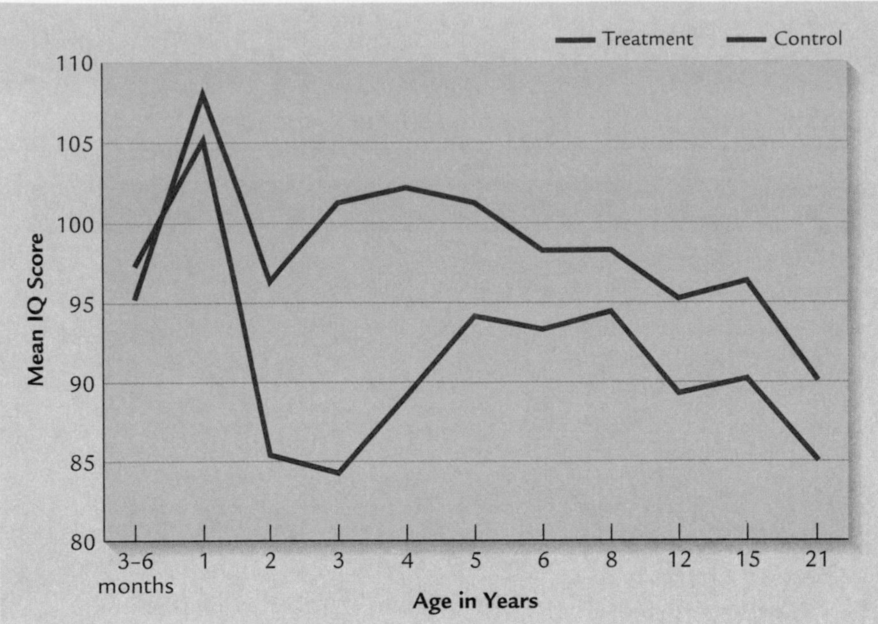

■ **FIGURE 5.6 IQ scores of treatment and control children from infancy to 21 years in the Carolina Abecedarian Project.** At 1 year, treatment children outperformed controls, an advantage consistently maintained through 21 years of age. The IQ scores of both groups declined gradually during the school years—a trend probably due to the damaging impact of poverty on mental development. (Adapted from Campbell et al., 2001.)

Ask Yourself

REVIEW

What probably accounts for the finding that habituation and recovery to visual stimuli predict later IQ better than does an infant mental test score?

APPLY

Fifteen-month-old Joey's developmental quotient (DQ) is 115. His mother wants to know exactly what this means and what she should do at home to support his mental development. How would you respond to her questions?

CONNECT

Using what you learned about brain development in Chapter 4, explain why intensive intervention for poverty-stricken children starting in the first 2 years has a greater long-term impact on IQ than intervention at a later age.

www.

Language Development

As perception and cognition improve during infancy, they pave the way for an extraordinary human achievement—language. On average, children say their first word at 12 months of age, with a range of about 8 to 18 months. Once words appear, language develops rapidly. Sometime between 1½ and 2 years, toddlers combine two words (Bloom, 1998). By age 6, they have a vocabulary of about 10,000 words, speak in elaborate sentences, and are skilled conversationalists.

How do children acquire language? Let's begin to address this question by examining several prominent theories, based on what we know about the beginnings of language in the first 2 years.

Three Theories of Language Development

In the 1950s, researchers did not take seriously the idea that very young children might be able to figure out important properties of the language they hear. As a result, the first two theories of how children acquire language were extreme views. One, *behaviorism*, regards language development as entirely due to environmental influences. The second, *nativism*, assumes that children are "prewired" to master the intricate rules of their language.

● **The Behaviorist Perspective.** Behaviorist B. F. Skinner (1957) proposed that language, like any other behavior, is acquired through *operant conditioning* (see Chapter 4, page 127). As the baby makes sounds, parents reinforce those that are most like words with smiles, hugs, and speech in return. For example, at 12 months, my older son, David, often babbled like this: "book-a-book-a-dook-a-dook-a-book-a-nook-a-book-aaa." One day while he babbled away, I held up his picture book and said, "Book!" Very soon, David was saying "book-aaa" in the presence of books.

Some behaviorists say children rely on *imitation* to rapidly acquire complex utterances, such as whole phrases and sentences (Moerk, 1992). And imitation can combine with reinforcement to promote language, as when a parent coaxes, "Say, 'I want a cookie,'" and delivers praise and a treat after the toddler responds, "Wanna cookie!"

Although reinforcement and imitation contribute to early language development, they are best viewed as supporting rather than fully explaining it. As Carolyn remarked one day, "It's amazing how creative Caitlin is with language. She combines words in ways she's never heard before, such as 'needle it' when she wants me to sew up her teddy bear and 'allgone outside' when she has to come in." Carolyn's observations are accurate: Young children create many novel utterances that are not reinforced by or copied from others.

● **The Nativist Perspective.** Linguist Noam Chomsky (1957) proposed a nativist theory that regards the young child's amazing language skill as etched into the structure of the human brain. Focusing on grammar, Chomsky reasoned that the rules of sentence organization are much too complex to be directly taught to or discovered by a young child. Instead, he argued, all children are born with a **language acquisition device (LAD),** an innate system that contains a set of rules common to all languages. It permits children, no matter which language they hear, to understand and speak in a rule-oriented fashion as soon as they pick up enough words.

Are children biologically primed to acquire language? Recall from Chapter 4 that newborn babies are remarkably sensitive to speech sounds and prefer to listen to the human voice. In addition, children the world over reach major language milestones in a similar sequence (Gleitman & Newport, 1996). And studies of isolated and abused children who experienced little human contact in childhood reveal lasting deficits in language, especially grammar and communication skills—evidence indicating that childhood is a *sensitive period* for language learning, although the precise boundary of that period is unclear (Curtiss, 1989). Taken together, these findings are consistent with Chomsky's idea of a biologically based language program.

At the same time, challenges to Chomsky's theory suggest that it, too, provides only a partial account of language development. First, researchers have had great difficulty identifying the single system of grammar that Chomsky believes underlies all languages (Maratsos, 1998; Tomasello, 1995). Second, children do not acquire language quite as quickly as nativist theory suggests. Their progress in mastering many sentence constructions is not immediate, but steady and gradual (Tager-Flusberg, 2001). This suggests that more learning and discovery are involved than Chomsky assumed.

© SOTOGRAPHS/LIAISON AGENCY

■ Infants are communicative beings from the start, as this interchange between a 3-month-old and her grandfather illustrates. How will this child accomplish the awesome task of becoming a fluent speaker of her native tongue within the next few years? Theorists disagree sharply on answers to this question.

● **The Interactionist Perspective.** In recent years, new ideas about language development have arisen, emphasizing *interactions* between inner capacities and environmental influences. Although several interactionist theories exist, all stress the social context of language learning. An active child, well endowed for acquiring language, observes and participates in social exchanges. From these experiences, children gradually discover the functions and regularities of language. According to the interactionist position, native capacity, a strong desire to interact with others, and a rich language and social environment combine to help children build a communicative system. And because genetic and environmental contributions vary across children, the interactionist perspective predicts individual differences in language learning (Bohannon & Bonvillian, 2001; Chapman, 2000).

Even among interactionists, debate continues over the precise nature of innate language abilities. Some theorists accept a modified view of Chomsky's position. They believe that children are primed to acquire language but that they form and refine hypotheses about its structure on the basis of experiences with language (Slobin, 1997). Others believe that children make sense of their complex language environments by applying powerful cognitive strategies rather than strategies specifically tuned to language (Bates, 1999; Tomasello & Brooks, 1999).

As we chart the course of early language growth, we will see a great deal of evidence that supports the interactionist position. But none of these theories has yet been fully tested. In reality, biology, cognition, and social experience may operate in different balances with respect to various aspects of language: pronunciation, vocabulary, grammar, and communication skills. Table 5.4 on page 166 provides an overview of early language milestones that we will take up in the next few sections.

Getting Ready to Talk

Before babies say their first word, they are preparing for language in many ways. They listen attentively to human speech and make speechlike sounds. And as adults, we can hardly help but respond.

● **Cooing and Babbling.** Around 2 months, babies begin to make vowel-like noises, called **cooing** because of their pleasant "oo" quality. Gradually, consonants are added, and around 4 months, **babbling** appears, in which infants repeat consonant–vowel combinations in long strings, such as "babababababa" or "nanananana."

The timing of early babbling seems to be due to maturation because babies everywhere start babbling at about the same age and produce a similar range of early sounds (Stoel-Gammon & Otomo, 1986). But for babbling to develop further, infants must be able to hear human speech. If a baby's hearing is impaired, these speechlike sounds are greatly delayed or, in the case of deaf infants, totally absent (Eilers & Oller, 1994; Oller, 2000).

As infants listen to spoken language, babbling expands to include a broader range of sounds. At around 7 months, it starts to include many sounds of mature spoken languages. And by 1 year, it contains the consonant–vowel and intonation

Table 5.4 Milestones of Language Development During the First Two Years

Approximate Age	Milestone
2 months	Infants coo, making pleasant vowel sounds.
4 months on	Infants babble, adding consonants to their cooing sounds and repeating syllables. By 7 months, babbling of hearing infants starts to include many sounds of mature spoken languages.
	Infants and parents establish joint attention, and parents often verbally label what the baby is looking at.
	Interaction between parents and baby includes turn-taking games, such as pat-a-cake and peekaboo. By 12 months, babies participate actively.
8–12 months	Babbling contains consonant–vowel and intonation patterns of the infant's language community.
	Infants begin using preverbal gestures, such as showing and pointing, to influence the behavior of others. Word comprehension first appears.
12 months	Toddlers say their first recognizable word.
18–24 months	Vocabulary expands from about 50 to 200 words.
20–26 months	Toddlers combine two words.

patterns of the infant's language community (Levitt & Utmann, 1992). Deaf infants exposed to sign language from birth babble with their hands in much the same way hearing infants do through speech (Petitto & Marentette, 1991). Furthermore, hearing babies of deaf, signing parents produce babblelike hand motions with the rhythmic patterns of natural language (Petitto et al., 2001). Infants' sensitivity to language rhythm, evident in both spoken and signed babbling, may help them discover and produce meaningful language units. And through babbling, babies seem to experiment with a great many sounds that can be blended into their first words.

● **Becoming a Communicator.** Besides responding to cooing and babbling, adults interact with infants in many other situations. Around 4 months, infants start to gaze in the same direction adults are looking, a skill that becomes more accurate between 12 and 15 months of age (Tomasello, 1999). Adults also follow the baby's line of vision and comment on what the infant sees, labeling the environment for the baby. Infants and toddlers who often experience this *joint attention* comprehend more language, produce meaningful gestures and words earlier, and show faster vocabulary development (Carpenter, Nagell, & Tomasello, 1998; Marcus et al., 1999).

Around 4 to 6 months, interaction between parent and baby begins to include *give-and-take,* as in turn-taking games, such as pat-a-cake and peekaboo. At first, the parent starts the game and the baby is an amused observer. Nevertheless, 4-month-olds are sensitive to the structure and timing of these interactions, smiling more to an organized than a disorganized peekaboo exchange (Rochat, Querido, & Striano, 1999). By 12 months, babies participate actively, trading roles with the parent. As they do so, they practice the turn-taking pattern of human conversation, a vital context for acquiring language and communication skills. Infants' play maturity and vocalizations during games predict advanced language progress between 1 and 2 years of age (Rome-Flanders & Cronk, 1995).

At the end of the first year, as infants become capable of intentional behavior, they use *preverbal gestures* to influence

■ This 15-month-old delights in playing peekaboo with her mother. As she participates, she practices the turn-taking pattern of human conversation.

© LAURA DWIGHT

■ This 14-month-old uses a preverbal gesture to attract her mother's attention to a fascinating sight. As the mother labels her daughter's pointing gesture, she promotes the transition to spoken language.

the behavior of others (Carpenter, Nagell, & Tomasello, 1998). For example, Caitlin held up a toy to show it and pointed to the cupboard when she wanted a cookie. Carolyn responded to her gestures and also labeled them ("Oh, you want a cookie!"). In this way, toddlers learn that using language leads to desired results. Soon they utter words along with their reaching and pointing gestures, the gestures recede, and spoken language is under way (Namy & Waxman, 1998).

First Words

In the middle of the first year, infants begin to understand word meanings. When 6-month-olds listened to the words "mommy" and "daddy" while looking at side-by-side videos of their parents, they looked longer at the video of the named parent (Tincoff & Jusczyk, 1999). First spoken words, around 1 year, build on the sensorimotor foundations Piaget described and on categories children form during their first 2 years. Usually they refer to important people ("Mama," "Dada"), objects that move ("car," "ball," "cat"), familiar actions ("bye-bye," "up," "more"), or outcomes of familiar actions ("dirty," "wet," "hot"). In their first 50 words, toddlers rarely name things that just *sit there,* like "table" or "vase" (Nelson, 1973).

Some early words are linked to specific cognitive achievements. For example, about the time toddlers master advanced object permanence problems, they use disappearance words, such as "all gone." And success and failure expressions, such as "There!" and "Uh-oh!", appear when toddlers can solve sensorimotor problems suddenly. According to one pair of researchers, "Children seem to be motivated to acquire words that are relevant to the particular cognitive problems they are working on at the moment" (Gopnik & Meltzoff, 1986, p. 1052).

Besides cognition, emotion influences early word learning. At first, when acquiring a new word for an object, person, or event, 1½-year-olds say it neutrally; they need to listen carefully to learn, and strong emotion diverts their attention. As words become better learned, toddlers integrate talking and expressing feelings (Bloom, 1998). "Shoe!" said one enthusiastic 22-month-old as her mother tied her shoelaces before an outing. At the end of the second year, children begin to label their emotions with words like "happy," "mad," and "sad"—a development we will consider further in Chapter 6.

When young children first learn words, they sometimes apply them too narrowly, an error called **underextension.** For example, at 16 months, Caitlin used "bear" to refer only to the worn and tattered bear that she carried around much of the day. A more common error is **overextension**—applying a word to a wider collection of objects and events than is appropriate. For example, Grace used "car" for buses, trains, trucks, and fire engines. Toddlers' overextensions reflect their sensitivity to categories. They apply a new word to a group of similar experiences, such as "car" to wheeled objects and "open" to opening a door, peeling fruit, and undoing shoelaces. This suggests that children sometimes overextend deliberately because they have difficulty recalling or have not acquired a suitable word (Bloom, 2000). As their vocabularies enlarge, overextensions disappear.

Overextensions illustrate another important feature of language development: the distinction between language *production* (the words children use) and language *comprehension* (the words children understand). Children overextend many more words in production than they do in comprehension. That is, a 2-year-old may refer to trucks, trains, and bikes as "car" but look at or point to these objects correctly when given their names (Naigles & Gelman, 1995). At all ages, comprehension develops ahead of production. This tells us that failure to say a word does not mean that toddlers do not understand it. If we rely only on what children say, we will underestimate their knowledge of language.

The Two-Word Utterance Phase

At first, toddlers add to their vocabularies slowly, at a rate of 1 to 3 words a month. Between 18 and 24 months, a spurt in vocabulary growth often takes place. As speed of identifying words in spoken sentences and memory and categorization improve, many children add 10 to 20 new words a week (Dapretto & Bjork, 2000; Fenson et al., 1994; Fernald, Swingley, & Pinto, 2001). When vocabulary approaches 200 words, toddlers start to combine two words, saying, for example, "Mommy shoe," "go car," and "more cookie." These two-word utterances are called **telegraphic speech** because, like a telegram, they leave out smaller and less important words. Children the world over use them to express an impressive variety of meanings.

Two-word speech is largely made up of simple formulas, such as "want + X" and "more + X," with many different words inserted in the X position. Although toddlers rarely make gross grammatical errors (such as saying "chair my" instead of "my chair"), they can be heard violating the rules. For example, at 20 months, Caitlin said "more hot" and "more read," combinations that are not acceptable in English grammar. The word-order regularities in toddlers' two-word utterances are usually copies of adult word pairings, as when the parent says, "That's *my book,*" or "How about *more sandwich*?" (Tomasello & Brooks, 1999). But it does not take long for children to figure out grammatical rules. As we will see in Chapter 7, the beginnings of grammar are in place by age 2½.

Individual and Cultural Differences

Each child's progress in acquiring language results from a complex blend of biological and environmental influences. For example, earlier we saw that Timmy's spoken language was delayed, in part because of Vanessa's tense, directive communication with him. But Timmy is also a boy, and many studies show that girls are ahead of boys in early vocabulary growth (Fenson et al., 1994). The most common biological explanation is girls' faster rate of physical maturation, believed to promote earlier development of the left cerebral hemisphere, where language is housed. But perhaps because of girls' slight language advantage, mothers also talk more to toddler-age girls than boys, so girls add vocabulary more quickly for both genetic and environmental reasons (Leaper, Anderson, & Sanders, 1998).

Besides the child's sex, personality makes a difference. Reserved, cautious toddlers often wait until they understand a great deal before trying to speak. When they finally do speak, their vocabularies grow rapidly (Nelson, 1973). In the week after her adoption, 16-month-old Grace spoke only a single Cambodian word. For the next 2 months, Grace listened to English conversation without speaking—a "silent period" typical of children beginning to acquire a second language (Saville-Troike, 1988). Around 18 months, words came quickly—first "Eli," then "doggie," "kitty," "Mama," "Dada," "book," "ball," "car," "cup," "clock," and "chicken," all within a single week.

Young children have unique styles of early language learning. Caitlin and Grace, like most toddlers, used a **referential style;** their early vocabularies consisted mainly of words that referred to objects. A smaller number of toddlers use an **expressive style;** compared with referential children, they produce many more pronouns and social formulas, such as "stop it," "thank you," and "I want it." These styles reflect early ideas about the functions of language. Grace, for example, thought words were for naming things. In contrast, expressive-style children believe words are for talking about people's feelings and needs. The vocabularies of referential-style children grow faster because all languages contain many more object labels than social phrases (Bates et al., 1994).

What accounts for a toddler's language style? Rapidly developing referential-style children often have an especially active interest in exploring objects. They also freely imitate their parents' eager naming of objects, and their parents imitate back—behaviors that support swift vocabulary growth by helping children remember new labels (Masur & Rodemaker, 1999). Expressive-style children tend to be highly sociable, and their parents more often use verbal routines ("How are you?" "It's no trouble") that support social relationships (Goldfield, 1987). The two language styles are also linked to culture. Whereas object words (nouns) are particularly common in the vocabularies of English-speaking toddlers, action words (verbs) are more numerous among Korean and Chinese toddlers. When caregivers' speech is examined in each culture, it reflects these differences (Choi & Gopnik, 1995; Tardif, Gelman, & Xu, 1999).

At what point should parents be concerned if their child does not talk or says very little? If a toddler's language is greatly delayed when compared with the norms in Table 5.4, then parents should consult the child's doctor or a speech and language therapist. Late babbling may be a sign of slow language development that can be prevented with early intervention (Oller et al., 1999). Some toddlers who do not follow simple directions or who, after age 2, have difficulty putting their thoughts into words may suffer from a hearing impairment or a language disorder that requires immediate treatment.

Supporting Early Language Development

There is little doubt that children are specially prepared for acquiring language, since no other species can develop as flexible and creative a capacity for communication as we can. Yet consistent with the interactionist view, a rich social environment builds on young children's natural readiness to speak their native tongue. The Caregiving Concerns table on the following page summarizes ways in which caregivers can consciously support early language learning. They also do so unconsciously—through a special style of speech.

Adults in many cultures speak to young children in **child-directed speech (CDS),** a form of communication made up of short sentences with high-pitched, exaggerated expression, clear pronunciation, distinct pauses between speech segments, and repetition of new words in a variety of contexts ("See the ball." "The ball bounced!") (Fernald et al., 1989; Kuhl, 2000).

■ This mother speaks to her baby in short, clearly pronounced sentences with high-pitched, exaggerated intonation. Adults in many countries use this form of language, called child-directed speech, with infants and toddlers. It eases the task of early language learning.

Supporting Early Language Learning

SUGGESTION	CONSEQUENCE
Respond to coos and babbles with speech sounds and words.	Encourages experimentation with sounds that can later be blended into first words. Provides experience with turn-taking pattern of human conversation.
Establish joint attention and comment on what child sees.	Predicts earlier onset of language and faster vocabulary development.
Play social games, such as pat-a-cake and peekaboo.	Provides experience with turn-taking pattern of human conversation. Permits pairing of words with actions they represent.
Engage toddlers in joint make-believe play.	Promotes all aspects of conversational dialogue.
Engage toddlers in frequent conversations.	Predicts faster early language development and academic competence during the school years.
Read to toddlers often, engaging them in dialogues about picture books.	Provides exposure to many aspects of language, including vocabulary, grammar, communicative conventions, and information about print and story structures.

Deaf parents use a similar style of communication when signing to their deaf babies (Masataka, 1996). CDS builds on several communicative strategies we have already considered: joint attention, turn-taking, and caregivers' sensitivity to children's preverbal gestures. Here is an example of Carolyn using CDS with 18-month-old Caitlin:

Caitlin: "Go car."

Carolyn: "Yes, time to go in the car. Where's your jacket?"

Caitlin: [*looks around, walks to the closet*] "Dacket!" [*pointing to her jacket*]

Carolyn: "There's that jacket! [*She helps Caitlin into the jacket.*] On it goes! Let's zip up. [*Zips up the jacket.*] Now, say bye-bye to Grace and Timmy."

Caitlin: "Bye-bye, G-ace."

Carolyn: "What about Timmy? Bye to Timmy?"

Caitlin: "Bye-bye, Te-te."

Carolyn: "Where's your bear?"

Caitlin: [*looks around*]

Carolyn: [*pointing*] "See? Go get the bear. By the sofa." [*Caitlin gets the bear.*]

From birth on, children prefer to listen to CDS over other kinds of adult talk, and by 5 months they are more emotionally responsive to it (Cooper & Aslin, 1994; Werker, Pegg, & McLeod, 1994). And parents constantly fine-tune it, adjusting the length and content of their utterances to fit their children's needs—adjustments that promote language comprehension and also permit toddlers to join in conversation (Murray, Johnson, & Peters, 1990).

Conversational give-and-take between parent and toddler is one of the best predictors of early language development and academic competence during the school years (Hart & Risley, 1995; Walker et al., 1994). Dialogues about picture books are especially effective. They expose children to great breadth of language and literacy knowledge, from vocabulary, grammar, and communication skills to information about print and story structures (Whitehurst & Lonigan, 1998).

Do social experiences that promote language development remind you of those that strengthen cognitive development in general? Notice how CDS and parent–child conversation create a *zone of proximal development* in which children's language expands. In contrast, impatience with and rejection of children's efforts to talk lead them to stop trying and result in immature language skills (Baumwell, Tamis-LeMonda, & Bornstein, 1997). In the next chapter, we will see that sensitivity to children's needs and capacities supports their emotional and social development as well.

Ask Yourself

REVIEW

Why is the interactionist perspective attractive to many investigators of language development? Cite evidence that supports it.

APPLY

Prepare a list of research-based recommendations for how to support language development during the first 2 years.

CONNECT

Cognition and language are interrelated. List examples of how cognition fosters language development. Next, list examples of how language fosters cognitive development.

www.

Summary

Piaget's Cognitive-Developmental Theory

According to Piaget, how do schemes change over the course of development?

- In Piaget's theory, by acting directly on the environment, children move through four stages in which psychological structures, or **schemes,** achieve a better fit with external reality.
- Schemes change in two ways. The first is through **adaptation,** which is made up of two complementary activities—**assimilation** and **accommodation.** The second is through **organization,** the internal rearrangement of schemes into a strongly interconnected cognitive system.

Describe the major cognitive achievements of the sensorimotor stage.

- Piaget's **sensorimotor stage** is divided into six substages. Through the **circular reaction,** the newborn baby's reflexes are gradually transformed into the more flexible action patterns of the older infant. During Substage 4, infants develop **intentional,** or **goal-directed, behavior** and begin to understand **object permanence.** By Substage 6, toddlers become capable of **mental representation,** as shown by sudden solutions to sensorimotor problems, mastery of object permanence problems involving invisible displacement, **deferred imitation,** and **make-believe play.**

What does recent research reveal about the accuracy of Piaget's sensorimotor stage?

- Many studies suggest that infants display certain understandings earlier than Piaget believed. Some awareness of object permanence, as revealed by the **violation-of-expectation method,** may be evident in the first few months. In addition, young infants display deferred imitation and analogical problem solving, which suggests that they are capable of mental representation in the first year.
- Today, researchers believe that newborns have more built-in equipment for making sense of their world than Piaget assumed, although they disagree on how much initial understanding infants have. According to the **core knowledge perspective,** infants begin life with core domains of thought that support early, rapid cognitive development. Overall, however, findings on early, ready-made knowledge are mixed. But there is broad agreement that many cognitive changes of infancy are continuous rather than stagelike and that various aspects of cognition develop unevenly rather than in an integrated fashion.

Information Processing

Describe the information-processing view of cognitive development and the general structure of the information-processing system.

- Information-processing researchers regard development as gradual and continuous and study many aspects of thinking. They want to know exactly what individuals of different ages do when faced with a task or problem.
- Most information-processing researchers assume that we hold information in three parts of the system, where **mental strategies** operate on it so that it can be retained and used efficiently. These parts are the **sensory register; working,** or **short-term, memory;** and **long-term memory.** To manage working memory's complex activities, the **central executive** directs the flow of information.

What changes in attention, memory, and categorization take place during the first 2 years?

- With age, infants attend to more aspects of the environment and take information in more quickly. In the second year, attention to novelty declines and sustained attention improves, especially during play with toys.
- As infants get older, they remember experiences longer. Young infants are capable of **recognition** memory, and by the end of the first year, they can **recall** past events. By the end of toddlerhood, recall for people, places, and objects is excellent. Both biology and social experience probably contribute to the emergence of **autobiographical memory.**
- During the first year, infants group stimuli into increasingly complex categories, and categorization shifts from a perceptual to a conceptual basis. By the second year, children become active categorizers, spontaneously sorting objects during play.

Describe contributions and limitations of the information-processing approach to our understanding of early cognitive development.

- Information-processing findings challenge Piaget's view of babies as purely sensorimotor beings who cannot mentally represent experiences. However, information processing has not yet provided a broad, comprehensive theory of children's thinking.

The Social Context of Early Cognitive Development

How does Vygotsky's concept of the zone of proximal development expand our understanding of early cognitive development?

- According to Vygotsky's sociocultural theory, complex mental activities originate in social interaction. Through the support and guidance of more skilled partners, infants master tasks within the **zone of proximal development**—that is, tasks just ahead of their current capacities. As early as the first 2 years, cultural variations in social experiences affect mental strategies.

Individual Differences in Early Mental Development

Describe the mental testing approach, the meaning of intelligence test scores, and the extent to which infant tests predict later performance.

- The mental testing approach measures intellectual development in an effort to predict future performance. **Intelligence quotients,** or **IQs,** compare an individual's test performance with that of a large, representative sample of same-age individuals.
- Infant tests consist largely of perceptual and motor responses; they predict later intelligence poorly. As a result, scores on infant tests are called **developmental quotients,** or **DQs,** rather than IQs. Speed of habituation and recovery to visual stimuli, which tap basic cognitive processes, are better predictors of future performance.

Discuss environmental influences on early mental development, including home, child care, and early intervention for at-risk infants and toddlers.

- Research with the **Home Observation for Measurement of the Environment (HOME)** shows that an organized, stimulating home environment and parental encouragement, involvement, and affection repeatedly predict early mental test scores. Although the HOME–IQ relationship is partly due to heredity, family living conditions do affect mental development.
- Quality of infant and toddler child care has a major impact on mental development. Standards for **developmentally appropriate practice** specify program characteristics that meet the developmental needs of young children. Intensive early intervention can prevent the gradual declines in intelligence and the poor academic performance of many poverty-stricken children. Findings of the Carolina Abecedarian Project reveal an IQ advantage still evident at 21 years of age and better school achievement, which translated into greater high school graduation rates for the treatment group than the control group.

Language Development

Describe three theories of language development, and indicate how much emphasis each places on innate abilities and environmental influences.

- Three theories provide different accounts of how young children develop language. According to the behaviorist perspective, parents train children in language skills through operant conditioning and imitation. In contrast, Chomsky's nativist view regards children as naturally endowed with a **language acquisition device (LAD)**. New interactionist theories suggest that innate abilities and a rich linguistic and social environment combine to promote language development.

Describe major milestones of language development in the first 2 years, individual differences, and ways adults can support infants' and toddlers' emerging capacities.

- During the first year, much preparation for language takes place. Infants begin **cooing** at 2 months and **babbling** at around 4 months. Adults encourage language progress by responding to infants' coos and babbles, establishing joint attention and labeling what babies see, playing turn-taking games, and acknowledging infants' preverbal gestures.
- In the middle of the first year, infants begin to understand word meanings. Around 12 months, toddlers say their first word. Young children often make errors of **underextension** and **overextension.** Between 18 and 24 months, a spurt in vocabulary growth often occurs, and two-word utterances called **telegraphic speech** appear. At all ages, language comprehension is ahead of production.
- Individual differences in early language development exist. Girls show faster progress than boys, and reserved, cautious toddlers may wait for a period of time before trying to speak. Most toddlers use a **referential style** of language learning, in which early words consist largely of names for objects. A few use an **expressive style,** in which pronouns and social formulas are common and vocabulary grows more slowly.
- Adults in many cultures speak to young children in **child-directed speech (CDS),** a simplified form of language that is well suited to their learning needs. Conversational give-and-take between parent and toddler is one of the best predictors of early language development and academic competence during the school years.

Important Terms and Concepts

accommodation (p. 144)
adaptation (p. 144)
assimilation (p. 144)
autobiographical memory (p. 156)
babbling (p. 165)
central executive (p. 153)
child-directed speech (CDS) (p. 168)
circular reaction (p. 145)
cooing (p. 165)
core knowledge perspective (p. 150)
deferred imitation (p. 147)
developmental quotient, or DQ (p. 160)
developmentally appropriate practice (p. 162)
expressive style of language learning (p. 168)
Home Observation for Measurement of the Environment (HOME) (p. 161)
intelligence quotient, or IQ (p. 160)
intentional, or goal-directed, behavior (p. 146)
language acquisition device (LAD) (p. 165)
long-term memory (p. 153)
make-believe play (p. 147)
mental representation (p. 147)
mental strategies (p. 153)
object permanence (p. 146)
organization (p. 145)
overextension (p. 167)
recall (p. 155)
recognition (p. 155)
referential style of language learning (p. 168)
scheme (p. 144)
sensorimotor stage (p. 144)
sensory register (p. 153)
telegraphic speech (p. 167)
underextension (p. 167)
violation-of-expectation method (p. 147)
working, or short-term, memory (p. 153)
zone of proximal development (p. 158)

FYI For Further Information and Help

Consult the Companion Website for *Development Through the Lifespan, Third Edition,* (www.ablongman.com/berk) where you will find the following resources for this chapter:

- **Chapter Objectives**
- **Flashcards** for studying important terms and concepts
- **Annotated Weblinks** to guide you in further research
- **Ask Yourself** questions, which you can answer and then check against a sample response
- **Suggested Readings**
- **Practice Test** with immediate scoring and feedback

6 Emotional and Social Development in Infancy and Toddlerhood

Erikson's Theory of Infant and Toddler Personality
Basic Trust versus Mistrust • Autonomy versus Shame and Doubt

Emotional Development
Development of Some Basic Emotions • Understanding and Responding to the Emotions of Others • Emergence of Self-Conscious Emotions • Beginnings of Emotional Self-Regulation

- A Lifespan Vista: Maternal Depression and Children's Development

Temperament and Development
The Structure of Temperament • Measuring Temperament • Stability of Temperament • Genetic Influences • Environmental Influences • Temperament and Child Rearing: The Goodness-of-Fit Model

- Biology & Environment: Biological Basis of Shyness and Sociability

Development of Attachment
Ethological Theory of Attachment • Measuring the Security of Attachment • Stability of Attachment and Cultural Variations • Factors That Affect Attachment Security • Multiple Attachments • Attachment and Later Development

- Social Issues: Does Child Care in Infancy Threaten Attachment Security and Later Adjustment?

Self-Development During the First Two Years
Self-Awareness • Categorizing the Self • Emergence of Self-Control

The mutual, ecstatic joy expressed by this mother and infant suggests that they have formed a deeply affectionate bond. The baby's sense of trust in his caregivers supports all aspects of early development.

As Caitlin reached 8 months of age, her parents noticed that she had become more fearful. One evening, when Carolyn and David left her with a babysitter, she wailed when they headed for the door—an experience she had accepted easily a few weeks earlier. Caitlin and Timmy's caregiver Ginette also observed an increasing wariness of strangers. When she turned to go to another room, both babies dropped their play to crawl after her. And a knock at the door from the mail carrier prompted them to cling to Ginette's legs and reach out to be picked up.

At the same time, each baby seemed more willful. An object removed from the hand at 5 months produced little response, but at 8 months Timmy resisted when his mother Vanessa took away a table knife he had managed to reach. He burst into angry screams and could not be consoled by the toys she offered in its place.

Monica and Kevin knew little about Grace's development during her first year, except that she had been deeply loved by her destitute, homeless mother. Separating from her, followed by a long journey to an unfamiliar home, left Grace in shock. At first she was extremely sad, turning away when Monica or Kevin picked her up. She did not smile for over a week.

But as Grace's new parents held her close, spoke gently, and satisfied her craving for food, Grace returned their affection. Two weeks after her arrival, her despondency gave way to a sunny, easygoing disposition. She burst into a wide grin, reached out at the sight of Monica and Kevin, and laughed at her brother Eli's funny faces. As her second birthday approached, she pointed to herself, exclaiming "Gwace!" and laid claim to treasured possessions. "Gwace's chicken!" she would announce at mealtimes, sucking the marrow from the drumstick, a practice she brought with her from Cambodia.

Taken together, Caitlin's, Timmy's, and Grace's reactions reflect two related aspects of personality development during the first 2 years: close ties to others and a sense of self. Our discussion begins with Erikson's psychosocial theory, which provides an overview of personality development during infancy and toddlerhood. Then we chart the course of emotional development. As we do so, we will discover why fear and anger became more apparent in Caitlin's and Timmy's range of emotions by the end of the first year. Our attention then turns to individual differences in temperament. We will examine biological and environmental contributions to these differences and their consequences for future development.

Next, we take up attachment to the caregiver, the child's first affectional tie. We will see how the feelings of security that grow out of this important bond provide a vital source of support for the child's sense of independence and expanding social relationships.

Finally, we focus on early self-development. By the end of toddlerhood, Grace recognized herself in mirrors and photographs, labeled herself as a girl, and showed the beginnings of self-control. "Don't touch!" she instructed herself one day as she resisted the desire to pull a lamp cord out of its socket. Cognitive advances combine with social experiences to produce these changes during the second year.

Erikson's Theory of Infant and Toddler Personality

Our discussion of major theories in Chapter 1 revealed that psychoanalytic theory is no longer in the mainstream of human development research. But one of its lasting contributions is its ability to capture the essence of personality development during each phase of life. Recall that Sigmund Freud, founder of the psychoanalytic movement, believed that psychological health and maladjustment could be traced to the early years—in particular, to the quality of the child's relationships with parents. Although Freud's limited concern with the channeling of instincts and his neglect of important experiences beyond infancy and early childhood came to be heavily criticized, the basic outlines of his theory were accepted and elaborated in several subsequent theories of personality development. The leader of these neo-Freudian perspectives is Erik Erikson's *psychosocial theory,* also introduced in Chapter 1.

Basic Trust versus Mistrust

Freud called the first year the *oral stage* and regarded gratification of the infant's need for food and oral stimulation as vital. Erikson accepted Freud's emphasis on feeding, but he expanded and enriched Freud's view. A healthy outcome during infancy, Erikson believed, does not depend on the *amount* of food or oral stimulation offered but rather on the *quality* of the caregiver's behavior. A mother who supports her baby's development relieves discomfort promptly and sensitively. For example, she holds the infant gently during feedings, patiently waits until the baby has had enough milk, and weans when the infant shows less interest in the breast or bottle.

Erikson recognized that no parent can be perfectly in tune with the baby's needs. Many factors affect her responsiveness—feelings of personal happiness, current life conditions (for example, additional young children in the family), and culturally valued child-rearing practices. But when the *balance of care* is sympathetic and loving, the psychological conflict of the first year—**basic trust versus mistrust**—is resolved on the positive side. The trusting infant expects the world to be good and gratifying, so he feels confident about venturing out and exploring it. The mistrustful baby cannot count on the kindness and compassion of others, so she protects herself by withdrawing from people and things around her.

Autonomy versus Shame and Doubt

In the second year, during Freud's *anal stage,* instinctual energies shift to the anal region of the body. Freud viewed toilet training, in which children must bring their anal impulses

■ According to Erikson, basic trust grows out of the quality of the early caregiving relationship. A parent who relieves the baby's discomfort promptly and holds the baby tenderly, during feeding and at other times, promotes basic trust—the feeling that the world is good and gratifying.

in line with social requirements, as crucial for personality development. Erikson agreed that the parent's manner of toilet training is essential for psychological health. But he viewed it as only one of many important experiences for newly walking, talking toddlers. Their familiar refrains—"No!" and "Do it myself!"—reveal that they have entered a period of budding selfhood. Toddlers want to decide for themselves—not just in toileting but in other situations as well. The great conflict of toddlerhood, **autonomy versus shame and doubt,** is resolved favorably when parents provide young children with suitable guidance and reasonable choices. A self-confident, secure 2-year-old has been encouraged not just to use the toilet but to eat with a spoon and to help pick up his toys. His parents do not criticize or attack him when he fails at these new skills. And they meet his assertions of independence with tolerance and understanding. For example, they grant him an extra 5 minutes to finish his play before leaving for the grocery store and wait patiently while he tries to zip his jacket.

According to Erikson, the parent who is over- or under-controlling in toileting is likely to be so in other aspects of the toddler's life. The outcome is a child who feels forced and shamed and doubts his ability to control his impulses and act competently on his own.

In sum, basic trust and autonomy grow out of warm, sensitive parenting and reasonable expectations for impulse control starting in the second year. If children emerge from the first few years without sufficient trust in caregivers and without a healthy sense of individuality, the seeds are sown for adjustment problems. Adults who have difficulty establishing intimate ties, who are overly dependent on a loved one, or who continually doubt their own ability to meet new challenges may not have fully mastered the tasks of trust and autonomy during infancy and toddlerhood.

Emotional Development

In the previous chapter, we considered babies' increasingly effective schemes for controlling the environment and ways that adults support cognitive and language development. Now let's focus on another aspect of infant and caregiver behavior: the exchange of emotions. Observe several infants and toddlers, noting the emotions each displays, the cues you rely on to interpret the baby's emotional state, and how caregivers respond. Researchers have conducted many such observations to find out how babies communicate their emotions and interpret those of others. They have discovered that emotions play a powerful role in organizing the attainments that Erikson regarded as so important: social relationships, exploration of the environment, and discovery of the self (Frijda, 2000; Izard, 1991; Saarni, Mumme, & Campos, 1998).

Because infants cannot describe their feelings, determining exactly which emotions they are experiencing is a challenge. Although vocalizations and body movements provide some information, facial expressions offer the most reliable cues. Cross-cultural evidence reveals that people around the world associate photographs of different facial expressions with emotions in the same way (Ekman & Friesen, 1972). These findings, which suggest that emotional expressions are built-in social signals, inspired researchers to analyze infants' facial patterns to determine the range of emotions they display at different ages.

Development of Some Basic Emotions

Basic emotions are universal in humans and other primates, have a long evolutionary history of promoting survival, and can be directly inferred from facial expressions. They include happiness, interest, surprise, fear, anger, sadness, and disgust. Do infants come into the world with the ability to express basic emotions? Although signs of some emotions are present, babies' earliest emotional life consists of little more than two global arousal states: attraction to pleasant stimulation and withdrawal from unpleasant stimulation. Over time, emotions become clear, well-organized signals (Camras, 1992; Fox, 1991).

Around 6 months, face, voice, and posture form organized patterns that vary meaningfully with environmental events. For example, Caitlin typically responded to her parents' playful interaction with a joyful face, pleasant cooing, and a relaxed posture, as if to say, "This is fun!" In contrast, an unresponsive parent often evokes a sad face, fussy vocalizations, and a drooping body (sending the message, "I'm despondent") or an angry face, crying, and "pick-me-up" gestures (as if to say, "Change this unpleasant event!") (Weinberg & Tronick, 1994; Yale et al., 1999). If parental depressive signals continue, they can profoundly disrupt emotional and social development (see the Lifespan Vista box on the following page). In sum, by the middle of the first year, emotional expressions are well organized and specific—and therefore tell us a great deal about the infant's internal state.

A Lifespan Vista

Maternal Depression and Children's Development

Approximately 8 to 10 percent of women experience chronic depression—mild to severe feelings of sadness and withdrawal that continue for months or years. Often the beginnings of this emotional state cannot be pinpointed; it simply becomes part of the person's daily life. In other instances, it emerges or strengthens after childbirth but fails to subside as the new mother adjusts to hormonal changes in her body and gains confidence in caring for her baby. Stella experienced this type, which is called *postpartum depression.* Although genetic makeup increases the risk of depressive illness, Stella's case shows that social and cultural factors are also involved (Swendsen & Mazure, 2000).

During Stella's pregnancy, her husband Kyle's lack of interest in the baby caused her to worry that having a child might be a mistake. Shortly after Lucy was born, Stella's mood plunged. She was anxious and weepy, overwhelmed by Lucy's needs, and angry that she no longer had control over her own schedule. When Stella approached Kyle about her own fatigue and his unwillingness to help with the baby, he snapped that she overreacted to every move he made. Stella's friends, who did not have children, stopped by once to see Lucy and did not call again.

Stella's depressed mood quickly affected her baby. In the weeks after birth, infants of depressed mothers sleep poorly, are less attentive to their surroundings, and have elevated stress hormone levels (Field, 1998). The more extreme the depression and the greater the number of stressors in a mother's life (such as marital discord, little or no social support, and poverty), the more the parent–child relationship suffers (Goodman et al., 1993). Stella, for example, rarely smiled and talked to Lucy, who responded to her mother's sad, vacant gaze by turning away, crying, and often looking sad or angry herself (Campbell, Cohn, & Meyers, 1995; Murray & Cooper, 1997). Each time this happened, Stella felt guilty and inadequate, and her depression deepened. By 6 months of age, Lucy showed emotional symptoms common in babies of depressed mothers—a negative, irritable mood and attachment difficulties (Martins & Gaffan, 2000).

When maternal depression persists, the parent–child relationship worsens. Depressed mothers view their infants more negatively than do independent observers (Hart, Field, & Roitfarb, 1999). And they use inconsistent discipline—sometimes lax, at other times too forceful. As we will see in later chapters, children who experience these maladaptive parenting practices often have serious adjustment problems. To avoid their parents' insensitivity, they sometimes withdraw into a depressive mood themselves. Or they mimic their parents' anger and become impulsive and antisocial (Conger, Patterson, & Ge, 1995; Murray et al., 1999).

Over time, the parenting just described leads children to develop a negative world view—one in which they lack confidence in themselves and perceive their parents and other people as threatening. Children who constantly feel in danger are likely to become overly aroused in stressful situations, easily losing control in the face of cognitive and social challenges (Cummings & Davies, 1994). Although children of depressed parents may inherit a tendency toward emotional and behavior problems, quality of parenting is a major factor in their adjustment.

Early treatment of maternal depression is vital to prevent the disorder from interfering with the parent–child relationship and harming children. Stella described her tearfulness, fatigue, and inability to comfort Lucy to her doctor. He referred her to a program for depressed mothers and their babies. A counselor worked with the family, helping Stella and Kyle with their marital problems and encouraging them to be more sensitive and patient with Lucy. At times, antidepressant medication is prescribed. In most cases of postpartum depression, short-term treatment is successful (Steinberg & Bellavance, 1999). When depressed mothers do not respond easily to treatment, a warm relationship with the father or another caregiver can safeguard children's development.

© LAURA DWIGHT/PHOTOEDIT

■ Depression disrupts parents' capacity to engage with children. This infant tries hard to get his despondent mother to react. If her unresponsiveness continues, the baby is likely to turn away, cry, and become negative and irritable. Over time, this disruption in the parent–child relationship leads to serious emotional and behavior problems.

Table 6.1 Milestones of Emotional Development During the First Two Years

Approximate Age	Milestone
Birth	Infants' emotions consist largely of two global arousal states: attraction to pleasant stimulation and withdrawal from unpleasant stimulation.
2–3 months	Infants engage in social smiling and respond in kind to adults' facial expressions.
3–4 months	Infants begin to laugh at very active stimuli.
6–8 months	Expressions of basic emotions are well organized and vary meaningfully with environmental events. Infants start to become angry more often and in a wider range of situations. Fear, especially stranger anxiety, begins to rise. Attachment to familiar caregivers is clearly evident, and separation anxiety appears. Infants use familiar caregivers as a secure base for exploration.
8–12 months	Infants perceive facial expressions as organized patterns, and meaningful understanding of them improves. Social referencing appears. Infants laugh at subtle elements of surprise.
18–24 months	Self-conscious emotions of shame, embarrassment, guilt, and pride appear. A vocabulary for talking about feelings develops rapidly, and emotional self-regulation improves. Toddlers appreciate that others' emotional reactions may differ from their own. First signs of empathy appear.

Three basic emotions—happiness, anger, and fear—have received the most research attention. Refer to Table 6.1 for an overview of changes in these emotions as well as others we will take up in this chapter.

● **Happiness.** Happiness—first in terms of blissful smiles and later through exuberant laughter—contributes to many aspects of development. Infants smile and laugh when they achieve new skills, expressing their delight in motor and cognitive mastery. The smile also encourages caregivers to be affectionate and stimulating, so the baby smiles even more. Happiness binds parent and baby into a warm, supportive relationship and fosters the infant's developing competencies.

During the early weeks, newborn babies smile when full, during sleep, and in response to gentle touches and sounds, such as stroking the skin, rocking, and the mother's soft, high-pitched voice. By the end of the first month, infants start to smile at interesting sights, but these must be dynamic and eye-catching, such as a bright object jumping suddenly across the baby's field of vision. Between 6 and 10 weeks, the human face evokes a broad grin called the **social smile** (Sroufe & Waters, 1976). These changes in smiling parallel the development of infant perceptual capacities—in particular, babies' increasing sensitivity to visual patterns, including the human face (see Chapter 4).

Laughter, which first occurs around 3 to 4 months, reflects faster processing of information than does smiling. But like smiling, the first laughs occur in response to very active stimuli, such as the parent saying playfully, "I'm gonna get you!" and kissing the baby's tummy. As infants understand more about their world, they laugh at events with subtler elements of surprise. At 10 months, Timmy chuckled as Vanessa played a silent game of peekaboo. At 1 year, he laughed heartily as she crawled on all fours and then walked like a penguin (Sroufe & Wunsch, 1972).

Around the middle of the first year, infants smile and laugh more when interacting with familiar people, a preference that strengthens the parent–child bond. Like adults, 10- to 12-month-olds have several smiles, which vary with context. They show a broad, "cheek-raised" smile in response to a parent's greeting; a reserved, muted smile in response to a friendly stranger; and a "mouth-open" smile during stimulating play (Dickson, Fogel, & Messinger, 1998).

● **Anger and Fear.** Newborn babies respond with generalized distress to a variety of unpleasant experiences, including hunger, painful medical procedures, changes in body temperature, and too much or too little stimulation. From 4 to 6 months into the second year, angry expressions increase in frequency and intensity. Older babies also show anger in a wider range of situations—for example, when an object is taken away, the caregiver leaves for a brief time, or they are put down for a nap (Camras et al., 1992; Stenberg & Campos, 1990).

Like anger, fear rises during the second half of the first year. Older infants often hesitate before playing with a new toy, and newly crawling infants soon show fear of heights (see Chapter 4). But the most frequent expression of fear is to unfamiliar adults, a response called **stranger anxiety.** Many infants and toddlers are quite wary of strangers, although the reaction does not always occur. It depends on several factors: the infant's temperament (some babies are generally more fearful), past experiences with strangers, and the current situation (Thompson & Limber, 1991). When an unfamiliar adult picks up the infant in a new setting, stranger anxiety is likely. But if the adult sits still while the baby moves around and a parent is nearby, infants often show positive and curious behavior (Horner, 1980). The stranger's style of interaction—expressing warmth, holding out an attractive toy, playing a familiar game, and approaching slowly rather than abruptly—reduces the baby's fear.

■ On an outing, this 1-year-old shows stranger anxiety, which increases during the second half of the first year. As infants move on their own, this rise in fear has adaptive value, increasing the chances that they will remain close to the parent and be protected from danger.

Why do angry and fearful reactions increase with age? Researchers believe that these emotions have survival value as babies begin to move on their own. Older infants can use the energy mobilized by anger to defend themselves or overcome obstacles (Izard & Ackerman, 2000). And fear keeps babies' enthusiasm for exploration in check, increasing the likelihood that they will remain close to the parent and be wary of unfamiliar people and objects. Anger and fear are also strong social signals that motivate caregivers to comfort a suffering infant.

Cognitive development plays an important role in infants' angry and fearful reactions, just as it does in expressions of happiness. Between 8 and 12 months, when (as Piaget pointed out) babies become capable of intentional behavior, they have a better understanding of the cause of their frustrations and know whom to get angry at. In the case of fear, improved memory permits older infants to distinguish familiar from unfamiliar events better than before.

Finally, culture can modify these emotions through infant-rearing practices. The maternal death rate is high among the Efe hunters and gatherers of Zaire, Africa. To ensure infant survival, a collective caregiving system exists in which Efe babies are passed from one adult to another. Consequently, Efe infants show little stranger anxiety (Tronick, Morelli, & Ivey, 1992). In contrast, in Israeli kibbutzim (cooperative agricultural settlements), living in an isolated community subject to terrorist attacks has led to widespread wariness of strangers. By the end of the first year, when infants look to others for cues about how to respond emotionally, kibbutz babies display far greater stranger anxiety than their city-reared counterparts (Saarni, Mumme, & Campos, 1998).

Understanding and Responding to the Emotions of Others

Infants' emotional expressions are closely tied to their ability to interpret the emotional cues of others. Already we have seen that within the first few months, babies match the feeling tone of the caregiver in face-to-face communication. Early on, infants detect others' emotions through a fairly automatic process of *emotional contagion,* just as we tend to feel happy or sad when we sense these emotions in others.

Between 7 and 10 months, infants perceive facial expressions as organized patterns, and they can match the emotion in a voice with the appropriate face of a speaking person (see Chapter 4). Responding to emotional expressions as organized wholes indicates that these signals have become meaningful to babies. As skill at detecting what others are looking at and reacting to improves, infants realize that an emotional expression not only has meaning but is also a meaningful reaction to a specific object or event (Moses et al., 2001; Tomasello, 1999a).

Once these understandings are in place, infants engage in **social referencing,** in which they actively seek emotional information from a trusted person in an uncertain situation. Many studies show that the caregiver's emotional expression (happy, angry, or fearful) influences whether a 1-year-old will be wary of strangers, play with an unfamiliar toy, or cross the deep side of the visual cliff (Repacholi, 1998; Sorce et al., 1985; Striano & Rochat, 2000).

Social referencing gives infants and toddlers a powerful means for learning. By responding to caregivers' emotional messages, they can avoid harmful situations, such as a shock from an electric outlet or a fall down a steep staircase, without first experiencing their unpleasant consequences. And parents can capitalize on social referencing to teach their young children, whose capacity to explore is rapidly expanding, how to react to a great many novel events. Social referencing also permits toddlers to compare their own assessments of events with those of others. By the middle of the second year, they appreciate that others' emotional reactions may differ from their own (Repacholi & Gopnik, 1997).

In sum, with the transition from infancy to toddlerhood, children move beyond simply reacting to the emotional messages of others. They use those signals to find out about others' internal states and preferences and to guide their own actions.

© CELIA ROBERTS/EARTH IMAGES

■ Self-conscious emotions appear at the end of the second year. This Guatemalan 2-year-old undoubtedly feels a sense of pride as she helps care for her elderly grandmother—an activity highly valued in her culture.

Emergence of Self-Conscious Emotions

Besides basic emotions, humans are capable of a second, higher-order set of feelings, including shame, embarrassment, guilt, envy, and pride. These are called **self-conscious emotions** because each involves injury to or enhancement of our sense of self. For example, when we are ashamed or embarrassed, we feel negatively about our behavior, and we want to retreat so others will no longer notice our failings. In contrast, pride reflects delight in the self's achievements, and we are inclined to tell others what we have accomplished (Saarni, Mumme, & Campos, 1998).

Self-conscious emotions appear in the middle of the second year, as the sense of self emerges. Shame and embarrassment can be seen as 18- to 24-month-olds lower their eyes, hang their heads, and hide their faces with their hands. Guilt-like reactions are also evident. One 22-month-old returned a toy she had grabbed and patted her upset playmate. Pride emerges around this time, and envy is present by age 3 (Barrett, 1998; Lewis et al., 1989).

Besides self-awareness, self-conscious emotions require an additional ingredient: adult instruction in *when* to feel proud, ashamed, or guilty. Parents begin this tutoring early when they say, "My, look at how far you can throw that ball!" or, "You should feel ashamed for grabbing that toy!"

As these comments indicate, self-conscious emotions play important roles in children's achievement-related and moral behaviors. The situations in which adults encourage these feelings vary from culture to culture. In most of the United States, children are taught to feel pride about personal achievement—throwing a ball the farthest, winning a game, and (later on) getting good grades. Among the Zuni Indians, shame and embarrassment are responses to purely individual success, whereas pride is evoked by generosity, helpfulness, and sharing (Benedict, 1934). In Japan, lack of concern for others—a parent, a teacher, or an employer—is cause for intense shame (Lewis, 1992).

Beginnings of Emotional Self-Regulation

Besides expressing a wider range of emotions, infants and toddlers begin to manage their emotional experiences. **Emotional self-regulation** refers to the strategies we use to adjust our emotional state to a comfortable level of intensity so we can accomplish our goals (Eisenberg et al., 1995; Thompson, 1994). If you drank a cup of coffee to wake up this morning, reminded yourself that an anxiety-provoking event would be over soon, or decided not to see a scary horror film, you were engaging in emotional self-regulation. A good start regulating emotion during the first 2 years contributes greatly to autonomy and mastery of cognitive and social skills (Crockenberg & Leerkes, 2000).

In the early months of life, infants have only a limited capacity to regulate their emotional states. Although they can turn away from unpleasant stimulation and can mouth and suck when their feelings get too intense, they are easily overwhelmed. They depend on the soothing interventions of caregivers—lifting the distressed baby to the shoulder, rocking, and talking softly.

Rapid development of the cerebral cortex increases the baby's tolerance for stimulation. Between 2 and 4 months, caregivers build on this capacity by initiating face-to-face play and attention to objects. In these interactions, parents arouse pleasure in the baby while adjusting the pace of their own behavior so the infant does not become overwhelmed and distressed. As a result, the baby's tolerance for stimulation increases further (Field, 1994). By 4 months, the ability to shift attention helps infants control emotion. Babies who more readily turn away from unpleasant events are less prone to distress (Axia, Bonichini, & Benini, 1999). At the end of the first year, crawling and walking enable infants to regulate feelings by approaching or retreating from various situations.

As caregivers help infants regulate their emotions, they contribute to the child's style of self-regulation. Parents who read and respond sympathetically to the baby's emotional cues have infants who are less fussy, more easily soothed, and more interested in exploration. In contrast, parents who wait to intervene until the infant has become extremely agitated reinforce the baby's rapid rise to intense distress (Eisenberg, Cumberland, & Spinrad, 1998). When caregivers fail to regulate stressful experiences for babies who cannot yet regulate them for themselves, brain structures that buffer stress may fail to develop properly, resulting in an anxious, reactive temperament (Nelson & Bosquet, 2000).

Caregivers also provide lessons in socially approved ways of expressing feelings. Beginning in the first few months, parents encourage infants to suppress negative emotion by often imitating their expressions of interest, happiness, and surprise and rarely imitating their expressions of anger and sadness. Infant boys get more of this training than do girls, in part because boys have a harder time regulating negative emotion. As a result, the well-known sex difference—females as emotionally expressive and males as emotionally controlled—is promoted at a tender age (Malatesta et al., 1986; Weinberg et al., 1999). Furthermore, collectivist cultures usually emphasize socially appropriate emotional behavior. Compared with North Americans, Japanese and Chinese adults discourage the expression of strong emotion in babies (Fogel, 1993; Kuchner, 1989). By the end of the first year, Chinese and Japanese infants smile and cry less than American infants (Camras et al., 1998).

In the second year, growth in representation and language leads to new ways of regulating emotions. A vocabulary for talking about feelings, such as "happy," "love," "surprised," "scary," "yucky," and "mad," develops rapidly after 18 months (Bretherton et al., 1986). Children of this age are not yet good at using language to comfort themselves (Grolnick, Bridges, & Connell, 1996). But once they can describe their internal states, they can guide caregivers to helping them. For example, while listening to a story about monsters, Grace whimpered, "Mommy, scary." Monica put the book down and gave Grace a comforting hug.

Ask Yourself

REVIEW

Why do many infants show stranger anxiety in the second half of the first year? What factors can increase wariness of strangers? What factors can decrease it?

APPLY

Dana is planning to meet her 10-month-old niece Laureen for the first time. How should Dana expect Laureen to react? How would you advise Dana to go about establishing a positive relationship with Laureen?

APPLY

At 14 months, Timmy danced joyfully to the tune "Old MacDonald" as several adults and children watched. At 20 months, he stopped dancing after a few steps, hiding his face behind his hands. What explains this change in Timmy's behavior?

CONNECT

How do babies of depressed mothers fare in development of emotional self-regulation? (See page 176.) What implications does their self-regulation competence have for handling cognitive and social challenges?

www.

Temperament and Development

Beginning in early infancy, Caitlin, Grace, and Timmy showed unique patterns of emotional responding. Caitlin's sociability was unmistakable to everyone who met her. She smiled and laughed while interacting with adults and readily approached other children during her second year. Meanwhile, Monica marveled at Grace's calm, relaxed disposition. At 19 months, she sat through a 2-hour family celebration at a restaurant, contented in her high chair. In contrast, Timmy was active and distractible. Vanessa found herself chasing him as he dropped one toy, moved on to the next, and climbed on chairs and tables.

When we describe one person as cheerful and "upbeat," another as active and energetic, and still others as calm, cautious, or prone to angry outbursts, we are referring to **temperament**—stable individual differences in quality and intensity of emotional reaction, activity level, attention, and emotional self-regulation (Rothbart & Bates, 1998). Researchers have become increasingly interested in temperamental differences among children because the psychological traits that make up temperament are believed to form the cornerstone of the adult personality.

The New York Longitudinal Study, initiated in 1956 by Alexander Thomas and Stella Chess, is the longest and most comprehensive study of temperament to date. A total of 141 children were followed from early infancy well into adulthood. Results showed that temperament increases the chances that a child will experience psychological problems or, alternatively, be protected from the effects of a highly stressful home life. However, Thomas and Chess (1977) also found that parenting practices can modify children's emotional styles considerably.

These findings inspired a growing body of research on temperament, including its stability, biological roots, and interaction with child-rearing experiences. Let's begin to explore these issues by looking at the structure, or makeup, of temperament and how it is measured.

The Structure of Temperament

Thomas and Chess's nine dimensions, listed in Table 6.2, served as the first influential model of temperament and inspired all others that followed. When detailed descriptions of infants' and children's behavior obtained from parental interviews were rated on these dimensions, certain characteristics clustered together, yielding three types of children:

- The **easy child** (40 percent of the sample) quickly establishes regular routines in infancy, is generally cheerful, and adapts easily to new experiences.

Table 6.2 Two Models of Temperament

Thomas and Chess		Rothbart	
Dimension	**Description**	**Dimension**	**Description**
Activity level	Ratio of active periods to inactive ones	Activity level	Level of gross motor activity
Rhythmicity	Regularity of body functions, such as sleep, wakefulness, hunger, and excretion	Soothability	Reduction of fussing, crying, and distress in response to caregiver's soothing
Distractibility	Degree to which stimulation from the environment alters behavior—for example, whether crying stops when a toy is offered	Attention span/ persistence	Duration of orienting or interest
Approach/ withdrawal	Response to a new object, food, or person	Fearful distress	Wariness and distress in response to intense or novel stimuli, including time to adjust to new situations
Adaptability	Ease with which child adapts to changes in the environment, such as sleeping or eating in a new place	Irritable distress	Extent of fussing, crying, and distress when desires are frustrated
Attention span and persistence	Amount of time devoted to an activity, such as watching a mobile or playing with a toy	Positive affect	Frequency of expression of happiness and pleasure
Intensity of reaction	Energy level of response, such as laughing, crying, talking, or gross motor activity		
Threshold of responsiveness	Intensity of stimulation required to evoke a response		
Quality of mood	Amount of friendly, joyful behavior as opposed to unpleasant, unfriendly behavior		

Sources: Left: Thomas & Chess, 1977. Right: Rothbart, 1981; Rothbart, Ahadi, & Evans, 2000.

- The **difficult child** (10 percent of the sample) is irregular in daily routines, is slow to accept new experiences, and tends to react negatively and intensely.
- The **slow-to-warm-up child** (15 percent of the sample) is inactive, shows mild, low-key reactions to environmental stimuli, is negative in mood, and adjusts slowly to new experiences.

Note that 35 percent of the children did not fit any of these categories. Instead, they showed unique blends of temperamental characteristics.

Of the three types, the difficult pattern has sparked the most interest, since it places children at high risk for adjustment problems—both anxious withdrawal and aggressive behavior in early and middle childhood (Bates, Wachs, & Emde, 1994; Thomas, Chess, & Birch, 1968). Compared with difficult children, slow-to-warm-up children do not present many problems in the early years. However, they tend to show excessive fearfulness and slow, constricted behavior in the late preschool and school years, when they are expected to respond actively and quickly in classrooms and peer groups (Chess & Thomas, 1984; Schmitz et al., 1999).

A second model of temperament, devised by Mary Rothbart (1981), is also shown in Table 6.2. It combines overlapping dimensions of Thomas and Chess and other researchers. For example, "distractibility" and "attention span and persistence," considered opposite ends of the same dimension, are called "attention span/persistence." It also includes a dimension not identified by Thomas and Chess—"irritable distress"—that taps emotional self-regulation. And it omits overly broad dimensions, such as "rhythmicity," "intensity of reaction," and "threshold of responsiveness" (Rothbart, Ahadi, & Evans, 2000). A child who is rhythmic in sleeping is not necessarily rhythmic in eating or bowel habits. And a child who smiles and laughs quickly and intensely is not necessarily quick and intense in fear or irritability. Overall, the characteristics shown in Table 6.2 provide a fairly complete picture of the temperamental traits most often studied.

Measuring Temperament

Temperament is often assessed through interviews or questionnaires given to parents. Behavior ratings by pediatricians, teachers, and others familiar with the child and direct observations by researchers have also been used. Parental reports have been emphasized because of their convenience and parents' depth of knowledge about the child. At the same time, information from parents has been criticized as being biased and subjective. Nevertheless, parent ratings are moderately related to observations of children's behavior (Mangelsdorf, Schoppe, & Buur, 2000). And parent perceptions are useful for understanding the way parents view and respond to their child.

Biology & Environment

Biological Basis of Shyness and Sociability

At age 4 months, Larry and Mitch visited the laboratory of Jerome Kagan, who observed their reactions to a variety of unfamiliar experiences. When exposed to new sights and sounds, such as a moving mobile decorated with colorful toys, Larry tensed his muscles, moved his arms and legs with agitation, and began to cry. Mitch's body remained relaxed and quiet, and he smiled and cooed pleasurably at the excitement around him.

Larry and Mitch returned to the laboratory as toddlers. This time, each experienced procedures designed to induce uncertainty. For example, electrodes were placed on their bodies and blood pressure cuffs on their arms to measure heart rate; toy robots, animals, and puppets moved before their eyes; and unfamiliar people entered and behaved in atypical ways or wore novel costumes. Larry whimpered and quickly withdrew. Mitch watched with interest, laughed, and approached the toys and strangers.

On a third visit, at age 4½, Larry barely talked or smiled during an interview with an unfamiliar adult. In contrast, Mitch spontaneously asked questions and communicated his pleasure at each intriguing activity. In a playroom with two unfamiliar peers, Larry pulled back. Mitch made friends quickly.

In longitudinal research on several hundred Caucasian children, Kagan (1998) found that about 20 percent of 4-month-old babies were easily upset by novelty (like Larry), whereas 40 percent were comfortable, even delighted, with new experiences (like Mitch). About 30 percent of these extreme groups retained their temperamental styles as they grew older (Kagan & Saudino, 2001; Kagan, Snidman, & Arcus, 1998). Those resembling Larry tended to become fearful, inhibited toddlers and preschoolers; those resembling Mitch developed into outgoing, uninhibited youngsters.

Physiological Correlates of Shyness and Sociability. Kagan believes that individual differences in arousal of the *amygdala,* an inner brain structure that controls avoidance reactions, contribute to these contrasting temperaments. In shy, inhibited children, novel stimuli easily excite the amygdala and its connections to the cerebral cortex and sympathetic nervous system, which prepares the body to act in the face of threat. The same level of stimulation evokes minimal neural excitation in highly sociable, uninhibited children. In support of this theory, several physiological responses of shy infants and children resemble those of highly timid animals and are known to be mediated by the amygdala:

- *Heart rate.* As early as the first few weeks of life, the heart rates of shy children are consistently higher than those of sociable youngsters, and they speed up further in response to unfamiliar events (Snidman et al., 1995).
- *Cortisol.* Saliva concentration of cortisol, a hormone that regulates blood pressure and is involved in resistance to stress, tends to be higher in shy than in sociable children (Gunnar & Nelson, 1994).
- *Pupil dilation, blood pressure, and skin surface temperature.* Compared with sociable children, shy children show

To explore the biological basis of temperament, researchers use physiological measures. Most efforts have focused on **inhibited,** or **shy, children,** who react negatively to and withdraw from novel stimuli (much like Thomas and Chess's slow-to-warm-up children), and **uninhibited,** or **sociable, children,** who react positively to and approach novel stimuli. As the Biology and Environment box above reveals, heart rate, hormone levels, and electrical brain-wave recordings in the frontal region of the cerebral cortex differentiate children with inhibited and uninhibited temperaments.

Stability of Temperament

It would be difficult to claim that temperament really exists if children's emotional styles were not stable over time. Indeed, many studies support the long-term stability of temperament. Infants and young children who score low or high on attention span, irritability, sociability, or shyness are likely to respond similarly when assessed again a few years later and, occasionally, even into the adult years (Caspi & Silva, 1995; Kochanska & Radke-Yarrow, 1992; Pedlow et al., 1993; Rothbart, Ahadi, & Evans, 2000; Ruff & Rothbart, 1996).

When the evidence as a whole is examined carefully, however, temperamental stability from one age period to the next is generally low to moderate (Putnam, Samson, & Rothbart, 2000). Although quite a few children remain the same, a good number have changed when assessed again. In fact, some characteristics, such as shyness and sociability, are stable over the long term only in children at the extremes—those who are very inhibited or very outgoing to begin with (Kagan & Saudino, 2001; Woodward et al., 2000).

Why is temperament not more stable? A major reason is that temperament itself develops with age. To illustrate, let's look at irritability and activity level. Recall from Chapter 3 that the early months are a period of fussing and crying for most babies. As infants can better regulate their attention and emotions, many who initially seemed irritable become calm and content. In the case of activity level, the meaning of the behavior changes. At first, an active, wriggling infant tends to be highly aroused and uncomfort-

greater pupil dilation, rise in blood pressure, and cooling of the fingertips when faced with novelty (Kagan et al., 1999).

Yet another physiological correlate of approach–withdrawal to people and objects is the pattern of brain waves in the frontal region of the cerebral cortex. Recall from Chapter 4 that the left hemisphere is specialized to respond with positive emotion, the right hemisphere with negative emotion. Shy infants and preschoolers show greater right than left frontal brain-wave activity; their sociable counterparts show the opposite pattern (Calkins, Fox, & Marshall, 1996; Fox, Calkins & Bell, 1994). Neural activity in the amygdala is transmitted to the frontal lobe and may influence these patterns.

Long-Term Consequences. According to Kagan (1998), extremely shy or sociable children inherit a physiology that biases them toward a particular temperamental style. Yet heritability research indicates that genes contribute only modestly to shyness and sociability. They share power with experience. When early inhibition persists, it leads to excessive cautiousness, social withdrawal, and loneliness (Caspi & Silva, 1995). At the same time, most inhibited infants and young children cope with novelty more effectively as they get older.

Child-rearing practices affect the chances that an emotionally reactive baby will become a fearful child. Warm, supportive parenting reduces the physiological reactivity of shy infants and preschoolers, whereas cold, intrusive parenting heightens anxiety and social reserve (Rubin, Burgess, & Hastings, 2002). In addition, when parents protect infants who dislike novelty from minor stresses, they make it harder for the child to overcome an urge to retreat from unfamiliar events. In contrast, parents who make appropriate demands for their baby to approach new experiences help the child overcome fear (Rubin et al., 1997). In sum, for children to develop at their best, parenting must be tailored to their temperaments—a theme we will encounter again in this and later chapters.

© ROBERT BRENNER/PHOTOEDIT

■ A strong physiological response to uncertain situations prompts this 2-year-old's withdrawal when a friend of her parents bends down to chat with her. Her mother's patient but insistent encouragement can modify her physiological reactivity and help her overcome her urge to retreat from unfamiliar events.

able, whereas an inactive baby is often alert and attentive. As infants begin to move on their own, the reverse is so! An active crawler is usually alert and interested in exploration, whereas a very inactive baby might be fearful and withdrawn.

These inconsistencies help us understand why long-term prediction of temperament is best achieved from the second year of life and after, when styles of responding are better established (Caspi, 1998; Lemery et al., 1999). At the same time, the changes shown by many children suggest that experience can modify biologically based temperamental traits (although children rarely change from one extreme to another—that is, a shy toddler practically never becomes highly sociable). With these ideas in mind, let's turn to genetic and environmental contributions to temperament and personality.

Genetic Influences

The word *temperament* implies a genetic foundation for individual differences in personality. Research shows that identical twins are more similar than fraternal twins across a wide range of temperamental and personality traits (Caspi, 1998; DiLalla, Kagan, & Reznick, 1994; Emde et al., 1992; Goldsmith et al., 1999). In Chapter 2, we noted that heritability estimates suggest a moderate role for heredity in temperament and personality: About half of the individual differences can be traced to differences in genetic makeup.

Consistent ethnic and sex differences in early temperament exist, again implying a role for heredity. Compared with Caucasian infants, Asian babies tend to be less active, less irritable, less vocal, more easily soothed when upset, and better at quieting themselves (Kagan et al., 1994; Lewis, Ramsay, & Kawakami, 1993). Grace's capacity to remain contentedly seated in her high chair through a long family dinner certainly fits with this evidence. And Timmy's high rate of activity is consistent with sex differences in temperament (Campbell & Eaton, 1999). From an early age, boys tend to be more active and daring and girls more anxious and timid—a difference reflected in boys' higher injury rates throughout childhood and adolescence.

Environmental Influences

Although genetic influences on temperament are clear, no study has shown that infants maintain their emotional styles without environmental supports. Instead, heredity and environment often combine to strengthen the stability of temperament, since a child's approach to the world affects the experiences to which she is exposed. To see how this works, let's take a second look at ethnic and sex differences in temperament.

Japanese mothers usually say that babies come into the world as independent beings who must learn to rely on their mothers through close physical contact. North American mothers are likely to believe just the opposite—that they must wean babies away from dependence into autonomy (Kojima, 1986). Consistent with these beliefs, Asian mothers interact gently, soothingly, and gesturally and (as we saw earlier) discourage strong emotion in their babies, whereas Caucasian mothers use a more active, stimulating, verbal approach (Rothbaum et al., 2000a). These behaviors enhance cultural differences in temperament.

A similar process seems to contribute to sex differences in temperament. Within the first 24 hours after birth (before they have had much experience with the baby), parents already perceive boys and girls differently. Sons are rated as larger, better coordinated, more alert, and stronger. Daughters are viewed as softer, more awkward, weaker, and more delicate (Stern & Karraker, 1989; Vogel et al., 1991). Gender-stereotyped beliefs carry over into the way parents treat their infants and toddlers. For example, parents more often encourage sons to be physically active and daughters to seek help and physical closeness (Ruble & Martin, 1998). These practices promote temperamental differences between boys and girls.

In families with several children, an additional influence on temperament is at work. Parents often look for and emphasize personality differences in their children. This is reflected in the comparisons parents make: "She's a lot more active," or "He's more sociable." In a study of identical-twin toddlers, mothers treated each twin differently, and this differential treatment predicted twin differences in psychological adjustment. The twin who received more warmth and less harshness was more positive in mood and social behavior (Deater-Deckard et al., 2001). Each child, in turn, evokes responses from caregivers that are consistent with parental views and with the child's actual temperamental style.

Besides different experiences within the family, as siblings get older they have unique experiences with teachers, peers, and others in their community that profoundly affect development (Caspi, 1998). And in middle childhood and adolescence, they often seek ways to differ from one another. For these reasons, both identical and fraternal twins tend to become more distinct in personality in adulthood (McCartney, Harris, & Bernieri, 1990). In sum, temperament and personality can be understood only in terms of complex interdependencies between genetic and environmental factors.

■ At birth, Chinese infants are calmer, more easily soothed when upset, and better at quieting themselves than are Caucasian infants. Although these differences may have biological roots, cultural variations in child rearing support them.

Temperament and Child Rearing: The Goodness-of-Fit Model

We have already indicated that the temperaments of many children change with age. This suggests that environments do not always sustain or intensify a child's existing temperament. If a child's disposition interferes with learning or getting along with others, adults must gently but consistently counteract the child's maladaptive behavior.

Thomas and Chess (1977) proposed a **goodness-of-fit model** to explain how temperament and environment can together produce favorable outcomes. Goodness of fit involves creating child-rearing environments that recognize each child's temperament while encouraging more adaptive functioning.

Goodness of fit helps explain why difficult children (who withdraw from new experiences and react negatively and intensely) are at high risk for later adjustment problems. These children, at least in Western middle-SES families, frequently experience parenting that fits poorly with their dispositions. As infants, they are far less likely to receive sensitive caregiving (van den Boom & Hoeksma, 1994). By the second year, parents of difficult children tend to resort to angry, punitive discipline, and the child reacts with defiance and disobedience. Then parents often behave inconsistently, rewarding the child's noncompliance by giving in to it, although they initially resisted (Lee & Bates, 1985). These practices maintain and even increase the child's irritable, conflict-ridden style. In contrast, when parents are positive and involved and engage in the sensitive, face-to-face play that helps infants regulate emotion, difficultness declines by age 2 (Feldman, Greenbaum, & Yirmiya, 1999).

Both difficult and shy children benefit from warm, accepting parenting that makes firm but reasonable demands for mastering new experiences. Yet goodness of fit depends in part on cultural values. For example, difficult children in low-SES Puerto Rican families are treated with sensitivity and patience; they are not at risk for adjustment problems (Gannon & Korn, 1983). In Western nations, shy children are regarded as socially incompetent—an attitude that discourages parents from helping them approach new situations. Yet in Chinese culture, adults evaluate shy children positively—as advanced in social maturity and understanding. In line with this view, in a study comparing Canadian and Chinese children, the Chinese children scored much higher in inhibition. Furthermore, Canadian mothers of shy children reported more protection and punishment and less acceptance and encouragement of achievement. Chinese mothers of shy children indicated just the opposite—less punishment and more acceptance and encouragement (Chen et al., 1998).

In cultures in which particular temperamental styles are linked to adjustment problems, a good match between rearing conditions and child temperament is best accomplished early, before unfavorable temperament–environment relationships have had a chance to produce maladjustment that is hard to undo. In the following sections, we will see that goodness of fit is also at the heart of infant–caregiver attachment. This first intimate relationship grows out of interaction between parent and baby, to which the emotional styles of both partners contribute.

Ask Yourself

REVIEW
Why is the stability of temperament only low to moderate?

REVIEW
How do genetic and environmental factors work together to promote a child's temperament? Cite examples from research.

APPLY
Eighteen-month-old, highly active Jake, who climbed out of his high chair, had a temper tantrum when his father made him sit at the table until the meal was finished. Using the concept of goodness of fit, suggest another way of handling Jake.

CONNECT
Do findings on ethnic and sex differences in temperament illustrate genetic–environmental correlation, discussed on page 69 of Chapter 2? Explain.

www.

Development of Attachment

Attachment is the strong, affectional tie we have with special people in our lives that leads us to experience pleasure and joy when we interact with them and to be comforted by their nearness during times of stress. By the second half of the first year, infants have become attached to familiar people who have responded to their needs. Watch babies of this age, and notice how they single out their parents for special attention. For example, when the mother enters the room, the baby breaks into a broad, friendly smile. When she picks him up, he pats her face, explores her hair, and snuggles against her. When he feels anxious or afraid, he crawls into her lap and clings closely.

Freud first suggested that the infant's emotional tie to the mother is the foundation for all later relationships. We will see shortly that research on the consequences of attachment is consistent with Freud's idea. But attachment has also been the subject of intense theoretical debate. Turn back to the description of Erikson's theory at the beginning of this chapter, and notice how the *psychoanalytic perspective* regards feeding as the primary context in which caregivers and babies build this close emotional bond. *Behaviorism,* too, emphasizes the importance of feeding, but for different reasons. According to a well-known behaviorist account, as the mother satisfies the baby's hunger, infants learn to prefer her soft caresses, warm smiles, and tender words of comfort because these events have been paired with tension relief.

Although feeding is an important context for building a close relationship, attachment does not depend on hunger satisfaction. In the 1950s, a famous experiment showed that rhesus monkeys reared with terry-cloth and wire-mesh "surrogate mothers" clung to the soft terry-cloth substitute, even though the wire-mesh "mother" held the bottle and infants had to climb on it to be fed (Harlow & Zimmerman, 1959). Similarly,

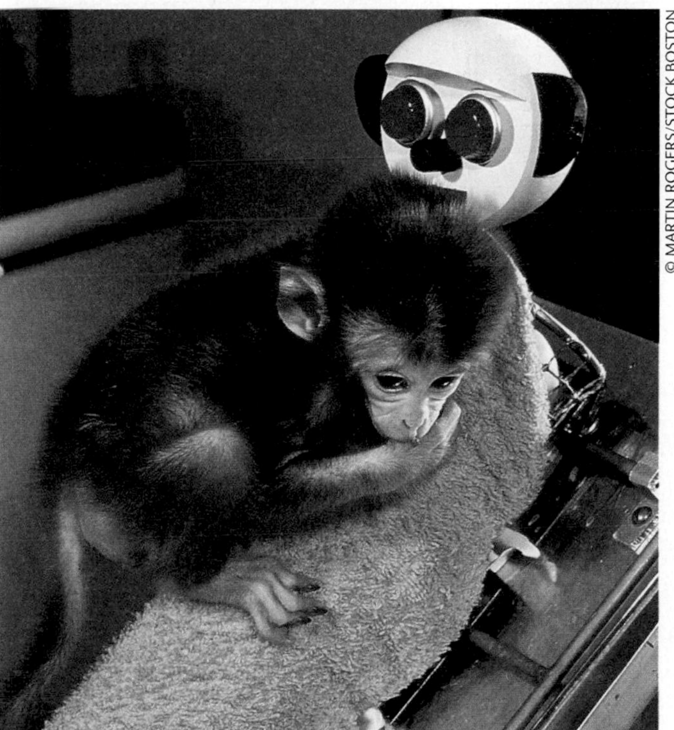

■ Baby monkeys reared with "surrogate mothers" from birth preferred to cling to a soft terry-cloth "mother" instead of a wire-mesh "mother" that held a bottle. These findings reveal that the drive-reduction explanation of attachment, which assumes that the mother–infant relationship is based on feeding, is incorrect.

human infants become attached to family members who seldom feed them, including fathers, siblings, and grandparents. And perhaps you have noticed that toddlers in Western cultures who sleep alone and experience frequent daytime separations from their parents sometimes develop strong emotional ties to cuddly objects, such as blankets and teddy bears. Yet such objects have never played a role in infant feeding!

Ethological Theory of Attachment

Today, **ethological theory of attachment,** which recognizes the infant's emotional tie to the caregiver as an evolved response that promotes survival, is the most widely accepted view. John Bowlby (1969), who first applied this idea to the infant–caregiver bond, was inspired by Konrad Lorenz's studies of imprinting in baby geese (see Chapter 1). He believed that the human infant, like the young of other animal species, is endowed with a set of built-in behaviors that help keep the parent nearby to protect the infant from danger and to provide support for exploring and mastering the environment (Waters & Cummings, 2000). Contact with the parent also ensures that the baby will be fed, but Bowlby pointed out that feeding is not the basis for attachment. Instead, the attachment bond has strong biological roots. It can best be understood in an evolutionary context in which survival of the species—through ensuring both safety and competence—is of utmost importance.

According to Bowlby, the infant's relationship with the parent begins as a set of innate signals that call the adult to the baby's side. Over time, a true affectional bond develops, which is supported by new cognitive and emotional capacities as well as by a history of warm, sensitive care. Attachment develops in four phases:

1. *The preattachment phase* (birth to 6 weeks). Built-in signals—grasping, smiling, crying, and gazing into the adult's eyes—help bring newborn babies into close contact with other humans. Once an adult responds, infants encourage her to remain nearby because closeness comforts them. Babies of this age recognize their own mother's smell and voice (see Chapter 3). But they are not yet attached to her, since they do not mind being left with an unfamiliar adult.
2. *The "attachment-in-the-making" phase* (6 weeks to 6–8 months). During this phase, infants respond differently to a familiar caregiver than to a stranger. For example, at 4 months, Timmy smiled, laughed, and babbled more freely when interacting with his mother and quieted more quickly when she picked him up. As infants learn that their own actions affect the behavior of those around them, they begin to develop a *sense of trust*—the expectation that the caregiver will respond when signaled. But even though they recognize the parent, babies still do not protest when separated from her.
3. *The phase of "clear-cut" attachment* (6–8 months to 18 months–2 years). Now attachment to the familiar caregiver is clearly evident. Babies display **separation anxiety,** becoming upset when the adult whom they have come to rely on leaves. Separation anxiety does not always occur; like stranger anxiety (see page 177), it depends on infant temperament and the current situation. But in many cultures, separation anxiety increases between 6 and 15 months (see Figure 6.1). Besides protesting the parent's departure, older infants and toddlers try hard to maintain her presence. They approach, follow, and climb on her in preference to others. And they use the familiar caregiver as a **secure base** or point from which to explore, venturing into the environment and then returning for emotional support.
4. *Formation of a reciprocal relationship* (18 months–2 years and on). By the end of the second year, rapid growth in representation and language permits toddlers to under-

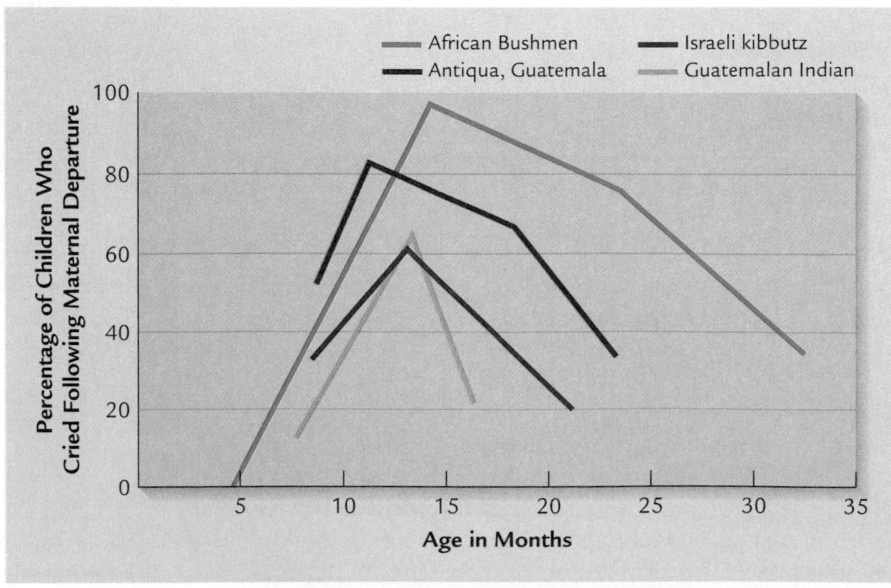

FIGURE 6.1 Development of separation anxiety. In cultures around the world, separation anxiety emerges in the second half of the first year, increases until about 15 months, and then declines. (From J. Kagan, R. B. Kearsley, & P. R. Zelazo, 1978, *Infancy: Its Place in Human Development*, Cambridge, MA: Harvard University Press, p. 107. Copyright 1978 by the President and Fellows of Harvard College. All rights reserved. Reprinted by permission.)

Table 6.3 Episodes in the Strange Situation

Episode	Events	Attachment behavior observed
1	Experimenter introduces parent and baby to playroom and then leaves.	
2	Parent is seated while baby plays with toys.	Parent as a secure base
3	Stranger enters, is seated, and talks to parent.	Reaction to unfamiliar adult
4	Parent leaves room. Stranger responds to baby and offers comfort if upset.	Separation anxiety
5	Parent returns, greets baby, and offers comfort if necessary. Stranger leaves room.	Reaction to reunion
6	Parent leaves room.	Separation anxiety
7	Stranger enters room and offers comfort.	Ability to be soothed by stranger
8	Parent returns, greets baby, offers comfort if necessary, and tries to reinterest baby in toys.	Reaction to reunion

Note: Episode 1 lasts about 30 seconds; each of the remaining episodes lasts about 3 minutes. Separation episodes are cut short if the baby becomes very upset. Reunion episodes are extended if the baby needs more time to calm down and return to play.
Source: Ainsworth et al., 1978.

stand some of the factors that influence the parent's coming and going and to predict her return. As a result, separation protest declines. Now children start to negotiate with the caregiver, using requests and persuasion to alter her goals. For example, at age 2, Caitlin asked Carolyn and David to read a story before leaving her with a baby-sitter. The extra time with her parents, along with a better understanding of where they were going ("to have dinner with Uncle Sean") and when they would be back ("right after you go to sleep"), helped Caitlin withstand her parents' absence.

According to Bowlby (1980), out of their experiences during these four phases, children construct an enduring affectional tie to the caregiver that they can use as a secure base in the parents' absence. This inner representation becomes a vital part of personality. It serves as an **internal working model,** or set of expectations about the availability of attachment figures and their likelihood of providing support during times of stress. This image becomes the model, or guide, for all future close relationships (Bretherton, 1992).

Measuring the Security of Attachment

Although virtually all family-reared babies become attached to a familiar caregiver by the second year, the quality of this relationship differs from child to child. A widely used technique for assessing the quality of attachment between 1 and 2 years of age is the **Strange Situation.** In designing it, Mary Ainsworth and her colleagues reasoned that securely attached infants and toddlers should use the parent as a secure base from which to explore an unfamiliar playroom. In addition, when the parent leaves, an unfamiliar adult should be less comforting than the parent. As summarized in Table 6.3, the Strange Situation takes the baby through eight short episodes in which brief separations from and reunions with the parent occur.

Observing the responses of infants to these episodes, researchers have identified a secure attachment pattern and three patterns of insecurity; a few babies cannot be classified (Ainsworth et al., 1978; Barnett & Vondra, 1999; Main & Solomon, 1990). Which pattern do you think Grace displayed after adjusting to her adoptive family? (See the description at the beginning of this chapter.)

- **Secure attachment.** These infants use the parent as a secure base from which to explore. When separated, they may or may not cry, but if they do, it is because the parent is absent and they prefer her to the stranger. When the parent returns, they actively seek contact, and their crying is reduced immediately. About 65 percent of North American infants show this pattern.
- **Avoidant attachment.** These infants seem unresponsive to the parent when she is present. When she leaves, they are usually not distressed, and they react to the stranger in much the same way as to the parent. During reunion, they avoid or are slow to greet the parent, and when picked up, they often fail to cling. About 20 percent of North American infants show this pattern.
- **Resistant attachment.** Before separation, these infants often seek closeness to the parent and often fail to explore. When she returns, they display angry, resistive behavior, sometimes hitting and pushing. Many continue to cry after being picked up and cannot be comforted easily. About 10 to 15 percent of North American infants show this pattern.
- **Disorganized–disoriented attachment.** This pattern reflects the greatest insecurity. At reunion, these infants show a variety of confused, contradictory behaviors.

They might look away while being held by the parent or approach her with flat, depressed emotion. A few cry out after having calmed down or display odd, frozen postures. About 5 to 10 percent of North American infants show this pattern.

Infants' reactions in the Strange Situation resemble their use of the parent as a secure base and their response to separation at home (Pederson & Moran, 1996; Pederson et al., 1998). For this reason, the procedure is a powerful tool for assessing attachment security.

Stability of Attachment and Cultural Variations

Research on the stability of attachment patterns between 1 and 2 years of age yields a wide range of findings (Thompson, 1998). But a close look at which infants change and which ones remain the same yields a more consistent picture. Quality of attachment is usually secure and stable for middle-SES babies experiencing favorable life conditions. And infants who move from insecurity to security typically have well-adjusted mothers with positive family and friendship ties. Perhaps many became parents before they were psychologically ready but, with social support, grew into the role. In contrast, for low-SES families with many daily stresses, attachment usually moves away from security or changes from one insecure pattern to another (Owen et al., 1984; Vaughn et al., 1979; Vondra, Hommerding, & Shaw, 1999).

These findings indicate that securely attached babies more often maintain their attachment status than do insecure babies. The exception to this trend is disorganized/disoriented attachment—an insecure pattern that is highly stable over the second year (Barnett, Ganiban, & Cicchetti, 1999; Hesse & Main, 2000). As we will see, many disorganized/disoriented infants experience extremely negative caregiving, which may disrupt emotional self-regulation so severely that the baby's confused behavior persists.

Furthermore, cross-cultural evidence indicates that attachment patterns may have to be interpreted differently in other cultures. For example, as Figure 6.2 reveals, German infants show considerably more avoidant attachment than American babies do. But German parents encourage their infants to be nonclingy and independent, so the baby's behavior may be an intended outcome of cultural beliefs and practices (Grossmann et al., 1985). An unusually high number of Japanese infants display a resistant response, but the reaction may not represent true insecurity. Japanese mothers rarely leave their babies in the care of unfamiliar people, so the Strange Situation probably creates far greater stress for them than for infants who frequently experience maternal separations (Takahashi, 1990). Also, Japanese parents value the infant clinginess and attention seeking that are part of resistant attachment, considering them to be normal indicators of infant closeness and dependence (Rothbaum et al., 2000b). Despite these cultural variations, the secure pattern is still the most common pattern of attachment in all societies studied to date (van IJzendoorn & Sagi, 1999).

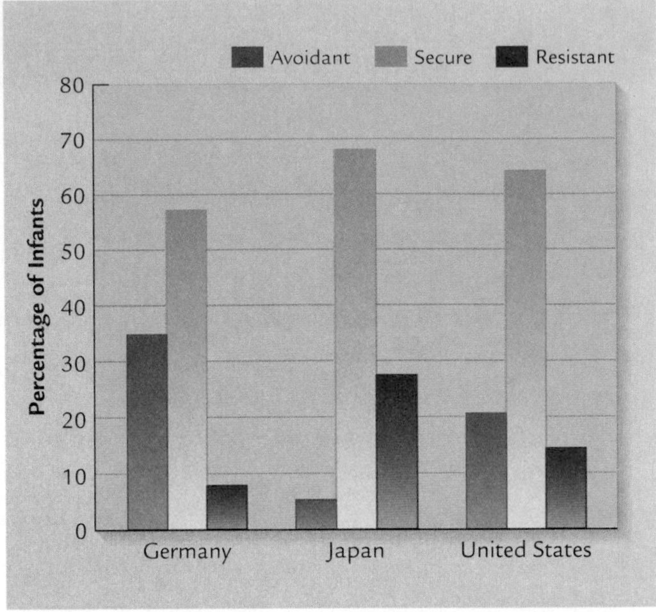

■ **FIGURE 6.2 A cross-cultural comparison of infants' reactions in the Strange Situation.** A high percentage of German babies seem avoidantly attached, whereas a substantial number of Japanese infants appear resistantly attached. Note that these responses may not reflect true insecurity. Instead, they are probably due to cultural differences in values and child-rearing practices. (Adapted from Thompson, 1998.)

Factors That Affect Attachment Security

What factors might influence attachment security? Researchers have looked closely at four important influences: (1) opportunity to establish a close relationship, (2) quality of caregiving, (3) the baby's characteristics, and (4) family context.

● **Opportunity for Attachment.** What happens when a baby does not have the opportunity to establish an affectional tie to a caregiver? In a series of studies, René Spitz (1946) observed institutionalized infants who had been given up by their mothers between 3 and 12 months of age. The babies were placed on a large ward where they shared a nurse with at least seven other babies. In contrast to the happy, outgoing behavior they had shown before separation, they wept and withdrew from their surroundings, lost weight, and had difficulty sleeping. If a consistent caregiver did not replace the mother, the depression deepened rapidly.

These institutionalized babies had emotional difficulties because they were prevented from forming a bond with one or a few adults (Rutter, 1996). Another study supports this conclusion. Researchers followed the development of infants in an institution with a good caregiver–child ratio and a rich selection of books and toys. However, staff turnover was so rapid that the average child had 50 different caregivers by age 4½! Many of these children became "late adoptees" who were placed in homes after age 4. Most developed deep ties with their adoptive parents, indicating that a first attachment bond can develop as late as 4 to 6 years of age (Tizard & Rees, 1975).

But these youngsters were more likely to display emotional and social problems, including an excessive desire for adult attention, "over-friendliness" to unfamiliar adults and peers, and few friendships. Adopted children who spent their first 8 months or more in deprived Romanian institutions often display these same difficulties (Hodges & Tizard, 1989; Zeanah, 2000). Although follow-ups into adulthood are necessary before we can be sure, these findings leave open the possibility that fully normal development depends on establishing close bonds with caregivers during the early years of life.

● **Quality of Caregiving.** Dozens of studies report that **sensitive caregiving**—responding promptly, consistently, and appropriately to infants and holding them tenderly and carefully—is moderately related to attachment security in diverse cultures (De Wolff & van IJzendoorn, 1997; Posada et al., 2002). In contrast, insecurely attached infants tend to have mothers who engage in less physical contact, handle them awkwardly, behave in a "routine" manner, and are sometimes negative, resentful, and rejecting (Ainsworth et al., 1978; Isabella, 1993; Pederson & Moran, 1996).

Also, in several studies, a special form of communication called **interactional synchrony** separated the experiences of secure and insecure babies. It is best described as a sensitively tuned "emotional dance," in which the caregiver responds to infant signals in a well-timed, rhythmic, appropriate fashion. In addition, both partners match emotional states, especially the positive ones (Isabella & Belsky, 1991; Kochanska, 1998).

© DAVID YOUNG-WOLFF/PHOTOEDIT

■ This mother and baby engage in a sensitively tuned form of communication called interactional synchrony in which they match emotional states, especially the positive ones. Interactional synchrony may support the development of secure attachment, but it does not characterize mother–infant interaction in all cultures.

Earlier we saw that sensitive face-to-face play, in which interactional synchrony occurs, helps infants regulate emotion. But moderate adult–infant coordination predicts attachment security, not "tight" coordination in which the adult responds to most infant cues (Jaffe et al., 2001). Perhaps warm, sensitive caregivers use a relaxed, flexible style of communication in which they comfortably accept and repair emotional mismatches, returning to a synchronous state. Furthermore, finely tuned, coordinated interaction does not characterize mother–infant interaction everywhere. Among the Gusii people of Kenya, mothers rarely cuddle, hug, or interact playfully with their babies, although they are very responsive to the needs of their infants. Most Gusii infants appear securely attached, using the mother as a secure base (LeVine et al., 1994). This suggests that attachment security depends on attentive caregiving but that its association with moment-by-moment contingent interaction is probably limited to certain cultures.

Compared with securely attached infants, avoidant babies tend to receive overly stimulating, intrusive care. Their mothers might, for example, talk energetically to them while they are looking away or falling asleep. By avoiding the mother, these infants try to escape from overwhelming interaction. Resistant infants often experience inconsistent care. Their mothers are unresponsive to infant signals. Yet when the baby begins to explore, these mothers interfere, shifting the infant's attention back to themselves. As a result, the baby is overly dependent as well as angry at the mother's lack of involvement (Cassidy & Berlin, 1994; Isabella & Belsky, 1991).

When caregiving is highly inadequate, it is a powerful predictor of disruptions in attachment. Child abuse and neglect (topics we will consider in Chapter 8) are associated with all three forms of attachment insecurity. Among maltreated infants, the most worrisome classification—disorganized–disoriented attachment—is especially high (Barnett, Ganiban, & Cicchetti, 1999). Depressed mothers and parents suffering from a traumatic event, such as loss of a loved one, also tend to promote the uncertain behaviors of this pattern (Teti et al., 1995; van IJzendoorn, 1995). Observations reveal that they often display frightening, contradictory, and unpleasant behaviors, such as looking scared, teasing the baby, holding the baby stiffly at a distance, or seeking reassurance from the upset child (Lyons-Ruth, Bronfman, & Parsons, 1999; Schuengel, Bakermans-Kranenburg, & van IJzendoorn, 1999).

● **Infant Characteristics.** Since attachment is the result of a *relationship* that builds between two partners, infant characteristics should affect how easily it is established. In Chapter 3, we saw that prematurity, birth complications, and newborn illness make caregiving more taxing. In stressed, poverty-stricken families, these difficulties are linked to attachment insecurity (Wille, 1991). But when parents have the time and patience to care for a baby with special needs and view their infants positively, at-risk newborns fare quite well in attachment security (Cox, Hopkins, & Hans, 2000; Pederson & Moran, 1995).

Social Issues

Does Child Care in Infancy Threaten Attachment Security and Later Adjustment?

Research suggests that infants placed in full-time child care before 12 months of age are more likely than infants who remain at home to display insecure attachment—especially avoidance—in the Strange Situation (Belsky, 1992). Does this mean that infants who experience daily separations from their employed parents and early placement in child care are at risk for developmental problems? A close look at the evidence reveals that we should be cautious about coming to this conclusion.

Attachment Quality. In studies reporting a child care–attachment association, the rate of insecurity is somewhat higher among child-care infants than non-child-care infants (36 versus 29 percent), but it nevertheless resembles the overall rate of insecurity reported for children in industrialized countries (Lamb, Sternberg, & Prodromidis, 1992). In fact, most infants of employed mothers are securely attached! Furthermore, not all investigations report a difference in attachment quality between child-care and home-reared infants (NICHD Early Child Care Research Network, 1997; Roggman et al., 1994).

Family Circumstances. We have seen that family conditions affect attachment security. Many employed women find the pressures of handling two full-time jobs (work and motherhood) stressful. Some respond less sensitively because they receive little caregiving assistance from the child's father and are fatigued and harried, thereby risking the infants' security (Stifter, Coulehan, & Fish, 1993). Other employed parents probably value and encourage their infants' independence. Or their babies are unfazed by the Strange Situation because they are used to separating from their parents. In these cases, the avoidant attachment pattern may represent healthy autonomy rather than insecurity (Lamb, 1998).

Quality and Extent of Child Care. Poor-quality child care and many hours in child care may contribute to a higher rate of insecure attachment among infants of employed mothers. In the U.S. National Institute of Child Health and Human Development (NICHD) Study of Early Child Care—the largest longitudinal study to date, including more than 1,300 infants and their families—child care alone did not contribute to attachment insecurity. But when babies were exposed to combined home and child-care risk factors—insensitive caregiving at home with insensitive caregiving in child care, long hours in child care, or more than one child-care arrangement—the rate of insecurity increased. Overall, mother–child interaction was more favorable when children attended higher-quality child care and were in child care for fewer hours (NICHD Early Childhood Research Network, 1997, 1999).

Furthermore, when the NICHD sample reached 3 years of age, a history of higher-quality child care predicted better social skills as rated by caregivers (NICHD Early Child Care Research Network, 2002b). At the same time, regardless of child-care quality, at age 4½ to 5, children averaging more than 30 child-care hours per week were rated by their mothers, caregivers, and kindergarten teachers as having more behavior problems, especially defiance, disobedience, and aggression (NICHD Early Childhood Research Network, 2001). This does not necessarily mean that child care causes behavior problems. Children prone to be aggressive may have parents who leave them in child care for longer hours.

Overall, findings of the NICHD Study indicate that parenting has a far stronger impact on preschoolers' problem behavior than does early, extensive child care (NICHD Early Child Care Research Network, 2002c). Indeed, having the opportunity to form a warm bond with a stable professional care-

Infants also vary in temperament, but its role in attachment security has been intensely debated. Some researchers believe that babies who are irritable and fearful may simply react to brief separations with intense anxiety, regardless of the parent's sensitivity to the baby (Kagan, 1998). Consistent with this view, emotionally reactive, difficult babies are more likely to develop later insecure attachments (Seifer et al., 1996; Vaughn & Bost, 1999).

But other evidence suggests that caregiving may be involved in the relationship between difficultness and attachment insecurity. In a study of disorganized–disoriented 1-year-olds, emotional reactivity increased sharply over the second year. Attachment disorganization was not caused by difficult temperament but rather seemed to promote it (Barnett, Ganiban, & Cicchetti, 1999). Furthermore, an intervention that taught mothers how to respond sensitively to their irritable 6-month-olds led to gains in maternal responsiveness and children's attachment security, exploration, cooperativeness, and sociability that were still present at 3½ years of age (van den Boom, 1995).

A major reason that temperament and other infant characteristics do not show strong relationships with attachment quality may be that their influence depends on goodness of fit. From this perspective, *many* child attributes can lead to secure attachment as long as the caregiver sensitively adjusts her behavior to fit the baby's needs (Seifer & Schiller, 1995). But when a parent's capacity to do so is strained—for example, by her own personality or stressful living conditions—then infants with illnesses, disabilities, and difficult temperaments are at risk for attachment problems.

giver seems to be particularly helpful to infants whose relationship with one or both parents is insecure. When followed into the preschool and early school years, such children show higher self-esteem and socially skilled behavior than their insecurely attached agemates who did not attend child care (Egeland & Hiester, 1995).

Conclusions. Taken together, research suggests that some infants may be at risk for attachment insecurity and adjustment problems due to inadequate child care, long hours in child care, and the joint pressures of full-time employment and parenthood experienced by their mothers. However, using this as evidence to justify a reduction in child-care services is inappropriate. When family incomes are limited or mothers who want to work are forced to stay at home, children's emotional security is not promoted. In a study carried out in Australia, first-time mothers with high career commitment and high levels of social support tended to return to work earlier in their baby's first year. Compared with other babies, infants of these career-committed mothers were more likely to be securely attached at 12 months of age (Harrison & Ungerer, 2002). They also benefited from Australia's government-subsidized high-quality child care, which is available to all families.

Consequently, it makes sense to increase the availability of high-quality child care, to provide paid employment leave so parents can limit the hours their children spend in child care (see pages 98–99), and to educate parents about the vital role of sensitive caregiving in early emotional development. Return to Chapter 5, page 163, to review signs of developmentally appropriate child care for infants and toddlers. For child care to foster attachment security and adjustment, the professional caregiver's relationship with the baby is vital. When caregiver–child ratios are generous, group sizes are small, and caregivers are educated about child development and child rearing, caregivers' interactions are more positive and children develop more favorably (NICHD Early Child Care Research Network, 2000b, 2002a). Child care with these characteristics can become part of an ecological system that relieves rather than intensifies parental and child stress, thereby promoting healthy development.

© B. MAHONEY/THE IMAGE WORKS

■ This child-care center meets rigorous, professionally established standards of quality. A generous caregiver–child ratio, a limited number of children in each room, an environment with appropriate equipment and toys, and training in child development enable caregivers to respond to infants' and toddlers' needs to be held, comforted, and stimulated.

● **Family Circumstances.** Timmy's parents divorced shortly after he was born, and his father moved to a distant city. Although Vanessa tried not to let his departure affect her caregiving, she became anxious and distracted. To make ends meet, she placed 1-month-old Timmy in Ginette's child-care home and began working 50- to 60-hour weeks. When Vanessa stayed late at the office, a baby-sitter picked Timmy up, gave him dinner, and put him to bed. Once or twice a week, Vanessa went to get Timmy. As he neared his first birthday, she noticed that the other children reached out, crawled, or ran to their parents. Timmy, in contrast, ignored Vanessa.

Timmy's behavior reflects a repeated finding: Job loss, a failing marriage, financial difficulties, and other stressors can undermine attachment by interfering with the sensitivity of parental care. Or they can affect babies' sense of security directly, by exposing them to angry adult interactions or unfavorable child-care arrangements (Thompson, 1998). But as the Social Issues box above indicates, placing infants and toddlers in child care does not necessarily affect the quality of attachment. The availability of social supports, especially assistance in caregiving, reduces stress and fosters attachment security. Ginette's sensitivity and the parenting advice that Ben, a psychologist, offered Vanessa were helpful. As Timmy turned 2, his relationship with his mother seemed warmer.

Parents bring to the family context a long history of attachment experiences, out of which they construct internal working models that they apply to the bonds established with their babies. Carolyn remembered her mother as deeply affectionate and caring and viewed her as a positive influence in her own parenting. Monica recalled her mother as tense and preoccupied and expressed regret that they had not had a closer relationship. Do these images of parenthood affect the quality of Caitlin's and Grace's attachments to their mothers?

To answer this question, researchers have assessed parents' internal working models by having them evaluate childhood memories of attachment experiences (George, Kaplan, & Main, 1985). In studies in several Western nations, parents who showed objectivity and balance in discussing their childhoods, regardless of whether they were positive or negative, tended to have securely attached infants and behaved sensitively toward them. In contrast, parents who dismissed the importance of early relationships or described them in angry, confused ways usually had insecurely attached babies and engaged in less sensitive caregiving (van IJzendoorn, 1995; Slade et al., 1999).

But we must be careful not to assume any direct transfer of parents' childhood experiences to quality of attachment with their own children. Internal working models are *reconstructed memories* affected by many factors, including relationship experiences over the life course, personality, and current life satisfaction. Longitudinal studies show that negative life events can weaken the link between an individual's own attachment security in infancy and a secure internal working model in adulthood. And insecurely attached babies who become adults with insecure internal working models often have lives that, based on adulthood self-reports, are filled with family crises (Waters et al., 2000; Weinfield, Sroufe, & Egeland, 2000).

In sum, our early rearing experiences do not destine us to become sensitive or insensitive parents. Rather, the way we view our childhoods—our ability to come to terms with negative events, to integrate new information into our working models, and to look back on our own parents in an understanding, forgiving way—is much more influential in how we rear our children than is the actual history of care we received (Main, 2000).

Multiple Attachments

We have already indicated that babies develop attachments to a variety of familiar people—not just mothers but also fathers, siblings, grandparents, and professional caregivers. Although Bowlby (1969) made room for multiple attachments in his theory, he believed that infants are predisposed to direct their attachment behaviors to a single special person, especially when they are distressed. For example, when an anxious, unhappy 1-year-old is permitted to choose between the mother and father for comfort and security, the infant generally chooses the mother (Lamb, 1997). This preference typically declines over the second year of life. An expanding world of attachments enriches the emotional and social lives of many babies.

● **Fathers.** Like mothers' sensitive caregiving, fathers' sensitivity predicts secure attachment—an effect that becomes stronger the more time they spend with their babies (van IJzendoorn & De Wolff, 1997). But as infancy progresses, mothers and fathers in many cultures—Australia, India, Israel, Italy, Japan, and the United States—relate to babies in different ways. Mothers devote more time to physical care and expressing affection. Fathers spend more time in playful interaction (Lamb, 1987; Roopnarine et al., 1990). Mothers and fathers also play differently with babies. Mothers more often provide toys, talk to infants, and engage in conventional games like pat-a-cake and peekaboo. In contrast, fathers tend to engage in more exciting, highly physical bouncing and lifting games, especially with their infant sons (Yogman, 1981).

However, this picture of "mother as caregiver" and "father as playmate" has changed in some families due to the revised work status of women. Employed mothers tend to engage in more playful stimulation of their babies than unemployed mothers, and their husbands are somewhat more involved in caregiving (Cox et al., 1992). When fathers are the primary caregivers, they retain their highly arousing play style (Lamb & Oppenheim, 1989). Such highly involved fathers are less gender-stereotyped in their beliefs; have sympathetic, friendly personalities; and regard parenthood as an especially enriching experience (Lamb, 1987; Levy-Shiff & Israelashvili, 1988).

Fathers' involvement with babies unfolds within a complex system of family attitudes and relationships. When mothers and fathers believe that men are capable of nurturing infants, fathers devote more time to caregiving (Beitel & Parke, 1998). A warm, gratifying marital relationship supports both parents' involvement with babies, but it is particularly important for fathers (Frosch, Mangelsdorf, & McHale, 2000; Grych & Clark, 1999). Cross-cultural support for this conclusion comes from the Aka hunters and gatherers of Central Africa, where fathers devote more time to infants than in any other known society. Aka fathers pick up and cuddle their babies at least five times more often than fathers in other African hunting-and-gathering societies. Aka couples have unusually cooperative and intimate relationships, sharing hunting, food preparation, and leisure activities. The more Aka parents are together, the greater the father's interaction with his baby (Hewlett, 1992).

● **Siblings.** Despite a declining family size, 80 percent of American children still grow up with at least one sibling. The arrival of a baby brother or sister is a difficult experience for most preschoolers, who quickly realize that now they must share their parents' attention and affection. They often become demanding and clingy for a time and engage in deliberate naughtiness. And their security of attachment typically declines, more so if they are over age 2 (old enough to feel threatened and displaced) and if the mother is under stress due to marital or psychological problems (Teti et al., 1996).

Yet resentment is only one feature of a rich emotional relationship that starts to build between siblings after a baby's birth. An older child can also be seen kissing, patting, and calling out, "Mom, he needs you," when the baby cries—signs of growing affection. By the end of the baby's first year, siblings typically spend much time together. Infants of this age are

■ This Aka father spends much time in close contact with his baby. In Aka society, husband and wife share many tasks of daily living and have an unusually cooperative and intimate relationship. Infants are generally within arm's reach of their fathers, who devote many hours to caregiving.

comforted by the presence of their preschool-age brother or sister during short parental absences (Stewart, 1983). And in the second year, they often imitate and actively join in play with the older child (Dunn, 1989).

Nevertheless, individual differences in sibling relationships appear shortly after a baby's birth. Temperament plays an important role. For example, conflict increases when one sibling is emotionally intense or highly active (Brody, Stoneman, & McCoy, 1994; Dunn, 1994). Parenting also makes a difference. Secure infant–mother attachment and parental warmth toward both children are related to positive sibling interaction (MacKinnon-Lewis et al., 1997; Volling & Belsky, 1992). Also, mothers who play with their children frequently and explain the toddler's wants and needs to the preschool sibling foster cooperative sibling ties. In contrast, maternal harshness and lack of involvement are linked to antagonistic sibling relationships (Howe, Aquan-Assee, & Bukowski, 2001).

The Caregiving Concerns table on page 194 suggests ways to promote positive relationships between babies and their preschool siblings. Siblings offer a rich social context in which young children learn and practice a wide range of skills, including affectionate caring, conflict resolution, and control of hostile and envious feelings.

■ Although the arrival of a baby brother or sister is a difficult experience for most preschoolers, a rich emotional relationship quickly builds between siblings. This toddler is already actively involved in play with her 4-year-old-brother, and both derive great pleasure from the interaction.

Attachment and Later Development

According to psychoanalytic and ethological theories, the inner feelings of affection and security that result from a healthy attachment relationship support all aspects of psychological development. Consistent with this view, an extensive longitudinal study found that preschool teachers viewed children who were securely attached as babies as high in self-esteem, socially competent, cooperative, and popular. In contrast, they viewed avoidantly attached agemates as isolated and disconnected and resistantly attached agemates as disruptive and difficult. Studied again at age 11 in summer camp, children who were secure as infants had more favorable relationships with peers and better social skills, as judged by camp counselors (Elicker, Englund, & Sroufe, 1992; Matas, Arend, & Sroufe, 1978; Shulman, Elicker, & Sroufe, 1994).

These findings have been taken by some researchers to mean that secure attachment in infancy causes improved cognitive,

Encouraging Affectionate Ties Between Infants and Their Preschool Siblings

SUGGESTION	DESCRIPTION
Spend extra time with the older child.	To minimize the older child's feelings of being deprived of affection and attention, set aside time to spend with her. Fathers can be especially helpful in this regard, planning special outings with the preschooler and taking over care of the baby so the mother can be with the older child.
Handle sibling misbehavior with patience.	Respond patiently to the older sibling's misbehavior and demands for attention, recognizing that these reactions are temporary. Give the preschooler opportunities to feel proud of being more grown-up than the baby. For example, encourage the older child to assist with feeding, bathing, dressing, and offering toys, and show appreciation for these efforts.
Discuss the baby's wants and needs.	By helping the older sibling understand the baby's point of view, parents can promote friendly, considerate behavior. Say, for example, "He's so little that he just can't wait to be fed" or "He's trying to reach his rattle and can't."

emotional, and social competence in later years. Yet more evidence is needed before we can be certain of this conclusion. In other longitudinal studies, secure infants sometimes develop more favorably than insecure infants and sometimes do not (Stams, Juffer, & van IJzendoorn, 2002; Lewis, 1997; Schneider, Atkinson, & Tardiff, 2001). Disorganized–disoriented attachment is an exception. It is consistently related to high hostility and aggression during the preschool and school years (Lyons-Ruth, 1996; Lyons-Ruth, Easterbrooks, & Cibelli, 1997).

Why, overall, is research on the long-term consequences of attachment quality unclear? Perhaps *continuity of caregiving* determines whether attachment security is linked to later development (Lamb et al., 1985; Thompson, 2000). When parents respond sensitively not just in infancy but also during later years, children are likely to develop favorably. In contrast, parents who react insensitively for a long time have children who establish lasting patterns of avoidant, resistive, or disorganized behavior and are at greater risk for academic, emotional, and social difficulties. At the same time, infants and young children are resilient beings. A child whose parental caregiving improves or who has compensating, affectionate ties outside the family is likely to fare well.

As we conclude our discussion of attachment, consider the diverse factors that affect the development of the parent–child bond. These include infant and parent characteristics, parents' relationship with each other, outside-the-family stressors, the availability of social supports, parents' views of their attachment history, and child-care arrangements. Although attachment builds within the warmth and intimacy of caregiver–infant interactions, it can be fully understood only from an ecological systems perspective (Cummings & Cummings, 2002). Return to Chapter 1, page 24, to review ecological systems theory. Notice how research confirms the contribution of each level of the environment to attachment security.

Ask Yourself

REVIEW

What factors explain stability in attachment pattern for some children and change for others? Are the same factors involved in the link between attachment in infancy and later development? Explain.

REVIEW

What contributions do quality of caregiving and infant characteristics make to attachment security? Which influence is more powerful, and why?

APPLY

What attachment pattern did Timmy display when Vanessa picked him up from child care, and what factors probably contributed to it? Will Timmy's insecurity necessarily compromise his development?

CONNECT

Review research on emotional self-regulation on page 179. How do the caregiving experiences of securely attached infants promote development of emotional self-regulation?

www.

Self-Development During the First Two Years

Infancy is a rich, formative period for the development of physical and social understanding. In Chapter 5, you learned that infants develop an appreciation of the permanence of objects. And in this chapter, we have seen that over the first year, infants recognize and respond appropriately to others' emotions and distinguish familiar from unfamiliar people. That both objects and people achieve an independent, stable existence for the infant implies that knowledge of the self as a separate, permanent entity emerges around this time.

Self-Awareness

After Caitlin's bath, Carolyn often held her in front of the bathroom mirror. As early as the first few months, Caitlin smiled and returned friendly behaviors to her image. At what age did she realize that the charming baby gazing and grinning back was really herself?

Emergence of the I-Self and Me-Self. To answer this question, researchers have exposed infants and toddlers to images of themselves in mirrors, on videotapes, and in photos. When shown two side-by-side video images of their kicking legs, one from their own perspective (camera behind the baby) and one from an observer's perspective (camera in front of the baby), 3-month-olds look longer at the observer's view (Rochat, 1998). This suggests that within the first few months, infants seem to have some sense of their own body as a distinct entity, since they have habituated to it, as indicated by their interest in novel images.

Researchers agree that the earliest aspect of the self to emerge is the **I-self,** or *sense of self as agent,* involving awareness that the self is separate from the surrounding world and can control its own thoughts and actions. According to many theorists, the beginnings of the I-self lie in infants' recognition that their own actions cause objects and people to react in predictable ways (Harter, 1998). In support of this idea, babies whose parents encourage them to explore and respond to their signals consistently and sensitively are advanced in constructing a sense of self as agent. For example, between 1 and 2 years of age, they display more complex self-related actions during make-believe play, such as making a doll labeled as the self take a drink or kiss a teddy bear (Pipp, Easterbrooks, & Harmon, 1992).

Then, as infants act on the environment, they notice different effects that help them sort out self from other people and objects (Rochat, 2001). For example, batting a mobile and seeing it swing in a pattern different from the infant's own actions informs the baby about the relation between self and physical world. Smiling and vocalizing at a caregiver who smiles and vocalizes back helps specify the relation between self and social world. And watching the movements of one's own hand provides still another kind of feedback—one under much more direct control than other people or objects.

A second aspect of the self is the **me-self,** a *sense of self as an object of knowledge and evaluation.* It consists of all qualities that make the self unique, including physical characteristics, possessions, and (as the child gets older) attitudes, beliefs, and personality traits. During the second year, toddlers start to construct a me-self; they become aware of the self's features. In one study, 9- to 24-month-olds were placed in front of a mirror. Then, under the pretext of wiping the baby's face, each mother was asked to rub red dye on her infant's nose. Younger infants touched the mirror as if the red mark had nothing to do with them. But by 15 months, toddlers rubbed their strange-looking red noses. They were keenly aware of their unique appearance (Lewis & Brooks-Gunn, 1979). By age 2, almost all children recognize themselves in photos and use their name or a personal pronoun ("I" or "me") to refer to themselves.

© LAURA DWIGHT/PHOTOEDIT

This 1-year-old notices the correspondence between her own movements and the movements of the image in the mirror, a cue that helps her figure out that the grinning baby is really herself.

Like the I-self, the me-self seems to be fostered by sensitive caregiving. Securely attached toddlers display more complex knowledge of their own and their parents' features (for example, by labeling body parts) than do their insecurely attached agemates (Pipp, Easterbrooks, & Brown, 1993).

Self-Awareness and Early Emotional and Social Development. Self-awareness quickly becomes a central part of children's emotional and social lives. Recall that self-conscious

■ Encouraging this toddler to help wipe up spilled milk fosters compliance and the beginnings of self-control. He joins in the clean-up task with an eager, willing spirit, which suggests that he is beginning to adopt the adult's directive as his own.

emotions depend on an emerging sense of self. As the self strengthens in the second year, toddlers show the beginnings of self-conscious behavior: bashfulness and embarrassment. Self-awareness also leads to the child's first efforts to appreciate another's perspective. It is accompanied by the first signs of **empathy**—the ability to understand another's emotional state and *feel with* that person, or respond emotionally in a similar way. For example, toddlers start to give to others what they themselves find comforting—a hug, a reassuring comment, or a favorite doll or blanket (Zahn-Waxler et al., 1992). At the same time, they demonstrate clearer awareness of how to upset others. One 18-month-old heard her mother comment to another adult, "Anny (sibling) is really frightened of spiders. In fact, there's a particular toy spider that we've got that she just hates" (Dunn, 1989, p. 107). The innocent-looking toddler ran to get the spider out of the toy box, returned, and pushed it in front of Anny's face!

Categorizing the Self

Once children have a me-self, they use their representational and language capacities to create a mental image of themselves. One of the first signs of this change is that toddlers compare themselves to other people. Between 18 and 30 months, children categorize themselves and others on the basis of age ("baby," "boy," or "man"), sex ("boy" versus "girl"), physical characteristics ("big," "strong"), and even goodness and badness ("I a good girl." "Tommy mean!") (Stipek, Gralinski, & Kopp, 1990). Toddlers' understanding of these social categories is limited, but they use this knowledge to organize their own behavior. For example, children's ability to label their own gender is associated with a sharp rise in gender-stereotyped responses (Fagot & Leinbach, 1989). As early as 18 months, toddlers select and play in a more involved way with toys that are stereotyped for their own gender—dolls and tea sets for girls, trucks and cars for boys. Then parents encourage these preferences by responding positively when toddlers display them (Fagot, Leinbach, & O'Boyle, 1992). As we will see in Chapter 8, gender-typed behavior increases dramatically in early childhood.

Emergence of Self-Control

Self-awareness also provides the foundation for **self-control,** the capacity to resist an impulse to engage in socially disapproved behavior. Self-control is essential for morality, another dimension of the self that will flourish during childhood and adolescence. To behave in a self-controlled fashion, children must have some ability to think of themselves as separate, autonomous beings who can direct their own actions. And they must have the representational and memory capacities to recall a caregiver's directive (such as "Caitlin, don't touch that light socket!") and apply it to their own behavior. The ability to shift attention from a captivating stimulus and focus on a less attractive alternative, supported by development of the frontal lobes of the cerebral cortex, is also essential (Rothbart & Bates, 1998).

As these capacities emerge, the first glimmerings of self-control appear between 12 and 18 months as **compliance.** Toddlers show clear awareness of caregivers' wishes and expectations and can obey simple requests and commands (Kaler & Kopp, 1990). And as every parent knows, they can also decide to do just the opposite! One way toddlers assert their autonomy is by resisting adult directives. But among toddlers who experience warm parenting, opposition is far less common than compliance with an eager, willing spirit, which suggests that the child is beginning to adopt the adult's directive as his

Helping Toddlers Develop Compliance and Self-Control

SUGGESTION	RATIONALE
Respond to the toddler with sensitivity and support.	Toddlers whose parents are sensitive and supportive are more compliant and self-controlled.
Provide advance notice when the toddler must stop an enjoyable activity.	Toddlers find it more difficult to stop a pleasant activity already under way than to wait before engaging in a desired action.
Offer many prompts and reminders.	Toddlers' ability to remember and comply with rules is limited; they need continuous adult oversight.
Respond to self-controlled behavior with verbal and physical approval.	Praise and hugs reinforce appropriate behavior, increasing its likelihood of occurring again.
Encourage sustained attention (see Chapter 5, pages 153–154).	Early sustained attention is related to self-control. Children who can shift attention from a captivating stimulus and focus on a less attractive alternative are better at controlling their impulses.
Support language development (see Chapter 5, pages 168–169).	Early language development is related to self-control. During the second year, children begin to use language to remind themselves of adult expectations.
Gradually increase rules in accord with the toddler's developing capacities.	As cognition and language improve, toddlers can follow more rules related to safety, respect for people and property, family routines, manners, and simple chores.

own (Kochanska, Tjebkes, & Forman, 1998). Compliance quickly leads to toddlers' first consciencelike verbalizations, as when Caitlin said, "No, can't" as she reached out for a light socket (Kochanska, 1993).

Around 18 months, the capacity for self-control appears, and it improves steadily into early childhood. In one study, toddlers were given three tasks that required them to resist temptation. In the first, they were asked not to touch an interesting toy telephone that was within arm's reach. In the second, raisins were hidden under cups, and they were instructed to wait until the adult said it was all right to pick up a cup and eat a raisin. In the third, they were told not to open a gift until the adult had finished her work. On all three problems, the ability to wait increased steadily between 18 and 30 months (Vaughn, Kopp, & Krakow, 1984).

Early, large individual differences in self-control remain modestly stable into middle childhood and adolescence (Shoda, Mischel, & Peake, 1990). Children who are advanced in sustained attention and language development are more self-controlled, so girls usually are ahead of boys (Cournoyer, Solomon, & Trudel, 1998; Rothbart, 1989). Already, some toddlers use verbal techniques, such as singing and talking to themselves, to keep from engaging in a prohibited act. In addition, mothers who are sensitive and supportive have toddlers who show greater gains in self-control (Kochanska, Murray, & Harlan, 2000). Such parenting seems to encourage as well as model patient, nonimpulsive behavior.

As self-control improves, mothers increase the rules they require toddlers to follow, from safety and respect for property and people to family routines, manners, and simple chores (Gralinski & Kopp, 1993). Still, toddlers' control over their own actions depends on constant parental oversight and reminders. To get Caitlin to stop playing so that she and her parents could go on an errand, several prompts ("Remember, we're going to go in just a minute") and gentle insistence were usually necessary. The Caregiving Concerns table above summarizes ways to help toddlers develop compliance and self-control.

As the second year of life drew to a close, Carolyn, Monica, and Vanessa were delighted at their children's readiness to learn the rules of social life. As we will see in Chapter 8, advances in cognition and language, along with parental warmth and reasonable maturity demands, lead children to make tremendous strides in this area during early childhood.

Ask Yourself

REVIEW

Why is insisting that infants comply with parental directives inappropriate? What competencies are necessary for the emergence of compliance and self-control?

CONNECT

What type of early parenting fosters the development of emotional self-regulation, attachment, and self-control? Why, in each instance, is it effective?

www.

Summary

Erikson's Theory of Infant and Toddler Personality

What personality changes take place during Erikson's stages of basic trust versus mistrust and autonomy versus shame and doubt?

- According to Erikson, warm, responsive caregiving leads infants to resolve the psychological conflict of **basic trust versus mistrust** on the positive side. During toddlerhood, the conflict of **autonomy versus shame and doubt** is resolved favorably when parents provide appropriate guidance and reasonable choices. If children emerge from the first few years without sufficient trust and autonomy, the seeds are sown for adjustment problems.

Emotional Development

Describe changes in happiness, anger, and fear over the first year, noting the adaptive function of each.

- During the first half-year, **basic emotions** become clear, well-organized signals. The **social smile** appears between 6 and 10 weeks, laughter around 3 to 4 months. Happiness strengthens the parent–child bond and reflects as well as supports physical and cognitive mastery. Anger and fear, especially in the form of **stranger anxiety,** increase in the second half of the first year. These reactions have survival value as infants' motor capacities improve.

Summarize changes during the first 2 years in understanding others' emotions, expression of self-conscious emotions, and emotional self-regulation.

- The ability to understand the feelings of others expands over the first year. Between 7 and 10 months, babies perceive facial expressions as organized patterns. Soon after, **social referencing** appears; infants actively seek emotional information from caregivers in uncertain situations. By the middle of the second year, infants appreciate that others' emotional reactions may differ from their own.

- During toddlerhood, self-awareness and adult instruction provide the foundation for **self-conscious emotions,** such as shame, embarrassment, and pride. Caregivers help infants with **emotional self-regulation** by relieving distress, engaging in stimulating play, and discouraging negative emotion. During the second year, growth in representation and language leads to more effective ways of regulating emotion.

Temperament and Development

What is temperament, and how is it measured?

- Infants differ greatly in **temperament,** or quality and intensity of emotion, activity level, attention, and emotional self-regulation. On the basis of parental descriptions of children's behavior, three patterns of temperament—the **easy child,** the **difficult child,** and the **slow-to-warm-up child**—were identified in the New York Longitudinal Study. Difficult children, especially, are likely to display adjustment problems.

- In addition to parental reports, behavior ratings, and direct observation of children, researchers use physiological measures. These assessments distinguish **inhibited,** or **shy, children** from **uninhibited,** or **sociable, children,** who may inherit a physiology that biases them toward a particular temperamental style.

Discuss the role of heredity and environment in the stability of temperament, including the goodness-of-fit model.

- Stability of temperament is generally low to moderate. Temperament has a genetic foundation, but child rearing and cultural beliefs and practices have much to do with maintaining or changing it. The **goodness-of-fit model** describes how temperament and environment work together to affect later development. Parenting practices that create a good fit with the child's temperament help difficult and shy children achieve more adaptive functioning.

Development of Attachment

Describe ethological theory of attachment and the development of attachment during the first 2 years.

- The most widely accepted perspective on development of **attachment** is **ethological theory.** It views babies as biologically prepared to contribute actively to ties established with their caregivers, which promote survival by ensuring both safety and competence.

- In early infancy, a set of built-in behaviors encourages the parent to remain close to the baby. Around 6 to 8 months, **separation anxiety** and use of the parent as a **secure base** indicate that a true attachment bond has formed. As representation and language develop, toddlers try to alter the parent's coming and going through requests and persuasion. Out of early caregiving experiences, children construct an **internal working model** that serves as a guide for all future close relationships.

Describe the Strange Situation, the four attachment patterns assessed by it, and factors that affect attachment security.

- The **Strange Situation** is a widely used technique for assessing the quality of attachment between 1 and 2 years of age. Four attachment patterns have been identified: **secure attachment, avoidant attachment, resistant attachment,** and **disorganized–disoriented attachment.** Securely attached babies in middle-SES families experiencing favorable life conditions more often maintain their attachment pattern than do insecure babies. An exception is the disorganized–disoriented pattern, which is highly stable. Cultural conditions must be considered in interpreting reactions to the Strange Situation.

- Attachment quality is influenced by the infant's opportunity to develop a close affectional tie with one or a few adults, **sensitive caregiving,** the fit between the baby's temperament and parenting practices, and family circumstances. In some (but not all) cultures, **interactional synchrony** characterizes the experiences of securely attached babies. Parents' internal working models are good predictors of infant attachment patterns. However, the transfer of parents' childhood experiences to quality of attachment with their own children is affected by many factors.

Discuss infants' attachments to fathers and siblings.

- Infants develop strong affectionate ties to fathers, whose sensitive caregiving predicts secure attachment. Fathers in a variety of cultures engage in more exciting, physical play with babies than do mothers. Early in the first year, infants begin to build rich emotional relationships with siblings that mix affection and caring with rivalry and resentment. Individual differences in the quality of sibling relationships are influenced by temperament and parenting practices.

Describe and interpret the relationship between secure attachment in infancy and cognitive, emotional, and social competence in childhood.

- Evidence for the impact of early attachment quality on cognitive, emotional, and social competence in later years is mixed. Continuity of caregiving may be the crucial factor that determines whether attachment security is linked to later development.

Self-Development During the First Two Years

Describe the development of self-awareness in infancy and toddlerhood, along with the emotional and social capacities it supports.

- The earliest aspect of the self to emerge is the **I-self,** a sense of self as agent. Its beginnings lie in infants' recognition that their own actions cause objects and people to react in predictable ways. During the second year, toddlers start to construct the **me-self,** a sense of self as an object of knowledge and evaluation. They become keenly aware of the self's visual appearance and, by age 2, use their name or a personal pronoun to refer to themselves.
- Self-awareness leads to toddlers' first efforts to appreciate another's perspective and to compare themselves to others. Social categories based on age, sex, physical characteristics, and goodness and badness are evident in toddlers' language. Self-awareness also provides the foundation for self-conscious emotions, **empathy, compliance,** and **self-control.**

Important Terms and Concepts

attachment (p. 185)
autonomy versus shame and doubt (p. 175)
avoidant attachment (p. 187)
basic emotions (p. 175)
basic trust versus mistrust (p. 174)
compliance (p. 196)
difficult child (p. 181)
disorganized–disoriented attachment (p. 187)
easy child (p. 180)
emotional self-regulation (p. 179)
empathy (p. 196)
ethological theory of attachment (p. 186)
goodness-of-fit model (p. 184)
I-self (p. 195)
inhibited, or shy, child (p. 182)
interactional synchrony (p. 189)
internal working model (p. 187)
me-self (p. 195)
resistant attachment (p. 187)
secure attachment (p. 187)
secure base (p. 186)
self-conscious emotions (p. 179)
self-control (p. 196)
sensitive caregiving (p. 189)
separation anxiety (p. 186)
slow-to-warm-up child (p. 181)
social referencing (p. 178)
social smile (p. 177)
Strange Situation (p. 187)
stranger anxiety (p. 177)
temperament (p. 180)
uninhibited, or sociable, child (p. 182)

FYI For Further Information and Help

Consult the Companion Website for *Development Through the Lifespan, Third Edition,* (www.ablongman.com/berk) where you will find the following resources for this chapter:

- **Chapter Objectives**
- **Flashcards** for studying important terms and concepts
- **Annotated Weblinks** to guide you in further research
- **Ask Yourself** questions, which you can answer and then check against a sample response
- **Suggested Readings**
- **Practice Test** with immediate scoring and feedback

Milestones

Development in Infancy and Toddlerhood

AGE	PHYSICAL	COGNITIVE	LANGUAGE	EMOTIONAL/SOCIAL
Birth–6 months	• Rapid height and weight gain. (114) • Reflexes decline. (101) • Sleep organized into a day-night schedule. (121) • Holds head up, rolls over, and reaches for objects. (130, 132) • Can be classically and operantly conditioned. (126–128) • Habituates to unchanging stimuli; recovers to novel stimuli. (128) • Hearing well developed; by the end of this period, displays greater sensitivity to speech sounds of own language. (133) • Depth and pattern perception emerge and improve. (134–138)	• Engages in immediate imitation and deferred imitation of adults' facial expressions. (128–129, 147) • Repeats chance behaviors leading to pleasurable and interesting results. (146) • Violation-of-expectation tasks suggest some awareness of object permanence. (147) • Attention becomes more efficient and flexible. (153–154) • Recognition memory for people, places, and objects improves. (154–155) • Forms perceptual categories based on objects' similar features. (155–157)	• Engages in cooing and babbling. (165) • Establishes joint attention with caregiver, who labels objects and events. (166) 	• Shows signs of almost all basic emotions (happiness, interest, surprise, fear, anger, sadness, disgust). (175–178) • Social smile and laughter emerge. (177) • Matches adults' emotional expressions during face-to-face interaction. (175, 178) • Emotional expressions become better organized and clearly tied to social events. (175) • I-self emerges. (195) 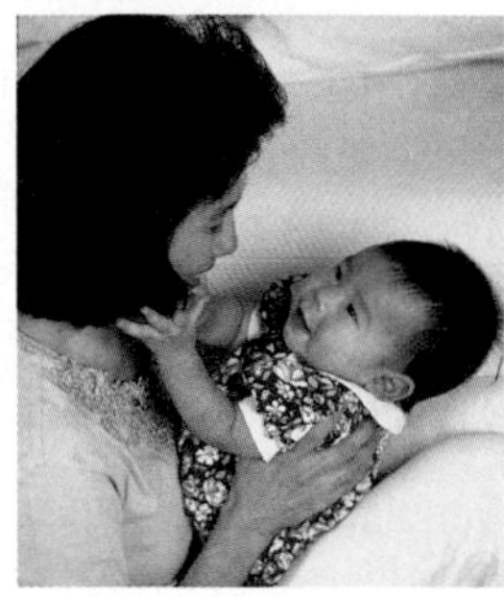
7–12 months	 • Sits alone, crawls, and walks. (130) • Shows refined pincer grasp. (132) • Perceives larger speech units crucial to understanding meaning. (133) • Depth and pattern perception improve further. (134–139)	• Combines sensorimotor schemes. (146) • Engages in intentional, or goal-directed, behavior. (146) • Finds object hidden in one place. (146) • Engages in deferred imitation of adults' actions with objects. (147) • Recall memory for people, places, and objects improves. (155) • Solves simple problems by analogy. (149–150) • Groups stimuli into wide range of meaningful categories. (155–157)	• Babbling expands to include sounds of spoken languages and the child's language community. (165-166) • Uses preverbal gestures (showing, pointing) to communicate. (166–167) 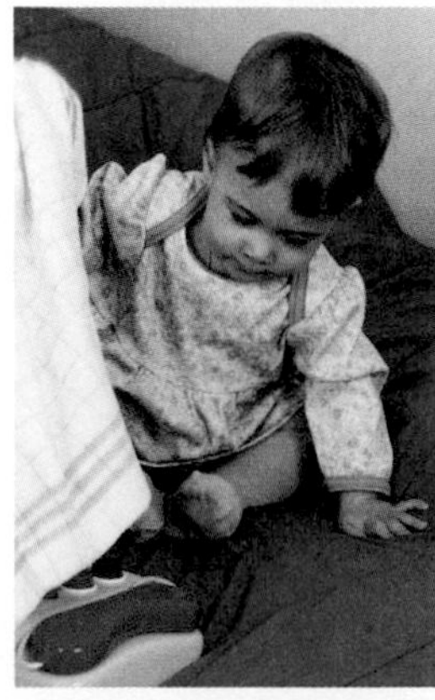	• Anger and fear increase in frequency and intensity. (177) • Stranger anxiety and separation anxiety appear. (177) • Uses caregiver as a secure base for exploration. (187) • Shows "clear-cut" attachment to familiar caregivers. (187) • Ability to detect the meaning of others' emotional expressions improves. (178) • Engages in social referencing. (178)

AGE	PHYSICAL	COGNITIVE	LANGUAGE	EMOTIONAL/SOCIAL
13–18 months	• Height and weight gain rapid, but not as great as in first year. (114) • Walking better coordinated. (130) • Manipulates small objects with improved coordination. (132) 	• Experiments with objects in a trial-and-error fashion. (146) • Finds object hidden in more than one place. (146) • Sorts objects into categories. (155–157) • Imitates actions across a change in context—for example, from home to laboratory. (149) • Sustained attention improves. (154)	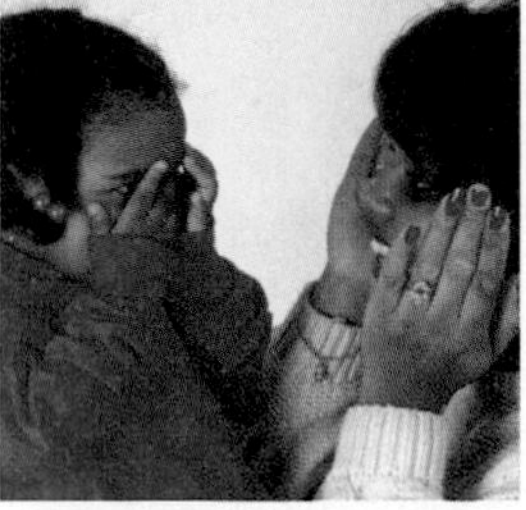 • Joint attention with caregiver becomes more accurate. (166) • Takes turns in games, such as pat-a-cake and peekaboo. (166) • Says first word. (167)	• Joins in play with familiar adults and siblings. (192–193) • Me-self emerges; recognizes image of self in mirror. (195) • Shows signs of empathy. (196) • Complies with simple directives. (196–197)
19–24 months	• Jumps, runs, and climbs. (130) • Manipulates small objects with good coordination. (132) 	• Solves sensorimotor problems suddenly. (147) • Finds object moved while out of sight. (146) • Engages in deferred imitation of actions an adult tries to produce, even if not fully realized. (147) • Engages in make-believe play. (158–159) • Sorts objects into categories more effectively. (155–157) • Recall memory for people, places, and objects improves further. (155)	• Vocabulary increases to 200 words. (166–167) • Combines two words. (167–168) 	• Self-conscious emotions (shame, embarrassment, guilt, and pride) emerge. (179) • Acquires a vocabulary of emotional terms. (180) • Begins using language to assist with emotional self-regulation. (180) • Begins to tolerate caregiver's absences more easily. (187) • Uses own name or personal pronoun to label self. (195) • Categorizes self and others on the basis of age, sex, physical characteristics, and goodness and badness. (196) • Shows gender-stereotyped toy choices. (196) • Self-control appears. (196–197)

Note: Numbers in parentheses indicate the page on which each milestone is discussed.

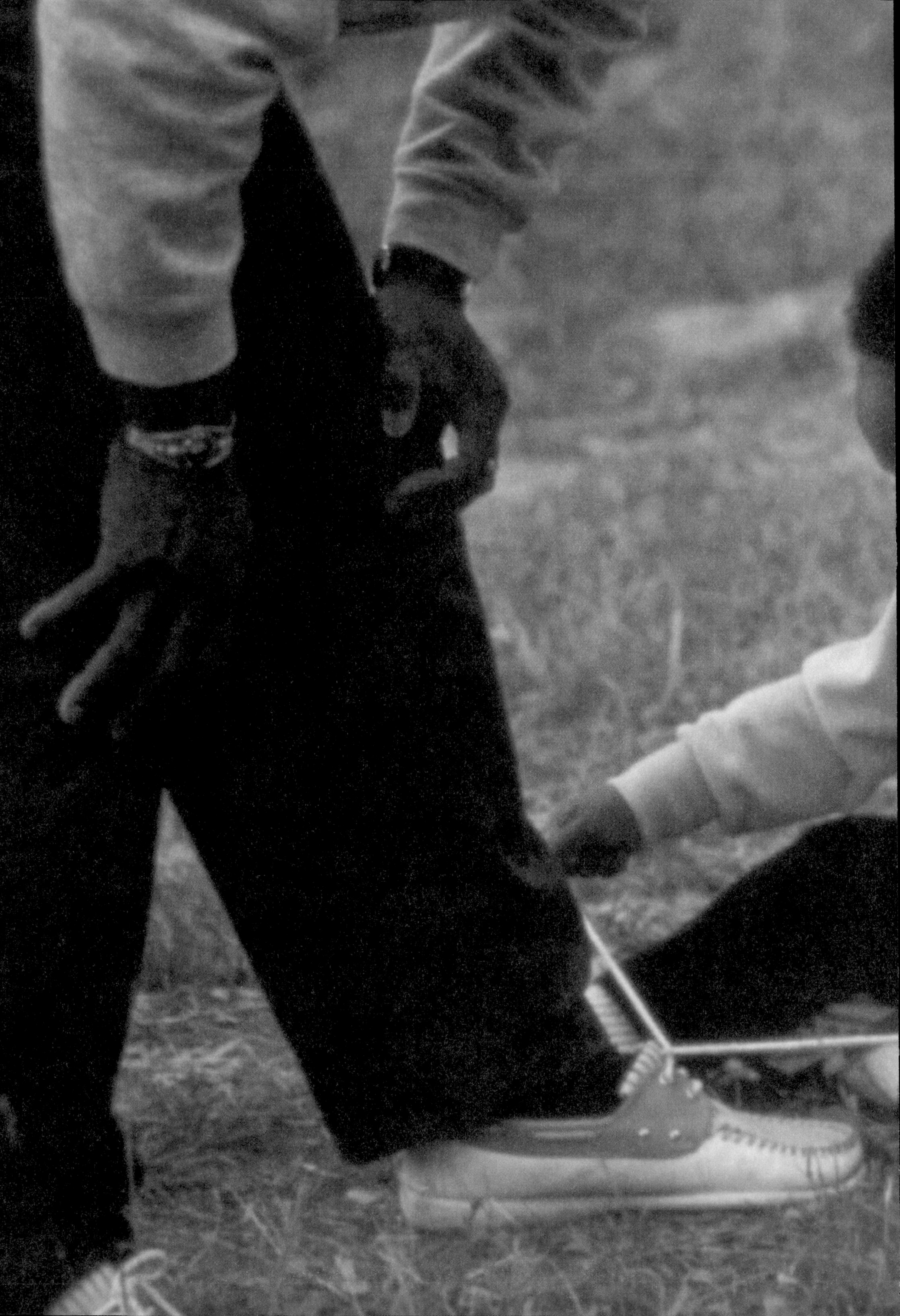

7

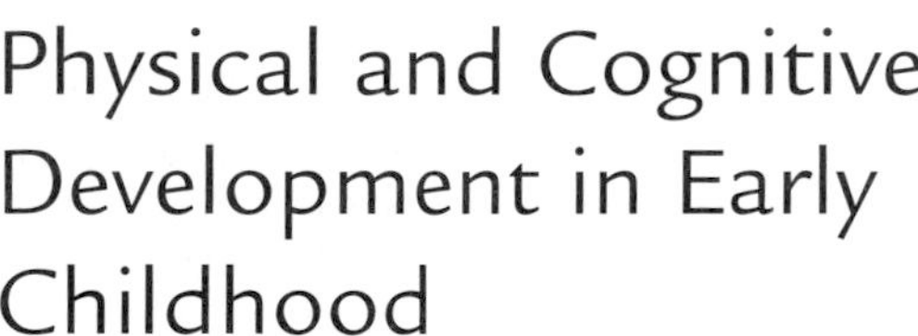

Physical and Cognitive Development in Early Childhood

© DORANNE JACOBSON

A transformed body and explosion of new motor and cognitive skills contribute to an expanding sense of competence in early childhood. Although it will take some time for this 3-year-old to master the intricate steps of shoe-tying, he is eager to get started.

Physical Development

Body Growth
Skeletal Growth • Asynchronies in Physical Growth

Brain Development
Handedness • Other Advances in Brain Development

Influences on Physical Growth and Health
Heredity and Hormones • Emotional Well-Being • Nutrition • Infectious Disease • Childhood Injuries

■ Social Issues: Chronic Middle Ear Infection in Early Childhood: Consequences for Development

Motor Development
Gross Motor Development • Fine Motor Development • Individual Differences in Motor Skills

Cognitive Development

Piaget's Theory: The Preoperational Stage
Advances in Mental Representation • Make-Believe Play • Limitations of Preoperational Thought • Follow-Up Research on Preoperational Thought • Evaluation of the Preoperational Stage • Piaget and Education

Vygotsky's Sociocultural Theory
Private Speech • Social Origins of Early Childhood Cognition • Vygotsky and Education • Evaluation of Vygotsky's Theory

■ Cultural Influences: Young Children's Daily Life in a Yucatec Mayan Village

Information Processing
Attention • Memory • The Young Child's Theory of Mind • Early Childhood Literacy • Young Children's Mathematical Reasoning

■ Biology & Environment: "Mindblindness" and Autism

Individual Differences in Mental Development
Home Environment and Mental Development • Preschool, Kindergarten, and Child Care • Educational Television

Language Development
Vocabulary • Grammar • Conversation • Supporting Language Learning in Early Childhood

For more than a decade, my fourth-floor office window overlooked the preschool and kindergarten play yard of our university laboratory school. On mild fall and spring mornings, the doors of the classrooms swung open, and sand table, woodworking bench, easels, and large blocks spilled out into a small, fenced courtyard. Around the side of the building was a grassy area with jungle gyms, swings, a small playhouse, and a flower garden planted by the children. Beyond it, I could see a circular path lined with tricycles and wagons. Each day, the setting was alive with activity.

The years from 2 to 6 are often called "the play years," and aptly so, since play blossoms during this time and supports every aspect of development. Our discussion of early childhood opens with the physical achievements of this period—growth in body size, improvements in motor coordination, and refinements in perception. We pay special attention to biological and environmental factors that support these changes, as well as to their intimate connection with other domains of development.

Then we explore the many facets of early childhood cognition, beginning with Piaget's preoperational stage. Recent research, along with Vygotsky's sociocultural theory and information processing, extends our understanding of preschoolers' cognitive competencies. Next, we turn to a variety of factors that contribute to early childhood mental development—the home environment, the quality of preschool and child care, and the many hours children spend watching television. Our chapter concludes with language development, the most awesome achievement of the preschool years.

The children whom I came to know well, first by watching from my office window and later by observing at close range in their classrooms, will join us as we trace the changes of early childhood. They will provide many examples of developmental trends and individual differences.

PHYSICAL DEVELOPMENT

Body Growth

The rapid increase in body size of the first 2 years tapers off into a slower growth pattern during early childhood. On average, children add 2 to 3 inches in height and about 5 pounds in weight each year. Boys continue to be slightly larger than girls. At the same time, the "baby fat" that began to decline in toddlerhood drops off further. The child gradually becomes thinner, although girls retain somewhat more body fat, and boys are slightly more muscular. As the torso lengthens and widens, internal organs tuck neatly inside, and the spine straightens. As Figure 7.1 shows, by age 5 the top-heavy, bowlegged, potbellied toddler has become a more streamlined, flat-tummied, longer-legged child with body proportions similar to those of adults. Consequently, posture and balance improve—changes that support the gains in motor coordination that we will take up later.

Individual differences in body size are even more apparent during early childhood than in infancy and toddlerhood. Looking down at the play yard one day, I watched 5-year-old Darryl speed around the bike path. At 48 inches tall and 55 pounds, he towered over his kindergarten classmates and was, as his mother put it, "off the growth charts" at the doctor's office. (The average North American 5-year-old boy is 43 inches tall and weighs 42 pounds.) Priti, an Asian-Indian child, was unusually small because of genetic factors linked to her cultural ancestry. Lynette and Hallie, two Caucasian children with impoverished home lives, were well below average for reasons we will discuss shortly.

Skeletal Growth

The skeletal changes that are under way in infancy continue throughout early childhood. Between ages 2 and 6, approximately 45 new *epiphyses,* or growth centers in which cartilage hardens into bone, emerge in various parts of the skeleton. Other epiphyses will appear in middle childhood. X-rays of these growth centers enable doctors to estimate children's *skeletal age,* or progress toward physical maturity (see Chapter 4, page 115). During early and middle childhood, information about skeletal age is helpful in diagnosing growth disorders.

Parents and children are especially aware of another aspect of skeletal growth: By the end of the preschool years, children start to lose their primary, or "baby," teeth. The age at which they do so is heavily influenced by genetic factors. For example, girls, who are ahead of boys in physical development, lose their primary teeth sooner. Environmental influences, especially prolonged malnutrition, can delay the appearance of permanent teeth.

Even though primary teeth are temporary, dental care is important. Diseased baby teeth can affect the health of permanent teeth. Brushing consistently, avoiding sugary foods, drinking fluoridated water, and getting topical fluoride treatments and sealants (plastic coatings that protect tooth surfaces) prevent cavities. Unfortunately, childhood tooth decay remains high, especially among low-SES children. An estimated 40 percent of North American 5-year-olds have at least some tooth decay. By the time they graduate from high school, about 80 percent of young people have decayed teeth (World Health Organization, 2001). Poor diet, lack of fluoridation in some communities, and inadequate health care are responsible.

Andy at 3 years

Andy at 4 years

Andy at 5 years

Andy at 5¾ years

Amy at 3 years

Amy at 3½ years

Amy at 4½ years

Amy at 5½ years

FIGURE 7.1 Body growth during early childhood.
Andy and Amy grew more slowly during the preschool years than they did in infancy and toddlerhood (see Chapter 4, page 115). By age 5, their bodies became more streamlined, flat-tummied, and longer legged. Boys continue to be slightly taller and heavier and more muscular than girls. But generally, the two sexes are similar in body proportions and physical capacities.

Asynchronies in Physical Growth

Body systems differ in their unique, carefully timed patterns of maturation. As Figure 7.2 on page 206 shows, physical growth is *asynchronous.* Body size (as measured by height and weight) and a variety of internal organs follow the **general growth curve,** which involves rapid growth during infancy, slower gains in early and middle childhood, and rapid growth again during adolescence. Yet there are exceptions to this trend. The genitals develop slowly from birth to age 4, change little throughout middle childhood, and then grow rapidly during adolescence. In contrast, the lymph glands grow at an astounding pace in infancy and childhood, and then their rate of growth declines in adolescence. The lymph system helps fight infection and assists with absorption of nutrients, thereby supporting children's health and survival (Malina & Bouchard, 1991).

Figure 7.2 on page 206 illustrates another growth trend with which you are already familiar: During the first few years, the brain grows more rapidly than any other part of the body. Let's look at some highlights of brain development during early childhood.

FIGURE 7.2 Growth of three different organ systems and tissues contrasted with the body's general growth. Growth is plotted in terms of percentage of change from birth to 20 years. Notice how growth of lymph tissue rises to nearly twice its adult level by the end of childhood. Then it declines. (Reprinted by permission of the publisher from J. M. Tanner, 1990, *Foetus into Man,* 2nd ed., Cambridge, MA: Harvard University Press, p. 16. Copyright © 1990 by J. M. Tanner. All rights reserved.)

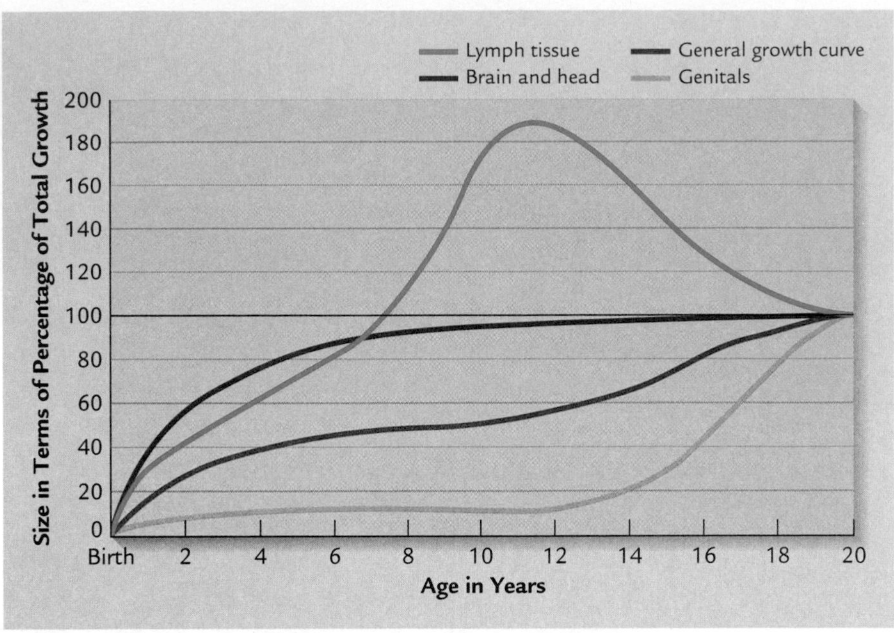

Brain Development

Between ages 2 and 6 years, the brain increases from 70 to 90 percent of its adult weight. Brain-imaging studies reveal that energy metabolism in the cerebral cortex reaches a peak around age 4. By this time, many cortical regions have overproduced synapses, resulting in a high energy need (Johnson, 1998). As *formation of synapses, myelinization,* and *synaptic pruning* continue (see Chapter 4), preschoolers improve in a wide variety of skills—physical coordination, perception, attention, memory, language, logical thinking, and imagination.

Recall that the cerebral cortex is made up of two *hemispheres.* Measures of neural activity in various cortical regions reveal especially rapid growth from 3 to 6 years in frontal-lobe areas devoted to planning and organizing behavior. Furthermore, for most children, the left hemisphere is especially active between 3 and 6 years and then levels off. In contrast, activity in the right hemisphere increases steadily throughout early and middle childhood (Thatcher, Walker, & Giudice, 1987; Thompson et al., 2000).

These findings fit nicely with what we know about several aspects of cognitive development. Language skills (typically housed in the left hemisphere) increase at an astonishing pace in early childhood, and they support children's increasing control over behavior. In contrast, spatial skills (such as finding one's way from place to place, drawing pictures, and recognizing geometric shapes) develop gradually over childhood and adolescence. Differences in rate of development between the two hemispheres suggest that they are continuing to *lateralize* (specialize in cognitive functions). Let's take a closer look at brain lateralization in early childhood by focusing on handedness.

Handedness

One morning on a visit to the preschool, I watched 3-year-old Moira as she drew pictures, worked puzzles, joined in snacktime, and played outside. Unlike most of her classmates, Moira does most things—drawing, eating, and zipping her jacket—with her left hand. But she uses her right hand for a few activities, such as throwing a ball. Hand preference is evident in 10 percent of 1-year-olds and strengthens during early

© LAURA DWIGHT

■ Twins usually lie in opposite orientations in the uterus during the prenatal period, which may explain why they are more likely to differ in handedness than ordinary siblings. Although left-handedness is associated with developmental problems, the large majority of left-handed children are normal, and some develop outstanding verbal and mathematical talents.

childhood. Ninety percent of 5-year-olds prefer one hand over the other (Oztürk et al., 1999).

A strong hand preference reflects the greater capacity of one side of the brain—the individual's **dominant cerebral hemisphere**—to carry out skilled motor action. Other important abilities may be located on the dominant side as well. In support of this idea, for right-handed people, who make up 90 percent of the population in Western nations, language is housed with hand control in the left hemisphere. For the remaining left-handed 10 percent, language is often shared between the hemispheres (Knecht et al., 2000). This indicates that the brains of left-handers tend to be less strongly lateralized than those of right-handers. Consistent with this idea, many left-handed individuals (like Moira) are *ambidextrous.* Although they prefer their left hand, they sometimes use their right hand skillfully as well (McManus et al., 1988).

Is handedness hereditary? Researchers disagree on this issue. Left-handed parents show only a weak tendency to have left-handed children. One genetic theory proposes that most children inherit a gene that *biases* them for right-handedness and a left-dominant cerebral hemisphere. However, that bias is not strong enough to overcome experiences that might sway children toward a left-hand preference (Annett, 2002).

Research confirms that experience can profoundly affect handedness. Both identical and fraternal twins are more likely than ordinary siblings to differ in handedness. The hand preference of each twin is related to body position in the uterus; twins usually lie in opposite orientations (Derom et al., 1996). According to one theory, the way most fetuses lie—turned toward the left—may promote greater postural control by the right side of the body (Previc, 1991). Also, wide cultural differences exist in rates of left-handedness. For example, in Tanzania, Africa, children are physically restrained and punished for favoring their left hand. Less than 1 percent of Tanzanians are left-handed (Provins, 1997).

Perhaps you have heard that left-handedness is more frequent among severely retarded and mentally ill people than it is in the general population. Although this is true, recall that when two variables are correlated, one does not necessarily cause the other. Atypical lateralization is probably not responsible for the problems of these individuals. Instead, they may have suffered early damage to the left hemisphere, which caused their disabilities and also led to a shift in handedness. In support of this idea, left-handedness is associated with prenatal and birth difficulties that can result in brain damage, including prolonged labor, prematurity, Rh incompatibility, and breech delivery (O'Callaghan et al., 1993; Powls et al., 1996).

Keep in mind, however, that only a small number of left-handers show developmental problems. In fact, unusual lateralization may have certain advantages. Left- and mixed-handed youngsters are more likely than their right-handed agemates to develop outstanding verbal and mathematical talents (Flannery & Liederman, 1995). More even distribution of cognitive functions across both hemispheres may be responsible.

Other Advances in Brain Development

Besides the cortex, other parts of the brain make strides during early childhood (see Figure 7.3). As we look at these changes, you will see that all involve establishing links between different parts of the brain, increasing the coordinated functioning of the central nervous system.

At the rear and base of the brain is the **cerebellum,** a structure that aids in balance and control of body movement. Fibers linking the cerebellum to the cerebral cortex begin to myelinate after birth, but they do not complete this process until about age 4 (Tanner, 1990). This change undoubtedly contributes to dramatic gains in motor control, so that by the end of the preschool years children can play hopscotch, pump a playground swing, and throw a ball with a well-organized set of movements.

The **reticular formation,** a structure of the brain that maintains alertness and consciousness, myelinates throughout childhood and into adolescence. Neurons in the reticular formation send out fibers to the frontal lobes of the cortex, contributing to improvements in sustained, controlled attention.

The **corpus callosum** is a large bundle of fibers that connects the two cortical hemispheres. Myelinization of the corpus callosum does not begin until the end of the first year of life. Between 3 and 6 years, it grows rapidly and then enlarges at a slower pace into adolescence (Thompson et al., 2000). The corpus callosum supports integration of many aspects of thinking, including perception, attention, memory, language, and problem solving. The more complex the task, the more crucial communication between the hemispheres becomes.

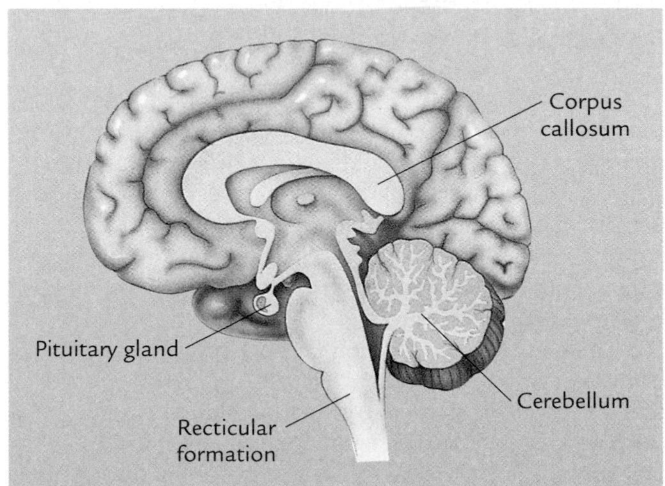

FIGURE 7.3 Cross section of the human brain, showing the location of the cerebellum, the reticular formation, and the corpus callosum. These structures undergo considerable development during early childhood. Also shown is the pituitary gland, which secretes hormones that control body growth (see page 208).

Ask Yourself

REVIEW
Cite evidence indicating that both heredity and environment contribute to handedness.

REVIEW
What aspects of brain development support the tremendous gains in language, thinking, and motor control of early childhood?

CONNECT
Explain why brain growth in early childhood involves not only an increase in neural connections but also a loss of synapses and cell death. Which assumption of the lifespan perspective do these processes illustrate? (If you need to review, see Chapter 1, pages 9–10, and Chapter 4, page 116.)

www.

Influences on Physical Growth and Health

In earlier chapters, we considered a wide variety of influences on physical growth during the prenatal period and infancy. As we discuss growth and health in early childhood, we will encounter some familiar themes. Although heredity remains important, environmental factors continue to play crucial roles. Emotional well-being, good nutrition, relative freedom from disease, and physical safety are essential.

Heredity and Hormones

The impact of heredity on physical growth is evident throughout childhood. Children's physical size and rate of growth are related to those of their parents (Malina & Bouchard, 1991). Genes influence growth by controlling the body's production of hormones. The **pituitary gland,** located at the base of the brain, plays a critical role by releasing two hormones that induce growth.

The first is **growth hormone (GH)**, which from birth on is necessary for development of all body tissues except the central nervous system and genitals. Children who lack GH reach an average mature height of only 4 feet, 4 inches. When treated with injections of GH starting at an early age, these GH-deficient children show catch-up growth and then grow at a normal rate, reaching a height much greater than they would have without treatment (Pasquino et al., 2001).

The second pituitary hormone affecting children's growth, **thyroid-stimulating hormone (TSH)**, stimulates the thyroid gland (located in the neck) to release *thyroxine,* which is necessary for normal development of the nerve cells of the brain and for GH to have its full impact on body size. Infants born with a deficiency of thyroxine must receive it at once or they will be mentally retarded. At later ages, children with too little thyroxine grow at a below-average rate. By then, the central nervous system is no longer affected because the most rapid period of brain development is complete. With prompt treatment, such children catch up in body growth and eventually reach normal size (Salerno et al., 2001).

Emotional Well-Being

In childhood as in infancy, emotional well-being can have a profound effect on growth and health. Preschoolers with very stressful home lives (due to divorce, financial difficulties, or a change in their parents' employment status) suffer more respiratory and intestinal illnesses and more unintentional injuries than others (Cohen & Herbert, 1996).

Extreme emotional deprivation can interfere with the production of GH and lead to **psychosocial dwarfism,** a growth disorder that appears between 2 and 15 years of age. Typical characteristics include very short stature, decreased GH secretion, immature skeletal age, and serious adjustment problems, which help distinguish psychosocial dwarfism from normal shortness (Doeker et al., 1999; Voss, Mulligan, & Betts, 1998). Lynette, the very small 4-year-old mentioned earlier in this chapter, was diagnosed with this condition. She had been placed in foster care after child welfare authorities discovered that she spent most of the day at home alone, unsupervised. She also may have been physically abused. When such children are removed from their emotionally inadequate environments, their GH levels quickly return to normal, and they grow rapidly. But if treatment is delayed, the dwarfism can be permanent.

Nutrition

With the transition to early childhood, many children become unpredictable and picky eaters. One father I know wistfully recalled his son's eager sampling of the cuisine at a Chinese restaurant during toddlerhood. "He ate rice, chicken chow mein, egg rolls, and more. Now, at age 3, the only thing he'll try is the ice cream!"

This decline in appetite is normal. It occurs because growth has slowed. Furthermore, preschoolers' wariness of new foods may be adaptive. By sticking to familiar foods, they are less likely to swallow dangerous substances when adults are not around to protect them (Birch & Fisher, 1995). Parents need not worry about variations in amount eaten from meal to meal. Preschoolers compensate for a meal in which they ate little with a later one in which they eat more (Hursti, 1999).

Even though they eat less, preschoolers need a high-quality diet. They require the same foods adults do—only smaller amounts. Fats, oils, and salt should be kept to a minimum because of their link to high blood pressure and heart disease in adulthood (Winkleby et al., 1999). Foods high in sugar should also be avoided. In addition to causing tooth decay, they lessen young children's appetite for healthy foods and increase their risk of overweight and obesity—a topic we will take up in Chapter 9.

© TOM MCCARTHY/PHOTOEDIT

■ Children's food tastes are trained by foods served often in their culture. Many Western children would refuse the spicy noodle dish these Japanese preschoolers are eating with enthusiasm.

The social environment powerfully influences young children's food preferences. Children tend to imitate the food choices of people they admire—adults as well as peers. For example, in Mexico, children often see family members delighting in the taste of peppery foods. Consequently, Mexican preschoolers enthusiastically eat chili peppers, whereas most North American children reject them (Birch, Zimmerman, & Hind, 1980). A pleasant mealtime climate also encourages healthy eating. Repeated exposure to a new food (without any direct pressure to eat it) increases children's acceptance (Sullivan & Birch, 1990). Sometimes parents bribe their children, saying, "Finish your vegetables, and you can have an extra cookie." This practice causes children to like the healthy food less and the treat more. Too much parental control over children's eating limits their opportunities to develop self-control (Birch, 1998).

Finally, as indicated in earlier chapters, many children in the United States and in developing countries are deprived of diets that support healthy growth. Five-year-old Hallie was bused to our laboratory preschool from a poor neighborhood. His mother's welfare check barely covered her rent, let alone food. Hallie's diet was deficient in protein as well as essential vitamins and minerals—iron (to prevent anemia), calcium (to support development of bones and teeth), vitamin A (to help maintain eyes, skin, and a variety of internal organs), and vitamin C (to facilitate iron absorption and wound healing). These are the most common dietary deficiencies of the preschool years (Kennedy, 1998). Not surprisingly, Hallie was thin, pale, and tired. By age 7, North American low-SES children are, on average, about 1 inch shorter than their economically advantaged counterparts (Yip, Scanlon, & Trowbridge, 1993).

Infectious Disease

Two weeks into the school year, I looked out my window and noticed that Hallie was absent from the play yard. Several weeks passed; still, I did not see him. When I asked Leslie, his preschool teacher, what had happened, she explained, "Hallie's been hospitalized with the measles. He's had a difficult time recovering—lost weight when there wasn't much to lose in the first place." In well-nourished children, ordinary childhood illnesses have no effect on physical growth. But when children are undernourished, disease interacts with malnutrition in a vicious spiral, and the consequences for physical growth can be severe.

● Infectious Disease and Malnutrition. Hallie's reaction to the measles is commonplace among children in developing nations, where a large proportion of the population lives in poverty. In these countries, many children do not receive a program of immunizations. Illnesses such as measles and chicken pox, which typically do not appear until after age 3 in industrialized nations, occur much earlier. Poor diet depresses the body's immune system, making children far more susceptible to disease. Of the 10 million annual worldwide deaths in children under age 5, 99 percent are in developing countries and 70 percent are due to infectious diseases (World Health Organization, 2000c).

Disease, in turn, is a major cause of malnutrition and, through it, hinders physical growth. Illness reduces appetite and limits the body's ability to absorb foods. These outcomes are especially severe in children with intestinal infections. In developing countries, diarrhea is widespread and increases in early childhood because of unsafe water and contaminated foods, leading to growth stunting and several million childhood deaths each year (Shann & Steinhoff, 1999).

Most growth retardation and deaths due to diarrhea can be prevented with nearly cost-free **oral rehydration therapy (ORT),** in which sick children are given a glucose, salt, and water solution that quickly replaces fluids the body loses. Since 1990, public health workers have taught nearly half the families in the developing world how to administer ORT. As a result, the lives of more than 1 million children are being saved annually (Victora et al., 2000).

● Immunization. In industrialized nations, childhood diseases have declined dramatically during the past half-century, largely due to widespread immunization of infants and young children. Hallie got the measles because, unlike his classmates from more advantaged homes, he did not receive a full program of immunizations. Overall, 24 percent of American preschoolers lack essential immunizations, a rate that rises to 40 percent for poverty-stricken preschoolers. These children do not receive full protection until age 5 or 6, when it is required for school entry (U.S. Department of Health and Human Services, 2002e). In contrast, fewer than 10 percent of preschoolers lack immunizations in Denmark and Norway, and fewer than 7 percent in Canada, the Netherlands, and Sweden (Bellamy, 2000; Health Canada, 2000c).

How have these countries managed to achieve higher rates of immunization than the United States? In earlier chapters, we noted that many children in the United States do not have access to the health care they need. In 1994, all medically uninsured American children were guaranteed free immunizations, a program that has led to a steady improvement in early childhood immunization rates.

Inability to pay for vaccines, however, is only one cause of inadequate immunization. Misconceptions also contribute—for example, the notion that vaccines do not work or that they weaken the immune system (Gellin, Maibach, & Marcuse, 2000). Furthermore, some parents have been influenced by media reports suggesting that the measles–mumps–rubella vaccine has contributed to a rise in number of children diagnosed with autism. Yet large-scale studies show no association between immunization and autism (Dales, Hammer, & Smith, 2001; DeStafano & Chen, 2001). Public education programs directed at increasing parental knowledge about the importance and safety of timely immunizations are badly needed.

A final point regarding communicable disease in early childhood deserves mention: Childhood illness rises with child-care attendance. On average, a child-care infant gets sick 9 to 10 times a year, a child-care preschooler 6 to 7 times. Diseases that spread most rapidly are diarrhea and respiratory infections—the illnesses most frequently suffered by young children. The risk that a respiratory infection will result in *otitis media,* or middle ear infection, is almost double that of children remaining at home (Uhari, Måntysaari, & Niemelä, 1996). To learn about the consequences of otitis media and how to prevent it, consult the Social Issues box on the following page.

Childhood Injuries

Three-year-old Tory caught my eye as I visited the preschool classroom one day. More than any other child, he had trouble sitting still and paying attention at storytime. Outside, he darted from one place to another, spending little time at any single activity. On a field trip to our campus museum, Tory ignored Leslie's directions and ran across the street without holding his partner's hand. Later in the year, I read in our local newspaper that Tory had narrowly escaped serious injury when he put his mother's car into gear while she was outside scraping its windows. The vehicle rolled through a guardrail and over the side of a 10-foot concrete underpass. There it hung until rescue workers arrived. Tory's mother was charged with failing to use a restraint seat for children under age 5.

Unintentional injuries—auto collisions, pedestrian accidents, drownings, poisonings, firearm wounds, burns, falls, and swallowing of foreign objects—are the leading cause of childhood mortality in industrialized countries (Roberts & DiGuiseppi, 1999). Figure 7.4 reveals that compared with other developed nations, the United States and Canada rank poorly in these largely preventable events. Nearly half of American and Canadian childhood deaths and three-fourths of adolescent deaths are due to injury (Children's Defense Fund, 2002; Health Canada, 2001a). And among injured children and youths who survive, thousands suffer pain, brain damage, and permanent physical disabilities.

Auto and traffic accidents, drownings, and burns are the most common injuries during early childhood. Motor vehicle collisions are by far the most frequent source of injury at all ages, ranking as the leading cause of death among children more than 1 year old.

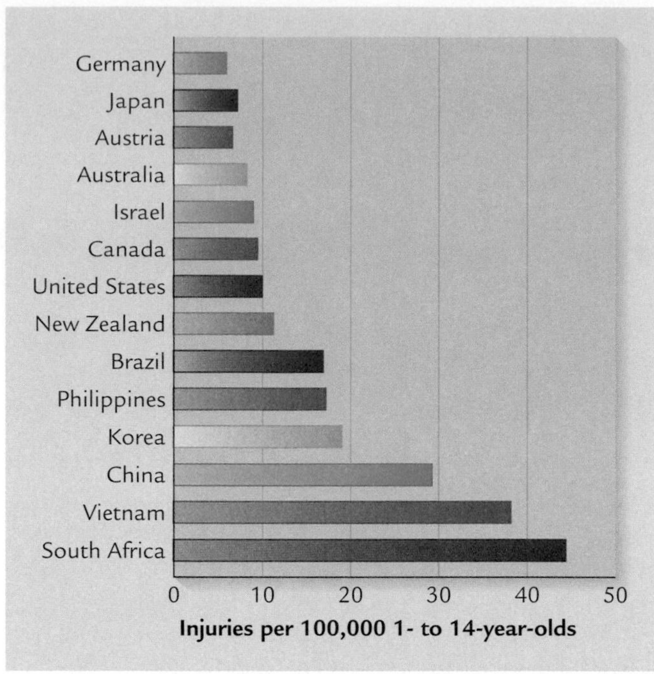

FIGURE 7.4 International rates of unintentional injury among 1- to 14-year-olds. Compared with other industrialized nations, the United States and Canada have high injury rates, due to widespread childhood poverty and shortages of high-quality child care. Injury rates are many times higher in developing nations, where poverty, rapid population growth, overcrowding in cities, and inadequate safety measures endanger children's lives. (Adapted from Safe Kids Worldwide, 2002; Health Canada, 2001a.)

● Factors Related to Childhood Injuries. We are used to thinking of childhood injuries as "accidental." But a close look reveals that meaningful causes underlie them, and we can, indeed, do something about them.

As Tory's case suggests, individual differences exist in the safety of children's behaviors. Because of their higher activity level and greater willingness to take risks during play, boys are more likely to be injured than girls (Laing & Logan, 1999). Temperamental characteristics—irritability, inattentiveness, and negative mood—are also related to childhood injuries. As we saw in Chapter 6, children with these traits present special child-rearing challenges. They are likely to protest when placed in auto seat restraints, to refuse to take a companion's hand when crossing the street, and to disobey even after repeated adult instruction and discipline (Matheny, 1991).

Poverty, low parental education, and more children in the home are also strongly associated with injury (Bradbury et al., 1999). Parents who must cope with many daily stresses often have little time and energy to monitor the safety of their youngsters. And their homes and neighborhoods pose further risks. Noise, crowding, and confusion characterize run-down, inner-city neighborhoods with few safe places to play (Kronenfeld & Glik, 1995).

Broad societal conditions also affect childhood injury. In developing nations, injury deaths before age 15 are five times higher than in developed nations and soon may exceed disease

Social Issues

Chronic Middle Ear Infection in Early Childhood: Consequences for Development

During his first year in child care, 18-month-old Alex caught five colds, had the flu on two occasions, and experienced *otitis media,* or repeated middle ear infection. Alex is not unusual. By age 3, more than 70 percent of North American children have had respiratory illnesses resulting in at least one bout of otitis media; 33 percent have had three or more bouts (Daly, Hunter, & Giebink, 1999). Although antibiotics eliminate the bacteria responsible for otitis media, they do not reduce fluid buildup in the middle ear, which causes mild to moderate hearing loss that can last for weeks or months.

The incidence of otitis media is greatest between 6 months and 3 years, when children are first acquiring language. Frequent and prolonged infections predict delayed language progress, reduced task persistence, social isolation in early childhood, and poorer academic performance after school entry (Miccio et al., 2001; Roberts et al., 2000; Rvachew et al., 1999).

How might otitis media disrupt language and academic progress? Difficulties in hearing speech sounds, particularly in noisy settings, may be responsible. Children with many bouts are less attentive to the speech of others and less persistent at tasks (Petinou et al., 2001; Roberts, Burchinal, & Campbell, 1994). Their distractibility may be due to repeated instances in which they could not make out what people around them were saying. When children have trouble paying attention, they may reduce the quality of others' interactions with them. In one study, mothers of preschoolers with frequent illnesses were less effective in teaching their child a task (Chase et al., 1995).

Current evidence strongly favors early prevention of otitis media, especially since the illness is so widespread. Crowded living conditions and exposure to cigarette smoke and other pollutants are linked to the disease—factors that probably account for its high incidence among low-SES children. In addition, child care creates opportunities for close contact, greatly increasing otitis media episodes.

© LAURA DWIGHT

■ High-quality child care reduces or eliminates language delays, social isolation, and later academic difficulties associated with frequent bouts of otitis media. These children profit from a verbally stimulating caregiver and a small group size, which ensures a relatively quiet environment where spoken language can be heard easily.

Negative outcomes of early otitis media can be prevented in the following ways:

- *Preventive doses of xylitol, a sweetener derived from birch bark.* A Finnish study revealed that children given a daily dose of xylitol in gum or syrup form show a 30 to 40 percent drop in otitis media compared with controls given gum or syrup without the sweetener. Xylitol appears to have natural, bacteria-fighting ingredients (Uhari, Kontiokari, & Niemelä, 1998). However, dosage must be carefully monitored, since too much xylitol can cause abdominal pain and diarrhea.
- *Frequent screening for the disease, followed by prompt medical intervention.* Plastic tubes that drain the inner ear are often used to treat chronic otitis media, although their effectiveness remains controversial.
- *Child-care settings that control infection.* Because infants and young children often put toys in their mouths, these objects should be rinsed frequently with a disinfectant. Spacious, well-ventilated rooms and small group sizes also limit the spread of disease.
- *Verbally stimulating adult–child interaction.* Developmental problems associated with otitis media are reduced or eliminated in responsive home environments and in high-quality child-care centers. When caregivers are verbally stimulating and keep noise to a minimum, children have more opportunities to hear spoken language (Roberts et al., 1998; Vernon-Feagans, Hurley, & Yont, 2002).

as the leading cause of childhood mortality (refer again to Figure 7.4). Poverty, rapid population growth, overcrowding in cities, and heavy road traffic combined with weak safety measures are major causes. Safety devices, such as car safety seats and bicycle helmets, are neither readily available nor affordable in most developing countries. To purchase a child safety seat requires more than 100 hours of wages in Vietnam, 53 hours in China, and only 2.5 hours in the United States (Safe Kids Worldwide, 2002).

Among developed nations, injury rates are high in the United States and Canada because of high child poverty rates, shortages of high-quality child care (to supervise children in their parents' absence), and—especially in the United States—a high rate of births to teenagers (who are neither psychologically nor financially ready to raise a child). But North American children from advantaged families are also at somewhat greater risk for injury than are children in European nations (Williams & Kotch, 1990). This indicates that besides reducing poverty and teenage pregnancy and upgrading the status of child care, additional steps must be taken to ensure children's safety.

● **Preventing Childhood Injuries.** Childhood injuries have many causes, so a variety of approaches are needed to control them (Tremblay & Peterson, 1999). Laws prevent many injuries by requiring car safety seats, child-resistant caps on medicine bottles, flameproof clothing, and fenced-in backyard swimming pools (the site of 90 percent of early childhood drownings).

Communities can help by modifying their physical environments. Inexpensive and widely available public transportation can reduce the time that children spend in cars. Playgrounds, a common site of injury, can be covered with protective surfaces, such as rubber matting, sand, and wood chips (Dowd, 1999). Free, easily installed window guards can be given to families in high-rise apartment buildings to prevent falls. And widespread media and information campaigns can inform parents and children about safety issues.

Nevertheless, even though they know better, many parents and children behave in ways that compromise safety. Preschoolers spontaneously recall only about half of their parents' home safety rules; they need prompting to remember others and supervision to ensure that they comply even with well-learned rules (Morrongiello, Midgett, & Shields, 2001). A variety of programs based on *behavior modification* (modeling and reinforcement) have improved safety practices. In one, counselors helped parents identify dangers in the home—fire hazards, objects that young children might swallow, poisons, firearms, and others. Then they demonstrated specific ways to eliminate the dangers (Tertinger, Greene, & Lutzker, 1984). Some interventions reward parents and children with prizes if the children arrive at child care or school restrained in car seats (Roberts, Alexander, & Knapp, 1990).

Efforts like these have been remarkably successful, yet their focus is fairly narrow—on decreasing specific environmental risks and risky behaviors (Peterson & Brown, 1994). Attention must also be paid to family conditions that can prevent childhood injury: relieving crowding in the home, providing social supports to ease parental stress, and teaching parents to use effective discipline, a topic we will take up in Chapter 8.

■ This Vietnamese mother and preschooler bike without helmets, and the child rides in an unsafe position. In developing countries, safety devices are neither readily available nor affordable, and overcrowding in cities and heavy road traffic result in childhood injury deaths five times higher than in developed nations.

Ask Yourself

REVIEW

Using research on malnutrition or unintentional injuries, show how physical growth and health in early childhood result from a continuous, complex interplay between heredity and environment.

APPLY

One day, Leslie prepared a new snack to serve at preschool: celery stuffed with ricotta cheese and pineapple. The first time she served it, few children touched it. How can Leslie encourage her students to accept the snack? What tactics should she avoid?

CONNECT

Using ecological systems theory, suggest ways to reduce childhood injuries by intervening in the microsystem, the mesosystem, and the macrosystem.

www.

Motor Development

Visit a playground at a neighborhood park, preschool, or child-care center, and observe several 2- to 6-year-olds. You will see that an explosion of new motor skills occurs in early childhood, each of which builds on the simpler movement patterns of toddlerhood.

The same principle that governs motor development during the first 2 years of life continues to operate during the preschool years. Children integrate previously acquired skills into more complex, *dynamic systems of action.* (Return to Chapter 4, page 130, if you need to review this concept.) Then they revise each new skill as their bodies grow larger and stronger, their central nervous systems develop, and their environments present new challenges.

Gross Motor Development

As children's bodies become more streamlined and less top-heavy, their center of gravity shifts downward, toward the trunk. As a result, balance improves greatly, paving the way for new motor skills involving large muscles of the body (Ulrich & Ulrich, 1985). By age 2, preschoolers' gaits become smooth and rhythmic—secure enough that soon they leave the ground, at first by running and later by jumping, hopping, galloping, and skipping.

As children become steadier on their feet, their arms and torsos are freed to experiment with new skills—throwing and catching balls, steering tricycles, and swinging on horizontal bars and rings. Then upper- and lower-body skills combine into more refined actions. Five- and 6-year-olds simultaneously steer and pedal a tricycle and flexibly move their whole body when throwing, catching, hopping, and jumping. By the end of the preschool years, all skills are performed with greater speed and endurance. Table 7.1 provides a closer look at gross motor development in early childhood.

Fine Motor Development

Like gross motor development, fine motor skills take a giant leap forward during the preschool years. Because control of the hands and fingers improves, young children put puzzles together, build with small blocks, cut and paste, and string beads. To parents, fine motor progress is most apparent in two areas: (1) children's care of their own bodies, and (2) the drawings and paintings that fill the walls at home, child care, and preschool.

● **Self-Help Skills.** As Table 7.1 shows, young children gradually become self-sufficient at dressing and feeding, although parents need to be patient about these abilities. When tired and in a hurry, young children often revert to eating with their fingers. And the 3-year-old who dresses himself sometimes ends up with his shirt on inside out, his pants on backwards, and his left snow boot on his right foot! Perhaps the most complex self-help skill of early childhood is shoe tying, mastered around age 6. Success requires a longer attention span, memory for an intricate series of hand movements, and the dexterity to perform them. Shoe tying illustrates the close connection between motor and cognitive development. Drawing and writing offer additional examples.

Table 7.1 Changes in Gross and Fine Motor Skills During Early Childhood

Age	Gross Motor Skills	Fine Motor Skills
2–3 years	Walks more rhythmically; hurried walk changes to run. Jumps, hops, throws, and catches with rigid upper body. Pushes riding toy with feet; little steering.	Puts on and removes simple items of clothing. Zips and unzips large zippers. Uses spoon effectively.
3–4 years	Walks up stairs, alternating feet, and downstairs, leading with one foot. Jumps and hops, flexing upper body. Throws and catches with slight involvement of upper body; still catches by trapping ball against chest. Pedals and steers tricycle.	Fastens and unfastens large buttons. Serves self food without assistance. Uses scissors. Copies vertical line and circle. Draws first picture of person, using tadpole image.
4–5 years	Walks downstairs, alternating feet. Runs more smoothly. Gallops and skips with one foot. Throws ball with increased body rotation and transfer of weight on feet; catches ball with hands. Rides tricycle rapidly, steers smoothly.	Uses fork effectively. Cuts with scissors following line. Copies triangle, cross, and some letters.
5–6 years	Increases running speed. Gallops more smoothly; engages in true skipping. Displays mature throwing and catching pattern. Rides bicycle with training wheels.	Uses knife to cut soft food. Ties shoes. Draws person with six parts. Copies some numbers and simple words.

Sources: Cratty, 1986; Getchell & Roberton, 1989; Newborg, Stock, & Wnek, 1984; Roberton, 1984.

● **From Scribbles to Pictures.** A variety of factors combine with fine motor control to influence the development of children's artful representations. These include cognitive advances—the realization that pictures can serve as symbols and gains in planning skills and spatial understanding, which result in a move from a focus on separate objects to a broader visual perspective (Golomb, 1992). The emphasis that the child's culture places on artistic expression also makes a difference.

Typically, drawing progresses through the following sequence:

1. *Scribbles.* Western children begin to draw during the second year. At first, gestures rather than the resulting scribbles contain the intended representation. For example, one 18-month-old took her crayon and hopped it around the page, explaining as she made a series of dots, "Rabbit goes hop-hop" (Winner, 1986).

2. *First representational forms.* By age 3, children's scribbles start to become pictures. Often this happens after they make a gesture with the crayon, notice that they have drawn a recognizable shape, and then decide to label it (Winner, 1986). Although few 3-year-olds spontaneously draw so others can tell what their picture represents, after an adult shows the child how pictures can be used to stand for objects, more children draw recognizable forms (Callaghan, 1999).

   A major milestone in drawing occurs when children use lines to represent the boundaries of objects. This enables 3- and 4-year-olds to draw their first picture of a person. Look at the tadpole image—a circular shape with lines attached—on the left in Figure 7.5. It is a universal one in which fine motor and cognitive limitations lead the preschooler to reduce the figure to the simplest form that still looks human.

3. *More realistic drawings.* Young children do not demand that a drawing be realistic. But as cognitive and fine motor skills improve, they learn to desire greater realism. As a result, they create more complex drawings, like the one on the right in Figure 7.5, by a 6-year-old child. These drawings contain more conventional figures, in which the head and body are differentiated. Still, older preschoolers' drawings contain perceptual distortions, since they have just begun to represent depth. Use of depth cues improves during middle childhood (Cox & Littlejohn, 1995). And instead of depicting objects separately (as in the drawing in Figure 7.5), older school-age children relate them to one another in an organized spatial arrangement (Case & Okamoto, 1996).

● **Cultural Variations in Children's Drawings.** In cultures with rich artistic traditions, children's drawings reflect the conventions of their culture and are more elaborate. In cultures with little interest in art, even older children and adolescents produce simple forms. The Jimi Valley is a remote region of Papua New Guinea with no indigenous pictorial art. Many Jimi children do not go to school and therefore have little opportunity to develop drawing skills. When a Western researcher asked nonschooled Jimi 10- to 15-year-olds to draw a

■ **FIGURE 7.5 Examples of young children's drawings.** The universal tadpolelike shape that children use to draw their first picture of a person is shown on the left. The tadpole soon becomes an anchor for greater detail as arms, fingers, toes, and facial features sprout from the basic shape. By the end of the preschool years, children produce more complex, differentiated pictures like the one on the right drawn by a 6-year-old child. (Tadpole drawings from H. Gardner, 1980, *Artful Scribbles: The Significance of Children's Drawings,* New York: Basic Books, p. 64. Reprinted with permission of Basic Books, a division of HarperCollins Publishers, Inc. Six-year-old's picture from E. Winner, August 1986, "Where Pelicans Kiss Seals," *Psychology Today, 20*[8], p. 35. Reprinted by permission of the author.)

FIGURE 7.6 Drawings produced by nonschooled 10- to 15-year-old children of the Jimi Valley of Papua New Guinea when asked to draw a human figure for the first time. Many produced nonrepresentational scribbles and shapes (a), "stick" figures (b), or "contour" figures (c). Compared with the Western tadpole form, the Jimi "stick" and "contour" figures emphasize the hands and feet. Otherwise, the drawings of these older children resemble those of young preschoolers. (From M. Martlew & K. J. Connolly, 1996, "Human Figure Drawings by Schooled and Unschooled Children in Papua New Guinea," *Child Development, 67,* pp. 2750–2751. © The Society for Research in Child Development, Inc. Adapted by permission.)

human figure for the first time, most produced nonrepresentational scribbles and shapes or simple "stick" or "contour" images (see Figure 7.6). These forms resemble preschoolers' and seem to be a universal beginning in drawing. Once children realize that lines must evoke human features, they find solutions to figure drawing that vary somewhat from culture to culture but, overall, follow the sequence described earlier.

● **Early Printing.** As preschoolers experiment with lines and shapes, notice print in storybooks, and observe people writing, they try to print letters and, later, words. Often the first word printed is the child's name. Initially, it may be represented by a single letter. "How do you make a *D?*" my older son, David, asked at age 3. When I printed a large uppercase *D,* he tried to copy. "*D* for David," he said as he wrote, quite satisfied with his backward, imperfect creation. A year later, David added several letters, and around age 5, he wrote his name clearly enough that others could read it. Like many children, David continued to reverse some letters in his printing until well into second grade. Not until they learn to read do children find it useful to distinguish between mirror-image forms, such as *b* and *d* and *p* and *q* (Bornstein, 1999; Casey, 1986).

Individual Differences in Motor Skills

We have discussed motor milestones largely in terms of the average age at which children reach them in Western nations, but, of course, wide individual differences occur. A child with a tall, muscular body tends to move more quickly and to acquire certain skills earlier than a short, stocky youngster. Researchers believe that body build contributes to the superior performance of African-American over Caucasian children in running and jumping. African-American youngsters tend to have longer limbs, so they have better leverage (Lee, 1980; Wakat, 1978).

Sex differences in motor skills are evident in early childhood. Boys are slightly ahead of girls in skills that emphasize force and power. By age 5, they can jump slightly farther, run slightly faster, and throw a ball much farther (about 5 feet farther). Girls have an edge in fine motor skills and in certain gross motor skills that require a combination of good balance and foot movement, such as hopping and skipping (Cratty, 1986; Thomas & French, 1985). Boys' greater muscle mass and (in the case of throwing) slightly longer forearms may contribute to their skill advantages. And girls' greater overall physical maturity may be partly responsible for their better balance and precision of movement.

From an early age, boys and girls are usually channeled into different physical activities. For example, fathers often play catch with their sons but seldom do so with their daughters. Baseballs and footballs are purchased for boys, jump ropes and hula hoops for girls. As children get older, sex differences in motor skills get larger, yet differences in physical capacity remain small until adolescence. These trends suggest that social pressures for boys to be active and physically skilled and for girls to play quietly at fine motor activities may exaggerate small, genetically based sex differences (Coakley, 1990; Greendorfer, Lewko, & Rosengren, 1996).

Children master the motor skills of early childhood as part of their everyday play. Aside from throwing (where direct instruction seems to make some difference), there is no evidence that preschoolers exposed to formal lessons are ahead in motor development. When children have play spaces appropriate for running, climbing, jumping, and throwing and have access to puzzles, construction sets, and art materials that promote manipulation, drawing, and writing, they respond eagerly to these challenges.

Finally, the social climate created by adults can enhance or dampen preschoolers' motor progress. When parents and teachers criticize a child's performance, push specific motor skills, or promote a competitive attitude, they risk undermining young children's self-confidence and, in turn, their motor progress (Kutner, 1993). Adults involved in young children's motor activities should focus on their having fun rather than on their learning the "correct" technique.

Ask Yourself

REVIEW

Describe typical changes in children's drawings during early childhood, along with factors that contribute to those differences.

REVIEW

Explain how physical and social factors jointly contribute to sex differences in motor skills during early childhood.

APPLY

Mabel and Chad want to do everything they can to support their 3-year-old daughter's athletic development. What advice would you give them?

www.

COGNITIVE DEVELOPMENT

One rainy morning, as I observed in our laboratory preschool, Leslie, the children's teacher, joined me at the back of the room to watch for a moment herself. "Preschoolers' minds are such a blend of logic, fantasy, and faulty reasoning," Leslie reflected. "Every day, I'm startled by the maturity and originality of what they say and do. Yet at other times, their thinking seems limited and inflexible."

Leslie's comments sum up the puzzling contradictions of early childhood cognition. Over the previous week, I had seen many examples as I followed the activities of 3-year-old Sammy. That day, I found him at the puzzle table, moments after a loud clash of thunder outside. Sammy looked up, startled, then turned to Leslie and pronounced, "The man turned on the thunder!" Leslie patiently explained that people can't turn thunder on or off. "Then a lady did it," Sammy stated with certainty.

In other respects, Sammy's cognitive skills seemed surprisingly advanced. At snack time, he accurately counted, "One, two, three, four!" and then got four cartons of milk, giving one to each child at his table. But when more than four children joined his snack group, Sammy's counting broke down. And some of his notions about quantity seemed as fantastic as his understanding of thunder. Across the snack table, Priti dumped out her raisins, and they scattered in front of her. "How come you got lots, and I only got this little bit?" asked Sammy, failing to realize that he had just as many; they were simply all bunched up in a tiny red box. Piaget's theory helps us understand Sammy's curious reasoning.

Piaget's Theory: The Preoperational Stage

As children move from the sensorimotor to the **preoperational stage,** which spans the years 2 to 7, the most obvious change is an extraordinary increase in representational, or symbolic, activity. Recall that infants and toddlers have some ability to represent their world. During early childhood, this capacity blossoms.

Advances in Mental Representation

Piaget acknowledged that language is our most flexible means of mental representation. By detaching thought from action, it permits far more efficient thinking than was possible earlier. When we think in words, we overcome the limits of our momentary experiences. We can deal with past, present, and future at once and combine concepts in unique ways, as when we think about a hungry caterpillar eating bananas or monsters flying through the forest at night.

Despite the power of language, Piaget did not believe that it plays a major role in cognitive development. Instead, he believed that sensorimotor activity leads to internal images of experience, which children then label with words (Piaget, 1936/1952). In support of Piaget's view, recall from Chapter 5 that the first words toddlers use have a strong sensorimotor basis. In addition, toddlers acquire an impressive range of categories long before they use words to label them (see page 155). Still, other theorists regard Piaget's account of the link between language and thought as incomplete, as we will see later in this chapter.

Make-Believe Play

Make-believe play is another excellent example of the development of representation during early childhood. Piaget believed that through pretending, young children practice and strengthen newly acquired representational schemes. Drawing on his ideas, several investigators have traced the development of make-believe during the preschool years.

Development of Make-Believe. One day, Sammy's 18-month-old brother, Dwayne, came to visit the classroom. Dwayne wandered around, picked up the receiver of a toy telephone, said, "Hi, Mommy," and then dropped it. In the housekeeping area, he found a cup, pretended to drink, and then toddled off again. In the meantime, Sammy joined a group of children in the block area for a space shuttle launch.

"That can be our control tower," he suggested to Vance and Lynette, pointing to a corner by a bookshelf. "Countdown!" Sammy announced, speaking into a small wooden block, his pretend walkie-talkie. "Five, six, two, four, one, blastoff!" Lynette made a doll push a pretend button, and the rocket was off!

A comparison of Dwayne's pretend with that of Sammy illustrates three important changes. Each reflects the preschool child's growing symbolic mastery:

- *Over time, play increasingly detaches from the real-life conditions associated with it.* In early pretending, toddlers use only realistic objects—for example, a toy telephone to talk into or a cup to drink from. Most of these first pretend acts imitate adults' actions and are not yet flexible. Children younger than age 2, for example, will pretend to drink from a cup but refuse to pretend a cup is a hat (Tomasello, Striano, & Rochat, 1999). As with drawings, at first preschoolers have difficulty understanding that an object with an obvious use can stand for another object (DeLoache & Smith, 1999).

  After age 2, children pretend with less realistic toys, such as a block standing in for a telephone receiver. And during the third year, they can flexibly imagine objects and events, without any support from the real world, as Sammy's imaginary control tower illustrates (Corrigan, 1987; O'Reilly, 1995).

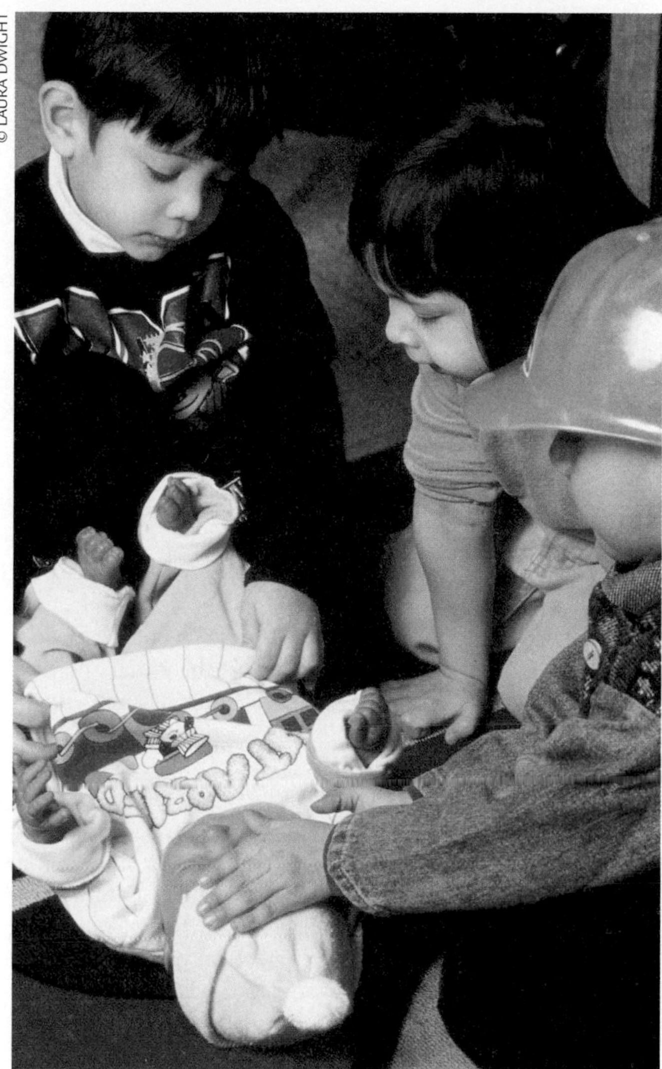

■ These 3- and 4-year-olds coordinate several make-believe roles as they jointly care for a sick baby. Sociodramatic play contributes to cognitive, emotional, and social development.

- *Play becomes less self-centered with age.* At first, make-believe is directed toward the self—for example, Dwayne pretends to feed only himself. A short time later, children direct pretend actions toward other objects, as when the child feeds a doll. And early in the third year, they become detached participants who make a doll feed itself or (in Lynette's case) push a button to launch a rocket. Make-believe gradually becomes less self-centered as children realize that agents and recipients of pretend actions can be independent of themselves (Corrigan, 1987; McCune, 1993).
- *Play gradually includes more complex scheme combinations.* For example, Dwayne can pretend to drink from a cup but he does not yet combine pouring and drinking. Later, children combine schemes with those of peers in **sociodramatic play,** the make-believe with others that is under way by age 2½ and increases rapidly during the next few years (Haight & Miller, 1993). Already, Sammy and his classmates create and coordinate several roles in an elaborate plot. By the end of early childhood, children have a sophisticated understanding of role relationships and story lines (Göncü, 1993).

The appearance of complex sociodramatic play indicates that children do not just represent their world; they are *aware* that make-believe is a representational activity—an understanding that increases between ages 4 and 8 (Lillard, 1998, 2001). Listen closely to preschoolers as they assign roles and negotiate make-believe plans: "You *pretend to be* the astronaut, I'll *act like* I'm operating the control tower!" In communicating about pretend, children think about their own and others' fanciful representations. This indicates that they have begun to reason about people's mental activities.

● **Benefits of Make-Believe.** Today, Piaget's view of make-believe as mere practice of representational schemes is regarded as too limited. Play not only reflects but also contributes to children's cognitive and social skills. Compared with social nonpretend activities (such as drawing or putting puzzles together), during sociodramatic play preschoolers' interactions last longer, show more involvement, draw larger numbers of children into the activity, and are more cooperative (Creasey, Jarvis, & Berk, 1998).

When we consider these findings, it is not surprising that preschoolers who spend more time at sociodramatic play are seen as more socially competent by their teachers (Connolly & Doyle, 1984). And many studies reveal that make-believe strengthens a wide variety of mental abilities, including attention, memory, logical reasoning, language and literacy, imagination, creativity, and the ability to reflect on one's own thinking and take another's perspective (Bergen & Mauer, 2000; Berk, 2001; Kavanaugh & Engel, 1998). We will return to the topic of early childhood play in this and the next chapter.

Limitations of Preoperational Thought

Aside from gains in representation, Piaget described preschoolers in terms of what they *cannot,* rather than *can,* understand (Beilin, 1992). He compared them to older, more competent children in the concrete operational stage, as the term *pre*operational suggests. According to Piaget, young children are not capable of *operations*—mental actions that obey logical rules. Instead, their thinking is rigid, limited to one aspect of a situation at a time, and strongly influenced by the way things appear at the moment.

● **Egocentrism.** For Piaget, the most serious deficiency of preoperational thinking, the one that underlies all others, is **egocentrism**—failure to distinguish the symbolic viewpoints of others from one's own. He believed that when children first mentally represent the world, they tend to focus on their own

■ **FIGURE 7.7 Piaget's three-mountains problem.** Each mountain is distinguished by its color and by its summit. One has a red cross, another a small house, and the third a snow-capped peak. Children at the preoperational stage respond egocentrically. They cannot select a picture that shows the mountains from the doll's perspective. Instead, they simply choose the photo that reflects their own vantage point.

viewpoint. Hence, they often assume that others perceive, think, and feel the same way they do.

Piaget's most convincing demonstration of egocentrism involves his *three-mountains problem,* described in Figure 7.7. Egocentrism, Piaget pointed out, shows up in other aspects of children's reasoning. Recall Sammy's firm insistence that someone must have turned on the thunder. Similarly, Piaget regarded egocentrism as responsible for preoperational children's **animistic thinking**—the belief that inanimate objects have lifelike qualities, such as thoughts, wishes, feelings, and intentions (Piaget, 1926/1930). The 3-year-old who charmingly explains that the sun is angry at the clouds and has chased them away is demonstrating this kind of reasoning. According to Piaget, because young children egocentrically assign human purposes to physical events, magical thinking is common during the preschool years.

Piaget argued that preschoolers' egocentric bias prevents them from *accommodating,* or reflecting on and revising their faulty reasoning in response to their physical and social worlds. But to appreciate their shortcomings fully, let's consider some additional tasks that Piaget gave to children.

● **Inability to Conserve.** Piaget's famous conservation tasks reveal a variety of deficiencies of preoperational thinking. **Conservation** refers to the idea that certain physical characteristics of objects remain the same, even when their outward appearance changes. At snack time, Priti and Sammy each had identical boxes of raisins, but after Priti spread hers out on the table, Sammy was convinced that she had more.

Another type of conservation task involves liquid. The child is shown two identical tall glasses of water and asked if they contain equal amounts. Once the child agrees that they do, the water in one glass is poured into a short, wide container, changing the appearance of the water but not its amount. Then the child is asked whether the amount of water is the same or has changed. Preoperational children think the quantity has changed. They explain, "There is less now because the water is way down here" (that is, its level is so low) or, "There is more now because it is all spread out." In Figure 7.8, you will find other conservation tasks that you can try with children.

Preoperational children's inability to conserve highlights several related aspects of their thinking. First, their understanding is *centered,* or characterized by **centration.** They focus on one aspect of a situation, neglecting other important features. In conservation of liquid, the child *centers* on the height of the water, failing to realize that all changes in height are compensated for by changes in width. Second, children are easily distracted by the perceptual appearance of objects. It *looks like* there is less water in the short, wide container, so it *must have* less water. Third, children treat the initial and final states of the water as unrelated events, ignoring the *dynamic transformation* (pouring of water) between them.

The most important illogical feature of preoperational thought is its **irreversibility,** an inability to mentally go through a series of steps in a problem and then reverse direction, returning to the starting point. *Reversibility* is part of every logical operation. After Priti spills her box of raisins, Sammy cannot reverse by thinking to himself, "I know that Priti doesn't have more raisins than I do. If we put them back in that little box, her raisins and my raisins would look just the same."

● **Lack of Hierarchical Classification.** Lack of logical operations leads preschoolers to have difficulty with **hierarchical classification**—the organization of objects into classes and subclasses on the basis of similarities and differences. Piaget's famous *class inclusion problem,* illustrated in Figure 7.9 on page 220, demonstrates this limitation. Preoperational children center on the overriding perceptual feature of yellow. They do not think reversibly by moving from the whole class (flowers) to the parts (yellow and blue) and back again.

Follow-Up Research on Preoperational Thought

Over the past two decades, researchers have challenged Piaget's account of a cognitively deficient preschooler. Many Piagetian problems contain unfamiliar elements or too many pieces of information for young children to handle at once. As a result, preschoolers' responses do not reflect their true abilities. Piaget also missed many naturally occurring instances of preschoolers' effective reasoning.

● **Egocentric, Animistic, and Magical Thinking.** Do young children really believe that a person standing elsewhere in a room sees the same thing they see? When researchers

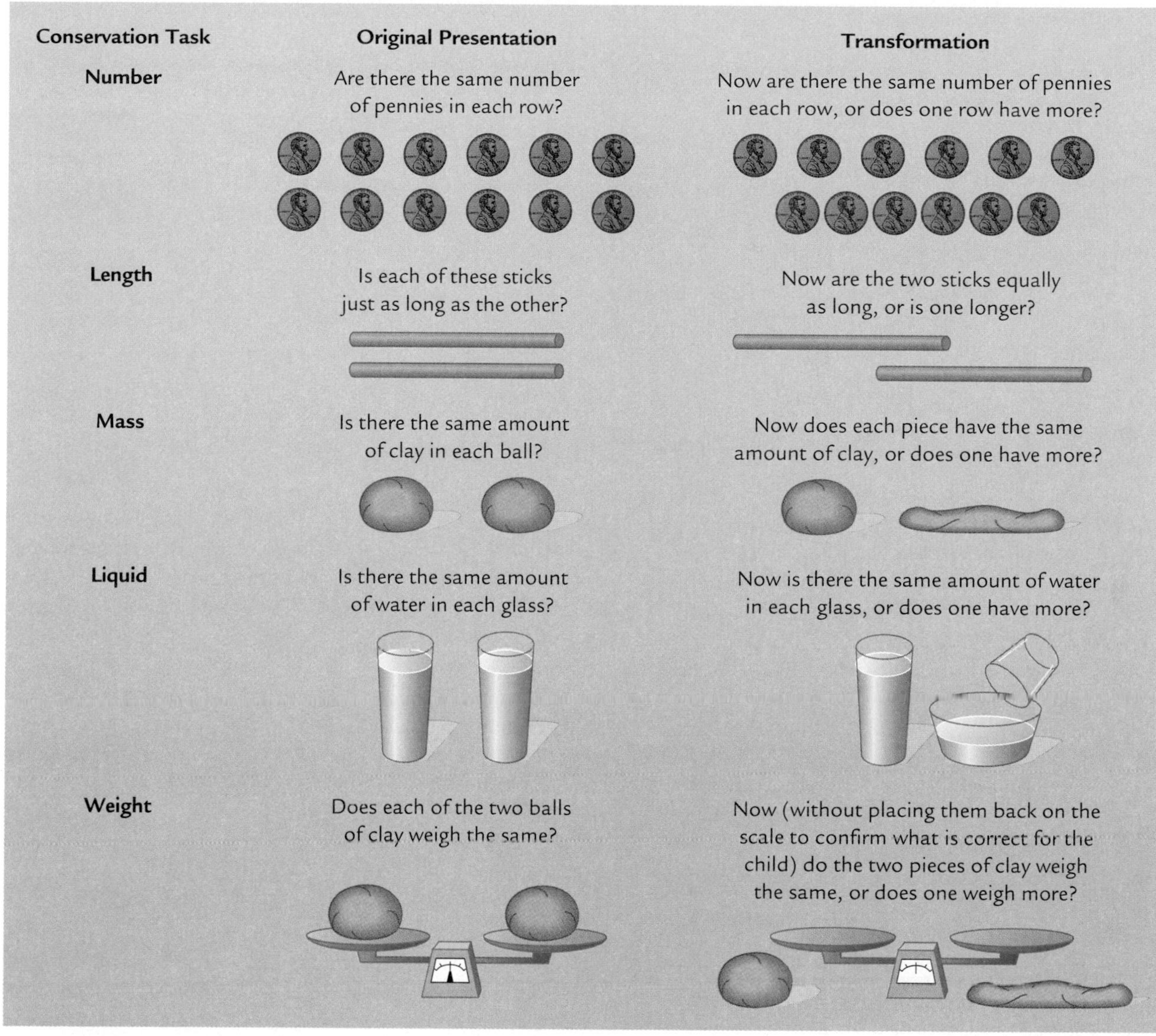

Conservation Task	Original Presentation	Transformation
Number	Are there the same number of pennies in each row?	Now are there the same number of pennies in each row, or does one row have more?
Length	Is each of these sticks just as long as the other?	Now are the two sticks equally as long, or is one longer?
Mass	Is there the same amount of clay in each ball?	Now does each piece have the same amount of clay, or does one have more?
Liquid	Is there the same amount of water in each glass?	Now is there the same amount of water in each glass, or does one have more?
Weight	Does each of the two balls of clay weigh the same?	Now (without placing them back on the scale to confirm what is correct for the child) do the two pieces of clay weigh the same, or does one weigh more?

FIGURE 7.8 Some Piagetian conservation tasks. Children at the preoperational stage cannot yet conserve.

change the nature of the three-mountains problem to include familiar objects and use methods other than picture selection (which is difficult even for 10-year-olds), 4-year-olds show clear awareness of others' vantage points (Borke, 1975; Newcombe & Huttenlocher, 1992).

Nonegocentric responses also appear in young children's conversations. For example, preschoolers adapt their speech to fit the needs of their listeners. Sammy uses shorter, simpler expressions when talking to his little brother, Dwayne, than to agemates or adults (Gelman & Shatz, 1978). Also, in describing objects, children do not use such words as "big" and "little" in a rigid, egocentric fashion. Instead, they *adjust* their descriptions, taking context into account. By age 3, children judge a 2-inch shoe as small when seen by itself (because it is much smaller than most shoes) but as big for a tiny, 5-inch doll (Ebeling & Gelman, 1994). However, in fairness, Piaget (1945/1951) in his later writings described preschoolers' egocentrism as a tendency rather than an inability. As we return to the topic of perspective taking in later chapters, we will see that it develops gradually through childhood and adolescence.

Piaget overestimated children's animistic beliefs because he asked children about objects with which they have little direct experience, such as the clouds, sun, and moon. Three-year-olds do make errors when questioned about certain vehicles, such as trains and airplanes. But these objects appear to be self-moving, a characteristic of almost all living things. And they have some lifelike features—for example, headlights that look like eyes (Massey & Gelman, 1988; Poulin-Dubois & Héroux, 1994). Preschoolers' responses result from incomplete knowledge about objects, not from a belief that inanimate objects are alive.

The same is true for other fantastic beliefs of the preschool years. Most 3- and 4-year-olds believe in the supernatural powers of fairies, goblins, and other enchanted creatures. But they deny that magic can alter their everyday experiences—for example, turn a picture into a real object

FIGURE 7.9 A Piagetian class inclusion problem. Children are shown 16 flowers, 4 of which are blue and 12 of which are yellow. Asked, "Are there more yellow flowers or flowers?" the preoperational child responds, "More yellow flowers," failing to realize that both yellow and blue flowers are included in the category "flowers."

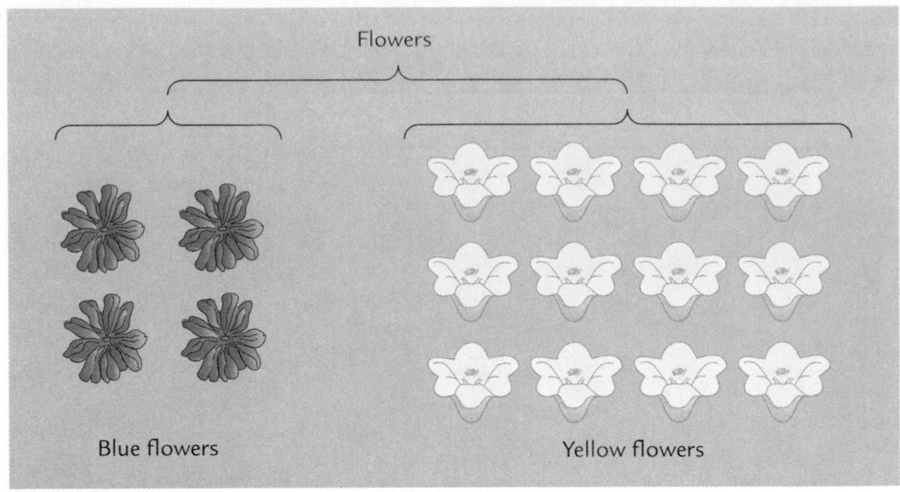

(Subbotsky, 1994). Instead, they think that magic accounts for events that they cannot explain (as in 3-year-old Sammy's magical explanation of thunder in the opening to this chapter) (Rosengren & Hickling, 2000).

Between 4 and 8 years, as familiarity with physical events and principles increases, children's magical beliefs decline. Children figure out who is really behind the activities of Santa Claus and the Tooth Fairy! They also realize that the antics of magicians are due to trickery, not special powers (Phelps & Woolley, 1994; Woolley et al., 1999). How quickly children give up certain fantastic ideas varies with religion and culture. For example, Jewish children express greater disbelief in Santa Claus and the Tooth Fairy than do their Christian agemates. Having been taught at home about the unreality of Santa, they seem to generalize this attitude to other mythical figures (Woolley, 1997).

■ Which of the children in this audience realize that a magician's powers depend on trickery? The younger children look surprised and bewildered. The older children think the magician's antics are funny. Between 4 and 8 years, as familiarity with physical events and principles increases, children's magical beliefs decline.

© J. SOHM/THE IMAGE WORKS

● **Illogical Thought.** Many studies have reexamined the illogical characteristics that Piaget saw in the preoperational stage. Results show that when preschoolers are given tasks that are simplified and relevant to their everyday lives, they do better than Piaget might have expected.

For example, when a conservation-of-number task is scaled down to include only three items instead of six or seven, 3-year-olds perform well (Gelman, 1972). And when preschoolers are asked carefully worded questions about what happens to substances (such as sugar) after they are dissolved in water, they give accurate explanations. Most 3- to 5-year-olds know that the substance is conserved—that it continues to exist, can be tasted, and makes the liquid heavier, even though it is invisible in the water (Au, Sidle, & Rollins, 1993; Rosen & Rozin, 1993).

Preschoolers' ability to reason about transformations is evident on other problems. For example, they can engage in impressive *reasoning by analogy* about physical changes. Presented with the picture-matching problem, *play dough is to cut-up play dough as apple is to ?,* even 3-year-olds choose the correct answer (a cut-up apple) from a set of alternatives, several of which share physical features with the right choice (a bitten apple, a cut-up loaf of bread) (Goswami, 1996). These findings indicate that preschoolers can overcome appearances and think logically about cause and effect in familiar contexts.

Finally, 3- and 4-year-olds use logical, causal expressions, such as *if–then* and *because,* with the same degree of accuracy as adults (McCabe & Peterson, 1988). Illogical reasoning seems to occur only when they grapple with unfamiliar topics, too much information, or contradictory facts, which they have trouble reconciling (Ruffman, 1999).

● **Categorization.** Although preschoolers have difficulty with Piagetian class inclusion tasks, their everyday knowledge is organized into nested categories at an early age. Recall that by the second half of the first year, children have formed a variety of global categories, such as furniture, animals, vehicles, and plants (Mandler, 1998). Notice that each of these categories includes objects that differ widely in perceptual features, challenging Piaget's assumption that preschoolers' thinking is governed by the way things appear. Indeed, 2- to 5-year-olds readily draw inferences about nonobservable characteristics that category members share (Keil & Lockhart, 1999). For example, after being told that a bird has warm blood and a stegosaurus (dinosaur) has cold blood, preschoolers infer that a pterodactyl (labeled a dinosaur) has cold blood, even though it closely resembles a bird.

Over the early preschool years, children's global categories differentiate. They form many *basic-level categories*—ones at an intermediate level of generality, such as "chairs," "tables," and "beds." By the third year, children easily move back and forth between basic-level categories and *general categories*, such as "furniture." Soon after, they break down basic-level categories into *subcategories*, such as "rocking chairs" and "desk chairs" (Johnson, Scott, & Mervis, 1997).

Preschoolers' expanding knowledge and their exposure to diverse examples of category members help them clarify the defining features of categories (Carmichael & Hayes, 2001). As they learn more about their world, they grasp ideas about underlying characteristics that category members share. For example, they realize that in addition to physical features, animals have internal organs and certain behaviors that determine their identity (Hirshfeld, 1995; Krascum & Andrews, 1998). Also, adults label and explain categories to children, and picture-book reading seems to be an especially rich context for doing so. While looking at books with their preschoolers, parents make such categorical statements as, "Penguins live at the South Pole, swim, and catch fish" (Gelman et al., 1998). The information adults provide helps guide children's inferences about categories.

In sum, preschoolers' category systems are not yet very complex. But the capacity to classify hierarchically is present in early childhood.

● **Appearance versus Reality.** So far, we have seen that preschoolers show remarkably advanced reasoning when presented with familiar situations and simplified problems. Yet in certain situations, young children are easily tricked by the outward appearance of things.

John Flavell and his colleagues presented children with objects that were disguised in various ways and asked what the items were, "really and truly." Preschoolers had difficulty with problems involving sights and sounds. When asked whether a white piece of paper placed behind a blue filter is "really and truly blue" or whether a can that sounds like a baby crying when turned over is "really and truly a baby," preschoolers often respond, "Yes!" Not until age 6 to 7 do children do well on these tasks (Flavell, 1993; Flavell, Green, & Flavell, 1987). Younger children's poor performance, however, is not due to a general difficulty in distinguishing appearance from reality, as Piaget suggested. Rather, these problems require a challenging form of representation: the ability to represent the true identity of an object in the face of a contradictory representation.

■ The capacity to categorize expands greatly in early childhood. Children form many categories based on underlying characteristics rather than perceptual features. Guided by knowledge that "dinosaurs have cold blood," this 4-year-old categorizes the pterodactyl (in the foreground) as a dinosaur rather than a bird, even though pterodactyls have wings and can fly.

How do children master appearance–reality distinctions? Make-believe play may be important. The more children engage in make-believe in their preschool classrooms, the better they distinguish the apparent identities of objects from their real identities (Schwebel, Rosen, & Singer, 1999). Experiencing the contrast between everyday and playful circumstances may help children realize that objects do not change their identity when their appearance changes.

Evaluation of the Preoperational Stage

Table 7.2 on page 222 provides an overview of the cognitive attainments of early childhood just considered. Compare them with Piaget's description of the preoperational child on pages 216–218. Although their reasoning is not as well developed as that of school-age children, when given scaled-down tasks based on familiar experiences, preschoolers show the beginnings of logical operations.

Table 7.2 Some Cognitive Attainments of the Preschool Years

Approximate Age	Cognitive Attainments
2–4 years	Shows a dramatic increase in representational activity, as reflected in the development of language, make-believe play, and categorization.
	Takes the perspective of others in simplified, familiar situations and in everyday, face-to-face communication.
	Distinguishes animate beings from inanimate objects; denies that magic can alter everyday experiences.
	Notices transformations, reverses thinking, and explains events logically in familiar contexts.
	Categorizes objects on the basis of common function and behavior (not just perceptual features) and devises ideas about underlying characteristics that category members share.
	Sorts familiar objects into hierarchically organized categories.
4–7 years	Becomes increasingly aware that make-believe (and other thought processes) are representational activities.
	Replaces magical beliefs about fairies, goblins, and events that violate expectations with plausible explanations.
	Shows improved ability to distinguish appearance from reality.

That preschoolers have some logical understanding suggests that they attain logical operations gradually. Over time, children rely on increasingly effective mental (as opposed to perceptual) approaches to solving problems. For example, children who cannot use counting to compare two sets of items do not conserve number (Sophian, 1995). Once preschoolers can count, they apply this skill to conservation-of-number tasks with only a few items. As counting improves, they extend the strategy to problems with more items. By age 6, they understand that number remains the same after a transformation, as long as nothing is added or taken away. Consequently, they no longer need to count to verify their answer (Klahr & MacWhinney, 1998).

That logical operations develop gradually poses a serious challenge to Piaget's stage concept, which assumes abrupt change toward logical reasoning around 6 or 7 years of age. Does a preoperational stage really exist? Some no longer think so. Recall from Chapter 5 that according to the information-processing perspective, children work out their understanding of each type of task separately. Their thought processes are regarded as basically the same at all ages—just present to a greater or lesser extent.

Other experts think the stage concept is still valid but must be modified. For example, some *neo-Piagetian theorists* combine Piaget's stage approach with the information-processing emphasis on task-specific change (Case, 1998; Halford, 1993). They believe that Piaget's strict stage definition must be transformed into a less tightly knit concept, one in which a related set of competencies develops over an extended time period, depending on brain development and specific experiences. These investigators point to findings indicating that as long as the complexity of tasks and children's exposure to them is carefully controlled, children approach those tasks in similar, stage-consistent ways (Case & Okamoto, 1996). For example, in drawing pictures, preschoolers depict objects separately, ignoring their spatial arrangement. In understanding stories, they grasp a single story line but have trouble with a main plot plus one or more subplots.

This flexible stage notion recognizes the unique qualities of early childhood thinking. At the same time, it provides a better account of why, to use Leslie's words, "preschoolers' minds are such a blend of logic, fantasy, and faulty reasoning."

Piaget and Education

Piaget's theory has had a major impact on education, especially during early childhood. Three educational principles derived from his theory continue to have a widespread influence on teacher training and classroom practices:

- *Discovery learning.* In a Piagetian classroom, children are encouraged to discover for themselves through spontaneous interaction with the environment. Instead of presenting ready-made knowledge verbally, teachers provide a rich variety of activities designed to promote exploration—art materials, puzzles, table games, dress-up clothing, building blocks, books, measuring tools, musical instruments, and more.
- *Sensitivity to children's readiness to learn.* A Piagetian classroom does not try to speed up development. Instead, Piaget believed that appropriate learning experiences build on children's current thinking. Teachers watch and listen to their students, introducing experiences that enable them to practice newly discovered schemes and that are likely to challenge their incorrect ways of viewing the world. But teachers do not impose new skills before children indicate they are interested and ready.
- *Acceptance of individual differences.* Piaget's theory assumes that all children go through the same sequence of develop-

ment, but at different rates. Therefore, teachers must plan activities for individual children and small groups rather than just for the whole class. In addition, teachers evaluate educational progress by comparing each child to that child's own previous development. They are less interested in how children measure up to normative standards, or the average performance of same-age peers.

Like his stages, educational applications of Piaget's theory have met with criticism. Perhaps the greatest challenge has to do with his insistence that young children learn mainly through acting on the environment. In the next section, we will see that young children also use language-based routes to knowledge, which Piaget de-emphasized.

Ask Yourself

REVIEW

Select two of the following features of Piaget's preoperational stage: egocentrism, a focus on perceptual appearances, difficulty reasoning about transformations, and lack of hierarchical classification. Cite findings that led Piaget to conclude that preschoolers are deficient in those ways. Then present evidence indicating that preschoolers are more capable thinkers than Piaget assumed.

APPLY

Brett's preschool teacher provides many opportunities for sociodramatic play. Brett's mother wonders whether Brett is learning anything from so much pretending. Using research findings, respond to her concern.

APPLY

At home, 4-year-old Will understands that his tricycle isn't alive and can't move by itself. Yet when Will went fishing with his family and his father asked, "Why do you think the river is flowing along?" Will responded, "Because it's alive and wants to." What explains this contradiction in Will's reasoning?

www.

Vygotsky's Sociocultural Theory

Piaget's de-emphasis on language as a source of cognitive development brought on yet another challenge, this time from Vygotsky's sociocultural theory, which stresses the social context of cognitive development. During early childhood, rapid growth in language broadens preschoolers' ability to participate in social dialogues with more knowledgeable individuals, who encourage them to master culturally important tasks. Soon children start to communicate with themselves in much the same way they converse with others. This greatly enhances the complexity of their thinking and their ability to control their own behavior. Let's see how this happens.

Private Speech

Watch preschoolers as they go about their daily activities, and you will see that they frequently talk out loud to themselves. For example, as Sammy worked a puzzle, he said, "Where's the red piece? Now, a blue one. No, it doesn't fit. Try it here."

Piaget (1923/1926) called these utterances *egocentric speech,* reflecting his belief that young children have difficulty taking the perspectives of others. For this reason, he said, their talk is often "talk for self" in which they run off thoughts in whatever form they happen to occur, regardless of whether a listener can understand. Piaget believed that cognitive maturity and certain social experiences—namely, disagreements with peers—eventually bring an end to egocentric speech. Through arguments with agemates, children repeatedly see that others hold viewpoints different from their own. As a result, egocentric speech declines.

Vygotsky (1934/1987) voiced a powerful objection to Piaget's conclusions. He reasoned that children speak to themselves for self-guidance. Because language helps children think about their mental activities and behavior and select courses of action, Vygotsky viewed it as the foundation for all higher cognitive processes, including controlled attention, deliberate memorization and recall, categorization, planning, problem solving, and self-reflection. As children get older and find tasks easier, their self-directed speech is internalized as silent, *inner speech*—the verbal dialogues we carry on with ourselves while thinking and acting in everyday situations.

Over the past three decades, almost all studies have supported Vygotsky's perspective. As a result, children's self-directed speech is now called **private speech** instead of egocentric speech. Research shows that children use more of it when tasks are difficult and they are confused about how to proceed. Also, as Vygotsky predicted, private speech goes underground with age, changing into whispers and silent lip movements (Duncan & Pratt, 1997; Patrick & Abravanel, 2000). Finally, children who freely use private speech during a challenging activity are more attentive and involved and do better than their less talkative agemates (Berk & Spuhl, 1995; Winsler, Diaz, & Montero, 1997).

Social Origins of Early Childhood Cognition

Where does private speech come from? Recall from Chapter 5 that Vygotsky believed children's learning takes place within a *zone of proximal development*—a range of tasks too difficult for the child to do alone but possible with the help of others. Consider the joint activity of Sammy and his mother, who assists him in putting together a difficult puzzle:

Sammy: "I can't get this one in." *[Tries to insert a piece in the wrong place]*

Mother: "Which piece might go down here?" *[Points to the bottom of the puzzle]*

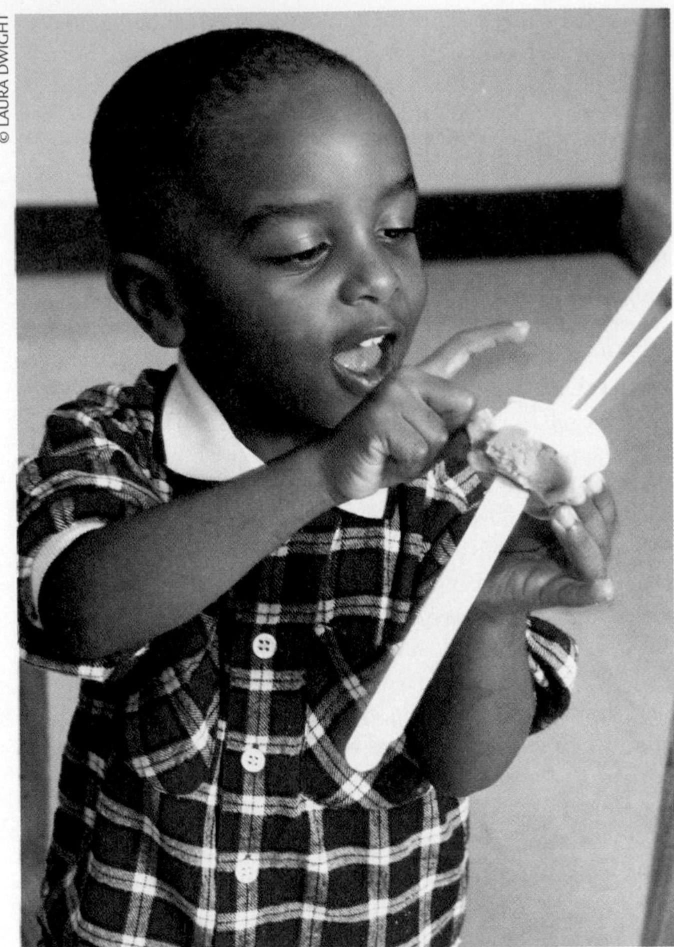

■ This 3-year-old makes a sculpture from play dough and plastic sticks with the aid of private speech. During the preschool years, children frequently talk to themselves as they play and tackle other challenging tasks. Research supports Vygotsky's theory that children use private speech to guide their own thinking and behavior.

Sammy: "His shoes." *[Looks for a piece resembling the clown's shoes but tries the wrong one]*

Mother: "Well, what piece looks like this shape?" *[Pointing again to the bottom of the puzzle]*

Sammy: "The brown one." *[Tries it, and it fits; then attempts another piece and looks at his mother]*

Mother: "Try turning it just a little." *[Gestures to show him]*

Sammy: "There!" *[Puts in several more pieces. His mother watches]*

Sammy's mother keeps the puzzle within his zone of proximal development, at a manageable level of difficulty. To do so, she engages in **scaffolding**—adjusting the support offered during a teaching session to fit the child's current level of performance. When the child has little notion of how to proceed, the adult uses direct instruction and breaks the task into manageable units. As the child's competence increases, effective scaffolders gradually and sensitively withdraw support, turning over responsibility to the child. Gradually, children take the language of these dialogues, make it part of their private speech, and use that speech to organize their independent efforts.

What evidence supports Vygotsky's ideas on the social origins of cognitive development? In several studies, parents who were effective scaffolders in teaching their child to solve challenging problems had children who used more private speech and were more successful when asked to do a similar task by themselves (Berk & Spuhl, 1995; Conner, Knight, & Cross, 1997; Winsler, Diaz, & Montero, 1997). Other research shows that although children benefit from working on tasks with same-age peers, their planning and problem solving show more improvement when their partner is either an "expert" peer (especially capable at the task) or an adult (Azmitia, 1988; Radziszewska & Rogoff, 1988). And peer disagreement (emphasized by Piaget) does not seem to be as important in fostering cognitive development as the extent to which children resolve differences of opinion and cooperate (Kobayashi, 1994; Tudge, 1992).

Vygotsky and Education

Piagetian and Vygotskian classrooms clearly have features in common. Both emphasize active participation and acceptance of individual differences. Yet a Vygotskian classroom goes beyond independent discovery. It promotes *assisted discovery.* Teachers guide children's learning with explanations, demonstrations, and verbal prompts, carefully tailoring their efforts to each child's zone of proximal development. Assisted discovery is also helped along by *peer collaboration* (Rogoff, 1998). Teachers group together classmates of differing abilities and encourage them to teach and help one another.

Vygotsky (1935/1978) saw make-believe play as the ideal social context for fostering cognitive development in early childhood. As children create imaginary situations, they learn to follow internal ideas and social rules rather than their immediate impulses. For example, a child pretending to go to sleep follows the rules of bedtime behavior. Another child imagining himself to be a father and a doll to be a child conforms to the rules of parental behavior. According to Vygotsky, make-believe play is a unique, broadly influential zone of proximal development in which children try out a wide variety of challenging activities and acquire many new competencies.

Turn back to page 217 to review findings that make-believe play enhances a diverse array of cognitive and social skills. Pretend play is also rich in private speech—a finding that supports its role in helping children bring action under the control of thought (Krafft & Berk, 1998). And in one study, preschoolers who engaged in more complex sociodramatic play showed greater gains in following classroom rules over a 4-month period (Elias & Berk, 2002).

© WILL FALLER

■ A Vygotskian classroom promotes assisted discovery. Teachers guide children's learning, tailoring their efforts to each child's zone of proximal development. They also promote peer collaboration, grouping together classmates of differing abilities and encouraging them to teach and help one another.

Ask Yourself

REVIEW
Describe features of social interaction that support children's cognitive development. How does such interaction create a zone of proximal development?

APPLY
Tanisha sees her 5-year-old son Toby talking out loud to himself while he plays. She wonders whether she should discourage this behavior. Use Vygotsky's theory to explain why Toby talks to himself. How would you advise Tanisha?

CONNECT
How is scaffolding involved in child-directed speech, discussed on page 168 of Chapter 5?

www.

Evaluation of Vygotsky's Theory

In granting social experience a fundamental role in cognitive development, Vygotsky's theory underscores the vital role of teaching and helps us understand the wide cultural variation in cognitive skills. It recognizes that children develop unique forms of thinking from engaging in activities that make up their culture's way of life. Nevertheless, Vygotsky's theory has not gone unchallenged. Verbal communication may not be the only means through which children's thinking develops—or even the most important means, in some cultures. When Western parents help their young children with challenging tasks, they assume much responsibility for children's motivation by frequently giving verbal instructions and conversing with the child. Their communication resembles the teaching that takes place in school, where their children will spend years preparing for adult life. But in cultures that place less emphasis on schooling and literacy, parents often expect children to take greater responsibility for acquiring new skills through keen observation and participation in community activities (Rogoff et al., 1993). Turn to the Cultural Influences box on page 226 for research that illustrates this difference.

Finally, Vygotsky's theory says little about how basic motor, perceptual, attention, memory, and problem-solving skills, discussed in Chapters 4 and 5, contribute to socially transmitted higher cognitive processes. For example, his theory does not address how these elementary capacities spark changes in children's social experiences, from which more advanced cognition springs (Moll, 1994). Piaget paid far more attention than did Vygotsky to the development of basic cognitive processes. It is intriguing to speculate about the broader theory that might exist today had Piaget and Vygotsky—the two twentieth-century giants of cognitive development—had a chance to meet and weave together their extraordinary accomplishments.

Information Processing

Return for a moment to the model of information processing discussed on page 154 of Chapter 5. Recall that information processing focuses on *mental strategies* that children use to transform stimuli flowing into their mental systems. During early childhood, advances in representation and in children's ability to guide their own behavior lead to more efficient ways of attending, manipulating information, and solving problems. Preschoolers also become more aware of their own mental life and begin to acquire academically relevant knowledge important for school success.

Attention

Parents and teachers are quick to notice that compared with school-age children, preschoolers spend relatively short times involved in tasks and are easily distracted. But recall from Chapter 5 that sustained attention improves in toddlerhood, a trend that continues over the preschool years, and fortunately so, since children will rely on this capacity greatly once they enter school.

During early childhood, children also become better at *planning*—thinking out a sequence of acts ahead of time and allocating their attention accordingly to reach a goal. As long as tasks are familiar and not too complex, preschoolers can generate and follow a plan. For example, they search for a lost object in a play yard systematically and exhaustively (Wellman, Somerville, & Haake, 1979). Still, planning has a long way to go. When asked to compare detailed pictures, preschoolers fail to search thoroughly. On complex tasks, they rarely decide what to do first and what to do next in an orderly

Cultural Influences

Young Children's Daily Life in a Yucatec Mayan Village

Conducting ethnographic research in a remote Mayan village of the Yucatan, Mexico, Suzanne Gaskins (1999) found that child-rearing values, daily activities and, consequently, 2- to 5-year-olds' competencies differed sharply from those of Western preschoolers. Yucatec Mayan adults are subsistence farmers. Men spend their days tending cornfields, aided by sons age 8 and older. Women oversee the household and yard, engaging in time-consuming meal preparation, clothes washing, and care of livestock and garden, assisted by daughters as well as sons not yet old enough to work in the fields.

In Yucatec Mayan culture, life is structured around adult work and religious and social events. Children join in these activities from the second year on. Adults make no effort to provide special experiences designed to satisfy children's interests or stimulate their development. When not participating with adults, children are expected to be independent. Even young children make many nonwork decisions for themselves—how much to sleep and eat, what to wear, when to bathe (as long as they do so every afternoon), and even when to start school.

© BERYL GOLDBERG

■ In Yucatec Mayan culture, adults rarely converse with children or scaffold their learning. And rather than engaging in make-believe, children join in the work of their community from an early age, spending many hours observing adults. This Mayan preschooler watches intently as her grandmother washes dishes. When the child begins to imitate adult tasks, she will be given additional responsibilities.

As a result, Mayan preschoolers spend much time at self-care and are highly competent at it. In contrast, their make-believe play is limited; when it occurs, it involves brief imitations of adult work or common scenes from adult life, organized and directed by older siblings. When not engaged in self-care or play, Mayan children watch others—for hours each day. By age 3, they can report the whereabouts and activities of all family members. At any moment, they may be called on to do a chore—fetch things from the house, run an errand, deliver a message, tend to livestock, or take care of toddler-age siblings.

Mayan parents rarely converse with children or scaffold their learning. Rather, when children imitate adult tasks, parents conclude that they are ready for more responsibility. Then they assign chores, selecting tasks the child can do with little help so that adult work is not disturbed. If a child cannot do a task, the adult takes over and the child observes, reengaging when able to contribute. This give-and-take occurs smoothly, with parent and child focused on the primary goal of getting the job done.

Cultural priorities and daily activities lead Mayan preschoolers' skills and behavior to differ sharply from those of their Western agemates. Expected to be independent and helpful, Mayan children seldom display attention-getting behaviors or ask others for something interesting to do. From an early age, they can sit quietly for long periods with little fussing—through a lengthy religious service or dance and even a 3-hour truck ride into town. And when an adult interrupts their activity and directs them to do a chore, Mayan children respond eagerly to a command that Western children frequently avoid or resent. By age 5, Mayan children spontaneously take responsibility for tasks beyond those assigned.

fashion. And even when young children do plan, they often fail to implement important steps (Friedman & Scholnick, 1997; Ruff & Rothbart, 1996).

Memory

Unlike infants and toddlers, preschoolers have the language skills to describe what they remember, and they can follow directions on simple memory tasks. As a result, memory becomes easier to study in early childhood.

● **Recognition and Recall.** Try showing a young child a set of 10 pictures or toys. Then mix them up with some unfamiliar items and ask the child to point to the ones in the original set. You will find that preschoolers' *recognition* memory (ability to tell whether a stimulus is the same as or similar to one they have seen before) is remarkably good. In fact, 4- and 5-year-olds perform nearly perfectly.

Now give the child a more demanding task. Keep the items out of view and ask the child to name the ones she saw. This requires *recall*—that the child generate a mental image of an absent stimulus. Young children's recall is much poorer

© SYRACUSE NEWSPAPERS/DAVID LASSMAN/THE IMAGE WORKS

■ This 4-year-old plays a memory game in which he must recognize whether he has previously seen a game piece that he pulled out of a bag held by his preschool teacher. During early childhood, recognition memory is excellent. In contrast, recall develops slowly because young children do not yet use memory strategies effectively.

than their recognition. At age 2, they can recall no more than one or two of the items, at age 4 only about three or four (Perlmutter, 1984).

Of course, recognition is much easier than recall for adults as well, but compared with adults' recall, children's recall is quite deficient. The reason is that young children are less effective at using **memory strategies,** deliberate mental activities that improve our chances of remembering. For example, when you want to retain information, you might *rehearse,* or repeat the items over and over again. Or you might *organize,* grouping together items that are alike so you can easily retrieve them by thinking of their similar characteristics.

Preschoolers do show the beginnings of memory strategies. When circumstances permit, they arrange items in space to aid their memories. In one study, an adult placed either an M&M or a wooden peg in each of 12 identical containers and handed them one by one to preschoolers, asking them to remember where the candy was hidden. By age 4, children put the candy containers in one place and the peg containers in another, a strategy that almost always led to perfect recall (DeLoache & Todd, 1988). But preschoolers do not yet rehearse or organize items into categories (for example, all the vehicles together, all the animals together) when asked to remember. Even when they are trained to do so, their memory performance rarely improves, and they do not apply these strategies in new situations (Gathercole, Adams, & Hitch, 1994; Miller & Seier, 1994).

Why do young children seldom use memory strategies? One reason is that strategies tax young children's limited working memories (Bjorklund & Coyle, 1995). Preschoolers have difficulty holding on to the to-be-learned information and applying a strategy at the same time.

● Memory for Everyday Experiences. Think about the difference in your recall of listlike information and your memory for everyday experiences. In remembering lists, you recall isolated pieces of information, reproducing them exactly as you originally learned them. In remembering everyday experiences, you recall complex, meaningful events.

Memory for Familiar Events. Like adults, preschoolers remember familiar, repeated events—what you do when you go to preschool or have dinner—in terms of **scripts,** general descriptions of what occurs and when it occurs in a particular situation. Young children's scripts begin as a structure of main acts. For example, when asked to tell what happens when you go to a restaurant, a 3-year-old might say, "You go in, get the food, eat, and then pay." Although children's first scripts contain only a few acts, they are almost always recalled in correct sequence (Bauer, 1997). With age, scripts become more elaborate, as in the following restaurant account given by a 5-year old child: "You go in. You can sit at a booth or a table. Then you tell the waitress what you want. You eat. If you want dessert, you can have some. Then you pay and go home" (Hudson, Fivush, & Kuebli, 1992).

Once formed, a script can be used to predict what will happen in the future. In this way, scripts help children organize and interpret repeated events. Children rely on them when listening to and telling stories. They also act out scripts in make-believe play as they pretend to go on a trip or play school. And scripts support children's earliest efforts at planning as they represent sequences of actions that lead to desired goals (Hudson, Sosa, & Shapiro, 1997).

Memory for One-Time Events. In Chapter 5, we considered a second type of everyday memory—*autobiographical memory,* or representations of personally meaningful, one-time events. As preschoolers' cognitive and conversational skills improve, their descriptions of special events become better organized, more detailed, and related to the larger context of their lives (Haden, Haine, & Fivush, 1997).

Adults use two styles for prompting children's autobiographical narratives. In the *elaborative style,* they ask varied questions, add information to children's statements, and volunteer their own recollections and evaluations of events. For example, after a field trip to the zoo, Leslie asked the children, "What was the first thing we did? Why weren't the parrots in their cages? I thought the lion was scary. What did you think?" In contrast, adults who use the *repetitive style* provide little information and ask the same short-answer questions over and over, as in, "Do you remember the zoo? What did we do at the zoo?" Preschoolers who experience the elaborative style produce more organized and detailed personal stories when followed up 1 to 2 years later (Farrant & Reese, 2000; Reese,

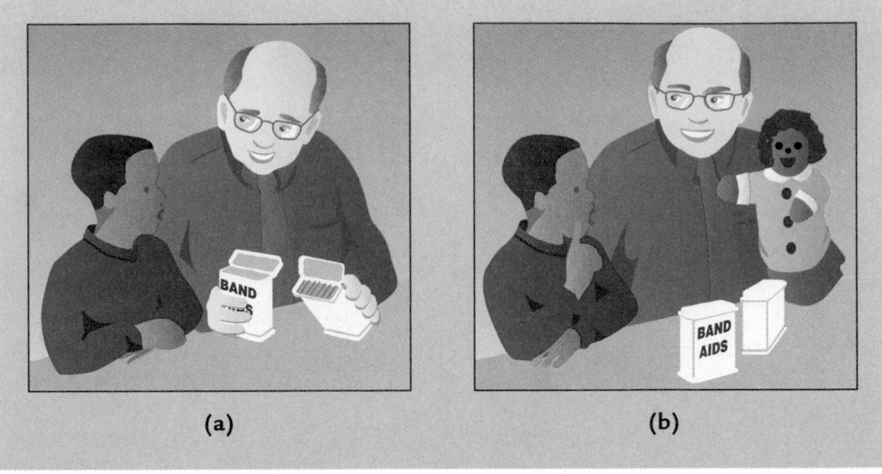

FIGURE 7.10 Example of a false-belief task. (a) An adult shows a child the contents of a Band-Aid box and an unmarked box. The Band-Aids are in the unmarked container. (b) The adult introduces the child to a hand puppet named Pam and asks the child to predict where Pam would look for the Band-Aids and to explain Pam's behavior. The task reveals whether children understand that without having seen that the Band-Aids are in the unmarked container, Pam will hold a false belief.

Haden, & Fivush, 1993). In line with Vygotsky's theory, social experiences aid children's memory skills.

The Young Child's Theory of Mind

As representation of the world, memory, and problem solving improve, children start to reflect on their own thought processes. They begin to construct a *theory of mind,* or coherent set of ideas about mental activities. This understanding is also called **metacognition.** The prefix *meta-,* meaning "beyond or higher," is included in the term because *metacognition* means "thinking about thought" (Flavell, 2000). As adults, we have a complex appreciation of our inner mental worlds, which we use to interpret our own and others' behavior and to improve our performance on various tasks. How early are children aware of their mental lives, and how complete and accurate is their knowledge?

• Awareness of Mental Life. Infants' and toddlers' capacity for joint attention, preverbal gestures, and social referencing suggests that they realize that people can share and influence each other's mental states. As 2-year-olds' language expands, "think," "remember," and "pretend" are among the first verbs in their vocabularies (Wellman, 1990). By age 3, children realize that thinking takes place inside their heads and that a person can think about something without seeing it, talking about it, or touching it (Flavell, Green, & Flavell, 1995). However, 2- to 3-year-olds have only a beginning grasp of the distinction between mental life and behavior. They think that people always behave in ways consistent with their *desires* and do not understand that less obvious mental states, such as *beliefs,* affect their actions.

Mastery of False Belief. Around age 4, children figure out that *both beliefs and desires* determine behavior. Dramatic evidence for this new understanding comes from games that test whether preschoolers realize that *false beliefs*—ones that do not represent reality accurately—can guide people's actions. For example: Show a child two small closed boxes, one a familiar Band-Aid box and the other a plain, unmarked box (see Figure 7.10). Then say, "Pick the box you think has the Band-Aids in it." Almost always, children pick the marked container. Next, open the boxes and show the child that in reality, the marked one is empty and the unmarked one contains Band-Aids. Finally, introduce the child to a hand puppet and explain, "Here's Pam. She has a cut. Where do you think she'll look for Band-Aids? Why would she look in there?" (Bartsch & Wellman, 1995). Only a handful of 3-year-olds but many 4-year-olds can explain Pam's false belief.

Children's grasp of false belief strengthens over the preschool years, becoming more secure by age 6 (Wellman, Cross, & Watson, 2001). During that time, it becomes a powerful tool for understanding oneself and others and a good predictor of social skills (Jenkins & Astington, 2000; Watson et al., 1999).

Factors Contributing to Preschoolers' Theory of Mind. How do children manage to develop a theory of mind at such a young age? Cognitive and language development, enabling children to reflect on thinking, is crucial. The ability to inhibit inappropriate responses, think flexibly, and plan predicts false-belief understanding (Carlson & Moses, 2001; Hughes, 1998). And a rich, mental-state vocabulary is helpful (de Villiers & de Villiers, 2000). Among the Quechua of the Peruvian highlands, adults refer to such states as "think" and "believe" indirectly, since their language lacks mental-state terms. Quechua children have difficulty with false-belief tasks for years after children in industrialized nations have mastered them (Vinden, 1996).

Social experience also promotes understanding of the mind. Preschoolers with older siblings are advanced in awareness of false belief, perhaps because relating to older siblings highlights the influence of beliefs on behavior—through teasing, trickery, and discussing feelings (Ruffman et al., 1998). Peer interaction also contributes. Three- and 4-year-olds who often engage in mental-state talk with friends, which occurs often during make-believe play, are ahead in understanding of false belief and other aspects of the mind (Harris & Leevers,

Biology & Environment

"Mindblindness" and Autism

Sidney stood at the water table in his preschool classroom, repeatedly filling a plastic cup and dumping out its contents. Dip-splash, dip-splash he went, until his teacher came over and redirected his actions. Without looking at his teacher's face, Sidney moved to a new repetitive pursuit: pouring water from one cup into another and back again. As other children entered the play space and conversed, Sidney hardly noticed. He rarely spoke, and when he did, he usually used words to get things he wanted, not to exchange ideas.

Sidney has *autism,* the most severe behavior disorder of childhood. The term *autism* means "absorbed in the self," an apt description of Sidney. Like other children with the disorder, Sidney is impaired in emotional expression and other nonverbal behaviors required for successful social interaction. In addition, his language is delayed and stereotyped; some autistic children do not speak at all. Sidney's interests, which focus on the physical world, are narrow and overly intense. For example, one day he sat for more than an hour making a toy ferris wheel go round and round.

Researchers agree that autism stems from abnormal brain functioning, usually due to genetic or prenatal environmental causes. Growing evidence suggests that one psychological factor involved is a severely deficient or absent theory of mind. Long after they reach the intellectual level of an average 4-year-old, children with autism have great difficulty with false-belief tasks. Most cannot attribute mental states to others or to themselves. Such words as "believe," "think," "know," and "pretend" are rarely part of their vocabularies (Happé, 1995; Yirmiya, Solomonica-Levi, & Shulman, 1996).

As early as the second year, autistic children show deficits in capacities believed to contribute to an understanding of mental life. For example, they less often establish joint attention, engage in social referencing, or imitate an adult's behaviors than normal children (Charman et al., 1997; Leekam, Lopez, & Moore, 2000). Furthermore, they are relatively insensitive to eye gaze as a cue to what a speaker is talking about (Baron-Cohen, Baldwin, & Crowson, 1997). Finally, autistic children engage in much less make-believe play than normally developing children and children with other developmental problems (Hughes, 1998).

Do these findings indicate that autism is due to impairment of an innate, core brain function, which leaves the child "mindblind" and therefore deficient in human sociability? Some researchers think so (Baron-Cohen, 2001; Scholl & Leslie, 2000). Another conjecture is that autism is due to a general memory deficit, which makes it hard to retain the parts of complex tasks (Bennetto, Pennington, & Rogers, 1996). Perhaps this explains autistic children's preoccupation with simple, repetitive acts. It also may contribute to their difficulty with tasks that require them to integrate several parts into a coherent whole to solve a problem (Jarrold et al., 2000; Yirmiya & Shulman, 1996). These memory and integration problems also interfere with understanding the social world, since social interaction takes place quickly and requires combining information from various sources.

At present, it is not clear which of these hypotheses is correct. Perhaps several biologically based cognitive deficits underlie the tragic social isolation of children like Sidney.

© WILL HART

■ This autistic girl does not take note of a speaker's gaze as a cue to what others are talking about. For this reason, the girl's teacher takes extra steps to capture her attention in a science lesson. Researchers disagree on whether autistic children's "mindblindness" is due to an impairment in an innate, core brain function, which leaves the child deficient in sociability, or to a general memory deficit, which makes it hard to retain the parts of complex tasks.

2000; Hughes & Dunn, 1998). Furthermore, interacting with more mature members of society is helpful. In a study of Greek preschoolers, daily contact with many adults and older children predicted mastery of false belief (Lewis et al., 1996). These encounters offer extra opportunities to observe different viewpoints and talk about inner states.

Core knowledge theorists (see Chapter 5, page 150) believe that to profit from the social experiences just described, children must be biologically prepared to develop a theory of mind. They claim that children with *autism,* who do not grasp false belief, are deficient in the brain mechanism that enables humans to detect mental states. See the Biology and Environment box above to find out more about the biological basis of reasoning about the mind.

● **Limitations of the Young Child's Understanding of Mental Life.** Although surprisingly advanced, preschoolers' awareness of mental activities is far from complete. For example,

3- and 4-year-olds are unaware that people continue to think while they are waiting or otherwise not doing something (Flavell, Green, & Flavell, 1993, 1995). They conclude that mental activity stops when there are no obvious external cues to indicate a person is thinking. Furthermore, children younger than age 5 pay little attention to the process of thinking. When questioned about subtle distinctions between mental states, such as "know" and "forget," they express confusion (Lyon & Flavell, 1994). And they believe that all events must be directly observed to be known. They do not understand that *mental inferences* can be a source of knowledge (Carpendale & Chandler, 1996).

These findings suggest that preschoolers view the mind as a passive container of information. Consequently, they greatly underestimate the amount of mental activity that goes on in people and are poor at inferring what people know or are thinking about. In contrast, older children view the mind as an active, constructive agent that selects and transforms information (Chandler & Carpendale, 1998; Flavell, 1999). We will consider this change further in Chapter 9 when we take up metacognition in middle childhood.

Early Childhood Literacy

One week, Leslie's pupils brought empty food boxes from home to place on shelves in the classroom. Soon a make-believe grocery store opened. Children labeled items with prices, made shopping lists, and wrote checks at the cash register. A sign at the entrance announced the daily specials: "APLS BNS 5¢" ("apples bananas 5¢").

As their grocery store play reveals, preschoolers understand a great deal about written language long before they learn to read or write in conventional ways. This is not surprising when we consider that children in industrialized nations live in a world filled with written symbols. Each day, they observe and participate in activities involving storybooks, calendars, lists, and signs. As part of these experiences, they try to figure out how written symbols convey meaning.

Young preschoolers search for units of written language as they "read" memorized versions of stories and recognize familiar signs, such as "PIZZA" at their favorite fast-food counter. But their early ideas about written language differ from ours. For example, many preschoolers think that a single letter stands for a whole word or that each letter in a person's signature represents a separate name. Gradually, children revise these ideas as their perceptual and cognitive capacities improve, as they encounter writing in many different contexts, and as adults help them with written communication.

Soon, preschoolers become aware of general characteristics of written language and create their own printlike symbols, as in the "story" and "grocery list" written by a 4-year-old in Figure 7.11. Eventually, children figure out that letters are parts of words and are linked to sounds in systematic ways, as you can see in the invented spellings that are typical between ages 5 and 7. At first, children rely heavily on sounds in the names of letters, as in "ADE LAFWTS KRMD NTU A LAVATR" ("eighty elephants crammed into a[n] elevator"). Over time, they grasp sound–letter correspondences and learn that some letters have more than one common sound (McGee & Richgels, 2000; Treiman et al., 1998).

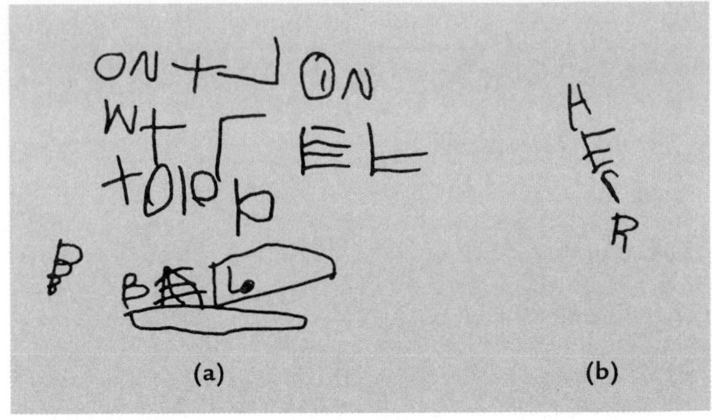

FIGURE 7.11 A story (a) and a grocery list (b) written by a 4-year-old child. This child's writing has many features of real print. It also reveals an awareness of different kinds of written expression. (From L. M. McGee & D. J. Richgels, 2000, *Literacy's Beginnings,* 3rd ed., Boston: Allyn and Bacon, p. 69. Reprinted by permission.)

The more informal literacy-related experiences young children have, the better prepared they will be to tackle the complex tasks involved in reading and writing. Adults can provide literacy-rich physical environments and encourage literacy-related play (Neuman, Copple, & Bredekamp, 2000; Roskos & Neuman, 1998). Storybook reading in which adults engage preschoolers in discussion and interpretation of story

■ Preschoolers acquire a great deal of literacy knowledge informally as they participate in everyday activities involving written symbols. They try to figure out how print conveys meaningful information, just as they strive to make sense of other aspects of their world.

© JIM PICKERELL

© MYRLEEN FERGUSON CATE/PHOTOEDIT

■ Counting on fingers is an early, spontaneous approach that children use to experiment with strategies for solving basic math facts. As they try out various routes to solution and select those that are efficient and accurate, answers become more strongly associated with problems. Soon children give up counting on fingers in favor of retrieving the right answer.

content is related to preschoolers' language and reading readiness scores, which predict success in school. And adult-supported writing activities that focus on connected discourse (preparing a letter or a story) also foster literacy progress (Purcell-Gates, 1996; Whitehurst & Lonigan, 1998).

Preschoolers from low-SES homes generally have far less access to storybooks than their higher-SES agemates. In a program that "flooded" child-care centers with children's books and provided training to caregivers on how to get 3- and 4-year-olds to spend time with books, children showed much greater gains in informal reading and writing knowledge than a control group (Neuman, 1999). Providing low-SES parents with children's books, along with guidance in how to stimulate literacy learning in preschoolers, greatly enhances literacy activities in the home (High et al., 2000).

Young Children's Mathematical Reasoning

Mathematical reasoning, like literacy, builds on a foundation of informally acquired knowledge. Between 14 and 16 months, toddlers display a beginning grasp of **ordinality,** or order relationships between quantities, such as three is more than two, and two is more than one. Soon they attach verbal labels (such as "lots," "little," "big," "small") to different amounts and sizes. And between ages 2 and 3, they begin to count. At first, counting is little more than a memorized routine, as in "Onetwothreefourfivesix!" Or children repeat a few number words while vaguely pointing toward objects (Fuson, 1992).

Soon, however, counting becomes more precise. Most 3- to 4-year-olds have established an accurate one-to-one correspondence between a short sequence of number words and the items they represent. Sometime between ages 4 and 5, they grasp the vital principle of **cardinality,** that the last number in a counting sequence indicates the quantity of items in a set (Bermejo, 1996). Mastery of cardinality increases the efficiency of children's counting. By age 4, children use counting to solve simple arithmetic problems. At first, their strategies are tied to the order of numbers as presented; when given 2 + 4, they count on from 2 (Ginsburg, Klein, & Starkey, 1998). Soon they experiment—for example, holding up 4 fingers on one hand, 2 on the other, then recognizing 6 as the total; or starting with the higher digit, 4, and counting on (Siegler, 1996). Gradually, children select the most efficient, accurate strategy—in this example, beginning with the higher digit. Then they generalize the strategy to subtraction, *counting down* to see how many items remain after some are taken away. With enough practice, children recall answers automatically.

The basic arithmetic knowledge just described emerges universally around the world. In homes and preschools where adults provide many occasions and requests for counting, comparing quantities and reasoning about number in meaningful situations, children master these basic understandings sooner (Geary, 1995). Then these concepts are solidly available as supports for the wide variety of mathematical skills children will be taught in school.

Ask Yourself

REVIEW

Describe a typical 4-year-old's understanding of mental activities, noting both strengths and limitations.

APPLY

Lena wonders why Gregor's preschool teacher, instead of teaching him literacy and math skills, permits him to spend much time playing. Gregor's teacher responds, "I *am* teaching him academics—through play." Why is play the best way for preschoolers to develop academically?

CONNECT

Cite evidence on the development of preschoolers' memory, theory of mind, and literacy and mathematical knowledge that is consistent with Vygotsky's theory.

www.

Individual Differences in Mental Development

Five-year-old Hallie sat in a testing room while Sarah gave him an intelligence test. Some of the questions Sarah asked were *verbal*. For example, she held out a picture of a shovel and said, "Tell me what this shows"—an item measuring vocabulary. Then she tested his memory by asking him to repeat sentences and lists of numbers back to her. Other tasks were *nonverbal* and largely assessed Hallie's spatial reasoning. Hallie copied designs with special blocks, figured out the pattern in a series of shapes, and indicated what a piece of paper folded and cut would look like when unfolded (Thorndike, Hagen, & Sattler, 1986).

Sarah was aware that Hallie came from an economically disadvantaged family. When low-SES and some ethnic minority preschoolers are faced with an unfamiliar adult who bombards them with questions, they sometimes become anxious. Also, such children may not define the testing situation in achievement terms. Often they look for attention and approval from the examiner and may settle for lower performance than their abilities allow. Sarah spent time playing with Hallie before she began testing, and she praised and encouraged him while testing was in progress. When these testing conditions are used, low-SES preschoolers improve in performance (Bracken, 2000).

Note that the questions Sarah asked Hallie tap knowledge and skills that not all children have had equal opportunity to learn. The issue of *cultural bias* in mental testing is a hotly debated topic that we will take up in Chapter 9. For now, keep in mind that intelligence tests do not sample all human abilities, and performance is affected by cultural and situational factors (Sternberg et al., 2000). Nevertheless, test scores are important because by age 5 to 6, they are good predictors of later intelligence and academic achievement, which are related to vocational success in industrialized societies. Let's see how the environments in which young children spend their days—home, preschool, and child care—affect mental test performance.

Home Environment and Mental Development

A special version of the *Home Observation for Measurement of the Environment (HOME)*, covered in Chapter 5, assesses aspects of 3- to 6-year-olds' home lives that support intellectual growth (see the Caregiving Concerns table below). Preschoolers who develop well intellectually have homes rich in toys and books. Their parents are warm and affectionate, stimulate language and academic knowledge, and arrange outings to places with interesting things to see and do. They also make reasonable demands for socially mature behavior—for example, that the child perform simple chores and behave courteously toward others. And these parents resolve conflicts with reason instead of physical force and punishment (Bradley & Caldwell, 1982; Espy, Molfese, & DiLalla, 2001; Roberts, Burchinal, & Durham, 1999).

As we saw in Chapter 2, these characteristics are less likely to be found in poverty-stricken families (Garrett, Ng'andu, & Ferron, 1994). When low-SES parents manage, despite daily pressures, to obtain high HOME scores, their preschoolers do substantially better on intelligence tests (Bradley & Caldwell,

Home Observation for Measurement of the Environment (HOME): Early Childhood Subscales

SUBSCALE	SAMPLE ITEM
Stimulation through toys, games, and reading material	Home includes toys that teach colors, sizes, and shapes.
Language stimulation	Parent teaches child about animals through books, games, and puzzles.
Organization of the physical environment	All visible rooms are reasonably clean and minimally cluttered.
Pride, affection, and warmth	Parent spontaneously praises child's qualities once or twice during observer's visit.
Stimulation of academic behavior	Child is encouraged to learn colors.
Modeling and encouragement of social maturity	Parent introduces interviewer to child.
Variety in daily stimulation	Family member takes child on one outing at least every other week (picnic, shopping).
Avoidance of physical punishment	Parent neither slaps nor spanks child during observer's visit.

Source: Bradley & Caldwell, 1979.

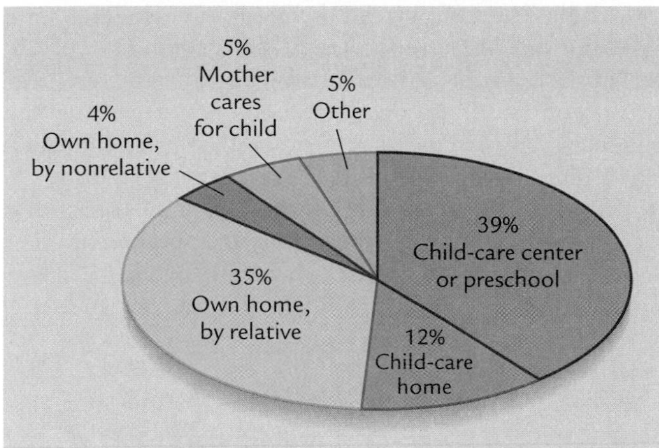

FIGURE 7.12 Who's minding America's preschoolers? The percentages refer to settings in which 3- and 4-year-olds spend most time while their parents are at work. Over one-fourth of 3- and 4-year-olds experience more than one type of child care, a fact not reflected in the chart. (Federal Interagency Forum on Child and Family Statistics, 2002.)

1982; Klebanov et al., 1998). These findings (as well as others we will discuss in Chapter 9) suggest that the home plays a major role in the generally poorer intellectual performance of low-SES children in comparison to their higher-SES peers.

Preschool, Kindergarten, and Child Care

Children between 2 and 6 spend even more time away from their homes and parents than infants and toddlers do. Over the last 30 years, the number of young children enrolled in preschool or child care has steadily increased. This trend is largely due to the dramatic rise in women's participation in the paid labor force. Currently, 64 percent of American and 68 percent of Canadian preschool-age children have mothers who are employed (Statistics Canada, 2002g; U.S. Bureau of the Census, 2002c). Figure 7.12 shows where American preschoolers spend their days while their parents are at work.

A *preschool* is a program with planned educational experiences aimed at enhancing the development of 2- to 5-year-olds. In contrast, *child care* identifies a variety of arrangements for supervising children of employed parents, ranging from care in someone else's or the child's own home to some type of center-based program. The line between preschool and child care is fuzzy. As Figure 7.12 indicates, parents often select a preschool as a child-care option. Many North American preschools—and public school kindergartens as well—have increased their hours from half to full days in response to the needs of employed parents (U.S. Department of Education, 2002b). At the same time, today we know that good child care is not simply a matter of keeping children safe and adequately fed. It should provide the same high-quality educational experiences that an effective preschool does, the only difference being that children attend for an extended day.

● Types of Preschool and Kindergarten. Preschool and kindergarten programs range along a continuum, from child-centered to teacher-directed. In **child-centered programs,** teachers provide a wide variety of activities from which children select, and much learning takes place through play. In contrast, in **academic programs,** teachers structure children's learning, teaching letters, numbers, colors, shapes, and other academic skills through formal lessons, often using repetition and drill.

Despite grave concern about the appropriateness of the approach, preschool and kindergarten teachers have felt increased pressure to stress formal academic training. Yet doing so undermines motivation and other aspects of emotional well-being. When preschoolers and kindergartners spend much time passively sitting and doing worksheets as opposed to being actively engaged in learning centers, they display more stress behaviors (such as wiggling and rocking), have less confidence in their abilities, and prefer less challenging tasks (Stipek et al., 1995). And follow-ups show that they have poorer study habits and achieve less well in grade school (Burts et al., 1992; Hart et al., 1998). These outcomes are strongest for low-SES children. Yet teachers tend to prefer an academic approach for economically disadvantaged children—a disturbing trend in view of its negative impact on motivation and learning (Stipek & Byler, 1997).

● Early Intervention for At-Risk Preschoolers. In the 1960s, when the United States launched a "War on Poverty," a wide variety of preschool intervention programs for economically disadvantaged children were initiated. Their underlying assumption was that learning problems are best treated early, before formal schooling begins. **Project Head Start,** begun by the U.S. federal government in 1965, is the most extensive of these programs. A typical Head Start center provides children with a year or two of preschool, along with nutritional and health services. Parent involvement is central to the Head Start philosophy. Parents serve on policy councils and contribute to program planning. They also work directly with children in classrooms, attend special programs on parenting and child development, and receive services directed at their own emotional, social, and vocational needs. Currently, more than 1,500 Head Start centers serve about 900,000 children (Head Start Bureau, 2002).

In 1995, Canada initiated **Aboriginal Head Start** for First Nations, Inuit, and Métis children younger than age 6, 60 percent of whom live in poverty. Like Project Head Start, the program provides children with preschool education and nutritional and health services and encourages parent involvement. Currently, Aboriginal Head Start has more than 80 sites and serves more than 3,000 children (Health Canada, 2000a).

Over two decades of research have established the long-term benefits of preschool intervention. The most important of these studies combined data from seven university-based interventions. Results showed that children who attended programs scored higher in IQ and school achievement than did controls during the first 2 to 3 years of elementary school. After that time, differences in test scores declined. Nevertheless, children who received intervention remained ahead into

■ This Inuit 4-year-old attends an Aboriginal Head Start program in Nunavut, Canada. Like American Head Start children, she receives preschool education and nutritional and health services, and her parents participate. In the classroom, she has many opportunities to engage in culturally meaningful activities. Here she plays with a model igloo.

adolescence on measures of real-life school adjustment. They were less likely to be placed in special education or retained in grade, and a greater number graduated from high school (Lazar & Darlington, 1982). A separate report on one program revealed benefits lasting into young adulthood—a reduction in delinquency and teenage pregnancy, greater likelihood of employment, and greater educational attainment, earnings, and marital stability at age 27 (Weikart, 1998).

Does the impact of outstanding university-based programs on school adjustment generalize to community-based Head Start programs? Outcomes are similar, although not as strong. Head Start preschoolers are more economically disadvantaged than children in university-based programs and hence have more severe learning problems. And quality of services is more variable across community-based programs (Barnett, 1998; Ramey, 1999).

A consistent finding is that gains in IQ and achievement scores from attending Head Start and other interventions do not last. These children typically enter inferior public schools in poverty-stricken neighborhoods, which undermine the benefits of preschool education (Schnur & Belanger, 2000). But Head Start children's favorable school adjustment is an impressive outcome. It may be partly due to program effects on parents. The more parents are involved in Head Start, the better their child-rearing practices and the more stimulating their home learning environments. These factors are positively related to preschoolers' year-end academic, language, and social skills and independence and task persistence in the classroom (Marcon, 1999; Parker et al., 1999).

Head Start is highly cost-effective. Program expenses are far less than the funds required to provide special education, treat delinquency, and support unemployed adults (Zigler & Styfco, 2001). Yet because of funding shortages, most poverty-stricken preschoolers in the United States and Canada do not receive services.

● **Child Care.** We have seen that high-quality early intervention can enhance the development of economically disadvantaged children. However, as noted in Chapter 5, much North American child care lacks quality. Preschoolers exposed to inadequate child care, regardless of family SES, score lower on measures of cognitive and social skills (Lamb, 1998; NICHD Early Child Care Research Network, 2002b).

What are the ingredients of high-quality child care during the preschool years? Large-scale studies of center- and home-based settings reveal that the following factors are important: group size (number of children in a single space), caregiver–child ratio, caregiver's educational preparation, and caregiver's personal commitment to learning about and caring for children. When these characteristics are favorable, adults are more verbally stimulating and sensitive to children's needs. Children, in turn, do especially well on measures of cognitive, language, and social skills—effects that persist into the early school years (Burchinal et al., 2000; Helburn, 1995; Peiser-Feinberg, 1999).

The Caregiving Concerns table on the following page summarizes characteristics of high-quality early childhood programs, based on standards for developmentally appropriate practice devised by the U.S. National Association for the Education of Young Children. These standards offer a set of worthy goals as the United States and Canada strive to upgrade child-care and educational services for young children.

Educational Television

Besides home and preschool, young children spend much time in another learning environment: television. The average 2- to 6-year-old watches TV from 1½ to 2 hours a day—a long time in a young child's life. During middle childhood, TV viewing increases to an average of 3½ hours a day for American children and 2½ hours a day for Canadian children (Comstock & Scharrer, 2001; Statistics Canada, 2001g). Low-SES children are more frequent viewers, perhaps because their parents are less able to pay for other entertainment. And if parents watch a lot of TV, their children do so as well (Huston & Wright, 1998).

Each afternoon, Sammy looked forward to his favorite TV program, *Sesame Street.* It uses lively visual and sound effects to stress basic literacy and number concepts and engaging puppet and human characters to teach general knowledge,

Signs of Developmentally Appropriate Early Childhood Programs

PROGRAM CHARACTERISTICS	SIGNS OF QUALITY
Physical setting	Indoor environment is clean, in good repair, and well ventilated. Classroom space is divided into richly equipped activity areas, including make-believe play, blocks, science, math, games, puzzles, books, art, and music. Fenced outdoor play space is equipped with swings, climbing equipment, tricycles, and sandbox.
Group size	In preschools and child-care centers, group size is no greater than 18 to 20 children with 2 teachers.
Caregiver–child ratio	In preschools and child-care centers, teacher is responsible for no more than 8 to 10 children. In family child-care homes, caregiver is responsible for no more than 6 children.
Daily activities	Most of the time, children work individually or in small groups. Children select many of their own activities and learn through experiences relevant to their own lives. Teachers facilitate children's involvement, accept individual differences, and adjust expectations to children's developing capacities.
Interactions among adults and children	Teachers move among groups and individuals, asking questions, offering suggestions, and adding more complex ideas. They use positive guidance techniques, such as modeling and encouraging expected behavior and redirecting children to more acceptable activities.
Teacher qualifications	Teachers have college-level specialized preparation in early childhood development, early childhood education, or a related field.
Relationships with parents	Parents are encouraged to observe and participate. Teachers talk frequently with parents about children's behavior and development.
Licensing and accreditation	Program is licensed by the state. If a preschool or child-care center, accreditation by the National Academy of Early Childhood Programs is evidence of an especially high-quality program. If a child-care home, accreditation by the National Association for Family Child Care is evidence of high-quality experiences for children.

Sources: Bredekamp & Copple, 1997; National Association for the Education of Young Children, 1998.

emotional and social understanding, and prosocial skills. Today, more than two-thirds of North American preschoolers watch *Sesame Street,* and it is broadcast in over 60 countries (Raugust, 1999).

The more children watch *Sesame Street,* the higher they score on tests and observations that measure the program's learning goals (Fisch, Truglio, & Cole, 1999). One study reported a link between preschool viewing of *Sesame Street* and other similar educational programs and getting higher grades, reading more books, and placing more value on achievement in high school (Anderson et al., 2001). In recent years, *Sesame Street* has reduced its rapid-paced, adlike format in favor of leisurely episodes with a clear story line (Truglio, 2000). Children's programs with slow-paced action and easy-to-follow narratives, such as *Mr. Rogers' Neighborhood* and *Barney and Friends,* lead to more elaborate make-believe play than those presenting quick, disconnected bits of information (Singer, 1999).

Does heavy TV viewing take children away from activities that promote cognitive development? Some evidence suggests that it does. The more preschool and school-age children watch TV—especially entertainment shows and cartoons—the less time they spend reading and interacting with others, and the weaker their academic skills (Huston et al., 1999; Wright et al., 2001). But television can support children's learning as long as viewing is not excessive and programs meet their developmental needs. We will look at the impact of TV on emotional and social development in the next chapter.

Ask Yourself

REVIEW

What findings indicate that child-centered rather than academic preschools and kindergartens are better suited to fostering young children's academic development?

APPLY

Senator Smith heard that IQ gains resulting from Head Start do not last, so he plans to vote against funding for the program. Write a letter to Senator Smith explaining why he should support Head Start.

www.

Language Development

Language is intimately related to virtually all the cognitive changes discussed in this chapter. Between ages 2 and 6, children make awesome and momentous advances in language. Their remarkable achievements, as well as mistakes along the way, indicate that they master their native tongue in an active, rule-oriented fashion.

Vocabulary

At age 2, Sammy had a vocabulary of 200 words. By age 6, he will have acquired around 10,000 words. To accomplish this feat, Sammy will learn about 5 new words each day (Anglin, 1993). How do children build their vocabularies so quickly? Researchers have discovered that they can connect a new word with an underlying concept after only a brief encounter, a process called **fast-mapping.**

Preschoolers fast-map some words more accurately and easily than others. Western children learn labels for objects especially rapidly because these refer to concrete items they know much about, and caregivers' speech often emphasizes names for things (Bloom, 1998). Soon action words ("go," "run," "broke") are added. However, Chinese-, Japanese-, and Korean-speaking children acquire verbs especially quickly. In these languages, nouns are often omitted from adult sentences, and verbs are stressed (Kim, McGregor, & Thompson, 2000; Tardif, Gelman, & Xu, 1999). Gradually, preschoolers add more modifiers ("red," "round," "sad"). If modifiers are related to one another in meaning, they take longer to learn. For example, 2-year-olds grasp the general distinction between "big" and "small," but not until age 3 to 5 are more refined differences between "tall" and "short," "high" and" "low," and "long" and "short" understood (Stevenson & Pollitt, 1987).

Preschoolers figure out the meanings of new words by contrasting them with words they already know. But exactly how they discover which concept each word picks out is not yet fully understood. Ellen Markman (1989, 1992) believes that in the early phases of vocabulary growth, children adopt a *principle of mutual exclusivity.* They assume that words refer to entirely separate (nonoverlapping) categories. Consistent with this idea, when 2-year-olds are told the names of two very different novel objects (a clip and a horn), they assign each label correctly (Waxman & Senghas, 1992).

However, sometimes adults call an object by more than one name. Children often draw on other aspects of language for help in these instances. According to one proposal, they figure out many word meanings by observing how words are used in the structure of sentences. Consider an adult who says, "This is a *citron* one," while showing the child a yellow car. Two- and 3-year-olds interpret a new word used as an adjective as referring to a property of the object (Hall & Graham, 1999).

Furthermore, preschoolers often use social cues to identify word meanings. In one study, an adult performed an action on an object and then used a new label while looking back and forth between the child and the object, as if to invite the child to play. Two-year-olds concluded that the label referred to the action, not the object (Tomasello & Akhtar, 1995). When no social cues or other information is available, children as young as 2 show remarkable flexibility in their word-learning strategies. They treat the new word as a second name for the object (Deák, 2000; Deák & Maratsos, 1998).

Once preschoolers have a sufficient vocabulary, they use words creatively to fill in for ones they have not learned. As early as age 2, children coin new words in systematic ways. For example, Sammy said "plant-man" for a gardener (created a compound word) and "crayoner" for a child using crayons (added the ending *-er*) (Clark, 1995). Preschoolers also extend language meanings through metaphor. For example, one 3-year-old used the expression "fire engine in my tummy" to describe a stomachache (Winner, 1988). The metaphors of young preschoolers involve concrete, sensory comparisons, such as "Clouds are pillows" and "Leaves are dancers." Once vocabulary and general knowledge expand, they also appreciate nonsensory comparisons, such as, "Friends are like magnets" (Karadsheh, 1991; Keil, 1986). Metaphors permit young children to communicate in especially vivid and memorable ways.

Grammar

Grammar refers to the way we combine words into meaningful phrases and sentences. Between ages 2 and 3, English-speaking children use simple sentences that follow a subject–verb–object word order. Children learning other languages adopt the word orders of the adult speech to which they are exposed (Maratsos, 1998). As they conform to word-order rules, preschoolers also make small additions and changes in words that enable us to express meanings flexibly and efficiently. For example, they add "-s" for plural ("cats"), use prepositions ("in" and "on"), and form various tenses of the verb "to be" ("is," "are," "were," "has been," "will"). All English-speaking children master these grammatical markers in a regular sequence, starting with the ones that involve the simplest meanings and structures (Brown, 1973; de Villiers & de Villiers, 1973).

By age 3½, children have acquired a great many grammatical rules, and they apply them so consistently that once in a while they overextend the rules to words that are exceptions, a type of error called **overregularization.** "My toy car *breaked*" and "We each have two *feets*" are expressions that appear between age 2 and 3 (Marcus et al., 1992; Marcus, 1995).

Between 3 and 6 years, children master even more complex grammatical structures, although they make predictable errors along the way. In asking questions, preschoolers are reluctant to let go of the subject–verb–object sequence so basic to the English language. At first, they use only a rising intonation and fail to invert the subject and verb, as in "Mommy baking cookies?" and "What you doing, Daddy?" (Stromswold, 1995). Because they cling to a consistent word order, they also have trouble with some passive sentences. When told, "The car was pushed by the truck," young preschoolers often make a toy car push a truck. By age 5, they understand expressions like these, but full mastery of the passive form is not complete until the end of middle childhood (Horgan, 1978).

Nevertheless, preschoolers' grasp of grammar is remarkable. By age 4 to 5, they form embedded sentences ("I think *he will come*"), tag questions ("Dad's going to be home soon, *isn't he?*"), and indirect objects ("He showed *his friend* the present"). As the preschool years draw to a close, children use most of the grammatical constructions of their language competently (Tager-Flusberg, 2001).

Conversation

Besides acquiring vocabulary and grammar, children must learn to engage in effective and appropriate communication with others. This practical, social side of language is called **pragmatics,** and preschoolers make considerable headway in mastering it.

At the beginning of early childhood, children are already skilled conversationalists. In face-to-face interaction, they take turns and respond appropriately to their partners' remarks (Pan & Snow, 1999). The number of turns over which children can sustain interaction and their ability to maintain a topic over time increase with age, but even 2-year-olds are capable of effective conversation.

By age 4, children already adjust their speech to fit the age, sex, and social status of their listeners. For example, in acting out different roles with hand puppets, they use more commands when playing socially dominant and male roles, such as teacher, doctor, and father. In contrast, they speak more politely and use more indirect requests when playing less dominant and female roles, such as student, patient, and mother (Anderson, 1992).

Preschoolers' conversational skills occasionally do break down. For example, have you tried talking on the telephone with a preschooler? Here is an excerpt of one 4-year-old's phone conversation with his grandfather:

Grandfather: "How old will you be?"
John: "Dis many." *[Holding up four fingers]*
Grandfather: "Huh?"
John: "Dis many." *[Again holding up four fingers]*(Warren & Tate, 1992, pp. 259–260)

Young children's conversations appear less mature in highly demanding situations in which they cannot see their listeners' reactions or rely on typical conversational aids, such as gestures and objects to talk about. However, when asked to tell a listener how to solve a simple puzzle, 3- to 6-year-olds' directions are more specific over the phone than in person, indicating that they realize more verbal description is necessary in the phone context (Cameron & Lee, 1997). Between ages 4 and 8, both conversing and giving directions over the phone improve greatly. Telephone talk provides yet another example of how preschoolers' competencies depend on the demands of the situation.

Supporting Language Learning in Early Childhood

How can adults foster preschoolers' language development? Interaction with more skilled speakers, which is so important during toddlerhood, remains vital in early childhood.

■ Conversational give-and-take with adults is vital for supporting children's language progress. Through it, children rapidly acquire vocabulary and grammar and learn how to engage in effective and appropriate communication with others.

Conversational give-and-take with adults, either at home or in preschool, is consistently related to general measures of language progress (Hart & Risley, 1995; Helburn, 1995).

Sensitive, caring adults use additional techniques that promote early language skills. When children use words incorrectly or communicate unclearly, they give helpful, explicit feedback, such as, "I can't tell which ball you want. Do you mean a large or small one or a red or green one?" At the same time, they do not overcorrect, especially when children make grammatical mistakes. Criticism discourages children from actively experimenting with language rules in ways that lead to new skills.

Instead, adults provide subtle, indirect feedback about grammar by using two strategies, often in combination: **expansions**—elaborating on children's speech, increasing its complexity; and **recasts**—restructuring inaccurate speech into correct form (Bohannon & Stanowicz, 1988). For example, if a child says, "I gotted new red shoes," the parent might respond, "Yes, you got a pair of new red shoes." However, some researchers question whether expansions and recasts are as important in children's mastery of grammar as mere exposure to a rich language environment. The techniques are not used in all cultures, and they do not consistently affect children's grammar (Strapp & Federico, 2000; Valian, 1996). Rather than eliminating errors, perhaps expansions and recasts model grammatical alternatives and encourage children to experiment with them.

Do the findings just described remind you once again of Vygotsky's theory? In language, as in other aspects of intellectual growth, parents and teachers gently prompt young children to take the next developmental step forward. Children strive to master language because they want to attain social connectedness to other people. Adults, in turn, respond to children's natural desire to become competent speakers by listening attentively, elaborating on what children say, modeling correct usage, and stimulating children to talk further. In the next chapter, we will see that this special combination of warmth and encouragement of mature behavior is at the heart of early childhood emotional and social development as well.

Ask Yourself

REVIEW

What can adults do to support language development in early childhood? Provide a list of recommendations, noting research that supports each.

APPLY

One day, Sammy's mother explained to him that the family would take a vacation in Miami. The next morning, Sammy emerged from his room with belongings spilling out of a suitcase and remarked, "I gotted my bag packed. When are we going to Your-ami?" What do Sammy's errors reveal about his approach to mastering language?

CONNECT

Explain how children's strategies for word learning support the interactionist perspective on language development, described on page 165 of Chapter 5.

www.

Summary

PHYSICAL DEVELOPMENT

Body Growth

Describe major trends in body growth during early childhood.

- Children grow more slowly in early childhood than they did in the first 2 years. Body fat declines, and children become longer and leaner. New growth centers appear in the skeleton, and by the end of early childhood, children start to lose their primary teeth.
- Different parts of the body grow at different rates. The **general growth curve** describes changes in body size—rapid during infancy, slower during early and middle childhood, rapid again in adolescence. Exceptions to this trend include the genitals, the lymph tissue, and the brain.

Brain Development

Describe brain development during early childhood.

- During early childhood, frontal-lobe areas of the cerebral cortex devoted to planning and organizing behavior develop rapidly. Also, the left cerebral hemisphere shows more neural activity than the right, supporting young children's expanding language skills.
- Hand preference strengthens during early and middle childhood, indicating that lateralization strengthens during this time. Handedness indicates an individual's **dominant cerebral hemisphere.** According to one theory, most children inherit a gene that biases them for right-handedness, but experience can sway children toward a left-handed preference. Although left-handedness is associated with developmental problems, the great majority of left-handed children are normal.
- During early childhood, connections are established among different brain structures. Fibers linking the **cerebellum** to the cerebral cortex myelinate, enhancing balance and motor control. The **reticular formation,** responsible for alertness and consciousness, and the **corpus callosum,** which connects the two cerebral hemispheres, also myelinate rapidly.

Influences on Physical Growth and Health

Explain how heredity influences physical growth.

- Heredity influences physical growth by controlling the release of hormones from the **pituitary gland.** Two hormones are especially influential: **growth hormone (GH)** and **thyroid-stimulating hormone (TSH).**

Describe the effects of emotional well-being, nutrition, and infectious disease on physical growth in early childhood.

- Emotional well-being continues to influence body growth in middle childhood. An emotionally inadequate home life can lead to **psychosocial dwarfism.**
- Preschoolers' slower growth rate causes their appetites to decline, and often they become picky eaters. Modeling by others, repeated exposure to new foods, and a positive emotional climate at mealtimes can promote healthy, varied eating in young children.
- Malnutrition can combine with infectious disease to undermine healthy growth. In developing countries, diarrhea is widespread and claims millions of young lives. Teaching families how to administer **oral rehydration therapy (ORT)** can prevent most of these deaths.
- Immunization rates are lower in the United States than in other industrialized nations because many economically disadvantaged children do not have access to health care.

What factors increase the risk of unintentional injuries, and how can childhood injuries be prevented?

- Unintentional injuries are the leading cause of childhood mortality in industrialized countries. Injury victims are more likely to be boys; to be temperamentally irritable, inattentive, and negative; and to be growing up in stressed, poverty-stricken inner-city families. Injury deaths are especially high in developing nations, where they may soon exceed disease as the leading cause of childhood deaths.
- A variety of approaches are needed to prevent childhood injuries, including reducing poverty and other sources of family stress; passing laws that promote child safety; creating safer home, travel, child-care, and play environments; improving public education; and changing parent and child behaviors.

Motor Development

Cite major milestones of gross and fine motor development in early childhood.

- During early childhood, the child's center of gravity shifts toward the trunk, and balance improves, paving the way for many gross motor achievements. Preschoolers run, jump, hop, gallop, eventually skip, throw and catch, and generally become better coordinated.
- Increasing control of the hands and fingers leads to dramatic improvements in fine motor skills. Preschoolers gradually become self-sufficient at dressing and using a knife and fork. By age 3, children's scribbles become pictures. With age, their drawings increase in complexity and realism and are greatly influenced by their culture's artistic traditions and by schooling. Preschoolers also try to print letters of the alphabet and, later, words.

Describe individual differences in preschoolers' motor skills.

- Body build and opportunity for physical play affect early childhood motor development. Sex differences that favor boys in skills requiring force and power and girls in skills requiring good balance and fine movements are partly genetic, but environmental pressures exaggerate them. Children master the motor skills of early childhood through informal play experiences.

COGNITIVE DEVELOPMENT

Piaget's Theory: The Preoperational Stage

Describe advances in mental representation and limitations of thinking during the preoperational stage.

- Rapid advances in mental representation, notably language and make-believe play, mark the beginning of Piaget's **preoperational stage.** With age, make-believe becomes increasingly complex, evolving into **sociodramatic play** with others. Preschoolers' make-believe supports many aspects of development.
- Aside from representation, Piaget described the young child in terms of deficits rather than strengths. Preoperational children are **egocentric**; they often fail to imagine the perspectives of others. Because egocentrism prevents children from reflecting on their own thinking and accommodating, it contributes to **animistic thinking, centration,** a focus on perceptual appearances, and **irreversibility.** These difficulties cause preschoolers to fail **conservation** and **hierarchical classification** tasks.

What are the implications of recent research for the accuracy of the preoperational stage?

- When young children are given simplified problems relevant to their everyday lives, their performance appears more mature than Piaget assumed. Preschoolers recognize differing perspectives, distinguish animate and inanimate objects, and reason by analogy about physical transformations. Also, their language reflects accurate causal reasoning and hierarchical classification, and they form many categories based on nonobvious features. Operational thinking develops gradually over the preschool years, a finding that challenges Piaget's stage concept.

What educational principles can be derived from Piaget's theory?

- A Piagetian classroom promotes discovery learning, sensitivity to children's readiness to learn, and acceptance of individual differences.

Vygotsky's Sociocultural Theory

Explain Vygotsky's perspective on the origins and significance of children's private speech, and describe applications of his theory to education.

- In contrast to Piaget, Vygotsky regarded language as the foundation for all higher cognitive processes. According to Vygotsky, **private speech,** or language used for self-guidance, emerges out of social communication as adults and more skilled peers help children master challenging tasks. Eventually private speech is internalized as inner, verbal thought. **Scaffolding** is a form of social interaction that promotes the transfer of cognitive processes to children.
- A Vygotskian classroom emphasizes assisted discovery. Verbal guidance from teachers and peer collaboration are vitally important. Make-believe play serves as a vital zone of proximal development that enhances many new competencies.

Information Processing

How do attention and memory change during early childhood?

- Attention gradually becomes more sustained, and planning improves. Nevertheless, compared with older children, preschoolers spend relatively short periods involved in tasks and are less systematic in planning.
- Young children's recognition memory is very accurate. Their recall for listlike information is much poorer than that of older children and adults because preschoolers use **memory strategies** less effectively. Like adults, preschoolers remember recurring everyday experiences in terms of **scripts,** which become more elaborate with age. As adults use an elaborative style of conversing with children about the past, their autobiographical memory becomes better organized and detailed.

Describe the young child's theory of mind.

- Preschoolers begin to construct a theory of mind, indicating that they are capable of **metacognition,** or thinking about thought. Around age 4, they understand that people can hold false beliefs. Many factors contribute to young children's appreciation of mental life, including cognitive and language development, having older siblings, interactions with friends, make-believe play, and contact with more mature members of society. Preschoolers' understanding of the mind is far from complete; they regard it as a passive container of information rather than as an active, constructive agent.

Summarize children's literacy and mathematical knowledge during early childhood.

- Children understand a great deal about literacy long before they read or write in conventional ways. Preschoolers gradually revise incorrect ideas about the meaning of written symbols as their perceptual and cognitive capacities improve, as they encounter writing in many different contexts, and as adults help them make sense of written information.

- In the second year, children have a beginning grasp of **ordinality,** which serves as the basis for more complex understandings. Soon they discover additional mathematical principles, including **cardinality,** and experiment with counting strategies, selecting the most efficient, accurate techniques. Both literacy and mathematical reasoning build on a foundation of informally acquired knowledge.

Individual Differences in Mental Development

Describe early childhood intelligence tests and the impact of home, educational programs, child care, and television on mental development in early childhood.

- Intelligence tests in early childhood sample a wide variety of verbal and nonverbal skills. By age 5 to 6, they are good predictors of later intelligence and academic achievement. Children growing up in warm, stimulating homes with parents who make reasonable demands for mature behavior score higher on mental tests.

- Preschool and kindergarten programs range along a continuum. In **child-centered programs,** much learning takes place through play. In **academic programs,** teachers train academic skills through formal lessons, often using repetition and drill. Stressing formal academic training, however, undermines young children's motivation and negatively influences later school achievement.

- **Project Head Start** is the largest federally funded preschool program for low-income children in the United States. In Canada, **Aboriginal Head Start** serves First Nations, Inuit, and Métis preschoolers. High-quality preschool intervention results in immediate test score gains and long-term improvements in school adjustment. The more parents are involved in Head Start, the higher children's year-end academic, language, and social skills. Regardless of SES, poor-quality child care undermines children's cognitive and social development.

- Children pick up many cognitive skills from educational television programs like *Sesame Street.* Programs with slow-paced action and easy-to-follow story lines foster more elaborate make-believe play. Heavy viewing of TV, especially entertainment shows and cartoons, takes children away from reading and interacting with others and is related to weaker academic skills.

Language Development

Trace the development of vocabulary, grammar, and conversational skills in early childhood.

- Supported by **fast-mapping,** children's vocabularies grow dramatically during early childhood. On hearing a new word, children contrast it with words they know and often assume the word refers to an entirely separate category. When adults call an object by more than one name, children observe how words are used in the structure of sentences or use social cues to figure out word meanings. Once preschoolers have sufficient vocabulary, they extend language meanings, coining new words and creating metaphors.

- Between ages 2 and 3, children adopt the basic word order of their language. As they master grammatical rules, they occasionally **overregularize,** or apply the rules to words that are exceptions. By the end of early childhood, children have acquired complex grammatical forms.

- **Pragmatics** refers to the practical, social side of language. In face-to-face interaction with peers, young preschoolers are already skilled conversationalists. By age 4, they adapt their speech to their listeners in culturally accepted ways. Preschoolers' communication skills appear less mature in highly demanding contexts.

Cite factors that support language learning in early childhood.

- Conversational give-and-take with more skilled speakers fosters language progress. Adults often provide explicit feedback on the clarity of children's language and indirect feedback about grammar through **expansions** and **recasts.** However, some researchers question the impact of these strategies on grammatical development. For this aspect of language, exposure to a rich language environment may be sufficient.

Important Terms and Concepts

Aboriginal Head Start (p. 233)
academic programs (p. 233)
animistic thinking (p. 218)
cardinality (p. 231)
centration (p. 218)
cerebellum (p. 207)
child-centered programs (p. 233)
conservation (p. 218)
corpus callosum (p. 207)
dominant cerebral hemisphere (p. 207)
egocentrism (p. 217)
expansions (p. 238)
fast-mapping (p. 236)
general growth curve (p. 205)
growth hormone (GH) (p. 208)
hierarchical classification (p. 218)
irreversibility (p. 218)
memory strategies (p. 227)
metacognition (p. 228)
oral rehydration therapy (ORT) (p. 209)
ordinality (p. 231)
overregularization (p. 236)
pituitary gland (p. 208)
pragmatics (p. 237)
preoperational stage (p. 216)
private speech (p. 223)
Project Head Start (p. 233)
psychosocial dwarfism (p. 208)
recasts (p. 238)
reticular formation (p. 207)
scaffolding (p. 224)
scripts (p. 227)
sociodramatic play (p. 217)
thyroid-stimulating hormone (TSH) (p. 208)

FYI For Further Information and Help

Consult the Companion Website for *Development Through the Lifespan, Third Edition,* (www.ablongman.com/berk) where you will find the following resources for this chapter:

- **Chapter Objectives**
- **Flashcards** for studying important terms and concepts
- **Annotated Weblinks** to guide you in further research
- **Ask Yourself** questions, which you can answer and then check against a sample response
- **Suggested Readings**
- **Practice Test** with immediate scoring and feedback

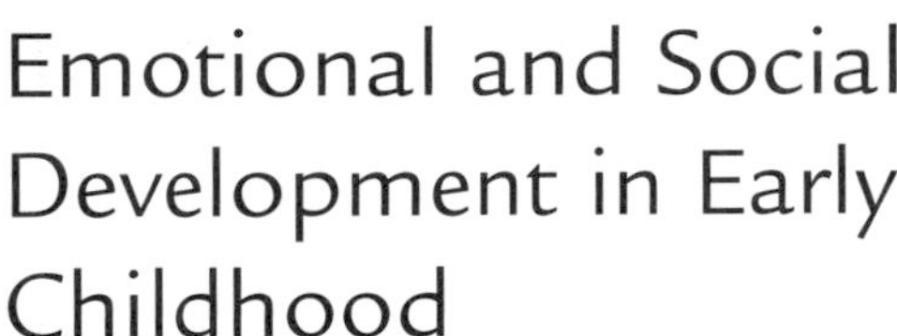

8 Emotional and Social Development in Early Childhood

Erikson's Theory: Initiative versus Guilt

Self-Understanding
Foundations of Self-Concept • Emergence of Self-Esteem

- Cultural Influences: Cultural Variations in Personal Storytelling: Implications for Early Self-Concept

Emotional Development
Understanding Emotion • Emotional Self-Regulation • Self-Conscious Emotions • Empathy

Peer Relations
Advances in Peer Sociability • First Friendships • Parental Influences on Early Peer Relations

Foundations of Morality
The Psychoanalytic Perspective • Social Learning Theory • The Cognitive-Developmental Perspective • The Other Side of Morality: Development of Aggression

- Biology & Environment: Temperament and the Development of Conscience in Young Children

Gender Typing
Gender-Stereotyped Beliefs and Behavior • Genetic Influences on Gender Typing • Environmental Influences on Gender Typing • Gender Identity • Reducing Gender Stereotyping in Young Children

- A Lifespan Vista: David: A Boy Who Was Reared as a Girl

Child Rearing and Emotional and Social Development
Child-Rearing Styles • Cultural Variations • Child Maltreatment

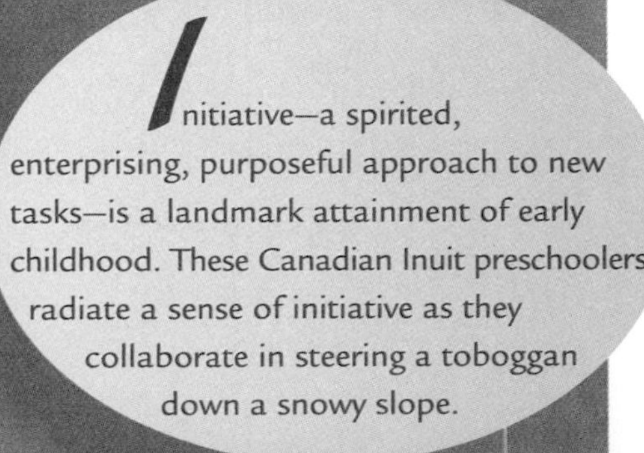

Initiative—a spirited, enterprising, purposeful approach to new tasks—is a landmark attainment of early childhood. These Canadian Inuit preschoolers radiate a sense of initiative as they collaborate in steering a toboggan down a snowy slope.

As the children in Leslie's classroom moved through the preschool years, their personalities took on clearer definition. By age 3, they voiced firm likes and dislikes as well as new ideas about themselves. "Stop bothering me," Sammy said to Mark, who tried to reach for Sammy's beanbag as Sammy aimed it toward the mouth of a large clown face. "See, I'm great at this game," Sammy announced with confidence, an attitude that kept him trying, even though he missed most of the throws.

The children's conversations also revealed their first notions about morality. Often they combined statements about right and wrong they had heard from adults with forceful attempts to defend their own desires. "You're 'posed to share," stated Mark, grabbing a beanbag out of Sammy's hand.

"I was here first! Gimme it back," demanded Sammy, who pushed Mark while reaching for the beanbag. The two boys struggled until Leslie intervened, provided an extra set of beanbags, and showed them how they could both play.

As Sammy's and Mark's interaction reveals, preschoolers quickly become complex social beings. Although arguments and aggression take place among all young children, cooperative exchanges are far more frequent. Between ages 2 and 6, first friendships form, in which children converse, act out complementary roles, and learn that their own desires for companionship and toys are best met when they consider the needs and interests of others.

The children's developing understanding of their social world was especially evident in the attention they gave to the dividing line between male and female. While Lynette and Karen cared for a sick baby doll in the housekeeping area, Sammy, Vance, and Mark transformed the block corner into a busy intersection. "Green light, go!" shouted police officer Sammy as Vance and Mark pushed large wooden cars and trucks across the floor. Already, the children preferred same-sex peers, and their play themes mirrored the gender stereotypes of their cultural community.

This chapter is devoted to the many facets of emotional and social development in early childhood. The theory of Erik Erikson provides an overview of personality change during the preschool years. Then we consider children's concepts of themselves, their insights into their social and moral worlds, their gender typing, and their increasing ability to manage their emotional and social behaviors. In the final sections of this chapter, we answer the question, What is effective child rearing? We also consider the complex conditions that support good parenting or lead it to break down, including the serious and widespread problems of child abuse and neglect.

Erikson's Theory: Initiative versus Guilt

Erikson (1950) described early childhood as a period of "vigorous unfolding." Once children have a sense of autonomy, they become less contrary than they were as toddlers. Their energies are freed for tackling the psychological conflict of the preschool years: **initiative versus guilt.** As the word *initiative* suggests, young children have a new sense of purposefulness. They are eager to tackle new tasks, join in activities with peers, and discover what they can do with the help of adults. And they also make strides in conscience development.

Erikson regarded play as a central means through which young children find out about themselves and their social world. Play permits preschoolers to try out new skills with little risk of criticism and failure. It also creates a small social organization of children who must cooperate to achieve common goals. Around the world, children act out family scenes and highly visible occupations—police officer, doctor, and nurse in Western societies, rabbit hunter and potter among the Hopi Indians, and hut builder and spear maker among the Baka of West Africa (Roopnarine et al., 1998).

Recall that Erikson's theory builds on Freud's psychosexual stages (see Chapter 1, page 16). In Freud's well-known Oedipus and Electra conflicts, to avoid punishment and maintain the affection of parents, children form a *superego,* or conscience, by *identifying* with the same-sex parent. That is, they take the parent's characteristics into their personality and, as a result, adopt the moral and gender-role standards of their society. Each time the child disobeys standards of conscience, painful feelings of guilt occur.

For Erikson, the negative outcome of early childhood is an overly strict superego that causes children to feel too much guilt because they have been threatened, criticized, and punished excessively by adults. When this happens, preschoolers' exuberant play and bold efforts to master new tasks break down.

Although Freud's Oedipus and Electra conflicts are no longer regarded as satisfactory explanations of conscience development, Erikson's image of initiative captures the diverse changes in young children's emotional and social lives. The preschool years are, indeed, a time when children develop a confident self-image, more effective control over their emo-

■ As this Mexican Indian folk healer administers herbal remedies to a patient, the healer's grandchildren watch and play at the art of healing. By acting out highly visible occupations in their culture during play, children acquire insight into what they can do and develop a sense of initiative.

tions, new social skills, the foundations of morality, and a clear sense of themselves as boy or girl. Let's look closely at each of these aspects of development.

Self-Understanding

During early childhood, new powers of representation permit children to reflect on themselves. Language enables them to talk about the *I-self*—their own subjective experience of being (Harter, 1998). In Chapter 7, we noted that preschoolers quickly acquire a vocabulary for talking about their inner mental lives and refine their understanding of mental states. As the I-self becomes more firmly established, children focus more intently on the *me-self*—knowledge and evaluation of the self's characteristics. They start to develop a **self-concept,** the set of attributes, abilities, attitudes, and values that an individual believes defines who he or she is.

Foundations of Self-Concept

Ask a 3- to 5-year-old to tell you about himself or herself, and you are likely to hear something like this: "I'm Tommy. See, I got this new red T-shirt. I'm 4 years old. I can brush my teeth, and I can wash my hair all by myself. I have a new Tinkertoy set, and I made this big, big tower." As these statements indicate, preschoolers' self-concepts are very concrete. Usually, they mention observable characteristics, such as their name, physical appearance, possessions, and everyday behaviors (Harter, 1996; Watson, 1990).

By age 3½, preschoolers also describe themselves in terms of typical emotions and attitudes, as in "I'm happy when I play with my friends" or "I don't like being with grown-ups" (Eder, 1989). This suggests that they have some awareness of their unique psychological characteristics. As further support for this budding grasp of personality, when given a trait label, such as "shy" or "mean," 4-year-olds infer appropriate motives and feelings. For example, they know that a shy person doesn't like to be with unfamiliar people (Heyman & Gelman, 1999). But preschoolers do not refer directly to traits when describing themselves. This capacity must wait for greater cognitive maturity.

In fact, very young preschoolers' concepts of themselves are so bound up with specific possessions and actions that they spend much time asserting their rights to objects, as Sammy did in the beanbag incident at the beginning of this chapter. The stronger children's self-definition, the more possessive they tend to be, claiming objects as "Mine!" (Fasig, 2000; Levine, 1983). These findings indicate that rather than a sign of selfishness, early struggles over objects seem to be a sign of developing selfhood, an effort to clarify boundaries between self and others.

A firmer sense of self also permits children to cooperate in resolving disputes over objects, playing games, and solving simple problems (Brownell & Carriger, 1990; Caplan et al., 1991). Accordingly, when trying to promote friendly peer interaction, parents and teachers can accept the young child's possessiveness as a sign of self-assertion ("Yes, that's your toy") and then encourage compromise ("but in a little while, would you give someone else a turn?"), rather than simply insisting on sharing.

Recall from Chapter 7 that adult–child conversations about personally experienced events contribute to the development of an autobiographical memory. Parents often use these discussions to impart evaluative information about the child's actions, as when they say, "You were a big boy when you did that!" Consequently, these narratives serve as a rich source of early self-knowledge and, as the Cultural Influences box on page 246 reveals, as a major means through which the young child's self-concept is imbued with cultural values.

Emergence of Self-Esteem

Another aspect of self-concept emerges in early childhood: **self-esteem,** the judgments we make about our own worth and the feelings associated with those judgments. Self-esteem ranks among the most important aspects of self-development, since evaluations of our own competencies affect our emotional experiences, future behavior, and long-term psychological adjustment. Take a moment to think about your

■ In their early self-descriptions, children usually emphasize observable characteristics, including physical appearance, possessions, and everyday behaviors. When asked to tell about herself, this 4-year-old might say proudly, "I can put my socks and shoes on all by myself!"

Cultural Influences

Cultural Variations in Personal Storytelling: Implications for Early Self-Concept

Preschoolers of many cultural backgrounds participate in personal storytelling with their parents. Striking cultural differences exist in parents' selection and interpretation of events in these early narratives, affecting the way children come to view themselves.

Peggy Miller and her colleagues spent hundreds of hours studying the storytelling practices of six middle-SES Irish-American families in Chicago and six middle-SES Chinese families in Taiwan. From extensive videotapes of adults' conversations with 2½-year-olds, the researchers identified personal stories and coded them for content, quality of their endings, and evaluation of the child (Miller, Fung, & Mintz, 1996; Miller et al., 1997).

Parents in both cultures discussed pleasurable holidays and family excursions about as often and in similar ways. Chinese parents, however, more often told lengthy stories about the child's misdeeds, such as using impolite language, writing on the wall, or playing in an overly rowdy way. These narratives were conveyed with warmth and caring, stressed the impact of misbehavior on others ("You made Mama lose face"), and often ended with direct teaching of proper behavior ("Saying dirty words is not good"). In the few instances in which Irish-American stories referred to transgressions, parents downplayed their seriousness, attributing them to the child's spunk and assertiveness.

Early narratives about the child seem to launch preschoolers' self-concepts on culturally distinct paths. Influenced by Confucian traditions of strict discipline and social obligations, Chinese parents integrated these values into their personal stories, affirming the importance of not disgracing the family and explicitly conveying expectations in the story's conclusion. Although Irish-American parents disciplined their children, they rarely dwelt on misdeeds in storytelling. Rather, they cast the child's shortcomings in a positive light, perhaps to encourage a positive sense of self. Hence, the Chinese child's self-image emphasizes obligations to others, whereas the American child's is more autonomous (Markus, Mullally, & Kitayama, 1997).

© PETER TURNLEY/CORBIS

■ This Chinese child on an outing with his mother listens as she speaks gently to him about proper behavior. Chinese parents often tell preschoolers stories about the child's misdeeds, stressing their negative impact on others. The Chinese child's self-concept, in turn, emphasizes social obligations.

own self-esteem. Besides a global appraisal of your worth as a person, you have a variety of separate self-judgments concerning how well you perform at different activities.

By age 4, preschoolers have several self-esteems, such as learning things well in school, trying hard at challenging tasks, making friends, and treating others kindly (Marsh, Craven, & Debus, 1998). However, their understanding is not as differentiated as that of older children and adults. And usually they rate their own ability as extremely high and underestimate the difficulty of tasks (Harter, 1990, 1998). Sammy's announcement that he was great at beanbag throwing despite his many misses is a typical self-evaluation during early childhood.

High self-esteem contributes greatly to preschoolers' initiative during a period in which they must master many new skills. Nevertheless, by age 4, some children give up easily when faced with a challenge, such as working a hard puzzle or building a tall block tower. They are discouraged after failure and conclude that they cannot do the task (Cain & Dweck, 1995; Smiley & Dweck, 1994). When these young nonpersisters are asked to act out with dolls an adult's reaction to failure, they often respond, "He's punished because he can't do the puzzle," or, "Daddy's mad and is going to spank her" (Burhans & Dweck, 1995). They are also likely to report that their parents would berate them for making small mistakes (Heyman, Dweck, & Cain, 1992). The Caregiving Concerns table on the following page suggests ways to avoid these self-defeating reactions and foster a healthy self-image in young children.

Fostering a Healthy Self-Image in Young Children

SUGGESTION	DESCRIPTION
Build a positive relationship.	Indicate that you want to be with the child by arranging times to be fully available. Listen without being judgmental, and express some of your own thoughts and feelings. Mutual sharing helps children feel valued.
Nurture success.	Adjust expectations appropriately, and provide assistance when asking the child to do something beyond his or her current limits. Accentuate the positive in the child's work or behavior. Promote self-motivation by emphasizing praise over concrete rewards. Instead of simply saying, "That's good," mention effort and specific accomplishments. Display the child's artwork and other products, pointing out increasing skill.
Foster the freedom to choose.	Choosing gives children a sense of responsibility and control over their own lives. Where children are not yet capable of deciding on their own, involve them in some aspect of the choice, such as when and in what order a task will be done.
Acknowledge the child's emotions.	Accept the child's strong feelings, and suggest constructive ways to handle them. When a child's negative emotion results from an affront to his or her self-esteem, offer sympathy and comfort along with a realistic appraisal of the situation so that the child feels supported and secure.
Use a warm, rational approach to child rearing.	The strategies discussed on pages 253 and 265 (induction and authoritative child rearing) promote self-confidence and self-control.

Source: Berne & Savary, 1993.

Emotional Development

Gains in representation, language, and self-concept support emotional development in early childhood. Between ages 2 and 6, children achieve a better understanding of their own and others' feelings, and their ability to regulate the expression of emotion improves. Self-development also contributes to a rise in *self-conscious emotions,* such as shame, embarrassment, guilt, envy, and pride.

Understanding Emotion

Preschoolers' vocabulary for talking about emotion expands rapidly. Early in the preschool years, they refer to causes, consequences, and behavioral signs of emotion, and over time, their understanding becomes more accurate and complex (Stein & Levine, 1999). By age 4 to 5, children correctly judge the causes of many basic emotions, as in "He's happy because he's swinging very high" or "He's sad because he misses his mother." However, they are likely to emphasize external factors over internal states as explanations—a balance that changes with age (Levine, 1995).

Preschoolers are also good at predicting what a playmate expressing a certain emotion might do next. Four-year-olds know that an angry child might hit someone and that a happy child is more likely to share (Russell, 1990). And they realize that thinking and feeling are interconnected—that a person reminded of a previous sad experience is likely to feel sad (Lagattuta, Wellman, & Flavell, 1997). Furthermore, they come up with effective ways to relieve others' negative feelings, such as hugging to reduce sadness (Fabes et al., 1988). Overall, preschoolers have an impressive ability to interpret, predict, and change others' feelings.

At the same time, in situations with conflicting cues about how a person is feeling, preschoolers have difficulty making sense of what is going on. For example, when asked what might be happening in a picture showing a happy-faced child with a broken bicycle, 4- and 5-year-olds tended to rely on the emotional expression: "He's happy because he likes to ride his bike." Older children more often reconciled the two cues: "He's happy because his father promised to help fix his broken bike" (Gnepp, 1983; Hoffner & Badzinski, 1989). As in their approach to Piagetian tasks, preschoolers focus on the most obvious aspect of a complex emotional situation to the neglect of other relevant information.

Preschoolers who are securely attached to their mothers are advanced in emotional understanding. Such mothers engage in richer conversations about emotions—interactions that help preschoolers accurately judge others' feelings (Denham, Zoller, & Couchoud, 1994; Laible & Thompson, 1998, 2000). As preschoolers learn more about emotion from interacting with adults, they engage in more "emotion talk" with siblings and friends, especially during sociodramatic play (Brown, Donelan-McCall, & Dunn, 1996; Hughes & Dunn, 1998). Make-believe, in turn, contributes to emotional understanding, especially when children play with siblings. The intense nature of the sibling relationship, combined with frequent acting out of feelings in make-believe, makes this an excellent context for learning about emotions.

Emotional Self-Regulation

Language also contributes to preschoolers' improved *emotional self-regulation.* By age 3 to 4, children verbalize a variety of strategies for adjusting their emotional arousal to a more comfortable level. For example, they know they can blunt emotions by restricting sensory input (covering your eyes or ears to block out a scary sight or sound), talking to themselves ("Mommy said she'll be back soon"), or changing their goals (deciding that you don't want to play anyway after being excluded from a game) (Thompson, 1990). Children's use of these strategies means fewer emotional outbursts over the preschool years.

Nevertheless, preschoolers' vivid imaginations, combined with their difficulty separating appearance from reality, make fears common in early childhood. The Caregiving Concerns table below lists ways that parents can help children manage them. As these interventions suggest, the social environment powerfully affects children's capacity to cope with stress. By watching adults handle their own feelings, preschoolers pick up strategies for regulating emotion. When parents have difficulty controlling anger and hostility, children have continuing problems with regulating emotion that seriously interfere with psychological adjustment (Eisenberg et al., 1999).

Besides parenting, temperament affects the development of emotional self-regulation. Children who experience negative emotion intensely find it harder to inhibit their feelings As early as the preschool years, they are more likely to respond with irritation to others' distress and to get along poorly with peers (Eisenberg et al., 1997; Walden, Lemerise, & Smith, 1999).

Self-Conscious Emotions

One morning in Leslie's classroom, a group of children crowded around for a bread-baking activity. Leslie asked them to wait patiently while she got a baking pan. In the meantime, Sammy reached to feel the dough, but the bowl tumbled over the side of the table. When Leslie returned, Sammy looked at her for a moment, covered his eyes with his hands, and said, "I did something bad." He was feeling ashamed and guilty.

As children's self-concepts become better developed, they experience *self-conscious emotions* more often—feelings that involve injury to or enhancement of their sense of self (see Chapter 6). By age 3, self-conscious emotions are clearly linked to self-evaluation (Lewis, 1995). Nevertheless, because preschoolers are still developing standards of excellence and conduct, they depend on adults' messages to know when to feel self-conscious emotions (Stipek, 1995). Parents who repeatedly give feedback about the worth of the child and her performance ("That's a bad job! I thought you were a good girl!") have children who experience self-conscious emotions intensely—more shame after failure and more pride after success. In contrast, parents who focus on how to improve performance ("You did it this way; you should have done it that way") induce moderate, more adaptive levels of shame and pride (Lewis, 1998).

Beginning in early childhood, intense shame is associated with feelings of personal inadequacy ("I'm stupid," "I'm a terri-

Helping Children Manage Common Fears of Childhood

FEAR	SUGGESTION
Monsters, ghosts, and darkness	Reduce exposure to frightening stories in books and on TV until the child is better able to sort out appearance from reality. Make a thorough "search" of the child's room for monsters, demonstrating that none are there. Leave a night-light burning, sit by the child's bed until the child falls asleep, and tuck in a favorite toy for protection.
Preschool or child care	If the child resists going to preschool but seems content once there, then the fear is probably separation. Under these circumstances, provide a sense of warmth and caring while gently encouraging independence. If the child fears being at preschool, try to find out what is frightening—the teacher, the children, or perhaps a crowded, noisy environment. Provide extra support by accompanying the child at first and gradually reducing the amount of time you are present.
Animals	Do not force the child to approach a dog, cat, or other animal that arouses fear. Let the child move at his or her own pace. Demonstrate how to hold and pet the animal, showing the child that when treated gently, the animal reacts in a friendly way. If the child is bigger than the animal, emphasize this: "You're so big. That kitty is probably afraid of you!"
Very intense fears	If a child's fear is very intense, persists for a long time, interferes with daily activities, and cannot be reduced in any of the ways just suggested, it has reached the level of a phobia. Sometimes phobias are linked to family problems, and special counseling is needed to reduce them. At other times, phobias diminish without treatment.

ble person") and with maladjustment—withdrawal and depression as well as intense anger and aggression at those who participated in the shame-evoking situation (Lindsay-Hartz, de Rivera, & Mascolo, 1995). In contrast, guilt—as long as it occurs in appropriate circumstances and shame does not accompany it—is related to good adjustment, perhaps because guilt helps children resist harmful impulses. And when children do transgress, guilt motivates them to repair the damage and behave more considerately (Ferguson et al., 1999; Tangney, 2001).

Empathy

Another emotional capacity—*empathy*—becomes more common in early childhood. It continues to serve as an important motivator of **prosocial,** or **altruistic, behavior**—actions that benefit another person without any expected reward for the self (Eisenberg & Fabes, 1998). Compared with toddlers, preschoolers rely more on words to communicate their empathic feelings, a change that indicates a more reflective level of empathy. And as the ability to take another's perspective improves, empathic responding increases.

Yet empathy, or *feeling with* another person and responding emotionally in a similar way, does not always yield acts of kindness and helpfulness. In some children, empathizing with an upset adult or peer escalates into personal distress. In trying to reduce these feelings, the child focuses on himself rather than the person in need. As a result, empathy does not give way to **sympathy**—feelings of concern or sorrow for another's plight.

Whether empathy occurs and prompts a personally distressed, self-focused response or sympathetic, prosocial behavior is related to temperament. Children who are sociable, assertive, and good at regulating emotion are more likely to help, share, and comfort others in distress. In contrast, children who are poor emotion regulators less often display sympathetic concern and prosocial behavior (Eisenberg et al., 1996, 1998). When faced with someone in need, these youngsters react with facial and physiological distress—frowning, lip biting, a rise in heart rate, and a sharp increase in brain-wave activity in the right cerebral hemisphere (which houses negative emotion), indicating that they are overwhelmed by their feelings (Miller et al., 1996; Pickens, Field, & Nawrocki, 2001).

As with emotional self-regulation, parenting affects empathy and sympathy. Parents who are warm and encouraging and show sensitive, sympathetic concern for their preschoolers have children who are likely to react in a concerned way to the distress of others—relationships that persist into adolescence and young adulthood (Eisenberg & McNally, 1993; Koestner, Franz, & Weinberger, 1990). In contrast, angry, punitive parenting disrupts empathy and sympathy at an early age. In one study, researchers observed physically abused preschoolers at a child-care center. Compared with nonabused agemates, they rarely showed signs of concern. Instead, they responded to a peer's unhappiness with fear, anger, and physical attacks (Klimes-Dougan & Kistner, 1990). The children's reactions resembled the behavior of their parents, since both responded insensitively to the suffering of others.

■ A 5-year-old empathizes with and comforts her 3-year-old brother, who has injured his finger. As young children's language skills expand and their ability to take the perspective of others improves, empathic responding increases and becomes an important motivator of prosocial, or altruistic, behavior.

Peer Relations

As children become increasingly self-aware, more effective at communicating, and better at understanding the thoughts and feelings of others, their skill at interacting with peers improves rapidly. Peers provide young children with learning experiences they can get in no other way. Because peers interact on an equal footing, they must assume responsibility for keeping a conversation going, cooperating, and setting goals in play. With peers, children form friendships—special relationships marked by attachment and common interests. Let's look at how peer interaction changes over the preschool years.

Advances in Peer Sociability

Mildred Parten (1932), one of the first to study peer sociability among 2- to 5-year-olds, noticed a dramatic rise with age in joint, interactive play. She concluded that social development proceeds in a three-step sequence. It begins with **nonsocial activity**—unoccupied, onlooker behavior and solitary play. Then it shifts to **parallel play,** in which a child plays near other children with similar materials but does not try to influence their behavior. At the highest level are two forms of true social interaction. One is **associative play,** in which children engage in separate activities, but they exchange toys and comment on one another's behavior. The other is **cooperative play,** a more advanced type of interaction in which children orient toward a common goal, such as acting out a make-believe theme or building a sand castle.

■ These children are engaged in parallel play. Although they sit side by side and use similar materials, they do not try to influence one another's behavior. Parallel play remains frequent and stable over the preschool years, accounting for as much of children's play as highly social, cooperative interaction.

● **Recent Evidence on Peer Sociability.** Longitudinal evidence indicates that these play forms emerge in the order suggested by Parten, but they do not form a developmental sequence in which later-appearing ones replace earlier ones (Howes & Matheson, 1992). Instead, all types coexist during the preschool years. Furthermore, although nonsocial activity declines with age, it is still the most frequent form among 3- to 4-year-olds. Even among kindergartners it continues to take up as much as a third of children's free-play time. Also, solitary and parallel play remain fairly stable from 3 to 6 years, accounting for as much of the young child's play as highly social, cooperative interaction.

■ As these children collaborate in operating a balance scale, they engage in an advanced form of interaction called cooperative play. With age, preschoolers more often cooperate to achieve a common goal.

We now understand that it is the *type,* rather than the amount, of solitary and parallel play that changes during early childhood. In studies of preschoolers' play in Taiwan and the United States, researchers rated the *cognitive maturity* of nonsocial, parallel, and cooperative play by applying the categories shown in Table 8.1. Within each of Parten's play types, older children displayed more cognitively mature behavior than younger children (Pan, 1994; Rubin, Watson, & Jambor, 1978).

Often parents wonder if a preschooler who spends much time playing alone is developing normally. Only *certain kinds* of nonsocial activity—aimless wandering, hovering near peers, and functional play involving immature, repetitive motor action—are cause for concern. Children who behave in these ways usually are temperamentally inhibited preschoolers who have not learned to regulate their high social fearfulness. Often their parents have overprotected them rather than encouraged them to approach other children (Burgess et al., 2001).

But not all preschoolers with low rates of peer interaction are socially anxious. To the contrary, most like to play by themselves, and their solitary activities are positive and constructive. Teachers encourage such play when they set out art materials, puzzles, and building toys. Children who spend much time in these activities are not maladjusted (Rubin & Coplan, 1998). They are bright children who, when they do play with peers, show socially skilled behavior.

● **Cultural Variations.** Peer sociability in collectivist societies takes different forms than in individualistic cultures. For example, children in India generally play in large groups that require high levels of cooperation. Much of their behavior during sociodramatic play and early games is imitative, occurs in unison, and involves close physical contact. In a game called Bhatto Bhatto, children act out a script about a trip to the market, touching each other's elbows and hands as they pretend to cut and share a tasty vegetable (Roopnarine et al., 1994).

Cultural beliefs about the importance of play also affect early peer associations. Adults who view play as mere entertainment are less likely to provide props and encourage pretend than those who value its cognitive and social benefits (Farver & Wimbarti, 1995). Korean-American parents, who emphasize task persistence as vital for learning, have preschoolers who spend less time than their Caucasian-American counterparts at joint make-believe and more time unoccupied and in parallel play (Farver, Kim, & Lee, 1995).

Return to the description of Yucatec Mayan preschoolers' pretending on page 226 of Chapter 7. Mayan parents do not promote children's play, and when it interferes with important cultural activities, they discourage it. Yet even though they spend little time pretending, Mayan children are socially competent (Gaskins, 2000). Perhaps Western-style sociodramatic play, with its elaborate materials and wide-ranging themes, is particularly important for social development in societies where child and adult worlds are distinct. It may be less crucial when children participate in adult activities from an early age.

Table 8.1 Developmental Sequence of Cognitive Play Categories

Play Category	Description	Examples
Functional play	Simple, repetitive motor movements with or without objects. Especially common during the first 2 years of life.	Running around a room, rolling a car back and forth, kneading clay with no intent to make something
Constructive play	Creating or constructing something. Especially common between 3 and 6 years.	Making a house out of toy blocks, drawing a picture, putting together a puzzle
Make-believe play	Acting out everyday and imaginary roles. Especially common between 2 and 6 years.	Playing house, school, or police officer; acting out storybook or television characters

Source: Rubin, Fein, & Vandenberg, 1983.

© DORANNE JACOBSON

■ These cousins at a family birthday party in a village in central India play an intricate hand-clapping game, called "Chapte," in which they clap in unison, clapping faster as the game proceeds. The game is played to a jingle with eleven verses, which take the girls through their lifespan and conclude with their turning into ghosts. The girls end the game by mimicking a scary ghost's antics. Their play reflects the value their culture places on group harmony.

First Friendships

As preschoolers interact, first friendships form that serve as important contexts for emotional and social development. Take a moment to jot down what the word *friendship* means to you. You probably thought of a mutual relationship involving companionship, sharing, understanding of thoughts and feelings, and caring for one another in times of need. In addition, mature friendships endure over time and survive occasional conflicts.

Preschoolers understand something about the uniqueness of friendship. They know that a friend is someone "who likes you" and with whom you spend a lot of time playing (Youniss, 1980). Yet their ideas about friendship are far from mature. Four- to 7-year-olds regard friendship as pleasurable play and sharing of toys. As yet, friendship does not have a long-term, enduring quality based on mutual trust (Selman, 1980). Indeed, Sammy could be heard declaring, "Mark's my best friend" on days when the boys got along well. But he would state just the opposite—"Mark, you're not my friend!"—when a dispute arose that was not quickly settled.

Nevertheless, interactions between young friends are unique. Preschoolers give twice as much reinforcement, in the form of greetings, praise, and compliance, to children they identify as friends, and they also receive more from them. Friends are also more emotionally expressive—talking, laughing, and looking at each other more often—than nonfriends (Hartup & Stevens, 1999; Vaughn et al., 2001). Furthermore, early childhood friendships offer social support. When children begin kindergarten with friends in their class or readily make new friends, they adjust to school more favorably (Ladd & Price, 1987; Ladd, Birch, & Buhs, 1999). Through friendships, children seem to integrate themselves into the learning environment in ways that foster both academic and social competence.

Parental Influences on Early Peer Relations

It is within the family that children first acquire skills for interacting with peers. Parents influence children's peer sociability both *directly*, through attempts to influence children's peer relations, and *indirectly*, through their child-rearing practices and play behaviors (Ladd & Pettit, 2002).

● **Direct Parental Influences.** Preschoolers whose parents frequently arrange informal peer play activities tend to have larger peer networks and be more socially skilled (Ladd,

© CREWS/THE IMAGE WORKS

■ Parents influence children's peer interaction skills by offering advice, guidance, and examples of how to behave. This father teaches his 3-year-old son how to offer a present as a guest at a birthday party.

LeSieur, & Profilet, 1993). In providing play opportunities, parents show children how to initiate their own peer contacts. Parents also offer guidance on how to act toward others. Their skillful suggestions for solving peer problems—such as managing conflict, discouraging teasing, and entering a play group—are associated with preschoolers' social competence and peer acceptance (Laird et al., 1994; Mize & Pettit, 1997),

● **Indirect Parental Influences.** Many parenting behaviors are aimed at promoting peer sociability but nevertheless spill over into peer relations. For example, secure attachments to parents are linked to more responsive, harmonious peer interactions (Bost et al., 1998; Schneider, Atkinson, & Tardif, 2001). The emotionally expressive and supportive communication that contributes to attachment security may be responsible. In several studies, highly involved, emotionally positive parent–child conversations and play were linked to children's prosocial behavior and positive peer relations (Clark & Ladd, 2000; Lindsey & Mize, 2000).

As early as the preschool years, some children have great difficulty with peer relations. In Leslie's classroom, Robbie was one of them. His demanding, aggressive behavior caused other children to dislike him. Wherever he happened to be, such comments as "Robbie ruined our block tower" and "Robbie hit me for no reason" could be heard. You will learn more about how parenting contributed to Robbie's peer problems as we take up moral development in the next section.

Ask Yourself

REVIEW
Among children who spend much time playing alone, what factors distinguish those who are likely to have adjustment difficulties from those who are well adjusted and socially skilled?

APPLY
Reread the description of Sammy and Mark's argument at the beginning of this chapter. On the basis of what you know about self-development, why was it a good idea for Leslie to resolve the dispute by providing an extra set of beanbags so both boys could play at once?

CONNECT
How does emotional self-regulation affect the development of empathy and sympathy? Why are these emotional capacities vital for positive peer relations?

CONNECT
Cite ways that parenting contributes to preschoolers' self-esteem, emotional self-regulation, self-conscious emotions, empathy and sympathy, and peer sociability. Do you see any patterns? Explain.

www.

Foundations of Morality

If you watch children's behavior and listen in on their conversations, you will find many examples of their developing moral sense. By age 2, they react with distress to acts that are aggressive or that otherwise might do harm, and they use words to evaluate behavior as "good" or "bad" (Kochanska, Casey, & Fukumoto, 1995).

Throughout the world, adults take note of this budding capacity to distinguish right from wrong. Some cultures have special words for it. The Utku Indians of Hudson Bay say the child develops *ihuma* (reason). The Fijians believe that *vakayalo* (sense) appears. In response, parents hold children more responsible for their behavior (Kagan, 1998). By the end of early childhood, children can state a great many moral rules, such as "You're not supposed to take things without asking" or "Tell the truth!" In addition, they argue over matters of justice, as when they say, "You sat there last time, so it's my turn," or, "It's not fair. He got more!"

All theories of moral development recognize that conscience begins to take shape in early childhood. And most

■ A teacher uses inductive discipline to explain to a child the impact of her transgression on others. Induction supports conscience development by clarifying how the child should behave, encouraging empathy and sympathetic concern, and permitting the child to grasp the reasons behind parental expectations.

agree that at first, the child's morality is *externally controlled* by adults. Gradually, it becomes regulated by *inner standards.* Truly moral individuals do not just do the right thing when authority figures are around. Instead, they have developed a compassionate concern for others and principles of good conduct, which they follow in a wide variety of situations.

Although points of agreement exist among major theories, each emphasizes a different aspect of morality. Psychoanalytic theory stresses the *emotional side* of conscience development—in particular, identification and guilt as motivators of good conduct. Social learning theory focuses on *moral behavior* and how it is learned through reinforcement and modeling. And the cognitive-developmental perspective emphasizes *thinking*—children's ability to reason about justice and fairness.

The Psychoanalytic Perspective

Recall from our discussion earlier in this chapter that according to Freud, young children form a *superego,* or conscience, by *identifying* with the same-sex parent, whose moral standards they adopt. Children obey the superego to avoid *guilt,* a painful emotion that arises each time they are tempted to misbehave. Moral development, Freud believed, is largely complete by 5 to 6 years of age, at the end of the phallic stage.

Today, most researchers disagree with Freud's account of conscience development. Notice how discipline promoting fear of punishment and loss of parental love is assumed to motivate conscience formation and moral behavior (Kochanska, 1993; Tellings, 1999). Yet children whose parents frequently use threats, commands, or physical force tend to violate standards frequently and feel little guilt after harming others. In the case of love withdrawal—for example, when a parent refuses to speak to or actually states a dislike for the child—children often respond with high levels of self-blame after misbehaving. They might think to themselves, "I'm no good," or "Nobody loves me." Eventually, these children may protect themselves from overwhelming feelings of guilt by denying the emotion when they do something wrong. So they, too, develop a weak conscience (Kochanska, 1991; Zahn-Waxler et al., 1990).

In contrast, a special type of discipline called **induction,** which helps the child notice feelings by pointing out the effects of the child's misbehavior on others, supports conscience formation. For example, a parent might say, "If you keep pushing him he'll fall down and cry," or, "She feels so sad because you won't give back her doll" (Hoffman, 2000). As long as the explanation matches the child's capacity to understand, induction is effective with children as young as 2 years of age. Preschoolers whose parents use it are more likely to make up for their misdeeds, and they more often display prosocial behavior (Zahn-Waxler, Radke-Yarrow, & King, 1979).

The success of induction may lie in its power to motivate children's active commitment to moral standards (Turiel, 1998). How does it do so? First, induction tells children how to behave so they can use this information in future situations. Second, by pointing out the impact of the child's actions on others, parents encourage empathy and sympathy, which promote prosocial behavior (Krevans & Gibbs, 1996). Third, providing children with reasons for changing their behavior encourages them to adopt moral standards because they make sense. In contrast, discipline that relies too heavily on threats of punishment or love withdrawal makes children so anxious and afraid that they cannot think clearly enough to figure out what they should do. As a result, these practices do not get children to internalize moral rules.

Furthermore, Freud's theory places a heavy burden on parents, who must ensure through their disciplinary practices that children develop an internalized conscience. Although good discipline is crucial, children's characteristics can affect the success of parenting techniques. Twin studies suggest a modest genetic contribution to empathy (Zahn-Waxler et al., 2001). A more empathic child requires less power assertion and is more responsive to induction. Turn to the Biology and Environment box on page 254 for recent findings on temperament and conscience development.

Finally, although little support exists for Freudian ideas about conscience development, Freud was correct that guilt is an important motivator of moral action. Inducing *empathy-based guilt* (expressions of personal responsibility and regret, such as "I'm sorry I hurt him") by explaining how the child's behavior is harmful is a way to influence children without using coercion. Empathy-based guilt reactions are consistently associated with stopping harmful actions, repairing damage, and acting prosocially (Baumeister, 1998). But guilt is not the only force that compels us to act morally. And contrary to what Freud believed, moral development is not complete by the end of early childhood. Instead, it is a far more gradual process, extending into adulthood.

Biology & Environment

Temperament and the Development of Conscience in Young Children

When her mother reprimanded her sharply for pouring water on the floor as she played in her bath, 3-year-old Katherine burst into tears. An anxious, sensitive child, Katherine was so distressed that it took her mother 10 minutes to calm her down. The next day, Katherine's mother watched as her next-door neighbor patiently asked her 3-year-old son not to pick tulips in the garden. Alex, an active, adventurous child, paid no attention. As he pulled at another tulip, Alex's mother grabbed him, scolded him harshly, and carried him inside. Alex responded by kicking, hitting, and screaming, "Let me down, let me down!"

What explains Katherine's and Alex's very different reactions to firm parental discipline? Grazyna Kochanska (1995) points out that children's temperaments affect the parenting practices that best promote responsibility and concern for others. She found that for temperamentally inhibited 2- and 3-year-olds, maternal gentle discipline—reasoning, polite requests, suggestions, and distractions—predicted conscience development at age 5, measured in terms of not cheating in games and completing stories about moral issues with prosocial themes (not taking someone else's toys, helping a child who is hurt). In contrast, for relatively fearless, impulsive children, mild disciplinary tactics showed no relationship to morality. Instead, a secure attachment with the mother predicted a mature conscience (Fowles & Kochanska, 2000; Kochanska, 1997).

According to Kochanska, inhibited children like Katherine, who are prone to anxiety, are easily overcome by intense discipline. Mild, patient tactics are sufficient to motivate them to adopt parental standards. But impulsive children, such as Alex, may not respond to gentle interventions with enough discomfort to promote a strong conscience. Yet frequent use of power-assertive methods is not effective either, since these techniques spark anger and resentment, which interfere with the child's processing of parental messages.

Why does secure attachment predict conscience development in nonanxious children? Kochanska suggests that when children are so low in anxiety that typically effective disciplinary practices fail, a close bond with the caregiver provides an alternative foundation for conscience formation. It motivates children who are unlikely to experience negative emotion to adopt parental rules as a means of preserving a spirit of affection and cooperation with the parent.

To foster early moral development, parents must tailor their child-rearing strategies to their child's temperament. In Katherine's case, a soft-spoken correction would probably be effective. For Alex, taking extra steps to build a warm, caring relationship during times when he behaves well is essential. Although Alex's parents need to use firmer and more frequent discipline than Katherine's do, emphasizing power assertion is counterproductive for both children. Do these findings remind you of the notion of *goodness of fit,* discussed in Chapter 6? Return to page 184 to review this idea.

Social Learning Theory

Social learning theory does not regard morality as a special human activity with a unique course of development. Instead, moral behavior is acquired just like any other set of responses: through reinforcement and modeling.

• The Importance of Modeling. Operant conditioning—following up children's good behavior with reinforcement in the form of approval, affection, and other rewards—is not enough for children to acquire moral responses. For a behavior to be reinforced, it must first occur spontaneously. Yet many prosocial acts, such as sharing, helping, or comforting an unhappy playmate, do not occur often enough at first for reinforcement to explain their rapid development in early childhood. Instead, social learning theorists believe that children largely learn to behave morally through *modeling*—by observing and imitating people who demonstrate appropriate behavior (Bandura, 1977; Grusec, 1988). Once children acquire a moral response, such as sharing or telling the truth, reinforcement in the form of praise increases its frequency (Mills & Grusec, 1989).

Many studies show that models who behave helpfully or generously increase young children's prosocial responses. The following characteristics of models affect children's willingness to imitate:

- *Warmth and responsiveness.* Preschoolers are more likely to copy the prosocial actions of an adult who is warm and responsive than those of an adult who is cold and distant (Yarrow, Scott, & Waxler, 1973). Warmth seems to make children more attentive and receptive to the model, and it is itself an example of a prosocial response.
- *Competence and power.* Children admire and therefore tend to select competent, powerful models to imitate—the reason they are especially willing to copy the behavior of older peers and adults (Bandura, 1977).
- *Consistency between assertions and behavior.* When models say one thing and do another—for example, announce that "it's important to help others" but rarely engage in helpful acts—children generally choose the most lenient standard of behavior that adults demonstrate (Mischel & Liebert, 1966).

Models are most influential during the preschool years. At the end of early childhood, children with a history of consistent exposure to caring adults tend to behave prosocially regardless of whether a model is present. By that time, they have internalized prosocial rules from repeated observations of and encouragement by others (Mussen & Eisenberg-Berg, 1977).

● **Effects of Punishment.** Many parents are aware that yelling at, slapping, or spanking children for misbehavior are ineffective disciplinary tactics. Sharp reprimands or physical force to restrain or move a child is justified when immediate obedience is necessary—for example, when a 3-year-old is about to run into the street. In fact, parents are most likely to use forceful methods under these conditions. When they wish to foster long-term goals, such as acting kindly toward others, they tend to rely on warmth and reasoning (Kuczynski, 1984). And parents often combine power assertion with reasoning in response to very serious transgressions, such as lying and stealing (Grusec & Goodnow, 1994).

When used frequently, however, punishment promotes only momentary compliance, not lasting changes in children's behavior. For example, Robbie's parents often punished, hitting, shouting, and criticizing. Robbie usually engaged in the unacceptable behavior again as soon as his parents were out of sight and he could get away with it. The more physical punishment children experience, the more likely they are to display depression, antisocial behavior, and poor academic performance in the future (Brezina, 1999; Gershoff, 2002).

Harsh punishment has undesirable side effects. First, when parents spank, they often do so in response to children's aggression (Holden, Coleman, & Schmidt, 1995). Yet the punishment itself models aggression! Second, children who are frequently punished soon learn to avoid the punishing adult. As a result, those adults have little opportunity to teach desirable behaviors. Finally, as punishment "works" to stop children's misbehavior temporarily, it offers immediate relief to adults, and they are reinforced for using coercive discipline. For this reason, a punitive adult is likely to punish with greater frequency over time, a course of action that can spiral into serious abuse.

In view of these findings, the widespread use of corporal punishment by North American parents is cause for concern. A survey of a nationally representative American sample of households revealed that although corporal punishment increases from infancy to age 5 and then declines, it is high at all ages (see Figure 8.1). Similarly, more than 70 percent of Canadian parents admit to having hit or spanked their children (Durrant, Broberg, & Rose-Krasnor, 2000; Straus & Stewart, 1999).

● **Alternatives to Harsh Punishment.** Alternatives to criticism, slaps, and spankings can reduce the undesirable side effects of punishment. A technique called **time out** involves removing children from the immediate setting—for example, by sending them to their rooms—until they are ready to act appropriately. Time out is useful when a child is out of control (Betz, 1994). It usually requires only a few minutes to change behavior, and it also offers a "cooling off" period for angry parents. Another approach is *withdrawal of privileges,* such as playing outside or watching a favorite TV program. Removing privileges may generate some resentment in children, but it allows parents to avoid harsh techniques that could easily intensify into violence.

When parents decide to use punishment, they can increase its effectiveness in three ways. The first is *consistency.* Permitting children to act inappropriately on some occasions but scolding them on others confuses children, and the unacceptable act persists (Acker & O'Leary, 1996). Second, a *warm parent–child relationship* is vital. Children of involved, caring parents find punishment especially unpleasant and want to regain parental warmth and approval as quickly as possible. Finally, *explanations*

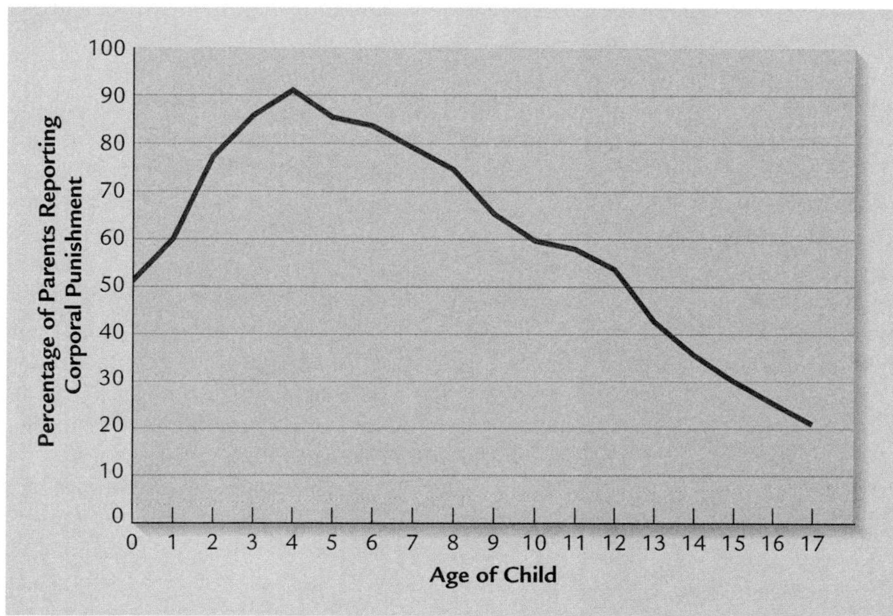

■ **FIGURE 8.1 Prevalence of corporal punishment by child's age.** Estimates are based on the percentage of American parents in a nationally representative sample of nearly 1,000 reporting one or more instances of spanking, slapping, pinching, shaking, or hitting with a hard object in the past year. Punishment increases sharply during early childhood and then declines, but it is high at all ages. (From M. A. Straus & J. H. Stewart, 1999, "Corporal Punishment by American Parents: National Data on Prevalence, Chronicity, Severity, and Duration, in Relation to Child and Family Characteristics," *Clinical Child and Family Psychology Review, 2,* p. 59. Adapted by permission of Kluwer Academic/Plenum Publishers and the author.)

■ Parents who engage in positive discipline encourage good conduct and reduce opportunities for misbehavior. This mother brought along plenty of quiet activities to keep her children occupied during a long train ride.

help children recall the misdeed and relate it to expectations for future behavior. Pairing reasons with mild punishment (such as time out) leads to a far greater reduction in misbehavior than using punishment alone (Larzelere et al., 1996).

● **Positive Discipline.** The most effective forms of discipline encourage good conduct—by building a positive relationship with the child, offering models of appropriate behavior, letting children know ahead of time how to act, and praising them when they behave well (Zahn-Waxler & Robinson, 1995). When preschoolers have positive and cooperative relationships with parents, they show firmer conscience development—in the form of responsible behavior, fair play in games, and consideration for others' welfare. These outcomes continue into the school years (Kochanska & Murray, 2000). Parent–child closeness leads children to want to heed parental demands because children feel a sense of commitment to the relationship.

Parents who use positive discipline also reduce opportunities for misbehavior. For example, on a long car trip, they bring along back-seat activities that relieve restlessness and boredom. At the supermarket, they engage preschoolers in conversation and encourage them to assist with shopping (Holden & West, 1989). Adults who help children acquire acceptable behaviors that they can use to replace forbidden acts greatly reduce the need for punishment.

The Cognitive-Developmental Perspective

The psychoanalytic and behaviorist approaches to morality focus on how children acquire ready-made standards of good conduct from adults. In contrast, the cognitive-developmental perspective regards children as *active thinkers* about social rules. As early as the preschool years, children make moral judgments, deciding what is right or wrong on the basis of concepts they construct about justice and fairness (Gibbs, 1991, 2003).

Young children have some well-developed ideas about morality. Three-year-olds know that a child who intentionally knocks a playmate off a swing is worse than one who does so accidentally (Yuill & Perner, 1988). By age 4, children can tell the difference between truthfulness and lying (Bussey, 1992). By the end of early childhood, children consider a person's intentions in evaluating lying. Influenced by collectivist values of social harmony and humility, Chinese children are more likely than Canadian children to judge lying favorably when an intention involves modesty—for example, when a child who generously picks up garbage in the school yard says, "I didn't do it." In contrast, both Chinese and Canadian children rate lying about antisocial acts as "very naughty" (Lee et al., 1997).

Furthermore, preschoolers distinguish *moral imperatives,* which protect people's rights and welfare, from two other types of action: *social conventions,* or customs such as table manners and dress styles; and *matters of personal choice,* which do not violate rights and are up to the individual (Nucci, 1996; Smetana, 1995). Three-year-olds judge moral violations (stealing an apple) as more wrong than social-conventional violations (eating ice cream with fingers) (Smetana & Braeges, 1990; Turiel, 1998). Preschoolers' concern with personal choice, conveyed through such statements as "I'm gonna wear *this* shirt," serves as the springboard for moral concepts of individual rights, which will expand greatly in adolescence (Killen & Smetana, 1999).

How do young children arrive at these distinctions? According to cognitive-developmental theorists, they do so by *actively make sense* of their experiences. They observe that after a moral offense, peers react emotionally, describe their own injury or loss, tell another child to stop, or retaliate (Arsenio & Fleiss, 1996). And an adult who intervenes is likely to call attention to the rights and feelings of the victim. In contrast, peers seldom react to violations of social convention. And in these situations, adults tend to demand obedience without explanation or point to the importance of obeying rules or keeping order (Turiel, Smetana, & Killen, 1991).

Although cognition and language support preschoolers' moral understanding, social experiences are vital. Disputes with siblings and peers over rights, possessions, and property give preschoolers opportunities to work out their first ideas about justice and fairness (Killen & Nucci, 1995). The way adults handle rule violations and discuss moral issues also helps children reason about morality. Children who are advanced in moral thinking have parents who adapt their communications about fighting, honesty, and ownership to what their children can understand, respect the child's opinion, and gently stimulate the child to think further, without being hostile or critical (Janssens & Dekovíc, 1997; Walker & Taylor, 1991a).

Preschoolers who are disliked by peers because of their aggressive approach to resolving conflict have trouble distinguishing moral rules from social conventions, and they violate both often (Sanderson & Siegal, 1988). Without special help, such children show long-term disruptions in moral development.

The Other Side of Morality: Development of Aggression

Beginning in late infancy, all children display aggression from time to time as they become better at identifying sources of anger and frustration. By the early preschool years, two forms of aggression emerge. The most common is **instrumental aggression,** in which children want an object, privilege, or space, and, in trying to get it, they push, shout at, or otherwise attack a person who is in the way. The other type, **hostile aggression,** is meant to hurt another person.

Hostile aggression comes in two varieties. The first is **overt aggression,** which harms others through physical injury or the threat of such injury—for example, hitting, kicking, or threatening to beat up a peer. The second is **relational aggression,** which damages another's peer relationships, as in social exclusion or rumor spreading. "Go away, I'm not your friend!" and "Don't play with Margie; she's a nerd" are examples.

Both the form of aggression and the way it is expressed change during early childhood. Physical aggression is gradually replaced by verbal aggression (Tremblay et al., 1999). And instrumental aggression declines as preschoolers learn to compromise over possessions. In contrast, hostile outbursts rise over early and middle childhood (Tremblay, 2000). Older children are better able to recognize malicious intentions and, as a result, more often retaliate in hostile ways.

On average, boys are more overtly aggressive than girls, a trend that appears in many cultures (Whiting & Edwards, 1988a). The sex difference is due in part to biology—in particular, to male sex hormones, or androgens. Androgens contribute to boys' greater physical activity, which may increase their opportunities for physically aggressive encounters (Collaer & Hines, 1995). At the same time, gender typing (a topic we will take up shortly) is important. As soon as 2-year-olds become dimly aware of gender stereotypes—that males and females are expected to behave differently—overt aggression drops off more sharply for girls than for boys (Fagot & Leinbach, 1989).

■ An occasional expression of aggression is normal in early childhood. This preschooler displays instrumental aggression as she grabs an attractive toy from a classmate. Instrumental aggression declines with age as children learn how to compromise and share.

© LAURA DWIGHT

But preschool and school-age girls are not less aggressive than boys! Instead, they are likely to express their hostility differently—through relational aggression (Crick, Casas, & Mosher, 1997; Crick, Casas, & Ku, 1999). When trying to harm a peer, children seem to do so in ways especially likely to thwart that child's social goals. Boys more often attack physically to block the dominance goals typical of boys. Girls resort to relational aggression because it interferes with the close, intimate bonds especially important to girls.

An occasional aggressive exchange between preschoolers is normal. But some young children—especially those who are impulsive, overactive, and disruptive—are at risk for lasting conduct problems (Brame, Nagin, & Tremblay, 2001; Coté et al., 2001). These negative outcomes, however, depend on child-rearing conditions.

● **The Family as Training Ground for Aggressive Behavior.** "I can't control him, he's impossible," complained Nadine, Robbie's mother, to Leslie one day. When Leslie asked if Robbie might be troubled by something going on at home, she discovered that Robbie's parents fought constantly and resorted to harsh, inconsistent discipline. The same child-rearing practices that undermine moral internalization—love withdrawal, power assertion, physical punishment, and inconsistency—are linked to aggression from early childhood through adolescence, in children of both sexes (Coie & Dodge, 1998; Stormshak et al., 2000).

Observations in families like Robbie's reveal that anger and punitiveness quickly create a conflict-ridden family atmosphere and an "out-of-control" child. The pattern begins with forceful discipline, which occurs more often with stressful life experiences, a parent's unstable personality, or a temperamentally difficult child. Once the parent threatens, criticizes, and punishes, the child whines, yells, and refuses until the parent "gives in." As these cycles become more frequent, they generate anxiety and irritability among other family members, who soon join in the hostile interactions (Patterson, 1997). Compared with siblings in typical families, preschool siblings who have critical, punitive parents are more verbally and physically aggressive to one another. Destructive sibling conflict, in turn, contributes to poor impulse control and antisocial behavior by the early school years (Garcia et al., 2000).

Because they are more active and impulsive and therefore harder to control, boys are more likely than girls to be targets of harsh, inconsistent discipline. Children who are products of these family processes soon view the world from a violent perspective, seeing hostile intent where it does not exist (Weiss et al., 1992). As a result, they make many unprovoked attacks. Soon they conclude that aggression "works" to control others. These cognitions contribute to the aggressive cycle (Egan, Monson, & Perry, 1998).

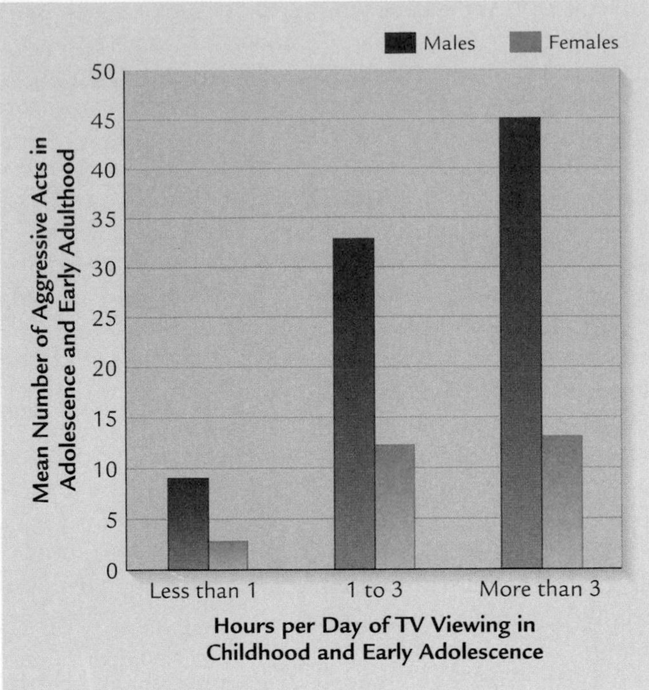

■ **FIGURE 8.2 Relationship of television viewing in childhood and early adolescence to aggressive acts in adolescence and early adulthood.** Interviews with more than 700 parents and youths revealed that the more TV watched in childhood and early adolescence, the greater the annual number of aggressive acts committed by the young person, as reported in follow-up interviews at ages 16 and 22. (Adapted from Johnson et al., 2002.)

Highly aggressive children tend to be rejected by peers, to fail in school, and (by adolescence) to seek out deviant peer groups, which lead them toward violent delinquency and adult criminality. We will consider this life-course path of antisocial activity in Chapter 12.

● **Television and Aggression.** In the United States, 57 percent of television programs between 6 A.M. and 11 P.M. contain violent scenes, often in the form of repeated aggressive acts that go unpunished. In fact, most TV violence does not show victims experiencing any serious harm, and few programs condemn violence or depict other ways of solving problems. Of all TV programs, children's cartoons are the most violent (Center for Communication and Social Policy, 1998). Although Canadian broadcasters follow a code that sharply restricts televised violence, Canadians devote two-thirds of their viewing time to American channels (Statistics Canada, 2001f).

Young children are especially likely to be influenced by television. One reason is that below age 8, children do not understand a great deal of what they see on TV. Because they have difficulty connecting separate scenes into a meaningful story line, they do not relate the actions of a TV character to motives or consequences (Collins et al., 1978). Young children also find it hard to separate true-to-life from fantasized television content. Not until age 7 do they fully realize that fictional characters do not retain the same roles in real life (Wright et al., 1994). These misunderstandings increase young children's willingness to uncritically accept and imitate what they see on TV.

Reviewers of thousands of studies have concluded that TV violence provides children with an extensive "how-to course in aggression" (Comstock & Scharrer, 1999; Slaby et al., 1995, p. 163). And a growing number of studies indicate that playing violent video and computer games has similar effects (Anderson & Bushman, 2001). Violent programming not only creates short-term difficulties in parent and peer relations but also has long-term effects. In three longitudinal studies, time spent watching TV in childhood and adolescence predicted aggressive behavior in early adulthood, after other factors linked to TV viewing (such as child and parent prior aggression, IQ, parental education, family income, and neighborhood crime) were controlled (see Figure 8.2) (Huesmann, 1986; Huesmann et al., 2003; Johnson et al., 2002). Highly aggressive youngsters have a greater appetite for violent TV. As they watch more, they become especially likely to resort to hostile ways of solving problems. But violent TV sparks hostile thoughts and behavior even in nonaggressive children; its impact is simply less intense (Bushman & Huesmann, 2001).

Furthermore, television violence "hardens" children to aggression, making them more willing to tolerate it in others. Heavy TV viewers believe that there is much more violence and danger in society, an effect that is especially strong for children who perceive televised aggression as relevant to their own lives (Donnerstein, Slaby, & Eron, 1994). As these responses indicate, violent television modifies children's attitudes toward social reality so they increasingly match what children see on TV.

The ease with which television can manipulate the beliefs and behavior of children has resulted in strong public pressures to improve its content. In the United States, the First Amendment right to free speech has hampered these efforts. Instead, broadcasters must rate television programs and manufacturers must build the V-Chip (also called the Violence Chip) into new TV sets so parents can block undesired violent and sexual material. Canada also mandates the V-Chip along with a program rating system. In addition, Canada's broadcasting code bans from children's shows realistic scenes of violence that minimize the consequences of violent acts and cartoons in which violence is the central theme. Further, violent programming intended for adults cannot be shown on Canadian channels before 9 P.M. (Canadian Broadcast Standards Council, 2002). Still, Canadian children can access violent TV fare on American channels.

At present, it is up to parents to regulate their children's TV exposure. The V-Chip is an incomplete solution, as the rating systems make TV shows more appealing to some youngsters (Cantor & Harrison, 1997). And as they get older, children may go to other homes to watch programs forbidden by their own parents. The Caregiving Concerns table on the following page lists some additional strategies parents can use to protect children from the dangers of excess TV viewing.

Regulating Children's TV Viewing

STRATEGY	DESCRIPTION
Limit TV viewing.	Avoid using TV as a baby-sitter. Provide clear rules that limit what children can watch—for example, an hour a day and only certain programs—and stick to the rules. Do not place a TV in the child's bedroom, where TV viewing is difficult to monitor.
Refrain from using TV as a reinforcer.	Do not use television to reward or punish children, a practice that increases its attractiveness.
Encourage child-appropriate viewing.	Encourage children to watch programs that are child-appropriate and prosocial.
Explain televised information to children.	As much as possible, watch with children, helping them understand what they see. When adults express disapproval of on-screen behavior and raise questions about the realism of televised information, they teach children to evaluate TV content rather than accept it uncritically.
Link televised content to everyday learning experiences.	Build on TV programs in constructive ways, encouraging children to move away from the set into active engagement with their surroundings. For example, a program on animals might spark a trip to the zoo or a visit to the library for books about animals.
Model good viewing practices.	Avoid excess television viewing yourself. Parental viewing patterns influence children's viewing patterns.

Source: Slaby et al., 1995.

● **Helping Children and Parents Control Aggression.** Treatment for aggressive children must begin early, before their antisocial behavior becomes so well practiced that it is difficult to change. Breaking the cycle of hostilities between family members and replacing it with effective interaction styles is crucial. Leslie suggested that Robbie's parents see a family therapist, who observed their inept practices and coached them in alternatives. They learned not to give in to Robbie, to pair commands with reasons, and to replace verbal insults and spankings with more effective punishments, such as time out and withdrawal of privileges (Patterson, 1982). The therapist also encouraged Robbie's parents to be warmer and to give him attention and approval for prosocial acts.

At the same time, Leslie began teaching Robbie more successful ways of relating to peers. As opportunities arose, she encouraged Robbie to talk about a playmate's feelings and express his own. This helped Robbie take the perspective of others, empathize, and feel sympathetic concern (Denham, 1998). Robbie also participated in *social problem-solving training.* Over several months, he met with Leslie and a small group of classmates to act out common conflicts with puppets, discuss effective and ineffective ways of resolving them, and practice successful strategies. Children who receive such training show gains in social competence still present several months later (Shure, 1997).

Finally, Robbie's parents got help with their marital problems. This, in addition to their improved ability to manage Robbie's behavior, greatly reduced tension and conflict in the household.

Ask Yourself

REVIEW

How does cognitive-developmental theory differ from psychoanalytic and social learning perspectives on moral development? How do preschoolers distinguish among moral imperatives, social conventions, and matters of personal choice? Why are these distinctions important for moral development?

REVIEW

Why are young children especially likely to imitate TV violence?

APPLY

Alice and Wayne want their two young children to develop a strong, internalized conscience and to become generous, caring individuals. List as many parenting practices as you can that would foster these goals.

APPLY

Suzanne has a difficult temperament, and her parents respond to her angry outbursts with harsh, inconsistent discipline. Explain why Suzanne is at risk for long-term difficulties in conscience development and peer relations.

www.

Gender Typing

The process of developing *gender roles,* or gender-linked preferences and behaviors valued by the larger society, is called **gender typing.** Early in the preschool years, gender typing is well under way. In Leslie's classroom, children tended to play and form friendships with peers of their own sex. Girls spent more time in the housekeeping, art, and reading corners, whereas boys gathered more often in spaces devoted to blocks, woodworking, and active play.

The same theories that provide accounts of morality have been used to explain gender-role development. *Social learning theory,* with its emphasis on modeling and reinforcement, and *cognitive-developmental theory,* with its focus on children as active thinkers about their social world, are major current approaches. We will see that neither is adequate by itself. Consequently, a third perspective that combines elements of both, called *gender schema theory,* has gained favor. In the following sections, we consider the early development of gender typing, along with genetic and environmental contributions to it.

Gender-Stereotyped Beliefs and Behavior

Even before children can label their own sex consistently, they stereotype their play world. When shown pairs of gender-stereotyped toys (vehicles and dolls), 18-month-olds look longer at one stereotyped for their own gender (Serbin et al., 2001). Recall from Chapter 6 that around age 2, children use such words as "boy" and "girl" and "lady" and "man" appropriately. As soon as gender categories are established, children sort out what they mean in terms of activities and behavior.

Preschoolers associate many toys, articles of clothing, tools, household items, games, occupations, and even colors (pink and blue) with one sex as opposed to the other (Ruble & Martin, 1998). And their actions fall in line with their beliefs—not only in play preferences but in personality traits as well. We have already seen that boys tend to be more active, assertive, and overtly aggressive, whereas girls tend to be more fearful, dependent, compliant, emotionally sensitive, and relationally aggressive (Geary, 1998; Eisenberg & Fabes, 1998; Feingold, 1994).

Over the preschool years, children's gender-stereotyped beliefs become stronger—so much so that they operate like blanket rules rather than flexible guidelines (Biernat, 1991; Martin, 1989). Once, when Leslie showed the children a picture of a Scottish bagpiper wearing a kilt, they insisted, "Men don't wear skirts!" During free play, they often exclaimed that girls can't be police officers and boys don't take care of babies. These one-sided ideas are a joint product of gender stereotyping in the environment and young children's cognitive limitations. Most preschoolers do not yet realize that characteristics *associated* with sex—activities, toys, occupations, hairstyle, and clothing—do not *determine* whether a person is male or female.

Genetic Influences on Gender Typing

The sex differences just described appear in many cultures around the world (Whiting & Edwards, 1988b). Certain of them—the preference for same-sex playmates as well as male activity level and overt aggression and female warmth and sensitivity—are also widespread among mammalian species (Beatty, 1992; de Waal, 1993). According to an evolutionary perspective, the adult life of our male ancestors was oriented toward competing for mates, that of our female ancestors toward rearing children. Therefore, males became genetically primed for dominance and females for intimacy and responsiveness. Evolutionary theorists claim that family and cultural forces can affect the intensity of biologically based sex differences, leading some individuals to be more gender-typed than others. But experience cannot eradicate those aspects of gender typing that served adaptive functions in human history (Geary, 1999; Maccoby, 2002).

Experiments with animals reveal that prenatally administered androgens (male sex hormones) increase active play and suppress maternal caregiving in many mammals. Eleanor Maccoby (1998) argues that hormones also affect human play styles, leading to rough, noisy movements among boys and calm, gentle actions among girls. Then, as children interact with peers, they choose partners whose interests and behaviors are compatible with their own. Over the preschool years, girls increasingly seek out other girls and like to play in pairs because of a common preference for quieter activities involving cooperative roles. And boys come to prefer larger-group play with other boys, who share a desire to run, climb, play-fight, compete, and build up and knock down (Benenson et al., 2001). At age 4, children already spend three times as much time with same-sex as with other-sex playmates. By age 6, this ratio has climbed to 11 to 1 (Maccoby & Jacklin, 1987).

Additional evidence for the role of biology in gender typing comes from a case study of a boy who experienced serious sexual-identity and adjustment problems because his biological makeup and sex of rearing were at odds. Turn to the Lifespan Vista box on the following page to find out about David's development. Note, also, that David's reflections on his upbringing caution us against minimizing the role of experience in gender typing. As we will see next, environmental forces build on genetic influences to promote children's awareness of and conformity to gender roles.

Environmental Influences on Gender Typing

A wealth of evidence reveals that family influences, encouragement by teachers and peers, and examples in the broader social environment combine to promote the vigorous gender typing of early childhood.

● **The Family.** Beginning at birth, parents hold different perceptions and expectations of their sons and daughters. Many parents state that they want their children to play with

A Lifespan Vista

David: A Boy Who Was Reared as a Girl

A happily married father of three children, 31-year-old David Reimer talks freely about his everyday life: his interest in auto mechanics, his problems at work, and the challenges of child rearing. But when asked about his first 15 years of life, he distances himself, speaking as if the child of his early life was another person. In essence, she was.

David—named Bruce at birth—underwent the first infant sex reassignment ever reported on a genetically and hormonally normal child. To find out about David's development, Milton Diamond and Keith Sigmundson (1997) intensively interviewed him and studied his medical and psychotherapy records. Later, John Colapinto (2001) extended this effort.

At age 8 months, Bruce's penis was accidentally severed during circumcision. His desperate parents soon heard about psychologist John Money's success in assigning a sex to children born with ambiguous genitals. They traveled from their home in Canada to Johns Hopkins University in Baltimore, where under Money's oversight, 22-month-old Bruce had surgery to remove his testicles and sculpt his genitals to look like those of a girl. The operation complete, Bruce's parents named their daughter Brenda.

Research on infants with ambiguous genitals indicates that a parent-chosen sex assignment usually works out (Zucker, 2001). But because of an imbalance in prenatal sex hormones, the organization of those children's central nervous systems might also be ambiguous, permitting development as either a male or a female. Brenda's upbringing, in contrast, was tragic. From the outset, she resisted her parents' efforts to steer her in a "feminine" direction.

Brian (Brenda's identical twin brother) recalled that Brenda looked like a delicate, pretty girl, but as soon as she moved or spoke, this impression evaporated. "She walked like a guy. Sat with her legs apart. She talked about guy things.... She played with my toys: Tinkertoys, dump trucks" (Colapinto, 2001, p. 57). Brian was quiet and gentle in personality. Brenda, in contrast, was a dominant, rough-and-tumble child who picked fights with other children and usually won. Former teachers and classmates agreed that Brenda was the more traditionally masculine of the two children.

At school, Brenda's boyish behavior led classmates to taunt and tease her. When she played with girls, she tried organizing large-group, active games, but they weren't interested. Uncomfortable as a girl and without friends, Brenda's behavior problems increased and her school performance worsened. During periodic medical follow-ups, she drew pictures of herself as a boy and refused additional surgery to create a vagina. Reflecting on Brenda's elementary school years, David explained that she realized she was not a girl and never would be.

As adolescence approached, Brenda's parents moved her from school to school and therapist to therapist, in an effort to help her fit in socially and accept a female identity. Brenda reacted with increased anxiety and insecurity, and conflict with her parents increased. At puberty, Brenda's shoulders broadened and her body added muscle, so her parents insisted that she begin estrogen therapy to feminize her appearance. Soon she grew breasts and added fat around her waist and hips. Repelled by her own feminizing shape, Brenda began overeating to hide it. Her classmates reacted to her confused appearance with stepped-up brutality.

At last, Brenda was transferred to a therapist who recognized her despair and encouraged her parents to tell her about her infancy. When Brenda was 14, her father explained the circumcision accident. David recalled reacting with relief. Deciding to return to his biological sex immediately, he chose for himself the name David, after the biblical lad who slew a giant and overcame adversity. David soon started injections of testosterone to masculinize his body, and he underwent surgery to remove his breasts and to construct a penis. Although his adolescence continued to be troubled, in his twenties he fell in love with Jane, a single mother of three children, and married her.

COURTESY OF DAVID REIMER

■ Because of a tragic medical accident when he was a baby, David Reimer underwent the first sex reassignment on a genetically and hormonally normal baby: He was reared as a girl. His case shows the overwhelming impact of biology on gender identity. David is pictured here as he is today—a happily married man and father of three children.

David's case confirms the impact of genetic sex and prenatal hormones on a person's sense of self as male or female. At the same time, his childhood highlights the importance of experience. David expressed outrage at adult encouragement of dependency in girls, having experienced it first hand. And he realized that had Brenda not been ostracized by peers for her "masculine" traits, she might have had an easier time reverting to her biological sex. In adulthood, David worked in a slaughterhouse with all male employees, who were extreme in their gender stereotyping. At one point he wondered—if he had had a typical childhood, would he have become like them? Of course, he can never know the answer to that question, but his case does clarify one issue: His gender reassignment failed because his male biology overwhelmingly demanded a consistent sexual identity.

"gender-appropriate" toys, and they also believe that boys and girls should be reared differently. Parents are likely to describe achievement, competition, and control of emotion as important for sons and warmth, "ladylike" behavior, and closely supervised activities as important for daughters (Brody, 1999; Turner & Gervai, 1995).

These beliefs carry over into actual parenting practices. Parents give toys that stress action and competition (such as guns, cars, tools, and footballs) to boys. They give toys that emphasize nurturance, cooperation, and physical attractiveness (dolls, tea sets, jewelry, and jump ropes) to girls (Leaper, 1994). Parents also actively reinforce independence in boys and closeness and dependency in girls. For example, they react more positively when a son plays with cars and trucks, demands attention, or tries to take toys from others. In contrast, they more often direct play activities, provide help, encourage participation in household tasks, and refer to emotions when interacting with a daughter (Fagot & Hagan, 1991; Kuebli, Butler, & Fivush, 1995; Leaper et al., 1995). Furthermore, mothers more often *label emotions* when talking to girls, thereby teaching them to "tune in" to others' feelings. In contrast, they more often *explain emotions,* noting causes and consequences, to boys—an approach that emphasizes why it is important to control the expression of emotion (Cervantes & Callanan, 1998).

These factors are major influences in children's gender-role learning, since parents who consciously avoid behaving in these ways have less gender-typed children (Weisner & Wilson-Mitchell, 1990). Other family members also contribute. For example, preschoolers with older, other-sex siblings are less gender-typed because they have many more opportunities to imitate and participate in "cross-gender" play (Rust et al., 2000). In any case, of the two sexes, boys are clearly the more gender-typed. One reason is that parents—particularly fathers—are far less tolerant of "cross-gender" behavior in their sons than in their daughters. They are more concerned if a boy acts like a sissy than if a girl acts like a tomboy (Gervai, Turner, & Hinde, 1995; Sandnabba & Ahlberg, 1999).

■ Parents and teachers can reduce preschoolers' gender stereotyping by modeling nonstereotyped behaviors and pointing out exceptions to stereotypes in their neighborhood and community. Perhaps because this boy has observed family members engaged in nonstereotyped activities, he enacts similar roles in his preschool classroom.

● **Teachers.** Besides parents, teachers encourage children's gender typing. Several times, Leslie caught herself responding in ways that furthered gender segregation and stereotyping in her classroom. One day, she called out, "Will the girls line up on one side and the boys on the other?" Then, as the class became noisy, she pleaded, "Boys, I wish you'd quiet down like the girls!"

As at home, girls get more encouragement to participate in adult-structured activities at preschool. They can frequently be seen clustered around the teacher, following directions in an activity. In contrast, boys more often choose areas of the classroom where teachers are minimally involved (Carpenter, 1983). As a result, boys and girls engage in very different social behaviors. Compliance and bids for help occur more often in adult-structured contexts, whereas assertiveness, leadership, and creative use of materials appear more often in unstructured pursuits.

● **Peers.** Children's same-sex peer groups strengthen gender-stereotyped beliefs and behavior. By age 3, same-sex peers positively reinforce one another for gender-typed play by praising, imitating, or joining in. In contrast, when preschoolers engage in "cross-gender" activities—for example, when boys play with dolls or girls with cars and trucks—peers criticize them. Boys are especially intolerant of "cross-gender" play in their male companions (Carter & McCloskey, 1984; Fagot, 1984). A boy who frequently crosses gender lines is likely to be ignored by other boys even when he does engage in "masculine" activities!

Children also develop different styles of social influence in gender-segregated peer groups. To get their way with male peers, boys more often rely on commands, threats, and physical force, whereas girls emphasize polite requests, persuasion, and acceptance. These tactics succeed with other girls but not with boys, who pay little attention to girls' gentle tactics (Leaper, 1994; Leaper, Tenenbaum, & Shaffer, 1999). Consequently, an additional reason that girls may stop interacting with boys is that they do not find it very rewarding to communicate with an unresponsive social partner.

● **The Broader Social Environment.** Finally, although children's everyday environments have changed to some degree, they continue to present many examples of gender-typed behavior—in occupations, leisure activities, entertainment TV, and achievements of men and women (Ruble & Martin, 1998). As we will see in the next section, children do more than imitate the many gender-linked responses they observe. They also start to view themselves and their environment in gender-biased ways, a perspective that can seriously restrict their interests, experiences, and skills.

Gender Identity

As adults, each of us has a **gender identity**—an image of oneself as relatively masculine or feminine in characteristics. By middle childhood, researchers can measure gender identity by asking children to rate themselves on personality traits. A child or adult with a "masculine" identity scores high on traditionally masculine items (such as ambitious, competitive, and self-sufficient) and low on traditionally feminine items (such as affectionate, cheerful, and soft-spoken). Someone with a "feminine" identity does just the reverse. Although most people view themselves in gender-typed terms, a substantial minority (especially females) have a gender identity called **androgyny,** scoring high on *both* masculine and feminine characteristics.

Gender identity is a good predictor of psychological adjustment. Masculine and androgynous children and adults have higher self-esteem, whereas feminine individuals often think poorly of themselves, perhaps because many of their traits are not highly valued by society (Alpert-Gillis & Connell, 1989; Boldizar, 1991). Also, androgynous individuals are more adaptable in behavior—for example, able to show masculine independence or feminine sensitivity, depending on the situation (Taylor & Hall, 1982). Research on androgyny shows that it is possible for children to acquire a mixture of positive qualities traditionally associated with each gender—an orientation that may best help them realize their potential.

● **Emergence of Gender Identity.** How do children develop a gender identity? Both social learning and cognitive-developmental answers exist. According to *social learning theory,* behavior comes before self-perceptions. Preschoolers first acquire gender-typed responses through modeling and reinforcement. Only then do they organize these behaviors into gender-linked ideas about themselves. In contrast, *cognitive-developmental theory* maintains that self-perceptions come before behavior. Over the preschool years, children first acquire a cognitive appreciation of the permanence of their sex. They develop **gender constancy,** the understanding that sex is biologically based and remains the same even if clothing, hairstyle, and play activities change. Then children use this idea to guide their behavior (Kohlberg, 1966).

Research indicates that gender constancy is not fully developed until the end of the preschool years, when children pass Piagetian conservation tasks (De Lisi & Gallagher, 1991). Shown a doll whose hairstyle and clothing are transformed before their eyes, children younger than age 6 typically indicate that the doll's sex has changed as well (McConaghy, 1979). And when asked such questions as "When you (a girl) grow up, could you ever be a daddy?" or "Could you be a boy if you wanted to?" young children freely answer yes (Slaby & Frey, 1975).

Because many young children in Western cultures do not see members of the opposite sex naked, they distinguish males and females using information they do have—the way each gender dresses and behaves. Although preschoolers who know about genital differences usually say a doll dressed in other-sex clothing is still the same sex, they do not justify their answer by referring to sex as an innate, unchanging quality of people (Szkrybalo & Ruble, 1999). This suggests that cognitive immaturity, not social experience, is responsible for preschoolers' difficulty grasping the permanence of sex.

Is cognitive-developmental theory correct that gender constancy is responsible for children's gender-typed behavior? Evidence for this assumption is weak. "Gender-appropriate" behavior appears so early in the preschool years that modeling and reinforcement must account for its initial appearance, as social learning theory suggests. At present, researchers disagree on just how gender constancy contributes to gender-role development. But they do know that once children begin to reflect on gender roles, their gender-typed self-images and behavior strengthen. Yet another theory shows how this happens.

● **Gender Schema Theory.** **Gender schema theory** is an information-processing approach to gender typing that combines social learning and cognitive-developmental features. It emphasizes that both environmental pressures and children's cognitions work together to shape gender-role development (Martin, 1993; Martin & Halverson, 1987). At an early age, children respond to instruction from others, picking up gender-typed preferences and behaviors. At the same time, they start to organize their experiences into *gender schemas,* or masculine and feminine categories, that they use to interpret their world. A young child who says, "Only boys can be doctors," or, "Cooking is a girl's job," already has some well-formed gender schemas. As soon as preschoolers can label their own sex, they select gender schemas consistent with it, applying those categories to themselves. As a result, their self-perceptions become gender-typed and serve as additional schemas that children use to process information and guide their own behavior.

Let's look at the example in Figure 8.3 on page 264 to see exactly how this network of gender schemas strengthens gender-typed preferences and behavior. Mandy has been taught that "dolls are for girls" and "trucks are for boys." She also knows that she is a girl. Mandy uses this information to make decisions about how to behave. Because her schemas lead her to conclude that "dolls are for me," when given a doll, she approaches it, explores it, and learns more about it. In contrast, on seeing a truck, she uses her gender schemas to conclude that "trucks are not for me" and responds by avoiding the "gender-inappropriate" toy. Gender schemas are so powerful that when children see others behaving in "gender-inconsistent" ways, they often cannot remember the information or distort it to make it "gender-consistent" (Liben & Signorella, 1993; Signorella & Liben, 1984).

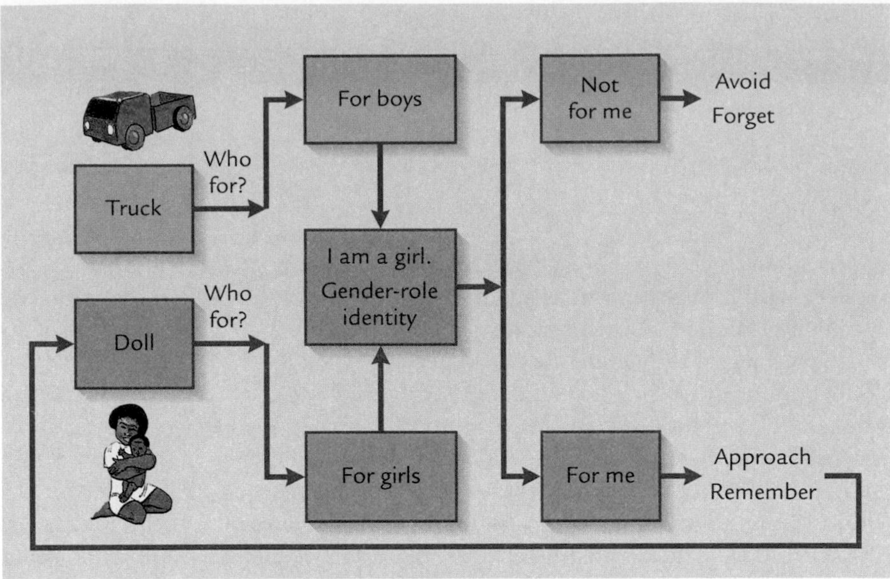

FIGURE 8.3 Impact of gender schemas on gender-typed preferences and behaviors. Mandy's network of gender schemas leads her to approach and explore "feminine" toys, such as dolls, and to avoid "masculine" toys, such as trucks. (From C. L. Martin & C. F. Halverson, Jr., 1981, "A Schematic Processing Model of Sex Typing and Stereotyping in Children," *Child Development, 52,* p. 1121. © The Society for Research in Child Development, Inc. Adapted by permission.)

Reducing Gender Stereotyping in Young Children

How can we help young children avoid developing rigid gender schemas that restrict their behavior and learning opportunities? No easy recipe exists for accomplishing this difficult task. Even children who grow up in homes and schools that minimize stereotyping will eventually encounter it in the media and in their communities. Consequently, children need experiences that counteract their readiness to absorb the extensive network of gender-linked associations that surrounds them.

Adults can begin by eliminating gender stereotyping from their own behavior and from the alternatives they provide for children. For example, mothers and fathers can take turns making dinner, bathing children, and driving the family car. They can provide sons and daughters with both trucks and dolls and pink and blue clothing. Teachers can make sure that all children spend time in both adult-structured and unstructured activities. Also, efforts can be made to shield children from television and other media presentations that portray rigid gender differences.

Once children notice the vast array of gender stereotypes in their society, parents and teachers can point out exceptions. For example, they can arrange for children to see men and women pursuing nontraditional careers. And they can reason with children, explaining that interests and skills, not sex, should determine a person's occupation and activities. Research shows that such reasoning is very effective in reducing children's tendency to view the world in a gender-biased fashion (Bigler & Liben, 1992). And as we will see in the next section, a rational approach to child rearing promotes healthy, adaptable functioning in many other areas as well.

Ask Yourself

REVIEW

Explain how gender stereotyping in the environment and young children's cognitive limitations contribute to rigid gender stereotyping in early childhood.

REVIEW

Cite research indicating that both biology and social experience influence gender-role adoption.

APPLY

When 4-year-old Roger was in the hospital, he was cared for by a male nurse named Jared. After Roger recovered, he told his friends about Dr. Jared. Using gender schema theory, explain why Roger remembered Jared as a doctor, not a nurse.

www.

Child Rearing and Emotional and Social Development

In this and previous chapters, we have seen how parents can foster children's competence—through warmth and sensitivity to children's needs, by serving as models and reinforcers of mature behavior, by using reasoning and inductive discipline, and by guiding and encouraging children's mastery of new skills. Now let's put these practices together into an overall view of effective parenting.

Child-Rearing Styles

Child-rearing styles are combinations of parenting behaviors that occur over a wide range of situations, creating an enduring child-rearing climate. In a landmark series of studies, Diana Baumrind gathered information on child rearing by watching parents interact with their preschoolers. Her findings, and those of others who have extended her work, reveal three features that consistently differentiate an *authoritative style* from less effective styles: (1) acceptance and involvement, (2) control, and (3) autonomy granting (Gray & Steinberg, 1999; Hart, Newell, & Olson, 2002). Table 8.2 shows how child-rearing styles differ in these features. Let's discuss each style in turn.

● **Authoritative Child Rearing.** The **authoritative style**—the most successful approach to child rearing—involves high acceptance and involvement, adaptive control techniques, and appropriate autonomy granting. Authoritative parents are warm, attentive, and sensitive to their child's needs. They establish an enjoyable, emotionally fulfilling parent–child relationship that draws the child into close connection. At the same time, authoritative parents exercise firm, reasonable control; they insist on mature behavior and give reasons for their expectations. Finally, authoritative parents engage in gradual, appropriate autonomy granting, allowing the child to make decisions in areas where he is ready to make choices (Kuczynski & Lollis, 2002; Russell, Mize, & Bissaker, 2002).

Throughout childhood and adolescence, authoritative parenting is linked to many aspects of competence. These include an upbeat mood, self-control, task persistence, cooperativeness, high self-esteem, social and moral maturity, and favorable school performance (Baumrind & Black, 1967; Herman et al., 1997; Luster & McAdoo, 1996; Steinberg, Darling, & Fletcher, 1995).

● **Authoritarian Child Rearing.** Parents who use an **authoritarian style** are low in acceptance and involvement, high in coercive control, and low in autonomy granting. Authoritarian parents appear cold and rejecting; they frequently degrade their child by putting her down. To exert control, they yell, command, and criticize. "Do it because I said so!" is the attitude of these parents. If the child disobeys, authoritarian parents resort to force and punishment. In addition, they make decisions for their child and expect the child to accept their word in an unquestioning manner. If the child does not, authoritarian parents resort to force and punishment.

Children of authoritarian parents are anxious and unhappy. When interacting with peers, they tend to react with hostility when frustrated. Boys, especially, show high rates of anger and defiance. Girls are dependent, lacking in exploration, and overwhelmed by challenging tasks (Baumrind, 1971; Hart et al., 2002, Nix et al., 1999).

● **Permissive Child Rearing.** The **permissive style** of child rearing is warm and accepting. But rather than being involved, such parents are overindulging or inattentive. Permissive parents engage in little control of their child's behavior. And instead of gradually granting autonomy, they allow children to make many of their own decisions at an age when they are not yet capable of doing so. Their children can eat meals and go to bed when they feel like it and watch as much television as they want. They do not have to learn good manners or do any household chores. Although some permissive parents truly believe that this approach is best, many others lack confidence in their ability to influence their child's behavior.

Children of permissive parents are impulsive, disobedient, and rebellious. They are also overly demanding and dependent on adults, and they show less persistence on tasks

Table 8.2 Features of Child-Rearing Styles

Child-Rearing Style	Acceptance and Involvement	Control	Autonomy Granting
Authoritative	Is warm, attentive, and sensitive to the child's needs	Makes reasonable demands for maturity, and consistently enforces and explains them	Permits the child to make decisions in accord with readiness
Authoritarian	Is cold and rejecting and frequently degrades the child	Makes many demands coercively by yelling, commanding, and criticizing	Makes decisions for the child Rarely listens to the child's point of view
Permissive	Is warm but overindulgent or inattentive	Makes few or no demands	Permits the child to make many decisions before the child is ready
Uninvolved	Is emotionally detached and withdrawn	Makes few or no demands	Is indifferent to the child's decision making and point of view

than children whose parents exert more control. The link between permissive parenting and dependent, nonachieving behavior is especially strong for boys (Barber & Olsen, 1997; Baumrind, 1971).

● **Uninvolved Parenting.** The **uninvolved style** combines low acceptance and involvement with little control and general indifference to autonomy granting. Often these parents are emotionally detached and depressed and so overwhelmed by life stress that they have little time and energy for children (Maccoby & Martin, 1983). At its extreme, uninvolved parenting is a form of child maltreatment called *neglect.* Especially when it begins early, it disrupts virtually all aspects of development, including attachment, cognition, and emotional and social skills (see Chapter 6, page 176).

● **What Makes Authoritative Child Rearing So Effective?** Like other correlational findings, the relationship between parenting and children's competence is open to interpretation. Perhaps parents of well-adjusted children are authoritative because their youngsters have especially cooperative, obedient dispositions. Children's characteristics do contribute to the ease with which parents can apply the authoritative style. An impulsive, noncompliant child makes it hard for parents to be warm, firm, and rational. But longitudinal research reveals that authoritative child rearing reduces difficult children's intense negative behavior, whereas parental coercion intensifies their difficultness (Stice & Barrera, 1995; Woodward, Taylor, & Dowdney, 1998).

Authoritative child rearing seems to create an emotional context for positive parental influence. First, warm, involved parents who are secure in the standards they hold for their children provide models of caring concern as well as confident, self-controlled behavior. Second, control that appears fair and reasonable to the child, not arbitrary, is far more likely to be complied with and internalized. Finally, authoritative parents make demands and engage in autonomy granting that fits with their children's ability to take responsibility for their own behavior. As a result, these parents let children know that they are competent individuals who can do things successfully for themselves, thereby fostering high self-esteem and cognitive and social maturity.

Cultural Variations

Despite broad agreement on the advantages of authoritative child rearing, ethnic groups often have distinct child-rearing beliefs and practices. For example, compared with Caucasian Americans, Chinese adults describe their parenting as more controlling (Berndt et al., 1993; Chao, 1994). They are more directive in teaching and scheduling their children's time, as a way of fostering self-control and high achievement (Huntsinger, Jose, & Larson, 1998). In most instances, Chinese parents combine control with high warmth. But when control becomes coercive, it is harmful in Chinese as well as Western cultures. In studies of Chinese children, authoritative parenting was positively related to cognitive and social competence, whereas authoritarian parenting predicted poorer academic achievement, increased aggression, and peer difficulties (Chen, Dong, & Zhou, 1997; Chen, Liu, & Li, 2000).

In Hispanic and Asian Pacific Island families, firm insistence on respect for parental authority, particularly that of the father, is paired with high parental involvement. Although at one time viewed as coercive, contemporary Hispanic fathers typically spend much time with their children and are warm and sensitive (García-Coll & Pachter, 2002; Jambunathan, Burts, & Pierce, 2000).

African-American mothers often expect immediate obedience, regarding strictness as important for promoting self-control and a watchful attitude in risky surroundings. Consistent with theses beliefs, low-SES African-American parents who use more controlling strategies tend to have more cognitively and socially competent children (Brody & Flor, 1998; Brody, Stoneman, & Flor, 1996). And in several studies, physical discipline predicted aggression only for Caucasian-American children, not for African-American children (Deater-Deckard & Dodge, 1997; Deater-Deckard et al., 1996). This does not mean that slaps and spankings are effective strategies. But it does suggest ethnic differences in how children view parental behavior that may modify its consequences. Most African-American parents who use "no-nonsense" discipline refrain from physical punishment and combine strictness with warmth and reasoning (Bluestone & Tamis-LeMonda, 1999; Petitt, Bates, & Dodge, 1998).

The cultural variations just considered remind us that child-rearing styles can be fully understood only in their larger ecological context. As we have seen in previous chapters, a great many factors contribute to good parenting: per-

■ African-American mothers often use strict, no-nonsense discipline combined with warmth and reasoning—an approach well suited to promoting self-control and a watchful attitude in risky surroundings.

sonal characteristics of the child and parent, SES, access to extended family and community supports, cultural values and practices, and public policies (Parke & Buriel, 1998).

As we turn to the topic of child maltreatment, our discussion will underscore, once again, that effective child rearing is sustained not just by mothers' and fathers' desire to be good parents. Almost all want to be. Unfortunately, when vital supports for parenting break down, children—as well as parents—can suffer terribly.

Child Maltreatment

Child maltreatment is as old as human history, but only recently has there been widespread acknowledgement that the problem exists and research aimed at understanding it. Perhaps public concern has increased because child maltreatment is especially common in large industrialized nations. In the year 2000, 880,000 American children (12 out of every 1,000) and 136,000 Canadian children (10 out of every 1,000) were identified as victims (Health Canada, 2002d; U.S. Department of Health and Human Services, 2002c). Because most cases go unreported, the true figures are much higher.

Child maltreatment takes the following forms:

- *Physical abuse:* assaults on children that produce pain, cuts, welts, bruises, burns, broken bones, and other injuries
- *Sexual abuse:* sexual comments, fondling, intercourse, and other forms of exploitation
- *Neglect:* living conditions in which children do not receive enough food, clothing, medical attention, or supervision
- *Psychological abuse:* Failure of caregivers to meet children's needs for affection and emotional support, and actions—such as ridicule, humiliation, or terrorizing—that damage children's cognitive, emotional, or social functioning

Some investigators regard psychological and sexual abuse as the most destructive forms. The rate of psychological abuse may be the highest, since it accompanies most other types. About 10 percent of confirmed maltreatment victims in the United States and Canada are sexually abused. And here again, many more children are affected, but they are too frightened to seek help or are pressured into silence. Although children of all ages are targets of sexual abuse, the largest number of victims are identified in middle childhood (U.S. Department of Health and Human Services, 2002c). Therefore, we will pay special attention to sexual abuse in Chapter 10.

• Origins of Child Maltreatment. Early findings suggested that child maltreatment was rooted in adult psychological disturbance (Kempe et al., 1962). But it soon became clear that although child maltreatment is more common among disturbed parents, a single "abusive personality type" does not exist. Sometimes even "normal" parents harm their children! Also, parents who were abused as children do not necessarily repeat the cycle with their own children (Buchanan, 1996; Simons et al., 1991).

For help in understanding child maltreatment, researchers turned to ecological systems theory (see Chapters 1 and 2). They discovered that many interacting variables—at the family, community, and cultural levels—promote child abuse and neglect. Table 8.3 summarizes factors associated with child maltreatment. The more of these risks that are present, the greater the likelihood that abuse or neglect will occur. Let's examine each set of influences in turn.

Table 8.3 Factors Related to Child Maltreatment

Factor	Description
Parent characteristics	Psychological disturbance; alcohol and drug abuse; history of abuse as a child; belief in harsh, physical discipline; desire to satisfy unmet emotional needs through the child; unreasonable expectations for child behavior; young age (most under 30); low educational level
Child characteristics	Premature or very sick baby; difficult temperament; inattentiveness and overactivity; other developmental problems
Family characteristics	Low income; poverty; homelessness; marital instability; social isolation; physical abuse of mother by husband or boyfriend; frequent moves; large families with closely spaced children; overcrowded living conditions; disorganized household; lack of steady employment; other signs of high life stress
Community	Characterized by social isolation; few parks, child-care centers, preschool programs, recreation centers, and churches to serve as family supports
Culture	Approval of physical force and violence as ways to solve problems

Sources: Cicchetti & Toth, 1998a.

The Family. Within the family, certain children—those whose characteristics make them more of a challenge to rear—are more likely to become targets of abuse. These include premature or very sick babies and children who are temperamentally difficult, are inattentive and overactive, or have other developmental problems (Kotch, Muller, & Blakely, 1999). But whether such children actually are maltreated depends on parents' characteristics.

Maltreating parents are less skillful than other parents in handling discipline confrontations. They also suffer from biased thinking about their child. For example, they often evaluate transgressions as worse than they are and attribute their child's misdeeds to a stubborn or bad disposition—perspectives that lead them to move quickly toward physical force (Milner, 1993; Rogosch et al., 1995).

Once abuse gets started, it quickly becomes part of a self-sustaining relationship. The small irritations to which abusive parents react—a fussy baby, a preschooler who knocks over her milk, or a child who will not mind immediately—soon become bigger ones. Then the harshness increases. By the preschool years, abusive and neglectful parents seldom interact with their children. When they do, they rarely express pleasure and affection; the communication is almost always negative (Wolff, 1999).

Most parents, however, have enough self-control not to respond to their child's misbehavior or developmental problems with abuse. Other factors must combine with these conditions to prompt an extreme parental response. Unmanageable parental stress is strongly associated with all forms of maltreatment. Abusive parents react to stressful situations with high emotional arousal. At the same time, low income, unemployment, marital conflict, overcrowded living conditions, frequent moves, and extreme household disorganization are common in abusive homes (Gelles, 1998; Kotch, Muller, & Blakely, 1999). These personal and situational conditions increase the chances that parents will be too overwhelmed to meet basic child-rearing responsibilities or will vent their frustrations by lashing out at their children.

The Community. The majority of abusive and neglectful parents are isolated from both formal and informal social supports. This social isolation has at least two causes. First, because of their own life histories, many of these parents have learned to mistrust and avoid others. They do not have the skills necessary for establishing and maintaining positive relationships with friends and relatives (Polansky et al., 1985). Second, maltreating parents are more likely to live in unstable, run-down neighborhoods that provide few links between family and community, such as parks, child-care centers, preschool programs, recreation centers, and churches (Coulton, Korbin, & Su, 1999). For these reasons, they lack "lifelines" to others and have no one to turn to for help during stressful times.

The Larger Culture. Cultural values, laws, and customs profoundly affect the chances that child maltreatment will occur when parents feel overburdened. Societies that view violence as an appropriate way to solve problems set the stage for child abuse. Although the United States and Canada have laws to protect children from maltreatment, our earlier consideration of physical punishment revealed that widespread support exists for use of physical force with children. In the United States, the Supreme Court has twice upheld the right of school officials to use corporal punishment. In countries that have policies or laws that prohibit physical punishment of children, such as Denmark, Norway, and Sweden, rates of child abuse are low (U.S. Department of State, 1999).

● **Consequences of Child Maltreatment.** The family circumstances of maltreated children impair the development of emotional self-regulation, empathy and sympathy, self-concept, social skills, and academic motivation. Over time, these youngsters show serious learning and adjustment problems, including academic failure, severe depression, aggressive behavior, peer difficulties, substance abuse, and delinquency (Bolger & Patterson, 2001; Shonk & Cicchetti, 2001).

How do these damaging consequences occur? Think back to our earlier discussion of hostile cycles of parent–child interaction, which are especially severe for abused children. Indeed, a family characteristic strongly associated with child abuse is spouse abuse (Margolin, 1998). Clearly, the home lives of abused children abound with opportunities to learn to use aggression as a way of solving problems.

Furthermore, demeaning parental messages, in which children are ridiculed, humiliated, rejected, or terrorized, result in low self-esteem, high anxiety, self-blame, aggression, and efforts to escape from extreme psychological pain—at times severe enough to lead to attempted suicide in adolescence (Wolfe, 1999). At school, maltreated children are serious discipline problems. Their noncompliance, poor motivation, and cognitive immaturity interfere with academic achievement—an outcome that further undermines their chances for life success (Margolin & Gordis, 2000).

Finally, the trauma of repeated abuse can lead to physiological changes, including abnormal brain-wave activity and heightened production of stress hormones (Ito et al., 1998; Cicchetti & Toth, 2000). These effects increase the chances that emotional self-regulation and adjustment problems will endure.

● **Preventing Child Maltreatment.** Because child maltreatment is embedded in families, communities, and society as a whole, efforts to prevent it must be directed at each of these levels. Many approaches have been suggested, including interventions that teach high-risk parents effective child-rearing strategies, high school child development courses that include direct experience with children, and broad social programs aimed at bettering economic conditions for low-SES families.

We have seen that providing social supports to families is very effective in easing parental stress. This approach sharply reduces child maltreatment as well. Research indicates that a trusting relationship with another person is the most important factor in preventing mothers with childhood histories of

■ Public service announcements help prevent child abuse by educating people about the problem and informing them of where to seek help. This poster reminds adults that degrading remarks can hit as hard as a fist.

abuse from repeating the cycle with their own youngsters (Egeland, Jacobvitz, & Sroufe, 1988). Parents Anonymous, an organization that has as its main goal helping child-abusing parents learn constructive parenting practices, does so largely through social supports. Its local chapters offer self-help group meetings, daily phone calls, and regular home visits to relieve social isolation and teach responsible child-rearing skills.

Even with intensive treatment, some adults persist in their abusive acts. An estimated 1,200 American children and 100 Canadian children die from maltreatment each year (Health Canada, 2002d; U.S. Department of Health and Human Services, 2002c). When parents are unlikely to change their behavior, taking the drastic step of separating parent from child and legally terminating parental rights is the only reasonable course of action.

Child maltreatment is a distressing and horrifying topic—a sad note on which to end our discussion of a period of childhood that is so full of excitement, awakening, and discovery. But there is reason to be optimistic. Great strides have been made over the past several decades in understanding and preventing child maltreatment.

Ask Yourself

REVIEW

Cite major differences between authoritative, authoritarian, and permissive child-rearing styles. Is the concept of authoritative parenting useful for understanding effective parenting across cultures? Explain.

REVIEW

Explain how the consequences of child maltreatment can increase the chances of further maltreatment and lead to lasting adjustment problems.

CONNECT

Which child-rearing style is most likely to be associated with use of inductive discipline, and why?

www.

Summary

Erikson's Theory: Initiative versus Guilt

What personality changes take place during Erikson's stage of initiative versus guilt?

■ Preschoolers develop a new sense of purposefulness as they grapple with the psychological conflict of **initiative versus guilt.** A healthy sense of initiative depends on exploring the social world through play, forming a conscience through identification with the same-sex parent, and receiving supportive child rearing. Erikson's image of initiative captures the emotional and social changes of early childhood.

Self-Understanding

Describe preschoolers' self-concepts and self-esteem.

■ Preschoolers' **self-concepts** largely consist of observable characteristics and typical emotions and attitudes. Their increasing self-awareness underlies struggles over objects as well as first efforts to cooperate.

■ During early childhood, **self-esteem** has already begun to differentiate into several self-judgments. Preschoolers' high self-esteem contributes to their mastery-oriented approach to the environment. However, even a little adult disapproval can undermine a young child's self-esteem and enthusiasm for learning.

Emotional Development

Cite changes in understanding and expression of emotion during early childhood, along with factors that influence those changes.

■ Young children have an impressive understanding of the causes, consequences, and behavioral signs of basic emotions. Secure attachment and conversations about emotions support their understanding. By age 3 to 4, children are aware of various strategies for emotional self-regulation.

Temperament and parental modeling influence preschoolers' capacity to handle negative emotion.

- As their self-concepts become better developed, preschoolers experience self-conscious emotions more often. Parental messages affect the situations in which self-conscious emotions occur and their intensity. Empathy also becomes more common. Temperament and parenting affect the extent to which empathy prompts **sympathy** and results in **prosocial,** or **altruistic, behavior.**

Peer Relations

Describe peer sociability and friendship in early childhood, along with cultural and parental influences on early peer relations.

- During early childhood, peer interaction increases. According to Parten, it moves from **nonsocial activity** to **parallel play** and then shifts to **associative** and **cooperative play.** However, preschoolers do not follow this straightforward developmental sequence. Despite increases in associative and cooperative play, solitary play and parallel play remain common. In collectivist societies, play occurs in large groups and is highly cooperative. Sociodramatic play seems especially important in societies where child and adult worlds are distinct.

- Preschoolers view friendship in concrete, activity-based terms. Their interactions with friends are especially positive and cooperative. Parents affect peer sociability both directly, through attempts to influence their child's peer relations, and indirectly, through their child-rearing practices. Secure attachment and emotionally positive parent–child conversations are linked to favorable peer interaction.

Foundations of Morality

What are the central features of psychoanalytic, social learning, and cognitive-developmental approaches to moral development?

- Psychoanalytic and social learning approaches to morality focus on how children acquire ready-made standards held by adults. In contrast to Freud's psychoanalytic theory, discipline using fear of punishment and loss of parental love does not foster conscience development. Instead, **induction** is far more effective in encouraging self-control and prosocial behavior.

- Social learning theory regards reinforcement and modeling as the basis for moral action. Effective adult models of morality are warm, powerful, and practice what they preach. Frequent, harsh punishment does not promote moral internalization and socially desirable behavior. Alternatives, such as **time out** and withdrawal of privileges, can help parents avoid the undesirable side effects of punishment. When parents use punishment, they can increase its effectiveness by being consistent, maintaining a warm relationship with the child, and offering explanations.

- The cognitive-developmental perspective views children as active thinkers about social rules. By age 4, children consider intentions in making moral judgments and distinguish truthfulness from lying. Preschoolers also distinguish moral imperatives from social conventions and matters of personal choice. Through sibling and peer interaction, children work out their first ideas about justice and fairness. Parents who discuss moral issues with their children help them reason about morality.

Describe the development of aggression in early childhood, including family and television as major influences.

- All children display aggression from time to time. During early childhood, **instrumental aggression** declines while **hostile aggression** increases. Two types of hostile aggression appear: **overt aggression,** more common among boys, and **relational aggression,** more common among girls.

- Ineffective discipline and a conflict-ridden family atmosphere promote and sustain aggression in children. Televised violence also triggers childhood aggression. Young children's limited understanding of TV content increases their willingness to accept uncritically and imitate what they see. Teaching parents effective child-rearing practices, providing children with social problem-solving training, reducing family hostility, and shielding children from violent TV can reduce aggressive behavior.

Gender Typing

Discuss genetic and environmental influences on preschoolers' gender-stereotyped beliefs and behavior.

- **Gender typing** is well under way in the preschool years. Preschoolers acquire many gender stereotypes and behaviors. Heredity, through prenatal hormones, contributes to boys' higher activity level and overt aggression and children's preference for same-sex playmates. At the same time, parents, teachers, peers, and the broader social environment encourage many gender-typed responses.

Describe and evaluate the accuracy of major theories that explain the emergence of gender identity.

- Although most people have a traditional **gender identity,** some are **androgynous,** combining both masculine and feminine characteristics. Masculine and androgynous identities are linked to better psychological adjustment.

- According to social learning theory, preschoolers first acquire gender-typed responses through modeling and reinforcement and then organize them into gender-linked ideas about themselves. Cognitive-developmental theory suggests that **gender constancy** must be mastered before children develop gender-typed behavior. In contrast to cognitive-developmental predictions, gender-role behavior is acquired long before gender constancy.

- **Gender schema theory** combines features of social learning and cognitive-developmental perspectives. As children acquire gender-stereotyped preferences and behaviors, they form masculine and feminine categories, or gender schemas, that they apply to themselves and use to interpret their world.

Child Rearing and Emotional and Social Development

Describe the impact of child-rearing styles on children's development, explain why authoritative parenting is effective, and note cultural variations in child rearing.

- Three features distinguish major **child-rearing styles:** (1) acceptance and involvement, (2) control, and (3) autonomy granting. Compared with the **authoritarian, permissive,** and **uninvolved styles,** the **authoritative style** promotes cognitive, emotional, and social competence. Warmth, explanations, and reasonable demands for mature behavior account for the effectiveness of the authoritative style. Certain ethnic groups, including Chinese, Hispanic, Asian Pacific Island, and African-American, rely on high levels of parental control. Research on Chinese children reveals that when such control is coercive, it impairs academic and social competence.

Discuss the multiple origins of child maltreatment, its consequences for development, and effective prevention.

- Child maltreatment is related to factors within the family, community, and larger culture. Child and parent characteristics often feed on one another to produce abusive behavior. Unmanageable parental stress and social isolation greatly increase the chances that abuse and neglect will occur. When a society approves of physical force as a means of solving problems, child abuse is promoted.

- Maltreated children are impaired in emotional self-regulation, empathy and sympathy, self-concept, social skills, and academic motivation. Over time, they show serious adjustment problems. Successful prevention of child maltreatment requires efforts at the family, community, and societal levels.

Important Terms and Concepts

androgyny (p. 263)
associative play (p. 249)
authoritarian style (p. 265)
authoritative style (p. 265)
child-rearing styles (p. 265)
cooperative play (p. 249)
gender constancy (p. 263)
gender identity (p. 263)
gender schema theory (p. 263)
gender typing (p. 260)
hostile aggression (p. 257)
induction (p. 253)
initiative versus guilt (p. 244)
instrumental aggression (p. 257)
nonsocial activity (p. 249)
overt aggression (p. 257)
parallel play (p. 249)
permissive style (p. 265)
prosocial, or altruistic, behavior (p. 249)
relational aggression (p. 257)
self-concept (p. 245)
self-esteem (p. 245)
sympathy (p. 249)
time out (p. 255)
uninvolved style (p. 266)

FYI For Further Information and Help

Consult the Companion Website for *Development Through the Lifespan, Third Edition,* (www.ablongman.com/berk) where you will find the following resources for this chapter:

- **Chapter Objectives**
- **Flashcards** for studying important terms and concepts
- **Annotated Weblinks** to guide you in further research
- **Ask Yourself** questions, which you can answer and then check against a sample response
- **Suggested Readings**
- **Practice Test** with immediate scoring and feedback

Milestones

Development in Early Childhood

AGE	PHYSICAL	COGNITIVE	LANGUAGE	EMOTIONAL/SOCIAL
2 years	• Slower gains in height and weight than in toddlerhood. (204) • Balance improves, walking becomes better coordinated. (204) • Running, jumping, hopping, throwing, and catching develop. (213) • Puts on and removes some items of clothing. (213) • Uses spoon effectively. (213)	• Make-believe becomes less dependent on realistic toys, less self-centered, and more complex. (216) • Can take the perspective of others in simplified situations. (218–219) • Recognition memory well developed. (226) • Aware of the difference between inner mental and outer physical events. (228)	• Vocabulary increases rapidly. (236) • Sentences follow basic word order of native language; adds grammatical markers. (236) • Displays effective conversational skills. (237) 	• Begins to develop self-concept and self-esteem. (245) • Cooperation and instrumental aggression appear. (244, 257) • Understands causes, consequences, behavioral signs of basic emotions. (247) • Empathy increases. (249) • Gender-stereotyped beliefs and behavior increase. (260)
3–4 years	 • Running, jumping, hopping, throwing, and catching become better coordinated. (213) • Galloping and one-foot skipping appear. (213) • Rides tricycle. (213) • Uses scissors, draws first picture of a person. (213–214) • Can tell the difference between writing and nonwriting. (215)	• Notices transformations, reverses thinking, and has a basic understanding of causality in familiar situations. (220) • Classifies familiar objects hierarchically. (221) • Uses private speech to guide behavior in challenging tasks. (223) • Attention becomes more sustained and planful. (225) • Uses scripts to recall familiar experiences. (227) • Understands that both beliefs and desires can determine behavior. (228) • Aware of some meaningful features of written language. (230) • Counts small numbers of objects and grasps cardinality. (231)	• Masters increasingly complex grammatical structures. (236) • Occasionally overextends grammatical rules to exceptions. (236) • Understands many culturally accepted ways of adjusting speech to fit the age, sex, and social status of speakers and listeners. (237) 	• Emotional self-regulation improves. (247) • Experiences self-conscious emotions more often. (248) • Nonsocial activity declines, and interactive play increases. (250) • Instrumental aggression declines, and hostile aggression increases. (257) • Forms first friendships. (251) • Distinguishes moral rules from social conventions and personal matters. (256) • Preference for same-sex playmates strengthens. (262) 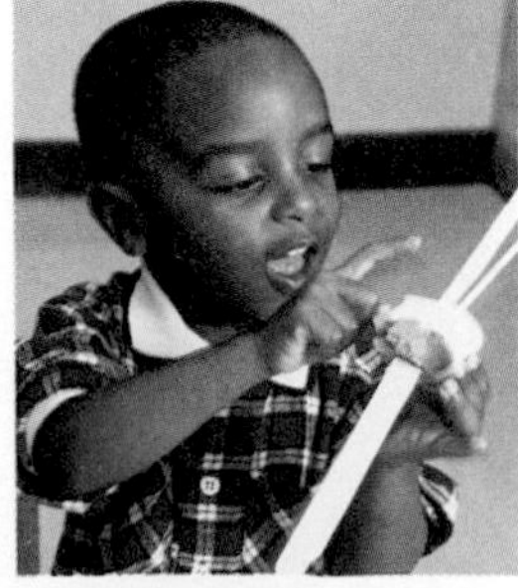

AGE	PHYSICAL	COGNITIVE	LANGUAGE	EMOTIONAL/SOCIAL
5–6 years	 • Body is streamlined and longer-legged with proportions similar to adults'. (204) • First permanent tooth erupts. (204) • Skipping appears. (213) • Shows mature throwing and catching patterns. 213) • Ties shoes, draws more complex pictures, writes name. (213, 215)	• Ability to distinguish appearance from reality improves. (220) • Attention continues to improve. (225) • Recall, scripted memory, and autobiographical memory improve. (226–227) • Understands that letters and sounds are linked in systematic ways. (230)  • Counts on and counts down, engaging in simple addition and subtraction. (231)	• Vocabulary reaches about 10,000 words. (236) • Uses many complex grammatical forms. (236)	• Ability to interpret and predict others' emotional reactions improves. (247)  • Relies more on language to express empathy. (249) • Has acquired many morally relevant rules and behaviors. (253–254) • Gender-stereotyped beliefs and behavior continue to increase. (260) • Understands gender constancy. (263)

Note: Numbers in parentheses indicate the page(s) on which each milestone is discussed.

About
Me!
Patchwork Patterns

9

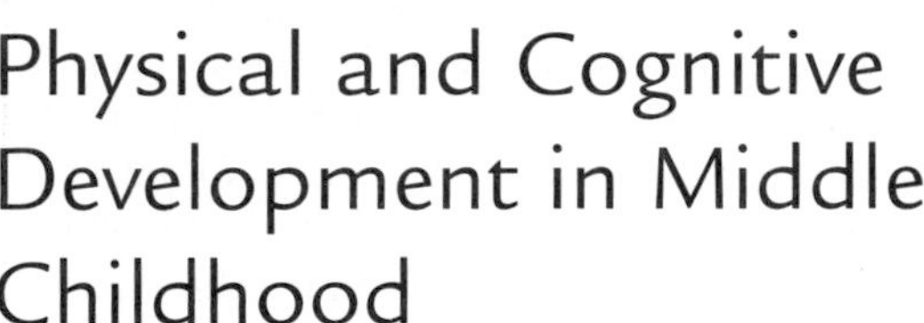

Physical and Cognitive Development in Middle Childhood

© FELICIA MARTINEZ/PHOTOEDIT

Physical Development

Body Growth

Common Health Problems
Vision and Hearing ● Malnutrition ● Obesity ● Illnesses ● Unintentional Injuries

Motor Development and Play
Gross Motor Development ● Fine Motor Development ● Sex Differences ● Games with Rules ● Physical Education

Cognitive Development

Piaget's Theory: The Concrete Operational Stage
Achievements of the Concrete Operational Stage ● Limitations of Concrete Operational Thought ● Recent Research on Concrete Operational Thought ● Evaluation of the Concrete Operational Stage

Information Processing
Attention ● Memory Strategies ● The Knowledge Base and Memory Performance ● Culture, Schooling, and Memory Strategies ● The School-Age Child's Theory of Mind ● Cognitive Self-Regulation ● Applications of Information Processing to Academic Learning

- Biology & Environment: Children with Attention-Deficit Hyperactivity Disorder

Individual Differences in Mental Development
Defining and Measuring Intelligence ● Recent Efforts to Define Intelligence ● Explaining Individual and Group Differences in IQ

Language Development
Vocabulary ● Grammar ● Pragmatics ● Learning Two Languages at a Time

Learning in School
Class Size ● Educational Philosophies ● Teacher–Student Interaction ● Grouping Practices ● Teaching Children with Special Needs ● How Well Educated Are North American Children?

- Social Issues: School Readiness and Grade Retention
- Cultural Influences: Education in Japan, Taiwan, and the United States

An improved capacity to remember, reason, and reflect on one's own thinking makes middle childhood a time of dramatic advances in academic learning and problem solving. Here, fourth graders collaborate in a quilt-making project.

"I'm on my way, Mom!" hollered 10-year-old Joey as he stuffed the last bite of toast into his mouth, slung his book bag over his shoulder, dashed out the door, jumped on his bike, and headed down the street for school. Joey's 8-year-old sister Lizzie followed, quickly kissing her mother good-bye and pedaling furiously until she caught up with Joey. Rena, the children's mother and one of my colleagues at the university, watched from the front porch as her son and daughter disappeared in the distance.

"They're branching out," Rena remarked to me over lunch that day, as she described the children's expanding activities and relationships. Homework, household chores, soccer teams, music lessons, scouting, friends at school and in the neighborhood, and Joey's new paper route were all part of the children's routine. "It seems as if the basics are all there; I don't have to monitor Joey and Lizzie constantly anymore. Although being a parent is still very challenging, it's more a matter of refinements—helping them become independent, competent, and productive individuals."

Joey and Lizzie have entered middle childhood, which spans the years from 6 to 11. Around the world, children of this age are assigned new responsibilities. Joey and Lizzie, like other children in industrialized nations, spend long hours in school. Indeed, middle childhood is often called the "school years," since its onset is marked by the start of formal schooling. In village and tribal cultures, the school may be a field or a jungle. But universally, mature members of society guide children of this age period toward more realistic tasks that increasingly resemble those they will perform as adults (Rogoff, 1996).

This chapter focuses on physical and cognitive development in middle childhood—changes that are less spectacular than those of earlier years. By age 6, the brain has reached 95 percent of its adult size, and the body continues to grow slowly. In this way, nature grants school-age children the mental powers to master challenging tasks as well as added time to learn before reaching physical maturity.

We begin by reviewing typical growth trends, special health concerns, and gains in motor skills. Then we return to Piaget's theory and the information-processing approach, which provide an overview of cognitive changes during the school years. Next, we examine the genetic and environmental roots of IQ scores, which often enter into important educational decisions. Our discussion continues with language, which blossoms further during middle childhood. Finally, we consider the importance of schools in children's learning and development.

PHYSICAL DEVELOPMENT

Body Growth

The rate of physical growth during the school years is an extension of the slow, regular pattern that characterized early childhood. At age 6, the average North American child weighs about 45 pounds and is 3½ feet tall. During the next few

FIGURE 9.1 Body growth during middle childhood. Andy and Amy continued the slow, regular pattern of growth that they showed in early childhood (see Chapter 7, page 205). But around age 9, Amy began to grow at a faster rate than Andy. At age 10½, she was taller, heavier, and more mature looking.

■ These 9-year-olds, who are taking a temperature reading for a science project, illustrate faster growth of the lower portion of the body in middle childhood. They appear longer-legged than they did as preschoolers, quickly growing out of their jeans and frequently needing larger shoes. As at earlier ages, they vary greatly in body size.

years, children add about 2 to 3 inches in height and 5 pounds in weight each year (see Figure 9.1). At ages 6 to 8, girls are slightly shorter and lighter than boys. By age 9, this trend reverses. Already, Rena noticed, Lizzie was starting to catch up with Joey in size as she approached the dramatic adolescent growth spurt, which takes place 2 years earlier in girls than in boys.

Because the lower portion of the body is growing fastest, Joey and Lizzie appeared longer-legged than they had in early childhood. Rena discovered that they grew out of their jeans more quickly than their jackets and frequently needed larger shoes. As in early childhood, girls have slightly more body fat and boys more muscle. After age 8, girls begin accumulating fat at a faster rate, and they will add even more during adolescence (Tanner, 1990).

During middle childhood, the bones of the body lengthen and broaden. However, ligaments are not yet firmly attached to bones. This, combined with increasing muscle strength, grants children unusual flexibility of movement as they turn cartwheels and perform handstands. As their bodies become stronger, many children experience a greater desire for physical exercise. Nighttime "growing pains"— stiffness and aches in the legs—are common as muscles adapt to an enlarging skeleton (Walco, 1997).

Between ages 6 and 12, all 20 primary teeth are replaced by permanent ones, with girls losing their teeth slightly earlier than boys. The first teeth to go are the lower and upper front teeth, giving many first and second graders a "toothless" smile. For a while, permanent teeth seem much too large. Growth of facial bones, especially the jaw and chin, gradually causes the child's face to lengthen and mouth to widen, accommodating the newly erupting teeth.

Common Health Problems

Children like Joey and Lizzie, who come from economically advantaged homes, appear to be at their healthiest in middle childhood, full of energy and play. The cumulative effects of good nutrition, combined with rapid development of the body's immune system, offer greater protection against disease. At the same time, growth in lung size permits more air to be exchanged with each breath, so children are better able to exercise vigorously without tiring.

Nevertheless, a variety of health problems do occur. We will see that many of them are more common among low-SES youngsters. Because economically disadvantaged American families often lack health insurance (see Chapter 7), many youngsters do not have regular access to a doctor. And a substantial number also lack such basic necessities as a comfortable home and regular meals. Not surprisingly, poverty continues to be a powerful predictor of ill health during the school years.

Vision and Hearing

The most common vision problem in middle childhood is *myopia*, or nearsightedness. By the end of the school years, nearly 25 percent of children are affected, a rate that rises to 60 percent by early adulthood. Heredity contributes to myopia, since identical twins are more likely to share the condition than fraternal twins (Pacella et al., 1999). Early biological trauma also can induce it. School-age children with low birth weights show an especially high rate, believed to result from immaturity of visual structures, slower eye growth, and a greater incidence of eye disease (O'Connor et al., 2002).

Parents often warn their children not to read in dim light or sit too close to the TV or computer screen, exclaiming, "You'll ruin your eyes." Their concern is well founded. Myopia progresses more rapidly during the school year, when children spend more time reading and doing other close work, than during the summer (Goss & Rainey, 1998). Furthermore, myopia is one of the few health conditions that increase with SES. For example, a dramatic rise in myopia among Hong Kong Chinese children has occurred in the past 50 years, during which the country changed from a largely illiterate to a highly educated society (Wu & Edwards, 1999). Fortunately, myopia can be corrected easily with glasses.

During middle childhood, the eustachian tube (canal that runs from the inner ear to the throat) becomes longer, narrower, and more slanted, preventing fluid and bacteria from traveling so easily from the mouth to the ear. As a result, *otitis media* (middle ear infection) becomes less frequent (see Chapter 7). Still, about 3 to 4 percent of the school-age population, and as many as 20 percent of low-SES children, develop permanent hearing loss from repeated infections (Daly, Hunter, & Giebink, 1999). Regular screening for both vision and hearing permits defects to be corrected before they lead to serious learning difficulties.

Malnutrition

School-age children need a well-balanced, plentiful diet to provide energy for successful learning in school and increased physical activity. Many youngsters are so focused on play, friendships, and new activities that they spend little time at the table. Also, the percentage of children eating dinner with their families drops sharply between ages 9 and 14 years, and family dinnertimes have waned in general over the past decade. Yet eating an evening meal with parents leads to a diet higher in fruits and vegetables and lower in fried foods and soft drinks (Gillman et al., 2000). Readily available, healthy between-meal snacks—cheese, fruit, raw vegetables, and peanut butter—also help meet nutritional needs in middle childhood.

As long as parents encourage healthy eating, mild nutritional deficits that result from the child's busy daily schedule have no impact on development. But as we have seen in earlier chapters, many poverty-stricken children in developing countries and in North America suffer from serious and prolonged malnutrition. By middle childhood, the effects are apparent in retarded physical growth, low intelligence test scores, poor motor coordination, inattention, and distractibility.

Unfortunately, when malnutrition persists from infancy or early childhood into the school years, permanent physical and mental damage usually results (Grantham-McGregor, Walker, & Chang, 2000). Prevention through government-sponsored food programs beginning in the early years and continuing throughout childhood and adolescence is necessary. In studies carried out in Egypt, Kenya, and Mexico, quality of food (protein, vitamin, and mineral content) was more important than quantity of food in predicting favorable cognitive development in middle childhood (Sigman, 1995; Wachs et al., 1995; Watkins & Pollitt, 1998).

Obesity

Mona, a very overweight child in Lizzie's class, often stood on the sidelines and watched during recess. When she did join in the children's games, she was slow and clumsy. On a daily basis, Mona was the target of unkind comments: "Move it, Tubs!" "No fatsoes allowed!" On most afternoons, she walked home from school by herself while the other children gathered in groups, talking, laughing, and chasing. Once at home, Mona sought comfort in high-calorie snacks, which promoted further weight gain.

Mona suffers from **obesity,** a greater-than-20-percent increase over average body weight, based on the child's age, sex, and physical build. During the past several decades, a rise in overweight and obesity has occurred in many Western nations, with a large increase in Canada, Denmark, Finland, Great Britain, New Zealand, and especially the United States. Today, 15 percent of Canadian and 25 percent of American children are obese (Tremblay & Willms, 2000; U.S. Department of Health and Human Services, 2002). Obesity is also becoming

© STEPHEN TRIMBLE

■ This Pima Indian medicine man of Arizona is very obese. By the time his two daughters reach adolescence, they are likely to follow in his footsteps. Because of a high-fat diet, the Pima residing in the southwestern United States have one of the highest rates of obesity in the world. In contrast, the Pima living in the remote Sierra Madre region of Mexico, who have a low-fat vegetarian diet, are average weight and rarely suffer from diabetes and its life-threatening complications.

common in developing nations, as urbanization shifts the population toward sedentary lifestyles and diets high in meats and refined foods (Troiana & Flegal, 1998).

Over 80 percent of affected children become overweight adults (Oken & Lightdale, 2000). Besides serious emotional and social difficulties, obese children are at risk for life-long health problems. High blood pressure and cholesterol levels and respiratory abnormalities begin to appear in the early school years, symptoms that are powerful predictors of heart disease, adult-onset diabetes, gallbladder disease, certain forms of cancer, and early death.

● **Causes of Obesity.** Not all children are equally at risk for becoming obese. Overweight children tend to have overweight parents, and identical twins are more likely to share the disorder than fraternal twins. But genetics accounts only for a tendency to gain weight (Salbe et al., 2002). One indication that environment is powerfully important is the consistent relation between low SES and obesity (Stunkard & Sørensen, 1993). Among factors responsible are lack of knowledge about healthy diet; a tendency to buy high-fat, low-cost foods; and family stress, which prompts overeating in some individuals.

Parental feeding practices contribute to childhood obesity. Some parents anxiously overfeed their infants and young children, interpreting almost all their discomforts as a desire for food. Others are overly controlling, constantly monitoring what their children eat. In either case, they fail to help children learn to regulate their own energy intake. Furthermore, parents of obese children often use high-fat, sugary foods to reward other behaviors—a practice that leads children to attach greater value to the treat (Birch & Fisher, 1995).

Because of these feeding experiences, obese children soon develop maladaptive eating habits (Johnson & Birch, 1994). They are more responsive than normal-weight individuals to external stimuli associated with food—taste, sight, smell, and time of day—and less responsive to internal hunger cues (Ballard et al., 1980). They also eat faster and chew their food less thoroughly, a behavior pattern that appears as early as 18 months of age (Drabman et al., 1979).

Furthermore, overweight children are less physically active than their normal-weight peers. This inactivity is both cause and consequence of their condition. Research indicates that the rise in childhood obesity is due in part to the many hours North American children spend watching television. In a study that tracked children's TV viewing over a 4-year period, those who watched more than 5 hours per day were more than 8 times likelier to become obese than those who watched 2 hours or less per day (see Figure 9.2) (Gortmaker et al., 1996). Television greatly reduces time devoted to physical exercise, and TV ads encourage children to eat fattening, unhealthy snacks. When researchers gave third and fourth graders 2 months of twice-weekly lessons in reducing TV viewing and videogame use, the children not only watched less but also lost weight (Robinson, 1999).

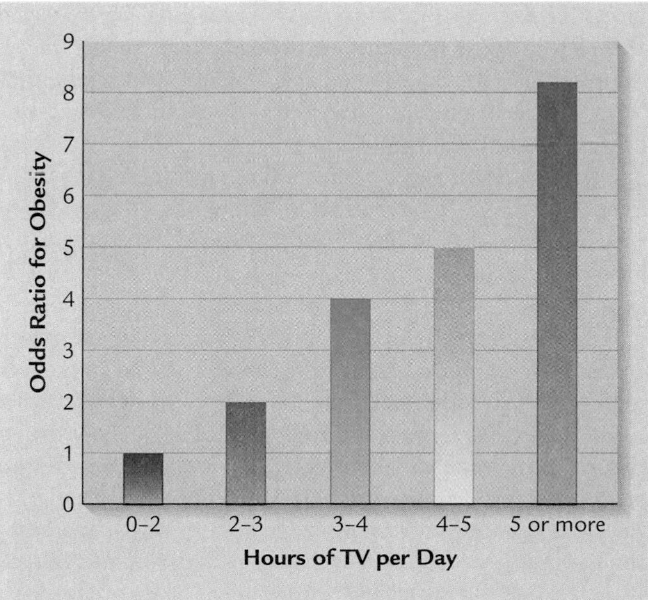

■ **FIGURE 9.2 Relationship between television viewing and development of childhood obesity.** Researchers tracked 10- to 15-year-olds' television viewing over a 4-year period. The more hours young people spent in front of the TV, the greater the likelihood that they became obese by the end of the study. (Adapted from Gortmaker et al., 1996.)

Finally, the broader food environment affects the incidence of obesity. The Pima Indians of Arizona, who recently changed from a traditional diet of plant foods to an affluent, high-fat diet, have one of the highest rates of obesity in the world. Compared with descendants of their ancestors living in the remote Sierra Madre region of Mexico, the Arizona Pima have body weights 50 percent higher. Half the population has diabetes (8 times the national average), and many are disabled by the disease in their twenties and thirties—blind, in wheelchairs, and on kidney dialysis (Gladwell, 1998; Ravussin et al., 1994). Although the Pima have a genetic susceptibility to overweight, it emerges only under Western dietary conditions.

● **Consequences of Obesity.** Unfortunately, physical attractiveness is a powerful predictor of social acceptance in Western societies. Both children and adults rate obese youngsters as unlikable, stereotyping them as lazy, sloppy, ugly, stupid, self-doubting, and deceitful (Kilpatrick & Sanders, 1978; Tiggemann & Anesbury, 2000). By middle childhood, obese children report feeling more depressed and display more behavior problems than their peers. Unhappiness and overeating contribute to one another, and the child remains overweight (Braet & Mervielde, 1997). As we will see in Chapter 13, these psychological consequences combine with continuing discrimination to result in reduced life chances in close relationships and employment.

● **Treating Obesity.** Childhood obesity is difficult to treat because it is often a family disorder. In Mona's case, the school nurse suggested that Mona and her obese mother enter a weight-loss program together. But Mona's mother, unhappily married for many years, had her own reasons for overeating. She rejected this idea, claiming that Mona would eventually lose weight on her own.

When parents decide to seek treatment for an obese child, long-term changes in body weight do occur. The most effective interventions are family based and focus on changing behaviors. In one study, both parent and child revised eating patterns, exercised daily, and reinforced each other with praise and points for progress, which they exchanged for special activities and times together. Follow-ups after 5 and 10 years showed that children maintained their weight loss more effectively than adults—a finding that underscores the importance of intervening at an early age. Furthermore, weight loss was greater when treatments focused on both dietary and lifestyle changes, including regular vigorous exercise (Epstein et al., 1990, 1994).

Schools can help reduce obesity by ensuring regular physical activity and serving healthier meals. The high-fat content of American school lunches and snacks can greatly affect body weight because children consume one-third of their daily caloric intake at school. In Singapore, school interventions consisting of nutrition education, low-fat food choices, and daily physical activity led child and adolescent obesity to decline from 14 to 11 percent (Schmitz & Jeffery, 2000).

Illnesses

Children experience a somewhat higher rate of illness during the first 2 years of elementary school than they will later, due to exposure to sick children and an immune system that is not yet mature. On average, illness causes children to miss about 11 days of school per year, but most absences can be traced to a few students with chronic health problems (Madan-Swain, Fredrick, & Wallander, 1999).

About 19 percent of North American children have chronic diseases and conditions (including physical disabilities). By far the most common—accounting for nearly one-third of childhood chronic illness and the most frequent cause of school absence and childhood hospitalization—is *asthma,* in which the bronchial tubes (passages that connect the throat and lungs) are highly sensitive (Newacheck & Halfon, 2000). In response to a variety of stimuli, such as cold weather, infection, exercise, pollution, allergies, and emotional stress, the bronchial tubes fill with mucus and contract, leading to coughing, wheezing, and serious breathing difficulties.

During the past three decades, the number of children with asthma has risen sharply. Today 8 percent of American and 12 percent of Canadian youngsters are affected (Health Canada, 1999b; U.S. Department of Health and Human Services, 2002e). Although heredity contributes to asthma, researchers believe that environmental factors are necessary to spark the illness. Boys, low-SES children, and children who were born underweight or whose parents smoke are at greatest risk (Chen, Matthews, & Boyce, 2002; Creer, 1998).

About 2 percent of North American youngsters have chronic illnesses that are more severe than asthma, such as sickle cell anemia, cystic fibrosis, diabetes, arthritis, cancer, and AIDS. Painful medical treatments, physical discomfort, and changes in appearance often disrupt the sick child's daily life, making it difficult to concentrate in school and causing withdrawal from peers. As the illness worsens, family stress increases. For these reasons, chronically ill children are at risk for academic, emotional, and social difficulties. A strong link between good family functioning and child well-being exists for chronically ill children, just as it does for physically healthy children (Barakat & Kazak, 1999). Interventions that foster positive family relationships help parent and child cope with the disease and improve children's adjustment. These include health education, counseling, social support, and disease-specific summer camps, which teach children self-help skills and give parents time off from the demands of caring for a chronically ill youngster.

Unintentional Injuries

As we conclude our discussion of threats to children's health during the school years, let's return for a moment to the topic of unintentional injuries (discussed in detail in Chapter 7). As Figure 9.3 shows, the frequency of injury fatalities increases from middle childhood into adolescence, with the rate for boys rising considerably above that for girls.

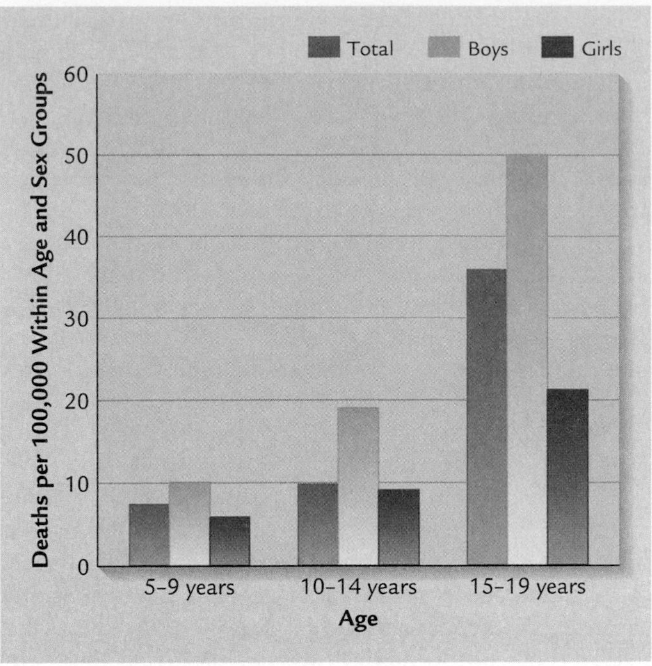

FIGURE 9.3 Rate of injury mortality in North America from middle childhood to adolescence. Injury fatalities increase with age, and the gap between boys and girls expands. Motor vehicle (passenger and pedestrian) accidents are the leading cause, with bicycle injuries next in line. American and Canadian injury rates are nearly identical. (From U.S. Department of Health and Human Services, 2002e.)

Motor vehicle accidents, involving children as passengers or pedestrians, continue to be the leading cause of injury, with bicycle accidents next in line (Health Canada, 1999b; U.S. Department of Health and Human Services, 2002e). Pedestrian injuries most often result from midblock dart-outs, bicycle accidents from disobeying traffic rules. Young school-age children are not yet good at thinking before they act, especially when many stimuli impinge on them at once (Tuchfarber, Zins, & Jason, 1997). They need frequent reminders, supervision, and prohibitions against venturing into busy traffic on their own.

As children range farther from home, safety education becomes especially important. School-based programs with lasting effects use extensive modeling and rehearsal of safety practices; give children feedback about their performance, along with praise and tangible rewards for acquiring safety skills; and provide occasional booster sessions (Zins et al., 1994). An important part of injury prevention is educating parents about children's age-related safety capacities, since parents often overestimate their child's safety knowledge and behavior (Rivara, 1995).

By middle childhood, the greatest risk-takers tend to be children whose parents do not act as safety-conscious models or who try to enforce rules with punitive or inconsistent discipline. As we saw in Chapter 8, these child-rearing techniques spark defiance in children and may actually promote high-risk behavior. Highly active boys remain particularly susceptible to injury. Compared with girls, boys judge risky play activities as less likely to result in injury, and they pay less attention to injury risk cues, such as a peer who looks hesitant or fearful (Morrongiello & Rennie, 1998). The greatest challenge for injury control programs is reaching these "more difficult to reach" youngsters and reducing the dangers to which they are exposed.

Motor Development and Play

Visit a park on a pleasant weekend afternoon, and watch several preschool and school-age children at play. You will see that gains in body size and muscle strength support improved motor coordination during middle childhood. In addition, greater cognitive and social maturity permit older children to use their new motor skills in more complex ways. A major change in children's play takes place at this time.

Gross Motor Development

During middle childhood, running, jumping, hopping, and ball skills become more refined. Third to sixth graders burst into sprints as they race across the playground, jump quickly over rotating ropes, engage in intricate patterns of hopscotch, kick and dribble soccer balls, bat at balls pitched by their classmates, and balance adeptly as they walk heel-to-toe across narrow ledges. These diverse skills reflect gains in four basic motor capacities:

- *Flexibility.* Compared with preschoolers, school-age children are physically more pliable and elastic, a difference that can be seen as children swing a bat, kick a ball, jump over a hurdle, or execute tumbling routines.
- *Balance.* Improved balance supports advances in many athletic skills, including running, hopping, skipping, throwing, kicking, and the rapid changes of direction required in many team sports.
- *Agility.* Quicker and more accurate movements are evident in the fancy footwork of jump rope and hopscotch, as well as in the forward, backward, and sideways motions older children use as they dodge opponents in tag and soccer.
- *Force.* Older youngsters can throw and kick a ball harder and propel themselves farther off the ground when running and jumping than they could at earlier ages (Cratty, 1986).

Although body growth contributes greatly to improved motor performance, more efficient information processing also plays an important role. Younger children often have difficulty with skills that require immediate responding, such as batting and dribbling. Steady gains in reaction time occur, with 11-year-olds responding twice as quickly as 5-year-olds. And the capacity to react only to relevant information increases (Band et al., 2000; Kail, 1993). Because 6- and 7-year-olds are seldom successful at batting a thrown ball, T-ball is more appropriate for them than baseball. And handball, four-square, and kickball should precede instruction in tennis, basketball, and football (Seefeldt, 1996).

Fine Motor Development

Fine motor development also improves over the school years. On rainy afternoons, Joey and Lizzie experimented with yo-yos, built model airplanes, and wove pot holders on small looms. Like many children, they took up musical instruments, which demand considerable fine motor control.

By age 6, most children can print the alphabet, their first and last names, and the numbers from 1 to 10 with reasonable clarity. However, their writing tends to be quite large because they use the entire arm to make strokes, rather than just the wrist and fingers. Children usually master uppercase letters first because their horizontal and vertical motions are easier to control than the small curves of the lowercase alphabet. Legibility of writing gradually increases as children produce more accurate letters with uniform height and spacing. These improvements prepare children for mastering cursive writing by third grade.

Children's drawings show dramatic gains in middle childhood. By the end of the preschool years, children can accurately copy many two-dimensional shapes, and they integrate these into their drawings. Some depth cues have also begun to appear, such as making distant objects smaller than near ones

■ Body size is sometimes the result of evolutionary adaptations to a particular climate. These boys of the Sudan, who live on the hot African plains, have long, lean physiques, which permit the body to cool easily.

■ **FIGURE 9.4 Increase in organization, detail, and depth cues in school-age children's drawings.** Compare both drawings to the one by a 6-year-old on page 214. In the drawing on the left, an 8-year-old depicts her family—father, mother, and three children. Notice how all parts are depicted in relation to one another, and the human figures are given much more detail. (The artist is your author, as a third grader. In the drawing, Laura can be found between her older sister and younger brother.) Integration of depth cues increases dramatically over the school years, as shown in the drawing on the right, by a 10-year-old artist from Singapore. Here, depth is indicated by overlapping objects, diagonal placement, and converging lines, as well as by making distant objects smaller than near ones.

(Braine et al., 1993). Around 9 to 10 years, the third dimension is clearly evident through overlapping objects, diagonal placement, and converging lines. Furthermore, as Figure 9.4 shows, school-age children not only depict objects in considerable detail but also relate them to one another as part of an organized whole (Case, 1998; Case & Okamoto, 1996).

Sex Differences

Sex differences in motor skills that appeared during the preschool years extend into middle childhood and, in some instances, become more pronounced. Girls remain ahead in the fine-motor area, including handwriting and drawing. They also continue to have an edge in skipping, jumping, and hopping, which depend on balance and agility. But boys outperform girls on all other gross motor skills, and in throwing and kicking, the gender gap is large (Cratty, 1986).

School-age boys' genetic advantage in muscle mass is not large enough to account for their gross motor superiority. Instead, environment plays a larger role. In a study of more than 800 elementary school students, children of both sexes viewed sports in a gender-stereotyped fashion—as more important for boys. And boys more often stated that it was vital to their parents that they participate in athletics. These attitudes affected children's self-confidence and behavior. Girls saw themselves as having less talent at sports, and by sixth grade they devoted less time to athletics than did their male classmates (Eccles & Harold, 1991; Eccles, Jacobs, & Harold, 1990). At the same time, girls and older school-age children regard boys' advantage in sports as unjust. They indicate, for example, that coaches should spend equal time with children of each sex and that female sports should command just as much media attention as male sports (Solomon & Bredemeier, 1999).

These findings indicate that extra measures must be taken to increase girls' participation, self-confidence, and sense of fair treatment in athletics. Educating parents about the minimal differences in school-age boys' and girls' physical capacities and sensitizing them to unfair biases against girls' athletic ability may prove helpful. In addition, greater emphasis on skill training for girls, along with increased attention to their athletic achievements, is likely to increase involvement. Middle childhood is a crucial time to take these steps because during the school years children start to discover what they are good at and make some definite skill commitments.

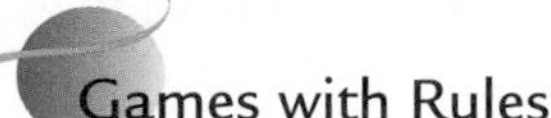

Games with Rules

The physical activities of school-age children reflect an important advance in the quality of their play: Games with rules become common. In cultures around the world, children engage in an enormous variety of informally organized games. Some are variants on popular sports, such as soccer, baseball, and basketball. Others are well-known childhood games, such as tag, jacks, and hopscotch. Children have also invented hun-

■ Is this Little League coach careful to encourage rather than criticize? To what extent does he emphasize teamwork, fair play, courtesy, and skill development over winning? These factors determine whether or not adult-organized sports are pleasurable, constructive experiences for children.

dreds of additional games, including red rover, statues, leapfrog, kick the can, and prisoner's base (Kirchner, 2000).

Gains in perspective-taking—in particular, children's increased ability to understand the roles of several players in a game—permit this transition to rule-oriented games. These play experiences contribute greatly to emotional and social development. Child-invented games usually rely on simple physical skills and a sizable element of luck. As a result, they rarely become contests of individual ability. Instead, they permit children to try out different styles of cooperating, competing, winning, and losing with little personal risk. Also, in their efforts to organize a game, children discover why rules are necessary and which ones work well. In fact, children often spend as much time working out the details of how a game should proceed as they spend playing the game itself! As we will see in Chapter 10, these experiences help children form more mature concepts of fairness and justice.

Today television, video games, and adult-organized sports (such as Little League Baseball and city soccer and hockey leagues) fill many hours that children used to devote to spontaneous play. Some researchers worry that adult-structured athletic activities, which mirror professional sports, are endangering children's development. So far, research indicates that for most children, these experiences do not result in psychological damage (Smoll & Smith, 1996).

But the arguments of critics are valid in certain cases. Children who join teams so early that the skills demanded are beyond their abilities soon lose interest (Bailey & Rasmussen, 1996). And coaches who criticize rather than encourage and who react angrily to defeat prompt intense anxiety in some children. Under these circumstances, weaker performers generally experience social ostracism (Strayer, Tofler, & Lapchick, 1998). Similarly, high parental pressure sets the stage for emotional difficulties and early athletic dropout, not elite performance (Marsh & Daigneault, 1999; Tofler, Knapp, & Drell, 1998). The Caregiving Concerns table on page 284 summarizes the pros and cons of adult-organized youth athletic leagues.

Physical Education

Physical activity supports many aspects of children's development—the health of their bodies, their sense of self-worth as active and capable beings, and the cognitive and social skills necessary for getting along with others. Yet physical education is not taught often enough in schools. The average American school-age child gets only 20 minutes of physical education a week. Canadian children fare better, at about 2 hours per week (Canadian Fitness & Lifestyle Research Institute, 2002; U.S. Department of Health and Human Services, 2000b). But in both nations, physical inactivity is pervasive. Among North American 5- to 17-year-olds, only about 40 percent of girls and 50 percent of boys are active enough for good health—that is, engage in at least 30 minute of vigorous aerobic activity and 1 hour of walking per day.

Besides offering more frequent physical education classes, many experts believe that schools should change the content of physical education programs. Training in competitive sports is often a high priority, but it is unlikely to reach the least physically fit youngsters, who draw back when an activity demands a high level of skill (Portman, 1995). Instead, programs should emphasize informal games that most children can perform well and individual exercise—walking, running, jumping, tumbling, and climbing. Furthermore, children of varying skill levels tend to sustain physical activity when teachers focus on each child's personal progress and contribution to team accomplishment (Whitehead & Corbin, 1997). Then physical education fosters a healthy sense of self while satisfying school-age children's need for relatedness.

■ Physical activity supports many aspects of children's development—the health of their bodies, their sense of self-worth as active and capable beings, and the cognitive and social skills necessary for getting along with others. These Mexican first graders follow their teacher's lead in an exercise routine during physical education class.

Pros and Cons of Adult-Organized Sports in Middle Childhood

PROS	CONS	RECOMMENDATIONS FOR COACHES AND PARENTS
Adult-structured athletics prepares children for realistic competition—the kind they may face as adults. Regularly scheduled games and practices ensure that children get plenty of exercise and fill free time that might otherwise be devoted to less constructive pursuits. Children get instruction in physical skills necessary for future success in athletics. Parents and children share an activity that both enjoy.	Adult involvement leads games to become overly competitive, placing too much pressure on children. When adults control the game, children learn little about leadership, followership, and fair play. When adults assign children to specific roles (such as catcher, first base), children lose the opportunity to experiment with rules and strategies. Highly structured, competitive sports are less fun than child-organized games; they resemble "work" more than "play."	Permit children to select from among appropriate activities the ones that suit them best. Do not push children into sports they do not enjoy. For children younger than age 9, emphasize basic skills, such as kicking, throwing, and batting, and simplified games that grant all participants adequate playing time. Permit children to progress at their own pace and to play for the fun of it, whether or not they become expert athletes. Adjust practice time to children's attention spans and need for unstructured time with peers, with family, and for homework. Two practices a week, each no longer than 30 minutes for younger school-age children and 60 minutes for older school-age children, are sufficient. Emphasize effort, skill gains, and teamwork rather than winning. Avoid criticism for errors and defeat, which promote anxiety and avoidance of athletics. Involve children in decisions about team rules. To strengthen desirable responses, reinforce compliance rather than punishing noncompliance.

Sources: Smith & Smoll, 1997; Strayer, Tofler, & Lapchick, 1998.

Physically fit children become more active adults who reap many benefits (Dennison et al., 1998). These include greater physical strength; resistance to many illnesses, from colds and flu to cancer, diabetes, and heart disease; enhanced psychological well-being; and a longer life.

Ask Yourself

REVIEW
What aspects of physical growth account for the long-legged appearance of many 8- to 12-year-olds?

REVIEW
Select one of the following health problems of middle childhood: myopia, obesity, asthma, or unintentional injuries. Explain how both genetic and environmental factors contribute to it.

APPLY
Joey complained to his mother that it wasn't fair that his younger sister Lizzie was almost as tall as he was. He worried that he wasn't growing fast enough. How should Rena respond to Joey's concern?

APPLY
On Saturdays, 8-year-old Gina gathers with friends at a city park to play kickball. Besides improved ball skills, what is she learning?

www.

COGNITIVE DEVELOPMENT

Finally!" Lizzie exclaimed the day she entered first grade. "Now I get to go to real school just like Joey!" Rena remembered how 6-year-old Lizzie had walked confidently into her classroom, pencils, crayons, and writing pad in hand, ready for a more disciplined approach to learning than she had experienced in early childhood.

Lizzie entered a whole new world of challenging mental activities. In a single morning, she and her classmates wrote in journals, met in reading groups, worked on addition and subtraction, and sorted leaves gathered on the playground for a science project. As Lizzie and Joey moved through the elementary school grades, they tackled increasingly complex tasks and became more accomplished at reading, writing, math skills, and general knowledge of the world. Cognitive development had prepared them for this new phase.

Piaget's Theory: The Concrete Operational Stage

When Lizzie visited my child development class as a 4-year-old, Piaget's conservation problems easily confused her (see Chapter 7, page 219). For example, she insisted that the amount of water had changed after it had been poured from a tall, narrow container into a short, wide one. At age 8, when Lizzie returned, these tasks were easy. "Of course it's the same," she exclaimed. "The water's shorter but it's also wider. Pour it back," she instructed the college student who was interviewing her. "You'll see, it's the same amount!"

Achievements of the Concrete Operational Stage

Lizzie has entered Piaget's **concrete operational stage,** which spans the years from 7 to 11. During this period, thought is far more logical, flexible, and organized than it was during early childhood.

● **Conservation.** The ability to pass *conservation tasks* provides clear evidence of *operations*—mental actions that obey logical rules. Notice how Lizzie is capable of **decentration,** focusing on several aspects of a problem and relating them, rather than centering on just one. Lizzie also demonstrates **reversibility,** the capacity to think through a series of steps and then mentally reverse direction, returning to the starting point. Recall from Chapter 7 that reversibility is part of every logical operation. It is solidly achieved in middle childhood.

● **Classification.** Between ages 7 and 10, children pass Piaget's *class inclusion problem* (see page 220). This indicates that they are more aware of classification hierarchies and can focus on relations between a general category and two specific categories at the same time—that is, three relations at once (Hodges & French, 1988; Ni, 1998). You can see this in children's play activities. Collections—stamps, coins, baseball cards, rocks, bottle caps, and more—become common in middle childhood. At age 10, Joey spent hours sorting and resorting his large box of baseball cards. At times he grouped them by league and team, at other times by playing position and batting average. He could separate the players into a variety of classes and subclasses and flexibly move back and forth between them.

■ An improved ability to categorize underlies children's interest in collecting objects during middle childhood. These older school-age children sort baseball cards into an elaborate structure of classes and subclasses.

● **Seriation.** The ability to order items along a quantitative dimension, such as length or weight, is called **seriation.** To test for it, Piaget asked children to arrange sticks of different lengths from shortest to longest. Older preschoolers can create the series, but they do so haphazardly. They put the sticks in a row but make many errors. In contrast, 6- to 7-year-olds are guided by an orderly plan. They create the series efficiently by beginning with the smallest stick, then moving to the next largest, and so on, until the ordering is complete.

The concrete operational child can also seriate mentally, an ability called **transitive inference.** In a well-known transitive inference problem, Piaget showed children pairings of differently colored sticks. From observing that Stick A is longer than Stick B and Stick B is longer than Stick C, children must make the mental inference that A is longer than C. Notice how this task, like Piaget's class inclusion task, requires children to integrate three relations at once—in this instance, A–B, B–C, and A–C. About half of 6-year-olds perform well on such problems, and performance improves considerably around age 8 (Andrews & Halford, 1998).

● **Spatial Reasoning.** Piaget found that school-age youngsters have a more accurate understanding of space. Let's take two examples: understanding of directions and maps.

Directions. When asked to name an object to the left or right of another person, 5- and 6-year-olds answer incorrectly; they apply their own frame of reference. Between 7 and 8 years, children start to perform *mental rotations,* in which they align the self's frame to match that of a person in a different orientation. As a result, they can identify left and right for positions they do not occupy (Roberts & Aman, 1993). Around 8 to 10 years, children can give clear, well-organized directions for how to get from one place to another by using a "mental walk" strategy in which they imagine another person's movements along a route. Without special prompting, 6-year-olds focus on the end point without describing exactly how to get there (Gauvain & Rogoff, 1989; Plumert et al., 1994).

Maps. Children's drawings of familiar large-scale spaces, such as their neighborhood or school, also change from early to middle childhood. Preschoolers and young school-age children display *landmarks* on the maps they draw, but their placement is fragmented. When asked to place stickers showing the location of desks and people on a map of their classroom, they perform better. But if the map is rotated relative to the orientation of the classroom, they have difficulty placing the stickers accurately (Liben & Downs, 1993).

During the school years, children's maps become more organized. They draw landmarks along an *organized route of travel*—an attainment that resembles their improved direction giving. By the end of middle childhood, children combine landmarks and routes into an *overall view of a large-scale space.* And they readily draw and read maps when the orientation of the map and the space it represents do not match (Liben, 1999).

■ In tribal and village societies, conservation is often delayed. These Vietnamese sisters gather firewood for their family. Although they have many opportunities to handle quantities, compared with their agemates in Western nations, they may seldom see two identical quantities arranged in different ways.

Limitations of Concrete Operational Thought

Although school-age children are far more capable problem solvers than they were during the preschool years, concrete operational thinking suffers from one important limitation. Children think in an organized, logical fashion only when dealing with concrete information they can perceive directly. Their mental operations work poorly with abstract ideas—ones not apparent in the real world.

Children's solutions to transitive inference problems provide a good illustration. When shown pairs of sticks of unequal length, Lizzie easily figured out that if Stick A is longer than Stick B and Stick B is longer than Stick C, then Stick A is longer than Stick C. But she had great difficulty with a hypothetical version of this task, such as "Susan is taller than Sally and Sally is taller than Mary. Who is the tallest?" Not until age 11 or 12 can children solve this problem easily.

That logical thought is at first tied to immediate situations helps account for a special feature of concrete operational reasoning. Perhaps you noticed that school-age children master Piaget's concrete operational tasks step by step, not all at once. For example, they usually grasp conservation of number before conservation of length, mass, and liquid. Piaget used the term **horizontal décalage** (meaning development within a stage) to describe this gradual mastery of logical concepts. The horizontal décalage is another indication of the concrete operational child's difficulty with abstractions. School-age children do not come up with the general logical principles and then apply them to all relevant situations. Instead, they seem to work out the logic of each problem separately.

Recent Research on Concrete Operational Thought

According to Piaget, brain maturation combined with experience in a rich and varied external world should lead children everywhere to reach the concrete operational stage. Yet recent evidence indicates that specific cultural and school practices have much to do with mastery of Piagetian tasks (Rogoff & Chavajay, 1995). The information-processing approach helps explain the gradual mastery of logical concepts in middle childhood.

● **The Impact of Culture and Schooling.** In tribal and village societies, conservation is often greatly delayed. For example, among the Hausa of Nigeria, who live in small agricultural settlements and rarely send their children to school, even the most basic conservation tasks—number, length, and liquid—are not understood until age 11 or later (Fahrmeier, 1978). This suggests that taking part in relevant everyday activities helps children master conservation and other Piagetian problems (Light & Perrett-Clermont, 1989). Joey and Lizzie, for example, think of fairness in terms of equal distribution—a value emphasized in their culture. They frequently divide materials, such as crayons, Halloween treats, and lemonade, equally among their friends. Because they often see the same quantity arranged in different ways, they grasp conservation early.

The very experience of going to school seems to promote mastery of Piagetian tasks. When children of the same age are tested, those who have been in school longer do better on transitive inference problems (Artman & Cahan, 1993). Opportunities to seriate objects, learn about order relations, and remember the parts of complex problems are probably responsible. Yet certain nonschool, informal experiences can also foster operational thought. Brazilian 6- to 9-year-old street vendors, who seldom attend school, do poorly on Piagetian class inclusion tasks but very well on versions relevant to street vending—for example, "If you have 4 units of mint chewing gum and 2 units of grape chewing gum, is it better to sell me the mint gum or [all] the gum?" (Ceci & Roazzi, 1994).

On the basis of findings like these, some investigators have concluded that the forms of logic required by Piagetian tasks do not emerge spontaneously but are heavily influenced by training, context, and cultural conditions. Does this view remind you of Vygotsky's sociocultural theory, which we discussed in earlier chapters?

● **An Information-Processing View of Concrete Operational Thought.** Piaget's notion of the horizontal décalage raises a familiar question about his theory: Is an abrupt stagewise transition to logical thought the best way to describe cognitive development in middle childhood?

Some *neo-Piagetian theorists* argue that the development of operational thinking can best be understood in terms of gains in information-processing capacity rather than a sudden shift to a new stage. For example, Robbie Case (1992, 1998) proposes that with practice, cognitive schemes demand less attention and become more automatic. This frees up space in *working memory* (see page 153) so children can focus on combining old schemes and generating new ones. For instance, the child confronted with water poured from one container to another recognizes that the height of the liquid changes. As this understanding becomes routine, the child notices that the width of the water changes as well. Soon children coordinate these observations, and they conserve liquid. Then, as this logical idea becomes well practiced, the child transfers it to more demanding situations.

Once the schemes of a Piagetian stage are sufficiently automatic, enough working memory is available to integrate them into an improved representation. As a result, children acquire *central conceptual structures,* networks of concepts and relations that permit them to think more effectively about a wide range of situations (Case, 1996, 1998). The central conceptual structures that emerge from integrating concrete operational schemes are highly efficient, abstract principles, which we will discuss in the context of formal operational thought in Chapter 11.

Case has applied his information-processing view to a wide variety of tasks, including solving arithmetic problems, understanding stories, drawing pictures, and interpreting social situations. In each, preschoolers' schemes focus on only one dimension. In understanding stories, for example, they grasp only a single story line. In drawing pictures, they depict objects separately. By the early school years, they coordinate two dimensions—two story lines in a single plot and drawings that show both the features of objects and their relationships. Around 9 to 11 years, central conceptual structures integrate multiple dimensions. Children tell coherent stories with a main plot and several subplots. And their drawings follow a set of rules for representing perspective and, therefore, include several points of reference, such as near, midway, and far.

According to Case, children show a horizontal décalage for two reasons. First, different forms of the same logical insight, such as the various conservation tasks, vary in their processing demands. Those acquired later require more working-memory resources. Second, children's experiences vary widely. A child who often tells stories but rarely draws pictures displays more advanced central conceptual structures in storytelling. Children who do not show the central conceptual structures expected for their age can usually be trained to attain them. And their improved understanding readily transfers to academic tasks (Case, Griffin, & Kelly, 2001). Consequently, the application of Case's neo-Piagetian theory to teaching is helping children who are behind in academic performance catch up and learn more effectively.

Evaluation of the Concrete Operational Stage

Piaget was correct that school-age youngsters approach a great many problems in systematic and rational ways not possible in early childhood. But whether this difference occurs because of *continuous* improvement in logical skills or *discontinuous* restructuring of children's thinking (as Piaget's stage idea assumes) is an issue that prompts much disagreement. Many researchers think that both types of change may be involved (Carey, 1999; Case, 1998; Fischer & Bidell, 1998). From early to middle childhood, children apply logical schemes to many more tasks. Yet in the process, their thought seems to undergo qualitative change—toward a comprehensive grasp of the underlying principles of logical thought. Piaget himself seems to have recognized this possibility in the very concept of the horizontal décalage. So perhaps some blend of Piagetian and information-processing ideas holds the greatest promise for understanding cognitive development in middle childhood.

Ask Yourself

REVIEW

Mastery of conservation provides one illustration of Piaget's horizontal décalage. Review the preceding sections and list additional examples that show that operational reasoning develops gradually.

REVIEW

Cite evidence that specific experiences influence children's mastery of concrete operational tasks.

CONNECT

Explain how advances in perspective taking contribute to school-age children's improved capacity to give directions and draw and use maps.

www.

Information Processing

In contrast to Piaget's focus on overall cognitive change, the information-processing perspective examines separate aspects of thinking. Attention and memory, which underlie every act of cognition, are central concerns in middle childhood, just as they were during infancy and the preschool years. Also, increased understanding of how

Biology & Environment

Children with Attention-Deficit Hyperactivity Disorder

While the other fifth graders worked quietly at their desks, Calvin squirmed in his seat, dropped his pencil, looked out the window, fiddled with his shoelaces, and talked out. "Hey Joey," he yelled over the top of several desks, "wanna play ball after school?" Joey and the other children weren't eager to play with Calvin. Out on the playground, Calvin was a poor listener and failed to follow the rules of the game. He had trouble taking turns at bat. In the outfield, he tossed his mitt up in the air and looked elsewhere when the ball came his way. Calvin's desk at school and his room at home were a chaotic mess. He often lost pencils, books, and other materials necessary for completing assignments.

Symptoms of ADHD. Calvin is one of 3 to 5 percent of school-age children with **attention-deficit hyperactivity disorder (ADHD),** a disorder involving inattention, impulsivity, and excessive motor activity that results in academic and social problems (American Psychiatric Association, 1994). Boys are diagnosed three to nine times more often than girls. However, many girls with ADHD may be overlooked because their symptoms are usually not as flagrant (Gaub & Carlson, 1997).

Children with ADHD cannot stay focused on a task for more than a few minutes. In addition, they often act impulsively, ignoring social rules and lashing out with hostility when frustrated. Many (but not all) are *hyperactive.* They charge through their days with excessive motor activity, exhausting parents and teachers and so irritating other children that they are quickly rejected by their classmates. ADHD youngsters have few friends; they are soundly rejected by their classmates. For a child to be diagnosed with ADHD, these symptoms must have appeared before age 7 as a persistent problem.

The intelligence of children with ADHD is normal, and they show no signs of serious emotional disturbance. According to one view that has amassed substantial research support, a common theme unifies ADHD symptoms: an impairment in inhibition, which makes it hard to delay action in favor of thought (Barkley, 1997, 1999). Consequently, such children do poorly on tasks requiring sustained attention, find it hard to ignore irrelevant information, and have difficulty with memory, planning, reasoning, and problem solving (Denckla, 1996).

Origins of ADHD. Heredity plays a major role in ADHD. The disorder runs in families, and identical twins share it more often than fraternal twins (Sherman, Iacono, & McGue, 1997). Children with ADHD also show abnormal brain functioning, including reduced electrical and blood-flow activity in the frontal lobes of the cerebral cortex and in other areas responsible for attention and inhibition of behavior (Giedd et al., 2001; Rapport & Chung, 2000). Several genes that affect neural communication have been implicated in the disorder (Biederman & Spencer, 2000; Quist & Kennedy, 2001).

At the same time, ADHD is associated with environmental factors. These children are more likely to come from homes in which marriages are unhappy

school-age children process information is being applied to their academic learning—in particular, to reading and mathematics.

Researchers believe that brain development contributes to two basic changes in information processing that facilitate the diverse aspect of thinking we are about to consider:

- *An increase in information-processing capacity.* Time needed to process information on a wide variety of cognitive tasks declines rapidly between ages 6 and 12 (Kail & Park, 1992, 1994). This suggests a biologically based gain in speed of thinking, possibly due to myelination and synaptic pruning in the brain (Kail, 2000). More efficient thinking increases information-processing capacity, since a faster thinker can hold on to and operate on more information at once.
- *Gains in cognitive inhibition.* **Cognitive inhibition**—the ability to control internal and external distracting stimuli—improves from infancy on. But great strides occur during middle childhood. Gains in cognitive inhibition are believed to be due to further development of the frontal lobes of the cerebral cortex (Dempster & Corkill, 1999). Individuals skilled at cognitive inhibition can prevent their minds from straying to irrelevant thoughts, a capacity that supports many information-processing skills.

Besides brain development, strategy use contributes to more effective information processing. As we will see, school-age children think far more strategically than preschoolers.

Attention

During middle childhood, attention changes in three ways. It becomes more selective, adaptable, and planful. First, children become better at deliberately attending to just those aspects of a situation that are relevant to their goals, ignoring other information. Researchers study this increasing selectivity of attention by introducing irrelevant stimuli into a task and seeing how well children attend to its central elements. Performance improves sharply between 6 and 9 years of age (Lin, Hsiao, & Chen, 1999; Smith et al., 1998).

and family stress is high (Bernier & Siegel, 1994). But a stressful home life rarely causes ADHD. Instead, the behaviors of these children can contribute to family problems, which intensify the child's preexisting difficulties. Furthermore, prenatal teratogens (particularly those involving long-term exposure, such as illegal drugs, alcohol, and cigarettes) are linked to inattention and hyperactivity (Milbeger et al., 1997).

Treating ADHD. Calvin's doctor eventually prescribed stimulant medication, the most common treatment for ADHD. As long as dosage is carefully regulated, these drugs reduce activity level and improve attention, academic performance, and peer relations for about 70 percent of children who take them (Greenhill, Halperin, & Abikoff, 1999). Stimulant medication seems to increase activity in the frontal lobes, thereby improving the child's capacity to sustain attention and to inhibit off-task and self-stimulating behavior.

Although stimulant medication is relatively safe, its impact is short-term. Drugs cannot teach children how to compensate for inattention and impulsivity. Combining medication with interventions that model and reinforce appropriate academic and social behavior seems to be the most effective treatment approach (Pelham, Wheeler, & Chronis, 1998). Family intervention is also important. Inattentive, overactive children strain the patience of parents, who are likely to react punitively and inconsistently in return—a child-rearing style that strengthens inappropriate behavior. Breaking this cycle is as important for children with ADHD as it is for the defiant, aggressive youngsters discussed in Chapter 8. In fact, at least 35 percent of the time, these two sets of behavior problems occur together (Lahey & Loeber, 1997).

Because ADHD can be a lifelong disorder, it often requires long-term therapy. Adults with ADHD need help structuring their environments, regulating negative emotion, selecting appropriate careers, and understanding their condition as a biological deficit rather than a character flaw.

© MICHAEL NEWMAN/PHOTOEDIT

■ This boy frequently engages in disruptive behavior, disturbing his classmates while they try to work. Children with ADHD have great difficulty staying on task and often act impulsively.

Second, older children flexibly adapt their attention to momentary requirements of situations. For example, when studying for a spelling test, 10-year-old Joey devoted most attention to words he knew least well. Lizzie was much less likely do so (Masur, McIntyre, & Flavell, 1973).

Finally, *planning* improves greatly in middle childhood (Scholnick, 1995). School-age children scan detailed pictures and written materials for similarities and differences more thoroughly than preschoolers. And on tasks with many parts, they make decisions about what to do first and what to do next in an orderly fashion. In one study, 5- to 9-year-olds were given lists of 25 items to obtain from a play grocery store. Before starting on a shopping trip, older children more often took time to scan the store, and they also followed shorter routes through the aisles (Szepkouski, Gauvain, & Carberry, 1994).

Children learn much about planning by collaborating on tasks with more expert planners. The demands of school tasks—and teachers' explanations of how to plan—also contribute to gains in planning. And parents can foster planning by encouraging it in everyday activities, from completing homework to loading dishes into the dishwasher. In one study of family interactions, discussions involving planning at ages 4 and 9 predicted adolescents' initiations of planning (Gauvain & Huard, 1999).

The selective, adaptable, and planful strategies just considered are crucial for success in school. Unfortunately, some children have great difficulty paying attention. See the Biology and Environment box above for a discussion of the serious learning and behavior problems of children with attention-deficit hyperactivity disorder.

Memory Strategies

As attention improves, so do *memory strategies,* the deliberate mental activities we use to store and retain information. When Lizzie had a list of things to learn, such as a phone number or the state capitals of the United States, she immediately used **rehearsal,** repeating the information. This memory strategy first appears in the early grade school years. Soon after, a second strategy becomes common: **organization**—grouping related items together (for example, all state capitals in the same part of the country), an approach that improves recall dramatically (Gathercole, 1998).

Memory strategies require time and effort to perfect. For example, 8-year-old Lizzie rehearsed in a piecemeal fashion. After being given the word *cat* in a list of items, she said, "Cat, cat, cat." In contrast, 10-year-old Joey combined previous words with each new item, saying, "Desk, man, yard, cat, cat" (Kunzinger, 1985). Joey also organized more skillfully, grouping items into fewer categories. In addition, he used organization in a wide range of memory tasks, whereas Lizzie used it only when categorical relations among items were obvious (Bjorklund et al., 1994). And Joey often combined several strategies—for example, organizing, stating the category names, and rehearsing (Coyle & Bjorklund, 1997). For all these reasons, Joey retained much more information.

By the end of middle childhood, children start to use **elaboration.** It involves creating a relationship, or shared meaning, between two or more pieces of information that are not members of the same category. For example, suppose the words *fish* and *pipe* are among those you must learn. You might generate a mental image of a fish smoking a pipe. Once children discover this memory technique, they find it so effective that it tends to replace other strategies. Elaboration develops late because it requires considerable mental effort and working-memory capacity. It becomes increasingly common during adolescence and early adulthood (Schneider & Pressley, 1997).

Because the strategies of organization and elaboration combine items into *meaningful chunks,* they permit children to hold on to much more information. As a result, the strategies further expand working memory. In addition, when children link a new item to information they already know, they can *retrieve* it easily by thinking of other items associated with it. As we will see in the next section, this is one reason that memory improves steadily during the school years.

The Knowledge Base and Memory Performance

During middle childhood, the long-term knowledge base grows larger and becomes organized into elaborate, hierarchically structured networks. This rapid growth of knowledge helps children use strategies and remember (Schneider, 1993). In other words, knowing more about a topic makes new information more meaningful and familiar so it is easier to store and retrieve.

To test this idea, researchers classified fourth graders as experts or novices in soccer knowledge. Then they gave both groups lists of soccer and nonsoccer items to learn. Experts remembered far more items on the soccer list (but not on the nonsoccer list) than nonexperts. And during recall, experts' listing of items was better organized, as indicated by clustering of items into categories (Schneider & Bjorklund, 1992). These findings suggest that highly knowledgeable children organize information in their area of expertise with little or no effort. Consequently, experts can devote more working-memory resources to using recalled information to reason and solve problems (Bjorklund & Douglas, 1997).

But knowledge is not the only important factor in children's strategic memory processing. Children who are expert in an area are usually highly motivated. As a result, they not only acquire knowledge more quickly but also *actively use what they know* to add more. In contrast, academically unsuccessful children fail to ask how previously stored information can clarify new material. This, in turn, interferes with the development of a broad knowledge base (Schneider & Bjorklund, 1998). So by the end of the school years, extensive knowledge and use of memory strategies are intimately related and support one another.

Culture, Schooling, and Memory Strategies

Think about situations in which the strategies of rehearsal, organization, and elaboration are useful. People usually employ these techniques when they need to remember information for its own sake. On many other occasions, they participate in daily activities that produce excellent memory as a natural by-product of the activity itself.

A repeated finding is that people in non-Western cultures who have no formal schooling do not use or benefit from instruction in memory strategies (Rogoff & Chavajay, 1995). Tasks that require children to recall isolated bits of information are common in classrooms, and they provide children with a great deal of motivation to use memory strategies. In fact, Western children get so much practice with this type of learning that they do not refine other techniques for remembering that rely on spatial location and arrangement of objects—cues

■ Among the Inupiaq people of northwestern Alaska, who still practice subsistence hunting and fishing, elders teach community responsibility and respect for the environment to children. As this Inupiaq girl assists her grandmother in the intricate art of weaving a fishing net, she demonstrates keen memory for information embedded in meaningful contexts. Yet on a list memory task of the kind often given in school, her performance may appear less sharp.

that are readily available in everyday life. Australian Aboriginal and Guatemalan Mayan children are considerably better at these memory skills (Kearins, 1981; Rogoff, 1986). Looked at in this way, the development of memory strategies is not just a matter of a more competent information-processing system. It is also a product of task demands and cultural circumstances.

The School-Age Child's Theory of Mind

During middle childhood, children's *theory of mind,* or set of ideas about mental activities, becomes more elaborate and refined. Recall from Chapter 7 that this awareness of thought is often called *metacognition.* School-age children's improved ability to reflect on their own mental life is another reason for their advances in thinking and problem solving.

Unlike preschoolers, who view the mind as a passive container of information, older children regard it as an active, constructive agent capable of selecting and transforming information (Kuhn, 2000). Consequently, they have a much better understanding of the process of thinking and the impact of psychological factors on performance. They know, for example, that mental inferences can be a source of knowledge and that doing well on a task depends on focusing attention—concentrating on it, wanting to do it, and not being tempted by anything else (Carpendale & Chandler, 1996; Miller & Bigi, 1979). They are also far more conscious of memory strategies, including which ones are likely to work best (Justice et al., 1997). Furthermore, they grasp the relationship between certain mental activities—for example, that remembering is crucial for understanding and that understanding strengthens memory (Schwanenflugel, Henderson, & Fabricius, 1998).

What promotes this reflective, process-oriented view of the mind? Perhaps children become aware of mental activities through quiet-time observation of their own thinking (Wellman & Hickling, 1994). Schooling may contribute as well. Asking children to keep their minds on what they are doing and to remember mental steps calls attention to the workings of the mind. And as children read, write, and solve math problems, they often use private speech, first out loud and then silently in their heads. As they "hear themselves think," they probably detect many aspects of mental life (Flavell, Green, & Flavell, 1995).

Once children are aware of the many factors that influence mental activity, they combine these into an integrated understanding. School-age children take account of *interactions* among variables—how age and motivation of the learner, effective use of strategies, and nature and difficulty of the task work together to affect cognitive performance (Wellman, 1990). In this way, metacognition truly becomes a comprehensive theory.

Cognitive Self-Regulation

Although metacognition expands, school-age children often have difficulty putting what they know about thinking into action. They are not yet good at **cognitive self-regulation,** the process of continuously monitoring progress toward a goal, checking outcomes, and redirecting unsuccessful efforts. For example, Lizzie knows she should group items when memorizing and that she should reread a complicated paragraph to make sure she understands. But she does not always do these things when working on an assignment.

To study cognitive self-regulation, researchers sometimes look at the impact that children's awareness of memory strategies has on how well they remember. By second grade, the more children know about memory strategies, the more they recall—a relationship that strengthens over middle childhood (Pierce & Lange, 2000).

Why does cognitive self-regulation develop gradually? Monitoring learning outcomes is cognitively demanding, requiring constant evaluation of effort and progress. By adolescence, self-regulation is a strong predictor of academic success (Joyner & Kurtz-Costes, 1997). Students who do well in school know when their learning is going well and when it is not. If they run up against obstacles, such as poor study conditions, a confusing text passage, or an unclear class presentation, they take steps to organize the learning environment, review the material, or seek other sources of support. This active, purposeful approach contrasts sharply with the passive orientation of students who do poorly (Zimmerman & Risemberg, 1997).

Parents and teachers can foster self-regulation by pointing out the special demands of tasks, suggesting effective strategies, and emphasizing the value of self-correction—practices that have a substantial effect on children's learning. In addition, explaining why strategies are effective encourages children to use them in new situations (Pressley, 1995).

Children who acquire effective self-regulatory skills develop confidence in their own ability—a belief that supports the use of self-regulation in the future (Zimmerman, 2002). Unfortunately, some children receive messages from parents and teachers that seriously undermine their academic self-esteem and self-regulatory skills. We will consider these *learned-helpless* youngsters, along with ways to help them, in Chapter 10.

Applications of Information Processing to Academic Learning

Fundamental discoveries about the development of information processing have been applied to children's learning of reading and mathematics. Researchers are identifying the cognitive ingredients of skilled performance, tracing their development, and distinguishing good from poor learners by pinpointing differences in cognitive skills. They hope, as a result, to design teaching methods that will help school-age children master these essential skills.

● **Reading.** While reading, we use a large number of skills at once, taxing all aspects of our information-processing systems. Joey and Lizzie must perceive single letters and letter combinations, translate them into speech sounds, hold chunks of text in working memory while interpreting their meaning, and combine the meanings of various parts of a text passage into an understandable whole. In fact, reading is so demanding that most or all of these skills must be done automatically (Perfetti, 1988). If one or more are poorly developed, they compete for space in our limited working memories, and reading performance declines.

Researchers do not yet know how children manage to acquire and combine all these varied skills into fluent reading. Consequently, psychologists and educators have been engaged in "great debate" about how to teach beginning reading. On one side are those who take a **whole-language approach.** They argue that reading should be taught in a way that parallels natural language learning. From the very beginning, children should be exposed to text in its complete form—stories, poems, letters, posters, and lists—so they can appreciate the communicative function of written language. According to these experts, as long as reading is kept whole and meaningful, children will be motivated to discover the specific skills they need (Goodman, 1986; Watson, 1989). On the other side of the debate are those who advocate a **basic-skills approach.** According to this view, children should be given simplified reading materials. At first, they should be coached on *phonics*—the basic rules for translating written symbols into sounds. Only later, after they have mastered these skills, should they get complex reading material (Rayner & Pollatsek, 1989).

Currently most experts believe that children learn best when they receive a mixture of both approaches. Kindergartners benefit from an emphasis on whole language, with gradual introduction of phonics (Jeynes & Littell, 2000). In first grade, teaching that includes phonics boosts reading achievement scores, especially for children from low-SES backgrounds at risk for reading difficulties (Rayner et al., 2001). And when teachers combine real reading and writing with teaching of basic skills and engage in other excellent teaching practices—encouraging children to tackle reading challenges and integrating reading into all school subjects—first graders show far greater literacy progress (Pressley et al., 2001).

Why might combining phonics with whole language make sense? Learning the basics—relationships between letters and sounds—enables children to decipher words they have never seen before. As this process becomes automatic, it releases working memory for the higher-level activities involved in comprehending the text's meaning. Yet if basic skills are overemphasized, children may lose sight of the goal of reading—understanding. Many teachers report cases of children who read aloud fluently but register little meaning. Such children have little knowledge of effective reading strategies—for example, that they must read more carefully if they will be tested on a passage. And they do not monitor their reading comprehension. Providing instruction aimed at increasing knowledge and use of reading strategies enhances reading performance from third grade on (Dickson et al., 1998).

● **Mathematics.** Mathematics teaching in elementary school builds on and greatly enriches children's informal knowledge of number concepts and counting. Written notation systems and formal computational procedures enhance children's ability to represent number and compute. Over the early elementary school years, children acquire basic math facts through a combination of frequent practice, reasoning about number concepts, and teaching that conveys effective strategies (Alibali, 1999; Canobi, Reeve, & Pattison, 1998). For example, when first graders realize that regardless of the order in which two sets are combined, they yield the same result ($2 + 6 = 8$ and $6 + 2 = 8$), they more often start with the higher digit (6) and count on (7, 8), a strategy that minimizes the work involved. Eventually children retrieve answers automatically and apply this knowledge to more complex problems.

■ Culture and language-based factors contribute to Asian children's skill at mathematics. The abacus supports these Japanese students' understanding of place value. Ones, tens, hundreds, and thousands are each represented by a different column of beads, and calculations are performed by moving the beads to different positions. As children become skilled at using the abacus, they learn to think in ways that facilitate solving complex arithmetic problems.

Arguments about how to teach mathematics resemble those in reading. Drill in computational skills is pitted against "number sense," or understanding. Yet once again, a blend of these two approaches is most beneficial. In learning math facts, poorly performing students spend little time experimenting with strategies but, instead, move quickly toward trying to retrieve answers from memory. Their responses are often wrong because they have not used strategies long enough to test which ones result in rapid, accurate solutions. By trying out strategies, good students grapple with underlying concepts and develop effective solution techniques (Siegler, 1996). This suggests that encouraging students to apply strategies and making sure they understand why certain strategies work well are vital for solid mastery of basic math.

In Asian countries, students receive a variety of supports for acquiring mathematical knowledge and often excel at math computation and reasoning. For example, use of the metric system helps Asian children grasp place value. The consistent structure of number words in Asian languages ("ten two" for 12, "ten three" for 13) also makes this idea clear (Ho & Fuson, 1998). Furthermore, Asian number words are shorter and more quickly pronounced. Therefore, more digits can be held in working memory at once, increasing the speed of thinking (Geary et al., 1996). Finally, as we will see later in this chapter, in Asian classrooms, much more time is spent exploring underlying math concepts and much less on drill and repetition.

Ask Yourself

REVIEW

Cite evidence that school-age children view the mind as an active, constructive agent.

REVIEW

Why is teaching both basic skills and understanding of concepts and strategies vital for children's solid mastery of reading and mathematics?

APPLY

After viewing a slide show on endangered species, second and fifth graders were asked to remember as many animals as they could. Explain why fifth graders recalled much more than second graders.

APPLY

Lizzie knows that if you have difficulty learning part of a task, you should devote extra attention to that part. But she plays each of her piano pieces from beginning to end instead of practicing the hard parts. Explain Lizzie's failure to apply what she knows.

www.

Individual Differences in Mental Development

During middle childhood, educators rely heavily on intelligence tests for assessing individual differences in mental development. Around age 6, IQ becomes more stable than it was at earlier ages, and it correlates well with academic achievement. Because IQ predicts school performance, it often enters into educational decisions. Do intelligence tests provide an accurate indication of the school-age child's ability to profit from academic instruction? Let's take a close look at this controversial issue.

Defining and Measuring Intelligence

Virtually all intelligence tests provide an overall score (the IQ), which represents *general intelligence* or reasoning ability, along with an array of separate scores measuring specific mental abilities. Intelligence is a collection of many capacities, not all of which are included on currently available tests. Test designers use a complicated statistical technique called *factor analysis* to identify the various abilities that intelligence tests measure. This procedure determines which sets of items on the test correlate strongly with one another. Those that do are assumed to measure a similar ability and, therefore, are designated as a separate factor. See Figure 9.5 on page 294 for items typically included in intelligence tests for children.

The intelligence tests given every so often in classrooms are *group administered tests.* They permit large numbers of students to be tested at once and require very little training of teachers who give them. Group tests are useful for instructional planning and for identifying children who require more extensive evaluation with *individually administered tests.* Unlike group tests, individually administered tests demand that the examiner have considerable training and experience. The examiner not only considers the child's answers but also observes the child's behavior, noting such responses as attentiveness to and interest in the tasks and wariness of the adult. These observations provide insight into whether the test score is accurate or underestimates the child's abilities. Two individual tests—the Stanford-Binet and the Wechsler—are often used to identify highly intelligent children and diagnose those with learning problems.

The *Stanford-Binet Intelligence Scale,* the modern descendant of Alfred Binet's first successful intelligence test, is appropriate for individuals between 2 years of age and adulthood. It measures general intelligence and four intellectual factors: verbal reasoning, quantitative reasoning, spatial reasoning, and short-term memory (Thorndike, Hagen, & Sattler, 1986). Within these factors are 15 subtests that permit a detailed analysis of each child's mental abilities. The verbal and quantitative factors emphasize culturally loaded, fact-oriented information, such as knowledge of vocabulary and sentence comprehension. The spatial factor is believed to be less culturally biased because it requires little in the way of specific information (see the spatial visualization item in Figure 9.5).

The *Wechsler Intelligence Scale for Children (WISC-III)* is the third edition of a widely used test for 6- through 16-year-olds. In addition to general intelligence, it assesses two broad intellectual factors: (1) verbal, which has six subtests, and (2) performance, which has five subtests. Performance items (see examples in Figure 9.5) require the child to arrange materials rather than talk to the examiner (Wechsler, 1991). They provided one of the first means through which non-English-speaking children and children with speech and language disorders could demonstrate their intellectual strengths. The WISC was also the first test to be standardized on children representing the total population of the United States, including

	TYPICAL VERBAL ITEMS
Vocabulary	Tell me what *carpet* means.
General Information	How many ounces make a pound? What day of the week comes right after Thursday?
Verbal Comprehension	Why are police officers needed?
Verbal Analogies	A rock is hard; a pillow is____.
Logical Reasoning	Five girls are sitting side by side on a bench. Jane is in the middle and Betty sits next to her on the right. Alice is beside Betty, and Dale is beside Ellen, who sits next to Jane. Who are sitting on the ends?
Number Series	Which number comes next in the series? **4 8 6 12 ___**

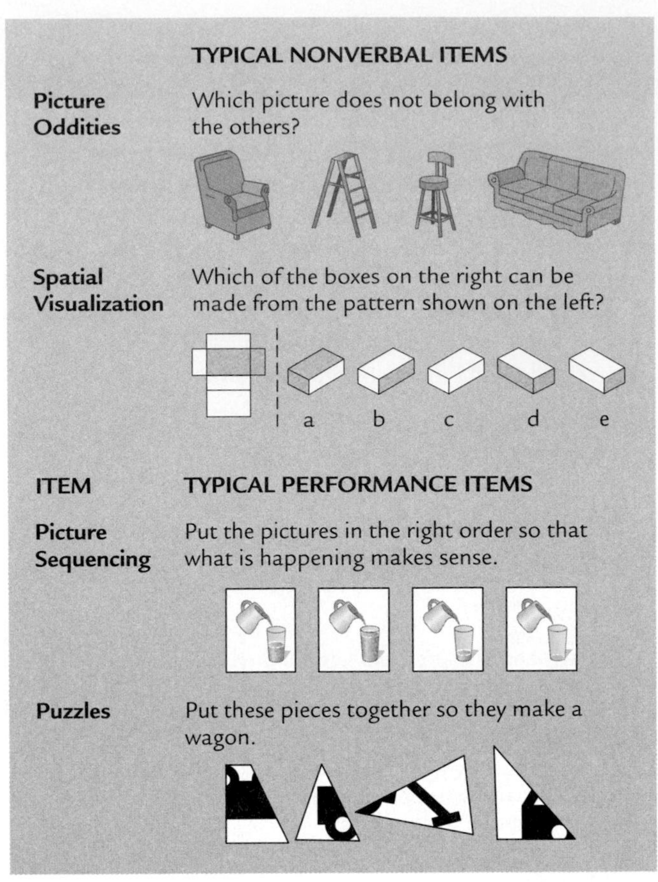

FIGURE 9.5 Test items like those on common intelligence tests for children. In contrast to verbal items, nonverbal items do not require reading or direct use of language. Performance items are also nonverbal, but they require the child to draw or construct something rather than merely give a correct answer. As a result, they appear only on individually administered intelligence tests. (Logical reasoning, picture oddities, and spatial visualization examples are adapted with permission of The Free Press, a Division of Simon & Shuster Adult Publishing Group and Thomson Publishing Services, from *Bias in Mental Testing* by Arthur R. Jensen. Copyright © 1980 by Arthur R. Jensen.)

ethnic minorities. It has been adapted for and standardized in Canada, where both English and French versions are available (Sarrazin, 1999; Wechsler, 1996).

Recent Efforts to Define Intelligence

Some researchers have combined the factor-analytic approach to defining intelligence with information processing. They believe that factors on intelligence tests have limited use unless we can identify the cognitive processes responsible for those factors. Once we understand the underlying basis of IQ, we will know much more about why a particular child does well or poorly and what capacities must be worked on to improve performance. These researchers conduct *componential analyses* of children's mental test scores. This means that they look for relationships between aspects (or components) of information processing and children's IQs.

Many studies reveal that speed of processing is related to IQ (Deary, 2000; Vernon et al., 2001). This suggests that individuals whose nervous systems function more efficiently, permitting them to take in and manipulate information quickly, have an edge in intellectual skills. But strategy use also predicts IQ, and it explains some of the association between response speed and intelligence (Miller & Vernon, 1992). Children who apply strategies effectively acquire more knowledge and can retrieve that knowledge rapidly—advantages that seem to carry over to performance on intelligence tests.

The componential approach has one major shortcoming: It regards intelligence as entirely due to causes within the child. Yet throughout this book, we have seen how cultural and situational factors affect children's cognitive skills. Robert Sternberg has expanded the componential approach into a comprehensive theory that regards intelligence as a product of both inner and outer forces.

● **Sternberg's Triarchic Theory of Intelligence.** As Figure 9.6 shows, Sternberg's (1985, 1997, 1999) **triarchic theory of intelligence** is made up of three interacting subtheories involving information-processing skills, experience with tasks, and contextual factors. The *componential subtheory* spells out the information-processing components that underlie intelligent behavior. You are already familiar with its main elements: strategy application, knowledge acquisition, metacognition, and self-regulation.

We apply these components in situations with which we have varying degrees of experience. The *experiential subtheory*

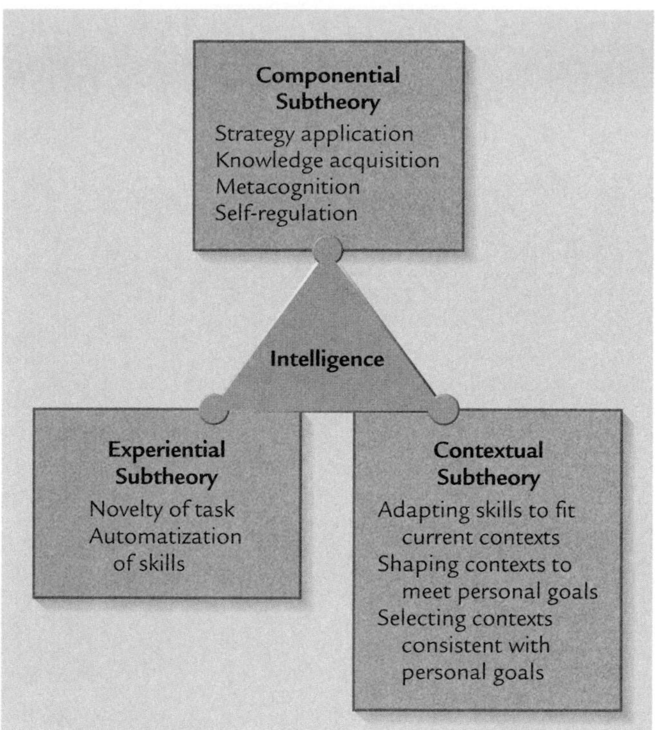

FIGURE 9.6 Sternberg's triarchic theory of intelligence.

states that highly intelligent individuals think more skillfully than others when faced with relatively novel tasks—ones at the edge of their understanding. When given a new task, the bright person learns rapidly, making strategies automatic so working memory is freed for more complex aspects of the situation. Consider the implications of this idea for measuring intelligence. To accurately compare people in brightness—in ability to deal with novelty and learn efficiently—all must be presented with equally unfamiliar test items. Otherwise, some will appear more intelligent because of their past experiences, not because they are more cognitively skilled.

This point brings us to the *contextual subtheory.* Intelligence is a goal-directed activity aimed at one or more of the following purposes: *adapting to an environment, shaping an environment, and selecting an environment.* Intelligent people skillfully *adapt* their thinking to fit with their personal desires and the demands of their everyday worlds. When they cannot adapt to a situation, they try to *shape,* or change, it to meet their needs. And if they cannot shape it, they *select* new contexts consistent with their goals. The contextual subtheory emphasizes that intelligent behavior is never culture-free. Because of their backgrounds, some individuals value behaviors required for success on intelligence tests, and they easily adapt to the tasks and testing conditions. Others with different life histories misinterpret the testing context or reject it because it does not suit their needs.

Sternberg's theory emphasizes the complexity of intelligent behavior. Children often use different abilities in academic tasks than in nonacademic everyday situations. Yet out-of-school, practical forms of intelligence are vital for life success, and they help explain why cultures vary widely in the behaviors they regard as intelligent (Sternberg et al., 2000). When ethnically diverse parents were asked for their view of an intelligent first grader, Caucasian Americans valued cognitive traits. In contrast, ethnic minorities (Cambodian, Filipino, Vietnamese, and Mexican immigrants) saw noncognitive capacities—motivation, self-management, and social skills—as particularly important (Okagaki & Sternberg, 1993). Clearly, Sternberg's ideas are relevant to the controversy surrounding cultural bias in intelligence testing, which we will address shortly.

● **Gardner's Theory of Multiple Intelligences.** In yet another view of how information-processing skills underlie intelligent behavior, Howard Gardner's (1983, 1993, 2000) **theory of multiple intelligences** defines intelligence in terms of distinct sets of processing operations that permit individuals to engage in a wide range of culturally valued activities. Dismissing the idea of general intelligence, Gardner proposes at least eight independent intelligences (see Table 9.1 on page 296).

Gardner believes that each intelligence has a unique biological basis, a distinct course of development, and different expert, or "end-state," performances. At the same time, he emphasizes that a lengthy process of education is required to transform any raw potential into a mature social role (Torff & Gardner, 1999). This means that cultural values and learning opportunities affect the extent to which a child's intellectual strengths are realized and the way they are expressed.

Gardner's list of abilities has yet to be firmly grounded in research. Neurological evidence for the independence of his abilities is weak. Furthermore, some exceptionally gifted individuals have abilities that are broad rather than limited to a particular domain (Goldsmith, 2000). And research with mental

© BOB DAEMMRICH/THE IMAGE WORKS

■ According to Gardner, children are capable of at least eight distinct intelligences. As these children classify wildflowers they collected during a walk through a forest and meadow, they enrich their naturalist intelligence.

Table 9.1 Gardner's Multiple Intelligences

Intelligence	Processing Operations	End-State Performance Possibilities
Linguistic	Sensitivity to the sounds, rhythms, and meaning of words and the functions of language	Poet, journalist
Logico-mathematical	Sensitivity to, and capacity to detect, logical or numerical patterns; ability to handle long chains of logical reasoning	Mathematician
Musical	Ability to produce and appreciate pitch, rhythm (or melody), and aesthetic quality of the forms of musical expressiveness	Instrumentalist, composer
Spatial	Ability to perceive the visual-spatial world accurately, to perform transformations on those perceptions, and to re-create aspects of visual experience in the absence of relevant stimuli	Sculptor, navigator
Bodily-kinesthetic	Ability to use the body skillfully for expressive as well as goal-directed purposes; ability to handle objects skillfully	Dancer, athlete
Naturalist	Ability to recognize and classify all varieties of animals, minerals, and plants	Biologist
Interpersonal	Ability to detect and respond appropriately to the moods, temperaments, motivations, and intentions of others	Therapist, salesperson
Intrapersonal	Ability to discriminate complex inner feelings and to use them to guide one's own behavior; knowledge of one's own strengths, weaknesses, desires, and intelligences	Person with detailed, accurate self-knowledge

Sources: Gardner, 1993, 1998a, 2000.

tests suggests that several of Gardner's intelligences (linguistic, logico-mathematical, and spatial) have common features. Nevertheless, Gardner's theory highlights other mental abilities not measured by intelligence tests. Consequently, his theory has been especially helpful in efforts to nurture children's talents, a topic we will discuss at the end of this chapter.

Explaining Individual and Group Differences in IQ

When we compare individuals in terms of academic achievement, years of education, and the status of their occupations, it quickly becomes clear that certain sectors of the population are advantaged over others. In trying to explain these differences, researchers have compared the IQ scores of ethnic and SES groups. American black children score, on average, 15 IQ points below American white children, although the difference has been shrinking (Hedges & Nowell, 1998; Loehlin, 2000).

The gap between middle-SES and low-SES children is about 9 points (Jensen & Figueroa, 1975). SES accounts for some, but not all, of the black–white IQ difference. When black children and white children are matched on family income, the black–white gap is reduced by a third (Jensen & Reynolds, 1982). Of course, considerable variation exists *within* each ethnic and SES group. Still, these group differences in IQ are large enough and of serious enough consequence that they cannot be ignored.

In the 1970s, the IQ nature–nurture controversy escalated after psychologist Arthur Jensen (1969) published a controversial article entitled, "How Much Can We Boost IQ and Scholastic Achievement?" Jensen's answer to this question was "not much." He argued that heredity is largely responsible for individual, ethnic, and SES variations in intelligence, a position he maintains (Jensen, 1998, 2001). Jensen's work was followed by an outpouring of responses and research studies. The controversy was rekindled in Richard Herrnstein and Charles Murray's (1994) *The Bell Curve.* Like Jensen, these authors concluded that the contribution of heredity to individual and SES differences in IQ is substantial. At the same time, they stated that the relative roles of heredity and environment in the black–white IQ gap remain unresolved. Let's look closely at some important evidence.

● **Nature versus Nurture.** In Chapter 2, we introduced the *heritability estimate.* Recall that heritabilities are obtained from *kinship studies,* which compare family members. The most powerful evidence on the role of heredity in IQ involves twin comparisons. The IQ scores of identical twins (who share all their genes) are more similar than those of fraternal twins (who are genetically no more alike than ordinary siblings). On the basis of this and other kinship evidence, researchers estimate that about half the differences in IQ among children can be traced to their genetic makeup.

However, recall that heritabilities risk overestimating genetic influences and underestimating environmental influences. Although these measures offer convincing evidence that genes contribute to IQ, disagreement persists over just how large the role of heredity really is (Grigorenko, 2000). And heritability estimates do not reveal the complex processes through which genes and experiences influence intelligence as children develop.

Compared with heritabilities, adoption studies offer a wider range of information. In one investigation, children of two extreme groups of biological mothers—those with IQs below 95 and those with IQs above 120—were adopted at birth by parents well above average in income and education. During the school years, children of the low-IQ biological mothers scored above average in IQ, indicating that test performance can be greatly improved by an advantaged home life. At the same time, they do not do as well as children of high-IQ biological mothers placed in similar adoptive families (Loehlin, Horn, & Willerman, 1997). Adoption research confirms that heredity and environment contribute jointly to IQ.

Some intriguing adoption research sheds light on the black–white IQ gap. African-American children placed in economically well-off homes during the first year of life score high on intelligence tests. In two such studies, adopted black children attained mean IQs of 110 and 117 by middle childhood—well above average, and 20 to 30 points higher than the typical scores of children growing up in low-income black communities (Moore, 1986; Scarr & Weinberg, 1983). Although the IQs of black adoptees declined in adolescence, they remained above the IQ average for low-SES African Americans (Weinberg, Scarr, & Waldman, 1992).

Adoption findings do not completely resolve questions about ethnic differences in IQ. Nevertheless, the IQ gains of black children "reared in the culture of the tests and schools" are consistent with a wealth of evidence that poverty severely depresses the intelligence of ethnic minority children (Nisbett, 1998).

■ This American Pueblo grandmother uses a collaborative style of communication with her granddaughters. They work smoothly and efficiently on the same aspect of a task—making bread—until it is complete. Ethnic minority parents with little education often communicate this way. Because their children are not accustomed to the hierarchical style of communication typical of classrooms, they may do poorly on tests and assignments.

● **Cultural Influences.** Jermaine, an African-American child in Lizzie's third-grade class, participated actively in class discussion and wrote complex, imaginative stories. Two years earlier, as a first grader, Jermaine had responded, "I don't know," to the simplest of questions, including "What's your name?" Fortunately, Jermaine's teacher understood his uneasiness. She helped him build a bridge between the learning style fostered by his cultural background and the style necessary for academic success. A growing body of evidence reveals that IQ scores are affected by specific learning experiences, including certain communication styles and knowledge.

Communication Styles. Ethnic minority families often foster unique communication skills that do not fit the expectations of most classrooms and testing situations. Shirley Brice Heath (1982, 1989), an anthropologist who spent many hours observing in low-SES black homes in a southeastern American city, found that black adults asked their children questions unlike those of white middle-SES families. From an early age, white parents ask knowledge-training questions, such as "What color is it?" and "What's this story about?" that resemble the questioning style of tests and classrooms. In contrast, the black parents asked only "real" questions—ones that they themselves could not answer. Often these were analogy questions ("What's that like?") or story-starter questions ("Didja hear Miss Sally this morning?") that called for elaborate responses about personal experiences and had no right answer.

These experiences lead low-SES black children to develop complex verbal skills at home, such as storytelling and exchanging quick-witted remarks. But their language emphasizes social and emotional topics rather than facts about the world (Blake, 1994). Not surprisingly, black children may be confused by the "objective" questions they encounter on tests and in classrooms and may withdraw into silence.

Furthermore, ethnic minority parents without extensive schooling often show a *collaborative style of communication* when completing tasks with children. They work together in a coordinated, fluid way, each focused on the same aspect of the problem. This pattern of adult–child engagement has been observed in Native American, Canadian Inuit, Hispanic, and Guatemalan Mayan cultures (Chavajay & Rogoff, 2002; Crago, Annahatak, & Ningiuruvik, 1993; Delgado-Gaitan, 1994). With increasing education, parents establish a *hierarchical style of communication,* like that of classrooms. The parent directs each child to carry out an aspect of the task, and children work independently. This sharp discontinuity between home and school communication practices may contribute to low-SES minority children's poorer test and school performance (Greenfield, Quiroz, & Raeff, 2000).

Test Content. Many researchers argue that IQ scores are affected by specific information acquired as part of majority-culture upbringing. Unfortunately, attempts to change tests by eliminating fact-oriented verbal tasks and relying only on spatial reasoning and performance items (believed to be less culturally loaded) have not raised the scores of low-SES minority children very much (Reynolds & Kaiser, 1990).

Yet even these nonverbal test items depend on learning opportunities. In one study, children's performance on a spatial reasoning task was related to their experience playing a popular but expensive game that (like the test items) required them to arrange blocks to duplicate a design as quickly as possible (Dirks, 1982). Playing video games also fosters success on spatial test items (Subrahmanyam & Greenfield, 1996). Low-income minority children, who often grow up in more "people-oriented" than "object-oriented" homes, may lack opportunities to use games and objects that promote certain intellectual skills.

Furthermore, the sheer amount of time a child spends in school is a strong predictor of IQ. When children of the same age who are in different grades are compared, those who have been in school longer score higher on intelligence tests (Ceci, 1991, 1999). Taken together, these findings indicate that children's exposure to the factual knowledge and ways of thinking valued in classrooms has a sizable impact on their intelligence test performance.

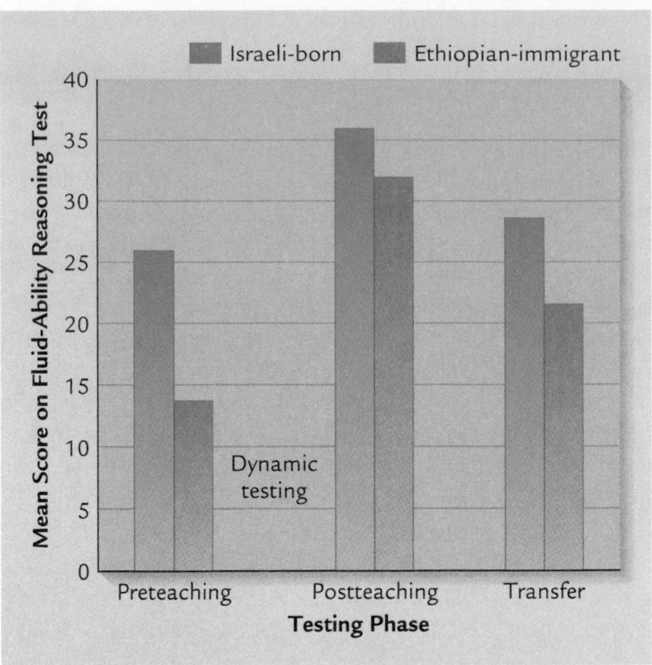

■ **FIGURE 9.7 Influence of dynamic testing on mental test scores of Ethiopian-immigrant and Israeli-born 6- and 7-year-olds.** Each child completed test items in a preteaching phase, a postteaching phase, and a transfer phase, in which they had to generalize their learning to new problems. After dynamic testing, Ethiopian and Israeli children's scores were nearly equal. Ethiopian children also transferred their learning to new test items, performing much better in the transfer than in the preteaching phase. (Adapted from Tzuriel & Kaufman, 1999.)

● **Reducing Cultural Bias in Intelligence Tests.** Although not all experts agree, many acknowledge that IQ scores can underestimate the intelligence of culturally different children. A special concern exists about incorrectly labeling minority children as slow learners and assigning them to remedial classes, which are far less stimulating than regular school experiences. Because of this danger, test scores need to be combined with assessments of adaptive behavior—children's ability to cope with the demands of their everyday environments. The child who does poorly on an IQ test yet plays a complex game on the playground or figures out how to rewire a broken TV is unlikely to be mentally deficient.

In addition, culturally relevant testing procedures enhance minority children's test performance. In one approach, called **dynamic testing,** an innovation consistent with Vygotsky's concept of the zone of proximal development, the adult introduces purposeful teaching into the testing situation to find out what the child can attain with social support. Many minority children perform more competently after adult assistance (Lidz, 2001; Tzuriel, 2001). In one study, Ethiopian 6- and 7-year-olds who had recently immigrated to Israel performed well below their Israeli-born agemates on spatial reasoning items. The Ethiopian children had little experience with this type of thinking. After several dynamic testing sessions in which the adult suggested effective strategies, the Ethiopian children's test scores increased sharply, nearly equaling those of Israeli-born children (see Figure 9.7). They also transferred their learning to new test items (Tzuriel & Kaufman, 1999).

■ Dynamic testing introduces purposeful teaching into the situation to find out what the child can attain with social support. This teacher assists a second grader in writing the alphabet. Many ethnic minority children perform more competently after adult assistance. And the approach helps identify the teaching style to which the child is most responsive.

In sum, intelligence tests are useful when interpreted carefully by examiners who are sensitive to cultural influences on test performance. And despite their limitations, IQ scores continue to be fairly accurate measures of school learning potential for the majority of Western children.

Ask Yourself

REVIEW

Using Sternberg's triarchic theory and Gardner's theory of multiple intelligences, explain the limitations of current mental tests in assessing the complexity of intelligence.

REVIEW

Summarize ethnic differences in IQ, and cite environmental factors that contribute to them.

APPLY

Desiree, an African-American child, was quiet and withdrawn while taking an intelligence test. Later she remarked to her mother, "I can't understand why that lady asked me all those questions. She's a grown-up. She must *know* what a ball and stove are for!" Explain Desiree's reaction. Why is her IQ score likely to underestimate her intelligence?

CONNECT

Explain how dynamic testing is consistent with Vygotsky's zone of proximal development and with scaffolding. (See Chapter 7, pages 223–224.)

www.

Language Development

Vocabulary, grammar, and pragmatics continue to develop in middle childhood, although less obviously than at earlier ages. In addition, school-age children's attitude toward language undergoes a fundamental shift. They develop *language awareness.*

Vocabulary

Because the average 6-year-old's vocabulary is already quite large, parents and teachers usually do not notice rapid gains during the school years. Between the start of elementary school and its completion, vocabulary increases fourfold, eventually reaching about 40,000 words. On average, children learn about 20 new words each day, a rate of growth exceeding that in early childhood. In addition to the word-learning strategies discussed in Chapter 7, school-age children add to their vocabularies by analyzing the structure of complex words. From "happy" and "decide," they quickly derive the meanings of "happiness" and "decision" (Anglin, 1993). Many more words are picked up from context, especially while reading (Nagy & Scott, 2000).

As their knowledge becomes better organized, school-age children think about and use words more precisely. Word definitions offer examples of this change. Five- and 6-year-olds give concrete descriptions that refer to functions or appearance—for example, *knife:* "when you're cutting carrots"; *bicycle:* "it's got wheels, a chain, and handlebars." By the end of elementary school, synonyms and explanations of categorical relationships appear—for example, *knife:* "something you could cut with. A saw is like a knife. It could also be a weapon" (Wehren, De Lisi, & Arnold, 1981). This advance reflects the older child's ability to deal with word meanings on an entirely verbal plane. Older children can add new words to their vocabulary simply by being given a definition.

School-age children's more reflective and analytical approach to language permits them to appreciate the multiple meanings of words. For example, they appreciate that many words, such as "cool" or "neat," have psychological as well as physical meanings: "What a cool shirt!" or "That movie was really neat!" This grasp of double meanings permits 8- to 10-year-olds to comprehend subtle metaphors, such as "sharp as a tack" and "spilling the beans" (Nippold, Taylor, & Baker, 1996; Winner, 1988). It also leads to a change in children's humor. Riddles and puns that go back and forth between different meanings of a key word are common.

Grammar

During the school years, mastery of complex grammatical constructions improves. For example, English-speaking children use the passive voice more frequently, and it expands from an abbreviated structure ("It broke") into full statements ("The glass was broken by Mary") (Horgan, 1978; Pinker, Lebeaux, & Frost, 1987). Although the passive form is challenging, language input makes a difference. When adults speak a language that emphasizes full passives, such as Inukitut (spoken by the Inuit people of Arctic Canada), children produce them sooner (Allen & Crago, 1996).

Another grammatical achievement of middle childhood is advanced understanding of infinitive phrases, such as the difference between "John is eager to please" and "John is easy to please" (Chomsky, 1969). Like gains in vocabulary, appreciation of these subtle grammatical distinctions is supported by an improved ability to analyze and reflect on language.

Pragmatics

Improvements in *pragmatics,* the communicative side of language, also take place. Children adapt to the needs of listeners in challenging communicative situations, such as describing one object among a group of very similar objects. Whereas preschoolers tend to give ambiguous descriptions, such as "the red one," school-age children are much more precise. They might say, "the round red one with stripes on it" (Deutsch & Pechmann, 1982).

Conversational strategies also become more refined. For example, older children are better at phrasing things to get their way. When faced with an adult who refuses to hand over a desired object, 9-year-olds, but not 5-year-olds, state their second requests more politely (Axia & Baroni, 1985). School-age children are also more sensitive than preschoolers to distinctions between what people say and what they mean. Lizzie, for example, knew that when her mother said, "The garbage is beginning to smell," she really meant, "Take that garbage out!" (Ackerman, 1978).

■ These American Pueblo children attend a bilingual education program in which they receive instruction in their native language and in English. In classrooms where both the first and second language are integrated into the curriculum, ethnic minority children are more involved in learning, participate more actively in class discussions, and acquire the second language more easily.

Learning Two Languages at a Time

Joey and Lizzie speak only one language, their native tongue of English. Yet throughout the world, many children grow up *bilingual,* learning two languages, and sometimes more than two, during childhood. An estimated 15 percent of American children—6 million in all—speak a language other than English at home (U.S. Bureau of the Census, 2002c). In Canada, where both English and French are official languages, 17 percent of the population—4.8 million in all—are French–English bilingual (Statistics Canada, 2001e). Most are young people, demonstrating the impact of Canada's school-based bilingual education programs.

● **Bilingual Development.** Children can become bilingual in two ways: (1) by acquiring both languages at the same time in early childhood, or (2) by learning a second language after mastering the first. Children of bilingual parents who teach them both languages in early childhood show no special problems with language development. They acquire normal native ability in the language of their surrounding community and good to native ability in the second language, depending on their exposure to it (Genessee, 2001). When children acquire a second language after they already speak a first language, they generally take 3 to 5 years to become as fluent as native-speaking agemates (Ramirez et al., 1991).

As with first-language development, a *sensitive period* for second-language development exists. Mastery must begin sometime in childhood for full development to occur. In a study of Chinese and Korean adults who had immigrated to the United States at varying ages, those who began mastering English between 3 and 7 years scored as well as native speakers on a test of grammar. As age of arrival in the United States increased, test scores gradually declined (Newport, 1991). But a precise age cutoff for a decline in second-language learning has not been established. Instead, a continuous age-related decrease occurs (Hakuta, Bialystok, & Wiley, 2003).

Research shows that bilingualism has positive consequences for development. Children who are fluent in two languages do better than others on tests of selective attention, analytical reasoning, concept formation, and cognitive flexibility (Bialystok, 1999, 2001). Also, bilingual children are advanced in ability to reflect on language. They are more aware that words are arbitrary symbols, more conscious of language structure and sounds, and better at noticing errors of grammar and meaning—capacities that enhance reading achievement (Bialystok & Herman, 1999; Campbell & Sais, 1995).

● **Bilingual Education.** The advantages of bilingualism provide strong justification for bilingual education programs in schools. In Canada, where both English and French are official languages, about 7 percent of elementary school students are enrolled in *language immersion programs,* in which English-speaking children are taught entirely in French for several years. The Canadian language immersion strategy is successful in developing children who are proficient in both languages (Harley & Jean, 1999; Holobow, Genessee, & Lambert, 1991).

Yet the question of how American ethnic minority children with limited English proficiency should be educated continues to be hotly debated. On one side are those who believe that time spent communicating in the child's native tongue detracts from English language achievement, which is crucial for success in school and work. On the other side are educators committed to truly *bilingual education*—developing minority children's native language while fostering mastery of English. Providing instruction in the native tongue lets minority children know that their heritage is respected. In addition, it prevents *semilingualism,* or inadequate proficiency in both languages. When minority children gradually lose the first language as a result of being taught the second, they end up limited in both languages for a time, a circumstance that leads to serious academic difficulties (Ovando & Collier, 1998). Semilingualism is believed to contribute to high rates of school failure and dropout among low-SES Hispanic youngsters, who make up nearly 50 percent of the American language-minority population.

At present, public opinion sides with the first of these two viewpoints. Many U.S. states have passed laws declaring English to be their official language, creating conditions in which schools have no obligation to teach minority students in languages other than English. Yet in classrooms where both languages are integrated into the curriculum, minority children are more involved in learning, participate more actively in class discussions, and acquire the second language more easily. In contrast, when teachers speak only in a language children can barely understand, minority children display frustration, boredom, and withdrawal (Crawford, 1997).

American supporters of English-only education often point to the success of Canadian language immersion programs, in which classroom lessons are conducted in the second language. Yet for American non-English-speaking minority children, whose native languages are not valued by the larger society, a truly bilingual educational approach seems necessary: one that promotes children's native-language skills while they learn English.

Ask Yourself

REVIEW

Cite examples of language progress that benefit from school-age children's greater language awareness.

REVIEW

Cite ways that bilingual education can contribute to ethnic minority children's cognitive and academic development.

APPLY

Ten-year-old Shana arrived home from soccer practice and commented, "I'm wiped out!" Megan, her 5-year-old sister, looked puzzled and asked, "What did'ya wipe out, Shana?" Explain Shana's and Megan's different understandings of this expression.

www.

Learning in School

Throughout this chapter, we have touched on evidence indicating that schools are vital forces in children's cognitive development. How do schools exert such a powerful influence? Research looking at schools as complex social systems—at their class size, educational philosophies, teacher–student interaction patterns, and larger cultural context—provides important insights into this question. As you read about these topics, refer to the Caregiving Concerns table on page 302, which summarizes characteristics of high-quality education in elementary school.

Class Size

As each school year began, Rena telephoned the principal's office and asked, "How large will Joey's and Lizzie's classes be?" Her concern is well founded. Class size influences children's learning. In a large field experiment, more than 6,000 Tennessee kindergartners were randomly assigned to three class types: small (13 to 17 students), regular (22 to 25 students), and regular with a teacher plus a full-time teacher's aide. These arrangements continued into third grade. Small-class students—especially minority children—scored higher in reading and math achievement and continued to do so after they returned to regular-size classes (Mosteller, 1995). Placing teacher's aides in regular-size classes had no impact. Instead, consistently being in small classes from kindergarten through third grade predicted substantially higher achievement from fourth through ninth grades (see Figure 9.8) (Nye, Hedges, & Konstantopoulos, 2001).

Why is small class size beneficial? Teachers of fewer children spend less time disciplining and more time giving individual

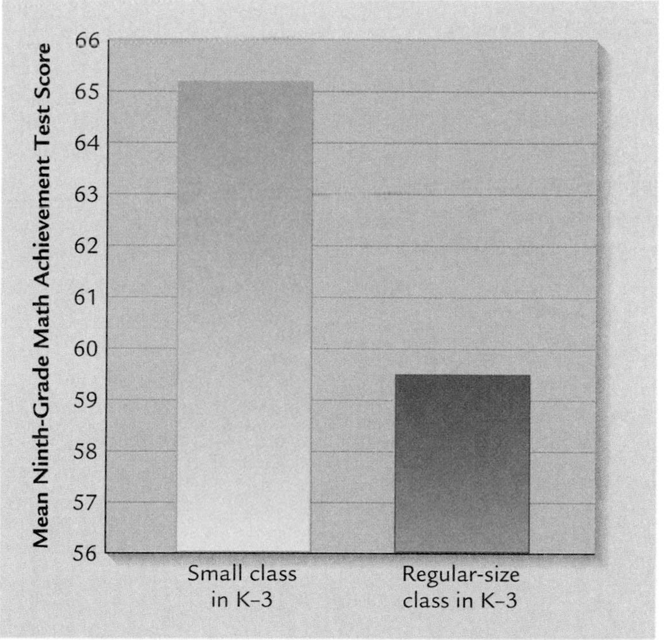

FIGURE 9.8 Impact of consistently being in small classes from kindergarten through third grade on ninth-grade math achievement. In this longitudinal study of thousands of Tennessee students, small class size in the early grades continued to be associated with higher academic achievement 6 years later, in ninth grade. Here, findings for math scores are shown. (Adapted from Nye, Hedges, & Konstantopoulos, 2001.)

Signs of High-Quality Education in Elementary School

CLASSROOM CHARACTERISTICS	SIGNS OF QUALITY
Class Size	Optimum class size is no larger than 18 children.
Physical Setting	Space is divided into richly equipped activity centers—for reading, writing, playing math or language games, exploring science, working on construction projects, using computers, and engaging in other academic pursuits. Spaces are used flexibly for individual and small-group activities and whole-class gatherings.
Curriculum	The curriculum helps children both achieve academic standards and make sense of their learning. Subjects are integrated so children apply knowledge in one area to others. The curriculum is implemented through activities responsive to children's interests, ideas, and everyday lives, including their cultural backgrounds.
Daily Activities	Teachers provide challenging activities that include opportunities for small-group and independent work. Groupings vary in size and makeup of children, depending on the activity and children's learning needs. Teachers encourage cooperative learning and guide children in attaining it.
Interactions Between Teachers and Children	Teachers foster each child's progress and use intellectually engaging strategies, including posing problems, asking thought-provoking questions, discussing ideas, and adding complexity to tasks. They also demonstrate, explain, coach, and assist in other ways, depending on each child's learning needs.
Evaluations of Progress	Teachers regularly evaluate children's progress through written observations and work samples, which they use to individualize teaching. They help children reflect on their work and decide how to improve it. They also seek information and perspectives from parents on how well children are learning and include parents' views in evaluations.
Relationship with Parents	Teachers forge partnerships with parents. They hold periodic conferences and encourage parents to visit the classroom anytime, to observe and volunteer.

Source: Bredekamp & Copple, 1997.

attention, and children's interactions with one another are more positive and cooperative. Also, children who learn in smaller groups show better concentration and higher-quality class participation and express more favorable attitudes toward school (Blatchford et al., 2001).

Educational Philosophies

Each teacher brings to the classroom an educational philosophy that plays a major role in children's learning. Two philosophical approaches have received the most research attention. They differ in what children are taught, the way they are believed to learn, and how their progress is evaluated.

● **Traditional versus Open Classrooms.** In a **traditional classroom,** the teacher is the sole authority for knowledge, rules, and decision making and does most of the talking. Students are relatively passive, listening, responding when called on, and completing teacher-assigned tasks. Their progress is evaluated by how well they keep pace with a uniform set of standards for their grade.

In an **open classroom,** students are viewed as active agents in their own development. The teacher assumes a flexible authority role, sharing decision making with students, who learn at their own pace. Students are evaluated in relation to their own prior development; comparisons with same-age students are less important. A glance inside an open classroom reveals richly equipped learning centers, small groups of students working on tasks they choose themselves, and a teacher responding to individual needs.

During the past few decades, education has swung back and forth between these two views. In the 1960s and early 1970s, open education gained in popularity, inspired by Piaget's vision of the child as an active, motivated learner. As concern over the academic progress of children and youths increased, a "back to basics" movement arose. Classrooms returned to traditional instruction, a style still prevalent today.

Although older children in traditional classrooms have a slight edge in academic achievement, open settings are associated with other benefits—gains in critical thinking, greater valuing of individual differences in classmates, and more positive attitudes toward school (Walberg, 1986). And, as noted in Chapter 7, when teacher-directed instruction is emphasized in preschool and kindergarten, it undermines academic motivation and achievement, especially among low-SES children. The heavy emphasis on knowledge absorption in many kindergarten and primary classrooms has contributed to a

Social Issues

School Readiness and Grade Retention

While waiting to pick up their sons from preschool, Susan and Vicky struck up a conversation about kindergarten enrollment. "Freddy will be 5 in August," Susan announced. "He's a month older than the cutoff date."

"But he'll be one of the youngest in the class," Vicky countered. "Better check into what kids have to do in kindergarten these days. Have you asked his teacher what she thinks?"

"Well," Susan admitted, "she did say Freddy was a bit young."

Since the 1980s, more parents have been delaying their child's kindergarten entry. Aware that boys lag behind girls in development, parents most often hold out sons whose birth dates are close to the cutoff date for enrolling in kindergarten. Is delaying kindergarten entry beneficial? Although some teachers and principals recommend it, research has not revealed any advantages. Younger children make just as much academic progress as older children in the same grade (Cameron & Wilson, 1990; Graue & DiPerna, 2000). And younger first graders reap academic gains from on-time enrollment; they outperform same-age children a year behind them in school (Stipek & Byler, 2001). Furthermore, delaying kindergarten entry does not seem to prevent or solve emotional and social difficulties. To the contrary, students who are older than the typical age for their grade show high rates of behavior problems—considerably higher than students who are young for their grade (Stipek, 2002).

A related dilemma concerns whether to retain a student who is not progressing well for a second year in kindergarten or in one of the primary grades. A wealth of research reveals no learning benefits and suggests negative consequences for motivation, self-esteem, peer relations, and attitudes toward school as early as kindergarten (Carlton & Winsler, 1999). In a Canadian study, students retained between kindergarten and second grade—regardless of the academic and social characteristics they brought to the situation—showed worsening academic performance, anxiety, and (among boys) disruptiveness throughout elementary school. These unfavorable trends did not characterize nonretained students (Pagani et al., 2001).

As an alternative to kindergarten retention, some school districts place poorly performing kindergarten children in a "transition" class—a waystation between kindergarten and first grade. Transition classes, however, are a form of homogeneous grouping. As with other "low groups," teachers may have reduced expectations and teach transition children in a less stimulating fashion than they do other children (Dornbusch, Glasgow, & Lin, 1996).

Each of the options just considered is based on the view that readiness for school largely results from biological maturation. An alternative perspective, based on Vygotsky's sociocultural theory, is that children acquire the knowledge, skills, and attitudes for school success through assistance from parents and teachers. The U.S. National Association for the Education of Young Children recommends that all children of legal age start kindergarten and be granted classroom experiences that foster their individual progress. Research shows that school readiness is not something to wait for; it can be cultivated.

■ Saying good-bye on the first day of school, this mother may wonder how ready her daughter is for classroom learning. Yet delaying kindergarten entry for a year has no demonstrated gains for academic or social development. Children of legal age to start school can best acquire the knowledge, skills, and attitudes for school success through on-time enrollment and assistance from parents and teachers.

growing trend among parents to delay their child's school entry. Traditional teaching practices may also increase the incidence of grade retention. See the Social Issues box above for research on these practices.

● **New Philosophical Directions.** The philosophies of some teachers fall in between traditional and open. They want to foster high achievement as well as critical thinking, positive social relationships, and excitement about learning. Approaches to elementary education, grounded in Vygotsky's sociocultural theory, represent this point of view. Vygotsky's emphasis on the social origins of higher cognitive processes has inspired the following educational themes:

- *Teachers and children as partners in learning.* A classroom rich in both teacher–child and child–child collaboration transfers culturally valued ways of thinking to children.
- *Experiences with many types of symbolic communication in meaningful activities.* As children master reading, writing, and quantitative reasoning, they become aware of their culture's communication systems, reflect on their own thinking, and bring it under voluntary control. (Can you identify research presented earlier in this chapter that supports this theme?)
- *Teaching adapted to each child's zone of proximal development.* Assistance that is responsive to current understandings but that encourages children to take the next step helps ensure that each child makes the best progress possible.

Let's take an example of a growing number of teaching approaches that have translated these ideas into action. In **reciprocal teaching,** a teacher and two to four students form a cooperative group and take turns leading dialogues on the content of a text passage. Within the dialogues, group members apply four cognitive strategies: questioning, summarizing, clarifying, and predicting. The technique was originally designed to improve reading comprehension in students achieving poorly. Because of its success, it has been adapted for other subjects and all school-age children (Palincsar & Herrenkohl, 1999).

The dialogue leader (at first the teacher, later a student) begins by *asking questions* about the content of the text passage. Students offer answers, raise additional questions, and, in case of disagreement, reread the original text. Next, the leader *summarizes* the passage, and children discuss the summary and *clarify* ideas that are unfamiliar to any group members. Finally, the leader encourages pupils to *predict* upcoming content based on prior knowledge and clues in the passage.

Elementary and junior high school students exposed to reciprocal teaching show impressive gains in reading comprehension compared with controls taught in other ways (Lederer, 2000; Rosenshine & Meister, 1994). Notice how reciprocal teaching creates a zone of proximal development in which children gradually assume more responsibility for comprehending text passages. Also, by collaborating with others, children forge group expectations for high-level thinking and acquire skills vital for learning and success in everyday life.

Teacher–Student Interaction

Elementary school students describe good teachers as caring, helpful, and stimulating—characteristics positively associated with learning (Daniels, Kalkman, & McCombs, 2001; Sanders & Jordan, 2000). Yet with respect to stimulation, a disappointing finding is that North American teachers emphasize rote, repetitive drill more than higher-level thinking, such as grappling with ideas and applying knowledge to new situations (Campbell, Hombo, & Mazzeo, 2000). In a study of fifth-grade social studies and math lessons, students were far more attentive when teachers encouraged high-level thinking (Stodolsky, 1988). And in a longitudinal investigation of middle-school students, those in more academically demanding classrooms showed better attendance and larger gains in math achievement over the next 2 years (Phillips, 1997).

Of course, teachers do not interact in the same way with all children. Well-behaved, high-achieving students typically get more encouragement and praise, whereas unruly students are often criticized and rarely called on in class discussions. When they seek special help or permission, their requests are usually denied (Good & Brophy, 1996).

Unfortunately, once teachers' attitudes toward students are established, they are in danger of becoming more extreme than is warranted by children's behavior. Of special concern are **educational self-fulfilling prophecies:** Children may adopt teachers' positive or negative views and start to live up to them. This effect is especially strong when teachers emphasize competition and publicly compare children (Weinstein et al., 1987).

When teachers hold inaccurate views, poorly achieving students are more affected (Madon, Jussim, & Eccles, 1997). High-achieving students have less room to improve when teachers think well of them, and they can fall back on their history of success when a teacher is critical. Low-achieving pupils' sensitivity to self-fulfilling prophecies enhances their school performance when teachers believe in them. But unfortunately, biased teacher judgments are usually slanted in a negative direction, so only rarely do poor achievers have a chance to reap these benefits.

Grouping Practices

In many schools, students are assigned to *homogeneous* groups or classes in which children of similar ability levels are taught together. Homogeneous grouping can be a potent source of self-fulfilling prophecies. Low-group students get more drill on basic facts and skills, engage in less discussion, and progress at a slower pace. Gradually, they show a drop in self-esteem and are viewed by themselves and others as "not smart." Not surprisingly, homogeneous grouping widens the gap between high and low achievers (Dornbusch, Glasgow, & Lin, 1996).

Partly because of these findings, some schools have increased the *heterogeneity* of classes by combining two or three adjacent grades. In *multigrade classrooms,* academic achievement, self-esteem, and attitudes toward school are usually more favorable than in the single-grade arrangement (Lloyd, 1999). Perhaps mixed-grade grouping decreases competition and promotes *cooperative learning,* in which small, heterogeneous groups of students share responsibility and consider one another's ideas while working toward common goals. When teachers explain, model, and have children role-play how to work together, cooperative learning promotes achievement across a wide range of school subjects (Gillies & Ashman, 1996, 1998).

© STEVE WARMOWSKI/*JOURNAL-COURIER*/THE IMAGE WORKS

■ Grouping practices can affect the quality of teachers' interactions with students. In this heterogeneous group, children receive warm, stimulating teaching that emphasizes high-level thinking. As a result, each student is attentive and involved in learning.

Teaching Children with Special Needs

We have seen that effective teachers flexibly adjust their teaching strategies to accommodate pupils with a wide range of characteristics. But such adjustments are increasingly difficult at the very low and high ends of the ability distribution. How do schools serve children with special learning needs?

● **Children with Learning Difficulties.** American and Canadian legislation mandates that schools place children requiring special supports for learning in the "least restrictive" (or closest to normal as possible) classroom environments that meet their educational needs. In **mainstreaming,** students with learning difficulties are placed in regular classrooms for part of the school day, a practice designed to prepare them better for participation in society. Largely due to parental pressures, mainstreaming has been extended to **full inclusion**—placement in regular classrooms full time.

© WILL HART

■ The student on the left, who has a learning disability, has been fully included in a regular classroom. Because his teacher takes special steps to encourage peer acceptance, individualizes instruction, minimizes comparisons with classmates, and promotes cooperative learning, this boy looks forward to school and is doing well.

Some mainstreamed students are **mildly mentally retarded**—children whose IQs fall between 55 and 70 and who also show problems in adaptive behavior, or skills of everyday living (American Psychiatric Association, 1994). But the largest number—5 to 10 percent of school-age children—have **learning disabilities,** great difficulty with one or more aspects of learning, usually reading. As a result, their achievement is considerably behind what would be expected on the basis of their IQ. The problems of these students cannot be traced to any obvious physical or emotional difficulty or to environmental disadvantage. Instead, subtle deficits in brain functioning seem to be involved (Kibby & Hynd, 2001). In most instances, the cause is unknown.

How effective is placement of these children in regular classes at providing appropriate academic experiences as well as integrated participation in classroom life? At present, the evidence is not positive on either of these points. Although some mainstreamed and fully included students benefit academically, many do not. Achievement gains depend on both the severity of the disability and the support services available in the regular classroom (Klingner et al., 1998; Waldron & McLeskey, 1998). Furthermore, children with disabilities are often rejected by regular-classroom peers. Students with mental retardation are overwhelmed by the social skills of their classmates; they cannot interact quickly or adeptly in a conversation or game. And the processing deficits of some children with learning disabilities lead to problems in social awareness and responsiveness (Gresham & MacMillan, 1997; Sridhar & Vaughn, 2001).

Does this mean that students with special needs cannot be served in regular classrooms? This extreme conclusion is not warranted. Often these children do best when they receive instruction in a resource room for part of the day and in the regular classroom for the remainder—an arrangement that the majority of school-age children with learning disabilities say they prefer (Vaughn & Klingner, 1998). In the resource room, a special education teacher works with students on an individual and small-group basis. Then, depending on their progress, children join regular classmates for different subjects and amounts of time.

Once children enter the regular classroom, special steps must to be taken to promote peer acceptance. Cooperative learning experiences in which children with learning difficulties and their classmates work together lead to friendly interaction and improved social acceptance (Siegel, 1996). Teachers can also prepare children for the arrival of a student with special needs—a process best begun early, before children have become less accepting of peers with disabilities (Okagaki et al., 1998).

● **Gifted Children.** In Joey and Lizzie's school, some children were **gifted,** displaying exceptional intellectual strengths. Their characteristics were diverse. In every grade, there were one or two students with IQ scores above 130, the standard definition of giftedness based on intelligence test performance (Gardner, 1998b). High-IQ children, as we have seen, are particularly quick at academic work. They have keen memories and an exceptional capacity to solve challenging academic

■ **FIGURE 9.9 Responses of an 8-year-old who scored high on a figural measure of divergent thinking.** This child was asked to make as many pictures as she could from the circles on the page. The titles she gave her drawings, from left to right, are as follows: "Dracula," "one-eyed monster," "pumpkin," "Hula-Hoop," "poster," "wheelchair," "earth," "moon," "planet," "movie camera," "sad face," "picture," "stoplight," "beach ball," "the letter O," "car," and "glasses." Tests of divergent thinking tap only one of the complex cognitive contributions to creativity. (Test form copyright © 1980 by Scholastic Testing Service, Inc. Reprinted by permission of Scholastic Testing Service, Inc., from *The Torrance Tests of Creative Thinking* by E. P. Torrance. Child's drawings reprinted by permission of Laura Berk.)

problems. Yet recognition that intelligence tests do not sample the entire range of human mental skills has led to an expanded conception of giftedness in schools.

Creativity and Talent. **Creativity** is the ability to produce work that is original yet appropriate—something others have not thought of that is useful in some way. High potential for creativity can result in a child being designated as gifted. Tests of creative capacity tap **divergent thinking**—the generation of multiple and unusual possibilities when faced with a task or problem. Divergent thinking contrasts sharply with **convergent thinking,** which involves arriving at a single correct answer and is emphasized on intelligence tests (Guilford, 1985).

Recognizing that highly creative children (like high-IQ children) are often better at some tasks than others, researchers have devised a variety of tests of divergent thinking (Runco, 1992; Torrance, 1988). A verbal measure might ask children to name uses for common objects (such as a newspaper). A figural measure might ask them to come up with drawings based on a circular motif (see Figure 9.9). A "real-world problem" measure requires students to suggest solutions to everyday problems. Responses to all these tests can be scored for the number of ideas generated and their originality.

Yet critics of these measures point out they are poor predictors of creative accomplishment in everyday life because they tap only one of the complex cognitive contributions to creativity (Cramond, 1994). Also involved are defining new and important problems, evaluating divergent ideas and choosing the most promising, and calling on relevant knowledge to understand and solve problems (Sternberg & Lubart, 1995).

Consider these additional ingredients, and you will see why people usually demonstrate creativity in only one or a few related areas. Even individuals designated as gifted by virtue of their high IQ often show uneven ability across academic subjects. Partly for this reason, definitions of giftedness have been extended to include **talent**—outstanding performance in a specific field. There is clear evidence that excellence in such endeavors as writing, mathematics, science, music, visual arts, athletics, and leadership has roots in specialized skills that first appear in childhood (Winner, 2000). Highly talented children are biologically prepared to master their domain of interest. And they display a passion for doing so.

At the same time, talent must be nurtured in a favorable environment. Studies of the backgrounds of talented children and highly accomplished adults often reveal parents who are warm and sensitive, who provide a stimulating home life, and who are devoted to developing their child's abilities. These parents are not driving and overly ambitious but, instead, are reasonably demanding (Winner, 1996). They arrange for caring teachers while the child is young and for more rigorous master teachers as the talent develops.

Extreme giftedness often results in social isolation. The highly driven, nonconforming, and independent styles of many gifted children and adolescents lead them to spend more time alone, partly because of their rich inner lives and partly because solitude is necessary for them to develop their talents. Still, gifted children desire gratifying peer relationships and some—girls more often than boys—try to hide their abilities to become better liked. Compared with their ordinary agemates, gifted youths, especially girls, report more emotional and social difficulties, including low self-esteem and depression (Gross, 1993; Winner, 2000).

Finally, whereas many talented youths become experts in their fields, few become highly creative. The skill involved in rapidly mastering an existing field is not the same as innovating in that field (Csikszentmihalyi, 1999). The world, however, needs both experts and creators.

Educating the Gifted. Although programs for the gifted exist in many schools, debate about their effectiveness usually focuses on factors irrelevant to giftedness—whether to provide enrichment in regular classrooms, to pull children out for special instruction (the most common practice), or to advance brighter students to a higher grade. Children of all ages fare well academically and socially within each of these models (Moon & Feldhusen, 1994). Yet the extent to which they foster creativity and talent depends on opportunities to acquire relevant skills.

Gardner's theory of multiple intelligences has inspired several model programs that provide enrichment to all students. Meaningful activities, each tapping a specific intelligence or set of intelligences, serve as contexts for assessing strengths and weaknesses and, on that basis, teaching new knowledge and original thinking (Gardner, 1993, 2000). For example, linguistic intelligence might be fostered through storytelling or playwriting, spatial intelligence through drawing, sculpting, or taking apart and reassembling objects.

Evidence is still needed on how effectively these programs nurture children's talents. But so far, they have succeeded in one way—by highlighting the strengths of some students who previously had been considered unexceptional or even at risk for school failure (Suzuki & Valencia, 1997). Consequently, they may be especially useful for identifying talented low-SES, ethnic minority children, who are underrepresented in school programs for the gifted.

	Country	Average Math Achievement Score
High-Performing Nations	Japan	557
	Korea, Republic of	547
	New Zealand	537
	Finland	536
	Australia	533
	Canada	**533**
	Switzerland	529
	United Kingdom	529
Intermediate-Performing Nations	Belgium	520
	France	517
	Austria	515
	Denmark	514
	Iceland	514
	Sweden	510
	Ireland	503
International Average = 500		
	Norway	499
	Czech Republic	498
	United States	**493**
	Germany	490
	Hungary	488
	Spain	476
	Poland	470
Low-Performing Nations	Italy	457
	Portugal	454
	Greece	447
	Luxembourg	446
	Mexico	387

FIGURE 9.10 Average mathematics scores of 15-year-olds by country. The Program for International Student Assessment assessed achievement in 27 nations. Japan, Korea, and Canada were among the top performers in mathematics, whereas the United States performed just below the international average. Similar outcomes occurred in reading and science. (Adapted from U.S. Department of Education, 2001a.)

How Well Educated Are North American Children?

Our discussion of schooling has largely focused on what teachers can do to support the education of children. Yet a great many factors, both within and outside schools, affect children's learning. Societal values, school resources, quality of teaching, and parental encouragement all play important roles. Nowhere are these multiple influences more apparent than when schooling is examined in cross-cultural perspective.

In international studies of reading, mathematics, and science achievement, young people in Hong Kong, Japan, Korea, and Taiwan are consistently among the top performers. Although they rank below students in Asian countries, Canadian students generally score high. American students, however, typically perform at the international average and sometimes below it (see Figure 9.10) (U.S. Department of Education, 2001a, 2001b).

Why do American children fall behind in academic accomplishment? According to international comparisons, instruction in the United States is not as challenging and focused as it is in other countries. In the Program for International Student Assessment, which assessed the academic achievement of 15-year-olds nearing the end of compulsory education in 27 nations, students were asked about their study habits. Compared with students in the top-achieving nations listed in Figure 9.10, many more American students reported studying by memorizing rather than relating information to previously acquired knowledge. And

Cultural Influences

Education in Japan, Taiwan, and the United States

Why do Asian children perform so well academically? Research examining societal, school, and family conditions in Japan, Taiwan, and the United States provides some answers.

Cultural Valuing of Academic Achievement. In Japan and Taiwan, natural resources are limited. Progress in science and technology is essential for economic well-being. Because a well-educated work force is necessary to meet this goal, children's mastery of academic skills is vital. Compared with Western countries, Japan, Taiwan, and other Asian nations invest more in education, including paying higher salaries to teachers (Rohlen, 1997). In the United States, attitudes toward academic achievement are far less unified. Many Americans believe it is more important for children to feel good about themselves and to explore various areas of knowledge than to perform well in school.

Emphasis on Effort. Japanese and Taiwanese parents and teachers believe that all children have the potential to master challenging academic tasks if they work hard enough. In contrast, American parents and teachers tend to regard native ability as the key to academic success (Stevenson, 1992). These differences in attitude may contribute to the fact that American parents are less inclined to encourage activities at home that might enhance school performance. Japanese and Taiwanese parents promote their children's commitment to academics, and they devote many more hours to helping with homework than American parents. As a result, Asian children spend more time studying and reading than children in the United States. When asked what factors influence academic success, Asian students most often say "studying hard." American students most often mention "having a good teacher" (Stevenson, Lee, & Mu, 2000).

Furthermore, effort takes on different meaning in Asian collectivist societies than in Western individualistic societies. Japanese and Chinese youths strive to achieve in school because effort is a moral obligation—part of one's responsibility to family and community. In contrast, American young people regard working hard academically as a matter of individual choice—of fulfilling personal goals (Bempchat & Drago-Severson, 1999).

High-Quality Education for All. Unlike American teachers, Japanese and Taiwanese teachers do not make early educational decisions on the basis of achievement. There are no separate ability groups or tracks in elementary school. Instead, all pupils receive the same nationally mandated, high-quality instruction. Academic lessons are particularly well organized and presented in ways that capture children's attention. Topics in mathematics are treated in greater depth, and there is

in-depth research on learning environments in top-performing Asian nations reveals that except for the influence of language on early counting skills (see page 293), Asian students do not start school with cognitive advantages over their North American peers (Geary, 1996). Instead, as the Cultural Influences box above explains, a variety of social forces combine to foster a strong national commitment to learning.

The Japanese and Taiwanese examples underscore that families, schools, and the larger society must work together to upgrade education. In the United States, more tax dollars are being invested in elementary and secondary education, and academic standards are being strengthened. In addition, many schools are working to increase parent involvement in children's education. Parents who create stimulating learning environments at home, monitor their child's academic progress, help with homework, and communicate often with teachers have children who consistently show superior academic progress (Christenson & Sheridan, 2001). The returns of these efforts can be seen in recent national assessments of educational progress (U.S Department of Education, 2002a). After two decades of decline, American students' overall academic achievement has risen, although not enough to enhance their standing internationally.

Ask Yourself

REVIEW

List educational practices that foster children's academic achievement and those that undermine it. For each practice, provide a brief explanation.

APPLY

Sandy wonders whether she should enroll her child in a multigrade classroom in which first, second, and third graders are taught together. How would you advise Sandy, and why?

APPLY

Carrie is a first grader with a learning disability. What steps can her school and teacher take to ensure that she develops at her best, academically and socially?

CONNECT

Review research on child-rearing styles on pages 265–266 of Chapter 8. What style do gifted children who realize their talents typically experience? Explain.

www.

less repetition of material taught the previous year.

More Time Devoted to Instruction. In Japan and Taiwan, the school year is over 50 days longer than in the United States. When one American elementary school experimented by adding 30 days to its school year, extended-year students scored higher in reading, general-knowledge, and math achievement than students in schools of similar quality with a traditional calendar (Frazier & Morrison, 1998).

Furthermore, on a day-to-day basis, Japanese and Taiwanese teachers devote much more time to academic pursuits. However, Asian schools are not regimented places, as many Americans believe. An 8-hour school day permits extra recesses and a longer lunch period, with plenty of time for play, social interaction, field trips, and extracurricular activities. Frequent breaks may increase children's capacity to learn (Pellegrini & Smith, 1998).

Communication Between Teachers and Parents. Japanese and Taiwanese teachers get to know their students especially well. They teach the same children for 2 or 3 years and make visits to the home once or twice a year. Continuous communication between teachers and parents takes place with the aid of small notebooks that children carry back and forth every day containing messages about assignments, academic performance, and behavior (Stevenson & Lee, 1990).

Do Japanese and Taiwanese children pay a price for the pressure placed on them to succeed? By high school, academic work often displaces other experiences, since Asian adolescents must pass a highly competitive entrance exam to gain admission to college. Yet Asian parenting practices that encourage academic competence do not undermine children's adjustment. In fact, high-performing Asian students are socially competent (Crystal et al., 1994; Huntsinger, Jose, & Larson, 1998).

■ Japanese children achieve considerably better than their American counterparts for a variety of reasons. A longer school day permits frequent alternation of academic instruction with pleasurable activity, an approach that fosters learning. During a break from academic subjects, these Japanese elementary-school children enjoy a class in the art of calligraphy.

Summary

PHYSICAL DEVELOPMENT

Body Growth

Describe major trends in body growth during middle childhood.

- Gains in body size during middle childhood extend the pattern of slow, regular growth established during the preschool years. Bones continue to lengthen and broaden, and all 20 primary teeth are replaced by permanent ones. By age 9, girls overtake boys in physical size.

Common Health Problems

What vision and hearing problems are common in middle childhood?

- During middle childhood, children from economically advantaged homes are at their healthiest, due to good nutrition and rapid development of the immune system. At the same time, a variety of health problems do occur, many of which are more prevalent among low-SES children.

- The most common vision problem in middle childhood is myopia, or nearsightedness. It is influenced by heredity, early biological trauma, and time spent reading and doing other close work. Myopia is one of the few health conditions that increase with SES. Because of untreated middle ear infections, some low-SES children develop permanent hearing loss during the school years.

Describe the causes and consequences of serious nutritional problems in middle childhood, giving special attention to obesity.

- Poverty-stricken children in developing countries and North America continue to suffer from malnutrition during middle childhood. When malnutrition is allowed to persist for many years, its negative impact on physical growth, intelligence, motor coordination, and attention is permanent.

- Overweight and obesity are growing problems in Western nations. Although heredity contributes to obesity, parental feeding practices, maladaptive eating habits, and lack of exercise also play important roles. Obese children are disliked by peers and adults and have serious adjustment problems. Family-based interventions in which both parents and children change their eating patterns and sedentary lifestyles are the most effective approaches to treatment.

What factors contribute to illness during the school years, and how can these health problems be reduced?

- Children experience more illnesses during the first 2 years of elementary school than later because of exposure to sick children and an immune system that is not yet mature. The most common cause of school absence and childhood hospitalization is asthma. Although heredity contributes to asthma, environmental factors—pollution, stressful home lives, and lack of access to good health care—have led the illness to increase.

- Children with more severe chronic illnesses are at risk for academic, emotional, and social difficulties. Positive family relationships predict favorable adjustment among chronically ill children.

Describe changes in unintentional injuries in middle childhood.

- Unintentional injuries increase over middle childhood and adolescence, especially for boys, with auto and bicycle accidents accounting for most of the rise. School-based programs that prevent injury teach safety practices through modeling and rehearsal and reward children for following them.

Motor Development and Play

Cite major changes in motor development and play during middle childhood.

- Gains in flexibility, balance, agility, and force contribute to school-age children's gross motor development. In addition, improvements in reaction time and in responding only to relevant information contribute to motor performance.

- Fine motor development also improves. Children's writing becomes more legible, and their drawings show increases in organization, detail, and representation of depth. Gender stereotypes, which affect parental encouragement, largely account for boys' superior performance on a wide range of gross motor skills.

- Games with rules become common during the school years. Children's spontaneous games support cognitive and social development. Many school-age youngsters are not physically fit. Frequent, high-quality physical education classes help ensure that all children have access to the benefits of regular exercise and play.

COGNITIVE DEVELOPMENT

Piaget's Theory: The Concrete Operational Stage

What are the major characteristics of concrete operational thought?

- During the **concrete operational stage,** children can reason logically about concrete, tangible information. Mastery of conservation indicates that children can **decenter** and **reverse** their thinking. They are also better at hierarchical classification and **seriation,** including **transitive inference.** School-age youngsters' spatial reasoning improves, as their understanding of distance and their ability to give directions reveal.

- Piaget used the term **horizontal décalage** to describe the school-age child's gradual mastery of logical concepts. Concrete operational thought is limited in that children have difficulty reasoning about abstract ideas.

Discuss recent research on concrete operational thought.

- Specific cultural practices, especially those associated with schooling, affect children's mastery of Piagetian tasks. Some researchers believe that the gradual development of operational thought can best be understood within an information-processing framework. According to Case's neo-Piagetian theory, with practice, cognitive schemes demand less attention, freeing up space in working memory for combining old schemes and generating new ones. Eventually, children consolidate schemes into highly efficient, central conceptual structures. On a wide variety of tasks, children move from a focus on only one dimension to coordinating two dimensions to integrating multiple dimensions.

Information Processing

Cite two basic changes in information processing, and describe the development of attention and memory in middle childhood.

- Brain development contributes to gains in information-processing capacity and cognitive inhibition. These changes facilitate many aspects of thinking.

- During the school years, attention becomes more selective, adaptable, and planful, and memory strategies also improve. **Rehearsal** appears first, followed by **organization** and then **elaboration.** The serious attentional difficulties of children with **attention-deficit hyperactivity disorder (ADHD)** lead to both academic and social problems.

- Development of the long-term knowledge base facilitates memory by making new information easier to store and retrieve. Children's motivation to use what they know also contributes to memory development. Memory strategies are promoted by learning activities in school.

Describe the school-age child's theory of mind and capacity to engage in self-regulation.

- Metacognition expands over middle childhood. School-age children regard the mind as an active, constructive agent, and they develop an integrated theory of mind. **Cognitive self-regulation**—putting what one knows about thinking into action—develops slowly over middle childhood and adolescence. It improves with instructions to monitor cognitive activity.

Discuss current controversies in teaching reading and mathematics to elementary school children.

- Skilled reading draws on all aspects of the information-processing system. Controversy exists over whether a **whole-language approach** or a **basic-skills approach** should be used to teach beginning reading. Research shows that a mixture of both is most effective. Teaching that blends practice in basic skills with conceptual understanding also is best in mathematics.

Individual Differences in Mental Development

Describe major approaches to defining and measuring intelligence.

- During the school years, IQ becomes more stable, and it correlates well with academic achievement. Most intelligence tests yield an overall score as well as scores for separate intellectual factors. The Stanford-Binet Intelligence Scale and the Wechsler Intelligence Scale for Children (WISC-III) are widely used individually administered intelligence tests.

- Aspects of information processing that are related to IQ include speed of thinking and effective strategy use. Sternberg's **triarchic theory** views intelligence as an interaction of information-processing skills, specific experiences, and contextual (or cultural) influences. According to Gardner's **theory of multiple intelligences,** at least eight mental abilities exist, each of which has a unique biological basis and distinct course of development. Gardner's theory has been especially helpful in understanding and nurturing children's talents.

Describe evidence indicating that both heredity and environment contribute to intelligence.

- Heritability estimates and adoption research indicate that intelligence is a product of both heredity and environment. Studies of black children adopted into well-to-do homes indicate that the black–white IQ gap is substantially influenced by environment.

- IQ scores are affected by specific learning experiences, including exposure to certain communication styles and to knowledge sampled by the test. Cultural bias in intelligence testing can lead test scores to underestimate minority children's intelligence. **Dynamic testing** helps many minority children perform more competently on mental tests.

Language Development

Describe changes in school-age children's vocabulary, grammar, and pragmatics, and cite the advantages of bilingualism for development.

- During middle childhood, vocabulary continues to grow rapidly, and children have a more precise and flexible understanding of word meanings. They also use complex grammatical constructions and conversational strategies. Language awareness contributes to school-age children's language progress.
- Mastery of a second language must begin in childhood for full development to occur. Bilingual children are advanced in cognitive development and language awareness. In Canada, language immersion programs are highly successful in making children proficient in both English and French. In the United States, bilingual education that combines instruction in the native tongue and in English supports ethnic minority children's academic learning.

Learning in School

Describe the impact of class size and educational philosophies on children's motivation and academic achievement.

- As class size drops, academic achievement improves. Older students in **traditional classrooms** have a slight edge in academic achievement. Those in **open classrooms** tend to be critical thinkers who respect individual differences and have more positive attitudes toward school.
- Vygotsky's sociocultural theory has inspired approaches to elementary education that emphasize teachers and children as partners in learning, rich experiences with symbolic communication, and teaching adapted to each child's zone of proximal development. **Reciprocal teaching** results in impressive gains in reading comprehension.

Discuss the role of teacher–student interaction and grouping practices in academic achievement.

- Instruction that encourages high-level thinking fosters children's interest and academic progress. **Educational self-fulfilling prophecies** are most likely to occur in classrooms that emphasize competition and public evaluation. Teachers' inaccurate views have a greater impact on poorly achieving students.

Under what conditions is placement of children with mild mental retardation and learning disabilities in regular classrooms successful?

- Pupils with **mild mental retardation** and **learning disabilities** are often placed in regular classrooms, usually through **mainstreaming** but also through **full inclusion.** The success of regular class placement depends on meeting individual learning needs and positive peer relations.

Describe the characteristics of gifted children and current efforts to meet their educational needs.

- **Giftedness** includes high IQ, **creativity,** and **talent.** Tests of creativity that tap **divergent** rather than **convergent thinking** focus on only one of the ingredients of creativity. Highly talented children are biologically prepared to master their domain of interest and have parents and teachers who nurture their extraordinary ability. Gifted children are best served by educational programs that build on their special strengths.

How well are North American children achieving compared with children in other industrialized nations?

- In international studies, young people in Asian nations are consistently top performers. Canadian students generally score high, whereas American students typically display an average or below-average performance. A strong commitment to learning in families and schools underlies the high academic success of Asian students.

Important Terms and Concepts

attention-deficit hyperactivity disorder (ADHD) (p. 288)
basic-skills approach (p. 292)
cognitive inhibition (p. 288)
cognitive self-regulation (p. 291)
concrete operational stage (p. 285)
convergent thinking (p. 306)
creativity (p. 306)
decentration (p. 285)
divergent thinking (p. 306)
dynamic testing (p. 298)
educational self-fulfilling prophecy (p. 304)
elaboration (p. 290)
full inclusion (p. 305)
gifted (p. 305)
horizontal décalage (p. 286)
learning disabilities (p. 305)
mainstreaming (p. 305)
mild mental retardation (p. 305)
obesity (p. 278)
open classroom (p. 302)
organization (p. 289)
reciprocal teaching (p. 304)
rehearsal (p. 289)
reversibility (p. 285)
seriation (p. 285)
talent (p. 306)
theory of multiple intelligences (p. 295)
traditional classroom (p. 302)
transitive inference (p. 285)
triarchic theory of intelligence (p. 294)
whole-language approach (p. 292)

FYI For Further Information and Help

Consult the Companion Website for *Development Through the Lifespan, Third Edition,* (www.ablongman.com/berk) where you will find the following resources for this chapter:

- **Chapter Objectives**
- **Flashcards** for studying important terms and concepts
- **Annotated Weblinks** to guide you in further research
- **Ask Yourself** questions, which you can answer and then check against a sample response
- **Suggested Readings**
- **Practice Test** with immediate scoring and feedback

© CHARLES GUPTON/CORBIS

10

Emotional and Social Development in Middle Childhood

Erikson's Theory: Industry versus Inferiority

Self-Understanding
Changes in Self-Concept • Development of Self-Esteem • Influences on Self-Esteem

Emotional Development
Self-Conscious Emotions • Emotional Understanding • Emotional Self-Regulation

Understanding Others: Perspective Taking

Moral Development
Learning About Justice Through Sharing • Changes in Moral and Social-Conventional Understanding

Peer Relations
Peer Groups • Friendships • Peer Acceptance

■ Biology & Environment: Bullies and Their Victims

Gender Typing
Gender-Stereotyped Beliefs • Gender Identity and Behavior • Cultural Influences on Gender Typing

Family Influences
Parent–Child Relationships • Siblings • Only Children • Divorce • Blended Families • Maternal Employment and Dual-Earner Families

Some Common Problems of Development
Fears and Anxieties • Child Sexual Abuse • Fostering Resiliency in Middle Childhood

■ A Lifespan Vista: Children of War

■ Social Issues: Children's Eyewitness Testimony

One late afternoon, Rena heard her son Joey dash through the front door, run upstairs, and call up his best friend Terry. "Terry, gotta talk to you," pleaded Joey, out of breath from running home. "Everything was going great until that word I got—'porcupine,'" remarked Joey, referring to the fifth-grade spelling bee at school that day. "Just my luck! *P-o-r-k,* that's how I spelled it! I can't believe it. Maybe I'm not so good at social studies," Joey confided, "but I *know* I'm better at spelling than that stuck-up Belinda Brown. Gosh, I knocked myself out studying those spelling lists. Then *she* got all the easy words. Did'ya see how snooty she acted after she won? If I *had* to lose, why couldn't it be to a nice person?"

Joey's conversation reflects a whole new constellation of emotional and social capacities. First, Joey shows evidence of *industriousness.* By entering the spelling bee, he energetically pursued meaningful achievement in his culture—a major change of the middle childhood years. At the same time, Joey's social understanding has expanded. He can size up strengths, weaknesses, and personality characteristics. Furthermore, friendship means something different to Joey than it did at younger ages. Terry is a best friend whom Joey counts on for understanding and emotional support.

We begin this chapter by returning to Erikson's theory for an overview of the personality changes of middle childhood. Then we take a close look at emotional and social development. We will see how, as children reason more effectively and spend more time in school and with peers, their views of themselves, others, and social relationships become more complex.

Although school-age children spend less time with parents than they did at younger ages, the family remains powerfully influential. Joey and Lizzie are growing up in a home profoundly affected by social change. Rena, their mother, has been employed since her children were preschoolers. In addition, Joey's and Lizzie's lives have been disrupted by parental divorce. Although family lifestyles are more diverse today than ever before, Joey's and Lizzie's experiences will help us appreciate that family functioning is far more important than family structure in ensuring children's well-being. Our chapter concludes with some common emotional problems of middle childhood.

Erikson's Theory: Industry versus Inferiority

According to Erikson (1950), the personality changes of the school years build on Freud's *latency stage* (see Chapter 1, page 16). Although Freud's theory is no longer widely accepted, children whose experiences have been positive enter middle childhood with the calm confidence Freud intended by the term *latency.* Their energies are redirected from the make-believe of early childhood into realistic accomplishment.

Erikson believed that the combination of adult expectations and children's drive toward mastery sets the stage for the psychological conflict of middle childhood: **industry versus inferiority,** which is resolved positively when experiences lead children to develop a sense of competence at useful skills and tasks. In cultures everywhere, improved physical and cognitive capacities mean that adults impose new demands. Children, in turn, are ready to meet these challenges and benefit from them.

In industrialized nations, the transition to middle childhood is marked by the beginning of formal schooling. With it comes literacy training, which prepares children for the vast array of specialized careers in complex societies. In school, children become aware of their own and others' unique capacities, learn the value of division of labor, and develop a sense of moral commitment and responsibility. The danger at

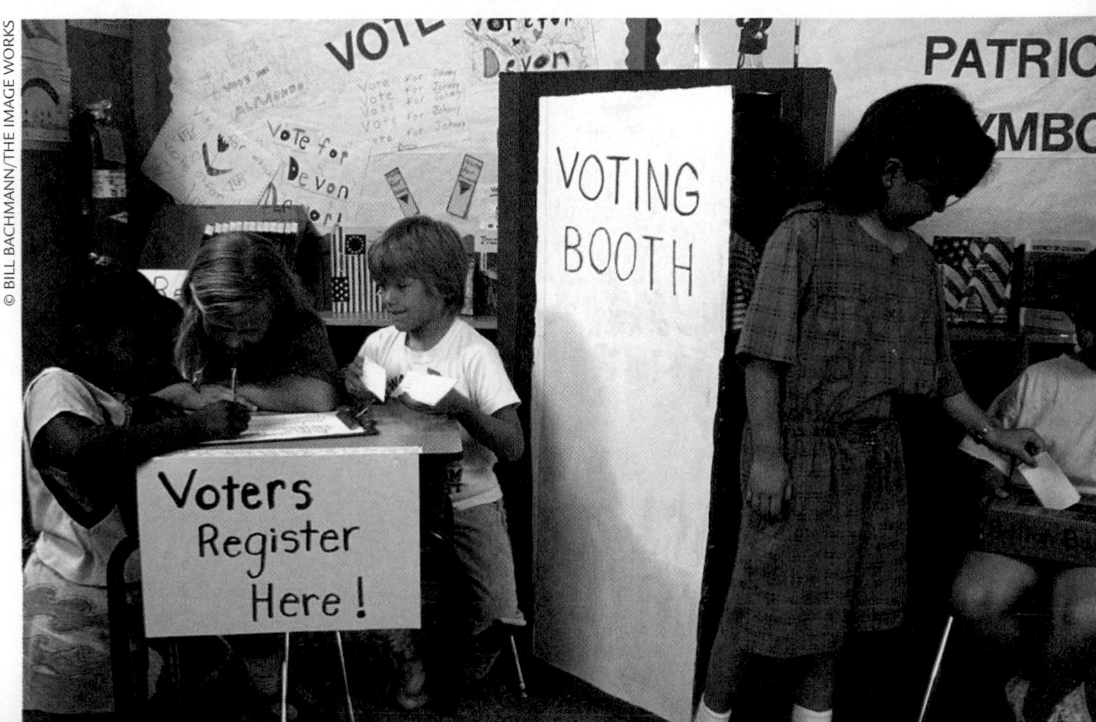

■ The industriousness of middle childhood involves mastery of useful skills and tasks. As these second graders conduct an election, they become aware of one another's unique capacities and come to view themselves as responsible, capable, and helpful.

© BILL BACHMANN/THE IMAGE WORKS

this stage is *inferiority,* reflected in the sad pessimism of children who have little confidence in their ability to do things well. This sense of inadequacy can develop when family life has not prepared children for school life or when experiences with teachers and peers are so negative that they destroy children's feelings of competence and mastery.

Erikson's sense of industry combines several developments of middle childhood: a positive but realistic self-concept, pride in accomplishment, moral responsibility, and cooperative participation with agemates. Let's see how these aspects of self and social relationships change over the school years.

Self-Understanding

Several transformations in self-understanding take place in middle childhood. First, children can describe themselves in terms of psychological traits. Second, they start to compare their own characteristics with those of their peers. Finally, they speculate about the causes of their strengths and weaknesses. These new ways of thinking about the self have a major impact on children's self-esteem.

Changes in Self-Concept

During the school years, children develop a much more refined *me-self,* or self-concept, organizing their observations of behaviors and internal states into general dispositions, with a major change taking place between ages 8 and 11. The following self-description from an 11-year-old reflects this change:

> My name is A. I'm a human being. I'm a girl. I'm a truthful person. I'm not pretty. I do so-so in my studies. I'm a very good cellist. I'm a very good pianist. I'm a little bit tall for my age. I like several boys. I like several girls. I'm old-fashioned. I play tennis. I am a very good swimmer. I try to be helpful. I'm always ready to be friends with anybody. Mostly I'm good, but I lose my temper. I'm not well-liked by some girls and boys. I don't know if I'm liked by boys or not. (Montemayor & Eisen, 1977, pp. 317–318)

Notice that instead of specific behaviors, this child emphasizes competencies, as in "I'm a very good cellist" (Damon & Hart, 1988). Also, she clearly describes her personality and mentions both positive and negative traits—"truthful" but "not pretty," "a good cellist [and] pianist" but only "so-so in my studies." Older school-age children are far less likely than younger children to describe themselves in all-or-none ways (Harter, 1996).

A major reason for these qualified self-descriptions is that school-age children often make **social comparisons,** judging their appearance, abilities, and behavior in relation to those of others. In commenting on the spelling bee, Joey expressed some thoughts about how good he was compared with his peers—better at spelling but not so good at social studies. Although 4- to 6-year-olds can compare their own performance to that of one peer, older children can compare multiple individuals, including themselves (Butler, 1998).

What factors are responsible for these revisions in self-concept? Cognitive development affects the changing *structure* of the self—children's ability to combine typical experiences and behaviors into psychological dispositions (Harter, 1999). The *content* of self-concept is a product of both cognitive capacities and feedback from others. Sociologist George Herbert Mead (1934) proposed that a well-organized psychological self emerges when the child's *I-self* adopts a view of the *me-self* that resembles others' attitudes toward the child. Mead's ideas indicate that *perspective-taking skills*—in particular, an improved ability to infer what other people are thinking—are crucial for the development of a self-concept based on personality traits. During the school years, children become better at "reading" the messages they receive from others and incorporating these into their self-definitions. As school-age children internalize others' expectations, they form an *ideal self* that they use to evaluate their *real self.* As we will see shortly, a large discrepancy between the two can greatly undermine self-esteem, leading to sadness, hopelessness, and depression.

During middle childhood, children look to more people for information about themselves as they enter a wider range of settings in school and community. This is reflected in children's frequent reference to social groups in their self-descriptions. "I'm a Boy Scout, a paper boy, and a Prairie City soccer player," Joey remarked when asked to describe himself. Gradually, as children move into adolescence, their sources of self-definition become more selective. Although parents remain influential, between ages 8 and 15 peers become more important. And over time, self-concept becomes increasingly vested in feedback from close friends (Oosterwegel & Oppenheimer, 1993).

Development of Self-Esteem

Recall that most preschoolers have very high self-esteem. As children move into middle childhood, they receive much more feedback about their performance in different activities compared with that of their peers. As a result, self-esteem differentiates, and it also adjusts to a more realistic level.

● **A Hierarchically Structured Self-Esteem.** Researchers have asked children to indicate the extent to which a variety of statements, such as, "I am good at homework" or "I'm usually the one chosen for games," are true of themselves. Findings reveal that by age 6 to 7, children have formed at least four self-esteems—academic competence, social competence, physical/athletic competence, and physical appearance—that become more refined with age (Marsh, 1990). Furthermore, the capacity to view the self in terms of stable dispositions permits school-age children to combine their separate self-evaluations into a general psychological image of themselves—an overall

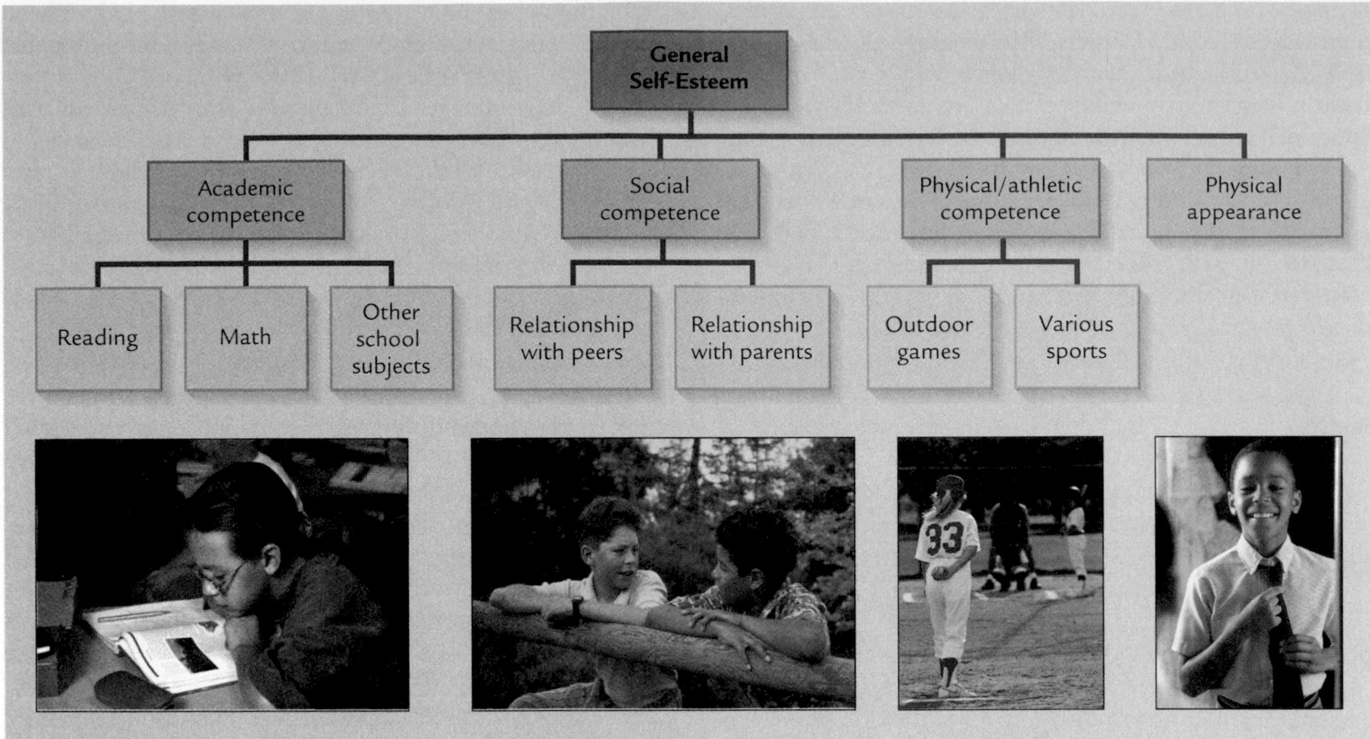

■ **FIGURE 10.1 Hierarchical structure of self-esteem in the mid-elementary school years.** From their experiences in different settings, children form at least four separate self-esteems: academic competence, social competence, physical/athletic competence, and physical appearance. These differentiate into additional self-evaluations and combine to form a general sense of self-esteem.

sense of self-esteem (Harter, 1998, 1999). As a result, self-esteem takes on the hierarchical structure shown in Figure 10.1.

Separate self-esteems, however, do not contribute equally to general self-esteem. Children attach greater importance to certain self-judgments. Although individual differences exist, during childhood and adolescence, perceived physical appearance correlates more strongly with overall self-worth than any other self-esteem factor (Hymel et al., 1999). The emphasis that society and the media place on appearance has major implications for young people's overall satisfaction with the self.

● **Changes in Level of Self-Esteem.** As children evaluate themselves in various areas, self-esteem drops during the first few years of elementary school (Marsh, Craven, & Debus, 1998; Wigfield et al., 1997). Typically, this decline is not great enough to be harmful. Most (but not all) children appraise their characteristics and competencies realistically while maintaining an attitude of self-acceptance and self-respect. Then, from fourth to sixth grade, self-esteem rises for the majority of youngsters, who feel especially good about their peer relationships and athletic capabilities (Twenge & Campbell, 2001; Zimmerman et al., 1997).

Influences on Self-Esteem

From middle childhood on, strong relationships exist between self-esteem and everyday behavior. Academic self-esteem predicts children's school achievement (Marsh, Smith, & Barnes, 1985). Children with high social self-esteem are better liked by peers (Harter, 1982). And as we saw in Chapter 9, boys come to believe they have more athletic talent than girls, and they are more advanced in a variety of physical skills. Furthermore, a profile of low self-esteem in all areas is linked to anxiety, depression, and increasing antisocial behavior (DuBois et al., 1999). What social influences might lead self-esteem to be high for some children and low for others?

● **Culture.** Cultural forces profoundly affect self-esteem. For example, an especially strong emphasis on social comparison in school may explain why Chinese and Japanese children score lower in self-esteem than North American children, despite their higher academic achievement (Chiu, 1992–1993; Hawkins, 1994). In Asian classrooms, competition is tough and achievement pressure is high. At the same time, Asian children less often call on social comparisons to promote their own self-esteem. Because their culture values modesty and social harmony, they tend to be reserved about judging themselves positively but generous in their praise of others (Heine & Lehman, 1995; Falbo et al., 1997).

Furthermore, a widely accepted cultural belief is that boys' overall sense of self-esteem is higher than girls', yet the difference is small. Girls may think less well of themselves because they internalize this negative cultural message (Kling et al., 1999). Compared with their Caucasian agemates, African-American children tend to have slightly higher self-esteem, perhaps because of warm, extended families and a strong sense of ethnic pride (Gray-Little & Hafdahl, 2000). And children and adolescents who attend schools or live in neighborhoods where their SES and ethnic groups are well represented feel a stronger sense of belonging and have fewer self-esteem problems (Gray-Little & Carels, 1997).

● **Child-Rearing Practices.** Children whose parents use an *authoritative* child-rearing style (see Chapter 8) feel especially good about themselves (Carolson, Uppal, & Prosser, 2000; Feiring & Taska, 1996). Warm, positive parenting lets children know that they are accepted as competent and worthwhile. And firm but appropriate expectations, backed up with explanations, seem to help children evaluate their own behavior against reasonable standards.

When parents help or make decisions for their youngsters when they do not need assistance, children often suffer from low self-esteem. These controlling parents communicate a sense of inadequacy to children (Pomerantz & Eaton, 2000). And overly tolerant, indulgent parenting is linked to unrealistically high self-esteem, which also undermines development. Children who feel superior to others tend to lash out at challenges to their overblown self-images and to have adjustment problems, including meanness and aggression (Hughes, Cavell, & Grossman, 1997).

Of special concern is that American cultural values have increasingly emphasized a focus on the self, perhaps leading parents to indulge children and boost their self-esteem too much. As Figure 10.2 illustrates, the self-esteem of American young people has risen sharply over the past few decades (Twenge & Campbell, 2001). Yet compared with previous generations, American youths are achieving less well and displaying more antisocial behavior.

Because the relationships between child rearing and self-esteem just described are correlational, we cannot really tell the extent to which parenting behaviors influence children's sense of self-worth. Research on the precise content of adults' messages to children has isolated factors that affect self-esteem. Let's see how these messages mold children's evaluations of themselves in achievement contexts.

● **Making Achievement-Related Attributions.** *Attributions* are our common, everyday explanations for the causes of behavior—our answers to the question "Why did I (or another person) do that?" Notice how Joey, in talking about the spelling bee at the beginning of this chapter, attributes his disappointing performance to *luck* (Belinda got all the easy words) and his usual success to *ability* (he *knows* he's a better speller than Belinda). Joey also appreciates that *effort* makes a difference; he "knocked himself out studying those spelling words."

Cognitive development permits school-age children to separate all these variables in explaining performance (Skinner, 1995). Those who are high in academic self-esteem make **mastery-oriented attributions,** crediting their successes to ability—a characteristic they can improve through trying hard and can count on when faced with new challenges. And they attribute failure to factors that can be changed and controlled, such as insufficient effort or a very difficult task (Heyman & Dweck, 1998). So whether these children succeed or fail, they take an industrious, persistent approach to learning.

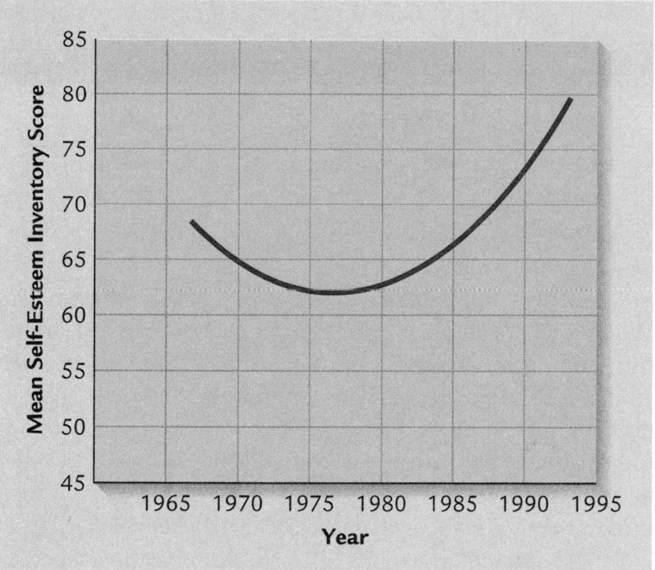

■ **FIGURE 10.2 Cohort effects for American junior-high students' self-esteem from 1965 to 1995.** Self-esteem dropped slightly during the late 1960s and 1970s. From 1980 on, a period of considerable public focus on boosting children's self-esteem, average self-esteem rose sharply. Self-esteem scores for younger and older students show a similar rise. (From J. M. Twenge & W. Keith Campbell, 2001, "Age and Birth Cohort Differences in Self-Esteem: A Cross-Temporal Meta-Analysis," *Personality and Social Psychology Review, 5,* p. 336. Adapted by permission.)

Unfortunately, children who develop **learned helplessness** attribute their failures, not their successes, to ability. When they succeed, they are likely to conclude that external factors, such as luck, are responsible. Furthermore, unlike their mastery-oriented counterparts, they have come to believe that ability is fixed and cannot be changed by trying hard. So when a task is difficult, these children experience an anxious loss of control—in Erikson's terms, a pervasive sense of inferiority. They give up before they have really tried (Elliott & Dweck, 1988).

Over time, the ability of learned-helpless children no longer predicts their performance. Because they fail to make the connection between effort and success, learned-helpless children do not develop the metacognitive and self-regulatory skills necessary for high achievement (see Chapter 9). Lack of effective learning strategies, reduced persistence, and a sense of being controlled by external forces sustain one another in a vicious cycle (Pomerantz & Saxon, 2001).

● **Influences on Achievement-Related Attributions.** What accounts for the different attributions of mastery-oriented and learned-helpless children? Adult communication plays a key role. Learned-helpless children have parents who set unusually high standards yet believe their child is not very capable and has to work harder than others to succeed. After these children fail, the parent might say, "You can't do that, can you? It's OK if you quit" (Hokoda & Fincham, 1995). And after the child succeeds, evaluating the child's traits, as in, "You're so smart," can promote helplessness. When used often, trait statements lead to a fixed view of ability, which encourages children to question their competence in the face of setbacks (Erdley et al., 1997).

Teachers' messages also affect children's attributions. When teachers are caring and helpful and emphasize learning over getting good grades, they tend to have mastery-oriented students (Anderman et al., 2001). In a study of 1,600 third to eighth graders, students who viewed their teachers as providing positive, supportive learning conditions worked harder and participated more in class—factors that predicted high achievement, which sustained children's belief in the role of effort. In contrast, students with unsupportive teachers regarded their performance as externally controlled (by teachers or luck). This predicted withdrawal from learning activities and declining achievement—outcomes that led children to doubt their ability (Skinner, Zimmer-Gembeck, & Connell, 1998).

Some children are especially likely to have their performance undermined by adult feedback. Despite their higher achievement, girls more often than boys blame their ability for poor performance. Girls also tend to receive messages from teachers and parents that their ability is at fault when they do not do well (Cole et al., 1999; Ruble & Martin, 1998). And in several studies, African-American and Mexican-American children received less favorable feedback from teachers (Irvine, 1986; Losey, 1995). Furthermore, when ethnic minority children observe that adults in their own family are not rewarded by society for their achievement efforts, they may give up themselves (Ogbu, 1997).

● **Supporting Children's Self-Esteem.** Attribution research suggests that at times, well-intended messages from adults undermine children's competence. *Attribution retraining* is an intervention that encourages learned-helpless children to believe that they can overcome failure by exerting more effort. Most often, children are given tasks hard enough so they will experience some failure. Then they get repeated feedback that helps them revise their attributions, such as, "You can do it if you try harder." Children are also taught to view their successes as due to both ability and effort rather than chance, by giving them additional feedback after they succeed, such as, "You're really good at this." Another approach is to encourage low-effort children to focus less on grades and more on mastering a task for its own sake (Ames, 1992). Instruction in metacognition and self-regulation is also helpful, to make up for learning lost in this area and to ensure that renewed effort will pay off (Borkowski & Muthukrisna, 1995).

To work well, attribution retraining is best begun early, before children's views of themselves become hard to change (Eccles, Wigfield, & Schiefele, 1998). An even better approach is to prevent learned helplessness, using the strategies summarized in the Caregiving Concerns table below.

Ways to Foster a Mastery-Oriented Approach to Learning

TECHNIQUE	DESCRIPTION
Provision of tasks	Select tasks that are meaningful, responsive to a diversity of student interests, and appropriately matched to current competence so that the child is challenged but not overwhelmed.
Parent and teacher encouragement	Communicate warmth, confidence in the child's abilities, the value of achievement, and the importance of effort in success. Model high effort in overcoming failure. (For teachers) Communicate often with parents, suggesting ways to foster children's effort and progress. (For parents) Monitor schoolwork; provide assistance that promotes knowledge of effective strategies and self-regulation.
Performance evaluations	Make evaluations private; avoid publicizing success or failure through wall posters, stars, privileges to "smart" children, and prizes for "best" performance. Stress individual progress and self-improvement.
School environment	Offer small classes, which permit teachers to provide individualized support for mastery. Provide for cooperative learning, in which children assist each other; avoid ability grouping, which makes evaluations of children's progress public. Accommodate individual and cultural differences in styles of learning. Create an atmosphere that sends a clear message that all students can learn.

Sources: Ames, 1992; Eccles, Wigfield, & Schiefele, 1998.

© MARY KATE DENNY/PHOTOEDIT

■ Repeated negative evaluations of their ability can cause children to develop learned helplessness—the belief that ability cannot be improved by trying hard. This learned-helpless boy seems to have concluded that he cannot improve. When faced with a challenging task, he is overwhelmed by negative thoughts and anxiety.

Ask Yourself

REVIEW

How does level of self-esteem change in middle childhood, and what accounts for these changes?

APPLY

Should parents promote children's self-esteem by telling them they're "smart" and "terrific"? Is it harmful if children do not feel good about everything they do? Why or why not?

CONNECT

What cognitive changes, described in Chapter 9, support the transition to a self-concept emphasizing competencies, personality traits, and social comparisons?

www.

Emotional Development

Greater self-awareness and social sensitivity support emotional development in middle childhood. Gains take place in experience of self-conscious emotions, understanding of emotional states, and emotional self-regulation.

Self-Conscious Emotions

In middle childhood, the self-conscious emotions of pride and guilt become clearly governed by personal responsibility. An adult need not be present for a new accomplishment to spark pride or for a transgression to arouse guilt (Harter & Whitesell, 1989). Also, children do not report guilt for any mishap, as they did at younger ages, but only for intentional wrongdoing, such as ignoring responsibilities, cheating, or lying (Ferguson, Stegge, & Damhuis, 1991).

Pride motivates children to take on further challenges. And guilt prompts them to make amends and strive for self-improvement as well. But harsh, insensitive reprimands from adults—such as, "Everyone else can do it! Why can't you?"—can lead to intense shame, which (as noted in Chapter 8) is particularly destructive. As children form an overall sense of self-esteem, they can take one or two unworthy acts to be the whole of self-worth, setting up maladaptive responses of high self-blame and passive retreat or intense anger at others who participated in the shame-evoking situation (Ferguson et al., 1999; Lindsay-Hartz, de Rivera, & Mascolo, 1995).

Emotional Understanding

School-age children's understanding of mental activity means that they are more likely to explain emotion by referring to internal states, such as happy or sad thoughts, than to external events—the focus of preschoolers (Flavell, Flavell, & Green, 2001). Older children are also more aware of the diversity of emotional experiences. Around age 8, children recognize that they can experience more than one emotion at a time, each of which may be positive or negative and differ in intensity (Wintre & Vallance, 1994). For example, recalling the birthday present he received from his grandmother, Joey reflected, "I was very happy that I got something but a little sad that I didn't get just what I wanted."

An appreciation of mixed emotions helps children realize that people's expressions may not reflect their true feelings (Saarni, 1997). It also fosters awareness of self-conscious emotions. For example, 8- and 9-year-olds understand that pride combines two sources of happiness—joy over accomplishment and joy that a significant person recognized that accomplishment (Harter, 1999). Furthermore, children of this age can reconcile contradictory facial and situational cues in figuring out another's feelings, whereas younger children rely only on the emotional expression (Hoffner & Badzinski, 1989).

As with self-understanding, gains in emotional understanding are supported by cognitive development and social experiences, especially adults' sensitivity to children's feelings and willingness to discuss emotions. Together, these factors lead to a rise in empathy as well (Ricard & Kamberk-Kilicci, 1995). As children move closer to adolescence, advances in perspective taking permit an empathic response not just to people's immediate distress, but also to their general life condition (Hoffman, 2000). As Joey and Lizzie imagined how people who are chronically ill or hungry feel and evoked those emotions in themselves, they gave part of their allowance to charity and joined in fundraising projects through school, church, and scouting.

Emotional Self-Regulation

Rapid gains in emotional self-regulation occur in middle childhood. As children engage in social comparison and care more about peer approval, they must learn to manage negative emotion that threatens their self-esteem.

By age 10, most children have an adaptive set of strategies for regulating emotion (Kliewer, Fearnow, & Miller, 1996). In situations where they have some control over an outcome (an anxiety-provoking test at the end of the week), they view problem solving and seeking social support as the best strategies. When outcomes are beyond their control (having received a bad grade), they opt for distraction or redefining the situation ("Things could be worse. There'll be another test."). Compared with preschoolers, school-age children more often use these internal strategies to manage emotion, due to an improved ability to reflect on their thoughts and feelings (Brenner & Salovey, 1997).

When emotional self-regulation has developed well, school-age children acquire a sense of *emotional self-efficacy*—a feeling of being in control of their emotional experience (Saarni, 1999).

■ Many children felt intensely fearful after witnessing on television the September 11, 2001, terrorist attack on the World Trade Center in New York City. At the suggestion of teachers, these children use an adaptive strategy to regulate their emotions. They offer sympathy to those directly harmed by attaching comforting messages to a flag their school made for families of victims. The flag was placed on public display.

This fosters a favorable self-image and an optimistic outlook, which further help children face emotional challenges.

Emotionally well-regulated children are generally upbeat in mood, more empathic and prosocial, and better liked by their peers. In contrast, poorly regulated children are overwhelmed by negative emotion, a response that interferes with prosocial behavior and peer acceptance (Eisenberg, Fabes, & Losoya, 1997). Recall from previous chapters that temperament and parenting influence emotional self-regulation. We will revisit these themes when we take up peer acceptance and children's ability to cope with stressful family circumstances.

Understanding Others: Perspective Taking

Already we have seen that middle childhood brings major advances in **perspective taking,** the capacity to imagine what other people may be thinking and feeling—changes that support self-concept and self-esteem, understanding of others, and a wide variety of social skills. Robert Selman's five-stage sequence describes changes in perspective-taking skill, based on children's and adolescents' responses to social dilemmas in which characters have differing information and opinions about an event.

As Table 10.1 indicates, at first children have only a limited idea of what other people might be thinking and feeling. Over time, they become more aware that people can interpret the same event quite differently. Soon, they can "step in another person's shoes" and reflect on how that person might regard their own thoughts, feelings, and behavior, as when they make statements like this: "I *thought you would think* I was just kidding when I said that." Finally, they can evaluate two people's perspectives simultaneously, at first from the vantage point of a disinterested spectator, and later by making reference to societal values. The following explanation reflects this advanced level: "I know why Joey hid the stray kitten in the basement, even though his mom was against keeping it. He believes in not hurting animals. If you put the kitten outside or give it to the pound, it might die."

Perspective taking varies greatly among children of the same age. Individual differences are due to cognitive maturity as well as experiences in which adults and peers explain their viewpoints, encouraging children to notice another's perspective (Dixon & Moore, 1990). Children with poor social skills—in particular, the angry, aggressive styles that we discussed in Chapter 8—have great difficulty imagining the thoughts and feelings of others. They often mistreat adults and peers without feeling the guilt and remorse prompted by awareness of another's viewpoint. Interventions that provide coaching and practice in perspective taking are helpful in reducing antisocial behavior and increasing empathy and prosocial responding (Chalmers & Townsend, 1990; Chandler, 1973).

Table 10.1 Selman's Stages of Perspective Taking

Stage	Approximate Age Range	Description
Level 0: Undifferentiated perspective taking	3–6	Children recognize that self and other can have different thoughts and feelings, but they frequently confuse the two.
Level 1: Social-informational perspective taking	4–9	Children understand that different perspectives may result because people have access to different information.
Level 2: Self-reflective perspective taking	7–12	Children can "step in another person's shoes" and view their own thoughts, feelings, and behavior from the other person's perspective. They also recognize that others can do the same.
Level 3: Third-party perspective taking	10–15	Children can step outside a two-person situation and imagine how the self and other are viewed from the point of view of a third, impartial party.
Level 4: Societal perspective taking	14–adult	Individuals understand that third-party perspective taking can be influenced by one or more systems of larger societal values.

Sources: Selman, 1976; Selman & Byrne, 1974.

Moral Development

Recall from Chapter 8 that preschoolers pick up a great many morally relevant behaviors through modeling and reinforcement. By middle childhood, they have had time to reflect on these experiences and internalize rules for good conduct, such as, "It's good to help others in trouble" or "It's wrong to take something that doesn't belong to you." This change leads children to become considerably more independent and trustworthy. They can take on many more responsibilities, from running errands at the supermarket to watching over younger siblings (Weisner, 1996). Of course, these advances take place only when children have had the consistent guidance and example of caring adults in their lives.

In Chapter 8, we also saw that children do not just copy their morality from others. As the cognitive-developmental approach emphasizes, they actively think about right and wrong. An expanding social world, the capacity to consider more information when reasoning, and perspective taking lead moral understanding to improve greatly in middle childhood.

Learning About Justice Through Sharing

In everyday life, children frequently experience situations that involve **distributive justice**—beliefs about how to divide material goods fairly. Heated discussions take place over how much weekly allowance is to be given to siblings of different ages, who has to sit where in the family car on a long trip, and in what way an eight-slice pizza is to be shared by six hungry playmates. William Damon (1977, 1988) has traced children's concepts of distributive justice over early and middle childhood.

■ These fourth-grade boys are figuring out how to divide a handful of penny candy fairly among themselves. Already, they have a well-developed sense of distributive justice.

Even 4-year-olds recognize the importance of sharing, but their reasons often seem self-serving: "I shared because if I didn't, she wouldn't play with me" or "I let her have some, but most are for me because I'm older." As children enter middle childhood, they express more mature notions of distributive justice. Their basis of reasoning follows an age-related, three-step sequence:

1. *Equality* (5 to 6 years). Children in the early school grades are intent on making sure that each person gets the same amount of a treasured resource, such as money, turns in a game, or a delicious treat.

2. *Merit* (6 to 7 years). A short time later, children say extra rewards should go to someone who has worked especially hard or otherwise performed in an exceptional way.

3. *Benevolence* (around 8 years). Finally, children recognize that special consideration should be given to those at a disadvantage—for example, that an extra amount might be given to a child who cannot produce as much or who does not get any allowance. Older children also adapt their basis of fairness to fit the situation, relying more on equality when interacting with strangers and more on benevolence when interacting with friends (McGillicuddy-De Lisi, Watkins, & Vinchur, 1994).

According to Damon (1988), the give-and-take of peer interaction makes children more sensitive to others' perspectives, and this, in turn, supports their developing ideas of justice. Advanced distributive justice reasoning, in turn, is associated with more effective social problem solving and a greater willingness to help and share with others (Blotner & Bearison, 1984; McNamee & Peterson, 1986).

Changes in Moral and Social-Conventional Understanding

As their ideas about justice advance, children clarify and link moral rules and social conventions. Over time, their understanding becomes more complex, taking into account an increasing number of variables.

School-age children, for example, distinguish social conventions with a clear *purpose* (not running in the school hallways to prevent injuries) from ones with no obvious justification (crossing a "forbidden" line on the playground). They regard violations of purposeful social conventions as closer to moral transgressions (Buchanan-Barrow & Barrett, 1998). They also realize that people's *intentions* and the *contexts* of their actions affect the moral implications of violating a social convention. In a Canadian study, 8- to 10-year-olds judged that because of a flag's symbolic value, burning it to express disapproval of a country or to start a cooking fire is worse than burning it accidentally. At the same time, 10-year-olds recognized that flag burning is a form of freedom of expression. Most agreed that in an unfair country, it would be acceptable (Helwig & Prencipe, 1999).

Children in Western and non-Western cultures use the same criteria to distinguish moral and social-conventional concerns (Nucci, Camino, & Sapiro, 1996; Tisak, 1995). When a directive is fair and caring, such as telling children to stop fighting or to share candy, school-age children view it as right, regardless of who states it—a principal, a teacher, or a child with no authority. This is true even for Korean children, whose culture places a high value on deference to authority. Korean 7- to 11-year-olds evaluate negatively a teacher's or principal's order to engage in immoral acts, such as stealing or refusing to share—a response that strengthens with age (Kim, 1998; Kim & Turiel, 1996).

Ask Yourself

REVIEW

How does emotional self-regulation improve in middle childhood? What implications do these changes have for children's self-esteem?

APPLY

Joey's fourth-grade class participated in a bowl-a-thon to raise money for a charity serving children with cancer. Explain how such activities can foster emotional development, perspective taking, and moral understanding.

CONNECT

Describe how older children's capacity to take more information into account affects each of the following: self-concept, emotional understanding, perspective taking, and moral understanding.

www.

Peer Relations

In middle childhood, the society of peers becomes an increasingly important context for development. Peer contact, as we have seen, contributes to perspective taking and understanding of self and others. These developments, in turn, enhance peer interaction, which becomes more prosocial over the school years. In line with this change, aggression declines, but the drop is greatest for physical attacks (Tremblay, 2000). As we will see, other types of hostile aggression continue as children form peer groups and start to distinguish "insiders" from "outsiders."

Peer Groups

Watch children in the school yard or neighborhood, and notice how groups of three to a dozen or more often gather. The organization of these collectives changes greatly with age. By the end of middle childhood, children display a strong desire for group belonging. They form **peer groups,** collectives that generate unique values and standards for behavior and a social structure of leaders and followers. Peer groups organize on the basis of proximity (being in the same classroom) and similarity in sex, ethnicity, and popularity (Cairns, Xie, & Leung, 1998).

The practices of these informal groups lead to a "peer culture" that typically consists of a specialized vocabulary, dress code, and place to "hang out" during leisure hours. For example, Joey formed a club with three other boys. They met in the treehouse in Joey's backyard and wore a "uniform" consisting of T-shirts, jeans, and tennis shoes. Calling themselves "the pack," the boys developed a secret handshake and chose Joey as their leader. Their activities included improving the club-

■ Peer groups first form in middle childhood. These girls have probably established a social structure of leader and followers as they gather for joint activities. Their body language suggests that they feel a strong sense of group belonging.

house, trading baseball cards, playing video games, and—just as important—keeping girls and adults out!

As children develop these exclusive associations, the codes of dress and behavior that grow out of them become more broadly influential. At school, children who deviate are often rebuffed. "Kissing up" to teachers, wearing the wrong kind of shirt or shoes, and tattling on classmates are grounds for critical glances and comments. These special customs bind peers together, creating a sense of group identity. Within the group, children acquire many social skills—cooperation, leadership, followership, and loyalty to collective goals. Through these experiences, children experiment with and learn about social organizations.

The beginning of peer group ties is also a time when some of the "nicest children begin to behave in the most awful way" (Redl, 1966). From third grade on, relational aggression—gossip, rumor spreading, and exclusion—rise among girls, who (because of gender-role expectations) express hostility in subtle, indirect ways (Crick & Grotpeter, 1995). Boys are more straightforward in their hostility toward the "out-group." Overt aggression, in the form of verbal insults and pranks—toilet-papering a front yard or ringing a doorbell and running away—occurs among small groups of boys, who provide one another with temporary social support for these mildly antisocial behaviors.

Unfortunately, peer groups often direct their hostilities toward their own members, excluding no-longer "respected" children. These cast-outs are profoundly wounded, and many find new group ties hard to establish. Their previous behavior, including expressed contempt for outsiders, reduces their chances of being included elsewhere. Excluded children often turn to other low-status peers for group belonging (Bagwell et al., 2001). As they join groups of children with poor social skills, they reduce their opportunities to learn socially competent behavior.

The school-age child's desire for group membership can also be satisfied through formal group ties—scouting, 4-H, religious youth groups, and other associations. Adult involvement holds in check the negative behaviors associated with children's informal peer groups. And as children work on joint projects and help in their communities, they gain in social and moral maturity (Killen & Nucci, 1995; Vandell & Shumow, 1999).

Friendships

Whereas peer groups provide children with insight into larger social structures, one-to-one friendships contribute to the development of trust and sensitivity. During the school years, friendship becomes more complex and psychologically based. Consider the following 8-year-old's ideas:

> *Why is Shelly your best friend?* Because she helps me when I'm sad, and she shares.... *What makes Shelly so special?* I've known her longer, I sit next to her and got to know her better.... *How come you like Shelly better than anyone else?* She's done the most for me. She never disagrees, she never eats in front of me, she never walks away when I'm crying, and she helps me with my homework.... *How do you get someone to like you?* ... If you're nice to [your friends], they'll be nice to you. (Damon, 1988, pp. 80–81)

As these responses show, friendship is no longer just a matter of engaging in the same activities. Instead, it is a mutually agreed-on relationship in which children like each other's

■ During middle childhood, concepts of friendship become more psychologically based. Although these boys both enjoy playing baseball, they want to spend time together because they like each other's personal qualities. Mutual trust is a defining feature of their friendship. Each child counts on the other for support and assistance.

personal qualities and respond to one another's needs and desires. And once a friendship forms, *trust* becomes its defining feature. School-age children state that a good friendship is based on acts of kindness that signify that each person can be counted on to support the other. Consequently, older children regard violations of trust, such as not helping when others need help, breaking promises, and gossiping behind the other's back, as serious breaches of friendship (Damon, 1977; Selman, 1980).

Because of these features, school-age children's friendships are more selective. Whereas preschoolers say they have lots of friends, by age 8 or 9, children name only a handful of good friends. Girls, especially, are more exclusive in their friendships because they demand greater closeness than boys (Markovitz, Benenson, & Dolensky, 2001). In addition, children tend to select friends like themselves in age, sex, race, ethnicity, and SES. Friends also resemble one another in personality (sociability, aggression), peer popularity, academic achievement, and prosocial behavior (Hartup, 1996). Note, however, that school and neighborhood characteristics affect friendship choices. For example, in integrated schools, 50 percent of students report at least one close other-race friend (DuBois & Hirsch, 1990).

Friendships remain fairly stable over middle childhood; most last for several years. Through them, children learn the importance of emotional commitment. They come to realize that close relationships can survive disagreements if friends are secure in their liking for one another (Rose & Asher, 1999). As a result, friendship provides an important context in which children learn to tolerate criticism and resolve disputes.

Yet the impact that friendships have on children's development depends on the nature of those friends. Children who bring kindness and compassion to their friendships strengthen each other's prosocial tendencies. When aggressive children make friends, the relationship often magnifies antisocial acts. The friendships of aggressive girls are high in exchange of private feelings but full of jealousy, conflict, and betrayal (Grotpeter & Crick, 1996). Among boys, aggressive friends' talk contains frequent coercive statements and attacks (Dishion, Andrews, & Crosby, 1995). These findings indicate that the social problems of aggressive children operate within their closest peer ties. As we will see next, these children are also at risk for rejection in the wider world of peers.

Peer Acceptance

Peer acceptance refers to likability—the extent to which a child is viewed by a group of agemates, such as classmates, as a worthy social partner. It differs from friendship in that it is not a mutual relationship. Rather, it is a one-sided perspective, involving the group's view of an individual. Nevertheless, some social skills that contribute to friendship also enhance peer acceptance. Consequently, better accepted children have more friends and better relationships with them (Gest, Graham-Bermann, & Hartup, 2001).

Researchers usually assess peer acceptance with self-report measures that ask classmates to evaluate one another's likability. Children's responses reveal four different categories: **popular children,** who get many positive votes; **rejected children,** who are actively disliked; **controversial children,** who get a large number of positive and negative votes; and **neglected children,** who are seldom chosen, either positively or negatively. About two-thirds of students in a typical elementary school classroom fit one of these categories (Coie, Dodge, & Coppotelli, 1982). The remaining one-third are *average* in peer acceptance; they do not receive extreme scores.

Peer acceptance is a powerful predictor of psychological adjustment. Rejected children, especially, are unhappy, alienated, poorly achieving children with low self-esteem. Both teachers and parents rate them as having a wide range of emotional and social problems. Peer rejection in middle childhood is also strongly associated with poor school performance, dropping out, antisocial behavior, and delinquency in adolescence and with criminality in young adulthood (Laird et al., 2001; Parker et al., 1995).

However, preceding influences—children's characteristics combined with parenting practices—may largely explain the link between peer acceptance and adjustment. School-age children with problems in peer relationships are more likely to have experienced family stress due to low income, insensitive child rearing, and coercive discipline (Woodward & Fergusson, 1999). Nevertheless, as we will see, rejected children evoke reactions from peers that contribute to their unfavorable development.

● **Determinants of Peer Acceptance.** What causes one child to be liked and another to be rejected? A wealth of research reveals that social behavior plays a powerful role.

Popular Children. Although most popular children are kind and considerate, a few are admired for their socially adept yet belligerent behavior. The large majority are **popular-prosocial children,** who combine academic and social competence. They perform well in school and communicate with peers in sensitive, friendly, and cooperative ways (Newcomb, Bukowski, & Pattee, 1993). In contrast, **popular-antisocial children** largely consist of "tough" boys who are athletically skilled but poor students. Although they are aggressive, their peers view them as "cool," perhaps because of their athletic ability and shrewd but devious social skills (Rodkin et al., 2000). Many are low-SES minority children who have concluded that they cannot succeed academically (Stormshak et al., 1999). Although their likability may offer some protection from future maladjustment, their poor school performance and antisocial behavior require intervention.

Rejected Children. Rejected children display a wide range of negative social behaviors. The largest subgroup, **rejected-aggressive children,** show high rates of conflict, hostility, and hyperactive, inattentive, and impulsive behavior. They are also deficient in perspective taking and regulation of negative

Biology & Environment

Bullies and Their Victims

Follow the activities of aggressive children over a school day, and you will see that they reserve their hostilities for certain peers. A particularly destructive form of interaction is **peer victimization,** in which certain children become frequent targets of verbal and physical attacks or other forms of abuse. What sustains these repeated assault–retreat cycles between pairs of children?

Research indicates that the majority of victims reinforce bullies by giving in to their demands, crying, assuming defensive postures, and failing to fight back. Victimized boys are passive when active behavior is expected. On the playground, they hang around chatting or wander on their own (Boulton, 1999). Biologically based traits—an inhibited temperament and a frail physical appearance—contribute to their behavior. But victimized children also have histories of resistant attachment; overly intrusive, controlling child rearing; and maternal overprotectiveness. These parenting behaviors prompt anxiety, low self-esteem, and dependency, resulting in a fearful demeanor that radiates vulnerability (Ladd & Ladd, 1998; Pepler & Craig, 2000).

About 10 percent of children and adolescents are harassed by aggressive agemates, who attack to achieve social status (Nansel et al., 2001). Peers expect these victims to give up desirable objects, show signs of distress, and fail to retaliate. In addition, children (especially those who are aggressive) feel little discomfort at the thought of causing pain and suffering to victims (Pellegrini, 2002).

Although bullies and victims are most often boys, at times they are girls who bombard a vulnerable classmate with relational hostility (Crick & Grotpeter, 1996). As early as kindergarten, victimization leads to a variety of adjustment difficulties, including depression, loneliness, low self-esteem, anxiety, and avoidance of school (Hawker & Boulton, 2000; Kochenderfer-Ladd & Wardrop, 2001).

Aggression and victimization are not polar opposites. A small number of extreme victims are also aggressive, picking fights or retaliating with relational aggression (Boulton & Smith, 1994; Crick & Bigbee, 1998). Perhaps these children foolishly provoke stronger agemates, who then prevail over them. Among rejected children, these bully/victims are the most despised, placing them at severe risk for maladjustment.

■ Children who are victimized by bullies have characteristics that make them easy targets. They are physically weak, rejected by peers, and afraid to defend themselves. Both temperament and child-rearing experiences contribute to their cowering behavior, which reinforces their attackers' abusive acts.

Interventions that change victimized children's negative opinions of themselves and that teach them to respond in nonreinforcing ways to their attackers are vital. Nevertheless, victimized children's behavior should not be taken to mean they are to blame for their abuse. Developing a school code against bullying, enlisting parents' assistance in changing both bullies' and victims' behavior, and moving aggressive children to another class or school can greatly reduce peer victimization (Olweus, 1995).

Another way to assist victimized children is to help them acquire the social skills needed to form and maintain a gratifying friendship. Anxious, withdrawn children who have a best friend seem better equipped to withstand peer attacks. They show fewer adjustment problems than victims with no close friends (Hodges et al., 1999).

emotion. For example, they tend to misinterpret the innocent behaviors of peers as hostile, to blame others for their social difficulties, and to act on their angry feelings (Coie & Dodge, 1998; Rubin et al., 1995). In contrast, **rejected-withdrawn children** are passive and socially awkward. These timid children are overwhelmed by social anxiety, hold negative expectations for how peers will treat them, and are very concerned about being scorned and attacked (Hart et al., 2000; Ladd & Burgess, 1999). Because of their inept, submissive style of interaction, rejected-withdrawn children are at risk for abuse by bullies (see the Biology and Environment box above).

Beginning in kindergarten, peers exclude rejected children. As a result, rejected children's classroom participation declines, their feelings of loneliness rise, their academic achievement falters, and they want to avoid school (Buhs & Ladd, 2001). Rejected children usually have few friends, and occasionally none.

Controversial and Neglected Children. Consistent with the mixed peer opinion they engender, controversial children display a blend of positive and negative social behaviors. Like rejected-aggressive children, they are hostile and disruptive, but they also engage in positive, prosocial acts. Even though some peers dislike them, they have qualities that protect them from social exclusion. As a result, they appear to be relatively happy and comfortable with their peer relationships (Newcomb, Bukowski, & Pattee, 1993).

Finally, perhaps the most surprising finding is that neglected children, once thought to be in need of treatment, are usually well adjusted. Although they engage in low rates of interaction, the majority are just as socially skilled as average children. They do not report feeling especially lonely or unhappy, and when they want to, they can break away from their usual pattern of playing by themselves (Harrist et al., 1997; Ladd & Burgess, 1999).

Neglected children remind us that there are other paths to emotional well-being besides an outgoing, gregarious personality style. Recall from Chapter 6 that in China, adults view restrained, cautious children as advanced in social maturity. Perhaps because shyness is consistent with a cultural emphasis on not standing out from the collective, it is associated with peer acceptance and social competence among Chinese 8- to 10-year-olds (Chen, 2002; Chen, Rubin, & Li, 1995).

● **Helping Rejected Children.** A variety of interventions exist to improve the peer relations and psychological adjustment of rejected children. Most involve coaching, modeling, and reinforcing positive social skills, such as how to begin interacting with a peer, cooperate in games, and respond to another child with friendly emotion and approval. Several social-skills training programs have produced lasting gains in social competence and peer acceptance (Asher & Rose, 1997).

Combining social-skills training with other treatments increases their effectiveness. Often rejected children are poor students, and their low academic self-esteem magnifies negative reactions to teachers and classmates. Intensive academic tutoring improves both their school achievement and social acceptance (O'Neil et al., 1997).

Still another approach involves training in perspective taking and social problem solving. Many rejected-aggressive children are unaware of their poor social skills and do not take responsibility for their social failures (Mrug, Hoza, & Gerdes, 2001). Rejected-withdrawn children, on the other hand, are likely to develop a learned-helpless approach to peer acceptance. They conclude, after repeated rebuffs, that they will never be liked (Rubin, Bukowski, & Parker, 1998). Both types of children need help attributing their peer difficulties to internal, changeable causes.

Finally, because rejected children's socially incompetent behaviors often originate in a poor fit between the child's temperament and parenting practices, interventions that focus on the child alone may not be sufficient. If the quality of parent–child interaction is not changed, children may soon return to their old behavior patterns.

Gender Typing

Children's understanding of gender roles broadens in middle childhood, and their gender-role identities (views of themselves as relatively masculine or feminine) change as well. We will see that development differs for boys and girls, and it can vary considerably across cultures.

Gender-Stereotyped Beliefs

During the school years, children extend the gender-stereotyped beliefs they acquired in early childhood. As they think more about people as personalities, they label some traits as more typical of one gender than the other. For example, they regard "tough," "aggressive," "rational," and "dominant" as masculine and "gentle," "sympathetic," and "dependent" as feminine—stereotyping that increases steadily with age (Serbin, Powlishta, & Gulko, 1993). Children derive these distinctions from observing gender differences as well as from adult treatment. Parents, for example, use more directive speech (telling the child what to do) with girls, less often encourage girls to make their own decisions, and less often praise girls for accomplishment (Leaper, Anderson, & Sanders, 1998; Pomerantz & Ruble, 1998).

Shortly after entering elementary school, children figure out which academic subjects and skill areas are "masculine" and which are "feminine." They regard reading, spelling, art, and music as more for girls and mathematics, athletics, and mechanical skills as more for boys (Eccles, Jacobs, & Harold, 1990; Jacobs & Weisz, 1994). These stereotypes influence children's preferences for and sense of competence at certain subjects. For example, boys feel more competent than girls at math and science, whereas girls feel more competent than boys at reading and spelling—even when children of equal skill level are compared (Andre et al., 1999; Freedman-Doan et al., 2000). As we will see in Chapter 11, these beliefs become realities for many young people in adolescence.

Although school-age children are aware of many stereotypes, they have a more open-minded view of what males and females *can do.* The ability to classify flexibly underlies this change. School-age children realize that a person can belong to more than one social category—for example, be a "boy" yet "like to play house" (Bigler, 1995). But acknowledging that people *can* cross gender lines does not mean that children always *approve* of doing so. Children and adults are fairly tolerant of girls' violations of gender roles. But they judge boys' violations ("playing with dolls" or "wearing a dress") harshly—as just as bad as a moral transgression (Levy, Taylor, & Gelman, 1995).

Gender Identity and Behavior

Boys' and girls' gender-role identities follow different paths in middle childhood. From third to sixth grade, boys strengthen their identification with "masculine" personality traits, whereas girls' identification with "feminine" traits declines. Girls begin to describe themselves as having some "other-gender" characteristics (Serbin, Powlishta, & Gulko, 1993). This difference is also evident in children's activities. Whereas boys usually stick to "masculine" pursuits, girls experiment with a wider range of options. Besides cooking, crafts, and baby-sitting, they join organized sports teams, take up science projects, and build forts in the backyard.

These changes are due to a mixture of cognitive and social forces. School-age children of both sexes are aware that society attaches greater prestige to "masculine" characteristics For example, they rate "masculine" occupations as having higher status than "feminine" occupations (Liben, Bigler, & Krogh, 2001). Messages from adults and peers are also influential. In Chapter 8 we saw that parents are especially concerned about boys' gender-role conformity. A tomboyish girl can make her way into boys' activities without losing status with her female peers, but a boy who hangs out with girls is likely to be ridiculed and rejected.

Cultural Influences on Gender Typing

Although the sex differences just described are typical in Western nations, they do not apply to children everywhere. Girls are less likely to experiment with "masculine" activities in cultures and subcultures in which the gap between male and female roles is especially wide. And when social and economic conditions make it necessary for boys to take over "feminine" tasks, their personalities and behavior are less stereotyped.

■ Boys of the Kinka tribe in Kenya are assigned "feminine" tasks—grinding corn and looking after younger siblings. Consequently, they are less likely to be gender stereotyped in personality traits than are most boys in other cultures.

For example, in Nyansongo, a small agricultural settlement in Kenya, mothers work 4 to 5 hours a day in the gardens. They assign the care of young children, the tending of the cooking fire, and the washing of dishes to older siblings. Because children of both sexes perform these duties, girls are relieved of total responsibility for "feminine" tasks and have more time to interact with agemates. Their greater freedom and independence lead them to score higher than girls of other village and tribal cultures in dominance, assertiveness, and playful roughhousing. In contrast, boys' caregiving responsibilities mean that they frequently engage in help giving and emotional support (Whiting & Edwards, 1988a).

Should these findings be taken to suggest that boys in Western cultures be assigned more "cross-gender" tasks? The consequences of doing so are not straightforward. Research shows that when fathers hold traditional gender-role beliefs and their sons engage in "feminine" housework, boys experience strain in the father–child relationship and judge themselves as less competent (McHale et al., 1990). So parental values may need to be consistent with task assignments for children to benefit.

Ask Yourself

REVIEW

Return to Chapter 8, page 263 and review the concept of androgyny. Which of the two sexes is more androgynous in middle childhood, and why?

REVIEW

How does friendship change in middle childhood?

APPLY

What changes in parent–child relationships are probably necessary to help rejected children?

CONNECT

Using your understanding of attributions, explain how rejected-aggressive and rejected-withdrawn children would probably account for their failure to gain peer acceptance. How might these attributions affect future social behavior?

www.

Family Influences

As children move into school, peer, and community contexts, the parent–child relationship changes. At the same time, children's well-being continues to depend on the quality of family interaction. In the following sections, we will see that contemporary changes in the American family—high rates of divorce, remarriage, and maternal employment—can have positive as well as negative effects on children. In later chapters, we will take up other family forms, including gay and lesbian families, never-married single-parent families, and the increasing numbers of grandparents rearing grandchildren.

Parent–Child Relationships

In middle childhood, the amount of time children spend with parents declines dramatically. The child's growing independence means that parents must deal with new issues. "I've struggled with how many chores to assign, how much allowance to give, whether their friends are good influences, and what to do about problems at school," Rena remarked. "And then there's the challenge of how to keep track of them when they're out of the house or even when they're home and I'm not there to see what's going on."

Although parents face new concerns, child rearing becomes easier for those who established an authoritative style during the early years. Reasoning works more effectively with school-age children because of their greater capacity for logical thinking and increased respect for parents' expert knowledge (Collins, Madsen, & Susman-Stillman, 2002). As children demonstrate that they can manage daily activities and responsibilities, effective parents gradually shift control from adult to child. This does not mean they let go entirely. Instead, they engage in **coregulation,** a transitional form of supervision in which they exercise general oversight while permitting children to be in charge of moment-by-moment decision making.

Coregulation grows out of a cooperative relationship between parent and child—one based on give-and-take and mutual respect. Parents must guide and monitor from a distance and effectively communicate expectations when they are with their children. And children must inform parents of their whereabouts, activities, and problems so parents can intervene when necessary (Maccoby, 1984). Coregulation supports and protects children while preparing them for adolescence, when they will make many important decisions themselves.

Although school-age children often press for greater independence, they know how much they need their parents' continuing support. In one study, fifth and sixth graders described parents as the most influential people in their lives. They often turned to mothers and fathers for affection, advice, enhancement of self-worth, and assistance with everyday problems (Furman & Buhrmester, 1992).

© ERIKA STONE

■ Although sibling rivalry tends to increase in middle childhood, siblings also provide one another with emotional support and help with difficult tasks.

Siblings

In addition to parents and friends, siblings are important sources of support for school-age children. Yet sibling rivalry tends to increase in middle childhood. As children participate in a wider range of activities, parents often compare siblings' traits and accomplishments. The child who gets less parental affection, more disapproval, or fewer material resources is likely to be resentful (Brody, Stoneman, & McCoy, 1994; Dunn, 1996).

When siblings are close in age and the same sex, parental comparisons are more frequent, resulting in more quarreling, antagonism, and poorer adjustment. This effect is particularly strong when parenting is cold or harsh (Feinberg & Hetherington, 2001). It is also strengthens when fathers prefer one child. Perhaps because fathers spend less time with children, their favoritism is more noticeable and triggers greater anger (Brody, Stoneman, & McCoy, 1992).

Siblings often take steps to reduce this rivalry by striving to be different from one another. For example, two brothers I know deliberately selected different athletic pursuits and musical instruments. If the older one did especially well at an activity, the younger one did not want to try it. Of course, parents can reduce these effects by making an effort not to compare children. But some feedback about their competencies is inevitable, and as siblings strive to win recognition for their own uniqueness, they shape important aspects of each other's development.

Although conflict rises, school-age siblings continue to rely on each other for companionship and assistance. When researchers asked siblings about shared daily activities, children mentioned that older siblings often helped younger siblings with academic and peer challenges. And both offered one another help with family issues (Tucker, McHale, & Crouter, 2001). Siblings whose parents are preoccupied and less involved with them sometimes fill in and become more supportive of one another (Bank, Patterson, & Reid, 1996).

Only Children

Although sibling relationships bring many benefits, they are not essential for healthy development. Contrary to popular belief, only children are not spoiled. Instead, they are as well adjusted as other children and advantaged in some respects. Children in one-child families score higher in self-esteem and achievement motivation. Consequently, they do better in school and attain higher levels of education (Falbo, 1992). One reason may be that only children have somewhat closer relationships with their parents, who exert more pressure for mastery and accomplishment.

Favorable development also characterizes only children in China, where a one-child family policy has been strictly enforced in urban areas for 2 decades to control overpopulation. Compared with agemates who have siblings, Chinese only children are advanced in cognitive development and academic achievement (Falbo & Poston, 1993; Jiao, Ji, & Jing, 1996). They also feel more emotionally secure, perhaps because government disapproval promotes tension in families with more than one child (Yang et al., 1995). Although many Chinese adults remain convinced that the one-child family policy breeds self-centered "little emperors," Chinese only children do not differ from children with siblings in social skills and peer acceptance (Chen, Rubin, & Li, 1995).

■ Limiting family size is a basic national policy in the People's Republic of China. In urban areas, the majority of couples have no more than one child.

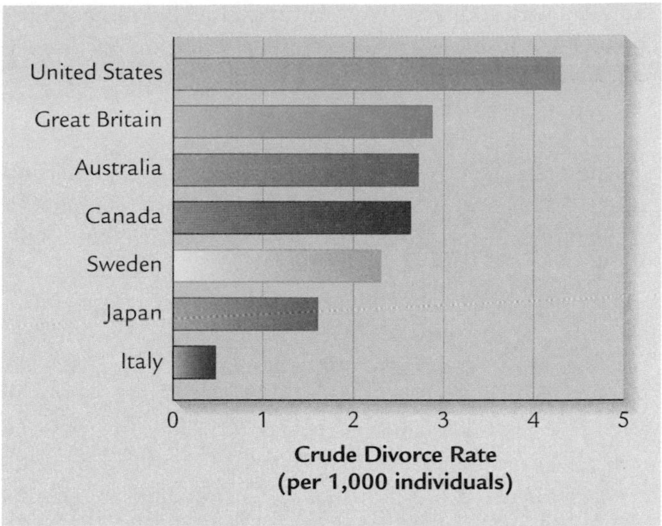

■ **FIGURE 10.3 Divorce rates in seven industrialized nations.** The U.S. divorce rate is the highest in the world, the Canadian divorce rate the fourth highest. (From Australian Bureau of Statistics, 2002; Statistics Canada, 2002c: U.S. Bureau of the Census, 2002c; United Nations, 1999.)

Divorce

Children's interactions with parents and siblings are affected by other aspects of family life. Joey and Lizzie's relationship, Rena told me, had been particularly negative only a few years before. Joey pushed, hit, and taunted Lizzie and called her names. Although she tried to retaliate, she was no match for Joey's larger size. The arguments usually ended with Lizzie running in tears to her mother. Joey's and Lizzie's fighting coincided with Rena and her husband's growing marital unhappiness. When Joey was 8 and Lizzie 5, their father, Drake, moved out.

The children were not alone in having to weather this traumatic event. Between 1960 and 1985, divorce rates in Western nations rose dramatically. Currently, the United States has the highest divorce rate in the world, Canada the fourth highest (see Figure 10.3). About 45 percent of American and 30 percent of Canadian marriages end in divorce; half involve children. At any given time, one-fourth of American children and one-fifth of Canadian children live in single-parent households. Although most reside with their mothers, the percentage in father-headed households has increased steadily, to about 12 percent in both nations (Hetherington & Stanley-Hagan, 2002; Statistics Canada, 2002c).

Children of divorce spend an average of 5 years in a single-parent home, or almost a third of childhood. For many, divorce leads to new family relationships. About two-thirds of divorced parents marry a second time. Half their children eventually experience a third major change—the end of a parent's second marriage (Hetherington, Bridges, & Insabella, 1998).

These figures reveal that divorce is not a single event in the lives of parents and children. Instead, it is a transition that leads to a variety of new living arrangements, accompanied by changes in housing, income, and family roles and responsibilities. Since the 1960s, many studies have reported that marital breakup is quite stressful for children (Amato & Booth, 2000). But the research also reveals great individual differences. How well children fare depends on many factors: the custodial parent's psychological health, the child's characteristics, and social supports within the family and surrounding community.

● **Immediate Consequences.** "Things were worst during the period Drake and I decided to separate," Rena reflected. "We fought over division of our belongings and the custody of the children, and the kids suffered. Sobbing, Lizzie told me she was 'sorry she made Daddy go away.' Joey kicked and threw things at home and didn't do his work at school. In the midst of everything, I could hardly deal with their problems. We had to sell the house; I couldn't afford it alone. And I needed a better-paying job."

Rena's description captures conditions in many newly divorced households. Family conflict often rises as parents try to settle disputes over children and possessions. Once one parent moves out, additional events threaten supportive interactions between parents and children. Mother-headed households typically experience a sharp drop in income. In Canada and the United States, the majority of single mothers with young children live in poverty, getting less than the full amount of child support from the absent father or none at all (Children's Defense Fund, 2002; Statistics Canada, 2002c). They often have to move to new housing for economic reasons, reducing supportive ties to neighbors and friends.

The transition from marriage to divorce typically leads to high maternal stress, depression, and anxiety and to a disorganized family situation (Hope, Power, & Rodgers, 1999; Marks & Lambert, 1998). "Meals and bedtimes were at all hours, the house didn't get cleaned, and I stopped taking Joey and Lizzie on weekend outings," said Rena. As children react with distress and anger to their less secure home lives, discipline may become harsh and inconsistent. Contact with noncustodial fathers often decreases over time (Hetherington & Kelly, 2002; Lamb, 1999). When fathers see their children only occasionally, they are inclined to be permissive and indulgent. This generally conflicts with the mother's style of parenting and makes her task of managing the child on a day-to-day basis even more difficult.

In view of these changes, it is not surprising that children experience painful emotional reactions. But the intensity of their feelings and the way they are expressed vary with the child's age, temperament, and sex.

Children's Age. Five-year-old Lizzie's fear that she had caused her father to leave is not unusual. The cognitive immaturity of preschool and early school-age children makes it difficult for them to grasp the reasons behind their parents' separation. Younger children often blame themselves and take the marital breakup as a sign that both parents may abandon them. They may whine and cling, displaying intense separation anxiety (Hetherington, 1989).

Older children are better able to understand the reasons behind their parents' divorce, which may reduce some of the pain they feel. Still, many school-age and adolescent youngsters react strongly, particularly when family conflict is high and supervision of children is low. Escaping into undesirable peer activities—such as running away, truancy, early sexual activity, and delinquency—and poor school achievement are common (Hetherington & Stanley-Hagan, 1999; Simons & Chao, 1996).

However, not all older children react this way. For some—especially the oldest child in the family—divorce can trigger more mature behavior. These youngsters may willingly take on extra burdens, such as household tasks, care and protection of younger siblings, and emotional support of a depressed, anxious mother. But if these demands are too great, older children may eventually become resentful and withdraw into some of the destructive behavior patterns just described (Hetherington, 1995, 1999a).

Children's Temperament and Sex. When temperamentally difficult children are exposed to stressful life events and inadequate parenting, their problems are magnified. In contrast, easy children are less often targets of parental anger and are also better at coping with adversity when it hits.

These findings help us understand sex differences in response to divorce. Girls sometimes respond as Lizzie did, with internalizing reactions, such as crying, self-criticism, and withdrawal. More often, they show demanding, attention-getting behavior. But in mother-custody families, boys typically experience more serious adjustment problems. Recall from Chapter 8 that boys are more active and noncompliant—behaviors that increase with exposure to parental conflict and inconsistent discipline. Coercive mother–child interaction and impulsive, defiant behavior on the part of sons are common in divorcing households (Hetherington & Kelly, 2002).

Perhaps because their behavior is so unruly, boys receive less emotional support from mothers, teachers, and peers. And as Joey's behavior toward Lizzie illustrates, the coercive cycles of interaction between boys and their divorced mothers soon spread to sibling relations (MacKinnon, 1989). These outcomes compound boys' difficulties. After divorce, children who are challenging to rear generally get worse (Hanson, 1999; Morrison & Coiro, 1999).

● **Long-Term Consequences.** Rena eventually found better-paying work and gained control over the daily operation of the household. Her own feelings of anger and rejection also declined. And after several meetings with a counselor, Rena and Drake realized the harmful impact of their quarreling on Joey and Lizzie. Drake visited regularly and handled Joey's unruliness with firmness and consistency. Soon Joey's school performance improved, his behavior problems subsided, and both children seemed calmer and happier.

Most children show improved adjustment by 2 years after divorce. Yet a few continue to have serious difficulties into early adulthood (Amato, 2000; Wolfinger, 2000). Boys and children with difficult temperaments are especially likely to drop out of school and display antisocial behavior. For both sexes, divorce is linked to problems with adolescent sexuality and with development of intimate ties. Young people who experience parental divorce—especially more than once—display high rates of adolescent parenthood and divorce in their adult lives (Booth, 1999; Hetherington, 1997).

The overriding factor in positive adjustment following divorce is effective parenting—in particular, how well the custodial parent handles stress and shields the child from conflict and the extent to which each parent uses authoritative child rearing (Amato & Gilbreth, 1999; Whiteside & Becker, 2000). In a study of 8- to 15-year-olds whose parents had divorced in the previous 2 years, children reporting high parental warmth and consistency of discipline had the fewest adjustment problems (Wolchik et al., 2000).

Fathers' involvement is also important. For girls, a good father–child relationship appears to protect against early sexual activity and unhappy romantic involvements. For boys, it seems to affect overall psychological well-being. In fact, several studies indicate that outcomes for sons are better when the father is the custodial parent (Clark-Stewart & Hayward, 1996; McLanahan, 1999). Fathers' greater economic security and image of authority seem to help them engage in effective parenting with sons. Furthermore, boys in father-custody families may benefit from greater involvement of both parents, since noncustodial mothers participate more in their children's lives than noncustodial fathers.

Although divorce is painful for children, remaining in a high-conflict intact family is much worse than making the transition to a low-conflict, single-parent household (Emery, 1999a; Hetherington, 1999b). When divorcing parents put aside their disagreements and support one another in their child-rearing roles, children have the best chance of growing up competent, stable, and happy. Caring extended family members, teachers, siblings, and friends also reduce the likelihood that divorce will result in long-term disruption (DeGarmo & Forgatch, 1999).

© PAUL BARTON/CORBIS

■ After parents divorce, children in mother-custody homes who also stay involved with their fathers fare better in development. Boys often adjust more favorably when the father is the custodial parent.

● **Divorce Mediation, Joint Custody, and Child Support.** Awareness that divorce is highly stressful for children and families has led to interventions aimed at helping families through this difficult time. One is **divorce mediation,** a series of meetings between divorcing adults and a trained professional aimed at reducing family conflict, including legal battles over property division and child custody. Research reveals that mediation increases out-of-court settlements, compliance with these agreements, cooperation between parents in child rearing, and feelings of well-being reported by divorcing parents and their children (Emery, 2001).

To encourage both parents to remain involved with children, courts today more often award **joint custody,** which grants the mother and father equal say in important decisions about the child's upbringing. In most instances, children reside with one parent and see the other on a fixed schedule, much like the typical sole-custody situation. But in other cases, parents share physical custody, and children move between homes and sometimes between schools and peer groups. These transitions introduce a new kind of instability that is especially hard on some children. The success of joint custody requires a cooperative relationship between divorcing parents (Emery, 1999a). If they continue to quarrel, it prolongs children's exposure to a hostile family atmosphere.

Finally, many single-parent families depend on child support from the absent parent to relieve financial strain. U.S. states and Canadian provinces have procedures for withholding wages from parents who fail to make these payments. Although child support is usually not enough to lift a single-parent family out of poverty, it can ease its burdens substantially. An added benefit is that noncustodial fathers are more likely to maintain contact with children if they pay child support (Garfinkel & McLanahan, 1995). The Caregiving Concerns table on page 332 summarizes ways to help children adjust to their parents' divorce.

Blended Families

"If you get married to Wendell, and Daddy gets married to Carol," Lizzie wondered aloud to Rena, "then I'll have two sisters and one more brother. And let's see, how many grandmothers and grandfathers? Gosh, a lot!" exclaimed Lizzie. "But what will I call them all?" she asked, looking worried.

Life in a single-parent family is often temporary. Many parents find a new partner within a few years. As Lizzie's comments indicate, entry into these *blended,* or *reconstituted, families* leads to a complex set of new relationships. For some children, this expanded family network is a positive turn of events that brings greater adult attention. But for most, it presents

Helping Children Adjust to Their Parents' Divorce

SUGGESTION	EXPLANATION
Shield children from conflict.	Witnessing intense parental conflict is very damaging to children. If one parent insists on expressing hostility, children fare better if the other parent does not respond in kind.
Provide children with as much continuity, familiarity, and predictability as possible.	Children adjust better during the period surrounding divorce when their lives have some stability—for example, the same school, bedroom, baby-sitter, and playmates and a dependable daily schedule.
Explain the divorce and tell children what to expect.	Children are more likely to develop fears of abandonment if they are not prepared for their parents' separation. They should be told that their mother and father will not be living together any more, which parent will be moving out, and when they will be able to see that parent. If possible, mother and father should explain the divorce together. Parents should provide a reason for the divorce that the child can understand and assure the child that he or she is not to blame.
Emphasize the permanence of the divorce.	Fantasies of parents getting back together can prevent children from accepting the reality of their current life. Children should be told that the divorce is final and that there is nothing they can do to change that fact.
Respond sympathetically to children's feelings.	Children need a supportive and understanding response to their feelings of sadness, fear, and anger. For children to adjust well, their painful emotions must be acknowledged, not denied or avoided.
Promote a continuing relationship with both parents.	When parents disentangle their lingering hostility toward the former spouse from the child's need for a continuing relationship with the other parent, children adjust well. Grandparents and other extended-family members can help by not taking sides.

Source: Teyber, 1992.

difficult adjustments. Stepparents often introduce new child-rearing practices, and having to switch to new rules and expectations can be stressful. In addition, children often regard steprelatives as "intruders." But how well they adapt is, once again, related to the overall quality of family functioning (Hetherington & Kelly, 2002). This often depends on which parent forms a new relationship and on the child's age and sex. As we will see, older children and girls seem to have the hardest time.

● **Mother–Stepfather Families.** The most frequent form of blended family is a mother–stepfather arrangement because mothers generally retain custody of the child. Boys usually adjust quickly. They welcome a stepfather who is warm and involved, who refrains from exerting his authority too quickly, and who offers relief from coercive mother–son cycles of interaction. Mothers' friction with sons also declines due to greater economic security, another adult to share household tasks, and an end to loneliness (Stevenson & Black, 1995). In contrast, girls adapt less favorably. Stepfathers disrupt the close ties many girls have established with their mothers, and girls often react to the new arrangement with sulky, resistant behavior (Bray, 1999).

Note, however, that age affects these findings. Older school-age and adolescent youngsters of both sexes display more irresponsible, acting-out behavior than peers not in stepfamilies (Hetherington & Stanley-Hagan, 2000). Often parents are warmer and more involved with their biological children than with their stepchildren. Older children are more likely to notice and challenge unfair treatment. And adolescents often view the new stepparent as a threat to their freedom, especially if they experienced little parental monitoring in the single-parent family.

● **Father–Stepmother Families.** Remarriage of noncustodial fathers often leads to reduced contact, as they tend to withdraw from their "previous" families, more so if they have daughters than sons (Hetherington, 1997). When fathers have custody, children typically react negatively to remarriage. One reason is that children living with fathers often start out with more problems. Perhaps the biological mother could no longer handle the unruly child (usually a boy), so the father and his new wife are faced with a youngster who has serious behavior problems. In other instances, the father is granted custody because of a very close relationship with the child, and his remarriage disrupts this bond (Buchanan, Maccoby, & Dornbusch, 1996).

Girls, especially, have a hard time getting along with their stepmothers. Sometimes (as just mentioned) this occurs because the girl's relationship with her father is threatened by the remarriage. In addition, girls often become entangled in loyalty conflicts between their two mother figures. But the longer girls live in father–stepmother households, the more positive their interaction with stepmothers becomes (Hetherington & Jodl, 1994). With time and patience they do adjust, and eventually girls benefit from the support of a second mother figure.

● **Support for Blended Families.** Family life education and therapy can help parents and children adapt to the complexities of blended families. Effective approaches encourage stepparents to move into their new roles gradually by first building a friendly relationship with the child. Only when a warm bond has formed is more active parenting possible (Ganong & Coleman, 2000). In addition, couples who form a "parenting coalition" through which they cooperate, provide consistency in child rearing, and limit loyalty conflicts have children who adjust more easily (Emery, 1999b).

Maternal Employment and Dual-Earner Families

Today, single and married mothers are in the labor market in nearly equal proportions, and more than three-fourths of those with school-age children are employed (Statistics Canada, 2002m; U.S. Bureau of the Census, 2002c). In Chapter 6, we saw that the impact of maternal employment on early development depends on the quality of child care and the continuing parent–child relationship. This same conclusion applies during later years.

● **Maternal Employment and Child Development.** Children of mothers who enjoy their work and remain committed to parenting show very favorable adjustment—higher self-esteem, more positive family and peer relations, less gender-stereotyped beliefs, and better grades in school. Girls, especially, profit from the image of female competence. Daughters of employed mothers perceive women's roles as involving more freedom of choice and satisfaction and are more achievement- and career-oriented (Hoffman, 2000).

These benefits undoubtedly result from parenting practices. Employed mothers who value their parenting role are more likely to use authoritative child rearing. Also, children in dual-earner households devote more daily hours to doing homework under parental guidance and participate more in household chores. And maternal employment leads fathers to take on greater child-care responsibility, with a small but increasing number staying home full-time (Gottfried, Gottfried, & Bathurst, 2002; Hoffman & Youngblade, 1999). More paternal contact with children is related to higher intelligence and achievement, mature social behavior, and a more flexible view of gender roles (Coltrane, 1996; Radin, 1994).

However, when employment places heavy demands on the mother's schedule, children are at risk for ineffective parenting. Working long hours and spending little time with school-age children are associated with less favorable adjustment (Moorehouse, 1991). In contrast, part-time employment and flexible work schedules seem to have benefits for children of all ages, probably because these arrangements prevent work–family role conflict, thereby helping parents meet children's needs (Frederiksen-Goldsen & Sharlach, 2000).

● **Support for Employed Parents and Their Families.** In dual-earner families, the husband's willingness to share responsibilities helps the mother engage in effective parenting. If the father helps very little or not at all, the mother carries a double load, at home and at work, leading to fatigue, distress, and little time and energy for children.

Employed mothers and dual-earner parents need assistance from work settings and communities in their child-rearing roles. Reduced work hours, flexible schedules, job sharing, and paid leave when children are ill help parents juggle the demands of work and child rearing. Equal pay and employment opportunities for women are also important. Because these policies enhance financial status and morale, they improve the way mothers feel and behave when they arrive home at the end of the working day.

● **Child Care for School-Age Children.** High-quality child care is vital for parents' peace of mind and children's well-being, even during middle childhood. Not all 5- to 13-year-olds, however, have round-the-clock supervision. An estimated 2.4 million in the United States and several hundred thousand in Canada are **self-care children,** who regularly look after themselves during after-school hours.

Research on these children reveals inconsistent findings. Some studies report that they suffer from low self-esteem, antisocial behavior, poor academic achievement, and fearfulness, whereas others show no such effects. Children's maturity and the way they spend their time seem to explain these contradictions. Among younger school-age children, those who spend more hours alone have more adjustment difficulties (Vandell &

© AMY ETRA/PHOTOEDIT

■ In this after-school program in Los Angeles, children spend time productively and enjoyably while their parents are at work. A community volunteer assists children with learning and completing homework. Children who attend such programs display better work habits, school grades, and peer relations.

Posner, 1999). As children become old enough to look after themselves, those who have a history of authoritative child rearing, are monitored from a distance by parental telephone calls, and have regular after-school chores appear responsible and well-adjusted. In contrast, children left to their own devices are more likely to bend to peer pressures and engage in antisocial behavior (Steinberg, 1986).

Before age 9 or 10, most children need supervision because they are not yet competent to deal with emergencies (Galambos & Maggs, 1991). Yet after-school programs serve only a minority of children in need and vary greatly in quality (Vandell, 1999). Programs with well-trained staffs, generous adult–child ratios, positive adult–child communication, and stimulating activities are linked to better social skills and emotional adjustment (Pierce, Hamm, & Vandell, 1999). And low-SES children who participate in "after-care" enrichment activities (scouting, music, or art lessons) show special benefits, including better work habits and school grades and fewer behavior problems (Posner & Vandell, 1994, 1999).

Ask Yourself

REVIEW

List findings from our discussion of the family that highlight the influence of fathers on children's development.

REVIEW

Describe and explain changes in sibling relationships during middle childhood.

APPLY

Steve and Marissa are in the midst of an acrimonious divorce. Their 9-year-old son Dennis has become hostile and defiant. How can Steve and Marissa help Dennis adjust?

CONNECT

How does each level in Bronfenbrenner's ecological systems theory—microsystem, mesosystem, exosystem, and macrosystem—contribute to the effects of maternal employment on children's development?

www.

Some Common Problems of Development

Throughout our discussion, we have considered a variety of stressful experiences that place children at risk for future problems. In the following sections, we touch on two more areas of concern: school-age children's fears and anxieties and the consequences of child sexual abuse. Finally, we sum up factors that help children cope effectively with stress.

Fears and Anxieties

Although fears of the dark, thunder and lightning, and supernatural beings (often stimulated by movies and television) persist into middle childhood, children's anxieties are also directed toward new concerns. As children begin to understand the realities of the wider world, the possibility of personal harm (being robbed, stabbed, or shot) and media events (war and disasters) often trouble them. Other common worries include academic failure, parents' health, physical injuries, and peer rejection (Muris et al., 2000; Silverman, La Greca, & Wasserstein, 1995).

Most children handle their fears constructively, by talking about them with parents, teachers, and friends and relying on the more sophisticated emotional self-regulation strategies that develop in middle childhood. Consequently, fears decline with age, especially for girls, who express more fears than boys throughout childhood and adolescence (Gullone & King, 1997).

From 10 to 20 percent of school-age children develop an intense, unmanageable anxiety of some kind (Barrios & Dell, 1998). For example, in **school phobia,** children feel severe apprehension about attending school, often accompanied by physical complaints (dizziness, nausea, stomachaches, and vomiting) that disappear once the child is allowed to stay home. About one-third of children with school phobia are 5- to 7-year-olds, most of whom do not fear school so much as separation from their mother. The difficulty often can be traced to parental overprotection and encouragement of dependency. Family therapy helps these children (Elliott, 1999).

Most cases of school phobia appear later, around 11 to 13, during the transition from middle childhood to adolescence. These children usually find a particular aspect of school frightening—an overcritical teacher, a school bully, or too much parental pressure for school success. Treating this form of school phobia may require a change in school environment or parenting practices. Firm insistence that the child return to school, along with training in how to cope with difficult situations, is also helpful (Blagg & Yule, 1996).

Severe childhood anxieties may arise from harsh living conditions. In inner-city ghettos and war-torn areas of the world, a great many children live in the midst of constant deprivation, chaos, and violence. As the Lifespan Vista box on the following page reveals, these youngsters are at risk for long-term emotional distress and behavior problems. Finally, as we saw in our discussion of child abuse in Chapter 8, too often violence and other destructive acts become part of adult–child relationships. During middle childhood, child sexual abuse increases.

Child Sexual Abuse

Until recently, child sexual abuse was viewed as a rare occurrence. When children came forward with it, adults rarely took their claims seriously. In the 1970s, efforts by professionals along with media attention caused child sexual abuse to be

A Lifespan Vista

Children of War

Violence stemming from ethnic and political tensions is increasingly being felt around the world. Children's experiences under armed conflict and terrorism are diverse. Some may participate in the fighting, either because they are forced or because they want to please adults. Others are kidnapped, assaulted, and tortured. Those who are bystanders often come under direct fire and may be killed or physically maimed for life. And many watch in horror as family members, friends, and neighbors flee, are wounded, or die. In the past decade, wars have left 4 to 5 million children physically disabled, 12 million homeless, and more than 1 million separated from their parents (Stichick, 2001). Half of all casualties of worldwide conflict are children.

When war and social crises are temporary, most children are easily comforted and do not show long-term emotional difficulties. But chronic danger requires children to make substantial adjustments, and their psychological functioning can be seriously impaired. Many children of war lose their sense of safety, become desensitized to violence, are haunted by terrifying intrusive memories, and build a pessimistic view of the future. Aggressive and antisocial behavior often increases (Muldoon & Cairns, 1999). These outcomes appear to be culturally universal, appearing among children from every war zone studied—from Iran, Bosnia, Rwanda, and Afghanistan to the West Bank and Gaza Strip (Garbarino, Andreas, & Vorrasi, 2002).

Parental affection and reassurance are the best protection against lasting problems. When parents offer security and serve as role models of calm emotional strength, most children can withstand even extreme war-related violence (Smith et al., 2001). Children who are separated from parents must rely on help from their communities. Preschool and school-age orphans in Eritrea who were placed in residential settings where they could form close emotional ties with at least one adult showed less emotional stress 5 years later than orphans placed in impersonal settings (Wolff & Fesseha, 1999). Educational programs are powerful safeguards, too, providing children with a sense of consistency in their lives along with teacher and peer supports.

The September 11, 2001, terrorist attack on the World Trade Center caused some American children to experience extreme wartime violence firsthand. Children in Public School 31 in Brooklyn, New York, for example, stared out windows as planes rushed toward the towers, were engulfed in flames, and crumbled. Many worried about the safety of family members, and some lost them. In the aftermath, most expressed intense fears—for example, that terrorists were in their neighborhoods and that planes flying overhead might smash into nearby buildings.

Unlike many war-traumatized children in the developing world, Public School 31 students received immediate intervention—a "trauma curriculum" in which they expressed their emotions through writing, drawing, and discussion and participated in experiences aimed at restoring trust and tolerance (Lagnado, 2001). Older children learned about the feelings of their Muslim classmates, the dire condition of children in Afghanistan, and ways to help victims as a means of overcoming a sense of helplessness.

When wartime drains families and communities of resources, international organizations must step in and help children. Efforts to preserve children's physical, psychological, and educational well-being may be the best way to stop transmission of violence to the next generation.

© AP/WIDE WORLD PHOTOS

■ These traumatized, displaced children eat in a refugee camp along the India–Pakistan border. Because of civil unrest, their families left their homes. Most have seen their neighborhoods severely damaged or destroyed and witnessed violence toward people they know. Without special support from caring adults, they may show lasting emotional problems.

recognized as a serious and widespread problem. About 90,000 cases in the United States and 14,000 cases in Canada were confirmed in the year 2000 (Health Canada, 2002d; U.S. Department of Health and Human Services, 2002c).

● **Characteristics of Abusers and Victims.** Sexual abuse is committed against children of both sexes but more often against girls. Most cases are reported in middle childhood, but sexual abuse also occurs at younger and older ages. For some victims, the abuse begins early in life and continues for many years (Trickett & Putnam, 1998).

Generally, the abuser is male—a parent or someone the parent knows well. Often it is a father, stepfather, or live-in boyfriend, somewhat less often an uncle or older brother. In a few instances, mothers are the offenders, more often with sons (Kolvin & Trowell, 1996). In the overwhelming majority of cases, the abuse is serious—vaginal or anal intercourse, oral–genital contact, fondling, and forced stimulation of the adult. Abusers make the child comply in a variety of distasteful ways, including deception, bribery, verbal intimidation, and physical force.

You may be wondering how any adult—especially a parent or close relative—could possibly violate a child sexually. Many offenders deny their own responsibility. They blame the abuse on the willing participation of a seductive youngster. Yet children are not capable of making a deliberate, informed decision to enter into a sexual relationship! Even at older ages, they are not free to say yes or no. Instead, the responsibility lies with abusers, who tend to have characteristics that predispose them toward sexual exploitation of children. They have great difficulty controlling their impulses and may suffer from alcohol and drug abuse. Often they pick out children who are unlikely to defend themselves—those who are physically weak, emotionally deprived, and socially isolated (Bolen, 2001).

Reported cases of child sexual abuse are strongly linked to poverty, marital instability, and the resulting weakening of family ties. Children who live in homes with a history of constantly changing characters—repeated marriages, separations, and new partners—are especially vulnerable. But middle-SES children in stable families are also victims, although their victimization is more likely to remain undetected (Gomez-Schwartz, Horowitz, & Cardarelli, 1990).

● **Consequences.** The adjustment problems of child sexual abuse victims are often severe. Depression, low self-esteem, mistrust of adults, and feelings of anger and hostility can persist for years after the abusive episodes. Younger children often react with sleep difficulties, loss of appetite, and generalized fearfulness. Adolescents may run away and show suicidal reactions, substance abuse, and delinquency (Feiring, Taska, & Lewis, 1999).

Sexually abused children frequently display sexual knowledge and behavior beyond their years. They have learned from their abusers that sexual overtures are acceptable ways to get attention and rewards. As they move toward early adulthood, abused young people become promiscuous. Women are likely to choose partners who abuse them and their children (Faller, 1990). As mothers, they often engage in irresponsible and coercive parenting, including child abuse and neglect (Pianta, Egeland, & Erickson, 1989). In these ways, the harmful impact of sexual abuse is transmitted to the next generation.

● **Prevention and Treatment.** Treating child sexual abuse is difficult. Since it typically appears in the midst of other serious family problems, long-term therapy with both children and parents is usually necessary (Olafson & Boat, 2000). The best way to reduce the suffering of victims is to prevent child sexual abuse from continuing. Today, courts are prosecuting abusers more vigorously and taking children's testimony more seriously (see the Social Issues box on the following page).

COURTESY OF THE NEW ZEALAND POLICE

■ In Keeping Ourselves Safe, New Zealand's national, school-based child abuse prevention program, teachers and police officers collaborate in teaching children to recognize abusive adult behaviors so they can take steps to protect themselves. Parents are informed about children's classroom learning experiences and encouraged to support and extend them at home.

Educational programs can teach children to recognize inappropriate sexual advances and show them where to go for help. Yet because of controversies over teaching children about sexual abuse, few schools offer these interventions. New Zealand is the only country in the world with a national, school-based prevention program targeting sexual abuse. In Keeping Ourselves Safe, children and adolescents learn that abusers are rarely strangers. Parent involvement ensures that home and school work together in teaching children self-protection skills. Evaluations reveal that virtually all New Zealand parents and children support the program and that it has helped many children avoid or report abuse (Briggs & Hawkins, 1996, 1999).

Fostering Resiliency in Middle Childhood

Throughout middle childhood—and other phases of development—children are confronted with challenging and sometimes threatening situations that require them to cope with psychological stress. In this and the previous chapter, we have considered such topics as chronic illness, learning disabilities, achievement expectations, divorce, and child sexual

Social Issues

Children's Eyewitness Testimony

Increasingly, children are being called on to testify in court cases involving child abuse and neglect, child custody, and other matters. Having to provide information on such topics can be difficult and traumatic. Almost always, children must report on highly stressful events, and they may have to speak against a parent or other relative to whom they feel loyal. In some family disputes, they may fear punishment for telling the truth. In addition, child witnesses are faced with an unfamiliar situation—at the very least an interview in the judge's chambers, and at most an open courtroom with judge, jury, spectators, and the possibility of unsympathetic cross-examination. Not surprisingly, these conditions can compromise the accuracy of children's recall.

Age Differences. Until recently, children younger than age 5 were rarely asked to testify, and not until age 10 were they assumed fully competent to do so. Yet as a result of societal reactions to rising rates of child abuse and difficulties in prosecuting perpetrators, legal requirements for child testimony have been relaxed in the United States and Canada. Children as young as age 3 frequently serve as witnesses (Ceci & Bruck, 1998).

Compared with preschoolers, school-age children are better at giving detailed descriptions of past experiences and making accurate inferences about others' motives and intentions. Older children are also more resistant to misleading questions of the sort asked by attorneys when they probe for more information or, in cross-examination, try to influence the content of the child's response (Bjorklund et al., 2000; Roebers & Schneider, 2001). Nevertheless, when properly questioned, even 3-year-olds can recall recent events accurately, including highly stressful ones (Peterson & Rideout, 1998).

Suggestibility. Court testimony, however, often involves repeated interviews. When adults lead children by suggesting incorrect facts ("He touched you there, didn't he?"), they increase the likelihood of incorrect reporting on the part of preschoolers and school-age children alike. Events that children fabricate in response to leading questions can be quite fantastic. In one study, after a visit to a doctor's office, children said yes to questions about events that not only never occurred but that implied abuse—"Did the doctor lick your knee?" "Did the nurse sit on top of you?" (Ornstein et al., 1997).

By the time children appear in court, it is weeks, months, or even years after the occurrence of the target events. When a long delay is combined with suggestions about what happened and with stereotyping of the accused ("He's in jail because he's been bad"), children can easily be misled into giving false information (Leichtman & Ceci, 1995). To ease the task of providing testimony, special interviewing methods have been devised for children. In many sexual abuse cases, anatomically correct dolls are used to prompt children's recall. Although this method helps older children provide more detail about experienced events, it increases the suggestibility of preschoolers, who report physical and sexual contact that never happened (Ceci & Bruck, 1998; Goodman et al., 1999).

Interventions. Adults must prepare child witnesses so they understand the courtroom process and know what to expect. In some places, "court schools" exist in which children are taken through the setting and given an opportunity to role-play court activities. As part of this process, children can be encouraged to admit not knowing an answer rather than guessing or going along with what an adult expects. At the same time, legal professionals must lessen the risk of suggestibility by limiting the number of times children are interviewed and asking questions in nonleading ways. And a warm, supportive interview tone fosters accurate recall, perhaps by decreasing children's fear so they feel freer to counter false suggestions (Ceci, Bruck, & Battin, 2000).

If a child is likely to experience emotional trauma or later punishment (in a family dispute), then courtroom procedures can be adapted to protect them. For example, they can testify over closed-circuit TV so they do not have to face an abuser. When it is not wise for a child to participate directly, expert witnesses can provide testimony that reports on the child's psychological condition and includes important elements of the child's story. But for such testimony to be worthwhile, witnesses must be impartial and trained in how to question to minimize false reporting (Bruck, Ceci, & Hembrooke, 1998).

abuse. Each taxes children's coping resources, creating serious risks for development.

At the same time, only a modest relationship exists between stressful life experiences and psychological disturbance in childhood (Garmezy, 1993). In our discussion in Chapter 3 of the long-term consequences of birth complications, we noted that some children manage to overcome the combined effects of birth trauma, poverty, and a deeply troubled family life. The same is true when we look at school difficulties, family transitions, and child maltreatment. Recall from Chapter 1 that three broad factors protect against maladjustment: (1) the child's personal characteristics, including an easy temperament and a mastery-oriented approach to new situations; (2) a warm, well-organized family life; and (3) an adult outside the immediate family who offers a support system and a positive coping model.

Any one of these ingredients of resiliency can account for why one child fares well and another poorly. Yet most of the time, personal and environmental factors are interconnected. Unfavorable life experiences increase the chances that children will act in ways that expose them to further hardship. And when negative conditions pile up, such as marital discord, poverty, crowded living conditions, neighborhood violence, and abuse, the rate of maladjustment multiplies (Farrington & Loeber, 2000; Wyman et al., 1999).

Throughout our discussion, we have seen many examples of how families, schools, communities, and society as a whole can enhance or undermine the school-age child's relationships and developing sense of competence. As the next two chapters will reveal, young people whose childhood experiences helped them learn to overcome obstacles, strive for self-direction, and respond considerately and sympathetically to others meet the challenges of adolescence—the transition to adulthood—quite well.

Ask Yourself

REVIEW

When children must testify in court cases, what factors increase the chances of accurate reporting?

REVIEW

List environmental factors that increase the risk of child sexual abuse.

APPLY

Claire told her 6-year-old daughter never to talk to strangers or take candy from them. Why will Claire's warning not protect her daughter from sexual abuse?

www.

Summary

Erikson's Theory: Industry versus Inferiority

What personality changes take place during Erikson's stage of industry versus inferiority?

- According to Erikson, children who successfully resolve the psychological conflict of **industry versus inferiority** develop the capacity for productive work, learn the value of division of labor, and develop a sense of moral commitment and responsibility.

Self-Understanding

Describe school-age children's self-concept and self-esteem, and discuss factors that affect their achievement-related attributions.

- During middle childhood, children's self-concepts include personality traits and **social comparisons.** Self-esteem differentiates further, becomes hierarchically organized, and declines over the early school years as children adjust their self-judgments to feedback from the environment. Authoritative parenting is linked to favorable self-esteem.
- Children with **mastery-oriented attributions** credit their successes to high ability and failures to insufficient effort. In contrast, children who receive negative feedback about their ability develop **learned helplessness,** attributing their successes to external factors, such as luck, and failures to low ability. Children who receive negative adult feedback about their ability and who encounter unsupportive teachers in school are likely to develop learned helplessness.

Emotional Development

Cite changes in understanding and expression of emotion in middle childhood.

- In middle childhood, the self-conscious emotions of pride and guilt become clearly governed by personal responsibility. Experiencing intense shame can shatter children's overall sense of self-esteem.
- School-age children recognize that people can experience more than one emotion at a time. They also attend to more cues in interpreting another's feelings, and empathy increases. By the end of middle childhood, most children have an adaptive set of strategies for regulating emotion. Emotionally well-regulated children are optimistic, prosocial, and well liked by peers.

Understanding Others: Perspective Taking

How does perspective taking change in middle childhood?

- **Perspective taking** improves greatly during the school years, as Selman's five-stage sequence indicates. Cognitive maturity and experiences in which adults and peers encourage children to take note of another's viewpoint support gains in perspective-taking skill. Good perspective takers have more positive social skills.

Moral Development

Describe changes in moral understanding during middle childhood.

- By middle childhood, children have internalized a wide variety of moral rules. Their concepts of **distributive justice** change, from equality to merit to benevolence. School-age children also clarify and link moral rules and social conventions. In judging the morality of rule violations, they take into account the purpose of the rule, people's intentions, and the context of their actions.

Peer Relations

How do peer sociability and friendship change in middle childhood?

- In middle childhood, peer interaction becomes more prosocial, and aggression declines. By the end of the school years, children organize themselves into **peer groups.** Friendships develop into mutual relationships based on trust. Children tend to select friends like themselves in many ways. Kind, compassionate friendships strengthen prosocial behavior. Friendships between aggressive children magnify antisocial acts.

Describe major categories of peer acceptance and ways to help rejected children.

- On measures of **peer acceptance, popular children** are liked by many agemates; **rejected children** are actively disliked; **controversial children** are both liked and disliked; and **neglected children** are seldom chosen, either positively or negatively.
- Two subtypes of popular children exist: **popular-prosocial children,** who are academically and socially competent, and **popular-antisocial children,** who are athletically skilled, poorly achieving, aggressive boys. Rejected children also divide into two subtypes: **rejected-aggressive children,** who show high rates of conflict and hostility, and **rejected-withdrawn children,** who are passive, socially awkward, and at risk for **peer victimization.** Rejected children often experience lasting adjustment difficulties. Helpful interventions include coaching in social skills; academic tutoring; training in perspective taking and social problem solving; teaching children to attribute peer difficulties to internal, changeable causes; and improving parent–child interaction.

Gender Typing

What changes in gender-stereotyped beliefs and gender identity take place during middle childhood?

- School-age children extend their awareness of gender stereotypes to personality characteristics and academic subjects. They also develop a more open-minded view of what males and females can do. Boys strengthen their identification with the masculine role, whereas girls often experiment with "other-gender" activities. Cultures shape gender-typed behavior through daily activities assigned to children.

Family Influences

How do parent–child communication and sibling relationships change in middle childhood?

- Effective parents of school-age youngsters engage in **coregulation,** exerting general oversight while permitting children to be in charge of moment-by-moment decision making. Coregulation depends on a cooperative relationship between parent and child.
- Sibling rivalry tends to increase as children participate in a wider range of activities and parents compare their traits and accomplishments. Only children are as well adjusted as children with siblings, and they do better in school and attain higher levels of education.

What factors influence children's adjustment to divorce and remarriage?

- Although painful emotional reactions accompany the period surrounding divorce, boys and children with difficult temperaments are more likely to show lasting problems in school performance and antisocial behavior. For children of both sexes, divorce is linked to problems with adolescent sexuality and development of intimate ties in adulthood.
- The overriding factor in positive adjustment following divorce is effective parenting. Contact with noncustodial fathers is important for children of both sexes, and outcomes for sons are better under father custody. Because **divorce mediation** helps parents resolve disputes and cooperate in parenting, it can help parents and children through the difficult period surrounding divorce. The success of **joint custody** depends on a cooperative relationship between divorcing parents.
- When divorced parents enter new relationships, girls, older children, and children in father–stepmother families display the greatest adjustment problems. Stepparents who move into their roles gradually and form a "parenting coalition" help children adjust.

How do maternal employment and life in dual-earner families affect school-age children?

- When mothers enjoy their work and remain committed to parenting, maternal employment is associated with higher self-esteem in children, more positive family and peer relations, less gender-stereotyped beliefs, and better grades in school. In dual-earner families, the father's willingness to share in household responsibilities is linked to many positive outcomes for children. Workplace supports, such as part-time employment, flexible schedules, and paid parental leave, help parents meet the demands of work and child rearing.
- **Self-care children** who are old enough to look after themselves, are monitored from a distance, and experience authoritative parenting appear responsible and well adjusted. Children in high-quality after-school programs reap academic and social benefits.

Some Common Problems of Development

Cite common fears and anxieties in middle childhood.

- School-age children's fears are directed toward new concerns, including physical safety, media events, academic performance, parents' health, and peer relations. Some children develop intense, unmanageable fears, such as **school phobia.** Severe anxiety can also result from harsh living conditions.

Discuss factors related to child sexual abuse and its consequences for children's development.

- Child sexual abuse is generally committed by male family members, more often against girls than boys. Abusers have characteristics that predispose them toward sexual exploitation of children. Reported cases are strongly associated with poverty and marital instability. Abused children often have severe adjustment problems.

Cite factors that foster resiliency in middle childhood.

- Overall, only a modest relationship exists between stressful life experiences and psychological disturbance in childhood. But when negative factors pile up, the rate of maladjustment is especially high. Personal characteristics of children; a warm, well-organized family life; and social supports outside the family are related to childhood resiliency in the face of stress.

Important Terms and Concepts

controversial children (p. 324)
coregulation (p. 328)
distributive justice (p. 321)
divorce mediation (p. 331)
industry versus inferiority (p. 314)
joint custody (p. 331)
learned helplessness (p. 317)
mastery-oriented attributions (p. 317)
neglected children (p. 324)
peer acceptance (p. 324)
peer group (p. 322)
peer victimization (p. 325)
perspective taking (p. 320)
popular children (p. 324)
popular-antisocial children (p. 324)
popular-prosocial children (p. 324)
rejected children (p. 324)
rejected-aggressive children (p. 324)
rejected-withdrawn children (p. 325)
school phobia (p. 334)
self-care children (p. 333)
social comparisons (p. 315)

FYI For Further Information and Help

Consult the Companion Website for *Development Through the Lifespan, Third Edition,* (www.ablongman.com/berk) where you will find the following resources for this chapter:

- **Chapter Objectives**
- **Flashcards** for studying important terms and concepts
- **Annotated Weblinks** to guide you in further research
- **Ask Yourself** questions, which you can answer and then check against a sample response
- **Suggested Readings**
- **Practice Test** with immediate scoring and feedback

Milestones

Development in Middle Childhood

AGE: **6–8 years**

PHYSICAL

- Slow gains in height and weight continue until adolescent growth spurt. (276–277)

- Gradual replacement of primary teeth by permanent teeth. (277)
- Writing becomes smaller and more legible. Letter reversals decline. (281)
- Drawings become more organized and detailed and include some depth cues. (281)
- Games with rules become common. (282)

COGNITIVE

- Thought becomes more logical, as shown by the ability to pass Piagetian conservation, class inclusion, and seriation problems. (285)
- Understanding of spatial concepts improves, as illustrated by ability to give clear, well-organized directions and to draw and read maps. (285)
- Attention becomes more selective, adaptable, and planful. (288)
- Uses memory strategies of rehearsal and organization. (289)
- Regards the mind as an active, constructive agent, capable of transforming information. (291)
- Awareness of memory strategies and the impact of psychological factors (attention, motivation) on task performance improves. (291)

LANGUAGE

- Vocabulary increases rapidly throughout middle childhood. (299)
- Word definitions are concrete, referring to functions and appearance. (299)
- Language awareness improves. (299)

EMOTIONAL/SOCIAL

- Self-concept begins to include personality traits and social comparisons. (315)
- Self-esteem differentiates, becomes hierarchically organized, and declines to a more realistic level. (315)
- Self-conscious emotions of pride and guilt are governed by personal responsibility. (319)
- Recognizes that individuals can experience more than one emotion at a time. (319)
- Attends to facial and situational cues in interpreting another's feelings. (319)
- Understands that access to different information often causes people to have different perspectives. (320)
- Becomes more responsible and independent. (328)

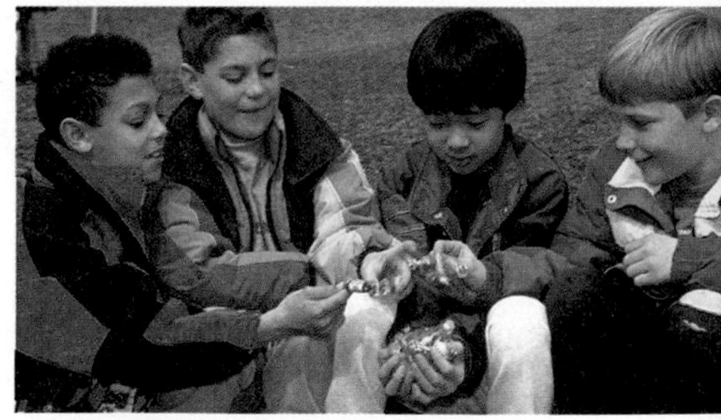

- Distributive justice reasoning changes from equality to merit to benevolence. (321)
- Peer interaction becomes more prosocial, and physical aggression declines. (322)

AGE	PHYSICAL	COGNITIVE	LANGUAGE	EMOTIONAL/SOCIAL
9–11 years	 • Adolescent growth spurt begins 2 years earlier in girls than in boys. (277) • Gross motor skills of running, jumping, throwing, catching, kicking, batting, and dribbling are executed more quickly and with better coordination. (281) • Reaction time improves, contributing to motor skill development. (281) • Representation of depth in drawings expands. (281–282)	• Logical thought remains tied to concrete situations. (286) • Piagetian tasks continue to be mastered in a step-by-step fashion. (286) • Memory strategies of rehearsal and organization become more effective. Begins to use elaboration. (289–290) • Applies several memory strategies at once. (290)  • Long-term knowledge base grows larger and becomes better organized. (290) • Cognitive self-regulation improves. (291)	• Word definitions emphasize synonyms and categorical relations. (299) • Grasps double meanings of words, as reflected in comprehension of metaphors and humor. (299)  • Use of complex grammatical constructions improves. (299) • Adapts messages to the needs of listeners in challenging communicative situations. (299) • Conversational strategies become more refined. (300)	• Self-esteem tends to rise. (316)  • Distinguishes ability, effort, and luck in attributions for success and failure. (317) • Has an adaptive set of strategies for regulating emotion. (320) • Can "step into another's shoes" and view the self from that person's perspective. (320) • Later, can view the relationship between self and other from the perspective of a third, impartial party. (320) • Appreciates the linkage between moral rules and social conventions. (322) • Peer groups emerge. (322) • Friendships are based on mutual trust. (323) • Becomes aware of more gender stereotypes, including personality traits and school subjects, but has a more flexible appreciation of what males and females can do. (326) • Sibling rivalry tends to increase. (328)

Note: Numbers in parentheses indicate the page(s) on which each milestone is discussed.

11

Physical and Cognitive Development in Adolescence

Physical Development

Conceptions of Adolescence
The Biological Perspective • The Social Perspective • A Balanced Point of View

Puberty: The Physical Transition to Adulthood
Hormonal Changes • Body Growth • Motor Development and Physical Activity • Sexual Maturation • Individual and Group Differences

The Psychological Impact of Pubertal Events
Reactions to Pubertal Changes • Pubertal Change, Emotion, and Social Behavior • Early versus Late Maturation

Health Issues
Nutritional Needs • Eating Disorders • Sexual Activity • Sexually Transmitted Disease • Adolescent Pregnancy and Parenthood • Substance Use and Abuse

- Social Issues: Homosexuality: Coming Out to Oneself and Others

Cognitive Development

Piaget's Theory: The Formal Operational Stage
Hypothetico-Deductive Reasoning • Propositional Thought • Recent Research on Formal Operational Thought

An Information-Processing View of Adolescent Cognitive Development
Scientific Reasoning: Coordinating Theory with Evidence • How Scientific Reasoning Develops

Consequences of Abstract Thought
Argumentativeness • Self-Consciousness and Self-Focusing • Idealism and Criticism • Planning and Decision Making

Sex Differences in Mental Abilities

- Biology & Environment: Sex Differences in Spatial Abilities

Learning in School
School Transitions • Academic Achievement • Dropping Out

- A Lifespan Vista: Extracurricular Activities: Contexts for Positive Youth Development

The dramatic physical and cognitive changes of adolescence make puberty both an exhilarating and apprehensive phase of the lifespan. Although their bodies are fully grown and sexually mature, these teenagers have many skills to acquire and hurdles to overcome before they are ready to assume adult roles.

On her eleventh birthday, Sabrina's friend Joyce gave her a surprise party, but Sabrina appeared somber during the celebration. Although Sabrina and Joyce had been close friends since third grade, their relationship was faltering. Sabrina was a head taller and some 20 pounds heavier than most of the other girls in her sixth-grade class. Her breasts were well developed, her hips and thighs had broadened, and she had begun to menstruate. In contrast, Joyce still had the short, lean, flat-chested body of a school-age child.

Ducking into the bathroom while Joyce and the other girls set the table for cake and ice cream, Sabrina looked herself over in the mirror and whispered, "Gosh, I feel so big and heavy." At church youth group on Sunday evenings, Sabrina broke away from Joyce and spent time with the eighth-grade girls, around whom she didn't feel so large and awkward.

Once every two weeks, parents gathered at Sabrina and Joyce's school for discussions about child-rearing concerns. Sabrina's Italian-American parents, Franca and Antonio, came whenever they could. "How you know they are becoming teenagers is this," volunteered Antonio. "The bedroom door is closed, and they want to be alone. Also, they contradict and disagree. I tell Sabrina, 'You have to go to Aunt Gina's for dinner with the family.' The next thing I know, she is arguing with me."

Sabrina has entered **adolescence,** the transition between childhood and adulthood. In contemporary societies, the skills young people must master are so complex and the choices confronting them so diverse that adolescence lasts for nearly a decade. But around the world, the basic tasks of this phase are much the same. Sabrina must accept her full-grown body, acquire adult ways of thinking, attain emotional and economic independence, develop more mature ways of relating to peers of both sexes, and construct an identity—a secure sense of who she is, sexually, morally, politically, and vocationally.

The beginning of adolescence is marked by **puberty,** a flood of biological events leading to an adult-sized body and sexual maturity. As Sabrina's reactions suggest, entry into adolescence can be a trying time, more so for some youngsters than for others. In this chapter, we trace the events of puberty and take up a variety of health concerns—nutrition, sexual activity, and serious health problems affecting teenagers who encounter difficulties on the path to maturity.

Adolescence brings with it the capacity for abstract thinking, which opens up new realms of learning. Teenagers can grasp complex scientific principles, grapple with social and political issues, and detect the hidden meaning of a poem or story. The second part of this chapter traces these extraordinary changes from both Piaget's and the information-processing perspective. Next we take a close look at findings that have attracted a great deal of public attention: sex differences in mental abilities. The final portion of this chapter is devoted to the primary setting in which adolescent thought takes shape: the school.

PHYSICAL DEVELOPMENT

Conceptions of Adolescence

Why is Sabrina self-conscious, argumentative, and in retreat from family activities? Historically, theorists explained the impact of puberty on psychological development by resorting to extremes—either a biological or a cultural explanation. Today, researchers realize that biological, social, and cultural forces jointly determine adolescent psychological change.

The Biological Perspective

Ask several parents of young children what they expect their sons and daughters to be like as teenagers. You will probably get answers like these: "Rebellious and reckless." "Full of rages and tempers" (Buchanan & Holmbeck, 1998). This widespread view dates back to the writings of eighteenth-century philosopher Jean-Jacques Rousseau (see Chapter 1). He believed that a natural outgrowth of the biological upheaval of puberty was heightened emotionality, conflict, and defiance of adults.

In the early twentieth century, this storm-and-stress perspective was picked up by major theorists. The most influential was G. Stanley Hall, whose view of development was grounded in Darwin's theory of evolution. Hall (1904) described adolescence as a cascade of instinctual passions, a phase of growth so turbulent that it resembled the period in which humans evolved from savages into civilized beings. Sigmund Freud, as well, emphasized the emotional storminess of the teenage years. He called adolescence the *genital stage,* a period in which instinctual drives reawaken and shift to the genital region of the body, resulting in psychological conflict and volatile, unpredictable behavior. Gradually, as adolescents find intimate partners, inner forces achieve a new, more mature harmony, and the stage concludes with marriage, birth, and child rearing. In this way, young people fulfill their biological destiny: sexual reproduction and survival of the species.

The Social Perspective

Recent research suggests that the notion of adolescence as a biologically determined period of storm and stress is greatly exaggerated. A number of problems, such as eating disorders, depression, suicide, and lawbreaking, occur more often in adolescence than earlier. But the overall rate of psychological disturbance rises only slightly (by about 2 percent) from childhood to adolescence, when it is the same as in the adult population—about 20 percent (Costello & Angold, 1995). Although some teenagers encounter serious difficulties, emotional turbulence is not a routine feature of adolescence.

The first researcher to point out the wide variability in adolescent adjustment was anthropologist Margaret Mead (1928). She traveled to the Pacific islands of Samoa and returned with a startling conclusion: Because of the culture's relaxed social relationships and openness toward sexuality, adolescence "is perhaps the pleasantest time the Samoan girl (or boy) will ever know" (p. 308).

Mead offered an alternative view in which the social environment is entirely responsible for the range of teenage experiences, from erratic and agitated to calm and stress-free. Yet this conclusion is just as extreme as the biological perspective it tried to replace! Later researchers found that Samoan adolescence was not as untroubled as Mead had assumed (Freeman, 1983). Still, Mead showed that greater attention must be paid to social and cultural influences for adolescent development to be understood.

A Balanced Point of View

Today we know that adolescence is a product of *both* biological and social forces. Biological changes are universal—found in all primates and all cultures. These internal stresses and the social expectations accompanying them—that the young person give up childish ways, develop new interpersonal relationships, and take on greater responsibility—are likely to prompt moments of uncertainty, self-doubt, and disappointment in all teenagers.

At the same time, the length of adolescence varies greatly from one culture to the next. Although simpler societies have a shorter transition to adulthood, adolescence is not absent (Weisfield, 1997). A study of 186 tribal and village cultures revealed that almost all had an intervening phase, however brief, between childhood and full assumption of adult roles (Schlegel & Barry, 1991).

In industrialized nations, where successful participation in economic life requires many years of education, adolescence is greatly extended. Young people face extra years of dependence on parents and postponement of sexual gratification as they prepare for a productive work life. We will see that the more the social environment supports young people in achieving adult responsibilities, the better they fare. For all the biological tensions and uncertainties about the future that modern teenagers feel, most are surprisingly good at negotiating this period of life. With this idea in mind, let's look closely at puberty, the dawning of adolescent development.

Puberty: The Physical Transition to Adulthood

The changes of puberty are dramatic and momentous. Within a few years, the body of the school-age child is transformed into that of a full-grown adult. Genetically influenced hormonal processes regulate puberty growth. Girls, who have been advanced in physical maturity since the prenatal period, reach puberty, on average, 2 years earlier than boys.

■ Sex differences in pubertal growth are obvious among these sixth graders. Although all the children are 11 or 12 years old, the girls are taller and more mature looking.

Hormonal Changes

The complex hormonal changes that underlie puberty take place gradually and are under way by age 8 or 9. Secretions of *growth hormone (GH)* and *thyroxine* (see Chapter 7, page 208) increase, leading to tremendous gains in body size and attainment of skeletal maturity.

Sexual maturation is controlled by the sex hormones. Although *estrogens* are thought of as female hormones and *androgens* as male hormones, both types are present in each sex but in different amounts. The boy's testes release large quantities of the androgen *testosterone,* which leads to muscle growth, body and facial hair, and other male sex characteristics. Androgens (especially testosterone for boys) also contribute to gains in body size. The testes secrete small amounts of estrogen as well—the reason that 50 percent of boys experience temporary breast enlargement (Larson, 1996).

Like androgens, estrogens contribute to pubertal growth in both sexes (Juul, 2001). And in girls, estrogens released by the ovaries cause the breasts, uterus, and vagina to mature and the body to take on feminine proportions. In addition, estrogens contribute to regulation of the menstrual cycle. *Adrenal androgens,* released from the adrenal glands on top of each kidney, influence the girl's height spurt and stimulate growth

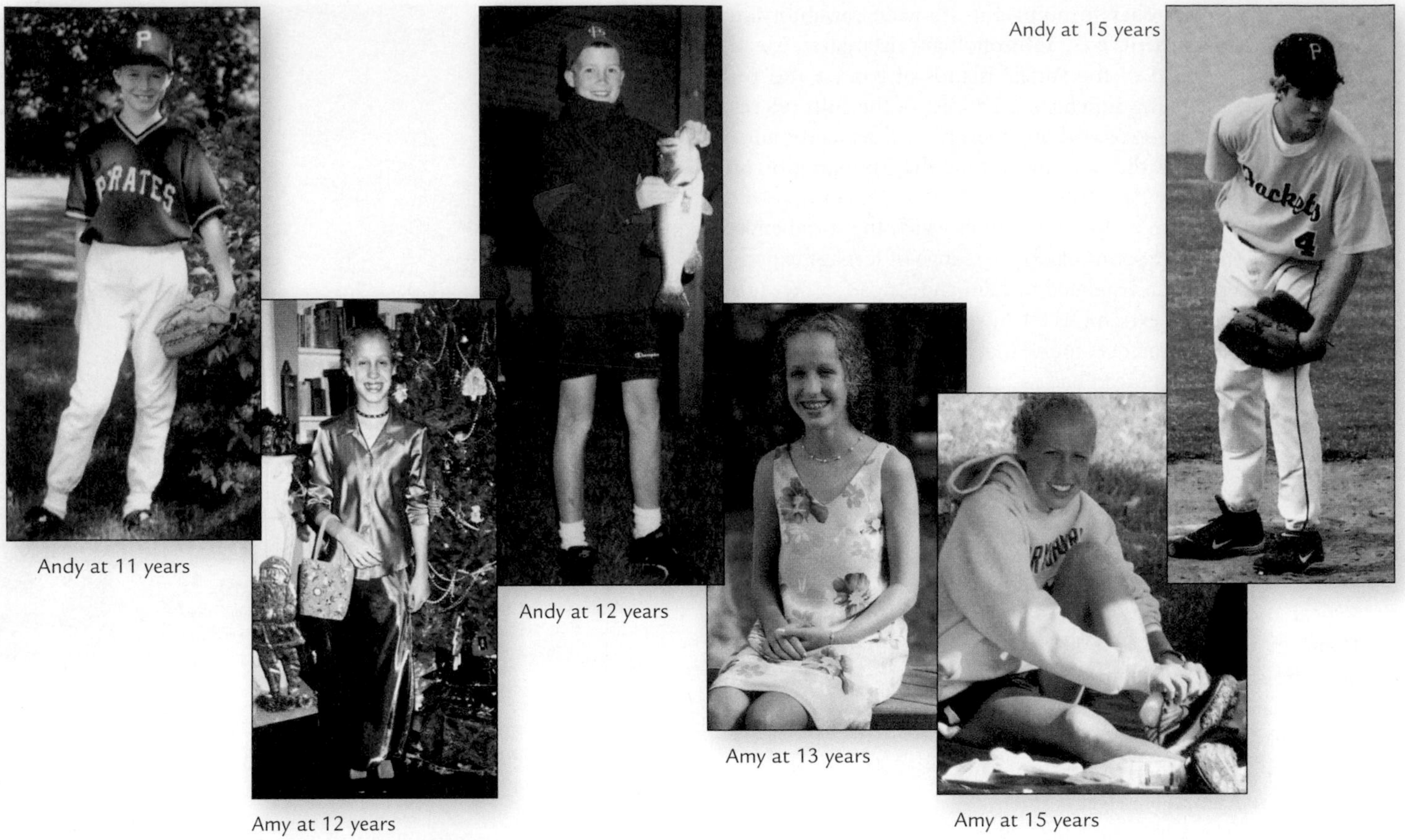

Andy at 11 years

Amy at 12 years

Andy at 12 years

Amy at 13 years

Amy at 15 years

Andy at 15 years

■ **FIGURE 11.1 Body growth during adolescence.** Because the pubertal growth spurt takes place earlier for girls than for boys, Amy reached her adult body size earlier than Andy. Rapid pubertal growth is accompanied by large sex differences in body proportions that were not present in middle childhood (see Chapter 9, page 276).

of underarm and pubic hair. They have little impact on boys, whose physical characteristics are mainly affected by androgen secretions from the testes.

As you can already tell, pubertal changes can be divided into two broad types: (1) overall body growth and (2) maturation of sexual characteristics. Although we will discuss these changes separately, they are interrelated. We have seen that the hormones responsible for sexual maturity also affect body growth; boys and girls differ in both aspects. In fact, puberty is the time of greatest sexual differentiation since prenatal life.

Body Growth

The first outward sign of puberty is the rapid gain in height and weight known as the **growth spurt.** On average, it is under way for North American girls shortly after age 10, for boys around age $12\frac{1}{2}$ (Malina, 1990). The girl is taller and heavier during early adolescence, but this advantage is short-lived. At age 14, she is surpassed by the typical boy, whose adolescent growth spurt has started, whereas hers is almost finished. Growth in body size is complete for most girls by age 16 and for boys by age $17\frac{1}{2}$, when the epiphyses at the ends of the long bones close completely (see Chapter 7, page 204). Altogether, adolescents add almost 10 inches in height and about 40 pounds in weight. Figure 11.1 illustrates pubertal changes in general body growth.

● **Body Proportions.** During adolescence, the cephalocaudal growth trend of infancy and childhood reverses. At first, the hands, legs, and feet accelerate, and then the torso, which accounts for most of the adolescent height gain (Sheehy et al., 1999). This pattern of development helps us understand why early adolescents often appear awkward and out of proportion—long-legged and with giant feet and hands.

Large sex differences in body proportions also appear, caused by the action of sex hormones on the skeleton. Boys' shoulders broaden relative to the hips, whereas girls' hips broaden relative to the shoulders and waist. Of course, boys also end up considerably larger than girls, and their legs are longer in relation to the rest of the body. The major reason is that boys have 2 extra years of preadolescent growth, when the legs are growing fastest (Graber, Petersen, & Brooks-Gunn, 1996).

● **Muscle–Fat Makeup and Other Internal Changes.** Compared with her later-developing girlfriends, Sabrina had accumulated much more fat, so she worried about her weight. Around age 8, girls start to add fat on their arms, legs, and trunk, a trend that accelerates between ages 11 and 16. In contrast, the arm and leg fat of adolescent boys decreases (Siervogel et al., 2000). Although both sexes gain in muscle, this increase is much greater for boys, who develop larger skeletal muscles, hearts, and lung capacity. Also, the number of red blood cells, and therefore the ability to carry oxygen from the lungs to the muscles, increases in boys but not in girls. Altogether, boys gain far more muscle strength than girls, a difference that contributes to boys' superior athletic performance during the teenage years (Ramos et al., 1998).

● **Changing States of Arousal.** Adolescence is a time of substantial change in sleep and wakefulness. On average, sleep declines from 10 hours in middle childhood to 7.5 to 8 hours in adolescence. Yet teenagers need almost as much sleep as they did during the school years—about 9.2 hours.

Adolescents go to bed much later than they did as children yet must get up early for school—before their sleep needs are satisfied. Biological changes may underlie this sleep "phase delay," as the tendency to stay up late strengthens with pubertal growth. But afternoon and evening activities, part-time jobs, and social pressures also contribute. Although most teenagers say they enjoy staying up late, they also complain of daytime sleepiness (Wolfson & Carskadon, 1998; Fins & Wohlgemuth, 2001). Sleep-deprived adolescents tend to achieve less well in school, more often suffer from depressed mood, and report irregular sleep schedules, which add to their daytime sleepiness and behavior problems (Link & Ancoli-Israel, 1995). Later school start times ease sleep loss but do not eliminate it (Kowalski & Allen, 1995).

Motor Development and Physical Activity

Puberty is accompanied by steady improvement in gross motor performance, but the pattern of change differs for boys and girls. Girls' gains are slow and gradual, leveling off by age 14. In contrast, boys show a dramatic spurt in strength, speed, and endurance that continues through the teenage years. By midadolescence, few girls perform as well as the average boy in running speed, broad jump, and throwing distance. And practically no boys score as low as the average girl (Malina & Bouchard, 1991).

Because girls and boys are no longer well matched physically, gender-segregated physical education usually begins in junior high school. At the same time, athletic options for both sexes expand. Many new sports are added to the curriculum—track and field, wrestling, tackle football, weight lifting, floor hockey, archery, tennis, and golf, to name just a few.

In 1972, the U.S. federal government required schools receiving public funds to provide equal opportunities for males and females in all educational programs, including athletics. Since then, high school girls' sports participation has increased greatly in both the United States and Canada, although it still falls short of boys' (see Figure 11.2). In Chapter 9, we saw that beginning at an early age, girls get less encouragement and recognition for athletic achievement, a pattern that persists into the teenage years.

The sex differences just described also characterize physical activity rates. Overall, 73 percent of American high school boys

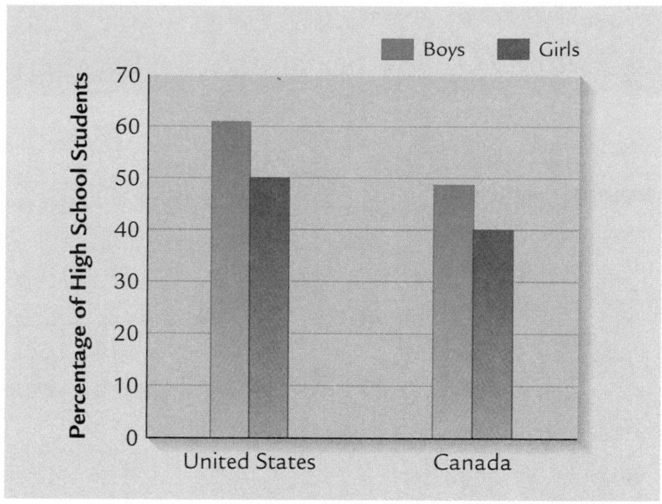

FIGURE 11.2 Involvement of American and Canadian high school students in school sports outside physical education classes. In both nations, more boys than girls participate. Still, girls' participation in high school sports has increased greatly over the past several decades. (From Canadian Fitness and Lifestyle Research Institute, 2001a; National Federation of State High School Associations, 2002.)

but only 53 percent of girls report regular vigorous physical activity (at least 20 minutes 3 days a week). About 22 percent of girls and 11 percent of boys who were physically active in ninth grade are no longer exercising regularly by twelfth grade. Physical activity rates for Canadian youths also decline over the teenage years (Canadian Fitness and Lifestyle Research Institute, 2001a; U.S. Department of Health and Human Services, 2002n).

Besides improving motor performance, sports and exercise influence cognitive and social development. Interschool and intramural athletics provide important lessons in competition, assertiveness, problem solving, and teamwork (Newcombe & Boyle, 1995). And regular, sustained physical activity has lifelong health benefits. Yet only 60 percent of American and 65 percent of Canadian secondary school students are enrolled in physical education (Canadian Fitness and Lifestyle Research Institute, 2001a; U.S. Department of Health and Human Services, 2002n). Required daily physical education, aimed at helping all teenagers find pleasure in sports and exercise, is a vital means of promoting adolescent physical and psychological well-being.

Sexual Maturation

Accompanying the rapid increase in body size are changes in physical features related to sexual functioning. Some, called **primary sexual characteristics,** involve the reproductive organs (ovaries, uterus, and vagina in females; penis, scrotum, and testes in males). Others, called **secondary sexual characteristics,** are visible on the outside of the body and serve as additional signs of sexual maturity (for example, breast development in females and the appearance of underarm and

Table 11.1 Average Age and Age Range of Major Pubertal Changes in North American Girls and Boys

Girls	Average Age	Age Range	Boys	Average Age	Age Range
Breasts begin to "bud"	10	(8–13)	Testes begin to enlarge	11.5	(9.5–13.5)
Height spurt begins	10	(8–13)	Pubic hair appears	12	(10–15)
Pubic hair appears	10.5	(8–14)	Penis begins to enlarge	12	(10.5–14.5)
Peak of strength spurt	11.6	(9.5–14)	Height spurt begins	12.5	(10.5–16)
Peak of height spurt	11.7	(10–13.5)	Spermarche (first ejaculation) occurs	13	(12–16)
Menarche (first menstruation) occurs	12.8	(10.5–15.5)	Peak of height spurt	14	(12.5–15.5)
Adult stature reached	13	(10–16)	Facial hair begins to grow	14	(12.5–15.5)
Breast growth completed	14	(10–16)	Voice begins to deepen	14	(12.5–15.5)
Pubic hair growth completed	14.5	(14–15)	Penis growth completed	14.5	(12.5–16)
			Peak of strength spurt	15.3	(13–17)
			Adult stature reached	15.5	(13.5–17.5)
			Pubic hair growth completed	15	(14–17)

Sources: Malina & Bouchard, 1991; Tanner, 1990.

pubic hair in both sexes). As Table 11.1 shows, these characteristics develop in a fairly standard sequence, although the ages at which each begins and is completed vary greatly.

● Sexual Maturation in Girls. Female puberty usually begins with the budding of the breasts and the growth spurt. **Menarche,** or first menstruation, typically happens around 12¾ years for North American girls, 13 for Europeans. But the age range is wide. Following menarche, pubic hair and breast development are completed, and underarm hair appears. Most girls take 3 to 4 years to complete this sequence, although this, too, can vary greatly, from 1½ to 5 years.

Notice in Table 11.1 how nature delays sexual maturity until the girl's body is large enough for childbearing; menarche occurs after the peak of the height spurt. As an extra measure of security, for 12 to 18 months following menarche, the menstrual cycle often takes place without an ovum being released from the ovaries (Tanner, 1990). However, this temporary period of sterility does not apply to all girls, and it cannot be counted on for protection against pregnancy.

● Sexual Maturation in Boys. The first sign of puberty in boys is the enlargement of the testes (glands that manufacture sperm), accompanied by changes in the texture and color of the scrotum. Pubic hair emerges a short time later, about the same time that the penis begins to enlarge (Graber, Petersen, & Brooks-Gunn, 1996).

Refer again to Table 11.1, and you will see that the growth spurt occurs much later in the sequence of pubertal events for boys than for girls. When it reaches its peak (at about age 14), enlargement of the testes and penis is nearly complete, and underarm hair appears soon after. Facial and body hair also emerge just after the peak in body growth and gradually increase for several years. Another landmark of male physical maturity is the deepening of the voice as the larynx enlarges and the vocal cords lengthen. (Girls' voices also deepen slightly.) Voice change usually takes place at the peak of the male growth spurt and is often not complete until puberty is over.

While the penis is growing, the prostate gland and seminal vesicles (which together produce semen, the fluid in which sperm are bathed) enlarge. Then, around age 13, **spermarche,** or first ejaculation, occurs (Jorgensen & Keiding, 1991). For a while, the semen contains few living sperm. So, like girls, boys have an initial period of reduced fertility.

Individual and Group Differences

Heredity contributes substantially to the timing of puberty, since identical twins generally reach menarche within a month or two of each other, whereas fraternal twins differ by about 12 months (Kaprio et al., 1995). Nutrition and exercise also contribute. In girls, a sharp rise in body weight and fat may trigger sexual maturation. Fat cells stimulate the ovaries and adrenal glands to produce sex hormones—which is probably why breast and pubic hair growth and menarche occur earlier for heavier and, especially, obese girls (Must & Strauss, 1999). In contrast, girls who begin serious athletic training at young ages or who eat very little (both of which reduce the percentage of body fat) often are delayed in sexual development (Rees, 1993).

Variations in pubertal growth also exist between regions of the world and between SES and ethnic groups. Heredity seems to play little role, since adolescents with very different genetic origins living under similarly advantaged conditions reach menarche at about the same average age (Morabia et al., 1998). Instead, physical health is largely responsible. In poverty-stricken regions where malnutrition and infectious disease are widespread, menarche is greatly delayed. In many parts of Africa, it does not occur until age 14 to 17. Girls from higher-income families reach menarche 6 to 18 months earlier than those living in economically disadvantaged homes. And African-American girls are ahead of Caucasian-American girls by 6 months in average age of menarche—a difference believed to be due to black girls' heavier body builds (Biro et al., 2001).

Early family experiences also seem to contribute to the timing of puberty. One theory suggests that humans have evolved to be sensitive to the emotional quality of their childhood environments. When children's safety and security are at risk, it is adaptive for them to reproduce early. In support of this view, several studies indicate that girls exposed to family conflict tend to reach menarche early, whereas those with warm family ties reach menarche relatively late (Ellis & Garber, 2000; Ellis et al., 1999; Moffitt et al., 1992).

Notice that in the research we have considered, threats to emotional health accelerate puberty, whereas threats to physical health delay it. A **secular trend,** or generational change, in pubertal timing lends added support to the role of physical well-being in pubertal development. In industrialized nations, age of menarche declined steadily from 1860 to 1970, by about 3 to 4 months per decade. Nutrition, health care, sanitation, and control of infectious disease improved greatly during this time. Since then, soaring rates of overweight and obesity have sustained this trend (Wattigney et al., 1999). A worrisome consequence is that girls who reach sexual maturity at ages 10 or 11 will feel pressure to act much older than they are. As we will see shortly, early maturing girls are at risk for unfavorable peer involvements, including sexual activity.

■ Because of improved health and nutrition, a secular trend in physical growth has taken place in industrialized nations. The adolescent girl in this photo is taller than her grandmother, mother, and aunt, and she probably reached menarche at an earlier age. Improved nutrition and health are responsible for gains in height and faster physical maturation from one generation to the next.

The Psychological Impact of Pubertal Events

Think back to your late elementary and junior high school days. As you reached puberty, how did your feelings about yourself and your relationships with others change? Research reveals that pubertal events affect the adolescent's self-image, mood, and interaction with parents and peers. Some of these outcomes are a response to dramatic physical change, regardless of when it occurs. Others have to do with the timing of pubertal maturation.

Reactions to Pubertal Changes

Two generations ago, menarche was often traumatic. Today, girls commonly react with "surprise," undoubtedly due to the sudden nature of the event. Otherwise, they typically report a mixture of positive and negative emotions. Yet wide individual differences exist that depend on prior knowledge and support from family members. Both are influenced by cultural attitudes toward puberty and sexuality.

For girls who have no advance information, menarche can be shocking and disturbing. In the 1950s, up to 50 percent were given no prior warning (Shainess, 1961). Today, few are uninformed. This shift is probably due to modern parents' greater willingness to discuss sexual matters with their youngsters and more widespread health education classes (Beausang & Razor, 2000; Brooks-Gunn, 1988b). Almost all girls get some information from their mothers. And girls whose fathers know about their daughters' pubertal changes adjust especially well. Perhaps a father's involvement reflects a family atmosphere that is highly understanding and accepting of physical and sexual matters (Brooks-Gunn & Ruble, 1980, 1983).

Like girls' reactions to menarche, boys' responses to spermarche reflect mixed feelings. Virtually all boys know about ejaculation ahead of time, but few get any information from parents. Usually they obtain it from reading material (Gaddis & Brooks-Gunn, 1985). Despite advance information, many boys say that their first ejaculation occurred earlier than they expected and that they were unprepared for it. As with girls, the better prepared boys feel, the more positively they react (Stein

■ Quinceañera, the traditional Hispanic fifteenth birthday celebration, is a rite of passage honoring a girl's journey from childhood to maturity and emphasizing the importance of family and community responsibility. It usually begins with a mass in which the priest blesses gifts presented to the girl, followed by a reception for family and friends. This girl receives a ring, symbolizing the unending circle of life and the emergence of her contributions to society.

& Reiser, 1994). In addition, whereas almost all girls eventually tell a friend that they are menstruating, far fewer boys tell anyone about spermarche (Downs & Fuller, 1991). Overall, boys get much less social support than girls for the physical changes of puberty. This suggests that boys might benefit, especially, from opportunities to ask questions and discuss feelings with a sympathetic parent or health professional.

The experience of puberty is affected by the larger cultural context. Many tribal and village societies celebrate puberty with a *rite of passage*—a community-wide initiation ceremony that marks an important change in privilege and responsibility. Consequently, young people know that pubertal changes are honored and valued in their culture. In contrast, Western societies grant little formal recognition to movement from childhood to adolescence or from adolescence to adulthood. Certain ethnic and religious ceremonies, such as the Jewish bar or bat mitzvah and the Quinceañera in Hispanic communities (celebrating a 15-year-old girl's journey to adulthood) resemble a rite of passage. But they usually do not lead to a meaningful change in social status.

Instead, Western adolescents are confronted with many ages at which they are granted partial adult status—for example, an age for starting employment, for driving, for leaving high school, for voting, and for drinking. In some contexts (on the highway and at work), they may be treated like adults. In others (at school and at home), they may still be regarded as children. The absence of a widely accepted marker of physical and social maturity makes the process of becoming an adult especially confusing.

Pubertal Change, Emotion, and Social Behavior

In the preceding section, we considered adolescents' reactions to their sexually maturing bodies. Puberty can also affect emotional state and social behavior. A common belief is that puberty has something to do with adolescent moodiness and the desire for greater physical and psychological separation from parents.

● **Adolescent Moodiness.** Although research reveals that higher pubertal hormone levels are linked to greater moodiness, these relationships are not strong (Buchanan, Eccles, & Becker, 1992). And we cannot be sure that a rise in pubertal hormones actually causes adolescent moodiness. What else might contribute to the common observation that adolescents are moody creatures? In several studies, the mood fluctuations of children, adolescents, and adults were tracked over a week by having them carry electronic pagers. At random intervals, they were beeped and asked to write down what they were doing, whom they were with, and how they felt.

As expected, adolescents reported less favorable moods than school-age children or adults (Csikszentmihalyi & Larson, 1984; Larson & Lampman-Petraitis, 1989). But negative moods were often linked to a greater number of negative life events, such as difficulties with parents, disciplinary actions at school, and breaking up with a boyfriend or girlfriend. Negative events increased steadily from childhood to adolescence, and teenagers also seemed to react to them with greater emotion than did children (Larson & Ham, 1993).

Furthermore, compared with the moods of adults, adolescents' feelings were less stable. They often varied from cheerful to sad and back again. But teenagers also moved from one situation to another more often, and their mood swings were strongly related to these changes. High points of their days were times spent with peers and in self-chosen leisure activities. Consequently, mood improved greatly on weekend evenings, especially at older ages as teenagers went out with friends and romantic partners (Larson & Richards, 1998). Low points tended to occur in adult-structured settings—class, job, school halls, school library, and religious services. Taken together, these findings suggest that situational factors join with hormonal influences to affect teenagers' moodiness. This explanation is consistent with the balanced view of biological and social forces described earlier in this chapter.

● **Parent–Child Relationships.** Sabrina's father noticed that as his children entered adolescence, their bedroom doors started to close, they resisted spending time with the family, and they became more argumentative. Sabrina and her mother squabbled over Sabrina's messy room ("Mom, it's *my* room. You don't have to live in it!") and her clothing purchases ("Sabrina, if you *buy* it, then *wear* it. Otherwise, you are wasting money!"). And Sabrina resisted the family's regular weekend visit to Aunt Gina's ("Why do I have to go *every* week?"). Many studies show that puberty is related to a rise in parent–child conflict. During this time, both parents and teenagers report feeling less close to one another (Laursen, Coy, & Collins, 1998; Steinberg & Morris, 2001). Frequency of conflict is surprisingly similar across American subcultures. It occurs about as often in families of European descent as in immigrant Chinese, Filipino, and Mexican families, whose traditions respect parental authority (Fuligni, 1998b).

Why should a youngster's new, more adultlike appearance trigger these disputes? The association may have some adaptive value. Among nonhuman primates, the young typically leave the family group around puberty. The same is true in many nonindustrialized cultures (Caine, 1986; Schlegel & Barry, 1991). Departure of young people discourages sexual relations between close blood relatives. But because children in industrialized societies remain economically dependent on parents long after they reach puberty, they cannot leave the family. Consequently, a modern substitute for physical departure seems to have emerged—psychological distancing.

As we will see later, adolescents' new powers of reasoning may also contribute to a rise in family tensions. In addition, friction rises because children have become physically mature and demand to be treated in adultlike ways. Parent–adolescent disagreements focus largely on mundane, day-to-day matters, such as driving, dating partners, and curfews (Adams & Laursen, 2001). But beneath these disputes are serious concerns—parental efforts to protect their teenagers from substance use, auto accidents, and early sex. The larger the gap between parents' and adolescents' views of teenagers' readiness to take on new responsibilities, the more quarreling (Dekovíc, Noom, & Meeus, 1997).

Most of these disputes are mild. In reality, parents and adolescents display both conflict and affection, and typically they agree on important values, such as honesty and education (Arnett, 1999). Although separation from parents is adaptive, both generations benefit from warm, protective family bonds throughout the lifespan.

Early versus Late Maturation

"All our children were early maturers," said Franca during the parents' discussion group. "The three boys were tall by age 12 or 13, but it was easier for them. They felt big and important. Sabrina was skinny as a little girl, but now she says she is too fat and wants to diet. She thinks about boys and doesn't concentrate on her schoolwork."

Findings of several studies match the experiences of Sabrina and her brothers. Both adults and peers viewed early maturing boys as relaxed, independent, self-confident, and physically attractive. Popular with agemates, they held many leadership positions in school and tended to be athletic stars. In contrast, late maturing boys were not well liked. Adults and peers viewed them as anxious, overly talkative, and attention-seeking (Brooks-Gunn, 1988a; Clausen, 1975; Jones, 1965). However, early maturing boys (despite being viewed as well

© BOB DAEMMRICH/STOCK BOSTON, LLC

■ These boys are all 13 years old, yet they differ in timing of pubertal maturation. The two early maturing boys in the center are probably popular, self-confident, athletic stars with a positive body image. The other two boys appear to be on-time maturers—about average in progress for their age.

adjusted) reported slightly more emotional stress than their later maturing agemates (Ge, Conger, & Elder, 2001).

In contrast, early maturing girls were unpopular, withdrawn, lacking in self-confidence, and anxious and held few leadership positions (Ge, Conger, & Elder, 1996; Jones & Mussen, 1958). In addition, they were more involved in deviant behavior (getting drunk, participating in early sexual activity) and achieved less well in school (Caspi et al., 1993; Dick et al., 2000). In contrast, their late maturing counterparts were especially well off—regarded as physically attractive, lively, sociable, and leaders at school.

Two factors largely account for these trends: (1) how closely the adolescent's body matches cultural ideals of physical attractiveness, and (2) how well young people "fit in" physically with their peers.

● The Role of Physical Attractiveness. Flip through the pages of your favorite popular magazine. You will see evidence of our society's view of an attractive female as thin and long-legged and of a good-looking male as tall, broad-shouldered, and muscular. The female image is a girlish shape that favors the late developer. The male image fits the early maturing boy.

A consistent finding is that early maturing girls report a less positive **body image**—conception of and attitude toward their physical appearance—than do their on-time and late maturing agemates. Among boys, the opposite occurs: Early maturation is linked to a positive body image, whereas late maturation predicts dissatisfaction with the physical self (Alsaker, 1995). The conclusions young people draw about their appearance strongly affect their self-esteem and psychological well-being (Usmiani & Daniluk, 1997).

● The Importance of Fitting in with Peers. Physical status in relation to peers also explains differences in adjustment between early and late maturers. From this perspective, early maturing girls and late maturing boys have difficulty because they fall at the extremes of physical development and feel "out of place" when with their agemates. Not surprisingly, adolescents feel most comfortable with peers who match their own level of biological maturity (Brooks-Gunn et al., 1986; Stattin & Magnusson, 1990).

Because few agemates of the same pubertal status are available, early maturing adolescents of both sexes seek older companions, sometimes with unfavorable consequences. Older peers often encourage them into activities they are not yet ready to handle emotionally, including sexual activity, drug and alcohol use, and minor delinquent acts. Perhaps because of involvements like these, early maturers of both sexes are emotionally stressed and show declines in academic performance (Caspi et al., 1993; Stattin & Magnusson, 1990).

Interestingly, school contexts can modify these maturational timing effects. In one study, early maturing sixth-grade girls felt better about themselves when they attended kindergarten through sixth-grade (K–6) schools rather than kindergarten through eighth-grade (K–8) schools, where they could mix with older adolescents. In the K–6 settings, they were relieved of pressures to adopt behaviors for which they were not ready (Blyth, Simmons, & Zakin, 1985). Similarly, a New Zealand study found that delinquency among early maturing girls was greatly reduced in all-girl schools, which limit opportunities to associate with norm-violating peers (most of whom are older boys) (Caspi et al., 1993).

● Long-Term Consequences. Do the effects of early and late maturation persist into adulthood? Long-term follow-ups show striking turnabouts in overall well-being. Many early maturing boys and late maturing girls, who had been so admired in adolescence, became rigid, conforming, and somewhat discontented adults. In contrast, late maturing boys and early maturing girls, who were stress-ridden as teenagers, often developed into adults who were independent, flexible, cognitively competent, and satisfied with the direction of their lives (Livson & Peshkin, 1980). Perhaps the confidence-inducing adolescence of early maturing boys and late maturing girls does not promote the coping skills needed to solve life's later problems. In contrast, the painful experiences associated with off-time pubertal growth may, in time, contribute to sharpened awareness, clarified goals, and greater stability.

Nevertheless, these long-term outcomes may not hold completely. In a Swedish study, achievement difficulties of early maturing girls persisted into young adulthood, in the form of lower educational attainment than their on-time and later maturing counterparts (Stattin & Magnusson, 1990). In countries with highly selective college entrance systems, perhaps it is harder for early maturers to recover from declines in school performance. Clearly, the effects of maturational timing involve a complex blend of biological, current social setting, and cultural factors.

Ask Yourself

REVIEW

Many people believe that the rising sexual passions of puberty cause teenage rebelliousness. Where did this belief originate? Explain why it is incorrect.

REVIEW

List factors that contribute to pubertal timing. Then summarize the consequences of early versus late maturation for adolescent development.

APPLY

Fourteen-year-old Chloe spends hours alone in her room and no longer wants to join in weekend family activities. Explain why Chloe's behavior is adaptive.

CONNECT

How might adolescent moodiness contribute to psychological distancing between parents and adolescents? (Hint: Think about bidirectional influences in parent-child relationships.)

www.

Health Issues

The arrival of puberty is accompanied by new health issues related to the young person's striving to meet physical and psychological needs. As adolescents are granted greater autonomy, their personal decision making becomes important, in health as well as other areas (Bearison, 1998). Yet none of the health concerns we are about to discuss can be traced to a single cause. Instead, biological, psychological, family, and cultural factors jointly contribute.

Nutritional Needs

When their sons reached puberty, Franca and Antonio reported a "vacuum cleaner effect" in the kitchen as the boys routinely emptied the refrigerator. Rapid body growth leads to a dramatic rise in food intake. During the growth spurt, boys require about 2,700 calories a day and much more protein than they did earlier, girls about 2,200 calories and somewhat less protein than boys because of their smaller size and muscle mass (Larson, 1996).

This increase in nutritional requirements comes at a time when the diets of many young people are the poorest. Of all age groups, adolescents are the most likely to consume empty calories. In a longitudinal study of Minnesota students, consumption of breakfast, fruits, vegetables, and milk declined after the transition to junior high school, whereas consumption of soft drinks and fast foods rose sharply (Lytle et al.,

2000). These eating habits are particularly harmful if they extend a lifelong pattern of poor nutrition, less serious if they are a temporary response to peer influences and a busy schedule.

The most common nutritional problem of adolescence is iron deficiency. A tired, irritable teenager may be suffering from anemia rather than unhappiness and should have a medical checkup. Most teenagers do not get enough calcium, and they are also deficient in riboflavin (vitamin B2) and magnesium, both of which support metabolism (Cavadini, Siega-Riz, & Popkin, 2000). And contrary to what many parents believe, obese children rarely outgrow their weight problem as teenagers (Berkowitz & Stunkard, 2002).

Adolescents, especially girls concerned about their weight, tend to be attracted to fad diets. Unfortunately, most are too limited in nutrients and calories to be healthy for fast-growing, active teenagers (Donatelle, 2003). When a young person wants to try a special diet, parents should insist that she first consult a doctor.

■ This anorexic girl's strict, self-imposed diet and obsession with strenuous exercise have led her to become painfully thin. Even so, her body image is so distorted that she probably regards herself as overweight.

Eating Disorders

Franca worried about Sabrina's desire to lose weight, explained to her that she was really quite average in build for an adolescent girl, and reminded Sabrina that her Italian ancestors thought a plump female body was more beautiful than a thin one. Girls who reach puberty early and who are very dissatisfied with their body image, and who grow up in families preoccupied with weight and thinness, are at risk for serious eating problems. Severe dieting is the strongest predictor of the onset of an eating disorder in adolescence (Patton et al., 1999). The two most serious are anorexia nervosa and bulimia nervosa.

● **Anorexia Nervosa.** **Anorexia nervosa** is a tragic eating disturbance in which young people starve themselves because of a compulsive fear of getting fat. About 1 percent of North American and Western European teenage girls are affected, a rate that increased sharply during the past half-century as a consequence of cultural admiration of female thinness. Asian-American, Caucasian-American, and Hispanic girls are at greater risk than African-American girls, who are more satisfied with their size and shape (Halpern et al., 1999; Rhea, 1999; Wildes, Emery, & Simons, 2001). Boys make up about 10 percent of anorexia cases. Half of these are homosexual or bisexual young people who are uncomfortable with a strong, muscular appearance (Robb & Dadson, 2002). Anorexia nervosa occurs equally often among SES groups (Rogers et al., 1997).

Anorexics have an extremely distorted body image. Even after they have become severely underweight, they conclude that they are too heavy. Most go on self-imposed diets so strict that they struggle to avoid eating in response to hunger. To enhance weight loss, they exercise strenuously.

In their attempt to reach "perfect" slimness, anorexics lose between 25 and 50 percent of their body weight. Because a normal menstrual cycle requires about 15 percent body fat, either menarche does not occur or menstrual periods stop. Malnutrition causes pale skin; brittle, discolored nails; fine, dark hairs all over the body; and extreme sensitivity to cold. If it is allowed to continue, the heart muscle can shrink, the kidneys can fail, and irreversible brain damage and loss of bone mass can occur. About 6 percent of anorexics die of the disorder, as a result of either physical complications or suicide (Schmidt, 2000).

Forces within the person, the family, and the larger culture give rise to anorexia nervosa. Identical twins share the disorder more often than fraternal twins, suggesting a genetic influence (Klump, Kaye, & Strober, 2001). We have also seen that the societal image of "thin is beautiful" contributes to the poorer body image of early maturing girls, who are at greatest risk for anorexia (Tyrka, Graber, & Brooks-Gunn, 2000). In addition, many anorexics have extremely high standards for their own behavior and performance, are emotionally inhibited, and avoid intimate ties outside the family. Consequently, these girls are excellent students who are responsible and well behaved—ideal daughters in many respects.

Yet parent–adolescent interactions reveal problems related to adolescent autonomy. Often mothers of these girls have high expectations for physical appearance, achievement, and social acceptance and are overprotective and controlling. Fathers tend to be emotionally distant. Instead of rebelling openly, anorexic girls do so indirectly—by fiercely pursuing perfection in achievement, respectable behavior, and thinness (Bruch, 2001). Nevertheless, whether maladaptive parent–child relationships precede the disorder, emerge as a response to it, or both is not yet clear.

Because anorexic girls typically deny that any problem exists, treating the disorder is difficult. Hospitalization is often necessary to prevent life-threatening malnutrition. Family therapy, aimed at changing parent–child interaction and expectations, is the most successful treatment (Gelbaugh et al., 2001). Still, only about 50 percent of anorexics fully recover (Fichter & Quadflieg, 1999).

● **Bulimia Nervosa.** When Sabrina's 16-year-old brother, Louis, brought his girlfriend, Cassie, to the house, Sabrina admired her good figure. "What willpower! Cassie hardly touches food," Sabrina thought to herself. "But what in the world is wrong with Cassie's teeth?"

Willpower was not the secret of Cassie's slender shape. When it came to food, she actually had great difficulty controlling herself. Cassie suffered from **bulimia nervosa,** an eating disorder in which young people (again, mainly girls, but homosexual and bisexual boys are vulnerable) engage in strict dieting and excessive exercise accompanied by binge eating, often followed by deliberate vomiting and purging with laxatives. When by herself, Cassie often felt lonely, unhappy, and anxious. She responded with eating rampages, consuming thousands of calories in an hour or two. The vomiting that followed eroded the enamel on Cassie's teeth. In some cases, life-threatening damage to the throat and stomach occurs.

Bulimia is more common than anorexia nervosa. About 2 to 3 percent of teenage girls are affected; only 5 percent have previously been anorexic. Twin studies show that bulimia, like anorexia, is influenced by heredity (Klump, Kaye, & Strober, 2001). Although bulimics share with anorexics a pathological fear of getting fat, they may have experienced their parents as disengaged and emotionally unavailable rather than controlling. One conjecture is that bulimics turn to food to compensate for feelings of emptiness resulting from lack of parental involvement (Attie & Brooks-Gunn, 1996).

Some bulimics, like anorexics, are perfectionists. Others lack self-control not just in eating but in other areas of their lives, engaging in petty shoplifting and alcohol abuse (Garner & Garfinkel, 1997). Bulimics differ from anorexics in that they usually feel depressed and guilty about their abnormal eating habits and are desperate to get help. As a result, bulimia is usually easier to treat than anorexia, using therapy focused on support groups, nutrition education, and revising eating habits and thoughts about food (Kaye et al., 2000).

Sexual Activity

Louis and Cassie hadn't planned to have intercourse; it "just happened." But before and after, a lot of things passed through their minds. Cassie had been dating Louis for 3 months, and she began to wonder, "Will he think I'm normal if I don't have sex with him? If he wants to and I say no, will I lose him?" Both young people knew their parents wouldn't approve. In fact, when Franca and Antonio noticed how attached Louis was to Cassie, they talked to him about the importance of waiting and the dangers of pregnancy. But that Friday evening, Louis and Cassie's feelings for each other seemed overwhelming. As things went farther and farther, Louis thought, "If I don't make a move, will she think I'm a wimp?" And Cassie had heard from one of her girlfriends that you couldn't get pregnant the first time.

With the arrival of puberty, hormonal changes—in particular, the production of androgens in young people of both sexes—lead to an increase in sex drive (Halpern, Udry, & Suchindran, 1997). As Louis and Cassie's inner thoughts reveal, adolescents become very concerned about how to manage sexuality in social relationships. New cognitive capacities involving perspective taking and self-reflection affect their efforts to do so. Yet like the eating behaviors we have just discussed, adolescent sexuality is heavily influenced by the young person's social context.

● **The Impact of Culture.** Think, for a moment, about when you first learned the "facts of life" and how you found out about them. In your family, was sex discussed openly or treated with secrecy? Exposure to sex, education about it, and efforts to limit the sexual curiosity of children and adolescents vary widely around the world.

Despite the publicity granted to the image of a sexually free modern adolescent, sexual attitudes in North America are relatively restrictive. Typically, parents give children little information about sex, discourage them from engaging in sex play, and rarely talk about sex in their presence. When young people become interested in sex, they seek information from friends, books, magazines, movies, the Internet, and television. On American prime-time TV, which adolescents watch the most, two-third of programs contain sexual content, and most depict partners as spontaneous, passionate, and with little commitment to each other. Characters are rarely shown taking steps to avoid pregnancy or sexually transmitted disease (Brown, 2002).

Consider the contradictory messages delivered by these sources. On one hand, adults emphasize that sex at a young age and outside of marriage is wrong. On the other hand, the broader social environment extols the excitement and romanticism of sex. American teenagers are left bewildered, poorly informed about sexual facts, and with little sound advice on how to conduct their sex lives responsibly.

● **Adolescent Sexual Attitudes and Behavior.** Although differences between subcultural groups exist, the sexual attitudes of American adolescents and adults have become more liberal over the past 30 years. Compared to a generation ago, more people believe that sexual intercourse before marriage is all right, as long as two people are emotionally committed to one another (Michael et al., 1994). Recently a slight swing back in the direction of conservative sexual beliefs has occurred, largely due to the risk of sexually transmitted disease (especially AIDS) and to teenage sexual abstinence programs sponsored by schools and religious organizations (Ali & Scelfo, 2002; Boonstra, 2002).

Trends in the sexual behavior of adolescents are consistent with their attitudes. The rate of premarital sex among American and Canadian young people rose for several decades but, since 1990, has declined (Dryburgh, 2001; U.S.

■ Adolescence is an especially important time for the development of sexuality. North American teenagers receive contradictory and confusing messages from the social environment about the appropriateness of sex. Although many adolescents are sexually active, typically they have had relations with only one or two partners in the past year. A considerable number, however, do not use contraception consistently and are at risk for unintended pregnancy and sexually transmitted disease.

Department of Health and Human Services, 2002n). Nevertheless, as Figure 11.3 illustrates, a substantial percentage of young people are sexually active quite early, by age 15.

Males tend to have their first intercourse earlier than females. And more youths in the United States than in Canada and other Western countries have sex before age 15. Yet timing of first intercourse provides only a limited picture of adolescent sexual behavior. About half of American and two-thirds of Canadian sexually active teenage boys have had relations with only one or two partners in the past year. Contrary to popular belief, a runaway sexual revolution does not characterize young people. In fact, American and Canadian rates of teenage sexual activity are about the same as those of Western European nations (Darroch, Frost, & Singh, 2001).

● **Characteristics of Sexually Active Adolescents.** Early and frequent teenage sexual activity is linked to personal, family, peer, and educational characteristics. These include early pubertal maturation, parental divorce, single-parent and stepfamily homes, large family size, little or no religious involvement, sexually active friends and older siblings, poor school performance, lower educational aspirations, and tendency to engage in norm-violating acts, including alcohol and drug use and delinquency (Kotchick et al., 2001).

Since many of these factors are associated with growing up in a low-income family, it is not surprising that early sexual activity is more common among young people from economically disadvantaged homes. In fact, the high rate of premarital intercourse among African-American teenagers—60 percent compared with 51 percent for all American young people—can largely be accounted for by widespread poverty in the black population (Darroch, Frost, & Singh, 2001).

● **Contraceptive Use.** Although adolescent contraceptive use has increased in recent years, 20 percent of American and 13 percent of Canadian sexually active teenagers are at risk for unintended pregnancy because they do not use contraception consistently (see Figure 11.4 on page 356) (Alan Guttmacher Institute, 2002a). Why do so many fail to take precautions? As we will see when we take up adolescent cognitive development, adolescents can consider many possibilities when faced with a problem. But they often fail to apply this reasoning to everyday situations. When asked to explain why they did not use contraception, they often give answers like these: "I was waiting until I had a steady boyfriend." "I wasn't planning to have sex."

One reason for these responses is that advances in perspective taking lead teenagers, for a time, to be extremely concerned about others' opinions of them. Recall how Cassie and Louis each worried about what the other would think if they decided not to have sex. Furthermore, in the midst of everyday social pressures, adolescents often overlook the consequences of engaging in risky behaviors (Beyth-Marom & Fischhoff, 1997).

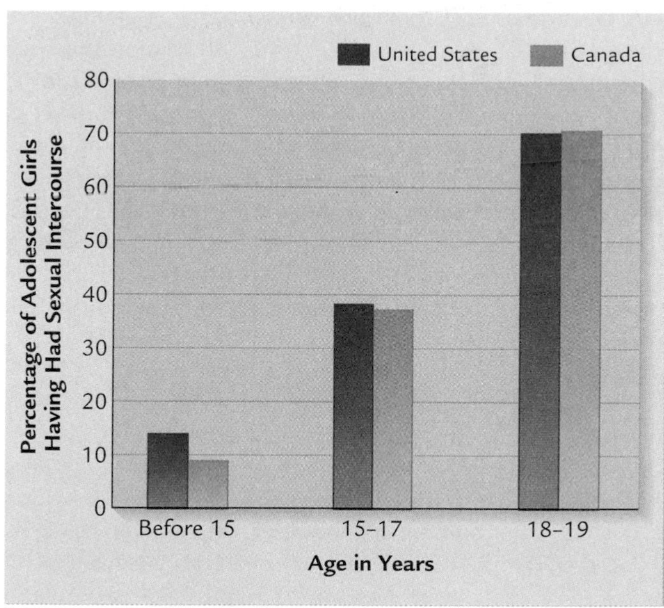

■ **FIGURE 11.3 Adolescent girls in the United States and Canada reporting ever having had sexual intercourse.** A greater percentage of American than of Canadian girls initiate sexual activity before age 15. Otherwise, rates of sexual activity are similar in the two countries and resemble those of Western European nations. Boys' sexual activity rates are 3 to 6 percent higher than the rates for girls shown here. (Adapted from Darroch, Frost, & Singh, 2001.)

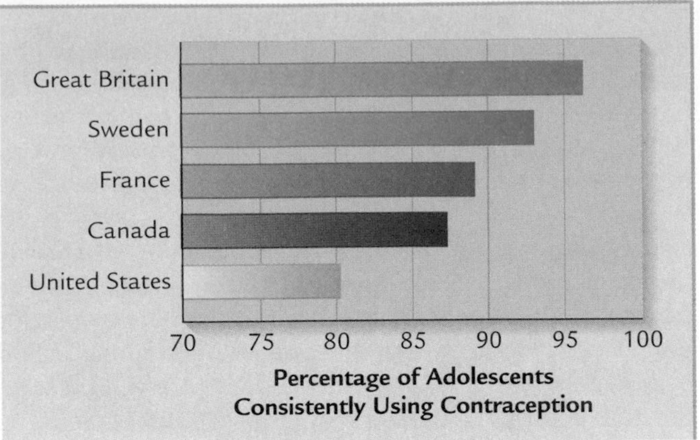

■ **FIGURE 11.4 Adolescent contraception use in five industrialized nations.** American sexually active teenagers are less likely to use contraception consistently than teenagers in other industrialized nations. Canadian adolescents, as well, fall below adolescents in Western Europe in contraceptive use. (Adapted from Darroch, Frost, & Singh, 2001.)

The social environment also contributes to teenagers' reluctance to use contraception. Those without the rewards of meaningful education and work are especially likely to engage in irresponsible sex, sometimes within relationships characterized by exploitation. About 12 percent of American girls and 5 percent of boys say they were forced to have intercourse. And among girls who voluntarily had sex, one-fourth indicate that they really did not want to do so (Alan Guttmacher Institute, 2002b; U.S. Department of Health and Human Services, 2000n).

In contrast, teenagers who report good relationships with parents and who talk openly with them about sex and contraception are more likely to use birth control (Whitaker & Miller, 2000). Unfortunately, many adolescents are too scared or embarrassed to ask parents questions. And too many leave sex education classes with incomplete or factually incorrect knowledge. Some do not know where to get birth control counseling and devices. When they do, they often worry that a doctor or family planning clinic might not keep their visits confidential (American Academy of Pediatrics, 1999).

● **Sexual Orientation.** Up to this point, our discussion has focused only on heterosexual behavior. About 3 to 6 percent of teenagers discover that they are lesbian or gay (see the Social Issues box on the following page). An as yet unknown but significant number are bisexual (Michael et al., 1994; Patterson, 1995). Adolescence is an equally crucial time for the sexual development of these young people, and societal attitudes, once again, loom large in how well they fare.

Heredity contributes importantly to homosexuality. Identical twins of both sexes are much more likely than fraternal twins to share a homosexual orientation. The same is true for biological as opposed to adoptive relatives (Bailey & Pillard, 1991; Bailey et al., 1993). Furthermore, male homosexuality tends to be more common on the maternal than on the paternal side of families. This suggests that it might be X-linked (see Chapter 2). Indeed, one gene-mapping study found that among 40 pairs of homosexual brothers, 33 (88 percent) had an identical segment of DNA on the X chromosome. One or several genes in that region might predispose males to become homosexual (Hamer et al., 1993).

How might heredity lead to homosexuality? According to some researchers, certain genes affect the level or impact of prenatal sex hormones, which modify brain structures in ways that induce homosexual feelings and behavior (Bailey et al., 1995; LeVay, 1993). Keep in mind, however, that both genetic and environmental factors can alter prenatal hormones. Girls exposed prenatally to very high levels of androgens or estrogens—because of either a genetic defect or drugs given to the mother to prevent miscarriage—are more likely to become homosexual or bisexual (Meyer-Bahlburg et al., 1995). Furthermore, homosexual men tend to have a later birth order and a higher-than-average number of older brothers. One controversial speculation is that mothers with several male children sometimes produce antibodies to androgens, which reduce the prenatal impact of male sex hormones on the brains of later-born boys (Blanchard et al., 1995; Blanchard & Bogaert, 1996).

Stereotypes and misconceptions about homosexuality continue to be widespread. For example, contrary to common belief, most homosexual adolescents are not "gender deviant" in dress or behavior. Furthermore, attraction to members of the same sex is not limited to gay and lesbian teenagers. Among heterosexual adolescents, about 18 percent of boys and 6 percent of girls report participating in at least one homosexual act (Braverman & Strasburger, 1993).

Sexually Transmitted Disease

Sexually active adolescents, both homosexual and heterosexual, are at risk for sexually transmitted disease (STD) (see Table 11.2 on page 358). Adolescents have the highest rates of STD of all age groups. Despite a recent decline in STD in the United States, one out of six sexually active teenagers contracts one of these illnesses each year—a rate much higher than that of Canada and other industrialized nations (U.S. Centers for Disease Control, 2001). When STD is left untreated, sterility and life-threatening complications can result. Teenagers in greatest danger of STD are the same ones who tend to engage in irresponsible sexual behavior: poverty-stricken young people who feel a sense of hopelessness about their lives (Darroch, Frost, & Singh, 2001).

By far the most serious STD is AIDS. Unlike Canada, where the incidence of AIDS among people younger than age 30 is low, one-fifth of U.S. AIDS cases occur between ages 20 and 29. Nearly all originate in adolescence, since AIDS symptoms typically take 8 to 10 years to develop. Drug-abusing adolescents who share needles and homosexual adolescents who have sex with HIV-positive partners account for most cases, but heterosexual spread of the disease has increased, especially among females. It is at least twice as easy for a male to

Social Issues

Homosexuality: Coming Out to Oneself and Others

Cultures vary as much in their acceptance of homosexuality as in their approval of extramarital sex. In the United States, homosexuals are stigmatized, as shown by the degrading language often used to describe them. This makes forming a sexual identity a much greater challenge for gay and lesbian youths than for their heterosexual counterparts.

Wide variations in sexual identity formation exist, depending on personal, family, and community factors. Yet interviews with homosexual adolescents and adults reveal that many (but not all) move through a three-phase sequence in coming out to themselves and others.

Feeling Different. Many gay men and lesbians say they felt different from other children when they were young (Savin-Williams, 1998). Typically, this first sense of their biologically determined sexual orientation appears between ages 6 and 12 and results from play interests more like those of the other gender (Mondimore, 1996). Boys may find that they are less interested in sports, drawn to quieter activities, and more emotionally sensitive than other boys, girls that they are more athletic and active than other girls.

Confusion. With the arrival of puberty, feeling different begins to include feeling sexually different. In research on ethnically diverse gay, lesbian, and bisexual youths, awareness of a same-sex attraction occurred, on average, between ages 11 and 12 for boys and 14 and 15 for girls, perhaps because social pressures toward heterosexuality are particularly intense for girls (Diamond, 1998; Herdt & Boxer, 1993). Realizing that homosexuality has personal relevance generally sparks confusion because most young people had assumed they were heterosexual like everyone else.

A few adolescents resolve their discomfort by crystallizing a gay or lesbian identity quickly, with a flash of insight into their sense of being different. But many experience an inner struggle, intensified by lack of role models and social support. Some throw themselves into activities they have come to associate with heterosexuality. Boys may go out for athletic teams; girls may drop softball and basketball in favor of dance. And homosexual youths typically try heterosexual dating, sometimes to hide their sexual orientation and at other times to develop intimacy skills that they later apply to same-sex relationships (Dubé, Savin-Williams, & Diamond, 2001). Those who are extremely troubled and guilt-ridden may escape into alcohol, drugs, and suicidal thinking.

© DONNA BINDER

■ This gay couple enjoys an evening at a high school prom. As long as friends and family members react with acceptance, coming out strengthens the young person's view of homosexuality as a valid, meaningful, and fulfilling identity.

Acceptance. The majority of gay and lesbian teenagers reach a point of accepting their homosexuality. Then they face another crossroad: whether to tell others. The most difficult disclosure is to parents, but many fear rejection by peers as well (Cohen & Savin-Williams, 1996). Powerful stigma against their sexual orientation lead some to decide that no disclosure is possible. As a result, they self-define but otherwise "pass" as heterosexual. In one study of gay adolescents, 85 percent said they tried concealment for a time (Newman & Muzzonigro, 1993).

Many homosexuals eventually acknowledge their sexual orientation publicly, usually by telling trusted friends first, then family members and acquaintances. When people react positively, coming out strengthens the young person's view of homosexuality as a valid, meaningful, and fulfilling identity. Contact with other gays and lesbians is important for reaching this phase, and changes in society permit many adolescents in urban areas to attain it earlier than they did a decade or two ago (Diamond, Savin-Williams, & Dubé, 1999). Gay and lesbian communities exist in large cities, along with specialized interest groups, social clubs, religious groups, newspapers, and periodicals. Small towns and rural areas remain difficult places to meet other homosexuals and to find a supportive environment. Teenagers in these locales have a special need for caring adults and peers who can help them find self- and social acceptance.

Gay and lesbian adolescents who succeed in coming out to themselves and others integrate their sexual orientation into a broader sense of identity, a process we will address in Chapter 12. As a result, they no longer need to focus so heavily on their homosexual self, and energy is freed for other aspects of psychological growth. In sum, coming out can foster many aspects of adolescent development, including self-esteem, psychological well-being, and relationships with family, friends, and co-workers.

Table 11.2 Most Common Sexually Transmitted Diseases of Adolescence

Disease	Reported Cases Among 15- to 19-Year-Olds (rate per 100,000) U.S.	Canada	Cause	Symptoms and Consequences	Treatment
Aids	20[a]	0.2[a]	Virus	Fever, weight loss, severe fatigue, swollen glands, and diarrhea. As the immune system weakens, severe pneumonias and cancers, especially on the skin, appear. Death due to other diseases usually occurs.	No cure; experimental drugs prolong life
Chlamydia	1,426	563	Bacteria	Discharge from the penis in males; painful itching, burning vaginal discharge, and dull pelvic pain in females. Often no symptoms. If left untreated, can lead to inflammation of the pelvic region, infertility, and sterility.	Antibiotic drugs
Cytomegalovirus	Unknown[b]		Virus of the herpes family	No symptoms in most cases. Sometimes a mild flu-like reaction. In pregnant women, can spread to the embryo or fetus and cause miscarriage or serious birth defects (see page 87).	None; usually disappears on its own
Genital warts	451	No data	Virus	Warts that typically grow near the vaginal opening in females, on the penis or scrotum in males. Can cause severe itching. Related to cancer of the cervix.	Removal of warts, but virus persists
Gonorrhea	500	59.4	Bacteria	Discharge from the penis or vagina, painful urination. Sometimes no symptoms. If left untreated, can spread to other regions of the body, resulting in such complications as infertility, sterility, blood poisoning, arthritis, and inflammation of the heart.	Antibiotic drugs
Herpes simplex 2 (genital herpes)	167	55.1	Virus	Fluid-filled blisters on the genitals, high fever, severe headache, and muscle aches and tenderness. No symptoms in a few people. In a pregnant woman, can spread to the embryo or fetus and cause birth defects (see pages 87–88).	No cure; can be controlled with drug treatment
Syphilis	1.9	0.2	Bacteria	Painless chancre (sore) at site of entry of germ and swollen glands, followed by rash, patchy hair loss, and sore throat within 1 week to 6 months. These symptoms disappear without treatment. Latent syphilis varies from no symptoms to damage to the brain, heart, and other organs after 5 to 20 years. In pregnant women, can spread to the embryo and fetus and cause birth defects.	Antibiotic drugs

[a]This figure includes both adolescents and young adults. For most U.S. cases, the virus is contracted in adolescence, and symptoms appear in early adulthood.

[b]Cytomegalovirus is the most common STD. Because there are no symptoms in most cases, its precise rate of occurrence is unknown. Half of the population or more may have had the virus sometime during their lives.

Sources: Health Canada, 2002e; U.S. Centers for Disease Control, 2001.

infect a female with any STD, including AIDS, as it is for a female to infect a male (U.S. Centers for Disease Control, 2001).

As the result of school course and media campaigns, over 90 percent of high school students are aware of basic facts about AIDS. But some hold false beliefs that put them at risk—for example, that birth control pills provide some protection (DiClemente, 1993). The Caregiving Concerns table on the following page lists strategies for preventing STD.

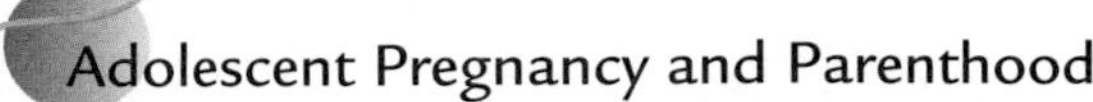

Adolescent Pregnancy and Parenthood

Cassie was lucky not to get pregnant after having sex with Louis, but some of her high school classmates weren't so fortunate. About 900,000 American teenage girls become pregnant annually, 30,000 of them younger than age 15. Despite a recent decline, the U.S. adolescent pregnancy rate is higher

Preventing Sexually Transmitted Disease

STRATEGY	DESCRIPTION
Know your partner well.	Take time to get to know your partner. Find out whether your partner has had sex with many people or has used injectable drugs.
Maintain mutual faithfulness.	For this strategy to work, neither partner can have an STD at the start of the relationship.
Do not use drugs.	Using a needle, syringe, or drug liquid previously used by others can spread STD. Alcohol, marijuana, or other illegal substances impair judgment, reducing your capacity to think clearly about the consequences of your behavior.
Always use a latex condom and vaginal contraceptive when having sex with a nonmarital partner.	Latex condoms give good (but not perfect) protection against STD by reducing the passage of bacteria and viruses. Vaginal contraceptives containing nonoxynol-9 can kill several kinds of STD microbes. They increase protection when combined with condom use.
Do not have sex with a person you know has an STD.	Even if you are protected by a condom, you risk contracting the disease. If either partner has engaged in behavior that might have caused HIV infection, a blood test must be administered and repeated at least 6 months after that behavior, since it takes time for the body to develop antibodies.
If you get an STD, inform all recent sexual partners.	Notifying people you may have exposed to STD permits them to get treatment before spreading the disease to others.

Source: Daugirdas, 1992.

than that of most other industrialized countries (see Figure 11.5). Although the Canadian rate is about half the U.S. rate, teenage pregnancy in Canada is nevertheless a problem. Three factors heighten the incidence of adolescent pregnancy: (1) effective sex education reaches too few teenagers, (2) convenient, low-cost contraceptive services for adolescents are scarce, and (3) many families live in poverty, which encourages young people to take risks.

Because 40 to 45 percent of American and Canadian teenage pregnancies end in abortion, the number of North American teenage births is actually lower than it was 30 years ago (Darroch, Frost, & Singh, 2001). But teenage parenthood is a much greater problem today because adolescents are far less likely to marry before childbirth. In 1960, only 15 percent of teenage births were to unmarried females, whereas today, 75 percent are (Coley & Chase-Lansdale, 1998). Increased social acceptance of single motherhood, along with the belief that a baby might fill a void in their lives, has meant that only a small number of girls give up their infants for adoption.

● **Correlates and Consequences of Adolescent Parenthood.** Becoming a parent is challenging and stressful for any person, but it is especially difficult for adolescents. Teenage parents have not yet established a clear sense of direction for their own lives. They have both life conditions and personal attributes that interfere with their ability to parent effectively (Jaffee et al., 2001).

Teenage mothers are many times more likely to be poor than are agemates who postpone childbearing. Their experiences often include low parental warmth and involvement; poor school performance; alcohol and drug use; adult models of unmarried parenthood, limited education, and unemployment; and residence in neighborhoods where other adolescents also display these risks (Scaramella et al., 1998). A high percentage of out-of-wedlock births are to members of low-income minorities, especially African-American, Native-American, Hispanic,

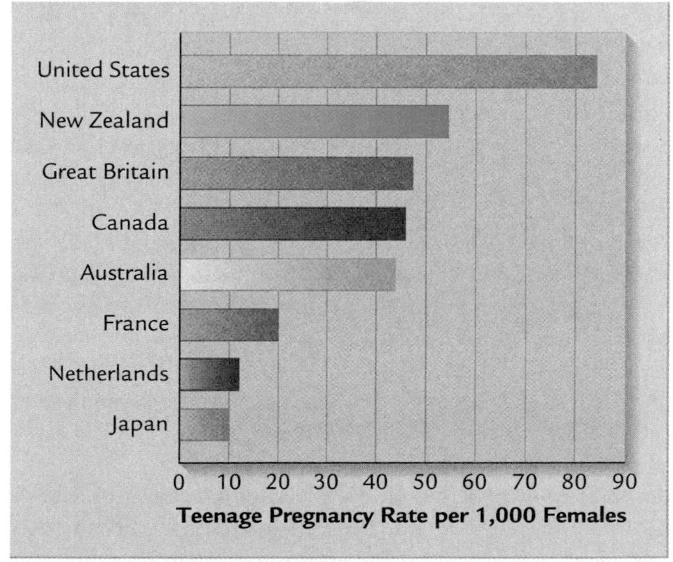

■ **FIGURE 11.5 Teenage pregnancy rate in eight industrialized nations.** (Adapted from Singh & Darroch, 2000.)

and Canadian Aboriginal teenagers. Many of these young people seem to turn to early parenthood as a way to move into adulthood when educational and career avenues are unavailable (Fagot et al., 1998).

The lives of pregnant teenagers are often troubled in many ways, and after the baby is born, their circumstances tend to worsen in at least three respects:

- *Educational attainment.* Giving birth before age 18 reduces the likelihood of finishing high school. Only 50 percent of American adolescent mothers graduate with either a diploma or a general equivalency degree (GED), compared with 96 percent of girls who wait to become parents (Hotz, McElroy, & Sanders, 1997).
- *Marital patterns.* Teenage motherhood reduces the chances of marriage. When these mothers do marry, they are more likely to divorce than their peers who delay childbearing (Moore et al., 1993). Consequently, teenage mothers spend more of their parenting years as single parents.
- *Economic circumstances.* Because of low educational attainment, marital instability, and poverty, many teenage mothers are on welfare. If they are employed, their limited education restricts them to unsatisfying, low-paid jobs. Adolescent fathers work more hours than their nonparent agemates in the years following their child's birth. Perhaps for this reason, they obtain less education and are also economically disadvantaged (Brien & Willis, 1997).

Because many pregnant teenage girls have inadequate diets, smoke, use alcohol and other drugs, and do not receive early prenatal care, their babies often experience prenatal and birth complications—especially low birth weight (Dell, 2001). And compared with adult mothers, adolescent mothers interact more negatively with the child's father, know less about child development, have unrealistically high expectations, perceive their infants as more difficult, and interact less effectively with them (Brooks-Gunn & Chase-Lansdale, 1995; Moore & Florsheim, 2001). Their children tend to score low on intelligence tests, achieve poorly in school, and engage in disruptive social behavior. And too often the cycle of adolescent pregnancy is repeated in the next generation (Jaffee et al., 2001; Moore, Morrison, & Green, 1997).

Still, how well adolescent parents and their children fare varies a great deal. If the adolescent finishes high school, avoids additional births, and finds a stable marriage partner, long-term disruptions in her own and her child's development are less severe. The small minority of young mothers who fail in all three of these areas face a life of continuing misfortune (Furstenberg, Brooks-Gunn, & Morgan, 1987).

● **Prevention Strategies.** Preventing teenage pregnancy means addressing the many factors underlying early sexual activity and lack of contraceptive use. Too often, sex education courses are given late in high school (after sexual activity has begun), last only a few sessions, and are limited to a catalogue of facts about anatomy and reproduction. Sex education that goes beyond this minimum does not encourage early sex, as some opponents claim. It does improve awareness of sexual facts—knowledge that is necessary for responsible sexual behavior (Katchadourian, 1990).

Knowledge, however, is not sufficient to influence teenagers' behavior. Sex education must also help them build a bridge between what they know and what they do. Today, more effective sex education programs have emerged with the following key elements:

- Teaching skills for handling sexual situations through creative discussion and role-playing techniques, which permit teenagers to confront sexual situations similar to those they will encounter in everyday life
- Promoting the value of abstinence to teenagers not yet sexually active
- Providing information about contraceptives and ready access to them

Sex education with these components can delay the initiation of sexual activity, increase contraceptive use, and reduce pregnancy rates (Aarons et al., 2000; Franklin et al., 1997).

The most controversial aspect of adolescent pregnancy prevention is providing access to contraceptives. Many adults argue that placing birth control pills or condoms in the hands of teenagers is equivalent to saying that early sex is OK. Yet in Western Europe, where school-based clinics offer contraceptives, teenage sexual activity is no higher than in North America, but pregnancy, childbirth, and abortion rates are much lower (Franklin & Corcoran, 2000).

Efforts to prevent adolescent pregnancy and parenthood must go beyond improving sex education to build social competence. In one study, researchers randomly assigned at-risk high school students either to a year-long community service class, called Teen Outreach, or to regular classroom experiences in health or social studies. In Teen Outreach, adolescents participated in at least 20 hours per week of volunteer work tailored to their interests. They returned to school for discussions that focused on enhancing their community service skills and their ability to cope with everyday challenges. At the end of the school year, pregnancy, school failure, and school suspension were substantially lower in the group enrolled in Teen Outreach, which fostered social skills, connectedness to the community, and self-respect (Allen et al., 1997).

Finally, teenagers who look forward to a promising future are far less likely to engage in early and irresponsible sex. By expanding educational, vocational, and employment opportunities, society can provide young people with good reasons to postpone childbearing.

● **Intervening with Adolescent Parents.** The most difficult and costly way to deal with adolescent parenthood is to wait until after it has happened. Young single mothers need health care for themselves and their children, encouragement to stay in school, job training, instruction in parenting and life

■ Early parenthood imposes lasting hardships on both generations—adolescent and newborn baby. Teenage fathers, like teenage mothers, suffer educationally and economically. This young father must work extra hours to pay for baby clothing and child care. His new burden will probably mean less opportunity for continued education and a lower income in the years to come.

management skills, and high-quality, affordable child care. Schools that provide these services reduce the incidence of low-birth-weight babies, increase mothers' educational success, and prevent additional childbearing (Seitz & Apfel, 1993, 1994).

Adolescent mothers also benefit from family relationships that are sensitive to their developmental needs. Older teenage mothers display more effective parenting when they establish their own residences with the help of relatives—an arrangement that grants the teenager a balance of autonomy and support. Independent living combined with high levels of grandparent assistance is associated with more effective parenting, warmer family ties, and children who develop more favorably (East & Felice, 1996).

Programs focusing on fathers are attempting to increase their financial and emotional commitment to the baby (Coley & Chase-Lansdale, 1998). Although almost half of young fathers visit their children during the first few years after birth, contact usually diminishes. But new laws that enforce child support payments may increase paternal responsibility and interaction. Teenage mothers who receive financial and child-care assistance from their child's father are less distressed and interact more favorably with their infants (Caldwell & Antonucci, 1997). And the fewer stressful life events teenage mothers experience, the more likely fathers are to stay involved and the better children's long-term adjustment (Cutrona et al., 1998; Furstenberg & Harris, 1993).

Substance Use and Abuse

At age 14, Louis took some cigarettes out of his uncle's pack, waited until he was alone in the house, and smoked. At an unchaperoned party, he and Cassie drank several cans of beer, largely because everyone else was doing it. Louis got little physical charge out of these experiences. He was a good student, was well liked by peers, and got along well with his parents. He had no need for drugs as an escape valve from daily life. But he knew of others at school for whom things were different—students who started with alcohol and cigarettes, moved to harder substances, and eventually were hooked.

In industrialized nations, teenage alcohol and drug use is pervasive. By age 14, 56 percent of American young people have tried cigarette smoking, 70 percent drinking, and 32 percent at least one illegal drug (usually marijuana). At the end of high school, 22 percent smoke cigarettes regularly, 60 percent have engaged in heavy drinking at least once, and over 50 percent have experimented with illegal drugs. About 30 percent have tried at least one highly addictive and toxic substance, such as amphetamines, cocaine, phencyclidine (PCP), or heroin. Canadian rates of teenage alcohol and drug use are similar (Adlaf & Paglia, 2001; U.S. Department of Health and Human Services, 2002i).

These figures represent an increase during the early 1990s, followed by a slight drop during the past few years. Why do so many young people subject themselves to the health risks of these substances? Part of the reason is cultural. Adolescents live in drug-dependent contexts. They see adults using caffeine to wake up in the morning, cigarettes to cope with daily hassles, a drink to calm down in the evening, and other remedies to relieve stress, depression, and physical illness. Reduced parental, school, and media focus on the hazards of drugs, followed by renewed public attention, may explain recent trends in adolescent drug taking.

For most young people, substance use simply reflects their intense curiosity about "adultlike" behaviors. The majority of teenagers dabble in alcohol, tobacco, and marijuana. These minimal *experimenters* are not headed for a life of decadence and addiction. Instead, they are psychologically healthy, curious young people (Shedler & Block, 1990). As Figure 11.6 on page 362 shows, tobacco and alcohol use is somewhat greater among European than American adolescents, perhaps because European adults more often smoke and drink. In contrast, illegal drug use is far more prevalent among American than European teenagers (Hibell, 2001). In the United States, a greater percentage of young people live in poverty, which is linked to family and peer contexts that promote illegal drug use.

Regardless of the type of drug, adolescent experimentation should not be taken lightly. Because most drugs impair perception and thought processes, a single heavy dose can lead

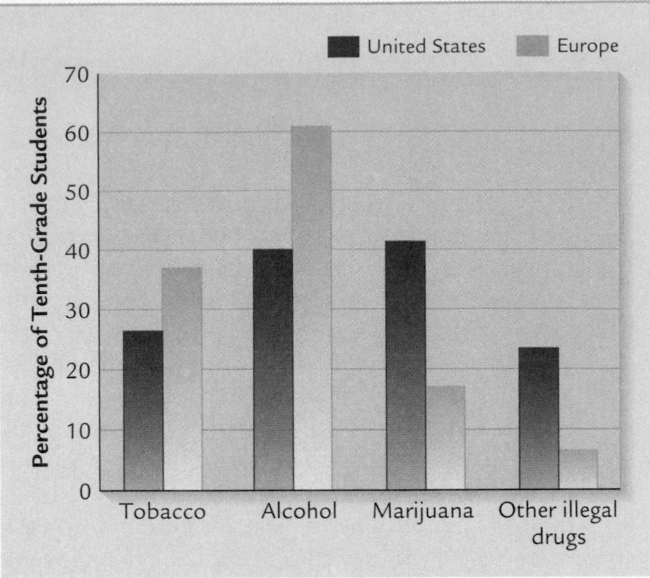

FIGURE 11.6 Tenth-grade students in the United States and Europe who have used various substances. Rates for tobacco and alcohol are based on any use in the past 30 days. Rates for marijuana and other illegal drugs are based on any lifetime use. Tobacco use and alcohol use are greater for European adolescents, whereas illegal drug use is greater for American adolescents. (Adapted from Hibell, 2001.)

to permanent injury or death. And a worrisome minority of teenagers move from substance *use* to *abuse*—taking drugs regularly, requiring increasing amounts to achieve the same effect, moving to harder substances, and using enough to impair their ability to meet school, work, and other responsibilities.

● **Correlates and Consequences of Adolescent Substance Abuse.** In contrast to experimenters, drug abusers are seriously troubled young people who express their unhappiness through antisocial acts. Their impulsive, disruptive style is often present in early childhood. And compared with other young people, their drug taking starts earlier and may have genetic roots (Chassin & Ritter, 2001). But a wide range of environmental factors also promote it. These include low SES, family mental health problems, parental and older sibling drug abuse, lack of parental warmth and involvement, physical and sexual abuse, and poor school performance. Peer encouragement—friends who use and provide access to drugs—predicts increased substance abuse and other antisocial behaviors (Kilpatrick et al., 2000; Patterson, Dishion, & Yoerger, 2000).

Adolescent substance abuse often has lifelong consequences. When teenagers depend on alcohol and hard drugs to deal with daily stresses, they fail to learn responsible decision-making skills and alternative coping techniques. These young people show serious adjustment problems, including depression and antisocial behavior (Luthar & Cushing, 1997). They often enter into marriage, parenthood, and the work world prematurely and fail at them—painful outcomes that encourage further addictive behavior.

● **Prevention and Treatment.** Programs that promote effective parenting (including monitoring of teenagers' activities) and that teach adolescents skills for resisting peer pressure reduce drug experimentation (Kosterman et al., 2000). But some drug taking seems inevitable. Therefore, interventions that teach teenagers about the dangers of drugs and that prevent them from harming themselves and others when they do experiment are essential. Many communities offer weekend on-call transportation services that any young person can contact for a safe ride home, with no questions asked.

Drug abuse, as we have seen, occurs for quite different reasons than occasional use. Therefore, different prevention strategies are required. One approach is to work with parents early, reducing family adversity and improving parenting skills, before children are old enough to become involved with drugs (Luthar, Cushing, & McMahon, 1997). Programs that teach at-risk teenagers strategies for handling life stressors and that build competence through community service reduce alcohol and drug use, just as they reduce teenage pregnancy (Richards-Colocino, McKenzie, & Newton, 1996).

When an adolescent becomes a drug abuser, hospitalization is often a necessary and even lifesaving first step. Once the young person is weaned from the drug, family and individual therapy to treat negative parent–child relationships, low self-esteem, anxiety, and impulsivity are generally needed. Academic and vocational training to improve life success makes a difference as well. Not much is known about the best way to treat adolescent drug abuse. Even the most comprehensive programs have alarmingly high relapse rates—from 35 to 85 percent (Gilvarry, 2000).

Ask Yourself

REVIEW
What unfavorable life experiences do teenagers who engage in early and frequent sexual activity and those who abuse drugs have in common?

REVIEW
Compare risk factors for anorexia nervosa and bulimia nervosa. How do treatments and outcomes differ for the two disorders?

APPLY
After 17-year-old Veronica gave birth to her second child, her parents told her they didn't have room for two babies. Veronica dropped out of school and moved in with her boyfriend. A few months later, he left. Why are Veronica and her children likely to experience long-term hardships?

CONNECT
List characteristics common to effective pregnancy and substance abuse prevention programs. Are these components well suited to fostering resiliency in at-risk adolescents? Explain. (Return to Chapter 1, page 10, if you need to review factors that promote resiliency.)

www.

COGNITIVE DEVELOPMENT

One mid-December evening, a knock at the front door announced the arrival of Franca and Antonio's oldest son Jules home for vacation after the fall semester of his sophomore year at college. Moments later, the family gathered around the kitchen table. "How did it all go, Jules?" inquired Antonio while passing out pieces of apple pie.

"Well, physics and philosophy were awesome. The last few weeks, our physics prof introduced us to Einstein's theory of relativity. Boggles my mind, it's so incredibly counterintuitive."

"Counter-what?" asked Sabrina, trying hard to follow the conversation.

"Counterintuitive. Unlike what you'd normally expect," explained Jules. "Imagine you're on a train, going unbelievably fast, like 160,000 miles a second. The faster you go approaching the speed of light, the slower time passes and the denser and heavier things get relative to on the ground. The theory revolutionized the way we think about time, space, matter—the entire universe."

Sabrina wrinkled her forehead, unable to comprehend Jules's otherworldly reasoning. "Time slows down when I'm bored, like right now, not on a train when I'm going somewhere exciting. No speeding train ever made me heavier, but this apple pie will if I eat any more of it," Sabrina announced, getting up and leaving the table.

Sixteen-year-old Louis reacted differently. "Totally cool, Jules. So what'd you do in philosophy?"

"It was a course in philosophy of technology. We studied the ethics of futuristic methods in human reproduction. For example, we argued the pros and cons of a world in which all embryos develop in artificial wombs."

"What do you mean?" asked Louis. "You order your kid at the lab?"

"That's right. I wrote my term paper on it. I had to evaluate it in terms of principles of justice and freedom. I can see some advantages but also lots of dangers. . . ."

As this conversation illustrates, adolescence brings with it vastly expanded powers of reasoning. At age 11, Sabrina finds it difficult to move beyond her firsthand experiences to a world of possibilities. Over the next few years, her thinking will take on the abstract qualities that characterize the cognition of her older brothers. Jules juggles variables in complex combinations and thinks about situations not easily detected in the real world or that do not exist at all. As a result, he can grasp complex scientific and mathematical principles and grapple with social and political issues. Compared with school-age children's thinking, adolescent thought is more enlightened, imaginative, and rational.

Piaget's Theory: The Formal Operational Stage

According to Piaget, around age 11 young people enter the **formal operational stage,** in which they develop the capacity for abstract, scientific thinking. Whereas concrete operational children can "operate on reality," formal operational adolescents can "operate on operations." In other words, they no longer require concrete things and events as objects of thought. Instead, they can come up with new, more general logical rules through internal reflection (Inhelder & Piaget, 1955/1958). Let's look at two major features of the formal operational stage.

Hypothetico-Deductive Reasoning

At adolescence, young people first become capable of **hypothetico-deductive reasoning.** When faced with a problem, they start with a *general theory* of all possible factors that might affect the outcome and *deduce* from it specific *hypotheses* (or predictions) about what might happen. Then they test these hypotheses in an orderly fashion to see which ones work in the real world. Notice how this form of problem solving begins with possibility and proceeds to reality. In contrast, concrete operational children start with reality—with the most obvious predictions about a situation. When these are not confirmed, they cannot think of alternatives and fail to solve the problem.

■ In Piaget's formal operational stage, adolescents engage in hypothetico-deductive reasoning. These high school students solve a complex scientific problem by thinking of all possible outcomes, not just the most obvious.

© WILL FALLER

■ **FIGURE 11.7 Piaget's pendulum problem.** Adolescents who engage in hypothetico-deductive reasoning think of all possibilities. Then they vary one factor at a time while holding all others constant. Soon they discover that the weight of the object, the height from which it is released, and how forcefully it is pushed have no effect on the speed with which the pendulum swings through its arc. Only string length makes a difference.

Adolescents' performance on Piaget's famous *pendulum problem* illustrates this new approach. Suppose we present several school-age children and adolescents with strings of different lengths, objects of different weights to attach to the strings, and a bar from which to hang the strings (see Figure 11.7). Then we ask each of them to figure out what influences the speed with which a pendulum swings through its arc.

Formal operational adolescents come up with four hypotheses: (1) the length of the string, (2) the weight of the object hung on it, (3) how high the object is raised before it is released, and (4) how forcefully the object is pushed. Then, by varying one factor at a time while holding all the others constant, they try out each possibility. Eventually they discover that only string length makes a difference.

In contrast, concrete operational children experiment unsystematically. They cannot separate the effects of each variable.They may test for the effect of string length without holding weight constant, comparing, for example, a short, light pendulum with a long, heavy one. Also, school-age children fail to notice variables that are not immediately suggested by the concrete materials of the task—the height at which and forcefulness with which the pendulum is released.

Propositional Thought

A second important characteristic of the formal operational stage is **propositional thought.** Adolescents can evaluate the logic of propositions (verbal statements) without referring to real-world circumstances. In contrast, children can evaluate the logic of statements only by considering them against concrete evidence in the real world.

In a study of propositional reasoning, a researcher showed children and adolescents a pile of poker chips and asked whether statements about the chips were true, false, or uncertain. In one condition, the researcher hid a chip in her hand and presented the following propositions:

"*Either* the chip in my hand is green *or* it is not green."

"The chip in my hand is green *and* it is not green."

In another condition, the experimenter held either a red or a green chip in full view and made the same statements.

School-age children focused on the concrete properties of the poker chips. When the chip was hidden from view, they replied that they were uncertain about both statements. When it was visible, they judged both statements to be true if the chip was green and false if it was red. In contrast, adolescents analyzed the logic of the statements. They understood that the "either–or" statement is always true and the "and" statement is always false, regardless of the poker chip's color (Osherson & Markman, 1975).

Although Piaget did not view language as playing a central role in children's cognitive development (see Chapter 7), he acknowledged it is more important in adolescence. Abstract thought requires language-based systems of representation that do not stand for real things, such as those in higher mathematics. Secondary school students use these systems in algebra and geometry. Formal operational thought also involves verbal reasoning about abstract concepts. Jules showed that he could think in this way when he pondered relationships among time, space, and matter in physics and wondered about justice and freedom in philosophy.

Recent Research on Formal Operational Thought

Research on formal operational thought poses questions similar to those we discussed with respect to Piaget's earlier stages: Does abstract thinking appear earlier than Piaget expected? Do all individuals reach formal operations during their teenage years?

● **Are Children Capable of Abstract Thinking?** School-age children show the glimmerings of hypothetico-deductive reasoning, but they are not as competent at it as adolescents. For example, in simplified situations—ones involving no more than two possible causal variables—6-year-olds understand that hypotheses must be confirmed by appropriate evidence (Ruffman et al., 1993). But school-age children cannot sort out evidence that bears on three or more variables at once. And as we will see when we take up information-processing research, children have difficulty explaining why a pattern of observations supports a hypothesis, even when they recognize the connection between the two.

School-age children's capacity for propositional thought is also limited. For example, they have great difficulty reasoning from premises that contradict reality or their own beliefs. Consider the following set of statements: "If dogs are bigger than elephants and elephants are bigger than mice, then dogs are bigger than mice." Children younger than 10 judge this reasoning to be false because not all the relations specified occur in real life (Moshman & Franks, 1986). They fail to grasp the *logical necessity* of propositional reasoning—that the validity of conclusions drawn from premises rests on the rules of logic, not on real-world confirmation.

As Piaget's theory indicates, people in Western nations become much better at analyzing the logic of propositions, irrespective of their content, around age 11. Propositional thought improves steadily over the adolescent years (Markovits & Vachon, 1989, 1990).

● Do All Individuals Reach the Formal Operational Stage? Try giving one or two of the formal operational tasks just described to your friends, and see how well they do. Even well-educated adults have difficulty with abstract reasoning! About 40 to 60 percent of college students fail Piaget's formal operational problems (Keating, 1979).

Why is it that so many college students, and adults in general, are not fully formal operational? The reason is that people are most likely to think abstractly in situations in which they have had extensive experience. This conclusion is supported by evidence that taking college courses leads to improvements in formal operational reasoning related to course content (Lehman & Nisbett, 1990). The physics student grasps Piaget's pendulum problem with ease. The English enthusiast excels at analyzing the themes of a Shakespeare play, whereas the history buff skillfully evaluates the causes and consequences of World War II.

Furthermore, in many village and tribal societies, formal operational tasks are not mastered at all (Cole, 1990; Gellatly, 1987). Piaget acknowledged that without the opportunity to solve hypothetical problems, people in some societies might not display formal operations. Still, these findings raise further questions about Piaget's stage sequence. Does the formal operational stage largely result from children's and adolescents' independent efforts to make sense of their world, as Piaget claimed? Or is it a culturally transmitted way of thinking that is specific to literate societies and taught in school? Just how do young people make the transition from concrete to formal operational thought? These issues have prompted many investigators to adopt an information-processing view.

An Information-Processing View of Adolescent Cognitive Development

Information-processing theorists agree with the broad outlines of Piaget's description of adolescent cognition (Case, 1992, 1998; Demetriou et al., 1993; Halford, 1993). However, they refer to a variety of specific mechanisms of cognitive change, each of which was discussed in previous chapters. Now let's draw them together:

- *Attention* becomes more focused on relevant information and better adapted to the changing demands of tasks.
- *Strategies* become more effective, improving storage, representation, and retrieval of information.
- *Knowledge* increases, easing strategy use.
- *Metacognition* (awareness of thought) expands, leading to new insights into effective strategies for acquiring information and solving problems.
- *Cognitive self-regulation* improves, yielding better moment-by-moment monitoring, evaluation, and redirection of thinking.
- *Processing capacity* increases due to the joint influence of brain development and the factors just mentioned on *speed of thinking* (Demetriou et al., 2002). As a result, space in working memory is freed so more information can be held at once and combined into highly efficient, abstract representations.

As we look at some influential findings from an information-processing perspective, we will see some of these mechanisms of change in action. And we will discover that researchers regard one of them—*metacognition*—as central to the development of abstract thought.

Scientific Reasoning: Coordinating Theory with Evidence

During a free moment in physical education class, Sabrina wondered why more of her tennis serves and returns seemed to pass the net and drop into her opponent's court when she used a particular brand of balls. "Maybe it's something about their color or size? Hmm, more likely it's their surface texture, which might affect their bounce," she thought to herself.

The heart of scientific reasoning is coordinating theories with evidence. Deanna Kuhn has conducted extensive research into the development of scientific reasoning, using problems that resemble Piaget's tasks, in that several variables might affect an outcome. In one series of studies, third, sixth, and ninth graders and adults were provided evidence, sometimes consistent with and sometimes conflicting with theories. Then they were questioned about the accuracy of each theory.

For example, participants were given a problem much like the one Sabrina posed. They were asked to theorize about which of several features of sports balls—size (large or small), color (light or dark), texture (rough or smooth), or presence or absence of ridges on its surface—influences the quality of a player's serve. Next, they were told about the theory of Mr. (or Ms.) S, who believes that the ball's size is important, and the theory of Mr. (or Ms.) C, who thinks color makes a difference. Finally, the interviewer presented evidence by placing balls with certain characteristics in two baskets labeled "good serve" and "bad serve" (see Figure 11.8 on page 366).

■ **FIGURE 11.8 Which features of these sports balls—size, color, surface texture, or presence or absence of ridges—influence the quality of a player's serve?** This set of evidence suggests that color might be important, since light-colored balls are largely in the good-serve basket and dark-colored balls in the bad-serve basket. But the same is true for texture! The good-serve basket has mostly smooth balls, the bad-serve basket rough balls. Since all light-colored balls are smooth and all dark-colored balls are rough, we cannot tell whether color or texture makes a difference. But we can conclude that size and presence or absence of ridges are not important because these features are equally represented in the good-serve and bad-serve baskets. (Adapted from Kuhn, Amsel, & O'Loughlin, 1988.)

The youngest participants often ignored conflicting evidence or distorted it in ways consistent with their preferred theory. Instead of viewing evidence as separate from and bearing on a theory, children often blended the two into a single representation of "the way things are." The ability to distinguish theory from evidence and to use logical rules to examine their relationship in complex, multivariable situations improves from childhood into adolescence and adulthood (Foltz, Overton, & Ricco, 1995; Kuhn et al., 1995; Schauble, 1996).

How Scientific Reasoning Develops

What factors support adolescents' skill at coordinating theory with evidence? Greater processing capacity, permitting a theory and the effects of several variables to be compared at once, is vital. Beyond this, *metacognition*—thinking about thought—is especially important (Kuhn, 1999; Moshman, 1999). Individuals must be able to represent the theory as an object of thought rather than as a mirror image of reality. And they must also set aside their own theoretical preference and consider what the evidence says as their sole basis for judgment.

How does skill in coordinating theory with evidence increase? Performance is strongly influenced by years of schooling (Kuhn, 1993). But even at advanced levels of education, scientific reasoning is rarely taught directly. Instead, in all subject-matter areas, students receive practice in setting aside their own experiences and beliefs to infer conclusions that follow from information given. Repeated opportunities to pit theory against evidence prompt adolescents to reflect on their current strategies, revise them, and become aware of the nature of logic (Kuhn et al., 1995; Moshman, 1998; Schauble, 1996).

Although much better at scientific reasoning than children, adolescents and adults continue to show a self-serving bias in their thinking. They apply logic more effectively to ideas they doubt than to those they favor. Reasoning scientifically, however, requires the metacognitive capacity to be objective rather than self-serving (Moshman, 1999). As we will see in Chapter 12, this flexible, open-minded approach is not just a cognitive milestone but a personality trait—one that assists teenagers greatly in forming an identity and in developing morally.

Adolescents develop formal operational thinking in a similar, step-by-step fashion on different types of tasks. In a series of studies, 10- to 20-year-olds were given sets of problems graded in difficulty. For example, one set consisted of quantitative-relational tasks like the pendulum problem on page 364. Another set contained propositional tasks like the poker chip problem on page 364. And still another set were causal-experimental tasks like the sports ball problem in Figure 11.8 (Demetriou, Efklides, & Platsidou, 1993; Demetriou et al., 1993; Demetriou et al., 1996).

In each type of task, adolescents mastered component skills in sequential order by expanding their metacognitive awareness. For example, on causal-experimental tasks, they first became aware of the many variables that could influence an outcome. This enabled them to formulate and test hypotheses. Over time, adolescents combined separate skills into a smoothly functioning system. They constructed a general model that could be applied to many instances of a given type of problem. In the researcher's words, young people seem to form a "hypercognitive system," or supersystem, that understands, organizes, and influences other aspects of cognition (Demetriou & Kazi, 2001).

Return to Chapter 9, page 287, and review Robbie Case's information-processing view of development. Does Case's concept of *central conceptual structures* remind you of the metacognitive advances just described? Piaget also underscored the role of metacognition in formal operational thought when he spoke of "operating on operations" (see page 363). However, information-processing findings reveal that scientific reasoning does not result from an abrupt, stagewise change, as Piaget believed. Instead, it develops gradually out of many specific experiences that require children and adolescents to match theories against evidence and reflect on and evaluate their thinking.

Consequences of Abstract Thought

The development of formal operations leads to dramatic revisions in the way adolescents see themselves, others, and the world in general. But just as adolescents are occasionally awkward in the use of their transformed bodies, they are initially faltering and clumsy in their abstract thinking. Parents and teachers must be careful not to mistake the many typical reactions of the teenage years—argumentativeness, self-concern, insensitive remarks, and indecisiveness—for anything other than inexperience with new reasoning powers. The Caregiving Concerns table below suggests ways to handle the everyday consequences of teenagers' newfound capacity for abstraction.

Argumentativeness

As adolescents acquire formal operations, they are motivated to use them. The once pliable school-age child becomes a feisty, argumentative teenager who can marshal facts and ideas to build a case (Elkind, 1994). "A simple, straightforward explanation used to be good enough to get Louis to obey," complained Antonio. "Now he wants a thousand reasons. And worse yet, he finds a way to contradict them all!"

As long as parent–child disagreements remain focused on principles and do not deteriorate into meaningless battles, they can promote development. Through discussions of family rules and practices, adolescents become more aware of their parents' values and the reasons behind them. Gradually, they come to see the validity of parental beliefs and adopt many as their own (Alessandri & Wozniak, 1987). Teenagers' capacity for effective argument also opens the door to intellectually stimulating pastimes, such as debate teams and endless bull sessions with friends over moral, ethical, and political concerns. By proposing, justifying, criticizing, and defending a variety of solutions, adolescents often move to a higher level of understanding (Moshman, 1998, 1999).

Self-Consciousness and Self-Focusing

Adolescents' ability to reflect on their own thoughts, combined with the physical and psychological changes they are undergoing, means that they start to think more about themselves. Piaget believed that a new form of egocentrism accompanies this stage: the inability to distinguish the abstract perspectives of self and other (Inhelder & Piaget, 1955/1958). As teenagers imagine what others must be thinking, two distorted images of the relation between self and other appear.

The first is called the **imaginary audience,** adolescents' belief that they are the focus of everyone else's attention and concern (Elkind & Bowen, 1979). Young teenagers regard themselves as always on stage. As a result, they become extremely self-conscious, often going to great lengths to avoid embarrassment. Sabrina, for example, woke up one Sunday morning with a large pimple on her chin. "I can't possibly go to church!" she cried. "Everyone will notice how ugly I look." The imaginary audience helps us understand the hours adolescents spend inspecting every detail of their appearance. It also accounts for their sensitivity to public criticism. To teenagers, who believe that everyone is monitoring their performance, a critical remark from a parent or teacher can be mortifying.

A second cognitive distortion is the **personal fable.** Because teenagers are so sure that others are observing and thinking about them, they develop an inflated opinion of their own

Ways to Handle the Consequences of Teenagers' New Capacity for Abstraction

CONSEQUENCE	SUGGESTION
Argumentativeness	During disagreements, remain calm, rational, and focused on principles. Express your point of view and the reasons behind it. Although adolescents may continue to challenge, explanations permit them to consider the validity of your beliefs at a later time.
Sensitivity to public criticism	Refrain from finding fault with the adolescent in front of others. If the matter is important, wait until you can speak to the teenager alone.
Exaggerated sense of personal uniqueness	Acknowledge the adolescent's unique characteristics. At opportune times, point out how you felt similarly as a young teenager, encouraging a more balanced perspective.
Idealism and criticism	Respond patiently to the adolescent's grand expectations and critical remarks. Point out positive features of targets, helping the teenager see that all worlds and people are blends of virtues and imperfections.
Difficulty making everyday decisions	Refrain from deciding for the adolescent. Offer patient reminders and diplomatic suggestions until he or she can make choices more confidently.

© DAVID YOUNG-WOLFF/PHOTOEDIT

■ These adolescents are acting for the camera. But the imaginary audience leads them to think that everyone is monitoring their performance at other times as well. Consequently, young teenagers are extremely self-conscious and go to great lengths to avoid embarrassment.

importance. They start to feel that they are special and unique. Many adolescents view themselves as reaching great heights of glory as well as sinking to unusual depths of despair—experiences that others could not possibly understand (Elkind, 1994). As one teenager wrote in her diary, "My parents' lives are so ordinary, so stuck in a rut. Mine will be different. I'll realize my hopes and ambitions." When combined with a sensation-seeking personality, the personal fable seems to contribute to adolescent risk taking by convincing teenagers of their invulnerability. In one study, young people with both high personal-fable and sensation-seeking scores took more sexual risks, more often used drugs, and committed more delinquent acts than their agemates (Greene et al., 2000).

The imaginary audience and personal fable are strongest during the transition to formal operations. They gradually decline as abstract reasoning becomes better established (Lapsley et al., 1988). Yet these distorted visions of the self may not represent a return to egocentrism. Instead, they seem to be an outgrowth of gains in perspective taking, which cause young teenagers to be very concerned with what others think (Vartanian & Powlishta, 1996). Adolescents also have emotional reasons for clinging to the idea that others are preoccupied with their appearance and behavior. Doing so helps them maintain a hold on important relationships as they struggle to separate from parents and establish an independent sense of self (Vartanian, 1997).

Idealism and Criticism

Because abstract thinking permits adolescents to go beyond the real to the possible, it opens up the world of the ideal and of perfection. Teenagers can imagine alternative family, religious, political, and moral systems, and they want to explore them. Doing so is part of investigating new realms of experience, developing larger social commitments, and defining their own values and preferences.

The idealism of teenagers leads them to construct grand visions of a perfect world—with no injustice, discrimination, or tasteless behavior. They do not make room for the shortcomings of everyday life. Adults, with their longer life experience, have a more realistic outlook. The disparity between adults' and teenagers' world views is often called the "generation gap," and it creates tension between parent and child. Aware of the perfect family against which their real parents and siblings do not measure up, adolescents may become fault-finding critics.

Yet overall, teenage idealism and criticism are advantageous. Once adolescents learn to see others as having both strengths and weaknesses, they have a much greater capacity to work constructively for social change and to form positive and lasting relationships (Elkind, 1994). Parents can help teenagers forge a better balance between the ideal and the real by tolerating their criticism while reminding the young person that all people are blends of virtues and imperfections.

Planning and Decision Making

Adolescents, who think more analytically, handle cognitive tasks more effectively than they did at younger ages. Given a homework assignment, they are far better at *cognitive self-regulation*—planning what to do first and what to do next, monitoring progress toward a goal, and redirecting actions that prove unsuccessful. For this reason, study skills improve from middle childhood into adolescence.

But when it comes to planning and decision making in everyday life, teenagers (especially young ones) often feel overwhelmed by the possibilities before them. Their efforts to choose among alternatives frequently break down, and they may resort to habit, act on impulse, or not make a decision at all (Elkind,

© JOHN LAWLOR/GETTY IMAGES/TAXI

■ This teenage girl looks as if she has nothing in common with her parents. Her idealistic image of a perfect world, to which her family cannot measure up, leads to a "generation gap," creating tension between parent and adolescent. Over time she will forge a better balance between the ideal and the real.

1994). On many mornings, for example, Sabrina tried on five or six outfits before leaving for school. Often she shouted from the bedroom, "Mom, what shall I wear?" Then, when Franca made a suggestion, Sabrina rejected it, opting for one of the two or three sweaters she had worn for weeks. Similarly, Louis procrastinated about registering for college entrance tests. When Franca mentioned that he was about to miss the deadline, Louis agonized over the forms, unable to decide when or where he wanted to take the test.

Everyday planning and decision making are challenging for teenagers because they have so many opportunities. When they were younger, adults usually specified their options, reducing the number of decisions that had to be made. As adolescents gather more experience, they make choices with greater confidence.

Ask Yourself

REVIEW

Using the concepts of hypothetico-deductive reasoning and propositional thought, illustrate the difference between school-age children's and adolescents' cognition.

REVIEW

Describe and cite examples of the role of metacognition in adolescent cognitive development.

APPLY

Thirteen-year-old Rosie had a crush on a boy who failed to return her affections. After her mother assured her that there would be other boys, Rosie snapped, "Mom! You don't know what it's like to be in love!" Which cognitive distortion of adolescence does Rosie's thinking illustrate? Explain.

CONNECT

Adolescent idealism and criticism, although troublesome for parents, are beneficial in the long run to both the developing individual and society. Explain why this is so.

www.

Sex Differences in Mental Abilities

Sex differences in intellectual performance have been studied since the beginning of this century, sparking almost as much controversy as the ethnic and SES differences in IQ considered in Chapter 9. Although boys and girls do not differ in general intelligence, they do vary in specific mental abilities. Throughout the school years, girls attain higher scores on reading and writing achievement tests and account for a lower percentage of children referred for remedial reading instruction (Campbell, Hombo, & Mazzeo, 2000; Halpern, 2000). Girls' continue to score slightly higher on tests of general verbal ability in adolescence (Hyde & Linn, 1988). Recall from Chapter 5 that girls show a biological advantage in earlier development of the left hemisphere of the cerebral cortex, where language is localized. Girls also receive more verbal stimulation from mothers in the preschool years. And after starting school, children view reading as a "feminine" subject.

By adolescence, boys do better than girls in mathematics, especially on tests of abstract reasoning, primarily complex word problems and geometry (Bielinski & Davison, 1998; Hedges & Nowell, 1995). Among mathematically talented youngsters, the gender gap appears much earlier, by second grade, and is large. In widely publicized research on more than 40,000 bright seventh and eighth graders invited to take the Scholastic Aptitude Test (SAT), boys outscored girls on the mathematics subtest year after year. Twice as many boys as girls had scores above 500; 13 times as many scored over 700 (Benbow & Stanley, 1983; Lubinski & Benbow, 1994).

Some researchers believe that the gender gap in mathematics—especially the tendency for many more boys to be extremely talented at math—is genetic. Accumulating evidence suggests that it is rooted in boys' biologically based superior spatial reasoning. See the Biology and Environment box on page 370 for a discussion of this issue.

Although heredity is involved, social pressures also contribute. Long before sex differences in math achievement appear, both boys and girls view math as a "masculine" subject. In addition, girls regard math as less useful for their future lives and more often blame their errors on lack of ability. These beliefs, in turn, lead girls to be less interested in math and less likely to consider math- or science-related careers (Byrnes & Takahira, 1993; Catsambis, 1994). The end result is that fewer girls than boys acquire abstract mathematical concepts and reasoning skills.

A positive sign, however, is that sex differences in cognitive abilities of all kinds have declined steadily over the past several decades. Paralleling this change is an increase in girls' enrollment in challenging math and science courses. Today, boys and girls reach advanced levels of high school math and science study in equal proportions—a crucial factor in reducing sex differences in knowledge and skill (Campbell, Hombo, & Mazzeo, 2000).

Clearly, extra steps must be taken to promote girls' interest in and confidence at math and science. When parents hold nonstereotyped gender-role values, daughters are less likely to show declines in math and science achievement at adolescence (Updegraff, McHale, & Crouter, 1996). In schools, teachers must demonstrate the relevance of math and science to everyday life. Girls, especially, respond positively to math and science taught from an applied, hands-on perspective (Eccles, 1994). At the same time, teachers must ensure that girls participate fully in hands-on group activities and are not reduced to a passive role by boys' more assertive style of peer interaction (Jovanovic & King, 1998). Finally, exposure to role models of successful women is likely to improve girls' belief in their capacity to do well at math and science.

Biology & Environment

Sex Differences in Spatial Abilities

Spatial abilities have become a key focus of researchers' efforts to explain sex differences in mathematical reasoning. Clear sex differences in spatial skills exist, but they occur only on certain tasks The gender gap favoring males is large for *mental rotation tasks,* in which individuals must rotate a three-dimensional figure rapidly and accurately inside their heads (see Figure 11.9). In addition, males do considerably better on *spatial perception tasks,* in which people must determine spatial relationships by considering the orientation of the surrounding environment. Sex differences on *spatial visualization tasks,* involving analysis of complex visual forms, are weak or nonexistent, perhaps because many strategies can be used to solve them. Both sexes may come up with effective procedures (Voyer, Voyer, & Bryden, 1995).

Sex differences in spatial abilities emerge in early childhood and persist throughout the lifespan (Levine et al., 1999). The pattern is consistent enough to suggest a biological explanation. One hypothesis is that heredity, perhaps through prenatal exposure to androgen hormones, enhances right hemispheric functioning, granting males a spatial advantage. (Recall that for most people, spatial skills are housed in the right hemisphere of the cerebral cortex.) Consistent with this idea, girls and women whose prenatal androgen levels were abnormally high show superior performance on spatial rotation tasks (Berenbaum, 2001). And people with severe prenatal deficits in either male or female hormones have difficulty with spatial reasoning (Hier & Crowley, 1982; Temple & Carney, 1995).

Research on hormone variations within normal range is less clear. Although many studies report a relationship between androgens and spatial abilities, not all do. And when a relationship is found, it is not always straightforward (Geary, 1998). But consistent with a biological contribution, boys show more rapid development of the right hemisphere of the cerebral cortex than girls (Breedlove, 1994).

Mental Rotation
Choose the responses that show the standard in a different orientation.
Standard Responses
1 2 3 4

Spatial Perception
Pick the tilted bottle that has a horizontal water line.
1 2 3 4

Spatial Visualization
Find the figure embedded in this complex shape.

FIGURE 11.9 Types of spatial tasks. Large sex differences favoring males appear on mental rotation, and males do considerably better than females on spatial perception. In contrast, sex differences on spatial visualization are weak or nonexistent. (From M. C. Linn & A. C. Petersen, 1985, "Emergence and Characterization of Sex Differences in Spatial Ability: A Meta-Analysis," *Child Development, 56,* pp. 1482, 1483, 1485. © The Society for Research in Child Development, Inc. Reprinted by permission.)

Why might a biologically based gender difference in spatial abilities exist? Evolutionary theorists point out that mental rotation skill predicts rapid, accurate map drawing and interpretation, in which males do better. During human evolution, the cognitive abilities of males became adapted for hunting, which required generating mental representations of large-scale spaces to find one's way (Moffat, Hampson, & Hatzipantelis, 1998).

Although biology is involved in males' superior spatial performance, experience also makes a difference. Children who engage in manipulative activities, such as block play, model-building, and carpentry, do better on spatial tasks (Baenninger & Newcombe, 1995). Furthermore, playing video games that require rapid mental rotation of visual images enhances spatial scores of boys and girls alike (Okagaki & Frensch, 1996; Subrahmanyam & Greenfield, 1996). Boys spend far more time at these pursuits than girls.

Do superior spatial skills contribute to the greater ease with which males solve complex math problems? Research indicates that they do (Casey et al., 1995). Yet in a study of high-ability college-bound adolescents, both mental rotation ability and self-confidence at doing math predicted higher scores on the math subtest of the SAT (Casey, Nuttall, & Pezaris, 1997). Boys are advantaged not only in mental rotation but also in math self-confidence, despite the fact that their grades are poorer than those of girls (Eccles et al., 1993b). In sum, biology and environment *jointly* determine variations in spatial and math performance—within and between the sexes.

Learning in School

In complex societies, adolescence coincides with entry into secondary school. Most young people move into either a middle or a junior high school and then into a high school. With each change, academic achievement becomes more serious, affecting college choices and job opportunities. In the following sections, we take up a variety of aspects of secondary school life.

School Transitions

When Sabrina started junior high, she left a small, intimate, self-contained sixth-grade classroom for a much larger school. "I don't know most of the kids in my classes," Sabrina complained to her mother at the end of the first week. "Besides, there's just too much homework. I get assignments in all my classes at once. I can't do all this!" she shouted, bursting into tears.

• Impact of School Transitions. As Sabrina's reactions suggest, school transitions can create adjustment problems. With each school change—from elementary to middle school or junior high and then to high school—adolescents' grades decline. The drop is partly due to tighter academic standards. At the same time, the transition to secondary school often brings with it less personal attention, more whole-class instruction, and less chance to participate in classroom decision making. In view of these changes, it is not surprising that students rate their junior-high learning experiences less favorably than their elementary school experiences (Wigfield & Eccles, 1994). They also report that their junior-high teachers care less about them, are less friendly, grade harder, and stress competition more. Consequently, many young people feel less academically competent (Anderman & Midgley, 1997).

Inevitably, students must readjust their feelings of self-confidence and self-worth as school becomes more impersonal and academic expectations are revised. A comprehensive study revealed that the timing of school transition is important, especially for girls (Simmons & Blyth, 1987). More than 300 adolescents living in a large Midwestern city were followed from sixth to tenth grade. Some were enrolled in schools with a 6–3–3 grade organization (a K–6 elementary school, a 3-year junior high, and a 3-year high school). These students made two school changes, one to junior high and one to high school. A comparison group attended schools with an 8–4 grade organization. They made only one school transition, from a K–8 elementary school to high school.

For the sample as a whole, grade point average dropped and feelings of anonymity increased after each transition. Participation in extracurricular activities declined more in the 6–3–3 than in the 8–4 arrangement, the drop being greater for girls. Furthermore, in 8–4 schools, school transition led to gains in self-esteem. In contrast, in 6–3–3 schools, sex differences in self-esteem were striking. Whereas boys remained stable, girls showed a sharp drop with each school change (see Figure 11.10).

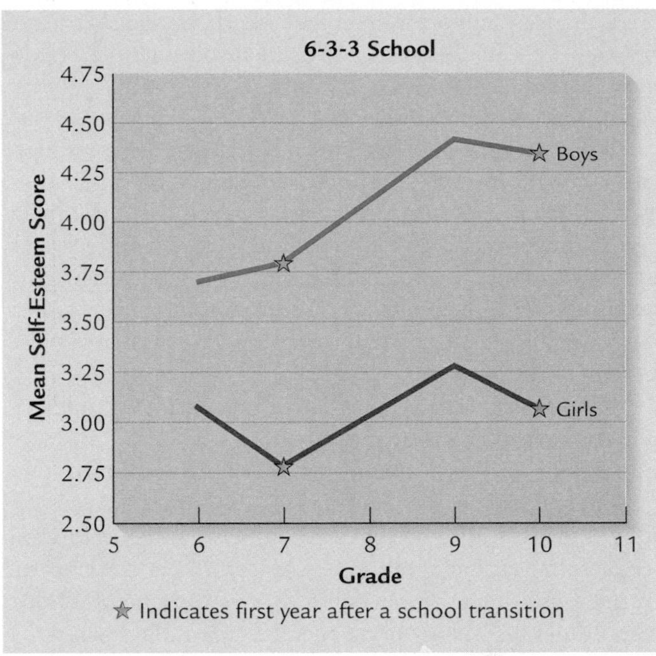

FIGURE 11.10 Self-esteem from sixth to tenth grade in 6–3–3 schools. In this longitudinal study of more than 300 adolescents, boys' self-esteem remained stable after school transition. In contrast, girls' self-esteem dropped sharply in the year after each school change. (Adapted from Simmons & Blyth, 1987.)

These findings show that the earlier school transition occurs, the more negative its impact. Girls in 6–3–3 schools fared poorest, the researchers argued, because movement to junior high tended to coincide with other life changes—namely, the onset of puberty and dating. Adolescents who face added strains, such as family disruption, parental unemployment, or learned helplessness on academic tasks, at the time they enter junior high are at greatest risk for emotional difficulties (Flanagan & Eccles, 1993; Rudolph et al., 2001).

Distressed young people whose school performance drops sharply often show a persisting pattern of poor self-esteem, motivation, and achievement (Roeser, Eccles, & Freedman-Doan, 1999). For some, school transition initiates a downward spiral in academic performance and school involvement that leads to dropping out (Eccles et al., 1997).

• Helping Adolescents Adjust to School Transitions. Consider the findings just reviewed, and you will see that school transitions often lead to environmental changes that fit poorly with adolescents' developmental needs. They disrupt close relationships with teachers at a time when adolescents need adult support. They emphasize competition during a period of heightened self-focusing. They reduce decision making and choice as the desire for autonomy increases. And they interfere with peer networks at a time of increased concern with peer acceptance.

Enhanced support from parents, teachers, and peers eases the strain of school transition. Parent involvement, monitoring, and gradual autonomy granting are associated with better adjustment after entering junior high (Grolnick et al., 2000). Because most students do better in an 8–4 school arrangement, school districts considering reorganization should seriously consider this plan. Also, smaller social units can be formed within large schools, permitting closer relations with teachers and peers. And students can also be assigned to classes with several familiar peers or a constant group of new peers—arrangements that promote emotional security and social support.

Finally, teenagers' perceptions of the sensitivity and flexibility of their school learning environments contribute substantially to their adjustment. When schools minimize competition and differential treatment by ability, students are less likely to feel angry and depressed, to be truant, and to show declines in academic values, self-esteem, and achievement during junior high school (Roeser, Eccles, & Sameroff, 2000). School rules that strike young people as fair rather than punitive also foster satisfaction with school life (Eccles et al., 1993a).

Academic Achievement

Adolescent achievement is the result of a long history of cumulative effects. Early on, positive educational environments, both family and school, lead to personal traits that support achievement—intelligence, confidence in one's own abilities, the desire to succeed, and high educational aspirations. Nevertheless, improving an unfavorable environment can help a poorly performing young person bounce back, opening the door to a more satisfying adult life. The Caregiving Concerns table below summarizes environmental factors that enhance achievement during the teenage years.

● Child-Rearing Practices. Authoritative parenting (which combines acceptance and involvement with reasonable demands for maturity and gradual autonomy granting) is linked to achievement in adolescence, just as it predicts mastery-oriented behavior during the childhood years. Research reveals that authoritative parenting is linked to higher grades for young people varying widely in SES. In contrast, authoritarian and permissive styles are associated with lower grades (Dornbusch et al., 1987; Steinberg, Darling, & Fletcher, 1995). Of all parenting approaches, an uninvolved style (low in both warmth and maturity demands) predicts the poorest grades and worsening school performance over time (Glasgow et al., 1997; Kaisa, Stattin, & Nurmi, 2000; Steinberg et al., 1994).

The relationship between authoritative parenting and adolescents' academic competence has been confirmed in countries with highly diverse value systems, including Argentina, Australia, China, Hong Kong, Pakistan, and Scotland (Steinberg, 2001). Why does authoritative parenting promote school success? In Chapter 8, we noted that authoritative parents adjust their expectations to children's capacity to take responsibility for their own behavior. Parents who engage in joint decision making with teenagers, permitting more autonomy with age, have youngsters who achieve especially well (Dornbusch et al., 1990). Open discussion accompanied by warmth and firmness makes adolescents feel competent, encourages constructive thinking and self-regulation, and increases awareness of the importance of doing well in school. These factors, in turn, are related to mastery-oriented attributions, effort, achievement, and high educational aspirations (Aunola, Stattin, & Nurmi, 2000; Trusty, 1999).

Factors That Support High Achievement During Adolescence

FACTOR	DESCRIPTION
Child-rearing practices	Authoritative parenting
	Joint parent–adolescent decision making
	Parent involvement in the adolescent's education
Peer influences	Peer valuing of and support for high achievement
School characteristics	Teachers who are warm and supportive, develop personal relationships with parents, and show them how to support their teenager's learning
	Learning activities that encourage high-level thinking
	Active student participation in learning activities and classroom decision making
Employment schedule	Job commitment limited to less than 15 hours per week
	High-quality vocational education for non-college-bound adolescents

© JOSE L. PELAEZ/CORBIS

■ This father is involved in his adolescent daughter's school career. Besides keeping tabs on her progress, he is probably in regular contact with the school. He sends a message to his daughter about the importance of education and teaches her how to solve academic problems and make wise educational decisions.

● **Parent–School Partnerships.** High-achieving young people typically have parents who keep tabs on their child's progress, communicate with teachers, and make sure that their child is enrolled in challenging, well-taught classes. These efforts are just as important during junior and senior high school as they were earlier (Epstein & Sanders, 2002). In a study of a nationally representative sample of more than 15,000 American adolescents, parents' school involvement in eighth grade strongly predicted students' grade point average in tenth grade, beyond the influence of SES and previous academic achievement. This relationship held for each ethnic group included—black, white, Native American, and Asian (Keith et al., 1998). Parents who are in frequent contact with the school send a message to their teenager about the value of education, promote wise educational decisions, and model constructive solutions to academic problems.

Parents living in low-income, high-risk neighborhoods face daily stresses that reduce the energy they have for school involvement. Yet schools could relieve some of this stress by forging stronger home–school links. Schools can do so through personal relationships with teachers, volunteer activities that tap parents' talents to increase the quality of school programs, and meaningful parental roles in school governance that ensure parents are invested in school goals (Epstein, 2001).

● **Peer Influences.** Peers play an important role in adolescent achievement, in a way that is related to both family and school. Teenagers whose parents value achievement generally choose friends who share those values (Berndt & Keefe, 1995). For example, when Sabrina began to make new friends in junior high, she often studied with her girlfriends. Each girl wanted to do well in school and reinforced the same desire in the others.

Peer support for high achievement also depends on the overall climate of the peer culture, which, in the case of ethnic minority youths, is powerfully affected by the surrounding social order. Low-SES minority adolescents often react against working hard, convinced that getting good grades will have little payoff for their future and may threaten their peer relationships and ethnic identity (Ogbu, 1997).

Yet not all economically disadvantaged minority students respond this way. A case study of six inner-city, poverty-stricken African-American adolescents who were high-achieving and optimistic about their futures revealed that they were intensely aware of oppression but believed in striving to alter their social position (O'Connor, 1997). How did they develop this sense of agency? Parents, relatives, and teachers had convinced them through discussion and example that injustice should not be tolerated and that together, blacks could overcome it—a perspective that facilitated academic motivation in the face of peer pressures against doing well in school and social obstacles to career success.

● **School Characteristics.** Adolescents need classroom environments that are responsive to their expanding powers of reasoning and their emotional and social needs. Without appropriate learning experiences, the potential for abstract thought is unlikely to be realized.

Classroom Learning Experiences. As noted earlier, the large, departmentalized school organization of most secondary schools leads many adolescents to report that their classes lack warmth and supportiveness—a circumstance that dampens their motivation (Eccles & Harold, 1996). Adolescents (not just children) need to form close relationships with teachers. As they begin to develop an identity beyond the family, they seek adult models other than their parents.

Of course, an important benefit of separate classes in each subject is that adolescents can be taught by experts, who are more likely to encourage high-level thinking—factors that promote school attendance and achievement (Phillips, 1997). But many secondary school classrooms do not consistently provide interesting, challenging teaching that stimulates abstract thought (Campbell, Hombo, & Mazzeo, 2000). Because of the uneven quality of instruction, many seniors graduate from high school deficient in basic academic skills. Although the achievement gap separating African-American and Hispanic students from white students has declined since the 1970s, mastery of reading, writing, mathematics, and science by American low-SES ethnic minority students remains disappointing (Campbell, Hombo, & Mazzeo, 2000). Too often these young people attend underfunded schools with run-down buildings, outdated equipment, and textbook shortages. In some, crime and discipline problems have become so overwhelming that addressing them receives more attention than learning and instruction. By junior high, large numbers of poverty-stricken minority students have been placed in low academic tracks, compounding their learning difficulties.

Tracking. Ability grouping, as we saw in Chapter 9, is detrimental during the elementary school years. At least into middle or junior high school, mixed-ability classes are desirable. Research shows that they do not stifle the more able students, and they have intellectual and social benefits for poorly performing youngsters (Oakes, Gamoran, & Page, 1992).

By high school, some grouping is unavoidable because certain aspects of education must dovetail with the young person's future educational and career plans. In the United States and Canada, high school students are counseled into college preparatory, vocational, or general education tracks. Unfortunately, this sorting tends to perpetuate educational inequalities of earlier years. Low-SES minority students are assigned in large numbers to noncollege tracks. Longitudinal research following the school performance of thousands of American students from eighth to twelfth grade revealed that assignment to a college track accelerates academic progress, whereas assignment to a vocational or general education track decelerates it (Hallinan & Kubitschek, 1999).

High school students are separated into academic and vocational tracks in virtually all industrialized nations. But in China, Japan, and most Western European nations, students take a national examination to determine their placement in high school. The outcome usually fixes future possibilities for the young person. In North America, educational decisions are more fluid. Students who are not assigned to a college preparatory track or who do poorly in high school can still get a college education. But by the adolescent years, SES differences in quality of education and academic achievement have already sorted students more drastically than is the case in other countries. In the end, many young people do not benefit from this more open system. Compared with many Western European nations, the United States and Canada have a higher percentage of young people who regard themselves as educational failures and drop out of high school (see Figure 11.11).

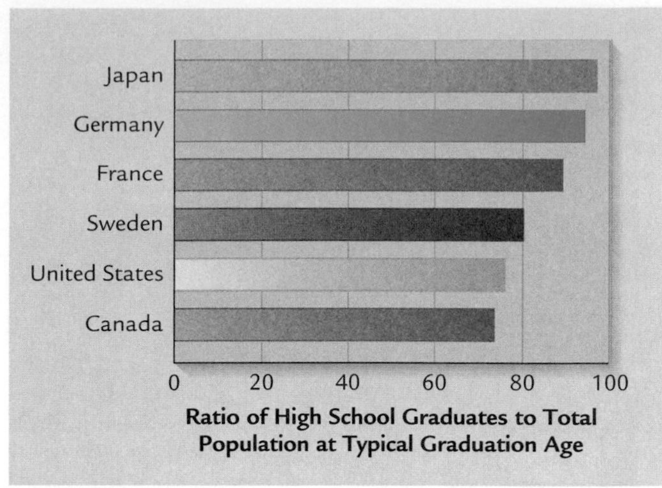

FIGURE 11.11 High school graduation rates in six industrialized nations. The United States and Canada rank below many other developed countries. (From Organization for Economic Cooperation and Development, 2000.)

● **Part-Time Work.** About half of American and Canadian high school students work part-time during the school year—a greater percentage than in other developed countries (Bowlby & McMullen, 2002; Children's Defense Fund, 2002). Most are middle-SES adolescents in pursuit of spending money rather than vocational exploration and training. Low-income teenagers who need to contribute to family income or to support themselves find it harder to get jobs (U.S. Department of Education, 2002b).

Furthermore, the jobs adolescents hold are limited to low-level repetitive tasks that provide little contact with adult supervisors. A heavy commitment to such jobs is harmful. Students who work more than 15 hours per week have poorer school attendance, lower grades, and less time for extracurricular activities. They also feel more distant from their parents and report more drug and alcohol use and delinquent acts (Barling, Rogers, & Kelloway, 1995; Kouvonen & Kivivuori, 2001). And perhaps because of the menial nature of their jobs, employed teenagers tend to be cynical about work life. Many admit to having stolen from their employers (Steinberg, Fegley, & Dornbusch, 1993).

When work experiences are specially designed to meet educational and vocational goals, outcomes are different. Participation in work–study programs is related to positive school and work attitudes and improved achievement among low-SES teenagers (Hamilton & Hamilton, 1999; Steinberg, 1984). Yet high-quality vocational preparation for non-college-bound American adolescents is scarce. Unlike some Western European nations, the United States and Canada have no widespread training system to prepare youths for skilled business and industrial occupations and manual trades (Heinz, 1999a). U.S. and Canadian federal, state, and provincial governments support some job-training programs. But most are too short to make a difference in the lives of poorly skilled adolescents, who need intensive training and academic remediation before they are ready to enter the job market. And at present, these programs serve only a small minority of young people who need assistance.

Dropping Out

Across the aisle from Louis in math class sat Norman, who daydreamed, crumpled his notes into his pocket after class, and rarely did his homework. On test days, he twirled a rabbit's foot for good luck but left most of the questions blank. Louis had been in school with Norman since fourth grade, but the two boys had little to do with one another. To Louis, who was quick at schoolwork, Norman seemed to live in another world. Once or twice each week, Norman cut class, and one spring day, he stopped coming altogether.

Norman is one of 11 percent of American and Canadian young people who leave high school without a diploma (Human Resources Development Canada, 2000; U.S. Department of Education, 2002b). The dropout rate is particularly high among low-SES ethnic minority youths, especially Hispanic and Canadian Aboriginal teenagers (see Figure 11.12).

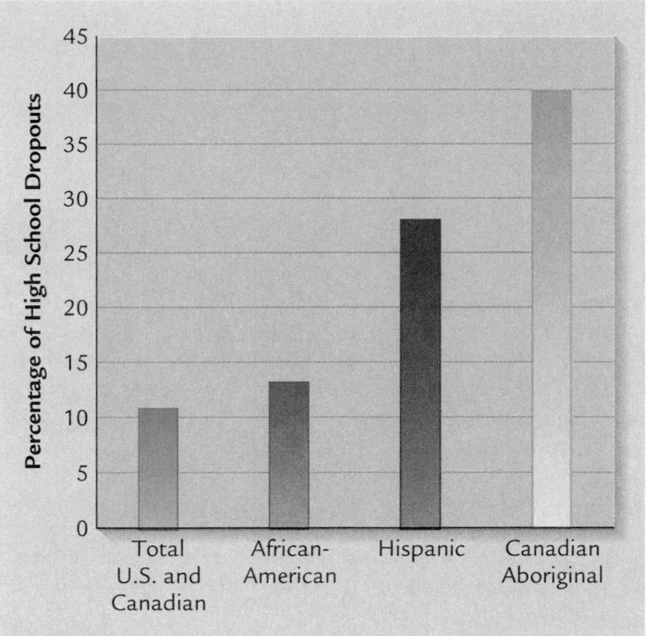

FIGURE 11.12 Percentage of high school dropouts by ethnicity, United States and Canada. Because many African-American, Hispanic, and Canadian Aboriginal teenagers come from low-income and poverty-stricken families, their dropout rates are above the national average. Rates for Hispanic and Canadian Aboriginal youths are especially high. (From Human Resources Development Canada, 2000; U.S. Department of Education, 2002b.)

The decision to leave school has dire consequences. Youths without upper secondary education have much lower literacy scores than high school graduates. As a result, American and Canadian high school dropouts lack skills valued by employers in today's knowledge-based economy. Consequently, as Figure 11.13 reveals, employment rates are much lower for dropouts than for high school graduates. And even when they are employed, dropouts are far more likely to remain in menial, low-paying jobs and to be out of work from time to time.

● **Factors Related to Dropping Out.** Although many dropouts achieve poorly and show high rates of norm-violating acts, a substantial number—like Norman—are young people with few behavior problems who experience academic difficulties and quietly disengage from school (Janosz et al., 2000; Newcomb et al., 2002). The pathway to dropping out starts early. Risk factors in first grade predict later dropout nearly as well as risk factors in secondary school (Alexander, Entwisle, & Kabbani, 2001).

Norman had a long history of marginal to failing school grades and low academic self-esteem. He gave up on tasks that presented the least bit of challenge and counted on luck—his rabbit's foot—to get by. As Norman got older, he attended class less regularly, paid little attention when he was there, and rarely did his homework. He didn't join any school clubs or participate in athletics. Because he was so uninvolved, few teachers or students got to know him well. The day Norman left, he felt alienated from all aspects of school life.

As with other dropouts, Norman's family background contributed to his problems. Compared with other students, even those with the same grade profile, dropouts are more likely to have parents who are less involved in their youngster's education. Many did not finish high school themselves and are unemployed or coping with the aftermath of divorce. When their youngsters bring home poor report cards, these parents are more likely to respond with punishment and anger—reactions that cause the young person to rebel further against academic work (Garnier, Stein, & Jacobs, 1997).

Academically marginal students who drop out often have school experiences that undermine their chances for success. Recent reports indicate that over 60 percent of adolescents in some American inner-city high schools do not graduate. Students in general education and vocational tracks, where teaching tends to be the least stimulating, are three times more likely to drop out than those in an a college preparatory track (U.S. Department of Education, 2002b).

● **Prevention Strategies.** Strategies for helping teenagers at risk for dropping out are diverse, but several common themes are related to success:

- *High-quality vocational training.* For many marginal students, the real-life nature of vocational education is more comfortable and effective than purely academic work. But to work well, vocational education must carefully integrate academic and job-related instruction so students see the relevance of what happens in the classroom to their future goals (Ianni & Orr, 1996).

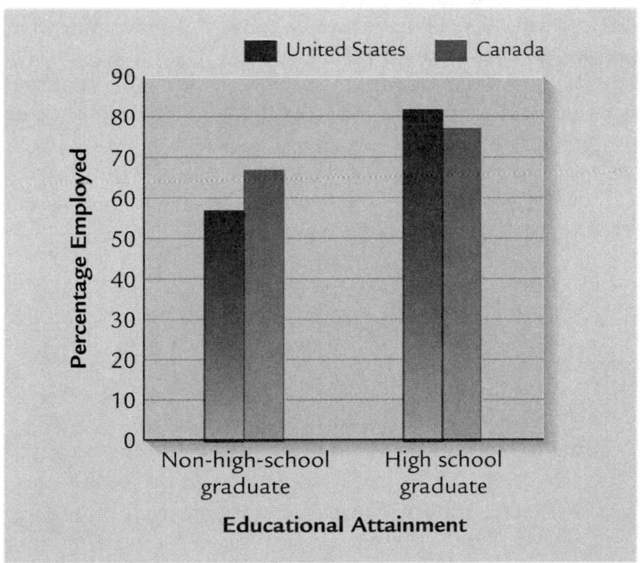

FIGURE 11.13 Employment rates of young adults by educational attainment, United States and Canada. Young people who do not graduate from high school are much less likely to get jobs than those with high school diplomas. (From Human Resources Development Canada, 2000; U.S. Department of Education, 2002b.)

A Lifespan Vista

Extracurricular Activities: Contexts for Positive Youth Development

The weekend before graduation, Terrell—a senior at an inner-city high school—attended a cast party celebrating the drama club's final performance of the year. That evening, Terrell had played a leading role in a production written and directed by club members. As Mrs. Meyer, the club's adviser, congratulated Terrell, he responded, "I loved this club. When I joined, I wasn't good at English and math and all that stuff, and I thought I couldn't do anything. Working on the sets and acting was so great—finding out that I could do these things well. Before, I wasn't secure with myself. Now I've got this boost of confidence."

Many studies show that high school extracurricular activities that focus on the arts, community service, and vocational development promote diverse academic and social skills and have a lasting positive impact on adjustment. Outcomes include improved academic performance, reduced antisocial behavior, more favorable self-esteem and initiative, greater peer acceptance, and increased concern for others (Mahoney, 2000; Sandstrom & Coie, 1999). The benefits of extracurricular involvement extend into adult life. After many factors had been controlled (including SES and academic performance), young people who were more involved in high school clubs and organizations achieved more in their occupations and engaged in more community service in their twenties and thirties (Berk, 1992).

How do extracurricular activities produce such wide-ranging benefits? Not just by giving young people something fun to do during leisure hours. In a Swedish study, adolescents who spent many afternoons and evenings in youth recreation centers offering such unstructured pastimes as pool, Ping-Pong, video games, and TV showed repeated and persisting antisocial behavior (Mahoney, Stattin, & Magnusson, 2001). In contrast, activities with a positive impact on development are highly structured, goal-oriented pursuits that require teenagers to take on challenging roles and responsibilities. In addition, such activities include caring and supportive interactions with peers and adults, who impose high expectations, help with problems, and serve as mentors (Roth et al., 1998).

Youths with academic, emotional, and social problems are especially likely to benefit from extracurricular participation. In a study of teenagers experiencing uninvolved parenting, those who engaged in extracurricular pursuits showed far lower levels of depressed mood. This outcome was strongest for adolescents reporting a trusting relationship with an activity adviser who validated their skills and encouraged them to do their best (Mahoney, Schweder, & Stattin, 2002). Furthermore, activity participation sometimes strengthens connectedness between parent and teenager as family members attend performances and exhibits or otherwise see the fruits of the young person's efforts (Mahoney & Magnuson, 2001).

Students seem to recognize the power of their extracurricular experiences to foster a smooth transition to adulthood. They report enjoyment, increased confidence, valuable relationships with adults, new friendships, and gains in setting goals, managing time, and working with others (Dworkin, Larson, & Hansen, 1993). Unfortunately, extracurricular activities are among the first aspects of school life to be cut during funding shortages. Yet a wealth of evidence indicates that these pursuits should be expanded, with special attempts made to reach academically and socially marginal young people.

■ A group of high school students gather after school to rehearse for a school play. Extracurricular involvement builds many competencies. Participants show improved academic performance, more self-esteem, greater peer acceptance, and increased concern for others. Benefits are greatest for youths with academic, emotional, and social problems. And they extend into adult life, enhancing occupational achievement and community service.

- *Remedial instruction and counseling that offer personalized attention.* Most potential dropouts need intensive remedial instruction in small classes that permit warm, caring teacher–student relationships to form. To overcome the negative psychological effects of repeated school failure, academic assistance must be combined with social support. One successful approach is to match at-risk students with retired adults, who serve as tutors, mentors, and role models in addressing academic and vocational needs (Lunenburg, 2000).

- *Efforts to address the many factors in students' lives related to leaving school early.* Programs that strengthen parent involvement, offer flexible work–study arrangements, and provide on-site child care for teenage mothers can make staying in school easier for at-risk adolescents.

- *Participation in extracurricular activities.* Another way of helping marginal students is to draw them into the community life of the school (Mahoney & Stattin, 2000). The most powerful influence on extracurricular involvement is small school size. In smaller high schools (500 to 700 students or less), a greater proportion of the student body is needed to staff and operate activities. Potential dropouts are far more likely to participate, feel needed, gain recognition for their abilities, and remain until graduation. Consult the Lifespan Vista box above for research indicating that extracurricular participation has a lasting favorable impact on development.

As we conclude our discussion of academic achievement, let's place the school dropout problem in historical perspective. Over the last half of the twentieth century, the percentage of American and Canadian adolescents completing high school increased steadily, from less than 50 percent to nearly 90 percent (Human Resources Development Canada, 2000; U.S. Department of Education, 2002b). During this same period, college attendance also rose. Today, nearly 40 percent of North American 18- to 24-year-olds are working toward college degrees—one of the highest rates in the world. Finally, although many dropouts get caught in a vicious cycle in which their lack of self-confidence and skills prevent them from seeking further education and training, about one-third return to finish their secondary education within a few years. And some extend their schooling further (Children's Defense Fund, 2002). As the end of adolescence approaches, many young people realize how essential education is for a rewarding job and a satisfying adult life.

Ask Yourself

REVIEW

List ways that parents can promote their adolescent's academic achievement, and explain why each is effective.

APPLY

Tanisha is finishing sixth grade. She could either continue in her current school through eighth grade or switch to a much larger junior high school. What would you suggest she do, and why?

CONNECT

How are educational practices that prevent school dropout similar to those that improve learning for adolescents in general?

www.

Summary

PHYSICAL DEVELOPMENT

Conceptions of Adolescence

How have conceptions of adolescence changed over the past century?

- **Adolescence,** initiated by **puberty,** is the period of transition between childhood and adulthood. Early biologically oriented theories viewed adolescence as an inevitable period of storm and stress. An alternative perspective regarded the social environment as entirely responsible for the wide variability in adolescent adjustment. Modern research shows that adolescence is a product of both biological and social forces.

Puberty: The Physical Transition to Adulthood

Describe pubertal changes in body size, proportions, sleep patterns, motor performance, and sexual maturity.

- Hormonal changes beginning in middle childhood initiate puberty, on the average, 2 years earlier for girls than boys. The first outward sign of puberty is the **growth spurt.** As the body enlarges, girls' hips and boys' shoulders broaden. Girls add more fat, boys more muscle.
- Adolescents go to bed much later than they did as children, a pattern that strengthens with pubertal growth. Sleep deprivation contributes to poorer achievement, depressed mood, and behavior problems.
- Pubertal changes lead gross motor performance to improve; boys show much larger gains than girls. Although girls' involvement in high school sports has increased, they continue to receive less athletic encouragement than boys. The number of adolescents participating in regular physical activity declines from ninth to twelfth grade, but at all grades, boys are considerably more active than girls.
- Sex hormones regulate changes in **primary** and **secondary sexual characteristics. Menarche** occurs late in the girl's sequence of pubertal events, following the rapid increase in body size. Among boys, as the sex organs and body enlarge and pubic and underarm hair appear, **spermarche** takes place.

What factors influence the timing of puberty?

- Heredity, nutrition, and overall physical health contribute to the timing of puberty. A **secular trend** toward earlier menarche has occurred in industrialized nations.

The Psychological Impact of Pubertal Events

Explain adolescents' reactions to the physical changes of puberty.

- Girls generally react to menarche with surprise and mixed emotions, but whether their feelings lean in a positive or a negative direction depends on prior knowledge and support from family members. Although boys usually know ahead of time about spermarche, they react with mixed feelings as well. Boys receive less social support for the physical changes of puberty than girls.
- Besides higher hormone levels, negative life events and adult-structured situations are associated with adolescents' negative moods. In contrast, teenagers feel upbeat when with peers and in self-chosen leisure activities. Puberty is accompanied by psychological distancing between parent and child. The reaction may be a modern substitute for physical departure from the family, which typically occurs at sexual maturity in primate species.

Describe the impact of maturational timing on adolescent adjustment, noting sex differences.

- Early maturing boys and late maturing girls, whose appearance closely matches cultural standards of physical attractiveness, have a more positive **body image** and usually adjust well in adolescence. In contrast, early maturing girls and late maturing boys, who fit in least well physically with peers, experience emotional and social difficulties.

Health Issues

Describe nutritional needs and cite factors related to serious eating disturbances during adolescence.

- As the body grows, nutritional requirements increase. Because of poor eating habits, many adolescents suffer from iron, vitamin, and mineral deficiencies.

- Girls who reach puberty early, who are very dissatisfied with their body images, and who grow up in homes where thinness is idealized are at increased risk for eating disorders. **Anorexia nervosa** tends to appear in girls with perfectionist, inhibited personalities; overprotective, controlling mothers; and emotionally distant fathers. The impulsive eating and purging of **bulimia nervosa** is associated with disengaged parenting. Some bulimics are perfectionists; others lack self-control not just in eating but also in other areas of their lives.

Discuss social and cultural influences on adolescent sexual attitudes and behavior.

- The hormonal changes of puberty lead to an increase in sex drive, but social factors affect how teenagers manage their sexuality. North America is fairly restrictive in its attitude toward adolescent sex. Young people receive contradictory messages from the social environment. Sexual attitudes and behavior of adolescents have become more liberal, with a slight swing back recently.

- Early and frequent sexual activity is linked to a variety of factors associated with economic disadvantage. Many sexually active teenagers do not practice contraception regularly. Adolescent cognitive processes and a lack of social support for responsible sexual behavior underlie the failure of these young people to protect themselves against pregnancy.

Describe factors involved in the development of homosexuality.

- About 3 to 6 percent of adolescents discover that they are lesbian or gay. Biological factors, including heredity and prenatal hormone levels, play an important role in homosexuality. Lesbian and gay teenagers face special problems in developing a positive sexual identity.

Discuss factors related to sexually transmitted disease and to teenage pregnancy and parenthood.

- Early sexual activity combined with inconsistent contraceptive use results in high rates of sexually transmitted disease (STD) among American adolescents. Many young adults with AIDS contracted the virus as teenagers.

- Adolescent pregnancy and parenthood rates are higher in the United States than in most other industrialized nations. Teenage pregnancy in Canada, though lower than in the United States, is still a problem. Adolescent parenthood is associated with school dropout, reduced chances of marriage, greater likelihood of divorce, and poverty—circumstances that jeopardize the well-being of both adolescent and newborn child.

- Improved sex education, access to contraceptives, programs that build social competence, and expanded educational and vocational opportunities help prevent early pregnancy. Adolescent mothers benefit from school programs that provide job training and child care and from family relationships that are sensitive to their needs. When teenage fathers stay involved, their children develop more favorably.

What personal and social factors are related to adolescent substance use and abuse?

- Teenage alcohol and drug use is widespread in industrialized nations. For most young people, drug experimentation reflects curiosity about these forbidden substances. The minority who move from use to abuse have serious psychological, family, and school problems. Programs that work with parents early to reduce family adversity and improve parenting skills and that build teenagers' competence help prevent substance abuse.

COGNITIVE DEVELOPMENT

Piaget's Theory: The Formal Operational Stage

What are the major characteristics of formal operational thought?

- During Piaget's **formal operational stage,** abstract thinking appears. Adolescents engage in **hypothetico-deductive reasoning.** When faced with a problem, they think of all possibilities and test them systematically. **Propositional thought** also develops. Young people can evaluate the logic of verbal statements, apart from real-world circumstances.

Discuss recent research on formal operational thought and its implications for the accuracy of Piaget's formal operational stage.

- Adolescents are capable of a much deeper grasp of scientific principles than school-age children. However, college students and adults think abstractly only in situations in which they have had extensive experience, and formal thought does not appear in many village and tribal cultures. These findings indicate that Piaget's highest stage is affected by specific, school-learning opportunities.

An Information-Processing View of Adolescent Cognitive Development

How do information-processing researchers account for the development of abstract reasoning?

- Information-processing researchers believe that a variety of specific mechanisms of change foster abstract thought, including improved attention, more effective strategies, greater knowledge, more effective cognitive self-regulation, gains in information-processing capacity, and, especially, advances in metacognition.

- Research on scientific reasoning indicates that the ability to coordinate theory with evidence improves during adolescence as young people solve increasingly complex problems and reflect on their thinking, acquiring more mature metacognitive understandings. Adolescents develop formal operational thinking in a similar, step-by-step fashion on different types of tasks, constructing general models that can be applied to many instances of a given type of problem.

Consequences of Abstract Thought

Describe typical reactions of adolescents that result from new abstract reasoning powers.

- Using their new cognitive powers, teenagers become more argumentative, idealistic, and critical. As they think more about themselves, two distorted images of the relation between self and other appear—the **imaginary audience** and the **personal fable.** Adolescents show gains in cognitive self-regulation, but they often have difficulty making decisions in everyday life.

Sex Differences in Mental Abilities

Describe sex differences in mental abilities at adolescence.

- During adolescence, the female advantage in verbal ability is slight. Boys do better in abstract mathematical reasoning. Among mathematically talented youngsters, the gender gap is large and stems, in part, from boys' biologically based superior spatial skills. At the same time, a variety of environmental factors, including childhood play activities, gender stereotyping of math as a "masculine" activity, and self-confidence and interest in doing math contribute to boys' better math performance.

Learning in School

Discuss the impact of school transitions on adolescent adjustment.

- School transitions in adolescence can be stressful. Girls experience more adjustment difficulties after the transition from elementary school to junior high because other life changes (puberty and the beginning of dating) tend to occur at the same time. Teenagers who must cope with added stresses—especially young people with both academic and mental health difficulties—are at greatest risk for increased adjustment problems following school transition.

Discuss family, peer, school, and employment influences on academic achievement during adolescence.

- Authoritative parenting and parents' school involvement promote high achievement. Teenagers with parents who encourage achievement are likely to choose friends who do the same. Warm, supportive learning environments with activities that emphasize high-level thinking enable adolescents to reach their academic potential.
- By high school, separate educational tracks that dovetail with students' future plans are necessary. Unfortunately, high school tracking in the United States and Canada usually extends the educational inequalities of earlier years.
- Spending more than 15 hours a week at a part-time job during the school year undermines adolescents' school performance and work-related attitudes. In contrast, work–study programs designed to meet both educational and vocational goals improve work attitudes and academic achievement among non-college-bound teenagers.

What factors are related to dropping out of school?

- Eleven percent of American and Canadian young people, many of whom are low-SES minority youths, leave high school without a diploma. Dropping out is the result of a long, gradual process of disengagement from school. Family and school influences combine to undermine the young person's chances for success. These include poor school performance, lack of parental support for academic achievement, and unstimulating teaching.

Important Terms and Concepts

adolescence (p. 344)
anorexia nervosa (p. 353)
body image (p. 351)
bulimia nervosa (p. 354)
formal operational stage (p. 363)
growth spurt (p. 346)
hypothetico-deductive reasoning (p. 363)
imaginary audience (p. 367)
menarche (p. 348)
personal fable (p. 367)
primary sexual characteristics (p. 347)
propositional thought (p. 354)
puberty (p. 344)
secondary sexual characteristics (p. 347)
secular trend (p. 349)
spermarche (p. 348)

FYI For Further Information and Help

Consult the Companion Website for *Development Through the Lifespan, Third Edition,* (www.ablongman.com/berk) where you will find the following resources for this chapter:

- **Chapter Objectives**
- **Flashcards** for studying important terms and concepts
- **Annotated Weblinks** to guide you in further research
- **Ask Yourself** questions, which you can answer and then check against a sample response
- **Suggested Readings**
- **Practice Test** with immediate scoring and feedback

THANK YO
VOLUNTEE
Hands On
MIAMI

© JEFF GREENBERG/PHOTOEDIT

As adolescents search for personally meaningful values to have faith in, they bring a sense of idealism and hopefulness to society. These teenagers are members of Hands On Miami, a volunteer organization devoted to improving the community. Among their activities are feeding the homeless, serving as companions to the elderly, restoring the natural environment, and tutoring at-risk children.

12

Emotional and Social Development in Adolescence

Erikson's Theory: Identity versus Identity Confusion

Self-Understanding
Changes in Self-Concept • Changes in Self-Esteem • Paths to Identity • Identity Status and Psychological Well-Being • Factors That Affect Identity Development

- Cultural Influences: Identity Development Among Ethnic Minority Adolescents

Moral Development
Piaget's Theory of Moral Development • Kohlberg's Extension of Piaget's Theory • Are There Sex Differences in Moral Reasoning? • Environmental Influences on Moral Reasoning • Moral Reasoning and Behavior

- Social Issues: Development of Civic Responsibility

Gender Typing

The Family
Parent–Child Relationships • Family Circumstances • Siblings

Peer Relations
Friendships • Cliques and Crowds • Dating • Peer Conformity

Problems of Development
Depression • Suicide • Delinquency

- A Lifespan Vista: Two Routes to Adolescent Delinquency

Louis sat on the grassy hillside overlooking the high school, waiting for his best friend, Darryl, to arrive from his fourth-period class. The two boys often met at noontime and then crossed the street to have lunch at a nearby hamburger stand.

Watching as hundreds of students poured onto the school grounds, Louis reflected on what he had learned in government class that day. "Suppose by chance I *had* been born in the People's Republic of China. I'd be sitting here, speaking a different language, being called by a different name, and thinking about the world in different ways. Gosh, I am who I am through some quirk of fate," Louis pondered.

Louis awoke from his thoughts with a start. Darryl was standing in front of him. "Hey, dreamer! I've been shouting and waving from the bottom of the hill for 5 minutes. How come you're so spaced out lately, Louis?" Darryl asked as they walked off.

"Oh, just wondering about stuff—like what I want, what I believe in. My older brother Jules—I envy him. He seems to know just where he's going. Most of the time, I'm up in the air about it. You ever feel that way?"

"Yeah, a lot," admitted Darryl, looking at Louis seriously as they approached the hamburger stand. "I often think, What am I really like? Who will I become?"

Louis and Darryl's introspective remarks are signs of a major reorganization of the self at adolescence: the development of identity. Both young people are attempting to formulate who they are—their personal values and the directions they will pursue in life. We begin this chapter with Erikson's account of identity development and the research it has stimulated on teenagers' thoughts and feelings about themselves. The quest for identity extends to many aspects of development. We will see how a sense of cultural belonging, moral understanding, and masculine and feminine self-images are refined during adolescence. And as parent–child relationships are revised and young people become increasingly independent of the family, friendships and peer networks become crucial contexts for bridging the gap between childhood and adulthood. Our chapter concludes with a discussion of several serious adjustment problems of adolescence: depression, suicide, and delinquency.

Erikson's Theory: Identity versus Identity Confusion

Erikson (1950, 1968) was the first to recognize **identity** as the major personality achievement of adolescence and as a crucial step toward becoming a productive, happy adult. Constructing an identity involves defining who you are, what you value, and the directions you choose to pursue in life. One expert described it as an explicit theory of oneself as a rational agent—one who acts on the basis of reason, takes responsibility for those actions, and can explain them (Moshman, 1999). This search for what is true and real about the self is the driving force behind many new commitments—to a sexual orientation; a vocation; interpersonal relationships; community involvement; ethnic group membership; and moral, political, religious, and cultural ideals.

Erikson called the psychological conflict of adolescence **identity versus identity confusion.** Successful outcomes of earlier stages pave the way to its positive resolution. Young people who reach adolescence with a weak sense of *trust* have trouble finding ideals to have faith in. Those with little *autonomy* or *initiative* do not engage in the active exploration required to choose among alternatives. And those who lack a sense of *industry* fail to select a vocation that matches their interests and skills.

Although the seeds of identity formation are planted early, not until adolescence do young people become absorbed in this task. According to Erikson, in complex societies, teenagers experience an *identity crisis*—a temporary period of confusion and distress as they experiment with alternatives before settling on values and goals. Adolescents who go through a process of inner soul-searching eventually arrive at a mature identity. They sift through characteristics that defined the self in childhood and combine them with new commitments. Then they mold these into a solid inner core that provides a sense of stability as they move through different roles in daily life. Once formed, identity continues to be refined in adulthood as people reevaluate earlier commitments and choices.

■ As these adolescents exchange opinions about a recent news event, they become more aware of a diversity of viewpoints. Exploring competing beliefs and values is essential for mature identity development—clarifying what is real and true about the self.

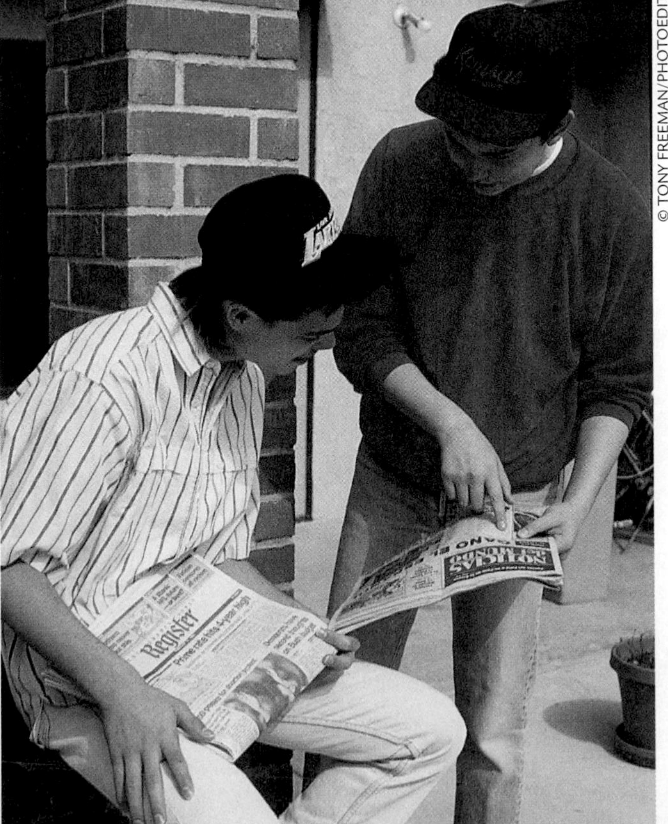

© TONY FREEMAN/PHOTOEDIT

Current theorists agree with Erikson that questioning of values, plans, and priorities is necessary for a mature identity, but they no longer refer to this process as a "crisis" (Grotevant, 1998). For some young people, identity development is traumatic and disturbing, but for most it is not. *Exploration* better describes the typical adolescent's gradual, uneventful approach to identity formation. By trying out various life possibilities and moving toward making enduring decisions, young people forge an organized self-structure (Arnett, 2000; Moshman, 1999).

Erikson described the negative outcome of adolescence as *identity confusion.* Some young people appear shallow and directionless, either because earlier conflicts have been resolved negatively or because society restricts their choices to ones that do not match their abilities and desires. As a result, they are unprepared for the psychological challenges of adulthood. For example, individuals find it difficult to risk the self-sharing involved in Erikson's young adult stage—*intimacy*—if they do not have a firm sense of self (an identity) to which they can return.

Does research support Erikson's ideas about identity development? In the following sections, we will see that adolescents go about the task of defining the self in ways that closely match Erikson's description.

Self-Understanding

During adolescence, cognitive changes transform the young person's vision of the self into a more complex, well-organized, and consistent picture. Changes in self-concept and self-esteem set the stage for development of a unified personal identity.

Changes in Self-Concept

Recall that by the end of middle childhood, children describe themselves in terms of personality traits. In early adolescence, they unify separate traits, such as "smart" and "talented," into more abstract descriptors, such as "intelligent." But these generalizations about the self are not interconnected, and often they are contradictory. For example, 12- to 14-year-olds might mention such opposing traits as "intelligent" and "airhead" or "shy" and "outgoing." These disparities result from social pressures to display different selves in different relationships—with parents, classmates, close friends, and romantic partners (Harter, 1998).

By middle to late adolescence, teenagers combine their traits into an organized system. And they begin to use qualifiers ("I have a *fairly* quick temper," "I'm not *thoroughly* honest"), revealing their awareness that psychological qualities often change from one situation to the next. Older adolescents also add integrating principles, which make sense out of formerly troublesome contradictions. For example, one young person remarked, "I'm very adaptable. When I'm around my friends, who think that what I say is important, I'm very talkative; but around my family I'm quiet because they're never interested enough to really listen to me" (Damon, 1990, p. 88).

Compared with school-age children, teenagers also place more emphasis on social virtues, such as being friendly, considerate, kind, and cooperative (Damon & Hart, 1988). Adolescents are very concerned with being viewed positively by others, and their statements about themselves reflect this concern. In addition, personal and moral values appear as key themes in older adolescents' self-concepts. As adolescents revise their views of themselves to include enduring beliefs and plans, they move toward the unity of self that is central to identity development.

Changes in Self-Esteem

Self-esteem, the evaluative side of self-concept, continues to differentiate during adolescence. To the self-evaluations of middle childhood—academic competence, social competence, physical/athletic competence, and physical appearance—are added several new dimensions: close friendship, romantic appeal, and job competence (Harter, 1990, 1999).

Level of self-esteem also changes. Except for temporary declines after school transitions for some adolescents (see Chapter 11, page 371), self-esteem is on the rise and remains high for the majority of young people (Twenge & Campbell, 2001; Zimmerman et al., 1997). Most of the time, becoming an adolescent leads to feelings of pride and self-confidence.

Still, young people vary widely in their self-esteem profiles. Whereas some evaluate themselves similarly in all areas,

■ In adolescence, self-esteem usually rises. This 14-year-old feels especially good about her athletic capabilities and peer relationships.

others are more satisfied in one or two than in others. A profile of all highly favorable self-evaluations is not more closely associated with adjustment than a profile that is generally positive. But teenagers who feel much better about their peer relations than their academic competence and family relations tend to have adjustment difficulties. And a profile of low self-regard in all areas is linked to anxiety, depression, and increasing antisocial behavior over time (DuBois et al., 1998, 1999).

All the same, the contexts in which young people find themselves can modify these group differences. Authoritative parenting predicts high self-esteem in adolescence, just as it did in childhood. And encouragement from teachers is linked to a favorable self-image as well (Carlson, Uppal, & Prosser, 2000; Steinberg et al., 1995). In contrast, when support from adults or peers is *conditional* (withheld unless the young person meets very high standards), teenagers frequently engage in behaviors they consider "false"—not representative of their true self. Adolescents who often "act phony" to protect themselves from disapproval suffer from low self-esteem and depression (Harter et al., 1996).

The larger social environment also influences self-esteem. Caucasian-American adolescents' self-esteem is less positive than that of African Americans, who benefit from warm, extended families and a strong sense of ethnic pride (Gray-Little & Hafdahl, 2000). Furthermore, most young people whose self-esteem drops in adolescence are girls (Eccles et al., 1999). Recall that girls worry more about their physical appearance and feel more insecure about their abilities. Finally, teenagers who attend schools or live in neighborhoods where their SES or ethnic group is well represented have fewer problems with self-esteem (Gray-Little & Carels, 1997). Schools and communities that accept the young person's cultural heritage support a positive sense of self-worth as well as a solid and secure personal identity.

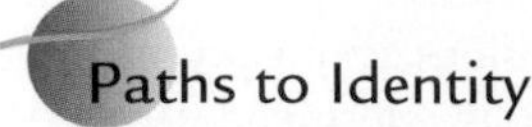

Paths to Identity

Adolescents' well-organized self-descriptions and differentiated sense of self-esteem provide the cognitive foundation for developing an identity. Using a clinical interviewing procedure, researchers group adolescents into four categories that reflect their progress in formulating a mature identity (Marcia, 1980). Table 12.1 describes and illustrates these *identity statuses:* **identity achievement, moratorium, identity foreclosure,** and **identity diffusion.**

Identity development follows many paths. Some adolescents remain in one status, whereas others experience many status transitions. And the pattern often varies across identity domains, such as sexual orientation, vocation, and religious and political values. Most young people change from "lower" statuses (foreclosure or diffusion) to higher statuses (moratorium or achievement) by the time they reach their twenties, but some move in the reverse direction (Kroger, 1995; Meeus, 1996).

Because college students have many opportunities to explore career options and lifestyles, they make more progress toward formulating an identity than they did in high school (Meeus et al., 1999). And in the years following college, many

Table 12.1 The Four Identity Statuses

Identity Status	Description	Example
Identity achievement	Having already explored alternatives, identity-achieved individuals are committed to a clearly formulated set of self-chosen values and goals. They feel a sense of psychological well-being, of sameness through time, and of knowing where they are going.	When asked how willing she would be to give up going into her chosen occupation if something better came along, Darla responded, "Well, I might, but I doubt it. I've thought long and hard about law as a career. I'm pretty certain it's for me."
Moratorium	*Moratorium* means "delay or holding pattern." These individuals have not yet made definite commitments. They are in the process of exploring—gathering information and trying out activities, with the desire to find values and goals to guide their lives.	When asked if he had ever had doubts about his religious beliefs, Ramon said, "Yes, I guess I'm going through that right now. I just don't see how there can be a God and yet so much evil in the world."
Identity foreclosure	Identity-foreclosed individuals have committed themselves to values and goals without exploring alternatives. They accept a ready-made identity that authority figures (usually parents but sometimes teachers, religious leaders, or romantic partners) have chosen for them.	When asked if she had ever reconsidered her political beliefs, Hillary answered, "No, not really, our family is pretty much in agreement on these things."
Identity diffusion	Identity-diffused individuals lack clear direction. They are not committed to values and goals or actively trying to reach them. They may have never explored alternatives or have found the task too threatening and overwhelming.	When asked about his attitude toward nontraditional gender roles, Joel responded, "Oh, I don't know. It doesn't make much difference to me. I can take it or leave it."

young people continue to obtain a broad range of life experiences before choosing a life course. Adolescents who go to work after high school graduation settle on a self-definition earlier than college-educated youths (Munro & Adams, 1977). But those who find it difficult to realize their occupational goals because of lack of training or vocational choices are at risk for identity diffusion (Archer, 1989).

At one time, researchers thought that adolescent girls postponed establishing an identity and instead focused on Erikson's next stage, intimacy development. Some girls do show more sophisticated reasoning than boys in identity areas related to intimacy, such as sexuality and family versus career priorities. Otherwise, adolescents of both sexes typically make progress on identity concerns before experiencing genuine intimacy in relationships (Kroger, 2000).

Identity Status and Psychological Well-Being

Identity achievement and moratorium are psychologically healthy routes to a mature self-definition, whereas foreclosure and diffusion are maladaptive. Young people who are identity achieved or actively exploring have a higher sense of self-esteem, are more likely to engage in abstract and critical thinking, report greater similarity between their ideal self and their real self, and are more advanced in moral reasoning (Josselson, 1994; Marcia et al., 1993).

Adolescents who get stuck in either foreclosure or diffusion have adjustment difficulties. Foreclosed individuals tend to be dogmatic, inflexible, and intolerant. Some regard any difference of opinion as a threat (Kroger, 1995). Most are afraid of rejection by people on whom they depend for affection and self-esteem. A few foreclosed teenagers who are alienated from their families and society may join cults or other extremist groups, uncritically adopting a way of life that is different from their past.

Long-term diffused teenagers are the least mature in identity development. They typically entrust themselves to luck or fate, have an "I don't care" attitude, and tend to go along with whatever the "crowd" is doing at the moment. As a result, they often experience time management and academic difficulties (Berzonsky & Kuk, 2000). And they are most likely to use drugs. Often at the heart of their apathy is a sense of hopelessness about the future (Archer & Waterman, 1990).

Factors That Affect Identity Development

Adolescent identity formation begins a lifelong process of refinement in personal commitments. Whenever the individual or the context changes, opportunities for reformulating identity exist (Yoder, 2000). A wide variety of factors influence identity development.

Personality—in particular, a flexible, open-minded approach to grappling with competing beliefs and values—fosters a mature identity. Adolescents who assume that absolute truth is always attainable tend to be foreclosed, whereas those who lack confidence about ever knowing anything with certainty are more often identity diffused. Adolescents who appreciate that they can use rational criteria to choose among alternatives are likely to be in a state of moratorium or identity achievement (Berzonsky & Kuk, 2000; Boyes & Chandler, 1992).

Recall from Chapter 6 that toddlers with a healthy sense of self have parents who provide both emotional support and freedom to explore. Similarly, when the family serves as a "secure base" from which teenagers can confidently move out into the wider world, identity development is enhanced. Adolescents who feel attached to their parents but also free to voice their own opinions tend to be identity achieved or in a state of moratorium (Grotevant & Cooper, 1998; Lapsley, Rice, & FitzGerald, 1990). Foreclosed teenagers usually have close bonds with parents, but they lack opportunities for healthy separation. And diffused young people report the lowest levels of warm, open communication at home (Papini, 1994).

As adolescents interact with a variety of peers, their exposure to ideas and values expands. Close friends assist each other in exploring options by providing emotional support and role models of identity development. In one study, adolescents' attachment to friends predicted exploration of careers and progress in choosing one (Felsman & Bluestein, 1999).

Schools and communities offering rich and varied opportunities for exploration also support identity development. Schools can help in many ways—through classrooms that promote high-level thinking, extracurricular activities that enable teenagers to take on responsible roles, teachers and counselors who encourage low-SES students to go to college, and vocational training programs that immerse young people in the real world of adult work (Cooper, 1998).

Finally, the larger cultural context and historical time period affect identity development. Among today's adolescents,

■ These Hispanic teenagers congregate in a city neighborhood where their ethnicity is well represented. Schools and communities that are accepting of the young person's cultural heritage promote high self-esteem.

Cultural Influences

Identity Development Among Ethnic Minority Adolescents

Most adolescents are aware of their cultural ancestry, and it is not a matter of intense concern for them. But for teenagers who are members of minority groups, **ethnic identity**—a sense of ethnic-group membership and attitudes and feelings associated with that membership—is central to the quest for identity, and it presents complex challenges. As they develop cognitively and become more sensitive to feedback from the social environment, minority youths become painfully aware that they are targets of discrimination and inequality. This discovery complicates their efforts to develop a sense of cultural belonging and a set of personally meaningful life goals.

Minority youths often feel caught between the standards of the larger society and the traditions of their culture of origin. In many immigrant families from collectivist cultures, adolescents' commitment to obeying their parents and fulfilling family obligations lessens the longer the family has been in the United States (Phinney, Ong, & Madden, 2000). Young people sometimes reject aspects of their ethnic background. In one study, Asian-American 15- to 17-year-olds were more likely than African Americans and Hispanics to hold negative attitudes toward their subcultural group (Phinney, 1989). Perhaps the absence of a social movement stressing ethnic pride of the kind available to black and Hispanic teenagers underlies this finding.

Some immigrant parents overly restrict their teenagers out of fear that assimilation into the larger society will undermine their cultural traditions, and their youngsters rebel. One Southeast-Asian refugee described his daughter's behavior: "She complains about going to the Lao temple on the weekend and instead joined a youth group in a neighborhood Christian church. She refused to wear traditional dress on the Lao New Year. The girl is setting a very bad example for her younger sisters and brothers" (Nidorf, 1985, pp. 422–423).

Other minority teenagers react to years of shattered self-esteem, school failure, and barriers to success in the mainstream by defining themselves in contrast to majority values. A Mexican-American teenager who had given up on school commented, "Mexicans don't have a chance to go on to college and make something of themselves." Another, responding to the question of what it takes to be a successful adult, pointed to his uncle, leader of a local gang, as an example (Matute-Bianche, 1986, pp. 250–251).

Because it is painful and confusing, minority high school students often dodge the task of forming an ethnic identity. Many are diffused or foreclosed on ethnic identity issues (Markstrom-Adams & Adams, 1995). But those who manage to deal with prejudice and discrimination proactively, through affirming their self-worth and disproving stereotypes of low achievement and antisocial behavior, generally feel strongly committed to their ethnic group (Phinney & Chavira, 1995).

How can society help minority adolescents resolve identity conflicts constructively? A variety of efforts are relevant, including

- promoting effective parenting, in which children and adolescents benefit from family ethnic pride yet are encouraged to explore the meaning of ethnicity in their own lives
- ensuring that schools respect minority youths' native languages, unique learning styles, and right to high-quality education, and
- fostering contact with peers of the same ethnicity along with respect between ethnic groups (García Coll & Magnuson, 1997)

© RUDI VON BRIEL/PHOTOEDIT

■ These East Indian adolescents dress in traditional costumes for a folk dancing demonstration at a town festival. When minority youths encounter respect for their cultural heritage in schools and communities, they are more likely to retain ethnic values and customs as an important part of their identities.

A secure ethnic identity is associated with higher self-esteem, optimism, a sense of mastery over the environment, and more positive attitudes toward one's ethnicity (Carlson, Uppal, & Prosser, 2000; Smith et al., 1999). But forming a **bicultural identity**—by exploring and adopting values from both the adolescent's subculture and the dominant culture—offers added benefits. Biculturally identified adolescents tend to be achieved in other areas of identity as well. And their relations with members of other ethnic groups are especially favorable (Phinney & Kohatsu, 1997). In sum, ethnic-identity achievement enhances many aspects of emotional and social development.

exploration and commitment take place earlier in vocational choice and gender-role preference than in religious and political values. Yet a generation ago, when the Vietnam War divided Americans, political beliefs took shape sooner (Archer, 1989). Societal forces also are responsible for the special problems that gay and lesbian youths (see Chapter 11) and some ethnic minority adolescents face in forming a secure identity (see the Cultural Influences box above). The Caregiving Concerns table on the following page summarizes ways that adults can support adolescents in their quest for identity.

Ways Adults Can Support Healthy Identity Development in Adolescence

STRATEGY	RATIONALE
Warm, open communication	Provides both emotional support and freedom to explore values and goals
Discussions at home and school that promote high-level thinking	Encourages rational and deliberate selection among competing beliefs and values
Opportunities to participate in extracurricular activities and vocational training programs	Permits young people to explore the real world of adult work
Opportunities to talk with adults and peers who have worked through identity questions	Offers models of identity achievement and advice on how to resolve identity concerns
Opportunities to explore ethnic heritage and learn about other cultures in an atmosphere of respect	Fosters identity achievement in all areas and ethnic tolerance, which supports the identity explorations of others

Ask Yourself

REVIEW
Return to the conversation between Louis and Darryl in the opening of this chapter. Which identity status best characterizes the two boys, and why?

REVIEW
Cite personal and contextual factors that contribute to identity development.

APPLY
Joe and Vera worry that their 18-year-old son Brad will waste time at college because he is unsure about his major and his career goals. Explain why Brad's uncertainty may be advantageous for his identity development.

CONNECT
Explain how changes in self-concept and self-esteem at adolescence pave the way for constructing an identity.

www.

Moral Development

Eleven-year-old Sabrina sat at the kitchen table reading the Sunday newspaper, her eyes wide with interest. "You gotta see this," she said to 16-year-old Louis, who sat munching cereal. Sabrina held up a page of large photos showing a 70-year-old woman standing in her home. The floor and furniture were piled with stacks of newspapers, cardboard boxes, tin cans, glass containers, food, and clothing. And the accompanying article described crumbling plaster on the walls, frozen pipes, and sinks, a toilet, and a furnace that no longer worked. The headline read: "Loretta Perry: My Life Is None of Their Business."

"Look what they're trying to do to this poor lady," exclaimed Sabrina. "They wanna throw her out of her house and tear it down! Those city inspectors must not care about anyone. Here it says, 'Mrs. Perry has devoted much of her life to doing favors for people.' Why doesn't someone help *her?*"

"Sabrina, you missed the point," Louis responded. "Mrs. Perry is violating 30 building code standards. The law says you're supposed to keep your house clean and in good repair."

"But Louis, she's old and she needs help. She says her life will be over if they destroy her home."

"The building inspectors aren't being mean, Sabrina. Mrs. Perry is stubborn. She refuses to obey the law. By not taking care of her house, she's not just a threat to herself. She's a danger to her neighbors, too. Suppose her house caught on fire. You can't live around other people and say your life is nobody's business."

"You don't just knock someone's home down," Sabrina replied angrily. "Where're her friends and neighbors in all this? Why aren't they over there fixing up that house? You're like those building inspectors, Louis. You've got no feelings!"

Louis and Sabrina's disagreement over Mrs. Perry's plight illustrates tremendous advances in moral understanding. Changes in cognition and social experience permit adolescents to better understand larger social structures—societal institutions and lawmaking systems—that govern moral responsibilities. As their grasp of social arrangements expands, adolescents construct new ideas about what ought to be done when the needs and desires of people conflict, and they move toward increasingly just, fair, and balanced solutions to moral problems.

Piaget's Theory of Moral Development

The most influential approach to moral development is Lawrence Kohlberg's cognitive-developmental perspective, which was inspired by Piaget's early work on the moral judgment of

■ This teenage boy's efforts to help a friend with disabilities experience a trip around the ice rink may be motivated by his understanding of the Golden Rule: "Do unto others as you would have others do unto you." A sophisticated grasp of reciprocity—expressing the same concern for others as for oneself—contributes to advances in moral understanding in older children and adolescents.

the child. Piaget (1932/1965) saw children as moving through two broad stages of moral understanding.

The first stage is **heteronomous morality,** which extends from about 5 to 10 years of age. The word *heteronomous* means "under the authority of another." As the term suggests, children of this stage view rules as handed down by authorities (God, parents, and teachers), as having a permanent existence, as unchangeable, and as requiring strict obedience. Also, in judging an act's wrongness, they focus on outcomes rather than intent to do harm. When asked to decide which child is naughtier—John, who accidentally breaks 15 cups while on his way to dinner, or Henry, who breaks 1 cup while stealing some jam—a 6- or 7-year-old chooses John.

According to Piaget, around age 10 children make the transition to the stage of **autonomous morality**—no longer viewing rules as fixed but seeing them as socially agreed-on principles that can be revised when there is a need to do so. In creating and changing rules, older children and adolescents use a standard of fairness called *ideal reciprocity.* They express the same concern for the welfare of others as they do for themselves. Most of us are familiar with ideal reciprocity in the form of the Golden Rule: "Do unto others as you would have others do unto you." Individuals at this stage realize that intentions, not just outcomes, should serve as the basis for judging behavior. Piaget believed that improvements in perspective taking, which result from cognitive development and opportunities to interact with peers, are responsible for this advance in moral judgment.

Think about Piaget's theory in light of what you learned about moral development in earlier chapters (return to pages 256 and 321–322). You will see that his account of young children as rigid, external, and focused on physical consequences underestimates their moral capacities. Nevertheless, Piaget's theory of morality, like his cognitive theory, does describe the general direction of moral development. Although children are more sophisticated moral thinkers than Piaget made them out to be, they are not as advanced as adolescents and adults. Over the past two decades, Piaget's groundbreaking work has been replaced by Kohlberg's more comprehensive theory, which regards moral development as extending beyond childhood into adolescence and adulthood in a six-stage sequence.

Kohlberg's Extension of Piaget's Theory

Kohlberg used a clinical interviewing procedure to study moral development. He gave children and adolescents *moral dilemmas*—stories that present a conflict between two moral values—and asked them what the main actor should do and why. The best known of these is the "Heinz dilemma," which presents a choice between the value of obeying the law (not stealing) and the value of human life (saving a dying person):

> In Europe a woman was near death from cancer. There was one drug that the doctors thought might save her. A druggist in the same town had discovered it, but he was charging ten times what the drug cost him to make. The sick woman's husband, Heinz, went to everyone he knew to borrow the money, but he could only get together half of what it cost. The druggist refused to sell it cheaper or let Heinz pay later. So Heinz got desperate and broke into the man's store to steal the drug for his wife. Should Heinz have done that? Why? (paraphrased from Colby et al., 1983, p. 77)

Kohlberg emphasized that it is *the way an individual reasons* about the dilemma, not *the content of the response* (whether or not to steal), that determines moral maturity. Individuals who believe Heinz should take the drug and those who think he should not can be found at each of Kohlberg's first four stages. At the two highest stages, moral reasoning and content are integrated into a coherent ethical system (Kohlberg, Levine, & Hewer, 1983). Given a choice between obeying the law and preserving individual rights, the most advanced moral thinkers support individual rights (in the Heinz dilemma, stealing the drug to save a life). Does this remind you of adolescents' effort to formulate a sound, well-organized set of personal values in identity development? According to some theorists, the development of identity and moral understanding are part of the same process (Blasi, 1994; Marcia, 1988).

● **Kohlberg's Stages of Moral Understanding.** Kohlberg organized his six stages into three general levels of moral development. He believed that moral understanding is promoted by the same factors Piaget thought were important for cognitive growth: (1) actively grappling with moral issues and

Table 12.2 Relations Among Kohlberg's Moral, Piaget's Cognitive, and Selman's Perspective-Taking Stages

Kohlberg's Moral Stage	Description	Piaget's Cognitive Stage	Selman's Perspective-Taking Stage[a]
Punishment and obedience orientation	Fear of authority and avoidance of punishment	Preoperational, early concrete operational	Social-informational
Instrumental purpose orientation	Satisfying personal needs	Concrete operational	Self-reflective
"Good boy–good girl" orientation	Maintaining the affection and approval of friends and relatives	Early formal operational	Third-party
Social-order-maintaining orientation	A duty to uphold laws and rules for their own sake	Formal operational	Societal
Social contract orientation	Fair procedures for changing laws to protect individual rights and the needs of the majority		
Universal ethical principle orientation	Abstract universal principles that are valid for all humanity		

[a]To review these stages, return to Chapter 10, page 321.

noticing weaknesses in one's current reasoning, and (2) gains in perspective taking, which permit individuals to resolve moral conflicts in more effective ways. As Table 12.2 shows, Kohlberg's moral stages are related to Piaget's cognitive and Selman's perspective-taking stages. As we examine Kohlberg's developmental sequence and illustrate it with responses to the Heinz dilemma, look for changes in perspective taking that each stage assumes.

The Preconventional Level. At the **preconventional level,** morality is externally controlled. As in Piaget's heteronomous stage, children accept the rules of authority figures and judge actions by their consequences. Behaviors that result in punishment are viewed as bad, and those that lead to rewards are seen as good.

Stage 1: The punishment and obedience orientation. Children at this stage find it difficult to consider two points of view in a moral dilemma. As a result, they ignore peoples' intentions and, instead, focus on fear of authority and avoidance of punishment as reasons for behaving morally.

> *Prostealing:* "If you let your wife die, you will get in trouble. You'll be blamed for not spending the money to help her and there'll be an investigation of you and the druggist for your wife's death." (Kohlberg, 1969, p. 381)

> *Antistealing:* "You shouldn't steal the drug because you'll be caught and sent to jail if you do. If you do get away, [you'd be scared that] the police would catch up with you any minute." (Kohlberg, 1969, p. 381)

Stage 2: The instrumental purpose orientation. Children become aware that people can have different perspectives in a moral dilemma, but this understanding is, at first, very concrete. They view right action as flowing from self-interest. Reciprocity is understood as equal exchange of favors—"you do this for me and I'll do that for you."

> *Prostealing:* "The druggist can do what he wants and Heinz can do what he wants to do. . . . But if Heinz decides to risk jail to save his wife, it's his life he's risking; he can do what he wants with it. And the same goes for the druggist; it's up to him to decide what he wants to do." (Rest, 1979, p. 26)

> *Antistealing:* "[Heinz] is running more risk than it's worth [to save a wife who is near death]." (Rest, 1979, p. 27)

The Conventional Level. At the **conventional level,** individuals continue to regard conformity to social rules as important, but not for reasons of self-interest. They believe that actively maintaining the current social system ensures positive relationships and societal order.

Stage 3: The "good boy–good girl" orientation, or the morality of interpersonal cooperation. The desire to obey rules because they promote social harmony first appears in the context of close personal ties. Stage 3 individuals want to maintain the affection and approval of friends and relatives by being a "good person"—trustworthy, loyal, respectful, helpful, and nice. The capacity to view a two-person relationship from the vantage point of an impartial, outside observer supports this new approach to morality. At this stage, the individual understands *ideal reciprocity,* expressed in the Golden Rule.

> *Prostealing:* "No one will think you're bad if you steal the drug, but your family will think you're an inhuman husband if you don't. If you let your wife die, you'll never be able to look anyone in the face again." (Kohlberg, 1969, p. 381)

> *Antistealing:* "It isn't just the druggist who will think you're a criminal, everyone else will too. After you steal it, you'll feel bad thinking how you've brought dishonor on your family and yourself." (Kohlberg, 1969, p. 381)

Stage 4: The social-order-maintaining orientation. At this stage, the individual takes into account a larger perspective—that of societal laws. Moral choices no longer depend on close ties to others. Instead, rules must be enforced in the same evenhanded fashion for everyone, and each member of society has a personal duty to uphold them. The Stage 4 individual believes that laws cannot be disobeyed under any circumstances because they are vital for ensuring societal order.

> *Prostealing:* "He should steal it. Heinz has a duty to protect his wife's life; it's a vow he took in marriage. But it's wrong to steal, so he would have to take the drug with the idea of paying the druggist for it and accepting the penalty for breaking the law later."

> *Antistealing:* "It's a natural thing for Heinz to want to save his wife, but. . . . Even if his wife is dying, it's still his duty as a citizen to obey the law. No one else is allowed to steal, why should he be? If everyone starts breaking the law in a jam, there'd be no civilization, just crime and violence." (Rest, 1979, p. 30)

The Postconventional or Principled Level. Individuals at the **postconventional** level move beyond unquestioning support for the laws and rules of their own society. They define morality in terms of abstract principles and values that apply to all situations and societies.

Stage 5: The social contract orientation. At Stage 5, individuals regard laws and rules as flexible instruments for furthering human purposes. They can imagine alternatives to their social order, and they emphasize fair procedures for interpreting and changing the law. When laws are consistent with individual rights and the interests of the majority, each person follows them because of a *social contract orientation*—free and willing participation in the system because it brings about more good for people than if it did not exist.

> *Prostealing:* "Although there is a law against stealing, the law wasn't meant to violate a person's right to life. Taking the drug does violate the law, but Heinz is justified in stealing in this instance. If Heinz is prosecuted for stealing, the law needs to be reinterpreted to take into account situations in which it goes against people's natural right to keep on living."

Stage 6: The universal ethical principle orientation. At this highest stage, right action is defined by self-chosen ethical principles of conscience that are valid for all humanity, regardless of law and social agreement. These values are abstract, not concrete moral rules like the Ten Commandments. Stage 6 individuals typically mention such principles as equal consideration of the claims of all human beings and respect for the worth and dignity of each person.

> *Prostealing:* "If Heinz does not do everything he can to save his wife, then he is putting some value higher than the value of life. It doesn't make sense to put respect for property above respect for life itself. [People] could live together without private property at all. Respect for human life and personality is absolute and accordingly [people] have a mutual duty to save one another from dying." (Rest, 1979, p. 37)

● **Research on Kohlberg's Stage Sequence.** Longitudinal studies provide the most convincing evidence for Kohlberg's stage sequence. With few exceptions, individuals move through the stages in the order that Kohlberg expected (Colby et al., 1983; Dawson, 2002; Walker & Taylor, 1991b). A striking finding is that moral development is very slow and gradual. Reasoning at Stages 1 and 2 decreases in early adolescence, whereas Stage 3 increases through midadolescence and then declines. Stage 4 reasoning rises over the teenage years until, by early adulthood, it is the typical response. Few people move beyond it to Stage 5. In fact, postconventional morality is so rare that there is no clear evidence that Kohlberg's Stage 6 actually follows Stage 5. The highest stage of moral reasoning is a matter of speculation and may represent a highly reflective endeavor achieved only by individuals with advanced education, usually with training in philosophy (Gibbs, 2003).

As you read the Heinz dilemma, you probably came up with your own solution to it. Now, try to think of a moral dilemma you recently faced. How did you solve it, and did your reasoning fall at the same stage as your thinking about Heinz and his dying wife? Real-life conflicts often elicit moral reasoning below a person's actual capacity because they bring out many practical considerations and mix cognition with intense emotion. In working through everyday moral challenges, people report feeling drained, confused, and torn by temptation. As one person observed, "It's a lot easier to be moral when you have nothing to lose" (Walker et al., 1999; Walker et al., 1995, p. 381).

The influence of situational factors on moral judgments suggests that like Piaget's cognitive stages, Kohlberg's moral stages are loosely organized. Rather than developing in a neat, stepwise fashion, people seem to draw on a range of moral responses that vary with context. With age, this range shifts upward as less mature moral reasoning is gradually replaced by more advanced moral thought.

Are There Sex Differences in Moral Reasoning?

As noted in the previous section, real-life moral dilemmas often highlight the contribution of emotion to moral judgment. Return once again to Sabrina's and Louis's moral discussion at the beginning of this section, and notice how her argument focuses on caring and commitment to others. Louis's approach is more impersonal. He looks at the dilemma of Loretta Perry in terms of competing rights and justice.

Carol Gilligan (1982) is the best-known figure among those who have argued that Kohlberg's theory does not adequately represent the morality of girls and women. She believes that feminine morality emphasizes an "ethic of care" that is devalued in Kohlberg's system. For example, Sabrina's reasoning falls at Stage 3 because she focuses on the importance of mutual trust and affection between people, whereas Louis's is at Stage 4 because he emphasizes the value of obeying the law to ensure societal order. According to Gilligan, a concern for others is a *different,* not a less valid, basis for moral judgment than a focus on impersonal rights.

Many studies have tested Gilligan's claim that Kohlberg's approach underestimates the moral maturity of females, and most do not support it (Turiel, 1998). On hypothetical dilemmas as well as everyday moral problems, adolescent and adult females display reasoning at the same stage as their male agemates or at higher stages. Also, themes of justice and caring appear in the responses of both sexes, and when girls do raise interpersonal concerns, they are not downscored in Kohlberg's system (Jadack et al., 1995; Kahn, 1992; Walker, 1995). These findings suggest that although Kohlberg emphasized justice rather than caring as the highest of moral ideals, his theory taps both sets of values.

Still, Gilligan makes a powerful claim that research on moral development has been limited by too much attention to rights and justice (a "masculine" ideal) and too little attention to care and responsiveness (a "feminine" ideal). Some evidence shows that although the morality of males and females taps both orientations, females do tend to stress care, whereas males either stress justice or use justice and care equally (Galotti, Kozberg, & Farmer, 1991; Garmon et al., 1996; Wark & Krebs, 1996).

The difference in emphasis appears most often on real-life rather than hypothetical dilemmas. Consequently, it may be largely a function of women's greater involvement in daily activities involving care and concern for others. In one study, American and Canadian 17- to 26-year-old females showed more complex reasoning about care issues than their male counterparts. But as Figure 12.1 shows, Norwegian males were just as advanced as Norwegian females in care-based understanding (Skoe, 1998). Perhaps Norwegian culture, which explicitly endorses gender equality, induces boys and men to think deeply about interpersonal obligations.

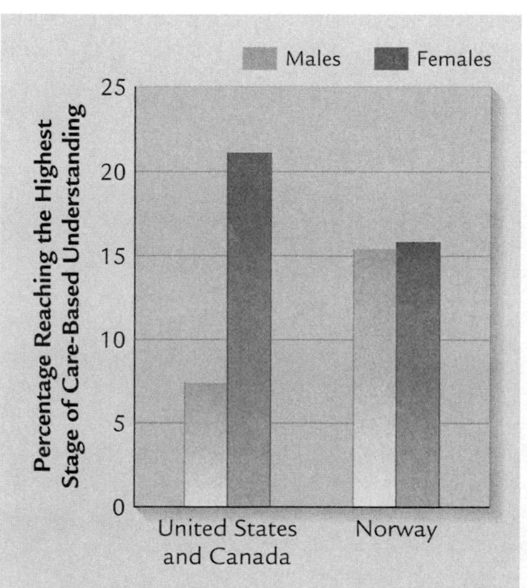

■ **FIGURE 12.1 Complex reasoning about care issues among U.S. and Canadian versus Norwegian males and females.** In this study of 17- to 26-year-olds, North American females scored much higher than males in complex care-based understanding. In contrast, Norwegian males and females displayed equally advanced care-based understanding. (Adapted from Skoe, 1998.)

Environmental Influences on Moral Reasoning

Change in moral stage is promoted by many environmental factors, including child-rearing practices, schooling, peer interaction, and aspects of culture. In line with Kohlberg's belief, growing evidence suggests that the way these experiences work is to present young people with cognitive challenges, which stimulate them to think about moral problems in more complex ways.

● **Child-Rearing Practices.** As in childhood, moral understanding in adolescence is fostered by warm parenting and discussion of moral concerns. Teenagers who gain most in moral development have parents who create a supportive atmosphere by listening sensitively, asking clarifying questions, and presenting higher-level reasoning. In contrast, parents who lecture, use threats, or make sarcastic remarks have youngsters who change little or not at all (Pratt et al., 1999; Walker & Taylor, 1991a).

● **Schooling.** Years of schooling completed is one of the most powerful predictors of moral understanding. Longitudinal research reveals that moral reasoning advances in late adolescence and early adulthood only as long as a person remains in college (Dawson, 2003; Speicher, 1994). Perhaps higher education has a strong impact on moral development because it introduces young people to social issues that extend beyond personal relationships to entire political or cultural groups. Consistent with this idea, college students who report more academic perspective-taking opportunities (for example, classes that emphasize open discussion of opinions) and who indicate that they have become more aware of social diversity tend to be advanced in moral reasoning (Mason & Gibbs, 1993).

● **Peer Interaction.** Research supports Piaget's belief that interaction among peers, who confront one another with differing viewpoints, promotes moral understanding. Adolescents who report more close friendships, who more often participate in conversations with their friends, and whom classmates view as leaders score higher in moral reasoning (Schonert-Reichl, 1999).

But peer interaction must have certain features to be effective. Consider, once again, Sabrina's and Louis's argument over the plight of Loretta Perry. Each teenager directly confronts and criticizes the other's statements, and emotionally intense expressions of disagreement occur. Peer discussions with these features lead to much greater stage change than do those in which adolescents state their opinions in a disorganized, uninvolved way (Berkowitz & Gibbs, 1983; Haan, Aerts, & Cooper,

■ Discussions about moral issues in which peers confront, critique, and attempt to clarify one another's statements lead to gains in moral understanding.

1985). Also note that Sabrina and Louis do not revise their ways of thinking after one discussion. It takes many peer interaction sessions over weeks or months to produce moral change.

● **Culture.** Young people in industrialized nations move through Kohlberg's stages more quickly and advance to a higher level than individuals in nonindustrialized, village societies, who rarely move beyond Stage 3. One explanation of these cultural differences focuses on the importance of larger social structures in advanced moral understanding. In village societies, moral cooperation is based on direct relations between people (Miller, 1997). Yet reasoning at Stages 4 to 6 depends on understanding the role of laws and government institutions in resolving moral conflict (Snarey, 1995).

In support of this view, in cultures where young people participate in the institutions of their society at early ages, moral reasoning is advanced. For example, on *kibbutzim,* small but technologically advanced agricultural settlements in Israel, children receive training in the governance of their community in middle childhood. By third grade, they mention more concerns about societal laws and rules when discussing moral conflicts than do Israeli city-reared or American children (Fuchs et al., 1986). During adolescence and young adulthood, a greater percentage of kibbutz than American individuals reach Kohlberg's Stages 4 and 5 (Snarey, Reimer, & Kohlberg, 1985).

A second possible reason for cultural variation is that responses to moral dilemmas in collectivist cultures (including village societies) are more other-directed than in Western Europe and North America (Miller, 1997). Consistent with this explanation, in village societies, moral statements that portray the individual as vitally connected to the social group are common. For example, one New Guinea village leader placed the blame for the Heinz dilemma on the entire community, stating, "If I were the judge, I would give him only light punishment because he asked everybody for help but nobody helped him" (Tietjen & Walker, 1985, p. 990).

Similarly, in research conducted in India, people less often held individuals accountable for moral transgressions. In their view, the self and social surroundings are inseparable, so solutions to moral dilemmas are the burden not of a single person but of the entire society (Miller & Bersoff, 1995). These findings raise the question of whether Kohlberg's highest level represents a culturally specific rather than a universal way of thinking—one limited to Western societies that emphasize individual rights and an appeal to an inner, private conscience.

Moral Reasoning and Behavior

According to Kohlberg, moral thought and action should come together at the higher levels of moral understanding. Mature moral thinkers realize that behaving in line with their beliefs is an important part of creating a just social world (Gibbs, 1995). Consistent with this idea, higher-stage adolescents more often act prosocially by helping, sharing, and defending victims of injustice (Carlo et al., 1996; Comunian & Gielan, 2000). They also less often engage in cheating, aggression, and other antisocial behaviors (Gregg, Gibbs, & Fuller, 1994; Taylor & Walker, 1997).

■ Young people growing up on Israeli kibbutzim receive training in the governance of their society at an early age. As a result, they understand the role of societal laws and rules in resolving moral conflict and are advanced in moral reasoning.

Social Issues

Development of Civic Responsibility

On Thanksgiving day, Jules, Louis, and Sabrina joined their parents at a soup kitchen, serving a holiday dinner to poverty-stricken people. Throughout the year, Sabrina volunteered on Saturday mornings at a nursing home, conversing with bedridden elders. In the months before a congressional election, all three young people attended special youth meetings with candidates, where they raised concerns. "What's your view on preserving our environment?" Louis asked. "How would you prevent the proposed tax cut from mostly benefiting the rich?" Jules chimed in. At school, Louis and his girlfriend Cassie formed an organization devoted to promoting ethnic and racial tolerance.

Already, these young people have a strong sense of civic responsibility—a complex capacity that combines cognition, emotion, and behavior. Civic responsibility involves *knowledge* of political issues and the means through which citizens can resolve differing views fairly; *feelings* of attachment to the community, of wanting to make a difference in its welfare; and *skills* for achieving civic goals, such as how to contact and question public officials and conduct meetings so all participants have a voice (Flanagan & Faison, 2001). New research reveals that family, school, and community experiences contribute to adolescents' civic responsibility.

Family Influences. Parents who bring up controversial issues and encourage their children to form opinions have teenagers who are more knowledgeable, more interested in civic issues, and better able to see them from more than one perspective (Santoloupo & Pratt, 1994). Also, adolescents who report that their families emphasize compassion for the less fortunate tend to hold socially responsible values. When asked what causes such social ills as unemployment, poverty, and homelessness, these teenagers more often mention situational and societal factors (lack of education, government policies, or the state of the economy) than individual factors (low intelligence or personal problems). Youths who endorse situational and societal causes, in turn, have more altruistic life goals, such as working to eradicate poverty or preserve the earth for future generations (Flanagan & Tucker, 1999).

© NANCY RICHMOND/THE IMAGE WORKS

■ During adolescence, young people develop a stronger sense of connection to their communities. These teenage volunteers gather food donations for the needy outside a grocery store on "Make a Difference Day." Involvement in community service grows out of a sense of civic responsibility, which is supported by family, school, and community experiences.

School and Community Influences. A democratic climate at school—one in which teachers hold high academic and moral standards for all students, express respect for students' ideas, and insist that students listen to and respect one another—fosters a sense of civic responsibility. Teenagers who say their teachers engage in these practices are more aware of political issues, better able to analyze them critically, and more committed to social causes (Flanagan & Faison, 2001).

Participation in extracurricular activities at school and in youth organizations is also associated with civic commitment that persists into adulthood (see Chapter 11, page 376). Two aspects of these involvements seem to account for their lasting impact. First, they introduce adolescents to the vision and skills required for mature civic engagement. Within clubs, teams, and other groups, young people see how their actions affect the wider school and community. They realize that collectively, they can achieve results greater than any one person can achieve alone. They also learn to work together, balancing strong convictions with compromise (Youniss, McLellan, & Yates, 1997). Second, while producing a weekly newspaper, participating in a dramatic production, or implementing a service project, young people explore political and moral ideals. Often they redefine themselves as having a responsibility to combat the misfortunes of others (Wheeler, 2002).

Current evidence points to growing self-interest and materialism among North American high school students (Rahn & Transue, 1998). Granting young people many opportunities to think in terms of *we* rather than *I* and to work with others toward a common good can combat this trend. The power of family, school, and community to promote civic responsibility may lie in discussions, educational practices, and activities that jointly foster moral thought, emotion, and behavior.

Yet even though a clear connection between advanced moral reasoning and action exists, it is only moderate. As we saw in earlier chapters, moral behavior is influenced by many factors besides cognition, including the emotions of empathy and guilt, individual differences in temperament, and a long history of experiences that affect moral decision making. Also, the degree to which morality is central to self-concept affects moral behavior. In a study of low-SES African-American and Hispanic teenagers, those who emphasized moral traits and goals in their self-descriptions displayed exceptional levels of community service. However, they did not differ from their agemates in moral reasoning (Hart & Fegley, 1995).

Researchers have yet to discover just how moral thought combines with other influences to foster moral action. Still, as the Social Issues box above reveals, promoting civic responsibility in young people can help them see the connection between their personal interests and the public interest—an insight that may foster all aspects of morality.

Gender Typing

As Sabrina entered adolescence, her thinking and behavior became more gender typed. For example, she began to place more emphasis on excelling in the traditionally feminine subjects of language, art, and music than in math and science. And when with peers, Sabrina worried about how she should walk, talk, eat, dress, laugh, and compete, judged according to accepted social standards for maleness and femaleness.

Early adolescence is a period of **gender intensification**—increased gender stereotyping of attitudes and behavior, and movement toward a more traditional gender identity (Basow & Rubin, 1999; Galambos, Almeida, & Petersen, 1990). Although gender intensification occurs in both sexes, it is stronger for girls, who feel less free to experiment with "other-gender" activities and behavior than they did in middle childhood (Huston & Alvarez, 1990).

What accounts for gender intensification? Biological, social, and cognitive factors are involved. Puberty magnifies sex differences in appearance, causing teenagers to spend more time thinking about themselves in gender-linked ways. Pubertal changes also prompt gender-typed pressures from others. Parents (especially those with traditional gender-role beliefs) may encourage "gender-appropriate" activities and behavior to a greater extent than they did in middle childhood (Crouter, Manke, & McHale, 1995). And when adolescents start to date, they often become more gender typed as a way of increasing their attractiveness (Maccoby, 1998). Finally, cognitive changes—in particular, greater concern with what others think—make young teenagers more responsive to gender-role expectations.

© MARY KATE DENNY/PHOTOEDIT

■ Early adolescence is a period of gender intensification. Puberty magnifies gender differences in appearance, causing teenagers to begin thinking about themselves in gender-linked ways. And when adolescents start to date, they often become more gender typed as a way of increasing their attractiveness to the other sex.

Gender intensification declines by middle to late adolescence, but not all young people move beyond it to the same degree. Teenagers who are encouraged to explore non-gender-typed options and to question the value of gender stereotypes for themselves and society are more likely to build an androgynous gender identity (see Chapter 8, page 263). Overall, androgynous adolescents tend to be psychologically healthier—more self-confident, more willing to speak their own mind, better liked by peers, and identity achieved (Dusek, 1987; Harter, 1998).

Ask Yourself

REVIEW

In our discussion of Kohlberg's theory, why were examples of both prostealing and antistealing responses to the Heinz dilemma presented for Stages 1 through 4, but only prostealing responses for Stages 5 and 6?

APPLY

Tam grew up in a small village culture, Lydia in a large industrial city. At age 15, Tam reasons at Kohlberg's Stage 2, Lydia at Stage 4. What factors might account for the difference?

CONNECT

What environmental factors that foster identity development are also likely to promote moral development?

www.

The Family

Franca and Antonio remember their son Louis's freshman year of high school as a difficult time. Because of a demanding project at work, Franca was away from home many evenings and weekends. Antonio took over in her absence, but when business declined at his hardware store, he, too, had less time for the family. That year, Louis and two friends used their computer know-how to crack the code of a long-distance telephone service. From the family basement, they made calls around the country. Louis's grades fell, and he often left the house without saying where he was going. Franca and Antonio began to feel uncomfortable about the long hours Louis spent in the basement and their lack of contact with him. Finally, when the telephone company traced the illegal calls to the family's phone number, Franca and Antonio knew they had cause for concern.

Development at adolescence involves striving for **autonomy**—a sense of oneself as a separate, self-governing individual. Autonomy first became a major issue in toddlerhood, but now it is revisited on a much higher plane. Teenagers strive to rely more on themselves and less on parents for guidance and decision making (Hill & Holmbeck, 1986; Steinberg & Silverberg, 1986). A major way that teenagers seek greater self-directedness is to shift away from family to peers, with whom they explore courses of action that depart from earlier patterns. Nevertheless, parent–child relationships remain vital for helping adolescents become autonomous, responsible individuals.

Parent–Child Relationships

Adolescents require freedom to experiment. Yet as Franca and Antonio's episode with Louis reveals, they also need guidance and, at times, protection from dangerous situations. Think back to what we said earlier about parent–child relationships that foster academic achievement (Chapter 11), identity formation, and moral maturity. Parental warmth and acceptance, combined with firm (but not overly restrictive) monitoring of the teenagers' activities, is related to many aspects of adolescent competence, including high self-esteem, self-reliance, academic achievement, and work orientation (Jacobson & Crockett, 2000; Steinberg et al., 1994). Note that these features make up the authoritative style that was adaptive in childhood as well. In contrast, parents who are coercive or psychologically controlling (for example, who belittle teenagers and discount their feelings) interfere with the development of autonomy. These tactics promote low self-esteem, depression, and antisocial behavior (Aquilino & Supple, 2001; Barber & Harmon, 2002).

Maintaining an authoritative style during adolescence involves special challenges and adjustments. In Chapter 11, we showed that puberty brings increased parent–child conflict, for both biological and cognitive reasons. Teenagers' improved ability to reason about social relationships adds to family tensions. Perhaps you can recall a time when you stopped viewing your parents as all-knowing and perfect and saw them as "just people." Once teenagers *de-idealize* their parents, they no longer bend as easily to parental authority as they did at earlier ages. They regard many matters—such as cleaning their rooms, coming and going from the household, and doing their schoolwork—as their own personal business, whereas parents continue to think of these as important social conventions—as shared concerns that foster family harmony (Nucci, 1996). Disagreements are harder to settle when parents and adolescents approach situations from such different perspectives.

In Chapter 2, we described the family as a *system* that must adapt to changes in its members. Many parents have reached their forties and are changing as well. While teenagers face a boundless future and a wide array of choices, their parents must come to terms with the fact that their own possibilities are narrowing. The pressures experienced by each generation act in opposition to one another (Holmbeck, 1996). Parents often can't understand why the adolescent wants to skip family activities to be with peers. And teenagers fail to appreciate that parents want the family to be together as often as possible because an important stage in adult life—parenthood—will soon be over.

In addition, parents and adolescents—particularly early adolescents—differ sharply on the appropriate age for granting certain responsibilities and privileges, such as control over clothing, school courses, and going out with friends (Collins et al., 1997). Parents typically say the young person is not ready for these signs of independence, whereas teenagers think they should have been granted long ago! Immigrant parents who place a high value on obedience to authority have great difficulty adapting to their teenagers' autonomy. Compared with nonimmigrant parents, they often react more strongly to adolescent disagreement. And when they do, their youngsters are more dissatisfied with daily life (Phinney & Ong, 2001).

■ A mother congratulates her 15-year-old daughter on winning a track meet event. Adolescent autonomy is effectively achieved in the context of warm parenting that relaxes control in accord with the young person's readiness for new responsibilities.

Throughout adolescence, the single most consistent predictor of mental health is the quality of the parent–child relationship (Steinberg & Silk, 2002). As teenagers move closer to adulthood, parents and children must balance connection and separation. In well-functioning families, teenagers remain attached to parents and seek their advice, but they do so in a context of greater freedom (Steinberg, 2001). The mild conflict that occurs along the way facilitates adolescent identity and autonomy by helping family members express and tolerate disagreement. Conflicts also inform parents of teenagers' changing needs and expectations, signaling that adjustments in the parent–child relationship are necessary. By middle to late adolescence, most parents and children achieve this mature, mutual relationship.

Family Circumstances

As Franca and Antonio's experience with Louis reminds us, difficulties at work as well as other life stresses can interfere with nurturant, involved child rearing and (in turn) with children's adjustment at any phase of development. However, maternal employment or a dual-earner family does not by itself reduce parental time with teenagers, nor is it harmful to adolescent development (Richards & Duckett, 1994). To the contrary, parents who are financially secure, invested in their work, and content with their marriages usually find it easier to grant teenagers appropriate autonomy (Seltzer & Ryff, 1994). When Franca and Antonio's work and financial pressures eased and they recognized Louis's need for more support and guidance, his problems subsided.

Less than 10 percent of families with adolescents have seriously troubled relationships—chronic and escalating levels of conflict and repeated arguments over serious issues. Of these, most have difficulties that began in childhood (Paikoff & Brooks-Gunn, 1991). Table 12.3 summarizes family circumstances considered in earlier chapters that pose challenges for adolescents. Teenagers who develop well despite family stresses continue to benefit from factors that fostered resiliency in earlier years: an appealing, easygoing disposition; a caring parent who combines warmth with high expectations; and (especially if parental supports are lacking) bonds to prosocial adults outside the family who care deeply about the adolescent's well-being (Masten, 2001).

Siblings

Like parent–child relationships, sibling interactions adapt to change at adolescence. As younger siblings mature and become more self-sufficient, they accept less direction from their older brothers and sisters. Consequently, the influence of older siblings declines. Sibling relationships also become less intense, in both positive and negative feelings. As teenagers become more involved in friendships and romantic relationships, they invest less time and energy in siblings, who are part of the family from which they are trying to establish autonomy (Hetherington, Henderson, & Reiss, 1999; Stocker & Dunn, 1994).

Despite a drop in companionship, attachment between siblings, like closeness to parents, remains strong for most young people. Brothers and sisters whose parents are warm and supportive and who established a positive bond in early childhood continue to display greater affection and caring during the teenage years (Bussell et al., 1999; Dunn, Slomkowski, & Beardsall, 1994).

Table 12.3 Family Circumstances with Implications for Adolescent Adjustment

Family Circumstance	To Review, Turn to . . .
Type of Family	
Adoptive	Chapter 2, pages 57–58
Divorced single-parent	Chapter 10, pages 330–331
Blended	Chapter 10, pages 331–333
Employed mother and dual-earner	Chapter 10, page 333
Family Conditions	
Child maltreatment	Chapter 8, page 268 Chapter 10, page 336
Economic hardship	Chapter 2, pages 61–62
Adolescent parenthood	Chapter 11, pages 359–360

Peer Relations

As adolescents spend less time with family members, peers become increasingly important. In industrialized nations, young people spend most of each weekday with agemates in school. Teenagers also spend much out-of-class time together, more in some cultures than others. For example, in one study American teenagers averaged 18 nonschool hours with peers, compared with 12 hours for Japanese and 9 hours for Taiwanese adolescents (Fuligni & Stevenson, 1995). Less demanding academic standards, which lead American youths to devote far less time than Asian youths to schoolwork, account for this difference. Schoolwork fills close to one-half of waking hours among Korean adolescents, one-third among Japanese adolescents, but only one-quarter among American adolescents (Larson & Verma, 1999).

In the following sections, we will see that adolescent peer relations can be both positive and negative. At their best, peers serve as critical bridges between the family and adult social roles.

Friendships

Adolescents report their most favorable moods when in the company of friends (Larson & Richards, 1991). The number of "best friends" declines from about four to six in early adolescence to one or two in adulthood (Hartup & Stevens, 1999). At the same time, the nature of the relationship changes. When asked about the meaning of friendship, teenagers stress two characteristics. The first, and most important, is *intimacy.* Adolescents seek psychological closeness, trust, and mutual understanding from their friends—the reason that self-disclosure (exchanges of private thoughts and feelings) between friends increases steadily over the adolescent years (see Figure 12.2). Second, more than younger children, teenagers want their

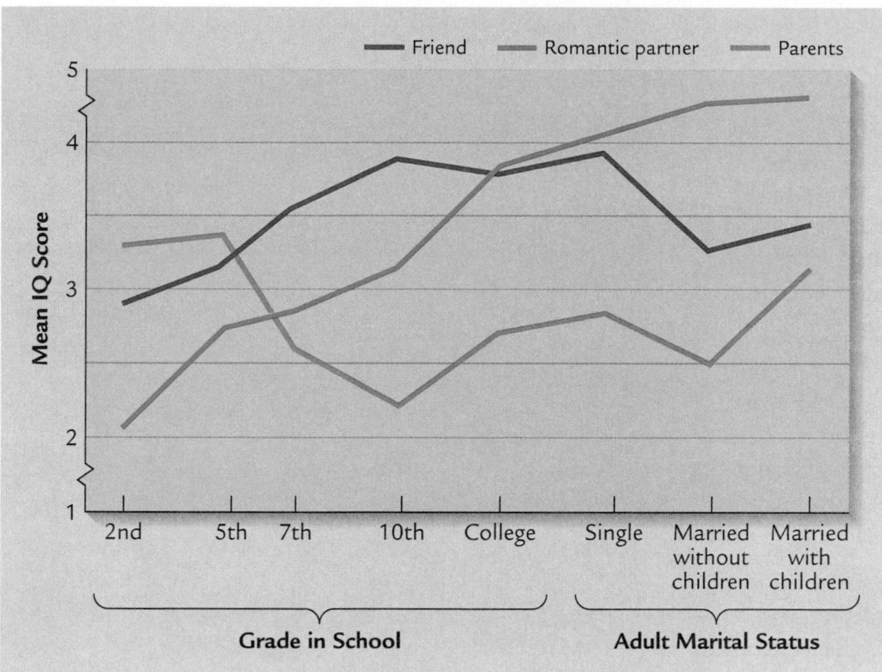

FIGURE 12.2 Age changes in reported self-disclosure to parents and peers, based on data from several studies. Self-disclosure to friends increases steadily during adolescence, reflecting intimacy as a major basis of friendship. Self-disclosure to romantic partners also rises. However, not until the college years does it surpass intimacy with friends. Self-disclosure to parents declines in early adolescence, a time of mild parent–child conflict. As family relationships readjust to the young person's increasing autonomy, self-disclosure to parents rises. (From D. Buhrmester, 1996, "Need Fulfillment, Interpersonal Competence, and the Developmental Contexts of Early Adolescent Friendship," in W. M. Bukowski, A. F. Newcomb, & W. W. Hartup, Eds., *The Company They Keep: Friendship During Childhood and Adolescence,* New York: Cambridge University Press, p. 168. Reprinted by permission.)

friends to be *loyal*—to stick up for them and not to leave them for somebody else (Buhrmester, 1996).

As frankness and faithfulness increase in friendships, teenagers get to know each other better as personalities. In addition to the many characteristics that school-age friends share (see Chapter 10, page 324), adolescent friends tend to be alike in identity status, educational aspirations, political beliefs, and willingness to try drugs and engage in lawbreaking acts. And they become more similar in these ways over time (Akers, Jones, & Coyl, 1998; Berndt & Keefe, 1995). At times, however, teenagers befriend agemates with differing attitudes and values, which permits them to explore new perspectives within the security of a compatible relationship. And during adolescence, cooperation and mutual affirmation between friends rise—changes that reflect greater effort and skill at preserving the relationship and increased sensitivity to a friend's needs and desires (Phillipsen, 1999).

Ask several adolescent girls and boys to describe their close friendships. You are likely to find a consistent sex difference. Emotional closeness is more common between girls than between boys (Markovitz, Benenson, & Dolensky, 2001). Girls frequently get together to "just talk, whereas boys more often gather in groups for an activity—usually sports and competitive games. When boys talk, they usually focus on their own or others' accomplishments (Buhrmester, 1996). Through friendship, girls typically focus on communal concerns, boys on achievement and status concerns. This does not mean that boys rarely form close friendship ties. They often do, but the quality of their friendships is more variable. The intimacy of boys' friendships is related to gender identity. Androgynous boys are just as likely as girls to form intimate same-sex ties, whereas boys with a "masculine" identity are less likely to do so (Jones & Dembo, 1989).

As long as adolescent friendships are not based on a common attraction to antisocial behavior, they are related to many aspects of psychological health and competence into early adulthood (Bagwell et al., 2001; Bukowski, 2001). The reasons are several.

- *Close friendships provide opportunities to explore the self and develop a deep understanding of another.* Through open, honest communication, adolescent friends become sensitive to each other's strengths and weaknesses, needs

During adolescence, intimacy and loyalty become defining features of friendship. Yet girls place a higher value on emotional closeness than do boys. Girls more often get together to "just talk," and they rate their friendships as higher in self-disclosure and emotional support.

and desires. This process supports the development of self-concept, perspective taking, and identity.

- *Close friendships provide a foundation for future intimate relationships.* Look again at Figure 12.2, and you will see that self-disclosure to friends precedes disclosure to romantic partners. Sexuality and romance are common topics of discussion between teenage friends—conversations that, along with the intimacy of friendship itself, may help adolescents establish and solve problems in romantic partnerships (Connolly & Goldberg, 1999).
- *Close friendships help young people deal with the stresses of adolescence.* Because supportive friendship enhances sensitivity to and concern for another, it increases the likelihood of empathy, sympathy, and prosocial behavior. As a result, anxiety and loneliness are reduced while self-esteem and sense of well-being are fostered.
- *Close friendships can improve attitudes toward and involvement in school.* When teenagers enjoy interacting with friends at school, perhaps they begin to view all aspects of school life more positively (Berndt & Keefe, 1995).

■ These members of a high school swim team form a crowd. To promote team spirit, they all paint themselves green before a competition. Unlike the more intimate clique, the larger, more loosely organized crowd grants adolescents an identity within the larger social structure of the school.

Cliques and Crowds

Friends also gather in *peer groups* (see Chapter 10), which become increasingly common and more tightly organized during adolescence. During the early teenage years, they are organized around **cliques,** small groups of about five to seven members who are good friends and, therefore, resemble one another in family background, attitudes, and values. At first, cliques are limited to same-sex members, but by midadolescence mixed groups become common. The cliques within a typical high school can be identified by their interests and social status, as the well-known "popular" and "unpopular" groups reveal (Cairns et al., 1995; Gillmore et al., 1997).

Often several cliques with similar values form a larger, more loosely organized group called a **crowd.** Unlike the more intimate clique, membership in a crowd is based on reputation and stereotype. It grants the adolescent an identity within the larger social structure of the school. Prominent crowds in a typical high school are the "brains," or nonathletes who enjoy academics; the "jocks," who are very involved in athletics; the "preppies," who are physically attractive, wear brand-name clothes, and are usually dating; the "partyers," who value socializing but care little about schoolwork; the "burnouts," who use drugs, skip school, and get into fights; and the "normals," average to good students who get along with peers in most other crowds (Kinney, 1999; Stone & Brown, 1999).

What influences the assortment of teenagers into cliques and crowds? In addition to personality and interests, family factors are important. In a study of 8,000 ninth to twelfth graders, adolescents who described their parents as authoritative were members of "brain," "jock," and "preppie" groups that accepted both the adult and peer reward systems of the school. In contrast, boys with permissive parents valued interpersonal relationships and aligned themselves with the "partyer" crowd. And teenagers who viewed their parents as uninvolved more often affiliated with "partyer" and "burnout" crowds, suggesting lack of identification with adult reward systems (Durbin et al., 1993).

These findings indicate that many peer-group values are extensions of values acquired at home. But once adolescents join a clique or crowd, it can modify their beliefs and behavior. However, the positive impact of having academically and socially skilled peers is greatest for teenagers whose own parents are authoritative. And the negative impact of having antisocial, drug-using friends is strongest for teenagers whose parents use less effective child-rearing styles (Mounts & Steinberg, 1995). In sum, family experiences affect the extent to which adolescents become like their peers over time.

In early adolescence, as interest in dating increases, boys' and girls' cliques come together. As mixed-sex cliques form and "hang out," they provide a supportive context for boys and girls to get to know each other. Cliques offer models for how to interact with the other sex and a chance to do so without having to be intimate. Gradually, the larger group divides into couples, several of whom spend time together, going to parties and movies. By late adolescence, boys and girls feel comfortable enough about approaching each other directly that the mixed-sex clique disappears (Connolly & Goldberg, 1999).

Crowds also decline in importance. As adolescents form personal values and goals, they no longer feel a need to broadcast, through dress, language, and preferred activities, who they are. And about half of young people switch crowds from tenth to twelfth grade, mostly in favorable directions. The "studious" and "normal" crowds grow and deviant crowds lose members as teenagers focus more on their future (Strouse, 1999). Both cliques and crowds serve vital functions. The clique provides a context for acquiring new social skills, and the crowd provides a temporary identity as adolescents separate from the family and construct a coherent sense of self.

■ As long as dating does not begin too soon, it extends the benefits of adolescent friendships. Besides fun and enjoyment, dating promotes sensitivity, empathy, and identity development as teenagers relate to someone whose needs differ from their own.

Dating

Although sexual interest is affected by the hormonal changes of puberty, the beginning of dating is regulated by cultural expectations. Western societies tolerate and even encourage romantic involvements between teens, which typically begin in junior high school (Larson, Clore, & Wood, 1999). When asked about their reasons for dating, younger teenagers more often mention recreation and achieving peer status. By late adolescence, as young people become ready for greater psychological intimacy, they look for someone who shares their interests, has clear goals for the future, and is likely to make a good permanent partner (Roscoe, Diana, & Brooks, 1987).

The achievement of intimacy in dating relationships typically lags behind that of friendships. Perhaps because communication between boys and girls remains stereotyped and shallow through midadolescence, early dating does not foster social maturity. Instead, it is related to drug use, delinquency, and poor academic achievement (Brown, Feiring, & Furman, 1999). These factors, along with a history of abusive family relationships, also are linked to dating violence (Pepler et al., 2002; Wolfe et al., 2001). Sticking with group activities, such as parties and dances, before becoming involved with a steady boyfriend or girlfriend is best for young teenagers.

Homosexual youths face special challenges in initiating and maintaining visible romances. Their first dating relationships seem to be short-lived and to involve little emotional commitment for reasons different from those of heterosexuals: They fear peer harassment and rejection. Recall from Chapter 11 that because of intense prejudice, homosexual adolescents often retreat into heterosexual dating. In addition, many have difficulty finding a same-sex partner because their homosexual peers have not yet come out. Often their first contacts with other sexual minority youths occur in support groups, where they are free to date publicly (Diamond, Savin-Williams, & Dubé, 1999). Homosexuals with a high proportion of same-sex romances are usually "out" to peers. Yet a sense of isolation from the larger peer world frequently means that dating partners place unreasonable demands on each other for fulfilling all social needs (Savin-Williams, 1996).

As long as it does not begin too soon, dating provides lessons in cooperation, etiquette, and dealing with people in a wider range of situations. As teenagers form a close emotional tie, sensitivity, empathy, social support, and identity development are enhanced (Connolly et al., 1999; Furman, 2002). About half of first romances do not survive high school graduation, and those that do become less satisfying (Shaver, Furman, & Buhrmester, 1985). Because young people are still forming their identities, those who like each other at one point in time often find that they have little in common later.

Peer Conformity

When Franca and Antonio discovered Louis's lawbreaking during his freshman year of high school, they began to worry (as many parents do) about the negative side of adolescent peer networks. Although conformity to peer pressure is greater during adolescence than in childhood or early adulthood, it is a complex process that varies with the young person's age and need for social approval and with the situation.

In one study of nearly 400 junior and senior high school students, adolescents felt greatest pressure to conform to the most obvious aspects of the peer culture—dress, grooming, and participation in social activities, such as dating and going to parties and school dances. Peer pressure to engage in proadult behavior, such as getting good grades and cooperating with parents, was also strong. Although pressure toward misconduct rose in early adolescence, it was low. Many teenagers said that their friends actively discouraged antisocial acts. These findings show that peers and parents often act in concert, toward desirable ends! Finally, peer pressures correlated only modestly with teenagers' actual values and behaviors (Brown, Lohr, & McClenahan, 1986).

Perhaps because of greater concern with what their friends think of them, early adolescents are more likely than younger or older individuals to give in to peer pressure (Brown, Clasen, & Eicher, 1986). Yet when parents and peers disagree, even young teenagers do not consistently rebel against the family. Instead, parents and peers differ in their spheres of greatest influence. Parents have more impact on teenagers' basic life values and educational plans (Steinberg, 2001). Peers are more influential in short-term day-to-day matters, such as dress, music, and choice of friends. Adolescents' personal characteristics also make a difference. Young people who feel competent and worthwhile are less likely to fall in line behind peers.

Finally, authoritative child rearing is related to resistance to unfavorable peer pressure. Teenagers whose parents are supportive and exert appropriate oversight respect their parents, an attitude that acts as an antidote to unfavorable peer pressure (Masten, 2001; Sim, 2000). In contrast, adolescents who experience extremes of parental behavior—either too much or too little control—tend to be highly peer oriented (Mason et al., 1996; Pettit et al., 1999). They more often rely on friends for advice about their personal lives and futures and are more willing to break their parents' rules, ignore their schoolwork, use drugs, commit delinquent acts, and engage in other problem behaviors.

Ask Yourself

REVIEW

What type of parenting fosters competence in adolescence? Explain why that parenting style is effective, and cite its many positive outcomes.

REVIEW

Cite the distinct, positive functions of friendships and group ties in adolescence. What factors lead some friendships and peer groups to have harmful consequences?

APPLY

Thirteen-year-old Mattie's parents are warm, firm in their expectations, and consistent in monitoring her activities. What type of crowd is Mattie likely to join, and why?

CONNECT

How might gender intensification contribute to the shallow quality of early adolescent dating relationships?

www.

Problems of Development

Although most young people move through adolescence with little difficulty, we have seen that some encounter major disruptions in development, such as premature parenthood, substance abuse, and school failure. Our discussion has also shown that psychological and behavior problems cannot be explained by any single factor. Instead, biological and psychological change, families, schools, peers, communities, and culture act together to produce particular outcomes. This theme is apparent in three additional problems of the teenage years: depression, suicide, and delinquency.

Depression

Depression—feeling sad, frustrated, and hopeless about life, accompanied by loss of pleasure in most activities and disturbances in sleep, appetite, concentration, and energy—is the most common psychological problem of adolescence. About 15 to 20 percent of teenagers have had one or more major depressive episodes, a rate comparable to that of adults. From 2 to 8 percent are chronically depressed—gloomy and self-critical for many months and sometimes years (Birmaher et al., 1996; Kessler et al., 1994). Depression is not absent in childhood. Yet as Figure 12.3 shows, depressive symptoms increase around the time of puberty. They occur twice as often in adolescent girls as in adolescent boys—a difference sustained throughout the lifespan (Nolen-Hoeksema, 2001).

■ Depression in teenagers should not be dismissed as a temporary side effect of puberty. Because adolescent depression can lead to long-term emotional problems, it deserves to be taken seriously. Without treatment, depressed teenagers have a high likelihood of becoming depressed adults.

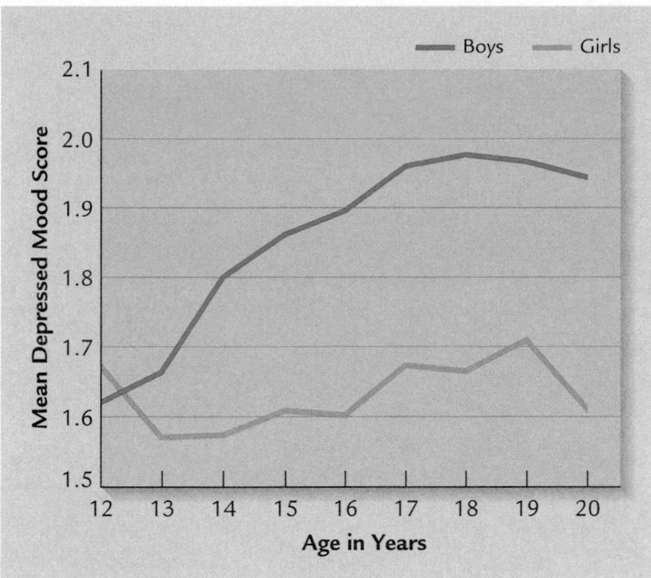

■ **FIGURE 12.3 Change in depressive symptoms from age 12 to 20 in a cross-sectional study of over 12,000 Norwegian adolescents.** Girls showed a more rapid rise in depression around the time of puberty than did boys. Similar trends occur in other industrialized nations. (From L. Wichstrøm, 1999, "The Emergence of Gender Difference in Depressed Mood During Adolescence: The Role of Intensified Gender Socialization," *Developmental Psychology, 35,* p. 237. Copyright © 1999 by the American Psychological Association. Reprinted by permission of the publisher and author.)

Depression prevents young people from mastering critical developmental tasks. Adolescent depression disrupts identity development. It is also associated with persistent anxiety, poor school performance, drug abuse, lawbreaking, and auto accidents. Without treatment, depressed teenagers are likely to become depressed adults with persistent problems in work, family life, and social life (Weissman et al., 1999).

Unfortunately, teachers and parents tend to minimize the seriousness of adolescents' depressive symptoms. Because of the popular stereotype of adolescence as a period of storm and stress, many adults interpret depression as just a passing phase. Depression also is hard to recognize in teenagers because they manifest it in diverse ways. Some engage in excessive brooding, worries about their health, and restless, undirected behavior. Others act it out by running away or behaving rebelliously. The first pattern is more typical of girls, the second more characteristic of boys (Gjerde, 1995).

● **Factors Related to Depression.** Combinations of biological and environmental factors lead to depression; the precise blend differs from one individual to the next. Kinship studies reveal that heredity plays an important role. Genes can induce depression by affecting the chemical balance in the brain, the development of brain regions involved in inhibiting negative emotion, or the body's hormonal response to stress (Cicchetti & Toth, 1998).

But experience can also activate depression, promoting any of the biological changes just described. Parents of depressed children and adolescents have a high incidence of depression and other psychological disorders. Although a genetic risk may be passed from parent to child, in earlier chapters we saw that depressed or otherwise stressed parents often engage in maladaptive parenting. As a result, their child's emotional self-regulation, attachment, and self-esteem may be impaired, with serious consequences for many cognitive and social skills (Garber, Braafladt, & Weiss, 1995; Garber et al., 1991).

Depressed youths usually display a learned-helpless attributional style (see Chapter 10) in which they view positive outcomes in school performance and peer relations as beyond their control. Consequently, numerous events can spark depression in a vulnerable young person—for example, failing at something important, parental divorce, the end of a close friendship or romantic partnership, or the challenges of school transition.

● **Sex Differences.** Why are girls more prone to depression? We know that biological changes associated with puberty are not responsible, since the gender gap is limited to industrialized nations. In developing countries, rates of depression are similar for males and females and occasionally higher in males (Culbertson, 1997). And even in developed nations, the size of the gender difference varies greatly. For example, it is smaller in China than in North America, perhaps because of decades of efforts by the Chinese government to eliminate gender inequalities (Greenberger et al., 2000).

Instead, stressful life events and gender-typed coping styles seem to be responsible. Early-maturing girls are especially prone to depression (see Chapter 11). And the gender intensification of early adolescence strengthens girls' passivity and dependency—maladaptive approaches to the tasks expected of teenagers in complex cultures. Consistent with this explanation, adolescents who identify strongly with "feminine" traits are more depressed, regardless of their sex (Wichstrøm, 1999). Girls who repeatedly feel overwhelmed develop an overly reactive physiological stress response and cope more poorly with challenges in the future (Nolen-Hoeksema, 2001). In this way, stressful experiences and stress reactivity feed on one another, sustaining depression.

Suicide

Profound depression can lead to suicidal thoughts, which all too often are translated into action. When a teenager tries to take his or her own life, or succeeds in doing so, depression is one of the factors that precedes it.

● **Factors Related to Adolescent Suicide.** The suicide rate increases over the lifespan. As Figure 12.4 on page 402 shows, it is lowest in childhood and highest in old age, but it jumps sharply at adolescence. Currently, suicide is the third-leading cause of death among American youths (after motor vehicle collisions and homicides) and the second-leading cause among Canadian youths (after motor vehicle collisions). Adolescent suicide has doubled in Canada and tripled

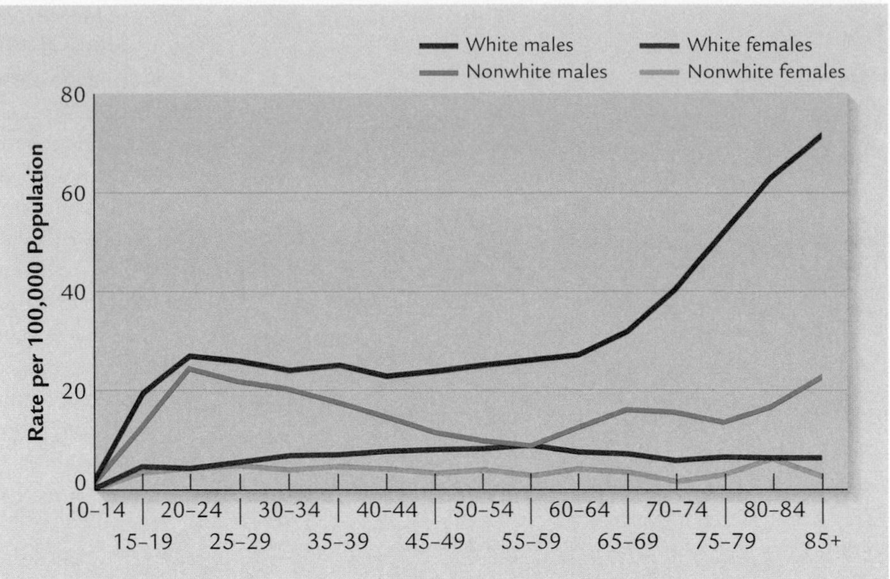

■ **FIGURE 12.4 Suicide rates over the lifespan in the United States.** Although teenagers do not commit suicide as often as adults and the aged, the suicide rate rises sharply from childhood to adolescence. Rates are greater for males than females and for white majority than nonwhite ethnic minority individuals. (From U.S. Bureau of the Census, 2002c.)

in the United States over the past 30 years; the greatest rise has occurred among young teenagers. At the same time, adolescent suicide varies widely among industrialized nations. Japan and most of Western Europe have low rates; Australia, Canada, and the United States have intermediate rates; and Estonia, Finland, and New Zealand have high rates (Johnson, Krug, & Potter, 2000). Although many theories exist, international differences remain unexplained.

Striking sex differences in suicidal behavior exist. The number of boys who kill themselves exceeds the number of girls by 4 or 5 to 1. This may seem surprising, given that girls show higher rates of depression. Yet the findings are not inconsistent. Girls make more unsuccessful suicide attempts and use methods from which they are more likely to be revived, such as a sleeping pill overdose. In contrast, boys tend to select more active techniques that lead to instant death, such as firearms or hanging. Gender-role expectations may be responsible. Less tolerance exists for feelings of helplessness and failed efforts in males than in females (Canetto & Sakinofsky, 1998).

Compared with the white majority, American ethnic minority teenagers, including African Americans and Hispanics, have lower suicide rates—a difference that increases in adulthood. Higher levels of support through extended families may be responsible. However, suicide among African-American adolescent males has recently risen; currently, it approaches that of Caucasian-American males. And Native American and Canadian Aboriginal youths commit suicide at rates two to seven times the national average. These trends are believed to be caused by high rates of profound family poverty, school failure, alcohol and drug use, and depression (Strickland, 1997). Gay, lesbian, and bisexual youths also are at high risk for suicide, making attempts three times more often than other adolescents. Those who have tried to kill themselves report more family conflict and peer victimization due to their sexual orientation (Hershberger, Pilkington, & D'Augelli, 1997).

Suicide tends to occur in two types of young people. In the first group are highly intelligent adolescents who are solitary, withdrawn, and unable to meet their own high standards or those of important people in their lives. A second, larger group shows antisocial tendencies. These young people express their despondency through bullying, fighting, stealing, and increased risk taking and drug abuse (Fergusson, Woodward, & Horwood, 2000). Besides being hostile and destructive toward others, they turn their anger and disappointment inward.

Parental emotional problems and family turmoil are common in the backgrounds of suicidal teenagers (Gould & Kramer, 2001). Their fragile self-esteem disintegrates in the face of stressful life events. Common circumstances just before a suicide include the breakup of an important peer relationship and the humiliation of having been caught engaging in irresponsible, antisocial acts.

Why is suicide on the rise in adolescence? Teenagers' improved ability to plan ahead seems to be involved. Although some act impulsively, many young people take purposeful steps toward killing themselves (McKeown et al., 1998). Other cognitive changes also contribute. Belief in the personal fable leads many depressed young people to conclude that no one could possibly understand the intense pain they feel. As a result, their despair, hopelessness, and isolation deepen. Warning signs of suicide are listed in Table 12.4.

● **Prevention and Treatment.** Picking up on the signals that a troubled teenager sends is a crucial first step in suicide prevention. Parents and teachers must be trained to see warning signs. Schools can help by providing sympathetic counselors, peer support groups, and information about telephone hot lines. Once a teenager takes steps toward suicide, staying with the young person, listening, and expressing sympathy and concern until professional help can be obtained are essential.

Table 12.4 Warning Signs of Suicide

Efforts to put personal affairs in order—smoothing over troubled relationships, giving away treasured possessions
Verbal cues—saying goodbye to family members and friends, making direct or indirect references to suicide ("I won't have to worry about these problems much longer"; "I wish I were dead")
Feelings of sadness, despondency, "not caring" anymore
Extreme fatigue, lack of energy, boredom
No desire to socialize; withdrawal from friends
Easily frustrated
Emotional outbursts—spells of crying or laughing, bursts of energy
Inability to concentrate, distractible
Decline in grades, absence from school, discipline problems
Neglect of personal appearance
Sleep change—loss of sleep or excessive sleepiness
Appetite change—eating more or less than usual
Physical complaints—stomachaches, backaches, headaches

Treatments for depressed and suicidal adolescents range from antidepressant medication to individual, family, and group therapy. Sometimes hospitalization is necessary to ensure the teenager's safety. Until the adolescent improves, parents are usually advised to remove weapons, knives, razors, scissors, and drugs from the home. On a broader scale, gun control legislation that limits adolescents' access to the most frequent and deadly suicide method in the United States would greatly reduce both the number of suicides and the high teenage homicide rate.

After a suicide, family and peer survivors need support to assist them in coping with grief, anger, and guilt for not having been able to help the victim. Teenage suicides often take place in clusters. When one occurs, it increases the likelihood of others among peers who knew the young person or heard about the death through the media (Grossman & Kruesi, 2000). In view of this trend, an especially watchful eye must be kept on vulnerable adolescents after a suicide happens. Restraint by journalists is also important, since publicity increases suicide contagion (Velting & Gould, 1997).

Delinquency

Juvenile delinquents are children or adolescents who engage in illegal acts. Although North American youth crime has declined since the mid-1990s, young people under age 18 continue to account for a substantial proportion of police arrests—about 17 percent the United States and 23 percent in Canada (Statistics Canada, 2002b; U.S. Department of Justice, 2002). When teenagers are asked directly and confidentially about lawbreaking, almost all admit that they are guilty of an offense of one sort or another (Farrington, 1987). Most of the time, they do not commit major crimes. Instead, they engage in petty stealing and disorderly conduct.

Both police arrests and self-reports show that delinquency rises over the early teenage years, remains high during middle adolescence, and then declines into young adulthood. What accounts for this trend? Recall that the desire for peer approval increases antisocial behavior among young teenagers. Over time, peers become less influential, moral reasoning matures, and young people enter social contexts (such as marriage, work, and career) that are less conducive to lawbreaking.

For most adolescents, a brush with the law does not forecast long-term antisocial behavior. But repeated arrests are cause for concern. Teenagers are responsible for 16 percent of violent crimes in the United States and 8 percent in Canada (Statistics Canada, 2002b; U.S. Department of Justice, 2002). A small percentage commit most of these crimes, developing into recurrent offenders. Some enter a life of crime. As the Lifespan Vista box on page 404 reveals, childhood-onset conduct problems are far more likely to persist than conduct problems emerging during the teenage years.

● Factors Related to Delinquency. In adolescence, the gender gap in overt aggression widens (Chesney-Lind, 2001). Depending on the estimate, about three to eight times as many boys as girls commit major crimes. Although SES and ethnicity are strong predictors of arrests, they are only mildly related to teenagers' self-reports of antisocial acts. The difference is due to the tendency to arrest, charge, and punish low-SES ethnic minority youths more often than their higher-SES white and Asian counterparts (Elliott, 1994).

Difficult temperament, low intelligence, poor school performance, peer rejection in childhood, and association with antisocial peers are linked to delinquency. How do these factors fit together? One of the most consistent findings about delinquent youths is that their families are low in warmth, high in conflict, and characterized by inconsistent discipline. Because parental divorce and remarriage often lead to these conditions, boys experiencing family transitions are especially prone to delinquency (Pagani et al., 1999). And youth crime peaks on weekdays between 2 and 8 o'clock, when many teenagers are unsupervised (U.S. Department of Justice, 2002).

Our discussion on page 257 of Chapter 8 explained how ineffective parenting can promote and sustain children's aggression. Boys are more likely than girls to be targets of angry, inconsistent discipline because they are more active and impulsive and therefore harder to control. When children extreme in these characteristics are exposed to inept parenting, aggression rises during childhood, is transformed into violent offenses by adolescence, and persists into adulthood (refer again to the Lifespan Vista box on page 404).

Delinquency increases under certain environmental conditions. Teenagers commit more crimes in poverty-stricken

A Lifespan Vista

Two Routes to Adolescent Delinquency

Persistent adolescent delinquency follows two paths of development, one with an onset of conduct problems in childhood, the second with an onset in adolescence. Research reveals that the early-onset type is far more likely to lead to a life-course pattern of aggression and criminality. The late-onset type usually does not persist beyond the transition to young adulthood (Farrington & Loeber, 2000).

Childhood-onset and adolescent-onset youths both engage in serious offenses; associate with deviant peers; participate in substance abuse, unsafe sex, and dangerous driving; and spend time in correctional facilities. Why does antisocial activity more often continue and escalate into violence in the first group than in the second? Longitudinal research extending from childhood into early adulthood sheds light on this question. So far, investigations have focused only on boys because of their greater delinquency.

Early-Onset Type. A difficult temperament distinguishes these boys; they are emotionally negative, restless, and willful as early as age 3. In addition, they show subtle deficits in cognitive functioning that seem to contribute to disruptions in the development of language, memory, and cognitive and emotional self-regulation (Loeber et al., 1999; Moffitt et al., 1996). Some have attention-deficit hyperactivity disorder (ADHD), which compounds their problems with learning and self-control (see Chapter 9, page 288) (White et al., 1996).

Yet these biological risks are not sufficient to sustain antisocial behavior, since most early-onset boys do not display serious delinquency followed by adult criminality. Among those who follow the life-course path, inept parenting transforms their undercontrolled style into hostility and defiance. As they fail academically and are rejected by peers, they befriend other deviant youths, who provide the attitudes and motivations for violent behavior (see Figure 12.5). Compared with their adolescent-onset counterparts, early-onset teenagers feel distant from their families and leave school early (Moffitt et al., 1996). Their limited cognitive and social skills result in high rates of unemployment, contributing further to their antisocial involvements.

Late-Onset Type. Other youths begin to display antisocial behavior around the time of puberty, gradually increasing their involvement. Their conduct problems arise from the peer context of early adolescence, not from biological deficits and a history of unfavorable development. For some, quality of parenting may decline for a time, perhaps due to family stresses or the challenges of disciplining an unruly teenager. When age brings gratifying adult privileges, they draw on prosocial skills mastered before adolescence and give up their antisocial ways (Moffitt et al., 1996).

A few late-onset youths, however, continue to engage in antisocial acts. The seriousness of their adolescent offenses seems to trap them in situations that close off opportunities for responsible behavior. In one study, finding a steady, well-paying job and entering a happy marriage led to a large reduction in repeat offending. In contrast, the longer antisocial young people spent in prison, the more likely they were to sustain a life of crime (Sampson & Laub, 1993).

These findings suggest a need for a fresh look at policies aimed at stopping youth crime. Keeping adolescent and young adult offenders locked up for many years disrupts their vocational and marital lives during a crucial period of development, committing them to a bleak future.

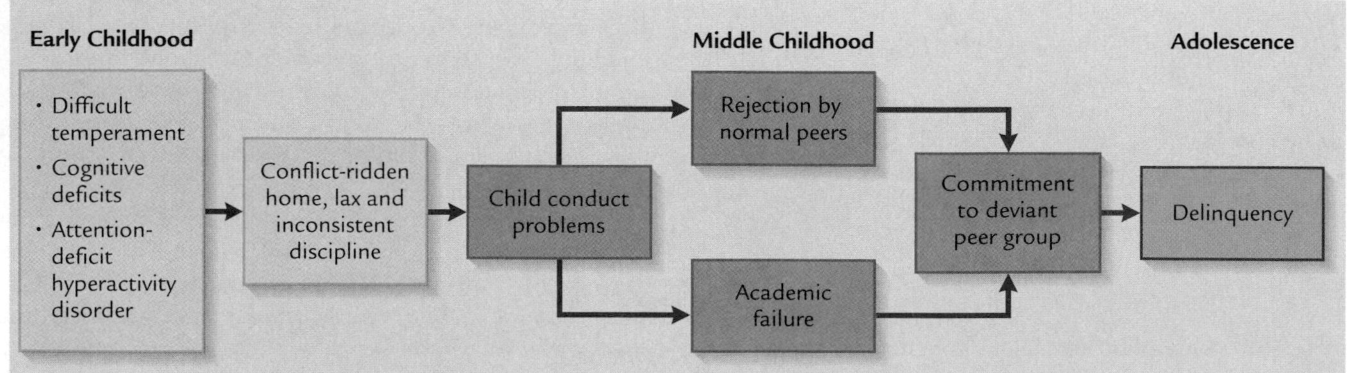

■ **FIGURE 12.5 Path to chronic delinquency for adolescents with childhood-onset antisocial behavior.** Difficult temperament and cognitive deficits characterize many of these youths in early childhood; some have attention-deficit hyperactivity disorder. Inept parenting transforms biologically based self-control difficulties into hostility and defiance. (Adapted from Patterson, DeBaryshe, & Ramsey, 1989.)

© BARBARA BURNES/PHOTO RESEARCHERS, INC.

■ Delinquency rises during the early teenage years, remains high during middle adolescence, and then declines. Although most of the time it involves petty stealing and disorderly conduct, a small percentage of young people engage in repeated, serious offenses and are at risk for a life of crime.

neighborhoods with limited recreational and employment opportunities and high adult criminality. In such neighborhoods, adolescents have easy access to deviant peers, drugs, and firearms and are especially likely to be recruited into antisocial gangs, whose members commit the vast majority of violent delinquent acts (Thornberry, 1998). Furthermore, schools in these locales typically fail to meet students' developmental needs. Large classes, weak instruction, and rigid rules are associated with higher rates of lawbreaking, even after other influences are controlled (Hawkins & Lam, 1987).

● **Prevention and Treatment.** Because delinquency has roots in childhood and results from events in several contexts, prevention must start early and take place at multiple levels. Authoritative parenting, high-quality teaching in schools, and communities with healthy economic and social conditions go a long way toward reducing adolescent criminality.

Treating serious offenders also requires an approach that recognizes the multiple determinants of delinquency. So far as possible, adolescents are best kept in their own homes and communities to increase the possibility that treatment changes will transfer to their daily lives. Many treatment models exist, including individual therapies, halfway houses, day treatment centers, special classrooms, work experience programs, and summer camps. Those that work best are lengthy and intensive and use problem-focused methods that teach cognitive and social skills needed to overcome family, peer, and school difficulties (Wasserman & Miller, 1998).

Yet even these multidimensional treatments can fall short if young people remain embedded in hostile home lives, antisocial peer groups, and violent neighborhoods. Intensive efforts to create nonaggressive environments—at the family, community, and cultural levels—are needed to support interventions for delinquent youths and to foster healthy development for all young people.

Ask Yourself

REVIEW

Why are adolescent girls at greater risk for depression and adolescent boys at greater risk for suicide?

APPLY

Throughout childhood, Mac had difficulty learning and picked fights with peers. At age 16, he was arrested for assault. Zeke had been a well-behaved child, but at age 16, he was arrested for property damage. Which boy is more likely to become a long-term offender, and why?

CONNECT

Reread the section on adolescent pregnancy and parenthood in Chapter 11. What factors does this problem have in common with suicide and delinquency?

www.

Summary

Erikson's Theory: Identity versus Identity Confusion

According to Erikson, what is the major personality achievement of adolescence?

■ Erikson's theory regards **identity** as the major personality achievement of adolescence. Young people who successfully resolve the psychological conflict of **identity versus identity confusion** construct a solid self-definition consisting of self-chosen values and goals.

Self-Understanding

Describe changes in self-concept and self-esteem during adolescence.

■ Cognitive changes lead adolescents' self-descriptions to become more organized and consistent, and personal and moral values appear as key themes. New dimensions of self-esteem are also added.

■ For most young people, self-esteem rises over the teenage years. Young people vary widely in their self-esteem profiles. Of those whose self-esteem drops, most are girls. Authoritative parenting and schools and neighborhoods that respect the young person's SES and ethnic group support positive self-esteem.

Describe the four identity statuses, along with factors that promote identity development.

■ In complex societies, a period of exploration is necessary to form a personally meaningful identity. **Identity achievement** and **moratorium** are psychologically healthy identity statuses. Long-term **identity foreclosure** and **identity diffusion** are related to adjustment difficulties.

- Adolescents who have a flexible, open-minded approach to grappling with competing beliefs and values and who feel attached to parents but free to voice their own opinions are likely to be advanced in identity development. Close friends assist young people in exploring options. Schools and communities that provide rich and varied opportunities also foster identity achievement. Ethnic minority youths who construct a **bicultural identity** are advantaged in many aspects of emotional and social development.

Moral Development

Describe Piaget's theory of moral development and Kohlberg's extension of it, and evaluate the accuracy of each.

- Piaget identified two stages of moral understanding: (1) **heteronomous morality,** in which moral rules are viewed as fixed dictates of authority figures, and (2) **autonomous morality,** in which rules are seen as flexible, socially agreed-on principles. Although Piaget's theory describes the general direction of moral development, it underestimates young children's moral capacities.

- According to Kohlberg, moral development is a gradual process that extends into adulthood. Moral reasoning advances through three levels, each of which contains two stages: (1) the **preconventional level,** in which morality is viewed as controlled by rewards, punishments, and the power of authority figures; (2) the **conventional level,** in which conformity to laws and rules is regarded as necessary to preserve positive human relationships and societal order; and (3) the **postconventional level,** in which individuals develop abstract, universal principles of justice. The influence of situational factors on moral judgment suggests that Kohlberg's moral stages are best viewed as a loosely organized sequence.

Evaluate claims that Kohlberg's theory does not adequately represent the morality of females.

- Kohlberg's theory does not underestimate the morality of females. Instead, justice and caring moralities coexist but vary in emphasis between males and females, especially when they are questioned about real-life moral dilemmas.

Describe influences on moral reasoning and the relationship of moral reasoning to behavior.

- Many experiences contribute to moral maturity, including warm, rational child-rearing practices, years of schooling, and peer discussions of moral issues. Young people in industrialized nations advance to higher levels of moral understanding than young people in village societies. Kohlberg's highest stages may represent a culturally specific rather than a universal morality.

- As individuals advance to higher stages, moral reasoning and behavior come closer together. Many other factors, including the emotions of empathy and guilt, the individual's history of morally relevant experiences, and the extent to which morality is central to self-concept, affect moral action.

Gender Typing

Why is early adolescence a period of gender intensification?

- **Gender intensification** occurs in early adolescence for several reasons. Physical and cognitive changes prompt young teenagers to view themselves in gender-linked ways, and gender-typed pressures from parents and peers increase. Teenagers who eventually build an androgynous gender identity show better psychological adjustment.

The Family

Discuss changes in parent–child and sibling relationships during adolescence.

- Effective parenting of adolescents requires an authoritative style that strikes a balance between connection and separation. Adapting family interaction to meet adolescents' need for **autonomy** is especially challenging. As teenagers de-idealize their parents, they often question parental authority. Because both adolescents and parents are undergoing major life transitions, they approach situations from different perspectives.

- Parents who are financially secure, invested in their work, and content with their marriages usually find it easier to grant teenagers support, guidance, and appropriate autonomy. When parents and adolescents have seriously troubled relationships, the difficulties usually began in childhood.

- Sibling relationships become less intense as adolescents separate from the family and turn toward peers. Still, attachment to siblings remains strong for most young people.

Peer Relations

Describe adolescent friendships, peer groups, and dating relationships and their consequences for development.

- During adolescence, the nature of friendship changes, moving toward greater intimacy and loyalty. Girls' friendships place greater emphasis on emotional closeness, boys' on status and mastery. As long as they are not based on attraction to antisocial behavior, adolescent friendships promote self-concept, perspective taking, identity, and the capacity for intimate relationships. They also help young people deal with stress and can foster improved attitudes toward school.

- Adolescent peer groups are organized into **cliques,** small groups of friends with similar attitudes, values, interests, and social status. Often several cliques form a larger, more loosely organized group called a **crowd** that grants the adolescent an identity within the larger social structure of the school. Parenting styles influence the assortment of teenagers into cliques and crowds. As mixed-sex cliques form, they provide a supportive context for boys and girls to get to know one another.

- Intimacy in dating relationships lags behind that of same-sex friendships. Because of prejudice, initiating and maintaining visible romances is especially challenging for homosexual youths. For all teenagers, first romances serve as practice for later, more mature bonds. They generally dissolve or become less satisfying after graduation from high school.

Discuss conformity to peer pressure in adolescence.

- Peer conformity is greater during adolescence than at younger or older ages. Young teenagers are most likely to give in to peer pressure for antisocial behavior. Yet most peer pressures are not in conflict with important adult values. Authoritative parenting is related to resistance to unfavorable peer pressure.

Problems of Development

What factors are related to adolescent depression and suicide?

- Depression is the most common psychological problem of the teenage years. Adolescents who are severely depressed are likely to remain so as adults. Heredity contributes to depression, but stressful life events are necessary to trigger it. Depression is more common in girls than in boys—a difference believed to be due to stressful life events and gender-typed coping styles.

- Profound depression often leads to suicidal thoughts. The suicide rate increases dramatically at adolescence. Boys account for most teenage suicides because their efforts usually succeed. Girls make more unsuccessful attempts. Teenagers at risk for suicide are sometimes intelligent, solitary, and withdrawn; more often, they are antisocial. Family turmoil is common in the backgrounds of suicidal adolescents, who react intensely to loss, failure, or humiliation.

Discuss factors related to delinquency.

- Although almost all teenagers engage in some delinquent activity, only a few are serious repeat offenders. Most of these are boys with a childhood history of conduct problems. Many factors are related to delinquency, but one of the most consistent is a family environment low in warmth, high in conflict, and characterized by inconsistent discipline. Poverty-stricken neighborhoods with high crime rates and schools that fail to meet adolescents' developmental needs also promote adolescent lawbreaking.

Important Terms and Concepts

autonomous morality (p. 388)
autonomy (p. 394)
bicultural identity (p. 386)
clique (p. 398)
conventional level (p. 389)
crowd (p. 398)
ethnic identity (p. 386)
gender intensification (p. 394)
heteronomous morality (p. 388)
identity (p. 382)
identity achievement (p. 384)
identity diffusion (p. 384)
identity foreclosure (p. 384)
identity versus identity confusion (p. 382)
moratorium (p. 384)
postconventional level (p. 390)
preconventional level (p. 389)

FYI For Further Information and Help

Consult the Companion Website for *Development Through the Lifespan, Third Edition,* (www.ablongman.com/berk) where you will find the following resources for this chapter:

- **Chapter Objectives**
- **Flashcards** for studying important terms and concepts
- **Annotated Weblinks** to guide you in further research
- **Ask Yourself** questions, which you can answer and then check against a sample response
- **Suggested Readings**
- **Practice Test** with immediate scoring and feedback

Milestones

Development in Adolescence

AGE

11–14 years

PHYSICAL

- If a girl, reaches peak of growth spurt. (348)
- If a girl, adds more body fat than muscle. (346)
- If a girl, starts to menstruate. (348)

- If a boy, begins growth spurt. (348)
- If a boy, starts to ejaculate seminal fluid. (348)
- Likely to become aware of sexual orientation. (356)
- If a girl, motor performance gradually increases and then levels off. (347)

COGNITIVE

- Becomes capable of formal operational reasoning. (363)
- Becomes better at coordinating theory with evidence. (365)
- Can argue more effectively. (367)
- Becomes more self-conscious and self-focused. (367)
- Becomes more idealistic and critical. (368)
- Metacognition and cognitive self-regulation continue to improve. (365)

EMOTIONAL/SOCIAL

- Moodiness and parent-child conflict increase. (350)
- Is likely to show increased gender stereotyping of attitudes and behavior. (394)
- Spends less time with parents and siblings. (394–396)
- Spends more time with peers. (396)

- Friendships are based on intimacy and loyalty. (396–397)
- Peer groups become organized around cliques. (398)
- Cliques with similar values form crowds. (398)
- Conformity to peer pressure increases. (399–400)

AGE

15–20 years

PHYSICAL

- If a girl, completes growth spurt. (348)
- If a boy, reaches peak and then completes growth spurt. (348)
- If a boy, voice deepens. (348)
- If a boy, adds muscle while body fat declines. (346–347)
- May have had sexual intercourse. (354)
- If a boy, motor performance increases dramatically. (347)

COGNITIVE

- Is likely to show formal operational reasoning on familiar tasks. (365)
- Masters the components of formal operational reasoning in sequential order on different types of tasks. (366)
- Becomes less self-conscious and self-focused. (368)
- Becomes better at everyday planning and decision making. (368)

EMOTIONAL/SOCIAL

- Combines features of the self into an organized self-concept. (383)
- Self-esteem differentiates further. (383)
- Self-esteem tends to rise. (383)
- Is likely to be searching for an identity. (384)
- Is likely to engage in societal perspective taking. (390)
- Is likely to have a conventional moral orientation. (389)

- Gender-stereotyped attitudes and behavior may decline. (394)
- Importance of cliques and crowds declines. (398)
- Has probably started dating. (399)

- Conformity to peer pressure may decline. (399–400)

Note: Numbers in parentheses indicate the page(s) on which each milestone is discussed.

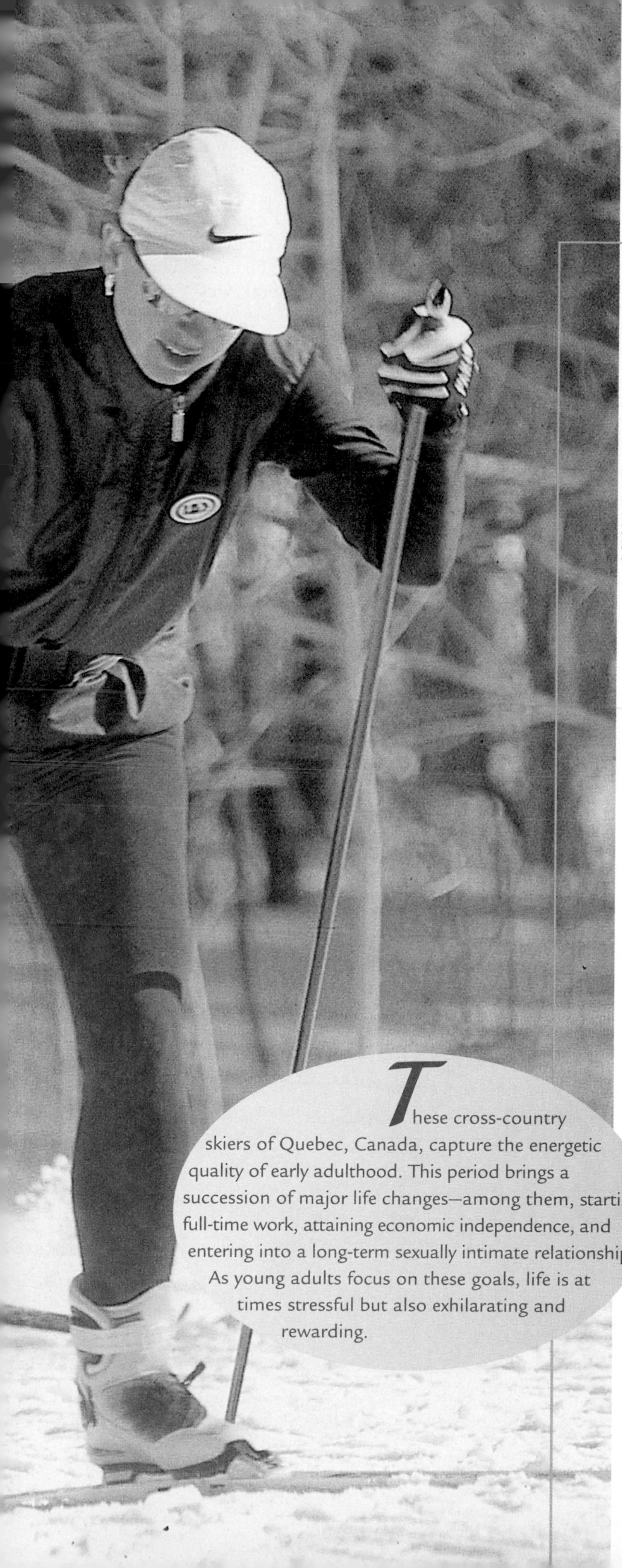

13

Physical and Cognitive Development in Early Adulthood

Physical Development

Biological Aging Begins in Early Adulthood
Aging at the Level of DNA and Body Cells • Aging at the Level of Organs and Tissues

Physical Changes
Cardiovascular and Respiratory Systems • Motor Performance • Immune System • Reproductive Capacity

Health and Fitness
Nutrition • Exercise • Substance Abuse • Sexuality • Psychological Stress

- A Lifespan Vista: The Obesity Epidemic: How Americans Became the Heaviest People in the World
- Social Issues: Sex Differences in Attitudes Toward Sexuality

Cognitive Development

Changes in the Structure of Thought
Perry's Theory • Schaie's Theory • Labouvie-Vief's Theory

Information Processing: Expertise and Creativity

Changes in Mental Abilities

The College Experience
Psychological Impact of Attending College • Dropping Out

Vocational Choice
Selecting a Vocation • Factors Influencing Vocational Choice • Vocational Preparation of Non-College-Bound Young Adults

- Cultural Influences: Work–Study Apprenticeships in Germany

These cross-country skiers of Quebec, Canada, capture the energetic quality of early adulthood. This period brings a succession of major life changes—among them, starting full-time work, attaining economic independence, and entering into a long-term sexually intimate relationship. As young adults focus on these goals, life is at times stressful but also exhilarating and rewarding.

The back seat and trunk piled high with belongings, 22-year-old Sharese hugged her mother and brother goodbye, jumped in the car, and headed toward the interstate with a sense of newfound independence mixed with apprehension. Three months earlier, the family had watched proudly as Sharese received her bachelor's degree in chemistry from a small university 40 miles from her home. Her college years had been a time of gradual release from economic and psychological dependence on her family. She returned home on weekends as often as she desired and lived there during the summer months. Her mother supplemented Sharese's loans with a monthly allowance. But this day marked a turning point. She was moving to her own apartment in a city 800 miles away, with plans to begin working on a master's degree. In charge of all her educational and living expenses, Sharese felt more self-sufficient than at any previous time in her life.

The college years were ones in which Sharese made important lifestyle changes and settled on a vocational direction. Overweight throughout high school, she lost twenty pounds during her freshman year, revised her diet, and began a regimen of exercise by joining the university's Ultimate Frisbee team. The sport helped her acquire healthier habits and leadership skills as team captain. A summer spent as a counselor at a camp for chronically ill children, combined with personal events we will take up later, convinced Sharese to apply her background in science to a career in public health.

Still, she wondered whether her choice was right. Two weeks before she was scheduled to leave, Sharese confided to her mother that she had doubts and might not go. Her mother advised, "Sharese, we never know ahead of time if the things we choose are going to suit us just right, and most times they aren't perfect. It's what we make of them—how we view and mold them—that turns a choice into a success." So Sharese embarked on her journey and found herself face to face with a multitude of exciting challenges and opportunities.

In this chapter, we take up the physical and cognitive sides of early adulthood—the period of the twenties and thirties. In Chapter 1, we emphasized that the adult years are difficult to divide into discrete periods, since the timing of important milestones varies greatly among individuals—much more so than in childhood and adolescence. But for most people, this first phase of adult life involves a common set of tasks: leaving home, completing education, beginning full-time work, attaining economic independence, establishing a long-term sexually and emotionally intimate relationship, and starting a family. As Sharese's conversation with her mother reveals, the momentous decisions of early adulthood inevitably lead to second thoughts and disappointments. But with the help of family, community, and societal contexts, most young adults make the best of wrong turns and solve problems successfully. These are energetic decades that, more than any phase, offer the potential for living to the fullest.

PHYSICAL DEVELOPMENT

In earlier chapters, we saw that throughout childhood and adolescence, the body grows larger and stronger, coordination improves, and sensory systems gather information more effectively. Once body structures reach maximum capacity and efficiency in the teens and twenties, **biological aging,** or **senescence,** begins—genetically influenced declines in the functioning of organs and systems that are universal in all members of our species (Cristofalo et al., 1999). However, like physical growth, biological aging is *asynchronous* (see Chapter 7, pages 204–206). Change varies widely across parts of the body, and some structures are not affected at all. In addition, individual differences are great—variation that the *lifespan perspective* helps us understand. Biological aging is influenced by a host of contextual factors, each of which can accelerate or slow age-related declines. These include the person's unique genetic makeup, lifestyle, living environment, and historical period (Arking, 1991). As a result, the physical changes of the adult years are, indeed, multidimensional and multidirectional.

In the following sections, we examine the process of biological aging. Then we turn to physical and motor changes already under way in early adulthood. Our discussion will show that biological aging is not fixed and immutable. Instead, it can be modified substantially through behavioral and environmental interventions. Indeed, that 25 to 30 years has been added over the past century to *average life expectancy* in industrialized nations (see Chapter 1, page 7) verifies the dramatic impact of improved nutrition, medical treatment, sanitation, and safety on length of life. We will take up life expectancy in greater depth in Chapter 17.

■ More than any other period of the lifespan, the decades of early adulthood—the twenties and the thirties—offer the potential for living to the fullest.

Biological Aging Begins in Early Adulthood

At an intercollegiate tournament, Sharese dashed across the playing field for hours, leaping into the air to catch Frisbees sailing her way. In her early twenties, she is at her peak in strength, endurance, sensory acuteness, and immune system responsiveness. Yet she will begin to age over the next 2 decades and will eventually show more noticeable declines as she moves into middle and late adulthood.

Biological aging is the combined result of many causes, some operating at the level of DNA, others at the level of cells, and still others at the level of tissues, organs, and whole organisms. Hundreds of theories exist, indicating that our understanding is still in an early stage (Cristofalo et al., 1999). One popular idea—the *"wear-and-tear" theory*—is that the body wears out from use. But unlike parts of a machine, worn-out parts of the body usually replace or repair themselves. Furthermore, no relationship exists between hard physical work and early death. To the contrary, vigorous exercise predicts a healthier and longer life (Paffenbarger, Blair, & Lee, 2001). We now know that this "wear-and-tear" theory is an oversimplification.

■ This 25-year-old student in a kick-boxing class demonstrates her agility and strength. Despite her overall health and exuberance, biological aging has already begun. Although physical declines are not yet noticeable, they will increase over time. However, young people can take steps to minimize negative physical changes through good health habits.

Aging at the Level of DNA and Body Cells

Current explanations of biological aging at the level of DNA and body cells are of two types: (1) those that emphasize the *programmed effects of specific genes,* and (2) those that emphasize the *cumulative effects of random events,* both internal and external, that damage genetic and cellular material. Support for both views exists, and a combination may eventually prove to be correct.

Genetically programmed aging receives some support from kinship studies indicating that longevity is a family trait. People whose parents had long lives tend to live longer themselves. And greater similarity exists in the lifespans of identical than fraternal twins. But the heritability of longevity is quite modest. In estimates based on diverse ethnic groups, it typically ranges from .15 to .25 (Kerber et al., 2001; Mitchell et al., 2001). Rather than inheriting longevity directly, people probably inherit one or more risk factors, which influence their chances of dying earlier or later.

One "genetic programming" theory proposes the existence of "aging genes" that control certain biological changes, such as menopause, gray hair, and deterioration of body cells. The strongest evidence for this view comes from research showing that human cells allowed to divide in the laboratory have a lifespan of 50 divisions plus or minus 10. With each duplication, a special type of DNA—called *telomeres,* located at the ends of chromosomes—shortens. Eventually, so little remains that the cells no longer duplicate at all. When researchers augment telomere activity through genetic engineering in the laboratory, they can extend the lifespan of human cells to some degree (Karlseder, Smogorzewska, & de Lange, 2002).

Cell death through telomere shortening characterizes only larger and longer-lived species. While causing cells to die, it also acts as a brake against disease-causing mutations, which increase as cells duplicate (Shay & Wright, 2001). According to an alternative, "random events" theory of biological aging, DNA in body cells is gradually damaged due to spontaneous or externally caused changes in DNA. As these mutations accumulate, cell repair and replacement become less efficient, or abnormal cancerous cells are produced. Animal studies confirm an increase in DNA breaks and deletions and damage to other cellular material with age. Among humans, similar evidence is accumulating (Wei & Lee, 2002).

One probable cause of age-related DNA and cellular abnormalities is the release of **free radicals,** naturally occurring, highly reactive chemicals that form in the presence of oxygen.

(Radiation and certain pollutants and drugs can trigger similar effects.) When oxygen molecules break down within the cell, the reaction strips away an electron, creating a free radical. As it seeks a replacement from its surroundings, it destroys nearby cellular material, including DNA, proteins, and fats essential for cell functioning. Free radicals are thought to be involved in more than sixty disorders of aging, including heart disease, cancer, cataracts, and arthritis (Miguel, 2001). Although our bodies produce substances that neutralize free radicals, some harm occurs and accumulates.

Some researchers believe that genes for longevity work by defending against free radicals. In this way, a programmed genetic response may limit random DNA and cellular deterioration. Foods rich in vitamins C and E and beta-carotene forestall free-radical damage as well.

Aging at the Level of Organs and Tissues

What consequences might the DNA and cellular deterioration just described have for the structure and functioning of organs and tissues? There are many possibilities. Among those with clear support is the **cross-linkage theory of aging.** Over time, protein fibers that make up the body's connective tissue form bonds, or links, with one another. When these normally separate fibers cross-link, tissue becomes less elastic, leading to many negative outcomes, including loss of flexibility in the skin and other organs, clouding of the lens of the eye, clogging of arteries, and damage to the kidneys. Like other aspects of aging, cross-linking can be reduced by external factors, including regular exercise and a vitamin-rich, low-fat diet (Schneider, 1992).

Gradual failure of the endocrine system, which is responsible for production and regulation of hormones, is yet another route to aging. An obvious example is decreased estrogen production in women, which culminates in menopause. Because hormones affect so many body functions, disruptions in the endocrine system can have widespread effects on health and survival. At present, scientists are studying the impact of key hormones on aging. Recent evidence indicates that a gradual drop in growth hormone (GH) is associated with loss of muscle and bone mass, addition of body fat, thinning of the skin, and decline in cardiovascular functioning. In adults with abnormally low levels of GH, hormone therapy can slow these symptoms, but it has serious side effects, including increased risk of cancer (Johannsson, Svensson, & Bengtsson, 2000; Sonntag et al., 2000). So far, diet and physical activity are safer ways to limit these aspects of biological aging.

Finally, declines in immune system functioning contribute to many conditions of aging. Among these are increased susceptibility to infectious disease, heightened risk of cancer, and changes in blood vessel walls associated with cardiovascular disease. Decreased vigor of the immune response seems to be genetically programmed, but other aging processes we have considered—such as weakening of the endocrine system—can intensify it (Malaguarnera et al., 2001). Indeed, combinations of theories—the ones just reviewed as well as others—are needed to explain the complexities of biological aging. With this in mind, let's look closely at physical signs and other characteristics of aging.

Physical Changes

During the twenties and thirties, changes in physical appearance and declines in body functioning are so gradual that many are hardly noticeable. Later, they will accelerate. The physical changes of aging are summarized in Table 13.1. We will examine several in detail, taking up others in later chapters. But before we begin, it is important to note that the data forming the basis for these trends are largely cross-sectional. Since younger cohorts have experienced better health care and nutrition, cross-sectional studies can exaggerate impairments associated with aging. Fortunately, longitudinal evidence is expanding, helping to correct this picture.

Cardiovascular and Respiratory Systems

During her first month in graduate school, Sharese pored over research articles on cardiovascular functioning. In her African-American extended family, her father, an uncle, and three aunts had died of heart attacks in their forties and fifties. The tragedies had prompted Sharese to worry about her own lifespan, reconsider her health-related behaviors, and enter the field of public health in hopes of finding ways to relieve health problems among black Americans. *Hypertension,* or high blood pressure, occurs 12 percent more often in the American black than in the American white population; the rate of death from heart disease among African Americans is 25 percent higher (U.S. Department of Health and Human Services, 2002h, 2002m).

Sharese was surprised to learn that there are fewer age-related changes in the heart than we might expect, in view of the fact that heart disease rises with age and is a leading cause of death throughout adulthood. In healthy individuals, the heart's ability to meet the body's oxygen requirements under typical conditions (as measured by heart rate in relation to volume of blood pumped) does not change during adulthood. Only during stressful exercise does performance of the heart decline with age. The change is due to a decrease in maximum heart rate and greater rigidity of the heart muscle. Consequently, the heart has difficulty delivering enough oxygen to the body during high activity and bouncing back from strain (Haywood & Getchell, 2001).

One of the most serious diseases of the cardiovascular system is *atherosclerosis,* in which heavy deposits of plaque containing cholesterol and fats collect on the walls of the main arteries. If it is present, it usually begins early in life, progresses during middle adulthood, and culminates in serious illness. Atherosclerosis is multiply determined, making it hard to separate the

Table 13.1 Physical Changes of Aging

Organ or System	Timing of Change	Description
Sensory		
Vision	From age 30	As the lens stiffens and thickens, ability to focus on close objects declines. Yellowing of the lens, weakening of muscles controlling the pupil, and clouding of the vitreous (gelatin-like substance that fills the eye) reduce light reaching the retina, impairing color discrimination and night vision. Visual acuity, or fineness of discrimination, decreases, with a sharp drop between ages 70 and 80.
Hearing	From age 30	Sensitivity to sound declines, especially at high frequencies but gradually extending to all frequencies. Change is more than twice as rapid for men as for women.
Taste	From age 60	Sensitivity to the four basic tastes—sweet, salty, sour, and bitter—is reduced. May be due to factors other than aging, since number and distribution of taste buds do not change.
Smell	From age 60	Loss of smell receptors reduces ability to detect and identify odors.
Touch	Gradual	Loss of touch receptors reduces sensitivity on the hands, particularly the fingertips.
Cardiovascular	Gradual	As the heart muscle becomes more rigid, maximum heart rate decreases, reducing the heart's ability to meet the body's oxygen requirements when stressed by exercise. As artery walls stiffen and accumulate plaque, blood flow to body cells is reduced.
Respiratory	Gradual	Under physical exertion, respiratory capacity decreases and breathing rate increases. Stiffening of connective tissue in the lungs and chest muscles makes it more difficult for the lungs to expand to full volume.
Immune	Gradual	Shrinking of the thymus limits maturation of T cells and disease-fighting capacity of B cells, impairing the immune response.
Muscular	Gradual	As nerves stimulating them die, fast-twitch muscle fibers (responsible for speed and explosive strength) decline in number and size to a greater extent than slow-twitch fibers (which support endurance). Tendons and ligaments (which transmit muscle action) stiffen, reducing speed and flexibility of movement.
Skeletal	Begins in the late 30s, accelerates in the 50s, slows in the 70s	Cartilage in the joints thins and cracks, leading bone ends beneath it to erode. New cells continue to be deposited on the outer layer of the bones, and mineral content of bone declines. The resulting broader but more porous bones weaken the skeleton and make it more vulnerable to fracture. Change is more rapid in women than in men.
Reproductive	In women, accelerates after age 35; in men, begins after age 40	Fertility problems (including difficulty conceiving and carrying a pregnancy to term) and risk of having a baby with a chromosomal disorder increase.
Nervous	From age 50	Brain weight declines as neurons lose water content and die, mostly in the cerebral cortex, and as ventricles (spaces) within the brain enlarge. Development of new synapses can, in part, compensate for the decline in number of neurons.
Skin	Gradual	Epidermis (outer layer) is held less tightly to the dermis (middle layer); fibers in the dermis and hypodermis (inner layer) thin; fat cells in the hypodermis decline. As a result, the skin becomes looser, less elastic, and wrinkled. Change is more rapid in women than in men.
Hair	From age 35	Grays and thins.
Height	From age 50	Loss of bone strength leads to collapse of disks in the spinal column, leading to a height loss of as much as 2 inches by the seventies and eighties.
Weight	Increases to age 50; declines from age 60	Weight change reflects a rise in fat and a decline in muscle and bone mineral. Since muscle and bone are heavier than fat, the resulting pattern is weight gain followed by loss. Body fat accumulates on the torso and decreases on the extremities.

Sources: Arking, 1991; Fabsitz, Sholinsky, & Carmelli, 1994; Pearson et al., 1995; Receputo et al., 1994; Whalley, 2001; Whitbourne, 1996.

contributions of biological aging from individual genetic and environmental influences. The complexity of causes is illustrated by animal research indicating that before puberty, a high-fat diet produces only fatty streaks on the artery walls (Olson, 2000). In sexually mature adults, however, it leads to serious plaque deposits. These findings suggest that sex hormones may heighten the insults of a high-fat diet.

As mentioned earlier, heart disease has decreased considerably since the middle of the twentieth century. An especially large drop has occurred during the last 15 years due to a deline in cigarette smoking, to improved diet and exercise among at-risk individuals, and to better medical detection and treatment of high blood pressure and cholesterol (U.S. Department of Health and Human Services, 2002h, 2002m). And as longitudinal research with several hundred thousand participants reveals, young and middle-aged adults at low risk for heart disease—defined by not smoking, low blood cholesterol, and normal blood pressure—show 40 to 60 percent reduced death rates over the next 16 to 22 years (see Figure 13.1) (Stamler et al., 1999). Later, when we consider health and fitness, we will see why heart attacks were so common in Sharese's family—and why they occur at especially high rates in the African-American population.

Like the heart, functioning of the lungs shows few age-related changes at rest, but during physical exertion, respiratory volume decreases and breathing rate increases with age. Maximum vital capacity (amount of air that can be forced in and out of the lungs) declines by 10 percent per decade after age 25 (Mahanran et al., 1999). Connective tissue in the lungs, chest muscles, and ribs stiffens with age, making it more difficult for the lungs to expand to full volume (Haywood & Getchell, 2001). Fortunately, under normal conditions, we use less than half our vital capacity; most of it is held in reserve. Nevertheless, aging of the lungs contributes to the fact that during heavy exercise, older adults find it more difficult to meet the body's oxygen needs.

Motor Performance

Declines in heart and lung functioning under conditions of exertion, combined with gradual muscle loss, lead to changes in motor performance. In ordinary people, the impact of biological aging on motor skills is difficult to separate from decreases in motivation and practice. Therefore, researchers study outstanding athletes, who try to attain their very best performance in real life. As long as athletes continue intensive training, their attainments at each age approach the limits of what is biologically possible (Ericsson, 1990).

Many studies show that athletic skill peaks between ages 20 and 30 and then declines. In several investigations, the mean ages for best performance of Olympic and professional athletes in a variety of sports were charted over time. Absolute performance in many events improved over the past century. Athletes continually set new world records, suggesting improved training methods. Nevertheless, ages of best performance remained relatively constant. As Figure 13.2 reveals, athletic tasks that require speed of limb movement, explosive strength, and gross body coordination—sprinting, jumping, and tennis—peak in the early twenties. Those that depend on endurance, arm–hand steadiness, and aiming—long-distance running, baseball, and golf—peak in the late twenties and early thirties. Because these tasks have to do with either stamina or precise motor control, they take longer to develop (Schulz & Curnow, 1988).

Research on outstanding athletes tells us that the upper biological limit of motor capacity is reached in the first part of

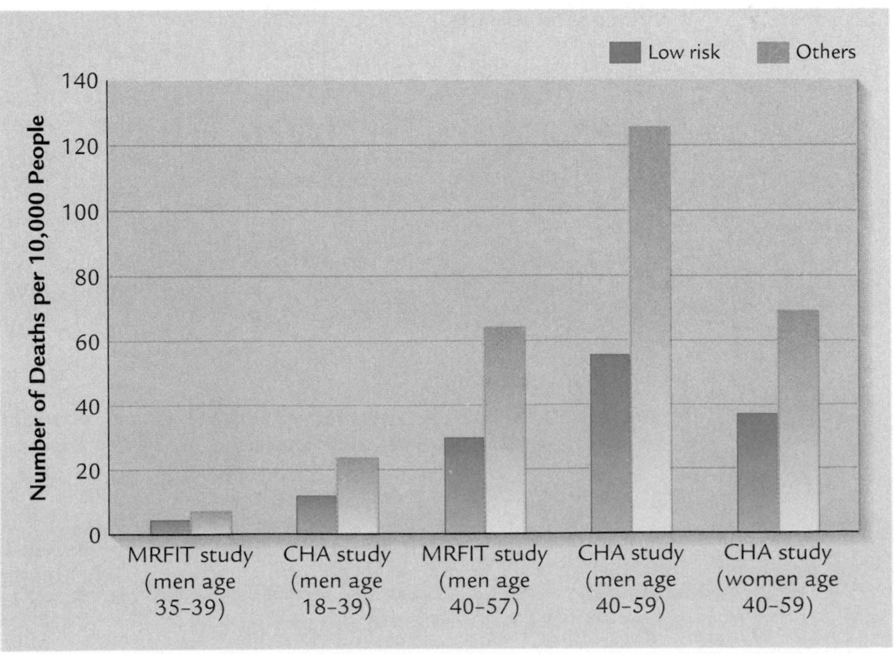

FIGURE 13.1 Death rates of individuals at low risk for heart disease compared with other individuals. Low risk for heart disease is defined by not smoking, low blood cholesterol, and normal blood pressure. Findings come from two longitudinal studies–the Multiple Risk Factor Intervention Trial (MRFIT) and the Chicago Heart Association Detection Project in Industry (CHA), in which five cohorts, totaling over 350,000 adults, were followed for 16 to 22 years. Those who did not smoke, whose total blood cholesterol was under 200 mg/dl, and whose systolic/diastolic blood pressure was 120/80 or lower showed death rates that were 40 to 60 percent lower than their same-sex agemates (Adapted from Stamler et al., 1999.)

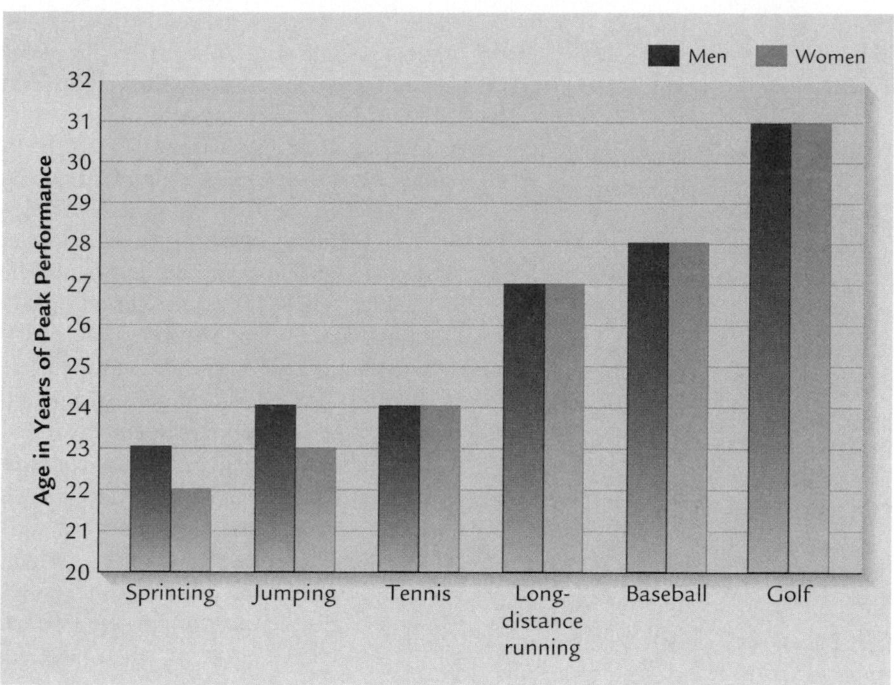

■ **FIGURE 13.2 Age of peak performance of Olympic and professional athletes in various sports.** For both men and women, sprinting, jumping, and tennis, which require speed of limb movement, explosive strength, and gross body coordination, peak in the early twenties. Long-distance running, baseball, and golf, which depend on endurance, arm–hand steadiness, and aiming, peak in the late twenties and early thirties. (Adapted from Schulz & Curnow, 1988.)

early adulthood. How quickly do athletic skills weaken in later years? Longitudinal research on master runners reveals that as long as practice continues, performance drops only slightly—about 2 percent per decade—from the twenties into the sixties and seventies (Hagberg et al., 1985; Pollock, Mengelkoch, & Graves, 1997). Indeed, sustained training leads to adaptations in body structures that minimize motor decline. For example, vital capacity is almost twice as great in older people actively participating in sports as in healthy controls without a history of exercise. Training also slows muscle loss, increases speed and force of muscle contraction, and leads fast-twitch muscle fibers to be converted into slow-twitch fibers, which support excellent long-distance running performance as well as other endurance skills (Trappe, 2001).

In sum, although athletic skill is at its best in early adulthood, only a small part of age-related decline is due to biological aging. Lower levels of performance in older healthy people largely reflect reduced capacities that result from adaptation to a less physically demanding lifestyle.

■ Athletic skill peaks between ages 20 and 30 and then declines. But as long as practice continues, performance drops only slightly—about 2 percent per decade—into the sixties and seventies. Here, older adults who have continued to train for many years keep pace with younger adults in a marathon.

Immune System

The immune response is the combined work of specialized cells that neutralize or destroy antigens (foreign substances) in the body. Two types of white blood cells play vital roles. *T cells,* which originate in the bone marrow and mature in the thymus (a small gland located in the upper part of the chest), attack antigens directly. *B cells,* manufactured in the bone marrow, secrete antibodies into the bloodstream that multiply, capture antigens, and permit the blood system to destroy them. Since receptors on their surfaces recognize only a single antigen, T and B cells come in great variety. They join with additional cells to produce immunity.

The capacity of the immune system to offer protection against disease increases through adolescence and declines after age 20. The trend is partly due to changes in the thymus, which is largest during the teenage years and shrinks until it is barely detectable by age 50. As a result, production of thymic hormones is reduced, and the thymus is less able to promote full maturity and differentiation of T cells. Since B cells release far more antibodies when T cells are present, the immune response is compromised further (Malaguarnera et al., 2001).

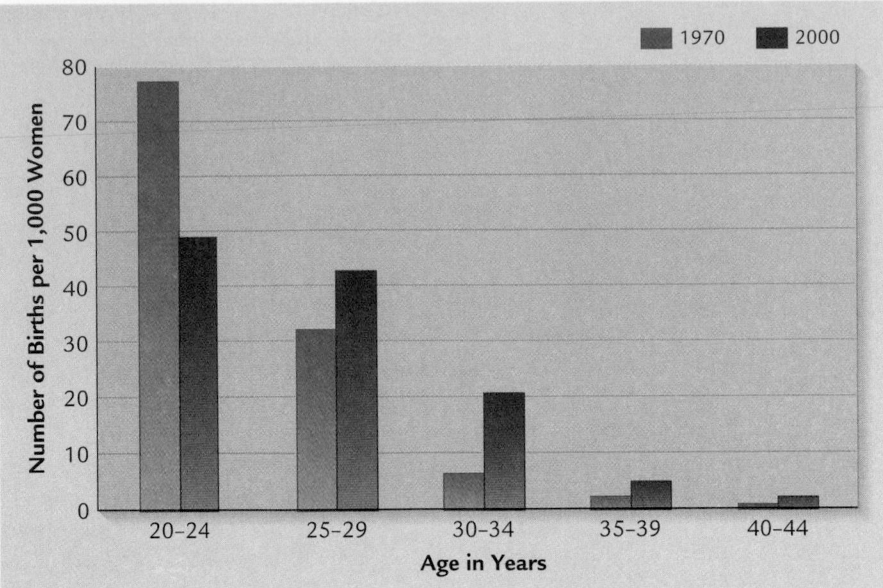

■ **FIGURE 13.3 First births to American women of different ages in 1970 and 2000.** The birth rate decreased during this period for women 20 to 24 years of age, whereas it increased for women 25 years of age and older. For women in their thirties, the birth rate more than doubled. Similar trends have occurred in Canada and other industrialized nations. (Adapted from U.S. Department of Health and Human Services, 2002g.)

Administering thymic hormones can help the aging body fight disease (Goya & Bolognani, 1999).

Withering of the thymus, however, is not the only reason that the body gradually becomes less effective in warding off illness. The immune system interacts with the nervous and endocrine systems. For example, stress can weaken the immune response. During final exams, Sharese was less resistant to colds. And in the month after her father died, she had great difficulty recovering from the flu. Divorce, caring for an ill aging parent, and chronic depression can also reduce immunity (Bernston et al., 1998). The link between stress and illness makes sense when we consider that stress hormones mobilize the body for action, whereas the immune response is fostered by reduced activity (Maier, Watkins, & Fleshner, 1994). But this also means that increased difficulty coping with physical and psychological stress can contribute to age-related declines in immune system functioning.

Reproductive Capacity

Sharese was born when her mother was in her early twenties—an age at which, a generation later, Sharese was still single and entering graduate school. Many people believe that pregnancy during the twenties is ideal, not only because the risks of miscarriage and chromosomal disorders are reduced (see Chapter 2) but also because younger parents have more energy to keep up with active children. However, as Figure 13.3 reveals, first births to women in their thirties have increased greatly over the past two decades. Many people are delaying childbearing until their education is complete, their careers are well established, and they know they can support a child.

Nevertheless, reproductive capacity does decline with age. Fertility problems among women increase from ages 15 to 50, with a dramatic rise in the mid-thirties. Between ages 25 and 34, nearly 14 percent are affected, a figure that climbs to 26 percent for 35- to 44-year-olds (U.S. Department of Health and Human Services, 2002c). The decline in female fertility is largely due to reduced quality of ova, since the uterus shows no consistent changes among women in their late 30s and 40s. In addition, some women in their later reproductive years have normal menstrual cycles but do not conceive because the reserve of ova in their ovaries is too low. A certain level of ova is necessary for conception in many mammals, including humans. This decline in reserve ova is the most important cause of the age-related drop in female fertility (Klein & Sauer, 2001).

Age also affects male reproductive capacity. Amount of semen and concentration of sperm in each ejaculation gradually decrease after age 40 (Murray & Meacham, 1993). Although there is no best time during adulthood to begin parenthood, individuals who postpone childbearing until their late thirties or their forties risk having fewer children than they desired or none at all. Reproductive technologies, such as donor insemination and in vitro fertilization, can help older couples bear children. But (like natural conception) their success rates drop sharply with age (see Chapter 2, page 54).

Ask Yourself

REVIEW

Cite examples of both programmed and random genetic influences on biological aging.

APPLY

Len noticed the term *free radicals* in an article on adult health in his local newspaper, but he doesn't understand how they contribute to biological aging. Explain the role of free radicals to Len.

APPLY

Penny is a long-distance runner for her college track team. She wonders what her running performance will be like 10 or 20 years from now. Describe factors that will affect Penny's long-term athletic skill.

CONNECT

How do genetic and environmental factors jointly contribute to age-related changes in cardiovascular, respiratory, and immune-system functioning?

www.

Health and Fitness

Figure 13.4 displays leading causes of death in early adulthood in the United States and Canada. Notice that death rates for all causes are lower for Canadians than Americans, due to Canada's national health insurance system and (in the case of homicide) strict gun registration, safety, and control policies, which ban possession of handguns and other firearms not designed for hunting. In later chapters, we will see that unintentional injury and homicide rates decline with age. In contrast, disease and physical disability rates rise. Biological aging clearly contributes to this trend. But we have already noted wide individual and group differences in physical changes that are linked to environmental risks and health-related behaviors.

SES variations in health over the lifespan reflect these influences. Income, education, and occupational status show a strong and continuous relationship with almost every disease and health indicator (Adler & Newman, 2002). Furthermore, when a representative sample of 3,600 Americans were asked about chronic illnesses and health-related limitations on their daily lives, SES differences were relatively small at 25 to 35 years, widened between ages 35 and 65, and contracted during old age (see Figure 13.5 on page 420). Longitudinal findings confirm that economically advantaged and well-educated individuals tend to sustain good health over most of their adult lives, whereas the health of lower-income individuals with limited education steadily declines (Lantz et al., 1998). SES differences in health-related circumstances and habits—stressful life events, crowding, pollution, diet, exercise, substance use, availability of supportive social relationships, and (in the United States) access of low-SES individuals to affordable health care—are largely responsible (Evans & Kantrowitz, 2002; Mulatu & Schooler, 2002).

SES disparities in health and mortality are larger in the United States than in Canada and other industrialized nations (Mackenbach, 2002; Ross et al., 2002). What explains this difference? Besides the lack of universal health insurance, American poverty-stricken families have lower incomes than their Canadian counterparts. In addition, SES groups are more likely to be segregated by neighborhood in American than in Canadian large cities, resulting in greater inequalities in environmental factors that affect health, such as housing, pollution, and community services.

These findings reveal, once again, that the living conditions that nations and communities provide combine with those that people create for themselves to affect physical aging. Because the incidence of health problems is much lower during the twenties and thirties than in succeeding decades, early adulthood is an excellent time to prevent later problems. In the following sections, we take up a variety of major health concerns—nutrition, exercise, substance abuse, sexuality, and psychological stress.

Nutrition

Bombarded with advertising claims and an extraordinary variety of food choices, adults find it increasingly difficult to make wise dietary decisions. An abundance of food, combined with a heavily scheduled life, means that most North Americans eat because they feel like it or because it is time to do so

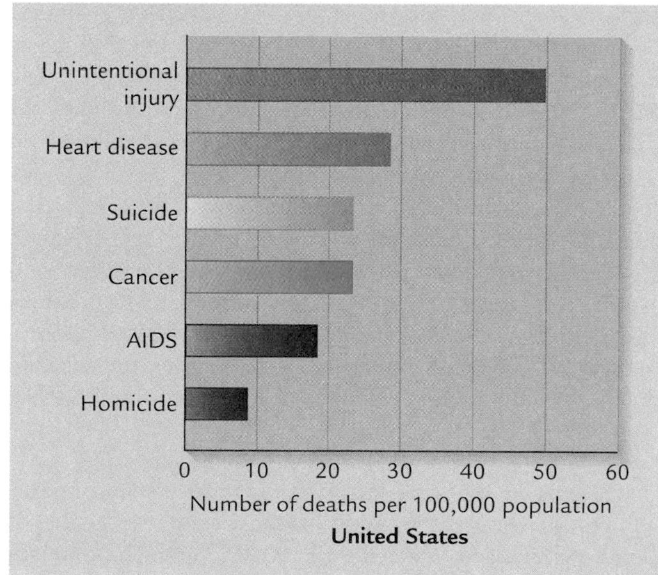

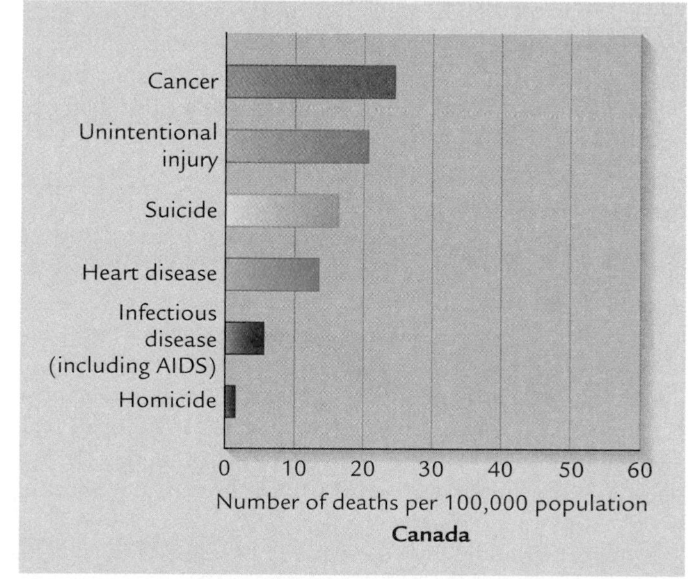

FIGURE 13.4 Leading causes of death between 25 and 44 years of age in the United States and Canada. Death rates are lower in Canada than in the United States, largely because all Canadians have government-funded health insurance. In addition, guns—involved in many homicides—are more stringently controlled in Canada than in the United States. (Adapted from Statistics Canada, 2002i; U.S. Bureau of the Census, 2002c.)

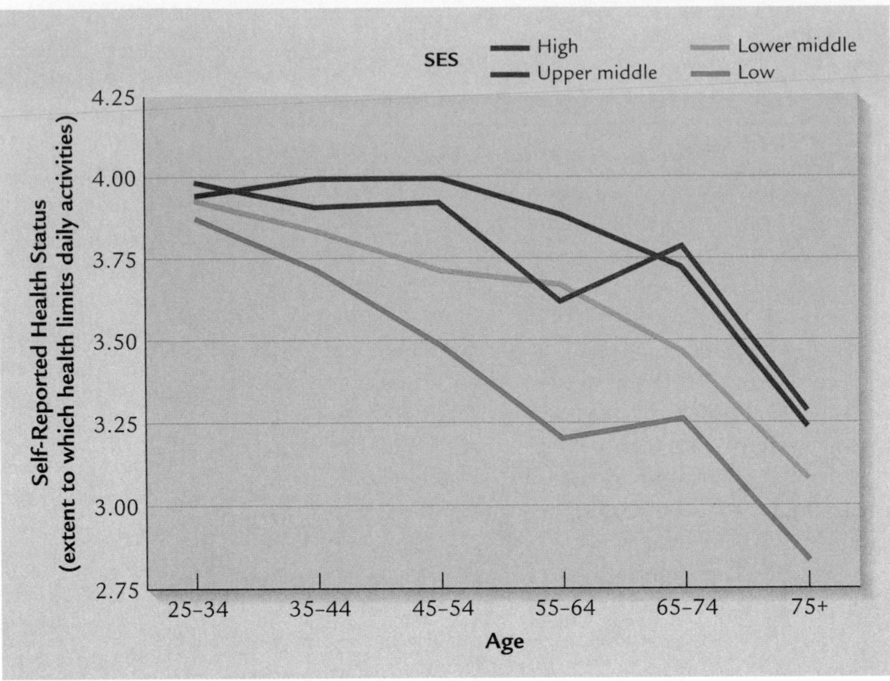

■ **FIGURE 13.5 Age differences in self-reported health status by socioeconomic status (SES).** Variations among SES groups are small at 25 to 34 years, increase from 35 to 64 years, and then contract. Environmental risks and poor health habits contribute to the earlier decline in health status among low-income and poverty-stricken adults. (From J. S. House et al., 1990, "Age, Socioeconomic Status, and Health," *Milbank Quarterly, 68,* p. 396. Reprinted by permission.)

rather than to maintain the body's functions (Donatelle, 2003). As a result, many eat the wrong types and amounts of food. Overweight and obesity and a high-fat diet are widespread nutritional problems with long-term consequences for health in adulthood.

● **Overweight and Obesity.** In Chapter 9, we noted that obesity (a greater than 20 percent increase over average body weight, based on age, sex, and physical build) has increased dramatically in many Western nations. Today, 20 percent of American adults (nearly 59 million people) and 14 percent of Canadian adults (3.3 million people) are affected. The incidence is especially high among low-SES ethnic minorities. Obesity rises to 24 percent among Hispanic adults and to 30 percent among African-American adults. The incidence among Native Americans and Canadian Aboriginal peoples is estimated to be more than 50 percent. In the United States and Western Europe, 5 to 7 percent more women than men suffer from obesity. In Canada, obesity rates for the two sexes are similar (Anand et al., 2001; Mokdad et al., 2001; U.S. Department of Health and Human Services, 2002j).

Overweight—a less extreme but nevertheless unhealthy condition—affects an additional 41 percent of Americans and 30 percent of Canadians. Add the rates of overweight and obesity together, and Americans emerge as the heaviest people in the world. Recall from Chapter 9 that overweight children are very likely to become overweight adults. But a substantial number of people (4 percent of men and 8 percent of women) show large weight gains in adulthood, most often between ages 25 and 40. And young adults who were already overweight or obese typically get heavier, leading obesity rates to rise steadily between ages 25 and 60 (Birmingham et al., 1999; Mokdad et al., 1999).

Causes and Consequences. As noted in Chapter 9, heredity plays a role in body weight; some people are more vulnerable to obesity than others. But environmental pressures underlie the rising rates of obesity in the United States and other industrialized nations. With the decline in need for physical labor in the home and workplace, our lives have become more sedentary. And the average number of calories and amount of sugar and fat that North Americans consume increased over most of the twentieth century, with a sharp rise after 1970 (see the Lifespan Vista box on pages 422–423).

Adding some weight between ages 25 and 50 is a normal part of aging, since **basal metabolic rate (BMR),** the amount of energy the body uses at complete rest, gradually declines as the number of active muscle cells (which create the greatest energy demand) drops off. But excess weight is strongly associated with serious health problems. These include high blood pressure, circulatory difficulties, atherosclerosis, stroke, adult-onset diabetes, liver and gallbladder disease, arthritis, sleep and digestive disorders, most forms of cancer, and early death (Calle et al., 2003; National Task Force on the Prevention and Treatment of Obesity, 2002). Furthermore, overweight adults suffer enormous social discrimination. They are less likely to find mates, to be rented apartments, to be given financial aid for college, and to be offered jobs (Allison & Pi-Sunyer, 1994; Gortmaker et al., 1993).

Treatment. Because obesity climbs in early and middle adulthood, treatment for adults should begin as soon as possible—preferably in the early twenties. Even moderate weight loss reduces health problems substantially (Robison et al., 1993). But successfully intervening in obesity is difficult. At present, 95 percent of individuals who start a weight-loss program return to their original weight within 5 years (Allison & Pi-Sunyer, 1994). The high rate of failure is partly due to limited knowledge of just how obesity disrupts the complex neural, hormonal, and metabolic factors that maintain a normal body-weight set point. Until more information is available, researchers are examining the characteristics of treatments and participants associated with greater success. The following elements promote lasting behavior change:

- *A well-balanced diet lower in calories and fat, plus exercise.* To lose weight, Sharese sharply reduced calories, sugar, and fat in her diet and exercised regularly. The precise balance of dietary protein, carbohydrates, and fats that best helps adults lose weight is currently a matter of heated debate. Although scores of diet books offer different recommendations, no clear-cut evidence exists for the long-term superiority of one approach over others (Aronne, 2001). Research does confirm that restricting calorie intake and fat (to no more than 20 to 30 percent of calories) and increasing physical activity is essential for reducing the impact of a genetic tendency to overweight (Rolls & Bell, 2000). In addition (as we will see shortly), exercise offers physical and psychological benefits that help prevent overeating. Yet less than half of American obese adults are advised by their doctors to control eating and to exercise (Mokdad et al., 2001).
- *Training participants to keep an accurate record of what they eat.* About 30 to 35 percent of obese people sincerely believe they eat less than they do, and from 25 to 45 percent report problems with binge eating (Wadden & Foster, 2000). When Sharese became aware of how often she ate when she was not hungry, she was better able to limit her food intake.
- *Social support.* Group or individual counseling and encouragement of friends and relatives help sustain weight-loss efforts by fostering self-esteem (Johnson, 2002). Once Sharese decided to act with the support of her family and a weight-loss counselor, she began to feel better about herself—walking and holding herself differently—even before the first pounds were shed.
- *Teaching problem-solving skills.* Acquiring cognitive and behavioral strategies for coping with tempting situations and periods of slowed progress is associated with long-term change. Weight-loss maintainers are more likely than individuals who relapse to be conscious of their behavior, to use social support, and to confront problems directly (Cooper & Fairburn, 2002).
- *Extended intervention.* Longer treatments (from 25 to 40 weeks) that include the components just listed grant people time to develop new habits.

Finally, nearly 40 percent of Americans and 33 percent of Canadians report that they are trying to lose weight (Mokdad et al., 2001). Although many dieters are overweight, about one-third are within normal range! Recall from Chapter 11 that the high value placed on thinness creates unrealistic expectations about desirable body weight and contributes to dangerous eating disorders of anorexia and bulimia, which remain common in early adulthood (see pages 353–354). Overweight people in weight-loss programs who set too-stringent weight goals are more likely to regain their lost pounds. And people of normal weight who diet often display "starvation syndrome" symptoms, including depression, anxiety, weakness, and preoccupation with food. A sensible body weight—neither too low nor too high—predicts physical and psychological health and longer life (Manson et al., 1995).

■ Overweight and obesity are serious health threats throughout the lifespan. This young woman participates in a weight-loss program in which she receives counseling and encouragement to sustain a low-fat diet and regular exercise. Her pleased expression conveys the benefits of attaining a healthy weight for self-esteem.

● **Dietary Fat.** After entering college, Sharese altered the diet of her childhood and adolescent years, sharply limiting red meat, eggs, butter, and fried foods. Public service announcements about the health risks of a high-fat diet have led the fat consumption of North Americans adults to drop slightly, but about 60 percent of adults eat too much. U.S. and Canadian national dietary recommendations include reducing dietary fat to 30 percent of total caloric intake, with no more than 10 percent made up of saturated fat (Health Canada, 1999a; U.S. Department of Health and Human Services, 2002f). Fat consumption is believed to play a role in breast cancer and (when it includes large amounts of red meat) is linked to colon cancer (Clifford & Kramer, 1993). But the main reasons for limiting dietary fat are the strong connection of total fat with obesity and of saturated fat with cardiovascular disease.

Fat is a type of molecule made up of carbon atoms dotted with hydrogen atoms. *Saturated fat* has as many hydrogen atoms as it can chemically hold, generally comes from meat and dairy products, and is solid at room temperature. *Unsaturated*

A Lifespan Vista

The Obesity Epidemic: How Americans Became the Heaviest People in the World

In the late 1980s, obesity in the United States started to soar. The maps in Figure 13.6 show how quickly it engulfed the nation. Today, the majority of Americans—61 percent—are either overweight or obese. The epidemic has spread to other Western nations, including Canada, which has a combined overweight and obesity rate of 44 percent. But no country matches the United States in prevalence of this life-threatening condition.

A Changing Food Environment and Lifestyle. Obesity has risen too quickly for a change in the genetic makeup of the population to be responsible. Instead, the following environmental and lifestyle factors have encouraged widespread, rapid weight gain:

- *Availability of cheap commercial fat and sugar.* In the 1970s, two massive changes in the U.S. food economy occurred: (1) the discovery and mass production of high-fructose corn syrup, a sweetener six times as sweet

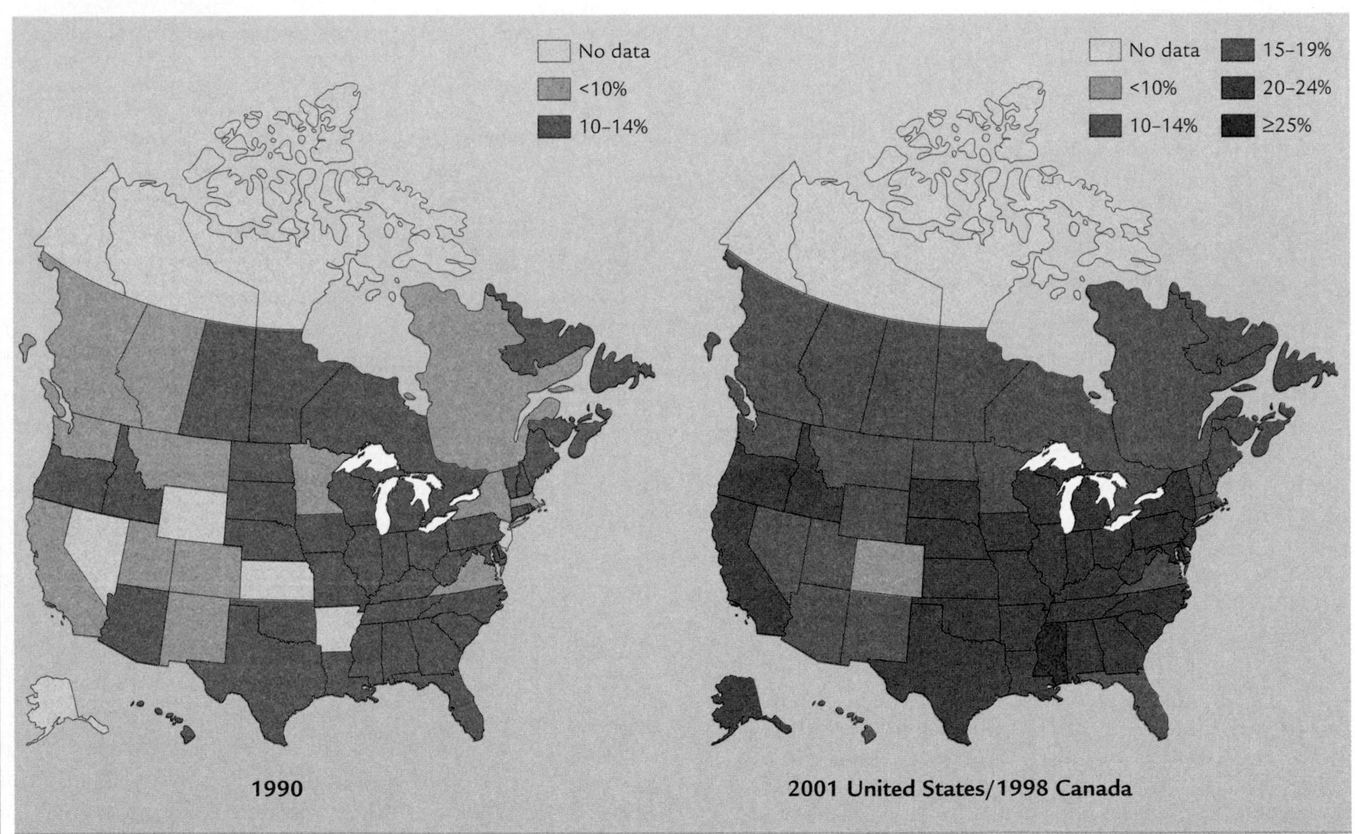

■ **FIGURE 13.6 Obesity trends among American and Canadian adults, 1990 and 1998–2001.** The darkening maps show that obesity has increased in both nations. But the rise during the past decade is especially extreme in the United States. (From A. Katzmarzyk, 2002, "The Canadian Obesity Epidemic, 1995–1998," *Canadian Medical Association Journal, 166,* pp. 1039–1040; Mokdad et al., 2001, "The Continuing Epidemics of Obesity and Diabetes in the United States," *Journal of the American Medical Association, 286,* p. 1198. Adapted by permission.)

fat has less than its full complement of hydrogen atoms and includes most liquid vegetable oils.

Moderate fat consumption is essential for normal body functioning. But when we consume too much, especially of the saturated variety, some is converted to cholesterol, which accumulates as plaque on the arterial walls in atherosclerosis. Earlier in this chapter, we noted that atherosclerosis is determined by multiple biological and environmental factors. But excess fat consumption (along with other societal conditions) is largely responsible for the high incidence of heart disease in the American black population, since black Africans have among the lowest rates of heart disease in the world (Begley, 1995).

and therefore far less expensive than ordinary sugar, and (2) the importing from Malaysia of large quantities of palm oil, both lower in cost and tastier than other vegetable oils because of its high saturated fat content. As food manufacturers relied on corn syrup and palm oil to make soft drinks and calorie-dense convenience foods, the production cost of these items dropped and their variety expanded. A new era of "cheap, abundant, and tasty calories had arrived" (Critser, 2003).

- *Portion supersizing.* Fast-food chains discovered a highly successful strategy for attracting customers: increasing portion sizes substantially and prices just a little for foods that had become relatively inexpensive to produce. Customers thronged to buy "value meals," jumbo burgers and burritos, pizza "by the foot," and 20-ounce Cokes. At McDonald's, a serving of French fries rose from 320 calories in the late 1970s to today's 610 calories (Critser, 2003). And research revealed that when presented with larger portions, preschoolers left the excess on their plates, but individuals age 5 and older devoured most of the food placed before them (Young & Nestle, 2002; Rolls, Engell, & Birch, 2000).
- *Increasingly busy lives.* Between the 1970s and 1990s, women entered the labor force in record numbers, and the average number of hours Americans worked rose by 350 per year (Schor, 2002). As time became scarce, eating out increased. In addition, Americans became frequent snackers, aided by an ever-increasing diversity of high-calorie snack foods on supermarket shelves. During this period, the number of calories Americans consumed away from home nearly doubled, and dietary fat increased from 19 to 38 percent. Overall, average daily food intake rose by almost 200 calories—enough to add an extra pound every 20 days (Nielsen & Popkin, 2003).
- *Declining rates of physical activity.* During the 1980s, physical activity, which had risen since the 1960s, started to fall as Americans spent more time working in sedentary jobs. At home, TV had become their major leisure pursuit—consuming, on average, about 4 hours per day and linked to weight gain in adults, just as in children (Woodring, 1998).
- *Misguided information about healthy weight, exercise, and diet.* As Americans got heavier, national health experts hesitated to inform them of their dire condition. In 1990, the federal government relaxed its recommendations on ideal weight. And recently, it advised that moderate activity is key to cardiovascular health, despite clear evidence that intense exercise is more protective (see page 424). Finally, an expanding health-advice industry dispensed a wealth of inconsistent messages, including the recommendation that Americans forget about dieting and focus on fitness instead (Molnar & Babbitt, 2000). Yet the overwhelming majority of overweight adults are not physically fit. And research repeatedly shows that the heavier people are, the greater the likelihood of illness and early death.

This conflicting information confused the public, especially low-SES adults among whom overweight and obesity are highest. In one survey, 37 percent of low-income African Americans agreed with the statement "I don't know if exercise is good or bad for me" (Airhihenbuwa et al., 1995).

Combatting the Obesity Epidemic. Obesity's toll on society is immense. It is responsible for 100 billion dollars in health expenditures and 280,000 premature deaths per year. Besides individual treatment, broad societal efforts are needed to combat obesity (Nestle & Jacobson, 2000). Suggestions include

- government funding to support massive public education about healthy eating and physical activity;
- a high priority placed on building parks and recreation centers in low-income neighborhoods, where overweight and obesity are highest;
- laws that mandate prominent posting of the calorie, sugar, and fat content of foods sold in restaurants, movie theaters, and convenience stores;
- a special tax on foods high in calories, sugar, or fat; and
- incentives to schools and workplaces for promoting healthy eating and daily exercise and for offering weight-management programs.

The best rule of thumb is to eat less fat of all kinds and to use unsaturated instead of saturated fat whenever possible. Furthermore (as we will see next), regular exercise can reduce the harmful influence of dietary fat because it creates chemical byproducts that help eliminate cholesterol from the body.

Exercise

Three times a week, over the noon hour, Sharese delighted in running, making her way to a wooded trail that cut through a picturesque area of the city. Regular exercise kept her fit and

© AP/WIDE WORLD PHOTOS

■ Regular exercise, through such activities as brisk walking, running, volleyball, basketball, aerobics, swimming, or cycling, can lead to a healthier, longer life. About 20 to 30 minutes of relatively vigorous use of large muscles of the body on most days can help protect against obesity, enhance immune functioning, and foster a positive outlook on life. These young mothers have limited child care, so they get their workouts with their babies.

slim. It also limited the number of respiratory illnesses she caught, compared to earlier days when she had been sedentary and overweight. And as Sharese explained to a friend one day, "Exercise gives me a positive outlook and calms me down. Afterward, I feel a burst of energy that gets me through the day. If I don't do it, I get tired in the afternoon."

Although most North Americans are aware of the health benefits of exercise, only 15 to 20 percent engage in at least moderate physical activity for 20 minutes or more at least five times a week. About 30 percent are minimally active or not active at all (Canadian Fitness and Lifestyle Research Institute, 2001b; U.S. Department of Health and Human Services, 2002f). More women than men are inactive. And inactivity is greater among low-SES adults—because they live in less safe neighborhoods, have a greater number of health problems, experience less social support for exercising regularly, and feel less personal control over their health (Grzywacz & Marks, 2001).

Besides reducing body fat and building muscle, exercise fosters resistance to disease. Frequent bouts of moderate-intensity exercise enhance the immune response, thereby lowering the risk of colds or flu and promoting faster recovery when these illnesses do strike (Nieman, 1994). Furthermore, in several longitudinal studies extending over 10 to 20 years, physical activity was linked to reduced incidence of cancer at all body sites except the skin, with the strongest findings for cancer of the rectum and colon (Albanes, Blair, & Taylor, 1989; Wannamethee, Shaper, & Macfarlane, 1993). Exercise also helps prevent adult-onset diabetes. Finally, physically active people are less likely to develop cardiovascular disease. If they do, it typically occurs at a later age and is less severe than among their inactive agemates (Bassey, 2000).

How does exercise help prevent the serious illnesses just mentioned? It may do so by reducing the incidence of obesity—a risk factor for heart disease, diabetes, and several forms of cancer. In addition, people who exercise probably adopt other healthful behaviors, thereby lowering the risk of diseases associated with high-fat diets, alcohol consumption, and smoking. Exercise can also have a direct impact on disease prevention. For example, in animal research, it inhibits growth of cancerous tumors beyond the impact of diet, body fat, and the immune response (Mackinnon, 1992). And it promotes cardiovascular functioning by strengthening the heart muscle, producing a form of "good cholesterol" (high-density lipoproteins) that helps remove "bad cholesterol" (low-density lipoproteins) from the artery walls, and decreasing blood pressure (Donatelle, 2003).

Yet another way that exercise may guard against illness is through its mental health benefits. Many studies show that physical activity reduces anxiety and depression, improves mood, and enhances alertness and energy (Kirkby & Lindner, 1998; Hassmén, Koivula, & Uutela, 2000). The impact of exercise on a "positive outlook," as Sharese expressed it, is most obvious just after a workout and can last for several hours (Chollar, 1995). The stress-reducing properties of exercise undoubtedly strengthen immunity to disease. And as physical activity enhances psychological well-being, it promotes self-esteem, ability to cope with stress, on-the-job productivity, and life satisfaction.

When we consider the evidence as a whole, it is not surprising that physical activity is associated with substantially lower death rates from all causes. The contribution of exercise to longevity cannot be accounted for by preexisting illness in inactive people because sedentary individuals who start to exercise live longer than those who remain inactive (Paffenbarger et al., 1993).

How much exercise is recommended for a healthier, happier, and longer life? Moderately intense physical activity—for example, 30 minutes of brisk walking—on most days leads to health benefits for previously inactive people. Adults who exercise at greater intensity—enough to build up a sweat—derive even greater protection against cardiovascular disease, diabetes, colon cancer, and obesity (Hu & Manson, 2001; Yu et al., 2003). Currently, the U.S. government recommends 30 minutes of moderate physical activity several times a week (U.S. Department of Health and Human Services, 2002f). Canadian recommendations are more stringent: at least 60 minutes of moderate activity per day, with time declining to 20 to 30 minutes as intensity increases to vigorous effort—for example, jogging, aerobics, or fast swimming (Health Canada, 2000d).

Substance Abuse

Despite an overall decrease in use of both legal and illegal substances in recent years, alcohol and drug dependency remain serious problems in early adulthood. They impair thinking, intensify psychological problems that underlie addiction, and increase the risk of unintentional injury and death. Return to Chapter 11, page 362 to review the personal and situational factors that lead individuals to become substance abusers in adolescence. The same characteristics are predictive in the adult years. Cigarette smoking and alcohol consumption are the most common substance disorders.

● **Cigarette Smoking.** Dissemination of information on the harmful effects of cigarette smoking has helped reduce its prevalence from 40 percent of North American adults in 1965 to about 25 percent in 2000 (Health Canada, 2002h; U.S. Department of Health and Human Services, 2002a). Still, smoking has declined very slowly.

Most of the drop is among college graduates; those who did not finish high school have changed very little. Furthermore, although more men than women smoke, the gender gap is much smaller today than in the past because smoking among young women who did not finish high school has increased sharply. Smoking among college students has also risen—for students of both sexes and of diverse ethnicities. More than 90 percent of men and 85 percent of women who smoke started before age 21 (Nordstrom et al., 2000; U.S. Department of Health and Human Services, 2001, 2002f). And the earlier people start smoking, the greater their daily cigarette consumption and likelihood of continuing, an important reason that preventive efforts with adolescents and young adults are vital.

The ingredients of cigarette smoke—nicotine, tar, carbon monoxide, and other chemicals—leave their damaging mark throughout the body. As the person inhales, delivery of oxygen to tissues is reduced, and heart rate and blood pressure rise. Over time, insufficient oxygen results in limited night vision, more rapid wrinkling of the skin, loss of bone mass, and a lower sperm count and higher rate of male sexual impotence (Margolin, Morrison, & Hulka, 1994; Tur, Yosipovitch, & Oren-Vulfs, 1992). More deadly outcomes include increased risk of heart attack, stroke, and cancer of the mouth, throat, larynx, esophagus, lungs, pancreas, kidneys, and bladder. Cigarette smoking is the single most important preventable cause of death in industrialized nations. One of every three young people who become regular smokers will die from a smoking-related disease (U.S. Department of Health and Human Services, 2002a).

The link between smoking and mortality is dose-related. The more cigarettes consumed, the greater the chance of premature death. At the same time, the benefits of quitting are great, including return of disease risks to nonsmoker levels within 3 to 8 years and a healthier living environment. Although millions of people have stopped smoking without help, those in treatment programs and those who use cessation aids (for example, nicotine gum, nasal spray, or patches, designed to reduce dependency gradually) often fail. After 1 year, 70 to 80 percent start smoking again (Brigham, 1998).

Unfortunately, too few treatments are long enough or teach skills for avoiding relapse. Sometimes a personal experience or the advice of a doctor is enough to energize quitting. My father, a heavy smoker in his twenties and thirties, sought treatment for chronic bronchitis (inflammation of the bronchial tubes, common among smokers). His doctor told him that if he did not stop smoking, he would not live to see his children grow up. That day marked his last cigarette. He died at age 80, having lived nearly 30 years longer than the average life expectancy for his year of birth.

● **Alcohol.** National surveys reveal that about 13 percent of men and 3 percent of women in the United States and Canada are heavy drinkers (Health Canada, 2002e; U.S. Department of Health and Human Services, 2002a). About one-third of heavy drinkers are *alcoholics*—people who cannot limit their alcohol use. In men, alcoholism usually begins in the teens and early twenties and worsens over the following decade. In women, its onset is typically later, in the twenties and thirties, and its course is more variable. Many alcoholics are also addicted to other drugs. About 80 percent are heavy cigarette smokers (Goodwin, 1997).

Twin studies support a genetic contribution to alcoholism (Tsuang et al., 2001). But whether a person comes to deal with life's problems through drinking is greatly affected by personal characteristics and circumstances, since half of hospitalized alcoholics do not have a family history of problem drinking (Hawkins, Catalano, & Miller, 1992). Alcoholism crosses SES and ethnic lines, but it is higher in some groups than others. For example, in cultures in which alcohol is a traditional part of religious or ceremonial activities, people are less likely to abuse it. Where access to alcohol is carefully controlled and is a sign of adulthood, dependency is more likely. Poverty and hopelessness also promote it (Donatelle, 2003).

© BILL ARON/PHOTOEDIT

■ In cultures where alcohol is a traditional part of religious or ceremonial activities, people are less likely to abuse it. Here, during a Jewish Passover service, participants say the blessing over the wine.

Alcohol acts as a depressant, impairing the ability of the brain to control thought and action. In a heavy drinker, it relieves anxiety but then induces it as the effects wear off, so the alcoholic drinks again. Chronic alcohol use does widespread physical damage. Its most well-known complication is liver disease, but it is also linked to cardiovascular disease, inflammation of the pancreas, irritation of the intestinal tract, bone marrow problems, disorders of the blood and joints, and some forms of cancer. Over time, alcohol causes brain damage, leading to confusion, apathy, inability to learn, and impaired memory (Brun & Andersson, 2001). The costs to society are enormous. About 40 percent of highway fatalities in the United States involve drivers who have been drinking. Nearly half of convicted felons are alcoholics, and about half of police activities in large cities involve alcohol-related offenses. Alcohol frequently plays a part in date rape, other forms of sexual coercion, and domestic violence (McKim, 2002).

The most successful treatments are comprehensive, combining personal and family counseling, group support, and aversion therapy (in which medication is used to produce a physically unpleasant reaction to alcohol, such as nausea and vomiting). Alcoholics Anonymous, a community support approach, helps many people exert greater control over their lives through the encouragement of others with similar problems. Nevertheless, breaking an addiction that has dominated a person's life is difficult; about 50 percent of alcoholics relapse within a few months (Volpicelli, 2001).

Sexuality

By the time their teenage years are over, more than 70 percent of young people have had sexual intercourse; by age 22, the figure rises to 90 percent (Darroch, Frost, & Singh, 2001; Michael et al., 1994). Compared with earlier generations, contemporary adults display a wider range of sexual choices and lifestyles, including cohabitation, marriage, extramarital experiences, and orientation toward a heterosexual or homosexual partner. In this chapter, we explore the attitudes, behaviors, and health concerns related to sexual activity becoming a regular event in young people's lives. In Chapter 14, we focus on the emotional side of close relationships.

● **Heterosexual Attitudes and Behavior.** One Friday evening, Sharese accompanied her roommate Heather to a young singles bar. Shortly after they arrived, two young men joined them. Faithful to her boyfriend Ernie, whom she met in college and who worked in another city, Sharese remained aloof for the next hour. In contrast, Heather was talkative and gave one of the men, Rich, her phone number. The next weekend, Heather went out with Rich. On the second date, they had intercourse, but the romance was short-lived. Within a few weeks, each went in search of a new partner. Aware of Heather's varied sex life, Sharese wondered whether her own was normal. Only after nearly a year of dating exclusively had she and Ernie slept together.

Since the 1950s, public display of sexuality in movies, newspapers, magazines, and books has steadily increased, fostering the impression that Americans are more sexually active than ever before. This escalation in sexual openness has prompted people whose sexual history falls short of popular images to question whether they might be missing something. What are modern adults' sexual attitudes and behaviors really like? Answers to this question were difficult to obtain until the completion of the National Health and Social Life Survey, the first in-depth study of Americans' sex lives based on a nationally representative sample. Nearly 4 out of 5 randomly chosen 18- to 59-year-olds agreed to participate—3,400 in all. Findings were remarkably similar to those of surveys conducted in France, Great Britain, and Finland (Laumann et al., 1994; Michael et al., 1994).

Recall from Chapter 11 that the sex lives of most teenagers do not dovetail with exciting media images. A similar picture emerges for adults in Western nations. Although their sexual practices are diverse, they are far less sexually active than we have come to believe. Couples like Sharese and Ernie are more typical (and satisfied) than couples like Heather and Rich.

Sexual partners, whether dating, cohabiting, or married, usually do not select each other arbitrarily. Instead, they tend to be alike in age (within 5 years), education, ethnicity, and (to a lesser extent) religion. In addition, most people who establish lasting relationships meet in conventional ways: Family members or friends introduce them, or they get to know each other at work, school, or social events where people similar to themselves congregate. Only 10 percent of adults who marry and 14 percent who decide to live together meet at bars, through personal ads, or on vacations. The powerful influence of social networks on sexual choice is adaptive. Sustaining an intimate relationship is easier when adults share interests and values and people they know approve of the match.

Consistent with popular belief, Americans today have more sexual partners than they did a generation ago. For example, one-third of adults over age 50 have had five or more partners in their lives, whereas half of 30- to 50-year-olds have accumulated that many in much less time. But when adults—either younger or older—are asked how many partners they have had in the past year, the usual reply (for 71 percent) is one.

What explains this trend toward more relationships in the context of sexual commitment? In the past, dating several partners was followed by marriage. Today, dating more often gives way to cohabitation, which leads either to marriage or to breakup. In addition, people are marrying later, and the divorce rate remains high. Together, these factors create more opportunities for new sexual partners. Still, surveys of college students reveal that almost all want to settle down with a mutually exclusive sexual partner at some point (Pedersen et al., 2002). In line with this goal, most people spend the majority of their lives with one partner. And only a small minority of Americans (3 percent), most of whom are men, report five or more partners in a single year. (See the Social Issues box on the following page for a discussion of sex differences in attitudes toward sexuality.)

How often do Americans have sex? Not nearly as frequently as the media would have us believe. Only one-third of 18- to 59-year-olds have intercourse as often as twice a week, another third have it a few times a month, and the remaining third have it a few times a year or not at all. Three factors affect frequency of sexual activity: age, whether people are co-

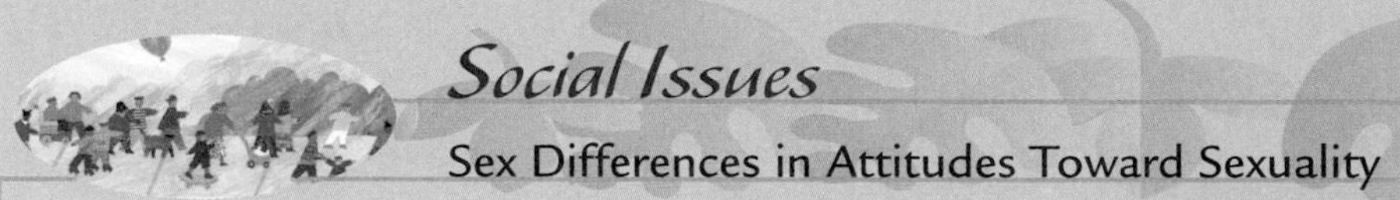

Social Issues

Sex Differences in Attitudes Toward Sexuality

Differences between men and women in sexual attitudes and behavior are widely assumed, and contemporary theories offer diverse explanations for them. For example, Nancy Chodorow, a feminist psychoanalytic theorist, and Carol Gilligan, a feminist theorist of moral development (see Chapter 12, page 390), believe that the emotional intensity of the infant–caregiver relationship is carried over into future intimate ties for girls. However, it is disrupted for boys as they form a "masculine" gender identity stressing independence and self-reliance (Chodorow, 1978; Gilligan, 1982).

According to an alternative evolutionary view, the desire to have children and ensure their survival powerfully shapes sexuality. Because sperm are plentiful and ova far less numerous, women must be more careful than men about selecting a partner with the commitment and resources needed to protect children's development (Bjorklund & Shackelford, 1999). Finally, social learning theory views sex differences as due to modeling and reinforcement of gender-role expectations (see Chapter 8, pages 260–262). Women receive more disapproval for having numerous partners and engaging in casual sex than do men, who are sometimes rewarded with admiration and social status.

Both small-scale studies and large-scale surveys confirm that women are more opposed to casual sex than men and are only half as likely as men to have engaged in it (Cubbins & Tanfer, 2000; Hyde & Oliver, 2000). However, when a small number of men with a great many sexual partners are excluded, contemporary men and women differ very little in average number of lifetime sexual partners (Chatterjee, Handcock, & Simonoff, 1995). Why is this so? From an evolutionary perspective, more effective contraception has permitted sexual activity with little risk of reproduction. Under these circumstances, women can have as many partners as men without endangering the welfare of their offspring.

©JON FEINGERSH/CORBIS

■ Because of more effective contraception, men and women differ much less in attitudes toward extramarital sex than they once did. And for both genders, satisfying sex is attained in the context of love, affection, and fidelity.

Still, when women complain that the men they meet are not interested in long-term commitments, their laments have a ring of truth. Many more men than women report that they are looking for sexual play and pleasure, with marriage and love not a part of the equation. And when infidelity occurs, men are more upset at the thought of their partner having sex with another person, women at the thought of their partner feeling affection for someone else—an outcome that could mean loss of investment in their offspring (Buss, 1999). These conflicting goals and attitudes are greatest for young adults. With age, both sexes become more similar, regarding a loving relationship as more central to sexuality (Schwartz & Rutter, 1998).

habiting or married, and how long the couple has been together. Younger, dating people have more partners, but this does not translate into more sex! Sexual activity increases through the twenties as people either cohabit or marry. Then, around age 30, it declines, even though hormone levels have not changed much. The demands of daily life—working, commuting, taking care of home and children—are probably responsible. Despite the common assumption that sexual practices vary greatly across social groups, the patterns just described are unaffected by education, SES, or ethnicity.

Although adults typically have only modest amounts of sex, most are happy with their sex lives. For those in committed relationships, feeling "extremely physically and emotionally satisfied" exceeds 80 percent; it rises to 88 percent for married couples. As number of sex partners increases, satisfaction declines sharply. These findings challenge the stereotypes of marriage as sexually dull and of people with many partners as having the "hottest" sex.

Furthermore, only a minority of adults—women more often than men—report persistent sexual problems. The two most frequent difficulties for women are lack of interest in sex (33 percent) and inability to achieve orgasm (24 percent); for men, climaxing too early (29 percent) and anxiety about performance (16 percent) are most common. Sexual difficulties are linked to low SES and psychological stress. They are also more common among people who are not married, have had more than five partners, and have experienced sexual abuse during childhood or (for women) sexual coercion in adulthood (Laumann, Paik, & Rosen, 1999). As these findings suggest, unfavorable personal relationships and sexual experiences increase the risk of sexual dysfunction.

But overall, a completely trouble-free physical experience is not essential for sexual happiness. Surveys of adults repeatedly show that satisfying sex involves more than technique; it is attained in the context of love, affection, and fidelity. In sum, happiness with partnered sex is linked to an emotionally fulfilling relationship, good mental health, and overall contentment with life (Bancroft, 2002; Michael et al., 1994).

● **Homosexual Attitudes and Behavior.** The tragic spread of AIDS has increased the visibility of homosexuals, reenergizing the gay liberation movement of the 1970s. Efforts at AIDS prevention, securing health care, and combating discrimination against people with AIDS have resulted in renewed attention to the civil rights of homosexuals. As a result, negative attitudes toward homosexuals have begun to change.

The majority of Americans support civil liberties and equal employment opportunities for gay men, lesbians, and bisexuals (Brooks, 2000). And although most Americans disapprove of sexual relations between two adults of the same sex, attitudes have become more accepting during the past decade, with about one-third of survey respondents saying that homosexuality is "not wrong at all" (Loftus, 2001). Homosexuals' political activism and greater openness about their sexual orientation have contributed to slow gains in acceptance. Exposure and interpersonal contact reduce negative attitudes (Epstein, 1999). Men judge homosexuals more harshly than do women, perhaps because heterosexual men are more concerned with conforming to prescribed gender roles (Kite & Whitley, 1998).

In the National Health and Social Life Survey, 2.8 percent of men and 1.4 percent of women identified themselves as homosexual or bisexual—figures similar to those of other recent national surveys conducted in the United States, France, and Great Britain (Black, Gates, & Saunders, 2000; Spira, 1992; Wellings et al., 1994). But an estimated 30 percent of same-sex couples do not report themselves as such in survey research. This unwillingness to answer questions, engendered by a climate of persecution, has limited what researchers have been able to find out about the sex lives of gay men and lesbians. The little evidence available indicates that homosexual sex follows many of the same rules as heterosexual sex: People tend to seek out partners similar in education and background to themselves; partners in committed relationships have sex more often and are more satisfied; and the overall frequency of sex is modest (Laumann et al., 1994; Michael et al., 1994).

Although homosexuals are few in number, they tend to live in large cities where many others share their sexual orientation and in college towns where attitudes are more accepting. Living in small communities where prejudice is intense and no social network exists through which to find compatible homosexual partners is isolating and lonely. People who identify themselves as gay or lesbian also tend to be well educated (Black, Gates, & Sanders, 2000). In the National Health and Social Life Survey, twice as many college-educated as high-school-educated men and 8 times as many college-educated as high-school-educated women stated they were homosexual. Although the reasons for these findings are not clear, they probably reflect greater social and sexual liberalism among the more highly educated and therefore greater willingness to disclose homosexuality.

● **Sexually Transmitted Disease.** A large number of North Americans—1 in 4 in the United States and 1 in 12 in Canada—are likely to contract a sexually transmitted disease (STD) some time in their lives (Health Canada, 2002e; U.S. Department of Health and Human Services, 1999b). (See Chapter 11, page 358, for a description of the most common STDs.) Although the incidence is highest in adolescence, STD continues to be prevalent in early adulthood. During the teens and twenties, people accumulate most of their sexual partners, and they often do not take appropriate precautions to prevent the spread of STD (see page 356). The overall rate of STD is higher among women than men, since it is at least twice as easy for a man to infect a woman with any STD, including AIDS, as it is for a woman to infect a man (Michael et al., 1994).

Although AIDS, the most deadly STD, remains concentrated among gay men and intravenous drug abusers, many homosexuals have responded to its spread by changing their sexual practices—limiting the number of partners with whom they have sex, choosing partners more carefully, and using latex condoms consistently and correctly. Heterosexuals at high risk due to a history of many partners have done the same. As a result, the number of infections is lower among gay and heterosexual men today than it was in the early 1980s. Still, AIDS remains the fifth-leading cause of death among American young adults (refer to Figure 13.5 on page 420). The incidence of HIV-positive adults is higher in the United States than in any other industrialized nations—2½ times the rate in Canada (Heuveline, 2002). AIDS is spreading most rapidly through heterosexual contact in poverty-stricken minority groups in which intravenous drug abuse is high and combines with poor health, inadequate education, high life stress, and hopelessness (Gayle, 2000). People overwhelmed by these problems are least likely to take preventive measures (Capaldi et al., 2002).

Yet AIDS can be contained and reduced—through sex education extending from childhood into adulthood and through access to health services, condoms, and clean needles and syringes for high-risk individuals (Stryker et al., 1995). In view of the dramatic rise in AIDS among women over the past two decades—from 7 percent to 19 percent of U.S. cases—a special need exists for female-controlled preventive measures (Hader et al., 2001). The recently developed female condom offers some promise, but its effectiveness and acceptability are not yet established.

● **Sexual Coercion.** STD is not the only serious health threat to accompany sexual activity. After a long day of classes, Sharese flipped on the TV and caught a talk show on sex without consent. Karen, a 25-year-old woman, described her husband Mike pushing, slapping, verbally insulting, and forcing her to have sex. "It was a control thing," Karen explained tearfully. "He complained that I wouldn't always do what he wanted. I was confused and blamed myself. I didn't leave because I was sure he'd come after me and get more violent."

One day, as Karen was speaking long distance to her mother on the phone, Mike grabbed the receiver and shouted, "She's not the woman I married! I'll kill her if she doesn't shape up!" Karen's parents realized the seriousness of her situation, arrived by plane to rescue her the next day, and helped her start divorce proceedings and get treatment.

An estimated 13 percent of women have endured *rape*, legally defined as intercourse by force, by threat of harm, or when the victim is incapable of giving consent (because of mental illness, mental retardation, or intoxication). Nearly half have experienced other forms of sexual aggression. The majority of victims (8 out of 10) are under age 30 (Humphrey & White, 2000). Women are vulnerable in the company of partners, acquaintances, and strangers. But most of the time their abusers are men they know well. Sexual coercion crosses SES and ethnic lines; people of all walks of life are offenders and victims. For example, in a national sample of over 6,000 American college students, 44 percent of women reported having experienced sexual coercion, and 19 percent of men said they had obtained sex through force (Koss, 1998).

Personal characteristics of the man with whom a woman is involved are far better predictors of her chances of becoming a victim than her own characteristics. Men who engage in sexual assault tend to believe in traditional gender roles, approve of violence against women, and accept rape myths ("Women want to be raped" or "Any healthy woman can resist if she really wants to"). Perpetrators also have difficulty interpreting women's social behavior accurately. They are likely to view friendliness as seductiveness, assertiveness as hostility, and resistance as desire. And they frequently deny their own responsibility, reasoning that "she brought it on herself" (Scully & Marolla, 1998). Furthermore, sexual abuse in childhood, promiscuity in adolescence, and alcohol abuse in adulthood are associated with sexually coercive behavior (Kalof, 2000; Senn et al., 2000).

Cultural forces—in particular, strong gender typing—contribute to sexual coercion. When men are taught from an early age to be dominant, competitive, and aggressive and women to be submissive, cooperative, and passive, the themes of rape are reinforced. Under these conditions, men may view a date not in terms of getting to know a partner but in terms of sexual conquest. And a husband may regard satisfaction of his sexual needs as his wife's duty, even when she is uninterested (Amaro, 1995). Societal acceptance of violence also sets the stage for rape, which typically occurs in relationships in which other forms of aggression are commonplace. Exposure to sexually aggressive pornography and other media images, which portray women desiring and enjoying the assault, also promote sexual coercion by dulling sensitivity to its harmful consequences (Donnerstein & Linz, 1998).

Consequences. Women's psychological reactions to rape resemble those of survivors of extreme trauma. Immediate responses include shock, confusion, withdrawal, and psychological numbing. These eventually give way to chronic fatigue, tension, disturbed sleep, depression, and suicidal thoughts (Resnick & Newton, 1992). When sexual coercion is ongoing, taking any action may seem dangerous (as it did to Karen), so the victim falls into a pattern of extreme passivity and fear (Goodman, Koss, & Russo, 1993). If she has a history of sexual abuse or received negative feedback after trying to tell someone, she is more likely to blame herself, which strengthens helpless reactions (Arata, 1999).

One-third to one-half of rape victims are physically injured. Some contract sexually transmitted diseases, and pregnancy results in about 5 percent of cases. Furthermore, women victimized by rape (and other crimes) report more symptoms of illness across almost all body systems. And they are more likely to engage in negative health behaviors, including smoking and alcohol use (Koss, Koss, & Woodruff, 1991).

Prevention and Treatment. Many rape victims are not as fortunate as Karen because their anxiety about provoking another attack keeps them from confiding even in trusted family members and friends. If they seek help for other problems, conflict over issues surrounding sexuality may lead a sensitive health professional to detect a possible rape. A variety of community services exist to help women take refuge from abusive partners, including safe houses, crisis hotlines, support groups, and legal assistance. Most, however, are underfunded and cannot reach out to everyone in need.

The trauma induced by rape is severe enough that therapy is important. In addition to individual treatment aimed at reducing anxiety and depression, many experts advocate

■ Psychological reactions to rape are severe enough that therapy is vital. In this rape crisis center, a professional counselor assists a victim in recovering from trauma through social support, validation of her experience, and safety planning.

group sessions because contact with other survivors is especially helpful in countering isolation and self-blame (Koss & Harvey, 1991; Neville & Heppner, 2002). Other critical features that foster recovery include

- *routine screening for victimization* when women seek health care to ensure referral to community services and protection from future harm;
- *validation of the experience,* by acknowledging that many other women have been physically and sexually assaulted by intimate partners; that such assaults lead to a wide range of persisting symptoms, are illegal and inappropriate, and should not be tolerated; and that the trauma can be overcome; and
- *safety planning,* even when the abuser is no longer present, to prevent recontact and reassault. This includes information about how to obtain police protection, legal intervention, a safe shelter, and other aid should a rape survivor find herself at risk again.

Finally, many steps can be taken at the level of the individual, the community, and society to prevent sexual coercion. Some are listed in the Caregiving Concerns table below.

● **Menstrual Cycle.** The menstrual cycle is central to women's lives, and it presents unique health concerns. Although almost all women experience some discomfort during menstruation, others have more severe difficulties. **Premenstrual syndrome (PMS)** refers to an array of physical and psychological symptoms that usually appear 6 to 10 days prior to menstruation. The most common are abdominal cramps, fluid retention, diarrhea, tender breasts, backache, headache, fatigue, tension, irritability, and depression; the precise combination varies from person to person. PMS is usually experienced for the first time after age 20. Nearly 40 percent of women have some form of it, usually mild. For 10 percent, PMS is severe enough to interfere with academic, occupational, and social functioning. PMS affects women of all SES levels and is a worldwide phenomenon—just as common in Italy and the Islamic nation of Bahrain as it is in the United States (Brody, 1992; American College of Obstetricians and Gynecologists, 2000).

The causes of PMS are not well established, but evidence for a genetic predisposition is accumulating. Identical twins are twice as likely as fraternal twins to share the syndrome (Freeman & Halbreich, 1998). PMS is related to hormonal changes that follow ovulation and precede menstruation. But since hormone therapy is not consistently effective, sensitivity of brain centers to these hormones, rather than the hormones themselves, may be responsible (Frackiewicz & Shiovitz, 2001). Common treatments include analgesics for pain, antidepressant medication, diuretics for fluid buildup, limiting caffeine intake (which can intensify symptoms), a low-fat, high-fiber diet, vitamin/mineral supplements, exercise, and other strategies for reducing stress. Although each of these approaches is helpful in certain cases, as yet no method has been successful in curing PMS.

Psychological Stress

A final health concern, threaded throughout previous sections, has such a broad impact that it merits a comment of its own. Psychological stress, measured in terms of adverse social conditions, negative life events, or daily hassles, is related to a wide variety of health outcomes. In addition to its association with many unhealthy behaviors, stress has clear physical consequences. For example, as it mobilizes the body for action, it elevates blood pressure. Chronic stress resulting from

Preventing Sexual Coercion

SUGGESTION	DESCRIPTION
Reduce gender stereotyping and gender inequalities.	The roots of sexual coercion lie in the historically subordinate role of women. Restricted educational and employment opportunities keep many women economically dependent on men and therefore poorly equipped to avoid partner violence.
Mandate treatment for men who physically or sexually assault women.	Ingredients of effective intervention include inducing personal responsibility for violent behavior; teaching social awareness, social skills, and anger management; and developing a support system to prevent future attacks.
Expand interventions for children and adolescents who have witnessed violence between adult caregivers.	Although most child witnesses to violence do not become involved in abusive relationships as adults, boys are at increased risk of assaulting their female partners and girls are at increased risk of becoming victims if they observed violence between their parents.
Teach women to take precautions that lower the risk of sexual assault.	Risk of sexual assault can be reduced by communicating sexual limits clearly to a date; developing neighborhood ties to other women; increasing the safety of the immediate environment (for example, installing deadbolt locks, checking the back seat of the car before entering); avoiding deserted areas; and not walking alone after dark.

Sources: Browne, 1993; Smith, 2002.

Fostering a Healthy Adult Life

SUGGESTION	DESCRIPTION
Engage in healthy eating behavior.	Educate yourself and those with whom you live about the makeup of a healthy diet. Eat in moderation, and learn to distinguish true hunger from eating due to boredom or stress.
Maintain a reasonable body weight.	If you need to lose weight, make a commitment to a lifelong change in the way you eat, not just a temporary diet. Select a sensible, well-balanced dietary plan, and exercise regularly.
Keep physically fit.	Choose a specific time to exercise, and stick with it. To help sustain physical activity and make it more enjoyable, exercise with your partner or a friend and encourage each other. Set reasonable expectations, and allow enough time to reach your fitness goals; many people become exercise dropouts because their expectations were too high.
Control alcohol intake, and do not smoke cigarettes.	Drink moderately or not at all. Do not allow yourself to feel you must drink to be accepted or to enjoy a social event. If you smoke, choose a time that is relatively stress-free to quit. Seek the support of your partner or a friend.
Engage in responsible sexual behavior.	Identify attitudes and behaviors that you need to change to develop a healthy intimate relationship. Educate yourself about sexual anatomy and functioning so you can make sound decisions about contraception and protect yourself against sexually transmitted disease.
Manage stress.	Seek a reasonable balance among work, family, and leisure. Become more aware of stressors, and identify effective ways of coping with them so you are better prepared when they arise. Engage in regular exercise, and find time each day for relaxation and quiet reflection.

Source: Donatelle, 2003.

economic hardship and inner-city living is consistently linked to hypertension, a relationship that contributes to the high incidence of heart disease in low-income groups, especially African Americans. Compared with higher-SES individuals, low-SES adults actually show a stronger cardiovascular response to stress (Carroll et al., 2000). Earlier we mentioned that psychological stress also interferes with immune system functioning, a link that may underlie its relationship to several forms of cancer. And by reducing digestive activity as blood flows to the brain, heart, and extremities, stress can cause gastrointestinal difficulties, including constipation, diarrhea, colitis, and ulcers (Donatelle, 2003).

The many challenging tasks of early adulthood make it a particularly stressful time of life. Young adults more often report feelings of depression than middle-aged people, many of whom have attained vocational success and financial security and are enjoying more free time as parenting responsibilities decline (Schieman, Gundy, & Taylor, 2001; Wade & Cairney, 1997). Also, as we will see in Chapters 15 and 16, middle-aged adults are better than young adults at coping with stress. Because of their longer life experience and greater sense of personal control over their lives, they are more likely to engage in effective problem solving when stressful conditions can be changed and to manage negative emotion when nothing can be done about an unpleasant situation (Lazarus, 1991).

In previous chapters, we repeatedly noted the stress-buffering effect of social support, which continues throughout life (Markides & Cooper, 1989). Helping stressed young adults establish and maintain satisfying social ties is as important a health intervention as any we have mentioned. Before we turn to the cognitive side of early adulthood, you may find it helpful to examine the Caregiving Concerns table above, which summarizes the many ways we can foster a healthy adult life.

Ask Yourself

REVIEW

List as many factors as you can that may have contributed to heart attacks and early death among Sharese's relatives.

REVIEW

Why are people in committed relationships likely to be more sexually active and satisfied than those who are dating several partners?

APPLY

Tom began going to a health club three days a week after work. Soon the pressures of his job convinced him that he no longer had time for so much exercise. Explain to Tom why he should keep up his exercise regimen, and suggest ways to fit it into his busy life.

CONNECT

Cite history-graded influences that have contributed to the obesity epidemic. (To review this aspect of the lifespan perspective, refer to Chapter 1, page 11.)

www.

COGNITIVE DEVELOPMENT

How does cognition change with the transition to adulthood? Lifespan theorists have examined this question from three familiar vantage points. First, they have proposed transformations in the structure of thought—new, qualitatively distinct ways of thinking that extend the cognitive-developmental changes of childhood and adolescence. Second, adulthood is a time of attaining advanced knowledge in a particular area, an accomplishment that has important implications for information processing and creativity. Finally, researchers have been interested in the extent to which the diverse mental abilities assessed by intelligence tests remain stable or change during the adult years.

Changes in the Structure of Thought

Sharese described her first year in graduate school as a "cognitive turning point." As part of her internship in a public health clinic, she observed firsthand the many factors that affect human health-related behaviors. For a time, she was intensely uncomfortable about the fact that clear-cut solutions to dilemmas in the everyday world were so hard to come by. "Working in this messy reality is so different from the problem solving I did in my undergraduate classes," she commented to her mother in a phone conversation one day.

Sharese's reflections agree with those of a variety of researchers who have studied **postformal thought**—cognitive development beyond Piaget's formal operational stage. Even Piaget (1967) acknowledged the possibility that important advances in thinking follow the attainment of formal operations. He observed that adolescents place excessive faith in abstract systems, preferring a logical, internally consistent but inaccurate perspective on the world to one that is vague, contradictory, and adapted to particular circumstances (see Chapter 11, page 368). To clarify how thinking is restructured in adulthood, let's look at some influential theories. Together, they show how educational, family, and occupational challenges spark increasingly flexible and practical ways of thinking.

Perry's Theory

William Perry (1970, 1981) interviewed students at the end of each of their four years of college, asking "what stood out" during the previous year. Responses indicated that cognitive perspectives changed as students experienced the complexities of university life and moved closer to adult roles—findings confirmed in recent research (Magolda, 2002; Moore, 2002).

Younger students regarded knowledge as made up of separate units (beliefs and propositions) whose truth could be determined by comparing them to abstract standards—standards that exist apart from the thinking person and his or her situation. As a result, they engaged in **dualistic thinking,** dividing information, values, and authority into right and wrong, good and bad, we and they. As one college freshman stated, "When I went to my first lecture, what the man said was just like God's word. I believe everything he said because he is a professor . . . and this is a respected position" (Perry, 1981, p. 81).

In contrast, older students were aware of a diversity of opinions on almost any topic. They moved toward **relativistic thinking,** viewing all knowledge as embedded in a framework of thought. Consequently, they gave up the possibility of absolute truth in favor of multiple truths, each relative to its context. Their thinking had become more flexible, tolerant, and realistic. A college senior reasoned, "Just seeing how [famous philosophers] fell short of an all-encompassing answer, [you realize] that ideas are really individualized. And you begin to have respect for how great their thought could be, without its being absolute" (p. 90). Relativistic thinking leads to the realization that commitment to one truth system is partly a matter of choice, since several frameworks may satisfy the criterion of internal logical consistency (Sinnott, 1998).

Perry's theory is based on a sample of highly educated young adults. He acknowledges that movement from dualism to relativism is probably limited to people confronted with the multiplicity of viewpoints typical of a college education. But the underlying theme of *adaptive cognition*—thought less constrained by the need to find one answer to a question and more responsive to its context—is also evident in two additional theories.

Schaie's Theory

According to K. Warner Schaie (1977/1978), it would be difficult for human cognition to exceed the complexity of Piaget's formal operational stage. But with entry into adulthood, the situations in which people must reason become more diverse. As a result, the goals of mental activity shift from acquiring knowledge to using it, as the following stage sequence reveals.

1. The **acquisitive stage** (childhood and adolescence). The first two decades of life are largely devoted to *knowledge acquisition.* As young people move from concrete to formal operational thought, they develop more powerful procedures for storing information, combining it, and drawing conclusions.

2. The **achieving stage** (early adulthood). In early adulthood, people must adapt their cognitive skills to situations (such as marriage and employment) that have profound implications for *achieving long-term goals.* As a result, they focus less on acquiring knowledge and more on applying it to everyday life. The problems encountered in work, intimate relationships, and child rearing, unlike those posed in classrooms, often do not have single cor-

■ Compared with adolescents, young adults focus less on acquiring knowledge and more on applying it to everyday life. Problems encountered at work, unlike those posed in classrooms or on intelligence tests, often do not have a single correct solution.

rect solutions. Yet how the individual handles them can affect the entire life course. To be successful, young adults must attend to both the problem and its context, not just the problem.

3. The **responsibility stage** (middle adulthood). As families and work lives become well established, *expansion of responsibilities to others* takes place on the job, in the community, and at home. As a result, cognition extends to situations involving social obligations. Maintaining a gratifying relationship with an intimate partner, staying involved in children's lives, assuming greater leadership on the job, and fulfilling community roles must be juggled simultaneously and effectively. The most advanced form of this type of thinking, called the **executive stage,** characterizes individuals whose responsibilities have become highly complex. People at the helm of large organizations—businesses, colleges and universities, and religious organizations—must monitor organizational progress and the activities of many people. This demands that they understand the dynamic forces that affect an elaborate social structure and combine information from many sources to make decisions.

4. The **reintegrative stage** (late adulthood). With retirement, people reexamine and *reintegrate their interests, attitudes, and values,* using them as a guide for maximizing quality of life. As the future shortens, the need to acquire knowledge and monitor decisions in terms of later consequences declines. Older adults move beyond asking, "What should I know?" and "How should I use what I know?" to asking, "Why should I know?" (Schaie & Willis, 2000, p. 179). Consequently, they are more selective about the circumstances in which they expend their cognitive energies, rarely wasting time on tasks that seem meaningless or irrelevant to their daily lives.

Labouvie-Vief's Theory

Gisella Labouvie-Vief's (1980, 1985) portrait of adult cognition echoes features of Schaie's theory. Adolescents, she points out, operate within a world of possibility. Adulthood involves movement from hypothetical to **pragmatic thought**, a structural advance in which logic becomes a tool for solving real-world problems.

According to Labouvie-Vief, the need to specialize motivates this change. As adults select and pursue one path out of many alternatives, they become more aware of the constraints of everyday life. And in the course of balancing various roles, they give up their earlier need to resolve contradictions. Instead, they accept inconsistencies as part of life and develop ways of thinking that thrive on imperfection and compromise. Sharese's friend Christy, a student and mother of two young children, illustrates:

> I've always been a feminist, and I wanted to remain true to my beliefs in family and career. But this is Gary's first year of teaching high school, and he's saddled with four preparations and coaching the school's basketball team. At least for now, I've had to settle for "give-and-take feminism"— going to school part-time and shouldering most of the responsibility for the kids while he gets used to his new job. Otherwise, we'd never make it financially.

Awareness of multiple truths, integration of logic with reality, and tolerance for the gap between the ideal and the real sum up qualitative transformations in thinking during the adult years (Sinnott, 1998). As we will see in the next section, adults' increasingly specialized and context-bound thought, although it closes off options, also opens new doors to higher levels of competence.

Information Processing: Expertise and Creativity

In Chapter 9, we noted that children's expanding knowledge improves their ability to remember new information related to what they already know. **Expertise**—acquisition of extensive knowledge in a field or endeavor—is supported by the specialization that begins with selecting a college major or an occupation, since it takes many years for a person to master any complex domain (Horn & Masunaga,

■ Expertise is supported by specialization. In accord with First Nation traditions, this member of the Alaskan Tsimshian Nation tribe carves a totem pole for a potlatch ceremony—a special group celebration of sharing food and resources, marking a birth or wedding or mourning the death of a leader. The sculptor's knowledge of his craft permits him to move from problem to solution quickly and effectively.

2000). Once expertise is attained, it has a profound impact on information processing.

Compared with novices, experts remember and reason more quickly and effectively. This makes sense, since the expert knows more domain-specific concepts and represents them in richer ways—at a deeper and more abstract level and as having more features that can be linked to other concepts. As a result, experts approach problems with underlying principles in mind, whereas novices' understanding is superficial. For example, a highly trained physicist notices when several problems deal with conservation of energy and can therefore be solved similarly. In contrast, a beginning physics student focuses only on surface features—whether the problem contains a disk, a pulley, or a coiled spring (Chi, Glaser, & Farr, 1988). Experts appear more cognitively mature because they can use what they know to arrive at solutions automatically—through quick and easy remembering. And when a problem is especially challenging, they are more likely to plan ahead, systematically analyzing and categorizing elements and selecting the best from many possibilities. The novice proceeds more by trial and error.

Besides enhancing problem solving, expertise is necessary for creativity. The creative products of adulthood differ from those of childhood in that they are not just original but also directed at a social or aesthetic need. Consequently, mature creativity requires a unique cognitive capacity—the ability to formulate new, culturally meaningful problems and to ask significant questions that have not been posed before. According to Patricia Arlin (1989), movement from *problem solving* to *problem finding* is a core feature of postformal thought evident in highly accomplished artists and scientists.

Case studies support the 10-year rule in development of master-level creativity—a decade between initial exposure to a field and sufficient expertise to produce a creative work (Feldman, 1999; Simonton, 2000). Furthermore, a century of research reveals that creative accomplishment rises in early adulthood, peaks in the late thirties or early forties, and gradually declines. However, exceptions exist. People off to an early start in creativity tend to peak and drop off sooner, whereas "late bloomers" reach their full stride at older ages. This suggests that creativity is more a function of "career age" than of chronological age. The course of creativity also varies across disciplines. For example, artists and musicians typically show an early rise in creativity, perhaps because they do not need extensive formal education before they begin to produce. Academic scholars and scientists usually display their achievements later and over a longer time because they must earn a higher degree and spend years doing research to make worthwhile contributions (Simonton, 1991).

Creativity is rooted in expertise, but not all experts are creative. Other qualities are needed, including an innovative thinking style, tolerance of ambiguity, a special drive to succeed, and a willingness to experiment and try again after failure (Csikszentmihalyi, 1999; Sternberg & Lubart, 1996). In addition, creativity requires time and energy. For women especially, it can be postponed or disrupted by child rearing, divorce, or an unsupportive partner (Vaillant & Vaillant, 1990). In sum, creativity is multiply determined. When personal and situational factors jointly promote it, creativity can continue for many decades, well into old age.

Changes in Mental Abilities

Intelligence tests are a convenient tool for assessing changes in a variety of mental abilities during the adult years. Factors included on tests for adults are similar to those for school-age children and adolescents (see pages 293–294). Think back to our discussion of postformal thought. It suggests that intelligence tests, well suited to measuring skills needed for success in school, are less adequate for assessing competencies relevant to many adults' everyday lives (Schaie & Willis, 1996). Consistent with this view, in a study of bank managers ranging in age from 24 to 59, verbal and spatial mental test scores bore no relation to managerial skill, based on job performance ratings and salary raises. But a measure of practical knowledge, which required participants to rate the quality of solutions to work-related problems, did predict outstanding on-the-job achievement (Colonia-Willner, 1998).

Nevertheless, research with mental tests sheds light on the widely held belief that intelligence declines in adulthood as structures in the brain deteriorate. Many cross-sectional studies show this pattern—a peak in intelligence test performance at age 35 followed by a steep drop into old age. But widespread

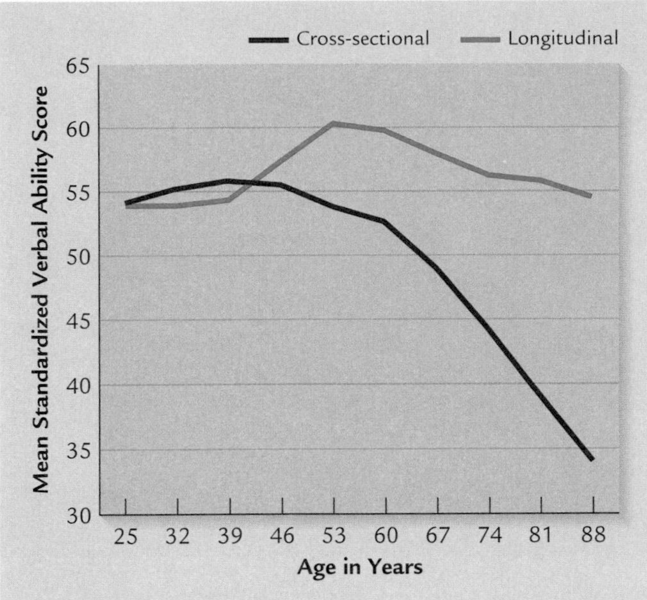

■ **FIGURE 13.7 Cross-sectional and longitudinal trends in verbal ability, illustrating cohort effects.** The steep cross-sectional decline is largely due to better health and education in younger generations. When adults are followed longitudinally, their verbal scores rise during early and middle adulthood and gradually decline during later years. However, this longitudinal trend does not hold for all abilities. For example, spatial skills begin declining earlier, as we will see in Chapter 15. (Adapted from K. W. Schaie, 1988, "Variability in Cognitive Functioning in the Elderly," in M. A. Bender, R. C. Leonard, & A. D. Woodhead, Eds., *Phenotypic Variation in Populations,* New York: Plenum, p. 201. Reprinted by permission.)

testing of young adults in college and in the armed services in the 1920s offered researchers a convenient opportunity to conduct longitudinal research, retesting people in middle adulthood. Results revealed an age-related increase! What explains this contradiction? To find out, K. Warner Schaie (1994, 1998) applied the *longitudinal-sequential design,* which combines longitudinal and cross-sectional approaches (see Chapter 1, page 35), in the Seattle Longitudinal Study.

In 1956, people ranging in age from 22 to 70 were tested cross-sectionally. Then, at regular intervals, longitudinal follow-ups were conducted and new samples added, yielding a total of 5,000 participants, five cross-sectional comparisons, and longitudinal data spanning 35 years. Findings on five mental abilities showed the typical cross-sectional drop after the mid-thirties. But longitudinal trends for those abilities revealed modest gains sustained into the fifties and early sixties, after which performance decreased very gradually.

Figure 13.7 illustrates Schaie's cross-sectional and longitudinal outcomes for just one intellectual factor—verbal ability. *Cohort effects* are largely responsible for the difference. In cross-sectional research, each new generation experienced better health and education than the one before it. Consequently, in Schaie's research and other similar but smaller-scale studies, younger test-takers score considerably higher than older test-takers (Kaufman, 2001). So a more accurate picture, obtained from longitudinal evidence, is steady improvement from early to middle adulthood in verbal skills and other abilities that depend on accumulated knowledge, followed by a decline late in life.

But not all mental abilities follow the trend just described. Those that depend very little on acquired knowledge but tap the capacity to detect relationships (such as spatial skills) decline earlier, perhaps starting in the mid-twenties—although the age at which performance begins to drop off is a matter of debate (Kaufman, 2001). Clearly, intellectual change in adulthood is a prime example of multidirectionality in development, an assumption of the lifespan perspective (see Chapter 1, page 9). In later chapters, we will consider factors that contribute to these contrasting patterns, as well as to vast individual differences in mental test performance that extend throughout adulthood.

Ask Yourself

REVIEW

How does expertise affect information processing? Why is expertise necessary for, but not the same as, creativity?

APPLY

For her human development course, Marcia wrote a paper in which she explained that behaviorists and psychoanalysts regard physical punishment as ineffective for different reasons. She concluded that both viewpoints are valid. Explain how Marcia's reasoning illustrates postformal thought.

APPLY

As a college sophomore, Steven took a course in decision making but didn't find it very interesting. At age 25, newly married and employed, he realized the value of what the professor had taught him. How do changes in the structure of thought help explain Steven's revised attitude?

www.

The College Experience

Looking back at the trajectory of their lives, many people view the college years as formative—more influential than any other period of adulthood. This is not surprising, since college serves as a "developmental testing ground," a time when full attention can be devoted to exploring alternative values, roles, and behaviors. To facilitate this exploration, college exposes students to a form of "culture shock"—encounters with new ideas and beliefs, new freedoms and opportunities, and new academic and social demands (Pascarella & Terenzini, 1991). About 70 percent of Canadian and 75 percent of American high school graduates enroll in an institution of higher education. Besides offering a route to a high-status career and its rewards, colleges and universities have a transforming impact on young people.

Psychological Impact of Attending College

A comprehensive review of thousands of studies revealed broad psychological changes from the freshman to the senior year of college (Pascarella & Terenzini, 1991). As Perry's theory revealed, students become better at reasoning about problems that have no clear solution and at identifying the strengths and weaknesses of opposing sides of complex issues. Their attitudes and values also broaden. They show increased interest in literature, the performing arts, and philosophical and historical issues and greater tolerance for ethnic and cultural diversity. And as noted in Chapter 12, college leaves its mark on moral reasoning by fostering concern with individual rights and human welfare. Finally, exposure to multiple world views encourages young people to look more closely at themselves. During the college years, students develop greater self-understanding, enhanced self-esteem, and a firmer sense of identity.

How do these interrelated changes come about? Type of college attended makes little difference, since cognitive growth is just as great at two-year community colleges as at four-year institutions (Bohr et al., 1994). Instead, the impact of college is jointly influenced by the person's involvement in academic and nonacademic activities and the richness and diversity of the campus setting. The more students interact with peers about course-related and non-course-related issues, the more they benefit. Residence hall living is one of the most consistent predictors of cognitive change because it maximizes involvement in the educational and social systems of the institution (Terenzini, Pascarella, & Blimling, 1999). Since over half of American college students commute, these findings underscore the importance of programs that integrate commuting students into out-of-class campus life. Quality of academic experiences also affects college outcomes. Psychological benefits increase with students' effort and willingness to participate in class and with challenging teaching that integrates learning in separate courses and that provides extensive contact with faculty (Franklin, 1995).

Dropping Out

A college degree channels people's postcollege lives, affecting their world view and opportunities in enduring ways. Yet 45 percent of students at two-year institutions and 26 percent of students at four-year institutions drop out, most within the first year and many within the first 6 weeks (American College Testing, 1998). The price paid is high, in lifelong earnings and personal development.

Academic factors are not the sole reason for leaving, since most young people who withdraw have the ability to succeed in the college to which they were admitted. Other personal and institutional characteristics play a larger role. Most entering freshmen have high hopes for college life but find the transition difficult. Those who have trouble adapting—because of lack of motivation, poor study skills, financial pressures, or emotional dependence on parents—quickly develop negative attitudes toward the college environment. Often these exit-prone students do not meet with their advisors or professors. At the same time, colleges that do little to provide encouragement and assistance to high-risk students have a higher percentage of dropouts (Moxley, Najor-Durack, & Dumbrigue, 2001).

© BILL ARON/PHOTOEDIT

■ Residential living enhances the beneficial effects of attending college on cognition, attitudes, and values. When students live on campus, they are more likely to experience the richness and diversity of the campus setting.

Factors involved in the decision to withdraw from college are usually not catastrophic; they are typical problems of early adulthood. Reaching out to students, especially during the early weeks and throughout the first year, is crucial. Young people who sense that they have entered a college community concerned about them as individuals are far more likely to persist to graduation.

Vocational Choice

Young adults—college-bound and non-college-bound alike—face a major life decision: the choice of a suitable work role. Being a productive worker calls for many of the same qualities as being an active citizen and a nurturant family member—good judgment, responsibility, dedication, and cooperation. How do young people make decisions about careers, and what influences their choices? What is the transition from school to work like, and what factors make it easy or difficult?

Selecting a Vocation

In societies with an abundance of career possibilities, occupational choice is a gradual process that begins long before adolescence. Major theorists view the young person as moving through several phases of vocational development (Ginzberg, 1972, 1988; Super, 1980, 1984):

1. The **fantasy period** (early and middle childhood). Young children gain insight into career options by fantasizing about them. However, their preferences are largely guided by familiarity, glamour, and excitement and bear little relation to the decisions they will eventually make.

2. The **tentative period** (early and middle adolescence). Between ages 11 and 17, adolescents start to think about careers in more complex ways. At first, they evaluate vocational options in terms of their interests. Later, as they become more aware of personal and educational requirements for different vocations, they take into account their *abilities and values.* "I like science and the process of discovery," Sharese thought to herself as she neared high school graduation. "But I'm also good with people, and I'd like to do something to help others. So maybe teaching or medicine would suit my needs."

3. The **realistic period** (late adolescence and early adulthood). By the early twenties, the economic and practical realities of adulthood are just around the corner, and young people start to narrow their options. At first, many do so through further *exploration,* gathering more information about possibilities that blend with their personal characteristics. Then they enter a final phase of *crystallization* in which they focus on a general vocational category. Within it, they experiment for a period of time before settling on a single occupation. As a college sophomore, Sharese pursued her interest in science, but she had not yet selected a major. Once she decided on chemistry, she considered whether to pursue teaching, medicine, or public health.

Factors Influencing Vocational Choice

Most, but not all, young people follow this pattern of vocational development. A few know from an early age just what they want to be and follow a direct path to a career goal. Some decide and later change their minds, and still others remain undecided for an extended period. College students are granted added time to explore various options. In contrast, the life conditions of many low-SES youths restrict their range of choices.

Consider how an occupational choice is made, and you will see that it is not just a rational process in which young people weigh abilities, interests, and values against career options. Like other developmental milestones, it is the result of a dynamic interaction between person and environment. A great many influences feed into the decision.

● **Personality.** People are attracted to occupations that complement their personalities. John Holland (1966, 1985) identified six personality types that affect vocational choice:

- The *investigative person,* who enjoys working with ideas and is likely to select a scientific occupation (for example, anthropologist, physicist, or engineer)
- The *social person,* who likes interacting with people and gravitates toward human services (counseling, social work, or teaching)
- The *realistic person,* who prefers real-world problems and work with objects and tends to choose a mechanical occupation (construction, plumbing, or surveying)
- The *artistic person,* who is emotional and high in need for individual expression and looks toward an artistic field (writing, music, or the visual arts)
- The *conventional person,* who likes well-structured tasks and values material possessions and social status—traits well suited to certain business fields (accounting, banking, or quality control)
- The *enterprising person,* who is adventurous, persuasive, and a strong leader and is drawn to sales and supervisory positions or politics

Research reveals a clear relationship between personality and vocational choice, but it is only moderate (Holland, 1985). Many people are blends of several personality types and can do well at more than one kind of occupation. Furthermore, career decisions are made in the context of family influences, educational opportunities, and current life circumstances. For example, Sharese's friend Christy scored high on Holland's "investigative" dimension. But when she married and had children early, she postponed her dream of becoming a college professor and chose a human services career that required fewer years of education. As Christy's case illustrates, personality takes us only partway in understanding vocational choice.

● **Family Influences.** Young people's vocational aspirations correlate strongly with the jobs of their parents. Individuals who grew up in higher-SES homes are more likely to select high-status, white-collar occupations, such as doctor, lawyer, scientist, and engineer. In contrast, those with low-income backgrounds tend to choose less prestigious, blue-collar careers—for example, plumber, construction worker, food service employee, and secretary. Parent–child similarity is partly a function of educational attainment. The single best predictor of occupational status is number of years of schooling completed (Featherman, 1980).

Family resemblance in occupational choice also comes about for other reasons. Higher-SES parents are more likely to give their children important information about the world of work and to have connections with people who can help the young person obtain a high-status position (Grotevant & Cooper, 1988). Parenting practices also shape work-related values. Recall from Chapter 2 that higher-SES parents tend to promote independence and curiosity, which are required in many high-status careers. Lower-SES parents, in contrast, are more likely to emphasize conformity and obedience. Eventually, young people choose careers that are compatible with these values. The jobs that appeal to them tend to be like those of their parents (Mortimer & Borman, 1988). Still, parents can foster higher aspirations. Parental pressure to do well in school and encouragement toward high-status occupations predict vocational attainment beyond SES (Bell et al., 1996).

● **Teachers.** Teachers also play a powerful role in career decisions. Young adults in careers requiring extensive educational preparation often report that teachers influenced their choice of a field of study (Reddin, 1997). College students have more opportunities to develop close relationships with teachers than do their non-college-educated counterparts, whose parents are more influential. These findings provide yet another reason to promote positive teacher–student relations, especially for high school students from low-SES families. The power of teachers as role models could serve as an important source of upward mobility for these young people.

● **Gender Stereotypes.** Over the past two decades, young men's career preferences have remained strongly gender-stereotyped, whereas young women have expressed increasing interest in occupations largely held by men (Gottfredson, 1996). Changes in gender-role attitudes along with the dramatic rise in employed mothers, who serve as career-oriented models for their daughters, are common explanations for women's growing attraction to nontraditional careers.

At the same time, women's progress in entering and excelling at male-dominated professions has been slow. As Table 13.2 shows, the percentage of women engineers, lawyers, and doctors increased between 1983 and 2000 in the United States, but it falls far short of equal representation. Women remain concentrated in less well-paid, traditionally feminine professions, such writing, social work, education, and nursing (U.S. Bureau of the Census, 2002). In virtually all fields, their achievements lag behind those of men, who write more books, make more discoveries, hold more positions of leadership, and produce more works of art.

Ability cannot account for these dramatic sex differences. Recall from Chapter 11 that the gender gap in cognitive performance of all kinds is small and is declining. Instead, gender-stereotyped messages play a key role. Although girls' grades are higher than boys', girls reach secondary school less confident of their abilities and more likely to underestimate their achievement (Wigfield et al., 2002). Between tenth and twelfth grade, the percentage of girls in programs for the gifted decreases. When girls were asked what discouraged them from continuing in such classes, parental and peer pressures and attitudes of teachers and counselors ranked high on their lists (Read, 1991; Winner, 1996).

During college, the career aspirations of academically talented females decline further. In one longitudinal study, high school valedictorians were followed over a 10-year period—through college and into the work world. By their sophomore year, young women showed a decline in estimates of their intelligence, whereas men did not. Women also shifted their expectations toward less demanding careers because of concerns about combining work with child rearing and unresolved questions about their ability (Arnold, 1994). Another study reported similar results. Educational aspirations of mathematically talented women declined considerably during college, as did the number majoring in the sciences (Benbow & Arjmand, 1990). And mathematically talented women who do select a science major more often choose medicine or other health professions—and less often choose engineering, math, or physical science—than their male counterparts (Benbow et al., 2000).

These findings reveal a pressing need for programs that sensitize high school and college personnel to the special problems women face in developing and maintaining high vocational aspirations and selecting nontraditional careers. Research shows that the aspirations of young women rise in response to career guidance that encourages them to set goals that match their abilities, interests, and values. Those who continue to achieve usually have four experiences in common:

- A college environment that values and supports the accomplishments of women and that attempts to enhance women's experiences in its curriculum
- Frequent interaction with faculty and professionals in their chosen fields
- The opportunity to test their abilities in a supportive environment
- Models of accomplished women who have successfully dealt with family–career role conflict (Arnold, 1994; Pascarella et al., 1997; Swanson & Fouad, 1999)

● **Access to Vocational Information.** Finally, many young people could profit from greater access to career information. In a longitudinal study of a nationally representative sample of over 1,200 American high school students followed for 5 years, youths of all SES levels and ethnicities were highly

Table 13.2 Percentage of Women in Various Professions in the United States, 1983 and 2000

Profession	1983	2000
Engineer	5.8	11.5
Lawyer	15.8	28.9
Doctor	15.8	24.5
Business executive	32.4	45.1[a]
Author, artist, entertainer	42.7	49.9
Social worker	64.3	71.4
Elementary or secondary school teacher	70.9	74.9
College or university professor	36.3	42.4
Librarian, museum curator	84.4	82.9
Registered nurse	95.8	92.9
Psychologist	57.1	64.9

Source: U.S. Bureau of the Census, 2002c.

[a]This percentage includes executives and managers at all levels. Women make up only 10 percent of senior management at large corporations, although that figure represents a threefold increase in the past decade.

© JIM CRAIGMYLE/CORBIS

■ Career aspirations of academically talented females often decline in adolescence and early adulthood. When college environments value and support the accomplishments of women and provide opportunities for them to test their abilities in a nurturing environment, they sustain high aspirations and usually achieve them. These young women have attained their initial career goals of securing well-paid positions with high responsibility in a corporate setting.

ambitious. Compared with previous generations, many more expected to graduate from college and enter professional jobs. But about half were unaware of the steps needed to reach their goal (Schneider & Stevenson, 1999). They had only sketchy knowledge of their chosen vocation, of the educational requirements to enter it, and of the future demand for it.

These high-ambition/low-knowledge young people were at risk of becoming "drifting dreamers" who failed to make strategic choices about how to invest their efforts wisely. Especially if they entered a community college with plans to transfer to a four-year institution, they frequently found that they did not have the high school prerequisites to take the courses they needed. And often they chose courses without checking to make sure that credit would transfer to a four-year college. Consequently, some remained in school for an unnecessarily long period of time, whereas others did not complete their educational plans—for both academic and economic reasons.

Four-year institutions must familiarize high school and community college students with the admissions process earlier, so these students can develop a better understanding of the link between their current course work and educational and vocational opportunities. And at both the high school and college levels, teachers and counselors must do a better job of helping young people learn about the work they are interested in—by introducing them to people in those jobs, encouraging participation in relevant extracurricular activities, and offering internships that provide firsthand experiences.

Vocational Preparation of Non-College-Bound Young Adults

Sharese's younger brother Leon graduated from high school in a vocational track. Like 25 percent of young people with a high school diploma, he had no plans to go to college. While in school, Leon held a part-time job selling candy at the local shopping mall. He hoped to work in data processing after graduation, but 6 months later he was still a part-time sales clerk at the candy store. Although Leon had filled out many job applications, he got no interviews or offers. He soon despaired of discovering any relationship between his schooling and a career.

Leon's inability to find a job other than the one he held as a student is typical for North American non-college-bound high school graduates. Although they are more likely to find employment than youths who drop out, they have fewer work opportunities than they did several decades ago. About 15 percent of Canadian and 20 percent of American recent high school graduates who do not continue their education are unemployed (Bowlby & McMullen, 2002; U.S. Department of Education, 2002b). When they do find work, most are limited to temporary, low-paid, unskilled jobs. In addition, they have few alternatives to turn to for vocational counseling and job placement as they make the transition from school to work (Shanahan, Mortimer, & Krüger, 2002).

North American employers regard the recent high school graduate as poorly prepared for a demanding, skilled occupation. Indeed, there is some truth to this conclusion. As noted in Chapter 11, unlike Western European nations, the United States and Canada do not have a vocational training system to serve non-college-bound youths. As a result, most graduate without work-related skills and experience a "floundering period" that lasts for several years (Grubb, 1999). Inspired by successful programs in Western Europe, youth apprenticeship strategies that coordinate on-the-job training with classroom instruction are being considered as an important dimension of educational reform. The Cultural Influences box on page 440 describes Germany's highly successful apprenticeship system.

Cultural Influences

Work–Study Apprenticeships in Germany

Rolf, an 18-year-old German vocational student, is an apprentice at Brandt, a large industrial firm known worldwide for its high-quality products. Like many German companies, Brandt has a well-developed apprenticeship program that includes a full-time professional training staff, a suite of classrooms, and a lab equipped with the latest learning aids. Apprentices move through more than 10 major divisions in the company that are carefully selected to meet their learning needs. Rolf has worked in purchasing, inventory, production, personnel, marketing, sales, and finance. Now in cost accounting, he assists Herr Stein, his supervisor, in designing a computerized inventory control system. Rolf draws a flowchart of the new system under the direction of Herr Stein, who explains that each part of the diagram will contain a set of procedures to be built into a computer program.

Rolf is involved in complex and challenging projects, guided by caring mentors who love their work and want to teach it to others. Two days a week, he attends the *Berufsschule*, a part-time vocational school. On the job, Rolf applies a wide range of academic skills, including reading, writing, problem solving, and logical thinking. His classroom learning is directly relevant to his daily life (Hamilton, 1990; Shanahan, Mortimer & Krüger, 2002).

Germany has one of the most successful apprenticeship systems in the world for preparing young people to enter modern business and industry. About two-thirds of youths participate, making it the most common form of secondary education. German adolescents who do not go to a *Gymnasium* (college-preparatory high school) usually complete full-time schooling by age 15 or 16, but education remains compulsory until age 18. They fill the two-year gap with part-time vocational schooling combined with apprenticeship. Students are trained for a wide range of blue- and white-collar occupations—more than 400, leading to over 20,000 specialized careers. Each apprenticeship is jointly planned by educators and employers. Apprentices who complete training and pass a qualifying examination are certified as skilled workers and earn union-set wages for that occupation. Businesses provide financial support for the program because they know it guarantees a competent, dedicated work force (Heinz, 1999b).

The German apprenticeship system offers a smooth and rewarding path from school to career for young people who do not enter higher education. Many apprentices are hired by the firms that trained them. Most others find jobs in the same occupation. For those who change careers, the apprentice certificate is a powerful credential. Employers view successful apprentices as responsible and capable workers. They are willing to invest in further training to adapt the individual's skills to other occupations. As a result, between ages 18 and 20, German young people establish themselves in well-paid careers with security and advancement possibilities that grant them dignity and full status as adults (Kerckhoff, 2002).

The success of the German system—and of similar systems in Austria, Denmark, Switzerland, and several East European countries—suggests that some kind of national apprenticeship program would improve the transition from school to work for young people in Canada and the United States. Nevertheless, implementing an apprenticeship system poses major challenges. Perhaps the greatest is overcoming the reluctance of employers to assume part of the responsibility for vocational training. Whereas North American companies are willing to provide training to college-educated workers, only about 10 percent of high school graduates receive company-sponsored training (Lewis et al., 1998).

Other barriers must also be addressed. These include creating institutional structures that ensure cooperation between schools and businesses and preventing low-income youths from being concentrated in the lowest-skilled apprenticeship placements—an obstacle that Germany itself has not yet fully overcome (Hamilton & Hamilton, 2000). Pilot apprenticeship projects are currently under way, in an effort to solve these problems and build bridges between learning and working.

■ A head chef explains food preparation to an apprentice in training. In Germany, high-quality vocational education combined with apprenticeship enables youths who do not go to college to enter well-paid careers. The success of apprenticeship systems in European countries suggests that a national apprenticeship program would improve the transition from school to work in the United States and Canada.

Bringing together the worlds of schooling and work offers many benefits. These include helping non-college-bound young people establish productive lives right after graduation, motivating at-risk youths to stay in school, and contributing to the nation's economic growth (Hamilton, 1993; Safyer, Leahy, & Colan, 1995).

Although vocational development is a lifelong process, adolescence and early adulthood are crucial periods for defining occupational goals and launching a career. Young people well prepared for an economically and personally satisfying work life are much more likely to become productive citizens, devoted family members, and contented adults. The support of families, schools, businesses, communities, and society as a whole can contribute greatly to a positive outcome. In Chapter 14, we will take up the challenges of establishing a career and integrating it with other life tasks.

Ask Yourself

REVIEW

What characteristics of students and of college environments contribute to favorable psychological changes during the college years?

APPLY

Jordan, a community college student, has little idea of the course work he needs to become a teacher. Why might the gap between Jordan's vocational goal and his knowledge have negative consequences for his educational and occupational attainment? What steps can colleges take to prevent students like Jordan from becoming "drifting dreamers"?

CONNECT

What have you learned in previous chapters about development of gender stereotypes and about sex differences in self-esteem that helps explain why women's progress in entering male-dominant professions has been slow? (See Chapter 10, page 326 and Chapter 12, page 384.)

www.

Summary

PHYSICAL DEVELOPMENT

Biological Aging Begins in Early Adulthood

Describe current theories of biological aging, including those at the level of DNA and body cells and those at the level of tissues and organs.

- Once body structures reach maximum capacity and efficiency in the teens and twenties, **biological aging,** or **senescence,** begins. The programmed effects of specific genes may control certain age-related biological changes in DNA and body cells. DNA may also be damaged as random mutations accumulate, leading to less efficient cell repair and replacement and to abnormal cancerous cells. Release of highly reactive **free radicals** is a likely cause of age-related DNA and cellular damage. Biological aging may result from a complex combination of programmed effects of specific genes and random events that cause cells to deteriorate.

- Genetic and cellular deterioration affects organs and tissues. The **cross-linkage theory** of aging suggests that over time, protein fibers form links and become less elastic, producing negative changes in many organs. Declines in the endocrine and immune systems may also contribute to aging.

Physical Changes

Describe the physical changes of aging, paying special attention to the cardiovascular and respiratory systems, motor performance, the immune system, and reproductive capacity.

- Gradual physical changes take place in early adulthood and later accelerate. Declines in heart and lung performance show up only during exercise. Heart disease is a leading cause of death in adults, although it has decreased considerably in the past half-century due to lifestyle changes and medical advances. Atherosclerosis is a serious, multiply determined cardiovascular disease involving fatty deposits on artery walls.

- Athletic skills requiring speed, strength, and gross body coordination peak in the early twenties; those requiring endurance, arm–hand steadiness, and aiming peak in the late twenties and early thirties. Less active lifestyles rather than biological aging account for most of the age-related decline in athletic skill and motor performance.

- The immune response strengthens through adolescence and declines after age 20. This trend is partly due to shrinking of the thymus gland. Increased difficulty coping with physical and psychological stress also contributes.

- After age 35, women's reproductive capacity declines dramatically due to reduced quality and quantity of ova. Men show a gradual decrease in amount of semen and concentration of sperm in each ejaculation after age 40.

Health and Fitness

Describe the impact of SES, nutrition, and exercise on health, and discuss obesity in adulthood.

- Economically advantaged and well-educated individuals tend to sustain good health over most of their adult lives, whereas the health of lower-income individuals with limited education declines. SES differences in health-related living conditions and habits are largely responsible.

- Many people in industrialized nations eat the wrong kinds and amounts of food. Obesity has increased dramatically in Western nations. Today, Americans are the most overweight people in the world. Low-SES ethnic minorities are most affected. Sedentary lifestyles and diets high in sugar and fat contribute to obesity, which is associated with serious health problems, social discrimination, and early death.

- Some weight gain between ages 25 and 50 results from a decrease in **basal metabolic rate (BMR),** but many young adults show large increases. Successful treatments for obesity typically involve a low-fat diet plus exercise, accurate recording of food consumption, social support, teaching problem-solving skills, and extended intervention.

- Regular exercise reduces body fat, builds muscle, helps prevent illness (including cardiovascular disease), and enhances psychological well-being. Physical activity is associated with substantially lower death rates from all causes. Moderately intense exercise on most days leads to health benefits, which increase with greater intensity of exercise.

What are the two most common substance disorders, and what health risks do they entail?

- Cigarette smoking and alcohol consumption are the two most common adult substance disorders. Most adults who smoke began before age 21. They are at increased risk for heart attack, stroke, numerous cancers, and premature death.

- About one-third of heavy drinkers suffer from alcoholism, to which both heredity and environment contribute. Alcohol is implicated in liver and cardiovascular disease, certain cancers, numerous other physical disorders, and social problems such as highway fatalities, crime, and sexual coercion.

Describe sexual attitudes and behavior of young adults, and discuss sexually transmitted diseases, sexual coercion, and premenstrual syndrome.

- Most adults are less sexually active than popular media images suggest, but compared with earlier generations, they display a wider range of sexual choices and lifestyles and have had more sexual partners. Nevertheless, most people spend the majority of their lives with one partner. Adults in committed relationships report high satisfaction with their sex lives. Only a minority of adults report persistent sexual problems—difficulties linked to low SES, stress, having many partners, and a history of childhood sexual abuse or adult sexual coercion.

- Negative attitudes toward homosexual men and women have begun to change, particularly in the area of equal employment opportunities. Homosexual relationships share many characteristics with heterosexual relationships, including similarity between partners in education and background, greater satisfaction in committed relationships, and modest frequency of sexual activity.

- Many North Americans are likely to contract a sexually transmitted disease (STD) during their lifetime; women are more vulnerable than men. Although AIDS remains concentrated among gay men and intravenous drug abusers, the incidence among homosexuals is declining. Currently, the disease is spreading most rapidly through heterosexual contact in poverty-stricken minority groups where drug abuse is high.

- Most rape victims have been harmed by men they know well. Men who commit sexual assault typically hold traditional gender roles, approve of violence against women, and have difficulty interpreting women's social behavior accurately. Cultural acceptance of strong gender typing and of violence contributes to sexual coercion. Victims react with extreme psychological trauma.

- Nearly 40 percent of women experience **premenstrual syndrome (PMS)**, usually in mild form. For some, PMS is severe enough to interfere with daily life. Evidence for a genetic predisposition to PMS is accumulating.

How does psychological stress affect health?

- Chronic psychological stress induces physical responses that contribute to heart disease, several types of cancer, and gastrointestinal problems. Because the many challenges of early adulthood make it a highly stressful time of life, interventions that help stressed young people form supportive social ties are especially important.

COGNITIVE DEVELOPMENT

Changes in the Structure of Thought

Describe characteristics of adult thought, and explain how thinking changes in adulthood.

- Cognitive development beyond Piaget's formal operational stage is known as **postformal thought.** Adult cognition typically reflects an awareness of multiple truths, integrates logic with reality, and tolerates the gap between the ideal and the real.

- According to Perry's theory, college students move from **dualistic thinking,** dividing information into right and wrong, to **relativistic thinking,** awareness of multiple truths.

- In Schaie's theory, as situations change in adulthood, the goals of mental activity shift from acquiring knowledge to using it in more complex ways. As a result, the **acquisitive stage** of childhood and adolescence gives way to the **achieving stage** in early adulthood, the **responsibility** and **executive stages** in middle adulthood, and the **reintegrative stage** in late adulthood.

- According to Labouvie-Vief's theory, the need to specialize motivates adults to progress from the ideal world of possibilities to **pragmatic thought,** which uses logic as a tool to solve real-world problems and accepts inconsistency, imperfection, and the need to compromise.

Information Processing: Expertise and Creativity

What roles do expertise and creativity play in adult thought?

- Specialization in college and in an occupation leads to **expertise,** which enhances problem solving and is necessary for creativity. Mature creativity involves formulating meaningful new problems and questions. Although creativity tends to rise in early adulthood and to peak in the late thirties or early forties, its development varies across disciplines and individuals. In addition to expertise, diverse personal and situational factors jointly promote creativity.

Changes in Mental Abilities

What do intelligence tests reveal about changes in adults' mental abilities over the life course?

- Cross-sectional studies show a peak in intelligence test performance around age 35 followed by a steep decline, whereas longitudinal studies reveal an age-related increase from early to middle adulthood. Applying the longitudinal-sequential design in the Seattle Longitudinal Study, Schaie found that the cross-sectional decline was largely due to cohort effects.

- Longitudinal evidence shows that verbal skills and other abilities that depend on acquired knowledge improve steadily into middle adulthood and do not decline until late in life. Abilities that involve detecting relationships, however, decline earlier.

The College Experience

Describe the impact of a college education on young people's lives, and discuss the problem of dropping out.

- Through involvement in academic programs and campus life, college students engage in exploration that produces gains in knowledge and reasoning ability, revised attitudes and values, enhanced self-esteem and self-knowledge, and preparation for a high-status career. Most young people who drop out do so during their freshman year and are struggling with typical problems of early adulthood. These students benefit from interventions that show concern for them as individuals.

Vocational Choice

Trace the development of vocational choice, and cite factors that influence it.

- Vocational choice moves through three phases: a **fantasy period,** in which children explore career options through play; a **tentative period,** in which teenagers weigh different careers against their interests, abilities, and values; and a **realistic period,** in which young people settle on a vocational category and then a specific occupation.

- Vocational choice is influenced by personality; parents' provision of educational opportunities, vocational information, and encouragement; close relationships with teachers; and access to vocational information in high schools and colleges. Women's progress in male-dominated professions has been slow, and their achievements lag behind those of men in virtually all fields. Gender-stereotyped messages about women's abilities prevent many from reaching their career potential.

What problems do American non-college-bound young people face in preparing for a vocation?

- North American non-college-bound high school graduates have fewer work opportunities today than they did several decades ago. Most are limited to low-paid, unskilled jobs, and too many are unemployed. To address their need for vocational training, youth apprenticeships inspired by those in Western Europe are being considered as an important dimension of educational reform in the United States.

Important Terms and Concepts

achieving stage (p. 432)
acquisitive stage (p. 432)
basal metabolic rate (BMR) (p. 420)
biological aging, or senescence (p. 412)
cross-linkage theory of aging (p. 414)
dualistic thinking (p. 432)
executive stage (p. 433)
expertise (p. 433)
fantasy period (p. 437)
free radicals (p. 413)
postformal thought (p. 432)
pragmatic thought (p. 433)
premenstrual syndrome (PMS) (p. 430)
realistic period (p. 437)
reintegrative stage (p. 433)
relativistic thinking (p. 432)
responsibility stage (p. 433)
tentative period (p. 437)

FYI For Further Information and Help

Consult the Companion Website for *Development Through the Lifespan, Third Edition,* (www.ablongman.com/berk) where you will find the following resources for this chapter:

- **Chapter Objectives**
- **Flashcards** for studying important terms and concepts
- **Annotated Weblinks** to guide you in further research
- **Ask Yourself** questions, which you can answer and then check against a sample response
- **Suggested Readings**
- **Practice Test** with immediate scoring and feedback

Emotional and Social Development in Early Adulthood

© YANG LIU/CORBIS

Erikson's Theory: Intimacy versus Isolation

Other Theories of Adult Psychosocial Development
Levinson's Seasons of Life • Vaillant's Adaptation to Life • Limitations of Levinson's and Vaillant's Theories • The Social Clock

Close Relationships
Romantic Love • Friendships • Loneliness

- A Lifespan Vista: Childhood Attachment Patterns and Adult Romantic Relationships

The Family Life Cycle
Leaving Home • Joining of Families in Marriage • Parenthood

- Social Issues: Partner Abuse
- Cultural Influences: A Global Perspective on Family Planning

The Diversity of Adult Lifestyles
Singlehood • Cohabitation • Childlessness • Divorce and Remarriage • Variant Styles of Parenthood

Career Development
Establishing a Career • Women and Ethnic Minorities • Combining Work and Family

The warm communication between this young couple shows that they are passionately in love. Yet many adjustments lie ahead. A lasting intimate partnership requires effort, compromise, sensitivity, and respect.

After completing her master's degree, Sharese returned to her hometown, where her marriage to Ernie would soon take place. A year-long engagement had preceded the wedding, during which Sharese had vacillated about whether to follow through. At times, she looked with envy at Heather, still unattached and free to pursue the career options before her.

Sharese also pondered the life circumstances of Christy and her husband, Gary—married their junior year in college and two children born within the next few years. Despite his good teaching performance, Gary's relationship with the high school principal deteriorated, and he quit his job at the end of his first year. Financial pressures and the demands of parenthood had put Christy's education and career plans on hold. Sharese wondered whether it was really possible to combine family and career.

Sharese's ambivalence intensified as her wedding approached. When Ernie asked why she was so agitated, she admitted that she didn't feel ready to marry. Ernie's admiration for Sharese had strengthened over their courtship, and he reassured her of his love. His career was launched, and at age 27, he looked forward to starting a family. Uncertain and conflicted, Sharese felt swept toward the altar. Relatives, friends, and gifts began to arrive. On the appointed day, she walked down the aisle.

In this chapter, we take up the emotional and social sides of early adulthood. Having achieved independence from family, young people still want and need close, affectionate ties. Yet like Sharese, they often fear losing their freedom. Once this struggle is resolved, the years from 20 to 40 lead to new family units and parenthood, accomplished in the context of diverse lifestyles. At the same time, young adults must master the skills and tasks of their chosen career. We will see that love and work are intertwined, and society expects success at both. In negotiating both arenas, young adults do more choosing, planning, and changing course than any other age group. When their decisions are in tune with themselves and their social and cultural worlds, they acquire many new competencies, and life is full and rewarding.

© CHIP HENDERSON/INDEX STOCK

■ To attain intimacy, young adults must give up some of their newfound independence and redefine their identity to include the values and interests of two people. The intense interaction of these marital partners may be part of their effort to reconcile their needs for independence and intimacy and establish a mutually gratifying close relationship.

Erikson's Theory: Intimacy versus Isolation

Erikson's contributions have energized the study of adult personality development. His vision has influenced all contemporary theories (McCrae & Costa, 1990). According to Erikson (1964), adults move through three stages, each bringing both opportunity and risk—"a turning point for better or worse" (p. 139). The psychological conflict of early adulthood is **intimacy versus isolation,** reflected in the young person's thoughts and feelings about making a permanent commitment to an intimate partner.

As Sharese's inner turmoil reveals, establishing a mutually gratifying close relationship is challenging. Most young adults have only recently attained economic independence from parents, and many are still involved in the quest for identity. Yet intimacy requires that they give up some of their newfound independence and redefine their identity to include the values and interests of two people. During their first year of marriage, Sharese separated from Ernie twice as she tried to reconcile her needs for independence and intimacy. Maturity involves balancing these forces. Without a sense of independence, people define themselves only in terms of their partner and sacrifice self-respect and initiative. Without intimacy, they face the negative outcome of Erikson's stage of early adulthood: loneliness and self-absorption. Ernie's patience and stability helped Sharese realize that marriage requires generosity and compromise but not total surrender of the self.

A secure sense of intimacy is also evident in other close relationships. For example, in friendships and work ties, young people who have achieved intimacy are cooperative, tolerant, and accepting of differences in background and values. Although they enjoy being with others, they also are comfortable when alone. People with a sense of isolation hesitate to form close ties because they fear loss of their own identity, tend to compete rather than cooperate, are not accepting of differences, and are easily threatened when others get too close (Hamachek, 1990). Erikson believed that successful resolution of intimacy versus isolation prepares the individual for the middle adulthood stage, which focuses on *generativity*—caring for the next generation and helping to improve society.

Research based on self-reports confirms that intimacy is a central concern of early adulthood (Ryff & Migdal, 1984; Whitbourne et al., 1992). But as noted previously, a fixed series of tasks tied neatly to age does not describe many adults. Childbearing and child rearing (aspects of generativity) usually occur in the twenties and thirties, and contributions to society through work also are under way. Furthermore, as we will see shortly, many combinations of marriage, children, and career exist, each with a unique pattern of timing and commitment (McAdams, de St. Aubin, & Logan, 1993; Weiland, 1993).

In sum, both intimacy and generativity emerge in early adulthood, with shifts in emphasis that differ among young people. Recognizing that Erikson's theory provides only a broad sketch of adult personality development, other theorists have expanded and modified his stage approach, adding detail and flexibility.

Other Theories of Adult Psychosocial Development

In the 1970s, growing interest in adult development led to several widely read books on the topic. Three of these volumes—Daniel Levinson's (1978) *The Seasons of a Man's Life* and George Vaillant's (1977, 2002) *Adaptation to Life* and *Aging Well*—present psychosocial theories in the tradition of Erikson. Each theory is summarized in Table 14.1.

Levinson's Seasons of Life

Looking for an underlying order in the life course, Levinson (1978) conducted in-depth biographical interviews with forty 35- to 45-year-old men from four occupational subgroups: hourly workers in industry, business executives, university biologists, and novelists. Later he interviewed forty-five women, also 35 to 45 years of age, from three subgroups: homemakers, business executives, and university professors. His results, and those of others, reveal a common path of change, within which men and women approach developmental tasks in somewhat different ways (Levinson, 1996; Roberts & Newton, 1987).

Like Erikson, Levinson (1978, 1996) conceived of development as a sequence of qualitatively distinct eras (stages or seasons). In each, biological and social forces introduce new psychological challenges. As Table 14.1 shows, each era begins with a *transition,* lasting about 5 years, which concludes the previous era and prepares the person for the next. Between transitions, people move into stable periods, lasting about 5 to 7 years, in which they build a life structure aimed at harmonizing inner personal and outer societal demands and, as a result, enhancing quality of life. Eventually people question the current structure, and a new transition ensues.

The **life structure,** a key concept in Levinson's theory, is the underlying design of a person's life, consisting of relationships with significant others—individuals, groups, and institutions. The life structure can have many components, but usually only a few, having to do with marriage/family and occupation, are central. However, wide individual differences exist in the weight of central and peripheral components.

Men's and women's accounts of their lives confirm Levinson's description of the life course. They also reveal that early adulthood is the era of "greatest energy and abundance, contradiction and stress" (Levinson, 1986, p. 5). These years can bring rich satisfactions in love, sexuality, family life, occupational advancement, and realization of life goals. But they also

Table 14.1 Stages of Adult Psychosocial Development

Period of Development	Erikson	Levinson	Vaillant
Early adulthood (20–40 years)	Intimacy versus isolation	Early adult transition: 17–22 years Entry life structure for early adulthood: 22–28 years Age 30 transition: 28–33 years Culminating life structure for early adulthood: 33–40 years	Intimacy Career consolidation
Middle adulthood (40–65 years)	Generativity versus stagnation	Midlife transition: 40–45 years Entry life structure for middle adulthood: 45–50 years Age 50 transition (50–55 years) Culminating life structure for middle adulthood (55–60 years)	Generativity Keeper of meanings
Late adulthood (65 years–death)	Ego integrity versus despair	Late adult transition (60–65 years) Late adulthood (65 years–death)	Ego integrity

■ To realize their dream, young adults form a relationship with a mentor, who fosters their occupational skills and knowledge of workplace values, customs, and characters. This experienced doctor advises two young interns over lunch in the hospital cafeteria.

involve serious decisions about marriage, children, work, and lifestyle before many people have the life experience to choose wisely.

● **Dreams and Mentors.** How do young adults cope with the opportunities and hazards of this period? Levinson found that during the early adult transition (17 to 22 years), most construct a *dream,* an image of themselves in the adult world that guides their decision making. The more specific the dream, the more purposeful the individual's structure building. For men, the dream usually emphasizes an independent achiever in an occupational role. In contrast, most career-oriented women display "split dreams" in which both marriage and career are prominent. Also, women's dreams tend to define the self in terms of relationships with husband, children, and colleagues. Men's dreams usually are more individualistic: They view significant others, especially wives, as vital supporters of their goals and less often see themselves as supporting others' goals.

Young adults also form a relationship with a mentor who facilitates realization of their dream. The mentor is generally several years older and experienced in the world the person seeks to enter. Most of the time, a senior colleague at work fills this role, occasionally a friend, neighbor, or relative. Mentors may act as teachers who enhance the person's occupational skills; guides who acquaint the person with values, customs, and characters in the occupational setting; and sponsors who foster the person's career advancement. As we will see when we take up vocational development, finding a supportive mentor is easier for men than for women.

According to Levinson, men oriented toward high-status careers spend their twenties acquiring professional skills, values, and credentials. Although some women follow this path, for many others career development extends into middle age. Men who serve as primary caregivers of their children also show a delayed career path (Levinson, 1978, 1996; Kogan & Vacha-Haase, 2002).

● **Age-30 Transition.** During the age-30 transition, young people reevaluate their life structure. Those who were preoccupied with their career and are still single usually focus on finding a life partner. However, men rarely reverse the relative priority of career and family, whereas career-oriented women sometimes do.

Women who had stressed marriage and motherhood often develop more individualistic goals. Recall that Christy had dreamed of becoming a professor. In her mid-thirties, she finally earned her doctoral degree and secured a college teaching position. Women also become conscious of aspects of their marriage that threaten to inhibit further development of the independent side of their dream. Married women tend to demand that their husbands recognize and accommodate their interests and aspirations beyond the home.

For men and women without satisfying relational or occupational accomplishments, the age-30 transition can be a crisis. For others who question whether they can create a meaningful life structure, it is a time of considerable conflict and instability.

● **Settling Down for Men, Continued Instability for Women.** To create the culminating life structure of early adulthood, men usually "settle down" by focusing on certain relationships and aspirations, setting others aside. In doing so, they try to establish a stable niche in society that is consistent with their values, whether those be wealth, power, prestige, artistic or scientific achievement, or forms of family or community participation. During his thirties, Sharese's husband

Ernie expanded his knowledge of real estate accounting, became a partner in his firm, coached his son's soccer team, and was elected treasurer of his church. He paid less attention to golf, travel, and playing the guitar than he had in his twenties.

"Settling down," however, does not accurately describe women's experiences during their thirties. Many remain unsettled because of the addition of an occupational or relationship commitment. When her two children were born, Sharese felt torn between her research position in the state health department and her family. She took 6 months off after the arrival of each baby. When she returned to work, she did not pursue attractive administrative openings because they required travel and time away from home. And shortly after Christy began teaching, she and Gary divorced. Becoming a single parent while starting her professional life introduced new strains. Not until middle age do many women attain the stability typical of men in their thirties—reaching career maturity and taking on more authority in the community (Levinson, 1996).

Vaillant's Adaptation to Life

Vaillant (1977) examined the development of nearly 250 men born in the 1920s, selected for study while they were students at a highly competitive liberal arts college and followed as many as possible over the lifespan. In college, the participants underwent extensive interviews. During each succeeding decade, they answered lengthy questionnaires. Then Vaillant (2002) interviewed the men at ages 47, 60, and 70 about work, family, and physical and mental health.

Other than denying a strict age-related schedule of change, Vaillant's theory is compatible with Levinson's. Both agree that quality of relationships with important people shape the life course. In studying how the men altered themselves and their social world to adapt to life, Vaillant confirmed Erikson's stages but filled gaps between them. Following a period in their twenties devoted to intimacy concerns, the men focused on career consolidation in their thirties, working hard and making the grade in their occupations. During their forties, they pulled back from individual achievement and became more generative—giving to and guiding others. In their fifties and sixties, they became "keepers of meaning," or guardians of their culture, expressing concern about the values of the new generation and the state of their society. Many felt a deep need to preserve and pass on cultural traditions by teaching others what they had learned from life experience (Vaillant & Koury, 1994). Finally, in their seventies, the men became more spiritual and reflective, contemplating the meaning of life and accepting its finiteness.

Although Vaillant initially studied only men, eventually he examined the development of a sample of bright, well-educated women who were participants in another lifelong study. His findings, and those of others, suggest that women undergo a series of changes similar to those just described (Block, 1971; Oden & Terman, 1968; Vaillant, 2002).

Limitations of Levinson's and Vaillant's Theories

Although psychosocial theorists express considerable consensus on adult development, their conclusions are largely based on interviews with people born in the first few decades of the twentieth century. The patterns identified by Levinson and Vaillant fit the life paths of Sharese, Ernie, Christy, and Gary. Nevertheless, those patterns might not apply as broadly to young people today as they did to past generations.

Two other factors limit the conclusions of these theorists. First, although non-college-educated, low-income adults were included in Levinson's sample, they were few in number, and low-SES women remain almost entirely uninvestigated. Yet SES can profoundly affect the life course. For example, Levinson's blue-collar workers rarely implemented an occupational dream. Perhaps because career advancement is less salient for them, low-SES men perceive "early adulthood" to end and "maturity" to arrive at a younger age than their higher-SES counterparts (Neugarten, 1979). Examining longitudinal archives on low-SES men who had grown up in the 1940s, Vaillant (1993) reported evidence for his stage sequence. Still, he acknowledged that the sample was limited.

Finally, Levinson's participants were middle aged when interviewed, and they might not have remembered all aspects of their early adulthoods accurately. In sum, studies of new generations—both men and women, of diverse SES and cultural backgrounds—are needed before we can conclude that the developmental sequences just described apply to most or all young people.

The Social Clock

In the previous section and in earlier parts of this book, we emphasized that changes in society from one generation to the next can affect the life course. Bernice Neugarten (1968a, 1979) pointed out that an important cultural and generational influence on adult development is the **social clock**—age-graded expectations for life events, such as beginning a first job, getting married, birth of the first child, buying a home, and retiring. All societies have timetables for accomplishing major developmental tasks. Being on time or off time can profoundly affect self-esteem because adults (like children and adolescents) make social comparisons, measuring the progress of their lives against those of their friends, siblings, and colleagues. Especially when evaluating family and occupational attainments, people often ask, "How am I doing for my age?"

A major source of personality change in adulthood is conformity to or departure from the social clock. In a study of college women born in the 1930s who were followed up at ages 27 and 43, researchers determined how closely participants followed a "feminine" social clock (marriage and parenthood in the early or mid-twenties) or a "masculine" social clock (entry into a high-status career and advancement by the late twenties). Those who

■ All societies have social clocks, or timetables for accomplishing major developmental tasks. Yet today, expectations for appropriate behavior are not as definite as they once were. This first-time mother has taken time to establish herself as a research scientist before having children. Is she departing from the social clock or establishing a new pattern?

started families on time became more responsible, self-controlled, tolerant, and caring but declined in self-esteem and felt more vulnerable as their lives progressed. Those who followed an occupational timetable typical for men became more dominant, sociable, independent, and intellectually effective, a trend also found in a cohort born a decade later (Vandewater & Stewart, 1997). Women not on a social clock—who had neither married nor begun a career by age 30—were doing especially poorly. They suffered from self-doubt, feelings of incompetence, and loneliness. One stated, "My future is a giant question mark" (Helson, 1992; Helson, Mitchell, & Moane, 1984, p. 1090).

As noted in Chapter 1, expectations for appropriate behavior during early, middle, and late adulthood are no longer as definite as they once were. When Australian college students were asked for the best age at which to accomplish various milestones of development, they mentioned later ages for marriage and grandparenthood and a broader age range for retirement than people did 30 years ago (Peterson, 1996). Still, many adults experience some psychological distress when they are substantially behind in timing of life events (Antonucci & Akiyama, 1997; Rook, Catalano, & Dooley, 1989). Following a social clock of some kind seems to foster confidence during early adulthood because it guarantees that young people will engage in the work of society, develop skills, and increase in understanding of the self and others (Helson, 1997; Helson & Moane, 1987). As Neugarten (1979) suggested, the stability of society depends on having people committed to social-clock patterns. With this in mind, let's take a closer look at how young men and women traverse the major tasks of young adulthood.

Ask Yourself

REVIEW
According to Levinson, how do the life structures of men and women differ?

APPLY
Using the concept of the social clock, explain why Sharese was so conflicted about getting married to Ernie after she finished graduate school.

CONNECT
Return to pages 27–30 in Chapter 1, and review common research methods for studying human development. Which method did Levinson and Vaillant use to chart the adult life course? What are the strengths and limitations of this method?

www.

Close Relationships

To establish an intimate tie to another person, people must find a partner, build an emotional bond, and sustain it over time. Although young adults are especially concerned with romantic love, the need for intimacy is also satisfied through other relationships that involve a mutual commitment—friends, siblings, and co-workers. Let's examine the multiple facets of intimacy.

Romantic Love

During her junior year of college, Sharese glanced around the room in government class, her eyes settling on Ernie, a senior and one of the top students. One weekend, Sharese and Ernie happened to attend the same party, where they struck up a conversation. Sharese discovered that Ernie was as warm and interesting as he had seemed from a distance. And Ernie found Sharese to be lively, intelligent, and attractive. By the end of the evening, the couple realized that they had similar opinions on important social issues and liked the same leisure activities. They began dating steadily. Four years later, they married.

Finding a partner with whom to share one's life is a major milestone of adult development, with profound consequences for self-concept and psychological well-being. It is also a complex process that unfolds over time and is affected by a variety of events, as Sharese and Ernie's relationship reveals.

● **Selecting a Mate.** Recall from Chapter 13 that intimate partners tend to meet in places where there are people of their own age, ethnicity, SES, and religion. Once in physical proximity, people usually select partners who resemble themselves in other ways—attitudes, personality, educational plans, intelligence, physical attractiveness, and even height (Keith & Schafer, 1991; Simpson & Harris, 1994). Romantic partners sometimes have complementary traits—for example, one more gregarious, the other more reserved; one more curious and sensation seeking, the other preferring consistency and routine. When this permits each person to satisfy personal preferences and goals, it contributes to compatibility (Dryer & Horowitz, 1997), But overall, little support exists for the idea that "opposites attract." In most ways, lovers resemble each other. And many studies confirm that the more similar they are, the more satisfied they tend to be with their relationship and the more likely they are to stay together (Caspi & Herbener, 1990; Eysenck & Wakefield, 1981; Richard, Wakefield, & Lewak, 1990).

Nevertheless, in choosing a long-term partner, men and women differ in the importance they place on certain characteristics. In research carried out in many countries, women assign greater weight to intelligence, ambition, financial status, and moral character, whereas men place more emphasis on physical attractiveness and domestic skills. In addition, women prefer a same-age or slightly older partner, men a younger partner (Stewart, Stinnett, & Rosenfeld, 2000; Kenrick et al., 1996; Pines, 1998).

Evolutionary theory helps us understand these findings. Recall from Chapter 13 that because their capacity to reproduce is limited, women seek a mate with traits that help ensure children's survival and well-being. Therefore, women are attuned to earning power and emotional commitment. In contrast, men look for a mate with traits that signal youth, health, sexual pleasure, and ability to give birth and care for offspring. As further evidence for this difference, men often want a relationship to move quickly toward physical intimacy. Women typically desire psychological intimacy—a longer time for partners to get to know each other—before sexual intercourse occurs (Buss, 2001).

According to an alternative, social learning perspective, gender roles profoundly influence criteria for mate selection. Beginning in childhood, men learn to be assertive and independent—behaviors that lead to success in the work world. Women acquire nurturant behaviors, which facilitate caregiving. Then each sex learns to value traits in the other that fit with a traditional division of labor (Eagly & Wood, 1999; Wood & Eagly, 2000). In support of this theory, in cultures and in younger generations experiencing greater gender equity, men and women are more alike in their mate preferences. For example, these men place more emphasis on their mate's financial prospects and less on her domestic skills. And both sexes care somewhat less about their mate's age relative to their own. Instead, they place a high value on caring and love—that is, on relationship satisfaction (Buss et al., 1990, 2001; Buunk et al., 2001).

Nevertheless, men continue to value physical attractiveness more than women do, and women continue to value earning capacity more than men do. Furthermore, these gender differences—along with gender similarity in desire for a caring, sensitive partner—also characterize homosexual men and women (Regan, Medina, & Joshi, 2001). In sum, both biological and social forces contribute to mate selection.

As the Lifespan Vista box on pages 452–453 reveals, young people's choice of an intimate partner and the quality of their relationship also are affected by memories of their early parent–child bond. Finally, for romance to lead to a lasting partnership, it must happen at the right time. Two people may be right for each other, but if one or both are not ready to marry in terms of their social clock, then the relationship is likely to dissolve.

● **The Components of Love.** What feelings and behaviors tell us that we are in love? According to Robert Sternberg's (1987, 1988) **triangular theory of love**, love has three components—intimacy, passion, and commitment—that shift in emphasis as romantic relationships develop. *Intimacy* is the emotional component. It involves warm, tender communication, expressions of concern about the other's well-being, and a desire for the partner to reciprocate. *Passion,* the desire for sexual activity and romance, is the physical- and psychological-arousal component. *Commitment* is the cognitive component. It leads partners to decide that they are in love and to maintain that love (Sternberg, 1987, 1988).

At the beginning of a relationship, **passionate love**—intense sexual attraction—is strong. Gradually, passion declines in favor of intimacy and commitment, which form the basis for **companionate love**—warm, trusting affection and caregiving (Fehr, 1994; Hatfield, 1988). Each aspect of love, however, helps sustain the relationship. Early passionate love is a strong predictor of whether partners keep dating. But without the quiet intimacy, predictability, and shared attitudes and values of companionate love, most romances eventually break up (Hendrick & Hendrick, 1992).

An ongoing relationship with a mate requires effort from both partners, as a study of newlyweds' feelings and behavior over the first year of marriage reveals. Husbands and wives gradually felt less "in love" and pleased with married life. A variety of factors contributed to this change. A sharp drop in time spent talking to one another and in doing things that brought each other pleasure (for example, saying "I love you" or making the other person laugh) occurred. In addition, although couples engaged in just as many joint activities at the beginning and end of the year, leisure pursuits gave way to household tasks and chores. Less pleasurable activities may have contributed to the decline in satisfaction (Huston, McHale, & Crouter, 1986).

But couples whose relationships endure generally report that they love each other more than they did earlier (Sprecher, 1999). In the transformation of romantic involvements from passionate to companionate, *commitment* may be the aspect of love that determines whether a relationship survives. Communicating that commitment—through warmth, sensitivity, caring, acceptance, and respect—can be of great benefit (Knapp & Taylor, 1994). For example, Sharese's doubts about getting married subsided largely because of Ernie's expressions

A Lifespan Vista

Childhood Attachment Patterns and Adult Romantic Relationships

Recall from Chapter 6 (page 187) that according to Bowlby's ethological theory of attachment, the early attachment bond leads to the construction of an *internal working model,* or set of expectations about attachment figures, that serves as a guide for close relationships throughout life. We also saw that adults' evaluations of their early attachment experiences are related to their parenting behaviors—specifically, to the quality of attachments they build with their children (see page 192). Additional evidence indicates that recollections of childhood attachment patterns strongly predict romantic relationships in adulthood.

In studies carried out in Australia, Israel, and the United States, researchers asked people to describe their early parental bonds (attachment history), their attitudes toward intimate relationships (internal working model), and their actual experiences with romantic partners. And in a few studies, investigators also observed couples' behaviors. Consistent with Bowlby's theory, adults' recall and interpretation of childhood attachment patterns were good indicators of internal working models and relationship experiences. (To review patterns of attachment, return to pages 187–188.)

Secure Attachment. Adults describing their attachment history as secure (warm, loving, and supportive) had internal working models that reflected this security. They viewed themselves as likable and easy to get to know, were comfortable with intimacy, and rarely worried about abandonment or someone getting too close to them. In line with these attitudes, they characterized their most important love relationship in terms of trust, happiness, and friendship (Cassidy, 2001). Furthermore, their behaviors toward their partner were supportive and their conflict resolution strategies constructive. They were also at ease in turning to their partner for comfort and assistance and reported mutually initiated, enjoyable sexual activity (Collins & Feeney, 2000; Creasey, 2002; Hazan & Zeifman, 1999; Roisman et al., 2002).

Avoidant Attachment. Adults who reported an avoidant attachment history (demanding, disrespectful, and critical parents) displayed internal working models that stressed independence, mistrust of love partners, and anxiety about people getting too close. They were convinced that others disliked them and that romantic love is hard to find and rarely lasts. Jealousy, emotional distance, and little enjoyment of physical contact pervaded their most important love relationship (Collins & Feeney, 2000). Avoidant adults often deny attachment needs through excessive work and brief sexual encounters and affairs (Feeney, 1998).

Resistant Attachment. Adults recalling a resistant attachment history (parents who responded unpredictably and unfairly) presented internal working models in which they sought to merge completely with another person and fall in love quickly (Cassidy, 2001). At the same time, they worried that their intense feelings would overwhelm others, who really did not love them and would not want to stay with them. Their most important love relationship was riddled with jealousy, emotional highs and lows, and desperation about whether the partner would return their affection (Feeney, 1999). Resistant adults are quick to express fear and anger, and they disclose information about themselves at inappropriate times (Brennan & Shaver, 1995).

Are adults' descriptions of their childhood attachment experiences accurate, or are they distorted or even completely invented? In several longitudinal studies, quality of parent–child interactions observed 6 to 23 years earlier were good predictors of internal working models in early adulthood (Allen & Hauser, 1996; Ogawa et al., 1997; Roisman et al., 2001). These findings suggest that adult

---

of commitment. In the most dramatic of these, he painted a large sign on her birthday and placed it in their front yard. It read, "I LOVE SHARESE." Sharese returned Ernie's sentiments, and the intimacy of their bond deepened.

Intimate partners who consistently express their commitment report higher-quality relationships (Duck, 1994). An important feature of their communication is constructive conflict resolution—raising issues gently and avoiding the escalation of negative interaction sparked by criticism, contempt, defensiveness, and stonewalling (Gottman et al., 1998; Schneewind & Gerhard, 2002). How men handle conflict is particularly important, since men tend to be less skilled at negotiating it than women, often avoiding discussion (Gayle, Preiss, & Allen, 2002). The Caregiving Concerns table on page 454 lists ways to help keep the embers of love aglow in a romantic partnership.

● **Culture and the Experience of Love.** Passion and intimacy, which form the basis for romantic love, became the dominant basis for marriage in twentieth-century Western nations as the value of individualism strengthened. From this vantage point, mature love is based on autonomy, appreciation of the partner's unique qualities, and intense emotion. Trying to satisfy dependency needs through an intimate bond is regarded as immature (Hatfield, 1993).

This Western view contrasts sharply with the perspectives of Eastern cultures, such as China and Japan. In Japanese, *amae,* or love, means "to depend on another's benevolence." Dependency throughout life is accepted and viewed positively. The traditional Chinese collectivist view defines the self in terms of role relationships. A Chinese man considers himself a son, a brother, a husband, and a father; he rarely thinks in

PHOTOS COURTESY OF ELIZABETH NAPOLITANO

■ How might the internal working model of the 2-year-old seated in her mother's lap (left) have influenced the relationship she forged as a young adult with her husband (center) and with her infant son (right)? Research indicates that early attachment pattern is one among several factors that predict the quality of later intimate ties. Characteristics of this woman's partner and her current life conditions are also influential.

recollections bear some resemblance to actual parent–child experiences. However, quality of attachment to parents is not the only factor that influences later internal working models and intimate ties. Characteristics of the partner and current life conditions also are important. In one study, adults with an inner sense of security fostered security in their partners as well as in their adolescent and young adult children (Cook, 2000).

In sum, negative parent–child experiences can be carried forward into adult relationships, predisposing people to believe that they are undeserving of love or that their intimate partners cannot be trusted. At the same time, internal working models are continuously "updated." When adults with unhappy love lives have a chance to form more satisfying intimate ties, they may revise their internal working models. As the new partner approaches the relationship with a secure state of mind and sensitive, supportive behavior, the insecure partner reappraises her expectations and responds in kind (Kobak & Hazan, 1991). This reciprocity creates a feedback loop through which a revised, more favorable internal working model, along with mutually gratifying interaction, is sustained over time.

terms of an independent self (Chu, 1985). Because feelings of affection are distributed across a broad social network, the intensity of any one relationship is reduced.

In choosing a mate, Chinese and Japanese young people are expected to consider obligations to others, especially parents. As one writer summarized, "An American asks, 'How does my heart feel?' A Chinese asks, 'What will other people say?'" (Hsu, 1981, p. 50). Consistent with this difference, college students of Asian heritage are less likely than those of American, Canadian, or European descent to endorse a view of love based on physical attraction and deep emotion. Instead they stress companionship and practical matters, such as similarity of background, career promise, and likelihood of being a good parent (Dion & Dion, 1993). Similarly, compared with American couples, dating couples in China report less passion but equally strong feelings of intimacy and commitment (Gao, 2001).

Friendships

Like romantic partners and childhood friends, adult friends are usually similar in age, sex, and SES—factors that contribute to common interests, experiences, and needs and therefore to the pleasure derived from the relationship. Friends offer many of the same benefits in adulthood as in earlier years. They enhance self-esteem through affirmation and acceptance and provide support during times of stress (Hartup & Stevens, 1999). Friends also make life more interesting by expanding social opportunities and access to knowledge and points of view.

Keeping Love Alive in a Romantic Partnership

SUGGESTION	DESCRIPTION
Make time for your relationship.	To foster relationship satisfaction and a sense of being "in love," plan regular times to be together.
Tell your partner of your love.	Express affection and caring, including the powerful words "I love you," at appropriate times. These messages increase perceptions of commitment and encourage your partner to respond in kind.
Be available to your partner in times of need.	Provide emotional support, giving of yourself when your partner is distressed.
Communicate constructively and positively about relationship problems.	When you or your partner is dissatisfied, suggest ways of overcoming difficulties, and ask your partner to collaborate in choosing and implementing a course of action. Avoid the four enemies of a gratifying, close relationship: criticism, contempt, defensiveness, and stonewalling.
Show an interest in important aspects of your partner's life.	Ask about your partner's work, friends, family, and hobbies, and express appreciation for his or her special abilities and achievements. In doing so, you grant your partner a sense of being valued.
Confide in your partner.	Share innermost feelings, keeping intimacy alive.
Forgive minor offenses, and try to understand major offenses.	Whenever possible, overcome feelings of anger through forgiveness. In this way, you acknowledge unjust behavior but avoid becoming preoccupied with it.

Sources: Donatelle, 2003; Harvey & Pauwels, 1999; Horowitz, McLaughlin, & White, 1997.

Trust, intimacy, and loyalty continue to be important in adult friendships, as they were in middle childhood and adolescence. Sharing thoughts and feelings is sometimes greater in friendship than in marriage, although commitment is less strong as friends come and go over the life course. Even so, some adult friendships continue for many years, at times throughout life. Friendship continuity is greater for women, who also see their friends more often, which helps sustain the relationship (Sherman, de Vries, & Lansford, 2000).

● **Same-Sex Friendships.** Throughout life, women continue to have more intimate same-sex friendships than men. When together, female friends say they prefer to "just talk," whereas male friends say they like to "do something," such as play sports, extending a pattern evident in childhood and adolescence (see Chapter 12, page 397). Men report barriers to intimacy with other men. For example, they indicate that they sometimes feel in competition with male friends and are therefore unwilling to disclose any weaknesses. They also worry that if they tell about themselves, their friends may not reciprocate (Reid & Fine, 1992). Because a balance of power and equality in social giving and taking is basic to a good friendship, women generally evaluate their same-sex friendships more positively than men do (Veniegas & Peplau, 1997).

Of course, individual differences in friendship quality exist. The longer-lasting men's friendships are, the closer they become and the more they involve disclosure of personal information (Sherman, de Vries, & Lansford, 2000). Furthermore, involvement in family roles affects reliance on friends. For single adults, friends are the preferred companions and confidants. As romantic ties form and people marry, young adults—especially men—direct more of their disclosures toward their partners (Carbery & Buhrmester, 1998). Still, friendships continue to be vital contexts for personal sharing throughout early adulthood. Turn back to Figure 12.2 on page 397 to view developmental trends in self-disclosure to romantic partners and friends.

● **Other-Sex Friendships.** Other-sex friendships are important to adults, although they occur less often and do not last as long as same-sex friendships. During the college years, these bonds are as common as romantic relationships. They decline after marriage for men but increase with age for women, who tend to form them in the workplace. Highly educated, employed women have the largest number of other-sex friends. Through these relationships, young adults often gain in companionship and self-esteem and learn a great deal about masculine and feminine styles of intimacy (Bleske & Buss, 2000). Because men confide especially easily in their female friends, such friendships offer them a unique opportunity to broaden their expressive capacity. And women sometimes say male friends offer objective points of view on problems and situations—perspectives not available from female friends (Monsour, 2002).

Sexual attraction is regulated more in other-sex friendships than in romantic ties. Many try to keep the relationship platonic to safeguard its integrity (Messman et al., 2000). Still, about half of college students engage in sexual activity with an other-sex friend whom they have no intention of dating. Men are more likely than women to feel sexually attracted to an other-sex friend (Kaplan & Keys, 1997). If these feelings persist, the relationship often changes into a romantic bond. Some friends sustain a platonic friendship with sexuality as part of it. Others find the platonic and sexual aspects incompatible, and the friendship disintegrates (Affifi & Faulkner, 2000). When a solid other-sex friendship does evolve into a romance, it may be more stable and enduring than a romantic relationship formed without a foundation in friendship (Hendrick & Hendrick, 1993).

● **Siblings as Friends.** Whereas intimacy is essential to friendship, commitment—willingness to maintain a relationship and care about the other—is the defining characteristic of family ties. As young people marry and invest less time in developing a romantic partnership, siblings—especially sisters whose earlier bond was positive—become more frequent companions than they were in adolescence. And often, friend and sibling roles merge. For example, Sharese described Heather's practical assistance—helping with moving and running errands during an illness—in kin terms: "She's like a sister to me. I can always turn to her." And adult sibling ties resemble friendships, in which the main concerns are staying in contact, offering social support, and enjoying being together (O'Connor, 1992). Relationships between same-sex siblings can be especially close. Despite rivalries and differences in interests that emerged in childhood, a shared background of experiences promotes similarity in values and perspectives and the possibility for deep mutual understanding.

Warm sibling relationships in adulthood are important sources of psychological well-being (Riggio, 2000). In Vaillant's (1977) study of well-educated men, a close sibling tie in early adulthood was the single best predictor of emotional health at age 65. In another investigation, one-fifth of married women identified a sister as a best friend (Oliker, 1989).

Loneliness

Young adults, who expect to form intimate ties at this time in their life, are at risk for **loneliness**—unhappiness resulting from a gap between the social relationships we currently have and those we desire. Adults may feel lonely because they do not have an intimate partner or because they lack gratifying friendships. Both situations give rise to similar emotions, but they are not interchangeable (Brehm, 1992). For example, even though she had several enjoyable friendships, Heather felt lonely from time to time because she was not dating someone she cared about. And although Sharese and Ernie were happily married, they felt lonely after moving to a new town where they did not know anyone.

Loneliness peaks during the late teens and early twenties, after which it declines steadily into the seventies. Figure 14.1 shows this trend, based on a large Canadian sample ranging in age from 13 to 80 (Rokach, 2001a). The rise in loneliness during early adulthood is understandable. Young people must constantly develop new relationships as they move through school and employment settings. Also, young adults may expect more from their intimate ties than older adults, who have learned to live with imperfections. With age, people become better at accepting loneliness and using it for positive ends—to sharpen awareness of their personal fears and needs (Rokach, 2001b).

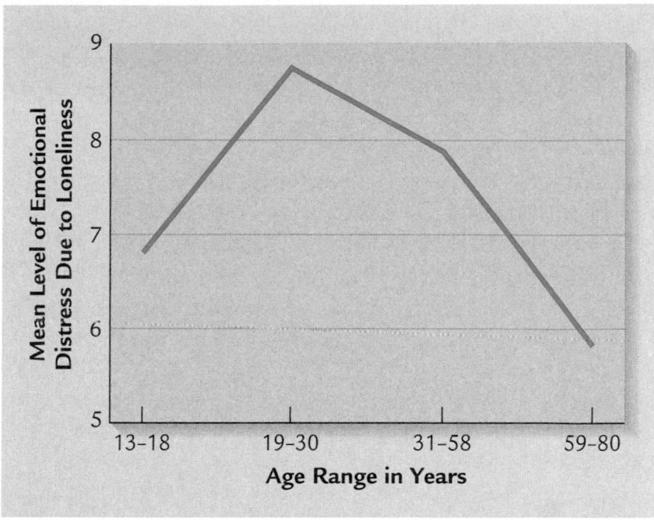

FIGURE 14.1 Changes in emotional distress due to loneliness from adolescence to late adulthood. More than 700 Canadian 13- to- 80-year-olds responded to a questionnaire assessing the extent to which they experienced emotional distress due to loneliness. Loneliness rose sharply from the teens to the twenties and then declined. (Adapted from Rokach, 2001a.)

Loneliness is at its peak during the late teens and early twenties. It usually stems from lack of an intimate partner or gratifying friendships. When extreme loneliness persists, it is associated with negative self-evaluations and socially unresponsive and insensitive behavior, which promote further isolation.

Adults are more likely to experience loneliness under certain conditions. Separated, divorced, or widowed adults are lonelier than their married, cohabiting, or single counterparts, suggesting that loneliness is intense after loss of an intimate tie. When not involved in a romantic relationship, men feel lonelier than women, perhaps because they have fewer alternatives for satisfying intimacy needs (Rubenstein & Shaver, 1982; Stroebe et al., 1996). And immigrants from collectivist cultures report higher levels of loneliness than people born in the United States and Canada (Rokach & Sharma, 1996). Leaving a large, close-knit family system for an individualistic society seems to prompt intense feelings of isolation.

Personal characteristics also contribute to loneliness, with shy, socially anxious people experiencing it more intensely (Bruch et al., 1989). When extreme loneliness persists, it is associated with a wide variety of self-defeating attitudes and behaviors. Lonely people evaluate themselves and others more negatively, tend to be socially unresponsive and insensitive, and are slow to develop intimacy because they are reluctant to tell others about themselves (Jones, 1990). The extent to which these responses are cause or consequence of loneliness is unclear, but once in place, they promote further isolation.

As long as loneliness is not overwhelming, it can motivate young people to take social risks and reach out to others. It can also encourage them to find ways to be comfortably alone and to use this time to understand themselves better. Much of healthy personality development involves striking this balance—between "satisfying relationships with others and a secure, internal base of satisfaction within ourselves" (Brehm, 1992, p. 345).

Ask Yourself

REVIEW

Cite differences between men and women in traits desired in a long-term partner. What findings indicate that *both* biological and social forces contribute to those differences?

REVIEW

Why does loneliness peak in early adulthood?

APPLY

After dating for 2 years, Mindy and Graham reported greater love and relationship satisfaction than during the first few months they had known each other. What features of communication probably contributed to a deepening of their bond, and why is it likely to endure?

APPLY

Claire and Tom, both married to other partners, got to know each other at work and occasionally have lunch together. What is each likely to gain from this other-sex friendship?

www.

The Family Life Cycle

For the majority of young people, the quest for intimacy leads to marriage. Their life course takes shape within the **family life cycle**—a sequence of phases that characterizes the development of most families around the world. In early adulthood, people typically live on their own, marry, and bear and rear children. As they become middle-aged and their children leave home, their parenting responsibilities diminish. Late adulthood brings retirement, growing old, and (mostly for women) death of one's spouse (Framo, 1994; McGoldrick, Heiman, & Carter, 1993).

However, we must be careful not to view the family life cycle as a fixed progression. Recall from Chapter 2 that the family is a dynamic system of relationships embedded in community, cultural, and historical contexts. Today, wide variations exist in the sequence and timing of family life-cycle phases. High rates of out-of-wedlock births, delayed childbearing, divorce, and remarriage are but a few illustrations. And some people—either voluntarily or involuntarily—do not experience some or all of the family life cycle.

Still, the family life-cycle model is useful. It offers an organized way of thinking about how the family system changes over time and the impact of each phase on the family unit and individuals within it.

Leaving Home

During her first semester of college, Sharese noticed a change in how she related to her mother. She found it more enjoyable to discuss daily experiences and life goals, sought advice and listened with greater openness, and expressed affection more freely. Over the next few years, Sharese's bedroom began to seem like a guest room. Looking around before she moved out permanently, she felt some nostalgia for the warmth and security of her childhood days, coupled with a sense of pride at being on her own.

Departure from the parental home is a major step toward assuming adult responsibilities. The average age of leaving has decreased in recent years as more young people live independently before marriage. In 1940, over 80 percent of North Americans in their twenties resided with their parents. Today, only about 50 percent do. Most industrialized nations show this trend (Goldscheider & Goldscheider, 1999; Statistics Canada, 2002p).

Timing of departure varies with the reason for leaving. Departures for education tend to occur at younger ages, those for full-time work and marriage later. Because the majority of North American young adults enroll in higher education, many leave home early, around age 18. Some young people who want to escape family friction also depart early (Stattin & Magnusson, 1996). And those from divorced, single-parent homes tend to be early leavers, perhaps because of family stress (Cooney & Mortimer, 1999). Compared with the previous generation, fewer young people leave home to marry, and more do so just to be "independent"—to express their adult status. But a difficult job market and high housing costs mean that many must take undesirable jobs or remain financially dependent on their parents (Lindsay, Almey, & Normand, 2002).

Nearly half of young adults return home for a brief time after initial leaving. Those who departed to marry are least likely to return. But single, independent living is a fragile arrangement. As people encounter unexpected twists and turns on the road to independence, the parental home becomes a safety net and base of operations for launching adult life. Failures in work or marriage can prompt a move back home. Also, young people who left because of family conflict usually return—largely because they were not ready for independent living. But most of the time, role transitions, such as the end of college or military service, bring people back. Contrary to popular belief, returning home is usually not a sign of weakness. Instead, it is a common event among unmarried adults (Graber & Brooks-Gunn, 1996).

Although most high school seniors expect to live on their own before marriage, the extent to which they do so varies with SES and ethnicity. Economically well-off young people are more likely to establish their own residence. Among African-American, Hispanic, and Canadian Aboriginal groups, poverty and a cultural tradition of extended family living lead to low rates of leaving home. Unmarried Asian young adults also tend to live with their parents. But the longer Asian families have lived in North America and thus been exposed to individualistic values, the more likely young people are to move out after finishing high school (Goldscheider & Goldscheider, 1999).

When young adults are prepared for independence, departure from the home is linked to more satisfying parent–child interaction and successful transition to adult roles. But leaving home very early can contribute to long-term disadvantage because it is associated with job seeking rather than education and with lack of parental financial and social support. Not surprisingly, non-college-bound youths who leave home early have less successful marriages and work lives (White, 1994).

Joining of Families in Marriage

Young adults delay marriage more today than a half-century ago. In 1950, the average age of first marriage was about 20 for women and 23 for men. In 2000, it was 25 and 27 in the United States and 27 and 29 in Canada. The number of first and second marriages has declined over the last few decades as more people remain single, cohabit, or do not remarry after divorce. Still, the United States and Canada remain cultures strongly committed to marriage. Nearly 90 percent of North Americans marry at least once in their lives. And at present, 59 percent of American adults and 47 percent of Canadian adults live together as married couples (Statistics Canada, 2002a; U.S. Bureau of the Census, 2002c).

Marriage is often regarded as the joining of two individuals. In actuality, it requires that two entire systems—the husband's and wife's families—adapt and overlap to create a new subsystem. Consequently, marriage presents couples with complex challenges. This is especially so today because husband–wife roles have only begun to move in the direction of a true partnership—educationally, occupationally, and in emotional connectedness.

Marital Roles. Their wedding and honeymoon over, Sharese and Ernie turned to a multitude of issues that they had previously decided individually or that their families of

Mixed marriages are more common today than in the past. This Kashmiri Hindu woman and American Catholic man combine their religious traditions on their wedding day. On the left, they light the Catholic Unity candle, which symbolizes the joining of their separate lives. Above, wedding guests bestow a Hindu blessing on the couple, showering them with flower petals.

origin had prescribed. They considered everyday matters—when and how to eat, sleep, talk, work, relax, have sex, and spend money. They also discussed which family traditions and rituals to retain and which to work out for themselves. And as they related to their social world as a couple, they modified relationships with parents, siblings, extended family, friends, and co-workers.

Recent alterations in the context of marriage, including changing gender roles and living farther away from family members, mean that contemporary couples must do more work to define their relationships. Although husbands and wives are usually similar in religious and ethnic background, "mixed" marriages occur more often today than they did in the past. Other-race unions account for 6 percent of the married population in the United States and 3 percent in Canada (Statistics Canada, 2003c; U.S. Bureau of the Census, 2002c). More young people also choose a mate of a different religion. For example, between one-third and one-half of North American Jews who marry today select a non-Jewish spouse (Weiner, 2002). When their backgrounds differ greatly, these couples face extra challenges in achieving a successful transition to married life.

Many contemporary couples live together before marriage, making it less of a turning point in the family life cycle than in the past. Still, defining marital roles can be difficult. Age of marriage is the most consistent predictor of marital stability. Young people who marry in their teens and early twenties are far more likely to divorce than those who marry at older ages (Heaton, 2002). Those who marry early may be running away from their own family or seeking the family they never had. Most have not developed a secure enough identity or sufficient independence for a mature marital bond. Both early marriage followed by childbirth and the reversal of family life-cycle events (childbirth before marriage) are more common among low-SES adults. This acceleration of family formation complicates adjustment to life as a couple (Leonard & Roberts, 1998).

Despite progress in the area of women's rights, **traditional marriages**, involving a clear division of husband's and wife's roles, still exist in Western nations. The man is the head of household; his primary responsibility is the economic well-being of his family. The woman devotes herself to caring for her husband and children and to creating a nurturant, comfortable home. However, traditional marriages have changed in recent decades. Although women make motherhood the top priority while their children are young, many return to the work force at a later date.

In **egalitarian marriages**, husband and wife relate as equals, and power and authority are shared. Both partners try to balance the time and energy they devote to their occupations, their children, and their relationship. The majority of well-educated, career-oriented women expect this form of marriage (Botkin, Weeks, & Morris, 2000). But women's employment has not had much impact on household division of labor. Men in dual-earner families participate more than those in single-earner families. But a recent international study revealed that women in the United States and Canada spend nearly twice as much time on housework as men do (see Figure 14.2). In Sweden, which places a high value on gender equality, men participate to a greater extent than in other nations. In contrast, men do very little housework in Japan, where gender roles remain traditional (Institute for Social Research, 2002). In sum, true equality in marriage is still rare, and couples who strive for it usually attain a form of marriage in between traditional and egalitarian.

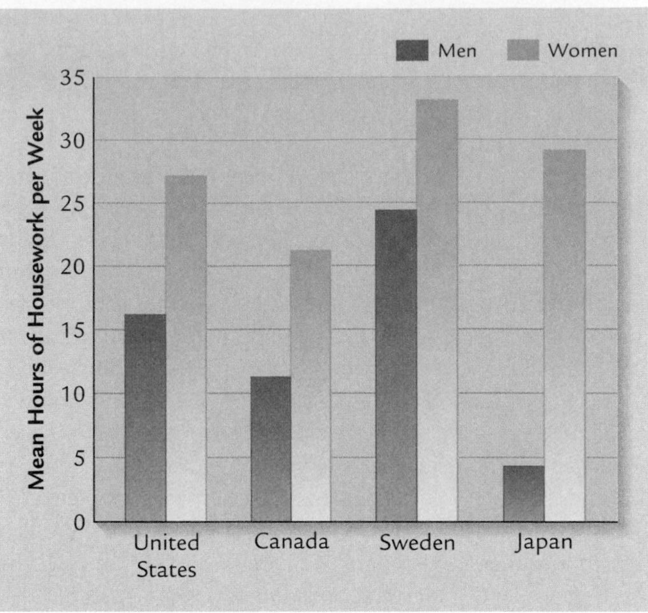

FIGURE 14.2 Average hours per week of housework reported by men and women in four nations. In each nation, women devote considerably more time than men to housework. Men's participation is greater in Sweden, which places a high value on gender equality. In Japan, where traditional gender roles prevail, men devote the least time to housework. (Data for the United States, Sweden, and Japan from Institute for Social Research, 2002; data for Canada derived from Statistics Canada, 2000l.)

● **Marital Satisfaction.** Despite its rocky beginnings, Sharese and Ernie's marriage grew to be especially happy. In contrast, Christy and Gary became increasingly discontented. What distinguishes marriages high in satisfaction from less successful partnerships? Differences between the two couples mirror the findings of a large body of research on personal and contextual factors, summarized in Table 14.2.

Christy and Gary had a brief courtship, married at a young age, had children early, and struggled financially. Gary's negative, critical personality led him to get along poorly with Christy's parents and to feel threatened when he and Christy disagreed. Christy tried her best to offer Gary encouragement and support, but her own needs for nurturance and individuality were not being met. Gary felt threatened by Christy's career aspirations. As she came closer to attaining them, the couple grew farther apart. In contrast, Sharese and Ernie married

Table 14.2 Factors Related to Marital Satisfaction

Factor	Happy Marriage	Unhappy Marriage
Family backgrounds	Partners similar in SES, education, religion, and age	Partners very different in SES, education, religion, and age
Age at marriage	After age 23	Before age 23
Length of courtship	At least 6 months	Less than 6 months
Timing of first pregnancy	After first year of marriage	Before or within first year of marriage
Relationship to extended family	Warm and positive	Negative; wish to maintain distance
Marital patterns in extended family	Stable	Unstable; frequent separations and divorces
Financial and employment status	Secure	Insecure
Personality characteristics	Emotionally positive; good conflict-resolution skills	Emotionally negative and impulsive; poor conflict-resolution skills

Note: The more factors present, the greater the likelihood of marital happiness or unhappiness.
Sources: Bradbury, Fincham, & Beach, 2000; Gottman et al., 1998; Sanders, Halford, & Behrens, 1999.

later, after their educations were complete. They postponed having children until their careers were under way and they had built a sense of togetherness that allowed each to thrive as an individual. Patience, caring, shared values, enjoyment of each other's company, and good conflict-resolution skills contributed to their compatibility.

Although the factors just described differentiate troubled from gratifying marital relationships, research also reveals clear sex differences in marital satisfaction. More men than women report being happily married (Dillaway & Broman, 2001; Kaslow, Hansson, & Lundblad, 1994). Simply being married is associated with gains in men's mental and physical health—an outcome that stems from enhanced feelings of attachment, belonging, and social support. For women, however, *relationship quality* has a greater impact on mental health. Consequently, women suffer more than men when a marriage has problems (Horowitz, White, & Howell-White, 1996; Horowitz, McLaughlin, & White, 1997). Women feel particularly dissatisfied when the demands of husband, children, housework, and career are overwhelming (Wilkie, Ferree, & Ratcliff, 1998). And although women are more willing to evaluate a relationship as unsatisfying and try to work on it, recall from our discussion of love that men often withdraw from conflict.

At their worst, marital relationships can become contexts for intense opposition, dominance–submission, and emotional and physical violence (Bradbury, Fincham, & Beach, 2000). As the Social Issues box on pages 460–461 explains, although women are more often targets of severe partner abuse, both men and women are perpetrators and victims.

● Marital Expectations and Myths. In a study in which fifty happily married couples were interviewed about their marriages, each participant reported good times and bad; none was happy all the time. Many admitted that there were moments when they wanted out, when they felt they had made a mistake. Clearly, marital happiness did not signify a "rose garden." Instead, it was grounded in mutual respect, pleasure and comfort in each other's company, and joint problem solving. All couples emphasized the need to reshape their relationship in response to new circumstances and to each partner's changing needs and desires (Wallerstein & Blakeslee, 1995).

Yet cultural expectations work against this view of marriage as an ongoing project requiring both partners' involvement and cooperation. Historically, women had little power in marriage and society; a wife's status came from her husband. This gender gap has such deep cultural roots that it continues to influence marriage today (Brooks, 1999). In a study of college students, more women than men said their partners should be superior to themselves, and more men than women said their partners should be inferior to themselves—in intelligence, education, vocational success, and income (Ganong & Coleman, 1992). Under these circumstances, women are likely to play down their abilities, sacrificing part of themselves. And men tend to limit themselves to the provider role rather than participating fully in family life.

Furthermore, many young people have a mythical image of marital bliss—one that is a far cry from reality. For example, a substantial number of college students endorse the following beliefs not supported by facts:

- A couple's satisfaction increases through the first year of marriage.
- The best single predictor of marital satisfaction is the quality of a couple's sex life.
- If my spouse loves me, he or she should instinctively know what I want and need to be happy.
- No matter how I behave, my spouse should love me simply because he or she is my spouse. (Larson, 1988, p. 5)

Social Issues

Partner Abuse

Violence in families is a widespread health and human rights issue, occurring in all cultures and SES groups. Often one form of domestic violence is linked to others. Recall the story of Karen in Chapter 13. Her husband Mike not only assaulted her sexually and physically but also abused her psychologically—isolating, humiliating, and demeaning her (Dutton et al., 2001). Violent adults also break their partner's favorite possessions, punch holes in walls, or throw objects. If children are present, they may become victims.

Partner abuse in which husbands are perpetrators and wives are physically injured is the type most likely to be reported to authorities. But many acts of family violence are not reported. When researchers ask North American couples about fights that led to acts of hostility, men and women report similar rates of assault (Straus, 1999; Zlotnick et al., 1998). For example, in a large national survey of Canadians, 8 percent of women and 7 percent of men indicated that they had been physically abused by a spouse within the past 5 years (Statistics Canada, 2001d). Women, however, are more likely to experience physical assaults that lead to serious injury—beatings, chokings, attempts to drown, and threats with guns. Men are more often targets of kicking, slapping, thrown objects, and threats with knives (Hoff, 2001).

Although self-defense is a frequent cause of domestic assault by women, men and women are equally likely to "strike first" (Carrado et al., 1996; Currie, 1999). "Getting my partner's attention," "gaining control," and "expressing anger" are reasons that partners typically give for abusing each other (Hamberger et al., 1997; Straus, 1999).

Factors Related to Partner Abuse. In abusive relationships, dominance–submission proceeds sometimes from husband to wife and at other times from wife to husband. In about half of cases, both partners are violent (Cook, 1997). Marvin's and Pat's relationship helps us understand how spouse abuse escalates. Shortly after their wedding, Pat began complaining about the demands of Marvin's work and insisted that he come home early to spend time with her. When he resisted, she hurled epithets, threw objects, and slapped him. One evening, Marvin became so angry at Pat's hostilities that he smashed a dish against the wall, threw his wedding ring at her, and left the house. The next morning, Pat apologized and promised not to attack again. But her outbursts became more frequent and desperate.

These violence-remorse cycles in which aggression escalates characterize many abusive relationships. Why do they occur? Personality and developmental history, family circumstances, and cultural factors combine to make partner abuse more likely (Dixon & Browne, 2003).

Many abusers are overly dependent on their spouses as well as jealous, possessive, and controlling. For example, the thought of Karen ever leaving induced such high anxiety in Mike that he monitored all her activities. Depression, anxiety, and low self-esteem also characterize abusers. And because they have great difficulty managing anger, trivial events—such as an unironed shirt or a late meal—can trigger abusive episodes (Guyer, 2000).

A high proportion of spouse abusers grew up in homes where parents engaged in hostile interactions, used coercive discipline, and were abusive toward their children (Bevan & Higgins, 2002; Reitzel-Jaffe & Wolfe, 2001). Perhaps this explains why conduct problems in childhood and violent delinquency in adolescence also predict partner abuse (Magdol et al., 1998). Adults with childhood exposure to domestic violence are not doomed to repeat it. But their parents provided them with negative expectations and behaviors that they often transfer to their close relation-

As these myths are overturned, couples react with disappointment, and marriage becomes less satisfying and more conflict-ridden. Interestingly, young people who hold a religious view of marriage as sacred are less likely to enter it with unrealistic expectations and better able to cope with disagreement. Perhaps because of their reverence for the marital bond, they are highly invested in forging a well-functioning relationship, engaging in more verbal collaboration and less conflict than other couples (Mahoney et al., 1999).

In view of its long-term implications, it is surprising that most couples spend little time before their wedding day reflecting on the decision to marry (McGoldrick, Heiman, & Carter, 1993). Courses in family life education in high schools and colleges can help dispel marital myths and promote better mate selection. And counseling aimed at helping couples discuss their desires openly and use positive, respectful conflict-resolution strategies are highly effective in easing adjustment to marriage and enhancing relationship quality (Christensen & Heavey, 1999).

Parenthood

In the past, the issue of whether to have children was, for many adults, "a biological given or an unavoidable cultural demand" (Michaels, 1988, p. 23). Today, in Western industrialized nations, it is a matter of true individual choice. Effective birth control techniques enable adults to avoid having chil-

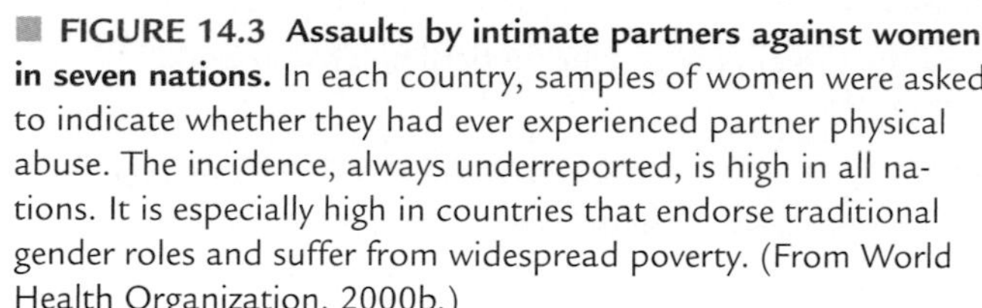

■ **FIGURE 14.3 Assaults by intimate partners against women in seven nations.** In each country, samples of women were asked to indicate whether they had ever experienced partner physical abuse. The incidence, always underreported, is high in all nations. It is especially high in countries that endorse traditional gender roles and suffer from widespread poverty. (From World Health Organization, 2000b.)

ships. Stressful life events, such as job loss or financial difficulties, increase the likelihood of partner abuse (Emery & Laumann-Billings, 1998). Because of widespread poverty, African Americans, Native Americans, and Canadian Aboriginal people report high rates of partner violence (Hoff, 2001; Statistics Canada, 2001d). Alcohol abuse also is related to it.

At a societal level, cultural norms that endorse male dominance and female submissiveness promote partner abuse (Stith & Farley, 1993). As Figure 14.3 shows, in countries that combine sanctioning of gender inequality with widespread poverty, partner violence against women is especially high, affecting as many as 40 to 50 percent of the female population (World Health Organization, 2000b).

Why don't people leave destructive intimate relationships before the abuse escalates? A variety of situational factors discourage them from doing so. A victimized wife may depend on her husband's earning power or fear even worse harm to herself or her children. Extreme assaults, including homicide, tend to occur after partner separation (Statistics Canada, 2001d). And victims of both sexes, but especially men, are deterred by the embarrassment of going to the police. Also, they may falsely believe that their partner will change (Straus, 1999).

Intervention and Treatment. Community services available to battered women include crisis telephone lines that provide anonymous counseling and support and shelters that offer protection and treatment (see pages 429–430). Because many women return to their abusive partners several times before making their final move, community agencies usually offer therapy to male batterers. Most rely on group sessions that confront rigid gender stereotyping; teach communication, problem solving, and anger control; and use social support to motivate change in behavior (Harway & Hansen, 1994).

Although existing treatments are far better than no treatment, most are too brief to pay sufficient attention to alcohol problems and relationship difficulties. Consequently, of the small number of perpetrators who agree to participate, at least half continue their violent behavior with the same or a new partner (Carden, 1994). At present, few interventions acknowledge that men also are victims. Yet ignoring their needs perpetuates domestic violence. When victims do not want to separate from a violent partner, a whole-family treatment approach that focuses on changing partner interaction and reducing high life stress is crucial.

dren in most instances. And changing cultural values allow people to remain childless with less fear of social criticism and rejection than was the case a generation or two ago.

In 1950, 78 percent of North American married couples were parents. Today, 70 percent bear children, and they tend to have their first child at a later age. Consistent with this pattern of delayed childbearing, family size in industrialized nations has declined. In 1950, the average number of children per couple was 3.1. Currently, it is 1.8 in the United States; 1.7 in Australia; 1.6 in Austria, Canada, Great Britain, and Sweden; 1.5 in the Netherlands and Japan; and 1.3 in Germany (Bellamy, 2000; Pearce, Cantisani, & Laihonen, 1999). Nevertheless, the vast majority of married people continue to embrace parenthood as one of life's most meaningful experiences. Why do they do so, and how do the challenges of child rearing affect the adult life course?

● **The Decision to Have Children.** The choice of parenthood is affected by a complex array of factors, including financial circumstances, personal and religious values, and health conditions. Women with traditional gender identities usually decide to have children. Whether a woman is employed has less impact on childbearing than her occupation. Women with high-status, demanding careers less often choose parenthood and, when they do, more often delay it than women with less time-consuming jobs (Barber, 2001; Tangri & Jenkins, 1997).

When asked about their desire to have children, couples mention a variety of advantages and disadvantages, listed in

Table 14.3 Advantages and Disadvantages of Parenthood Mentioned by Contemporary Couples

Advantages	Disadvantages
Giving and receiving warmth and affection	Loss of freedom, being tied down
Experiencing the stimulation and fun that children add to life	Financial strain
Being accepted as a responsible and mature member of the community	Role overload—not enough time for both family and work responsibilities
Experiencing new growth and learning opportunities that add meaning to life	Interference with mother's employment opportunities
Having someone carry on after one's own death	Worries over children's health, safety, and well-being
Gaining a sense of accomplishment and creativity from helping children grow	Risks of bringing up children in a world plagued by crime, war, and pollution
Learning to become less selfish and to sacrifice	Reduced time to spend with partner
Having offspring who help with parents' work or add their own income to the family's resources	Loss of privacy
	Fear that children will turn out badly, through no fault of one's own

Source: Cowan & Cowan, 2000.

Table 14.3. Although some ethnic and regional differences exist, reasons for having children that are most important to all groups include the desire for a warm, affectionate relationship and the stimulation and fun that children provide. Also frequently mentioned are growth and learning experiences that children bring to the lives of adults, the desire to have someone carry on after one's own death, and the feelings of accomplishment and creativity that come from helping children grow (Cowan & Cowan, 1992, 2000).

Most young adults are also aware that having children means years of extra burdens and responsibilities. When asked about the disadvantages of parenthood, they mention "loss of freedom" most often, followed by "financial strain." Indeed, the cost of child rearing is a major factor in modern family planning. According to a conservative estimate, today's new parents will spend about $260,000 to rear a child from birth through four years of college (U.S. Department of Labor, 2003). Finally, many adults worry greatly about family–work conflict—not having enough time to meet both child-rearing and job responsibilities (Hochschild, 1997).

Greater freedom to choose whether and when to have children makes family planning more challenging today than in the past. As each partner expects to have equal say, childbearing often becomes a matter of delicate negotiation (Cowan & Cowan, 2000). Yet careful weighing of the pros and cons of parenthood means that many more couples are making informed and personally meaningful choices—a trend that should increase the chances that they will have children when ready and that their own lives will be enriched by their decision.

● **Transition to Parenthood.** The early weeks after a baby enters the family are full of profound changes: disrupted sleep schedules, new caregiving and household tasks, less time for couples to devote to each other, and added financial responsibilities. In addition, the roles of husband and wife often become more traditional, even for couples like Sharese and Ernie who are strongly committed to gender equality and were used to sharing household tasks (Cowan & Cowan, 1997; Huston & Vangelisti, 1995).

For most new parents, however, the arrival of a baby does not cause significant marital strain. Marriages that are gratifying and supportive tend to remain so and resemble childless marriages in overall happiness (Feeney et al., 2001; Miller, 2000). In contrast, troubled marriages usually become more distressed after a baby is born. In a study of newlyweds who were interviewed annually for 6 years, the husband's affection, expression of "we-ness" (values and goals similar to his wife's), and awareness of his wife's daily life predicted mothers' stable or increasing marital satisfaction after childbirth. In contrast, the husband's negativity and the couple's out-of-control conflict predicted a drop in mothers' satisfaction (Shapiro, Gottman, & Carrere, 2000).

Also, the larger the difference in men's and women's caregiving responsibilities, the greater the decline in marital satisfaction after childbirth, especially for women—with negative consequences for parent–infant interaction. In contrast, sharing caregiving predicts greater parental happiness and sensitivity to the baby (Feldman, 2000; Feeney et al., 2001). Postponing childbearing until the late twenties or thirties, as more couples are doing today, eases the transition to parenthood. Waiting permits couples to pursue occupational goals and gain life experience. Under these circumstances, men are more enthusiastic about becoming fathers and therefore more willing to participate. And women whose careers are under way are more likely to encourage their husbands to share housework and child care (Coltrane, 1990).

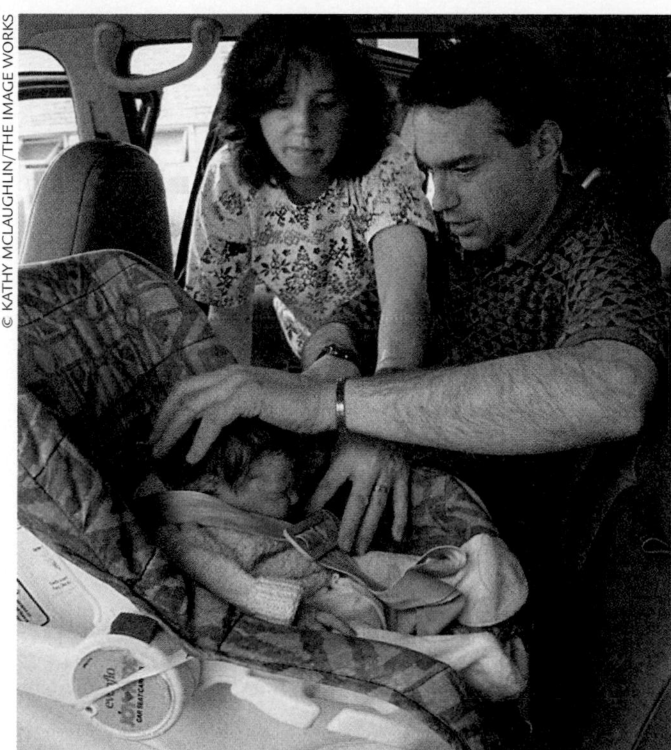

■ These parents settle their newborn daughter into her car seat for the trip home from the hospital. The transition to parenthood can enrich a warm, gratifying marriage or worsen a tense, unhappy marriage. Sharing caregiving tasks enhances marital satisfaction and is related to both parents' sensitivity toward their new baby.

In many non-Western cultures, childbirth has little or no impact on marital satisfaction. In these societies, family life and parenthood are central for women, and husband–wife division of labor is not questioned. In addition, the extended family typically assists with child rearing (Levy-Shiff, 1994). In Western nations, however, the trend toward gender equality and isolation of the nuclear family unit leads marital and parenting roles to be closely linked. A change in happiness in one role is often accompanied by a similar change in the other (Rogers & White, 1998).

Special interventions exist to ease the transition to parenthood. For those who are not at high risk for problems, couples' groups led by counselors are highly effective. In one program, first-time expectant couples gathered once a week for 6 months to discuss their dreams for the family and changes in relationships sparked by the baby's arrival. Eighteen months after the program ended, participating fathers described themselves as more involved with their children than did fathers in a no-intervention condition. Perhaps because of fathers' caregiving assistance, participating mothers maintained their prebirth satisfaction with family and work roles. Three years after the birth, the marriages of all participating couples were intact and just as happy as they had been before parenthood. In contrast, 15 percent of couples receiving no intervention had divorced (Cowan & Cowan, 1997). For high-risk parents struggling with poverty or the birth of a child with disabilities, interventions must be more intensive, focusing on enhancing social support and parenting skills (Cowan & Cowan, 1995).

● **Additional Births.** The same array of factors that affect the decision to become parents also influences family size. Besides more effective birth control, a major reason that couples in industrialized nations have fewer children today than in the past is women's increased career orientation. Also, more couples get divorced before their child-rearing plans are complete.

Overall, a smaller family size enhances parent–child interaction. Parents of fewer children are more patient and less punitive. They also have more time to devote to each child's activities, schoolwork, and other special needs. Furthermore, in smaller families, siblings are more likely to be widely spaced (born more than 2 years apart), which adds to the attention and resources parents can invest in each other and in each child. Together, these findings may account for the fact that small family size is linked to greater marital satisfaction and physically healthier children who score higher in IQ and do better in school (Downey, 1995; Powell & Steelman, 1993).

However, large families are usually less well off economically. Crowded housing, inadequate nutrition, and stressed and poorly educated parents seem to contribute to the negative relationship between family size and well-being. When well-educated, high-income parents have large families, unfavorable outcomes are eliminated (Guo & VanWey, 1999). As the Cultural Influences box on page 464 indicates, education and family planning are closely linked. Both are vital for women's and children's quality of life.

● **Families with Young Children.** A year after the birth of their first child, Sharese and Ernie received a phone call from Heather, who asked how well they liked parenthood: "Is it a joy, a dilemma, a stressful experience—how would you describe it?"

Chuckling, Sharese and Ernie responded in unison, "All of the above!"

In today's complex world, men and women are less certain about how to rear children than they were in previous generations. Clarifying child-rearing values and implementing them in warm, involved, and appropriately demanding ways are crucial for the welfare of the next generation and society. Yet cultures do not always place a high priority on parenting, as indicated by the lack of many societal supports for children and families (see Chapter 2, page 64). Furthermore, changing family forms mean that the lives of today's parents differ substantially from those of past generations.

In previous chapters, we discussed a wide variety of influences on child-rearing styles—personal characteristics of children and parents, SES, ethnicity, and more. The couple's relationship is also vital. Parents who work together as a "coparenting team," cooperating and showing solidarity and respect for each other in parenting roles, are more likely to feel

Cultural Influences

A Global Perspective on Family Planning

Approximately one-fifth of the world's population—one billion people—live in extreme poverty, the majority in slums and shantytowns of developing countries. If current trends in population growth continue, the number of poor will quadruple within the next 60 to 70 years (United Nations, 1998). Poverty and rapid population growth are intertwined: Poverty leads to high birthrates, and rising birthrates heighten poverty and deprivation. Why is this so?

First, in poor regions of the world where child death rates are high, parents have more children to compensate for the fact that some will certainly die. Second, lack of status, education, and opportunities for women, characteristic of most nonindustrialized societies, restrict life choices to early marriage and prolonged childbearing. Third, in regions with few basic services and labor-saving technologies, families often depend on children to help in the fields and at home. Fourth, poverty is associated with absence of family-planning services, which causes birthrates to remain high even when people realize the advantages of smaller families (Bulatao, 1998). And finally, lack of hope for the future is a major obstacle to life planning in general and family planning in particular.

As a country's population grows, poverty worsens. The number of people seeking work expands more rapidly than available jobs, and a new generation of unemployed or underemployed parents emerges. Basic resources, including food, water, land, and fuel, are in shorter supply, and health and educational services are increasingly strained. As a result, overcrowding in urban areas—along with malnutrition, disease, illiteracy, and hopelessness—spreads. A circuit forms through which poverty and high birthrates perpetuate one another.

Two interrelated strategies are vital for intervening in this cycle:

- Emphasizing education and literacy, particularly for girls. Years of schooling strongly predicts small family size. Because women with more education have better life opportunities, they are more likely to marry at a later age and take advantage of family-planning services. As a result, they have fewer, more widely spaced, and healthier children (Caldwell, 1999).
- Making family-planning services available to all who want them, in ways compatible with each country's cultural and religious traditions. During the past half-century, use of birth control by married women in developing nations increased from 10 to 50 percent, contributing to an overall decline in birthrate from 6.1 to 3.3 (United Nations, 1998).

Still, the unmet need for family planning remains high. About 100 million women in the developing world want to limit family size or increase spacing between births (DaVanzo & Adamson, 1998). Yet those who are minimally literate or illiterate have difficulty accessing family-planning information. Their knowledge about effective birth control is often incorrect (Gazmararian, Parker, & Baker, 1999). Hence, they have many unintended pregnancies. As Figure 14.4 shows, the world's population continues to increase at an astounding pace because of high birthrates in poverty-stricken countries. In Nigeria, for example, the average woman will give birth to 6.5 children in her lifetime.

Education combined with family planning leads to substantial declines in birthrates and to resulting improvements in life chances for both mothers and children. These benefits carry over to future generations.

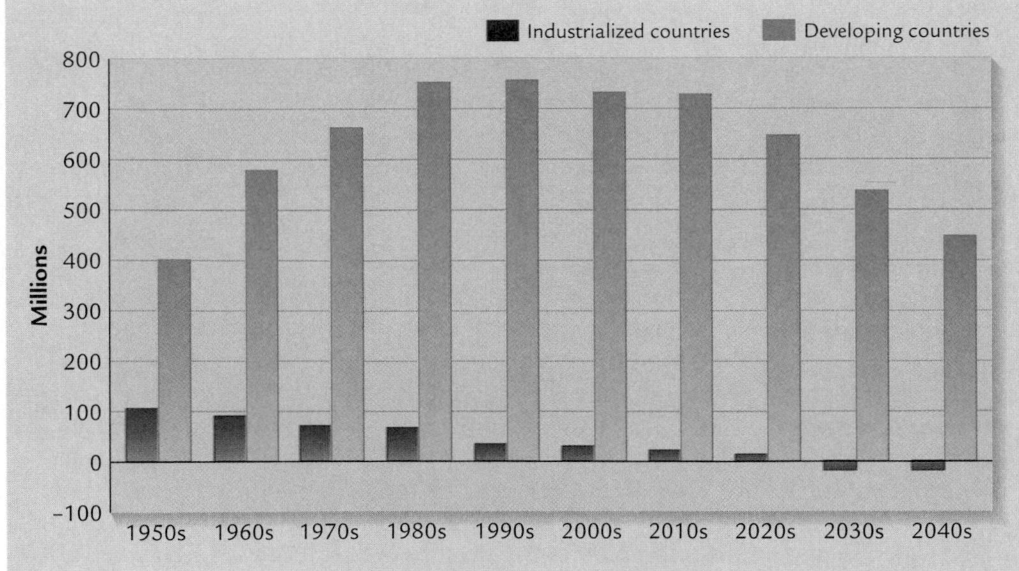

FIGURE 14.4 Population increases by decade in industrialized and developing countries, projected through the first half of the twenty-first century. Although birthrates are declining, the downward trend is very recent for developing nations. The world's population is still growing, with almost all this growth concentrated in developing countries. In the 2030s, the population in industrialized countries will start to decrease, whereas that of developing nations will continue to swell by more than 500 million people. (From DaVanzo & Adamson, 2000.)

© TONY FREEMAN/PHOTOEDIT

■ Despite its many challenges, rearing young children is a powerful source of adult development. Parents report that it expands their emotional capacities and enriches their lives. Involved parents say that child rearing helped them become more sensitive, tolerant, self-confident, and responsible.

competent as parents, use effective child-rearing practices, and have children who are developing well (McHale et al., 2002). When parents forge this coparenting alliance early, within the first few months after childbirth, it is more likely to persist (Fivaz-Depeursinge & Corboz-Warnery, 1999).

For employed parents, a major struggle during this phase of family life is finding good child care. The younger the child, the greater parents' sense of risk and difficulty finding the help they need (Lamb, 1998). When competent, convenient, affordable child care is not available, it usually leads to additional pressures on the woman. Either she must curtail or give up her career, or she must endure unhappy children, missed workdays, and constant searches for new arrangements.

Despite its many challenges, rearing young children is a powerful source of adult development. Parents report that it expands their emotional capacities and enriches their lives. For example, Ernie remarked that through sharing in child rearing, he felt "rounded out" as a person. Other involved parents say that parenthood helped them tune in to others' feelings and needs and become more tolerant, self-confident, and responsible (Coltrane, 1990).

● **Families with Adolescents.** Adolescence brings sharp changes in parental roles. In Chapters 11 and 12, we noted that parents must establish a revised relationship with their adolescent youngsters—blending guidance with freedom and gradually loosening control. As adolescents gain in autonomy and explore values and goals in their search for identity, parents often complain that their teenager is too focused on peers and no longer seems to care about the family. Heightened parent–child bickering over mundane issues takes a toll, especially on mothers, who do most of the negotiating with teenagers.

Overall, children seem to navigate the challenges of adolescence more easily than parents, many of whom report a dip in marital and life satisfaction. More people seek family therapy during this period of the family life cycle than during any other (Steinberg & Silk, 2002; Young, 1991).

● **Parent Education.** In the past, family life changed little from one generation to the next, and adults learned what they needed to know about parenting through modeling and direct experience. Today's world confronts adults with a host of factors that impinge on their ability to succeed as parents.

Contemporary parents eagerly seek information on child rearing through popular books. New mothers regard these sources as particularly valuable, second in importance only to their doctors (Deutsch et al., 1988). Special parent education courses have also emerged, designed to help parents clarify child-rearing values, improve family communication, understand how children develop, and apply more effective parenting strategies. Although their formats vary, many of these programs yield positive outcomes, including improved parent–child interaction, more flexible parenting attitudes, and heightened awareness by parents of their role as educators of their children (Shumow, 1998; Smith, Perou, & Lesesne, 2002). Yet another benefit is social support—opportunities to discuss concerns with experts and other dedicated parents, who share the view that no job is more important to the future of society than child rearing.

Ask Yourself

REVIEW
Why do more men than women report being happily married?

REVIEW
What strategies can couples use to ease the transition to parenthood?

APPLY
After her wedding, Sharese was convinced she had made a mistake. Cite factors that sustained her marriage and that led it to become especially happy.

CONNECT
What aspects of adolescent development make rearing teenagers a stressful time for parents, resulting in a dip in marital and life satisfaction? (See Chapter 11, pages 350–351, 367–368, and Chapter 12, page 395.)

www.

The Diversity of Adult Lifestyles

The current array of adult lifestyles arose in the 1960s, a decade in which young people began to question the conventional wisdom of previous generations. Many asked, How can I find happiness? What kinds of commitments should I make to live a full and rewarding life? As the public became more accepting of diverse lifestyles, choices seemed more available—among them, staying single, cohabiting, remaining childless, and divorcing.

Today, nontraditional family options have penetrated the North American mainstream. Many adults experience not just one, but several. As we consider these variations in the following sections, we will see that some adults make a deliberate decision to adopt a lifestyle, whereas others drift into it. The lifestyle may be imposed by society, as is the case for cohabiting homosexual couples, who cannot marry legally. Or people may decide on a certain lifestyle because they feel pushed away from another, such as a marriage gone sour. In sum, the adoption of a lifestyle can be within or beyond the person's control.

Singlehood

On finishing her education, Heather joined the Peace Corps and spent 5 years in Africa. Although open to a long-term relationship, she had only fleeting romances. When she returned to the United States, she accepted an executive position with an insurance company. Professional challenge and travel preoccupied her. At age 35, she reflected on her life while having lunch with Sharese: "I was open to marriage, but after my career took off, it would have interfered. Now I'm so used to independence that I question whether I could adjust to living with another person. I like being able to pick up and go where I want, when I want, without having to ask anyone or think about caring for anyone. But there's a trade-off: I sleep alone, eat most of my meals alone, and spend a lot of my leisure time alone."

Singlehood—not living with an intimate partner—has increased in recent years, especially among young adults. For example, rates of never-married North American 30- to 34-year-olds have risen sixfold since 1970, to about 30 percent of males and 20 percent of females. Besides more people marrying later or not at all, divorce has added to the numbers of single adults. In view of these trends, it is likely that most North Americans will spend a substantial part of their adult lives single, and a growing minority—about 8 to 10 percent—will stay that way (Statistics Canada, 2003d; U.S. Bureau of the Census, 2002c).

Because they marry later, more young adult men than women are single. But women are far more likely than men to remain single for many years or their entire life. With age, fewer men are available with characteristics that most women seek in a mate—the same age or older, equally or better educated, and professionally successful. Men find partners more easily, since they can select from a large pool of younger unmarried women. Because of the tendency for women to "marry up" and men to "marry down," men in blue-collar occupations and women in prestigious careers are overrepresented among singles after age 30.

Ethnic differences also exist. For example, the percentage of never-married African Americans is nearly twice as great as that of Caucasian Americans in early adulthood. As we will see later, high unemployment among black men interferes with marriage. But many African Americans eventually marry in their late thirties and forties, a period in which black and white marriage rates come closer together (Cherlin, 1992; U.S. Bureau of the Census, 2002c).

Singlehood is a multifaceted experience with different meanings. At one extreme are people who choose it deliberately, at the other people who regard themselves as single because of circumstances beyond their control. Most, like Heather, are in the middle—adults who wanted to marry but made choices that took them in a different direction. In interview studies of never-married women, some said they focused on occupational goals instead of marriage. Others reported that they found singlehood preferable to the disappointing relationships they had with men. And still others commented that they just did not meet "the right person" (Dalton, 1992; Lewis, 2000).

Of the various advantages of singlehood, those mentioned most are freedom and mobility. But singles also recognize drawbacks—loneliness, the dating grind, limited sexual and social life, reduced sense of security, and feelings of exclusion from the world of married couples (Chasteen, 1994). Single men have more physical and mental health problems than single women, who usually come to terms with their lifestyle. The greater social support available to women through intimate same-sex friendships is partly responsible. In addition, never-married men are more likely to have conflict-ridden family backgrounds and personal characteristics that contribute to both their singlehood and their adjustment difficulties (Buunk & van Driel, 1989).

Many single people go through a stressful period in their late twenties, when most of their friends get married. The mid-thirties is another trying time for single women, due to the approaching biological deadline for childbearing. A few decide to become parents through adoption, artificial insemination, or a love affair.

Cohabitation

Cohabitation refers to the lifestyle of unmarried couples who have an intimate, sexual relationship and share a residence. Until the 1960s, cohabitation in Western nations was largely limited to low-SES adults. Since then, it has increased in all groups, with an especially dramatic rise among well-educated,

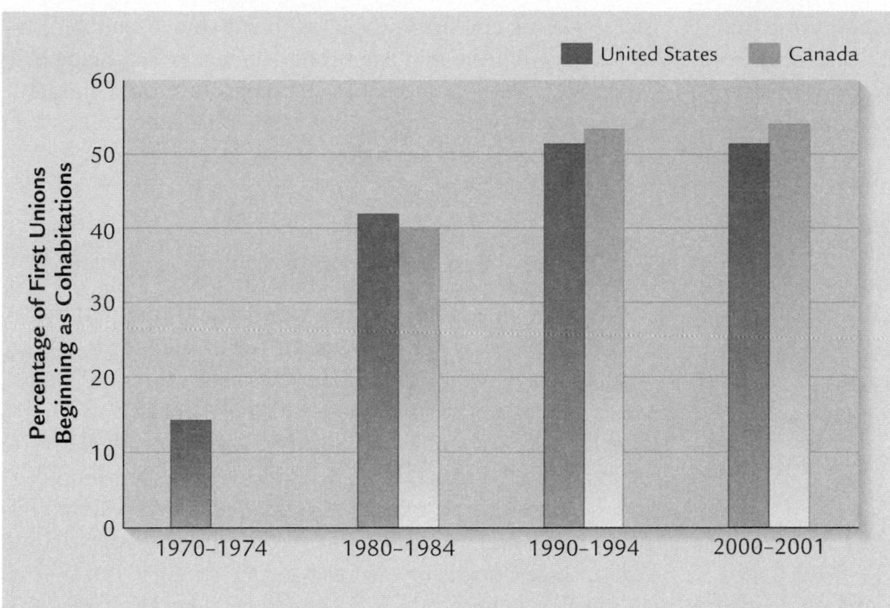

FIGURE 14.5 Generational increase in first conjugal unions beginning as cohabitations in the United States and Canada. Compared with a generation ago, North American young people are far more likely to choose cohabitation as a way of forming their first conjugal union. (1970–1974 data are available only for the United States.) (From Bumpass & Lu, 2000; Statistics Canada, 2002a; Turcotte & Bélanger, 1997; U.S. Bureau of the Census, 2002c.)

economically advantaged young people. As Figure 14.5 shows, today's North American young adults are much more likely than those of a generation ago to choose cohabitation as a way of forming their first conjugal union. Among people in their twenties, cohabitation is now the preferred mode of entry into a committed intimate partnership, with more than 50 percent of couples choosing it (U.S. Bureau of the Census, 2002c; Statistics Canada, 2002a). Rates of cohabitation are even higher among adults with failed marriages. Half of cohabiting relationships in the United States and Canada involve at least one partner who is separated or divorced; one-third of these households include children (Cohan & Kleinbaum, 2002).

Like singlehood, cohabitation has different meanings. For some, it serves as *preparation for marriage*—a time to test the relationship and get used to living together. For others, it is an *alternative to marriage*—an arrangement that offers the rewards of sexual intimacy and companionship along with the possibility of easy departure if satisfaction declines. In view of this variation, it is not surprising that cohabiters differ greatly in the extent to which they share money and possessions and take responsibility for each other's children.

Although North Americans are more open to cohabitation than in the past, their attitudes are not yet as positive as those of Western Europeans. In the Netherlands, Norway, and Sweden, cohabitation is thoroughly integrated into society. From 70 to 90 percent of young people cohabit in their first intimate partnership, and cohabiters are nearly as devoted to one another as married people (Kaslow, Hansson, & Lundblad, 1994; Ramsøy, 1994). Whereas about 50 percent of cohabiting unions in North America break up within 2 years, only 6 to 16 percent dissolve in Western Europe (Brown, 2000; Kiernan, 2002) When they decide to marry, Dutch, Norwegian, and Swedish cohabiters more often do so to legalize their relationships, especially for the sake of children. North American cohabiters typically marry to confirm their love and commitment—sentiments that Western Europeans attach to cohabitation.

Furthermore, American and Canadian couples who cohabit before marriage are more prone to divorce than married couples who did not cohabit, an association less strong or not present in Western European nations (Kiernan, 2001, 2002; Krishnan, 1998). In North America, people who cohabit tend to have less conventional values. They have had more sexual partners and are more politically liberal, less religious, and more androgynous. In addition, a larger number have parents who divorced (Axinn & Barber, 1997; Cunningham & Antill, 1994).

These personal characteristics may contribute to the negative outcomes associated with cohabitation. In addition, the cohabitation experience plays a role. Compared with married people, cohabiters are less likely to pool finances or jointly own a house. In addition, formerly cohabiting married couples display more conflict-ridden interaction (Cohan & Kleinbaum, 2002). Perhaps the open-ended nature of the cohabiting relationship reduces motivation to develop effective conflict-resolution skills. As cohabiters carry this poor communication into marriage, it undermines marital satisfaction. Finally, a history of parental divorce may increase cohabiters' willingness to dissolve a union when it becomes less satisfying.

Certain couples, however, are exceptions to the trends just described. People who cohabit after separation or divorce often test a new relationship carefully to prevent another failure, especially when children are involved. As a result, they cohabit longer and are less likely to move toward marriage (Smock & Gupta, 2002). Similarly, cohabitation is often an alternative to marriage among low-SES couples. Many regard their earning power as too uncertain for marriage and continue living together, sometimes giving birth to children and marrying when their financial status improves (Jayakody & Cabrera, 2002). Finally, homosexual couples report strong commitment—as intense as that of married people. When their relationships become difficult, they end more often than marriages only because there are fewer barriers to separating, including children in common, financial dependence on a partner, and concerns about the costs of divorce (Kurdek, 1998).

Clearly, cohabitation has advantages and drawbacks. For people not ready for marriage, it combines the rewards of a close relationship with the opportunity to remain partially uncommitted. Although most North American couples cohabit to avoid legal obligations, they can encounter difficulties because they do not have them (Mahoney, 2002). Bitter fights over property, money, rental contracts, and responsibility for children are the rule rather than the exception when unmarried couples split up.

Childlessness

At work, Sharese got to know Beatrice and Daniel. Married for 7 years and in their mid-thirties, they did not have children and were not planning any. To Sharese, their relationship seemed especially caring and affectionate. "At first, we were open to becoming parents," Beatrice explained, "but eventually we decided to focus on our marriage."

Some people are *involuntarily* childless because they did not find a partner with whom to share parenthood or their efforts at fertility treatments did not succeed. Beatrice and Daniel are in another category—men and women who are *voluntarily* childless.

How many North American couples choose not to have children is a matter of dispute. Some researchers claim the rate has been low for several decades—between 3 and 6 percent. Others say it rose in the 1980s and early 1990s to about 10 to 15 percent (Jacobson & Heaton, 1989; Morell, 1994). These differing reports may be due to the fact that voluntary childlessness is not always a permanent condition. A few people decide early that they do not want to be parents and stick to these plans. But most, like Beatrice and Daniel, make their decision after they are married and have developed a lifestyle they do not want to give up. Later, some change their minds.

Besides marital satisfaction and freedom from child-care responsibilities, common reasons for not having children include the woman's career and economic security. Consistent with these motives, the voluntarily childless are usually college educated, have prestigious occupations, and are highly committed to their work. Many were only or first-born children whose parents encouraged achievement and independence. In cultures that negatively stereotype childlessness, it is not surprising that voluntarily childless women are more self-reliant and assertive (Houseknecht, 1987; Morell, 1994).

Voluntarily childless adults are just as content with their lives as parents who have warm relationships with their children. In contrast, adults who cannot overcome infertility and parents whose children have serious problems are likely to be dissatisfied and depressed (Connidis & McMullin, 1993; Nichols & Pace-Nichols, 2000). These findings challenge the stereotype of childless people as lonely and unfulfilled. Instead, they indicate that parenthood enhances well-being only when the parent–child relationship is positive. And childlessness interferes with adjustment and life satisfaction only when it is beyond a person's control.

Divorce and Remarriage

Divorce rates have declined slightly since the mid-1980s, partly due to rising age of marriage, which is linked to greater financial stability and marital satisfaction (Heaton, 2002). Still, nearly half of American and one-third of Canadian marriages dissolve. Most divorces occur within 7 years of marriage, so many involve young children. Divorces are also common during the transition to midlife, when people have teenage children—a period (as noted earlier) of low marital satisfaction. About three-fourths of divorced people remarry. But marital failure is even greater during the first few years of second marriages—7 percent above that for first marriages. Afterward, the divorce rates for first and second marriages are similar (Coleman, Ganong, & Fine, 2000; Wu & Penning, 1997).

● **Factors Related to Divorce.** Why do so many marriages fail? As Christy and Gary's divorce illustrates, the most obvious reason is a disrupted husband–wife relationship. Christy and Gary did not argue more than Sharese and Ernie. But their problem-solving style was ineffective. When Christy raised concerns, Gary reacted with resentment, anger, and retreat—a demand–withdraw pattern found in many partners who split up. Another typical style involves little conflict, but partners increasingly lead separate lives because they have different expectations of family life and few shared interests, activities, or friends (Gottman & Levenson, 2000).

What problems underlie these maladaptive communication patterns? In a 9-year longitudinal study, researchers telephoned a U.S. national sample of 2,000 married people, asking about marital problems and following up 3, 6, and 9 years later to find out who had separated or divorced (Amato & Rogers, 1997). Wives reported more problems than husbands, with the gender difference largely involving the wife's emotions, such as anger, hurt feelings, and moodiness. Husbands seemed to have difficulty sensing their wife's distress, which contributed to her view of the marriage as unhappy. Regardless of which spouse reported the problem or was judged responsible for it, the strongest predictors of divorce during the following decade were infidelity, spending money foolishly, drinking or using drugs, expressing jealousy, engaging in irritating habits, and moodiness.

Younger age at marriage, not attending religious services, being previously divorced, and having parents who had divorced increased the chances of divorce, in part because these background factors were linked to marital difficulties. For example, couples who married at younger ages were more likely to report infidelity and jealousy. Low religious involvement may raise the odds of divorce by subtracting an influential context for instilling positive marital attitudes and behaviors. And parental divorce elevates divorce in the next generation in part because it reduces commitment to the norm of lifelong marriage (Amato & DeBoer, 2001). As a result, partners are more likely to engage in infidelity and other inconsiderate behaviors and less likely to try to work through relationship difficulties.

Economically disadvantaged couples who suffer multiple life stresses are especially likely to split up (Amato, 2000). But Christy's case represents another trend—rising marital breakup among well-educated, economically independent women. Women are twice as likely as men to initiate divorce (Rice, 1994).

● **Consequences of Divorce.** When Sharese heard that Christy and Gary's marriage had dissolved, she remarked that it seemed as if "someone had died." Her description is fitting, since divorce involves the loss of a way of life and therefore part of the self sustained by that way of life. As a result, it carries with it opportunities for both positive and negative change.

Immediately after separation, both men and women are depressed and anxious and display impulsive behavior. For most, these reactions subside within 2 years. Women who were in traditional marriages and who organized their identities around their husbands have an especially hard time. As one divorcée remarked, "I used to be Mrs. John Jones, the bank manager's wife. Now I'm Mary Jones. Who is Mary Jones?" (Hetherington, Law, & O'Connor, 1994, p. 216). Noncustodial fathers often feel disoriented and rootless as a result of decreased contact with their children. Others distract themselves through a frenzy of social activity (Cherlin, 1992).

Finding a new partner contributes most to the life satisfaction of divorced adults. But it is more crucial for men, who are better adjusted in the context of marriage than on their own. Despite loneliness and reduced income (see Chapter 10), most divorced women prefer their new life to an unhappy marriage. For example, Christy developed new skills and a sense of self-reliance that might not have emerged had she remained married to Gary. However, a few women—especially those who are anxious and fearful or who remain strongly attached to their ex-spouses—show a drop in self-esteem, become depressed, and tend to form repeated unsuccessful relationships (Amato, 2000; Ganong & Coleman, 1994). Job training, continued education, career advancement, and social support from family and friends play vital roles in the economic and psychological well-being of divorced women (DeGarmo & Forgatch, 1997).

● **Remarriage.** On average, people remarry within 4 years of divorce, men somewhat faster than women. For several reasons, remarriages are especially vulnerable to breakup. First, although people often remarry for love, practical matters—financial security, help in rearing children, relief from loneliness, and social acceptance—figure more heavily into a second marriage than a first. These concerns do not provide a sound footing for a lasting partnership. Second, some people transfer the negative patterns of interaction and problem solving learned in their first marriage to the second. Third, people who have already had a failed marriage are more likely to view divorce as an acceptable solution when marital difficulties resurface. And finally, remarried couples experience more stress from stepfamily situations (Bray, 1999; Coleman, Ganong, & Fine, 2000). Recall from Chapter 10 that adults in blended families have few societal guidelines for how to relate to their steprelatives, including stepchildren. As we will see in the next section, stepparent–stepchild ties are powerful predictors of marital happiness.

Divorce and remarriage, like other adult lifestyles, lead to diverse outcomes. It generally takes 3 to 5 years for blended families to develop the connectedness and comfort of intact biological families (Ihinger-Tallman & Pasley, 1997). Family life education, couples counseling, and group therapy can help divorced and remarried adults adapt to the complexities of their new circumstances (Forgatch, Patterson, & Ray, 1996).

Variant Styles of Parenthood

Diverse family forms result in varied styles of parenthood. Among these are a growing number of cohabiting and remarried parents, never-married parents, and gay and lesbian parents. Each type of family presents unique challenges to parenting competence and adult psychological well-being.

● **Stepparents.** Whether stepchildren live in the household or visit only occasionally, stepparents are in a difficult position. Since the parent–child tie predates the blended family, the stepparent enters as an outsider. Too often, stepparents move into their new role too quickly. Because they do not have a warm attachment bond to build on, their discipline is usually ineffective. Stepparents frequently criticize the biological parent for being too lenient. The parent, in turn, tends to view the stepparent as too harsh. These differences can divide the couple (Ganong, Coleman, & Fine, 1995). Remarried parents typically report higher levels of tension and disagreement than first-marriage parents, most centering on child-rearing issues. Because of more opportunities for conflict, relationship quality is poorer when both adults have children from prior marriages than when only one does (Coleman, Ganong, & Fine, 2000).

Stepmothers, especially, are likely to experience conflict. Expected to be in charge of family relationships, they quickly find that stepparent–stepchild ties do not develop instantly. Often mothers are jealous, uncooperative, and possessive of their children following divorce. Even when their husbands do not have custody, stepmothers feel stressed (MacDonald & DeMaris, 1996). As stepchildren go in and out of the home, stepmothers compare life with and without resistant children, and many prefer life without them. No matter how hard a stepmother tries to build a close parent–child bond, her efforts are probably doomed to failure in the short run.

Stepfathers with children of their own have an easier time. They establish positive ties with stepchildren relatively quickly, perhaps because they are experienced in building warm parent–child ties and feel less pressure than stepmothers to plunge into parenting. As stepchildren recognize stepfathers' efforts to connect with them through enjoyable activities, they generally respond favorably (Ganong et al., 1999). Stepfathers without biological children, however, are new to child rearing. If they have unrealistic expectations or their wives push them into the father role, their interactions with stepchildren can be troublesome. After making several overtures that are ignored or rebuffed, they often withdraw from parenting (Hetherington & Clingempeel, 1992).

A caring husband–wife relationship, the cooperation of the biological parent, and the willingness of children to accept their parent's new spouse are crucial for stepparent adjustment. Because stepparent–stepchild bonds are hard to establish, the divorce rate is higher for couples with stepchildren than for those without them (Bray, 1999).

● **Never-Married Single Parents.** About 10 percent of American children and 5 percent of Canadian children live with a single parent who has never married. Of these, about 90 percent are mothers, 10 percent fathers (U.S. Bureau of the Census, 2002c; Vanier Institute of the Family, 2002b). Earlier we mentioned that single adults occasionally decide to become parents on their own. Births to women in high-status occupations who have not married by their thirties have increased. However, they are still few in number, and little is known about how these mothers and their children fare.

■ For most never-married African-American mothers, marriage follows birth of the first child by several years and is not necessarily to the child's father. But never-married parenthood greatly increases financial hardship. When never-married mothers find ways to overcome poverty, they experience less stress and their children develop more favorably.

In the United States, the largest group of never-married parents is African-American young women. Over 60 percent of births to black mothers in their twenties are to women without a partner, compared with 13 percent of births to white women (U.S. Bureau of the Census, 2002c). African-American women postpone marriage more and childbirth less than women in other American ethnic groups. Job loss, persisting unemployment, and consequent inability of many black men to support a family have contributed to the number of African-American never-married, single-mother families.

Never-married black mothers tap the extended family, especially their own mothers, for help in caring for children (Gasden, 1999). For just over one-third, marriage occurs within 9 years after birth of the first child, not necessarily to the child's biological father (Wu, Bumpass, & Musick, 2001). These couples function much like other first-marriage parents. Their children are often unaware that the father is a stepfather, and parents do not report the child-rearing difficulties typically associated with blended families (Ganong & Coleman, 1994).

Still, for low-SES women, never-married parenthood generally increases financial hardship. About 47 percent of white and 59 percent of black mothers have a second child while unmarried. And they are far less likely to receive paternal child support payments than are divorced mothers (Hill, 1997; Wu, Bumpass, & Musick, 2001). U.S. welfare policies currently mandate that parents find work to sustain welfare benefits. In states where the combination of wages and income supplements is sufficient to lift families out of poverty, children gain in academic performance and social behavior, perhaps because of reduced parental stress. When welfare programs require parents to take low-paid jobs without counteracting poverty, they do not improve parents' or children's well-being (Morris, 2002)

Finally, children of never-married mothers who lack the involvement of a father achieve more poorly in school and display more antisocial behavior than children in low-SES, first-marriage families. These adjustment problems make life more difficult for mothers (Coley, 1998). Strengthening social support, education, and employment opportunities for low-SES parents would encourage marriage as well as help unmarried-mother families.

● **Gay and Lesbian Parents.** Several million American gay men and lesbians are parents, most through previous heterosexual marriages, a few through adoption or reproductive technologies (Patterson, 2002). In the past, laws assuming that homosexuals could not be adequate parents led those who divorced a heterosexual partner to lose custody of their children. Today, some states hold that sexual orientation is irrelevant to custody. In others, fierce prejudice against homosexual parents still prevails.

Research on homosexual parents and children is limited and based on small samples. Nevertheless, findings consistently indicate that gay and lesbian parents are as committed to and effective at child rearing as heterosexual parents (Patterson, 2001). Some evidence suggests that gay fathers are more consistent in setting limits and more responsive to their children's needs than heterosexual fathers, perhaps because gay men's less traditional gender identity fosters involvement with children (Bigner & Jacobsen, 1989). In lesbian families, quality of mother–child interaction is as positive as in heterosexual families. And children of lesbian mothers regard their mother's partner as very much a parent (Brewaeys et al., 1997). Whether born to or adopted by their parents or conceived through donor insemination, children in homosexual families are as well adjusted as other children. Also, the large majority are heterosexual (Allen & Burrell, 1996; Chan, Raboy, & Patterson, 1998; Golombok & Tasker, 1996).

When extended-family members have difficulty accepting them, homosexual mothers and fathers often build "families of choice" through friends, who assume the roles of relatives. But most of the time, parents of gays and lesbians cannot endure a permanent rift (Hare, 1994). With time, interactions between homosexual parents and their families of origin become more positive and supportive.

Homosexual couples' joint involvement in parenting varies with the way children entered the family. When partners choose parenthood through adoption or reproductive technologies, they report more even division of child-care and household tasks and greater relationship satisfaction than heterosexual couples do (Chan, Rabo, & Patterson, 1998). When children resulted from a previous heterosexual relationship, the biological parent typically assumes a larger parenting role (Hare & Richards, 1993).

Overall, families headed by homosexuals can be distinguished from other families only by issues related to living in a nonsupportive society. A major concern of gay and lesbian parents is that their children will be stigmatized by their parents' sexual orientation.

© TOM MCKITTERICK

■ Homosexual parents are as committed to and effective at child rearing as are heterosexual parents. Their children are well adjusted, and the large majority develop a heterosexual orientation.

Ask Yourself

REVIEW

What factors affect the impact of childlessness on adult adjustment?

REVIEW

Why is never-married single parenthood especially high among African Americans? What conditions affect parent and child well-being in these families?

APPLY

After dating for a year, Wanda and Scott decided to live together. Their parents worried that cohabitation would reduce Wanda and Scott's chances for a successful marriage. Is this fear justified? Why or why not?

CONNECT

Return to Chapter 10, pages 330–331, and review the impact of divorce and remarriage on children and adolescents. How do those findings resemble outcomes for adults? What might account for the similarities?

www.

Career Development

Besides family life, vocational life is a vital domain of social development in early adulthood. After choosing an occupation, young people must learn how to perform its tasks well, get along with co-workers, respond to authority, and protect their own interests. When work experiences go well, adults develop new competencies, feel a sense of personal accomplishment, make new friends, and become financially independent and secure. And as we have already seen, for women especially but also for men who support their partners' career development, aspirations and accomplishments in the workplace and the family are interwoven.

Establishing a Career

Our earlier discussion of Levinson's and Vaillant's theories highlighted diverse paths and timetables for career development. Consider, once again, the wide variations among Sharese, Ernie, Christy, and Gary in establishing their careers. As is typical for men, Ernie's and Gary's career lives were long and *continuous,* beginning after completion of formal education and ending with retirement. Like many women, Sharese and Christy had *discontinuous* career paths—ones that were interrupted or deferred by child rearing and other family needs (Betz, 1993). Furthermore, not all people embark on the vocation of their dreams. For example, although half of young people aspire to professional occupations, only 20 percent attain them (U.S. Bureau of the Census, 2002c).

Even for those who enter their chosen field, initial experiences can be discouraging. At the health department, Sharese discovered that committee meetings and paperwork consumed much of her day. Because each grant proposal and research project had a deadline, the pressure of productivity weighed heavily on her. Adjusting to unanticipated disappointments in salary, supervisors, and co-workers is difficult. As new workers become aware of the gap between their expectations and reality, resignations are common. On average, people in their twenties move to a new job every 2 years; five or six changes are not unusual (Petersen & Gonzales, 1999).

After a period of evaluation and adjustment, young adults generally settle into their work. In careers with opportunities for promotion, high aspirations must often be revised downward, since the structure of most work settings resembles a pyramid, with few high-level executive and supervisory jobs. In a longitudinal study of over 400 AT&T lower-level male managers, the importance of work in men's lives varied with career advancement and age (Howard & Bray, 1988). For men who advanced very little, "work disengagement" occurred early; family, recreation, and community service assumed greater importance by the early thirties. Men with average levels of career success emphasized nonwork roles at a later age. In contrast, men who were highly successful became more involved in their jobs over time. Although the desire for advancement tends to decline with age, most people still seek challenges and find satisfaction in their work roles.

Besides opportunity, personal characteristics affect career progress. As we will see in the next section, a *sense of self-efficacy*—belief in one's own ability to be successful—affects career choice and development. (Return to Chapter 1, page 18, if you need to review this concept.) Young people who are very anxious about the possibility of making mistakes or failing tend to set career aspirations that are either too high or too low. When they encounter obstacles, they quickly conclude that career tasks are too hard and give up (Albert & Luzzo, 1999). As a result, they achieve far less than their abilities would permit.

Recall from our discussion of Levinson's theory that career success often depends on the quality of a mentoring relationship. Access to an effective mentor—a person with advanced experience and knowledge who is emotionally invested in the junior person's development and who fosters a bond of trust—is jointly affected by the availability of willing people and the individual's capacity to select an appropriate individual (Crosby, 1998). Interestingly, the best mentors usually are not top executives, who tend to be preoccupied and therefore less helpful and sympathetic (Seligman, 1994). Most of the time, young adults fare better with a mentor lower on the corporate ladder.

Women and Ethnic Minorities

Although women and ethnic minorities have penetrated nearly all professions, they attain less than they otherwise would if their talents were developed to the fullest. Women in general—and those who are members of economically disadvantaged minorities in particular—remain concentrated in occupations that offer little opportunity for advancement, and they are underrepresented in executive and managerial roles (see Chapter 13, page 438). And although the overall gap between men's and women's earnings is not as large today as 2 decades ago, it remains considerable. For every dollar earned by a man, the average American woman earns 78 cents, the average Canadian woman 82 cents (Harkness & Waldfogel, 2002). Gender disparities in career development are largely responsible. Men and women with similar work experience and job status differ much less in income (Venable, 2002).

Especially for women in traditionally feminine occupations, career planning is often short-term and subject to change. Many enter and exit the labor market several times, to rear children and move with their families. Between ages 18 and 34, the typical woman has been out of the labor force 27 percent of the time, in contrast to 11 percent for the typical man (U.S. Department of Labor, 2002a, 2002b). Time away from a career greatly hinders advancement. Furthermore, women's low self-efficacy with respect to male-dominated fields limits not only their occupational choices but also their career progress. Women who pursue nontraditional careers usually have "masculine" qualities—high achievement orientation, self-reliance, and belief that their efforts will result in success (Petersen & Gonzales, 1999).

Even when women enter high-status professions, few move into top-level management positions. Gender-stereotyped images of women as followers rather than leaders and competing work and family obligations slow their advancement. Singlehood or late marriage and few or no children are strongly associated with career achievement in women—a relationship that does not hold for men (Betz & Fitzgerald, 1987). Furthermore, since men dominate high-status fields, women are less available to serve as mentors. Although amount of mentor support is similar in same-sex and other-sex mentoring relationships, women with female mentors tend to be more productive (Goldstein, 1979; O'Neill, Horton, & Crosby, 1999). Perhaps female mentors are more likely to be perceived as role models and to provide guidance on the unique problems women encounter in the workplace.

Despite laws guaranteeing equal opportunity, racial bias in the labor market remains strong. In one study, researchers responded to over 1,300 help-wanted newspaper ads with fictitious résumés, some containing higher qualifications and some lower qualifications. Half the résumés were assigned a very white-sounding name (Emily Walsh or Brendan Baker) and half a very black-sounding name (Lakisha Washington or Jamal Jones). At all job levels, from clerical work to top management, résumés with white names evoked 50 percent more callbacks than résumés with black names. And although whites received substantially more callbacks in response to high-quality than to low-quality résumés, having a high-quality résumé made little difference for blacks (see Figure 14.6). As the researchers noted, "Discrimination appears to bite

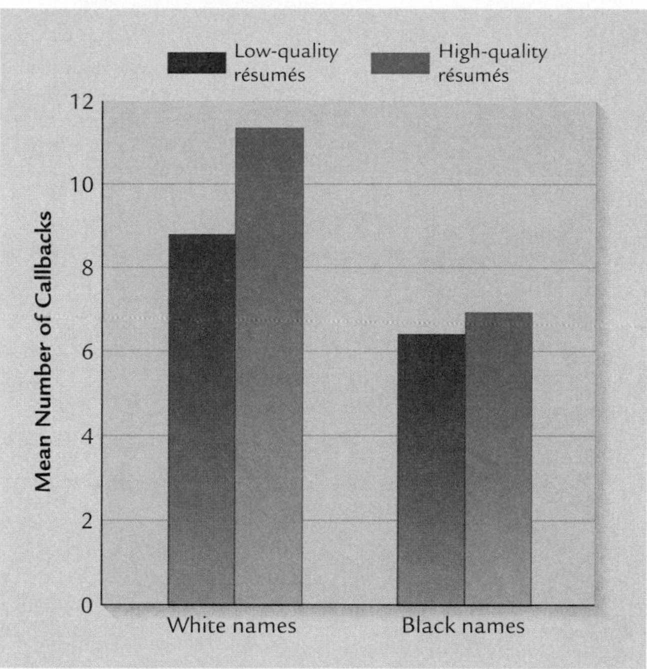

FIGURE 14.6 Relationship of ethnicity of job applicant's name to employer callbacks. Researchers responded to help-wanted newspaper ads with fictitious résumés, some having white-sounding names and others black-sounding names. Résumés with "white names" evoked many more callbacks than résumés with black names. When résumés were high in quality, callbacks to whites increased, but those to blacks showed little change. (Adapted from Bertrand & Mullainathan, 2002.)

twice, making it harder for African Americans to find a job and to improve their employability" (Bertrand & Mullainathan, 2002, p. 3).

Ethnic minority women face special challenges in realizing their career potential because they must surmount combined gender and racial discrimination. Those who succeed often display an unusually high sense of self-efficacy, attacking problems head-on despite repeated obstacles to achievement (Byars & Hackett, 1998). In an interview study of African-American women who had become leaders in diverse fields, all reported intense persistence, fueled by supportive relationships with other women, including teachers, colleagues, and friends who countered their sense of professional isolation. Many described their mothers as inspiring role models who had set high standards for them. Others mentioned support from their African-American communities, stating that a deep sense of connection to their people had empowered them (Richie et al., 1997).

Despite obstacles to success, young and middle-aged women reporting the highest levels of life satisfaction generally have developed rewarding careers (Burke, 2001; Sears & Barbie, 1977). This finding suggests that some of the discontent that married women frequently express may not be due to marriage per se, but rather to lack of a gratifying work life. Consistent with this idea, most young women prefer to blend work and family (Barnett & Rivers, 1996). For women in financially stressed families, this is usually not a choice; it is a necessity.

Combining Work and Family

Whether women work because they want to or have to (or both), the dominant family form today is the **dual-earner marriage,** in which both husband and wife are employed. Most dual-earner couples are also parents, since the majority of women with children are in the work force (see Chapter 10). In about one-third of these families, moderate to severe conflict occurs over trying to meet both work and family responsibilities (Gilbert & Brownson, 1998; Phillips & Imhoff, 1997).

What are the main sources of strain in dual-earner marriages? When Sharese returned to her job after her children were born, she felt a sense of *role overload,* or conflict between work and family responsibilities. Not only did she have a demanding career, but (like most employed women) she shouldered most of the household and child-care tasks. And both Sharese and Ernie felt torn between the desire to excel at their jobs and the desire to spend more time with each other, their children, and their friends and relatives. Role overload is linked to a rise in psychological stress, poorer marital relations, less effective parenting, and child behavior problems (Crouter et al., 1999; Perry-Jenkins, Repetti, & Crouter, 2000).

Role overload is greater for women, especially for those in low-status work roles with rigid schedules and little autonomy (Marshall, 1997). Couples in prestigious careers have

Famed contemporary author Terry McMillan uses the theme of women's struggles in her best-selling novels. Persistence in the face of gender and racial discrimination, a passion for writing, a mother who encouraged and inspired her, and a vision of connection to others, especially other African-American women, underlie her success.

■ Like most women in dual-earner marriages, this mother takes responsibility for the majority of child-rearing and household tasks. Women are more likely than men to experience role overload, or conflict between work and family responsibilities.

more control over both work and family domains. For example, Sharese and Ernie devised ways to spend more time with their children. They picked them up at child care early one day a week, compensating by doing certain occupational tasks on evenings and weekends. Like other career-oriented mothers, Sharese coped with role pressures by setting priorities. She decreased the amount of time she spent on household chores, not child rearing (Institute for Social Research, 2002).

Having two careers in one family usually means that certain career decisions become more complex. A move to a new job can mean vocational sacrifices for one partner. Usually this is the wife, since a decision in favor of the husband's career (typically further along and better paid) is more likely to maximize family income. One solution to the geographical limitations of the dual-earner marriage is to live apart. Although more couples are doing this, the strain of separation and risk of divorce are high.

Clearly, dual-earner marriages pose difficulties. But when couples cooperate to surmount these, they profit greatly from involvement in both work and family roles. Besides higher earnings and a better standard of living, a major advantage is women's self-fulfillment and improved well-being (Swanson et al., 1997). Ernie took great pride in Sharese's accomplishments, which contributed to his view of her as an interesting, self-confident, and capable helpmate in life. Multiple roles also granted both young people expanded contexts for experiencing success and greater similarity in everyday experiences, which fostered gratifying communication (Barnett & Hyde, 2001).

In sum, a challenging, rewarding occupation in the context of a supportive spouse can strengthen a marriage and promote adult development. Under other circumstances—for example, when a woman tries to combine a low-status, low-paying job with marriage to a disapproving man—the toll on physical and mental health can be severe. The Caregiving Concerns table on the following page lists strategies that help dual-earner couples combine work and family roles in ways that promote mastery and pleasure in both spheres of life.

Ask Yourself

REVIEW

Why do professionally accomplished women, especially those who are members of economically disadvantaged minorities, typically display high self-efficacy?

APPLY

Heather climbed the career ladder of her company quickly, reaching a top-level executive position by her early thirties. In contrast, Sharese and Christy did not attain managerial roles in early adulthood. What factors account for this disparity in career progress?

CONNECT

Work and family life are inseparably intertwined. Explain how this is so in early adulthood.

www.

Strategies That Help Dual-Earner Couples Combine Work and Family Roles

STRATEGY	DESCRIPTION
Devise a plan for sharing household tasks.	As soon as possible in the relationship, discuss division of household responsibilities. Decide who does a particular chore on the basis of who has the needed skill and time, not on the basis of gender. Schedule regular times to rediscuss your plan to fit changing family circumstances.
Begin sharing child care right after the baby's arrival.	For fathers, strive to spend equal time with the baby early. For mothers, refrain from imposing your standards on your partner. Instead, share the role of "child-rearing expert" by discussing parenting values and concerns often. Attend a parent education course together.
Talk over conflicts about decision making and responsibilities.	Face conflict through communication. Clarify your feelings and needs and express them to your partner. Listen and try to understand your partner's point of view. Then be willing to negotiate and compromise.
Establish a balance between work and family.	Critically evaluate the time you devote to work in view of your values and priorities. If it is too much, cut back.
Make sure your relationship receives regular loving care and attention.	See the Caregiving Concerns table on page 454.
Press for workplace and public policies that assist dual-earner couples.	Difficulties faced by dual-earner couples are partly due to lack of workplace and societal supports. Encourage your employer to provide benefits that help combine work and family roles, such as flexible work hours; parental leave with pay; and onsite high-quality, affordable child care. Communicate with lawmakers and other citizens about improving public policies for children and families.

Summary

Erikson's Theory: Intimacy versus Isolation

According to Erikson, what personality changes take place during early adulthood?

- In Erikson's theory, young adults must resolve the conflict of **intimacy versus isolation,** balancing independence and intimacy as they form a close relationship with a partner. Research confirms that intimacy is a central concern of early adulthood. Young people also focus on aspects of generativity, including contributions to society through work and child rearing.

Other Theories of Adult Psychosocial Development

Describe Levinson's and Vaillant's psychosocial theories of adult personality development.

- Levinson described a series of eras, each consisting of a transition and a stable period, in which people revise their **life structure.** Young adults usually construct a dream, typically involving career for men and both marriage and career for women, and form a relationship with a mentor to help them realize their dream. In their thirties, they focus on aspects of their lives that have received less attention. Men settle down, whereas many women remain unsettled into middle adulthood.

- Vaillant refined Erikson's stages, portraying the twenties as devoted to intimacy, the thirties to career consolidation, the forties to guiding others, and the fifties to cultural and philosophical values.

What is the social clock, and how does it affect personality in adulthood?

- Although societal expectations have become less rigid, conformity to or departure from the **social clock,** the culturally determined timetable for major life events, can be a major source of personality change in adulthood. Following a social clock grants confidence to young adults, whereas deviating from it can bring psychological distress.

Close Relationships

Describe factors that affect mate selection and the role of romantic love in the young adult's quest for intimacy.

- Establishing an intimate bond is a major milestone of adult development. Romantic partners tend to resemble one another in age, ethnicity, SES, religion, and various personal and physical attributes.

- According to evolutionary theory, women seek a mate with traits that help ensure children's survival, whereas men look for characteristics that signal sexual pleasure and ability to bear offspring. An alternative, social learning perspective emphasizes that gender roles profoundly influence criteria for mate selection. Research suggests that both biological and social forces are involved.

- According to the **triangular theory of love,** the balance among passion, intimacy, and commitment changes as romantic relationships move from the intense sexual attraction of **passionate love** toward more settled

companionate love. Commitment is key to a satisfying, enduring relationship. The Western emphasis on romantic love in mate selection does not characterize all cultures.

Describe adult friendships and sibling relationships and the role of loneliness in adult development.

- Adult friendships have characteristics and benefits similar to earlier friendships and are based on trust, intimacy, and loyalty. Women's same-sex friendships tend to be more intimate than men's. Other-sex friendships are beneficial in adulthood but less common and enduring than same-sex friendships. Siblings become more frequent companions in early adulthood than they were in adolescence, often taking on the characteristics of friendship, especially between same-sex siblings.

- Young adults are vulnerable to **loneliness,** which declines with age as they form satisfying intimate ties. As long as it is not too intense, loneliness can encourage young people to reach out to others and better understand themselves.

The Family Life Cycle

Trace phases of the family life cycle that are prominent in early adulthood, and cite factors that influence these phases today.

- Wide variations exist in the sequence and timing of phases of the **family life cycle.** Leaving home is a major step in assuming adult responsibilities. A large percentage of teenagers depart relatively early when they go to college. SES and ethnicity influence the likelihood that a young person will live independently before marriage. Returning home for a period of time is common among unmarried young adults.

- Nearly 90 percent of North Americans marry, although most do so later than in the past. Both **traditional marriages** and **egalitarian marriages** are affected by women's participation in the work force. Husbands and wives today must work harder to define their marital roles. Even in dual-earner families, North American women spend nearly twice as much time on housework as men do.

- Men tend to be happier and healthier in marriage than women, who suffer more when a relationship has problems. Many young people enter marriage with unrealistic expectations. Even happy marriages have ups and downs and require adaptability from both partners.

- Effective birth control techniques and changing cultural values make childbearing a matter of choice in Western industrialized nations. Although most couples become parents, they do so at a later age and have fewer children than in the past.

- The arrival of children requires couples to adjust to increased responsibilities, less time for each other, and more traditional roles. Marriages that are gratifying and supportive tend to remain so after childbirth, whereas troubled marriages usually become more distressed. Sharing care of the baby is linked to greater parental happiness and positive parent–infant interaction. Unlike many non-Western cultures, marital and parenting roles are closely linked in Western nations. A change in happiness in one role is usually accompanied by a similar change in the other.

- Challenges facing families with young children include inadequate preparation for parenthood, lack of societal supports for child rearing, the need for cooperation in the marital relationship, and difficulties in finding good child care. Couples who work together as "coparenting teams" are more likely to feel competent as parents and to be sensitive to their children.

- In families with adolescents, parents must establish new relationships with their increasingly autonomous teenagers, blending guidance with freedom and gradually loosening control. Marital satisfaction often declines during in this phase. Parent education programs can help parents clarify their child-rearing values and use more effective strategies.

The Diversity of Adult Lifestyles

Discuss the diversity of adult lifestyles, focusing on singlehood, cohabitation, and childlessness.

- Singlehood has risen in recent years due to a trend toward later marriage and a high divorce rate. Women with high-status careers and men in blue-collar occupations are most likely to remain single. Women tend to adjust more favorably to singlehood than do men.

- **Cohabitation** has risen dramatically, especially among well-educated, economically advantaged young adults, for whom it is the preferred mode of entry into a committed intimate partnership. Cohabitation rates are especially high among separated and divorced adults. Compared with their Western European counterparts, North American cohabiters tend to be less conventional in values and behavior and less committed to their partner, and their subsequent marriages are more likely to fail.

- Voluntarily childless adults tend to be well educated and career oriented and are just as satisfied with their lives as parents who have good relationships with their children. But when childlessness is beyond a person's control, it interferes with adjustment and life satisfaction.

Discuss today's high rates of divorce and remarriage, and cite factors that contribute to them.

- Nearly half of American and one-third of Canadian marriages dissolve, often while children are at home. About three-fourths of divorced people remarry, many of whom divorce again. Maladaptive communication patterns, family poverty, and the changing status of women contribute to divorce by heightening marital difficulties.

- Finding a new partner is important to many divorced adults, especially men. Remarriages break up for several reasons, including the prominence of practical concerns rather than love in the decision to remarry, the persistence of negative styles of communication, the acceptance of divorce as a solution to marital difficulties, and problems adjusting to a stepfamily.

Discuss the challenges associated with variant styles of parenthood, including stepparents, never-married parents, and gay and lesbian parents.

- Establishing stepparent–stepchild ties is difficult, especially for stepmothers and for stepfathers without children of their own. A caring husband–wife relationship, the cooperation of the biological parent, and children's acceptance are crucial for stepparent adjustment.

- Never-married single parenthood is especially high among low-income African-American women in their twenties. Unemployment among black men contributes to this trend. Although these mothers often receive help from extended family members, they find it difficult to overcome poverty.

- Gay and lesbian parents are as loving and effective at child rearing as heterosexual parents. Furthermore, children reared by homosexual parents are as well adjusted as those reared by heterosexual parents. Families headed by homosexuals face difficulties related to living in an unsupportive society.

Career Development

Discuss patterns of career development, and cite difficulties faced by women, ethnic minorities, and couples seeking to combine work and family.

- Men's career paths are usually continuous, whereas women's are often discontinuous because of child rearing and other family needs. After adjusting to the realities of the work world, young adults settle into an occupation. Their progress is affected by opportunities for promotion in their chosen occupations, by personal characteristics such as self-efficacy, and by access to an effective mentor.
- Women and ethnic minorities have penetrated nearly all professions but have made limited progress in career advancement. Women are hampered by time away from the labor market, low self-efficacy with respect to traditionally male-dominated fields, gender stereotypes, conflict between work and family obligations, and difficulties in finding a suitable mentor. Racial bias in the labor market remains strong, and ethnic minority women who succeed display an unusually high sense of self-efficacy.
- Couples, and particularly women, in **dual-earner marriages** often experience role overload and make vocational sacrifices to further their spouse's career. When dual-earner couples cooperate to surmount difficulties, they benefit from higher earnings, a better standard of living, and women's self-fulfillment and improved well-being.

Important Terms and Concepts

cohabitation (p. 466)
companionate love (p. 451)
dual-earner marriage (p. 473)
egalitarian marriage (p. 458)
family life cycle (p. 456)
intimacy versus isolation (p. 446)
life structure (p. 447)
loneliness (p. 455)
passionate love (p. 451)
social clock (p. 449)
traditional marriage (p. 458)
triangular theory of love (p. 451)

FYI For Further Information and Help

Consult the Companion Website for *Development Through the Lifespan, Third Edition*, (www.ablongman.com/berk) where you will find the following resources for this chapter:

- **Chapter Objectives**
- **Flashcards** for studying important terms and concepts
- **Annotated Weblinks** to guide you in further research
- **Ask Yourself** questions, which you can answer and then check against a sample response
- **Suggested Readings**
- **Practice Test** with immediate scoring and feedback

Milestones

Development in Early Adulthood

AGE

20–30 years

PHYSICAL

- Athletic skills that require speed of limb movement, explosive strength, and gross-motor coordination peak early in this decade and then decline. (416)
- Athletic skills that depend on endurance, arm–hand steadiness, and aiming peak at the end of this decade and then decline. (416)

- Declines in touch sensitivity; respiratory, cardiovascular, and immune system functioning; and elasticity of the skin begin and continue throughout adulthood. (415)
- As basal metabolic rate declines, gradual weight gain begins in the middle of this decade and continues through middle adulthood. (420)
- Sexual activity increases. (427)

COGNITIVE

- If college educated, dualistic thinking (dividing information, values, and authority into right and wrong) declines in favor of relativistic thinking (viewing all knowledge as embedded in a framework of thought). (432)
- Narrows vocational options and settles on a specific career. (436–437)
- With entry into marriage and employment situations, focuses less on acquiring knowledge and more on applying it to everyday life. (432)

- Develops expertise (acquisition of extensive knowledge in a field of endeavor), which enhances problem solving. (433)
- Creativity (generating useful original products) increases. (434)
- Steady improvement in mental abilities that depend on accumulated knowledge through middle adulthood. (434–435)

EMOTIONAL/SOCIAL

- Leaves home permanently. (456)
- Strives to make a permanent commitment to an intimate partner. (450, 457)
- Usually constructs a dream, an image of the self in the adult world that guides decision making. (448)
- Usually forms a relationship with a mentor, who facilitates realization of the dream. (448)

- If in a high-status career, acquires professional skills, values, and credentials (for women, may be delayed and take longer). (471–472)
- Begins to develop mutually gratifying adult friendships and work ties. (453–454)
- May cohabit, marry, and bear children. (457, 461, 466–467)
- Sibling relationships become more companionate. (455)
- As people move in and out of relationships, loneliness peaks early in this decade and then declines steadily throughout adulthood. (455)

AGE	PHYSICAL	COGNITIVE	EMOTIONAL/SOCIAL
30–40 years	• Declines in vision, hearing, and the skeletal system begin and continue throughout adulthood. (415)  • In women, fertility problems increase sharply in the middle of this decade. (418) • Hair begins to gray and thin in the middle of this decade. (415) • Sexual activity declines, probably due to the demands of daily life. (427)	 • As family and work lives expand, the capacity to juggle many responsibilities improves. (433)  • Creativity (generating useful original products) often peaks. (434)	• Reevaluates life structure and tries to change components that are inadequate. (448) • Establishes a stable niche within society through family, occupation, and community activities (for women, career maturity and authority in the community may be delayed). (448–449) 

Note: Numbers in parentheses indicate the page(s) on which each milestone is discussed.

Middle adulthood is a time of narrowing life options, but it brings compensating gains. Expertise—a wealth of accumulated knowledge that supports a high level of performance in a vocational or leisure pursuit—reaches its height. This husband and wife paleontologist-paleobotanist team bring their expertise to bear on a dig site in the Red Deer River badlands in Alberta, Canada.

15

Physical and Cognitive Development in Middle Adulthood

Physical Development

Physical Changes
Vision • Hearing • Skin • Muscle–Fat Makeup • Skeleton • Reproductive System

■ Biology & Environment: Anti-Aging Effects of Dietary Calorie Restriction: Relevant to Humans?

■ Cultural Influences: Menopause as a Biocultural Event

Health and Fitness
Sexuality • Illness and Disability • Hostility and Anger

Adapting to the Physical Challenges of Midlife
Stress Management • Exercise • An Optimistic Outlook • Gender and Aging: A Double Standard

Cognitive Development

Changes in Mental Abilities
Crystallized and Fluid Intelligence • Individual and Group Differences

Information Processing
Speed of Processing • Attention • Memory • Practical Problem Solving and Expertise • Creativity • Information Processing in Context

Vocational Life and Cognitive Development

Adult Learners: Becoming a College Student in Midlife
Characteristics of Returning Students • Supporting Returning Students

On a snowy December evening, Devin and Trisha sat down to read the holiday cards piled high on the kitchen counter. Devin's 55th birthday had just passed; Trisha would reach 48 in a few weeks. During the past year, they had celebrated their 24th wedding anniversary. These milestones, along with greetings from friends who sent annual updates about their lives, brought the changes of midlife into bold relief.

Instead of new births, children starting school, or a first promotion at work, notations on the cards sounded new themes. Jewel's recap of the past year reflected a growing awareness of a finite lifespan, one in which time had become more precious. She wrote,

> My mood has been lighter ever since my birthday. There was some burden I laid down by turning 49. My mother passed away when she was 48, so it all feels like a gift now. Blessed be!

George and Anya reported on their son's graduation from law school and their daughter Michelle's first year of university. The house empty of children, George wrote,

> Anya is filling the gap created by the children's departure by returning to college for a nursing degree. After enrolling this fall, she was surprised to find herself in the same psychology class as Michelle. At first, Anya was worried about handling the academic work, but after a semester of success, she's feeling more confident.

Tim's message reflected continuing robust health, acceptance of physical change, and a new burden: caring for aging parents—a firm reminder of the limits of the lifespan:

> I used to be a good basketball player in college, but I've recently noticed that my 20-year-old nephew Brent can dribble and shoot circles around me. It must be my age! I ran our city marathon in September, coming in seventh in the over-50 division. Brent ran, too, but he opted out a few miles short of the finish line to get some pizza while I pressed on. That must be my age, too!
>
> The saddest news is that my dad had a bad stroke. His mind is clear, but his body is partially paralyzed. It's really upsetting because he was getting to enjoy the computer I gave him, and it was so upbeat to talk with him about it in the months before the stroke.

Middle age begins around age 40 and ends at about 60. This phase is marked by narrowing of life options and a shrinking future as children leave home and career paths become more determined. In other ways, middle adulthood is hard to define, since wide variations in attitudes and behaviors exist. Some individuals seem physically and mentally young at age 60—active and optimistic, having attained a sense of serenity and stability. Others feel old at age 35 or 40—as if their lives had peaked and were on a downhill course.

Yet another reason that middle age eludes clear definition is that it is a product of contemporary times. Before the twentieth century, only a brief interval existed between the tasks of early adulthood and those of old age. For example, women often became widows by their mid-fifties, before their youngest child left home. And harsh living conditions led people to accept a ravaged body as a natural part of life. As life expectancy—and with it, health and vigor—increased over the past century, adults became more aware of their own aging and mortality.

In this chapter, we trace physical and cognitive development during the fifth and sixth decades of life. In both domains, we will encounter not just progressive declines but also

© JUDY GELLES/STOCK BOSTON, LLC.

■ Middle adulthood is difficult to define, since it encompasses wide variations in attitudes and behaviors. These middle-aged individuals seem physically and mentally young—active and optimistic. As volunteers with Habitat for Humanity, they build a house for a needy family.

sustained performance and compensating gains. As in earlier chapters, we will see that change occurs in manifold ways. Besides heredity and biological aging, our personal approach to passing years combines with family, community, and cultural contexts to affect the way we age.

PHYSICAL DEVELOPMENT

Physical development in midlife is a continuation of the gradual changes under way in early adulthood. Yet enough time has passed that a look in the mirror or at family photos prompts awareness of an older body, even in the most vigorous adults. Hair grays and thins, new lines appear on the face, and a less youthful, fuller body shape is evident. During midlife, most individuals begin to experience life-threatening health episodes—if not in themselves, then in their partners and friends. And a change in time orientation, from "years since birth" to "years left to live," adds to consciousness of aging (Neugarten, 1968b).

These factors lead to a revised physical self-image, which often emphasizes fewer hoped-for gains and more feared declines. Prominent concerns among 40- to 60-year-olds include getting a fatal disease, being too ill to maintain independence, and losing mental capacities. Unfortunately, many middle-aged adults fail to mention realistic alternatives as central life goals—becoming more physically fit and developing into a healthy, energetic older adult (Hooker & Kaus, 1994).

As we examine physical changes and health issues of middle adulthood, we will see that certain aspects of aging cannot be controlled. Yet many positive outcomes can be attained and feared outcomes avoided. There is much we can do to promote physical vigor and good health in midlife.

Physical Changes

One morning as they dressed before work, Trisha commented half-jokingly to Devin, "I just let the mirror get dusty; then I can't see the wrinkles and gray hairs very well." As she caught sight of her image, her tone became more serious. "I'm certainly not happy about my weight. Look at this fat—it just doesn't want to go! I need to get back to some regular exercise, adjust my life to fit it in." Devin responded by glancing down soberly at his own enlarged midriff.

At breakfast, Devin took his glasses on and off and squinted while reading the paper. "Trish—what's the eye doctor's phone number? I need to get these bifocals adjusted again." As they conversed between the kitchen and the adjoining den, Devin sometimes asked Trisha to repeat herself. And he turned the radio and TV volume up so that Trisha frequently queried, "Does it need to be so loud?" Devin, it seemed, couldn't hear as clearly as before.

In the following sections, we look closely at the major physical changes of midlife—those especially salient to Devin and Trisha and some others that occur in the reproductive system and skeleton. As we do so, you may find it helpful to refer back to Table 13.1 on page 415, which provides a summary.

Vision

By the forties, difficulty reading small print is common, due to growth in size of the lens combined with weakening of the muscle that enables the eye to adjust its focus (accommodate) to nearby objects. As new fibers appear on the surface of the lens, they compress older fibers toward the center, creating a thicker, denser, less pliable structure that eventually cannot be transformed at all. By age 50, the accommodative ability of the lens is one-sixth of what it was at age 20. Around age 60, the lens loses its capacity to adjust to objects at varying distances entirely, a condition called **presbyopia** (meaning, literally, "old eyes"). Corrective lenses for reading, which for nearsighted people take the form of bifocals, ease this problem. Because of the enlarging lens, the eye rapidly becomes more farsighted between ages 40 and 60 (Kalsi, Heron, & Charman, 2001).

A second set of changes limits ability to see in dim light, which declines at twice the rate of daylight vision (Jackson & Owsley, 2000). Throughout adulthood, the size of the pupil shrinks and the lens yellows. In addition, starting at age 40, the vitreous (transparent gelatinlike substance that fills the eye) develops opaque areas, reducing the amount of light reaching the retina. Changes in the lens and vitreous also cause light to scatter within the eye, increasing sensitivity to glare. As a college student, Devin had enjoyed driving at night. Now he

■ Because the lens of the eye gradually loses its ability to adjust its focus to nearby objects, difficulty reading small print is a common experience by the forties. Corrective lenses for reading, which for nearsighted people take the form of bifocals, ease this problem.

sometimes had trouble making out signs and moving objects. And his vision was more disrupted by bright light sources, such as the headlights of oncoming cars (Owsley et al., 1998).

Finally, yellowing of the lens and increasing density of the vitreous limit color discrimination, especially at the green–blue–violet end of the spectrum (Kraft & Werner, 1999). As a result, from time to time Devin had to ask whether his sport coat, tie, and socks matched. Besides structural changes in the eye, neural changes in the visual system occur. Gradual loss of light- and color-receptor cells in the retina (rods and cones) and of neurons in the optic nerve (the pathway between the retina and the cerebral cortex) contributes to visual declines (Pardhan et al., 1996).

After age 40, adults are at increased risk of **glaucoma,** a disease in which pressure builds up within the eye due to poor fluid drainage, damaging the optic nerve. Glaucoma often progresses without noticeable symptoms and is a leading cause of blindness among older adults. The disease runs in families, with African Americans at especially high risk (Lee & Bailey, 2003). A glaucoma test, involving an air puff to detect fluid pressure, should be part of eye exams starting in midlife. Drugs that promote release of fluid and surgery to open blocked drainage channels prevent vision loss.

Hearing

An estimated 14 percent of North American adults between ages 45 and 64 have hearing loss. Adult-onset hearing impairments account for many of these cases. Although some conditions run in families and may be hereditary, most are age-related, a condition called **presbycusis** (meaning "old hearing") (Adams & Marano, 1995).

As we age, inner-ear structures that transform mechanical sound waves into neural impulses deteriorate due to natural cell death or reduced blood supply caused by atherosclerosis. Processing of neural messages in the auditory cortex also declines. The first sign is a noticeable hearing loss at high frequencies, around 50 years of age. Gradually, the impairment extends to all frequencies so that late in life, human speech becomes more difficult to make out, although hearing loss remains greatest for high tones. Still, throughout middle adulthood, most people hear reasonably well across a wide frequency range, suggesting that severe hearing problems are caused by factors other than biological aging (Fozard & Gordon-Salant, 2001).

Men's hearing declines earlier and at a faster rate than women's, a difference thought to be due to exposure to intense noise in some male-dominated occupations (Wiley et al., 1998). Many North American production workers are or have been exposed to noise levels that induce hearing declines similar to those associated with aging (Willott, Chisolm, & Lister, 2001). Government regulations requiring industries to implement safeguards, including noise monitoring, provision of earplugs, and regular hearing tests, have greatly reduced hearing damage. Still, some employers do not fully comply with safety measures (Sataloff & Sataloff, 2001).

Most middle-aged and elderly people with hearing difficulties benefit from sound amplification with hearing aids. Once perception of the human voice is affected, speaking to the person patiently, clearly, and with good eye contact aids understanding.

Skin

Our skin consists of three layers: (1) the *epidermis,* or outer protective layer, where new skin cells are constantly produced; (2) the *dermis,* or middle supportive layer, which consists of connective tissue that stretches and bounces back, giving the skin flexibility; and (3) the *hypodermis,* an inner fatty layer that adds to the soft lines and shape of the skin. As we age, the epidermis becomes less firmly attached to the dermis, fibers in the dermis thin, and fat in the hypodermis diminishes.

These changes lead the skin to wrinkle and loosen. In the thirties, lines develop on the forehead as a result of smiling, furrowing the brow, and other facial expressions. In the forties, these become more pronounced, and "crow's-feet" appear around the eyes. Gradually, the skin loses elasticity and begins to sag, especially on the face, arms, and legs. After age 50, "age spots," collections of pigment under the skin, increase. Blood vessels in the skin become more visible as the fatty layer thins.

Because sun exposure hastens wrinkling and spotting, individuals who have spent much time outdoors without proper skin protection look older than their contemporaries. And partly because the dermis of women is not as thick as that of men, women's skin ages more quickly (Whitbourne, 1996, 2001).

Muscle–Fat Makeup

As Trisha and Devin make clear, weight gain—often referred to as "middle-age spread"—is a concern to both men and women. A common pattern of change is an increase in body fat and a loss of lean body mass (muscle and bone). The rise in fat largely affects the torso and occurs as fatty deposits within the body cavity; as noted earlier, fat beneath the skin on the limbs declines. On average, the size of the abdomen increases 6 to 16 percent in men, 25 to 35 percent in women from early through middle adulthood (Whitbourne, 1996). Sex differences in fat distribution also appear. Men accumulate more on the back and upper abdomen, women around the waist and upper arms. Muscle mass declines very gradually in the forties and fifties, largely due to atrophy of fast-twitch fibers, responsible for speed and explosive strength.

Yet, as indicated in Chapter 13, large weight gain and loss of muscle power are not inevitable. A low-fat diet, with gradual reduction in caloric intake to adjust for the age-related decline in basal metabolic rate (see page 420), helps maintain a steady, healthy weight. In nonhuman mammals, greater dietary restraint dramatically increases longevity while sustaining health and vitality. Currently, researchers are trying to uncover the biological mechanisms involved and to determine whether these beneficial outcomes apply to humans as well (see the Biology and Environment box on the following page.)

Biology & Environment

Anti-Aging Effects of Dietary Calorie Restriction: Relevant to Humans?

For nearly 70 years, scientists have known that dietary calorie restriction in nonhuman animals slows aging while maintaining good health and body functions. Rats and mice fed 30 to 40 percent fewer calories than in their typical diets beginning early in life show a reduced incidence of age-related diseases and live more than 50 percent longer. Mild to moderate calorie restriction begun after rodents reach physical maturity also slows aging and extends longevity, although to a lesser extent (Roth et al., 2000). Others studies reveal similar effects of dietary restriction in water fleas, spiders, and fish.

Primate Research. Why does calorie restriction promote a healthier, longer life in diverse species? Would primates, and especially humans, also benefit? To answer these questions, researchers are tracking health indicators in rhesus monkeys and squirrel monkeys ranging from 1 to 20 years of age, after placing some on regimens of 30 percent reduced calories, supplemented with vitamins and minerals to ensure a healthy diet. More than a decade of longitudinal findings reveals that, compared with freely eating controls, dietary-restricted monkeys are smaller but not overly thin. And they accumulate body fat differently—less on the torso, a type of fat distribution that reduces middle-aged humans' risk of heart disease.

Calorie-restricted monkeys also have a lower body temperature and basal metabolic rate—changes that suggest they shift physiological processes away from growth to life-maintaining functions. Consequently, like calorie-restricted rodents, they seem better able to withstand severe physical stress, such as surgery and infectious disease, perhaps because their bodies more easily repair damaged cells (Weindruch et al., 2001). Furthermore, calorie-restricted monkeys sustain youthful levels of the endocrine hormone DHEAS longer. When DHEAS drops in aging primates, cardiovascular disease, diabetes, cancer, and immune-system disorders rise (Lane et al., 1997).

Of all physiological processes mediating the benefits of calorie restriction, reduced blood glucose and insulin levels, signaling improved carbohydrate metabolism, seem to be the most powerful (Roth et al., 2000). Efficient blood glucose regulation protects against diabetes and cardiovascular disease as primates age. Lower blood pressure and cholesterol and a high ratio of "good" to "bad" cholesterol in calorie-restricted monkeys strengthen these antidisease effects (Roth, Ingram, & Lane, 2001).

The earlier monkeys are placed on calorie restriction, the more favorable its physiological consequences. Because none of the monkeys have yet reached old age, we do not know whether calorie restriction grants them an exceptionally long lifespan (Roberts et al., 2001). But so far, early death rates for calorie-restricted monkeys have been lower than for fully fed controls.

Human Parallels. The laboratory experimental strategies used to study calorie restriction in animals cannot be applied to humans. But two natural experiments indicate that people who limit their food intake may benefit similarly.

Compared with mainland Japanese citizens, residents of the island of Okinawa consume an average of 20 percent fewer calories, while maintaining a healthy diet. Okinawans' restricted diet is associated with a 2- to 40-fold increase over other regions of Japan in seniors living to age 100 (Kagawa, 1978).

In the early 1990s, eight American men and women ranging in age from 25 to 67 entered the Biosphere, an enclosed three-acre natural habitat, where they lived and raised their own food for 2 years. Eating 1,800 calories a day and engaging in 3 to 4 hours of daily heavy labor, the Biospherians experienced physiological changes similar to those observed in calorie-restricted rodents and monkeys, including reduced blood glucose, insulin, cholesterol, and blood pressure and a stronger immune-system response (Walford et al., 1999).

Despite these promising findings, medical and ethical considerations (including the risks of malnutrition and insufficient energy to meet the requirements of daily life) argue against attempting to get people to limit their food intake substantially. Furthermore, the great majority of people would be unwilling to maintain a reduced diet for most of their lifespan, even if it meant more healthy years. For these reasons, scientists have begun to explore *calorie-restriction mimetics*—agents, such as natural food substances, herbs, and vigorous exercise regimens—that might yield the same benefits as calorie restriction, without dieting (Butler et al., 2002; Poehlman et al., 2001). At present, these investigations are in the early stages.

■ This resident of the Biosphere, an enclosed three-acre natural habitat where a group of men and women lived and raised their own food for 2 years, examines the roots of plants that will soon be harvested. The Biospherians consumed a low-calorie diet and engaged in regular heavy labor. They showed physiological changes—including reduced blood glucose, insulin, cholesterol, blood pressure, and bouts of illness—that seem to lengthen life in calorie-restricted nonhuman animals.

Furthermore, continued exercise offsets both excess weight and muscle loss. Within the same individual, strength varies between often-used and little-used muscles (Arking, 1991). And consider Devin's 57-year-old friend Tim, who for years has ridden his bike to and from work and jogged on weekends, averaging 1 hour of vigorous activity per day. Like many endurance athletes, he maintained the same weight and muscular physique throughout early and middle adulthood (Horber et al., 1996).

Skeleton

As new cells accumulate on their outer layers, the bones broaden, but their mineral content declines so they become more porous. This leads to a gradual loss in bone mass that begins in the late thirties and accelerates in the fifties, especially among women. Women's reserve of bone minerals is lower than men's to begin with. And following menopause, the favorable impact of estrogen on bone mineral absorption is lost. Reduction in bone density during adulthood is substantial—about 8 to 12 percent for men and 20 to 30 percent for women (Arking, 1991).

Loss of bone strength causes the disks in the spinal column to collapse. Consequently, height may drop by as much as 1 inch by age 60, a change that will hasten thereafter. In addition, the weakened bones cannot support as much load, and they fracture more easily and heal more slowly. A healthy lifestyle—including exercise, adequate calcium and vitamin D intake, and avoidance of smoking and heavy alcohol consumption—can slow bone loss in postmenopausal women by as much as 30 to 50 percent (Dawson-Hughes et al., 1995; Reid, 1996).

When bone loss is very great, it leads to a debilitating disorder called *osteoporosis.* We will take up this condition shortly when we consider illness and disability.

Reproductive System

The midlife transition in which fertility declines is called the **climacteric.** It differs markedly between the two sexes because it brings an end to reproductive capacity in women, whereas fertility in men diminishes but is retained.

● **Reproductive Changes in Women.** The changes involved in women's climacteric occur gradually over a 10-year period, during which the production of estrogen drops. As a result, the number of days in a woman's monthly cycle shortens, from about 28 in her twenties and thirties to perhaps 23 by her late forties. Her cycles also become more irregular. In some, ova are not released; when they are, more are defective (see Chapter 2, page 51). The climacteric concludes with **menopause,** the end of menstruation and reproductive capacity. This occurs, on average, at age 51 among North American, European, and East Asian women, although the age range is large—from 42 to 58. Women who smoke and who have not borne children tend to reach menopause earlier (Avis, Crawford, & Johannes, 2002).

Following menopause, estrogen declines further. This causes the reproductive organs to shrink in size, the genitals to be less easily stimulated, and the vagina to lubricate more slowly during arousal. The drop in estrogen also contributes to other physical changes, including decreased elasticity of the skin and loss of bone mass. And estrogen's ability to slow accumulation of plaque on the walls of the arteries is lost.

The period leading up to and following menopause is often accompanied by emotional and physical symptoms, such as mood fluctuations and *hot flashes*—periodic sensations of warmth accompanied by a rise in body temperature; redness in the face, neck, and chest; followed by sweating (Bromberger et al., 2001). Hot flashes affect about 75 percent of women in Western industrialized nations and are a common cause of midlife sleeplessness and irritability.

To reduce the discomforts of menopause and to protect women from other impairments due to estrogen loss, many doctors recommend **hormone therapy,** or low daily doses of estrogen. Hormone therapy comes in two types: (1) estrogen alone, or *estrogen replacement therapy (ERT),* for women who have had hysterectomies (surgical removal of the uterus), and (2) estrogen plus progesterone, or *hormone replacement therapy (HRT),* for other women. Taking estrogen increases the chances of cancer of the endometrium (lining of the uterus), and combining estrogen with progesterone lessens this risk.

Hormone therapy is highly successful at counteracting menopausal symptoms, such as hot flashes, mood swings, and vaginal dryness. It also offers some protection against bone deterioration and colon cancer (Nelson et al., 2002). Nevertheless, a large-scale experiment, in which more than 16,000 women randomly assigned to take HRT or a sugar pill were followed for 5 years, revealed two negative outcomes. First, although HRT was expected to reduce cardiovascular disease, it actually caused a mild increase in heart attacks, stroke, and blood clots. Second, HRT taken for more than 4 years slightly elevated the incidence of breast cancer (Women's Health Initiative, 2002). Furthermore, an additional experiment involving 4,500 65-to 79-year-olds indicated that HRT slightly elevated the risk of mild cognitive declines and nearly doubled the risk of Alzheimer's disease and other dementias (Rapp et al., 2003; Shumaker et al., 2003).

In view of these findings, women and their doctors should make decisions about hormone therapy carefully. Those with cardiovascular disease or who have a family history of breast cancer are generally advised against it. A recent study showed that a relatively safe migraine-headache medication, called gabapentin, substantially reduces hot flashes, perhaps by acting on the brain's temperature regulation center. Several new antidepressant drugs have similar effects (Guttuso et al., 2003). Alternative medications are also available to protect the bones, although their long-term safety is not yet clear.

According to some experts, the menopausal symptoms are best treated through lifestyle changes—a good diet, regular exercise, and avoidance of smoking (Willett, Colditz, & Stampfer, 2000). Compared with North American, European,

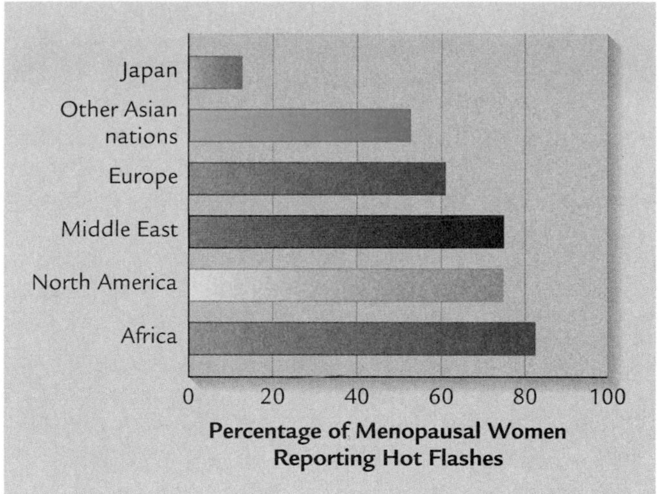

FIGURE 15.1 Percentage of menopausal women in different regions of the world reporting hot flashes. Findings are derived from interviews with large samples in each region. Women in Asian nations, especially Japanese women, are less likely to suffer from hot flashes, perhaps because they eat soy-based foods, a rich source of plant estrogen. See the Cultural Influences box on page 488 for additional evidence on the low rates of menopausal symptoms among Japanese women. (Adapted from Obermeyer, 2000.)

African, and Middle Eastern women, fewer Asian women report menopausal complaints, including hot flashes (see Figure 15.1) (Obermeyer, 2000). Asian diets, which are low in fat and high in soy-based foods, a rich source of plant estrogen, may be largely responsible.

● **Women's Psychological Reactions to Menopause.** How do women react to menopause—a clear-cut signal that their childbearing years are over? The answer lies in how they interpret the event in relation to their past and future lives.

For Jewel, who had wanted marriage and family but never attained these goals, menopause was traumatic. Her sense of physical competence was still bound up with the ability to have children. Discomfort from physical symptoms—hot flashes, headaches, and loss of sleep—can also make menopause a difficult time. And as the climacteric evokes concern about an aging body in a society that values a youthful appearance, some women respond with disappointment to a loss of sex appeal.

At the same time, many women find menopause to be little or no trouble and regard it as a new beginning (George, 2002). They do not want more children, and they are relieved to see their menstrual periods end and no longer have to worry about birth control. More highly educated, career-oriented women with fulfilling lives outside the home usually have more positive attitudes toward menopause than women with less education (Theisen et al., 1995). And the little evidence we have suggests that African-American and Mexican-American women have a generally favorable view, a trend that is stronger for Mexican Americans who have not yet adopted the language (and perhaps certain beliefs) of the larger society (Bell, 1995; Holmes-Rovner et al., 1996).

The wide variation in physical symptoms and attitudes indicates that menopause is more than just a hormonal event; it is also affected by societal beliefs and practices. For a cross-cultural look at women's experience of menopause, see the Cultural Influences box on page 488.

● **Reproductive Changes in Men.** Men also experience a climacteric, but the change is limited to a decrease in quantity of semen and sperm after age 40. However, sperm continue to be produced throughout life, and men in their nineties have fathered children. Although testosterone production declines gradually with age, the change is minimal in healthy men who continue to engage in sexual activity, which stimulates cells that release testosterone (Hermann et al., 2000). Consequently, no male counterpart to menopause exists.

Nevertheless, because of reduced blood flow to and changes in connective tissue in the penis, more stimulation is required for an erection, and it may be harder to maintain. The inability to attain an erection when desired can occur at any age, but it becomes more common in midlife, affecting about 25 percent of men by age 60 (Blanker et al., 2001). An episode or two of impotence is not serious, but frequent bouts can lead some men to fear that their sex life is over and undermine their self-image. Stress, alcohol abuse, cardiovascular disease, and other illnesses can contribute to the problem (Lue, 2000). Viagra, a drug that increases blood flow to the penis, offers temporary relief from erectile dysfunction. Publicity surrounding the drug has prompted open discussion of men's sexuality and has encouraged more men to seek help for sexual problems. Most cases of impotence respond well to medical or psychological treatment.

Ask Yourself

REVIEW

Cite cultural and gender-role influences on the experience of menopause.

APPLY

At age 42, Stan began to wear bifocals, and over the next 10 years, he required an adjustment to his corrective lenses almost every year. What physical changes account for Stan's repeated need for new eyewear?

APPLY

Between ages 40 and 50, Nancy gained 20 pounds. She also had trouble opening tightly closed jars, and her calf muscles ached after she climbed a flight of stairs. Nancy thought, "Exchanging muscle for fat must be an inevitable part of aging." Is Nancy correct? Why or why not?

www.

Cultural Influences

Menopause as a Biocultural Event

Biology and culture join forces to influence women's response to menopause, making it a *biocultural event.* In Western industrialized nations, menopause is "medicalized"—assumed to be a syndrome requiring treatment. Many women experience physical and emotional symptoms. The more symptoms they report, the more negative their attitude toward menopause tends to be (Theisen et al., 1995).

Yet change the circumstances in which menopause is evaluated, and attitudes change as well. In one study, nearly 600 men and women between ages 19 and 85 described their view of menopause in one of three contexts—as a medical problem, as a life transition, or as a symbol of aging. The medical context evoked many more negative statements than the other contexts (Gannon & Ekstrom, 1993).

Research in non-Western cultures reveals that middle-aged women's social status also affects the experience of menopause. In societies where older women are respected and the mother-in-law and grandmother roles bring new privileges and responsibilities, complaints about menopausal symptoms are rare (Patterson & Lynch, 1988). Perhaps partly for this reason, women in Asian nations seldom report discomforts. And when they do, their symptoms usually differ from those of Western women.

For example, although they rarely complain of hot flashes, a small number of Japanese women report shoulder stiffness, back pain, headaches, and fatigue. In midlife, a Japanese woman attains peak respect and responsibility. Typically her days are filled with monitoring the household economy, attending to grandchildren, caring for dependent parents-in-law, and part-time employment (Lock & Kaufert, 2001). The rare woman who experiences menopausal distress seems to interpret it in light of these socially valued commitments. Neither Japanese women nor their doctors consider menopause to be a significant marker of female middle age. Rather, midlife is viewed as an extended period of "socially recognized, productive maturity" (Menon, 2002, p. 58).

A comparison of rural Mayan women of the Yucatan with rural Greek women on the island of Evia reveals additional biocultural influences on the menopausal experience (Beyene, 1992). In both societies, old age is a time of increased status, and menopause brings freedom from child rearing and more time for leisure activities. Otherwise, Mayan and Greek women differ greatly.

Mayan women marry at age 13 or 14. By 35 to 40, they have given birth to many children but rarely menstruated, due to repeated pregnancies and breastfeeding. Because they are eager for childbearing to end, they welcome menopause, describing it with such phrases as "being happy" and "free like a young girl again." None report hot flashes or any other symptoms.

Like North Americans, rural Greek women use birth control to limit family size. And most report hot flashes and sweating at menopause. But they regard these as temporary discomforts that will stop on their own, not as medical symptoms requiring treatment. When asked what they do about hot flashes, the Greek women reply, "Pay no attention," "Go outside for fresh air," and "Throw off the covers at night."

Does frequency of childbearing affect menopausal symptoms, as this contrast between Mayan and Greek women suggests? More research is needed to be sure. At the same time, the difference between North American and Greek women in attitudes toward and management of hot flashes is striking. This—along with other cross-cultural findings—highlights the impact of culture on menopausal experiences.

■ For these rural Mayan women of the Yucatan, old age is a time of increased status, and menopause brings freedom. After decades of childbearing, Mayan women welcome menopause, describing it as "being happy" and "free like a young girl again."

Health and Fitness

In midlife, about 75 percent of North Americans rate their health as either "excellent" or "good"—still a large majority, but nevertheless fewer than in early adulthood, when the figure is 95 percent (Statistics Canada, 1999b; U.S. Department of Health and Human Services, 2002f). Whereas younger people usually attribute health complaints to temporary infections, middle-aged adults more often point to chronic diseases. Visits to the doctor's office and hospital stays become more frequent. As we will see shortly, among middle-aged adults who rate their health unfavorably, men are more likely to suffer from fatal illnesses, women from nonfatal, limiting health problems.

Our discussion takes up sexuality as a positive indicator of health, in addition to typical negative indicators—major diseases and disabling conditions. Before we begin, it is important to note that our understanding of health in middle and late adulthood is limited by insufficient research on women and ethnic minorities. Most studies of illness risk factors, prevention, and treatment have been carried out on men. Fortunately, this situation is changing. For example, the Women's Health Initiative represents a 15-year commitment by the U.S. federal government to study the impact of various lifestyle and medical prevention strategies on the health of more than 164,000 postmenopausal women of all ethnic groups and SES levels. Recent findings on cardiovascular disease and breast cancer risks associated with hormone therapy, discussed earlier, resulted from this effort.

Sexuality

Frequency of sexual activity among married couples tends to decline in middle adulthood, but the drop is only slight and is unrelated to the woman's menopausal status (Avis, Crawford, & Johannes, 2002). Longitudinal research reveals that stability of sexual activity is far more typical than dramatic change. Couples who have sex often in early adulthood continue to do so in midlife. And the best predictor of sexual frequency is marital happiness, an association that is probably bidirectional (Edwards & Booth, 1994). Sex is more likely to occur in the context of a good marriage, and couples who have sex often probably view their relationship more positively.

Nevertheless, *intensity* of sexual response declines somewhat in midlife due to physical changes of the climacteric. Both men and women take longer to feel aroused and to reach orgasm (Bartlik & Goldstein, 2001; Gelfand, 2000). If partners perceive each other as less attractive, this may contribute to a drop in sexual desire. Yet in the context of a positive outlook, sexual activity can become more satisfying. Devin and Trisha, for example, viewed each other's aging bodies with acceptance and affection—as a sign of their enduring and deepening relationship. And with greater freedom from the demands of work and family, their sex life became more spontaneous.

Finally, when surveys include both married and unmarried people, a striking gender difference in age-related sexual activity appears. The proportion of men with no sexual partners in the previous year increases only slightly, from 8 percent in the thirties to 11 percent in the late fifties. In contrast, the rise for women is dramatic, from 9 percent to 30 percent—a gender gap that becomes even greater in late adulthood (Avis, 2000). Opportunity, not desire, is responsible. A higher male mortality rate and the value women place on affection and continuity in sexual relations make partners less available to them. When we look at the evidence as a whole, sexual activity in midlife, as in earlier periods, is the combined result of biological, psychological, and social forces.

Illness and Disability

As Figure 15.2 on page 490 shows, cancer and cardiovascular disease are the leading causes of death in middle age. Unintentional injuries continue to be a major health threat, although they occur at a lower rate than in young adulthood, largely because motor vehicle collisions decline. Despite a rise in vision problems, older adults' many years of driving experience and greater cautiousness may reduce these deaths. In contrast, falls resulting in bone fractures and death nearly double from early to middle adulthood (Statistics Canada, 2002i; U.S. Department of Health and Human Services, 2002f).

Middle-aged men are equally prone to cardiovascular disease and cancer, and overall, men are more vulnerable than women to most health problems. Among middle-aged women, cancer is by far the leading cause of death (refer again to Figure 15.2). As in earlier decades, economic disadvantage is a strong predictor of poor health and premature death in midlife. And largely because of more severe poverty and lack of universal health insurance, the United States continues to exceed Canada in death rates from all major causes. Finally, as we take a closer look at illness and disability in the following sections, we will encounter yet another familiar theme: the close connection between emotional and physical well-being. Personality traits that magnify stress—especially hostility and anger—are serious threats to health in midlife.

● **Cancer.** The death rate due to cancer multiplies 10-fold from early to middle adulthood. In midlife, it is responsible for about one-third of all deaths in the United States and one-half in Canada. Although the incidence of many types of cancer is currently leveling off or declining, cancer mortality was on the rise for many decades, largely because of a dramatic increase in lung cancer due to cigarette smoking. In the last 10 years, lung cancer dropped in men; 50 percent fewer smoke today than in the 1950s. In contrast, lung cancer has increased in women, many of whom took up smoking after World War II (Greenlee, 2000).

Cancer occurs when a cell's genetic program is disrupted, leading to uncontrolled growth and spread of abnormal cells

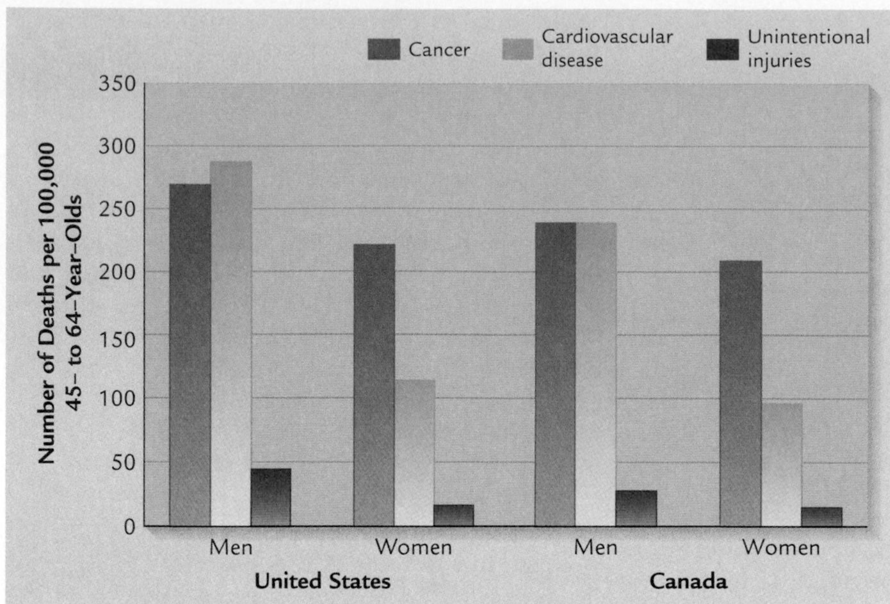

FIGURE 15.2 Leading causes of death between 45 and 64 years of age in the United States and Canada. As in early adulthood, death rates are lower in Canada than in the United States, due to Canada's universal government-funded health care and less extreme poverty. Men are more vulnerable than women to each leading cause of death. Deaths rates due to cancer and cardiovascular disease are similar for men. Cancer is by far the leading killer of women. (Adapted from Statistics Canada, 2002i; U.S. Bureau of the Census, 2002c.)

that crowd out normal tissues and organs. Why does this happen? Recall from Chapter 13 that according to one theory, random error in duplication of body cells increases with age, either due to release of free radicals or breakdown of the immune system. External agents may also initiate or intensify this process. Damage to a gene called p53, which keeps cells with defective DNA from multiplying, is involved in 60 percent of cancers, including bladder, breast, cervix, liver, lung, prostate, and skin. New cancer therapies targeting the p53 gene, in the early stages of research, show promise in counteracting tumor growth (Fisher, 2001).

Furthermore, an inherited proneness to certain cancers exists. For example, many patients with familial breast cancer who respond poorly to treatment lack a particular tumor-suppressing gene (either BRCA1 or BRCA2). Genetic screening for these mutations is available, permitting prevention efforts to begin early (Coughlin, Khoury, & Steinberg, 1999). But even though women with one of these defective genes are at much greater risk for breast cancer than other women, only 35 to 50 percent actually develop the disease. Other genes and lifestyle factors, not yet well understood, modify their risk (Rebbeck, 2002). Consequently, genetic testing for BRCA1 or BRCA2 raises ethical concerns, such as heightening women's fears of future illness and engaging in medical interventions when outcomes are uncertain (see Chapter 2, page 56).

Overall, a complex interaction of heredity, biological aging, and environment contributes to cancer. Figure 15.3 shows the most common types and the incidence of each.

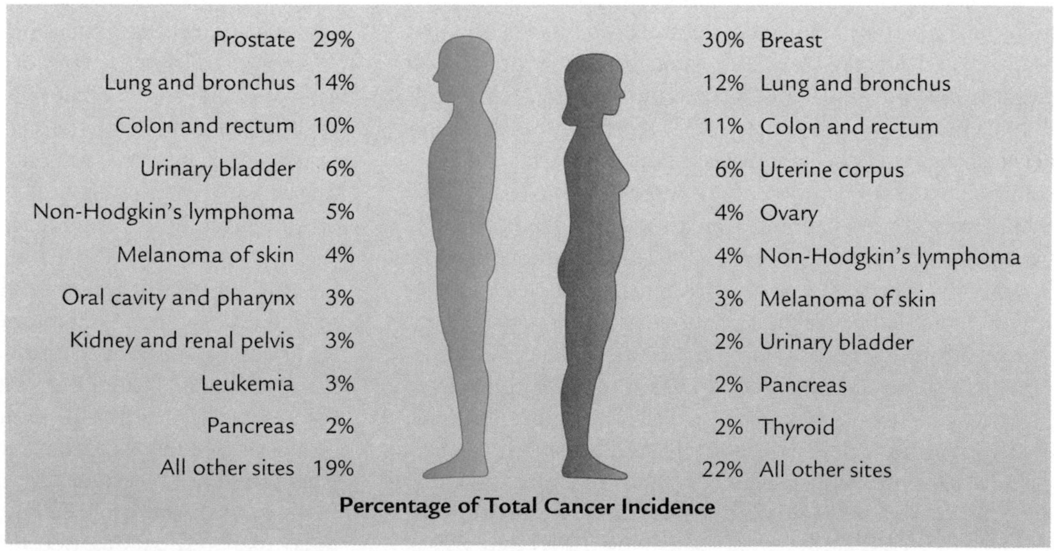

FIGURE 15.3 Cancer incidence among men and women in the United States. Although cancer rates are lower in Canada than the United States, the relative incidence of different types of cancers is similar in the two countries. (From R. T. Greenlee, 2000, "Cancer Statistics, 2000," *CA: A Cancer Journal for Clinicians, 50,* p. 16. Reprinted by permission.)

Reducing the Incidence of Cancer and Cancer Deaths

INTERVENTION	DESCRIPTION
Know the seven warning signs of cancer.	Change in bowel or bladder habits, sore that does not heal, unusual bleeding or discharge, thickening or lump in a breast or elsewhere in your body, indigestion or swallowing difficulty, obvious change in a wart or mole, nagging cough or hoarseness. If you have any of these signs, consult your doctor immediately.
Do self-examination.	Women should self-examine the breasts and men the testicles for lumps and other changes once a month. If detected early, breast and testicular cancers usually can be cured.
Schedule regular medical checkups and cancer-screening tests.	Regular medical checkups, mammograms every 1 to 2 years and Pap tests every year for women, and other screening tests for both sexes increase early detection and cure.
Avoid tobacco.	Cigarette smoking causes 90 percent of lung cancer deaths and 30 percent of all cancer deaths. Smokeless (chewing) tobacco increases risk of cancer of the mouth, larynx, throat, and esophagus.
Avoid sun exposure.	Sun exposure causes many cases of skin cancer. When in the sun for an extended time, use a sun blocker and cover exposed skin.
Avoid unnecessary X-ray exposure.	Excessive exposure to X-rays increases cancer risk. Most medical X-rays are adjusted to deliver the lowest possible dose but should not be used unnecessarily.
Avoid exposure to industrial chemicals and other pollutants.	Exposure to nickel, chromate, asbestos, vinyl chloride, and other industrial agents increases risk of various cancers.
Weigh the benefits versus risks of hormone therapy.	Because estrogen replacement increases risk of uterine and breast cancer and estrogen/progesterone replacement may further elevate risk of breast cancer, carefully consider hormone therapy with your doctor.
Maintain a healthy diet.	Avoid too much dietary fat and salt-cured, smoked, and nitrite-cured foods; eat vegetables and foods rich in fiber and vitamins A and C.

Source: Greenlee, 2000; Holland, 1998.

Illness and mortality rates are higher for certain groups of people than for others. For cancers that affect both sexes, men are generally more vulnerable than women. The difference may be due to genetic makeup, exposure to cancer-causing agents as a result of lifestyle or occupation, and a tendency to delay going to the doctor (Greenlee, 2000).

Many people fear cancer because they believe it is incurable. Yet 40 percent of people with cancer are cured (free of the disease for 5 or more years). Breast cancer is the leading type for women, prostate cancer for men. Lung cancer—largely preventable through avoiding tobacco—ranks second for both sexes, followed closely by colon and rectal cancer. Scheduling annual medical checkups that screen for these and other forms of cancer and taking the additional steps listed in the Caregiving Concerns table above can reduce cancer illness and death rates considerably. Although the relationship of SES to cancer varies with site (for example, lung and stomach cancers are linked to lower SES, breast and prostate cancers to higher SES), cancer survival drops sharply as SES decreases (Balfour & Kaplan, 1998). Poorer medical care and reduced ability to fight the disease, due to inadequate diet, high life stress, and lack of social support, underlie this trend.

Surviving cancer is a triumph, but it also brings emotional challenges. During cancer treatment, relationships focus on the illness. Afterward, they must refocus on health and full participation in daily life. Unfortunately, stigmas associated with cancer exist. Friends, family, and co-workers may need reminders that cancer is not contagious and that research shows cancer survivors to be just as productive on the job as other people (Kornblith, 1998).

● **Cardiovascular Disease.** Despite a decline during the last few decades (see Chapter 13), each year about 28 percent of middle-aged Americans and Canadians who die succumb to cardiovascular disease (Statistics Canada, 2002i; U.S. Bureau of the Census, 2002c). We associate cardiovascular disease with heart attacks, but for Devin (like many middle-aged and older adults) the condition was discovered during an annual checkup. His doctor detected high blood pressure, high blood cholesterol, and *atherosclerosis*—a buildup of plaque in his coronary arteries, which encircle the heart and provide its muscles with oxygen and nutrients. These indicators of cardiovascular disease are known as "silent killers" because they often have no symptoms.

When symptoms *are* evident, they take different forms. The most extreme is a *heart attack*—blockage of normal blood supply to an area of the heart. It is usually brought on by a blood clot in one or more plaque-filled coronary arteries. Intense pain results as muscle in the affected region dies. A heart attack is a medical emergency; over 50 percent of victims die before reaching the hospital, another 15 percent during treatment, and an additional 10 percent over the next few years (American Heart Association, 2002). Among other less extreme symptoms of cardiovascular disease are *arrhythmia,* or irregular heartbeat. When it persists, it can prevent the heart from pumping enough blood and result in faintness. It can also allow clots to form within the heart's chambers, which may break loose and travel to the brain. In some individuals, indigestion-like pain or crushing chest pain, called *angina pectoris,* reveals an oxygen-deprived heart.

Today, cardiovascular disease can be treated in many ways—including coronary bypass surgery, medication, and pacemakers to regulate heart rhythm. To relieve arterial blockage, Devin had *angioplasty,* a procedure in which a surgeon threaded a needle-thin catheter into his arteries and inflated a balloon at its tip, which flattened fatty deposits to allow blood to flow more freely. At the same time, his doctor warned that unless Devin took other measures to reduce his risk, the arteries would clog again within a year. As the Caregiving Concerns table below indicates, adults can do much to prevent heart disease or slow its progress.

Of course, some risks, such as heredity, advanced age, and being male, cannot be changed. But cardiovascular disease is so disabling and deadly that people must be alert for it where it is least expected. A special concern is accurate diagnosis in women. Because men account for over 70 percent of cases in middle adulthood, doctors often view a heart condition as a "male problem." Consequently, they frequently overlook women's symptoms, which tend to be milder, more often taking the form of angina than a full-blown heart attack (Roger et al., 2000). In one study, researchers had male and female actors present identical symptoms of angina to a sample of over 700 doctors. The doctors were far less likely to suspect heart problems in women—especially African-American women, who are at greater risk for heart disease than Caucasian-American women. This combined gender–racial bias greatly

Reducing the Risk of Heart Attack

INTERVENTION	RISK REDUCTION
Quit smoking.	→ 70% 5 years after quitting, up to 70 percent lower risk compared to current smokers
Reduce blood cholesterol level.	→ 60% 2 to 3 percent decline in risk for each 1 percent reduction in blood cholesterol. Reductions in cholesterol average 10 percent with diet therapy and can exceed 20 percent with drug therapy.
Treat high blood pressure.	→ 60% Combined diet and drug therapy can lower blood pressure substantially, leading to as much as a 60 percent risk reduction.
Maintain ideal weight.	→ 55% Up to 55 percent lower risk for people who maintain ideal body weight compared to those who are obese
Exercise regularly.	→ 45% 45 percent lower risk for people who maintain an active rather than a sedentary lifestyle
Drink an occasional glass of wine or beer.[a]	→ 45% Up to 45 percent lower risk for people who consume small-to-moderate amounts of alcohol; believed to promote high-density lipoproteins, a form of "good cholesterol" that reduces "bad cholesterol"
Take low-dose aspirin.	→ 33% Up to 33 percent lower risk for people who take 162 mg (½ tablet) daily or every other day; reduces the likelihood of blood clots (should be medically recommended; long-term use can have serious side effects)
Reduce hostility and other forms of psychological stress.	Extent of risk reduction not yet known.

Source: American Heart Association, 2002; Harpaz et al., 1996; Rimm, 1996.

[a]Recall from Chapter 13 that heavy alcohol use increases the risk of cardiovascular disease.

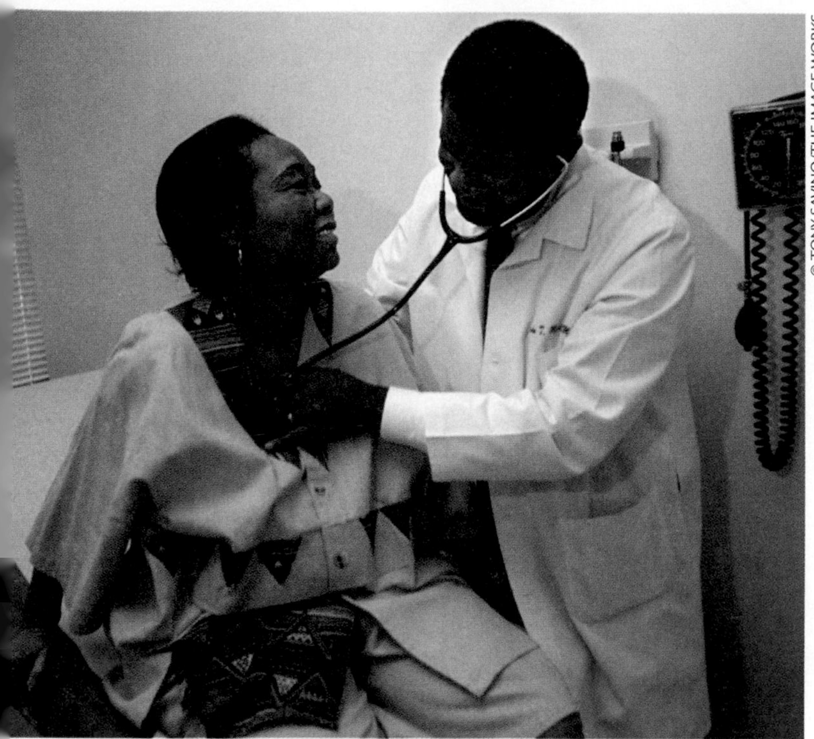

■ Accurate diagnosis of cardiovascular disease is of special concern in women. Because doctors often view a heart condition as a "male problem," they tend to overlook women's symptoms. African-American women, who are at increased risk for heart disease, are particularly likely to experience a lack of appropriate follow-up tests.

reduced the extent to which African-American women were referred for a standard follow-up test (Schulman et al., 1999).

● **Osteoporosis.** When age-related bone loss is severe, a condition called **osteoporosis** develops. In the largest study of osteoporosis conducted to date, consisting of more than 200,000 postmenopausal middle-aged women, 40 percent had bone density levels low enough to be of concern, and 7 percent were diagnosed with the disorder (Chestnut, 2001). Osteoporosis affects the majority of people of both sexes over age 70 (Donatelle, 2003). Although we associate it with a slumped-over posture, a shuffling gait, and a "dowager's hump" in the upper back, this extreme is rare. Because the bones gradually become more porous over many years, osteoporosis may not be evident until fractures—typically in the spine, hips, and wrist—occur or are discovered through X-rays.

A major factor related to osteoporosis is the decline in estrogen associated with menopause. In middle and late adulthood, women lose about 50 percent of their bone mass, about half of which is lost in the first 10 years following menopause. Because of estrogen loss, the earlier a woman reaches menopause, the greater her chances of developing the disease. Heredity plays an important role, as identical twins are more likely than fraternal twins to share the disorder, and a family history increases risk (Stewart & Ralston, 2000). People with thin, small-framed bodies are more likely to be affected because they typically attain a lower peak bone mass in adolescence. In contrast, higher bone density makes African Americans less susceptible than Asian Americans, Caucasians, and Hispanics. An unhealthy lifestyle can also contribute to osteoporosis, through a calcium-deficient diet or physical inactivity, both of which reduce bone mass. In addition, cigarette smoking and heavy caffeine and alcohol use are linked to the disease, although how these substances exert their effects is not known (Donatelle, 2003; Dowd, 2001).

Osteoporosis affects more than 25 million Americans and 1.4 million Canadians and greatly magnifies the risk of bone fractures (Osteoporosis Society of Canada, 2002; U.S. Department of Health and Human Services, 2002). When major breaks (such as the hip) occur, 12 to 20 percent of patients die within a year (Center et al., 1999). To treat the disorder, doctors recommend a diet enriched with calcium and vitamin D (which promotes calcium absorption), weight-bearing exercise (walking rather than swimming), and hormone therapy (for women not at high risk for breast and uterine cancer and cardiovascular disease) or other bone-strengthening medications. Each of these interventions helps the bones regain mineral content. Prevention early in life is an even better course of action. Increasing calcium and vitamin D intake and engaging in regular exercise in childhood, adolescence, and early adulthood reduce lifelong risk by maximizing peak bone density (NIH Consensus Development Panel, 2001).

Hostility and Anger

Each time Trisha's sister Dottie called, she seemed like a powder keg ready to explode. Dottie was critical of her boss at work and dissatisfied with the way Trisha, a lawyer, had handled the family's affairs after their father died. All conversations ended the same way—with Dottie making demeaning, hurtful remarks as her anger rose to the surface. "Any lawyer knows that, Trisha; how could you be so stupid! I should have called a *real* lawyer." "You and Devin are so stuck in your privileged lives that you can't think of anyone else. You don't know what work *is.*"

Trisha listened as long as she could bear. Then she warned, "Dottie, if you continue, I'm going to hang up... Dottie, I'm ending this right now!"

Off the telephone, Dottie's life was full of health-related complaints and problems. At age 53, she had high blood pressure, difficulty sleeping, and back pain. During the past 5 years, she had been hospitalized five times—twice for treatment of digestive problems, twice for an irregular heartbeat, and once for a benign tumor on her thyroid gland. Trisha often wondered whether Dottie's personal style was partly responsible for her physical condition.

That hostility and anger might have negative effects on health is a centuries-old idea. Several decades ago, researchers first tested this notion by identifying 35- to 59-year-old men who displayed the **Type A behavior pattern**—extreme competitiveness, ambition, impatience, hostility, angry outbursts, and a sense of time pressure. They found that within the next

© BOB DAEMMRICH

■ Is this angry man headed for an early death? Recent evidence pinpoints hostility as the "toxic" ingredient in the Type A behavior pattern. Expressed hostility, especially in the form of frequent angry outbursts and rude, disagreeable behavior, leads to greater cardiovascular arousal, health complaints, and illness.

8 years, Type As were more than twice as likely as Type Bs (people with a more relaxed disposition) to develop heart disease (Rosenman et al., 1975).

Later studies, however, often failed to confirm these results, perhaps because Type A is a mix of behaviors, only one or two of which affect health. Current evidence pinpoints hostility as the "toxic" ingredient of Type A, since isolating it from global Type A consistently predicts heart disease and other health problems in both men and women (Aldwin et al., 2001; Low et al., 1998; Williams, 2000). The risk of atherosclerosis is $2\frac{1}{2}$ times higher in adults scoring high than low on hostility measures (Iribarren et al., 2000).

Recent findings suggest that *expressed hostility* in particular—frequent angry outbursts; rude, disagreeable behavior; critical and condescending nonverbal cues during social interaction, including glares; expressions of contempt and disgust; and a hard, insistent voice—leads to greater cardiovascular arousal, health complaints, and illness (Chesney et al., 1997). As people get angry, heart rate, blood pressure, and stress hormones rise—physiological reactions that escalate until the body's response is extreme.

Of course, people who are repeatedly enraged are more likely to be depressed and dissatisfied with their lives, to lack social supports, and to engage in unhealthy behaviors (Raynor et al., 2002). But hostility predicts health problems even after such factors as smoking, alcohol consumption, overweight, general unhappiness, and negative life events are controlled. And since men score higher in hostility than women (Dottie is an exception), emotional style may contribute to the sex differences in heart disease described earlier (Stoney & Engebretson, 1994).

To preserve her health, should Dottie bottle up her hostility instead of expressing it? Repeatedly suppressing anger—seething inside while looking calm on the outside—is also associated with an increase in heart disease, and it may pose other as yet unknown health risks (Julkunen, 1996). As we will see shortly, a better alternative is to develop effective ways of handling stress and conflict.

Adapting to the Physical Challenges of Midlife

Middle adulthood is often a productive period in which people attain their greatest accomplishments and life satisfactions. Nevertheless, it takes considerable stamina to cope with the full array of changes this phase can bring. Devin responded to his expanding waistline and cardiovascular symptoms by leaving his desk twice a week to attend a low-impact aerobics class and by reducing job-related stress through daily 10-minute meditation sessions. Aware of her sister Dottie's difficulties, Trisha resolved to handle her own hostile feelings more adaptively. And her generally optimistic outlook enabled her to cope successfully with the physical changes of midlife, the pressures of her legal career, and Devin's chronic illness.

Stress Management

Turn back to Chapter 13, page 430, and review the negative consequences of psychological stress on the cardiovascular, immune, and gastrointestinal systems. As adults encounter problems at home and at work, daily hassles can add up to a serious stress load. Stress management is important at any age for a more satisfying life. In middle adulthood, it can limit the age-related rise in illness and, when disease strikes, reduce its severity.

The Caregiving Concerns table on the following page summarizes effective ways to reduce stress. Although many stressors cannot be eliminated, people can change how they handle some and view others. At work, Trisha focused on problems she could control—not on her boss's irritability, but on ways to delegate routine tasks to her staff so she could focus on problems that required her knowledge and skills. When Dottie's phone calls came, Trisha learned to distinguish normal emotional reactions from unreasonable self-blame. She stopped interpreting Dottie's anger as a sign of her own incompetence and, instead, reminded herself of Dottie's difficult temperament and hard life.

Managing Stress

TECHNIQUE	DESCRIPTION
Reevaluate the situation.	Learn to tell the difference between normal reactions and those based on irrational beliefs.
Focus on events you can control.	Don't worry about things you cannot change or that may never happen; focus on strategies for handling events under your control.
View life as fluid.	Expect change and accept it as inevitable; then many unanticipated changes will have less emotional impact.
Consider alternatives.	Don't rush into action; think before you act.
Set reasonable goals for yourself.	Aim high, but be realistic in terms of your capacities, motivation, and the situation.
Exercise regularly.	A physically fit person can handle stress better, both physically and emotionally.
Master relaxation techniques.	Relaxation helps refocus energies and reduce the physical discomfort of stress. Classes and self-help books teach these techniques.
Use constructive approaches to anger reduction.	In addition to the techniques listed above, seek a delay in responding ("Let me check into that and get back to you"); use mentally distracting behaviors (counting to 10 backwards) and self-instruction (a covert "Stop!") to control anger arousal; then engage in calm, self-controlled problem solving ("I should call him rather than confront him personally").
Seek social support.	Friends, family members, co-workers, and organized support groups can offer information, assistance, and suggestions for handling stressful situations.

And greater life experience helped her accept change as inevitable. Consequently, she was better equipped to deal with the jolt of sudden events, such as Devin's hospitalization for treatment of heart disease.

As Trisha's efforts illustrate, people use two general strategies to cope with stress. In *problem-centered coping,* they appraise the situation as changeable, identify the difficulty, and decide what to do about it. If problem solving does not work, people engage in *emotion-centered coping,* which is internal, private, and aimed at controlling distress when there is little we can do about a situation (Lazarus & Lazarus, 1994). Longitudinal research shows that adults who effectively reduce stress use a mixture of problem-centered and emotion-centered techniques, depending on the situation (Zakowski et al., 2001). And their approach is deliberate, thoughtful, and respectful of both themselves and others. In contrast, ineffective coping is largely emotion-centered and either impulsive or escapist (Lazarus, 1991, 1999).

Constructive approaches to anger reduction are a vital health intervention (refer again to the Caregiving Concerns table). In terms of problem-centered coping, teaching people to be assertive rather than hostile and to negotiate rather than explode interrupts the intense physiological response that intervenes between psychological stress and illness. If reasonable communication is not possible, it is best to delay responding by leaving a provocative situation, as Trisha did when she told Dottie that one more insult would cause her to hang up the phone (Deffenbacher, 1994).

As noted in Chapter 13, people tend to cope with stress more effectively as they move from early to middle adulthood. Many are more realistic about their ability to change situations than they were earlier. And midlifers may be more skilled at anticipating stressful events and taking steps to avoid them (Aldwin & Levenson, 2002). Furthermore, when middle-aged adults surmount a highly stressful experience, they often report lasting personal benefits. Some describe a sense of disbelief at what they accomplished under extremely trying conditions and a greater sense of mastery. And after a serious illness and brush with death, changed values and perspectives—such as rediscovery of the importance of health-promoting behaviors and of family relationships—are common (Aldwin, Sutton, & Lachman, 1996). In this way, managing intense stress can serve as a context for positive development.

Still, some people have difficulty handling the challenges of midlife, and communities provide fewer social supports for them than for young adults and senior citizens. For example, Jewel had little knowledge of what to expect during the climacteric. "It would have helped to have a support group so I could have learned about menopause and dealt with it more easily," she commented in a phone call to Trisha. Community programs addressing other midlife concerns, such as those of adult learners returning to college and caregivers of elderly parents, can also reduce stress during this period.

Exercise

Regular exercise, as noted in Chapter 13, has a wide variety of physical and psychological benefits—among them, equipping adults to handle stress more effectively. On the way to his first aerobics class, Devin wondered, "Can starting to exercise at age 50 counteract years of physical inactivity?" Devin's question is important, since more than half of American and Canadian middle-aged adults are sedentary. Of those who begin an exercise program, 50 percent discontinue within the first 6 months. And among those who stay active, fewer than 20 percent exercise at levels that lead to health benefits (Canadian Fitness and Lifestyle Research Institute, 2001b; U.S. Department of Health and Human Services, 2002f).

■ The exercise format most likely to lead to sustained physical activity in midlife varies with the beginning exerciser's characteristics. Overweight adults do better with a home-based program planned by a consultant than with group classes, where they may find it difficult to keep up with the pace of the class. This man works out at a health club with a personal trainer, who also recommends a daily, home-based routine.

A person beginning to exercise in midlife must overcome initial barriers and obstacles along the way, such as lack of time and energy, inconvenience, and work conflicts. *Self-efficacy*—belief in one's ability to succeed—is just as vital in adopting, maintaining, and exerting oneself in an exercise regimen as it is in career progress (see Chapter 14). An important outcome of starting an exercise program is that it leads sedentary adults to gain in self-efficacy, which promotes physical activity all the more (McAuley & Blissmer, 2000). Enhanced physical fitness, in turn, prompts middle-aged adults to feel better about their physical selves. Over time, their physical self-esteem—sense of body conditioning and attractiveness—rises (McAuley, Mihalko, & Bane, 1997).

The exercise format that works best varies with the beginning exerciser's characteristics. Normal-weight adults are more likely to stick with group classes than are overweight adults, who may strain to keep up with the pace of the class and feel embarrassed in front of others. Overweight people do better in a home-based program planned by a consultant who individualizes the routine (King, 2001). In contrast, adults with highly stressful lives are more likely to persist in group classes than in a home-based program, which requires them to find time to exercise. Without a scheduled exercise period and the face-to-face support of others, stress can become a major barrier to behavior change (King et al., 1997). Yet when stressed people manage to sustain a home-based program, it substantially reduces stress—more so than the group format (King, Taylor, & Haskell, 1993). Perhaps succeeding on their own helps stressed adults gain better control over their lives.

Accessible, attractive, and safe exercise environments—parks, walking and biking trails, and community recreation centers—and frequent opportunities to observe others using them also promote physical activity (King et al., 2000). And when researchers placed signs at choice points for taking the stairs or an escalator, more people took the stairs and continued to do so for several months after the signs were removed (Blamey, Mutrie, & Aitchison, 1995). Finally, Australia has experimented with TV public service announcements aimed at promoting physical activity—an approach that increased walking for exercise among older adults (Owen et al., 1995).

An Optimistic Outlook

Our ability to handle the inevitable changes of life depends in part on personality strengths. What type of individual is likely to cope with stress adaptively, thereby reducing its impact on illness and mortality? Searching for answers to this question, researchers have studied a set of three personal qualities—control, commitment, and challenge—that, together, they call **hardiness** (Maddi, 1999).

Trisha fit the pattern of a hardy individual. First, she regarded most experiences as *controllable*. "You can't stop all bad things from happening," she advised Jewel after hearing about her menopausal symptoms, "but you can try to do something about them." Second, Trisha displayed a *committed*, involved approach to daily activities, finding interest and meaning in almost all of them. Finally, she viewed change as a *challenge*—a normal part of life and a chance for personal growth.

Research shows that hardiness influences the extent to which people appraise stressful situations as manageable, interesting, and enjoyable. These positive appraisals, in turn, predict health-promoting behaviors, tendency to seek social support, and fewer physical symptoms (Florian, Mikulincer, & Taubman, 1995; Wiebe & Williams, 1992). Furthermore, high-hardy individuals are likely to use active, problem-centered coping strategies in situations they can control. In contrast, low-hardy people more often use emotion-centered and avoidant coping strategies—for example, saying, "I wish I could change how I feel," denying that the stressful event occurred, or eating and drinking to forget about it (Maddi & Hightower, 1999; Soderstrom et al., 2000).

In some studies, hardiness-related positive appraisals were associated with lower physiological arousal to stress—a major means by which hardiness is believed to protect against illness. The link between hardiness and reduced physiological reactivity does not always appear, perhaps because a hardy person's active coping sometimes leads to greater arousal. But over time, this increase may be offset by a calm that comes with effective stress management (Wiebe & Williams, 1992). In support of this idea, hardiness is associated with higher blood levels of cortisol, a hormone that regulates blood pressure and is involved in resistance to stress. And the optimistic component of hardiness predicts lower average blood pressure, measured over a 3-day period (Räikkönen et al., 1999; Zorilla, DeRubeis, & Redei, 1995).

In this and previous chapters, we have seen that many factors act as stress-resistant resources—heredity, diet, exercise, social support, coping strategies, and more. Research on hardiness adds yet another ingredient: a generally optimistic outlook and zest for life.

Gender and Aging: A Double Standard

Earlier in this chapter, we mentioned that negative stereotypes of aging lead many middle-aged adults to fear physical changes. These stereotypes are more likely to be applied to women than to men, yielding a double standard. Despite the fact that many women in midlife say they have "hit their stride"—feel assertive, confident, versatile, and capable of resolving life's problems—people often rate them as less attractive and as having more negative characteristics than middle-aged men. In some studies, aging men actually gain slightly in positive judgments of appearance, maturity, and power, whereas aging women show a decline. And the sex of the person doing the rating makes a difference: Compared with women, men judge an aging female much more harshly (Kogan & Mills, 1992).

These effects appear more often when people rate photos as opposed to verbal descriptions of men and women. Consequently, researchers believe that the ideal of a sexually attractive woman—smooth skin, good muscle tone, and lustrous hair—is at the heart of the double standard of aging. Recall from Chapter 14 that women prefer same-age or slightly older sexual partners, whereas men prefer younger partners. To explain why, evolutionary theory points to sex differences in reproductive capacity, whereas social learning theory emphasizes gender roles (see page 451). Some evidence suggests that the end of a woman's ability to bear children contributes to negative judgments, especially by men (Marcus-Newhall, Thompson, & Thomas, 2001). Yet societal forces exaggerate this view. For example, when media ads include middle-aged people, they usually are male executives, fathers, and grandfathers, who are images of competence and security. And the cosmetics industry offers many products designed to hide signs of aging for women, but far fewer for men.

At one time in our evolutionary history, a double standard of aging may have been adaptive. Today, as many couples limit childbearing and devote more time to career and leisure pursuits, it has become irrelevant. Consequently, new surveys suggest that the double standard of aging is declining—that more people are viewing middle age as a potentially upbeat, satisfying time for both genders (Menon, 2002). Models of older women, whose lives are full of intimacy, accomplishment, hope, and imagination, are promoting acceptance of physical aging and a new vision of growing older—one that emphasizes gracefulness, fulfillment, and inner strength.

Ask Yourself

REVIEW

Cite evidence that biological aging, individual heredity, and environmental factors contribute to osteoporosis.

APPLY

When Cara complained of chest pains to Dr. Furrow, he decided to "wait and see" before conducting additional tests. In contrast, Dr. Furrow gave Bill a battery of tests aimed at detecting cardiovascular disease, even though he did not complain of symptoms. What might account for Dr. Furrow's different approach to Cara than to Bill?

APPLY

Because his assistant misplaced some files, Tom lost a client to a competitor. Tom felt his anger building to the breaking point. Explain why Tom's emotional response is unhealthy. How can he best cope with this stressful experience?

CONNECT

According to the lifespan perspective, development is multidimensional—affected by biological, psychological, and social forces. Using examples, show how this assumption characterizes health at midlife.

www.

COGNITIVE DEVELOPMENT

In middle adulthood, the cognitive demands of everyday life extend to new and, at times, more challenging situations. Consider a typical day in the lives of Devin and Trisha. Recently appointed dean of faculty at a small college, Devin was at his desk by 7 A.M. In between strategic-planning meetings, he reviewed files of applicants for new positions, worked on the coming year's budget, and spoke at an alumni luncheon. In the meantime, Trisha prepared for a civil trial, participated in jury selection at the courthouse, and then returned to her firm to join the other top lawyers for a conference about management issues. That evening, Trisha and Devin advised their 20-year-old son, Mark, who had dropped by to discuss his uncertainty over whether to change his college major. By 7:30 P.M., Trisha was off to an evening meeting of the local school board. And Devin left for a biweekly gathering of an amateur quartet, in which he played the cello.

Recall from Chapter 13 that Schaie characterized middle adulthood as the *responsibility stage*—a time when expansion of responsibilities takes place on the job, in the community, and at home. To juggle diverse roles and perform effectively, Devin and Trisha called on a wide array of intellectual abilities, including accumulated knowledge, verbal fluency, memory, rapid analysis of information, reasoning, problem solving, and expertise in their areas of specialization. What changes in thinking take place in middle adulthood? How does vocational life—a major arena in which cognition is expressed—influence intellectual skills? And what can be done to support the rising tide of adults who are returning to college in hopes of enhancing their knowledge and quality of life? Let's see what research has to say about these topics.

■ Valecia Crisafulli, the coordinator of Illinois Main Street, an organization run by Illinois citizens to revitalize downtown business districts, speaks to a full audience at a Jacksonville Main Street information meeting. She exemplifies Schaie's responsibility stage—a time when responsibilities expand on the job, in the community, and at home.

Changes in Mental Abilities

At age 50, when he occasionally couldn't recall a name or had to pause in the middle of a lecture or speech to think about what to say next, Devin wondered to himself, "Are these the first signs of an aging mind?" Twenty years earlier, Devin took little notice of the same mental events. His current questioning stems from a widely held stereotype—a view of the aging mind as on a path of inevitable decline. The majority of aging research has focused on deficits—because they are cause for concern—while neglecting cognitive stability and gains (Salthouse, 1991a).

In Chapter 13, we noted that some apparent decrements in cognitive aging result from weaknesses in the research itself. In cross-sectional studies, *cohort effects,* in which older participants are less well educated and in poorer health, can paint an inaccurate picture of both the timing and the extent of decline. Also, the tests given may tap abilities less often used by older individuals, whose lives no longer require that they learn information for its own sake but, instead, that they build knowledge and skills that help them tackle current, real-world problems.

As we examine changes in thinking in middle adulthood, we will revisit the theme of diversity in development. Different aspects of cognitive functioning show different patterns of change. Although declines occur in some areas, most people display cognitive competence, especially in familiar contexts, and some attain outstanding accomplishment. Overall, the evidence supports an optimistic view of adult cognitive potential.

The changes we are about to consider bring into bold relief core assumptions of the lifespan perspective: development as *multidimensional,* or the combined result of biological, psychological, and social forces; development as *multidirectional,* or the joint expression of growth and decline, with the precise mix varying across abilities and individuals; and development as *plastic,* or open to change, depending on how a person's biological and environmental history combines with current life conditions. Before we begin, you may find it helpful to return to Chapter 1, pages 9–10, to review these ideas.

Crystallized and Fluid Intelligence

Many studies report consistent age-related trends in two broad mental abilities. Each includes a diverse array of specific intellectual factors tapped by intelligence tests.

The first of these broad abilities is **crystallized intelligence.** It refers to skills that depend on accumulated knowledge and experience, good judgment, and mastery of social

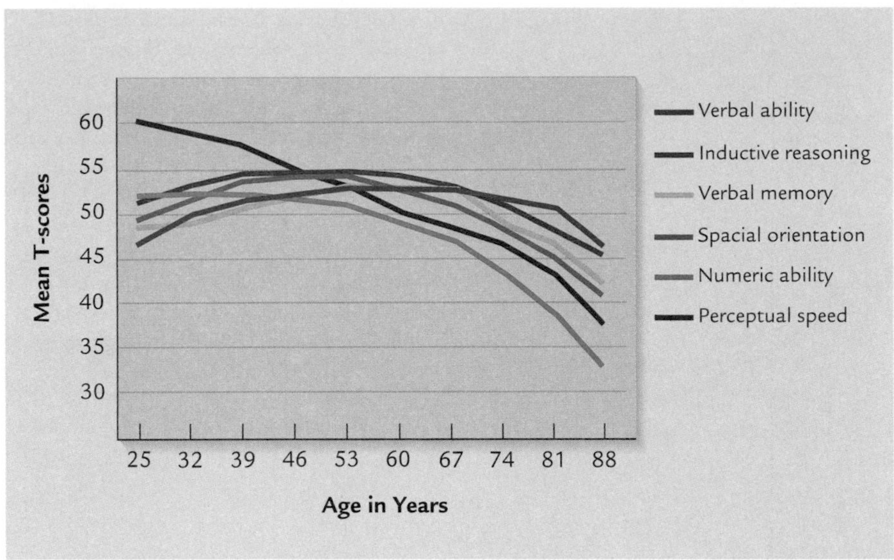

FIGURE 15.4 Longitudinal trends in six mental abilities, from the Seattle Longitudinal Study. In five abilities, modest gains occurred into the fifties and early sixties, followed by gradual declines. The sixth ability—perceptual speed—dropped steadily from the twenties to the late eighties. And late in life, fluid factors (spatial orientation, perceptual speed, and numeric ability) showed greater decrements than crystallized factors (verbal ability, inductive reasoning, and verbal memory.) Still, these findings are more optimistic than other investigations, which show an earlier and steeper decline in fluid IQ. (From K. W. Schaie, 1994, "The Course of Adult Intellectual Development," *American Psychologist, 49,* p. 308. Copyright © 1994 by the American Psychological Association. Reprinted by permission.)

conventions. Together, these capacities represent abilities acquired because they are valued by the individual's culture. Devin made use of crystallized intelligence when he expressed himself articulately at the alumni luncheon and suggested effective ways to save money in budget planning. On intelligence tests, vocabulary, general information, verbal analogy, and logical reasoning items measure crystallized intelligence. (Turn back to Chapter 9, page 294, to see examples.)

In contrast, **fluid intelligence** depends more heavily on basic information-processing skills—the ability to detect relationships among stimuli, the speed with which we can analyze information, and the capacity of working memory. Fluid intelligence often works with crystallized intelligence to support effective reasoning, abstraction, and problem solving. But fluid intelligence is believed to be influenced more by conditions in the brain and learning unique to the individual, less by culture (Horn & Noll, 1997). Intelligence test items that reflect fluid abilities include number series, spatial visualization, and picture sequencing. (Refer again to page 294 for examples.)

Many studies show that crystallized intelligence increases steadily though middle adulthood, whereas fluid intelligence begins to decline in the twenties. These trends have been found repeatedly in cross-sectional comparisons in which the education and health status of younger and older participants are similar, thereby largely correcting for cohort effects (Horn & Donaldson, 1980; Horn, Donaldson, & Engstrom, 1981; Kaufman & Horn, 1996). The rise in crystallized abilities makes sense, since adults are constantly adding to their knowledge and skills at work, at home, and in leisure activities. In addition, many crystallized skills are practiced almost daily. But the early decline in fluid intelligence identified in cross-sectional research is a matter of intense debate.

● Schaie's Seattle Longitudinal Study.

Recall Schaie's Seattle Longitudinal Study, introduced in Chapter 13, which revealed modest gains in five mental abilities from early to middle adulthood that were sustained until the fifties and early sixties, after which performance gradually declined. Figure 15.4 shows these trends. The five abilities—verbal ability, inductive reasoning, verbal memory, spatial orientation, and numeric ability—tap both crystallized and fluid skills. Their paths of change show that midlife is the period of peak performance on some of the most complex mental abilities. According to these findings, middle-aged adults appear intellectually "in their prime," not "over the hill" as stereotypes of aging would have us believe (Willis & Schaie, 1999).

A sixth ability is also shown in Figure 15.4: perceptual speed, a fluid skill in which participants must, for example, identify during a time limit which of five shapes is identical to a model or whether pairs of multidigit numbers are the same or different (Schaie, 1994, 1996, 1998). Perceptual speed dropped steadily from the twenties to the late eighties, a finding that fits with a wealth of research indicating that cognitive processing slows as people get older. And notice in Figure 15.4 how late in life, fluid factors (spatial orientation, numeric ability, and perceptual speed) show greater decrements than the crystallized factors (verbal ability, inductive reasoning, and verbal memory).

● Kaufman's Research on Verbal and Performance IQ.

Some researchers argue that to be certain about changes in various mental abilities across the lifespan, we need more sensitive mental tests than the one used in the Seattle Longitudinal Study, which was group administered and did not include people older than 18 in its standardization sample. Using a sample of nearly 2,500 mentally and physically healthy 16- to 85-year-olds representing the ethnic and SES diversity of the United States, Alan Kaufman (2001) examined cross-sectional trends in performance on the Wechsler Adult Intelligence Scale (WAIS). Like the Wechsler Intelligence Scales for Children (see Chapter 9, page 293), the WAIS measures two broad intellectual factors: *verbal,* which largely taps crystallized

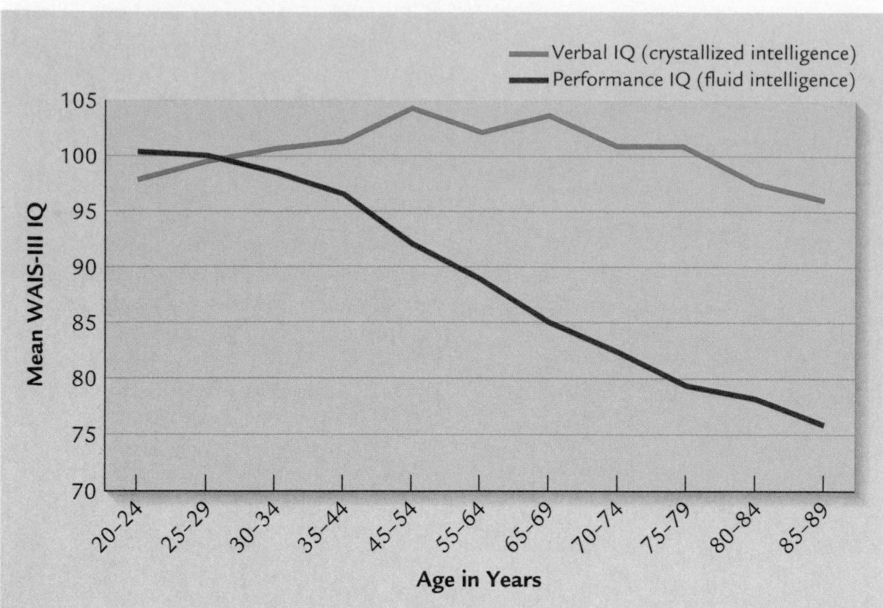

FIGURE 15.5 Cross-sectional trends in verbal and performance IQ on the Wechsler Adult Intelligence Scale, after controlling for education to minimize cohort effects. Verbal IQ, which largely taps crystallized intelligence, peaked in midlife, between ages 45 and 54, and then declined only slightly into the eighties. Performance IQ, which largely taps fluid intelligence, declined from the early twenties, showing an earlier and steeper drop-off than indicated by fluid factors in the Seattle Longitudinal Study. (Adapted from Kaufman, 2001.)

intelligence; and *performance,* which is made up of nonverbal reasoning tasks and largely taps fluid intelligence.

Controlling for education, Kaufman found that verbal IQ peaked in midlife, between ages 45 and 54, and then declined only slightly into the eighties. Confirming the optimism of the Seattle Longitudinal Study, 80-year-olds remained verbally adept, scoring as well as people in their twenties! Performance IQ, in contrast, showed an earlier and steeper decline than indicated by the Seattle fluid factors. As Figure 15.5 reveals, it reached its high point in the twenties and then dropped, precipitously after age 45. In a smaller longitudinal investigation, Kaufman reported similar patterns of change (Kaufman, 2001). Overall, his findings verify the results of early research—that crystallized intelligence is maintained into old age, whereas fluid intelligence is highly vulnerable to biological aging. Indeed, WAIS performance IQ dropped more than 20 points from the twenties to the eighties.

● **Explaining Changes in Mental Abilities.** Some theorists believe that a general slowing of central nervous system functioning underlies nearly all age-related declines in cognitive performance (Salthouse, 1993, 1996). Many studies offer at least partial support for this idea. For example, Kaufman (2001) reported that scores on WAIS speeded tasks mirror the sharp decline in performance IQ shown in Figure 15.5. Researchers have also discovered other important changes in information processing, some of which may be triggered by declines in speed.

But before we turn to this evidence, let's clarify why research reveals gains followed by stability in crystallized abilities, despite a much earlier decline in fluid intelligence, or basic information-processing skills. First, the decrease in basic processing, while substantial after age 45, may not be great enough to affect many well-practiced performances until quite late in life. Second, as we will see shortly, adults often find ways to compensate for cognitive weaknesses by drawing on their cognitive strengths. Finally, as people discover that they are no longer as good as they once were at certain tasks, they accommodate, shifting to activities that depend less on cognitive efficiency and more on accumulated knowledge. The basketball player becomes a coach, the quick-witted salesperson a manager (Salthouse, 1991b).

Individual and Group Differences

Hidden beneath the age trends just described are large individual differences. Some adults, because of illness or unfavorable environments, decline intellectually much earlier than others. And a considerable number show full functioning, even in fluid abilities, at an advanced age (Schaie, 1989).

Adults who use their intellectual skills seem to maintain them longer. In the Seattle Longitudinal Study, declines were delayed for people with above-average education, highly complex occupations, and stimulating leisure pursuits that included reading, traveling, attending cultural events, and participating in clubs and professional organizations. People with flexible personalities, lasting marriages (especially to a cognitively high-functioning partner), and absence of cardiovascular and other chronic diseases were also likely to maintain mental abilities well into late adulthood. And being economically well off was linked to favorable cognitive development, undoubtedly because SES is associated with many of the factors just mentioned (Schaie, 1996, 2000).

Several sex differences also emerged, consistent with those obtained in childhood and adolescence. In early and middle adulthood, women outperformed men on verbal tasks and perceptual speed. In contrast, men excelled at spatial skills (Maitland et al., 2000). Overall, however, changes in mental abilities over the adult years were remarkably similar for the two sexes—findings that defy the stereotype that older women are less competent than older men.

Furthermore, cohort effects were evident. For example, when the baby boom generation, currently middle-aged, was compared with the previous generation at the same age, the baby boom cohort performed substantially better on verbal memory, inductive reasoning, and spatial orientation (Willis & Schaie, 1999). These generational gains reflect advances in education, technology, environmental stimulation, and health care. They are expected to continue in the twenty-first century because today's children also attain higher mental test scores than children of past generations (Flynn, 1999). Interestingly, numeric ability declined in later-born cohorts in the Seattle Longitudinal Study. The recent generation's reliance on calculators and computers seems to have compromised its calculation skills!

Finally, adults who maintained higher levels of perceptual speed tended to be advantaged on other mental abilities. As we turn now to information processing in midlife, we will see why a decrement in speed of processing affects other aspects of cognitive functioning.

■ Compared with his 20-year-old son, this man in his fifties reacts less swiftly in playing a computer game. Yet older adults can make up for slowing of cognitive processing through practice, experience, and a compensatory approach that permits them to prepare responses in advance.

Information Processing

Information-processing researchers interested in adult development usually use the model of the mental system introduced in Chapter 5 (see page 154) to guide their exploration of different aspects of thinking. As processing speed slows, certain aspects of attention and memory decline. Yet midlife is also a time of great expansion in cognitive competence as adults apply their vast knowledge and life experience to problem solving in the everyday world.

Speed of Processing

One day, Devin looked over the shoulder of his 20-year-old son, Mark, who was playing a computer game. Mark responded to multiple cues on the screen in rapid-fire fashion. "Let me try it," suggested Devin, who practiced over several days but remained well behind Mark in performance. And on a family holiday in Australia, Mark adjusted quickly to driving on the left side of the road. After a week, Trisha and Devin still felt confused at intersections, where they had to respond rapidly to lights, cross traffic, and oncoming vehicles.

These real-life experiences fit with laboratory findings. On simple reaction-time tasks (pushing a button in response to a light) and complex reaction-time tasks (pushing a left-hand button to a blue light, a right-hand button to a yellow light), response time increases steadily from the early twenties into the nineties. Older adults are increasingly disadvantaged as situations requiring rapid responding become more complex. The decline in speed is small, under 1 second in most studies, but it is nevertheless of practical significance (Fozard et al., 1994; Salthouse, 2000).

What causes this age-related slowing of cognitive processing? Although researchers agree that changes in the brain are probably responsible, they disagree on the precise explanation. According to the **neural network view**, as neurons in the brain die, breaks in neural networks occur. The brain adapts by forming bypasses—new synaptic connections that go around the breaks but are less efficient (Cerella, 1990). A second hypothesis, the **information-loss view**, suggests that older adults experience greater loss of information as it moves through the cognitive system. As a result, the whole system must slow down to inspect and interpret the information. Imagine making a photocopy and using it to make another copy. Each time we do this, the copy becomes less clear. Similarly, with each step of thinking, information degrades. The older the adult, the more exaggerated this effect. Because complex tasks have more processing steps, they are more affected by information loss (Myerson et al., 1990).

Yet many older adults perform complex, familiar tasks with considerable efficiency. Devin, for example, played a Mozart quartet on his cello with great speed and dexterity, keeping up with three other players who were 10 years his junior. How did he manage? Compared with the other players, he more often looked ahead in the score. This compensatory approach permitted him to prepare a response in advance, thereby minimizing the importance of speed. In one study, researchers asked 19- to 72-year-olds to do transcription typing and also tested their reaction time. Although reaction time slowed with age, speed of typing did not change. Like Devin, older individuals looked further ahead in the transcription, anticipating their next keystrokes (Salthouse, 1984).

Practice and experience can also compensate for impairments in processing speed. Devin's many years of playing the cello and his familiarity with the Mozart quartet undoubtedly supported his ability to play swiftly and fluidly. Because older adults find ways to compensate for cognitive slowing on familiar tasks, their reaction time is considerably better on verbal items (indicating as quickly as possible whether a string of letters forms a word) than on nonverbal items (responding to a light or other signal) (Hultsch, MacDonald, & Dixon, 2002; Wingfield & Stine-Morrow, 2000).

Attention

Studies of attention focus on how much information adults can take into their mental systems at once; the extent to which they can attend selectively, ignoring irrelevant information; and the ease with which they can adapt their attention, switching from one task to another as the situation demands. When Dottie telephoned, Trisha sometimes tried to prepare dinner or continue working on a legal brief while talking on the phone. With age, she found it harder to engage in two activities simultaneously. Consistent with Trisha's experience, laboratory research reveals that sustaining two complex tasks at once becomes more challenging with age (Kausler, 1991; Madden & Plude, 1993). An age-related decrement also occurs in the ability to attend to relevant information and to switch back and forth between mental operations, such as judging one of a pair of numbers as "odd or even" on some trials and as "more or less" on others (Kray & Lindenberger, 2000; Salthouse et al., 1998).

These declines in attention might be due to the slowdown in information processing described earlier, which limits the amount of information a person can attend to at once (Plude & Hoyer, 1985). Reduced speed of processing may also contribute to a related finding: a decrement with age in the ability to combine many pieces of visual information into a meaningful pattern. When the mind inspects stimuli slowly, they are more likely to remain disconnected. This problem, in turn, can intensify attentional difficulties (Plude & Doussard-Roosevelt, 1989). Without a coherent pattern to serve as a guide, *cognitive inhibition*—resisting interference from irrelevant information—is harder (Hasher, Zacks, & May, 1999). Consequently, at times older adults appear distractible, inappropriately captured by a thought or a feature of the environment and unable to turn back to the task at hand.

But once again, adults can compensate for these changes. People highly experienced in attending to critical information and performing several tasks at once, such as air traffic controllers and pilots, know exactly what to look for. As a result, they show smaller age-related attentional declines (Morrow et al., 1994; Tsang & Shaner, 1998). Furthermore, practice can improve the ability to divide attention between two tasks, selectively focus on relevant information, and switch back and forth between mental operations. When older adults receive training in these skills, their performance improves as much as that of younger adults, although training does not close the gap between age groups (Kausler, 1994; Kramer, Hahn, & Gopher, 1998).

Memory

Memory is crucially important for all aspects of information processing—an important reason that we place great value on a good memory in middle and late adulthood. From the twenties into the sixties, the amount of information people can retain in working memory diminishes. Whether given lists of words or digits or meaningful prose passages to learn, middle-aged and older adults recall less than young adults, although memory for prose suffers less than memory for list items (Hultsch et al., 1998; Salthouse & Skovronek, 1992).

This change is largely due to a decline in use of memory strategies on these tasks. Older individuals rehearse less than younger individuals—a difference believed to be due to a slower rate of thinking. Older people cannot repeat new information to themselves as quickly as younger people (Salthouse & Babcock, 1991).

Memory strategies of organization and elaboration are also applied less often and less effectively with age. (See Chapter 9, pages 289–290, if you need to review these strategies.) Both require people to link incoming information with already stored information. One reason older adults are less likely to use organization and elaboration is that they find it harder to retrieve information from long-term memory that would help them recall. For example, given a list of words containing *parrot* and *blue jay,* they don't immediately access the category "bird," even though they know it well (Hultsch et al., 1998). Why does this happen? Greater difficulty keeping one's attention on relevant as opposed to irrelevant information seems to be involved. As irrelevant stimuli take up space in working memory, less is available for the memory task at hand (Salthouse & Meinz, 1995).

But we must keep in mind that the memory tasks given by researchers require strategies that many adults seldom use and may not be motivated to use, since most are not in school (see Chapter 9, page 290). When a word list has a strong category-based structure, older adults organize as well as younger adults do (Small et al., 1999). And when instructed to organize or elaborate, middle-aged and older people willingly do so, and their performance improves. Furthermore, tasks can be designed to help older people compensate for age-related declines in working memory. For example, we can slow the pace at which information is presented so that adults have enough time to process it, or we can cue the link between new and previously stored information ("To learn these words, try thinking of the category 'bird'") (Hay & Jacoby, 1999; Kausler, 1994).

When we consider the variety of memory skills we call on in daily life, the decrements just described are limited in scope. General *factual knowledge* (such as historical events), *procedural knowledge* (such as how to drive a car, ride a bike, or solve a math problem), and knowledge related to one's occupation either remain unchanged or increase into midlife (Baltes, Dittmann-Kohli, & Dixon, 1984). Middle-aged people who have trouble recalling something often draw on decades of accumulated *metacognitive knowledge* about how to maximize performance—reviewing major points before an important presentation, organizing notes and files so information can be found quickly, and parking the car in the same area of the parking lot each day. Metacognition shows no decline with age (Berg, 2000).

In sum, age-related changes in memory vary widely across tasks and individuals as people use their cognitive capacities to meet the requirements of their everyday worlds. This may remind you of Sternberg's *triarchic theory of intelligence,* described in Chapter 9—in particular, his *contextual subtheory* (see page 295). It emphasizes that intelligent people adapt their information-processing skills to fit with their personal desires and the demands of their environments. To understand memory development (and other aspects of cognition) in adulthood, we must view it in context. As we turn to problem solving, expertise, and creativity, we will encounter this theme again.

© LYRIC OPERA OF CHICAGO ARCHIVES

■ As its general director, Ardis Krainik (right) saved the Lyric Opera of Chicago from disaster, erasing the opera's deficit and restoring its sagging reputation. Her creative achievements depended on her ability to reconcile the company's subjective, artistic needs with its objective, financial needs. Drawing on her life experience, she dealt with a complex practical problem in a unique way. She is shown here with Patricia Ryan of Lyric Opera's Executive Committee and world-renowned tenor Plácido Domingo.

Practical Problem Solving and Expertise

One evening, as Devin and Trisha sat in the balcony of the Chicago Opera House awaiting curtain time, a figure appeared on the stage to announce that 67-year-old Ardis Krainik, general director and "life force" of the opera company, had died. A hush fell over the theater. Soon, members of the audience turned to one another, asking about the woman who had made the opera company into one of the world's greatest.

Starting as a chorus singer and clerk typist, Ardis rose rapidly through the ranks, becoming assistant to the director and developing a reputation for tireless work and unmatched organizational skill. When the opera company fell deeply into debt, the board of directors turned to Ardis to save it from disaster. As newly appointed general director, she erased the deficit within a year and restored the company's sagging reputation. She charmed donors into making large contributions, attracted world-class singers, and filled the house to near capacity. On her office wall hung a sign she had received as a gift. It read, "Wonder Woman" (Rhein, 1997).

As Ardis's story illustrates, middle-aged adults have special opportunities to display continued cognitive growth in the realm of **practical problem solving,** which requires people to size up real-world situations and analyze how best to achieve goals that have a high degree of uncertainty. Gains in *expertise*—an extensive, highly organized, and integrated knowledge base that can be used to support a high level of performance—help us understand why practical problem solving takes a leap forward.

The development of expertise is under way in early adulthood. But it reaches its height in midlife, leading to highly efficient and effective approaches to solving problems that are organized around abstract principles and intuitive judgments. Saturated with experience, the expert intuitively feels when an approach to a problem will work and when it will not. This rapid, implicit application of knowledge is the result of years of learning and experience. It cannot be assessed by laboratory tasks or mental tests that do not call on this knowledge (Ackerman, 2000; Wagner, 2000).

Expertise is not just the province of the highly educated and of those who rise to the top of administrative ladders. It can emerge in any field of endeavor. In a study of food service workers, researchers identified the diverse ingredients of expert performance in terms of physical skills (strength and dexterity); technical knowledge (of menu items, ordering, and food presentation); organizational skills (a sense of priority, anticipating customer needs); and social skills (confident presentation and a pleasant, polished manner). Next, 20- to 60-year-olds with fewer than 2 to more than 10 years of experience were evaluated on these qualities. Although physical strength and dexterity declined with age, job knowledge and practice increased. Compared to younger adults with similar years of experience, middle-aged employees performed more competently, serving customers in especially adept, attentive ways (Perlmutter, Kaplan, & Nyquist, 1990).

Age-related advantages are also evident in solutions to everyday problems (Denney, 1990; Denney & Pearce, 1989). Consider the following dilemma:

> What would you do if you had a landlord who refused to make some expensive repairs you want done because he or she thinks they are too costly?
>
> (a) Try to make the repairs yourself.
>
> (b) Try to understand your landlord's view and decide whether they are necessary repairs.
>
> (c) Try to get someone to settle the dispute between you and your landlord.
>
> (d) Accept the situation and don't dwell on it.
> (Cornelius & Caspi, 1987, p. 146)

On tasks like these, middle-aged and older adults select better strategies (as rated by independent judges) than young adults. In the preceding example, the preferred choice is (b), a problem-centered approach that involves seeking information and using it to guide action (Cornelius & Caspi, 1987). As response (b) illustrates, from middle age on, adults place greater emphasis on thinking through a practical problem—trying to understand it better, to interpret it from different perspectives, and to solve it through logical analysis. Perhaps for this reason, middle-aged and older adults are more likely to use a broader range of everyday problem-solving strategies and to combine them to suit the situation (Blanchard-Fields, Chen, & Norris, 1997).

■ In middle adulthood, creativity becomes more thoughtful, integrative, and altruistic. Oren Lyons, spiritual leader of the Turtle Clan, Onondaga Nation native people of New York, puts finishing touches on his well-known painting, "Tree of Peace," which captures these qualities of midlife creativity.

Creativity

As noted in Chapter 13, creative accomplishment peaks in the late thirties or early forties and then declines, with some variation across individuals and disciplines. But like problem solving, the *quality* of creativity may change with advancing age—in at least three ways.

First, youthful creativity in literature and the arts is often spontaneous and intensely emotional. In contrast, creative works produced after age 40 appear more deliberately thoughtful. Perhaps for this reason, poets produce their most frequently cited works at younger ages than do authors of fiction and nonfiction. Poetry depends more on language play and "hot" expression of feelings, whereas story- and book-length works require extensive planning and molding (Cohen-Shalev, 1986).

Second, with age, creators typically shift from generating unusual products to combining extensive knowledge and experience into unique ways of thinking (Abra, 1989; Sasser-Coen, 1993). Creative works by older adults more often sum up or integrate ideas. Consequently, mature academics typically devote less energy to new discoveries in favor of writing memoirs, histories of their field, and other reflective works. Also, in older creators' novels, scholarly writings, paintings, and musical compositions, living with old age and facing death are common themes (Beckerman, 1990; Sternberg & Lubart, 2001).

Finally, creativity in middle adulthood reflects a transition from a largely egocentric concern with self-expression to more altruistic goals. As the middle-aged person overcomes the youthful illusion that "life can last forever," the desire to contribute to humanity and enrich the lives of others increases (Dacey, 1989).

Taken together, these changes may contribute to the midlife decline in overall creative output. In reality, however, creativity emerges in new forms.

Information Processing in Context

In sum, cognitive gains in middle adulthood are especially likely in areas involving experience-based buildup and transformation of knowledge and skills. Consider the evidence just reviewed, and you will see that processing speed varies with the situation. When given challenging real-world problems related to their expertise, middle-aged adults are likely to win out in both efficiency and excellence of thinking. Furthermore, when researchers' laboratory tasks and test items are relevant to the real-life endeavors of intelligent, cognitively active adults, older people respond as quickly and competently as their younger counterparts do!

In a study of 30- to 70-year-old professors at the University of California, Berkeley, performance on reaction-time tasks and tasks requiring recall of unrelated information revealed typical age-related declines. But on complex tasks involving planning, organization, and recall of meaningful information, the professors showed no decrements with age (Shimamura et al., 1995). Years of applying these cognitive strategies in their work lives, which required constant integration of new information with a large existing knowledge base, seemed to counteract the effects of aging

By midlife, people's past and current experiences vary enormously—more so than in previous decades. Therefore, thinking in middle adulthood is characterized by an increase in specialization; people branch out in a multitude of directions. Yet to reach their cognitive potential, middle-aged adults must have opportunities for continued growth. Let's see how vocational and educational environments can support cognition in middle adulthood.

Ask Yourself

REVIEW

Cite research findings on changes in crystallized and fluid intelligence during adulthood. What accounts for these differing trends?

REVIEW

How does age-related slowing of cognitive processing affect attention and memory in midlife? What can older adults do to compensate for these declines?

APPLY

Asked about hiring older waiters, one restaurant manager replied, "They are my best employees. . . . I do not know what you mean by slowness. They are the fastest ones on the floor" (Perlmutter, Kaplan, & Nyquist, 1990, p. 189). Why does this manager find older employees desirable, despite the age-related decline in speed of processing?

CONNECT

In what aspects of cognition did Devin decline, and in what aspects did he gain? How do changes in Devin's thinking reflect assumptions of the lifespan perspective?

www.

Vocational Life and Cognitive Development

Vocational settings are vital contexts for maintaining old skills and learning new ones. Yet they vary in the extent to which they are cognitively stimulating and promote autonomy. At times, work environments carry negative stereotypes of age-related problem-solving and decision-making skills that may result in older employees being assigned less challenging work.

Recall from Chapter 13 that cognitive and personality characteristics affect occupational choice. Devin, for example, chose college teaching and administration because he enjoyed reading and writing, framing new ideas, and helping others learn. Once a person is immersed in a job, it influences cognition. In a study of over 600 American men representing a wide range of occupations, researchers asked participants about the task complexity and self-direction of their jobs. During this interview, they also assessed cognitive flexibility, based on logical reasoning, awareness of both sides of an issue, and independence of judgment. A decade later, the job and cognitive variables were remeasured, permitting a look at their effects on each other. As expected, cognitively flexible men sought work that offered challenge and autonomy. But complex work also led to gains in cognitive flexibility! In other words, the relationship between vocational life and cognition was reciprocal (Kohn & Schooler, 1978).

These same findings emerged in large-scale studies carried out in Japan and Poland—cultures quite different from the United States (Kohn et al., 1990; Kohn & Slomczynski, 1990). In each nation, having a stimulating, nonroutine job helped explain the relationship between SES and flexible, abstract thinking. Furthermore, learning on the job generalizes to other realms of life. People who do intellectually demanding work seek out stimulating leisure pursuits, which also foster cognitive flexibility (Kohn et al., 2000). Flexible thinkers come to value self-direction, both for themselves and for their children. Consequently, they are likely to pass their cognitive preferences to the next generation.

Is the impact of a challenging job on cognitive growth greatest for young adults, who are in the early phase of career development? Research shows that it is not. People in their fifties and early sixties gain as much as those in their twenties and thirties. The relationship also holds for people of different generations and, therefore, widely varying life experiences (Avolio & Sosik, 1999; Miller, Slomczynski, & Kohn, 1985). Once again, we are reminded of the plasticity of development. Cognitive flexibility is responsive to vocational experience well into middle adulthood, and perhaps beyond. This means that more jobs should be designed to promote intellectual stimulation and challenge—factors linked to higher cognitive functioning later in the lifespan.

Adult Learners: Becoming a Student in Midlife

Adults are returning to undergraduate and graduate study in record numbers. During the past quarter-century, students over age 25 in North American colleges and universities increased from 28 to 40 percent, with an especially sharp rise in those over age 35 (Statistics Canada, 2001b; U.S. Department of Education, 2002b). Their reasons for enrolling are diverse—a career change, a better income, self-enrichment, a sense of personal achievement, or simply a degree. Life transitions often trigger a return to formal education, as was the case for Devin and Trisha's friend Anya, who entered a nursing program after her last child left home. A divorce, widowhood, a job layoff, or a youngest child going to school for the first time are other transitions that commonly precede reentry (Bradburn, Moen, & Dempster-McClain, 1995).

Characteristics of Returning Students

Women are the majority—about 58 percent—of adult learners. As Anya's fear of not being able to handle class work suggests (see page 482), first-year reentry women report feeling more self-conscious, inadequate, and hesitant to talk in class than either returning men or traditional-age students (under age 25) (Wilke & Thompson, 1993). Their anxiety is partly due to not having practiced academic learning for many years. It is also prompted by stereotypes of aging, since returning women are often convinced that traditional-age students are smarter.

Role demands outside the educational setting—from spouses, children, other family members, friends, and employers—pull many returning women in several, often conflicting, directions. Those reporting high psychological stress typically have career rather than enrichment goals, young children, limited financial resources, and nonsupportive husbands (Novak & Thacker, 1991; Padula & Miller, 1999). As a classmate told Anya one day, "I tried keeping the book open and reading, cooking, and talking to the kids. It didn't work. They felt I was ignoring them." Because of multiple demands on their time, mature-age women tend to take fewer credits, experience more interruptions in their academic programs, and progress at a slower pace than mature-age men. Family responsibilities are their most frequent reason for not finishing their education (Robertson, 1991).

Supporting Returning Students

As these findings suggest, social supports for returning students can make the difference between continuing in school and dropping out. Adult students need family members and friends who encourage their efforts and help them find time for uninterrupted study. Anya's classmate explained, "My doubts subsided when my husband volunteered, "I can cook dinner and do the laundry. You take your books and do what you need to do.'" Institutional services for returning students are also essential. Personal relationships with faculty, peer networks enabling adults to get to know one another, conveniently scheduled evening and Saturday classes, online courses, and financial aid for part-time students (many of whom are returning adults) increase the chances of academic success.

Although nontraditional students rarely require assistance in settling on career goals, they report a strong desire for help in choosing the most appropriate courses and in exploring jobs related to their talents (Luzzo, 1999). Academic advising and professional internship opportunities responsive to their needs are vital for their career development.

Low-income and ethnic minority students often need special assistance, such as academic tutoring, sessions in con-

© BILL ARON/PHOTOEDIT

■ This mature woman's return to college may have been sparked by a life transition, such as divorce, widowhood, or job layoff. At first, adult learners often question their ability to be successful at academic work. When family members, friends, and educational institutions are supportive, returning students reap great personal benefits and do well academically.

Facilitating Adult Reentry to College

SOURCE OF SUPPORT	DESCRIPTION
Partner and children	Value and encourage educational efforts. Assist with household tasks to permit time for uninterrupted study.
Extended family and friends	Value and encourage educational efforts.
Educational institution	Provide orientation programs and literature that inform adult students about services and social supports. Provide counseling and intervention addressing academic weaknesses, self-doubts about success, and matching courses to career goals. Facilitate peer networks through regular meetings or phone contact. Promote personal relationships with faculty. Encourage active engagement and discussion in classes and integration of course content with real-life experiences. Offer evening, Saturday, and off-campus classes and online courses. Provide financial aid for part-time students. Initiate campaigns to recruit returning students, including those from low-income families and ethnic minority groups. Assist students with young children in finding child care arrangements, and provide on-campus child care.
Workplace	Value and encourage educational efforts. Accommodate work time to class schedules.

fidence building and assertiveness, and help adjusting to styles of learning that are at odds with their cultural background. One Chinese returning student noted that she found criticizing ideas and arguing with her professors difficult because Chinese students are taught to respect, not to disagree with, their teachers.

The Caregiving Concerns table above suggests ways to facilitate adult reentry to college. When support systems are in place, most returning students reap great personal benefits and do well academically. They especially value forming new relationships, sharing opinions and experiences, and relating subject matter to their own lives. And their greater ability to integrate knowledge leads to an enhanced appreciation of classroom experiences and assignments. Yet another benefit of adult students in college classes is intergenerational contact. As younger students observe the capacities and talents of older individuals, unfavorable stereotypes of aging decline.

In previous chapters, we have seen that education transforms development. It continues to do so in middle adulthood. After finishing her degree, Anya secured a position as a parish nurse with creative opportunities to counsel members of a large congregation about health concerns. Education granted her new life options, financial rewards, and higher self-esteem as she reevaluated her own competencies (Redding & Dowling, 1992). Sometimes, the revised values and self-reliance forged by education spark change in other spheres of life, such as a divorce or a new intimate partnership (Esterberg, Moen, & Dempster-McClain, 1994). In this way too, returning to school can powerfully reshape the life course.

Ask Yourself

REVIEW

Considering the influence of vocational and educational experiences on midlife cognitive development, evaluate the saying "You can't teach an old dog new tricks."

APPLY

Marcella completed one year of college in her twenties. Now, at age 42, she has returned to earn a degree. Plan a set of experiences for Marcella's first semester that will increase her chances of success.

CONNECT

Most high-level government and corporate positions are held by middle-aged and older adults rather than by young adults. What cognitive capacities enable mature adults to perform these jobs well?

www.

Summary

PHYSICAL DEVELOPMENT

Physical Changes

Describe the physical changes of middle adulthood, paying special attention to vision, hearing, the skin, muscle–fat makeup, and the skeleton.

- The gradual physical changes begun in early adulthood continue in midlife, contributing to a revised physical self-image, which often emphasizes fewer hoped-for gains and more feared declines.
- Vision is affected by **presbyopia,** or loss of the accommodative ability of the lens, reduced ability to see in dim light, and diminished color discrimination. After age 40, adults are at increased risk of **glaucoma.** Hearing loss, called **presbycusis,** first affects detection of high frequencies and then spreads to other tones. Presbycusis begins earlier and is more rapid for men, due to intense noise in some male-dominated occupations.
- The skin wrinkles, loosens, and starts to develop age spots, especially in women and in people exposed to the sun. Muscle mass declines and fat deposits increase, with men and women developing different patterns of fat distribution. A low-fat diet and continued exercise can offset excess weight and muscle loss.
- Bone density declines in both sexes, but to a greater extent in women, especially after menopause. Loss in height and bone fractures can result.

Describe reproductive changes in women during middle adulthood, and discuss women's psychological reactions to menopause.

- The **climacteric** in women occurs gradually over a 10-year period as estrogen production drops and concludes with **menopause.** Many doctors recommend **hormone therapy** to reduce the discomforts of menopause and to protect women from other impairments due to estrogen loss. However, hormone therapy remains controversial because of an increased risk of certain cancers and cardiovascular disease.
- Menopause is a biocultural event—affected by hormonal changes as well as societal beliefs and practices. Physical symptoms and psychological reactions vary widely. Whether women find menopause traumatic or liberating depends on how they interpret it in relation to their past and future lives.

Describe reproductive changes in men during middle adulthood.

- Although men also experience a climacteric, their reproductive capacity merely declines rather than ends. Therefore, no male counterpart to menopause exists. Occasional episodes of impotence are more common in midlife but usually can be treated successfully.

Health and Fitness

Discuss sexuality in middle adulthood and its association with psychological well-being.

- Frequency of sexual activity among married couples remains fairly stable in midlife, dropping only slightly, and is associated with marital happiness. Intensity of sexual response declines more, due to physical changes of the climacteric. Many more women than men are without sexual partners, a trend that continues into late adulthood.

Discuss cancer, cardiovascular disease, and osteoporosis, noting risk factors and interventions.

- The death rate from cancer increases 10-fold from early to middle adulthood. Cancer is the leading killer of middle-aged women. A complex interaction of heredity, biological aging, and environment contributes to cancer. Today 40 percent of people with cancer are cured. Annual screenings and various preventive steps (such as not smoking) can reduce the incidence of cancer and cancer deaths.
- Although cardiovascular disease has declined in recent decades, it remains a major cause of death in middle adulthood, especially among men. Symptoms include high blood pressure, high blood cholesterol, atherosclerosis, heart attack, arrhythmia, and angina pectoris. Diet, exercise, drug therapy, and stress reduction can reduce risks and aid in treatment. A special concern is accurate diagnosis in women.
- **Osteoporosis** affects 7 percent of postmenopausal women, and another 40 percent have bone density levels low enough to be of concern. Weight-bearing exercise, calcium and vitamin D, and hormone therapy or other bone-strengthening medications can help prevent and treat osteoporosis.

Discuss the association of hostility and anger with heart disease and other health problems.

- Hostility is the component of the **Type A behavior pattern** that predicts heart disease and other health problems, largely due to physiological arousal associated with anger. Because inhibiting the expression of emotion is also related to health problems, a better alternative is to develop effective ways of handling stress and conflict.

Adapting to the Physical Challenges of Midlife

Discuss the benefits of stress management, exercise, and an optimistic outlook in dealing effectively with the physical challenges of midlife.

- The changes and responsibilities of middle adulthood can cause psychological stress, with negative consequences for the cardiovascular, immune, and gastrointestinal systems. Effective stress management includes both problem-centered and emotion-centered coping, depending on the situation; constructive approaches to anger reduction; and social support. In middle adulthood, people tend to cope with stress more effectively, and when they do, they often report lasting personal benefits.
- Regular exercise confers many physical and psychological advantages, making it worthwhile for sedentary middle-aged people to begin exercising. Developing a sense of self-efficacy and choosing an exercise format that fits the individual's characteristics (home-based for overweight adults, group-based for highly stressed adults) increase the chances that a beginner will stick with an exercise regimen.
- **Hardiness** is made up of three personal qualities: control, commitment, and challenge. By inducing a generally optimistic outlook on life, hardiness helps people cope with stress adaptively.

Explain the double standard of aging.

- Negative stereotypes of aging discourage older adults of both sexes. Yet middle-aged women are more likely to be viewed unfavorably, especially by men. Although this double standard of aging may have been adaptive at an earlier time in our evolutionary history, it is irrelevant in an era of limited childbearing and greater involvement in career and leisure pursuits. New surveys suggest that the double standard of aging is declining.

COGNITIVE DEVELOPMENT

Changes in Mental Abilities

Describe changes in crystallized and fluid intelligence in middle adulthood, and discuss individual and group differences in intellectual development.

- Cross-sectional and longitudinal research reveals gains from early to middle adulthood in skills that tap **crystallized intelligence** (which depends on accumulated knowledge and experience). In contrast, **fluid intelligence** (which depends more on basic information-processing skills) starts to decline in the twenties and continues to do so throughout adulthood. In the Seattle Longitudinal Study, perceptual speed follows this pattern, but trends for other fluid factors suggest a more optimistic picture. Research with the Wechsler Adult Intelligence Scale (WAIS), however, confirms the early drop-off in fluid intelligence, which intensifies in midlife.
- Large individual differences among middle-aged adults remind us that intellectual development is multidimensional, multidirectional, and plastic. Some people, become of illness or unfavorable environments, show early intellectual declines. Those who use their intellectual skills are more likely to retain them. Stimulating occupations and leisure pursuits, good health, lasting marriages, flexible personalities, and economic advantage are linked to favorable cognitive development.
- In early and middle adulthood, women outperform men on verbal tasks and perceptual speed, whereas men excel at spatial skills. Generational gains in certain intellectual skills reflect advances in education, technology, environmental stimulation, and health care.

Information Processing

How does information processing change in midlife?

- Speed of cognitive processing slows with age, a change explained by either the **neural network view** or the **information-loss view.** Slower processing speed makes it harder for middle-aged people to divide their attention, focus on relevant stimuli, and switch from one task to another as the situation demands. Cognitive inhibition becomes more difficult, at times prompting distractibility.
- Adults in midlife retain less information in working memory, largely due to a decline in use of memory strategies. Training, practice, improved design of tasks, and metacognitive knowledge enable middle-aged and older adults to compensate for decrements in processing speed, attention, and memory.

Discuss the development of practical problem solving, expertise, and creativity in middle adulthood.

- Middle-aged adults in all walks of life often become good at **practical problem solving,** largely due to development of expertise. In midlife, creativity becomes more deliberately thoughtful. It also shifts from generating unusual products to integrating ideas, and from concern with self-expression to more altruistic goals.

Vocational Life and Cognitive Development

Describe the relationship between vocational life and cognitive development.

- At all ages and in different cultures, the relationship between vocational life and cognitive development is reciprocal. Stimulating, complex work and flexible, abstract, autonomous thinking support one another.

Adult Learners: Becoming a College Student in Midlife

Discuss the challenges that adults face in returning to college, ways to support returning students, and benefits of earning a degree in midlife.

- Often motivated to return to college by life transitions, women make up the majority of the growing number of adult students. Returning students must cope with a lack of recent practice at academic work, stereotypes of aging, and multiple role demands. Low-income and ethnic minority students need special assistance.
- Social support from family and friends and institutional services suited to their needs can help returning students succeed. Further education results in enhanced competencies, new relationships, intergenerational contact, and reshaped life paths.

Important Terms and Concepts

climacteric (p. 486)
crystallized intelligence (p. 498)
fluid intelligence (p. 499)
glaucoma (p. 484)
hardiness (p. 496)
hormone therapy (p. 486)
information-loss view (p. 501)
menopause (p. 486)
neural network view (p. 501)
osteoporosis (p. 493)
practical problem solving (p. 503)
presbycusis (p. 484)
presbyopia (p. 483)
Type A behavior pattern (p. 493)

FYI For Further Information and Help

Consult the Companion Website for *Development Through the Lifespan, Third Edition,* (www.ablongman.com/berk) where you will find the following resources for this chapter:

- **Chapter Objectives**
- **Flashcards** for studying important terms and concepts
- **Annotated Weblinks** to guide you in further research
- **Ask Yourself** questions, which you can answer and then check against a sample response
- **Suggested Readings**
- **Practice Test** with immediate scoring and feedback

Emotional and Social Development in Middle Adulthood

© NORBERT SCHAEFER/CORBIS

Erikson's Theory: Generativity versus Stagnation

Other Theories of Psychosocial Development in Midlife
Levinson's Seasons of Life • Vaillant's Adaptation to Life • Is There a Midlife Crisis? • Stage or Life Events Approach

- A Lifespan Vista: Generative Adults Tell Their Life Stories

Stability and Change in Self-Concept and Personality
Possible Selves • Self-Acceptance, Autonomy, and Environmental Mastery • Coping Strategies • Gender Identity • Individual Differences in Personality Traits

- Biology & Environment: What Factors Promote Psychological Well-Being in Midlife?

Relationships at Midlife
Marriage and Divorce • Changing Parent–Child Relationships • Grandparenthood • Middle-Aged Children and Their Aging Parents • Siblings • Friendships • Relationships Across Generations

- Social Issues: Grandparents Rearing Grandchildren: The Skipped-Generation Family

Vocational Life
Job Satisfaction • Career Development • Career Change at Midlife • Unemployment • Planning for Retirement

The sense of well-being that stems from generativity—giving to and guiding the next generation—is keenly apparent in this grandfather's involved, contented expression as he encourages a future chess player. For those who experience it, grandparenthood is a major milestone of middle adulthood.

On a weekend in which Devin, Trisha, and their 24-year-old son, Mark, were vacationing, the two middle-aged parents knocked on Mark's hotel room door. "Your dad and I are going off to see a crafts exhibit," Trisha explained. "No problem with your staying behind," she offered, recalling Mark's antipathy toward attending such events as an adolescent. "We'll be back around 12 for lunch."

"That exhibit sounds great!" Mark replied. "I'll meet you in the lobby. We've got so little time together as it is."

"Sometimes I forget he's an adult!" exclaimed Trisha as she and Devin returned to their room to grab their coats. "It's been great to have Mark with us these few days, like spending time with a good friend."

During their forties and fifties, Trisha and Devin built on earlier strengths and intensified their commitment to leaving a legacy for those who would come after them. As Mark graduated from college, took his first job, fell in love, and married, they felt a sense of pride at having escorted a member of the next generation into responsible adult roles. Family activities, which had declined during Mark's adolescent and college years, now increased as Trisha and Devin related to their son and daughter-in-law not just as kin but as enjoyable adult companions. Challenging work and more time for community involvement, leisure pursuits, and each other contributed to a richly diverse and gratifying time of life.

The years of midlife were not as smooth for two of Trisha and Devin's friends. Fearing that she might grow old alone, Jewel frantically pursued her quest for an intimate partner. She attended singles events, registered with dating services, and traveled in hopes of meeting a like-minded companion. "I can't stand the thought of turning 50. I look like an old bag with big circles under my eyes," she lamented in a letter to Trisha. In other ways, Jewel's life had compensating satisfactions—friendships that had grown more meaningful, a warm relationship with a nephew and niece, and a successful consulting business.

Tim, Devin's best friend from graduate school, had been divorced for over 15 years. Recently, he had met Elena, for whom he cared deeply. But Elena was in the midst of major life changes. Besides her own divorce, she was dealing with a troubled daughter, a change in careers, and a move away from the city that served as a constant reminder of her unhappy past. Whereas Tim had reached the peak of his career and was ready to enjoy life, Elena wanted to recapture much of what she had missed in earlier decades—not just a satisfying intimate relationship, but opportunities to realize her talents. "I don't know where I fit into Elena's plans," Tim wondered aloud to Trisha in a phone conversation.

With the arrival of middle adulthood, half or more of the lifespan is over. Increasing awareness of limited time ahead prompts adults to reevaluate the meaning of their lives and reach out to future generations. As we will see, most middle-aged people make modest adjustments in their outlook and daily lives. A few experience profound inner turbulence and initiate major changes, often in an effort to make up for lost time. Besides advancing years, family and work transitions contribute greatly to emotional and social development.

Erikson's Theory: Generativity versus Stagnation

Erikson's psychological conflict of midlife is called **generativity versus stagnation.** Generativity involves reaching out to others in ways that give to and guide the next generation. Recall from Chapter 14 that generativity is under way in early adulthood, typically through childbearing and child rearing and establishing a niche in the occupational world. It expands greatly in midlife. At this stage, commitment extends beyond oneself (identity) and one's life partner (intimacy) to a larger group—family, community, or society. The generative adult combines the need for self-expression with the need for communion, integrating personal goals with the welfare of the larger social world (Wrightsman, 1994). The resulting strength is the capacity to care for others in a broader way than in previous stages.

Erikson (1950) selected the term *generativity* to encompass everything generated that can outlive the self and ensure society's continuity and improvement: children, ideas, products, and works of art. Although parenting is a major means of realizing generativity, some people, because of misfortune or special gifts, do not express it through their own children. Adults can be generative in other family relationships (as Jewel was with her nephew and niece), as mentors in the workplace, in volunteer endeavors, and through many forms of productivity and creativity.

■ Generativity expands greatly in midlife. It involves reaching out to others in ways that give to and guide the next generation. Here an election supervisor assists a young volunteer at a center established to field calls from poll workers trying to resolve problems ranging from malfunctioning voting machines to voters showing up at the wrong precincts.

Look closely at what we have said so far, and you will see that generativity brings together personal desires and cultural demands. On the personal side, middle-aged adults feel a need to be needed; they want to attain symbolic immortality—that is, make a contribution that will survive their death (Kotre, 1984, 1999; McAdams, Hart, & Maruna, 1998). On the cultural side, society imposes a social clock for generativity in midlife, requiring adults to take responsibility for the next generation through their roles as parents, teachers, mentors, leaders, and coordinators (McAdams, 2001). And according to Erikson, a culture's "belief in the species"—the conviction that life is good and worthwhile, even in the face of human destructiveness and deprivation—is a major motivator of generative action. Without this optimistic world view, people would not have any hope of improving humanity.

The negative outcome of this stage is *stagnation*. Erikson recognized that once people attain certain life goals, such as marriage, children, and career success, they may become self-centered and self-indulgent. Adults with a sense of stagnation cannot contribute to society's welfare because they place their own comfort and security above challenge and sacrifice (Hamachek, 1990). Their self-absorption is expressed in many ways—through lack of involvement with and concern for young people (including their own children), through a focus on what they can get from others rather than what they can give, and through taking little interest in being productive at work, developing their talents, or bettering the world in other ways.

Researchers study generativity in diverse ways. Some assess personality traits, such as assertiveness, nurturance, and responsibility. Others ask people to rate themselves on generative characteristics and report on generative activities. Still others look for generative themes in people's narrative descriptions of themselves (Keyes & Ryff, 1998a; McAdams et al., 1997; Ryff & Migdal, 1984). Regardless of the method used, generativity consistently increases in midlife. For example, in longitudinal and cross-sectional studies of college-educated women, self-rated generativity rose steadily from the thirties to the fifties (see Figure 16.1). At the same time, participants expressed greater awareness of aging, increased security with their identities, and a stronger sense of competence (Stewart, Ostrove, & Helson, 2001; Zucker, Ostrove, & Stewart, 2002). And as the Lifespan Vista box on page 514 illustrates, generativity is a major, unifying theme in middle-aged adults' life stories.

Just as Erikson's theory suggests, highly generative people appear especially well adjusted—low in anxiety and depression and high in self-acceptance and life satisfaction (Ackerman, Zuroff, & Moskowitz, 2000; Grossbaum & Bates, 2002; Stewart & Vandewater, 1998). They are also more open to differing viewpoints, possess leadership qualities, desire more from work than financial rewards, and care greatly about the welfare of their children, their partner, their aging parents, and the wider society (Peterson, 2002; Peterson, Smirles, & Wentworth, 1997). Furthermore, generativity is associated with greater involvement in political activities, including voting, campaigning, and contacting public officials, especially among adults for whom political participation was central to identity in earlier years (Cole & Stewart, 1996).

Although the findings just described characterize adults of all backgrounds, group differences in contexts for generativity exist. Having children seems to foster men's generative development more than women's. In two studies, fathers scored higher in generativity than childless men (McAdams & de St. Aubin, 1992; Snarey et al., 1987). In contrast, motherhood is unrelated to women's generativity scores. Perhaps parenting awakens in men a tender, caring attitude toward the next generation that women have opportunities to develop in other ways. Finally, compared with Caucasians, African Americans more often engage in certain types of generativity. They are more involved in religious groups and activities, offer more social support, and are more likely to view themselves as role models and sources of wisdom for their children (Hart et al., 2001). A life history of strong support from church and extended family may strengthen these generative values and actions.

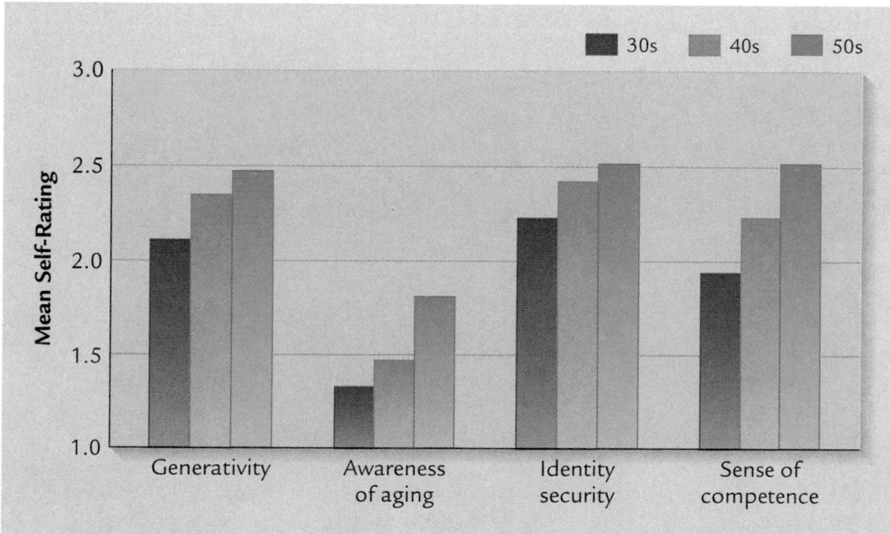

FIGURE 16.1 Age-related changes in self-rated generativity, awareness of aging, identity security, and sense of competence. In a longitudinal study of college-educated women, self-rated generativity increased from the thirties to the fifties, as did awareness of aging. The rise in generativity was accompanied by other indicators of psychological health—greater security with one's identity and sense of competence. (Adapted from Stewart, Ostrove, & Helson, 2001.)

A Lifespan Vista

Generative Adults Tell Their Life Stories

In research aimed at understanding how highly generative adults make sense of their lives, Dan McAdams and his colleagues (1997, 2001) interviewed two groups of midlifers: those who often behave generatively and those who seldom do. Participants were asked to relate their life stories, including a high point, a low point, a turning point, and important scenes from childhood, adolescence, and adulthood. Analyses of story lines and themes revealed that adults high and low in generativity reconstruct their past and anticipate their future in strikingly different ways.

Narratives of highly generative people usually contained an orderly sequence of events that the researchers called a *commitment story,* in which adults give to others as a means of giving back to family, community, and society (McAdams, 1993). The generative storyteller typically describes an early special advantage (such as a good family or a special talent), along with early awareness of the suffering of others. This clash between blessing and suffering motivates the person to view the self as "called," or committed, to being good to others. In commitment stories, the *theme of redemption* is prominent. Highly generative adults frequently describe scenes in which extremely negative life events, involving frustration, failure, loss, or death, are followed by good outcomes—personal renewal, improvement, and enlightenment. That is, bad scenes are redeemed, or made better, by what follows.

Consider a story episode related by Diana, a 49-year-old fourth-grade teacher. Born in a small town to a minister and his wife, Diana was a favorite among the parishioners, who showered her with attention and love. At age 8, however, her life descended to its low point. She watched in horror as her younger brother ran into the street and was hit by a car; he died later that day. After her brother's death, Diana sensed her father's anguish and tried to be the "son" he lost, although her efforts failed. The scene ends on an upbeat note, with Diana marrying a man who forged a warm bond with her father and who became accepted "as his own son." One of Diana's life goals was to improve her teaching, because "I'd like to give something back . . . to grow and help others grow" (McAdams et al., 1997, p. 689). Her interview overflowed with expressions of generative commitment.

Whereas highly generative adults tell stories in which bad scenes turn good, less generative adults relate stories with *themes of contamination,* in which good scenes turn bad. For example, a child gets a treasured musical instrument, which is immediately stolen. A good first year of college turns sour when a professor grades unfairly. And a young woman loses weight, looks good, but can't overcome her low self-esteem.

Why are generativity and life-story redemption events, in which bad gives way to good, connected? First, some adults may view their generative activities as a way to redeem negative aspects of their lives. In a study of the life stories of ex-convicts who turned away from crime, many spoke of a strong desire to do good works as penance for their transgressions (Maruna, 1997). Second, generativity seems to entail the conviction that the imperfections of today can be transformed into a better tomorrow. Through guiding and giving to the next generation, mature adults increase the chances that the mistakes of the past will not happen again. Finally, interpreting one's own life in terms of redemption offers hope that hard work will lead to future benefits—an expectation that may sustain generative efforts of all kinds, from rearing children to advancing communities and societies (Kotre, 1999).

Life stories offer insight into how people imbue their lives with meaning and purpose. Adults high and low in generativity do not differ in the number of positive and negative events included in their narratives. Instead, they interpret those events differently. Commitment stories, filled with redemption, involve a way of thinking about the self that fosters a caring, compassionate approach to others. The more redemptive events adults include in their life stories, the higher their life satisfaction, self-esteem, and certainty that the challenges of life are meaningful, manageable, and rewarding (McAdams, 2001). Researchers still have much to learn about factors that lead people to view good as emerging from adversity.

■ Highly generative adults often tell commitment stories with themes of redemption, in which negative events in their lives are redeemed, or made better, by good outcomes. Consequently they feel "called," or committed, to helping others. Perhaps redemptive experiences contributed to the enthusiasm and patience with which this Plains Indian man of South Dakota teaches younger members of his culture the arts of ceremonial dress and dance.

Other Theories of Psychosocial Development in Midlife

Recall that Erikson's theory provides only a broad sketch of adult personality development. For a closer look at psychosocial change in midlife, let's revisit Levinson's and Vaillant's theories, which were introduced in Chapter 14.

Levinson's Seasons of Life

Return to page 447 to review Levinson's eras (stages or seasons), and notice that like early adulthood, middle adulthood begins with a transitional period (age 40 to 45), followed by the building of an entry life structure (age 45 to 50). Then this structure is evaluated and revised (age 50 to 55), resulting in a culminating life structure (age 55 to 60). Among the adults Levinson (1978, 1996) interviewed, the majority displayed these phases. But because of gender stereotypes and differences in opportunity, men's and women's experiences were somewhat different.

● Midlife Transition. Around age 40, people evaluate their success in meeting early adulthood goals. Realizing that from now on, more time will lie behind than ahead, they regard the remaining years as increasingly precious. Consequently, some people make drastic revisions in family and occupational components of the life structure: divorcing, remarrying, changing careers, or displaying enhanced creativity. Others make smaller changes while staying in the same marriage, surroundings, occupation, and workplace.

Whether these years bring a gust of wind or a storm, most people turn inward for a time, focusing on personally meaningful living (Neugarten, 1968b). Part of the reason is that for many middle-aged adults, only limited career advancement and personal growth remain possible. Others are disappointed in not having fully realized their early adulthood dream and want to find a more satisfying path before it is too late. Even people who have reached their goals ask, What good are these accomplishments to others, to society, and to myself?

According to Levinson, for middle-aged adults to reassess their relation to themselves and the external world, they must confront four developmental tasks, summarized in Table 16.1. Each requires the person to reconcile two opposing tendencies within the self, thereby attaining greater internal harmony. Let's see how this happens.

● Modifying the Life Structure: Gender Similarities and Differences. At midlife, adults must give up certain youthful qualities, find age-appropriate ways to express other qualities, and accept being older, thereby creating a youth–age balance more in tune with their time of life. Physical changes, personal encounters with illness, and aging parents intensify this task, and it often triggers reassessment of what is important. Because of the double standard of aging (see Chapter 15), women find it harder than men to accept being older. For Jewel, the stereotypical image of an older woman prompted a desperate fear of becoming unattractive and unlovable. She tried numerous remedies, from skin creams to a face lift, to maintain her youth. Women are more likely than men to perceive themselves as younger than their chronological age. And as Figure 16.2 on page 516 indicates, especially for women, the gap between subjective and objective age widens over time (Montepare & Lachman, 1989).

Table 16.1 Levinson's Four Developmental Tasks of Middle Adulthood

Task	Description
Young–Old	The middle-aged person must seek new ways of being both young and old. This means giving up certain youthful qualities, retaining and transforming others, and finding positive meaning in being older.
Destruction–Creation	With greater awareness of mortality, the middle-aged person focuses on ways he or she has acted destructively and others have done the same. Past hurtful acts toward parents, intimate partners, children, friends, and rivals are countered by a strong desire to become more creative—by making products of value to the self and others and participating in activities that advance human welfare.
Masculinity–Femininity	The middle-aged person must come to terms with masculine and feminine parts of the self, creating a better balance. For men, this means becoming more empathic and caring; for women, it often means becoming more autonomous, dominant, and assertive.
Engagement–Separateness	The middle-aged person must create a better balance between engagement with the external world and separateness. For men, this generally means pulling back from ambition and achievement and becoming more in touch with the self. Women who have devoted themselves to child rearing or who have unfulfilling jobs typically move in the other direction—toward greater involvement in the work world and wider community.

Sources: Levinson, 1978, 1996.

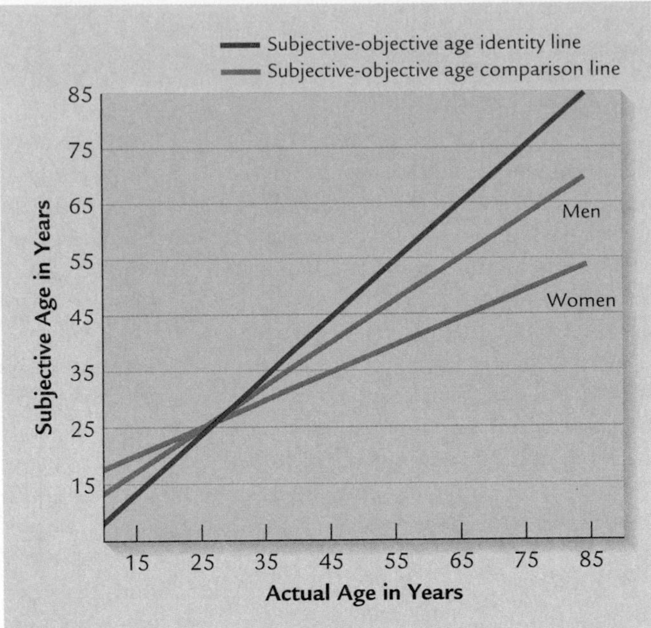

FIGURE 16.2 Relationship between subjective and objective age across the lifespan. After age 25, men and women report perceiving themselves as younger than they actually are. Especially for women, the gap between subjective and objective age widens over time. Because of the double standard of aging, women find it harder to accept being older. (Adapted from J. Montepare & M. Lachman, 1989, "'You're Only as Old as You Feel': Self-Perceptions of Age, Fears of Aging, and Life Satisfaction from Adolescence to Old Age," *Psychology and Aging, 4,* p. 75. Copyright © 1989 by the American Psychological Association. Reprinted by permission.)

As middle-aged adults confront their own mortality and the actual or impending death of agemates, they become more aware of ways people can act destructively—to parents, intimate partners, children, friends, and co-workers. Countering this force is a desire to strengthen life-affirming aspects of the self by advancing human welfare and, as a result, leaving a legacy for future generations. The image of a legacy, which flourishes in midlife, can be satisfied in many ways—through charitable gifts, creative products, volunteer service, or mentoring young people.

Middle age is also a time when people must reconcile masculine and feminine parts of the self. For men, this means being more accepting of "feminine" traits of nurturance and caring, which enhance close relationships and compassionate exercise of authority in the workplace. For women, it generally means being more open to "masculine" characteristics of autonomy, dominance, and assertiveness (Gilligan, 1982; Harris, Ellicott, & Holmes, 1986). Recall from Chapter 8 that people who combine masculine and feminine traits have an androgynous gender identity. Later we will see that androgyny is associated with many favorable personality traits.

Finally, midlife requires a middle ground between engagement with the external world and separateness. Many men must reduce their concern with ambition and achievement and attend more fully to the self. Some women, who have had active, successful careers, must do so as well. But after a period of self-reflection, those who devoted their early adulthood to child rearing or an unfulfilling job often feel compelled to move in the other direction (Levinson, 1996). At age 48, for example, Elena left her position as a reporter for a small-town newspaper, pursued an advanced degree in creative writing, eventually accepted a college teaching position, and began writing a novel. As Tim looked inward, he recognized his overwhelming desire for a gratifying romantic partnership. By scaling back his own career, he realized could grant Elena the time and space she needed to build a rewarding work life—and that doing so might deepen their attachment to one another.

● **The Life Structure in Social Context.** Rebuilding the life structure depends on supportive social contexts. When poverty, unemployment, and lack of a respected place in society dominate the life course, energies are directed toward survival rather than pursuit of a satisfying life structure. Even adults whose jobs are secure and who live in pleasant neighborhoods may find that employment conditions place too much emphasis on productivity and profit and too little on the meaning of work, thereby restricting possibilities for growth. In her early forties, Trisha left a large law firm for a small practice because of constant pressures to bring in high-fee clients and little acknowledgment of her efforts.

Opportunities for advancement permit realization of the early adulthood dream, thereby easing the transition to middle adulthood. Yet they are far less available to women than to men. Individuals of both sexes in blue-collar jobs also have few possibilities for promotion. The industrial workers in Levinson's (1978) sample made whatever vocational adjustments they could—becoming active union members, shop stewards, or mentors of younger workers. Many found compensating rewards in moving to the senior generation of their families.

Vaillant's Adaptation to Life

Because Levinson interviewed 35- to 45-year-olds, his findings cannot tell us about psychosocial change in the fifties. In Vaillant's (1977, 2002) longitudinal research on well-educated men and women, participants were followed past the half-century mark. Recall from Chapter 14 that they became "keepers of meaning," or guardians of their culture (see page 449). Adults in their late forties and fifties carry peak responsibility for the functioning of society. Vaillant reported that the most successful and best adjusted entered a calmer, quieter time of life, an outcome verified in other research (Whitbourne & Weinstock, 1979). "Passing the torch"—concern that the positive aspects of their culture survive—became a major preoccupation.

In societies around the world, older people are guardians of traditions, laws, and cultural values. This stabilizing force holds in check too rapid change sparked by the questioning and challenging of adolescents and young adults. As people move toward the end of middle age, they focus on longer-term, less personal goals, such as the state of human relations in their society. And they become more philosophical, accepting the fact that not all problems can be solved in their lifetime.

Is There a Midlife Crisis?

Levinson (1978, 1996) reported that most men and women in his samples experienced substantial inner turmoil during the transition to middle adulthood. Yet Vaillant (1977) saw few examples of crisis. Instead, change was typically slow and steady. These contrasting findings raise the question of how much personal upheaval actually accompanies entry to midlife. Are self-doubt and stress especially great during the forties, and do they prompt major restructuring of the personality, as the term **midlife crisis** implies?

Think about the reactions of Trisha, Devin, Jewel, Tim, and Elena to middle adulthood. Trisha and Devin moved easily into this period. In contrast, Jewel, Tim, and Elena displayed greater questioning of their situations and sought alternative life paths. Similarly, research suggests wide individual differences in response to midlife. Overall, changes for men are more likely to occur in the early forties (in accord with Levinson's timetable). Those for women may be postponed to the late forties and fifties, until a reduction in parenting responsibilities gives them time and freedom to confront personal issues (Harris, Ellicott, & Holmes, 1986; Mercer, Nichols, & Doyle, 1989). In addition, the direction of change for women is more variable than for men. It depends in part on whether they focused as young adults on a "feminine" social clock (marriage and motherhood), a "masculine" social clock (high-status career), or a combination of the two (Helson & Roberts, 1994; Stewart & Vandewater, 1992).

But sharp disruption and agitation are more the exception than the rule. For example, Elena had considered both a divorce and a new career long before she initiated these changes. In her thirties, she separated from her husband, later reconciled, and told him of her desire to return to school, which he firmly opposed. She put her own life on hold because of her daughter's academic and emotional difficulties and her husband's resistance. In a survey of more than 700 adults, only one-fourth reported experiencing a midlife crisis. When asked what they meant by the term, the participants used a much looser definition than that of researchers. Some reported a crisis well before age 40, others well after age 50. And most attributed it not to age but rather to challenging life events (Wethington, 2000).

Furthermore, midlife is sometimes experienced as relief rather than crisis. In Chapter 15, we noted that women often welcome menopause because it liberates them from unwanted pregnancy and the need to use birth control. In a study of well-known contemporary novelists, many expressed "feeling safe at last in the middle years," free of anxieties about the adequacy of their performance. Midlife brought a sense of confidence and accomplishment and a more insightful view of

■ These sisters responded to the midlife transition by opening a new restaurant and cooking school in a small Colorado town. Yet their career change was probably not the result of a crisis. Most middle-aged adults who decide to change careers plan carefully and choose a second career related to the kind of work they did and enjoyed in their first career.

difficulties—that even the bleakest event is an episode that will eventually pass (Gullette, 1988).

In sum, concern with mortality and life evaluation is common during middle age, but only a minority of people drastically alter their life structure (Lachman & James, 1997). The few who are in crisis typically have had early adulthoods in which gender roles, family pressures, or low income and poverty severely limited their ability to fulfill personal needs and goals, at home or in the wider world (McAdams, 1988).

Stage or Life Events Approach

If crisis and major restructuring in midlife are rare, is middle adulthood actually a *stage* of development, as Erikson's, Levinson's, and Vaillant's theories indicate? Some researchers believe the midadult transition is not stagelike. Instead, they regard it as simply an adaptation to normative life events, such as children growing up, reaching the crest of a career, and impending retirement (McCrae & Costa, 1990).

Yet recall from earlier chapters that life events are no longer as age-graded as they were in the past. Their timing is so variable that they cannot be the sole cause of midlife change. Furthermore, in several studies, people were asked to trace their thoughts, feelings, attitudes, and hopes during early and middle adulthood. Psychosocial change, in terms of personal disruption followed by reassessment, coincided with both family life cycle events and chronological age. For this reason, most experts regard adaptation during midlife as the combined result of growing older and social experiences (Ellicott, 1985; Schroots & Birren, 1990; Whitbourne et al., 1992). Return to our discussion of generativity and the midlife transition, and notice how both factors are involved.

Finally, in describing their lives, the large majority of middle-aged people report troubling moments that prompt new understandings and goals—a finding consistent with the stage approach. As we look closely at emotional and social development in middle adulthood, we will see that this period, like others, is characterized by *both continuity and stagewise change.* With this in mind, let's turn to the diverse inner concerns and outer experiences that contribute to psychological well-being and decision making in midlife.

Ask Yourself

REVIEW

What personal and cultural forces motivate generativity? Why is it vital for favorable adjustment in midlife?

APPLY

After years of experiencing little personal growth at work, 42-year-old Mel looked for a new job. When he received an attractive offer in another city, he felt torn between leaving close friends and a long-awaited career opportunity. After several weeks of soul searching, he took the new job. Was Mel's dilemma a midlife crisis? Why or why not?

CONNECT

Review the concept of androgyny on page 263 of Chapter 8. Why might acceptance of both "masculine" and "feminine" parts of the self foster favorable adjustment in middle adulthood?

www.

Stability and Change in Self-Concept and Personality

Midlife changes in self-concept and personality reflect growing awareness of a finite lifespan, longer life experience, and generative concerns. At the same time, certain aspects of personality remain stable, revealing that individual differences established during earlier phases persist.

Possible Selves

On a business trip, Jewel found a spare afternoon to visit Trisha. The two women sat in a coffee shop, reminiscing about the past and thinking about the future. "It's been tough living on my own and building the business," Jewel commented. "What I hope for is to become better at my work, to be more community-oriented, and to stay healthy and available to my friends. Of course, I don't want to grow old alone, but if I don't find that special person, I suppose I can take comfort in the fact that I'll never have to face divorce or widowhood."

Jewel is discussing **possible selves,** future-oriented representations of what one hopes to become and what one is afraid of becoming. Possible selves are the temporal dimension of self-concept—what the individual is striving for and attempting to avoid. Lifespan researchers regard these hopes and fears as just as vital for explaining behavior as people's views of their current characteristics. Indeed, possible selves may be an especially strong motivator of action in midlife, as more meaning becomes attached to time. Some researchers speculate that as we age, we rely less on social comparisons in judging our self-worth and more on temporal comparisons—how well we are doing in relation to what we had planned for ourselves (Suls & Mullen, 1982).

Throughout adulthood, people's descriptions of their current selves show considerable stability. A 30-year-old who says he is cooperative, competent, outgoing, or successful is likely to report a similar picture at a later age. But reports of possible selves change greatly. Adults in their early twenties mention many possible selves, and they are lofty and idealistic—being "perfectly happy," "rich and famous," "healthy throughout life," and not being "down and out" or "a person who does nothing important." With age, possible selves become fewer in number and more modest and concrete. Most middle-aged people no longer desire to be the best and the most successful. Instead, they are largely concerned with performance of roles and responsibilities already begun—"being competent at work," "being a good husband and father," or "putting my children through the colleges of their choice" and not being "in poor health" or "without enough money to meet my daily needs" (Cross & Markus, 1991; Ryff, 1991).

What explains these shifts in possible selves? Because the future no longer holds limitless opportunities, adults adjust their hopes and fears so they can preserve mental health. They must maintain a sense of unachieved possibility to stay motivated, yet they must still manage to feel good about themselves and their lives in spite of disappointments (Lachman & Bertrand, 2002). For example, Jewel no longer desired to be an executive in a large company, as she had in her twenties. Instead, she wanted to grow in her current occupation. And although she feared loneliness in old age, she reminded herself that marriage could also lead to equally negative outcomes, such as divorce and widowhood—possibilities that made not having attained an important interpersonal goal easier to bear.

Unlike current self-concept, which is constantly responsive to others' feedback, possible selves (although influenced by others) can be defined and redefined by the individual, as needed. Consequently, they permit affirmation of the self, even when things are not going well (Cross & Markus, 1991). Researchers believe that possible selves may be the key to continued well-being in adulthood, as people revise these future images to achieve a better match between desired and achieved goals.

Many studies reveal that the self-esteem of middle-aged and older individuals equals or surpasses that of younger individuals, perhaps because of the protective role of possible selves (Bengston, Reedy, & Gordon, 1985).

Self-Acceptance, Autonomy, and Environmental Mastery

An evolving mix of competencies and experiences leads certain personality traits to change in middle adulthood. One of the most consistent findings is a rise in introspection as people contemplate the second half of life. Middle-aged adults tend to offer more complex, integrated descriptions of themselves than do younger and older individuals (Labouvie-Vief et al., 1995), And many have reshaped contexts to suit their personal needs and values.

These developments undoubtedly contribute to other gains in personal functioning. In a study of well-educated adults ranging in age from the late teens into the seventies, three traits increased from early to middle adulthood and then leveled off. The first was *self-acceptance.* More than young adults, middle-aged people acknowledged and accepted both their good and bad qualities and felt positively about themselves and life. Second, they saw themselves as more *autonomous*—less concerned about expectations and evaluations of others and more concerned with following self-chosen standards. Third, they regarded themselves as high in *environmental mastery*—capable of managing a complex array of tasks easily and effectively (Ryff, 1991).

In Chapter 15, we noted that midlife brings gains in expertise and practical problem solving. These cognitive changes may support the confidence, initiative, and decisiveness of this period. Overall, midlife is a time of increased comfort with the self, independence, assertiveness, and commitment to personal values—outcomes apparent not just in cross-sectional research, but in longitudinal studies as well (Helson & Wink, 1992; Helson, Jones, & Kwan, 2002; Mitchell & Helson, 1990). Perhaps because of these personal attributes, people sometimes refer to middle age as "the prime of life." Although individual differences exist (see the Biology and Environment box on pages 520–521), the middle-adulthood years are a time when many people report feeling especially happy and functioning at their best.

Notions of happiness, however, are not the same in all cultures. For example, a comparison of Korean and North American adults in their fifties revealed that the Koreans reported lower levels of psychological well-being, largely because they were less willing to endorse individualistic traits, such as self-acceptance and autonomy, as characteristic of themselves (Keyes & Ryff, 1998b). Consistent with their collectivist orientation, Koreans' highest well-being scores were on positive relations with others. They viewed personal fulfillment as achieved through family, especially the success of children. North American adults also regarded family relations as relevant to well-being. But they placed greater emphasis on their own traits and accomplishments than on their children's.

Coping Strategies

In Chapter 15, we discussed the importance of stress management in the prevention of illness. It is also vital for psychological well-being. Recall that midlife brings an increase in effective coping strategies. Middle-aged individuals are more likely to look for the "silver lining" or positive side of a difficult situation, postpone action to permit evaluation of alternatives, anticipate and plan ways to handle future discomforts, and use humor to express ideas and feelings without offending others (Diehl, Coyle, & Labouvie-Vief, 1996). Younger individuals more often engage in denial of troubling emotions, acting out (temper outbursts), avoidance (sleep, substance use), and blaming others. Notice how effective coping blends problem-centered and emotion-centered strategies,

■ Greater confidence accompanies gains in expertise and practical problem solving in middle adulthood. This middle-aged drama coach exudes a sense of mastery as he demonstrates improvisational acting to a class of high school students attending a summer camp for the performing arts.

Biology & Environment

What Factors Promote Psychological Well-Being in Midlife?

For Trisha and Devin, midlife brought contentment and high life satisfaction. But the road to happiness was rockier for Jewel, Tim, and Elena. What factors contribute to variations in psychological well-being at midlife? Consistent with the lifespan perspective, biological, psychological, and social forces are involved, and their effects are interwoven.

Good Health and Exercise. Good health affects energy and zest for life. Adults of any age who rate their health as good to excellent are more likely to feel positively about their life circumstances. But during middle and late adulthood, taking steps to improve health and prevent disability becomes a better predictor of psychological well-being. In a study of over 3,000 20- to 64-year-olds, engaging in regular exercise—walking, dancing, jogging, or swimming—was more strongly associated with self-rated health and a positive outlook in older than in younger individuals (Ransford & Palisi, 1996). Middle-aged adults who maintain an exercise regimen are likely to perceive themselves as particularly active for their age and, therefore, to feel a special sense of accomplishment. And exercise may reduce feelings of vulnerability to illness that increase with age, conveying extra psychological benefits. Fear of disease and disability is one of the strongest contributors to poor psychological well-being at midlife (Barsky, Cleary, & Klerman, 1992).

Sense of Control and Personal Life Investment. Middle-aged adults who report a high sense of control over events in various aspects of their lives—health, family, and work—report more favorable psychological well-being. A sense of control fosters self-efficacy, a belief in one's ability to surmount challenges. Consequently, it helps sustain a positive outlook in the face of health, family, and work difficulties (Bandura, 1997; Smith et al., 2000).

But beyond feeling in control, personal life investment—firm commitment to goals and involvement in pursuit of those goals—adds to mental health and life satisfaction (Staudinger, Fleeson, & Baltes, 1999). According to Mihaly Csikszentmihalyi, a vital wellspring of happiness is *flow*—the psychological state of being so engrossed in a demanding, meaningful activity that one loses all sense of time and self-awareness. People describe flow as the height of enjoyment, even as an ecstatic state. The more they experience it, the more they judge their lives to be gratifying (Nakamura & Csikszentmihalyi, 2002). Although flow is common in people engaged in creative endeavors, many others report it—students who love studying, employees who like their jobs, adults involved in challenging leisure pursuits, and parents and grandparents engaged in pleasurable learning activities with children. Flow depends on perseverance and skill at complex endeavors that offer potential for growth. These qualities are well developed in middle adulthood.

Positive Social Relationships. Developing gratifying social ties is closely linked to midlife psychological well-being. In a longitudinal study of 90 men selected for good physical and mental health as college students and followed over 32 years, most maintained their physical health status. A good mentor relationship in early adulthood (which fosters high career achievement) and favorable peer ties were among the best predictors of well-being in late forties and early fifties (Westermeyer, 1998). In a survey of college alumni, those who preferred occupational prestige and high income to close friends were twice as likely as other respondents to describe themselves as "fairly" or "very" unhappy (Perkins, 1991, as cited by Myers, 2000).

A Good Marriage. Although friendships and positive relationships with co-workers are important, a good marriage boosts psychological well-being

whereas ineffective coping is largely emotion-centered (see Chapter 15, page 495).

Why might effective coping increase in middle adulthood? Other personality changes seem to support it. In one study, complex, integrated self-descriptions—which increase in midlife, indicating an improved ability to reflect on strengths and weaknesses and blend them into an organized picture—predicted good coping strategies (Labouvie-Vief & Diehl, 2000). Greater confidence at handling life's problems may also contribute. In a longitudinal investigation of well-educated women, taking initiative to overcome difficult times in early adulthood predicted advanced self-understanding, social and moral maturity, and high life satisfaction at age 43 (Helson & Roberts, 1994). Overall, these findings suggest that years of experience in overcoming stress promote enhanced self-knowledge, which joins with life experience to foster more sophisticated, flexible coping during middle age.

Gender Identity

In her forties and early fifties, Trisha appeared more assertive at work, speaking out more freely at meetings and taking a leadership role when a team of lawyers worked on an especially complex case. She was also more dominant in family relationships, expressing her preferences and points of view to even more. In the longitudinal study just described, the role of marriage as a marker of mental health increased with age, becoming a powerful predictor by age 50 (Westermeyer, 1998). Although women's adjustment is more sensitive to the quality of the marital relationship than men's (see Chapter 14), marriage bestows considerable benefits on both genders. Longitudinal studies tracking people as they moved in and out of intimate relationships suggest that marriage actually brings about well-being.

For example, when interviews with over 13,000 adults were repeated 5 years later, people who remained married reported greater happiness than those who remained single. Those who separated or divorced became less happy, reporting considerable depression. Men and women who married for the first time experienced a sharp increase in happiness, those who entered their second marriage a modest increase (Marks & Lambert, 1998).

The link between marriage and well-being is similar in many nations, suggesting that marriage changes people's behavior in ways that make them better off (Diener et al., 2000). Married partners monitor each other's health and offer care in times of illness. They also earn and save more money than single people. Higher income is modestly linked to psychological well-being (Myers, 2000; Waite, 1999). Furthermore, sexual satisfaction predicts mental health, and married couples have more satisfying sex lives than unmarried couples and singles (see Chapter 14). Of course, not everyone is better off married. But for many people, the long-term commitment that marriage entails fosters mental health.

Mastery of Multiple Roles. Finally, success in handling multiple roles is linked to psychological well-being. Women are generally happier today than they used to be because they now reap satisfactions not just from family relationships but also from vocational achievements. In a study of nearly 300 middle-aged women, researchers asked about feelings of competence and control in four roles: wife, mother, caregiver of an impaired parent, and employee. Participants experienced higher levels of mastery in their work roles than in any other. But competence and control in all four roles predicted life satisfaction and reduced depression (Christensen, Stephens, & Townsend, 1998). Women who occupied several roles—in work and family arenas—seemed to benefit from added opportunities to enhance their sense of mastery.

■ Personal life investment adds to mental health and life satisfaction during middle adulthood. This middle-aged adult scaling the vertical surface of a mountain looks as if she is experiencing flow—the psychological state of being so engrossed in a demanding, meaningful activity that one loses all sense of time and self-awareness. The more people experience flow, the more they judge their lives to be gratifying.

her husband and son more readily than she had 10 or 15 years earlier. In contrast, Devin's sense of empathy and caring became more apparent, and he was less assertive and more accommodating to Trisha's wishes than before.

Many studies report an increase in "masculine" traits in women and "feminine" traits in men across middle age (Huyck, 1990; James et al., 1995). Women become more confident, self-sufficient, and forceful, men more emotionally sensitive, caring, considerate, and dependent. These trends appear in cross-sectional and longitudinal research, in people varying in SES, and in highly diverse cultures—not just Western industrialized nations but also village societies such as the Mayans of Guatemala, the Navajo of the United States, and the Druze of the Middle East (Fry, 1985; Gutmann, 1977; Turner, 1982). Consistent with Levinson's theory, gender identity in midlife becomes more androgynous—a mixture of both "masculine" and "feminine" characteristics.

Although the existence of these changes is well accepted, explanations for them are controversial. According to a well-known evolutionary view called **parental imperative theory,** identification with traditional gender roles is maintained during the active parenting years to help ensure the survival of children. Men become more goal-oriented, whereas women emphasize nurturance. After children reach adulthood, parents are free to express the "other-gender" side of their personalities (Gutmann & Huyck, 1994). A related idea is that the

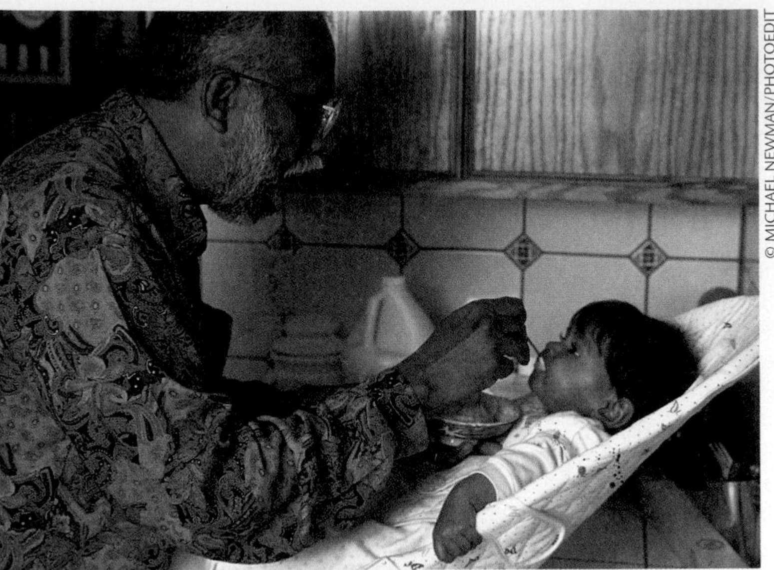
© MICHAEL NEWMAN/PHOTOEDIT

■ Gender identity becomes more androgynous in midlife—a mixture of both "masculine" and "feminine" characteristics. As this grandfather illustrates, men typically become more emotionally sensitive and nurturant.

decline in sex hormones associated with aging may contribute to androgyny in later life (Rossi, 1980).

But these biological accounts have been criticized. Think back to what you learned in earlier chapters, and you will see that both warmth and assertiveness (in the form of firmness and consistency) are necessary for parents to rear children effectively. Furthermore, although children's departure from the home is related to men's openness to the "feminine" side of their personalities, it is less clearly linked to a rise in "masculine" traits among women (Huyck, 1996, 1998). In longitudinal research, college-educated women in the labor force became more independent by their early forties, regardless of whether they had children; those who were homemakers did not. Women attaining high status at work gained most in dominance, assertiveness, and outspokenness by their early fifties (Helson & Picano, 1990; Wink & Helson, 1993). Finally, androgyny is not associated with menopause—a finding at odds with a hormonal explanation of gender-identity change (Helson & Wink, 1992).

Besides reduced parenting responsibilities, other demands and experiences of midlife may prompt a more androgynous orientation. For example, among men, a need to enrich a marital relationship after children have departed, along with reduced opportunities for career advancement, may be involved in the awakening of emotionally sensitive traits. Compared with men, women are far more likely to face economic and social disadvantages. A greater number remain divorced, are widowed, and encounter discrimination in the workplace. Self-reliance and assertiveness are vital for coping with these circumstances.

In sum, androgyny in midlife results from a complex combination of social roles and life conditions. In Chapter 8, we noted that androgyny predicts high self-esteem. In adulthood, it is also associated with advanced moral reasoning and psychosocial maturity (Prager & Bailey, 1985; Waterman & Whitbourne, 1982). People who do not integrate the masculine and feminine sides of their personalities tend to have mental health problems, perhaps because they are unable to adapt flexibly to the challenges of aging (Huyck, 1996).

Individual Differences in Personality Traits

Although Trisha and Jewel both became more self-assured and assertive in midlife, in other respects they differed. Trisha had always been more organized and hardworking, whereas Jewel was more gregarious and fun-loving. Once, the two women traveled together. At the end of each day, Trisha was disappointed if she had not kept to a schedule and visited every tourist attraction. Jewel liked to "play it by ear"—wandering through streets and stopping to talk with shopkeepers and residents.

In previous sections, we considered personality changes common to many middle-aged adults, but stable individual differences also exist. The hundreds of personality traits on which people differ have been organized into five basic factors, called the **"big five" personality traits:** neuroticism, extroversion, openness to experience, agreeableness, and conscientiousness. Table 16.2 provides a description of each. Notice that Trisha is high in conscientiousness, whereas Jewel is high in extroversion (Costa & McCrae, 1994; McCrae & Costa, 1990).

Cross-sectional studies of Canadian and American men and women reveal that neuroticism, extroversion, and openness to experience show modest declines from the teenage years through middle age, whereas agreeableness and conscientiousness increase—changes that reflect "settling down" and greater maturity. Similar trends have been identified in many other countries varying widely in cultural traditions—for example, Germany, Italy, Japan, Russia, and South Korea (Costa et al., 2000; McCrae et al., 2000). The consistency of these cross-cultural findings has led some researchers to conclude that adult personality change is genetically influenced. As support for a genetic view, they note that individual differences in the "big five" traits are large and highly stable. A person who scores high or low at one age is likely to do the same at another, over intervals ranging from 3 to 30 years (Costa & McCrae, 1994). In a reanalysis of more than 150 longitudinal studies including more than 50,000 participants, personality-trait stability increased during early and middle adulthood, reaching a peak in the decade of the fifties (see Figure 16.3) (Roberts & DelVecchio, 2000).

How can there be high stability in personality traits, yet significant changes in aspects of personality discussed earlier? Studies of the "big five" traits include very large samples. Typically, they do not examine the impact of a host of contextual factors—including life events, the social clock, and cultural values—that shape aspirations, goals, and expectations for appropriate behavior (Caspi & Roberts, 2001). Look

Table 16.2 The "Big Five" Personality Traits

Trait	Description
Neuroticism	Individuals who are high on this trait are worrying, temperamental, self-pitying, self-conscious, emotional, and vulnerable. Individuals who are low are calm, even-tempered, self-content, comfortable, unemotional, and hardy.
Extroversion	Individuals who are high on this trait are affectionate, talkative, active, fun-loving, and passionate. Individuals who are low are reserved, quiet, passive, sober, and emotionally unreactive.
Openness to experience	Individuals who are high on this trait are imaginative, creative, original, curious, and liberal. Individuals who are low are down-to-earth, uncreative, conventional, uncurious, and conservative.
Agreeableness	Individuals who are high on this trait are soft-hearted, trusting, generous, acquiescent, lenient, and good-natured. Individuals who are low are ruthless, suspicious, stingy, antagonistic, critical, and irritable.
Conscientiousness	Individuals who are high on this trait are conscientious, hardworking, well organized, punctual, ambitious, and persevering. Individuals who are low are negligent, lazy, disorganized, late, aimless, and nonpersistent.

Source: McCrae & Costa, 1990.

closely at the traits in Table 16.2, and you will see that they differ from the attributes considered in previous sections. They do not take into account motivations, preferred tasks, and coping styles. And they do not consider how certain aspects of personality, such as masculinity and femininity, are integrated (Block, 1995; Helson & Stewart, 1994). Theorists concerned with change due to experience focus on the way personal needs and life events induce new strategies and goals. In contrast, theorists who emphasize stability due to heredity measure personality traits on which individuals can easily be compared and that are present at any time of life (Rosenberg, Rosenberg, & Farrell, 1999).

Perhaps we can resolve the apparent contradiction by thinking of adults as changing in overall organization and integration of personality, but doing so on a foundation of basic, enduring dispositions that support a coherent sense of self as people adapt to changing life circumstances. When more than 2,000 individuals in their forties were asked to reflect on their personalities during the previous 6 years, 52 percent said they had "stayed the same," 39 percent said they had "changed a little," and 9 percent said they had "changed a lot" (Herbst et al., 2000). Once again, these findings contradict a view of middle adulthood as a period of great turmoil and change. But they also underscore that personality remains an "open system," responsive to the pressures of life experiences. Indeed, certain midlife personality changes may strengthen trait consistency! Improved self-understanding, self-acceptance, and skill at handling challenging situations may result in less need to modify basic personality dispositions over time (Caspi & Roberts, 2001).

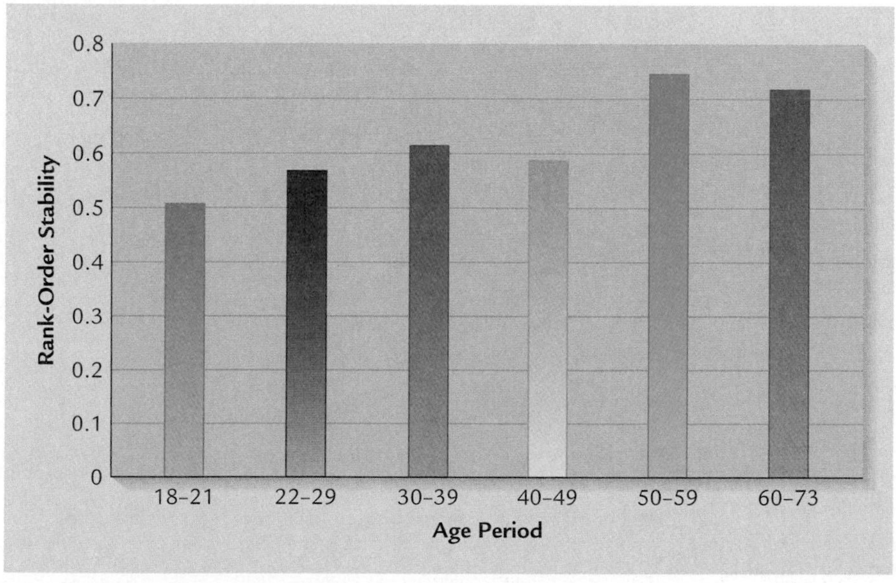

FIGURE 16.3 Increase in stability of the "big five" personality traits across early and middle adulthood. In a reanalysis of more than 150 longitudinal studies, stability of personality traits was strong at all ages; adults who scored high or low in certain traits largely retained that pattern when retested. Stability reached a peak in the fifties. Nevetheless, at no age was stability perfect, indicating that personality remains an "open system." (Adapted from Roberts & DelVecchio, 2000.)

Ask Yourself

REVIEW
Summarize personality changes at midlife. How can these changes be reconciled with increasing stability of the "big five" personality traits, peaking in the fifties?

APPLY
Around age 40, Luellen no longer thought about becoming a performing pianist. Instead, she decided to concentrate on accompanying other musicians and expanding her studio of young students. How do Luellen's plans reflect midlife changes in possible selves?

APPLY
Jeff, age 46, suggested to his wife, Julia, that they set aside time once a year to discuss their relationship—both positive aspects and ways to improve. Julia reacted with surprise, since Jeff had never before expressed interest in working on their marriage. What developments at midlife probably fostered this new concern?

CONNECT
List cognitive gains that typically occur during middle adulthood. (See Chapter 15, pages 503–505.) How might they support midlife personality changes?

www.

Relationships at Midlife

The emotional and social changes of midlife take place within a complex web of family relationships and friendships. Although a few middle-aged people live alone, the vast majority—9 out of 10 in the United States and Canada—live in families, most with a spouse (Statistics Canada, 2001c; U.S. Bureau of the Census, 2002c).

The middle adulthood phase of the family life cycle is often referred to as "launching children and moving on." At one time it was called the "empty nest," but this phrase implies a negative transition, especially for women. When adults devote themselves entirely to their children, the end of active parenting can trigger feelings of emptiness and regret. But for many people, middle adulthood is a liberating time, offering a sense of completion and an opportunity to strengthen existing ties and build new ones.

A century ago, most parents reared children for almost all of their active adulthood. Due to a declining birthrate and a longer life expectancy, contemporary parents launch children about 20 years before retirement and then seek other rewarding activities. Because of the lengthening of this period, it is marked by the greatest number of exits and entries of family members. As adult children leave home and marry, middle-aged people must adapt to new roles of parent-in-law and grandparent. At the same time, they must establish a different type of relationship with their aging parents, who may become ill or infirm and die. Let's see how ties within and beyond the family change during this time of life.

Marriage and Divorce

Although not all couples are financially comfortable, middle-aged households are well off economically compared with other age groups. North Americans between 45 and 54 have the highest average annual income (Statistics Canada, 2002j; U.S. Bureau of the Census, 2002c). Partly because of increased financial security, and because the time between ddeparture of the last child and retirement is so long, the contemporary social view of marriage in midlife is one of expansion and new horizons.

These forces strengthen the need to review and adjust the marital relationship. For Devin and Trisha, this shift was gradual. By middle age, their marriage had permitted satisfaction of family and individual needs, endured many changes, and culminated in deeper feelings of love. Elena's marriage, in contrast, became more conflict-ridden as her teenage daughter's problems introduced added strains and as departure of children made marital difficulties more obvious. Tim's failed marriage revealed yet another pattern. With passing years, both love expressed and number of problems declined. As less happened in the relationship, either good or bad, the couple had little to keep them together (McCullough & Rutenberg, 1989).

© RHODA SIDNEY/PHOTOEDIT

■ For many middle-aged couples, having forged a relationship that permits satisfaction of both family and individual needs results in deeper feelings of love.

As the Biology and Environment box on pages 520–521 revealed, marital satisfaction is a strong predictor of midlife psychological well-being. Middle-aged men who have focused only on career often realize the limited nature of their pursuits. At the same time, their wives may insist on a more gratifying relationship. In addition, children fully engaged in adult roles remind middle-aged parents that they are in the latter part of their lives, so many decide that the time for improving their marriages is now (Berman & Napier, 2000).

As in early adulthood, divorce is one way of resolving an unsatisfactory marriage in midlife. Although most divorces occur within 5 to 10 years of marriage, about 10 percent take place after 20 years or more (Statistics Canada, 2002d; U.S. Department of Health and Human Services, 2002l). Divorce at any age takes a heavy psychological toll, but midlifers seem to adapt more easily than younger people. A survey of more than 13,000 Americans revealed that following divorce, middle-aged men and women reported less decline in psychological well-being than their younger counterparts (Marks & Lambert, 1998). Midlife gains in practical problem solving and effective coping strategies may reduce the stressful impact of divorce.

A substantial number of midlife divorces occur to people who have had one or more previous unsuccessful marriages, since the divorce rate is more than twice as great among remarried couples as among couples in first marriages. Highly educated middle-aged adults are more likely to divorce, probably because their more comfortable economic circumstances make it easier to leave an unhappy marriage (Wu & Penning, 1997). Nevertheless, for many women, marital breakup—especially when it is repeated—severely reduces standard of living. In a study of over five thousand 30- to 44-year-old women followed over 15 years, divorce led to a 39 percent drop in average income. Many African-American women opted for separation rather than divorce, probably because of the high cost of divorce proceedings. Women who separated had the highest poverty rate—both before the marital transition (27 percent) and afterward (57 percent). Neither separated nor divorced women escaped economic disadvantage after adjusting to their new life circumstances. Even 7 or 8 years after their marriages ended, the high rate of poverty persisted (Morgan, 1991).

As these findings reveal, marital breakup, in midlife and earlier, is a strong contributor to the **feminization of poverty**—a trend in which women who support themselves or their families have become the majority of the adult poverty population, regardless of age and ethnic group. Because of weak public policies safeguarding families (see Chapter 2), the gender gap in poverty is higher in the United States and Canada than in other Western industrialized nations (Paquet, 2002; U.S. Bureau of the Census, 2002c).

What do recently divorced middle-aged people say about why their marriages ended? In one study, women mentioned communication problems most often, followed by husband's substance abuse, husband's physical and verbal abuse, and their own desire for autonomy. Notice how these responses dwell on neutral circumstances (communication) and the husband's faults. When women attribute divorce to a cause that is more self-accusing ("my husband's lack of interest in me"), they show poorer adjustment (Davis & Aron, 1989).

Longitudinal evidence reveals that middle-aged women who weather divorce successfully tend to become more tolerant, comfortable with uncertainty, nonconforming, and self-reliant in personality—factors believed to be fostered by divorce-forced independence. As in earlier periods, divorce represents both a time of trauma and a time of growth (Marks & Lambert, 1998; Rockwell, Elder, & Ross, 1979). Little is known about long-term adjustment following divorce among middle-aged men, perhaps because most enter new relationships and remarry within a short time.

Changing Parent–Child Relationships

Parents' positive relationships with their grown children are the result of a gradual process of "letting go," starting in childhood, gaining momentum in adolescence, and culminating in children's independent living. As mentioned earlier, most middle-aged parents adjust well to the launching phase of the family life cycle; only a minority have difficulty. Investment in nonparental relationships and roles, children's characteristics, parents' marital and economic circumstances, and cultural forces affect the extent to which this transition is expansive and rewarding or sad and distressing.

After moving their son Mark into his college dormitory at the start of his freshman year, Devin and Trisha felt a twinge of nostalgia. On the way home, they recalled his birth, first day of school, and high school graduation and commented on their suddenly tranquil household. Beyond this, they returned to rewarding careers and community participation and delighted in having more time for each other. Parents who have developed gratifying alternative activities typically welcome their children's adult status. A strong work orientation, especially, predicts gains in life satisfaction after children depart from the home (Seltzer & Ryff, 1994; Silverberg, 1996).

Regardless of whether they reside with parents, adolescent and young adult children who are "off-time" in development—not showing expected signs of independence and accomplishment—can prompt parental strain (Aquilino, 1996; Ryff et al., 1994). Consider Elena, whose daughter was frequently truant from high school and in danger of not graduating. The need for greater parental oversight and guidance caused anxiety and unhappiness for Elena, who was ready to complete the active parenting phase and focus on her own personal and vocational development (Raup & Myers, 1989).

However, wide variations exist in the social clock for launching children. Recall from Chapter 13 that many young people from low-SES homes and with cultural traditions of extended-family living do not leave home early. In the southern European countries of Greece, Italy, and Spain, parents often actively delay their children's departure, so many more young adults reside with their parents than do so in other Western nations (Cordon, 1997). In Italy, adult children's relationships with parents are usually positive, making living at

© DAVID SIMSON/STOCK BOSTON, LLC.

■ Cultures set quite different social clocks for the middle-aged phase of the life cycle: "launching children and moving on." In Italy, it is not unusual for children in their thirties to be living with their parents, who actively encourage them to stay. Italian mothers are admired and respected when their grown children live at home.

home attractive. And middle-aged mothers are held in higher esteem when their adult children live with them, even past age 30 (Scabini & Cigoli, 1997).

The end of parent–child coresidence is accompanied by a substantial decline in parental authority. Devin and Trisha no longer knew of Mark's daily comings and goings or expected him to inform them. Nevertheless, Mark telephoned at regular intervals to report on events in his life and seek advice about major decisions. Although the parental role changes, its continuation is important to middle-aged adults. Departure of children is a relatively minor event when parent–child contact and affection are sustained. When it results in little or no communication, parents' life satisfaction declines (White, 1994). In a large longitudinal study of New Zealand families, parents' relations with adult children were linked to quality of parenting years before. Parents who had been warm and supportive in middle childhood and adolescence were more likely to experience contact and closeness with their child in early adulthood (Belsky et al., 2001).

Throughout middle adulthood, parents continue to give more assistance to children than they receive, especially while children are unmarried or when they face difficulties, such as marital breakup or unemployment (Zarit & Eggebeen, 2002). Providing emotional and financial support while children get their lives under way is related to midlife psychological well-being. Due to disrupted relationships and economic need, divorced and remarried parents are less likely than parents in first marriages to offer adult children support, and they are also less content (Marks, 1995). Children's psychological adjustment is more important than their educational and occupational success for midlife parental adjustment (Ryff, Schmutte, & Lee, 1996). When young adult children are relatively free of emotional and social problems, they relate more positively to their parents. Favorable adaptation to the launching phase depends on feeling successful as a parent and on not being estranged from one's children.

When children marry, parents face additional challenges in enlarging the family network to include in-laws. Difficulties occur when parents do not approve of their child's partner or when the young couple adopts a way of life inconsistent with the values of the family of origin. But when warm, supportive relationships endure, intimacy between parents and children increases over the adult years, with great benefits for parents' life satisfaction (Ryff, Singer, & Seltzer, 2002). Once young adults strike out on their own, members of the middle generation, especially mothers, usually take on the role of **kin-keeper,** gathering the family for celebrations and making sure everyone stays in touch.

As children become adults, parents expect a mature relationship with them, marked by tranquility and contentment. Yet many factors—on the child's and the adult's side—affect whether that goal is achieved. The Caregiving Concerns table on the following page suggests ways middle-aged parents can increase the chances that bonds with adult children will be loving and rewarding and serve as contexts for personal growth.

Grandparenthood

Two years after Mark married, Devin and Trisha were thrilled to learn that a granddaughter was on the way. Although the stereotypical image of grandparents as very elderly persists, on average American adults become grandparents in their mid- to late forties, Canadian adults in their late forties to early fifties (Rosenthal & Gladstone, 2000; Szinovacz, 1998). A longer life expectancy means that adults will spend as much as one-third of their lifespan in the grandparent role.

● **Meanings of Grandparenthood.** Why did Trisha and Devin, like many people their age, greet the announcement of a grandchild with such enthusiasm? Grandparenthood is a highly significant milestone to most who experience it. When asked about its meaning, people generally mention one or more of the following gratifications:

- Valued elder—being perceived as a wise, helpful person
- Immortality through descendants—leaving behind not just one but two generations after death
- Reinvolvement with personal past—being able to pass family history and values to a new generation
- Indulgence—having fun with children without major child-rearing responsibilities (Kivnick, 1983; Miller & Cavanaugh, 1990)

● **Grandparent–Grandchild Relationships.** Grandparents' styles of relating to grandchildren vary as widely as the meanings they derive from their new role. The grandparent's and grandchild's age and sex make a difference. When their granddaughter was young, Trisha and Devin enjoyed an affectionate, playful relationship with her. As she got older, she

Ways Middle-Aged Parents Can Promote Positive Ties with Their Adult Children

SUGGESTION	DESCRIPTION
Emphasize positive communication.	Let adult children know of your respect, support, and interest. This not only communicates affection but also permits conflict to be handled in a constructive context.
Avoid unnecessary comments that are a holdover from childhood.	Adult children, like younger children, appreciate an age-appropriate relationship. Comments that have to do with safety, eating, and cleanliness (for example, "Be careful on the freeway," "Don't eat those foods," and "Make sure you wear a sweater on a cold day") annoy adult children and can stifle communication.
Accept the possibility that some cultural values and practices and aspects of lifestyle will be modified in the next generation.	In constructing a personal identity, most adult children have gone through a process of evaluating the meaning of cultural values and practices for their own lives. Traditions and lifestyles cannot be imposed on adult children.
When an adult child encounters difficulties, resist the urge to "fix" things.	Accept the fact that no meaningful change can take place without the willing cooperation of the adult child. Stepping in and taking over communicates a lack of confidence and respect. Find out whether the adult child wants your help, advice, and decision-making skills.
Be clear about your own needs and preferences.	When it is difficult to arrange for a visit, baby-sit, or provide other assistance, say so and negotiate a reasonable compromise rather than letting resentment build.

Source: Toder, 1994.

looked to them for information and advice in addition to warmth and caring. By the time their granddaughter reached adolescence, Trisha and Devin had become role models, family historians, and conveyers of social, vocational, and religious values (Hurme, 1991). Typically, relationships are closer between grandparents and grandchildren of the same sex and, especially, between maternal grandmothers and granddaughters—a pattern found in many countries (Smith, 1991). Grandmothers also report higher satisfaction with the grandparent role than grandfathers, perhaps because it is an important means through which middle-aged women satisfy their kinkeeping function (Somary & Stricker, 1998; Thomas, 1989).

Nearness of residence made Trisha and Devin's pleasurable interaction with their granddaughter possible. Grandparents who live far from young grandchildren usually have more distant relationships with them, appearing mainly on holidays, birthdays, and other formal occasions but otherwise having little contact. Despite high family mobility in Western industrialized nations, most grandparents live within 30 minutes of at least one grandchild, which makes regular visits possible (Werner, 1991). And even when grandparents reside far away, a strong desire to affect the development of grandchildren can motivate them to become involved in grandchildren's lives. As grandchildren get older, distance has less impact. Instead, the extent to which the adolescent or young adult grandchild believes the grandparent values contact is a good predictor of a close bond (Brussoni & Boon, 1998).

SES and ethnicity are additional influences on grandparent–grandchild ties. In higher-income families, the grandparent role is not central to family maintenance and survival. For this reason, it is fairly unstructured and takes many forms. In contrast, grandparents perform essential activities in low-income families. For example, many single parents live with their families of origin, where grandparents' financial and caregiving assistance reduces the impact of poverty. Compared with grandchildren in intact families, grandchildren in single-parent and stepparent families report engaging in more diverse, higher-quality activities with their grandparents (Kennedy & Kennedy, 1993). As children experience the stress of family transition, bonds with grandparents take on increasing importance.

In some cultures, grandparents are absorbed into an extended-family household and become actively involved in child rearing. When a Chinese, Korean, or Mexican-American maternal grandmother is a homemaker, she is the preferred caregiver while parents of young children are at work (Kamo, 1998; Williams & Torrez, 1998). Similarly, involvement in child care is high among Native American and Canadian Aboriginal grandparents. In the absence of a biological grandparent, an unrelated elder may be integrated into the family to serve as a mentor and disciplinarian for children (Werner, 1991). (See Chapter 2 page 63, for a description of the grandmother's role in the African-American extended family.)

● **Recent Trends in Grandparenting.** Increasingly, grandparents have stepped in as primary caregivers in the face of serious family problems. As the Social Issues box on page 528 reveals, a rising number of North American children live apart from their parents in grandparent-headed households. Despite their willingness to help and their competence at child

Social Issues

Grandparents Rearing Grandchildren: The Skipped-Generation Family

Nearly 1.5 million American and 56,000 Canadian children—1 to 2 percent of the child population—live with grandparents but apart from parents, in **skipped-generation families** (Statistics Canada, 2003a; U.S. Bureau of the Census, 2002c). The number of grandparents rearing grandchildren has increased over the past decade. The arrangement occurs in all ethnic groups, though more often in African-American, Hispanic, and Canadian Aboriginal families than in Caucasian families (Downey et al., 2002; Szinovacz, 1998).

The following grandmother's account illustrates the circumstances in which skipped-generation families form:

> I think [the child's mother] is doing an awful lot of drugs. I don't know if it's serious drugs, but enough that she wasn't doing a good job. They were doing an awful lot of partying, and I just finally went to her one day and said, "Your life is messed up, let me take the baby for a while so you can get your life together." And it was just because we wanted her [the granddaughter] out of there. She had become our main priority. (Jendrek, 1994, p. 209)

In about half of skipped-generation families, grandparents step in because substance-abuse problems prevent parents, usually the child's mother, from engaging in competent child rearing. In most other instances, parental emotional or physical illness is involved (Pruchno & McKenney, 2000; Weber & Waldrop, 2000). At times, child welfare authorities, out of a preference for placing the child with relatives rather than in a foster home, approach the grandparent, who assumes temporary or permanent legal custody. In most instances, grandparents offer their assistance, sometimes with and sometimes without legal responsibility. Most say they took action to protect the child only when the parents' situation became intolerable.

Because the skipped-generation family structure is not freely chosen, many custodial grandparents face highly stressful life circumstances. The absent parents' adjustment difficulties and uncertain involvement with the child strain family relationships (Hirshorn, Van Meter, & Brown, 2000). And unfavorable child-rearing experiences have left their mark on the children, who show high rates of learning difficulties, depression, and antisocial behavior (Pinson-Millburn et al., 1996). In addition, these youngsters introduce financial burdens into households, many of which are already low-income. Grandparents struggle with daily dilemmas—wanting to be grandparents, not parents; wanting the parent to be present in the child's life but fearing for the child's well-being if the parent returns and does not provide good care. Child-rearing tasks mean that grandparents have less time for spouses, friends, and leisure when they had expected to have more time. Many report feeling tired, emotionally drained, and depressed (Giarrusso et al., 2000; Hayslip et al., 2002).

■ Vera Saunders, age 71, shown here with her 17-year-old grandson who lives with her, is tired but proud. In middle age, she took custody of three grandchildren. She takes special pleasure in the achievements of her 25-year-old granddaughter, who won a scholarship to Yale University and became a medical doctor. "I took her when she was only 3 days old," Vera recalls, "because her mother had a nervous breakdown." Grandparents in skipped-generation families face shattered dreams of freedom and relaxation and significant financial burdens. To Vera, the sacrifices were definitely worthwhile.

Skipped-generation families have a tremendous need for social and financial support. Despite the burdens, these grandparents seem to realize their widespread image as "silent saviors." Often they forge close emotional bonds with their grandchildren (Fuller-Thomson & Minkler, 2000). And a survey of a large, representative sample of American families revealed that compared with children in divorced, single-parent families or in blended families, children reared by grandparents were better behaved in school, were less susceptible to physical illness, and were doing just as well academically (Solomon & Marx, 1995). These outcomes foster compensating feelings of accomplishment and pride in grandparents. Despite great hardship, many custodial grandparents report joy from sharing children's lives and helping them grow. And some view the rearing of grandchildren as a "second chance"—an opportunity to make up for earlier, unfavorable parenting experiences and "do it right" (Minkler & Roe, 1993; Waldrop & Weber, 2001).

© TONY FREEMAN/PHOTOEDIT

■ In some subcultures, grandparents are absorbed into an extended family household and become actively involved in child rearing. When a Chinese, Korean, or Mexican-American maternal grandmother is a homemaker, she is the preferred caregiver while parents of young children are at work.

rearing, grandparents who take full responsibility for young children experience considerable emotional and financial strain. They need much more assistance from community and government agencies than is currently available.

In most families, parents serve as gatekeepers of grandparents' contact with grandchildren. When grandparents and parents get along poorly, the grandparent–grandchild tie usually suffers. After marital breakup, for example, grandparents on the custodial parent's side have more frequent contact with grandchildren than grandparents on the noncustodial side (Johnson, 1998). A growing concern among grandparents is maintaining contact with grandchildren after parental divorce. Currently, all 50 U.S. states and the Canadian province of Québec permit grandparents to seek legal visitation judgments. Yet grandparent visitation rights can add to family difficulties. When parents are divorcing, the behavior of grandparents varies greatly, from constructive help to entanglement in parental battles. Unfortunately, intense conflict usually lies behind the legal petitions of grandparents who fail to work out visitation informally (Smith & Drew, 2002). Consequently, courts are wise to exercise restraint in awarding grandparents visitation privileges.

When family relationships are positive, grandparenthood provides an important means of fulfilling personal and societal needs in midlife and beyond. Typically, grandparents are a frequent source of pleasure, support, and knowledge for children, adolescents, and young adults. They also provide the young with firsthand experience in how older people think and function. In return, grandchildren become deeply attached to grandparents and keep them abreast of social change. Clearly, grandparenthood is a vital context for sharing between generations.

Middle-Aged Children and Their Aging Parents

Compared with earlier generations, today's adults spend more years not only as parents and grandparents, but also as children of aging parents. The percentage of North American middle-aged people with living parents has risen dramatically—from 10 percent in 1900 to 50 percent in 2000 (U.S. Bureau of the Census, 2002c; Vanier Institute of the Family, 2002a). A longer life expectancy means that adult children and their parents are increasingly likely to grow old together. What are middle-aged children's relationships with their aging parents like? And how does life change for adult children when an aging parent's health declines?

● **Frequency and Quality of Contact.** A widespread myth is that adults of past generations were more devoted to their aging parents than are adults of the present generation. Although adult children spend less time in close proximity to their parents, the reason is not neglect or isolation. Fewer aging adults live with younger generations now than in the past because of a desire to be independent, made possible by gains in health and financial security. Nevertheless, approximately two-thirds of older adults in the United States and Canada live close to at least one of their children, and frequency of contact is high through both visits and telephone calls (Rosenthal & Gladstone, 2000; U.S. Bureau of the Census, 2002c). Proximity increases with age. Elders who move usually do so in the direction of kin, and younger people tend to move in the direction of their aging parents.

Middle age is a time when adults reassess relationships with their parents, just as they rethink other close ties (Helson & Moane, 1987). Many adult children become more appreciative of their parents' strengths and generosity. Trisha, for example, marveled at her parents' fortitude in rearing three college-educated children despite limited income. And she recalled her mother's sound advice just before her marriage to Devin nearly three decades earlier: "Build a life together but also forge your own life. You'll be happier." Trisha had called on that advice at several turning points, and it had influenced her decisions.

In the non-Western world, older adults most often live with their married children. Chinese, Japanese, and Korean elderly, for example, generally reside with the eldest son and his wife and children (Kamo, 1998; Youn et al., 1999). Regardless of whether coresidence and daily contact are typical, relationship quality varies widely. Usually, patterns established earlier persist; positive parent–child ties remain so, as do conflict-ridden interactions.

Help exchanged between adult children and their aging parents is responsive to past and current family circumstances. The closer family ties were when children were growing up, the more help given and received (Whitbeck, Hoyt, & Huck, 1994). Also, parents give more to unmarried children and to children with disabilities. Similarly, children give more to widowed parents and parents in poor health. At the same time, a shift in helping occurs over the adult years. Parent-to-child advice, household aid, gift giving, and financial assistance decline, whereas child-to-parent help of various kinds increases (Rossi & Rossi, 1990; Zarit & Eggebeen, 2002). Even when the early parent–child relationship was emotionally distant, adult children offer more support as their parents age, out of a sense of altruism and family duty (Silverstein et al., 2002).

● Caring for Aging Parents. The burden of caring for aging parents can be great. In Chapter 2, we noted that the family structure has become more "top-heavy," with more generations alive, but fewer younger members as birthrates have declined. This means that more than one older family member is likely to need assistance, with fewer younger adults available to provide it. About 20 percent of midlifers in the United States and Canada are involved in caring for an aging parent with a chronic illness or disability (Takamura & Williams, 2002; Vanier Institute of the Family, 2002a). Today's middle-aged adults with ill or frail parents often face competing demands of children (some of whom are under age 18 and still at home) and employment. They are called the **sandwich generation** because they are "sandwiched," or squeezed, between the needs of aging parents and financially dependent children.

When an aging parent's spouse cannot provide care, adult daughters are the next most likely relatives to do so (see Figure 16.4). Even when the spouse is available, adult children—again, usually daughters—often pitch in as needed. Why are women more often elected as caregivers? Families turn to the person who seems most available—living nearby and with fewer commitments regarded as interfering with the ability to assist. These unstated rules, in addition to parents' preference for same-sex caregivers (aging mothers live longer), lead more women to fill the role. Daughters also feel a stronger sense of obligation to care for aging parents than do sons (Stein et al., 1998). About 50 percent of women caregivers are employed; another 10 to 30 percent quit their jobs to provide care. And time devoted to caring for a disabled aging parent is substantial, averaging 20 hours per week (Nichols & Junk, 1997; Takamura & Williams, 2002).

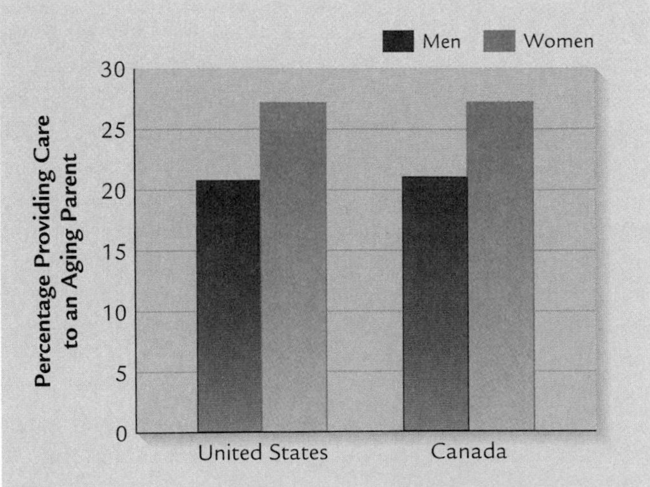

FIGURE 16.4 Percentage of American and Canadian 45- to 54-year-olds providing care to an aging parent. In both nations, more women than men provide care. Still, men make a substantial contribution to care of aging parents. (Adapted from Takamura & Williams, 2002; Vanier Institute of the Family, 2002a.)

Nevertheless, as Figure 16.4 shows, men make a substantial contribution to care of aging parents—one that should not be overlooked. Tim, for example, looked in on his father, a recent stroke victim, every evening, reading to him, running errands, making household repairs, and taking care of finances. His sister, however, provided more hands-on care in her own home—cooking, feeding, and bathing. The care sons and daughters provide tends to be divided along gender-role lines. About 10 percent of the time, sons become primary caregivers, generally when no other family member can do so, and are heavily involved in basic-care tasks (Campbell & Martin-Matthews, 2000, Harris, 1998).

As adults move from early to later middle age, the sex difference in parental caregiving declines. Perhaps as men reduce their vocational commitments and feel less need to conform to a "masculine" gender role, they grow more able and willing to provide basic care (Marks, 1996). At the same time, parental caregiving may contribute to men's greater openness to the "feminine" side of their personalities. A man who cared for his mother, severely impaired by Alzheimer's disease, commented on how the experience altered his outlook:

> Having to do personal care, becoming a male nurse, was a great adjustment. It was so difficult to do these tasks; things a man, a son, is not supposed to do. But, I had to alter, since charity must come before maintaining a selfish, conventional view. I have definitely modified my views on conventional expectations. (Hirsch, 1996, p. 112)

Although most adult children help willingly, caring for a chronically ill or disabled parent is highly stressful. Some people regard it as similar to caring for a young child, but it is radically different. The need for parental care typically occurs suddenly, after a heart attack, fall, stroke, or diagnosis of cancer, leaving little time for preparation. Whereas children become increasingly independent, the parent usually gets worse and the caregiving task as well as its cost escalates. "One of the most difficult aspects is the emotional strain of being such a close observer of my father's physical and mental decline," Tim explained to Devin and Trisha. Tim also felt a sense of grief over the loss of a cherished relationship, as his father no longer seemed to be his former self. Because duration of caregiving is uncertain, caregivers often feel they no longer have control over their lives (Gatz, Bengtson, & Blum, 1990).

Adults who share a household with ill parents—about 10 percent of caregivers—experience the most stress (Marks, 1996). In addition to the factors just mentioned, a parent and child who have lived separately for years usually dislike moving in together. Conflicts are likely to arise over routines and lifestyles. But the greatest source of stress is problem behavior, especially for caregivers of parents who have deteriorated mentally. Tim's sister reported that their father would wake during the night, ask repetitive questions, follow her around the house, and become agitated and combative.

Parental caregiving often has emotional and physical health consequences. It leads to role overload (conflict among employment, spouse, parent, and elder care roles), high job absenteeism, exhaustion, inability to concentrate, feelings of hostility, anxiety about aging, and rates of depression as high as 30 to 50 percent (Ingersoll-Dayton, Neal, & Hammer, 2001; Stephens et al., 1997, 2001). In collectivist cultures where adult children feel an especially strong sense of obligation to care for aging parents, caregiver involvement is generally higher. In a study of Korean, Korean-American, and Caucasian-American caregivers of parents with mental disabilities, the Korean and Korean Americans reported the highest levels of family obligation and care burden—and also the highest levels of anxiety and depression (Youn et al., 1999).

Social support is highly effective in reducing caregiver stress. Tim's encouragement, assistance, and willingness to listen helped his sister cope with in-home care of their father. As a result, she found satisfactions and rewards in it. Despite having more time to care for an ill parent, women who quit work generally fare poorly, probably because of social isolation and financial strain (Pohl et al., 1994). Positive experiences at work can actually reduce the stress of parental care as caregivers bring a favorable self-evaluation and a positive mood home with them (Stephens & Franks, 1999).

In Denmark, Sweden, and Japan, a government-sponsored home helper system eases the burden of parental care by making specially trained nonfamily caregivers available, based on the elder's needs (Blomberg, Edebalk, & Petersson, 2000; Yamanoi, 1993). In the United States and Canada, in-home care by a nonfamily caregiver is generally not an option because of its high cost and limited availability outside large cities; only 10 to 20 percent of family caregivers arrange for in-home assistance (Family Caregiver Alliance, 2002). And unless they must, few people want to place their parents in formal care, such as nursing homes, which also are expensive. At present, North American adult children have limited choices for how to provide parental care. The Caregiving Concerns table below summarizes ways to relieve the stress of caring for an aging parent—at the individual, family, community, and societal levels.

Siblings

As Tim's relationship with his sister reveals, siblings are ideally suited to provide social support. Nevertheless, a survey of a large sample of ethnically diverse Americans revealed that sibling contact and support decline from early to middle adulthood, rebounding only after age 70 for siblings living near one another (White, 2001). Decreased midlife contact is probably due to the demands of middle-aged adults' diverse roles. However, most adult siblings report getting together or talking on the phone at least monthly (Antonucci, Akiyama, & Merline, 2002).

Despite reduced contact, many siblings feel closer in midlife, often in response to major life events (Stewart et al., 2001). Launching and marriage of children seem to prompt siblings to think more about one another. As Tim commented, "It helped our relationship when my sister's children were out of the house and married. I'm sure she cared about me. I think she just didn't have time!" Parental illness can have a profound impact on sibling ties. Brothers and sisters who previously had little to do with one another find themselves in touch about parental care. When parents die, adult children realize they

Relieving the Stress of Caring for an Aging Parent

STRATEGY	DESCRIPTION
Use effective coping strategies.	Use problem-centered coping to manage the parent's behavior and caregiving tasks. Delegate responsibilities to other family members, seek assistance from friends and neighbors, and recognize the parent's limits while calling on capacities the parent does have. Use emotion-centered coping to reinterpret the situation in a positive way, such as emphasizing the opportunity it offers for personal growth and for giving to parents in the last years of their lives. Avoid denial of anger, depression, and anxiety in response to the caregiver work burden, which heightens stress.
Seek social support.	Confide in family members and friends about the stress of caregiving, seeking their encouragement and help. So far as possible, avoid quitting work to care for an ill parent, because doing so leads to social isolation and loss of financial resources.
Make use of community resources.	Contact community organizations to seek information and assistance, in the form of in-home respite help, home-delivered meals, transportation, and adult day care.
Press for workplace and public policies that relieve the emotional and financial burdens of caring for an aging parent.	Encourage your employer to provide elder care benefits, such as flexible work hours and caregiver leave without pay. Communicate with lawmakers and other citizens about the need for additional government funding to help pay for elder care. Emphasize the need for improved health insurance plans that reduce the financial strain of elder care on middle- and low-income families.

have become the oldest generation and must look to one another to sustain family ties (Gold, 1996). As in early adulthood, sister–sister relationships are closer than sister–brother and brother–brother ties, a trend apparent in many industrialized nations (Cicerelli, 1995).

Although the trend is toward closer relationships, not all sibling bonds improve. Recall Trisha's negative encounters with her sister, Dottie (see Chapter 15, page 493). Yet Dottie's difficult temperament had made her hard to get along with since childhood. Her temper flared when their father died and problems arose over family finances. When siblings do not help with parental caregiving, the child shouldering the burden can unleash powerful negative feelings (Merrill, 1997). As one expert expressed it, "As siblings grow older, good relationships [often] become better and rotten relationships get worse" (Moyer, 1992, p. 57).

Ethnic background also affects the quality of sibling ties. In one study, Italian-American siblings (especially sisters) had much greater contact and warmer relationships than white Protestant siblings (Johnson, 1985). The researchers speculated that strong parental authority in Italian immigrant families led siblings to turn to one another for support early in life. These warm bonds persisted into adulthood.

In industrialized nations, sibling relationships are voluntary. In village societies, they are generally involuntary and basic to family functioning. For example, among Asian Pacific Islanders, family social life is organized around strong brother–sister attachments. A brother–sister pair is often treated as a unit in exchange marriages with another family. After marriage, brothers are expected to protect sisters, and sisters serve as spiritual mentors to brothers. Families not only include biological siblings but also grant other relatives, such as cousins, the status of brother or sister (Cicerelli, 1995). This leads to an unusually large network of sibling support throughout life.

In village societies, cultural norms reduce sibling conflict, thereby ensuring family cooperation (Weisner, 1993). In industrialized nations, promoting positive sibling interaction in childhood is vital for warm, supportive sibling bonds in later years.

Friendships

As family responsibilities declined, Devin had more time to spend with friends in middle age. On Friday afternoons, he met several male friends at a coffee house, and they chatted for a couple of hours. However, the majority of Devin's friendships were couple-based—relationships he shared with Trisha. Compared with Devin, Trisha more often got together with friends on her own (Blieszner & Adams, 1992).

Characteristics of middle-age friendships are a continuation of trends discussed in Chapter 14. At all ages, men are less expressive with friends than are women. Men tend to talk about sports, politics, and business, whereas women focus on feelings and life problems. For this reason, when Trisha and Devin gathered with friends, men often congregated in one area, women in another (Fox, Gibbs, & Auerbach, 1985). Women report a greater number of close friends and say they both receive and provide their friends with more emotional support (Antonucci, 1994).

© BILLY E. BARNES/PHOTOEDIT

■ In communicating with friends, men typically talk about sports, politics, and business, whereas women focus on feelings and life problems. Because of gender differences in interests, when couples get together, men usually interact with men and women with women.

Nevertheless, for both sexes, number of friends declines with age, probably because people become less willing to invest in nonfamily ties unless they are very rewarding (Carbery & Buhrmester, 1998). As selectivity of friendship increases, older adults express more complex ideas about friendship. They also try harder to get along with friends (Antonucci & Akiyama, 1995). Having chosen a friend, middle-aged people attach great value to the relationship and take extra steps to protect it.

By midlife, family relationships and friendships support different aspects of psychological well-being. Family ties protect against serious threats and losses, offering security within a long-term time frame. In contrast, friendships serve as current sources of pleasure and satisfaction, with women benefiting somewhat more than men (Antonucci, Akiyama, & Merline, 2002). As middle-aged couples renew their sense of companionship, they may combine the best of family and friendship. Indeed, research indicates that viewing a spouse as a best friend contributes greatly to marital happiness (Bengtson, Rosenthal, & Burton, 1990).

Relationships Across Generations

A widely recognized aspiration of families, communities, and society as a whole is that each generation leave the next better off. For more than 2 centuries, North Americans have achieved this goal. For example, Trisha's high-school-educated parents did everything possible to support her desire to earn a professional degree. And Devin and Trisha reared their son under financially more comfortable circumstances than they had experienced as children.

Yet Trisha and Devin often wondered, as many North Americans do, whether the current generation of young people would be as well off as the previous one. And on the basis of high rates of marital breakup, single-parent families, and youth crime, they questioned whether younger people would continue to respect and care for their elders. Trisha and Devin also worried that improved financial conditions among the elderly, in comparison to the high poverty rate among families with children (see Chapter 2), might fuel intergenerational resentment.

A survey of a large, nationally representative sample of American 18- to 90-year-olds addressed the question of whether solidarity between generations is eroding. Findings showed that despite the social changes just mentioned, supportive ties among younger and older individuals remain strong (Bengtson & Harootyan, 1994; Lawton, Silverstein, & Bengtson, 1994). At the family level, the large majority (90 percent) of adult children reported feeling close to an aging parent. In fact, adult children described closer ties to their aging parents than aging parents felt toward their adult children (although 80 percent of these ties were strong). And in line with the high rate of help giving noted earlier in this chapter, younger adults expressed a deep sense of commitment toward their parents—more than their parents expected! Furthermore, adult children's willingness to grant time and assistance to aging parents is unaffected by parents' wealth (Wong, Kitayama, & Soldo, 1999).

Young, middle-aged, and older survey respondents reported a wide range of volunteering in their communities. Middle-aged adults frequently devoted time to children and adolescents—usually through religious activities and academic tutoring programs. Older adults, by contrast, more often provided volunteer assistance to the elderly. But they were more likely to do so if they had warm relationships with their adult children. More than 70 percent of the sample described giving some kind of informal assistance to relatives, friends and neighbors, and other community residents (Harootyan & Vorek, 1994).

Finally, most participants expressed no resentment about government benefits to other age groups. Instead, a *norm of equity* characterized their responses. When people believed the needs of one generation were unmet, they expressed dissatisfaction about government benefits to another age group. Only rarely did they say their own age group was neediest and should get more (Schlesinger & Kronebusch, 1994). Very young adults (18- to 24-year-olds) are more likely than other age groups to view government programs serving the elderly as too costly. But having had frequent contact with grandparents increases their endorsement of these benefits (Silverstein & Parrott, 1997).

In sum, despite public concern over intergenerational conflicts, most people are not guided by self-interest. To the contrary, their values, opinions, and behaviors reveal "hidden bridges" between generations.

Ask Yourself

REVIEW

How do age, sex, proximity, and culture affect grandparent–grandchild ties?

APPLY

Raylene and her brother Walter live in the same city as their aging mother, Elsie. When Elsie could no longer live independently, Raylene took primary responsibility for her care. What factors probably contributed to Raylene's involvement in caregiving and Walter's lesser role?

APPLY

As a young adult, Daniel maintained close friendship ties with six college classmates. At age 45, he continued to see only two of them. What explains Daniel's reduced circle of friends in midlife?

CONNECT

Cite evidence that early family relationships affect middle-aged adults' bonds with adult children, aging parents, and siblings.

www.

Vocational Life

We have already seen that the midlife transition typically involves vocational adjustments. For Devin, it resulted in a move up the career ladder to a demanding administrative post as college dean. Trisha reoriented her career from a large to a small law firm, where she felt her efforts were appreciated. Recall from Chapter 15 that after her oldest child left home, Anya earned a college degree and entered the work force for the first time. Jewel strengthened her commitment to an already successful business, whereas Elena changed careers. Finally, Tim reduced his career obligations as he prepared for retirement.

Work continues to be a salient aspect of identity and self-esteem in middle adulthood. More so than in earlier or later years, people attempt to increase the personal meaning and self-direction of their vocational lives (Levinson, 1978, 1996). At the same time, certain aspects of job performance improve. Older employees have lower rates of absenteeism, turnover, and accidents and show no change in work productivity (Warr, 1994). Consequently, the value of an older employee ought to be equal to, and possibly even greater than, that of a younger employee.

The post–World War II baby boom, along with the elimination of mandatory retirement age in most industrialized nations, means that the number of older workers will rise dramatically over the next few decades. Yet a favorable transition from adult worker to older worker is hindered by negative stereotypes of aging—incorrect beliefs about limited learning capacity, slower decision making, and resistance to change and supervision (Sterns & Huyck, 2001). Furthermore, gender discrimination continues to restrict the career attainments of many women. Let's take a close look at middle-aged work life.

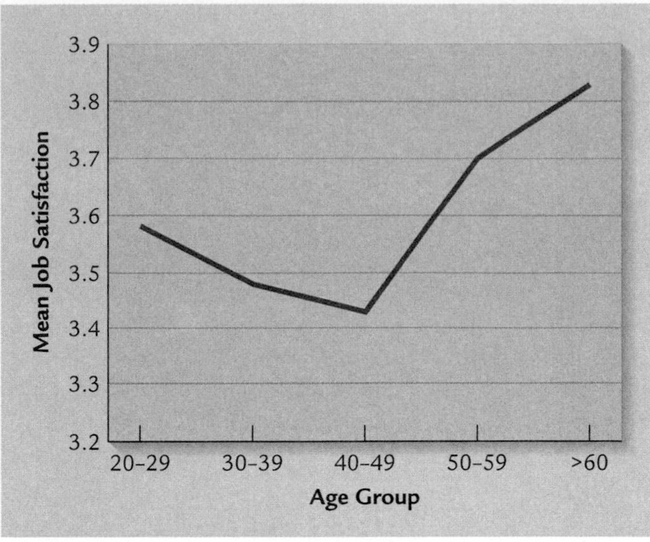

FIGURE 16.5 Age-related change in job satisfaction. In this study of more than 2,000 university employees at all levels, from secretary to university president, job satisfaction dropped slightly in early adulthood as people encountered some discouraging experiences (see Chapter 14). In middle age, job satisfaction showed a steady rise. (From W. A. Hochwarter et al., 2001, "A Note on the Nonlinearity of the Age-Job-Satisfaction Relationship," *Journal of Applied Social Psychology, 31,* p. 1232. Reprinted by permission.)

Job Satisfaction

Job satisfaction has both psychological and economic significance. If people are dissatisfied at work, the consequences include strikes, grievances, absenteeism, and turnover, all of which are costly to employers.

Research shows that job satisfaction increases in midlife at all occupational levels, from executives to hourly workers (see Figure 16.5). The relationship is weaker for women than for men, probably because women's reduced chances for advancement result in a sense of unfairness. It is also weaker for blue-collar than for white-collar workers, perhaps because blue-collar workers have less control over their own work schedules and activities (Fotinatos-Ventouratos & Cooper, 1998; Avolio & Sosik, 1999). When different aspects of jobs are considered, intrinsic satisfaction—happiness with the work itself—shows a strong age-related gain. Extrinsic satisfaction—contentment with supervision, pay, and promotions—changes very little (Hochwarter et al., 2001).

What explains the rise in job satisfaction during middle adulthood? A broader time perspective probably contributes. "I recall complaining about how much I had to do when I first started teaching," remarked Devin. "Since then, I've seen a lot of hard times. From my current vantage point, I can tell a big problem from a trivial one." Moving out of unrewarding work roles, as Trisha did, can also boost morale. Key characteristics that predict job well-being include involvement in decision making, reasonable workloads, and good physical working conditions. Older people may have greater access to jobs that are attractive in these ways. Finally, having fewer alternative positions into which they can move, older workers generally reduce their career aspirations. As the perceived gap between actual and possible achievements declines, work involvement increases (Warr, 1992).

Although emotional engagement with work is usually seen as psychologically healthy, it can also result in **burnout**—a condition in which long-term job stress leads to mental exhaustion, a sense of loss of personal control, and feelings of reduced accomplishment. Burnout occurs more often in the helping professions, including health care, human services, and teaching, which place high emotional demands on employees (Zapf et al., 2001). Although people in interpersonally demanding jobs are as psychologically healthy as other people, sometimes a worker's dedication exceeds his or her coping skills, especially in an unsupportive work environment. Burnout is associated with excessive work assignments for available time and lack of encouragement and feedback from supervisors. It tends to occur more often in North America than in Western Europe, perhaps because of North Americans' greater achievement orientation (Maslach, Schaufeli, & Leiter, 2001)

Burnout is a serious occupational hazard because it is linked to absenteeism, turnover, poor job performance, and impaired health (Wright & Bonett, 1997). Employers can prevent burnout by making sure workloads are reasonable, providing opportunities for workers to take time out from stressful situations, limiting hours of stressful work, and offering social support. Interventions that enlist employees' participation in designing higher-quality work environments show promise for increasing work engagement and effectiveness and reducing burnout (Leiter & Maslach, 2000). Other proposals include provisions for work at home, which would respond to the needs of some people for less time pressure and noise.

Career Development

After several years as a parish nurse, Anya felt a need for additional training to do her job better. Trisha appreciated her firm's generous support of workshop and course attendance, which helped her keep abreast of new legal developments. And as college dean, Devin took several seminars each year on management effectiveness and instructional technology. As Anya, Trisha, and Devin reveal, career development is vital throughout work life.

● **Job Training.** Anya's 35-year-old supervisor, Roy, was surprised when she asked for time off to upgrade her skills. "You're in your fifties," he commented insensitively. "What're you going to do with so much new information at this point in your life?"

Although Roy's narrow-minded attitude is usually unspoken, it is all too common among managers—even some who are older themselves! Research suggests that training and on-the-job career counseling are less available to older workers. And when career development activities are offered, older employees may be less likely to volunteer for them (Cleveland & Shore, 1992; Salthouse & Maurer, 1996). What influences willingness to engage in job training and updating? Characteristics of the person and the work environment make a difference.

On the person side, the degree to which an individual wants to change is important. With age, growth needs decline somewhat in favor of security needs. Consequently, learning and challenge may have less intrinsic value to many older workers. Perhaps for this reason, older employees depend more on co-worker and supervisor encouragement for vocational development. Yet we have just seen that they are less likely to have supportive supervisors. Furthermore, negative stereotypes of aging reduce older workers' self-efficacy, or confidence that they can renew and expand their skills—another reason they may not volunteer for training experiences. Self-efficacy is a powerful predictor of employees' efforts to improve career-relevant skills (Maurer, 2001; Maurer & Tarulli, 1994).

Challenging tasks facilitate updating. An employee given work that requires new learning has to obtain that learning to complete the assignment. Unfortunately, older workers sometimes receive more routine tasks than younger workers. Therefore, some of their reduced motivation to engage in career-relevant learning may be due to the type of assignments they receive. Interaction among co-workers can also have a profound impact. Within project teams, people similar in age communicate more often. Age-balanced work groups (with more than one person in each age range) foster on-the-job learning because communication is a source of support as well as a means of acquiring job-relevant information (Zenger & Lawrence, 1989).

Computer networks facilitate such team interaction and also enhance sharing of information in large companies. In this way, they help employees at lower levels develop a broader perspective on their contribution to company goals and feel a part of the larger organization (Avolio & Sosik, 1999). Furthermore, expansion of computer technology compels older workers to

■ This project team, made up of two young adults and two middle-aged adults, works on a challenging assignment in an advertising agency. Because team members similar in age communicate more, age-balanced work groups foster on-the-job learning and enhanced performance.

© AP/WIDE WORLD

■ Women and ethnic minorities face a glass ceiling—an invisible barrier to advancement up the career ladder—in major corporate and government organizations. A few, such as Nancy Pelosi, Democratic leader of the United States Congress and the first woman to lead a major American political party, succeed in shattering the glass ceiling. Here, she discusses controversial medical malpractice legislation with a man who had his limbs amputated after two doctors failed to diagnose an acute infection.

upgrade their computer literacy. This, in turn, helps them keep pace with changing aspects of their jobs.

● **Gender and Ethnicity: The Glass Ceiling.** In her thirties, Jewel became a company president by starting her own business. As a woman, she decided that her chances of rising to a top flight executive position in a large corporation were so slim that she didn't even try. Recall from Chapter 14 that women and ethnic minorities rarely move into high-level management jobs. At major North American firms, only 5 percent of senior executives are women. Fewer than 1 percent are members of ethnic minorities (Globalist, 2003).

Women and ethnic minorities face a **glass ceiling,** or invisible barrier to advancement up the corporate ladder. Contrary to popular belief, their low numbers cannot be attributed to poor management skills. A survey of employees at six large American firms revealed that compared to men, women managers were rated as more effective and satisfying to work for and as more likely to motivate extra effort. Characteristics that distinguished them from their male counterparts were charisma, inspiration, and considerateness (Bass & Avolio, 1994). Modern businesses realize that the best managers must not only display “masculine” authority and decisiveness but also build consensus through broad participation in decision making—an approach that requires “feminine” qualities of caring and collaboration.

Why is there a glass ceiling? Management is an art and skill that must be taught. Yet women and members of ethnic minorities have less access to mentors, role models, and informal networks that serve as training routes. Also, the majority of chief executives acknowledge that they spend less money on formal training programs for their female employees. The reasons given are stereotyped doubts about women's career commitment and ability to become strong managers. Furthermore, challenging, high-risk, high-visibility assignments that require leadership and open the door to advancement, such as startup ventures, international experience, and troubleshooting, are rarely granted to women and minorities (Barr, 1996; Lyness & Thompson, 1997). And when women do demonstrate qualities linked to leadership and advancement, such as assertiveness, confidence, forcefulness, and ambitiousness, they encounter prejudice because they deviate from traditional gender roles. They are evaluated more negatively than men with these traits (Carli & Eagly, 2000; Eagly & Karau, 2002).

Like Jewel, many women have shattered the glass ceiling by going around it. Nearly twice as many female as male middle managers quit their jobs in large corporations, largely because of lack of career opportunities (Stroh, Brett, & Reilly, 1996). Today, there are 4 million American women who, like Jewel, have started their own businesses—a number six times that of men (Mergenhagen, 1996). When women and ethnic minorities leave the corporate world to further their careers, companies not only lose valuable talent but also fail to address the leadership needs of an increasingly diversified work force.

Career Change at Midlife

Although most people remain in the same vocation through middle age, career change does occur, as Elena's shift from journalism to teaching and creative writing illustrates. Recall that circumstances at home and at work motivated Elena's decision to pursue a new vocation. Like other career changers, she wanted a more satisfying life—a goal she attained by ending an unhappy marriage and initiating a long-awaited vocational move at the same time.

As noted earlier, midlife career changes are usually not radical; they typically involve leaving one line of work for a related one. Elena sought a more stimulating, involving job. But other people move in the reverse direction—to careers that are more relaxing, free of painful decisions, and less demanding in terms of responsibility for others (Juntunen, Wegner, & Matthews, 2002). The decision to change is often difficult. The individual must weigh years invested in one set of skills, current income, and job security against present frustrations and hoped-for gains from a new vocation.

When an extreme career shift occurs, it usually signals a personal crisis (Young & Rodgers, 1997). In a study of professionals who abandoned their well-paid, prestigious positions for routine, poorly paid, semiskilled work, nonwork problems influenced the break with an established career. For example, an eminent 55-year-old TV producer became a school bus driver, and a New York banker become a waiter in a ski resort. Each responded to feelings of personal meaninglessness by escaping from family conflict, difficult relationships with colleagues, and work that had become unsatisfying into a freer, more independent lifestyle (Sarason, 1977).

Unemployment

Devin and Trisha's friend George worked in a corporate retirement-planning office, counseling retirees on how to enjoy leisure time or find new work. When he lost his job at age 54, he had to apply his counseling skills to himself. George found unemployment to be a culture shock. For the first 2 weeks, he spent most of his days in bed, didn't shave, and drank heavily. Even after this initial phase, George was depressed and had frequent bouts of illness.

As companies downsize and jobs are eliminated, the majority of people affected are middle-aged and older. Although unemployment is difficult at any time, middle-aged workers show a sharper decline in physical and mental health than their younger counterparts (Kulik, 2000). Older workers affected by layoffs remain without work for a longer time, suffering substantial income loss (Couch, 1998). In addition, people over age 40 who must reestablish occupational security find themselves "off-time" in terms of the social clock. Consequently, job loss can disrupt major tasks of midlife, including generativity and reappraisal of life goals and accomplishments (Broomhall & Winefield, 1990). Finally, having been more involved in and committed to an occupation, the older employed worker has also lost something of greater value.

After a despondent period, George began to follow the advice he had given his clients. He made a list of what he really liked to do, what he didn't want to do again, and the risks he could take given his current financial and life circumstances. He formed his own small business and continued to advise retirees, write articles, and give speeches on all aspects of retirement, working from home in a T-shirt instead of a business suit. Effective problem-centered coping strategies enabled George to reestablish a satisfying work life (Kinicki, Prussia, & McKee-Ryan, 2000).

Social support is vital for reducing stress and reassuring middle-aged job seekers of their worth. However, not all forms of social support work equally well. Recognition of the person's abilities and communication with others who share interests and values help the most (Mallinckrodt & Fretz, 1988). Both are experiences that occur often in relations with co-workers.

People who have lost their jobs in midlife usually do not duplicate the status and pay of their previous positions. As they search, they encounter age discrimination and find that they are overqualified for many openings. Counseling that focuses on financial planning, reducing feelings of humiliation due to the stigma of unemployment, and encouraging personal flexibility can help people find alternative, gratifying work roles.

Planning for Retirement

One evening, Devin and Trisha met George and Anya for dinner. Halfway through the meal, Devin inquired, "George, you're an expert on the topic. Tell us what you and Anya are going to do about retirement. Are you planning to close down your business or work part-time? Do you think you'll stay here or move out of town?"

Three or four generations ago, the two couples would not have had this conversation. In 1900, about 70 percent of North American men age 65 and over were in the labor force. By 1970, the figure had dropped to 27 percent. In 2000, it was 9 percent in Canada and 17 percent in the United States (Statistics Canada, 2002h; U.S. Bureau of the Census, 2002c). Retirement is no longer a privilege reserved for the wealthy. Government-sponsored retirement benefits began in Canada in 1927 and in the United States in 1935. Today, the U.S. and Canadian federal governments pay social security benefits to the majority of the aged. Others are covered by employer-based private pension plans (Chappell et al., 2003; Meyer & Bellas, 1995).

Most workers report looking forward to retirement, and an increasing number are leaving full-time work in midlife. The average age of retirement declined during the past 2 decades. Currently, it is age 62 in Canada and the United States and hovers between 60 and 63 in other Western nations (Statistics Canada, 2002h; U.S. Bureau of the Census, 2002c). This means that increasing numbers of people spend up to one-fourth of their lives in retirement.

Retirement is a lengthy, complex process that begins as soon as the middle-aged person first thinks about it (Kim & Moen, 2002b). Planning is important, since retirement leads to a loss of two important work-related rewards—income and status—and to change in many other aspects of life. Like other life transitions, retirement is often stressful.

"Retirement planning helps you evaluate your options, learn about the availability of resources, and prepare emotionally for the changes ahead," Devin and Trisha heard George explain when they attended one of his retirement seminars. Up to 40 percent of middle-aged people engage in no concrete retirement planning, yet research consistently shows that planning results in better retirement adjustment and satisfaction (Pery, 1995; Quick & Moen, 1998).

The Caregiving Concerns table on page 538 lists the variety of issues addressed in a typical retirement preparation program. Financial planning is vital, since income typically drops by 50 percent. Although more people are involved in financial planning than in other forms of preparation, even those who attend financial education programs often fail to look closely at their financial well-being and to make wise decisions

Ingredients of Effective Retirement Planning

ISSUE	DESCRIPTION
Finances	Ideally, financial planning for retirement should start with the first paycheck; at a minimum, it should begin 10 to 15 years before retirement, since most people spend more than 20 years retired.
Fitness	Starting a fitness program in middle age is important because good health is crucial for well-being in retirement.
Role adjustment	Retirement is harder for people who strongly identify with their work role. Preparing for a radical role adjustment reduces stress.
Where to live	The pros and cons of moving should be considered carefully because where one lives affects access to health care, friends, family, recreation, entertainment, and part-time employment.
Leisure activities	A retiree typically gains an additional 50 hours per week of free time. Careful planning of what to do with that time has a major impact on psychological well-being.
Health insurance	Finding out about government-sponsored health insurance options helps protect quality of life after retirement.
Legal affairs	The preretirement period is an excellent time to finalize a will and begin estate planning.

Source: Pery, 1995.

(Hershey et al., 1998). Many could benefit from an expert's financial analysis and counsel.

Retirement leads to ways of spending time that are largely guided by what one wants to do, not by what one has to do. Individuals who have not thought carefully about how to fill this time may find their sense of purpose in life seriously threatened. Research in the United States and Canada reveals that planning for an active life has an even greater impact on happiness after retirement than financial planning. Undoubtedly this is because participation in activities promotes many factors essential for psychological well-being, including a structured time schedule, social contact, and self-esteem (MacEwen et al., 1995; Ostling & Kelloway, 1992). Carefully considering whether or not to relocate at retirement is related to an active life, since it affects access to family, friends, recreation, entertainment, and part-time work.

Devin retired at age 62, George at age 66. Although several years younger, Trisha and Anya—like many married women—coordinated their retirements with those of their husbands (Ruhm, 1996). In good health but without an intimate partner to share her life, Jewel kept her consulting business going until age 75. Tim took early retirement and moved to be near Elena, where he devoted himself to public service—tutoring second graders in a public school, transporting inner-city children to museums, and coaching after-school and weekend youth sports. For Tim, like many executives, retirement offered the first opportunity to pay attention to the world around him.

Unfortunately, less educated people with lower lifetime earnings are least likely to attend retirement preparation programs, yet they stand to benefit the most. And compared with men, women do less planning for retirement, often depending on their husband's preparations—a finding that may change as women increasingly become equal, rather than secondary, family earners (Han & Moen, 1999). Employers must take extra steps to encourage lower-paid workers and women to participate in planning activities (Gibson & Burns, 1991). In addition, enhancing retirement adjustment among the economically disadvantaged depends on access to better health care, vocational training, and jobs at early ages. Clearly, a lifetime of opportunities and experiences affects the transition to retirement. In Chapter 18, we will consider the decision to retire and retirement adjustment in greater detail.

Ask Yourself

REVIEW
What factors lead job satisfaction to increase with age?

APPLY
An executive asks you what his large corporation might do to promote advancement of women and ethnic minorities to upper management positions. What would you recommend?

CONNECT
Supervisors sometimes assign only the more routine tasks to older workers, believing that they can no longer handle complex assignments. Cite evidence presented in this and the previous chapter that shows this assumption is incorrect.

www.

Summary

Erikson's Theory: Generativity versus Stagnation

According to Erikson, how does personality change in middle age?

- Generativity begins in early adulthood but expands greatly as middle-aged adults face Erikson's psychological conflict of **generativity versus stagnation.** Highly generative people find fulfillment as they make contributions to society through parenthood, other family relationships, the workplace, volunteer activities, and many forms of productivity and creativity. Cultural demands and inner desires jointly shape adults' generative activities. Highly generative people are especially well adjusted.

Other Theories of Psychosocial Development in Midlife

Describe Levinson's and Vaillant's views of psychosocial development in middle adulthood, and discuss similarities and differences between men and women.

- According to Levinson, middle-aged adults reassess their relation to themselves and the external world. They must confront four developmental tasks, each of which requires them to reconcile two opposing tendencies within the self: young–old, destruction–creation, masculinity–femininity, and engagement–separateness.
- Because of the double standard of aging, women have more trouble than men accepting being older. Men may adopt "feminine" traits of nurturance and caring, whereas women may take on "masculine" traits of autonomy, dominance, and assertiveness. Men and successful career-oriented women often reduce their focus on ambition and achievement. Women who have devoted themselves to child rearing or an unfulfilling job typically increase their involvement in work and the community.
- According to Vaillant, middle-aged adults become guardians of their culture. In their late forties and fifties, they carry peak responsibility for the functioning of society.

Does the term midlife crisis *fit most people's experience of middle adulthood?*

- Wide individual differences exist in response to midlife. Only a minority of people experience a **midlife crisis** in which intense self-doubt and inner turmoil lead them to make drastic changes in their personal lives and careers.

Characterize middle adulthood using a life events approach and a stage approach.

- Both continuity and stagewise change characterize emotional and social development in middle adulthood. Some changes are adaptations to external events (such as the family life cycle), but these events are less age-graded than in the past. At the same time, the majority of middle-aged adults report stagelike development, involving troubling moments that prompt new understandings and goals.

Stability and Change in Self-Concept and Personality

Describe changes in self-concept and personality in middle adulthood.

- **Possible selves** become fewer in number as well as more modest and concrete as middle-aged individuals adjust their hopes and fears to the circumstances of their lives. Revising possible selves enables adults to maintain self-esteem and stay motivated.
- Adults become more introspective and in touch with themselves in midlife. Self-acceptance, autonomy, and environmental mastery increase, leading some people to consider middle age the "prime of life." In addition, coping strategies become more effective as middle-aged adults develop greater confidence in their ability to handle life's problems.

Describe changes in gender identity in midlife.

- Both men and women become more androgynous in middle adulthood. Biological explanations, such as **parental imperative theory,** have been criticized, and evidence for them is mixed. A complex combination of social roles and life conditions is probably responsible for midlife changes in gender identity.

Discuss stability and change in the "big five" personality traits in adulthood.

- Among the **"big five" personality traits,** neuroticism, extroversion, and openness to experience show modest declines during adulthood, whereas agreeableness and conscientiousness increase. At the same time, individual differences in the "big five" traits are large and highly stable. Although adults change in overall organization and integration of personality, they do so on a foundation of basic, enduring dispositions.

Relationships at Midlife

Describe the middle adulthood phase of the family life cycle, and discuss midlife relationships with a marriage partner, adult children, grandchildren, and aging parents.

- "Launching children and moving on" is the newest and longest phase of the family life cycle. Middle-aged adults must adapt to many entries and exits of family members as their own children leave, marry, and produce grandchildren and as their own parents age and die.
- The changes of midlife prompt many adults to focus on improving their marriages. When divorce occurs, middle-aged adults seem to adapt more easily than younger people. For women, marital breakup usually brings significant economic disadvantage, contributing to the **feminization of poverty.**
- Most middle-aged parents adjust well to the launching phase of the family life cycle, especially if they have a strong work orientation and if parent–child contact and affection are sustained. As children marry and bring in-laws into the family network, middle-aged parents, especially mothers, often become **kinkeepers.**
- When family relationships are positive, grandparenthood is an important means of fulfilling personal and societal needs. In low-income families and in some subcultures, grandparents perform essential activities, including financial assistance and child care. Increasingly, grandparents have become primary caregivers when

serious family problems exist, resulting in **skipped-generation families.**

- Middle-aged adults reassess their relationships with aging parents, often becoming more appreciative. Yet quality of the early parent–child relationship—positive or conflict-ridden—usually persists, affecting help given and received.

- Adults caught between the demands of ill or frail parents and financially dependent children are called the **sandwich generation.** The burden of caring for aging parents falls most heavily on adult daughters, but both men and women participate. As adults move from early to later middle age, the sex difference in parental caregiving declines. Social support is highly effective in relieving caregiver stress.

Describe midlife sibling relationships and friendships, and discuss relationships across generations.

- Sibling contact and support decline from early to middle adulthood, probably because of the demands of middle-aged adults' diverse roles. However, most middle-aged siblings tend to feel closer, often in response to major life events, such as launching and marriage of children and illness and death of parents. Sister–sister ties are typically closest in industrialized nations, where sibling relationships are voluntary. In nonindustrialized societies, where sibling relationships are basic to family functioning, other sibling attachments (such as brother–sister) may be unusually strong.

- In midlife, friendships are fewer, more selective, and more deeply valued. Men continue to be less expressive with their friends than are women, who have more close friendships. Viewing a spouse as a best friend can contribute greatly to marital happiness.

- Recent research reveals strong supportive ties across generations in families and communities. Most people do not resent government benefits for other age groups.

Vocational Life

Discuss job satisfaction and career development in middle adulthood, paying special attention to sex differences and experiences of ethnic minorities.

- Vocational readjustments are common as middle-aged people seek to increase the personal meaning and self-direction of their work lives. Certain aspects of job performance improve. Overall job satisfaction increases at all occupational levels, though less so for women than for men.

- **Burnout** is a serious occupational hazard, especially in helping professions. Burnout can be prevented by ensuring reasonable workloads, limiting hours of stressful work, and providing workers with social support. Interventions that enlist employees' participation in designing higher-quality work environments also seem helpful.

- Career development is vital throughout work life. Yet older workers engage in less job training because of negative stereotypes of aging, which impair self-efficacy; lack of encouragement from supervisors; and more routine work assignments.

- Women and ethnic minorities face a **glass ceiling** because of reduced access to formal and informal management training and prejudice against women who demonstrate qualities linked to leadership and advancement. Many women further their careers by leaving the corporate world, often to start their own businesses.

Discuss career change and unemployment in middle adulthood.

- Most middle-aged people remain in the same vocation. Those who change careers usually leave one line of work for a related one. Radical career change often signals a personal crisis.

- Unemployment is especially difficult for middle-aged adults, who make up the majority of workers affected by corporate downsizing and layoffs. Social support is vital for reducing stress. Counseling can help midlife job seekers find alternative, gratifying work roles, but these rarely duplicate the status and pay of their previous positions.

Discuss the importance of planning for retirement, noting various issues that middle-aged adults should address.

- An increasing number of American workers are retiring from full-time work in midlife. Planning for retirement is important because the changes it brings—loss of income and status and increase in free time—are often stressful. Besides financial planning, planning for an active life is vital and has an especially strong impact on happiness following retirement. Employers must take extra steps to encourage lower-paid workers and women to participate in retirement preparation programs.

Important Terms and Concepts

"big five" personality traits (p. 522)
burnout (p. 534)
feminization of poverty (p. 525)
generativity versus stagnation (p. 512)
glass ceiling (p. 536)
kinkeeper (p. 526)
midlife crisis (p. 517)
parental imperative theory (p. 521)
possible selves (p. 518)
sandwich generation (p. 530)
skipped-generation family (p. 528)

FYI For Further Information and Help

Consult the Companion Website for *Development Through the Lifespan, Third Edition,* (www.ablongman.com/berk) where you will find the following resources for this chapter:

- **Chapter Objectives**
- **Flashcards** for studying important terms and concepts
- **Annotated Weblinks** to guide you in further research
- **Ask Yourself** questions, which you can answer and then check against a sample response
- **Suggested Readings**
- **Practice Test** with immediate scoring and feedback

Milestones

Development in Middle Adulthood

AGE

40-50 years

PHYSICAL

- Accommodative ability of the lens of the eye, ability to see in dim light, and color discrimination decline; sensitivity to glare increases. (483)
- Hearing loss at high frequencies occurs. (484)
- Hair grays and thins. (483)
- Lines on the face become more pronounced, and skin loses elasticity and begins to sag. (484)
- Weight gain continues, accompanied by a rise in fatty deposits in the torso, whereas fat beneath the skin declines. (484)
- Loss of lean body mass (muscle and bone) occurs. (484)

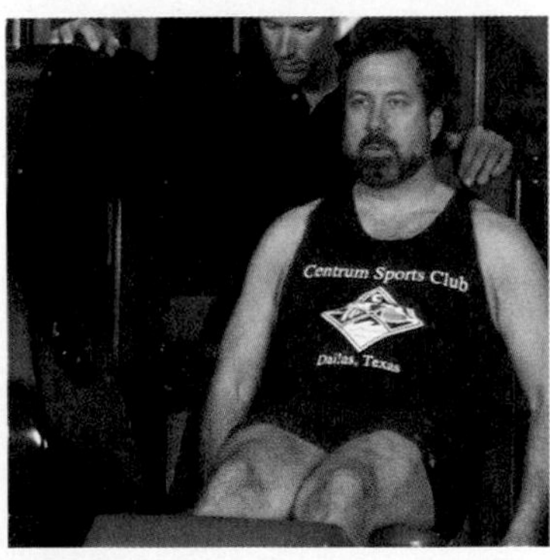

- In women, production of estrogen drops, leading to shortening and irregularity of the menstrual cycle. (486)
- For men, quantity of semen and sperm declines. (487)
- Intensity of sexual response declines, but sexual activity drops only slightly. (489)
- Rates of cancer and cardiovascular disease increase, more for men than for women. (489–492)

COGNITIVE

- Consciousness of aging increases. (483)
- Crystallized intelligence increases; fluid intelligence declines. (498–499)
- Processing speed declines; adults compensate through practice and experience. (501)

- On complex tasks, ability to divide and control attention declines; adults compensate through practice and experience. (502)
- Amount of information retained in working memory declines, largely due to reduced use of memory strategies. (502)
- Retrieving information from long-term memory becomes more difficult. (502-503)
- General factual knowledge, procedural knowledge, and knowledge related to one's occupation remain unchanged or increase. (503)
- Gains in practical problem solving and expertise occur. (503)
- Creativity focuses on integrating ideas and becomes more altruistic. (504)
- If in an occupation offering challenge and autonomy, shows gains in cognitive flexibility. (505)

EMOTIONAL/SOCIAL

- Generativity increases. (512)
- Focuses more on personally meaningful living. (515)

- Possible selves become fewer in number and more modest and concrete. (518)
- Introspection increases as people contemplate the second half of life. (519)
- Self-acceptance, autonomy, and environmental mastery increase. (519)
- Coping strategies become more effective. (519)
- Gender identity becomes more androgynous: "masculine" traits increase in women, "feminine" traits in men. (521)
- May launch children. (524)
- May enlarge the family network to include in-laws. (526)
- May become a kinkeeper, especially if a mother. (526)
- May care for a parent with a disability or chronic illness. (530)
- Siblings may feel closer. (531)
- Number of friends generally declines. (532)
- Job satisfaction increases. (534)

AGE	PHYSICAL	COGNITIVE	EMOTIONAL/SOCIAL
50-60 years	• Lens of the eye loses its accommodative ability entirely. (483) • Hearing loss extends to all frequencies but remains greatest for highest tones. (484) • Skin continues to wrinkle and sag, and "age spots" appear. (484) • Menopause occurs. (486) • Continued loss of bone mass, accelerating especially for women after menopause and leading to high rates of osteoporosis. (493) 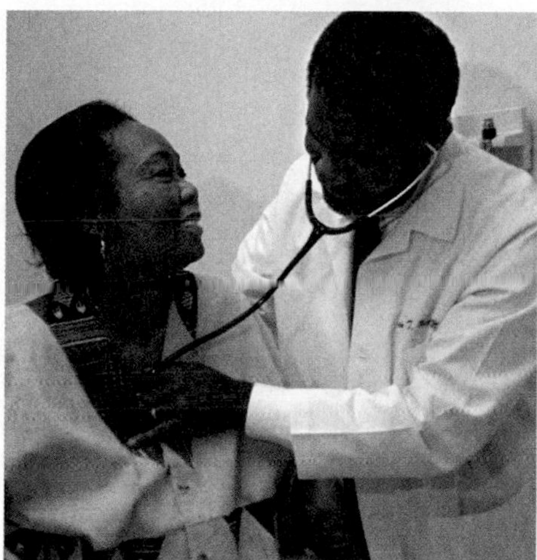 • Due to collapse of disks in the spinal column, height may drop by as much as 1 inch. (486)	• Changes in cognition described on the previous page continue. 	• Emotional and social changes described on the previous page continue. • May become a grandparent. (526) • Parent-to-child help-giving declines, and child-to-parent help-giving increases. (530) • May retire. (537) 

Note: Numbers in parentheses indicate the page(s) on which each milestone is discussed.

17

Physical and Cognitive Development in Late Adulthood

Physical Development

Life Expectancy
Variations in Life Expectancy ● Life Expectancy in Late Adulthood ● Maximum Lifespan

- A Lifespan Vista: What Can We Learn About Aging from Centenarians?

Physical Changes
Nervous System ● Sensory Systems ● Cardiovascular and Respiratory Systems ● Immune System ● Sleep ● Physical Appearance and Mobility ● Adapting to Physical Changes of Late Adulthood

- Cultural Influences: Cultural Variations in the Experience of Aging

Health, Fitness, and Disability
Nutrition and Exercise ● Sexuality ● Physical Disabilities ● Mental Disabilities ● Health Care

- Social Issues: Interventions for Caregivers of Elders with Dementia

Cognitive Development

Memory
Deliberate versus Automatic Memory ● Associative Memory ● Remote Memory ● Prospective Memory

Language Processing

Problem Solving

Wisdom

Factors Related to Cognitive Change

Cognitive Interventions

Lifelong Learning
Types of Programs ● Benefits of Continuing Education

Although elders do not see and hear as well or move and think as quickly as they did at younger ages, they find ways to sustain high levels of functioning. And their extensive life experience enhances their ability to solve human problems. Consequently, important leadership positions are often reserved for the old.

At age 64, Walt gave up his photography business and looked forward to leisure years ahead with 60-year-old Ruth, who retired from her position as a social worker at the same time. This culminating phase of Walt's and Ruth's lives was filled with volunteer work, golfing three times a week, and joint summer vacations with Walt's older brother Dick and his wife, Goldie. Walt also took up activities he had always loved but had little time to pursue—writing poems and short stories, attending theater performances, enrolling in a class on world politics, and cultivating a garden that became the envy of the neighborhood. Ruth read voraciously, served on the board of directors of an adoption agency, and had more time to visit her sister Ida in a nearby city.

Over the next 20 years, Walt's and Ruth's energy and vitality were an inspiration to everyone who met them. Their warmth, concern for others, and generosity with their time led not just their own children and grandchildren, but also nieces, nephews, children of their friends, and former co-workers, to seek them out. On weekends, their home was alive with visitors.

Then, in their early eighties, the couple's lives changed profoundly. Walt had surgery to treat an enlarged, cancerous prostate gland and within 3 months was hospitalized again after a heart attack. He lingered for 6 weeks with Ruth at his side, and then died. Ruth's grieving was interrupted by the need to care for Ida. Alert and spry at age 78, Ida deteriorated mentally in her seventy-ninth year, despite otherwise being in excellent physical health. In the meantime, Ruth's arthritis worsened and her vision and hearing weakened.

As Ruth turned 85, certain activities had become more difficult—but not impossible. "It just takes a little adjustment!" Ruth exclaimed in her usual, upbeat manner. Reading was harder, so she checked out "talking books" from her local library. Her gait was slower and her eyesight less reliable, so she hesitated to go out alone. When her daughter and family took Ruth to dinner, the conversation moved so quickly in the noisy restaurant that Ruth felt overwhelmed and said little. But interacting with her on a one-to-one basis revealed that she was far from passive and withdrawn! In a calm environment, she showed the same intelligence, wit, and astute insights that she had displayed all her life.

Late adulthood stretches from age 60 to the end of the lifespan. Unfortunately, popular images of old age fail to capture the quality of these final decades. Instead, many myths prevail—that the elderly are feeble, senile, and sick; that they have little contact with their families, who set them aside in nursing homes; that they are no longer able to learn; and that they have entered a phase of deterioration and dependency. Young people who have little contact with older adults are often surprised that elders like Walt and Ruth even exist—active and involved in the world around them.

As we trace physical and cognitive development in old age, we will see that the balance of gains and declines shifts as death approaches. But the typical 60-year-old in industrialized nations can anticipate nearly 2 healthy, rewarding decades before this shift has meaning for everyday life. And as Ruth illustrates, even after older adults become frail, many find ways to surmount physical and cognitive challenges.

Late adulthood is best viewed as an extension of earlier periods, not a break with them. As long as social and cultural contexts grant elders support, respect, and purpose in life, these years are a time of continued potential.

PHYSICAL DEVELOPMENT

If you were to guess the ages of older people on the basis of their appearance, chances are that you would frequently be wrong. Indeed, we often remark that an older person "looks young" or "looks old" for his or her age—a statement acknowledging that chronological age is an imperfect indicator of **functional age,** or actual competence and performance. Because people age biologically at different rates, experts distinguish between the **young-old elderly,** who appear physically young for their advanced years, and the **old-old elderly,** who appear frail

■ How old are these elders? How old do they look and feel? Because people age biologically at different rates, the 76-year-old woman on the right appears younger than her 74-year-old sister on the left.

and show signs of decline.[1] According to this functional distinction, it is possible for an 80-year-old to be young-old and a 65-year-old to be old-old (Neugarten & Neugarten, 1987). Yet even these labels do not fully capture the wide variation in biological aging. Recall from Chapter 13 that within each person, change differs across parts of the body. For example, Ruth became infirm physically but remained active mentally, whereas Ida was physically fit for her age but found it hard to carry on a conversation, keep appointments, and complete familiar tasks.

So much variation exists between and within individuals that as yet, researchers have not been able to identify any biological measure that predicts the overall rate at which an elderly person will age (Hayflick, 1994). But we do have estimates of how much longer older adults can expect to live, and our knowledge of factors affecting longevity in late adulthood has increased rapidly.

Life Expectancy

"I wonder how many years I have left," Ruth asked herself each time a major life event, such as retirement and widowhood, occurred. Dramatic gains in **average life expectancy**—the number of years that an individual born in a particular year can expect to live—provide powerful support for the multiplicity of factors considered in previous chapters that slow biological aging, including improved nutrition, medical treatment, sanitation, and safety. Recall from Chapter 1 that a North American baby born in 1900 had an average life expectancy of just under 50 years. In 2000, this figure reached 77 in the United States (74 for men and 80 for women) and 79 in Canada (76 for men and 82 for women). Twentieth-century gains in life expectancy were so extraordinary that they equaled those of the previous 5,000 years! Steady declines in infant mortality (see Chapter 3) are a major contributor to longer life expectancy. But death rates among adults have decreased as well. For example, heart disease, the leading cause of overall adult death in North America, has dropped by nearly 50 percent in the past 30 years, due to declines in risk factors (such as high blood pressure and smoking) and, mostly, advances in medical treatment (Kuulasmaa et al., 2000; Statistics Canada, 2002n; U.S. Bureau of the Census, 2002c).

Variations in Life Expectancy

Consistent group differences in life expectancy underscore the joint contribution of heredity and environment to biological aging. As the figures just given reveal, on average, women can look forward to 4 to 7 more years of life than men—a difference found in almost all cultures. The female life expectancy advantage also characterizes several animal species, including rats, mice, and dogs (Shock, 1977). The protective value of the female's extra X chromosome (see Chapter 2) is believed to be responsible. Yet since the early 1970s, the gender gap in life expectancy has narrowed in industrialized nations (Newman & Brach, 2001). Because men are at higher risk for disease and early death, they reap somewhat larger generational gains from positive lifestyle changes and new medical discoveries.

Life expectancy varies substantially with SES, ethnicity, and nationality. For example, an American white child born in the year 2000 is likely to live 5 to 7 years longer than an African-American child and 4 to 5 years longer than a Native-American child. Similarly, in regions of Canada with a greater than 20 percent Aboriginal population, average life expectancy is 5 to 15 years below that for the nation as a whole (Statistics Canada, 1999a; U.S. Bureau of the Census, 2002c). These differences can be accounted for by higher rates of infant mortality, unintentional injuries, life-threatening disease, poverty-linked stress, and (in the United States) violent death in low-SES minority groups.

Length of life—and even more important, *quality of life* in old age—can be predicted by a country's health care, housing, and social services, along with lifestyle factors. When researchers estimate **active lifespan,** the number of years of vigorous, healthy life an individual born in a particular year can expect, Japan ranks first, Canada twelfth, and the United States a disappointing twenty-fourth (see Figure 17.1 on page 548). Japan's low rate of heart disease, linked to its low-fat diet, in combination with favorable health care and other policies for the aged, account for its leading status. Because the United States falls short of other industrialized nations in all these respects, Americans spend somewhat more time disabled and die earlier than elders in most other developed countries. In developing nations with widespread poverty, malnutrition, disease, and armed conflict, average life expectancy hovers around 50 years, and active lifespan is even shorter—44 in Haiti, 38 in Afghanistan, 33 in Rwanda, and 26 in Sierra Leone (World Health Organization, 2000a).

Life Expectancy in Late Adulthood

Although poverty-stricken groups lag behind the economically advantaged, the number of people age 65 and older has risen dramatically in the industrialized world. From the beginning to the end of the twentieth century, the elderly increased from 4 percent to nearly 13 percent of the North American population—a figure expected to reach 23 percent by the middle of the twenty-first century. The fastest-growing segment of senior citizens is the 85-and-older group, who currently make up 1 percent of Americans and 1.4 percent of Canadians. By the year 2020, the proportion of seniors age 85 and older will double, and by 2050 it will quadruple (Statistics Canada, 2002k; U.S. Bureau of the Census, 2002c).

[1] In the popular literature on aging, this distinction has been confused with chronological age: young-old referring to age 65 to 75, old-old to 75 to 85, and oldest-old to 85 to 99 (see, for example, Safire, 1997). In Western nations, persistence of the incorrect belief that age determines function is probably due to stereotypes of aging.

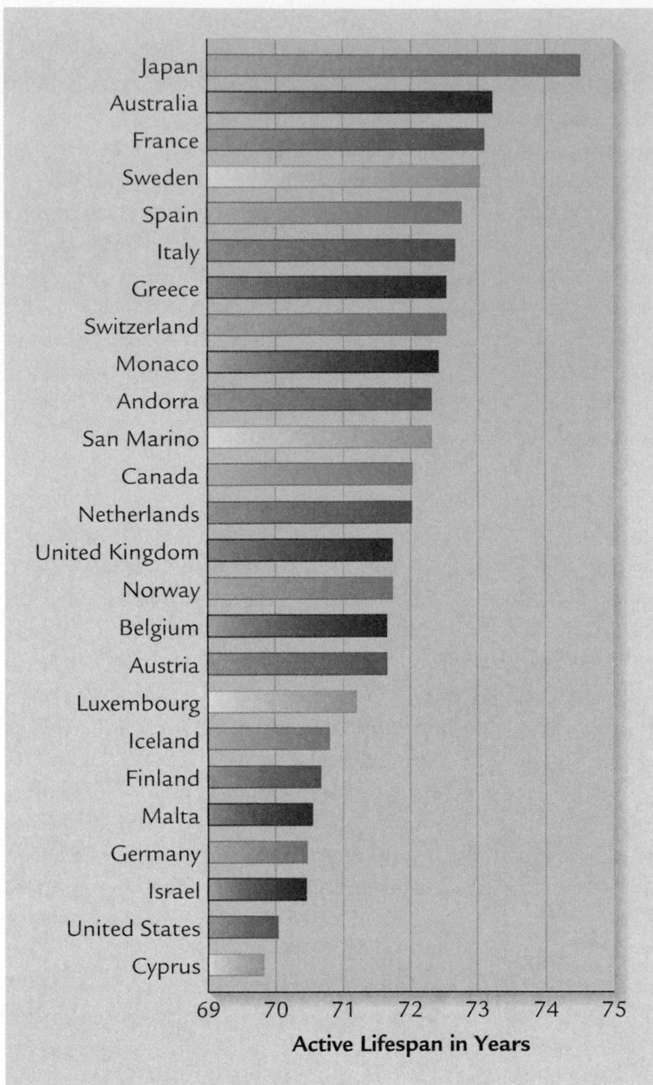

■ **FIGURE 17.1 Active lifespan in twenty-five nations.** Lifestyle factors and favorable health care and other policies for the aged contribute to active lifespan. Japan ranks first, Canada twelfth, and the United States a disappointing twenty-fourth. (Adapted from World Health Organization, 2000a.)

People reaching age 65 in the early twenty-first century can look forward, on average, to 18 more years of life in the United States and 19 more in Canada. As at earlier ages, life expectancy continues to be greater for women than for men. Today, the 65- to 69-year age group consists of 111 women for every 100 men; for people age 85 and older, this number climbs to 160 (Health Canada, 2002b; U.S. Department of Health and Human Services, 2002c). Discrepancies like these occur in all developed countries. They are not always present in the developing world because of high death rates of women during childbirth (Hayflick, 1994).

Although women outnumber men by a greater margin, differences in average life expectancy between the sexes decline as elders advance in age. A North American newborn girl can expect to live about 7 to 8 years longer than a newborn boy. At age 65, the difference narrows to about 3½ years; at age 85, to just over 1 year. Over age 100, the gender gap in life expectancy disappears. Similarly, differences in rates of chronic illness and in life expectancy between higher-SES whites and low-SES ethnic minorities decline with age. Around age 85, a **life expectancy crossover** occurs, in that surviving members of low-SES ethnic minority groups live longer than members of the white majority. Figure 17.2 illustrates this finding for African Americans. Researchers speculate that only the biologically sturdiest males and members of low-SES groups survive into very old age (Barer, 1994).

Of course, average life expectancy does not tell us how enjoyable living to a ripe old age is likely to be. Most North Americans age 65 and older are capable of living independent, productive lives, although with age, those needing assistance increase. After age 80, about 35 percent have difficulty carrying out activities of daily living independently (U. S. Department of Health and Human Services, 2002f; Statistics Canada, 2002n). Throughout this book, we have seen that genetic and environmental factors jointly affect aging. With respect to heredity, identical twins typically die within 3 years of each other, whereas fraternal twins of the same sex differ by more than 6 years. Also, longevity runs in families. People with long-lived ancestors tend to survive longer and to be physically healthier in old age. And when both parents survive to age 70 or older, the chances that their children will live to 90 or 100 are double that of the general population (Hayflick, 1994; Mitchell et al., 2001).

At the same time, evidence from twin studies suggests that once people pass 75 to 80 years, the contribution of heredity to length of life decreases in favor of environmental factors—a healthy diet; normal body weight; regular exercise; little or no tobacco, alcohol, and drug use; an optimistic outlook; low psychological stress; and social support (McGue et al., 1993; Rowe & Kahn, 1998). The study of centenarians—people who cross the 100-year mark—offers special insights into how biological, psychological, and social influences work together to promote a long, satisfying life. Refer to the Lifespan Vista box on pages 550–551 to find out more about the survival and successful adaptation of these longest-lived individuals.

Maximum Lifespan

Finally, perhaps you are wondering: What is the **maximum lifespan,** or the genetic limit to length of life for a person free of external risk factors? According to current estimates, it varies between 70 and 110 for most people, with 85 about average (Harman, 2002). The oldest verified age to which an individual has lived is 122 years.

Do these figures reflect the upper bound of human longevity, or can our lifespans be extended further? At present, scientists disagree on answers to this question. Some believe that about 85 or 90 years is as much as most humans can expect, since gains in average life expectancy are largely the result of reducing health risks in the first 20 or 30 years. Expected life for

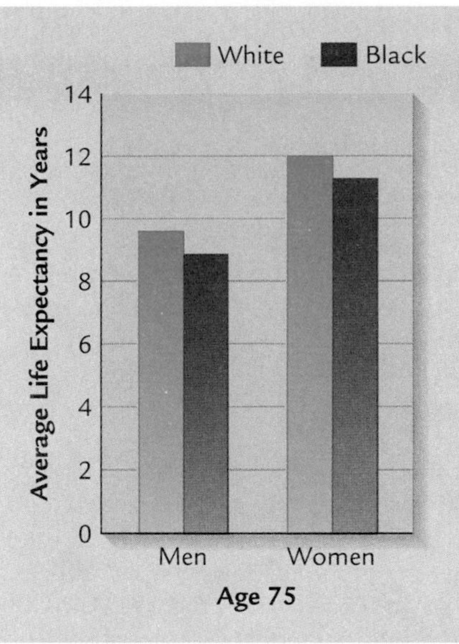

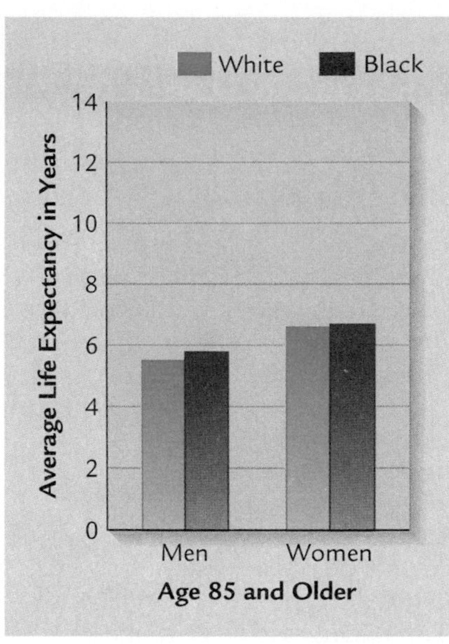

FIGURE 17.2 Average life expectancy for Caucasians and African Americans at ages 75 and 85, illustrating the life expectancy crossover. Below age 85, life expectancy is greater for American whites than blacks. At age 85 and older, this trend reverses; average life expectancy is greater for blacks than whites. Because only the biologically sturdiest members of low-SES minority groups survive into very old age, their life expectancy is slightly longer than that of their higher-SES counterparts. Notice, also, how the gender gap in life expectancy declines with age. (Adapted from U.S. Bureau of the Census, 2002c.)

people age 65 and older has increased very little—only about 5 months—over the past decade (U.S. Department of Health and Human Services, 2002f). Others think we have not yet identified the human genetic limit because the lifespans of several species have been stretched in the laboratory—through selective breeding, genetic engineering, and (as we saw in Chapter 15) dietary calorie restriction.

The possibility of similar achievements in humans raises another issue: *Should* the lifespan be increased as far as possible? In considering this question, many people respond that quality, not just quantity, of life is the important goal—that is, doing everything possible to extend active lifespan. Most experts agree that only after reducing the high rates of preventable illness and disability among low-SES individuals and wiping out age-related diseases does it make sense to invest in lengthening the maximum lifespan.

Physical Changes

The programmed effects of specific genes and the random cellular events believed to underlie biological aging (see Chapter 13) make physical declines more apparent in late adulthood. More organs and systems of the body are affected. Nevertheless, most body structures can last into our eighties and beyond, if we take good care of them. For an overview of the physical changes we are about to discuss, return to Table 13.1 on page 415.

Nervous System

On a routine office visit, 80-year-old Ruth responded to her doctor's query about how she was getting along by explaining, "During the last two days, I forgot the name of the family that just moved in next door, couldn't recall where I had put a pile of bills, and had trouble finding the right words to explain to a delivery service how to get to my house. Am I losing my mind?" Ruth asked anxiously.

"You're much too sharp for that," Dr. Wiley responded. "Ruth, if you were losing your mind, you wouldn't be so concerned about forgetting." Ruth also wondered why extremes of hot and cold weather felt more uncomfortable than in earlier years. And she required more time to coordinate a series of movements and had become less sure of her balance.

Aging of the nervous system affects a wide range of complex thoughts and activities. Although brain weight declines throughout adulthood, the loss becomes greater after age 60 and may amount to as much as 5 to 10 percent by age 80, due to death of neurons and enlargement of ventricles (spaces) within the brain (Vinters, 2001). Neuron loss occurs throughout the cerebral cortex but at different rates in different regions. In the visual, auditory, and motor areas, as many as 50 percent of neurons die. In contrast, parts of the cortex (such as the frontal lobes) that are responsible for integration of information, judgment, and reflective thought show less change. Besides the cortex, the cerebellum (which controls balance and coordination) loses neurons—in all, about 25 percent. Glial cells, which myelinate neural fibers, decrease as well, contributing to diminished efficiency of the central nervous system (Whitbourne, 1999, 2001).

But the brain can overcome some of these declines. In several studies, growth of neural fibers in the brains of older adults unaffected by illness took place at the same rate as in middle-aged people. Aging neurons established new synapses after other neurons had degenerated (Flood & Coleman, 1988). Furthermore, the aging cerebral cortex can, to some degree, generate new neurons (Gould et al., 1999). And brain-imaging research reveals that compared with younger adults, elders who do well on memory tasks sometimes show more widely distributed activity across areas of the cerebral cortex.

A Lifespan Vista

What Can We Learn About Aging from Centenarians?

Born in 1893 in Croatia, Dr. Bogdan Stojic continued to practice medicine until he became a centenarian. After retiring at age 101, he spent his days reading, writing, going for walks, and enjoying his many friends. During his long life, he surmounted many difficult times. In his youth, he fought in three wars—the Balkan War against Turkey, the Serbo-Bulgarian War, and World War I. As a middle-aged soldier during World War II, he was taken prisoner by the Nazis, served as a medic in a prisoner-of-war camp, and narrowly escaped a death sentence. After the war, he entered private practice. At age 68, he and his wife packed their belongings and traveled thousands of miles, settling in Australia to live close to their daughter. There Dr. Stojic continued to treat patients for another 33 years.

About living a long and healthy life, Dr. Stojic reflected, "My parents were 90, but I never thought about it—how old I'd live. It just happened. . . . don't smoke; don't drink too much. . . . My father and mother were very good. They spoke with us as if we were grown-up people when we were 10 years of age . . . [and] my marriage has been very good." About his current trouble walking, he noted optimistically, "I'll be treated. Soon, I hope I will walk without the stick." He continues to have goals—among them, contributing to a cure for Alzheimer's disease (Deveson, 1994, pp. 218, 222–223).

Due to stereotypes of aging and researchers' focus on the very old with the heaviest burden of illness and disability, popular images of the most senior members of the human species are ones of extreme frailty. Yet the past 40 years have seen a tenfold increase in centenarians in the industrialized world—a trend expected to accelerate. Currently, American centenarians number about 72,000, and Canadian centenarians about 3,800. In 2050, these figures are expected to be thirteen times greater (Statistics Canada, 2002k; U.S. Department of Health and Human Services, 2002b).

Women centenarians greatly outnumber men—by about 4 to 1. About 60 to 70 percent have physical and mental impairments that interfere with independent functioning. But the rest lead active, autonomous lives (Hagberg et al., 2001; Silver, Jilinskaia, & Perls, 2001). These robust centenarians are of special interest, since they represent the ultimate potential of the human species. What are they like? To find out, several longitudinal studies of centenarians have been initiated. Results reveal that they are diverse in years of education (none to postgraduate), economic well-being (very poor to very rich), and ethnicity. At the same time, their physical condition and life stories reveal common threads.

Health. Longevity runs in centenarians' families, suggesting a genetically based survival advantage. These elders usually have grandparents, parents, and siblings who reached very old age. And their children, most of whom are in their 70s and 80s, appear physically young for their age (Perls et al., 2002).

The majority of robust centenarians have escaped age-related chronic illnesses, such as cardiovascular disease, cancer, diabetes, and dementia. Genetic testing reveals a low incidence of genes associated with immune-deficiency disorders and Alzheimer's disease. Consistent with these findings, robust centenarians typically have efficiently functioning immune systems, and after-death examinations reveal few brain abnormalities (Silver & Perls, 2000). Furthermore, compared with the general population, about four times as many centenarian women gave birth to healthy children after age 40 (Perls et al., 2000). Late childbearing may indicate that the woman's reproductive system is aging slowly—and that the rest of her is doing so as well.

This suggests that one way older adults compensate for neuron loss is to call on additional brain regions to support cognitive processing (Grady & Craik, 2000).

The autonomic nervous system, involved in many life-support functions, also performs less well in old age. For example, Ruth's reduced tolerance for hot weather was due to decreased sweating. And her body found it more difficult to raise its core temperature during cold exposure. For these reasons, the elderly are at risk during heat waves and cold spells. However, among physically fit elders who are free of disease, these declines are mild (Whitbourne, 2001). The autonomic nervous system also releases higher levels of stress hormones into the bloodstream than it did at younger ages, perhaps in an effort to arouse body tissues that have become less responsive to these hormones over the years (Whitbourne, 1999). Later we will see that this change may contribute to decreased immunity and to sleep problems among older adults.

Sensory Systems

Changes in sensory functioning become increasingly noticeable in late life. Older adults see and hear less well, and taste, smell, and touch sensitivity may also decline. As Figure 17.3 on page 552 shows, hearing impairments are far more common than visual impairments, and they extend trends described for middle adulthood, in that many more men than women are affected.

As a group, robust centenarians are of average or slender build and practice moderation in eating. Many have most or all of their own teeth—another sign of unusual physical health. Despite heavy tobacco use in their generation, the large majority have never smoked. And most report lifelong physical activity extending past age 100 (Kropf & Pugh, 1995).

Personality. In personality, these very senior citizens appear highly optimistic. Instead of dwelling on fears and tragedies, they focus on a better tomorrow. The less anxious and fearful centenarians are, the more favorable their self-rated health and their psychological well-being (Quinn et al., 1999). In a study in which robust centenarians retook personality tests after 18 months, they reported more fatigue and depression, perhaps in response to increased frailty at the very end of their lives. But they also scored higher in toughmindedness, independence, emotional security, and openness to experience—traits that may be vital for surviving beyond 100 (Martin, Long, & Poon, 2002). When asked about contributors to their longevity, these extremely long-lived elders often mention close family bonds and a long and happy marriage. An unusually large percentage of centenarian men—about one-fourth—are still married (Velkoff, 2000).

Activities. Robust centenarians have a history of community involvement—working for just causes that are central to their growth and happiness. Their current activities often include stimulating work, leisure pursuits, and learning, which may help sustain their good cognition and life satisfaction (Samuelsson et al., 1997). Writing letters, poems, plays, and memoirs; making speeches; teaching music lessons and Sunday school; nursing the sick; chopping wood; selling merchandise, bonds, and insurance; painting; practicing medicine; and preaching sermons are among robust centenarians' varied involvements. In several cases, illiterate centenarians learned to read and write. One of the most impressive was a 105-year-old woman who enrolled in classes four nights a week. Within a short time, she could read road signs, newspaper headlines, and sections of the Bible (Beard, 1991).

Robust centenarians are often regarded as rare curiosities who do not represent the general population. As their numbers increase, they are likely to be viewed less as exceptions and more as people for whom normal development is at its best. These independent, mentally alert, fulfilled 100-year-olds illustrate how a healthy lifestyle, personal resourcefulness, and close ties to family and community can build on biological strengths, thereby pushing the limits of the active lifespan.

■ Dr. Bogdan Stojic, centenarian, defies stereotypes of the very old as extremely frail. He practiced medicine until age 101.

● **Vision.** In Chapter 15 (see pages 483–484), we noted that structural changes in the eye make it harder to focus on nearby objects, see in dim light, and perceive color. In late adulthood, vision diminishes further. For example, the cornea (clear covering of the eye) becomes more translucent and scatters light, which blurs images and increases sensitivity to glare. The lens continues to yellow, leading to further impairment in color discrimination. From middle to old age, cloudy areas in the lens called **cataracts** increase, resulting in foggy vision and (without surgery) eventual blindness. The number of individuals with cataracts increases tenfold from middle to late adulthood; 25 percent of people in their seventies and 50 percent of those in their eighties are affected (Lindsay, 1999; U.S. Bureau of the Census, 2002c). Besides biological aging, heredity, sun exposure, and certain diseases (such as diabetes) increase the risk of cataracts (Hammond et al., 2000). Fortunately, removal of the lens and replacement with an artificial lens implant or corrective eyewear are highly successful in restoring vision.

Impaired eyesight in late adulthood largely results from a reduction in light reaching the retina (caused by yellowing of the lens, shrinking of the pupil, and clouding of the vitreous) and from cell loss in the retina and optic nerve (refer again to Chapter 15). Dark adaptation—moving from a brightly lit to a dim environment—is harder, making entering a movie theater after the show has started a challenge. In addition, depth perception is less reliable, since binocular vision (the brain's ability to combine images received from both eyes) declines (Brabyn et al., 2001). And visual acuity (fineness of discrimination)

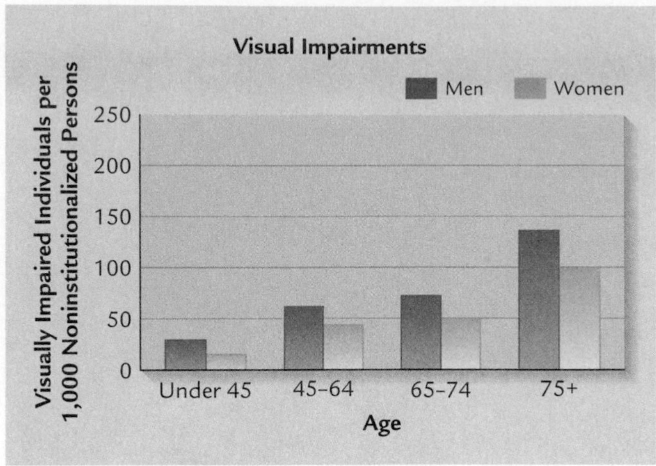

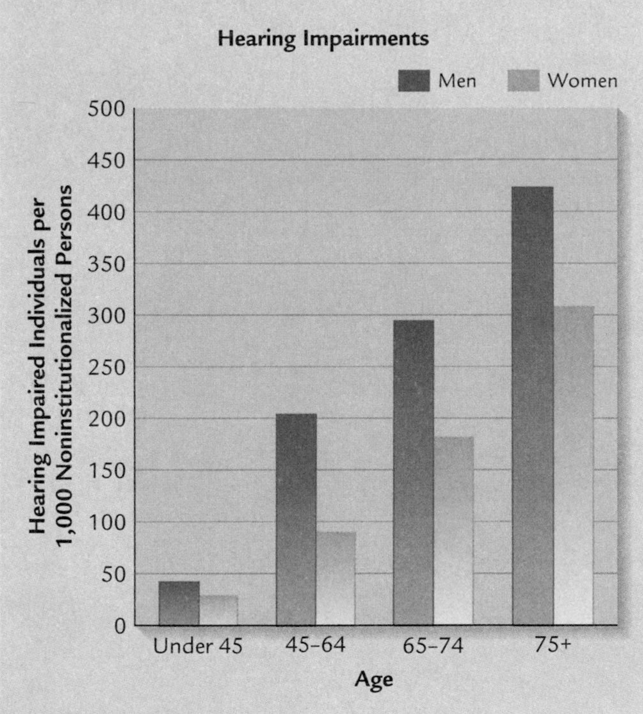

FIGURE 17.3 Rates of visual and hearing impairments among American men and women by age. Hearing impairments are far more common than visual impairments in aging individuals, and many more men than women are affected. Nevertheless, a visual impairment is more likely than a hearing impairment to interfere with self-care. Trends are similar in Canada, although the overall incidence of sensory impairments is slightly lower. (Adapted from U.S. Bureau of the Census, 2002c.)

worsens, with a sharp drop after age 70 (Fozard & Gordon-Salant, 2001).

When light-sensitive cells in the *macula,* or central region of the retina, break down, older adults may develop **macular degeneration,** in which central vision blurs and gradually is lost. Macular degeneration is the leading cause of blindness among older adults. If diagnosed early, it can sometimes be treated with laser therapy (Mayo Clinic, 2000). Several large-scale studies suggest that a diet rich in green, leafy vegetables reduces the risk of macular degeneration and slows vision loss in those with the condition. These sources of vitamins A, C, and E and carotenoids (yellow and red plant pigments) are believed to exert their beneficial effects by protecting cells in the macula from free-radical damage. Vitamin supplements are helpful in reducing the incidence of both cataracts and macular degeneration (Jacques, 1999; Sackett & Schenning, 2002).

Visual difficulties have a profound impact on older people's self-confidence and everyday behavior. Ruth gave up driving in her mid-seventies, and she worried greatly when Walt had trouble making decisions behind the wheel. He found it hard to shift focus between the road and the dashboard and strained to make out pedestrians at dusk and at night (Owsley et al., 1998). On foot, older adults' problems with depth perception and dark adaptation increase their chances of stumbling.

When vision loss is extensive, it can affect leisure pursuits and be very isolating. Ruth could no longer enjoy museums, movies, playing bridge, and working crossword puzzles. Her poor vision also meant that she had to depend on others for help with housekeeping and shopping. Still, even among people age 85 and older, only 30 percent experience visual impairment severe enough to interfere with daily living (U.S. Department of Health and Human Services, 2002f). As in other aspects of aging, wide individual differences in visual capacity exist.

● **Hearing.** At a Thanksgiving gathering, 85-year-old Ruth had trouble hearing. "Mom, this is Leona, Joe's cousin. I'd like you to meet her," said Ruth's daughter Sybil. In the clamor of boisterous children, banging dishes, television sounds, and nearby verbal exchanges, Ruth didn't catch Leona's name or her relationship to Sybil's husband, Joe.

"Tell me your name again," Ruth asked. "This is a mighty busy place right now, and it's hard for me to hear. Let's go into the next room so we can speak a bit."

Reduced blood supply and natural cell death in the inner ear and auditory cortex, discussed in Chapter 15, along with stiffening of membranes (such as the eardrum), cause hearing to decline in late adulthood. Decrements are greatest at high frequencies, although detection of soft sounds diminishes throughout the frequency range (see page 484). In addition, responsiveness to startling noises lessens, and discriminating complex tone patterns becomes harder (Fitzgibbons & Gordon-Salant, 1998; Ford et al., 1995; Stevens et al., 1998).

Although hearing loss has less impact on self-care than vision loss, it affects safety and enjoyment of life. In the din of traffic on city streets, 80-year-old Ruth didn't always interpret warnings correctly—both spoken ("Watch it, don't step out yet") and nonspoken (the beep of a horn or a siren). And when she turned up the radio or television volume, she sometimes missed the ring of the telephone or a knock at the door.

Of all hearing difficulties, the age-related decline in speech perception has the greatest impact on life satisfaction. Ability to detect the content and emotionally expressive features of conversation declines after age 70, a difficulty that worsens in noisy settings (Schneider et al., 2000; Villaume et al., 1997). Although Ruth used problem-centered coping to increase her chances of hearing conversation, she wasn't always successful. And sometimes people were inconsiderate. For example, on a dinner outing, Joe raised his voice impatiently when Ruth asked him to repeat himself. And he said to Sybil in Ruth's presence, "Be honest, Syb, Ruth's going deaf, isn't she?" This time, Ruth heard every word. At the family's Thanksgiving reunion, fewer relatives took time to talk with Ruth, and she felt some pangs of loneliness.

As with vision, most elders do not suffer from hearing loss great enough to disrupt their daily lives. Of those who do, compensating with a hearing aid and minimizing background noise are helpful. Furthermore, recall from Chapter 4 (page 000) that beginning at birth, our perception is *intermodal* (combines information from more than one sensory system). By attending to facial expressions, gestures, and lip movements, older adults can use vision to help interpret the spoken word. When family members and others speak in quiet environments, older people are far more likely to convey an image of alertness and competence than of reduced sensitivity to the surrounding world.

● **Taste and Smell.** Walt's brother Dick was a heavy smoker. In his sixties, he poured salt and pepper over his food, stirred extra spoonfuls of sugar into his coffee, and satisfied his fondness for spicy foods by asking for "extra hot" in Mexican and Indian restaurants.

Dick's reduced sensitivity to the four basic tastes—sweet, salty, sour, and bitter—is evident in many adults after age 60. In addition, older adults have greater difficulty recognizing familiar foods by taste alone (Mojet, Christ-Hazelhoff, & Heidema, 2001; Stevens et al., 1995). But no change in the number or distribution of taste buds takes place late in life, so the drop in taste sensitivity just described may be due to factors other than aging. Smoking, dentures, medications, and environmental pollutants can affect taste perception. When taste is harder to detect, food is less enjoyable, increasing the likelihood of deficiencies in the elderly person's diet. Flavor additives can help make food more attractive to older adults (Drewnowski & Shultz, 2001).

Smell contributes to enjoyment of food and also has a self-protective function. An aging person who has difficulty detecting rancid food, gas fumes, or smoke may be in a life-threatening situation. A decrease in the number of smell receptors after age 60 contributes to declines in odor sensitivity. Older adults are less accurate at linking odors with such descriptors as *floral, musky, fruity,* or *sweet.* Researchers believe that odor perception becomes distorted in late adulthood, a change that may promote complaints that "food no longer smells and tastes right" (Russell et al., 1993). But other factors may lead this decline to appear greater than it actually is. For example, some older adults have difficulty recalling odor labels, and this makes odor recognition tasks harder (Larsson & Bäckman, 1998). Furthermore, practicing odor discriminations affects their maintenance (Engen, 1982). Walt had been a wine enthusiast all his life. In his early eighties, he could still distinguish the aromas of fine wines.

● **Touch.** Touch sensitivity is especially crucial for certain adults, such as the severely visually impaired who must read in Braille and people who make fine judgments about texture in their occupations or leisure pursuits—for example, in art and handicraft activities. To measure touch perception, researchers determine how close two stimuli on the skin must be before they are perceived as one. Findings indicate that aging brings a sharp decline on the hands, especially the fingertips, and less of a drop on the arms and lips. Decreased touch sensitivity may be due to loss of touch receptors in certain regions of the skin and slowing of blood circulation to the extremities. After age 70, nearly all elderly are affected (Stevens & Cruz, 1996).

© MIKE YAMASHITA/WOODFIN CAMP & ASSOCIATES

■ Aging brings a decline in touch sensitivity, with the hands and, especially, the fingertips most affected. Still, many older adults who enjoy art and handicraft activities draw on previously acquired expertise to sustain a high level of performance.

If we consider the spacing among elements of Braille letters, 45 percent of middle-aged and 100 percent of elderly people should have difficulty interpreting them (Stevens, Foulke, & Patterson, 1996). Just as the sighted need new visual aids for reading as they age, so the visually impaired may need new tactile aids—an important consideration in responding to diversity in the aging population.

Cardiovascular and Respiratory Systems

Aging of the cardiovascular and respiratory systems proceeds gradually during early and middle adulthood, usually without notice. Signs of change are more apparent in late adulthood, and they prompt concern among aging individuals, who know these organ systems are vital for quality and length of life. In their sixties, Ruth and Walt noticed that they felt more physically stressed after running to catch a departing bus or to reach the far side of the street before the light changed.

As the years pass, the heart muscle becomes more rigid, and some of its cells die while others enlarge, leading the walls of the left ventricle (the largest heart chamber, from which blood is pumped to the body) to thicken. In addition, artery walls stiffen and accumulate some plaque (cholesterol and fats) due to normal aging (much more if the person has atherosclerosis). Finally, the heart muscle becomes less responsive to signals from pacemaker cells within the heart, which initiate each contraction (Whitbourne, 1996, 1999).

The combined result of these changes is that the heart pumps with less force, maximum heart rate decreases, and blood flow throughout the circulatory system slows. This means that sufficient oxygen may not be delivered to body tissues during high physical activity. (Recall from Chapter 13 that a healthy heart supports typical levels of exertion well into old age.)

Changes in the respiratory system compound the reduced oxygenation just described. Because lung tissue gradually loses its elasticity, between ages 25 and 80, vital capacity (amount of air that can be forced in and out of the lungs) is reduced by half. As a result, the lungs fill and empty less efficiently, causing the blood to absorb less oxygen and give off less carbon dioxide. This explains why older people increase their breathing rate more and feel more out of breath when exercising.

Cardiovascular and respiratory deficiencies are more extreme in people who have smoked cigarettes throughout their lives, failed to reduce dietary fat, or had many years of exposure to environmental pollutants. We have seen in previous chapters that exercise is a powerful means of slowing cardiovascular aging. Exercise also facilitates respiratory functioning, as we will see when we discuss health and fitness in a later section.

■ Functioning of the cardiovascular and respiratory systems declines in late adulthood. But regular physical activity is a powerful means of slowing these changes. These elderly Havana residents exercise in a neighborhood park as part of a national exercise program. Cuba requires all able-bodied elders to participate in a physical fitness routine for 45 minutes, 5 days a week. Cuba boasts an average life expectancy of 75 years, the highest in Latin America.

Immune System

As the immune system ages, T cells, which attack antigens (foreign substances) directly, become less effective (see Chapter 13, page 417). In addition, the immune system is more likely to malfunction by turning against normal body tissues in an **autoimmune response.** A less competent immune system can increase the elderly person's risk for a variety of illnesses, including infectious diseases (such as the flu), cardiovascular disease, certain forms of cancer, and a variety of autoimmune disorders, such as rheumatoid arthritis and diabetes. But an age-related decline in immune functioning is not the cause of most illnesses among the elderly. It merely permits disease to progress, whereas a stronger immune reaction would have stamped out the disease agent (Pawelec et al., 1999; Weksler, 1995).

Older adults vary greatly in immunity. A few have sturdy immune systems that continue to respond nearly as well as in early adulthood. But the responses of most range from partial to profound loss of function (Pawelec et al., 1999a). The strength of the aging person's immune system seems to be a sign of overall physical vigor. Certain immune indicators, such as high T cell activity, predict survival over the next 2 years in very old people (Wikby et al., 1998).

In Chapter 13, we emphasized that other physical changes contribute to impaired immune functioning. Recall that stress hormones undermine immunity. With age, the autonomic nervous system releases higher levels of these into the bloodstream (refer back to page 560). New findings suggest that a healthy diet and exercise help protect the immune response in old age, whereas obesity aggravates the age-related decline (Bogden & Louria, 1999; Moriguchi et al., 1995; Shephard & Shek, 1995).

Sleep

When Walt climbed into bed at night, he usually lay awake for a half-hour to an hour before falling asleep, remaining in a drowsy state longer than when he was younger. During the night, he spent less time in the deepest phase of NREM sleep (see Chapter 3, page 102) and awoke several times. Again, he sometimes lay awake for a half-hour or more before drifting back to sleep.

Older adults require about the same total sleep time as younger adults: around 7 hours per night. Yet as people age, they have more difficulty falling asleep, staying asleep, and sleeping deeply—a trend that begins after age 30 for men and after age 50 for women. The timing of sleep tends to change as well, toward earlier bedtime and earlier morning awakenings (Hoch et al., 1997). Changes in brain structures controlling sleep and higher levels of stress hormones in the bloodstream, which have an alerting effect on the central nervous system, are believed to be responsible (Whitbourne, 1996).

Until age 70 or 80, men experience more sleep disturbances than women for several reasons. First, enlargement of the prostate gland, which occurs in almost all aging men, constricts the urethra (the tube draining the bladder) and leads to a need to urinate more often, including during the night. Second, men are more prone to **sleep apnea,** a condition in which breathing ceases for 10 seconds or longer, resulting in many brief awakenings. The incidence of sleep apnea in elderly men is high; 30 to 50 percent have twenty or more episodes per night. Finally, periodic rapid movement of the legs sometimes accompanies sleep apnea but also occurs at other times of night. Called "restless legs," these movements may be due to muscle tension, reduced circulation, or age-related changes in motor areas of the brain. Although common among the elderly and not dangerous, they do disrupt sleep (Martin, Shochat, & Ancoli-Israel, 2000).

Older adults often express concern about sleep difficulties. From 25 to 35 percent report some degree of insomnia (Roberts, Shema, & Kaplan, 1999). Poor sleep can feed on itself. For example, Walt's nighttime wakings led to daytime fatigue and short naps, which made it harder to fall asleep the following evening. And because Walt expected to have trouble sleeping, he worried about it, which also interfered with sleep.

Fortunately, there are many ways to foster restful sleep, such as establishing a consistent bedtime and waking time, exercising regularly, and using the bedroom only for sleep (not for eating, reading, or watching TV). Explaining that even very healthy older adults have trouble sleeping lets people know that age-related changes in the sleep–wake pattern are normal. The elderly receive more prescription sedatives for sleep complaints than do 40- to 60-year-olds. Used briefly, these drugs can help relieve temporary insomnia. But long-term medication can make matters worse by increasing the frequency and severity of sleep apnea and by inducing rebound insomnia after the drug is discontinued (Riedel & Lichstein, 2000). Finally, discomfort due to an enlarged prostate, including frequent urination at night, can be corrected with new laser surgical procedures that relieve symptoms without complications (Aho & Gilling, 2003).

Physical Appearance and Mobility

The inner physical declines we have considered are accompanied by many outward signs of growing older—involving the skin, hair, facial structure, and body build. In earlier chapters, we saw that changes leading to an aged appearance are under way as early as the twenties and thirties. Because these changes occur gradually, older adults may not notice their elderly appearance until its arrival is obvious. Each year during their summer travels, Walt and Ruth observed that Dick's and Goldie's skin appeared more wrinkled. Their hair turned from gray to white as all pigment was lost, and their bodies were rounder and their arms and legs thinner. When they returned home, Walt and Ruth also were more aware that they themselves had aged.

Creasing and sagging of the skin, described in Chapter 15, extends into old age. In addition, oil glands that lubricate the skin become less active, leading to dryness and roughness. "Age spots" increase; in some elderly individuals, the arms, backs of the hands, and face may be dotted with these pigmented marks. Moles and other small skin growths may also appear. Blood vessels can be seen beneath the more transparent skin, which has largely lost its layer of fatty support (Whitbourne, 1999, 2001). This loss further limits the older adult's ability to adapt to hot and cold temperatures.

The face is especially likely to show these effects because it is frequently exposed to the sun, which accelerates aging. Other facial changes occur: The nose and ears broaden as new cells are deposited on the outer layer of the skeleton. Teeth may be yellowed, cracked, and chipped, and gums may recede; with better dental care, these outcomes are likely to be less pronounced in future generations. As hair follicles under the skin's surface die, hair on the head thins in both sexes, and the scalp may be visible. In men with hereditary pattern baldness, follicles do not die but, instead, begin to produce fine, downy hair (Whitbourne, 1996).

Body build changes as well. Height continues to decline, especially in women, as loss of bone mineral content leads to further collapse of the spinal column. Weight generally drops after age 60 due to additional loss in lean body mass (bone density and muscle), which is heavier than the fat deposits accumulating on the torso.

Several factors affect mobility. The first is muscle strength, which generally declines at a faster rate in late adulthood than in middle age. On the average, by 60 to 70 years of age, 10 to 20 percent of muscle power has been lost, a figure that climbs to 30 to 50 percent after ages 70 to 80 (Whitbourne, 1996, 2001). Second, bone strength deteriorates because of reduced bone mass, and tiny cracks in response to stress weaken the bones further. Third, strength and flexibility of the joints and the tendons and ligaments (which connect muscle to bone) diminish. In her eighties, Ruth's reduced ability to support her body, flex her limbs, and rotate her hips made walking at a steady, moderate pace, climbing stairs, and rising from a chair difficult.

In Chapter 13, we noted that endurance athletes who continue training throughout adulthood retain their muscular physiques and much of their strength into their sixties and seventies. These especially active individuals lose fast-twitch muscle fibers like other aging individuals, but they compensate by strengthening remaining slow-twitch fibers so they work more efficiently. Careful planning of an exercise program can also enhance joint flexibility and range of movement. When Ruth complained of joint stiffness and pain, her doctor pointed out that certain rhythmic and flexing exercises, as well as exercises that strengthen muscles supporting the joints, could be helpful.

Adapting to Physical Changes of Late Adulthood

Great diversity exists in older adults' adaptation to the physical changes of aging. Dick and Goldie took advantage of an enormous industry designed to stave off the appearance of old age, including cosmetics, wigs, and plastic surgery. In contrast, Ruth and Walt gave little thought to their thinning white hair and wrinkled skin. Their identities were not as bound up with their appearance as with their ability to remain actively engaged in their surroundings.

People vary in the aspects of physical aging that matter most to them. And because parts of the body age at different rates, older adults' sense of physical aging is multidimensional: They feel older in some domains than in others. Compared with Dick and Goldie, Ruth and Walt approached aging with a more positive outlook and greater peace of mind. They did not try to hang on to a youthful identity but, instead, resolved to intervene in those aspects of aging that could be changed and to accept those that could not.

Research shows that the most obvious, outward signs of aging—graying hair, facial wrinkles, and baldness—bear no relationship to sensory, cognitive, and motor functioning or to longevity (Schnohr et al., 1998). In contrast, neurological, sensory, cardiovascular, respiratory, immune-system, and skeletal and muscular health predict cognitive performance and both quality and length of later life (Anstey, Luszcz, & Sanchez, 2001; Korten et al., 1999). Furthermore, people can do more to prevent declines in the functioning of these internal body systems than they can do to prevent gray hair and baldness!

● **Effective Coping Strategies.** Think back to our discussion of problem-centered and emotion-centered coping in Chapter 15. It applies here as well. As Walt and Ruth prevented and compensated for age-related changes through diet, exercise, environmental adjustments, and an active, stimulating lifestyle, they felt a sense of personal control over their fates. This prompted additional positive coping and improved physical functioning.

In one study, elders who generated new, problem-centered coping strategies for dealing with vision loss—such as relying on sound to recognize people, acting more cautiously in risky situations, and exchanging ideas with other visually impaired individuals—adapted more favorably to everyday challenges and felt less depressed (Brennan & Cardinali, 2000). In contrast, older adults who avoid confronting age-related declines—who think these are inevitable and uncontrollable—tend to be passive when faced with them and to report more physical and psychological adjustment difficulties (Whitbourne & Primus, 1996).

● **Assistive Technology.** A rapidly expanding **assistive technology,** or array of devices that permit people with disabilities to improve their functioning, is available to help older people cope with physical declines. Computers are the greatest source of these innovative products. People with sensory impairments can use special software to enlarge text or have it read aloud. Phones that can be dialed and answered by voice commands help elders who have difficulty pushing buttons or getting to the phone from across the room. And for elders who take multiple medications, a tiny computer chip, called a "smart cap," can be placed on medicine bottles. It beeps on a programmed schedule to remind the older person to take the drug and tracks how many and at what time pills have been taken.

Architects are also designing "smart homes" that include a variety of features that promote safety and mobility. For example, sensors in floors can activate room lights when an older person gets up at night to use the bathroom, thereby preventing injuries. Another remarkable device is a harness attached to a track in the ceiling, which carries people with reduced mobility from room to room (Hooyman & Kiyak, 2002). At present, "smart home" systems are beyond the means of most elders. But as the older population increases, future housing may be designed to permit easy and cost-effective installation.

In the United States and Canada, government-sponsored health care coverage for assistive technology is largely limited to essential medical equipment. Sweden's health care system, in contrast, covers many assistive devices that promote function and safety (Stone, Staisey, & Sonn, 1991). In this way, Sweden helps older adults compensate for impaired abilities and remain as independent as possible.

■ Assistive technology, permitting older adults with physical declines to improve their functioning, is rapidly expanding. In this "smart" bedroom, an elder who cannot walk sits in a hoist attached to an overhead track. Using a system of color-coded strings, he gets out of bed and moves about the room.

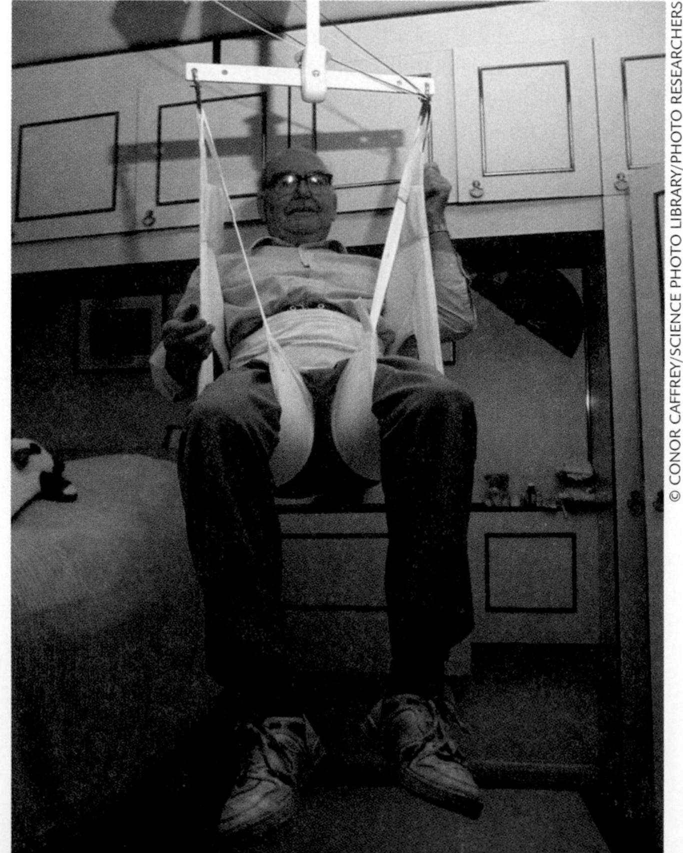

© HINATA HAGA/HAGA/THE IMAGE WORKS

■ In cultures where the elderly are treated with deference and respect, an aging appearance may be a source of pride. This Maori bride of New Zealand seeks a blessing from the groom's grandmother.

● **Overcoming Stereotypes of Aging.** Stereotypes of late adulthood, which view "deterioration as inevitable," are widespread in Western nations. Overcoming this pessimistic picture is vital for helping people adapt favorably to late-life physical changes. In a survey of older adults diverse in SES, many reported experiencing prejudice and discrimination. For example, from 30 to 40 percent had been ignored, talked down to, or assumed to be unable to hear or understand well because of their age (Palmore, 2001).

Like gender stereotypes, aging stereotypes often operate automatically, without awareness; people "see" elders in stereotypical ways, even when they appear otherwise. As seniors encounter these negative messages, they are likely to accept them as true of the elderly population, and of themselves. In several studies, researchers exposed older adults to words associated with either negative aging stereotypes ("decrepit," "confused") or positive aging stereotypes ("sage," "enlightened"). Those in the negative-stereotype condition showed a more intense physiological response to stress and worse handwriting, memory performance, self-efficacy, and will to live (Levy & Banaji, 2002; Levy et al., 2000).

As these findings indicate, negative stereotypes of aging have a stressful, disorganizing impact on elders' functioning, whereas positive stereotypes reduce stress and foster competence. In a longitudinal investigation, people with positive self-perceptions of aging—who, for example, agreed with such statements as "As I get older, things are better than I thought they'd be"— lived, on average, 7½ years longer than those with negative self-perceptions. This survival advantage remained after gender, SES, loneliness, and physical health status had been controlled (Levy et al., 2002).

In cultures where images of aging place less emphasis on physical decline and the elderly are treated with deference and respect, an aging appearance can be a source of pride. For example, the closest word to "elder" in the native language of the Inuit of Canada is *isumataq,* or "one who knows things"—a high status that begins when a couple becomes head of the extended family unit. When Inuit older adults in a small community on Victoria Island were asked for their thoughts on aging well, they mentioned attitudes—a positive approach to life, interest in transmitting cultural knowledge to young people, and community involvement—nearly twice as often as physical health (Collings, 2001).

In Japan, a ritual called *kanreki* recognizes the older person's release from the responsibilities of middle age, new freedoms and competencies, and senior place in the family and society. Japanese extended families in the United States and Canada often plan the kanreki as a surprise sixtieth birthday party, incorporating elements of both the traditional ritual (such as dress) and the Western birthday (a special cake) (Doi, 1991; Gelfand, 1994). Cultural valuing of aging prompts a welcoming approach to late adulthood, including some of its physical transitions. (See also the Cultural Influences box on page 558.)

Although declines are inevitable, physical aging can be viewed with either optimism or pessimism. As Walt commented, "You can think of your glass as half full or half empty." Today, a wealth of research supports the "half full" alternative. In the next section, we will encounter additional examples.

Ask Yourself

REVIEW

Cite examples of how older adults can compensate for age-related physical declines.

APPLY

Sixty-five-year-old Herman inspected his thinning hair in the mirror. "The best way to adjust to this is to learn to like it," he thought. "I remember reading that bald older men are regarded as leaders." What type of coping is Herman using, and why is it effective?

CONNECT

Reread the story of Dr. Stojic, a robust centenarian, in the Lifespan Vista box on pages 550–551. What aspects of his life history are consistent with research findings on factors that contribute to a long and healthy life?

CONNECT

Review gender schema theory on page 263 of Chapter 8. How do aging stereotypes similarly affect elders' self-perceptions and behavior?

www.

Cultural Influences

Cultural Variations in the Experience of Aging

In a study of diverse communities around the world, a team of anthropologists found that older people fare best when they retain social status and opportunities for community participation, even after they become frail (Fry et al., 1997; Keith et al., 1994). When elders are excluded from important social roles and infirmity brings separation from the community, aging leads to reduced psychological well-being.

Consider the Herero, a pastoral people of Botswana, Africa. Older adults who are strong and active spend their days just as younger adults do, tending the cattle and performing other chores. When elders decline physically, they retain positions of seniority and are treated with respect. A status hierarchy makes the oldest man and his wife village leaders. They are responsible for preserving the sacred flame of the ancestors, who remain significant family members after death. Children are sent to live in the homes of frail elders to provide care—an assignment that is a source of great pride and prestige.

Old age is also a gratifying time of life in Momence, Illinois, a small, working-class farming and manufacturing town. The population is highly stable, so elders are granted positions of authority because of their length of residence and intimate knowledge of the community. Town, church, and club leaders tend to be older, and past leaders are included in decision making. Frail elders remain part of community life in a less direct way. Because they are embedded in family, neighborhood, and church networks that have persisted for many years, other citizens often inquire about them, visit them, and monitor their condition.

In Swarthmore, a middle-SES Philadelphia suburb, life for elders is less certain. Although 25 percent of residents are over age 60, most moved to the community after retirement, and their adult children live elsewhere. As a result, older people are granted no special status due to seniority, and they occupy few valued community roles. Unlike the Herero and the citizens of Momence, Swarthmore elders spend much time in age-segregated settings, such as bridge games and church groups for seniors. Townspeople tend to equate "being older" with age—60 and beyond. In contrast, the Herero and the residents of Momence seldom refer to older adults in terms of their age. Rather, they mention knowledge and social position.

Among the Herero and in Momence, neither age nor physical decline threatens community ties. In Swarthmore, being old limits integration into community life, and frailty has profound, negative consequences. It typically brings social isolation.

© JACOB HALASKA/INDEX STOCK IMAGERY

■ Among the Herero people of Botswana, Africa, elders continue to participate actively in their community until advanced old age. When these Herero women decline physically, children will be assigned the prestigious task of living in their homes to provide care. And the oldest man and woman will be named leaders of the village.

Health, Fitness, and Disability

At Walt and Ruth's fiftieth wedding anniversary, 77-year-old Walt thanked a roomful of well-wishers for joining in the celebration. Then he announced emotionally, "I'm so grateful Ruth and I are in good health and still able to give to our family, friends, and community."

As Walt's remarks affirm, health is central to psychological well-being in late life. When researchers ask the elderly about their possible selves (see Chapter 16, page 518), number of hoped-for physical selves declines with age and number of feared physical selves increases. Despite this realistic response to physical changes, older adults are generally optimistic about their health. Because they judge themselves against same-age peers, the majority rate their health favorably (see Figure 17.4) (Statistics Canada, 2002n; U.S. Department of Health and Human Services, 2002k). As with respect to protecting their health, elders' sense of self-efficacy is as high as

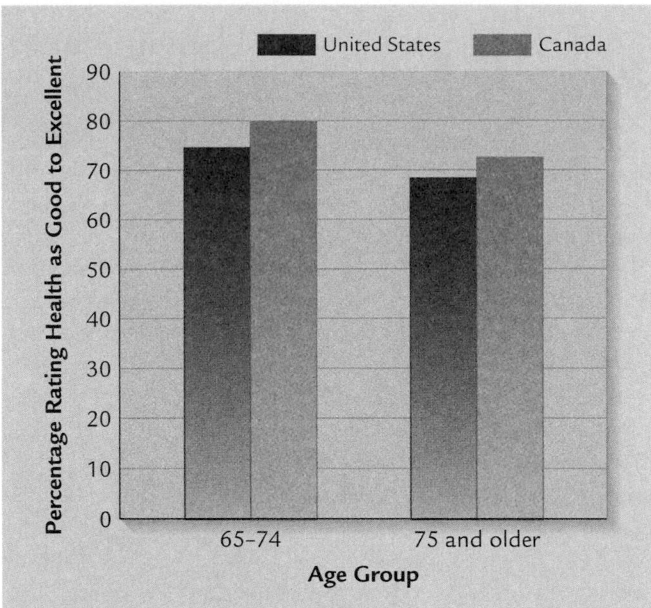

FIGURE 17.4 Percentage of older Americans and Canadians who rate their health as good to excellent. The majority of elders in both countries are optimistic about their health, even at age 75 and older. More Canadians than Americans evaluate their health favorably—judgments that dovetail with Canadian elders' overall better health and longer active lifespan; see Figure 17.1 on page 548. (Adapted from Statistics Canada, 2000n; U.S. Department of Health and Human Services, 2002f.)

that of young adults and higher than that of middle-aged people (Frazier, 2002; Hooker, 1992).

The more optimistic elders are about their capacity to cope with physical challenges, the better they are at overcoming threats to health, which promotes further optimism and continued health-enhancing behaviors (Hooker & Kaus, 1992; Schwarzer, 1999). Longitudinal research reveals that disability need not follow a path toward further disability and dependency. In several studies, 12 to 20 percent of elders with disabilities showed improvement 2 to 6 years later (Ostir et al., 1999). Furthermore, good health permits older adults to remain socially active, thereby fostering psychological well-being (Heidrich & Ryff, 1993). In this way, physical and mental health are intimately related in late life.

As mentioned earlier, SES and ethnic variations in health diminish in late adulthood. Nevertheless, before age 85, SES continues to predict physical functioning (Berkman & Gurland, 1998). African-American and Hispanic elderly (one-fifth of whom live in poverty) remain at greater risk for certain health problems (see Table 17.1). Native-American and Canadian Aboriginal older adults are even worse off. The majority are poor, and chronic health conditions are so widespread that in the United States, the federal government grants Native Americans special health benefits. These begin as early as age 45, reflecting a much harder and shorter lifespan (Gelfand, 1994). Unfortunately, low-SES elders are less likely than their higher-SES counterparts to seek medical treatment. When they do, they often do not comply with the doctor's directions because they are less likely to believe they can control their health and that treatment will work (Hopper, 1993). A high sense of self-efficacy does not apply to these older adults, further impairing their physical condition.

The sex differences noted in Chapter 15, with men more prone to fatal diseases and women to non-life-threatening disabling conditions, extend into late adulthood. By very old age (85 and beyond), women are more impaired than men because only the sturdiest men have survived (Arber & Cooper, 1999). In addition, with fewer physical limitations, older men are better able to remain independent and to engage in exercise, hobbies, and involvement in the social world, all of which promote better health.

Widespread health-related optimism among the elderly means that substantial inroads into preventing disability can be made even in the last few decades of life. Yet the persistence of poverty and negative lifestyle factors means that as longevity increases, diseases of old age may rise as well, straining the nation's health care resources. Ideally, as life expectancy extends, we want the average period of diminished vigor before death to decrease—a public health goal called the **compression of morbidity.** As we look closely at health, fitness, and disability in late adulthood, we will expand our discussion

Table 17.1 Poverty Rates and Health Problems Among Elderly Ethnic Minorities

Ethnic Minority	Poverty Rate Age 65 And Over	Health Problems Greater than in the General Population of Elderly
African-American	23%	Cardiovascular disease, a variety of cancers, diabetes
Hispanic	20%	Cardiovascular disease, diabetes
Native-American	Over 80%	Diabetes, kidney disease, liver disease, tuberculosis, hearing and vision impairments
Canadian Aboriginal	Over 63%	Cardiovascular disease, diabetes, liver disease, tuberculosis

Sources: MacMillan et al., 1996; Health Canada, 2002i; U.S. Bureau of the Census, 2000c.

of health promotion in earlier chapters, taking up additional ways to reach this objective.

Nutrition and Exercise

The physical changes of late life lead to an increased need for certain nutrients—calcium and vitamin D to protect the bones; zinc and vitamins B_6, C, and E to protect the immune system; and vitamins A, C, and E to prevent free radicals (see Chapter 13, page 413). Yet declines in physical activity, in the senses of taste and smell, and in ease of chewing (because of deteriorating teeth) can reduce the quantity and quality of food eaten (Morley, 2001). Furthermore, the aging digestive system has greater difficulty absorbing certain nutrients, such as protein, calcium, and vitamin D. And older adults who live alone may have problems shopping or cooking and may feel less like eating when by themselves at mealtimes. Together, these physical and environmental conditions increase the risk of dietary deficiencies, which affect 10 to 25 percent of elders (High, 2001).

Older adults who take vitamin–mineral supplements show improved health and physical functioning. In one study, a daily vitamin–mineral tablet resulted in an enhanced immune response and a 50 percent drop in days of infectious illness (Chandra, 1992). In another study, elderly men in a weight-bearing exercise class who took a daily protein energy supplement showed a larger increase in muscle mass than did exercising classmates who did not take the supplement (Meredith et al., 1992). And as discussed earlier, dietary supplements containing vitamins A, C, and E help prevent macular degeneration. But their role in protecting against cardiovascular disease is still uncertain (Watkins et al., 2000). And although malnutrition depresses cognitive functioning at all ages, various nutrients and herbs identified as "cognitive enhancers"—including vitamin E and ginkgo biloba—show weak effects at best (Gold, Cahill, & Wenk, 2002; McDaniel, Maier, & Einstein, 2002).

Besides nutrition, exercise continues to be a powerful health intervention. Sedentary healthy older adults up to age 80 who begin endurance training (walking, cycling, aerobic dance) show gains in vital capacity that compare favorably with those of much younger individuals (Stratton et al., 1994). And weight-bearing exercise begun in late adulthood—even as late as age 90—promotes muscle size and strength, blood flow to muscles, and ability of muscles to extract oxygen from blood. This translates into improved walking speed, balance, posture, and ability to carry out everyday activities, such as opening a stubborn jar lid, carrying an armload of groceries, or lifting a 30-pound grandchild (Goldberg, Dengel, & Hagberg, 1996; Pyka et al., 1994).

Exercise also increases blood circulation to the brain, which helps preserve brain structures and behavioral capacities. Brain scans show that compared with sedentary elders, those who are physically fit experienced less tissue loss—in both neurons and glial cells—in diverse areas of the cerebral cortex (Colcombe et al., 2003). Consistent with these findings, exercise reduces cognitive declines. In a large study of Canadians 65 years and older, those who engaged in regular physical activity were less likely to be cognitively impaired when reassessed 5 years later (Laurin et al., 2001).

Although good nutrition and physical activity are most beneficial when they are lifelong, it is never too late to change. Beginning in his sixties and until his death at age 94, Walt's Uncle Louie played 1 to 2 hours of tennis a day and went ballroom dancing three nights a week. Exercise led Louie to sustain a high sense of physical self-esteem. As a dancer, he dressed nattily and moved gracefully. He often commented on how dance and other sports could transform an older person's appearance from dowdy to elegant, expressing the beauty of the inner self.

Elders who come to value the intrinsic benefits of exercise—feeling stronger, healthier, and more energetic—are likely to engage in it regularly (Caserta & Gillett, 1998). Yet lack of awareness of the health benefits of exercise and expected discomforts from engaging in it are major barriers to getting older people to take up a fitness routine; 75 percent of men and 80 percent of women are not active enough (Stewart et al., 2001). Furthermore, many with chronic disease symptoms think "taking it easy" is the best treatment and that exercise actually will do harm. In planning exercise programs for older adults, it is important to instill a sense of control over the aging process—by stressing the health-enhancing rewards of physical activity and by changing negative beliefs that interfere with sustained effort (Lachman et al., 1997). Active seniors can serve as positive role models and sources of encouragement.

Sexuality

When Walt turned 60, he asked his 90-year-old Uncle Louie at what age sexual desire and activity cease, if they do. Walt's question stemmed from a widely held myth that sex drive disappears among the elderly (Hillman, 2000). "It's especially important to be reasonably rested and patient during sex," Louie explained to Walt. "I can't do it as often, and it's a quieter experience than it was in my youth, but my sexual interest has never gone away. Rachella and I have led a happy intimate life, and it's still that way."

Although virtually all cross-sectional studies report a decline in sexual desire and frequency of sexual activity in older people, this trend may be exaggerated by cohort effects. A new generation of elders, accustomed to viewing sexuality positively, will probably be more sexually active. Furthermore, longitudinal evidence indicates that most healthy older married couples report continued, regular sexual enjoyment. The same generalization we discussed for midlife applies to late life: Good sex in the past predicts good sex in the future. Among unmarried people over age 65, about 70 percent of men and 50 percent of women manage to have sex from time to time (Levy, 1994).

Too often, intercourse is used as the only measure of sexual activity—a circumstance that promotes a narrow view of pleasurable sex. Even at the most advanced ages, there is more to sexuality than the sex act itself—feeling sensual, enjoying close companionship, and being loved and wanted (Hodson &

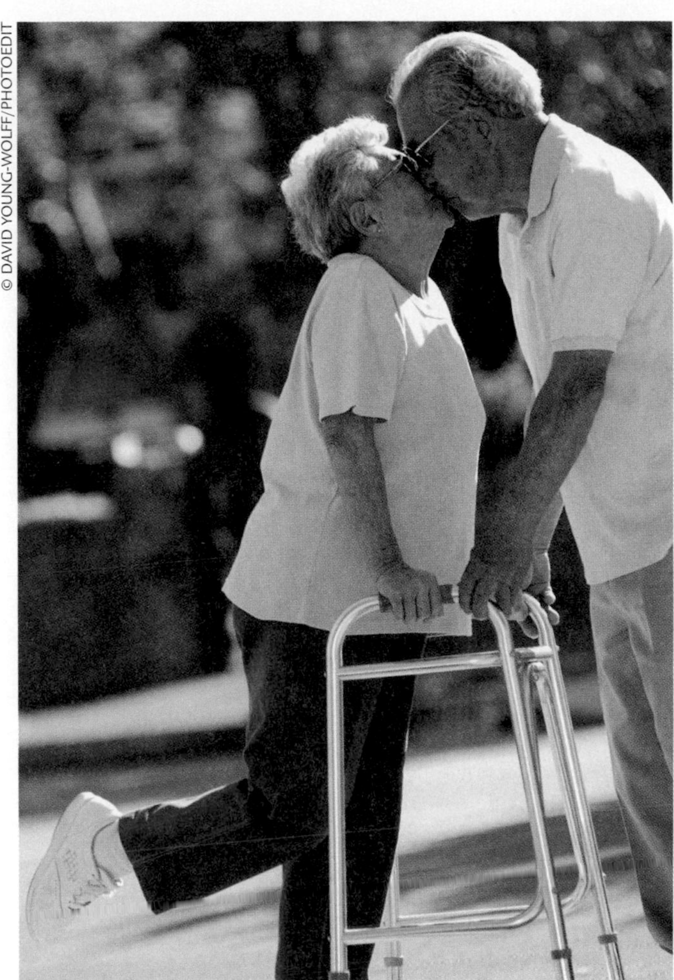
© DAVID YOUNG-WOLFF/PHOTOEDIT

■ Most healthy older couples report continued, regular sexual enjoyment. And even at the most advanced ages, there is more to sexuality than the sex act itself—feeling sensual, enjoying close companionship, and being loved and wanted.

Skeen, 1994). Both older men and older women report that the male partner is usually the one who ceases to interact sexually. In a culture that emphasizes an erection as necessary for being sexual, a man may withdraw from all erotic activity when he finds that erections are harder to achieve and more time must elapse between them (Pedersen, 1998).

Disabilities that disrupt blood flow to the penis—most often, disorders of the autonomic nervous system, cardiovascular disease, and diabetes—are responsible for dampening sexuality in older men. Cigarette smoking, excessive alcohol intake, and a variety of prescription medications also lead to diminished sexual performance. Among women, poor health and absence of a partner are major factors that reduce sexual activity (Kellett, 2000). Because the sex ratio increasingly favors females, aging heterosexual women have fewer and fewer opportunities for sexual encounters.

In most tribal and village cultures, sexual activity among elders is expected, and men and women engage in sex regularly until very late in life (Winn & Newton, 1982). In Western nations, sex in old age often meets with disapproval. Educational programs that inform older adults about normal, age-related changes in sexual functioning and that foster a view of sex as extending throughout adulthood promote positive sexual attitudes (Hillman & Stricker, 1994). In nursing homes, education for caregivers is vital for ensuring residents' rights to privacy and other living conditions that permit sexual expression (Levy, 1994).

Physical Disabilities

Illness and disability climb as the end of the lifespan approaches. If you compare the death rates shown in Figure 17.5 on page 562 with those in Figure 15.2 on page 490, you will see that cardiovascular disease and cancer—illnesses we discussed in Chapter 15—increase dramatically from mid- to late life and remain the leading causes of death. As before, death rates from cardiovascular disease and cancer are higher for men than for women, although the sex difference declines with advancing age (Lindsay, 1999; U.S. Bureau of the Census, 2002c).

Respiratory diseases also rise sharply in late adulthood. Among them is *emphysema,* caused by extreme loss of elasticity in lung tissue, which results in serious breathing difficulty. Although a few cases of emphysema are inherited, most result from long-term cigarette smoking. In addition, as the longest-lived people escape chronic diseases or weaken because of them, the immune system eventually encounters an infection it cannot fight. Consequently, many of the very old succumb to one of the more than fifty lung inflammations classified as *pneumonia.* Doctors recommend that people age 65 and older be vaccinated against the most common type.

Stroke is the fourth most common killer among the aged. It is caused by hemorrhage or blockage of blood flow in the brain and is a major cause of disability in late adulthood and death after age 75. Other diseases are less frequent killers, but they limit older adults' ability to live fully and independently. Osteoporosis, discussed in Chapter 15 (see page 493), continues to rise in late adulthood; recall that it affects the majority of men and women after age 70. Yet another bone disorder—*arthritis*—adds to the physical limitations of many elders. And *adult-onset diabetes* and *unintentional injuries* also multiply in late adulthood. In the following sections, we take up these latter three conditions.

Finally, an important point must be kept in mind as we discuss physical and mental disabilities of late adulthood: None are a normal part of the aging process, even though their incidence increases with age. In other words, just because certain physical and mental disabilities are *related to age* does not mean that they are *caused by aging.* To clarify this distinction, some experts distinguish between **primary aging** (another term for *biological aging*), or genetically influenced declines that affect all members of our species and take place even in the context of overall good health, and **secondary aging,** declines due to hereditary defects and negative environmental influences, such as poor diet, lack of exercise, disease, substance abuse, environmental pollution, and psychological stress.

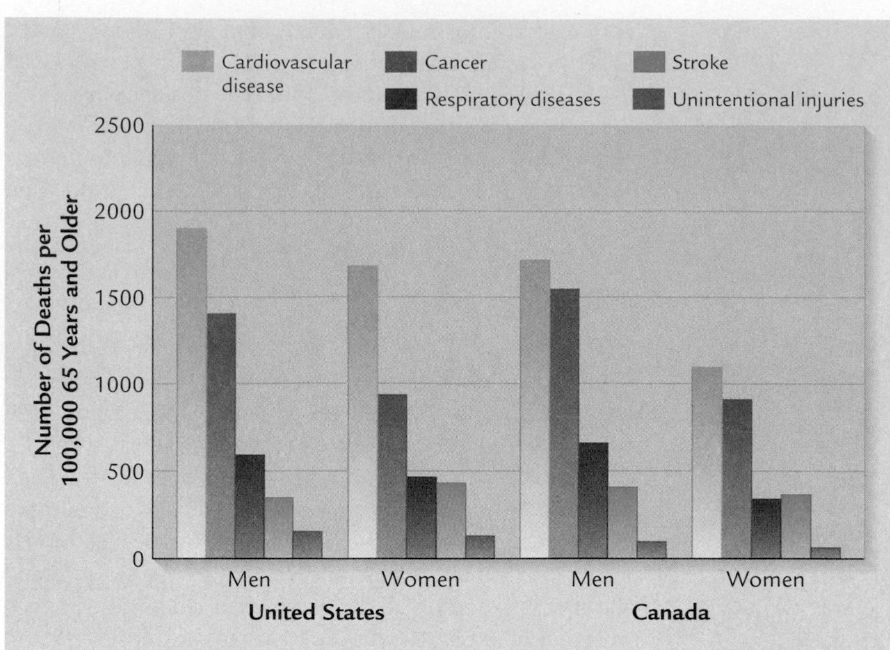

FIGURE 17.5 Leading causes of death among people age 65 and older in the United States and Canada. In late adulthood, cardiovascular disease is the leading cause of death among men and women, followed by cancer. Respiratory diseases and stroke also claim the lives of many elders. Notice that death rates for all causes, and especially cardiovascular disease, are lower among Canadian than American women. Cardiovascular disease among men shows a similar trend. Several factors might underlie these differences, including higher rates of certain risk factors, such as obesity, among American adults and better prevention and treatment of disease in Canada because of its generous, government-supported health care system. (Adapted from Statistics Canada, 2002n; U.S. Bureau of the Census, 2002c.)

Throughout this book, we have seen that it is difficult to distinguish primary from secondary aging. But unless we try to do so, we are in danger of magnifying decrements and promoting a false, stereotyped view of aging as illness and disease (Lemme, 2002). All the disabilities we are about to discuss are extreme conditions that fall within *secondary aging.*

● **Arthritis.** Beginning in her fifties, Ruth felt a slight morning stiffness in her neck, back, hips, and knees. In her sixties, she developed bony lumps on the end joints of her fingers. As the years passed, she experienced periodic joint swelling and some loss of flexibility—changes that affected her ability to move quickly and easily.

Arthritis, a condition of inflamed, painful, stiff, and sometimes swollen joints and muscles, becomes more common in late adulthood. It occurs in several forms. Ruth has **osteoarthritis,** the most common type, which involves deteriorating cartilage on the ends of bones of frequently used joints. Otherwise known as "wear-and-tear arthritis" or "degenerative joint disease," it is one of the few age-related disabilities in which years of use makes a difference. Although a genetic proneness seems to exist, the disease usually does not appear until the forties and fifties. In frequently used joints, cartilage on the ends of the bones, which reduces friction during movement, gradually deteriorates. Or obesity places abnormal pressure on the joints and damages cartilage. Almost all older adults show some osteoarthritis on X-rays, although wide individual differences in severity exist (Hinton, Moody, & Davis, 2002).

Unlike osteoarthritis, which is limited to certain joints, **rheumatoid arthritis** involves the whole body. An autoimmune response leads to inflammation of connective tissue, particularly the membranes that line the joints. The result is overall stiffness, inflammation, and aching. Tissue in the cartilage tends to grow, damaging surrounding ligaments, muscles, and bones. The result is deformed joints and often serious loss of mobility. Sometimes other organs, such as the heart and lungs, are affected.

Disability due to arthritis affects 45 percent of American and 34 percent of Canadian men over age 65 and rises modestly with age. Among North American women, the incidence is higher and increases sharply with age; about 52 percent of 65- to 84-year-olds and 70 percent of 85-year-olds are affected (Lindsay, 1999; U.S. Bureau of the Census, 2002c). The reason for the sex difference is not clear. Although rheumatoid arthritis can strike at any age, it rises after menopause. But unlike osteoporosis, it does not respond to hormone therapy, so something other than estrogen loss must be involved. Rheumatoid arthritis may be due to a late-appearing genetic defect in the immune system; twin studies reveal a strong hereditary contribution. However, identical twins differ widely in disease severity, indicating that as yet unknown environmental factors make a difference. Early treatment with new, powerful anti-inflammatory medications helps slow the course of the disease (Lee & Weinblatt, 2001).

Managing arthritis requires a balance of rest when the disease flares, pain relief, and physical activity involving gentle stretching of all muscles to maintain mobility. Twice a week, 84-year-old Ruth attended a water-based exercise class. Within 2 months, her symptoms lessened, and she no longer needed a walker (Strange, 1996). Weight loss in obese people is also helpful.

Although osteoarthritis responds to treatment more easily than rheumatoid arthritis, the course of each varies greatly. With proper medication, joint protection, and lifestyle changes, many people with either form of the illness lead long, productive lives. If hip or knee joints are badly damaged or deformed, they can be surgically rebuilt or replaced with plastic or metal devices.

● **Adult-Onset Diabetes.** After we eat a meal, the body breaks down the food, releasing glucose (the primary energy source for cell activity) into the bloodstream. Insulin, produced by the pancreas, keeps the blood concentration of glucose within set limits by stimulating muscle and fat cells to absorb it. When this balance system fails, either because not enough insulin is produced or because body cells become insensitive to it, *adult-onset diabetes* (otherwise known as *diabetes mellitus*) results. Over time, abnormally high blood glucose damages the blood vessels, increasing the risk of stroke, heart attack, circulatory problems in the legs, and injury to the eyes, kidneys, and nerves.

From middle to late adulthood, the incidence of adult-onset diabetes doubles, affecting 10 percent of the elderly in the United States and Canada (Lindsay, 1999; U.S. Bureau of the Census, 2002c). Diabetes runs in families, suggesting that heredity is involved. But inactivity and abdominal fat deposits greatly increase the risk. Higher rates of adult-onset diabetes are found among African-American, Mexican-American, Native-American, and Canadian Aboriginal minorities for genetic as well as environmental reasons, including high-fat diets and obesity associated with poverty. The Pima Indians of Arizona have the highest rate of diabetes in the world (Quinn, 2001). Return to Chapter 9, page 279, to review how nature and nurture combine to induce the disease, which affects 50 percent of Pima adults.

Treating adult-onset diabetes requires lifestyle changes, including a carefully controlled diet, regular exercise, and weight loss (Goldberg, Dengel, & Hagberg, 1996). By promoting glucose absorption and reducing abdominal fat, physical activity lessens disease symptoms.

■ Declines in vision and reaction time combined with greater cautiousness and indecisiveness increase elders' risk of auto and pedestrian accidents. Confusing intersections with traffic lights that do not provide enough time to cross the street also heighten the chances of a tragedy.

● **Unintentional Injuries.** At age 65 and older, the death rate from unintentional injuries is at an all-time high—more than twice as great as in adolescence and young adulthood. Motor vehicle collisions and falls are largely responsible.

Motor Vehicle Accidents. Motor vehicle collisions account for only one-fourth of injury mortality in late life, compared with one-half in middle adulthood. But a look at individual drivers tells a different story. Older adults have higher rates of traffic violations, accidents, and fatalities per mile driven than any other age group, with the exception of drivers under age 25 (Messinger-Rapport & Rader, 2000). The high rate of injury persists, even though many elders limit their driving after noticing that their ability to drive safely is slipping. Women are more likely to take these preventive steps. Deaths due to injuries—motor vehicle and otherwise—continue to be much higher for men than for women in late adulthood (Lindsay, 2000; U.S. Bureau of the Census, 2002c).

Recall that visual declines led Walt to have difficulty seeing the dashboard and identifying pedestrians at night. Most information used in driving is visual. The greater elders' visual processing difficulties, the higher their rate of moving violations and crashes (Owsley & McGwin, 1999). Compared with young drivers, the elderly are less likely to drive quickly and recklessly but more likely to fail to heed signs, yield the right of way, and turn appropriately. They often try to compensate for their difficulties by being more cautious. But slowed reaction time and indecisiveness can pose hazards as well. Recall from Chapter 15 that with age, adults find it harder to attend selectively, engage in two activities at once, and switch back and forth between tasks—abilities essential for safe driving (De Raedt & Ponjaert-Kristoffersen, 2000).

The elderly are also at risk on foot, accounting for more than 30 percent of all pedestrian deaths (Transport Canada, 2001; U.S. Bureau of the Census, 2002c). Confusing intersections and crossing signals that do not allow older people enough time to get to the other side of the street are often involved.

Falls. On one occasion, Ruth fell down the basement steps and lay there with a broken ankle until Walt arrived home an hour later. Ruth's tumble represents the leading type of accident among the elderly. About 30 percent of adults over age 65 and 40 percent over age 80 have experienced a fall within the last year. Declines in vision, hearing, and mobility, making it harder to avoid hazards and keep one's balance, increase the risk of falling in late adulthood (Fuller, 2000). Because of weakened bones and difficulty breaking a fall, serious injury results about 10 percent of the time. Among the most common is hip fracture. It increases twentyfold from age 65 to 85 and is associated with a 12 to 20 percent increase in mortality. Of those who survive, half never regain the ability to walk without assistance (Simoneau & Leibowitz, 1996).

Preventing Unintentional Injury in Late Adulthood

SUGGESTION	DESCRIPTION
Motor Vehicle Collisions and Pedestrian Accidents	
Modify driving behavior in accord with visual and other limitations.	Drive fewer miles; reduce or eliminate driving during rush hour, at night, or in bad weather.
Modify pedestrian behavior in accord with visual and other limitations.	Wear light-colored clothing at night; allow extra time to cross streets; walk with a companion.
Attend training classes for older drivers; if not available, press for them in your community.	Practice tracking vehicles and pedestrians in dim light, judging vehicle speed, and reading signs and dashboard displays; review rules of the road.
Falls	
Schedule regular medical checkups.	Eye exams to ensure that corrective lenses are up to date; physical exams to identify health risks that increase the chances of falling; review of medications for effects on attention and coordination
Engage in regular exercise.	Strength and balance training to promote coordination and counteract fear of falling
Use walking aids when necessary.	Canes and walkers to compensate for poor balance and unsteady gait
Improve safety of the living environment.	Extra lighting in dim areas, such as entrances, hallways, and staircases; handrails in hallways and grab bars in bathrooms; loose rugs secured to floor or moved; furniture and other objects arranged so they are not obstacles
Be alert and plan ahead in risky situations.	Watch for slippery pavement; carry a flashlight at night; allow extra time to cross streets; become familiar with new settings before moving about freely.

Sources: Fuller, 2000; Messinger-Rapport & Rader, 2000.

Falling can also impair health indirectly—by promoting fear of falling. Almost half of older adults who have fallen admit that they purposefully avoid activities because they are afraid of falling again (Tinetti, Speechley, & Ginter, 1988). In this way, a fall can limit mobility and social contact, undermining both physical and psychological well-being. Although an active lifestyle may expose the elderly to more situations that can cause a fall, the health benefits of activity far outweigh the risk of serious injury due to falling.

Preventing Unintentional Injuries. Many steps can be taken to reduce unintentional injury in late adulthood. Designing motor vehicles and street signs to accommodate the visual needs of the elderly is a goal for the future. In the meantime, training that enhances visual and cognitive skills essential for safe driving and that helps older adults avoid high-risk situations (such as busy intersections and rush hour) can save lives.

Similarly, efforts to prevent falls must address risks within the person and the environment—through corrective eyewear, strength and balance training, and improved safety in homes and communities. The Caregiving Concerns table above summarizes ways to protect the elderly from injury.

Mental Disabilities

Normal age-related cell death in the brain, described earlier in this chapter, does not lead to loss of ability to engage in everyday activities. But when cell death and structural and chemical abnormalities are profound, serious deterioration of mental and motor functions occurs.

Dementia refers to a set of disorders occurring almost entirely in old age in which many aspects of thought and behavior are so impaired that everyday activities are disrupted. Dementia rises sharply with age, striking adults of both sexes about equally (Liesi et al., 2001). Approximately 1 percent of people in their sixties are affected, a rate that increases steadily with age, rising sharply after age 75 until it reaches about 50 percent for people age 85 and older—trends that apply to Canada, the United States, and other Western nations (Launer et al., 1999; Pfizer, 2002). Although dementia rates are similar across SES and most ethnic groups, African Americans are at higher risk—a finding we will take up shortly (Alzheimer's Association, 2002).

About a dozen types of dementia have been identified. Some are reversible with proper treatment, but most are irreversible and incurable. A few forms, such as Parkinson's disease,[2] involve deterioration in subcortical brain regions (primitive structures below the cortex) that often extends to the cerebral cortex and, in many instances, involves brain abnormalities resembling Alzheimer's disease. Some researchers be-

[2]In Parkinson's disease, neurons in the part of the brain that controls muscle movements deteriorate. Symptoms include tremors, shuffling gait, loss of facial expression, rigidity of limbs, difficulty maintaining balance, and stooped posture.

lieve that Parkinson's and Alzheimer's are related (Lieberman, 2002). But in the large majority of dementia cases, subcortical brain regions are intact, and only progressive damage to the cerebral cortex occurs. *Cortical dementia* comes in two varieties: Alzheimer's disease and cerebrovascular dementia.

● **Alzheimer's Disease.** When Ruth took 79-year-old Ida to the ballet, an occasion that the two sisters looked forward to each year, Ida's behavior was different than before. Having forgotten the engagement, she reacted with anger at Ruth for arriving unannounced at her door. After Ida calmed down, she got lost driving to the theater in familiar parts of town, insisting that she knew her way perfectly well. Settled in the balcony, Ida talked loudly and dug noisily in her purse as the lights dimmed and the music began.

"Shhhhhh," responded a dozen voices from the surrounding seats.

"It's just the music!" Ida snapped at full volume. "You can talk all you want until the dancing starts." Ruth was astonished and embarrassed at the behavior of her once socially sensitive sister.

Six months later, Ida was diagnosed with **Alzheimer's disease,** the most common form of dementia, in which structural and chemical brain deterioration is associated with gradual loss of many aspects of thought and behavior. Alzheimer's accounts for 60 percent of all dementia cases and, at older ages, for an even higher percentage. Approximately 8 to 10 percent of people over age 65 have the disorder. Of those over age 80, close to 50 percent are affected (Heinz & Blass, 2002; Launer et al., 1999; Pfizer, 2002). Each year, about 5 percent of all deaths among the elderly—63,000 in the United States and 6,500 in Canada—involve Alzheimer's (Wilkins et al., 2000). This makes the disease a leading cause of late-life mortality, although on death certificates, more immediate causes, such as infection and respiratory failure, are usually listed.

Symptoms and Course of the Disease. Severe memory problems are often the earliest symptom—forgetting names, dates, appointments, familiar routes of travel, or the need to turn off the stove in the kitchen. At first, recent memory is most impaired, but as serious disorientation sets in, recall of distant events and such basic facts as time, date, and place evaporates. Faulty judgment puts the person in danger. For example, Ida insisted on driving after she was no longer capable of doing so. Personality changes occur—loss of spontaneity and sparkle, anxiety in response to uncertainties created by mental problems, aggressive outbursts, reduced initiative, and social withdrawal. Depression often appears in the early phase of Alzheimer's and other forms of dementia and seems to be part of the disease process (Espiritu et al., 2001; Mulsant & Ganguli, 1999). However, depression may worsen as the older adult reacts to disturbing mental changes.

As the disease progresses, skilled and purposeful movements disintegrate. When Ruth took Ida into her home, she had to help her dress, bathe, eat, brush her teeth, and (eventually) walk and use the bathroom. Ruth also found that Ida's sleep was disrupted by delusions and imaginary fears. She often awoke in the night and banged on the wall, insisting that it was dinnertime. Sometimes she cried out that someone was choking her. Over time, Ida lost the ability to comprehend and produce speech. And when her brain ceased to process information, she could no longer recognize objects and familiar people. In the final months, Ida became increasingly vulnerable to infections, lapsed into a coma, and died.

The course of Alzheimer's varies greatly, from 1 to as many as 15 years. The average is about 6 to 7 years (National Institute on Aging, 2000).

Brain Deterioration. A diagnosis of Alzheimer's disease is made through exclusion, after other causes of dementia have been ruled out by a physical examination and psychological testing—an approach that is more than 90 percent accurate. To confirm Alzheimer's, doctors inspect the brain after death for a set of abnormalities that either cause or result from the disease (Heinz & Blass, 2002). However, new developments in brain-imaging techniques, which yield three-dimensional pictures of brain volume and activity, predict in nearly 90 percent of cases whether elders who do not yet show symptoms receive an after-death confirmation of Alzheimer's (Silverman et al., 2001). Brain-imaging procedures offer hope of very early diagnosis, opening the door to more successful interventions.

Two major structural changes in the cerebral cortex, especially in memory and reasoning areas, are associated with Alzheimer's. Inside neurons, **neurofibrillary tangles** appear—bundles of twisted threads that are the product of collapsed neural structures. Outside neurons, **amyloid plaques,** dense deposits of a deteriorated protein called *amyloid,* surrounded by clumps of dead nerve cells, develop. Amyloid appears in

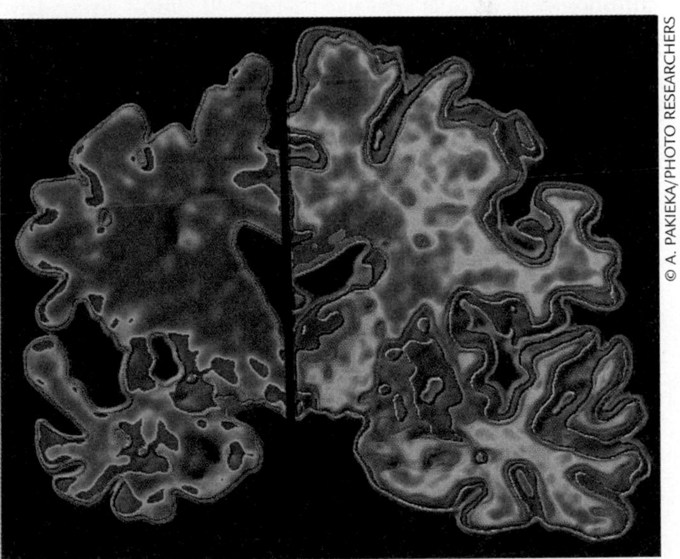

■ These computerized images compare a brain scan of an Alzheimer's patient (left) with one of a healthy older adult (right). The Alzheimer's brain is shrunken, due to massive degeneration and death of neurons. In addition, activity and blood flow (marked by yellow and green color coding in the right scan) in the Alzheimer's brain is greatly reduced. Such images offer hope of early diagnosis and, therefore, more effective interventions.

tissue with reduced immunity. As it breaks down, it may destroy surrounding neurons and their communication networks, perhaps by causing inflammation in the brain, by generating free radicals, or by reducing blood flow (Knowles et al., 1999). Although some neurofibrillary tangles and amyloid plaques are present in the brains of normal middle-aged and older people and increase with age, they are far more abundant and widely distributed in Alzheimer's victims (Munoz & Feldman, 2000).

Accompanying massive death of neurons and resulting shrinkage of brain volume are chemical changes—lowered levels of *neurotransmitters* necessary for communication between neurons. Alzheimer's involves destruction of neurons that release the neurotransmitter acetylcholine, which transports messages between distant areas of the brain. As a result, perception, memory, reasoning, and judgment are disrupted further (National Institute on Aging, 2000). A drop in serotonin, a neurotransmitter that regulates arousal and mood, may contribute to sleep disturbances, aggressive outbursts, and depression (Lanctot et al., 2002; Mintzer, 2001).

Risk Factors. Alzheimer's disease comes in two types: *familial,* which runs in families, and *sporadic,* which has no obvious family history. Familial Alzheimer's generally has an early onset (before age 65) and progresses more rapidly than the later-appearing, sporadic Alzheimer's. Researchers have identified genes on chromosomes 1, 14, and 21 that are linked to familial Alzheimer's. In each case, the abnormal gene is dominant; if it is present in only one of the pair of genes inherited from parents, the person will develop early-onset Alzheimer's (Heinz & Blass, 2002; National Institute on Aging, 2000). Recall that chromosome 21 is involved in Down syndrome. Individuals with this chromosomal disorder who live past age 40 almost always have the brain abnormalities and symptoms of Alzheimer's.

Heredity also plays a role in sporadic Alzheimer's, through spontaneous mutations. People with this form of the disease often have an abnormal gene on chromosome 19, which results in excess levels of *ApoE4,* a blood protein that carries cholesterol throughout the body. A high blood concentration of ApoE4 is linked to amyloid plaque formation in Alzheimer's victims (Heinz & Blass, 2002; Holtzman et al., 2000). The abnormal ApoE4 gene is the most commonly known risk factor for this form of Alzheimer's; it is present in 50 percent of cases. Genetic testing has revealed other regions of the genome with as yet unidentified genes that seem to make a contribution to the disease as great as or greater than that of the ApoE4 gene (Blacker et al., 2003; Scott et al., 2002).

Nevertheless, many sporadic Alzheimer's victims show no currently known genetic marker. Besides heredity, the roles of toxic substances and viruses, defects in the blood–brain barrier (which protects the brain from harmful agents), vitamin–mineral deficiencies (especially folate), excess dietary fat, and cardiovascular disease are being explored. Head injury, which can cause brain damage and deterioration of the amyloid protein, is clearly linked to Alzheimer's (Plassman et al., 2000). Although elevated levels of aluminum have been found in neurofibrillary tangles and amyloid plaques of Alzheimer's victims, the role of environmental exposure to aluminum remains uncertain (Heinz & Blass, 2002).

Alzheimer's disease probably results from different combinations of genetic and environmental factors, each of which leads to a somewhat different course of the disease. The high incidence of Alzheimer's and other forms of dementia among African-American elderly illustrates the complexity of potential causes. Compared with African Americans, Yoruba village dwellers of Nigeria show a much lower Alzheimer's incidence, and only a weak association between the ApoE4 gene and the disease. Some investigators speculate that intermarriage with Caucasians heightened genetic risk among African Americans and that environmental factors translated that risk into reality (Hendrie, 2001; Hendrie et al., 2001). Whereas the Yoruba of Nigeria eat a low-fat diet, the African-American diet is high in fat. Eating fatty foods may elevate the chances that the ApoE4 gene will lead to Alzheimer's (Notkola et al., 1998). And even for African Americans without the ApoE4 gene, a high-fat diet is risky. The more dietary fat consumed, the greater the incidence of Alzheimer's (Evans et al., 2000).

Protective Factors. Among factors that protect against Alzheimer's are vitamin C and E supplements and anti-inflammatory drugs, such as aspirin, ibuprofen, and the steroid prednisone (Engelhart et al., 2002; Heinz & Blass, 2002). Vitamins probably exert beneficial effects by lessening free-radical damage. And anti-inflammatory medications may slow the course of the disease by reducing inflammation of brain tissue caused by amyloid and other abnormalities. Although hormone therapy was previously believed to offer women protection against Alzheimer's, recent rigorously designed research yielded contrary findings. The estrogen/progesterone combination (HRT) actually heightened risk of the disease (See Chapter 15, page 486).

Other protective factors are education and an active lifestyle. Compared with their better-educated counterparts, elders with little education show more that double the rate of Alzheimer's (Qiu et al., 2001). Some researchers speculate that education leads to more synaptic connections, which act as a *cognitive reserve,* equipping the aging brain with greater tolerance for injury before it crosses the threshold into mental disability. Finally, late-life engagement in social and leisure activities reduces the risk of Alzheimer's and of dementia in general, perhaps by stimulating synaptic growth and thereby preserving cognitive functioning (Wang et al., 2002).

Helping Alzheimer's Victims and Their Caregivers. As Ida's Alzheimer's worsened, the doctor prescribed a mild sedative and an antidepressant to help control her behavior. Drugs that limit breakdown of the neurotransmitter acetylcholine also reduce dementia symptoms (Winblad et al., 2001).

But with no cure available, family interventions ensure the best adjustment possible for the Alzheimer's victim, spouse, and other relatives. Dementia caregivers devote substantially more time to caregiving and experience more stress than do people caring for elders with disabilities (Ory et al., 2000).

They need assistance and encouragement from extended-family members, friends, and community agencies. The Social Issues box on page 568 describes a variety of helpful interventions for family caregivers. In addition to these strategies, avoiding dramatic changes in living conditions, such as moving to a new location, rearranging furniture, or modifying daily routines, helps elders with Alzheimer's disease feel as secure as possible in a cognitive world that is gradually disintegrating.

● Cerebrovascular Dementia. In **cerebrovascular dementia,** a series of strokes leaves areas of dead brain cells, producing step-by-step degeneration of mental ability, with each step occurring abruptly after a stroke. About 5 to 10 percent of all cases of dementia are cerebrovascular, and about 10 percent are due to a combination of Alzheimer's and repeated strokes (Corey-Bloom, 2000).

Heredity affects cerebrovascular dementia indirectly, through high blood pressure, cardiovascular disease, and diabetes, each of which increases the risk of stroke. But many environmental influences, including cigarette smoking, heavy alcohol use, high salt intake, very low dietary protein, obesity, inactivity, and stress, also heighten stroke risk, so cerebrovascular dementia results from a combination of genetic and environmental forces.

Because of their susceptibility to cardiovascular disease, more men than women have cerebrovascular dementia by their late sixties. Women are not at great risk until after age 75 (Sachdev, Brodaty, & Looi, 1999). The disease also varies among countries. For example, deaths due to stroke are high in Japan. Although a low-fat diet reduces Japanese adults' risk of cardiovascular disease, high intake of alcohol and salt and a diet very low in animal protein increase the risk of stroke. As Japanese consumption of alcohol and salt declined and intake of meat rose in recent decades, the rates of cerebrovascular dementia and stroke-caused deaths dropped. However, they remain higher than in other developed nations (Goldman & Takahashi, 1996; Myers, 1996).

Although Japan presents a unique, contradictory picture (a culture where cardiovascular disease is low and stroke is high), in most cases cerebrovascular dementia is caused by atherosclerosis. Prevention is the only effective way to stop the disease. The incidence of cerebrovascular dementia has dropped in the last 2 decades, probably as a result of the decline in heart disease and more effective stroke prevention methods (Elkind & Sacco, 1998). Signs that a stroke might be coming are weakness, tingling, or numbness in an arm, a leg, or the face; sudden vision loss or double vision; speech difficulty; and severe dizziness and imbalance. Doctors may prescribe drugs to reduce the tendency of the blood to clot. Once strokes occur, paralysis and loss of speech, vision, coordination, memory, and other mental abilities are common.

● Misdiagnosed and Reversible Dementia. Careful diagnosis of dementia is crucial because other disorders can be mistaken for it. And some forms of dementia can be treated and a few reversed.

Depression is the disorder most often misdiagnosed as dementia. The depressed (but not demented) older adult is likely to exaggerate his or her mental difficulties, whereas the demented person minimizes them and is not fully aware of cognitive declines. Fewer than 1 percent of people over age 65 are severely depressed and another 2 percent are moderately depressed—rates that are lower than for young and middle-aged adults (King & Markus, 2000). Most older people have probably learned how to adjust what they expect from life, resulting in fewer feelings of worthlessness. However, as we will see in Chapter 18, depression rises with age. It is often related to physical illness and pain and can lead to cognitive deterioration (Magni & Frisoni, 1996). As at younger ages, the support of family members and friends; antidepressant medication; and individual, family, and group therapy can help relieve depression. However, the elderly in general and low-SES, ethnic minorities in particular are unlikely to seek mental health services. This increases the chances that depression will deepen and be confused with dementia (Padgett et al., 1994).

The older we get, the more likely we are to be taking drugs that might have side effects resembling dementia. For example, some medications for coughs, diarrhea, and nausea inhibit the neurotransmitter acetylcholine, leading to Alzheimer's-like symptoms. Because tolerance for drugs decreases with age, these reactions intensify in late adulthood. In addition, some diseases can cause temporary memory loss and mental symptoms, especially among the elderly, who often become confused and withdrawn when ill. Treatment of the underlying illness relieves the problem. Finally, environmental changes and social isolation can trigger mental declines (Gruetzner, 1992). When supportive ties are restored, cognitive functioning usually bounces back.

Health Care

Health care professionals and lawmakers in industrialized nations worry about the economic consequences of rapid increase in the elderly population. Rising government-supported health care costs and demand for certain health care services, particularly long-term care, are of greatest concern.

● Cost of Health Care for the Elderly. Adults age 65 and older, who make up 12 percent of the North American population, account for 30 percent of government health care expenditures in the United States and for 42 percent in Canada (Health Canada, 2002f; U.S. Bureau of the Census, 2002c). According to current estimates, the cost of government-sponsored health insurance, or Medicare, for the elderly is expected to double by the year 2020 and to nearly triple by 2040 as the "baby boom" generation reaches late adulthood and average life expectancy extends further (Bodenheimer, 1999).

Medicare expenses rise steeply with age. People age 75 and older receive, on average, 70 percent more benefits than younger senior citizens. Most of this increase is due to a need for long-term care—in hospitals and nursing homes—because

Social Issues

Interventions for Caregivers of Elders with Dementia

Margaret, wife and caregiver of a 71-year-old Alzheimer's patient, sent a desperate plea to an advice columnist at her local newspaper: "My husband can't feed or bathe himself, or speak to anyone or ask for assistance. I must constantly anticipate his needs and try to meet them. Please help me. I'm at the end of my rope."

The effects of Alzheimer's disease are devastating not just to elderly victims but also to family members who provide care with little or no outside assistance. Caregiving under these conditions has been called the "36-hour day" because of its constant demands. Although the majority of home caregivers are middle-aged, an estimated 15 to 25 percent are elders themselves who care for a spouse or an aging parent (Chappell et al., 2003; Family Caregiver Alliance, 2001). The burdens of such care are high, and family members who exceed their caregiving capacities suffer greatly in physical and mental health (Lund, 1993a).

Interventions designed to support family caregivers are available in most communities, although they need to be expanded and made more cost-effective. Those that work best begin before caregivers become overwhelmed and address multiple needs: knowledge, coping strategies, caregiving skills, and respite.

Knowledge. Virtually all interventions try to enhance knowledge about the disease, caregiving challenges, and available community assistance. Although knowledge usually is delivered through classes, in one innovative approach, computers were installed in caregivers' homes, enabling them to access a database with wide-ranging information on caregiving (Brennan, Moore, & Smyth, 1991). Gains in knowledge, however, must be combined with other approaches to improve caregivers' well-being.

Coping Strategies. Many interventions teach caregivers everyday problem-solving strategies for managing the dependent elder's behavior, along with techniques for dealing with negative thoughts and feelings, such as resentment for having to provide constant care. Modes of delivery include support groups, individual therapy, and "coping with frustration" classes. In one study, interveners assessed caregivers' current coping strategies and provided individualized training in more effective techniques. Compared with no-intervention controls, trained caregivers felt less burdened and depressed, and their patients engaged in fewer disturbing behaviors—gains still evident 3 months after the intervention (Marriott et al., 2000).

Caregiving Skills. Caregivers benefit from lessons in how to communicate with elders who can no longer handle everyday tasks—for example, distracting rather than scolding when the person asks the same question over and over; responding patiently with reminders and lists when the person blames others for memory problems; and introducing pleasant activities, such as music and slow-paced children's TV programs, that relieve agitation. Interventions that teach communication skills reduce elders' troublesome behavior and, as a result, may boost caregivers' sense of self-efficacy and reports of positive emotion (Bourgeois et al., 1997).

Respite. Caregivers usually say that *respite*—time away from providing care—is the assistance they most desire (Shope et al., 1993). Yet even when communities offer respite services, such as adult day care or temporary placement in a care facility, caregivers are reluctant to use them because of cost, worries about the elder's adjustment, and guilt. Yet respite at least twice a week for several hours improves physical and mental health by permitting caregivers to maintain friendships, engage in enjoyable activities, and sustain a balanced life (Lund & Wright, 2001; Zarit et al., 1998).

Recognizing caregivers' need for relief from unrelenting in-home care, one group of researchers devised a unique tool called Video Respite—a series of videotapes suited to the interests of Alzheimer's patients that provide caregivers with a half-hour to an hour break. On each tape, a professional actor conducts a friendly, slow-paced, simple conversation about familiar experiences, people, and objects, pausing occasionally for the impaired elder to respond. Evaluations of the videos show that they not only capture the attention of people with Alzheimer's but also reduce problem behaviors, such as wandering, agitation, repeated questioning, and aggression (Lund et al., 1995).

No single magic bullet exists for eliminating caregiver stress. But interventions that include most of the ingredients just considered, that begin early in the caregiving process, that continue for many weeks or months, and that are tailored to caregivers' individual needs make a substantial difference in caregivers' lives. And they usually delay placement of dementia patients in institutions as well (Kennet, Burgio, & Schultz, 2000).

■ This elderly woman cares for her husband, who has Alzheimer's disease, full time. Although the task has compensating rewards, it is physically demanding and emotionally draining. A great need exists for interventions that provide caregivers with essential knowledge about the elder's illness, teach coping strategies and caregiving skills, and provide respite from unrelenting caregiving responsibilities.

of the age-related rise in disabling chronic diseases and acute illnesses. Because U.S. Medicare funds only about half of older adults' medical needs, American elders spend nearly five times the percentage of their annual incomes on health care as do Canadian elders—19 versus 4 percent (Lindsay, 1999; Crystal et al., 2000). But as we will see, Medicare in both nations provides far less support for long-term care than elders with severe disabilities need.

● **Long-Term Care.** When Ida moved into Ruth's home, Ruth promised that she'd never place Ida in an institution. But as Ida's condition worsened and Ruth faced health problems of her own, she couldn't keep her word. Ida needed round-the-clock monitoring. Reluctantly, Ruth placed her in a nursing home.

As Figure 17.6 reveals, advancing age is strongly associated with use of long-term care services, especially nursing homes. Among disorders of aging, dementia, especially Alzheimer's disease, most often leads to nursing home placement, followed by hip fracture (Agüero-Torres et al., 2001; Rockwood, Stolee, & Dowell, 1996). Greater use of nursing homes is also prompted by loss of informal caregiving support through widowhood—which mostly affects women—and aging of adult children and other relatives.

Overall, only 4.5 percent of Americans and 6 percent of Canadians age 65 and older are institutionalized, a rate about half that of other industrialized nations, such as the Netherlands and Sweden. These countries provide more generous public financing of institutional care. In the United States and Canada, unless nursing home placement follows hospitalization for an acute illness, older adults must pay for it until their resources are exhausted. At that point, Medicaid (health insurance for the poor) takes over in the United States and Medicare in Canada. Consequently, the largest users of nursing homes are people with very low and high incomes (Torrey, 1992). Middle-income elderly and their families are more likely to try to protect their savings from being drained by high nursing home costs.

Nursing home use also varies across ethnic groups. For example, among people age 75 and older, Caucasian Americans are one-and-a-half times more likely to be institutionalized than African Americans. Large, closely knit extended families mean that over 70 percent of African-American elders do not live alone and over one-third reside with their adult children (Gibson & Jackson, 1992). Similarly, Asian, Hispanic, Native-American and Canadian Aboriginal elders use nursing homes less often than Caucasian Americans due to families' strong sense of caregiving responsibility (Gabrel, 2000; Lindsay, 1999). Nevertheless, families provide at least 60 to 80 percent of all long-term care in Australia, Canada, New Zealand, the United States, and Western Europe. As we have seen in this and the previous chapter, families of diverse ethnic and SES backgrounds willingly step in to care for elders in times of need.

To reduce institutionalized care of the elderly and its associated high cost, some experts have advocated alternatives, such as publicly funded in-home help for family caregivers (see Chapter 16, page 531). Another option rapidly becoming more widely available is *assisted living*—a homelike housing arrangement for seniors who require more care than can be provided at home but less than is usually provided in nursing homes. Assisted living is a cost-effective alternative to nursing homes that prevents unnecessary institutionalization (Maddox, 2001). It also can enhance residents' autonomy, social life, community involvement, and life satisfaction—potential benefits that we will take up in Chapter 18. In Denmark, the combination of a government-sponsored home-helper system and expansion of assisted-living housing resulted in a 30 percent reduction in the need for nursing home beds over the past 15 years. At the same time, the Danish government saved money: Public expenditures for long-term care declined by 8 percent (Stuart & Weinrich, 2001).

When nursing home placement is the best choice, steps can be taken to improve the quality of services. For example, the Netherlands has established separate facilities for patients with mental and physical disabilities because each has distinctly different needs. And every elderly person, no matter how disabled, benefits from opportunities to maintain existing strengths and acquire new skills, which (as we will see in the next section) can compensate for declines. Among institutionalized elderly, health, sense of personal control, gratifying

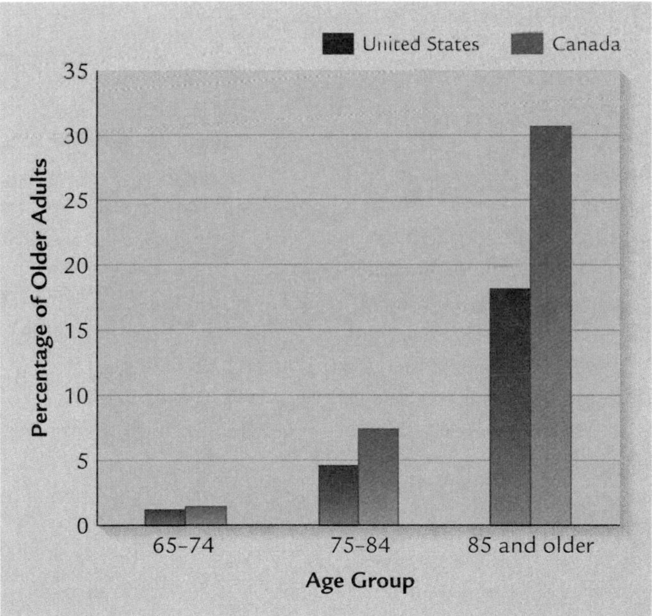

■ **FIGURE 17.6 Increase with age in nursing home care in the United States and Canada.** In both nations, placement in nursing homes rises steeply with age. Canadian elders are generally financially better off than American elders (see Chapter 2, page 65), which helps explain the higher percentages of Canadians in the two older age groups who reside in nursing homes. Another related factor is Canada's longer average life expectancy. (Adapted from Statistics Canada, 2002n; U.S. Department of Health and Human Services, 2002k.)

© NANCY SIESEL/THE NEW YORK TIMES

■ Assisted living is a rapidly expanding alternative to high-cost nursing home care. Once a week, mothers of toddlers take their children to visit residents of this warmly appointed assisted-living facility for people with mild dementia. Elders recall how they cared for their own children, and the babies bring joy into their lives.

social relationships, and meaningful and enjoyable daily activities strongly predict life satisfaction (Logsdon, 2000). These aspects of living are vital for all older people, wherever they live.

Ask Yourself

REVIEW

Cite evidence that both genetic and environmental factors contribute to Alzheimer's disease and cerebrovascular dementia.

REVIEW

Explain how depression can combine with physical illness and disability to promote cognitive deterioration in the elderly. Should cognitive declines due to physical limitations and depression be called dementia? Explain.

APPLY

Marissa complained to a counselor that at age 68, her husband, Wendell, stopped initiating sex and no longer cuddled her. Why might Wendell have ceased to interact sexually? What interventions—both medical and educational—could be helpful to Marissa and Wendell?

CONNECT

Explain how each level of ecological systems theory (Chapter 1, pages 24–25) contributes to caregiver well-being and quality of home care for elders with dementia.

www.

COGNITIVE DEVELOPMENT

When Ruth complained to her doctor about difficulties with memory and verbal expression, she voiced common concerns about cognitive functioning in late adulthood. Decline in speed of processing, under way throughout the adult years, is believed to affect many aspects of cognition in old age. In Chapter 15, we noted that reduced efficiency of thinking compromises attention, amount of information that can be held in working memory, use of memory strategies, and retrieval from long-term memory. These decrements continue in the final decades of life.

Recall that the more a mental ability depends on fluid intelligence (biologically based information-processing skills), the earlier it starts to decline. In contrast, mental abilities that rely on crystallized intelligence (culturally based knowledge) are sustained for a longer time. (Return to Figure 15.5 on page 500 to review these trends.) But maintenance of crystallized intelligence depends on continued opportunities to enhance cognitive skills. When these are available, crystallized abilities—general information and expertise in specific endeavors—can offset losses in fluid intelligence.

Look again at Figure 15.5 on page 500. In advanced old age, decrements in fluid intelligence eventually limit what people can accomplish with the help of cultural supports, including a rich background of experience, knowledge of how to remember and solve problems, and a stimulating daily life. Consequently, crystallized intelligence shows a mild decline (Baltes, 1997; Kaufman, 2001).

Overall, loss outweighs improvement and maintenance as people move closer to the end of life. But plasticity of development is still possible (Baltes & Carstensen, 1996). Research reveals large individual differences in cognitive functioning among the elderly—greater than at any other time of life (Hultsch & Dixon, 1990; Morse, 1993). Besides fuller expression of genetic and lifestyle influences, increased freedom to pursue self-chosen courses of action—some that enhance and others that undermine cognitive skills—may be responsible.

How can older adults make the most of their cognitive resources? According to one view, elders who sustain high levels of functioning engage in **selective optimization with compensation.** That is, they narrow their goals, *selecting* personally valued activities as a way of *optimizing* (or maximizing) returns from their diminishing energy. They also come up with new ways of *compensating* for losses (Baltes, 1997; Freund & Baltes, 2000). One day, Ruth and Walt watched a public television rebroadcast of an interview with 80-year-old concert pianist Arthur Rubinstein, who was asked how he managed to sustain such extraordinary piano playing, despite his advanced age. Rubinstein used each of the strategies just mentioned. He stated that he was *selective;* he played fewer pieces. This enabled him to *optimize* his energy; he could practice each piece more. Finally, he developed new, *compensatory* techniques for a decline in playing speed. For example, before a fast passage, he played extra slowly, so the fast section appeared to his audience to move more quickly.

© GLENN A. BAKER ARCHIVES/REDFERNS

■ World-renowned concert pianist Arthur Rubenstein, performing before a sold-out audience in New York City at age 89, sustained his extraordinary piano playing through *selective optimization with compensation.* He was selective, so he could optimize his energy. He compensated for a decline in playing speed by playing extra slowly before a fast passage, so the fast section appeared to listeners to move more quickly.

As we review major changes in memory, language processing, and problem solving, we will consider ways that older adults can optimize and compensate in the face of declines. We will also see that certain abilities that depend on extensive life experience, not processing efficiency, are sustained or increase in old age. Last, we take up programs that recognize the elderly as lifelong learners empowered by new knowledge, just as they were at earlier periods of development.

Memory

As older adults take in information more slowly and find it harder to apply strategies, inhibit irrelevant information, and retrieve relevant knowledge from long-term memory, the chances of memory failure increase (Bäckman, Small, & Wahlin, 2001; Persad et al., 2002). A reduced capacity to hold material in working memory while operating on it means that memory problems are especially evident on complex tasks.

Deliberate versus Automatic Memory

"Ruth, you know that movie we saw—with the little 5-year-old boy who did such a wonderful acting job. I'd like to suggest it to Dick and Goldie. Do you recall the name of it?" asked Walt.

"I can't think of it, Walt. We've seen several movies lately, and that one just doesn't ring a bell. Which theater was it at, and who'd we go with? Tell me more about the little boy, and maybe I'll think of it."

Although all of us have had memory failures like this from time to time, difficulties with recall rise in old age. When Ruth and Walt watched the movie, their slower cognitive processing meant that they retained fewer details. And because their working memories could hold less at once, they attended poorly to context—where they saw the movie and who went with them (Craik & Jacoby, 1996; Wegesin et al., 2000). When we try to remember, context serves as an important retrieval cue. Because older adults take in less about a stimulus and its context, they sometimes cannot distinguish an experienced event from one they imagined (Rybash & Hrubi-Bopp, 2000).

A few days later, Ruth saw a clip from the movie on TV, and she recognized it immediately. Compared with recall, recognition memory suffers less in late adulthood because a multitude of environmental supports for remembering are present (Craik & Jennings, 1992). Age-related memory declines are largely limited to tasks that require deliberate processing. Because recognition is a fairly automatic type of memory that demands little mental effort, performance does not change much in old age.

Consider another automatic form of memory called **implicit memory,** or memory without conscious awareness. In a typical implicit memory task, you would be asked to fill in a word fragment (such as t k) after being shown a list of words. You would probably complete the sequence with a word you had just seen (*task*) rather than other words (*took* or *teak*). Notice that you engaged in recall without trying to do so.

Age differences in implicit memory are much smaller than in explicit, or deliberate, memory. When memory depends on familiarity rather than on conscious use of strategies, it is spared from impairment in old age (Davis, Trussell, & Klebe, 2001; Titov & Knight, 1997). The memory problems elders report—for names of people, places where they put important objects, directions for getting from one place to another, and (as we will see later) appointments and medication schedules—all place substantial demands on their more limited working memories.

Associative Memory

One way of characterizing the memory deficits just described is in terms of a general, age-related decline in binding information into complex memories. Researchers call this an **associative memory deficit,** or difficulty creating and retrieving links between pieces of information—for example, two items or an item and its context, such as Ruth trying to remember the name of the movie with the child actor or where she had seen the movie.

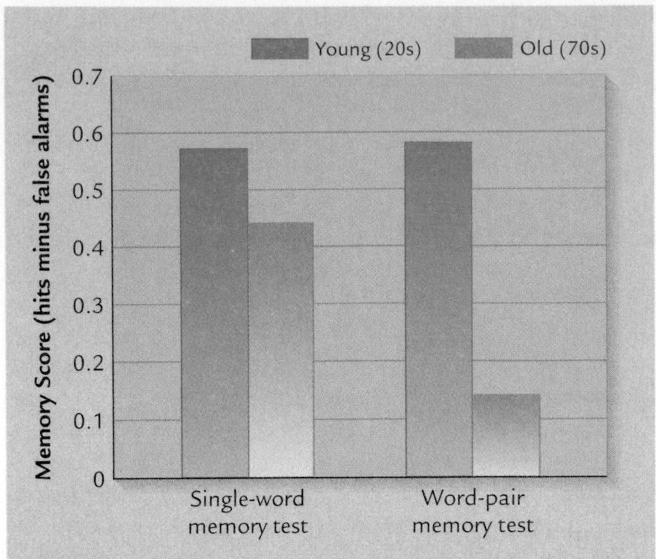

FIGURE 17.7 Young and old adults' performance on single-word and word-pair memory tests, supporting an associative memory deficit in late adulthood. After studying pairs of unrelated words, some adults were asked to identify single words they had seen. Others were asked to identify word pairs they had seen. Old adults performed almost as well as young adults on the single-word memory test. But they did far worse on the word-pair memory test. These findings support an associative memory deficit in late adulthood. (Adapted from Naveh-Benjamin, 2000.)

To find out whether older adults have greater difficulty with associative memory than younger adults, researchers showed them pairs of unrelated words (such as table–overcoat, sandwich–radio), asking that they study the pairs for an upcoming memory test. During the test, one group of participants was given a page of *single words,* some that had appeared in the study phase and some that had not, and asked to circle words they had studied. The other group was given a page of *word pairs,* some intact from the study phase (table–overcoat) and some that had been rearranged (overcoat–radio), and asked to circle pairs they had studied. As Figure 17.7 shows, older adults did almost as well as younger adults on the single-word memory test. But they performed far worse on the word-pair test—findings that support an associative memory deficit (Naveh-Benjamin, 2000).

Notice that the memory tasks in this study relied on recognition. Most of the time, recognition memory requires only that we recognize single pieces of information, so elders perform well. When researchers complicate recognition tasks by making them depend on associative memory, elders have difficulty. This suggests that providing older adults with more cues for remembering in any situation that requires linking pieces of information would enhance their memory performance. For example, to associate names with faces, elders profit from mention of relevant facts about those individuals. Enhancing the meaningfulness of names improves recall (Schmidt et al., 1999).

Remote Memory

Although older people often say that their **remote memory,** or very long-term recall, is clearer than their memory for recent events, research does not support this conclusion. In several studies, adults ranging in age from their twenties to their seventies were asked to recall names of grade school teachers and high school classmates and Spanish vocabulary from high school—information very well learned early in life. Memory declined rapidly for the first 3 to 6 years, then changed little for the next 20 years. After that, additional modest forgetting occurred (Bahrick, 1984; Bahrick, Bahrick, & Wittlinger, 1975).

How about *autobiographical memory,* or memory for personally meaningful events, such as what you did on your first date or how you celebrated your college graduation? To test for this type of memory, researchers typically give a series of words (such as *book, machine, sorry, surprised*) and ask adults to report a personal memory cued by each. People between 50 and 90 recall both remote and recent events more frequently than intermediate events, with recent events mentioned most often. Among remote events recalled, most happened between ages 10 and 30 (see Figure 17.8) (Jansari & Parkin, 1996; Rubin & Schulkind, 1997a).

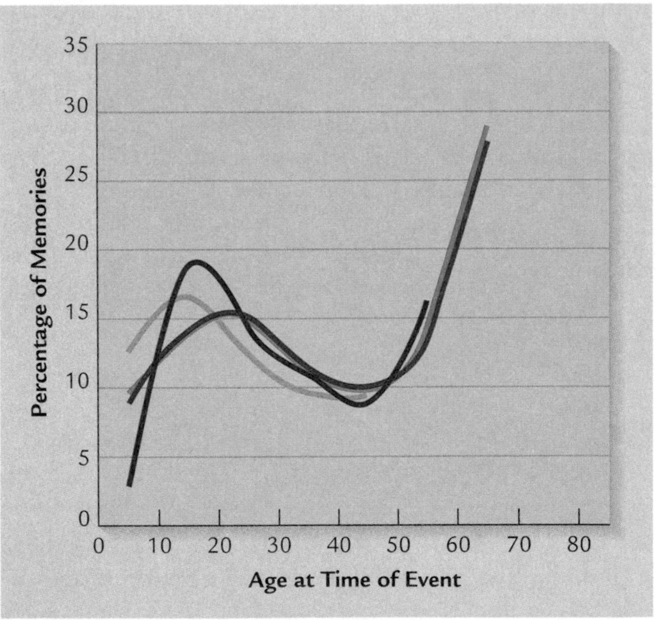

FIGURE 17.8 Distribution of older adults' autobiographical memories by reported age at time of the event. In the four studies of 50- to 90-year-olds represented here, later events were remembered better than early events. Among early events, most of those recalled occurred between ages 10 and 30. (Adapted from D. C. Rubin, T. A. Rahhal, & L. W. Poon, 1998, "Things Learned in Early Adulthood Are Remembered Best," *Memory and Cognition, 26,* p. 4. Copyright © 1998 by the Psychonomic Society, Inc. Reprinted by permission.)

Why do older adults recall their adolescent and early adulthood experiences more readily than their mid-adulthood lives? Perhaps youthful events are remembered best because they occurred during a period of rapid life change filled with novel experiences—ones that stand out from the humdrum of daily life. Adolescence and early adulthood are also times of identity development, when many personally significant experiences occur. Recall from Chapter 5 that such events are likely to become part of the individual's life story and to be long-lasting. Even public events linked to this period—World Series winners, Academy Award winners, and current events—are especially salient to elders (Rubin, Rahhal, & Poon, 1998).

Nevertheless, older adults recall recent personal experiences more readily than remote ones, probably because of interference produced by years of additional experience (Kausler, 1994). As we accumulate more memories, some inevitably resemble others. As a result, certain early memories become less clear than they once were.

Prospective Memory

Elderly people often complain that they have become more absentminded about daily events. Because Ruth and Walt knew they were more likely to forget an appointment, they often asked about it repeatedly. "Sybil, what time is our dinner engagement?" Walt said several times during the 2 days before the event. His questioning was not a sign of dementia. He simply wanted to be sure to remember an important date.

So far, we have considered various aspects of *retrospective memory* (or remembrance of things past). **Prospective memory** refers to remembering to engage in planned actions in the future. The amount of mental effort required determines whether older adults have trouble with prospective memory. Remembering the dinner date was challenging for Walt because he typically ate dinner with his daughter on Thursday evenings at 6 P.M., but this time, dinner was set for Tuesday at 7:15 P.M.

In the laboratory, older adults do better on *event-based* than on *time-based* prospective memory tasks. In an event-based task, an event (such as a certain word appearing on a computer screen) serves as a cue for remembering to do something (pressing a key) while the participant engages in an ongoing activity (reading paragraphs). As long as the event-based task is not complex, older adults do as well as younger adults. But when researchers introduce extra memory demands (press the key when any one of four cues appears), older adults' performance decreases (Einstein et al., 1997; Einstein et al., 1992, 2000). In time-based tasks, the adult must engage in an action after a certain time interval has elapsed, without any obvious external cue (for example, pressing a key every 10 minutes). Time-based prospective memory requires considerable initiative to keep the planned action in mind, and declines in late adulthood are large (Park et al., 1997; West & Craik, 1999).

But difficulties with prospective memory that are evident in the laboratory do not always appear in real life, since older adults are very good at setting up reminders for themselves, such as a buzzer ringing in the kitchen or a note tacked up in a prominent location. When trying to remember a future activity, younger adults rely more on strategies like rehearsal, older adults on external aids to memory (Dixon, de Frias, & Bäckman, 2001; Marsh, Hicks, & Landau, 1998). In this way, the elderly compensate for their reduced-capacity working memories and the challenge of dividing attention between what they are doing now and what they must do in the future.

Language Processing

Language and memory skills are closely related. In language comprehension (understanding the meaning of spoken or written prose), we recollect what we have heard or read without conscious awareness. Like implicit memory, language comprehension changes very little in late life as long as conversational partners do not speak very quickly and elders are given enough time to process written text accurately (Brébion, Smith, & Ehrlich, 1997; Hultsch et al., 1998; Stine-Morrow & Miller, 1999). In one longitudinal study, 55- to 70-year-olds showed a slight increase over a 6-year period in their recall of written stories! And into their eighties and nineties, elders made good use of story organization to help them recall both main ideas and details (Small et al., 1999).

In contrast to language comprehension, two aspects of language production show age-related losses. The first is retrieving words from long-term memory. When conversing with others, Ruth and Walt sometimes had difficulty coming up with just the right word to convey their thoughts. Consequently, their speech contained more pronouns and other unclear references than it did at younger ages. They also spoke more slowly and paused more often, partly because they needed time to search their memories for certain words (MacKay & Abrams, 1996).

Second, planning what to say and how to say it is harder in late adulthood. As a result, Walt and Ruth displayed more hesitations, false starts, word repetitions, and sentence fragments as they aged. Their statements were also less well organized than before (Kemper, Kynette, & Norman, 1992).

What explains these changes? Once again, age-related limits on working memory seem to be responsible. Because less information can be held at once, the elderly have difficulty coordinating the multiple tasks required to produce speech. As a result, they sometimes have trouble remembering the nonverbal information they want to communicate, putting it into words, and conveying it in a coherent fashion.

As with memory, older adults develop compensatory techniques for their language production problems. For

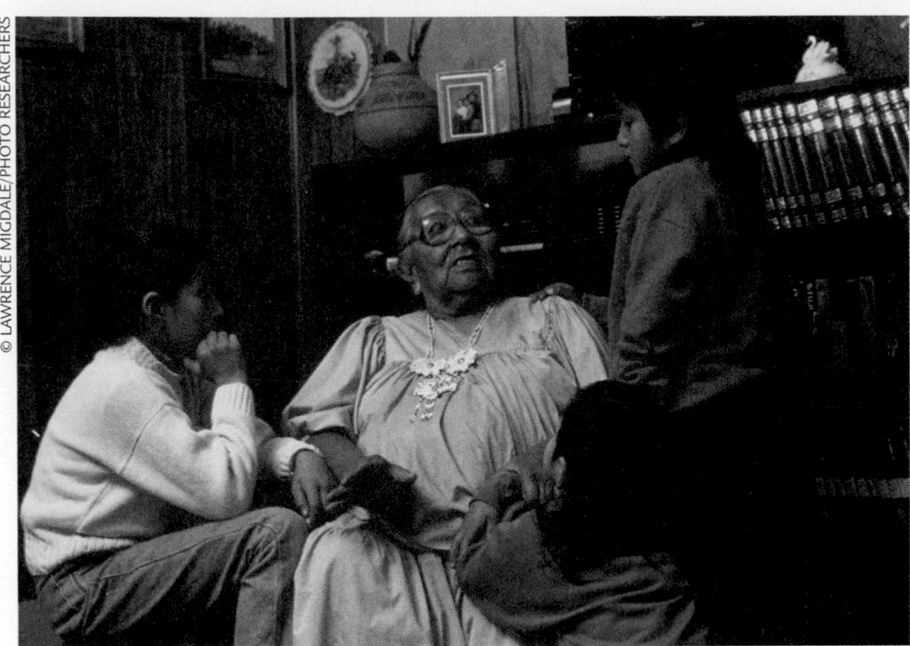

■ As this senior of the Apache people of Arizona tells stories to young members of her community, she is likely to enrich her tales with inferences and moral lessons grounded in extensive life experience.

example, they simplify their grammatical structures so they can devote more effort to retrieving words and organizing their thoughts. In this way, they convey their message in more sentences, sacrificing efficiency for greater clarity (Kemper, Thompson, & Marquis, 2001).

The elderly also compensate by representing information they want to communicate in terms of gist rather than details (Jepson & Labouvie-Vief, 1992). For example, when Walt told his granddaughter Marci fairy tales, he left out many concrete facts. Instead, he included personal inferences and a moral lesson—elements that appear less often in the storytelling of younger adults. Here is Walt's rendition of *Sleeping Beauty:* "An evil fairy condemns Sleeping Beauty to death. But a kind fairy changes the curse from death to sleep. Then a handsome prince awakens the girl with a kiss. So you see, Marci, both good and bad exist in the world. The bad things instill in us the need to think of and care for others."

Older adults often make the best of their limited working memories by extracting the essence of a message. Then they enrich it with symbolic interpretations, drawing on their extensive life experience.

Problem Solving

Problem solving is yet another cognitive skill that illustrates how aging brings not only deterioration but also important adaptive changes. In late adulthood, traditional problem solving, which lacks a real-life context (as in playing Twenty Questions), shows declines. Older adults' memory limitations make it hard to keep all relevant facts in mind when dealing with a complex hypothetical problem (Sinnott, 1989).

Yet the problematic situations the elderly encounter in everyday life and care about differ from hypothetical problems devised by researchers. Elders' everyday problems also differ from those experienced at earlier ages. Being retired, most do not have to deal with problems in the workplace. Even the social problems they confront at home may be reduced. Their children are typically grown and living on their own, and their marriages have endured long enough to have fewer difficulties (Berg et al., 1998). Instead, major concerns involve managing activities of daily living, such as preparing nutritious meals, handling finances, and attending to health concerns. Surveys in Germany and the United States reveal that older adults spend a third to half of a typical day on these issues (Willis, 1996).

How do the elderly solve problems of daily living? Their strategies extend the adaptive problem solving of midlife. As long as they perceive problems as under their control, elders are active and effective in solving them (Haught et al., 2000). Perhaps because older adults often conclude that they cannot change other people, they are likely to avoid interpersonal conflicts (Blanchard-Fields, Chen, & Norris, 1997). But as we will see in Chapter 18, this strategy often makes sense for elders, who are highly motivated to conserve energy and limit stress.

The health arena, especially salient for elders, illustrates the adaptiveness of everyday problem solving in late adulthood (Sansone & Berg, 1993). Older adults make quicker decisions about whether they are ill and seek medical care sooner. In contrast, young and middle-aged adults are more likely to adopt a "wait and see" approach in favor of gathering more facts, even when the health problem is serious

(Leventhal et al., 1993; Meyer, Russo, & Talbot, 1995). This swift response of the elderly is interesting in view of their slower cognitive processing. Perhaps years of experience in coping with illnesses enable them to draw on extensive personal knowledge and move ahead with greater certainty. Acting decisively when faced with health risks is sensible in old age.

Finally, compared with younger married couples, older married couples are more likely to collaborate in everyday problem solving, and researchers judge their jointly generated strategies as highly effective (Meegan & Berg, 2002). In solving problems together, older couples seem to compensate for each other's moments of cognitive difficulty, yielding enhanced accomplishment of life tasks.

Wisdom

We have seen that a wealth of life experience enhances the storytelling and problem solving of the elderly. It also underlies another capacity believed to reach its height in old age: **wisdom.** When researchers ask people to describe wisdom, most mention breadth and depth of practical knowledge, ability to reflect on and apply that knowledge in ways that make life more bearable and worthwhile, emotional maturity, including the ability to listen, evaluate, and give advice; and the altruistic form of creativity discussed in Chapter 15 that involves contributing to humanity and enriching others' lives. One group of researchers summed up the multiple cognitive and personality traits that make up wisdom as "expertise in the conduct and meaning of life" (Baltes & Staudinger, 2000, p. 124; Staudinger et al., 1998).

During her college years, Ruth and Walt's granddaughter Marci telephoned with a pressing personal dilemma. Ruth's advice reflected the features of wisdom just mentioned. Unsure whether her love for her boyfriend Ken would endure, Marci began to date another student after Ken moved to another city to attend medical school. "I can't stand being pulled in two directions. I'm thinking of calling Ken and telling him about Steve," she exclaimed. "Do you think I should?"

"This is not a good time, Marci," Ruth advised. "You'll break Ken's heart before you've had a chance to size up your feelings for Steve. And you said Ken's taking some important exams in two weeks. If you tell him now and he's distraught, it could affect the rest of his life."

Wisdom—whether applied to personal problems or to community, national, and international concerns—requires the "pinnacle of insight into the human condition" (Baltes & Staudinger, 2000). Not surprisingly, cultures around the world assume that age and wisdom go together. In village and tribal societies, the most important social positions, such as chieftain and shaman (religious leader), are reserved for the old. Similarly, in industrialized nations, people over age 60 are chief executive officers of large corporations, high-level religious leaders, members of legislatures, and supreme court justices. What explains this widespread trend? According to an evolutionary view, the genetic program of our species grants health, fitness, and strength to the young. Culture tames this

■ Former U.S. President Jimmy Carter exemplifies wisdom, applied on an international scale. Here he greets Prime Minister P. J. Patterson on a visit to Jamaica, just before Jamaican elections that gave Patterson a third term in office. For decades, Carter has championed democracy, peaceful conflict resolution, and human rights around the world. For example, he often serves as an unbiased observer in elections in developing countries. In 2002, Carter was awarded the Nobel Peace Prize for his highly successful efforts.

youthful advantage in physical power with the insights of the old, ensuring balance and interdependence between generations (Assmann, 1994; Csikszentmihalyi & Rathunde, 1990).

Few efforts have been made to investigate the development of wisdom. In one series of studies, adults ranging in age from 20 to 89 responded to uncertain real-life situations—for example, what to consider and do if a good friend is about to commit suicide or if, after reflecting on your life, you discover that you have not achieved your goals (Staudinger, 1999). Responses were rated for five ingredients of wisdom:

- knowledge about fundamental concerns of life, including human nature, social relations, and emotions
- effective strategies for applying that knowledge to making life decisions, handling conflict, and giving advice
- a view of people that considers the multiple demands of their life contexts
- a concern with ultimate human values, such as the common good, as well as respect for individual differences in values, and
- awareness and management of the uncertainties of life—that many problems have no perfect solution.

Results revealed that age was no guarantee of wisdom. A small number of adults of diverse ages ranked among the wise. But type of life experience made a difference. People in human-service careers who had extensive training and practice in grappling with human problems tended to attain high wisdom scores (Smith, Staudinger, & Baltes, 1994; Staudinger, Smith, & Baltes, 1992). And when age and relevant life experiences were considered together, more older than younger people scored in the top 20 percent. Consistent with this finding, a panel of citizens asked to nominate public figures high in wisdom mostly selected older adults ranging in age from 50 to 70, with an average of 64 (Baltes et al., 1995).

In addition to age and life experience, having faced and overcome adversity appears to be an important contributor to late-life wisdom. In a longitudinal study of people who were young adults during the Great Depression of the 1930s, those who experienced economic hardship and surmounted it scored especially high in wisdom nearly 40 years later, as indicated by thoughtful interview responses to life events, insights into their own motives and behavior, and warmth and compassion (Ardelt, 1998). Stress-related growth may be one of several as yet unknown paths to the development of wisdom.

Older adults with the cognitive, reflective, and emotional qualities that make up wisdom tend to be better educated and physically healthier and to forge more positive relations with others. And in two investigations, wisdom predicted life satisfaction more strongly than did objective life conditions, such as physical health, SES, and quality of living environment (Ardelt, 1997, 2000). Wise elders seem to flourish, even when faced with physical and cognitive challenges. This suggests that finding ways to promote wisdom would be a powerful means of both contributing to human welfare and fostering a gratifying old age. Currently, researchers are devising strategies for teaching young people to think about life questions and make decisions wisely, starting in adolescence (Sternberg, 2001).

Factors Related to Cognitive Change

As in middle adulthood, a mentally active life—above-average education, stimulating leisure pursuits, community participation, and a flexible personality—predicts maintenance of mental abilities into advanced old age. To illustrate, recall our discussion of age-related declines in prospective memory. In one study, well-educated elders high in verbal ability and involved in their communities performed as well as young adults on an event-based prospective memory task. In contrast, elders who had little more than a high school diploma and were low in verbal ability did poorly (Cherry & LeCompte, 1999).

As noted earlier in this chapter, health status becomes an increasingly strong predictor of intellectual performance in late adulthood. A wide variety of chronic conditions, including vision and hearing impairments, cardiovascular disease, osteoporosis, and arthritis, are associated with cognitive declines (Anstey & Christensen, 2000; Baltes, Lindenberger, & Staudinger, 1998). But we must be cautious in interpreting this link between physical and cognitive deterioration. The relationship may be exaggerated by the fact that brighter adults are more likely to engage in health-protective behaviors, which postpone the onset of serious disease.

Retirement also affects cognitive change, both positively and negatively. When people leave routine jobs for stimulating leisure activities, outcomes are favorable. In contrast, retiring from a highly complex job without developing challenging substitutes accelerates intellectual declines (Schaie, 1996).

After age 75, cognitive decrements are related to distance to death rather than to chronological age (Small & Bäckman, 1999). In the year before Walt died, those close to him noticed that he had become less active and more withdrawn. In the company of friends, he talked and moved less. At home, he spent more time looking out the window instead of immersing himself in creative writing and gardening.

Terminal decline refers to a steady, marked decrease in cognitive functioning prior to death. Researchers are not certain if it is limited to a few aspects of intelligence or if it affects all aspects, signifying general deterioration. Furthermore, studies vary greatly in estimated length of terminal decline. Some report that it lasts only 1 to 3 years, others that it extends for as much as 14 years. The average is about 5 years

(Bosworth & Schaie, 1999; Hassing et al., 2002; Maier & Smith, 1999).

Perhaps the reason for these conflicting findings is that there are different kinds of terminal decline. One type might arise from disease processes. Another might be part of a general biological breakdown due to normal aging (Berg, 1996). At present, all we know for sure is that an extended, steep falloff in cognitive functioning is a sign of loss of vitality and impending death.

Cognitive Interventions

For most of late adulthood, cognitive declines are gradual. Although aging of the brain contributes to them, recall from our earlier discussion that the brain can compensate by growing new neural fibers. Furthermore, some cognitive decrements may be due to disuse of particular skills rather than biological aging (Dixon & Hultsch, 1999). If plasticity of development is possible in old age, then interventions that train the elderly in cognitive strategies should at least partially reverse the age-related declines we have discussed.

The Adult Development and Enrichment Project (ADEPT) is the most extensive cognitive intervention program conducted to date. By using participants in the Seattle Longitudinal Study (see Chapter 13, page 435, and Chapter 15, page 499), researchers were able to do what no other investigation had yet done: assess the effects of cognitive training on long-term development (Schaie, 1996).

Intervention began with adults over age 64, some of whom had maintained their scores on two mental abilities (inductive reasoning and spatial orientation) over the previous 14 years and others who had shown declines. After just five 1-hour training sessions in relevant cognitive skills, two-thirds of the participants improved their performance. The gains for decliners were dramatic. Forty percent returned to the level at which they had been functioning 14 years earlier! A follow-up after 7 years revealed that although the scores of trained adults dropped somewhat, they were still doing better than untrained controls. Finally, "booster" training at this time led to further gains, although these were not as large as the earlier gains.

In other short-term studies, training resulted in improvements in memory and problem solving among the elderly (Kotler-Cope & Camp, 1990; Willis & Schaie, 1994). Clearly, a wide range of cognitive skills can be enhanced in old age. A vital goal is to transfer intervention from the laboratory to the community, weaving it into the daily experiences of elderly people. As we will see in the next section, a promising approach is to provide older adults with well-designed, highly interesting learning experiences in which cognitive training is an integral part.

Lifelong Learning

Think about the competencies that older adults need to live in our complex, changing world. They are the same as those of younger people—communicating effectively through spoken and written systems; locating information, sorting through it, and selecting what is needed; using math strategies such as estimation; planning and organizing activities, including making good use of time and resources; mastering new technologies; and understanding past and current events and the relevance of each to their own lives. The elderly also need to develop new, problem-centered coping strategies—ways to sustain health and operate their households efficiently and safely—and self-employment skills for those who must continue their work lives.

Because of better health and earlier retirement, participation of the elderly in continuing education has increased substantially over the past few decades. Successful programs include a wide variety of offerings responsive to the diversity of senior citizens and teaching methods suited to their developmental needs.

Types of Programs

One summer, Walt and Ruth attended an Elderhostel at a nearby university. After moving into a dormitory room, they joined 30 other senior citizens for 2 weeks of morning lectures on Shakespeare, afternoon visits to points of interest; and evening performances of plays at a nearby Shakespeare festival.

Elderhostel programs (in Canada, recently renamed Routes to Learning) attract over a quarter million North American older adults annually. Local educational institutions serve as hosts, combining stimulating 1- to 3-week courses taught by experts with recreational pursuits. Some programs make use of community resources through classes on local ecology or folk life. Others involve travel abroad. Still others focus on innovative topics and experiences—writing one's own life story, discussing contemporary films with screenwriters, whitewater rafting, Chinese painting and calligraphy, and acquiring French language skills.

Similar educational programs have sprung up around the world. Originating in France, the University of the Third Age[3] offers community-sponsored courses for elders that vary widely in content and style of presentation. These include open lectures, access to established university courses, workshops on special topics, excursions, and physical health programs. The model has spread to many countries. In Australia and Great Britain, elders often do the teaching, based on the

[3]The term *third age* refers to the period after the second age of midlife, when older people are freed of work and parenting responsibilities and have more time to invest in lifelong learning.

© AP/WIDE WORLD PHOTOS

■ Because older adults are healthier and retire earlier than in previous generations, they often continue their education, expanding their knowledge and skills. These seniors, ranging in age from 72 to 80, call themselves the Silver Stringers Computer Group. They created the website in the foreground, overturning the prevailing assumption that aging people and new technology do not mix.

idea that experts of all kinds retire (Swindell & Thompson, 1995). Some programs foster intergenerational relations and community involvement. In Austria, for example, training in academic subjects is offered to grandparents so they can help grandchildren, in whose education they are encouraged to participate.

In many Western nations, including the United States and Canada, universities offer tuition-free course enrollment to older adults. But participants in the programs mentioned so far tend to be active, well educated, and financially well off (Abraham, 1998). Much less is available for elders with little education and limited income. Community senior centers with inexpensive offerings related to everyday living attract more low-SES people than programs such as Elderhostel (Knox, 1993). Regardless of course content and which seniors attend, using the techniques summarized in the Caregiving Concerns table below increases the effectiveness of instruction.

Benefits of Continuing Education

Elderly participants in continuing education report a rich array of benefits. These include learning new facts, understanding new ideas in many disciplines, making new friends, and de-

Increasing the Effectiveness of Instruction for Older Adults

TECHNIQUE	DESCRIPTION
Provide a positive learning environment.	Some elders have internalized negative stereotypes of their abilities and come to the learning environment with low self-esteem. A supportive group atmosphere, in which the instructor is viewed as a colleague, helps convince older adults that they can learn.
Allow ample time to learn new information.	Rate of learning varies widely among older adults, and some master new material at a fairly slow rate. Presenting information over several sessions or allowing for self-paced instruction aids mastery.
Present information in a well-organized fashion.	Older adults do not organize information as effectively as younger adults. Material that is outlined, presented, and then summarized enhances memory and understanding. Digressions make a presentation harder to comprehend.
Give information related to elders' experiences.	Relating new material to what elders have already learned, by drawing on their experiences and giving many vivid examples, enhances recall.

Source: Thompson, 1992.

veloping a broader perspective on the world (Long & Zoller-Hodges, 1995). Furthermore, seniors come to see themselves differently. Many arrive with deeply ingrained stereotypes of aging, which they abandon when they realize that adults in their seventies and eighties—including themselves—can still engage in complex learning. In Elderhostel courses, participants with the least education report learning the most, an argument for recruiting less economically privileged people into these programs (Brady, 1984).

The educational needs of seniors are likely to be given greater attention in coming decades, as their numbers grow and they assert their right to lifelong learning. Once this happens, false stereotypes, such as "the elderly are too old to learn" and "education is for the young," are likely to weaken and, perhaps, disappear.

Ask Yourself

REVIEW

When Ruth couldn't recall which movie Walt was thinking about, she asked several questions: "Which theater was it at, and who'd we go with? Tell me more about the little boy [in the movie]." Which memory deficits of aging is Ruth trying to overcome?

REVIEW

Describe cognitive functions that are maintained or improve in late adulthood. What aspects of aging contribute to them?

APPLY

Estelle complained that she had forgotten two recent hair appointments and sometimes had trouble finding the right words to convey her thoughts. What cognitive changes account for Estelle's difficulties? What can she do to compensate?

CONNECT

What processes in brain development contribute to the success of elders' efforts to compensate for cognitive declines? (See pages 549–550.)

www.

Summary

PHYSICAL DEVELOPMENT

Life Expectancy

Distinguish between chronological age and functional age, and discuss changes in life expectancy over the past century.

- Vastly different rates of aging are apparent in late adulthood. In terms of **functional age,** older adults of any chronological age may be **young-old** or **old-old,** depending on their physical condition.
- Twentieth-century gains in **average life expectancy** were extraordinary, confirming that biological aging can be modified by environmental factors. As a result, the number of people age 65 and older has risen dramatically in the industrialized world. Length of life and, even more important, **active lifespan,** can be predicted by a country's health care, housing, and social services, along with lifestyle factors.
- With advancing age, women outnumber men by a greater margin, but sex differences in average life expectancy decline. Differences between higher-SES whites and low-SES ethnic minorities also diminish until age 85, when a **life expectancy crossover** occurs.
- Longevity runs in families, but environmental factors become increasingly important with age. Scientists disagree on whether **maximum lifespan** can be extended beyond an average of about 85 years.

Physical Changes

Describe changes in the nervous system and the senses in late adulthood.

- Loss of neurons occurs throughout the cerebral cortex, with the visual, auditory, and motor areas most affected. However, the brain compensates by forming new synapses. The autonomic nervous system functions less well and releases more stress hormones.
- Older adults tend to suffer impaired eyesight, and **cataracts** and **macular degeneration** may occur. Visual deficits affect elders' self-confidence and everyday behavior and can be very isolating.
- Hearing difficulties are more common than visual difficulties, especially in men. Impaired speech perception has the greatest impact on life satisfaction.
- Taste and odor sensitivity decline, making food less appealing. Touch sensitivity also deteriorates, particularly on the fingertips.

Describe cardiovascular, respiratory, and immune system changes in late adulthood.

- Reduced capacity of the cardiovascular and respiratory systems becomes more apparent in late adulthood. As at earlier ages, not smoking, reducing dietary fat, avoiding environmental pollutants, and exercising can slow the effects of aging on these systems.
- The immune system functions less effectively in late life, permitting diseases to progress and making **autoimmune responses** more likely.

Discuss sleep difficulties in late adulthood.

- Older adults find it harder to fall asleep, stay asleep, and sleep deeply. Until age 70 or 80, men have more trouble sleeping than women, due to enlargement of the prostate leading to a need to urinate often, **sleep apnea,** and "restless legs."

Describe changes in physical appearance and mobility in late adulthood.

- Outward signs of aging, such as white hair, wrinkled and sagging skin, age spots, and decreased height and weight, become more noticeable. Mobility diminishes as muscle and bone strength and joint flexibility decline.

- Problem-centered coping strategies yield improved physical functioning in the elderly. Also, a rapidly expanding **assistive technology** is available to help older people cope with physical declines. Negative stereotypes of aging in Western society make adapting to late-life physical changes more difficult. Older people fare best when they have status and opportunities for social participation, even when they are frail.

Health, Fitness, and Disability

Discuss health and fitness in late life, paying special attention to nutrition, exercise, and sexuality.

- Most elders are optimistic about their health and, with respect to protecting it, have a high sense of self-efficacy. Low-SES ethnic minority elders remain at greater risk for certain health problems and are less likely to believe they can control their health.

- As in early adulthood, in late life men are more prone to fatal diseases and women to disabling conditions. By very old age, women are more impaired than surviving men. Poverty and negative lifestyle factors make the public health goal of **compression of morbidity** hard to attain.

- Because risk of dietary deficiencies increases, vitamin–mineral supplements are beneficial. Exercise continues to be a powerful health intervention, even when begun in late adulthood.

- Compared with other parts of the body, the reproductive organs undergo minimal change in late adulthood. Sexual desire and sexual activity decline but need not disappear.

Discuss physical disabilities common in late adulthood.

- Illness and disability increase toward the end of life. Cardiovascular disease, cancer, stroke, and emphysema claim many lives. Because of declines in immune system functioning, vaccination against the most common type of pneumonia is advisable.

- **Secondary aging,** rather than **primary aging,** is responsible for many disabilities in late adulthood. **Osteoarthritis** and **rheumatoid arthritis** are widespread among older adults, especially women. Adult-onset diabetes also increases.

- Motor vehicle collisions per mile driven, pedestrian accidents, and falls increase in late adulthood. Changes in elders' lifestyles and environments can help prevent these causes of unintentional injury.

Discuss mental disabilities common in late adulthood.

- **Alzheimer's disease** is the most common form of **dementia.** Often starting with severe memory problems, it brings personality changes, depression, loss of ability to comprehend and produce speech, disintegration of purposeful movements, and death. Underlying these changes are abundant **neurofibrillary tangles** and **amyloid plaques** and lowered neurotransmitter levels in the brain.

- Familial Alzheimer's generally has an early onset, progresses rapidly, and is linked to dominant genes on chromosomes 1, 14, and 21. Sporadic Alzheimer's may also be hereditary; people with this form often have an abnormal ApoE4 gene on chromosome 19. However, many sporadic Alzheimer's victims show no currently known genetic marker. Diverse environmental factors, including head injuries and a high-fat diet, increase the risk of Alzheimer's.

- Heredity contributes to **cerebrovascular dementia** indirectly, through high blood pressure, cardiovascular disease, and diabetes. Many environmental influences also heighten stroke risk. Because of their greater susceptibility to cardiovascular disease, men are affected more than women.

- Treatable problems, such as depression, side effects of medication, and reactions to social isolation, can be mistaken for dementia. Therefore, careful diagnosis is essential.

Discuss health care issues that affect senior citizens.

- Only a small percentage of North American seniors are institutionalized, a rate about half that of other industrialized nations that provide more generous public financing of nursing home care than the United States and Canada. Family members provide most long-term care, especially among ethnic minorities with closely knit extended families. Publicly funded in-home help and assisted-living arrangements can reduce nursing home placement of the elderly and its associated high cost.

COGNITIVE DEVELOPMENT

Describe overall changes in cognitive functioning in late adulthood.

- Individual differences in cognitive functioning are greater in late adulthood than at any other time of life. Although both fluid intelligence and crystallized intelligence decline in advanced old age, plasticity of development is still possible. Older adults can make the most of their cognitive resources through **selective optimization with compensation.**

Memory

How does memory change in late life?

- Age-related limitations on working memory make memory difficulties more apparent on tasks that are complex and require deliberate processing. Automatic forms of memory, such as recognition and **implicit memory,** are largely spared. In general, an **associative memory deficit,** or difficulty creating and retrieving links between pieces of information, seems to characterize older adults' memory problems.

- Contrary to what older people sometimes report, **remote memory** is not clearer than recent memory. Autobiographical memory is best for recent experiences, followed by personally meaningful events that happened between ages 10 and 30, a period of rapid life change and identity development. In the laboratory, older adults do better on event-based than on time-based **prospective memory** tasks. In everyday life, they compensate for declines in prospective memory by using external memory aids.

Language Processing

Describe changes in language processing in late adulthood.

- Although language comprehension changes little in late life, two aspects of language production—finding the right words and planning what to say and how to say it—show age-related losses. Older adults compensate by simplifying their grammatical structures and communicating gist rather than details.

Problem Solving

How does problem solving change in late life?

- Traditional problem solving declines in late adulthood. In everyday problem solving, older adults extend the adaptive strategies developed in midlife. In matters of health, elders often respond more decisively than younger people, perhaps because of more experience in coping with illness. And older married couples are more likely to collaborate in everyday problem solving, generating highly effective strategies.

Wisdom

What capacities make up wisdom, and how is it affected by age and life experience?

- **Wisdom** involves extensive practical knowledge, ability to reflect on and apply that knowledge in ways that make life more bearable and worthwhile, emotional maturity, and altruistic creativity. When age and life experience in grappling with human problems are combined, more older than younger people rank among the wise.

- Having faced and overcome adversity seems to foster late-life wisdom. And elders who score higher in wisdom tend to be better educated and physically healthier and to forge more positive relations with others. Wisdom is a strong predictor of life satisfaction in late adulthood.

Factors Related to Cognitive Change

List factors related to cognitive change in late adulthood.

- Mentally active people are likely to maintain their cognitive abilities into advanced old age. A wide array of chronic conditions are associated with cognitive decline. Retirement can bring about either positive or negative changes. As death approaches, **terminal decline**—a steady, marked decrease in cognitive functioning—often occurs.

Cognitive Interventions

Can cognitive interventions help older adults sustain their mental abilities?

- The Adult Development and Enrichment Project (ADEPT) and various short-term studies show that training can enhance cognitive skills in older adults, including those who have suffered declines.

Lifelong Learning

Discuss types of continuing education and benefits of such programs in late life.

- Better health and earlier retirement permit increasing numbers of older people to continue their education through university courses, community offerings, and programs such as Elderhostel. Those who participate are enriched by new knowledge, new friends, a broader perspective on the world, and an image of themselves as more competent. Unfortunately, fewer continuing-education opportunities are available to low-SES seniors.

Important Terms and Concepts

active lifespan (p. 547)
Alzheimer's disease (p. 565)
amyloid plaques (p. 565)
assistive technology (p. 556)
associative memory deficit (p. 571)
autoimmune response (p. 554)
average life expectancy (p. 547)
cataracts (p. 551)
cerebrovascular dementia (p. 567)
compression of morbidity (p. 559)
dementia (p. 564)
functional age (p. 546)
implicit memory (p. 571)
life expectancy crossover (p. 548)
macular degeneration (p. 552)
maximum lifespan (p. 548)
neurofibrillary tangles (p. 565)
old-old (p. 546)
osteoarthritis (p. 562)
primary aging (p. 561)
prospective memory (p. 573)
remote memory (p. 572)
rheumatoid arthritis (p. 562)
secondary aging (p. 561)
selective optimization with compensation (p. 570)
sleep apnea (p. 555)
terminal decline (p. 576)
wisdom (p. 575)
young-old (p. 546)

FYI For Further Information and Help

Consult the Companion Website for *Development Through the Lifespan, Third Edition,* (www.ablongman.com/berk) where you will find the following resources for this chapter:

- **Chapter Objectives**
- **Flashcards** for studying important terms and concepts
- **Annotated Weblinks** to guide you in further research
- **Ask Yourself** questions, which you can answer and then check against a sample response
- **Suggested Readings**
- **Practice Test** with immediate scoring and feedback

18

© DAVID TURNLEY/CORBIS

Emotional and Social Development in Late Adulthood

Erikson's Theory: Ego Integrity versus Despair

Other Theories of Psychosocial Development in Late Adulthood
Peck's Theory: Three Tasks of Ego Integrity • Labouvie-Vief's Theory: Emotional Expertise • Reminiscence and Life Review

■ A Lifespan Vista: World War II Refugee and Evacuee Children Look Back from the Vantage Point of Old Age

Stability and Change in Self-Concept and Personality
Secure and Multifaceted Self-Concept • Agreeableness, Sociability, and Acceptance of Change • Spirituality and Religiosity

Individual Differences in Psychological Well-Being
Control versus Dependency • Health • Negative Life Changes • Social Support and Social Interaction

■ Social Issues: Elder Suicide

A Changing Social World
Social Theories of Aging • Social Contexts of Aging: Communities, Neighborhoods, and Housing

■ Biology & Environment: Aging, Time Perception, and Social Goals

Relationships in Late Adulthood
Marriage • Gay and Lesbian Partnerships • Divorce and Remarriage • Widowhood • Never-Married, Childless Older Adults • Siblings • Friendships • Relationships with Adult Children • Relationships with Adult Grandchildren and Great-Grandchildren • Elder Maltreatment

Retirement and Leisure
The Decision to Retire • Adjustment to Retirement • Leisure Activities

Successful Aging

This senior couple of Italy enjoys a day's outing in the countryside. When a marital relationship is positive, older adults engage in more joint leisure activities than they did at earlier ages, and their love deepens.

With Ruth at his side, Walt disclosed to guests on their fiftieth anniversary, "Even when there were hard times, it's the period of life I was in at the moment that I always liked the most. I adored playing baseball as a kid and learning the photography business when I was in my twenties. And I recall our wedding, the most memorable day of all," Walt continued, glancing affectionately at Ruth. "Then came the Depression, when professional picture taking was a luxury few people could afford. But we found ways to have fun without money, such as singing in the church choir and acting in community theater. A short time later, Sybil was born. It meant so much to me to be a father—and now a grandfather and a great-grandfather. Looking back at my parents and grandparents and forward at Sybil, Marci, and Marci's son, Jamel, I feel a sense of unity with past and future generations."

Walt and Ruth greeted old age with calm acceptance, grateful for the gift of long life and loved ones. Yet not all older adults find peace of mind in these final decades. Walt's brother Dick was contentious and critical, often over petty issues, at other times over major disappointments. "Goldie, why did you serve cheesecake? No one eats cheesecake on birthdays" and "You know the reason we've got these financial worries? Uncle Louie wouldn't lend me the money to keep the bakery going, so I *had* to retire."

A mix of gains and losses characterizes these twilight years, extending the multidirectionality of development begun early in life. On one hand, old age is an Indian summer—a time of pleasure and tranquility. Children are grown, life's work is nearly done, and responsibilities are lightened. On the other hand, it brings concerns about declining physical functions, unwelcome loneliness, and the growing specter of imminent death.

In this chapter, we consider how older adults reconcile these opposing forces. Although some are weary and discontented, most traverse this phase with poise and calm composure. They attach deeper significance to life and reap great benefits from family and friendship bonds, leisure activities, and community involvement. We will see how personal attributes and life history combine with home, neighborhood, community, and societal conditions to mold emotional and social development in late life.

■ In late adulthood, people come to terms with their lives. Elders who arrive at a sense of integrity feel whole, complete, and satisfied with their achievements. Erik Erikson and his wife Joan provided ideal role models for Erikson's final stage. They grew old gracefully and could often be seen walking hand in hand, deeply in love.

Erikson's Theory: Ego Integrity versus Despair

The final psychological conflict of Erikson's (1950) theory, **ego integrity versus despair,** involves coming to terms with one's life. Adults who arrive at a sense of integrity feel whole, complete, and satisfied with their achievements. They have adapted to the mix of triumphs and disappointments that are an inevitable part of love relationships, child rearing, work, friendships, and community participation. They realize that the paths they followed, abandoned, and never selected were necessary for fashioning a meaningful life course.

The capacity to view one's life in the larger context of all humanity—as the chance combination of one person and one segment in history—contributes to the serenity and contentment that accompany integrity. "These last few decades have been the happiest," Walt murmured as he clasped Ruth's hand, only weeks before the heart attack that would end his life. He was at peace with himself, his wife, and his children, having accepted his life course as something that had to be the way it was. In a study of people ranging in age from 17 to 82, increased age was associated with greater psychosocial maturity, measured in terms of striving for generativity and ego integrity in everyday behavior. Generativity and ego integrity, in turn, largely accounted for the link between age and psychological well-being (Sheldon & Kasser, 2001). Just as Erikson's theory indicates, the psychosocial maturity of these later years seems to bring increased happiness.

One day, as he scanned the newspaper, Walt pondered, "I keep reading these percentages: 1 out of 5 people will get heart disease, 1 out of 3 will get cancer. But in truth, 1 out of 1 will die. We are all mortal and must accept this fate." The year before, Walt had given his granddaughter, Marci, his collection of prized photos, which had absorbed him for over half a century. With the realization that the integrity of one's own life is part of an extended chain of human existence, death loses its sting (Vaillant, 1994, 2002).

The negative outcome of this stage, despair, occurs when elders feel they have made many wrong decisions, yet time is too short to find an alternate route to integrity. Without another chance, the despairing person finds it hard to accept that death is near and is overwhelmed with bitterness, defeat, and hopelessness. According to Erikson, these attitudes are often expressed as anger and contempt for others, which disguise contempt for oneself. Dick's argumentative, faultfinding behavior, tendency to blame others for his personal failures, and regretful view of his own life reflect this deep sense of despair.

Other Theories of Psychosocial Development in Late Adulthood

As with Erikson's stages of early and middle adulthood, other theorists have clarified and refined his vision of late adulthood, specifying the tasks and thought processes that contribute to a sense of ego integrity. All agree that successful development in the later years involves greater integration and deepening of the personality.

Peck's Theory: Three Tasks of Ego Integrity

According to Robert Peck (1968), Erikson's conflict of ego integrity versus despair comprises three distinct tasks. Each must be resolved for integrity to develop:

- *Ego differentiation versus work-role preoccupation.* This task results from retirement. It requires aging people who have invested heavily in their careers to find other ways of affirming their self-worth. The person striving for integrity must differentiate a set of family, friendship, and community roles that are equally as satisfying as work life.
- *Body transcendence versus body preoccupation.* Old age brings declines in appearance, physical capacities, and resistance to disease. For people whose psychological well-being is heavily bound up with the state of their bodies, late adulthood can be a difficult time. Older adults need to *transcend* physical limitations by emphasizing cognitive and social powers, which offer alternative, compensating rewards.
- *Ego transcendence versus ego preoccupation.* Whereas middle-aged people realize that life is finite, the elderly are reminded of the certainty of death as siblings, friends, and peers in their community die. They must find a constructive way of facing this reality—through investing in a longer future than their own lifespan. Although the generative years of early and middle adulthood prepare people for a satisfying old age, attaining ego integrity requires a continuing effort to make life more secure, meaningful, and gratifying for those who will go on after one dies.

In Peck's theory, ego integrity requires that older adults move beyond their life's work, their bodies, and their separate identities. Drawing on wisdom, they must reach out to others in ways that make the world better. Their success is apparent in their inner state of contentment and positive impact on others.

Labouvie-Vief's Theory: Emotional Expertise

In Chapter 13, we discussed Gisella Labouvie-Vief's theory of cognitive change, noting how in early adulthood abstract thought becomes *pragmatic*—a tool for solving real-world problems. Labouvie-Vief has also explored the development of adults' reasoning about emotion. She believes that older and more psychologically mature individuals are more in touch with their feelings. They describe emotional reactions in more complex and personalized ways, perhaps as a result of reflecting on life experiences, realizing that every moment and event may be one of the last, and using coping strategies that permit fuller acknowledgment of emotion.

In one study, participants from adolescence to old age were asked to relate personal experiences in which they were happy, angry, fearful, and sad and to indicate how they knew they felt the emotion. Younger people explained feelings technically, as though they were observing them from the outside. For example, they were likely to say, "My adrenaline was high," or, "My heart rate increased." They also emphasized how they *should* feel rather than how they did feel. In contrast, older adults gave vivid descriptions that viewed both mind and body as contributing to feeling states and that integrated subjective and objective aspects of emotion. Here is one:

> You have sunshine in your heart. During the wedding the candles were glowing. And that's just how I felt. I was glowing too. It was kind of dull outside. But that isn't how I felt. Everybody in the church felt like they were glowing. It was that kind of feeling. (Labouvie-Vief, DeVoe, & Bulka, 1989, p. 429)

Older adults' emotional perceptiveness assists them in separating emotional interpretations from objective aspects of situations more effectively than younger adults do. Consequently, their coping strategies often include making sure their own biases are fully understood before deciding on a course of action (Blanchard-Fields, 1997; Labouvie-Vief et al.,

■ These parents of the bride and groom dance at a wedding reception in a small city in Hungary. How will they capture their elation in words? Elders describe emotional reactions in more complex and personalized ways.

1995). Elders also acknowledge periods of intense rumination and are better able to interpret negative events in a positive light—factors that may contribute to the self-acceptance inherent in ego integrity. In sum, a significant late-life psychosocial attainment is becoming expert at processing emotional information and engaging in emotional self-regulation (Labouvie-Vief & Diehl, 1999; Lawton, 2001b).

Reminiscence and Life Review

When we think of the elderly, we often picture them engaged in **reminiscence**—telling stories about people and events from their past and reporting associated thoughts and feelings. Indeed, the image of a reminiscing elder is so widespread that it ranks among the negative stereotypes of aging. A common view is that older people live in the past to escape the realities of a shortened future and the nearness of death. Researchers do not yet have a full understanding of why older people reminisce more often than younger people do (Boden & Bielby, 1983; Lamme & Baars, 1993). But current theory and research indicate that reflecting on the past can be positive and adaptive.

Return to Walt's commentary about major events in his life at the beginning of this chapter. Walt engaged in a special form of reminiscence called **life review,** in which the person calls up, reflects on, and reconsiders past experiences, contemplating their meaning with the goal of achieving greater self-understanding. According to Robert Butler (1968), most older adults engage in life review as part of attaining ego integrity, preventing despair, and accepting the end of life. Butler's ideas have been so influential that many therapists encourage the elderly to engage in life-review reminiscence. In one study, homebound 61- to 99-year-olds who participated in a counselor-led life-review process showed gains in life satisfaction still apparent a year later. This did not occur for participants who simply enjoyed friendly visits or who received no treatment (Haight, 1992).

Although life review often prompts self-awareness and self-respect, many elders do not spend much time evaluating their pasts. When asked to do so, 30 percent of one sample claimed that the best part of life was "right now"! Middle age also received high marks, whereas childhood and adolescence ranked as less satisfying (Field, 1997). These findings clearly contradict the widespread belief that older adults wish to be young again. As the Lifespan Vista box on pages 588–589 illustrates, recollections of distant events are colored by many factors, including personality, the timing and nature of the events, and subsequent life experiences.

Besides life review, reminiscence serves other purposes. When it is *self-focused,* engaged in to reduce boredom and revive bitter events, it is linked to adjustment problems. Despairing elders tend to ruminate often, dwelling on painful content

■ Reflecting on the past can be positive and adaptive. Many older adults engage in reminiscence and life review as part of attaining ego integrity. And often they talk about the past to teach younger people about family and cultural history—an activity that makes life richer and more rewarding.

from the past, which sustains their negative emotion (Cully, LaVoie, & Gfeller, 2001; Webster & McCall, 1999). In contrast, extraverted elders more often engage in *other-focused* reminiscence—directed at social goals, such as solidifying family and friendship ties and reliving relationships with lost loved ones. And at times, older adults—especially those who score high in openness to experience—engage in *knowledge-based* reminiscence, drawing on their past for effective problem-solving strategies and for teaching younger people. These socially engaged, mentally stimulating forms of reminiscence help make life rich and rewarding (Cappeliez & O'Rourke, 2002). Perhaps because of their strong storytelling traditions, African-American and Chinese immigrant elders are more likely than their Caucasian counterparts to use reminiscence to teach others about the past (Merriam, 1993; Webster, 2002).

Finally, reminiscence—for young and old alike—often occurs during times of life transition (Parker, 1995). The older adult who has recently retired, been widowed, or moved to a new residence may turn to the past to sustain a sense of personal continuity. Indeed, elders who feel their current lives are adrift reminisce especially often to recapture a sense of meaning and purpose (Fry, 1995). As long as they do not get stuck in mulling over unresolved difficulties, reminiscence probably helps them adjust.

Stability and Change in Self-Concept and Personality

Longitudinal research reveals continuing stability of the "big five" personality traits from mid- to late life (see Chapter 16, page 523). Yet consider the ingredients of ego integrity: wholeness, contentment, and image of the self as part of a larger world order. These attributes are reflected in several significant late-life changes in both self-concept and personality.

Secure and Multifaceted Self-Concept

Older adults have accumulated a lifetime of self-knowledge, leading to more secure and complex conceptions of themselves than at earlier ages (Labouvie-Vief & Diehl, 1999). Ruth, for example, knew with certainty that she was good at counseling others, growing a flower garden, giving dinner parties, budgeting money, and figuring out who could be trusted and who couldn't. At the same time, she commented wistfully that she couldn't get around the city as easily as before.

The firmness and multifaceted nature of Ruth's self-concept permitted her to compensate for lack of skill in domains she had never tried, did not master, or could no longer perform as well as before. Consequently, it allowed for self-acceptance—a key feature of integrity. In a study of old (70 to 84 years) and very old (85 to 103 years) German elders asked to respond to the question "Who am I?", participants mentioned a broad spectrum of life domains, including hobbies, interests, social participation, family, health, and personality traits. Adults of both ages expressed more positive than negative self-evaluations, although a slight increase in negative comments occurred in the older group. Positive, multifaceted self-definitions predicted psychological well-being (Freund & Smith, 1999).

As the future shortens, most elders, into their eighties and nineties, continue to mention hoped-for selves in the areas of good health, relationships, and social responsibility and are very active in pursuing them. These grant older adults goals in life and a sense of further development (Frazier, 2002; Markus & Herzog, 1992). Even in advanced old age, when some capacities decline, the majority of older adults retain a coherent sense of self. They regard themselves as very much the same person they have always been (Troll & Skaff, 1997).

Agreeableness, Sociability, and Acceptance of Change

During late adulthood, shifts in three personality characteristics take place—changes that, once again, defy stereotypes of the elderly. Old age is not a time in which the personality inevitably becomes rigid and morale declines. Instead, a flexible, optimistic approach to life is common.

Rating open-ended interviews with elders in their sixties, and again when they reached their eighties and nineties, researchers found that scores on adjectives that make up *agreeableness*—*generous, acquiescent,* and *good-natured*—were higher on the second occasion than the first for over one-third of the sample. These qualities seem to characterize people who have come to terms with life despite its imperfections. However, participants showed a slight dip in *sociability* as they aged (Field & Millsap, 1991). Perhaps this reflects a narrowing of social contacts as people become more selective about relationships and as family members and friends die—trends we will take up in a later section.

A third, related development is greater *acceptance of change*—an attribute the elderly frequently mention as important to psychological well-being (Ryff, 1989). That many older adults adjust well to change is evident in what they say when asked about dissatisfactions in their lives. They often respond that they are not unhappy about anything! A capacity to accept life's twists and turns, many of which are beyond one's control, is vital for positive functioning in late adulthood. Most elders are resilient; they bounce back in the face of adversity, especially if they tended to do so earlier in their lives.

Spirituality and Religiosity

How do older adults manage to accept declines and losses yet feel whole and complete and anticipate death with calm composure? One possibility is that they develop a more mature

A Lifespan Vista

World War II Refugee and Evacuee Children Look Back from the Vantage Point of Old Age

How do elders look back on a traumatic period that occurred early in their lives? To find out, Glen Palmer (1999) interviewed older adults who had come to Australia just before or during World War II as Jewish refugees from Austria, Germany, and Poland or as evacuees from Great Britain. All had been younger than age 16 when their parents put them on boats, giving them up in anguish to protect them. Most parents of the Austrian, Polish, and German children were murdered by the Nazis. After kissing them good-bye, these children never saw their parents again. In contrast, the British evacuees returned to their homeland and families at the end of the war.

Looking back, these elders were largely supportive of their parents' decision to send them away. For the refugees, it meant that their lives had been spared (see Chapter 10, page 335). Beyond this, recollections and assessments of the lifelong impact of their experiences varied. Among factors that made a difference were quality of care, age at separation, and temperament.

Care situations included relatives, foster homes, group homes, and boarding schools. For children placed with families, being loved and wanted led to new attachments and pleasant memories, often at the expense of former bonds. As one elder noted, "If you send a child away in the formative years, you take the risk of losing the affection of that child." When relatives and foster families were unkind or outright rejecting, painful memories of an "emotional limbo" emerged. Without new attachments, these children clung to fading images of their families, idealized their parents, and longed to return to them. Separation and inadequate care had a more profound, lasting impact on temperamentally withdrawn, anxious individuals than on outgoing individuals, who were better able to develop supportive social ties beyond the family.

Compared with interviewees who were children at the time of separation, those who were adolescents were less affected by the absence of close bonds with adults. "I knew in my mind that I had plenty of family support, even though they weren't around," one Polish refugee remarked. On the threshold of adulthood, these young people had little desire for substitute parents. Instead, they fared best in group homes and boarding schools that provided appropriate oversight, balancing security with autonomy. Often their warmest memories were of bonds established with one another. In contrast, when children were placed in group homes and denied family attachments, their sadness was still apparent in old age: "There was a desolation because there was no love. . . . You need to be able to say to a child, 'You're my special one.'"

By the end of the war, the majority of interviewees had become young adults. The British evacuees were reunited with their families, but rarely with "happily ever after" endings. Children and parents had changed during the intervening years. As one interviewee put it, "It was easier to go than to come back." Still, reunion was vital, since its absence usually brings a lifetime of unresolved grief.

For most Polish and German refugees, loss of immediate family was total. More than a half-century later, their searching and sorrow continue. One commented, "The Holocaust pursues us daily. It is like having a flea in your head." This suggests that rather than healing with the passage of time, the pain of childhood loss and trauma can intensify in later life. Reunions with other refugees helped soothe persisting wounds as elders shared experiences with those best able to understand their continued suffering.

Although many refugees and evacuees remained deeply affected by their childhood experiences, few were maladjusted. To the contrary, most found turning points through families, careers, or other interests and led productive, fulfilling lives. When late-life adversities arose, some drew on inner strengths developed during early years of separation. As one Holocaust survivor commented after her husband died, "[W]hen you've been through what I've been through, you can survive anything."

Nevertheless, these victims of genocide—and other similarly traumatized individuals around the world—probably cannot make a complete recovery. None of the refugees returned to live in Europe, although some visited their countries of birth. Rarely were these journeys comfortable. Almost always, they brought heartaches to the surface—yet another reminder that ethnic and political violence has pervasive effects throughout the lifespan.

---

sense of spirituality. That is, they actively seek a higher meaning for life, knowing that it will end in the foreseeable future. Spirituality is not the same as religion. A transcendent sense of truth and beauty can be found in art, nature, and relationships with others. But religion provides many people with beliefs, symbols, and rituals that guide this quest for meaning.

Older adults attach great value to religious beliefs and behaviors. According to a national survey, 76 percent of Americans age 65 and older say that religion is very important in their lives, and 16 percent describe it as fairly important. Over half attend religious services weekly, nearly two-thirds watch religious TV programs, and about one-fourth pray at least three times a day (Princeton Religion Research Center, 1994). Canadian elders are not as involved in organized religion as their American counterparts. Nevertheless, 37 percent attend religious services or other religious functions at least

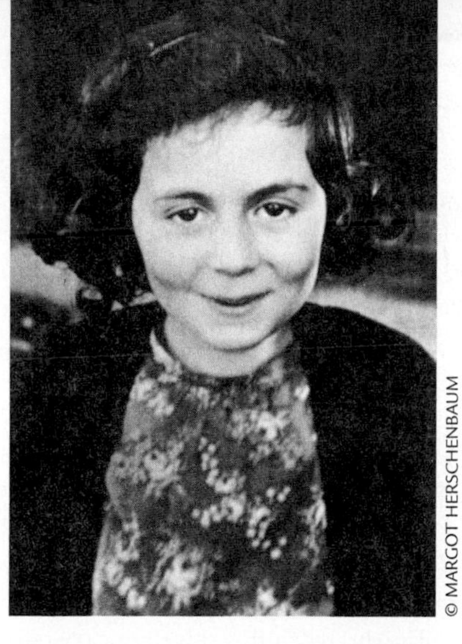

■ Above are the German Jewish children, pictured on the deck of the *Orama*, which sailed for Australia in June of 1939, just before the outbreak of World War II. Margot, standing third from left, reflected on her childhood from the vantage point of old age: "Leaving wasn't that traumatic. I mean when you leave for a holiday you leave your parents. . . . I didn't know then that I wasn't going to see them again. . . . I don't have happy recollections of any particular incident in Australia." Anne, shown on the far left, was a British evacuee who was reunited with her parents as a young adult at the end of the war. She recalled, "The war years were not always easy or happy, but it was far worse to return to a family I had missed for so long and to find we had grown so far apart."

once a week, and half attend at least once a month (Lindsay, 1999).

Although declining health and transportation difficulties reduce organized religious participation in advanced old age, informal religious activities remain prominent in the lives of today's elders (Ainlay, Singleton, & Swigert, 1992). Longitudinal research suggests that religious involvement is fairly stable throughout adulthood. Even so, the need for religious affiliation may rise among the elderly. In one study, over a third of older people said they would devote more time to religion if they could (DeGenova, 1992). And in longitudinal research extending from early to late adulthood, spirituality rose substantially after the mid-fifties (Wink & Dillon, 2002). Physical declines, more time to reflect on the meaning of life, and concern with issues related to immortality may motivate elders to search for inner spiritual resources.

■ Older adults attach great value to religious beliefs and behavior, and their faith may advance to a high level. This elderly Tibetan woman is deeply engrossed in a daily ritual of Buddhist prayers.

Furthermore, spirituality and faith may advance to a higher level in late life—away from prescribed beliefs to a more reflective approach that emphasizes links to others and is at ease with mystery and uncertainty (Birren, 1990). In his theory of the development of faith, James Fowler (1981) posits five stages (summarized in Table 18.1), which have been confirmed in several studies. Notice how adults who reach Stage 4 become aware of their own belief system as one of many possible world views, contemplate the deeper significance of religious symbols and rituals, and open themselves to other religious perspectives as sources of inspiration (McFadden, 1996). For example, as a complement to his Catholicism, Walt became intensely interested in Buddhism, especially its focus on attaining perfect peace and happiness by mastering thoughts and feelings, never harming others, and resisting attachment to worldly objects.

Involvement in both organized and informal religious activities is especially high among low-SES ethnic minority elders, including African-American, Hispanic, Native-American, and Canadian Aboriginal groups. In African-American communities, churches not only provide contexts for deriving meaning from life but also are centers for education, health, social welfare, and political activities aimed at improving life conditions. African-American elders look to religion as a powerful resource for social support beyond the family and for the inner strength to withstand daily stresses and physical impairments (Arm-

Table 18.1 Fowler's Stages of Faith Development

Stage of Faith	Period of Development	Description
1. Intuitive-projective	3–7 years	Children's fantasy and imitation lead them to be powerfully influenced by stories, moods, and behaviors demonstrating the faith of adults. They become aware of right and wrong actions.
2. Mythic-literal	7–11 years	Children begin to internalize the stories, beliefs, and observances of their religious community, which they take literally. For example, they often hold concrete images of God living on top of the world and watching over everybody.
3. Synthetic-conventional	Adolescence	Adolescents have a coherent set of deeply felt beliefs and values, which provides a basis for identity. They have not yet examined this ideology systematically.
4. Individuative-reflective	Adulthood	Adults who reach this stage critically reflect on their beliefs and values, recognizing that their world view is only one of many possible world views. They actively shape a personal ideology, forming and re-forming it over time. About religious rituals and symbols, they ask, "What does this really mean?"
5. Conjunctive	Late Adulthood	The few people who attain this stage form an enlarged vision of an all-inclusive human community. They act to bring it about by standing up against persecution and injustice and by promoting a common good that serves the needs of diverse groups. Great religious leaders, such as Mahatma Gandhi and Martin Luther King, Jr., illustrate conjunctive faith.

Source: Fowler, 1981.

strong & Crowther, 2002; Husaini, Blasi, & Miller, 1999). Asked about her philosophy of life, an African-American 65-year-old reveals how faith enabled her to do more than survive. As she tells it, she has much to be thankful for:

> We've had lots of misfortunes ... but we always knowed that it could be worse.... I know somedays I get up I'd be stiff and my knees aching and my back is aching and my head is hurting and I can get up and go in the bathroom. I say, "Thank you Lord because I have the activities of my limbs...." And then we get a meal on the table and we have ... at least a portion of health and strength.... [I] thank the Lord for things being as well as they are. (Nye, 1993, p. 109)

Among elders of all backgrounds, religious involvement has many benefits. These include outcomes as diverse as exercising, feeling close to family and friends, and psychological well-being (Fry, 2001; Levin & Chatters, 1998). In one study, religious participation was a strong predictor of better physical functioning over a 12-year period, beyond the effects of optimism, health practices, and social support (Idler & Kasl, 1997). And in other longitudinal research, organized and informal religious activities predicted longer survival, after family background, health, social, and psychological factors known to affect mortality were controlled (Helm et al., 2000; Strawbridge et al., 2001).

Sex differences in religious involvement and spirituality are evident throughout adulthood. Women are more likely than men to be church or synagogue members, to engage in religious activities, and to report a personal quest for connectedness with a higher power (Levin, Taylor, & Chatters, 1994; Wink & Dillon, 2002). Women's greater poverty, widowhood, and participation in caregiving, including caring for chronically ill family members, expose them to higher levels of stress and anxiety. As with ethnic minorities, they turn to religion for social support and for a larger vision of community that places life's challenges in perspective.

Individual Differences in Psychological Well-Being

In this and the previous chapter, we have seen that most adults adapt well to old age. Yet a few feel dependent, incompetent, and worthless. Identifying personal and environmental influences on late-life psychological well-being is vital for designing interventions that foster positive adjustment.

Control versus Dependency

As Ruth's eyesight, hearing, and mobility declined in her eighties, Sybil visited daily to assist with self-care and household tasks. During the hours mother and daughter were together, Sybil interacted most often with Ruth when she asked for help with activities of daily living. When Ruth handled tasks on her own, Sybil usually withdrew.

Observations of people interacting with older adults in both private homes and institutions reveal two highly predictable, complementary behavior patterns. In the first, called the **dependency–support script,** dependent behaviors are attended to immediately. In the second, called the **independence–ignore script,** independent behaviors are mostly ignored. Notice how these sequences reinforce dependent behavior at the expense of independent behavior, regardless of the older person's competencies. Even a self-reliant elder like Ruth did not always resist Sybil's unnecessary help because it brought about social contact (Baltes, 1995, 1996).

Among elders who experience no difficulty with daily activities, opportunities to interact with others are related to high satisfaction with everyday life. In contrast, among elders who have trouble performing daily activities, social contact is linked to a less positive everyday existence (Lang & Baltes, 1997). This suggests that social interaction while assisting elders with physical care, household chores, and errands is often not meaningful and rewarding, but rather demeaning and unpleasant. Consider these typical reactions of care recipients to a spouse's help with daily activities: "felt dependent," "felt indebted," "felt like a weak, incapable person" (Newsom, 1999).

Longitudinal research shows that negative reactions to caregiving can foster persisting depression (Newsom & Schultz,

■ Is this son encouraging his mother's dependency by helping her with her grocery shopping? The answer lies in whether she assumes personal control over her dependency. Dependency can be adaptive if it permits older people to conserve their strength and invest in highly valued activities.

© CHRIS STEELE-PERKINS/MAGNUM PHOTOS

Social Issues

Elder Suicide

When 65-year-old Abe's wife died, he withdrew from life. Living far from his two daughters, he spent his nonworking days alone, watching television and reading mystery novels. As grandchildren were born, Abe visited his daughters' homes from time to time. When he did, he carried his despondent behavior with him. "Look at my new pajamas, Grandpa!" Abe's 6-year-old grandson Tony exclaimed on one occasion. Abe didn't acknowledge the little boy.

Abe retired after arthritis made walking difficult. With more empty days, his depression deepened. Gradually, he developed painful digestive difficulties, but he refused to go to the doctor. "Don't need to," he said abruptly when one of his daughters begged him to get medical attention. Answering her invitation to Tony's tenth birthday party, Abe wrote, "Maybe—if I'm still around next month. By the way, when I go, I want my body cremated." Two weeks later, Abe died from an intestinal blockage. His body was found in the living room chair where he habitually spent his days. Although it may surprise you, Abe's self-destructive acts are a form of suicide.

Factors Related to Elder Suicide. Return to Figure 12.4 on page 402, and notice that suicide peaks in late life. It climbs during the elder years, reaching its highest rate among people age 75 and older. Although the incidence of suicide varies among nations, older adults are at increased risk around the world (World Health Organization, 2002).

Recall from Chapter 12 that the suicide rate is much higher among adolescent males than females. This sex difference persists throughout the lifespan. In the United States and Canada, 5 times as many elderly men as women take their own lives (Lindsay, 1999; U.S. Bureau of the Census, 2002c). Furthermore, compared with the white majority, most North American ethnic minority elders have low suicide rates.

What explains these trends? Despite the lifelong pattern of higher rates of depression and more suicide attempts among females, elderly women's closer ties to family and friends, greater willingness to seek social support, and religiosity prevent many from taking their own lives. High levels of social support through extended families and church affiliations may also prevent suicide among ethnic minorities. And among certain minorities, such as Alaskan Natives, deep respect for and reliance on older adults to teach cultural traditions foster self-esteem and social integration. This reduces elder suicide, making it nonexistent after age 80 (Kettl, 1998).

As in earlier years, the method favored by elder males (firearms) offers less chance of revival than that favored by elder females (poisoning or drug overdose). Nevertheless, failed suicides are much rarer in old age than in adolescence. The ratio of attempts to completions for the young is as high as 300 to 1; for the elderly, it is 4 to 1 or lower (Conwell, Duberstein, & Caine, 2002). When elders decide to die, they seem especially determined to succeed.

Underreporting of suicides probably occurs at all ages, but it is more common in old age. Medical examiners are less likely to pursue suicide as a cause of death when a person is old. And many elders, like Abe, engage in indirect self-destructive acts rarely classified as suicide, such as deciding not to go to a doctor when ill and refusing to eat or take prescribed medications. Among institutionalized elders, these efforts to hasten death are widespread (Kennedy & Tanenbaum, 2000). Consequently, elder suicide is an even

1998). But whether assistance from others undermines well-being is a function of many factors, including the quality of help, the caregiver–elder relationship, and the social and cultural context in which helping occurs. Why do family members and other caregivers often respond in ways that promote excessive dependency in old age? A stereotype of the elderly as passive and incompetent appears to be responsible. Older adults seem well aware of others' low expectations for them. They frequently attribute their dependency to overresponsive social partners (Wahl, 1991).

In Western societies, which place a high value on independence, many elders fear becoming dependent on others (Frazier, 2002). They often say that when they can no longer care for themselves, it is time to die. Does this mean we should encourage elders to be as independent as possible? According to Mary Baltes (1996), this alternative is as counterproductive as promoting passivity and incompetence. Aging brings diminished energy at a time when people confront many challenging developmental tasks. Dependency can be adaptive if it permits older people to conserve their strength by investing it in highly valued activities, using a set of strategies we considered in Chapter 17: *selective optimization with compensation.*

Health

As noted in Chapter 16, health is a powerful predictor of psychological well-being in late adulthood. Physical declines and chronic disease can be highly stressful, leading to a sense of loss of personal control—a major factor in adult mental health. Furthermore, physical illness resulting in disability is among the strongest risk factors for late-life depression (Geerlings et al., 2001). Although fewer older than young and middle-aged adults are depressed (see Chapter 17), profound

larger problem than official statistics indicate.

Two types of events prompt suicide in late life. Losses—retirement from a highly valued occupation, widowhood, and social isolation—place elders who have difficulty coping with change at risk for persistent depression. A second set of risks are chronic and terminal illnesses that severely reduce physical functioning or cause intense pain (Conwell et al., 2001). As comfort and quality of life diminish, feelings of hopelessness and helplessness deepen. Very old people, especially men, are particularly likely to take their own lives under these conditions. The chances are even greater when a sick elder is socially isolated—living alone or in a nursing home with high staff turnover and minimal caregiver support.

Prevention and Treatment. Warning signs of suicide in late adulthood overlap with those at earlier ages. They include efforts to put personal affairs in order, statements about dying, despondency, and sleep and appetite changes. Family members, friends, and caregivers must also watch for indirect self-destructive acts, such as refusing food or medical treatment, which are unique to old age. Too often, people in close touch with the elderly have difficulty recognizing symptoms, assuming incorrectly that they are a "natural" consequence of aging. More than 70 percent of older suicide victims visited their doctors within a month of taking their lives, and 30 percent did so the same week (Pearson & Brown, 2000). Yet their suicidal risk was not recognized.

When suicidal elders are depressed, the most effective treatment combines antidepressant medication with therapy, including help in coping with role transitions, such as retirement, widowhood, and dependency brought about by illness. Distorted ways of thinking, such as "I'm old, and nothing can be done about my problems," must be countered and revised. Meeting with the family to find ways to reduce loneliness and desperation is also helpful.

Although youth suicide has risen (see Chapter 12, page 401), elder suicide has declined during the past 50 years, due to increased economic security among older adults, improved medical care and social services, and more favorable cultural attitudes toward retirement. Communities are beginning to recognize the importance of additional preventive steps, such as telephone hot lines with trained volunteers who provide emotional support and agencies that arrange for regular home visitors or "buddy system" phone calls. In institutions, providing residents with privacy, autonomy, and space helps prevent self-destructive behavior (Conwell & Duberstein, 2001).

Finally, elder suicide raises a controversial ethical issue: Do people with incurable illnesses have the right to take their own lives? We will take up this topic in Chapter 19.

© JON FEINGERSH/CORBIS

■ As comfort and quality of life diminish, feelings of hopelessness and helplessness deepen. Under these circumstances, very old people are especially likely to take their own lives.

feelings of hopelessness rise with age as physical disability and consequent social isolation increase (Roberts et al., 1997).

The relationship between physical and mental health problems can become a vicious cycle, each intensifying the other. At times, the rapid decline of a sick elder is the result of despondency and "giving up" (Penninx et al., 2000). This downward spiral can be hastened by a move to a nursing home, requiring the older person to adjust to distance from family and friends and to a new self-definition as "a person who can survive only in an institution." In the month after admission, many residents deteriorate rapidly and become severely depressed. The stress of illness together with institutionalization is associated with heightened health problems and mortality (Tobin, 1989).

Depression in old age is often lethal. People age 65 and older have the highest suicide rate of all age groups (see the Social Issues box above). What factors enable elders like Ruth to surmount the physical impairment–depression relationship, remaining optimistic and content? Personal characteristics discussed in earlier chapters—effective coping and a sense of self-efficacy—are vitally important. But for frail elders to display these attributes, families and caregivers must grant them autonomy by avoiding the dependency–support script. When older adults remain in charge of personally important areas of their lives, they retain essential aspects of their identity in the face of change and report a more favorable outlook on their past and future (Brandtstädter & Rothermund, 1994).

Negative Life Changes

Ruth lost Walt to a heart attack, cared for Ida as her Alzheimer's symptoms worsened, and faced health problems of her own—all within a span of a few years. Elders are at risk for a variety of negative life changes—death of spouse, siblings, and friends; illness and physical disabilities; declining income; and greater dependency. Negative life changes are difficult for all people. But these events may actually evoke less stress and depression in older than in younger adults (Gatz, Kasl-Godley, & Karel, 1996). Many elders have learned to cope with hard times and to accept loss as part of human existence.

Still, when negative changes pile up, they test the coping skills of older adults (Krasij, Arensman, & Spinhoven, 2002). In very old age, such changes are greater for women than for men. Women over age 75 are far less likely to be married, more often have lower incomes, and suffer from more illnesses—especially ones that restrict mobility. Furthermore, elderly women more often say that others depend on them for emotional support. This means that their social relations, even in very old age, are more often a source of stress. And because of failing health, older women may not be able to meet others' needs for caregiving, with negative consequences for their self-esteem. Not surprisingly, women of very advanced age report a lower sense of psychological well-being than men (Pinquart & Sörensen, 2001).

Social Support and Social Interaction

In late adulthood, social support continues to play a powerful role in reducing stress, thereby promoting physical health and psychological well-being. Social support increases the odds of living longer (Liang et al., 1999; Seeman et al., 1993). And it may help explain the relationship of religious participation to survival, noted earlier. Most of the time, elders receive informal assistance from family members—first from their spouse or, if none exists, from children, and then from siblings. If these individuals are not available, other relatives and friends may step in.

Nevertheless, many older adults place such high value on independence that they do not want a great deal of support from people close to them unless they can reciprocate. When assistance is excessive or cannot be returned, it often results in psychological distress (Liang, Krause, & Bennett, 2001). Perhaps for this reason, adult children express a deeper sense of obligation toward their aging parents than their parents expect from them (see Chapter 16, page 533). Formal support—a paid home helper or agency-provided services—as a complement to informal assistance not only helps relieve caregiving burdens but also spares elders from feeling overly dependent in their close relationships (Krause, 1990).

Ethnic minority elders do not readily accept formal assistance. But they are more willing to do so when home helpers are connected to a familiar neighborhood organization, especially the church. Although African-American seniors say they rely more on their families than on the church for assistance, those with support and meaningful roles in both contexts score highest in mental health (Coke, 1992; Walls & Zarit, 1991). Support from religious congregants has psychological benefits for elders of all backgrounds, perhaps because recipients feel that it is motivated by genuine care and concern, not just obligation. Also, the warm atmosphere of religious organizations fosters a sense of social acceptance and belonging (Krause, 2001).

Having an extraverted personality is also linked to high morale in old age (Adkins, Martin, & Poon, 1996). Sociable elders are more likely to take advantage of opportunities to interact with others, thereby reducing loneliness and depression and fostering self-esteem and life satisfaction. But as we will see in the next section, supportive communication in old age has little to do with quantity of contact. Instead, high-quality relationships, involving expressions of kindness, encouragement, and respect, have the greatest impact on mental health in late life.

Overall, for social support to foster well-being, elders need to assume personal control of it. This means consciously giving up primary control in some areas to remain in control of other, highly valued pursuits. For example, although she could handle the dressing, financial matters, shopping, and food preparation, Ruth permitted her daughter Sybil to assist with these activities so she would have more stamina for pleasurable reading. To optimize her energies, Ruth selected certain domains in which to become dependent. This permitted her to compensate for poor eyesight by taking extra time to use a magnifying glass while reading or to listen to a book on tape. In this way, social support granted Ruth autonomy—a means for managing her own aging.

When we intervene with older adults, we must ask ourselves, What kind of assistance are we providing? Help that is not wanted or needed or that exaggerates weaknesses can undermine mental health. It can also accelerate physical disability if existing skills fall into disuse. In contrast, help that frees up energy for endeavors that are personally satisfying and that lead to growth enhances elders' quality of life.

Ask Yourself

REVIEW

Many elders adapt effectively to negative life changes. List personal and environmental factors that facilitate this generally positive outcome.

REVIEW

Although involvement in religion is fairly stable throughout adulthood, it takes on greater meaning and is linked to many positive outcomes in old age. Explain why, drawing on late-life physical and psychosocial development.

APPLY

At age 80, Miriam took a long time to get dressed. Joan, her home helper, suggested, "Wait until I arrive before dressing. Then I can help you and it won't take so long." What impact is Joan's approach likely to have on Miriam's personality? What alternative approach to helping Miriam would you recommend?

CONNECT

Fowler's stages of faith, summarized in Table 18.1 on page 590, were inspired by Kohlberg's theory of moral development (see Chapter 12, pages 388–390). Cite similarities between Kohberg's and Fowler's stages.

www.

A Changing Social World

Walt and Ruth's outgoingness led many family members and friends to seek them out, and they often reciprocated. In contrast, Dick's stubborn nature meant that his and Goldie's network of social ties was far more restricted, as it had been for many years.

In late adulthood, extroverts (like Walt and Ruth) continue to interact with a wider range of people than do introverts and people (like Dick) with poor social skills. Nevertheless, both cross-sectional and longitudinal research reveals that size of social networks and therefore amount of social interaction decline for virtually everyone (Carstensen, 1992; Lang, Staudinger, & Carstensen, 1998). This finding presents a curious paradox. If social interaction and social support are necessary for mental health, how is it possible for elders to interact less yet be generally satisfied with life and less depressed than younger adults?

Social Theories of Aging

Social theories of aging offer explanations for the decline in social interaction just described. Two prominent perspectives—disengagement theory and activity theory—interpret it in opposite ways. A recent approach—socioemotional selectivity theory—accounts for a wider range of findings on social contacts in old age.

● **Disengagement Theory.** According to **disengagement theory,** mutual withdrawal between elders and society takes place in anticipation of death (Cumming & Henry, 1961). Older people decrease their activity levels and interact less frequently, becoming more preoccupied with their inner lives. At the same time, society frees elders from employment and family responsibilities. The result is viewed as beneficial for both sides. Elders are granted a life of tranquility. And once they disengage, their deaths are less disruptive to society.

However, think back to our discussion of wisdom in Chapter 17. Because of their long life experience, older adults in many cultures move into new positions of prestige and power. Clearly, not everyone disengages! Even after retirement, some people sustain certain aspects of their work, and others develop new, rewarding roles in the community and in leisure. In tribal and village societies, most elders continue to hold important social positions (Luborsky & McMullen, 1999). Consequently, when old people disengage, it may not represent their personal preference. Instead, it may be due to a failure of the social world to provide opportunities for engagement. The more social opportunities elders report, the more strongly they believe they can create worthwhile social experiences for themselves (Lang, Featherman, & Nesselroade, 1997). (Return to the Cultural Influences box in Chapter 17, page 558, for some striking community variations in meaningful social roles available to elders.)

■ Young people seek advice and learn about the land and history of their people from this Navajo elder. Because of his long life experience, he is granted increased prestige and power in his cultural community. The greater social responsibility of many older adults presents a major challenge to disengagement theory of aging.

As we will see shortly, older adults' retreat from interaction is more complex than disengagement theory implies. Instead of disengaging from all social ties, they let go of unsatisfying contacts and maintain satisfying ones. And sometimes, they put up with less than satisfying relationships to remain engaged! For example, Ruth reluctantly agreed to travel with Dick and Goldie because she wanted to share the experience with Walt. But she often complained about Dick's insensitive behavior.

● **Activity Theory.** Attempting to overcome the flaws of disengagement theory, **activity theory** states that social barriers to engagement, not the desires of elders, cause declining rates of interaction. When older people lose certain roles (for example, through retirement or widowhood), they do their best to find others in an effort to stay active and busy. According to this view, arranging conditions that permit elders to remain engaged in roles and relationships is vital for life satisfaction (Maddox, 1963).

Although many people seek alternative sources of meaning and gratification in response to social losses, activity theory fails to acknowledge any psychological change in old age. Many studies show that merely offering elders opportunities for social contact does not lead to greater social activity. Indeed, the majority do not take advantage of such opportunities. In nursing homes, for example, social partners are abundant, but social interaction is very low, even among the healthiest residents—a circumstance that we will examine when we discuss housing arrangements for the elderly. Finally, especially troubling for activity theory is the repeated finding that when health status is controlled, elders who have larger social networks and engage in more activities are not necessarily happier (Lee & Markides, 1990; Ritchey, Ritchey, &

Dietz, 2001). Instead, recall that quality, not quantity, of relationships predicts psychological well-being in old age.

• Socioemotional Selectivity Theory. A more recent approach asserts that our social networks become more selective as we age. According to **socioemotional selectivity theory,** social interaction does not decline suddenly in late adulthood. Rather, it extends lifelong selection processes. In middle adulthood, marital relationships deepen, siblings feel closer, and number of friendships declines. In old age, contacts with family and long-term friends are sustained until the eighties, when they diminish gradually in favor of a few very close relationships. In contrast, contacts with acquaintances and willingness to form new social ties fall off steeply from middle through late adulthood (see Figure 18.1) (Fung, Carstensen, & Lang, 2001; Lang, Staudinger, & Carstensen, 1998).

What explains these changes? Socioemotional selectivity theory states that physical and psychological aspects of aging lead to changes in the functions of social interaction. Consider the reasons you interact with members of your social network. At times, you approach them to get information. At other times, you seek affirmation of your uniqueness and worth as a person. And you also choose social partners to regulate emotion, approaching those who evoke positive feelings and avoiding those who make you feel sad, angry, or uncomfortable. In old age, the information-gathering and self-affirming functions become less significant. Because older adults have gathered a lifetime of information, there are fewer people with knowledge they desire. And elders realize it is risky to approach people they do not know for self-affirmation. Stereotypes of aging increase the odds of receiving a condescending, hostile, or indifferent response.

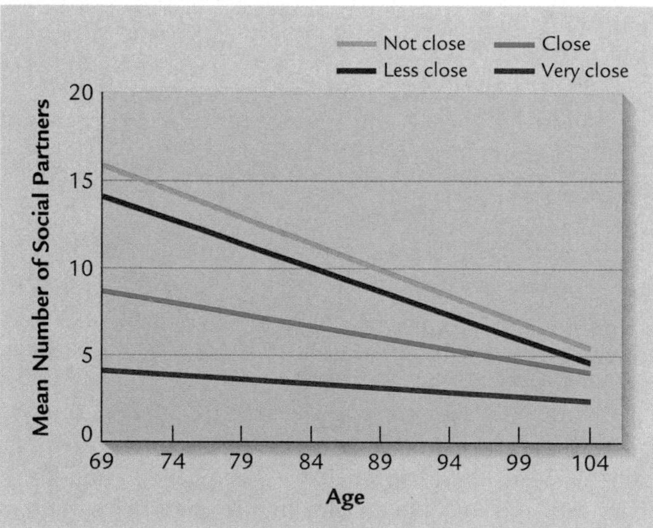

■ **FIGURE 18.1 Age-related change in number of social partners varying in closeness.** In a study of elders ranging in age from 69 to 104, the number of "not close" and "less close" partners fell off steeply with age, whereas the number of "close" and "very close" partners declined minimally and gradually. (Adapted from F. R. Lang, U. M. Staudinger, & L. L. Carstensen, 1998, "Perspectives on Socioemotional Selectivity in Late Life: How Personality and Social Context Do (and Do Not) Make a Difference," *Journal of Gerontology, 53B,* p. 24. Copyright © 1998 by the Gerontological Society of America. Reprinted by permission.)

Instead, elders emphasize the emotion-regulating function of interaction. Physical fragility makes it more important to avoid stress and maintain emotional equilibrium. In one study, younger and older adults were asked to categorize their social partners. Younger people more often used information seeking and future contact as the basis for sorting. In contrast, older people stressed anticipated feelings (Frederickson & Carstensen, 1990). They appeared highly motivated to approach pleasant relationships and avoid unpleasant ones.

Interacting mostly with relatives and friends makes it more likely that elders' self-concepts and emotional equilibrium will be preserved. Although their social networks are smaller, more older adults than younger adults are happy with their current number of friends (Lansford, Sherman, & Antonucci, 1998). Indeed, as the Biology and Environment box on the following page reveals, people's perception of time is strongly linked to their social goals. When time to live is limited, adults of all ages place more emphasis on the emotional quality of their social experiences.

In sum, socioemotional selectivity theory views reduced-quantity but high-quality social ties in old age as due to changing life conditions. As a result, elders much prefer partners with whom they have developed pleasurable, rewarding ties.

Social Contexts of Aging: Communities, Neighborhoods, and Housing

Elders live in contexts—both physical and social—that affect their social experiences and, consequently, their development and adjustment. Communities, neighborhoods, and housing arrangements vary in the extent to which they enable aging residents to satisfy their social needs.

• Communities and Neighborhoods. About half of American and three-fourths of Canadian ethnic minority older adults live in cities, compared with only one-third of Caucasians. The majority of senior citizens reside in suburbs, where they moved earlier in their lives and usually remain after retirement. Suburban elders have higher incomes and report better health than inner-city elders do. Inner-city elders, however, are better off in terms of transportation and proximity to social services, and they are not as disadvantaged in terms of health, income, and availability of services as the one-fourth of American and one-third of Canadian seniors who live in small towns and rural areas. In addition, small-town and rural elderly are less likely to live near their children, who often leave the community in early adulthood (Lindsay, 1999; Statistics Canada, 2003b; U.S. Bureau of the Census, 2002c).

Yet small-town and rural elderly compensate for distance from family members and social services by interacting more with neighbors and friends. Positive aspects of smaller

Biology & Environment

Aging, Time Perception, and Social Goals

With whom would you spend time if you knew you would soon be moving away from your community? When asked this question, young people typically choose close friends and relatives, with whom they have enduring, positive relationships. Their response resembles that of older people. Elders view time as precious and perceive it as flying by rapidly (Kennedy, Fung, & Carstensen, 2001). Increasingly aware that it is "running out," they don't waste it on unlikely future payoffs. Instead, they opt for emotionally gratifying social experiences in the here and now.

Socioemotional selectivity theory underscores that our time perspective plays a crucial role in the social goals we select and pursue (Carstensen, Isaacowitz, & Charles, 1999). People who perceive future time as limited search for meaningful social experiences. Yet tests of the theory documenting an age-related rise in time spent with familiar, highly rewarding social partners do not permit us to separate the influence of age from that of time orientation on people's social goals. Is time perspective really at the heart of elders' focus on old friends and family members as desired social partners?

To find out, Laura Carstensen and her colleagues uncoupled age from time. In a study that held age constant but varied time left in life, the researchers compared three groups of men in their late thirties. The first group was HIV-negative, the second HIV-positive without AIDS symptoms, and the third HIV-positive and actively experiencing deadly symptoms of AIDS (Carstensen & Fredrickson, 1998). When asked to categorize a variety of potential social partners, each successive group increasingly emphasized the emotional rewards of the relationship. Men with AIDS symptoms—the group with the least time left—focused nearly exclusively on the emotional quality of social ties, just as very old people do.

In another investigation, the researchers permitted age to vary but held time perspective constant by giving participants a hypothetical situation in which their future had expanded. People between ages 11 and 92 were asked to imagine that they had just received a telephone call from their doctor, who told them of a new medical breakthrough that would add 20 years to their life (Fung, Carstensen, & Lutz, 1999). Under these conditions, older people's strong bias for familiar, emotionally close social partners disappeared! Their social preferences were just as diverse as those of younger people.

In further research, the investigators examined the social goals of young and old adults in Taiwan and Mainland China. In both cultures, the elderly expressed a stronger desire than young people for familiar social partners likely to provide emotionally satisfying interaction. But this preference was particularly strong among the Mainland Chinese, who perceived future time as more limited. In line with this perception, average life expectancy is 7 years shorter in Mainland China than in Taiwan (Fung, Lai, & Ng, 2001).

■ This senior delights in spending time with his adult granddaughter. When time is limited, people search for meaningful social experiences. Elders' focus on family members and close friends as desired social partners is consistent with the preferences of people of any age who are confronted with a shortened time to live.

In sum, the social preferences of old age are an active adaptation to shrinking longevity. When our days are numbered, present-oriented goals—social connectedness, social support, and emotional depth in relationships—become high priorities.

communities—stability of residents, shared values and lifestyles, willingness to exchange social support, and frequent social visits as country people "drop in" on one another—foster gratifying relationships. And many suburban and rural communities have responded to elder residents' needs by developing transportation programs (such as special buses and vans) to take elders to health and social services, senior centers, and shopping centers.

Both urban and rural older adults report greater life satisfaction when many senior citizens reside in their neighborhood and are available as like-minded companions. Presence of family is not as crucial, so long as neighbors and nearby friends provide social support (Lawton et al., 1999). This does not mean that neighbors replace family relationships. But elders are content as long as their children and other relatives arrange occasional visits (Hooyman & Kiyak, 2002).

© A. RAMSEY/WOODFIN CAMP & ASSOCIATES

■ These men gather regularly to play Chinese chess in Monterey Park, California. In both urban and rural areas, older adults report greater life satisfaction when many senior citizens reside in their neighborhood and are available as like-minded companions.

Compared with older adults in urban areas, those in quiet neighborhoods in small and mid-sized communities are more satisfied with life. The major reason is that smaller communities have lower crime rates (Scheidt & Windley, 1985; Statistics Canada, 2003b). As we will see next, fear of crime has profound, negative consequences for elders' sense of security and comfort.

● **Victimization and Fear of Crime.** Walt and Ruth's single-family house stood in an urban neighborhood, five blocks from the business district where Walt's photography shop had been prior to his retirement. When leaving home for more than a few hours, Walt and Ruth telephoned their next-door neighbor and asked her to keep an eye on the property. As the neighborhood aged, some homes fell into disrepair, and the population became more transient. Although shops were open Thursday and Friday evenings, Walt and Ruth saved their errands for bright daylight hours. Although they had never been victimized, crime was on their minds and affected their behavior. In a large, nationally representative Canadian survey, 41 percent of adults age 65 and older said they felt unsafe walking in their neighborhoods, compared with 24 percent of younger adults (Lindsay, 1999)

Media attention has led to a widely held belief that crime against the elderly is common. In reality, older adults are less often targets of crime, especially violent crime, than other age groups. However, in urban areas, purse snatching and pickpocketing are more often committed against elders (especially women) than younger people, probably because perpetrators feel that they can easily overpower older and female victims (U.S. Department of Justice, 2002). A single incident can strike intense anxiety into the hearts of seniors, given its financial consequences for those with low incomes and its potential for physical injury.

For frail older adults living alone and in inner-city areas, fear of crime is sometimes greater than worries about income, health, and housing. These elders feel especially vulnerable because they believe assistance will not be available should they need it. Their fear restricts their activities and undermines morale (Joseph, 1997; Thompson & Krause, 1998). Neighborhood Watch and other programs that encourage residents to look out for one another reduce fear and feelings of isolation from the community. Some cities have established special police units to investigate and prevent crimes against seniors. They improve life satisfaction (Zevitz & Gurnack, 1991).

● **Housing Arrangements.** Overwhelmingly, older adults in Western industrialized nations want to stay in the neighborhoods where they spent their adult lives, and the large majority—90 percent—remain in or near their old home. In the United States and Canada, fewer than 5 percent relocate to other communities (Che-Alford & Stevenson, 1998; U.S. Department of Health and Human Services, 2002k). These moves are usually motivated by a desire to live closer to children or, among the more economically advantaged and healthy, a desire for a more temperate climate and a place to pursue leisure interests.

Most elder relocations occur within the same town or city; are prompted by declining health, widowhood, or disability; and increase with age (Chappell et al., 2003). As we look at housing arrangements for older adults, we will see that the more a setting deviates from home life, the harder it is for elders to adjust.

Ordinary Homes. For the majority of elders, who are not physically impaired, staying in their own homes affords the greatest possible personal control—freedom to arrange space and schedule daily events as one chooses. More elders in the United States, Canada, and other Western nations live on their own today than ever before—a trend due to improved health and economic well-being (Lindsay, 1999; U.S. Department of Health and Human Services, 2002f). But when health and mobility problems appear, independent living poses risks. Most homes are designed for younger people. They are seldom modified to suit the physical capacities of their elder residents. And living alone in ill health is linked to social isolation and loneliness (Victor et al., 2000).

When Ruth reached her mid-eighties, Sybil begged her to move into her home. Like many adult children of Southern, Central, and Eastern European descent (Greek, Italian, Polish, and others), Sybil felt an especially strong obligation to care for her frail mother. Older adults of these cultural backgrounds, as well as African Americans, Asians, Hispanics, Native Americans, and Canadian Aboriginals, more often live in extended families (Gabrel, 2000; Hays & George, 2002).

Yet increasing numbers of ethnic minority elders want to live on their own, although poverty often prevents them from doing so. For example, two decades ago, 75 percent of Korean-American older adults were living with their children, whereas today just over 50 percent are (Yoo & Sung, 1997). With sufficient income to keep her home, Ruth refused to move in with Sybil. Why do many elders react this way, even after health

© CATHERINE KARNOW/WOODFIN CAMP & ASSOCIATES

■ Because of improved health and economic well-being, today more older adults than ever before live in their own homes.

problems accumulate? As sites of memorable life events, the home strengthens continuity with the past, sustaining elders' sense of identity in the face of physical declines and social losses. And it permits older adults to adapt to their surroundings in familiar, comfortable ways (Atchley, 1999). Elders also value their independence, privacy, and network of nearby friends and neighbors.

During the past half-century, the number of unmarried, divorced, and widowed elders living alone has risen dramatically. Currently, 33 percent of American and 29 percent of Canadian elders live by themselves, a figure that rises to nearly 50 percent for those age 85 and older. This trend is evident in all segments of the elderly population. However, it is less pronounced among men, who are far more likely than women to be living with a spouse into advanced old age (Lindsay, 1999; U.S. Bureau of the Census, 2002c).

Over 40 percent of American and 38 percent of Canadian elders who live alone are poverty-stricken—rates many times greater than among elderly couples. The large majority (more than 70 percent) are widowed women. Some arrived at old age poor because of lower earnings in earlier years. Others became poor for the first time, often because they outlived a spouse who suffered a lengthy, costly illness. With age, their financial status worsens as their assets shrink and their own health care costs rise (Law Commission of Canada, 2001; Vartanian & McNamara, 2002). Under these conditions, isolation, loneliness, and depression can pile up. Poverty among lone elderly women is deeper in the United States than Canada because of less generous government-sponsored income and health benefits. Still, in both nations, elderly women living on their own are worse off than their counterparts in Western Europe (Hardy & Hazelrigg, 1993). Consequently, feminization of poverty deepens in old age.

Residential Communities. About 6 to 8 percent of North American senior citizens live in residential communities, which come in great variety (Pynoos & Golant, 1996). Housing developments for the aged, either single-dwelling or apartment complexes, differ from ordinary homes only in that they have been modified to suit elders' capacities (featuring, for example, single-level living space and grab bars in bathrooms). Some are federally subsidized units for the elderly poor, but most are privately developed retirement villages with adjoining recreational facilities. For elders who need more help with everyday tasks, *assisted living* arrangements are available (see Chapter 17, page 569). **Congregate housing**—an increasingly popular long-term care option—provides a variety of support services, including meals in a common dining room, along with watchful oversight of residents with physical and mental disabilities. **Life care communities** offer a range of housing alternatives, from independent or congregate housing to full nursing home care. For a large initial payment and additional monthly fees, life care guarantees that elders' changing needs will be met in one place as they age.

Unlike Ruth and Walt, who remained in their own home, Dick and Goldie entered congregate housing in their late sixties. For Dick, the move was a positive turn of events that permitted him to relate to peers on the basis of their current life together, setting aside past failures in the outside world. Dick found gratifying leisure pursuits—leading an exercise class, organizing a charity drive with Goldie, and using his skills as a baker to make cakes for birthday and anniversary celebrations.

Studies of diverse residential communities for the aged reveal that they can have positive effects on physical and mental

■ Older adults enjoy an evening sing-along in a common living area of their retirement community. When elders in congregate housing have shared values and goals, live in a facility small enough to promote frequent communication, and have access to meaningful and pleasurable roles and activities, they express high life satisfaction.

© CATRINA GENOVESE/INDEX STOCK IMAGERY

health. A specially designed physical space and care on an as-needed basis help elders overcome mobility limitations, thereby permitting greater social participation (Fonda, Clipp, & Maddox, 2002). And in societies where old age leads to reduced status, age-segregated living is gratifying to most elders who choose it. It may open up useful roles and leadership opportunities, which result in a more vigorous social life (Ball et al., 2000). A study of congregate housing in Israel revealed that the more older adults perceived the environment as socially supportive, the more they provided assistance to other residents (Litwin, 1998). Congregate housing appears to be well suited to promoting mutually supportive relationships among elder residents.

Yet a collection of elders does not guarantee a comfortable, content community. Shared values and goals among residents with similar backgrounds, a small enough facility to promote frequent communication, and availability of meaningful roles enhance life satisfaction. When older adults feel socially integrated into the setting, they are more likely to consider it their home. Those who remain distant and reclusive are unlikely to characterize their apartment or room as home, citing lack of warmth and like-minded companions as the reason (Young, 1998).

Nursing Homes. The small percentage of North Americans age 65 and older who live in nursing homes experience the most extreme restriction of autonomy. As noted in Chapter 17, sense of personal control and gratifying social relationships are as vital for the mental health of nursing home residents as they are for community-dwelling elders.

Potential social partners are abundant in nursing homes, but interaction is low. To regulate emotion in social interaction (so important to elders), personal control over social experiences is vital. Yet nursing home residents have little opportunity to choose their social partners, and timing of contact is generally determined by the nursing home staff rather than by elders' preferences. Social withdrawal is an adaptive response to these often overcrowded, hospital-like settings. Interaction with people in the outside world predicts nursing home residents' life satisfaction; interaction within the institution does not (Baltes, Wahl, & Reichert, 1992). Not surprisingly, nursing home residents with physical but not mental impairments are far more depressed and anxious than their community-dwelling counterparts (Guildner et al., 2001).

Designing nursing homes to be more homelike could do much to increase residents' sense of security and control over their social experiences. North American nursing homes are often packed with residents and institutional in their operation, largely because most are operated for profit. In contrast, European facilities are liberally supported by public funds. Residents live in private suites or small apartments furnished in part with their own belongings. Specially adapted parks and gardens draw residents away from passive activities, such as TV viewing, into outdoor communal spaces. When an elder's condition worsens, caregivers modify the existing space rather than move the individual to more medically oriented quarters (Horgas, Wilms, & Baltes, 1998; Schwarz, 1996). In this way, they preserve the person's identity, sense of place, and social relationships as much as possible.

Ask Yourself

REVIEW

Cite features of neighborhoods and residential communities that enhance elders' life satisfaction.

APPLY

Sam lives by himself in the same home he has occupied for over 30 years. His adult children can't understand why he won't move across town to a modern apartment. Why does Sam prefer to stay where he is?

APPLY

Vera, a nursing home resident, speaks to her adult children and to a close friend on the phone every day. In contrast, she seldom attends nursing home social events or interacts with her roommate. Using socioemotional selectivity theory, explain Vera's behavior.

CONNECT

According to socioemotional selectivity theory, when time is limited, adults focus on the emotional quality of their social experiences. How might this emphasis foster emotional expertise, discussed on pages 585–586?

www.

Relationships in Late Adulthood

The **social convoy** is an influential model of changes in our social networks as we move through life. Picture yourself in the midst of a cluster of ships traveling together, granting one another safety and support. Ships in the inner circle represent people closest to you, such as a spouse, best friend, parent, or child. Those less close but still important travel on the outside. With age, ships exchange places in the convoy, and some drift off while others join the procession (Antonucci, 1990). But as long as the convoy continues to exist, you adapt positively.

In the following sections, we examine the ways elders with diverse lifestyles sustain social networks of family members and friends—an effort that fosters personal continuity and security in the face of major life changes (Atchley, 1999). We will see that as ties are lost, older adults draw others closer and even add replacements, although not at the rate they did at younger ages. Tragically, for some older adults the social convoy breaks down. We will also explore the circumstances in which elders experience abuse and neglect at the hands of those close to them.

Marriage

Even with high divorce rates, 1 in every 4 or 5 first marriages in North America is expected to survive for at least 50 years. Walt's comment to Ruth that "the last few decades have been the happiest" characterizes the attitudes and behaviors of many elderly couples who have spent their adult lives together. Marital satisfaction rises from middle to late adulthood, when it is at its peak (Goodman, 1999; Levenson, Carstensen, & Gottman, 1993). Several changes in life circumstance and couples' communication underlie this trend.

First, perceptions of fairness in the relationship increase as men participate more in household tasks after retirement. For elders who experienced little social pressure for gender equality in their youth, division of labor in the home still reflects traditional roles (Condie, 1989; Vinick & Ekerdt, 1991). Men take on more home maintenance projects, whereas women's duties—cooking, cleaning, laundry, and shopping—continue as before. Among adults retiring today, "feminine" tasks are more equally shared than they were during work life (Kulik, 2001). In either case, men's increased involvement in caring for the home results in a greater sense of equity in marriage than before.

© DAVID YOUNG-WOLFF/PHOTOEDIT

■ Marital satisfaction peaks in late adulthood. Older couples have more time to enjoy each other's company. And because of their expertise at understanding and regulating emotion, they interact more positively and express more affection.

Second, with extra time together, the majority of couples engage in more joint leisure activities. Ruth and Walt walked, worked in the garden, played golf, and took frequent day trips. In interviews with a diverse sample of retired couples, women often stated that more time with their husbands enhanced marital closeness (Vinick & Ekerdt, 1991).

Finally, greater emotional understanding and emphasis on regulating emotion in relationships lead to more positive interactions. Couples married for at least 35 years resolve conflicts in ways that are less negative and more affectionate than do middle-aged couples. Even in unhappy marriages, elders are less likely to let disagreements escalate into expressions of anger and resentment (Carstensen, Gottman, & Levenson, 1995; Carstensen, Isaacowitz, & Charles, 1999). For example, when Dick complained about Goldie's cooking, Goldie tried to appease him: "All right, Dick, next birthday I won't make cheesecake." And when Goldie brought up Dick's bickering and criticism, Dick usually said, "I know, dear," and retreated to another room. As in other relationships, the elderly protect themselves from stress by molding marital ties to make them as pleasant as possible.

When marital dissatisfaction is present, it continues to take a greater toll on women than on men. Recall from Chapter 14 that women tend to confront marital problems and try to solve them. In old age, the energy expended is especially taxing on their physical and mental health. Husbands, in contrast, often protect themselves by withdrawing, as they did in their twenties and thirties (Levenson, Carstensen, & Gottman, 1993).

Gay and Lesbian Partnerships

Elderly gays and lesbians in long-term partnerships have sustained their relationships through a historical period of hostility and discrimination toward homosexuals. Nevertheless, most report happy, highly fulfilling relationships, pointing to their partner as their most important source of social support. And compared with homosexual elders who live alone, homosexual partners rate their physical and mental health more favorably (Grossman, Daugelli, & Hershberger, 2000; Wojciechowski, 1998).

A lifetime of effective coping with an oppressive social environment may have strengthened homosexuals' skill at dealing with late-life physical and social changes, thereby contributing to a satisfying partnernship (Gabbay & Wahler, 2002). And greater gender-role flexibility enables gay and lesbian couples to adapt easily to sharing household tasks following retirement. Furthermore, because of imagined or real strain in family relationships when they told others about their homosexuality, gays and lesbians less often assume that family members will provide support in old age. Consequently, many have forged strong friendships to replace or build on family ties (Kimmel, 2002). Homosexual couples with gratifying friendship networks report high life satisfaction and less fear of aging (Slusher, Mayer, & Dunkle, 1996).

These supportive social contexts and positive feelings probably enhance their happiness as aging couples.

Nevertheless, because of continuing prejudice and lack of social recognition of their partnerships, aging gays and lesbians face unique challenges (Woolf, 2001). Health care systems are often unresponsive to their unique needs. And if their loved one becomes frail or ill, partners be welcome in hospitals or nursing homes or be allowed to participate in health care decisions—an issue we will return to in Chapter 19. These circumstances can make late-life declines and losses especially painful.

Divorce and Remarriage

When Walt's uncle Louie was 61, he divorced his wife Sandra, to whom he had been married for 17 years. Although she knew the marriage was far from perfect, Sandra had lived with Louie long enough that the divorce came as a shock. A year later, Louie married Rachella, a divorcée who shared his enthusiasm for sports and dance.

Couples who divorce in late adulthood constitute a very small proportion of all divorces in any given year—less than 1 percent. But the divorce rate among people age 65 and older is increasing as new generations of elders become more accepting of marital breakup and as the divorce risk rises for second and subsequent marriages. When asked about the reasons for divorce, elderly men typically mention lack of shared interests and activities, whereas women frequently cite their partner's refusal to communicate and emotional distance. "We never talked. I felt isolated," Sandra said (Weingarten, 1988).

Compared with younger adults, long-time married elders have given their adult lives to the relationship. Following divorce, they find it harder to separate their identity from that of their former spouse, and they suffer more from a sense of personal failure. Relationships with family and friends shift at a time when close bonds are crucial for psychological well-being. Women suffer most from late-life divorce because they are more likely than men to spend their remaining years living alone. The financial consequences are severe—greater than for widowhood because many accumulated assets are lost in property settlements (Miller, Hemesath, & Nelson, 1997).

In younger individuals, divorce often leads to greater awareness of and resolve to change negative patterns of behavior. In contrast, self-criticism in divorced elders heightens guilt and depression because their self-worth depends more on past than on future accomplishments. Louie and Sandra blamed each other. "I was always miserable with Sandra," Louie claimed, even though the couple's earlier days had been reasonably happy. Blaming the partner may distort the marital history, but it is a common coping strategy that enables older adults to preserve integrity and self-esteem (Weingarten, 1989).

Remarriage rates are low in late adulthood and decline with age, although they are considerably higher among divorced than widowed elders. Older men's opportunities for remarriage are far greater than women's. Nevertheless, the gender gap in elder remarriage is much smaller after divorce than after widowhood (see Figure 18.2). Perhaps because their previous relationship was disappointing, divorcées find it easier than widows to enter a new relationship. Also, divorced older women may be more motivated to remarry because of their more extreme economic circumstances. Finally, some divorced elders (like Louie and Rachella) leave their marriages only after a new bond is forming (Huyck, 1995).

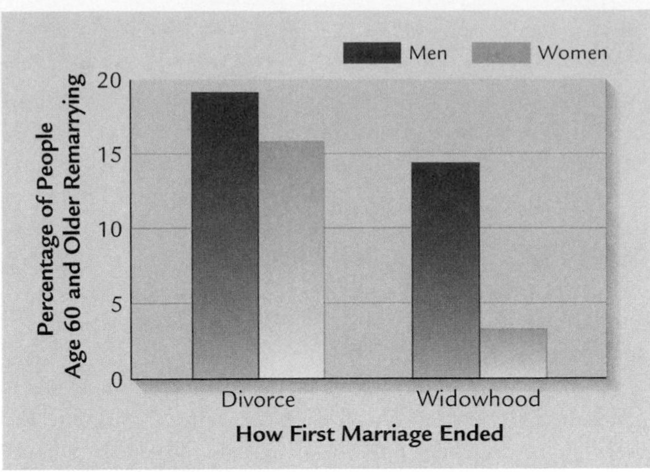

■ **FIGURE 18.2 Rates of marriage after divorce and widowhood among American men and women age 65 and older.** More divorced than widowed elders remarry. In addition, the gender gap in late-life remarriage is much smaller after divorce than after widowhood. (From U.S. Bureau of the Census, 2002c.)

Compared with younger people who remarry, elders who do so enter more stable relationships, as their divorce rate is much lower. In Louie and Rachella's case, the second marriage lasted for 32 years! Perhaps late-life marriages are more successful because they involve a better balance of romantic with practical concerns. Older couples who remarry are very satisfied with their new relationships, although men continue to be more content than women (Brubaker, 1985). With fewer potential mates, perhaps women who remarry in late life must settle for less desirable partners.

Widowhood

Walt died shortly after Ruth turned 80. Like over 70 percent of widowed elders, Ruth described the loss of her spouse as the most stressful event of her life (Lund, Caserta, & Dimond, 1993). She felt lonely, anxious, and depressed for several months after the funeral.

Widows make up one-third of the elderly population in the United States and Canada. Because women live longer than men and are less likely to remarry, nearly 50 percent of women age 65 and older are widowed, whereas only 13 percent of men are. Ethnic minorities with high rates of poverty and chronic disease are more likely to be widowed (Lindsay, 1999; U.S. Bureau of the Census, 2002c).

Earlier we mentioned that most widows and widowers live alone rather than in extended families, a trend that is stronger for whites than for ethnic minorities. Although they are less well off financially than married elders, most want to retain control over their time and living space and to avoid disagreements with their adult children. When widowed elders relocate because they cannot make mortgage payments or keep up their homes, they usually move closer to family rather than into the same residence (Lopata, 1996).

The greatest problem for recently widowed elders is profound loneliness (Lund, 1993b). At the same time, wide variation in adaptation exists, with age, social support, and personality making a difference. Elders have fewer lasting problems than younger individuals who are widowed, probably because death in later life is expected and viewed as less unfair (Stroebe & Stroebe, 1993). And most widowed elders—especially those with outgoing personalities and high self-esteem—are resilient in the face of loneliness (Moore & Stratton, 2002; van Baarsen, 2002). They try to maintain social relationships that were important before the spouse's death and report that relatives and friends respond in kind, contacting them at least as often as before (Utz et al., 2002). Also, the stronger elders' sense of self-efficacy in handling tasks of daily living, the more favorably they adjust (Fry, 2001).

Widowed individuals must reorganize their lives, reconstructing an identity that is separate from the deceased spouse. The task is harder for wives whose roles depended on their husbands' than for those who developed rewarding roles of their own. But overall, men show more physical and mental health problems and greater risk of mortality (Ferraro, 2001). Most relied on their wives for social connectedness and household tasks. And overall, men tend to be less involved in religious activities—a vital source of social support and inner strength (Lee et al., 2001). African-American widowers, however, report less depression than Caucasian widowers, perhaps because of greater support from extended family and church (Balaswamy & Richardson, 2001).

Sex differences in the experience of widowhood contribute to men's higher remarriage rate. Women's kinkeeper role (see Chapter 16, page 526) and ability to form close friendships may lead them to feel less need to remarry. In addition, because many elderly women share the widowed state, they probably offer one another helpful advice and sympathy. In contrast, men often lack skills for maintaining family relationships, forming emotionally satisfying ties outside marriage, and handling the chores of their deceased wives.

Although widowhood affects lifestyle, most widowed elders fare well within a few years, resembling their married counterparts in psychological well-being. Those who arrive at this traumatic event with high self-esteem and a sense of purpose in life are better prepared to deal with loss of a spouse. About 15 to 25 percent have long-term difficulties (Lund & Caserta, 2001). The Caregiving Concerns table on page 604 suggests a variety of ways to foster adaptation to widowhood in old age.

Never-Married, Childless Older Adults

Shortly after Ruth and Walt's marriage in their twenties, Ruth's father died. Her sister Ida continued to live with and care for their mother, who was in ill health until she died 16 years later. When, at age 25, Ida received a marriage proposal, she responded, "I can't marry anybody while my mother is still living. I'm expected to look after her." Ida's decision was not unusual for a daughter of her day. She never married or had children.

About 5 percent of older North Americans have remained unmarried and childless throughout their lives. Almost all are conscious of being different from the norm, but most have developed alternative meaningful relationships. Ida, for example, formed a strong bond with a neighbor's son. In his childhood,

© ELENA ROOROID/PHOTOEDIT

■ These women share the experience of widowhood and offer one another companionship, social support, advice, and sympathy. Men have a harder time adjusting because most relied on their wives for social connectedness and household tasks.

Fostering Adaptation to Widowhood in Late Adulthood

SUGGESTION	DESCRIPTION
Family and Friends	
Social support and interaction	Social support and interaction must extend beyond the grieving period to ongoing assistance and caring relationships. Family members and friends can help most by making support available while encouraging the widowed elder to use effective coping strategies.
Community	
Senior centers	Senior centers offer communal meals and other social activities, enabling widowed and other elders to connect with people in similar circumstances and to access other community resources, such as listings of part-time employment and available housing.
Support groups	Support groups can be found in senior centers, churches, and other agencies. Besides new relationships, they offer an accepting atmosphere for coming to terms with loss, effective role models, and assistance with developing skills for daily living.
Religious activities	Involvement in church, synagogue, or mosque can help relieve the loneliness associated with loss of a spouse and offer social support, new relationships, and meaningful roles.
Volunteer activities	One of the best ways for widowed elders to find meaningful roles is through volunteer activities. Some are sponsored by formal service organizations, such as the Red Cross or the Retired and Senior Volunteer Program. Other volunteer programs exist in hospitals, senior centers, schools, and charitable organizations.

she provided emotional support and financial assistance, which helped him overcome a stressful home life. He included Ida in significant family events and visited her regularly until she died. Other nonmarried elders also speak of the centrality of younger people—often nieces and nephews—in their social networks and of influencing them in enduring ways (Rubinstein et al., 1991). In addition, same-sex friendships are key in never-married elderly women's lives. These tend to be unusually close and often involve joint travel, periods of co-residence, and associations with one another's extended families.

In a large, nationally representative sample of Americans over age 70, childless men without marital partners were far more likely than childless women to feel lonely (Zhang & Hayward, 2001). Never-married elderly women report a level of life satisfaction equivalent to that of married elders and greater than that of divorcées and recently widowed elders. Only when they agree with the stereotype that "life is empty without a partner," or when they cannot maintain social contacts because of declining health, do they report feeling lonely (Dykstra, 1995; Rubinstein, 1987). These single women often state that they avoided many problems associated with being a wife and mother, and they view their enhanced friendships as an advantage of not marrying. At the same time, they realize that friendships are not the same as blood ties when it comes to caregiving in old age.

Most unmarried, childless elders say that support is available from friends and family members, and men often rely on a sister for help (Wenger, 2001; Wu & Pollard, 1998). Never-married elderly women are more likely than other older adults to have relatives and nonrelatives living in their households, with whom they form mutually supportive relationships (Stull & Scarisbrick-Hauser, 1989).

Siblings

Nearly 80 percent of Americans over age 60 have at least one living sibling. Most elder siblings live within 100 miles of each other, communicate regularly, and visit at least several times a year. In one study, 77 percent of a sample of Canadian older adults considered at least one sibling to be a close friend (Connidis, 1989). Both men and women perceive bonds with sisters to be closer than bonds with brothers. Perhaps because of women's greater emotional expressiveness and nurturance, the closer the tie to a sister, the higher elders' psychological well-being (Cicerelli, 1989; O'Bryant, 1988b).

Elderly siblings in industrialized nations are more likely to socialize than to provide each other with direct assistance because most older adults turn to their spouse and children before they turn to their siblings. Nevertheless, siblings seem to be an important "insurance policy" in late adulthood. Figure 18.3 shows the extent to which, in a large, nationally representative American survey, individuals ranging in age from 16 to 85 reported giving or receiving aid from a sibling. As we saw in early chapters, sibling support rises in early adulthood and then declines in middle adulthood. After age 70, it increases for siblings living within 25 miles of each other (White, 2001). Most elders say they would turn to a sibling for help in a crisis, less often in other situations (Connidis, 1994).

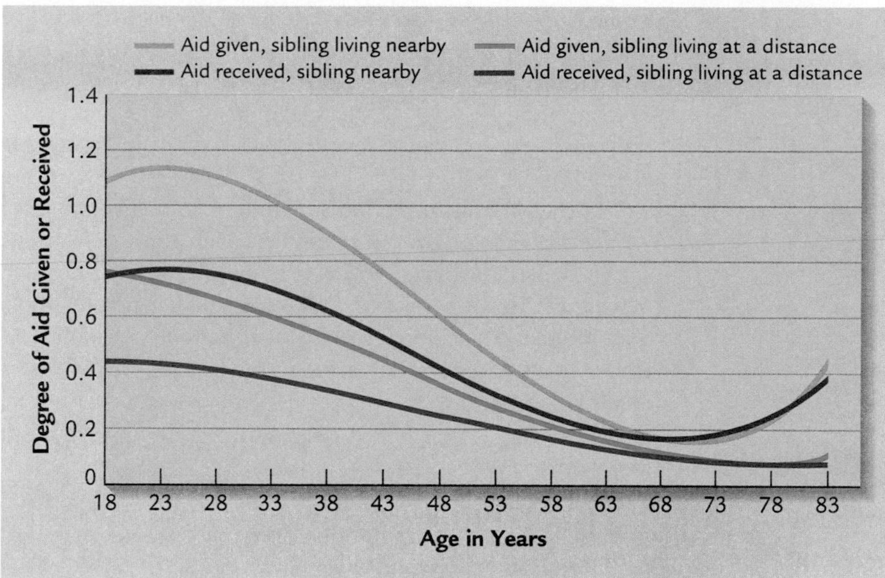

■ **FIGURE 18.3 Age-related change in aid given or received from a sibling.** In a large, nationally representative American survey, adults reported a rise in sibling aid in early adulthood, a decline during middle adulthood, and then a rise after age 70 for siblings living near one another (within 25 miles). In late life, siblings seem to be an important "insurance policy" when help is not available from a spouse or child. (Adapted from L. White, 2001, "Sibling Relationships Over the Life Course: A Panel Analysis," *Journal of Marriage and the Family, 63,* p. 564. Reprinted by permission.)

Widowed and never-married elders have more contacts with siblings, perhaps because they have fewer competing family relationships. They are also more likely to receive sibling support during illness (Connidis & Campbell, 1995). For example, when Ida's Alzheimer's symptoms worsened, Ruth came to her aid. Although Ida had many friends, Ruth was her only living relative.

Because siblings share a long and unique history, joint reminiscing about earlier times increases in late adulthood (Cicerelli, 1995). Walt and Dick often talked about their boyhood days, evoking the warmth of early family life. These discussions helped them appreciate the lifelong significance of the sibling bond and contributed to a sense of family continuity and harmony—important aspects of ego integrity.

Friendships

As family responsibilities and vocational pressures lessen, friendships take on increasing importance. Having friends is an especially strong predictor of mental health among the elderly (Blieszner & Adams, 1992; Nussbaum, 1994). In one study, retired adults were paged at random intervals and asked to write down what they were doing, whom they were with, and how they felt (much like the adolescents described on page 350 of Chapter 11). They reported more favorable experiences with friends than with family members, a difference partly due to the many pleasurable leisure activities shared with friends. But unique qualities of friendship interaction—openness, spontaneity, mutual caring, and common interests—seemed especially influential (Larson, Mannell, & Zuzanek, 1986).

● **Functions of Elder Friendships.** The diverse functions of friendship in late adulthood clarify its profound significance:

- *Intimacy and companionship are basic to meaningful elder friendships.* As Ida and her best friend Rosie took walks, went shopping, or visited each other, the two women disclosed their deepest sources of happiness and worry. They also engaged in pleasurable conversation, laughed, and had fun (Crohan & Antonucci, 1989). When elders are asked to describe qualities of their close friendships, their responses reveal that mutual interests, feelings of belongingness, and opportunities to express feelings and confide in another sustain these bonds over time (Field, 1999).
- *Elderly women mention acceptance as a primary aspect of close friendship.* Late-life friends shield one another from others' negative judgments about their capabilities and worth as a person, which frequently stem from stereotypes of aging (Adams, 1985–1986). "Where's your cane, Rosie?" Ida asked when the two women were about to leave for a restaurant. "Come on, don't be self-conscious. When y'get one of those 'you're finished' looks from someone, just remember: In the Greek village where my mother grew up, there was no separation between generations, so the young ones got used to wrinkled skin and weak knees and recognized older women as the wise ones. Why, they were midwives, matchmakers, experts in herbal medicine; they knew about everything!" (Deveson, 1994).
- *Friendships link elderly people to the larger community.* For elders who cannot go out as often, interactions between friends can keep them abreast of events in the wider world (Peterson, 1989). "Rosie," Ida reported, "did you know that the Thompson girl was named high school valedictorian . . . and the business community is putting its support behind Jesse for mayor?" Friends can also open up new experiences that older adults might not take part in alone. Often a first trip to a senior citizens' center takes place within the context of friendship (Nussbaum, 1994).

© STEPHEN AGRICOLA/STOCK BOSTON

■ Having friends is a strong predictor of mental health among the elderly. Even after declining health restricts their mobility, many elders find ways to sustain ties with friends, who offer intimacy and companionship, links to the larger community, and social support in the face of loss.

- *Friendships help protect elders from the psychological consequences of loss.* Older adults in declining health who remain in contact with friends through phone calls and visits show improved psychological well-being. Similarly, when close relatives die, friends offer compensating social supports (Newsom & Schulz, 1996).

● **Characteristics of Elder Friendships.** Although older adults prefer familiar, established relationships over new ones, friendship formation continues throughout life. Ties to old and dear friends who live far away are maintained, but practical restrictions promote more frequent interaction with friends in the immediate environment. With age, elders report that the friends they feel closest to are fewer in number and live in the same community. Nevertheless, close elder friendships are not affected by patterns of daily life, such as participation in social groups or volunteerism (Dugan & Kivett, 1998). Many elders find ways to sustain links with friends, even after declining health restricts their mobility.

As in earlier years, elders tend to choose friends whose age, sex, race, ethnicity, and values are like their own. However, as agemates die, the very old report more intergenerational friendships (Johnson & Troll, 1994). In her eighties, Ruth spent time with Margaret, a 55-year-old widow she met while serving on the board of directors of an adoption agency. Two or three times a month, Margaret came to Ruth's home for tea and lively conversation.

Sex differences in friendship, discussed in previous chapters, extend into late adulthood. Women are more likely to have intimate friends; men depend on their wives and, to a lesser extent, their sisters for warm, open communication. Also, older women have more **secondary friends**—people who are not intimates but with whom they spend time occasionally, such as a group that meets for lunch, bridge, or museum tours. Through these associates, elders meet new people, remain socially involved, and gain in psychological well-being (Gupta & Korte, 1994).

In elder friendships, affection and emotional support are both given and received to maintain balance in the relationship (Ikkink & Tilburg, 1998). Although friends call on each other for help with tasks of daily living, they generally do so only in emergencies or for occasional, limited assistance. As elders avoid excessive dependency on friends, they register their own autonomy.

Relationships with Adult Children

About 80 percent of older adults in North America are parents of living children, most of whom are middle aged. In Chapter 16, we noted that exchanges of help vary with the closeness of the parent–child bond and with the parent's and adult child's needs. Recall, also, that over time, parent-to-child help declines, whereas child-to-parent assistance increases. Elders and their adult children are often in touch, even when they live far from each other. Nevertheless, as with other ties, quality rather than quantity of these interactions affects older adults' life satisfaction. As people grow older, children usually continue to provide rich rewards, including love, companionship, and stimulation. These warm bonds reduce the negative impact of physical impairments and other losses (such as death of a spouse) on psychological well-being. Alternatively, conflict or unhappiness with adult children contributes to poor physical and mental health (Peterson, 1989; Silverstein & Bengtson, 1991).

Although aging parents and adult children in Western nations provide each other with various forms of help, the level of assistance is typically modest. In interviews with a nationally representative American sample, elders reported that exchanges of advice were most common. Only one-fifth said their children had assisted with household tasks and transportation within the past month. To avoid dependency, older parents expect more emotional support than practical assistance and usually do not seek help from children in the absence of pressing need (Eggebeen, 1992). Moderate support, with opportunities to reciprocate, is psychologically beneficial to elders. Extensive support that cannot be returned is linked to poor psychological well-being (Davey & Eggebeen, 1998; Silverstein, Chen, & Heller, 1996).

Sex differences in older parent–adult child interaction are evident. As kinkeepers, adult daughters are primary agents of their aging parents' family contacts—telephoning, writing letters, and arranging visits more often than sons. Mother–daughter ties are particularly warm (Suitor et al., 1995). Currently, many divorced, noncustodial fathers are reaching late adulthood. Closeness of the parent–child tie affects adult children's sense of obligation to assist an aging parent who had not resided with them for much of their childhood (Ganong & Coleman, 1998). Consequently, in the next few decades, the number of elderly men with limited family contact may increase.

As social networks shrink in size, relationships with adult children become more important sources of family involvement. Elders 85 years and older with children have substantially more contacts with relatives than do those without children (Johnson & Troll, 1992). Why is this so? Consider Ruth, whose daughter, Sybil, linked her to grandchildren, great-grandchildren, and relatives by marriage. When childless elders reach their eighties, siblings, other same-age relatives, and close friends may have become frail or died and hence may no longer be available as companions.

Relationships with Adult Grandchildren and Great-Grandchildren

Older adults with adult grandchildren and great-grandchildren benefit from a wider potential network of support. Ruth and Walt saw their granddaughter, Marci, and their great-grandson, Jamel, at family gatherings. At other times, Marci telephoned, visited, and sent greeting cards, expressing deep affection for her aging grandparents.

In developed nations, slightly more than half of elders over age 65 have a grandchild who is at least 18 years old (Farkas & Hogan, 1994). In the few studies available on grandparent–adult grandchild relationships, the overwhelming majority of grandchildren felt obligated to assist grandparents in need. Grandparents expected affection (but not practical help) from grandchildren, and in most cases they received it. They regarded the adult grandchild tie as very gratifying and as a vital link between themselves and the future (Langer, 1990).

Still, grandparent–adult grandchild relationships vary greatly. Degree of grandparent involvement during childhood strongly predicts the quality of the current relationship. And often, the tie with one grandchild is "special," characterized by more frequent contact, mutual expressions of affection, and enjoyable times together—factors that enhance elders' psychological well-being (Fingerman, 1998). However, longitudinal research reveals that as grandparents and grandchildren move through life, contact declines. Many grandchildren establish distant homes and become immersed in work and family roles that compete for time with extended-family members.

But despite less contact, grandparents' affection for their adult grandchildren strengthens with age, usually exceeding grandchildren's expressed closeness toward their grandparents (which is still strong) (Giarrusso et al., 2001; Harwood, 2001; Silverstein & Long, 1998). This difference in emotional investment reflects each generation's distinct needs and goals—adult grandchildren in establishing independent lives, grandparents in preserving closeness of family relationships and continuity of values across generations. When with their adult

■ Grandparents' affection for their adult grandchildren strengthens with age, even though contact declines as grandchildren establish distant homes and become immersed in work and family roles. Grandchildren become increasingly important sources of emotional meaning for elders in the final decade or two of life.

grandchildren, grandparents frequently convey information and engage in activities related to family cultural heritage—telling stories about earlier times and discussing family traditions and customs (Wiscott & Kopera-Frye, 2000). Grandchildren become increasingly important sources of emotional meaning for elders in the last decade or two of life.

About 40 percent of older adults have great-grandchildren (Hooyman & Kiyak, 2002). Most describe their new role as limited and a sign of advancing age. Nevertheless, they welcome it with enthusiasm, commenting that it reaffirms the continuance of their families. Parents mediate great-grandchild contact, just as they did contact with young grandchildren (see Chapter 16).

Elder Maltreatment

Although the majority of older adults enjoy positive relationships with family members, friends, and professional caregivers, some suffer maltreatment at the hands of these individuals. Recent media attention has led elder maltreatment to become a serious public concern in Western nations.

Reports from Australia, Canada, Finland, Great Britain, the Netherlands, and the United States reveal surprisingly similar rates of maltreatment: about 3 to 7 percent of all elders, or 1.8 million in the United States and 280,000 in Canada (Pavlik et al., 2001; Wolf, 2000). Elder maltreatment crosses ethnic lines, although it is lower in Asian, Hispanic, Native-American, and Canadian Aboriginal groups with strong traditions of respect for and obligation to the aged (Rittman, Kuzmeskus, & Flum, 2000). Yet all figures underestimate the actual incidence, because most acts take place in private and victims are often unable or unwilling to complain.

Elder maltreatment takes the following forms:

- *Physical abuse*—intentional infliction of pain, discomfort, or injury, through hitting, cutting, burning, physical force, restraint, sexual assault, and other acts.
- *Physical neglect*—intentional or unintentional failure to fulfill caregiving obligations, which results in lack of food, medication, or health services or in the elderly person being left alone or isolated.
- *Psychological abuse*—verbal assaults (such as name calling), humiliation (being treated as a child), and intimidation (threats of isolation or placement in a nursing home).
- *Financial abuse*—illegal or improper exploitation of the elder's property or financial resources, through theft or use without the elder's consent.

Although these four types are widely recognized, a common definition of elder maltreatment does not exist. As with child maltreatment (see Chapter 8, page 267), an agreed-on definition is vital for finding out why elder maltreatment occurs and preventing it. A person with multiple bruises and head injuries has surely been abused, but maltreatment is harder to detect in other circumstances. For example, when a frail elder is left alone, how much time must elapse to constitute neglect? And when a relative demands that an elder give back a treasured gift received years earlier and the elder complies in fear of an angry outburst, has financial abuse occurred?

Reported cases suggest that financial abuse is the most common form, followed by psychological abuse and neglect. Often several types occur in combination (Neale et al., 1996; Peake, Oelschlager, & Kearns, 2000). The perpetrator is usually a person the older adult loves, trusts, and depends on for care and assistance. Most abusers are family members—spouses (usually men), followed by children of both sexes and then by other relatives. Some are friends, neighbors, and in-home caregivers (Hornick, McDonald, & Robertson, 1992). Abuse in nursing homes is a major concern. In one survey, one-third of nurses indicated that they had observed it, and 10 percent admitted that they had engaged in it (Wilber & McNeilly, 2001).

● **Risk Factors.** Characteristics of the victim, the abuser, their relationship, and its social context are related to the incidence and severity of elder maltreatment. The more of the following risk factors that are present, the greater the likelihood that abuse and neglect will occur.

Dependency of the Victim. Very old, frail, and mentally and physically impaired elders are more vulnerable to maltreatment (Dyer et al., 2000). This does not mean that declines in functioning cause abuse; most older adults with disabilities do not experience it. Rather, when other conditions are ripe for maltreatment, elders with severe disabilities are least able to protect themselves. Those with physical or cognitive impairments may also have personality traits that make them vulnerable—a tendency to lash out when angry or frustrated, a passive or avoidant approach to handling problems, and a low sense of self-efficacy (Comijs et al., 1999).

Dependency of the Perpetrator. Many abusers are dependent, emotionally or financially, on their victims. This dependency, experienced as powerlessness, can lead to aggressive, exploitative behavior. Often the perpetrator–victim relationship is one of mutual dependency. The abuser needs the older person for money or housing, and the older person needs the abuser for assistance with everyday tasks or to relieve loneliness (Kingston & Reay, 1996).

Psychological Disturbance and Stress of the Perpetrator. Abusers are more likely than other caregivers to have psychological problems and to be dependent on alcohol or other drugs. Often they are socially isolated, have difficulties at work, or are unemployed, with resulting financial worries. These factors increase the likelihood that they will lash out when caregiving is highly demanding or the behavior of an elder with dementia is irritating or hard to manage.

History of Family Violence. Elder abuse is often part of a long history of family violence. Adults who were abused as children are at increased risk of harming elders (Reay & Browne, 2001). And in Chapter 8, we showed how aggressive cycles between family members can easily become self-sustaining, leading to the development of individuals who cope with anger through hostility toward others (Buttell, 1999). In many instances, elder abuse may be an extension of years of partner abuse (see Chapter 14, page 460).

Institutional Conditions. Elder maltreatment is more likely to occur in nursing homes that are run down and overcrowded and that have staff shortages, minimal staff supervision, high staff turnover, and few visitors (Glendenning, 1993). When highly stressful work conditions combine with minimal oversight of caregiving quality, the stage is set for abuse and neglect.

● **Preventing Elder Maltreatment.** Preventing elder maltreatment by family members is especially challenging. Victims may fear retribution, wish to protect abusers who are spouses, sons, or daughters, or feel embarrassed that they could not control the situation. And they may have been intimidated into silence or not know whom to contact for help (Henderson, Buchanan, & Fisher, 2002). Once abuse is discovered, intervention involves immediate protection and provision of unmet needs for the elder and of mental health services and social support for the caregiver.

Prevention programs offer caregivers counseling, education, and respite services, such as elder day care and in-home help. Sometimes trained volunteer "buddies" make visits to the home to combat social isolation among elders and assist them with problem solving to avoid further harm. Support groups help seniors identify abusive acts, practice appropriate responses, and form new relationships. And agencies that provide informal financial services to older adults who are unable to manage on their own, such as writing and cashing checks and holding valuables in a safe, reduce financial abuse (Peake, Oelschlager, & Kearns, 2000).

When elder abuse is extreme, legal action is the best way to shield elders from abusers, yet it seldom happens. Many victims are reluctant to initiate court procedures or, because of mental impairments, cannot do so (Griffiths, Roberts, & Williams, 1993). In these instances, social service professionals must help caregivers rethink their role, even if it means that the aging person might be institutionalized. In nursing homes, abuse and neglect can be prevented by improving staff selection, training, and working conditions.

Combating elder maltreatment also requires efforts at the level of the larger society. Public education to encourage reporting of suspected cases and improved understanding of the needs of older people is vital. As part of this effort, elders benefit from information on how to recognize abusive treatment and where to go for help (Wilber & McNeilly, 2002). Finally, countering negative stereotypes of aging reduces maltreatment, since recognition of elders' dignity, individuality, and autonomy is incompatible with acts of physical and psychological harm.

Ask Yourself

REVIEW

What factors lead marital satisfaction to increase in late adulthood?

REVIEW

Why is adjustment to late-life divorce usually more difficult for women, and adjustment to widowhood more difficult for men?

CONNECT

Mae, age 51, lost her job, could no longer afford to pay rent, and moved in with her 78-year-old widowed mother, Beryl, who welcomed Mae's companionship. Mae grew depressed and drank heavily. When Beryl complained about Mae's failure to look for work, Mae pushed and slapped her. Explain why this mother–daughter relationship led to elder abuse.

CONNECT

After reviewing sections on sibling relationships in Chapter 14 (page 455) and Chapter 16 (page 531), explain life-course trends in sibling support, shown in Figure 18.3 on page 605.

www.

Retirement and Leisure

In Chapter 16, we noted that the period of retirement has lengthened due to increased life expectancy and a steady decline in average age of retirement—trends occurring in all Western industrialized nations. These changes have also led to a blurring of the distinction between work and retirement. Because mandatory retirement no longer exists for most workers in Western countries, older adults have more choices about when to retire and how they spend their time. The retirement process may include a planning period that lasts for years, the decision itself, diverse acts of retiring, and continuous adjustment and readjustment of activities for the rest of the life course (Atchley, 1988; Mutchler et al., 1997).

Some older adults retire gradually by cutting down their hours and responsibilities, taking part-time *bridge jobs* that serve as transitions between full-time career and retirement. Others give up their jobs but later return to work both to support themselves because of limited financial resources and to introduce interest and challenge into the retirement years. Recent estimates indicate that 30 to 40 percent of North American retirees re-enter the labor force within 1 year after retirement in some capacity, usually part-time. However, the likelihood of return to work declines with age (Lindsay, 1999; Marshall, Clarke, & Ballantyne, 2001; Quinn, 1999).

In the following sections, we examine factors that affect the decision to retire, happiness during the retirement years, and leisure pursuits. We will see that the process of retirement and retired life reflect an increasingly diverse retired population.

The Decision to Retire

When Walt and Ruth retired, both had worked for enough years to be eligible for comfortable income-replacement benefits—Walt's through the government-sponsored Social Security program, Ruth's through a private pension plan. In addition, Walt and Ruth had planned for retirement (see Chapter 16, pages 537–538) and decided when they would leave the work force. They wanted to retire early enough to pursue leisure activities while they were both in good health and could enjoy them together. In contrast, Walt's brother Dick was forced to retire as the operating costs of his bakery rose and his clientele dropped off. His wife Goldie kept her part-time job as a bookkeeper to help cover their living expenses.

Affordability of retirement is usually the first consideration in the decision to retire. Yet even in the face of some economic concerns, many preretirees favor letting go of a steady work life for alternative, personally meaningful work or leisure activities. As one retired automobile worker said, "I was working since I was 10 years old. I thought of all the years I've been working, and I wanted a rest." Exceptions to this favorable outlook are people like Dick—forced into retirement or anticipating serious financial difficulties (Bossé, Spiro, & Kressin, 1996).

Retire
Adequate retirement benefits
Compelling leisure interests
Low work commitment
Declining health
Spouse retiring
Routine, boring job

Continue Working
Limited or no retirement benefits
Few leisure interests
High work commitment
Good health
Spouse working
Flexible work schedule
Pleasant, stimulating work environment

LEFT: © DAVID YOUNG WOLFF/PHOTOEDIT
RIGHT: © LARRY DALE GORDON/GETTY IMAGES

FIGURE 18.4 Personal and workplace factors that influence the decision to retire.

Figure 18.4 summarizes personal and workplace factors in addition to income that influence the decision to retire. People in good health, for whom vocational life is central to self-esteem, and whose work environments are pleasant and stimulating are likely to keep on working. For these reasons, individuals in professional occupations usually retire later than those in blue-collar or clerical jobs (Moen, 1996; Moen et al., 2000). Self-employed elders also stay with their jobs longer, probably because they can flexibly adapt their working hours to changing needs. In contrast, people in declining health; who are engaged in routine, boring work; and who have compelling leisure interests often opt for retirement.

Societal factors also affect retirement decisions. When many younger, less costly workers are available to replace older workers, industries are likely to offer added incentives for people to retire, such as increments to pension plans and earlier benefits—a trend that has contributed to a rising number of retirements before age 62 in Western nations (Sterns & Gray, 1999). But when concern increases about the burden on younger generations of a rising population of retirees, eligibility for retirement benefits may be postponed to a later age.

Retirement decisions vary with gender and ethnicity. On average, women retire earlier than men, largely because family events—a husband's retirement or the need to care for an ill spouse or parent—play larger roles in their decisions (Chappell et al., 2003; Smith & Moen, 1998). However, women in or near poverty are an exception. Many find themselves without the financial resources to retire and must continue working into old age. This trend is especially pronounced among African-American women, who are more likely to have minimal retirement benefits and to be caring for other family members (Choi, 1994; Flippen & Tienda, 2000).

In other Western nations, higher minimum pension benefits make retirement feasible for the economically disadvantaged. Denmark, France, Germany, Finland, and Sweden have gradual retirement programs in which older employees reduce their work hours, receive a partial pension to make up income loss, and continue to accrue pension benefits. This approach not only strengthens financial security but introduces a transitional phase that fosters retirement planning (Reday-Mulvey, 2000). In addition, some countries have retirement policies sensitive to women's more interrupted work lives. In Canada, France, and Germany, for example, time devoted to child rearing is given some credit when figuring retirement benefits (O'Grady-LeShane & Williamson, 1992).

In sum, individual preferences shape retirement decisions. At the same time, older adults' opportunities and limitations greatly affect their choices.

Adjustment to Retirement

Because retirement involves giving up roles that are a vital part of identity and self-esteem, it usually is assumed to be a stressful process that contributes to declines in physical and mental health. Yet consider Dick, who reacted to the closing of his bakery with anxiety and depression. His adjustment difficulties were not very different from those of younger people experiencing job loss (see Chapter 16, page 537). Also, recall that Dick had a cranky, disagreeable personality. In this respect, his psychological well-being after retirement was similar to what it had been before!

We must be careful not to assume a cause-and-effect relationship each time retirement and unfavorable reactions are paired. For example, a wealth of evidence confirms that physical health problems lead elders to retire, rather than the reverse. And for most people, mental health is fairly stable from the pre- to postretirement years, with little change prompted by retirement itself (Ross & Drentea, 1998).

The widely held belief that retirement inevitably leads to adjustment problems is contradicted by countless research findings indicating that most people adapt well. They describe themselves as active and socially involved—major determinants of retirement satisfaction. Still, when elders are asked about the stresses of retirement, about 30 percent mention some adjustment difficulties (Bossé et al., 1990).

Workplace factors—especially, financial worries and having to give up one's job—predict stress following retirement. In addition, pressures at work make a difference. Moving out of a high-stress job is associated with positive adaptation to retirement, whereas leaving a pleasant, low-stress job or a highly satisfying job before one is ready is linked to slightly greater difficulties (Gray & Citera, 1996). And especially for women, a continuous work life leading to consistency between career expectations and actual achievements enhances retirement quality (Quick & Moen, 1998).

Among psychological factors, a sense of personal control over life events, including deciding to retire for internally motivated reasons (to do other things), is strongly linked to retirement satisfaction (Kim & Moen, 2002c; Quick & Moen, 1998). At the same time, those who find it hard to give up the predictable schedule and social contacts of the work setting experience discomfort with their less structured way of life. Overall, however, well-educated people in high-status careers adjust more favorably, perhaps because the satisfactions derived from challenging, meaningful work readily transfer to nonwork pursuits (Kim & Moen, 2002a).

Social support reduces stress associated with retirement, just as it helps relieve the stress of other major life events. Although size of the social network typically shrinks as relationships with co-workers decline, quality of social support (the number of people elders "can count on") remains fairly stable. In Dick's case, entering congregate housing eased a difficult postretirement period. It led to new friends and rewarding leisure pursuits, some of which he shared with Goldie. Besides friends, spouses are a vital source of support in fostering retirement adjustment. The number of leisure activities couples enjoy together predicts retirement satisfaction (Reeves & Darville, 1994).

Finally, earlier in this chapter we noted that marital happiness tends to rise after retirement. When a couple's relationship is positive, it can buffer the uncertainty of retirement. And retirement can enhance marital satisfaction by granting husband and wife more time for companionship (Kim & Moen, 2002a). Consequently, a good marriage not only promotes adjustment to retirement but also benefits from the greater freedom of the retirement years. Return to Chapter 16, page 538, for ways adults can plan ahead to increase the chances of a favorable transition to retirement.

Leisure Activities

With retirement, most older adults have more time for leisure pursuits than ever before. After a "honeymoon period" of trying out new activities, many find that leisure interests and skills do not develop suddenly. Instead, meaningful leisure pursuits are usually formed earlier and sustained or expanded during retirement (Mannell, 1999). For example, Walt's fondness for writing, theater, and gardening dated back to his youth. And Ruth's strong focus on her vocation of social work led her to become an avid community volunteer. The stability of leisure activities over the lifespan suggests that the best preparation for leisure in late life is to develop rewarding interests at a young age.

Involvement in leisure activities is related to better physical and mental health and reduced mortality (Cutler & Hendricks, 2001). But simply participating does not explain this relationship. Instead, elders select leisure pursuits because they permit self-expression, new achievements, the rewards of helping others, or pleasurable social interaction. These factors account for gains in well-being (Guinn, 1999).

With age, the frequency and variety of leisure pursuits tend to decline, and travel, outdoor recreation, and exercise are especially likely to drop off. After age 75, mobility limitations lead leisure activities to become more sedentary and home-based (Armstrong & Morgan, 1998; Strain et al., 2002). Elders in residential communities participate more than those in ordinary homes because activities are conveniently available. But regardless of living arrangements, older adults do not spend much time in programs designed just for them. Rather, they choose activities on the basis of whether they are personally gratifying. Partly for this reason, organized activities in community senior centers attract only about 15 percent of elders who live nearby (Krout, Cutler, & Coward, 1990). Nevertheless, these structured opportunities are important for elders with limited incomes and for those who lack daily companionship.

■ Older adults contribute vitally to their communities through volunteerism, especially if they were used to volunteering earlier in their lives. This senior spends time with a hospitalized young patient who has Lou Gehrig's disease. The paralyzing illness does not affect the patient's mind, so the woman visits daily to talk, read, and offer comfort.

Older adults make a vital contribution to society through volunteer work—in hospitals, senior centers, schools, charitable organizations, and other community settings. Younger, better-educated, and financially secure elders with social interests are more likely to volunteer, and women do so more often than men. However, volunteer work is seldom begun in late adulthood. Like other leisure pursuits, it originates earlier, usually during the individual's working years. Nevertheless, nonvolunteers are especially receptive to volunteer activities in the first 2 years after retiring—a prime time to recruit them into these personally rewarding and socially useful pursuits (Caro & Bass, 1997).

Finally, when Walt and Ruth got together with Dick and Goldie, the two couples often discussed politics. Older adults report greater awareness of and interest in public affairs and vote at a higher rate than any age group. And their political knowledge shows no sign of decline, even in late old age. After retiring, elders have more time to read and watch TV, through which they keep abreast of current events. They also have a major stake in political debates over policies central to their welfare. But elders' political concerns are far broader than those that serve their own age group, and their voting behavior is not driven by self-interest (Binstock & Quadagno, 2001). Instead, their political involvement may stem from a deep desire for a safer, more secure world for future generations.

Successful Aging

Walt, Ruth, Dick, Goldie, and Ida, and the research findings they illustrate, reveal great diversity in development during the final decades of life. Walt and Ruth fit contemporary experts' view of **successful aging,** in which gains are maximized and losses minimized. Both remained in reasonably good health until advanced old age, coped well with negative life changes, and enjoyed a happy intimate partnership, other close relationships, and daily lives filled with gratifying activities. Ida, too, was a successful ager until the onset of Alzheimer's symptoms overwhelmed her ability to manage life's challenges. As a single adult, she built a rich social network that sustained her into old age, despite the hardship of having spent many years caring for her ailing mother. In contrast, Dick and Goldie reacted with despondency to physical aging and other losses (such as Dick's forced retirement). And Dick's angry outbursts restricted their social contacts, although the couple's move to congregate housing eventually led to an improved social life.

Successful agers are people for whom growth, vitality, and striving limit and, at times, overcome physical, cognitive, and social declines. Researchers want to know more about their characteristics and development so they can help more seniors age well. Yet theorists disagree on the precise ingredients of a satisfying old age. Some focus on easily measurable outcomes, such as excellent cardiovascular functioning, absence of disability, superior cognitive performance, and creative achievements. Yet this view has been heavily criticized (Baltes & Carstensen, 1996). Not everyone can become an outstanding athlete, an innovative scientist, or a talented artist. And many older adults do not want to keep on accomplishing and producing—the main markers of success in Western nations. Each of us is limited by our genetic potential as it combines with a lifetime of environments we encounter and select for ourselves. And outcomes valued in one culture may not be valued in others.

Recent views of successful aging have turned away from specific achievements toward processes people use to reach personally valued goals (Freund & Baltes, 1998; Lund, 1998). This perspective avoids identifying one set of standards as "successful." Instead, it focuses on how people minimize losses while maximizing gains. In recent research on three samples of adults followed over the lifespan, George Vaillant looked at how various life-course factors contributed to late-life physical and psychological well-being. His findings revealed that factors people could control to some degree (such as health habits, coping strategies, marital stability, and years of education) far outweighed uncontrollable factors (parental SES, family warmth in childhood, early physical health, and longevity of family members) in predicting a happy, active old age (Vaillant & Mukamal, 2001).

Consider the following description of one participant, who in childhood had experienced low SES, parental discord, a depressed mother, and seven siblings crowded into a tenement apartment. Despite these early perils, he became happily married and, through the GI bill, earned an accounting degree. At 70, he was aging well:

> Anthony Pirelli may have been *ill* considering his heart attack and open-heart surgery, but he did not feel *sick.* He was physically active as ever, and he continued to play tennis. Asked what he missed about his work, he exulted, "I'm so busy doing other things that I don't have time to miss work.... Life is not boring for me." He did not smoke or abuse alcohol; he loved his wife; he used mature [coping strategies]; he obtained 14 years of education; he watched his waistline; and he exercised regularly. (Adapted from Vaillant, 2002, pp. 12, 305)

Vaillant concluded, "The past often predicts but never determines our old age" (p. 12). Successful aging is an expression of remarkable resiliency during this final phase of the lifespan.

Take a moment to list the many ways, discussed in this and the previous chapter, that older adults realize their goals. Here are the most important ones, with page references so you can review:

- Optimism and sense of self-efficacy in improving health and physical functioning (pages 558–559)
- Selective optimization with compensation to make the most of limited physical energies and cognitive resources (pages 570 and 594)
- Strengthening of self-concept, which promotes self-acceptance and pursuit of hoped-for possible selves (page 587)

- Enhanced emotional understanding and emotional self-regulation, which supports meaningful, rewarding social ties (pages 585–586)
- Acceptance of change, which fosters life satisfaction (page 587)
- A mature sense of spirituality and faith, permitting anticipation of death with calmness and composure (pages 587–591)
- Personal control over domains of dependency and independence (pages 591–592, 594)
- High-quality relationships, which offer social support and pleasurable companionship (pages 596–597)

Successful aging is facilitated by societal contexts that permit elders to manage life changes effectively. Older adults need well-funded social security plans, good health care, safe housing, and diverse social services. (See, for example, the description of the U.S. Area Agencies on Aging in Chapter 2, page 65.) Yet inadequate funding and difficulties reaching rural communities mean that many older adults' needs remain unmet. And isolated elders with little education often do not know how to access available assistance. Furthermore, the American Medicare system of sharing health care costs with senior citizens strains their financial resources. And housing that adjusts to changes in elders' capacities, permitting them to age in familiar surroundings without disruptive and disorienting moves, is available only to the economically well-off in the United States and Canada.

Besides improving policies that meet older adults' basic needs, new future-oriented approaches must prepare for increased aging of the population. More emphasis on lifelong learning for workers of all ages would help people maintain and even increase skills as they grow older. Also, reforms that prepare for expected growth in the number of frail elders are vital, including affordable help for family caregivers, adapted housing, and sensitive nursing home care.

All these changes involve recognizing, supporting, and enhancing the contributions that senior citizens make to society—both the elders of today and those of tomorrow. A nation that takes care of its senior citizens and grants them a multitude of opportunities for personal growth maximizes the chances that each of us, when our time comes to be old, will age successfully.

Ask Yourself

REVIEW

What psychological and contextual factors predict favorable adjustment to retirement?

REVIEW

Explain how both personal capacities and social conditions foster successful aging.

APPLY

Nate, happily married to Gladys, adjusted well to retirement. He also found that his marriage became even happier. How can a good marriage ease the transition to retirement? How can retirement enhance marital satisfaction?

CONNECT

Leisure interests and skills usually form early and persist over the lifespan. After referring to earlier parts of this book, cite examples of childhood, adolescent, and early adulthood experiences that are likely to foster meaningful leisure pursuits after retirement.

www.

Summary

Erikson's Theory: Ego Integrity versus Despair

According to Erikson, how does personality change in late adulthood?

- The final psychological conflict of Erikson's theory, **ego integrity versus despair,** involves coming to terms with one's life. Adults who arrive at a sense of integrity feel whole and satisfied with their achievements. Despair occurs when elders feel they have made many wrong decisions, yet time is too short for change.

Other Theories of Psychosocial Development in Late Adulthood

Describe Peck's and Labouvie-Vief's views of development in late adulthood, and discuss the functions of reminiscence and life review in older adults' lives.

- According to Robert Peck, the conflict of ego integrity versus despair comprises three distinct tasks: (1) ego differentiation versus work-role preoccupation; (2) body transcendence versus body preoccupation; and (3) ego transcendence versus ego preoccupation.
- Gisella Labouvie-Vief addresses the development of adults' reasoning about emotion, pointing out that older, more psychologically mature individuals are more in touch with their feelings. They become expert at processing emotional information and engaging in emotional self-regulation.
- Researchers do not yet fully understand why older people engage in **reminiscence** more often than younger people. In a special

form of reminiscence called **life review,** elders call up, reflect on, and reconsider past experiences with the goal of achieving greater self-understanding. Reminiscence may also serve social goals, such as strengthening close relationships. And at times, elders use it to identify effective problem-solving strategies and to teach younger people. Sometimes, reminiscence helps elders cope with life transitions.

Stability and Change in Self-Concept and Personality

Cite stable and changing aspects of self-concept and personality in late adulthood.

- The "big five" personality traits show continuing stability from mid- to late life. Elders have accumulated a lifetime of self-knowledge, leading to more secure and complex self-concepts than at earlier ages.
- During late adulthood, shifts in three personality traits take place. Agreeableness and acceptance of change tend to rise, whereas sociability dips slightly.

Discuss spirituality and religiosity in late adulthood, and trace the development of faith.

- Although organized religious participation declines, informal religious activities remain common in late adulthood. Religious involvement is especially high among low-SES ethnic minority elders and women and is linked to many positive outcomes, including physical functioning, psychological well-being, and longer survival.
- Faith and spirituality may advance to a higher level in late adulthood, away from prescribed beliefs to a more reflective approach that is at ease with uncertainty.

Individual Differences in Psychological Well-Being

Discuss individual differences in psychological well-being as older adults respond to increased dependency, declining health, and negative life changes.

- Friends, family members, and caregivers often promote excessive dependency in elders. In patterns of behavior called the **dependency–support script** and the **independence–ignore script,** older adults' dependency behaviors are attended to immediately and their independent behaviors are ignored. For dependency to foster well-being, elders need to assume personal control over it.
- Health is a powerful predictor of psychological well-being in late adulthood. The relationship between physical and mental health problems can become a vicious cycle, each intensifying the other. Physical declines often lead to depression, and if hopelessness deepens, it can be lethal. Older adults have the highest suicide rate of all age groups.
- Although elders are at risk for a variety of negative life changes, these events actually evoke less stress and depression in elders than in younger adults. Many seniors have learned to cope with hard times.

Describe the role of social support and social interaction in promoting physical health and psychological well-being in late adulthood.

- In late adulthood, social support from family, friends, and paid helpers reduces stress, thereby promoting physical health and psychological well-being. At the same time, older adults do not want a great deal of support from people close to them unless they can reciprocate. Having an extraverted personality, which fosters interaction with others, is linked to high morale in old age. However, high-quality relationships have the greatest impact on mental health.

A Changing Social World

Describe social theories of aging, including disengagement theory, activity theory, and socioemotional selectivity theory.

- **Disengagement theory** holds that social interaction declines because of mutual withdrawal between elders and society in anticipation of death. However, not everyone disengages, and elders' retreat from interaction is more complex than this theory implies.
- According to **activity theory,** social barriers to engagement, not the desires of elders, cause declining rates of interaction. Yet offering older adults opportunities for social contact does not guarantee greater social activity.
- **Socioemotional selectivity theory** states that social networks become more selective as we age. As older adults emphasize the emotion-regulating function of interaction, they tend to limit their contacts to familiar partners with whom they have developed pleasurable, rewarding relationships.

How do communities, neighborhoods, and housing arrangements affect elders' social lives and adjustment?

- Elders residing in suburbs are better off in terms of income and health, whereas elders in inner cities have easier access to social services. Those in small towns and rural areas are least well off in these ways. Older adults who have many senior citizens in their neighborhood and who live in small and mid-sized communities report greater life satisfaction.
- For elders living alone and in inner-city areas, fear of crime is sometimes greater than worries about income, health, and housing. It restricts their activities and seriously undermines their morale.
- The housing arrangement that offers seniors the greatest personal control is their own home. But when health and mobility problems appear, independent living poses risks. Many older adults who live alone, especially widowed women, are poverty-stricken and suffer from unmet needs.
- Most residential communities for senior citizens are privately developed retirement villages with adjoining recreational facilities. Among assisted living arrangements, **congregate housing** offers a variety of support services, including meals in a common dining room. **Life care communities** offer a range of options, from independent or congregate housing to full nursing home care. A sense of community in planned housing enhances life satisfaction.
- The small number of North Americans who live in nursing homes experience extreme restriction of autonomy. Typically, social interaction among residents is low. Designing homelike nursing homes could enhance elders' sense of control over their social experiences.

Relationships in Late Adulthood

Describe changes in social relationships in late adulthood, including marriage, divorce, remarriage, and widowhood, and discuss never-married, childless older adults.

- As we move through life, a **social convoy,** or cluster of family members and friends, provides safety and support. With age, some bonds become closer, others more distant, while still others are added or drift away. To preserve personal continuity and security in the face of life changes, elders do their best to sustain social networks of family members and friends.
- Marital satisfaction rises from middle to late adulthood as perceptions of fairness in the relationship increase, couples engage in joint leisure activities, and communication becomes more positive. Most gay and lesbian elders also report happy, highly fulfilling relationships.
- When divorce occurs, stress is higher for older than for younger adults. Although remarriage rates are low in late adulthood, those who do remarry enter into more stable relationships. Because of greater financial hardship and less likelihood of remarrying, women suffer more than men from late-life divorce.
- Wide variation exists in adaptation to widowhood, with age, social support, and personality making a difference. Elders fare better than younger individuals. Efforts to

maintain social ties, an outgoing personality, high self-esteem, and a sense of self-efficacy in handling tasks of daily living foster adjustment. Women—especially those who developed rewarding roles outside the marital relationship—fare better than men.

- Most older adults who remain unmarried and childless throughout their lives develop alternative meaningful relationships. At the same time, men more often feel lonely than women. And unmarried, childless elders frequently worry about how care will be provided in old age if it is needed.

How do sibling relationships and friendships change in late life?

- In late adulthood, social support from siblings increases, especially when siblings live nearby. Bonds with sisters are closer than bonds with brothers. Because siblings share a long and unique history, joint reminiscing increases in late adulthood, contributing to a sense of family continuity and harmony.

- Friendships serve a variety of functions in late adulthood: intimacy and companionship, acceptance, a link to the larger community, and protection from the psychological consequences of loss. Women are more likely than men to have both intimate friends and **secondary friends**—people with whom they spend time occasionally.

Describe older adults' relationships with adult children, adult grandchildren, and great-grandchildren.

- Elders and their adult children are often in touch, typically exchanging advice rather than direct assistance. Moderate support, with opportunities to reciprocate, is psychologically beneficial to elders. Adult daughters are primary agents of elderly parents' family contacts.

- Seniors with adult grandchildren and great-grandchildren benefit from a wider network of support. Most often, grandparents expect and receive affection rather than practical help from their grandchildren. Grandparent–grandchild contact declines over time, but grandparents' affection for their adult grandchildren remains high.

Discuss elder maltreatment, including risk factors and strategies for prevention.

- Some elders suffer maltreatment at the hands of family members, friends, or professional caregivers. Risk factors include a dependent perpetrator–victim relationship; perpetrator psychological disturbance and stress; a history of family violence; and overcrowded nursing homes with staff shortages and turnover.

- Elder-abuse prevention programs offer caregivers counseling, education, and respite services. Elders benefit from trained volunteers and support groups that help them avoid future harm. Societal efforts, including public education to encourage reporting of suspected cases and improved understanding of the needs of older people, are also vital.

Retirement and Leisure

Discuss the decision to retire, adjustment to retirement, and involvement in leisure activities.

- The decision to retire depends on affordability, health status, opportunities to pursue meaningful activities, societal factors such as early retirement benefits, gender, and ethnicity. On average, women retire earlier than men because family events play larger roles in their decisions. However, those near poverty may need to continue working, a trend especially pronounced among African Americans.

- Health status; financial stability; a sense of personal control over life events, including the retirement decision; characteristics of the workplace; satisfactions derived from work; social support; and marital happiness are among factors that affect adjustment to retirement.

- Engaging in meaningful and pleasurable leisure activities is related to physical and mental health and to reduced mortality. The best way to prepare for leisure experiences in late life is to develop rewarding interests at a young age.

Successful Aging

Discuss the meaning of successful aging.

- Elders who experience **successful aging** have developed many ways to minimize losses and maximize gains. Social contexts that permit elders to manage life changes effectively foster successful aging. These include well-funded social security plans, good health care, safe housing that adjusts to changes in elders' capacities, social services, and opportunities for lifelong learning. Reforms are also needed to ensure the well-being of frail elders through affordable in-home help, adapted housing, and sensitive nursing home care.

Important Terms and Concepts

activity theory (p. 595)
congregate housing (p. 599)
dependency–support script (p. 591)
disengagement theory (p. 595)
ego integrity versus despair (p. 584)
independence–ignore script (p. 591)
life care communities (p. 599)
life review (p. 586)
reminiscence (p. 586)
secondary friends (p. 606)
social convoy (p. 600)
socioemotional selectivity theory (p. 596)
successful aging (p. 612)

FYI For Further Information and Help

Consult the Companion Website for *Development Through the Lifespan, Third Edition,* (www.ablongman.com/berk) where you will find the following resources for this chapter:

- **Chapter Objectives**
- **Flashcards** for studying important terms and concepts
- **Annotated Weblinks** to guide you in further research
- **Ask Yourself** questions, which you can answer and then check against a sample response
- **Suggested Readings**
- **Practice Test** with immediate scoring and feedback

Milestones

Development in Late Adulthood

AGE

60–80 years

PHYSICAL

- Neurons die at a faster rate, but the brain compensates through growth of new synapses. (549)
- Autonomic nervous system performs less well, impairing adaptation to hot and cold weather. (550)
- Declines in vision continue, in terms of increased sensitivity to glare and impaired color discrimination, dark adaptation, depth perception, and visual acuity. (551–552)

- Declines in hearing continue throughout the frequency range. (552)
- Taste and odor sensitivity may decline. (553)
- Touch sensitivity declines on the hands, particularly the fingertips, less so on the arms. (553)

- Declines in cardiovascular and respiratory functioning lead to greater physical stress during exercise. (553)
- Aging of the immune system increases risk for a variety of illnesses. (554)
- Sleep difficulties increase, especially for men. (554–555)

COGNITIVE

- Processing speed continues to decline; crystallized abilities are largely sustained. (570)
- Amount of information that can be retained in working memory diminishes further; memory problems are greatest on tasks requiring deliberate processing and associative memory. (571)
- Modest forgetting of remote memories occurs. (572)
- Use of external aids for prospective memory increases. (573)
- Retrieving words from long-term memory and planning what to say and how to say it become more difficult. (573)
- Information is more likely to be remembered in terms of gist than details. (574)
- Traditional problem solving declines; everyday problem solving remains adaptive. (574)

- May hold one of the most important positions in society, such as chief executive officer, religious leader, or Supreme Court justice. (575–576)

EMOTIONAL/SOCIAL

- Comes to terms with life, developing ego integrity. (584)
- Describes emotional reactions in more complex and personalized ways; improves in emotional self-regulation. (585–586)
- May engage in reminiscence and life review. (586)
- Self-concept strengthens, becoming more secure and complex. (587)
- Agreeableness and acceptance of change increase. (587)
- Faith and spirituality may advance to a higher level. (587–590)

- Size of social network and amount of social interaction decline. (595)
- Selects social partners on the basis of emotion, approaching pleasant relationships and avoiding unpleasant ones. (596)
- Marital satisfaction increases. (601)

- May be widowed. (602)
- Sibling closeness and support may increase. (601)
- Number of friends generally declines. (606)

AGE	PHYSICAL	COGNITIVE	EMOTIONAL/SOCIAL
60–80 years *(continued)*	• Graying and thinning of the hair continue; the skin wrinkles further and becomes more transparent as it loses its fatty layer of support (554) • Height and weight (due to loss of lean body mass) decline. (555) • Loss of bone mass leads to rising rates of osteoporosis. (555, 562)  • Intensity of sexual response and sexual activity decline, although most healthy married couples report regular sexual enjoyment. (560)	• May excel at wisdom. (575–576) • Can improve a wide range of cognitive skills through training. (577) 	• May become a great-grandparent. (607)  • May retire. (609) • More likely to be knowledgeable about politics and to vote. (612)
80 years and older	• Physical changes described above continue. • Mobility diminishes, due to loss of muscle and bone strength and joint flexibility. (555) 	• Cognitive changes described above continue.  • Fluid abilities decline further; crystallized abilities drop as well. (570)	• Emotional and social changes described above continue. • As relatives and friends die, may develop friendships with younger people. (606) • Relationships with adult children become more important. (607) • Frequency and variety of leisure activities decline. (611) 

Note: Numbers in parentheses indicate the page(s) on which each milestone is discussed.

19

Death, Dying, and Bereavement

© ASAP LTD./INDEX STOCK

How We Die
Physical Changes ● Defining Death ● Death with Dignity

Understanding of and Attitudes Toward Death
Childhood ● Adolescence ● Adulthood ● Death Anxiety

Thinking and Emotions of Dying People
Do Stages of Dying Exist? ● Contextual Influences on Adaptations to Dying

A Place to Die
Home ● Hospital ● The Hospice Approach

The Right to Die
Passive Euthanasia ● Voluntary Active Euthanasia ● Assisted Suicide

■ Social Issues: Voluntary Active Euthanasia: Lessons from Australia and the Netherlands

Bereavement: Coping with the Death of a Loved One
Grief Process ● Personal and Situational Variations ● Bereavement Interventions

■ Cultural Influences: Cultural Variations in Mourning Behavior

Death Education

All cultures have special rituals for celebrating the end of life and helping the bereaved cope with profound loss. Here members of a Druze community in Israel reach out to touch the coffin of a respected leader.

As every life is unique, so each death is unique. The final forces of the human spirit separate themselves from the body in manifold ways.

My mother Sofie's death was the culmination of a 5-year battle against cancer. In her last months, the disease invaded organs throughout her body, attacking the lungs in its final fury. She withered slowly, having been granted the mixed blessing of time to prepare against certain knowledge that death was just around the corner. My father, Philip, lived another 18 years. At age 80, he was outwardly healthy, active, and about to depart on a long-awaited vacation when a heart attack snuffed out his life suddenly, without time for last words or deathbed reconciliations.

As I set to work on this chapter, my 65-year-old neighbor Nicholas gambled for a higher quality of life. To be eligible for a kidney transplant, he elected bypass surgery to strengthen his heart. Doctors warned that his body might not withstand the operation. But Nicholas knew that without taking a chance, he would live only a few years, in debilitated condition. Shortly after the surgery, infection set in, traveling throughout his system and so weakening him that only extreme measures—a respirator to sustain breathing and powerful drugs to elevate his fading blood pressure—could keep him alive.

"Come on, Dad, you can do it," encouraged Nicholas's daughter Sasha, standing by his bedside and stroking his hand. But Nicholas could not. After 2 months in intensive care, he experienced brain seizures and slipped into a coma. Three doctors met with his wife, Giselle, to tell her there was no hope. She asked them to disconnect the respirator, and within half an hour Nicholas drifted away.

Death is essential for the survival of our species. We die so that our own children and the children of others may live. When it comes to this fate, nature treats humankind, with all its unique capabilities, just as it treats every other living creature (Nuland, 1993). As hard as it is to accept the reality that we too will die, our greatest solace lies in the knowledge that death is part of ongoing life.

In this chapter, we address the culmination of lifespan development. Over the past century, technology has provided us with so much to keep death at bay that many people regard it as a forbidden topic. But pressing social and economic dilemmas that are an outgrowth of the dramatic increase in life expectancy are forcing us to attend to life's end—its quality, its timing, and ways to help people adjust to their own and others' final leave-taking.

Our discussion addresses the physical changes of dying; understanding of and attitudes toward death in childhood, adolescence, and adulthood; the thoughts and feelings of people as they stand face to face with death; hopelessly ill patients' right to die; and coping with the death of a loved one. The experiences of Sofie, Philip, Nicholas, their families, and others illustrate how each person's life history joins with social and cultural contexts to shape death and dying, lending great diversity to this universal experience.

How We Die

Our vast literature on death is largely aimed at helping people cope with the emotional trauma of dying and its aftermath. Few people are aware of the physical aspects of death, since opportunities to witness death are less available than in previous generations. Today, the large majority of people in industrialized nations die in hospitals, where doctors and nurses, not loved ones, typically attend their last moments. Nevertheless, many people want to know how we die, either to anticipate their own end or grasp what is happening to a dying loved one.

Physical Changes

My father's fatal heart attack came suddenly during the night. On being told the news, I longed for reassurance that his death had been swift and without suffering.

When asked how they would like to die, most people say they want "death with dignity"—either a quick, agony-free end during sleep or a clear-minded final few moments in which they can say farewell and review their lives. In reality, death is the culmination of a straightforward biological process. In about 20 percent of people, it is gentle—especially when narcotic drugs ease pain and mask the destructive events taking place (Nuland, 1993). But most of the time it is not.

Recall that unintentional injuries are the leading cause of death in childhood and adolescence, cardiovascular disease and cancer in adulthood. Of the one-quarter of people in industrialized nations who die suddenly, within a few hours of experiencing symptoms, 80 to 90 percent are victims of heart attacks (Nuland, 1993). My yearning for a painless death for my father was probably not fulfilled. Undoubtedly he felt the sharp, crushing sensation of a heart deprived of oxygen. As his heart twitched uncontrollably (called *fibrillation*) or stopped entirely, blood circulation slowed and ceased, and he was thrust into unconsciousness. A brain starved of oxygen for more than 2 to 4 minutes is irreversibly damaged—an outcome indicated by the pupils of the eyes becoming unresponsive to light and widening into large, black circles. Other oxygen-deprived organs stop functioning as well.

Death is long and drawn out for three-fourths of people—many more than in times past, due to life-saving medical technology (Benoliel & Degner, 1995). They succumb in many different ways. Of those with heart disease, most have congestive heart failure, the cause of Nicholas's death. His scarred heart could no longer contract with the force needed to deliver enough oxygen to his tissues. As it tried harder, its muscle weakened further. Without sufficient blood pressure, fluid backed up in Nicholas's lungs. This hampered his breathing and created ideal conditions for inhaled bacteria to multiply, enter the bloodstream, and run rampant in his system, leading many organs to fail.

Cancer also chooses diverse paths to inflict its damage. When it *metastasizes,* bits of tumor travel through the bloodstream and implant and grow in vital organs—perhaps the liver, the kidneys, the intestinal tract, the lungs, or the brain—disrupting their functioning. Medication made my mother's final days as comfortable as possible, granting a relatively easy death. But the preceding weeks involved physical suffering, including impaired breathing and digestion and turning and twisting to find a comfortable position in bed.

In general, dying takes place in three phases:

1. The **agonal phase:** The Greek word *agon* means "struggle." Here *agonal* refers to gasps and muscle spasms during the first moments in which the body can no longer sustain life.
2. **Clinical death:** A short interval follows in which heartbeat, circulation, breathing, and brain functioning stop but resuscitation is still possible.
3. **Mortality:** The individual passes into permanent death. Within a few hours, the newly lifeless being appears shrunken, not at all like the person he or she was when alive.

Defining Death

Think about what we have said so far, and you will see that death is not an event that happens at a single point in time. Rather, it is a process in which organs stop functioning in a sequence that varies from person to person. Because the dividing line between life and death is fuzzy, societies need a definition of death to help doctors decide when life-saving measures should be terminated, to signal survivors that they must begin to grieve their loss and reorganize their lives, and to establish when donated organs can be removed.

Several decades ago, loss of heartbeat and respiration signified death. But these criteria are no longer adequate, since resuscitation techniques frequently permit vital signs to be restored. Today, **brain death,** irreversible cessation of all activity in the brain and the brain stem (which controls reflexes), is used in most industrialized nations.

But not all countries accept this standard. In Japan, for example, doctors rely on traditional criteria (absence of heartbeat and respiration) that fit with Japanese laypeople's views. This approach stands in the way of a national organ transplant program, since few organs can be salvaged from bodies without artificially maintaining vital signs. Buddhist, Confucian, and Shinto beliefs about death, which stress ancestor worship and time for the spirit to leave the corpse, may be partly responsible for Japan's discomfort with brain death and organ donation. Marring the body to harvest organs violates respect for the deceased. Today, Japanese law allows people who want to be organ donors to choose the standard of brain death, as long as their families do not object (Morioka, 2001). Otherwise, they are considered to be alive until their heart stops beating.

Furthermore, in a great many cases the brain death standard does not solve the dilemma of when to halt treatment. Consider Nicholas, who, though not brain dead, had entered a **persistent vegetative state,** in which the cerebral cortex no longer registered electrical activity but the brain stem remained active. Doctors were certain they could not restore consciousness or body movement. Because thousands of North Americans are in a persistent vegetative state, with health care costs totaling many millions of dollars annually, some experts believe that absence of activity in the cerebral cortex should be sufficient to declare a person dead. Others, however, point to a few cases in which patients who had been vegetative for months regained consciousness, although usually with very limited functioning (Jennett, 2002). In still other instances of illness, a fully conscious but suffering person refuses life-saving

■ Japanese Buddhist monks perform a fire ritual, in which they light hundreds of prayer sticks, symbolizing a fire within their hearts that drives out confusion while retaining wisdom. Buddhist, Confucian, and Shinto beliefs about death, which stress ancestor worship, may be partly responsible for Japanese people's discomfort with organ donation.

measures—an issue we will consider later when we take up the right to die.

Death with Dignity

Our brief look at the process of dying indicates that nature rarely delivers the idealized, easy end most people want, and medical science cannot guarantee it. Therefore, the greatest dignity in death is in the integrity of the life that precedes it—an integrity we can foster by the way we communicate with and care for the dying person.

First, we can assure the majority of dying people, who succumb gradually, that we will support them through their physical and psychological distress. We can do everything possible to provide the utmost in humane and compassionate care. And we can treat them with esteem and respect—for example, by taking interest in those aspects of their lives that they most value and by addressing their greatest concerns (Chochinov, 2002).

Second, we can be candid about death's certainty. Unless people are aware that they are dying and understand (so far as possible) the likely circumstances of their death, they cannot plan for end-of-life care and decision making and share the sentiments that bring closure to relationships they hold most dear. Because Sofie knew how and when her death would probably take place, she chose a time when she, Philip, and her children could express what their lives had meant to one another. Among those precious bedside exchanges was Sofie's memorable last wish that Philip remarry after her death so he would not live out his final years alone. Openness about impending death granted Sofie a final generative act, helped her let go of the person closest to her, and offered comfort as she faced death.

© AP/WIDE WORLD PHOTOS

■ We can best ensure death with dignity by offering the dying person care, affection, companionship, and maximum personal control over this final phase of life. This teenager was diagnosed with lymphoblastic leukemia, a highly aggressive cancer of the lymph system. She poses with Kona, one of the dolphins she played with in Hawaii, thanks to the Make-a-Wish Foundation, which makes dreams come true for children and adolescents with life-threatening illnesses.

Finally, doctors and nurses can help dying people learn enough about their condition to make reasoned choices about whether to fight on or say no to further treatment. An understanding of how the normal body works simplifies comprehension of how disease affects it—education that can begin as early as the childhood years.

In sum, although the conditions of illness often do not permit a graceful, serene death, we can ensure the most dignified exit possible by offering the dying person care, affection, and companionship; the truth about diagnosis; and the maximum personal control over this final phase of life. These are essential ingredients of a "good death," and we will revisit them throughout this chapter.

Understanding of and Attitudes Toward Death

A century ago, deaths most often occurred at home. People of all ages, including children, helped with care of the dying family member and were present at the moment of death. They saw their loved one buried on family property or in the local cemetery, where the grave could be visited regularly. Because infant and childhood mortality rates were high, all people were likely to know someone the same age as or younger than themselves who had died. And it was common for children to experience the death of a parent.

Compared with earlier generations, today's children and adolescents in developed nations are insulated from death. Residents of inner-city ghettos, where violence is part of everyday existence, are exceptions. But overall, more young people reach adulthood without having experienced the death of someone they know well. When they do, professionals in hospitals and funeral homes take care of most tasks that involve confronting death directly (Wass, 1995).

This distance from death undoubtedly contributes to a sense of uneasiness about it. Despite frequent images of death in television shows, movies, and news reports of accidents, murders, wars, and natural disasters, we live in a death-denying culture. Adults are often reluctant to talk about death with children and adolescents. And a variety of substitute expressions, such as "passing away," "going out," and "departing," permit us to avoid acknowledging it candidly. In the following sections, we examine the development of conceptions of and attitudes toward death, along with ways to foster increased understanding and acceptance.

Childhood

Five-year-old Miriam arrived at our university laboratory preschool the day after her dog Pepper died. Leslie, her teacher, noticed that instead of joining the other children as she usually did, Miriam stood by herself, looking anxious and unhappy. "What's wrong, Miriam?" Leslie asked.

"Daddy said Pepper had a sick tummy. He fell asleep and died." For a moment, Miriam looked hopeful. "When I get home, Pepper might be up."

Leslie answered directly, "No, Pepper won't get up again. He's not asleep. He's dead, and that means he can't sleep, eat, run, or play anymore."

Miriam wandered off. Later, she returned and confessed to Leslie, "I chased Pepper too hard," tears streaming from her eyes.

Leslie put her arm around Miriam. "Pepper didn't die because you chased him. He was very old and very sick," she explained.

Over the next few days, Miriam asked many questions: "When I go to sleep, will I die?" "Can a tummy ache make you die?" "Will Mommy and Daddy die?" "Can Pepper see me?" "Does Pepper feel better now?" (Corr, 1997).

● **Development of the Death Concept.** A realistic understanding of death is based on three ideas:

1. **Permanence:** Once a living thing dies, it cannot be brought back to life.
2. **Universality:** All living things eventually die.
3. **Nonfunctionality:** All living functions, including thought, feeling, movement, and body processes, cease at death.

Without clear explanations, preschoolers rely on egocentric and magical thinking to make sense of death. They may believe, as Miriam did, that they are responsible for a relative's or pet's death. And they can easily arrive at incorrect conclusions—in Miriam's case, that sleeping or a stomachache can cause someone to die.

Preschoolers grasp the three components of the death concept in the order just given, and most master them between ages 7 and 10 (Kenyon, 2001). *Permanence* is the first and most easily understood idea. When Leslie explained that Pepper would not get up again, Miriam accepted this fact quickly, perhaps because she had seen it in other, less emotionally charged situations, such as the dead butterflies and beetles that she picked up and inspected while playing outside (Furman, 1990). Appreciation of *universality* comes slightly later. At first, children think that certain people do not die, especially those with whom they have close emotional ties or who are like themselves—other children. Finally, *nonfunctionality* is the most difficult component for children to grasp. Many preschoolers view dead things as retaining living capacities. When they first comprehend nonfunctionality, they do so in terms of its most visible aspects, such as heartbeat and breathing. Only later do they understand that thinking, feeling, and dreaming cease as well (Lazar & Torney-Purta, 1991; Speece & Brent, 1996).

● **Individual and Cultural Variations.** Although children usually have an accurate conception of death by middle childhood, there are wide individual differences. Experiences with death make a difference (Speece & Brent, 1996). For example, terminally ill children under age 6 often have a well-developed concept of death. If parents and health professionals have not been forthright with them, children find out that they are deathly ill in other ways—through nonverbal communication, eavesdropping, and talking with other child patients (O'Halloran & Altmaier, 1996). Children growing up on Israeli kibbutzim (agricultural settlements) who have experienced terrorist attacks, family members' departure on army tours to high-tension areas, and parental anxiety about safety express an accurate understanding of death by age 6 (Mahon, Goldberg, & Washington, 1999).

Ethnic variations suggest that religious teachings also affect children's understanding. A comparison of four ethnic groups in Israel revealed that Druze and Moslem children's death concepts lagged behind those of Christian and Jewish children

© AP/WIDE WORLD PHOTOS

■ Experiences with death often prompt a more accurate conception of death. These Palestinian children pray over the flag-draped body of an elderly member of their community during his funeral in Gaza.

(Florian & Kravetz, 1985). The Druze emphasis on reincarnation and the greater religiosity of both the Druze and Muslim groups may underlie these findings. Religious teachings seem to have an especially strong impact on children's grasp of the permanence of death. For example, children of Southern Baptist families, who believe in an afterlife, are less likely to endorse permanence than are children of Unitarian families, who focus on the here and now—peace and justice in today's world (Candy-Gibbs, Sharp, & Petrun, 1985).

● **Enhancing Children's Understanding.** Parents often worry that discussing death candidly with children will fuel their fears, but this is not so. Instead, children with a good grasp of the facts of death have an easier time accepting it (Essa & Murray, 1994). Direct explanations, like Leslie's, that fit the child's cognitive maturity, work best.

When adults use clichés or make misleading statements, children may take these literally and react with confusion. For example, after a parent said to her 5-year-old daughter, "Grandpa went on a long trip," the child wondered, "Why didn't he take me?" In another instance, a father whose wife died of cancer told his 9-year-old son that his mother "was sick," explaining nothing else about the illness. The father was surprised when, 10 months later, the boy caught the flu and became terribly afraid of dying (Wolfelt, 1997). Sometimes children ask very difficult questions, such as "Will I die? Will you die?" Parents can be truthful as well as comforting by taking advantage of children's sense of time. They can say something like "Not for many, many years. First I'm going to enjoy you as a grown-up and be a grandparent."

Discussions with children should also be culturally sensitive. Rather than presenting scientific evidence as counteracting religious beliefs, parents and teachers can assist children in blending the two sources of knowledge. As children get older, they often combine an accurate conception of death with spiritual and philosophical views, which offer solace during times of bereavement (Cuddy-Casey & Orvaschel, 1997). As we will see later, open, honest discussions with children not only contribute to a realistic understanding of death but also facilitate grieving after a child has experienced a loss.

■ This adolescent knows that death happens to everyone and can occur at any time, but his risk taking suggests otherwise. Teenagers' difficulty integrating logic with reality extends to their understanding of death.

Adolescence

Adolescents can easily voice the permanence, universality, and nonfunctionality of death, but their understanding is largely limited to the realm of possibility. Recall that teenagers have difficulty integrating logical insights with the realities of everyday life (Corr, 1995). Consequently, their understanding of death is not yet fully mature.

● **The Gap Between Logic and Reality.** Teenagers' difficulty applying what they know about death is evident in both their reasoning and their behavior. Although they can explain the permanence and nonfunctionality of death, they are attracted to alternatives. Consequently, the dividing line between life and death actually becomes less sharp in adolescence than it was in childhood. For example, teenagers often describe death as an enduring abstract state, such as "darkness," "eternal light," "transition," or "nothingness" (Brent et al., 1996). They also formulate personal theories about life after death. Besides images of heaven and hell influenced by their religious background, they speculate about reincarnation, transmigration of souls, and spiritual survival on earth or at another level (Noppe & Noppe, 1997).

Although mortality in adolescence is low compared with that in infancy and adulthood, teenage deaths are typically sudden and human-induced; unintentional injuries, homicide, and suicide are leading causes. Adolescents are clearly aware that death happens to everyone and can occur at any time. But this does not make them more safety-conscious. Rather, their high-risk activities suggest that they do not take death personally. When asked to comment on death, teenagers often make statements like these: "Why dwell on it? There is nothing I can do about it." (Wass, 1991, p. 27). In one study, the lower adolescent boys' consciousness of personal mortality, the more they engaged in risky behavior (Word, 1996).

What explains teenagers' difficulty integrating logic with reality in the domain of death? First, adolescence is a period of rapid growth and onset of reproductive capacity—attainments that are the opposite of death! Second, recall the adolescent personal fable, which leads teenagers to be so wrapped up in their own uniqueness that they can conclude they are beyond the reach of death. Finally, as teenagers construct a personal identity and experience their first freely chosen love relationships, they may be strongly attracted to romantic notions of death, which challenge logic (Noppe & Noppe, 1991, 1996). Not until early adulthood are young people capable of the relativistic thinking needed to reconcile these conflicting ideas (see Chapter 13, page 432).

● **Enhancing Adolescents' Understanding.** By encouraging adolescents to discuss concerns about death, adults can help them build a bridge between death as a logical concept and their personal experiences. In Chapter 12, we noted that teenagers with authoritative parents are more likely to turn to adults for guidance on important issues. In one study of 12- to 15-year-olds, most wanted to talk with parents rather than peers about the "meaning of life and death" and "what happens when you die." But 60 percent thought that their parents would not be genuinely interested, and the majority of parents felt inadequately prepared for the task (McNeil, 1986).

Taking up adolescents' thoughts and feelings about death can be part of everyday conversation, sparked by a news report or the death of an acquaintance. Parents can capitalize on these moments to express their own views, listen closely, accept teenagers' feelings, and correct misconceptions. Such mutual sharing deepens bonds of love and provides the basis for further exploration when the need arises. The Caregiving Concerns table below suggests ways to discuss concerns about death with children and adolescents.

Adulthood

In early adulthood, many people brush aside thoughts of death (Gresser, Wong, & Reker, 1987). This avoidance may be prompted by death anxiety, which we will consider in the next section. Alternatively, it may be due to relative disinterest in death-related issues, given that young adults typically do not know very many people who have died and (like adolescents) think of their own death as a long way off.

In Chapters 15 and 16, we described midlife as a time of stock taking in which people begin to view the lifespan in terms of time left to live, focusing on tasks to be completed. Middle-aged people no longer have a vague conception of their own death. They know that in the not-too-distant future, it will be their turn to grow old and die.

In late adulthood, adults think and talk more about death because it is much closer. Increasing evidence of mortality comes from physical changes, higher rates of illness and disability, and loss of relatives and friends (see Chapter 17). Compared with middle-aged people, older adults spend more time pondering the process and circumstances of dying than the state of death. Nearness to death seems to lead to a practical concern with how and when it might happen (de Vries, Bluck, & Birren, 1993; Kastenbaum, 2001).

Finally, although we have traced age-related changes, keep in mind that large individual differences exist. Some people

Discussing Concerns About Death with Children and Adolescents

SUGGESTION	DESCRIPTION
Take the lead.	Be alert to the child's or adolescent's nonverbal behaviors, bringing up the subject sympathetically, especially after a death-related situation has occurred.
Listen perceptively.	Grant full attention to the child or adolescent and the feelings underlying his or her words. When adults pretend to listen while thinking about other things, young people quickly pick up this sign of indifference and withdraw their confidence.
Acknowledge feelings.	Accept the child's or adolescent's emotions as real and important; avoid being judgmental. For example, paraphrase sentiments you detect, such as, "I see you're very puzzled about that. Let's talk more about it."
Provide factual information in a candid, culturally sensitive fashion.	For children who do not yet have a realistic understanding of the permanence, universality, and nonfunctionality of death, provide simple, direct, and accurate explanations. Avoid misleading statements, such as, "Grandpa went away on a long trip." Do not contradict the young person's religious beliefs. Rather, assist him or her in blending scientific with religious sources of knowledge.
Engage in joint problem solving.	When questions do not have easy answers, such as, "Where does your soul go when you die?", convey your belief in the young person's worth by indicating that you do not want to impose a point of view but rather to help him or her come to personally satisfying conclusions. To questions you cannot answer, say, "I don't know." Such honesty shows a willingness to generate and evaluate solutions jointly.

focus on life and death issues early on, whereas others are less reflective, moving into advanced old age without giving these matters much attention.

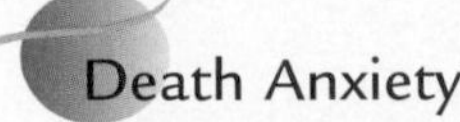

Death Anxiety

As you read the following statements, do you find yourself agreeing, disagreeing, or reacting neutrally?

"Never feeling anything again after I die upsets me."

"I hate the idea that I will be helpless after I die."

"The total isolation of death is frightening to me."

"The feeling that I will be missing out on so much after I die disturbs me." (Thorson & Powell, 1994, pp. 38–39)

Items like these appear on questionnaires used to measure **death anxiety**—fear and apprehension of death. Even people who are very accepting of the reality of death may be afraid of it (Firestone, 1994).

What predicts whether thoughts of our own demise trigger intense distress, relative calm, or something in between? To answer this question, researchers measure both general death anxiety and a variety of specific factors—fear of no longer existing, loss of control, a painful death, the body decaying, being separated from loved ones, the unknown, and others (Neimeyer, 1994). Findings reveal large individual and cultural variations in aspects of death that arouse fear. For example, in a study of devout Islamic Saudi Arabians, certain factors that appear repeatedly in the responses of Westerners, such as fear of the body decaying and of the unknown, were entirely absent (Long, 1985).

■ Death anxiety declines in old age, and this 81-year-old of the Netherlands seems to have very little! She shows off her custom-made bookshelf, which will eventually become her coffin. She explained that she had the piece of furniture made because "it's a waste to use a coffin just for burial." The pillow on the top shelf will support her head after she dies.

© MICHEL PORRO/GETTY IMAGES

Among Westerners, spirituality—a sense of life's meaning—seems to be more important than religious commitment in limiting death anxiety (Rasmussen & Johnson, 1994). Having a well-developed personal philosophy of death also reduces fear. In one study, both devout Christians and devout atheists viewed death as less threatening than did people with ambivalent religious views (Moore, 1992). Death anxiety is especially low among adults with deep faith in some form of higher force or being—faith that may or may not be influenced by religion (Cicirelli, 1999, 2002; Rasmussen & Johnson, 1994).

From what you have learned about adult psychosocial development, how do you think death anxiety might change with age? If you predicted it would decline, reaching its lowest level in late adulthood, you are correct (see Figure 19.1) (Thorson & Powell, 2000; Tomer, Eliason, & Smith, 2000). This age-related drop has been found in many cultures and ethnic groups. Recall from Chapter 18 that the attainment of ego integrity and a more mature sense of spirituality reduce fear of death. Older adults have also had more time to develop *symbolic immortality*—the belief that one will continue to live on through one's children or through one's work or personal influence (see Chapter 16, page 513). In a study of Israeli adults, symbolic immortality predicted reduced fear of death, especially among those with secure attachments (Florian & Mikulincer, 1998). Gratifying, close interpersonal ties seem to help people develop a sense of symbolic immortality. And people who view death as an opportunity to pass a legacy to future generations are less likely to fear it (Cicirelli, 2001).

Regardless of age, women appear more anxious about death than men do (refer again to Figure 19.1). This difference appears in both Eastern and Western cultures (Cicerelli, 1998; Tomer, Eliason, & Smith, 2000). Perhaps women are more likely to admit and men more likely to avoid troubled feelings about mortality—an explanation consistent with females' greater emotional expressiveness throughout the lifespan.

Experiencing some anxiety about death is normal. But like other fears, when it is very intense, it can undermine effective adjustment. Although physical health in adulthood is not related to death anxiety, mental health clearly is. People who are depressed or generally anxious are likely to have more severe death concerns (Neimeyer & Van Brunt, 1995). A large gap between their actual and ideal self-concepts leaves these individuals with a sense of incompleteness when they contemplate death. In contrast, self-confident individuals with a clear sense of purpose and control over their lives express less fear of death (Cicerelli, 1999).

Death anxiety is largely limited to adolescence and adulthood. Children rarely display it unless they live in dangerous neighborhoods or war-torn areas where they are in constant danger (see Chapter 10, page 335). Terminally ill children are

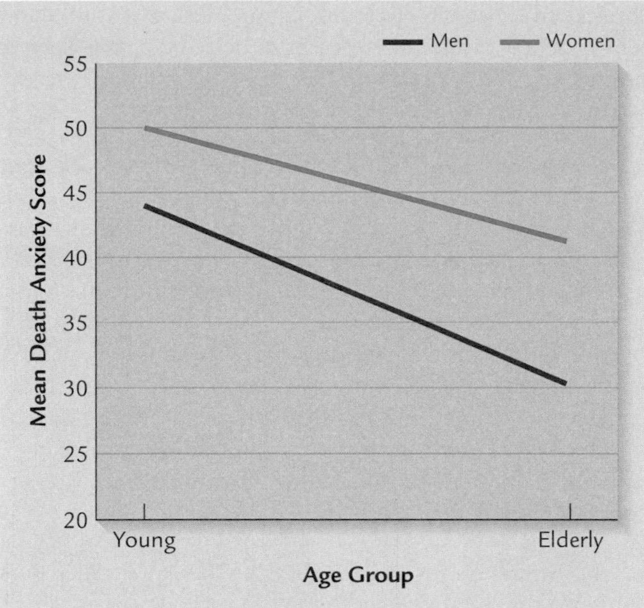

FIGURE 19.1 Relationship of age and gender to death anxiety. In this study comparing young and elderly adults, death anxiety declined with age. At both ages, women expressed greater fear of death than men. Many other studies show similar findings. (Adapted from Tomer, Eliason, & Smith, 2000.)

also at risk for high death anxiety. Compared with other same-age patients, children with cancer express more destructive thoughts and negative feelings about death (Malone, 1982). For those whose parents make the mistake of not telling them they are going to die, loneliness and death anxiety can be extreme (O'Halloran & Altmaier, 1996).

Ask Yourself

REVIEW

Explain why older adults think and talk more about death than do younger people but nevertheless feel less anxious about it.

APPLY

When 4-year-old Chloe's aunt died, Chloe asked, "Where's Aunt Susie?" Her mother explained, "Aunt Susie is taking a long, peaceful sleep." For the next 2 weeks, Chloe refused to go to bed, and, when finally coaxed into her room, she lay awake for hours. Explain the likely reason for Chloe's behavior, and suggest a better way of answering Chloe's question.

CONNECT

How do advances in cognition contribute to adolescents' concepts of death? (Refer to Chapter 11, pages 363–368.)

www.

Thinking and Emotions of Dying People

In the year before her death, Sofie did everything possible to surmount her illness. In between treatments to control the cancer, she tested her strength. She continued to teach high school, traveled to visit her children, cultivated a garden, and took weekend excursions with Philip. Hope pervaded Sofie's approach to her deadly condition, and she spoke often about the disease—so much so that her friends wondered how she could confront it so directly.

As Sofie deteriorated physically, she moved in and out of a range of mental and emotional states. She was frustrated, and at times angry and depressed, about her inability to keep on fighting. I recall her lamenting anxiously on a day when she was in pain, "I'm sick, so very sick! I'm trying so hard, but I can't keep on." Once she asked when my husband and I, who were newly married, would have children. "If only I could live long enough to hold them in my arms!" she cried. In the last week, she appeared tired but free of struggle. Occasionally, she spoke of her love for us and commented on the beauty of the hills outside her window. But mostly, she looked and listened rather than actively participated in conversation. One afternoon, she fell permanently unconscious.

Do Stages of Dying Exist?

As dying people move closer to death, are their reactions predictable? Do they go through a series of changes that are the same for everyone, or are their thoughts and feelings unique?

What is this dying man—a husband and father—thinking about during his final weeks? Dying people move in and out of a range of mental and emotional states affected by many personal and situational factors.

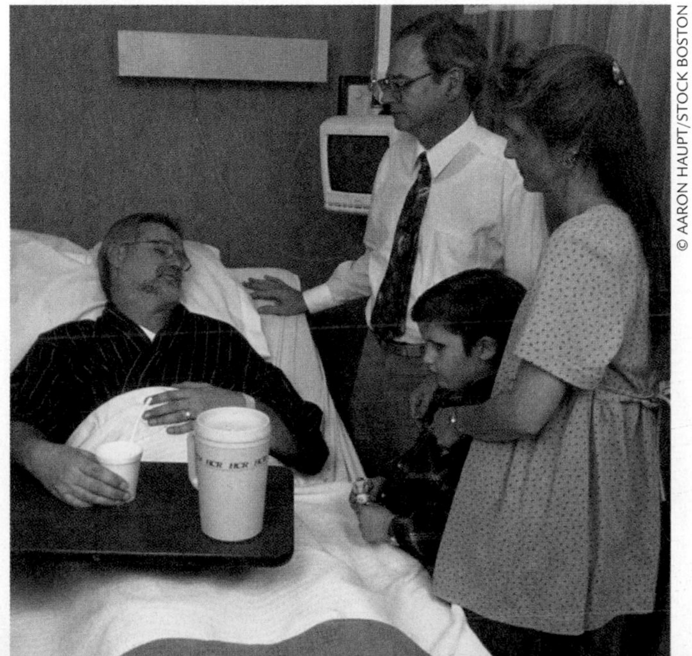

• Kübler-Ross's Theory. Although her theory has been heavily criticized, Elisabeth Kübler-Ross (1969) is credited with awakening society's sensitivity to the psychological needs of dying patients. From interviews with over 200 terminally ill people, she devised a theory of five typical responses—initially proposed as stages—to the prospect of death and the ordeal of dying. According to Kübler-Ross, when family members and health professionals understand these responses, they are in a better position to provide compassionate support.

Denial. On learning of the terminal illness, the person denies its seriousness to escape from the prospect of death. While the patient still feels reasonably well, denial is self-protective. It allows the individual to invest in rewarding activities and to deal with the illness at his or her own pace. In reality, most people move in and out of denial, making great plans one day and acknowledging that death is near the next (Smith, 1993b).

Kübler-Ross recommends accepting the dying patient's denial as a coping strategy he or she may need initially. But family members and health professionals should not act in ways that prolong denial by distorting the truth about the person's condition. In doing so, they prevent the dying person from adjusting to impending death and hinder necessary arrangements—for social support, for bringing closure to relationships, for what should and should not be done medically once the person is near death, and for giving away possessions.

Anger. Recognition that time is short promotes anger at having to die without being given a chance to do all one wants to do. Family members and health professionals are often targets of the patient's rage, resentment, and envy of those who will go on living. Still, they must tolerate rather than lash out at the patient's behavior, recognizing that the underlying cause is the unfairness of death.

Bargaining. Recognition inevitability of death, the terminally ill person attempts to forestall it by bargaining for extra time—a deal he or she may try to strike with family members, friends, doctors, nurses, or God. Listening sympathetically is the best response to these efforts to sustain hope, as one doctor did to the pleas of a young AIDS-stricken father to live long enough to dance with his 8-year-old daughter at her wedding (Selwyn, 1996). At times, bargains are altruistic acts. Witness the following request that Tony, a 15-year-old leukemia patient, expressed to his mother:

> I don't want to die yet. Gerry [youngest brother] is only 3 and not old enough to understand. If I could live just one more year, I could explain it to him myself and he will understand. Three is just too young. (Komp, 1996, pp. 69–70)

Although many dying patients' bargains are unrealistic and impossible to fulfill, Tony lived for exactly one year—a gift to those who survived him.

Depression. When denial, anger, and bargaining fail to postpone the course of illness, the person becomes depressed about the loss of his or her life. Kübler-Ross regards depression as necessary preparation for the last stage, acceptance. Unfortunately, many experiences associated with dying, including physical and mental deterioration, pain, lack of control, and being hooked to machines, intensify despondency (Maier & Newman, 1995). Health care that responds humanely to the patient's wishes can limit hopelessness and despair.

Acceptance. Most people who reach acceptance, a state of peace and quiet about upcoming death, do so only in the last weeks or days. The weakened patient yields to death, with its release from pain and anxiety. Disengagement from all but a few family members, friends, and caregivers occurs. Some dying people, in an attempt to pull away from all they have loved, withdraw into themselves for long periods of time. "I'm getting my mental and emotional house in order," one patient explained (Samarel, 1995, p. 101). As in the other stages, people who have reached acceptance usually maintain some hope of living, if only a flicker. When hope is entirely gone, Kübler-Ross observed, death follows quickly.

• Evaluation of Kübler-Ross's Theory. Kübler-Ross cautioned that her five stages should not be viewed as a fixed sequence and that not all people display each response—warnings that might have been better heeded had she not called them stages. Too often her theory has been interpreted simplistically, as the series of steps a "normal" dying person follows. Some health professionals, unaware of diversity in dying experiences, have insensitively tried to push patients through Kübler-Ross's sequence. And caregivers, through callousness or ignorance, can too easily dismiss a dying patient's legitimate complaints about treatment as "just what you would expect in Stage 2" (Corr, 1993; Kastenbaum, 2001).

Research confirms that, in line with Kübler-Ross's observations, dying people are more likely to display denial after learning of their condition and acceptance shortly before death (Kalish, 1985). But rather than stages, the five reactions Kübler-Ross observed are best viewed as coping strategies that anyone may call on in the face of threat. Furthermore, her list is much too limited. Dying people react in many additional ways—for example, through efforts to conquer the illness, as Sofie displayed; through an overwhelming need to control what happens to their bodies during the dying process; and through acts of generosity and caring, as Tony's concern for his 3-year-old brother, Gerry, illustrates.

As these examples suggest, the most serious drawback to Kübler-Ross's theory is that in it, dying patients' thoughts and feelings are removed from the contexts that grant them meaning. As we will see next, people's adaptations to impending death can be understood only in relation to the multidimensional influences that have contributed to their life course and that also shape this final phase.

Contextual Influences on Adaptations to Dying

From the moment of her diagnosis, Sofie spent little time denying the deadliness of her disease. Instead, she met it head on, just as she had dealt with other challenges of life. Her impassioned plea to hold her grandchildren in her arms was less a bargain with fate than an expression of profound defeat that on the threshold of late adulthood, she would not live to enjoy its rewards. At the end, her quiet, withdrawn demeanor was probably resignation, not acceptance. She had been a person with a fighting spirit all her life, unwilling to give in to challenge.

According to recent theorists, a single strategy, such as acceptance, is not best for every dying patient. Instead, an **appropriate death** is one that makes sense in terms of the individual's pattern of living and values and, at the same time, preserves or restores significant relationships and is as free of suffering as possible (Worden, 2000; Samarel, 1995; Weisman, 1984). When asked about a good death, most patients are clear about what, ideally, they would like to happen and mention the following goals:

- Maintaining a sense of identity, or inner continuity with one's past
- Clarifying the meaning of one's life and death
- Maintaining and enhancing relationships
- Achieving a sense of control over the time that remains
- Confronting and preparing for death (Steinhauser et al., 2000)

Research reveals that biological, psychological, and social and cultural forces affect people's coping with dying and, therefore, the extent to which they attain these goals. Let's look at some important influences on how people fare.

● Nature of the Disease. The course of the illness and its symptoms affect the dying person's reactions. For example, the extended nature of Sofie's illness and her doctor's initial optimism about achieving a remission undoubtedly contributed to her efforts to try to conquer the disease. During the final month, when cancer had spread to Sofie's lungs and she could not catch her breath, she was agitated and fearful until oxygen and medication relieved her uncertainty about being able to breathe. In contrast, Nicholas's weakened heart and failing kidneys so depleted his strength that he responded only with passivity.

Because of the toll of the disease, about one-fourth of cancer patients experience severe depression—reactions distinct from the sadness, grief, and worry that typically accompany the dying process. Profound depression amplifies pain and impairs the patient's capacity for pleasure, meaning, and connection. Consequently, it requires immediate treatment—through therapy, antidepressant medication, and patient and family education (Block, 2000).

● Personality and Coping Style. Understanding the way individuals view stressful life events and have coped with them in the past helps us appreciate the way they manage the dying process. In one study, terminally ill patients were asked to describe their image of dying. Each regarded it differently—for example, as a responsibility, an insurmountable obstacle, a punishment, or an act of courage. These meanings helped explain their responses to their worsening illness (Paige, 1980, as reported by Samarel, 1995). Poorly adjusted individuals—those with conflict-ridden relationships and many disappointments in life—are usually more distressed (Kastenbaum, 2001).

● Family Members' and Health Professionals' Behavior. Earlier we noted that a candid approach, in which everyone close to and caring for the dying person acknowledges the terminal illness, is best. Yet this also introduces the burden of participating in the work of dying with the patient—bringing relationships to closure, reflecting on life, and dealing with fears and regrets.

Because some people find it hard to engage in these tasks, they pretend that the disease is not as bad as it is. In patients inclined toward denial, a game of mutual pretending can be set in motion, in which all participants know the patient is dying but act as though it were not so. The game softens psychological pain for the moment but makes dying much more difficult (Samarel, 1995). At other times, the patient comes to suspect what he or she has not been told. In one instance, a terminally ill child flew into a rage because his doctor and a nurse spoke to him in ways that denied the fact that he would not grow up. Trying to get the child to cooperate with a medical procedure, the doctor said,

> "I thought you would understand, Sandy. You told me once you wanted to be a doctor."
>
> He screamed back, "I'm not going to be anything!" and threw an empty syringe at her.
>
> The nurse standing nearby asked, "What are you going to be?"
>
> "A ghost," said Sandy, and turned away from them. (Bluebond-Langner, 1977, p. 59)

The behavior of health professionals impeded Sandy's efforts to form a realistic time perspective and intensified his anger at the injustice of his premature death.

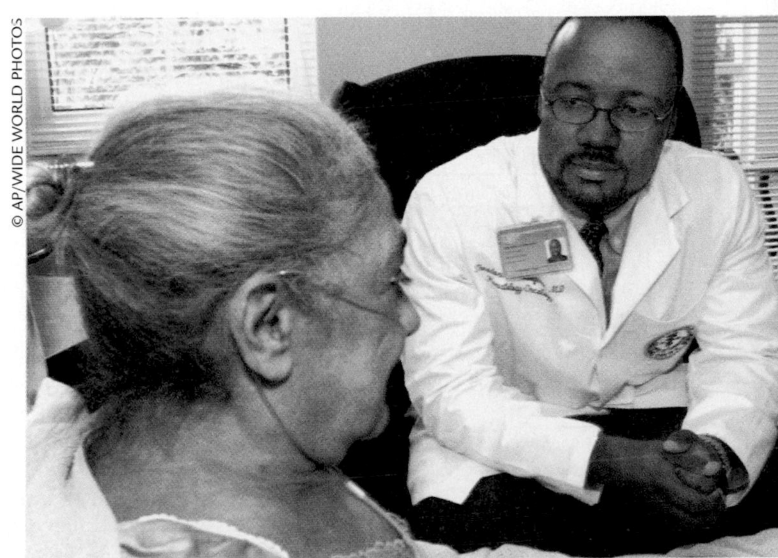

■ When doctors communicate openly and sensitively with terminally ill patients, they help them take steps to prepare for death by bringing relationships to closure, reflecting on life, and dealing with fears and regrets.

Even when doctors want to inform patients of their prognosis, they may encounter resistance, especially within certain ethnic groups. Withholding information is common in Southern and Eastern Europe, Central and South America, much of Asia, and the Middle East. In recent surveys, 52 percent of Mexican Americans and 65 percent of Korean Americans stated that patients should not be told the truth about a terminal illness, believing that doing so hastens death (Blackhall et al., 2001; Blackhall et al., 1995). In these instances, providing information is complex. When a family insists that a patient not be told, the doctor can make an offer of information to the patient and, if the patient refuses, ask who should receive information and make health care decisions. The patient's preference can be honored and reassessed at regular intervals (Zane & Yeh, 2002).

Care of the terminally ill is demanding and stressful. In one study, nurses trained to respond to the psychological needs of dying patients and their families consistently provided empathetic and supportive care. Keys to their success were staff meetings aimed at strengthening interpersonal skills; day-to-day mutual support among staff; and development of a personal philosophy of living and dying, which ensured that each nurse did not feel especially threatened by his or her own death (Samarel, 1991). Research indicates that highly death-anxious health professionals are more likely to delay informing terminal patients of their diagnoses and to prolong life-saving treatment when death is near (Eggerman & Dustin, 1985; Schulz & Aderman, 1979).

Social support from family members also affects adaptation to dying. Dying patients who feel they have much unfinished business to attend to are more anxious about impending death. But family contact reduces their sense of urgency to prolong life, perhaps because it permits patients to work through at least some incomplete tasks (Mutran et al., 1997).

Effective communication with the dying person is honest, fostering a trusting relationship. At the same time, it is oriented toward maintaining hope. Dying patients often move through a hope trajectory—at first, hope for a cure; later, hope for prolonging life; and finally, hope for a peaceful death with as few burdens as possible (Fanslow, 1981). Once patients near death stop expressing hope, those close to them must accept. Family members who find letting go very difficult may benefit from expert, sensitive guidance. The Caregiving Concerns table below offers suggestions for communicating with the dying.

● **Spirituality, Religion, and Culture.** Earlier we noted that a strong sense of spirituality reduces fear of death. Informal reports from health professionals suggest that this is as true for dying patients as for people in general. One experienced nurse commented,

> At the end, those [patients] with a faith—it doesn't really matter in what, but a faith in something—find it easier. Not always, but as a rule. I've seen people with faith panic and I've seen those without faith accept it [death]. But, as a rule, it's much easier for those with faith. (Samarel, 1991, pp. 64–65)

Vastly different cultural attitudes, guided by religious beliefs, also shape people's dying experiences. Buddhism, widely practiced in China, India, and Southeast Asia, fosters acceptance of death. By reading sutras (teachings of Buddha) to the dying person to calm the mind and emphasizing that dying leads to rebirth, Buddhists believe that it is possible to reach Nirvana, a state beyond the world of suffering (Yeung, 1996). In many Native-American groups, death is met with stoic self-control, an approach taught at an early age through stories that emphasize a circular, rather than linear, relationship be-

Communicating with Dying People

SUGGESTION	DESCRIPTION
Be truthful about the diagnosis and course of the disease.	Be honest about what the future is likely to hold, thereby permitting the dying person to bring closure to his or her life by expressing sentiments and wishes and participating in decisions about treatment.
Listen perceptively and acknowledge feelings.	Be truly present, focusing full attention on what the dying person has to say and accepting the patient's feelings. Patients who sense another's presence and concern are more likely to relax physically and emotionally and express themselves.
Maintain realistic hope.	Assist the dying person in maintaining hope by encouraging him or her to focus on a realistic goal that might yet be achieved—for example, resolution of a troubled relationship or special moments with a loved one. Knowing the dying person's hope, family members and health professionals can often help fulfill it.
Assist in the final transition.	Assure the dying person that he or she is not alone, offering a sympathetic touch, a caring thought, or just a calm presence. Some patients who struggle may benefit from being granted permission to die—the message that giving up and letting go is all right.

Sources: Benoliel & Degner, 1995; Samarel, 1995.

tween life and death and the importance of making way for others (Cox, 2002). The Inuit of Canada encourage the dying to find comfort in current social relationships rather than in an afterlife, which they describe mainly to reduce worry about it (Mills & Slobodin, 1994). And for African Americans, a dying loved one signals a crisis that unites family members in caregiving. The terminally ill person remains an active and vital force within the family until he or she no longer can carry out this role—an attitude of respect that undoubtedly eases the dying process (Sullivan, 1995).

In sum, dying prompts a multitude of thoughts, emotions, and coping strategies. Which ones are selected and emphasized depends on a wide array of contextual influences and differs greatly across individuals. Does this remind you of vital aspects of the lifespan perspective—that development is multidimensional and multidirectional? These assumptions are just as relevant to this final phase as they are to each earlier period.

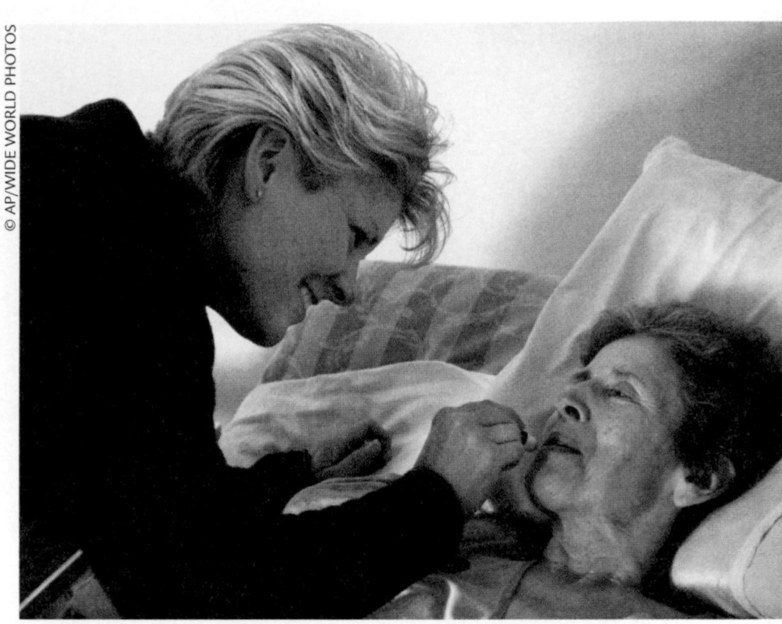

© AP/WIDE WORLD PHOTOS

■ Hospice helps patients and family members prepare for death in ways that are satisfying to them. An interdisciplinary team of professionals focuses on protecting quality of life rather than extending life. This daughter caringly applies lip balm to her dying mother's lips. Their facial expressions convey close, warm communication during these last few weeks of the mother's life.

A Place to Die

Unlike times past, when most deaths occurred at home, today about 70 percent in Canada and 80 percent in the United States take place in hospitals (Peres, 2002; Wilson, 2002). In the large, impersonal hospital environment, meeting the human needs of dying patients and their families is secondary, not because professionals lack concern, but because the work to be done focuses on saving lives. A dying patient represents a failure.

In the 1960s, a death awareness movement arose as a reaction to the death-avoiding practices of hospitals—complicated machinery hooked to patients with no chance of survival and lack of communication with dying patients. It led to medical care suited to the needs of dying people and to hospice programs, which have spread throughout the world. Let's visit each of these settings for dying.

Home

Had Sofie and Nicholas been asked where they wanted to die, undoubtedly each would have responded, "At home"—the preference of about 70 to 80 percent of North Americans (Peres, 2002; Ratner et al., 2001; Wilson, 2002). The reason is clear; the home offers an atmosphere of intimacy and loving care in which the terminally ill person is unlikely to feel abandoned or humiliated by physical decline or dependence on others.

However, only about one-fourth of Canadians and one-fifth of Americans experience home death (Mezey et al., 2002; Wilson, 2002). And it is important not to romanticize dying at home. Because of dramatic improvements in medicine, dying people tend to be sicker or much older than they used to be. Consequently, their bodies may be extremely frail, making ordinary activities—eating, sleeping, taking a pill, toileting, and bathing—major ordeals (Ellingson & Fuller, 2001). Health problems of elderly spouses; work and other responsibilities of family members; and the physical, psychological, and financial strain of providing home care can make it difficult to honor a terminally ill person's wish to die at home.

For many people, the chance to be with the dying person until the very end is a rewarding trade-off for the high demands of caregiving. But the advantages and disadvantages of home death should be carefully weighed before undertaking it. Adequate support for the caregiver is essential. A home health aide is often necessary—a service (as we will see shortly) that hospice programs have made more accessible. When family relationships are conflict-ridden, a dying patient introduces additional strains, negating the benefits of home death. Finally, even with professional help, most homes are poorly equipped to handle the medical and comfort-care needs of the dying. For all these reasons, family members continue to report more psychological stress 10 months after a home death than do family members whose loved one died elsewhere (Addington-Hall, 2000).

Hospital

Hospital dying takes many forms. Each pattern is affected by the physical state of the dying person, the hospital unit in which it takes place, and the goal and quality of care.

Sudden deaths, due to injury or critical illness, typically occur in emergency rooms. Doctors and nurses must evaluate the problem and take action quickly. Little time is available for

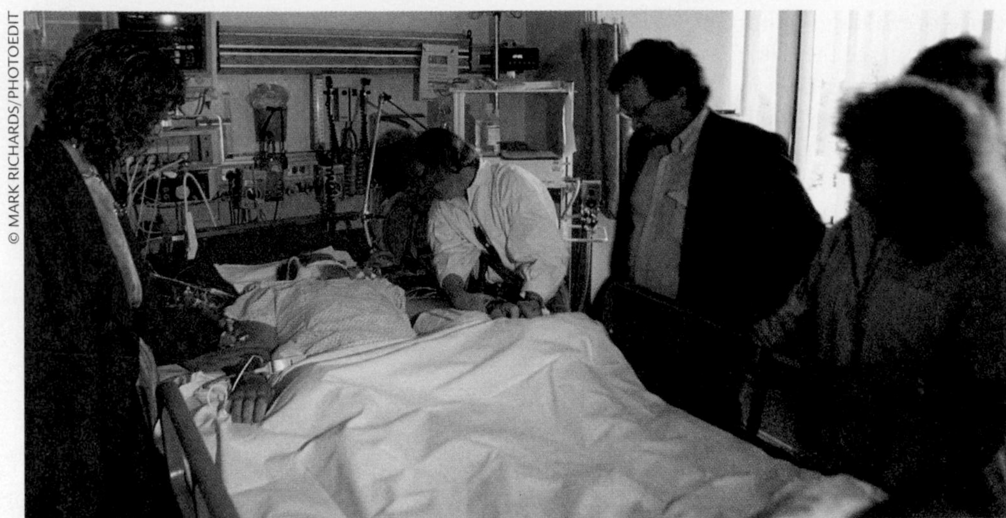
© MARK RICHARDS/PHOTOEDIT

■ In this hospital intensive care unit, family members visit a dying patient at a scheduled time. Monitoring the patient's condition and responding medically in case of an emergency supersedes privacy and communication. Dying in intensive care is a depersonalizing experience unique to technologically sophisticated societies. Some patients linger, hooked to machines, for months.

contact with family members. When staff break the news of death in a sympathetic manner and provide explanations, family members are grateful. Otherwise, feelings of anger, frustration, and confusion can add to their grief (Benoliel & Degner, 1995). Crisis intervention services are needed to help survivors cope with sudden death, and some hospitals are doing a better job of providing them.

Nicholas died on an intensive care ward organized to prevent death in patients whose condition can worsen quickly. Privacy and communication with the family were secondary to monitoring his condition. To prevent disruption of nurses' activities, Giselle and Sasha could be at Nicholas's side only at scheduled times. Dying in intensive care is an experience unique to technologically sophisticated societies. It is especially depersonalizing for patients like Nicholas, who linger between life and death while hooked to machines for months.

Cancer patients, who account for most cases of prolonged dying, typically die in general or specialized cancer care hospital units. When hospitalized for a long time, they reach out for help with physical and emotional needs, usually with mixed success. In these hospital settings, as in intensive care, a conflict of values is apparent (Hanson, Danis, & Garrett, 1997). The tasks associated with dying must be performed efficiently so that all patients can be served and health professionals are not drained emotionally by repeated attachments and separations.

A recent national report revealed that only 14 percent of American hospitals have comprehensive treatment programs aimed at easing physical, emotional, and spiritual suffering at the end of life. Although pain management programs are more common, existing in 42 percent of hospitals, the majority do not have them. And less than 1 percent of doctors and nurses are specially trained in managing pain in dying people. In sum, the report concluded, many people die in painful, frightening, and depersonalizing hospital conditions, without their wishes being met (Peres, 2002; Zwillich, 2002). The hospice approach aims to reduce these profound caregiving failures.

The Hospice Approach

In medieval times, a *hospice* was a place where travelers could find rest and shelter. In the nineteenth and twentieth centuries, the word referred to homes for dying patients. Today, **hospice** is not a place but a comprehensive program of support services for terminally ill people and their families. It aims to provide a caring community sensitive to the dying person's needs so patients and family members can prepare for death in ways that are satisfying to them. Quality of life is central to the hospice approach. Here are its main features:

- The patient and family as a unit of care
- Emphasis on meeting the patient's physical, emotional, social, and spiritual needs, including controlling pain, retaining dignity and self-worth, and feeling cared for and loved
- Care provided by an interdisciplinary team: the patient's doctor, a nurse or nurse's aide, a chaplain, a counselor or social worker, and a pharmacist
- The patient kept at home or in an inpatient setting with a homelike atmosphere where coordination of care is possible
- Focus on improving the quality of remaining life with **palliative,** or **comfort, care** that relieves pain and other symptoms (such as nausea, breathing difficulties, insomnia, and depression), aimed at protecting the patient's quality of life rather than prolonging life
- In addition to regularly scheduled home care visits, on-call services available 24 hours a day, 7 days a week
- Follow-up bereavement services offered to families in the year after a death

Because hospice care is a philosophy, not a facility, it can be applied in diverse ways. In Great Britain, care in a special inpatient unit, sometimes associated with a hospital, is typical. Home care has been emphasized in the United States, where nearly 80 percent of hospice patients die in their own residence. In Canada, half of hospice care is home-based and half is hospital-based (Gardner, 2001; National Hospice and Palliative Care Organization, 2002). But hospice programs everywhere have expanded to include a continuum of care, from home to inpatient options, including hospitals and nursing homes. Central to the hospice approach is that the dying person and his or her family be offered choices that guarantee an appropriate death. Some programs offer hospice day care, which permits caregivers to continue working or be relieved of the stresses of long-term care (Mahoney, 1998). Contact with others facing terminal illness is a supportive byproduct of many hospice arrangements.

Currently, the United States has over 3,000 hospices serving over 540,000 terminally ill patients annually, and Canada has 650 hospice and palliative care organizations, also reaching thousands of patients. About 50 percent of patients dying of cancer and over 30 percent dying of AIDS choose hospice. Programs also admit patients suffering from a variety of other terminal illnesses (Gardner, 2001; National Hospice and Palliative Care Organization, 2002). Because hospice care is a cost-effective alternative to expensive life-saving treatments, government health care benefits cover it in both the United States and Canada, making it affordable for most dying patients and their families. Hospices also serve dying children—a tragedy so devastating that social support and bereavement intervention are vital (Wolfe, 1997).

Research reveals that besides reducing patient physical suffering, hospice care contributes to improved family functioning. The majority of patients and families report improved coping, increased social support, and increased feelings of family closeness after receiving hospice services (Godkin et al., 1984). In one study, family members experiencing hospice participated more fully in the funeral (choosing music, delivering a eulogy), and also scored higher in psychological well-being 1 to 2 years following their loved one's death, than did nonhospice family members (Ragow-O'Brien, Hayslip, & Guarnaccia, 2000).

As a long-range goal, hospice organizations are striving for broader acceptance of their patient- and family-centered approach. At present, the majority of North Americans are unfamiliar with the philosophy, although when it is described to them, nearly 90 percent say it is the type of end-of-life care they want (Gardner, 2001; National Hospice and Palliative Care Organization, 2002). Recently, Canada launched a web-based hospice outreach service, called the Canadian Virtual Hospice, to support patients, families, and those providing care with information, resources, and connections to others with similar concerns. A special strength of the website is its accessibility to people in all geographic regions, including underserved rural and remote areas (Mars Hill Group, 2002). Finally, culturally sensitive approaches are needed to reach more ethnic minority patients, who participate in hospice at much lower rates than white patients (Crawley et al., 2000).

Ask Yourself

REVIEW

How does hospice help terminally ill patients attain an appropriate death?

REVIEW

Why is the stage notion an inaccurate account of terminally ill patients' mental and emotional reactions as they move closer to death?

APPLY

When 5-year-old Timmy's kidney failure was diagnosed as terminal, his parents could not accept the tragic news. Their hospital visits became shorter, and they evaded his anxious questions. Eventually, Timmy blamed himself. He died with little physical pain but alone, and his parents suffered prolonged guilt. Explain how hospice care could have helped Timmy and his family.

CONNECT

Reread the description of Sofie's mental and emotional reactions to dying on page 627. Then review the story of Sofie's life on page 4. How were Sofie's responses consistent with her personality and lifelong style of coping with adversity?

www.

The Right to Die

In 1976, the parents of Karen Ann Quinlan, a young woman who took drugs at a party and fell into an irreversible coma, sued to have her respirator turned off. The New Jersey Supreme Court, calling on Karen's right to privacy and the power of her parents as guardians, complied with this request. Although Karen was expected to die quickly, she breathed independently, continued to be fed intravenously, and lived another 10 years in a persistent vegetative state.

In 1983, 25-year-old Nancy Cruzan fell asleep at the wheel and was thrown from her car. After resuscitation, her heart and lungs worked but her brain was badly damaged. Like Karen, Nancy lay in a persistent vegetative state. Her parents wanted to end her meaningless existence by halting artificial feeding—a preference consistent with Nancy's statement to a friend a year earlier. But the Missouri courts refused, claiming that no written evidence of Nancy's wishes existed. The U.S. Supreme Court upheld this ruling, but it encouraged the state of Missouri to reconsider. Eventually, Nancy's doctor agreed that it was in her best interests to terminate treatment, and a county judge honored this request. She died a week later, nearly 8 years after losing consciousness.

Before the 1950s, medical science could do little to extend the lives of terminally ill patients, so the right to die was of

Table 19.1 Forms of Euthanasia

Form	Description
Voluntary passive euthanasia	At the patient's request, the doctor withholds or withdraws treatment, thereby permitting the patient to die naturally. For example, the doctor does not perform surgery or administer medication that could prolong life, or the doctor turns off the respirator of a patient who cannot breathe independently.
Voluntary active euthanasia	The doctor ends a suffering patient's life at the patient's request—for example, by administering a lethal dose of drugs.
Assisted suicide	The doctor helps a suffering patient take his or her own life. For example, the doctor enables the patient to swallow or inject a lethal dose of drugs.
Involuntary active euthanasia	The doctor ends a suffering patient's life without the patient's permission. For example, without obtaining the patient's consent, the doctor administers a lethal dose of drugs.

little concern. Today, medical advances mean that the same procedures that preserve life can prolong inevitable death, diminishing both quality of life and personal dignity.

The Quinlan and Cruzan cases brought right-to-die issues to the forefront of public attention, sparking over 40 U.S. states to pass legislation honoring patients' desires concerning withdrawal of treatment in cases of terminal illness and, in a few states, in cases of a persistent vegetative state. But in the United States, and in Canada as well, no uniform right-to-die policy exists, and controversy persists over how to handle the diverse circumstances in which patients and families make requests (Filene, 1998).

Euthanasia is the practice of ending the life of a person suffering from an incurable condition. Its various forms are summarized in Table 19.1. As we will see in the following sections, public acceptance of euthanasia is high, except when it involves ending the life of an anguished, terminally ill patient without his or her permission.

Passive Euthanasia

In **passive euthanasia,** life-sustaining treatment is withheld or withdrawn, permitting a patient to die naturally. Should Nancy Cruzan have been allowed to die sooner? Was it right for Nicholas's doctors to turn off his respirator at Giselle's request? Consider an Alzheimer's victim, whose disease has progressed to the point where he has lost all awareness and body functions. Should life support be withheld?

Recent polls reveal that the majority of people answer yes to these questions. When there is no hope of recovery, more than 80 percent of Americans and Canadians support the patient's or family members' right to end treatment (Singer et al., 1995; Woodman, 1998). In 1986, the American Medical Association endorsed withdrawing all forms of treatment from the terminally ill when death is imminent and from those in a permanent vegetative state. Consequently, passive euthanasia is widely practiced as part of ordinary medical procedure, in which doctors exercise professional judgment.

Still, a minority of citizens do not endorse passive euthanasia. Religious denomination has surprisingly little effect on people's opinions. For example, most Catholics hold favorable views, despite slow official church acceptance because of fears that passive euthanasia might be a first step toward government-approved mercy killing. However, ethnicity makes a difference: Nearly twice as many African Americans as Caucasian Americans desire all medical intervention possible, regardless of the patient's condition (Hopp & Duffy, 2000). Perhaps this reluctance to forgo treatment reflects strong cultural beliefs about respecting and preserving life.

Because of controversial court cases, some doctors and health care institutions are unwilling to end treatment without legal protection. In the absence of national consensus on passive euthanasia, people can best ensure that their wishes will be followed by preparing an **advance medical directive**—a written statement of desired medical treatment should they become incurably ill. The majority of U.S. states recognize two

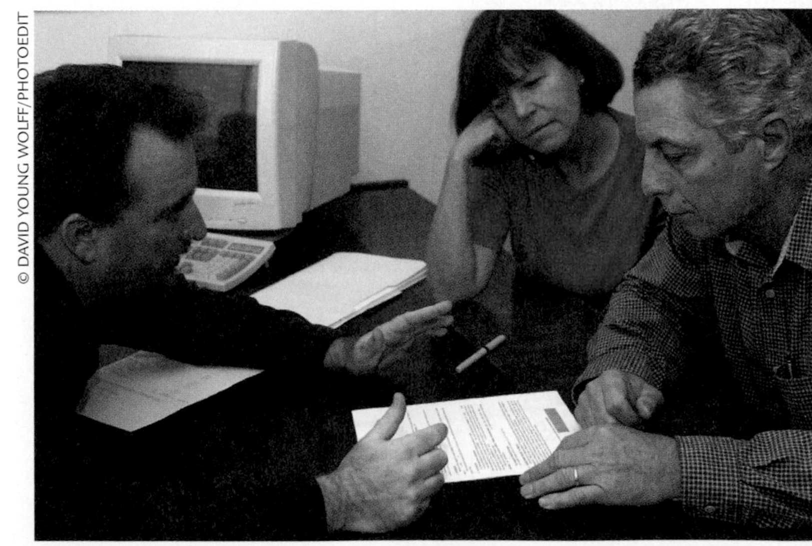

■ This couple talks about an advance medical directive with a hospital chaplain. Appointing, through a durable power of attorney, a trusted spokesperson to make health care decisions is the most flexible approach to ensuring that one's desires will be granted because it offers latitude for dealing with unexpected situations.

types of advance directives: a *living will* and *a durable power of attorney for health care.* Most Canadian provinces, and a few U.S. states, combine them into one document (Gunter-Hunt et al., 2002; Joint Centre for Bioethics, 2002).

In a **living will,** people specify the treatments they do or do not want in case of a terminal illness, coma, or other near-death situation (see Figure 19.2). For example, a person might state that without reasonable expectation of recovery, he or she should not be kept alive through medical intervention of any kind. In addition, living wills sometimes specify that pain-relieving medication be given, even though this may shorten life. In Sofie's case, her doctor administered a powerful narcotic to relieve labored breathing and quiet her fear of suffocation. The narcotic suppressed respiration, causing death to occur hours or days earlier than if the medication had not been prescribed, but without distress. Such palliative care is accepted as appropriate and ethical medical practice.

Although living wills help ensure personal control, they do not guarantee it. Recognition of living wills is usually limited to patients who are terminally ill or are otherwise expected to die shortly. Only a few U.S. states cover people in a persistent vegetative state due to injury or elders who linger with many chronic problems, including Alzheimer's disease—because these conditions are not classified as terminal. Furthermore, even when terminally ill patients have living wills, doctors often do not follow them for a variety of reasons, including fear of lawsuits, their own moral beliefs, and failure to inquire about patients' directives (Lawton, 2001a; Lens & Pollack, 2000).

LIVING WILL

THIS DECLARATION is made this _____ day of_______________, 20 _____.

I, _________________, being of sound mind, willfully and voluntarily make known my desires that my moment of death shall not be artificially postponed. If at any time I should have an incurable and irreversible injury, disease, or illness judged to be a terminal condition by my attending physician who has personally examined me and has determined that my death is imminent except for death delaying procedures. I direct that such procedures which would only prolong the dying process be withheld or withdrawn, and that I be permitted to die naturally with only the administration of medication, sustenance, or the performance of any medical procedure deemed necessary by my attending physician to provide me with comfort care.

In the absence of my ability to give directions regarding the use of such death delaying procedures, it is my intention that this declaration shall be honored by my family and physician as the final expression of my legal right to refuse medical or surgical treatment and accept the consequences from such refusal.

Signed: ____________________

City, County and State of Residence: ____________________

____

____

The declarant is personally known to me and I believe him or her to be of sound mind. I saw the declarant sign the declaration in my presence (or the declarant acknowledged in my presence that he or she had signed the declaration) and I signed the declaration as a witness in the presence of the declarant. At the date of this instrument, I am not entitled to any portion of the estate of the declarant according to the laws of intestate succession or, to the best of my knowledge and belief, under any will of declarant or other instrument taking effect at declarant's death, or directly financially responsible for declarant's medical care.

Witness: ____________________

Witness: ____________________

■ **FIGURE 19.2 Example of a living will. This document is legal in the State of Illinois.** Each person completing a living will should use a form specific to the U.S. state or Canadian province in which he or she resides because laws vary widely. (Courtesy of Office of the Attorney General, State of Illinois.)

Because living wills cannot anticipate all future medical conditions and can easily be ignored, a second form of advance directive has become common. The **durable power of attorney for health care** authorizes appointment of another person (usually, although not always, a family member) to make health care decisions on one's behalf. It generally requires only a short signed and witnessed statement like this.

> I hereby appoint [name] to act for me and in my name (in any way I could act in person) to make any and all decisions for me concerning my personal care, medical treatment, hospitalization, and health care and to require, withhold, or withdraw any type of medical treatment or procedure, even though my death may ensue. (Courtesy of Office of the Attorney General, State of Illinois.)

The durable power of attorney for health care is more flexible than the living will because it permits a trusted spokesperson to confer with the doctor as medical circumstances arise. Because authority to speak for the patient is not limited to terminal illnesses, more latitude exists for dealing with unexpected situations (Jasper, 1996). And in gay and lesbian and other close relationships not sanctioned by law, the durable power of attorney can ensure the partner's role in advocating for the patient's health care needs and in decision making.

Whether or not a person supports passive euthanasia, it is important to have a living will, durable power of attorney, or both, since most deaths occur in hospitals. Yet fewer than 20 percent of North

Americans have executed such documents, perhaps because of widespread uneasiness about bringing up the topic of death, especially with relatives (Humphrey & Clement, 1998). To encourage people to make decisions about potential treatment while they are able, U.S. federal law now requires all medical facilities receiving federal funds to provide information at admission about state laws and institutional policies on patients' rights and advance directives.

Health care professionals, unclear about a patient's intent and fearing liability, will probably decide to continue treatment regardless of cost and a person's prior oral statements. Perhaps for this reason, some U.S. states and Canadian provinces permit appointment of a *health care proxy,* or substitute decision maker, if a patient failed to provide an advance medical directive while competent (Jasper, 1996; Joint Centre for Bioethics, 2002). Proxies are an important means of covering children and adolescents, who cannot legally execute advance medical directives.

Voluntary Active Euthanasia

In recent years, the right-to-die debate has shifted from withdrawal of treatment for the hopelessly ill to more active alternatives. In **voluntary active euthanasia,** doctors or others act directly, at a patient's request, to end suffering before a natural end to life. The practice is a form of mercy killing and is a criminal offense in most countries, including Canada and almost all U.S. states (Scherer & Simon, 1999). But support for voluntary active euthanasia is growing. As Figure 19.3 shows, about 70 to 90 percent of people in Western nations approve of it (Caddell & Newton, 1995; Gallup Canada, 1997). When doctors engage in it, American and Canadian judges are usually lenient, granting suspended sentences or probation—a trend reflecting rising public interest in self-determination in death as in life.

Nevertheless, attempts to legalize voluntary active euthanasia have prompted heated controversy. Supporters believe that it represents the most compassionate option for terminally ill people in severe pain. Opponents stress the moral difference between "letting die" and "killing" and point out that at times, even very sick patients recover. They also argue that involving doctors in taking the lives of suffering patients may impair people's trust in health professionals. Finally, a fear exists that legalizing this practice—even when strictly monitored to make sure it does not arise out of depression, loneliness, coercion, or a desire to diminish the burden of illness on others—could lead to a broadening of euthanasia. Initially limited to the terminally ill, it might be applied involuntarily to the frail, demented, or disabled—outcomes that most people find unacceptable and immoral (Kerridge & Mitchell, 1996).

Will legalizing voluntary active euthanasia lead us down a "slippery slope" to the killing of vulnerable people who did not ask to die? The Social Issues box on the following page presents lessons from the Australian state of the Northern Territory, where a law allowing voluntary active euthanasia was passed in 1996, and from the Netherlands, where doctors have practiced it for years and it was recently legalized.

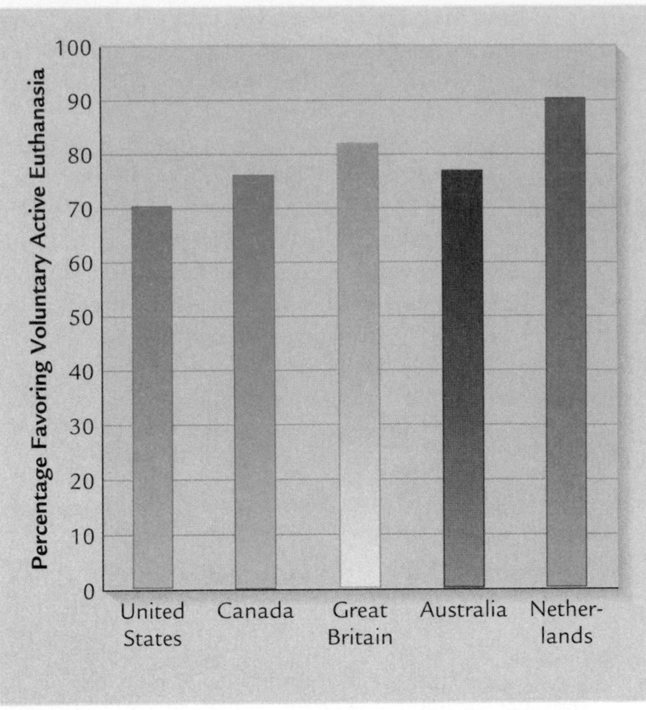

FIGURE 19.3 Public opinion favoring voluntary active euthanasia in five nations. A struggle exists between public opinion, which has increasingly favored voluntary active euthanasia over the past 30 years, and legal statutes, which prohibit it. The majority of people in Western nations believe that a hopelessly ill, suffering patient who asks for a lethal injection should be granted that request. Public support for voluntary active euthanasia is highest in the Netherlands—the only nation in the world where the practice is legal under certain conditions. (From Gallup Canada, 1997; van der Maas, Pijnenborg, & van Delden, 1995; Scherer & Simon, 1999.)

Assisted Suicide

After checking Diane's blood count, Dr. Timothy Quill gently broke the news: leukemia. If she were to have any hope of survival, a strenuous course of treatment with only a 25 percent success rate would have to begin immediately. Convinced that she would suffer unspeakably from side effects and lack of control over her body, Diane chose not to undergo chemotherapy and a bone marrow transplant.

Dr. Quill made sure that Diane understood her options. As he adjusted to her decision, she opened another issue: Diane wanted no part of a lingering death. She calmly insisted that when the time came, she desired to take her own life in the least painful way possible—a choice she had discussed with her husband and son, who respected it. Realizing that Diane could get the most out of the time she had left only if her fears of prolonged pain were allayed, Dr. Quill granted her request for sleeping pills, making sure she knew the amounts needed for both sleep and suicide.

Diane's next few months were busy and fulfilling. Her son took leave from college to be with her, and her husband

Social Issues

Voluntary Active Euthanasia: Lessons from Australia and the Netherlands

In 1996, Australia's Northern Territory passed legislation allowing a terminally ill patient of sound mind and suffering from pain or other distress to ask a doctor to end his or her life. Two other doctors had to agree that the patient could not be cured, and a psychiatrist had to confirm the absence of treatable depression.

In the months after its passage, four deaths occurred under the Northern Territory euthanasia statute, and it was heavily criticized. The Aborigines, valuing harmony and balance with nature, regarded it as culturally inappropriate. Their leaders claimed the law would discourage Aboriginal elders, many of whom have experienced a lifetime of persecution at the hands of European settlers, from seeking medical care (Fleming, 2000). Others considered the law to be a national issue because patients traveled from other states to make use of it.

In 1997, the Northern Territory legislation was overturned by the Australian Parliament, which claimed that assemblies do not have the right to legislate intentional killing. But the episode placed Australia at the center of the euthanasia debate, and the issue continues to spark high passions across the country. Opponents question the mental condition of patients seeking euthanasia and worry about error and abuse of the practice. Supporters emphasize compassion and mercy and the right of individuals to control the course of their own lives. June Burns, a 59-year-old Australian woman with bladder cancer who participated in TV ads that documented the course of her illness, responded, "If I were a dog, they would have put me down by now. I feel life is very precious and . . . I wish I could go on, but I can't and I'd like to die with dignity" (British Broadcasting Corporation, 1999).

For the past several decades, doctors in the Netherlands have engaged in voluntary active euthanasia without criminal prosecution. In 2001, the practice became legal under the following conditions: when physical or mental suffering is severe, with no prospect of relief; when no doubt exists about the patient's desire to die; when the patient's decision is voluntary, well-informed, and stable over time; when all other options for care have been exhausted or refused; and when another doctor has been consulted.

■ Australian Aborigines' valuing of harmony with nature is evident in the natural images that permeate their artwork. The Aborigines regarded the Northern Territory's legislation permitting voluntary active euthanasia as unnatural and culturally inappropriate. Their leaders also worried that the law would discourage Aboriginal elders from seeking medical care. It was eventually overturned by the Australian Parliament, which opposed legalizing intentional killing. The issue continues to spark heated debate across Australia.

Over 50 percent of Dutch doctors say they perform euthanasia, with cancer patients being the majority of their cases. Despite safeguards, both voluntary and involuntary euthanasia have occurred. Many doctors acknowledge occasionally suggesting euthanasia to patients. And in hundreds of cases annually, they admit actively causing death when a patient did not ask for it. Justifications included the impossibility of treating pain, a low quality of life, and terminal patients not dying after withdrawal of treatment (Hendin, 1999). Although doctors judged most of these patients incompetent to decide for themselves, an estimated 37 percent were competent. When asked the reason for not discussing the decision with competent patients, doctors usually said they had done so previously; they did not voice a need to check how the patient felt currently. Voluntary active euthanasia in the Netherlands has risen steadily over the past decade. Currently, about 3,200 people die this way annually—2.4 percent of all deaths (Hendin, 2002).

The Northern Territory and Dutch examples reveal that legalizing voluntary active euthanasia can spark both the fear and the reality of death without consent. Nevertheless, terminally ill individuals in severe pain continue to plead for such laws. Probably all would agree that when doctors feel compelled by respect for self-determination and relief of suffering to assist a patient in dying, they should be subject to the most stringent professional and legal oversight possible.

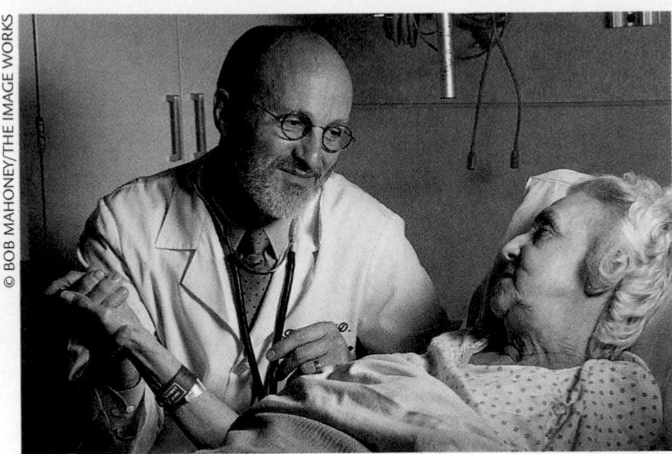

© BOB MAHONEY/THE IMAGE WORKS

■ In a prestigious medical journal, Dr. Timothy Quill explained how and why he assisted a terminally ill patient in taking her own life. With the majority of North Americans favoring it, assisted suicide may become a more widespread, unofficial practice in the United States and Canada.

worked at home as much as possible. Gradually, bone pain, fatigue, and fever set in. Saying good-bye to her family and friends, Diane asked to be alone for an hour, took a lethal dose of medication, and died at home (Quill, 1991).

Assisting a suicide is illegal in Canada and in many, but not all, U.S. states. In Western Europe, doctor-assisted suicide is legal in Belgium, Germany, the Netherlands, and Switzerland and is tacitly accepted in many other countries (Hill, 2003; Scherer & Simon, 1999). In North America, Oregon is unique in having passed a law, called the Death with Dignity Act, that explicitly allows physicians to prescribe drugs so terminally ill patients can end their lives. To get a prescription, patients must have two doctors agree that they have less than 6 months to live and must request the drugs several times. Consistent with its policy of leaving assisted suicide in the hands of the states, the U.S. Supreme Court refused to review a challenge to the Oregon statute but also upheld the right of other states to ban assisted suicide. In a 1997 ballot, Oregon residents voted to retain their assisted-suicide law.

As with voluntary active euthanasia, strong public support for doctor-assisted suicide exists. About 75 percent of Americans and 70 percent of Canadians approve of the practice (Angus Reid Poll, 1995; Scherer & Simon, 1999). A survey of nearly 1,000 terminally ill U.S. patients ranging in age from 22 to 109 revealed that although they largely endorsed assisted suicide, only one-tenth seriously considered it for themselves. In a follow-up, many of those had changed their minds (Emanuel, Fairclough, & Emanuel, 2000). People who say they would choose assisted suicide if they were terminally ill tend to be higher in SES and more often Caucasian than African American (Cicerelli, 1997).

The number of Oregon residents dying by assisted suicide has increased since passage of the law, from 16 in 1998 to 38 in 2002. Still, assisted suicide accounts for only one-tenth of 1 percent of Oregon deaths. Ten times as many terminally ill people initiate the qualification process for assisted suicide as actually engage in it. Hundreds of others have discussed it with their doctors, and thousands of Oregonians say they find comfort in knowing the option is available should they suffer while dying (Hedberg, Hopkins, & Kohn, 2003). Oregon hospice nurses and social workers who had cared for patients asking for assisted suicide rated desire to control the circumstances of death, desire to die at home, and belief that continuing to live is pointless as the most important reasons for these requests (Ganzini et al., 2002).

Public interest in assisted suicide was sparked by Dr. Jack Kevorkian's "suicide machines," which permitted terminally ill patients, after brief counseling, to self-administer lethal drugs and carbon monoxide. Dr. Kevorkian participated in more than 50 such deaths. Less publicity surrounded Dr. Quill's decision to assist Diane—a patient he knew well after serving for years as her personal doctor. After he told her story in a prestigious medical journal, reactions were mixed, as they currently are toward Oregon's suicide law. Some view doctors who help suffering people who want die as compassionate and respectful of patients' personal choices. Others oppose assisted suicide on religious and moral grounds or believe that the role of doctors should be limited to saving, not taking, lives.

Nevertheless, with the majority of citizens favoring it, assisted suicide may become a more widespread practice in North America. Yet grave dilemmas, like those we discussed for euthanasia, surround it. Analyzing the practice, one group of medical experts concluded that it is warranted only when the following conditions are met:

- The patient requests assisted suicide repeatedly and freely and is suffering intolerably, with no satisfactory options.
- The doctor thoroughly explores comfort-care alternatives with the patient.
- The practice is consistent with the doctor's fundamental values. (If not, the doctor should recommend transfer of care.)
- Even when the doctor and patient agree that there is no other acceptable choice, independent monitoring is crucial for preventing abuse. (Stevens, 1997)

Juries have seldom returned guilty verdicts in cases involving doctor-assisted suicide. Yet in April 1999, Kevorkian—the most vigorous American proponent of assisted suicide—was sentenced to 10 to 25 years in prison for second-degree murder and delivery of a controlled substance. Seven months earlier, he had given a terminally ill man a lethal injection, videotaped the death, and permitted the event to be broadcast on the CBS television program *60 Minutes.* Then he dared prosecutors to charge him. The murder indictment prevented Kevorkian from introducing evidence indicating that the man wanted to kill himself. Such evidence would have been permissible had the charge been assisted suicide or voluntary active euthanasia.

Previously we noted that public opinion favors voluntary active euthanasia over assisted suicide. Yet in assisted suicide, the final act is solely the patient's, reducing the possibility of coercion. For this reason, some experts believe that legalizing assisted suicide is preferable to legalizing voluntary active euthanasia. However, in an atmosphere of high family caregiving burdens and intense pressure to contain health care costs (see Chapter 17), legalizing either practice poses risks. The American and Canadian Medical Associations oppose both voluntary active euthanasia and assisted suicide. Helping incurable, suffering patients who yearn for death is a profound moral and legal problem.

Ask Yourself

REVIEW

Why do the majority of North Americans support voluntary active euthanasia and assisted suicide? What benefits and risks does legalizing each practice pose?

APPLY

Thinking ahead to the day she dies, Noreen imagines a peaceful scene in which she says good-bye to loved ones. What social and medical practices are likely to increase Noreen's chances of dying in the manner she desires?

APPLY

If he should ever fall terminally ill, Ramon is certain that he wants doctors to halt life-saving treatment. To best ensure that his wish will be granted, what should Ramon do?

www.

Bereavement: Coping with the Death of a Loved One

Loss is an inevitable part of existence throughout the lifespan. Even when change takes place for the better, we let go of some aspects of experience so we can embrace others. Consequently, our development prepares us for profound loss.

Bereavement is the experience of losing a loved one by death. The root of this word means "to be robbed," suggesting unjust and injurious stealing of something valuable. Consistent with this image, we respond to loss with **grief**—intense physical and psychological distress. When we say someone is grief-stricken, we imply that his or her total way of being is affected.

Because grief can be overwhelming, cultures have devised ways of helping their members move beyond it to deal with the life changes demanded by death of a loved one. **Mourning** is the culturally specified expression of the bereaved person's thoughts and feelings. Customs—such as gathering with family and friends, dressing in black, attending the funeral, and observing a prescribed mourning period with special rituals—vary greatly among societies and ethnic groups. But all have in common the goal of helping people work through their grief and learn to live in a world that does not include the deceased.

Clearly, grief and mourning are closely linked. In fact, we often use the two words interchangeably in everyday language. Let's look closely at how people respond to the death of a loved one.

Grief Process

Many theorists studying a wide range of types of loss and bereaved individuals—both children and adults—have concluded that grieving usually takes place in three phases, each characterized by a different set of responses (Bowlby, 1980; Parkes & Weiss, 1983; Rando, 1995). In reality, however, people vary greatly in behavior and timing and often move back and forth between these reactions. Consequently, rather than a sequence of steps that results in grief resolution, a more accurate account is a roller-coaster ride, with many ups and downs and gradual recovery over time (Lund, 1996; Lund, Caserta, & Dimond, 1986).

- **Avoidance.** On hearing the news, the survivor experiences shock followed by disbelief, which may last from hours to weeks. A numbed feeling serves as "emotional anesthesia" while the bereaved person begins to experience painful awareness of the loss. The loved one has been a part of daily life for so long that his or her death is too much to comprehend immediately.

- **Confrontation.** As the mourner confronts the reality of the loss, grief is experienced most intensely. The bereaved person often experiences a cascade of emotional reactions, including anxiety, sadness, protest, anger, helplessness, frustration, abandonment, and yearning for the loved one. Obsessively reviewing the circumstances of death, asking how it might have been prevented, and searching for meaning in the loss are common. In addition, the grief-stricken person may be absentminded, unable to concentrate, and preoccupied with thoughts of the deceased and may experience loss of sleep and appetite. At times, self-destructive behaviors, such as taking drugs or driving too fast, occur. Most of these responses are symptoms of depression—an invariable component of grieving.

Although confrontation is difficult, each pang of grief that results from an unmet wish to be reunited with the deceased brings the mourner closer to the realization that the loved one is gone. After hundreds, perhaps thousands, of these painful moments, the bereaved person comprehends that a cherished relationship must be transformed from a physical presence to an inner representation.

● **Restoration.** At first, confronting the loss preoccupies most bereaved individuals. But they must also deal with other stressors that are secondary outcomes of the death—overcoming loneliness by reaching out to others; mastering tasks (such as finances or cooking) that the deceased had performed; reorganizing daily life without the loved one; and revising their identity from "spouse" to "widow" or from "parent" to "parent of a deceased child." According to a recent perspective, called the **dual-process model of coping with loss,** effective coping requires people to oscillate between dealing with the emotional consequences of loss and attending to life changes, which—when handled successfully—have restorative, or healing, effects (Stroebe & Schut, 1999, 2001). Moving back and forth offers temporary distraction and relief from painful grieving. Much research indicates that confronting grief without relief has severe negative consequences for physical and mental health (Rimé et al., 1998).

As grief subsides, emotional energies increasingly shift toward life-restoring pursuits—meeting everyday responsibilities, investing in new activities and goals, and strengthening old ties and building new relationships. On certain days, such as family celebrations or the anniversary of death, grief reactions may resurface and require attention, but they do not interfere with a healthy, positive approach to life. How long does grieving last? No single answer can be given. Sometimes confrontation continues for a few months, at other times for several years. An occasional upsurge of grief may persist for a lifetime and is a common response to losing a much-loved spouse, partner, child, or friend (Shuchter & Zisook, 1995).

Personal and Situational Variations

Like dying, grieving is affected by many factors. Once again, personality, coping style, and religious and cultural background influence adjustment. Sex differences are also evident. Compared with women, men typically express distress and depression less directly and seek social support less readily (Doka & Martin, 2000; Stroebe, 1998). Furthermore, the quality of the mourner's relationship with the deceased is important. For example, an end to a loving, fulfilling bond may lead to anguished grieving, but it is unlikely to leave the residue of guilt and regret that often follows dissolution of a conflict-ridden, ambivalent tie.

Circumstances surrounding the death—whether it is sudden and unanticipated or follows a prolonged illness—also shape mourners' responses. The nature of the lost relationship and when the death occurs in the life course make a difference as well.

● **Sudden, Unanticipated versus Prolonged, Expected Deaths.** Sudden, unexpected deaths are usually the result of murder, suicide, war, accident, or natural disaster. In these instances, avoidance may be especially pronounced and confrontation highly traumatic because shock and disbelief are extreme. In a survey of a representative sample of 18- to 45-year-old adults in a large American city, the trauma most often reported as prompting an intense, debilitating stress reaction was the sudden, unanticipated death of a loved one, an event experienced by 60 percent of the participants (Breslau et al., 1998). In contrast, during prolonged dying, the bereaved person has had time to engage in **anticipatory grieving**—acknowledging that the loss is inevitable and preparing emotionally for it.

© PAUL CONKLIN/PHOTOEDIT

■ Sudden, unanticipated deaths often prompt an intense, debilitating stress reaction. Here teenagers express their feelings for a peer victim of an automobile accident by constructing a shrine at the site of the crash. Shock and disbelief make confronting the reality of this avoidable, premature death especially traumatic.

Adjusting to a death is made easier when the survivor understands the reasons for it. Without explanations that make sense, mourners often remain anxious and confused. This barrier to confronting loss is tragically apparent in cases of sudden infant death syndrome (SIDS), in which doctors cannot tell parents exactly why their apparently healthy baby died (see Chapter 3, page 103). That death seems "senseless" also complicates grieving after suicides, terrorist attacks, school and drive-by shootings, and hurricanes, earthquakes, and other natural disasters. In Western societies, people tend to believe that momentous events should be comprehensible and nonrandom. When a death is sudden and unexpected, it can threaten basic assumptions about a just, benevolent, and controllable world (Gluhoski & Wortman, 1996).

Suicide, particularly that of a young person, is especially hard to bear. Compared with survivors of other sudden deaths, people grieving a suicidal loss are more likely to blame themselves for what happened, concluding that they contributed to or could have prevented it. In one study, college students who had survived a suicide scored higher than those who had experienced any other type of loss in feelings of shame, a sense of rejection by the deceased, and self-destructive behavior, perhaps because of persisting feelings of responsibility for what had happened (Silverman, Range, & Overholser, 1994). Typically, recovery from grief after a suicide is quite prolonged (Thompson & Range, 1992).

● **Parents Grieving the Loss of a Child.** The death of a child, whether unexpected or foreseen, is the most difficult loss an adult can face (Stillion, 1995). It brings special grieving problems because children are extensions of parents' feelings about themselves—the focus of hopes and dreams, including parents' sense of immortality. Also, since children depend on, admire, and appreciate their parents in a deeply gratifying way, they are an unmatched source of love. Finally, the unnaturalness of a child's death complicates it. Children are not supposed to die before their parents.

Parents who have lost a child often report considerable distress many years later, along with frequent thoughts of the deceased (Rubin & Malkinson, 2001). The guilt triggered by outliving their child frequently becomes a tremendous burden, even when parents "know" better. For example, a psychologist who understood that his daughter's cancer was not hereditary nevertheless said to a therapist, "Her genes allowed her to develop cancer. I gave her her genes. Therefore, I killed my daughter" (Rando, 1991b, p. 239).

Although a child's death sometimes leads to marital breakup, this is likely to happen only when the relationship was unsatisfactory before. If parents can reorganize the family system and reestablish a sense of life's meaning through valuing the lost child's impact on their lives and investing in other children and activities, then the result can be firmer family commitments and personal growth (Wheeler, 2001).

● **Children and Adolescents Grieving the Loss of a Parent.** The loss of an attachment figure has long-standing consequences for children. When a parent dies, children's basic sense of security and being cared for is threatened. Death of a sibling not only deprives children of a close emotional tie but also informs them, often for the first time, of their own vulnerability.

Children grieving a family loss describe frequent crying, trouble concentrating in school, sleep difficulties, headaches, and other physical symptoms several months to years after a death. And clinical studies reveal that persistent, mild depression, anxiety, and angry outbursts are common (Dowdney, 2000; Silverman & Worden, 1992). At the same time, many children say they actively maintained mental contact with their dead parent or sibling, dreaming about and speaking to them regularly. In a follow-up 7 to 9 years after sibling loss, thinking about the deceased brother or sister at least once a day was common (Martinson, Davies, & McClowry, 1987; Silverman & Nickman, 1996). These images seem to facilitate coping with loss and are sometimes reported by bereaved adults as well.

Cognitive development contributes to the ability to grieve. For example, children with an immature understanding of death may believe the dead parent left voluntarily, perhaps in anger, and that the other parent may also disappear. For these reasons, young children need careful, repeated explanations assuring them that the parent did not want to die and was not angry at them (Christ, Siegel, & Christ, 2002). Keeping the truth from children isolates them and often leads to profound regrets. One 8-year-old who learned only a half-hour in advance that his sick brother was dying reflected, "If only I'd known, I could have said good-bye."

Regardless of children's level of understanding, honesty, affection, and reassurance help them tolerate painful feelings of loss. Grief-stricken school-age children are usually more willing than adolescents to confide in parents. To appear normal, teenagers tend to keep their grieving from both adults and peers. Consequently, they are more likely than children to become depressed or to escape from grief through acting-out behavior (Corr & Balk, 1996).

● **Adults Grieving the Loss of an Intimate Partner.** Recall from Chapter 18 that after the death of a spouse, adaptation to widowhood varies greatly, with age, social support, and personality making a difference. After a period of intense grieving, most widowed elders in Western nations fare well (see pages 602–603 to review). Younger individuals display more negative outcomes (Stroebe & Stroebe, 1993). Loss of a spouse or partner in early or middle adulthood is a nonnormative event that profoundly disrupts life plans. Older widows and widowers have many more contemporaries in similar circumstances. And most have already attained important life goals or adjusted to the fact that some goals will not be attained.

In addition to dealing with feelings of loss, young and middle-aged widows and widowers often must assume a greater role in comforting others, especially children. They also face the stresses of single parenthood and rapid shrinking of the social network established during their life as a couple (Lopata, 1996). Younger widows frequently report that the most valuable support they received came from their mothers, many of whom were also widowed (Bankoff, 1983).

The death of an intimate partner in a gay or lesbian relationship presents unique challenges. When relatives limit or bar the partner from participating in funeral services, they profoundly disrupt the process of grieving (Housel, 1995). Fortunately, gay and lesbian communities provide helpful alternative support, in the form of memorial services and other rituals.

Cultural Influences

Cultural Variations in Mourning Behavior

The ceremonies that commemorated Sofie's and Nicholas's deaths—the first Jewish, the second Quaker—were strikingly different. Yet they served common goals: providing a setting in which loss could be shared and mobilizing the community in support of the bereaved.

At the funeral home, Sofie's body was washed and shrouded, a Jewish ritual signifying return to a state of purity. Then it was placed in a plain wooden (not metal) coffin, so as not to impede the natural process of decomposition. To underscore the finality of death, Jewish tradition does not permit viewing of the body; it remains in a closed coffin. Traditionally, the coffin is not left alone until burial; the community maintains a day-and-night vigil.

To return the body quickly to the life-giving earth from which it sprang, Sofie's funeral was scheduled 3 days after death, as soon as relatives could gather. As the service began, Sofie's husband and children symbolized their anguish by cutting a black ribbon and pinning it to their clothing. The rabbi recited psalms of comfort, followed by a eulogy in which he related memories of Sofie that family members had shared with him. The service continued at the graveside. Once the coffin had been lowered into the ground, relatives and friends took turns shoveling earth onto it, each participating in the irrevocable act of burial. The service concluded with the "homecoming" prayer called *Kaddish,* which affirms life while accepting death.

The family returned home to light a memorial candle, which burned throughout *shiva,* the 7-day period of mourning (Hazell, 2001). A meal of consolation prepared by others followed, creating a warm feeling of community. Jewish custom prescribes that after 30 days, life should gradually return to normal. When a parent dies, the mourning period is extended to 12 months.

In the tradition of Quaker simplicity, Nicholas's death did not require elaborate preparation of the body, a casket, or a hearse to carry him to a cemetery. He was cremated promptly. During the next week, relatives and close friends gathered with Giselle and Sasha at their home. Together, they planned a memorial service uniquely suited to celebrating Nicholas's life.

On the appointed day, people who had known Nicholas sat in chairs arranged in concentric circles. Standing at the center, a clerk of the Friends (Quaker) Meeting extended a welcome and explained to newcomers the custom of worshipping silently, with those who feel moved to speak rising and sharing thoughts and feelings at any time. During the next hour, many people offered personal statements about Nicholas or read poems and selections from Scripture. Giselle and Sasha provided concluding comments. Then everyone joined hands to close the service, and a reception for the family followed.

Variations in mourning behavior are vast—both within and across societies. For most Jews and Christians, extensive ritual accompanies a funeral and burial. In contrast, the Quaker memorial service is among the least ritualized. In some groups, grief is expressed freely. For example, venting deep emotion is often part of African-American funerals, especially in the southern United States. Christians and Jews of European descent are usually restrained in their display of sorrow (Sullivan, 1995). And in some societies, letting any feeling show is actively discouraged. The Balinese of Indonesia must remain calm in the face of death if the gods are to hear their prayers. With the help of supporters who joke, tease, and distract, mourners work hard to keep their composure, although the Balinese acknowledge the existence of underlying grief (Rosenblatt, 1993).

Religions also render accounts of the aftermath of death—teachings that console both dying and bereaved individuals. Jewish tradition affirms personal survival, although its form is not clearly specified. Greater emphasis is placed on the survival of a people—living on by granting life and care to others. Unlike other Christian faiths, Quakers focus almost exclusively on the here and now—"salvation by character" through working for peace, justice, and a loving community. Little attention is given to fear of hell or hope of heaven. The religions of tribal and village cultures typically include elabo-

● **Bereavement Overload.** When a person experiences several deaths at once or in close succession, *bereavement overload* can occur. Multiple losses deplete the coping resources of even well-adjusted people, leaving them emotionally overwhelmed and unable to resolve their grief. For many young adults, especially members of the gay community who have lost partners and friends, AIDS presents this challenge. In a study of over 700 homosexual men, those experiencing two or more losses in close succession reported more distress, suicidal thoughts, and substance use than did those with only a single loss (Martin & Dean, 1993). Fear of discrimination may prevent homosexuals overwhelmed by grief from seeking treatment, especially if they are unaware of psychological services, including counseling and social support networks, in the gay community (Springer & Lease, 2000),

Because old age often brings the death of spouse, siblings, and friends in close succession, elders are also at risk for bereavement overload. But recall from Chapter 18 that compared

rate beliefs about ancestor spirits and the afterlife and customs designed to ease the journey of the deceased to the spiritual world.

By announcing that a death has occurred, ensuring social support, commemorating the deceased, and conveying a philosophy of life after death, funerals and memorial services are of great assistance to the bereaved. Yet some evidence indicates that these customs may be on the decline in Western nations (Fulton, 1995). In large cities, more deaths are being followed by disposal of the body, without a public or religious observance of any kind. The trend is worrisome, for it deprives grief-stricken people of community assurance of love and concern.

In the current cyber age, a new ritual has arisen: memorials to the deceased on the Internet. These website "cemeteries" offer benefits not available through traditional funerals: postings whenever bereaved individuals feel ready to convey their thoughts and feelings, creation of tributes at little or no cost, and continuous, easy access to the memorial. Rather than using the traditional obituary format, most creators of web tributes choose to tell personalized stories, highlighting a laugh, a favorite joke, or a touching moment (Roberts & Vidal,1999–2000). Besides celebrating the deceased, some survivors use web memorials to grieve openly, others to converse with the lost loved one. Web cemeteries also provide a means for people excluded from traditional death rituals to engage in public mourning. As one bereaved friend expressed:

> It's been 7 years since you've been gone. For 7 years I've dealt quite poorly I might add, with the fact there was no funeral, no wake, no grave to visit you at. Your parents whisked you away. Your body was cremated, and I didn't even get to say "good-bye" . . .

Another captured the unusual nature of this highly flexible medium for mourning:

> I wish I could maintain contact with you, to keep alive the vivid memories of your impact on my life. Perhaps I can do this by sharing my memories in this seemingly unconventional way. Because I cannot visit your grave today, I use this means to tell you how much you are loved.

■ Venting deep emotion often is part of African-American funerals, particularly in the southern United States. In this Baptist funeral service, a family member weeps openly.

with young people, older adults are often better equipped to handle these losses. They know that decline and death are expected in late adulthood, and they have had a lifetime of experience through which to develop effective coping strategies.

Finally, the recent rash of random murders in American schools and the televised replays of the September 11, 2001, terrorist attacks have led more North Americans to witness unexpected, violent deaths. For survivors whose loved ones died on September 11, including an estimated 15,000 children who lost a parent, images of devastation and horror often block efforts to come to terms with loss. Children and adolescents suffer profoundly—from intense shock, prolonged grief, frequent mental replays of the vicious attack and gruesome outcomes, and fear of the settings in which those events occurred (Nader, 2002; Webb, 2002). The greater the exposure to the catastrophic death scene, the more severe these reactions (Nader et al., 1990).

Funerals and other bereavement rituals, illustrated in the Cultural Influences box above, assist mourners of all ages in

■ In the aftermath of the March 2000 Columbine High School shooting, these students wear ribbons in memory of slain victims and are overcome as they stand before a memorial in a city park. Because of the rash of random murders in American schools, more children and adolescents have witnessed the unexpected, violent deaths of teachers and classmates. They are at risk for bereavement overload—prolonged, overwhelming grief that is difficult to resolve.

resolving grief with the help of family and friends. Bereaved individuals who remain preoccupied with loss and who have difficulty resuming interest in everyday activities benefit from special interventions designed to help them adjust.

Bereavement Interventions

Sympathy and understanding are sufficient for most people to undertake the tasks necessary to recover from grief (see the Caregiving Concerns table below). Yet effective support is often difficult to provide, and relatives and friends could benefit from training in how to respond. Sometimes they give advice aimed at hastening recovery, an approach that most bereaved people dislike (Lehman, Ellard, & Wortman, 1986). At other times, mourners are too overcome to acknowledge well-meaning interactions, causing others to withdraw (Stylianos & Vachon, 1993). Listening patiently and "just being there" are among the best ways to help.

Bereavement interventions typically encourage people to draw on their existing social network, while providing them with additional social support. Self-help groups that bring together mourners who have experienced the same type of loss are highly effective in reducing stress (Lieberman, 1993). In one program for recently widowed older adults, group meetings also included training in tasks of daily living and good health practices. This added focus greatly enhanced participants' sense of self-efficacy in managing their own lives (Caserta, Lund, & Rice, 1999).

Interventions for children and adolescents following violent deaths must protect them from unnecessary reexposure, assist parents and teachers with their own distress so they can

Suggestions for Resolving Grief After a Loved One Dies

SUGGESTION	DESCRIPTION
Give yourself permission to feel the loss.	Permit yourself to confront all thoughts and emotions associated with the death. Make a conscious decision to overcome your grief, recognizing that this will take time.
Accept social support.	In the early part of grief, let others reach out to you by making meals, running errands, and keeping you company. Be assertive; ask for what you need so people who would like to help will know what to do.
Be realistic about the course of grieving.	Expect to have some negative and intense reactions, such as feeling anguished, sad, and angry, that last from weeks to months and may occasionally resurface years after the death. There is no one way to grieve, so find the best way for you.
Remember the deceased.	Review your relationship to and experiences with the deceased, permitting yourself to see that you can no longer be with him or her as before. Form a new relationship based on memories, keeping it alive through photographs, commemorative donations, prayers, and other symbols and actions.
At the appropriate time, invest in new activities and relationships.	Determine which roles you must give up and which ones you must assume as a consequence of the death, and take deliberate steps to incorporate these into your life. Set small goals at first, such as a night at the movies, a dinner date with a friend, or a week's vacation.

effectively offer comfort, and be culturally sensitive. After shootings on the grounds of one school, administrators arranged for a ceremony that removed "angry dead souls." Only then would the large number of Vietnamese Buddhist students return to their classrooms (Nader, Dubrow, & Stamm, 1999).

At times, mourning does not proceed smoothly. Sudden, violent, and unexplainable deaths; the loss of a child; a death that the mourner feels he or she could have prevented; or an ambivalent or dependent relationship with the deceased make it harder for bereaved people to overcome their loss. In these instances, *grief therapy,* or individual counseling with a specially trained professional, is sometimes helpful (Neimeyer, 2000). One approach is to assist bereaved adults in finding some value in the grieving experience—for example, gaining insight into the meaning of relationships, discovering their own capacity to cope with adversity, and crystallizing a sense of purpose in their lives (Davis et al., 2000). In a study tracking middle-aged adults' adjustment after a family member's death, the ability to see benefits for one's own life was strongly related to well-being 13 and 18 months later (Davis, Nolen-Hoeksema, & Larson, 1998).

Death Education

Preparatory steps can be taken to help people of all ages cope with death more effectively. The death awareness movement that sparked increased sensitivity to the needs of dying patients has also led to the rise of death education. Courses in death, dying, and bereavement are now a familiar part of offerings in colleges and universities. Instruction has been integrated into the professional training of doctors, nurses, psychologists, social workers, and other helping professionals. It can also be found in adult education programs in many communities. And it has filtered down to a few elementary and secondary schools.

Death education at all levels has the following goals:

- Increasing students' understanding of the physical and psychological changes that accompany dying
- Helping students learn how to cope with the death of a loved one
- Preparing students to be informed consumers of medical and funeral services
- Promoting understanding of social and ethical issues involving death

Courses vary widely in format. Some focus on conveying information, whereas others are experiential and include many activities—role playing, discussions with the terminally ill, visits to mortuaries and cemeteries, and personal awareness exercises. Research reveals that using a lecture style leads to gains in knowledge, but it often leaves students more uncomfortable about death than when they entered. In contrast, experiential programs that help people confront their own mortality are less likely to heighten death anxiety and may sometimes reduce it (Durlak & Riesenberg, 1991; Maglio & Robinson, 1994). These findings suggest that to reach students cognitively and emotionally, death educators need to be more than well informed. They must be sensitive and responsive, skilled at communication, and able to help people deal with distress.

Whether acquired in the classroom or in our daily lives, our thoughts and feelings about death are forged through interactions with others. Becoming more aware of how we die and our own mortality, we encounter our greatest loss, but we also gain. Dying people have at times confided in those close to them that awareness of the limits of their lifespan permitted them to dispense with superficial distractions and wasted energies and focus on what is truly important in their lives. As one AIDS patient summed up, "[It's] kind of like life, just speeded up"—an accelerated process in which, over a period of weeks to months, one grapples with issues that normally would have taken years or decades to resolve (Selwyn, 1996, p. 36). Applying this lesson to ourselves, we learn that by being in touch with death and dying, we can live ever more fully.

Ask Yourself

REVIEW
What circumstances are likely to induce bereavement overload? Cite examples.

REVIEW
Explain how death intervention can help people cope with death more effectively.

CONNECT
Compare grieving individuals' reactions with terminally ill patients' thoughts and feelings as they move closer to death, described on page 628. Can a dying person's reactions be viewed as a form of grieving? Explain.

CONNECT
Cite examples of age-graded, history-graded, and nonnormative influences on the grieving process, referring to research findings on pages 640–643. If you need to review these contexts for lifespan development, return to Chapter 1, pages 10–12.

www.

Summary

How We Die

Describe the physical changes of dying, along with their implications for defining death and the meaning of death with dignity.

- Of the one-quarter of people in industrialized nations who die suddenly, the overwhelming majority are victims of heart attacks. Death is long and drawn out for three-fourths of people—many more than in times past, due to life-saving medical technology.

- In general, dying takes place in three phases: the **agonal phase,** or first moments in which the body can no longer sustain life; **clinical death,** a short interval in which resuscitation is still possible; and **mortality,** or permanent death.

- In most industrialized nations, **brain death** is accepted as the definition of death. However, the thousands of patients who remain in a **persistent vegetative state** reveal that the brain death standard does not always solve the dilemma of when to halt treatment of the incurably ill.

- Most people will not experience an easy death. Therefore, we can best ensure death with dignity by supporting dying patients through their physical and psychological distress, being candid about death's certainty, and helping them learn enough about their condition to make reasoned choices about treatment.

Understanding of and Attitudes Toward Death

Discuss age-related changes in conceptions of and attitudes toward death, and cite factors that influence death anxiety.

- We live in a death-denying culture. Compared with earlier generations, more young people reach adulthood having had little contact with death, and adults are often reluctant to bring up the topic.

- Between ages 7 and 10, most children grasp the three components of the death concept in the following order: **permanence, universality,** and **nonfunctionality**. Experiences with death and religious teachings affect children's understanding. When parents talk about death candidly, children usually have a good grasp of the facts of death and an easier time accepting it.

- Adolescents often fail to apply their understanding of death to everyday life. Aware that death happens to everyone and can occur at any time, teenagers nevertheless are attracted to alternative views, are high risk takers, and do not take death personally. By discussing concerns about death with adolescents, adults can help them build a bridge between death as a logical concept and their personal experiences.

- As people pass from early to middle adulthood, they become more conscious of the finiteness of their lives. Compared with middle-aged adults, elders focus more on practical matters of how and when death might happen.

- Large individual and cultural variations in **death anxiety** exist. Overall, fear of death declines with age, reaching its lowest level in late adulthood and in adults with deep faith in some form of higher being. Women seem more anxious about death than men. And people with mental health problems express more severe death anxiety.

Thinking and Emotions of Dying People

Describe and evaluate Kübler-Ross's stage theory, citing factors that influence the responses of dying patients.

- According to Elisabeth Kübler-Ross's theory, dying people typically express five responses, initially proposed as stages: denial, anger, bargaining, depression, and acceptance. However, these responses do not occur in fixed sequence, and dying people often display other coping strategies.

- An **appropriate death** is one that makes sense in terms of the individual's pattern of living and values, preserves or restores significant relationships, and is as free of suffering as possible. A host of contextual variables—nature of the disease; personality and coping style; family members' and health professionals' truthfulness and sensitivity; and spirituality, religion, and cultural background—affect the way people respond to their own dying and, therefore, the extent to which they attain an appropriate death.

A Place to Die

Evaluate the extent to which homes, hospitals, and the hospice approach meet the needs of dying people and their families.

- Although the overwhelming majority of people want to die at home, caring for a dying patient is highly stressful. Even with professional help, most homes are poorly equipped to handle the medical and comfort-care needs of the dying.

- In hospitals, sudden deaths typically occur in emergency rooms. Sympathetic explanations from staff can reduce family members' anger, frustration, and confusion. Intensive care is especially depersonalizing for patients who are hooked to machines for months. Even in general or specialized cancer care units, emphasis on efficiency usually interferes with an appropriate death. Most hospitals do not have comprehensive treatment programs aimed at easing suffering at the end of life.

- Whether a person dies at home or in a hospital, the **hospice** approach strives to meet the dying person's physical, emotional, social, and spiritual needs and emphasizes quality of life over life-prolonging measures through **palliative,** or **comfort, care.** In the year after death, bereavement services are offered to families. Besides reducing patients' physical suffering, hospice care contributes to improved family functioning and better psychological well-being among family survivors 1 to 2 years following their loved one's death.

The Right to Die

Discuss controversies surrounding euthanasia and assisted suicide.

- The same medical procedures that preserve life can prolong inevitable death, diminishing quality of life and personal dignity. **Euthanasia**—ending the life of a person suffering from an incurable condition—takes various forms.

- **Passive euthanasia,** withholding or withdrawing life-sustaining treatment from a hopelessly ill patient, is widely accepted and practiced. But in the absence of consensus on end-of-life health care, people can best ensure that their wishes will be followed by preparing an **advance medical directive.** A **living will** contains instructions for treatment. A more flexible approach is the **durable power of attorney for health care,** which authorizes appointment of another person to make health care decisions on one's behalf. When no advance medical directive exists, some U.S. states and Canadian provinces permit appointment of a health care proxy.

- Public support for **voluntary active euthanasia,** in which doctors or others comply with a suffering patient's request to die before a natural end to life, is growing. Never-theless, the practice has sparked heated controversy, fueled by fears that it will undermine trust in health professionals and lead us down a "slippery slope" to the killing of vulnerable people who did not ask to die.

- Slightly less public consensus exists for assisted suicide. But because the final act is solely the patient's, some experts believe that legalizing assisted suicide is preferable to legalizing voluntary active euthanasia.

Bereavement: Coping with the Death of a Loved One

Describe the phases of grieving, factors that underlie individual variations, and bereavement interventions.

- **Bereavement** refers to the experience of losing a loved one by death, **grief** to the intense physical and psychological distress that accompanies loss. **Mourning** customs are culturally prescribed expressions of thoughts and feelings designed to help people work through their grief.
- Although many theorists regard grieving as taking place in orderly phases of avoidance, confrontation, and restoration, a more accurate account is a roller-coaster ride, with many ups and downs and gradual recovery over time. According to the **dual-process model of coping with loss,** effective coping involves oscillating between dealing with the emotional consequences of loss and attending to life changes, which offer temporary distraction and relief from painful grieving.
- Like dying, grieving is affected by many personal and situational factors. After a sudden, unanticipated death, avoidance may be especially pronounced and confrontation highly traumatic because shock and disbelief are extreme. In contrast, a prolonged, expected death grants the bereaved person time to engage in **anticipatory grieving.**
- When a parent loses a child or a child loses a parent or sibling, grieving is generally very intense and prolonged. Loss of a spouse or partner in early or middle adulthood profoundly disrupts life plans. Younger widowed individuals usually display more negative outcomes than widowed elders. When relatives limit or bar a homosexual partner's participation in funeral services, they profoundly disrupt the process of grieving.
- People who experience several deaths at once or in close succession may suffer from bereavement overload. Among those at risk are young people who have lost partners and friends to AIDS, the elderly, and people who have witnessed unexpected, violent deaths.
- Sympathy and understanding are sufficient for most people to recover from grief. Self-help groups can provide extra social support. When bereaved individuals find it very hard to overcome loss, grief therapy is sometimes helpful.

Death Education

Explain how death education can help people cope with death more effectively.

- Today, instruction in death, dying, and bereavement can be found in colleges and universities; training programs for doctors, nurses, and helping professionals; adult education programs; and a few elementary and secondary schools. Courses are more likely to reach students cognitively and emotionally when they include an experiential component.

Important Terms and Concepts

advance medical directive (p. 634)
agonal phase (p. 621)
anticipatory grieving (p. 640)
appropriate death (p. 629)
bereavement (p. 639)
brain death (p. 621)
clinical death (p. 621)
death anxiety (p. 626)
dual-process model of coping with loss (p. 640)
durable power of attorney for health care (p. 635)
euthanasia (p. 634)
grief (p. 639)
hospice (p. 632)
living will (p. 635)
mortality (p. 621)
mourning (p. 639)
nonfunctionality (p. 623)
palliative, or comfort, care (p. 632)
passive euthanasia (p. 634)
permanence (p. 623)
persistent vegetative state (p. 621)
universality (p. 623)
voluntary active euthanasia (p. 636)

FYI For Further Information and Help

Consult the Companion Website for *Development Through the Lifespan, Third Edition,* (www.ablongman.com/berk) where you will find the following resources for this chapter:

- **Chapter Objectives**
- **Flashcards** for studying important terms and concepts
- **Annotated Weblinks** to guide you in further research
- **Ask Yourself** questions, which you can answer and then check against a sample response
- **Suggested Readings**
- **Practice Test** with immediate scoring and feedback

Glossary

Aboriginal Head Start A Canadian federal program that provides First Nations, Inuit, and Métis children younger than age 6 with preschool education and nutritional and health services and that encourages parent involvement in program planning and children's learning. (p. 233)

academic programs Preschools and kindergartens in which teachers structure children's learning, teaching academic skills through formal lessons, often using repetition and drill. Distinguished from *child-centered programs.* (p. 233)

accommodation In Piaget's theory, that part of adaptation in which new schemes are created and old ones adjusted to produce a better fit with the environment. Distinguished from *assimilation.* (p. 144)

achieving stage Schaie's stage of early adulthood in which people adapt their cognitive skills to situations that have profound implications for achieving long-term goals. Consequently, they focus less on acquiring knowledge and more on applying it to everyday life. (p. 432)

acquisitive stage Schaie's stage of childhood and adolescence in which the goal of mental activity is knowledge acquisition. (p. 432)

active lifespan The number of years of vigorous, healthy life. Distinguished from *average life expectancy* and *maximum lifespan.* (p. 547)

activity theory A social theory of aging that states that the decline in social interaction in late adulthood is due to failure of the social environment to offer opportunities for social contact, not the desires of elders. Distinguished from *disengagement theory* and *socioemotional selectivity theory.* (p. 595)

adaptation In Piaget's theory, the process of building schemes through direct interaction with the environment. Made up of two complementary processes: *assimilation* and *accommodation.* (p. 144)

adolescence The transition between childhood and adulthood. Begins with puberty and involves accepting one's full-grown body, acquiring adult ways of thinking, attaining emotional and economic independence, developing more mature ways of relating to peers of both sexes, and constructing an identity. (p. 344)

advance medical directive A written statement of desired medical treatment should a person become incurably ill. (p. 634)

age of viability The age at which the fetus can first survive if born early. Occurs sometime between 22 and 26 weeks. (p.80)

age-graded influences Influences on lifespan development that are strongly related to age and therefore fairly predictable in when they occur and how long they last. (p. 10)

agonal phase The phase of dying in which gasps and muscle spasms occur during the first moments in which the body can no longer sustain life. Distinguished from *clinical death* and *mortality.* (p. 621)

Alzheimer's disease The most common form of dementia, in which structural and chemical deterioration in the brain is associated with gradual loss of many aspects of thought and behavior, including memory, skilled and purposeful movements, and comprehension and production of speech. (p. 565)

amnion The inner membrane that forms a protective covering around the prenatal organism. (p. 78)

amyloid plaques A structural change in the brain associated with Alzheimer's disease in which deposits of the protein amyloid are surrounded by clumps of dead nerve cells, which amyloid appears to destroy. (p. 565)

androgyny A type of gender identity in which the person scores high on both masculine and feminine personality characteristics. (p. 263)

animistic thinking The belief that inanimate objects have lifelike qualities, such as thoughts, wishes, feelings, and intentions. (p. 218)

anorexia nervosa An eating disorder in which individuals (usually females) starve themselves because of a compulsive fear of getting fat. (p. 353)

anoxia Inadequate oxygen supply. (p. 95)

anticipatory grieving Before a prolonged, expected death, acknowledging that the loss is inevitable and preparing emotionally for it. (p. 640)

Apgar Scale A rating used to assess the newborn baby's physical condition immediately after birth. (p. 93)

applied behavior analysis A set of practical procedures that combine conditioning and modeling to eliminate undesirable behaviors and increase socially acceptable responses. (p. 19)

appropriate death A death that makes sense in terms of the individual's pattern of living and values and, at the same time, preserves or restores significant relationships and is as free of suffering as possible. (p. 629)

assimilation In Piaget's theory, that part of adaptation in which the external world is interpreted in terms of current schemes. Distinguished from *accommodation.* (p. 144)

assistive technology An array of devices that permit people with disabilities, including older adults, to improve their functioning. (p. 556)

associative memory deficit Difficulty creating and retrieving links between pieces of information—for example, two items or an item and its context. (p. 571)

associative play A form of true social participation in which children are engaged in separate activities, but they interact by exchanging toys and comment on one another's behavior. Distinguished from *nonsocial activity, parallel play,* and *cooperative play.* (p. 249)

attachment The strong, affectional tie that humans have with special people in their lives that leads them to feel pleasure and joy when interacting with them and to be comforted by their nearness during times of stress. (p. 185)

attention-deficit hyperactivity disorder (ADHD) A childhood disorder involving inattentiveness, impulsivity, and excessive motor activity. Often leads to academic failure and social problems. (p. 288)

authoritarian style A child-rearing style that is low in acceptance and involvement, is high in coercive control, and restricts rather than grants autonomy. Distinguished from *authoritative, permissive,* and *uninvolved styles.* (p. 265)

authoritative style A child-rearing style that is high in acceptance and involvement, emphasizes firm control with explanations, and includes gradual, appropriate autonomy granting. Distinguished from *authoritarian, permissive , and uninvolved styles.* (p. 265)

autobiographical memory Representations of special, one-time events that are long lasting because they are imbued with personal meaning. (p. 156)

autoimmune response An abnormal response of the immune system in which it turns against normal body tissues. (p. 554)

autonomous morality Piaget's second stage of moral development, in which children view rules as flexible, socially agreed-on principles that can be revised when there is a need to do so. Begins around age 10. (p. 388)

autonomy At adolescence, a sense of oneself as a separate, self-governing individual. Involves relying more on oneself and less on parents for direction and guidance and engaging in careful, well-reasoned decision making. (p. 394)

autonomy versus shame and doubt In Erikson's theory, the psychological conflict of toddlerhood, which is resolved positively if parents provide young children with suitable guidance and appropriate choices. (p. 175)

autosomes The 22 matching chromosome pairs in each human cell. (p. 46)

average life expectancy The number of years an individual born in a particular year can expect to live. Distinguished from *maximum lifespan* and *active lifespan.* (p. 547)

avoidant attachment The quality of insecure attachment characterizing infants who are usually not distressed by parental separation and who avoid the parent when she returns. Distinguished from *secure, resistant,* and *disorganized/disoriented attachment.* (p. 187)

babbling Repetition of consonant–vowel combinations in long strings, beginning around 4 months of age. (p. 165)

basal metabolic rate (BMR) The amount of energy the body uses at complete rest. (p. 420)

basic emotions Emotions that are universal in humans and other primates, have a long evolutionary history of promoting survival, and can be directly inferred from facial expressions. Includes happiness, interest, surprise, fear, anger, sadness, and disgust. (p. 175)

basic trust versus mistrust In Erikson's theory, the psychological conflict of infancy, which is resolved positively if caregiving, especially during feeding, is sympathetic and loving. (p. 174)

basic-skills approach An approach to beginning reading instruction that emphasizes training in phonics—the basic rules for translating written symbols into sounds—and simplified reading materials. Distinguished from *whole-language approach.* (p. 292)

behavior modification Procedures that combine conditioning and modeling to eliminate undesirable behaviors and increase desirable responses. (p. 19)

behaviorism An approach that views directly observable events—stimuli and responses—as the appropriate focus of study and the development of behavior as taking place through classical and operant conditioning. (p. 18)

bereavement The experience of losing a loved one by death. (p. 639)

bicultural identity The identity constructed by adolescents who explore and adopt values from both their subculture and the dominant culture. (p. 386)

"big five" personality traits Five basic factors, into which hundreds of personality traits have been organized: neuroticism, extroversion, openness to experience, agreeableness, and conscientiousness. (p. 522)

biological aging, or **senescence** Genetically influenced, age-related declines in the functioning of organs and systems that are universal in all members of our species. Sometimes called *primary aging.* (p. 412)

body image Conception of and attitude toward one's physical appearance. (p. 351)

brain death Irreversible cessation of all activity in the brain and the brain stem. The definition of death accepted in most industrialized nations. (p. 621)

brain plasticity The ability of other parts of the brain to take over functions of damaged regions. (p. 118)

breech position A position of the baby in the uterus that would cause the buttocks or feet to be delivered first. (p. 95)

bulimia nervosa An eating disorder in which individuals (mainly females) engage in strict dieting and excessive exercise accompanied by binge eating, often followed by deliberate vomiting and purging with laxatives. (p. 354)

burnout A condition in which long-term job stress leads to mental exhaustion, a sense of loss of personal control, and feelings of reduced accomplishment. (p. 534)

canalization The tendency of heredity to restrict the development of some characteristics to just one or a few outcomes. (p. 69)

cardinality The mathematical principle that the last number in a counting sequence indicates the quantity of items in the set. (p. 231)

carrier A heterozygous individual who can pass a recessive gene to his or her children. (p. 46)

cataracts Cloudy areas in the lens of the eye that increase from middle to old age, resulting in foggy vision and (without surgery) eventual blindness. (p. 551)

central executive The conscious part of working memory that directs the flow of information through the mental system by deciding what to attend to, coordinating incoming information with information already in the system, and selecting, applying, and monitoring strategies. (p. 153)

centration The tendency to focus on one aspect of a situation and neglect other important features. Distinguished from *decentration.* (p. 218)

cephalocaudal trend An organized pattern of physical growth and motor control that proceeds from head to tail. Distinguished from *proximodistal trend.* (p. 115)

cerebellum A brain structure that aids in balance and control of body movements. (p. 207)

cerebral cortex The largest structure of the human brain, which accounts for the highly developed intelligence of the human species. (p. 117)

cerebrovascular dementia A form of dementia in which a series of strokes leaves dead brain cells, producing step-by-step degeneration of mental ability, with each step occurring abruptly after a stroke. (p. 567)

cesarean delivery A surgical delivery in which the doctor makes an incision in the mother's abdomen and lifts the baby out of the uterus. (p. 96)

child-centered programs Preschools and kindergartens in which teachers provide a wide variety of activities from which children select, and much learning takes place through play. Distinguished from *academic programs.* (p. 233)

child-directed speech A form of language used by adults to speak to infants and toddlers that consists of short sentences with high-pitched, exaggerated expression, clear pronunciation, distinct pauses between speech segments, and repetition of new words in a variety of contexts. (p. 168)

child-rearing styles Combinations of parenting behaviors that occur in a wide range of situations, thereby creating an enduring child-rearing climate. (p. 265)

chorion The outer membrane that forms a protective covering around the prenatal organism. It sends out tiny fingerlike villi, from which the placenta begins to emerge. (p. 78)

chromosomes Rodlike structures in the cell nucleus that store and transmit genetic information. (p. 44)

chronosystem In ecological systems theory, temporal changes in environments, which produce new conditions that affect development. These changes can be imposed externally or arise from within the organism, since people select, modify, and create many of their own settings and experiences. (p. 25)

circular reaction In Piaget's theory, a means of building schemes in which infants try to repeat a chance event caused by their own motor activity. (p. 145)

classical conditioning A form of learning that involves associating a neutral stimulus with a stimulus that leads to a reflexive response. Once that baby's nervous system makes the connection between the two stimuli, the new stimulus produces the behavior by itself. (p. 126)

climacteric Midlife transition in which fertility declines. Brings an end to reproductive capacity in women and diminished fertility in men. (p. 486)

clinical death The phase of dying in which heartbeat, circulation, breathing, and brain functioning stop, but resuscitation is still possible. Distinguished from *agonal phase* and *mortality.* (p. 621)

clinical interview A method that uses a flexible, conversational style to probe for the participant's point of view. (p. 28)

clinical, or **case study, method** A method in which the researcher tries to understand the unique person by combining interview data, observations, and sometimes test scores. (p. 29)

clique A small group of about five to seven members who are good friends and, therefore, resemble one another in family background, attitudes, and values. (p. 398)

codominance A pattern of inheritance in which both genes, in a heterozygous combination, are expressed. (p. 47)

cognitive inhibition The ability to control internal and external distracting stimuli, preventing them from capturing attention and cluttering working memory with irrelevant information. (p. 288)

cognitive self-regulation The process of continuously monitoring progress toward a goal, checking outcomes, and redirecting unsuccessful efforts. (p. 291)

cognitive-developmental theory An approach introduced by Piaget that views children as actively building psychological structures as they explore and manipulate their world and cognitive development as taking place in stages. (p. 19)

cohabitation The lifestyle of unmarried couples who have an intimate, sexual relationship and share a residence. (p. 466)

cohort effects The effects of history-graded influences on research findings: People born in one period of time are influenced by particular historical and cultural conditions. (p. 35)

collectivist societies Societies in which people define themselves as part of a group and stress group over individual goals. Distinguished from *individualistic societies.* (p. 63)

companionate love Love based on warm, trusting affection and caregiving. Distinguished from *passionate love.* (p. 451)

compliance Voluntary obedience to requests and commands. (p. 196)

compression of morbidity The public health goal of reducing the average period of diminished vigor before death as life expectancy extends. So far, persistence of poverty and negative lifestyle factors have interfered with progress toward this goal. (p. 559)

concordance rate The percentage of instances in which both twins show a trait when it is present in one twin. Used to study the role of heredity in emotional and behavior disorders, which can be judged as either present or absent. (p. 67)

concrete operational stage Piaget's third stage, during which thought is logical, flexible, and organized in its application to concrete information. However, the capacity for abstract thinking is not yet present. Spans the years from 7 to 11. (p. 285)

conditioned response (CR) In classical conditioning, an new response produced by a conditioned stimulus (CS) that resembles the unconditioned, or reflexive, response (UCR). (p. 127)

conditioned stimulus (CS) In classical conditioning, a neutral stimulus that through pairing with an unconditioned stimulus (UCS) leads to a new response (CR). (p. 127)

congregate housing Housing for the elderly that provides a variety of support services, including meals in a common dining room, along with watchful oversight of elders with physical and mental disabilities (p. 599)

conservation The understanding that certain physical characteristics of objects remain the same, even when their outward appearance changes. (p. 218)

contexts Unique combinations of personal and environmental circumstances that can result in markedly different paths of development. (p. 6)

continuous development A view that regards development as a process of gradually augmenting the same types of skills that were there to begin with. Distinguished from *discontinuous development.* (p. 6)

contrast sensitivity A general principle accounting for early pattern preferences, which states that if babies can detect a difference in contrast between two patterns, they will prefer the one with more contrast. (p. 135)

controversial children Children who get a large number of positive and negative votes on self-report measures of peer acceptance. Distinguished from *popular, neglected,* and *rejected* children. (p. 324)

conventional level Kohlberg's second level of moral development, in which moral understanding is based on conforming to social rules to ensure positive human relationships and societal order. (p. 389)

convergent thinking The generation of a single correct answer to a problem. The type of cognition emphasized on intelligence tests. Distinguished from *divergent thinking.* (p. 306)

cooing Pleasant vowel-like noises made by infants beginning around 2 months of age. (p. 165)

cooperative play A form of true social participation in which children's actions are directed toward a common goal. Distinguished from *nonsocial activity, parallel play,* and *associative play.* (p. 249)

core knowledge perspective A view that assumes infants begin life with innate knowledge systems, or core domains of thought, each of which permits a read grasp of new related information and therefore supports early, rapid development of certain aspects of cognition. (p. 150)

coregulation A transitional form of supervision in which parents exercise general oversight, while permitting children to be in charge of moment-by-moment decision making. (p. 328)

corpus callosum A large bundle of fibers that connects the two hemispheres of the brain. (p. 328)

correlation coefficient A number, ranging from +1.00 to –1.00, that describes the strength and direction of the relationship between two variables. The size of the number shows the strength of the relationship. The sign of the number (+ or -) refers to the direction of the relationship. (p. 32)

correlational design A research design that gathers information without altering participants' experiences and examines relationships between variables. Cannot determine cause and effect. (p. 31)

creativity The ability to produce work that is original yet appropriate—something others have not thought of that is useful in some way. (p. 306)

cross-linkage theory of aging A theory of biological aging asserting that the formation of bonds, or links, between normally separate protein fibers causes the body's connective tissue to become less elastic over time, leading to many negative physical consequences. (p. 414)

cross-sectional design A research design in which groups of participants of different ages are studied at the same point in time. Distinguished from *longitudinal design.* (p. 35)

crowd A large, loosely organized group consisting of several cliques with similar normative characteristics. (p. 398)

crystallized intelligence Intellectual skills that depend on accumulated knowledge and experience, good judgment, and mastery of social conventions. Together, these capacities represent abilities acquired because they are valued by the individual's culture. Distinguished from *fluid intelligence.* (p. 498)

death anxiety Fear and apprehension of death. (p. 626)

decentration The ability to focus on several aspects of a problem at once and relate them. Distinguished from *centration.* (p. 285)

deferred imitation The ability to remember and copy the behavior of models who are not present. (p. 147)

dementia A set of disorders occurring almost entirely in old age in which many aspects of thought and behavior are so impaired that everyday activities are disrupted. (p. 564)

deoxyribonucleic acid (DNA) Long, double-stranded molecules that make up chromosomes. (p. 44)

dependency–support script A typical pattern of interaction in which elders' dependency behaviors are attended to immediately, thereby reinforcing those behaviors. Distinguished from *independence–ignore script.* (p. 591)

dependent variable The variable the researcher expects to be influenced by the independent variable in an experiment. Distinguished from *independent* variable. (p. 33)

developmental quotient, or **DQ** A score on an infant intelligence test, based primarily on perceptual and motor responses. Computed in the same manner as an IQ. (p. 160)

developmentally appropriate practice A set of standards devised by the National Association for the Education of Young Children that specify program characteristics that meet the developmental and individual needs of young children of varying ages, based on current research and the consensus of experts. (p. 162)

differentiation theory The view that perceptual development involves the detection of increasingly fine-grained, invariant features in the environment. (p. 139)

difficult child A child whose temperament is such that he or she is irregular in daily routines, is slow to accept new experiences, and tends to react negatively and intensely. Distinguished from *easy child* and *slow-to-warm-up child.* (p. 181)

discontinuous development A view in which new and different ways of interpreting and responding to the world emerge at particular time periods. Distinguished from *continuous development.* (p. 6)

disengagement theory A social theory of aging that states that the decline in social interaction in late adulthood is due to mutual withdrawal between elders and society in anticipation of death. Distinguished from *activity theory* and *socioemotional selectivity theory.* (p. 595)

disorganized-disoriented attachment The quality of insecure attachment characterizing infants who respond in a confused, contradictory fashion when reunited with the parent. Distinguished from *secure, avoidant,* and *resistant attachment.* (p. 187)

distributive justice Beliefs about how to divide resources fairly. (p. 321)

divergent thinking The generation of multiple and unusual possibilities when faced with a task or problem. Associated with creativity. Distinguished from *convergent thinking.* (p. 306)

divorce mediation A series of meetings between divorcing adults and a trained professional, who tries to help them settle disputes. Aimed at reducing family conflict during the period surrounding divorce. (p. 331)

dominant cerebral hemisphere The hemisphere of the brain responsible for skilled motor action. The left hemisphere is dominant in right-handed individuals. In left-handed individuals, the right hemisphere may be dominant, or motor and language skills may be shared between the hemispheres. (p. 207)

dominant–recessive inheritance A pattern of inheritance in which, under heterozygous conditions, the influence of only one gene is apparent. (p. 46)

dual-earner marriage A family form in which both husband and wife are employed. (p. 473)

dual-process model of coping with loss A perspective that assumes that effective coping with loss requires people to oscillate between dealing with the emotional consequences of loss and attending to life changes, which—when handled successfully—have restorative, or healing, effects. (p. 640)

dualistic thinking In Perry's theory, the cognitive approach of younger college students, who search for absolute truth and therefore divide information, values, and authority into right and wrong, good and bad, we and they. Distinguished from *relativistic thinking.* (p. 432)

durable power of attorney for health care A written statement that authorizes appointment of another person (usually, although not always, a family member) to make health care decisions on one's behalf in case of incompetence. (p. 635)

dynamic systems theory of motor development In motor development, combinations of previously acquired abilities that lead to more advanced ways of exploring and controlling the environment. Each new skill is a joint product of central nervous system development, movement possibilities of the body, environmental supports for the skill, and the goal the child has in mind. (p. 130)

dynamic testing An approach to testing consistent with Vygotsky's concept of the zone of proximal development, in which purposeful teaching is introduced into the testing situation to find out what the child can attain with social support. (p. 298)

easy child A child whose temperament is such that he or she quickly establishes regular routines in infancy, is generally cheerful, and adapts easily to new experiences. Distinguished from *difficult child* and *slow-to-warm-up child.* (p. 180)

ecological systems theory Bronfenbrenner's approach, which views the person as developing within a complex system of relationships affected by multiple levels of the environment, from immediate settings of family and school to broad cultural values and programs. (p. 24)

educational self-fulfilling prophecy The idea that children may adopt teachers' positive or negative attitudes toward them and start to live up to these views. (p. 304)

egalitarian marriage A form of marriage in which husband and wife share power and authority. Both try to balance the time and energy they devote to the workplace, the children, and their relationship. Distinguished from *traditional marriage.* (p. 458)

ego integrity versus despair In Erikson's theory, the psychological conflict of late adulthood, which is resolved positively when elders feel whole, complete, and satisfied with their achievements, having accepted their life course as something that had to be the way it was. (p. 584)

egocentrism Failure to distinguish the symbolic viewpoints of others from one's own. (p. 217)

elaboration The memory strategy of creating a relation between two or more items that are not members of the same category. (p. 290)

embryo The prenatal organism from 2 to 8 weeks after conception, during which time the foundations of all body structures and internal organs are laid down. (p. 79)

emotional self-regulation Strategies for adjusting our emotional state to a comfortable level of intensity so we can accomplish our goals. (p. 179)

empathy The ability to understand another's emotional state and feel with that person, or respond emotionally in a similar way. (p. 196)

epigenesis Development resulting from ongoing, bidirectional exchanges between heredity and all levels of the environment. (p. 71)

ethnic identity An enduring aspect of the self that includes a sense of ethnic group membership and attitudes and feelings associated with that membership. (p. 386)

ethnography A method by which the researcher attempts to understand the unique values and social processes of a culture or a distinct social group, achieving this goal through participant observation. The researcher lives with the cultural community for a period of months or years, participating in all aspects of its daily life. (p. 30)

ethological theory of attachment A theory formulated by Bowlby, which views the infant's emotional tie to the mother as an evolved response that promotes survival. (p. 186)

ethology An approach concerned with the adaptive, or survival, value of behavior and its evolutionary history. (p. 22)

euthanasia The practice of ending the life of a person suffering from an incurable condition. (p. 634)

evolutionary developmental psychology A new area of research that seeks to understand the adaptive value of species-wide cognitive, emotional, and social competencies as those competencies change with age. (p. 22)

executive stage In Schaie's theory, a more advanced form of the responsibility stage that characterizes people at the helm of large organizations, whose responsibilities have become highly complex. (p. 433)

exosystem In ecological systems theory, settings that do not contain children but that affect their experiences in immediate settings. (p. 25)

expansions Adult responses that elaborate on children's speech, increasing its complexity. (p. 238)

experimental design A research design in which the investigator randomly assigns participants to two or more treatment conditions. Permits inferences about cause and effect. (p. 33)

expertise Acquisition of extensive knowledge in a field or endeavor, supported by the specialization that begins with selecting a college major or an occupation in early adulthood. (p. 433)

expressive style of language learning A style of early language learning in which toddlers frequently produce pronouns and social formulas, such as "stop it," "thank you," and "I want it." They use language mainly to talk about the feelings and needs of themselves and other people. Distinguished from *referential style.* (p. 168)

extended family household A household in which three or more generations live together. (p. 62)

family life cycle A sequence of phases that characterizes the development of most families around the world. In early adulthood, people typically live on their own, marry, and bear and rear children. During middle age, their parenting responsibilities diminish. Late adulthood brings retirement, growing old, and (mostly for women) death of one's spouse. (p. 456)

fantasy period Period of vocational development in which young children fantasize about career options through make-believe play Distinguished from *tentative period* and *realistic period.* (p. 437)

fast-mapping Connecting a new word with an underlying concept after only a brief encounter. (p. 236)

feminization of poverty A trend in which women who support themselves or their families have become the majority of the adult poverty population, regardless of age and ethnic group. (p. 525)

fetal alcohol effects (FAE) The condition of children who display some but not all of the defects of fetal alcohol syndrome. Usually their mothers drank alcohol in smaller quantities during pregnancy than did mothers of children with fetal alcohol syndrome (FAS). (p. 86)

fetal alcohol syndrome (FAS) A set of defects that results when women consume large amounts of alcohol during most or all of pregnancy. Includes mental retardation; impaired motor coordination, attention, memory, and language; overactivity; slow physical growth, and facial abnormalities. (p. 86)

fetal monitors Electronic instruments that track the baby's heart rate during labor. (p. 95)

fetus The prenatal organism from the beginning of the third month to the end of pregnancy, during which time completion of body structures and dramatic growth in size takes place. (p. 79)

fluid intelligence Intellectual skills that largely depend on basic information-processing skills—the speed with which we can analyze information, the capacity of working memory, and the ability to detect relationships among stimuli. Distinguished from *crystallized intelligence.* (p. 499)

formal operational stage Piaget's final stage, in which adolescents develop the capacity for abstract, scientific thinking. Begins around 11 years of age. (p. 363)

fraternal, or **dizygotic, twins** Twins resulting from the release and fertilization of two ova. They are genetically no more alike than ordinary siblings. Distinguished from *identical,* or *monozygotic, twins.* (p. 46)

free radicals Naturally occurring, highly reactive chemicals that form in the presence of oxygen and destroy cellular material, including DNA, proteins, and fats essential for cell functioning. (p. 413)

full inclusion Placement of pupils with learning difficulties in regular classrooms for the entire school day. (p. 305)

functional age Actual competence and performance of an older adult (as distinguished from chronological age). (p. 546)

gametes Human sperm and ova, which contain half as many chromosomes as a regular body cell. (p. 45)

gender constancy The understanding that sex is biologically based and remains the same even if clothing, hairstyle, and play activities change. (p. 263)

gender identity An image of oneself as relatively masculine or feminine in characteristics. (p. 263)

gender intensification Increased gender stereotyping of attitudes and behavior. Occurs in early adolescence. (p. 394)

gender schema theory An information-processing approach to gender typing that combines social learning and cognitive-developmental features to explain how environmental pressures and children's cognitions work together to shape gender-role development. (p. 263)

gender typing The process of developing gender roles, or gender-linked preferences and behaviors valued by the larger society. (p. 260)

gene A segment of a DNA molecule that contains instructions for production of various proteins that contribute to growth and functioning of the body. (p. 44)

general growth curve A curve that represents overall changes in body size—rapid growth during infancy, slower gains in early and middle childhood, and rapid growth once more during adolescence. (p. 205)

generativity versus stagnation In Erikson's theory, the psychological conflict of midlife, which is resolved positively if the adult can integrate personal goals with the welfare of the larger social environment. The resulting strength is the capacity to give to and guide the next generation. (p. 512)

genetic counseling A communication process designed to help couples assess their chances of giving birth to a baby with a hereditary disorder. (p. 52)

genetic imprinting A pattern of inheritance in which genes are imprinted, or chemically marked, in such a way that one member of the pair (either the mother's or the father's) is activated, regardless of its makeup. (p. 49)

genetic–environmental correlation The idea that heredity influences the environments to which people are exposed. (p.69)

genotype The genetic makeup of the individual. Distinguished from *phenotype.* (p. 44)

giftedness Exceptional intellectual ability. Includes high IQ, high creativity, and specialized talent. (p. 305)

glass ceiling Invisible barrier, faced by women and ethnic minorities, to advancement up the corporate ladder. (p. 536)

glaucoma A disease in which pressure builds up within the eye due to poor fluid drainage, damaging the optic nerve. A leading cause of blindness among older adults. (p. 484)

glial cells Cells serving the function of myelination. (p. 116)

goodness of fit An effective match between child-rearing practices and a child's temperament, leading to favorable adjustment. (p. 184)

grief Intense physical and psychological distress following to loss of a loved one. (p. 639)

growth hormone (GH) A pituitary hormone that from birth on is necessary for development of all body tissues except the central nervous system and genitals. (p. 208)

growth spurt Rapid gain in height and weight during adolescence. (p. 346)

habituation A gradual reduction in the strength of a response as the result of repetitive stimulation. (p. 128)

hardiness A set of three personal qualities—control, commitment, and challenge—that help people cope with stress adaptively, thereby reducing its impact on illness and mortality. (p. 496)

heritability estimate A statistic that measures the extent to which individual differences in complex traits, such as intelligence or personality, in a specific population are due to genetic factors. (p. 67)

heteronomous morality Piaget's first stage of moral development, in which children view moral rules as permanent features of the external world that are handed down by authorities and cannot be changed. Extends from about 5 to 10 years of age. (p. 388)

heterozygous Having two different genes at the same place on a pair of chromosomes. Distinguished from *homozygous*. (p. 46)

hierarchical classification The organization of objects into classes and subclasses on the basis of similarities and differences between the groups. (p. 218)

history-graded influences Influences on lifespan development that are unique to a particular historical era and explain why people born around the same time (called a *cohort*) tend to be alike in ways that set them apart from people born at other times. (p. 11)

Home Observation for Measurement of the Environment (HOME) A checklist for gathering information about the quality of children's home lives through observation and parental interview. (p. 161)

homozygous Having two identical genes at the same place on a pair of chromosomes. Distinguished from *heterozygous*. (p. 46)

horizontal décalage Development within a Piagetian stage. Gradual mastery of logical concepts during the concrete operational stage is an example. (p. 286)

hormone therapy Low daily doses of estrogen, aimed at reducing the discomforts of menopause and protecting women from other impairments due to estrogen loss, such as bone deterioration. Comes in two types: (1) estrogen alone, or *estrogen replacement therapy (ERT)*, for women who have had hysterectomies (surgical removal of the uterus), and (2) estrogen plus progesterone, or *hormone replacement therapy (HRT)*, for other women. (p. 486)

hospice A comprehensive program of support services that focuses on meeting terminally ill patients' physical, emotional, social, and spiritual needs and that offers follow-up bereavement services to families. (p. 632)

hostile aggression Aggression intended to harm another individual. Distinguished from *instrumental aggression*. (p. 257)

human development A field of study devoted to understanding constancy and change throughout the lifespan. (p. 5)

hypothetico-deductive reasoning A formal operational problem-solving strategy in which adolescents begin with a general theory of all possible factors that could affect an outcome in a problem and deduce specific hypotheses, which they test systematically. (p. 363)

I-self A sense of self as subject, or agent, who is separate from but attends to and acts on objects and other people. Distinguished from *me-self*. (p. 195)

identical, or **monozygotic, twins** Twins that result when a zygote, during the early stages of cell duplication, divides in two. They have the same genetic makeup. Distinguished from *fraternal*, or *dizygotic, twins*. (p. 46)

identity A well-organized conception of the self, made up of values, beliefs, and goals to which the individual is solidly committed. (p. 382)

identity achievement The identity status of individuals who have explored and committed themselves to self-chosen values and occupational goals. Distinguished from *moratorium, identity foreclosure*, and *identity diffusion*. (p. 384)

identity diffusion The identity status of individuals who do not have firm commitments to values and goals and are not actively trying to reach them. Distinguished from *identity achievement, moratorium*, and *identity foreclosure*. (p. 384)

identity foreclosure The identity status of individuals who have accepted ready-made values and goals that authority figures have chosen for them. Distinguished from *identity achievement, moratorium*, and *identity diffusion*. (p. 384)

identity versus identity confusion In Erikson's theory, the psychological conflict of adolescence, which is resolved positively when adolescents attain an identity after a period of exploration and inner soul-searching. (p. 382)

imaginary audience Adolescents' belief that they are the focus of everyone else's attention and concern. (p. 367)

imitation Learning by copying the behavior of another person. Also called *modeling* or *observational learning*. (p. 128)

implantation Attachment of the blastocyst to the uterine lining 7 to 9 days after fertilization. (p. 78)

implicit memory Memory without conscious awareness. (p. 571)

independence–ignore script A typical pattern of interaction in which elders' independent behaviors are mostly ignored, thereby leading them to occur less often. Distinguished from *dependency–support script*. (p. 591)

independent variable In an experiment, the variable the investigator expects to cause changes in another variable and that the researcher manipulates by randomly assigning participants to treatment conditions. Distinguished from *dependent variable*. (p. 33)

individualistic societies Societies in which people think of themselves as separate entities and are largely concerned with their own personal needs. Distinguished from *collectivist societies*. (p. 63)

induction A type of discipline in which the effects of the child's misbehavior on others are communicated to the child. (p. 253)

industry versus inferiority In Erikson's theory, the psychological conflict of middle childhood, which is resolved positively when experiences lead children to develop a sense of competence at useful skills and tasks. (p. 314)

infant mortality The number of deaths in the first year of life per 1,000 live births. (p. 98)

information processing An approach that views the human mind as a symbol-manipulating system through which information flows, that often uses flowcharts to map the precise series of steps individuals use to solve problems and complete tasks, and that regards cognitive development as a continuous process. (p. 21)

information-loss view A view that attributes age-related slowing of cognitive processing to greater loss of information as it moves through the system. As a result, the whole system must slow down to inspect and interpret the information. Distinguished from *neural network view*. (p. 501)

inhibited, or **shy, child** A child whose temperament is such that he or she reacts negatively to and withdraws from novel stimuli. Distinguished from *uninhibited*, or *sociable, child*. (p. 182)

initiative versus guilt In Erikson's theory, the psychological conflict of early childhood, which is resolved positively through play experiences that foster a healthy sense of initiative and through development of a superego, or conscience, that is not overly strict and guilt-ridden. (p. 244)

instrumental aggression Aggression aimed at obtaining an object, privilege, or space with no deliberate intent to harm another person. Distinguished from *hostile aggression*. (p. 257)

intelligence quotient, or **IQ** A score that permits an individual's performance on an intelligence test to be compared to the performances of other individuals of the same age. (p. 160)

intentional, or **goal-directed, behavior** A sequence of actions in which schemes are deliberately combined to solve a problem. (p. 146)

interactional synchrony A sensitively tuned "emotional dance," in which the caregiver responds to infant signals in a well-timed, appropriate fashion and both partners match emotional states, especially the positive ones. (p. 189)

intermodal perception Perception that combines information from more than one modality, or sensory system. (p. 138)

internal working model A set of expectations derived from early caregiving experiences concerning the availability of attachment figures and their likelihood of providing support during times of stress. Becomes a model, or guide, for all future close relationships. (p. 187)

intimacy versus isolation In Erikson's theory, the psychological conflict of young adulthood, which is resolved positively when young adults give up some of their newfound independence and make a permanent commitment to an intimate partner. (p. 446)

invariant features Features that remain stable in a constantly changing perceptual world. (p. 139)

irreversibility The inability to mentally go through a series of steps in a problem and then reverse direction, returning to the starting point. Distinguished from *reversibility.* (p. 218)

joint custody A child custody arrangement following divorce in which the court grants each parent equal say in important decisions about the child's upbringing. (p. 331)

kinkeeper Role assumed by members of the middle generation, especially mothers, who take responsibility for gathering the family for celebrations and making sure everyone stays in touch. (p. 526)

kinship studies Studies comparing the characteristics of family members to determine the importance of heredity in complex human characteristics. (p. 67)

kwashiorkor A disease usually appearing between 1 and 3 years of age that is caused by a diet low in protein. Symptoms include an enlarged belly, swollen feet, hair loss, skin rash, and irritable, listless behavior. (p. 125)

language acquisition device (LAD) In Chomsky's theory, a biologically based innate system that permits children, no matter which language they hear, to understand and speak in a rule-oriented fashion as soon as they have picked up enough words. (p. 165)

lanugo A white, downy hair that covers the entire body of the fetus, helping the vernix stick to the skin. (p. 80)

lateralization Specialization of functions of the two hemispheres of the cerebral cortex. (p. 118)

learned helplessness Attributions that credit success to external factors, such as luck, and failure to low ability. Leads to anxious loss of control in the face of challenging tasks. Distinguished from *mastery-oriented attributions.* (p. 317)

learning disabilities Specific learning disorders that lead children to achieve poorly in school, despite an average or above-average IQ. Believed to be due to faulty brain functioning. (p. 305)

life care communities Housing for the elderly that offers a range of alternatives, from independent or congregate housing to full nursing home care. For a large initial payment and additional monthly fees, guarantees that elders' needs will be met in one place as they age. (p. 599)

life expectancy crossover An age-related reversal in life expectancy of sectors of the population. For example, members of ethnic minorities who survive to age 85 live longer than members of the white majority. (p. 548)

life review The process of calling up, reflecting on, and reconsidering past experiences, contemplating their meaning with the goal of achieving greater self-understanding. (p. 586)

life structure In Levinson's theory, the underlying pattern or design of a person's life at a given time. Consists of relationships with significant others (the most important of which have to do with marriage/family and occupation) that are reorganized during each period of adult development. (p. 447)

lifespan perspective A balanced perspective that assumes development is lifelong, multidimensional and multidirectional, highly plastic, and embedded in multiple contexts. (p. 8)

living will A written statement that specifies the treatments a person does or does not want in case of a terminal illness, coma, or other near-death situation. (p. 635)

loneliness Feelings of unhappiness that result from a gap between actual and desired social relationships. (p. 455)

longitudinal design A research design in which in which one group of participants is studied repeatedly at different ages. Distinguished from *cross-sectional design.* (p. 34)

longitudinal-sequential design A research design with both longitudinal and cross-sectional components in which groups of participants born in different years are followed over time. (p. 35)

long-term memory In information processing, the part of the mental system that contains our permanent knowledge base. (p. 153)

macrosystem In ecological systems theory, the values, laws, customs, and resources of a culture that influence experiences and interactions at inner levels of the environment. (p. 25)

macular degeneration Blurring and eventual loss of central vision due to a break-down of light-sensitive cells in the macula, or central region of the retina. (p. 552)

mainstreaming The integration of pupils with learning difficulties into regular classrooms for part of the school day. (p. 305)

make-believe play A type of play in which children pretend, acting out everyday and imaginary activities. (p. 147)

marasmus A disease usually appearing in the first year of life that is caused by a diet low in all essential nutrients. Leads to a wasted condition of the body. (p. 125)

mastery-oriented attributions Attributions that credit success to high ability and failure to insufficient effort. Leads to high self-esteem and a willingness to approach challenging tasks. Distinguished from *learned helplessness.* (p. 317)

maturation A genetically determined, naturally unfolding course of growth. (p. 13)

maximum lifespan The genetic limit to length of life for a person free of external risk factors. Distinguished from *average life expectancy* and *active lifespan.* (p. 548)

me-self A reflective observer who treats the self as an object of knowledge and evaluation. Distinguished from *I-self.* (p. 195)

meiosis The process of cell division through which gametes are formed and in which the number of chromosomes in each cell is halved. (p. 45)

memory strategies Deliberate mental activities that improve the likelihood of remembering. (p. 227)

menarche First menstruation. (p. 348)

menopause The end of menstruation and, therefore, reproductive capacity in women. Occurs, on the average, at age 51 among North American women, although the age range is large—from 42 to 58. (p. 495)

mental representation Internal depiction of information that the mind can manipulate The most powerful mental representations are images and concepts. (p. 147)

mental strategies In information processing, procedures that operate on and transform information, increasing the chances that we will retain information and use it efficiently. (p. 153)

mesosystem In ecological systems theory, connections between children's immediate settings. (p. 25)

metacognition Thinking about thought; awareness of mental activities. (p. 228)

microsystem In ecological systems theory, the activities and interaction patterns in the person's immediate surroundings. (p. 24)

midlife crisis Inner turmoil, self-doubt, and major restructuring of the personality during the transition to middle adulthood. Characterizes the experiences of only a minority of adults. (p. 517)

mild mental retardation Substantially below-average intellectual functioning, resulting in an IQ between 55 and 70 and problems in adaptive behavior, or skills of everyday living. (p. 305)

mitosis The process of cell duplication, in which each new cell receives an exact copy of the original chromosomes. (p. 45)

moratorium The identity status of individuals who are exploring alternatives in an effort to find values and goals to guide their life. Distinguished from *identity achievement, identity foreclosure,* and *identity diffusion.* (p. 384)

mortality The phase of dying in which the individual passes into permanent death. Distinguished from *agonal phase* and *clinical death.* (p. 629)

mourning The culturally specified expression of the bereaved person's thoughts and feelings through funerals and other rituals. (p. 621)

mutation A sudden but permanent change in a segment of DNA. (p. 49)

myelination A process in which neural fibers are coated with an insulating fatty sheath called *myelin* that improves the efficiency of message transfer. (p. 116)

natural, or **prepared, childbirth** An approach designed to reduce pain and medical intervention and to make childbirth a rewarding experience for parents. (p. 94)

naturalistic observation A method in which the researcher goes into the natural environment to observe the behavior of interest. Distinguished from *structured observation.* (p. 27)

nature– nurture controversy Disagreement among theorists about whether genetic or environmental factors are the most important determinants of development and behavior. (p. 7)

neglected children Children who are seldom chosen, either positively or negatively, on self-report measures of peer acceptance. Distinguished from *popular, rejected,* and *controversial children.* (p. 324)

Neonatal Behavioral Assessment Scale (NBAS) A test developed to assess the behavior of the infant during the newborn period. (p. 107)

neural network view A view that attributes age-related slowing of cognitive processing to breaks in neural networks as neurons die. The brain forms bypasses—new synaptic connections that go around the breaks but are less efficient. Distinguished from *information-loss view.* (p. 501)

neural tube The primitive spinal cord that develops from the ectoderm, the top of which swells to form the brain. (p. 79)

neurofibrillary tangles A structural change in the brain associated with Alzheimer's disease in which abnormal bundles of threads run through the body of the neuron and into fibers establishing synaptic connections with other neurons. (p. 565)

neurons Nerve cells that store and transmit information. (p. 116)

niche-picking A type of genetic–environmental correlation in which individuals actively choose environments that complement their heredity. (p. 69)

noble savage Rousseau's view of the child as naturally endowed with an innate plan for orderly, healthy growth. (p. 13)

nonfunctionality The component of the death concept specifying that all living functions, including thought, feeling, movement, and body processes, cease at death. (p. 623)

nonnormative influences Influences on lifespan development that are irregular, in that they happen to just one or a few individuals and do not follow a predictable timetable. (p. 11)

nonorganic failure to thrive A growth disorder usually present by 18 months of age that is caused by lack of affection and stimulation. (p. 126)

non-rapid-eye-movement (NREM) sleep A "regular" sleep state in which the body is quiet and heart rate, breathing, and brain wave activity are slow and regular. Distinguished from *rapid-eye-movement (REM) sleep.* (p. 102)

nonsocial activity Unoccupied, onlooker behavior and solitary play. Distinguished from *parallel, associative,* and *cooperative play.* (p. 249)

normative approach An approach in which measures of behavior are taken on large numbers of individuals and age related averages are computed to represent typical development. (p. 14)

obesity A greater-than-20-percent increase over average body weight, based on the individual's age, sex, and physical build. (p. 278)

object permanence The understanding that objects continue to exist when they are out of sight. (p. 146)

old-old Showing signs of physical decline in old age. Distinguished from *young-old.* (p. 546)

open classroom An elementary school classroom based on the educational philosophy that children are active agents in their own development. The teacher assumes a flexible authority role, sharing decision making with students, who learn at their own pace. Students are evaluated in relation to their own prior development. Distinguished from *traditional classroom.* (p. 302)

operant conditioning A form of learning in which a spontaneous behavior is followed by a stimulus that changes the probability that the behavior will occur again. (p. 127)

oral rehydration therapy (ORT) A treatment for diarrhea, in which sick children are given a glucose, salt, and water solution that quickly replaces fluids the body loses. (p. 209)

ordinality A principle specifying order relationships between quantities, such as three is more than two and two is more than one. (p. 231)

organization In Piaget's theory, the internal rearrangement and linking together of schemes so that they form a strongly interconnected cognitive system. In information processing, the memory strategy of grouping together related items. (p. 145; p. 289)

osteoarthritis A form of arthritis characterized by deteriorating cartilage on the ends of bones of frequently used joints. Leads to swelling, stiffness, and loss of flexibility. Otherwise known as "wear-and-tear" arthritis. Distinguished from *rheumatoid arthritis.* (p. 562)

osteoporosis A severe version of age-related bone loss. Porous bones are easily fractured and when very extreme, lead to a slumped-over posture, a shuffling gait, and a "dowager's hump" in the upper back. (p. 493)

overextension An early vocabulary error in which a word is applied too broadly, to a wider collection of objects and events than is appropriate. Distinguished from *underextension.* (p. 167)

overregularization Application of regular grammatical rules to words that are exceptions. (p. 236)

overt aggression A form of hostile aggression that harms others through physical injury or the threat of such injury—for example, hitting, kicking, or threatening to beat up a peer. (p. 257)

palliative, or **comfort, care** Care for terminally ill, suffering patients that relieves pain and other symptoms (such as breathing difficulties, insomnia, and depression), aimed at protecting the patient's quality of life rather than prolonging life. (p. 632)

parallel play A form of limited social participation in which the child plays near other children with similar materials but does not interact with them. Distinguished from *nonsocial, associative,* and *cooperative play.* (p. 249)

parental imperative theory A theory that claims that traditional gender roles are maintained during the active parenting years to help ensure the survival of children. After children reach adulthood, parents are free to express the "other-gender" side of their personalities. (p. 521)

passionate love Love based on intense sexual attraction. Distinguished from *companionate love.* (p. 451)

passive euthanasia The practice of withholding or withdrawing life-sustaining treatment, permitting a patient to die naturally. Distinguished from *voluntary active euthanasia.* (p. 634)

peer acceptance Likability, or the extent to which a child is viewed by a group of agemates (such as classmates) as a worthy social partner. (p. 324)

peer group Collectives that generate unique values and standards for behavior and a social structure of leaders and followers. (p. 322)

peer victimization A destructive form of peer interaction in which certain children become frequent targets of verbal and physical attacks or other forms of abuse. (p. 325)

permanence The component of the death concept specifying that once a living thing dies, it cannot be brought back to life. (p. 623)

permissive style A child-rearing style that is high in acceptance but overindulging or inattentive, low in control, and lenient rather than appropriate in autonomy granting. Distinguished from *authoritative, authoritarian,* and *uninvolved styles.* (p. 265)

persistent vegetative state A state produced by absence of brain wave activity in the cortex, in which the person is unconscious, displays no voluntary movements, and has no hope of recovery. (p. 621)

personal fable Adolescents' belief that they are special and unique. Leads them to conclude that others cannot possibly understand their thoughts and feelings and may promote a sense of invulnerability to danger. (p. 367)

perspective taking The capacity to imagine what other people may be thinking and feeling. (p. 320)

phenotype The individual's physical and behavioral characteristics, which are determined by both genetic and environmental factors. Distinguished from *genotype.* (p. 44)

pituitary gland A gland located near the base of the brain that releases hormones that induce physical growth. (p. 208)

placenta The organ that separates the mother's bloodstream from the embryo or fetal bloodstream but permits exchange of nutrients and waste products. (p. 78)

polygenic inheritance A pattern of inheritance in which many genes determine a characteristic. (p. 50)

popular children Children who get many positive votes on self-report measures of peer acceptance. Distinguished from *rejected, controversial,* and *neglected children.* (p. 324)

popular-prosocial children A subgroup of popular children who combine academic and social competence. Distinguished from *popular-antisocial children.* (p. 324)

popular-antisocial children A subgroup of popular children largely made up of "tough" boys who are athletically skilled, aggressive, and poor students. Distinguished from *popular-prosocial children.* (p. 324)

possible selves Future-oriented representations of what one hopes to become and is afraid of becoming. The temporal dimension of self-concept. (p. 518)

postconventional level Kohlberg's highest level of moral development, in which individuals define morality in terms of abstract principles and values that apply to all situations and societies. (p. 390)

postformal thought Cognitive development beyond Piaget's formal operational stage. (p. 432)

practical problem solving Problem solving that requires people to size up real-world situations and analyze how best to achieve goals that have a high degree of uncertainty. (p. 503)

pragmatic thought In Labouvie-Vief's theory, adult thought in which logic becomes a tool to solve real-world problems and inconsistencies and imperfections are accepted. (p. 433)

pragmatics The practical, social side of language that is concerned with how to engage in effective and appropriate communication with others. (p. 237)

preconventional level Kohlberg's first level of moral development, in which moral understanding is based on rewards, punishments, and the power of authority figures. (p. 389)

preformationism Medieval view of the child as a miniature adult. (p. 13)

premenstrual syndrome (PMS) An array of physical and psychological symptoms that usually appear 6 to 10 days prior to menstruation. The most common are abdominal cramps, fluid retention, diarrhea, tender breasts, backache, headache, fatigue, tension, irritability, and depression. (p. 430)

prenatal diagnostic methods Medical procedures that permit detection of developmental problems before birth. (p. 52)

preoperational stage Piaget's second stage, in which rapid growth in representation takes place. However, thought is not yet logical. Spans the years from 2 to 7. (p. 216)

presbycusis Age-related hearing impairments that involve a sharp loss at high frequencies around age 50, which gradually extends to all frequencies. (p. 484)

presbyopia Condition of aging in which, around age 60, the lens of the eye loses its capacity to accommodate entirely to nearby objects. (p. 483)

preterm Infants born several weeks or more before their due date. (p. 97)

primary aging Genetically influenced age-related declines in the functioning of organs and systems that affect all members of our species and take place even in the context of overall good health. Also called *biological aging.* Distinguished from *secondary aging.* (p. 561)

primary sexual characteristics Physical features that involve the reproductive organs directly (ovaries, uterus, and vagina in females; penis, scrotum, and testes in males). Distinguished from *secondary sexual characteristics.* (p. 347)

principle of mutual exclusivity The assumption by children in the early stages of vocabulary growth that words mark entirely separate (nonoverlapping) categories. (p. 236)

private speech Self-directed speech that children often use to plan and guide their own behavior. (p. 223)

Project Head Start A U.S. federal program that provides poverty-stricken children with a year or two of preschool along with nutritional and health services and that encourages parent involvement in program planning and children's learning. (p. 233)

propositional thought A type of formal operational reasoning in which adolescents evaluate the logic of verbal statements without referring to real-world circumstances. (p. 354)

prosocial, or **altruistic, behavior** Actions that benefit another person without any expected reward for the self. (p. 249)

prospective memory Recall that involves remembering to engage in planned actions at an appropriate time in the future. (p. 573)

proximodistal trend An organized pattern of physical growth and motor control that proceeds from the center of the body outward. Distinguished from *cephalocaudal trend.* (p. 115)

psychoanalytic perspective An approach to personality development introduced by Freud that assumes children move through a series of stages in which they confront conflicts between biological drives and social expectations. The way these conflicts are resolved determines psychological adjustment. (p. 15)

psychosexual theory Freud's theory, which emphasizes that how parents manage children's sexual and aggressive drives during the first few years is crucial for healthy personality development. (p. 15)

psychosocial dwarfism A growth disorder observed between 2 and 15 years of age. Characterized by very short stature, decreased GH secretion, immature skeletal age, and serious adjustment problems. Caused by emotional deprivation. (p. 208)

psychosocial theory Erikson's theory, which emphasizes that at each Freudian stage, individuals not only develop a unique personality, but also acquire attitudes and skills that help them become active, contributing members of their society. (p. 17)

puberty Biological changes at adolescence that lead to an adult-sized body and sexual maturity. (p. 344)

public policies Laws and government programs designed to improve current conditions. (p. 63)

punishment In operant conditioning, removing a desirable stimulus or presenting an unpleasant one to decrease the occurrence of a response. (p. 127)

random assignment An evenhanded procedure for assigning participants to treatment groups, such as drawing numbers out of a hat or flipping a coin. Increases the chances that participants' characteristics will be equally distributed across treatment conditions in an experiment. (p. 33)

range of reaction Each person's unique, genetically determined response to a range of environmental conditions. (p. 68)

rapid-eye-movement (REM) sleep An "irregular" sleep state in which brain wave activity is similar to that of the waking state; eyes dart beneath the lids, heart rate, blood pressure, and breathing are uneven, and slight muscle twitches occur. Distinguished from *non-rapid-eye-movement (NREM) sleep.* (p. 102)

realistic period Period of vocational development in which older adolescents and young adults focus on a general vocational category and, slightly later, settle on a single occupation. Distinguished from *fantasy period* and *tentative period.* (p. 437)

recall A type of memory that involves remembering a stimulus that is not present. (p. 155)

recasts Adult responses that restructure children's incorrect speech into a more mature form. (p. 238)

reciprocal teaching An approach to teaching based on Vygotsky's theory in which a teacher and two to four students form a collaborative learning group and take turns leading dialogues on the content of a text passage, using four cognitive strategies: questioning, summarizing, clarifying, and predicting. Creates a zone of proximal development in which reading comprehension improves. (p. 304)

recognition A type of memory that involves noticing whether a stimulus is identical or similar to one previously experienced. (p. 155)

recovery Following habituation, an increase in responsiveness to a new stimulus. (p. 128)

referential style of language learning A style of early language learning in which toddlers produce many words that refer to objects. They use language mainly to name things. Distinguished from *expressive style of language.* (p. 168)

reflex An inborn, automatic response to a particular form of stimulation. (p. 100)

rehearsal The memory strategy of repeating information. (p. 289)

reinforcer In operant conditioning, a stimulus that increases the occurrence of a response. (p. 127)

reintegrative stage Schaie's stage of late adulthood, in which people reexamine and reintegrate their interests, attitudes, and values, using them as a guide for maximizing quality of life as they retire and the future shortens. (p. 433)

rejected children Children who are actively disliked and get many negative votes on self-report measures of peer acceptance. Distinguished from *popular, controversial,* and *neglected children.* (p. 324)

rejected-aggressive children A subgroup of rejected children who engage in high rates of conflict, hostility, and hyperactive, inattentive, and impulsive behavior. Distinguished from *rejected-withdrawn children.* (p. 324)

rejected-withdrawn children A subgroup of rejected children who are passive and socially awkward. Distinguished from *rejected-aggressive children.* (p. 325)

relational aggression A form of hostile aggression that damages another's peer relationships, as in social exclusion or rumor spreading. (p. 257)

relativistic thinking In Perry's theory, the cognitive approach of older college students, who favor multiple truths, each relative to its context of evaluation. Distinguished from *dualistic thinking.* (p. 432)

reminiscence The process of telling stories about people and events from the past and reporting associated thoughts and feelings. (p. 586)

remote memory Recall of events that happened long ago. (p. 572)

resiliency The ability to adapt effectively in the face of threats to development. (p. 10)

resistant attachment The quality of insecure attachment characterizing infants who remain close to the parent before departure and display angry, resistive behavior when she returns. Distinguished from *secure, avoidant,* and *disorganized/disoriented attachment.* (p. 187)

respiratory distress syndrome A disorder of preterm infants in which the lungs are so immature that the air sacs collapse, causing serious breathing difficulties. (p. 97)

responsibility stage Schaie's stage of middle adulthood in which people adapt cognitive skills to an expansion of social responsibilities on the job, in the community, and at home. (p. 433)

reticular formation A brain structure that maintains alertness and consciousness. (p. 207)

reversibility The ability to think through a series of steps in a problem and then mentally reverse direction, returning to the starting point. Distinguished from *irreversibility.* (p. 285)

Rh factor A protein that, when present in the fetus's blood but not in the mother's, can cause the mother to build up antibodies. If these return to the fetus's system, they destroy red blood cells, reducing the oxygen supply to organs and tissues. (p. 89)

rheumatoid arthritis A form of arthritis in which the immune system attacks the body, resulting in inflammation of connective tissue, particularly the membranes that line the joints. The result is stiffness, inflammation, aching, deformed joints, and serious loss of mobility. Distinguished from *osteoarthritis.* (p. 562)

sandwich generation Today's middle-aged adults, who are "sandwiched," or squeezed, between the needs of ill or frail parents and financially dependent children. (p. 530)

scaffolding Adjusting the quality of support during a teaching session to fit the child's current level of performance. Direct instruction is offered when a task is new; less help is provided as competence increases. (p. 224)

scheme In Piaget's theory, a specific structure, or organized way of making sense of experience, that changes with age. (p. 144)

school phobia Severe apprehension about attending school, often accompanied by physical complaints that disappear once the child is allowed to remain home. (p. 334)

scripts General descriptions of what occurs and when it occurs in a particular situation. A basic means through which children organize and interpret their everyday experiences. (p. 227)

secondary aging Declines due to hereditary defects and environmental influences, such as poor diet, lack of exercise, substance abuse, environmental pollution, and psychological stress. Distinguished from *primary aging.* (p. 561)

secondary friends People who are not intimates but with whom the individual spends time occasionally, such as a group that meets for lunch, bridge, or museum tours. (p. 606)

secondary sexual characteristics Features visible on the outside of the body that serve as signs of sexual maturity but do not involve the reproductive organs (for example, breast development in females, appearance of underarm and pubic hair in both sexes). Distinguished from *primary sexual characteristics.* (p. 347)

secular trend in physical growth Change in body size and rate of growth from one generation to the next. For example, in industrialized nations, age of menarche declined from 1860 to 1970, signifying faster physical maturation in modern young people. (p. 349)

secure attachment The quality of attachment characterizing infants who are distressed by parental separation and easily comforted by the parent when she returns. Distinguished from *avoidant, resistant,* and *disorganized/disoriented attachment.* (p. 187)

secure base Infants' use of the familiar caregiver as a point from which to explore the environment and return for emotional support. (p. 186)

selective optimization with compensation A set of strategies that permits the elderly to sustain high levels of functioning. They *select* personally valued activities as a way of *optimizing* returns from their diminishing energies and come up with new ways of *compensating* for losses. (p. 570)

self-care children Children who look after themselves while their parents are at work. (p. 333)

self-concept The sum total of attributes, abilities, attitudes, and values that an individual believes defines who he or she is. (p. 245)

self-conscious emotions Emotions that involve injury to or enhancement of the sense of self. Examples are shame, embarrassment, guilt, envy, and pride. (p. 179)

self-control The capacity to resist an impulse to engage in socially disapproved behavior. (p. 196)

self-esteem An aspect of self-concept that involves judgments about one's own worth and the feelings associated with those judgments. (p. 245)

sensitive caregiving Caregiving involving prompt, consistent, and appropriate responding to infant signals. (p. 189)

sensitive period A time span that is optimal for certain capacities to emerge and in which the individual is especially responsive to environmental influences. (p. 22)

sensorimotor stage Piaget's first stage, during which infants and toddlers "think" with their eyes, ears, hands, and other sensorimotor equipment. Spans the first 2 years of life. (p. 144)

sensory register In information processing, that part of the mental system in which sights and sounds are held briefly before they decay or are transferred to working, or short-term, memory. (p. 153)

separation anxiety An infant's distressed reaction to the departure of the familiar caregiver. (p. 186)

seriation The ability to order items along a quantitative dimension, such as length or weight. (p. 285)

sex chromosomes The twenty-third pair of chromosomes, which determines the sex of the individual. In females, called *XX*; in males, called *XY*. (p. 46)

skipped-generation family A family structure in which children live with grandparents but apart from parents. (p. 528)

sleep apnea A condition during sleep in which breathing ceases for 10 seconds or longer, resulting in many brief awakenings. (p. 555)

slow-to-warm-up child A child whose temperament is such that he or she is inactive, shows mild, low-key reactions to environmental stimuli, is negative in mood, and adjusts slowly when faced with new experiences. Distinguished from *easy child* and *difficult child.* (p. 181)

small for date Infants whose birth weight is below their expected weight when length of pregnancy is taken into account. Some are full term; others are preterm infants who are especially underweight. (p. 97)

social clock Age-graded expectations for life events, such as beginning a first job, getting married, birth of the first child, buying a home, and retiring. (p. 449)

social comparisons Judgments of one's own abilities, behavior, appearance, and other characteristics in relation to those of others. (p. 315)

social convoy A model of age-related changes in social networks, which views the individual within a cluster of relationships moving through life. Close ties are in the inner circle, less close ties on the outside. With age, people change places in the convoy, new ties are added, and some are lost entirely. (p. 600)

social learning theory An approach that emphasizes the role of modeling, or observational learning, in the development of behavior. (p. 18)

social referencing Relying on a trusted person's emotional reaction to decide how to respond in an uncertain situation. (p. 178)

social smile The smile evoked by the stimulus of the human face. First appears between 6 and 10 weeks. (p. 177)

sociocultural theory Vygotsky's theory, in which children acquire the ways of thinking and behaving that make up a community's culture through cooperative dialogues with more knowledgeable members of society. (p. 23)

sociodramatic play The make-believe play with others that first appears around age 21⁄2 and increases rapidly until 4 to 5 years. (p. 217)

socioeconomic status (SES) A measure of a family's social position and economic well-being that combines three interrelated, but not completely overlapping, variables: (1) years of education and (2) the prestige of and skill required by one's job, both of which measure social status; and (3) income, which measures economic status. (p. 59)

socioemotional selectivity theory A social theory of aging that states that the decline in social interaction in late adulthood is due to physical and psychological changes, which lead elders to emphasize the emotion-regulating function of interaction. Consequently, they prefer familiar partners with whom they have developed pleasurable relationships. Distinguished from *disengagement theory* and *activity theory.* (p. 596)

spermarche First ejaculation of seminal fluid. (p. 348)

stage A qualitative change in thinking, feeling, and behaving that characterizes a specific period of development. (p. 6)

states of arousal Different degrees of sleep and wakefulness. (p. 102)

Strange Situation A procedure involving short separations from and reunions with the parent that assesses the quality of the attachment bond. (p. 187)

stranger anxiety The infant's expression of fear in response to unfamiliar adults. Appears in many babies after 6 months of age. (p. 177)

structured interview An interview method in which each participant is asked the same questions in the same way. (p. 29)

structured observation A method in which the investigator sets up a cue for the behavior of interest and observes it in a laboratory. Distinguished from *naturalistic observation.* (p. 27)

subculture A group of people with beliefs and customs that differ from those of the larger culture. (p. 62)

successful aging Aging in which gains are maximized and losses minimized. (p. 612)

sudden infant death syndrome (SIDS) The unexpected death, usually during the night, of an infant under 1 year of age that remains unexplained after thorough investigation. (p. 103)

sympathy Feelings of concern or sorrow for another's plight. (p. 249)

synapses The gaps between neurons, across which chemical messages are sent. (p. 116)

synaptic pruning Loss of connective fibers by seldom-stimulated neurons, thereby returning them to an uncommitted state so they can support the development of future skills. (p. 116)

tabula rasa Locke's view of the child as a blank slate whose character is shaped by experience. (p. 13)

talent Outstanding performance in a specific field. (p. 306)

telegraphic speech Toddlers' two-word utterances that, like a telegram, leave out smaller and less important words. (p. 167)

temperament Stable individual differences in quality and intensity of emotional reaction. (p. 180)

tentative period Period of vocational development in which adolescents weigh vocational options against their interests, abilities, and values. Distinguished from *fantasy period* and *realistic period.* (p. 437)

teratogen Any environmental agent that causes damage during the prenatal period. (p. 81)

terminal decline A steady, marked decrease in cognitive functioning prior to death. (p. 576)

theory An orderly, integrated set of statements that describes, explains, and predicts behavior. (p. 5)

theory of multiple intelligences Gardner's theory, which identifies eight independent intelligences on the basis of distinct sets of processing operations that permit individuals to engage in a wide range of culturally valued activities. (p. 295)

thyroid-stimulating hormone (TSH) A pituitary hormone that stimulates the thyroid gland to release thyroxine, which is necessary for normal brain development and body growth. (p. 208)

time out A form of mild punishment in which children are removed from the immediate setting until they are ready to act appropriately. (p. 255)

traditional classroom An elementary school classroom based on the educational philosophy that the teacher is the sole authority for knowledge, rules, and decision making. Students are relatively passive, listening, responding when called on, and completing teacher-assigned tasks. Their progress is evaluated by how well they keep up with a uniform set of standards for their grade. Distinguished from *open classroom.* (p. 302)

traditional marriage A form of marriage involving clear division of husband's and wife's roles. The man is the head of household and economic provider. The woman devotes herself to caring for her husband and children and creating a nurturant, comfortable home. Distinguished from *egalitarian marriage.* (p. 458)

transition Climax of the first stage of labor, in which the frequency and strength of contractions are at their peak and the cervix opens completely. (p. 91)

transitive inference The ability to seriate—or order items along a quantitative dimension—mentally. (p. 285)

triarchic theory of intelligence Sternberg's theory, which states that information-processing skills, prior experience with tasks, and contextual (or cultural) factors interact to determine intelligent behavior. (p. 294)

triangular theory of love Sternberg's view of love as having three components—intimacy, passion, and commitment—that shift in emphasis as romantic relationships develop. (p. 451)

trimesters Three equal time periods in the prenatal period, each of which lasts three months. (p. 79)

Type A behavior pattern A behavior pattern consisting of extreme competitiveness, ambition, impatience, hostility, angry outbursts, and a sense of time pressure. (p. 493)

umbilical cord The long cord connecting the prenatal organism to the placenta that delivers nutrients and removes waste products. (p. 78)

unconditioned response (UCR) In classical conditioning, a reflexive response that is produced by an unconditioned stimulus (UCS). (p. 126)

unconditioned stimulus (UCS) In classical conditioning, a stimulus that leads to a reflexive response. (p. 126)

underextension An early vocabulary error in which a word is applied too narrowly, to a smaller number of objects and events than is appropriate. Distinguished from *overextension.* (p. 167)

uninhibited, or sociable, child A child whose temperament is such that he or she displays positive emotion to and approaches novel stimuli. Distinguished from *inhibited,* or *shy, child.* (p. 182)

uninvolved style A child-rearing style that combines low acceptance and involvement with little control and effort to grant autonomy. Reflects minimal commitment to parenting. Distinguished from *authoritative, authoritarian,* and *permissive styles.* (p. 266)

universality The component of the death concept specifying that all living things eventually die. (p. 623)

vernix A white, cheeselike substance covering the fetus and preventing the skin from chapping due to constant exposure to the amniotic fluid. (p. 80)

violation-of-expectation method A method in which researchers habituate infants to a physical event and then determine whether they recover to (look longer at) a possible event (a variation of the first event that conforms to physical laws) or an impossible event (a variation that violates physical laws). Recovery to the impossible event suggests that the infant is surprised at a deviation from reality and is aware of that aspect of the physical world. (p. 147)

visual acuity Fineness of visual discrimination. (p. 106)

voluntary active euthanasia The practice of ending a patient's suffering, at the patient's request, before a natural end to life. A form of mercy killing. Distinguished from *passive euthanasia.* (p. 636)

whole-language approach An approach to beginning reading instruction that parallels children's natural language learning and keeps reading materials whole and meaningful. Distinguished from *basic-skills approach.* (p. 292)

wisdom A form of cognition that combines breadth and depth of practical knowledge; ability to reflect on and apply that knowledge in ways that make life more bearable and worthwhile; emotional maturity, including the ability to listen, evaluate, and give advice; and altruistic creativity—contributing to humanity and enriching others' lives. (p. 575)

working, or **short-term, memory** In information processing, the conscious part of the mental system, where we actively "work" on a limited amount of information to ensure that it will be retained. (p. 153)

X-linked inheritance A pattern of inheritance in which a recessive gene is carried on the X chromosome. Males are more likely to be affected. (p. 49)

young-old Appearing physically young for one's advanced years. Distinguished from *old-old.* (p. 546)

zone of proximal development In Vygotsky's theory, a range of tasks that the child cannot yet handle alone but can do with the help of more skilled partners. (p. 158)

zygote The newly fertilized cell formed by the union of sperm and ovum at conception. (p. 45)

References

Aarons, S. J., Jenkins, R. R., Raine, T. R., El-Khorazaty, M. N., Woodward, K. M., Williams, R. L., Clark, M. C., & Wingrove, B. K. (2000). Postponing sexual intercourse among urban junior high school students—a randomized controlled trial. *Journal of Adolescent Health, 27,* 236–247.

Abbott, S. (1992). Holding on and pushing away: Comparative perspectives on an eastern Kentucky child-rearing practice. *Ethos, 20,* 33–65.

Abra, J. (1989). Changes in creativity with age: Data, explanations, and further predictions. *International Journal of Aging and Human Development, 28,* 105–126.

Abraham, S. (1998). Satisfaction of participants in university-administered elderhostel programs. *Educational Gerontology, 24,* 529–536.

Abramovitch, R., Freedman, J. L., Henry, K., & Van Brunschot, M. (1995). Children's capacity to agree to psychological research: Knowledge of risks and benefits and voluntariness. *Ethics and Behavior, 5,* 25–48.

Achenbach, T. M., Phares, V., Howell, C. T., Rauh, V. A., & Nurcombe, B. (1990). Seven-year outcome of the Vermont program for low-birthweight infants. *Child Development, 61,* 1672–1681.

Acker, M. M., & O'Leary, S. G. (1996). Inconsistency of mothers' feedback and toddlers' misbehavior and negative affect. *Journal of Abnormal Child Psychology, 24,* 703–714.

Ackerman, B. P. (1978). Children's understanding of speech acts in unconventional frames. *Child Development, 49,* 311–318.

Ackerman, P. L. (2000). Domain-specific knowledge as the "dark matter" of adult intelligence: Personality and interest correlates. *Journal of Gerontology, 55B,* P69–P84.

Ackerman, S., Zuroff, D. C., & Moskowitz, D. S. (2000). Generativity in midlife and young adults: Links to agency, communion, and subjective well-being. *International Journal of Aging and Human Development, 50,* 17–41.

Adams, P. F., & Marano, M. A. (1995). Current estimates from the National Health Interview Survey, 1994. National Center for Health Statistics. *Vital Health Statistics, 10*(193), 83–84.

Adams, R., & Laursen, B. (2001). The organization and dynamics of adolescent conflict with parents and friends. *Journal of Marriage and the Family, 63,* 97–110.

Adams, R. G. (1985–1986). Emotional closeness and physical distance between friends: Implications for elderly women living in age-segregated and age-integrated settings. *International Journal of Aging and Human Development, 22,* 55–76.

Adams, R. J., Courage, M. L., & Mercer, M. E. (1994). Systematic measurement of human neonatal color vision. *Vision Research, 34,* 1691–1701.

Addington-Hall, J. (2000). Do home deaths increase distress in bereavement? *Palliative Medicine, 14,* 161–162.

Adkins, G., Martin, P., & Poon, L. W. (1996). Personality traits and states as predictors of subjective well-being in centenarians, octogenarians, and sexagenarians. *Psychology and Aging, 11,* 408–416.

Adlaf, E. M., & Paglia, A. (2001). *Drug use among Ontario students, 1997–2001: Findings from the OSDUS. CAMH Research Document Series No. 10.* Toronto: Centre for Adiction and Mental Health.

Adler, N. E., & Newman, K. (2002). Socioeconomic disparities in health: Pathways and policies. *Health Affairs, 21,* 60–76.

Adolph, K. E. (1997). Learning in the development of infant locomotion. *Monographs of the Society for Research in Child Development, 62*(3, Serial No. 251).

Adolph, K. E. (2000). Specificity of learning: Why infants fall over a veritable cliff. *Psychological Science, 11,* 290–295.

Adolph, K. E. A., Vereijken, B., & Denny, M. A. (1998). Learning to crawl. *Child Development, 69,* 1299–1312.

Affifi, W. A., & Faulkner, S. L. (2000). On being "just friends": The frequency and impact of sexual activity in cross-sex friendships. *Journal of Social and Personal Relationships, 17,* 205–222.

Agüero-Torres, H., von Strauss, E., Viitanen, M., Winblad, B., & Fratiglioni, L. (2001). Institutionalization in the elderly: The role of chronic diseases and dementia. Cross-sectional and longitudinal data from a population-based study. *Journal of Clinical Epidemiology, 54,* 795–801.

Aguiar, A., & Baillargeon, R. (1999). 2.5-month-old infants' reasoning about when objects should and should not be occluded. *Cognitive Psychology, 39,* 116–157.

Aguiar, A., & Baillargeon, R. (2002). Developments in young infants' reasoning about occluded objects. *Cognitive Psychology, 45,* 267–336.

Aho, T. F., & Gilling, P. J. (2003). Laser therapy for benign prostatic hyperplasia: A review of recent developments. *Current Opinion in Urology, 13,* 39–44.

Ainlay, S. C., Singleton, R., & Swigert, V. L. (1992). Aging and religious participation: Reconsidering the effects of health. *Journal for the Scientific Study of Religion, 31,* 175–188.

Ainsworth, M. D. S., Blehar, M. C., Waters, E., & Wall, S. (1978). *Patterns of attachment.* Hillsdale, NJ: Erlbaum.

Airhihenbuwa, C. A., Kumanyika, S., Agurs, T. D., & Lowe, A. (1995). Perceptions and beliefs about exercise, rest, and health among African Americans. *American Journal of Health Promotion, 9,* 426–429.

Akers, J. F., Jones, R. M., & Coyl, D. D. (1998). Adolescent friendship pairs: Similarities in identity status development, behaviors, attitudes, and intentions. *Journal of Adolescent Research, 13,* 178–201.

Akshoomoff, N. A., & Stiles, J. (1995). Developmental trends in visuospatial analysis and planning: I. Copying a complex figure. *Neuropsychology, 9,* 378–389.

Akshoomoff, N. A., Feroleto, C. C., Doyle, R. E., & Stiles, J. (2001). The impact of early unilateral brain injury on perceptual organization and visual memory. *Neuropsychologia, 40,* 539–561.

Alan Guttmacher Institute. (2002a). Teen pregnancy: Trends and lessons learned. Retrieved from http://www.agi-usa.org/pubs/ib_1-02.html

Alan Guttmacher Institute. (2002b). Teen sex and pregnancy. Retrieved from http://www.agi-usa.org/pubs/ib_teen_sex.html

Albanes, D., Blair, A., & Taylor, P. R. (1989). Physical activity and risk of cancer in the NHANES I population. *American Journal of Public Health, 79,* 744–750.

Albert, K. A., & Luzzo, D. A. (1999). The role of perceived barriers in career development: A social-cognitive perspective. *Journal of Counseling and Development, 77,* 431–436.

Aldwin, C. M., & Levenson, M. (2002). Stress, coping, and health at midlife: A developmental perspective. In M. E. Lachman (Ed.), *Handbook of midlife development* (pp. 188–214). New York: Wiley.

Aldwin, C. M., Spiro, A., III, Levenson, M. R., & Cupertino, A. P. (2001). Longitudinal findings from The Normative Aging Study: III. Personality, individual health trajectories, and mortality. *Psychology and Aging, 16,* 450–465.

Aldwin, C. M., Sutton, K. J., & Lachman, M. (1996). The development of coping resources in adulthood. *Journal of Personality, 64,* 91–113.

Alessandri, S. M., Bendersky, M., & Lewis, M. (1998). Cognitive functioning in 8- to 18-month-old drug-exposed infants. *Developmental Psychology, 34,* 565–573.

Alessandri, S. M., & Wozniak, R. H. (1987). The child's awareness of parental beliefs concerning the child: A developmental study. *Child Development, 58,* 316–323.

Alexander, J. M., Lucas, M. J., Ramin, S. M., McIntire, D. D., & Leveno, K. J. (1998). The course of labor with and without epidural analgesia. *American Journal of Obstetrics and Gynecology, 178,* 516–520.

Alexander, K. L., Entwisle, D. R., & Kabbani, N. S. (2001). The dropout process in life course perspective: Early risk factors at home and school. *Teachers College Record, 103,* 760–822.

Ali, L., & Scelfo, J. (2002, December 9). Choosing virginity. *Newsweek,* pp. 60–65.

Alibali, M. W. (1999). How children change their minds: Strategy change can be gradual or abrupt. *Developmental Psychology, 35,* 127–145.

Allen, J. P., & Hauser, S. T. (1996). Autonomy and relatedness in adolescent–family interactions as predictors of young adults' states of mind regarding attachment. *Development and Psychopathology, 8,* 793–809.

Allen, J. P., Philliber, S., Herrling, S., & Kuperminc, G. P. (1997). Preventing teen pregnancy and academic failure: Experimental evaluation of a developmentally based approach. *Child Development, 64,* 729–742.

Allen, M., & Burrell, N. (1996). Comparing the impact of homosexual and heterosexual parents on children: Meta-analysis of existing research. *Journal of Homosexuality, 32,* 19–35.

Allen, S. E. M., & Crago, M. B. (1996). Early passive acquisition in Inukitut. *Journal of Child Language, 23,* 129–156.

Allison, D. B., & Pi-Sunyer, X. (1994, May–June). Fleshing out obesity. *The Sciences, 34*(3), 38–43.

Allison, J. A., & Wrightsman, L. S. (1993). *Rape: The misunderstood crime.* Newbury Park, CA: Sage.

Alpert-Gillis, L. J., & Connell, J. P. (1989). Gender and sex-role influences on children's self-esteem. *Journal of Personality, 57,* 97–114.

Alsaker, F. D. (1995). Timing of puberty and reactions to pubertal changes. In M. Rutter (Ed.), *Psychosocial disturbances in young people* (pp. 37–82). New York: Cambridge University Press.

Alzheimer's Association. (2002). *African Americans and Alzheimer's disease: The silent epidemic.* Chicago: Author.

Amaro, H. (1995). Love, sex and power: Considering women's realities in HIV prevention. *American Psychologist, 50,* 437–447.

Amato, P. R. (2000). The consequences of divorce for adults and children. *Journal of Marriage and the Family, 62,* 1269–1287.

Amato, P. R., & Booth, A. (1996). A prospective study of divorce and parent–child relationships. *Journal of Marriage and the Family, 58,* 356–365.

Amato, P. R, & Booth, A. (2000). *A generation at risk: Growing up in an era of family upheaval.* Cambridge, MA: Harvard University Press.

Amato, P. R., & DeBoer, D. D. (2001). The transmission of marital instability across generations: Relationship skills or commitment to marriage? *Journal of Marriage and the Family, 63,* 1038–1051.

Amato, P. R., & Gilbreth, J. (1999). Nonresident fathers and children's well-being: A meta-analysis. *Journal of Marriage and the Family, 61,* 557–573.

Amato, P. R., & Rogers, S. J. (1997). A longitudinal study of marital problems and subsequent divorce. *Journal of Marriage and the Family, 59,* 612–624.

American Academy of Pediatrics. (1999). Contraception and adolescents. *Pediatrics, 104,* 1161–1166.

American Academy of Pediatrics. (2000). Changing concepts of sudden infant death syndrome: Implications for infant sleeping environment and sleep position. *Pediatrics, 105,* 650–656.

American College of Obstetricians and Gynecologists. (2000). Premenstrual syndrome: Clinical management guidelines for obstetrican-gynecologists. *ACOG Practice Bulletin, 15,* 1–9.

American College Testing. (1998). New low for college graduation rate, but dropout picture brighter. *ACT Newsroom.* Retrieved from http://www.act.org/news/releases/1998

American Heart Association. (2002). Heart attack and angina statistics. Retrieved from http://www.americanheart.org

American Psychiatric Association. (1994). *Diagnostic and statistical manual of mental disorders* (4th ed.). Washington, DC: Author.

American Psychological Association. (1994). *Publication manual of the American Psychological Association.* Washington, DC: Author.

Ames, C. (1992). Classrooms: Goals, structures, and student motivation. *Journal of Educational Psychology, 84,* 261–271.

Ames, E. (1997). *The development of Romanian orphanage children adopted to Canada (Final Report to the National Welfare Grants Program).* Burnaby, British Columbia: Simon Fraser University.

Anand, S. S., Yusuf, S., Jacobs, R., Davis, A. D., Yi, Q., & Gerstein, H. (2001). Risk factors, atherosclerosis, and cardiovascular disease among Aboriginal people in Canada: The study of health assessment and risk evaluation in Aboriginal peoples (SHARE-AP). *Lancet, 358,* 1147–1153.

Anderman, E. M., Eccles, J. S., Yoon, K. S., Roeser, R., Wigfield, A., & Blumenfeld, P. (2001). Learning to value mathematics and reading: Relations to mastery and performance-oriented instructional practices. *Contemporary Educational Psychology, 26,* 76–95.

Anderman, E. M., & Midgley, C. (1997). Changes in achievement goal orientations, perceived academic competence, and grades across the transition to middle-level schools. *Contemporary Educational Psychology, 22,* 269–298.

Anderson, C. A., & Bushman, B. J. (2001). Effects of violent video games on aggressive behavior, aggressive cognition, aggressive affect, physiological arousal, and prosocial behavior: A meta-analytic review of the scientific literature. *Psychological Science, 12,* 353–359.

Anderson, D. M., Huston, A. C., Schmitt, K. L., Linebarger, D. L., & Wright, J. C. (2001). Early childhood television viewing and adolescent behavior. *Monographs of the Society for Research in Child Development, 66*(1, Serial No. 264).

Anderson, E. (1992). *Speaking with style: The sociolinguistic skills of children.* London: Routledge.

Andersson, B.-E. (1989). Effects of public day care—A longitudinal study. *Child Development, 60,* 857–866.

Andersson, B.-E. (1992). Effects of day care on cognitive and socio-emotional competence of thirteen-year-old Swedish schoolchildren. *Child Development, 63,* 20–36.

Andersson, S. W., Bengtsson, C., Hallberg, L., Lapidus, L., Niklasson, A., Wallgren, A., & Huthén, L. (2001). Cancer risk in Swedish women: The relation to size at birth. *British Journal of Cancer, 84,* 1193–1198.

Andre, T., Whigham, M., Hendrickson, A., & Chambers, S. (1999). Competence beliefs, positive affect, and gender stereotypes of elementary students and their parents about science versus other school subjects. *Journal of Research in Science Teaching, 36,* 719–747.

Andrews, G., & Halford, G. S. (1998). Children's ability to make transitive inferences: The importance of premise integration and structural complexity. *Cognitive Development, 13,* 479–513.

Andrews, L. B., & Elster, N. (2000). Regulating reproductive technologies. *Journal of Legal Medicine, 21,* 35–65.

Anglin, J. M. (1993). Vocabulary development: A morphological analysis. *Monographs of the Society for Research in Child Development, 58*(10, Serial No. 238).

Angus Reid Poll. (1995). Euthanasia in Canada. Retrieved from http://www.rights.org/deathnet/Angus_Reid.html

Anisfeld, M., Turkewitz, G., Rose, S. A., Rosenberg, F. R., Shelber, F. J., Couturier-Fagan, D. A., Ger, J. S., & Sommer, I. (2001). No compelling evidence that newborns imitate oral gestures. *Infancy, 2,* 111–122.

Annett, M. (2002). *Handedness and brain asymmetry: The right shift theory.* Hove, U.K.: Psychology Press.

Anslow, P. (1998). Birth asphyxia. *European Journal of Radiology, 26,* 148–153.

Anstey, K., & Christensen, H. (2000). Education, activity, health, blood pressure, and apolipoprotein E as predictors of cognitive change in old age: A review. *Gerontology, 46,* 163–177.

Anstey, K. J., Luszcz, M. A., & Sanchez, L. (2001). Two-year decline in vision but not hearing is associated with memory decline in very old adults in a population-based sample. *Gerontology, 47,* 289–293.

Antonucci, T. (1990). Social supports and social relationships. In R. Binstock & L. K. George (Eds.), *Handbook of aging and the social sciences* (3rd ed., pp. 205–227). New York: Academic Press.

Antonucci, T., & Akiyama, H. (1997). Concern with others at midlife: Care, comfort, or compromise? In M. E. Lachman & J. B. James (Eds.), *Multiple paths of midlife development* (pp. 145–169). Chicago: University of Chicago Press.

Antonucci, T. C. (1994). A life-span view of women's social relations. In B. F. Turner & L. E. Troll (Eds.), *Women growing older* (pp. 239–269). Thousand Oaks, CA: Sage.

Antonucci, T. C., & Akiyama, H. (1995). Convoys of social relations: Family and friendships within a life span context. In R. Blieszner & V. H. Bedford (Eds.), *Handbook of aging and the family* (pp. 355–371). Westport, CT: Greenwood Press.

Antonucci, T. C., Akiyama, H., & Merline, A. (2002). Dynamics of social relationships in midlife. In M. E. Lachman (Ed.), *Handbook of midlife development* (pp. 571–598). New York: Wiley.

Apgar, V. (1953). A proposal for a new method of evaluation in the newborn infant. *Current Research in Anesthesia and Analgesia, 32,* 260–267.

Aquilino, W. S. (1996). The returning adult child and parental experience at midlife. In C. D. Ryff & M. M. Seltzer (Eds.), *The parental experience in midlife* (pp. 423–458). Chicago: University of Chicago Press.

Aquilino, W. S., & Supple, A. J. (2001). Long-term effects of parenting practices during adolescence on well-being outcomes in young adulthood. *Journal of Family Issues, 22,* 289–308.

Arata, C. M. (1999). Coping with rape: The roles of prior sexual abuse and attributions of blame. *Journal of Interpersonal Violence, 14,* 62–78.

Arber, S., & Cooper, H. (1999). Gender differences in health in later life: The new paradox? *Social Science and Medicine, 48,* 61–76.

Archer, S. L. (1989). The status of identity: Reflections on the need for intervention. *Journal of Adolescence, 12,* 345–359.

Archer, S. L., & Waterman, A. S. (1990). Varieties of identity diffusions and foreclosures: An exploration of subcategories of the identity statuses. *Journal of Adolescent Research, 5,* 96–111.

Ardelt, M. (1997). Wisdom and life satisfaction in old age. *Journal of Gerontology, 52B,* P15–P27.

Ardelt, M. (1998). Social crisis and individual growth: The long-term effects of the Great Depression. *Journal of Aging Studies, 12,* 291–314.

Ardelt, M. (2000). Intellectual versus wisdom-related knowledge: The case for a different kind of learning in the later years of life. *Educational Gerontology, 26,* 771–789.

Arking, R. (1991). *Biology of aging.* Englewood Cliffs, NJ: Prentice Hall.

Arlin, P. K. (1989). Problem solving and problem finding in young artists and young scientists. In M. L. Commons, J. D. Sinnott, F. A. Richards, & C. Armon (Eds.), *Adult development: Vol 1. Comparisons and applications of developmental models* (pp. 197–216). New York: Praeger.

Armstrong, G. K., & Morgan, K. (1998). Stability and change in levels of habitual physical activity in later life. *Age and Ageing, 27*(Suppl. 3), 17–23.

Armstrong, K. L., Quinn, R. A., & Dadds, M. R. (1994). The sleep patterns of normal children. *Medical Journal of Australia, 161,* 202–206.

Armstrong, T. D., & Crowther, M. R. (2002). Spirituality among older African Americans. *Journal of Adult Development, 9,* 3–12.

Arnett, J. J. (1999). Adolescent storm and stress reconsidered. *American Psychologist, 54,* 317–326.

Arnett, J. J. (2000). Emerging adulthood: A theory of development from the late teens through the twenties. *American Psychologist, 55,* 469–480.

Arnold, K. (1994). The Illinois Valedictorian Project: Early adult careers of academically talented male and female high school students. In R. F. Subotnik & K. D. Arnold (Eds.), *Beyond Terman: Contemporary longitudinal studies of giftedness and talent* (pp. 24–51). Norwood, NJ: Ablex.

Aronne, L. J. (2001). Epidemiology, morbidity, and treatment of overweight and obesity. *Journal of Clinical Psychiatry, 62,* 13–22.

Aronson, M., Hagberg, B., & Gillberg, C. (1997). Attention deficits and autistic spectrum problems in children exposed to alcohol during gestation: A follow-up study. *Developmental Medicine and Child Neurology, 39,* 583–587.

Arora, S., McJunkin, C., Wehrer, J., & Kuhn, P. (2000). Major factors influencing breastfeeding rates: Mother's perception of father's attitude and milk supply. *Pediatrics, 106,* e67.

Arsenio, W., & Fleiss, K. (1996). Typical and behaviourally disruptive children's understanding of the emotional consequences of socio-moral events. *British Journal of Developmental Psychology, 14,* 173–186.

Artman, L., & Cahan, S. (1993). Schooling and the development of transitive inference. *Developmental Psychology, 29,* 753–759.

Asakawa, K. (2001). Family socialization practices and their effects on the internationalization of educational values for Asian and white American adolescents. *Applied Developmental Science, 5,* 184–194.

Asher, S. R., & Rose, A. J. (1997). Promoting children's social-emotional adjustment with peers. In P. Salovey & D. J. Sluyter (Eds.), *Emotional development and emotional intelligence* (pp. 193–195). New York: Basic Books.

Ashley-Koch, A., Robinson, H., Glicksman, A. E., Nolin, S. L., Schwartz, C. E., & Brown, W. T. (1998). Examination of factors associated with instability of the FMR1 CGG repeat. *American Journal of Human Genetics, 63,* 776–785.

Aslin, R. N. (1993). Perception of visual direction in human infants. In C. E. Granrud (Ed.), *Visual perception and cognition in infancy* (pp. 91–119). Hillsdale, NJ: Erlbaum.

Aslin, R. N. (2000). Why take the cog out of infant cognition? *Infancy, 1,* 463–470.

Aslin, R. N., Jusczyk, P. W., & Pisoni, D. B. (1998). Speech and auditory processing during infancy: Constraints on and precursors to language. In D. Kuhn & R. S. Siegler (Eds.), *Handbook of child psychology: Vol. 2. Cognition, perception, and language* (5th ed., pp. 147–198). New York: Wiley.

Assmann, A. (1994). Wholesome knowledge: Concepts of wisdom in a historical and cross-cultural perspective. In D. L. Featherman, R. M. Lerner, & M. Perlmutter (Eds.), *Lifespan development and behavior* (pp. 187–224). Hillsdale, NJ: Erlbaum.

Astley, S. J., Clarren, S. K., Little, R. E., Sampson, P. D., & Daling, J. R. (1992). Analysis of facial shape in children gestationally exposed to marijuana, alcohol, and/or cocaine. *Pediatrics, 89,* 67–77.

Atchley, R. (1988). *Social forces and aging: An introduction to social gerontology.* Belmont, CA: Wadsworth.

Atchley, R. C. (1999). *Continuity and adaptation in aging.* Baltimore: Johns Hopkins University Press.

Atkinson, R. C., & Shiffrin, R. M. (1968). Human memory: A proposed system and its control processes. In K. W. Spence & J. T. Spence (Eds.), *Advances in the psychology of learning and motivation* (Vol. 2, pp. 90–195). New York: Academic Press.

Attie, I., & Brooks-Gunn, J. (1996). The development of eating regulation across the life span. In D. Cicchetti & D. J. Cohen (Eds.), *Developmental psychology: Vol. 2. Risk, disorder, and adaptation* (pp. 332–368). New York: Wiley.

Au, T. K., Sidle, A. L., & Rollins, K. B. (1993). Developing an intuitive understanding of conservation and contamination: Invisible particles as a plausible mechanism. *Developmental Psychology, 29,* 286–299.

Aunola, K., Stattin, H., & Nurmi, J.-E. (2000). Parenting styles and adolescents' achievement strategies. *Journal of Adolescence, 23,* 205–222.

Australian Bureau of Statistics. (2002). Divorce rates. Retrieved from http://www.abus.gov.au

Avis, N. E. (2000). Sexual function and aging in men and women: Community and population-based studies. *Journal of Gender-Specific Medicine, 3,* 37–41.

Avis, N. E., Crawford, S., & Johannes, C. B. (2002). Menopause. In G. M. Wingood & R. J. DeClemente (Eds.), *Handbook of women's sexual and reproductive health* (pp. 367–391). New York: Kluwer.

Avolio, B. J., & Sosik, J. J. (1999). A lifespan framework for assessing the impact of work on white-collar workers. In S. L. Willis & J. D. Reid (Eds.), *Life in the middle* (pp. 249–274). San Diego, CA: Academic Press.

Axia, G., & Baroni, R. (1985). Linguistic politeness at different age levels. *Child Development, 56,* 918–927.

Axia, G., Bonichini, S., & Benini, F. (1999). Attention and reaction to distress in infancy: A longitudinal study. *Developmental Psychology, 35,* 500–504.

Axinn, W. G., & Barber, J. S. (1997). Living arrangements and family formation attitudes in early adulthood. *Journal of Marriage and the Family, 59,* 595–611.

Azmitia, M. (1988). Peer interaction and problem solving: When are two heads better than one? *Child Development, 59,* 87–96.

Bach-y-Rita, P. (2001). Theoretical and practical considerations in the restoration of function after stroke. *Topics in Stroke Rehabilitation, 8,* 1–15.

Bäckman, L., Small, B. J., & Wahlin, A. (2001). Aging and memory: Cognitive and biological perspectives. In J. E. Birren & K. W. Schaie (Eds.), *Handbook of the psychology of aging* (5th ed., pp. 186–214). San Diego: Academic Press.

Baddeley, A. (1993). Working memory and conscious awareness. In A. F. Collins, S. E. Gathercole, M. A. Conway, & P. E. Morris (Eds.), *Theories of memory* (pp. 11–28). Hove, U.K.: Erlbaum.

Baddeley, A. (2000). Short-term and working memory. In E. Tulving & R. I. M. Craik (Eds.), *The Oxford handbook of memory* (pp. 77–92). New York: Oxford University Press.

Baenninger, M., & Newcombe, N. (1995). Environmental input to the development of sex-related differences in spatial and mathematical ability. *Learning and Individual Differences, 7,* 363–379.

Bagwell, C. L., Schmidt, M. E., Newcomb, A. F., & Bukowski, W. M. (2001). Friendship and peer rejection as predictors of adult adjustment. In D. W. Nangle & C. A. Erdley (Eds.), *The role of friendship in psychological adjustment* (pp. 25–49). San Francisco: Jossey-Bass.

Bahrick, H. P. (1984). Semantic memory content in permastore: Fifty years of memory for Spanish learned in school. *Journal of Experimental Psychology: General, 113,* 1–29.

Bahrick, H. P., Bahrick, P. O., & Wittlinger, R. P. (1975). Fifty years of memory for names and faces: A cross-sectional approach. *Journal of Experimental Psychology: General, 104,* 54–75.

Bahrick, L. E. (1992). Infants' perceptual differentiation of amodal and modality-specific audio-visual relations. *Journal of Experimental Child Psychology, 53,* 180–199.

Bahrick, L. E. (2001). Increasing specificity in perceptual development: Infants' detection of nested levels of multimodal stimulation. *Journal of Experimental Child Psychology, 79,* 253–270.

Bahrick, L. E., Moss, L., & Fadil, C. (1996). Development of visual self-recognition in infancy. *Ecological Psychology, 8,* 189–208.

Bahrick, L. E., Netto, D., & Hernandez-Reif, M. (1998). Intermodal perception of adult and child faces and voices by infants. *Child Development, 69,* 1263–1275.

Bai, D. L., & Bertenthal, B. I. (1992). Locomotor status and the development of spatial search skills. *Child Development, 63,* 215–226.

Bailey, D. A., & Rasmussen, R. L. (1996). Sport and the child: Physiological and skeletal issues. In F. L. Smoll & R. E. Smith (Eds.), *Children and youth in sport: A biopsychological perspective* (pp. 187–199). Dubuque, IA: Brown & Benchmark.

Bailey, J. M., Bobrow, D., Wolfe, M., & Mikach, S. (1995). Sexual orientation of adult sons of gay fathers. *Developmental Psychology, 31,* 124–129.

Bailey, J. M., & Pillard, R. C. (1991). A genetic study of male sexual orientation. *Archives of General Psychology, 43,* 808–812.

Bailey, J. M., Pillard, R. C., Neale, M. C., & Agyei, Y. (1993). Heritable factors influence sexual orientation in women. *Archives of General Psychiatry, 50,* 217–223.

Baillargeon, R. (1994). Physical reasoning in infancy. In M. S. Gazzaniga (Ed.), *The cognitive neurosciences* (pp. 181–204). Cambridge, MA: MIT Press.

Baillargeon, R. (2000). Reply to Bogartz, Shinskey, and Schilling; Schilling; and Cashon and Cohen. *Infancy, 1,* 447–462.

Baillargeon, R., & DeVos, J. (1991). Object permanence in young infants: Further evidence. *Child Development, 62,* 1227–1246.

Balaswamy, S., & Richardson, V. E. (2001). The cumulative effects of life event, personal, and social resources on subjective well-being of elderly widowers. *International Journal of Aging and Human Development, 53,* 311–327.

Balfour, J. L., & Kaplan, G. A. (1998). Social class/socioeconomic factors. In J. C. Holland (Ed.), *Psycho-oncology* (pp. 78–90). New York: Oxford University Press.

Ball, M. M., Whittington, F. J., Perkins, M. M., Patterson, V. L., Hollingsworth, C., King, S. V., & Combs, B. L. (2000). Quality of life in assisted living facilities: Viewpoints of residents. *Journal of Applied Gerontology, 19,* 304–325.

Ballard, B. D., Gipson, M. T. Guttenberg, W., & Ramsey, K. (1980). Palatability of food as a factor influencing obese and normal- weight children's eating habits. *Behavior Research and Therapy, 18,* 598–600.

Baltes, M. M. (1995, February). Dependency in old age: Gains and losses. *Psychological Science, 4*(1), 14–19.

Baltes, M. M. (1996). *The many faces of dependency in old age.* New York: Cambridge University Press.

Baltes, M. M., & Carstensen, L. L. (1996). The process of successful ageing. *Ageing and Society, 16,* 397–422.

Baltes, M. M., Wahl, H.-W., & Reichert, M. (1992). Successful aging in long-term care institutions. In K. W. Schaie & M. P. Lawton (Eds.), *Annual review of gerontology and geriatrics* (pp. 311–337). New York: Springer.

Baltes, P. B. (1997). On the incomplete architecture of human ontogeny: Selection, optimization, and compensation as foundation of developmental theory. *American Psychologist, 52,* 366–380.

Baltes, P. B., Dittmann-Kohli, F., & Dixon, R. A. (1984). New perspectives on the development of intelligence in adulthood: Toward a dual-process conception and a model of selective optimization with compensation. In P. B. Baltes & O. G. Brim, Jr. (Eds.), *Life-span development and behavior* (Vol. 6, pp. 33–76). San Diego, CA: Academic Press.

Baltes, P. B., Lindenberger, U., & Staudinger, U. M. (1998). Life-span theory in developmental psychology. In R. M. Lerner (Ed.), *Handbook of child psychology: Vol. 1. Theoretical models of human development* (5th ed., pp. 1029–1143). New York: Wiley.

Baltes, P. B., & Staudinger, U. M. (2000). Wisdom: A metaheuristic (pragmatic) to orchestrate mind and virtue toward excellence. *American Psychologist, 55,* 122–136.

Baltes, P. B., Staudinger, U. M., Maercker, A., & Smith, J. (1995). People nominated as wise: A comparative study of wisdom-related knowledge. *Psychology and Aging, 10,* 155–166.

Bancroft, J. (2002). The medicalization of female sexual dysfunction: The need for caution. *Archives of Sexual Behavior, 31,* 451–455.

Band, G. P. H., van der Molen, M. W., Overtoom, C. C. E., & Verbaten, M. N. (2000). The ability to activate and inhibit speeded responses: Separate developmental trends. *Journal of Experimental Child Psychology, 75,* 263–290.

Bandura, A. (1977). *Social learning theory.* Englewood Cliffs, NJ: Prentice-Hall.

Bandura, A. (1989). Social cognitive theory. In R. Vasta (Ed.), *Annals of child development* (Vol. 6, pp. 1–60). Greenwich, CT: JAI Press.

Bandura, A. (1992). Perceived self-efficacy in cognitive development and functioning. *Educational Psychologist, 28,* 117–118.

Bandura, A. (1997). *Self-efficacy: The exercise of control.* New York: Freeman.

Bandura, A. (1999). Social cognitive theory of personality. In L. A. Pervin (Ed.), *Handbook of personality: Theory and research* (2nd ed., pp. 154–196). New York: Guilford.

Banish, M. T. (1998). Integration of information between the cerebral hemispheres. *Current Directions in Psychological Science, 7,* 32–37.

Banish, M. T., & Heller, W. (1998). Evolving perspectives on lateralization of function. *Current Directions in Psychological Science, 7,* 1–2.

Bank, L., Patterson, G. R., & Reid, J. B. (1996). Negative sibling interaction as a predictor of later adjustment problems in adolescent and young adult males. In G. H. Brody (Ed.), *Sibling relationships: Their causes and consequences* (pp. 197–229). Norwood, NJ: Ablex.

Bankoff, E. A. (1983). Aged parents and their widowed daughters: A support relationship. *Journal of Gerontology, 38,* 226–230.

Banks, M. S. (1980). The development of visual accommodation during early infancy. *Child Development, 51,* 646–666.

Banks, M. S., & Bennett, P. J. (1988). Optical and photoreceptor immaturities limit the spacial and chromatic vision of human neonates. *Journal of the Optical Society of America, 5,* 2059–2097.

Banks, M. S., & Ginsburg, A. P. (1985). Early visual preferences: A review and new theoretical treatment. In H. W. Reese (Ed.), *Advances in child development and behavior* (Vol. 19, pp. 207–246). New York: Academic Press.

Banta, D. H., & Thacker, S. B. (2001). Historical controversy in health technology assessment: The case of electronic fetal monitoring. *Obstetrical and Gynecological Survey, 56,* 707–719.

Barakat, L. P., & Kazak, A. E. (1999). Family issues. In R. T. Brown (Ed.), *Cognitive aspects of chronic illness in children* (pp. 333–354). New York: Guilford.

Barber, B. K., & Harmon, E. L.. (2002). Violating the self: Parental psychological control of children and adolescents. In B. K. Barber (Ed.), *Parental psychological control of children and adolescents* (pp. 15–52). Washington, DC: American Psychological Association.

Barber, B. K., & Olsen, J. A. (1997). Socialization in context: Connection, regulation, and autonomy in the family, school, and neighborhood, and with peers. *Journal of Adolescent Research, 12,* 287–315.

Barber, J. S. (2001). Ideational influences on the transition to parenthood: Attitudes toward childbearing and competing alternatives. *Social Psychology Quarterly, 64,* 101–127.

Barer, B. M. (1994). Men and women aging differently. *International Journal of Aging and Human Development, 38,* 29–40.

Barker, D. (2002). Fetal programming of coronary heart disease. *Trends in Endocrinology and Metabolism, 13,* 364.

Barker, D. J. P. (1994). *Mothers, babies, and disease in later life.* London: British Medical Journal Publishing.

Barker, D. J. P. (1999) Fetal origins of cardiovascular disease. *Annals of Medicine, 21*(Suppl.1), 3–6.

Barker, R. G. (1955). *Midwest and its children.* Stanford, CA: Stanford University Press.

Barkley, R. A. (1997). Behavioral inhibition, sustained attention, and executive functions: Constructing a unifying theory of ADHD. *Psychological Bulletin, 121,* 65–94.

Barkley, R. A. (1999). Theories of attention-deficit/hyperactivity disorder. In H. C. Quay & A. E. Hogan (Eds.), *Handbook of disruptive behavior disorders* (pp. 295–313). New York: Kluwer.

Barling, J., Rogers, K., & Kelloway, K. (1995). Some effects of teenagers' part-time employment: The quantity and quality of work make the differences. *Journal of Organizational Behavior, 16,* 143–154.

Barnett, D., Ganiban, J., & Cicchetti, D. (1999). Maltreatment, negative expressivity, and the development of Type D attachments from 12 to 24 months of age. In J. I Vondra & D. Barnett (Eds.), Atypical attachment in infancy and early childhood among children at developmental risk. *Monographs of the Society for Research in Child Development, 64*(3, Serial No. 258), 97–118.

Barnett, D., & Vondra, J. I. (1999). Atypical patterns of early attachment: Theory, research, and current directions. In J. I Vondra & D. Barnett (Eds.), Atypical attachment in infancy and early childhood among children at developmental risk. *Monographs of the Society for Research in Child Development, 64*(3, Serial No. 258), 1–24.

Barnett, R. C., & Hyde, J. S. (2001). Women, men, work, family: An expansionist theory. *American Psychologist, 56,* 781–796.

Barnett, R. C., & Rivers, C. (1996). *She works/he works.* San Francisco: Harper.

Barnett, W. S. (1998). Long-term cognitive and academic effects of early childhood education on children in poverty. *Preventive Medicine, 27,* 204–207.

Baron-Cohen, S. (2001). Theory of mind and autism: A review. In L. M. Glidden (Ed.), *International review of research in mental retardation: Autism* (Vol. 23, pp. 169–184). San Diego: Academic Press.

Baron-Cohen, S., Baldwin, D. A., & Crowson, M. (1997). Do children with autism use the speaker's direction of gaze strategy to crack the code of language? *Child Development, 68,* 48–57.

Barr, H. M., Streissguth, A. P., Darby, B. L., & Sampson, P. D. (1990). Prenatal exposure to alcohol, caffeine, tobacco, and aspirin. Effects on fine and gross motor performance in 4-year-old children. *Developmental Psychology, 26,* 339–348.

Barr, R. G. (2001). "Colic" is something infants do, rather than a condition they "have": A developmental approach to crying phenomena patterns, pacification and (patho)genesis. In R. G. Barr, I. St. James-Roberts, & M. R. Keefe (Eds.), *New evidence on unexplained infant crying* (pp. 87–104). St. Louis: Johnson & Johnson Pediatric Institute.

Barr, R. G., & Gunnar, M. (2000). Colic: The 'transient responsivity' hypothesis. In R. G. Barr, B. Hopkins, & J. A. Green (Eds.), *Crying as a sign, a symptom, and a signal* (pp. 41–66). Cambridge, U.K.: Cambridge University Press.

Barr, S. (1996, September). Up against the glass. *Management Review, 85*(9), 12–17.

Barratt, M. S., Roach, M. A., & Leavitt, L. A. (1996). The impact of low-risk prematurity on maternal behaviour and toddler outcomes. *International Journal of Behavioral Development, 19,* 581–602.

Barrett, K. C. (1998). The origins of guilt in early childhood. In J. Bybee (Ed.), *Guilt and children* (pp. 75–90). San Diego: Academic Press.

Barrios, B. A., & Dell, S. L. (1998). Fears and anxieties. In E. J. Mash & R. A. Barkley (Eds.), *Treatment of childhood disorders* (2nd ed., pp. 249–337). New York: Guilford.

Barsky, A. J., Cleary, P. D., & Klerman, G. L. (1992). Determinants of perceived health status of medical outpatients. *Social Science and Medicine, 34,* 1147–1154.

Bartlik, B., & Goldstein, M. Z. (2001). Men's sexual health after midlife. *Practical Geriatrics, 52,* 291–306.

Bartrip, J., Morton, J., & de Schonen, S. (2001). Responses to mother's face in 3-week- to 5-month-old infants. *British Journal of Developmental Psychology, 19,* 219–232.

Bartsch, K., & Wellman, H. (1995). *Children talk about the mind.* New York: Oxford University Press.

Basow, S. A., & Rubin, L. R. (1999). Gender influences on adolescent development. In N. G. Johnson & M. C. Roberts (Eds.), *Beyond appearance: A new look at adolescent girls* (pp. 25–52). Washington, DC: American Psychological Association.

Bass, B. M., & Avolio, B. J. (1994). Shatter the glass ceiling: Women may make better managers. *Human Resource Management, 33,* 549–560.

Bassey, E. J. (2000). The benefits of exercise for the health of older people. *Reviews in Clinical Gerontology, 10,* 17–31.

Bates, E. (1999). Plasticity, localization, and language development. In S. H. Broman & J. M. Fletcher (Eds.), *The changing nervous system: Neurobehavioral consequences of early brain disorders* (pp. 214–247). New York: Oxford University Press.

Bates, E., Marchman, V., Thal, D., Fenson, L., Dale, P., Reznick, J. S., Reilly, J., & Hartung, J. (1994). Developmental and stylistic variation in the composition of early vocabulary. *Journal of Child Language, 21,* 85–123.

Bates, J. E., Wachs, T. D., & Emde, R. N. (1994). Toward practical uses for biological concepts. In J. E. Bates & T. D. Wachs (Eds.), *Temperament: Individual differences at the interface of biology and behavior* (pp. 275–306). Washington, DC: American Psychological Association.

Bauer, P. J. (1997). Development of memory in early childhood. In N. Cowan (Ed.), *The development of memory in childhood* (pp. 83–111). Hove, U.K.: Psychology Press.

Baumeister, R. F. (1998). Inducing guilt. In J. Bybee (Ed.), *Guilt and children* (pp. 185–213). San Diego: Academic Press.

Baumrind, D. (1971). Current patterns of parental authority. *Developmental Psychology Monograph, 4*(No. 1, Pt. 2).

Baumrind, D., & Black, A. E. (1967). Socialization practices associated with dimension of competence in preschool boys and girls. *Child Development, 38,* 291–327.

Baumwell, L., Tamis-LeMonda, C. S., & Bornstein, M. H. (1997). Maternal verbal sensitivity and child language comprehension. *Infant Behavior and Development, 20,* 247–258.

Bayer, A., & Tadd, W. (2000). Unjustified exclusion of elderly people from studies submitted to research ethics committee for approval: Descriptive study. *British Medical Journal, 321,* 992–993.

Bayley, N. (1969). *Bayley Scales of Infant Development.* New York: Psychological Corporation.

Bayley, N. (1993). *Bayley Scales of Infant Development* (2nd ed.). San Antonio, TX: Psychological Corporation.

Beard, B. B. (1991). *Centenarians: The new generation.* New York: Greenwood Press.

Bearison, D. J. (1998). Pediatric psychology and children's medical problems. In I. G. Sigel & K. A. Renninger (Eds.), *Handbook of child psychology: Vol. 4. Child psychology in practice* (5th ed., pp. 635–711). New York: Wiley.

Beatty, W. W. (1992). Gonadal hormones and sex differences in nonreproductive behaviors. In A. A. Gerall, H. Moltz, & I. L. Ward (Eds.), *Handbook of behavioral neurobiology: Vol. 11. Sexual differentiation* (pp. 85–128). New York: Plenum.

Beausang, C. C., & Razor, A. G. (2000). Young Western women's experiences of menarche and menstruation. *Health Care for Women International, 21,* 517–528.

Beckerman, M. B. (1990). Leos Janácek and "the late style" in music. *Gerontologist, 30,* 632–635.

Beckwith, L., & Sigman, M. D. (1995). Preventive interventions in infancy. *Child and Adolescent Psychiatric Clinics of North America, 4,* 683–700.

Begley, S. (1995, February 13). Three is not enough. *Newsweek,* pp. 67–69.

Behrman, R. E., Kliegman, R. M., & Arvin, A. M. (Eds.). (1996). *Nelson textbook of pediatrics* (15th ed.). Philadelphia: Saunders.

Behrman, R. E., Kliegman, R. M., & Jenson, H. B. (2000). *Nelson textbook of pediatrics* (16th ed.). Philadelphia: Saunders.

Beilin, H. (1992). Piaget's enduring contribution to developmental psychology. *Developmental Psychology, 28,* 191–204.

Beitel, A. H., & Parke, R. D. (1998). Paternal involvement in infancy: The role of maternal and paternal attitudes. *Journal of Family Psychology, 12,* 268–288.

Bell, K. L., Allen, J. P., Hauser, S. T., & O'Connor, T. G. (1996). Family factors and young adult transitions: Educational attainment and occupational prestige. In J. A. Graber, J. Brooks-Gunn, & A. C. Petersen (Eds.), *Transitions through adolescence: Interpersonal domains and context* (pp. 345–366). Mahwah, NJ: Erlbaum.

Bell, M. A. (1998). Frontal lobe function during infancy: Implications for the development of cognition and attention. In J. E. Richards (Ed.), *Cognitive neuroscience of attention: A developmental perspective* (pp. 327–362). Mahwah, NJ: Erlbaum.

Bell, M. A., & Fox, N. A. (1994). Brain development over the first year of life: Relations between EEG frequency and coherence and cognitive and affective behaviors. In G. Dawson & K. W. Fischer (Eds.), *Human behavior and the developing brain* (pp. 314–345). New York: Guilford.

Bell, M. A., & Fox, N. A. (1996). Crawling experience is related to changes in cortical organization during infancy: Evidence from EEG coherence. *Developmental Psychobiology, 29,* 551–561.

Bell, M. L. (1995). Attitudes toward menopause among Mexican American women. *Health Care for Women International, 16,* 425–435.

Bellamy, C. (1998). *The state of the world's children 1998.* New York: Oxford University Press (in cooperation with UNICEF).

Bellamy, C. (2000). *The state of the world's children 2000.* New York: Oxford University Press (in cooperation with UNICEF).

Belsky, J. (1992). Consequences of child care for children's development: A deconstructionist view. In A. Booth (Ed.), *Child care in the 1990s: Trends and consequences* (pp. 83–85). Hillsdale, NJ: Erlbaum.

Belsky, J., Jaffee, S., Hsieh, K., & Silva, P. A. (2001) Child-rearing antecedents of intergenerational relations in young adulthood: A prospective study. *Developmental Psychology, 37,* 801–813.

Bempechat, J., & Drago-Severson, E. (1999). Cross-national differences in academic achievement: Beyond etic conceptions of children's understandings. *Review of Educational Research, 69,* 287–314.

Benbow, C. P., & Arjmand, O. (1990). Predictors of high academic achievement in mathematics and science by mathematically talented students: A longitudinal study. *Journal of Educational Psychology, 82*, 430–441.

Benbow, C. P., Lubinski, D., Shea, D. L., & Eftekhara-Sanjani, H. (2000). Sex differences in mathematical reasoning ability at age 13: Their status 20 years later. *Psychological Science, 11*, 474–480.

Benbow, C. P., & Stanley, J. C. (1983). Sex differences in mathematical reasoning: More facts. *Science, 222*, 1029–1031.

Bench, R. J., Collyer, Y., Mentz, L., & Wilson, I. (1976). Studies in infant behavioural audiometry: I. Neonates. *Audiology, 15*, 85–105.

Bender, S. L., Word, C. O., DiClemente, R. J., Crittenden, M. R., Persaud, N. A., & Ponton, L. (1995). The developmental implications of prenatal and/or postnatal crack cocaine exposure in preschool children: A preliminary report. *Developmental and Behavioral Pediatrics, 16*, 418–424.

Benedict, R. (1934). Anthropology and the abnormal. *Journal of Genetic Psychology, 10*, 59–82.

Benenson, J. F., Nicholson, C., Waite, A., Roy, R., & Simpson, A. (2001). The influence of group size on children's competitive behavior. *Child Development, 72*, 921–928.

Bengtson, V. L., & Harootyan, R. A. (1994). *Intergenerational linkages: Hidden connections in American society.* New York: Springer.

Bengston, V. L., Reedy, M. N., & Gordon, C. (1985). Aging and self-conceptions: Personality processes and social contexts. In J. E. Birren & K. W. Schaie (Eds.), *Handbook of the psychology of aging* (pp. 544–593). New York: Van Nostrand Reinhold.

Bengtson, V. L., Rosenthal, C. L., & Burton, L. (1990). Families and aging: Diversity and heterogeneity. In R. H. Binstock & L. K. George (Eds.), *Handbook of aging and the social sciences* (3rd ed., pp. 263–287). San Diego: Academic Press.

Benoliel, J. Q., & Degner, L. F. (1995). Institutional dying: A convergence of cultural values, technology, and social organization. In H. Wass & R. A. Neimeyer (Eds.), *Dying: Facing the facts* (pp. 117–162). Washington, DC: Taylor and Francis.

Berenbaum, S. A. (2001). Cognitive function in congenital adrenal hyperplasia. *Endocrinology and Metabolism Clinics of North America, 30*, 173–192.

Berg, C. A. (2000). Intellectual development in adulthood. In R. J. Sternberg (Ed.), *Handbook of intelligence* (pp. 117–137). New York: Cambridge University Press.

Berg, C. A., Strough, J., Calderone, K. S., Sansone, C., & Weir, C. (1998). The role of problem definitions in understanding age and context effects on strategies for solving everyday problems. *Psychology and Aging, 13*, 29–44.

Berg, S. (1996). Aging, behavior, and terminal decline. In J. E. Birren & K. W. Schaie (Eds.), *Handbook of the psychology of aging* (4th ed., pp. 323–337). San Diego: Academic Press.

Bergen, D., & Mauer, D. (2000). Symbolic play, phonological awareness, and literacy skills at three age levels. In K. A. Roskos & J. F. Christie (Eds.), *Play and literacy in early childhood: Research from multiple perspectives* (pp. 45–62). Mahwah, NJ: Erlbaum.

Berk, L. E. (1985). Relationship of caregiver education to child-oriented attitudes, job satisfaction, and behaviors toward children. *Child Care Quarterly, 14*, 103–129.

Berk, L. E. (1992). The extracurriculum. In P. W. Jackson (Ed.), *Handbook of research on curriculum* (pp. 1003–1043). New York: Macmillan.

Berk, L. E. (2001). *Awakening children's minds: How parents and teachers can make a difference.* New York: Oxford University Press.

Berk, L. E., & Spuhl, S. (1995). Maternal interaction, private speech, and task performance in preschool children. *Early Childhood Research Quarterly, 10*, 145–169.

Berkman, C. S., & Gurland, B. J. (1998). The relationship between ethnoracial group and functional level in older persons. *Ethnicity and Health, 3*, 175–188.

Berkowitz, M. W., & Gibbs, J. C. (1983). Measuring the developmental features of moral discussion. *Merrill-Palmer Quarterly, 29*, 399–410.

Berkowitz, R. J., & Stunkard, A. J. (2002). Development of childhood obesity. In T. A. Wadden & A. J. Stunkard (Eds.), *Handbook of obesity treatment* (pp. 515–531). New York: Guilford.

Berkus, M. D., Langer, O., Samueloff, A., Xanakis, E. M., & Field, N. T. (1999). Electronic fetal monitoring: What's reassuring? *Acta Obstetrica et Gynecologica Scandinavica, 78*, 15–21.

Berlin, L. J., Brooks-Gunn, J., McCarton, C., & McCormick, M. C. (1998). The effectiveness of early intervention: Examining risk factors and pathways to enhanced development. *Preventive Medicine, 27*, 238–245.

Berman, E., & Napier, A. Y. (2000). The midlife family: Dealing with adolescents, young adults, and the marriage in transition. In W. C. Nichols, M. A. Pace-Nichols, D. S. Becvar, & A. Y. Napier (Eds.), *Handbook of family development and intervention* (pp. 208–234). New York: Wiley.

Berman, P. W. (1980). Are women more responsive than men to the young? A review of developmental and situational variables. *Psychological Bulletin, 88*, 668–695.

Bermejo, V. (1996). Cardinality development and counting. *Developmental Psychology, 32*, 263–268.

Berndt, T. J., Cheung, P. C., Lau, S., Hau, K.-T., & Lew, W. J. F. (1993). Perceptions of parenting in mainland China, Taiwan, and Hong Kong: Sex differences and societal differences. *Developmental Psychology, 29*, 156–164.

Berndt, T. J., & Keefe, K. (1995). Friends' influence on adolescents' adjustment to school. *Child Development, 66*, 1312–1329.

Berne, P. H., & Savary, L. M. (1993). *Building self-esteem in children.* New York: Continuum.

Bernier, J. C., & Siegel, D. H. (1994). Attention-deficit hyperactivity disorder: A family ecological systems perspective. *Families in Society, 75*, 142–150.

Bernston, G. G., Malarkey, W. B., Kiecolt-Glaser, J. K., Sheridan, J. F., & Poehlmann, K. M. (1998). Autonomic, neuroendocrine, and immune responses to psychological stress: the reactivity hypothesis. *Annals of the New York Academy of Sciences, 840*, 664–673.

Bertenthal, B. I. (1993). Infants' perception of biomechanical motions: Instrinsic image and knowledge-based constraints. In C. Granrud (Ed.), *Visual perception and cognition in infancy* (pp. 175–214). Hillsdale, NJ: Erlbaum.

Bertenthal, B. I., Campos, J. J., & Barrett, K. (1984). Self-produced locomotion: An organizer of emotional, cognitive, and social development in infancy. In R. Emde & R. Harmon (Eds.), *Continuities and discontinuities in development* (pp. 174–210). New York: Plenum.

Bertrand, M., & Mullainathan, S. (2002). *Are Emily and Brendan more employable than Lakisha and Jamal? A field experiment on labor market discrimination.* Unpublished manuscript, University of Chicago.

Berzonsky, M. D., & Kuk, L. S. (2000). Identity status, identity processing style, and the transition to university. *Journal of Adolescent Research, 15*, 81–98.

Betz, C. (1994, March). Beyond time-out: Tips from a teacher. *Young Children, 49*(3), 10–14.

Betz, N. E. (1993). Women's career development. In F. L. Denmark & M. A. Paludi (Eds.), *Psychology of women* (pp. 627–684). Westport, CT: Greenwood Press.

Betz, N. E., & Fitzgerald, L. F. (1987). *The career psychology of women.* New York: Academic Press.

Bevan, E., & Higgins, D. J. (2002). Is domestic violence learned? The contribution of five forms of child maltreatment to men's violence and adjustment. *Journal of Family Violence, 17*, 223–245.

Beyene, Y. (1992). Menopause: A biocultural event. In A. J. Dan & L. L. Lewis (Eds.), *Menstrual health in women's lives* (pp. 169–177). Urbana, IL: University of Illinois Press.

Beyth-Marom, R., & Fischhoff, B. (1997). Adolescents' decisions about risks: A cognitive perspective. In J. Schulenberg, J. L. Maggs, & K. Hurrelmann (Eds.), *Health risks and developmental transitions during adolescence* (pp. 110–135). New York: Cambridge University Press.

Bhatt, R. S., Rovee-Collier, C., & Weiner, S. (1994). Developmental changes in the interface between perception and memory retrieval. *Developmental Psychology, 30*, 151–162.

Bialystok, E. (1999). Cognitive complexity and attentional control in the bilingual mind. *Child Development, 70*, 636–644.

Bialystok, E. (2001). *Bilingualism in development: Language, literacy, and cognition.* New York: Cambridge University Press.

Bialystok, E., & Herman, J. (1999). Does bilingualism matter for early literacy? *Language and Cognition, 2*, 35–44.

Bianco, A., Stone, J., Lynch, L., Lapinski, R., Berkowitz, G., & Berkowitz, R. L. (1996). Pregnancy outcome at age 40 and older. *Obstetrics and Gynecology, 87*, 917–922.

Biederman, J., & Spencer, T. J. (2000). Genetics of childhood disorders: XIX, ADHD, part 3: Is ADHD a noradrenergic disorder? *Journal of the American Academy of Child and Adolescent Psychiatry, 39*, 1330–1333.

Bielinski, J., & Davison, M. L. (1998). Gender differences by item difficulty interactions in multiple-choice mathematics items. *American Educational Research Journal, 35*, 455–476.

Biernat, M. (1991). Gender stereotypes and the relationship between masculinity and femininity: A developmental analysis. *Journal of Personality and Social Psychology, 61,* 351–365.

Bigelow, A. (1992). Locomotion and search behavior in blind infants. *Infant Behavior and Development, 15,* 179–189.

Bigler, R. S. (1995). The role of classification skill in moderating environmental influences on children's gender stereotyping: A study of the functional use of gender in the classroom. *Child Development, 66,* 1072–1087.

Bigler, R. S., & Liben, L. S. (1992). Cognitive mechanisms in children's gender stereotyping: Theoretical and educational implications of a cognitive-based intervention. *Child Development, 63,* 1351–1363.

Bigner, J. J., & Jacobsen, R. B. (1989). Parenting behaviors of homosexual and heterosexual fathers. *Journal of Homosexuality, 18,* 173–186.

Bijeljac-Babic, R., Bertoncini, J., & Mehler, J. (1993). How do 4-day-old infants categorize multisyllable utterances? *Developmental Psychology, 29,* 711–721.

Binstock, R. H., & Quadagno, J. (2001). Aging and politics. In R. H. Binstock & L. K. George (Eds.), *Handbook of aging and the social sciences* (5th ed., pp. 333–351). San Diego, CA: Academic Press.

Birch, E. E. (1993). Stereopsis in infants and its developmental relation to visual acuity. In K. Simons (Ed.), *Early visual development: Normal and abnormal* (pp. 224–236). New York: Oxford University Press.

Birch, L. L. (1998). Psychological influences on the childhood diet. *Journal of Nutrition, 128,* 407S–410S.

Birch, L. L., & Fisher, J. A. (1995). Appetite and eating behavior in children. *Pediatric Clinics of North America, 42,* 931–953.

Birch, L. L., Zimmerman, S., & Hind, H. (1980). The influence of social–affective context on preschool children's food preferences. *Child Development, 51,* 856–861.

Biringen, Z., Emde, R. N., Campos, J. J., & Appelbaum, M. I. (1995). Affective reorganization in the infant, the mother, and the dyad: The role of upright locomotion and its timing. *Child Development, 66,* 499–514.

Birmaher, B., Ryan, N., Williamson, D., Brent, D., & Kaufman, J. (1996). Childhood and adolescent depression: A review of the past 10 years. Part II. *Journal of the American Academy of Child and Adolescent Psychiatry, 35,* 1575–1583.

Birmingham, C. L., Muller, J. L., Palepu, A., Spinelli, J. J., & Anis, A. H. (1999). The cost of obesity in Canada. *Canadian Medical Association Journal, 160,* 483–438.

Biro, F. M., McMahon, R. P., Striegel-Moore, R., Crawford, P. B., Obarzanek, E., & Morrison, J. A. (2001). Impact of timing of pubertal maturation on growth in black and white female adolescents: The National Heart, Lung, and Blood Institute Growth and Health Study. *Journal of Pediatrics, 138,* 636–643.

Birren, J. E. (1990). Spiritual maturity in psychological development. In J. J. Seeber (Ed.), *Spiritual maturity in later years* (pp. 41–53). New York: Haworth.

Bjorklund, D. F., Cassel, W. S., Bjorklund, B. R., Brown, R. D., Park, C. L., Ernst, K., & Owen, F. A. (2000). Social demand characteristics in children's and adults' eyewitness memory and suggestibility: The effect of different interviewers on free recall and recognition. *Applied Cognitive Psychology, 14,* 421–423.

Bjorklund, D. F., & Coyle, T. R. (1995). Utilization deficiencies in the development of memory strategies. In F. E. Weinert & W. Schneider (Eds.), *Research on memory development: State of the art and future directions* (pp. 161–180). Hillsdale, NJ: Erlbaum.

Bjorklund, D. F., & Douglas, R. N. (1997). The development of memory strategies. In N. Cowan (Ed.), *The development of memory in childhood* (pp. 83–111). Hove, U.K.: Psychology Press.

Bjorklund, D. F., & Pellegrini, A. D. (2000). Child development and evolutionary psychology. *Child Development, 71,* 1687–1708.

Bjorklund, D. F., Schneider, W., Cassel, W. S., & Ashley, E. (1994). Training and extension of a memory strategy: Evidence for utilization deficiencies in high- and low-IQ children. *Child Development, 65,* 951–965.

Bjorklund, D. F., & Shackelford, T. K. (1999). Differences in parental investment contribute to important differences between men and women. *Current Directions in Psychological Science, 8,* 86–89.

Black, D., Gates, G., & Sanders, S. (2000). Demographics of the gay and lesbian population in the United States: Evidence from available systematic data sources. *Demography, 37,* 139–154.

Blacker, D., Bertram, L., Saunders, A. J., Moscarillo, T. J., Albert, M. S., & Wiener, H. (2003). Results of a high-resolution genome screen of 437 Alzheimer's disease families. *Human Molecular Genetics, 12,* 23–92.

Blackhall, L. J., Frank, G., Murphy, S., & Michel, V. (2001). Bioethics in a different tongue: The case of truth-telling. *Journal of Urban Health, 78,* 59–71.

Blackhall, L. J., Murphy, S. T., Frank, G., Michel, V., & Azen, S. (1995). Ethnicity and attitudes toward patient autonomy. *Journal of the American Medical Association, 274,* 820–825.

Blagg, N., & Yule, W. (1996). School phobia. In T. H. Ollendick, N. J. King, & W. Yule (Eds.), *International handbook of phobic and anxiety disorders in children and adolescents* (pp. 169–186). New York: Plenum.

Blake, I. K. (1994). Language development and socialization in young African-American children. In P. M. Greenfield & R. R. Cocking (Eds.), *Cross-cultural roots of minority child development* (pp. 167–195). Hillsdale, NJ: Erlbaum.

Blamey, A., Mutrie, N., & Aitchison, T. (1995). Health promotion by encouraged use of stairs. *British Medical Journal, 311,* 289–290.

Blanchard, R., & Bogaert, A. F. (1996). Homosexuality in men and number of older brothers. *American Journal of Psychiatry, 153,* 27–31.

Blanchard, R., Zucker, K. J., Bradley, S. J., & Hume, C. S. (1995). Birth order and sibling sex ratio in homosexual male adolescents and probably prehomosexual feminine boys. *Developmental Psychology, 31,* 22–30.

Blanchard-Fields, F. (1997). The role of emotion in social cognition across the adult life span. In K. W. Schaie & M. P. Lawton (Eds.), *Annual review of gerontology and geriatrics* (Vol. 17, pp. 325–352). New York: Springer.

Blanchard-Fields, F., Chen, Y., & Norris, L. (1997). Everyday problem solving across the adult life span: Influence of domain specificity and cognitive appraisal. *Psychology and Aging, 12,* 684–693.

Blanker, M. H., Bosch, J. L., Groeneveld, F. P., Bohnen, A. M., Prins, A., Thomas, S., & Hop, W. C. (2001). Erectile and ejaculatory dysfunction in a community-based sample of men 50 to 78 years old: Prevalence, concern, and relation to sexual activity. *Urology, 57,* 763–768.

Blasi, A. (1994). Moral identity: Its role in moral functioning. In B. Puka (Ed.), *Fundamental research in moral development: A compendium* (Vol. 2, pp. 123–167). New York: Garland.

Blass, E. M. (1999). Savoring sucrose and suckling milk: Easing pain, saving calories, and learning about mother. In M. Lewis & D. Ramsay (Eds.), *Soothing and stress* (pp. 79–107). Mahwah, NJ: Erlbaum.

Blass, E. M., Ganchrow, J. R., & Steiner, J. E. (1984). Classical conditioning in newborn humans 2–48 hours of age. *Infant Behavior and Development, 7,* 223–235.

Blatchford, P., Baines, E., Kutnick, P., & Martin, C. (2001). Classroom contexts: Connections between class size and within-class grouping. *British Journal of Educational Psychology, 71,* 283–302.

Bleske, A. L., & Buss, D. M. (2000). Can men and women be just friends? *Personal Relationships, 7,* 131–151.

Blieszner, R., & Adams, R. G. (1992). *Adult friendship.* Newbury Park, CA: Sage.

Block, J. (1971). *Lives through time.* Berkeley, CA: Bancroft.

Block, J. (1995). A contrarian view of the five-factor approach to personality description. *Psychological Bulletin, 117,* 187–215.

Block, S. D. (2000). Assessment and management of depression in the terminally ill. *Annals of Internal Medicine, 132,* 209–218.

Blomberg, S., Edebalk, P. G., & Petersson, J. (2000). The withdrawal of the welfare state: Elderly care in Sweden in the 1990s. *European Journal of Social Work, 3,* 151–163.

Bloom, L. (1998). Language acquisition in its developmental context. In D. Kuhn & R. S. Siegler (Eds.), *Handbook of child psychology: Vol. 2. Cognition, perception, and language* (5th ed., pp. 309–370). New York: Wiley.

Bloom, L. (2000). The intentionality model of language development: How to learn a word, any word. In R. Golinkoff, K. Hirsh-Pasek, N. Akhtar, L. Bloom, G. Hollich, L. Smith, M. Tomasello, & A. Woodward (Eds.), *Becoming a word learner: A debate on lexical acquisition.* New York: Oxford University Press.

Blotner, R., & Bearison, D. J. (1984). Developmental consistencies in socio-moral knowledge: Justice reasoning and altruistic behavior. *Merrill-Palmer Quarterly, 30,* 349–367.

Bluebond-Langner, M. (1977). Meanings of death to children. In H. Feifel (Ed.), *New meanings of death* (pp. 47–66). New York: McGraw-Hill.

Bluestone, C., & Tamis-LeMonda, C. S. (1999). Correlates of parenting styles in predominantly working- and middle-class African American mothers. *Journal of Marriage and the Family, 61,* 881–893.Blum, M. (2000). Embryos and the new reproductive technologies. Retrieved from http://www.surrogacy.com/legals/embryotech.html

Blum, N. J., & Carey, W. B. (1996). Sleep problems among infants and young children. *Pediatrics in Review, 17,* 87–93.

Blumberg, M. S., & Lucas, D. E. (1996). A developmental and component analysis of active sleep. *Developmental Psychobiology, 29,* 1–22.

Blumenstein, P., & Schwartz, P. (1983). *American couples.* New York: Morrow.

Blumenthal, H. T. (2001). Milestone or genomania? The relevance of the Human Genome Project to biological aging and the age-related diseases. *Journal of Gerontology, 56A,* M529–537.

Blyth, D. A., Simmons, R. G., & Zakin, D. F. (1985). Satisfaction with body image for early adolescent females: The impact of pubertal timing within different school environments. *Journal of Youth and Adolescence, 14,* 207–225.

Bock, G. R., & Goode, J. A. (Eds.). (1996). *Genetics of criminal and antisocial behavior, Ciba Foundation Symposium 194.* Chicester, U.K.: Wiley.

Boden, D., & Bielby, D. D. V. (1983). The past as resource: A conversational analysis of elderly talk. *Human Development, 26,* 308–319.

Bodenheimer, T. (1999). Questions and answers about Medicare. *International Journal of Health Services, 29,* 519–523.

Bogartz, R. S., Shinskey, J. L., & Schilling, T. H. (2000). Object permanence in five-and-a-half-month-old infants. *Infancy, 1,* 403–428.

Bogden, J. D., & Louria, D. B. (1999). Aging and the immune system: The role of micronutrient nutrition. *Nutrition, 15,* 593–595.

Bohannon, J. N., III, & Bonvillian, J. D. (2001). Theoretical approaches to language acquisition. In J. Berko Gleason (Ed.), *The development of language* (5th ed., pp. 254–314). Boston: Allyn and Bacon.

Bohannon, J. N., III, & Stanowicz, L. (1988). The issue of negative evidence: Adult responses to children's language errors. *Developmental Psychology, 24,* 684–689.

Bohman, M. (1996) Predispositions to criminality: Swedish adoption studies in retrospect. In G. R. Bock & J. A. Goode (Eds.), *Genetics of criminal and antisocial behavior, Ciba Foundation Symposium 194* (pp. 99–114). Chichester, U.K.: Wiley.

Bohman, M., & Sigvardsson, S. (1990). Outcome in adoption: Lessons from longitudinal studies. In D. M. Brodzinsky & M. D. Schechter (Eds.), *The psychology of adoption* (pp. 93–106). New York: Oxford University Press.

Bohr, L., Pascarella, E., Nora, A., Zusman, B., Jacobs, M., Desler, M., & Bulakowski, C. (1994). Cognitive effects of two-year and four-year institutions: A preliminary study. *Community College Review, 22,* 4–11.

Boldizar, J. P. (1991). Assessing sex typing and androgyny in children: The children's sex role inventory. *Developmental Psychology, 27,* 505–515.

Bolen, R. M. (2001). *Child sexual abuse.* New York: Kluwer Academic.

Bolger, K. E., & Patterson, C. J. (2001). Developmental pathways from child maltreatment to peer rejection. *Child Development, 72,* 549–568.

Bookstein, F. L., Sampson, P. D., Connor, P. D., & Streissguth, A. P. (2002). Midline corpus callosum is a neuroanatomical focus of fetal alcohol damage. *Anatomical Record, 269,* 162–174.

Boonstra, H. (2002). Teen pregnancy: Trends and lessons learned. *Guttmacher Report on Public Policy, 5*(1), 1–10.

Booth, A. (1999). Causes and consequences of divorce: Reflections on recent research. In R. A. Thompson & P. R. Amato (Eds.), *The postdivorce family: Children, parenting, and society* (pp. 29–48). Thousand Oaks, CA: Sage.

Borke, H. (1975). Piaget's mountains revisited: Changes in the egocentric landscape. *Developmental Psychology, 11,* 240–243.

Borkowski, J. G., & Muthukrisna, N. (1995). Learning environments and skill generalization: How contexts facilitate regulatory processes and efficacy beliefs. In F. Weinert & W. Schneider (Eds.), *Memory performances and competence: Issues in growth and development* (pp. 283–300). Hillsdale, NJ: Erlbaum.

Bornstein, M. H. (1989). Sensitive periods in development: Structural characteristics and causal interpretations. *Psychological Bulletin, 105,* 179–197.

Bornstein, M. H. (1999). Perception across the life span. In M. H. Bornstein & M. E. Lamb (Eds.), *Developmental psychology: An advanced textbook* (pp. 155–209). Mahwah, NJ: Erlbaum.

Bornstein, M. H., Haynes, O. M., Pascual, L., Painter, K. M., & Galperin, C. (1999). Play in two societies: Pervasiveness of process, specificity of structure. *Child Development, 70,* 317–331.

Bornstein, M. H., Vibbert, M., Tal, J., & O'Donnell, K. (1992). Toddler language and play in the second year: Stability, covariation, and influences of parenting. *First Language, 12,* 323–338.

Borst, C. G. (1995). *Catching babies: The professionalization of childbirth, 1870–1920.* Cambridge, MA: Harvard University Press.

Borstelmann, L. J. (1983). Children before psychology: Ideas about children from antiquity to the late 1800s. In W. Kessen (Ed.), *Handbook of child psychology: Vol. 1. History, theory, and methods* (pp. 1–40). New York: Wiley.

Bortolus, R., Parazzini, F., Chatenoud, L., Benzi, G., Bianchi, M. M., & Marini, A. (1999). The epidemiology of multiple births. *Human Reproduction Update, 5,* 179–187.

Bossé, R., Aldwin, C. M., Levenson, M. R., Workman-Daniels, K., & Ekerdt, D. J. (1990). Differences in social support among retirees and workers: Findings from the Normative Aging Study. *Psychology and Aging, 5,* 41–47.

Bossé, R., Spiro, A., III, & Kressin, N. R. (1996). The psychology of retirement. In R. T. Woods (Ed.), *Handbook of the clinical psychology of aging* (pp. 141–157). Chicester, U.K.: Wiley.

Bost, K. K., Vaughn, B. E., Washington, W. N., Cielinski, K. L., & Bradbard, M. R. (1998). Social competence, social support, and attachment: Demarcation of construct domains, measurement, and paths of influence for preschool children attending Head Start. *Child Development, 69,* 192–218.

Boston Women's Health Book Collective. (1992). *The new Our Bodies, Ourselves: A book by and for women* (2nd ed.). New York: Simon & Schuster.

Bosworth, H. B., & Schaie, K. W. (1999). Survival effects in cognitive function, cognitive style, and sociodemographic variables. *Experimental Aging Research, 25,* 121–140.

Botkin, D. R., Weeks, M. O., & Morris, J. E. (2000). Changing marriage role expectations: 1961–1996. *Sex Roles, 42,* 933–942.

Bouchard, T. J., Jr., Lykken, D. T., McGue, M., Segal, N. L., & Tellegen, A. (1990). Sources of human psychological differences: The Minnesota Study of Twins Reared Apart. *Science, 250,* 223–228.

Boukydis, C. F. Z., & Burgess, R. L. (1982). Adult physiological response to infant cries: Effects of temperament of infant, parental status and gender. *Child Development, 53,* 1291–1298.

Boukydis, C. F. Z., & Lester, B. M. (1998). Infant crying, risk status and social support in families of preterm and term infants. *Early Development and Parenting, 7,* 31–39.

Boulton, M. J. (1999). Concurrent and longitudinal relations between children's playground behavior and social preference, victimization, and bullying. *Child Development, 70,* 944–954.

Boulton, M. J., & Smith, P. K. (1994). Bully/victim problems in middle-school children: Stability, self-perceived competence, peer perceptions and peer acceptance. *British Journal of Developmental Psychology, 12,* 315–329.

Bourgeois, M., Burgio, L., Schulz, R., Beach, S., & Palmer, B. (1997). Modifying repetitive verbalization of community dwelling patients with AD. *Gerontologist, 37,* 30–39.

Bowlby, J. (1969). *Attachment and loss: Vol. 1. Attachment.* New York: Basic Books.

Bowlby, J. (1979). *The making and breaking of affectional bonds.* London: Tavistock.

Bowlby, J. (1980). *Attachment and loss: Vol. 3. Loss: Sadness and depression.* New York: Basic Books.

Bowlby, J. W., & McMullen, K. (2002). *At a crossroads: First results for the 18- to 20-year-old cohort of the Youth in Transition Survey*. Ottawa: Human Resources Development Canada.

Boyer, K., & Diamond, A. (1992). Development of memory for temporal order in infants and young children. In A. Diamond (Ed.), *Development and neural bases of higher cognitive function* (pp. 267–317). New York: New York Academy of Sciences.

Boyes, M. C., & Chandler, M. (1992). Cognitive development, epistemic doubt, and identity formation in adolescence. *Journal of Youth and Adolescence, 21,* 277–304.

Brackbill, Y., McManus, K., & Woodward, L. (1985). *Medication in maternity: Infant exposure and maternal information.* Ann Arbor: University of Michigan Press.

Bracken, B. A. (2000). Maximizing construct relevant assessment:

The optimal preschool testing situation. In B. A. Bracken (Ed.), *The psychoeducational assessment of preschool children* (3rd ed., pp. 33–44). Upper Saddle River, NJ: Prentice-Hall.

Bradburn, E. M., Moen, P., & Dempster-McClain, D. (1995). Women's return to school following the transition to motherhood. *Social Forces, 73,* 1517–1551.

Bradbury, K., Janicke, D. M., Riley, A. W., & Finney, J. W. (1999). Predictors of unintentional injuries to school-age children seen in pediatric primary care. *Journal of Pediatric Psychology, 24,* 423–433.

Bradbury, T. N., Fincham, F. D., & Beach, S. R. H. (2000). Research on the nature and determinants of marital satisfaction: A decade in review. *Journal of Marriage and the Family, 62,* 964–980.

Bradley, R. H., & Caldwell, B. M. (1979). Home Observation for Measurement of the Environment: A revision of the preschool scale. *American Journal of Mental Deficiency, 84,* 235–244.

Bradley, R. H., & Caldwell, B. M. (1982). The consistency of the home environment and its relation to child development. *International Journal of Behavioral Development, 5,* 445–465.

Bradley, R. H., Caldwell, B. M., Rock, S. L., Ramey, C. T., Barnard, D. E., Gray, C., Hammond, M. A., Mitchell, S., Gottfried, A., Siegel, L., & Johnson, D. L. (1989). Home environment and cognitive development in the first 3 years of life: A collaborative study involving six sites and three ethnic groups in North America. *Developmental Psychology, 25,* 217–235.

Bradley, R. H., Whiteside, L., Mundfrom, D. J., Casey, P. H., Kelleher, K. J., & Pope, S. K. (1994). Early indications of resilience and their relation to experiences in the home environments of low birthweight, premature children living in poverty. *Child Development, 65,* 346–360.

Brady, E. M. (1984). Demographic and educational correlates of self-reported learning among older students. *Educational Gerontology, 10,* 27–38.

Braet, C., & Mervielde, I. (1997). Psychological aspects of childhood obesity: A controlled study in a clinical and nonclinical sample. *Journal of Pediatric Psychology, 22,* 59–71.

Braine, L. G., Schauble, L., Kugelmass, S., & Winter, A. (1993). Representation of depth by children: Spatial strategies and lateral biases. *Developmental Psychology, 29,* 466–479.

Brame, B., Nagin, D. S., & Tremblay, R. E. (2001). Developmental trajectories of physical aggression from school entry to late adolescence. *Journal of Child Psychology and Psychiatry, 42,* 503–512.

Brandtstädter, J., & Rothermund, K. (1994). Self-percepts of control in middle and later adulthood: Buffering losses by rescaling goals. *Psychology and Aging, 9,* 265–273.

Bratt, R. G. (2002). Housing: The foundation of family life. In F. Jacobs, D. Wertlieb, & R. M. Lerner (Eds.), *Handbook of applied developmental science* (Vol. 2, pp. 445–468). Thousand Oaks, CA: Sage.

Braverman, P. K., & Strasburger, V. C. (1993). Adolescent sexual activity. *Clinical Pediatrics, 32,* 658–668.

Bray, J. H. (1999). From marriage to remarriage and beyond: Findings from the Developmental Issues in Stepfamilies Research Project. In E. M. Hetherington (Ed.), *Coping with divorce, single parenting, and remarriage: A risk and resiliency perspective* (pp. 295–319). Mahwah, NJ: Erlbaum.

Braybyn, J., Schneck, M., Haegerstrom-Portnoy, G., & Lott, L. (2001). The Smith-Kettlewell Institute (SKI) longitudinal study of vision function and its impact among the elderly: An overview. *Optometry and Vision Science, 78,* 264–269.

Brazelton, T. B., Koslowski, B., & Tronick, E. (1976). Neonatal behavior among urban Zambians and Americans. *Journal of the American Academy of Child Psychiatry, 15,* 97–107.

Brazelton, T. B., & Nugent, J. K. (1995). *Neonatal Behavioral Assessment Scale.* London, Mac Keith Press.

Brazelton, T. B., Nugent, J. K., & Lester, B. M. (1987). Neonatal Behavioral Assessment Scale. In J. D. Osofsky (Ed.), *Handbook of infant development* (2nd ed., pp. 780–817). New York: Wiley.

Brébion, G., Smith, M. J., & Ehrlich, M.-F. (1997). Working memory and aging: Deficit or strategy differences? *Aging, Neuropsychology, and Cognition, 4,* 58–73.

Bredekamp, S., & Copple, C. (Eds.). (1997). *Developmentally appropriate practice in early childhood programs* (rev. ed.). Washington, DC: National Association for the Education of Young Children.

Breedlove, S. M. (1994). Sexual differentiation of the human nervous system. *Annual Review of Psychology, 45,* 389–418.

Brehm, S. S. (1992). *Intimate relationships* (2nd ed.). New York: McGraw-Hill.

Bremner, J. G. (1998). From perception to action: The early development of knowledge. In F. Simion & G. Butterworth (Eds.), *Development of sensory, motor, and cognitive capacities in early infancy* (pp. 239–255). East Sussex, U.K.: Psychology Press.

Brennan, K. A., & Shaver, P. R. (1995). Dimensions of adult attachment, affect regulation, and romantic relationship functioning. *Personality and Social Psychology Bulletin, 21,* 267–283.

Brennan, M., & Cardinali, G. (2000). The use of preexisting and novel coping strategies in adapting to age-related vision loss. *Gerontologist, 40,* 327–334.

Brennan, P. F., Moore, S. M., & Smyth, K. A. (1991). ComputerLink: Electronic support for the home caregiver. *Advances in Nursing Science, 13,* 14–27.

Brennan, W. M., Ames, E. W., & Moore, R. W. (1966). Age differences in infants' attention to patterns of different complexities. *Science, 151,* 354–356.

Brenner, E., & Salovey, P. (1997). Emotional regulation during childhood: Developmental, interpersonal, and individual considerations. In P. Salovey & D. Sluyter (Eds.), *Emotional literacy and emotional development* (pp. 168–192). New York: Basic Books.

Brenner, R. A., Simons-Morton, B. G., Bhaskar, B., Revenis, M., Das, A., & Clemens, J. D. (2003). Infant–parent bed sharing in an inner-city population. *Archives of Pediatrics and Adolescent Medicine, 157,* 33–39.

Brent, R. L. (1999). Utilization of developmental basic science principles in the evaluation of reproductive risks from pre- and postconception environmental radiation exposures. *Teratology, 59,* 182–204.

Brent, S. B., Speece, M. W., Lin, C., Dong, Q., & Yang, C. (1996). The development of the concept of death among Chinese and U.S. children 3–17 years of age: From binary to "fuzzy" concepts? *Omega, 33,* 67–83.

Breslau, N., Kessler, R. C., Chilcoat, H. D., Schultz, L. R., Davis, G. C., & Andreski, P. (1998). Trauma and posttraumatic stress disorder in the community: The 1996 Detroit Area Survey of Trauma. *Archives of General Psychiatry, 55,* 626–632.

Bretherton, I. (1992). The origins of attachment theory: John Bowlby and Mary Ainsworth. *Developmental Psychology, 28,* 759–775.

Bretherton, I., Fritz, J., Zahn-Waxler, C., & Ridgeway, D. (1986). Learning to talk about emotions: A functionalist perspective. *Child Development, 57,* 529–548.

Brewaeys, A., Ponjaert, I., Van Hall, E. V., & Golombok, S. (1997). Donor insemination: Child development and family functioning in lesbian mother families. *Human Reproduction, 12,* 1349–1359.

Brezina, T. (1999). Teenage violence toward parents as an adaptation to family strain: Evidence from a national survey of male adolescents. *Youth and Society, 30,* 416–444.

Brien, M. J., & Willis, R. J. (1997). Costs and consequences for the fathers. In R. A. Maynard (Ed.), *Kids having kids* (pp. 95–144). Washington, DC: Urban Institute.

Briggs, F., & Hawkins, R. (1996). *Keeping ourselves safe: Who benefits?* Wellington, NZ: New Zealand Council for Educational Research.

Briggs, F., & Hawkins, R. (1999). The importance of parent involvement in child-protection curricula. In L. E. Berk (Ed.), *Landscapes of development* (pp. 321–335). Belmont, CA: Wadsworth.

Brigham, J. (1998). *Dying to quit: Why we smoke and how we stop.* Washington, DC: National Academy Press.

British Broadcasting Corporation. (1999, May 11). *Euthanasia special report: Lessons from down under.* London: Author.

Broberg, A. G., Wessels, H., Lamb, M. E., & Hwang, C. P. (1997). Effects of day care on the development of cognitive abilities in 8-year-olds: A longitudinal study. *Developmental Psychology, 33,* 62–69.

Brody, G. H., & Flor, D. L. (1998). Maternal resources, parenting practices, and child competence in rural, single-parent African American families. *Child Development, 69,* 803–816.

Brody, G. H., Stoneman, Z., & Flor, D. (1996). Parental religiosity, family processes, and youth competence in rural, two-parent African-American families. *Developmental Psychology, 32,* 696–706.

Brody, G. H., Stoneman, Z., & McCoy, J. K. (1992). Associations of maternal and paternal direct and differential behavior with sibling relationships: Contemporaneous and longitudinal analyses. *Child Development, 63,* 82–92.

Brody, G. H., Stoneman, Z., & McCoy, J. K. (1994). Forecasting sibling relationships in early adolescence from child temperaments and family processes in middle childhood. *Child Development, 65,* 771–784.

Brody, J. E. (1992, November 11). PMS is a worldwide phenomenon. *The New York Times,* p. C14.

Brody, L. (1999). *Gender, emotion, and the family.* Cambridge, MA: Harvard University Press.

Brody, N. (1997a). *Intelligence* (2nd ed.). San Diego: Academic Press.

Brody, N. (1997b). Intelligence, schooling, and society. *American Psychologist, 52,* 1046–1050.

Bromberger, J. T., Meyer, P. M., Kravitz, H. M., Sommer, B., Cordal, A., & Powell, L. (2001). Psychologic distress and natural menopause: A multiethnic community study. *American Journal of Public Health, 91,* 1435–1442.

Bronfenbrenner, U. (1989). Ecological systems theory. In R. Vasta (Ed.), *Annals of child development* (Vol. 6, pp. 187–251). Greenwich, CT: JAI Press.

Bronfenbrenner, U. (1995). The bioecological model from a life course perspective: Reflections of a participant observer. In P. Moen, G. H. Elder, Jr., & K. Lüscher (Eds.), *Examining lives in context* (pp. 599–618). Washington, DC: American Psychological Association.

Bronfenbrenner, U., & Ceci, S. J. (1994). Nature–nurture reconceptualized in developmental perspective: A bioecological model. *Psychological Review, 101,* 568–586.

Bronfenbrenner, U., & Morris, P. A. (1998). The ecology of developmental processes. In R. M. Lerner (Ed.), *Handbook of child psychology: Vol. 1. Theoretical models of human development* (5th ed., pp. 535–584). New York: Wiley.

Bronson, G. W. (1991). Infant differences in rate of visual encoding. *Child Development, 62,* 44–54.

Brooks, C. (2000). Civil rights, liberalism, and the suppression of a Republican political realignment in the United States, 1972 to 1996. *American Sociological Review, 65,* 483–505.

Brooks, D., & Barth, R. P. (1999). Adult transracial and inracial adoptees: Effects of race, gender, adoptive family structure, and placement history on adjustment outcomes. *American Journal of Orthopsychiatry, 69,* 87–99.

Brooks, G. R. (1999). The role of gender in marital dysfunction. In R. M. Eisler & M. Hersen (Eds.), *Handbook of gender, culture, and health* (pp. 449–470). Mahwah, NJ: Erlbaum.

Brooks-Gunn, J. (1988a). Antecedents and consequences of variations in girls' maturational timing. *Journal of Adolescent Health Care, 9,* 365–373.

Brooks-Gunn, J. (1988b). The impact of puberty and sexual activity upon the health and education of adolescent girls and boys. *Peabody Journal of Education, 64,* 88–113.

Brooks-Gunn, J., & Chase-Lansdale, P. L. (1995). Adolescent parenthood. In M. H. Bornstein (Ed.), *Handbook of parenting: Vol. 3. Status and social conditions of parenting* (pp. 113–149). Mahwah, NJ: Erlbaum.

Brooks-Gunn, J., & Duncan, G. J. (1997). The effects of poverty on children. *Future of Children, 7,* 55–71.

Brooks-Gunn, J., McCarton, C. M., Casey, P. H., McCormick, M. C., Bauer, C. R., Bernbaum, J. C., Tyson, J., Swanson, M., Bennett, F. C., Scott, D. T., Tonascia, J., & Meinert, C. L. (1994). Early intervention in low-birth-weight premature infants. *Journal of the American Medical Association, 272,* 1257–1262.

Brooks-Gunn, J., & Ruble, D. N. (1980). Menarche: The interaction of physiology, cultural, and social factors. In A. J. Dan, E. A. Graham, & C. P. Beecher (Eds.), *The menstrual cycle: A synthesis of interdisciplinary research* (pp. 141–159). New York: Springer-Verlag.

Brooks-Gunn, J., & Ruble, D. N. (1983). The experience of menarche from a developmental perspective. In J. Brooks-Gunn & A. C. Petersen (Eds.), *Girls at puberty* (pp. 155–177). New York: Plenum.

Brooks-Gunn, J., Warren, M. P., Samelson, M., & Fox, R. (1986). Physical similarity of and disclosure of menarcheal status to friends: Effects of grade and pubertal status. *Journal of Early Adolescence, 6,* 3–14.

Broomhall, H. S., & Winefield, A. H. (1990). A comparison of the affective well-being of young and middle-aged unemployed men matched for length of employment. *British Journal of Medical Psychology, 63,* 43–52.

Brown, B. B., Clasen, D., & Eicher, S. (1986). Perceptions of peer pressure, peer conformity dispositions, and self-reported behavior among adolescents. *Developmental Psychology, 22,* 521–530.

Brown, B. B., Feiring, C., & Furman, W. (1999). Missing the love boat: Why researchers have shied away from adolescent romance. In W. Furman, B. B. Brown, & C. Feiring (Eds.), *The development of romantic relationships in adolescence* (pp. 1–16). New York: Cambridge University Press.

Brown, B. B., Lohr, M. J., & McClenahan, E. L. (1986). Early adolescents' perceptions of peer pressure. *Journal of Early Adolescence, 6,* 139–154.

Brown, J. D. (2002). Mass media influences on sexuality. *Journal of Sex Research, 39,* 42–45.

Brown, J. R., Donelan-McCall, N., & Dunn, J. (1996). Why talk about mental states? The significance of children's conversations with friends, siblings, and mothers. *Child Development, 67,* 836–849.

Brown, R. W. (1973). *A first language: The early stages.* Cambridge, MA: Harvard University Press.

Brown, S. L. (2000). Union transitions among cohabitors: The significance of relationship assessments and expectations. *Journal of Marriage and the Family, 62,* 833–846.

Browne, A. (1993). Violence against women by male partners: Prevalence, outcomes and policy implications. *American Psychologist, 48,* 1077–1087.

Brownell, C. A., & Carriger, M. S. (1990). Changes in cooperation and self-other differentiation during the second year. *Child Development, 61,* 1164–1174.

Brubaker, T. (1985). *Later life families.* Beverly Hills, CA: Sage.

Bruce, B., & Wilfley, D. (1996). Binge eating among the overweight population: A serious and prevalent problem. *Journal of the American Dietetic Association, 96,* 58–61.

Bruce, D., Dolan, A., & Phillips-Grant, K. (2000). On the transition from childhood amnesia to recall of personal memories. *Psychological Science, 11,* 360–364.

Bruch, H. (2001). *The golden cage: The enigma of anorexia nervosa.* Cambridge, MA: Harvard University Press.

Bruch, M. A., Gorsky, J. M., Collins, T. M., & Berger, P. A. (1989). Shyness and sociability examined: A multicomponent analysis. *Journal of Personality and Social Psychology, 57,* 904–915.

Bruck, M., Ceci, S. J., & Hembrooke, H. (1998). Reliability and credibility of young children's reports. *American Psychologist, 53,* 136–151.

Brun, A., & Andersson, J. (2001). Frontal dysfunction and frontal cortical synapse loss in alcoholism: The main cause of alcohol dementia? *Dementia and Geriatric Cognitive Disorders, 12,* 289–294.

Brussoni, M. J., and Boon, S. D. (1998). Grandparental impact in young adults' relationships with their closest grandparents: The role of relationship strength and emotional closeness. *International Journal of Aging and Human Development, 45,* 267–286.

Buchanan, A. (1996). *Cycles of child maltreatment.* Chichester, U.K.: Wiley.

Buchanan, C. M., Eccles, J. S., & Becker, J. B. (1992). Are adolescents the victims of raging hormones? Evidence for activational effects of hormones on moods and behavior at adolescence. *Psychological Bulletin, 111,* 62–107.

Buchanan, C. M., & Holmbeck, G. N. (1998). Measuring beliefs about adolescent personality and behavior. *Journal of Youth and Adolescence, 27,* 609–629.

Buchanan, C. M., Maccoby, E. E., & Dornbusch, S. M. (1996). *Adolescents after divorce.* Cambridge, MA: Harvard University Press.

Buchanan-Barrow, E., & Barrett, M. (1998). Children's rule discrimination within the context of the school. *British Journal of Developmental Psychology, 16,* 539–551.

Buekens, P., Kotelchuck, M., Blondel, B., Kristensen, F. B., Chen, J.-H., & Masuy-Stroobant, G. (1993). A comparison of prenatal care use in the United States and Europe. *American Journal of Public Health, 83,* 31–36.

Buhrmester, D. (1996). Need fulfillment, interpersonal competence, and the developmental contexts of early adolescent friendship. In W. M. Bukowski, A. F. Newcomb, & W. W. Hartup (Eds.), *The company they keep: Friendship during childhood and adolescence* (pp. 158–185). New York: Cambridge University Press.

Buhrmester, D., & Furman, W. (1990). Perceptions of sibling relationships during middle childhood and adolescence. *Child Development, 61,* 1387–1398.

Buhs, E. S., & Ladd, G. W. (2001). Peer rejection as antecedent of young children's school adjustment: An examination of mediating processes. *Developmental Psychology, 37,* 550–560.

Bukowski, W. M. (2001). Friendship and the worlds of childhood. In D. W. Nangle & C. A. Erdley (Eds.), *The role of friendship in psychological adjustment* (pp. 93–105). San Francisco: Jossey-Bass.

Bulatao, R. A. (1998). *The value of family planning programs in developing countries.* New York: Rand Corporation.

Bumpass, L. L., & Lu, H. H. (2000). Trends in cohabitation and implications for children's family contexts in the United States. *Population Studies, 54,* 29–41.

Burchinal, M. R., Peisner-Feinberg, E., Bryant, D. M., & Clifford, R. (2000). Children's social and cognitive development and child-care quality: Testing for differential associations related to poverty, gender, or ethnicity. *Applied Developmental Science, 4,* 149–165.

Burgess, K. B., Rubin, K. H., Chea, C. S. L., & Nelson, L. J. (2001). Behavioral inhibition, social withdrawal, and parenting. In R. Crozier & L. Alden (Eds.), *International handbook of social anxiety* (pp. 137–158). New York: Wiley.

Burhans, K. K., & Dweck, C. S. (1995). Helplessness in early childhood: The role of contingent worth. *Child Development, 66,* 1719–1738.

Burke, R. J. (2001). Organizational values, work experiences, and satisfactions among managerial and professional women. *Journal of Management Development, 20,* 346–354.

Burke, W., Atkins, D., & Gwin, M. (2002). Genetic test evaluation: Information needs of clinicians, policy makers, and the public. *American Journal of Epidemiology, 156,* 311-318.

Burts, D.C., Hart, C. H., Charlesworth, R., Fleege, P. O., Mosely, J., & Thomasson, R. H. (1992). Observed activities and stress behaviors of children in developmentally appropriate and inappropriate kindergarten classrooms. *Early Childhood Research Quarterly, 7,* 297–318.

Bushman, B. J., & Huesmann, L. R. (2001). Effects of televised violence of aggression. In D. G. Singer & J. L. Singer (Eds.), *Handbook of children and the media* (pp. 223–254). Thousand Oaks, CA: Sage.

Bushnell, E. W., & Boudreau, J. P. (1993). Motor development and the mind: The potential role of motor abilities as a determinant of aspects of perceptual development. *Child Development, 64,* 1005–1021.

Buss, D. M. (1999). *Evolutionary psychology: The new science of the mind.* Boston: Allyn and Bacon.

Buss, D. M. (2001). The strategies of human mating. In P. W. Sherman & J. Alcock (Eds.), *Exploring animal behavior: Readings from American Scientist* (3rd ed., pp. 240–251). Sunderland, MA: Sinauer Associates.

Buss, D. M., Abbott, M., Angleitner, A., & Asherian, A. (1990). International preferences in selecting mates: A study of 37 cultures. *Journal of Cross-Cultural Psychology, 21,* 5–47.

Buss, D. M., Shackelford, T. K., Kirkpatrick, L. A., & Larsen, R. J. (2001). A half century of mate preferences: The cultural evolution of values. *Journal of Marriage and the Family, 63,* 491–503.

Bussell, D. A., Neiderhiser, J. M., Pike, A., Plomin, R., Simmens, S., Howe, G. W., Hetherington, E. M., Carroll, E., & Reiss, D. (1999). Adolescents' relationships to siblings and mothers: A multivariate genetic analysis. *Developmental Psychology, 35,* 1248–1259.

Bussey, K. (1992). Lying and truthfulness: Children's definitions, standards, and evaluative reactions. *Child Development, 63,* 129–137.

Butler, R. (1998). Age trends in the use of social and temporal comparison for self-evaluation: Examination of a novel developmental hypothesis. *Child Development, 69,* 1054–1073.

Butler, R. N. (1968). The life review: An interpretation of reminiscence in the aged. In B. Neugarten (Ed.), *Middle age and aging* (pp. 486–496). Chicago: University of Chicago Press.

Butler, R. N., Fossel, M., Harman, M., Heward, C. B., Olshansky, S. J., & Perls, T. T. (2002). Is there an antiaging medicine? *Journal of Gerontology, 57A,* B333–B338.

Buttell, F. P. (1999). The relationship between spouse abuse and the maltreatment of dementia sufferers by their caregivers. *American Journal of Alzheimer's Disease, 14,* 230–232.

Buunk, B. P., Dijkstra, P., Kenrick, D. T., & Warntjes, A. (2001). Age preferences for mates as related to gender, own age, and involvement level. *Evolution and Human Behavior, 22,* 241–250.

Buunk, B. P., & van Driel, B. (1989). *Variant lifestyles and relationships.* Newbury Park, CA: Sage.

Byars, A. M., & Hackett, G. (1998). Applications of social cognitive theory to the career development of women of color. *Applied and Preventive Psychology, 7,* 255–267.

Byrnes, J. P., & Takahira, S. (1993). Explaining gender differences on SAT-math items. *Developmental Psychology, 29,* 805–810.

Caddell, D. P., & Newton, R. R. (1995). Euthanasia: American attitudes toward the physician's role. *Social Science and Medicine, 40,* 1671–1681.

Cadoret, R. J., Cain, C. A., & Crowe, R. R. (1983). Evidence for gene–environment interaction in the development of adolescent antisocial behavior. *Behavior Genetics, 13,* 301–310.

Cain, K. M., & Dweck, C. S. (1995). The relation between motivational patterns and achievement cognitions through the elementary school years. *Merrill-Palmer Quarterly, 41,* 25–52.

Caine, N. (1986). Behavior during puberty and adolescence. In G. Mitchell & J. Erwin (Eds.), *Comparative primate biology: Vol. 2A. Behavior, conservation, and ecology* (pp. 327–361). New York: Liss.

Cairns, R. B. (1998). The making of developmental psychology. In R. M. Lerner (Ed.), *Handbook of child psychology: Vol. 1. Theoretical models of human development* (5th ed., pp. 25–105). New York: Wiley.

Cairns, R. B., Leung, M.-C., Buchanan, L., & Cairns, B. D. (1995). Friendships and social networks in childhood and adolescence: Fluidity, reliability, and interrelations. *Child Development, 66,* 1330–1345.

Cairns, R., Xie, H., & Leung, M.-C. (1998). The popularity of friendship and the neglect of social networks: Toward a new balance. In W. M. Bukowski & A. H. Cillessen (Eds.), *Sociometry then and now: Building on six decades of measuring children's experiences with the peer group* (pp. 25–53). San Francisco: Jossey-Bass.

Caldwell, B. M., & Bradley, R. H. (1994). Environmental issues in developmental follow-up research. In S. L. Friedman & H. C. Haywood (Eds.), *Developmental follow-up* (pp. 235–256). San Diego: Academic Press.

Caldwell, C. H., & Antonucci, T. C. (1997). Childbearing during adolescence: Mental health risks and opportunities. In J. Schulenberg, J. L. Maggs, & K. Hurrelmann (Eds.), *Health risks and developmental transitions during adolescence* (pp. 220–245). New York: Cambridge University Press.

Caldwell, J. (1999). Paths to lower fertility. *British Medical Journal, 319,* 985–987.

Calkins, S. D., Fox, N. A., & Marshall, T. R. (1996). Behavioral and physiological antecedents of inhibited and uninhibited behavior. *Child Development, 67,* 523–540.

Callaghan, T. C. (1999). Early understanding and production of graphic symbols. *Child Development, 70,* 1314–1324.

Calle, E. E., Rodriguez, C., Walker-Thurmond, K., & Thun, M. J. (2003). Overweight, obesity, and mortality from cancer in a prospectively studied cohort of U.S. adults. *New England Journal of Medicine, 348,* 1625–1638.

Cameron, C. A., & Lee, K. (1997). The development of children's telephone communication. *Journal of Applied Developmental Psychology, 18,* 55–70.

Cameron, M. B., & Wilson, B. J. (1990). The effects of chronological age, gender, and delay of entry on academic achievement and retention: Implications for academic redshirting. *Psychology in the Schools, 27,* 260–263.

Campbell, D. W., & Eaton, W. O. (1999). Sex differences in the activity level of infants. *Infant and Child Development, 8,* 1–17.

Campbell, F. A., Pungello, E. P., Miller-Johnson, S., Burchinal, M., & Ramey, C. T. (2001). The development of cognitive and academic abilities: Growth curves from an early childhood educational experiment. *Developmental Psychology, 37,* 231–242.

Campbell, F. A., & Ramey, C. T. (1995). Cognitive and school outcomes for high-risk African-American students at middle adolescence: Positive effects of early intervention. *American Educational Research Journal, 32,* 743–772.

Campbell, J. R., Hombo, C. M., & Mazzeo, J. (2000). *NAEP 1999: Trends in academic progress.* Washington, DC: U.S. Department of Education.

Campbell, L. D., & Martin-Matthews, A. (2000). Caring sons: Exploring men's involvement in filial care. *Canadian Journal on Aging, 19,* 57–97.

Campbell, R., & Sais, E. (1995). Accelerated metalinguistic phonological awareness in bilingual children. *British Journal of Developmental Psychology, 13,* 61–68.

Campbell, S. B., Cohn, J. F., & Meyers, T. (1995). Depression in first-time mothers: Mother–infant interaction and depression chronicity. *Developmental Psychology, 31,* 349–357.

Campos, J. J., Anderson, D. I., Barbu-Roth, M. A., Hubbard, E. M., Hertenstein, J. J., & Witherington, D. (2000). Travel broadens the mind. *Infancy, 1,* 149–219.

Campos, J. J., Kermoian, R., & Zumbahlen, M. R. (1992). Socioemotional transformation in the family system following infant crawling onset. In N. Eisenberg & R. A. Fabes (Eds.), *New directions for child development* (No. 55, pp. 25–40). San Francisco: Jossey-Bass.

Campos, R. G. (1989). Soothing pain-elicited distress in infants with swaddling and pacifiers. *Child Development, 60,* 781–792.

Camras, L. A. (1992). Expressive development and basic emotions. *Cognition and Emotion, 6,* 267–283.

Camras, L. A., Oster, H., Campos, J., Campos, R., Ujie, T., Miyake, K., Wang, L., & Meng, Z. (1998). Production of emotional and facial expressions in European American, Japanese, and Chinese infants. *Developmental Psychology, 34,* 616–628.

Camras, L. A., Oster, H., Campos, J. J., Miyake, K., & Bradshaw, D. (1992). Japanese and American infants' responses to arm restraint. *Developmental Psychology, 28,* 578–583.

Canadian Broadcast Standards Council. (2002). Canada deals with media violence. Retrieved from http://www.cbsc.ca/english/canada.htm

Canadian Fitness and Lifestyle Research Institute. (2001a). Survey of physical activity in Canadian schools. Retrieved from http://www.cflri.ca/cflri/cflri.html

Canadian Fitness and Lifestyle Research Institute. (2001b). 2001 physical activity monitor. Retrieved from http://www.cflri.ca/cflri/pa/surveys/2001survey/2001survey.html

Canadian Fitness and Lifestyle Research Institute. (2002). 2000 physical activity monitor. Retrieved from http://www.cflri.ca/cflri/resources/pub.php#2001capacity

Canadian National Council of Welfare. (2002). *Child poverty profile, 2000.* Ottawa: Canadian Ministry of Public Works and Government Services.

Candy-Gibbs, S., Sharp, K., & Petrun, C. (1985). The effects of age, object, and cultural/religious background on children's concepts of death. *Omega, 154,* 329–345.

Canetto, S. S., & Sakinofsky, I. (1998). The gender paradox in suicide. *Suicide and Life-Threatening Behavior, 28,* 1–23.

Canobi, K. H., Reeve, R. A., & Pattison, P. E. (1998). The role of conceptual understanding in children's addition problem solving. *Developmental Psychology, 34,* 882–891.

Cantor, J., & Harrison, K. (1997). Ratings and advisories for television programming: University of Wisconsin, Madison study. In Mediascope (Ed.), *National Television Violence Study: Scientific Papers 1994–1995.* Studio City, CA: Author.

Capaldi, D. M., Stoolmiller, M., Clark, S., & Owen, L. D. (2002). Heterosexual risk behaviors in at-risk young men from early adolescence to young adulthood: Prevalence, prediction, and association with STD contraction. *Developmental Psychology, 38,* 394–406.

Caplan, M., Vespo, J., Pedersen, J., & Hay, D. F. (1991). Conflict and its resolution in small groups of one- and two-year-olds. *Child Development, 62,* 1513–1524.

Cappeliez, P., & O'Rourke, N. (2002). Personality traits and existential concerns as predictors of the functions of reminiscence in older adults. *Journal of Gerontology, 57B,* P116–P123.

Carbery, J., & Buhrmester, D. (1998). Friendship and need fulfillment during three phases of young adulthood. *Journal of Social and Personal Relationships, 15,* 393–409.

Carden, A. D. (1994). Wife abuse and the wife abuser: Review and recommendations. *Counseling Psychologist, 22,* 539–582.

Carey, S. (1999). Sources of conceptual change. In E. K. Scholnick, K. Nelson, S. A. Gelman, & P. H. Miller (Eds.), *Conceptual development: Piaget's legacy* (pp. 293–326). Mahwah, NJ: Erlbaum.

Carey, S., & Markman, E. M. (1999). Cognitive development. In B. M. Bly & D. E. Rumelhart (Eds.), *Cognitive science* (pp. 201–254). San Diego: Academic Press.

Carli, L. L., & Eagly, A. H. (2000). Gender effects on influence and emergent leadership. In G. N. Powell (Ed.), *Handbook of gender in organizations* (pp. 203–222). Newbury Park, CA: Sage.

Carlo, G., Koller, S. H., Eisenberg, N., Da Silva, M., & Frohlich, C. (1996). A cross-national study on the relations among prosocial moral reasoning, gender role orientations, and prosocial behaviors. *Developmental Psychology, 32,* 231–240.

Carlson, C., Uppal, S., & Prosser, E. (2000). Ethnic differences in processes contributing to the self-esteem of early adolescent girls. *Journal of Early Adolescence, 20,* 44–67.

Carlson, K. J., Eisenstat, S. A., & Ziporyn, T. (1996). *The Harvard guide to women's health.* Cambridge, MA: Harvard University Press.

Carlson, S. M., & Moses, L. J. (2001). Individual differences in inhibitory control and children's theory of mind. *Child Development, 72,* 1032–1053.

Carlton, M. P., & Winsler, A. (1999). School readiness: The need for a paradigm shift. *School Psychology Review, 28,* 338–352.

Carmichael, C. A., & Hayes, B. K. (2001). Prior knowledge and exemplar encoding in children's concept acquisition. *Child Development, 72,* 1071–1090.

Carmichael, S. L., & Shaw, G M. (2000). Maternal life stress and congenital abnormalities. *Epidemiology, 11,* 30–35.

Caro, F. G., & Bass, S. A. (1997). Receptivity to volunteering in the immediate postretirement period. *Journal of Applied Gerontology, 16,* 427–441.

Caroll, D., Harrison, L. K., Johnston, D. W., Ford, G., Hunt, K., Der, G., & West, P. (2000). Cardiovascular reactions to psychological stress: The influence of demographic variables. *Journal of Epidemiological Community Health, 54,* 876-877.

Carolson, C., Uppal, S., & Prosser, E. C. (2000). Ethnic differences in processes contributing to the self-esteem of early adolescent girls. *Journal of Early Adolescence, 20,* 44–67.

Carpendale, J. I., & Chandler, M. J. (1996). On the distinction between false belief understanding and subscribing to an interpretive theory of mind. *Child Development, 67,* 1686–1706.

Carpenter, C. J. (1983). Activity structure and play: Implications for socialization. In M. Liss (Eds.), *Social and cognitive skills: Sex roles and children's play* (pp. 117–145). New York: Academic Press.

Carpenter, M., Nagell, K., & Tomasello, M. (1998). Social cognition, joint attention, and communicative competence. *Monographs of the Society for Research in Child Development, 63*(4, Serial No. 255).

Carrado, M., George, M. J., Loxam, E., Jones, L., & Templar, D. (1996). Aggression in British heterosexual relationships: A descriptive analysis. *Aggressive Behavior, 22,* 401–415.

Carstensen, L. L. (1992). Selectivity theory: Social activity in life-span context. In K. W. Schaie & M. P. Lawton (Eds.), *Annual review of gerontology and geriatrics* (pp. 195–217). New York: Springer.

Carstensen, L. L., & Fredrickson, B. F. (1998). Socioemotional selectivity in healthy older people and younger people living with the human immunodeficiency virus: The centrality of emotion when the future is constrained. *Health Psychology, 17,* 1–10.

Carstensen, L. L., Gottman, J. M., & Levenson, R. W. (1995). Emotional behavior in long-term marriages. *Psychology and Aging, 10,* 140–149.

Carstensen, L. L., Isaacowitz, D. M., & Charles, S. T. (1999). Taking time seriously: A theory of socioemotional selectivity. *American Psychologist, 54,* 165–181.

Carta, J. J., Atwater, J. B., Greenwood, C. R., McConnell, S. R., & McEvoy, M. A. (2001). Effects of cumulative prenatal substance exposure and environmental risks on children's developmental trajectories. *Journal of Clinical Psychology, 30,* 327–337.

Carter, D. B., & McCloskey, L. A. (1984). Peers and the maintenance of sex-typed behavior: The development of children's conceptions of cross-gender behavior in their peers. *Social Cognition, 2,* 294–314.

Carus, F. A. (1808). *Psychologie. Zweiter Teil: Specialpsychologie.* Leipzig: Barth & Kummer.

Case, R. (1992). *The mind's staircase: Exploring the conceptual underpinnings of children's thought and knowledge.* Hillsdale, NJ: Erlbaum.

Case, R. (1996). Introduction: Reconceptualizing the nature of children's conceptual structures and their development in middle childhood. In R. Case & Y. Okamoto (Eds.), The role of central conceptual structures in the development of children's thought. *Monographs of the Society for Research in Child Development, 246*(61, Serial No. 246), pp. 1–26.

Case, R. (1998). The development of central conceptual structures. In D. Kuhn & R. Siegler (Eds.), *Handbook of child psychology: Vol. 2. Cognition, perception, and language* (5th ed., pp. 745–800). New York: Wiley.

Case, R., Griffin, S., & Kelly, W. M. (2001). Socioeconomic differences in children's early cognitive development and their readiness for schooling. In S. L. Golbeck (Ed.), *Psychological perspectives on early education* (pp. 37–63). Mahwah, NJ: Erlbaum.

Case, R., & Okamoto, Y. (Eds.). (1996). The role of central conceptual structures in the development of children's thought. *Monographs of the Society for Research in Child Development, 61*(1–2, Serial No. 246).

Caserta, M. S., & Gillett, P. A. (1998). Older women's feelings about exercise and their adherence to an aerobic regimen over time. *Gerontologist, 38,* 602–609.

Caserta, M. S., Lund, D. A., & Rice, S. J. (1999). Pathfinders: A self-care and health education program for older widows and widowers. *Gerontologist, 39,* 615–620.

Casey, M. B. (1986). Individual differences in selective attention among prereaders: A key to mirror-image confusions.

Developmental Psychology, 22, 824–831.

Casey, M. B., Nuttall, R. L., & Pezaris, E. (1997). Mediators of gender differences in mathematics college entrance test scores: A comparison of spatial skills with internalized beliefs and anxieties. *Developmental Psychology, 33,* 669–680.

Casey, M. B., Nuttall, R., Pezaris, E., & Benbow, C. P. (1995). The influence of spatial ability on gender differences in mathematics collect entrance test scores across diverse samples. *Developmental Psychology, 31,* 697–705.

Cashon, C. H., & Cohen, L. B. (2000). Eight-month-old infants' perceptions of possible and impossible events. *Infancy, 1,* 429–446.

Caspi, A. (1998). Personality development across the life course. In N. Eisenberg (Ed.), *Handbook of child psychology: Vol. 3. Social, emotional, and personality development* (5th ed., pp. 311–388). New York: Wiley.

Caspi, A. (2000). The child is father of the man: Personality continuities from childhood to adulthood. *Journal of Personality and Social Psychology, 78,* 158–172.

Caspi, A., Elder, G. H., Jr., & Bem, D. J. (1987). Moving against the world: Life-course patterns of explosive children. *Developmental Psychology, 23,* 308–313.

Caspi, A., Elder, G. H., Jr., & Bem, D. J. (1988). Moving away from the world: Life-course patterns of shy children. *Developmental Psychology, 24,* 824–831.

Caspi, A., & Herbener, E. (1990). Continuity and change: Assortative marriage and the consistency of personality in adulthood. *Journal of Personality and Social Psychology, 58,* 250–258.

Caspi, A., Lynam, D., Moffitt, T. E., & Silva, P. A. (1993). Unraveling girls' delinquency: Biological, dispositional, and contextual contributions to adolescent misbehavior. *Developmental Psychology, 29,* 19–30.

Caspi, A., & Roberts, B. W. (2001). Personality development across the life course: The argument for change and continuity. *Psychological Inquiry, 12,* 49–66.

Caspi, A., & Silva, P. A. (1995). Temperamental qualities at age three predict personality traits in young adulthood: Longitudinal evidence from a birth cohort. *Child Development, 66,* 486–498.

Cassel, C. K. (1988). Ethical issues in the conduct of research in long-term care. *Gerontologist, 28*(Suppl.), 90–96.

Cassidy, J. (2001). Adult romantic attachments: A developmental perspective on individual differences. *Review of General Psychology, 4,* 111–131.

Cassidy, J., & Berlin, L. J. (1994). The insecure/ambivalent pattern of attachment: Theory and research. *Child Development, 65,* 971–991.

Catsambis, S. (1994). The path to math: Gender and racial-ethnic differences in mathematics participation from middle school to high school. *Sociology of Education, 67,* 199–215.

Cavadini, C., Siega-Riz, A. M., & Popkin, B. M. (2000). U.S. adolescent food intake trends from 1965 to 1996. *Archives of Diseases in Childhood, 83,* 18–24.

Cavazzana-Calvo, M., Hacein-Bey, S., de Saint Basile, G., et al. (2000). Gene therapy of human severe combined immunodeficiency (SCID)-X1 disease. *Science, 288,* 669–672.

Ceci, S. J. (1991). How much does schooling influence general intelligence and its cognitive components? A reassessment of the evidence. *Developmental Psychology, 27,* 703–722.

Ceci, S. J. (1999). Schooling and intelligence. In S. J. Ceci & W. M. Williams (Eds.), *The nature–nurture debate: The essential readings* (pp. 168–175). Oxford, U.K.: Blackwell.

Ceci, S. J., & Bruck, M. (1998). Children's testimony: Applied and basic issues. In I. Sigel & K. A. Renninger (Eds.), *Handbook of child psychology: Vol. 4. Child psychology in practice* (5th ed., pp. 713–774). New York: Wiley.

Ceci, S. J., Bruck, M., & Battin, D. B. (2000). The suggestibility of children's testimony. In D. F. Bjorklund (Ed.), *False-memory creation in children and adults* (pp. 169–201). Mahwah, NJ: Erlbaum.

Ceci, S. J., & Roazzi, A. (1994). The effects of context on cognition: Postcards from Brazil. In R. J. Sternberg (Ed.), *Mind in context* (pp. 74–101). New York: Cambridge University Press.

Center, J. R., Nguyen, T. V., Schneider, D., Sambrook, P. N., & Eisman, J. A. (1999). Mortality after all major types of osteoporotic fracture in men and women: An observational study. *Lancet, 353,* 878–882.

Center for Communication and Social Policy. (Ed.). (1998). *National Television Violence Study* (Vol. 2). Newbury Park, CA: Sage.

Cerella, J. (1990). Aging and information processing rate. In J. E. Birren & K. W. Schaie (Eds.), *Handbook of the psychology of aging* (3rd ed.), pp. 201–221). San Diego: Academic Press.

Cernoch, J. M., & Porter, R. H. (1985). Recognition of maternal axillary odors by infants. *Child Development 56,* 1593–1598.

Cervantes, C. A., & Callanan, M. A. (1998). Labels and explanations in mother–child emotion talk: Age and gender differentiation. *Developmental Psychology, 34,* 88–98.

Chalmers, J. B., & Townsend, M. A. R. (1990). The effects of training in social perspective taking on socially maladjusted girls. *Child Development, 61,* 178–190.

Chan, R. W., Raboy, B., & Patterson, C. J. (1998). Psychosocial adjustment among children conceived via donor insemination by lesbian and heterosexual mothers. *Child Development, 69,* 443–457.

Chandler, M. J. (1973). Egocentrism and antisocial behavior: The assessment and training of social perspective-taking skills. *Developmental Psychology, 9,* 326–332.

Chandler, M. J., & Carpendale, J. I. (1998). Inching toward a mature theory of mind. In M. Ferrari & R. J. Sternberg (Eds.), *Self-awareness: Its nature and development* (pp. 148–190). New York: Guilford.

Chandra, R. K. (1991). Interactions between early nutrition and the immune system. In *Ciba Foundation Symposium No. 156* (pp. 77–92). Chichester, U.K.: Wiley.

Chandra, R. K. (1992). Effect of vitamin and trace-element supplementation on immune responses and infection in elderly subjects. *Lancet, 340,* 1124–1127.

Chao, R. K. (1994). Beyond parental control and authoritarian parenting style: Understanding Chinese parenting through the cultural notion of training. *Child Development, 65,* 1111–1119.

Chapman, R. S. (2000). Children's language learning: An interactionist perspective. *Journal of Child Psychology and Psychiatry, 41,* 33–54.

Chapman, R. S., & Hesketh, L. J. (2000). Behavioral phenotype of individuals with Down syndrome. *Mental Retardation and Developmental Disabilities Research Reviews, 6,* 84–95.

Chappell, N., Gee, E., McDonald, L., & Stones, M. (2003). *Aging in contemporary Canada.* Toronto: Pearson Education Canada.

Charman, T., Swettenham, J., Baron-Cohen, S., Cox, A., Baird, G., & Drew, A. (1997). Infants with autism: An investigation of empathy, pretend play, joint attention, and imitation. *Developmental Psychology, 33,* 781–789.

Chase, C., Teele, D. W., Klein, J. O., & Rosner, B. A. (1995). Behavioral sequelae of otitis media for infants at one year of age and their mothers. In D. J. Lim, C. D. Bluestone, J. O. Klein, J. D. Nelson, & P. L. Ogra (Eds.), *Recent advances in otitis media.* Hamilton, Ontario: Decker.

Chase-Lansdale, P. L., Brooks-Gunn, J., & Zamsky, E. S. (1994). Young African-American multigenerational families in poverty: Quality of mothering and grandmothering. *Child Development, 65,* 373–393.

Chase-Lansdale, P. L., Gordon, R., Brooks-Gunn, J., & Klebanov, P. K. (1997). Neighorhood and family influences on the intellectual and behavioral competence of preschool and early school-age children. In J. Brooks-Gunn, G. Duncan, & J. L. Aber (Eds.), *Neighborhood poverty: Context and consequences for development* (pp. 79–118). New York: Russell Sage Foundation.

Chassin, L., & Ritter, J. (2001). Vulnerability to substance use disorders in childhood and adolescence. In R. E. Ingram & J. M. Price (Eds.), *Vulnerability to psychopathology: Risk across the lifespan* (pp. 107–134). New York: Guilford.

Chasteen, A. L. (1994). "The world around me": The environment and single women. *Sex Roles, 31,* 309–328.

Chatterjee, S., Handcock, M. S., & Simonoff, J. S. (1995). *A casebook for a first course in statistics and data analysis.* New York: Wiley.

Chavajay, P., & Rogoff, B. (1999). Cultural variation in management of attention by children and their caregivers. *Developmental Psychology, 35,* 1079–1090.

Chavajay, P., & Rogoff, B. (2002). Schooling and traditional collaborative social organization of problem solving by Mayan mothers and children. *Developmental Psychology, 38,* 55–66.

Che-Alford, J., & Stevenson, K. (1998, Spring). Older Canadians on the move. *Canadian Social Trends,* pp. 15–18. Ottawa: Statistics Canada.

Chen, E., Matthews, K. A., & Boyce, W. T. (2002). Socioeconomic differences in children's health: How and why do these relationships change with age? *Psychological Bulletin, 128,* 295–329.

Chen, X. (2002). Peer relationships and networks and socioemotional adjustment: A Chinese perspective. In B. Cairns & T. Farmer (Eds.), *Social networks from a developmental perspective.* New York: Cambridge University Press.

Chen, X., Dong, Q., & Zhou, H. (1997). Authoritative and authoritarian parenting practices and social and school performance in Chinese children. *International Journal of Behavioral Development, 21,* 855–873.

Chen, X., Hastings, P. D., Rubin, K. H., Chen, H., Cen, G., & Stewart, S. L. (1998). Childrearing attitudes and behavioral inhibition in Chinese and Canadian toddlers: A cross-cultural study. *Developmental Psychology, 34,* 677–686.

Chen, X., Liu, M., & Li, D. (2000). Parental warmth, control, and indulgence and their relations to adjustment in Chinese children: A longitudinal study. *Journal of Family Psychology, 14,* 401–419.

Chen, X., Rubin, K. H., & Li, B. (1994). Only children and sibling children in urban China: A re-examination. *International Journal of Behavioral Development, 17,* 413–421.

Chen, X., Rubin, K. H., & Li, Z. (1995). Social functioning and adjustment in Chinese children: A longitudinal study. *Developmental Psychology, 31,* 531–539.

Chen, Y.-C., Yu, M.-L., Rogan, W., Gladen, B., & Hsu, C.-C. (1994). A 6-year follow-up of behavior and activity disorders in the Taiwan Yu-cheng children. *American Journal of Public Health, 84,* 415–421.

Chen, Y.-J., & Hsu, C.-C. (1994). Effects of prenatal exposure to PCBs on the neurological function of children: A neuropsychological and neurophysiological study. *Developmental Medicine and Child Neurology, 36,* 312–320.

Chen, Z., Sanchez, R. P., & Campbell, T. (1997). From beyond to within their grasp: The rudiments of analogical problem solving in 10- to 13-month-olds. *Developmental Psychology, 33,* 790–801.

Cherlin, A. J. (1992). *Marriage, divorce, remarriage* (rev. ed.). Cambridge, MA: Harvard University Press.

Cherney, S. S. (1994). Home environmental influences on general cognitive ability. In J. C. DeFries, R. Plomin, & D. W. Fulker (Eds.), *Nature and nurture during middle childhood* (pp. 262–280). Cambridge, MA: Blackwell.

Cherry, K. E., & LeCompte, D. C. (1999). Age and individual differences influence prospective memory. *Psychology and Aging, 14,* 60–76.

Chesney, M. A., Ekman, P., Friesen, W. V., Black, G. W., & Hecker, M. H. L. (1997). Type A behavior pattern: Facial behavior and speech components. In P. Ekman & E. L. Rosenberg (Eds.), *What the face reveals* (pp. 453–468). New York: Oxford University Press.

Chesney-Lind, M. (2001). Girls, violence, and delinquency: Popular myths and persistent problems. In S. O. White (Ed.), *Handbook of youth and justice* (pp. 135–158). New York: Kluwer Academic.

Chess, S., & Thomas, A. (1984). *Origins and evolution of behavior disorders.* New York: Brunner/Mazel.

Chestnut, C. H., III. (2001). Osteoporosis, an underdiagnosed disease. *Journal of the American Medical Association, 286,* 2865–2866.

Chi, M. T. H., Glaser, R., & Farr, M. J. (Eds.). (1988). *The nature of expertise.* Hillsdale, NJ: Erlbaum.

Children's Defense Fund. (2002). *The state of America's children: Yearbook 2002.* Washington, DC: Author.

Children's Defense Fund (2003, April). Number of black children in extreme poverty hits record high [Online]. Available: www.childrensdefense.org/release030430.php

Childs, C. P., & Greenfield, P. M. (1982). Informal modes of learning and teaching: The case of Zinacanteco weaving. In N. Warren (Ed.), *Advances in cross-cultural psychology* (Vol. 2, pp. 269–316). London: Academic Press.

Chisholm, J. S. (1989). Biology, culture, and the development of temperament: A Navajo example. In J. K. Nugent, B. M. Lester, & T. B. Brazelton (Eds.), *Biology, culture, and development* (Vol. 1, pp. 341–364). Norwood, NJ: Ablex.

Chiu, L.-H. (1992–1993). Self-esteem in American and Chinese (Taiwanese) children. *Current Psychology: Research and Reviews, 11,* 309–313.

Chochinov, H. M. (2002). Dignity-conserving care: A new model for palliative care. Helping the patient feel valued. *Journal of the American Medical Association, 287,* 2253–2260.

Chodirker, B., Cadrin, C., Davies, G., Summers, A., Wilson, R., Winsor, E., & Young, D. (2001). Genetic indications for prenatal diagnosis. *Journal of the Society of Obstetricians and Gynaecologists of Canada, 23,* 525–531.

Chodorow, N. (1978). *The reproduction of mothering.* Berkeley: University of California Press.

Choi, N. G. (1994). Racial differences in timing and factors associated with retirement. *Journal of Sociology and Social Welfare, 21,* 31–52.

Choi, S., & Gopnik, A. (1995). Early acquisition of verbs in Korean: A cross-linguistic study. *Journal of Child Language, 22,* 497–529.

Chollar, S. (1995, June). The psychological benefits of exercise. *American Health, 14*(5), 72–75.

Chomsky, C. (1969). *The acquisition of syntax in children from five to ten.* Cambridge, MA: MIT Press.

Chomsky, N. (1957). *Syntactic structures.* The Hague: Mouton.

Christ, G. H., Siegel, K., & Christ, A. E. (2002). "It never really hit me . . . until it actually happened." *Journal of the American Medical Association, 288,* 1269–1278.

Christensen, A., & Heavey, C. L. (1999). Interventions for couples. *Annual Review of Psychology, 50,* 165–190.

Christensen, K. A., Stephens, M. A. P., & Townsend, A. L. (1998). Mastery in women's multiple roles and well-being: Adult daughters providing care to impaired parents. *Health Psychology, 17,* 163–171.

Christenson, S. L., & Sheridan, S. M. (2001). *Schools and families.* New York: Guilford.

Chu, G. C. (1985). The changing concept of self in contemporary China. In A. J. Marsella, G. DeVos, & F. L. K. Hsu (Eds.), *Culture and self: Asian and Western perspectives* (pp. 252–277). London: Tavistock.

Cicchetti, D., & Aber, J. L. (1986). Early precursors of later depression: An organizational perspective. In L. P. Lipsitt & C. Rovee-Collier (Eds.), *Advances in infancy research* (Vol. 4, pp. 87–137). Norwood, NJ: Ablex.

Cicchetti, D., & Toth, S. L. (1998). The development of depression in children and adolescents. *American Psychologist, 53,* 221–241.

Cicchetti, D., & Toth, S. L. (2000). Developmental processes in maltreated children. In. D. J. Hansen (Ed.), *Nebraska Symposium on Motivation* (Vol. 46, pp. 85–160). Lincoln: University of Nebraska Press.

Cicerelli, V. G. (1989). Feelings of attachment to siblings and well-being in later life. *Psychology and Aging, 4,* 211–216.

Cicerelli, V. G. (1995). *Sibling relationships across the life span.* New York: Plenum.

Cicerelli, V. G. (1997). Relationship of psychosocial and background variables to older adults' end-of-life decisions. *Psychology and Aging, 12,* 72–83.

Cicerelli, V. G. (1998). Personal meanings of death in relation to fear of death. *Death Studies, 22,* 713–733.

Cicirelli, V. G. (1999). Personality and demographic factors in older adults' fear of death. *Gerontologist, 39,* 569–579.

Cicirelli, V. G. (2001). Personal meanings of death in older adults and young adults in relation to their fears of death. *Death Studies, 25,* 663–683.

Cicirelli, V. G. (2002). *Older adults' views on death.* New York: Springer.

Clark, E. V. (1995). The lexicon and syntax. In J. L. Miller & P. D. Eimas (Eds.), *Speech, language, and communication* (pp. 303–337). San Diego: Academic Press.

Clark, K. E., & Ladd, G. W. (2000). Connectedness and autonomy support in parent–child relationships: Links to children's socioemotional orientation and peer relationships. *Developmental Psychology, 36,* 485–498.

Clark, R., Hyde, J. S., Essex, M. J., & Klein, M. H. (1997). Length of maternity leave and quality of mother–infant interaction. *Child Development, 68,* 364–383.

Clarke-Stewart, K. A. (1998). Historical shifts and underlying themes in ideas about rearing young children in the United States: Where have we been? Where are we going? *Early Development and Parenting, 7,* 101–117.

Clarke-Stewart, K. A., & Hayward, C. (1996). Advantages of father custody and contact for the psychological well-being of school-age children. *Journal of Applied Developmental Psychology, 17,* 239–270.

Clausen, J. A. (1975). The social meaning of differential physical and sexual maturation. In S. E. Dragastin & G. H. Elder (Eds.), *Adolescence in the life cycle: Psychological change and the social context* (pp. 25–47). New York: Halsted.

Cleveland, J., & Shore, L. (1992). Self- and supervisory perspectives on age and work attitudes and performance. *Journal of Applied Psychology, 77,* 469–484.

Clifford, C., & Kramer, B. (1993). Diet as risk and therapy for cancer. *Medical Clinics of North America, 77,* 725–744.

Clifton, R. K., Perris, E., & Bullinger, A. (1991). Infants' perception of auditory space. *Developmental Psychology, 27,* 161–171.

Clifton, R. K., Rochat, P., Robin, D. J., & Berthier, N. E. (1994). Multimodal perception in the control of infant reaching.

Journal of Experimental Psychology: Human Perception and Performance, 20, 876–886.

Coakley, J. (1990). *Sport and society: Issues and controversies* (4th ed.). St. Louis: Mosby.

Cohan, C. L., & Kleinbaum, S. (2002). Toward a greater understanding of the cohabitation effect: Premarital cohabitation and marital communication. *Journal of Marriage and the Family, 64,* 180–192.

Cohen, K. M., & Savin-Williams, R. C. (1996). Developmental perspectives on coming out to self and others. *The lives of lesbians, gays, and bisexuals: Children to adults* (pp. 113–151). Ft. Worth, TX: Harcourt Brace.

Cohen, S., & Herbert, T. B. (1996). Health psychology: Psychological factors and physical disease from the perspective of human psychoneuroimmunology. *Annual Review of Psychology, 47,* 113–142.

Cohen, S., & Williamson, G. M. (1991). Stress and infectious disease in humans. *Psychological Bulletin, 109,* 5–24.

Cohen-Shalev, A. (1986). Artistic creativity across the adult life span: An alternative approach. *Interchange, 17*(4), 1–16.

Coie, J. D., & Dodge, K. A. (1998). *Aggression and antisocial behavior.* In N. Eisenberg (Ed.), *Handbook of child psychology: Vol. 3. Social, emotional, and personality development* (5th ed., pp. 779–862). New York: Wiley.

Coie, J. D., Dodge, K. A., & Coppotelli, H. (1982). Dimensions and types of social status: A cross-age perspective. *Developmental Psychology, 18,* 557–570.

Coke, M. M. (1992). Correlates of life satisfaction among elderly African Americans. *Journal of Gerontology, 47,* P316–P320.

Colapinto, J. (2001). *As nature made him: The boy who was raised as a girl.* New York: Perennial.

Colby, A., Kohlberg, L., Gibbs, J., & Lieberman, M. (1983). A longitudinal study of moral judgment. *Monographs of the Society for Research in Child Development, 48*(1–2, Serial No. 200).

Colcombe, S. J., Erickson, K. I., Raz, N., Webb, A. G., Cohen, N. J., & McAuley, E. (2003). Aerobic fitness reduces brain tissue loss in aging humans. *Journal of Gerontology, 58,* M176–M180.

Cole, D. A., Martin, J. M., Peeke, L. A., Seroczynski, A. D., & Fier, J. (1999). Children's over- and underestimation of academic competence: A longitudinal study of gender differences, depression, and anxiety. *Child Development, 70,* 459–473.

Cole, E. R., & Stewart, A. J. (1996). Meanings of political participation among black and white women: Political identity and social responsibility. *Journal of Personality and Social Psychology, 71,* 130–140.

Cole, M. (1990). Cognitive development and formal schooling: The evidence from cross-cultural research. In L. C. Moll (Ed.), *Vygotsky and education* (pp. 89–110). New York: Cambridge University Press.

Coleman, M., Ganong, L., & Fine, M. (2000). Reinvestigating remarriage: Another decade of progress. *Journal of Marriage and the Family, 62,* 1288–1307.

Coley, R. L. (1998). Children's socialization experiences and functioning in single-mother households: The importance of fathers and other men. *Child Development, 69,* 219–230.

Coley, R. L., & Chase-Lansdale, P. L. (1998). Adolescent pregnancy and parenthood: Recent evidence and future directions. *American Psychologist, 53,* 152–166.

Collaer, M. L., & Hines, M. (1995). Human behavioral sex differences: A role for gonadal hormones during early development? *Psychological Bulletin, 118,* 55–107.

Collie, R., & Hayne, H. (1999). Deferred imitation by 6- and 9-month-old infants: More evidence for declarative memory. *Developmental Psychobiology, 35,* 83–90.

Collings, P. (2001). "If you got everything, it's good enough": Perspectives on successful aging in a Canadian Inuit community. *Journal of Cross-Cultural Gerontology, 16,* 127–155.

Collins, F. S., & McKusick, V. A. (2001). Implications of the human genome project for medical science. *Journal of the American Medical Association, 285,* 540–544.

Collins, J. A. (1994). Reproductive technology—the price of progress. *New England Journal of Medicine, 331,* 270–271.

Collins, N. L., & Feeney, B. C. (2000). A safe haven: An attachment theory perspective on support-seeking and caregiving in intimate relationships. *Journal of Personality and Social Psychology, 78,* 1053–1073.

Collins, W. A., Laursen, B., Mortensen, N., Luebker, C., & Ferreira, M. (1997). Conflict processes and transitions in parent and peer relationships: Implications for autonomy and regulation. *Journal of Adolescent Research, 12,* 178–198.

Collins, W. A., Maccoby, E. E., Steinberg, L., Hetherington, E. M., & Bornstein, M. H. (2000). Contemporary research on parenting: The case for nature and nurture. *American Psychologist, 55,* 218–232.

Collins, W. A., Madsen, S. D., & Susman-Stillman, A. (2002). Parenting during middle childhood. In M. H. Bornstein (Ed.), *Handbook of parenting: Vol. 1* (2nd ed., pp. 73–101). Mahwah, NJ: Erlbaum.

Collins, W. A., Wellman, H., Keniston, A. H., & Westby, S. D. (1978). Age-related aspects of comprehension and inference from a televised dramatic narrative. *Child Development, 49,* 389–399.

Colman, L. L., & Colman, A. D. (1991). *Pregnancy: The psychological experience.* New York: Noonday Press.

Colombo, J. (1995). On the neural mechanism underlying developmental and individual differences in visual fixation in infancy. *Developmental Review, 15,* 97–135.

Colonia-Willner, R. (1998). Practical intelligence at work: Relationship between aging and cognitive efficiency among managers in a bank environment. *Psychology and Aging, 13,* 45–57.

Coltrane, S. (1990). Birth timing and the division of labor in dual-earner families. *Journal of Family Issues, 11,* 157–181.

Coltrane, S. (1996). *Family man.* New York: Oxford University Press.

Comijs, H. C., Jonker, C., van Tilberg, W., and Smit, J. H. (1999). Hostility and coping capacity as risk factors of elder mistreatment. *Social Psychiatry and Psychiatric Epidemiology, 34,* 48–52.

Comstock, G. A., & Scharrer, E. (1999). *Television: What's on, who's watching, and what it means.* San Diego: Academic Press.

Comstock, G. A., & Scharrer, E. (2001). The use of television and other film-related media. In D. G. Singer & J. L. Singer (Eds.), *Handbook of children and the media* (pp. 47–72). Thousand Oaks, CA: Sage.

Comunian, A. L., & Gielen, U. P. (2000). Sociomoral reflection and prosocial and antisocial behavior: Two Italian studies. *Psychological Reports, 87,* 161–175.

Condie, S. J. (1989). Older married couples. In S. J. Bahr & E. T. Peterson (Eds.), *Aging and the family* (pp. 143–158). Lexington, MA: Lexington Books.

Conger, R. D., Patterson, G. R., & Ge, X. (1995). It takes two to replicate: A mediational model for the impact of parents' stress on adolescent adjustment. *Child Development, 66,* 80–97.

Conner, D. B., Knight, D. K., & Cross, D. R. (1997). Mothers' and fathers' scaffolding of their 2-year-olds during problem-solving and literacy interactions. *British Journal of Developmental Psychology, 15,* 323–338.

Connidis, I. A. (1989). Siblings as friends in later life. *American Behavioral Scientist, 33,* 81–93.

Connidis, I. A. (1994). Sibling support in older age. *Journal of Gerontology, 49,* S309–S317.

Connidis, I. A., & Campbell, L. D. (1995). Closeness, confiding, and contact among siblings in middle and late adulthood. *Journal of Family Issues, 16,* 722–745.

Connidis, I. A., & McMullin, J. A. (1993). To have or have not: Parent status and the subjective well-being of older men and women. *The Gerontologist, 33,* 630–636.

Connolly, J., Craig, W., Goldberg, A., & Pepler, D. (1999). Conceptions of cross-sex friendships and romantic relationships in early adolescence. *Journal of Youth and Adolescence, 28,* 481–494.

Connolly, J., & Goldberg, A. (1999). Romantic relationships in adolescence: The role of friends and peers in their emergence and development. In W. Furman, B. B. Brown, & C. Feiring (Eds.), *The development of romantic relationships in adolescence* (pp. 266–290). New York: Cambridge University Press.

Connor, P. D., Sampson, P. D., Bookstein, F. L., Barr, H. M., & Streissguth, A. P. (2001). Direct and indirect effects of prenatal alcohol damage on executive function. *Developmental Neuropsychology, 18,* 331–354.

Connors, L. J., & Epstein, J. L. (1996). Parent and school partnerships. In M. H. Bornstein (Ed.), *Handbook of parenting: Vol. 4. Applied and practical parenting* (pp. 437–458). Mahwah, NJ: Erlbaum.

Conti-Ramsden, G., & Pérez-Pereira, M. (1999). Conversational interactions between mothers and their infants who are congenitally blind, have low vision, or are sighted. *Journal of Visual Impairment and Blindness, 93,* 691–703.

Conwell, Y., & Duberstein, P. R. (2001). Suicide in elders. *Annals of the New York Academy of Sciences, 932,* 132–150.

Conwell, Y., Duberstein, P. R., & Caine, E. D. (2002). Risk factors for suicide in later life. *Biological Psychiatry, 52,* 193–204.

Conwell, Y., Lyness, J. M., Duberstein, P., Cox, C., Seidlitz, L., DiGiorgio, A., & Caine, E. D. (2001). Completed suicide among older patients in primary care practices: A controlled study. *Journal of the American Geriatrics Society, 48,* 23–29.

Cook, P. W. (1997). *Abused men: the hidden side of domestic violence.* Westport, CT: Praeger.

Cook, W. L. (2000). Understanding attachment security in family context. *Journal of Personality and Social Psychology, 78,* 285–294.

Cooney, T. M., & Mortimer, J. T. (1999). Family structure differences in the timing of leaving home: Exploring mediating factors. *Journal of Research on Adolescence, 9,* 367–393.

Cooper, C. R. (1998). *The weaving of maturity: Cultural perspectives on adolescent development.* New York: Oxford University Press.

Cooper, R. P., & Aslin, R. N. (1994). Developmental differences in infant attention to the spectral properties of infant-directed speech. *Child Development, 65,* 1663–1677.

Cooper, S., & Glazer, E. S. (1999). *Choosing assisted reproduction: Social, emotional, and ethical considerations.* New York: Dimensions.

Cooper, Z., & Fairburn, C. G. (2002). Cognitive-behavioral treatment of obesity. In T. A. Wadden & A. J. Stunkard (Eds.), *Handbook of obesity treatment* (3rd ed., pp. 465–479). New York: Guilford.

Cordon, J. A. F. (1997). Youth residential independence and autonomy: A comparative study. *Journal of Family Issues, 18,* 576–807.

Corey-Bloom, J. (2002). Dementia. In S. K. Whitbourne (Ed.), *Psychopathology in later adulthood* (pp. 217–243). New York: Wiley.

Cornelius, M. D., Day, N. L., Richardson, G. A., & Taylor, P. M. (1999). Epidemiology of substance abuse during pregnancy. In P. J. Ott & R. E. Tarter (Eds.), *Sourcebook on substance abuse: Etiology, epidemiology, assessment, and treatment* (pp. 1–13). Boston: Allyn and Bacon.

Cornelius, M. D., Ryan, C. M., Day, N. L., Goldschmidt, L., & Willford, J. A. (2001). Prenatal tobacco effects on neuropsychological outcomes among preadolescents. *Developmental and Behavioral Pediatrics, 22,* 217–225.

Cornelius, S. W., & Caspi, A. (1987). Everyday problem solving in adulthood and old age. *Psychology and Aging, 2,* 144–153.

Corr, C. A. (1993). Coping with dying: Lessons that we should and should not learn from the work of Elisabeth Kübler-Ross. *Death Studies, 17,* 69–83.

Corr, C. A. (1995). Entering into adolescent understandings of death. In E. A. Grollman (Eds.), *Bereaved children and teens* (pp. 21–35). Boston: Beacon Press.

Corr, C. A., & Balk, D. E. (1996). *Handbook of adolescent death and bereavement.* New York: Springer.

Corrigan, R. (1987). A developmental sequence of actor–object pretend play in young children. *Merrill-Palmer Quarterly, 33,* 87–106.

Cosden, M., Peerson, S., & Elliott, K. (1997). Effects of prenatal drug exposure on birth outcomes and early child development. *Journal of Drug Issues, 27,* 525–539.

Costa, P. T., Jr., & McCrae, R. R. (1994). Set like plaster? Evidence for the stability of adult personality. In T. F. Heatherton & J. L. Weinberger (Eds.), *Can personality change?* (pp. 21–40). Washington, DC: American Psychological Association.

Costa, P. T., Jr., McCrae, R. R., Martin, T. A., Oryol, V. E., Senin, I. G., & Rukavishnikow, A. A. (2000). Personality development from adolescence through adulthood: Further cross-cultural comparisons of age differences. In V. J. Molfese & D. L. Molfese (Eds.), *Temperament and personality development across the life span* (pp. 235–252). Mahwah, NJ: Erlbaum.

Costello, E. J., & Angold, A. (1995). Developmental epidemiology. In D. Cicchetti & D. Cohen (Eds.), *Developmental psychopathology: Vol. 1. Theory and method* (pp. 23–56). New York: Wiley.

Coté, S., Zoccolillo, M., Tremblay, R., Nagin, D., & Vitaro, F. (2001). Predicting girls' conduct disorder in adolescence from childhood trajectories of disruptive behaviors. *Journal of the American Academy of Child and Adolescent Psychiatry, 40,* 678–684.

Couch, K. A. (1998). Late life job displacement. *Gerontologist, 38,* 7–17.

Coughlin, S. S., Khoury, M. J., & Steinberg, K. K. (1999). BRCA1 and BRCA2 gene mutations and risk of breast cancer: Public health perspectives. *American Journal of Preventive Medicine, 16,* 91–98.

Coulton, C. J., Korbin, J. E., & Su, M. (1999). Neighborhoods and child maltreatment: A multi-level study. *Child Abuse and Neglect, 23,* 1019–1040.

Courage, M. L., & Howe, M. L. (2002). From infant to child: The dynamics of cognitive change in the second year of life. *Psychological Bulletin, 128,* 250–277.

Cournoyer, M., Solomon, C. R., & Trudel, M. (1998). I speak then I expect: Language and self-control in the young child at home. *Canadian Journal of Behavioural Science,* 30, 69–81.

Cowan, C. P., & Cowan, P. A. (1992). *When partners become parents.* New York: Basic Books.

Cowan, C. P., & Cowan, P. A. (1995). Interventions to ease the transition to parenthood: Why they are needed and what they can do. *Family Relations, 44,* 412–423.

Cowan, C. P., & Cowan, P. A. (1997). Working with couples during stressful transitions. In S. Dreman (Ed.), *The family on the threshold of the 21st century* (pp. 17–47). Mahwah, NJ: Erlbaum.

Cowan, C. P., & Cowan, P. A. (2000). Working with couples during stressful transitions. In S. Dreman (Ed.), *The family on the threshold of the 21st century* (pp. 17–47). Mahwah, NJ: Erlbaum.

Cowan, P. A., Powell, D., & Cowan, C. P. (1998). Parenting interventions: A family systems perspective. In I. E. Sigel & K. A. Renninger (Eds.), *Handbook of child psychology: Vol. 4. Child psychology in practice* (5th ed., pp. 3–72). New York: Wiley.

Cox, G. (2002). The Native American patient. In R. B. Gilbert (Ed.), *Health care and spirituality: Listening, assessing, caring* (pp. 107–127). Amityville, NY: Baywood.

Cox, K., & Schwartz, J. D. (1990). *The well-informed patient's guide to caesarean births.* New York: Dell.

Cox, M., & Littlejohn, K. (1995). Children's use of converging obliques in their perspective drawings. *Educational Psychology, 15,* 127–139.

Cox, M. J., Owen, M. T., Henderson, V. K., & Margand, N. A. (1992). Prediction of infant–father and infant–mother attachment. *Developmental Psychology, 28,* 474–483.

Cox, M. J., Paley, B., & Harter, K. (2001). Interparental conflict and parent–child relationships. In J. H. Grych & F. D. Fincham (Eds.), *Interparental conflict and child development: Theory, research, and applications* (pp. 249–272). New York: Cambridge University Press.

Cox, S. M., Hopkins, J., & Hans, S. L. (2000). Attachment in preterm infants and their mothers: Neonatal risk status and maternal representations. *Infant Mental Health Journal, 21,* 464–480.

Coyle, T. R., & Bjorklund, D. F. (1997). Age differences in, and consequences of, multiple- and variable-strategy use on a multi-trial sort-recall task. *Developmental Psychology, 33,* 372–380.

Crago, M. B., Annahatak, B., & Ningiuruvik, L. (1993). Changing patterns of language socialization in Inuit homes. *Anthropology and Education Quarterly, 24,* 205–223.

Craik, F. I. M., & Jacoby, L. L. (1996). Aging and memory: Implications for skilled performance. In W. A. Rogers, A. D. Fisk, & N. Walker (Eds.), *Aging and skilled performance* (pp. 113–137). Mahwah, NJ: Erlbaum.

Craik, F. I. M., & Jennings, J. M. (1992). Human memory. In F. I. M. Craik & T. A. Salthouse (Eds.), *Handbook of aging and cognition* (pp. 51–110). Hillsdale, NJ: Erlbaum.

Crair, M. C., Gillespie, D. C., & Stryker, M. P. (1998). The role of visual experience in the development of columns in cat visual cortex. *Science, 279,* 566–570.

Cramond, B. (1994). The Torrance Tests of Creative Thinking: From design through establishment of predictive validity. In R. F. Subotnik & K. D. Arnold (Eds.), *Beyond Terman: Contemporary longitudinal studies of giftedness and talent* (pp. 229–254). Norwood, NJ: Ablex.

Cratty, B. J. (1986). *Perceptual and motor development in infants and children* (3rd ed.), Englewood Cliffs, NJ: Prentice-Hall.

Crawford, J. (1997). *Best evidence: Research foundations of the bilingual education act.* Washington, DC: National Clearinghouse for Bilingual Education.

Crawley, L., Payne, R., Bolden, J., Payne, T., Washington, P., & Williams, S. (2000). Palliative and end-of-life care in the African American community. *Journal of the American Medical Association, 284,* 2518–2521.

Creasey, G. (2002). Associations between working models of attachment and conflict management behavior in romantic couples. *Journal of Counseling Psychology, 49,* 365–375.

Creasey, G. L., Jarvis, P. A., & Berk, L. E. (1998). Play and social competence. In O. N. Saracho & B. Spodek (Eds.), *Multiple perspectives on play in early childhood education* (pp. 116–143). Albany: State University of New York Press.

Creer, T. L. (1998). Childhood asthma. In T. H. Ollendick & M. Hersen (Eds.), *Handbook of child psychopathology* (3rd ed., pp. 395–415). New York: Plenum.

Crick, N. R., & Bigbee, M. A. (1998). Relational and overt forms of peer victimization: A multi-informant approach. *Journal of Consulting and Clinical Psychology, 66,* 337–347.

Crick, N. R., Casas, J. F., & Ku, H.-C. (1999). Relational and physical forms of peer victimization in preschool. *Developmental Psychology, 35,* 376–385.

Crick, N. R., Casas, J. F., & Mosher, M. (1997). Relational and overt aggression in preschool. *Developmental Psychology, 33,* 579–588.

Crick, N. R., & Grotpeter, J. K. (1995). Relational aggression, gender, and social-psychological adjustment. *Child Development, 66,* 710–722.

Crick, N. R., & Grotpeter, J. K. (1996). Children's treatment by peers: Victims of relational and overt aggression. *Development and Psychopathology, 8,* 367–380.

Cristofalo, V. J., Tresini, M., Francis, M. K., & Volker, C. (1999). Biological theories of senescence. In V. L. Bengtson & K. W. Schaie (Eds.), *Handbook of theories of aging* (pp. 98–112). New York: Springer.

Critser, G. (2003). *Fat land.* Boston: Houghton Mifflin.

Crockenberg, S., & Leerkes, E. (2000). Infant social and emotional development in family context. In C. H. Zeanah, Jr., *Handbook of infant mental health* (2nd ed., pp. 60–90). New York: Guilford.

Crohan, S. E., & Antonucci, T. C. (1989). Friends as a source of social support in old age. In R. G. Adams & R. Blieszner (Eds.), *Older adult friendship* (pp. 129–146). Newbury Park, CA: Sage.

Crosby, F. J. (1998). The developing literature on developmental relationships. In A. J. Murrell, F. J. Crosby, & R. J. Ely (Eds.), *Mentoring dilemmas* (pp. 3–20). Mahwah, NJ: Erlbaum.

Cross, S., & Markus, H. (1991). Possible selves across the life span. *Human Development, 34,* 230–255.

Crouter, A. C., Bumpass, M. F., Maguire, M. C., & McHale, S. M. (1999). Linking parents' work pressure and adolescents' well-being: Insights into dynamics in dual earner families. *Developmental Psychology, 35,* 1453–1461.

Crouter, A. C., Manke, B. A., & McHale, S. M. (1995). The family context of gender intensification in early adolescence. *Child Development, 66,* 317–329.

Crystal, D. S., Chen, C., Fuligni, A. J., Stevenson, H. W., Hsu, C.-C., Ko, H-J., Kitamura, S., & Kimura, S. (1994). Psychological maladjustment and academic achievement: A cross-cultural study of Japanese, Chinese, and American high school students. *Child Development, 65,* 738–753.

Crystal, S., Johnson, R. W., Harman, J., Sambamoorthi, U., & Kumar, R. (2000). Out-of-pocket health care costs among older Americans. *Journal of Gerontology, 55B,* S51–S62.

Csikszentmihalyi, M. (1999). Implications of a systems perspective for the study of creativity. In R. J. Sternberg (Ed.), *Handbook of creativity* (pp. 313–335). Cambridge, U.K.: Cambridge University Press.

Csikszentmihalyi, M., & Larson, R. (1984). *Being adolescent: Conflict and growth in the teenage years.* New York: Basic Books.

Csikszentmihalyi, M., & Rathunde, K. (1990). The psychology of wisdom: An evolutionary interpretation. In R. J. Sternberg (Ed.), *Wisdom: Its nature, origins, and development* (pp. 25–51). New York: Cambridge University Press.

Cubbins, L. A., & Tanfer, K. (2000). The influence of gender on sex: A study of men's and women's self-reported high-risk sexual behavior. *Archives of Sexual Behavior, 29,* 229–257.

Cuddy-Casey, M., & Orvaschel, H. (1997). Children's understanding of death in relation to child suicidality and homicidality. *Clinical Psychology Review, 17,* 33–45.

Culbertson, F. M. (1997). Depression and gender: An international review. *American Psychologist, 52,* 25–51.

Cully, J. A., LaVoie, D., & Gfeller, J. D. (2001). Reminiscence, personality, and psychological functioning in older adults. *Gerontologist, 41,* 89–95.

Culnane, M., Fowler, M. G., Lee, S. S., McSherry, G., Brady, M., & O'Donnell, K. (1999). Lack of long-term effects of in utero exposure to zidovudine among uninfected children born to HIV-infected women. *Journal of the American Medical Association, 281,* 151–157.

Cumming, E., & Henry, W. E. (1961). *Growing old: The process of disengagement.* New York: Basic Books.

Cummings, E. M., & Cummings, J. S. (2002). Parenting and attachment. In M. H. Bornstein (Ed.), *Handbook of parenting* (2nd ed., pp. 35–58). Mahwah, NJ: Erlbaum.

Cummings, E. M., & Davies, P. T. (1994). Maternal depression and child development. *Journal of Child Psychology and Psychiatry, 35,* 73–112.

Cunningham, J. D., & Antill, J. K. (1994). Cohabitation and marriage: Retrospective and predictive comparisons. *Journal of Social and Personal Relationships, 11,* 77–93.

Currie, D. H. (1999). Violent men or violent women? Whose definition counts? In R. K. Bergen (Ed.), *Issues in intimate violence* (pp. 97–111). Thousand Oaks, CA: Sage.

Curtin, S. C., & Park, M. M. (1999). Trends in the attendant, place, and timing of births and in the use of obstetric interventions: United States, 1989–1997. *National Vital Statistics Report, 47*(27), 1–12.

Curtiss, S. (1989). The independence and task-specificity of language. In M. H. Bornstein & J. S. Bruner (Eds.), *Interaction in human development* (pp. 105–137). Hillsdale, NJ: Erlbaum.

Cutler, S. J., & Hendricks, J. (2001). Emerging social trends. In R. H. Binstock & L. K. George (Eds.), *Handbook of aging and the social sciences* (5th ed., pp. 462–480). San Diego, CA: Academic Press.

Cutrona, C. E., Hessling, R. M., Bacon, P. L., & Russell, D. W. (1998). Predictors and correlates of continuing involvement with the baby's father among adolescent mothers. *Journal of Family Psychology, 12,* 369–387.

D'Agostino, J. A., & Clifford, P. (1998). Neurodevelopmental consequences associated with the premature neonate. *AACN Clinical Issues, 9,* 11–24.

Dacey, J. S. (1989). Peak periods of creative growth across the life span. *Journal of Creative Behavior, 23,* 224–248.

Dahl, R.E., Scher, M. S., Williamson, D. E., Robles, N., & Day, N. (1995). A longitudinal study of prenatal marijuana use: Effects on sleep and arousal at age 3 years. *Archives of Pediatric and Adolescent Medicine, 149,* 145-150.

Dales, L., Hammer, S. J., & Smith, N. J. (2001). Time trends in autism and MMR immunization coverage in California. *Journal of the American Medical Association, 285,* 1183–1185.

Dalton, S. T. (1992). Lived experience of never-married women. *Issues in Mental Health Nursing, 13,* 69–80.

Daly, K. A., Hunter, L. L., & Giebink, G. S. (1999). Chronic otitis media with effusion. *Pediatrics in Review, 20,* 85–93.

Damon, W. (1977). *The social world of the child.* San Francisco: Jossey-Bass.

Damon, W. (1988). *The moral child.* New York: Free Press.

Damon, W. (1990). Self-concept, adolescent. In R. M. Lerner, A. C. Petersen, & J. Brooks-Gunn (Eds.), *The encyclopedia of adolescence* (Vol. 2, pp. 87–91). New York: Garland.

Damon, W., & Hart, D. (1988). *Self-understanding in childhood and adolescence.* New York: Cambridge University Press.

Daniels, D. H., Kalkman, D. L., & McCombs, B. L. (2001). Young children's perspectives on learning and teacher practices in different classroom contexts: Implications for motivation. *Early Education and Development, 12,* 253–273.

Dannemiller, J. L., & Stephens, B. R. (1988). A critical test of infant pattern preference models. *Child Development, 59,* 210–216.

Dapretto, M., & Bjork, E. L. (2000). The development of word retrieval abilities in the second year and its relation to early vocabulary growth. *Child Development, 71,* 635–648.

Darroch, J. E., Frost, J. J., & Singh, S. (2001). *Teenage sexual and reproductive behavior in developed countries: Can more progress be made?* New York: Alan Guttmacher Institute.

Darton-Hill, L., & Coyne, E. T. (1998). Feast and famine: Socioeconomic disparities in global nutrition and health. *Public Health and Nutrition, 1,* 23–31.

Darwin, C. (1936). *On the origin of species by means of natural selection.* New York: Modern Library. (Original work published 1859)

Datta-Bhutada, S., Johnson, H. L., & Rosen, T. S. (1998). Intrauterine cocaine and crack exposure: Neonatal outcome. *Journal of Perinatology, 18,* 183–188.

Daugirdas, J. T. (1992). *Sexually transmitted diseases.* Hinsdale, IL: Medtext.

DaVanzo, J., & Adamson, D. M. (2000). *Family planning in developing countries: An unfinished success story.* New York: Rand Corporation.

Davey, A., & Eggebeen, D. J. (1998). Patterns of intergenerational exchange and mental health. *Journal of Gerontology, 53B,* P86–P95.

Davidson, R. J. (1994). Asymmetric brain function, affective style, and psychopathology: The role of early experience and plasticity. *Development and Psychopathology, 6,* 741–758.

Davies, T., Howell, R. T., & Gardner, A. (2001). *Human genetics.* New York: Oxford University Press.

Davis, B., & Aron, A. (1989). Perceived causes of divorce and postdivorce adjustment among recently divorced midlife women. *Journal of Divorce, 12,* 41–55.

Davis, C. G., Nolen-Hoeksema, S., & Larson, J. (1998). Making sense of loss and benefiting from the experience: Two construals of meaning. *Journal of Personality and Social Psychology, 75,* 561–574.

Davis, C. G., Wortman, C. B., Lehman, D. R., & Silver, R. C. (2000). Searching for meaning in loss: Are clinical assumptions correct? *Death Studies, 24,* 497–540.

Davis, H. P., Trussell, L. H., & Klebe, K. J. (2001). A ten-year longitudinal examination of repetition priming, incidental recall, free recall, and recognition in young and elderly. *Brain and Cognition, 46,* 99–104.

Dawson, T. L. (2002). New tools, new insights: Kohlberg's moral judgment stages revisited. *International Journal of Behavioral Development, 26,* 154–166.

Dawson-Hughes, B., Harris, S. S., Krall, E. A., Dallal, G. E., Falconer, G., & Green, C. L. (1995). Rates of bone loss in postmenopausal women randomly assigned to one of two dosages of vitamin D. *American Journal of Clinical Nutrition, 61,* 1140–1145.

Day, N. L., Leach, S. L., Richardson, G. A., Cornelius, M. D., Robles, N., & Larkby, C. (2002). Prenatal alcohol exposure predicts continued deficits in offspring size at 14 years of age. *Alcoholism: Clinical and Experimental Research, 26,* 1584–1591.

Deák, G. O. (2000). Hunting the fox of word learning: Why "constraints" fail to capture it. *Developmental Review, 20,* 29–80.

Deák, G. O., & Maratsos, M. (1998). On having complex representations of things: Preschoolers use multiple words for objects and people. *Developmental Psychology, 34,* 224–240.

Deary, I. J. (2000). Simple information processing and intelligence. In R. J. Sternberg (Ed.), *Handbook of intelligence* (pp. 267–284). New York: Cambridge University Press.

Deater-Deckard, K., & Dodge, K. A. (1997). Externalizing behavior problems and discipline revisited: Nonlinear effects and variation by culture, context, and gender. *Psychological Inquiry, 8,* 161–175.

Deater-Deckard, K., Dodge, K. A., Bates, J. E., & Petit, G. S. (1996). Physical discipline among African American and European American mothers: Links to children's externalizing behaviors. *Developmental Psychology, 32,* 1065–1072.

Deater-Deckard, K., Pike, A., Petrill, S. A., Cutting, A. L., Hughes, C., & O'Connor, T. G. (2001). Nonshared environmental processes in social-emotional development: An observational study of identical twin differences in the preschool period. *Developmental Science, 4,* F1–F6.

DeCasper, A. J., & Spence, M. J. (1986). Prenatal maternal speech influences newborns' perception of speech sounds. *Infant Behavior and Development, 9,* 133–150.

Deffenbacher, J. L. (1994). Anger reduction: Issues, assessment, and intervention strategies. In A. W. Siegman & T. W. Smith (Eds.), *Anger, hostility, and the heart* (pp. 239–269). Hillsdale, NJ: Erlbaum.

DeGarmo, D. S., & Forgatch, M. S. (1997). Determinants of observed confidant support for divorced mothers. *Journal of Personality and Social Psychology, 72,* 336–345.

DeGarmo, D. S., & Forgatch, M. S. (1999). Contexts as predictors of changing maternal parenting practices in diverse family structures: A social interactional perspective of risk and resilience. In E. M. Hetherington (Ed.), *Coping with divorce, single parenting, and remarriage: A risk and resiliency perspective* (pp. 227–252). Mahwah, NJ: Erlbaum.

DeGenova, M. K. (1992). If you had your life to live over again: What would you do differently? *International Journal of Aging and Human Development, 34,* 135–143.

Dejin-Karlsson, E., Hanson, B. S., Estergren, P.-O., Sjoeberg, N.-O., & Marsal, K. (1998). Does passive smoking in early pregnancy increase the risk of small-for-gestational-age infants? *American Journal of Public Health, 88,* 1523–1527.

Dekovíc, M., Noom, M. J., & Meeus, W. (1997). Expectations regarding development during adolescence: Parent and adolescent perceptions. *Journal of Youth and Adolescence, 26,* 253–271.

Delgado-Gaitan, C. (1994). Socializing young children in Mexican-American families: An intergenerational perspective. In P. Greenfield & R. Cocking (Eds.), *Cross-cultural roots of minority child development* (p. 55–86). Hillsdale, NJ: Erlbaum.

De Lisi, R., & Gallagher, A. M. (1991). Understanding gender stability and constancy in Argentinean children. *Merrill-Palmer Quarterly, 37,* 483–502.

Dell, D. L. (2001). Adolescent pregnancy. In N. L. Stotland & D. E. Stewart (Eds.), *Psychological aspects of women's health care* (pp. 95–116). Washington, DC: American Psychiatric Association.

DeLoache, J. S., & Smith, C. M. (1999). Early symbolic representation. In I. E. Sigel (Ed.), *Development of mental representation* (pp. 61–86). Mahwah, NJ: Erlbaum.

DeLoache, J. S., & Todd, C. M. (1988). Young children's use of spatial categorization as a mnemonic strategy. *Journal of Experimental Child Psychology, 46,* 1–20.

Demetriou, A., & Kazi, S. (2001). *Unity and modularity in the mind and the self: Studies on the relationships between self-awareness, personality, and intellectual development from childhood to adolescence.* London: Routledge.

Demetriou, A., Christou, C., Spanoudis, G., & Platsidou, M. (2002). The development of mental processing: Efficiency, working memory, and thinking. *Monographs of the Society for Research in Child Development, 67*(1, Serial No. 268).

Demetriou, A., Efklides, A., Papadaki, M., Papantoniou, G., & Economou, A. (1993). Structure and development of causal–experimental thought: From early adolescence to youth. *Developmental Psychology, 29,* 480–497.

Demetriou, A., Efklides, A., & Platsidou, M. (1993). The architecture and dynamics of developing mind. *Monographs of the Society for Research in Child Development, 58*(No. 5–6, Serial No. 234).

Demetriou, A., Pachaury, A., Metallidou, Y., & Kazi, S. (1996). Universals and specificities in the structure and development of quantitative-relational throught: A cross-cultural study in Greece and India. *International Journal of Behavioral Development, 19,* 255–290.

Dempster, F. N., & Corkill, A. J. (1999). Interference and inhibition in cognition and behavior: Unifying themes for educational psychology. *Educational Psychology Review, 11,* 1–88.

Denckla, M. B. (1996). Biological correlates of learning and attention: What is relevant to learning disability and attention-deficit hyperactivity disorder? *Developmental and Behavioral Pediatrics, 17,* 114–119.

Denham, S. (1998). *Emotional development in young children.* New York: Guilford.

Denham, S. A., Zoller, D., & Couchoud, E. (1994). Socialization of preschoolers' emotion understanding. *Developmental Psychology, 30,* 928–936.

Denner, J., Cooper, C. R., Lopez, E. M., & Dunbar, N. (1999). Beyond "giving science away": How university–community partnerships inform youth programs, research, and policy. *Social Policy Report of the Society for Research in Child Development, 13*(1).

Denney, N. W. (1990). Adult age differences in traditional and practical problem solving. *Advances in Psychology, 72,* 329–349.

Denney, N. W., & Pearce, K A. (1989). A developmental study of practical problem solving in adults. *Psychology and Aging, 4,* 438–442.

Dennis, W. (1960). Causes of retardation among institutionalized children: Iran. *Journal of Genetic Psychology, 96,* 47–59.

Dennison, B. A., Straus, J. H., Mellits, D., & Charney, E. (1998). Childhood physical fitness tests: Predictor of adult physical activity levels? *Pediatrics, 82,* 342–350.

De Raedt, R., & Ponjaert-Kristoffersen, I. (2000). The relationship between cognitive/neuropsychological factors and car driving performance in older adults. *Journal of the American Geriatrics Society, 48,* 1664–1668.

Derom, C., Thiery, E., Vlietinck, R., Loos, R., & Derom, R. (1996). Handedness in twins according to zygosity and chorion type: A preliminary report. *Behavior Genetics, 26,* 407–408.

DeStafano, F., & Chen, R. T. (2001). Autism and measles-mumps-rubella vaccination: Controversy laid to rest? *CNS Drugs, 15,* 831–837.

Deutsch, F. M., Ruble, D. N., Fleming, A., Brooks-Gunn, J., & Stangor, C. (1988). Information-seeking and maternal self-definition during the transition to motherhood. *Journal of Personality and Social Psychology, 55,* 420–431.

Deutsch, W., & Pechmann, T. (1982). Social interaction and the development of definite descriptions. *Cognition, 11,* 159–184.

Deveson, A. (1994). *Coming of age: Twenty-one interviews about growing older.* Newham, Australia: Scribe.

de Villiers, J. G., & de Villiers, P. A. (1973). A cross-sectional study of the acquisition of grammatical morphemes in child speech. *Journal of Psycholinguistic Research, 2,* 267–278.

de Villiers, J. G., & de Villiers, P. A. (2000). Linguistic determinism and the understanding of false beliefs. In P. Metchell & K. J. Riggs (Eds.), *Children's reasoning and the mind* (pp. 87–99). Hove, U.K.: Psychology Press.

de Vries, B., Bluck, S., & Birren, J. E. (1993). The understanding of death and dying in a life-span perspective. *Gerontologist, 33,* 366–372.

de Waal, F. B. M. (1993). Sex differences in chimpanzee (and human) behavior: A matter of social values? In M. Hechter, L. Nadel, & R. E. Michod (Eds.), *The origin of values* (pp. 285–303). New York: Aldine de Gruyter.

de Waal, F. B. M. (1999). The end of nature versus nurture. *Scientific American, 281*(6), 94–99.

Dewey, K. G. (2001). Nutrition, growth, and complementary feeding of the breastfed infant. *Pediatric Clinics of North America, 48,* 87–104.

De Wolff, M. S., & van IJzendoorn, M. H. (1997). Sensitivity and attachment: A meta-analysis on parental antecedents of infant attachment. *Child Development, 68,* 571–591.

Dewsbury, D. A. (1992). Comparative psychology and ethology: A reassessment. *American Psychologist, 47,* 208–215.

Diamond, A., Cruttenden, L., & Neiderman, D. (1994). AB with multiple wells: 1. Why are multiple wells sometimes easier than two wells? 2. Memory or memory + inhibition. *Developmental Psychology, 30,* 192–205.

Diamond, L. M. (1998). Development of sexual orientation among adolescent and young adult women. *Developmental Psychology, 34,* 1085–1095.

Diamond, L. M., Savin-Williams, R. C., & Dubé, E. M. (1999). Sex, dating, passionate friendships, and romance: Intimate peer relations among lesbian, gay, and bisexual adolescents. In W. Furman, B. B. Brown, & C. Feiring, (Eds.), *The development of romantic relationships in adolescence* (pp. 175–210). New York: Cambridge University Press.

Diamond, M., & Hopson, J. (1999). *Magic trees of the mind.* New York: Plume.

Diamond, M., & Sigmundson, H. K. (1999). Sex reassignment at birth. In S. J. Ceci & W. M. Williams (Eds.), *The nature–nurture debate* (pp. 55–75). Malden, MA: Blackwell.

Dick, D. M., Rose, R. J., Viken, R. J., & Kaprio, J. (2000). Pubertal timing and substance use: Associations between and within families across late adolescence. *Developmental Psychology, 36,* 180–189.

Dick-Read, G. (1959). *Childbirth without fear.* New York: Harper & Row.

Dickson, K. L., Fogel, A., & Messinger, D. (1998). The development of emotion from a social process view. In M. F. Mascolo (Ed.), *What develops in emotional development?* (pp. 253–271). New York: Plenum.

Dickson, S. V., Collins, V. L., Simmons, D. C., & Dameenui, E. J. (1998). Metacognitive strategies: Research bases. In D. C. Simmons & E. J. Kameenui (Eds.), *What reading research tells us about children with diverse learning needs: Bases and basics* (pp. 295–360). Mahwah, NJ: Erlbaum.

DiClemente, R. J. (1993). Preventing HIV/AIDS among adolescents. *Journal of the American Medical Association, 270,* 760–762.

Diehl, M., Coyle, N., & Labouvie-Vief, G. (1996). Age and sex differences in strategies of coping and defense across the life span. *Psychology and Aging, 11,* 127–139.

Diener, E., Gohm, C. L., Suh, E., & Oishi, S. (2000). Similarity of the relations between marital status and subjective well-being across cultures. *Journal of Cross-Cultural Psychology, 31,* 419–436.

Dietrich, K. N. (1999). Environmental toxicants and child development. In H. Tager-Flusberg (Ed.), *Neurodevelopmental disorders* (pp. 469–490). Boston: MIT Press.

Dietrich, K. N., Berger, O. G,. & Succop, P. A. (1993). Lead exposure and the motor developmental status of urban six-year-old children in the Cincinnati Prospective Study. *Pediatrics, 91,* 301–307.

DiLalla, L. F., Kagan, J., & Reznick, J. S. (1994). Genetic etiology of behavioral inhibition among 2-year-old children. *Infant Behavior and Development,* 17, 405–412.

Dildy, G. A., Jackson, G. M., Fowers, G. K., Oshiro, B. T., Varner, M. W., & Clark, S. L. (1996). Very advanced maternal age. Pregnancy after age 45. *American Journal of Obstetrics and Gynecology, 175,* 668–674.

Dillaway, H., & Broman, C. (2001). Race, class, and gender differences in marital satisfaction and divisions of household labor among dual-earner couples. *Journal of Family Issues, 22,* 309–327.

DiNitto, D. M. (2002). *Social welfare: Politics and public policy* (2nd ed.). Boston: Allyn and Bacon.

Dion, K. L., & Dion, K. K. (1993). Gender and ethnocultural comparisons in styles of love. *Journal of Social Issues, 49,* 53–69.

DiPietro, J. A., Hodgson, D. M., Costigan, K. A., & Hilton, S. C. (1996a). Fetal neurobehavioral development. *Child Development, 67,* 2553–2567.

DiPietro, J. A., Hodgson, D. M., Costigan, K. A., & Johnson, T. R. B. (1996b). Fetal antecedents of infant temperament. *Child Development, 67,* 2568–2583.

Dirks, J. (1982). The effect of a commercial game on children's Block Design scores on the WISC–R test. *Intelligence, 6,* 109–123.

Dishion, T. J., Andrews, D. W., & Crosby, L. (1995). Antisocial boys and their friends in early adolescence: Relationship characteristics, quality, and interactional processes. *Child Development, 66,* 139–151.

Dixon, J. A., & Moore, C. F. (1990). The development of perspective taking: Understanding differences in information and weighting. *Child Development, 61,* 1502–1513.

Dixon, L., & Browne, K. (2003). The heterogeneity of spouse abuse: A review. *Aggression and Violent Behavior, 8,* 107–130.

Dixon, R. A., de Frias, C. M., & Bäckman, L. (2001). Characteristics of self-reported memory compensation in older adults. *Journal of Clinical and Experimental Neuropsychology, 23,* 650–661.

Dixon, R. A., & Hultsch, D. F. (1999). Intelligence and cognitive potential in late life. In J. C. Cavanaugh & S. K. Whitbourne (Eds.), *Gerontology: An interdisciplinary perspective* (pp. 213–237). New York: Oxford University Press.

Dixon, R. A., & Lerner, R. M. (1999). History and systems in developmental psychology. In M. H. Bornstein & M. E. Lamb (Eds.), *Developmental psychology: An advanced textbook* (4th ed., pp. 3–46). Mahwah, NJ: Erlbaum.

Dodge, K. A., Pettit, G. S., & Bates, J. E. (1994). Socialization mediators of the relation between socioeconomic status and child conduct problems. *Child Development, 65,* 649–655.

Doeker, B., Simic-Schleicher, A., Hauffa, B. P., & Andler, W. (1999). Psychosozialer Kleinwuchs maskiert als Wachstumshormonmangel [Psychosocially stunted growth masked as growth hormone deficiency]. *Klinisch Padiatrie, 211,* 394–398.

Doherty, G., Lero, D. S., Goelman, H., Tougas, J., & LaGrange, A. (2000). *You bet I care! Caring and learning environments: Quality in regulated family child care across Canada.* Guelph, ON: Centre for Families, Work and Well-Being, University of Guelph.

Doi, M. (1991). A transformation of ritual: The Nisei 60th birthday. *Journal of Cross-Cultural Gerontology, 6,* 153–161.

Doka, K., & Martin, T. (2000). Take it like a man: Masculine response to loss. In D. A. Lund (Ed.), *Men coping with grief* (pp. 37–47). Amityville, NY: Baywood.

Donatelle, R. J. (2003). *Health: The basics.* San Francisco: Benjamin Cummings.

Dondi, M., Simion, F., & Caltran, G. (1999). Can newborns discriminate between their own cry and the cry of another newborn infant? *Developmental Psychology, 35,* 418–426.

Donnerstein, E., & Linz, D. (1998). Mass media, sexual violence, and male viewers. In M. E. Odem & J. Clay-Warner (Eds.*), Confronting rape and sexual assault* (pp.181–198). Wilmington, DE: Scholarly Resources.

Donnerstein, E., Slaby, R. G., & Eron, L. D. (1994). The mass media and youth aggression. In L. D. Eron, J. H. Gentry, & P. Schlegel (Eds.), *Reason to hope: A psychosocial perspective on violence and youth* (pp. 219–250). Washington, DC: American Psychological Association.

Dornbusch, S. M., Glasgow, K. L., & Lin, I.-C. (1996). The social structure of schooling. *Annual Review of Psychology, 47,* 401–427.

Dornbusch, S. M., Ritter, P. L., Liederman, P. H., Roberts, D. F., & Fraleigh, M. J. (1987). The relation of parenting style to adolescent school performance. *Child Development, 58,* 1244–1257.

Dornbusch, S. M., Ritter, P. L., Mont-Reynaud, R., & Chen, Z. (1990). Family decision making and academic performance in a diverse high school population. *Journal of Adolescent Research, 5,* 143–160.

Dorris, M. (1989). *The broken cord.* New York: Harper & Row.

Dowd, M. D. (1999). Childhood injury prevention at home and play. *Current Opinion in Pediatrics, 11,* 578–582.

Dowd, R. (2001). Role of calcium, vitamin D, and other essential nutrients in the prevention and treatment of osteoporosis. *Nursing Clinics of North America, 36,* 417–431.

Dowdney, L. (2000). Annotation: Childhood bereavement following parental death. *Journal of Child Psychology and Psychiatry and Allied Disciplines, 41,* 819–830.

Downey, B., Armstrong, K., Lindsay, G., & Dorey, D. (2002). *Finding our way: A sexual and reproductive sourcebook for Aboriginal communities.* Ottawa: Aboriginal Nurses Association of Canada.

Downey, D. B. (1995). When bigger is not better: Family size, parental resources, and children's educational performance. *American Sociological Review, 60,* 746–761.

Downs, A. C., & Fuller, M. J. (1991). Recollections of spermarche: An exploratory investigation. *Current Psychology: Research and Reviews, 10,* 93–102.

Drabman, R. S., Cordua, G. D., Hammer, D., Jarvie, G. J., & Horton, W. (1979). Developmental trends in eating rates of normal and overweight preschool children. *Child Development, 50,* 211–216.

Drewnowski, A., & Shultz, J. M. (2001). Impact of aging on eating behaviors, food choices, nutrition, and health status. *Journal of Nutrition, Health, and Aging, 5,* 75–79.

Drotar, D., Pallotta, J., & Eckerle, D. (1994). A prospective study of family environments of children hospitalized for nonorganic failure-to-thrive. *Developmental and Behavioral Pediatrics, 15,* 78–85.

Druker, B. J., & Lydon, N. B. (2000). Lessons learned from the development of an ABL tyrosine kinase inhibitor for chronic myelogenous leukemia. *Journal of Clinical Investigation, 105,* 3–7.

Dryburgh, H. (2001). Teenage pregnancy. *Health Reports, 12*(1), Statistics Canada, Cat. No. 82–003.

Dryer, D. C., & Horowitz, L. M. (1997). When do opposites attract? Interpersonal complementarity versus similarity. *Journal of Personality and Social Psychology, 72,* 592–603.

Dubé, E. M., Savin-Williams, R. C., & Diamond, L. M. (2001). Intimacy development, gender, and ethnicity among sexual-minority youths. In A. R. D'Augelli & C. J. Patterson (Eds.), *Lesbian, gay, and bisexual identities and youth* (pp. 129–152). New York: Oxford University Press.

DuBois, D. L., Bull, C. A., Sherman, M. D., & Roberts, M. (1998). Self-esteem and adjustment in early adolescence: A social-contextual perspective. *Journal of Youth and Adolescence, 27,* 557–583.

DuBois, D. L., Felner, R. D., Brand, S., & George, G. R. (1999). Profiles of self-esteem in early adolescence: Identification and investigation of adaptive correlates. *American Journal of Community Psychology, 27,* 899–932.

DuBois, D. L., & Hirsch, B. J. (1990). School and neighborhood friendship patterns of blacks and whites in early adolescence. *Child Development, 61,* 524–536.

Duck, S. (1994). *Meaningful relationships.* Thousand Oaks, CA: Sage.

Dugan, E., & Kivett, V. R. (1998). Implementing the Adams and Blieszner conceptual model: Predicting interactive friendship processes of older adults. *Journal of Social and Personal Relationships, 15,* 607–622.

Duncan, R. M., & Pratt, M. W. (1997). Microgenetic change in the quantity and quality of preschoolers' private speech. *International Journal of Behavioral Development, 20,* 367–383.

Duniz, M., Scheer, P. J., Trojovsky, A., Kaschnitz, W., Kvas, E., & Macari, S. (1996). *European Child and Adolescent Psychiatry, 5,* 93–100.

Dunn, J. (1989). Siblings and the development of social understanding in early childhood. In P. G. Zukow (Ed.), *Sibling interaction across cultures* (pp. 106–116). New York: Springer-Verlag.

Dunn, J. (1994). Temperament, siblings, and the development of relationships. In W. B. Carey & S. C. McDevitt (Eds.), *Prevention and early intervention* (pp. 50–58). New York: Brunner/Mazel.

Dunn, J. (1996). Sibling relationships and perceived self-competence: Patterns of stability between childhood and early adolescence. In A. J. Sameroff & M. M. Haith (Eds.), *The five to seven year shift* (pp. 253–270). Chicago: University of Chicago Press.

Dunn, J., Slomkowski, C., & Beardsall, L. (1994). Sibling relationships from the preschool period through middle childhood and early adolescence. *Developmental Psychology, 30,* 315–324.

Durbin, D. L., Darling, N., Steinberg, L., & Brown, B. B. (1993). Parenting style and peer group membership among European-American adolescents. *Journal of Research on Adolescence, 3,* 87–100.

Durlak, J. A., & Riesenberg, L. A. (1991). The impact of death education. *Death Studies, 15,* 39–58.

Durrant, J., Broberg, A., & Rose-Krasnor, L. (2000). Predicting use of physical punishment during mother–child conflicts in Sweden and Canada. In P. Hastings & C. Piotrowski (Eds.), *Conflict as a context for understanding maternal beliefs about child rearing and children's misbehavior: New directions for child development.* San Francisco: Jossey-Bass.

Dusek, J. B. (1987). Sex roles and adjustment. In D. B. Carter (Ed.), *Current conceptions of sex roles and sex typing* (pp. 211–222). New York: Praeger.

Dutton, D. G., Landolt, M. A., Starzomski, A., & Bodnarchuk, M. (2001). Validation of the propensity for abusiveness scale in diverse male populations. *Journal of Family Violence, 16,* 59–73.

Dworkin, J. B., Larson, R., & Hansen, D. (1993). Adolescents' accounts of growth experiences in youth activities. *Journal of Youth and Adolescence, 32,* 17–26.

Dybing, E., & Sanner, T. (1999). Passive smoking, sudden infant death syndrome (SIDS), and childhood infections. *Human Experimental Toxicology, 18,* 202–205.

Dyer, C. B., Pavlik, V. N., Murphy, K. P., & Hyman, D. J. (2000). The high prevalence of depression and dementia in elder abuse and neglect. *Journal of the American Geriatrics Society, 48,* 205–208.

Dye-White, E. (1986). Environmental hazards in the work setting: Their effect on women of child-bearing age. *American Association of Occupational Health and Nursing Journal, 34,* 76–78.

Dykstra, P. A. (1995). Loneliness among the never and formerly married: The importance of supportive friendships and a desire for independence. *Journal of Gerontology, 50B,* S321–S329.

Eacott, M. J. (1999). Memory for the events of early childhood. *Current Directions in Psychological Science, 8,* 46–48.

Eagly, A. H., & Karau, S. J. (2002). Role congruity theory of prejudice toward female leaders. *Psychological Review, 109,* 573–598.

Eagly, A. H., & Wood, W. (1999). The origins of sex differences in human behavior: Evolved dispositions versus social roles. *American Psychologist, 54,* 408–423.

East, P. L., & Felice, M. E. (1996). *Adolescent pregnancy and parenting: Findings from a racially diverse sample.* Mahwah, NJ: Erlbaum.

Ebeling, K. S., & Gelman, S. A. (1994). Children's use of context in interpreting "big" and "little." *Child Development, 65,* 1178–1192.

Eberhart-Phillips, J. E., Frederick, P. D., & Baron, R. C. (1993). Measles in pregnancy: A descriptive study of 58 cases. *Obstetrics and Gynecology, 82,* 797–801.

Eccles, J., Barber, B., Jozefowicz, D., Malenchuk, O., & Vida, M. (1999). Self-evaluations of competence, task values, and self-esteem. In N. G. Johnson & M. C. Roberts (Eds.), *Beyond appearance: A new look at adolescent girls* (pp. 53–83). Washington, DC: American Psychological Association.

Eccles, J. S. (1994). Understanding women's educational and occupational choices: Applying the Eccles et al. model of achievement-related choices. *Psychology of Women Quarterly, 18,* 585–609.

Eccles, J. S., Early, D., Frasier, K., Belansky, E., & McCarthy, K. (1997). The relation of connection, regulation, and support for autonomy to adolescents' functioning. *Journal of Adolescent Research, 12,* 263–286.

Eccles, J. S., & Harold, R. D. (1991). Gender differences in sport involvement: Applying the Eccles' expectancy-value model. *Journal of Applied Sport Psychology, 3,* 7–35.

Eccles, J. S., & Harold, R. D. (1996). Family involvement in children's and adolescents' schooling. In A. Booth & J. F. Dunn (Eds.), *Family–school links: How do they affect educational outcomes?* (pp. 3–34). Mahwah, NJ: Erlbaum.

Eccles, J. S., Jacobs, J. E., & Harold, R. D. (1990). Gender-role stereotypes, expectancy effects, and parents' role in the socialization of gender differences in self-perceptions and skill acquisition. *Journal of Social Issues, 46,* 183–201.

Eccles, J. S., Lord, S., Roeser, R. W., Barber, B., & Josefowicz-Hernandez, D. (1997). The association of school transitions in early adolescence with developmental trajectories through high school. In J. Schulenberg, J. Maggs, & K. Hurrelmann (Eds.), *Health risks and developmental transitions during adolescence* (pp. 283–320). New York: Cambridge University Press.

Eccles, J. S., Midgley, C., Wigfield, A., Buchanan, C. M., Reuman, D., Flanagan, C., & MacIver, D. (1993a). Development during adolescence: The impact of stage–environment fit on young adolescents' experiences in schools and in families. *American Psychologist, 48,* 90–101.

Eccles, J. S., Wigfield, A., Harold, R., & Blumenfeld, P. B. (1993b). Age and gender differences in children's self- and task perceptions during elementary school. *Child Development, 64,* 830–847.

Eccles, J. S., Wigfield, A., & Schiefele, U. (1998). Motivation to succeed. In N. Eisenberg (Ed.), *Handbook of child psychology: Vol. 3. Social, emotional, and personality development* (5th ed., pp. 1017–1095). New York: Wiley.

Eder, R. A. (1989). The emergent personologist: The structure and content of 3½, 5½ and 7½-year-olds' concepts of themselves and other persons. *Child Development, 60,* 1218–1228.

Edwards, J. N., & Booth, A. (1994). Sexuality, marriage, and well-being: The middle years. In A. S. Rossi (Ed.), *Sexuality across the life course* (pp. 233–259). Chicago: University of Chicago Press.

Egan, S. K., Monson, T. C., & Perry, D. G. (1998). Social-cognitive influences on change in aggression over time. *Developmental Psychology, 34,* 996–1006.

Egeland, B., & Hiester, M. (1995). The long-term consequences of infant day-care and mother–infant attachment. *Child Development, 66,* 474–485.

Egeland, B., Jacobvitz, D., & Sroufe, L. A. (1988). Breaking the cycle of abuse. *Child Development, 59,* 1080–1088.

Eggebeen, D. J. (1992). Parent–child support in aging American families. *Generations, 16*(3), 45–49.

Eggerman, S., & Dustin, D. (1985). Death orientation and communication with the terminally ill. *Omega, 16,* 255–265.

Eiben, B., Hammans, W., Hansen, S., Trawicki, W., Osthelder, B., Stelzer, A., Jaspers, K.-D., & Goebel, R. (1997). On the complication risk of early amniocentesis versus standard amniocentesis. *Fetal Diagnosis and Therapy, 12,* 140–144.

Eiden, R. D., & Reifman, A. (1996). Effects of Brazelton demonstrations on later parenting: A meta-analysis. *Journal of Pediatric Psychology, 21,* 857–868.

Eilers, R. E., & Oller, D. K. (1994). Infant vocalizations and the early diagnosis of severe hearing impairment. *Journal of Pediatrics, 124,* 199–203.

Einstein, G. O., Holland, L. J., McDaniel, M. A., & Guynn, M. J. (1992). Age-related deficits in prospective memory: The influence of task complexity. *Psychology and Aging, 7,* 471–478.

Einstein, G. O., Smith, R. E., McDaniel, M. A., & Shaw, P. (1997). Aging and prospective memory: The influence of increased task demands at encoding and retrieval. *Psychology and Aging, 12,* 479–488.

Eisenberg, L. (1999). Experience, brain, and behavior: The importance of a head start. *Pediatrics, 103,* 1031–1035.

Eisenberg, N. (1998). Introduction. In N. Eisenberg (Ed.), *Handbook of child psychology: Vol. 3. Social, emotional, and personality development* (5th ed., pp. 1–24). New York: Wiley.

Eisenberg, N., Cumberland, A., & Spinrad, T. L. (1998). Parental socialization of emotion. *Psychological Inquiry, 9,* 241–273.

Eisenberg, N., & Fabes, R. A. (1998). Prosocial development. In N. Eisenberg (Ed.), *Handbook of child psychology: Vol. 3. Social, emotional, and personality development* (5th ed., pp. 701–778). New York: Wiley.

Eisenberg, N., Fabes, R. A., Carlo, G., Speer, A. L., Switzer, G., Karbon, M., & Troyer, D. (1993). The relations of empathy-related emotions and maternal practices to children's comforting behavior. *Journal of Experimental Child Psychology, 55,* 131–150.

Eisenberg, N., Fabes, R. A., & Losoya, S. (1997). Emotional responding: Regulation, social correlates, and socialization. In P. Alovey & D. J. Sluyter (Eds.), *Emotional development and emotional intelligence* (pp. 129–162). New York: Basic Books.

Eisenberg, N., Fabes, R. A., Murphy, B., Karbon, M., Smith, M., & Maszk, P. (1996). The relations of children's dispositional empathy-related responding to their emotionality, regulation, and social functioning. *Developmental Psychology, 32,* 195–209.

Eisenberg, N., Fabes, R. A., Murphy, B., Maszk, P., Smith, M., & Karbon, M. (1995). The role of emotionality and regulation in children's social functioning: A longitudinal study. *Child Development, 66,* 1360–1384.

Eisenberg, N., Fabes, R. A., Shepard, S. A., Guthrie, I., Murphy, B. C., & Reiser, M. (1999). Parental reactions to children's negative emotions: Longitudinal relations to quality of children's social functioning. *Child Development, 70,* 513–534.

Eisenberg, N., Fabes, R. A., Shepard, S. A., Murphy, B. C., Guthrie, I. K., Jones, S., Friedman, J., Poulin, R., & Maszk, P. (1997). Contemporaneous and longitudinal prediction of children's social functioning from regulation and emotionality. *Child Development, 68,* 642–664.

Eisenberg, N., Fabes, R. A., Shepard, S. A., Murphy, B. C., Jones, S., & Guthrie, I. K. (1998). Contemporaneous and longitudinal prediction of children's sympathy from dispositional regulation and emotionality. *Developmental Psychology, 34,* 910–924.

Eisenberg, N., & McNally, S. (1993). Socialization and mothers' and adolescents' empathy-related characteristics. *Journal of Research on Adolescence, 3,* 171–191.

Ekman, P., & Friesen, W. (1972). Constants across culture in the face of emotion. *Journal of Personality and Social Psychology, 17,* 124–129.

Elardo, R., & Bradley, R. H. (1981). The Home Observation for Measurement of the Environment (HOME) Scale: A review of research. *Developmental Review, 1,* 113–145.

Elder, G., Jr. (1998). The life course and human development. In R. M. Lerner (Ed.), *Handbook of child psychology: Vol. 1. Theoretical models of human development* (5th ed., pp. 939–991). New York: Wiley.

Elder, G. H., Jr. (1999). *Children of the Great Depression* (25th anniversary ed.). Boulder, CO: Westview.

Elder, G. H., Jr., & Caspi, A. (1988). Human development and social change: An emerging perspective on the life course. In N. Bolger, A. Caspi, G. Downey, & M. Moorehouse (Eds.), *Persons in context: Developmental processes* (pp. 77–113). Cambridge, U.K.: Cambridge University Press.

Elder, G. H., Jr., & Hareven, T. K. (1993). Rising above life's disadvantage: From the Great Depression to war. In G. H. Elder, Jr., J. Modell, & R. D. Parke (Eds.), *Children in time and place* (pp. 47–72). Cambridge, U.K.: Cambridge University Press.

Elder, G. H., Jr., Liker, J. K., & Cross, C. E. (1984). Parent–child behavior in the Great Depression: Life course and intergenerational influences. In P. B. Baltes & O. G. Brim (Eds.), *Life-span development and behavior* (Vol. 6, pp. 109–158). New York: Academic Press.

Elder, G. H., Jr., Van Nguyen, T., & Caspi, A. (1985). Linking family hardship to children's lives. *Child Development, 56,* 361–375.

Elias, C. L., & Berk, L. E. (2002). Self-regulation in young children: Is there a role for sociodramatic play? *Early Childhood Research Quarterly, 17,* 1–17.

Elicker, J., Englund, M., & Sroufe, L. A. (1992). Predicting peer competence and peer-relationships in childhood from early parent–child relationships. In R. D. Parke & G. W. Ladd (Eds.), *Family–peer relationships: Modes of linkage* (pp. 77–106). Hillsdale, NJ: Erlbaum.

Elkind, D. (1994). *A sympathetic understanding of the child: Birth to sixteen* (3rd ed.). Boston: Allyn and Bacon.

Elkind, D., & Bowen, R. (1979). Imaginary audience behavior in children and adolescents. *Developmental Psychology, 15,* 33–44.

Elkind, M. S., & Sacco, R. L. (1998). Stroke risk factors and stroke prevention. *Seminars in Neurology, 18,* 429–440.

Ellicott, A. M. (1985). Psychosocial changes as a function of family-cycle phase. *Human Development, 28,* 270–274.

Ellingson, S., & Fuller, J. D. (2001). A good death? Finding a balance between the interests of patients and caregivers. In M. B. Holstein & P. B. Mitzen (Eds.), *Ethics in community-based elder care* (pp. 200–207). New York: Springer.

Elliott, D. S. (1994). Serious violent offenders: Onset, developmental course, and termination. *Criminology, 32,* 1–21.

Elliott, D. S., Wilson, W. J., Huizinga, D., Sampson, R. J., Elliott, A., & Rankin, B. (1996). The effects of neighborhood disadvantage on adolescent development. *Journal of Research in Crime and Delinquency, 33,* 389–426.

Elliott, E. S., & Dweck, C. S. (1988). Goals: An approach to motivation and achievement. *Journal of Personality and Social Psychology, 54,* 5–12.

Elliott, J. G. (1999). School refusal: Issues of conceptualisation, assessment, and treatment. *Journal of Child Psychology and Psychiatry and Allied Disciplines, 40,* 1001–1012.

Ellis, B. J., & Garber, J. (2000). Psychosocial antecedents of variation in girls' pubertal timing: Maternal depression, stepfather presence, and marital and family stress. *Child Development, 71,* 485–501.

Ellis, B. J., McFadyen-Ketchum, S., Dodge, K. A., Pettit, G. S., & Bates, J. E. (1999). Quality of early family relationships and individual differences in the timing of pubertal maturation in girls: A longitudinal test of an evolutionary model. *Journal of Personality and Social Psychology, 77,* 933–952.

Elman, J. L., Bates, E. A., Johnson, M. H., Karmiloff-Smith, A., Parisi, D., & Plunkett, K. (1996). *Rethinking innateness: A connectionist perspective on development.* Cambridge, MA: MIT Press.

El-Sheikh, M., Cummings, E. M., & Reiter, S. (1996). Preschoolers' responses to ongoing interadult conflict: The role of prior exposure to resolved versus unresolved arguments. *Journal of Abnormal Child Psychology, 24,* 665–679.

Emanuel, E. J., Fairclough, D. L., & Emanuel, L. L. (2000). Attitudes and desires related to euthanasia and physician-assisted suicide among terminally ill patients and their caregivers. *Journal of the*

American Medical Association, 284, 2460–2468.

Emde, R. N. (1992). Individual meaning and increasing complexity: Contributions of Sigmund Freud and René Spitz to developmental psychology. *Developmental Psychology, 28,* 347–359.

Emde, R. N., Plomin, R., Robinson, J., Corley, R., DeFries, J., Fulker, D. W., Reznick, J. S., Campos, J., Kagan, J., & Zahn-Waxler, C. (1992). Temperament, emotion, and cognition at fourteen months: The MacArthur Longitudinal Twin Study. *Child Development, 63,* 1437–1455.

Emery, R. E. (1999a). Postdivorce family life for children: An overview of research and some implications for policy. In R. A. Thompson & P. R. Amato (Eds.), *The postdivorce family: Children, parenting, and society* (pp. 3–27). Thousand Oaks, CA: Sage.

Emery, R. E. (1999b). Psychological interventions for separated and divorced families. In E. M. Hetherington (Ed.), *Coping with divorce, single parenting, and remarriage: A risk and resiliency perspective* (pp. 253–271). Mahwah, NJ: Erlbaum.

Emery, R. E. (2001). Interparental conflict and social policy. In J. H. Grych & F. D. Fincham (2001). *Interparental conflict and child development: Theory, research, and applications* (pp. 417–439). New York: Cambridge University Press.

Emery, R. E., & Laumann-Billings, L. (1998). An overview of the nature, causes, and consequences of abusive family relationships: Toward differentiating maltreatment and violence. *American Psychologist, 53,* 121–135.

Emory, E. K., Schlackman, L. J., & Fiano, K. (1996). Drug–hormone interactions on neurobehavioral responses in human neonates. *Infant Behavior and Development, 19,* 213–220.

Engelhart, M. J., Geerlings, M. I., Ruitenberg, A., van Swieten, J. C., Hofman, A., Witteman, J. C., & Breteler, M. M. (2002). Dietary intake of antioxidants and risk of Alzheimer's disease. *Journal of the American Medical Association, 26,* 3223–3229.

Engen, T. (1982). *The perception of odors.* New York: Academic Press.

Epstein, J. L. (2001). *School, family, and community partnerships: Preparing educators and improving schools.* Bolder, CO: Westview.

Epstein, J. L., & Sanders, M. G. (2002). Family, school, and community partnerships. In M. H. Bornstein (Ed.), *Handbook of parenting: Vol. 5* (2nd ed., pp. 407–437). Mahwah, NJ: Erlbaum.

Epstein, L. H., McCurley, J., Wing, R. R., & Valoski, A. (1990). Five-year follow-up of family-based treatments for childhood obesity. *Journal of Consulting and Clinical Psychology, 58,* 661–664.

Epstein, L. H., McKenzie, S. J., Valoski, A., Klein, K. R., & Wing, R. R. (1994). Effects of mastery criteria and contingent reinforcement for family-based child weight control. *Addictive Behaviors, 19,* 135–145.

Epstein, S. (1999). Gay and lesbian movements in the United States: Dilemmas of identity, diversity, and political strategy. In B. D. Adam, J. W. Duyvendak, & A. Krouwel (Eds.), *The global emergence of gay and lesbian politics: National imprints of a worldwide movement* (pp. 30–90). Philadelphia: Temple University Press.

Erdley, C. A., Cain, K. M., Loomis, C. C., Dumas-Hines, F., & Dweck, C. S. (1997). Relations among children's social goals, implicit personality theories, and responses to social failure. *Developmental Psychology, 33,* 263–272.

Erel, O., & Burman, B. (1995). Interrelatedness of marital relation and parent–child relations: A meta-analytic review. *Psychological Bulletin, 118,* 108–132.

Ericsson, K. A. (1990). Peak performance and age: An examination of peak performance in sports. In P. B. Baltes & M. M. Baltes (Eds.), *Successful aging* (pp. 164–196). Cambridge, U.K.: Cambridge University Press.

Erikson, E. H. (1950). *Childhood and society.* New York: Norton.

Erikson, E. H. (1964). *Insight and responsibility.* New York: Norton.

Erikson, E. H. (1968). *Identity, youth, and crisis.* New York: Norton.

Eriksson, J. G., Forsén, T., Tuomilehto, J., Osmond, C., & Barker, D. J. (2001). Early growth and coronary heart disease in later life: A longitudinal study. *British Medical Journal, 322,* 949–953.

Espiritu, D. A. V., Rashid, H., Mast, B. T., Fitzgerald, J., Steinberg, J., & Lichtenberg, P. A. (2001). Depression, cognitive impairment and function in Alzheimer's disease. *International Journal of Geriatric Psychiatry, 16,* 1098–1103.

Espy, K. A., Kaufmann, P. M., & Glisky, M. L. (1999). Neuropsychological function in toddlers exposed to cocaine in utero: A preliminary study. *Developmental Neurospsychology, 15,* 447–460.

Espy, K. A., Molfese, V. J., & DiLalla, L. F. (2001). Effects of environmental measures on intelligence in young children: Growth curve modeling of longitudinal data. *Merrill-Palmer Quarterly, 47,* 42–73.

Essa, E. L., & Murray, C. I. (1994). Young children's understanding and experience with death. *Young Children, 49*(4), 74–81.

Esterberg, K. G., Moen, P., & Dempster-McClain, D. (1994). Transition to divorce: A life-course approach to women's marital duration and dissolution. *Sociological Quarterly, 35,* 289–307.

Evans, G. W., & Kantrowitz, E. (2002). Socioeconomic status and health: The potential role of environmental risk exposure. *Annual Review of Public Health, 23,* 303–331.

Evans, G. W., Maxwell, L. E., & Hart, B. (1999). Parental language and verbal responsiveness to children in crowded homes. *Developmental Psychology, 35,* 1020–1023.

Evans, R. M., Emsley, C. L., Gao, S., Sahota, A., Farlow, M. R., & Hendrie, H. C. (2000). Serum cholesterol, ApoE gene type and the risk of Alzheimer's disease: A population-based study of African Americans. *Neurology, 54,* 95–99.

Everman, D. B., & Cassidy, S. B. (2000). Genetics of childhood disorders: XII. Genomic imprinting: Breaking the rules. *Journal of the American Academy of Child and Adolescent Psychiatry, 38,* 386–389.

Eysenck, H. J., & Wakefield, J. A. (1981). Psychological factors as predictors of marital satisfaction. *Advances in Behaviour Research and Therapy, 3,* 151–192.

Fabes, R. A., Eisenberg, N., McCormick, S. E., & Wilson, M. S. (1988). Preschoolers' attributions of the situational determinants of others' naturally occurring emotions. *Developmental Psychology, 24,* 376–385.

Fabsitz, R. R., Sholinsky, P., & Carmelli, D. (1994). Genetic influences on adult weight gain and maximum body mass index in male twins. *American Journal of Epidemiology, 140,* 711–720.

Fagan, J. F., III. (1973). Infants' delayed recognition memory and forgetting. *Journal of Experimental Child Psychology, 16,* 424–450.

Fagan, J. F., III, & Singer, L. T. (1979). The role of simple feature differences in infants' recognition of faces. *Infant Behavior and Development, 2,* 39–45.

Fagard, J., & Pezé, A. (1997). Age changes in interlimb coupling and the development of bimanual coordination. *Journal of Motor Behavior, 29,* 199–208.

Fagot, B. I. (1984). The child's expectations of differences in adult male and female interactions. *Sex Roles, 11,* 593–600.

Fagot, B. I., & Hagan, R. I. (1991). Observations of parent reactions to sex-stereotyped behaviors: Age and sex effects. *Child Development, 62,* 617–628.

Fagot, B. I., & Leinbach, M. D. (1989). The young child's gender schema: Environmental input, internal organization. *Child Development, 60,* 663–672.

Fagot, B. I., Leinbach, M. D., & O'Boyle, C. (1992). Gender labeling, gender stereotyping, and parenting behaviors. *Developmental Psychology, 28,* 225–230.

Fagot, B. I., Pears, K. C., Apaldi, D. M., Crosby, L., & Lee, C. S. (1998). Becoming an adolescent father: Precursors and parenting. *Developmental Psychology, 34,* 1209–1219.

Fahrmeier, E. D. (1978). The development of concrete operations among the Hausa. *Journal of Cross-Cultural Psychology, 9,* 23–44.

Falbo, T. (1992). Social norms and the one-child family: Clinical and policy implications. In F. Boer & J. Dunn (Eds.), *Children's sibling relationships* (pp. 71–82). Hillsdale, NJ: Erlbaum.

Falbo, T., & Poston, D. L., Jr. (1993). The academic, personality, and physical outcomes of only children in China. *Child Development, 64,* 18–35.

Falbo, T., Poston, D. L., Jr., Triscari, R. S., & Zhang, X. (1997). Self-enhancing illusions among Chinese schoolchildren. *Journal of Cross-Cultural Psychology, 28,* 172–191.

Fall, C. H., Stein, C. E., Kumaran, K., Cox, V., Osmond, C., Barker, D. J., & Hales, C. N. (1998). Size at birth, maternal weight, and type 2 diabetes in South India. *Diabetic Medicine, 15,* 220–227.

Faller, K. C. (1990). *Understanding child sexual maltreatment.* Newbury Park, CA: Sage.

Family Caregiver Alliance. (2001). Fact sheet: Selected caregiver statistics. Retrieved from http://www.caregiver.org/factsheets/selected_caregiver_statisticsC.html

Family Caregiver Alliance. (2002). *Fact sheet: Selected caregiver statistics* [Online]. Available: www.nlm.nih.gov/medlineplus/caregivers.html

Fanslow, C. A. (1981). Death: A natural facet of the life continuum. In D. Krieger (Ed.), *Foundations for holistic health nursing practices: The renaissance nurse* (pp. 249–272). Philadelphia: Lippincott.

Fantz, R. L. (1961, May). The origin of form perception. *Scientific American, 204* (5), 66–72.

Farkas, J., & Hogan, D. (1994). The demography of changing intergenerational relationships. In V. L. Bengtson, K. W. Schaie, & L. M. Burton (Eds.), *Adult intergenerational relations: Effects of societal change* (pp. 1–19). New York: Springer.

Farrant, K., & Reese, E. (2000). Maternal style and children's participation in reminiscing: Stepping stones in children's autobiological memory development. *Journal of Cognition and Development, 1,* 193–225.

Farrington, D. P. (1987). Epidemiology. In H. C. Quay (Ed.), *Handbook of juvenile delinquency* (pp. 33–61). New York: Wiley.

Farrington, D. P., & Loeber, R. (2000). Epidemiology of juvenile violence. *Juvenile Violence, 9,* 733–748.

Farver, J. M. (1993). Cultural differences in scaffolding pretend play: A comparison of American and Mexican mother–child and sibling–child pairs. In K. MacDonald (Ed.), *Parent–child play* (pp. 349–366). Albany, NY: SUNY Press.

Farver, J. M., & Branstetter, W. H. (1994). Preschoolers' prosocial responses to their peers' distress. *Developmental Psychology, 30,* 334–341.

Farver, J. M., Kim, Y. K., & Lee, Y. (1995). Cultural differences in Korean- and Anglo-American preschoolers' social interaction and play behaviors. *Child Development, 66,* 1088–1099.

Farver, J., & Wimbarti, S. (1995). Indonesian toddlers' social play with their mothers and older siblings. *Child Development, 66,* 1493–1503.

Fasig, L. G. (2000). Toddlers' understanding of ownership: Implications for self-concept development. *Social Development, 9,* 370–382.

Fasouliotis, S. J., & Schenker, J. G. (2000). Ethics and assisted reproduction. *European Journal of Obstetrics, Gynecology, and Reproductive Biology, 90,* 171–180.

Fattibene, P., Mazzei, F., Nuccetelli, C., & Risica, S. (1999). Prenatal exposure to ionizing radiation: Sources, effects, and regulatory aspects. *Acta Paediatrica, 88,* 693–702.

Featherman, D. (1980). Schooling and occupational careers: Constancy and change in worldly success. In O. Brim, Jr., & J. Kagan (Eds.), *Constancy and change in human development* (pp. 675–738). Cambridge, MA: Harvard University Press.

Federal Interagency Forum on Child and Family Statistics. (2002). *America's children: Key national indicators of well-being: 2002.* Washington, DC: U.S. Government Printing Office.

Feeney, J. A. (1998). Adult attachment and relationship-centered anxiety: Responses to physical and emotional distancing. In J. A. Simpson & W. S. Rholes (Eds.), *Attachment theory and close relationships* (pp. 189–218). New York: Guilford.

Feeney, J. A. (1999). Adult romantic attachment and couple relationships. In J. Cassidy & P. R. Shaver (Eds.), *Handbook of attachment* (pp. 355–377). New York: Guilford.

Feeney, J. A., Hohaus, L., Noller, P., & Alexander, R. P. (2001). *Becoming parents: Exploring the bonds between mothers, fathers, and their infants.* New York: Cambridge University Press.

Fehr, B. (1994). Prototype based assessment of laypeoples' views of love. *Personal Relationships, 1,* 309–331.

Feigenson, L., Carey, S., & Spelke, E. (2002). Infants' discrimination of number vs. continuous extent. *Cognitive Psychology, 44,* 33–66

Feinberg, M. E., & Hetherington, E. M. (2001). Differential parenting as a within-family variable. *Journal of Family Psychology, 51,* 22–37.

Feingold, A. (1994). Gender differences in personality: A meta-analysis. *Psychological Bulletin, 116,* 429–456.

Feiring, C., & Taska, L. S. (1996). Family self-concept: Ideas on its meaning. In B. Bracken (Ed.), *Handbook of self-concept* (pp. 317–373). New York: Wiley.

Feiring, C., Taska, L., & Lewis, M. (1999). Age and gender differences in children's and adolescents' adaptation to sexual abuse. *Child Abuse and Neglect, 23,* 115–128.

Feldman, D. H. (1999). The development of creativity. In R. J. Sternberg (Ed.), *Handbook of creativity* (pp. 169–186). Cambridge, U.K.: Cambridge University Press.

Feldman, R. (2000). Parents' convergence on sharing and marital satisfaction, father involvement, and parent–child relationship in the transition to parenthood. *Infant Mental Health Journal, 21,* 176–191.

Feldman, R., Greenbaum, C. W., & Yirmiya, N. (1999). Mother–infant affect synchrony as an antecedent of the emergence of self-control. *Developmental Psychology, 35,* 223–231.

Felsman, D. E., & Blustein, D. L. (1999). The role of peer relatedness in late adolescent career development. *Journal of Vocational Behavior, 54,* 279–295.

Fenson, L., Dale, P. S., Reznick, J. S., Bates, E., Thal, D. J., & Pethick, S. J. (1994). Variability in early communicative development. *Monographs of the Society for Research in Child Development, 59*(5, Serial No. 242).

Fergus, M. A. (1995, March 19). 99 years and . . . a joy to be around. *The Pantagraph,* pp. C1–C2.

Ferguson, T. J., Stegge, H., & Damhuis, I. (1991). Children's understanding of guilt and shame. *Child Development, 62,* 827–839.

Ferguson, T. J., Stegge, H., Miller, E. R., & Olsen, M. E. (1999). Guilt, shame, and symptoms in children. *Developmental Psychology, 35,* 347–357.

Fergusson, D. M., & Woodward, L. J. (1999). Breast-feeding and later psychosocial adjustment. *Paediatric and Perinatal Epidemiology, 13,* 144–157.

Fergusson, D. M., Woodward, L. J., & Horwood, L. J. (2000). Risk factors and life processes associated with the onset of suicidal behaviour during adolescence and early adulthood. *Psychological Medicine, 30,* 23–39.

Fernald, A., Swingley, D., & Pinto, J. P. (2001). When half a word is not enough: Infants can recognize spoken words using partial phonetic information. *Child Development, 72,* 1003–1015.

Fernald, A., Taeschner, T., Dunn, J., Papousek, M., Boyssen-Bardies, B., & Fukui, I. (1989). A cross-language study of prosodic modifications in mothers' and fathers' speech to preverbal infants. *Journal of Child Language, 16,* 477–502.

Fernald, L. C., & Grantham-McGregor, S. M. (1998). Stress response in school-age children who have been growth-retarded since early childhood. *American Journal of Clinical Nutrition, 68,* 691–698.

Fernandes, O., Sabharwal, M., Smiley, T., Pastuszak, A., Koren, G., & Einarson, T. (1998). Moderate to heavy caffeine consumption during pregnancy and relationship to spontaneous abortion and abnormal fetal growth: A meta-analysis. *Reproductive Toxicology, 12,* 435–444.

Ferraro, K. F. (2001). Aging and role transitions. In R. H. Binstock & L. K. George (Eds.), *Handbook of aging and the social sciences* (pp. 313–330). San Diego: Academic Press.

Ficca, G., Fagioli, I., Giganti, F., & Salzarulo, P. (1999). Spontaneous awakenings from sleep in the first year of life. *Early Human Development, 55,* 219–228.

Fichter, M. M., & Quadflieg, N. (1999). Six-year course and outcome of anorexia nervosa. *International Journal of Eating Disorders, 16,* 359–385.

Field, D. (1997). "Looking back, what period of your life brought you the most satisfaction?" *International Journal of Aging and Human Development, 45,* 169–194.

Field, D. (1999). Stability of older women's friendships: A commentary on Roberto. *International Journal of Aging and Human Development, 48,* 81–83.

Field, D., & Millsap, R. E. (1991). Personality in advanced old age: Continuity or change? *Journal of Gerontology, 46,* 299–308.

Field, T. (1998). Maternal depression effects on infants and early interventions. *Preventive Medicine, 27,* 200–203.

Field, T. (2001). Massage therapy facilitates weight gain in preterm infants. *Current Directions in Psychological Science, 10,* 51–54.

Field, T. M. (1994). The effects of mother's physical and emotional unavailability on emotion regulation. In N. A. Fox (Ed.), The development of emotion regulation: Biological and behavioral considerations. *Monographs of the Society for Research in Child Development, 59*(2–3, Serial No. 240).

Field, T. M. (1998). Massage therapy effects. *American Psychologist, 53,* 1270–1281.

Field, T. M., Schanberg, S. M., Scafidi, F., Bauer, C. R., Vega-Lahr, N., Garcia, R. Nystrom, J., & Kuhn, C. M. (1986). Effects of tactile/kinesthetic stimulation on preterm neonates. *Pediatrics, 77,* 654–658.

Field, T. M., Woodson, R., Greenberg, R., & Cohen, D. (1982). Discrimination and imitation of facial expressions by neonates. *Science, 218,* 179–181.

Fiese, B. (1990). Playful relationships: A contextual analysis of mother-toddler interaction and symbolic play. *Child Development, 61,* 1648–1656.

Filene, P. G. (1998). *In the arms of others: A cultural history of the right-to-die in America.* New York: Ivan R. Dee.

Fingerman, K. L. (1998). The good, the bad, and the worrisome: Emotional complexities in grandparents' experiences with individual grandchildren. *Family Relations, 47,* 403–414.

Fins, A. I., & Wohlgemuth, W. K. (2001). Sleep disorders in children and adolescents. In H. Orvaschel & J. Faust (Eds.), *Handbook of conceptualization and treatment of child psychopathology* (pp. 437–448). Amsterdam: Pergamon.

Firestone, R. W. (1994). Psychological defenses against death anxiety. In R. A. Neimeyer (Ed.), *Death anxiety handbook* (pp. 217–241). Washington, DC: Taylor & Francis.

Fisch, S. M., Truglio, R. T., & Cole, C. F. (1999). The impact of *Sesame Street* on preschool children: A review and synthesis of 30 years' research. *Media Psychology, 1,* 165–190.

Fischer, K. W., & Bidell, T. R. (1998). Dynamic development of psychological structures in action and thought. In R. M. Lerner (Ed.), *Handbook of child psychology: Vol. 1. Theoretical models of human development* (5th ed., pp. 467–562). New York: Wiley.

Fischer, K. W., & Rose, S. P. (1995, Fall). Concurrent cycles in the dynamic development of brain and behavior. *SRCD Newsletter,* pp. 3–4, 15–16.

Fisher, C. B. (1993, Winter). Integrating science and ethics in research with high-risk children and youth. *Social Policy Report of the Society for Research in Child Development, 4*(4).

Fisher, D. E. (2001). The p53 tumor suppressor: Critical regulator of life and death in cancer. *Apoptosis, 6,* 7–15.

Fitzgibbons, P. J., & Gordon-Salant, S. (1998). Auditory temporal order perception in younger and older adults. *Journal of Speech, Language, and Hearing Research, 41,* 1052–1060.

Fivaz-Depeursinge, E., & Corboz-Warnery, A. (1999). *The primary triangle: A developmental systems view of mothers, fathers, and infants.* New York: Basic Books.

Flake, A., Roncarolo, M., Puck, J. M., Almeidaporada, G., Evins, M. I., Johnson, M. P., Abella, E. M., Harrison, D. D., & Zanjani, E. D. (1996). Treatment of X-linked severe combined immunodeficiency by in utero transplantation of paternal bone marrow. *New England Journal of Medicine, 335,* 1806–1810.

Flamm, B. L., & Quilligan, E. J. (Eds.). (1995). *Cesarean section: Guidelines for appropriate utilization.* New York: Springer-Verlag.

Flanagan, C. A., & Eccles, J. S. (1993). Changes in parents' work status and adolescents' adjustment at school. *Child Development, 64,* 246–257.

Flanagan, C. A., & Faison, N. (2001). Youth civic development: Implications of research for social policy and programs. *Social Policy Report of the Society for Research in Child Development, 15*(1).

Flanagan, C. A., & Tucker, C. J. (1999). Adolescents' explanations for political issues: Concordance with their views of self and society. *Developmental Psychology, 35,* 1198–1209.

Flannery, K. A., & Liederman, J. (1995). Is there really a syndrome involving the co-occurrence of neurodevelopmental disorder, talent, non-right handedness and immune disorder among children? *Cortex, 31,* 503–515.

Flavell, J. H. (1993). The development of children's understanding of false belief and the appearance–reality distinction. *International Journal of Psychology, 28,* 595–604.

Flavell, J. H. (1999). Cognitive development: Children's knowledge about the mind. *Annual Review of Psychology, 50,* 21–45.

Flavell, J. H. (2000). Development of children's knowledge about the mental world. *International Journal of Behavioral Development, 24,* 15–23.

Flavell, J. H., Flavell, E. R., & Green, F. L. (2001). Development of children's understanding of connections between thinking and feeling. *Psychological Science, 12,* 430–432.

Flavell, J. H., Green, F. L., & Flavell, E. R. (1987). Development of knowledge about the appearance–reality distinction. *Monographs of the Society for Research in Child Development, 51*(1, Serial No. 212).

Flavell, J. H., Green, F. L., & Flavell, E. R. (1993). Children's understanding of the stream of consciousness. *Child Development, 64,* 387–398.

Flavell, J. H., Green, F. L., & Flavell, E. R. (1995). Young children's knowledge about thinking. *Monographs of the Society for Research in Child Development, 60*(1, Serial No. 243).

Flavell, J. H., Miller, P. H., & Miller, S A. (2002). *Cognitive development* (4th ed.). Upper Saddle River, NJ: Prentice-Hall.

Fleming, J. I. (2000). Death, dying, and euthanasia: Australia versus the Northern Territory. *Issues in Law and Medicine, 15,* 291–305.

Flippen, C., & Tienda, M. (2000). Pathways to retirement: Patterns of labor force participation and labor market exit among the pre-retirement population by race, Hispanic origins, and sex. *Journal of Gerontology, 55B,* S14–S27.

Floccia, C., Christophe, A., & Bertoncini, J. (1997). High-amplitude sucking and newborns: The quest for underlying mechanisms. *Journal of Experimental Child Psychology, 64,* 175–198.

Flood, D. G., & Coleman, P. D. (1988). Cell type heterogeneity of changes in dendritic extent in the hippocampal region of the human brain in normal aging and in Alzheimer's disease. In T. L. Petit & G. O. Ivy (Ed.), *Neural plasticity: A lifespan approach* (pp. 265–281). New York: Alan R. Liss.

Florian, V., & Kravetz, S. (1985). Children's concepts of death: A cross-cultural comparison among Muslims, Druze, Christians, and Jews in Israel. *Journal of Cross-Cultural Psychology, 16,* 174–179.

Florian, V., & Mikulincer, M. (1998). Symbolic immortality and the management of the terror of death: The moderating role of attachment style. *Journal of Personality and Social Psychology, 74,* 725–734.

Florian, V., Mikulincer, M., & Taubman, O. (1995). Does hardiness contribute to mental health during a stressful real-life situation? The roles of appraisal and coping. *Journal of Personality and Social Psychology, 68,* 687–695.

Flynn, J. R. (1999). Searching for justice: The discovery of IQ gains over time. *American Psychologist, 54,* 5–20.

Fogel, A. (1993). *Developing through relationships: Origins of communication, self and culture.* New York: Harvester Wheatsheaf.

Foltz, C., Overton, W. F., & Ricco, R. B. (1995). Proof construction: Adolescent development from inductive to deductive problem-solving strategies. *Journal of Experimental Child Psychology, 59,* 179–195.

Fonda, S. J., Clipp, E. C., & Maddox, G. L. (2002). Patterns in functioning among residents of an affordable assisted living housing facility. *Gerontologist, 42,* 178–187.

Ford, J. M., Roth, W. T., Isaacks, B. G., White, P. M., Hood, S. H., & Pfefferbaum, A. (1995). *Biological Psychology, 39,* 57–80.

Forgatch, M. S., Patterson, G. R., & Ray, J. A. (1996). Divorce and boys' adjustment problems: Two paths with a single model. In E. M. Hetherington (Ed.), *Stress, coping, and resiliency in children and the family* (pp. 67–105). Hillsdale, NJ: Erlbaum.

Forsén, T., Eriksson, J., Tuomilehto, J., Reunanen, A., Osmond, C., & Barker, D. (2000). The fetal and childhood growth of persons who develop type 2 diabetes. *Annals of Internal Medicine, 133,* 176–182.

Fotinatos-Ventouratos, R., & Cooper, C. L. (1998). Social class differences and occupational stress. *International Journal of Stress Management, 5,* 211–222.

Fowler, J. W. (1981). *Stages of faith.* San Francisco: Harper & Row.

Fowles, D. C., & Kochanska, G. (2000). Temperament as a moderator of pathways to conscience in children: The contribution of electrodermal activity. *Psychophysiology, 37,* 788–795.

Fox, M., Gibbs, M., & Auerbach, D. (1985). Age and gender dimensions of friendship. *Psychology of Women Quarterly, 9,* 489–501.

Fox, N. A. (1991). If it's not left, it's right: Electroencephalograph asymmetry and the development of emotion. *American Psychologist, 46,* 863–872.

Fox, N. A., Calkins, S. D., & Bell, M. A. (1994). Neural plasticity and development in the first two years of life: Evidence from cognitive and socioemotional domains of research. *Development and Psychopathology, 6,* 677–696.

Fox, N. A., & Davidson, R. J. (1986). Taste-elicited changes in facial signs of emotion and the asymmetry of brain electrical activity in newborn infants. *Neuropsychologia, 24,* 417–422.

Fozard, J. L., & Gordon-Salant, S. (2001). Changes in vision and hearing with aging. In J. E. Birren & K. W. Schaie (Eds.), *Handbook of the psychology of aging* (pp. 241–266). San Diego: Academic Press.

Fozard, J. L., Vercruyssen, M., Reynolds, S. L., Hancocke, P. A., & Quilter, R. E. (1994). Age differences and changes in reaction time: The Baltimore Longitudinal Study of Aging. *Journal of Gerontology, 49,* P179–P189.

Frackiewicz, E. J., & Shiovitz, T. M. (2001). Evaluation and management of premenstrual syndrome and premenstrual dysphoric disorder. *Journal of the American Pharmaceutical Association, 41,* 437–447.

Framo, J. L. (1994). The family life cycle: Impressions. *Contemporary Family Therapy, 16,* 87–117.

Franco, M. C., Danias, A. P., Akamine, E. H., Kawamoto, E. M., Fortes, Z. B., Scavone, C., Tostes, R. C., Carvalho, M. H., & Nigro, D. (2002). Enhanced oxidative stress as a potential mechanism underlying the programming of hypertension in utero. *Journal of Cardiovascular Pharmacology, 40,* 501–509.

Franco, P., Chabanski, S., Szliwowski, H., Dramaiz, M., & Kahn, A. (2000). Influence of maternal smoking on autonomic nervous system in healthy infants. *Pediatric Research, 47,* 215–220.

Franklin, C., & Corcoran, J. (2000). Preventing adolescent pregnancy: A review of programs and practices. *Social Work, 45,* 40–52.

Franklin, C., Grant, D., Corcoran, J., O'Dell-Miller, P., & Bultman, L. (1997). Effectiveness of prevention programs for adolescent pregnancy: A meta-analysis. *Journal of Marriage and the Family, 59,* 551–567.

Franklin, M. (1995). The effects of differential college environments on academic learning and student perceptions of cognitive development. *Research in Higher Education, 36,* 127–153.

Frazier, J. A., & Morrison, F. J. (1998). The influence of extended-year schooling on growth of achievement and perceived competence in early elementary school. *Child Development, 69,* 495–517.

Frazier, L. D. (2002). Perceptions of control over health: Implications for sense of self in healthy and ill older adults. In S. P. Shohov (Ed.), *Advances in psychology research* (Vol. 10, pp. 145–163). Huntington, NY: Nova Science Publishers.

Frederickson, B. L., & Carstensen, L. L. (1990). Relationship classification using grade of membership analysis: A typology of sibling relationships in later life. *Journal of Gerontology, 45,* 343–351.

Frederiksen-Goldsen, K. I., & Sharlach, A. E. (2000). *Families and work: New directions in the twenty-first century.* New York: Oxford University Press.

Freedman Doan, C., Wigfield, A., Eccles, J. S., Blumenfeld, P., Arbreton, A., & Harold, R. D. (2000). What am I best at? Grade and gender differences in children's beliefs about ability improvement. *Journal of Applied Developmental Psychology, 21,* 379–402.

Freeman, D. (1983). *Margaret Mead and Samoa: The making and unmaking of an anthropological myth.* Cambridge, MA: Harvard University Press.

Freeman, E. W., & Halbreich, U. (1998). Premenstrual syndromes. *Psychopharmacological Bulletin, 34,* 291–295.

Freud, S. (1973). *An outline of psychoanalysis.* London: Hogarth. (Original work published 1938)

Freud, S. (1974). *The ego and the id.* London: Hogarth. (Original work published 1923)

Freund, A. M., & Baltes, P. B. (1998). Selection, optimization, and compensation as strategies of life management: Correlations with subjective indicators of successful aging. *Psychology and Aging, 13,* 531–543.

Freund, A. M., & Baltes, P. B. (2000). The orchestration of selection, optimization and compensation: An action-theoretical conceptualization of a theory of developmental regulation. In W. J. Perrig & A. Grob (Eds.), *Control of human behavior, mental processes, and consciousness* (pp. 35–58). Mahwah, NJ: Erlbaum.

Freund, A. M., & Smith, J. (1999). Content and function of the self-definition in old and very old age. *Journal of Gerontology, 54B,* P55–P67.

Frick, J. E., Colombo, J., & Saxon, T. F. (1999). Individual and developmental differences in disengagement of fixation in early infancy. *Child Development, 70,* 537–548.

Fried, P. A. (1993). Prenatal exposure to tobacco and marijuana: Effects during pregnancy, infancy, and early childhood. *Clinical Obstetrics and Gynecology, 36,* 319–337.

Fried, P. A., & Makin, J. E. (1987). Neonatal behavioral correlates of prenatal exposure to marijuana, cigarettes, and alcohol in a low risk population. *Neurobehavioral Toxicology and Teratology, 9,* 1–7.

Fried, P. A., Watkinson, B., & Gray, R. (1999). Growth from birth to early adolescence in offspring prenatally exposed to cigarettes and marijuana. *Neurotoxicology and Teratology, 21,* 513–525.

Friedman, J. M. (1996). *The effects of drugs on the fetus and nursing infant: A handbook for health care professionals.* Baltimore: Johns Hopkins University Press.

Friedman, S. L., & Scholnick, E. K. (1997). An evolving "blueprint" for planning: Psychological requirements, task characteristics, and social–cultural influences. In S. L. Friedman & E. K. Scholnick (Eds.), *The developmental psychology of planning: Why, how, and when do we plan?* (pp. 3–22). Mahwah, NJ: Erlbaum.

Frijda, N. (2000). The psychologist's point of view. In M. Lewis & J. M. Haviland-Jones (Eds.), *Handbook of emotions* (pp. 59–74). New York: Guilford.

Frodi, A. (1985). When empathy fails: Aversive infant crying and child abuse. In B. M. Lester & C. F. Z. Boukydis (Eds.), *Infant crying: Theoretical and research perspectives* (pp. 263–277). New York: Plenum.

Frosch, C. A., Mangelsdorf, S. C., & McHale, J. L. (2000). Marital behavior and security of preschooler–parent attachment relationships. *Journal of Family Psychology, 14,* 144–161.

Fry, C., Dickerson-Putman, J., Draper, P., Ikels, C., Keith, J., Glascock, A. P., & Harpending, H. C. (1997). Culture and the meaning of a good old age. In J. Sokolovsky (Ed.), *The cultural context of aging: Worldwide perspectives* (2nd ed., pp. 99–124). New York: Bergin & Garvey.

Fry, C. L. (1985). Culture, behavior, and aging in the comparative perspective. In J. E. Birren & K. W. Schaie (Eds.), *Handbook of the psychology of aging* (2nd ed., pp. 216–244). New York: Van Nostrand Reinhold.

Fry, P. M. (1995). Individual differences in reminiscence among older adults: Predictors of frequency and pleasantness ratings of reminiscence activity. In J. Hendricks (Ed.), *The meaning of reminiscence and life review* (pp. 83–97). Amityville, NY: Baywood.

Fry, P. M. (2001). Predictors of health-related quality of life perspectives, self-esteem, and life satisfactions of older adults following spousal loss: An 18-month follow-up study of widows and widowers. *Gerontologist, 41,* 787–798.

Fuchs, I., Eisenberg, N., Hertz-Lazarowitz, R., & Sharabany, R. (1986). Kibbutz, Israeli city, and American children's moral reasoning about prosocial moral conflicts. *Merrill-Palmer Quarterly, 32,* 37–50.

Fuligni, A. J. (1997). The academic achievement of adolescents from immigrant families: The roles of family background, attitudes, and behavior. *Child Development, 68,* 261–273.

Fuligni, A. J. (1998a). Authority, autonomy, and parent–adolescent conflict and cohesion: A study of adolescents from Mexican, Chinese, Filipino, and European backgrounds. *Developmental Psychology, 34,* 782–792.

Fuligni, A. J. (1998b). The adjustment of children from immigrant families. *Current Directions in Psychological Science, 7,* 99–103.

Fuligni, A. J., Burton, L., Marshall, S., Perez-Febles, A., Yarrington, J., Kirsh, L.B., & Merriwether-DeVries, C. (1999). Attitudes toward family obligations among American adolescents with Asian, Latin American, and European backgrounds. *Child Development, 70,* 1030–1044.

Fuligni, A. J., & Stevenson, H. W. (1995). Time use and mathematics achievement among American, Chinese, and Japanese high school students. *Child Development, 66,* 830–842.

Fuller, G. F. (2000). Falls in the elderly. *American Family Physician, 61,* 2159–2168.

Fuller-Thomson, E., & Minkler, M. (2000). The mental and physical health of grandmothers who are raising their grandchildren. *Journal of Mental Health and Aging, 6,* 311–323.

Fulton, R. (1995). The contemporary funeral: Functional or dysfunctional? In H. Wass & R. A. Neimeyer (Eds.), *Dying: Facing the facts* (3rd ed., pp. 185–209). Washington, DC: Taylor and Francis.

Fung, H. H., Carstensen, L. L., & Lang, F. R. (2001). Age-related patterns in social networks among European Americans and African Americans: Implications for socioemotional selectivity across the life span. *International Journal of Aging and Human Development, 52,* 185–206.

Fung, H. H., Carstensen, L. L., & Lutz, A. (1999). The influence of time on social preferences: Implications for life-span development. *Psychology and Aging, 14,* 595–604.

Fung, H. H., Lai, P., & Ng, R. (2001). Age differences in social preferences among Taiwanese and mainland Chinese: The role of perceived time. *Psychology and Aging, 16,* 351–356.

Furman, E. (1990, November). Plant a potato—learn about life (and death). *Young Children, 46*(1), 15–20.

Furman, W. (2002). The emerging field of adolescent romantic relationships. *Current Directions in Psychological Science, 11,* 177–180.

Furman, W., & Buhrmester, D. (1992). Age and sex differences in perceptions of networks of personal relationships. *Child Development, 63,* 103–115.

Furstenberg, F. F., Jr., Brooks-Gunn, J., & Morgan, S. P. (1987). *Adolescent mothers and their children in later life.* Cambridge, U.K.: Cambridge University Press.

Furstenberg, F. F., Jr., & Harris, K. M. (1993). When and why fathers matter: Impact of father involvement on children of adolescent mothers. In R. I. Lerman & T. J. Ooms (Eds.), *Young unwed fathers* (pp. 117–138). Philadelphia: Temple University Press.

Fuson, K. C. (1992). Research on learning and teaching addition and subtraction of whole numbers. In G. Leinhardt, R. T. Putnam, & R. A. Hattrup (Eds.), *The analysis of arithmetic for mathematics teaching* (pp. 53–187). Hillsdale, NJ: Erlbaum.

Gabbay, S. G., & Wahler, J. J. (2002). Lesbian aging: Review of a growing literature. *Journal of Gay and Lesbian Social Services, 14,* 1–21.

Gabrel, C. S. (2000). *Advance data from Vital and Health Statistics of the Centers for Disease Control and Prevention.* Washington, DC: U.S. Department of Health and Human Services.

Gaddis, A., & Brooks-Gunn, J. (1985). The male experience of pubertal change. *Journal of Youth and Adolescence, 14,* 61–69.

Galambos, N. L., Almeida, D. M., & Petersen, A. C. (1990). Masculinity, femininity, and sex role attitudes in early adolescence: Exploring gender intensification. *Child Development, 61,* 1905–1914.

Galambos, N. L., & Maggs, J. L. (1991). Children in self-care: Figures, facts, and fiction. In J. V. Lerner & N. L. Galambos (Eds.), *Employed mothers and their children* (pp. 131–157). New York: Garland.

Gale, G., & VandenBerg, K. A. (1998). Kangaroo care. *Neonatal Network, 17*(5), 69–71.

Gallard, B. C., Taylor, B. J., & Bolton, D. P. (2002). Prone versus supine sleep position: A review of the physiological studies in SIDS research. *Journal of Paediatric Child Health, 38,* 332–338.

Galler, J. R., Ramsey, C. F., Morley, D. S., Archer, E., & Salt, P. (1990). The long-term effects of early kwashiorkor compared with marasmus. IV. Performance on the National High School Entrance Examination. *Pediatric Research, 28,* 235–239.

Galler, J. R., Ramsey, F., & Solimano, G. (1985a). A follow-up study of the effects of early malnutrition on subsequent development: I. Physical growth and sexual maturation during adolescence. *Pediatric Research, 19,* 518–523.

Galler, J. R., Ramsey, F., & Solimano, G. (1985b). A follow-up study of the effects of early malnutrition on subsequent development: II. Fine motor skills in adolescence. *Pediatric Research, 19,* 524–527.

Gallup Canada. (1997, June). The Gallup Poll—Canadians voice their opinions on doctor-assisted suicide. Retrieved from http://www.web.apc.org/dwd/index.html

Galotti, K. M., Kozberg, S. F., & Farmer, M. C. (1991). Gender and developmental differences in adolescents' conceptions of moral reasoning. *Journal of Youth and Adolescence, 20,* 13–30.

Gannon, L., & Ekstrom, B. (1993). Attitudes toward menopause: The influence of sociocultural paradigms. *Psychology of Women Quarterly, 17,* 275–288.

Gannon, S., & Korn, S. J. (1983). Temperament, cultural variation, and behavior disorder in preschool children. *Child Psychiatry and Human Development, 13,* 203–212.

Ganong, L. H., & Coleman, M. (1992). Gender differences in expectations of self and future partner. *Journal of Family Issues, 13,* 55–64.

Ganong, L. H., & Coleman, M. (1994). *Remarried family relationships.* Thousand Oaks, CA: Sage.

Ganong, L. H., & Coleman, M. (1998). Attitudes regarding filial responsibilities to help elderly divorced parents and stepparents. *Journal of Aging Studies, 12,* 271–290.

Ganong, L. H., & Coleman, M. (2000). Remarried families. In C. Hendrick & S. S. Hendrick (Eds.), *Close relationships* (pp. 155–168). Thousand Oaks, CA: Sage.

Ganong, L. H., Coleman, M., & Fine, M. (1995). Remarriage and stepfamilies. In W. Burr & R. Day (Eds.), *Advanced family science.* Provo, UT: Brigham Young University Press.

Ganong, L., Coleman, M., Fine, M., & Martin, P. (1999). Stepparents' affinity-seeking and affinity-maintaining strategies with stepchildren. *Journal of Family Issues, 20,* 299–327.

Ganzini, L., Harvath, T. A., Jackson, A., Goy, E. R., Miller, L. L., & Delorit, M. A. (2002). Experiences of Oregon nurses and social workers with hospice patients who requested assistance with suicide. *New England Journal of Medicine, 347,* 582–588.

Gao, G. (2001). Intimacy, passion, and commitment in Chinese and U.S. American romantic relationships. *International Journal of Intercultural Relations, 25,* 329–342.

Garbarino, J., & Kostelny, K. (1993). Neighborhood and community influences on parenting. In T. Luster & L. Okagaki (Eds.), *Parenting: An ecological perspective* (pp. 203–226). Hillsdale, NJ: Erlbaum.

Garbarino, J., Andreas, J. B., & Vorrasi, J. A. (2002). Beyond the body count: Moderating the effects of war on children's long-term adaptation. In F. Jacobs, D. Wertlieb, & R. M. Lerner (Eds.), *Handbook of developmental science* (Vol. 2, pp. 137–158). Thousand Oaks, CA: Sage.

Garber, J., Braafladt, N., & Weiss, B. (1995). Affect regulation in depressed and nondepressed children and young adolescents. *Development and Psychopathology, 7,* 93–115.

Garber, J., Quiggle, N., Panak, W., & Dodge, K. (1991). Aggression and depression in children: Comorbidity, specificity, and social cognitive processing. In D. Cicchetti & S. L. Toth (Eds.), *Rochester Symposium on Developmental Psychopathology: Vol. 2. Internalizing and externalizing expressions of dysfunction* (pp. 225–264). Hillsdale, NJ: Erlbaum.

Garcia, M. M., Shaw, D. S., Winslow, E. B., & Yaggi, K. E. (2000). Destructive sibling conflict and the development of conduct problems in young boys. *Developmental Psychology, 36,* 44–53.

García-Coll, C., & Magnuson, K. (1997). The psychological experience of immigration: A developmental perspective. In A. Booth, A. C. Crouter, & N. Landale (Eds.), *Immigration and the family* (pp. 91–131). Mahwah, NJ: Erlbaum.

García-Coll, C., & Pachter, L. M. (2002). Ethnic and minority parenting. In M. H. Bornstein (Ed.), *Handbook of parenting* (2nd ed., Vol. 4, pp. 1–20). Mahwah, NJ: Erlbuam.

Gardner, E. (2001). Making every day count: Hospice palliative care in Canada. *Transition Magazine, 31*(1). Retrieved from http://www.vifamily.ca/tm/311/1.htm

Gardner, H. (1980). *Artful scribbles: The significance of children's drawings.* New York: Basic Books.

Gardner, H. (1983). *Frames of mind: The theory of multiple intelligences.* New York: Basic Books.

Gardner, H. (1993). *Multiple intelligences: The theory in practice.* New York: Basic Books.

Gardner, H. E. (1998a). Are there additional intelligences? The case of the naturalist, spiritual, and existential intelligences. In J. Kane (Ed.), *Educational information and transformation.* Upper Saddle River, NJ: Prentice-Hall.

Gardner, H. E. (1998b). Extraordinary cognitive achievements (ECA): A symbol systems approach. In R. M. Lerner (Ed.), *Handbook of child psychology: Vol. 1. Theoretical models of human development* (5th ed., pp. 415–466). New York: Wiley.

Gardner, H. E. (2000). *Intelligence reframed: Multiple intelligences for the twenty-first century.* New York: Basic Books.

Garfinkel, I., & McLanahan, S. (1995). The effects of child support reform on child well-being. In P. L. Chase-Lansdale & J. Brooks-Gunn (Eds.), *Escape from poverty: What makes a difference for children?* (pp. 211–238). New York: Cambridge University Press.

Garmezy, N. (1993). Children in poverty: Resilience despite risk. *Psychiatry, 56,* 127–136.

Garmon, L. C., Basinger, K. S., Gregg, V. R., & Gibbs, J. C. (1996). Gender differences in stage and expression of moral judgment. *Merrill-Palmer Quarterly, 42,* 418–437.

Garner, D. M., & Garfinkel, P. E. (Eds.). (1997). *Handbook of treatment for eating disorders* (2nd ed.). New York: Guilford.

Garnier, H. E., Stein, J. A., & Jacobs, J. K. (1997). The process of dropping out of high school: A 19-year perspective. *American Educational Research Journal, 34,* 395–410.

Garrett, P., Ng'andu, N., & Ferron, J. (1994). Poverty experiences of young children and the quality of their home environments. *Child Development, 65,* 331–345.

Gasden, V. (1999). Black families in intergenerational and cultural perspective. In M. E. Lamb (Ed.), *Parenting and child development in "nontraditional" families* (pp. 221–246). Mahwah, NJ: Erlbaum.

Gaskins, S. (1999). Children's daily lives in a Mayan village: A case study of culturally constructed roles and activities. In R. Göncü (Ed.), *Children's engagement in the world: Sociocultural perspectives* (pp. 25–61). Cambridge, U.K.: Cambridge University Press.

Gaskins, S. (2000). Children's daily activities in a Mayan village: A culturally grounded description. *Cross-Cultural Research, 34,* 375–389.

Gathercole, S. E. (1998). The development of memory. *Journal of Child Psychology and Psychiatry, 39,* 3–27.

Gathercole, S. E., Adams, A.-M., & Hitch, G. (1994). Do young children rehearse? An individual-differences analysis. *Memory and Cognition, 22,* 201–207.

Gatz, M., Bengtson, V. L., & Blum, M. J. (1990). Caregiving families. In J. E. Birren & K. W. Schaie (Eds.), *Handbook of the psychology of aging* (3rd ed., pp. 404–426). San Diego, CA: Academic Press.

Gatz, M., Kasl-Godley, J. E., & Karel, M. J. (1996). Aging and mental disorders. In J. E. Birren & K. W. Schaie (Eds.), *Handbook of the psychology of aging* (pp. 365–382). Sand Diego: Academic Press.

Gaub, M., & Carlson, C. L. (1997). Gender differences in ADHD: A meta-analysis and critical review. *Journal of the American Academy of Child and Adolescent Psychiatry, 36,* 1036–1045.

Gauvain, M. (1999). Cognitive development in social and cultural context. *Current Directions in Psychological Science, 7,* 188–192.

Gauvain, M., & Huard, R. D. (1999). Family interaction, parenting style, and the development of planning: A longitudinal analysis using archival data. *Journal of Family Psychology, 13,* 75–92.

Gauvain, M., & Rogoff, B. (1989). Ways of speaking about space: The development of children's skill in communicating spatial knowledge. *Cognitive Development, 4,* 295–307.

Gayle, B. M., Preiss, R. W., & Allen, M. (2002). A meta-analytic interpretation of nonintimate interpersonal conflict. In M. Allen & R. W. Preiss (Eds.), *Interpersonal communication research: Advances through meta-analysis* (pp. 345–368). Mahwah, NJ: Erlbaum.

Gayle, H. (2000). An overview of the global HIV/AIDS epidemic, with a focus on the United States. *AIDS 2000, 14*(Suppl. 2), S8–S17.

Gazmararian, J. A., Parker, R. M., & Baker, D. W. (1999). Reading skills and family planning knowledge and practices in a low-income managed-care population. *Obstetrics and Gynecology, 93,* 239–244.

Ge, X., Conger, R. D., & Elder, G. H., Jr. (1996). Coming of age too early: Pubertal influences on girls' vulnerability to psychological distress. *Child Development, 67,* 3386–3400.

Ge, X., Conger, R. D., & Elder, G. H., Jr. (2001). The relation between puberty and psychological distress in adolescent boys. *Journal of Research on Adolescence, 11,* 49–70.

Geary, D. C. (1994). *Children's mathematical development.* Washington, DC: American Psychological Association.

Geary, D. C. (1995). *Children's mathematical development: Research and practical applications.* Washington, DC: American Psychological Association.

Geary, D. C. (1996). International differences in mathematics achievement: The nature, causes, and consequences. *Current Directions in Psychological Science, 5,* 133–137.

Geary, D. C. (1998). *Male, female: The evolution of human sex differences.* Washington, DC: American Psychological Association.

Geary, D. C. (1999). Evolution and developmental sex differences. *Current Directions in Psychological Science, 8,* 115–120.

Geary, D. C., & Bjorklund, D. F. (2000). Evolutionary developmental psychology. *Child Development, 71,* 57–65.

Geary, D. C., Bow-Thomas, C. C., Liu, F., & Siegler, R. S. (1996). Development of arithmetical competencies in Chinese and American children: Influence of age, language, and schooling. *Child Development, 67,* 2022–2044.

Geerlings, S. W., Beekman, A. T. F., Deeg, D. J. H., Twisk, J. W. R., & van Tilburg, W. (2001). The longitudinal effect of depression on functional limitations and disability in older adults: An eight-wave prospective community-based study. *Psychological Medicine, 31,* 1361–1371.

Gelbaugh, S., Ramos, M., Soucar, E., & Urena, R. (2001). Therapy for anorexia nervosa. *Journal of the American Academy of Child and Adolescent Psychiatry, 40,* 129–130.

Gelfand, D. E. (1994). *Aging and ethnicity.* New York: Springer.

Gelfand, M. M. (2000). Sexuality among older women. *Journal of Women's Health and Gender-Based Medicine, 9*(Suppl.1), S15–S20.

Gellatly, A. R. H. (1987). Acquisition of a concept of logical necessity. *Human Development, 30,* 32–47.

Gelles, R. J. (1998). The youngest victims. Violence toward children. In R. Bergen & R. Kennedy (Eds.), *Issues in intimate violence* (pp. 5–24). Thousand Oaks, CA: Sage.

Gellin, B. G., Maibach, E. W., & Marcuse, E. K. (2000). Do parents understand immunizations? A national telephone survey. *Pediatrics, 106,* 1097–1102.

Gelman, R. (1972). Logical capacity of very young children: Number invariance rules. *Child Development, 43,* 75–90.

Gelman, R., & Shatz, M. (1978). Appropriate speech adjustments: The operation of conversational constraints on talk to two-year-olds. In M. Lewis & L. A. Rosenblum (Eds.), *Interaction, conversation, and the development of language* (pp. 27–61). New York: Wiley.

Gelman, S. A., Coley, J. D., Rosengren, K. S., Hartman, E., & Pappas, A. (1998). Beyond labeling: The role of maternal input in the acquisition of richly structured categories. *Monographs of the Society for Research in Child Development, 63*(1, Serial No. 253).

Genessee, F. (2001). Portrait of the bilingual child. In V. Cook (Ed.), *Portraits of the second language user* (pp. 170–196). Clevedon, U.K.: Multilingual Matters.

George, C., Kaplan, N., & Main, M. (1985). *The Adult Attachment Interview.* Unpublished manuscript, University of California at Berkeley.

George, S. A. (2002). The menopause experience: A woman's perspective. *Journal of Obstetric, Gynecologic, and Neonatal Nursing, 31,* 71–85.

Gershoff, E. T. (2002). Corporal punishment by parents and associated child behaviors and experiences: A meta-analytic and theoretical review. *Psychological Bulletin, 128,* 539–579.

Gervai, J., Turner, P. J., & Hinde, R. A. (1995). Gender-related behaviour, attitudes, and personality in parents of young children in England and Hungary. *International Journal of Behavioral Development, 18,* 105–126.

Geschwind, D. H., Boone, K. B., Miller, B. L., & Swerdloff, R. S. (2000). Neurobehavioral phenotype of Klinefelter syndrome. *Mental Retardation and Developmental Disabilities Research Reviews, 6,* 107–116.

Gesell, A. (1933). Maturation and patterning of behavior. In C. Murchison (Ed.), *A handbook of child psychology.* Worcester, MA: Clark University Press.

Gesell, A., & Ilg, F. L. (1949a). The child from five to ten. In A. Gesell & F. Ilg (Eds.), *Child development* (pp. 394–454). New York: Harper & Row. (Original work published 1946)

Gesell, A., & Ilg, F. L. (1949b). The infant and child in the culture of today. In A. Gesell & F. Ilg (Eds.), *Child development* (pp. 1–393). New York: Harper & Row. (Original work published 1943)

Gest, S. D., Graham-Bermann, S. A., & Hartup, W. W. (2001). Peer experience: Common and unique features of number of friendships, social network, centrality, and socioeconomic status. *Social Development, 10,* 23–40.

Getchell, N., & Roberton, M. A. (1989). Whole body stiffness as a function of developmental level in children's hopping. *Developmental Psychology, 25,* 920–928.

Ghim, H. R. (1990). Evidence for perceptual organization in infants: Perception of subjective contours by young infants. *Infant Behavior and Development, 13,* 221–248.

Giarrusso, R., Feng, D., Silverstein, M., & Bengtson, V. L. (2001). Grandparent–adult grandchild affection and consensus. *Journal of Family Issues, 22,* 456–477.

Giarrusso, R., Feng, D., Silverstein, M., & Marenco, A. (2000). Primary and secondary stressors of grandparents raising grandchildren: Evidence from a national survey. *Journal of Mental Health and Aging, 6,* 291–310.

Gibbons, A. (1998). Which of our genes make us human? *Science, 281,* 1432–1434.

Gibbs, J. C. (1995). The cognitive developmental perspective. In W. M. Kurtines & J. L. Gewirtz (Eds.), *Moral development: An introduction* (pp. 27–48). Boston: Allyn and Bacon.

Gibbs, J. C. (2003). *Moral development and reality: Beyond the theories of Kohlberg and Hoffman.* Thousand Oaks, CA: Sage.

Gibson, E. J. (1970). The development of perception as an adaptive process. *American Scientist, 58,* 98–107.

Gibson, E. J. (2000). Perceptual learning in development: Some basic concepts. *Ecological Psychology, 12,* 295–302.

Gibson, E. J., & Walk, R. D. (1960). The "visual cliff." *Scientific American, 202,* 64–71.

Gibson, F. L., Ungerer, J. A., McMahon, C. A., Leslie, G. I., & Saunders, D. M. (2000). The mother–child relationship following in vitro fertilization (IVF): Infant attachment, responsivity, and maternal sensitivity. *Journal of Child Psychology and Psychiatry, 41,* 1015–1023.

Gibson, J. J. (1979). *The ecological approach to visual perception.* Boston: Houghton Mifflin.

Gibson, R. C., & Burns, C. J. (1991, Winter). The health, labor force, and retirement experiences of aging minorities. *Generations, 15*(4), 31–35.

Gibson, R. C., & Jackson, J. S. (1992). The black oldest old: Health, functioning, and informal support. In R. M. Suzman, D. P. Willis, & K. G. Manton (Eds.), *The oldest old* (pp. 321–340). New York: Oxford University Press.

Giedd, J. N., Blumenthal, J., Molloy, E., & Castellanos, F. X. (2001). Brain imaging of attention deficit/hyperactivity disorder: Brain mechanisms and life outcomes. *Annals of the New York Academy of Sciences* (Vol. 931, pp. 33–49). New York: New York Academy of Sciences.

Gilbert, L. A., & Brownson, C. (1998). Current perspectives on women's multiple roles. *Journal of Career Assessment, 6,* 433–448.

Gilbert-Barness, E. (2000). Maternal caffeine and its effect on the fetus. *American Journal of Medical Genetics, 93,* 253.

Gillies, R. M., & Ashman, A. F. (1996). Teaching collaborative skills to primary school children in classroom-based workgroups. *Learning and Instruction, 6,* 187–200.

Gillies, R. M., & Ashman, A. F. (1998). Behavior and interactions of children in cooperative groups in lower and middle elementary school grades. *Journal of Educational Psychology, 90,* 746–757.

Gilligan, C. F. (1982). *In a different voice.* Cambridge, MA: Harvard University Press.

Gillman, M. W., Rifas-Shiman, S. L., Frazier, A. L., Rockett, H. R. H., Camargo, C. A., Jr., Field, A. E., Berkey, C. S., & Colditz, G. A. (2000). Family dinner and diet quality among older children and adolescents. *Archives of Family Medicine, 9,* 235–240.

Gillmore, M. R., Hawkins, J. D., Day, L. E., & Catalano, R. F. (1997). Friendship and deviance: New evidence on an old controversy. *Journal of Early Adolescence, 16.*

Gilvarry, E. (2000). Substance abuse in young people. *Journal of Child Psychology and Psychiatry, 41,* 55–80.

Ginsburg, H. P. (1997). *Entering the child's mind: The clinical interview in psychological research and practice.* New York: Cambridge University Press.

Ginsburg, H. P., Klein, A., & Starkey, P. (1998). The development of children's mathematical thinking: Connecting research with practice. In I. E. Sigel & K. A. Renninger (Eds.), *Handbook of child psychology: Vol. 4. Cognition, perception, and language* (5th ed., pp. 401–476). New York: Wiley.

Ginzberg, E. (1972). Toward a theory of occupational choice: A restatement. *Vocational Guidance Quarterly, 20,* 169–176.

Ginzberg, E. (1988). Toward a theory of occupational choice. *Career Development Quarterly, 36,* 358–363.

Giusti, R. M., Iwamoto, K., & Hatch, E. E. (1995). Diethylstilbestrol revisited: A review of the long-term health effects. *Annals of Internal Medicine, 122,* 778–788.

Gjerde, P. F. (1995). Alternative pathways to chronic depressive symptoms in young adults: Gender differences in developmental trajectories. *Child Development, 66,* 1277–1300.

Gladwell, M. (1998, February 2). The Pima paradox. *The New Yorker,* pp. 44–57.

Glasgow, K. L., Dornbusch, S. M., Troyer, L., Steinberg, L., & Ritter, P. L. (1997). Parenting styles, adolescents' attributions, and educational outcomes in nine heterogeneous high schools. *Child Development, 68,* 507–523.

Gleitman, L. R., & Newport, E. (1996). *The invention of language by children.* Cambridge, MA: MIT Press.

Glendenning, F. (1993). What is elder abuse and neglect? In P. Decalmer & F. Glendenning (Eds.), *The mistreatment of elderly people* (pp. 1–34). London: Sage.

Globalist. (2003, February). What is the male–female ratio like in the working world? Retrieved from http://www.theglobalist.com/DBWeb/StoryId.aspx?StoryId=1201

Glosten, B. (1998). Controversies in obstetric anesthesia. *Anesthesia and Analgesia, 428*(Suppl. 32–8), 32–38.

Gluhoski, V. L., & Wortman, C. B. (1996). The impact of trauma on world views. *Journal of Social and Clinical Psychology, 15,* 417–429.

Gnepp, J. (1983). Children's social sensitivity: Inferring emotions from conflicting cues. *Developmental Psychology, 19,* 805–814.

Godfrey, K. M., & Barker, D. J. (2000). Fetal nutrition and adult disease. *American Journal of Clinical Nutrition, 71,* 1344S–1352S.

Godfrey, K. M., & Barker, D. J. (2001). Fetal programming and adult health. *Public Health Nutrition, 4,* 611–624.

Godkin, M., Krant, M., & Doster, N. (1984). The impact of hospice care on families. *International Journal of Psychiatry in Medicine, 13,* 153–165.

Goelman, H., Doherty, G., Lero, D., LaGrange, A., & Tougas, J. (2000). *You bet I care! Caring and learning environments: Quality in child care centers across Canada.* Guelph, Ontario: Centre for Families, Work and Well-Being, University of Guelph.

Goering, J. (Ed.). 2003. *Choosing a better life? How public housing tenants selected a HUD experiment to improve their lives and those of their children: The Moving to Opportunity Demonstration Program.* Washington, DC: Urban Institute Press.

Gold, D. T. (1996). Continuities and discontinuities in sibling relationships across the life span. In V. L. Bengtson (Ed.), *Adulthood and aging: Research on continuities and discontinuities* (pp. 228–243). New York: Springer.

Gold, P. E., Cahill, L., & Wenk, G. L. (2002). Ginkgo biloba: A cognitive enhancer? *Psychological Science in the Public Interest, 3,* 3–11.

Goldberg, A. P., Dengel, D. R., & Hagberg, J. M. (1996). Exercise physiology and aging. In E. L. Schneider & J. W. Rowe (Eds.), *Handbook of the biology of aging* (pp. 331–354). San Diego: Academic Press.

Goldfield, B. A. (1987). The contributions of child and caregiver to referential and expressive language. *Applied Psycholinguistics, 8,* 267–280.

Goldman, N., & Takahashi, S. (1996). Old-age mortality in Japan: Demographic and epidemiological perspectives. In G. Caselli & A. D. Lopez (Eds.), *Health and mortality among elderly populations* (pp. 157–181). New York: Oxford University Press.

Goldscheider, F., & Goldscheider, C. (1999). *The changing transition to adulthood: Leaving and returning home.* Thousand Oaks, CA: Sage.

Goldsmith, H. H., Lemery, K. S., Buss, K. A., & Campos, J. J. (1999). Genetic analyses of focal aspects of infant temperament. *Developmental Psychology, 35,* 972–985.

Goldsmith, L. T. (2000). Tracking trajectories of talent: Child prodigies growing up. In R. C. Friedman & B. M. Shore (Eds.), *Talents unfolding: Cognition and development* (pp. 89–122). Washington, DC: American Psychological Association.

Goldstein, E. (1979). Effect of same-sex and cross-sex role models on the subsequent academic productivity of scholars. *American Psychologist, 34,* 407–410.

Golomb, C. (1992). *The child's creation of a pictorial world.* Berkeley: University of California Press.

Golombok, S., MacCallum, F., & Goodman, E. (2001). The "test-tube" generation: Parent–child relationships and the psychological well-being of in vitro fertilization children at adolescence. *Child Development, 72,* 599–608.

Golombok, S., & Tasker, F. L. (1996). Do parents influence the sexual orientation of their children? Findings from a longitudinal study of lesbian families. *Developmental Psychology, 32,* 3–11.

Golub, M. S. (1996). Labor analgesia and infant brain development. *Pharmacology, Biochemistry and Behavior, 55,* 619–628.

Gomez-Schwartz, B., Horowitz, J. M., & Cardarelli, A. P. (1990). *Child sexual abuse: Initial effects.* Newbury Park, CA: Sage.

Göncü, A. (1993). Development of intersubjectivity in the dyadic play of preschoolers. *Early Childhood Research Quarterly, 8,* 99–116.

Gonzales, N. A., Cauce, A. M., Friedman, R. J., & Mason, C. A. (1996). Family, peer, and neighborhood influences on academic achievement among African-American adolescents: One-year prospective effects. *American Journal of Community Psychology, 24,* 365–387.

Good, T. L., & Brophy, J. E. (1996). *Looking in classrooms* (7th ed.). New York: Addison-Wesley.

Goodlet, C. R., & Johnson, T. B. (1999). Temporal windows of vulnerability within the third trimester equivalent: Why "knowing when" matters. In J. H. Hannigan, L. P. Spear, N. P. Spear, & C. R. Goodlet (Eds.), *Alcohol and alcoholism: Effects on brain and development* (pp. 59–91). Mahwah, NJ: Erlbaum.

Goodlin-Jones, B. L., Burnham, M. M., & Anders, T. F. (2000). Sleep and sleep disturbances: Regulatory processes in infancy. In A. J. Sameroff, M. Lewis, & S. M. Miller (Eds.), *Handbook of developmental psychology* (2nd ed., pp. 309–325). New York: Kluwer.

Goodman, C. (1999). Intimacy and autonomy in long-term marriage. *Journal of Gerontological Social Work, 32,* 83–97.

Goodman, G. S., Quas, J. A., Bulkley, J., & Shapiro, C. (1999). Innovations for child witnesses: A national survey. *Psychology, Public Policy, and Law, 5,* 255, 281.

Goodman, K. S. (1986). *What's whole in whole language?* Portsmouth, NH: Heinemann.

Goodman, L. A., Koss, M. P., & Russo, N. F. (1993). Violence against women: Physical and mental health effects. *Applied and Preventive Psychology, 2,* 79–89.

Goodman, S. H., Brogan, D., Lynch, M. E., & Fielding, B. (1993). Social and emotional competence in children of depressed mothers. *Child Development, 64,* 516–531.

Goodwin, D. W. (1997). Alcohol: Clinical aspects. In J. H. Lowinson, P. Ruiz, & R. B. Millman (Eds.), *Substance abuse* (3rd ed.). Baltimore: Lippincott, Williams & Wilkins.

Gopnik, A., & Meltzoff, A. N. (1986). Relations between semantic and cognitive development in the one-word stage: The specificity hypothesis. *Child Development, 57,* 1040–1053.

Gopnik, A., & Meltzoff, A. N. (1987). The development of categorization in the second year and its relation to other cognitive and linguistic developments. *Child Development, 58,* 1523–1531.

Gormally, S., Barr, R G., Wertheim, L., Alkawaf, R., Calinoiu, N., & Young, S. N. (2001). Contact and nutrient caregiving effects on newborn infant pain responses. *Developmental Medicine and Child Neurology, 43,* 28–38.

Gortmaker, S. L., Must, A., Perrin, J. M., Sobol, A. M., & Dietz, W.

H. (1993). Social and economic consequences of overweight in adolescence and young adulthood. *New England Journal of Medicine, 329,* 1008–1012.

Gortmaker, S. L., Must, A., Sobol, A. M., Peterson, K., Colditz, G. A., & Dietz, W. H. (1996). Television viewing as a cause of increasing obesity among children in the United States, 1986–1990. *Archives of Pediatric and Adolescent Medicine, 150,* 356–362.

Goss, D. A., & Rainey, B. B. (1998). Relation of childhood myopia progression rates to time of year. *Journal of the American Optometric Association, 69,* 262–266.

Goswami, U. (1996). Analogical reasoning and cognitive development. In H. Reese (Ed.), *Advances in child development and behavior* (Vol. 26, pp. 91–138). New York: Academic Press.

Gott, V. L. (1998). Antoine Marfan and his syndrome: One hundred years later. *Maryland Medical Journal, 47,* 247–252.

Gottesman, I. I. (1963). Genetic aspects of intelligent behavior. In N. Ellis (Ed.), *Handbook of mental deficiency* (pp. 253–296). New York: McGraw-Hill.

Gottesman, I. I. (1991). *Schizophrenia genetics: The origins of madness.* New York: Freeman.

Gottfredson, L. S. (1996). Godfredson's theory of circumscription and compromise. In D. Brown & L. Brooks (Eds.), *Career choice and development* (3rd ed.). San Francisco: Jossey-Bass.

Gottfried, A. E., Gottfried, A. W., & Bathurst, K. (2002). Maternal and dual-earner employment status and parenting. In M. H. Bornstein (Ed.), *Handbook of parenting. Vol. 2: Biology and ecology of parenting* (2nd ed., pp. 207–229). Mahwah, NJ: Erlbaum.

Gottlieb, G. (1998). Normally occurring environmental and behavioral influences on gene activity: From central dogma to probabilistic epigenesis. *Psychological Review, 105,* 792–802.

Gottlieb, G. (2000). Environmental and behavioral influences on gene activity. *Current Directions in Psychological Science, 9,* 93–97.

Gottlieb, G. (2002). *Individual development and evolution: The genesis of novel behavior.* New York: Oxford University Press.

Gottman, J., Coan, J., Carrere, S., & Swanson, C. (1998). Predicting marital happiness and stability from newlywed interactions. *Journal of Marriage and the Family, 60,* 5–22.

Gottman, J. M., & Levenson, R. W. (2000). The timing of divorce: Predicting when a couple will divorce over a 14-year period. *Journal of Marriage and the Family, 62,* 737–745.

Gould, E., Reeves, A. J., Graziano, M. S. A., & Gross, C. G. (1999). Neurogenesis in the neocortex of adult primates. *Science, 286,* 548–552.

Gould, J. L., & Keeton, W. T. (1997). *Biological science.* New York: Norton.

Gould, M. S., & Kramer, R. A. (2001). Youth suicide prevention. *Suicide and Life-Threatening Behavior, 31,* 6–31.

Goya, R. G., & Bolognani, F. (1999). Homeostatis, thymic hormones, and aging. *Gerontology, 45,* 174–178.

Graber, J. A., & Brooks-Gunn, J. (1996). Expectations for and precursors to leaving home in young women. In J. A. Graber & J. S. Dubas (Eds.), *New directions for child development* (No. 71, pp. 21–38). San Francisco: Jossey-Bass.

Graber, J. A., Petersen, A. C., & Brooks-Gunn, J. (1996). Pubertal processes: Methods, measures, and models. In J. A. Graber, J. Brooks-Gunn, & A. C. Petersen (Eds.), *Transitions through adolescence* (pp. 23–53). Mahwah, NJ: Erlbaum.

Grady, C. L., & Craik, F. I M. (2000). Changes in memory processing with age. *Current Opinion in Neurobiology, 10,* 224–231.

Gralinski, J. H., & Kopp, C. B. (1993). Everyday rules for behavior: Mothers' requests to young children. *Developmental Psychology, 29,* 573–584.

Granot, M., Spitzer, A., Aroian, K. J., Ravid, C., Tamir, B., & Noam, R. (1996). Pregnancy and delivery practices and beliefs of Ethiopian immigrant women in Israel. *Western Journal of Nursing Research, 18,* 299–313.

Grantham-McGregor, S., Powell, C., Walker, S., Chang, S., & Fletcher, P. (1994). The long-term follow-up of severely malnourished children who participated in an intervention program. *Child Development, 65,* 428–439.

Grantham-McGregor, S., Schofield, W., & Powell, C. (1987). Development of severely malnourished children who received psychosocial stimulation: Six-year follow-up. *Pediatrics, 79,* 247–254.

Grantham-McGregor, S. M., Walker, S. P., & Chang, S. (2000). Nutritional deficiencies and later behavioral development. *Proceedings of the Nutrition Society, 59,* 47–54.

Graue, M. E., & DiPerna, J. (2000). Redshirting and early retention: Who gets the "gift of time" and what are its outcomes? *American Educational Research Journal, 37,* 509–534.

Gray, J. H., & Citera, M. (1996). *A meta-analytical review of the relation between work attitudes and retirement attitudes.* Paper presented at the annual meeting of the Society for Industrial and Organizational Psychology, San Diego, CA.

Gray, M. R., & Steinberg, L. (1999). Unpacking authoritative parenting: Reassessing a multidimensional construct. *Journal of Marriage and the Family, 61,* 574–587.

Gray-Little, B., & Carels, R. (1997). The effects of racial and socioeconomic consonance on self-esteem and achievement in elementary, junior high, and high school students. *Journal of Research on Adolescence, 7,* 109–131.

Gray-Little, B., & Hafdahl, A. R. (2000). Factors influencing racial comparisons of self-esteem: A quantitative review. *Psychological Bulletin, 126,* 26–54.

Green, G. E., Irwin, J. R., & Gustafson, G. E. (2000). Acoustic cry analysis, neonatal status and long term developmental outcomes. In R. G. Barr, B. Hopkins, & J. A. Green (Eds.), *Crying as a sign, a symptom, and a signal* (pp. 137–156). Cambridge, U.K.: Cambridge University Press.

Greenberger, E., Chen, C., Tally, S. R., & Dong, Q. (2000). Family, peer, and individual correlates of depressive symptomatology among U.S. and Chinese adolescents. *Journal of Counseling and Clinical Psychology, 68,* 209–219.

Greenberger, E., O'Neil, R., & Nagel, S.K. (1994). Linking workplace and homeplace: Relations between the nature of adults' work and their parenting behavior. *Developmental Psychology, 30,* 990–1002.

Greendorfer, S. L., Lewko, J. H., & Rosengren, K. S. (1996). Family and gender-based socialization of children and adolescents. In F. L. Smoll & R. E. Smith (Eds.), *Children and youth in sport: A biopsychological perspective* (pp. 89–111). Dubuque, IA: Brown & Benchmark.

Greene, K., Krcmar, M., Walters, L. H., Rubin, D. L., Hale, J., & Hale, L. (2000). Targeting adolescent risk-taking behaviors: The contributions of egocentrism and sensation-seeking. *Journal of Adolescence, 23,* 439–461.

Greenfield, P. (1992, June). *Notes and references for developmental psychology.* Conference on Making Basic Texts in Psychology More Culture-Inclusive and Culture-Sensitive, Western Washington University, Bellingham, WA.

Greenfield, P. M. (1994). Independence and interdependence as developmental scripts: Implications for theory, research, and practice. In P. M. Greenfield & R. R. Cocking (Eds.), *Cross-cultural roots of minority child development* (pp. 1–37). Hillsdale, NJ: Erlbaum.

Greenfield, P. M., Quiroz, B., & Raeff, C. (2000). Cross-cultural conflict and harmony in the social construction of the child. In S. Harkness, C. Raeff, & C. M. Super (Eds.), *Variability in the social construction of the child* (pp. 93–108). San Francisco: Jossey-Bass.

Greenhill, L. L., Halperin, J. M., & Abikoff, H. (1999). Stimulant medications. *Journal of the American Academy of Child and Adolescent Psychiatry, 38,* 503–512.

Greenlee, R. T. (2000). Cancer Statistics, 2000. *CA—A Cancer Journal for Clinicians, 50,* 7–33.

Greenough, W. T., & Black, J. E. (1992). Induction of brain structure by experience: Substrates for cognitive development. In M. R. Gunnar & C. A. Nelson (Eds.), *Minnesota Symposia on Child Psychology* (pp. 155–200). Hillsdale, NJ: Erlbaum.

Greenough, W. T., Wallace, C. S., Alcantara, A. A., Anderson, B. J., Hawrylak, N., Sirevaag, A. M., Weiler, I. J., & Withers, G. S. (1993). Experience affects the structure of neurons, glia, and blood vessels. In N. J. Anastasiow & S. Harel (Eds.), *At-risk infants: Interventions, family, and research* (pp. 175–185). Baltimore: Paul H. Brookes.

Gregg, V., Gibbs, J. C., & Fuller, D. (1994). Patterns of developmental delay in moral judgment by male and female delinquents. *Merrill-Palmer Quarterly, 40,* 538–553.

Gresham, F. M., & MacMillan, D. L. (1997). Social competence and affective characteristics of students with mild disabilities. *Review of Educational Research, 67,* 377–415.

Gresser, G., Wong, P., & Reker, G. (1987). Death attitudes across the life-span: The development and validation of the Death Attitude Profile (DAP). *Omega, 18,* 113–128.

Griffiths, A., Roberts, G., & Williams, J. (1993). Elder abuse and the law. In P. Decalmer & F. Glendenning (Eds.), *The mistreatment of elderly people* (pp. 62–75). London: Sage.

Grigorenko, E. L. (2000). Heritability and intelligence. In R. J. Sternberg (Ed.), *Handbook of intelligence* (pp. 53–91). Cambridge, U.K.: Cambridge University Press.

Grody, W. W. (1999). Cystic fibrosis: Molecular diagnosis, population screening, and public policy. *Archives of Pathology and Laboratory Medicine, 123,* 1041–1046.

Grolnick, W. S., Bridges, L. J., & Connell, J. P. (1996). Emotion regulation in two-year-olds: Strategies and emotional expression in four contexts. *Child Development, 67,* 928–941.

Grolnick, W. S., Kurowski, C. O., Dunlap, K. G., & Hevey, C. (2000). Parental resources and the transition to junior high. *Journal of Research on Adolescence, 10,* 466–488.

Groome, L. J., Swiber, M. J., Atterbury, J. L., Bentz, L. S., & Holland, S. B. (1997). Similarities and differences in behavioral state organization during sleep periods in the perinatal infant before and after birth. *Child Development, 68,* 1–11.

Gross, M. (1993). *Exceptionally gifted children.* London: Routledge.

Grossbaum, M. F., & Bates, G. W. (2002). Correlates of psychological well-being at midlife: The role of generativity, agency and communion, and narrative themes. *International Journal of Behavioral Development, 26,* 120–127.

Grossman, A. H., Daugelli, A. R., & Hershberger, S. L. (2000). Social support networks of lesbian, gay, and bisexual adults 60 years of age and older. *Journal of Gerontology, 55B,* P171–179.

Grossman, J. A., & Kruesi, M. J. P. (2000). Innovative approaches to youth suicide prevention: An update of issues and research findings. In R. W. Maris, S. S. Canetto, J. L. McIntosh, & M. M. Silverman (Eds.), *Review of Suicidology, 2000* (pp. 170–201). New York: Guilford.

Grossmann, K., Grossmann, K. E., Spangler, G., Suess, G., & Unzner, L. (1985). Maternal sensitivity and newborns' orientation responses as related to quality of attachment in Northern Germany. In I. Bretherton & E. Waters (Eds.), Growing points of attachment theory and research. *Monographs of the Society for Research in Child Development, 50*(1–2, Serial No. 209).

Grotevant, H. D. (1998). Adolescent development in family contexts. In N. Eisenberg (Ed.), *Handbook of child psychology: Vol. 3. Social, emotional, and personality development* (5th ed., pp. 1097–1149). New York: Wiley.

Grotevant, H. D., & Cooper, C. R. (1998). Individuality and connectedness in adolescent development: Review and prospects for research on identity, relationships, and context. In E. Skoe & A. von der Lippe (Eds.), *Personality development in adolescence* (pp. 3–37). London: Routledge & Kegan Paul.

Grotevant, H. D., & Kohler, J. K. (1999). Adoptive families. In M. E. Lamb (Ed.), *Parenting and child development in "nontraditional" families* (pp. 161–190). Mahwah, NJ: Erlbaum.

Grotpeter, J. K., & Crick, N. R. (1996). Relational aggression, overt aggression, and friendship. *Child Development, 67,* 2328–2338.

Grubb, W. N. (1999). The subbaccalaureate labor market in the United States: Challenges for the school-to-work transition. In W. R. Heinz (Ed.), *From education to work: Cross-national perspectives* (pp. 171–193). New York: Cambridge University Press.

Gruetzner, H. (1992). *Alzheimer's.* New York: Wiley.

Grusec, J. E. (1988). *Social development: History, theory, and research.* New York: Springer.

Grusec, J. E., & Goodnow, J. J. (1994). Impact of parental discipline methods on the child's internalization of values: A reconceptualization of current points of view. *Developmental Psychology, 30,* 4–19.

Grych, J. H., & Clark, R. (1999). Maternal employment and development of the father–infant relationship in the first year. *Developmental Psychology, 35,* 893–903.

Grzywacz, J. G., & Marks, N. F. (2001). Social inequalities and exercise during adulthood: Toward an ecological perspective. *Journal of Health and Social Behavior, 42,* 202–220.

Guildner, S. H., Loeb, S., Morris, D., Penrod, J., Bramlett, M., Johnston, L., & Schlotzhauer, P. (2001). A comparison of life satisfaction and mood in nursing home residents and community-dwelling elders. *Archives of Psychiatric Nursing, 15,* 232–240.

Guilford, J. P. (1985). The structure-of-intellect model. In B. B. Wolman (Ed.), *Handbook of intelligence* (pp. 225–266). New York: Wiley.

Guinn, B. (1999). Leisure behavior motivation and the life satisfaction of retired persons. *Activities, Adaptation, and Aging, 23,* 13–20.

Gullette, M. M. (1988). *Safe at last in the middle years.* Berkeley: University of California Press.

Gullone, E., & King, N. J. (1997). Three-year follow-up of normal fear in children and adolescents aged 7 to 18 years. *British Journal of Developmental Psychology, 15,* 97–111.

Gunnar, M. R. (2001). Effects of early deprivation: Findings from orphanage-reared infants and children. In C. A. Nelson & M. Luciana (Eds.), *Handbook of developmental cognitive neuroscience* (pp. 617–629). Cambridge, MA: MIT Press.

Gunnar, M. R., & Nelson, C. A. (1994). Event-related potentials in year-old infants: Relations with emotionality and cortisol. *Child Development, 65,* 80–94.

Gunter-Hunt, G., Mahoney, J. E., & Sieger, C. E. (2002). A comparison of state advance directive documents. *Gerontologist, 42,* 51–60.

Guo, G., & VanWey, L. K. (1999). Sibship size and intellectual development: Is the relationship causal? *American Sociological Review, 64,* 169–187.

Gupta, V., & Korte, C. (1994). The effects of a confidant and a peer group on the well-being of single elders. *International Journal of Aging and Human Development, 39,* 293–302.

Gustafson, G. E., Green, J. A., & Cleland, J. W. (1994). Robustness of individual identity in the cries of human infants. *Developmental Psychobiology, 27,* 1–9.

Gustafson, G. E., Wood, R. M., & Green, J. A. (2000). Can we hear the causes of infants' crying? In R. G. Barr & B. Hopkins (Eds.), *Crying as a sign, a symptom, and a signal: Clinical, emotional, and developmental aspects of infant and toddler crying* (p.p. 8–22). New York: Cambridge University Press.

Gutmann, D. (1977). The cross-cultural perspective: Notes toward a comparative psychology of aging. In J. E. Birren & K. W. Schaie (Eds.), *Handbook of the psychology of aging* (pp. 302–326). New York: Van Nostrand Reinhold.

Gutmann, D. L., & Huyck, M. H. (1994). Development and pathology in post-parental men: A community study. In E. Thompson, Jr. (Ed.), *Older men's lives* (pp. 65–84). Thousand Oaks, CA: Sage.

Guttuso, T., Jr., Kurlan, R., McDermott, M. P., & Kieburtz, K. (2003). Gabapentin's effects on hot flashes in postmenopausal women: A randomized controlled trial. *Obstetrics and Gynecology, 101,* 337–345.

Guyer, C. G., II. (2000). Spouse abuse. In F. W. Kaslow (Ed.), *Handbook of couple and family forensics* (pp. 206–234). New York: Wiley.

Gwiazda, J., & Birch, E. E. (2001). Perceptual development: Vision. In E. B. Goldstein (Ed.), *Blackwell handbook of perception* (pp. 636–668). Oxford, U.K.: Blackwell.

Haan, N., Aerts, E., & Cooper, B. (1985). *On moral grounds: The search for practical morality.* New York: New York University Press.

Hack, M., Wright, L. L., Shankaran, S., & Tyson, J. E. (1995). Very low birth weight outcomes of the National Institute of Child Health and Human Development Neonatal Network, November 1989 to October 1990. *American Journal of Obstetrics and Gynecology, 172,* 457–464.

Hack, M. B., Taylor, H. G., Klein, N., Eiben, R., Schatschneider, C., & Mercuri-Minich, N. (1994). School-age outcomes in children with birth weights under 750 g. *New England Journal of Medicine, 331,* 753–759.

Haddad, F. F., Yeatman, T. J., Shivers, S. C., & Reintgen, D. S. (1999). The Human Genome Project: A dream becoming a reality. *Surgery, 125,* 575–580.

Haden, C. A., Haine, R. A., & Fivush, R. (1997). Developing narrative structure in parent–child reminiscing across the preschool years. *Developmental Psychology, 33,* 295–307.

Hader, S. L., Smith, D. K., Moore, J. S., & Holmberg, S. D. (2001). HIV infection in women in the United States. *Journal of the American Medical Association, 285,* 1186–1192.

Hagberg, B., Alfredson, B. B., Poon, L. W., & Homma, A. (2001). Cognitive functioning in centenarians: A coordinated analysis of results from three countries. *Journal of Gerontology, 56B,* P141–P151.

Hagberg, J. M., Allen, W. K., Seals, D. R., Hurley, B. F., Ehsani, A. A., & Holloszy, J. O. (1985). A hemodynamic comparison of young and older endurance athletes during exercise. *Journal of Applied Physiology, 58,* 2041–2046.

Hagekull, B., Bohlin, G., & Rydell, A. (1997). Maternal sensitivity, infant temperament, and the development of early feeding problems. *Infant Mental Health Journal, 18,* 92–106.

Hager, M. (1996, December 23). The cancer killer. *Newsweek,* pp. 42–47.

Haight, B. K. (1992). Long-term effects of a structured life review process. *Journal of Gerontology, 47,* P312–P315.

Haight, W. L., & Miller, P. J. (1993). *Pretending at home: Early development in a sociocultural context.* Albany, NY: State University of New York Press.

Haith, M. M. (1999). Some thoughts about claims for innate knowledge and infant physical reasoning. *Developmental Science, 2,* 153–156.

Haith, M. M., & Benson, J. B. (1998). Infant cognition. In D. Kuhn & R. S. Siegler (Eds.), *Handbook of child psychology: Vol. 2. Cognition, perception, and language* (5th ed., pp. 199–254). New York: Wiley.

Hakuta, K., Bialystok, E., & Wiley, E. (2003). Critical evidence: A test of the critical-period hypothesis for second-language acquisitions. *Psychological Science, 14,* 31–38.

Halford, G. S. (1993). *Children's understanding: The development of mental models.* Hillsdale, NJ: Erlbaum.

Hall, D. G., & Graham, S. A. (1999). Lexical form class information guides word-to-object mapping in preschoolers. *Child Development, 70,* 78–91.

Hall, G. S. (1904). *Adolescence.* New York: Appleton.

Hall, S. S. (1996, January/February). Short like me. *Health,* pp. 98–106.

Hallett, M. (2000). Brain plasticity and recovery from hemiplegia. *Journal of Medical Speech-Language Pathology, 9,* 107–115.

Halliday, J. L., Watson, L. F., Lumley, J., Danks, D. M., & Sheffield, L. S. (1995). New estimates of Down syndrome risks of chorionic villus sampling, amniocentesis, and live birth in women of advanced maternal age from a uniquely defined population. *Prenatal Diagnosis, 15,* 455–465.

Hallinan, M. T., & Kubitschek, W. N. (1999). Curriculum differentiation and high school achievement. *Social Psychology of Education, 3,* 41–62.

Halpern, C. T., Udry, J. R., Campbell, B., & Suchindran, C. (1999). Effects of body fat on weight concerns, dating, and sexual activity: A longitudinal analysis of black and white adolescent girls. *Developmental Psychology, 35,* 721–736.

Halpern, C. T., Udry, J. R., & Suchindran, C. (1997). Testosterone predicts initiation of coitus in adolescent females. *Psychosomatic Medicine, 59,* 161–171.

Halpern, D. F. (1992). *Sex differences in cognitive abilities* (2nd ed.). Hillsdale, NJ: Erlbaum.

Halpern, D. F. (1997). Sex differences in intelligence. *American Psychologist, 52,* 1091–1102.

Halpern, D. F. (2000). *Sex differences in cognitive abilities* (3rd ed.). Mahwah, NJ: Erlbaum.

Halpern, L. F., MacLean, W. E., & Baumeister, A. A. (1995). Infant sleep–wake characteristics: Relation to neurological status and the prediction of developmental outcome. *Developmental Review, 15,* 255–291.

Hamachek, D. (1990). Evaluating self-concept and ego status in Erikson's last three psychosocial stages. *Journal of Counseling and Development, 68,* 677–683.

Hamberger, L. K., Lohr, J. M., Bonge, D., & Tolin, D. (1997). An empirical classification of motivations for domestic violence. *Violence Against Women, 3,* 401–423.

Hamer, D. H., Hu, S., Magnuson, V. L., Hu, N., & Pattatucci, A. M. L. (1993). A linkage between DNA markers on the X chromosome and male sexual orientation. *Science, 261,* 321–327.

Hamilton, S. F. (1990). *Apprenticeship for adulthood: Preparing youth for the future.* New York: Free Press.

Hamilton, S. F. (1993). Prospects for an American-style youth apprenticeship system. *Educational Researcher, 22*(3), 11–16.

Hamilton, S. F., & Hamilton, M. A. (1999). Creating new pathways to adulthood by adapting German apprenticeship in the United States. In W. R. Heinz (Ed.), *From education to work: Cross-national perspectives* (pp. 194–213). New York: Cambridge University Press.

Hamilton, S. F., & Hamilton, M. A. (2000). Research, intervention, and social change: Improving adolescents' career opportunities. In L. J. Crockett & R. K. Silbereisen (Eds.), *Negotiating adolescence in times of social change* (pp. 267–283). New York: Cambridge University Press.

Hammond, C. J., Snieder, H., Spector, T. D., & Gilbert, C. E. (2000). Genetic and environmental factors in age-related nuclear cataracts in monozygotic and dizygotic twins. *New England Journal of Medicine, 342,* 1786–1790.

Han, J. J., Leichtman, M. D., & Wang, Q. (1998). Autobiographical memory in Korean, Chinese, and American children. *Developmental Psychology, 34,* 701–713.

Han, S. K., & Moen, P. (1999). Clocking out: Temporal patterning of retirement. *American Journal of Sociology, 105,* 191–236.

Hanson, L. C., Danis, M., & Garrett, J. (1995). What is wrong with end-of-life care? Opinions of bereaved family members. *Journal of the American Geriatric Society, 45,* 1339–1344.

Hanson, T. L. (1999). Does parental conflict explain why divorce is negatively associated with child welfare? *Social Forces, 77,* 1283–1316.

Happé, F. G. E. (1995). The role of age and verbal ability in the theory of mind task performance of subjects with autism. *Child Development, 66,* 843–855.

Hare, J. (1994). Concerns and issues faced by families headed by a lesbian couple. *Families in Society, 43,* 27–35.

Hare, J., & Richards, L. (1993). Children raised by lesbian couples: Does context of birth affect father and partner involvement? *Family Relations, 42,* 249–255.

Harkness, S., & Waldfogel, J. (2002). The family gap in pay: Evidence from seven industrialized countries. Retrieved from http://www.sticerd.lse.ac.uk/Case

Harley, B., & Jean, G. (1999). Vocabulary skills of French immersion students in their second language. *Zeitschrift für Interkulturellen Fremdsprachenunterricht, 4*(2). Retrieved from http://www.ualberta.ca

Harlow, H. F., & Zimmerman, R. (1959). Affectional responses in the infant monkey. *Science, 130,* 421–432.

Harman, D. (2002). Aging: Overview. *Annals of the New York Academy of Sciences, 959,* 1–21.

Harootyan, R. A., & Vorek, R. E. (1994). Volunteering, helping and gift giving in families and communities. In V. L. Bengtson & R. A. Harootyan (Eds.), *Intergenerational linkages: Hidden connections in American society* (pp. 77–111). New York: Springer.

Harpaz, D., Benderly, M., Goldbourt, U., Kishan, Y., & Behar, S. (1996). Effect of aspirin on mortality in women with symptomatic or silent myocardial ischemia. *American Journal of Cardiology, 78,* 1215–1219.

Harris, G. (1997). Development of taste perception and appetite regulation. In G. Bremner, A. Slater, & G. Butterworth (Eds.), *Infant development: Recent advances* (pp. 9–30). East Sussex, U.K.: Erlbaum.

Harris, J. R. (1998). *The nurture assumption: Why children turn out the way they do.* New York: Free Press.

Harris, K. M. (2000). The health status and risk behavior of adolescents in immigrant families. In D. J. Hernandez (Ed.), *Children of immigrants: Health, adjustment, and public assistance.* Washington, DC: National Academy Press.

Harris, P. B. (1998). Listening to caregiving sons: Misunderstood realities. *Gerontologist, 38,* 342–352.

Harris, P. L., & Leevers, H. J. (2000). Reasoning from false premises. In P. Mitchell & K. J. Riggs (Eds.), *Children's reasoning and the mind* (pp. 67–99). Hove, U.K.: Psychology Press.

Harris, R. L., Ellicott, A. M., & Holmes, D. S. (1986). The timing of psychosocial transitions and changes in women's lives: An examination of women aged 45 to 60. *Journal of Personality and Social Psychology, 51,* 409–416.

Harrison, A. O., Wilson, M. N., Pine, C. J., Chan, S. Q., & Buriel, R. (1994). Family ecologies of ethnic minority children. In G. Handel & G. G. Whitchurch (Eds.), *The psychosocial interior of the family* (pp. 187–210). New York: Aldine de Gruyter.

Harrison, L. J., & Ungerer, J. A. (2002). Maternal employment and infant–mother attachment security at 12 months postpartum. *Developmental Psychology, 38,* 758–773.

Harrist, A. W., Zaia, A. F., Bates, J. E., Dodge, K. A., & Pettit, G. S. (1997). Subtypes of social withdrawal in early childhood: Sociometric status and social–cognitive differences across four years. *Child Development, 68,* 278–294.

Hart, B., & Risley, T. R. (1995). *Meaningful differences in the everyday experience of young American children.* Baltimore: Paul H. Brookes.

Hart, C. H., Burts, D. C., Durland, M. A., Charlesworth, R., DeWolf, M., & Fleege, P. O. (1998). Stress behaviors and activity type participation of preschoolers in more and less developmentally appropriate classrooms: SES and sex differences. *Journal of Research in Childhood Education, 13,* 176–196.

Hart, C. H., Nelson, D. A., Robinson, C. C., Olsen, S. F., McNeilly-Choque, M. K., Porter, C. L., & Mckee, T. R. (2002). Russian parenting styles and family processes: Linkages with subtypes of victimization and aggression. In K. A. Kerns, J. M. Contreras, & A. M. Neal-Barnett (Eds.), *Family and peers: Linking two social worlds.* Westport, CT: Praeger.

Hart, C. H., Newell, L. D., & Olsen, S. F. (2002). Parenting skills and social/communicative competence in childhood. In J. O.

Greene & B. R. Burleson (Eds.), *Handbook of communication and social interaction skill.* Hillsdale, NJ: Erlbaum.

Hart, C. H., Yang, C., Nelson, L. J., Robinson, C. C., Olsen, J. A., Nelson, D. A., Porter, C. L., Jin, S., Olsen, S. F., & Wu, P. (2000). Peer acceptance in early childhood and subtypes of socially withdrawn behavior in China, Russia, and the United States. *International Journal of Behavioral Development, 24,* 73–81.

Hart, D., & Fegley, S. (1995). Prosocial behavior and caring in adolescence: Relations to self-understanding and social judgment. *Child Development, 66,* 1346–1359.

Hart, H. M., McAdams, D. P., Hirsch, B. J., & Bauer, J. J. (2001). Generativity and social involvement among African Americans and white adults. *Journal of Research in Personality, 35,* 208–230.

Hart, S., Field, T., & Roitfarb, M. (1999). Depressed mothers' assessments of their neonates' behaviors. *Infant Mental Health Journal, 20,* 200–210.

Harter, S. (1982). The perceived competence scale for children. *Child Development, 53,* 87–97.

Harter, S. (1990). Issues in the assessment of the self-concept of children and adolescents. In A. LaGreca (Ed.), *Through the eyes of a child* (pp. 292–325). Boston: Allyn and Bacon.

Harter, S. (1996). Developmental changes in self-understanding across the 5 to 7 shift. In A. J. Sameroff & M. M. Haith (Eds.), *The five to seven year shift* (pp. 207–236). Chicago: University of Chicago Press.

Harter, S. (1998). The development of self-representations. In N. Eisenberg (Ed.), *Handbook of child psychology: Vol. 3. Social, emotional, and personality development* (5th ed., pp. 553–618). New York: Wiley.

Harter, S. (1999). *The construction of self: A developmental perspective.* New York: Guilford.

Harter, S., Marold, D. B., Whitesell, N. R., & Cobbs, G. (1996). A model of the effects of parent and peer support on adolescent false self-behavior. *Child Development, 67,* 360–374.

Harter, S., & Whitesell, N. (1989). Developmental changes in children's understanding of simple, multiple, and blended emotion concepts. In C. Saarni & P. Harris (Eds.), *Children's understanding of emotion* (pp. 81–116). Cambridge, U.K.: Cambridge University Press.

Hartshorn, K., Rovee-Collier, C., Gerhardstein, P., Bhatt, R. S., Wondoloski, T. L., Klein, P., Gilch, J., Wurtzel, N., & Campos-de-Carvalho, M. (1998). The ontogeny of long-term memory over the first year-and-a-half of life. *Developmental Psychobiology, 32,* 69–89.

Hartup, W. W. (1996). The company they keep: Friendships and their developmental significance. *Child Development, 67,* 1–13.

Hartup, W. W., & Stevens, N. (1999). Friendships and adaptation across the life span. *Current Directions in Psychological Science, 8,* 76–79.

Harvey, J. H., & Pauwels, B. G. (1999). Recent developments in close-relationships theory. *Current Directions in Psychological Science, 8,* 93–99.

Harway, M., & Hansen, M. (1994). *Spouse abuse.* Sarasota, FL: Professional Resource Press.

Harwood, J. (2001). Comparing grandchildren's and grandparents' stake in their relationship. *International Journal of Aging and Human Development, 53,* 195–210.

Hasher, L., Zacks, R. T., & May, C. P. (1999). Inhibitory control, circadian arousal, and age. In D. Gopher & A. Koriat (Eds.), *Attention and performance* (Vol. 17, pp. 653–675). Cambridge, MA: MIT Press.

Hassing, L. B., Johansson, B., Berg, S., Nilsson, S. E., Pedersen, N. L., Hofer, S. M., & McClearn, G. (2002). Terminal decline and markers of cerebro- and cardiovascular disease: Findings from a longitudinal study of the oldest old. *Journal of Gerontology, 57B,* P268–P276.

Hassmén, P., Koivula, N., & Uutela, A. (2000). Physical exercise and psychological well-being: A population study in Finland. *Preventive Medicine, 30,* 17–25.

Hatch, M. C., Shu, X.-O., McLean, D. E., Levin, B., Begg, M., Reuss, L., & Susser, M. (1993). Maternal exercise during pregnancy, physical fitness, and fetal growth. *American Journal of Epidemiology, 137,* 1105–1114.

Hatfield, E. (1988). Passionate and companionate love. In R. J. Sternberg & M. L. Barnes (Eds.), *The psychology of love* (pp. 191–217). New Haven, CT: Yale University Press.

Hatfield, E. (1993). *Love, sex, and intimacy: Their psychology, biology, and history.* New York: HarperCollins.

Hatton, D. D., Bailey, D. B., Jr., Burchinal, M. R., & Ferrell, K. A. (1997). Developmental growth curves of preschool children with vision impairments. *Child Development, 68,* 788–806.

Haught, P. A., Hill, L. A., Nardi, A. H., & Walls, R. T. (2000). Perceived ability and level of education as predictors of traditional and practical adult problem solving. *Experimental Aging Research, 36,* 89–101.

Hausfather, A., Toharia, A., LaRoche, C., & Engelsmann, F. (1997). Effects of age of entry, day-care quality, and family characteristics on preschool behavior. *Journal of Child Psychology and Psychiatry, 38,* 441–448.

Hauth, J. C., Goldenberg, R. L., Parker, C. R., Cutter, G. R., & Cliver, S. P. (1995). Low-dose aspirin—Lack of association with an increase in abruptio placentae or perinatal mortality. *Obstetrics and Gynecology, 85,* 1055–1058.

Hawker, D. S. J., & Boulton, M. J. (2000). Twenty years' research on peer victimization and psychosocial maladjustment: A meta-analytic review of cross-sectional studies. *Journal of Child Psychology and Psychiatry, 41,* 441–455.

Hawkins, J. D., Catalano, R. F., & Miller, J. Y. (1992). Risk and protective factors for alcohol and other drug problems in adolescence and early adulthood: Implications for substance abuse prevention. *Psychological Bulletin, 112,* 64–105.

Hawkins, J. D., & Lam, T. (1987). Teacher practices, social development, and delinquency. In J. D. Burchard & S. N. Burchard (Eds.), *Prevention of delinquent behavior* (pp. 241–274). Newbury Park, CA: Sage.

Hawkins, J. N. (1994). Issues of motivation in Asian education. In H. F. O'Neil, Jr., & M. Drillings (Eds.), *Motivation: Theory and research* (pp. 101–115). Hillsdale, NJ: Erlbaum.

Hay, J. F., & Jacoby, L. L. (1999). Separating habit and recollection in young and older adults: Effects of elaborative processing and distinctiveness. *Psychology and Aging, 14,* 122–134.

Hayflick, L. (1994). *How and why we age.* New York: Ballantine.

Hayne, H. (2002). Thoughts from the crib: Meltzoff and Moore (1994) alter our views of mental representation during infancy. *Infant Behavior and Development, 25,* 62–64.

Hayne, H., Boniface, J., & Barr, R. (2000). The development of declarative memory in human infants: Age-related changes in deferred imitation. *Behavioral Neuroscience, 114,* 77–83.

Hayne, H., Rovee-Collier, C., & Perris, E. E. (1987). Categorization and memory retrieval by three-month-olds. *Child Development, 58,* 750–767.

Hays, J. C., & George, L. K. (2002). The life-course trajectory toward living alone: Racial differences. *Research on Aging, 24,* 283–307.

Hayslip, B., Emick, M. A., Henderson, C. E., & Elias, K. (2002). Temporal variations in the experience of custodial grandparenting: A short-term longitudinal study. *Journal of Applied Gerontology, 21,* 139–156.

Haywood, K. M., & Getchell, N. (2001). *Life span motor development* (3rd ed.). Champaign, IL: Human Kinetics.

Hazan, C., & Zeifman D. (1999). Pair-bonds as attachments: Evaluating the evidence. In J. Cassidy & P. R. Shaver (Eds.), *Handbook of attachment* (pp. 336–354). New York: Guilford.

Hazell, L. V. (2001). Multicultural aftercare issues. In O. D. Weeks & C. Johnson (Eds.), *When all the friends have gone: Guide for aftercare providers* (pp. 57–71). Amityville, NY: Baywood.

Head Start Bureau. (2002). 2000 Head Start fact sheet. Retrieved from http://www2.acf.dhhs.gov/programs/hsb/research/00_hsfs.htm

Health Canada. (1999a). Canadian dietary guidelines, recommendations and standards. Retrieved from http://www.sfu.ca

Health Canada. (1999b). Measuring up: A health surveillance update on Canadian children and youth. Retrieved from http://www.hc-sc.gc.ca/pphb-dgspsp/publicat/meas-haut

Health Canada. (2000a). *Aboriginal Head Start in urban and northern communities.* Ottawa: Minister of Health.

Health Canada. (2000b). Canadian perinatal health report 2000. Ottawa: Minister of Public Works and Government Services. Retrieved from http://www.hc-sc.gc.ca/pphb-dgspsp/publicat/cphr-rspc00/pdf/cphr00e.pdf

Health Canada. (2000c). Paediatrics and child health: Canadian Report on Immunization, 1997. Retrieved from http://www.hc-sc.gc.ca

Health Canada. (2000d). Physical activity guide. Retrieved from http://www.hc-sc.gc.ca/hppb/paguide

Health Canada. (2001a). Childhood injury: Deaths and hospitalizations in Canada. Retrieved from http://www.hc-sc.gc.ca

Health Canada. (2001b). Prenatal and postpartum women and tobacco. Ottawa: Tobacco Control Programme. Retrieved from http://www.hc-sc.gc.ca/hecs-sesc/tobacco/pdf/prenatal.pdf

Health Canada. (2001c). Sudden infant death syndrome (SIDS). Retrieved from http://www.hc-sc.gc.ca/english/media/releases/2001/2001_113ebk.htm

Health Canada. (2002a). Alcohol and pregnancy. Retrieved from http://www.hc-sc.gc.ca/pphb-dgspsp/rhs-ssg/factshts/alcprg_e.html

Health Canada. (2002b). Breast-feeding. *Perinatal Surveillance System.* Retrieved from http://www.hc-sc.gc.ca/pphb-dgspsp/rhs-ssg/factshts/brstfd_e.html

Health Canada. (2002c). The Canada Prenatal Nutrition Program. Retrieved from http://www.hc-sc.gc.ca/hppb/childhood-youth/cbp/cpnp/index.html

Health Canada. (2002d). Child maltreatment in Canada: Incidence and characteristics. Retrieved from http://www.hc-sc.gc.ca

Health Canada. (2002e). Disease surveillance online. Retrieved from http://www.hc-sc.gc.ca/pphb-dgspsp/dsol-smed/index.html

Health Canada. (2002f). Health expenditures in Canada by age and sex. Retrieved from http://www.hc-sc.gc.ca/english/care/expenditures/exp_age_sex.html

Health Canada. (2002g). *Healthy Canadians: A federal report on comparable health indicators.* Retrieved from http://www.hc-sc.gc.ca/iacb-dgiac/arad-draa/english/accountability/indicators/html#high

Health Canada. (2002h). Smoking among adults. Retrieved from http://www.hc-sc.gc.ca

Health Canada. (2002i). Statistical report on the health of Canadians 1999. Retrieved from http://www.statcan.ca:80/english/freepub/82-570-XIE/partb.htm

Heath, S. B. (1982). Questioning at home and at school: A comparative study. In G. Spindler (Ed.), *Doing the ethnography of schooling: Educational anthropology in action* (pp. 102–127). New York: Holt.

Heath, S. B. (1989). Oral and literate traditions among black Americans living in poverty. *American Psychologist, 44,* 367–373.

Heath, S. B. (1990). The children of Trackton's children: Spoken and written in social change. In J. Stigler, G. Herdt, & R. A. Shweder (Eds.), *Cultural psychology: Essays on comparative human development* (pp. 496–519). New York: Cambridge University Press.

Heaton, T. B. (2002). Factors contributing to increasing marital stability in the United States. *Journal of Family Issues, 23,* 392–409.

Hedberg, K., Hopkins, D., & Kohn, M. (2003). Five years of legal physician-assisted suicide in Oregon. *New England Journal of Medicine, 348,* 961–964.

Hedges, L. V., & Nowell, A. (1995). Sex differences in mental scores, variability, and numbers of high-scoring individuals. *Science, 269,* 41–45.

Hedges, L. V., & Nowell, A. (1998). Black–white test score convergence since 1995. In C. Jencks & M. Phillips (Eds.), *The black–white test score gap* (pp. 149–181). Washington, DC: Brookings Institution.

Hediger, M. L., Overpeck, M. D., Ruan, W. J., & Troendle, J. F. (2002). Birthweight and gestational age effects on motor and social development. *Paediatric and Perinatal Epidemiology, 16,* 33–46.

Heidrich, S. M., & Ryff, C. D. (1993). Physical and mental health in later life: The self-system as mediator. *Psychology and Aging, 8,* 327–338.

Heine, S. J., & Lehman, D. R. (1995). Cultural variation in unrealistic optimism: Does the West feel more invulnerable than the East? *Journal of Personality and Social Psychology, 68,* 595–607.

Heinl, T. (1983). *The baby massage book.* London: Coventure.

Heinz, A., & Blass, J. P. (2002). *Alzheimer's disease: A status report for 2002.* New York: American Council on Science and Health.

Heinz, W. R. (1999a). Introduction: Transitions to employment in a cross-national perspective. In W. R. Heinz (Ed.), *From education to work: Cross-national perspectives* (pp. 1–21). New York: Cambridge University Press.

Heinz, W. R. (1999b). Job entry patterns in life-course perspective. In W. R. Heinz (Ed.), *From education to work: Cross-national perspectives* (pp. 214–231). New York: Cambridge University Press.

Helburn, S. W. (Ed.). (1995). *Cost, quality and child outcomes in child care centers.* Denver: University of Colorado.

Helm, H. M., Hays, J. C., Flint, E. P., Koenig, H. G., & Blazer, D. G. (2000). Does private religious activity prolong survival? A six-year follow-up study of 3,851 older adults. *Journal of Gerontology, 55A,* M400–M405.

Helson, R. (1992). Women's difficult times and the rewriting of the life story. *Psychology of Women Quarterly, 16,* 331–347.

Helson, R. (1997). The self in middle age. In M. E. Lachman & J. B. James (Eds.), *Multiple paths of midlife development* (pp. 21–43). Chicago: University of Chicago Press.

Helson, R., Jones, C. J., & Kwan, V. S. Y. (2002). Personality change over 40 years of adulthood: Hierarchical linear modeling analyses of two longitudinal samples. *Journal of Personality and Social Psychology, 83,* 752–766.

Helson, R., Mitchell, V., & Moane, G. (1984). Personality and patterns of adherence and nonadherence to the social clock. *Journal of Personality and Social Psychology, 46,* 1079–1096.

Helson, R., & Moane, G. (1987). Personality change in women from college to midlife. *Journal of Personality and Social Psychology, 53,* 176–186.

Helson, R., & Picano, J. (1990). Is the traditional role bad for women? *Journal of Personality and Social Psychology, 59,* 311–320.

Helson, R., & Roberts, B. W. (1994). Ego development and personality change in adulthood. *Journal of Personality and Social Psychology, 66,* 911–920.

Helson, R., & Stewart, A. (1994). Personality change in adulthood. In T. F. Heatherton & J. L. Weinberger (Eds.), *Can personality change?* (pp. 201–225). Washington, DC: American Psychological Association.

Helson, R., & Wink, P. (1992). Personality change in women from the early 40s to the early 50s. *Psychology and Aging, 7,* 46–55.

Helwig, C. C., & Prencipe, A. (1999). Children's judgments of flags and flag-burning. *Child Development, 70,* 132–143.

Henderson, D., Buchanan, J. A., & Fisher, J. E. (2002). Violence and the elderly population: Issues for prevention. In P. A. Schewe (Ed.), *Preventing violence in relationships: Interventions across the lifespan* (pp. 223–245). Washington DC: American Psychological Association.

Hendin, H. (1999). Suicide, assisted suicide, and euthanasia. In D. G. Jacobs (Ed.), *The Harvard Medical School guide to suicide assessment and intervention* (pp. 540–560). San Francisco: Jossey-Bass.

Hendin, H. (2002). The Dutch experience. In D. Foley & H. Hendin (Eds.), *The case against assisted suicide: For the right to end-of-life care* (pp. 97–121). Baltimore: Johns Hopkins University Press.

Hendrick, S. S., & Hendrick, C. (1992). *Romantic love.* Newbury Park, CA: Sage.

Hendrick, S. S., & Hendrick, C. (1993). Lovers as friends. *Journal of Social and Personal Relationships, 10,* 459–466.

Hendrie, H. C., Ogunniyi, A., Hall, K. S., Baiyewu, O., Unverzagt, F. W., & Gureje, O. (2001). The incidence of dementia and AD in two communities: Yoruba residing in Ibadan, Nigeria, and African Americans residing in Indianapolis, USA. *Journal of the American Medical Association, 285,* 739–747.

Hendrie, H. H. (2001). Exploration of environmental and genetic risk factors for Alzheimer's disease: The value of cross-cultural studies. *Current Directions in Psychological Science, 10,* 98–101.

Herbst, J. H., McCrae, R. R., Costa, P. T., Jr., Feaganes, J. R., & Siegler, I. C. (2000). Self-perceptions of stability and change in personality at midlife: The UNC Alumni Heart Study. *Assessment, 7,* 379–388.

Herdt, G., & Boxer, A. M. (1993). *Children of horizons: How gay and lesbian teens are leading a new way out of the closet.* Boston: Beacon Press.

Herman, M. R., Dornbusch, S. M., Herron, M. C., & Herting, J. R. (1997). The influence of family regulation, connection, and psychological autonomy on six measures of adolescent functioning. *Journal of Adolescent Research, 12,* 34–67.

Hermann, M., Untergasser, G., Rumpold, H., & Berger, P. (2000). Aging of the male reproductive system. *Experimental Gerontology, 35,* 1267–1279.

Herrnstein, R. J., & Murray, C. (1994). *The bell curve.* New York: Free Press.

Hershberger, S. L., Pilkington, N. W., & D'Augelli, A. R. (1997). Predictors of suicide attempts among gay, lesbian, and bisexual youth. *Journal of Adolescent Research, 12,* 477–497.

Hershey, D. A., Walsh, D. A., Brougham, R., & Carter, S. (1998). Challenges of training pre-retirees to make sound financial planning decisions. *Educational Gerontology, 24,* 447–470.

Hespos, S. J., & Baillargeon, R. (2001). Reasoning about containment events in very young infants. *Cognition, 78,* 207–245.

Hesse, E., & Main, M. (2000). Disorganized infant, child, and adult attachment: Collapse in behavioral and attentional strategies. *Journal of the American Psychoanalytic Association, 48,* 1097–1127.

Hetherington, E. M. (1989). Coping with family transitions:

Winners, losers, and survivors. *Child Development, 60,* 1–14.

Hetherington, E. M. (1995, March). *The changing American family and the well-being of children.* Master lecture presented at the biennial meeting of the Society for Research in Child Development, Indianapolis.

Hetherington, E. M. (1997). Teenaged childbearing and divorce. In S. Luthar, J. A. Burack, D. Cicchetti, & J. Weisz (Eds.), *Developmental psychopathology: Perspectives on adjustment, risk, and disorders* (pp. 350–373). Cambridge, U.K.: Cambridge University Press.

Hetherington, E. M. (1999a). Should we stay together for the sake of the children? In E. M. Hetherington (ed.), *Coping with divorce, single-parenting, and remarriage: A risk and resiliency perspective* (pp. 93–116). Hillsdale, NJ: Erlbaum.

Hetherington, E. M. (1999b). Social capital and the development of youth from nondivorced, divorced, and remarried families. In A. Collins (Ed.), *Minnesota Symposia on Child Psychology* (Vol. 29). Hillsdale, NJ: Erlbaum.

Hetherington, E. M., Bridges, M., & Insabella, G. M. (1998). What matters? What does not? Five perspectives on the association between marital transitions and children's adjustment. *American Psychologist, 53,* 167–184.

Hetherington, E. M., & Clingempeel, W. G. (1992). Coping with marital transitions: A family systems perspective. *Monographs of the Society for Research in Child Development, 57*(2–3, Serial No. 227).

Hetherington, E. M., Henderson, S. H., & Reiss, D. (1999). Adolescent siblings in stepfamilies: Family functioning and adolescent adjustment. *Monographs of the Society for Research in Child Development, 64*(4, Serial No. 259).

Hetherington, E. M., & Jodl, K. M. (1994). Stepfamilies as settings for child development. In A. Booth & J. Dunn (Eds.), *Stepfamilies: Who benefits? Who does not?* (pp. 55–79). Hillsdale, NJ: Erlbaum.

Hetherington, E. M., & Kelly, J. (2002). *For better or for worse: Divorce reconsidered.* New York: Norton.

Hetherington, E. M., Law, T. C., & O'Connor, T. G. (1994). Divorce: Challenges, changes, and new chances. In F. Walsh (Ed.), *Normal family processes* (2nd ed., pp. 208–234). New York: Guilford.

Hetherington, E. M., & Stanley-Hagan, M. (1999). The adjustment of children with divorced parents: A risk and resiliency perspective. *Journal of Child Psychology and Psychiatry, 40,* 129–140.

Hetherington, E. M., & Stanley-Hagan, M. (2000). Diversity among stepfamilies. In D. H. Demo, K. R. Allen, & M. A. Fine (Eds.), *Handbook of family diversity* (pp. 173–196). New York: Oxford University Press.

Hetherington, E. M., & Stanley-Hagan, M. (2002). Parenting in divorced and remarried families. In M. H. Bornstein (Ed.), *Handbook of parenting* (2nd ed., Vol. 3, pp. 287–315). Mahwah, NJ: Erlbaum.

Hetherington, S. E. (1990). A controlled study of the effect of prepared childbirth classes on obstetric outcomes. *Birth, 17,* 86–90.

Heuveline, P. (2002). An international comparison of adolescent and young adult morality. *Annals of the American Association of Political and Social Science, 580,* 172–200.

Hewlett, B. S. (1992). Husband–wife reciprocity and the father–infant relationship among Aka pygmies. In B. S. Hewlett (Ed.), *Father–child relations: Cultural and biosocial contexts* (pp. 153–176). New York: Aldine de Gruyter.

Heyman, G. D., & Dweck, C. S. (1998). Children's thinking about traits: Implications for judgments of the self and others. *Child Development, 69,* 391–403.

Heyman, G. D., Dweck, C. S., & Cain, K. M. (1992). Young children's vulnerability to self-blame and helplessness: Relationship to beliefs about goodness. *Child Development, 63,* 401–415.

Heyman, G. D., & Gelman, S. A. (1999). The use of trait labels in making psychological inferences. *Child Development, 70,* 604–619.

Hibell, B. (2001). *European School Survey Project on Alcohol and Drugs.* Stockholm: Swedish Council for Information on Alcohol and Other Drugs.

Hicks, P. (1997, January). The impact of aging on public policy. *OECD Observer,* No. 203, 19–21.

Hier, D. B., & Crowley, W. F. (1982). Spatial ability in androgen-deficient men. *New England Journal of Medicine, 302,* 1202–1205.

High, D. M., & Doole, M. M. (1995). Ethical and legal issues in conducting research involving elderly subjects. *Behavioral Sciences and the Law, 13,* 319–335.

High, K. P. (2001). Nutritional strategies to boost immunity and prevent infection in elderly individuals. *Aging and Infectious Diseases, 33,* 1892–1900.

High, P. C., LaGasse, L., Becker, S., Ahlgren, I., & Gardner, A. (2000). Literacy promotion in primary care pediatrics: Can we make a difference? *Pediatrics, 105,* 927–934.

Hill, D. (2003). Europe: When dying seems better than living. New York: Radio Free Europe. Retrieved from http://www.rferl. org/nca/features/2003/01/22012003154227.as

Hill, J. P., & Holmbeck, G. (1986). Attachment and autonomy during adolescence. In G. Whitehurst (Ed.), *Annals of child development* (Vol. 3, pp. 145–189). Greenwich, CT: JAI Press.

Hill, R. B. (1997). Social welfare policies and African American families. In H. P. McAdoo (Ed.), *Black families* (3rd ed., pp. 349–364). Thousand Oaks, CA: Sage.

Hillman, J. L. (2000). *Clinical perspectives on elderly sexuality.* New York: Kluwer Academic.

Hillman, J. L., & Stricker, G. (1994). A linkage of knowledge and attitudes toward elderly sexuality: Not necessarily a uniform relationship. *Gerontologist, 34,* 256–260.

Hinde, R. A. (1989). Ethological and relationships approaches. In R. Vasta (Ed.), *Annals of child development* (Vol. 6, pp. 251–285). Greenwich, CT: JAI Press.

Hines, S., & Bennett, F. (1996). Effectiveness of early intervention for children with Down syndrome. *Mental Retardation and Developmental Disabilities Research Reviews, 2,* 96–101.

Hinton, R., Moody, R. L., & Davis, A. W. (2002). Osteoarthritis: Diagnosis and therapeutic considerations. *American Family Physician, 65,* 841–848.

Hirsch, C. (1996). Understanding the influence of gender role identity on the assumption of family caregiving roles by men. *International Journal of Aging and Human Development, 42,* 103–121.

Hirschfeld, L. A. (1995). Do children have a theory of race? *Cognition, 54,* 209–252.

Hirshorn, B. A., Van Meter, J. V., & Brown, D. R. (2000). When grandparents raise grandchildren due to substance abuse: Responding to a uniquely destabilizing factor. In B. Hayslip, Jr., & R. Goldberg-Glen (Eds.), *Grandparents raising grandchildren: Theoretical, empirical, and clinical perspectives* (pp. 269–288). New York: Springer.

Ho, C. S.-H., & Fuson, K. C. (1998). Children's knowledge of teen quantities as tens and ones: Comparisons of Chinese, British, and American kindergartners. *Journal of Educational Psychology, 90,* 536–544.

Hoch, C. C., Dew, M. A., Reynolds, C. F., III, Buysse, D. J., Nowell, P. D., Monk, T. H., Mazumdar, S., Borland, M. D., Miewald, J., & Kupfer, D. J. (1997). Longitudinal changes in diary- and laboratory-based sleep measures in healthy "old old" and "young old" subjects: A three-year follow-up. *Sleep, 20,* 192–202.

Hochschild, A. R. (1997). *The time bind: When work becomes home and home becomes work.* New York: Metropolitan Books.

Hochwarter, W. A., Ferris, G. R., Perrewe, P. L., Witt, L. A., & Kiewitz, C. (2001). A note on the nonlinearity of the age–job satisfaction relationship. *Journal of Applied Social Psychology, 31,* 1223–1237.

Hodapp, R. M. (1996). Down syndrome: Developmental, psychiatric, and management issues. *Child and Adolescent Psychiatric Clinics of North America, 5,* 881–894.

Hodges, E. V. E., Boivin, M., Vitaro, F., & Bukowski, W. M. (1999). The power of friendship: Protection against an escalating cycle of peer victimization. *Developmental Psychology, 35,* 94–101.

Hodges, J., & Tizard, B. (1989). Social and family relationships of ex-institutional adolescents. *Journal of Child Psychology and Psychiatry, 30,* 77–97.

Hodges, R. M., & French, L. A. (1988). The effect of class and collection labels on cardinality, class-inclusion, and number conservation tasks. *Child Development, 59,* 1387–1396.

Hodson, D. S., & Skeen, P. (1994). Sexuality and aging: The hammerlock of myths. *Journal of Applied Gerontology, 13,* 219–235.

Hoff, B. (2001). *Full report of the prevalence, incidence, and consequences of violence against women.* Washington, DC: U.S. Department of Justice.

Hoff, E., Laursen, B., & Tardif, T. (2002). Socioeconomic status and parenting. In M. H. Bornstein (Ed.), *Handbook of parenting* (pp. 231–252). Mahwah, NJ: Erlbaum.

Hoffman, L. W. (2000). Maternal employment: Effects of social context. In R. D. Taylor & M. C. Wang (Eds.), *Resilience across contexts: Family, work, culture, and community* (pp. 147–176). Mahwah, NJ: Erlbaum.

Hoffman, L. W., & Youngblade, L. M. (1999). *Mothers at work: Effects on children's well-being.* New

York: Cambridge University Press.

Hoffman, M. L. (2000). *Empathy and moral development.* New York: Cambridge University Press.

Hoffman, S., & Hatch, M. C. (1996). Stress, social support and pregnancy outcome: A reassessment based on research. *Paediatric and Perinatal Epidemiology, 10,* 380–405.

Hoffner, C., & Badzinski, D. M. (1989). Children's integration of facial and situational cues to emotion. *Child Development, 60,* 411–422.

Hokoda, A., & Fincham, F. D. (1995). Origins of children's helpless and mastery achievement patterns in the family. *Journal of Educational Psychology, 87,* 375–385.

Holden, G. W., Coleman, S. M., & Schmidt, K. L. (1995). Why 3-year-old children get spanked: Determinants as reported by college-educated mothers. *Merrill-Palmer Quarterly, 41,* 431–452.

Holden, G. W., & West, M. J. (1989). Proximate regulation by mothers: A demonstration of how differing styles affect young children's behavior. *Child Development, 60,* 64–69.

Holland, J. C. (Ed.). (1998). *Psycho-oncology.* New York: Oxford University Press.

Holland, J. L. (1966). *The psychology of vocational choice.* Waltham, MA: Blaisdell.

Holland, J. L. (1985). *Making vocational choices: A theory of vocational personalities and work environments.* Englewood Cliffs, NJ: Prentice-Hall.

Holmbeck, G. N. (1996). A model of family relational transformations during the transition to adolescence: Parent–adolescent conflict and adaptation. In J. A. Graber, J. Brooks-Gunn, & A. C. Petersen (Eds.), *Transitions through adolescence* (pp. 167–199). Mahwah, NJ: Erlbaum.

Holmes-Rovner, M., Rovner, D. R., Padonu, G., Talarczyk, G., Kroll, J., Rothert, M., & Breer, L. (1996). African-American women's attitudes and expectations of menopause. *American Journal of Preventive Medicine, 12,* 420–423.

Holobow, N., Genessee, F., & Lambert, W. (1991). The effectiveness of a foreign language immersion program for children from different ethnic and social class backgrounds: Report 2. *Applied Psycholinguistics, 12,* 179–198.

Holtzman, D. M., Bales, K. R., Tendova, T., Fagan, A. M., Parsadanian, M., & Sartorius, L. J. (2000). Apoliprotein E isoform-dependent amyloid deposition and neuritic degeneration in a model of Alzheimer's disease. *Proceedings of the National Academy of Sciences, 97,* 2892–2897.

Hood, B. M., Atkinson, J., & Braddick, O. J. (1998). Selection-for-action and the development of orienting and visual attention. In J. E. Richards (Ed.), *Cognitive neuroscience of attention: A developmental perspective* (pp. 219–251). Mahwah, NJ: Erlbaum.

Hooker, K. (1992). Possible selves and perceived health in older adults and college students. *Journal of Gerontology, 47,* P85–P89.

Hooker, K., & Kaus, C. R. (1992). Possible selves and health behaviors in later life. *Journal of Aging and Health, 4,* 390–411.

Hooker, K., & Kaus, C. R. (1994). Health-related possible selves in young and middle adulthood. *Psychology and Aging, 9,* 126–133.

Hooyman, N. B., & Kiyak, H. A. (2002). *Social gerontology* (6th ed.). Boston: Allyn and Bacon.

Hope, S., Power, C., & Rodgers, B. (1999). Does financial hardship account for elevated psychological distress in lone mothers? *Social Science and Medicine, 29,* 381–389.

Hopkins, B., & Butterworth, G. (1997). Dynamical systems approaches to the development of action. In G. Bremner, A. Slater, & G. Butterworth (Eds.), *Infant development: Recent advances* (pp. 75–100). East Sussex, U.K.: Psychology Press.

Hopkins, B., & Westra, T. (1988). Maternal handling and motor development: An intracultural study. *Genetic, Social and General Psychology Monographs, 14,* 377–420.

Hopp, F. P., & Duffy, S. A. (2000). Racial variations in end-of-life care. *Journal of the American Geriatrics Society, 48,* 658–663.

Hopper, S. V. (1993). The influence of ethnicity on the health of older women. *Clinics in Geriatric Medicine, 9,* 231–259.

Horber, F. F., Kohler, S. A., Lippuner, K., & Jaeger, P. (1996). Effect of regular physical training on age-associated alteration of body composition in men. *European Journal of Clinical Investigation, 26,* 279–285.

Horgan, D. (1978). The development of the full passive. *Journal of Child Language, 5,* 65–80.

Horgas, A. L., Wilms, H., & Baltes, M. M. (1998). Daily life in very old age: Everyday activities as expression of successful living. *Gerontologist, 38,* 556–568.

Horn, J. L., & Donaldson, G. (1980). Cognitive development in adulthood. In O. G. Brim, Jr., & J. Kagan (Eds.), *Constancy and change in human development* (pp. 445–529). Cambridge, MA: Harvard University Press.

Horn, J. L., Donaldson, G., & Engstrom, R. (1981). Apprehension, memory, and fluid intelligence decline through the "vital years" of adulthood. *Research on Aging, 3,* 33–84.

Horn, J. L., & Masunaga, H. (2000). New directions for research into aging and intelligence: The development of expertise. In T. J. Perfect & E. A. Maylor (Eds.), *Models of cognitive aging* (pp. 125–159). New York: Oxford University Press.

Horn, J. L., & Noll, J. (1997). Human cognitive capabilities: Gf–Gc theory. In D. P. Flanagan, J. L., Genshaft, & P. L. Harrison (Eds.), *Beyond traditional intellectual assessment* (pp. 53–91). New York: Guilford.

Horn, J. M. (1983). The Texas Adoption Project: Adopted children and their intellectual resemblance to biological and adoptive parents. *Child Development, 54,* 268–275.

Horne, R. S., Sly, D. J., Cranage, S. M., Chau, B., & Adamson, T. M. (2000). Effects of prematurity on arousal from sleep in the newborn infant. *Pediatric Research, 47,* 468–474.

Horner, T. M. (1980). Two methods of studying stranger reactivity in infants: A review. *Journal of Child Psychology and Psychiatry, 21,* 203–219.

Hornick, J. P., McDonald, L., & Robertson, G. B. (1992). Elder abuse in Canada and the United States: Prevalence, legal, and service issues. In R. D. Peters, R. J. McMahon, & V. L. Quinsey (Eds.), *Aggression and violence throughout the life span* (pp. 301–335). Newbury Park, CA: Sage.

Horowitz, A. V., McLaughlin, J., & White, H. R. (1997). How the negative and positive aspects of partner relationships affect the mental health of young married people. *Journal of Health and Social Behavior, 39,* 124–136.

Horowitz, A. V., White, H. R., & Howell-White, S. (1996). Becoming married and mental health: A longitudinal study of a cohort of young adults. *Journal of Marriage and the Family, 58,* 895–907.

Horowitz, F. D. (1992). John B. Watson's legacy: Learning and environment. *Developmental Psychology, 28,* 360–367.

Hotz, V. J., McElroy, S. W., & Sanders, S. G. (1997). The costs and consequences of teenage childbearing for mothers. In R. A. Maynard (Ed.), *Kids having kids* (pp. 55–94). Washington, DC: Urban Institute.

House, J. S., Kessler, R. C., Herzog, A. R., Mero, R. P., Kinney, A. M., & Breslow, M. J. (1990). Age, socioeconomic status, and health. *Milbank Quarterly, 68,* 383–411.

Houseknecht, S. K. (1987). Voluntary childlessness. In M. B. Sussman & S. K. Steinmetz (Eds.), *Handbook of marriage and the family* (pp. 369–392). New York: Plenum.

Housel, D. A. (1995). Spirituality and death and dying from a gay perspective. In J. K. Parry & A. S. Ryan (Eds.), *A cross-cultural look at death, dying, and religion* (pp. 117–130). Chicago: Nelson-Hall.

Howard, A., & Bray, D. W. (1988). *Managerial lives in transition: Advancing age and changing times.* New York: Guilford Press.

Howe, M. L. (2003). Memories from the cradle. *Current Directions in Psychological Science, 12,* 62–65.

Howe, N., Aquan-Assee, J., & Bukowski, W. M. (2001). Predicting sibling relations over time: Synchrony between maternal management styles and sibling relationship quality. *Merrill-Palmer Quarterly, 47,* 121–141.

Howes, C., & Matheson, C. C. (1992). Sequences in the development of competent play with peers: Social and social pretend play. *Developmental Psychology, 28,* 961–974.

Hsu, F. L. K. (1981). *Americans and Chinese: Passage to difference* (3rd ed.). Honolulu: University of Hawaii Press.

Hu, F. B., & Manson, J. E. (2001). Diet, lifestyle, and the risk of type 2 diabetes mellitus in women. *New England Journal of Medicine, 345,* 790–797.

Hudson, J. A., Fivush, R., & Kuebli, J. (1992). Scripts and episodes: The development of event memory. *Applied Cognitive Psychology, 6,* 483–505.

Hudson, J. A., Sosa, B. B., & Shapiro, L. R. (1997). Scripts and plans: The development of preschool children's event knowledge and event planning. In S. L. Friedman & E. K. Scholnick (Eds.), *The developmental psychology of planning* (pp. 77–102). Mahwah, NJ: Erlbaum.

Huesmann, L. R. (1986). Psychological processes promoting the relation between exposure to media violence and aggressive behavior by the viewer. *Journal of Social Issues, 42,* 125–139.

Huesmann, L. R., Moise-Titus, J., Podolski, C. & Eron, L. D. (2003). Longitudinal relations between children's exposure to TV violence and their aggressive and violent behavior in young adulthood: 1977–1992. *Developmental Psychology, 39,* 201–221.

Hughes, C. (1998). Finding your marbles: Does preschoolers' strategic behavior predict later understanding of mind? *Developmental Psychology, 34,* 1326–1339.

Hughes, C., & Dunn, J. (1998). Understanding mind and emotion: Longitudinal associations with mental-state talk between young friends. *Developmental Psychology, 34,* 1026–1037.

Hughes, J. N., Cavell, T. A., & Grossman, P. B. (1997). A positive view of self: Risk or protection for aggressive children? *Development and Psychopathology, 9,* 75–94.

Hultsch, D. F., & Dixon, R. A. (1990). Learning and memory in aging. In J. E. Birren & K. W. Schaie (Eds.), *Handbook of the psychology of aging* (3rd ed., pp. 258–274). San Diego: Academic Press.

Hultsch, D. F., Hertzog, C., Dixon, R. A., & Small, B. J. (1998). *Memory change in the aged.* New York: Cambridge University Press.

Hultsch, D. F., MacDonald, S. W. S., & Dixon, R. A. (2002). Variability in reaction time performance of younger and older adults. *Journal of Gerontology, 57B,* P101–P115.

Human Resources Development Canada. (2000). *Dropping out of high school: Definitions and costs.* Hull, Quebec: Author.

Humphrey, D., & Clement, M. (1998). *Freedom to die: People, politics, and the right-to-die movement.* New York: St. Martin's Press.

Humphrey, J. A., & White, J. W. (2000). Women's vulnerability to sexual assault from adolescence to young adulthood. *Journal of Adolescent Health, 27,* 419–424.

Humphrey, T. (1978). Function of the nervous system during prenatal life. In U. Stave (Ed.), *Perinatal physiology* (pp. 651–683). New York: Plenum.

Hunt, E., Streissguth, A. P., Kerr, B., & Olson, H. C. (1995). Mothers' alcohol consumption during pregnancy: Effects on spatial-visual reasoning in 14-year-old children. *Psychological Science, 6,* 339–342.

Hunter, M., Salter-Ling, N., & Glover, L. (2000). Donor insemination: Telling children about their origins. *Child Psychology and Psychiatry Review, 5,* 157–163.

Huntsinger, C. S., Jose, P. E., & Larson, S. L. (1998). Do parent practices to encourage academic competence influence the social adjustment of young European American and Chinese American Children? *Developmental Psychology, 34,* 747–756.

Hurme, H. (1991). Dimensions of the grandparent role in Finland. In P. K. Smith (Ed.), *The psychology of grandparenthood: An international perspective* (pp. 19–31). London: Routledge.

Hursti, U. K. (1999). Factors influencing children's food choice. *Annals of Medicine, 31, 26–32.*

Husaini, B. A., Blasi, A. J., & Miller, O. (1999). Does public and private religiosity have a moderating effect on depression? A bi-racial study of elders in the American south. *International Journal of Aging and Human Development, 48,* 63–72.

Huston, A. C., & Alvarez, M. M. (1990). The socialization context of gender role development in early adolescence. In R. Montemayor, G. R. Adams, & T. P. Gullotta (Eds.), *From childhood to adolescence: A transitional period?* (pp. 156–179). Newbury Park, CA: Sage.

Huston, A. C., & Wright, J. C. (1998). Mass media and children's development. In I. E. Siegel & K. A. Renninger (Eds.), *Handbook of child psychology: Vol. 4. Child psychology in practice* (5th ed., pp. 999–1058). New York: Wiley.

Huston, A. C., Wright, J. C., Marquis, J., & Green, S. B. (1999). How young children spend their time: Television and other activities. *Developmental Psychology, 35,* 912–925.

Huston, P., McHale, S., & Crouter, A. (1986). When the honeymoon's over: Changes in the marriage relationship over the first year. In R. Gilmour & S. Duck (Eds.), *The emerging field of personal relationships* (pp. 109–132). Hillsdale, NJ: Erlbaum.

Huston, T. L., & Vangelisti, A. L. (1995). How parenthood affects marriage. In M. A. Fitzpatrick & A. L. Vangelisti (Eds.), *Explaining family interactions* (pp. 147–176). Thousand Oaks, CA: Sage.

Huttenlocher, P. R. (1994). Synaptogenesis in the human cerebral cortex. In G. Dawson & K. W. Fischer (Eds.), *Human behavior and the developing brain* (pp. 137–152). New York: Guilford.

Huyck, M. H. (1990). Gender differences in aging. In J. E. Birren & K. W. Schaie (Eds.), *Handbook of the psychology of aging* (3rd ed., pp. 124–134). New York: Academic Press.

Huyck, M. H. (1995). Marriage and close relationships of the marital kind. In R. Blieszner & V. H. Bedford (Eds.), *Handbook of aging and the family* (pp. 181–200). Westport, CT: Greenwood Press.

Huyck, M. H. (1996). Continuities and discontinuities in gender identity. In V. L. Bengtson (Ed.), *Adulthood and aging* (pp. 98–121). New York: Springer-Verlag.

Huyck, M. H. (1998). Gender roles and gender identity in midlife. In S. L. Willis & J. D. Reid (Eds.), *Life in the middle* (pp. 209–232). San Diego: Academic Press.

Hyde, J. S., Klein, M. H., Essex, M. J., & Clark, R. (1995). Maternity leave and women's mental health. *Psychology of Women Quarterly, 19,* 257–285.

Hyde, J. S., & Linn, M. C. (1988). Gender differences in verbal ability: A meta-analysis. *Psychological Bulletin, 104,* 53–69.

Hyde, J. S., & Oliver, M. B. (2000). Gender differences in sexuality: Results from meta-analysis. In C. B. Travis & J. W. White (Eds.), *Sexuality, society, and feminism* (pp. 57–77). Washington, DC: American Psychological Association.

Hymel, S., LeMare, L., Ditner, E., & Woody, E. Z. (1999). Assessing self-concept in children: Variations across self-concept domains. *Merrill-Palmer Quarterly, 45,* 602–623.

Ianni, F. A. J., & Orr, M. T. (1996). Dropping out. In J. A. Graber, J. Brooks-Gunn, & A. C. Petersen (Eds.), *Transitions through adolescence: Interpersonal domains and context* (pp. 285–322). Mahwah, NJ: Erlbaum

Idler, E. L., & Kasl, S. V. (1997). Religion among disabled and nondisabled persons I: Cross-sectional patterns in health practices, social activities, and well-being. *Journal of Gerontology, 52B,* S294–S305.

Ihinger-Tallman, M., & Pasley, K. (1997). Stepfamilies in 1984 and today—A scholarly perspective. *Marriage and Family Review, 26,* 19–40.

Ikkink, K. K., & van Tilburg, T. (1998). Do older adults' network members continue to provide instrumental support in unbalanced relationships? *Journal of Social and Personal Relationships, 15,* 59–75.

Ingersoll-Dayton, B., Neal, M. B., & Hammer, L. B. (2001). Aging parents helping adult children: The experience of the sandwiched generation. *Family Relations, 50,* 262–271.

Inhelder, B., & Piaget, J. (1958). *The growth of logical thinking from childhood to adolescence: An essay on the construction of formal operational structures.* New York: Basic Books. (Original work published 1955)

Institute for Social Research. (2002). U.S. husbands do more housework. Ann Arbor: Author. Retrieved from: http://www.newswise.com/articles/2002/3/timeuse.umi.html

Iribarren, C., Sidney, S., Bild, D. E., Liu, K., Markowvitz, J. H., Roseman, J. M., & Matthews, K. (2000). Association of hostility with coronary artery calcification in young adults. *Journal of the American Medical Association, 283,* 2546–2551.

Irvine, J. J. (1986). Teacher–student interactions: Effects of student race, sex, and grade level. *Journal of Educational Psychology, 78,* 14–21.

Isabella, R. (1993). Origins of attachment: Maternal interactive behavior across the first year. *Child Development, 64,* 605–621.

Isabella, R., & Belsky, J. (1991). Interactional synchrony and the origins of infant–mother attachment: A replication study. *Child Development, 62,* 373–384.

Ismail, M. A., Nagib, N., Ismail, T., & Sibils, L. A. (1999). Comparison of vaginal and cesarean section delivery for fetuses in breech presentation. *Journal of Perinatal Medicine, 27,* 339–351.

Ito, Y., Teicher, M. H., Glod, C. A., & Ackerman, E. (1998). Preliminary evidence for aberrant cortical development in abused children: A quantitative EEG study. *Journal of Neuropsychiatry and Clinical Neuroscience, 10,* 298–307.

Izard, C. E. (1991). *The psychology of emotions.* New York: Plenum.

Izard, C. E., & Ackerman, B. P. (2000). Motivational, organizational, and regulatory functions of discrete emotions. In M. Lewis, & J. M. Haviland-Jones (Ed.), *Handbook of emotions* (2nd ed., pp. 253–264). New York: Guilford.

Jackson, G. R., & Owsley, C. (2000). Scotopic sensitivity during adulthood. *Vision Research, 40,* 2467–2473.

Jacobs, J. E., & Weisz, V. (1994). Gender stereotypes: Implications for gifted education. *Roeper Review, 16,* 152–155.

Jacobson, C. K., & Heaton, T. B. (1989). Voluntary childlessness among American men and women in the late 1980s. *Social Biology, 38,* 79–93.

Jacobson, J. L., Jacobson, S. W., Fein, G., Schwartz, P. M., & Dowler, J. (1984). Prenatal exposure to an environmental toxin: A test of the multiple effects model. *Developmental Psychology, 20,* 523–532.

Jacobson, J. L., Jacobson, S. W., Padgett, R. J., Brumitt, G. A., & Billings, R. L. (1992). Effects of prenatal PCB exposure on cognitive processing efficiency and sustained attention. *Developmental Psychology, 28,* 297–306.

Jacobson, K. C., & Crockett, L. J. (2000). Parental monitoring and adolescent adjustment: An ecological perspective. *Journal of Research on Adolescence, 10,* 65–97.

Jacobson, S. W. (1998). Specificity of neurobehavioral outcomes associated with prenatal alcohol exposure. *Alcoholism: Clinical and Experimental Research, 22,* 313–320.

Jacques, P. F. (1999). The potential preventive effects of vitamins for cataract and age-related macular degeneration. *International Journal of Vitamin Nutrition Research, 69,* 198–205.

Jadack, R. A., Hyde, J. S., Moore, C. F., & Keller, M. L. (1995). Moral reasoning about sexually transmitted diseases. *Child Development, 66,* 167–177.

Jaffe, J., Beebe, B., Feldstein, S., Crown, C. L., & Jasnow, M. D. (2001). Rhythms of dialogue in infancy. *Monographs of the Society for Research in Child Development, 66*(2, Serial No. 265).

Jaffee, S., Caspi, A., Moffitt, T. E., Belsky, J., & Silva, P. (2001). Why are children born to teen mothers at risk for adverse outcomes in young adulthood? Results of a 20-year longitudinal study. *Development and Psychopathology, 13,* 377–397.

Jain, A., Concat, J., & Leventhal, J. M. (2002). How good is the evidence linking breastfeeding and intelligence? *Pediatrics, 109,* 1044–1053.

Jambunathan, S., Burts, D. C., & Pierce, S. (2000). Comparisons of parenting attitudes among five ethnic groups in the United States. *Journal of Comparative Family Studies, 31,* 395–406.

James, D. (1998). Recent advances in fetal medicine. *British Medical Journal, 316,* 1580–1583.

James, J. B., Lewkowicz, C., Libhaber, J., & Lachman, M. (1995). Rethinking the gender identity crossover hypothesis: A test of a new model. *Sex Roles, 32,* 185–207.

Janosz, M., Le Blanc, M., Boulerice, B., & Tremblay, R. E. (2000). Predicting different types of school dropouts: A typological approach with two longitudinal samples. *Journal of Educational Psychology, 92,* 171–190.

Jansari, A., & Parkin, A. J. (1996). Things that go bump in your life: Explaining the reminiscence bump in autobiographical memory. *Psychology and Aging, 11,* 85–91.

Janssens, J. M. A. M., & Dekovíc, M. (1997). Child rearing, prosocial moral reasoning, and prosocial behaviour. *International Journal of Behavioral Development, 20,* 509–527.

Jarrold, C., Butler, D. W., Cottington, E. M., & Jimenez, F. (2000). Linking theory of mind and central coherence bias in autism and in the general population. *Developmental Psychology, 36,* 126–138.

Jasper, M. (1996). *The right to die.* Dobbs Ferry, NY: Oceana Publications.

Jayakody, R., & Cabrera, N. (2002). What are the choices for low-income families? Cohabitation, marriage, and remaining single. In A. Booth & A. C. Crouter (Eds.), *Just living together* (pp. 85–96). Mahwah, NJ: Erlbaum.

Jendrek, M. P. (1994). Grandparents who parent their grandchildren: Circumstances and decisions. *Gerontologist, 34,* 206–216.

Jenkins, J. M., & Astington, J. W. (2000). Theory of mind and social behavior: Causal models tested in a longitudinal study. *Merrill-Palmer Quarterly, 46,* 203–220.

Jennett, B. (2002). The vegetative state. *Journal of Neurology and Neurosurgical Psychiatry, 73,* 355–356.

Jensen, A. R. (1969). How much can we boost IQ and scholastic achievement? *Harvard Educational Review, 39,* 1–123.

Jensen, A. R. (1980). *Bias in mental testing.* New York: Free Press.

Jensen, A. R. (1985). The nature of the black–white difference on various psychometric tests: Spearman's hypothesis. *Behavioral and Brain Sciences, 8,* 193–219.

Jensen, A. R. (1998). *The g factor: The science of mental ability.* New York: Praeger.

Jensen, A. R. (2001). Spearman's hypothesis. In J. M. Collis & S. Messick (Eds.), *Intelligence and personality: Bridging the gap in theory and measurement* (pp. 3–24). Mahwah, NJ: Erlbaum.

Jensen, A. R., & Figueroa, R. A. (1975). Forward and backward digit-span interaction with race and IQ: Predictions from Jensen's theory. *Journal of Educational Psychology, 67,* 882–893.

Jensen, A. R., & Reynolds, C. R. (1982). Race, social class and ability patterns on the WISC-R. *Personality and Individual Differences, 3,* 423–438.

Jepson, K. L., & Labouvie-Vief, G. (1992). Symbolic processing of youth and elders. In R. L. West and J. D. Sinnott (Eds.), *Everyday memory and aging* (pp. 124–137). New York: Springer.

Jessor, R. (1996). Ethnographic methods in contemporary perspective. In R. Jessor, A. Colby, & R. A. Shweder (Eds.), *Ethnography and human development* (pp. 3–14). Chicago: University of Chicago Press.

Jeynes, W. H., & Littell, S. W. (2000). A meta-analysis of studies examining the effect of whole language instruction on the literacy of low-SES students. *Elementary School Journal, 101,* 21–33.

Jiao, S., Ji, G., & Jing, Q. (1996). Cognitive development of Chinese urban only children and children with siblings. *Child Development, 67,* 387–395.

Jimeniz-Sanchez, G., Childs, B., & Valle, D. (2001). Human disease genes. *Nature, 409,* 853–855.

Johannsson, G., Svensson, J., & Bengtsson, B.-A. (2000). Growth hormone and aging. *Growth Hormone and IGF Research 2000, Supplement B,* S25–S30.

Johnson, C. (2002). Obesity, weight management, and self-esteem. In T. A. Wadden & A. J. Stunkard (Eds.), *Handbook of obesity treatment* (pp. 480–493). New York: Guilford.

Johnson, C. L. (1985). *Growing up and growing old in Italian-American families.* New Brunswick, NJ: Rutgers University Press.

Johnson, C. L. (1998). Effects of adult children's divorce on grandparenthood. In M. E. Szinovacz (Ed.), *Handbook on grandparenthood* (pp. 87–96). Westport, CT: Greenwood Press.

Johnson, C. L., & Troll, L. (1992). Family functioning in late life. *Journal of Gerontology, 47,* S66–S72.

Johnson, C. L., & Troll, L. E. (1994). Constraints and facilitators to friendships in late life. *Gerontologist, 34,* 79–87.

Johnson, D. E. (2000). Medical and developmental sequelae of early childhood institutionalization in Eastern European adoptees. In C. A. Nelson (Ed.), *Minnesota symposia on child psychology* (Vol. 31, pp. 113–162). Mahwah, NJ: Erlbaum.

Johnson, G. R., Krug, E. G., & Potter, L. B. (2000). Suicide among adolescents and young adults: A cross-national comparison of 34 countries. *Suicide and Life-Threatening Behavior, 30,* 74–82.

Johnson, J. G., Cohen, P., Smailes, E. M., Kasen, S., & Brook, J. S. (2002). Television viewing and aggressive behavior during adolescence and adulthood. *Science, 295,* 2468–2471.

Johnson, K. E., Scott, P., & Mervis, C. B. (1997). Development of children's understanding of basic–subordinate inclusion relations. *Developmental Psychology, 33,* 745–763.

Johnson, M. H. (1995). The inhibition of automatic saccades in early infancy. *Developmental Psychobiology, 28,* 281–291.

Johnson, M. H. (1998). The neural basis of cognitive development. In D. Kuhn & R. S. Siegler (Ed.), *Handbook of child psychology: Vol. 2. Cognition, perception, and language* (5th ed., pp. 1–49). New York: Wiley.

Johnson, M. H. (1999). Ontogenetic constraints on neural and behavioral plasticity: Evidence from imprinting and face processing. *Canadian Journal of Experimental Psychology, 55,* 77–90.

Johnson, M. H. (2001). The development and neural basis of face recognition: Comment and speculation. *Infant and Child Development, 10,* 31–33.

Johnson, S. L., & Birch, L. L. (1994). Parents' and children's adiposity and eating style. *Pediatrics, 94,* 653–661.

Joint Centre for Bioethics. (2002). Is a living will legal in Canada? Toronto: University of Toronto. Retrieved from http://www.utoronto.ca/jcb/_lwdisclaimer/canchap2.htm

Jones, G. P., & Dembo, M. H. (1989). Age and sex role differences in intimate friendships during childhood and adolescence. *Merrill-Palmer Quarterly, 35,* 445–462.

Jones, J. L., Lopez, A. Wilson, M., Schulkin, J., & Gibbs, R. (2001). Congenital toxoplasmosis: A review. *Obstetrical and Gynecological Survey, 56,* 296–305.

Jones, M. C. (1965). Psychological correlates of somatic development. *Child Development, 36,* 899–911.

Jones, M. C., & Mussen, P. H. (1958). Self-conceptions, motivations, and interpersonal attitudes of early- and late-maturing girls. *Child Development, 29,* 491–501.

Jones, W. H. (1990). Loneliness and social exclusion. *Journal of Social and Clinical Psychology, 9,* 214–220.

Jongbloet, P. H., Zielhuis, G. A., Groenewoud, H. M., & Pasker-De Jong, P. C. (2001). The secular trends in male:female ratio at birth in postwar industrialized countries. *Environmental Health Perspectives, 109,* 749–752.

Jordan, B. (1993). *Birth in four cultures.* Prospect Heights, IL: Waveland.

Jorgensen, K. M. (1999). Pain assessment and management in the newborn infant. *Journal of PeriAnesthesia Nursing, 14,* 349–356.

Jorgensen, M., & Keiding, K. (1991). Estimation of spermarche from longitudinal spermaturia data. *Biometrics, 47,* 177–193.

Joseph, J. (1997). Fear of crime among black elderly. *Journal of Black Studies, 27,* 698–717.

Josselson, R. (1994). The theory of identity development and the question of intervention. In S. L. Archer (Ed.), *Interventions for adolescent identity development* (pp. 12–25). Thousand Oaks, CA: Sage.

Jovanovic, J., & King, S. S. (1998). Boys and girls in the performance-based science classroom: Who's doing the performing? *American Educational Research Journal, 35,* 477–496.

Joyner, M. H., & Kurtz-Costes, B. (1997). Metamemory development. In W. Schneider & F. E. Weinert (Eds.), *Memory performance and competencies: Issues in growth and development* (pp. 275–300). Hillsdale, NJ: Erlbaum.

Julkunen, J. (1996). Suppressing your anger: Good manners, bad health? In C. D. Spielberger & I. G. Sarason (Eds.), *Stress and emotion: Anxiety, anger, and curiosity* (Vol. 16, pp. 227–240). Washington, DC: Taylor & Francis.

Juntunen, C. L., Wegner, K. E., & Matthews, L. G. (2002). Promoting positive career change in midlife. In C. L. Juntunen & D. R. Atkinson (Eds.), *Counseling across the lifespan* (pp. 329–347). Thousand Oaks, CA: Sage.

Jusczyk, P. (1995). Language acquisition: Speech sounds and phonological development. In J. L. Miller & P. D. Eimas (Eds.), *Handbook of perception and cognition: Vol. 2. Speech, language, and communication* (pp. 263–301). Orlando, FL: Academic Press.

Jusczyk, P. W. (2001). In the beginning, was the word.... In F. Lacerda & C. von Hofsten (Eds.), *Emerging cognitive abilities in early infancy* (pp. 173–192). Mahwah, NJ: Erlbaum.

Jusczyk, P. W., & Aslin, R. N. (1995). Infants' detection of the sound patterns of words in fluent speech. *Cognitive Psychology, 29,* 1–23.

Jusczyk, P. W., & Hohne, E. A. (1997). Infants' memory for spoken words. *Science, 277,* 1984–1986.

Jusczyk, P. W., Houston, D. M., & Newsome, M. (1999). The beginnings of word segmentation in English-learning infants. *Cognitive Psychology, 39,* 159–207.

Justice, E. M., Baker-Ward, L., Gupta, S., & Jannings, L. R. (1997). Means to the goal of remembering: Developmental changes in awareness of strategy use–performance relations. *Journal of Experimental Child Psychology, 65,* 293–314.

Juul, A. (2001). The effects of oestrogens on linear bone growth. *Human Reproduction Update, 7,* 303–313.

Kagan, J. (1998). Biology and the child. In N. Eisenberg (Ed.), *Handbook of child psychology: Vol. 3. Social, emotional, and personality development* (5th ed., pp. 177–236). New York: Wiley.

Kagan, J., Arcus, D., Snidman, N., Feng, W. Y. Hendler, J., & Greene, S. (1994). Reactivity in infants: A cross-national comparison. *Developmental Psychology, 30,* 342–345.

Kagan, J., Kearsley, R. B., & Zelazo, P. R. (1978). *Infancy: Its place in human development.* Cambridge, MA: Harvard University Press.

Kagan, J., & Saudino, K. J. (2001). Behavioral inhibition and related temperaments. In R. N. Emde & J. K. Hewitt (Eds.), *Infancy to early childhood: Genetic and environmental influences on developmental change* (pp. 111–119). New York: Oxford University Press.

Kagan, J., Snidman, N., & Arcus, D. (1998). Childhood derivatives of high and low reactivity in infancy. *Child Development, 69,* 1483–1493.

Kagan, J., Snidman, N., Zentner, M., & Peterson, E. (1999). Infant temperament and anxious symptoms in school-age children. *Development and Psychopathology, 11,* 209–224.

Kagawa, Y. (1978). Impact of westernization on the nutrition of Japanese: Changes in physique, cancer, longevity, and centenarians. *Preventive Medicine, 7,* 205–217.

Kahn, P. H., Jr. (1992). Children's obligatory and discretionary moral judgments. *Child Development, 63,* 416–430.

Kail, R. (1993). Processing time decreases globally at an exponential rate during childhood and adolescence. *Journal of Experimental Child Psychology, 57,* 281–291.

Kail, R. (2000). Speed of information processing: Developmental change and links to intelligence. *Journal of School Psychology, 38,* 51–61.

Kail, R., & Park, Y. (1992). Global developmental change in processing time. *Merrill-Palmer Quarterly, 38,* 525–541.

Kail, R., & Park, Y. (1994). Processing time, articulation time, and memory span. *Journal of Experimental Child Psychology, 57,* 281–291.

Kaisa, A., Stattin, H., & Nurmi, J. (2000). Parenting styles and adolescents' achievement strategies. *Journal of Adolescence, 23,* 205–222.

Kaler, S. B., & Kopp, C. B. (1990). Compliance and comprehension in very young toddlers. *Child Development, 61,* 1997–2003.

Kalish, R. A. (1985). The social context of death and dying. In R. H. Binstock & E. Shanas (Eds.), *Handbook of aging and the social sciences* (2nd ed., pp. 149–170). New York: Van Nostrand Reinhold.

Kalof, L. (2000). Ethnic differences in female sexual victimization. *Sexuality and Culture, 2,* 75–97.

Kalsi, M., Heron, G., & Charman, W. N. (2001). Changes in the static accommodation response with age. *Ophthalmic and Physiological Optics, 21,* 77–84.

Kamerman, S. B. (1993). International perspectives on child care policies and programs. *Pediatrics, 91,* 248–252.

Kamerman, S. B. (2000). From maternity to parental leave policies: Women's health, employment, and child and family well-being. *Journal of the American Medical Women's Association,* 55, 96–99.

Kamo, Y. (1998). Asian grandparents. In M. E. Szinovacz (Ed.), *Handbook on grandparenthood* (pp. 97–112). Westport, CT: Greenwood Press.

Kandall, S. R., & Gaines, J. (1991). Maternal substance use and subsequent sudden infant death syndrome (SIDS) in offspring. *Neurotoxicology and Teratology, 13,* 235–240.

Kao, G. (2000). Psychological well-being and educational achievement among immigrant youth. In D. J. Hernandez (Ed.), *Children of immigrants: Health, adjustment, and public assistance.* Washington, DC: National Academy Press.

Kao, G., & Tienda, M. (1995). Optimism and achievement: The educational performance of immigrant youth. *Social Science Quarterly, 76,* 1–19.

Kaplan, D. L., & Keys, C. B. (1997). Sex and relationship variables as predictors of sexual attraction in cross-sex platonic friendships between young heterosexual adults. *Journal of Social and Personal Relationships, 14,* 191–206.

Kaprio, J., Rimpela, A., Winter, T., Viken, R. J., Pimpela, M., & Rose, R. J. (1995). Common genetic influence on BMI and age at menarche. *Human Biology, 67,* 739–753.

Karadsheh, R. (1991, April). *This room is a junkyard!: Children's comprehension of metaphorical language.* Paper presented at the biennial meeting of the Society for Research in Child Development, Seattle, WA.

Karlseder, J., Smogorzewska, A., & de Lange, T. (2002). Senescence induced by altered telomere state, not telomere loss. *Science, 295,* 2446–2449.

Karmiloff-Smith, A. (1992). *Beyond modularity: A developmental perspective on cognitive science.* Cambridge, MA: MIT Press.

Kaslow, F. W., Hansson, K., & Lundblad, A. (1994). Long-term marriages in Sweden: And some comparisons with similar couples in the United States. *Contemporary Family Therapy, 16,* 521–537.

Kastenbaum, R. J. (2001). *Death, society, and human experience* (7th ed.). Boston: Allyn and Bacon.

Katchadourian, H. (1990). Sexuality. In S. S. Feldman & G. R. Elliott (Eds.), *At the threshold: The developing adolescent* (pp. 330–351). Cambridge, MA: Harvard University Press.

Katzmarzyk, A. (2002). The Canadian obesity epidemic, 1995–1998. *Canadian Medical Association Journal, 166,* 1039–1040.

Kaufman, A. S. (2001). WAIS-III IQs, Horn's theory, and generational changes from young adulthood to old age. *Intelligence, 29,* 131-167.

Kaufman, A. S., & Horn, J. L. (1996). Age changes on tests of fluid and crystallized intelligence for females and males on the Kaufman Adolescent and Adult Intelligence Test (KAIT) at ages 17 to 94 years. *Archives of Clinical Neuropsychology, 11,* 97–121.

Kausler, D. H. (1991). *Experimental psychology, cognition, and human aging.* New York: Springer.

Kausler, D. H. (1994). *Learning and memory in normal aging.* San Diego: Academic Press.

Kavanaugh, R. D., & Engel, S. (1998). The development of pretense and narrative in early childhood. In O. N. Saracho & B. Spodek (Eds.), *Multiple perspectives on play in early childhood education* (pp. 80–99). Albany:

State University of New York Press.

Kay, M. A., Manno, C. S., Ragni, M. V., Larson, P. J., Couto, L. B., McClelland, A., et al. (2000). Evidence for gene transfer and expression of factor IX in haemophilia B patients treated with an AAV vector. *Nature Genetics, 24,* 257–261.

Kaye, K., & Marcus, J. (1981). Infant imitation: The sensory-motor agenda. *Developmental Psychology, 17,* 258–265.

Kaye, W. H., Klump, K. L., Frank, G. K. W., & Strober, M. (2000). Anorexia and bulimia nervosa. *Annual Review of Medicine, 51,* 299–313.

Kearins, J. M. (1981). Visual spatial memory in Australian aboriginal children of desert regions. *Cognitive Psychology, 13,* 434–460.

Keating, D. (1979). Adolescent thinking. In J. Adelson (Ed.), *Handbook of adolescent psychology* (pp. 211–246). New York: Wiley.

Keil, F. C. (1986). Conceptual domains and the acquisition of metaphor. *Cognitive Development, 1,* 73–96.

Keil, F. C., & Lockhart, K. L. (1999). Explanatory understanding in conceptual development. In E. K. Scholnick, K. Nelson, S. A. Gelman, & P. H. Miller (Eds.), *Conceptual development: Piaget's legacy* (pp. 103–130). Mahwah, NJ: Erlbaum.

Keith, J., Fry, C. L., Glascock, A. P., Ikels, C., Dickerson-Putman, J., Harpending, H. C., & Draper, P. (1994). *The aging experience: Diversity and commonality across cultures.* Thousand Oaks, CA: Sage.

Keith, P. M., & Schafer, R. B. (1991). *Relationships and well-being over the life stages.* New York: Praeger.

Keith, T. Z., Keith, P. B., Quirk, K. J., Sperduto, J., Santillo, S., & Killings, S. (1998). Longitudinal effects of parent involvement on high school grades: Similarities and differences across gender and ethnic groups. *Journal of School Psychology, 36,* 335–363.

Kellett, J. M. (2000). Older adult sexuality. In L. T. Szuchman & F. Muscarella (Eds.), *Psychological perspectives on human sexuality* (pp. 355–379). New York: Wiley.

Kelly, S. J., Day, N., & Streissguth, A. P. (2000). Effects of prenatal alcohol exposure on social behavior in humans and other species. *Neurotoxicology and Teratology, 22,* 143–149.

Kempe, C. H., Silverman, B. F., Steele, P. W., Droegemueller, P. W., & Silver, H. K. (1962). The battered-child syndrome. *Journal of the American Medical Association, 181,* 17–24.

Kemper, S., Kynette, D., & Norman, S. (1992). Age differences in spoken language. In R. L. West & J. D. Sinnott (Eds.), *Everyday memory and aging* (pp. 138–152). New York: Springer-Verlag.

Kemper, S., Thompson, M., & Marquis, J. (2001). Longitudinal change in language production: Effects of aging and dementia on grammatical complexity and prepositional content. *Psychology and Aging, 16,* 600–614.

Kennedy, C. M. (1998). Childhood nutrition. *Annual Review of Nursing Research, 16,* 3–38.

Kennedy, G. E., & Kennedy, C. E. (1993). Grandparents: A special resource for children in stepfamilies. *Journal of Divorce and Remarriage, 19,* 45–68.

Kennedy, G. J., & Tanenbaum, S. (2000). Suicide and aging: International perspectives. *Psychiatric Quarterly, 71,* 345–362.

Kennedy, Q., Fung, H. H., & Carstensen, L. L. (2001). Aging, time estimation, and emotion. In S. H. McFadden & R. C. Atchley (Eds.), *Aging and the meaning of time: A multidisciplinary exploration* (pp. 51–70). New York: Springer.

Kennell, J., Klaus, M., McGrath, S., Robertson, S., & Hinkley, C. (1991). Continuous emotional support during labor in a U.S. hospital. *Journal of the American Medical Association, 265,* 2197–2201.

Kennet, J., Burgio, L., & Schultz, R. (2000). Interventions for in-home caregivers: A review of research 1990 to present. In R. Schulz (Ed.), *Handbook on dementia caregiving* (pp. 61–126). New York: Springer.

Kenrick, D. T., Gabrielidis, C., Keefe, R. C., & Cornelius, J. S. (1996). Adolescents' age preferences for dating partners: Support for an evolutionary model of life-history strategies. *Child Development, 67,* 1499–1511.

Kenyon, B. L. (2001). Current research in children's conceptions of death: A critical review. *Omega, 43,* 63–91.

Kerber, R. A., O'Brien, E., Smith, K. R., & Cawthon, R. M. (2001). Familial excess longevity in Utah genealogies. *Journals of Gerontology, 567,* B130–B139.

Kerckhoff, A. C. (2002). The transition from school to work. In J. T. Mortimer & R. Larson (Eds.), *The changing adolescent experience* (pp. 52–87). New York: Cambridge University Press.

Kerridge, I. H., & Mitchell, K. R. (1996). The legislation of active voluntary euthanasia in Australia: Will the slippery slope prove fatal? *Journal of Medical Ethics, 22,* 273–278.

Kessler, R., McGonagle, K., Zhao, S., Nelson, C., Hughes, M., Eshleman, S., Wittchen, H., & Kendler, K. (1994). Lifetime and 12-month prevalence of DSM-III-R psychiatric disorders in the United States: Results from the national comorbidity survey. *Archives of General Psychiatry, 51,* 8–19.

Kettl, P. (1998). Alaska Native suicide: Lessons for elder suicide. *International Psychogeriatrics, 10,* 205–211.

Keyes, C. L. M., & Ryff, C. D. (1998a). Generativity in adult lives: Social structural contours and quality of life consequences. In D. P. McAdams & E. de St. Aubin (Eds.), *Generativity and adult development* (pp. 227–257). Washington, DC: American Psychological Association.

Keyes, C. L. M., & Ryff, C. D. (1998b). Psychological well-being in midlife. In S. L. Willis & J. D. Reid (Eds.), *Life in the middle* (pp. 161–180). San Diego: Academic Press.

Kibby, M. Y., & Hynd, G. W. (2001). Neurobiological basis of learning disabilities. In D. P. Hallahan & B. K. Keogh (Eds.), *Research and global perspectives in learning disabilities* (pp. 25–42). Mahwah, NJ: Erlbaum.

Kiernan, K. (2001). European perspectives on nonmarital childbearing. In L. L. Wu & B. Wolfe (Eds.), *Out of wedlock: Causes and consequences of nonmarital fertility* (pp. 77–108). New York: Russell Sage Foundation.

Kiernan, K. (2002). Cohabitation in Western Europe: Trends, issues, and implications. In A. Booth & A. C. Crouter (Eds.), *Just living together* (pp. 3–32). Mahwah, NJ: Erlbaum.

Killen, M., & Nucci, L. P. (1995). Morality, autonomy, and social conflict. In M. Killen & D. Hart (Eds.), *Morality in everyday life: Developmental perspectives* (pp. 52–86). Cambridge, U.K.: Cambridge University Press.

Killen, M., & Smetana, J. G. (1999). Social interactions in preschool classrooms and the development of young children's conceptions of the personal. *Child Development, 70,* 486–501.

Kilpatrick, D. G., Acierno, R., Saunders, B., Resnick, H. S., Best, C. L., & Schnurr, P. P. (2000). Risk factors for adolescent substance abuse and dependence: Data from a national sample. *Journal of Consulting and Clinical Psychology, 68,* 19–30.

Kilpatrick, S. W., & Sanders, D. M. (1978). Body image stereotypes: A developmental comparison. *Journal of Genetic Psychology, 132,* 87–95.

Kim, J. E., & Moen, P. (2002a). Is retirement good or bad for subjective well-being? *Current Directions in Psychological Science, 10,* 83–86.

Kim, J. E., & Moen, P. (2002b). Moving into retirement: Preparation and transitions in late midlife. In M. E. Lachman (Ed.), *Handbook of midlife development* (pp. 487–527). New York: Wiley.

Kim, J. E., & Moen, P. (2002c). Retirement transitions, gender, and psychological well-being: A life-course, ecological model. *Journal of Gerontology, 57B,* P212–P222.

Kim, J. M. (1998). Korean children's concepts of adult and peer authority and moral reasoning. *Developmental Psychology, 34,* 947–955.

Kim, J. M., & Turiel, E. (1996). Korean children's concepts of adult and peer authority. *Social Development, 5,* 310–329.

Kim, M., McGregor, K. K., & Thompson, C. K. (2000). Early lexical development in English- and Korean-speaking children: Language-general and language-specific patterns. *Journal of Child Language, 27,* 225–254.

Kimmel, D. C. (2002). Aging and sexual orientation. In B. E. Jones & M. J. Hill (Eds.), *Mental health issues in lesbian, gay, bisexual, and transgender communities. (Review of psychiatry, Vol. 21,* pp. 17–36). Washington, DC: American Psychiatric Publishing.

Kimmel, D. C., & Moody, H. R. (1990). Ethical issues in gerontological research and services. In J. E. Birren & K. W. Schaie (Eds.), *Handbook of the psychology of aging* (3rd ed., pp. 489–501). San Diego, CA: Academic Press.

King, A. C. (2001). Interventions to promote physical activity by older adults. *Journal of Gerontology, 56A,* 36A–46A.

King, A. C., Castro, C., Wilcox, S., Eyler, A. A., Sallis, J. F., & Brownson, R. C. (2000). Personal and environmental factors associated with physical inactivity among different racial–ethnic groups of U.S. middle-aged and older-aged women. *Health Psychology, 19,* 354–364.

King, A. C., Kiernan, M., Oman, R. F., Kraemer, H., Hull, M., & Ahn, D. (1997). Can we identify who will adhere to long-term physical activity? Signal detection methodology as a potential aid to clinical decision making. *Health Psychology, 16,* 380–389.

King, A. C., Taylor, C. B., & Haskell, W. L. (1993). Effects of differing intensities and formats of 12 months of exercise training on psychological outcomes in older adults. *Health Psychology, 12,* 292–300.

King, D. A., & Markus, H. E. (2000). Mood disorders in older adults. In S. K. Whitbourne (Ed.), *Psychopathology in later adulthood* (pp. 141–172). New York: Wiley.

Kingston, P., & Reay, A. (1996). Elder abuse and neglect. In R. T. Woods (Ed.), *Handbook of the clinical psychology of ageing* (pp. 423–438). Chichester, U.K.: Wiley.

Kinicki, A. J., Prussia, G. E., & McKee-Ryan, F. M. (2000). A panel study of coping with involuntary job loss. *Academy of Management Journal, 43,* 90–100.

Kinney, D. (1999). From "headbangers" to "hippies": Delineating adolescents' active attempts to form an alternative peer culture. In J. A. McLellan & M. J. V. Pugh (Eds.), *The role of peer groups in adolescent social identity: Exploring the importance of stability and change* (pp. 21–35). San Francisco: Jossey-Bass.

Kirchner, G. (2000). *Children's games from around the world.* Boston: Allyn and Bacon.

Kirkby, R. J., & Lindner, H. (1998). Exercise is linked to reductions in anxiety but not premenstrual syndrome in women with prospectively-assessed symptoms. *Psychology, Health and Medicine, 3,* 211–222.

Kisilevsky, B. S., & Low, J. A. (1998). Human fetal behvior: 100 years of study. *Developmental Review, 18,* 1–29.

Kite, M. E., & Whitley, B. E., Jr. (1998). Do heterosexual women and men differ in their attitudes toward homosexuality? In G. M. Herek (Ed.), *Stigma and sexual orientation* (pp. 39–61). Thousand Oaks, CA: Sage.

Kivnick, H. Q. (1983). Dimensions of grandparenthood meaning: Deductive conceptualization and empirical derivation. *Journal of Personality and Social Psychology, 44,* 1056–1068.

Klahr, D., & MacWhinney, B. (1998). Information processing. In D. Kuhn & R. S. Siegler (Eds.), *Handbook of child psychology: Vol. 2. Cognition, perception, and language* (5th ed., pp. 631–678). New York: Wiley.

Klebanov, P. K., Brooks-Gunn, J., McCarton, C., & McCormick, M. C. (1998). The contribution of neighborhood and family income to developmental test scores over the first three years of life. *Child Development, 69,* 1420–1436.

Klein, J., & Sauer, M. V. (2001). Assessing fertility in women of advanced reproductive age. *American Journal of Obstetrics and Gynecology, 185,* 758–770.

Klein, P. J., & Meltzoff, A. N. (1999). Long-term memory, forgetting, and deferred imitation in 12-month-old infants. *Developmental Science, 2,* 102–113.

Klesges, L., M., Johnson, K. C., Ward, K. D., & Barnard, M. (2001). Smoking cessation in pregnant women. *Obstetrics and Gynecology Clinics of North America, 28,* 269–282.

Kliewer, W., Fearnow, M. D., & Miller, P. A. (1996). Coping socialization in middle childhood: Tests of maternal and paternal influences. *Child Development, 67,* 2339–2357.

Klimes-Dougan, B., & Kistner, J. (1990). Physically abused preschoolers' responses to peers' distress. *Developmental Psychology, 26,* 599–602.

Kling, K. C., Hyde, J. S., Showers, C. J., & Buswell, B. N. (1999). Gender differences in self-esteem: A meta-analysis. *Psychological Bulletin, 125,* 470–500.

Klingner, J. K., Vaughn, S., Hughes, M. T., Schumm, J. S., & Elbaum, B. (1998). Outcomes for students with and without learning disabilities in inclusive classrooms. *Learning Disabilities Research and Practice, 13,* 153–161.

Klump, K. L., Kaye, W. H., & Strober, M. (2001). The evolving foundations of eating disorders. *Psychiatric Clinics of North America, 24,* 215–225.

Knapp, M. L., & Taylor, E. H. (1994). Commitment and its communication in romantic relationships. In A. L. Weber & J. H. Harvey (Eds.), *Perspectives on close relationships* (pp. 153–175). Boston: Allyn and Bacon.

Knecht, S., Draeger, B., Deppe, M., Bobe, L., Lohmann, H., Floeel, A., Ringelstein, E.-B., & Henningsen, H. (2000). Handedness and hemispheric language dominance in healthy humans. *Brain, 123,* 2512–2518.

Knobloch, H., & Pasamanick, B. (Eds.). (1974). *Gesell and Amatruda's Developmental Diagnosis.* Hagerstown, MD: Harper & Row.

Knoers, N., van den Ouweland, A., Dreesen, J., Verdijk, M., Monnens, L. S., & van Oost, B. A. (1993). Nephrogenic diabetes insipidus: Identification of the genetic defect. *Pediatric Nephrology, 7,* 685–688.

Knowles, R. B., Wyart, C., Buldyrev, S. V., Cruz, L., Urbanc, B., Hasselmo, M. E., Stanley, H. E., & Hyman, B. T. (1999). Plaque-induced neurite abnormalities: Implications of disruption of neuronal networks in Alzheimer's disease. *Proceedings of the National Academy of Sciences, 96,* 5274–5279.

Knox, A. B. (1993). *Strengthening adult and continuing education.* San Francisco: Jossey-Bass.

Kobak, R. R., & Hazan, C. (1991). Attachment in marriage: Effects of security and accuracy of working models. *Journal of Personality and Social Psychology, 60,* 861–869.

Kobayashi, Y. (1994). Conceptual acquisition and change through social interaction. *Human Development, 37,* 233–241.

Kochanska, G. (1991). Socialization and temperament in the development of guilt and conscience. *Child Development, 62,* 1379–1392.

Kochanska, G. (1993). Toward a synthesis of parental socialization and child temperament in early development of conscience. *Child Development, 64,* 325–347.

Kochanska, G. (1995). Children's temperament, mothers' discipline, and security of attachment: Multiple pathways to emerging internalization. *Child Development, 66,* 597–615.

Kochanska, G. (1997). Multiple pathways to conscience for children with different temperaments: From toddlerhood to age 5. *Developmental Psychology, 33,* 228–240.

Kochanska, G. (1998). Mother–child relationship, child fearfulness, and emerging attachment: A short-term longitudinal study. *Developmental Psychology, 34,* 480–490.

Kochanska, G., Casey, R. J., & Fukumoto, A. (1995). Toddlers' sensitivity to standard violations. *Child Development, 66,* 643–656.

Kochanska, G., & Murray, K. T. (2000). Mother–child mutually responsive orientation and conscience development: From toddler to early school age. *Child Development, 71,* 417–431.

Kochanska, G., Murray, K. T., & Harlan, E. T. (2000). Effortful control in early childhood: Continuity and change, antecedents, and implications for social development. *Developmental Psychology, 36,* 220–232.

Kochanska, G., & Radke-Yarrow, M. (1992). Inhibition in toddlerhood and the dynamics of the child's interaction with an unfamiliar peer at age five. *Child Development, 63,* 325–335.

Kochanska, G., Tjebkes, T. L., & Forman, D. R. (1998). Children's emerging regulation of conduct: Restraint, compliance, and internalization from infancy to the second year. *Child Development, 69,* 1378–1389.

Kochenderfer-Ladd, B., & Wardrop, J. L. (2001). Chronicity and instability of children's peer victimization experiences as predictors of loneliness and social satisfaction trajectories. *Child Development, 72,* 134–151.

Koestner, R., Franz, C., & Weinberger, J. (1990). The family origins of empathic concern: A 26-year longitudinal study. *Journal of Personality and Social Psychology, 58,* 709–717.

Koff, T. H., & Park, R. W. (1999). *Aging public policy: Bonding the generations* (2nd ed.). Amityville, NY: Baywood.

Kogan, L. R., & Vacha-Haase, T. (2002). Supporting adaptation to new family roles in middle age. In C. L. Juntunen & D. R. Atkinson (Eds.), *Counseling across the lifespan* (pp. 299–347). Thousand Oaks, CA: Sage.

Kogan, N., & Mills, M. (1992). Gender influences on age cognitions and preferences: Sociocultural or sociobiological? *Psychology and Aging, 7,* 98–106.

Kohen, D., Hunter, T., Pence, A., & Goelman, H. (2000). The Victoria Day Care Research Project: Overview of a longitudinal study of child care and human development in Canada. *Canadian Journal of Research in Early Childhood Education, 8,* 49–54.

Kohlberg, L. (1966). A cognitive-developmental analysis of children's sex-role concepts and attitudes. In E. E. Maccoby (Ed.), *The development of sex differences* (pp. 82–173). Stanford, CA: Stanford University Press.

Kohlberg, L. (1969). Stage and sequence: The cognitive-developmental approach to socialization. In D. A. Goslin (Ed.), *Handbook of socialization theory and research* (pp. 347–480). Chicago: Rand McNally.

Kohlberg, L., Levine, C., & Hewer, A. (1983). *Moral stages: A current formulation and a response to critics.* Basel, Switzerland: Karger.

Kohlendorfer, U., Kiechl, S., & Sperl, W. (1998). Sudden infant death syndrome: Risk factor profiles for distinct subgroups. *American Journal of Epidemiology, 147,* 960–968.

Kohn, M. L., Naoi, A., Schoenbach, C., Schooler, C., & Slomczynski, K. M. (1990). Position in the class structure and psychological functioning in the United States,

Japan, and Poland. *American Journal of Sociology, 95,* 964–1008.

Kohn, M. L., & Schooler, C. (1978). The reciprocal effects of the substantive complexity of work and intellectual flexibility: A longitudinal assessment. *American Journal of Sociology, 84,* 24–52.

Kohn, M. L., & Slomczynski, D. M. (1990). *Social structure and self-direction: A comparative analysis of the United States and Poland.* Cambridge, MA: Blackwell.

Kohn, M. L., Zaborowski, W., Janicka, K., Mach, B. W., Khmelko, V., Slomczynski, K. M., Heyman, C., & Podobnik, B. (2000). *Social Psychology Quarterly, 63,* 187–208.

Kojima, H. (1986). Childrearing concepts as a belief–value system of the society and the individual. In H. Steveson, H. Azuma, & K. Hakuta (Eds.), *Child development and education in Japan* (pp. 39–54). New York: Freeman.

Kolb, B., & Gibb, R. (2001). Early brain injury, plasticity, and behavior. In C. A. Nelson & M. Luciana (Eds.), *Handbook of developmental cognitive neuroscience* (pp. 175–190). Cambridge, MA: MIT Press.

Kolominsky, Y., Igumnov, S., & Drozdovitch, V. (1999). The psychological development of children from Belarus exposed in the prenatal period to radiation from the Chernobyl atomic power plant. *Journal of Child Psychology and Psychiatry, 40,* 299–305.

Kolvin, I., & Trowell, J. (1996). Child sexual abuse. In I. Rosen (Ed.), *Sexual deviation* (3rd ed., pp. 337–360). Oxford, U.K.: Oxford University Press.

Komp, D. M. (1996). The changing face of death in children. In H. M. Spiro, M. G. M. Curnen, & L. P. Wandel (Eds.), *Facing death: Where culture, religion, and medicine meet* (pp. 66–76). New Haven: Yale University Press.

Kopp, C. B. (1994). Infant assessment. In C. B. Fisher & R. M. Lerner (Eds.), *Applied developmental psychology* (pp. 265–293). New York: McGraw-Hill.

Kornblith, A. B. (1998). Psychosocial adaptation of cancer survivors. In J. C. Holland (Ed.), *Psycho-oncology* (pp. 78–90). New York: Oxford University Press.

Korten, A. E., Jorm, A. F., Jiao, Z., Letenneur, L., Jacomb, P. A., Henderson, A. S., Christensen, H., & Rodgers, B. (1999). Health, cognitive, and psychosocial factors as predictors of mortality in an elderly community sample. *Journal of Epidemiology and Community Health, 53,* 83–88.

Koss, M. P. (1998). Hidden rape: Sexual aggression and victimization in a national sample of students in higher education. In M. E. Odem & J. Clay-Warner (Eds.), *Confronting rape and sexual assault* (pp. 51–69). Wilmington, DE: Scholarly Resources.

Koss, M. P., & Harvey, M. (1991). *The rape victim: Clinical and community interventions.* Newbury Park, CA: Sage.

Koss, M. P., Koss, P., & Woodruff, W. J. (1991). Deleterious effects of criminal victimization on women's health and medical utilization. *Archives of Internal Medicine, 151,* 342–357.

Kosterman, R., Hawkins, J. D., Guo, J., Catalano, R. F, & Abbott, R. D. (2000). The dynamics of alcohol and marijuana initiation: Patterns and predictors of first use in adolescence. *American Journal of Public Health, 90,* 360–366.

Kotch, J. B., Muller, G. O., & Blakely, C. H. (1999). Understanding the origins and incidence of spousal violence in North America. In T. P. Gullotta & S. J. McElhaney (Eds.), *Violence in homes and communities* (pp. 1–38). Thousand Oaks, CA: Sage.

Kotchick, B. A., Shaffer, A., Forehand, R., & Miller, K. S. (2001). Adolescent sexual risk behavior: A multi-system perspective. *Clinical Psychology Review, 21,* 493–519.

Kotler-Cope, S., & Camp, C. J. (1990). Memory interventions in aging populations. In E. A. Lovelace (Ed.), *Aging and cognition* (pp. 231–261). Amsterdam: North-Holland.

Kotre, J. (1984). *Outliving the self: Generativity and the interpretation of lives.* Baltimore: Johns Hopkins University Press.

Kotre, J. (1999). *Make it count: How to generate a legacy that gives meaning to your life.* New York: Free Press.

Kouvonen, A., & Kivivuori, J. (2001). Part-time jobs, delinquency, and victimization among Finnish adolescents. *Journal of Scandinavian Studies in Criminology and Crime Prevention, 2,* 191–212.

Kowalski, N., & Allen, R. (1995). School sleep lag is less but persists with a very late starting high school. *Sleep Research, 24,* 124.

Kraaij, V., Arensman, E., & Spinhoven, P. (2002). Negative life events and depression in elderly persons: A meta-analysis. *Journal of Gerontology, 57B,* P87–P94.

Kraemer, H. C., Yesavage, J. A., Taylor, J. L., & Kupfer, D. (2000). How can we learn about developmental processes from cross-sectional studies, or can we? *American Journal of Psychiatry, 157,* 163–171.

Krafft, K., & Berk, L. E. (1998). Private speech in two preschools: Significance of open-ended activities and make-believe play for verbal self-regulation. *Early Childhood Research Quarterly, 13,* 637–658.

Kraft, J. M., & Werner, J. S. (1999). Aging and the saturation of colors. 2. Scaling of color appearance. *Journal of the Optical Society of America, 16,* 231–235.

Kramer, A. F., Hahn, S., & Gopher, D. (1998). Task coordination and aging: Explorations of executive control processes in the task switching paradigm. *Acta Psychologica, 101,* 339–378.

Krascum, R. M., & Andrews, S. (1998). The effects of theories on children's acquisition of family-resemblance categories. *Child Development, 69,* 333–346.

Krause, N. (1990). Perceived health problems, formal/informal support, and life satisfaction among older adults. *Journal of Gerontology, 45,* S193–S205.

Krause, N. (2001). Social support. In R. H. Binstock & L. K. George (Eds.), *Handbook of aging and the social sciences* (5th ed., pp. 272–294). San Diego, CA: Academic Press.

Kray, J., & Lindenberger, U. (2000). Adult age differences in task switching. *Psychology and Aging, 15,* 126–147.

Krevans, J., & Gibbs, J. C. (1996). Parents' use of inductive discipline: Relations to children's empathy and prosocial behavior. *Child Development, 67,* 3263–3277.

Krishnan, V. (1998). Premarital cohabitation and marital disruption. *Journal of Divorce and Remarriage, 28,* 157–170.

Kroger, J. (1995). The differentiation of "firm" and "developmental" foreclosure identity statuses: A longitudinal study. *Journal of Adolescent Research, 10,* 317–337.

Kroger, J. (2000). *Identity development: Adolescence through adulthood.* Thousand Oaks, CA: Sage.

Kronenfeld, J. J., & Glik, D. C. (1995). Unintentional injury: A major health problem for young children and youth. *Journal of Family and Economic Issues, 16,* 365–393.

Kropf, N. P., & Pugh, K. L. (1995). Beyond life expectancy: Social work with centenarians. *Journal of Gerontological Social Work, 23,* 121–137.

Krout, J., Cutler, S. J., & Coward, R. T. (1990). Correlates of senior center participation: A national analysis. *Gerontologist, 30,* 72–79.

Kübler-Ross, E. (1969). *On death and dying.* New York: Macmillan.

Kuchner, J. (1989, April). *Chinese-American and European-American mothers and infants: Cultural influences in the first three months of life.* Paper presented at the biennial meeting of the Society for Research in Child Development, Kansas City, MO.

Kuczynski, L. (1984). Socialization goals and mother–child interaction: Strategies for long-term and short-term compliance. *Developmental Psychology, 20,* 1061–1073.

Kuczynski, L., & Lollis, S. (2002). Four foundations for a dynamic model of parenting. In J. R. M. Gerris (Eds.), *Dynamics of parenting.* Hillsdale, NJ: Erlbaum.

Kuebli, J., Butler, S., & Fivush, R. (1995). Mother–child talk about past emotions: Relations of maternal language and child gender over time. *Cognition and Emotion, 9,* 265–283.

Kuhl, P. K. (2000). A new view of language acquisition. *Proceedings of the National Academy of Sciences, 97,* 11850–11857.

Kuhl, P. K., Williams, K. A., Lacerda, F., Stevens, K. N., & Lindblom, B. (1992). Linguistic experience alters phonetic perception in infants by 6 months of age. *Science, 255,* 606–608.

Kuhn, D. (1993). Connecting scientific and informal reasoning. *Merrill-Palmer Quarterly, 39,* 74–103.

Kuhn, D. (1999). Metacognitive development. *Current Directions in Psychological Science, 9,* 178–181.

Kuhn, D. (2000). Theory of mind, metacognition, and reasoning: A life-span perspective. In P. Mitchell & K. J. Riggs (Eds.), *Children's reasoning and the mind* (pp. 301–326). Hove, U.K.: Psychology Press.

Kuhn, D., Amsel, E., & O'Loughlin, M. (1988). *The development of scientific thinking skills.* Orlando, FL: Academic Press.

Kuhn, D., Garcia-Mila, M., Zohar, A., & Andersen, C. (1995). Strategies of knowledge acquisition. *Monographs of the Society for Research in Child Development, 60*(245, Serial No. 4).

Kulik, K. (2001). Marital relationships in late adulthood: Synchronous versus asynchronous couples. *International Journal of Aging and Human Development, 52,* 323–339.

Kulik, L. (2000). Jobless men and women: A comparative analysis of job search intensity, attitudes toward unemployment, and related outcomes. *Journal of Occupational and Organizational Psychology, 73,* 487–500.

Kunzinger, E. L., III. (1985). A short-term longitudinal study of memorial development during early grade school. *Developmental Psychology, 21,* 642–646.

Kurdek, L. A. (1998). Relationship outcomes and their predictors: Longitudinal evidence from heterosexual married, gay cohabiting, and lesbian cohabiting couples. *Journal of Marriage and the Family, 60,* 553–568.

Kutner, L. (1993, June). Getting physical. *Parents,* pp. 96–98.

Kuulasmaa, K., Tunstall-Pedoe, H., Dobson, A., Fortman, S., Sans, S., & Tolonen, H. (2000). Estimation of contribution of changes in classic risk factors to trends in coronary-event rates across the WHO MONICA Project populations. *Lancet, 355,* 675–685.

Labouvie-Vief, G. (1980). Beyond formal operations: Uses and limits of pure logic in life-span development. *Human Development, 23,* 141–160.

Labouvie-Vief, G. (1985). Logic and self-regulation from youth to maturity: A model. In M. Commons, F. Richards, & C. Armon (Eds.), *Beyond formal operations: Late adolescent and adult cognitive development* (pp. 158–180). New York: Praeger.

Labouvie-Vief, G., Chiodo, L. M., Goguen, L. A., Diehl, M., & Orwoll, L. (1995). Representations of self across the life span. *Psychology and Aging, 10,* 404–415.

Labouvie-Vief, G., DeVoe, M., & Bulka, D. (1989). Speaking about feelings: Conceptions of emotion across the life span. *Psychology and Aging, 4,* 425–437.

Labouvie-Vief, G., & Diehl, M. (1999). Self and personality development. In J. C. Kavanaugh & S. K. Whitbourne (Eds.), *Gerontology: An interdisciplinary perspective* (pp. 238–268). New York: Oxford University Press.

Labouvie-Vief, G., & Diehl, M. (2000). Cognitive complexity and cognitive-affective integration: Related or separate domains of adult development? *Psychology and Aging, 15,* 490–504.

Labouvie-Vief, G., Diehl, M., Chiodo, L. M., & Coyle, N. (1995). Representations of self and parents across the life span. *Journal of Adult Development, 2,* 207–222.

Lachman, M. E., & Bertrand, R. M. (2002). Personality and self in midlife. In M. E. Lachman (Ed.), *Handbook of midlife development* (pp. 279–309). New York: Wiley.

Lachman, M. E., & James, J. B. (1997). Charting the course of midlife development: An overview. In M. E. Lachman & J. B. James (Eds.), *Multiple paths of midlife development* (pp. 1–17). Chicago: University of Chicago Press.

Lachman, M. E., Jette, A., Tennstedt, S., Howland, J., Harris, B. A., & Peterson, E. (1997). A cognitive-behavioural model for promoting regular physical activity in older adults. *Pyschology, Health and Medicine, 2,* 251–261.

Ladd, G. W., Birch, S. H., & Buhs, E. S. (1999). Children's social and scholastic lives in kindergarten: Related spheres of influence? *Child Development, 70,* 1373–1400.

Ladd, G. W., & Burgess, K. B. (1999). Charting the relationship trajectories of aggressive, withdrawn, and aggressive/withdrawn children during early grade school. *Child Development, 70,* 910–929.

Ladd, G. W., & Ladd, B. K. (1998). Parenting behaviors and parent–child relationships: Correlates of peer victimization in kindergarten? *Developmental Psychology, 34,* 1450–1458.

Ladd, G. W., LeSieur, K., & Profilet, S. M. (1993). Direct parental influences on young children's peer relations. In S. Duck (Ed.), *Learning about relationships* (Vol. 2, pp. 152–183). London: Sage.

Ladd, G. W., & Pettit, G. S. (2002). Parenting and the development of children's peer relationships. In M. Bornstein (Ed.), *Handbook of parenting* (2nd ed.). Mahwah, NJ: Erlbaum.

Ladd, G. W., & Price, J. M. (1987). Predicting children's social and school adjustment following the transition from preschool to kindergarten. *Child Development, 58,* 1168–1189.

Lagattuta, K. H., Wellman, H. M., & Flavell, J. H. (1997). Preschoolers' understanding of the link between thinking and feeling: Cognitive cuing and emotional change. *Child Development, 68,* 1081–1104.

Lagercrantz, H., & Slotkin, T. A. (1986). The "stress" of being born. *Scientific American, 254,* 100–107.

Lagnado, L. (2001, November 2). Kids confront Trade Center trauma. *Wall Street Journal,* pp. B1, B6.

Lahey, B. B., & Loeber, R. (1997). Attention-deficit/hyperactivity disorder, oppositional defiant disorder, conduct disorder, and adult antisocial behavior: A life span perspective. In D. M. Stoff, J. Breiling, & J. D. Maser (Eds.), *Handbook of antisocial behavior* (pp. 51–59). New York: Wiley.

Laible, D. J., & Thompson, R. A. (1998). Attachment and emotional understanding in preschool children. *Developmental Psychology, 34,* 1038–1045.

Laible, D. J., & Thompson, R. A. (2000). Mother–child discourse, attachment security, shared positive affect, and early conscience development. *Child Development, 71,* 1424–1440.

Laing, G. J., & Logan, S. (1999). Patterns of unintentional injury in childhood and their relation to socioeconomic factors. *Public Health, 113,* 291–294.

Laird, R. D., Jordan, K. Y., Dodge, K. A., Pettit, G. S., & Bates, J. E. (2001). Peer rejection in childhood, involvement with antisocial peers in early adolescence, and the development of externalizing behavior problems. *Development and Psychopathology, 13,* 337–354.

Laird, R. D., Pettit, G. S., Mize, J., & Lindsey, E. (1994). Mother–child conversations about peers: Contributions to competence. *Family Relations, 43,* 425–432.

Lamaze, F. (1958). *Painless childbirth.* London: Burke.

Lamb, M. E. (1987). *The father's role: Cross-cultural perspectives.* Hillsdale, NJ: Erlbaum.

Lamb, M. E. (1997). The development of father–infant relationships. In M. E. Lamb (Ed.), *The role of the father in child development* (3rd ed., pp. 104–120). New York: Wiley.

Lamb, M. E. (1998). Nonparental child care: Context, quality, correlates, and consequences. In I. E. Sigel & K. A. Renninger (Eds.), *Handbook of child psychology: Vol. 4. Child psychology in practice* (5th ed., pp. 73–133). New York: Wiley.

Lamb, M. E. (1999). Noncustodial fathers and their impact on the children of divorce. In R. A. Thompson & P. R. Amato (Eds.), *The postdivorce family: Children, parenting, and society* (pp. 105–125). Thousand Oaks, CA: Sage.

Lamb, M. E., & Oppenheim, D. (1989). Fatherhood and father–child relationships: Five years of research. In S. H. Cath, A. Gurwitt, & L. Gunsberg (Eds.), *Fathers and their families* (pp. 11–26). Hillsdale, NJ: Erlbaum.

Lamb, M. E., Sternberg, K. J., & Prodromidis, M. (1992). Nonmaternal care and the security of infant–mother attachment: A reanalysis of the data. *Infant Behavior and Development, 15,* 71–83.

Lamb, M. E., Thompson, R. A., Gardner, W., Charnov, E. L., & Connell, J. P. (1985). Infant–mother attachment: The origins and developmental significance of individual differences in the Strange Situation: Its study and biological interpretation. *Behavioral and Brain Sciences, 7,* 127–147.

Lamme, S., & Baars, J. (1993). Including social factors in the analysis of reminiscence in elderly individuals. *International Journal of Aging and Human Development 37,* 297–311.

Lampl, M. (1993). Evidence of saltatory growth in infancy. *American Journal of Human Biology, 5,* 641–652.

Lampl, M., Veldhuis, J. D., & Johnson, M. L. (1992). Saltation and stasis: A model of human growth. *Science, 258,* 801–803.

Lanctot, K. L., Herrmann, N., Eryavec, G., van Reekum, R., Reed, K., & Naranjo, C. A. (2002). Central serotonergic activity is related to the aggressive behaviors of Alzheimer's disease. *Neuropsychopharmacology, 27,* 646–654.

Landry, S. H., & Whitney, J. A. (1996). The impact of prenatal cocaine exposure: Studies of the developing infant. *Seminars in Perinatology, 20,* 99–106.

Lane, M. A., Ingram, D. K., Ball, S. S., & Roth, G. S. (1997). Dehydroepiandorsterone sulfate: A biomarker of primate aging slowed by caloric restriction. *Journal of Clinical and Endocrinology Metabolism, 82,* 2093–2096.

Lang, F. R., & Baltes, M. M. (1997). Being with people and being alone in later life: Costs and benefits for everyday functioning. *International Journal of Behavioral Development, 21,* 729–749.

Lang, F. R., Featherman, D. L., & Nesselroade, J. R. (1997). Social self-efficacy and short-term variability in social relationships: The MacArthur Successful Aging Studies. *Psychology and Aging, 12,* 657–666.

Lang, F. R., Staudinger, U. M., & Carstensen, L. L. (1998). Perspectives on socioemotional selectivity in late life: How personality and social context do (and do not) make a difference. *Journal of Gerontology, 53B,* P21–P30.

Langer, N. (1990). Grandparents and adult grandchildren: What do they do for one another? *International Journal of Aging and Human Development, 31,* 101–110.

Lansford, J. E., Sherman, A. M., & Antonucci, T. C. (1998). Satisfaction with social networks: An examination of socioemotional selectivity theory across cohorts.

Psychology and Aging, 13, 544–552.

Lantz, P. M., House, J. S., Lepkowski, J. M., Williams, D. R., Mero, R. P., & Chen, J. (1998). Socioeconomic factors, health behaviors, and mortality. *Journal of the American Medical Association, 279,* 1703–1708.

Lapsley, D. K., Jackson, S., Rice, K., & Shadid, G. (1988). Self-monitoring and the "new look" at the imaginary audience and personal fable: An ego-developmental analysis. *Journal of Adolescent Research, 3,* 17–31.

Lapsley, D. K., Rice, K. G., & FitzGerald, D. P. (1990). Adolescent attachment, identity, and adjustment to college: Implications for the continuity of adaptation hypothesis. *Journal of Counseling and Development, 68,* 561–565.

Larson, D. E. (1996). *Mayo Clinic family health book.* New York: Morrow.

Larson, J. H. (1988). The marriage quiz: College students' beliefs in selected myths about marriage. *Family Relations, 37,* 3–11.

Larson, R., & Ham, M. (1993). Stress and "storm and stress" in early adolescence: The relationship of negative events with dysphoric affect. *Developmental Psychology, 29,* 130–140.

Larson, R., & Lampman-Petraitis, C. (1989). Daily emotional states as reported by children and adolescents. *Child Development, 60,* 1250–1260.

Larson, R., Mannell, R., & Zuzanek, J. (1986). Daily well-being of older adults with friends and family. *Psychology and Aging, 1,* 117–126.

Larson, R., & Richards, M. (1998). Waiting for the weekend: Friday and Saturday night as the emotional climax of the week. In A. C. Crouter & R. Larson (Eds.), *Temporal rhythms in adolescence: Clocks, calendars, and the coordination of daily life* (pp. 37–51). San Francisco: Jossey-Bass.

Larson, R., & Richards, M. H. (1991). Daily companionship in late childhood and early adolescence: Changing developmental contexts. *Child Development, 62,* 284–300.

Larson, R. W., Clore, G. L., & Wood, G. A. (1999). The emotions of romantic relationships: Do they wreak havoc on adolescents? In W. Furman, B. B. Brown, & C. Feiring (Eds.), *The development of romantic relationships in adolescence* (pp. 19–49). New York: Cambridge University Press.

Larson, R. W., & Verma, S. (1999). How children and adolescents spend time across the world: Work, play, and developmental opportunities. *Psychological Bulletin, 125,* 701–736.

Larsson, M., & Bäckman, L. (1998). Modality memory across the adult life span: Evidence for selective age-related olfactory deficits. *Experimental Aging Research, 24,* 63–82.

Larzelere, R. E., Schneider, W. N., Larson, D. B., & Pike, P. L. (1996). The effects of discipline responses in delaying toddler misbehavior recurrences. *Child and Family Behavior Therapy, 18,* 35–57.

Latz, S., Wolf, A. W., & Lozoff, B. (1999). Sleep practices and problems in young children in Japan and the United States. *Archives of Pediatric and Adolescent Medicine, 153,* 339–346.

Laucht, M., Esser, G., & Schmidt, M. H. (1997). Developmental outcome of infants born with biological and psychosocial risks. *Journal of Child Psychology and Psychiatry, 38,* 843–853.

Laumann, E. O., Gagnon, J. H., Michael, R. T., & Michaels, S. (1994). *The social organization of sexuality.* Chicago: University of Chicago Press.

Laumann, E. O., Paik, A., & Rosen, R. C. (1999). Sexual dysfunction in the United States: Prevalence and predictors. *Journal of the American Medical Association, 281,* 537–544.

Launer, L. J., Andersen, K., Dewey, M. E., Letenneur, L., Ott, A., & Amaducci, L. A. (1999). Incidence Research and Work Groups. Rates and risk factors for dementia and Alzheimer's disease: Results from EURODEM pooled analyses. *Neurology, 52,* 78–84.

Laurin, D., Verreault, R., Lindsay, J., MacPherson, K., & Rockwood, K. (2001). Physical activity and risk of cognitive impairment and dementia in elderly persons. *Archives of Neurology, 58,* 498–504.

Laursen, B., Coy, K., & Collins, W. A. (1998). Reconsidering changes in parent–child conflict across adolescence: A meta-analysis. *Child Development, 69,* 817–832.

Law Commission of Canada. (2001). A fact sheet on the economics of aging in Canada. Retrieved from http://www.lcc.gc.ca/en/themes/pr/oa/mcgregor/chap03.asp

Lawton, L., Silverstein, M., & Bengtson, V. L. (1994). Solidarity between generations in families. In V. L. Bengtson & R. A. Harootyan (Eds.), *Intergenerational linkages: Hidden connections in American society* (pp. 19–42). New York: Springer-Verlag.

Lawton, M. P. (1980). Environment and aging. Monterey, CA: Brooks/Cole.

Lawton, M. P. (2001a). *Annual review of gerontology and geriatrics: Vol. 20. Focus on the end of life: Scientific and social issues.* New York: Springer.

Lawton, M. P. (2001b). Emotion in later life. *Current Directions in Psychological Science, 10,* 120–123.

Lawton, M. P., Winter, L., Kleban, M. H., & Ruckdeschel, K. (1999). Affect and quality of life. *Journal of Aging and Health, 11,* 169–198.

Lazar, A., & Torney-Purta, J. (1991). The development of the subconcepts of death in young children: A short-term longitudinal study. *Child Development, 62,* 1321–1333.

Lazar, I., & Darlington, R. (1982). Lasting effects of early education: A report from the Consortium for Longitudinal Studies. *Monographs of the Society for Research in Child Development, 47*(2–3, Serial No. 195).

Lazarus, R. S. (1991). *Emotion and adaptation.* New York: Oxford University Press.

Lazarus, R. S. (1999). *Stress and emotion: A new synthesis.* New York: Springer.

Lazarus, R. S., & Lazarus, B. N. (1994). *Passion and reason.* New York: Oxford University Press.

Leach, C. E. A., Blair, P. S., Fleming, P. J., Smith, I. J., Platt, M. W., & Berry, P. J. (1999). Epidemiology of SIDS and explained sudden infant deaths. *Pediatrics, 104,* e43.

Leaper, C. (1994). Exploring the correlates and consequences of gender segregation: Social relationships in childhood, adolescence, and adulthood. In C. Leaper (Ed.), *New directions for child development* (No. 65, pp. 67–86). San Francisco: Jossey-Bass.

Leaper, C., Anderson, K. J., & Sanders, P. (1998). Moderators of gender effects on parents' talk to their children: A meta-analysis. *Developmental Psychology, 34,* 3–27.

Leaper, C., Leve, L., Strasser, T., & Schwartz, R. (1995). Mother–child communication sequences: Play activity, child gender, and marital status effects. *Merrill-Palmer Quarterly, 41,* 307–327.

Leaper, C., Tenenbaum, H. R., & Shaffer, T. G. (1999). Communication patterns of African-American girls and boys from low-income, urban backgrounds. *Child Development, 70,* 1489–1503.

Lederer, J. M. (2000). Reciprocal teaching of social studies in inclusive elementary classrooms. *Journal of Learning Disabilities, 33,* 91–106.

Lee, A. M. (1980). Child-rearing practices and motor performance of Black and White children. *Research Quarterly for Exercise and Sport, 51,* 494–500.

Lee, C. L., & Bates, J. E. (1985). Mother–child interaction at age two years and perceived difficult temperament. *Child Development, 56,* 1314–1325.

Lee, D. J., & Markides, K. S. (1990). Activity and morality among aged persons over an eight-year period. *Journal of Gerontology, 45,* S39–S42.

Lee, D. M., & Weinblatt, M. E. (2001). Rheumatoid arthritis. *Lancet, 358,* 903–911.

Lee, G. R., DeMaris, A., Bavin, S., & Sullivan, R. (2001). Gender differences in the depressive effect of widowhood in later life. *Journal of Gerontology, 56B,* S56–S61.

Lee, J., & Bailey, G. (2003). Glaucoma. Retrieved from http://www.allaboutvision.com/conditions/glaucoma.htm

Lee, K., Cameron, C., Xu, F., Fu, G., & Board, J. (1997). Chinese and Canadian children's evaluations of lying and truth telling: Similarities and differences in the context of pro- and antisocial behaviors. *Child Development, 68,* 924–934.

Lee, S. H., Ewert, D. P., Frederick, P. D., & Mascola, L. (1992). Resurgence of congenital rubella syndrome in the 1990s. *Journal of the American Medical Association, 267,* 2616–2620.

Leekam, S. R., Lopez, B., & Moore, C. (2000). Attention and joint attention in preschool children with autism. *Developmental Psychology, 36,* 261–273.

Lehman, D., Ellard, J., & Wortman, C. (1986). Social support for the bereaved: Recipients' and providers' perspectives on what is helpful. *Journal of Consulting and Clinical Psychology, 54,* 438–446.

Lehman, D. R., & Nisbett, R. E. (1990). A longitudinal study of the effects of undergraduate training on reasoning. *Developmental Psychology, 26,* 952–960.

Leichtman, M. D., & Ceci, S. J. (1995). The effect of stereotypes and suggestions on preschoolers' reports. *Developmental Psychology, 31,* 568–578.

Leiter, M. P., & Maslach, C. (2000). *Preventing burnout and building engagement: A complete program for organizational renewal.* San Francisco: Jossey-Bass.

Lemery, K. S., Goldsmith, H. H., Klinnert, M. D., & Mrazek, D. A. (1999). Developmental models of infant and childhood temperament. *Developmental Psychology, 35,* 189–204.

Lemme, B. H. (2002). *Development in adulthood* (3rd ed.). Boston: Allyn and Bacon.

Lens, V., & Pollack, D. (2000). Advance directives: Legal remedies and psychosocial interventions. *Death Studies, 24,* 377–399.

Leonard, K. E., & Roberts, L. J. (1998). Marital aggression, quality, and stability in the first year of marriage: Findings from the Buffalo Newlywed Study. In T. N. Bradbury (Ed.), *The developmental course of marital dysfunction* (pp. 44–73). New York: Cambridge University Press.

Lerner, R. M., Fisher, C. B., & Weinberg, R. A. (2000). Toward a science for and of the people: Promoting civil society through the application of developmental science. *Child Development, 71,* 11–20.

Lester, B. M. (1985). Introduction: There's more to crying than meets the ear. In B. M. Lester & C. F. Z. Boukydis (Eds.), *Infant crying* (pp. 1–27). New York: Plenum.

Lester, B. M. (2000). Prenatal cocaine exposure and child outcome: A model for the study of the infant at risk. *Israel Journal of Psychiatry and Related Sciences, 37,* 223–235.

Lester, B. M., & Dreher, M. (1989). Effects of marijuana use during pregnancy on newborn cry. *Child Development, 60,* 765–771.

LeVay, S. (1993). *The sexual brain.* Cambridge, MA: MIT Press.

Levenson, R. W., Carstensen, L. L., & Gottman, J. M. (1993). Long-term marriage: Age, gender, and satisfaction. *Psychology and Aging, 8,* 301–313.

Leventhal, E. A., Leventhal, H., Schaefer, P. M., & Easterling, D. (1993). Conservation of energy, uncertainty reduction, and swift utilization of medical care among the elderly. *Journal of Gerontology, 48,* P78–P86.

Levin, J. S., & Chatters, L. M. (1998). Religion, health, and psychological well-being in older adults. *Journal of Aging and Health, 10,* 504–531.

Levin, J. S., Taylor, R. J., & Chatters, L. M. (1994). Race and gender differences in religiosity among older adults: Findings from four national surveys. *Journal of Gerontology, 49,* S137–S145.

Levine, L. E. (1983). Mine: Self-definition in 2-year-old boys. *Developmental Psychology, 19,* 544–549.

Levine, L. J. (1995). Young children's understanding of the causes of anger and sadness. *Child Development, 66,* 697–709.

LeVine, R. A., Dixon, S., LeVine, S., Richman, A., Leiderman, P. H., Keefer, C. H., & Brazelton, T. B. (1994). *Child care and culture: Lessons from Africa.* New York: Cambridge University Press.

Levine, S. C., Huttenlocher, J., Taylor, A., & Langrock, A. (1999). Early sex differences in spatial skill. *Developmental Psychology, 35,* 940–949.

Levinson, D. J. (1978). *The seasons of a man's life.* New York: Knopf.

Levinson, D. J. (1986). A conception of adult development. *American Psychologist, 41,* 3–13.

Levinson, D. J. (1996). *The seasons of a woman's life.* New York: Knopf.

Levitt, A. G., & Utmann, J. G. A. (1992). From babbling towards the sound systems of English and French: A longitudinal two-case study. *Journal of Child Language, 19,* 19–40.

Levtzion-Korach, O., Tennenbaum, A., Schnitzer, R., & Ornoy, A. (2000). Early motor development of blind children. *Journal of Paediatric and Child Health, 36,* 226–229.

Levy, B. R., & Banaji, M. R. (2002). Implicit ageism. In T. D. Nelson (Ed.), *Ageism: Stereotyping and prejudice against older persons* (pp. 49–75). Cambridge, MA: MIT Press.

Levy, B. R., Hausdorff, J., Hencke, R., & Wei, J. Y. (2000). Reducing cardiovascular stress with positive self-stereotypes of aging. *Journal of Gerontology, 55B,* P205–P213.

Levy, B. R., Slade, M. D., Kunkel, S. R., & Kasl, S. V. (2002). Longevity increased by positive self-perceptions of aging. *Journal of Personality and Social Psychology, 83,* 261–270.

Levy, G. D., Taylor, M. G., & Gelman, S. A. (1995). Traditional and evaluative aspects of flexibility in gender roles, social conventions, moral rules, and physical laws. *Child Development, 66,* 515–531.

Levy, J. A. (1994). Sex and sexuality in later life stages. In A. S. Rossi (Ed.), *Sexuality across the life course* (pp. 287–309). Chicago: University of Chicago Press.

Levy-Shiff, R. (1994). Individual and contextual correlates of marital change across the transition to parenthood. *Developmental Psychology, 30,* 591–601.

Levy-Shiff, R. (2001). Psychological adjustment of adoptees in adulthood: Family environment and adoption-related correlates. *International Journal of Behavioral Development, 25,* 97–104.

Levy-Shiff, R., & Israelashvili, R. (1988). Antecedents of fathering: Some further exploration. *Developmental Psychology, 24,* 434–440.

Lewis, C., Freeman, N. H., Kyriadidou, C., Maridakikassotaki, K., & Berridge, D. M. (1996). Social influences on false belief access—specific sibling influences or general apprenticeship? *Child Development, 67,* 2930–2947.

Lewis, K. G. (2000). *With or without a man: Single women taking control of their lives.* New York: Bull Publishing.

Lewis, M. (1992). *Shame: The exposed self.* New York: Free Press.

Lewis, M. (1995). Embarrassment: The emotion of self-exposure and evaluation. In J. P. Tangney & K. W. Fischer (Eds.), *Self-conscious emotions* (pp. 198–218). New York: Guilford Press.

Lewis, M. (1997). *Altering fate: Why the past does not predict the future.* New York: Guilford.

Lewis, M. (1998). Emotional competence and development. In D. Pushkar, W. M. Bukowski, A. E. Schwartzman, E. M. Stack, & D. R. White (Eds.), *Improving competence across the lifespan* (pp. 27–36). New York: Plenum.

Lewis, M., & Brooks-Gunn, J. (1979). *Social cognition and the acquisition of self.* New York: Plenum.

Lewis, M., Ramsay, D. S., & Kawakami, K. (1993). Differences between Japanese infants and Caucasian American infants in behavioral and cortisol response to inoculation. *Child Development, 64,* 1722–1731.

Lewis, M., Sullivan, M. W., Stanger, C., & Weiss, M. (1989). Self development and self-conscious emotions. *Child Development, 60,* 146–156.

Lewis, T., Stone, J., III, Shipley, W., & Madzar, S. (1998). The transition from school to work: An examination of the literature. *Youth and Society, 29,* 259–292.

Lewkowicz, D. J. (1996). Infants' response to the audible and visible properties of the human face. I. Role of lexical syntactic context, temporal synchrony, gender, and manner of speech. *Developmental Psychology, 32,* 347–366.

Liang, J., Bennett, J. M., Krause, N. M., Chang, M., Lin, S., Chuang, Y. L., & Wo, S. (1999). Stress, social relationships, and old age mortality in Taiwan. *Journal of Clinical Epidemiology, 52,* 983–995.

Liang, J., Krause, N. M., & Bennett, J. M. (2001). Social exchange and well-being: Is giving better than receiving? *Psychology and Aging, 16,* 511–523.

Liaw, F., & Brooks-Gunn, J. (1993). Patterns of low-birth-weight children's cognitive development. *Developmental Psychology, 29,* 1024–1035.

Liben, L. S. (1999). Developing an understanding of external spatial representations. In I. E. Sigel (Ed.), *Development of mental representation* (pp. 297–321). Mahwah, NJ: Erlbaum.

Liben, L. S., Bigler, R. S., & Krogh, H. R. (2001). Pink and blue collar jobs: Children's judgments of job status and job aspirations in relation to sex of worker. *Journal of Experimental Child Psychology, 79,* 346–363.

Liben, L. S., & Downs, R. M. (1993). Understanding person–space–map relations: Cartographic and developmental perspectives. *Developmental Psychology, 29,* 739–752.

Liben, L. S., & Signorella, M. L. (1993). Gender-schematic processing in children: The role of initial interpretations of stimuli. *Developmental Psychology, 29,* 141–149.

Lickliter, R., & Bahrick, L. E. (2000). The development of infant intersensory perception: Advantages of a comparative convergent-operations approach. *Psychological Bulletin, 126,* 260–280.

Lidz, C. S. (2001). Multicultural issues and dynamic assessment. In L. A. Suzuki & J. G. Ponterotto (Eds.), *Handbook of multicultural assessment: Clinical, psychological, and educational applications* (2nd ed., pp. 523–539). San Francisco: Jossey-Bass.

Lieberman, A. (2002). Dementia in Parkinson's disease. Retrieved from http://www.parkinson.org/pddement.htm

Lieberman, M. A. (1993). Bereavement selfhelp groups: A review of conceptual and methodological issues. In M. S. Stroebe, W. Stroebe, & R. O. Hansson (Eds.), *Handbook of bereavement* (pp. 427–453). New York: Cambridge University Press.

Liesi, E. H., Scherr, P. A., McCann, J. J., Beckett, L. A., & Evans, D. A. (2001). Is the risk of developing Alzheimer's disease greater for women than for men? *American Journal of Epidemiology, 153,* 132–136.

Light, P., & Perrett-Clermont, A.-N. (1989). Social context effects in learning and testing. In A. Gellatly, D. Rogers, & J. Sloboda (Eds.), *Cognition and social worlds* (pp. 99–112). Oxford, U.K.: Clarendon Press.

Lillard, A. S. (1998). Playing with a theory of mind. In O. N. Saracho & B. Spodek (Eds.), *Multiple perspectives on play in early childhood education* (pp. 11–33). Albany: State University of New York Press.

Lillard, A. S. (2001). Pretending, understanding pretense, and understanding minds. In S. Reifel (Ed.), *Play and culture studies* (Vol. 3). Norwood, NJ: Ablex.

Lin, C. C., Hsiao, C. K., & Chen, W. J. (1999). Development of sustained attention assessed using the continuous performance test among children 6–15 years. *Journal of Abnormal Child Psychology, 27,* 403–412.

Lindsay, C. (1999). *A portrait of seniors in Canada* (3rd ed.). Ottawa: Statistics Canada.

Lindsay, C., Almey, M., & Normand, J. (2002). *Youth in Canada.* Retrieved from http://www.statcan.ca/english/IPS/Data/85-511-XPE.htm

Lindsay-Hartz, J., de Rivera, J., & Mascolo, M. F. (1995). Differentiating guilt and shame and their effects on motivation. In J. P. Tangney & K. W. Fischer (Eds.), *Self-conscious emotions* (pp. 274–300). New York: Guilford.

Lindsey, E. W., & Mize, J. (2000). Parent–child physical and pretense play: Links to children's social competence. *Merrill-Palmer Quarterly, 46,* 565–591.

Link, S. C., & Ancoli-Israel, S. (1995). Sleep and the teenager. *Sleep Research, 24a,* 184.

Linn, M. C., & Petersen, A. C. (1985). Emergence and characterization of sex differences in spatial ability: A meta-analysis. *Child Development, 56,* 1479–1498.

Lissens, W., & Sermon, K. (1997). Preimplantation genetic diagnosis—current status and new developments. *Human Reproduction, 12,* 1756–1761.

Liston, R., Crane, J., Hamilton, E., Hughes, O., Kuling, S., & MacKinnon, C. (2002). Fetal health surveillance during labour. *Journal of Obstetrics and Gynecology of Canada, 24,* 250–276.

Litovsky, R. Y., & Ashmead, D. H. (1997). Development of binaural and spatial hearing in infants and children. In R. H. Gilkey & T. R. Anderson (Eds.), *Binaural and spatial hearing in real and virtual environments* (pp. 571–592). Mahwah, NJ: Erlbaum.

Litwin, H. (1998). The provision of informal support by elderly people residing in assisted living facilities. *Gerontologist, 38,* 239–246.

Livson, N., & Peshkin, H. (1980). Perspectives on adolescence from longitudinal research. In J. Adelson (Ed.), *Handbook of adolescent psychology* (pp. 47–98). New York: Wiley.

Lloyd, L. (1999). Multi-age classes and high ability students. *Review of Educational Research, 69,* 187–212.

Lock, M., & Kaufert, P. (2001). Menopause, local biologies, and cultures of aging. *American Journal of Human Biology, 13,* 494–504.

Locke, J. (1892). Some thoughts concerning education. In R. H. Quick (Ed.), *Locke on education* (pp. 1–236). Cambridge, U.K.: Cambridge University Press. (Original work published 1690)

Lockhart, R. S., & Craik, F. I. M. (1990). Levels of processing: A retrospective commentary on a framework for memory research. *Canadian Journal of Psychology, 44,* 87–112.

Loeber, R. L., Farrington, D. P., Stouthamer-Loeber, M., Moffitt, T. E., & Caspi, A. (1999). The development of male offending: Key findings from the first decade of the Pittsburgh Youth Study. *Studies on Crime and Crime Prevention, 8,* 245–263.

Loehlin, J. C. (2000). Group differences in intelligence. In R. J. Sternberg (Ed.), *Handbook of intelligence* (pp. 176–193). New York: Cambridge University Press.

Loehlin, J. C., Horn, J. M., & Willerman, L. (1997). Heredity, environment, and IQ in the Texas Adoption Project. In R. J. Sternberg & E. L. Grigorenko (Eds.), *Intelligence, heredity, and environment* (pp. 105–125). New York: Cambridge University Press.

Loehlin, J. C., Willerman, L., & Horn, J. M. (1988). Human behavior genetics. *Annual Review of Psychology, 38,* 101–133.

Loftus, J. (2001). America's liberalization in attitudes toward homosexuality, 1973 to 1998. *American Sociological Review, 66,* 762–782.

Logsdon, R. G. (2000). *Enhancing quality of life in long term care: A comprehensive guide.* New York: Hatherleigh Press.

Long, D. D. (1985). A cross-cultural examination of fears of death among Saudi Arabians. *Omega, 16,* 43–50.

Long, H. B., & Zoller-Hodges, D. (1995). Outcomes of Elderhostel participation. *Educational Gerontology,* 21, 113–127.

Lopata, H. Z. (1996). *Current widowhood: Myths and realities.* Thousand Oaks, CA: Sage.

Lorenz, K. (1952). *King Solomon's ring.* New York: Crowell.

Lorenz, K. Z. (1943). Die angeborenen Formen möglicher Erfahrung. *Zeitschrift für Tierpsychologie, 5,* 235–409.

Losey, K. M. (1995). Mexican-American students and classroom interaction: An overview and critique. *Review of Educational Research, 65,* 283–318.

Low, K. G., Fleisher, C., Colman, R., Dionne, A., Casey, G., & Legendre, S. (1998). Psychosocial variables, age, and angiographically determined coronary artery disease in women. *Annals of Behavioral Medicine, 20,* 221–226.

Lozoff, B., Wolf, A., Latz, S., & Paludetto, R. (1995, March). *Cosleeping in Japan, Italy, and the U.S.: Autonomy versus interpersonal relatedness.* Paper presented at the biennial meeting of the Society for Research in Child Development, Indianapolis.

Lubinski, D., & Benbow, C. P. (1994). The study of mathematically precocious youth: The first three decades of a planned 50-year study of intellectual talent. In R. F. Subotnik & K. D. Arnold (Eds.), *Beyond Terman: Contemporary longitudinal studies of giftedness and talent* (pp. 255–281). Norwood, NJ: Ablex.

Luborsky, M. R., & McMullen, K. (1999). Culture and aging. In J. C. Kavanaugh & S. K. Whitbourne (Eds.), *Gerontology: An interdisciplinary perspective* (pp. 65–90). New York: Oxford University Press.

Ludemann, P. M. (1991). Generalized discrimination of positive facial expressions by seven- and ten-month-old infants. *Child Development, 62,* 55–67.

Lue, T. F. (2000). Erectile dysfunction. *New England Journal of Medicine, 342,* 1802–1813.

Lund, D. A. (1993a). Caregiving. In R. Kastenbaum (Ed.), *Encyclopedia of adult development* (pp. 57–63). Phoenix, AZ: Oryx Press.

Lund, D. A. (1993b). Widowhood: The coping response. In R. Kastenbaum (Ed.), *Encyclopedia of adult development* (pp. 537–541). Phoenix, AZ: Oryx Press.

Lund, D. A. (1996). Bereavement and loss. In J. E. Birren (Ed.), *Encyclopedia of gerontology* (pp. 173–183). San Diego: Academic Press.

Lund, D. A. (1998). Statements and perspectives from leaders in the field of aging in Utah. In *Utah sourcebook on aging.* Salt Lake City: Empire Publishing.

Lund, D. A., & Caserta, M. S. (2001). When the unexpected happens: Husbands coping with the deaths of their wives. In D. Lund (Ed.), *Men coping with grief* (pp. 147–166). Amityville, NY: Baywood.

Lund, D. A., Caserta, M. S., & Dimond, M. F. (1986). Gender differences through two years of bereavement among the elderly. *Gerontologist, 26,* 314–320.

Lund, D. A., Caserta, M. S., & Dimond, M. F. (1993). The course of spousal bereavement in later life. In M. S. Stroebe, W. Stroebe, & R. O. Hansson (Eds.), *Handbook of bereavement* (pp. 240–245). New York: Cambridge University Press.

Lund, D. A., Hill, R. D., Caserta, M. S., & Wright, S. D. (1995). Video RespiteTM: An innovative resource for family, professional cargivers, and persons with dementia. *Gerontologist, 35,* 683–687.

Lund, D. A., & Wright, S. D. (2001). Respite services: Enhancing the quality of daily life for caregivers and persons with dementia. Retrieved from http://www.nurs.utah.edu/Gerontology

Lunenburg, F. C. (2000). America's hope: Making schools work for all children. *Journal of Instructional Psychology, 27,* 39–46.

Luster, T., & McAdoo, H. (1996). Family and child influences on educational attainment: A secondary analysis of the High/Scope Perry Preschool data. *Developmental Psychology, 32,* 26–39.

Luthar, S. S., & Cushing, G. (1997). Substance use and personal adjustment among disadvantaged teenagers: A six-month prospective study. *Journal of Youth and Adolescence, 26,* 353–372.

Luthar, S. S., Cushing, T. J., & McMahon, T. J. (1997). Interdisciplinary interface: Developmental principles brought to substance abuse research. In S. S. Luthar, J. A. Burack, D. Cicchetti, & J. R. Weisz (Eds.), *Developmental psychopathology* (pp. 437–456). Cambridge, U.K.: Cambridge University Press.

Lutz, D. J., & Sternberg, R. J. (1999). Cognitive development. In M. H. Bornstein & M. E. Lamb (Eds.), *Developmental psychology: An advanced textbook* (4th ed., pp. 275–311). Mahwah, NJ: Erlbaum.

Luzzo, D. A. (1999). Identifying the career decision-making needs of nontraditional college students. *Journal of Counseling and Development, 77,* 135–140.

Lyness, K., & Thompson, D. (1997). Above the glass ceiling? A comparison of matched samples of female and male executives. *Journal of Applied Psychology, 82,* 359–375.

Lyon, T. D., & Flavell, J. H. (1994). Young children's understanding of "remember" and "forget." *Child Development, 65,* 1357–1371.

Lyons-Ruth, K. (1996). Attachment relationships among children with aggressive behavior problems: The role of disorganized early attachment patterns. *Journal of Consulting and Clinical Psychology, 64,* 64–73.

Lyons-Ruth, K., Bronfman, E., & Parsons, E. (1999). Maternal frightened, frightening, or atypical behavior and disorganized infant attachment patterns. *Monographs of the Society for Research in Child Development, 64*(3, Serial No. 258), 67–96.

Lyons-Ruth, K., Easterbrooks, A., & Cibelli, C. (1997). Infant attachment strategies, infant mental lag, and maternal depressive symptoms: Predictors of internalizing and externalizing problems at age 7. *Developmental Psychology, 33,* 681–692.

Lytle, L. A., Seifert, S., Greenstein, J., & McGovern, P. (2000). How do children's eating patterns and food choices change over time? Results from a cohort study. *American Journal of Health Promotion, 14,* 222–228.

Lytton, H., & Gallagher, L. (2002). Parenting twins and the genetics of parenting. In M. H. Bornstein (Ed.), *Handbook of parenting* (Vol. 1, pp. 227–253). Mahwah, NJ: Erlbaum.

Maccoby, E. E. (1984). Socialization and developmental change. *Child Development, 55,* 317–328.

Maccoby, E. E. (1998). *The two sexes: Growing up apart, coming together.* Cambridge, MA: Belknap.

Maccoby, E. E. (2002). Gender and group process: A developmental perspective. *Current Directions in Psychological Science, 11,* 54–58.

Maccoby, E. E., & Jacklin, C. N. (1987). Gender segregation in childhood. In E. H. Reese (Ed.), *Advances in child development and behavior* (Vol. 20, pp. 239–287). New York: Academic Press.

Maccoby, E. E., & Martin, J. A. (1983). Socialization in the context of the family: Parent–child interaction. In E. M. Hetherington (Ed.), *Handbook of child psychology: Vol. 4. Socialization, personality, and social development* (4th ed., pp. 1–101). New York: Wiley.

MacDonald, W. L., & DeMaris, A. (1996). The effects of stepparent's gender and new biological children. *Journal of Family Issues, 17,* 5–25.

MacDorman, M. F., & Atkinson, J. O. (1999). Infant mortality statistics from the 1997 period linked birth/infant death data set. *National Vital Statistics Report, 47*(23), 1–23.

MacEwen, K. E., Barling, J., Kelloway, E. K., & Higginbottom, S. F. (1995). Predicting retirement anxiety: The roles of parental socialization and personal planning. *Journal of Social Psychology, 135,* 203–213.

Machenbach, J. P. (2002). Income inequality and population health. *British Medical Journal, 324,* 1–2.

MacKay, D. G., & Abrams, L. (1996). Language, memory, and aging: Distributed deficits and the structure of new-versus-old connections. In J. E. Birren & K. W. Schaie (Eds.), *Handbook of the psychology of aging* (pp. 251–265). San Diego: Academic Press.

Mackey, M. C. (1995). Women's evaluation of their childbirth performance. *Maternal–Child Nursing Journal, 23,* 57–72.

MacKinnon, C. E. (1989). An observational investigation of sibling interactions in married and divorced families. *Developmental Psychology, 25,* 36–44.

Mackinnon, L. T. (1992). *Exercise and immunology.* Champaign, IL: Human Kinetics.

MacKinnon-Lewis, C., Starnes, R., Volling, B., & Johnson, S. (1997). Perceptions of parenting as predictors of boys' sibling and peer relations. *Developmental Psychology, 33,* 1024–1031.

MacMillan, H. L., MacMillan, A. B., Offord, D. R., & Dingle, J. L. (1996). Aboriginal health—Canada. *Canadian Medical Association Journal, 155,* 1569–1578.

Madan-Swain, A., Fredrick, L. D., & Wallander, J. L. (1999). Returning to school after a serious illness or injury. In R. T. Brown (Ed.), *Cognitive aspects of chronic illness in children* (pp. 312–332). New York: Guilford.

Madden, D. J., & Plude, D. J. (1993). Selective preservation of selective attention. In J. Cerella & J. M. Rybash (Eds.), *Adult information processing: Limits on loss* (pp. 273–300). San Diego: Academic Press.

Maddi, S. R. (1999). The personality construct of hardiness: I. Effects on experiencing, coping, and strain. *Consulting Psychology Journal: Practice and Research, 51,* 83–94.

Maddi, S. R., & Hightower, M. (1999). Hardiness and optimism as expressed in coping patterns. *Consulting Psychology Journal: Practice and Research, 51,* 95–105.

Maddox, G. L. (1963). Activity and morale: A longitudinal study of selected elderly subjects. *Social Forces, 42,* 195–204.

Maddox, G. L. (2001). Housing and living arrangements. In R. H. Binstock & L. K. George (Eds.), *Handbook of aging and the social sciences* (5th ed., pp. 426–443). San Diego: Academic Press.

Madon, S., Jussim, L., & Eccles, J. (1997). In search of the powerful self-fulfilling prophecy. *Journal of Personality and Social Psychology, 72,* 791–809.

Magdol, L., Moffitt, T. E., Caspi, A., & Silva, P. A. (1998). Developmental antecedents of partner abuse: A prospective-longitudinal study. *Journal of Abnormal Psychology, 107,* 375–389.

Maglio, C. J., & Robinson, S. E. (1994). The effects of death education on death anxiety: A meta-analysis. *Omega, 29,* 319–335.

Magni, E., & Frisoni, G. B. (1996). Depression and somatic symptoms in the elderly: The role of cognitive function. *International Journal of Geriatric Psychiatry, 11,* 517–522.

Magnuson, K. A., & Duncan, G. J. (2002). Parents in poverty. In M. H. Bornstein (Ed.), *Handbook of parenting* (Vol. 4, pp. 95–122). Mahwah, NJ: Erlbaum.

Magnusson, D., & Stattin, H. (1998). Person-context interaction theories. In R. M. Lerner (Ed.), *Handbook of child psychology: Vol. 1. Theoretical models of human development* (5th ed., pp. 685–759). New York: Wiley.

Magolda, M. B. B. (2002). Epistemological reflection: The evolution of epistemological assumptions from age 18 to 30. In B. K. Hofer & P. R. Pintrich (Eds.), *Personal epistemology* (pp. 89–102). Mahwah, NJ: Erlbaum.

Mahanran, L. G., Bauman, P. A., Kalman, D., Skolnik, H., & Pele, S. M. (1999). Master athletes: Factors affecting performance. *Sports Medicine, 28,* 273–285.

Mahon, M. M., Goldberg, E. Z., & Washington, S. K. (1999). Concept of death in a sample of Israeli kibbutz children. *Death Studies, 23,* 43–59.

Mahoney, A., Pargament, K. I., Jewell, T., Swank, A. B., Scott, E., Emery, E., & Rye, M. (1999). Marriage and the spiritual realm: The role of proximal and distal religious constructs in marital functioning. *Journal of Family Psychology, 13,* 321–338.

Mahoney, J. (1998). An update on efforts by the hospice community and the National Hospice Organization to improve access to quality hospice care. *Hospital Journal, 13,* 139–144.

Mahoney, J. L. (2000). Participation in school extracurricular activities as a moderator in the development of antisocial patterns. *Child Development, 71,* 502–516.

Mahoney, J. L., & Magnuson, D. (2001). Parent participation in community activities and the persistence of criminality. *Development and Psychopathology, 13,* 123–139.

Mahoney, J. L., & Stattin, H. (2000). Leisure activities and antisocial behavior: The role of structure and social context. *Journal of Adolescence, 23,* 113–127.

Mahoney, J. L., Schweder, A. E., & Stattin, H. (2002). Structured after-school activities as a moderator of depressed mood for adolescents with detached relations to their parents. *Journal of Community Psychology, 30,* 69–86.

Mahoney, J. L., Stattin, H., & Magnusson, D. (2001). Youth recreation centre participation and criminal offending: A 20-year longitudinal study of Swedish boys. *International Journal of Behavioral Development, 25,* 509–520.

Mahoney, M. M. (2002). The economic rights and responsibilities of unmarried cohabitants. In A. Booth & A. C. Crouter (Eds.), *Just living together* (pp. 247–254). Mahwah, NJ: Erlbaum.

Maier, D. M., & Newman, M. J. (1995). Legal and psychological considerations in the development of a euthanasia statute for adults in the United States. *Behavioral Sciences and the Law, 13,* 3–25.

Maier, H., & Smith, J. (1999). Psychological predictors of mortality in old age. *Journal of Gerontology, 54B,* P44–P54.

Maier, S. F., Watkins, L. R., & Fleshner, M. (1994). Psychoneuroimmunology: The interface between behavior, brain, and immunity. *American Psychologist, 49,* 1004–1017.

Main, M. (2000). The organized categories of infant, child, and adult attachment: Flexible vs. inflexible attention under attachment-related stress. *Journal of the American Psychoanalytic Association, 48,* 1055–1096.

Main, M., & Solomon, J. (1990). Procedures for identifying infants as disorganized/disoriented during the Ainsworth Strange Situation. In M. Greenberg, D. Cicchetti, & M. Cummings (Eds.), *Attachment in the preschool years: Theory, research, and intervention* (pp. 121–160). Chicago: University of Chicago Press.

Maitland, S. B., Intrieri, R. C., Schaie, K. W., & Willis, S. L. (2000). Gender differences and changes in cognitive abilities across the adult life span. *Aging, Neuropsychology, and Cognition, 7,* 32–53.

Makin, J., Fried, P. A., & Watkinson, B. (1991). A comparison of active and passive smoking during pregnancy: Long-term effects. *Neurotoxicology and Teratology, 13,* 5–12.

Malaguarnera, L., Ferlito, L., Imbesi, R. M., Gulizia, G. S., Di Mauro, S., Maugeri, D., Malaguarnera, M., & Messina, A. (2001). Immunosenescence: A review. *Archives of Gerontology and Geriatrics, 32,* 1–14.

Malatesta, C. Z., Grigoryev, P., Lamb, C., Albin, M., & Culver, C. (1986). Emotion socialization and expressive development in preterm and full-term infants. *Child Development, 57,* 316–330.

Malina, R. M. (1990). Physical growth and performance during the transitional years (9–16). In R. Montemayor, G. R. Adams, & T. P. Gullotta (Eds.), *From childhood to adolescence: A transitional period?* (pp. 41–62). Newbury Park, CA: Sage.

Malina, R. M., & Bouchard, C. (1991). *Growth, maturation, and physical activity.* Champaign, IL: Human Kinetics.

Mallinckrodt, B., & Fretz, B. R. (1988). Social support and the impact of job loss on older professionals. *Journal of Counseling Psychology, 35,* 281–286.

Malloy, M. H., & Hoffman, H. J. (1995). Prematurity, sudden infant death syndrome, and age of death. *Pediatrics, 96,* 464–471.

Malone, M. M. (1982). Consciousness of dying and projective fantasy of young children with malignant disease. *Developmental and Behavioral Pediatrics, 3,* 55–60.

Maloni, J. A., Cheng, C. Y., Liebl, C. P., & Maier, J. S. (1996). Transforming prenatal care: Reflections on the past and present with implications for the future. *Journal of Obstetrics, Gynecology, and Neonatal Nursing, 25,* 17–23.

Mandler, J. M. (1998). Representation. In D. Kuhn & R. S. Siegler (Eds.), *Handbook of child psychology: Vol. 2. Cognition, perception, and language* (5th ed., pp. 255–308). New York: Wiley.

Mandler, J. M. (2000). What global-before-basic trend? Comment on perceptually based approaches to early categorization. *Infancy, 1,* 99–110.

Mandler, J. M., & McDonough, L. (1993). Concept formation in infancy. *Cognitive Development, 8,* 291–318.

Mandler, J. M., & McDonough, L. (1996). Drinking and driving don't mix: Inductive generalization in infancy. *Cognition, 59,* 307–335.

Mandler, J. M., & McDonough, L. (1998). On developing a knowledge base in infancy. *Developmental Psychology, 34,* 1274–1288.

Mange, E. J., & Mange, A. P. (1998). *Basic human genetics* (2nd ed.). Sunderland, MA: Sinauer Associates.

Mangelsdorf, S. C., Schoppe, S. J., & Burr, H. (2000). The meaning of parental reports: A contextual approach to the study of temperament and behavior problems. In V. J. Molfese & D. L. Molfese (Eds.), *Temperament and personality across the life span* (pp. 121–140). Mahwah, NJ: Erlbaum.

Mannell, R. C. (1999). Older adults, leisure, and wellness. *Journal of Leisurability, 26*(2), 3–10.

Manson, J. E., Willett, W. C., Stampfer, M. J., Colditz, G. A., Hunter, D. J., Hankinson, S. E., Hennekens, C. H., & Speizer, F. E. (1995). Body weight and mortality among women. *New England Journal of Medicine, 333,* 678–685.

Maratsos, M. (1998). The acquisition of grammar. In D. Kuhn & R. S. Siegler (Eds.), *Handbook of child psychology: Vol. 2. Cognition, perception, and language* (5th ed., pp. 421–466). New York: Wiley.

Marcia, J. E. (1980). Identity in adolescence. In J. Adelson (Ed.), *Handbook of adolescent psychology* (pp. 159–187). New York: Wiley.

Marcia, J. E. (1988). Common processes underlying ego identity, cognitive/moral development, and individuation. In D. K. Lapsley & F. P. Clark (Eds.), *Self, ego, and identity* (pp. 211–225). New York: Springer-Verlag.

Marcia, J. E., Waterman, A. S., Matteson, D. R., Archer, S. L., & Orlofsky, J. L. (1993). *Ego identity: A handbook for psychosocial research.* New York: Springer-Verlag.

Marcon, R. A. (1999). Positive relationships between parent–school involvement and public school inner-city preschoolers' development and academic performance. *School Psychology Review, 28,* 395–412.

Marcus, G. F. (1995). Children's overregularization of English plurals: A quantitative analysis. *Journal of Child Language, 22,* 447–459.

Marcus, G. F., Pinker, S., Ullman, M., Hollander, M., Rosen, T. J., & Xu, F. (1992). Overregularization in language acquisition. *Monographs of the Society for Research in Child Development, 57*(4, Serial No. 228).

Marcus, G. F., Vijayan, S., Rao, S. B., & Vishton, P. M. (1999). Rule learning by seven-month-old infants. *Science, 283,* 77–80.

Marcus-Newhall, A., Thompson, S., & Thomas, C. (2001). Examining a gender stereotype: Menopausal women. *Journal of Applied Social Psychology, 31,* 698–719.

Margolin, B. H., Morrison, H. I., & Hulka, B. S. (1994). Cigarette smoking and sperm density: A meta-analysis. *Fertility and Sterility, 61,* 35–43.

Margolin, G. (1998). Effects of domestic violence on children. In P. K. Trickett & C. J. Schellenbach (Eds.), *Violence against children in the family and community* (pp. 57–102). Washington, DC: American Psychological Association.

Margolin, G., & Gordis, E. B. (2000). The effects of family and community violence on children. *Annual Review of Psychology, 51,* 445–479.

Markides, K. S., & Cooper, C. L. (1989). Aging, stress, social support and health: An overview. In K. S. Markides & C. L. Cooper (Eds.), *Aging, stress and health* (pp. 1–10). Chicester: Wiley.

Markman, E. M. (1989). *Categorization and naming in children.* Cambridge, MA: MIT Press.

Markman, E. M. (1992). Constraints on word learning: Speculations about their nature, origins, and domain specificity. In M. R. Gunnar & M. P. Maratsos (Eds.), *Minnesota Symposia on Child Psychology* (Vol. 25, pp. 59–101). Hillsdale, NJ: Erlbaum.

Markovits, H., Benenson, J., & Dolensky, E. (2001). Evidence that children and adolescents have internal models of peer interactions that are gender differentiated. *Child Development, 72,* 879–886.

Markovits, H., & Vachon, R. (1989). Reasoning with contrary-to-fact propositions. *Journal of Experimental Child Psychology, 47,* 398–412.

Markovits, H., & Vachon, R. (1990). Conditional reasoning, representation, and level of abstraction. *Developmental Psychology, 26,* 942–951.

Marks, N. (1995). Midlife marital status differences in social support relationships with adult children and psychological well-being. *Journal of Family Issues, 16,* 5–28.

Marks, N. F. (1996). Caregiving across the lifespan: National prevalence and predictors. *Family Relations, 45,* 27–36.

Marks, N. F., & Lambert, J. D. (1998). Marital status continuity and change among young and midlife adults. *Journal of Family Issues, 19,* 652–686.

Markstrom-Adams, C., & Adams, G. R. (1995). Gender, ethnic group, and grade differences in psychosocial functioning during middle adolescence? *Journal of Youth and Adolescence, 24,* 397–417.

Markus, H. R., & Herzog, A. R. (1992). The role of self-concept in aging. In K. W. Schaie & M. P. Lawton (Eds.), *Annual review of gerontology and geriatrics* (pp. 110–143). New York: Springer.

Markus, H. R., Mullally, P. R., & Kitayama, S. (1997). Selfways: Diversity in modes of cultural participation. In U. Neisser & D. Jopling (Eds.), *The conceptual self in context* (pp. 13–61). New York: Cambridge University Press.

Marlier, L., & Schaal, B. (1997). La perception de la familiarité olfactive chez le nouveau-né: Influence différentielle du mode d'alimentation? [The perception of olfactory familiarity in the neonate: Differential influence of the mode of feeding?] *Enfance, 1,* 47–61.

Marlier, L., Schaal, B., & Soussignan, R. (1998). Neonatal responsiveness to the odor of amniotic and lacteal fluids: A test of perinatal chemosensory continuity. *Child Development, 69,* 611–623.

Marriott, A., Donaldson, C., Tarrier, N., & Burns, A. (2000). Effectiveness of cognitive-behavioural family intervention in reducing the burden of care in carers of patients with Alzheimer's disease. *British Journal of Psychiatry, 176,* 557–562.

Mars Hill Group. (2002). Needs assessment for the Canadian virtual hospice. Retrieved from http://www. canadianvirtualhospice.ca/survey/home.html

Marsh, H. W. (1990). The structure of academic self-concept: The Marsh/Shavelson model. *Journal of Educational Psychology, 82,* 623–636.

Marsh, H. W., Craven, R., & Debus, R. (1998). Structure, stability, and development of young children's self-concepts: A multicohort–multioccasion study. *Child Development, 69,* 1030–1053.

Marsh, H. W., Smith, I. D., & Barnes, J. (1985). Multidimensional self-concepts: Relations with sex and academic achievement. *Journal of Educational Psychology, 77,* 581–596.

Marsh, J. S., & Daigneault, J. P. (1999). The young athlete. *Current Opinion in Pediatrics, 11,* 84–88.

Marsh, R. L., Hicks, J., & Landau, J. D. (1998). An investigation of everyday prospective memory. *Memory and Cognition, 26,* 633–643.

Marshall, N. L. (1997). Combining work and family. In S. J. Gallant, G. P. Keita, & R. Royak-Schaler (Eds.), *Health care for women* (pp. 163–174). Washington, DC: American Psychological Association.

Marshall, V. W., Clarke, P. J., & Ballantyne, P. J. (2001). Instability in the retirement transition: Effects on health and well-being in a Canadian study. *Research on Aging, 23,* 379–409.

Marshall-Baker, A., Lickliter, R. & Cooper, R. P. (1998). Prolonged exposure to a visual pattern may promote behavioral organization in preterm infants. *Journal of Perinatal and Neonatal Nursing, 12,* 50–62.

Martin, C. L. (1989). Children's use of gender-related information in making social judgments. *Developmental Psychology, 25,* 80–88.

Martin, C. L. (1993). New directions for investigating children's gender knowledge. *Developmental Review, 13,* 184–204.

Martin, C. L., & Halverson, C. F. (1981). A schematic processing model of sex typing and stereotyping in children. *Child Development, 52,* 1119–1134.

Martin, C. L., & Halverson, C. F. (1987). The role of cognition in sex role acquisition. In D. B. Carter (Ed.), *Current conceptions of sex roles and sex typing: Theory and research* (pp. 123–137). New York: Praeger.

Martin, J., Shochat, T., & Ancoli-Israel, S. (2000). Assessment and treatment of sleep disturbances in older adults. *Clinical Psychology Review, 20,* 783–805.

Martin, J. A. (1981). A longitudinal study of the consequences of early mother–infant interaction: A microanalytic approach. *Monographs of the Society for Research in Child Development, 46*(3, Serial No. 190).

Martin, J. E., & Dean, L. (1993). Bereavement following death from AIDS: Unique problems, reactions, and special needs. In M. S. Stroebe, W. Stroebe, & R. O. Hansson (Eds.), *Handbook of bereavement* (pp. 317–330). Cambridge, U.K.: Cambridge University Press.

Martin, P., Long, M. V., & Poon, L. W. (2002). Age changes and differences in personality traits and states of the old and very old. *Journal of Gerontology, 57B,* P144–P152.

Martins, C., & Gaffan, E. A. (2000). Effects of maternal depression on patterns of infant–mother attachment: A meta-analytic investigation. *Journal of Child Psychology and Psychiatry, 41,* 737–746.

Martinson, I. M., Davies, E., & McClowry, S. G. (1987). The long-term effect of sibling death on self-concept. *Journal of Pediatric Nursing, 2,* 227–235.

Martlew, M., & Connolly, K. J. (1996). Human figure drawings by schooled and unschooled children in Papua New Guinea. *Child Development, 67,* 2743–2762.

Martyn, C. N., Barker, D. J. P., & Osmond, C. (1996). Mothers' pelvic size, fetal growth, and death from stroke and coronary heart disease in men in the UK. *Lancet, 348,* 1264–1268.

Maruna, S. (1997). Going straight: Desistance from crime and life narratives of reform. In A. Lieblich & R. Josselson (Eds.), *The narrative study of lives* (pp. 59–93). Thousand Oaks, CA: Sage.

Masataka, N. (1996). Perception of motherese in a signed language by 6-month-old deaf infants. *Developmental Psychology, 32,* 874–879.

Maslach, C., Schaufeli, W. B., & Leiter, M. P. (2001). Job burnout. *Annual Review of Psychology, 52,* 397–422.

Mason, C. A., Cauce, A. M., Gonzales, N., & Hiraga, Y. (1996). Neither too sweet nor too sour: Problem peers, maternal control, and problem behavior in African-American adolescents. *Child Development, 67,* 2115–2130.

Mason, M. G., & Gibbs, J. C. (1993). Social perspective taking and moral judgment among college students. *Journal of Adolescent Research, 8,* 109–123.

Massey, C. M., & Gelman, R. (1988). Preschoolers' ability to decide whether a photographed unfamiliar object can move itself. *Developmental Psychology, 24,* 307–317.

Masten, A. S. (2001). Ordinary magic: Resilience processes in development. *American Psychologist, 56,* 227–238.

Masten, A. S., Hubbard, J. J., Gest, S. D., Tellegen, A., Garmezy, N., & Ramirez, M. (1999). Adaptation in the context of adversity: Pathways to resilience and maladaptation from childhood to late adolescence. *Development and Psychopathology, 11,* 143–169.

Mastropieri, D., & Turkewitz, G. (1999). Prenatal experience and neonatal responsiveness to vocal expression of emotion. *Developmental Psychobiology, 35,* 204–214.

Masur, E. F., McIntyre, C. W., & Flavell, J. H. (1973). Developmental changes in apportionment of study time among items in a multi-trial free recall task. *Journal of Experimental Child Psychology, 15,* 237–246.

Masur, E. F., & Rodemaker, J. E. (1999). Mothers' and infants' spontaneous vocal, verbal, and action imitation during the second year. *Merrill-Palmer Quarterly, 45,* 392–412.

Matas, L., Arend, R., & Sroufe, L. A. (1978). Continuity of adaptation in the second year: The relationship between quality of attachment and later competence. *Child Development, 49,* 547–556.

Matheny, A. P., Jr. (1991). Children's unintentional injuries and gender: Differentiation and psychosocial aspects. *Children's Environment Quarterly, 8,* 51–61.

Matthews, T. J. (2001). Smoking during pregnancy in the 1990s. *National Vital Statistics Reports 49*(7). Retrieved from http://www.cdc.gov/nchs/releases/01news/smokpreg.htm

Mattson, S. N., Riley, E. P., Delis, D. C., & Jones, K. L. (1998). Neuropsychological comparison of alcohol-exposed children with or without physical features of fetal alcohol syndrome. *Neuropsychology, 12,* 146–153.

Matute-Bianchi, M. E. (1986). Ethnic identities and patterns of school success and failure among Mexican-descent and Japanese-American students in a California high school: An ethnographic analysis. *American Journal of Education, 95,* 233–255.

Maurer, T. J. (2001). Career-relevant learning and development, worker age, and beliefs about self-efficacy for development. *Journal of Management, 27,* 123–140.

Maurer, T., & Tarulli, B. (1994). Perceived environment, perceived outcome, and person variables in relationship to voluntary development activity by employees. *Journal of Applied Psychology, 79,* 3–14.

Mayes, L. C., & Bornstein, M. H. (1997). Attention regulation in infants born at risk: Prematurity and prenatal cocaine exposure. In J. A. Burack & J. T. Enns (Eds.), *Attention, development, and psychopathology* (pp. 97–122). New York: Guilford.

Mayes, L. C., Bornstein, M. H., Chawarska, K., & Haynes, O. M. (1996). Impaired regulation of arousal in 3-month-old infants exposed prenatally to cocaine and other drugs. *Development and Psychopathology, 8,* 29–42.

Mayes, L. C., & Zigler, E. (1992). An observational study of the affective concomitants of mastery in infants. *Journal of Child Psychology and Psychiatry, 33,* 659–667.

Maylor, E., & Valentine, T. (1992). Linear and nonlinear effects of aging on categorizing and naming faces. *Psychology and Aging, 7,* 317–323.

Mayo Clinic. (2000, March). Age-related macular degeneration: Who gets it and what you can do about it. *Women's Healthsource, 4*(3), 1–2.

McAdams, D. (1993). *The stories we live by: Personal myths and the making of the self.* New York: William Morrow.

McAdams, D., Reynolds, J., Lewis, M., Patten, A. H., & Bowman, P. J. (2001). When bad things turn good and good things turn bad: Sequences of redemption and contamination in life narrative and their relation to psychosocial adaptation in midlife adults and children. *Personality and Social Psychology Bulletin, 27,* 474–485.

McAdams, D. P. (1988). *Power, intimacy, and the life story.* New York: Guilford.

McAdams, D. P. (2001). Generativity in midlife. In M. E. Lachman (Ed.), *Handbook of midlife development* (pp. 395–443). New York: Wiley.

McAdams, D. P., & de St. Aubin, E. (1992). A theory of generativity and its assessment through self-report, behavioral acts, and narrative themes in autobiography. *Journal of Personality and Social Psychology, 62,* 1003–1015.

McAdams, D. P., de St. Aubin, E., & Logan, R. L. (1993). Generativity among young, midlife, and older adults. *Psychology and Aging, 8,* 221–230.

McAdams, D. P., Diamond, A., de St. Aubin, E., & Mansfield, E. (1997). Stories of commitment: The psychosocial construction of generative lives. *Journal of Personality and Social Psychology, 72,* 678–694.

McAdams, D. P., Hart, H. M., & Maruna, S. (1998). The anatomy of generativity. In D. P. McAdams & E. de St. Aubin (Eds.), *Generativity and adult development* (pp. 7–43). Washington, DC: American Psychological Association.

McAuley, E., & Blissmer, B. (2000). Self-efficacy determinants and consequences of physical activity. *Exercise and Sport Sciences Reviews, 28,* 85–88.

McAuley, E., Mihalko, S. L., & Bane, S. M. (1997). Exercise and self-esteem in middle-aged adults: Multidimensional relationships and physical fitness and self-efficacy influences. *Journal of Behavioral Medicine, 20,* 67–83.

McCabe, A. E., & Peterson, C. (1988). A comparison of adults' versus children's spontaneous use of *because* and *so. Journal of Genetic Psychology, 149,* 257–268.

McCall, R. B. (1993). Developmental functions for general mental performance. In D. K. Detterman (Ed.), *Current topics in human intelligence* (Vol. 3, pp. 3–29). Norwood, NJ: Ablex.

McCall, R. B., & Carriger, M. S. (1993). A meta-analysis of infant habituation and recognition memory performance as predictors of later IQ. *Child Development, 64,* 57–79.

McCartney, K., Harris, M. J., & Bernieri, F. (1990). Growing up and growing apart: A developmental meta-analysis of twin studies. *Psychological Bulletin, 107,* 226–237.

McCarton, C. (1998). Behavioral outcomes in low birth weight infants. *Pediatrics, 102,* 1293–1297.

McCarton, C. M., Brooks-Gunn, J., Wallace, I. F., Bauer, C. R., Bennett, F. C., Bernbaum, J. C., Broyles, R. S., Casey, P. H., McCormick, M. C., Scott, D. T., Tyson, J., Tonascia, J., and Meinert, C. L. (1997). Results at age 8 years of early intervention for low-birth-weight premature infants: The infant health and development program. *Journal of the American Medical Association, 277,* 126–132.

McCarty, M. E., & Ashmead, D. H. (1999). Visual control of reaching and grasping in infants. *Developmental Psychology, 35,* 620–631.

McClearn, G. E., Johansson, B., Berg, S., & Pedersen, N. L. (1997). Substantial genetic influence on cognitive abilities in twins 80 or more years old. *Science, 276,* 1560–1563.

McConaghy, M. J. (1979). Gender permanence and the genital basis of gender: Stages in the development of constancy of gender identity. *Child Development, 50,* 1223–1226.

McCrae, R. R., & Costa, P. T., Jr. (1990). *Personality in adulthood.* New York: Guilford.

McCrae, R. R., Costa, P. T., Jr., Ostendorf, F., Angleitner, A., Hrebickov, M., & Avia, M. D. (2000). Nature over nurture: Temperament, personality, and life span development. *Journal of Personality and Social Psychology, 78,* 173–186.

McCullough, P., & Rutenberg, S. (1989). Launching children and moving on. In B. Carter & M. McGoldrick (Eds.), *The changing family life cycle* (pp. 285–309). Boston: Allyn and Bacon.

McCune, L. (1993). The development of play as the development of consciousness. In M. H. Bornstein & A. O'Reilly (Eds.), *New directions for child development* (No. 59, pp. 67–79). San Francisco: Jossey-Bass.

McDaniel, M. A., Maier, S. F., & Einstein, G. O. (2002). "Brain-specific" nutrients: A memory cure? *Psychological Science in the Public Interest, 3,* 12–38.

McDonough, L. (1999). Early declarative memory for location. *British Journal of Developmental Psychology, 17,* 381–402.

McFadden, S. H. (1996). Religion, spirituality, and aging. In J. E. Birren & K. W. Schaie (Eds.), *Handbook of the psychology of aging* (pp. 162–177). San Diego: Academic Press.

McGee, G. (1997). Legislating gestation. *Human Reproduction, 12,* 407–408.

McGee, L. M., & Richgels, D. J. (2000). *Literacy's beginnings* (3rd ed.). Boston: Allyn and Bacon.

McGillicuddy-De Lisi, A. V., Watkins, C., & Vinchur, A. J. (1994). The effect of relationship on children's distributive justice reasoning. *Child Development, 65,* 1694–1700.

McGoldrick, M., Heiman, M., & Carter, B. (1993). The changing family life cycle: A perspective on normalcy. In F. Walsh (Ed.), *Normal family processes* (pp. 405–443). New York: Guilford.

McGue, M., & Bouchard, T. J., Jr. (1998). Genetic and environmental influences on human behavioral differences. *Annual Review of Neuroscience, 21,* 1–24.

McGue, M., & Christensen, K. (2002). The heritability of level and rate-of-change in cognitive functioning in Danish twins aged 70 years and older. *Experimental Aging Research, 28,* 435–451.

McGue, M., Vaupel, J. W., Holm, N., & Harvald, B. (1993). Longevity is moderately heritable in a sample of Danish twins born 1870–1880. *Journal of Gerontology, 48,* B237–B244.

McGuffin, P., & Sargeant, M. P. (1991). Major affective disorder. In P. McGuffin & R. Murray (Eds.), *The new genetics of mental illness* (pp. 165–181). London: Butterworth-Heinemann.

McHale, J., Khazan, I., Erera, P., Rotman, T., DeCourcey, W., & McConnell, M. (2002). Coparenting in diverse family systems. In M. H. Bornstein (Ed.), *Handbook of parenting: Vol. 3* (2nd ed., pp. 75–107). Mahwah, NJ: Erlbaum.

McHale, S. M., Bartko, W. T., Crouter, A. C., & Perry-Jenkins, M. (1990). Children's housework and psychosocial functioning: The mediating effects of parents' sex-role behaviors and attitudes. *Child Development, 61,* 1413–1426.

McKenna, J. J. (2001). Why we never ask "Is it safe for infants to sleep alone?" *Academy of Breast Feeding Medicine News and Views, 7*(4), 32, 38.

McKenna, J. J. (2002, October 2). Personal communication.

McKeown, R. E., Garrison, C. Z., Cuffe, S. P., Waller, J. L., Jackson, K. L., & Addy, C. L. (1998). Incidence and predictors of suicidal behaviors in a longitudinal sample of young adolescents. *Journal of the American Academy of Child and Adolescent Psychiatry, 37,* 612–619.

McKim, W. A. (2002). *Drugs and behavior* (5th ed.). Upper Saddle River, NJ: Prentice-Hall.

McKusick, V. A. (1998). *Mendelian inheritance in man: A catalog of human genes and genetic disorders.* Baltimore: Johns Hopkins University Press.

McLanahan, S. (1999). Father absence and the welfare of children. In E. M. Hetherington (Ed.), *Coping with divorce, single parenting, and remarriage: A risk and resiliency perspective* (pp. 117–145). Mahwah, NJ: Erlbaum.

McLean, D. F., Timajchy, K. H., Wingo, P. A., & Floyd, R. L. (1993). Psychosocial measurement: Implications of the study of preterm delivery in black women. *American Journal of Preventive Medicine, 9,* 39–81.

McLeod, J. D., & Shanahan, M. J. (1996). Trajectories of poverty and children's mental health. *Journal of Health and Social Behavior, 37,* 207–220.

McLoyd, V. C. (1998). Children in poverty: Development, public policy, and practice. In I. Sigel & A. Renninger (Eds.), *Handbook of child psychology: Vol. 4. Child psychology in practice* (5th ed., pp. 135–208). New York: Wiley.

McManus, I. C., Sik, G., Cole, D. R., Mellon, A. F., Wong, J., & Kloss, J. (1988). The development of handedness in children. *British Journal of Developmental Psychology, 6,* 257–273.

McNamee, S., & Peterson, J. (1986). Young children's distributive justice reasoning, behavior, and role taking: Their consistency and relationship. *Journal of Genetic Psychology, 146,* 399–404.

McNeil, J. N. (1986). Talking about death: Adolescents, parents, and peers. In C. A. Corr & J. N. McNeil (Eds.), *Adolescence and death* (pp. 185–201). New York: Springer.

MCR Vitamin Study Research Group. (1991). Prevention of neural tube defects: Results of the Medical Research Council Vitamin Study. *Lancet, 338,* 131–137.

Mead, G. H. (1934). *Mind, self, and society.* Chicago: University of Chicago Press.

Mead, M. (1928). *Coming of age in Samoa.* Ann Arbor, MI: Morrow.

Mead, M., & Newton, N. (1967). Cultural patterning of perinatal behavior. In S. Richardson & A. Guttmacher (Eds.), *Childbearing: Its social and psychological aspects* (pp. 142–244). Baltimore: Williams & Wilkins.

Meegan, S. P., & Berg, C. A. (2002). Contexts, functions, forms, and processes of collaborative everyday problem solving in older adulthood. *International Journal of Behavioral Development, 26,* 6–15.

Meeus, W. (1996). Studies on identity development in adolescence: An overview of research and some new data. *Journal of Youth and Adolescence, 25,* 569–598.

Meeus, W., Iedema, J., Helsen, M., & Vollebergh, W. (1999). Patterns of adolescent identity development: Review of literature and longitudinal analysis. *Developmental Review, 19,* 419–461.

Mehlmadrona, L., & Madrona, M. M. (1997). Physician- and midwife attended home births—effects of breech, twin, and post-dates outcome data on mortality rates. *Journal of Nurse-Midwifery, 42,* 91–98.

Meltzoff, A. N. (1990). Towards a developmental cognitive science. *Annals of the New York Academy of Sciences, 608,* 1–37.

Meltzoff, A. N. (1995). Understanding the intentions of others: Re-enactment of intended acts by 18-month-old children. *Developmental Psychology, 31,* 838–850.

Meltzoff, A. N., & Kuhl, P. K. (1994). Faces and speech: Intermodal processing of biologically relevant signals in infants and adults. In D. J. Lewkowicz & R. Lickliter (Eds.), *The development of intersensory perception* (pp. 335–369). Hillsdale, NJ: Erlbaum.

Meltzoff, A. N., & Moore, M. K. (1977). Imitation of facial and manual gestures by human neonates. *Science, 198,* 75–78.

Meltzoff, A. N., & Moore, M. K. (1994). Imitation, memory, and the representation of persons. *Infant Behavior and Development, 17,* 83–99.

Meltzoff, A. N., & Moore, M. K. (1998). Object representation, identity, and the paradox of early permanence: Steps toward a new framework. *Infant Behavior and Development, 21,* 201–235.

Meltzoff, A. N., & Moore, M. K. (1999). Persons and representations: Why infant imitation is important for theories of human development. In J. Nadel & G. Butterworth (Eds.), *Imitation in infancy* (pp. 9–35). Cambridge, U.K.: Cambridge University Press.

Mennella, J. A., & Beauchamp, G. K. (1998). Early flavor experiences: Research update. *Nutrition Reviews, 56,* 205–211.

Menon, U. (2002). Middle adulthood in cultural perspective: The imagined and the experienced in three cultures. In M. E. Lachman (Ed.), *Handbook of midlife development* (pp. 40–74). New York: Wiley.

Mercer, R. T., Nichols, E. G., & Doyle, G. C. (1989). *Transitions in a woman's life: Major life events in developmental context.* New York: Springer-Verlag.

Meredith, C. N., Frontera, W. R., O'Reilly, K. P., & Evans, W. J. (1992). Body composition in elderly men: Effect of dietary modification during strength training. *Journal of the American Geriatrics Society, 40,* 155–162.

Mergenhagen, P. (1996). Her own boss. *American Demographics, 18,* 36–41.

Merriam, S. B. (1993). The uses of reminiscence in older adulthood. *Educational Gerontology, 8,* 275–290.

Merrill, D. M. (1997). *Caring for elderly parents.* Westport, CT: Auburn House.

Messinger-Rapport, B. J., & Rader, E. (2000). High risk on the highway: How to identify and treat the impaired older driver. *Geriatrics, 55,* 32–45.

Messman, S. J., Canary, D. J., & Hause, K. S. (2000). Motives to remain platonic, equity, and the use of maintenance strategies in opposite-sex friendships. *Journal of Social and Personal Relationships, 17,* 67–94.

Meyer, B. J. F., Russo, C., & Talbot, A. (1995). Discourse comprehension and problem solving: Decisions about the treatment of breast cancer by women across the lifespan. *Psychology and Aging, 10,* 84–103.

Meyer, M. H., & Bellas, M. L. (1995). U.S. old-age policy and the family. In R. Blieszner & V. H. Bedford (Eds.), *Handbook of aging and the family* (pp. 263– 283). Westport, CT: Greenwood Press.

Meyer-Bahlburg, H. F. L., Ehrhardt, A. A., Rosen, L. R., Gruen, R. S., Veridiano, N. P., Vann, F. H., & Neuwalder, H. F. (1995). Prenatal estrogens and the development of homosexual orientation. *Developmental Psychology, 31,* 12–21.

Meyers, C., Adam, R., Dungan, J., & Prenger, V. (1997). Aneuploidy in twin gestations: When is maternal age advanced? *Obstetrics and Gynecology, 89,* 248–251.

Mezey, M., Dubler, N. N., Mitty, E., & Brody, A. A. (2002). What impact do setting and transitions have on the quality of life at the end of life and the quality of the dying process? *Gerontologist, 42*(Special Issue III), 54–76.

Miccio, A., Yont, K. M., Clemons, H. L., & Vernon-Feagans, L. (2001). Otitis media and the acquisition of consonants. In F. Windsor & M. L. Kelly (Eds.), *Investigations in clinical phonetics and linguistics* (pp. 429–435). Mahwah, NJ: Erlbaum.

Michael, R. T., Gagnon, J. H., Laumann, E. O., & Kolata, G. (1994). *Sex in America.* Boston: Little, Brown.

Michaels, G. Y. (1988). Motivational factors in the decision and timing of pregnancy. In G. Y. Michaels & W. A. Goldberg (Eds.), *The transition to parenthood: Current theory and research* (pp. 23–61). New York: Cambridge University Press.

Michels, K. B., Trichopoulos, D., Robins, J. M., Rosner, B. A., Manson, J. E., Hunter, D. J., Colditz, G. A., Hankinson, S. E., Speizer, F. E., & Willett, W. C. (1996). Birthweight as a risk factor for breast cancer. *Lancet, 348,* 1542–1546.

Miguel, J. (2001). Nutrition and aging. *Public Health Nutrition, 4,* 1385–1388.

Milberger, S., Biederman, J., Faraone, S. V., Guite, J., & Tsuang, M. T. (1997). Pregnancy, delivery and infancy complications and attention deficit hyperactivity disorder: Issues of gene–environment interaction. *Biological Psychiatry, 41,* 65–75.

Milgram, N. A., & Palti, G. (1993). Psychosocial characteristics of resilient children. *Journal of Research in Personality, 27,* 207–221.

Miller, B. C., Fan, X., Christensen, M., Grotevant, H. D., & van Dulmen, M. (2000). Comparisons of adopted and nonadopted adolescents in a large, nationally representative sample. *Child Development, 71,* 1458–1473.

Miller, J. G. (1997). Culture and self: Uncovering the cultural grounding of psychological theory. In J. G. Snodgrass & R. L. Thompson (Eds.), *Annals of the New York Academy of Sciences* (Vol. 18, pp. 217–231). New York: New York Academy of Sciences.

Miller, J. G., & Bersoff, D. M. (1995). Development in the context of everyday family relationships: Culture, interpersonal morality, and adaptation. In M. Killen & D. Hart (Eds.), *Morality in everyday life: Developmental perspectives* (pp. 259–282). Cambridge: Cambridge University Press.

Miller, J., Slomczynski, K. M., & Kohn, M. L. (1985). Continuity of learning-generalization: The effect of job on men's intellective process in the United States and Poland. *American Journal of Sociology, 91,* 593–615.

Miller, L. T., & Vernon, P. A. (1992). The general factor in short-term memory, intelligence, and reaction time. *Intelligence, 16,* 5–29.

Miller, L. T., & Vernon, P. A. (1997). Developmental changes in speed of information processing in young children. *Developmental Psychology, 33,* 549–554.

Miller, P. A., Eisenberg, N., Fabes, R. A., & Shell, R. (1996). Relations of moral reasoning and vicarious emotion to young children's prosocial behavior toward peers and adults. *Developmental Psychology, 32,* 210–219.

Miller, P. H., & Bigi, L. (1979). The development of children's understanding of attention. *Merrill-Palmer Quarterly, 25,* 235–250.

Miller, P. H., & Seier, W. L. (1994). Strategy utilization deficiencies in children: When, where, and why. In H. W. Reese (Ed.), *Advances in child development and behavior* (Vol. 24, pp. 107–156). New York: Academic Press.

Miller, P. J., Fung, H., & Mintz, J. (1996). Self-construction through narrative practices: A Chinese and American comparison of early socialization. *Ethos, 24,* 1–44.

Miller, P. J., Wiley, A. R., Fung, H., & Liang, C.-H. (1997). Personal storytelling as a medium of socialization in Chinese and American families. *Child Development, 68,* 557–568.

Miller, R. B. (2000). Do children make a marriage unhappy? *Family Science Review, 13,* 60–73.

Miller, R. B., Hemesath, K., & Nelson, B. (1997). Marriage in middle and later life. In T. D. Hargrave & S. M. Hanna (Eds.), *The aging family* (pp. 178–198). New York: Brunner/Mazel.

Miller, S. S., & Cavanaugh, J. C. (1990). The meaning of grandparenthood and its relationship to demographic, relationship, and social participation variables. *Journal of Gerontology, 45,* P244–P246.

Mills, A., & Slobodin, R. (Eds.). (1994). *Amerindian rebirth: Reincarnation belief among North American Indians and Inuit.* Toronto: University of Toronto Press.

Mills, R., Coffey-Corina, S., & Neville, H. J. (1997). Language comprehension and cerebral specialization from 13 to 20 months. *Developmental Neuropsychology, 13,* 397–445.

Mills, R., & Grusec, J. (1989). Cognitive, affective, and behavioral consequences of praising altruism. *Merrill-Palmer Quarterly, 35,* 299–326.

Milner, J. S. (1993). Social information processing and physical child abuse. *Clinical Psychology Review, 13,* 275–294.

Minde, K. (2000). Prematurity and serious medical conditions in infancy: Implications for development, behavior, and intervention. In C. H. Zeanah, Jr. (Ed.), *Handbook of infant mental health* (pp. 176–194). New York: Guilford.

Minkler, M., & Roe, K. M. (1993). *Grandmothers as caregivers: Raising children of the crack cocaine epidemic.* Newbury Park, CA: Sage.

Mintzer, J. E. (2001). Underlying mechanisms of psychosis and aggression in patients with Alzheimer's disease. *Journal of Clinical Psychiatry, 62*(Suppl. 21), 23–25.

Mischel, W., & Liebert, R. M. (1966). Effects of discrepancies between observed and imposed reward criteria on their acquisition and transmission. *Journal of Personality and Social Psychology, 3,* 45–53.

Mitchell, B. D., Hsueh, W. C., King, T. M., Pollin, T. I., Sorkin, J., Agarwala, R., Schäffer, A. A., & Shuldiner, A. R. (2001). Heritability of life span in the Old Order Amish. *American Journal of Medical Genetics, 102,* 346–352.

Mitchell, V., & Helson, R. (1990). Women's prime of life. *Psychology of Women Quarterly, 14,* 451–470.

Mize, J., & Pettit, G. S. (1997). Mothers' social coaching, mother–child relationship style, and children's peer competence: Is the medium the message? *Child Development, 68,* 312–332.

Moen, P. (1996). Gender, age, and the life course. In R. H. Binstock & L. K. George (Eds.), *Handbook of aging and the social sciences* (pp. 171–187). San Diego: Academic Press.

Moen, P., Fields, V., Quick, H. E., & Hofmeister, H. (2000). A life-course approach to retirement and social integration. In K. Pillemer, P. Moen, E. Wthington, & N. Glasgow (Eds.), *Social integration in the second half of life* (pp. 75–107). Baltimore: Johns Hopkins University Press.

Moerk, E. L. (1992). *A first language taught and learned.* Baltimore: Paul H. Brookes.

Moffat, S. D., Hampson, E., & Hatzipantelis, M. (1998). Navigation is a "virtual" maze: Sex differences and correlation with psychometric measures of spatial ability in humans. *Evolution and Human Behavior, 19*(2), 73–87.

Moffitt, T. E., Caspi, A., Belsky, J., & Silva, P. A. (1992). Childhood experience and onset of menarche: A test of a sociobiological model. *Child Development, 63,* 47–58.

Moffitt, T. E., Caspi, A., Dickson, N., Silva, P., & Stanton, W. (1996). Childhood-onset versus adolescent-onset antisocial conduct problems in males: Natural history from ages 3 to 18 years. *Development and Psychopathology, 8,* 399–424.

Mogford-Bevan, K. (1999). Twins and their language development. In A. C. Sandbank (Ed.), *Twin and triplet psychology.* New York: Routledge.

Mogilner, A., Grossman, J. A., Ribary, U., Joliot, M., Volkmann, J., Rapaport, D., Beasley, R. W., & Linas, R. R. (1993). Somatosensory cortical plasticity in adult humans revealed by magnetoencephalography. *Proceedings of the National Academy of Sciences, 90,* 3593–3597.

Mojet, J., Christ-Hazelhof, E., & Heidema, J. (2001). Taste perception with age: Generic or specific losses in threshold sensitivity to the five basic tastes? *Chemical Senses, 26,* 845–860.

Mokdad, A. H., Bowman, B. A., Ford, E. S., Vinicor, F., Marks, J. S., & Koplan, J. P. (2001). The continuing epidemics of obesity and diabetes in the United States. *Journal of the American Medical Association, 286,* 1195–1200.

Mokdad, A. H., Serdula, M. K., Dietz, W. H., Bowman, B. A., Marks, J. S., & Koplan, J. P. (1999). The spread of the obesity epidemic in the United States, 1991–1998. *Journal of the American Medical Association, 282,* 1519–1522.

Moll, I. (1994). Reclaiming the natural line in Vygotsky's theory of cognitive development. *Human Development, 37,* 333–342.

Molnar, J., & Babbitt, B. (2000). *You don't have to be thin to win.* New York: Villard.

Mondimore, F. M. (1996). *A natural history of homosexuality.* Baltimore: Johns Hopkins University Press.

Mondloch, C. J., Lewis, T., Budreau, D. R., Maurer, D., Dannemillier, J. L., Stephens, B. R., & Kleiner-Gathercoal, K. A. (1999). Face perception during early infancy. *Psychological Science, 10,* 419–422.

Money, J. (1993). Specific neurocognitional impairments associated with Turner (45,X) and Klinefelter (47,XXY) syndromes: A review. *Social Biology, 40,* 147–151.

Monk, C., Fifer, W. P., Myers, M. M., Sloan, R. P., Trien, L., & Hurtado, A. (2000). Maternal stress responses and anxiety during pregnancy: Effects on fetal heart rate. *Developmental Psychobiology, 36,* 67–77.

Monsour, M. (2002). *Women and men as friends.* Mahwah, NJ: Erlbaum.

Montepare, J., & Lachman, M. (1989). "You're only as old as you feel": Self-perceptions of age, fears of aging, and life satisfaction from adolescence to old age. *Psychology and Aging, 4,* 73–78.

Moon, C., Cooper, R. P., & Fifer, W. P. (1993). Two-day-old infants prefer their native language. *Infant Behavior and Development, 16,* 495–500.

Moon, S. M., & Feldhusen, J. F. (1994). The Program for Academic and Creative Enrichment (PACE): A follow-up study ten years later. In R. F. Subotnik & K. D. Arnold (Eds.), *Beyond Terman: Contemporary longitudinal studies of giftedness and talent* (pp. 375–400). Norwood, NJ: Ablex.

Moore, A., & Stratton, D. C. (2002). *Resilient widowers.* New York: Springer.

Moore, D. R., & Florsheim, P. (2001). Interpersonal processes and psychopathology among expectant and nonexpectant adolescent couples. *Journal of Consulting and Clinical Psychology, 69,* 101–113.

Moore, E. G. J. (1986). Family socialization and the IQ test performance of traditionally and transracially adopted black children. *Developmental Psychology, 22,* 317–326.

Moore, K. A., Morrison, D. R., & Green, A. D. (1997). Effects on the children born to adolescent mothers. In R. A. Maynard (Ed.), *Kids having kids* (pp. 145–180). Washington, DC: Urban Institute.

Moore, K. A., Myers, D. E., Morrison, D. R., Nord, C. W., Brown, B., & Edmonston, B. (1993). Age at first childbirth and later poverty. *Journal of Research on Adolescence, 3,* 393–422.

Moore, K. L., & Persaud, T. V. N. (1998). *Before we are born* (5th ed.). Philadelphia: Saunders.

Moore, M. K. (1992). An empirical investigation of the relationship between religiosity and death concern. *Dissertation Abstracts International, 53*(2–A), 527.

Moore, M. K., & Meltzoff, A. N. (1999). New findings on object permanence: A developmental difference between two types of occlusion. *British Journal of Developmental Psychology, 17,* 563–584.

Moore, W. S. (2002). Understanding learning in a postmodern world: Reconsidering the Perry scheme of ethical and intellectual development. In B. K. Hofer & P. R. Pintrich (Eds.), *Personal epistemology* (pp. 17–36). Mahwah, NJ: Erlbaum.

Moorehouse, M. J. (1991). Linking maternal employment patterns to mother–child activities and children's school competence. *Developmental Psychology, 27,* 295–303.

Morabia, A., Costanza, M. C., & the World Health Organization Collaborative Study of Neoplasia and Steroid Contraceptives. (1998). International variability in ages at menarche, first live birth, and menopause. *American Journal of Epidemiology, 148,* 1195–1205.

Morell, C. M. (1994). *Unwomanly conduct: The challenges of intentional childlessness.* New York: Routledge.

Morelli, G., Rogoff, B., Oppenheim, D., & Goldsmith, D. (1992). Cultural variation in infants' sleeping arrangements: Questions of independence. *Developmental Psychology, 28,* 604–613.

Morgan, L. A. (1991). *After marriage ends: Economic consequences for midlife women.* Newbury Park, CA: Sage.

Morgane, P. J., Austin-LaFrance, R., Bronzino, J., Tonkiss, J., Diaz-Cintra, S., Cintra, L., Kemper, T., & Galler, J. R. (1993). Prenatal malnutrition and development of the brain. *Neuroscience and Biobehavioral Reviews, 17,* 91–128.

Moriguchi, S., Oonishi, K., Kato, M., & Kishino, Y. (1995). Obesity is a risk factor for deteriorating cellular immune functions decreased with aging. *Nutrition Research, 15,* 151–160.

Morioka, M. (2001). Reconsidering brain death: A lesson from Japan's fifteen years of experience. *Hastings Center Report, 31*(4), 41–46.

Morley, J. E. (2001). Decreased food intake with aging. *Journal of Gerontology, 56A,* 81–88.

Morris, P. A. (2002). The effects of welfare reform policies on children. *Social Policy Report of the Society for Research in Child Development, 16*(1).

Morrison, D. R., & Coiro, M. J. (1999). Parental conflict and marital disruption: Do children benefit when high-conflict marriages are dissolved? *Journal of Marriage and the Family, 61,* 626–637.

Morrison, F. E., Griffith, E. M., & Alberts, D. M. (1997). Nature–nurture in the classroom: Entrance age, school readiness, and learning in children. *Developmental Psychology, 33,* 254–262.

Morrongiello, B. A. (1986). Infants' perception of multiple-group auditory patterns. *Infant Behavior and Development, 9,* 307–319.

Morrongiello, B. A., Fenwick, K. D., & Chance, G. (1998). Crossmodal learning in newborn infants: Inferences about properties of auditory-visual events. *Infant Behavior and Development, 21,* 543–554.

Morrongiello, B. A., Midgett, C., & Shields, R. (2001). Don't run with scissors: Young children's knowledge of home safety rules. *Journal of Pediatric Psychology, 26,* 105–115.

Morrongiello, B. A., & Rennie, H. (1998). Why do boys engage in more risk taking than girls? The role of attributions, beliefs, and risk appraisals. *Journal of Pediatric Psychology, 23,* 33–43.

Morrow, D., Leirer, V., Altieri, P., & Fitzsimmons, C. (1994). When expertise reduces age differences in performance. *Psychology and Aging, 9,* 134–148.

Morse, C. K. (1993). Does variability increase with age? An archival study of cognitive measures. *Psychology and Aging, 8,* 156–164.

Mortensen, E. L., Michaelsen, K. F., Sanders, S. A., & Reinisch, J. M. (2002). The association between duration of breastfeeding and adult intelligence. *Journal of the American Medical Association, 287,* 2365–2371.

Mortimer, J. T., & Borman, K. M. (Eds.). (1988). *Work experience and psychological development through the lifespan.* Boulder, CO: Westview.

Moses, L. J., Baldwin, D. A., Rosicky, J. G., & Tidball, G. (2001). Evidence for referential understanding in the emotions domain at twelve and eighteen months. *Child Development, 72,* 718–735.

Moshman, D. (1998). Identity as a theory of oneself. *Genetic Epistemologist, 26*(3), 1–9.

Moshman, D. (1999). *Adolescent psychological development: Rationality, morality, and identity.* Mahwah, NJ: Erlbaum.

Moshman, D., & Franks, B. A. (1986). Development of the concept of inferential validity. *Child Development, 57,* 153–165.

Mosko, S., Richard, C., & McKenna, J. J. (1997a). Infant arousals during mother–infant bed sharing: Implications for infant sleep and sudden infant death syndrome research. *Pediatrics, 100,* 841–849.

Mosko, S., Richard, C., & McKenna, J. J. (1997b). Maternal sleep and arousals during bedsharing with

infants. *Sleep, 20,* 142–150.
Mosteller, F. (1995, Fall). The Tennessee Study of Class Size in the Early School Grades. *Future of Children, 5*(2), 113–127.
Mounts, N. S., & Steinberg, L. (1995). An ecological analysis of peer influence on adolescent grade point average and drug use. *Developmental Psychology, 31,* 915–922.
Moxley, D. P., Najor-Durack, A., & Dumbrigue, C. (2001). *Keeping students in higher education.* London: Kogan Page.
Moyer, M. S. (1992). Sibling relationships among older adults. *Generations, 16*(3), 55–58.
Mrug, S., Hoza, B., & Gerdes, A. C. (2001). Children with attention-deficit/hyperactivity disorder: Peer relationships and peer-oriented interventions. In D. W. Nangle & C. A. Erdley (Eds.), *The role of friendship in psychological adjustment* (pp. 51–77). San Francisco: Jossey-Bass.
Mulatu, M. S., & Schooler, C. (2002). Causal connections between socioeconomic status and health: Reciprocal effects and mediating mechanisms. *Journal of Health and Social Behavior, 43,* 22–41.
Muldoon, O., & Cairns, E. (1999). Children, young people, and war: Learning to cope. In E. Frydenberg (Ed.), *Learning to cope: Developing as a person in complex societies* (pp. 322–337). New York: Oxford University Press.
Mullen, M. K. (1994). Earliest recollections of childhood: A demographic analysis. *Cognition, 52,* 55–79.
Muller, F., Rebiff, M., Taillandier, A., Qury, J. F., & Mornet, E. (2000). Parental origin of the extra chromosome in prenatally diagnosed fetal trisomy. *Human Genetics, 106,* 340–344.
Mulsant, B. H., & Ganguli, M. (1999). Epidemiology and diagnosis of depression in late life. *Journal of Clinical Psychiatry, 60*(Suppl. 20), 9–15.
Munoz, D. G., & Feldman, H. (2000). Causes of Alzheimer's disease. *Canadian Medical Association Journal, 162,* 65–72.
Munro, G., & Adams, G. R. (1977). Ego identity formation in college students and working youth. *Developmental Psychology, 13,* 523–524.
Muret-Wagstaff, S., & Moore, S. G. (1989). The Hmong in America: Infant behavior and rearing practices. In J. K. Nugent, B. M. Lester, & T. B. Brazelton (Eds.), *Biology, culture, and development* (Vol. 1, pp. 319–339). Norwood, NJ: Ablex.
Muris, P., Merckelbach, H., Gadet, B., & Moulaert, V. (2000). Fears, worries, and scary dreams in 4- to 12-year-old children: Their content, developmental pattern, and origins. *Journal of Clinical Child Psychology, 29,* 43–52.
Murray, A. D. (1985). Aversiveness is in the mind of the beholder. In B. M. Lester & C. F. Z. Boukydis (Eds.), *Infant crying* (pp. 217–239). New York: Plenum.
Murray, A. D., Johnson, J., & Peters, J. (1990). Fine-tuning of utterance length to preverbal infants: Effects on later language development. *Journal of Child Language, 17,* 511–525.
Murray, L., & Cooper, P. J. (1997). Postpartum depression and child development. *Psychological Medicine, 27,* 253–260.
Murray, L., Sinclair, D., Cooper, P., Ducournau, P., & Turner, P. (1999). The socioemotional development of 5-year-old children of postnatally depressed mothers. *Journal of Child Psychology and Psychiatry, 8,* 1259–1271.
Murray, M. J., & Meacham, R. B. (1993). The effect of age on male reproductive function. *World Journal of Urology, 11,* 137–140.
Mussen, P., & Eisenberg-Berg, N. (1977). *Roots of caring, sharing, and helping.* San Francisco: Freeman.
Must, A., & Strauss, R. S. (1999). Risks and consequences of childhood and adolescent obesity. *International Journal of Obesity, 23,* S2–S11.
Mutchler, J. E., Burr, J. A., Pienta, A. M., & Massagli, M. P. (1997). Pathways to labor force exit: Work transitions and instability. *Journal of Gerontology, 52B,* S4–S12.
Mutran, E. J., Danis, M., Bratton, K. A., Sudha, S., & Hanson, L. (1997). Attitudes of the critically ill toward prolonging life: The role of social support. *Gerontologist, 37,* 192–199.
Myers, D. G. (2000). The funds, friends, and faith of happy people. *American Psychologist, 55,* 56–67.
Myers, G. C. (1996). Comparative mortality trends among older persons in developed countries. In G. Caselli & A. D. Lopez (Eds.), *Health and mortality among elderly populations* (pp. 87–111). Oxford, U.K.: Clarendon Press.
Myerson, J., Hale, S., Wagstaff, D., Poon, L. W., & Smith, G. A. (1990). The information-loss model: A mathematical theory of age-related cognitive slowing. *Psychological Review, 97,* 475–487.
Nader, K. (2002). Treating children after violence in schools and communities. In N. B. Webb (Ed.), *Helping bereaved children: A handbook for practitioners* (pp. 214–244). New York: Guilford.
Nader, K., Dubrow, N., & Stamm, B. H. (1999). *Honoring differences: Cultural issues in the treatment of trauma and loss.* Washington, DC: Taylor & Francis.
Nader, K. O., Pynoos, R., Rairbanks, L., & Frederick, C. (1990). Childen's PTSD reactions one year after a sniper attack at their school. *American Journal of Psychiatry, 147,* 1526–1530.
Nagy, W. E., & Scott, J. A. (2000). Vocabulary processes. In M. L. Kamil & P. B. Mosenthal (Eds.), *Handbook of reading research* (Vol. 3, pp. 269–284). Mahwah, NJ: Erlbaum.
Naigles, L. G., & Gelman, S. A. (1995). Overextensions in comprehension and production revisited: Preferential-looking in a study of dog, cat, and cow. *Journal of Child Language, 22,* 19–46.
Nakamura, J., & Csikszentmihalyi, M. (2002). The concept of flow. In C. R. Snyder & S. J. Lopez (Eds.), *Handbook of positive psychology* (pp. 89–105). New York: Oxford University Press.
Namy, L. L., & Waxman, S. R. (1998). Words and gestures: Infants' interpretations of different forms of symbolic reference. *Child Development, 69,* 295–308.
Nánez, J., Sr., & Yonas, A. (1994). Effects of luminance and texture motion on infant defensive reactions to optical collision. *Infant Behavior and Development, 17,* 165–174.
Nansel, T. R., Overpeck, M., Pila, R. S., Ruan, W. J., Simons-Morton, B., & Scheidt, P. (2001). Bullying behaviors among U.S. youth: Prevalence and association with psychosocial adjustment. *Journal of the American Medical Association, 285,* 2094–2100.
National Association for the Education of Young Children. (1998). *Accreditation criteria and procedures of the National Academy of Early Childhood Programs* (2nd ed.). Washington, DC: Author.
National Federation of State High School Associations. (2002). *High school athletic participation survey.* Kansas City, MO: Author.
National Hospice and Palliative Care Organization. (2002). *Facts and figures on hospice care in America.* Alexandria, VA: Author.
National Institute for Child Health and Development (NICHD), Early Child Care Research Network. (1997). The effects of infant child care on infant–mother attachment security: Results of the NICHD Study of Early Child Care. *Child Development, 68,* 860–879.
National Institute of Child Health and Human Development (NICHD), Early Child Care Research Network. (1999). Child care and mother–child interaction in the first 3 years of life. *Developmental Psychology, 35,* 1399–1413.
National Institute of Child Health and Human Development (NICHD), Early Child Care Research Network. (2000a). Characteristics and quality of child care for toddlers and preschoolers. *Applied Developmental Science, 4,* 116–135.
National Institute of Child Health and Human Development (NICHD), Early Child Care Research Network. (2000b). The relation of child care to cognitive and language development. *Child Development, 71,* 960–980.
National Institute of Child Health and Human Development (NICHD), Early Child Care Research Network. (2001, April). *Early child care and children's development prior to school entry.* Symposium presented at the biennial meeting of the Society for Research in Child Development, Minneapolis, MN.
National Institute of Child Health and Human Development (NICHD), Early Child Care Research Network. (2002a). Child-care structure → process → outcome: Direct and indirect effects of child-care quality on young children's development. *Psychological Science, 13,* 199–206.
National Institute of Child Health and Human Development (NICHD), Early Child Care Research Network. (2002b). The interaction of child care and family risk in relation to child development at 24 and 36 months. *Applied Developmental Science, 6,* 144–156.
National Institute of Child Health and Human Development (NICHD), Early Child Care Research Network. (2002c). Parenting and family influences when children are in child care: Results from the NICHD Study of Early Child Care. In J. G. Borkowski & S. L. Ramey (Eds.), *Parenting and the child's world* (pp. 99–123). Mahwah, NJ: Erlbaum.
National Institute on Aging. (2000). *Progress report on Alzheimer's disease 2000.* Washington, DC: U.S. Government Printing Office.
National Task Force on the Prevention and Treatment of Obesity. (2002). Medical care for obese patients: Advice for health care professionals. *American Family Physician, 65,* 81–88.

Navarrete, C., Martinez, I., & Salamanca, F. (1994). Paternal line of transmission in chorea of Huntington with very early onset. *Genetic Counseling, 5,* 175–178.

Naveh-Benjamin, M. (2000). Adult age differences in memory performance: Tests of an associative deficit hypothesis. *Journal of Experimental Psychology: Learning, Memory, and Cognition, 26,* 1170–1187.

Neale, A. V., Hwalek, M. A., Goodrich, C. S., & Quinn, K. M. (1996). The Illinois Elder Abuse System: Program description and administrative findings. *Gerontologist, 36,* 502–511.

Neimeyer, R. A. (2000). Searching for the meaning of meaning: Grief therapy and the process of reconstruction. *Death Studies, 24,* 541–558.

Neimeyer, R. A. (Ed.). (1994). *Death anxiety handbook.* Washington, DC: Taylor & Francis.

Neimeyer, R. A., & Van Brunt, D. (1995). Death anxiety. In H. Waas & R. A. Neimeyer (Eds.), *Dying: Facing the facts* (3rd ed., pp. 49–88). Washington, DC: Taylor & Francis.

Nelson, C. A. (2000). Neural plasticity and human development: The role of early experience sculpting memory systems. *Developmental Science, 3,* 115–130.

Nelson, C. A. (2002). Neural devel opment and lifelong plasticity. In R. M. Lerner, F. Jacobs, & D. Wertlieb (Eds.), *Handbook of applied developmental science* (Vol. 1, pp. 31–60). Thousand Oaks, CA: Sage.

Nelson, C. A., & Bosquet, M. (2000). Neurobiology of fetal and infant development: Implications for infant mental health. In C. H. Zeanah, Jr. (Ed.), *Handbook of infant mental health* (2nd ed., pp. 37–59). New York: Guilford.

Nelson, E. A. S., Schiefenhoevel, W., & Haimerl, F. (2000). Child care practices in nonindustrialized societies. *Pediatrics, 105,* e75.

Nelson, H. D., Humphrey, L. L., Nygren, P., Teutsch, S. M., & Allan, J. D. (2002). Postmenopausal hormone replacement therapy: Scientific review. *Journal of the American Medical Association, 288,* 872–881.

Nelson, K. (1973). Structure and strategy in learning to talk. *Monographs of the Society for Research in Child Development, 38*(1–2, Serial No. 149).

Nelson, K. (1993). The psychological and social origins of autobiographical memory. *Psychological Science, 1,* 1–8.

Nestle, M., & Jacobson, M. F. (2000). Halting the obesity epidemic: A public health policy approach. *Public Health Reports, 115,* 12–24.

Neugarten, B. L. (1968a). Adult personality: Toward a psychology of the life cycle. In B. Neugarten (Ed.), *Middle age and aging* (pp. 137–147). Chicago: University of Chicago Press.

Neugarten, B. L. (1968b). The awareness of middle aging. In B. L. Neugarten (Ed.), *Middle age and aging* (pp. 93–98). Chicago: University of Chicago Press.

Neugarten, B. L. (1979). Time, age, and the life cycle. *American Journal of Psychiatry, 136,* 887–894.

Neugarten, B., & Neugarten, D. (1987, May). The changing meanings of age. *Psychology Today, 21*(5), 29–33.

Neuman, S. B. (1999). Books make a difference: A study of access to literacy. *Reading Research Quarterly, 34,* 286–311.

Neuman, S. B., Copple, C., & Bredekamp, S. (2000). *Learning to read and write: Developmentally appropriate practices for young children.* Washington, DC: National Association for the Education of Young Children.

Neville, H. A., & Heppner, M. J. (2002). Prevention and treatment of violence against women: An examination of sexual assault. In C. L. Juntunen & D. R. Atkinson (Eds.), *Counseling across the lifespan: Prevention and treatment* (pp. 261–277). Thousand Oaks, CA: Sage.

Neville, H. J., & Bruer, J. T. (2001). Language processing: How experience affects brain organization. In D. B. Bailey, Jr., J. T. Bruer, F. J. Symons, & J. W. Lichtman (Eds.), *Critical thinking about critical periods* (pp. 151–172). Baltimore: Paul H. Brookes.

Newacheck, P. W., & Halfon, N. (2000). Prevalence, impact, and trends in childhood disability due to asthma. *Archives of Pediatric and Adolescent Medicine, 154,* 287–293.

Newborg, J., Stock, J. R., & Wnek, L. (1984). *Batelle Developmental Inventory.* Allen, TX: LINC Associates.

Newcomb, A. F., Bukowski, W. M., & Pattee, L. (1993). Children's peer relations: A meta-analytic review of popular, rejected, neglected, controversial, and average sociometric status. *Psychological Bulletin, 113,* 99–128.

Newcomb, M. D., Abbott, R. D., Catalano, R. F., Hawkins, J. D., Battin-Pearson, S., & Hill, K. (2002). Mediational and deviance theories of late high school failure: Process roles of structural strains, academic competence, and general versus specific problem behavior. *Journal of Counseling Psychology, 49,* 172–186.

Newcombe, N., & Huttenlocher, J. (1992). Children's early ability to solve perspective-taking problems. *Developmental Psychology, 28,* 635–643.

Newcombe, P. A., & Boyle, G. J. (1995). High school students' sports personalities: Variations across participation level, gender, type of sport, and success. *International Journal of Sports Psychology, 26,* 277–294.

Newman, A. B., & Brach, J. S. (2001). Gender gap in longevity and disability in older persons. *Epidemiologic Reviews, 23,* 343–350.

Newman, B. S., & Muzzonigro, P. G. (1993). The effects of traditional family values on the coming out process of gay male adolescents. *Adolescence, 28,* 213–226.

Newman, C., Atkinson, J., & Braddick, O. (2001). The development of reaching and looking preferences in infants to objects of different sizes. *Developmental Psychology, 37,* 561–572.

Newnham, J. P., Evans, S. F., Michael, C. A., Stanley, F. J., & Landau, L. I. (1993). Effects of frequent ultrasound during pregnancy: A randomized control trial. *Lancet, 342,* 887–890.

Newport, E. L. (1991). Contrasting conceptions of the critical period for language. In S. Cary & R. Gelman (Eds.), *The epigenesis of mind: Essays on biology and cognition* (pp. 111–130). Hillsdale, NJ: Erlbaum.

Newsom, J. T. (1999). Another side to caregiving: Negative reactions to being helped. *Current Directions in Psychological Science, 8,* 183–187.

Newsom, J. T., & Schulz, R. (1996). Social support as a mediator in the relation between functional status and quality of life in older adults. *Psychology and Aging, 11,* 34–44.

Ni, Y. (1998). Cognitive structure, content knowledge, and classificatory reasoning. *Journal of Genetic Psychology, 159,* 280–296.

Nichols, L. S., & Junk, V. W. (1997). The sandwich generation: Dependency, proximity, and task assistance needs of parents. *Journal of Family and Economic Issues, 18,* 299–326.

Nichols, W. C., & Pace-Nichols, M. A. (2000). Childless married couples. In W. C. Nichols, M. A. Pace-Nichols, D. S. Becvar, & A. Y. Napier (Eds.), *Handbook of family development and prevention* (pp. 171–188). New York: Wiley.

Nidorf, J. F. (1985). Mental health and refugee youths: A model for diagnostic training. In T. C. Owen (Ed.), *Southeast Asian mental health: Treatment, prevention, services, training, and research* (pp. 391–427). Washington, DC: National Institute of Mental Health.

Nielsen, S. J., & Popkin, B. M. (2003). Patterns and trends in food portion sizes. *Journal of the American Medical Association, 289,* 450–453.

Nieman, D. (1994). Exercise: Immunity from respiratory infections. *Swimming Technique, 31*(2), 38–43.

NIH Consensus Development Panel on Osteoporosis Prevention, Diagnosis, and Therapy. (2001). Osteoporosis prevention, diagnosis, and therapy. *Journal of the American Medical Association, 285,* 785–795.

Nilsson, L., & Hamberger, L. (1990). *A child is born.* New York: Delacorte.

Nippold, M. A., Taylor, C. L., & Baker, J. M. (1996). Idiom understanding in Australian youth: A cross-cultural comparison. *Journal of Speech and Hearing Research, 39,* 442–447.

Nisbett, R. E. (1998). Race, genetics, and IQ. In C. Jencks & M. Phillips (Eds.), *The black–white test score gap* (pp. 86–102). Washington, DC: Brookings Institution.

Nix, R. L., Pinderhughes, E. E., Dodge, K. A., Bates, J. E., Pettit, G. S., & McFadyen-Ketchum, S. A. (1999). The relation between mothers' hostile attribution tendencies and children's externalizing behavior problems: The mediating role of mothers' harsh discipline practices. *Child Development, 70,* 896–909.

Nolen-Hoeksema, S. (2001, January). *Why women are more prone to depression than men.* Address presented at the 23rd Annual National Institute on the Teaching of Psychology, St. Petersburg Beach, Florida.

Noppe, I. C., & Noppe, L. D. (1997). Evolving meanings of death during early, middle, and later adolescence. *Death Studies, 21,* 253–275.

Noppe, L. D., & Noppe, I. C. (1991). Dialectical themes in adolescent conceptions of death. *Journal of Adolescent Research, 6,* 28–42.

Noppe, L. D., & Noppe, I. C. (1996). Ambiguity in adolescent understanding of death. In C. A. Corr & D. E. Balk (Eds.), *Handbook of adolescent death and bereavement* (pp. 25–41). New York: Springer.

Nordstrom, B. L., Kinnunen, T. U., Krall, C. H., & Et, E. A. (2000). Predictors of continued smoking over 25 years of follow-up in the normative aging study. *American Journal of Public Health, 90,* 404–406.

Notkola, I. L., Sulkava, R., Pekkanen, J., Erkinjuntti, T., Ehnholm, C., Kivinen, P., Tuomilehto, J., & Nissinen, A. (1998). Serum total cholesterol, apoliprotein E epsilon 4 allele, and Alzheimer's disease. *Neuroepidemiology, 17,* 14–20.

Nourse, C. B., & Butler, K. M. (1998). Perinatal transmission of HIV and diagnosis of HIV infection in infants: A review. *Irish Journal of Medical Science, 167,* 28–32.

Novak, M., & Thacker, C. (1991). Satisfaction and strain among middle-aged women who return to school: Replication and extension of findings in a Canadian context. *Educational Gerontology, 17,* 323–342.

Nucci, L. P. (1996). Morality and the personal sphere of action. In E. Reed, E. Turiel, & T. Brown (Eds.), *Values and knowledge* (pp. 41–60). Hillsdale, NJ: Erlbaum.

Nucci, L. P., Camino, C., & Sapiro, C. M. (1996). Social class effects on Northeastern Brazilian children's conceptions of areas of personal choice and social regulation. *Child Development, 67,* 1223–1242.

Nuckolls, K. B., Cassel, J., & Kaplan, B. H. (1972). Psychosocial assets, life crisis, and the prognosis of pregnancy. *American Journal of Epidemiology, 95,* 431–441.

Nuland, S. B. (1993). *How we die.* New York: Random House.

Nussbaum, J. F. (1994). Friendship in older adulthood. In M. L. Hummer, J. M. Wiemann, & J. F. Nussbaum (Eds.), *Interpersonal communication in older adulthood* (pp. 209–225). Thousand Oaks, CA: Sage.

Nye, B., Hedges, L. V., & Konstantopoulos, S. (2001). Are effects of small classes cumulative? Evidence from a Tennessee experiment. *Journal of Educational Research, 94,* 336–345.

Nye, W. P. (1993). Amazing grace: Religion and identity among elderly black individuals. *International Journal of Aging and Human Development, 36,* 103–114.

Oakes, J., Gamoran, A., & Page, R. N. (1992). Curriculum differentiation: Opportunities, outcomes, and meanings. In P. W. Jackson (Ed.), *Handbook of research on curriculum* (pp. 570–608). New York: Macmillan.

Oakes, L. M., Coppage, D. J., & Dingel, A. (1997). By land or by sea: The role of perceptual similarity in infants' categorization of animals. *Developmental Psychology, 33,* 396–407.

Obermeyer, C. M. (2000). Menopause across cultures: A review of the evidence. *Menopause, 7,* 184–192.

O'Bryant, S. L. (1988). Sibling support and older widows' well-being. *Journal of Marriage and the Family, 50,* 173–183.

O'Callaghan, M. J., Burn, Y. R., Mohay, H. A., Rogers, Y., & Tudehope, D. I. (1993). The prevalence and origins of left hand preference in high risk infants, and its implications for intellectual, motor, and behavioral performance at four and six years. *Cortex, 29,* 617–627.

O'Connor, A. R., Stephenson, T., Johnson, A., Tobin, M. J., Ratib, S., Ng, Y., & Fielder, A. R. (2002). Long-term ophthalmic outcome of low birth weight children with and without retinopathy of prematurity. *Pediatrics, 109,* 12–18.

O'Connor, C. (1997). Dispositions toward (collective) struggle and educational resilience in the inner city: A case analysis of six African-American high school students. *American Educational Research Journal, 34,* 593–629.

O'Connor, P. (1992). *Friendships between women.* New York: Guilford.

Oden, M. H., & Terman, L. M. (1968). The fulfillment of promise—40-year follow-up of the Terman gifted group. *Genetic Psychology Monographs, 77,* 3–93.

Oden, S. (2000). How researchers can support community efforts for change: Illustrations from two case studies. *Applied Developmental Science, 4,* 28–37.

Ogawa, J. R., Sroufe, L. A., Weinfield, N. S., Carlson, E. A., & Egeland, B. (1997). Development and the fragmented self: Longitudinal study of dissociative symptomatology in a nonclinical sample. *Development and Psychopathology, 9,* 855–879.

Ogbu, J. U. (1997). Understanding the school performance of urban blacks: Some essential background knowledge. In H. J. Walberg, O. Reyes, & R. P. Weissberg (Ed.), *Children and youth: Interdisciplinary perspectives* (pp. 190–222). Thousand Oaks, CA: Sage.

O'Grady-LeShane, R., & Williamson, J. B. (1992). Family provisions in old-age pensions. In M. E. Szinovacz, D. J. Ekerdt, & B. H. Vinick (Eds.), *Families and retirement* (pp. 64–77). Newbury Park, CA: Sage.

O'Halloran, C. M., & Altmaier, E. M. (1996). Awareness of death among children: Does a life-threatening illness alter the process of discovery? *Journal of Counseling and Development, 74,* 259–262.

Okagaki, L., Diamond, K. E., Kontos, S. J., & Hestenes, L. L. (1998). Correlates of young children's interactions with classmates with disabilities. *Early Childhood Research Quarterly, 13,* 67–86.

Okagaki, L., & Frensch, P. A. (1996). Effects of video game playing on measures of spatial performance: Gender effects in late adolescence. In P. M. Greenfield & R. R. Cocking (Eds.), *Interacting with video* (pp. 115–140). Norwood, NJ: Ablex.

Okagaki, L., & Sternberg, R. J. (1993). Parental beliefs and children's school performance. *Child Development, 64,* 36–56.

Oken, E., & Lightdale, J. R. (2000). Updates in pediatric nutrition. *Current Opinion in Pediatrics, 12,* 282–290.

Olafson, E., & Boat, B. W. (2000). Long-term management of the sexually abused child: Considerations and challenges. In R. M. Reece (Ed.), *Treatment of child abuse: Common ground for mental health, medical, and legal practitioners* (pp. 14–35). Baltimore: Johns Hopkins University Press.

Oliker, S. J. (1989). *Best friends and marriage: Exchange among women.* Berkeley: University of California Press.

Oller, D. K. (2000). *The emergence of the speech capacity.* Mahwah, NJ: Erlbaum.

Oller, D. K., Eilers, R. E., Neal, A. R., & Schwartz, H. K. (1999). Precursors to speech in infancy: The prediction of speech and language disorders. *Journal of Communication Disorders, 32,* 223–245.

Olsen, O. (1997). Meta-analysis of the safety of home birth. *Birth: Issues in Perinatal Care, 24,* 4–13.

Olson, R. E. (2000). Is it wise to restrict fat in the diets of children? *Journal of the American Dietetic Association, 100,* 28–32.

Olweus, D. (1995). Bullying or peer abuse at school: Facts and intervention. *Current Directions in Psychological Science, 4,* 196–200.

Ondrusek, N., Abramovitch, R., Pencharz, P., & Koren, G. (1998). Empirical examination of the ability of children to consent to clinical research. *Journal of Medical Ethics, 24,* 158–165.

O'Neill, R. M., Horton, S., & Crosby, F. J. (1999). *Mentoring dilemmas* (pp. 63–80). Mahwah, NJ: Erlbaum.

O'Neill, R., Welsh, M., Parke, R. D., Wang, S., & Strand, C. (1997). A longitudinal assessment of the academic correlates of early peer acceptance and rejection. *Journal of Clinical Child Psychology, 26,* 290–303.

Oosterwegel, A., & Oppenheimer, L. (1993). *The self-system: Developmental changes between and within self-concepts.* Hillsdale, NJ: Erlbaum.

Organization for Economic Cooperation and Development. (2000). *Education at a Glance: OECD Indicators.* Paris: Author.

O'Reilly, A. W. (1995). Using representations: Comprehension and production of actions with imagined objects. *Child Development, 66,* 999–1010.

O'Reilly, A. W., & Bornstein, M. H. (1993). Caregiver–child interaction in play. In M. H. Bornstein & A. W. O'Reilly (Eds.), *New directions for child development* (No. 59, pp. 55–66). San Francisco: Jossey-Bass.

Ory, M. G., Yee, J. L., Tennstedt, S. L., & Schulz, R. (2000). The extent and impact of dementia care: Unique challenges experienced by family caregivers. In R. Schulz (Ed.), *Handbook on dementia caregiving* (pp. 1–32). New York: Springer.

Osgood, N. J., Brant, B. A., & Lipman, A. (1991). *Suicide among the elderly in long-term care facilities.* New York: Greenwood Press.

Osherson, D. N., & Markman, E. M. (1975). Language and the ability to evaluate contradictions and tautologies. *Cognition, 2,* 213–226.

Osmond, C., & Barker, D. J. (2000). Fetal, infant, and childhood growth are predictors of coronary heart disease, diabetes, and hypertension in adult men and women. *Environmental Health Perspectives, 108,* 545–553.

Osteoporosis Society of Canada. (2002). About osteoporosis. Retrieved from http://www.osteoporosis.ca/english/about%20osteoporosis/default.asp?s=1

Ostir, G. V., Carlson, J. E., Black, S. A., Rudkin, L., Goodwin, J. S., & Markides, K. S. (1999). Disability in older adults 1: Prevalence, causes, and consequences. *Behavioral Medicine, 24,* 147–156.

Ostling, I., & Kelloway, E. K. (1992, June). *Predictors of life satisfaction in retirement: A mediational model.* Paper presented at the annual conference of the Canadian Psychological Association, Quebec City.

Ostrea, E. M., Jr., Ostrea, A. R., & Simpson, P. M. (1997). Mortality within the first 2 years in infants exposed to cocaine, opiate, or cannabinoid during gestation. *Pediatrics, 100,* 79–83.

Ovando, C. J., & Collier, V. P. (1998). *Bilingual and ESL classrooms: Teaching in multicultural contexts.* Boston: McGraw-Hill.

Overgaard, C., & Knudsen, A. (1999). Pain-relieving effect of sucrose in newborns during heel prick. *Biology of the Neonate, 75,* 279–284.

Owen, M. T., Easterbrooks, M. A., Chase-Lansdale, L., & Goldberg, W. A. (1984). The relation between maternal employment status and the stability of attachments to mother and father. *Child Development, 55,* 1894–1901.

Owen, N., Bauman, A., Booth, M., Oldenburg, B., & Magnus, P. (1995). Serial mass-media campaigns to promote physical activity: Reinforcing or redundant? *American Journal of Public Health, 85,* 244–248.

Owsley, C., Ball, K., McGwin, G., Jr., Sloane, M. E., Roenker, D. L., White, M. F., & Overley, E. T. (1998). Visual processing impairment and risk of motor vehicle crash among older adults. *Journal of the American Medical Association, 279,* 1083–1088.

Owsley, C., & McGwin, G., Jr. (1999). Vision impairment and driving. *Survey of Ophthalmology, 43,* 535–550.

Öztürk, C., Durmazlar, N., Ural, B., Karaagaoglu, E., Yalaz, K., & Anlar, B. (1999). Hand and eye preference in normal preschool children. *Clinical Pediatrics, 38,* 677–680.

Pacella, R., McLellan, M., Grice, K., Del Bono, E. A., Wiggs, J. L., & Gwiazda, J. E. (1999). Role of genetic factors in the etiology of juvenile-onset myopia based on a longitudinal study of refractive error. *Optometry and Vision Science, 76,* 381–386.

Padgett, D. K., Patrick, C., Bruns, B. J., & Schlesinger, H. J. (1994). Women and outpatient mental health services: Use by black, Hispanic, and white women in a national insured population. *Journal of Mental Health Administration, 2,* 347–360.

Padula, M. A., & Miller, D. L. (1999). Understanding graduate women's reentry experiences. *Psychology of Women Quarterly, 23,* 327–343.

Paffenbarger, R. S., Jr., Blair, S. N., & Lee, I. M. (2001). A history of physical activity, cardiovascular health and longevity: The scientific contributions of Jeremy N. Morris. *International Journal of Epidemiology, 30,* 1184–1192.

Paffenbarger, R. S., Jr., Hyde, R. T., Wing, A. L., Lee, I.-M., Jung, D. L., & Kampert, J. B. (1993). The association of changes in physical-activity level and other lifestyle characteristics with mortality among men. *New England Journal of Medicine, 329,* 538–545.

Pagani, L., Boulerice, B., Vitaro, F., & Tremblay, E. (1999). Effects of poverty on academic failure and delinquency in boys: A change and process model approach. *Journal of Child Psychology and Psychiatry, 40,* 1209–1219.

Pagani, L., Tremblay, R. E., Vitaro, F., Boulerice, B., & McDuff, P. (2001). Effects of grade retention on academic performance and behavioral development. *Development and Psychopathology, 13,* 297–315.

Paikoff, R. L., & Brooks-Gunn, J. (1991). Do parent–child relationships change during puberty? *Psychological Bulletin, 110,* 47–66.

Palincsar, A. S., & Herrenkohl, L. R. (1999). Designing collaborative contexts: Lessons from three research programs. In A. M. O'Donnell & A. King (Eds.), *Cognitive perspectives on peer learning. The Rutgers Invitational Symposium on Education Series* (pp. 151–177). Mahwah, NJ: Erlbaum.

Palmer, G. (1999). *Reluctant refugee.* Sydney: Kangaroo Press.

Palmore, E. (2001). The ageism survey: First findings. *Gerontologist, 41,* 572–575.

Palta, M., Sadek-Badawi, M., Evans, M., Weinstein, M. R., & McGuinness, G. (2000). Functional assessment of a multicenter very low-birth-weight cohort at age 5 years. *Archives of Pediatric and Adolescent Medicine, 154,* 23–30.

Pan, B. A., & Snow, C. E. (1999). The development of conversation and discourse skills. In M. Barrett (Ed.), *The development of language* (pp. 229–249). Hove, U.K.: Psychology Press.

Pan, H. W. (1994). Children's play in Taiwan. In J. L. Roopnarine, J. E. Johnson, & F. H. Hooper (Eds.), *Children's play in diverse cultures* (pp. 31–50). Albany, NY: SUNY Press.

Papini, D. R. (1994). Family interventions. In S. L. Archer (Ed.), *Interventions for adolescent identity development* (pp. 47–61). Thousand Oaks, CA: Sage.

Papousek, M., & Papousek, H. (1996). Infantile persistent crying, state regulation, and interaction with parents: A systems view. In M. H. Bornstein & J. L. Genevro (Eds.), *Child development and behavioral pediatrics* (pp. 11–33). Mahwah, NJ: Erlbaum.

Paquet, B. (2002). *Low-income cutoffs from 1992–2001 and low-income measures from 1991–2000.* Ottawa: Minister of Industry.

Pardhan, S., Gilchrist, J., Elliott, D. B., & Beh, G. K. (1996). A comparison of sampling efficiency and internal noise level in young and old subjects. *Vision Research, 36,* 1641–1648.

Park, D. C., Hertzog, C., Kidder, D. P., Morrell, R. W., & Mayhorn, C. B. (1997). Effect of age on event-based and time-based prospective memory. *Psychology and Aging, 12,* 314–327.

Parke, R. D., & Buriel, R. (1998). Socialization in the family: Ethnic and ecological perspectives. In N. Eisenberg (Ed.), *Handbook of child psychology: Vol. 3. Social, emotional, and personality development* (pp. 463–552). New York: Wiley.

Parker, F. L., Boak, A. Y., Griffin, K. W., Ripple, C., & Peay, L. (1999). Parent–child relationship, home learning environment, and school readiness. *School Psychology Review, 28,* 413–425.

Parker, J. G., Rubin, K. H., Price, J., & DeRosier, M. E. (1995). Peer relationships, child development, and adjustment: A developmental psychopathology perspective. In D. Cicchetti & D. Cohen (Eds.), *Developmental psychopathology: Vol. 2. Risk, disorder, and adaptation* (pp. 96–161). New York: Wiley.

Parker, R. G. (1995). Reminiscence: A continuity theory framework. *Gerontologist, 35,* 515–525.

Parkes, C. M., & Weiss, R. S. (1983). *Recovery from bereavement.* New York: Basic Books.

Parks, W. (1996). Human immunodeficiency virus. In R. D. Behrman, R. M. Kliegman, & A. M. Arvin (Eds.), *Nelson textbook of pediatrics* (15th ed., pp. 916–919). Philadelphia: Saunders.

Parmelee, P. A., & Lawton, M. P. (1990). The design of special environments for the aged. In J. E. Birren & K. W. Schaie (Eds.), *Handbook of the psychology of aging* (3rd ed., pp. 464–488). San Diego, CA: Academic Press.

Parten, M. (1932). Social participation among preschool children. *Journal of Abnormal and Social Psychology, 27,* 243–269.

Pascalis, O., de Haan, M., & Nelson, C. A. (1998). Long-term recognition memory for faces assessed by visual paired comparison in 3- and 6-month-old infants. *Journal of Experimental Psychology: Learning, Memory, and Cognition, 24,* 249–260.

Pascarella, E. T., & Terenzini, P. T. (1991). *How college affects students.* San Francisco: Jossey-Bass.

Pascarella, E. T., Whitt, E. J., Edison, M. I., Nora, A., Hagecdorn, L. S., Yeager, P. M., & Terenzini, P. T. (1997). Women's perceptions of a "chilly climate" and their cognitive outcomes during the first year of college. *Journal of College Student Development, 38,* 109–124.

Pasquino, A. M., Albanese, A., Bozzola, M., Butler, G. E., Buzi, F., & Cherubini, V. (2001). Idiopathic short stature. *Journal of Pediatric Endocrinology and Metabolism, 14,* 967–972.

Patrick, E., & Abravanel, E. (2000). The self-regulatory nature of preschool children's private speech in a naturalistic setting. *Applied Psycholinguistics, 21,* 45–61.

Patterson, C. J. (2001). Lesbian and gay parenting. Retrieved from http://www.apa.org/pi/parent.html

Patterson, C. J. (2002). Lesbian and gay parenthood. In M. H. Bornstein (Ed.), *Handbook of parenting* (Vol. 3, pp. 317–338). Mahwah, NJ: Erlbaum.

Patterson, G. R. (1982). *Coercive family processes.* Eugene, OR: Castilia Press.

Patterson, G. R. (1995). Coercion—A basis for early age of onset for arrest. In J. McCord (Ed.), *Coercion and punishment in long-term perspective* (pp. 81–105). New York: Cambridge University Press.

Patterson, G. R. (1997). Performance models for parenting: A social interactional perspective. In J. E. Grusec & L. Kuczynski (Eds.), *Parenting and children's internalization of values* (pp. 193–226). New York: Wiley.

Patterson, G. R., Dishion, T. J., & Yoerger, K. (2000). Adolescent growth in new forms of problem behavior: Macro- and micro-peer dynamics. *Prevention Science, 1,* 3–13.

Patterson, G. R., Reid, J. B., & Dishion, T. J. (1992). *Antisocial boys.* Eugene, OR: Castalia.

Patterson, M. M., & Lynch, A. Q. (1988). Menopause: Salient issues for counselors. *Journal of Counseling and Development, 67,* 185–188.

Patton, G. C., Selzer, R., Coffey, C., Carlin, J. B., & Wolfe, R. (1999). Onset of adolescent eating disorders: Population based cohort study over 3 years. *British Medical Journal, 318,* 765–768.

Pavlik, V. N., Hyman, D. J., Festa, N. A., & Dyer, C. B. (2001). Quantifying the problem of abuse and neglect in adults—analysis of a statwide database. *Journal of the American Geriatrics Society, 49,* 45–48.

Pawelec, G., Wagner, W., Adibzadeh, M., & Engel, A. (1999). T cell immunosenescence in vitro and in vivo. *Experimental Gerontology, 34,* 419–429.

Peake, T. H., Oelschlager, J. R., & Kearns, D. F. (2000). Elder abuse: Families, systems, causes, and interventions. In F. W. Kaslow (Ed.), *Handbook of couple and family forensics* (pp. 400–425). New York: Wiley.

Pearce, D., Cantisani, G., & Laihonen, A. (1999). Changes in fertility and family sizes in Europe. *Population Trends, 95,* 33–40.

Pearson, J. D., Morell, C. H., Gordon-Salant, S., Brant, L. J., Metter, E. J., Klein, L. L., & Fozard, J. L. (1995). Gender differences in a longitudinal study of age-associated hearing loss. *Journal of the Acoustical Society of America, 97,* 1196–1205.

Pearson, J. L., & Brown, G. K. (2000). Suicide prevention in late life: Directions for science and practice. *Clinical Psychology Review, 20,* 685–705.

Pebody, R. G., Edmunds, W. J., Conyn-van Spaendonck, M., Olin, P., Berbers, G., & Rebiere, I. (2000). The seroepidemiology of rubella in western Europe. *Epidemiology and Infections, 125,* 347–357.

Peck, R. C. (1968). Psychological developments in the second half of life. In B. L. Neugarten (Ed.), *Middle age and aging* (pp. 88–92). Chicago: University of Chicago Press.

Pedersen, J. B. (1998). Sexuality and aging. In I. H. Nordhus, G. R. VandenBos, S. Berg, & P. Fromholt (Eds.), *Clinical Geropsychology* (pp. 141–145). Washington, DC: American Psychological Association.

Pedersen, W. C., Miller, L. C., Putcha-Bhagavatula, A. D., & Yang, Y. (2002). Evolved sex differences in the number of partners desired? The long and the short of it. *Psychological Science, 13,* 157–161.

Pederson, D. R., Gleason, K. E., Moran, G., & Bento, S. (1998). Maternal attachment representations, maternal sensitivity, and the infant–mother attachment relationship. *Developmental Psychology, 34,* 925–933.

Pederson, D. R., & Moran, G. (1995). A categorical description of infant–mother relationships in the home and its relation to Q-sort measures of infant–mother interaction. In E. Waters, B. E.Vaughn, G. Posada, & K. Kondo-Ikemura (Eds.), Caregiving, cultural, and cognitive perspectives on secure-base behavior and working models: New growing points of attachment theory and research. *Monographs of the Society for Research in Child Development, 60*(2–3, Serial No. 244).

Pederson, D. R., & Moran, G. (1996). Expressions of the attachment relationship outside of the Strange Situation. *Child Development, 67,* 915–927.

Pedlow, R., Sanson, A., Prior, M., & Oberklaid, F. (1993). Stability of maternally reported temperament from infancy to 8 years. *Developmental Psychology, 29,* 998–1007.

Peiser-Feinberg, E. S. (1999). *The children of the Cost, Quality, and Outcomes Study go to school.* Chapel Hill, NC: University of North Carolina.

Pelham, W. E., Jr., & Hoza, B. (1996). Intensive treatment: A summer treatment program for children with ADHD. In E. D. Hibbs & P. S. Jensen (Eds.), *Psychosocial treatments for child and adolescent disorders: Empirically based strategies for clinical practice* (pp. 311–340). Washington, DC: American Psychological Association.

Pelham, W. E., Jr., Wheeler, T., & Chronis, A. (1998). Empirically supported psychosocial treatments for attention deficit hyperactivity disorder. *Clinical Child Psychology, 27,* 190–205.

Pellegrini, A. D. (2002). Bullying, victimization, and sexual harassment during the transition to middle school. *Educational Psychologist, 37,* 151–163.

Pellegrini, A. D., & Smith, P. K. (1998). Physical activity play: The nature and function of a neglected aspect of play. *Child Development, 69,* 557–598.

Penninx, B. W., Deeg, D. J., van Eijk, J. T., Beekman, A. T., & Guralnik, J. M. (2000). Changes in depression and physical decline in older adults: A longitudinal perspective. *Journal of Affective Disorders, 61,* 1–12.

Peplau, L. A. (1991). Lesbian and gay relationships. In J. C. Gonsiorek & J. D. Weinrich (Eds.), *Homosexuality* (pp. 177–196). Newbury Park, CA: Sage.

Pepler, D. J., & Craig, W. (2000). Making a difference in bullying. Retrieved from http://www.yorku.ca/lamarsh/Making%20a%20Difference%20in%20Bullying.pdf

Pepler, D. J., Craig, W. M., Connolly, J., & Henderson, K. (2002). Bullying, sexual harassment, dating violence, and substance use among adolescents. In C. Wekerle & A. Wall (Eds.), *The violence and addiction equation* (pp. 153–168). New York: Brunner-Routledge.

Peres, J. R. (2002). *Means to a better end: A report on dying in America today.* Princeton, NJ: Robert Wood Johnson Foundation.

Perfetti, C. A. (1988). Verbal efficiency in reading ability. In M. Daneman, G. E. MacKinnon, & T. G. Waller (Eds.), *Reading research: Advances in theory and practice* (Vol. 6, pp. 109–143). San Diego, CA: Academic Press.

Perie, M., Sherman, J. D., Phillips, G., & Riggan, M. (2000). Elementary and secondary education: An international perspective. *Education Statistics Quarterly.* Retrieved from http://nces.ed.gov/pubs2000/quarterly/summer/5int/q51.html

Perkins, H. W. (1991). Religious commitment, yuppie values, and well-being in post-collegiate life. *Review of Religious Research, 32,* 244–251.

Perkins, K. (1993). Working-class women and retirement. *Journal of Gerontological Social Work, 20,* 129–146.

Perleth, C., & Heller, K. A. (1994). The Munich Longitudinal Study of Giftedness. In R. F. Subotnik & K. D. Arnold (Eds.), *Beyond Terman: Contemporary studies of giftedness and talent* (pp. 77–114). Norwood, NJ: Ablex.

Perlmutter, M. (1984). Continuities and discontinuities in early human memory: Paradigms, processes, and performances. In R. V. Kail, Jr., & N. R. Spear (Eds.), *Comparative perspectives on the development of memory* (pp. 253–287). Hillsdale, NJ: Erlbaum.

Perlmutter, M., Kaplan, M., & Nyquist, L. (1990). Development of adaptive competence in adulthood. *Human Development, 33,* 185–197.

Perls, T., Levenson, R., Regan, M., & Puca, A. (2002). What does it take to live to 100? *Mechanisms of Ageing and Development, 123,* 231–242.

Perls, T., Terry, D. F., Silver, M., Shea, M., Bowen, J., & Joyce, E. (2000). Centenarians and the genetics of longevity. *Results and Problems in Cell Differentiation, 29,* 1–20.

Perls, T. T. (1995, January). The oldest old. *Scientific American, 272*(1), 70–75.

Perry, D. G., Perry, L. C., & Weiss, R. J. (1989). Sex differences in the consequences that children anticipate for aggression. *Developmental Psychology, 25,* 312–319.

Perry, D. G., Williard, J. C., & Perry, L. C. (1990). Peers' perceptions of the consequences that victimized children provide aggressors. *Child Development, 61,* 1310–1325.

Perry, W. G., Jr. (1970). *Forms of intellectual and ethical development in the college years.* New York: Holt, Rinehart & Winston.

Perry, W. G., Jr. (1981). Cognitive and ethical growth. In A. Chickering (Ed.), *The modern American college* (pp. 76–116). San Francisco: Jossey-Bass.

Perry-Jenkins, M., Repetti, R. L., & Crouter, A. C. (2000). Work and family in the 1990s. *Journal of Marriage and the Family, 62,* 981–998.

Persad, C. C., Abeles, N., Zacks, R. T., & Denburg, N. L. (2002). Inhibitory changes after age 60 and their relationship to measures of attention and memory. *Journal of Gerontology, 57B,* P223–P232.

Persson, I., Weiderpass, E., Bergkvist, L., Bergstrom, R., & Schairer, C. (1999). Risks of breast and endometrial cancer after estrogen and progestin replacement therapy. *Cancer Causes and Control, 10,* 253–260.

Pery, F. (1995, November–December). Careers: Retirement planning essentials. *Healthcare Executive, 10*(6), 42–43.

Peshkin, A. (1978). *Growing up American: Schooling and the survival of the community.* Chicago: University of Chicago Press.

Peshkin, A. (1997). *Places of memory: Whiteman's schools and native American communities.* Mahwah, NJ: Erlbaum.

Petersen, N., & Gonzales, R. C. (1999). *Career counseling models for diverse populations.* Belmont, CA: Wadsworth.

Peterson, B. E. (2002). Longitudinal analysis of midlife generativity, intergenerational roles, and caregiving. *Psychology and Aging, 17,* 161–168.

Peterson, B. E., & Klohnen, E. C. (1995). Realization of generativity in two samples of women at midlife. *Psychology and Aging, 10,* 20–29.

Peterson, B. E., Smirles, K. A., & Wentworth, P. A. (1997). Generativity and authoritarianism: Implications for personality, political involvement, and parenting. *Journal of Personality and Social Psychology, 72,* 1202–1216.

Peterson, C., & Rideout, R. (1998). Memory for medical emergencies experienced by 1- and 2-year-olds. *Developmental Psychology, 34,* 1059–1072.

Peterson, C. C. (1996). The ticking of the social clock: Adults' beliefs about the timing of transition events. *International Journal of Aging and Human Development, 42,* 189–203.

Peterson, L. (1989). Latchkey children's preparation for self-care:

Overestimated, underrehearsed, and unsafe. *Journal of Clinical Child Psychology, 18,* 36–43.

Peterson, L., & Brown, D. (1994). Integrating child injury and abuse–neglect research: Common histories, etiologies, and solutions. *Psychological Bulletin, 116,* 293–315.

Petinou, K. C., Schwartz, R. G., Gravel, J. S., & Raphael, L. J. (2001). A preliminary account of phonological and morphological perception in young children with and without otitis media. *International Journal of Language and Communication Disorders, 36,* 21–42.

Petitto, L. A., Holowka, S., Sergio, L. E., & Ostry, D. (2001, September 6). Language rhythms in babies' hand movements. *Nature, 413,* 35–36.

Petitto, L. A., & Marentette, P. F. (1991). Babbling in the manual mode: Evidence for the ontogeny of language. *Science, 251,* 1493–1496.

Pettit, G. S., Bates, J. E., & Dodge, K. A. (1998). Supportive parenting, ecological context, and children's adjustment: A seven-year longitudinal study. *Child Development, 68,* 908–923.

Pettit, G. S., Bates, J. E., Dodge, K. A., & Meece, D. W. (1999). The impact of after-school peer contact on early adolescent externalizing problems is moderated by parental monitoring, perceived neighborhood safety, and prior adjustment. *Child Development, 70,* 768–778.

Pfizer. (2002). *New facts about dementia in older adults.* Groton, CT: Author.

Phelps, K. E., & Woolley, J. D. (1994). The form and function of young children's magical beliefs. *Developmental Psychology, 30,* 385–394.

Phillips, D. A., Voran, M., Kisker, E., Howes, C., & Whitebook, M. (1994). Child care for children in poverty: Opportunity or inequity? *Child Development, 65,* 472–492.

Phillips, D. A., & Zimmerman, M. (1990). The developmental course of perceived competence and incompetence among competent children. In R. Sternberg & J. Kolligian (Eds.), *Competence considered* (pp. 41–66). New Haven, CT: Yale University Press.

Phillips, M. (1997). What makes schools effective? A comparison of the relationships of communitarian climate and academic climate to mathematics achievement and attendance during middle school. *American Educational Research Journal, 34,* 633–662.

Phillips, O. P., & Elias, S. (1993). Prenatal genetic counseling issues in women of advanced reproductive age. *Journal of Women's Health, 2,* 1–5.

Phillips, S. D., & Imhoff, A. R. (1997). Women and career development: A decade of research. *Annual Review of Psychology, 48,* 31–59.

Phillipsen, L. C. (1999). Associations between age, gender, and group acceptance and three components of friendship quality. *Journal of Early Adolescence, 19,* 438–464.

Phinney, J., & Ong, A. (2001). *Family obligations and life satisfaction among adolescents from immigrant and nonimmigrant families: Direct and moderated effects.* Unpublished manuscript, California State University, Los Angeles.

Phinney, J. S. (1989). Stages of ethnic identity development in minority group adolescents. *Journal of Early Adolescence, 9,* 34–49.

Phinney, J. S. (1993). Multiple group identities: Differentiation, conflict, and integration. In J. Kroger (Ed.), *Discussions on ego identity* (pp. 47–73). Hillsdale, NJ: Erlbaum.

Phinney, J. S., & Chavira, V. (1995). Parental ethnic socialization and adolescent outcomes in ethnic minority families. *Journal of Research on Adolescence, 5,* 31–53.

Phinney, J. S., & Kohatsu, E. L. (1997). Ethnic and racial identity development and mental health. In J. Schulenberg, J. L. Maggs, & K. Hurrelmann (Eds.), *Health risks and developmental transitions during adolescence* (pp. 420–443). Cambridge, U.K.: Cambridge University Press.

Phinney, J. S., Ong, A., & Madden, T. (2000). Cultural values and intergenerational value discrepancies in immigrant and nonimmigrant families. *Child Development, 71,* 528–539.

Piaget, J. (1926). *The language and thought of the child.* New York: Harcourt, Brace & World. (Original work published 1923)

Piaget, J. (1930). The child's conception of the world. New York: Harcourt, Brace, & World. (Original work published 1926)

Piaget, J. (1950). *The psychology of intelligence.* New York: International Universities Press.

Piaget, J. (1951). *Play, dreams, and imitation in childhood.* New York: Norton. (Original work published 1945)

Piaget, J. (1952). *The origins of intelligence in children.* New York: International Universities Press. (Original work published 1936)

Piaget, J. (1965). *The moral judgment of the child.* New York: Free Press. (Original work published 1932)

Piaget, J. (1967). *Six psychological studies.* New York: Vintage.

Piaget, J. (1971). *Biology and knowledge.* Chicago: University of Chicago Press.

Piaget, J. (1985). *The equilibration of cognitive structures: The central problem of intellectual development.* Chicago: University of Chicago Press.

Pianta, R., Egeland, B., & Erickson, M. F. (1989). The antecedents of maltreatment: Results of the Mother–Child Interaction Research Project. In D. Cicchetti & V. Carlson (Eds.), *Child maltreatment* (pp. 203–253). New York: Cambridge University Press.

Piatt, B. (1993). *Only English? Law and language policy in the United States.* Albuquerque: University of New Mexico Press.

Pickens, J., Field, T., & Nawrocki, T. (2001). Frontal EEG asymmetry in response to emotional vignettes in preschool age children. *International Journal of Behavioral Development, 25,* 105–112.

Pickering, L. K., Granoff, D. M., Erickson, J. R., Mason, M. L., & Cordle, C. T. (1998). Modulation of the immune system by human milk and infant formula containing nucleotides. *Pediatrics, 101,* 242–249.

Pierce, K. M., Hamm, J. V., & Vandell, D. L. (1999). Experiences in after-school programs and children's adjustment in first-grade classrooms. *Child Development, 70,* 756–767.

Pierce, S. H., & Lange, G. (2000). Relationships among metamemory, motivation and memory performance in young school-age children. *British Journal of Developmental Psychology, 18,* 121–135.

Pierce, W. D., & Epling, W. F. (1995). *Behavior analysis and learning.* Englewood Cliffs, NJ: Prentice-Hall.

Pietz, J., Dunckelmann, R., Rupp, A., Rating, D., Meinck, H. M., Schmidt, H., & Bremer, H. J. (1998). Neurological outcome in adult patients with early-treated phenylketonuria. *European Journal of Pediatrics, 157,* 824–830.

Pilkington, C. L., & Piersel, W. C. (1991). School phobia: A critical analysis of the separation anxiety theory and an alternative conceptualization. *Psychology in the Schools, 28,* 290–303.

Pillemer, K. A., & Finkelhor, D. (1988). The prevalence of elder abuse: A random sample survey. *Gerontologist, 28,* 51–57.

Pillow, B. H. (1995). Two trends in the development of conceptual perspective taking. An elaboration of the passive–active hypothesis. *International Journal of Behavioral Development, 18,* 649–676.

Pinderhughes, E. E., Dodge, K. A., Bates, J. E., Pettit, G. S., & Zelli, A. (2000). Discipline responses: Influences of parents' socioeconomic status, ethnicity, beliefs about parenting, stress, and cognitive-emotional processes. *Journal of Family Psychology, 14,* 380–400.

Pines, A. M. (1998). A prospective study of personality and gender differences in romantic attraction. *Personality and Individual Differences, 25,* 147–157.

Pinker, S., Lebeaux, D. S., & Frost, L. A. (1987). Productivity and constraints in the acquisition of the passive. *Cognition, 26,* 195–267.

Pinquart, M., & Sörensen, S. (2001). Gender differences in self-concept and psychological well-being in old age: A meta-analysis. *Journal of Gerontology, 56B,* P195–P213.

Pinson-Milburn, N. M., Fabian, E. S., Schlossberg, N. K., & Pyle, M. (1996). Grandparents raising grandchildren. *Journal of Counseling and Development, 74,* 548–554.

Pipes, P. L. (1996). *Nutrition in infancy and childhood* (6th ed.). St. Louis: Mosby.

Pipp, S., Easterbrooks, M. A., & Brown, S. R. (1993). Attachment status and complexity of infants' self- and other-knowledge when tested with mother and father. *Social Development, 2,* 1–14.

Pipp, S., Easterbrooks, M. A., & Harmon, R. J. (1992). The relation between attachment and knowledge of self and mother in one-year-old infants to three-year-old infants. *Child Development, 63,* 738–750.

Pivarnik, J. M. (1998). Potential effects of maternal physical activity on birth weight: Brief review. *Medicine and Science in Sports and Exercise, 30,* 407–414.

Plassman, B. L., & Breitner, C. S. (1996). Recent advances in the genetics of Alzheimer's disease and vascular dementia with an emphasis on gene–environment interactions. *Journal of the American Genetics Society, 44,* 1242–1250.

Plassman, B. L., Havlik, R. J., Steffens, D. C., Helms, M. J., Newman, T. N., & Drosdick, D. (2000). Documented head injury in early adulthood and risk of Alzheimer's disease and other dementias. *Neurology, 55,* 1158–1166.

Plessinger, M. A., & Woods, J. R., Jr. (1998). Cocaine in pregnancy: Recent data on maternal and fetal risks. *Substance Abuse in Pregnancy, 25,* 99–118.

Plomin, R. (1994a). The Emanuel Miller Memorial Lecture 1993: Genetic research and identification of environmental influences. *Journal of Child Psychology and Psychiatry, 35,* 817–834.

Plomin, R. (1994b). *Genetics and experience: The interplay between nature and nurture.* Thousand Oaks, CA: Sage.

Plomin, R. (1994c). Nature, nurture, and social development. *Social Development, 3,* 37–53.

Plomin, R., Reiss, D., Hetherington, E. M., & Howe, G. W. (1994). Nature and nurture: Genetic contributions to measures of the family environment. *Developmental Psychology, 30,* 32–43.

Plucker, J. A., Callahan, C. M., & Tomchin, E. M. (1996). Wherefore art thou, multiple intelligences? Alternative assessments for identifying talent in ethnically diverse and low income students. *Gifted Child Quarterly, 40,* 81–92.

Plude, D. J., & Doussard-Roosevelt, J. A. (1989). Aging, selective attention, and feature integration. *Psychology and Aging, 4,* 98–105.

Plude, D. J., & Hoyer, W. J. (1985). Attention and performance: Identifying and localizing age deficits. In N. Charness (Ed.), *Aging and human performance* (pp. 47–99). Chichester, England: Wiley.

Plumert, J. M., Pick, H. L., Jr., Marks, R. A., Kintsch, A. S., & Wegesin, D. (1994). Locating objects and communicating about locations: Organizational differences in children's searching and direction-giving. *Developmental Psychology, 30,* 443–453.

Podrouzek, W., & Furrow, D. (1988). Preschoolers' use of eye contact while speaking: The influence of sex, age, and conversational partner. *Journal of Psycholinguistic Research, 17,* 89–93.

Poehlman, E. T., Turturro, A., Bodkin, N., Cefalu, W., Heymsfield, S., Holloszy, J., & Kemnitz, J. (2001). Caloric restriction mimetics: Physical activity and body composition changes. *Journals of Gerontology, 56A,* 45–54.

Pohl, J. M., Given, C. W., Collins, C. E., & Given, B. A. (1994). Social vulnerability and reactions to caregiving in daughters and daughters-in-law caring for disabled aging parents. *Health Care for Women International, 15,* 385–395.

Pohl, R. (2001). *Homelessness in Canada: Part 1—An introduction.* Ottawa: Innercity Ministries.

Pohl, R. (2002). *Poverty in Canada.* Ottawa: Innercity Ministries.

Polansky, N. A., Gaudin, J. M., Ammons, P. W., & Davis, K. B. (1985). The psychological ecology of the neglectful mother. *Child Abuse and Neglect, 9,* 265–275.

Polka, L., & Werker, J. F. (1994). Developmental changes in perception of non-native vowel contrasts. *Journal of Experimental Psychology: Human Perception and Performance, 20,* 421–435.

Pollitt, E. (1996). A reconceptualization of the effects of undernutrition on children's biological, psychosocial, and behavioral development. *Social Policy Report of the Society for Research in Child Development, 10*(5).

Pollock, G. H. (1987). The mourning–liberation process in health and disease. *Psychiatric Clinics of North America, 10,* 345–354.

Pollock, L. (1987). *A lasting relationship: Parents and children over three centuries.* Hanover, NH: University Press of New England.

Pollock, M. L., Mengelkoch, L. J., & Graves, J. E. (1997). Twenty-year follow-up of aerobic power and body composition of older track athletes. *Journal of Applied Physiology, 82,* 1508–1516.

Polloway, E. A., Patton, J. R., Smith, T. E. C., & Buck, G. H. (1997). Mental retardation and learning disabilities: Conceptual and applied issues. *Journal of Learning Disabilities, 30,* 297–308.

Pomerantz, E. M., & Eaton, M. M. (2000). Developmental differences in children's conceptions of parental control: "They love me, but they make me feel incompetent." *Merrill-Palmer Quarterly, 46,* 140–167.

Pomerantz, E. M., & Ruble, D. N. (1998). The multidimensional nature of control: Implications for the development of sex differences in self-evaluation. In J. Heckhausen & C. S. Dweck (Eds.), *Motivation and self-regulation across the lifespan* (pp. 159–184). New York: Cambridge University Press.

Pomerantz, E. M., & Saxon, J. L. (2001). Conceptions of ability as stable and self-evaluative processes: A longitudinal examination. *Child Development, 72,* 152–173.

Popkin, B. M. (1994). The nutrition transition in low-income countries: An emerging crisis. *Nutrition Review, 52,* 285–298.

Popkin, B. M., & Doak, C. M. (1998). The obesity epidemic is a worldwide phenomenon. *Nutrition Reviews, 56,* 106–114.

Popkin, B. M., Richards, M. K., & Montiero, C. A. (1996). Stunting is associated with overweight in children of four nations that are undergoing the nutrition transition. *Journal of Nutrition, 126,* 3009–3016.

Porter, R. H., Makin, J. W., Davis, L. B., & Christensen, K. M. (1992). An assessment of the salient olfactory environment of formula-fed infants. *Physiology and Behavior, 50,* 907–911.

Porter, R. H., & Winberg, J. (1999). Unique salience of maternal breast odors for newborn infants. *Neuroscience and Biobehavioral Reviews, 23,* 439–449.

Portman, P. A. (1995). Who is having fun in physical education classes? Experiences of sixth-grade students in elementary and middle schools. *Journal of Teaching in Physical Education, 14,* 445–453.

Posada, G., Jacobs, A., Richmond, M. K., Carbonell, O. A., Alzate, G., Bustamante, M. R., & Quiceno, J. (2002). Maternal caregiving and infant security in two cultures. *Developmental Psychology, 38,* 67–78.

Posner, J. K., & Vandell, D. L. (1994). Low-income children's after-school care: Are there beneficial effects of after-school programs? *Child Development, 58,* 568–595.

Posner, J. K., & Vandell, D. L. (1999). After-school activities and the development of low-income urban children: A longitudinal study. *Developmental Psychology, 35,* 868–879.

Poulin-Dubois, D., & Héroux, G. (1994). Movement and children's attributions of life properties. *International Journal of Behavioral Development, 17,* 329–347.

Poulin-Dubois, D., Serbin, L. A., Kenyon, B., & Derbyshire, A. (1994). Infants' intermodal knowledge about gender. *Developmental Psychology, 30,* 436–442.

Powell, B., & Steelman, L. C. (1993). The educational benefits of being spaced out: Sibship density and educational progress. *American Sociological Review, 58,* 367–381.

Powell, D. R. (1986, March). Parent education and support programs. *Young Children, 41*(3), 47–53.

Powers, S. I., Hauser, S. T., & Kilner, L. A. (1989). Adolescent mental health. *American Psychologist, 44,* 200–208.

Powls, A., Botting, N., Cooke, R. W. I., & Marlow, N. (1996). Handedness in very-low-birthweight (VLBW) children at 12 years of age: Relation to perinatal and outcome variables. *Developmental Medicine and Child Neurology, 38,* 594–602.

Prado, L. M., & Markman, H. J. (1998). Unearthing the seeds of marital distress: What we have learned from married and remarried couples. In M. J. Cox & J. Brooks-Gunn (Eds.), *Conflict and cohesion in families* (pp. 51–85). Mahwah, NJ: Erlbaum.

Prager, K. J., & Bailey, J. M. (1985). Androgyny, ego development, and psychological crisis resolution. *Sex Roles, 13,* 525–535.

Pratt, M. W., Arnold, M. L., Pratt, A. T., & Diessner, R. (1999). Predicting adolescent moral reasoning from family climate: A longitudinal study. *Journal of Early Adolescence, 19,* 148–175.

Prechtl, H. F. R., & Beintema, D. (1965). *The neurological examination of the full-term newborn infant.* London: Heinemann Medical Books.

Preisler, G. M. (1991). Early patterns of interaction between blind infants and their sighted mothers. *Child: Care, Health and Development, 17,* 65–90.

Preisler, G. M. (1993). A descriptive study of blind children in nurseries with sighted children. *Child: Care, Health and Development, 19,* 295–315.

Pressley, M. (1995). More about the development of self-regulation: Complex, long-term, and thoroughly social. *Educational Psychologist, 30,* 207–212.

Pressley, M., Wharton-McDonald, R., Allington, R., Block, C. C., Morrow, L., Tracey, D., Baker, K., Brooks, G., Cronin, J., Nelson, E., & Woo, D. (2001). A study of effective first-grade literacy instruction. *Scientific Studies of Reading, 5,* 35–58.

Previc, F. H. (1991). A general theory concerning the prenatal origins of cerebral lateralization. *Psychological Review, 98,* 299–334.

Price, J. L., & Morris, J. C. (1999). Tangles and plaques in nondemented aging and "preclinical" Alzheimer's disease. *Annals of Neurology, 45,* 358–368.

Princeton Religion Research Center. (1994). *Religion in America.* Princeton, NJ: Gallup Poll.

Prinz, P. N., Vitiello, M. V., Raskind, M. A., & Thorpy, M. J. (1990). Geriatrics: Sleep disorders and aging. *New England Journal of Medicine, 323,* 520–526.

Privacy Commissioner of Canada. (2002). Genetic testing must not trade away privacy. Retrieved from http://www.privcom.gc.ca.

Prochaska, J. O., DiClemente, C. C., & Norcross, J. C. (1992). In search of how people change: Applications to addictive behaviors. *American Psychologist, 47,* 1102–1114.

Provins, K. A. (1997). Handedness and speech: A critical reappraisal of the role of genetic and environmental factors in the cerebral lateralization of function. *Psychological Review, 104,* 554–571.

Pruchno, R., & McKenney, D. (2000). The effects of custodial and coresident households on the mental health of grandmothers. *Journal of Mental Health and Aging, 6,* 291–310.

Prysak, M., Lorenz, R. P., & Kisly, A. (1995). Pregnancy outcome in nulliparous women 35 years and older. *Obstetrics and Gynecology, 85,* 65–70.

Pungello, E. P., & Kurtz-Costes, B. (1999). Why and how working women choose child care: A review with a focus on infancy. *Developmental Review, 19,* 31–96.

Purcell, D. W., & Hicks, D. W. (1996). Lesbian sexuality. In R. P. Cabaj & T. S. Stein (Eds.), *Textbook of homosexuality and mental health* (pp. 763–782). Washington, DC: American Psychiatric Press.

Purcell-Gates, V. (1996). Stories, coupons, and the TV Guide: Relationships between home literacy experiences and emergent literacy knowledge. *Reading Research Quarterly, 31,* 406–428.

Putnam, S. P., Samson, A. V., & Rothbart, M. K. (2000). Child temperament and parenting. In V. J. Molfese & D. L. Molfese (Eds.), *Temperament and personality across the life span* (pp. 255–277). Mahwah, NJ: Erlbaum.

Pyeritz, R. E. (1998). Sex: What we make of it. *Journal of the American Medical Association, 279,* 269.

Pyka, G., Lindenberger, E., Charette, S., & Marcus, R. (1994). Muscle strength and fiber adaptations to a year-long resistance training program in elderly men and women. *Journal of Gerontology, 49,* M22–M27.

Pynoos, J., & Golant, S. (1996). Housing and living arrangements for the elderly. In R. H. Binstock & L. K. George (Eds.), *Handbook of aging and the social sciences* (pp. 303–324). San Diego: Academic Press.

Qiu, C., Bäckman, L., Winblad, B., Agüero-Torres, H., & Fratiglioni, L. (2001). The influence of education on clinically diagnosed dementia incidence and mortality data from the Kungsholmen Project. *Archives of Neurology, 58,* 2034–2039.

Quadagno, J., & Hardy, M. (1996). Work and retirement. In R. H. Binstock & L. K. George (Eds.), *Handbook of aging and the social sciences* (pp. 325–345). San Diego: Academic Press.

Quick, H. E., & Moen, P. (1998). Gender, employment, and retirement quality: A life course approach to the differential experiences of men and women. *Journal of Occupational Health Psychology, 3,* 44–64.

Quill, T. E. (1991). Death and dignity: A case of individualized decision making. *New England Journal of Medicine, 324,* 691–694.

Quinn, J. F. (1999). Retirement patterns and bridge jobs in the 1990s. *EBRI Issue Brief, 206,* 1–22.

Quinn, J. F., & Kozy, M. (1996). The role of bridge jobs in the retirement transition: Gender, race, and ethnicity. *Gerontologist, 36,* 363–372.

Quinn, L. (2001). Type 2 diabetes. *Nursing Clinics of North America, 36,* 175–192.

Quinn, M. E., Johnson, M. A., Poon, L. W., & Martin, P. (1999). Psychosocial correlates of subjective health in sexagenarians, octogenarians, and centenarians. *Issues in Mental Health Nursing, 20,* 151–171.

Quinn, P. C., Johnson, M. H., Mareschal, D., Rakison, D. H., & Younger, B. A. (2000). Understanding early categorization: One process or two? *Infancy, 1,* 111–122.

Quinn, T. M., & Adzick, N. S. (1997). Fetal surgery. *Obstetrics and Gynecology Clinics of North America, 24,* 143–157.

Quint, J. C., Box, J. M., & Polit, D. F. (1997). *New Chance: Final report on a comprehensive program for disadvantaged young mothers and their children.* New York: Manpower Demonstration Research Corporation.

Quintero, R. A., Puder, K. S., & Cotton, D. B. (1993). Embryoscopy and fetoscopy. *Obstetrics and Gynecology Clinics of North America, 20,* 563–581.

Quist, J. F., & Kennedy, J. L. (2001). Genetics of childhood disorders: XXIII. ADHD, part 7: The serotonin system. *Journal of the American Academy of Child and Adolescent Psychiatry, 40,* 253–256.

Quyen, G. T., Bird, H. R., Davies, M., Hoven, C., Cohen, P., Jensen, P. S., & Goodman, S. (1998). Adverse life events and resilience. *Journal of the American Academy of Child and Adolescent Psychiatry, 37,* 1191–1200.

Radin, N. (1994). Primary caregiving fathers in intact families. In A. E. Gottfried & A. W. Gottfried (Eds.), *Redefining families: Implications for children's development* (pp. 11–54). New York: Plenum.

Radziszewska, B., & Rogoff, B. (1988). Influence of adult and peer collaboration on the development of children's planning skills. *Developmental Psychology, 24,* 840–848.

Rafferty, Y. (1995). The legal rights and educational problems of homeless children and youth. *Educational Evaluation and Policy Analysis, 17,* 39–61.

Ragow-O'Brien, D., Hayslip, B., Jr., & Guarnaccia, C. A. (2000). The impact of hospice on attitudes toward funerals and subsequent bereavement adjustment. *Omega, 41,* 291–305.

Rahn, W. M., & Transue, J. E. (1998). Social trust and value change: The decline of social capital in American youth, 1976–1995. *Political Psychology, 19,* 545–565.

Räihä, N. C. R., & Axelsson, I. E. (1995). Protein nutrition in infancy. *Pediatric Clinics of North America, 42,* 745–763.

Räikkönen, K., Matthews, K. A., Flory, J. D., Owens, J. F., & Gump, B. B. (1999). Effects of optimism, pessimism, and trait anxiety on ambulatory blood pressure and mood during everyday life. *Journal of Personality and Social Psychology, 76,* 104–113.

Raisler, J. (1999). Breast-feeding and infant illness: A dose-response relationship? *American Journal of Public Health, 89,* 25–30.

Rakison, D. H., & Butterworth, G. E. (1998). Infants' use of object parts in early categorization. *Developmental Psychology, 34,* 49–62.

Raley, R. K. (1996). A shortage of marriageable men? A note on the role of cohabitation in black–white differences in marriage rates. *American Sociological Review, 61,* 973–983.

Ramey, C. T., Campbell, F. A., & Ramey, S. L. (1999). Early intervention: Successful pathways to improving intellectual development. *Developmental Neuropsychology, 16,* 385–392.

Ramey, C. T., & Ramey, S. L. (1998). Early intervention and early experience. *American Psychologist, 53,* 109–120.

Ramey, S. L. (1999). Head Start and preschool education: Toward continued improvement. *American Psychologist, 54,* 344–346.

Ramey, S. L., & Ramey, C. T. (1999). Early experience and early intervention for children "at risk" for developmental delay and mental retardation. *Mental Retardation and Developmental Disabilities, 5,* 1–10.

Ramirez, J. D., Yuen, S. D., Ramey, D. R., & Pasta, D. (1991). *Longitudinal study of structured English immersion strategy, early exit and late-exit transitional bilingual education programs for language minority: Final report* (Vols. 1 & 2). San Mateo, CA: Aguirre International.

Ramos, E., Frontera, W. R., Llorpart, A., & Feliciano, D. (1998). Muscle strength and hormonal levels in adolescents: Gender related differences. *International Journal of Sports Medicine, 19,* 526–531.

Ramsøy, N. R. (1994). Non-marital cohabitation and change in norms: The case of Norway. *Acta Sociologica, 37,* 23–37.

Rando, T. A. (1991a). *How to go on living when someone you love dies.* New York: Bantam.

Rando, T. A. (1991b). Parental adjustment to the loss of a child. In D. Papadatou & C. Papadatos (Eds.), *Children and death* (pp. 233–253). New York: Hemisphere.

Rando, T. A. (1995). Grief and mourning: Accommodating to loss. In H. Wass & R. A. Neimeyer (Eds.), *Dying: Facing the facts* (3rd ed., pp. 211–241). Washington, DC: Taylor & Francis.

Rank, M. R. (2000). Socialization of socioeconomic status. In W. C. Nichols & M. A. Pace-Nichols (Eds.), *Handbook of family development and intervention* (pp. 129–142). New York: Wiley.

Ransford, H. E., & Palisi, B. J. (1996). Aerobic exercise, subjective health and psychological well-being within age and gender subgroups. *Social Science and Medicine, 42,* 1555–1559.

Raphael, B., Middleton, W., Martinek, N., & Misso, V. (1993). Counseling and therapy of the bereaved. In M. S. Stroebe, W. Stroebe, & R. O. Hansson (Eds.), *Handbook of bereavement* (pp. 427–453). New York: Cambridge University Press.

Rapp, S. R., Espeland, M. A., Shumaker, S. A., Henderson, V. W., Brunner, R. L., & Manson, J. E. (2003). Effect of estrogen plus progestin on global cognitive function in postmenopausal women: The Women's Health Initiative Memory Study: A randomized controlled trial. *Journal of the American Medical Association, 289,* 2663–2672.

Rappaport, L. (1993). The treatment of nocturnal enuresis—Where are we now? *Pediatrics, 92,* 465–466.

Rapport, M. D., & Chung, K.-M. (2000). Attention deficit hyperactivity disorder. In M. Hersen & R. T. Ammerman (Eds.),

Advanced abnormal child psychology (2nd ed., pp. 413–440). Mahwah, NJ: Erlbaum.

Rasmussen, C. H., & Johnson, M. E. (1994). Spirituality and religiosity: Relative relationships to death anxiety. *Omega, 29,* 313–318.

Rast, M., & Meltzoff, A. N. (1995). Memory and representation in young children with Down syndrome: Exploring deferred imitation and object permanence. *Development and Psychopathology, 7,* 393–407.

Ratner, E., Norlander, L., & McSteen, K. (2001). Death at home following a targeted advance-care planning process at home: The kitchen table discussion. *Journal of the American Geriatrics Society, 49,* 778–781.

Ratner, N., & Bruner, J. S. (1978). Social exchange and the acquisition of language. *Journal of Child Language, 5,* 391–402.

Raugust, K. (1999, October). Can you tell me how to get to Sesamestrasse? *Animation World Magazine,* Issue 4.7. Retrieved from http://www.awn.com/mag/issue4.07/4.07pages/raugustctw.php3

Raup, J. L., & Myers, J. E. (1989). The empty nest syndrome: Myth or reality? *Journal of Counseling and Human Development, 68,* 180–183.

Ravussin, E., Valencia, M. E., Esparza, J., Bennett, P. H., & Schulz, L. O. (1994). Effects of a traditional lifestyle on obesity in Pima Indians. *Diabetes Care, 17,* 1067–1074.

Rayner, K., Foorman, B. R., Perfetti, C. A., Pesetsky, D., & Seidenberg, M. S. (2001). How psychological science informs the teaching of reading. *Psychological Science in the Public Interest, 2,* 31–74.

Rayner, K., & Pollatsek, A. (1989). *The psychology of reading.* Englewood Cliffs, NJ: Prentice-Hall.

Raynor, D. A., Pogue-Geile, M. F., Kamarck, T. W., McCaffery, J. M., & Manuck, S. B. (2002). Covariation of psychosocial characteristics associated with cardiovascular disease: Genetic and environmental influences. *Psychosomatic Medicine, 64,* 191–203.

Read, C. R. (1991). Achievement and career choices: Comparisons of males and females. *Roeper Review, 13,* 188–193.

Reay, A. M., & Browne, K. D. (2001). Risk factor characteristics in carers who physically abuse or neglect their elderly dependents. *Aging and Mental Health, 5,* 56–62.

Rebbeck, T. R. (2002). Inherited predisposition and breast cancer: Modifiers of BRCA1/2-associated breast cancer risk. *Environmental and Molecular Mutagenesis, 39,* 228–234.

Receputo, G., Di Stefano, S., Fornaro, D., Malaguarnera, M., & Motta, L. (1994). Comparison of tactile sensitivity in a group of elderly and young adults and children using a new instrument called a "Tangoceptometer." *Archives of Gerontology and Geriatrics, 18,* 207–214.

Reday-Mulvey, G. (2000). Gradual retirement in Europe. *Journal of Aging and Social Policy, 11,* 49–60.

Reddin, J. (1997). High-achieving women: Career development patterns. In H. S. Farmer (Ed.), *Diversity and women's career development* (pp. 95–126). Thousand Oaks, CA: Sage.

Redding, N. P., & Dowling, W. D. (1992). Rites of passage among women reentering higher education. *Adult Education Quarterly, 42,* 221–236.

Redl, F. (1966). *When we deal with children.* New York: The Free Press.

Rees, M. (1993). Menarche when and why? *Lancet, 342,* 1375–1376.

Reese, E., Haden, C. A., & Fivush, R. (1993). Mother–child conversations about the past: Relationships of style and memory over time. *Cognitive Development, 8,* 403–430.

Reese, E., Haden, C. A., & Fivush, R. (1996). Mothers, fathers, daughters, sons: Gender differences in autobiographical reminiscing. *Research on Language and Social Interaction, 29,* 27–56.

Reeves, J. B., & Darville, R. L. (1994). Social contact patterns and satisfaction with retirement of women in dual-career/ earner families. *International Journal of Aging and Human Development, 39,* 163–175.

Regan, P. C., Medina, R., Joshi, A. (2001). Partner preferences among homosexual men and women: What is desirable in a sex partner is not necessarily desirable in a romantic partner. *Social Behavior and Personality, 29,* 625–634.

Reid, H. M., & Fine, A. (1992). Self-disclosure in men's friendships: Variations associated with intimate relations. In P. M. Nardi (Ed.), *Men's friendships* (pp. 153–171). Newbury Park, CA: Sage.

Reid, I. R. (1996). Therapy of osteoporosis—calcium, vitamin D and exercise. *American Journal of the Medical Sciences, 312,* 278–286.

Reifler, B. V. (1994). Depression: Diagnosis and comorbidity. In L. S. Schneider, C. F. Reynolds, III, B. D. Lebowitz, & A. J. Friedhoff (Eds.), *Diagnosis and treatment of depression in late life* (pp. 55–59). Washington, DC: American Psychiatric Press.

Reiger, D. A., Boyd, J. H., Burke, J. D., Rae, D. S., Myers, J. K., Dramer, M., Robins, L. N., George, L. K., Karno, M., & Locke, B. Z. (1988). One-month prevalence of mental disorders in the United States. *Archives of General Psychiatry, 45,* 977–986.

Reilly, J. S., Bates, E. A., & Marchman, V. A. (1998). Narrative discourse in children with early focal brain injury. *Brain and Language, 61,* 335–375.

Reiman, E. M., Caselli, R. J., Yun, L. S., Chen, K., Bandy, D., Minoshima, S., Thibodeau, S. N., & Osborne, D. (1996). Preclinical evidence of Alzheimer's disease in persons homozygous for the epsilon 4 allele for apolipoprotein E. *New England Journal of Medicine, 334,* 752–758.

Reinke, B. J., Holmes, D. S., & Harris, R. L. (1985). The timing of psychosocial changes in women's lives: The years 25 to 45. *Journal of Personality and Social Psychology, 48,* 1353-1364.

Reis, M., & Nahmiash, D. (1995). When seniors are abused: An intervention model. *Gerontologist, 35,* 666–671.

Reiser, J., Yonas, A., & Wikner, K. (1976). Radial localization of odors by human neonates. *Child Development, 47,* 856–859.

Reisman, J. E. (1987). Touch, motion, and proprioception. In P. Salapatek & L. Cohen (Eds.), Handbook of infant perception: Vol. 1. *From sensation to perception* (pp. 265–303). Orlando, FL: Academic Press.

Reitzel-Jaffe, D., & Wolfe, D. A. (2001). Predictors of relationship abuse among young men. *Journal of Interpersonal Violence, 16,* 99–115.

Renninger, K. A. (1998). Developmental psychology and instruction: Issues from and for practice. In I. Sigel & K. A. Renninger (Eds.), *Handbook of child psychology: Vol. 4. Child psychology and practice* (pp. 211–274). New York: Wiley.

Repacholi, B. M. (1998). Infants' use of attentional cues to identify the referent of another person's emotional expression. *Developmental Psychology, 33,* 12–21.

Repacholi, B. M., & Gopnik, A. (1997). Early reasoning about desires: Evidence from 14- and 18-month-olds. *Developmental Psychology, 33,* 12–21.

Resnick, H. S., & Newton, T. (1992). Assessment and treatment of post-traumatic stress disorder in adult survivors of sexual assault. In D. Fox (Ed.), *Treating PTSD: Procedures for combat veterans, battered women, adult and child sexual assaults* (pp. 99–126). New York: Guilford.

Resnick, M. B., Gueorguieva, R. V., Carter, R. L., Ariet, M., Sun, Y., Roth, J., Bucciarelli, R. L., Curran, J. S., & Mahan, C. S. (1999). The impact of low birth weight, perinatal conditions, and sociodemographic factors on educational outcome in kindergarten. *Pediatrics, 104,* e74.

Rest, J. R. (1979). *Development in judging moral issues.* Minneapolis: University of Minnesota Press.

Rest, J. R., & Narvaez, D. (1991). The college experience and moral development. In W. M. Kurtines & J. L. Gewirtz (Eds.), *Handbook of moral behavior and development* (Vol. 2, pp. 229–245). Hillsdale, NJ: Erlbaum.

Reynolds, A. J., & Temple, J. A. (1998). Extended early childhood intervention and school achievement: Age thirteen findings from the Chicago Longitudinal Study. *Child Development, 69,* 231–246.

Reynolds, C. R., & Kaiser, S. M. (1990). Test bias in psychological assessment. In T. B. Gutkin & C. R. Reynolds (Eds.), *The handbook of school psychology* (pp. 487–525). New York: Wiley.

Reznick, J. S., & Goldfield, B. A. (1992). Rapid change in lexical development in comprehension and production. *Developmental Psychology, 28,* 406–413.

Rhea, D. J. (1999). Eating disorder behaviors of ethnically diverse urban female adolescent athletes and non-athletes. *Journal of Adolescence, 22,* 379–388.

Rhein, J. von (1997, January 19). Ardis Krainik, Lyric Opera's life force, dies. *Chicago Tribune,* pp. 1, 16.

Ricard, M., & Kamberk-Kilicci, M. (1995). Children's empathic responses to emotional complexity. *International Journal of Behavioral Development, 18,* 211–225.

Ricciardelli, L. A. (1992). Bilingualism and cognitive development: Relation to threshold theory. *Journal of Psycholinguistic Research, 21,* 301–316.

Riccio, C. A., Hynd, G. W., Cohen, M. J., & Gonzalez, J. J. (1993). Neurological basis of attention deficit hyperactivity disorder. *Exceptional Children, 60,* 118–124.

Rice, F. P. (1999). *The adolescent: Development, relationships, and culture* (8th ed.). Boston: Allyn and Bacon.

Rice, J. K. (1994). Reconsidering research on divorce, family life cycle, and the meaning of family. *Psychology of Women Quarterly, 18,* 559–584.

Richard, L. S., Wakefield, J. A., & Lewak, R. (1990). Similarity of personality variables as predictors of marital satisfaction: A Minnesota Multiphasic Personality Inventory (MMPI) item analysis. *Personality and Individual Differences, 11,* 39–43.

Richards, J. E., & Holley, F. B. (1999). Infant attention and the development of smooth pursuit tracking. *Developmental Psychology, 35,* 856–867.

Richards, M. H., & Duckett, E. (1994). The relationship of maternal employment to early adolescent daily experience with and without parents. *Child Development, 65,* 225–236.

Richards-Colocino, N., McKenzie, P., & Newton, R. R. (1996). Project Success: Comprehensive intervention services for middle school high-risk youth. *Journal of Adolescent Research, 11,* 130–163.

Richardson, G. A. (1998). Prenatal cocaine exposure: A longitudinal study of development. In J. A. Harvey & B. E. Kosofsky (Eds.), *Annals of the New York Academy of Sciences* (Vol. 846, pp. 144–152). New York: New York Academy of Sciences.

Richardson, G. A., Hamel, S. C., Goldschmidt, L., & Day, N. L. (1996). The effects of prenatal cocaine use on neonatal neurobehavioral status. *Neurotoxicology and Teratology, 18,* 519–528.

Richardson, S. A., Koller, H., & Katz, M. (1986). Factors leading to differences in the school performance of boys and girls. *Developmental and Behavioral Pediatrics, 7,* 49–55.

Rich-Edwards, J. W., Colditz, G. A., Stampfer, M. J., Willett, W. C., Gillman, M. W., Hennekens, C. H., Speizer, F. E., & Manson, J. E. (1999). Birthweight and the risk for type 2 diabetes mellitus in adult women. *Annals of Internal Medicine, 130,* 278–284.

Rich-Edwards, J. W., Stampfer, M. J., Manson, J. E., Rosner, B., Hankinson, S. E., Colditz, G. A., Willett, W. C., & Hennekens, C. H. (1997). Birth weight and risk of cardiovascular disease in a cohort of women followed up since 1976. *British Medical Journal, 315,* 396–400.

Richgels, D. J., McGee, L. M, & Slaton, E. A. (1989). Teaching expository text structure in reading and writing. In K. D. Muth (Ed.), *Children's comprehension of text* (pp. 167–184). Newark, DE: International Reading Association.

Richie, B. S., Fassinger, R. E., Linn, S. G., Johnson, J., Prosser, J., & Robinson, S. (1997). Persistence, connection, and passion: A qualitative study of the career development of highly achieving African American–black and white women. *Journal of Counseling Psychology, 44,* 133–148.

Rickel, A. U., & Becker, E. (1997). *Keeping children from harm's way.* Washington, DC: American Psychological Association.

Riedel, B. W., & Lichstein, K. L. (2000). Insomnia in older adults. In S. K. Whitbourne (Ed.), *Psychopathology in later adulthood* (pp. 299–322). New York: Wiley.

Riggio, H. R. (2000). Measuring attitudes toward adult sibling relationships: The lifespan sibling relationship scale. *Journal of Social and Personal Relationships, 17,* 707–728.

Rigler, S. K. (1999). Preventing falls in older adults. *Hospital Practice, 34,* 117–120.

Rimé, B., Finkenauer, C., Luminet, O., Zech, E., & Philippot, P. (1998). Social sharing of emotion: New evidence and new questions. In W. Stroebe & M. Hewstone (Eds.), *European review of social psychology* (Vol. 9). Chichester, U.K.: Wiley.

Rimm, E. B. (1996). Alcohol consumption and coronary heart disease: Good habits may be more important than just good wine. *American Journal of Epidemiology, 143,* 1094–1098.

Riordan, J., Gross, A., Angeron, J., Drumwiede, B., & Melin, J. (2000). The effect of labor pain relief medication on neonatal suckling and breastfeeding duration. *Journal of Human Lactation, 16,* 7–12.

Ritchey, L. H., Ritchey, P. N., & Dietz, B. E. (2001). Clarifying the measurement of activity. *Activities, Adaptation, and Aging, 26,* 1–21.

Ritchie, K. (1998). Mental health of the oldest old: The relevance of centenarian studies to psychogeriatric research. *International Psychogeriatrics, 10,* 7–9.

Rittman, M., Kuzmeskus, L. B., & Flum, M. A. (2000). A synthesis of current knowledge on minority elder abuse. In T. Tatara (Ed.), *Understanding elder abuse in minority populations* (pp. 221–238). Philadelphia: Brunner/Mazel.

Rivara, F. P. (1995). Developmental and behavioral issues in childhood injury prevention. *Developmental and Behavioral Pediatrics, 16,* 362–370.

Rivera, S. M., Wakeley, A., & Langer, J. (1999). The drawbridge phenomenon: Representational reasoning or perceptual preference? *Developmental Psychology, 35,* 427–435.

Roazzi, A., & Bryant, P. (1997). Explicitness and conservation: Social class differences. *International Journal of Behavioral Development, 21,* 51–70.

Robb, A. S., & Dadson, M. J. (2002). Eating disorders in males. *Child and Adolescent Psychiatric Clinics of North America, 11,* 399–418.

Roberto, K. A., & Kimboko, P. J. (1989). Friendships in later life: Definitions and maintenance patterns. *International Journal of Aging and Human Development, 28,* 9–19.

Roberton, M. A. (1984). Changing motor patterns during childhood. In J. R. Thomas (Ed.), *Motor development during childhood and adolescence* (pp. 48–90). Minneapolis: Burgess.

Roberts, B. W., & DelVecchio, W. E. (2000). The rank-order consistency of personality traits from childhood to old age: A quantitative review of longitudinal studies. *Psychological Bulletin, 126,* 3–25.

Roberts, I., & DiGuiseppi, C. (1999). Injury prevention. *Archives of Disease in Childhood, 81,* 200–201.

Roberts, J. E., Burchinal, M. R., & Campbell, F. (1994). Otitis media in early childhood and patterns of intellectual development and later academic performance. *Journal of Pediatric Psychology, 19,* 347–367.

Roberts, J. E., Burchinal, M. R., & Durham, M. (1999). Parents' report of vocabulary and grammatical development of American preschoolers: Child and environment associations. *Child Development, 70,* 92–106.

Roberts, J. E., Burchinal, M. R., Jackson, S. C., Hooper, S. R., Roush, J., Mundy, M., Neebe, E. C., & Zeisel, S. A. (2000). Otitis media in childhood in relation to preschool language and school readiness skills among black children. *Pediatrics, 106,* 725–735.

Roberts, J. E., Burchinal, M. R., Zeisel, S. A., Neebe, E. C., Hooper, S. R., Roush, J., Bryant, D., Mundy, M., & Henderson, F. W. (1998). Otitis media, the caregiving environment, and language and cognitive outcomes at 2 years. *Pediatrics, 102,* 346–354.

Roberts, M. C., Alexander, K., & Knapp, L. G. (1990). Motivating children to use safety belts: A program combining rewards and "flash for life." *Journal of Community Psychology, 18,* 110–119.

Roberts, P., & Newton, P. M. (1987). Levinsonian studies of women's adult development. *Psychology and Aging, 2,* 154–163.

Roberts, P., & Vidal, L. A. (1999–2000). Perpetual care in cyberspace: A portrait of memorials on the Web. *Omega, 40,* 521–545.

Roberts, R. E., Kaplan, G. A., Shema, S. J., & Strawbridge, W. J. (1997). Prevalence and correlates of depression in an aging cohort: The Alameda County Study. *Journal of Gerontology, 52B,* S252–S258.

Roberts, R. E., Shema, S. J., & Kaplan, G. A. (1999). Prospective data on sleep complaints and associated risk factors in an older cohort. *Psychosomatic Medicine, 61,* 188–196.

Roberts, R. J., Jr., & Aman, C. J. (1993). Developmental differences in giving directions: Spatial frames of reference and mental rotation. *Child Development, 64,* 1258–1270.

Roberts, S. B., Pi-Sunyer, X., Kuller, L., Lane, M., Ellison, P., Prior, J. C., & Shapses, S. (2001). Physiologic effects of lowering caloric intake in nonhuman primates and nonobese humans. *Journals of Gerontology, 56A,* 66–75.

Roberts, W., & Strayer, J. (1996). Empathy, emotional expressiveness, and prosocial behavior. *Child Development, 67,* 449–470.

Robertson, D. L. (1991). Gender differences in the academic progress of adult undergraduates: Patterns and policy implications. *Journal of College Student Development, 32,* 490–496.

Robin, D. J., Berthier, N. E., & Clifton, R. K. (1996). Infants' predictive reaching for moving objects in the dark. *Developmental Psychology, 32,* 824–835.

Robinson, E. J., & Mitchell, P. (1994). Young children's false-belief reasoning: Interpretation of messages is not easier than the classic task. *Developmental Psychology, 30,* 67–72.

Robinson, J., & Godbey, G. (1997). *Time for life.* College Park: Pennsylvania State University.

Robinson, T. N. (1999). Reducing children's television viewing to prevent obesity. *Journal of the American Medical Association, 282,* 1561–1567.

Robinson, T. N., Killen, J. D., Litt, I. F., Hammer, L. D., Wilson, D. M., Haydel, K. F., Hayward, C., & Taylor, C. B. (1996). Ethnicity and body dissatisfaction: Are Hispanic and Asian girls at increased risk for eating disorders? *Journal of Adolescent Health, 19,* 384–393.

Robison, J. I., Hoerr, S. L., Strandmark, J., & Mavis, B. (1993). Obesity, weight loss, and health. *Journal of the American Dietetic Association, 93,* 445–449.

Rochat, P. (1998). Self-perception and action in infancy. *Experimental Brain Research, 123,* 102–109.

Rochat, P. (2001). The infant's world. Cambridge, MA: Harvard University Press.

Rochat, P., & Goubet, N. (1995). Development of sitting and reaching in 5- to 6-month-old infants. *Infant Behavior and Development, 18,* 53–68.

Rochat, P., Querido, J. G., & Striano, T. (1999). Emerging sensitivity to the timing and structure of proto-conversation. *Developmental Psychology, 35,* 950–957.

Roche, A. F. (1979). Secular trends in stature, weight, and maturation. In A. F. Roche (Ed.), Secular trends in human growth, maturation, and development. *Monographs of the Society for Research in Child Development, 44*(3–4, Serial No. 179).

Rockwell, R. C., Elder, G. H., & Ross, D. J. (1979). Psychological patterns in marital timing and divorce. *Social Psychology Quarterly, 42,* 399–404.

Rockwood, K., Stolee, P., & Dowell, I. (1996). Factors associated with institutionalization of older people in Canada: Testing a multifactorial definition of frailty. *Journal of the American Geriatrics Society, 44,* 578–582.

Rodkin, P. C., Farmer, T. W., Pearl, R., & Van Acker, R. (2000). Heterogeneity of popular boys: Antisocial and prosocial configurations. *Developmental Psychology, 36,* 14–24.

Roebers, C. M., & Schneider, W. (2001). Individual differences in children's eyewitness recall: The influence of intelligence and shyness. *Applied Developmental Science, 5,* 9–20.

Roebuck, T. M., Mattson, S. N., & Riley, E. P. (1999). Prenatal exposure to alcohol: Effects on brain structure and neuropsychological functioning. In J. H. Hannigan & L. P. Spear (Eds.), *Alcohol and alcoholism: Effects on brain and development* (pp. 1–16). Mahwah, NJ: Erlbaum.

Roeser, R. W., Eccles, J. S., & Freedman-Doan, C. (1999). Academic functioning and mental health in adolescence: Patterns, progressions, and routes from childhood. *Journal of Adolescent Research, 14,* 135–174.

Roeser, R. W., Eccles, J. S., & Sameroff, A. J. (2000). School as a context of early adolescents' academic and social-emotional development: A summary of research findings. *Elementary School Journal, 100,* 443–471.

Roffwarg, H. P., Muzio, J. N., & Dement, W. C. (1966). Ontogenetic development of the human sleep–dream cycle. *Science, 152,* 604–619.

Roger, V. L., Farkouh, M. E., Weston, S. A., Reeder, G. S., Jacobsen, S. J., Zinsmeister, A. R., Yawn, B. P., Kopeky, S. L., & Gabriel, S. E. (2000). Sex differences in evaluation and outcome of unstable angina. *Journal of the American Medical Association, 283,* 646–652.

Rogers, C., & Shiff, M. (1996). Early versus late prenatal care in New Mexico: Barriers and motivators. *Birth, 23,* 26–30.

Rogers, L., Resnick, M. D., Mitchell, J. E., & Blum, R. W. (1997). The relationship between socioeconomic status and eating disordered behaviors in a community sample of adolescent girls. *International Journal of Eating Disorders, 22,* 15–23.

Rogers, S. J., & White, L. K. (1998). Satisfaction with parenting: The role of marital happiness, family structure, and parents' gender. *Journal of Marriage and the Family, 60,* 293–308.

Roggman, L. A., Langlois, J. H., Hubbs-Tait, L., & Rieser-Danner, L. A. (1994). Infant day-care, attachment, and the "file drawer problem." *Child Development, 65,* 1429–1443.

Rogoff, B. (1986). The development of strategic use of context in spatial memory. In M. Perlmutter (Ed.), *Perspectives on intellectual development* (pp. 107–123). Hillsdale, NJ: Erlbaum.

Rogoff, B. (1996). Developmental transitions in children's participation in sociocultural activities. In A. J. Sameroff & M. M. Haith (Eds.), *The five to seven year shift: The age of reason and responsibility* (pp. 273–294). Chicago: University of Chicago Press.

Rogoff, B. (1998). Cognition as a collaborative process. In D. Kuhn & R. S. Siegler (Eds.), *Handbook of child psychology: Vol. 2. Cognition, perception, and language* (5th ed., pp. 679–744). New York: Wiley.

Rogoff, B., & Chavajay, P. (1995). What's become of research on the cultural basis of cognitive development? *American Psychologist, 50,* 859–877.

Rogoff, B., Malkin, C., & Gilbride, K. (1984). Interaction with babies as guidance in development. In B. Rogoff & J. V. Wertsch (Eds.), *New directions for child development* (No. 23, pp. 31–44). San Francisco: Jossey-Bass.

Rogoff, B., Mosier, C., Mistry, J., & Göncü, A. (1993). Toddlers' guided participation with their caregivers in cultural activity. In E. A. Forman, N. Minick, & C. A. Stone (Eds.), *Contexts for learning* (pp. 230–253). New York: Oxford University Press.

Rogosch, F., Cicchetti, D., Shields, A., & Toth, S. L. (1995). Parenting dysfunction in child maltreatment. In M. H. Bornstein (Ed.), *Handbook of parenting* (Vol. 4, pp. 127–159). Hillsdale, NJ: Erlbaum.

Rogow, S. (1988). *Helping the visually impaired child with developmental problems: Effective practice in home, school, and community.* New York: Teachers College Press.

Rohlen, T. P. (1997). Differences that make a difference: Explaining Japan's success. In W. K. Cumings & P. G. Altbach (Eds.), *The challenge of Eastern Asian education: Implications for America* (pp. 223–248). Albany, NY: SUNY Press.

Roisman, G. I., Madsen, S. D., Hennighausen, K. H., Sroufe, L. A., & Collins, W. A. (2001). The coherence of dyadic behavior across parent–child and romantic relationships as mediated by the internalized representation of experience. *Attachment and Human Development, 3,* 156–172.

Roisman, G. I., Padron, E., Sroufe, L. A., & Egeland, B. (2002). Earned-secure attachment status in retrospect and prospect. *Child Development, 73,* 1204–1219.

Rokach, A. (2001a). Perceived causes of loneliness in adulthood. *Journal of Social Behavior and Personality, 15,* 67–84.

Rokach, A. (2001b). Strategies of coping with loneliness throughout the lifespan. *Current Psychology: Developmental, Learning, Personality, Social, 20,* 3–18.

Rokach, A., & Sharma, M. (1996). The loneliness experience in a cultural context. *Journal of Social Behavior and Personality, 11,* 827–839.

Rolls, B. J., & Bell, E. A. (2000). Dietary approaches to the treatment of obesity. *Medical Clinics of North America, 84,* 401–418.

Rolls, B. J., Engell, D., & Birch, L. (2000). Serving portion size influences 5-year-old but not 3-year-old children's food intakes. *Journal of the American Dietetic Association, 100,* 232–234.

Romans, S. M., Roeltgen, D. P., Kushner, H., & Ross, J. L. (1997). Executive function in girls with Turner's syndrome. *Developmental Neuropsychology, 13,* 23–40.

Rome-Flanders, T., & Cronk, C. (1995). A longitudinal study of infant vocalizations during mother–infant games. *Journal of Child Language, 22,* 259–274.

Rook, K. S., Catalano, R., & Dooley, D. (1989). The timing of major life events: Effects of departing from the social clock. *American Journal of Community Psychology, 17,* 233–258.

Roopnarine, J. L., Hossain, Z., Gill, P., & Brophy, H. (1994). Play in the East Indian context. In J. L. Roopnarine, J. E. Johnson, & F. H. Hooper (Eds.), *Children's play in diverse cultures* (pp. 9–30). Albany: State University of New York Press.

Roopnarine, J. L., Lasker, J., Sacks, M., & Stores, M. (1998). The cultural contexts of children's play. In O. N. Saracho & B. Spodek (Eds.), *Multiple perspectives on play in early childhood education* (pp. 194–219). Albany: State University of New York Press.

Roopnarine, J. L., Talukder, E., Jain, D., Joshi, P., & Srivastav, P. (1990). Characteristics of holding, patterns of play, and social behaviors between parents and infants in New Delhi, India. *Developmental Psychology, 26,* 667–673.

Roscoe, B., Diana, M. S., & Brooks, R. H. (1987). Early, middle, and late adolescents' views on dating and factors influencing partner selection. *Adolescence, 22,* 59–68.

Rose, A. J., & Asher, S. R. (1999). Children's goals and strategies in response to conflicts within a friendship. *Developmental Psychology, 35,* 69–79.

Rose, R. J. (1995) Genes and human behavior. *Annual Review of Psychology, 46,* 625–654.

Rose, S. A., & Feldman, J. F. (1997). Memory and speed: Their role in the relation of infant information processing to later IQ. *Child Development, 68,* 610–620.

Rose, S. A., Jankowski, J. J., & Senior, G. J. (1997). Infants' recognition of contour-deleted figures. *Journal of Experimental Psychology: Human Perception and Performance, 23,* 1206–1216.

Rosen, A. B., & Rozin, P. (1993). Now you see it, now you don't: The preschool child's conception of invisible particles in the context of dissolving. *Developmental Psychology, 29,* 300–311.

Rosen, W. D., Adamson, L. B., & Bakeman, R. (1992). An experimental investigation of infant social referencing: Mothers' messages and gender differences. *Developmental Psychology, 28,* 1172–1178.

Rosenberg, S. (1988). Self and others: Studies in social personality and autobiography. In L. Berkowitz (Ed.), *Advances in experimental social psychology* (Vol. 21, pp. 56–96). New York: Academic Press.

Rosenberg, S. D., Rosenberg, H. J., & Farrell, M. P. (1999). The midlife crisis revisited. In S. L. Willis & J. D. Reid (Eds.), *Life in the middle* (pp. 47–73). San Diego: Academic Press.

Rosenblatt, P. C. (1993). Cross-cultural variation in the experience, expression, and understanding of grief. In D. P. Irish, K. F. Lundquist, & V. J. Nelsen (Eds.), *Ethnic variations in dying, death, and grief* (pp. 13–19). Washington, DC: Taylor & Francis.

Rosengren, K. S., & Hickling, A. K. (2000). The development of children's thinking about possible events and plausible mechanisms. In K. S. Rosengren, C. N. Johnson, & P. L. Harris (Eds.), *Imagining the impossible* (pp. 75–98). Cambridge, U.K.: Cambridge University Press.

Rosenman, R. H., Brand, R. J., Jenkins, C. D., Friedman, M., Strauss, R., & Wurm, M. (1975). Coronary heart disease in the Western Collaborative Group Study: Final follow-up experience of 8½ years. *Journal of the American Medical Association, 223,* 872–877.

Rosenshine, B., & Meister, C. (1994). Reciprocal teaching: A review of nineteen experimental studies. *Review of Educational Research, 64,* 479–530.

Rosenthal, C. J., & Gladstone, J. (2000). *Grandparenthood in Canada.* Ottawa: Vanier Institute of the Family.

Rosenthal, J. A. (1992). *Special-needs adoption: A study of intact families.* New York: Praeger.

Roskos, K., & Neuman, S. B. (1998). Play as an opportunity for literacy. In O. N. Saracho & B. Spodek (Eds.), *Multiple perspectives on play in early childhood education* (pp. 100–115). Albany: State University of New York Press.

Ross, C. E., & Drentea, P. (1998). Consequences of retirement activities for distress and the sense of personal control. *Journal of Health and Social Behavior, 39,* 317–334.

Ross, J., Zinn, A., & McCauley, E. (2000). Neurodevelopmental and psychosocial aspects of Turner syndrome. *Mental Retardation and Developmental Disabilities Research Review, 6,* 135–141.

Ross, N. A., Wolfson, M. C., Dunn, J. R., Berthelot, J., Kaplan, G. A., & Lynch, J. W. (2000). Relation between income inequality and mortality in Canada and in the United States: Cross-sectional assessment using census data and vital statistics. *British Medical Journal, 320,* 898–902.

Rossi, A. S. (1980). Life-span theories and women's lives. *Signs: Journal of Women in Culture and Society, 6,* 4–32.

Rossi, A. S., & Rossi, P. H. (1990). *Of human bonding: Parent–child relations across the life course.* New York: Aldine de Gruyter.

Roth, G. S., Ingram, D. K., Black, A., & Lane, M. A. (2000). Effects of reduced energy intake on the biology of aging: The primate model. *European Journal of Clinical Nutrition, 54*(Suppl. 3), S15–S20.

Roth, G. S., Ingram, D. K., & Lane, M. A. (2001). Caloric restriction in primates and relevance to humans. *Annals of the New York Academy of Sciences, 928,* 305–315.

Roth, J., Brooks-Gunn, J., Murray, L., & Foster, W. (1998). Promoting healthy adolescents: Synthesis of youth development program evaluations. *Journal of Research on Adolescence, 8,* 423–459.

Rothbart, M. K. (1981). Measurement of temperament in infancy. *Child Development, 52,* 569–578.

Rothbart, M. K. (1989). Temperament and development. In G. A. Kohnstamm, J. A. Bates, & M. K. Rothbart (Eds.), *Temperament in childhood* (pp. 59–73). New York: Wiley.

Rothbart, M. K., Ahadi, S. A., & Evans, D. E. (2000). Temperament and personality: Origins and outcome. *Journal of Personality and Social Psychology, 78,* 122–135.

Rothbart, M. K., & Bates, J. E. (1998). Temperament. In N. Eisenberg (Ed.), *Handbook of child psychology: Vol. 3. Social, emotional, and personality development* (5th ed., pp. 105–176). New York: Wiley.

Rothbart, M. K., Derryberry, D., & Posner, M. I. (1994). A psychobiological approach to the development of temperament. In J. E. Bates & T. D. Wachs (Eds.), *Temperament: Individual differences at the interface of biology and behavior* (pp. 83–116). Washington, DC: American Psychological Association.

Rothbart, M. K., & Mauro, J. A. (1990). Questionnaire approaches to the study of infant temperament. In J. W. Fagen & J. Colombo (Eds.), *Individual differences in infancy: Reliability, stability and prediction* (pp. 411–429). Hillsdale, NJ: Erlbaum.

Rothbaum, F., Pott, M., Azuma, H., Miyake, K., & Weisz, J. (2000a). The development of close relationships in Japan and the United States: Paths of symbiotic harmony and generative tension. *Child Development, 71,* 1121–1142.

Rothbaum, F., Weisz, J., Pott, M., Miyake, K., & Morelli, G. (2000b). Attachment and culture: Security in the United States and Japan. *American Psychologist, 55,* 1093–1104.

Roughan, P. A., Kaiser, F. E., & Morley, J. E. (1993). Sexuality and the older woman. *Clinics in Geriatric Medicine, 9,* 87–106.

Rousseau, J. J. (1955). *Emile.* New York: Dutton. (Original work published 1762)

Rovee-Collier, C. (1996). Shifting the focus from what to why. *Infant Behavior and Development, 19,* 385–400.

Rovee-Collier, C. (2001). Information pickup by infants: What is it, and how can we tell? *Journal of Experimental Child Psychology, 78,* 35–49.

Rovee-Collier, C., & Barr, R. (2001). Infant learning and memory. In G. Bremner (Ed.), *Blackwell handbook of infant development* (pp. 139–168). Malden, MA: Blackwell.

Rovee-Collier, C. K. (1987). Learning and memory. In J. D. Osofsky (Ed.), *Handbook of infant development* (2nd ed., pp. 98–148). New York: Wiley.

Rovee-Collier, C. K. (1999). The development of infant memory. *Current Directions in Psychological Science, 8,* 80–85.

Rovee-Collier, C. K., & Bhatt, R. S. (1993). Evidence of long-term memory in infancy. *Annals of Child Development, 9,* 1–45.

Rowe, D. C. (1994). *The limits of family influence: Genes, experience, and behavior.* New York: Guilford.

Rowe, J. W., & Kahn, R. L. (1998). *Successful aging.* New York: Random House.

Royal College of Obstetricians and Gynecologists. (1997, October). *Report of the panel to review fetal pain.* London: Author.

Rubenstein, C. M., & Shaver, P. (1982). *In search of intimacy.* New York: Delacorte Press.

Rubin, D. C., Rahhal, T. A., & Poon, L. W. (1998). Things learned in early adulthood are remembered best. *Memory and Cognition, 26,* 3–19.

Rubin, D. C., & Schulkind, M. D. (1997a). Distribution of important and word-cued autobiographical memories in 20-, 35-, and 70-year-old adults. *Psychology and Aging, 12,* 524–535.

Rubin, D. C., & Schulkind, M. D. (1997b). The distribution of autobiographical memories across the lifespan. *Memory and Cognition, 25,* 859–866.

Rubin, K., Bukowski, W., & Parker, J. G. (1998). Peer interactions, relationships, and groups. In N. Eisenberg (Ed.), *Handbook of child psychology: Vol. 3. Social, emotional, and personality development* (5th ed., pp. 619–700). New York: Wiley.

Rubin, K. H., Burgess, K. B., & Hastings, P. D. (2002). Stability and social-behavioral consequences of toddlers' inhibited temperament and parenting behaviors. *Child Development, 73,* 483–495.

Rubin, K. H., & Coplan, R. J. (1998). Social and nonsocial play in childhood: An individual differences perspective. In O. N. Saracho & B. Spodek (Eds.), *Multiple perspectives on play in early childhood education* (pp. 144–170). Albany: State University of New York Press.

Rubin, K. H., Coplan, R. J., Fox, N. A., & Calkins, S. (1995). Emotionality, emotion regulation, and preschoolers' social adaptation. *Development and Psychopathology, 7,* 49–62.

Rubin, K. H., Fein, G. G., & Vandenberg, B. (1983). Play. In E. M. Hetherington (Ed.), *Handbook of child psychology: Vol. 4. Socialization, personality, and social development* (4th ed., pp. 693–744). New York: Wiley.

Rubin, K. H., Hastings, P. D., Stewart, S. L., Henderson, H. A., & Chen, X. (1997). The consistency and concomitants of inhibition: Some of the children, all of the time. *Child Development, 68,* 467–483.

Rubin, K. H., Watson, K. S., & Jambor, T. W. (1978). Free-play behaviors in preschool and kindergarten children. *Child Development, 49,* 539–536.

Rubin, S. S., & Malkinson, R. (2001). Parental response to child loss across the life cycle: Clinical and research perspectives. In M. S. Stroebe, R. O. Hansson, W. Stroebe, & H. Schut (Eds.), *Handbook of bereavement research* (pp. 219–240). Washington, DC: American Psychological Association.

Rubinowitz, L. S., & Rosenbaum, J. E. (2000). *Crossing the class and color lines: From public housing to white suburbia.* Chicago: University of Chicago Press.

Rubinstein, R. L. (1987). Never married elderly as a social type: Re-evaluating some images. *Gerontologist, 27,* 108–113.

Rubinstein, R. L., Alexander, B. B., Goodman, M., & Luborsky, M. (1991). Key relationships of never married, childless older women: A cultural analysis. *Journal of Gerontology, 46,* S270–S277.

Ruble, D. N., & Frey, K. S. (1991). Changing patterns of comparative behavior as skills are acquired: A functional model of self-evaluation. In J. Suls & T. A. Wills (Eds.), *Social comparison: Contemporary theory and research* (pp. 70–112). Hillsdale, NJ: Erlbaum.

Ruble, D. N., & Martin, C. L. (1998). Gender development. In N. Eisenberg (Ed.), *Handbook of child psychology: Vol. 3. Social, emotional, and personality development* (5th ed., pp. 933–1016). New York: Wiley.

Rudolph, D. K., Lambert, S. F., Clark, A. G., & Kurlakowsky, K. D. (2001). Negotiating the transition to middle school: The role of self-regulatory processes. *Child Development, 72,* 929–946.

Rudolph, K. D., & Hammen, C. (1999). Age and gender as determinants of stress exposure, generation, and reactions in youngsters: A transactional perspective. *Child Development, 70,* 660–677.

Ruff, H. A., & Lawson, K. R. (1990). Development of sustained, focused attention in young children during free play. *Developmental Psychology, 26,* 85–93.

Ruff, H. A., Lawson, K. R., Parrinello, R., & Weissberg, R. (1990). Long-term stability of individual differences in sustained attention in the early years. *Child Development, 61,* 60–75.

Ruff, H. A., & Rothbart, M. K. (1996). *Attention in early development.* New York: Oxford University Press.

Ruffman, T. (1999). Children's understanding of logical inconsistency. *Child Development, 70,* 887–895.

Ruffman, T., & Langman, L. (2002). Infants' reaching in a multi-well A not B task. *Infant Behavior and Development, 25,* 237–246.

Ruffman, T., Perner, J., Naito, M., Parkin, L., & Clements, W. A. (1998). Older (but not younger) siblings facilitate false belief understanding. *Developmental Psychology, 34,* 161–174.

Ruffman, T., Perner, J., Olson, D. R., & Doherty, M. (1993). Reflecting on scientific thinking: Children's understanding of the hypothesis–evidence relation. *Child Development, 64,* 1617–1636.

Ruhm, C. J. (1996). Gender differences in employment behavior during late middle age. *Journal of Gerontology, 51B,* S11–S17.

Rumbaut, R. G. (1997). Ties that bind: Immigration and immigrant families in the United States. In A. Booth, A. C. Crouter, & N. Landale (Eds.), *Immigration and the family: Research and policy on U.S. immigrants* (pp. 3–46). Mahwah, NJ: Erlbaum.

Rumberger, R. W. (1990). Second chance for high school dropouts: Dropout recovery programs in the United States. In D. Inbar (Ed.), *Second chance in education: An interdisciplinary and international perspective* (pp. 227–250). Philadelphia: Falmer.

Runco, M. A. (1992). Children's divergent thinking and creative ideation. *Developmental Review, 12,* 233–264.

Russell, A., Mize, J., & Bissaker, K. (2002). Parent–child relationships. In P. K. Smith & C. H. Hart (Eds.), *Handbook of childhood social development.* Oxford, U.K.: Blackwell.

Russell, J. A. (1990). The preschooler's understanding of the causes and consequences of emotion. *Child Development, 61,* 1872–1881.

Russell, M. J., Cummings, B. J., Proffitt, B. F., Wysocki, C. J., Gilbert, A. N., & Cotman, C. W. (1993). Life span changes in the verbal categorization of odors. *Journal of Gerontology, 48,* P49–P53.

Russell, R. J. H., & Wells, P. A. (1994). Predictors of happiness in married couples. *Personality and Individual Differences, 17,* 313–321.

Rust, J., Golombok, S., Hines, M., Johnston, K., Golding, J., & the ALSPAC Study Team. (2000). The role of brothers and sisters in the gender development of preschool children. *Journal of Experimental Child Psychology, 77,* 292–303.

Rutter, M. (1987). Psychosocial resilience and protective mechanisms. *American Journal of Orthopsychiatry, 57,* 316–331.

Rutter, M. (1996). Maternal deprivation. In M. H. Bornstein (Ed.), *Handbook of parenting: Vol. 4. Applied and practical parenting* (pp. 3–31). Mahwah, NJ: Erlbaum.

Rutter, M., & the English and Romanian Adoptees Study Team. (1998). Developmental catch-up, and deficit, following adoption after severe global early privation. *Journal of Child Psychology and Psychiatry, 39,* 465–476.

Rvachew, S., Slawinski, E., Williams, M., & Green, C. L. (1999). The impact of early onset otitis media on babbling and early language development. *Journal of the Acoustical Society of America, 105,* 467–475.

Rybash, J. M., & Hrubi-Bopp, K. L. (2000). Isolating the neural mechanisms of age-related changes in human working memory. *Nature Neuroscience, 3,* 509–515.

Ryff, C. D. (1989). In the eye of the beholder: Views of psychological well-being among middle-aged and older adults. *Psychology and Aging, 4,* 195–210.

Ryff, C. D. (1991). Possible selves in adulthood and old age: A tale of shifting horizons. *Psychology and Aging, 6,* 286–295.

Ryff, C. D., Lee, Y. H., Essex, M. J., & Schmutte, P. S. (1994). My children and me: Midlife evaluations of grown children and of self. *Psychology and Aging, 9,* 195–205.

Ryff, C. D., & Migdal, S. (1984). Intimacy and generativity: Self-perceived transitions. *Signs: Journal of Women in Culture and Society, 9,* 470–481.

Ryff, C. D., Schmutte, P. S., & Lee, Y. H. (1996). How children turn out: Implications for parental self-evaluation. In C. D. Ryff & M. M. Seltzer (Eds.), *The parental experience in midlife* (pp. 383–422). Chicago: University of Chicago Press.

Ryff, C. D., Singer, B. H., & Seltzer, M. M. (2002). Pathways through challenge: Implications for well-being and health. In L. Pulkkinen & A. Caspi (Eds.), *Paths to successful development* (pp. 302–328). Cambridge, U.K.: Cambridge University Press.

Ryynänen, M., Kirkinen, P., Mannermaa, A., & Saarikoski, S. (1995). Carrier diagnosis of the fragile X syndrome—A challenge in antenatal clinics. *American Journal of Obstetrics and Gynecology, 172,* 1236–1239.

Saarni, C. (1993). Socialization of emotion. In M. Lewis & J. M. Haviland (Eds.), *Handbook of emotions* (pp. 435–446). New York: Guilford.

Saarni, C. (1997). Emotional competence and self-regulation in childhood. In P. Salovey & D. J. Sluyter (Eds.), *Emotional development and emotional intelligence* (pp. 35–66). New York: Basic Books.

Saarni, C. (1999). *The development of emotional competence.* New York: Guilford.

Saarni, C., Mumme, D. L., & Campos, J. J. (1998). Emotional development: Action, communication, and understanding. In N. Eisenberg (Ed.), *Handbook of child psychology: Vol. 3. Social, emotional, and personality development* (5th ed., pp. 237–309). New York: Wiley.

Sachdev, P. S., Brodaty, H., & Looi, J. C. L. (1999). Vascular dementia: Diagnosis, management, and possible prevention. *Medical Journal of Australia, 170,* 81–85.

Sackett, C. S., & Schenning, S. (2002). The age-related eye disease study: The results of the clinical trial. *Insight, 27,* 5–7.

Sacks, C. H., & Mergendoller, J. R. (1997). The relationship between teachers' theoretical orientation toward reading and student outcomes in kindergarten children with different initial reading abilities. *American Educational Research Journal, 34,* 721–739.

Sadeh, A. (1997). Sleep and melatonin in infants: A preliminary study. *Sleep, 20,* 185–191.

Sadler, T. W. (1995). *Langman's medical embryology* (7th ed.). Baltimore: Williams & Wilkins.

Sadler, T. W. (2000). *Langman's medical embryology* (8th ed.). Baltimore: Williams & Wilkins.

Safe Kids Worldwide. (2002). Childhood injury worldwide: Meeting the challenge. Retrieved from http://www.safekidsworldwide.org

Saffran, J. R., Aslin, R. N., & Newport, E. L. (1996). Statistical learning by 8-month-old infants. *Science, 27,* 1926–1928.

Safire, W. (1997, March 9). The young old. *New York Times Magazine,* p. 14.

Safman, P. C. (1988). Women from special populations: The challenge of reentry. In L. H. Lewis (Ed.), *Addressing the needs of returning women* (pp. 79–94). San Francisco: Jossey-Bass.

Safyer, A. W., Leahy, B. H., & Colan, N. B. (1995). The impact of work on adolescent development. *Families in Society, 76,* 38–45.

Sahni, R., Schulze, K. F., Stefanski, M., Myers, M. M., & Fifer, W. P. (1995). Methodological issues in coding sleep states in immature infants. *Developmental Psychobiology, 28,* 85–101.

Salapatek, P. (1975). Pattern perception in early infancy. In L. B. Cohen & P. Salapatek (Eds.), *Infant perception: From sensation to cognition* (pp. 133–248). New York: Academic Press.

Salbe, A. D., Weyer, C., Lindsay, R. S., Ravussin, E., & Tataranni, P. A. (2002). Assessing risk factors for obesity between childhood and adolescence: I. Birth weight, childhood adiposity, parental obesity, insulin, and leptin. *Pediatrics, 110,* 299–306.

Salerno, M., Micillo, M., Di Maio, S., Capalbo, D., Ferri, P., & Lettiero, T. (2001). Longitudinal growth, sexual maturation and final height in patients with congenital hypothyroidism detected by neonatal screening. *European Journal of Endocrinology, 145,* 377–383.

Salthouse, T. A. (1984). Effects of age and skill in typing. *Journal of Experimental Psychology: General, 113,* 345–371.

Salthouse, T. A. (1985). Speed of behavior and its implications for cognition. In J. E. Birren & K. W. Schaie (Eds.), *Handbook of the psychology of aging* (2nd ed., pp. 400–426). New York: Van Nostrand Reinhold.

Salthouse, T. A. (1991a). Cognitive facets of aging well. *Generations, 51*(1), 35–38.

Salthouse, T. A. (1991b). *Theoretical perspectives in cognitive aging.* Hillsdale, NJ: Erlbaum.

Salthouse, T. A. (1993). Speed mediation of adult age differences in cognition. *Developmental Psychology, 29,* 722–738.

Salthouse, T. A. (1996). Constraints on theories of cognitive aging. *Psychonomic Bulletin and Review, 3,* 287–299.

Salthouse, T. A. (2000). Aging and measures of processing speed. *Biological Psychology, 54,* 35–54.

Salthouse, T. A., & Babcock, R. L. (1991). Decomposing adult age differences in working memory. *Developmental Psychology, 27,* 763–776.

Salthouse, T. A., Fristoe, N., McGuthry, K. E., & Hambrick, D. Z. (1998). Relation of task switching to speed, age, and fluid intelligence. *Psychology and Aging, 13,* 445–461.

Salthouse, T. A., & Maurer, T. J. (1996). Aging, job performance, and career development. In J. E. Birren & K. W. Schaie (Eds.), *Handbook of the psychology of aging* (pp. 353–364). San Diego, CA: Academic Press.

Salthouse, T. A., & Meinz, E. J. (1995). Aging, inhibition, working memory, and speed. *Journal of Gerontology, 50,* P297–P306.

Salthouse, T. A., & Skovronek, E. (1992). Within-context assessment of working memory. *Journal of Gerontology, 47,* P110–P129.

Samarel, N. (1991). *Caring for life and death.* Washington, DC: Hemisphere.

Samarel, N. (1995). The dying process. In H. Wass & R. A. Neimeyer (Eds.), *Dying: Facing the facts* (3rd ed., pp. 89–116). Washington, DC: Taylor & Francis.

Sameroff, A. J., Seifer, R., Baldwin, A., & Baldwin, C. (1993). Stability of intelligence from preschool to adolescence: The influence of social and family risk factors. *Child Development, 64,* 80–97.

Sampson, R. J., & Laub, J. H. (1993). *Crime in the making: Pathways and turning points through life.* Cambridge, MA: Harvard University Press.

Samuels, N., & Samuels, M. (1996). *The new well pregnancy book.* New York: Summitt.

Samuelsson, S. M., Alfredson, B. B., Hagberg, B., Anonymous, Nordbeck, B., Brun, A., Gustafson, L., & Risberg, J. (1997). The Swedish Centenarian Study: A multidisciplinary study of five consecutive cohorts at the age of 100. *International Journal of Aging and Human Development, 45,* 223–253.

Sandberg, D. E., Brook, A. E., & Campos, S. P. (1994). Short stature: A psychosocial burden requiring growth hormone therapy? *Pediatrics, 94,* 832–840.

Sanders, M. G., & Jordan, W. J. (2000). Student–teacher relations and academic achievement in high school. In M. G. Sanders (Ed.), *Schooling students placed at risk: Research, policy, and practice in the education of poor and minority adolescents* (pp. 65–82). Mahwah, NJ: Erlbaum.

Sanders, M. R., Halford, W. K., & Behrens, B. C. (1999). Parental divorce and premarital couple communication. *Journal of Family Psychology, 13,* 60–74.

Sanderson, J. A., & Siegal, M. (1988). Conceptions of moral and social rules in rejected and nonrejected preschoolers. *Journal of Clinical Child Psychology, 17,* 66–72.

Sandman, C. A., Wadhwa, P., Hetrick, W., Porto, M., & Peeke, H. V. S. (1997). Human fetal heart rate dishabituation between thirty and thirty-two weeks gestation. *Child Development, 68,* 1031–1040.

Sandnabba, N. K., & Ahlberg, C. (1999). Parents' attitudes and expectations about children's cross-gender behavior. *Sex Roles, 40,* 249–263.

Sandqvist, K. (1992). Sweden's sex-role scheme and commitment to gender equality. In S. Lewis, D. N. Izraeli, & H. Hottsmans (Eds.), *Dual-earner families: International perspective.* London: Sage.

Sandstrom, M. J., & Coie, J. D. (1999). A developmental perspective on peer rejection: Mechanisms of stability and change. *Child Development, 70,* 955–966.

Sanford, J. P. (1985). *Comprehension-level tasks in secondary classrooms.* Austin: Research and Development Center for Teacher Education, University of Texas at Austin.

Sankar, A. (1993). Images of home death and the elderly patient: Romantic versus real. *Generations, 27*(2), 59–63.

Sankaranarayanan, K. (1998). Ionizing radiation and genetic risks IX. Estimates of the frequencies of Mendelian diseases and spontaneous mutation rates in human populations: A 1998 perspective. *Mutation Research, 411,* 129–178.

Sansavini, A., Bertoncini, J., & Giovanelli, G. (1997). Newborns discriminate the rhythm of multisyllabic stressed words. *Developmental Psychology, 33,* 3–11.

Sanson, A. V., Pedlow, R., Cann, W., Prior, M., & Oberklaid, F. (1996). Shyness ratings: Stability and correlates in early childhood. *International Journal of Behavioural Development, 19,* 705–724.

Sansone, C., & Berg, C. A. (1993). Adapting to the environment across the life span: Different process or different inputs? *International Journal of Behavioral Development, 16,* 215–241.

Santoloupo, S., & Pratt, M. (1994). Age, gender, and parenting style variations in mother–adolescent dialogues and adolescent reasoning about political issues. *Journal of Adolescent Research, 9,* 241–261.

Sarason, S. B. (1977). *Work, aging, and social change.* New York: Free Press.

Sarrazin, G. (1999). WISC-III, *Échelle d'intelligence de Wechsler pour Enfants troisième Édition, adaptation canadienne-française, Manuel d'administration.* Toronto: Psychological Corporation.

Sasser-Coen, J. A. (1993). Qualitative changes in creativity in the second half of life: A life-span developmental perspective. *Journal of Creative Behavior, 27,* 18–27.

Sataloff, R. T., & Sataloff, J. (2001). A monumental achievement. *Occupational Health and Safety, 70,* 122–125.

Satz, P. (1993). Brain reserve capacity on symptom onset after brain injury: A formulation and review of evidence for threshold theory. *Neuropsychology, 7,* 273–295.

Savage, A. R., Petersen, M. B., Pettay, D., Taft, L., Allran, K., Freeman, S. B., Karadima, G., Avramopoulos, D., Torfs, C., Mikkelsen, M., & Hassold, T. J. (1998). Elucidating the mechanisms of paternal non-disjunction of chromosome 21 in humans. *Human Molecular Genetics, 7,* 1221–1227.

Saville-Troike, M. (1988). Private speech: Evidence for second language learning strategies during the 'silent' period. *Journal of Child Language, 15,* 567–590.

Savin-Williams, R. C. (1996). Dating and romantic relationships among gay, lesbian, and bisexual youths. In R. C. Savin-Williams & K. M. Cohen (Eds.), *The lives of lesbians, gays, and bisexuals* (pp. 166–180). Fort Worth: Harcourt Brace.

Savin-Williams, R. C. (1998).... *And then I became gay: Young men's stories.* New York: Routledge.

Saxe, G. B. (1988, August–September). Candy selling and math learning. *Educational Researcher, 17(6),* 14–21.

Scabini, E., & Cigoli, V. (1997). Young adult families: An evolutionary slowdown or a breakdown in the generational transition? *Journal of Family Issues, 18,* 608–626.

Scaramella, L. V., Conger, R. D., Simons, R. L., & Whitbeck, L. B. (1998). Predicting risk for pregnancy by late adolescence: A social contextual perspective. *Developmental Psychology, 34,* 1233–1245.

Scarr, S. (1985). Constructing psychology: Making facts and fables for our times. *American Psychologist, 40,* 499–512.

Scarr, S. (1996). Individuality and community: The contrasting role of the state in family life in the United States and Sweden. *Scandinavian Journal of Psychology, 37,* 93–102.

Scarr, S. (1997). Behavior-genetic and socialization theories of intelligence: Truce and reconciliation. In R. J. Sternberg & E. L. Grigorenko (Eds.), *Intelligence, heredity, and environment* (pp. 3–41). New York: Cambridge University Press.

Scarr, S. (1998). American child care today. *American Psychologist, 53,* 95–108.

Scarr, S., & McCartney, K. (1983). How people make their own environments: A theory of genotype environment effects. *Child Development, 54,* 424–435.

Scarr, S., Phillips, D. A., & McCartney, K. (1990). Facts, fantasies, and the future of child care in America. *Psychological Science, 1,* 26–35.

Scarr, S., & Weinberg, R. A. (1983). The Minnesota Adoption Studies:

Genetic differences and malleability. *Child Development, 54,* 260–267.
Schaffer, J., & Kral, R. (1988). Adoptive families. In C. S. Chilman, E. W. Nunnally, & F. M. Cox (Eds.), *Variant family forms* (pp. 165–184). Newbury Park, CA: Sage.
Schaie, K. W. (1977/1978). Toward a stage theory of adult cognitive development. *Aging and Human Development, 8,* 129–138.
Schaie, K. W. (1988). Variability in cognitive functioning in the elderly. In M. A. Bender, R. C. Leonard, & A. D. Woodhead (Eds.), *Phenotypic variation in populations* (p. 201). New York: Plenum.
Schaie, K. W. (1989). Individual differences in rate of cognitive change in adulthood. In V. L. Bengtson & K. W. Schaie (Eds.), *The course of later life: Research and reflections* (pp. 68–83). New York: Springer.
Schaie, K. W. (1994). The course of adult intellectual development. *American Psychologist, 49,* 304–313.
Schaie, K. W. (1996). *Intellectual development in adulthood: The Seattle Longitudinal Study.* New York: Cambridge University Press.
Schaie, K. W. (1998). The Seattle Longitudinal Studies of Adult Intelligence. In M. P. Lawton & T. A. Salthouse (Eds.), *Essential papers on the psychology of aging* (pp. 263–271). New York: New York University Press.
Schaie, K. W. (2000). The impact of longitudinal studies on understanding development from young adulthood to old age. *International Journal of Behavioral Development, 24,* 257–266.
Schaie, K. W., & Willis, S. L. (1996). *Adult development and aging* (4th ed.). Boston: Addison-Wesley.
Schaie, K. W., & Willis, S. L. (2000). A stage theory model of adult cognitive development revisited. In R. L. Rubinstein & M. Moss (Eds.), *The many dimensions of aging* (pp. 175–193). New York: Springer.
Schauble, L. (1996). The development of scientific reasoning in knowledge-rich contexts. *Developmental Psychology, 32,* 102–119.
Scheidt, R. J., & Windley, P. G. (1985). The ecology of aging. In J. E. Birren & K. W. Schaie (Eds.), *Handbook of the psychology of aging* (pp. 245–258). New York: Van Nostrand Reinhold.
Scher, A., Tirosh, E., Jaffe, M., Rubin, L., Sadeh, A., & Lavie, P. (1995). Sleep patterns of infants and young children in Israel. *International Journal of Behavioral Development, 18,* 701–711.
Scherer, J. M., & Simon, R. J. (1999). *Euthanasia and the right to die: A comparative view.* Lanham, MD: Rowman & Littlefield.
Schieman, S., Gundy, V., & Taylor, K. (2001). Status, role, and resource explanations for age patterns in psychological distress. *Journal of Health and Social Behavior, 42,* 80–96.
Schlaud, M., Eberhard, C., Trumann, B., Kleemann, W. J., Poets, C. F., Tietze, K. W., & Schwartz, F. W. (1999). Prevalence and determinants of prone sleeping position in infants: Results for two cross-sectional studies on risk factors for SIDS in Germany. *American Journal of Epidemiology, 150,* 51–57.
Schlegel, A., & Barry, H., III (1991). *Adolescence: An anthropological inquiry.* New York: Free Press.
Schlesinger, M., & Kronebusch, K. (1994). The sources of intergenerational burdens and tensions. In V. L. Bengtson & R. A. Harootyan (Eds.), *Intergenerational linkages: Hidden connections in American society* (pp. 185–209). New York: Springer.
Schmidt, I. W., Berg, I. J., Deelman, B. G., & Pelemans, W. (1999). Memory training for remembering names in older adults. *Clinical Gerontologist, 20,* 57–73.
Schmidt, U. (2000). Eating disorders. In D. Kohen (Ed.), *Women and mental health* (pp. 174–197). London: Routledge.
Schmitz, M. K. H., & Jeffery, R. W. (2000). Public health interventions for the prevention and treatment of obesity. *Medical Clinics of North America, 84,* 491–512.
Schmitz, S., Fulker, D. W., Plomin, R., Zahn-Waxler, C., Emde, R. N., & DeFries, J. C. (1999). Temperament and problem behaviour during early childhood. *International Journal of Behavioural Development, 23,* 333–355.
Schneewind, K. A., & Gerhard, A. (2002). Relationship personality, conflict resolution, and marital satisfaction in the first 5 years of marriage. *Family Relations, 51,* 63–71.
Schneider, B., & Stevenson, D. (1999). *The ambitious generation: America's teenagers, motivated but directionless.* New Haven, CT: Yale University Press.
Schneider, B. A., Daneman, M., Murphy, D. R., & See, S. K. (2000). Listening to discourse in distracting settings: The effects of aging. *Psychology and Aging, 15,* 110–125.
Schneider, B. H., Atkinson, L., & Tardif, C. (2001). Child–parent attachment and children's peer relations: A quantitative review. *Developmental Psychology, 37,* 86–100.
Schneider, E. L. (1992). Biological theories of aging. *Generations, 16*(4), 7–10.
Schneider, W. (1993). Domain-specific knowledge and memory performance in children. *Educational Psychology Review, 5,* 257–274.
Schneider, W., & Bjorklund, D. F. (1992). Expertise, aptitude, and strategic remembering. *Child Development, 63,* 461–473.
Schneider, W., & Bjorklund, D. F. (1998). Memory. In D. Kuhn & R. S. Siegler (Eds.), *Handbook of child psychology: Vol. 2. Cognition, perception, and language* (5th ed., pp. 467–521). New York: Wiley.
Schneider, W., & Pressley, M. (1997). *Memory development between two and twenty* (2nd ed.). Mahwah, NJ: Erlbaum.
Schnohr, P., Nyboe, J., Lange, P., & Jensen, G. (1998). Longevity and gray hair, baldness, facial wrinkles, and arcus senilis in 13,000 men and women: The Copenhagen City Heart Study. *Journal of Gerontology, 53,* M347–350.
Schnur, E., & Belanger, S. (2000). What works in Head Start. In M. P. Kluger & G. Alexander (Eds.), *What works in child welfare* (pp. 277–284). Washington, DC: Child Welfare League of America.
Scholl, B. J., & Leslie, A. M. (2000). Minds, modules, and meta-analysis. *Child Development, 72,* 696–701.
Scholl, T. O., Heidiger, M. L., & Belsky, D. H. (1996). Prenatal care and maternal health during adolescent pregnancy: A review and meta-analysis. *Journal of Adolescent Health, 15,* 444–456.
Scholnick, E. K. (1995, Fall). Knowing and constructing plans. *SRCD Newsletter,* pp. 1–2, 17.
Schonert-Reichl, K. A. (1999). Relations of peer acceptance, friendship adjustment, and social behavior to moral reasoning during early adolescence. *Journal of Early Adolescence, 19,* 249–279.
Schonfeld, A. M., Mattson, S. N., Lang, A. R., Delis, D. C., & Riley, E. P. (2001). Verbal and nonverbal fluency in children with heavy prenatal alcohol exposure. *Journal of Studies on Alcohol, 62,* 239–246.
Schor, J. B. (2002). Time crunch among American parents. In S. A. Hewlett, N. Rankin, & C. West (Eds.), *Taking parenting public* (pp. 83–102). Boston: Rowman & Littlefield.
Schothorst, P. F., & van Engeland, H. (1996). Long-term behavioral sequelae of prematurity. *Journal of the American Academy of Child and Adolescent Psychiatry, 35,* 175–183.
Schroots, J., & Birren, J. (1990). Concept of time and aging in science. In J. E. Birren & K. W. Schaie (Eds.), *Handbook of the psychology of aging* (3rd ed., pp. 45–64). San Diego: Academic Press.
Schuengel, G., Bakermans-Kranenburg, M. J., & van IJzendoorn, M. H. (1999). Attachment and loss: Frightening maternal behavior linking unresolved loss and disorganized infant attachment. *Journal of Consulting and Clinical Psychology, 67,* 54–63.
Schull, W. J., & Otake, M. (1999). Cognitive function and prenatal exposure to ionizing radiation. *Teratology, 59,* 222–226.
Schulman, J. D., & Black, S. H. (1997). Screening for Huntington disease and certain other dominantly inherited disorders: A case for preimplantation genetic testing. *Journal of Medical Screening, 4,* 58–59.
Schulman, K. A., Berlin, J. A., Harless, W., Kerner, J. F., Sistrunk, S., & Gersh, B. J. (1999). The effect of race and sex on physicians' recommendations for cardiac catheterization. *New England Journal of Medicine, 340,* 618–626.
Schulz, R., & Aderman, D. (1979). Physicians' death anxiety and patient outcomes. *Omega, 9,* 327–332.
Schulz, R., & Curnow, C. (1988). Peak performance and age among superathletes: Track and field, swimming, baseball, tennis, and golf. *Journal of Gerontology, 43,* P113–P120.
Schwanenflugel, P. J., Henderson, R. L., & Fabricius, W. V. (1998). Developing organization of mental verbs and theory of mind in middle childhood: Evidence from extensions. *Developmental Psychology, 34,* 512–524.
Schwartz, P., & Rutter, V. (1998). *The gender of sexuality.* Thousand Oaks, CA: Pine Forge.
Schwarz, B. (1996). *Nursing home design: Consequences of employing the medical model.* New York: Garland.
Schwarzer, R. (1999). Self-regulatory processes in the adoption and maintenance of health behaviors. *Journal of Health Psychology, 4,* 115–127.
Schwebel, D. C., Rosen, C. S., & Singer, J. L. (1999). Preschoolers' pretend play and theory of mind:

The role of jointly constructed pretense. *British Journal of Developmental Psychology, 17,* 333–348.

Schwebel, M., Maher, C. A., & Fagley, N. S. (1990). Introduction: The social role in promoting cognitive growth over the life span. In M. Schwebel, C. A. Maher, & N. S. Fagley (Eds.), *Promoting cognitive growth over the life span* (pp. 1–20). Hillsdale, NJ: Erlbaum.

Scott, W. K., Vance, J. M., Haines, J. L., & Pericak-Vance, M. A. (2002). Linkage of Parkinsonism and Alzheimer's disease with Lewy body pathology to chromosome 12. *Annals of Neurology, 52,* 524.

Scully, D., & Marolla, J. (1998). "Riding the bull at Gilley's": Convicted rapists describe the rewards of rape. In M. E. Odem & J. Clay-Warner (Eds.), *Confronting rape and sexual assault* (pp. 181–198). Wilmington, DE: Scholarly Resources.

Sears, P. S., & Barbie, A. H. (1977). Career and life satisfaction among Terman's gifted women. In J. C. Stanley, W. George, & C. Solano (Eds.), *The gifted and creative: Fifty year perspective* (pp. 154–172). Baltimore: Johns Hopkins University Press.

Seccombe, K. (2002). "Beating the odds" versus "changing the odds": Poverty, resilience, and family policy. *Journal of Marriage and the Family, 64,* 384–394.

Seefeldt, V. (1996). The concept of readiness applied to the acquisition of motor skills. In F. L. Smoll & R. E. Smith (Eds.), *Children and youth in sport: A biopsychological perspective* (pp. 49–56). Dubuque, IA: Brown & Benchmark.

Seeman, T. E., Berkman, L. F., Kohout, F., Lacroix, A., Glynn, R., & Blazer, D. (1993). Intercommunity variations in the association between social ties and mortality in the elderly. *Annals of Epidemiology,* 3, 325–335.

Seifer, R., & Schiller, M. (1995). The role of parenting sensitivity, infant temperament, and dyadic interaction in attachment theory and assessment. In E. Waters, B. E. Vaughn, G. Posada, & K. Kondo-Ikemura (Eds.), Caregiving, cultural, and cognitive perspectives on secure-base behavior and working models: New growing points of attachment theory and research. *Monographs of the Society for Research in Child Development, 60*(2–3, Serial No. 244).

Seifer, R., Schiller, M., Sameroff, A. J., Resnick, S., & Riordan, K. (1996). Attachment, maternal sensitivity, and infant temperament during the first year of life. *Developmental Psychology, 32,* 12–25.

Seitz, V., & Apfel, N. H. (1993). Adolescent mothers and repeated childbearing: Effects of a school-based intervention program. *American Journal of Orthopsychiatry, 63,* 572–581.

Seitz, V., & Apfel, N. H. (1994). Effects of a school for pregnant students on the incidence of low-birthweight deliveries. *Child Development, 65,* 666–676.

Seligman, L. (1994). *Developmental career counseling and assessment* (2nd ed.). Thousand Oaks, CA: Sage.

Seligman, M. E. P. (1975). *Helplessness: On depression, development, and death.* San Francisco: Freeman.

Selikowitz, M. (1997). *Down syndrome: The facts* (2nd ed.). Oxford, U.K.: Oxford University Press.

Selman, R. L. (1976). Social-cognitive understanding: A guide to educational and clinical practice. In T. Lickona (Ed.), *Moral development and behavior: Theory, research, and social issues* (pp. 299–316). New York: Holt, Rinehart and Winston.

Selman, R. L. (1980). *The growth of interpersonal understanding.* New York: Academic Press.

Selman, R. L., & Byrne, D. F. (1974). A structural-developmental analysis of levels of role taking in middle childhood. *Child Development, 45,* 803–806.

Seltzer, M. M., & Ryff, C. D. (1994). Parenting across the life span: The normative and nonnormative cases. In D. L. Featherman, R. M. Lerner, & M. Perlmutter (Eds.), *Lifespan development and behavior* (pp. 1–40). Hillsdale, NJ: Erlbaum.

Selwyn, P. A. (1996). Before their time: A clinician's reflections on death and AIDS. In H. M. Spiro, M. G. M. Curnen, & L. P. Wandel (Eds.), *Facing death: Where culture, religion, and medicine meet* (pp. 33–37). New Haven, CT: Yale University Press.

Sen, M. G., Yonas, A., & Knill, D. C. (2001). Development of infants' sensitivity to surface contour information for spatial layout. *Perception, 30,* 167–176.

Senn, C. Y., Desmarais, S., Verberg, N., & Wood, E. (2000). Predicting coercive sexual behavior across the lifespan in a random sample of Canadian men. *Journal of Social and Personal Relationships, 17,* 95–113.

Serbin, L. A., Poulin-Dubois, D., Colburne, K. A., Sen, M. G., & Eichstedt, J. A. (2001). Gender stereotyping in infancy: Visual preferences for and knowledge of gender-stereotyped toys in the second year. *International Journal of Behavioral Development, 25,* 7–15.

Serbin, L. A., Powlishta, K. K., & Gulko, J. (1993). The development of sex typing in middle childhood. *Monographs of the Society for Research in Child Development, 58*(2, Serial No. 232).

Seward, R. R., Yeats, D. E., & Zottarelli, L. K. (2002). Parental leave and father involvement in child care: Sweden and the United States. *Journal of Comparative Family Studies, 33,* 387–399.

Shainess, N. (1961). A re-evaluation of some aspects of femininity through a study of menstruation: A preliminary report. *Comparative Psychiatry, 2,* 20–26.

Shanahan, M. J., Mortimer, J. T., & Krüger, H. (2002). Adolescence and adult work in the twenty-first century. *Journal of Research on Adolescence, 12,* 99–120.

Shann, F., & Steinhoff, M. C. (1999). Vaccines for children in rich and poor countries. *Paediatrics, 354,* 7–11

Shapiro, A. E., Gottman, J. M., & Carrere, S. (2000). The baby and the marriage: Identifying factors that buffer against decline in marital satisfaction after the first baby arrives. *Journal of Family Psychology, 14,* 59–70.

Sharma, A. R., McCue, M. K., & Benson, P. L. (1998). The psychological adjustment of United States adopted adolescents and their nonadopted siblings. *Child Development, 69,* 791–802.

Shaver, P., Furman, W., & Buhrmester, D. (1985). Transition to college: Network changes, social skills, and loneliness. In S. Duck & D. Perlman (Eds.), *Understanding personal relationships: An interdisciplinary approach* (pp. 193–219). London: Sage.

Shay, J. W., & Wright, W. E. (2001). When do telomeres matter? *Science, 291,* 839–840.

Shedler, J., & Block, J. (1990). Adolescent drug use and psychological health: A longitudinal inquiry. *American Psychologist, 45,* 612–630.

Sheehy, A., Gasser, T., Molinari, L., & Largo, R. H. (1999). An analysis of variance of the pubertal and midgrowth spurts for length and width. *Annals of Human Biology, 26,* 309–331.

Sheldon, K. M., & Kasser, T. (2001). Getting older, getting better? Personal strivings and psychological maturity across the life span. *Developmental Psychology, 37,* 491–501.

Shephard, R. J., & Shek, P. N. (1995). Exercise, aging and immune function. *International Journal of Sports Medicine, 16,* 1–6.

Sherman, A. M., de Vries, B., & Lansford, J. E. (2000). Friendship in childhood and adulthood: Lessons across the life span. *International Journal of Aging and Human Development, 51,* 31–51.

Sherman, D. K., Iacono, W. G., & McGue, M. K. (1997). Attention-deficit hyperactivity disorder dimensions: A twin study of inattention and impulsivity–hyperactivity. *Journal of the American Academy of Child and Adolescent Psychiatry, 36,* 745–753.

Sherrill, C. L., & Pinderhughes, E. E. (1999). Conceptions of family and adoption among older adoptees. *Adoption Quarterly, 2,* 21–48.

Shields, P. J., & Rovee-Collier, C. K. (1992). Long-term memory for context-specific category information at six months. *Child Development, 63,* 245–259.

Shiloh, S. (1996). Genetic counseling: A developing area of interest for psychologists. *Professional Psychology: Research and Practice, 27,* 475–486.

Shimamura, A. P., Berry, J. M., Mangels, J. A., Rusting, C. L., & Jurica, P. J. (1995). Memory and cognitive abilities in university professors: Evidence for sucessful aging. *Psychological Science, 6,* 271–277.

Shock, N. W. (1977). Biological theories of aging. In J. E. Birren & K. W. Schaie (Eds.), *Handbook of the psychology of aging* (pp. 103–115). New York: Van Nostrand Reinhold.

Shoda, Y., Mischel, W., & Peake, P. K. (1999). Predicting adolescent cognitive and self-regulatory competencies from preschool delay of gratification: Identifying diagnostic conditions. *Developmental Psychology, 26,* 978–986.

Shonk, S. M., & Cicchetti, D. (2001). Maltreatment, competency deficits, and risk for academic and behavioral maladjustment. *Developmental Psychology, 37,* 3–17.

Shope, J. T., Holmes, S. B., Sharpe, P. A., & Goodman, C. (1993). Services for persons with dementia and their families., A survey of information and referral agencies in Michigan. *Gerontologist, 33,* 529–533.

Shuchter, S. R., & Zisook, S. (1995). The course of normal grief. In M. S. Stroebe, W. Stroebe, & R. O. Hansson (Eds.), *Handbook of bereavement* (pp. 44–61). Cambridge: Cambridge University Press.

Shulman, S., Elicker, J., & Sroufe, A. (1994). Stages of friendship growth in preadolescence as related to attachment history. *Journal of Social and Personal Relationships, 11,* 341–361.

Shumaker, S. A., Legault, C., Thal, L., Wallace, R. B., Ockene, J. K., & Hendrix, S. L. (2003). Estrogen plus progestin and the incidence of dementia and mild cognitive impairment in postmenopausal women: The Women's Health Initiative Memory Study: A randomized controlled trial. *Journal of the American Medical Association, 289,* 2651–2662.

Shumow, L. (1998). Contributions of parent education to adult development. In C. M. Smith & T. Pourchot (Eds.), *Adult learning and development: Perspectives from educational psychology* (pp. 239–255). Mahwah, NJ: Erlbaum.

Shure, M. B. (1997). Interpersonal cognitive problem solving: Primary prevention of early high-risk behaviors in the preschool and primary years. In G. W. Albee & T. P. Gullotta (Eds.), *Primary prevention works* (pp. 167–188). Thousand Oaks, CA: Sage.

Shweder, R. A. (1996). True ethnography: The lore, the law, and the lure. In R. Jessor, A. Colby, & R. A. Shweder (Eds.), *Ethnography and human development* (pp. 15–52). Chicago: University of Chicago Press.

Shweder, R. A., Goodnow, J., Hatano, G., LeVine, R. A., Markus, H., & Miller, P. (1998). The cultural psychology of development: One mind, many mentalities. In R. M. Lerner (Ed.), *Handbook of child psychology: Vol. 1. Theoretical models of human development* (5th ed., pp. 865–937). New York: Wiley.

Siegel, B. (1996, Spring). Is the emperor wearing clothes? Social policy and the empirical support for full inclusion of children with disabilities in the preschool and early elementary school grades. *Social Policy Report of the Society for Research in Child Development, 10*(2–3), 2–17.

Siegler, R. S. (1996). *Emerging minds: The process of change in children's thinking.* New York: Oxford University Press.

Siegler, R. S. (1998). *Children's thinking* (3rd ed.). Upper Saddle River, NJ: Prentice-Hall.

Siervogel, R. M., Maynard, L. M., Wisemandle, W. A., Roche, A. F., Guo, S. S., Chumlea, W. C., & Towne, B. (2000). Annual changes in total body fat and fat-free mass in children from 8 to 18 years in relation to changes in body mass index: The Fels Longitudinal Study. *Annals of the New York Academy of Science, 904,* 420–423.

Sigman, M. (1995). Nutrition and child development: More food for thought. *Current Directions in Psychological Science, 4,* 52–55.

Sigman, M. (1999). Developmental deficits in children with Down syndrome. In H. Tager-Flusberg (Ed.), *Neurodevelopmental disorders: Developmental cognitive neuroscience* (pp. 179–195). Cambridge, MA: MIT Press.

Sigman, M., Cohen, S. E., & Beckwith, L. (1997). Why does infant attention predict adolescent intelligence? *Infant Behavior and Development, 20,* 133–140.

Signorella, M., & Liben, L. S. (1984). Recall and reconstruction of gender-related pictures: Effects of attitude, task difficulty, and age. *Child Development, 55,* 393–405.

Silver, M. H., Jilinskaia, E., & Perls, T. T. (2001). Cognitive functional status of age-confirmed centenarians in a population-based study. *Journal of Gerontology, 56B,* P134–P140.

Silver, M. H., & Perls, T. T. (2000). Is dementia the price of a long life? An optimistic report from centenarians. *Journal of Geriatric Psychiatry, 33,* 71–79.

Silverberg, S. B. (1996). Parents' well-being at their children's transition to adolescence. In C. D. Ryff & M. M. Seltzer (Eds.), *The parental experience in midlife* (pp. 215–254). Chicago: University of Chicago Press.

Silverman, D. H. S., Small, G. W., Chang, C. Y., Lu, C. S., Kung de Aburto, M. A., Chen, W., et al. (2001). Positron emission tomography in evaluation of dementia: Regional brain metabolism and long-term outcome. *Journal of the American Medical Association, 286,* 2120–2127.

Silverman, E., Range, L., & Overholser, J. (1994). Bereavement from suicide as compared to other forms of bereavement. *Omega, 30,* 41–51.

Silverman, P. R., & Nickman, S. L. (1996). Children's construction of their dead parents. In D. Klass, P. R. Silverman, & S. L. Nickman (Ed.), *Continuing bonds: New understandings of grief* (pp. 73–86). Washington, DC: Taylor & Francis.

Silverman, P. R., & Worden, J. M. (1992). Children's reactions in the early months after the death of a parent. *American Journal of Orthopsychiatry, 62,* 93–104.

Silverman, W. K., La Greca, A. M., & Wasserstein, S. (1995). What do children worry about? Worries and their relation to anxiety. *Child Development, 66,* 671–686.

Silverstein, M., & Bengtson, V. L. (1991). Do close parent–child relations reduce the mortality risk of older parents? *Journal of Health and Social Behavior, 32,* 382–395.

Silverstein, M., Chen, X., & Heller, K. (1996). Too much of a good thing? Intergenerational social support and the psychological well-being of older parents. *Journal of Marriage and the Family, 58,* 970–982.

Silverstein, M., Conroy, S., Wang, H., Giarrusso, R., & Bengtson, V. L. (2002). Reciprocity in parent–child relations over the adult life course. *Journal of Gerontology, 57B,* S3–S13.

Silverstein, M., & Long, J. D. (1998). Trajectories of grandparents' perceived solidarity with adult grandchildren: A growth curve analysis over 23 years. *Journal of Marriage and the Family, 60,* 912–923.

Silverstein, M., & Parrott, T. M. (1997). Attitudes toward public support of the elderly: Does early involvement with grandparents moderate generational tensions? *Research on Aging, 19,* 108–132.

Sim, T. N. (2000). Adolescent psychosocial competence: The importance and role of regard for parents. *Journal of Research on Adolescence, 10,* 49–64.

Simmons, R. G., & Blyth, D. A. (1987). *Moving into adolescence.* New York: Aldine de Gruyter.

Simoneau, G. G., & Leibowitz, H. W. (1996). Posture, gait, and falls. In J. Birren & K. W. Schaie (Eds.), *Handbook of the psychology of aging* (4th ed., pp. 204–217). San Diego: Academic Press.

Simons, R. L., & Chao, W. (1996). Conduct problems. In R. L. Simons & Associates (Eds.), *Understanding differences between divorced and intact families* (pp. 125–143). Thousand Oaks, CA: Sage.

Simons, R. L., Whitbeck, L. B., Conger, R. D., & Wu, C.-I. (1991). Intergenerational transmission of harsh parenting. *Developmental Psychology, 27,* 159–171.

Simonton, D. K. (1991). Creative productivity through the adult years. *Generations, 15*(2), 13–16.

Simonton, D. K. (2000). Creativity: Cognitive, personal, developmental, and social aspects. *American Psychologist, 55,* 151–158.

Simpson, J. A., & Harris, B. A. (1994). Interpersonal attraction. In A. L. Weber & J. H. Harvey (Eds.), *Perspectives on close relationships* (pp. 45–66). Boston: Allyn and Bacon.

Simpson, J. M. (2001). Infant stress and sleep deprivation as an aetiological basis for the sudden infant death syndrome. *Early Human Development, 61,* 1–43.

Singer, D. G. (1999). Imaginative play and television: Factors in a child's development. In J. A. Singer & P. Salovey (Eds.), *At play in the fields of consciousness: Essays in honor of Jerome L. Singer* (pp. 303–326). Mahwah, NJ: Erlbaum.

Singer, P. A., Choudhry, S., Armstrong, J., Meslin, E. M., & Lowry, F. H. (1995). Public opinion regarding end-of-life decisions: Influence of prognosis, practice, and process. *Social Science and Medicine, 41,* 1517–1521.

Singh, S., & Darroch, J. E. (2000). Adolescent pregnancy and childbearing: Levels and trends in developed countries. *Family Planning Perspectives, 32,* 14–23.

Sinnott, J. D. (1989). A model for solution of ill-structured problems: Implications for everyday and abstract problem solving. In J. D. Sinnott (Ed.), *Everyday problem solving: Theory and applications* (pp. 72–99). New York: Praeger.

Sinnott, J. D. (1998). *The development of logic in adulthood: Postformal thought and its applications.* New York: Plenum.

Skinner, B. F. (1957). *Verbal behavior.* New York: Appleton-Century-Crofts.

Skinner, B. F. (1983). Intellectual self-management in old age. *American Psychologist, 38,* 239–244.

Skinner, E. A. (1995). *Perceived control, motivation, and coping.* Thousand Oaks, CA: Sage.

Skinner, E. A., Zimmer-Gembeck, M. J., & Connell, J. P. (1998). Individual differences and the development of perceived control. *Monographs of the Society for Research in Child Development, 63*(2–3, Serial No. 254).

Skoe, E. S. A. (1998). The ethic of care: Issues in moral development. In E. E. A. Skoe & A. L. von der Lippe (Eds.), *Personality development in adolescence* (pp. 143–171). London: Routledge.

Slaby, R. G., & Frey, K. S. (1975). Development of gender constancy and selective attention to same-sex models. *Child Development, 46,* 849–856.

Slaby, R. G., Roedell, W. C., Arezzo, D., & Hendrix, K. (1995). *Early violence prevention.* Washington, DC: National Association for the Education of Young Children.

Slade, A., Belsky, J., Aber, J. L., & Phelps, J. L. (1999). Mothers' representations of their relationships with their toddlers: Links to adult attachment and observed mothering. *Developmental Psychology, 35,* 611–619.

Slater, A., Brown, E., Mattock, A., & Bornstein, M. H. (1996). Continuity and change in habituation in the first 4 months from birth. *Journal of Reproductive and Infant Psychology, 14,* 187–194.

Slater, A., Quinn, P. C., Brown, E., & Hayes, R. (1999). Intermodal perception at birth: Intersensory redundancy guides newborn infants' learning of arbitrary auditory–visual pairings. *Developmental Science, 2,* 333–338.

Slobin, D. I. (Ed.). (1997). *The crosslinguistic study of language acquisition: Vol. 5. Expanding the contexts* (pp. 265–324). Mahwah, NJ: Erlbaum.

Slusher, M. P., Mayer, C. J., & Dunkle, R. E. (1996). Gays and lesbians older and wiser (GLOW): A support group for older gay people. *Gerontologist, 36,* 118–123.

Small, B. J., & Bäckman, L. (1997). Cognitive correlates of mortality: Evidence from a population-based sample of very old adults. *Psychology and Aging, 12,* 309–313.

Small, B. J., Dixon, R. A., Hultsch, D. F., & Hertzog, C. (1999). Longitudinal changes in quantitative and qualitative indicators of word and story recall in young–old and old–old adults. *Journal of Gerontology, 54B,* P107–P115.

Small, M. (1998). *Our babies, ourselves.* New York: Anchor.

Smetana, J. G. (1995). Morality in context: Abstractions, ambiguities, and applications. In R. Vasta (Ed.), *Annals of child development* (Vol. 10, p. 83–130). London: Jessica Kingsley.

Smetana, J. G., & Braeges, J. L. (1990). The development of toddlers' moral and conventional judgments. *Merrill-Palmer Quarterly, 36,* 329–346.

Smiley, P. A., & Dweck, C. S. (1994). Individual differences in achievement goals among young children. *Child Development, 65,* 1723–1743.

Smith, A. E., Jussim, L., Eccles, J., VanNoy, M., Madon, S., & Palumbo, P. (1998). Self-fulfilling prophecies, perceptual biases, and accuracy at the individual and group levels. *Journal of Experimental Social Psychology, 34,* 530–561.

Smith, C., Perou, R., & Lesesne, C. (2002). Parent education. M. H. Bornstein (Ed.), *Handbook of parenting.* (Vol. 4, pp. 389–410). Mahwah, NJ: Erlbaum.

Smith, D. B., & Moen, P. (1998). Spousal influence on retirement: His, her, and their perceptions. *Journal of Marriage and the Family, 60,* 734–744.

Smith, D. C. (1993). The terminally ill patient's right to be in denial. *Omega, 27,* 115–121.

Smith, E. P., Walker, K., Fields, L., Brookins, C. C., & Seay, R. C. (1999). Ethnic identity and its relationship to self-esteem, perceived efficacy, and prosocial attitudes in early adolescence. *Journal of Adolescence, 22,* 867–880.

Smith, G. C., Kohn, S. J., Savage-Stevens, S. E., Finch, J. J., Ingate, R., & Lim, Y. (2000). The effects of interpersonal and personal agency on perceived control and psychological well-being in adulthood. *Gerontologist, 40,* 458–468.

Smith, J., & Baltes, P. B. (1999). Life-span perspectives on development. In M. H. Bornstein & M. E. Lamb (Eds.), *Developmental psychology: An advanced textbook* (4th ed., pp. 275–311). Mahwah, NJ: Erlbaum.

Smith, J., & Prior, M. (1995). Temperament and stress resilience in school-age children: A within-families study. *Journal of the American Academy of Child and Adolescent Psychiatry, 34,* 168–179.

Smith, J., Staudinger, U. M., & Baltes, P. B. (1994). Occupational settings facilitating wisdom-related knowledge: The sample case of clinical psychologists. *Journal of Consulting and Clinical Psychology, 66,* 989–999.

Smith, K. E., Landry, S. H., Swank, P. R., Baldwin, C. D., Denson, S. E., & Wildin, S. (1996). The relation of medical risk and maternal stimulation with preterm infants' development of cognitive, language and daily living skills. *Journal of Child Psychology and Psychiatry, 37,* 855–864.

Smith, L. B., Thelen, E., Titzer, R., & McLin, D. (1999). Knowing in the context of acting: The task dynamics of the A-not-B error. *Psychological Review, 106,* 235–260.

Smith, M. (Ed.). (2002). *Sex without consent.* New York: New York University Press.

Smith, M. L., Klim, P., & Hanley, W. B. (2000). Executive function in school-aged children with phenylketonuria. *Journal of Developmental and Physical Disabilities, 12,* 317–332.

Smith, P. (1991). Introduction: The study of grandparenthood. In P. K. Smith (Ed.), *The psychology of grandparenthood: An international perspective* (pp. 1–16). London: Routledge.

Smith, P., Perrin, S., Yule, W., & Rabe-Hesketh, S. (2001). War exposure and maternal reactions in the psychological adjustment of children from Bosnia-Hercegovina. *Journal of Child Psychology and Psychiatry and Allied Disciplines, 42,* 395–404.

Smith, P. K., & Drew, L. M. (2002). Grandparenthood. In M. H. Bornstein (Ed.), *Handbook of parenting, Vol. 3* (2nd ed., pp. 141–172). Mahwah, NJ: Erlbaum.

Smith, R. E., & Smoll, F. L. (1997). Coaching the coaches: Youth sports as a scientific and applied behavior setting. *Current Directions in Psychological Science, 6,* 16–21.

Smock, P. J., & Gupta, S. (2002). *Cohabitation in contemporary North America.* In A. Booth & A. C. Crouter (Eds.), *Just living together* (pp. 53–84). Mahwah, NJ: Erlbaum.

Smoll, F. L., & Smith, R. E. (Eds.). (1996). *Children and youth in sport: A biopsychological perspective.* Dubuque, IA: Brown & Benchmark.

Snarey, J. (1995). In a communitarian voice: The sociological expansion of Kohlbergian theory, research, and practice. In W. M. Kurtines & J. L. Gewirtz (Eds.), *Moral development: An introduction* (pp. 109–134). Boston: Allyn and Bacon.

Snarey, J., Son, L., Kuehne, V. S., Hauser, S., & Vaillant, G. (1987). The role of parenting in men's psychosocial development: A longitudinal study of early adulthood infertility and midlife generativity. *Developmental Psychology, 23,* 593–603.

Snarey, J. R., Reimer, J., & Kohlberg, L. (1985). The development of social–moral reasoning among kibbutz adolescents: A longitudinal cross-cultural study. *Developmental Psychology, 21,* 3–17.

Snidman, N., Kagan, J., Riordan, L., & Shannon, D. C. (1995). Cardiac function and behavioral reactivity. *Psychophysiology, 32,* 199–207.

Society for Research in Child Development (1993). Ethical standards for research with children. In *Directory of Members* (pp. 337–339). Ann Arbor, MI: Author.

Soderstrom, M., Dolbier, C., Leiferman, J., & Steinhardt, M. (2000). The relationship of hardiness, coping strategies, and perceived stress to symptoms of illness. *Journal of Behavioral Medicine, 23,* 311–328.

Soken, H. H., & Pick, A. D. (1992). Intermodal perception of happy and angry expressive behaviors by seven-month-old infants. *Child Development, 63,* 787–795.

Solomon, G. B., & Bredemeier, B. J. L. (1999). Children's moral conceptions of gender stratification in sport. *International Journal of Sport Psychology, 30,* 350–368.

Solomon, J. C., & Marx, J. (1995). "To grandmother's house we go": Health and school adjustment of children raised solely by grandparents. *Gerontologist, 35,* 386–394.

Somary, K., & Stricker, G. (1998). Becoming a grandparent: A longitudinal study of expectations and early experiences as a function of sex and lineage. *Gerontologist, 38,* 53–61.

Sonntag, W. E., Lynch, C., Thornton, P., Khan, A., Bennett, S., & Ingram, R. (2000). The effects of growth hormone and IGF-1 deficiency on cerebrovascular and brain ageing. *Journal of Anatomy, 197,* 575–585.

Sophian, C. (1995). Representation and reasoning in early numerical development: Counting, conservation, and comparisons between sets. *Child Development, 66,* 559–577.

Sorce, J., Emde, R., Campos, J., & Klinnert, M. (1985). Maternal emotional signaling: Its effect on the visual cliff behavior of 1-year-olds. *Developmental Psychology, 21,* 195–200.

Sosa, R., Kennell, J., Klaus, M., Robertson, S., & Urrutia, J. (1980). The effect of a supportive companion on perinatal problems, length of labor, and mother–infant interaction. *New England Journal of Medicine, 303,* 597–600.

Speece, M. W., & Brent, S. B. (1996). The development of children's understanding of death. In C. A. Corr & D. M. Corr (Eds.), *Handbook of childhood death and bereavement* (pp. 29–50). New York: Springer.

Speicher, B. (1994). Family patterns of moral judgment during adolescence and early adulthood. *Developmental Psychology, 30,* 624–632.

Spelke, E. S. (1987). The development of intermodal perception. In P. Salapatek & L. Cohen (Eds.), *Handbook of infant perception: Vol. 2. From perception to cognition* (pp. 233–273). Orlando, FL: Academic Press.

Spelke, E. S., & Newport, E. L. (1998). Nativism, empiricism, and the development of knowledge. In R. M. Lerner (Ed.), *Handbook of child psychology: Vol. 1. Theoretical models of human development* (5th ed., pp. 199–254). New York: Wiley.

Spence, M. J., & DeCasper, A. J. (1987). Prenatal experience with low-frequency maternal voice sounds influences neonatal perception of maternal voice samples. *Infant Behavior and Development, 10,* 133–142.

Spencer, J. P., Verejiken, B., Diedrich, F. J., & Thelen, E. (2000). Posture and the emergence of manual skills. *Developmental Science, 3,* 216–233.

Spira, A. (1992). *Les comportements sexuels en France.* Paris: La documentation Française.

Spitz, R. A. (1946). Anaclitic depression. *Psychoanalytic Study of the Child, 2,* 313–342.

Spock, B., & Parker, S. J. (1998). *Dr. Spock's baby and child care* (7th ed.). New York: Pocket Books.

Sprecher, S. (1999). "I love you more today than yesterday": Romantic partners' perceptions of changes in love and related affect over time. *Journal of Personality and Social Psychology, 76,* 46–53.

Springer, C. A., & Lease, S. H. (2000). The impact of multiple AIDS-related bereavement in the gay male population. *Journal of Counseling and Development, 78,* 297–304.

Sridhar, D., & Vaughn, S. (2001). Social functioning of students with learning disabilities. In D. P. Hallahan & B. K. Keogh (Eds.), *Research and global perspectives in learning disabilities* (pp. 65–91). Mahwah, NJ: Erlbaum.

Sroufe, L. A., Egeland, B., & Kreutzer, T. (1990). The fate of early experience following developmental change: Longitudinal approaches to individual adaptation. *Child Development, 61,* 1363–1373.

Sroufe, L. A., & Waters, E. (1976). The ontogenesis of smiling and laughter: A perspective on the organization of development in infancy. *Psychological Review, 83,* 173–189.

Sroufe, L. A., & Wunsch, J. P. (1972). The development of laughter in the first year of life. *Child Development, 43,* 1324–1344.

Stamler, J., Stamler, R., Neaton, J. D., Wentworth, D., Daviglus, M. L., Garside, D., Dyer, A. R., Liu, K. A., & Greenland, P. (1999). Low risk-factor profile and long-term cardiovascular and noncardiovascular mortality and life expectancy: Findings for five large cohorts of young adult and middle-aged men and women. *Journal of the American Medical Association, 282,* 2012–2018.

Stams, G. J. M., Juffer, F., & van IJzendoorn, M. H. (2002). Maternal sensitivity, infant attachment, and temperament in early childhood predict adjustment in middle childhood: The case of adopted children and their biologically unrelated parents. *Developmental Psychology, 38,* 806–821.

Standley, J. M. (1998). The effect of music and multimodal stimulation on responses of premature infants in neonatal intensive care. *Pediatric Nursing, 24,* 532–538.

Stanley, B., & Seiber, J. E. (Eds.). (1992). *Social research on children and adolescents: Ethical issues.* Newbury Park, CA: Sage.

Statistics Canada. (1999a, Winter). Life expectancy. *Health Reports, 11*(3).

Statistics Canada. (1999b). Statistical report on the health of Canadians. Retrieved from http://www.hc-sc.gc.ca/hppb/phdd/report/stat/report.html

Statistics Canada. (2000). *Immigrant Youth in Canada.* Retrieved from http://www.ccsd.ca

Statistics Canada. (2001a). Births: 1999. Retrieved from http://www.statcan.ca/Daily/English/011210/d011210b.htm

Statistics Canada. (2001b). *Education in Canada.* Ottawa: Statistics Canada.

Statistics Canada. (2001c). *Families and household living arrangements: Highlight tables 2001.* Ottawa: Census Operations Division.

Statistics Canada. (2001d). *Family violence in Canada: A statistical profile.* Ottawa: National Clearinghouse on Family Violence.

Statistics Canada. (2001e). Population by knowledge of official languages, showing age groups, for Canada, provinces, and territories, 1996 Census. Retrieved from http://www.statcan.ca/english/census96/dec2/off.htm

Statistics Canada. (2001f, January 25). Television viewing. *The Daily.* Retrieved from http://www.statcan.ca/Daily/English/010125/d010125a.htm

Statistics Canada. (2001g). *Television viewing data bank.* Ottawa: Author.

Statistics Canada. (2002a). *Changing conjugal life in Canada.* Ottawa: Statistics Canada.

Statistics Canada. (2002b). Crime in Canada. *Juristat, 22*(6). Retrieved from http://www.statcan.ca

Statistics Canada. (2002c). Divorces. Retrieved from http://www.statcan.ca

Statistics Canada. (2002d, December 2). Divorces. *The Daily.* Retrieved from http://www.statcan.ca/Daily/English/021202/d021202f.htm

Statistics Canada. (2002f). Family studies kit. Retrieved from http://www.statcan.ca/english/kits/Family/intro.htm

Statistics Canada. (2002g). Family studies kit. Retrieved from http://www.statcan.ca/english/kits/Family/pdf/ch3_3e.pdf

Statistics Canada. (2002h). *Labour force historical review.* Catalogue No. 71F0004. Ottawa: Author.

Statistics Canada. (2002i). *Mortality, summary list of causes—shelf tables.* Ottawa: Health Statistics Division.

Statistics Canada. (2002j). Number of income recipients and their average income in constant dollars by sex and age groups, for Canada, provinces and territories. Retrieved from http://www.statcan.ca/english/census96/may12/t1.htm

Statistics Canada. (2002k). Population. Retrieved from http://www.statcan.ca

Statistics Canada. (2002l). Population 15 years and over by hours spent on unpaid housework. Retrieved from http://www.statcan.ca/english/Pgdb/famil56a.htm

Statistics Canada. (2002m). Robert Glossup on the Canadian family. Retrieved from http://www.statcan.ca/english/ads/11-008-XPE/family.html

Statistics Canada. (2002n). Statistical report on the health of Canadians. Retrieved from http://www.statcan.ca:80/english/freepub/82-570-XIE/partb.htm

Statistics Canada. (2002o, January 23). Youth in transition survey, 2000. *The Daily.* Retrieved from http://www.statcan.ca/Daily/English/000928/d000928b.htm

Statistics Canada (2002p). 2001 Census [Online]. Available: www12.statcan.ca/English/census01/products/analytic/companion/fam/provs.cfm

Statistics Canada. (2003a). Age groups, number of grandparents, and sex for grandchildren living with grandparents with no parent present, 2001. Retrieved from http://www12.statcan.ca/English/census01

Statistics Canada. (2003b). Aging in rural communities and small towns. *Expression, 9*(1). Retrieved from http://www.hc-sc.gc.ca/seniors-aines/pubs/expression/9-1/exp-9-1-e.htm

Statistics Canada. (2003c). Canada's ethnocultural portrait: The changing mosaic. Retrieved from http://www12.statcan.ca/English/census01/products/analytic/companion/etoimm/Canada.cfm

Statistics Canada. (2003d). Population 15 years and over by marital status, showing selected age groups and sex, for Canada, provinces, and territories. Retrieved from http://www.statcan.ca/english/census96/oct14/mar1.htm

Stattin, H., & Magnusson, D. (1996). Leaving home at an early age among females. In J. A. Graber & J. S. Dubas (Eds.), *New directions for child development* (No. 71, pp. 53–69). San Francisco: Jossey-Bass.

Stattin, H., & Magnusson, D. (1990). *Pubertal maturation in female development.* Hillsdale, NJ: Erlbaum.

Staub, E. (1996). Cultural–societal roots of violence. *American Psychologist, 51,* 117–132.

Staudinger, U. M. (1999). Older and wiser? Integrating results on the relationship between age and wisdom-related peformance. *International Journal of Behavioral Development, 23,* 641–664.

Staudinger, U. M., Fleeson, W., & Baltes, P. B. (1999). Predictors of subjective physical health and global well-being: Similarities and differences between the United States and Germany. *Journal of Personality and Social Psychology, 76,* 305–319.

Staudinger, U. M., Maciel, A. G., Smith, J., & Baltes, P. B. (1998). What predicts wisdom-related performance? A first look at personality, intelligence, and facilitative experiential contexts. *European Journal of Personality, 12,* 1–17.

Staudinger, U. M., Smith, J., & Baltes, P. B. (1992). Wisdom-related knowledge in a life-review task: Age differences and the role of professional specialization. *Psychology and Aging, 7,* 271–281.

Stein, C. H., Wemmerus, V. A., Ward, M., Gaines, M. E., Freeberg, A. L., & Jewell, T. C. (1998). "Because they're my parents": An intergenerational study of felt obligation and parental caregiving. *Journal of Marriage and the Family, 60,* 611–622.

Stein, J. H., & Reiser, L. W. (1994). A study of white middle-class adolescent boys' responses to "semenarche" (the first ejaculation). *Journal of Youth and Adolescence, 23,* 373–384.

Stein, N., & Levine, L. J. (1999). The early emergence of emotional understanding and appraisal: Implications for theories of development. In T. Dalgleish & M. J. Power (Eds.), *Handbook of cognition and emotion* (pp. 383–408). Chichester, U.K.: Wiley.

Steinberg, L. (1984). The varieties and effects of work during adolescence. In M. Lamb, A. Brown, & B. Rogoff (Eds.), *Advances in developmental psychology* (pp. 1–37). Hillsdale, NJ: Erlbaum.

Steinberg, L. (2001). We know some things: Parent–adolescent relationships in retrospect and prospect. *Journal of Research on Adolescence, 11,* 1–19.

Steinberg, L., Darling, N. E., & Fletcher, A. C. (1995). Authoritative parenting and adolescent development: An ecological journey. In P. Moen, G. H. Elder, Jr., & K. Luscher (Eds.), *Examining lives in context* (pp. 423–466). Washington, DC: American Psychological Association.

Steinberg, L., Fletcher, A., & Darling, N. (1994). Parental monitoring and peer influences on adolescent substance use. *Pediatrics, 93,* 1060–1064.

Steinberg, L., & Morris, A. S. (2001). Adolescent development. *Annual Review of Psychology, 52,* 83–110.

Steinberg, L., & Silk, J. S. (2002). Parenting adolescents. In M. H. Bornstein (Ed.), *Handbook of parenting* (Vol. 1, pp. 103–134). Mahwah, NJ: Erlbaum.

Steinberg, L., & Silverberg, S. (1986). The vicissitudes of autonomy in early adolescence. *Child Development, 57,* 841–851.

Steinberg, L. D. (1986). Latchkey children and susceptibility to peer pressure: An ecological analysis. *Developmental Psychology, 22,* 433–439.

Steinberg, L. D., Darling, N. E., & Fletcher, A. C. (1995). Authoritative parenting and adolescent development: An ecological journey. In P. Moen, G. H. Elder, Jr., & K. Luscher (Eds.), *Examining lives in context* (pp. 423–466). Washington, DC: American Psychological Association.

Steinberg, L. D., Fegley, S., & Dornbusch, S. (1993). Negative impact of part-time work on adolescent adjustment: Evidence from a longitudinal study. *Developmental Psychology, 29,* 171–180.

Steinberg, L. D., Lamborn, S. D., Darling, N., Mounts, N. S., & Dornbusch, S. M. (1994). Over-time changes in adjustment and competence among adolescents from authoritative, authoritarian, indulgent, and neglectful families. *Child Development, 65,* 754–770.

Steinberg, S., & Bellavance, F. (1999). Characteristics and treatment of women with antenatal and postpartum depression. *International Journal of Psychiatry and Medicine, 29,* 209–233.

Steiner, J. E. (1979). Human facial expression in response to taste and smell stimulation. In H. W. Reese & L. P. Lipsitt (Eds.), *Advances in child development and behavior* (Vol. 13, pp. 257–295). New York: Academic Press.

Steinhauser, K. E., Clipp, E. C., McNeilly, M., Christakis, N. A., McIntyre, L. M., & Tulsky, J. A. (2000). In search of a good death: Observations of patients, families, and providers. *Annals of Internal Medicine, 132,* 825–832.

Stenberg, C., & Campos, J. (1990). The development of anger expressions in infancy. In N. Stein, B. Leventhal, & T. Trabasso (Eds.), *Psychological and biological approaches to emotion* (pp. 247–282). Hillsdale, NJ: Erlbaum.

Stephens, M. A. P., & Franks, M. M. (1999). Parent care in the context of women's multiple roles. *Current Directions in Psychological Science, 8,* 149–152.

Stephens, M. A. P., Franks, M. M., & Atienza, A. A. (1997). Where two roles intersect: Spillover between parent care and employment. *Psychology and Aging, 12,* 30–37.

Stephens, M. A. P., Townsend, A. L., Martire, L. M., & Druley, A. (2001). Balancing parent care with other roles: Interrole conflict of adult daughter caregivers. *Journal of Gerontology, 56B,* P24–P34.

Stern, M., & Karraker, K. H. (1989). Sex stereotyping of infants: A review of gender labeling studies. *Sex Roles, 20,* 501–522.

Sternberg, R. J. (1985). *Beyond IQ: A triarchic theory of human intelligence.* New York: Cambridge University Press.

Sternberg, R. J. (1987). Liking versus loving: A comparative evaluation of theories. *Psychological Bulletin, 102,* 331–345.

Sternberg, R. J. (1988). Triangulating love. In R. J. Sternberg & M. L. Barnes (Eds.), *The psychology of love* (pp. 119–138). New Haven, CT: Yale University Press.

Sternberg, R. J. (1997). *Successful intelligence.* New York: Plume.

Sternberg, R. J. (1999). A triarchic approach to understanding and assessment of intelligence in multicultural populations. *Journal of School Psychology, 37,* 145–159.

Sternberg, R. J. (2001). Why schools should teach for wisdom: The balance theory of wisdom in educational settings. *Educational Psychologist, 36,* 227–245.

Sternberg, R. J., Forsythe, G. B., Hedlund, J., Horvath, J. A., Wagner, R. K., Williams, W. M., Snook, S. A., & Grigorenko, E. L. (2000). *Practical intelligence in everyday life.* Cambridge, U.K.: Cambridge University Press.

Sternberg, R. J., & Lubart, T. I. (1995). *Defying the crowd.* New York: Basic Books.

Sternberg, R. J., & Lubart, T. I. (1996). Investing in creativity. *American Psychologist, 51,* 677–688.

Sternberg, R. J., & Lubart, T. I. (2001). Wisdom and creativity. In J. E. Birren & K. W. Schaie (Eds.), *Handbook of the psychology of aging* (pp. 500–522). San Diego: Academic Press.

Sterns, H. L., & Gray, J. H. (1999). Work, leisure, and retirement. In J. C. Cavanaugh & S. K. Whitbourne (Eds.), *Gerontology: An interdisciplinary perspective* (pp. 355–390). New York: Oxford University Press.

Sterns, H. L., & Huyck, M. H. (2001). The role of work in midlife. In M. E. Lachman (Ed.), *Handbook of midlife development* (pp. 447–486). New York: Wiley.

Stettler, N., Zemel, B. S., Kumanyika, S., & Sallings, V. A. (2002). Infant weight gain and childhood overweight status in a multicenter, cohort study. *Pediatrics, 109,* 194–199.

Stevens, J. C., & Cruz, L. A. (1996). Spatial acuity of touch: Ubiquitous decline with aging revealed by repeated threshold testing. *Somatosensory and Motor Research, 13,* 1–10.

Stevens, J. C., Cruz, L. A., Hoffman, J. M., & Patterson, M. Q. (1995). Taste sensitivity and aging: High incidence of decline revealed by repeated threshold measures. *Chemical Senses, 20,* 451–459.

Stevens, J. C., Cruz, L. A., Marks, L. E., & Lakatos, S. (1998). A multimodal assessment of sensory thresholds in aging. *Journal of Gerontology, 53B,* P263–P272.

Stevens, J. C., Foulke, E., & Patterson, M. Q. (1996). Tactile acuity, aging, and Braille reading in long-term blind adults. *Journal of Experimental Psychology: Applied, 2,* 91–106.

Stevens, M. L. T. (1997, March–April). What Quinlan can tell Kevorkian about the right to die. *The Humanist, 57*(2), 10–14.

Stevenson, H. W. (1992, December). Learning from Asian schools. *Scientific American, 267*(6), 32–38.

Stevenson, H. W., & Lee, S.-Y. (1990). Contexts of achievement: A study of American, Chinese, and Japanese children. *Monographs of the Society for Research in Child Development, 55*(1–2, Serial No. 221).

Stevenson, H. W., Lee, S., & Mu, X. (2000). Successful achievement in mathematics: China and the United States. In C. F. M. van Lieshout & P. G. Heymans (Eds.), *Developing talent across the life-span* (pp. 167–183). Philadelphia: Psychology Press.

Stevenson, M. R., & Black, K. N. (1995). *How divorce affects offspring: A research approach.* Dubuque, IA: Brown & Benchmark.

Stevenson, R., & Pollitt, C. (1987). The acquisition of temporal terms. *Journal of Child Language, 14,* 533–545.

Steward, D. K. (2001). Behavioral characteristics of infants with nonorganic failure to thrive during a play interaction. *American Journal of Maternal Child Nursing, 26,* 79–85.

Stewart, A. J., Ostrove, J. M., & Helson, R. (2001). Middle aging in women: Patterns of personality change from the 30s to the 50s. *Journal of Adult Development, 8,* 23–37.

Stewart, A. J., & Vandewater, E. A. (1992, August). *Combining tough and tender methods to study women's lives.* Paper presented at the annual meeting of the American Psychological Association, Washington, DC.

Stewart, A. J., & Vandewater, E. A. (1998). The course of generativity. In D. P. McAdams & E. de St. Aubin (Eds.), *Generativity and adult development* (pp. 75–100). Washington, DC: American Psychological Association.

Stewart, A. L., Verboncoeur, C. J., McLellan, B. Y., Gillis, D. E., Rush, S., & Mills, K. M. (2001). Physical activity outcomes of CHAMPS II: A physical activity promotion program for older adults. *Journal of Gerontology, 56A,* M465–M470.

Stewart, P., Reihman, J., Lonky, E., Darvill, T., & Pagano, J. (2000). Prenatal PCB exposure and neonatal behavioral assessment scale (NBAS) performance. *Neurotoxicology and Teratology, 22,* 21–29.

Stewart, R. B. (1983). Sibling attachment relationships: Child–infant interactions in the Strange Situation. *Developmental Psychology, 19,* 192–199.

Stewart, R. B., Kozak, A. L., Tingley, L. M., Goddard, J. M., Blake, E. M., & Cassel, W. A. (2001). Adult sibling relationships: Validation of a typology. *Personal Relationships, 8,* 299–324.

Stewart, S., Stinnett, H., & Rosenfeld, L. B. (2000). Sex differences in desired characteristics of short-term and long-term relationship partners. *Journal of Social and Personal Relationships, 17,* 843–853.

Stewart, T. L., & Ralston, S. H. (2000). Role of genetic factors in the pathogenesis of osteoporosis. *Journal of Endocrinology, 166,* 235–245.

Stice, E., & Barrera, M., Jr. (1995). A longitudinal examination of the reciprocal relations between perceived parenting and adolescents' substance use and externalizing behaviors. *Developmental Psychology, 31,* 322–334.

Stichick, T. (2001). The psychosocial impact of armed conflict on children. *Child and Adolescent Psychiatric Clinics of North America, 10,* 797–814.

Stifter, C. A., Coulehan, C. M., & Fish, M. (1993). Linking employment to attachment: The mediating effects of maternal separation anxiety and interactive behavior. *Child Development, 64,* 1451–1460.

Stiles, J. (2001a). Neural plasticity in cognitive development. *Developmental Neuropsychology, 18,* 237–272.

Stiles, J. (2001b). Spatial cognitive development. In C. A. Nelson & M. Luciana (Eds.), *Handbook of developmental cognitive neuroscience* (pp. 399–414). Cambridge, MA: MIT Press.

Stiles, J., Bates, E. A., Thai, D., Trauner, D., & Reilly, J. (1998). Linguistic, cognitive, and affective development in children with pre- and perinatal focal brain injury: A ten-year overview from the San Diego Longitudinal Project. In C. Rovee-Collier, L. P. Lipsitt, & H. Hayne (Eds.), *Advances in infancy research* (pp. 131–163). Stamford, CT: Ablex.

Stillion, J. M. (1995). Death in the lives of adults: Responding to the tolling of the bell. In H. Wass & R. A. Neimeyer (Eds.), *Dying: Facing the facts* (pp. 303–322). New York: Taylor & Francis.

Stine-Morrow, E. A. L., & Miller, L. M. S. (1999). Basic cognitive processes. In J. C. Cavanaugh & S. K. Whitbourne (Eds.), *Gerontology: An interdisciplinary perspective* (pp.186–212). New York: Oxford University Press.

Stipek, D. (1995). The development of pride and shame in toddlers. In J. P. Tangney & K. W. Fischer (Eds.), *Self-conscious emotions* (pp. 237–252). New York: Guilford.

Stipek, D. (2002). At what age should children enter kindergarten? A question for policy makers and parents. *Social Policy Report of the Society for Research in Child Development, 16*(3).

Stipek, D. J., & Byler, P. (1997). Early childhood education teachers: Do they practice what they preach? *Early Childhood Research Quarterly, 12,* 305–326.

Stipek, D., & Byler, P. (2001). Academic achievement and social behaviors associated with age of entry into kindergarten. *Journal of Applied Developmental Psychology, 22,* 175–189.

Stipek, D. J., Feiler, R., Daniels, D., & Milburn, S. (1995). Effects of different instructional approaches on young children's achievement and motivation. *Child Development, 66,* 209–223.

Stipek, D. J., Gralinski, J. H., & Kopp, C. B. (1990). Self-concept development in the toddler years. *Developmental Psychology, 26,* 972–977.

Stith, S. M., & Farley, S. C. (1993). A predictive model of male spousal violence. *Journal of Family Violence, 8,* 183–201.

Stoch, M. B., Smythe, P. M., Moodie, A. D., & Bradshaw, D. (1982). Psychosocial outcome and CT findings after growth undernourishment during infancy: A 20-year developmental study. *Developmental Medicine and Child Neurology, 24,* 419–436.

Stocker, C., & Dunn, J. (1994). Sibling relationships in childhood and adolescence. In J. C. DeFries, R. Plomin, & D. W. Fulker (Eds.), *Nature and nurture in middle childhood* (pp. 214–232). Cambridge, MA: Blackwell.

Stodolsky, S. S. (1988). *The subject matters.* Chicago: University of Chicago Press.

Stoel-Gammon, C., & Otomo, K. (1986). Babbling development of hearing-impaired and normal hearing subjects. *Journal of Speech and Hearing Disorders, 51,* 33–41.

Stone, M. R., & Brown, B. B. (1999). Identity claims and projections: Descriptions of self and crowds in secondary school. In J. A. McLellan & M. J. V. Pugh (Eds.), *The role of peer groups in adolescent social identity: Exploring the importance of stability and change* (pp. 7–20). San Francisco: Jossey-Bass.

Stone, R. G., Staisey, N., & Sonn, U. (1991). Systems for delivery of assistive equipment to elders in Canada, Sweden, and the United States. *International Journal of Technology and Aging, 4,* 129–140.

Stoney, C. M., & Engebretson, T. O. (1994). Anger and hostility: Potential mediators of the gender difference in coronary artery disease. In A. W. Siegman & T. W. Smith (Eds.), *Anger, hostility, and the heart* (pp. 215–238). Hillsdale, NJ: Erlbaum.

Stormshak, E. A., Bierman, K. L., Bruschi, C., Dodge, K. A., & Coie, J. D. (1999). The relation between behavior problems and peer preference in different classroom contexts. *Child Development, 70,* 169–182.

Stormshak, E. A., Bierman, K. L., McMahon, R. J., Lengua, L. J., & the Conduct Problems Prevention Research Group. (2000). Parenting practices and child disruptive behavior problems in early elementary school. *Journal of Clinical Child Psychology, 29,* 17–29.

Strain, L. A., Grabusic, C. C., Searle, M. S., & Dunn, N. J. (2002). Continuing and ceasing leisure activities in later life: A longitudinal study. *Gerontologist, 42,* 217–223.

Strange, C. J. (1996). *Coping with arthritis in its many forms.* Washington, DC: U.S. Government Printing Office. (Reprint from *FDA Consumer Magazine.)*

Strapp, C. M., & Federico, A. (2000). Imitations and repetitions: What do children say following recasts? *First Language, 20,* 273–290.

Stratton, J. R., Levy, W. C., Cereueira, M. D., Schwartz, R. S., & Abrass, I. B. (1994). Cardiovascular responses to exercise: Effects of aging and exercise training in healthy men. *Circulation, 89,* 1648–1655.

Straus, M. A. (1999). The controversy over domestic violence by women: A methodological, theoretical, and sociology of science analysis. In X. B. Arriaga & S. Oskamp (Eds.), *Violence in intimate relationships* (pp.17–44). Thousand Oaks, CA: Sage.

Straus, M. A., & Stewart, J. H. (1999). Corporal punishment by American parents: National data on prevalence, chronicity, severity, and duration, in relation to child and family characteristics. *Clinical Child and Family Psychology Review, 2,* 55–70.

Strawbridge, W. J., Shema, S. J., Cohen, R. D., & Kaplan, G. A. (2001). Religious attendance increases survival by improving and maintaining good health behaviors. *Annals of Behavioral Medicine, 23,* 68–74.

Strayer, B. K., Tofler, I. R., & Lapchick, R. (1998). A developmental overview of child and youth sports in society. *Child and Adolescent Psychiatric Clinics of North America, 7,* 697–719.

Streissguth, A. P., Barr, H. M., Sampson, P. D., & Bookstein, F. L. (1994). Prenatal alcohol and offspring development: The first fourteen years. *Drug and Alcohol Dependence, 36,* 89–99.

Streissguth, A. P., Barr, H. M., Sampson, P. D., Darby, B. L., & Martin, D. C. (1989). IQ at age 4 in relation to maternal alcohol use and smoking during pregnancy. *Developmental Psychology, 25,* 3–11.

Streissguth, A. P., Treder, R., Barr, H. M., Shepard, T., Bleyer, W. A., Sampson, P. D., & Martin, D. G. (1987). Aspirin and acetaminophen use by pregnant women and subsequent child IQ and attention decrements. *Teratology, 35,* 211–219.

Striano, T., & Rochat, P. (2000). Emergence of selective social referencing in infancy. *Infancy, 1,* 253–264.

Strickland, C. J. (1997). Suicide among American Indian, Alaskan Native, and Canadian Aboriginal youth: Advancing the research agenda. *International Journal of Mental Health, 25,* 11–32.

Stroebe, M. (1998). New directions in bereavement research: Exploration of gender differences. *Palliative Medicine, 12,* 5–12.

Stroebe, M., & Schut, H. (1999). The dual process model of coping with bereavement: Rationale and description. *Death Studies, 23,* 197–224.

Stroebe, M., & Schut, H. (2001). Models of coping with bereavement: A review. In M. S. Stroebe, R. O. Hansson, W. Stroebe, & H. Schut (Eds.), *Handbook of bereavement research* (pp. 375–403). Washington, DC: American Psychological Association.

Stroebe, W., & Stroebe, M. S. (1993). Determinants of adjustment to bereavement in younger widows and widowers. In M. S. Stroebe, W. Stroebe, & R. O. Hansson (Eds.), *Handbook of bereavement* (pp. 208–226). New York: Cambridge University Press.

Stroebe, W., Stroebe, M., Abakoumkin, G., & Schut, H. (1996). The role of loneliness and social support in adjustment to loss: A test of attachment versus stress theory. *Journal of Personality and Social Psychology, 70,* 1241–1249.

Stroh, L. K., Brett, J. M., & Reilly, A. H. (1996). Family structure, glass ceiling, and traditional explanations for the differential rate of turnover of female and male managers. *Journal of Vocational Behavior, 49,* 99–118.

Stromswold, K. (1995). The acquisition of subject and object questions. *Language Acquisition, 4,* 5–48.

Strouse, D. L. (1999). Adolescent crowd orientations: A social and temporal analysis. In J. A. McLellan & M. J. V. Pugh (Eds.), *The role of peer groups in adolescent social identity: Exploring the importance of stability and change* (pp. 37–54). San Francisco: Jossey-Bass.

Stryker, J., Coates, T. J., DeCarlo, P., Haynes-Sanstad, K., Shriver, M.,

& Makadon, H. J. (1995). Prevention of HIV infection: Looking back, looking ahead. *Journal of the American Medical Association, 273,* 1143–1148.

Stuart, M., & Weinrich, M. (2001). Home- and community-based long-term care: Lessons from Denmark. *Gerontologist, 41,* 474–480.

Stull, D., & Scarisbrick-Hauser, A. (1989). Never-married elderly. *Research on Aging, 11,* 124–139.

Stunkard, A. J., & Sørensen, T. I. A. (1993). Obesity and socioeconomic status—a complex relation. *New England Journal of Medicine, 329,* 1036–1037.

Stunkard, A. J., Sørenson, T. I. A., Hanis, C., Teasdale, T. W., Chakraborty, R., Schull, W. J., & Schulsinger, F. (1986). An adoption study of human obesity. *New England Journal of Medicine, 314,* 193–198.

Stylianos, S. K., & Vachon, M. L. S. (1993). The role of social support in bereavement. In M. S. Stroebe, W. Stroebe, & R. O. Hansson (Eds.), *Handbook of bereavement* (pp. 397–410). New York: Cambridge University Press.

Suarez-Orozco, C., & Suarez-Orozco, M. M. (1995). *Transformation: Immigration, family life, and achievement motivation among Latino adolescents.* Stanford, CA: Stanford University Press.

Subbotsky, E. V. (1994). Early rationality and magical thinking in preschoolers: Space and time. *British Journal of Developmental Psychology, 12,* 97–108.

Subrahmanyam, K., & Greenfield, P. M. (1996). Effect of video game practice on spatial skills in girls and boys. In P. M. Greenfield & R. R. Cocking (Eds.), *Interacting with video* (pp. 95–114). Norwood, NJ: Ablex.

Suitor, J. J., Pillemer, K., Keeton, S., & Robison, J. (1995). Aging parents and aging children: Determinants of relationships quality. In R. Blieszner & V. H. Bedford (Eds.), *Handbook of aging and the family* (pp. 223–242). Westport, CT: Greenwood Press.

Sullivan, M. A. (1995). May the circle be unbroken: The African-American experience of death, dying, and spirituality. In J. K. Parry & A. S. Ryan (Eds.), *A cross-cultural look at death, dying, and religion* (pp. 160–171). Chicago: Nelson-Hall.

Sullivan, S. A., & Birch, L. L. (1990). Pass the sugar, pass the salt: Experience dictates preference. *Developmental Psychology, 26,* 546–551.

Suls, J., & Mullen, B. (1982). From the cradle to the grave: Comparison and self-evaluation across the life span. In J. Suls (Ed.), *Psychological perspectives on the self* (Vol. 1, pp. 97–128). Hillsdale, NJ: Erlbaum.

Sundell, H. (2001). Why does maternal smoke exposure increase the risk of sudden infant death syndrome? *Acta Paediatrica, 90,* 718–720.

Super, C. M. (1981). Behavioral development in infancy. In R. H. Monroe, R. L. Monroe, & B. B. Whiting (Eds.), *Handbook of cross-cultural human development* (pp. 181–270). New York: Garland.

Super, D. (1980). A life-span, life-space approach to career development. *Journal of Vocational Behavior, 16,* 282–298.

Super, D. (1984). Career and life development. In D. Brown & L. Brooks (Eds.), *Career choice and development* (pp. 192–234). San Francisco: Jossey-Bass.

Sureau, C. (1997). Trials and tribulations of surrogacy: From surrogacy to parenthood. *Human Reproduction, 12,* 410–411.

Sutcliffe, A. G. (2002). Health risks in babies born after assisted reproduction. *British Medical Journal, 325,* 117–118.

Suzuki, L. A., & Valencia, R. R. (1997). Race–ethnicity and measured intelligence: Educational implications. *American Psychologist, 52,* 1103–1114.

Swanson, J. L., & Fouad, N. A. (1999). Applying theories of person–environment fit to the transition from school to work. *Career Development Quarterly, 47,* 337–347.

Swanson, N. G., Piotrkowski, C. S., Keita, G. P., & Becker, A. B. (1997). Occupational stress and women's health. In S. J. Gallant, G. P. Keita, & R. Royak-Schaler (Eds.), *Health care for women* (pp. 147–159). Washington, DC: American Psychological Association.

Swendsen, J. D., & Mazure, C. M. (2000). Life stress as a risk factor for postpartum depression: Current research and methodological issues. *Clinical Psychology—Science and Practice, 7,* 17–31.

Swindell, R., & Thompson, J. (1995). An international perspective on the University of the Third Age. *Educational Gerontology, 21,* 429–447.

Szepkouski, G. M., Gauvain, M., & Carberry, M. (1994). The development of planning skills in children with and without mental retardation. *Journal of Applied Developmental Psychology, 15,* 187–206.

Szinovacz, M. (1998). Grandparents today: A demographic profile. *Gerontologist, 38,* 37–52.

Szkrybalo, J., & Ruble, D. N. (1999). "God made me a girl": Sex-category constancy judgments and explanations revisited. *Developmental Psychology, 35,* 392–402.

Taddio, A., Katz, J., Ilersich, A. L., & Koren, G. (1997). Effect of neonatal circumcision on pain response during subsequent routine vaccination. *Lancet, 349,* 599–603.

Tager-Flusberg, H. (2001). Putting words together: Morphology and syntax in the preschool years. In J. Berko Gleason (Ed.), *The development of language* (4th ed., pp. 159–209). Boston: Allyn and Bacon.

Takahashi, K. (1990). Are the key assumptions of the "Strange Situation" procedure universal? A view from Japanese research. *Human Development, 33,* 23–30.

Takamura, J., & Williams, B. (2002). *Informal caregiving: Compassion in action.* Arlington, TX: Arc of the United States.

Tamis-LeMonda, C. S., & Bornstein, M. H. (1989). Habituation and maternal encouragement of attention in infancy as predictors of toddler language, play, and representational competence. *Child Development, 60,* 738–751.

Tamis-LeMonda, C. S., & Bornstein, M. H. (1994). Specificity in mother–toddler language–play relations across the second year. *Developmental Psychology, 30,* 283-292.

Tangney, J. P. (2001). Constructive and destructive aspects of shame and guilt. In A. C. Bohart & D. J. Stipek (Eds.), *Constructive and destructive behavior* (pp. 127–145). Washington, DC: American Psychological Association.

Tangri, S. S., & Jenkins, S. R. (1997). Why expecting conflict is good. *Sex Roles, 36,* 725–746.

Tanner, J. M. (1990). *Foetus into man* (2nd ed.). Cambridge, MA: Harvard University Press.

Tanner, J. M., Healy, M., & Cameron, N. (2001). *Assessment of skeletal maturity and prediction of adult height* (3rd ed.). Philadelphia: Saunders.

Tardif, T., Gelman, S. A., & Xu, F. (1999). Putting the "noun bias" in context: A comparison of English and Mandarin. *Child Development, 70,* 620–635.

Taylor, J. H., & Walker, L. J. (1997). Moral climate and the development of moral reasoning: The effects of dyadic discussions between young offenders. *Journal of Moral Education, 26,* 21–43.

Taylor, M. C., & Hall, J. A. (1982). Psychological androgyny: Theories, methods, and conclusions. *Psychological Bulletin, 92,* 347–366.

Taylor, R. D., & Roberts, D. (1995). Kinship support and maternal and adolescent well-being in economically disadvantaged African-American families. *Child Development, 66,* 1585–1597.

Taylor, R. L. (2000). Diversity within African-American families. In D. H. Demo & K. R. Allen (Eds.), *Handbook of family diversity* (pp. 232–251). New York: Oxford University Press.

Teller, D. Y. (1997). First glances: The vision of infants. *Investigative Ophthalmology and Visual Science, 38,* 2183–2203.

Teller, D. Y. (1998). Spatial and temporal aspects of infant color vision. *Vision Research, 38,* 3275–3282.

Tellings, A. (1999). Psychoanalytical and genetic-structuralistic approaches to moral development: Incompatible views? *Psychoanalytic Review, 86,* 903–914.

Temple, C. M., & Carney, R. A. (1995). Patterns of spatial functioning in Turner's syndrome. *Cortex, 31,* 109–118.

Terenzini, P. T., Pascarella, E. T., & Blimling, G. S. (1999). Students' out-of-class experiences and their influence on learning and cognitive development: A literature review. *Journal of College Student Development, 40,* 610–623.

Terestchenko, N. Y., Lyaginskaya, A. M., & Burtzeva, L. I. (1991). Stochastic, non-stochastic effects and some population-genetic characteristics in children of the critical group in period of basic organogenesis. In *The scientific and practical aspects of preservation of health of the people exposed to radiation influence as a result of the accident at the Chernobyl atomic power station* (in Russian) (pp. 73–74). Minsk: Publishing House of Belarussian Committee "Chernobyl Children."

Tertinger, D. A., Greene, B. F., & Lutzker, J. R. (1984). Home safety: Development and validation of one component of an ecobehavioral treatment program for abused and neglected children. *Journal of Applied Behavior Analysis, 17,* 159–174.

Tetens, J. N. (1777). *Philosophische Versuche über die menschliche Natur und ihre Entwicklung.* Leipzig: Weidmanns Erben & Reich.

Teti, D. M., Gelfand, D. M., Messinger, D. S., & Isabella, R. (1995). Maternal depression and the quality of early attachment: An examination of infants, preschoolers, and their mothers. *Developmental Psychology, 31,* 364–376.

Teti, D. M., Saken, J. W., Kucera, E., & Corns, K. M. (1996). And baby makes four: Predictors of attachment security among preschool-age firstborns during the transition to siblinghood. *Child Development, 67,* 579–596.

Teyber, E. (1992). *Helping children cope with divorce.* New York: Lexington Books.

Thacker, S. B., Stroup, D., & Chang, M. (2001). Continuous electronic heart rate monitoring for fetal assessment during labor. *Cochrane Database of Systematic Reviews, 2001*(2), CD000063.

Thatcher, R. W., Lyon, G. R., Rumsey, J., & Krasnegor, J. (1996). *Developmental neuroimaging.* San Diego, CA: Academic Press.

Thatcher, R. W., Walker, R. A., & Giudice, S. (1987). Human cerebral hemispheres develop at different rates and ages. *Science, 236,* 1110–1113.

Theisen, S. C., Mansfield, P. K., Seery, B. L., & Voda, A. (1995). Predictors of midlife women's attitudes toward menopause. *Health Values, 19,* 22–31.

Thelen, E. (1989). The (re)discovery of motor development: Learning new things from an old field. *Developmental Psychology, 25,* 946–949.

Thelen, E. (2001). Dynamic mechanisms of change in early perceptual–motor development. In J. L. McClelland & R. S. Siegler (Eds.), *Mechanisms of cognitive development: Behavioral and neural perspectives* (pp. 161–184). Mahwah, NJ: Erlbaum.

Thelen, E., & Adolph, K. E. (1992). Arnold Gesell: The paradox of nature and nurture. *Developmental Psychology, 28,* 368–380.

Thelen, E., Fisher, D. M., & Ridley-Johnson, R. (1984). The relationship between physical growth and a newborn reflex. *Infant Behavior and Development, 7,* 479–493.

Thelen, E., & Smith, L. B. (1998). Dynamic systems theories. In R. M. Lerner (Ed.), *Handbook of child psychology: Vol. 1. Theoretical models of human development* (5th ed., pp. 563–634). New York: Wiley.

Thomas, A., & Chess, S. (1977). *Temperament and development.* New York: Brunner/Mazel.

Thomas, A., Chess, S., & Birch, H. G. (1968). *Temperament and behavior disorders in children.* New York: New York University Press.

Thomas, J. L. (1989). Gender and perceptions of grandparenthood. *International Journal of Aging and Human Development, 29,* 269–282.

Thomas, J. R., & French, K. E. (1985). Gender differences across age in motor performance: A meta-analysis. *Psychological Bulletin, 98,* 260–282.

Thomas, R. M. (2000). *Comparing theories of child development* (5th ed.). Belmont, CA: Wadsworth.

Thompson, D. N. (1992). Applications of psychological research for the instruction of elderly adults. In R. L. West & J. D. Sinnott (Eds.), *Everyday memory and aging* (pp. 173–181). New York: Springer-Verlag.

Thompson, E. E., & Krause, N. (1998). Living alone and neighborhood characteristics as predictors of social support in late life. *Journal of Gerontology, 53,* S354–S364.

Thompson, K., & Range, L. (1992). Bereavement following suicide and other deaths: Why support attempts fail. *Omega, 26,* 61–70.

Thompson, P. M., Giedd, J. N., Woods, R. P., MacDonald, D., Evans, A. C., & Toga, A. W. (2000). Growth patterns in the developing brain detected by using continuum mechanical tensor maps. *Nature, 404,* 190–192.

Thompson, R. A. (1990). On emotion and self-regulation. In R. A. Thompson (Ed.), *Nebraska Symposia on Motivation* (Vol. 36, pp. 383–483). Lincoln: University of Nebraska Press.

Thompson, R. A. (1992). Developmental changes in research risk and benefit: A changing calculus of concerns. In B. Stanley & J. E. Sieber (Eds.), *Social research on children and adolescents: Ethical issues* (pp. 31–64). Newbury Park, CA: Sage.

Thompson, R. A. (1998). Early sociopersonality development. In N. Eisenberg (Ed.), *Handbook of child psychology: Vol. 3. Social, emotional, and personality development* (5th ed., pp. 25–104). New York: Wiley.

Thompson, R. A. (2000). The legacy of early attachments. *Child Development, 71,* 145–152.

Thompson, R. A., & Leger, D. W. (1999). From squalls to calls: The cry as a developing socioemotional signal. In B. Lester, J. Newman, & F. Pedersen (Eds.), *Biological and social aspects of infant crying.* New York: Plenum.

Thompson, R. A., & Limber, S. (1991). "Social anxiety" in infancy: Stranger wariness and separation distress. In H. Leitenberg (Ed.), *Handbook of social and evaluation anxiety* (pp. 85–137). New York: Plenum.

Thompson, R. A., & Nelson C. A. (2001). Developmental science and the media. *American Psychologist, 56,* 5–15.

Thornberry, T. P. (1998). Membership in youth gangs and involvement in serious and violent juvenile offending. In R. Loeber & D. P. Farrington (Eds.), *Serious and violent juvenile offenders: Risk factors and successful interventions* (pp. 147–166). Thousand Oaks, CA: Sage.

Thorndike, R. L., Hagen, E. P., & Sattler, J. M. (1986). *The Stanford-Binet Intelligence Scale: Guide for administering and scoring* (4th ed.). Chicago: Riverside Publishing.

Thornton, S. (1999). Creating conditions for cognitive change: The interaction between task structures and specific strategies. *Child Development, 70,* 588–603.

Thorson, J. A., & Powell, F. C. (1994). A Revised Death Anxiety Scale. In R. A. Neimeyer (Ed.), *Death anxiety handbook* (pp. 31–43). Washington, DC: Taylor & Francis.

Thorson, J. A., & Powell, F. C. (2000). Death anxiety in younger and older adults. In A. Tomer (Ed.), *Death attitudes and the older adult: Theories, concepts, and applications* (pp. 123–136). Philadelphia: Taylor & Francis.

Tienari, P., Wynne, L. C., Moring, J., & Lahti, I. (1994). The Finnish adoptive family study of schizophrenia: Implications for family research. *British Journal of Psychiatry, 164,* 20–26.

Tietjen, A., & Walker, L. (1985). Moral reasoning and leadership among men in a Papua New Guinea village. *Developmental Psychology, 21,* 982–992.

Tiggemann, M., & Anesbury, T. (2000). Negative stereotyping of obesity in children: The role of controllability beliefs. *Journal of Applied Social Psychology, 30,* 1977–1993.

Tincoff, R., & Jusczyk, P. W. (1999). Some beginnings of word comprehension in 6-month-olds. *Psychological Science, 10,* 172–175.

Tinetti, M. E., Speechley, M., & Ginter, S. F. (1988). Risk factors for falls among elderly persons living in the community. *New England Journal of Medicine, 319,* 1701–1707.

Tisak, M. S. (1995). Domains of social reasoning and beyond. In R. Vasta (Ed.), *Annals of child development* (Vol. 11, pp. 95–130). London: Jessica Kingsley.

Titov, N., & Knight, R. G. (1997). Adult age differences in controlled and automatic memory processing. *Psychology and Aging, 12,* 565–573.

Tizard, B., & Rees, J. (1975). The effect of early institutional rearing on the behaviour problems and affectional relationships of four-year-old children. *Journal of Child Psychology and Psychiatry, 16,* 61–73.

Tobin, S. S. (1989). The effects of institutionalization. In K. S. Markides & C. L. Cooper (Eds.), *Aging, stress and health* (pp. 139–164). Chichester: Wiley.

Toder, F. A. (1994). *Your kids are grown: Moving on with and without them.* New York: Plenum.

Tofler, I. R., Knapp, P. K., & Drell, M. J. (1998). The achievement by proxy spectrum in youth sports: Historical perspective and clinical approach to pressured and high-achieving children and adolescents. *Child and Adolescent Psychiatric Clinics of North America, 7,* 803–820.

Tomasello, M. (1995). Language is not an instinct. *Cognitive Development, 10,* 131–156.

Tomasello, M. (1999a). Having intentions, understanding intentions, and understanding communicative intentions. In P. D. Zelazo, J. W. Astington, & J. Wilde (Eds.), *Developing theories of intention: Social understanding and self-control* (pp. 63–75). Mahwah, NJ: Erlbaum.

Tomasello, M. (1999b). The human adaptation for culture. *Annual Review of Anthropology, 28,* 509–529.

Tomasello, M., & Akhtar, N. (1995). Two-year-olds use pragmatic cues to differentiate reference to objects and actions. *Cognitive Development, 10,* 201–224.

Tomasello, M., & Brooks, P. (1999). Early syntactic development: A construction grammar approach. In M. Barrett (Ed.), *The development of language* (pp. 161–190). Philadelphia: Psychology Press.

Tomasello, M., Striano, T., & Rochat, P. (1999). Do young children use objects as symbols? *British Journal of Developmental Psychology, 17,* 563–584.

Tomer, A., Eliason, G., & Smith, J. (2000). Beliefs about the self, life, and death: Testing aspects of a comprehensive model of death anxiety and death attitudes. *Death attitudes and the older adult: Theories, concepts, and applications* (pp. 109–122). Philadelphia: Taylor & Francis.

Tong, S., Caddy, D., & Short, R. V. (1997). Use of dizygotic to monozygotic twinning ratio as a measure of fertility. *Lancet, 349,* 843–845.

Torff, B., & Gardner, H. (1999). The vertical mind—The case for multiple intelligences. In M. Anderson (Ed.), *The development of intelligence* (pp. 139–159). Hove, U.K.: Psychology Press.

Torrance, E. P. (1988). The nature of creativity as manifest in its testing. In R. J. Sternberg (Ed.), *The nature of creativity: Contemporary psychological perspectives* (pp. 43–75). New York: Cambridge University Press.

Torrey, B. B. (1992). Sharing increasing costs on declining income: The visible dilemma of the invisible aged. In R. M. Suzman, D. P. Willis, & K. G. Manton (Eds.), *The oldest old* (pp. 381–393). New York: Oxford University Press.

Transport Canada. (2001). Pedestrian fatalities and injuries, 1988–1997. Retrieved from http://www.tc.gc.ca/roadsafety/tp2436/rs200101/en/menu.htm

Trappe, S. (2001). Master athletes. *International Journal of Sport Nutrition and Exercise Metabolism, 11,* S196–S207.

Trasti, N., Vik, T., Jacobson, G., & Bakketeig, L. S. (1999). Smoking in pregnancy and children's mental and motor development at age 1 and 5 years. *Early Human Development, 55,* 137–147.

Treiman, R., Tincoff, R., Rodriguez, K., Mouzaki, A., & Francis, D. J. (1998). The foundations of literacy: Learning the sounds of letters. *Child Development, 69,* 1524–1540.

Tremblay, G. C., & Peterson, L. (1999). Prevention of childhood injury: Clinical and public policy challenges. *Clinical Psychology Review, 19,* 415–434.

Tremblay, M. S., & Willms, J. D. (2000). Secular trends in the body mass index of Canadian children. *Canadian Medical Association Journal, 163,* 1429–1433.

Tremblay, R. E. (2000). The development of aggressive behaviour during childhood: What have we learned in the past century? *International Journal of Behavioral Development, 24,* 129–141.

Tremblay, R. E., Japel, C., Perusse, D., Voivin, M., Zoccolillo, M., Montplaisir, J., & McDuff, P. (1999). The search for the age of "onset" of physical aggression: Rousseau and Bandura revisited. *Criminal Behavior and Mental Health, 9,* 8–23.

Trent, K., & Harlan, S. L. (1994). Teenage mothers in nuclear and extended households. *Journal of Family Issues, 15,* 309–337.

Triandis, H. C. (1995). *Individualism and collectivism.* Boulder, CO: Westview Press.

Triandis, H. C. (1998, May). *Cross-cultural versus cultural psychology: A synthesis?* Colloquium presented at Illinois Wesleyan University, Bloomington, IL.

Trickett, P. K., & Putnam, F. W. (1998). Developmental consequences of child sexual abuse. In P. K. Trickett & C. J. Schellenbach (Eds.), *Violence against children in the family and community* (pp. 39–56). Washington, DC: American Psychological Association.

Troiana, R. P., & Flegal, K. M. (1998). Overweight children and adolescents: Description, epidemiology, and demographics. *Pediatrics, 101,* 497–504.

Troll, L. E., & Skaff, M. M. (1997). Perceived continuity of self in very old age. *Psychology and Aging, 12,* 162–169.

Tronick, E., Morelli, G., & Ivey, P. (1992). The Efe forager infant and toddler's pattern of social relationships: Multiple and simultaneous. *Developmental Psychology, 28,* 568–577.

Tronick, E. Z., Thomas, R. B., & Daltabuit, M. (1994). The Quechua manta pouch: A caretaking practice for buffering the Peruvian infant against the multiple stressors of high altitude. *Child Development, 65,* 1005–1013.

Tröster, H., & Brambring, M. (1992). Early social-emotional development in blind infants. *Child: Care, Health and Development, 18,* 207–227.

Truglio, R. (2000, April). *Research guides "Sesame Street."* Public lecture presented as part of the Consider the Children program, Illinois State University, Normal, IL.

Trusty, J. (1999). Effects of eighth-grade parental involvement on late adolescents' educational expectations. *Journal of Research and Development in Education, 32,* 224–233.

Tsang, P. S., & Shaner, T. L. (1998). Age, attention, expertise, and time-sharing performance. *Psychology and Aging, 13,* 323–347.

Tsuang, M. T., Bar, J. L., Harley, R. M., & Lyons, M. J. (2001). The Harvard Twin Study of Substance Abuse: What we have learned. *Harvard Review of Psychiatry, 9,* 267–279.

Tuchfarber, B. S., Zins, J. E., & Jason, L. A. (1997). Prevention and control of injuries. In R. Weissberg, T. P. Gullotta, R. L. Hampton, B. A. Ryan, & G. R. Adams (Eds.), *Enhancing children's wellness* (pp. 250–277). Thousand Oaks, CA: Sage.

Tucker, C. J., McHale, S. M., & Crouter, A. C. (2001). Conditions of sibling support in adolescence. *Journal of Family Psychology, 15,* 254–271.

Tudge, J. R. H. (1992). Processes and consequences of peer collaboration: A Vygotskian analysis. *Child Development, 63,* 1364–1397.

Tudge, J. R. H., Hogan, D. M., Snezhkova, I. A., Kulakova, N. N., & Etz, K. E. (2000). Parents' child-rearing values and beliefs in the United States and Russia: The impact of culture and social class. *Infant and Child Development, 9,* 105–121.

Tur, E., Yosipovitch, F., & Oren-Vulfs, S. (1992). Chronic and acute effects of cigarette smoking on skin blood flow. *Angiology, 43,* 328–335.

Turcotte, P., & Bélanger, A. (1997). *The dynamics of formation and dissolution of first common-law unions in Canada.* Ottawa: Statistics Canada.

Turiel, E. (1998). The development of morality. In N. Eisenberg (Ed.), *Handbook of child psychology: Vol. 3. Social, emotional, and personality development* (5th ed., pp. 863–932). New York: Wiley.

Turiel, E., Smetana, J. G., & Killen, M. (1991). Social contexts in social cognitive development. In W. M. Kurtines & J. L. Gewirtz (Eds.), *Handbook of moral behavior and development* (Vol. 2, pp. 307–332). Hillsdale, NJ: Erlbaum.

Turner, B. F. (1982). Sex-related differences in aging. In B. B. Wolman (Ed.), *Handbook of developmental psychology* (pp. 912–936). Englewood Cliffs, NJ: Prentice-Hall.

Turner, P. J., & Gervai, J. (1995). A multidimensional study of gender typing in preschool children and their parents: Personality, attitudes, preferences, behavior, and cultural differences. *British Journal of Developmental Psychology, 11,* 323–342.

Twenge, J. M., & Campbell, W. K. (2001). Age and birth cohort differences in self-esteem: A cross-temporal meta-analysis. *Personality and Social Psychology Review, 5,* 321–344.

Tyrka, A. R., Graber, J. A., & Brooks-Gunn, J. (2000). The development of disordered eating: Correlates and predictors of eating problems in the context of adolescence. In A. J. Sameroff & M. Lewis (Eds.), *Handbook of developmental psychopathology* (2nd ed., pp. 607–624). New York: Kluwer.

Tzuriel, D. (2001). *Dynamic assessment of young children.* Dordrecht, Netherlands: Kluwer.

Tzuriel, D., & Kaufman, R. (1999). Mediated learning and cognitive modifiability: Dynamic assessment of young Ethiopian immigrant children to Israel. *Journal of Cross-Cultural Psychology, 30,* 359–380.

Uhari, M., Kontiokari, T., & Niemelä, M. (1998). A novel use of xylitol sugar in preventing acute otitis media. *Pediatrics, 102,* 879–884.

Uhari, M., Mäntysaari, & Niemelä, M. (1996). A meta-analytic review of the risk factors for acute otitis media. *Clinical Infectious Diseases, 22,* 1079–1083.

Ulrich, B. D., & Ulrich, D. A. (1985). The role of balancing in performance of fundamental motor skills in 3-, 4-, and 5-year-old children. In J. E. Clark & J. H. Humphrey (Eds.), *Motor development* (Vol. 1, pp. 87–98). Princeton, NJ: Princeton Books.

United Nations. (1998). *World population prospects: The 1998 revision* (Vol. 1). New York: Author.

United Nations. (1999). *World social situation in the 1990s.* New York: Author.

United Nations Children's Fund. (2000). *Child poverty in rich nations.* Florence, Italy: Innocenti Research Centre.

United Nations Development Programme. (2002). *Human development report 2002.* New York: Oxford University Press.

U.S. Bureau of the Census. (2002a). *IDB Summary Demographic Data.* Retrieved from http://www.census.gov/ipc/www/idbsum.html

U.S. Bureau of the Census. (2002b). International data base. Retrieved from http://www.census.gov/ipc/www/idbnew.html

U.S. Bureau of the Census. (2002c). *Statistical abstract of the United States* (122nd ed.). Washington, DC: U.S. Government Printing Office.

U.S. Centers for Disease Control. (2001). STD surveillance. Retrieved from http://www.cdc.gov/std/stats/TOC2001.htm

U.S. Department of Education. (2001a). Program for International Student Assessment (PISA) 2000 highlights. Retrieved from http://nces.ed.gov/surveys/pisa/2000highlights.asp

U.S. Department of Education. (2001b). *Pursuing excellence: A study of U.S. twelfth-grade mathematics and science achievement in international context.* Washington, DC: U.S. Government Printing Office.

U.S. Department of Education. (2002a). *The condition of education 2002 in brief.* Washington, DC: National Center for Education Statistics.

U.S. Department of Education. (2002b). *Digest of education statistics 2001.* Washington, DC: U.S. Government Printing Office.

U.S. Department of Health and Human Services. (1999a). Assisted reproductive technology success rates. Retrieved from http://www.cdc.gov/nccdphp/drh/art99/99nation.htm

U.S. Department of Health and Human Services. (1999b). An introduction to sexually transmitted diseases. Retrieved from http://www.niaid.nih.gov/factsheets/stdinfo.htm

U.S. Department of Health and Human Services. (2000a). *Breastfeeding: HHS blueprint for action.* Washington, DC: U.S. Government Printing Office.

U.S. Department of Health and Human Services. (2000b). *Promoting better health for young people through physical activity and sports.* Washington, DC: U.S. Government Printing Office.

U.S. Department of Health and Human Services. (2000c). *Vital statistics of the United States.* Washington, DC: U.S. Government Printing Office.

U.S. Department of Health and Human Services. (2001). Women and smoking: A report of the Surgeon General—2001. Retrieved from http://www.cdc.gov/tobacco/sgr/sgr_forwomen

U.S. Department of Health and Human Services. (2002a, April 12). Annual smoking-attributable mortality, years of potential life lost, and economic costs—United States, 1995–1999. *Morbidity and Mortality Weekly Report, 51*(14), 300–303.

U.S. Department of Health and Human Services. (2002b). *Centenarians in the United States.* Washington, DC: U.S. Government Printing Office.

U.S. Department of Health and Human Services. (2002c). *Child maltreatment 2000.* Washington, DC: U.S. Government Printing Office.

U.S. Department of Health and Human Services. (2002d). Fetal alcohol syndrome. Retrieved from http://www.cdc.gov/ncbddd/fas/fasask.htm

U.S. Department of Health and Human Services. (2002e). *Health, United States, 1999–2000 and injury chartbook.* Washington, DC: U.S. Government Printing Office.

U.S. Department of Health and Human Services. (2002f). *Health, United States, 2002 with chartbook on trends in the health of Americans.* Hyattsville, MD: National Center for Health Statistics.

U.S. Department of Health and Human Services. (2002g). *Impaired fecundity by age and selected characteristics.* Washington, DC: U.S. Government Printing Office.

U.S. Department of Health and Human Services. (2002h). *Men and heart disease: An atlas of racial and ethnic disparities in mortality.* Washington, DC: U.S. Government Printing Office.

U.S. Department of Health and Human Services. (2002i). *National survey results on drug use from the Monitoring the Future Study: Vol. 1. Secondary school students.* Washington, DC: U.S. Government Printing Office.

U.S. Department of Health and Human Services (2002j). Obesity still on the rise, new data show. *Health and Human Services News.* Retrieved from http://www.cdc.gov/nchs/releases/02news/obesityonrise.htm

U.S. Department of Health and Human Services. (2002k). Profile of older Americans 2001. Retrieved from http://www.aoa.dhhs.gov/aoa/stats/profile/2001/default.htm

U.S. Department of Health and Human Services. (2002l). *Vital statistics of the United States.* Washington, DC: U.S. Government Printing Office.

U.S. Department of Health and Human Services. (2002m). *Women and heart disease: An atlas of racial and ethnic disparities in mortality.* Washington, DC: U.S. Government Printing Office.

U.S. Department of Health and Human Services. (2002n, June 28). Youth risk behavior surveillance—2001. *Morbidity and Mortality Weekly Report, 51*(SS04), 1–64.

U.S. Department of Justice. (2002). *Crime in the United States.* Washington, DC: U.S. Government Printing Office.

U.S. Department of Labor. (2002a). National longitudinal surveys of young men and older men. Retrieved from http://www.bls.gov

U.S. Department of Labor. (2002b). National longitudinal surveys of young women and mature women. Retrieved from http://www.bls.gov

U.S. Department of Labor, Bureau of Labor Statistics. (2003, February). *Consumer Price Index Monthly Labor Review, 125*(2).

U.S. Department of State. (1999). Country reports on human rights practices. Retrieved from http://www.state.gov/global/humanrights

Updegraff, K. A., McHale, S. M., & Crouter, A. C. (1996). Gender roles in marriage: What do they mean for girls' and boys' school achievement? *Journal of Youth and Adolescence, 25,* 73–88.

Uribe, F. M. T., LeVine, R. A., & LeVine, S. E. (1994). Maternal behavior in a Mexican community: The changing environments of children. In P. M. Greenfield & R. R. Cocking (Eds.), *Cross-cultural roots of minority child development* (pp. 41–54). Hillsdale, NJ: Erlbaum.

Usmiani, S., & Daniluk, J. (1997). Mothers and their adolescent daughters: Relationship between self-esteem, gender role identity, and body image. *Journal of Youth and Adolescence, 26,* 45–60.

Utz, R. L., Carr, D., Nesse, R., & Wortman, C. B. (2002). The effect of widowhood on older adults' social participation: An evaluation of activity, disengagement, and continuity theories. *Gerontologist, 42,* 522–533.

Vaillant, G. E. (1977). *Adaptation to life.* Boston: Little, Brown.

Vaillant, G. E. (1993). *The wisdom of the ego.* Cambridge, MA: Harvard University Press.

Vaillant, G. E. (1994). "Successful aging" and psychosocial well-being. In E. H. Thompson, Jr. (Ed.), *Older men's lives* (pp. 22–41). Thousand Oaks, CA: Sage.

Vaillant, G. E. (2002). *Aging well.* Boston: Little, Brown.

Vaillant, G. E., & Koury, S. H. (1994). Late midlife development. In G. H. Pollock & S. I. Greenspan (Eds.), *The course of life* (pp. 1–22). Madison, CT: International Universities Press.

Vaillant, G. E., & Mukamal, K. (2001). Successful aging. *American Journal of Psychiatry, 158,* 839–847.

Vaillant, G. E., & Vaillant, C. O. (1990). Determinants and consequences of creativity in a cohort of gifted women. *Psychology of Women Quarterly, 14,* 607–616.

Valdés, G. (1998). The world outside and inside schools: Language and immigrant children. *Educational Researcher, 27*(6), 4–18.

Valian, V. V. (1996). *Parental replies: Linguistic status and didactic roles.* Cambridge, MA: MIT Press.

van Baarsen, B. (2002). Theroies on coping with loss: The impact of social support and self-esteem on adjustment to emotional and social loneliness following a partner's death in later life. *Journal of Gerontology, 57B,* S33–S42.

Vandell, D. L. (1999). When school is out: Analysis and recommendations. *The Future of Children, 9*(2). Retrieved from http://www.futureofchidlren.org

Vandell, D. L., & Posner, J. K. (1999). Conceptualization and measurement of children's after-school environments. In S. L. Friedman & T. D. Wachs (Eds.), *Measuring environment across the life span* (pp. 167–196). Washington, DC: American Psychological Association.

Vandell, D. L., & Shumow, L. (1999). After-school child care programs. *Future of Children, 9*(2), 64–80.

van den Boom, D. C. (1995). Do first-year intervention effects endure? Follow-up during toddlerhood of a sample of Dutch irritable infants. *Child Development, 66,* 1798–1816.

van der Maas, P. J., Pijnenborg, L., & van Delden, J. J. M. (1995). Changes in Dutch opinions on active euthanasia, 1966 through 1991. *Journal of the American Medical Association, 273,* 1411–1414.

Vandewater, E. A., & Stewart, A. J. (1997). Women's career commitment patterns and personality development. In M. E. Lachman & J. B. James (Eds.), *Multiple paths of midlife development* (pp. 375–410). Chicago: University of Chicago Press.

Vanier Institute of the Family. (2002a). Profiling Canada's families II. Retrieved from http://www.vifamily.ca/profiling/notes.htm

Vanier Institute of the Family. (2002b). Who are Canada's lone parents? Retrieved from http://www.tifamily.ca/profiling/parti31.htm

van IJzendoorn, M. H. (1995). Adult attachment representations, parental responsiveness, and infant attachment: A meta-analysis on the predictive validity of the Adult Attachment Interview. *Psychological Bulletin, 117,* 387–403.

van IJzendoorn, M. H., & De Wolff, M. S. (1997). In search of the absent father—meta-analyses of infant–father attachment: A rejoinder to our discussants. *Child Development, 68,* 604–609.

van IJzendoorn, M. H., & Sagi, A. (1999). Cross-cultural patterns of attachment. In J. Cassidy & P. R. Shaver (Eds.), *Handbook of attachment: Theory, research, and clinical applications* (pp. 713–734). New York: Guilford.

Varendi, H., Christensson, K., Porter, R. H., & Winberg, J. (1998). Soothing effect of amniotic fluid smell in newborn infants. *Early Human Development, 51,* 47–55.

Vartanian, L. R. (1997). Separation–individuation, social support, and adolescent egocentrism: An exploratory study. *Journal of Early Adolescence, 17,* 245–270.

Vartanian, L. R., & Powlishta, K. K. (1996). A longitudinal examination of the social-cognitive foundations of adolescent egocentrism. *Journal of Early Adolescence, 16,* 157–178.

Vartanian, T. P., & McNamara, J. M. (2002). Older women in poverty: The impact of midlife factors. *Journal of Marriage and the Family, 64,* 532–547.

Vatten, L. J., Maehle, B. O., Lund, N. T., Treti, S., Hsieh, C. C., Trichopoulos, D., & Stuver, S. O. (2002). Birth weight as a predictor of breast cancer: A case-control study in Norway. *British Journal of Cancer, 86,* 89–91.

Vaughn, B. E., & Bost, K. K. (1999). Attachment and temperament: Redundant, independent, or interacting influences on interpersonal adaptation and personality development? In J. Cassidy & P. Shaver (Eds.), *Handbook of attachment: Theory, research, and clinical applications* (pp. 265–286). New York: Guilford.

Vaughn, B. E., Colvin, T. N., Azria, M. R., Caya, L., & Krzysik, L. (2001). Dyadic analyses of friendship in a sample of preschool-age children attending Head Start: Correspondence between measures and implications for social competence. *Child Development, 72,* 862–878.

Vaughn, B. E., Egeland, B. R., Sroufe, L. A., & Waters, E. (1979). Individual differences in infant–mother attachment at twelve and eighteen months: Stability and change in families under stress. *Child Development, 50,* 971–975.

Vaughn, B. E., Kopp, C. B., & Krakow, J. B. (1984). The emergence and consolidation of self-control from eighteen to thirty months of age: Normative trends and individual differences. *Child Development, 55,* 990–1004.

Vaughn, S., & Klingner, J. K. (1998). Students' perceptions of inclusion and resource room settings. *Journal of Special Education, 32,* 79–88.

Velkoff, V. (2000, January–March). Centenarians in the United States, 1990 and beyond. *Statistical Bulletin, U.S. Bureau of the Census.* Washington, DC: U.S. Government Printing Office.

Velting, D. M., & Gould, M. (1997). Suicide contagion. In R. Maris, S. Canetto, & M. M. Silverman (Eds.), *Review of suicidology, 1997* (pp. 96–137). New York: Guilford.

Venable, D. (2002). *The wage gap myth.* Dallas: National Center for Policy Analysis.

Veniegas, R., & Peplau, L. A. (1997). Power and the quality of same-sex friendships. *Psychology of Women Quarterly, 21,* 279–297.

Vernon, P. A., Wickett, J. C., Bazana, G., & Stelmack, R. M. (2001). The neuropsychology and psychophysiology of human intelligence. In R. J. Sternberg (Ed.), *Handbook of intelligence* (pp. 245–264). Cambridge, U.K.: Cambridge University Press.

Vernon-Feagans, L., Hurley, M., & Yont, K. (2002). The effect of otitis media and daycare quality on mother/child bookreading and language use at 48 months of age. *Journal of Applied Developmental Psychology, 23,* 113–133.

Victor, C., Scambler, S., Bond, J., & Bowling, A. (2000). Being alone in later life: Loneliness, social isolation, and living alone. *Reviews in Clinical Gerontology, 10,* 407–417.

Victora, C. G., Bryce, J., Fontaine, O., & Monasch, R. (2000). Reducing deaths from diarrhea through oral rehydration therapy. *Bulletin of the World Health Organization, 78,* 1246–1255.

Villaume, W. A., Brown, M. H., Darling, R., Richardson, D., Hawk, R., Henry, D. M., & Reid, T. (1997). Presbycusis and conversation: Elderly interactants adjusting to multiple hearing losses. *Research on Language and Social Interaction, 30,* 235–262.

Vinden, P. G. (1996). Junín Quechua children's understanding of mind. *Child Development, 67,* 1707–1716.

Vinick, B. H., & Ekerdt, D. J. (1991). Retirement: What happens to husband–wife relationships? *Journal of Geriatric Psychiatry, 24,* 23–40.

Vinters, H. V. (2001). Aging and the human nervous system. In J. E. Birren & K. W. Schaie (Eds.), *Handbook of the psychology of aging* (pp. 135–160). San Diego: Academic Press.

Vogel, D. A., Lake, M. A., Evans, S., & Karraker, H. (1991). Children's and adults' sex-stereotyped perceptions of infants. *Sex Roles, 24,* 605–616.

Volling, B. L., & Belsky, J. (1992). Contribution of mother–child and father–child relationships to the quality of sibling interaction: A longitudinal study. *Child Development, 63,* 1209–1222.

Volpicelli, J. R. (2001). Alcohol abuse and alcoholism. *Journal of Clinical Psychiatry, 62*(Suppl. 20), 4–10.

Vondra, J. I., Hommerding, K. D., & Shaw, D. S. (1999). Stability and change in infant attachment in a low-income sample. In J. I. Vondra & D. Barnett (Eds.), Atypical attachment in infancy and early childhood among children at developmental risk. *Monographs of the Society for Research in Child Development, 64*(3, Serial No. 258), 119–144.

von Hofsten, C., & Rosander, K. (1998). The establishment of gaze control in early infancy. In S. Simion & S. G. Butterworth (Eds.), *The development of sensory, motor and cognitive capacities in early infancy* (pp. 49–66). Hove, U.K.: Psychology Press.

Voss, L. D., Mulligan, J., & Betts, P. R. (1998). Short stature at school entry—an index of social deprivation? (The Wessex Growth Study). *Child: Care, Health and Development, 24,* 145–156.

Vostanis, P., Grattan, E., & Cumella, S. (1997). Psychosocial functioning of homeless children. *Journal of the American Academy of Child and Adolescent Psychiatry, 36,* 881–889.

Voyer, D., Voyer, S., & Bryden, M. P. (1995). Magnitude of sex differences in spatial abilities: A meta-analysis and consideration of critical variables. *Psychological Bulletin, 117,* 250–270

Vygotsky, L. S. (1978). *Mind in society: The development of higher mental processes.* Cambridge, MA: Harvard University Press. (Original works published 1930, 1933, and 1935)

Vygotsky, L. S. (1987). Thinking and speech. In R. W. Rieber, & A. S. Carton (Eds.), & N. Minick (Trans.), *The collected works of L. S. Vygotsky: Vol. 1. Problems of general psychology* (pp. 37–285). New York: Plenum. (Original work published 1934)

Wachs, T. D. (1995). Relation of mild-to-moderate malnutrition to human development: Correlational studies. *Journal of Nutrition Supplement, 125,* 22455–22545.

Wachs, T. D. (1999). The what, why, and how of temperament: A piece of the action. In L. Balter & C. S. Tamis-LeMonda (Eds.), *Child psychology: A handbook of contemporary issues* (pp. 23-44). Philadelphia: Psychology Press.

Wachs, T. D. (2000). *Necessary but not sufficient: The respective roles of single and multiple influences on individual development.* Washington, DC: American Psychological Association.

Wachs, T. D., Bishry, Z., Moussa, W., Yunis, F., McCabe, G., Harrison, G., Swefi, I., Kirksey, A., Galal, O., Jerome, N., & Shaheen, F. (1995). Nutritional intake and context as predictors of cognition and adaptive behavior of Egyptian school-age children. *International Journal of Behavioral Development, 18,* 425–450.

Wadden, T. A., & Foster, G. D. (2000). Behavioral treatment of obesity. *Medical Clinics of North America, 84,* 441–461.

Waddington, C. H. (1957). *The strategy of the genes.* London: Allen & Unwin.

Wade, T. J., & Cairney, J. (1997). Age and depression in a nationally representative sample of Canadians: A preliminary look at the National Population Health Survey. *Canadian Journal of Public Health, 88,* 297–302.

Wagner, R. K. (2000). Practical intelligence. In R. J. Sternberg (Ed.), *Handbook of intelligence* (pp. 380–395). New York: Cambridge University Press.

Wahl, H.-W. (1991). Dependence in the elderly from an interactional point of view: Verbal and observational data. *Psychology and Aging, 6,* 238–246.

Wahlsten, D. (1994). The intelligence of heritability. *Canadian Psychology, 35,* 244–259.

Waite, L. J. (1999, July). *Debunking the marriage myth: It works for women, too.* Paper presented at the annual Smart Marriages Conference, Washington, DC.

Wakat, D. K. (1978). Physiological factors of race and sex in sport. In L. K. Bunker & R. J. Rotella (Eds.), *Sport psychology: From theory to practice* (pp. 194–209). Charlotte, VA: University of Virginia. (Proceedings of the 1978 Sport Psychology Institute)

Wakeley, A., Rivera, S., & Langer, J. (2000). Can young infants add and subtract? *Child Development, 71,* 1477–1720.

Walberg, H. J. (1986). Synthesis of research on teaching. In M. C. Wittrock (Ed.), *Handbook of research on teaching* (3rd ed., pp. 214–229). New York: Macmillan.

Walco, G. A. (1997). Growing pains. *Developmental and Behavioral Pediatrics, 18,* 107–108.

Walden, T., Lemerise, E., & Smith, M. C. (1999). Friendship and popularity in preschool classrooms. *Early Education and Development, 10,* 351–371.

Waldron, N. L., & McLeskey, J. (1998). The effects of an inclusive school program on students with mild and severe learning disabilities. *Exceptional Children, 64,* 395–405.

Waldrop, D. P., & Weber, J. A. (2001). From grandparent to caregiver: The stress and satisfaction of raising grandchildren. *Families in Society, 82,* 461–472.

Walford, R. L., Mock, D., MacCallum, T., & Laseter, J. L. (1999). Physiologic changes in humans subjected to severe, selective calorie restriction for two years in Biosphere 2: Health, aging, and toxicological perspectives. *Toxicological Sciences, 52*(Suppl.), 61–65.

Walker, A., Rosenberg, M., & Balaban-Gil, K. (1999). Neurodevelopmental and neurobehavioral sequelae of selected substances of abuse and psychiatric medications in utero. *Neurological Disorders: Developmental and Behavioral Sequelae, 8,* 845–867.

Walker, D., Greenwood, C., Hart, B., & Carta, J. (1994). Prediction of school outcomes based on early language production and socioeconomic factors. *Child Development, 65,* 606–621.

Walker, L. (1995). Sexism in Kohlberg's moral psychology? In W. M. Kurtines & J. L. Gewirtz (Eds.), *Moral development: An introduction* (pp. 83–107). Boston: Allyn and Bacon.

Walker, L. J., & Taylor, J. H. (1991a). Family interactions and the development of moral reasoning. *Child Development, 62,* 264–283.

Walker, L. J., & Taylor, J. H. (1991b). Stage transitions in moral reasoning: A longitudinal study of developmental processes. *Developmental Psychology, 27,* 330–337.

Walker, L. J., Pitts, R. C., Hennig, K. H., & Matsuba, M. K. (1999). Reasoning about morality and real-life moral problems. In M. Killen & D. Hart (Eds.), *Morality in everyday life* (pp. 371–407). New York: Cambridge University Press.

Wallerstein, J. S., & Blakeslee, S. (1995). *The good marriage.* Boston: Houghton Mifflin.

Walls, C. T., & Zarit, S. H. (1991). Informal support from black churches and the well-being of elderly blacks. *Gerontologist, 31,* 490–495.

Wang, H. X., Karp, A., Winblad, B., & Fratiglioni, L. (2002). Late-life engagement in social and leisure activities is associated with a decreased risk of dementia: A longitudinal study from the Kungsholmen project. *American Journal of Epidemiology, 155,* 1081–1087.

Wannamethee, G., Shaper, A. G., & Macfarlane, P. W. (1993). Heart rate, physical activity, and mortality from cancer and other noncardiovascular diseases. *American Journal of Epidemiology, 137,* 735–748.

Wapner, R. J. (1997). Chorionic villus sampling. *Obstetrics and Gynecology Clinics of North America, 24,* 83–110.

Wark, G. R., & Krebs, D. L. (1996). Gender and dilemma differences in real-life moral judgment. *Developmental Psychology, 32,* 220–230.

Warr, P. B. (1992). Age and occupational well-being. *Psychology and Aging, 7,* 37–45.

Warr, P. B. (1994). Age and employment. In M. D. Dunnette, L. Hough, & H. Triandis (Eds.), *Handbook of industrial and organizational psychology* (pp. 485–550). Palo Alto, CA: Consulting Psychologists Press.

Warren, A. R., & Tate, C. S. (1992). Egocentrism in children's telephone conversations. In R. M. Diaz & L. E. Berk (Eds.), *Private speech: From social interaction to self-regulation* (pp. 245–264). Hillsdale, NJ: Erlbaum.

Warren, D. H. (1994). *Blindness and children: An individual difference approach.* New York: Cambridge University Press.

Wass, H. (1991). Helping children cope with death. In D. Papadatou & C. Papadatos (Eds.), *Children and death* (pp. 11–32). New York: Hemisphere.

Wass, H. (1995). Death in the lives of children and adolescents. In H. Wass & R. A. Neimeyer (Eds.), *Dying: Facing the facts* (3rd ed., pp. 269–301). Washington, DC: Taylor & Francis.

Wasserman, G., Graziano, J. H., Factor-Litvak, P., Popovac, D., Morina, N., & Musabegovic, A. (1994). Consequences of lead exposure and iron supplementation on childhood development at age 4 years. *Neurotoxicology and Teratology, 16,* 233–240.

Wasserman, G. A., Liu, X., Pine, D. S., & Graziano, J. H. (2001). Contribution of maternal smoking during pregnancy and lead exposure to early childhood behavior problems. *Neurotoxicology and Teratology, 23,* 13–21.

Wasserman, G. A., & Miller, L. S. (1998). The prevention of serious and violent juvenile offending. In R. Loeber & D. P. Farrington (Eds.), *Serious and violent juvenile offenders* (pp. 197–247). Thousand Oaks, CA: Sage.

Waterman, A. S., & Whitbourne, S. K. (1982). Androgyny and psychosocial development among college students and adults. *Journal of Personality, 50,* 121–133.

Waters, E., & Cummings, E. M. (2000). A secure base from which to explore close relationships. *Child Development, 71,* 164–172.

Waters, E., Merrick, S., Treboux, D., Crowell, J., & Albersheim, L. (2000). Attachment security in infancy and early adulthood: A twenty-year longitudinal study. *Child Development, 71,* 684–689.

Watkins, M. L., Erickson, J. D., Thun, M. J., Mulinare, J., & Heath, C. W., Jr. (2000). Multivitamin use and mortality in a large prospective study. *American Journal of Epidemiology, 152,* 149–162.

Watkins, W. E., & Pollitt, E. (1998). Iron deficiency and cognition among school-age children. In S. G. McGregor (Ed.), *Recent advances in research on the effects of health and nutrition on children's development and school achievement in the Third World.* Washington, DC: Pan American Health Organization.

Watson, A. C., Nixon, C. L., Wilson, A., & Capage, L. (1999). Social interaction skills and theory of mind in young children. *Developmental Psychology, 35,* 386–391.

Watson, D. J. (1989). Defining and describing whole language. *Elementary School Journal, 90,* 129–141.

Watson, J. B., & Raynor, R. (1920). Conditioned emotional reactions. *Journal of Experimental Psychology, 3,* 1–14.

Watson, M. (1990). Aspects of self development as reflected in children's role playing. In D. Cicchetti & M. Beeghly (Eds.), *The self in transition: Infancy to childhood* (pp. 281–307). Chicago: University of Chicago Press.

Wattigney, W. A., Srinivasan, S. R., Chen, W., Greenlund, K. J., & Berenson, G. S. (1999). Secular trend of earlier onset of menarche with increasing obesity in black and white girls: The Bogalusa Heart Study. *Ethnicity and Disease, 9,* 181–189.

Waxman, S. R. (1995). Words as invitations to form categories: Evidence from 12- to 13-month-old infants. *Cognitive Psychology, 29,* 254–302.

Waxman, S. R., & Senghas, A. (1992). Relations among word meanings in early lexical development. *Developmental Psychology, 28,* 862–873.

Webb, N. B. (2002). September 11, 2001. In N. B. Webb (Ed.), *Helping bereaved children: A handbook for practitioners* (pp. 365–384). New York: Guilford.

Weber, J. A., & Waldrop, D. P. (2000). Grandparents raising grandchildren: Families in transition. *Journal of Gerontological Social Work, 33,* 27–46.

Webster, J. D. (2002). Reminiscence function in adulthood: Age, ethnic, and family dynamics correlates. In J. D. Webster & B. K. Haight (Eds.), *Critical advances in reminiscence work* (pp. 140–142). New York: Springer.

Webster, J. D., & McCall, M. E. (1999). Reminiscence functions across adulthood: A replication and extension. *Journal of Adult Development, 6,* 73–85.

Wechsler, D. (1991). *Manual for the Wechsler Intelligence Test for Children–III.* New York: Psychological Corporation.

Wechsler, D. (1996). *Canadian supplement manual for the WISC-III.* Toronto: Psychological Corporation.

Wegesin, D. J., Jacobs, D. M., Zubin, N. R., & Ventura, P. R. (2000). Source memory and encoding strategy in normal aging. *Journal of Clinical and Experimental Neuropsychology, 22,* 455–464.

Wehren, A., De Lisi, R., & Arnold, M. (1981). The development of noun definition. *Journal of Child Language, 8,* 165–175.

Wei, Y. H., & Lee, H. C. (2002). Oxidative stress, mitochondrial DNA mutation, and impairment of antioxidant enzymes in aging. *Experimental Biology and Medicine, 227,* 671–682.

Weikart, D. P. (1998). Changing early childhood development through educational intervention. *Preventive Medicine, 27,* 233–237.

Weiland, S. (1993). Erik Erikson: Ages, stages, and stories. *Generations, 17*(2), 17–22.

Weinberg, M. K., & Tronick, E. Z. (1994). Beyond the face: An empirical study of infant affective configurations of facial, vocal, gestural, and regulatory behaviors. *Child Development, 65,* 1503–1515.

Weinberg, M. K., Tronick, E. Z., Cohn, J. F., & Olson, K. L. (1999). Gender differences in emotional expressivity and self-regulation during early infancy. *Developmental Psychology, 35,* 175–188.

Weinberg, R. A., Scarr, S., & Waldman, I. D. (1992). The Minnesota transracial adoption study: A follow-up of IQ test performance at adolescence. *Intelligence, 16,* 117–135.

Weindruch, R., Keenan, K. P., Carney, J. M., Fernandes, G., Feuers, R. J., & Floyd, R. A. (2001). Caloric restriction mimetics: Metabolic interventions. *Journal of Gerontology, 56A,* 20–33.

Weiner, J. (2002). Jews accept intermarriage in growing numbers. Retrieved from http://www.interfaithfamily.com/article/issue48/weiner.phtml

Weinfeild, N. S., Sroufe, L. A., & Egeland, B. (2000). Attachment from infancy to early adulthood in a high-risk sample: Continuity, discontinuity, and their correlates. *Child Development, 71,* 695–702.

Weingarten, H. R. (1988). Late life divorce and the life review.

Journal of Gerontological Social Work, 12(3–4), 83–97.

Weingarten, H. R. (1989). The impact of late life divorce: A conceptual and empirical study. *Journal of Divorce, 12,* 21–38.

Weinstein, R. S., Marshall, H. H., Sharp, L., & Botkin, M. (1987). Pygmalion and the student: Age and classroom differences in children's awareness of teacher expectations. *Child Development, 58,* 1079–1093.

Weisfield, G. E. (1997). Puberty rites as clues to the nature of human adolescence. *Cross-Cultural Research, 31,* 27–54.

Weisman, A. D. (1984). *The coping capacity: On the nature of being mortal.* New York: Sciences Press.

Weisner, T. S. (1993). Ethnographic and ecocultural perspectives on sibling relationships. In Z. Stoneman & P. W. Berman (Eds.), *The effects of mental retardation, disability, and illness on sibling relationships* (pp. 51–83). Baltimore: Paul H. Brookes.

Weisner, T. S. (1996). The 5 to 7 year transition as an ecocultural project. In A. J. Sameroff & M. M. Haith (Eds.), *The five to seven year shift* (pp. 295–326). Chicago: University of Chicago Press.

Weisner, T. S., & Wilson-Mitchell, J. E. (1990). Nonconventional family life-styles and sex typing in six-year-olds. *Child Development, 61,* 1915–1933.

Weiss, B., Dodge, K. A., Bates, J. E., & Pettit, G. S. (1992). Some consequences of early harsh discipline: Child aggression and a maladaptive social information processing style. *Child Development, 63,* 1321–1335.

Weissman, M., Wolk, S., Goldstein, R. B., Moreau, D., Adams, P., & Greenwald, S. (1999). Depressed adolescents grown up. *Journal of the American Medical Association, 281,* 1707–1713.

Weksler, M. E. (1995). Immune senescence: Deficiency or dysregulation. *Nutrition Reviews, 53,* S3–S7.

Wellings, K., Field, J., Johnson, A., & Wadsworth, J. (1994). *Sexual behavior in Britain: The National Survey of Sexual Attitudes and Lifestyles.* New York: Penguin.

Wellman, H. M. (1990). *The child's theory of mind.* Cambridge, MA: MIT Press.

Wellman, H. M., Cross, D., & Watson, J. (2001). Meta-analysis of theory-of-mind development: The truth about false belief. *Child Development, 72,* 655–684.

Wellman, H. M., & Hickling, A. K. (1994). The mind's "I": Children's conception of the mind as an active agent. *Child Development, 65,* 1564–1580.

Wellman, H. M., Somerville, S. C., & Haake, R. J. (1979). Development of search procedures in real-life spatial environments. *Developmental Psychology, 15,* 530–542.

Wendland-Carro, J., Piccinini, C. A., & Millar, W. S. (1999). The role of an early intervention on enhancing the quality of mother-infant interaction. *Child Development, 70,* 713–731.

Wenger, G. C. (2001). Ageing without children: Rural Wales. *Journal of Cross-Cultural Gerontology, 16,* 79–109.

Wentworth, N., Benson, J. B., & Haith, M. M. (2000). The development of infants' reaches for stationary and moving targets. *Child Development, 71,* 576–601.

Werker, J. F., Pegg, J. E., & McLeod, P. J. (1994). A cross-language investigation of infant preference for infant-directed communication. *Infant Behavior and Development, 17,* 323–333.

Werner, E. (2001). *Journeys from childhood to midlife: Risk, resilience, and recovery.* Ithaca, NY: Cornell University Press.

Werner, E. E. (1989, April). Children of the garden island. *Scientific American, 260*(4), 106–111.

Werner, E. E. (1991). Grandparent–grandchild relationships amongst U.S. ethnic groups. In P. K. Smith (Ed.), *The psychology of grandparenthood: An international perspective* (pp. 68–82). London: Routledge.

Werner, E. E. (1993). Risk, resilience, and recovery: Perspectives from the Kauai Longitudinal Study. *Development and Psychopathology, 5,* 503–515.

Werner, E. E., & Smith, R. S. (1982). *Vulnerable but invincible.* New York: McGraw-Hill.

Werner, E. E., & Smith, R. S. (1992). *Overcoming the odds: High-risk children from birth to adulthood.* Ithaca, NY: Cornell University Press.

Werner, E. E., & Smith, R. S. (2001). *Journeys from childhood to midlife: Risk, resilience, and recovery.* Ithaca, NY: Cornell University Press.

Wertsch, J. V., & Tulviste, P. (1992). L. S. Vygotsky and contemporary developmental psychology. *Developmental Psychology, 28,* 548–557.

West, R. L., & Craik, F. I. M. (1999). Age-related decline in prospective memory: The roles of cue accessibility and cue sensitivity. *Psychology and Aging, 14,* 264–272.

Westen, D., & Gabbard, G. O. (1999). Psychoanalytic approaches to personality. In L. A. Pervin & O. P. John (Eds.), *Handbook of personality: Theory and research* (2nd ed.). New York: Guilford.

Westermeyer, J. F. (1998). Predictors and characteristics of mental health among men at midlife: A 32-year longitudinal study. *American Journal of Orthopsychiatry, 68,* 265–273.

Wethington, E. (2000). Expecting stress: Americans and the "midlife crisis." *Motivation and Emotion, 24,* 85–103.

Whalley, L. (2001). *The aging brain.* London: Weidenfeld & Nicolson.

Wheeler, I. (2001). Parental bereavement: The crisis of meaning. *Death Studies, 25,* 51–66.

Wheeler, T., Barker, D. J. P., & O'Brien, P. M. S. (1999). *Fetal programming: Influences on development and disease in later life.* London: RCOG Press.

Wheeler, W. (2002). Youth leadership for development: Civic activism as a component of youth development programming and a strategy for strengthening civil society. In R. M. Lerner, F. Jacobs, & D. Wertlieb (Eds.), *Handbook of applied developmental science* (Vol. 2, pp. 491–506). Thousand Oaks, CA: Sage.

Whitaker, D. J., & Miller, K. S. (2000). Parent–adolescent discussions about sex and condoms: Impact on peer influences of sexual risk behavior. *Journal of Adolescent Research, 15,* 251–273.

Whitbeck, L., Hoyt, D. R., & Huck, S. M. (1994). Early family relationships, intergenerational solidarity, and support provided to parents by their adult children. *Journal of Gerontology, 49,* 585–594.

Whitbourne, S. K. (1996). *The aging individual: Physical and psychological perspectives.* New York: Springer.

Whitbourne, S. K. (1999). Physical changes. In J. C. Kavanaugh & S. K. Whitbourne (Eds.), *Gerontology: An interdisciplinary perspective* (pp. 33–64). New York: Oxford University Press.

Whitbourne, S. K. (2001). The physical aging process in midlife: Interactions with psychological and sociocultural factors. In M. E. Lachman (Ed.), *Handbook of midlife development* (pp. 109–155). New York: Wiley.

Whitbourne, S. K., & Primus, L. (1996). Identity, physical. In J. E. Birren (Ed.), *Encyclopedia of aging* (pp. 733–742). San Diego: Academic Press.

Whitbourne, S. K., & Weinstock, C. S. (1979). *Adult development: The differentiation of experience.* New York: Holt, Rinehart & Winston.

Whitbourne, S. K., Zuschlag, M. K., Elliot, L. B., & Waterman, A. S. (1992). Psychosocial development in adulthood: A 22-year sequential study. *Journal of Personality and Social Psychology, 63,* 260–271.

White, B., & Held, R. (1966). Plasticity of sensorimotor development in the human infant. In J. F. Rosenblith & W. Allinsmith (Eds.), *The causes of behavior* (pp. 60–70). Boston: Allyn and Bacon.

White, J. L., Moffitt, T. E., Caspi, A., Bartusch, D. J., Needles, D. J., & Stouthamer-Loeber, M. (1996). Measuring impulsivity and examining its relationship to delinquency. *Journal of Abnormal Psychology, 103,* 192–205.

White, L. (2001). Sibling relationships over the life course: A panel analysis. *Journal of Marriage and the Family, 63,* 555–568.

White, L. K. (1994). Coresidence and leaving home: Young adults and their parents. *Annual Review of Sociology, 20,* 81–102.

Whitehead, J. R., & Corbin, C. B. (1997). Self-esteem in children and youth: The role of sport and physical education. In K. R. Fox (Ed.), *The physical self: From motivation to well-being* (pp. 175–204). Champaign, IL: Human Kinetics.

Whitehurst, G. J., & Lonigan, C. J. (1998). Child development and emergent literacy. *Child Development, 69,* 848–872.

Whiteside, M. F., & Becker, B. J. (2000). Parental factors and the young child's postdivorce adjustment: A meta-analysis with implications for parenting arrangements. *Journal of Family Psychology, 14,* 5–26.

Whiting, B., & Edwards, C. P. (1988a). *Children in different worlds.* Cambridge, MA: Harvard University Press.

Whiting, B., & Edwards, C. P. (1988b). A cross-cultural analysis of sex differences in the behavior of children aged 3 through 11. In G. Handel (Ed.), *Childhood socialization* (pp. 281–297). New York: Aldine de Gruyter.

Whitney, M. P., & Thoman, E. B. (1994). Sleep in premature and full-term infants from 24-hour home recordings. *Infant Behavior and Development, 17,* 223–234.

Wichstrøm, L. (1999). The emergence of gender difference in depressed mood: The role of intensified gender socialization.

Developmental Psychology, 35, 232–245.

Wiebe, D. J., & Williams, P. G. (1992). Hardiness and health: A social psychophysiological perspective on stress and adaptation. *Journal of Social and Clinical Psychology, 11,* 238–262.

Wigfield, A., Battle, A., Keller, L. B., & Eccles, J. S. (2002). Sex differences in motivation, self-concept, career aspiration, and career choice: Implications for cognitive development. In A. McGillicuddy-De Lisi & R. De Lisi (Eds.), *Biology, society, and behavior: The development of sex differences in cognition* (pp. 93–124). Westport, CT: Ablex.

Wigfield, A., & Eccles, J. S. (1994). Children's competence beliefs, achievement values, and general self-esteem change across elementary and middle school. *Journal of Early Adolescence, 14,* 107–138.

Wigfield, A., Eccles, J. S., Yoon, K. S., Harold, R. D., Arbreton, A. J., Freedman-Doan, C., & Blumenfeld, P. C. (1997). Changes in children's competence beliefs and subjective task values across the elementary school years: A three-year study. *Journal of Educational Psychology, 89,* 451–469.

Wikby, A., Maxson, P., Olsson, J., Johansson, B., & Ferguson, F. G. (1998). Changes in CD8 and CD4 lymphocyte subsets, T cell proliferation responses and non-survival in the very old: The Swedish longitudinal OCTO-immune study. *Mechanisms of Ageing and Development, 102,* 187–198.

Wilber, K. H., & McNeilly, D. P. (2001). Elder abuse and victimization. In J. E. Birren (Ed.), *Handbook of the psychology of aging* (pp. 569–591). San Diego: Academic Press.

Wilcox, A. J., Weinberg, C. R., & Baird, D. D. (1995). Timing of sexual intercourse in relation to ovulation: Effects on the probability of conception, survival of the pregnancy, and sex of the baby. *New England Journal of Medicine, 333,* 1517–1519.

Wildes, J. E., Emery, R. E., & Simons, A. D. (2001). The roles of ethnicity and culture in the development of eating disturbance and body dissatisfaction: A meta-analytic review. *Clinical Psychology Review, 21,* 521–551.

Wiley, T. L., Cruickshanks, K. J., Nondahl, D. M., Tweed, T. S., Klein, R., & Klein, B. E. K. (1998). Aging and high-frequency hearing sensitivity. *Journal of Speech, Language, and Hearing Research, 41,* 1061–1072.

Wilke, C. J., & Thompson, C. A. (1993). First-year reentry women's perceptions of their classroom experiences. *Journal of the Freshman Year Experience, 5,* 69–90.

Wilkie, J. R., Ferree, M. M., & Ratcliff, K. S. (1998). Gender and fairness: Marital satisfaction and two-earner couples. *Journal of Marriage and the Family, 60,* 577–594.

Wilkins, K., Parsons, G. F., Gentleman, J. F., & Forbes, W. F. (2000). *Chronic diseases in Canada, 20*(3), 1–17. Retrieved from http://www. hc-sc.gc.ca/pphb-dgspsp/publicat/cdic-mcc/20-1/e_e.html#tab1

Willatts, P. (1999). Development of means–end behavior in young infants: Pulling a support to retrieve a distant object. *Developmental Psychology, 35,* 651–667.

Wille, D. E. (1991). Relation of preterm birth with quality of infant–mother attachment at one year. *Infant Behavior and Development, 14,* 227–240.

Willett, W. C., Colditz, G., & Stampfer, M. (2000). Postmenopausal estrogens opposed, unopposed, or none of the above. *Journal of the American Medical Association, 283,* 485–491.

Williams, B. C., & Kotch, J. B. (1990). Excess injury mortality among children in the United States: Comparison of recent international statistics. *Pediatrics, 86*(6, Pt. 2), 1067–1073.

Williams, N., & Torrez, D. J. (1998). Grandparenthood among Hispanics. In M. E. Szinovacz (Ed.), *Handbook on grandparenthood* (pp. 87–96). Westport, CT: Greenwood Press.

Williams, R. B. (2000). Psychological factors, health, and disease: The impact of aging and the life cycle. In S. B. Manuck, R. Jennings, B. S. Rabin, & A. Baum (Eds.), *Behavior, health, and aging* (pp. 135–151). Mahwah, NJ: Erlbaum.

Willinger, M., Ko, C.-W., Hoffman, H. J., Kessler, R. C., & Corwin, M. J. (2003). Trends in infant bed sharing in the United States. *Archives of Pediatrics and Adolescent Medicine, 157,* 43–49.

Willis, S. (1996). Everyday problem solving. In J. E. Birren & K. W. Schaie (Eds.), *Handbook of the psychology of aging* (4th ed., pp. 287–307). San Diego: Academic Press.

Willis, S. L., & Schaie, K. W. (1994). Cognitive training in the normal elderly. In F. Forette, Y. Christen, & F. Boller (Eds.), *Plasticité cérébrale et stimulation cognitive* [Cerebral plasticity and cognitive stimulation] (pp. 91–113). Paris: Fondation Nationale de Gérontologie.

Willis, S. L., & Schaie, K. W. (1999). Intellectual functioning in midlife. In S. L. Willis & J. D. Reid (Eds.), *Life in the middle* (pp. 105–146). San Diego: Academic Press.

Willner, J. P. (1998). Reproductive genetics and today's patient options: Prenatal diagnosis. *Mount Sinai Journal of Medicine, 65,* 173–177.

Willott, J. F., Chisolm, T. H., & Lister, J. L. (2001). Modulation of presbycusis: Current status and future directions. *Audiology and Neuro-Otology, 6,* 231–249.

Wilson, D. M. (2002). Addressing myths about end-of-life care: Research into the use of acute care hospitals over the last five years of life. *Journal of Palliative Care, 18,* 29–38.

Wilson, M. N., Greene-Bates, C., McKim, L., Simmons, T. A., Curry-El, J., & Hinton, I. D. (1995). African American family life: The dynamics of interactions, relationships, and roles. In M. N. Wilson (Ed.), *African American family life: Its structural and ecological aspects* (pp. 5–21). San Francisco: Jossey-Bass.

Winblad, B., Engedal, K., Soineinen, H., Berhey, F., Waldemar, G., & Wimo, A. (2001). A 1-year, randomized, placebo-controlled study of donepezil in patients with mild to moderate AD. *Neurology, 57,* 489–495.

Wingfield, A., & Stine-Morrow, E. A. L. (2000). Language and speech. In F. I. M. Craik & T. A. Salthouse (Eds.), *Handbook of aging and cognition* (2nd ed., pp. 359–416). Mahwah, NJ: Erlbaum.

Wink, P., & Dillon, M. (2002). Spiritual development across the adult life course: Findings from a longitudinal study. *Journal of Adult Development, 9,* 79–94.

Wink, P., & Helson, R. (1993). Personality change in women and their partners. *Journal of Personality and Social Psychology, 65,* 597–605.

Winkleby, M. A., Robinson, T. N., Sundquist, J., & Kraemer, H. C. (1999). Ethnic variation in cardiovascular disease risk factors among children and young adults. *Journal of the American Medical Association, 281,* 1006–1013.

Winn, R., & Newton, N. (1982). Sexual activity in aging: A study of 106 cultures. *Archives of Sexual Behavior, 11,* 283–298.

Winner, E. (1986, August). Where pelicans kiss seals. *Psychology Today, 20*(8), 25–35.

Winner, E. (1988). *The point of words: Children's understanding of metaphor and irony.* Cambridge, MA: Harvard University Press.

Winner, E. (1996). *Gifted children: Myths and realities.* New York: Basic Books.

Winner, E. (2000). The origins and ends of giftedness. *American Psychologist, 55,* 159–169.

Winsler, A., Diaz, R. M., & Montero, I. (1997). The role of private speech in the transition from collaborative to independent task performance in young children. *Early Childhood Research Quarterly, 12,* 59–79.

Wintre, M. G., & Vallance, D. D. (1994). A developmental sequence in the comprehension of emotions: Intensity, multiple emotions, and valence. *Developmental Psychology, 30,* 509–514.

Wiscott, R., & Kopera-Frye, K. (2000). Sharing of culture: Adult grandchildrens' perceptions of intergenerational relations. *International Journal of Aging and Human Development, 5,* 199–215.

Wojciechowski, W. C. (1998). Issues in caring for older lesbians. *Journal of Gerontological Nursing, 24,* 28-33.

Wolchik, S. A., Wilcox, K. L., Tein, J.-Y., & Sandler, I. N. (2000). Maternal acceptance and consistency of discipline as buffers of divorce stressors on children's psychological adjustment problems. *Journal of Abnormal Child Psychology, 28,* 87–102.

Wolf, R. S. (2000). Elder abuse. In V. B. Van Hasselt & M. Hersen (Eds.), *Aggression and violence* (pp. 135–151). Boston: Allyn and Bacon.

Wolfe, D. A., (1999). *Child abuse* (2nd ed.). Thousand Oaks, CA: Sage.

Wolfe, D. A., Scott, K., Wekerle, C., & Pittman, A. (2001). Child maltreatment: Risk of adjustment problems and dating violence in adolescence. *Journal of the American Academy of Child and Adolescent Psychiatry, 40,* 282–289.

Wolfe, J. J. (1997). Hospice support for families facing multiple deaths of children. *American Journal of Hospital Palliative Care, 14,* 224–227.

Wolfelt, A. D. (1997). Death and grief in the school setting. In T. N. Fairchild (Ed.), *Crisis intervention strategies for school-based helpers* (pp. 199–244). Springfield, IL: Charles C. Thomas.

Wolff, P. H. (1966). The causes, controls and organization of behavior in the neonate. *Psychological Issues, 5*(1, Serial No. 17).

Wolff, P. H., & Fesseha, G. (1999). The orphans of Eritrea: A five-year follow-up study. *Journal of Child Psychology and Psychiatry and Allied Disciplines, 40,* 1231–1237.

Wolfinger, N. H. (2000). Beyond the intergenerational transmission of divorce: Do people replicate the patterns of marital instability they grew up with? *Journal of Family Issues, 21,* 1061–1086.

Wolfson, A. R., & Carskadon, M. A. (1998). Sleep schedules and daytime functioning in adolescents. *Child Development, 69,* 875–887.

Wolpe, J., & Plaud, J. J. (1997). Pavlov's contributions to behavior therapy: The obvious and not so obvious. *American Psychologist, 52,* 966–972.

Women's Health Initiative. (2002). Risks and benefits of estrogen plus progestin in healthy postmenopausal women: Principal results from the Women's Health Initiative randomized control trial. *Journal of the American Medical Association, 288,* 321–333.

Wong, R., Kitayama, K. E., & Soldo, B. J. (1999). Ethnic differences in time transfers from adult children to elderly parents: Unobserved heterogeneity across families? *Research on Aging, 21,* 144–175.

Wood, W., & Eagly, A. H. (2000). Once again, the origins of sex differences. *American Psychologist, 55,* 1062–1063.

Wood. S. (2001). Interview. *Frontline*. Retrieved from http://www.pbs.org/wgbh/pages/frontline/shows/fertility/interviews/wood.html

Woodman, S. (1998). *Last rights: The struggle over the right to die.* New York: Plenum.

Woodring, B. C. (1998). Relationship of physical activity and television watching with body weight and level of fatness: Results from the Third National Health and Nutrition Survey. *Journal of Child and Family Nursing, 1,* 78–79.

Woodward, L., Taylor, E., & Dowdney, L. (1998). The parenting and family functioning of children with hyperactivity. *Journal of Child Psychology and Psychiatry, 39,* 161–169.

Woodward, L. J., & Fergusson, D. M. (1999). Childhood peer relationship problems and psychosocial adjustment in late adolescence. *Journal of Abnormal Child Psychology, 27,* e87.

Woodward, S. A., Lenzenweger, M. F., Kagan, J., Snidman, N., & Arcus, D. (2000). Taxonic structure of infant reactivity: Evidence from a taxometric perspective. *Psychological Science, 11,* 296–301.

Woolf, L. M. (2001). Gay and lesbian aging. *SIECUS Report, 30,* 16–21.

Woolley, J. D. (1997). Thinking about fantasy: Are children fundamentally different thinkers and believers from adults? *Child Development, 68,* 991–1011.

Woolley, J. D., Phelps, K. E., Davis, D. L., & Mandell, D. J. (1999). Where theories of mind meet magic: The development of children's beliefs about wishing. *Child Development, 70,* 571–587.

Wooster, D. M. (1999). Assessment of nonorganic failure to thrive. *Infant–Toddler Intervention, 9,* 353–371.

Wooster, D. M. (2000). Intervention for nonorganic failure to thrive. *Transdisciplinary Journal, 10,* 37–45.

Word, S. (1996). Mortality awareness and risk-taking in late adolescence. *Death Studies, 20,* 133–148.

Worden, J. W. (2000). Toward an appropriate death. In T. A. Rando (Ed.), *Clinical dimensions of anticipatory mourning* (pp. 267–277). Champaign, IL: Research Press.

World Health Organization. (2000a). *Healthy life expectancy rankings.* Geneva: Author.

World Health Organization. (2000b). Violence against women information pack. Retrieved from http://www.who.int/frh-whd/VAW/infopack/English

World Health Organization. (2000c). *The world health report, 2000.* Geneva: Author.

World Health Organization. (2001, November 30). Oral Health County/Area Profile Program. Retrieved from http://www.whocollab.od.mah.se/amro/canada/data/canadacar.html

World Health Organization. (2002). *World health statistics annual 2002.* Geneva: Author.

Wright, J. C., Huston, A. C., Murphy, K. C., St. Peters, M., Pinon, M., Scantlin, R., & Kotler, J. (2001). The relations of early television viewing to school readiness and vocabulary of children from low-income families: The Early Window Project. *Child Development, 72,* 1347–1366.

Wright, J. C., Huston, A. C., Reitz, A. L., & Piemyat, S. (1994). Young children'sperceptions of television reality: Determinants and developmental differences. *Developmental Psychology, 30,* 229–239.

Wright, J. W. (Ed.). (1999). *The universal almanac 1999.* Kansas City: Andrews and McMeel.

Wright, T. A., & Bonett, D. G. (1997). The contribution of burnout to work performance. *Journal of Organizational Behavior, 18,* 491–499.

Wrightsman, L. S. (1994). *Adult personality development: Vol. 1. Theories and concepts.* Thousand Oaks, CA: Sage.

Wu, L. L., Bumpass, L. L., & Musick, K. (2001). Historical and life course trajectories of nonmarital childbearing. In L. L. Wu & B. Wolfe (Eds.), *Out of wedlock: Causes and consequences of nonmarital fertility* (pp. 3–48). New York: Russell Sage Foundation.

Wu, M. M., & Edwards, M. H. (1999). The effect of having myopic parents: An analysis of myopia in three generations. *Optometry and Vision Science, 76,* 387–392.

Wu, Z., & Penning, M. J. (1997). Marital instability after midlife. *Journal of Family Issues, 18,* 459–478.

Wu, Z., & Pollard, M. S. (1998). Social support among unmarried childless elderly persons. *Journal of Gerontology, 53B,* S324–S335.

Wyman, P. A., Cowen, E. L., Work, W. C., Hoyt-Meyers, L., Magnus, K. B., & Fagen, D. B. (1999). Caregiving and developmental factors differentiating young at-risk urban children showing resilient versus stress-affected outcomes: A replication and extension. *Child Development, 70,* 645–659.

Wynn, K. (1992). Addition and subtraction by human infants. *Nature, 358,* 749–750.

Wynn, K. (2002). Do infants have numerical expectations or just perceptual preferences? Comment. *Developmental Science, 5,* 207–209.

Wynn, K., Bloom, P., & Chiang, W.-C. (2002). Enumeration of collective entities by 5-month-old infants. *Cognition, 83,* B55–B62.

Yale, M. E., Messinger, D. S., Cobo-Lewis, A. B., Oller, D. K., & Eilers, R. E. (1999). An event-based analysis of the coordination of early infant vocalizations and facial actions. *Developmental Psychology, 35,* 505–513.

Yamanoi, K. (1993). Care for the elderly in Sweden and Japan. Retrieved from http://www.wao.or.jp/yamanoi/report/lunds/index.htm

Yang, B., Ollendick, T. H., Dong, Q., Xia, Y., & Lin, L. (1995). Only children and children with siblings in the People's Republic of China: Levels of fear, anxiety, and depression. *Child Development, 66,* 1301–1311.

Yang, E. Y., Flake, A. W., & Adzick, N. S. (1999). Prospects for fetal gene therapy. *Seminars in Perinatology, 23,* 524–534.

Yarrow, M. R., Scott, P. M., & Waxler, C. Z. (1973). Learning concern for others. *Developmental Psychology, 8,* 240–260.

Yates, W. R., Cadoret, R. J., & Troughton, E. P. (1999). The Iowa adoption studies: Methods and results. In M. C. LaBuda & E. L. Grigorenko (Eds.), *On the way to individuality: Current methodological issues in behavioral genetics* (pp. 95–125). Commack, NY: Nova Science Publishers.

Yeung, W. (1996). Buddhism, death, and dying. In J. K. Parry & A. S. Ryan (Eds.), *A cross-cultural look at death, dying, and religion* (pp. 74–83). Chicago: Nelson-Hall.

Yip, R., Scanlon, K., & Trowbridge, F. (1993). Trends and patterns in height and weight status of low-income U.S. children. *Critical Reviews in Food Science and Nutrition, 33,* 409–421.

Yirmiya, N., & Shulman, C. (1996). Seriation, conservation, and theory of mind abilities in individuals with autism, individuals with mental retardation, and normally developing children. *Child Development, 67,* 2045–2059.

Yirmiya, N., Solomonica-Levi, D., & Shulman, C. (1996). The ability to manipulate behavior and to understand manipulation of beliefs: A comparion of individuals with autism, mental retardation, and normal development. *Developmental Psychology, 32,* 62–69.

Yoder, A. E. (2000). Barriers to ego identity status formation: A contextual qualification of Marcia's identity status paradigm. *Journal of Adolescence, 23,* 95–106.

Yogman, M. W. (1981). Development of the father–infant relationship. In H. Fitzgerald, B. Lester, & M. W. Yogman (Eds.), *Theory and research in behavioral pediatrics* (Vol. 1, pp. 221–279). New York: Plenum.

Yonas, A., Granrud, E. C., Arterberry, M. E., & Hanson, B. L. (1986). Infants' distance perception from linear perspective and texture gradients. *Infant Behavior and Development, 9,* 247–256.

Yoo, S. H., & Sung, K.-T. (1997). Elderly Koreans' tendency to live independently from their adult children: Adaptation to cultural differences in America. *Journal of Cross-Cultural Gerontology, 12,* 225–244.

Youn, G., Knight, B. G., Jeon, H., & Benton, D. (1999). Differences in familism values and caregiving outcomes among Korean, Korean American, and White American dementia caregivers. *Psychology and Aging, 14,* 355–364.

Young, H. M. (1998). Moving to congregate housing: The last chosen home. *Journal of Aging Studies, 12,* 149–165.

Young, J. B., & Rodgers, R. F. (1997). A model of radical career change in the context of psychosocial development. *Journal of Career Assessment, 5,* 167–172.

Young, L. R., & Nestle, M. (2002). The contribution of expanding portion sizes to the U.S. obesity epidemic. *American Journal of Public Health, 92,* 246–249.

Young, P. (1991). Families with adolescents. In F. H. Brown (Ed.), *Reweaving the family tapestry* (pp. 131–168). New York: Norton.

Youniss, J. (1980). *Parents and peers in social development: A Piagetian-Sullivan perspective.* Chicago: University of Chicago Press.

Youniss, J., McLellan, J. A., & Yates, M. (1997). What we know about engendering civic identity. *American Behavioral Scientist, 40,* 620–631.

Yu, S., Yarnell, J. W. G., Sweetnam, P. M., & Murray, L. (2003). What level of physical activity protects against premature cardiovascular death? *The Caerphilly Study. Heart, 89,* 502–506.

Yuill, N., & Perner, J. (1988). Intentionality and knowledge in children's judgments of actor's responsibility and recipient's emotional reaction. *Developmental Psychology, 24,* 358–365.

Zafeiriou, D. I. (2000). Plantar grasp reflex in high-risk infants during the first year of life. *Pediatric Neurology, 22,* 75–76.

Zahn-Waxler, C., Kochanska, G., Krupnick, J., & McKnew, D. (1990). Patterns of guilt in children of depressed and well mothers. *Developmental Psychology, 26,* 51–59.

Zahn-Waxler, C., Radke-Yarrow, M., & King, R. M. (1979). Child-rearing and children's prosocial initiations toward victims of distress. *Child Development, 50,* 319–330.

Zahn-Waxler, C., Radke-Yarrow, M., Wagner, E., & Chapman, M. (1992). Development of concern for others. *Developmental Psychology, 28,* 126–136.

Zahn-Waxler, C., & Robinson, J. (1995). Empathy and guilt: Early origins of feelings of responsibility. In J. P. Tangney & K. W. Fischer (Eds.), *Self-conscious emotions* (pp. 143–173). New York: Guilford.

Zahn-Waxler, C., Schiro, K., Robinson, J. L., Emde, R. N., & Schmitz, S. (2001). Empathy and prosocial patterns in young MZ and DZ twins: Development and genetic and environmental influences. In R. N. Emde & J. K. Hewitt (Eds.), *Infancy to early childhood: Genetic and environmental influences on developmental change* (pp. 141–162). New York: Oxford University Press.

Zakowski, S. G., Hall, M. H., Klein, L. C., & Baum, A. (2001). Appraised control, coping, and stress in a community sample: A test of the goodness-of-fit hypothesis. *Annals of Behavioral Medicine, 23,* 158–165.

Zane, N., & Yeh, M. (2002). The use of culturally based variables in assessment: Studies on loss of face. In K. Kurasaki, S. Okazaki, & S. Sue (Eds.), *Asian American mental health: Assessment theories and methods* (pp. 123–138). Dordrecht, Netherlands: Kluwer Academic.

Zapf, D., Seifert, C., Schmutte, B., Mertini, H., & Hotz, M. (2001). Emotion work and job stressors and their effects on burnout. *Psychology and Health, 16,* 527–545.

Zarit, S. H., & Eggebeen, D. J. (2002). Parent–child relationships in adulthood and later years. In M. H. Bornstein (Ed.), *Handbook of parenting, Vol. 1* (2nd ed., pp. 135–161). Mahwah, NJ: Erlbaum.

Zarit, S. H., Stephens, M. A. P., Townsend, A., & Greene, R. (1998). Stress reduction for family caregivers: Effects of adult day care use. *Journal of Gerontology, 53B,* S267–S277.

Zeanah, C. H. (2000). Disturbances of attachment in young children adopted from institutions. *Developmental and Behavioral Pediatrics, 21,* 230–236.

Zelazo, N. A., Zelazo, P. R., Cohen, K. M., & Zelazo, P. D. (1993). Specificity of practice effects on elementary neuromotor patterns. *Developmental Psychology, 29,* 686–691.

Zelazo, P. R. (1983). The development of walking: New findings on old assumptions. *Journal of Motor Behavior, 2,* 99–137.

Zenger, T., & Lawrence, B. (1989). Organizational demography: The differential effects of age and tenure distributions on technical communication. *Academy of Management Journal, 32,* 353–376.

Zeskind, P. S., & Barr, R. G. (1997). Acoustic characteristics of naturally occurring cries of infants with "colic." *Child Development, 68,* 394–403.

Zevitz, R. G., & Gurnack, A. M. (1991). Factors related to elderly crime victims' satisfaction with police service: The impact of Milwaukee's Gray Squad. *Gerontologist, 31,* 92–101.

Zhang, Z., & Hayward, M. D. (2001). Childlessness and the psychological well-being of older persons. *Journal of Gerontology, 56B,* S311–S320.

Zhou, M., & Bankston, C. L. (1998). *Growing up American: How Vietnamese children adapt to life in the United States.* New York: Russell Sage Foundation.

Zigler, E., & Styfco, S. J. (2001). Can early childhood intervention prevent delinquency? A real possibility. In A. C. bohart & D. J. Stipek (Eds.), *Constructive and destructive behavior: Implications for family, school, and society* (pp. 231–248). Washington, DC: American Psychological Association.

Zigler, E. F., & Finn-Stevenson, M. (1999). Applied developmental psychology. In M. H. Bornstein & M. E. Lamb (Eds.), *Developmental psychology: An advanced textbook* (4th ed., pp. 555–598). Mahwah, NJ: Erlbaum.

Zigler, E. F., & Gilman, E. (1998). The legacy of Jean Piaget. In G. A. Kimble & M. Wertheimer (Eds.), *Portraits of pioneers in psychology* (Vol. 3, pp. 145–160). Washington, DC: American Psychological Association.

Zigler, E. F., & Hall, N. W. (2000). *Child development and social policy: Theory and applications.* New York: McGraw-Hill.

Zimmerman, B. J. (2002). Achieving academic excellence: A self-regulatory perspective. In M. Ferrari (Ed.), *The pursuit of excellence through education* (pp. 85–110). Mahwah, NJ: Erlbaum.

Zimmerman, B. J., & Risemberg, R. (1997). Self-regulatory dimensions of academic learning and motivation. In G. D. Phye (Ed.), *Handbook of academic learning: Construction of knowledge* (pp. 105–125). San Diego: Academic Press.

Zimmerman, M. A., & Arunkumar, R. (1994). Resiliency research: Implications for schools and policy. *Social Policy Report of the Society for Research in Child Development, 8*(4).

Zimmerman, M. A., Copeland, L. A., Shope, J. T., & Dielman, T. E. (1997). A longitudinal study of self-esteem: Implications for adolescent development. *Journal of Youth and Adolescence, 26,* 117–141.

Zins, J. E., Garcia, V. F., Tuchfarber, B. S., Clark, K. M., & Laurence, S. C. (1994). Preventing injury in children and adolescents. In R. J. Simeonsson (Ed.), *Risk, resilience, and prevention: Promoting the well-being of all children* (pp. 183–202). Baltimore: Paul H. Brookes.

Zlotnick, C., Kohn, R., Peterson, J., & Pearlstein, T. (1998). Partner physical victimization in a national sample of American families: Relationship to psychological functioning, psychosocial factors, and gender. *Journal of Interpersonal Violence, 13,* 156–166.

Zorilla, E. P., DeRubeis, R. J., & Redei, E. (1995). High self-esteem, hardiness and affective stability are associated with higher basal pituitary–adrenal hormone levels. *Psychoneuroendocrinology, 20,* 591–601.

Zucker, A. N., Ostrove, J. M., & Stewart, A. J. (2002). College-educated women's personality development in adulthood: Perceptions and age differences. *Psychology and Aging, 17,* 236–244.

Zucker, K. J. (2001). Biological influences on psychosexual differentiation. In R. K. Unger (Ed.), *Handbook of the psychology of women and gender* (pp. 101–115). New York: Wiley.

Zwillich, T. (2002, November 19). U.S. system fails patients' dying wishes. *Reuters Health.* Retrieved from http://www.dwd.org/fss/news/reut.11.19.02.asp

Name Index

Italic *n* following page number indicates source for an illustration or figure.

A
Aarons, S. J., 360
Abbott, S., 122
Aber, J. L., 128
Abikoff, H., 289
Abra, J., 504
Abraham, S., 578
Abramovitch, R., 38
Abrams, L., 573
Abravanel, E., 223
Achenbach, T. M., 98
Acker, M. M., 255
Ackerman, B. P., 178, 300
Ackerman, P. L., 503
Ackerman, S., 513
Adams, A. M., 227
Adams, G. R., 385, 386
Adams, P. F., 484
Adams, R., 351
Adams, R. G., 532, 605
Adams, R. J., 107
Adamson, D. M., 464, 464*n*
Addington-Hall, J., 631
Aderman, D., 630
Adkins, G., 594
Adlaf, E. M., 361
Adler, N. E., 419
Adolph, K. E., 15, 131, 135, 139
Adzick, N. S., 56
Aerts, E., 391
Affifi, W. A., 455
Agüero-Torres, H., 569
Aguiar, A., 148
Ahadi, S. A., 181, 181*n*, 182
Ahlberg, C., 262
Aho, T. F., 555
Ainlay, S. C., 589
Ainsworth, M. D. S., 187, 187*n*, 189
Airhihenbuwa, C. A., 423
Aitchison, T., 496
Akers, J. F., 397
Akhtar, N., 236
Akiyama, H., 450, 531, 532
Akshoomoff, N. A., 120
Alan Guttmacher Institute, 355, 356
Albanes, D., 424
Albert, K. A., 472
Alberts, D. M., 303
Aldwin, C. M., 494, 495
Alessandri, S. M., 83, 367
Alexander, J. M., 95
Alexander, K., 212
Alexander, K. L., 375
Ali, L., 354
Alibali, M. W., 292
Allen, J. P., 360, 452
Allen, M., 452, 471
Allen, R., 347
Allen, S. E. M., 299
Allison, D. B., 420
Almeida, D. M., 394
Almey, M., 457
Alpert-Gillis, L. J., 263
Alsaker, F. D., 351
Altmaier, E. M., 623, 627
Alvarez, M. M., 394
Alzheimer's Association, 564
Aman, C. J., 285
Amaro, H., 429
Amato, P. R., 59, 330, 331, 468, 469
American Academy of Pediatrics, 356
American College of Obstetricians and Gynecologists, 430
American College Testing, 436
American Heart Association, 492, 492*n*
American Psychiatric Association, 288, 305
American Psychological Association, 37, 38*n*
Ames, C., 318
Ames, E., 119
Ames, E. W., 135
Amsel, E., 366*n*
Anand, S. S., 420
Ancoli-Israel, S., 347, 555
Anderman, E. M., 318, 371
Anders, T. F., 102
Anderson, C. A., 258
Anderson, D. M., 84–85, 235
Anderson, R. J., 168, 326
Andersson, B.-E., 162, 237
Andersson, J., 426
Andre, T., 326
Andreas, J. B., 335
Andrews, D. W., 324
Andrews, G., 285
Andrews, L. B., 55
Andrews, S., 221
Anesbury, T., 279
Anglin, J. M., 236, 299
Angold, A., 344
Angus Reid Poll, 638
Anisfeld, M., 128
Annahatak, B., 297
Annett, M., 206
Anslow, P., 95
Anstey, K. J., 556, 576
Antill, J. K., 467
Antonucci, T. C., 361, 450, 531, 532, 596, 600, 605
Apfel, N. H., 361
Apgar, V., 93, 93*n*
Aquan-Assee, J., 193
Aquilino, W. S., 395, 525
Arata, C. M., 429
Arber, S., 559
Archer, S. L., 385, 386
Arcus, D., 182
Ardelt, M., 576
Arend, R., 193
Arensman, E., 594
Ariès, P., 13
Arjmand, O., 438
Arking, R., 412, 415*n*, 486
Arlin, P. K., 19, 20, 434
Armstrong, G. K., 611
Armstrong, K. L., 121
Armstrong, T. D., 590
Arnett, J. J., 351, 383
Arnold, K., 438
Arnold, M., 299
Aron, A., 525
Aronne, L. J., 421
Aronson, M., 86
Arora, S., 124
Arsenio, W., 256
Artman, L., 286
Arunkumar, R., 11
Arvin, A. M., 48*n*
Asakawa, K., 31
Asher, S. R., 324, 326
Ashley-Koch, A., 49
Ashman, A. F., 304
Ashmead, D. H., 132, 133
Aslin, R. N., 106, 107, 133, 148, 169
Assmann, A., 576
Astington, J. W., 228
Astley, S. J., 86
Atchley, R. C., 599, 600, 609
Atkins, D., 56
Atkinson, J., 132, 154
Atkinson, J. O., 103
Atkinson, L., 194, 252
Atkinson, R. C., 21
Attie, I., 354
Au, T. K., 220
Auerbach, D., 532
Aunola, K., 372
Australian Bureau of Statistics, 329*n*
Avis, N. E., 486, 489
Avolio, B. J., 505, 534, 535, 536
Axia, G., 179, 300
Axinn, W. G., 467
Azmitia, M., 224

B
Baars, J., 586
Babbitt, B., 423
Babcock, R. L., 502
Bach-y-Rita, P., 121
Bäckman, L., 553, 571, 573, 576
Baddeley, A., 153
Badzinski, D. M., 247, 319
Baenninger, M., 370
Bagwell, C. L., 323, 397
Bahrick, H. P., 572
Bahrick, L. E., 139, 155
Bahrick, P. O., 572
Bai, D. L., 135
Bailey, D. A., 283
Bailey, G., 484
Bailey, J. M., 356, 522
Baillargeon, R., 148, 148*n*, 151
Baird, D. D., 77
Baker, D. W., 464
Baker, J. M., 299
Bakermans-Kranenburg, M. J., 189
Balaban-Gil, K., 83, 84
Balaswamy, S., 603
Baldwin, D. A., 229
Balfour, J. L., 491
Balk, D. E., 641
Ball, M. M., 600
Ballantyne, P. J., 609
Ballard, B. D., 279
Baltes, M. M., 570, 591, 592, 600, 612
Baltes, P. B., 8, 9, 23, 503, 520, 570, 575, 576, 612
Banaji, M. R., 557
Bancroft, J., 428
Band, G. P. H., 281
Bandura, A., 18, 254, 520
Bane, S. M., 496
Banish, M. T., 118
Bank, L., 328
Bankoff, E. A., 641
Banks, M. S., 106, 135, 135*n*
Bankston, C. L., 30, 31
Banta, D. H., 95
Barakat, L. P., 280
Barber, B. K., 266, 395
Barber, J. S., 461, 467
Barbie, A. H., 473
Barer, B. M., 548
Barker, D. J. P., 84, 88
Barker, R. G., 62
Barkley, R. A., 288
Barling, J., 374
Barnes, J., 316
Barnett, D., 187, 188, 189, 190
Barnett, R. C., 473, 474
Barnett, W. S., 234
Baron, R. C., 87
Baron-Cohen, S., 229
Baroni, R., 300
Barr, H. M., 85
Barr, R., 149, 156
Barr, R. G., 104, 105
Barr, S., 536
Barratt, M. S., 97
Barrera, M., Jr., 266
Barrett, K., 134
Barrett, K. C., 179
Barrett, M., 322
Barrios, B. A., 334
Barry, H., III, 345, 351
Barsky, A. J., 520
Barth, R. P., 57
Bartlik, B., 489
Bartrip, J., 138
Bartsch, K., 228
Basow, S. A., 394
Bass, B. M., 536
Bass, S. A., 612
Bassey, E. J., 424
Bates, E. A., 119, 120, 165, 168
Bates, G. W., 513
Bates, J. E., 58, 67, 180, 181, 184, 196, 266
Bathurst, K., 25, 333
Battin, D. B., 337
Bauer, P. J., 227
Baumeister, A. A., 103
Baumeister, R. F., 253
Baumrind, D., 265
Baumwell, L., 169
Bayer, A., 38
Bayley, N., 130*n*, 160
Beach, S. R. H., 459, 459*n*
Beard, B. B., 551
Beardsall, L., 35, 396
Bearison, D. J., 322, 352
Beatty, W. W., 260
Beauchamp, G. K., 106
Beausang, C. C., 349
Becker, B. J., 331
Becker, E., 62
Becker, J. B., 350
Beckerman, M. B., 504
Beckwith, L., 161
Begley, S., 422
Behrens, B. C., 459*n*
Behrman, R. E., 48*n*, 88
Beilin, H., 217
Beintema, D., 101*n*
Beitel, A. H., 192
Bélanger, A., 467*n*
Belanger, S., 234
Bell, E. A., 421
Bell, K. L., 438
Bell, M. A., 119, 135, 149, 183
Bell, M. L., 487
Bellamy, C., 125, 209, 461
Bellas, M. L., 537
Bellavance, 176
Belsky, D. H., 88
Belsky, J., 189, 190, 193, 526
Bem, D. J., 34
Bempchat, J., 308
Benbow, C. P., 369, 438
Bench, R. J., 106
Bender, S. L., 83
Bendersky, M., 83
Benedict, R., 179
Benenson, J., 324, 397
Benenson, J. F., 260
Bengtson, V. L., 519, 530, 533, 606
Bengtsson, B.-A., 414
Benini, F., 179
Bennett, F., 51
Bennett, J. M., 594
Bennett, P. J., 106
Bennetto, L., 229
Benoliel, J. Q., 620, 630*n*, 632
Benson, J. B., 132, 139, 150
Benson, P. L., 57
Berenbaum, S. A., 370
Berg, C. A., 503, 574, 575
Berg, S., 577
Bergen, D., 217
Berger, O. G., 87
Berk, L. E., 23, 27, 217, 223, 224, 376
Berkman, C. S., 559
Berkowitz, M. W., 391
Berkowitz, R. J., 353
Berkus, M. D., 95
Berlin, L. J., 99, 189
Berman, E., 525
Berman, P. W., 93
Bermejo, V., 231
Berndt, T. J., 266, 373, 397, 398
Berne, P. H., 247
Bernier, J. C., 289
Bernieri, F., 184
Bernston, G. G., 418
Bersoff, D. M., 392
Bertenthal, B. I., 134, 135, 137
Bertoncini, J., 106, 127
Bertrand, M., 473, 473*n*
Bertrand, R. M., 518
Berzonsky, M. D., 385
Betts, P. R., 208
Betz, C., 255
Betz, N. E., 471, 472
Bevan, E., 460
Beyene, Y., 488
Beyth-Marom, R., 355

Bhatt, R. S., 155, 155*n*
Bialystok, E., 300
Bianco, A., 89
Bidell, T. R., 20, 119, 287
Biederman, J., 288
Bielby, D. D. V., 586
Bielinski, J., 369
Biernat, M., 260
Bigbee, M. A., 325
Bigelow, A., 136
Bigi, L., 291
Bigler, R. S., 264, 326, 327
Bigner, J. J., 471
Bijeljac-Babic, R., 106
Binstock, R. H., 612
Birch, E. E., 106, 133, 134, 135
Birch, H. G., 181
Birch, L. L., 125, 208, 209, 279, 423
Birch, S. H., 251
Biringen, Z., 129
Birmaher, B., 400
Birmingham, C. L., 420
Biro, F. M., 349
Birren, J. E., 12, 517, 590, 625
Bissaker, K., 265
Bjork, E. L., 167
Bjorklund, D. F., 23, 152, 227, 290, 427
Black, A. E., 265
Black, D., 428
Black, J. E., 119
Black, K. N., 332
Black, S. H., 48*n*
Blacker, D., 566
Blackhall, L. J., 630
Blagg, N., 334
Blair, A., 424
Blair, S. N., 403
Blake, I. K., 297
Blakely, C. H., 268
Blakeslee, S., 459
Blamey, A., 496
Blanchard, R., 356
Blanchard-Fields, F., 504, 574, 585
Blanker, M. H., 487
Blasi, A., 388
Blasi, A. J., 591
Blass, E. M., 105, 127
Blass, J. P., 565, 566
Blatchford, P., 302
Bleske, A. L., 454
Blieszner, R., 532, 605
Blimling, G. S., 436
Blissmer, B., 496
Block, J., 361, 449, 523
Block, S. D., 629
Blomberg, S., 531
Bloom, L., 164, 167, 236
Bloom, P., 152
Blotner, R., 322
Bluck, S., 625
Bluebond-Langner, M., 629
Bluestein, D. L., 385
Bluestone, C., 266
Blum, M. J., 54, 530
Blum, N. J., 121
Blumberg, M. S., 103
Blumenthal, H. T., 56
Blyth, D. A., 352, 371, 371*n*
Boat, B. W., 336
Bock, G. R., 67
Boden, D., 586
Bodenheimer, T., 567
Bogaert, A. F., 356
Bogartz, R. S., 148
Bogden, J. D., 554
Bohannon, J. N., III, 165, 238
Bohlin, G., 126
Bohman, M., 57, 57*n*, 70
Bohr, L., 436
Boldizar, J. P., 263
Bolen, R. M., 336
Bolger, K. E., 268
Bolognani, F., 418
Bolton, D. P., 103
Bonett, D. G., 534
Bonichini, S., 179
Boniface, J., 149
Bonvillian, J. D., 165
Bookstein, F. L., 86
Boon, S. D., 527
Boonstra, H., 354
Booth, A., 59, 330, 331, 489
Borke, H., 219
Borkowski, J. G., 318
Borman, K. M., 437
Bornstein, M. H., 22, 32, 96, 158, 159, 169, 215
Borst, C. G., 93
Borstelmann, L. J., 13
Bortolus, R., 46, 46*n*
Bosquet, M., 118, 179
Bossé, R., 609, 611
Bost, K. K., 190, 252
Bosworth, H. B., 577
Botkin, D. R., 458
Bouchard, C., 205, 208, 347, 348*n*
Bouchard, T. J., Jr., 69
Boudreau, J. P., 134
Boukydis, C. F. Z., 104, 105
Boulton, M. J., 325
Bourgeois, M., 568
Bowen, R., 367
Bowlby, J., 7, 22, 186, 187, 192, 639
Bowlby, J. W., 374, 439
Boxer, A. M., 357
Boyce, W. T., 280
Boyer, K., 156
Boyes, M. C., 385
Boyle, G. J., 347
Braafladt, N., 401
Brabyn, J., 552
Brach, J. S., 547
Brackbill, Y., 95
Bracken, B. A., 232
Bradburn, E. M., 506
Bradbury, K., 211
Bradbury, T. N., 459, 459*n*
Braddick, O. J., 132, 154
Bradley, R. H., 99, 161, 161*n*, 162, 232
Brady, E. M., 578
Braeges, J. L., 256
Braet, C., 279
Braine, L. G., 282
Brambring, M., 137
Brame, B., 257
Brandtstädter, J., 593
Branstetter, W. H., 27
Bratt, R. G., 61
Braverman, P. K., 356
Bray, D. W., 472
Bray, J. H., 332, 469, 470
Brazelton, T. B., 107
Brébion, G., 573
Bredekamp, S., 163, 163*n*, 230, 235, 302
Bredemeier, B. J. L., 282
Breedlove, S. M., 370
Brehm, S. S., 455, 456
Brennan, K. A., 452
Brennan, M., 556
Brennan, P. F., 568
Brennan, W. M., 135
Brenner, E., 320
Brenner, R. A., 122
Brent, R. L., 50
Brent, S. B., 623, 624
Breslau, N., 640
Bretherton, I., 180, 187
Brett, J. M., 536
Brewaeys, A., 471
Brezina, T., 255
Bridges, L. J., 180
Bridges, M., 329
Brien, M. J., 360
Briggs, F., 336
Brigham, J., 425
British Broadcasting Corporation, 637
Broberg, A., 255
Broberg, A. G., 162
Brodaty, H., 567
Brody, G. H., 193, 266, 328
Brody, J. E., 430
Brody, L., 262
Brody, N., 60, 67, 162, 430
Broman, C., 459
Bromberger, J. T., 486
Bronfenbrenner, U., 24, 58, 68
Bronfman, E., 189
Bronson, G. W., 136
Brooks, C., 428
Brooks, D., 57
Brooks, G. R., 459
Brooks, P., 165, 168
Brooks, R. H., 399
Brooks-Gunn, J., 60, 63, 97, 99, 195, 346, 349, 351, 352, 353, 354, 360, 396, 457
Broomhall, H. S., 537
Brophy, J. E., 304
Brown, B. B., 398, 399, 400
Brown, D., 212
Brown, D. R., 528
Brown, G. K., 593
Brown, J. D., 354
Brown, J. R., 247
Brown, R. W., 236
Brown, S. L., 467
Brown, S. R., 195
Browne, K. D., 460, 608
Brownell, C. A., 245
Brownson, C., 473
Brubaker, T., 602
Bruce, D., 156
Bruch, H., 353
Bruch, M. A., 456
Bruck, M., 337
Bruer, J. T., 118
Brun, A., 426
Brussoni, M. J., 527
Bryden, M. P., 370
Buchanan, A., 267
Buchanan, C. M., 332, 344, 350
Buchanan, J. A., 608
Buchanan-Barrow, E., 322
Buekens, P., 98
Buhrmester, D., 35, 328, 397, 397*n*, 399, 454, 532
Buhs, E. S., 251, 325
Bukowski, W. M., 193, 324, 326, 397
Bulatao, R. A., 464
Bulka, D., 585
Bullinger, A., 133
Bumpass, L. L., 467*n*, 470
Burchinal, M. R., 161, 210, 232, 234
Burgess, K. B., 183, 250, 325, 326
Burgess, R. L., 104
Burgio, L., 568
Burhans, K. K., 246
Buriel, R., 58, 267
Burke, R. J., 473
Burke, W., 56
Burman, B., 59
Burnham, M. M., 102
Burns, C. J., 538
Burrell, N., 471
Burton, L., 533
Burts, D. C., 233, 266
Burtzeva, L. I., 86
Bushman, B. J., 258
Bushnell, E. W., 134
Buss, D. M., 427, 451, 454
Bussell, D. A., 396
Bussey, K., 256
Butler, K. M., 87
Butler, R., 315
Butler, R. N., 485, 586
Butler, S., 262
Buttell, F. P., 608
Butterworth, G. E., 131, 155
Buunk, B. P., 451, 466
Buur, H., 181
Byars, A. M., 473
Byler, P., 233, 303
Byrne, D. F., 321*n*
Byrnes, J. P., 369

C

Cabrera, N., 467
Caddell, D. P., 636
Caddy, D., 46
Cadoret, R. J., 70, 70*n*
Cahan, S., 286
Cahill, L., 560
Cain, C. A., 70*n*
Cain, K. M., 246
Caine, E. D., 592
Caine, N., 351
Cairney, J., 431
Cairns, E., 335
Cairns, R. B., 17, 322, 398
Caldwell, B. M., 161, 232
Caldwell, C. H., 361
Caldwell, J., 464
Calkins, S. D., 183
Callahan, D., 214
Callanan, M. A., 262
Calle, E. E., 420
Caltran, G., 104
Cameron, C. A., 237
Cameron, M. B., 303
Cameron, N., 115
Camino, C., 322
Camp, C. J., 577
Campbell, D. W., 183
Campbell, F. A., 163, 164*n*, 210
Campbell, J. R., 304, 369, 373
Campbell, L. D., 530, 605
Campbell, R., 300
Campbell, S. B., 176
Campbell, T., 150, 150*n*
Campbell, W. K., 316, 317, 317*n*, 383
Campos, J. J., 129, 134, 135, 175, 177, 178, 179
Campos, R. G., 105
Camras, L. A., 175, 177, 180
Canadian Broadcast Standards Council, 258
Canadian Fitness and Lifestyle Research Institute, 283, 347, 347*n*, 424, 496
Canadian National Council of Welfare, 60
Candy-Gibbs, S., 624
Canetto, S. S., 402
Canobi, K. H., 292
Cantisani, G., 461
Cantor, J., 258
Capaldi, D. M., 428
Caplan, M., 245
Cappeliez, P., 587
Carberry, M., 289
Carbery, J., 454, 532
Cardarelli, A. P., 336
Carden, A. D., 461
Cardinali, G., 556
Carels, R., 317, 384
Carey, S., 146, 150, 152, 287
Carey, W. B., 121
Carli, L. L., 536
Carlo, G., 392
Carlson, C., 384, 386
Carlson, C. L., 288
Carlson, K. J., 90
Carlson, S. M., 228
Carlton, M. P., 303
Carmelli, D., 415*n*
Carmichael, C. A., 221
Carmichael, S. L., 89
Carney, R. A., 370
Caro, F. G., 612
Carolson, C., 317
Carpendale, J. I., 230, 291
Carpenter, C. J., 262
Carpenter, M., 166, 167
Carrado, M., 460
Carrere, S., 462
Carriger, M. S., 161, 245
Carroll, D., 431
Carskadon, M. A., 347
Carstensen, L. L., 570, 595, 596, 596*n*, 597, 601, 612
Carta, J. J., 84
Carter, B., 456, 460
Carter, D. B., 262
Carus, F. A., 14
Casas, J. F., 257
Case, R., 20, 153, 214, 222, 282, 287, 365
Caserta, M. S., 560, 568, 602, 603, 639, 644
Casey, M. B., 215, 370
Casey, R. J., 252
Cashon, C. H., 148
Caspi, A., 34, 35, 36, 182, 183, 184, 351, 352, 451, 504, 522, 523
Cassel, C. K., 38*n*, 39
Cassel, J., 89
Cassidy, J., 189, 452
Cassidy, S. B., 49
Catalano, R. F., 425, 450
Catsambis, S., 369
Cavadini, C., 353
Cavanaugh, J. C., 526
Cavazzana-Calvo, M., 56
Cavell, T. A., 317
Ceci, S. J., 68, 286, 298, 337
Center, J. R., 493
Center for Communication and Social Policy, 258
Centers for Disease Control, 204
Cerella, J., 501
Cernoch, J. M., 106
Cervantes, C. A., 262
Chalmers, J. B., 320
Chan, R. W., 54, 471
Chance, G., 139
Chandler, M. J., 230, 291, 320, 385
Chandra, R. K., 88, 560
Chang, M., 95
Chang, S., 278
Chao, R. K., 266
Chao, W., 330
Chapman, R. S., 50, 165
Chappell, N., 537, 568, 598, 610
Charles, S. T., 601
Charman, T., 229
Charman, W. N., 483
Chase-Lansdale, P. L., 63, 89, 162, 210, 359, 360, 361
Chassin, L., 362
Chasteen, A. L., 466
Chatterjee, S., 427
Chatters, L. M., 591
Chavajay, P., 159, 286, 297
Chavira, V., 386
Che-Alford, J., 598

Chen, E., 280
Chen, R. T., 210
Chen, W. J., 288
Chen, X., 29, 185, 266, 326, 329, 606
Chen, Y., 504, 574
Chen, Y.-C., 87
Chen, Y.-J., 87
Chen, Z., 150, 150*n*
Cherlin, A. J., 466, 469
Cherney, S. S., 162
Cherry, K. E., 576
Chesney, M. A., 494
Chesney-Lind, M., 403
Chess, S., 7, 180, 181, 181*n*, 184
Chestnut, C. H., III, 493
Chi, M. T. H., 434
Chiang, W.-C., 152
Children's Defense Fund, 60, 64, 88, 90, 96, 126, 162, 163, 211, 330, 374, 377
Childs, B., 53
Childs, C. P., 24
Chisholm, J. S., 107
Chisolm, T. H., 484
Chiu, L. H., 316
Chochinov, H. M., 622
Chodirker, B., 48*n*
Chodorow, N., 427
Choi, N. G., 610
Choi, S., 168
Chollar, S., 424
Chomsky, C., 299
Chomsky, N., 165
Christ, A. E., 641
Christ, G. H., 641
Christ-Hazelhoff, E., 553
Christensen, A., 460
Christensen, H., 576
Christensen, K., 67
Christensen, K. A., 521
Christenson, S. L., 308
Christophe, A., 127
Chronis, A., 289
Chu, G. C., 453
Chung, K.-M., 288
Cibelli, C., 194
Cicchetti, D., 128, 188, 189, 190, 267*n*, 268, 401
Cicerelli, V. G., 532, 604, 605, 626, 638
Cigoli, V., 526
Citera, M., 611
Clark, E. V., 236
Clark, K. E., 252
Clark, R., 99, 192
Clarke, P. J., 609
Clarke-Stewart, K. A., 13, 331
Clasen, D., 400
Clausen, J. A., 351
Cleary, P. D., 520
Cleland, J. W., 104
Clement, M., 636
Cleveland, J., 535
Clifford, C., 421
Clifford, P., 96*n*
Clifton, R. K., 132, 133
Clingempeel, W. G., 470
Clipp, E. C., 600
Clore, G. L., 399
Coakley, J., 215
Coatsworth, J. D., 10
Coffey-Corina, S., 119
Cohan, C. L., 467
Cohen, K. M., 357
Cohen, L. B., 148
Cohen, S., 89, 208
Cohen, S. E., 161
Cohen-Shalev, A., 504
Cohn, J. F., 176
Coie, J. D., 257, 324, 325, 376
Coiro, M. J., 330
Coke, M. M., 594
Colan, N. B., 441
Colapinto, J., 261
Colby, A., 388, 390
Colcombe, S. J., 560
Colditz, G., 486
Cole, C. F., 235
Cole, D. A., 318
Cole, E. R., 513
Cole, M., 365
Coleman, M., 333, 459, 468, 469, 470, 607
Coleman, P. D., 549
Coleman, S. M., 255
Coley, R. L., 89, 359, 361, 470
Collaer, M. L., 257
Collie, R., 149
Collier, V. P., 301
Collings, P., 557
Collins, F. S., 56
Collins, N. L., 452
Collins, W. A., 24, 68, 257, 328, 351, 395
Colman, A. D., 108
Colman, L. L., 108
Colombo, J., 161
Colonia-Willner, R., 434
Coltrane, S., 333, 462, 465
Columbo, J., 154
Comijs, H. C., 608
Comunian, A. L., 392
Comstock, G. A., 234, 258
Concat, J., 125
Condie, S. J., 601
Conel, L., 117*n*
Conger, R. D., 176, 351
Connell, J. P., 180, 263, 318
Conner, D. B., 224
Connidis, I. A., 468, 604, 605
Connolly, J., 398, 399
Connolly, J. A., 217
Connolly, K. J., 215*n*
Connor, P. D., 86
Connors, L. J., 25
Conti-Ramsden, G., 137
Conwell, Y., 592, 593
Cook, P. W., 460
Cook, W. L., 453
Cooney, T. M., 456
Cooper, B., 391
Cooper, C. L., 431, 534
Cooper, C. R., 385, 437
Cooper, H., 559
Cooper, P. J., 176
Cooper, R. P., 97, 106, 169
Cooper, S., 54
Cooper, Z., 421
Coplan, R. J., 7, 250
Coppage, D. J., 155
Copple, C., 163, 163*n*, 230, 235, 302
Coppotelli, H., 324
Corbin, C. B., 283
Corboz-Warnery, A., 465
Corcoran, J., 360
Cordon, J. A. F., 525
Corey-Bloom, J., 567
Corkhill, A. J., 288
Cornelius, M. D., 83, 85
Cornelius, S. W., 504
Corr, C. A., 623, 624, 628, 641
Corrigan, R., 216, 217
Cosden, M., 83
Costa, P. T., Jr., 446, 517, 522, 523*n*
Costello, E. J., 344
Coté, S., 257
Cotton, D. B., 52*n*
Couch, K. A., 537
Couchoud, E., 247
Coughlin, S. S., 490
Coulehan, C. M., 190
Coulton, C. J., 268
Courage, M. L., 107, 157
Cournoyer, M., 197
Cowan, C. P., 25, 462, 462*n*, 463
Cowan, P. A., 25, 462, 462*n*, 463
Coward, R. T., 611
Cox, K., 96
Cox, M., 214
Cox, M. J., 59, 192, 631
Cox, S. M., 189
Coy, K., 351
Coyl, D. D., 397
Coyle, N., 519
Coyle, T. R., 227, 290
Coyne, E. T., 123
Crago, M. B., 297, 299
Craig, W., 325
Craik, F. I. M., 21, 550, 571, 573
Crair, M. C., 119
Cramond, B., 306
Cratty, B. J., 213*n*, 215, 281, 282
Craven, R., 246, 316
Crawford, J., 301
Crawford, S., 486, 489
Crawley, L., 633
Creasey, G., 452
Creasey, G. L., 217
Creer, T. L., 280
Crick, N. R., 257, 323, 324, 325
Cristofalo, V. J., 412, 413
Critser, G., 423
Crockenberg, S., 179
Crockett, L. J., 395
Crohan, S. E., 605
Cronk, C., 166
Crosby, F. J., 472
Crosby, L., 324
Cross, C. E., 36
Cross, D. R., 224, 228
Cross, S., 518
Crouter, A. C., 328, 369, 394, 451, 473
Crowe, R. R., 70*n*
Crowley, W. F., 370
Crowson, M., 229
Crowther, M. R., 590
Cruttenden, L., 149
Cruz, L. A., 553
Crystal, D. S., 309
Crystal, S., 569
Csikszentmihalyi, M., 307, 350, 434, 520, 576
Cubbins, L. A., 427
Cuddy-Casey, M., 624
Culbertson, F. M., 401
Cully, J. A., 587
Culnane, M., 87
Cumberland, A., 179
Cumella, S., 61
Cummings, E. M., 33, 176, 186, 194, 595
Cummings, J. S., 194
Cunningham, J. D., 467
Curnow, C., 416, 417*n*
Currie, D. H., 460
Curtin, S. C., 94
Curtiss, S., 165
Cushing, G., 362
Cushing, T. J., 362
Cutler, S. J., 611
Cutrona, C. E., 361

D

Dacey, J. S., 504
Dadds, M. R., 121
Dadson, M. J., 353
D'Agostino, J. A., 96*n*
Dahl, R. E., 84
Daigneault, J. P., 283
Dales, L., 210
Daltabuit, M., 104
Dalton, S. T., 466
Daly, K. A., 210, 277
Damhuis, I., 319
Damon, W., 315, 321, 322, 323, 324, 383
Daniels, D. H., 304
Daniluk, J., 351
Danis, M., 632
Dannemiller, J. L., 138
Dapretto, M., 167
Darling, N. E., 265, 372
Darlington, R., 234
Darroch, J. E., 64*n*, 355, 355*n*, 356, 356*n*, 359, 359*n*, 426
Darton-Hill, L., 123
Darville, R. L., 611
Darwin, C., 14
Datta-Bhutada, S., 83
Daugelli, A. R., 402, 601
Daugirdas, J. T., 359*n*
DaVanzo, J., 464, 464*n*
Davey, A., 606
Davidson, R. J., 118
Davies, E., 641
Davies, P. T., 176
Davies, T., 45
Davis, A. W., 562
Davis, B., 525
Davis, C. G., 645
Davis, H. P., 571
Davison, M. L., 369
Dawson, T. L., 390, 391
Dawson-Hughes, B., 486
Day, N. L., 86
de Frias, C. M., 573
de Haan, M., 154
de Lange, T., 413
De Lisi, R., 263, 299
De Raedt, R., 563
de Rivera, J., 249, 319
de St. Aubin, E., 447, 513
de Villiers, J. G., 228, 236
de Villiers, P. A., 228, 236
de Vries, B., 454, 625
de Waal, F. B. M., 7, 260
De Wolff, M. S., 189, 192
Deák, G. O., 236
Dean, L., 642
Deary, I. J., 294
Deater-Deckard, K., 184, 266
DeBaryshe, B. D., 404*n*
DeBoer, D. D., 469
Debus, R., 246, 316
DeCasper, A. J., 81, 106
Deffenbacher, J. L., 495
DeGarmo, D. S., 331, 469
DeGenova, M. K., 589
Degner, L. F., 620, 630*n*, 632
Dejin-Karlsson, E., 85
Dekovíc, M., 256, 351
Delgado-Gaitan, C., 297
Dell, D. L., 360
Dell, S. L., 334
DeLoache, J. S., 216, 227
DelVecchio, W. E., 522, 523*n*
DeMaris, A., 469
Dembo, M. H., 397
Dement, W. C., 102
Demetriou, A., 365, 366
Dempster, F. N., 288
Dempster-McClain, D., 506, 507
Denckla, M. B., 288
Dengel, D. R., 560, 563
Denham, S. A., 247, 259
Denner, J., 66
Denney, N. W., 504
Dennis, W., 131
Denny, M. A., 131
Derom, C., 206
DeRubeis, R. J., 497
deSchonen, S., 138
DeStafano, R., 210
Deutsch, F. M., 465
Deutsch, W., 300
Deveson, A., 550, 605
DeVoe, M., 585
DeVos, J., 148, 148*n*
Dewey, K. G., 124
Dewsbury, D. A., 22
Diamond, A., 149, 156
Diamond, L. M., 357, 399
Diamond, M., 116, 261
Diana, M. S., 399
Diaz, R. M., 223, 224
Dick, D. M., 351
Dick-Read, G., 94
Dickson, K. L., 177
Dickson, S. V., 292
DiClemente, R. J., 358
Diehl, M., 519, 520, 586, 587
Diener, E., 521
Dietrich, K. N., 87
Dietz, B. E., 595
DiGuiseppi, C., 211
DiLalla, L. F., 161, 183
Dildy, G. A., 89
Dillaway, H., 459
Dillon, M., 589, 591
Dimond, M. F., 568, 602, 639
Dingel, A., 155
DiNitto, D. M., 64
Dion, K. K., 453
Dion, K. L., 453
DiPerna, J., 303
DiPietro, J. A., 80, 103
Dirks, J., 298
Dishion, T. J., 324, 362
Dittmann-Kohli, F., 503
Dixon, J. A., 330
Dixon, L., 460
Dixon, R. A., 14, 502, 503, 570, 573, 577
Dodge, K. A., 58, 257, 266, 324, 325
Doeker, B., 208
Doherty, G., 162
Doi, M., 557
Doka, K., 640
Dolan, A., 156
Dolensky, E., 324, 397
Donaldson, G., 499
Donatelle, R. J., 353, 420, 424, 425, 431, 431*n*, 454*n*, 493
Dondi, M., 104
Donelan-McCall, N., 247
Dong, Q., 266
Donnerstein, E., 258, 429
Doole, M. M., 38
Dooley, D., 450
Dornbusch, S., 374
Dornbusch, S. M., 332, 372
Dorris, M., 85
Douglas, R. N., 290
Doussard-Roosevelt, J. A., 502
Dowd, M. D., 212
Dowd, R., 493
Dowdney, L., 266, 641
Dowell, I., 569
Dowling, W. D., 507
Downey, B., 528
Downey, D. B., 463
Downs, A. C., 350
Downs, R. M., 286
Doyle, A. B., 217
Doyle, G. C., 517
Drabman, R. S., 279
Drago-Severson, E., 308
Dreher, M., 84
Drell, M. J., 283
Drentea, P., 610
Drew, L. M., 529
Drewnowski, A., 553
Drotar, D., 126
Drozdovitch, V., 86
Druker, B. J., 56

Dryburgh, H., 354
Dryer, D. C., 451
Dubé, E. M., 357, 399
Duberstein, P. R., 592, 593
DuBois, D. L., 316, 324, 384
Dubrow, N., 645
Duck, S., 452
Duckett, E., 396
Duffy, S. A., 634
Dugan, E., 606
Dumbrigue, C., 436
Duncan, G. J., 60, 61
Duncan, R. M., 223
Duniz, M., 126
Dunkle, R. E., 601
Dunn, J., 35, 193, 196, 229, 247, 328, 396
Durbin, D. L., 398
Durham, M., 161, 232
Durlak, J. A., 645
Durrant, J., 255
Dusek, J. B., 394
Dustin, D., 630
Dutton, D. G., 460
Dweck, C. S., 246, 317
Dworkin, J. B., 376
Dybing, E., 103
Dye-White, E., 87
Dyer, C. B., 608
Dykstra, P. A., 604

E

Eacott, M. J., 156
Eagly, A. H., 451, 536
East, P. L., 361
Easterbrooks, A., 194
Easterbrooks, M. A., 195
Eaton, M. M., 317
Eaton, W. O., 183
Ebeling, K. S., 219
Eberhart-Phillips, J. E., 87
Eccles, J. S., 62, 282, 304, 318, 326, 350, 369, 370, 371, 372, 373, 384
Eckerle, D., 126
Edebalk, P. G., 531
Eder, R. A., 245
Edwards, C. P., 257, 260, 327
Edwards, J. N., 489
Edwards, M. H., 277
Efklides, A., 366
Egan, S. K., 257
Egeland, B., 7, 191, 192, 269, 336
Eggebeen, D. J., 526, 530, 606
Eggerman, S., 630
Ehrlich, M.-F., 573
Eiben, B., 52*n*
Eicher, S., 400
Eiden, R. D., 107
Eilers, R. E., 165
Einstein, G. O., 560, 573
Eisen, M., 315
Eisenberg, N., 27, 68, 116, 179, 248, 249, 260, 320
Eisenberg-Berg, N., 255
Eisenstat, S. A., 90
Ekerdt, D. J., 601
Ekman, P., 175
Ekstrom, B., 488
El-Sheikh, M., 33
Elardo, R., 161, 161*n*
Elder, G., 36
Elder, G. H., Jr., 5, 34, 36, 351, 525
Elias, C. L., 224
Eliason, G., 626, 627*n*
Elicker, J., 193
Elkind, D., 367, 368
Elkind, M. S., 567
Ellard, J., 644
Ellicott, A. M., 516, 517
Ellinson, S., 631
Elliott, D. S., 61, 403
Elliott, E. S., 317
Elliott, J. G., 334
Elliott, K., 83
Ellis, B. J., 349
Elman, J. L., 150
Elster, N., 55
Emanuel, E. J., 638
Emanuel, L. L., 638
Emde, R. N., 17, 181, 183
Emery, R. E., 25, 331, 333, 353, 461
Emory, E. K., 95
Engebretson, T. O., 494
Engel, S., 217
Engelhart, M. J., 566
Engell, D., 423
Engen, T., 553
Englund, M., 193
Engstrom, R., 499
Entwisle, D. R., 375
Epling, W. F., 19
Epstein, J. L., 25, 373
Epstein, L. H., 279
Epstein, S., 428
Erdley, C. A., 318
Erel, O., 59
Erickson, M. F., 336
Ericsson, K. A., 416
Erikson, E. H., 17, 244, 314, 382, 446, 512, 584
Eriksson, J. G., 84
Eron, L. D., 258
Espiritu, D. A. V., 565
Espy, K. A., 83, 161, 232
Essa, E. L., 624
Esser, G., 100
Esterberg, K. G., 507
Evans, D. E., 181, 181*n*, 182
Evans, G. W., 162, 419
Evans, R. M., 566
Everman, D. B., 49
Eysenck, H. J., 451

F

Fabes, R. A., 247, 249, 260, 320
Fabricius, W. V., 291
Fabsitz, R. R., 415*n*
Fagan, J. F., III, 128*n*, 154
Fagard, J., 132
Fagley, N. S., 24
Fagot, B. I., 196, 257, 262, 360
Fahrmeier, E. D., 286
Fairburn, C. G., 421
Fairclough, D. L., 638
Faison, N., 393
Falbo, T., 316, 329
Fall, C. H., 84
Faller, K. C., 336
Family Caregiver Alliance, 531, 568
Fanslow, C. A., 630
Fantz, R. L., 135
Farkas, J., 607
Farley, S. C., 461
Farmer, M. C., 391
Farr, M. J., 434
Farrant, K., 227
Farrell, M. P., 523
Farrington, D. P., 338, 403, 404
Farver, J., 250
Farver, J. M., 27, 159, 250
Fasig, L. G., 245
Fasouliotis, S. J., 55
Fattibene, P., 86
Faulkner, S. L., 455
Fearnow, M. D., 320
Featherman, D. L., 437, 595
Federal Interagency Forum on Child and Family Statistics, 233
Federico, A., 238
Feeney, B. C., 452
Feeney, J. A., 452, 462
Fegley, S., 374, 393
Fehr, B., 451
Feigenson, L., 152
Fein, G. G., 251*n*
Feinberg, M. E., 328
Feingold, A., 260
Feiring, C., 317, 336, 399
Feldhusen, J. F., 307
Feldman, D. H., 434
Feldman, H., 566
Feldman, J. F., 161
Feldman, R., 184, 462
Felice, M. E., 361
Felsman, D. E., 385
Fenson, L., 167, 168
Fenwick, K. D., 139
Fergus, M. A., 62
Ferguson, T. J., 249, 319
Fergusson, D. M., 124, 324, 402
Fernald, A., 167, 168
Fernald, L. C., 126
Fernandes, O., 83
Ferraro, K. F., 603
Ferree, M. M., 459
Ferron, J., 232
Fesseha, G., 335
Fiano, K., 95
Ficca, G., 121
Fichter, A. I., 354
Field, D., 586, 587, 605
Field, T. M., 98, 128, 129*n*, 176, 179, 249
Fiese, B., 162
Fifer, W. P., 106
Figueroa, R. A., 296
Filene, P. G., 634
Fincham, F. D., 318, 459, 459*n*
Fine, A., 454
Fine, M., 468, 469
Fingerman, K. L., 607
Finn-Stevenson, M., 64
Fins, A. I., 347
Firestone, R. W., 626
Fisch, S. M., 235
Fischer, K. W., 20, 119, 287
Fischhoff, B., 355
Fish, M., 190
Fisher, C. B., 27, 38
Fisher, D. E., 490
Fisher, D. M., 101, 101*n*
Fisher, J. A., 125, 208, 279
Fisher, J. E., 608
FitzGerald, D. P., 385
Fitzgerald, L. F., 472
Fitzgibbons, P. J., 552
Fivaz-Depeursinge, E., 465
Fivush, R., 156, 227–228, 262
Flake, A. W., 53, 56
Flamm, B. L., 96
Flanagan, C. A., 371, 393
Flannery, K. A., 207
Flavell, E. R., 221, 228, 230, 291, 319
Flavell, J. H., 151, 221, 228, 230, 247, 289, 291, 319
Fleeson, W., 520
Flegal, K. M., 278
Fleiss, K., 256
Fleming, J. I., 637
Fleshner, M., 418
Fletcher, A. C., 265, 372
Flippen, C., 610
Floccia, C., 127
Flood, D. G., 549
Flor, D. L., 266
Florian, V., 497, 624, 626
Florsheim, P., 360
Flum, M. A., 608
Flynn, J. R., 501
Fogel, A., 177, 180
Foltz, C., 366
Fonda, S. J., 600
Ford, J. M., 552
Forgatch, M. S., 331, 469
Forman, D. R., 197
Forsén, T., 84
Foster, G. D., 421
Fotinatos-Ventouratos, R., 534
Fouad, N. A., 438
Foulke, E., 553
Fowler, J. W., 590, 590*n*
Fowles, D. C., 254
Fox, M., 532
Fox, N. A., 118, 119, 135, 175, 183
Fozard, J. L., 484, 501, 552
Frackiewicz, E. J., 430
Framo, J. L., 456
Franco, M. C., 84
Franco, P., 84
Franklin, C., 360
Franklin, M., 436
Franks, B. A., 365
Franks, M. M., 531
Franz, C., 249
Frazier, J. A., 309
Frazier, L. D., 559, 587, 592
Frederick, P. D., 87
Fredericksen-Goldsen, K. I., 333
Frederickson, B. L., 596
Fredrick, L. D., 280
Fredrickson, B. F., 597
Freedman-Doan, C., 326, 371
Freeman, D., 345
Freeman, E. W., 430
French, K. E., 215
French, L. A., 285
Frensch, P. A., 370
Fretz, B. R., 537
Freud, S., 16
Freund, A. M., 9, 570, 587, 612
Frey, K. S., 263
Frick, J. E., 154
Fried, P. A., 84, 85
Friedman, J. M., 83, 85
Friedman, S. L., 226
Friesen, W., 175
Frijda, N., 175
Frisoni, G. B., 567
Frodi, A., 105
Frosch, C. A., 192
Frost, J. J., 355, 355*n*, 356, 356*n*, 359, 426
Frost, L. A., 299
Fry, C., 558
Fry, C. L., 521
Fry, P. M., 587, 591, 603
Fuchs, I., 392
Fukumoto, A., 252
Fuligni, A. J., 30, 31, 351, 369
Fuller, D., 392
Fuller, G. F., 563, 564*n*
Fuller, J. D., 631
Fuller, M. J., 350
Fuller-Thomson, E., 528
Fulton, R., 643
Fung, H., 246
Fung, H. H., 596, 597
Furman, E., 623
Furman, W., 35, 328, 399
Furstenberg, F. F., Jr., 360, 361
Fuson, K. C., 231, 293

G

Gabbard, G. D., 17
Gabbay, S. G., 601
Gabrel, C. S., 569, 598
Gaddis, A., 349
Gaffan, E. A., 176
Gaines, J., 103
Galambos, N. L., 334, 394
Gale, G., 98
Gallagher, A. M., 263
Gallagher, L., 46
Gallard, B. C., 103
Galler, J. R., 125, 126
Gallup Canada, 636, 636*n*
Galotti, K. M., 391
Gamoran, A., 374
Ganchrow, J. R., 127
Ganguli, M., 565
Ganiban, J., 188, 189, 190
Gannon, L., 488
Gannon, S., 184
Ganong, L. H., 333, 459, 468, 469, 470, 607
Ganzini, L., 638
Gao, G., 453
Garbarino, J., 61, 335
Garber, J., 349, 401
García, M. M., 257
García-Coll, C., 266, 386
Gardner, A., 45
Gardner, E., 633
Gardner, H. E., 29, 214*n*, 295, 296*n*, 305, 307
Garfinkel, I., 331
Garfinkel, P. E., 354
Garmezy, N., 10, 337
Garmon, L. C., 391
Garner, D. M., 354
Garnier, H. E., 375
Garrett, J., 632
Garrett, P., 232
Gasden, V., 470
Gaskins, S., 226, 250
Gates, G., 428
Gathercole, S. E., 227, 290
Gatz, M., 530, 594
Gaub, M., 288
Gauvain, M., 24, 285, 289
Gayle, B. M., 452
Gayle, H., 428
Gazmararian, J. A., 464
Ge, X., 176, 351
Geary, D. C., 21, 23, 152, 231, 260, 293, 308, 370
Geerlings, S. W., 592
Gelbaugh, S., 354
Gelfand, D. E., 557, 559
Gelfand, M. M., 489
Gellatly, A. R. H., 365
Gelles, R. J., 268
Gellin, B. G., 210
Gelman, R., 219, 220
Gelman, S. A., 167, 168, 219, 221, 236, 245, 326
Genessee, F., 300
George, C., 192
George, L. K., 598
George, S. A., 487
Gerdes, A. C., 326
Gerhard, A., 452
Gershoff, E. T., 255
Gervai, J., 262
Geschwind, D. H., 51
Gesell, A., 14, 15
Gest, S. D., 324
Getchell, N., 213*n*, 414, 416
Gfeller, J. D., 587
Ghim, H. R., 137, 138*n*
Giarrusso, R., 528, 607
Gibb, R., 121
Gibbons, A., 45
Gibbs, J. C., 253, 256, 390, 391, 392
Gibbs, M., 532
Gibson, E. J., 134, 139
Gibson, F. L., 54
Gibson, J. J., 139
Gibson, R. C., 538, 569
Giebink, G. S., 210, 277
Giedd, J. N., 288
Gielan, U. P., 392

Gilbert, L. A., 473
Gilbert-Barness, E., 83
Gilbreth, J., 331
Gilbride, K., 158
Gillberg, C., 86
Gillespie, D. C., 119
Gillett, P. A., 560
Gillies, R. M., 304
Gilligan, C. F., 390, 427, 516
Gilling, P. J., 555
Gillman, M. W., 278
Gillmore, M. R., 398
Gilman, E., 19
Gilvarry, E., 362
Ginsburg, A. P., 135
Ginsburg, H. P., 28, 231
Ginter, S. F., 564
Ginzberg, E., 437
Giovanelli, G., 106
Giudice, S., 206
Giusti, R. M., 83
Gjerde, P. F., 401
Gladstone, J., 526, 529
Gladwell, M., 279
Glaser, R., 434
Glasgow, K. L., 372
Glazer, E. S., 54
Gleitman, L. R., 165
Glendenning, F., 608
Glik, D. C., 211
Glisky, M. L., 83
Globalist, 536
Glosten, B., 95
Glover, L., 55
Gluhoski, V. L., 641
Gnepp, J., 247
Godfrey, K. M., 84
Godkin, M., 633
Goelman, H., 64, 162
Goering, J., 61
Golant, S., 599
Gold, D. T., 532
Gold, P. E., 560
Goldberg, A., 398, 399
Goldberg, A. P., 560, 563
Goldberg, E. Z., 623
Goldfield, B. A., 168
Goldman, N., 567
Goldscheider, C., 456, 457
Goldscheider, F., 456, 457
Goldsmith, H. H., 183
Goldsmith, L. T., 29, 295
Goldstein, E., 472
Goldstein, M. Z., 489
Golomb, C., 214
Golombok, S., 54, 471
Golub, M. S., 95
Gomez-Schwartz, B., 336
Göncü, A., 217
Gonzales, N. A., 61
Gonzales, R. C., 472
Good, T. L., 304
Goode, J. A., 67
Goodlet, C. R., 86
Goodlin-Jones, B. L., 102
Goodman, C., 601
Goodman, E., 54
Goodman, G. S., 337
Goodman, K. S., 292
Goodman, L. A., 176, 429
Goodman, S. H., 176
Goodnow, J. J., 255
Goodwin, D. W., 425
Gopher, D., 502
Gopnik, A., 155, 167, 168, 178
Gordis, E. B., 268
Gordon, C., 519
Gordon-Salant, S., 484, 552
Gormally, S., 105
Gortmaker, S. L., 279, 279*n*, 420
Goss, D. A., 277
Goswami, U., 150, 220
Gott, V. L., 48*n*
Gottesman, I. I., 67, 68
Gottfredson, L. S., 438
Gottfried, A. E., 25, 333
Gottfried, A. W., 25, 333
Gottlieb, G., 71, 71*n*
Gottman, J. M., 452, 459*n*, 462, 468, 601
Goubet, N., 132
Gould, E., 549
Gould, J. L., 45
Gould, M. S., 402, 403
Goya, R. G., 418
Graber, J. A., 346, 353, 457
Grady, C. L., 550
Graham, S. A., 236
Graham-Bermann, S. A., 324
Gralinski, J. H., 196, 197
Granot, M., 94
Grantham-McGregor, S. M., 88, 126, 278
Grattan, E., 61
Graves, J. E., 417
Gray, J. H., 610, 611
Gray, M. R., 265
Gray, R., 84
Gray-Little, B., 317, 384
Green, A. D., 360
Green, F. L., 221, 228, 230, 291, 319
Green, G. E., 105
Green, J. A., 104
Greenbaum, C. W., 184
Greenberger, E., 60, 401
Greendorfer, S. L., 215
Greene, B. F., 212
Greene, K., 368
Greenfield, P. M., 23, 24, 131, 297, 298, 370
Greenhill, L. L., 289
Greenlee, R. T., 489, 490*n*, 491, 491*n*
Greenough, W. T., 116, 119
Gregg, V., 392
Gresham, F. M., 305
Gresser, G., 625
Griffin, S., 287
Griffiths, A., 609
Grigorenko, E. L., 296
Grody, W. W., 48*n*
Grolnick, W. S., 180, 372
Groome, L. J., 103
Gross, M., 307
Grossbaum, M. F., 513
Grossman, A. H., 601
Grossman, J. A., 403
Grossman, P. B., 317
Grossmann, K., 188
Grotevant, H. D., 57, 383, 385, 437
Grotpeter, J. K., 323, 324, 325
Grubb, W. N., 439
Gruetzner, H., 567
Grusec, J. E., 254
Grych, J. H., 192
Grzywacz, J. G., 424
Guarnaccia, C. A., 633
Guildner, S. H., 600
Guilford, J. P., 306
Guinn, B., 611
Gulko, J., 326, 327
Gullette, M. M., 517
Gullone, E., 334
Gundy, V., 431
Gunnar, M. R., 105, 119, 182
Gunter-Hunt, G., 635
Guo, G., 463
Gupta, S., 467
Gupta, V., 606
Gurland, B. J., 559
Gurnack, A. M., 598
Gustafson, G. E., 104, 105
Gutmann, D. L., 521
Guttuso, T., Jr., 486
Guyer, C. G., II, 460
Gwiazda, J., 106, 133, 135
Gwinn, M., 56

H

Haake, R. J., 225
Haan, N., 391
Hack, M. B., 96
Hackett, G., 473
Haddad, F. F., 56
Haden, C. A., 156, 227–228
Hader, S. L., 428
Hafdahl, A. R., 317, 384
Hagan, R. I., 262
Hagberg, B., 86, 550
Hagberg, J. M., 417, 560, 563
Hagekull, B., 126
Hagen, E. P., 232, 293
Hahn, S., 502
Haight, B. K., 586
Haight, W. L., 158, 217
Haimerl, F., 122
Haine, R. A., 227
Haith, M. M., 132, 139, 147, 150, 152
Hakuta, K., 300
Halbreich, U., 430
Halfon, N., 280
Halford, G. S., 222, 285, 365
Halford, W. K., 459*n*
Hall, D. G., 236
Hall, G. S., 14, 344
Hall, J. A., 263
Hall, N. W., 60, 62
Hallet, M., 121
Halliday, J. L., 51
Hallinan, M. T., 374
Halperin, J. M., 289
Halpern, C. T., 353, 354
Halpern, D. F., 49, 369
Halpern, L. F., 103
Halverson, C. F., 263, 264*n*
Ham, M., 350
Hamachek, D., 447, 513
Hamberger, L., 77*n*, 79, 80
Hamberger, L. K., 460
Hamer, D. H., 356
Hamilton, M. A., 374, 441
Hamilton, S. F., 374, 441
Hamm, J. V., 334
Hammer, L. B., 531
Hammer, S. J., 210
Hammond, C. J., 551
Hampson, E., 370
Han, J. J., 156
Han, S. K., 538
Handcock, M. S., 427
Hanley, W. B., 47
Hans, S. L., 189
Hansen, D., 376
Hansen, M., 461
Hanson, L. C., 632
Hanson, T. L., 330
Hansson, K., 459, 467
Happé, F. G. E., 229
Hardy, M. A., 599
Hare, J., 471
Hareven, T. K., 36
Harkness, S., 472
Harlan, E. T., 197, 256
Harlan, S. L., 63
Harley, B., 300
Harlow, H. F., 185
Harman, D., 548
Harmon, R. J., 195
Harmon, E. L., 395
Harold, R. D., 62, 282, 326, 373
Harootyan, R. A., 533
Harpaz, D., 492*n*
Harris, G., 106
Harris, J. R., 70
Harris, K. M., 30, 361
Harris, M. J., 184
Harris, P. B., 530
Harris, P. L., 228
Harris, R. L., 516, 517
Harrison, A. O., 62
Harrison, K., 258
Harrison, L. J., 191
Harrist, A. W., 326
Hart, 233
Hart, B., 161, 162, 169, 238
Hart, C. H., 233, 265, 325
Hart, D., 315, 383, 393
Hart, H. M., 513
Hart, S., 176
Harter, K., 59
Harter, S., 195, 245, 246, 315, 316, 319, 383, 384, 394
Hartshorn, K., 155
Hartup, W. W., 251, 324, 396, 453
Harvey, J. H., 454*n*
Harvey, M., 430
Harway, M., 461
Harwood, J., 607
Hasher, L., 502
Haskell, W. L., 496
Hassing, L. B., 577
Hassmén, P., 424
Hastings, P. D., 183
Hatch, E. E., 83
Hatch, M. C., 88, 89
Hatfield, E., 451, 452
Hatton, D. D., 136
Hatzipantelis, M., 370
Haught, P. A., 574
Hauser, S. T., 452
Hausfather, A., 162
Hauth, J. C., 83
Hawker, D. S. J., 325
Hawkins, J. D., 405, 425
Hawkins, J. N., 316
Hawkins, R., 336
Hay, J. F., 503
Hayes, B. K., 221
Hayflick, L., 547, 548
Hayne, H., 128, 149, 155*n*
Hays, J. C., 598
Hayslip, B., Jr., 528, 633
Hayward, C., 331
Hayward, M. D., 604
Haywood, K. M., 414, 416
Hazan, C., 452, 453
Hazell, L. V., 642
Hazelrigg, L. E., 599
Head Start Bureau, 233
Health Canada, 84, 86, 89, 96, 103, 124, 209, 211, 211*n*, 233, 267, 269, 280, 335, 358*n*, 421, 424, 425, 428, 548, 559, 567
Healy, M., 115
Heath, S. B., 60, 62, 297
Heaton, T. B., 458, 468
Heavey, C. L., 460
Hedberg, K., 638
Hedges, L. V., 296, 301, 301*n*, 369
Hediger, M. L., 97
Heese, E., 188
Heidema, J., 553
Heidiger, M. L., 88
Heidrich, S. M., 559
Heiman, M., 456, 460
Heine, S. J., 316
Heinl, T., 105
Heinz, A., 565, 566
Heinz, W. R., 374, 441
Helburn, S. W., 162, 234, 238
Held, R., 132
Heller, K., 606
Heller, W., 118
Helm, H. M., 591
Helson, R., 450, 513, 513*n*, 517, 519, 520, 522, 523, 529
Helwig, C. C., 322
Hembrooke, H., 337
Hemesath, K., 602
Henderson, D., 608
Henderson, R. L., 291
Henderson, S. H., 396
Hendin, H., 637
Hendrick, C., 451, 455
Hendrick, S. S., 451, 455
Hendricks, J., 611
Hendrie, H. C., 566
Hendrie, H. H., 566
Henry, W. E., 595
Heppner, M. J., 430
Herbener, E., 451
Herbert, T. B., 208
Herbst, J. H., 523
Herdt, G., 357
Herman, J., 300
Herman, M. R., 265
Hermann, M., 487
Hernandez-Reif, M., 139, 155
Heron, G., 483
Héroux, G., 219
Herrenkohl, L. R., 304
Herrnstein, R. J., 296
Hershberger, S. L., 402, 601
Hershey, D. A., 538
Herzog, A. R., 587
Hesketh, L. J., 50
Hespos, S. J., 151
Hetherington, E. M., 25, 59, 328, 329, 330, 331, 332–333, 396, 469, 470
Hetherington, S. E., 94
Heuveline, P., 428
Hewer, A., 388
Hewlett, B. S., 192
Heyman, G. D., 245, 246, 317
Hibell, B., 361, 362*n*
Hickling, A. K., 220, 291
Hicks, J., 573
Hier, D. B., 370
Hiester, M., 191
Higgins, D. J., 460
High, D. M., 38
High, K. P., 560
High, P. C., 231
Hightower, M., 497
Hill, D., 638
Hill, J. P., 394
Hill, R. B., 470
Hillman, J. L., 560, 5610
Hind, H., 208
Hinde, R. A., 22, 262
Hines, M., 257
Hines, S., 51
Hinton, R., 562
Hirsch, B. J., 324
Hirsch, C., 530
Hirschfeld, L. A., 221
Hirshorn, B. A., 528
Hitch, G., 227
Ho, C. S.-H., 293
Hoch, C. C., 555
Hochschild, A. R., 462
Hochwarter, W. A., 534, 534*n*
Hodapp, R. M., 50
Hodges, E. V. E., 325
Hodges, J., 189
Hodges, R. M., 285
Hodson, D. S., 560
Hoeksma, J. B., 184
Hoff, B., 461
Hoff, E., 59
Hoffman, H. J., 103
Hoffman, L. W., 253, 320, 333
Hoffman, S., 89

Hoffner, C., 247, 319
Hogan, D., 607
Hohne, E. A., 133
Hokoda, A., 318
Holden, G. W., 255, 256
Holland, J. C., 491*n*
Holland, J. L., 437
Holley, F. B., 154
Holmbeck, G. N., 344, 394, 395
Holmes, D. S., 516, 517
Holmes-Rovner, M., 487
Holobow, N., 300
Holtzman, D. M., 566
Hombo, C. M., 304, 369, 373
Hommerding, K. D., 188
Hood, B. M., 154
Hooker, K., 483, 559
Hooyman, N. R., 61, 65, 556, 607
Hope, S., 330
Hopkins, B., 131, 131*n*, 132
Hopkins, D., 638
Hopkins, J., 189
Hopp, F. D., 634
Hopper, S. V., 559
Hopson, J., 116
Horber, F. F., 486
Horgan, D., 237, 299
Horgas, A. L., 600
Horn, J. L., 433, 499
Horn, J. M., 67, 69, 297
Horne, R. S., 103
Horner, T. M., 177
Hornick, J. P., 608
Horowitz, A. V., 454*n*, 459
Horowitz, F. D., 18
Horowitz, J. M., 336
Horowitz, L. M., 451
Horton, S., 472
Horwood, L. J., 402
Hotz, V. J., 360
House, J. S., 420*n*
Houseknecht, S. K., 468
Housel, D. A., 641
Houston, D. M., 133
Howard, A., 472
Howe, M. L., 156, 157
Howe, N., 193
Howell, R. T., 45
Howell-White, S., 459
Howes, C., 250
Hoyer, W. J., 502
Hoyt, D. R., 529
Hoza, B., 326
Hrubi-Bopp, K. L., 571
Hsiao, C. K., 288
Hsu, C.-C., 87
Hsu, F. L. K., 453
Hu, F. B., 424
Huard, R. D., 289
Huck, S. M., 529
Hudson, J. A., 227
Huesmann, L. R., 258
Hughes, C., 228, 229, 247
Hughes, J. N., 317
Hulka, B. S., 425
Hultsch, D. F., 502, 570, 573, 577
Human Resources Development Canada, 374, 375*n*, 377
Humphrey, D., 636
Humphrey, J. A., 429
Humphrey, T., 105
Hunt, E., 86
Hunter, L. L., 210, 277
Hunter, M., 55
Huntsinger, C. S., 266, 309
Hurley, M., 211
Hurme, H., 527
Hursti, U. K., 208
Husaini, B. A., 591
Huston, A. C., 234, 235, 394
Huston, P., 451
Huston, T. L., 462
Huttenlocher, J., 219
Huttenlocher, P. R., 116
Huyck, M. H., 521, 522, 534, 602
Hyde, J. S., 99, 369, 427, 474
Hymel, S., 316
Hynd, G. W., 305

I

Iacono, W. G., 288
Ianni, F. A. J., 375
Idler, E. L., 591
Igumnov, S., 86
Ihinger-Tallman, M., 469
Ikkink, K. K., 606
Ilg, F. L., 15
Imhoff, A. R., 473
Ingersoll-Dayton, B., 531
Ingram, D. K., 485
Inhelder, B., 363, 367
Insabella, G. M., 329
Institute for Social Research, 458, 458*n*, 474
Iribarren, C., 494
Irvine, J. J., 318
Irwin, J. R., 105
Isaacowitz, D. M., 601
Isabella, R., 189
Ismail, M. A., 96
Israelashvili, R., 192
Ito, Y., 268
Ivey, P., 178
Iwamoto, K., 83
Izard, C. E., 175, 178

J

Jacklin, C. N., 260
Jackson, G. R., 483
Jackson, J. S., 569
Jacobs, J. E., 282, 326
Jacobs, J. K., 375
Jacobsen, R. B., 471
Jacobson, C. K., 468
Jacobson, J. L., 87
Jacobson, K. C., 395
Jacobson, M. F., 423
Jacobson, S. W., 87
Jacobvitz, D., 269
Jacoby, L. L., 503, 571
Jacques, P. F., 552
Jadack, R. A., 391
Jaffe, J., 189
Jaffee, S., 359, 360
Jain, A., 125
Jambor, T. W., 250
Jambunanthan, S., 266
James, D., 53
James, J. B., 517, 521
Jankowski, J. J., 136, 138*n*
Janosz, M., 375
Jansari, A., 572
Janssens, J. M. A. M., 256
Jarrold, C., 229
Jarvis, P. A., 217
Jason, L. A., 280
Jasper, M., 635, 636
Jayakody, R., 467
Jean, G., 300
Jeffery, R. W., 279
Jendrek, M. P., 528
Jenkins, J. M., 228
Jenkins, S. R., 461
Jennett, B., 621
Jennings, J. M., 571
Jensen, A. R., 68, 294*n*, 296
Jenson, H. B., 88
Jepson, K. L., 574
Jessor, R., 30
Jeynes, W. H., 292
Ji, G., 329
Jiao, S., 329
Jilinskaia, E., 550
Jimeniz-Sanchez, G., 53
Jing, Q., 329
Jodl, K. M., 332–333
Johannes, C. B., 489
Johannsson, G., 414
Johnson, C., 421
Johnson, C. B., 486
Johnson, C. L., 116, 529, 532, 606, 607
Johnson, G. R., 402
Johnson, H. L., 83
Johnson, J., 169
Johnson, J. G., 258, 258*n*
Johnson, K. E., 221
Johnson, M. E., 626
Johnson, M. H., 119, 134, 137, 138*n*, 206
Johnson, M. L., 114
Johnson, S. L., 279
Johnson, T. B., 86
Joint Centre for Bioethics, 636
Jones, C. J., 519
Jones, G. P., 397
Jones, J. L., 88
Jones, M. C., 351
Jones, R. M., 397
Jones, W. H., 456
Jongbloet, P. H., 49
Jordan, B., 93
Jordan, W. J., 304
Jorgensen, K. M., 105
Jorgensen, M., 348
Jose, P. E., 266, 309
Joseph, J., 598
Joshi, A., 451
Josselson, R., 385
Jovanovic, J., 369
Joyner, M. H., 291
Juffer, F., 57, 194
Julkunen, J., 494
Junk, V. W., 530
Juntunen, C. L., 536
Jusczyk, P. W., 106, 133, 167
Jussim, L., 304
Justice, E. M., 291
Juul, A., 345

K

Kabbani, N. S., 375
Kagan, J., 182, 183, 186*n*, 190, 252
Kagawa, Y., 485
Kahn, P. H., Jr., 391
Kahn, R. L., 548
Kail, R., 281, 288
Kaisa, A., 372
Kaiser, S. M., 298
Kaler, S. B., 196
Kalish, R. A., 628
Kalkman, D. L., 304
Kalof, L., 429
Kalsi, M., 483
Kamberk-Kilicci, M., 320
Kamerman, S. B., 98, 99
Kamo, Y., 527, 529
Kandall, S. R., 103
Kantrowitz, E., 419
Kao, G., 30, 30*n*
Kaplan, B. H., 89
Kaplan, D. L., 455
Kaplan, G. A., 491, 555
Kaplan, M., 503
Kaplan, N., 192
Kaprio, J., 348
Karadsheh, R., 236
Karau, S. J., 536
Karel, M. J., 594
Karlseder, J., 413
Karmiloff-Smith, A., 150
Karraker, K. H., 184
Kasl, S. V., 591
Kasl-Godley, J. E., 594
Kaslow, F. W., 459, 467
Kasser, T., 584
Kastenbaum, R. J., 625, 628, 629
Katchadourian, H., 360
Katzmarzyk, A., 422*n*
Kaufert, P., 488
Kaufman, A. S., 435, 499, 500, 500*n*, 570
Kaufman, R., 298, 298*n*
Kaufmann, P. M., 83
Kaus, C. R., 483, 559
Kausler, D. H., 502, 503, 573
Kavanaugh, R. D., 217
Kawakami, K., 183
Kay, M. A., 56
Kaye, K., 146
Kaye, W. H., 353, 354
Kazak, A. E., 280
Kazi, S., 366
Kearins, J. M., 291
Kearns, D. F., 608, 609
Kearsley, R. B., 186*n*
Keating, D., 365
Keefe, K., 373, 397, 398
Keeton, W. T., 45
Keiding, K., 348
Keil, F. C., 221, 236
Keith, J., 558
Keith, P. M., 451
Keith, T. Z., 373
Kellett, J. M., 561
Kelloway, E. K., 538
Kelloway, K., 374
Kelly, J., 330, 332
Kelly, S. J., 86
Kelly, W. M., 287
Kempe, C. H., 267
Kemper, S., 573, 574
Kennedy, C. E., 527
Kennedy, C. M., 209
Kennedy, G. E., 527
Kennedy, G. J., 592
Kennedy, J. L., 288
Kennedy, Q., 597
Kennell, J., 94
Kennet, J., 568
Kenrick, D. T., 451
Kenyon, B. L., 623
Kerber, R. A., 403
Kerckhoff, A. C., 441
Kermoian, R., 129
Kerridge, I. H., 636
Kessler, R., 400
Kettl, P., 592
Keyes, C. L. M., 503, 519
Keys, C. B., 455
Khoury, M. J., 490
Kibby, M. Y., 305
Kiechl, S., 103
Kiernan, K., 467
Killen, M., 256, 323
Kilpatrick, D. G., 362
Kilpatrick, S. W., 279
Kim, J. E., 537, 611
Kim, J. M., 322
Kim, M., 236
Kim, Y. K., 250
Kimmel, D. C., 38, 601
King, A. C., 496
King, D. A., 567
King, N. J., 334
King, R. M., 253
King, S. S., 369
Kingston, P., 608
Kinicki, A. J., 537
Kinney, D., 398
Kirchner, G., 283
Kirkby, R. J., 424
Kisilevsky, B. S., 81
Kisly, A., 89
Kistner, J., 249
Kitayama, K. E., 533
Kitayama, S., 246
Kite, M. E., 428
Kivett, V. R., 606
Kivivuori, J., 374
Kivnick, H. Q., 526
Kiyak, H. A., 61, 65, 556, 607
Klahr, D., 21, 153, 222
Klebanov, P. K., 161, 162, 233
Klebe, K. J., 571
Klein, A., 231
Klein, J., 408
Klein, P. J., 149
Kleinbaum, S., 467
Klerman, G. L., 520
Klesges, L., 84
Kliegman, R. M., 48*n*, 88
Kliewer, W., 320
Klim, P., 47
Klimes-Dougan, B., 249
Kling, K. C., 316
Klingner, J. K., 305
Klump, K. L., 353, 354
Knapp, L. G., 212
Knapp, M. L., 451
Knapp, P. K., 283
Knecht, S., 206
Knight, D. K., 224
Knight, R. G., 571
Knill, D. C., 134
Knobloch, H., 101*n*
Knoers, N., 48*n*
Knowles, R. B., 566
Knox, A. B., 578
Knudsen, A., 105
Kobak, R. R., 453
Kobayashi, Y., 224
Kochanska, G., 182, 189, 197, 253, 254, 256
Kochenderfer-Ladd, B., 325
Koestner, R., 249
Koff, T. H., 65
Kogan, L. R., 448
Kogan, N., 497
Kohatsu, E. L., 386
Kohen, D., 162
Kohlberg, L., 263, 388, 389, 392
Kohlendorfer, U., 103
Kohler, J. K., 57
Kohn, M., 638
Kohn, M. L., 505
Koivula, N., 424
Kojima, H., 184
Kolb, B., 121
Kolominsky, Y., 86
Kolvin, I., 336
Komp, D. M., 628
Konstantopoulos, S., 301, 301*n*
Kontiokari, T., 210
Kopera-Frye, K., 607
Kopp, C. B., 161, 196, 197
Korbin, J. E., 268
Korn, S. J., 184
Kornblith, A. B., 491
Korte, C., 606
Korten, A. E., 556
Koslowski, B., 107
Koss, M. P., 429, 430
Koss, P., 429
Kostelny, K., 61
Kosterman, R., 362
Kotch, J. B., 212, 268
Kotchick, B. A., 355
Kotler-Cope, S., 577
Kotre, J., 513, 514
Koury, S. H., 449
Kouvonen, A., 374
Kowalski, N., 347
Kozberg, S. F., 391
Krasij, V., 594
Kraemer, H. C., 35

Krafft, K., 224
Kraft, J. M., 484
Krakow, J. B., 197
Kral, R., 57
Kramer, A. F., 502
Kramer, B., 421
Kramer, R. A., 402
Krascum, R. M., 221
Krause, N., 594, 598
Krause, N. M., 594
Kravetz, S., 624
Kray, J., 502
Krebs, D. L., 391
Kressin, N. R., 609
Kreutzer, T., 7
Krevans, J., 253
Krishnan, V., 467
Kroger, J., 384, 385
Krogh, H. R., 327
Kronebusch, K., 533
Kronenfeld, J. J., 211
Kropf, N. P., 551
Krout, J., 611
Kruesi, M. J. P., 403
Krug, E. G., 402
Krüger, H., 439, 441
Ku, H.-C., 257
Kubitschek, W. N., 374
Kübler-Ross, E., 628
Kuchner, J., 180
Kuczynski, L., 255, 265
Kuebli, J., 227, 262
Kuhl, P. K., 128, 133, 168
Kuhn, D., 291, 366, 366*n*
Kuk, L. S., 385
Kulik, L., 537, 601
Kunzinger, E. L., III., 290
Kurdek, L. A., 468
Kurtz-Costes, B., 162, 291
Kutner, L., 215
Kuulasmaa, K., 547
Kuzmeskus, L. B., 608
Kynette, D., 573

L

La Greca, A. M., 334
Labouvie-Vief, G., 20, 433, 519, 520, 574, 585, 586, 587
Lachman, M., 495, 515, 516*n*
Lachman, M. E., 517, 518, 560
Ladd, B. K., 325
Ladd, G. W., 251–252, 252, 325, 326
Lagattuta, K. H., 247
Lagercrantz, H., 91
Lagnado, L., 335
Lahey, B. B., 289
Lai, P., 597
Laible, D. J., 247
Laihonen, A., 461
Laing, G. J., 211
Laird, R. D., 252, 324
Lam, T., 405
Lamaze, F., 94
Lamb, M. E., 162, 190, 192, 194, 234, 330, 465
Lambert, J. D., 330, 521, 525
Lambert, W., 300
Lamme, S., 586
Lampl, M., 114
Lampman-Petraitis, C., 350
Lanctot, K. L., 566
Landau, J. D., 573
Landry, S. H., 83
Lane, M. A., 485
Lang, F. R., 591, 595, 596, 596*n*
Lange, G., 291
Langer, J., 148, 152
Langer, N., 607
Langman, L., 149
Lansford, J. E., 454, 596
Lantz, P. M., 419
Lapchick, R., 283, 284
Lapsley, D. K., 368, 385
Larson, D. E., 345, 352
Larson, J., 645
Larson, J. H., 459
Larson, R., 350, 376, 396, 605
Larson, R. W., 399
Larson, S. L., 266, 309
Larsson, M., 553
Larzelere, R. E., 256
Latz, S., 122
Laub, J. H., 404
Laucht, M., 100
Laumann, E. O., 426, 427, 428
Laumann-Billings, L., 25, 461
Launer, L. J., 564, 565
Laurin, D., 560
Laursen, B., 59, 351
LaVoie, D., 587
Law, T. C., 469
Law Commission of Canada, 599
Lawrence, B., 535
Lawson, K. R., 154
Lawton, L., 533
Lawton, M. P., 61, 62, 586, 597, 635
Lazar, A., 623
Lazar, I., 234
Lazarus, B. N., 495
Lazarus, R. S., 431, 495
Leach, C. E. A., 103
Leahy, B. H., 441
Leaper, C., 168, 262, 326
Lease, S. H., 642
Leavitt, L. A., 97
Lebeaux, D. S., 299
LeCompte, D. C., 576
Lederer, J. M., 304
Lee, A. M., 215
Lee, C. L., 184
Lee, D. J., 595
Lee, D. M., 562
Lee, G. R., 603
Lee, H. C., 413
Lee, I. M., 403
Lee, J., 484
Lee, K., 237, 256
Lee, S. H., 87
Lee, S. Y., 308, 309
Lee, Y., 250
Lee, Y. H., 526
Leekam, S. R., 229
Leerkes, E., 179
Leevers, H. J., 228–229
Leger, D. W., 104
Lehman, D., 644
Lehman, D. R., 316, 365
Leibowitz, H. W., 564
Leichtman, M. D., 156, 337
Leinbach, M. D., 196, 257
Leiter, M. P., 534
Lemerise, E., 248
Lemery, K. S., 183
Lemme, B. H., 10, 562
Lens, V., 635
Leonard, K. E., 458
Lerner, R. M., 14, 27
Lesesne, C., 465
LeSieur, K., 251–252
Leslie, A. M., 229
Lester, B. M., 83, 84, 105, 107
Leung, M.-C., 322
LeVay, S., 356
Levenson, M., 495
Levenson, R. W., 468, 601
Leventhal, E. A., 575
Leventhal, J. M., 125
Levin, J. S., 591
Levine, C., 388
Levine, L. E., 245
Levine, L. J., 247
LeVine, R. A., 30, 60, 189
Levine, S. C., 370
LeVine, S. E., 60
Levinson, D. J., 18, 447, 448, 449, 515, 515*n*, 516, 517, 534
Levitt, A. G., 166
Levtzion-Korach. O., 136
Levy, B. R., 557
Levy, G. D., 326
Levy, J. A., 560, 561
Levy-Shiff, R., 57, 192, 463
Lewak, R., 451
Lewis, C., 229
Lewis, K. G., 466
Lewis, M., 83, 179, 183, 194, 195, 248, 336
Lewis, T., 440
Lewko, J. H., 215
Lewkowicz, D. J., 139
Li, D., 266
Li, Z., 326, 329
Liang, J., 594
Liaw, F., 97
Liben, L. S., 263, 264, 286, 327
Lichstein, K. L., 555
Lickliter, R., 97, 139
Lidz, C. S., 298
Lieberman, A., 565
Lieberman, M. A., 644
Liebert, R. M., 254
Liederman, J., 207
Liesi, E. H., 564
Light, P., 286
Lightdale, J. R., 278
Liker, J. K., 36
Lillard, A. S., 217
Limber, S., 177
Lin, C. C., 288
Lindenberger, U., 8, 502, 576
Lindner, H., 424
Lindsay, C., 457, 551, 561, 562, 563, 569, 589, 592, 596, 598, 599, 602, 609
Lindsay-Hartz, J., 249, 319
Lindsey, E. W., 252
Link, S. C., 347
Linn, M. C., 369, 370
Linz, D., 429
Lissens, W., 52*n*
Lister, J. L., 484
Liston, R., 95
Litovsky, R. Y., 133
Littell, S. W., 292
Littlejohn, K., 214
Litwin, H., 600
Liu, M., 266
Livson, N., 352
Lloyd, L., 304
Lock, M., 488
Locke, J., 13
Lockhart, K. L., 221
Lockhart, R. S., 21
Loeber, R., 289, 338, 404
Loeber, R. L., 404
Loehlin, J. C., 67, 69, 296, 297
Loftus, J., 428
Logan, R. L., 447
Logan, S., 211
Logsdon, R. G., 570
Lohr, M. J., 400
Lollis, S., 265
Long, D. D., 626
Long, H. B., 578
Long, J. D., 607
Long, M. V., 551
Lonigan, C. J., 169, 231
Looi, J. C. L., 567
Lopata, H. Z., 603, 641
Lopez, B., 229
Lorenz, K. Z., 22, 93
Lorenz, R. P., 89
Losey, K. M., 318
Losoya, S., 320
Louria, D. B., 554
Low, J. A., 81
Low, K. G., 494
Lozoff, B., 122
Lu, H. H., 467*n*
Lubart, T. I., 306, 434, 504
Lubinski, D., 369
Luborsky, M. R., 595
Lucas, D. E., 103
Ludemann, P. M., 138
Lue, T. F., 487
Lund, D. A., 568, 602, 603, 612, 639, 644
Lundblad, A., 459, 467
Lunenburg, F. C., 376
Luster, T., 265
Luszcz, M. A., 556
Luthar, S. S., 362
Lutz, A., 597
Lutz, D. J., 21
Lutzker, J. R., 212
Luzzo, D. A., 472, 506
Lyaginskaya, A. M., 86
Lydon, N. B., 56
Lynch, A. Q., 488
Lyness, K., 536
Lyon, T. D., 230
Lyons-Ruth, K., 189, 194
Lytle, L. A., 352
Lytton, H., 46

M

Maccoby, E. E., 260, 266, 328, 332, 394
MacDonald, S. W. S., 502
MacDonald, W. L., 469
MacDorman, M. F., 103
MacEwen, K. E., 538
Macfarlane, P. W., 424
MacKay, D. G., 573
Mackenbach, J. P., 419
Mackey, M. C., 94
MacKinnon, C. E., 330
Mackinnon, L. T., 424
MacKinnon-Lewis, C., 193
MacLean, W. E., 103
MacMillan, D. L., 305
MacMillan, H. L., 559
MacWhinney, B., 21, 153, 222
Madan-Swain, A., 280
Madden, D. J., 502
Madden, T., 386
Maddi, S. R., 496, 497
Maddox, G. L., 569, 595, 600
Madon, S., 304
Madrona, M. M., 95
Madsen, S. D., 328
Maggs, J. L., 334
Maglio, C. J., 645
Magni, E., 567
Magnuson, D., 376
Magnuson, K. A., 61, 386
Magnusson, D., 37, 352, 376, 456
Magolda, M. B. B., 432
Mahanran, L. G., 416
Maher, C. A., 24
Mahon, M. M., 623
Mahoney, A., 460
Mahoney, J., 633
Mahoney, J. L., 376
Mahoney, M. M., 468
Maibach, E. W., 209
Maier, D. M., 628
Maier, H., 577
Maier, S. F., 418, 560
Main, M., 187, 188, 192
Maitland, S. B., 500
Makin, J., 85
Makin, J. E., 84
Malaguarnera, L., 414, 417
Malatesta, C. Z., 180
Malina, R. M., 205, 208, 346, 347, 348*n*
Malkin, C., 158
Malkinson, R., 641
Mallinckrodt, B., 537
Malloy, M. H., 103
Malone, M. M., 627
Maloni, J. A., 90
Mandler, J. M., 140, 155, 157, 221
Mange, A. P., 46*n*
Mange, E. J., 46*n*
Mangelsdorf, S. C., 181, 192
Manke, B. A., 394
Mannell, R. C., 605, 611
Manson, J. E., 421, 424
Måntysaari, K., 210
Marano, M. A., 484
Maratsos, M., 165, 236
Marchman, V. A., 120
Marcia, J. E., 384, 385, 388
Marcon, R. A., 234
Marcus, G. F., 133, 166, 236
Marcus, J., 146
Marcus-Newhall, A., 497
Marcuse, E. K., 209
Marentette, P. F., 166
Margolin, B. H., 425
Margolin, G., 268
Markides, K. S., 431, 595
Markman, E. M., 146, 150, 236, 237, 364
Markovits, H., 324, 365, 397
Marks, N., 526
Marks, N. F., 330, 424, 521, 525, 530
Markstrom-Adams, C., 386
Markus, H. E., 567
Markus, H. R., 246, 518, 587
Marlier, L., 106
Marolla, J., 429
Marquis, J., 574
Marriott, A., 568
Mars Hill Group, 633
Marsh, H. W., 246, 315, 316
Marsh, J. S., 283
Marsh, R. L., 573
Marshall, N. L., 473
Marshall, T. R., 183
Marshall, V. W., 609
Marshall-Baker, A., 97
Martin, C. L., 184, 260, 262, 263, 264*n*, 318
Martin, J., 555
Martin, J. A., 32, 266
Martin, J. E., 642
Martin, P., 551, 594
Martin, T., 640
Martin-Matthews, A., 530
Martinez, I., 49
Martins, C., 176
Martinson, I. M., 641
Martlew, M., 215*n*
Martyn, C. N., 84
Maruna, S., 513, 514
Marx, J., 528
Masataka, N., 168
Mascolo, M. F., 249, 319
Maslach, C., 534
Mason, C. A., 400
Mason, M. G., 391
Massey, C. M., 219
Masten, A. S., 10, 396, 400
Mastropieri, D., 106
Masunaga, H., 433
Masur, E. F., 168, 289
Matas, L., 193
Matheny, A. P., Jr., 211
Matheson, C. C., 250
Matthews, K. A., 280

Matthews, L. G., 536
Matthews, T. J., 84
Mattson, S. N., 86
Matute-Bianchi, M. E., 386
Mauer, D., 217
Maurer, T. J., 535
Maxwell, L. E., 162
May, C. P., 502
Mayer, C. J., 601
Mayes, L. C., 83, 96, 129
Maylor, E., 14
Mayo Clinic, 552
Mazure, C. M., 176
Mazzeo, J., 304, 369, 373
McAdams, D. P., 447, 513, 514, 517
McAdoo, H., 265
McAuley, E., 496
McCabe, A. E., 220
McCall, M. E., 587
McCall, R. B., 160, 161
McCallum, F., 54
McCartney, K., 69, 184
McCarton, C. M., 99
McCarty, M. E., 132
McCauley, E., 51
McClearn, G. E., 67
McClenahan, E. L., 400
McCloskey, L. A., 262
McClowry, S. G., 641
McCombs, B. L., 304
McConaghy, M. J., 263
McCoy, J. K., 193, 328
McCrae, R. R., 446, 517, 522, 523*n*
McCue, M. K., 57
McCullough, P., 524
McCune, L., 217
McDaniel, M. A., 560
McDonald, L., 608
McDonough, L., 149, 155, 157
McElroy, S. W., 360
McFadden, S. H., 590
McGee, G., 55
McGee, L. M., 230, 230*n*
McGillicuddy-De Lisi, A. V., 322
McGoldrick, M., 456, 460
McGregor, K. K., 236
McGue, M. K., 67, 69, 288, 548
McGuffin, P., 67
McGwin, G., Jr., 563
McHale, J., 465
McHale, J. L., 192
McHale, S. M., 327, 328, 369, 394, 451
McIntyre, C. W., 289
McKee-Ryan, F. M., 537
McKenna, J. J., 122
McKenney, D., 528
McKenzie, P., 362
McKeown, R. E., 402
McKim, W. A., 426
McKusick, V. A., 47*n*, 48*n*, 49, 56
McLanaham, S., 331
McLaughlin, J., 454*n*, 459
McLean, D. F., 89
McLellan, J. A., 393
McLeod, J. D., 61
McLeod, P. J., 169
McLeskey, J., 305
McLoyd, V. C., 60
McMahon, T. J., 362
McManus, I. C., 206
McManus, K., 95
McMullen, K., 374, 439, 595
McMullin, J. A., 468
McNally, S., 249
McNamara, J. M., 599
McNamee, S., 322
McNeil, J. N., 625
McNeilly, D. P., 609
MCR Vitamin Study Research Group, 88
Meacham, R. B., 418
Mead, G. H., 315
Mead, M., 93, 345
Medina, R., 451
Meegan, S. P., 575
Meeus, W., 351, 384
Mehler, J., 106
Mehlmadrona, L., 95
Meinz, E. J., 502
Meister, C., 304
Meltzoff, A. N., 128, 129*n*, 138, 148, 149, 152, 155, 167
Mengelkoch, L. J., 417
Mennella, J. A., 106
Menon, U., 488, 497
Mercer, M. E., 107
Mercer, R. T., 517
Meredith, C. N., 560
Mergenhagen, P., 536
Merline, A., 531, 532
Merriam, S. B., 587
Merrill, D. M., 532
Mervielde, I., 279
Mervis, C. B., 221
Messinger, D., 177
Messinger-Rapport, B. J., 563, 564*n*
Messman, S. J., 454
Meyer, B. J. F., 575
Meyer, M. H., 537
Meyer-Bahlburg, H. F. L., 356
Meyers, C., 51, 52
Meyers, T., 176
Mezey, M., 631
Miccio, A., 210
Michael, R. T., 354, 356, 426, 428
Michaels, G. Y., 460
Michaels, S., 85, 85*n*
Midgett, C., 212
Midgley, C., 371
Migdal, S., 447, 513
Miguel, J., 419
Mihalko, S. L., 496
Mikulincer, M., 497, 626
Milberger, S., 289
Milgram, N. A., 10
Millar, W. S., 107
Miller, B. C., 57, 462
Miller, D. L., 506
Miller, J., 505
Miller, J. G., 392
Miller, J. Y., 425
Miller, K. S., 356
Miller, L. M. S., 573
Miller, L. S., 405
Miller, L. T., 153, 294
Miller, O., 591
Miller, P. A., 249, 320
Miller, P. H., 151, 227, 291
Miller, P. J., 158, 217, 246
Miller, R. B., 602
Miller, S. A., 151
Miller, S. S., 526
Mills, A., 631
Mills, M., 497
Mills, R., 119, 254
Millsap, R. E., 587
Milner, J. S., 268
Minde, K., 96
Minkler, M., 528
Mintz, J., 246
Mintzer, J. E., 566
Mischel, W., 197, 254
Mitchell, B. D., 403, 548
Mitchell, K. R., 636
Mitchell, V., 450, 519
Mize, J., 252, 265
Moane, G., 450, 529
Moen, P., 506, 507, 537, 538, 610, 611
Moerk, E. L., 164
Moffat, S. D., 370
Moffitt, T. E., 349, 404
Mogford-Bevan, K., 46
Mogilner, A., 120
Mojet, J., 553
Mokdad, A. H., 420, 421, 422*n*
Molfese, V. J., 161, 232
Moll, I., 225
Molnar, J., 423
Mondimore, F. M., 357
Mondloch, C. J., 137, 138*n*
Money, J., 51
Monk, C., 89
Monson, T. C., 257
Monsour, M., 454
Montemayor, R., 315
Montepare, J., 515, 516*n*
Montero, I., 223, 224
Montiero, C. A., 125
Moody, H. R., 38
Moody, R. L., 562
Moon, C., 106
Moon, S. M., 307
Moore, A., 603
Moore, C., 229
Moore, C. F., 320
Moore, D. R., 360, 626
Moore, E. G. J., 297
Moore, K. A., 360
Moore, K. L., 46, 52*n*, 53*n*, 76*n*, 77*n*, 78, 80, 82, 82*n*, 116
Moore, M. K., 128, 129*n*, 148, 149, 152, 626
Moore, R. W., 135
Moore, S. G., 107
Moore, S. M., 568
Moore, W. S., 432
Moorehouse, M. J., 333
Morabia, A., 349
Moran, G., 188, 189
Morell, C. M., 468
Morelli, G., 122, 178
Morgan, K., 611
Morgan, L. A., 525
Morgan, S. P., 360
Morgane, P. J., 88
Moriguchi, S., 554
Morioka, M., 621
Morley, J. E., 560
Morris, A. S., 351
Morris, J. E., 458
Morris, P. A., 24, 68, 470
Morrison, D. R., 330, 360
Morrison, F. J., 309
Morrison, H. I., 425
Morrongiello, B. A., 133, 139, 212, 281
Morrow, D., 502
Morse, C. K., 570
Mortensen, E. L., 125
Mortimer, J. T., 437, 439, 441, 456
Morton, J., 138
Moses, L. J., 178, 228
Mosher, M., 257
Moshman, D., 365, 366, 367, 382, 383
Mosko, S., 122
Moskowitz, D. S., 513
Mosteller, F., 301
Mounts, N. S., 398
Moxley, D. P., 436
Moyer, M. S., 532
Mrug, S., 326
Mu, X., 308
Mukamal, K., 612
Mulatu, M. S., 419
Muldoon, O., 335
Mullainathan, S., 473, 473*n*
Mullally, P. R., 246
Mullen, B., 518
Mullen, M. K., 156
Muller, F., 51
Muller, G. O., 268
Mulligan, J., 208
Mulsant, B. H., 565
Mumme, D. L., 175, 178, 179
Munoz, D. G., 566
Munro, G., 385
Muret-Wagstaff, S., 107
Muris, P., 334
Murray, A. D., 104, 169
Murray, C., 296
Murray, C. I., 624
Murray, K. T., 197, 256
Murray, L., 176
Murray, M. J., 418
Musick, K., 470
Mussen, P. H., 255, 351
Must, A., 348
Mutchler, J. E., 609
Muthukrisna, N., 318
Mutran, E. J., 630
Mutrie, N., 496
Muzio, J. N., 102
Muzzonigro, P. G., 357
Myers, D. G., 520, 521
Myers, G. C., 567
Myers, J. E., 525
Myerson, J., 501

N

Nader, K. O., 643, 645
Nagel, S. K., 60
Nagell, K., 166, 167
Nagin, D. S., 257
Nagy, W. E., 299
Naigles, L. G., 167
Najor-Durack, A., 436
Nakamura, J., 520
Namy, L. L., 167
Nánez, J., Sr., 134
Nansel, T. R., 325
Napier, A. Y., 525
National Association for the Education of Young Children, 163, 163*n*, 235
National Federation of State High School Associations, 347, 347*n*
National Hospice and Palliative Care Organization, 633
National Institute on Aging, 565, 566
National Task Force on the Prevention and Treatment of Obesity, 420
Navarrete, C., 49
Naveh-Benjamin, M., 572, 572*n*
Nawrocki, T., 249
NcNeilly, D. P., 608
Neal, M. B., 531
Neale, A. V., 608
Neiderman, D., 149
Neimeyer, R. A., 626, 645
Nelson, B., 602
Nelson, C. A., 117*n*, 118, 119, 122, 154, 179, 182, 486
Nelson, H. D., 7
Nelson, K., 156, 167, 168
Nelville, H. J., 118
Nesselroade, J. R., 595
Nestle, M., 423
Netto, D., 139, 155
Neugarten, B. L., 449, 450, 483, 515, 547
Neugarten, D., 547
Neuman, S. B., 230, 231
Neville, H. A., 430
Neville, H. J., 119
Newacheck, P. W., 280
Newborg, J., 213*n*
Newcomb, A. F., 324, 326
Newcomb, M. D., 375
Newcombe, N., 219, 370
Newcombe, P. A., 347
Newell, L. D., 265
Newman, A. B., 547
Newman, B. S., 357
Newman, C., 132
Newman, K., 419
Newman, M. J., 628
Newnham, J. P., 52*n*
Newport, E., 165
Newport, E. L., 133, 150, 300
Newsom, J. T., 591, 606
Newsome, M., 133
Newton, N., 93, 561
Newton, P. M., 447
Newton, R. R., 362, 636
Newton, T., 429
Ng, R., 597
Ng'andu, N., 232
Ni, Y., 285
NICHD Early Child Care Research Network, 64, 162, 190, 191, 234
Nichols, E. G., 517
Nichols, L. S., 530
Nichols, W. C., 468
Nickman, S. L., 641
Nidorf, J. F., 386
Nielsen, S. J., 423
Nieman, D., 424
Niemelä, M., 210
NIH Consensus Development Panel, 493
Nilsson, L., 77*n*, 79, 80
Ningiuruvik, L., 297
Nippold, M. A., 299
Nisbett, R. E., 297, 365
Nix, R. L., 265
Nolen-Hoeksema, S., 400, 401, 645
Noll, J., 499
Noom, M. J., 351
Noppe, I. C., 624, 625
Noppe, L. D., 624, 625
Nordstrom, B. L., 425
Norman, S., 573
Normand, J., 457
Norris, L., 504, 574
Notkola, I. L., 566
Nourse, C. B., 87
Novak, M., 506
Nowell, A., 296, 369
Nucci, L. P., 256, 322, 323, 395
Nuckolls, K. B., 89
Nugent, J. K., 107
Nuland, S. B., 620
Nurmi, J., 372
Nurmi, J.-E., 372
Nussbaum, J. F., 605
Nuttall, R. L., 370
Nye, B., 301, 301*n*
Nye, W. P., 591
Nyquist, L., 503

O

Oakes, J., 374
Oakes, L. M., 155
Obermeyer, C. M., 487, 487*n*
O'Boyle, C., 196
O'Brien, P. M. S., 84
O'Bryant, S. L., 604
O'Callaghan, M. J., 207
O'Connor, A. R., 277
O'Connor, C., 373
O'Connor, P., 455
O'Connor, T. G., 469
Oden, M. H., 449
Oden, S., 28
Oelschlager, J. R., 608, 609

Ogawa, J. R., 452
Ogbu, J. U., 318, 373
O'Grady-LeShane, R., 610
O'Halloran, C. M., 623, 627
Okagaki, L., 295, 305, 370
Okamoto, Y., 214, 222, 282
Oken, E., 278
Olafson, E., 336
O'Leary, S. G., 255
Oliker, S. J., 455
Oliver, M. B., 427
Oller, D. K., 165, 168
O'Loughlin, M., 366*n*
Olsen, J. A., 266
Olsen, O., 95
Olson, R. E., 416
Olson, S. F., 265
Olweus, D., 325
Ondrusek, N., 38
O'Neil, R., 60
O'Neill, R., 326
O'Neill, R. M., 472
Ong, A., 386, 395
Oosterwegel, A., 315
Openheimer, L., 315
Oppenheim, D., 192
O'Reilly, A. W., 158, 216
Oren-Vulfs, S., 425
Organization for Economic Cooperation and Development, 374*n*
Ornstein, P. A., 337
O'Rourke, N., 587
Orr, M. T., 375
Orvaschel, H., 624
Ory, M. G., 564
Osherson, D. N., 364
Osmond, C., 84
Osteoporosis Society of Canada, 493
Ostir, G. V., 559
Ostling, I., 538
Ostrea, A. R., 83
Ostrea, E. M., Jr., 83
Ostrove, J. M., 513, 513*n*
Otake, M., 86
Otomo, K., 165
Ovando, C. J., 301
Overgaard, C., 105
Overholser, J., 641
Overton, W. F., 366
Owen, M. T., 188
Owen, N., 496
Owsley, C., 483, 484, 552, 563
Oztürk, 207

P

Pace-Nichols, M. A., 468
Pacella, R., 277
Pachter, L. M., 266
Padgett, D. K., 567
Padula, M. A., 506
Paffenbarger, R. S., 403, 424
Pagani, L., 303, 403
Page, R. N., 374
Paglia, A., 361
Paik, A., 427
Paikoff, R. L., 396
Paley, B., 59
Palincsar, A. S., 304
Palisi, B. J., 520
Pallotta, J., 126
Palmer, G., 588
Palmore, E., 556
Palta, M., 96
Palti, G., 10
Pan, B. A., 237
Pan, H. W., 250
Papini, D. R., 385
Papousek, H., 105
Papousek, M., 105
Paquet, B., 525
Pardhan, S., 484
Park, D. C., 573
Park, M. M., 94
Park, R. W., 65
Park, Y., 288
Parke, R. D., 58, 192, 267
Parker, F. L., 234
Parker, J. G., 324, 326
Parker, R. G., 587
Parker, R. M., 464
Parker, S. J., 122
Parkes, C. M., 639
Parkin, A. J., 572
Parks, W., 87
Parmelee, P. A., 61, 62
Parrott, T. M., 533
Parsons, E., 189
Parten, M., 249
Pasamanick, B., 101*n*
Pascalis, O., 154
Pascarella, E. T., 435, 436, 438
Pasley, K., 469
Pasquino, A. M., 208
Patrick, E., 223
Pattee, L., 324, 326
Patterson, C. J., 54, 268, 356, 470, 471
Patterson, G. R., 176, 257, 259, 328, 356, 362, 404*n*, 469
Patterson, M. M., 488
Patterson, M. Q., 553
Pattison, P. E., 292
Patton, G. C., 353
Pauwels, B. G., 454*n*
Pavlik, V. N., 607
Pawelec, G., 554
Peake, P. K., 197
Peake, T. H., 608, 609
Pearce, D., 461
Pearce, K A., 504
Pearson, J. D., 415*n*
Pearson, J. L., 593
Pebody, R. G., 87
Pechmann, T., 300
Peck, R. C., 585
Pedersen, J. B., 561
Pedersen, W. C., 426
Pederson, D. R., 188, 189
Pedlow, R., 182
Peerson, S., 83
Pegg, J. E., 169
Peiser-Feinberg, E. S., 234
Pelham, W. E., Jr., 289
Pellegrini, A. D., 23, 309, 325
Penning, M. J., 468, 525
Pennington, B. F., 229
Penninx, B. W., 593
Peplau, L. A., 454
Pepler, D. J., 325, 399
Peres, J. R., 631, 632
Pérez-Pereira, M., 136
Perfetti, C. A., 292
Perie, M., 64*n*
Perkins, H. W., 520
Perlmutter, M., 227, 503
Perls, T. T., 550
Perner, J., 256
Perou, R., 465
Perrett-Clermont, A.-N., 286
Perris, E. E., 133, 155*n*
Perry, D. G., 257
Perry, W. G., Jr., 432
Perry-Jenkins, M., 473
Persad, C. C., 571
Persaud, T. V. N., 46, 52*n*, 53*n*, 76*n*, 77*n*, 78, 80, 82, 82*n*, 116
Pery, F., 537, 538*n*
Peshkin, A., 30
Peshkin, H., 352
Peters, J., 169
Petersen, A. C., 346, 370, 394
Petersen, N., 472
Peterson, B. E., 513
Peterson, C., 220, 337
Peterson, C. C., 450
Peterson, J., 322, 531
Peterson, L., 212, 605, 606
Petinou, K. C., 210
Petitto, L. A., 166
Petrun, C., 624
Pettit, G. S., 58, 251, 252, 266, 400
Pezaris, E., 370
Pezé, A., 132
Pfizer, 564, 565
Phelps, K. E., 220
Phillips, M., 304, 373
Phillips, S. D., 473
Phillips-Grant, K., 156
Phillipsen, L. C., 397
Phinney, J. S., 386, 395
Pi-Sunyer, X., 420
Piaget, J., 19, 28, 145, 146, 159, 216, 218, 223, 363, 367, 388, 432
Pianta, R., 336
Picano, J., 522
Piccinini, C. A., 107
Pick, A. D., 139
Pickens, J., 249
Pickering, L. K., 124
Pierce, K. M., 334
Pierce, S., 266
Pierce, S. H., 291
Pierce, W. D., 19
Pietz, J., 47
Pijnenborg, L., 636*n*
Pilkington, M. W., 102
Pillard, R. C., 356
Pinderhughes, E. E., 57, 60
Pines, A. M., 451
Pinker, S., 299
Pinquart, M., 594
Pinson-Milburn, N. M., 528
Pinto, J. P., 167
Pipes, P. L., 123
Pipp, S., 195
Pisoni, D. B., 106
Pivarnik, J. M., 88
Plassman, B. L., 566
Platsidou, M., 366
Plaud, J. J., 19
Plessinger, M. A., 83
Plomin, R., 67, 68, 69
Plude, D. J., 502
Plumert, J. M., 285
Poehlman, E. T., 485*n*
Pohl, J. M., 531
Pohl, R., 61, 62
Polansky, N. A., 268
Polka, L., 133
Pollard, M. S., 604
Pollatsek, A., 292
Pollitt, C., 236
Pollitt, E., 88, 278
Pollock, D., 635
Pollock, M. L., 417
Pomerantz, E. M., 317, 326
Ponjaert-Kristoffersen, I., 563
Poon, L. W., 551, 572*n*, 573, 594
Popkin, B. M., 125, 353, 423
Porter, R. H., 106
Portman, P. A., 283
Posada, G., 189
Posner, J. K., 61, 333–334, 334
Poston, D. L., Jr., 329
Potter, L. B., 402
Poulin-Dubois, D., 155, 219
Powell, B., 463
Powell, C., 88
Powell, D., 25
Powell, F. C., 626
Power, C., 330
Powlishta, K. K., 326, 327, 368
Powls, A., 207
Prager, K. J., 522
Pratt, M., 393
Pratt, M. W., 223, 391
Prechtl, H. F. R., 101*n*
Preisler, G. M., 136
Preiss, R. W., 452
Prencipe, A., 322
Pressley, M., 290, 291, 292
Previc, F. H., 206
Price, J. M., 251
Primus, L., 556
Princeton Religion Research Center, 588
Prior, M., 10, 11
Privacy Commissioner of Canada, 56
Prodromidis, M., 190
Profilet, S. M., 251–252
Prosser, E., 384, 386
Prosser, E. C., 317
Provins, K. A., 207
Pruchno, R., 528
Prussia, G. E., 537
Prysak, M., 89
Puder, K. S., 52*n*
Pugh, K. L., 551
Pungello, E. P., 162
Purcell-Gates, V., 231
Putnam, F. W., 336
Putnam, S. P., 182
Pyeritz, R. E., 49
Pyka, G., 560
Pynoos, J., 599

Q

Qiu, C., 566
Quadagno, J., 612
Quadflieg, N., 354
Quan, V. S. Y., 519
Querido, J. G., 166
Quick, H. E., 537, 611
Quill, T. E., 638
Quilligan, E. J., 96
Quinn, J. F., 609
Quinn, L., 563
Quinn, M. E., 551
Quinn, P. C., 157
Quinn, R. A., 121
Quintero, R. A., 52*n*
Quiroz, B., 297
Quist, J. F., 288
Quyen, G. T., 11

R

Raboy, B., 54, 471
Rader, E., 563, 564*n*
Radin, N., 333
Radke-Yarrow, M., 182, 253
Radziszewska, B., 224
Raeff, C., 297
Ragow-O'Brien, D., 633
Rahhal, T. A., 572*n*, 573
Rahn, W. M., 393
Räikkönen, K., 497
Rainey, B. B., 277
Raisler, J., 124
Rakison, D. H., 155
Ralston, S. H., 493
Ramey, C. T., 163
Ramey, S. L., 163, 234
Ramirez, J. D., 300
Ramos, E., 347
Ramsay, D. S., 183
Ramsey, E., 404*n*
Ramsey, F., 125, 126
Ramsøy, N. R., 467
Rando, T. A., 639, 641
Range, L., 641
Rank, M. R., 59
Ransford, H. E., 520
Rapp, S. R., 486
Rapport, M. D., 288
Rasmussen, C. H., 626
Rasmussen, R. L., 283
Ratcliff, K. S., 459
Rathunde, K., 576
Ratner, E., 631
Raugust, K., 235
Raup, J. L., 525
Ravussin, E., 279
Ray, J. A., 469
Rayner, K., 292
Raynor, D. A., 494
Raynor, R., 18
Razor, A. G., 349
Read, C. R., 438
Reay, A., 608
Reay, A. M., 608
Rebbeck, T. R., 490
Receputo, G., 415*n*
Reday-Mulvey, G., 610
Reddin, J., 438
Redding, N. P., 507
Redei, E., 497
Redl, F., 323
Reedy, M. N., 519
Rees, J., 189
Rees, M., 348
Reese, E., 156, 227–228
Reeves, J. B., 611
Regan, P. C., 451
Reichert, M., 600
Reid, H. M., 454
Reid, I. R., 486
Reid, J. B., 328
Reifman, A., 107
Reilly, A. H., 536
Reilly, J. S., 120
Reimer, J., 392
Reiser, J., 106
Reiser, L. W., 350
Reisman, J. E., 105
Reiss, D., 396
Reiter, S., 33
Reitzel-Jaffe, D., 460
Reker, G., 625
Rennie, H., 281
Repacholi, B. M., 178
Repetti, R. L., 473
Resnick, H. S., 429
Resnick, M. B., 100
Rest, J. R., 389, 390
Reynolds, C. R., 296, 298
Reznick, J. S., 183
Rhea, D. J., 353
Rhein, J. von, 503
Ricard, M., 320
Ricco, R. B., 366
Rice, J. K., 469
Rice, K. G., 385
Rice, S. J., 644
Rich-Edwards, J. W., 84
Richard, C., 122
Richard, L. S., 451
Richards, J. E., 154
Richards, L., 471
Richards, M. H., 350, 396
Richards, M. K., 125
Richards-Colocino, N., 362
Richardson, G. A., 83
Richardson, V. E., 603
Richgels, D. J., 230, 230*n*
Richie, B. S., 473
Rickel, A. U., 62
Rideout, R., 337
Ridley-Johnson, R., 101, 101*n*
Riedel, B. W., 555
Riesenberg, L. A., 645
Riggio, H. R., 455
Riley, E. P., 86
Rimé, B., 640
Rimm, E. B., 492*n*

Riordan, J., 95
Risemberg, R., 291
Risley, T. R., 161, 169, 238
Ritchey, L. H., 595
Ritchey, P. N., 595
Ritter, J., 362
Rittman, M., 608
Rivara, F. P., 280
Rivera, S. M., 148, 152
Rivers, C., 473
Roach, M. A., 97
Roazzi, A., 286
Robb, A. S., 353
Roberton, M. A., 213*n*
Roberts, B. W., 517, 520, 522, 523, 523*n*
Roberts, D., 63
Roberts, G., 609
Roberts, I., 211
Roberts, J. E., 161, 210, 232
Roberts, L. J., 458
Roberts, M. C., 212
Roberts, P., 447, 643
Roberts, R. E., 555, 593
Roberts, R. J., Jr., 285
Roberts, S. B., 485
Robertson, D. L., 506
Robertson, G. B., 608
Robinson, J., 256
Robinson, S. E., 645
Robinson, T. N., 279
Robison, J. I., 420
Rochat, P., 132, 166, 178, 195, 216
Rockwell, R. C., 525
Rockwood, K., 569
Rodemaker, J. E., 168
Rodgers, B., 330
Rodgers, R. F., 536
Rodkin, P. C., 324
Roe, K. M., 528
Roebers, C. M., 337
Roebuck, T. M., 86
Roeser, R. W., 371, 372
Roffwarg, H. P., 102
Roger, V. L., 492
Rogers, C., 90
Rogers, K., 374
Rogers, L., 353
Rogers, S. J., 229, 463, 468
Roggman, L. A., 190
Rogoff, B., 24, 157, 158, 159, 224, 225, 276, 285, 286, 291, 297
Rogosch, F., 268
Rohlen, T. P., 308
Roisman, A., 452
Roisman, G. I., 452
Roitfarb, M., 176
Rokach, A., 455, 455*n*, 456
Rollins, K. B., 220
Rolls, B. J., 421, 423
Rome-Flanders, T., 166
Rook, K. S., 450
Roopnarine, J. L., 244, 250
Rosander, K., 134
Roscoe, B., 399
Rose, A. J., 324, 326
Rose, S. A., 136, 138*n*, 161
Rose, S. P., 119
Rose-Krasnor, L., 255
Rosen, A. B., 220
Rosen, C. S., 221
Rosen, R. C., 427
Rosen, T. S., 83
Rosenbaum, J. E., 61
Rosenberg, H. J., 523
Rosenberg, S. D., 523
Rosenblatt, P. C., 642
Rosenburg, M., 83, 84
Rosenfeld, L. B., 451
Rosengren, K. S., 215, 220
Rosenman, R. H., 494
Rosenshine, B., 304
Rosenthal, C. J., 526, 529
Rosenthal, C. L., 533
Roskos, K., 230
Ross, C. E., 610
Ross, D. J., 525
Ross, J., 51
Ross, N. A., 419
Rossi, A. S., 522, 530
Rossi, P. H., 530
Roth, G. S., 485
Roth, J., 376
Rothbart, M. K., 67, 154, 180, 181, 181*n*, 182, 196, 197, 226
Rothbaum, F., 184, 188
Rothermund, K., 593
Rousseau, J. J., 13
Rovee-Collier, C. K., 128, 154, 155, 155*n*, 156
Rowe, D. C., 70
Rowe, J. W., 548
Royal College of Obstetricians and Gynecologists, 81
Rozin, P., 220
Rubenstein, C. M., 456
Rubin, D. C., 572, 572*n*, 573
Rubin, K. H., 7, 183, 250, 251*n*, 325, 326, 329
Rubin, L. R., 394
Rubin, S. S., 641
Rubinowitz, L. S., 61
Rubinstein, R. L., 604
Ruble, D. N., 184, 260, 262, 263, 318, 326, 349
Rudolph, D. K., 371
Ruff, H. A., 154, 182, 226
Ruffman, T., 149, 220, 228, 364
Ruhm, C. J., 538
Rumbaut, R. G., 30
Runco, M. A., 306
Russell, A., 265
Russell, J. A., 247
Russo, C., 575
Russo, N. F., 429
Rust, J., 262
Rutenberg, S., 524
Rutter, M., 10, 188
Rutter, V., 427
Rvachew, S., 210
Rybash, J. M., 571
Rydell, A., 126
Ryff, C. D., 396, 447, 503, 513, 518, 519, 525, 526, 559, 587

S

Saarni, C., 175, 178, 179, 319, 320
Sacco, R. L., 567
Sachdev, P. S., 567
Sackett, C. S., 552
Sadeh, A., 121
Sadler, T. W., 78
Safe Kids Worldwide, 210*n*, 212
Saffran, J. R., 133
Safire, W., 547*n*
Safyer, A. W., 441
Sagi, A., 188
Sahni, R., 103
Sais, E., 300
Sakinofsky, I., 402
Salamanca, F., 49
Salapatek, P., 135*n*, 137*n*
Salbe, A. D., 278
Salerno, M., 208
Salovey, P., 320
Salter-Ling, N., 55
Salthouse, T. A., 498, 500, 501, 502, 535
Samarel, N., 628, 629, 630, 630*n*
Sameroff, A. J., 372
Sampson, R. J., 404
Samson, A. V., 182
Samuels, M., 96
Samuels, N., 96
Samuelsson, S. M., 551
Sanchez, L., 556
Sanchez, R. P., 150, 150*n*
Sanders, D. M., 279
Sanders, M. G., 304, 373
Sanders, M. R., 459*n*
Sanders, P., 168, 326
Sanders, S. G., 360
Sanderson, J. A., 256
Sandman, C. A., 128
Sandnabba, N. K., 262
Sandstrom, M. J., 376
Sanner, T., 103
Sansavini, A., 106
Sansone, C., 574
Santoloupo, S., 393
Sapiro, C. M., 322
Sarason, S. B., 537
Sargeant, M. P., 67
Sarrazin, G., 294
Sasser-Coen, J. A., 504
Sataloff, J., 484
Sataloff, R. T., 484
Sattler, J. M., 232, 293
Saudino, K. J., 182
Sauer, M. V., 408
Saunders, S., 428
Savage, A. R., 51
Savary, L. M., 247
Saville-Troike, M., 168
Savin-Williams, R. C., 357, 399
Saxe, G. B., 24
Saxon, J. L., 317
Saxon, T. F., 154
Scabini, E., 526
Scanlon, K., 209
Scaramella, L. V., 359
Scarisbrick-Hauser, A., 604
Scarr, S., 62, 67, 69, 297
Scelfo, J., 354
Schaal, B., 106
Schafer, R. B., 451
Schaffer, J., 57
Schaie, K. W., 10, 432, 433, 434, 435, 435*n*, 499, 499*n*, 500, 501, 576, 577
Scharrer, E., 234, 258
Schauble, L., 366
Schaufeli, W. B., 534
Scheidt, R. J., 598
Schenker, J. G., 55
Schenning, S., 552
Scher, A., 121
Scherer, J. M., 636, 636*n*, 638
Schiefele, U., 318
Schiefenhoevel, W., 122
Schieman, S., 431
Schiller, M., 190
Schlackman, L. J., 95
Schlaud, M., 103
Schlegel, A., 345, 351
Schlesinger, M., 533
Schmidt, I. W., 572
Schmidt, K. L., 255
Schmidt, M. H., 100
Schmidt, U., 353
Schmitz, M. K. H., 279
Schmitz, S., 181
Schmutte, P. S., 526
Schneewind, K. A., 452
Schneider, B., 439
Schneider, B. A., 552
Schneider, B. H., 194, 252
Schneider, E. L., 414
Schneider, W., 290, 337
Schnohr, P., 556
Schnur, E., 234
Schofield, W., 88
Scholastic Testing Service, 306*n*
Scholl, B. J., 229
Scholl, T. O., 88
Scholnick, E. K., 226, 289
Schonert-Reichl, K. A., 391
Schonfeld, A. M., 86
Schooler, C., 419, 505
Schoppe, S. J., 181
Schor, J. B., 423
Schothorst, P. F., 97
Schroots, J., 12, 517
Schuengel, G., 189
Schulkind, M. D., 572
Schull, W. J., 86
Schulman, J. D., 48*n*
Schulman, K. A., 493
Schultz, R., 568
Schulz, R., 416, 417*n*, 591, 606, 630
Schut, H., 640
Schwanenflugel, P. J., 291
Schwartz, J. D., 96
Schwartz, P., 427
Schwarz, B., 600
Schwarzer, R., 559
Schwebel, D. C., 221
Schwebel, M., 24
Schweder, A. E., 376
Scott, J. A., 299
Scott, P., 221
Scott, P. M., 33, 253
Scott, W. K., 566
Scully, D., 429
Sears, P. S., 473
Seccombe, K., 11
Seefeldt, V., 281
Seeman, T. E., 594
Seiber, J. E., 38
Seier, W. L., 227
Seifer, R., 190
Seitz, V., 361
Seligman, L., 472
Seligman, M. E. P., 128
Selikowitz, M., 50
Selman, R. L., 251, 321*n*, 324
Seltzer, M. M., 396, 525, 526
Selwyn, P. A., 628, 645
Sen, M. G., 134
Senghas, A., 236
Senior, G. J., 137, 138*n*
Senn, C. Y., 429
Serbin, L. A., 260, 326, 327
Sermon, K., 52*n*
Seward, R. R., 98
Shackelford, T. K., 427
Shaffer, T. G., 262
Shainess, N., 349
Shanahan, M. J., 61, 439, 441
Shaner, T. L., 502
Shann, F., 209
Shaper, A. G., 424
Shapiro, A. E., 462
Shapiro, L. R., 227
Sharlach, A. E., 333
Sharma, A. R., 57
Sharma, M., 456
Sharp, K., 624
Shatz, M., 219
Shaver, P. R., 399, 452, 456
Shaw, D. S., 188
Shaw, G. M., 89
Shay, J. W., 413
Shedler, J., 361
Sheehy, A., 346
Shek, P. N., 554
Sheldon, K. M., 584
Shema, S. J., 555
Shephard, R. J., 554
Sheridan, S. M., 308
Sherman, A. M., 454, 596
Sherman, D. K., 288
Sherrill, C. L., 57
Shields, P. J., 128
Shields, R., 212
Shiff, M., 90
Shiffrin, R. M., 21
Shilling, T. H., 148
Shiloh, S., 52
Shimamura, A. P., 505
Shinskey, J. L., 148
Shiovitz, T. M., 430
Shochat, T., 555
Shock, N. W., 547
Shoda, Y., 197
Sholinsky, P., 415*n*
Shonk, S. M., 268
Shope, J. T., 568
Shore, L., 535
Short, R. V., 46
Shuchter, S. R., 640
Shulman, C., 229
Shulman, S., 193
Shultz, J. M., 553
Shumaker, S. A., 486
Shumow, L., 323, 334, 465
Shure, M. B., 259
Shweder, R. A., 7, 30
Sidle, A. L., 220
Siega-Riz, A. M., 353
Siegal, M., 256
Siegel, B., 305
Siegel, D. H., 289
Siegel, K., 641
Siegler, R. S., 21, 153, 231, 292
Siervogel, R. M., 346
Sigman, M., 50, 161, 278
Sigmundson, H. K., 261
Signorella, M. L., 263
Sigvardsson, S., 57, 57*n*
Silk, J. S., 395, 465
Silva, P. A., 182, 183
Silver, M. H., 550
Silverberg, S. B., 394, 525
Silverman, D. H. S., 565
Silverman, E., 641
Silverman, P. R., 641
Silverman, W. K., 334
Silverstein, M., 530, 533, 606, 607
Sim, T. N., 400
Simion, F., 104
Simmons, R. G., 352, 371, 371*n*
Simon, R. J., 636, 636*n*, 638
Simoneau, G. G., 564
Simonoff, J. S., 427
Simons, A. D., 353
Simons, R. L., 267, 330
Simonton, D. K., 434
Simpson, J. M., 103
Simpson, P. M., 83
Singer, B. H., 526
Singer, D. G., 235
Singer, J. L., 221
Singer, L. T., 128*n*
Singer, P. A., 634
Singh, S., 64*n*, 355, 355*n*, 356, 356*n*, 359, 359*n*, 426
Singleton, R., 589
Sinnott, J. D., 432, 433, 574
Skaff, M. M., 587
Skeen, P., 560
Skinner, B. F., 9, 164
Skinner, E. A., 317, 318
Skoe, E. S. A., 391, 391*n*
Skovronek, E., 502
Slaby, R. G., 258, 259, 263
Slade, A., 192
Slater, A., 139, 154
Slobin, D. I., 165
Slobodin, R., 631
Slomczynski, D. M., 505
Slomczynski, K. M., 505
Slomkowski, C., 35, 396

Slotkin, T. A., 91
Slusher, M. P., 601
Small, B. J., 502, 571, 573, 576
Small, M., 104
Smetana, J. G., 256
Smiley, P. A., 246
Smirles, K. A., 513
Smith, A. E., 288
Smith, B. A., 98
Smith, C., 465
Smith, C. M., 216
Smith, D. B., 610
Smith, D. C., 628
Smith, E. P., 149, 386
Smith, G. C., 520
Smith, I. D., 316
Smith, J., 8, 10, 11, 23, 576, 577, 587, 626, 627*n*
Smith, K. E., 98
Smith, L. B., 131, 157
Smith, M. C., 248
Smith, M. J., 573
Smith, M. L., 47
Smith, N. J., 210
Smith, P., 335
Smith, P. K., 309, 325, 527, 529
Smith, R. E., 283, 284
Smith, R. S., 7, 10, 100
Smock, P. J., 467
Smogorzewska, A., 413
Smoll, F. L., 283, 284
Smyth, K. A., 568
Snarey, J. R., 392, 513
Snidman, N., 182
Snow, C. E., 237
Society for Research in Child Development, 38, 38*n*
Soderstrom, M., 497
Soken, H. H., 139
Soldo, B. J., 533
Solimano, G., 125, 126
Solomon, B. B., 282
Solomon, C. R., 197
Solomon, J., 187
Solomon, J. C., 528
Solomonica-Levi, D., 229
Somary, K., 527
Somerville, S. C., 225
Sonn, U., 556
Sonntag, W. E., 414
Sophian, C., 222
Sorce, J., 178
Sörensen, S., 594
Sørensen, T. I. A., 278
Sosa, B. B., 227
Sosa, R., 94
Sosik, J. J., 505, 534, 535
Soussignan, R., 106
Speece, M. W., 623
Speechley, M., 564
Speicher, B., 391
Spelke, E. S., 138, 150, 152
Spence, M. J., 81, 106
Spencer, J. P., 132
Spencer, T. J., 288
Sperl, W., 103
Spinhoven, P., 594
Spinrad, T. L., 179
Spira, A., 428
Spiro, A., III, 609
Spitz, R. A., 188
Spock, B., 122
Sprecher, S., 451
Springer, C. A., 642
Spuhl, S., 223, 224
Sridhar, D., 305
Sroufe, A., 193
Sroufe, L. A., 7, 177, 192, 193, 269
Staisey, N., 556
Stamler, J., 416, 416*n*
Stamm, B. H., 645
Stampfer, M., 486
Stams, G. J. M., 57, 194
Standley, J. M., 97
Stanley, B., 38
Stanley, J. C., 369
Stanley-Hagan, M., 25, 329, 330, 332
Stanowicz, L., 238
Starkey, P., 231
Statistics Canada, 7, 30, 64, 96, 162, 233, 234, 258, 300, 329, 329*n*, 330, 333, 403, 419*n*, 456, 457, 458, 458*n*, 460, 461, 466, 467, 467*n*, 489, 490*n*, 491, 506, 524, 525, 528, 537, 547, 548, 550, 558, 562*n*, 569*n*, 596, 598
Stattin, H., 37, 352, 372, 376, 456
Staudinger, U. M., 8, 520, 575, 576, 595, 596, 596*n*
Steelman, L. C., 463
Stegge, H., 319
Stein, C. H., 530
Stein, J. A., 375
Stein, J. H., 350
Stein, N., 247
Steinberg, K. K., 490
Steinberg, L. D., 265, 334, 351, 372, 374, 384, 394, 395, 398, 400, 465
Steinberg, S., 176
Steiner, J. E., 106, 127
Steinhauser, K. E., 629
Steinhoff, M. C., 209
Stenberg, C., 177
Stephens, B. R., 138
Stephens, M. A. P., 521, 531
Stern, M., 184
Sternberg, K. J., 190
Sternberg, R. J., 21, 232, 294, 295, 306, 434, 451, 504, 576
Sterns, H. L., 534, 610
Stettler, N., 125
Stevens, J. C., 552, 553
Stevens, M. L. T., 638
Stevens, N., 251, 396, 453
Stevenson, D., 439
Stevenson, H. W., 308, 309, 369
Stevenson, K., 598
Stevenson, M. R., 332
Stevenson, R., 236
Steward, D. K., 126
Stewart, A. J., 450, 513, 513*n*, 517, 523
Stewart, A. L., 560
Stewart, J. H., 255*n*
Stewart, P., 87
Stewart, R. B., 193, 531
Stewart, S., 451
Stewart, T. L., 493
Stice, E., 266
Stichick, T., 335
Stifter, C. A., 190
Stiles, J., 116, 120, 121*n*
Stillion, J. M., 641
Stine-Morrow, E. A. L., 502, 573
Stinnett, H., 451
Stipek, D. J., 196, 233, 248
Stith, S. M., 461
Stoch, M. B., 125
Stock, J. R., 213*n*
Stocker, C., 396
Stodolsky, S. S., 304
Stoel-Gammon, C., 165
Stolee, P., 569
Stone, M. R., 398
Stone, R. G., 556
Stoneman, Z., 193, 328
Stoney, C. M., 494
Stormshak, E. A., 58, 257, 324
Strain, L. A., 611
Strange, C. J., 562
Strapp, C. M., 238
Strasburger, V. C., 356
Stratton, D. C., 603
Stratton, J. R., 560
Straus, M. A., 255, 255*n*, 460, 461
Strauss, R. S., 348
Strawbridge, W. J., 591
Strayer, B. K., 283, 284
Streissguth, A. P., 83, 85, 86
Striano, T., 166, 178, 216
Stricker, G., 527, 5610
Strickland, C. J., 402
Strober, M., 353, 354
Stroebe, M. S., 603, 640, 641
Stroebe, W., 456, 603, 641
Stroh, L. K., 536
Stromswold, K., 237
Stroup, D., 95
Strouse, D. L., 399
Stryker, J., 428
Stryker, M. P., 119
Stuart, M., 569
Stull, D., 604
Stunkard, A. J., 123, 278, 353
Styfco, S. J., 234
Stylianos, S. K., 644
Su, M., 268
Suarez-Orozco, C., 30
Suarez-Orozco, M. M., 30
Subbotsky, E. V., 220
Subrahmanyam, K., 298, 370
Succop, P. A., 87
Suchindran, C., 354
Suitor, J. J., 606
Sullivan, M. A., 631, 642
Sullivan, S. A., 208
Suls, J., 518
Sundell, H., 103
Sung, K.-T., 598
Super, C. M., 131
Super, D., 437
Supple, A. J., 395
Sureau, C., 54
Susman-Stillman, A., 328
Sutcliffe, A. G., 52*n*, 54
Sutton, K. J., 495
Suzuki, L. A., 307
Svensson, J., 414
Swanson, J. L., 438
Swanson, N. G., 474
Swendsen, J. D., 176
Swigert, V. L., 589
Swindell, R., 577
Swingley, D., 167
Szepkouski, G. M., 289
Szinovacz, M., 526, 528
Szkrybalo, J., 263

T

Tadd, W., 38
Taddio, A., 106
Tager-Flusberg, H., 165, 237
Takahashi, K., 122, 188
Takahashi, S., 567
Takahira, S., 369
Takamura, J., 530, 530*n*
Talbot, A., 575
Tamis-LeMonda, C. S., 32, 158, 169, 266
Tanenbaum, S., 592
Tanfer, K., 427
Tangney, J. P., 249
Tangri, S. S., 461
Tanner, J. M., 114, 115, 123, 205*n*, 207, 277, 348, 348*n*
Tardif, C., 194, 252
Tardif, T., 168, 236
Tardiff, T., 59
Tarulli, B., 535
Taska, L. S., 317, 336
Tasker, F. L., 471
Tate, C. S., 237
Taubman, O., 497
Taylor, B. J., 103
Taylor, C. B., 496
Taylor, C. L., 299
Taylor, E., 266
Taylor, E. H., 451
Taylor, J. H., 256, 390, 391, 392
Taylor, K., 431
Taylor, M. C., 263
Taylor, M. G., 326
Taylor, P. R., 424
Taylor, R. D., 63
Taylor, R. J., 591
Taylor, R. L., 63
Teller, D. Y., 133, 135
Tellings, A., 253
Temple, C. M., 370
Tenenbaum, H. R., 262
Terenzini, P. T., 435, 436
Terestchenko, N. Y., 86
Terman, L. M., 449
Tertinger, D. A., 212
Tetens, J. N., 14
Teti, D. M., 189, 192
Teyber, E., 332
Thacker, C., 506
Thacker, S. B., 95
Thatcher, R. W., 117, 206
Theisen, S. C., 487, 488
Thelen, E., 15, 101, 101*n*, 130, 131, 132, 157
Thoman, E. B., 121
Thomas, A., 7, 180, 181, 181*n*, 184
Thomas, C., 497
Thomas, J. L., 527
Thomas, J. R., 215
Thomas, R. B., 104
Thomas, R. M., 17
Thompson, C. A., 506
Thompson, C. K., 236
Thompson, D., 536
Thompson, D. N., 579
Thompson, E. E., 598
Thompson, J., 577
Thompson, K., 641
Thompson, M., 574
Thompson, P. M., 118, 194, 206, 207
Thompson, R. A., 38, 104, 117*n*, 177, 179, 188, 188*n*, 191, 247, 248
Thompson, S., 497
Thornberry, T. P., 405
Thorndike, R. L., 232, 293
Thornton, S., 21, 21*n*
Thorson, J. A., 626
Tienari, P., 70
Tienda, M., 30, 610
Tietjen, A., 392
Tiggemann, M., 279
Tincoff, R., 167
Tinetti, M. E., 564
Tisak, M. S., 322
Titov, N., 571
Tizard, B., 189
Tjebkes, T. L., 197
Tobin, S. S., 593
Todd, C. M., 227
Toder, F. A., 527*n*
Tofler, I. R., 283, 284
Tomasello, M., 165, 166, 167, 168, 178, 216, 236
Tomer, A., 626, 627*n*
Tong, S., 46
Torff, B., 295
Torney-Purta, J., 623
Torrance, E. P., 306
Torrey, B. B., 569
Torrez, D. J., 527
Toth, S. L., 267*n*, 268, 401
Townsend, A. L., 521
Townsend, M. A. R., 320
Transport Canada, 563
Transue, J. E., 393
Trappe, S., 417
Trasti, N., 85
Tremblay, G. C., 211
Tremblay, M. S., 278
Tremblay, R. E., 257, 322
Trent, K., 63
Triandis, H. C., 29, 63
Trickett, P. K., 336
Troiana, R. P., 278
Troll, L. E., 587, 606, 607
Tronick, E. Z., 104, 107, 175, 178
Tröster, H., 136
Troughton, E. P., 70
Trowbridge, F., 209
Trowell, J., 336
Trudel, M., 197
Truglio, R. T., 235
Trussell, L. H., 571
Trusty, J., 372
Tsang, P. S., 502
Tsuang, M. T., 425
Tuchfarber, B. S., 280
Tucker, C. J., 328, 393
Tudge, J. R. H., 59, 224
Tulviste, P., 23, 157
Tur, E., 425
Turcotte, P., 467*n*
Turiel, E., 253, 256, 322, 391
Turkewitz, G., 106
Turner, B. F., 521
Turner, P. J., 262
Twenge, J. M., 316, 317, 317*n*, 383
Tyrka, A. R., 353
Tzuriel, D., 298, 298*n*

U

Udry, J. R., 354
Uhari, M., 210
Ulrich, B. D., 213
Ulrich, D. A., 213
Ungerer, J. A., 191
United Nations, 329*n*, 464
United Nations Children's Fund, 64*n*
United Nations Development Programme, 64*n*, 65*n*
Updegraff, K. A., 369
Uppal, S., 317, 384, 386
Uribe, F. M. T., 60
U.S. Bureau of the Census, 7, 55, 60, 61, 64*n*, 65, 96, 98, 99*n*, 233, 233*n*, 300, 329*n*, 333, 402*n*, 419*n*, 438, 438*n*, 457, 458, 466, 467, 467*n*, 470, 471, 490*n*, 491, 524, 525, 528, 529, 537, 547, 549*n*, 551, 552*n*, 559*n*, 561, 562, 562*n*, 563, 567, 592, 596, 599, 602, 602*n*
U.S. Centers for Disease Control, 356, 358, 358*n*
U.S. Department of Education, 233, 307, 307*n*, 309, 374, 375, 375*n*, 377, 439, 506
U.S. Department of Health and Human Services, 54, 86, 124, 209, 267, 269, 278, 280, 280*n*, 283, 335, 347, 355, 356, 361, 414, 416, 418, 418*n*, 420, 421, 424, 425, 428, 489, 493, 496, 525, 548, 549, 550, 552, 558, 559*n*, 569*n*, 598

U.S. Department of Justice, 403, 598
U.S. Department of Labor, 462, 472
U.S. Department of State, 268
Usmiani, S., 351
Utmann, J. G. A., 166
Utz, R. L., 603
Uutela, A., 424

V

Vacha-Haase, T., 448
Vachon, M. L. S., 644
Vachon, R., 365
Vaillant, C. O., 434
Vaillant, G. E., 18, 434, 449, 455, 516, 517, 585, 612
Valdés, G., 30
Valencia, R. R., 307
Valentine, T., 14
Valian, V. V., 238
Vallance, D. D., 319
Valle, D., 53
van Baarsen, B., 603
Van Brunt, D., 626
van Delden, J. J. M., 636*n*
van den Boom, D. C., 184, 190
van der Maas, P. J., 636*n*
van Driel, B., 466
van Engeland, H., 97
van IJzendoorn, M. H., 57, 188, 189, 192, 194
Van Meter, J. V., 528
Van Nguyen, T., 36
van Tilburg, T., 606
Vandell, D. L., 61, 323, 333–334, 334
Vandenberg, B., 251*n*
Vandenberg, K. A., 98
Vandewater, E. A., 450, 513, 517
Vangelisti, A. L., 462
Vanier Institute of the Family, 470, 529, 530, 530*n*
VanWey, L. K., 463
Varendi, H., 106
Vartanian, L. R., 368
Vartanian, T. P., 599
Vatten, L. J., 85
Vaughn, B. E., 188, 190, 197, 251
Vaughn, S., 305
Veldhuis, J. D., 114
Velkoff, V., 551
Velting, D. M., 403
Venable, D., 472
Veniegas, R., 454
Vereijken, B., 131
Verma, S., 396
Vernon, P. A., 153, 294
Vernon-Feagans, L., 211
Victor, C., 598
Victoria, G. C., 209
Vidal, L. A., 643
Villaume, W. A., 552
Vinchur, A. J., 322
Vinden, P. G., 228
Vinick, B. H., 601
Vinters, H. V., 549
Vogel, D. A., 184
Volling, B. L., 193
Volpicelli, J. R., 426
von Hofsten, C., 134
Vondra, J. I., 187, 188
Vorek, R. E., 533
Vorrasi, J. A., 335
Voss, L. D., 208
Vostanis, P., 61
Voyer, D., 370
Voyer, S., 370
Vygotsky, L. S., 23, 223, 224

W

Wachs, T. D., 7, 68, 126, 181, 278
Wadden, T. A., 421
Waddington, C. H., 69
Wade, T. J., 431
Wagner, R. K., 503
Wahl, H.-W., 592, 600
Wahler, J. J., 601
Wahlin, A., 571
Wahlsten, D., 68*n*, 69
Waite, L. J., 521
Wakat, D. K., 215
Wakefield, J. A., 451
Wakeley, A., 148, 152
Walberg, H. J., 302
Walco, G. A., 277
Walden, T., 248
Waldfogel, J., 472
Waldman, I. D., 297
Waldron, N. L., 305
Waldrop, D. P., 528
Walford, R. L., 485
Walk, R. D., 134
Walker, A., 83, 84
Walker, D., 60, 169
Walker, L. J., 256, 390, 391, 392
Walker, R. A., 206
Walker, S. P., 278
Wallander, J. L., 280
Wallerstein, J. S., 459
Walls, C. T., 594
Wang, H. X., 566
Wang, Q., 156
Wannamethee, G., 424
Wapner, R. J., 52*n*
Wardrop, J. L., 325
Wark, G. R., 391
Warr, P. B., 534
Warren, A. R., 237
Warren, D. H., 136
Washington, S. K., 623
Wass, H., 622, 624
Wasserman, G. A., 85, 87, 405
Wasserstein, S., 334
Waterman, A. S., 385, 522
Waters, E., 177, 186, 192
Watkins, C., 322
Watkins, L. R., 418
Watkins, M. L., 560
Watkins, W. E., 278
Watkinson, B., 84, 85
Watson, A. C., 228
Watson, D. J., 292
Watson, J., 228
Watson, J. B., 18
Watson, K. S., 250
Watson, M., 245
Wattigney, W. A., 349
Waxler, C. Z., 33, 254
Waxman, S. R., 157, 167, 236
Webb, N. B., 643
Weber, J. A., 528
Weber, L. A., 528
Webster, J. D., 587
Wechsler, D., 293, 294
Weeks, M. D., 458
Wegesin, D. J., 571
Wegner, K. E., 536
Wehren, A., 299
Wei, Y. H., 413
Weikart, D. P., 234
Weiland, S., 447
Weinberg, C. R., 77
Weinberg, M. K., 175, 180
Weinberg, R. A., 27, 67, 297
Weinberger, J., 249
Weinblatt, M. E., 562
Weindruch, R., 485
Weiner, J., 458
Weiner, S., 155*n*
Weinfield, N. S., 192
Weingarten, H. R., 602
Weinrich, M., 569
Weinstein, R. S., 304
Weinstock, C. S., 516
Weisfield, G. E., 345
Weisman, A. D., 629
Weisner, T. S., 262, 321, 532
Weiss, B., 257, 401
Weiss, R. S., 639
Weissman, M., 401
Weisz, V., 326
Weksler, M. E., 554
Wellings, K., 428
Wellman, H. M., 225, 228, 247, 291
Wendland-Carro, J., 107
Wenger, G. C., 604
Wenk, G. L., 560
Wentworth, N., 132
Wentworth, P. A., 513
Werker, J. F., 133, 169
Werner, E. E., 7, 10, 100, 527
Werner, J. S., 484
Wertsch, J. V., 23, 157
West, M. J., 256
West, R. L., 573
Westen, D., 17
Westermeyer, J. F., 520, 521
Westra, T., 131*n*, 132
Wethington, E., 517
Whalley, L., 415*n*
Wheeler, I., 641
Wheeler, T., 84, 289
Wheeler, W., 393
Whitaker, D. J., 356
Whitbeck, L., 529
Whitbourne, S. K., 35, 415*n*, 447, 484, 516, 517, 522, 549, 550, 554, 555, 556
White, B., 132
White, H. R., 454*n*, 459
White, J. L., 404
White, J. W., 429
White, L., 531, 604, 605*n*
White, L. K., 457, 463, 526
White, S. H., 15
Whitehead, J. R., 283
Whitehurst, G. J., 169, 231
Whitesell, N., 319
Whiteside, M. F., 331
Whiting, B., 257, 260, 327
Whitley, B. E., Jr., 428
Whitney, J. A., 83
Whitney, M. P., 121
Wichstrøm, L., 401, 401*n*
Wiebe, D. J., 497
Wigfield, A., 316, 318, 371, 438
Wikby, A., 554
Wikner, K., 106
Wilber, K. H., 608, 609
Wilcox, A. J., 77
Wildes, J. E., 353
Wiley, E., 300
Wiley, T. L., 484
Wilke, C. J., 506
Wilkie, J. R., 459
Wilkins, K., 565
Willatts, P., 149
Wille, D. E., 189
Willerman, L., 67, 69, 297
Willett, W. C., 486
Williams, B., 530, 530*n*
Williams, B. C., 211
Williams, J., 609
Williams, N., 527
Williams, P. G., 497
Williams, R. B., 494
Williamson, G. M., 89
Williamson, J. B., 610
Willinger, M., 122
Willis, R. J., 360
Willis, S. L., 433, 434, 499, 501, 574, 577
Willms, J. D., 278
Willner, J. P., 52*n*
Willott, J. F., 484
Wilms, H., 600
Wilson, B. J., 303
Wilson, D. M., 631
Wilson, M. N., 63
Wilson-Mitchell, J. E., 262
Wimbarti, S., 159, 250
Winberg, J., 106
Winblad, B., 566
Windley, P. G., 598
Winefield, A. H., 537
Wingfield, A., 502
Wink, P., 519, 522, 589, 591
Winkleby, M. M., 208
Winn, R., 561
Winner, E., 214, 214*n*, 236, 299, 306, 307, 438
Winsler, A., 223, 224, 303
Wintre, M. G., 319
Wiscott, R., 607
Wittlinger, R. P., 572
Wnek, L., 213*n*
Wohlgemuth, W. K., 347
Wojciechowski, W. C., 601
Wolchick, S. A., 331
Wolf, A. W., 122
Wolf, R. S., 607
Wolfe, D. A., 268, 399, 460
Wolfe, J. J., 633
Wolfelt, A. D., 624
Wolff, P. F., 268, 335
Wolff, P. H., 102*n*
Wolfinger, N. H., 330
Wolfson, A. R., 347
Wolpe, J., 19
Women's Health Initiative, 486
Wong, P., 625
Wong, R., 533
Wood, G. A., 399
Wood, R. M., 104
Wood, S., 55
Wood, W., 451
Woodard, L., 266
Woodman, S., 634
Woodring, B. C., 423
Woodruff, W. J., 429
Woods, J. R., Jr., 83
Woodward, L., 95
Woodward, L. J., 124, 324, 402
Woodward, S. A., 182
Woolf, L. M., 602
Woolley, J. D., 220
Wooster, D. M., 126
Word, S., 624
Worden, J. M., 641
World Health Organization, 204, 209, 461*n*, 547, 548*n*, 592
Wortman, C. B., 641, 644
Wozniak, R. H., 367
Wright, J. C., 234, 235, 257
Wright, J. W, 61
Wright, S. D., 568
Wright, T. A., 534
Wright, W. E., 413
Wrightsman, L. S., 512
Wu, L. L., 470
Wu, M. M., 277
Wu, Z., 468, 525, 604
Wunsch, J. P., 177
Wyman, P. A., 11, 338
Wynn, K., 152, 152*n*

X

Xie, H., 322
Xu, F., 168, 236

Y

Yale, M. E., 175
Yamanoi, K., 531
Yang, B., 329
Yang, E. Y., 56
Yarrow, M. R., 33, 254
Yates, M., 393
Yates, W. R., 70
Yeats, D. E., 98
Yeh, M., 630
Yeung, W., 630
Yip, R., 209
Yirmiya, N., 184, 229
Yoder, A. E., 385
Yoerger, K., 362
Yogman, M. W., 192
Yonas, A., 106, 134
Yont, K., 211
Yoo, S. H., 598
Yosipovitch, F., 425
Youn, G., 529, 531
Young, H. M., 600
Young, J. B., 536
Young, L. R., 423
Young, P., 465
Youngblade, L. M., 333
Youniss, J., 251, 393
Yu, S., 424
Yuill, N., 256
Yule, W., 334

Z

Zacks, R. T., 502
Zafeiriou, D. I., 101
Zahn-Waxler, C., 196, 253, 256
Zakin, D. F., 352
Zakowski, S. G., 495
Zamsky, E. S., 63
Zane, N., 630
Zapf, D., 534
Zarit, S. H., 526, 530, 568, 594
Zeanah, C. H., 189
Zeifman, D., 452
Zelazo, N. A., 101
Zelazo, P. R., 101, 186*n*
Zenger, T., 535
Zeskin, P. S., 105
Zevitz, R. G., 598
Zhang, Z., 604
Zhou, H., 266
Zhou, M., 30, 31
Zigler, E. F., 19, 60, 62, 64, 129, 234
Zimmer-Gembeck, M. J., 318
Zimmerman, B. J., 291
Zimmerman, M. A., 11, 316, 383
Zimmerman, R., 185
Zimmerman, S., 208
Zinn, A., 51
Zins, J. E., 280
Ziporyn, T., 90
Zisook, S., 640
Zlotnick, C., 460
Zoller, D., 247
Zoller-Hodges, D., 578
Zorilla, E. P., 497
Zottarelli, K., 98
Zucker, A. N., 513
Zucker, K. J., 261
Zumbahlen, M. R., 129
Zuroff, D. C., 513
Zuzanek, J., 605
Zwillich, T., 632

Subject Index

A

AA (Alcoholics Anonymous), 426
AARP (American Association of Retired Persons), 66
AARP Public Policy Agenda, 66
A–B search, 146, 147, 149, 150
Abacus, 292
Abilities. *See specific ability or skill (e.g., Categorization abilities)*
Ability grouping in schools, 304, 374
Aboriginal Head Start, 233–234
Aboriginals, Canadian. *See* Canadian Aboriginals
Aborigines, Australian
  voluntary active euthanasia, 637
Abortion, and teenage pregnancies, 359
Absenteeism, school. *See* Dropping out of school; School attendance
Absenteeism, work, 534
Abstract thinking
  adolescence, 344, 363–369
  argumentativeness and, 367
  brain growth spurts and, 119
  caregiving guidelines, 367
  cognitive-developmental theory of, 363–365
  consequences of, 367–369
  criticism and, 367, 368
  cultural influences on, 365
  decision making and, 367, 368–369
  idealism and, 367, 368
  identity development and, 385
  individual differences in, 365
  information processing theories of, 365–366
  Kohlberg's stages of moral reasoning and, 390
  planning and, 368–369
  self-consciousness and self-focusing and, 367–368
Abuse
  alcohol. *See* Alcohol use and abuse
  children. *See* Child maltreatment
  drugs. *See* Drug use and abuse
  elderly. *See* Elder maltreatment
  spouse. *See* Partner abuse
Abusive parenting. *See* Child maltreatment
Academic achievement
  ability grouping in schools and, 304
  ADHD and, 288–289
  adolescence, 372–374
  child maltreatment and, 268
  child-rearing practices and, 265, 372
  classroom and school environment and, 373–374
  cognitive self-regulation and, 291
  cross-cultural research, 307–309
  cultural influences on, 308
  delinquent behavior and, 403, 404
  educational enrichment and, 163
  educational philosophies and, 302
  employment and, 374
  family size and, 463
  Head Start programs and, 233–234
  homelessness and, 61
  immigrant youths, 30
  mainstreaming of students with learning difficulties and, 305
  maternal employment and, 333
  only children, 329
  otitis media and, 210
  parent-school partnerships and, 373
  peer acceptance and, 324
  peer relations and, 373
  readiness for school and, 303
  school transitions and, 371
  self-care children, 333
  self-esteem and, 315–316
  sexual activity of adolescents and, 355
  skipped-generation families, 528
  sleep-deprivation and, 347
  socioeconomic status and, 60, 373
  teacher–student interaction and, 304
Academic learning. *See* Education; Higher education; Schools and schooling; Teacher *entries*
Academic preschools, 233
  developmentally appropriate programs, signs of, 235
Academic self-esteem, 315–316
Academic subjects. *See also specific subjects (e.g., Mathematics; Reading)*
  gender stereotyping of, 326
Acceptance
  friendships in late adulthood and, 605
  peers. *See* Peer acceptance
  self. *See* Self-acceptance
  stage of dying, 628
Acceptance of change, as personality trait, 587
  successful aging and, 613
Access Program, 28
Access to vocational information, 439
Accidents. *See* Injuries; Motor vehicle accidents
Accommodation, in cognitive-developmental theory, 144, 218
Acetylcholine, in Alzheimer's disease, 566, 567
Achievement. *See also* Expertise; Mastery
  academic. *See* Academic achievement
  identity achievement. *See* Identity achievement
  maternal employment and, 333
  maturational timing and achievement difficulties of girls in Sweden, 352
  motivation. *See* Motivation
  peer acceptance and, 324
  self-conscious emotions and, 248
  vocational. *See* Vocational achievement
Achievement-related attributions, 317–318
Achieving stage of cognitive development (Schaie), 432–433
Acquired immune deficiency syndrome (AIDS), 280
  adolescence, 356, 358
  bereavement overload and, 642
  early adulthood, death caused by, 419
  homosexuality and, 428
  hospice care and, 633
  prenatal development and, 87
  prenatal transmission of, 87
  sexual attitudes, effect on, 354
Acquisitive stage of cognitive development (Schaie), 432
Action words, in early vocabulary, 236
Active genetic–environmental correlation, 69
Active lifespan, 547, 548
Activities. *See* Athletics; Extracurricular activities; Games; Leisure activities; Play; Toys
Activities of daily living in late adulthood, 591–592
Activity level, as temperament dimension, 181
Activity theory of aging, 595–596
Adaptability, as temperament dimension, 181
Adaptation. *See also* Coping strategies
  aging process and, 556–557
  cognitive-developmental theory, 19, 144–145
  Darwin's theory of survival of fittest, 14
  dying, to, 627–631
  evolutionary developmental psychology and, 22–23
  families adapting to change, 59
  immigrant youths, 30–31
  midlife transition, 517
  newborn's adaptation to labor and delivery, 91–92
  resiliency of children, 10–11
Adaptation to Life (Vaillant), 447
Adaptive cognition, 432
Adaptive problem solving in late adulthood, 574
Addiction. *See* Alcohol use and abuse; Drug use and abuse; Smoking
ADEPT (Adult Development and Enrichment Project), 577
ADHD (attention-deficit hyperactivity disorder), 288–289, 404
Adjustment. *See also* Adaptation
  biological aging, to, 515
  breastfeeding, to, 108
  divorce, to, 332, 469, 525, 602
  newborns, family's adjustment to, 108
  parenthood, to, 462–463
  retirement, to, 610–611
Adjustment problems. *See also* Psychological disturbances
  ADHD and, 288–289
  aggression and, 259
  blended families, 469–470
  child maltreatment and, 268
  divorce, following, 330–331, 469, 602
  homosexuality in adolescence and, 357
  identity development and, 385
  maternal depression and, 176
  menopause and, 487
  midlife transition and, 517
  obesity and, 279
  psychosocial dwarfism, 208
  retirement and, 610–611
  school transitions and, 371–372
  sexual abuse victims, 336
  widowhood, following, 603
Adolescence, 343–407. *See also* Puberty; Sexual maturation
  abstract thinking in, 344, 363–369
  academic achievement in, 372–374
  alcohol use and abuse in, 361–362
  athletics in, 347
  autonomy in, 394–395
  bereavement in, 641
  biological perspective of, 344, 345
  body composition in, 346–347
  body growth in, 346–347
  body proportions in, 346
  childbearing in. *See* Teenage pregnancy and parenthood
  child-rearing practices in, 395–396. *See also* Parent–adolescent relationship
  cognitive development in, 363–377. *See also* Formal operational stage of cognitive development
    milestones, 408–409
  cognitive self-regulation in, 368
  contraception use in, 355–356, 360
  cross-gender behavior in, 394
  crowds, membership in, 398–399
  cultural influences on, 245
  dating in, 399
  death concept, development of, 624–625
  defined, 344
  delinquency in. *See* Delinquency
  depression in, 344, 400–401
  drug use and abuse in, 361–362
  eating disorders in, 344, 353–354
  emotional and social development in, 381–407
    milestones, 408–409
  employment in, 372, 374
  family influences on, 394–396
  friendships in, 396–398
  gender identity in, 394
  gender stereotyping in, 394
  gender typing in, 394
  growth spurts in, 346
  hair growth in, 345, 348
  Hall's description of, 344
  health issues in, 352–362
  height in, 346
  homosexuality in, 356, 357
  information processing in, 365–366
  injuries in, 620
  job-training programs, 374
  milestones in, 408–409
  moral development in, 387–393
  motor development in, 347
  muscle development in, 346–347
  nutrition in, 352–353
  overview, 8

parental relationships in. *See* Parent–adolescent relationship
peer groups in, 398–399
peer relations in, 396–400
perspective taking in, 368
physical development in, 344–362
milestones, 408–409
physical education in, 347
pregnancy in. *See* Teenage pregnancy and parenthood
public policies concerning, 64
Rousseau's description of, 344
school learning in, 371–377
self-concept in, 383
self-development in, 383–387
self-esteem in, 383–384
sexuality in, 354–356
sexually transmitted diseases in, 356, 358
sibling relationships in, 396
sleep in, 347
smoking in, 361–362
social perspective on, 344–345
states of arousal in, 347
substance use and abuse in, 361–362
suicide in, 344, 401–403
voice changes in, 348
weight in, 346
Adoption, 56–57
attachment and, 189
homosexual parents and, 471
Adoption studies
antisocial behavior, 70
intelligence, 297
mental illness, 67
schizophrenia, 70
Adrenal androgens, 345–346
Adult children. *See* Aging parents; Parent-adult-child relationship
Adult development, 7. *See also* Human development; *specific periods (e.g., Early adulthood; Middle adulthood)*
brain plasticity and, 120–121
diversity of lifestyles during, 466–471
early philosophies of, 13–14
neighborhood influences on, 61
postformal thought and, 20
prenatal environment and health in later life, 84–85
research on, 5
Adult Development and Enrichment Project (ADEPT), 577
Adult learners, 506–507
benefits gained, 507
characteristics of, 506
late adulthood, 577–579
reasons for becoming, 506
support of, 506–507
Adult-onset diabetes, 561, 563. *See also* Diabetes
Advance medical directives, 634–636
African Americans
academic achievement among, 373
achievement-related attributions among, 318
aging parents, caring for and contact with, 598
anorexia nervosa among, 353
bereavement rituals among, 642
body construction and motor development among, 215
child-rearing practices of, 266
cosleeping arrangements in infancy and childhood, 122
dementia incidence, 564, 566
diabetes incidence among, 563
drop out rate among, 375
dying, adaptation to, 631
euthanasia, response to, 634
extended family households of, 62, 63, 470
health in late adulthood, 559
heart disease among, 414, 422, 431, 492–493
hypertension among, 414
identity development in adolescence, 386
infant mortality among, 98
institutionalization of elderly, 569
IQ scores of, 296, 297
life expectancy of, 547
low birth weight among, 98
marital age of, 466
menopause, women's attitude toward, 487
obesity in adulthood of, 420
osteoporosis, vulnerability to, 493
out-of-wedlock births among, 359, 470
physical development in infancy and toddlerhood, 114, 115
poverty and, 60
reminiscence in late adulthood among, 587
retirement of, 610
self-esteem among, 316–317
self-esteem in adolescence, 384
sexual activity of adolescents, 355
sickle cell anemia among, 48
social support and interaction among elderly, 594
spiritual and religious involvement in late adulthood, 590–591
suicide rate, 402
widowhood, adjustment to, 603
young adults leaving home, 457
African societies. *See also specific entries (e.g., Aka of Central Africa; Kenya)*
extended family households in, 63
sickle cell anemia among, 49
After-school programs, 61, 334
Age. *See also* Milestones
athletics, age of peak performance, 416–417
blended families, adjustment to and, 332
"career age," 434
children's response to divorce and, 330
fetal viability, 80
functional age, 546–547
grandparenthood, average age of, 526
immune response, decline of, 417
leaving home, average age of, 456–457, 525–526
marital. *See* Marital age
maternal. *See* Maternal age
menarche. *See* Menarche
paternal. *See* Paternal age
pubertal changes, 346
retirement, average age of, 537
skeletal age, 115
social clock and, 449–450
spermarche, 348
stages of development. *See* Stage concept of development
subjective versus objective, 515–516
wisdom and, 575–576
Age 30 transition, in Levinson's theory of adult development, 448
Age discrimination. *See* Aging stereotyping
Age of viability, 80
Age spots, 484, 555
Age-graded influences, 10–11
Agency, sense of, in infancy and toddlerhood, 195
Aggression. *See also* Antisocial behavior; Difficult child, as temperament type; Violence
androgens and, 256, 260
attachment security and, 194
child-rearing practices and, 257
early childhood, 256–259
family influences on, 257
gender typing and, 256–257
hostile aggression, 257
instrumental aggression, 257
intervention programs for, 259
middle childhood, 323
modeling of, 254
overt aggression, 256, 323
peer relations and, 259
peer victimization, 325
perspective taking and, 320
rejected-aggressive children, 324–325
relational aggression, 256, 257, 323
sexual. *See* Sexual coercion
television and, 258–259
Agility. *See also* Motor development
middle childhood, 281
Aging. *See also* Biological aging; Late adulthood
activity theory of, 595–596
Area Agencies on Aging, 65
cross-linkage theory of aging, 414
disengagement theory of, 595
low-calorie diets and, 485
parents. *See* Aging parents
primary aging, 561–562. *See also* Biological aging
secondary aging, 561–562
social theories of, 595–596
socioemotional selectivity theory of aging, 596, 597
stereotyping. *See* Aging stereotyping
successful aging, 612–613
"wear-and-tear" theory of, 413
"Aging genes," 413
Aging parents, 529–531, 606–607
Alzheimer's disease victims, caring for, 566–567, 568
caring for, 530–531, 533, 591–592, 594, 598
long-term care. *See* Long-term care
relationships with adult children, 529–530
sibling relationships and, 531–532
Aging stereotyping, 7, 547, 550
adult learners and, 506
cognitive ability, 498
dependency reinforcement and, 592
elder maltreatment and, 609
lifelong learning and, 578
overcoming, 557
physical attractiveness, 497
reminiscence and, 586
sexuality and, 560
social interaction and, 596
vocational settings, 505, 535
Aging Well (Vaillant), 447
Agonal phase of dying, 621
Agreeableness, as personality trait, 522–523, 587
AIDS. *See* Acquired immune deficiency syndrome (AIDS)
Aka of Central Africa, father-infant relationships among, 192
Alaskan Native Americans, elder suicide among, 592
Albinism, 47
Alcohol use and abuse
adolescence, 361–362
bulimia and, 354
divorce and, 525
early adulthood, 425–426, 431
Fetal Alcohol Effects, 86
Fetal Alcohol Syndrome, 86
heart disease and, 492
homosexuality in adolescence and, 357
impotence and, 487
life expectancy and, 548
osteoporosis and, 493
partner abuse and, 461
prenatal development and, 85–86
sexual abuse victims, 336
sexuality in late adulthood and, 561
strokes and, 567
Alcoholics Anonymous (AA), 426
Alcoholism, 425, 426
Alienation. *See also* Isolation; Loneliness
identity foreclosure and, 385
Allergies. *See also* Asthma
genetic inheritance, 49
Altruistic behavior. *See* Prosocial behavior
Aluminum, and Alzheimer's disease, 566
Alzheimer's disease, 565–567
brain deterioration in, 565–566
caregivers and, 566–567, 568
incidence of, 565
intervention programs, 566–567, 568
protective factors, 566
risk factors, 566
symptoms and course of, 565
AMA. *See* American Medical Association
Ambidexterity, 206, 207
American Association of Retired Persons (AARP), 66
American Medical Association
passive euthanasia, endorsement of, 634
voluntary active euthanasia and assisted suicide, response to, 639
American Psychological Association, ethical guidelines for research, 37
Amnesia, infantile, 156
Amniocentesis, 52, 53
Amnion, 78
Amniotic fluid, 78
Amodal relations, detection of, 139
Amphetamines. *See also* Drug use and abuse
teenage use, 361
Amygdala, and shyness versus sociability, 182
Amyloid plaques, in Alzheimer's disease, 566
Anal stage of development (Freud), 16, 174–175
Analgesics
childbirth, during, 95
dying and, 620
living will specifications, 635
pain management programs, 632
premenstrual syndrome, 430
Analogical problem solving, 149–150
early childhood, 220
Androgens, 345
aggression and, 256, 260
gender typing and, 260
spatial abilities and, 370
Androgynous traits, 263, 394, 516. *See also* Cross-gender reasoning and behavior
middle adulthood gender identity and, 521–522

Anemia
  adolescence, 353
  Cooley's, 48
  sickle cell, 47–49, 280
Anencephaly, folic acid and, 88
Anesthesia
  childbirth, during, 95
  newborns and, 105–106
Anger
  Alzheimer's disease and, 565, 566
  dying, stage of, 628
  health and, 493–494
  heart disease and, 493–494
  infancy and toddlerhood, 177–178
  parents' anger, effect on children, 33
  perspective taking and, 320
  reducing, 495
  sexual abuse victims, 336
Anger stage of dying, 628
Angina pectoris, 492
Angioplasty, 492
Animals. *See also* Ethology
  brain plasticity studies, 121
  caregiving behaviors after birth, 97–98
  cloning, 55
  dietary calorie restriction studies, 485
  fears of, caregiving concerns, 248
  sensory deprivation studies, 119
  sex differences in life expectancy, 547
  smell responses in, 106
Animistic thinking, 218–220
Anorexia nervosa, 353–354, 421
A-not-B search error, 146, 149
Anoxia, 95
Antibodies. *See also* Immune system
  fetus, transmission to, 81
  Rh factor. *See* Rh factor
Anticipatory grieving, 640
Antidepressants, 486, 567, 692
Antisocial behavior. *See also* Aggression; Delinquency
  delinquency and, 404
  extracurricular activities and, 376
  genetic–environmental correlation, 70
  peer acceptance and, 324
  peer conformity and, 400
  self-care children, 333
  self-esteem and, 384
  substance abuse and, 362
  suicide and, 402
Anxiety. *See also* Fear; Stress
  adult learners, 506
  alcohol abuse and, 426
  Alzheimer's disease and, 565
  child maltreatment and, 268
  death anxiety, 626–627
  dieting and, 421
  divorce and, 469
  exercise and, 424
  long-term care residents, 600
  middle childhood, 334
  peer victimization and, 325
  prenatal development and, 89
  self-esteem and, 384
  separation anxiety, 186
  stranger anxiety, 177–178
Apathy, in identity diffusion, 385
Apgar Scale, 93
  SIDS and, 103
ApoE4, and Alzheimer's disease, 566
Appalachians, cosleeping arrangements in infancy and childhood, 122
Appearance and attractiveness. *See also* Body image; Dress and grooming
  aging stereotyping, 497
  late adulthood, 555
  maturational timing and, 351–352
  newborns, 92–93
  self-esteem and, 315–316
Appearance-reality distinction, 221
Appetite. *See* Eating habits; Nutrition
Applied field of human development, 5. *See also* Public policies; Social issues
Apprenticeship programs. *See* Work-study programs
Approach-withdrawal, as temperament dimension, 181
Appropriate death, 629
Area Agencies on Aging, 65
Argumentativeness, and abstract thinking, 367
Arithmetic. *See* Mathematics
Arizona Pima Indians, diabetes incidence among, 563
Armed conflict. *See* War and armed conflict
Arms. *See* Limbs; Skeletal development
Arousal. *See* States of arousal
Arrhythmia, 492
Artery walls. *See* Atherosclerosis; Cardiovascular system
Arthritis, 280, 561, 562
  cognitive deterioration and, 576
  free radicals and, 414
  obesity and, 420
Artistic expression. *See also* Creativity; Drawing
  early childhood, 213–215
Artistic personality type, in vocational choice, 437
Asian Americans
  aging parents, caring for, 598
  anorexia nervosa among, 353
  elder maltreatment, incidence of, 607–608
  extended family households of, 62
  identity development in adolescence, 386
  institutionalization of elderly, 569
  osteoporosis, vulnerability to, 493
  physical development in infancy and toddlerhood, 114–115
  young adults leaving home, 457
Asian Pacific Islanders. *See* Pacific Islanders
Asian societies. *See also specific countries or cultures*
  academic achievement among, 307–309
  mathematics learning among, 293
  newborn behavior and child-rearing practices, 107
  temperament of infants, 183, 184
Aspirin, prenatal development and maternal use of, 83
Assimilation, in cognitive-developmental theory, 144
Assisted discovery learning, 224
Assisted living, 569, 599
Assisted suicide, 636, 638–639
Assistive technology, 556
Associative memory deficit, 571–572
Associative play, 249
Asthma, 280
  genetic inheritance, 49
Asynchronicity
  biological aging, 412
  physical development, 204–206
Atheism, and death anxiety, 626
Atherosclerosis, 414, 416, 491
  cerebrovascular dementia caused by, 567
  fat consumption and, 422
  obesity and, 420
Athletics. *See also* Exercise; Extracurricular activities; Physical education
  adolescence, 347
  adult-structured, 283, 284
  aging, effects of, 416–417
  caregiving concerns, 284
  gender stereotyping in, 282
  menarche and, 348
  self-esteem and, 315–316
  sex differences in performance and participation, 347
At-risk children and adolescents, resiliency of, 10–11
Attachment, 5, 185–194
  adoption and, 189
  adult romantic relationships and childhood attachment patterns, 452–453
  avoidant attachment, 187
  behaviorist theories of, 185
  Bowlby's theory of, 22, 186–187, 192
  caregiving quality and, 189, 190, 194
  child care and, 190–191
  child maltreatment and, 189
  cultural differences in, 188
  defined, 185
  disorganized–disoriented attachment, 187–188
  ethological theory of, 186–187, 193
  factors affecting, 188–192
  fathers and, 192
  home environment and, 191–192
  infant characteristics and, 189–190
  internal working model of, 187, 191–192, 452
  later development and, 193–194
  longitudinal studies of, 193–194
  measurement of attachment security, 187–188
  multiple attachments, 192–193
  objects of, 185–186
  opportunity for, 188–189
  peer victimization and, 325
  phases of, 186–187
  psychoanalytic theories of, 185, 193
  reproductive technologies and, 54
  resistant attachment, 187
  secure attachment, 187
  siblings and, 192–193, 194
  stability of, 188
  theories of, 185–187
  World War II refugee and evacuee children, 588
Attachment-in-the making phase, 186
Attention
  ADHD and, 288–289
  adolescence, 365
  early childhood, 225–226
  emotional self-regulation and, 179
  habituation research, 154
  infancy and toddlerhood, 153–154
  information processing theories, 153–154
  joint attention. *See* Joint attention
  malnutrition and, 278
  middle adulthood, 502
  middle childhood, 288–289
  self-control and, 197
Attention from parents. *See* Parent *entries*
Attention span, as temperament dimension, 181
Attention-deficit hyperactivity disorder (ADHD), 288–289
  delinquent behavior and, 404
Attractiveness. *See* Appearance and attractiveness
Attribution retraining, 318
Attributions. *See also* Personality traits
  achievement-related attributions, 317–318
  defined, 317
  mastery-oriented attributions, 317–318
Atypical lateralization of brain, 207
Auditory. *See* Ears; Hearing
Australia
  Aboriginals, memory skills of, 291
  elder maltreatment, incidence of, 607
  elderly poverty rate, 65
  institutionalization of elderly, 569
  lifelong learning programs, 577
  reproductive technologies, 55
  suicide rate, 402
  teenage pregnancy rate, 359
  voluntary active euthanasia in, 636, 637
  World War II refugee and evacuee children, 588–589
Austria
  lifelong learning programs, 577–578
  World War II refugee and evacuee children, 588–589
Authoritarian parenting, 265
  academic achievement in adolescence and, 372
Authoritative parenting, 265, 266
  academic achievement in adolescence and, 372
  adolescent development and, 395
  maternal employment and, 333
  peer conformity and, 400
  peer groups in adolescence and, 398
  self-care children and, 334
  self-esteem and, 317, 384
Autism, 229
  fragile X syndrome and, 49
  immunization and, 209
Auto accidents. *See* Motor vehicle accidents
Autobiographical memory, 156, 227, 245, 572–573
Autoimmune response, 554
  rheumatoid arthritis and, 562
Automatic memory, 571
Autonomic nervous system. *See also* Nervous system
  aging process, 550
  sexuality in late adulthood and, 561
Autonomous morality, 388
Autonomy. *See also* Independent living
  adolescent development and, 394–395, 396
  anorexia nervosa and, 353
  divorce and, 525
  late adulthood, 591–592
  long-term care residents, 600
  middle adulthood, 519
  psychosocial stages of development, 17, 174–175
  school transitions and, 372
  successful aging and, 613
Autonomy versus shame and doubt, 17, 174–175
Autosomal diseases, 48
Autosomes, 46
  genetic imprinting, 49
Average life expectancy, 412, 547
Aversion therapy, 426
Avoidance, as phase of grieving, 639
Avoidant attachment, 187
  adult romantic relationships and childhood attachment patterns,

452
later development and, 193
mother-child relationship and, 189

B

B cells, 417
Babbling, in language development, 165–166
Babies. *See* Childbirth; Infancy and toddlerhood; Newborns; Pregnancy; Prenatal development
Babinski reflex, 101
Baby and Child Care (Spock), 15, 122
"Baby boom" generation, and health care costs, 567
Baby fat, 114, 204
Baby teeth, 204
"Back to basics" movement, 302
Bacterial diseases. *See* Infectious diseases
Balance
aging process, 560
early childhood, 212
middle childhood, 281
sex differences in, 215
Balance of care, in psychosocial theory of development, 174
Baldness, 555
Balinese, bereavement rituals, 642
Ball skills
adolescence, 347
early childhood, 213
sex differences in, 215
Bandura's social learning theory. *See* Social learning theory (Bandura)
Baptists, death concept among, 624
Bar mitzvah, 350
Bargaining, as stage of dying, 628
Barney and Friends, 235
Basal metabolism rate (BMR), 420
Bases, in DNA, 44
Basic emotions, 175, 177–178
Basic issues in development, 5–7
Basic trust versus mistrust, 17, 174
Basic-level categories, 221
Basic-skills approach to reading, 292
Bat mitzvah, 350
Battered spouses. *See* Partner abuse
Bayley Scales of Infant Development, 160
Behavior, types of. *See specific entries (e.g., Antisocial behavior; Prosocial behavior)*
Behavior disorders. *See* Adjustment problems; Antisocial behavior; Delinquency; Psychological disturbances
Behavior modification, 19
childhood injuries, preventing, 212
Behavioral assessment of newborns, 107
Behaviorism, 18–19, 26. *See also* Classical conditioning; Operant conditioning
attachment theories, 185
language development theories, 164
Locke as forerunner of, 13
moral development theories, 253–256
Belgium, assisted suicide laws, 638
Beliefs. *See also* Bias; Spirituality and religiosity; Values
false beliefs, children's recognition of, 228
Bell-shaped curve, in intelligence test scores, 160
Benevolence, in distributive justice reasoning, 322
Bereavement, 639–649. *See also* Death and dying; Widowhood
anticipatory grieving, 640
caregiving concerns, 644
child's death and, 640–641
counseling, 645
cultural influences on, 639, 642–643
defined, 639
dual-process model of coping with loss, 640
educational programs, 645
individual differences in, 640–644
interventions, 644–645
intimate partner's death, 641
overload, 642–644
parent's death and, 641
phases of grieving, 639–640
prolonged, expected death and, 640–641
rituals, 642–643
sibling's death and, 641
sudden, unanticipated death and, 640–641
Bereavement overload, 642–644
Bhatto Bhatto, 250
Bias. *See also* Aging stereotyping; Gender stereotyping
cancer victims, against, 491
homosexuality, against, 428, 470, 602
identity foreclosure and intolerance, 385
intelligence tests, bias in, 232, 293, 297–299
juvenile justice system, 403
obesity, against, 279, 420
self-serving bias, 366
Bicultural identity, 386
Bicycle helmets, 211
Bicycle skills in early childhood, 213
Bidirectional influences
environmental influences on gene expression, 71
teratogens, 82
Bidirectional relationships
ecological systems theory, 24
family systems theory, 58
"Big five" personality traits, 522–523, 587
Bilingual education, 300–301
Bilingualism, 300–301
Binet's intelligence test. *See* Stanford-Binet Intelligence Scale
Binge eating. *See* Bulimia
Binocular depth cues, 134
aging process, 551
Bioecological model (Bronfenbrenner), 24
Biological aging, 412–418. *See also* Aging *entries;* Late adulthood
adjustment to, 515
"aging genes" theory, 413
coping strategies, 556–557
cross-linkage theory of, 414
DNA and body cells, 413–414, 489–490
free radicals and, 413–414
individual differences in, 546–547
low-calorie diets and, 485
organs and tissues, 414
physical changes of, 414–418, 549–557
"random events" theory, 413
secondary aging versus primary aging, 561–562
theories of, 413–414
"wear-and-tear" theory, 413
Biological influences. *See* Genetic *entries;* Inheritance; Nature–nurture controversy
Biosphere, 485
Birth. *See* Childbirth; Conception; Newborns; Pregnancy; Prenatal *entries*
Birth complications. *See also* Birth defects; Low birth weight; Preterm infants
long-term consequences of, 99–100
Birth control. *See* Contraception
Birth defects. *See also* Chromosomal abnormalities
drug use during pregnancy and, 82–84
environmental pollution and, 87
infectious diseases during pregnancy and, 87–88
left-handedness and, 207
nutrition during pregnancy and, 88
radiation during pregnancy and, 86–87
Rh blood incompatibility and, 89
stress during pregnancy and, 89
teratogens and, 81–88
Birth order, and sexual orientation, 356
Birth rates, 464
sex differences in, 49
Birth weight. *See also* Low birth weight
average, 92
health in later life and, 84–85
Bisexuality, 428
Blacks. *See* African Americans; African societies
"Blank slate," 13
Blastocyst, 76, 77, 78
Blended families, 331–333, 469–470
homosexual parents, 471
Blindness. *See* Visual impairment
Blood pressure. *See also* Hypertension
anger's effect on, 493–494
calorie restrictions and, 485
prenatal health care, 90
shyness versus sociability, 182–183
smoking, effects of, 425
stress and, 430–431
Blood tests
genetic counseling, 52
maternal blood analysis, 52
phenylketonuria, 47
Blood type
dominant–recessive inheritance, 47
incompatible blood types
cesarean delivery and, 96
prenatal development and, 89
BMR (basal metabolic rate), 420
Bodily-kinesthetic intelligence (Gardner), 296
Body composition
adolescence, 346–347
aging process, 555
early childhood, 204
infancy and toddlerhood, 114–115
late adulthood, 555
middle adulthood, 484, 486
middle childhood, 277
Body construction, and motor development, 215
Body fat. *See* Fat, body
Body growth. *See* Growth *entries;* Physical development; Prenatal development; Puberty; Sexual maturation
Body hair growth, 345, 348
Body image. *See also* Appearance and attractiveness
anorexia nervosa and, 353
defined, 351
late adulthood, Peck's theory of developmental tasks, 585
maturational timing and, 351–352
Body proportion
adolescence, 346
early childhood, 204
infancy and toddlerhood, 115–116
Body size. *See* Body proportion; Height; Weight
Body transcendence versus body preoccupation (Peck), 585
Body weight. *See* Obesity; Weight
Bonding. *See* Attachment
Bones. *See* Osteoporosis; Skeletal *entries*
Botswana !Kung, infant caregiving, 104
Bottle-feeding, 123–125
Bowlby's theory of attachment, 22, 186–187, 192
Braille, touch sensitivity and, 553
Brain activity. *See also* States of arousal
shyness versus sociability, 182–183
sleep and, 102–103
Brain damage
AIDS in infancy and, 87
alcohol abuse and, 426
brain plasticity studies and, 120–121
crying and, 105
environmental pollution and, 87
infectious diseases during pregnancy and, 87
reflexes and, 101
sleep cycles and, 103
small-for-date infants, 97
Brain death, 621
Brain development. *See also* Nervous system; *specific entries (e.g., Cerebral cortex; Neurons)*
cognitive development and, 288
early childhood, 206–207
emotional self-regulation and, 179
exercise and, 560
growth curve, 205–206
growth spurts, 119
infancy and toddlerhood, 116–123
infantile amnesia and, 156
late adulthood, 549–550
milestones in, 117
prenatal development, 80, 82
sensitive periods, 119–120
Brain lateralization. *See* Lateralization of brain
Brain plasticity, 118–119, 120–121
Brain-imaging techniques for Alzheimer's disease, 565–566
Brain-wave activity. *See* Brain activity; States of arousal
Brazil, mathematical learning among child candy sellers, 24
BRCA1 and BRCA2, 490
Breast cancer, 490, 491
estrogen replacement therapy and, 486
fat consumption and, 421
genetic testing, 56
high birth weight and, 84–85
Breast development, 345, 347, 348
Breastfeeding, 123–125
adjustments to, 108
caregiving concerns, 124
cosleeping arrangements and, 122
rooting reflex and, 100
smell responses in newborns and, 106
sucking reflex and, 101
Yurok Indians, 17
Breech position, 95
cesarean delivery, 96
Bridge jobs, 609

British. *See* England; Great Britain
Bronfenbrenner's ecological systems theory, 24–25, 26, 58–59
Buddhism, 590
beliefs about death, 621
dying, adaptation to, 630
Bulimia, 354, 421
Bullies, 325. *See also* Aggression; Antisocial behavior
Burial ceremonies, 642–643
Burnout, job, 534–535

C
Caffeine
osteoporosis and, 493
premenstrual syndrome and, 430
prenatal development and, 83
Calcium
adolescence, deficiencies in, 353
bone loss and, 486
osteoporosis and, 493
Caloric intake, 423
Calorie-restricted diets, 485
Calorie-restriction mimetics, 485
Camps, summer
chronic illness in childhood, 280
delinquency intervention programs, 405
Canada
adolescent crime rate, 403
adolescent school sports participation, 347
average family size, 461
breastfeeding rate, 123
centenarians, 550
cesarean delivery, 96
child care in, 162
child maltreatment, incidence of, 267, 335
child poverty rate, 60
childhood asthma rate, 280
childhood unintentional injuries, incidence of, 210
divorce rate, 329, 468
elder maltreatment, incidence of, 607
elderly dementia rate, 565
elderly disability rates, 562
elderly in nursing homes, 569
elderly poverty rates, 65, 559
euthanasia laws, 638, 538
grandparent visitation rights, Québec, 529
grandparenthood, average age of, 526
health care policy, 64, 569
high school dropout rate, 64, 374
homelessness in, 60–61
hospice care, 633
immigrant youths, 30
immunization in, 209
independent living in late adulthood, 599
infant mortality rate, 98
institutionalization of elderly in, 569
language immersion programs, 301
leading causes of death, adulthood, 419, 490, 562
leaving and returning home, young adults, 456
legislation for children with learning disabilities, 305
life expectancy, 7, 547
low birth weight, incidence of, 96
male birth rate, 49
marriage rate, 457
maternal employment rate, 233, 333
national dietary recommendations, 421
national physical activity recommendations, 424
obesity rate, 278, 422
Old Age Security Program, 65
parental leave policy, 98
partner abuse, rates of, 460–461
poverty rate, 60
Prenatal Nutrition Program, 88–89
public policies re children, youth, and families, 64
reproductive technologies, 55
retirement benefits for women, 610
retirement, average age of, 537
self-care children, incidence of, 333
sexually transmitted disease rates, 356, 358
sibling relationships in late adulthood, 604
social security system, 64–65
suicide rate, 401–402
teenage physical fitness and activity rates, 347, 424
teenage pregnancy rate, 359
teenage sexual activity rate, 354–355
television policies and standards, 258
Canadian Aboriginals
aging parents, caring for, 598
communication style, 297
diabetes incidence, 563
drop out rate, 374–375
dying, adaptation to, 631
elder maltreatment, incidence of, 607–608
extended family households of, 62
grandparent-grandchildren relationships, 527
health in late adulthood, 559
institutionalization of elderly, 569
life expectancy among, 547
low birth weight, 98
obesity in adulthood, 420
out of wedlock births, 360
poverty and, 60
spiritual and religious involvement in late adulthood, 590
suicide rate, 402
valuing of elders, 557
young adults leaving home, 457
Canadian Medical Association
voluntary active euthanasia and assisted suicide, response to, 639
Canadian Virtual Hospice, 633
Canalization, 69, 70
Cancer, 280, 489–491. *See also specific entries (e.g., Breast cancer; Lung cancer)*
alcohol abuse and, 426
childhood, in, 86
death, as cause of, 419, 489, 561, 562, 620, 621
death anxiety and, 627
depression and, 629
diethylstilbestrol and, 83
DNA markers, 53
early adulthood, death caused by, 419
estrogen replacement therapy and, 486
exercise and, 424
fat consumption and, 421
free radicals and, 414, 490
genetic testing, 56
high birth weight and, 84–85
hospice care, 633
hospital dying, 632
incidence of, 489–490
late adulthood, cause of death in, 561, 562
metastasis, 621
middle adulthood, cause of death in, 489
obesity and, 278, 420
radiation exposure and, 50, 86
reducing incidence of and deaths from, 491
smoking and, 84, 425
stress and, 431
Car accidents. *See* Motor vehicle accidents
Car safety seats, 211
Carbon monoxide, and prenatal exposure to nicotine, 85
Cardinality, early childhood grasp of, 231
Cardiovascular system. *See also* Heart disease
aging of, 414–416, 553–554
anger's effect on, 493–494
estrogen decrease and, 486
exercise and, 424, 560
prenatal development of, 79, 82
SIDS and, 103
smoking, effects of, 425
Career. *See* Employment; Job *entries;* Vocational *entries*
"Career age," 434
Career consolidation, Vaillant's stage development, 447, 449
Caregiving
aging parents, for, 530–531, 533, 591–592, 594, 598
Alzheimer's disease victims, for, 566–567, 568
children. *See* Attachment; Child-rearing practices; Father *entries;* Mother *entries;* Parent *entries*
long-term care. *See* Long-term care
Caregiving concerns guidelines
abstract thinking, 367
adult learners, supporting, 507
aging parents, stress of caring for, 531
bereavement, 644
cancer, reducing incidence of and deaths from, 491
communicating with dying people, 630
compliance and self-control, 197
crying, 104–105
death concept, development of, 625
divorce, adjustment to, 332
dual-earner households, 475
education programs in late adulthood, 579
fears in early childhood, 248
heart attack, reducing risk of, 492
high-quality education, signs of, 302
identity development, 387
injury prevention in late adulthood, 564
pregnancy, 90
romantic relationships, keeping love alive in, 454
self-concept, 247
sexual coercion, 430
sexually transmitted diseases, 359
sibling relationship, 194
suicide risk factors, 403
television, 259
widowhood, 604
Carolina Abecedarian Project, 163–164
Carrier tests, 52
Carriers of genetic traits, 46
Carus's philosophy of adult development, 13–14
Case studies. *See* Clinical method of research
Cataracts, 551
dominant–recessive inheritance, 47
Down syndrome and, 50
free radicals and, 414
Catching skills in early childhood, 213
Catch-up growth, 123
hormone treatment and, 208
marasmic children, 125
nonorganic failure to thrive, 126
Categorization abilities. *See also* Organization
concrete operational stage of cognitive development, 285
information processing theory, 155, 157
preoperational stage of cognitive development, 218, 221
self-categorizing, 196
Catell's crystallized versus fluid intelligence, 498–500, 570
Catholicism, 590
euthanasia, response to, 634
Caucasian Americans
aging parents, caregivers of, 531
anorexia nervosa among, 353
euthanasia, response to, 634
institutionalization of elderly, 569
life expectancy of, 547
osteoporosis, vulnerability to, 493
self-esteem in adolescence, 384
suicide rate, 402
widowhood, adjustment to, 603
Causal-experimental tasks, and abstract thinking, 366
Cells, 44–46. *See also specific cells (e.g., Glial cells; T cells)*
aging process, 413–414, 490
division of (meiosis), 45, 50, 51
duplication of (mitosis), 45, 50
sex cells, 45–46. *See also* Ovum; Sperm
Centenarians, 548, 550–551
Center of gravity, changes in, 212
Central conceptual structures
adolescence, 366
middle childhood, 287
Central Europeans, caring for aging parents, 598
Central executive, in information processing, 153, 154
Central nervous system. *See* Brain *entries;* Nervous system
Centration thinking, 218
Cephalocaudal trend, 115–116
adolescence, 346
motor development, 130
Cerebellum, 207
neuron loss, in aging process, 549
Cerebral cortex. *See also* Brain *entries*
defined, 117
dominant hemisphere, 206
functions of, 117–118
lateralization of, 118–119, 206
neuron loss, in aging process, 549
postnatal development, 117–119
prenatal development, 80
Cerebral palsy, 95
Cerebrovascular dementia, 567
Cervix, 76–77
dilation and effacement in childbirth, 91, 92
Cesarean delivery, 96
Change. *See also* Adaptation; Adjustment; Transitions
acceptance of change, as personality trait, 587, 613
stability versus change, in development, 7

Characteristics. *See* Personality traits; Temperament
Chemical pollutants. *See* Environmental pollutants
Chernobyl, 86–87
Chicken pox, 209
prenatal development and, 87
Child abuse. *See* Child maltreatment
Child care, 465
attachment and, 190–191
cognitive development and, 162, 233, 234
developmentally appropriate practices, 162, 163, 235
early childhood, 233, 234
educational enrichment and, 163
fears of, caregiving concerns, 248
high-quality programs, signs of, 162, 163, 235
illness and, 209
infancy and toddlerhood, 162, 163
injury mortality rate and, 211
middle childhood, 333–334
NICHD study of early child care, 190–191
otitis media and, 210
public policies, 64, 162
quality of care, 190–191
social issues, 190–191
Sweden, 161
Child custody. *See* Custody of children
Child development. *See also* Human development; *specific periods (e.g., Infancy and toddlerhood; Early childhood)*
Darwin's theory of, 14
early philosophies of, 13
Hall's theory of, 14–15
neighborhood influences on, 61
public policies, 64, 65
research ethics, 37–39
Child maltreatment, 267–269
attachment and, 189
community influences on, 61, 268
consequences of, 268
crying of newborns and, 105
cultural influences on, 268
empathy expressions in abused children, 249
eyewitness testimony by children, 336, 337
family dynamics and, 268
forms of, 267
incidence of, 267
intervention programs, 268–269
origins of, 267–268
premature (preterm) infants, 268
preterm infants, 97
prevention of, 268–269
sexual abuse, 334–336
Child neglect, 266, 267. *See also* Child maltreatment
Child support, 330
public policies, 64
Child visitation (following divorce), 331
grandparents' rights, 529
Childbearing. *See* Conception; Pregnancy; Prenatal *entries*
Childbirth, 91–100. *See also* Birth *entries;* Newborns
age of. *See* Maternal age
approaches to, 93–95
breech position, 95
cesarean delivery, 96
fetal monitoring, 95
freestanding birth centers, 93
home delivery, 93, 94–95
medical interventions, 93, 95–96
medications during, 95
natural (prepared) childbirth, 93–94
parental leave, 98–99
stages of, 91, 92
Child-centered preschools, 233
developmentally appropriate programs, signs of, 235
Child-directed speech, 168–169
Childhood mortality. *See* Death and dying; Infant mortality
Childlessness, 468
late adulthood and, 603–604
vocational development of women and, 472
Child–parent relationship. *See* Attachment; Father *entries;* Mother *entries;* Parent *entries*
Child-rearing costs, 462
Child-rearing practices, 463, 465. *See also* Attachment; Discipline; Families; Father *entries;* Home environment; Mother *entries;* Parent *entries*
academic achievement in adolescence and, 372
achievement-related attributions and, 317–318
ADHD and, 288–289
adolescent development and, 395–396
aggression and, 257
American values of, 62–63
athletics, caregiving concerns, 284
attachment and, 189, 190, 194
authoritarian style, 265
authoritative style. *See* Authoritative parenting
birth complications and, 100
blended families, 469–470
cognitive development and, 232
coregulation, 328
cosleeping arrangements and, 122
cultural influences on, 226, 266–267
death concept, development of, 624, 625
delinquent behavior and, 403, 404
depression and, 401
difficult children and, 184
divorce, adjustment of children to and, 330–331
dropping out of school and, 375
dual-earner families, 473–474
early childhood, 264–269
early intervention programs, 234
eating habits and, 208–209
education about, 465
emotional self-regulation and, 179–180
empathy development and, 249
extended family households and, 63
fear management in children, 248
gender identity and, 521–522
gender typing and, 260–262
goodness-of-fit model, 184–185, 190, 254
marital relationship and, 59, 463, 465
maternal employment and, 333
middle childhood, 328
moral development and, 391
nonorganic failure to thrive and, 126
peer groups in adolescence and, 398
peer relations and, 251–252
peer victimization and, 325
permissive style, 265–266, 372, 398
popular literature about, 15
preterm infants, 97–98
Puritan practices, 13
research on, 5, 265–266
self-care children and, 333–334
self-control and, 197
self-esteem and, 317
self-image of child and, 246
sexual abuse victims as parents, 336
shyness versus sociability and, 183
sibling relationships and, 328
situational influences on, 266–267
skipped-generation families, 528
socioeconomic status and, 59–60
teenage parenthood, 360–361
temperament and, 184
uninvolved style, 265, 266, 398
World War II refugee and evacuee children, 588
Child-rearing styles, 265–266
Children's Defense Fund, 65
Chinese
aging parents, relationships with, 529
child-rearing practices, 266
dying, adaptation to, 630
emotional expression among, 180
national examinations, 374
only children, 329
parental leave, 98
romantic relationships, 452–453
self-esteem among, 316
shyness, valuing of, 185
Chinese Americans. *See also* Asian Americans
grandparent-grandchildren relationships, 527
reminiscence in late adulthood among, 587
Chlamydia, 358
Cholesterol, 422, 491, 492
calorie restrictions and, 485
exercise and, 424
heart disease and, 554
Chomsky's theory of language development, 165
Chorion, 78
Chorionic villus sampling, 52, 53, 78*n*
Christianity. *See also specific faith*
bereavement rituals, 642
death anxiety and, 626
death concept, development of, 623–624
Chromosomal abnormalities, 50–51
genetic counseling, 52
identification of DNA markers, 53
prenatal diagnosis and fetal medicine, 52–53, 56
Chromosomes, 44–50
Alzheimer's disease risk and, 566
mapping of, 53
sex chromosomes, 46, 49, 51
X chromosome, 46, 49, 547
Y chromosome, 46, 49
Chronic illness
childhood, 280
cognitive deterioration and, 576
cognitive development in middle adulthood and, 500
life expectancy and, 550
psychological well-being and, 592–593
Chronological age. *See* Age
Chronosystems, in ecological systems theory, 24, 25, 59
Church attendance. *See* Spirituality and religiosity
Cigarettes. *See* Smoking
Circular reactions, in sensorimotor stage of cognitive development, 145–147
Circumcision, 105–106
Cities, influence on development, 61–62
Civil responsibility, development of, 393
Class inclusion problems, 218, 220, 285
Classical conditioning, 18, 126–127
Classification. *See also* Categorization abilities
concrete operational stage of cognitive development, 285
hierarchical, 218
Classrooms. *See* Schools and schooling; Teacher *entries*
Clear-cut attachment phase, 186
Cleft palate, stress during pregnancy and, 89
Climacteric, 486–487, 488. *See also* Menopause
Clinical death, 621
Clinical interviews, 28–29, 32
Piaget's use of, 19
Clinical method of research, 29, 32
psychoanalytic theorists' use of, 17
Cliques, 398–399
Cloning, 55
Close relationships. *See* Friendships; Intimacy; Marriage
Coaching. *See* Athletics
Cocaine. *See also* Drug use and abuse
prenatal development and, 83–84
teenage use, 361
Codominant pattern of inheritance, 47, 49
Coercion, sexual. *See* Sexual coercion
Cognition. *See also* Cognitive development
abstract. *See* Abstract thinking
adaptive, 432
dying people's thoughts, 627–631
metacognition. *See* Metacognition
Cognitive development. *See also* Intelligence *entries;* Language development; Learning; Mental retardation
achieving stage, 432–433
acquisitive stage, 432
adolescence, 363–377
milestones, 408–409
alcohol use during pregnancy and, 85–86
athletics and, 347
bereavement and, 641
bilingualism and, 300–301
brain development and, 288
brain growth spurts and, 119
child care and, 162, 233, 234
child maltreatment and, 268
core knowledge perspective, 150–151
cultural influences on, 286–287
delinquent behavior and, 404
Down syndrome, 50–51
early adulthood, 432–441
milestones, 478–479
early childhood, 216–238
milestones, 272–273
early intervention programs, 162–164, 233–234, 235
emotional development and, 178
emotional self-regulation and, 320
environmental influences on, 161–162, 232–235
environmental pollution and, 87
executive stage of, 433
fetal alcohol syndrome and, 86
gender identity and, 263
gender stereotyping and, 260
higher education and, 436
home environment and, 232–233
identity development and, 385
individual differences in, 160–164, 232–235, 293–299, 500–501, 570

infancy and toddlerhood, 143–170
milestones, 200–201
information processing theories. *See* Information processing theories
Labouvie-Vief's theory of postformal thought, 433
language development and, 167, 364
late adulthood, 10, 433, 570–579
milestones, 616–617
lateralization of brain and, 206
low birth weight and, 96
make-believe play and, 158–159
middle adulthood, 433, 498–507
milestones, 542–543
middle childhood, 284–309
milestones, 340–341
milestones in
adolescence, 222, 408–409
early adulthood, 478–479
early childhood, 222, 272–273
infancy and toddlerhood, 151, 200–201
late adulthood, 616–617
middle adulthood, 542–543
middle childhood, 340–341
nonorganic failure to thrive and, 126
nutrition during pregnancy and, 88
overview, 9
Perry's theory of postformal thought, 432
Piaget's theory of. *See* Cognitive-developmental theory (Piaget)
play and, 250
postformal thought. *See* Postformal thought
pragmatic thought, 433
reintegrative stage, 433
responsibility stage, 433, 498
Schaie's theory of postformal thought, 432–433, 435, 499
selective optimization with compensation, 570, 592, 612
self-concept and, 315
sex chromosome disorders and, 51
sex differences in, 369–370
small-for-date infants, 97
social-cognitive learning theory, 18–19
sociocultural theory of. *See* Sociocultural theory of cognitive development (Vygotsky)
sociodramatic play and, 217
socioeconomic status and, 60
television and, 234–235
terminal decline, 576–577
vocational life and, 505
voluntary reaching and, 132
Cognitive disequilibrium, 144–145
Cognitive equilibrium, 19, 144–145
Cognitive flexibility, 505
Cognitive inhibition
middle adulthood, 502
middle childhood, 288
Cognitive interventions. *See* Intervention programs
Cognitive maturity. *See* Cognitive development
Cognitive psychology. *See* Information processing theories
Cognitive reserve, in aging process, 566
Cognitive self-regulation
adolescence, 365, 368
middle childhood, 291
Cognitive-developmental theory (Piaget), 19–20, 26, 144–151
accommodation in, 144, 218
adaptation in, 19, 144–145
assimilation in, 144
disequilibrium in, 144–145
educational principles based on, 222–223, 302
equilibrium in, 19, 144–145
evaluation of, 157
gender identity and, 260, 263
key concepts, 144–145
make-believe play and, 158
methods of study, 19
moral development, 256, 321, 387–388, 389
organization in, 145
schemes in, 144, 217
stages of, 19, 20. *See also specific stages (e.g., Formal operational stage of cognitive development; Sensorimotor stage of cognitive development)*
Cohabitation, 466–468
sexual activity and, 426–427
Cohort, 11
Cohort effects, 35–36
intelligence studies, 435
mental abilities, 498, 501
sexuality in late adulthood, 560
Colic, 105
Collaborative communication style, 297
Collectivist societies, 62–63
academic achievement, valuing of, 308
aging parents, caregivers of, 531
emotional expression among, 180
make-believe play and, 159
moral development among, 392
psychological well-being in middle adulthood among, 519
romantic relationships in, 452
College education. *See* Higher education
College-preparatory education, 374
Colon cancer, 491
exercise and, 424
fat consumption and, 421
genetic testing, 56
Color vision, 484
postnatal development, 107
Columbine High School shooting, bereavement after, 644
Comfort care, in dying, 632
Coming out, 357
Commitment
generativity stage of development, 512, 514
hardiness and, 496–497
love, component of, 451–452
political activism and, 513
psychological well-being in middle adulthood and, 520
Communication. *See also* Language development; Speech
dying people, with, 630
parent-teacher interactions, 62
teacher–student interaction, 302, 304
Communication styles
intelligence and, 297
sex differences in, 262
Community influences. *See also* Cultural influences
child maltreatment, 268
childhood injuries, preventing, 212
civil responsibility, 393
delinquent behavior, 403–405
elder social interactions and, 596–598
family systems theory, 61–62
identity development, 385
immigrant youths, 30–31
resiliency of children, 11
Community senior centers. *See* Senior centers
Community service. *See also* Volunteering
extracurricular activities and, 376
life expectancy and, 551
Community-researcher partnership, 28
Commuting college students, 436
Companionate love, 451
Companionship. *See* Peer *entries*
Comparisons, social, 315
Compassion. *See also* Empathy; Prosocial behavior
perspective taking and, 320
Compensatory strategies
selective optimization with compensation, 570, 592, 612
Compliance. *See also* Conformity
caregiving concerns, 197
early childhood, 226
family systems research, 58
infancy and toddlerhood, 196–197
maternal behavior and child's compliance with parental demands, 32
Componential analysis of intelligence, 294
Componential subtheory of intelligence (Sternberg), 294
Comprehension versus production of language, 167
Compression of morbidity, 559
Computers
assistive technology, 556
games. *See* Video and computer games
job training and, 535–536
Websites. *See* Websites
Conception, 45–46. *See also* Fertility; Pregnancy; Prenatal *entries*
breastfeeding's effect on, 123
caregiving concerns, 56
genetic counseling, 52
menarche and, 348
prenatal diagnosis and fetal medicine, 52–53, 56
reproductive technologies, 54–55
spermarche and, 348
Conceptual categorization, 155, 157
Concordance rates, 67–68
Concrete operational stage of cognitive development, 19, 20, 284–287
classification in, 285
conservation in, 285
defined, 285
evaluation of theory, 287
limitations of, 286
research on, 286–287
seriation in, 285
spatial reasoning in, 285–286
Conditional support, and self-esteem, 384
Conditioned response (CR), 127
Conditioned stimulus (CS), 127
Conditioning. *See* Classical conditioning; Operant conditioning
Condoms. *See* Contraception
Conduct problems. *See* Adjustment problems; Antisocial behavior; Delinquency
Confidence. *See* Self-confidence
Confidentiality. *See* Privacy
Conflict. *See also* Aggression; Violence
marital. *See* Divorce
peer acceptance and, 324
sibling relationships, 108, 328
Conflict resolution. *See* Social problem solving
Conformity
adolescent peer conformity, 399–400
social conventions, to, 256, 322. *See also* Moral development
Confrontation, as phase of grieving, 639
Confucian beliefs about death, 621
Confusion
appearance versus reality confusion in preoperational stage, 221
identity confusion, psychosocial stage of development, 17, 382–383
Congestive heart failure. *See* Heart disease
Congregate housing for elderly, 599–600
adjustment to retirement and, 611
Conjunctive stage of faith development (Fowler), 590
Conscience. *See* Moral *entries;* Superego
Conscientiousness, as personality trait, 522–523
Consent to research, 38
Conservation
concrete operational stage of cognitive development, 285
preoperational stage of cognitive development, 218, 219, 220, 222
Constructive play, 251
Contamination themes, in life stories, 514
Contexts for development, 6–7, 10–12, 22. *See also specific influences (e.g., Cultural influences; Home environment)*
Contextual subtheory of intelligence (Sternberg), 295, 503
Continuing education
death education, 645
late adulthood, 577–579, 613
vocational development, 535–536
Continuous course of development, 6
cognitive development in early and middle childhood, 287
Locke's *tabula rasa,* 13
stance of major developmental theories, 26
vocational development, 471
Contraception
adolescents' use of, 355–356, 360
family planning, 464
sexual attitudes and behavior and, 427
teenage pregnancy rate and, 360
Contractions in childbirth, 91–92
Contrast sensitivity, 135
Control. *See also* Self-control
late adulthood, control versus dependency, 591–592
psychological well-being in middle adulthood and, 520
Controversial children, and peer acceptance, 324, 326
Conventional level of moral development, 389–390
Conventional personality type, in vocational choice, 437
Convergent thinking, 306
Conversation. *See* Pragmatic development; Speech
Cooing, in language development, 165–166
Cooley's anemia, 48
Cooperation. *See also* Compliance; Prosocial behavior
dating and, 399
morality of interpersonal cooperation, 389
Cooperative family structures, 62
Cooperative learning, 304
mainstreaming of students with learning difficulties and, 305

Cooperative play, 249–250
Coordination of secondary circular reactions, in sensorimotor stage of cognitive development, 145, 146
Coping strategies, 494–496
aging parents, caregivers of, 531
aging process and, 556–557
caregivers of elders with dementia and, 568
divorce in late adulthood and, 602
dual-process model of coping with loss, 640
dying, adaptation to, 628, 629
emotion-centered coping, 495, 519–520, 556
friendships and, 398
gender-typed coping styles and depression, 401
hardiness traits, 496–497
hearing loss, 552–553
late adulthood, negative life changes and, 593–594
maturational timing and, 352
middle adulthood, 519–520
obesity treatment and, 421
problem-centered coping. *See* Problem-centered coping
resilient children, 336–338
successful aging and, 612–613
teenage substance abuse and, 362
unemployment, 537
Core knowledge perspective, 150–152
metacognition and, 229
Coregulation, in supervision of children, 328
Cornea. *See* Eyes
Corporal punishment, 254–255, 266, 268
Corpus callosum, 207
Corpus luteum, 77
Correlation, genetic–environmental, 69–70, 162
Correlation coefficients, 32
Correlational research design, 31–33, 34
Cortex. *See* Cerebral cortex
Cortical dementia, 565–567
Cortisol
hardiness and, 497
shyness versus sociability, 182
Cosleeping arrangements, 122
Costs
child-rearing costs, 462
health care costs of elderly, 567, 569, 613
Counseling
anorexia nervosa, 354
blended families, for, 333
bulimia, 354
chronic illness in childhood, 280
delinquency prevention and treatment, 405
depression, 176, 567, 629
divorce, following, 469
drop out prevention programs, 376
elder maltreatment prevention and treatment, 608
genetic counseling, 52
genetic testing, 56
grief therapy, 645
life-review reminiscence, 586
parent–adolescent relationship and, 465
parenthood, transition to, 463
partner abuse, 461
postpartum depression, 176
prenatal surgery, about, 53
remarriage, following, 469
sexual abuse victims, 336
sexual coercion victims, 429–430
substance abuse, 362, 426
suicide prevention, 402–403, 593
unemployment, following, 537
Counting skills, 231
Course of development, 6–7
stance of major developmental theories, 26
Court proceedings. *See* Legal proceedings
CR (conditioned response), 127
Crawling. *See also* Motor development
depth perception and, 134–135
Creativity
flow and, 520
gifted students, 306–307
information processing theories of, 433–434
middle adulthood, 504, 512, 515
Crime. *See also* Delinquency; Violence
alcohol abuse and, 426
delinquency and adult criminality, 404
elderly victims, 598
Critical thinking
identity development and, 385
teaching techniques to encourage, 302
Criticism, and abstract thinking, 367, 368
Cross-age relationships. *See* Intergenerational relationships
Cross-cultural research, 29–30. *See also* Cultural influences
academic achievement, 307–309
adolescence, period of, 345
cognitive flexibility and vocation, 505
mathematical reasoning studies, 231
motor development, 131–132
sociocultural theory, 23–24
Cross-gender reasoning and behavior. *See also* Androgynous traits
adolescence, 394
cultural influences on, 327
early childhood, 262
middle childhood, 326
Cross-generation relationships. *See* Intergenerational relationships
Cross-linkage theory of aging, 414
Cross-sectional research, 34, 35. *See also* Cohort effects
"big five" personality traits, 522–523
crystallized and fluid intelligence, changes in, 499
gender identity in middle adulthood, 521
intelligence in adulthood, 434–435
self-concept in middle adulthood, 519
sexuality in late adulthood, 560
Crowds, adolescent membership in, 398–399
Crow's-feet, 484
Cruzan case, and right to die, 633
Crying in newborns, 103–105
Crystallized intelligence (Cattell), 498–500, 570
CS (conditioned stimulus), 127
Cults, identity development and, 385
Cultural bias in intelligence tests, 232, 293, 297–299
Cultural influences, 62–66. *See also* Cross-cultural research; Ethnicity and race; *specific entries (e.g., African societies; Native Americans)*
abstract thinking, 365
academic achievement, 307–309
adolescence, 345
aging parents, caring for, 598
aging process, coping with, 557–558
alcohol abuse, 425
attachment, 188
bereavement, 639, 642–643
child maltreatment, 268
childbirth, 93
child-rearing practices, 226, 266–267
cognitive development, 286–287
death anxiety, 626
death concept, development of, 623–624
dying, adaptation to, 630–631
elder maltreatment, incidence of, 607–608
emotional development, 178
family planning, 464
gender typing, 327
grandparent–grandchild relationships, 527
hand preference, 206–207
hospice care, 633
identity development, 385–386
infant sleeping arrangements, 121–123
institutionalization of elderly, 569
intelligence, 297–299
make-believe play in infancy and toddlerhood, 159
memory development, 290–291
menopause, attitudes toward, 488
moral development, 322, 391, 392
motor development, 131–132
newborn behavior and child-rearing practices, 107
parent–child relationship, 192
parenthood and marital satisfaction, 463
partner abuse, 461
peer relations in adolescence, 396
personal storytelling, 246
play, 250
psychosocial stages of development, 17
pubertal changes, responses to, 350
romantic relationships, 452–453
self-concept, 246
self-conscious emotions, expression of, 179
self-esteem, 316–317
separation anxiety, 186
sexual activity, 354
sociability, 250
social clock, in adult development, 449–450
sociocultural theory. *See* Sociocultural theory of cognitive development (Vygotsky)
subcultures, 62
substance use and abuse, 361
temperament, 184
young adults leaving home, 457, 525–526
Cultural meanings, 30
Cursive. *See* Writing
Custody of children, 331
homosexuality and, 470
Cystic fibrosis, 48, 280
DNA markers, identification, 53
Cytomegalovirus, 358
prenatal development and, 87–88
Cytoplasm, 45

D

Dark adaptation, aging process and, 551
Darwin's theory of evolution, 14
Date rape. *See* Sexual coercion
Dating. *See also* Romantic relationships
adolescence, 399
early adulthood, 426–428
Dating violence, 399
Day care. *See* Child care
Day treatment centers
delinquency intervention programs, 405
Deafness. *See* Hearing loss
Death and dying, 619–645. *See also* Bereavement; Euthanasia; Infant mortality; Suicide
adolescent injuries as cause of death, 620
adolescent understanding of and attitudes toward, 624–625
adult understanding of and attitudes toward, 625–626
advance medical directives, 634–636
agonal phase, 621
AIDS as cause of death, 432
Alzheimer's disease as cause of death, 565
anorexia nervosa as cause of death, 353
anxiety about death, 626–627
appropriate death, 629
brain death, 621
caregiving guidelines, 625
child maltreatment causing death, 269
childhood understanding of and attitudes toward, 623–624
clinical death, 621
comfort care, 632
communicating with dying people, 630
concept of death, development of, 623–626
contextual influences on adaptation to dying, 628–631
defining death, 621–622
disease course and, 629
durable power of attorney for health care, 635–636
early adulthood, leading causes of death in, 419, 620
early childhood injuries as cause of death, 211, 620
education about death, 645
family members' behavior and, 629–630
health professionals' behavior and, 629–630
home deaths, 631
hospice care, 632–633
hospital dying, 631–632
individual differences in adaptation to dying, 629–631
intimate partner's death, 641
Kübler-Ross's stages of dying, 628
late adulthood, leading causes of death in, 561, 562
living wills, 635–636
middle adulthood, leading causes of death in, 489, 490, 620
mortality phase, 621
obesity as cause of death, 420, 423
parental death, 532, 641
persistent vegetative state, 621
personality traits and, 629
physical changes in, 620–621
right to die, 633–639
smoking as cause of death, 425
spirituality and, 624, 626, 630–631
stages of dying, 627–628
terminal decline, 576–577

thoughts and emotions of dying people, 627–631
understanding of and attitudes toward, 622–627
Death anxiety, 626–627
educational programs and, 645
health care professionals, in, 630
Death concept, development of, 623–626
Death education, 645
Death with dignity, 620, 622
Oregon law, 638
Debriefing in research, 39
Decentration thinking, in cognitive-developmental theory, 285
Deception in research, 39
Decision making
abstract thinking and, 367, 368–369
parenthood, entry into, 461–462
retirement, entering, 609–610, 611
Deferred imitation, 147, 149
Deformities. *See* Birth defects; Chromosomal abnormalities; *specific entries (e.g., Cleft palate)*
Degenerative joint disease, 562
Deliberate memory, 571
Delinquency, 344, 403–405
child maltreatment and, 268
depression and, 401
divorce and, 330
early-onset type, 404
extended family households and, 63
factors related to, 403–405
Head Start programs and, 234
incidence of, 403
late-onset type, 404
partner abuse and, 460
peer acceptance and, 324
prevention and treatment, 405
sexual abuse victims, 336
Delivery. *See* Childbirth
Dementia, 564–567
Alzheimer's disease, 565–567
cerebrovascular, 567
misdiagnosis, 567
Denial, as stage of dying, 628
Denmark
home health care for aging parents, 531
immunization in, 209
institutionalization of elderly, 569
male birth rate, 49
obesity rate, 278
retirement programs, 610
Dental care, 204
Dental development. *See* Teeth
Dentures, taste perception and, 553
Deoxyribonucleic acid (DNA), 44–45
aging and, 413–414, 490
identification of markers, 53
Dependency
elder maltreatment risk factor, 608
late adulthood, 591–592
peer victimization and, 325
successful aging and, 613
Dependency–support script, 591, 593
Dependent variables, in experimental research design, 33
Depression. *See also* Suicide
adolescence, 344, 400–401
aging parents, caregivers of, 531
Alzheimer's disease and, 565, 566
cancer and, 629
child maltreatment and, 268
death anxiety and, 626
dementia misdiagnosis and, 567
dieting and, 421
divorce and, 469, 602
dying, stage of, 628
exercise and, 424
extracurricular activities and, 376
genetic factors, 67
homosexuality in adolescence and, 357
immune response and, 418
late adulthood, 567, 591–593
long-term care residents, 600
maternal depression, 176, 189
obesity and, 279
peer victimization and, 325
postpartum depression, 176
premenstrual syndrome and, 430
self-esteem and, 384
sexual abuse victims, 336
sexual coercion and, 429
sleep-deprivation and, 347
substance abuse and, 362
Depression era, consequences on lifespan development, 36, 576
Depression stage of dying, 628
Deprivation. *See also* Malnutrition; Poverty; Sensitive periods; Socioeconomic influences
nonorganic failure to thrive, 126
sensory deprivation studies of animals, 119
stimulus deprivation, 119
Depth perception, 134–138, 139
aging process, 551–552
drawing and depth cues, 281
Dermis, 484
DES (diethylstilbestrol), 83
Design of research. *See* Research designs
"Designer" babies, 55
Despair. *See also* Depression
life review and, 586–587
Peck's theory of developmental tasks, 585
psychosocial stage of development, 17, 447, 584–585
Destruction versus creation, in midlife development, 515
Determinants of development, 5, 6–8. *See also* Environmental influences; Genetic *entries;* Nature–nurture controversy
stance of major developmental theories, 26
Developing countries
average life expectancy, 547
breastfeeding versus bottlefeeding, 123
malnutrition in, 125
Development. *See* Human development; *specific entries (e.g., Cognitive development; Physical development)*
Developmental quotients (DQs), 160
Developmental research designs, 34–37
Developmentally appropriate practices, 162, 163, 235
DHEAS, 485
Diabetes, 280
calorie restrictions and, 485
cataracts and, 551
cerebrovascular dementia and, 567
exercise and, 424
genetic inheritance, 48
late adulthood, 561, 562, 563
low birth weight and, 84
obesity and, 278, 420
prenatal health care, 89
sexuality in late adulthood and, 561
Diarrhea, 209
Diet. *See* Eating habits; Malnutrition; Nutrition
Dietary fat. *See* Fat, dietary
Dietary supplements. *See* Food supplement programs; Vitamins and minerals
Diethylstilbestrol (DES), 83
Dieting, 421, 485
anorexia nervosa and, 353
Differentiation theory of perceptual development, 139–140
Difficult child, as temperament type, 181
child-rearing practices and, 184
Diffusion, in identity development. *See* Identity diffusion
Digestive system. *See* Gastrointestinal system
Dignity, dying with. *See* Death with dignity
Dilation and effacement of cervix, 91, 92
Dimples, dominant–recessive inheritance, 47
Directions, in concrete operational stage of cognitive development, 285
Disability
learning. *See* Learning disabilities
mental. *See* Dementia; Psychological disturbances
physical. *See* Physical disability
Disadvantaged. *See* Poverty; Socioeconomic influences
Discipline. *See also* Child-rearing practices; Punishment (disciplinary)
aggression in children and, 257
divorce's effect on, 330
inductive, 253
moral development and, 252–253
positive discipline, 255–256
temperament and, 254
time out, 255
Discontinuous course of development, 6
cognitive development in early and middle childhood, 287
Rousseau's "noble savage," 13
stance of major developmental theories, 26
vocational development, 471
Discovery learning, 222
assisted discovery learning, 224
Discrimination. *See* Bias; Ethnicity and race; Socioeconomic status (SES)
Diseases. *See also* Health care; Immune system; Infectious diseases; *specific diseases (e.g., Cancer; Diabetes; Heart disease)*
autosomal, 48
breastfeeding and, 123–124
cognitive development in middle adulthood and, 500
dominant–recessive inheritance, 47, 48
early childhood development and, 209
fetal medicine, 52–53, 56
genetic counseling, 52
identification of DNA markers, 53
impotence and, 487
late adulthood, 561–564
life expectancy and, 547, 550
middle childhood, 280
otitis media, 209, 210
prenatal development and, 87–88
prenatal diagnosis, 52–53, 56
X-linked, 48, 54
Disengagement theory of aging, 595
Disequilibrium, in cognitive-developmental theory, 144–145
Disorganized–disoriented attachment, 187–188
later development and, 194
mother-child relationship and, 189, 190
Disposition. *See* Personality traits; Temperament
Distractibility, as temperament dimension, 181
Distress. *See also* Anxiety; Fear; Stress
temperament dimension, as, 181
Distributive justice, 321–322
Divergent thinking, 306
Diversity of lifestyles in adult development, 466–471
Division of cells, 45
chromosomal abnormalities, 50, 51
Divorce, 329–331, 468–469. *See also* Single-parent families
adjustment to, 332, 469, 525, 602
age of child and response to divorce, 330
child-rearing practices and, 330–331
cohabitating relationships after, 467
consequences of, 469
custody arrangements. *See* Custody of children
economic impact of, 525, 602
emotional impact on children, 59
factors related to, 468–469
grandparent–grandchildren relationships and, 529
immediate consequences of, 330, 602
immune response and, 418
incidence of, 329, 457, 468, 525, 602
late adulthood, 602
loneliness and, 456
long-term consequences of, 330–331, 525
marital age and, 458
middle adulthood, 524–525
parental bereavement and, 641
poverty and, 60
psychological well-being in middle adulthood and, 521
reasons for, 525, 602
sex of child and response to divorce, 330
sexual activity of adolescents and, 355
temperament of child and response to divorce, 330
visitation rights. *See* Child visitation (following divorce)
Divorce mediation, 331
Dizygotic twins. *See* Fraternal twins
DNA. *See* Deoxyribonucleic acid
Domains of development, 8, 9. *See also* *specific domains (e.g., Cognitive development; Physical development)*
Domestic chores. *See* Household chores
Domestic violence. *See also* Child maltreatment; Elder maltreatment; Partner abuse
alcohol abuse and, 426
elder maltreatment risk factor, 608
Dominant cerebral hemisphere, 206
Dominant–recessive inheritance, 46–47, 48, 49
Donor insemination, 54
Donor ova and sperm, 54–55
Double-jointedness, dominant–recessive inheritance, 47
Doubt. *See also* Anxiety
psychosocial stage of development, 17, 174–175

Down syndrome, 50–51, 566
Downsizing, 537
Drawing
  cultural influences, 214–215
  early childhood, 213–215
  middle childhood, 281–282
Dreams, in Levinson's theory of adult development, 448
Dress and grooming. *See also* Appearance and attractiveness
  adolescent peer conformity and, 399
  Alzheimer's disease and, 565
  early childhood motor development, 213
  peer culture and, 322–323
  peer groups, 398
Drinking. *See* Alcohol use and abuse
Driving accidents. *See* Motor vehicle accidents
Dropping out of college, 436
Dropping out of school, 374–377
  factors related to, 375
  incidence of, 374–375
  peer acceptance and, 324
  prevention strategies, 375–377
Drug education programs, 362
Drug use and abuse. *See also* Medications
  adolescence, 361–362
  AIDS and, 432
  child maltreatment and, 268
  depression and, 401
  divorce and, 525
  early adulthood, 424–425
  homosexuality in adolescence and, 357
  identity development and, 385
  life expectancy and, 548
  prenatal development and maternal use, 82–83
  sexual abuse victims, 336
  SIDS and drug use during pregnancy, 103
Druze, death concept among, 623–624
Dual-earner households, 333–334, 458, 473–475. *See also* Maternal employment
  adolescent development and, 396
  caregiving guidelines, 475
Dualistic thinking, 432
Dual-process model of coping with loss, 640
Duchenne muscular dystrophy, 48
  DNA markers, identification, 53
Duplication of cells, 45, 50
Durable power of attorney for health care, 635–636
Dutch. *See* Netherlands
Dwarfism, psychosocial, 208
Dying. *See* Death and dying
Dynamic systems theory
  cognitive development, 157
  motor development, 130–131, 212
Dynamic testing, 298
Dynamic transformations, 218

E

Ear. *See* Ears; Hearing
Ear infections, 209, 210, 277
Early adulthood, 8, 411–477
  alcohol use and abuse in, 425–426, 431
  cognitive development in, 432–441
    milestones, 478–479
  dating in, 426–428
  death concept, development of, 625–626
  death in, leading causes of, 419, 620
  drug use and abuse in, 424–425
  eating habits in, 421, 431
  emotional and social development in, 445–477
    milestones, 478–479
  exercise in, 423–424
  family life in, 456–465
  friendships in, 453–455
  health issues in, 419–431
  homosexuality in, 428
  information processing in, 433–434
  injuries in, 419
  intelligence tests in, 434–435
  Levinson's theory of development in, 447–449
  loneliness in, 455–456
  marriage in, 446
  milestones in, 478–479
  nutrition in, 419–423
  obesity in, 420–421
  parental relationships in, 456–457
  peer relations in, 450–456
  physical development in, 412–431
    milestones, 478–479
  sexuality in, 426–430, 431
  sexually transmitted diseases in, 428
  sibling relationships in, 455
  smoking in, 425, 431
  stress in, 430–431
  substance abuse in, 424–426
  Vaillant's theory of development in, 449
  vocational development in, 436–441, 471–474
  vocational education in, 439–440
  weight in, 431
Early childhood, 8, 203–271
  academic training in, 233
  aggression in, 256–259
  artistic expression in, 213–215
  attention in, 225–226
  bereavement in, 641
  body composition in, 204
  body proportions in, 204
  brain development in, 206–207
  brain plasticity in, 120
  child care in, 233, 234
  child-rearing practices in, 264–269
  cognitive development in, 216–238. *See also* Preoperational stage of cognitive development
    milestones, 272–273
  cross-gender reasoning and behavior in, 262
  death concept, development of, 623–624
  diseases in, 209
  drawing in, 213–215
  eating habits in, 208–209
  emotional and social development in, 244–271
    milestones, 272–273
  emotional self-regulation in, 248
  emotions in, 247–249
  empathy in, 249
  fears in, 248
  fine motor development in, 213–215
  friendships in, 251
  gender identity in, 263–264
  gender stereotyping in, 260, 264
  gender typing in, 260–264
  grammar in, 236–237
  gross motor development in, 212–213
  guilt in, 248–249
  height in, 204
  information processing in, 225–231
  injuries in, 209, 211–212, 620
  intelligence tests in, 232
  intervention programs in, 233–234, 235
  language development in, 216, 236–238
    milestones, 272–273
  make-believe play in, 251
  malnutrition in, 209
  mathematical reasoning in, 231
  memory in, 226–228
  metacognition in, 228–230
  milestones in, 272–273
  moral development in, 252–259
  motor development in, 212–215
  muscle development in, 204
  nutrition in, 208–209
  peer relations in, 249–252
  physical development in, 204–215
    milestones, 272–273
  planning in, 225–226
  pragmatic development in, 237
  preschools. *See* Preschools
  pride in, 248–249
  reading in, 230–231
  recall memory in, 226–227
  recognition memory in, 226–227
  self-concept in, 245, 246
  self-conscious emotions in, 248–249
  self-development in, 245–247
  self-esteem in, 245–247
  self-help skills in, 213
  shame in, 248–249
  skeletal age in, 204
  skeletal development in, 204
  sociocultural theory of, 223–225
  speech in, 237
  spelling in, 230–231
  taste in, 208
  vocabulary in, 236
  walking in, 212–213
  weight in, 204
  writing in, 213–214, 215, 230–231
Early experiences, 7
  psychosexual theory of, 16
Early intervention programs
  cognitive development, 162–164
  Down syndrome and, 50
  early childhood, 233–234
Early learning centers, 120
Earnings gap, 472
Ears. *See also* Hearing
  aging process, 552–553, 555
  infections in, 209, 210, 277
  prenatal development of, 79, 82
Eastern Europeans, caring for aging parents, 598
Easy child, as temperament type, 180–181
Eating disorders, 421
  adolescence, 344, 353–354
Eating habits. *See also* Nutrition
  adolescence, 352–353
  early adulthood, 421, 431
  early childhood, 208–209
  late adulthood, 553, 560
  low-calorie diets, 485
  obesity and, 278–279, 420, 421, 422–423
Eclampsia, 90
Ecological systems theory (Bronfenbrenner), 24–25, 26, 58–59
  attachment and, 194
Ecological transitions, 25
Economic influences. *See* Poverty; Socioeconomic influences; Socioeconomic status (SES)
Ectoderm, formation of, 79
Education. *See also* Learning; Schools and schooling; Teacher *entries; specific subjects (e.g., Mathematics; Reading)*
  Alzheimer's disease and, 566
  bereavement, 645
  bilingual education, 300–301
  child-rearing practices, 465
  cognitive development and, 576
  cognitive-developmental principles of, 222–223, 302
  college. *See* Higher education
  college-preparatory, 374
  continuing. *See* Continuing education
  death education, 645
  drug education programs, 362
  early childhood academics, 233
  elder maltreatment prevention and treatment, 608, 609
  family life education. *See* Family life education
  family size and, 464
  health education. *See* Health education
  information processing principles of, 291–293
  late adulthood, 577–579
  National Association for the Education of Young Children, 303
  parent education, 465
  physical. *See* Physical education
  public policies, 64
  remedial education, in drop out prevention programs, 376
  sex education. *See* Sex education
  sexual abuse prevention, 336
  SIDS public education programs, 103
  sociocultural principles of, 224, 304
  socioeconomic status and, 60
  vocational. *See* Vocational education
Educational attainment. *See also* Higher education
  employment and, 375, 439
  socioeconomic status and, 60
  teenage pregnancies and, 360
  vocational development and, 437
Educational philosophies, 302–304
  cognitive-developmental principles, 20, 222–223, 302
  information processing principles, 291–293
  sociocultural principles, 224, 304
Educational self-fulfilling prophecies, 304
Educational television, 234–235
Efe of Zaire, stranger anxiety among infants, 178
Effacement of cervix, 91, 92
Egalitarian marriages, 458
Ego, 15–16
Ego differentiation versus work-role preoccupation (Peck), 585
Ego integrity
  death anxiety and, 626
  life review and, 586–587
  Peck's theory of developmental tasks, 585
  psychosocial stage of development, 17, 447, 584–585
  self-concept and, 587
  Vaillant's stage development, 447
Ego integrity versus despair, 17, 447, 584–585
Ego transcendence versus ego preoccupation (Peck), 585
Egocentric speech, 223

- Egocentrism. *See also* Perspective taking
  - concept of death and, 623
  - formal operational stage of cognitive development, 367
  - preoperational stage of cognitive development, 217–218
- Ejaculation, first, 348, 349–350
- Elaboration, as memory strategy, 290, 502
- Elder maltreatment, 607–609
  - forms of, 608
  - incidence of, 607–608
  - prevention and treatment, 608–609
  - risk factors, 608
- Elder suicide, 592–593
- Elderhostel programs, 577, 578
- Elderly. *See* Aging *entries;* Late adulthood
- Electra conflict, 16, 244
- Elementary schools. *See* Schools and schooling
- Embarrassment
  - adolescent self-consciousness, 367–368
  - infancy and toddlerhood, 178–179
- Embryo, period of prenatal development, 77, 79
  - teratogens in, 82
- Embryonic disk, 76, 78
- Emotional and social development, 8, 9. *See also* Emotions; Moral development; Personality traits; Self *entries;* Social *entries;* Temperament
  - activity theory of aging, 595–596
  - adolescence, 381–407
    - milestones, 408–409
  - athletics and, 347
  - attachment. *See* Attachment
  - autism and, 229
  - child maltreatment and, 268
  - child-rearing practices and, 264–269
  - disengagement theory of aging, 595
  - divorce's effect on, 59
  - early adulthood, 445–477
    - milestones, 478–479
  - early childhood, 244–271
    - milestones, 272–273
  - ethological theory, 22
  - friendships and, 397–398
  - infancy and toddlerhood, 173–197
    - milestones, 200–201
  - Labouvie-Vief's theory of emotional reasoning, 585–586
  - late adulthood, 583–615
    - milestones, 616–617
  - Levinson's theory of. *See* Levinson's theory of adult development
  - middle adulthood, 511–541
    - milestones, 542–543
  - middle childhood, 313–339
    - milestones, 340–341
  - milestones in, 177
    - adolescence, 408–409
    - early adulthood, 478–479
    - early childhood, 272–273
    - infancy and toddlerhood, 200–201
    - late adulthood, 616–617
    - middle adulthood, 542–543
    - middle childhood, 340–341
  - motor development and, 129
  - neighborhood influences on, 61
  - nonorganic failure to thrive and, 126
  - Peck's theory of, in late adulthood, 585
  - peer relations. *See* Peer relations
  - play and, 244
  - psychosexual theory of. *See* Psychosexual theory of development (Freud)
  - psychosocial theory of. *See* Psychosocial theory of development (Erikson)
  - pubertal changes and, 350–351
  - reflexes contributing to, 101
  - responding to others' emotions, 178, 247, 319–320
  - self-awareness and, 195–196
  - self-conscious emotions. *See* Self-conscious emotions
  - self-regulation of emotions. *See* Emotional self-regulation
  - social learning theory of. *See* Social learning theory (Bandura)
  - sociodramatic play and, 217
  - socioemotional selectivity theory of aging, 596, 597
  - systematic observation studies of social development, 27
  - understanding of emotions, 178, 247–248, 319–320, 585–586, 613
  - Vaillant's theory of. *See* Vaillant's theory of adult development
- Emotional contagion, 178
- Emotional disturbances. *See* Psychological disturbances
- Emotional expertise, 585–586
- Emotional expression
  - health risks of expression and suppression, 494
  - sex differences in, 180
- Emotional neglect of children, 267. *See also* Child maltreatment
- Emotional self-efficacy, middle childhood, 320
- Emotional self-regulation. *See also* Self-control
  - anxiety management and, 334
  - child maltreatment and, 268
  - early childhood, 248
  - infancy and toddlerhood, 179–180
  - late adulthood, 601
  - middle childhood, 320
  - successful aging and, 613
- Emotional stress. *See* Stress
- Emotional styles. *See* Temperament
- Emotional well-being. *See* Psychological well-being
- Emotion-centered coping, 495, 519–520
  - aging process and, 556
- Emotions. *See also* Emotional and social development; *specific entries (e.g., Anger; Fear)*
  - adolescence, 344, 350
  - basic emotions, 175, 177–178
  - bereavement. *See* Bereavement
  - dying, stages of, 628
  - dying people, 627–631
  - early childhood, 247–249
  - individual differences in. *See* Temperament
  - infancy and toddlerhood, 175, 177–178
  - Labouvie-Vief's theory of emotional reasoning, 585–586
  - late adulthood, 585–586
  - middle childhood, 319–320
  - self-conscious. *See* Self-conscious emotions
  - self-regulation. *See* Emotional self-regulation
- Empathy
  - defined, 249
  - early childhood, 249
  - moral development and, 253
  - perspective taking and, 320
  - self-awareness and, 196
- Empathy-based guilt, 253
- Emphysema, 561, 562
- Employment. *See also* Job *entries;* Unemployment; Vocational *entries*
  - adolescence, 372, 374
  - bridge jobs, 609
  - delinquency and, 404
  - dual-earner families, 458, 473–474
  - educational attainment and, 375, 439
  - extracurricular activities and, 376
  - Head Start programs and, 234
  - identity development and, 385
  - marital satisfaction and, 459
  - maternal employment. *See* Maternal employment
  - part-time. *See* Part-time work
  - retirees returning to employment, 609
  - teenage pregnancies and, 360
- "Empty nest" syndrome, 524, 525
- Endocrine system. *See* Hormones
- Endoderm, formation of, 79
- Endurance, and aging process, 416–417, 555, 560
- Engagement versus separateness, in midlife development, 515, 516
- England. *See also* Great Britain
  - childbirth practices, 94
  - reproductive technologies, 55
- Enrichment programs. *See* Early intervention programs; Gifted children
- Enterprising personality type, in vocational choice, 437
- Environmental hazards. *See also* Environmental pollutants; Radiation
  - conception, caregiving concerns, 56
  - mutation of genes, 49–50
  - sperm counts and, 49
- Environmental influences, 7, 58–71. *See also* Nature–nurture controversy; *specific influences (e.g., Cultural influences; Home environment)*
  - aging, 548
  - Alzheimer's disease, 566
  - asthma, 280
  - canalization, 69, 70
  - cognitive development, 161–162, 232–235
  - depression, 401
  - gender typing, 260–263
  - gene expression, 70–71
  - moral development, 391–392
  - motor development, 282
  - obesity in adulthood, 420
  - physical development, 123–126, 208–212
  - prenatal development, 81–90
  - relationship between heredity and environment, 66–71
  - SIDS, 103
  - strokes, 567
  - temperament, 184
  - tooth development, 204
- Environmental mastery, in middle adulthood, 519
- Environmental pollutants
  - Alzheimer's disease and, 566
  - cancer and, 491
  - heart disease and, 554
  - prenatal development and, 87
  - taste perception and, 553
- Envy, 178–179
- Epidermis, 484
- Epigenesis, 71
- Epigenetic framework, 71
- Epiphyses, 204
  - adolescence, 346
- Equality, in distributive justice reasoning, 321
- Equilibrium, in cognitive-developmental theory, 19, 144–145
- Equilibrium, physical. *See* Balance
- Equity, norm of, 533
- Erikson's psychosocial theory. *See* Psychosocial theory of development (Erikson)
- Eritrea, orphans created by war, 335
- ERT. *See* Estrogen replacement therapy
- Estonia, suicide rate, 402
- Estrogen replacement therapy (ERT), 486
- Estrogens, 345. *See also* Sex Hormones
  - menopause and, 486
  - premenstrual syndrome and, 430
- Ethics
  - genetic testing, 56
  - reproductive technologies, 54–55
  - research, 37–39
- "Ethics of care," in moral reasoning, 390–391
- Ethnic identity, 386
  - immigrant youths, 31
- Ethnic pride, 384
- Ethnicity and race. *See also* Cultural influences; *specific entries (e.g., African Americans; Hispanics)*
  - adoption, transracial and transcultural, 57
  - adult learners and, 506–507
  - bicultural identity, 386
  - death concept, development of, 623–624
  - delinquent behavior, 403
  - diabetes incidence, 563
  - drop out rates, 374–375
  - elder maltreatment, incidence of, 607–608
  - euthanasia, response to, 634
  - fraternal twinning and, 46
  - grandparent-grandchildren relationships and, 527
  - health in late adulthood and, 559
  - health studies related to ethnic minorities, 489
  - heritability estimates and, 68, 296
  - identity development and, 386
  - independent living in late adulthood and, 598
  - infant mortality and, 98
  - institutionalization of elderly and, 569
  - IQ scores and, 296, 297
  - life expectancy and, 547, 548
  - mixed marriages, 458
  - osteoporosis, vulnerability to, 493
  - physical development and, 114–115, 204
  - poverty and, 60
  - retirement decisions and, 610
  - self-esteem and, 318, 384
  - sibling relationships in middle adulthood and, 532
  - spiritual and religious involvement in late adulthood and, 590
  - suicide and, 402, 592
  - temperament and, 183
  - vocational advancement and, 536
  - widowhood, incidence of, 602
- Ethnography, 30, 32

Mayans' daily life in early childhood, study of, 226
Ethology, 22. *See also* Evolutionary theories
attachment theory and, 186–187, 193
Etiquette, and dating in adolescence, 399
Europe. *See also specific countries or regions (e.g., Sweden; Western Europe)*
long-term care facilities, 600
reproductive technologies, 55
Eustachian tube, 277
Euthanasia, 633–639
assisted suicide, 636, 638–639
defined, 634
forms of, 634
passive, 634–636
voluntary active, 636, 637
Evacuee children of World War II, 588–589
Event-based prospective memory tasks, 573
Evidence
eyewitness testimony by children, 336, 337
scientific reasoning and, 355–356
Evocative genetic–environmental correlation, 69
Evolutionary developmental psychology, 22–23, 26
Evolutionary theories. *See also* Ethology
Darwin, 14
gender typing and, 260
mate selection characteristics and, 451, 497
parental imperative theory, 521–522
sexual attitudes and behavior and, 427
wisdom, valuing of, 575–576
Executive stage of cognitive development (Schaie), 433
Exercise. *See also* Athletics; Physical education
adolescence, 347
anorexia nervosa and, 353
arthritis and, 562
bone loss and, 486
brain development and, 560
cardiovascular health and, 554
diabetes and, 563
early adulthood, 423–424
heart disease and, 492
immune response and, 554
joint flexibility and, 555
late adulthood, 560
life expectancy and, 414, 548, 551
middle adulthood, 486, 496
obesity and, 125, 279, 421, 423
osteoporosis and, 493
physical education programs, 283–284
premenstrual syndrome and, 430
prenatal development and, 88
psychological well-being in middle adulthood and, 520
respiratory health and, 554
sleep and, 555
Exosystems, 24, 25, 61
Expansions, in language development, 238
Expectations
marriage, in, 459–460
self-fulfilling prophecy in teacher–student interactions, 304
Experiential subtheory of intelligence (Sternberg), 294–295
Experimental research design, 33, 34
Expertise. *See also* Achievement; Mastery
emotional expertise, 585–586
information processing theories, 433–434
middle adulthood, 503–504
wisdom and, 575
Explicit memory, 571
Expressive style of language learning, 168
Extended family households, 62, 63
never-married single parents and, 470
young adults leaving home and, 457, 525
Extinction, in classical conditioning, 127
Extracurricular activities. *See also* Athletics
after-school programs, 61, 334
benefits of, 376
civil responsibility and, 393
drop out prevention programs, 376–377
Extramarital sex. *See also* Sexual attitudes and behavior
adolescent premarital sex, 354–355
divorce and, 468–469
Extroversion, as personality trait, 522–523
late adulthood and, 594
Eye blink reflex, 101
Eyes. *See also* Vision
aging process, 483–484, 551–552
diabetes and, 563
postnatal development of, 106
prenatal development of, 79, 82
Eyewitness testimony by children, 336, 337

F

Face perception, 135–138, 139
emotional development and, 177
Facial hair development, 345, 348
Facial structure, aging process, 555
Factor analysis of intelligence test components, 293
Factual knowledge, in middle adulthood, 503
FAE (fetal alcohol effects), 86
Failure to thrive (nonorganic), 126
Fairness. *See* Justice; Prosocial behavior
Faith. *See* Spirituality and religiosity
Fallopian tubes, 76–77. *See also* Reproductive system
Falls, injuries from in late adulthood, 563–564
False beliefs, children's recognition of, 228
Familial Alzheimer's, 566
Families, 58–59. *See also* Child-rearing practices; Grandparent *entries;* Home environment; Parent *entries;* Sibling relationship
adapting to change, 59
adoptive. *See* Adoption
blended. *See* Blended families
child care. *See* Child care
cultural values and, 62
divorce. *See* Divorce
dual-earner households, 458, 473–474
dying member, response to, 629–630
early adulthood, 456–465
intergenerational relationships, 533
kinkeeper role. *See* Kinkeeper role
life expectancy and, 548, 550
maternal employment. *See* Maternal employment
newborns, adjustments to, 108
one-child families, 329
poverty's effect on, 60–61
public policies, 64
reconstituted. *See* Blended families
single-parent. *See* Single-parent families
size of. *See* Family size
socioeconomic status and, 59–60
stepfamilies. *See* Blended families
young adults leaving home, 456–457, 525
"Families of choice," 471
Family backgrounds, and marital satisfaction, 459
Family history
adoption and, 57
fraternal twinning and, 46
genetic counseling, 52
Family influences, 58–59. *See also* Home environment
aggression in children, 257
child maltreatment, 268
civil responsibility, 393
delinquency, 403, 404
direct influences, 58–59
gender typing, 260–262
immigrant youths, 30–31
indirect influences, 59
vocational choice, 437–438
Family life cycle, 456
Family life education, 460. *See also* Parent education
blended families, for, 333
depression among those with cancer, 629
divorce, following, 469
remarriage, following, 469
Family planning, 464
Family size, 463
decline in, 463
family dynamics, effect on, 59
injury proneness and, 211
only children, 329
poverty and, 464
sexual activity of adolescents and, 355
socioeconomic status and, 59
temperament and, 184
vocational development of women and, 472
Family stress. *See* Home environment; Stress
Family systems, 58–59, 456
Family therapy. *See* Counseling
Family trees, in genetic counseling, 52
Family violence. *See* Child maltreatment; Domestic violence; Elder maltreatment; Partner abuse
Fantasy period, of vocational development, 437
Fantasy play. *See* Make-believe play
FAS (fetal alcohol syndrome), 86
Fast food, and obesity, 423
Fast-mapping, in vocabulary development, 236
Fat, body. *See also* Obesity
adolescence, 346–347
baby fat, 114, 204
early childhood, 204
menarche and, 348
middle adulthood, 484, 486
middle childhood, 277
prenatal development, 81
Fat, dietary, 421–423. *See also* Nutrition
Alzheimer's disease and, 566
diabetes and, 563
exercise and, 424
heart disease and, 554
Father-child relationship. *See also* Parent *entries*
cultural influences on, 192
divorce and, 329, 330, 331
incest, 336
infant, 192
maternal employment and, 333
remarriage and, 332
socioeconomic status and, 59
teenage parenthood, 361
Father-headed households. *See* Single-parent families
Father-stepmother families, 332–333
Fatness. *See* Obesity
Favoritism, and sibling rivalry, 328
Fear. *See also* Anxiety
death anxiety, 626–627
early childhood, caregiving concerns, 248
elderly's fear of crime, 598
falling, fear of in late adulthood, 564
infancy and toddlerhood, 177–178
middle childhood, 334
self-care children, 333
sexual abuse victims, 336
sexual coercion and, 429
Fearful distress, as temperament dimension, 181
Feeding practices. *See also* Eating habits; Nutrition
bottle-feeding, 123–125
breastfeeding. *See* Breastfeeding
early childhood motor development, 213
obesity and, 278–279
Females. *See* Gender *entries;* Mother *entries;* Sex differences
Feminine morality (Gilligan), 390–391, 427
Feminine social clock, in adult development, 449–450, 517
Femininity. *See* Gender *entries*
Feminization of poverty, 525, 599
Fertility. *See also* Conception
aging, effects of, 418
pubertal development and, 348
reproductive technologies, 54–55
Fertility drugs, 46
Fetal Alcohol Effects (FAE), 86
Fetal Alcohol Syndrome (FAS), 86
Fetal medicine, 52–53, 56
Fetal monitors, 95
Fetoscopy, 52
Fetus, period of prenatal development, 77, 79–81
teratogens in, 82
Fibrillation, 620
Field experiments, 33
Fijians, moral development among, 252
Financial abuse of elderly, 608. *See also* Elder maltreatment
Financial planning for retirement, 537–538
Fine motor development
early childhood, 213–215
infancy and toddlerhood, 130, 132
middle childhood, 281–282
sex differences in, 215
Finland
adoption study of schizophrenia, 70
elder maltreatment, incidence of, 607
male birth rate, 49
obesity rate, 278
retirement programs, 610
suicide rate, 402

Firearms
  adolescent suicide and, 403
  deaths caused by, in early adulthood, 419
First Nations. *See* Canadian Aboriginals
First words, 164, 167
Fitness. *See* Exercise; Health; Nutrition
Flexibility
  cognitive flexibility, 505
  joint flexibility and exercise, 555
  middle childhood, 281
Flow, and psychological well-being in middle adulthood, 520
Fluid intelligence (Cattell), 498–500, 570
Fluoridated water, and dental health, 204
Folic acid, prenatal care and, 88
Food. *See* Eating *entries;* Feeding practices; Malnutrition; Nutrition; Vitamins and minerals
Food supplement programs, 278
  prenatal care, 88–89
Force, in middle childhood, 281
Foreclosure, in identity development. *See* Identity foreclosure
Foreign adoptions, 57
Formal operational stage of cognitive development, 19, 20, 363–365, 366
  egocentrism in, 367
  hypothetico-deductive reasoning in, 363–364
  propositional thought in, 364–365
  research on, 364–365
Formulas (bottle-feeding), 123–125
Foster care, grandparents providing, 528
4-H clubs, 323
Fractures
  falls in late adulthood and, 564
  long-term care placement and, 569
  osteoporosis and, 493
Fragile X syndrome, 49
France
  elderly poverty rate, 65
  reproductive technologies, 55
  retirement programs, 610
  teenage pregnancy rate, 359
  University of the Third Age, 577
Fraternal twins, 46. *See also* Twin studies; Twins
  life expectancy, 548
Free radicals, 560
  aging and, 414, 490
Freestanding birth centers, 93
Freud's psychosexual theory. *See* Psychosexual theory of development (Freud)
Friendships. *See also* Peer relations
  adolescence, 396–398
  cliques, 398–399
  early adulthood, 453–455
  early childhood, 251
  late adulthood, 605–606
  middle adulthood, 532–533
  middle childhood, 323–324
  other-sex friendships, 454–455
  same-sex friendships, 454
  secondary friends, 606
  sex differences in, 324
  siblings as friends, 455
  singlehood and, 604
Frontal lobes, 117–118
Full inclusion (in schools), 305
Functional age, 546–547
Functional play, 251
Funerals, 642–643

G
Gabapentin, 486
Games. *See also* Athletics; Play; Toys
  language development and, 166, 168–169
  middle childhood, 282–283
  organized games with rules, 282–283
Gametes, 45, 46. *See also* Ovum; Sperm
Gangs, youth, 405. *See also* Delinquency
Gardner's multiple intelligences theory, 295–296, 307
Gastrointestinal system
  late adulthood, 560
  obesity and disorders of, 420
  stress and disorders of, 431
Gay men. *See* Homosexuality
Gender appropriate. *See* Gender stereotyping
Gender constancy, 263
Gender differences. *See* Sex differences
Gender gap. *See* Sex differences
Gender identity
  adolescence, 394
  cognitive-developmental theory of, 260, 263
  defined, 263
  early childhood, 263–264
  emergence of, 263
  friendships and, 397
  gender schema theory of, 260, 263–264
  middle adulthood, 516, 520–522
  middle childhood, 326–327
  parental imperative theory of, 521–522
  self-esteem and, 263
  sex differences in development, 326–327
  social learning theory of, 260, 263
  theories of, 260, 263–264
Gender intensification, 394
Gender schema theory, 260, 263–264
Gender stereotyping. *See also* Gender typing
  academic subjects, 326
  adolescence, 394
  aging and, 497, 515–516
  early childhood, 260, 264
  father–infant relationship and, 192
  infancy and toddlerhood, 184, 196
  maternal employment and, 333
  middle childhood, 326
  motor development and, 215
  partner abuse and, 461
  reducing, 264
  sports and, 282
  toys and, 196
  vocational development and, 438, 472, 536
Gender typing. *See also* Gender identity; Gender stereotyping; Sex differences
  adolescence, 394
  aggression and, 256–257
  child-rearing practices and, 260–262
  cultural influences on, 327
  defined, 260
  depression and gender-typed coping styles, 401
  early childhood, 260–264
  environmental influences on, 260–263
  family influences on, 260–262
  genetic influences on, 260, 261
  marriage and, 458
  middle childhood, 326–327
  peer relations and, 262
  sex reassignment, 261
  sexual coercion and, 429
  sibling relationships and, 262
  social environment's influence on, 262–263
  teachers influencing, 262
Gender-linked preferences. *See* Gender typing
Gender-role adoption. *See* Gender stereotyping
Gender-role identity. *See* Gender identity
Gene expression, environmental influences on, 70–71
Gene therapy, 53, 56
General categories, 221
General growth curve, 205–206
Generation gap, 368
Generational differences. *See* Cohort effects
Generativity
  life stories and, 514
  psychosocial stage of development, 17, 447, 512–514, 584
  Vaillant's stage development, 447, 449
Generativity versus stagnation, 17, 447, 512–514, 584
Genes, 44. *See also* Genetic *entries*
  "aging genes," 413
  cloning, 55
  dominant–recessive inheritance, 46–47, 48, 49
  identification of, 53
  mutation. *See* Mutation of genes
Genetic code, 44–45
Genetic counseling, 52, 57
Genetic disorders. *See* Chromosomal abnormalities
Genetic engineering, 53
Genetic imprinting, 49
Genetic screening for cancer, 490
Genetic testing, 56, 57
Genetic–environmental correlation, 69–70, 162
Genetics, 44–51. *See also* Chromosomal abnormalities; Chromosomes; Inheritance; Nature–nurture controversy
  ADHD and, 288
  aging and, 412, 413, 548
  alcohol abuse and, 425
  Alzheimer's disease and, 566
  anorexia nervosa and, 353
  arthritis and, 562
  autism and, 229
  bulimia and, 354
  canalization, 69, 70
  cancer and, 490
  cerebrovascular dementia and, 567
  depression and, 401
  diabetes and, 563
  family studies of. *See* Kinship studies; Twin studies
  gender typing and, 260, 261
  handedness and, 206
  heritability estimates, 67, 68, 296
  homosexuality and, 356
  intelligence and, 296–297
  obesity and, 278
  osteoporosis and, 493
  physical development and, 123, 208
  relationship between heredity and environment, 66–71
  shyness versus sociability, 183
  temperament and, 183
Genital herpes, 358
Genital stage of development (Freud), 16, 344
Genital warts, 358
Genitals, 76–77. *See also* Reproductive system
  abnormalities of, in DES daughters and sons, 83
  ambiguous genitals, 261
  cancer of, in DES daughters and sons, 83
  growth curve, 205–206
  middle adulthood, 486
  prenatal development, 79, 82, 83
  pubertal development of, 348
Genome Project, 53, 56
Genotypes, 44
German measles, prenatal development and, 87
Germany
  assisted suicide laws, 638
  attachment style of infants, 188
  elderly poverty rate, 65
  male birth rate, 49
  parental leave, 98–99
  retirement programs, 610
  work-study apprenticeships, 440
  World War II refugee and evacuee children, 588–589
Gesell's theory of child development, 14–15
Gestures, preverbal, 166–167
GH. *See* Growth hormone
Gifted children, 305–307, 448
  case study of, 29
Gilligan's theory of feminine morality, 390–391, 427
Glare, sensitivity to, 483–484
Glass ceiling, 536
Glaucoma, 484
Glial cells, 116
  aging process, 549
  prenatal development of, 80
Glucose. *See* Diabetes; Sugar
Goal-directed behavior, in sensorimotor stage of cognitive development, 146
Golden Rule, 388, 389
Gonorrhea, 358. *See also* Sexually transmitted diseases (STDs)
"Good boy-good girl" orientation in moral development, 389
Goodness-of-fit model of child-rearing, 184–185
  attachment and, 190
  moral development and, 254
Government policies. *See* Public policies
Grade advancement, 307
Grade retention, 303
Grade structures of schools, 371–372
Grammar
  early childhood, 236–237
  middle childhood, 299
Grandparent-grandchild relationship, 526–527
  adult grandchildren, 607
Grandparent-headed households, 527–529
Grandparenthood, 526–529, 607
  average age of, 526
  child-rearing, participation in, 62, 63
  extended family households and, 63
  impact of grandparents on child's development, 59
  meanings of, 526
  relationships with adult grandchildren, 607
  relationships with grandchildren, 526–527
  skipped-generation families, 528
Grandparents' visitation rights, 529

Grasp reflex and skills, 132
Great Britain. *See also* England
  elder maltreatment, incidence of, 607
  hospice care, 632–633
  lifelong learning programs, 577
  obesity rate, 278
  teenage pregnancy rate, 359
  World War II refugee and evacuee children, 588–589
Great Depression, consequences on lifespan development, 36, 576
Great-grandparenthood, 607
Greece
  women's attitude toward menopause, 488
  young adults leaving home, 525
Grief, 639. *See also* Bereavement
Grief therapy, 645
Grooming. *See* Dress and grooming
Gross motor development
  early childhood, 212–213
  infancy and toddlerhood, 130
  middle childhood, 281
  sex differences in, 215
Group differences. *See* Ethnicity and race; Sex differences; Socio-economic status (SES)
Group-administered intelligence tests, 293
Grouping in schools, by ability, 304, 374
Groups
  peers. *See* Peer groups
  religious groups for youths, 323
  support. *See* Support groups
"Growing pains," 277
Growth. *See* Body *entries;* Physical development
Growth, catch-up. *See* Catch-up growth
Growth centers, 204
Growth curves, 115, 205–206
Growth hormone (GH), 208, 345
  aging and, 414
  psychosocial dwarfism and, 208
Growth spurts
  adolescence, 346
  brain development, 119
  infancy and toddlerhood, 114
Guatemalan Mayans. *See* Mayans
Guidance Study, 34, 36
Guilt
  divorce in late adulthood and, 602
  early childhood, 248–249
  empathy-based guilt, 253
  infancy and toddlerhood, 178–179
  middle childhood, 319
  moral development and, 252–253
  parental bereavement and, 641
  perspective taking and, 320
  psychosocial stage of development, 17, 244
Gun control, 419
Guns. *See* Firearms
Gusii of Kenya
  infant caregiving, 104
  mother–child relationship and attachment, 189

H

Habituation/recovery response
  attention and memory studies, 154–155
  infant learning and, 128
  intelligence predictions based on, 161
  object permanence studies, 147–148
Hair
  adolescence, facial and body hair growth, 345, 348
  aging process, 415, 555
  dominant–recessive inheritance, 47
Halfway houses, delinquency intervention programs, 405
Hall's theory of child development, 14–15
  adolescence and, 344
Hand preference, 206–207
Happiness, 177
Hardiness, 496–497
  life expectancy and, 551
Harsh punishment. *See* Child maltreatment; Punishment (disciplinary)
Hausa of Nigeria, accomplishment of conservation tasks, 286
Hawaii, Kauai study of birth complications, 99–100
Head Start programs, 233–234
Health. *See also* Diseases; Nutrition
  adolescence, 352–362
  anger and, 493–494
  cognitive deterioration and, 576
  early adulthood, 419–431
  generativity and, 513
  job burnout and, 534
  late adulthood, 558–570
    problem solving and, 574–575
  life expectancy and, 548, 550–551
  marriage, influence on, 459
  middle adulthood, 489–494
  prenatal environment and health in later life, 84–85
  psychological well-being and, 520, 592–593
  retirement decisions and, 610
  self-reported health status, age and social class differences, 419
Health care. *See also* Diseases; Medications; Surgery; *specific procedures (e.g., Blood tests; Ultrasound)*
  advance medical directives, 634–636
  Alzheimer's disease, 566–567
  arthritis, 562
  assistive technology, 556
  cancer screenings, 491
  cataracts, 551
  childbirth. *See* Childbirth
  comfort care, 632
  conception, caregiving concerns, 56
  cost of, for elderly, 567, 569, 613
  dental care, 204
  depression, 567, 629
  diabetes, 563
  durable power of attorney for, 635–636
  euthanasia. *See* Euthanasia
  genetic counseling, 52
  hearing aids, 553
  hearing screenings, 277–278
  heart disease, 492
  hospice care, 632–633
  immunization, 209
  late adulthood, 559, 567, 569–570
  life expectancy and, 547
  living wills, 635–636
  macular degeneration, 552
  menopause, 486–487
  Native Americans, government benefits in adulthood, 559
  osteoporosis, 493
  otitis media, 210
  pain management programs, 632
  passive euthanasia, 634–636
  prenatal. *See* Prenatal health care
  prenatal diagnosis and fetal medicine, 52–53, 56
  preterm infants, 97–98
  public policies, comparison of U.S. and Canada with other nations, 64
  reproductive technologies, 54–55
  vision screenings, 277–278
  voluntary active euthanasia, 636, 637
Health care costs, for elderly, 567, 569, 613
Health care professionals
  dying patients, response to, 629–630
  job burnout, 534
Health care proxies, 636
Health education. *See also* Sex education
  chronic illness in childhood, 280
  middle childhood, 282–284
Health insurance, 209. *See also* Medicaid; Medicare
  prenatal health care and, 90
  public policies, 64
Hearing. *See also* Ears
  aging process, 415, 550, 552–553
  dominant–recessive inheritance, 47
  infancy and toddlerhood, 133
  intermodal perception, 138–139
  language development and, 166, 168
  middle adulthood, 484
  neuron loss in auditory area of cerebral cortex, 549
  newborns' capacities, 106
  presbycusis, 484
Hearing aids, 484
Hearing loss
  adult onset, 484
  cognitive deterioration and, 576
  middle childhood, 277–278
  sign language, 118
Heart. *See* Cardiovascular system
Heart attacks, 492
Heart disease, 491–493. *See also* Hypertension
  African Americans, 414
  alcohol abuse and, 425
  Alzheimer's disease and, 566
  anger and, 493–494
  caregiving concerns, 492
  cerebrovascular dementia and, 567
  cognitive deterioration and, 576
  death, as cause of, 416, 547, 620
    early adulthood, 419
    late adulthood, 561, 562
    middle adulthood, 489, 490
  diabetes and, 563
  early adulthood, death caused by, 419
  exercise and, 424
  fat consumption and, 421
  free radicals and, 414
  late adulthood, cause of death in, 561, 562
  low birth weight, 84
  middle adulthood, cause of death in, 489, 490
  obesity and, 278, 420
  reducing risk of, 492
  sexuality in late adulthood and, 561
  smoking and, 425
  stress and, 431
  Type A behavior pattern and, 493–494
  vitamin supplements in late adulthood and, 560
Heart rate, and shyness versus sociability, 182
Height
  adolescence, 346
  aging process, 415, 555
  early childhood, 204
  infancy and toddlerhood, 114–115
  middle adulthood, 486
  middle childhood, 276–277
  psychosocial dwarfism, 208
"Heinz dilemma" research, 388–390
Helpfulness. *See also* Prosocial behavior
  modeling by child care provider, effect of, 33
Helping professions, job burnout, 534
Helplessness, learned, 291, 317–318
Hemispheres of brain. *See* Lateralization of brain
Hemophilia, 48
  gene therapy, 56
Hereditary pattern baldness, 555
Heredity. *See* Genetic *entries;* Inheritance; Nature–nurture controversy
Herero of Botswana, aging process among, 558
Heritability estimates, 67, 296
  limitations, 68
Heroin. *See also* Drug use and abuse
  prenatal development and, 83
  teenage use, 361
Herpes viruses
  cesarean delivery, 96
  prenatal development and, 87–88
  simplex, 358
Heteronomous morality, 388
Heterosexual attitudes and behavior, 426–428
Heterozygous genes, 46
Hierarchical classification
  concrete operational stage of cognitive development, 285
  preoperational stage of cognitive development, 218
Hierarchical communication style, 297
Hierarchically structured self-esteem, 315–316
High birth weight, and cancer in adulthood, 84–85
High blood pressure. *See* Hypertension
High school valedictorians, vocational choices among, 438
High schools. *See* Schools and schooling
High-density lipoproteins, 424. *See also* Cholesterol
Higher education, 435–436
  adult learners. *See* Adult learners
  attendance rate, 435
  cognitive development in middle adulthood and, 500
  dropping out of college, 436
  identity development and, 384–385
  late adulthood, 577–578
  moral development and, 390, 391
  program to foster educational aspirations, 28
  psychological impact of, 436
High-income households. *See* Socioeconomic influences; Socioeconomic status (SES)
Hip fractures
  falls in late adulthood and, 564
  long-term care placement and, 569
Hiroshima, bombing of, 86
Hispanics. *See also* Mexican Americans; Mexico
  academic achievement among, 373
  aging parents, caring for, 598
  anorexia nervosa among, 353
  child-rearing practices, 266
  communication style, 297
  drop out rate, 374–375
  elder maltreatment, incidence of, 607–608
  extended family households of, 62

health in late adulthood, 559
identity development in adolescence, 386
institutionalization of elderly, 569
obesity in adulthood, 420
out-of-wedlock births, 359
poverty and, 60
Quinceañera, adolescent initiation ceremony, 350
spiritual and religious involvement in late adulthood, 590
suicide rate, 402
young adults leaving home, 457
Historical foundations of development, 13–15
History-graded influences, 11, 35. *See also* Cohort effects
HIV. *See* Acquired immune deficiency syndrome (AIDS)
Holocaust, refugee children from, 588–589
HOME. *See* Home Observation for Measurement of the Environment (HOME)
Home births, 93, 94–95
Home deaths, 631
Home environment. *See also* Families; Parent *entries;* Sibling relationship
ADHD and, 288–289
aggression in children and, 257
attachment and, 191–192
birth complications, long-term consequences of and, 100
child maltreatment and, 268
cognitive development and, 161–162, 232–233
delinquent behavior and, 403
divorce's effect on, 330
dropping out of school and, 375
injury proneness and, 211
late adulthood, 598–600
leaving home, 456–457, 525
otitis media and, 210
resilient children, 337
suicide and, 402
talent, influence on, 306
Home health care, 594
aging parents, caring for, 531
dying patients, 631
government financing, for elderly, 569
Home Observation for Measurement of the Environment (HOME), 161–162
early childhood subscales, 232
infancy and toddler subscales, 161
Homelessness, 60–61. *See also* Poverty
Homicide deaths, in early adulthood, 419
Homosexuality
adolescents, 356, 357
anorexia nervosa and, 353
cohabitating relationships, 467–468
coming out, 357
dating relationships in adolescence, 399
death of partner, 641
early adulthood, 428
incidence of, 428
late-adulthood partnerships, 601–602
parenthood and, 470–471
sexual attitudes and behavior, 428
suicide and, 402
Homozygous genes, 46
Hong Kong. *See also* Chinese
academic achievement of students, 307
myopia among, 277
Hope trajectory of dying patients, 630
Hoped-for selves. *See* Possible selves
Hopping, in early childhood, 212–213
Horizontal décalage, 286, 287
Hormone replacement therapy (HRT), 486
Alzheimer's disease and, 566
cancer and, 491
osteoporosis and, 493
Hormone therapy, 486. *See also* Estrogen replacement therapy (ERT); Hormone replacement therapy (HRT)
Hormones. *See also specific hormones (e.g., Cortisol; Estrogens; Stress hormones)*
aging and, 414
depression and, 401
gender typing and, 260, 261
puberty, 345–346
sex drive and, 354
Hospice care, 632–633
Hospital births, 93, 95–96
Hospital dying, 631–632
Hostile aggression, 257
Hostility. *See* Aggression; Anger; Antisocial behavior
Hot flashes, 486
Hours of work, and obesity, 423
Household chores
dual-earner families, 333, 473–474
late adulthood, 601
marital roles and, 458
Housing
assisted living, 569, 599
assistive technology, 556
homelessness, 60–61
late adulthood, 598–600, 613
nursing homes. *See* Long-term care
HRT (hormone replacement therapy), 486. *See also* Estrogen replacement therapy (ERT)
Human development
basic issues, 5–7
contexts for, 6–7, 10–12
course of development, 6–7
stance of major developmental theories, 26
definition, 5
determinants of, 5, 6–8. *See also* Environmental influences; Genetic *entries;* Nature–nurture controversy
stance of major developmental theories, 26
domains of, 8, 9. *See also specific domains (e.g., Cognitive development; Physical development)*
field of study, 5
historical foundations of, 13–15
major periods of, 8. *See also specific period (e.g., Adolescence; Late adulthood)*
overview of lifespan, 8. *See also specific periods (e.g., Prenatal development; Early childhood; Middle adulthood)*
research. *See* Research
theories. *See* Theories
Human Genome Project, 53, 56
Human immunodeficiency virus (HIV). *See* Acquired immune deficiency syndrome (AIDS)
Humor, and language development, 299
Huntington disease, 47, 48, 49
DNA markers, identification, 53
Husbands. *See* Marital *entries;* Marriage
Hyaline membrane disease, 97
Hyperactivity. *See* Attention-deficit hyperactivity disorder (ADHD)
Hypertension, 491, 492
African Americans, 414
cerebrovascular dementia and, 567
obesity and, 278, 420
stress and, 431
Hypodermis, 484
Hypothesis, 27
Hypothetico-deductive reasoning, 363–364

I

Id, 15–16
Ideal reciprocity, in moral reasoning, 388, 389
Ideal self, 315, 385
Idealism, and abstract thinking, 367, 368
Identical twins, 46. *See also* Twin studies; Twins
life expectancy, 548
Identification with same-sex parent, 244
moral development and, 252
Identity. *See also* Gender identity; Self *entries*
bicultural identity, 386
caregiving guidelines, 387
dating and, 399
defined, 382
ethnic identity, 386
factors affecting development, 385–386
friendships and, 397, 398
paths to, 384–385
psychological well-being and, 385
psychosocial stage of development, 17, 382–383
stages of development, longitudinal-sequential study of, 35–36
Identity achievement, 384–385
androgynous traits and, 394
Identity confusion, psychosocial stage of development, 17, 382–383
Identity crisis, 382–383
Identity diffusion, 384–385
ethnic identity and, 386
Identity foreclosure, 384–385
ethnic identity and, 386
Identity versus identity confusion, 17, 382–383
Illegal drugs. *See* Drug use and abuse
Illinois, Momence, experience of aging in, 558
Illiteracy. *See* Literacy
Illnesses. *See* Diseases
Illogical thought in preoperational stage of cognitive development, 220
Imaginary audience, and abstract thinking, 367
Imitation, 18–19. *See also* Modeling
autism and, 229
deferred imitation, 147, 149
early childhood, 226
infant learning and, 128–129
language development and, 164
Immigrants
adaptation by immigrant youths, 30–31
loneliness of, 456
Immortality, symbolic, 626
Immune system
aging and, 414, 415, 417–418, 490, 554
exercise and, 424
life expectancy and, 550
malnutrition and, 209
middle childhood, 280
nutrition during pregnancy and, 88
prenatal development, 81
stress and, 431
vitamin supplements in late adulthood and, 560
Immunization, 209
Implantation of blastocyst, 76, 77, 78
Implicit memory, 571
Impotence, 487
smoking and, 425
Impoverishment. *See* Deprivation; Poverty
Imprinting, ethological, 22
attachment and, 186
Imprinting, genetic, 49
Impulsivity. *See* Emotional self-regulation; Self-control
In vitro fertilization, 54–55
fraternal twinning and, 46
preimplantation genetic diagnosis, 52
Incest, 336
Income. *See* Poverty; Socioeconomic status (SES)
Independence. *See* Autonomy
Independence–ignore script, 591
Independent living
centenarians, 550
late adulthood, 548, 569, 598–599
widowhood, following, 603
young adults leaving home, 456–457
Independent variables, experimental research design, 33
India
dying, adaptation to, 630
moral development in, 392
play and games in early childhood, 250
Individual differences, 7
abstract thinking, 365
biological aging, 546–547
cognitive development, 160–164, 232–235, 293–299, 500–501, 570
death anxiety, 626
dying, adaptation to, 629–631
emotions. *See* Temperament
injury proneness, 211
language development, 168
learning, 222–223
motor development, 215
perspective taking, 320
physical development, 114–115, 204
psychological well-being in late adulthood, 591–594
pubertal development, 348–349
reminiscence in late adulthood, 586–587
self-control, 197
sibling relationships, 193
temperament. *See* Temperament
Individualistic societies, 62–63
academic achievement, valuing of, 308
psychological well-being in middle adulthood among, 519
romantic relationships in, 452
Individually administered intelligence tests, 293
Individuative-reflective stage of faith development (Fowler), 590
Indonesia
Balinese bereavement rituals, 642
make-believe play in, 159
Induction, 253
Inductive reasoning, in middle adulthood, 499
Industry versus inferiority, 17, 314–315
Infancy and toddlerhood, 8, 113–197. *See also* Newborns

AIDS in, 87
anger in, 177–178
attachment in. *See* Attachment
attention in, 153–154
body proportion in, 115–116
brain development in, 116–123
brain plasticity in, 120
child care in, 162, 163
cognitive development in, 143–170. *See also* Sensorimotor stage of cognitive development
milestones, 200–201
compliance in, 196–197
cosleeping arrangements in, 122
death in. *See* Infant mortality
embarrassment in, 178–179
emotional and social development in, 173–197
milestones, 200–201
emotional self-regulation in, 179–180
emotions in, 175, 177–178
envy in, 178–179
father-infant relationship, 192
fears in, 177–178
fine motor development in, 130, 132
gender stereotyping in, 184, 196
gross motor development in, 130
growth spurts in, 114
guilt in, 178–179
happiness in, 177
hearing in, 133
height in, 114–115
information processing in, 153–157
intelligence tests in, 160–161
language development in, 164–170
milestones, 200–201
laughter in, 177
learning capacities in, 126–129
low birth weight. *See* Low birth weight
make-believe play in, 158–159
malnutrition in, 125–126
memory in, 154–155
milestones in, 200–201
motor development in, 129–133
muscle development in, 114
nutrition in, 123–126
parental relationships in. *See* Parent-infant relationship
perceptual development in, 133–140
personality development theories of, 174–179
physical development in, 113–140
milestones, 200–201
preterm. *See* Preterm infants
pride in, 178–179
psychoanalytic theories of, 174–179
recall memory in, 155
recognition memory in, 155
self-awareness in, 195–196
self-conscious emotions in, 178–179
self-control in, 196–197
self-development in, 174–175, 194–197
sense of agency in, 195
shame in, 178–179
skeletal development in, 115
sleep in, 120–123
smiling in, 177
sociocultural theory of, 157–159
states of arousal in, 120–123
temperament in, 180–185
vision in, 133–138
visual acuity in, 133
vocabulary in, 164, 167–168
weight in, 114–115
Infant Health and Development Project, 98–99
Infant intelligence tests, 160–161
Infant mortality, 98–99
AIDS and, 87
declines in, 547
drug use during pregnancy and, 83
home delivery and, 95
public policies, comparison of U.S. and Canada with other nations, 64
Rh blood incompatibility and, 89
small-for-date infants, 97
smoking during pregnancy and, 84
sudden infant death syndrome. *See* Sudden infant death syndrome (SIDS)
Infant–caregiver attachment. *See* Attachment
Infantile amnesia, 156
Infantile autism. *See* Autism
Infant-parent relationship. *See* Attachment; Parent-infant relationship
Infant-rearing practices. *See* Child-rearing practices
Infectious diseases. *See also* Diseases; *specific diseases (e.g., Acquired immune deficiency syndrome (AIDS))*
aging process and, 554
early childhood development and, 209
immunization, 209
prenatal development and, 87–88
pubertal development and, 349
vitamin supplements in late adulthood and, 560
Inferences
children's recognition of, 230
transitive, 285
Inferiority
learned helplessness and, 317
psychosocial stage of development, 17, 314–315
Infertility, 468
reproductive technologies, 54–55
Infidelity. *See* Extramarital sex
Infinitive phrases, in language development, 299
Information processing theories, 21–22, 26. *See also* Attention; Memory; Metacognition
adolescence, 365–366
capacity, 288, 365, 366
categorization in, 155, 157
creativity in, 433–434
early adulthood, 433–434
early childhood, 225–231
education principles based on, 291–293
evaluation of, 157
expertise in, 433–434
gender schema theory, 260, 263–264
infancy and toddlerhood, 153–157
intelligence, 294
mental strategies in, 153
middle adulthood, 501–505
middle childhood, 287–293
motor development and, 281
self-regulation in, 291
sensory register in, 153, 154
speed of processing. *See* Speed of processing
Information-loss view, of information processing, 501
Informed consent of research participants, 38–39
Inheritance, 7. *See also* Genetic *entries*
canalization, 69, 70
codominant pattern of, 47, 49
concordance rates, 67–68
dominant–recessive, 46–47, 48, 49
mutation of genes. *See* Mutation of genes
polygenic, 50, 66
prenatal diagnostic methods, 52–53, 56
X-linked inheritance, 49, 50
Inhibited (shy) temperament, 182–183
discipline and, 254
Initiative versus guilt, 17, 244
Injuries
adolescence, 620
caregiving concerns, 564
death, as cause of, 211, 620
early adulthood, 419
early childhood, 209, 211–212, 620
late adulthood, 561, 563–564
middle adulthood, 489, 490
middle childhood, 280–281
prevention, 211–212, 280–281, 564
In-law relationships, 526
Inner speech, 223
Insecurity. *See* Attachment; Self-confidence
Insemination. *See* Conception
Insomnia. *See* Sleep disorders
Institutionalization
elderly. *See* Long-term care
infant attachment and, 188–189
motor development and, 131, 132
stimulus deprivation, 119
Instruction. *See* Education; Learning; Schools and schooling; Teacher *entries; specific subjects (e.g., Mathematics; Reading)*
Instrumental aggression, 257
Instrumental purpose orientation in moral development, 389
Insulin. *See* Diabetes
Insurance. *See* Health insurance
Integrity. *See* Ego integrity; Moral *entries*
Intellectual development. *See* Cognitive development
Intelligence. *See also* Cognitive development
ADHD and, 288
alcohol use during pregnancy and, 86
brain growth spurts and, 119
crystallized intelligence, 498–500, 570
cultural influences on, 297–299
definitions of, 293–296
delinquent behavior and, 403
drug use during pregnancy and, 83
environmental influences on, 296–297
fluid intelligence, 498–500, 570
genetic factors, 66, 296–297
habituation-recovery response as predictor of, 161
kinship studies, 67
language customs and, 297
malnutrition and, 126
maternal employment and, 333
measurement of. *See* Intelligence quotients (IQs); Intelligence tests
middle adulthood, 498–501
multiple intelligences theory (Gardner), 295–296, 307
reaction range, 68–69
small-for-date infants, 97
terminal decline, 576–577
theories of, 294–296
triarchic theory of (Sternberg), 294–295, 503
Intelligence quotients (IQs), 160, 293
componential analyses of, 294
computation and distribution of, 160
early intervention programs and, 163
ethnic differences in, 296, 297
family size and, 463
gifted students, 305
Head Start programs and, 233–234
HOME scores and, 161
malnutrition and, 278
predicting later performance from, 160–161
socioeconomic differences in, 296, 297
stability of, 160
Intelligence tests
computing scores, 160
convergent thinking and, 306
cultural bias in, 232, 293, 297–299
divergent thinking and, 306
dynamic testing, 298
early adulthood, 434–435
early childhood, 232
factor analysis of components, 293
familiarity with test content, effect on performance, 297–298
group administered, 293
individually administered, 293
infant intelligence tests, 160–161
middle adulthood, 499–500
middle childhood, 293–294
scores on. *See* Intelligence quotients (IQs)
Stanford-Binet Intelligence Scale, 15, 293
typical items, 294
Wechsler Intelligence Scales, 293–294, 499–500
Intensity of reaction, as temperament dimension, 181
Intentional behavior, in sensorimotor stage of cognitive development, 146–147
Interactional synchrony, in attachment, 189
Interactionist theories of language development, 165
Interactions
marital. *See* Marital relationship
parent–child. *See* Father *entries;* Mother *entries;* Parent *entries*
peer. *See* Peer *entries*
sibling. *See* Sibling relationship
teacher–student. *See* Teacher–student interaction
Interdisciplinary study of development, 5
Intergenerational relationships, 533, 577
friendships in late adulthood and, 606
Intermodal perception, 138–139
hearing loss and, 553
Internal working model, in attachment theory, 187, 191–192
adult romantic relationships and childhood attachment patterns, 452
Internalization. *See* Superego
Internet. *See* Websites
Interpersonal behavior. *See* Families; Friendships; Marital relationship; Peer *entries;* Sibling relationship
Interpersonal cooperation, morality of, 389
Interpersonal intelligence (Gardner), 296
Interracial marriages, 458

Intervention programs. *See also* Early intervention programs
  ADHD, for, 289
  aggression, for, 259
  alcohol abuse, for, 426
  Alzheimer's disease, for, 566–567, 568
  anorexia nervosa, for, 353–354
  bereavement, for, 644–645
  bulimia, for, 354
  child maltreatment, for, 268–269
  childhood injuries, preventing, 211–212
  chronic illness in childhood, 280
  cognitive deterioration in late adulthood, for, 577
  delinquency prevention and treatment, 405
  dropping out of school, for, 375–377
  educational enrichment for children living in poverty, 163
  elder maltreatment prevention and treatment, 608–609
  food supplement programs, 88–89, 278
  newborns, NBAS-based interventions, 107
  obesity, for, 279, 420–421, 423
  partner abuse, 461
  peer acceptance and, 326
  peer victimization, 325
  perspective taking, for, 320
  postpartum depression, 176
  prenatal malnutrition and, 88–89
  preterm infants, for, 97–99
  school transitions, for, 371–372
  self-esteem, for, 318
  sexual abuse victims, for, 336
  sexual coercion victims, for, 429–430
  SIDS public education programs, 103
  social services for elderly, 65
  stress management in midlife, 495–496
  suicide, for, 402–403, 593
  teenage pregnancy and childbearing, for, 360–361
  teenage substance use and abuse, for, 362
Interview research methods, 28–29
Intestinal infections, 209
Intimacy. *See also* Peer relations; Romantic relationships
  adolescence, 383, 396, 398, 399
  dating relationships, 399
  friendships and, 454
  friendships in late adulthood and, 605
  loneliness and, 456
  love, component of, 451
  psychosocial stage of development, 17, 446–447
  sex differences in, 385
  stages of development, longitudinal-sequential study of, 35–36
  Vaillant's stage development, 447, 449
Intimacy versus isolation, 17, 446–447
Intrapersonal intelligence (Gardner), 296
Introspection, 519
Intuitive-projective stage of faith development (Fowler), 590
Inuits. *See* Canadian Aboriginals
Invariant features, in differentiation theory, 139
Investigative personality type, in vocational choice, 437
Invisible displacement, 147
Ionizing radiation. *See* Radiation
IQs. *See* Intelligence quotients
Iron deficiency in adolescence, 353
Irreversibility in thinking, 218
Irritable distress, as temperament dimension, 181
I-self
  emergence of, 195
  middle childhood, 315
Islam
  death anxiety and, 626
  death concept, development of, 623–624
Isolation. *See also* Loneliness
  dementia-like symptoms and, 567
  otitis media and, 210
  psychosocial stage of development, 17, 446–447
  visual impairment and, 552
  widowhood, following, 603
Israel
  congregate housing for elderly, 600
  death anxiety study, 626
  death concept among children, 623–624
Israeli kibbutzim
  death concept among children, 623
  moral development among children, 392
  stranger anxiety among infants, 178
Italian Americans, sibling relationships in middle adulthood, 532
Italy
  reproductive technologies, 55
  young adults leaving home, 525–526

J
Jamaican West Indians, motor development in infancy, 131–132
Japanese
  academic achievement of students, 307, 308–309
  aging parents, relationships with, 529
  attachment style of infants, 188
  bombing of Hiroshima and Nagasaki, 86
  cesarean delivery, 96
  cognitive flexibility and vocation among, 505
  cosleeping arrangements in infancy and childhood, 122
  death, definition of, 621
  emotional expression among, 180
  home health care for aging parents, 531
  infant caregiving, 104
  kanreki ritual in aging process, 557
  life expectancy among, 547
  menopause, women's attitude toward, 488
  national examinations, 374
  peer relations in adolescence, 396
  romantic relationships, 452–453
  self-conscious emotions among, 179
  self-esteem among, 316
  strokes, incidence of death caused by, 567
  suicide rate, 402
  teenage pregnancy rate, 359
  temperament of infants, 184
Jarara of South America, childbirth practices, 93
Jealousy
  divorce and, 468
  sibling relationships, 108, 328
Jews. *See also* Israel; Judaism
  mixed marriages, 458
  Tay-Sachs disease, 48
  World War II refugee and evacuee children, 588–589
Job. *See* Employment; Vocational *entries*
Job burnout, 534–535
Job performance, 534
Job satisfaction, 472, 534–535
Job training programs. *See also* Vocational education; Work-study programs
  adolescents, 374
  middle adulthood, 535–536
Joint attention
  autism and, 229
  language development and, 166, 168
Joint custody of children, 331
Joints
  arthritis. *See* Arthritis
  dominant–recessive inheritance, 47
  exercise and, 555
Judaism. *See also* Israel; Jews
  adolescent rites of passage, 350
  bereavement rituals, 642
  death concept, development of, 623–624
Jumping
  adolescence, 347
  early childhood, 212–213
  sex differences in, 215
Justice
  distributive, 321–322
  legal proceedings. *See* Legal proceedings
  moral reasoning and, 391
Juvenile delinquency. *See* Delinquency
Juvenile justice system, bias in, 403
Juveniles. *See* Adolescence

K
Kaddish, 642
"Kangaroo baby care," 98
Kanreki ritual in aging process, 557
Karyotype, 44
Kauai study on long-term consequences of birth complications, 99–100
Keeper of meanings, Vaillant's stage of development, 447, 449, 516–517
Keeping Ourselves Safe (sexual abuse prevention program), 336
Kenya
  Gusii
    infant caregiving, 104
    mother-child relationship and attachment, 189
  Kinka tribe, gender typing in, 327
  Kipsigis, motor development in infancy, 131
  Nyansongo, gender typing among, 327
Kevorkian's "suicide machines," 638
Kibbutzim. *See* Israeli Kibbutzim
Kindergarten, delay in entering, 303
Kinka tribe in Kenya, gender typing in, 327
Kinkeeper role, 526
  aging parents and, 606
  widowhood and, 603
Kinship studies, 67. *See also* Twin studies
  depression, 401
  heritability estimates and, 296
Kipsigis of Kenya, motor development in infancy, 131
Klinefelter syndrome, 51
Knowledge. *See also* Cognitive development; Education; Intelligence *entries;* Learning
  adolescence, 365
  caregivers of elders with dementia and, 568
  core knowledge perspective, 150–151, 229
  factual knowledge, in middle adulthood, 503
  memory and, 290
  metacognitive. *See* Metacognition
  procedural knowledge, in middle adulthood, 503
Knowledge-based reminiscence, 587
Kohlberg's theory of moral development, 388–390
Korean Americans
  aging parents, caregivers of, 531
  grandparent-grandchildren relationships, 527
  independent living in late adulthood, 598
  play in early childhood, 250
Koreans
  academic achievement of students, 307
  aging parents, relationships with, 529, 531
  moral development in, 322
Kübler-Ross's stages of dying, 628
!Kung of Botswana, infant caregiving, 104
Kwashiorkor, 125–126

L
Labeling
  categorization and, 221
  object labeling, in vocabulary development, 236
Labor. *See* Employment; Job *entries;* Vocational *entries*
Labor, birthing. *See* Childbirth
Laboratory experiments, 33
Labouvie-Vief's theories
  emotional expertise, 585–586
  postformal thought, 433
Language, second, 300–301
Language acquisition device (LAD), 165
Language customs and intelligence, 297
Language development. *See also* Speech; *specific entries (e.g., Grammar; Reading; Vocabulary)*
  autism and, 229
  behaviorist theories of, 164
  bilingualism and, 300–301
  brain growth spurts and, 119
  brain plasticity and, 120
  caregiving concerns, 169
  child-directed speech and, 168–169
  cognitive development and, 364
  cooing and babbling, 165–166
  delays in, 168
  Down syndrome and, 50
  early childhood, 216, 236–238
    milestones, 272–273
  emotional self-regulation and, 180, 248
  expressive style of learning, 168
  first words, 164, 167
  hearing loss and, 166, 168
  imitation and, 164
  individual differences in, 168
  infancy and toddlerhood, 164–170
    milestones, 200–201
  interactionist theories of, 165
  late adulthood, 573–574
  lateralization of brain and, 206
  middle adulthood, 499
  middle childhood, 299–301
    milestones, 340–341
  milestones in, 166
    early childhood, 272–273
    infancy and toddlerhood, 200–201
    middle childhood, 340–341
  nativist theory of, 165

operant conditioning and, 164
otitis media and, 210
overextension in, 167
prelinguistic development, 165–166
production versus comprehension, 167
referential style of learning, 168
self-control and, 197
sensitive periods, 300
sex chromosome disorders and, 51
sex differences in, 168
socioeconomic status and, 60
supporting language learning, 168–169, 237–238
telegraphic speech in, 167–168
two-word utterance phase, 167–168
underextension in, 167
visual impairment and, 136
Language immersion programs, 300
Lanugo, formation of, 80
Late adulthood, 7, 8, 545–615
adaptation to physical changes in, 556–557
aging process. *See* Aging *entries;* Biological aging
appearance in, 555
autonomy in, 591–592
brain development in, 549
cardiovascular system in, 553–554
centenarians, 550–551
childlessness in, 603–604
cognitive development in, 10, 433, 570–579
milestones, 616–617
continuing education in, 577–579, 613
death concept, development of, 625–626
death in, leading causes of, 561, 562
dependency in, 591–592
depression in, 567, 591–593
diabetes in, 561, 562, 563
diseases in, 561–564
divorce in, 602
early philosophies of, 13–14
eating habits in, 553, 560
education in, 577–579
emotional and social development in, 583–615
milestones, 616–617
emotional self-regulation in, 601
emotions in, 585–586
exercise in, 560
falls, injuries from, 563–564
friendships in, 605–606
health care in, 559, 567, 569–570
health issues in, 558–570
problem solving and, 574–575
health perceptions in, 558–559
hearing in, 552–553
higher education in, 577–579
home environment in, 598–600
immune system in, 554
independent living in, 548, 569, 598–599
injuries in, 561, 562, 563–564
language processing in, 573–574
leisure activities in, 611–612
life expectancy in, 547–548
long-term memory in, 571, 572–573
marriage in, 601
memory in, 14, 549–550, 570, 571–573, 577
mental health in, 564–567
milestones, 616–617
mobility in, 555
motor vehicle accidents in, 563, 564
neighborhood influences on, 61
never-married in, 603–604
nutrition in, 560
parent-adult-child relationship. *See* Aging parents; Parent-adult-child relationship
personality traits in, 587–591
physical changes in, 549–557
physical development in, 546–570
milestones, 616–617
political activities in, 612
population, percentage of, 547
poverty in, 65
problem solving in, 574–575, 577
psychological disturbances in, 564–567
public policies, 64–65
reaction time in, 563
recall memory in, 571
recognition memory in, 571, 572
relationships in, 600–609
relocation during, 598
remarriage in, 602
research about, 5, 37–39
respiratory system in, 553–554
retirement. *See* Retirement
self-concept in, 587–591
sexuality in, 560–561
sibling relationships in, 604–605
singlehood in, 603–604
sleep in, 554–555
smell in, 553
social support and interaction in, 594
spirituality and religiosity in, 587–591
suicide in, 592–593
taste in, 553
time perception in, 597
touch in, 553
victimization in, 598
vision in, 551–552
volunteering in, 612
wisdom in, 575–576
working memory in, 571
Latency stage of development (Freud), 16, 314
Lateralization of brain, 118–119, 206–207
Latin Americans. *See* Hispanics; Mexican Americans
Laughter in infancy and toddlerhood, 177
Layoffs, 537
Lead poisoning, and prenatal development, 87
Lean body mass. *See* Body composition; Fat, body
Learned helplessness, 291, 317–318
depression and, 401
Learning. *See also* Cognitive development; *specific forms (e.g., Imitation; Operant conditioning)*
academic. *See* Education; Schools and schooling; Teacher *entries; specific subjects (e.g., Mathematics; Reading)*
achievement in. *See* Academic achievement
assisted discovery, 224
cooperative learning, 304
discovery learning, 222
early learning centers, 120
evolutionary developmental psychology and, 23
individual differences in, 222–223
infancy and toddlerhood, capacities in, 126–129
life expectancy and, 551
lifelong learning. *See* Continuing education
observational. *See* Modeling
peer collaboration, 224
readiness to learn, 222
scaffolding, 224, 226
social learning theory. *See* Social learning theory (Bandura)
social referencing and, 178
Learning disabilities
adoptive children, 57
attention-deficit hyperactivity disorder, 288–289
child maltreatment and, 268
education of students, 305
full inclusion, 305
sex differences in, 49
Left-handedness, 206–207
lateralization of brain and, 118
Legal issues. *See also* Ethics; Public policies
childhood injuries, preventing, 211
genetic testing, 56
reproductive technologies, 54–55
Legal proceedings
assisted suicide, 638
child support, 331
divorce mediation, 331
elder maltreatment, 609
eyewitness testimony by children, 336, 337
grandparents' visitation rights, 529
juvenile justice system, 403
right to die, 633–634, 638
Legs. *See* Limbs; Skeletal development
Leisure activities
Alzheimer's disease and, 566
cognitive development and, 500, 576
late adulthood, 611–612
life expectancy and, 551
visual impairment and, 552
Lens of eye
aging process, 551
development of, 106
Lesbian women. *See* Homosexuality
Levinson's theory of adult development
early adulthood, 447–449
middle adulthood, 515–516
Licensing of child care facilities, 163
Life care communities, 599
Life events approach to midlife transition, 517–518
Life expectancy, 7, 412, 547–549
centenarians, 550–551
compression of morbidity, 559
Down syndrome and, 50
family dynamics, effect on, 59
gains in, 547
health care costs and, 567
late adulthood and, 547–548
maximum lifespan, 548–549
social goals and, 597
social support and, 594
spiritual and religious involvement in late adulthood and, 591
variations in, 547
Life expectancy crossover, 548, 549
Life review, 586–587
World War II refugee and evacuee children, 588–589
Life structures, in Levinson's theory of adult development, 447–448, 516
Lifelong learning. *See* Continuing education
Lifespan, 8. *See also specific periods (e.g., Prenatal development; Early childhood; Middle adulthood)*
Lifespan, active, 547, 548
Lifespan, maximum, 548–549
Lifespan perspective, 7–12, 26. *See also* Cross-sectional research; Longitudinal research
assumptions of, 8–12
development of, 7–8
lifelong aspect of development, 8
multidimensional and multi-directional aspect, 9, 498
multiple contexts for development, 10–12, 22
plasticity in development, 9–10, 570, 577
Lifestyle diversity in adult development, 466–471
Light, fetal response to, 80
Likability. *See* Peer acceptance
Limbs. *See also* Skeletal development
prenatal development of, 79, 82
"restless legs" during sleep, 555
Linguistic development. *See* Language development
Linguistic intelligence (Gardner), 296
Linguistic knowledge, in core knowledge perspective, 151
Lipoproteins, 424. *See also* Cholesterol
Liquor. *See* Alcohol use and abuse
Literacy. *See also* Language development; Reading; Writing
centenarians achieving, 551
early childhood, 230–231
family size and, 464
Liver disease, and alcohol abuse, 425
Living wills, 635–636
Locke's philosophy of child development, 13
Locomotion. *See* Motor development; Walking
Logico-mathematical intelligence (Gardner), 296
Loneliness. *See also* Isolation
early adulthood, 455–456
peer victimization and, 325
widowhood, following, 603
Longevity. *See* Life expectancy
Longitudinal research, 34–35
aggression and television violence, 258
Alzheimer's disease and ERT, 566
athletic performance, 417
"big five" personality traits, 587
birth complications, long-term consequences of, 99–100
centenarians, 550–551
cognitive flexibility and vocation, 505
coping strategies in middle adulthood, 519–520
delinquency, 404
disability, 559
divorce, factors related to, 468
divorce, women's adjustment to, 525
exercise and cancer, 424
gender identity in middle adulthood, 521
grandparent-grandchild relationship, 607
health status in adulthood and socioeconomic staus (SES), 419
intelligence in adulthood, 435
intelligence quotients, stability of, 160
mental abilities, changes in, 499–500
moral reasoning, 390
New York Longitudinal Study on temperament, 180
Oakland Growth Study, 36
religious involvement in adulthood, 589

  self-concept in middle adulthood, 519
  self-esteem and school transitions, 371
  sexuality in late adulthood, 560
  Vaillant's stages of adult development, applicability to women, 449
  vocational choice of women, 438
  work satisfaction and importance of work, 472
Longitudinal-sequential research, 34, 35, 37
  intelligence in adulthood, 435
Long-term care, 569–570, 600
  elder maltreatment and, 609
  Medicare costs of, 569
  placement of aging parents in, 531
  psychological well-being and, 600
  sexuality of residents, 561
  social interaction among residents, 595, 600
Long-term memory, 153, 154. *See also* Memory
  language processing and, 573
  late adulthood, 571, 572–573
  middle adulthood, 502–503
Lorenz's study of imprinting, 22, 186
Loss
  bone loss, 486
  death. *See* Bereavement; Death and dying
  hearing loss. *See* Hearing loss
  neuron loss, 549
  primary (baby) teeth, 204
  vision loss. *See* Visual impairment
Love relationships. *See* Marriage; Romantic relationships
Low birth weight, 96–99. *See also* Preterm infants
  adult health and, 84
  drug use during pregnancy and, 83–84
  environmental pollution and, 87
  infant mortality and, 98
  infectious diseases during pregnancy and, 87
  SIDS and, 103
  smoking during pregnancy and, 84–85
  stress during pregnancy and, 89
  teenage pregnancies, 360
Low-calorie diets, 485
Low-density lipoproteins, 424. *See also* Cholesterol
Low-income households. *See* Poverty; Socioeconomic influences; Socioeconomic status (SES)
Loyalty, in friendships, 397, 454
Lung cancer, 489, 491
Lungs. *See* Respiratory system
Lymph system, growth curve, 205–206

M

Macrosystems, 24, 25, 58, 62
Macular degeneration, 552
  vitamin supplements and, 560
Magical thinking, 218–220
  concept of death and, 623
Magnesium, deficiencies in adolescence, 353
Mainstreaming (in schools), 305
Major periods of development, 8. *See also specific period (e.g., Adolescence; Late adulthood)*
Make-believe play, 147
  advantages of, 217
  autism and, 229
  cultural influences on, 159
  early childhood, 251
  emotional and social development and, 247–248
  infancy and toddlerhood, 147, 158–159
  language development and, 168
  preoperational stage of cognitive development, 216–217, 221
Maladjustment. *See* Adjustment problems
Malaria, 49
Males. *See* Father *entries;* Gender *entries;* Sex differences
Malnutrition. *See also* Nutrition
  anorexia nervosa and, 353
  diseases and, 209
  early childhood, 209
  infancy and toddlerhood, 125–126
  late adulthood, 560
  life expectancy and, 547
  middle childhood, 278
  prenatal malnutrition, 88–89
  small-for-date infants, 97
  tooth development, effect on, 204
Maltreatment. *See* Child maltreatment; Elder maltreatment; Partner abuse
Management positions, glass ceiling in, 536
Mapping of chromosomes, 53
Maps, in concrete operational stage of cognitive development, 285–286
Marasmus, 125–126
Marfan syndrome, 48
  DNA markers, identification, 53
Marijuana. *See also* Drug use and abuse
  prenatal development and, 84
  teenage use, 361
Marital age, 457, 466
  divorce and, 458
  marital satisfaction and, 459
  vocational development of women and, 472
Marital relationship. *See also* Divorce; Marital satisfaction; Marriage; Partner abuse
  aggression in children and, 257
  child maltreatment and, 268
  child-rearing practices and, 59, 463, 465
  intimacy versus independence, 446
  sexual abuse and, 336
Marital roles, 457–458
  parenthood and, 462
Marital satisfaction, 458–459
  childlessness and, 468
  family size and, 463
  friendship between spouses and, 533
  late adulthood, 601
  middle adulthood, 525
  parenthood and, 462–463
  retirement and, 611
  sexual activity and, 489
Marriage, 457–460. *See also* Marital *entries;* Remarriage; Romantic relationships
  age of. *See* Marital age
  cognitive development in middle adulthood and, 500
  death of partner, 641
  divorce. *See* Divorce
  dual-earner households, 458, 473–474
  early adulthood, 446
  egalitarian marriages, 458
  expectations and myths in, 459–460
  incidence of, 457
  in-law relationships, 526
  late adulthood, 601
  life expectancy and, 551
  loneliness in, 456
  middle adulthood, 524–525
  psychological well-being in middle adulthood and, 520–521
  roles in, 457–458
    parenthood and, 462
  satisfaction in. *See* Marital satisfaction
  selection of mate, 450–451, 466, 497
  sexual activity in, 426–427, 489
  teenage pregnancies and, 360
  traditional marriages, 458
Masculine social clock, in adult development, 449–450, 517
Masculinity. *See* Gender *entries*
Masculinity versus femininity, in midlife development, 515, 516
Massage of preterm infants, 98
Mastery. *See also* Achievement; Expertise
  environmental mastery, in middle adulthood, 519
  psychological well-being in middle adulthood and, 521
Mastery-oriented attributions, 317–318
Materialism, among youth, 393
Maternal age
  delay in childbearing, 418, 462
  Down syndrome and, 51
  fraternal twinning and, 46
  genetic counseling, 52
  prenatal development and, 89
  prenatal diagnosis and, 52
Maternal depression, 176
  attachment and, 189
Maternal employment, 333–334
  adolescent development and, 396
  attachment and, 190–191
  child care. *See* Child care
  child development and, 333
  incidence of, 333
  percentage of preschool-age children with employed mothers, 233
  support for employed mothers, 333
  vocational choice of daughters and, 438
Maternal leave, 98–99, 333
Maternal relationship. *See* Mother *entries;* Parent *entries*
Mathematical reasoning in early childhood, 231
Mathematics
  class size and mathematics achievement, 301
  cross-cultural research on academic achievement, 307–309
  gender stereotyping in academic subject preferences, 326
  information processing principles of, 292–293
  logico-mathematical intelligence, 296
  "number sense" and, 292
  sex differences in, 369, 370, 438
Maturation
  brain. *See* Brain development
  cognitive. *See* Cognitive development
  concept of development, 13
  emotional and social. *See* Emotional and social development
  moral. *See* Moral development
  motor skills. *See* Motor development
  physical. *See* Physical development
  sexual. *See* Sexual maturation
Maturational timing, 351–352
Maximum lifespan, 548–549
Mayans
  childbirth practices, 93
  communication style, 297
  cosleeping arrangements in infancy and childhood, 122
  daily life in early childhood, 226
  memory skills of, 291
  menopause, women's attitude toward, 488
  play in early childhood, 250
Measles, 209
Media
  crimes against elderly, reporting of, 598
  safety campaigns, 212
  sexual attitudes and, 426
Mediation of divorce, 331
Medicaid, 569
  hospice care, 633
Medical care. *See* Health care
Medical insurance. *See* Health insurance
Medicare, 64–65, 567, 569, 613
  hospice care, 633
Medications. *See also* Analgesics; Anesthesia; Drug use and abuse
  ADHD, 289
  AIDS in infancy, 87
  Alzheimer's disease and, 566
  assistive technology, 556
  childbirth, during, 95
  dementia-like symptoms and, 567
  depression, 176, 567, 629
  fertility drugs, 46
  heart disease, 492
  postpartum depression, 176
  premenstrual syndrome, 430
  prenatal development and maternal use of, 82–83
  sexuality in late adulthood and, 561
  sleep disorders, 555
  suicide prevention, 403, 593
  taste perception and, 553
Medieval philosophy of childhood, 13
Meiosis, 45
  chromosomal abnormalities, 50, 51
Melatonin, 121
Memorial services, 642–643
Memory. *See also* Long-term memory; Working memory
  alcohol abuse and, 426
  Alzheimer's disease and, 565, 566
  associative memory, 571–572
  autism and, 229
  autobiographical. *See* Autobiographical memory
  cultural influences on, 290–291
  deliberate versus automatic, 571
  early childhood, 226–228
  everyday events and, 227–228
  explicit memory, 571
  eyewitness testimony by children, 337
  habituation research, 154–155
  implicit memory, 571
  infancy and toddlerhood, 154–155
  infantile amnesia, 156
  knowledge base and, 290
  language processing in late adulthood and, 573
  late adulthood, 14, 549–550, 570, 571–573, 577
  middle adulthood, 502–503
  middle childhood, 289–291
  operant conditioning studies, 154–155
  prospective memory, 573
  recall ability. *See* Recall memory
  recognition ability. *See* Recognition memory

remote memory, 572–573
research on, 290
retrospective memory, 573
schooling influencing, 290–291
scripts in, 227
self-control and, 197
Memory strategies
early childhood, 227
middle adulthood, 502–503
middle childhood, 289–291
Men. *See* Father *entries;* Gender *entries;* Sex differences
Menarche, 348–349
anorexia nervosa and, 353
reactions to, 349–350
Menopause, 486–487, 488
osteoporosis and, 493
reproductive technologies and, 55
Menstrual cycle, 77, 430
anorexia nervosa and, 353
estrogens and, 345
menarche. *See* Menarche
menopause. *See* Menopause
middle adulthood, 486–487, 488
Mental development. *See* Cognitive development; Intelligence *entries*
Mental health. *See* Adjustment problems; Counseling; Psychological disturbances; Psychological well-being; *specific entries (e.g., Anxiety; Depression; Stress)*
Mental inferences, children's recognition of, 230
Mental representation
preoperational stage of cognitive development, 216
sensorimotor stage of cognitive development, 145, 147, 149–150
Mental retardation
Down syndrome, 50–51
environmental pollution and, 87
fetal alcohol syndrome and, 86
fragile X syndrome and, 49
infectious diseases during pregnancy and, 87
left-handedness and, 207
mainstreaming of students, 305
phenylketonuria and, 47
Rh blood incompatibility and, 89
sex chromosome disorders and, 51
sex differences in, 49
thyroxine deficiency and, 208
Mental rotations
concrete operational stage of cognitive development, 285
sex differences in, 370
Mental strategies
adolescence, 365
information processing, 153
Mental testing movement, 15. *See also* Intelligence tests
"Mental walk," in spatial reasoning, 285
Mentors
educational aspirations, 28
generativity stage of development, 512
glass ceiling and, 536
Levinson's theory of adult development, 448
vocational success and, 472
women's vocational development and, 472
Mercury, prenatal development and, 87
Mercy killing, 627, 636, 637, 639. *See also* Euthanasia
Merit, in distributive justice reasoning, 322
Me-self
emergence of, 195
middle childhood, 315
Mesoderm, formation of, 79
Mesosystems, 24, 25, 61
Metacognition
adolescence, 365, 366
autism and, 229
early childhood, 228–230
learned helplessness and, 317
middle adulthood, 503
middle childhood, 291
Metaphors, in language development, 236, 299
Metastasis of cancer, 621
Methadone. *See also* Drug use and abuse
prenatal development and, 83
Methods of research. *See* Research methods
Métis. *See* Canadian Aboriginals
Metric system, 293
Mexican Americans. *See also* Hispanics
achievement-related attributions, 318
diabetes incidence, 563
grandparent-grandchildren relationships, 527
identity development in adolescence, 386
menopause, women's attitude toward, 487
Mexico
make-believe play in, 159
Zinacanteco Indians. *See* Zinacanteco Indians
Microsystems, 24–25, 58
Middle adulthood, 8, 481–541
attention in, 502
autonomy in, 519
body composition in, 484, 486
cognitive development in, 433, 498–507
milestones, 542–543
coping strategies in, 519–520
creativity in, 504, 511, 515
death concept, development of, 625–626
death in, leading causes of, 489, 490, 620
divorce in, 524–525
emotional and social development in, 511–541
milestones, 542–543
environmental mastery in, 519
exercise in, 486, 496
expertise in, 503–504
factual knowledge in, 503
friendships in, 532–533
gender identity in, 516, 520–522
gender stereotypes of, 497
health issues in, 489–494
hearing in, 484
height in, 486
information processing in, 501–505
injuries in, 489, 490
intelligence in, 498–501
intelligence tests in, 499–500
job performance in, 534
job training programs in, 535–536
Levinson's theory of, 515–516
long-term memory in, 502–503
marriage in, 524–525
memory in, 502–503
menstrual cycle in, 486–487, 488
metacognition in, 503
milestones in, 542–543
motivation in, 518–519
muscle development in, 484, 486
parenthood in, 524
personality traits in, 518–524
physical development in, 482–497
milestones, 542–543
political activities in, 513
practical problem solving in, 503–504
procedural knowledge in, 503
psychological well-being in, 520–521
reaction time in, 501–502, 505
relationships in, 524–533
reproductive system in, 486–488
retirement planning in, 537–538
self-concept in, 518–524
self-development in, 518–524
self-esteem in, 519
sexuality in, 489
sibling relationships in, 531–532
skeletal development in, 486
social support in, 495–496
speed of processing in, 501–502, 505
stress in, 494–496
unemployment in, 537
Vaillant's theory of, 516–517
vision in, 483–484
vocational achievement, 536
vocational development in, 505, 533–538
weight in, 484
working memory in, 502–503
Middle childhood, 8, 275–339
anxiety in, 334
attention in, 288–289
balance in, 281
bereavement in, 641
body composition in, 277
child care in, 333–334
child-rearing practices in, 328
cognitive development in, 284–309. *See also* Concrete operational stage of cognitive development
milestones, 340–341
cognitive self-regulation in, 291
cross-gender reasoning and behavior in, 326
death concept, development of, 623–624
diseases in, 280
drawing in, 281–282
emotional and social development in, 313–339
milestones, 340–341
emotional self-regulation in, 320
emotions in, 319–320
family influences in, 327–334
fears in, 334
fine motor development in, 281–282
friendships in, 323–324
games in, 282–283
gender identity in, 326–327
gender stereotyping in, 326
gender typing in, 326–327
grammar in, 299
gross motor development in, 281
guilt in, 319
health problems in, 277–281
hearing loss in, 277–278
height in, 276–277
immune system in, 280
information processing in, 287–293
injuries in, 280–281
intelligence tests in, 293–294
language development in, 299–301
milestones, 340–341
malnutrition in, 278
memory in, 289–291
metacognition in, 291
milestones in, 340–341
moral development in, 321–322
motor development in, 281–284
muscle development in, 277
nutrition in, 278
obesity in, 278–279
parent–child relationship in, 328
peer acceptance, 324–326
peer groups in, 322–323
peer relations in, 322–326
perspective taking in, 320–321
phobias in, 334
physical development in, 276–284
milestones, 340–341
physical education in, 283–284
planning in, 289
pragmatic development in, 299–300
pride in, 319
reaction time in, 281
resiliency in, 336–338
self-concept in, 315
self-conscious emotions in, 319
self-development in, 315–318
self-esteem in, 315–318
sibling relationships in, 328
skeletal development in, 277
teeth in, 277
vision in, 277–278
vocabulary in, 299
weight in, 276–277
working memory in, 287
writing in, 281
Middle Eastern societies. *See specific entries (e.g., Saudi Arabia)*
Middle schools. *See* Schools and schooling
Middle-age spread, 484
Middle-income households. *See* Socioeconomic influences; Socioeconomic status (SES)
Midlife crisis, 517
Midlife transition
crisis in, 517
Levinson's theory of adult development, 515
life events approach to, 517–518
Mid-twentieth century theories of development, 15–20
Midwives, 94–95
Mild mental retardation. *See also* Mental retardation
mainstreaming of students, 305
Milestones. *See also* Age
adolescence, 408–409
brain development, 117
cognitive development
adolescence, 222, 408–409
early adulthood, 478–479
early childhood, 222, 272–273
infancy and toddlerhood, 151, 200–201
late adulthood, 616–617
middle adulthood, 542–543
middle childhood, 340–341
early adulthood, 478–479
early childhood, 272–273
emotional and social development, 177
adolescence, 408–409
early adulthood, 478–479
early childhood, 272–273
infancy and toddlerhood, 200–201
late adulthood, 616–617
middle adulthood, 542–543
middle childhood, 340–341
infancy and toddlerhood, 200–201
language development, 166
early childhood, 272–273
infancy and toddlerhood, 200–201
middle childhood, 340–341

late adulthood, 616–617
middle adulthood, 542–543
middle childhood, 340–341
physical development
adolescence, 408–409
early adulthood, 478–479
early childhood, 272–273
infancy and toddlerhood, 200–201
late adulthood, 616–617
middle adulthood, 542–543
middle childhood, 340–341
prenatal development, 77
Mind, theory of. *See* Metacognition
"Mindblindness." *See* Autism
Minerals. *See* Vitamins and minerals
Minorities. *See* Ethnicity and race; *specific entries (e.g., African Americans; Hispanics)*
Miscarriage
chromosomal abnormalities and, 51
diethylstilbestrol and, 83
genetic counseling and, 52
infectious diseases during pregnancy and, 87
radiation exposure and, 50, 86
Rh blood incompatibility and, 89
sex differences in, 49
smoking during pregnancy and, 84
stress during pregnancy and, 89
Missouri, right-to-die case, 633
Mr. Rogers' Neighborhood, 235
Mistrust. *See also* Trust
psychosocial stage of development, 17, 171
Mitosis, 45
chromosomal abnormalities, 50
Mixed marriages, 458
Mixed-sex cliques, 398–399
Mobility. *See* Motor development
Modeling, 18–19. *See also* Imitation
childhood injury prevention programs, 212
field experiment on effect of modeling by child care provider, 33
moral development and, 253–254
peer acceptance intervention programs, 326
Modified experimental research designs, 33–34
Moles, and aging process, 555
Momence, Illinois, experience of aging in, 558
Monkeys
attachment experiment with, 185–186
dietary calorie restriction studies, 485
Monozygotic twins. *See* Identical twins
Mood quality, as temperament dimension, 181
Moodiness. *See also* Depression
adolescence, 350
Moral development. *See also* Aggression; Moral reasoning; Prosocial behavior; Self-control
adolescence, 387–393
autonomous morality, 388
child-rearing practices and, 265, 391
civil responsibility and, 393
cognitive-developmental theory of, 256, 321, 387–388, 389
cultural influences on, 322, 391, 392
distributive justice in, 321–322
early childhood, 252–259
environmental influences on, 391–392
Gilligan's theory of feminine morality, 390–391, 427
heteronomous morality, 388
Kohlberg's theory of, 388–390
middle childhood, 321–322
peer groups and, 323
peer relations and, 391–392
psychoanalytic theories of, 252–253
schooling and, 391
self-conscious emotions and, 249
self-control and, 196
sex differences in, 390–391
social learning theories of, 253–256
temperament and, 254
theories of, 252–256
Moral dilemma research, 388–390
Moral imperatives, 256
Moral reasoning. *See also* Moral development
androgyny and, 522
behavior and, 392–393
early childhood, 256
higher education and, 436
identity development and, 385
Kohlberg's stages of, 388–390
Moral rules, 256, 322
Kohlberg's stages of moral reasoning and, 389, 390
Morality of interpersonal cooperation, 389
Moratorium, in identity development, 384–385
Morbidity, compression of, 559
Moro reflex, 101
Mortality. *See* Death and dying; Infant mortality; Life expectancy
Mortality phase of dying, 621
Mosaic pattern of chromosomes, 50
Mother-child relationship. *See also* Maternal *entries;* Parent *entries*
divorce and, 329, 330
incest, 336
remarriage and, 332
teenage parenthood, 361
Mother-headed households. *See* Single-parent families
Mother-infant relationship. *See also* Parent *entries*
attachment and, 189. *See also* Attachment
interactional synchrony in, 189
Mother-stepfather families, 332
Motion. *See* Motor development; Walking
Motivation
child maltreatment and, 268
educational philosophies and, 302
extracurricular activities and, 376
grade retention and, 303
middle adulthood, 518–519
only children, 329
Motor development. *See also* Sensorimotor stage of cognitive development; *specific activities (e.g., Crawling; Drawing; Walking)*
adolescence, 347
aging process, 416–417, 555
alcohol use during pregnancy and, 86
athletics and, 347
body construction and, 215
brain development and, 207
canalization, 69
cultural variations in, 131–132
depth perception and, 134–135
drug use during pregnancy and, 83
dynamic systems theory of, 130–131, 212
early childhood, 212–215
fine. *See* Fine motor development
gross. *See* Gross motor development
individual differences in, 215
infancy and toddlerhood, 129–133
late adulthood, 555
malnutrition and, 126, 278
middle childhood, 281–284
perceptual development and, 139
rate of, 130
sequence of, 130
sex differences in, 215, 347
stepping reflex and, 100–101
visual impairment and, 136
voluntary reaching in, 132
Motor vehicle accidents
alcohol abuse and, 426
depression and, 401
early adulthood, cause of death in, 419
early childhood, cause of injuries in, 211
late adulthood, 563, 564
middle adulthood, cause of death in, 489, 490
middle childhood, cause of injuries in, 280
Mourning, 639, 642–643. *See also* Bereavement
Movement. *See* Motor development
Mr. Rogers' Neighborhood, 235
Multicultural research, 29–30
Multidimensional aspect of development, 9, 498
Multidirectional aspect of development, 9, 498
Multiple attachments, 192–193
Multiple births, 46. *See also* Twins
Multiple intelligences theory (Gardner), 295–296, 307
Multiple roles, mastery of in middle adulthood, 521
Mumps, 209
prenatal development and, 87
Muscle development
adolescence, 346–347
aging process, 415, 555
early childhood, 204
exercise and, 560
infancy and toddlerhood, 114
middle adulthood, 484, 486
middle childhood, 277
prenatal development, 79
SIDS and, 103
vitamin supplements in late adulthood and, 560
Muscle-fat makeup. *See* Body composition; Fat, body
Muscular dystrophy, 48
DNA markers, identification, 53
Musical intelligence (Gardner), 296
Muslims. *See* Islam
Mutation of genes, 49–50
aging and, 413, 489–490
Mutual exclusivity principle, in vocabulary development, 236
Myelin, 116
Myelination, 116–117, 206, 207
aging process and, 549
malnutrition and, 126
Myopia, 277
Mythic-literal stage of faith development (Fowler), 590
Myths, marital, 459–460

N

Nagasaki, bombing of, 86
National Association for the Education of Young Children, 303
National Health and Social Life Survey, 426, 428
National Institute of Child Health and Human Development (NICHD) Study of Early Child Care, 190–191
Native Americans. *See also specific tribes*
aging parents, caring for, 598
alcohol abuse among, 86
communication style, 297
diabetes incidence, 563
dying, adaptation to, 630–631
elder maltreatment, incidence of, 607–608
extended family households of, 62
grandparent-grandchildren relationships, 527
health in late adulthood, 559
institutionalization of elderly, 569
life expectancy among, 547
newborn behavior and child-rearing practices, 107
obesity in adulthood, 420
out-of-wedlock births, 359
poverty and, 60
spiritual and religious involvement in late adulthood, 590
suicide rate, 402
Nativist theory of language development, 165
Natural childbirth, 93–94
Natural experiments, 33–34
Natural selection, Darwin's theory of, 14
Naturalist intelligence (Gardner), 296
Naturalistic observation research method, 27, 32
Nature–nurture controversy, 7, 66–71. *See also* Environmental influences; Genetic *entries*
canalization, 69, 70
combined point of view. *See* Lifespan perspective
environmental influences on gene expression, 70–71
epigenesis, 71
genetic–environmental correlation, 69–70, 162
intelligence, 296–297
Locke's *tabula rasa,* 13
reaction range, 68–69, 70
resiliency of children and, 10–11
Rousseau's "noble savage," 13
stance of major developmental theories, 26
sudden infant death syndrome, 103
NBAS (Neonatal Behavioral Assessment Scale), 107
Near-sightedness, 277
Neglect
children, 266, 267. *See also* Child maltreatment
elderly, 608. *See also* Elder maltreatment
Neglected children, and peer acceptance, 324, 326
Neighborhood influences, 61. *See also* Community influences
late adulthood, 596–598
Neighborhood Watch, 598
Neo-Freudians. *See* Psychoanalytic theories; Psychosocial theory of development (Erikson)
Neonatal Behavioral Assessment Scale (NBAS), 107
Neonatal mortality, 98. *See also* Infant mortality
Neonates. *See* Newborns

Nervous system. *See also* Brain *entries*
aging of, 415, 549–550
alcohol abuse and, 426
exercise and, 560
mental decline and, 500
prenatal development of, 79, 82
SIDS and, 103
Netherlands
assisted suicide laws, 638
cesarean delivery, 96
childbirth practices, 94
cohabitation, attitude toward, 467
elder maltreatment, incidence of, 607
elderly poverty rate, 65
immunization in, 209
institutionalization of elderly in, 569
male birth rate, 49
prenatal health care, 98
teenage pregnancy rate, 359
voluntary active euthanasia in, 636, 637
Neural network view, of information processing, 501
Neural tube, 79, 116
Neurofibrillary tangles, in Alzheimer's disease, 566
Neurons, 116–117
Alzheimer's disease and, 566
death of, in aging process, 549
prenatal development of, 80
Neuroticism, as personality trait, 522–523
Neurotransmitters
Alzheimer's disease and, 566
Neutral stimulus, 127
Never-married, 466
late adulthood and, 603–604
parenthood and, 470
sibling relationships and, 605
New Jersey, right-to-die case, 633
New York Longitudinal Study, 180
New Zealand
institutionalization of elderly, 569
obesity rate, 278
parent–adult child relationship, 526
reproductive technologies, 55
sexual abuse prevention program, 336
suicide rate, 402
teenage pregnancy rate, 359
Newborns. *See also* Infancy and toddlerhood
appearance of, 92–93
behavioral assessment, 107
breastfeeding. *See* Breastfeeding
capacities of, 100–107
death of. *See* Infant mortality
family's adjustment to, 108
low birth weight. *See* Low birth weight
physical assessment, 93
preterm. *See* Preterm infants
reflexes of, 100–101
sensory capacities, 105–107
"small for date," 97
states of arousal in, 102–105
NICHD study of early child care, 190–191
Niche-picking, 69
Nicotine. *See* Smoking
Nigeria
birth rate in, 464
Yoruba village dwellers, dementia incidence among, 566
Night vision, 483–484
"Noble savage" view of child development, 13
Noise pollution, hearing loss and, 484
Nonfunctionality, as component of death concept, 623
Nonnormative influences, 11–12
Nonorganic failure to thrive, 126
Nonprescription drugs. *See* Medications
Non-rapid-eye-movement (NREM) sleep
aging process and, 554
newborns, 102–103
Nonsocial activity, 249, 250
Norm of equity, 533
Normal curve, in intelligence test scores, 160
Normative, defined, 11
Normative approach to child study, 14–15
Norms, in intelligence test scores, 160
Norway
cohabitation, attitude toward, 467
immunization in, 209
male birth rate, 49
Nose. *See also* Smell
aging process, 555
prenatal development of, 79
NREM sleep. *See* Non-rapid-eye-movement (NREM) sleep
Nuclear bombs and accidents, 86
Nucleus, of cell, 44, 45
Numbers. *See* Mathematics
Numerical knowledge, in core knowledge perspective, 151, 152
Nurse-midwives, 94–95
Nursing homes. *See* Long-term care
Nurture. *See* Environmental influences; Nature–nurture controversy
Nutrition. *See also* Eating habits; Malnutrition; Vitamins and minerals
adolescence, 352–353
cancer and, 491
diabetes and, 563
early adulthood, 419–423
early childhood, 208–209
fraternal twinning and, 46
heart disease and, 416, 554
immune response and, 554
infancy and toddlerhood, 123–126
late adulthood, 560
life expectancy and, 548, 551
macular degeneration and, 552
middle childhood, 278
obesity and, 125
osteoporosis and, 493
phenylketonuria and, 47
prenatal development and, 88–89
pubertal development and, 349
respiratory disease and, 554
strokes and, 567
Nyansongo of Kenya, gender typing among, 327

O

Oakland Growth Study, 36
Obedience. *See also* Compliance; Discipline
punishment/obedience orientation in moral development, 389
Obesity
arthritis and, 562
causes of, 278–279, 420, 422–423
chubby babies and, 125
consequences of, 279, 420
defined, 278
diabetes and, 563
early adulthood, 420–421
epidemic of, 422–423
exercise and, 424
fat consumption and, 421, 422–423
immune response and, 554
incidence of, 278
middle childhood, 278–279
strokes and, 567
treatment of, 279, 420–421, 423
Object labeling, in vocabulary development, 236
Object permanence
language development and, 167
research on, 148–149
sensorimotor stage of cognitive development, 146, 147
Object sorting, 155
preoperational stage of cognitive development, 221
Object words, in language development, 236
Object-hiding tasks, in sensorimotor stage of cognitive development, 146
Objects
attachment objects, 185–186
categorization. *See* Categorization abilities
conservation principle. *See* Conservation
possessiveness and, 245
self-concept and, 245
seriation, 285
Object-search behaviors, and brain growth spurts, 119
Observation research methods, 27–28, 29
Observational learning, 18. *See also* Imitation; Modeling
Occupations. *See* Employment; Job *entries;* Vocational *entries*
Odor. *See* Smell
Oedipus conflict, 16, 244
Older adults. *See* Late adulthood
Old-old elderly, 546–547
One-child families, 329
On-the-job training, 535–536. *See also* Work-study programs
Open classrooms, 302–303
Openness to experience, as personality trait, 522–523
Operant conditioning, 18, 127–128
language development and, 164
memory studies, 154–155
moral development and, 253–254
Opposite-sex friendships, 454–455
Optimism
ethnic identity and, 386
generativity and, 513
hardiness and, 496–497
health perceptions in late adulthood, 559
late adulthood, 587
life expectancy and, 548, 551
spiritual and religious involvement in late adulthood and, 591
successful aging and, 612
Oral rehydration therapy (ORT), 209
Oral stage of development (Freud), 16, 174
Ordinality, early childhood grasp of, 231
Oregon, death with dignity law, 638
Organ donation, 621
Organization
cognitive-developmental theory, 145
memory strategy, 227, 289–290, 502
Organized games with rules, 282–283
Organs. *See also specific entries (e.g., Cardiovascular system; Respiratory system)*
aging of, 414
prenatal development of, 79
ORT (oral rehydration therapy), 209
Osteoarthritis, 562
Osteoporosis, 486, 493, 561, 562
cognitive deterioration and, 576
Other-focused reminiscence, 587
Other-sex friendships, 454–455
Otitis media, 209, 210, 277
Out-of-wedlock births, 470
teenage pregnancies, 359–360
"Outreach" research, 27
Ovaries, 76. *See also* Reproductive system
Overextension, in language development, 167
Overregularization, in grammar development, 236
Overt aggression, 256, 323
Over-the-counter drugs. *See* Medications
Overweight. *See* Obesity; Weight
Ovulation, 77. *See also* Menarche; Menstrual cycle
Ovum, 45–46
aging process, 51
conception, 76–77
donor banks, 54–55
Oxygen deprivation during childbirth, 92
Oxygenation, aging process and, 554

P

Pacific Islanders
childbirth practices, 93
child-rearing practices, 266
sibling relationships among, 532
Pain
fetal response to, 81
newborns' response to, 105–106
Pain management programs, 632
Pain-relievers. *See* Analgesics
Palliative care, in dying, 632
Palmar grasp reflex, 101, 132
Parallel play, 249, 250
Parent education, 465. *See also* Family life education
preterm infants, 98–99
Parent–adolescent relationship, 395–396, 465, 525–526. *See also* Child-rearing practices; Families; Home environment
abstract thinking, handling consequences of, 367
anorexia nervosa and, 353–354
conflict in, 350–351
extended family households and, 63
identity development and, 385, 386
Parent–adult child relationship
aging parents. *See* Aging parents
early adulthood, 456–457
leaving home, 456–457, 525–526
middle adulthood, 525–526, 533
promoting positive ties with adult children, 527
return home, of young adults, 457
Parental imperative theory, 521–522
Parental leave, 98–99, 333
Parent–child relationship, 463, 465. *See also* Child-rearing practices; Families; Father-child relationship; Home environment; Mother-child relationship
adoption and, 57
blended families, 469–470
divorce and, 329, 330, 331
extended family households and, 63
homosexuality and, 471
make-believe play and, 158–159

  middle childhood, 328
  only children, 329
  reproductive technologies and, 54
  resiliency of children and, 10–11
  schools, involvement with, 308, 309
  socioeconomic status and, 59–60
  visual impairment and, 136
Parenthood, 460–465. *See also* Child-rearing practices; Families; Father *entries;* Mother *entries*
  additional births, 463
  adjustment to, 462–463
  advantages and disadvantages of, 461–462, 465
  bereaved parents, 640, 641
  costs of child rearing, 462
  decision to have children, 461–462
  family planning, 464
  gender identity and, 521–522
  generativity and, 512, 513
  homosexual parents, 470–471
  middle adulthood, 524
  never-married parents, 470
  single parenting. *See* Single-parent families
  teenage parenting. *See* Teenage pregnancy and parenthood
  variant styles of, 469–471
Parent–infant relationship. *See also* Child-rearing practices; Families; Home environment
  attachment, 185–194. *See also* Attachment
  cosleeping arrangements, 122
  father–infant, 192
  mother–infant, 189
Parenting practices. *See* Child-rearing practices
Parent-in-law relationships, 526
Parents Anonymous, 269
Parent-school partnerships, and academic achievement, 373
Parent-teacher interactions, 62, 373
Parkinson's disease, 564–565
Participant observation, 30
Partner abuse, 460–461
  child maltreatment risk factor, 268
  divorce and, 525
  factors related to, 460–461
  intervention and treatment, 461
Part-time work
  adolescents, 374
  bridge jobs, 609
  maternal employment, 333
Passionate love, 451
Passive euthanasia, 634–636
Passive genetic–environmental correlation, 69
Passive smoking, and prenatal development, 85
Paternal age
  chromosomal abnormalities and, 51
  delay in parenthood, 462
  Down syndrome and, 51
Paternal leave, 333
Paternal relationship. *See* Father *entries;* Parent *entries*
Pattern perception, 135–138, 139
Pavlov's theory of classical conditioning, 18, 126–127
PCBs, prenatal development and, 87
PCP. *See also* Drug use and abuse
  teenage use, 361
Peck's theory of psychosocial development, 585
Pedaling, in early childhood, 213
Pedestrian accidents, in late adulthood, 563
Pedigrees, in genetic counseling, 52
Peer acceptance, 324–326
  androgynous traits and, 394
  delinquent behavior and, 403, 404
Peer collaboration, 224
Peer conformity, 399–400
Peer culture, 322–323
Peer groups
  adolescence, 398–399
  defined, 322
  delinquent behavior and, 404
  middle childhood, 322–323
Peer pressure, 399
Peer relations. *See also* Friendships; Peer groups; Play
  academic achievement in adolescence and, 373
  ADHD and, 288
  adolescence, 396–400
  adult learners, 506
  aggression and, 259
  child maltreatment and, 268
  early adulthood, 450–456
  early childhood, 249–252
  ethnic identity and, 386
  exercise groups, 496
  gender typing and, 262
  grade retention and, 303
  identity development and, 385
  intergenerational relationships, 533, 577
  mainstreaming of students with learning difficulties and, 305
  maternal employment and, 333
  maturational timing and, 352
  middle childhood, 322–326
  moral development and, 256, 322, 391–392
  parental influences on, 251–252
  psychological well-being in middle adulthood and, 520
  self-concept and, 315
  self-esteem and, 316
  substance abuse and, 362
  World War II refugee and evacuee children, 588
Peer sociability. *See* Sociability
Peer victimization, 325
Pendulum problem, and abstract thinking, 363, 365
Penis. *See* Genitals; Reproductive system
Pennsylvania, Swarthmore, experience of aging in, 558
Pension plans, 537–538
People's Republic of China. *See* Chinese
Perceptual categorization, 155, 157
Perceptual development. *See also* *specific senses (e.g., Hearing; Touch; Vision)*
  aging process, 415
  canalization, 69
  differentiation theory of, 139–140
  infancy and toddlerhood, 133–140
  intermodal perception, 138–139
  late adulthood, 550–553
  newborns, 105–107
Perceptual speed in middle adulthood, 499, 501
Permanence
  death concept, component of, 623
  object permanence. *See* Object permanence
Permissive parenting, 265–266
  academic achievement in adolescence and, 372
  peer groups in adolescence and, 398
Perry's theory of postformal thought, 432
Persistence
  child-rearing practices and, 265
  early childhood, 246
  learned helplessness and, 317
  otitis media and, 210
  temperament dimension, 181
Persistent vegetative state, 621
Personal choice, and moral development, 256
Personal fables, and abstract thinking, 367–368
  adolescent concept of death and, 625
Personal life investment in middle adulthood, 520
Personal standards, 18–19
Personal storytelling, 246
Personality development. *See* Emotional and social development; Self *entries*
Personality traits. *See also* Temperament; *specific traits (e.g., Aggression; Shyness)*
  Alzheimer's disease and, 565
  "big five" personality traits, 522–523, 587
  cognitive development and, 500, 576
  cohabiting couples, 467
  dying, adaptation to and, 629
  extreme personality styles, longitudinal study of, 34–35
  generativity and, 513
  genetic factors, 66
  hardiness, 496–497
  heritability research, 67
  identity development and, 385–386
  individual differences in, 522–523
  late adulthood, 587–591
  life expectancy and, 551
  marital satisfaction and, 459
  middle adulthood, 518–524
  partner abuse and, 460
  resiliency of children, 10
  social clock, conformity to and, 449–450
  stability of, 522–523
  vocational choice and, 437
  widowhood, adjustment to and, 603
Perspective taking
  adolescence, 355, 368
  defined, 320–321
  distributive justice and, 322
  friendships and, 398
  games and, 283
  middle childhood, 320–321
  moral development and, 389, 391
  peer acceptance intervention programs, 326
  self-concept and, 315
  Selman's stages of, 320, 321
  social skills and, 320
Peruvian Quechua, swaddling of infants, 104
Pesticides. *See also* Environmental pollutants
  sperm count and, 49
p53 gene, cancer and, 490
Phallic stage of development (Freud), 16, 244, 252
Pharmaceuticals. *See* Medications
Phencyclidine (PCP). *See also* Drug use and abuse
  teenage use, 361
Phenotypes, 44
Phenylalanine, 47
Phenylketonuria (PKU), 47, 48, 49
Philadelphia, experience of aging in, 558
Philosophies. *See also* Theories
  adult development, early philosophies of, 13–14
  child development, early philosophies of, 13
  educational. *See* Educational philosophies
Phobias. *See also* Fear
  early childhood, 248
  middle childhood, 334
Phonics, 293
Physical abuse
  children, 267. *See also* Child maltreatment
  elderly, 608. *See also* Elder maltreatment
Physical activities. *See* Athletics; Exercise; Extracurricular activities; Games; Play
Physical appearance. *See* Appearance and attractiveness; Body image
Physical deformities. *See* Birth defects; Chromosomal abnormalities; *specific entries (e.g., Cleft palate)*
Physical development, 8, 9. *See also* Body *entries;* Growth *entries;* Motor development; Perceptual development; Skeletal *entries; specific entries (e.g., Brain development; Height; Weight)*
  adolescence, 344–362. *See also* Puberty; Sexual maturation
    milestones, 408–409
  asynchronies in, 204–206
  Down syndrome, 50–51
  early adulthood, 412–431
    milestones, 478–479
  early childhood, 204–215
    milestones, 272–273
  environmental influences on, 123–126, 208–212
  ethnicity and race differences in, 114–115
  fetal alcohol syndrome and, 86
  genetic influences on, 123, 208
  individual differences in, 114–115
  infancy and toddlerhood, 113–140
    milestones, 200–201
  late adulthood, 546–570
    milestones, 616–617
  malnutrition and, 278
  middle adulthood, 482–497
    milestones, 542–543
  middle childhood, 276–284
    milestones, 340–341
  milestones in
    adolescence, 408–409
    early adulthood, 478–479
    early childhood, 272–273
    infancy and toddlerhood, 200–201
    late adulthood, 616–617
    middle adulthood, 542–543
    middle childhood, 340–341
  nutrition and, 123–126
  patterns of growth, 115–116
  prenatal. *See* Prenatal development
  psychological well-being and, 126
  radiation during pregnancy and, 86
  sex differences in, 114–115, 346–347
  sexual activity of adolescents and, 355
  SIDS and, 103
Physical disability
  arthritis as cause of, 562
  assistive technology, 556
  late adulthood, 551–564

longitudinal research on, 559
Physical education. *See also* Exercise
adolescence, 347
gender segregation, 347
middle childhood, 283–284
Physical knowledge, in core knowledge perspective, 150–151
Physical neglect
children, 267. *See also* Child maltreatment
elderly, 608. *See also* Elder maltreatment
Physical well-being. *See* Health
Physical/athletic self-esteem, 315–316
Physiological measures of temperament, 182–183
Piaget's cognitive-developmental theory. *See* Cognitive-developmental theory (Piaget)
Pictorial depth cues, 134
Picture books, and language development, 169
Pima Indians of Arizona, diabetes incidence, 563
Pincer grasp, 132
Pituitary gland, 208
PKU. *See* Phenylketonuria
Placenta, 78, 91, 92
Planning
adolescence, 368–369
early childhood, 225–226
middle childhood, 289
retirement, 537–538
Plaque
amyloid plaques, in Alzheimer's disease, 566
arteries. *See* Atherosclerosis
Plasticity in development, 7, 9–10, 570, 577
Plasticity of brain, 118–119, 120–121
Play. *See also* Games; Make-believe play; Peer relations; Toys
associative play, 249
child care facilities, 163
cognitive development and, 285
constructive play, 251
cooperative play, 249–250
cultural influences, 250
developmental sequence, 249–250
emotional and social development and, 244
functional play, 251
parallel play, 249, 250
sociodramatic play, 217
spatial abilities and, 370
visual impairment and, 136
"Play years," 204
Playgrounds, injury on, 212
PMS (premenstrual syndrome), 430
Pneumonia, mortality rate among elderly, 561, 562
Poland
cognitive flexibility and vocation, 505
World War II refugee and evacuee children, 588–589
Political activities
late adulthood, 612
middle adulthood, 513
Pollution. *See* Environmental pollutants
Polychlorinated biphenyls (PCBs), prenatal development and, 87
Polygenic inheritance, 50, 66
Popular children, and peer acceptance, 324
Popular-antisocial children, and peer acceptance, 324
Popular-prosocial children, and peer acceptance, 324
Population increases, 464
Positive affect, as temperament dimension, 181
Positive discipline, 255–256
Possessiveness, and self-definition, 245
Possible selves, 518–519
late adulthood, 558, 587
successful aging and, 612
Postconventional level of moral development, 390
Postformal thought, 20, 432–433
higher education and, 436
Schaie's theory of. *See* Schaie's theory of postformal thought
Postmenopause. *See* Menopause
Postpartum depression, 176
Poverty. *See also* Socioeconomic influences
child maltreatment and, 268
divorce and, 525
educational enrichment and, 163
elderly, among, 65
family functioning, effect on, 60–61
family planning and, 464
feminization of, 525, 599
Great Depression, consequences on lifespan development, 36, 576
health in late adulthood, 559
homelessness, 60–61
increase in poverty rate, 60
independent living in late adulthood and, 599
infant mortality and, 98
infectious diseases and, 209
injury mortality and, 211
life expectancy and, 547
low birth weight and, 96
malnutrition and, 278
partner abuse and, 461
physical development and, 209
prenatal health care and, 90
prenatal malnutrition and, 88–89
public policies, comparison of U.S. and Canada with other nations, 64
sexual abuse and, 336
sexually transmitted diseases and, 356
teenage pregnancy rate and, 359
"War on Poverty," 233
Power of attorney for health care, 635–636
Practical problem solving
late adulthood, 574–575
middle adulthood, 503–504
Practice effects, 35
Pragmatic development
early childhood, 237
middle childhood, 299–300
Pragmatic thought, 433
Preattachment phase, 186
Preconventional level of moral development, 389
Preformationism, 13
Pregnancy. *See also* Childbirth; Conception; Contraception; Prenatal *entries*
adolescence. *See* Teenage pregnancy and parenthood
age of. *See* Maternal age
caregiving concerns, 90
drug use during, 83–84
first pregnancy and marital satisfaction, 459
infectious diseases during, 87–88
miscarriage. *See* Miscarriage
nutrition during, 88–89
radiation during, 86
sexual coercion resulting in, 429
smoking during, 84–85
stress during, 89
Preimplantation genetic diagnosis, 52
Prejudice. *See* Bias
Premarital sex. *See also* Sexual attitudes and behavior
adolescence, 354–355
Premature academic training, 233
Premature infants. *See* Preterm infants
Premenstrual syndrome (PMS), 430
Prenatal development, 8, 76–90. *See also* Conception; Pregnancy
age of viability, 80
AIDS and, 87
environmental influences on, 81–90
milestones in, 77
nutrition and, 88–89
overview, 8
phases of, 77–81
trimesters, 79–81
Prenatal diagnostic methods, 52–53, 56
Prenatal health care, 89–90, 98
teenage pregnancies, 360
Prenatal malnutrition, 88–89
small-for-date infants, 97
Prenatal surgery, 53
Preoperational stage of cognitive development, 19, 20, 216–223
accommodation in, 218
animistic thinking in, 218–220
appearance versus reality confusion in, 221
categorization abilities, 218, 221
centration thinking in, 218
conservation in, 218, 219, 222
egocentrism in, 217–218
evaluation of theory, 221–222
hierarchical classification in, 218
illogical thought in, 220
irreversible thinking in, 218
language in, 216
limitations of, 217–218
magical thinking in, 218–220
make-believe play in, 216–217, 221
mental representation in, 216
research on, 218–221
state-versus-transformation thinking in, 218
Prepared childbirth, 93–94
Prereaching movements in infants, 132
Presbycusis, 484
Presbyopia, 483
Preschoolers. *See* Early childhood
Preschools, 233–234
fears of, caregiving concerns, 248
sex differences in activities, 262
Prescription drugs. *See* Medications
Pretending. *See* Make-believe play
Preterm infants, 96–99. *See also* Low birth weight
caregiving for, 98–99
child maltreatment and, 97, 268
defined, 97
drug use during pregnancy and, 83
environmental pollution and, 87
interventions for, 97–99
medical interventions, 97–99
SIDS and, 103
"small for date" versus preterm, 97
smoking during pregnancy and, 84
stress during pregnancy and, 89
Prevention programs. *See* Intervention programs
Preverbal gestures, 166–167
Pride
adolescence, 383
early childhood, 248–249
infancy and toddlerhood, 178–179
middle childhood, 319
Primary aging, 561–562. *See also* Biological aging
Primary (baby) teeth, 204
Primary circular reactions, in sensorimotor stage of cognitive development, 145, 146
Primary sexual characteristics, 347
Principle of mutual exclusivity, in vocabulary development, 236
Principled level of moral development, 390
Privacy
genetic testing results, 56
research participants' rights, 38
Private speech, 223
Privileges, withdrawal as disciplinary technique, 255
Problem finding, in postformal thought, 434
Problem solving
analogical, 149–150, 220
class inclusion problems, 218, 220, 285
late adulthood, 574–575, 577
pendulum problem, 363, 365
postformal thought and, 434
practical problem solving, 503–504, 574–575
social. *See* Social problem solving
three-mountains problem, 218, 219
Problem-centered coping, 495, 519–520
aging process and, 556
hardiness and, 497
hearing loss, for, 552–553
unemployment, response to, 537
Problems, emotional. *See* Adjustment problems; Psychological disturbances; Stress
Procedural knowledge, in middle adulthood, 503
Processing capacity
adolescence, 365, 366
middle childhood, 288
Processing speed. *See* Speed of processing
Production versus comprehension of language, 167
Program for International Student Assessment, 307
Project Head Start, 233–234
Prolonged death, and bereavement, 640–641
Propositional thought, 364–365
Prosocial behavior, 249. *See also* Empathy
creativity in middle adulthood and, 504
modeling of, 253–254
moral reasoning and, 392
peer acceptance and, 324
perspective taking and, 320
Prospective memory, 573
Prostate. *See also* Reproductive system
cancer of, 491
sleep disorders and, 555
Proteins, genetic, 45
modification, 56
Proteomics, 56
Protestants. *See specific denomination*
Proxies, for health care, 636
Proximal development, zone of. *See* Zone of proximal development, in sociocultural theory

Proximodistal trend, 115
motor development, 130
Psychoanalytic theories, 15–18, 26. *See also specific theories (e.g., Psychosocial theory of development (Erikson))*
attachment and, 185, 193
infancy and toddlerhood, 174–175
moral development and, 252–253
Psychological abuse
children, 267. *See also* Child maltreatment
elderly, 608. *See also* Elder maltreatment
Psychological development. *See* Emotional and social development; Moral development; Self *entries;* Social *entries*
Psychological disturbances. *See also* Adjustment problems; *specific entries (e.g., Depression; Schizophrenia)*
adolescence, 344
adoptive children, 57
Alzheimer's disease and, 565
child maltreatment and, 268
elder maltreatment risk factor, 608
genetic–environmental correlation, 70
homelessness and, 61
late adulthood, 564–567
left-handedness and, 207
peer acceptance and, 324
sex differences in, 49
war, effect on children, 335
Psychological knowledge, in core knowledge perspective, 151
Psychological stress. *See* Stress
Psychological well-being
dependency and, 591–592
divorce and, 525
ego integrity and, 584
exercise and, 424
extended family households and, 63
family relationships in middle adulthood and, 532–533
friendships and, 398, 532, 605–606
generativity and, 513, 584
health and, 559, 592–593
immigrant youths, 30
individual differences, in late adulthood, 591–594
leisure activities and, 611
life expectancy and, 551
long-term care residents, 600
marital satisfaction and, 525
marriage, influence on, 459
middle adulthood, 520–521
negative life changes in late adulthood and, 593–594
parent-adult-child relationship and, 526
parenthood and, 462–463
physical development and, 126, 208
possible selves and, 518
retirement and, 538, 610
sexual activity and, 489
sibling relationships in adulthood and, 455
singlehood and, 604
social support and, 594
widowhood and, 603
Psychosexual theory of development (Freud), 15–16, 17, 174–175. *See also specific stages (e.g., Anal stage of development; Oral stage of development)*
Erikson's expansion of. *See* Psychosocial theory of development (Erikson)
moral development and, 252, 253
Psychosocial development. *See* Emotional and social development; Self *entries;* Social *entries*
Psychosocial dwarfism, 208
Psychosocial theory of development (Erikson), 16–18
autonomy versus shame and doubt, 174–175
basic trust versus mistrust, 174
ego integrity versus despair, 584–585
generativity versus stagnation, 512–514, 584
identity versus identity confusion, 382–383
industry versus inferiority, 314–315
initiative versus guilt, 244
intimacy versus isolation, 446–447
longitudinal-sequential study of stages of, 35–36
overview of stages of, 17, 447
Psychotherapy. *See* Counseling
Puberty, 344, 345–352. *See also* Adolescence; Sex hormones; Sexual maturation
early vs. late maturation, 351–352
group differences in, 348–349
hormonal changes, 345–346
individual differences in, 348–349
physical development in, 345–349
psychological impact of, 349–352
Pubic hair development, 346, 348
Public policies, 63–65. *See also* Intervention programs; Social issues
advance medical directives, 634–636
age-group benefits, public support of, 533
aging parents, caring for, 531
assistive technology, 556
bilingual education, 301
child care, 162
children, youth, and families, 64, 65
defined, 63
early intervention programs for cognitive development, 162–164
elder maltreatment prevention and treatment, 609
elderly, 64–65
Head Start programs, 233–234
health care costs for elderly, 567, 569, 613
health studies involving women and ethnic minorities, 489
immunization, 209
job-training programs, 374
reproductive technologies, 55
right to die, 633
social services for elderly, 613
surrogate motherhood, 55
vocational education, 439, 441
Public television, and children's programming, 234–235
Puerto Ricans
difficult children, child-rearing practices, 184
Pukapukans of Pacific Islands, childbirth practices, 93
Punishment (disciplinary), 58–59
aggression and, 257
child maltreatment and, 268
corporal punishment, 254–255, 266, 268
moral development and, 254–255
principles of effective punishment, 255–256
socioeconomic status and, 59–60
Punishment (operant conditioning), 18, 127–128
Punishment/obedience orientation in moral development, 389
Puns, in language development, 299
Pupils. *See* Academic achievement; Schools and schooling; Teacher–student interaction
Purging. *See* Bulimia
Puritan child-rearing practices, 13
Pyloric stenosis, stress during pregnancy and, 89

Q
Quakers, bereavement rituals, 642
Quality of mood, as temperament dimension, 181
Quantitative-relational tasks, and abstract thinking, 366
Quechua of Peru, swaddling of infants, 104
Quinceañera, adolescent initiation ceremony, 350
Quinlan case, and right to die, 633

R
Race. *See* Ethnicity and race
Radiation
cancer and, 491
mutation of genes and, 50
prenatal development and, 86
Rage. *See* Anger
Random assignment, in experimental research design, 33
Range of reaction, 68–69, 70
Rape. *See* Sexual coercion
Rapid-eye-movement (REM) sleep
infants, 121
newborns, 102–103
Reaching, voluntary, 132
Reaction intensity, as temperament dimension, 181
Reaction range, 68–69, 70
Reaction time
late adulthood, 563
middle adulthood, 501–502, 505
middle childhood, 281
Readiness to learn, 222
Reading
basic-skills approach, 292
class size and reading achievement, 301
early childhood, 230–231
gender stereotyping in academic subject preferences, 326
information processing principles of, 291–292
language development and, 169
sex differences in, 369
whole-language approach, 292
Real self, 315, 385
Realistic period, of vocational development, 437
Realistic personality type, in vocational choice, 437
Reasoning
abstract. *See* Abstract thinking
cross-gender. *See* Cross-gender reasoning and behavior
distributive justice, 321–322
emotional (Labouvie-Vief), 585–586
hypothetico-deductive reasoning, 363–364
mathematical reasoning in early childhood, 231
moral. *See* Moral reasoning
scientific reasoning, 365–366
spatial. *See* Spatial reasoning and skills
Recall memory
early childhood, 226–227
infancy and toddlerhood, 155
late adulthood, 571
Recasts, in language development, 238
Recessive characteristics, 46–47, 48, 49
Reciprocal relationships
ecological systems theory, 24
ethological theory of attachment, 186–187
Reciprocal teaching, 304
Reciprocity, in moral reasoning, 388, 389
Recognition memory
early childhood, 226–227
infancy and toddlerhood, 155
late adulthood, 571, 572
Reconstituted families. *See* Blended families
Reconstructed memories, and attachment, 192
Recovery, 128. *See also* Habituation-recovery response
Recreational drugs. *See* Drug use and abuse
Rectal cancer, 491
Redemption themes, in life stories, 514
Referential style of language learning, 168
Reflexes
defined, 100
Neonatal Behavioral Assessment Scale (NBAS), 107
newborns' capacities, 100–101
SIDS and, 103
Reflexive schemes, in sensorimotor stage of cognitive development, 146, 147
Refugee children of World War II, 588–589
Rehearsal, as memory strategy, 289–290, 502
Reinforcement
childhood injury prevention programs, 212
food as reinforcer, 279
moral development and, 254
peer acceptance intervention programs, 326
Reinforcement (operant conditioning), 18, 19, 127–128
Reintegrative stage of cognitive development (Schaie), 433
Rejected children, and peer acceptance, 324–325
Rejected-aggressive children, 324–325
Rejected-withdrawn children, 325
Relational aggression, 256, 257, 323
peer victimization and, 325
Relationships. *See specific entries (e.g., Marital relationship; Parent–child relationship; Peer relations)*
Relativistic thinking, 432
concept of death and, 625
higher education and, 436
Religion. *See* Spirituality and religiosity
Relocation
late adulthood, 598
widowhood, following, 603
young adults leaving home, 456–457, 525–526
REM sleep
infants, 121
newborns, 102–103

Remarriage, 331–333, 468, 469. *See also* Blended families
divorce following, 469, 525, 602
incidence of, 457
late adulthood, 602
reasons for, 469
Remedial education, in drop out prevention programs, 376
Reminiscence, 586–587
sibling relationships and, 605
Remote memory, 572–573
Representation, mental. *See* Mental representation
Reproduction. *See* Conception; Fertility; Pregnancy; Sexual maturation
Reproductive choices, 51–57
Reproductive system, 76–77. *See also* Genitals
aging of, 415, 418
DES and reproductive abnormalities, 83
female, 76–77
male, 77
middle adulthood, 486–488
pubertal development, 347–348
Reproductive technologies, 54–55, 418
homosexual parents, 471
Research, 5, 27–39. *See also specific topics or studies (e.g. Kinship studies; Twin studies)*
applications of, 5. *See also* Public policies; Social issues
community-researcher partnership, 28
ethics in research, 37–39
historical foundations of, 13–15
rights of participants, 38–39
Research designs, 31–37. *See also specific designs (e.g., Correlational research design; Experimental research design; Longitudinal research)*
defined, 27
developmental research designs, 34–37
general research designs, 31–34
Research methods, 27–30. *See also specific methods (e.g., Clinical method of research; Systematic observation research methods)*
defined, 27
strengths and limitations, 32
Research rights of participants, 38–39
Residential communities for elderly, 599–600
leisure activities in, 611
Residential hall living in college, 436
Resilient children, 10–11
birth complications, long-term consequences of and, 99–100
middle childhood, 336–338
Resistance to temptation. *See* Self-control
Resistant attachment, 187
adult romantic relationships and childhood attachment patterns, 452–453
mother-child relationship and, 189
Resource rooms, in schools, 305
Respiratory distress syndrome, 97
Respiratory system
aging process, 414–416, 553–554
emphysema, 561, 562
pneumonia, 561, 562
prenatal development, 79, 80
preterm infants, 97
SIDS and, 103
sleep apnea, 555
smoking, effects of, 425
Respite services
caregivers of elders with dementia and, 568
elder maltreatment prevention and treatment, 608
Response
autoimmune. *See* Autoimmune response
conditioned, 127
habituation-recovery. *See* Habituation-recovery response
unconditioned, 126–127
Responsibility stage of cognitive development (Schaie), 433, 498
Responsiveness threshold, as temperament dimension, 181
"Restless legs" during sleep, 555
Restoration, as phase of grieving, 640
Retardation. *See* Mental retardation
Reticular formation, 207
Retina, 106. *See also* Eyes
Retirement, 609–611. *See also* Leisure activities
adjustment to, 610–611
American Association of Retired Persons (AARP), 66
average age of, 537
bridge jobs, 609
cognitive development and, 576
decision to retire, 609–610, 611
planning for, 537–538
returning to employment, 609
Retirement planning, 537–538
Retirement villages, 599
Retrieval of information. *See* Long-term memory; Memory
Retrospective memory, 573
Returning students. *See* Adult learners
Reversibility in thinking, 285
Rh factor
dominant–recessive inheritance, 47
incompatibility
cesarean delivery, 96
prenatal development and, 89
Rheumatoid arthritis, 562
Rhythmicity, as temperament dimension, 181
Riboflavin, deficiencies in adolescence, 353
Riddles, in language development, 299
Right to die, 633–639
Right-handedness, 118, 206–207
Rights of research participants, 38–39
Rites of passage, 350
Rituals
adolescent rites of passage, 350
bereavement, 642–643
kanreki ritual in aging process, 557
Role models. *See* Modeling
Roles
gender-role adoption. *See* Gender stereotyping
gender-role identity. *See* Gender identity
kinkeeper role. *See* Kinkeeper role
marital roles, 457–458, 462
multiple roles, mastery in middle adulthood, 521
work-role preoccupation, versus ego differentiation, 585
Romantic relationships, 450–453. *See also* Marriage
childhood attachment patterns and, 452–453
components of love, 451–452
cultural influences on experience of love, 452–453
keeping love alive in, 454
selection of mate, 450–451, 497
Rooting reflex, 100, 101
Rotations, mental. *See* Mental rotations
Rousseau's philosophy of child development, 13
adolescence and, 344
Routes to Learning programs, 577
Rubella, 209
prenatal development and, 87
Rules
moral. *See* Moral rules
organized games with rules, 282–283
Running
adolescence, 347
early childhood, 212–213
sex differences in, 215
Running away
divorce of parents and, 330
sexual abuse victims, 336
Rural areas. *See* Community influences; Neighborhood influences

S

Safety devices, preventing childhood injuries, 211
Salt, and strokes, 567
Same-sex friendships, 454
Same-sex parent, identification with. *See* Identification with same-sex parent
Sandwich generation, 530
SAT (Scholastic Aptitude Test)
sex differences in scores, 369
Satisfaction
job satisfaction, 472, 534–535
marital. *See* Marital satisfaction
Saturated fat, 421
Saudi Arabia, death anxiety among Islamic, 626
Scaffolding, 224, 226
Schaie's theory of postformal thought, 432–433, 435, 499
Schemes, in cognitive-developmental theory, 144
make-believe play and, 217
Schizophrenia
genetic factors, 67
genetic–environmental correlation, 70
Scholastic achievement. *See* Academic achievement
Scholastic Aptitude Test (SAT)
sex differences in scores, 369
School achievement. *See* Academic achievement
School attendance. *See also* Dropping out of school
divorce and, 330
homelessness and, 61
School lunches, 279
School phobia, 334
School readiness, 303
School transitions, 371–372
School year, 309
School-age children. *See* Middle childhood
Schools and schooling, 301–309, 371–377. *See also* Education *entries;* Teacher *entries; specific subjects (e.g., Mathematics; Reading)*
ability grouping in, 304, 374
achievement in. *See* Academic achievement
after-school programs, 61, 334
"back to basics" movement, 302
bilingual education, 300–301
civil responsibility, development of, 393
class size, 301–302
cognitive development influenced by, 286–287
corporal punishment, 268
dropping out. *See* Dropping out of school
educational self-fulfilling prophecies, 304
ethnic identity and, 386
family life education courses, 460
friendships and school adjustments, 398
full inclusion, 305
gifted children. *See* Gifted children
grade advancement, 307
grade retention, 303
grade structures of, 371–372
identity development and, 385
language immersion programs, 300
mainstreaming of students with learning difficulties, 305
meals provided by, 279
memory strategies, effect on, 290–291
memory strategies and, 290
moral development and, 391
obesity and, 279
open classrooms, 302–303
parental involvement in, 308, 309
peer victimization in, 325
physical education programs. *See* Physical education
premature academic training, 233
preschools. *See* Preschools
readiness to attend, 303
resource rooms for students with learning difficulties, 305
scientific reasoning and, 366
self-esteem and, 317
sex education programs. *See* Sex education
special needs students, 305–307
starting times, 347
tracking students, 374
traditional classrooms, 302–303
transition classes, 303
"trauma curriculum," 335
truancy. *See* School attendance
work-study programs, 374
Science, sex differences in involvement in, 369, 438
Scientific reasoning, 365–366
Scientific studies. *See* Research *entries*
Scouting, 323
Scribbles, 213–214
Scripts, in memory development, 227
Scrotum, 77. *See also* Genitals
Seasons of a Man's Life, The (Levinson), 447
Seattle Longitudinal Study, 435, 499, 500, 577
Second language, 300–301
Secondary aging, 561–562
Secondary circular reactions, in sensorimotor stage of cognitive development, 145, 146
Secondary friends, 606
Secondary schools. *See* Schools and schooling
Secondary sexual characteristics, 347–348
Secondhand smoke, and prenatal development, 85

Secular trend in pubertal development, 349
Secure attachment, 187
adult romantic relationships and childhood attachment patterns, 452
discipline and, 254
later development and, 193
Secure base, for attachment, 186
Selective breeding, 55
Selective optimization with compensation, 570, 592, 612
Self-acceptance
late adulthood, 584, 587
middle adulthood, 519
successful aging and, 612
Self-awareness
coping strategies in middle adulthood and, 520
emotional and social development and, 195–196
infancy and toddlerhood, 195–196
life review and, 586
self-categorizing and, 196
self-control and, 196
self-recognition and, 195
Self-blame. *See* Guilt
Self-care children, 333–334
Self-categorizing, 196
Self-concept. *See also* Gender identity; Identity; Self-esteem
adolescence, 383
caregiving concerns, 247
child maltreatment and, 268
cultural influences, 246
death anxiety and, 626
defined, 245
early childhood, 245, 246
friendships and, 398
late adulthood, 587–591
life review and, 586
middle adulthood, 518–524
middle childhood, 315
moral reasoning and, 393
personal storytelling and, 246
possible selves, 518–519
social interaction and, 596
successful aging and, 612
Self-confidence
adolescence, 383
androgynous traits and, 394
mental abilities and, 369, 438
Self-conscious emotions. *See also* *specific emotions (e.g., Empathy; Guilt; Shame)*
early childhood, 248–249
infancy and toddlerhood, 178–179
middle childhood, 319
Self-consciousness, and abstract thinking, 367
Self-control. *See also* Compliance; Self-regulation
caregiving concerns, 197
infancy and toddlerhood, 196–197
Self-definition. *See* Self-concept
Self-development. *See also* Identity
adolescence, 383–387
early childhood, 245–247
infancy and toddlerhood, 174–175, 194–197
make-believe play and, 217
middle adulthood, 518–524
middle childhood, 315–318
Self-disclosure, and friendships, 397, 398
Self-efficacy, 18–19
bereavement intervention programs and, 644
emotional self-efficacy in middle childhood, 320
exercise and, 496
health protection in late adulthood and, 558–559
political activism and, 513
psychological well-being in middle adulthood and, 520
vocational development and, 472, 535
Self-employment, and decision to retire, 610
Self-esteem
ability grouping in schools and, 304
achievement-related attributions and, 317–318
adolescence, 383–384
androgyny and, 522
caregiving concerns, 247
changes in level of, 316
child maltreatment and, 268
child-rearing practices and, 265, 317
cultural influences on, 316–317
defined, 245
divorce and, 469
early childhood, 245–247
ethnic identity and, 386
exercise and, 424, 560
friendships and, 398, 454, 455
gender identity and, 263
grade retention and, 303
hierarchically structured, 315–316
identity development and, 385
immigrant youths, 30
influences on, 316–318
intervention programs, 318
loneliness and, 456
mainstreaming of students with learning difficulties and, 305
maternal employment and, 333
middle adulthood, 519
middle childhood, 315–318
only children, 329
peer acceptance and, 326
peer victimization and, 325
school transitions and, 371
self-care children, 333
sexual abuse victims, 336
Self-focused reminiscence, 586–587
Self-focusing, and abstract thinking, 367
Self-fulfilling prophecy, in teacher–student interaction, 304
Self-help groups. *See* Support groups
Self-help skills
chronic illness in childhood, 280
early childhood, 213, 226
Self-image. *See* Self-concept
Self-interest, among youth, 393
Self-recognition, 195
Self-reflective perspective taking, 321
Self-regulation. *See also* Self-control
cognitive. *See* Cognitive self-regulation
emotional. *See* Emotional self-regulation
learned helplessness and, 317
middle childhood, 291
Self-report research methods, 28–29, 32
Self-serving bias, 366
Self-understanding. *See* Self-awareness; Self-concept
Self-worth. *See* Self-concept; Self-esteem
Selman's stages of perspective taking, 320, 321
Semantic development. *See* Language development; Vocabulary
Semilingualism, 300–301
Senescence. *See* Biological aging
Senior centers, 578
organized activities of, 611
widowhood, support following, 604
Senior citizens. *See* Late adulthood
Sense of agency in infancy and toddlerhood, 195
Sense of self. *See* Self *entries*
Senses. *See* Perceptual development; *specific senses (e.g., Hearing; Touch; Vision)*
Sensitive caregiving, in attachment, 189, 190
Sensitive periods, 22
brain development, 119–120
language development, 300
prenatal development, 82
Sensorimotor stage of cognitive development, 19, 20, 144, 145–147
circular reactions in, 145–147
evaluation of, 150–151
language development and, 167
research on, 147–150
substages, 145–147
Sensory perception. *See* Perceptual development; *specific senses (e.g., Hearing; Touch; Vision)*
Sensory register, in information processing, 153, 154
Separation, marital. *See* Divorce
Separation anxiety, 186
September 11 attack on World Trade Center, 335, 643
Sequential research designs, 35, 37
Seriation, in concrete operational stage of cognitive development, 285
Serotonin, in Alzheimer's disease, 566
SES. *See* Socioeconomic status
Sesame Street, 234–235
Settling down, in Levinson's theory of adult development, 448–449
Sex. *See* Gender *entries;* Sexuality
differences in. *See* Sex differences
Sex cells, 45–46. *See also* Ovum; Sperm
Sex chromosomes, 46
abnormalities of, 51
genetic imprinting, 49
Sex differences. *See also* Gender stereotyping; Gender typing
ADHD diagnosis, 288
adult learners, 506
aggression, 256–257, 323
aging parents, relationship with, 530, 606–607
alcohol abuse, 425
anger, 494
arthritis, 562
asthma, 280
athletic performance and participation, 347
autobiographical memory, 156
bereavement, 640
blended families, adjustment to, 332
bone loss, 493
cancer, vulnerability to, 489–491
centenarians, 550
cerebrovascular dementia, 567
cognitive development, 369–370, 500
communication styles, 262
creative period in vocations, 434
cross-gender behavior, 262, 394
death anxiety, 626
delinquent behavior, 403
depression, 400, 401
divorce of parents, response to, 330
emotional and social development in early adulthood, 448, 449
emotional expression, 180
engagement versus separateness, in midlife development, 516
exercise, 424
extreme personality styles, longitudinal study of, 35
fat distribution, 484
fine motor development, 282
friendships, 324, 397, 454, 532, 606
gender identity development, 326–327
gender intensification, 394
genetic code and, 49
grandparent-grandchildren relationships, 527
gross motor development, 282
health in late adulthood, 559
health in middle adulthood, 489
hearing loss, 484, 550
heart disease, 492–493, 494, 561
identity development, 385
independent living in late adulthood, 599
injuries, 211, 280–281
intimacy development, 385
language development, 168
life expectancy, 547, 548, 550
loneliness, 456
marital expectations, 459
marital satisfaction, 459, 601
mate selection, characteristics valued, 451, 497
mathematical ability, 369
maturational timing, reactions and responses to, 351–352
midlife transition, 517
moral development, 390–391
motor development, 215, 282, 347
motor vehicle accidents, 563
newborn length and weight, 92
obesity in adulthood, 420
parent–child relationship, 192
physical development, 114–115, 204, 276–277, 346–347
poverty, 60
preschool activities, 262
reading ability, 369
remarriage in late adulthood, 602
retirement, 610
science courses, interest in, 369
self-control and, 197
self-esteem, 316, 384
sexual activity, 355, 489, 561
sexual attitudes and behavior, 427
sexual maturation, 348
sexually transmitted diseases, 428
sibling relationships, 532, 604
singlehood, adjustment to, 466
skeletal age, 115
sleep disorders, 555
spatial reasoning and skills, 370
spiritual and religious involvement in late adulthood, 591
suicide rate, 402, 592
temperament, 183, 184
vocational achievement, 536
vocational development, 438, 472
volunteering in late adulthood, 612
widowhood, 602, 603
X-linked inheritance, 49
Sex education, 336, 360
AIDS and, 432
contraceptive use and, 356
late adulthood, 561
teenage pregnancy rate and, 360
Sex hormones, 345
aggression and, 256
heart disease and, 416
Sex reassignment, 261

Sex stereotyping. *See* Gender stereotyping
Sex typing. *See* Gender typing
Sexism. *See* Gender stereotyping
Sex-role identity. *See* Gender identity
Sexual abstinence programs, 354
Sexual abuse of children, 267, 334–336. *See also* Child maltreatment
Sexual attitudes and behavior
  adolescence, 354–355
  cultural influences on, 354
  divorce of parents and, 331
  early adulthood, 426–428
  friendships and, 455–456
  heterosexual, 426–428
  homosexual, 428
  late adulthood, 560–561
  problems in sexual interest and behavior, 427
  psychological well-being in middle adulthood and, 521
  sex differences in, 427
  sexual abuse victims, adult relationships of, 336
Sexual characteristics, primary and secondary, 347–348
Sexual coercion, 428–430
  alcohol abuse and, 425, 426
  caregiving guidelines, 430
  characteristics of perpetrators, 429
  consequences of, 429
  incidence of, 429
  prevention and treatment, 429–430
Sexual dysfunction, 427
  impotence, 425, 487
Sexual intercourse. *See* Sexual attitudes and behavior
Sexual maturation, 45–46, 347–348. *See also* Puberty
Sexual orientation
  bisexuality, 428
  heterosexual attitudes and behavior, 426–428
  homosexuality. *See* Homosexuality
Sexuality. *See also* Sexual attitudes and behavior; Sexual maturation
  adolescence, 354–356
  early adulthood, 426–430, 431
  late adulthood, 560–561
  middle adulthood, 489
Sexually transmitted diseases (STDs)
  adolescence, 356, 358
  early adulthood, 428
  prevention, 359
  sexual attitudes, effect on, 354
  sexual coercion resulting in, 429
Shame
  early childhood, 248–249
  infancy and toddlerhood, 178–179
  psychosocial stage of development, 17, 174–175
Sharing. *See* Prosocial behavior
Shinto beliefs about death, 621
Shiva, 642
Shoe-tying, motor development and, 213
Shoplifting, bulimia and, 354
Short-term memory. *See* Working memory
Shyness
  biological basis of, 182–183
  contexts of development and, 6–7
  cultural differences in response to, 184–185
  genetic influences on, 183
  loneliness and, 456
Sibling relationship
  adolescence, 396
  attachment and, 192–193, 194
  caregiving concerns, 194
  cross-sectional study of, 35
  death of sibling and bereavement, 641
  development of, 193
  early adulthood, 455
  friendships among siblings, 455
  gender typing and, 262
  individual differences in, 193
  late adulthood, 604–605
  make-believe play and, 159
  middle adulthood, 531–532
  middle childhood, 328
  newborns, adjustments to, 108
  parental comparisons and, 328
  support given or received by sibling over lifespan, 604, 605
  temperament and, 193
Sibling rivalry, 328
Sickle cell anemia, 47–49, 280
SIDS. *See* Sudden infant death syndrome
Sight. *See* Eyes; Vision
Sign language, 118
Singlehood, 466. *See also* Widowhood
  late adulthood and, 603–604
  parenthood and, 470
  sibling relationships and, 605
  vocational development of women and, 472
Single-parent families, 329–330. *See also* Divorce
  child-rearing practices in, 330–331
  extended family households and, 63, 470
  grandparent-grandchildren relationships and, 527
  never-married parenthood, 359, 470
  poverty and, 61, 62
Size of body. *See* Body proportion; Height; Weight
Size of family. *See* Family size
Skeletal age, 115
  early childhood, 204
  psychosocial dwarfism, 208
Skeletal development
  adolescence, 345
  aging process, 415, 555
  early childhood, 204
  estrogen decrease and, 486
  infancy and toddlerhood, 115
  middle adulthood, 486
  middle childhood, 277
  osteoporosis and, 493
  prenatal development, 79
Skin
  aging process, 415, 484, 555
  dominant–recessive inheritance, 47
  estrogen decrease and, 486
  smoking, effects of, 425
Skinner's operant conditioning theory. *See* Operant conditioning
Skipped-generation families, 528
Sleep
  adolescence, 347
  aging and, 554–555
  cosleeping arrangements, 122
  infancy and toddlerhood, 120–123
  newborns, 102–103
  "phase delay," 347
  SIDS and, 103
Sleep apnea, 555
Sleep disorders
  Alzheimer's disease and, 565, 566
  estrogen decrease and, 486
  late adulthood, 555
  obesity and, 420
  sexual abuse victims, 336
  sexual coercion and, 429
Slow-to-warm-up child, as temperament type, 181
Small towns, influence on development, 62
Small-for-date infants, 97
"Smart home" systems, 556
Smell. *See also* Nose
  aging process, 415, 553
  newborns' response to, 106
Smiling in infancy and toddlerhood, 177
Smoking
  adolescence, 361–362
  cancer and, 489
  early adulthood, 425, 431
  emphysema caused by, 561
  heart disease and, 492, 554
  life expectancy and, 551
  osteoporosis and, 493
  otitis media and, 210
  prenatal development and, 84–85
  sexuality in late adulthood and, 561
  SIDS and, 103
  strokes and, 567
  taste perception and, 553
Sociability
  biological basis of, 182–183
  cultural influences, 250
  early childhood, 249–250
  late adulthood, 587, 594
  parental influences on, 251–252
Social acceptance. *See* Peer acceptance
Social class. *See* Socioeconomic status (SES)
Social clock, 449–450
  generativity stage of development and, 513
  midlife transition and, 517
  unemployment and, 537
  young adults leaving home, 525–526
Social comparisons, in self-concept, 315
Social conflict. *See* Social problem solving
Social contract orientation in moral development, 390
Social conventions, and moral development, 256, 322
Social convoy, 600
Social development. *See* Emotional and social development
Social goals, 597
Social interactions. *See* Social support; *specific entries (e.g., Friendships; Peer relations)*
Social isolation. *See* Isolation
Social issues. *See also* Intervention programs; Public policies
  advocacy for children and aged, 65–66
  child care and attachment, 190–191
  child maltreatment. *See* Child maltreatment
  civil responsibility, development of, 393
  community-researcher partnership, 28
  elder suicide, 592–593
  eyewitness testimony by children, 337
  genetic testing, 56
  infant mortality, 98–99
  prenatal health care, 98–99
  reproductive technologies, 54–55
  sex education, 360
  skipped-generation families, 528
  voluntary active euthanasia, 636, 637
  war, effect on children, 335
Social learning theory (Bandura), 18–19, 26
  gender identity and, 260, 263
  mate selection characteristics and, 451
  moral development and, 254–256
  sexual attitudes and behavior and, 427
Social personality type, in vocational choice, 437
Social policy. *See* Public policies; Social issues
Social problem solving. *See also* Social skills
  peer acceptance intervention programs, 326
  perspective taking and, 320
  training in, 259
Social referencing, 178
  autism and, 229
Social Security, 64–65, 537
Social self-esteem, 315–316
Social services. *See* Intervention programs; Social issues
Social skills. *See also* Social problem solving
  attachment security and, 193–194
  intervention programs, 326
  peer acceptance and, 326
  peer groups and, 323
  perspective taking and, 320
Social smiles, 177
Social support
  adult learners and, 506–507
  aging parents, caregivers of, 531
  Alzheimer's disease caregivers, 567
  bereavement and, 644
  child maltreatment and, 268–269
  childbirth, during, 93–94
  chronic illness in childhood, 280
  depression and, 567
  divorce, following, 469
  dying, adaptation to and, 630
  extended family households and, 63
  homosexuality in adolescence and, 357
  late adulthood, 594
  life expectancy and, 548
  middle adulthood, 495–496
  obesity treatment and, 421
  partner abuse and, 461
  pubertal changes and, 350
  resilient children and, 11, 337
  retirement and, 611
  sibling relationships as, 328, 531
  skipped-generation families, 528
  spiritual and religious involvement in late adulthood and, 591
  stress and, 431
  successful aging and, 613
  unemployment and, 537
  widowhood and, 603, 604
Social systems perspective, 58–59
Social theories of aging, 595–596
Social-cognitive learning theory (Bandura), 18–19. *See also* Social learning theory (Bandura)
Social-informational perspective taking, 321
Social-order-maintaining orientation in moral development, 390
Societal perspective taking, 321
Sociocultural theory of cognitive development (Vygotsky), 23–24

early childhood, 223–225
educational principles based on, 224, 304
evaluation of, 225
infancy and toddlerhood, 157–159
make-believe play and, 158–159
private speech, 223
research on, 158–159
zone of proximal development, 158, 169, 223, 304
Sociodramatic play, 217, 250
Socioeconomic influences. *See also* Poverty; Socioeconomic status (SES)
delinquent behavior, 403, 405
divorce, 469
Great Depression, consequences on lifespan development, 36, 576
long-term care use, 569
maternal employment, response to, 333
retirement, 609–610
sexual abuse, 336
sexual activity of adolescents, 355
suicide, 402
Socioeconomic status (SES), 59–60. *See also* Poverty
academic achievement and, 60, 373
defined, 59
delinquent behavior and, 403
dropping out of school and, 374
exercise and, 424
family functioning and, 59–60
grandparent-grandchildren relationships and, 527
health in late adulthood and, 559
health status in adulthood and, 419
IQ scores and, 296, 297
life expectancy and, 419, 547, 548
obesity in adulthood and, 420
pubertal development and, 349
school tracking and, 374
self-esteem and, 317, 384
spiritual involvement in late adulthood and, 590
vocational choice and family socioeconomic level, 437
young adults leaving home and, 457, 525
Socioemotional selectivity theory of aging, 596, 597
Soothability, as temperament dimension, 181
Sound. *See also* Hearing
fetal response to, 80, 81
Southeast Asia, adaptation to dying, 630
Southern Europeans, caring for aging parents, 598
Spain, young adults leaving home, 525
Spanking. *See* Corporal punishment
Spatial intelligence (Gardner), 296
Spatial perception, sex differences in, 370
Spatial reasoning and skills
brain plasticity and, 120, 121
concrete operational stage of cognitive development, 285–286
lateralization of brain and, 206
middle adulthood, 499
sex differences in, 370
visual impairment and, 136
Spatial visualization, sex differences in, 370
Special needs students, 305–307
Special Supplemental Food Program for Women, Infants, and Children, 88
Speech. *See also* Language development
aging process and speech perception, 552–553
Alzheimer's disease and, 565
child-directed, 168–169
early childhood, 237
egocentric, 223
infants' response to, 133
newborns' response to, 106
private, 223
telegraphic, 167–168
Speed, perceptual, in middle adulthood, 499
Speed of processing
adolescence, 365
late adulthood, 570
middle adulthood, 501–502, 505
Spelling in early childhood, 230–231
Sperm, 45
conception, 76–77
donor banks, 54–55
pesticides and, 49
Spermarche, 348
reactions to, 349–350
Spina bifida, folic acid and, 88
Spirituality and religiosity. *See also specific faiths (e.g., Buddhism; Judaism)*
adolescent rites of passage, 350
adolescent understanding of death and, 624
bereavement rituals, 642–643
death, definition of and, 621
death anxiety and, 626
divorce and, 468–469
dying and, 630–631
euthanasia and, 634
late adulthood, 587–591
mixed marriages, 458
stages of faith development, 590
successful aging and, 613
widowhood, support following, 604
youth groups, 323
Sporadic Alzheimer's, 566
Sports. *See* Athletics; Extracurricular activities
Spouse. *See* Marital *entries;* Marriage
Spouse abuse. *See* Partner abuse
Stability
attachment patterns, 188
change versus, in development, 7
IQ scores, 160
personality traits, 522–523
temperament, 182–183
Stage concept of development, 6
cognitive development (Schaie), 432–433
cognitive-developmental theory, 19, 20. *See also specific stages (e.g., Formal operational stage of cognitive development; Sensorimotor stage of cognitive development)*
dying, 627–628
faith development, 590
moral reasoning, 388–390
perspective taking, 320, 321
psychosexual theory, 16. *See also specific stages (e.g., Anal stage of development; Oral stage of development)*
psychosocial theory, 17. *See also specific stages (e.g., Autonomy; Trust)*
Rousseau's philosophy of, 13
Stagnation, as psychosocial stage of development, 17, 447, 512–514
Standard of living. *See* Socioeconomic status (SES)
Stanford-Binet Intelligence Scale, 15, 293
Starvation. *See* Malnutrition
"Starvation syndrome," 421
State of America's Children, The, 65
States of arousal
adolescence, 347
infancy and toddlerhood, 120–123
newborns, 102–105
State-versus-transformation thinking, in preoperational stage of cognitive development, 218
Statistics
concordance rates, 67–68
correlation coefficients, 32
factor analysis, 293
heritability estimates, 67, 68
STDs. *See* Sexually transmitted diseases
Steering skills in early childhood, 213
Stepfamilies. *See* Blended families
Stepping reflex, 100–101
Stereotyping. *See* Aging stereotyping; Gender stereotyping
Sternberg's triarchic theory of intelligence, 294–295, 503
Stimulant medications, for ADHD, 289
Stimulation
brain development and, 116, 119–120
emotional self-regulation and, 179
preterm infants, 97–98
temperament and, 184
voluntary reaching and, 132
Stimulus
conditioned, 127
habituation to. *See* Habituation-recovery response
neutral, 127
unconditioned, 126–127
Strange Situation technique to measure attachment, 187, 188, 190
Stranger anxiety, 177–178
Strategies
coping. *See* Coping strategies
memory. *See* Memory strategies
mental strategies, in information processing, 153
prevention. *See* Intervention programs
Stress
adult learners, 506
aggression in children and family stress, 257
aging parents, caring for, 530–531
aging stereotypes and, 557
attachment and family stressors, 190, 191
caregiving guidelines, 495
child maltreatment and parental stress, 268
depression and, 401
divorce and, 469
dual-earner families, 473–474
early adulthood, 430–431
elder maltreatment risk factor, 608
exercise and, 424
extended family households and, 63
hardiness and, 496–497
heart disease and, 492
immune response and, 418
impotence and, 487
job burnout, 534–535
life expectancy and, 548
middle adulthood, 494–496
parents' anger, effect on children, 33
partner abuse and, 461
physical development and, 89, 208
poverty and, 60
prenatal development and, 89
prenatal health care and, 90
resilient children, 10–11, 336–338
retirement and, 611
socioeconomic status and, 59–60
strokes and, 567
unemployment and, 537
Stress hormones
aging process, 550
anger's effect on, 494
childbirth, during, 92
immune response and, 418, 554
prenatal development and, 89
sleep and, 555
Stress management, 494–496. *See also* Coping strategies
Stroke, 561, 562
brain plasticity and, 120–121
cerebrovascular dementia resulting from, 567
diabetes and, 563
low birth weight and, 84
warning signs of, 567
Structured interviews, 29, 32
Structured observation research method, 27, 32
Students. *See* Academic achievement; Adult learners; Schools and schooling; Teacher–student interaction
Subcategories, 221
Subcultures, 62
Substance abuse. *See* Alcohol use and abuse; Drug use and abuse; Smoking
Suburbs. *See* Community influences
Successful aging, 612–613
Sucking reflex, 101
Sudden death, and bereavement, 640–641
Sudden infant death syndrome (SIDS), 103
bereavement and, 640
cosleeping arrangements and, 122
Sugar
diabetes. *See* Diabetes
obesity and, 420, 422–423
Suicide
adolescence, 344, 401–403
assisted suicide, 636, 638–639
bereavement and, 641
child maltreatment and, 269
early adulthood, death caused by, 419
factors related to, 401–402
homosexuality in adolescence and, 357
incidence of, over lifespan, 401–402, 592
late adulthood, 592–593
prevention and treatment, 402–403, 593
sex differences in, 402, 592
sexual abuse victims, 336
sexual coercion and, 429
warning signs of, 402–403, 593
Summer camps
chronic illness in childhood, 280
delinquency intervention programs, 405
Sun exposure
aging of skin and, 484, 555
cancer and, 491
cataracts and, 551

Superego
  psychosexual theory of, 15–16, 244, 252
  psychosocial theory of, 17
Support groups. *See also* Intervention programs; Social support
  Alcoholics Anonymous, 426
  bereavement, 644
  elder maltreatment prevention and treatment, 609
  widowhood and, 604
Suppression of emotion, health risks of, 494
Supreme Court, U.S. *See* U.S. Supreme Court
Surgery. *See also* Health care
  cesarean delivery, 96
  heart disease, 492
  joint replacements, 562
  prenatal surgery, 53
  prostate enlargement, 555
Surrogate motherhood, 54–55
Surrogate mothering, attachment experiment with monkeys, 185–186
Survival. *See* Life expectancy
Survival of fittest, Darwin's theory of, 14
Swaddling, 104
Swarthmore, Pennsylvania, experience of aging in, 558
Sweden
  adoption study of antisocial behavior, 70
  assistive technology, 556
  child care in, 162
  childbirth practices, 94
  cohabitation, attitude toward, 467
  divorce rate, 329
  elderly poverty rate, 65
  home health care for aging parents, 531
  immunization in, 209
  institutionalization of elderly in, 569
  maturational timing and achievement difficulties of girls, 352
  parental leave, 98
  reproductive technologies, 55
  retirement programs, 610
Swimming reflex, 101
Switzerland
  assisted suicide laws, 638
Symbolic immortality, 626
Sympathy, 249. *See also* Empathy; Prosocial behavior
Synapses, 116–117, 206
  aging process, 549
  brain growth spurts and, 119
Synaptic pruning, 116, 119, 206
Synthetic-conventional stage of faith development (Fowler), 590
Syphilis, 358. *See also* Sexually transmitted diseases (STDs)
Systematic observation research methods, 27–28, 32

T

T cells, 417
  aging process, 554
Tabula rasa, 13
Taiwan. *See also* Chinese
  academic achievement of students, 307, 308–309
  peer relations in adolescence, 396
Talent, in gifted students, 306
Talking. *See* Language development; Speech
Tanzania, hand preference in, 207
Taste
  aging, effects of, 415, 553
  early childhood, 208
  newborns' response to, 106
Tay-Sachs disease, 48, 51, 57
Teacher–student interaction, 304
  encouraging critical thinking by students, 302
  self-fulfilling prophecy in, 304
Teacher's aides, 301
Teachers and teaching techniques. *See also* Education; Learning; Schools and schooling; *specific subjects (e.g., Mathematics; Reading)*
  achievement-related attributions and, 318
  civil responsibility, development of, 393
  critical thinking, encouraging, 302
  gender typing, influence on, 262
  open classrooms, 302–303
  parent-teacher interactions, 62
  philosophies of education. *See* Educational philosophies
  reciprocal teaching, 304
  self-fulfilling prophecies, 304
  traditional classrooms, 302–303
  vocational choice, influences on, 438
Technology. *See also* Computers
  assistive technology, 556
  reproductive technologies. *See* Reproductive technologies
Teen Outreach, 360
Teenage pregnancy and parenthood, 358–361
  consequences of, 360
  correlates of, 350–360
  extended family households and, 63
  Head Start programs and, 234
  incidence of, 358–359
  injury mortality rate and, 211
  intervention programs, 360–361
  prevention strategies, 360
  public policies, comparison of U.S. and Canada with other nations, 64
Teenagers. *See* Adolescence
Teeth
  aging process, 555
  bulimia and, 354
  life expectancy and, 551
  loss of primary (baby) teeth, 204
  middle childhood, 277
  prenatal development, 82
Telegraphic speech, 167–168
Television
  aggression and, 258–259
  caregiving concerns, 259
  cognitive development and, 234–235
  educational television, 234–235
  influence on children, 257–259
  obesity and television viewing, 279
  religious programs, 588
  sexual attitudes and, 354
  V-Chip, 258
Telomeres, 413
Temperament. *See also* Personality traits; *specific traits (e.g., Anxiety; Shyness)*
  attachment and, 190
  child maltreatment and, 268
  child-rearing practices and, 184–185
  defined, 180
  delinquent behavior and, 403, 404
  dimensions of, 181
  divorce of parents, response to and, 330
  empathy and, 249
  environmental influences on, 184
  genetic influences on, 183
  heritability research, 67
  infancy and toddlerhood, 180–185
  injury proneness and, 211
  loneliness and, 456
  measurement of, 181–182
  models of, 181
  moral development and, 254
  New York Longitudinal Study of, 180
  peer victimization and, 325
  prenatal development, 80
  resilient children, 10, 337
  sibling relationships and, 193
  stability of, 182–183
  structure of, 180–181
Temperature
  aging process and response to, 550
  baby fat and, 114
  newborns' response to, 103
Tentative period, of vocational development, 437
Teratogens, 81–88
  alcohol, 85–86
  drugs, illegal, 83–84
  drugs, prescription and nonprescription, 82–83
  environmental pollution, 87
  infectious diseases, 87–88
  radiation, 86–87
  tobacco, 84–85
Terminal decline, 576–577
Terminal illness. *See* Death and dying
Tertiary circular reactions, in sensorimotor stage of cognitive development, 145, 146–147
"Test tube" babies, 54–55
Testes, 77. *See also* Genitals
Testimony by children, 336, 337
Testosterone, 345, 487. *See also* Sex hormones
Tests. *See specific tests (e.g., Blood tests; Intelligence tests)*
Tetens's philosophy of adult development, 13–14
Thalidomide, 82–83
Theories, 5–7. *See also specific theories (e.g., Behaviorism; Lifespan perspective)*
  aging, 413–414, 595–596
  attachment, 185–187
  comparing and evaluating, 26–27
  course of development, 6–7
  defined, 5
  determinants of development, 6–8
  gender identity, 260, 263–264
  historical foundations, 13–15
  intelligence, 294–296
  language development, 164–165
  mid-twentieth century theories, 15–20
  moral development, 252–256
  purpose of, 5
  recent perspectives, 20–25
  scientific reasoning and, 365–366
  scientific verification of, 5, 27. *See also* Research *entries*
Theory of mind. *See* Metacognition
Therapy. *See* Counseling
Thinking. *See* Cognition; Reasoning; *specific entries (e.g., Abstract thinking; Relativistic thinking)*
Third age, 577*n*
Third parties, ecological systems theory, 59
Third-party perspective taking, 321
Three-mountains problem, 218, 219
Threshold of responsiveness, as temperament dimension, 181
Throwing
  adolescence, 347
  early childhood, 213
  sex differences in, 215
Thymus, effects of aging, 417–418
Thyroid-stimulating hormone (TSH), 208
Thyroxine, 208, 345
Time out, as disciplinary technique, 255
Time perception in late adulthood, 597
Time-based prospective memory tasks, 573
Timidity. *See* Anxiety; Fear; Shyness
Tobacco. *See* Smoking
Toddlerhood. *See* Infancy and toddlerhood
Toilet training, 16, 175
Tonic neck reflex, 101
Tooth. *See* Teeth
"Top-heavy" family structures, 59, 530
Touch
  aging, effects of, 415, 553
  Braille and touch sensitivity, 553
  intermodal perception, 138–139
  newborns' response to, 105–106
  prenatal development, 79
  preterm infants, 97–98
Towns, influence on development, 61–62
Toxemia, 90
Toxoplasmosis, prenatal development and, 88
Toys. *See also* Games; Play
  child care facilities, 163
  gender stereotyping and, 196
Tracking in schools, 304, 374
Traditional classrooms, 302–303
Traditional marriages, 458
Training. *See* Education; Intervention programs; Schools and schooling
  vocational. *See* Vocational education
Traits. *See also specific traits (e.g., Aggression; Shyness)*
  androgynous. *See* Androgynous traits
  dominant–recessive, 46–47, 48, 49
  personality. *See* Personality traits
  polygenic, 50, 66
Transformations
  dynamic transformations, 218
  state-versus-transformation thinking, 218
Transition classes, 303
Transitions. *See also* Adaptation; Adjustment problems; Change
  Levinson's theory of adult development, 447, 448, 515
  midlife. *See* Midlife transition
  parenthood, to, 462–463
  reminiscence and, 587
  school transitions, 371–372
Transitive inference, in concrete operational stage of cognitive development, 285
Transracial and transcultural adoption, 57
Trauma. *See* Stress; *specific experience (e.g., Sexual coercion; War and armed conflict)*
"Trauma curriculum," 335
Treatment programs. *See* Counseling; Intervention programs
Trials, legal. *See* Legal proceedings
Triangular theory of love, 451

Triarchic theory of intelligence (Sternberg), 294–295, 503
Tribal and village societies. *See also specific* societies
  abstract thinking among, 365
  adolescent periods among, 345
  adolescent rites of passage, 350
  bereavement rituals, 642–643
  childbirth practices, 93
  contexts of development and, 6–7
  elders, valuing of, 595
  gender typing among, 327
  infant caregiving, 104
  moral development among, 392
  sexuality in late adulthood, 561
  sibling relationships, 532
  transition from childhood to adulthood, 345
  wisdom, valuing of, 575
Trimesters, in prenatal development, 77, 79–81
Triple X syndrome, 51
Trophoblast, 78
Truancy. *See* Dropping out of school; School attendance
Trust
  friendships and, 324, 454
  psychosocial stage of development, 17, 174
  sexual abuse victims, 336
Turner syndrome, 51
TV. *See* Television
Twentieth-century theories of development, 15–20
Twin studies. *See also* Kinship studies
  alcohol abuse, 425
  anorexia nervosa, 353
  bulimia, 354
  concordance rate, 67–68
  hand preference, 206
  heritability estimates, 67, 296
  IQ scores, 296
  life expectancy, 548
  niche-picking and, 69
  pubertal development, 348
  sexual orientation, 356
  temperament and personality, 183, 184
Twins, 46
  ADHD and, 288
  life expectancy among, 413, 548
  obesity among, 278
Two-word utterance phase, in language development, 167–168
Type A behavior pattern, 493–494
Tyrosine, 47

U

UCR (unconditioned response), 126–127
UCS (unconditioned stimulus), 126–127
Ukraine, Chernobyl nuclear accident, 86–87
Ulcers, stress and, 431
Ulnar grasp, 132
Ultrasound, 52
Umbilical cord, 78
Unconditioned response (UCR), 126–127
Unconditioned stimulus (UCS), 126–127
Underarm hair development, 346, 348
Underextension, in language development, 167
Underweight babies. *See* Low birth weight
Undifferentiated perspective taking, 321
Unemployment
  education and, 439
  middle adulthood, 537
Uninhibited (sociable) temperament, 182–183
Uninvolved parenting, 265, 266
  peer groups in adolescence and, 398
Unitarians, death concept among, 624
Universal ethical principle orientation in moral development, 390
Universality, as component of death concept, 623
University education. *See* Higher education
University of the Third Age, 577
Unmarried. *See* Never-married; Singlehood; Widowhood
Unsaturated fat, 421–422
Urban areas, influence on development, 61–62. *See also* Community influences; Neighborhood influences
U.S. Supreme Court
  corporal punishment decisions, 268
  right-to-die decisions, 633, 638
Uterus, 76–77. *See also* Reproductive system
Utku Indians, moral development among, 252

V

Vaccination, 209
Vagina. *See* Genitals; Reproductive system
Vaillant's theory of adult development, 447
  early adulthood, 449
  middle adulthood, 516–517
Valedictorians, vocational choices among, 438
Values. *See also* Bias; Cultural influences; Moral development; Spirituality and religiosity
  child maltreatment and cultural values, 268
  child-rearing practices and American values, 62–63
  elders, valuing of, 595
  immigrant youths, 31
  mate selection, characteristics valued, 450–451, 497
  shyness, valuing of, 184–185
  socioeconomic status and family values, 59
  wisdom, valuing of, 575–576
Variables, in experimental research design, 33
V-Chip, 258
Vegetative state, persistent, 621
Verbal abilities. *See* Language development; Speech
Verbal memory, in middle adulthood, 499
Vernix, formation of, 80
Victimization. *See also* Violence
  elderly, 598
  peer victimization, 325
Video and computer games
  spatial abilities and, 370
  violence in, 258
Village societies. *See* Tribal and village societies
Violation-of-expectation method of research, 147–148, 150, 152
Violence. *See also* Aggression
  child maltreatment. *See* Child maltreatment
  childhood anxieties and, 334, 335
  community influences on, 61
  cultural values and, 268
  dating violence, 399
  domestic. *See* Domestic violence
  elderly, against, 598. *See also* Elder maltreatment
  partner abuse. *See* Partner abuse
  September 11 attack on World Trade Center, 335, 643
  television programming and, 258–259
  video and computer games, 258
  war. *See* War and armed conflict
Violence Chip, 258
Viral disease. *See* Infectious diseases
Vision. *See also* Eyes; Visual impairment
  aging, effects of, 415, 551–552, 553
  color vision, 484
  depth perception, 134–138, 139
  dominant–recessive inheritance, 47
  face perception, 135–138, 177
  glaucoma, 484
  infancy and toddlerhood, 133–138
  intermodal perception, 138–139
  middle adulthood, 483–484
  middle childhood, 277–278
  myopia, 277
  neuron loss in visual area of cerebral cortex, 549
  newborns' capacities, 106–107
  night vision, 483–484
  pattern perception, 135–138
  presbyopia, 483
  smoking, effects of, 425
Visitation rights of grandparents, 529
Visual acuity
  aging process, 552
  defined, 106
  infancy and toddlerhood, 133
  newborns' capacities, 106
Visual cliff studies, 134
Visual impairment
  Braille letters and touch sensitivity, 553
  cataracts leading to, 551
  cognitive deterioration and, 576
  infant development and, 136–137
  macular degeneration leading to, 552
Vitamins and minerals. *See also* Nutrition; *specific entries (e.g., Calcium)*
  adolescence, deficiencies in, 353
  Alzheimer's disease and, 566
  free-radical damage and, 414
  late adulthood, 560
  macular degeneration and, 552
  osteoporosis and, 493
  premenstrual syndrome and, 430
Vitreous. *See* Eyes
Vocabulary
  early childhood, 236
  infancy and toddlerhood, 164, 167–168
  maternal behavior and development of, 32
  middle childhood, 299
  peer culture and, 322
  peer groups, 398
Vocational achievement, 472
  education and, 439
  middle adulthood, 536
  sex- and ethnic-related differences in, 472–473, 536
Vocational development. *See also* Employment; Maternal employment; Unemployment
  access to vocational information and, 439
  burnout, 534–535
  changes in vocation, 472, 517, 536–537
  childbearing decisions and, 461
  childlessness and, 468
  cognitive development and, 500, 505
  continuing education, 535–536
  creative period in, 434
  divorce and, 469
  early adulthood, 436–441, 471–474
  ethnic minorities, 472–473, 536
  factors influencing, 437–439
  family influences on, 437–438
  gender stereotypes and, 438
  glass ceiling, 536
  job-training programs, 374
  mentor relationships. *See* Mentors
  middle adulthood, 505, 533–538
  on-the-job training, 535–536
  performance on job, 534
  personality types and, 437
  satisfaction with vocation, 472, 534–535
  selection of vocation, 436–439, 471–472
  teachers' influences on, 438
  women, 438, 472, 473, 536
Vocational education. *See also* Work-study programs
  cultural influences, 440
  drop out prevention programs, 375
  early adulthood, 439–440
  tracking in schools, 374
  work-study programs, 374
Voice changes in adolescence, 348
Voluntary active euthanasia, 636, 637
Voluntary reaching, 132
Volunteering, 512, 516
  civil responsibility, development of, 393
  generativity stage of development, 512
  intergenerational relationships and, 533
  late adulthood, 612
  widowhood, following, 604
Vygotsky's sociocultural theory. *See* Sociocultural theory of cognitive development (Vygotsky)

W

WAIS (Wechsler Adult Intelligence Scale), 293–294
Walking. *See also* Motor development
  crawling and depth perception, 134–135
  early childhood, 212–213
  stepping reflex, 100–101
War and armed conflict
  children, effect on, 335
  life expectancy and, 547
  World War II. *See* World War II
"War on Poverty," 233
Water, fluoridated, 204
Watson's theory. *See* Behaviorism
"Wear-and-tear" theory of aging, 413
Websites
  hospice care, 633
  memorials to deceased, 643
Wechsler Intelligence Scales, 293–294, 499–500
Weight. *See also* Obesity
  adolescence, 346
  aging, effects of, 415, 555
  birth weight, 92. *See also* Low birth weight
  early adulthood, 431

early childhood, 204
genetic influences on, 123
heart disease and, 492
infancy and toddlerhood, 114–115
life expectancy and, 548
middle adulthood, 484
middle childhood, 276–277
Welfare programs. *See* Intervention programs; Medicaid; Medicare; Public policies
Well-being. *See* Health; Psychological well-being
West Indians of Jamaica, motor development in infancy, 131–132
Western Europe. *See also specific countries*
assisted suicide laws, 638
cohabitation, attitude toward, 467
institutionalization of elderly in, 569
national examinations, 374
prenatal health care, 98
social security systems, 64
suicide rate, 402
vocational education, 374
Whites. *See* Caucasian Americans
Whole-language approach to reading, 292
Widowhood, 602–603
bereavement and, 641
caregiving guidelines, 604
independent living in late adulthood, 599
loneliness and, 456
poverty and, 60
sibling relationships and, 605
WISC-III (Wechsler Intelligence Scale for Children-III), 293–294
Wisdom, 9, 575–576
Withdrawal of privileges, as disciplinary technique, 255
Withdrawal reflex, 101
Wives. *See* Marital *entries;* Marriage
Women. *See* Gender *entries;* Mother *entries;* Sex differences
Women's Health Initiative, 489
Words. *See* Language development; Speech
Work. *See* Employment; Job *entries;* Vocational *entries*
Working memory, 153, 154. *See also* Memory
fluid intelligence and, 499
language processing and, 573
late adulthood, 571, 573
middle adulthood, 502–503
middle childhood, 287
Working mothers. *See* Maternal employment
Work-role preoccupation versus ego differentiation (Peck), 585
Work-study programs, 374. *See also* Vocational education
delinquency intervention programs, 405
Germany, 440
World Trade Center, terrorist attack on, 335, 643
World War II
bombing of Hiroshima and Nagasaki, 86
consequences on lifespan development, 36
refugee and evacuee children, 588–589
Writing
early childhood, 213–214, 215, 230–231
early printing, 215
middle childhood, 281
scribbles, 213–214

X

X chromosome, 46, 49
fragile X syndrome, 49
life expectancy and, 547
X-linked diseases, 48, 54
X-linked inheritance, 49, 50
XO syndrome, 51
X-rays. *See also* Radiation
skeletal age, determining, 115
XXX syndrome, 51
XXY syndrome, 51
Xylitol, 210
XYY syndrome, 51

Y

Y chromosome, 46, 49
Yolk sac, 78
Young adults. *See* Early adulthood
Young-old elderly, 546–547
Youth. *See* Adolescence
Youth gangs, 405. *See also* Delinquency
Youth organizations
civil responsibility and, 393
religious groups, 323
Youth-age balance, in midlife development, 515
Yurok Indians, breastfeeding among, 17

Z

Zaire, Efe tribe, stranger anxiety among infants, 178
Zambia, newborn behavior and child-rearing practices, 107
Zidovudine (ZDZ), 87
Zinacanteco Indians
motor development in infancy, 131
social interaction and learning among, 24
Zone of proximal development, in sociocultural theory, 158, 169, 223
reciprocal teaching and, 304
Zuni Indians, self-conscious emotions among, 179
Zygote, 45, 76, 77
Zygote, period of prenatal development, 78
teratogens in, 82